Editorial Board

Editors-in-Chief

Ernst Knobil

Jimmy D. Neill

Associate Editors

Eli Y. Adashi
Department of Obstetrics and Gynecology
University of Utah
Salt Lake City, Utah

Fuller W. Bazer
Albert B. Alkek Institute of Biosciences and Technology
Texas A&M University
Houston & College Station, Texas

Ian P. Callard
Department of Biology
Boston University
Boston, Massachusetts

Kenneth G. Davey
Department of Biology
York University
North York, Ontario, Canada

Claude Desjardins
College of Medicine
University of Illinois at Chicago
Chicago, Illinois

Marc E. Freeman
Department of Biological Science
Florida State University
Tallahassee, Florida

Michael J. K. Harper
Conrad Program
Arlington, Virginia

Paul Licht
College of Letters and Science
University of California, Berkeley
Berkeley, California

John S. Pearse
Institute of Marine Sciences
University of California, Santa Cruz
Santa Cruz, California

Donald W. Pfaff
Rockefeller University
New York, New York

Mary Lake Polan
Department of Gynecology and Obstetrics
Stanford University School of Medicine
Stanford, California

Jerome F. Strauss, III
Department of Obstetrics and Gynecology
University of Pennsylvania
Philadelphia, Pennsylvania

Peter Thomas
University of Texas Marine Science Institute
Port Aransas, Texas

H. Allen Tucker
Department of Animal Science
Michigan State University
East Lansing, Michigan

John C. Wingfield
Department of Zoology
University of Washington
Seattle, Washington

Encyclopedia of REPRODUCTION

Volume 3   M–Pri

Encyclopedia of REPRODUCTION

Volume 3 M–Pr

Editors-in-Chief

Ernst Knobil

H. Wayne Hightower Professor in the Medical Sciences
and
Ashbel Smith Professor,
University of Texas Health Sciences Center
Houston, Texas

Jimmy D. Neill

Distinguished Professor
University of Alabama at Birmingham

ACADEMIC PRESS

San Diego   London   Boston   New York   Sydney   Tokyo   Toronto

This book is printed on acid-free paper.

Copyright © 1998 by ACADEMIC PRESS

All Rights Reserved.
No part of this publication may be reproduced or transmitted in any form or by any means, electronic or mechanical, including photocopy, recording, or any information storage and retrieval system, without permission in writing from the publisher.

Academic Press
a division of Harcourt Brace & Company
525 B Street, Suite 1900, San Diego, California 92101-4495, USA
http://www.apnet.com

Academic Press Limited
24-28 Oval Road, London NW1 7DX, UK
http://www.hbuk.co.uk/ap/

Library of Congress Catalog Card Number: 98-84463

International Standard Book Number: 0-12-227020-7 (set)
International Standard Book Number: 0-12-227021-5 (vol. 1)
International Standard Book Number: 0-12-227022-3 (vol. 2)
International Standard Book Number: 0-12-227023-1 (vol. 3)
International Standard Book Number: 0-12-227024-X (vol. 4)

PRINTED IN THE UNITED STATES OF AMERICA
98 99 00 01 02 03 MM 9 8 7 6 5 4 3 2 1

Contents

Contents by Subject Area   xxiii
Preface   xxxiii
Guide to the Encyclopedia   xxxv

M

Magnocellular System   1
R. John Bicknell

Male Reproductive Disorders   5
Ivan Damjanov

Male Reproductive System, Amphibians   10
Riccardo Pierantoni

Male Reproductive System, Birds   15
Timothy R. Birkhead

Male Reproductive System, Fish   20
Florence Le Gac, Maurice Loir

Male Reproductive System, Human   30
John F. Redman

Male Reproductive System, Insects   41
Cedric Gillott

Male Reproductive System, Nonhuman Mammals   49
Larry Johnson, Gabrielle U. Falk, Genevieve E. Spoede

Male Reproductive System, Reptiles   60
Daniel H. Gist

Mammary Gland Development   71
Jose Russo, Irma H. Russo

Mammary Gland, Overview   81
Isabel A. Forsyth

Marine Invertebrate Larvae   89
Craig M. Young

Marine Invertebrates, Modes of Reproduction in   98
Jan A. Pechenik

Marsupials   104
Marilyn B. Renfree, Geoffrey Shaw

Mate Choice, Overview   115
Patricia Adair Gowaty

Mating Behaviors, Insects   129
William H. Cade

Mating Behaviors, Mammals   137
Michael J. Baum

Mating Behaviors, Invertebrates Other Than Insects   141
Janet Leonard

Median Eminence   151
Ann-Judith Silverman

Meiosis   160
V. Polanski, Jacek Z. Kubiak

Meiotic Cell Cycle, Oocytes   168
Nava Dekel

Menarche   177
Harry Hatasaka

Menopause   183
Lawrence C. Udoff, Eli Y. Adashi

Menstrual Cycle   189
Sarah L. Berga

Menstrual Disorders   195
Isaac Schiff, Shimon Segal

Menstruation Linda R. Nelson	200
Metamorphosis, Insects Fred Nijhout	205
Microtinae (Voles) Theresa M. Lee	209
Migration, Amphibians Raymond D. Semlitsch, Travis J. Ryan	221
Migration, Birds Peter Berthold	228
Migration, Fish Graham Young	234
Migration, Insects Kenneth Wilson	244
Migration, Reptiles David W. Owens	251
Milk, Composition and Synthesis Harold M. Farrell, Jr.	256
Milk Ejection Jonathan B. Wakerley	264
Mollusca A. Saber M. Saleuddin	276
Molt and Nuptial Color Christopher W. Thompson	283
Monotremes Mervyn Griffiths	295
Morning Sickness and Hyperemesis Gravidarum Marc Jackson, Marcelo F. Noguera	302
Myriapoda Richard L. Hoffman	308

N

Naked Mole-Rats Christopher G. Faulkes	321
Nematodes and Related Phyla Denis J. Wright	326
Nematomorpha Andreas Schmidt-Rhaesa	333
Nemertea James M. Turbeville	341
Nervus Terminalis Marlene Schwanzel-Fukuda	350
Nesting, Birds Joanna Burger	357
Neuroendocrine Systems George Fink	366
Neurohypophysial Hormones Hans H. Zingg	384
Neuropeptides Andrés F. Negro-Vilar, Brian J. Arey, Francisco J. López	391
Neurosecretion Harold Gainer	400
Neurotransmitters William F. Ganong	405
Nutritional Factors and Reproduction Gary L. Williams	412
Nutritional Factors and Lactation Michael J. VandeHaar	422

O

Obstetric Anesthesia Brett B. Gutsche, David L. Hepner	443
Olfaction and Reproduction Dietrich L. Meyer, Rakesh K. Rastogi	445
Onychophora Michael T. Ghiselin	456
Oocyte and Embryo Transport Horacio B. Croxatto, Manuel Villalón, Luis Velasquez	459
Oocyte, Mammalian Anne Byskov, Maria Strömstedt	468
Oocyte Maturation and Spawning in Starfish Takeo Kishimoto	481

Oogenesis, in Mammals 488
Roger G. Gosden, Helen M. Picton

Oogenesis, in
Nonmammalian Vertebrates 498
Charles A. Lessman

Oostatins, Folliculostatins, and
Antigonadotropins, Insects 509
Terry S. Adams

Opossums 515
John D. Harder, Leslie M. Jackson

Orchitis 524
Wolfgang Weidner, W. Krause

Orgasm 528
Kevin E. McKenna

Orthonectida 532
John S. Pearse

Osteoporosis 533
Claude D. Arnaud, E. Bruce Roe

Ovarian Cancer 542
Mark K. Dodson, Jason L. Johnson

Ovarian Cycle, Mammals 547
Michel Ferin

Ovarian Cycle, Teleost Fish 552
Izhar A. Khan, Peter Thomas

Ovarian Cycles and Follicle Development
in Birds 564
Alan L. Johnson

Ovarian Function in the Perimenopause 574
Elizabeth B. Connell

Ovarian Hormones, Overview 578
Shao-Yao Ying, Zhong Zhang

Ovarian Innervation 583
Sergio Ojeda, Gregory A. Dissen

Ovary, Overview 590
Janice Bahr, Humphrey H. C. Yao

Oviposition in Molluscs 598
Jeffrey L. Ram

Ovulation 605
Lawrence L. Espey

Ovulation and Oviposition, Insects 615
Marc J. Klowden

Oxytocics 620
Laird Wilson, Jr.,
Ramkrishna Mehendale

Oxytocin 630
Carol Sue Carter,
A. Courtney DeVries

P

Pampiniform Plexus 635
Brian P. Setchell

Parasites and Reproduction 638
Jack J. O'Brien

Parasitoids 646
Nancy E. Beckage

Paraspermatozoa 656
Alan N. Hodgson

Parental Behavior, Arthropods 668
Gary A. Polis, Joseph D. Barnes,
Andrew N. Beld, C. Todd Jackson

Parental Behavior, Birds 674
John D. Buntin

Parental Behavior, Mammals 684
Michael Numan

Parthenogenesis and Natural Clones 695
Robert C. Vrijenhoek

Parturition, Nonhuman Mammals 703
Michael Fields, Anna-Riitta Fuchs

Pelvic Inflammatory Disease (PID) 716
Thomas E. Snyder

Pelvic Nerve 725
Karen J. Berkley

Pelvimetry 731
Samuel Parry, Mark A. Morgan

Penis 739
Paul F. Schelhammer,
Gerald H. Jordan

Pheromones, Fish 748
Norman Stacey, Peter Sorensen

Pheromones, Insects Jeremy N. McNeil, Johanne DeLisle, Claude Everaerts	755	Placental Steroidogenesis in Primate Pregnancy Eugene D. Albrecht, Gerald J. Pepe	889
Pheromones, Mammals John G. Vandenbergh	764	Placozoa Vicki Buchsbaum Pearse	898
Phoronida Russel L. Zimmer	770	Platyhelminthes Seth Tyler	901
Photoperiodism, Vertebrates Randy J. Nelson	779	PMS (Premenstrual Syndrome) Ellen W. Freeman	908
Pigeons Richard F. Johnston	789	Poecilogony Glenys D. Gibson	917
Pigs Rodney D. Geisert	792	Polycystic Ovary Syndrome Richard S. Legro	925
Pineal Gland, Melatonin Biosynthesis and Secretion Stuart E. Dryer	800	Polyspermy R. H. F. Hunter	930
Pineal Gland, Regulatory Function Fred W. Turek	802	Porifera Paul E. Fell	938
Pituitary Gland, in Fish Martin P. Schreibman, Lucia Magliulo-Cepriano	812	Postdate (Postterm) Pregnancy Brian J. Koos, Jennifer Claman	946
Pituitary Gland, Overview Béla Halász	823	Postpartum Depression Joseph F. Mortola	954
Placenta and Placental Analogs in Elasmobranchs William C. Hamlett	831	Prader-Willi Syndrome Shahab S. Minassian	959
Placenta and Placental Analogs in Reptiles and Amphibians Daniel G. Blackburn	840	Preeclampsia/Eclampsia James N. Martin, Everett F. Magann	964
Placenta: Implantation and Development Kurt Benirschke	848	Pregnancy in Dogs and Cats Patrick W. Concannon, John Verstegen	970
Placental and Decidual Protein Hormones, Human Stuart Handwerger	855	Pregnancy in Farm Animals Troy L. Ott	980
		Pregnancy in Humans, Overview Carmen L. Regan	986
Placental Gas Exchange Lawrence D. Longo	863	Pregnancy in Other Mammals Lloyd L. Anderson	992
Placental Lactogens Daniel I. H. Linzer	874	Pregnancy, Maintenance of Fuller W. Bazer	1002
Placental Nutrient Transport Colin P. Sibley	881	Pregnancy, Maternal Recognition of Thomas E. Spencer	1006

Pregnancy, Metabolic Changes in William W. Hay, Jr.	1016	**Priapulida** Christian Lemburg, Andreas Schmidt-Rhaesa	1053
Prenatal Genetic Screening Deborah A. Driscoll	1027	**Primates, Nonhuman** Bill Lasley, Susan Shideler	1058
Preterm Labor and Delivery Steve N. Caritis, Douglas A. Woelkers	1037	**Primordial Germ Cells** Peter J. Donovan	1064

Contents of Other Volumes

VOLUME 1

Contents by Subject Area
Preface
Guide to the Encyclopedia

A

Abortion
Steven J. Sondheimer

Acanthocephala
D. W. T. Crompton

Acrosome Reaction
Gregory S. Kopf

Activin and Activin Receptors
Ralph H. Schwall

Adrenal Androgens
Collin B. Smikle

Adrenal Hyperplasia, Congenital Virilizing
Peter A. Lee, Selma F. Witchel

Adrenarche
Frank Gonzalez

Aedes aegypti
Alexander S. Raikhel, Thomas W. Sappington

Aggressive Behavior
David A. Edwards

Agnatha
Stacia A. Sower, Aubrey Gorbman

Allantochorion (Chorioallantois)
Fuller W. Bazer

Allantoic Fluid
Fuller W. Bazer

Allantois
Fuller W. Bazer

Allatostatins
Stephen S. Tobe

Altitude, Effects on Humans
Lorna G. Moore

Altricial and Precocial Development in Birds
F. M. Anne McNabb

Altruism in Insect Reproduction
Laurence Packer

Altruistic Behavior, Vertebrates
R. Haven Wiley

Amenorrhea
Sarah Berga

Amniocentesis
Nancy C. Rose, Tzazil Ayala

Amniotic Fluid
Robert A. Brace

Amphibian Ovarian Cycles
Alberta M. Polzonetti-Magni

Amphibian Reproduction, Overview
Marvalee H. Wake

Androgen Inhibitors/Antiandrogens
Vivian L. Fuh, Elizabeth Stoner

Androgen Insensitivity Syndromes
James Aiman

Androgens
Bernard Robaire

Androgens, Effects in Birds
Cheryl F. Harding

Androgens, Effects in Mammals
Shalender Bhasin

Androgens, Subavian Species
Kyle W. Selcer, Jeffrey W. Clemens

Andrology: Origins and Scope
Philip Troen

Annelida
Damhnait McHugh

Anterior Pituitary
Robert B. Page

Antiestrogens
Donald P. McDonnell

Antiprogestins
Irving M. Spitz

Apgar Score
Joseph Dancis,
Karen D. Hendricks-Muñoz

Aphids
Jim Hardie

Aplysia
Stephen Arch

Apoptosis (Cell Death)
Paul F. Terranova,
Christopher C. Taylor

Armadillo
Richard D. Peppler

Aromatization
Alan J. Conley, Karen W. Walters

Artificial Insemination, in Animals
William L. Flowers

Artificial Insemination, in Humans
Douglas T. Carrell, Deborah Cartmill

Asexual Reproduction
Kerstin Wasson

Autonomic Nervous System
and Reproduction
Karen J. Berkley

Avian Reproduction, Overview
Tony D. Williams

Avian Reproductive System,
Developmental Endocrinology
James E. Woods,
Robert C. Thommes

B

Benign Prostatic Hyperplasia (BPH)
Kevin T. McVary, John T. Grayhack

Birds, Diversity of
Scott V. Edwards

Blastocyst
Richard M. Schultz

Blood–Testis Barrier
Brian P. Setchell

Brachiopoda
Stephen A. Stricker

Breast Cancer
Laura Esserman, Hope Wallace

Breast Disorders
Stefanie S. Jeffrey, Diana O. Cua

Breastfeeding
Carla D. Harris, Eloise D. Clawson

Breeding Strategies for
Domestic Animals
George R. Foxcroft

Brood Parasitism in Birds
Alfred M. Dufty, Jr.

Bruce Effect
Peter Brennan

Bryozoa (Ectoprocta)
Robert Woollacott

C

Caenorhabditis elegans
Dave Pilgrim

Captive Breeding of Wildlife
Barbara S. Durrant

Cardiovascular Adaptation
to Pregnancy
Margaret K. McLaughlin,
Robin F. Gandley

Castration, Effects in Humans (Male)
Harald H. J. Hoekstra,
Mels F. van Driel,
Pax H. B. Willemse

Castration, Effects in
Nonhuman Mammals (Female)
Sandra J. Legan

Castration, Effects in Nonhuman
Mammals (Male)
Graeme B. Martin, David R. Lindsay

Cats
David E. Wildt, Janine L. Brown,
William F. Swanson

Cattle
Roy Fogwell

CBG (Corticosteroid-Binding Globulin)
Geoffrey L. Hammond

Cephalochordata
M. Dale Stokes

Cervical Cancer
James A. Roberts, B. Hannah Ortiz

Cervix
Kamran S. Moghissi

Cesarean Delivery
Linda J. Heffner

Chaetognatha
George L. Shinn

Chelicerate Arthropods
W. Reuben Kaufman

Chickens, Control of Reproduction in
Peter J. Sharp

Choriocarcinoma
Anthony C. Evans, Jr.

Chorionic Gonadotropin, Human
Robert E. Canfield, Stephen Birken,
John O'Connor, Leslie Lobel

Chorionic Gonadotropins,
Nonhuman Mammals
Marylynn Barkley,
Mary B. Zelinski-Wooten

Circadian Rhythms
Fred W. Turek

Circannual Rhythms
Irving Zucker, Brian J. Prendergast

Circumcision
M. Sean Esplin

Clitoris
Ursula Kuhnle

Cloning Mammals by Nuclear Transfer
Randall S. Prather

Cnidaria
Daphne Gail Fautin

Conjugation in Ciliates
Peter J. Bruns

Contraceptive Methods and
Devices, Female
Malcolm Potts, Diana S. Wolfe

Contraceptive Methods and
Devices, Male
William J. Bremner,
M. Cristina Meriggiola

Copulation, Mammals
Robert L. Meisel

Corpora Lutea of
Nonmammalian Species
Giovanni Chieffi,
Gabriella Chieffi Baccari

Corpus Allatum
Stephen S. Tobe

Corpus Cardiacum, Insects
Barry G. Loughton

Corpus Luteum (CL)
Gordon D. Niswender,
Jennifer J. Juengel, Eric W. McIntush

Corpus Luteum of Pregnancy
Richard L. Stouffer

Corpus Luteum Peptides
O. David Sherwood, Phillip A. Fields

Cotyledonary Placenta
Stephen P. Ford

Cricetidae (Hamsters and Lemmings)
Bruce D. Goldman,
David A. Freeman

Critical Period, Estrous Cycle
Lewis C. Krey

Crustacea
Hans Laufer, Matthew Landau

Cryopreservation of Embryos
C. Matthew Peterson,
Akiyasu Mizukami,
Douglas T. Carrell

Cryopreservation of Sperm
Rupert P. Amann

Cryptorchidism
George W. Kaplan, Irene M. McAleer

Ctenophora
George I. Matsumoto

Cycliophora
Peter Funch,
Reinhardt Møbjerg Kristensen

Cytokines
Sarah A. Robertson

D

Decidua
Linda C. Giudice, Juan C. Irwin

Deciduoma
Geula Gibori, Yan Gu

Deer
Edward D. Plotka

DHEA (Dehydroepiandrosterone)
John E. Nestler, Nancy Pahle

Diapause
David L. Denlinger, Seiji Tanaka

Dihydrotestosterone
Shutsung Liao, Richard A. Hiipakka

Diploptera punctata
Barbara Stay

Discoidal Placenta
John J. Rasweiler, IV,
Nilima K. Badwaik

Dogs
Cheryl S. Asa

Dorsal Bodies in Mollusca
A. Saber M. Saleuddin

Drosophila
Marla B. Sokolowski, John Ewer

E

Ecdysiotropins and Ecdysiostatins
Henry H. Hagedorn

Ecdysteroids
Henry H. Hagedorn

Echinodermata
Maria Byrne

Ectopic Pregnancy
Christos Coutifaris

Egg, Avian
Carol M. Vleck

Egg Coverings, Insects
Michael P. Kambysellis,
Lukas Margaritis,
Elysse M. Craddock

Eicosanoids
William J. Silvia

Ejaculation
Kevin E. McKenna

Elasmobranch Reproduction
Thomas J. Koob

Elephants
Keith Hodges, Cheryl Niemuller,
Janine Brown

Embryogenesis, Mammalian
Carol A. Burdsal

Embryo Transfer
George E. Seidel, Jr.

Endogenous Opioids
John A. Russell, C. H. Brown,
R. W. Caron

Endometriosis
Camran Nezhat, Farr Nezhat,
Ceana Nezhat

Endometrium
Linda C. Giudice

Endotheliochorial Placentation
Vibeke Dantzer

Energetics of Reproduction
Cynthia Carey

Energy Balance, Effects on Reproduction
George N. Wade

Environmental Estrogens
Stephen H. Safe

VOLUME 2

Contents by Subject Area
Preface
Guide to the Encyclopedia

E

Epididymis
Trevor G. Cooper

Epitheliochorial Placentation
Vibeke Dantzer

Equine Chorionic Gonadotropin
Janet F. Roser

Erection
George J. Christ

Erythroblastosis Fetalis
Donald J. Dudley, G. Marc Jackson

Estrogen Action, Behavior
Lynda Uphouse, Sharmin Maswood

Estrogen Action, Bone
Robert Marcus

Estrogen Action, Breast
Serdar E. Bulun, Evan R. Simpson

Estrogen Action on the Female Reproductive Tract
Kenneth S. Korach, Jonathan Lindzey

Estrogen Effects and Receptors, Subavian Species
Marina Paolucci, Noemi Custodia, Ian P. Callard

Estrogen, Effects in Birds
Barney A. Schlinger, Colin J. Saldanha

Estrogen Replacement Therapy
Rogerio A. Lobo

Estrogen Secretion, Regulation of
Koji Yoshinaga

Estrogens, Overview
Carolyn L. Smith

Estrous Cycle
Neena B. Schwartz, Signe M. Kilen

Estrus
Janice E. Thornton, Patricia D. Finn

Eunuchoidism
Victor Y. Fujimoto, Michael R. Soules

F

Fallopian Tube
Michael P. Diamond, Diaa M. El-Mowafi

Family Planning
Allan Rosenfield, Victoria L. Dunning

Female Reproductive Disorders, Overview
Robert W. Rebar

Female Reproductive System, Amphibians
Rakesh K. Rastogi, Luisa Iela

Female Reproductive System, Birds
Patricia Johnson

Female Reproductive System, Fish
Martin A. Connaughton, Katsumi Aida

Female Reproductive System, Humans
Bruce R. Carr

Female Reproductive System, Insects
Erwin Huebner

Female Reproductive System, Nonhuman Mammals
Robert A. Dailey

Female Reproductive System, Reptiles
Valentine A. Lance

Female Sterilization
Richard M. Soderstrom

Fertility and Fecundity
Philip J. Dziuk

Fertilization
Gerald Schatten

Fetal Adrenals
Robert B. Jaffe, Samuel Mesiano

Fetal Alcohol Syndrome
Nancy C. Rose, David M. Stamilio

Fetal Anomalies
E. Albert Reece, Arnon Wiznitzer

Fetal Diagnosis, Invasive
Carl P. Weiner, Celeste Sheppard

Fetal Growth and Development
William W. Hay, Jr.

Fetal Hormones
Theresa M. Siler-Khodr

Fetal Lung Development
Philip L. Ballard, Susan Guttentag

Fetal Membranes
Jerome F. Strauss, III, Erdal Budak

Fetal Monitoring and Testing
Iraj Forouzan

Fetal-Placental Unit
Bruce R. Carr

Fetal Surgery
N. Scott Adzick, Theresa M. Quinn

α-Fetoprotein and Triple Screening
Jacob A. Canick

Fetus, Overview
Timothy A. Cudd

Fish, Modes of Reproduction in
Rudolf Reinboth

Follicular Atresia
J. Yeh, G. D. Chen, R. H. Oliver

Follicular Development
Bradley J. Van Voorhis

Follicular Steroidogenesis
Bradley J. Van Voorhis

Follistatin
David M. Robertson, Christopher Gilfillan

Freemartin
Sherrill E. Echternkamp

FSH (Follicle-Stimulating Hormone)
Leo E. Reichert, Jr.

G

Galactorrhea
Howard A. Zacur

Gametes, Overview
James M. Robl, Rafael A. Fissore

Gene Transfer, Sperm-Mediated
Jorge A. Piedrahita, Jagdeece J. Ramsoondar

Genitalia
Mary Min-chin Lee

Germ Layers
Jonathan J. H. Pearce

Global Zones and Reproduction
Franklin Bronson

Gnathostomulida
Wolfgang Sterrer

GnRH (Gonadotropin-Releasing Hormone)
P. Michael Conn, L. Jennes, J. A. Janovick

GnRH Pulse Generator
Jon E. Levine

Gonadogenesis, Female
Michael K. Skinner, Jeffrey A. Parrott

Gonadogenesis, Male
Mary Min-chin Lee, Jose Teixeira

Gonadotropes
Gwen V. Childs

Gonadotropin Biosynthesis
Raymond Counis

Gonadotropin Receptors
David Puett

Gonadotropin Secretion, Control of
Charles A. Blake

Gonadotropins, Overview
M. Ram Sairam

Graafian Follicle
J. Yeh, G. D. Chen, R. H. Oliver

Granulosa Cells
Kenneth H. H. Wong, Eli Y. Adashi

Growth Factors
Asgerally T. Fazleabas, J. Julie Kim

Guinea Pig, Female
Reinhold J. Hutz, Amanda L. Trewin

Gynecomastia
Samuel Smith

Hemichordata
Gary M. King

Hemochorial Placentation
Jerome F. Strauss, III, Erdal Budak

Hemocoelic Insemination
Kenneth G. Davey

Hermaphroditism
Kerstin Wasson

Hirsutism
William J. Butler

HIV Infection and AIDS
Penelope J. Hitchcock,
Kearston Schmidt

Homosexuality
Vivienne Cass

Hormonal Contraception
Malcolm Potts, Claire Norris

Hormonal Control of the Reproductive
Tract, Subavian Vertebrates
Ian P. Callard, Vicki Abrams-Motz,
Georgia Giannoukos, Lisa A. Sorbera

Hormone Receptors, Overview
David Puett, Adviye Ergul

Hormones and Reproductive
Behaviors, Fish
Harold H. Zakon, Kent D. Dunlap

Hormones of Pregnancy
Glen E. Hofmann

Horses
Dan C. Sharp

Human Placental Lactogen (Human
Chorionic Somatomammotropin)
Michael Freemark

Hybridization
Michael L. Arnold

Hydra
Vicki J. Martin

Hyenas
Stephen E. Glickman,
Christine M. Drea,
Elizabeth M. Coscia

Hyperprolactinemia
Howard A. Zacur

Hypogonadism
Andrew J. Friedman

Hypophysectomy
Donald C. Johnson

Hypopituitarism
William W. Hurd

Hypospadias
Steven G. Docimo, Ranjiv Mathews

Hypothalamic–Hypophysial Complex
(Pituitary Portal System)
Robert B. Page

Hypoxia, Effect on Reproduction
Charles A. Ducsay

Hysterectomy
Howard T. Sharp

IGF (Insulin-like Growth Factor)
James M. Hammond

Immunocytochemistry
William C. Okulicz

Immunology of Reproduction
Joan S. Hunt, Peter M. Johnson

Implantation
Daniel D. Carson

Impotence
Irwin Goldstein, Lawrence S. Hakim,
Ajay Nehra

Infections in Pregnancy
Jack Ludmir

Infertility
Lawrence C. Udoff, Eli Y. Adashi

Inhibin
Ralph H. Schwall

Insect Accessory Glands
Cedric Gillott

Insect Reproduction, Overview
Kenneth G. Davey

Interferons
Troy L. Ott

Internal Fertilization in Birds and Mammals
Sally D. Perreault, John D. Kirby

Interrenal Gland, Stress Response and Reproduction
Wilfrid Hanke

Intersexuality in Mammals
R. H. F. Hunter

Intrauterine Growth Restriction and Mechanisms of Fetal Growth
Victor K. M. Han

Intrauterine Position Phenomenon
Frederick S. vom Saal,
Mertice M. Clark,
Bennett G. Galef, Jr.,
Lee C. Drickamer,
John G. Vandenbergh

***In Vitro* Fertilization**
Lewis C. Krey, Alan S. Berkeley

J

Juvenile Hormone
Gerard R. Wyatt

K

Kallmann's Syndrome
Lisa M. Halvorson

Kamptozoa (Entoprocta)
Kerstin Wasson

Kinorhyncha
Birger Neuhaus

Klinefelter's Syndrome
Fady I. Sharara

L

Labor and Delivery, Human
Peter W. Nathanielsz,
Gordon C. S. Smith

Lactational Amenorrhea
Amy Banulis, William D. Schlaff

Lactational Anestrus
Jeffrey S. Stevenson

Lactation, Human
Margaret C. Neville

Lactation, Nonhuman
Robert J. Collier

Lactogenesis
R. Michael Akers

Lactotrophs
Tom E. Porter

Leiomyoma
Linda C. Giudice, Salli Tazuke

Leukemia Inhibitory Factor
Colin L. Stewart

Leydig and Sertoli Cells, Nonmammalian
Jeffrey Pudney

Leydig Cells
Matthew P. Hardy,
Benson T. Akingbemi, Ren-Shan Ge

LH (Luteinizing Hormone)
George R. Bousfield

Local Control Systems in Reproduction
John A. McCracken

Locusts
Cedric Gillott

Lordosis
Donald W. Pfaff

Luteinization
Anthony J. Zeleznik

Luteolysis
John A. McCracken

Luteotropic Hormones
Gilbert S. Greenwald

Lymphokines
Wenbin Tuo, Fuller W. Bazer,
Wendy C. Brown

VOLUME 4

Contents by Subject Area
Preface
Guide to the Encyclopedia

P

Progesterone Actions on Behavior
Anne M. Etgen

Progesterone Actions on Reproductive Tract
Francesco J. DeMayo,
Cindee R. Funk

Progesterone Effects and Receptors, Subavian Species
Marina Paolucci, Noemi Custodia,
Ian P. Callard

Progestins
Thomas Burris

Prolactin, Overview
Nadine Binart, Vincent Goffin,
Christopher J. Ormandy,
Paul A. Kelly

Prolactin, Actions of
James A. Rillema

Prolactin Inhibitory Factors
Michael Selmanoff

Prolactin, in Nonmammalian Vertebrates
E. Gordon Grau, Gregory M. Weber

Prolactin Secretion, Regulation of
György Nagy, Pal Gööz,
Katalin M. Horváth, Béla E. Tóth

Prostate Cancer
James M. Kozlowski

Prostate Gland
Chung Lee, Lynn Janulis

Prostate-Specific Antigen
Joseph E. Oesterling,
Ricardo Beduschi

Protein Hormones of Primate Pregnancy
Gerald J. Pepe, Eugene D. Albrecht

Protozoa
O. Roger Anderson

Pseudocyesis
Joseph F. Mortola

Pseudopregnancy
Mary S. Erskine

Puberty Acceleration
John G. Vandenbergh

Puberty, in Humans
Thomas A. Klein

Puberty, in Nonhuman Primates
Tony M. Plant

Puberty, in Nonprimate Mammals
Douglas L. Foster,
Francis J. P. Ebling

Puberty, Precocious
Leo Plouffe, Jr.

Puerperal Infections
John W. Riggs, Jorge D. Blanco

Puerperium
Harish M. Sehdev

R

Rabbits
Josephine B. Miller

Radioimmunoassay
Terry M. Nett, Jennifer M. Malvey

Receptors for Hormones, Overview
Kevin J. Catt

Reflex (Induced) Ovulation
Arnold L. Goodman

Regulation of Sertoli Cells
Michael D. Griswold

Relaxin, Mammalian
Russell V. Anthony

Relaxin, Nonmammalian
Thomas J. Koob

Reproductive Senescence, Human
Charles V. Mobbs

Reproductive Senescence, Nonhuman Mammals
Anne N. Hirshfield, Jodi A. Flaws

Reproductive Technologies, Overview
Alan O. Trounson

Reproductive Toxicology
Robert E. Chapin

Reptilian Reproduction, Overview
David Crews

Reptilian Reproductive Cycles
Valentine A. Lance

Respiratory Distress Syndrome
Rebecca A. Simmons

Rhodnius prolixus
Kenneth G. Davey

Rhombozoa
John S. Pearse

Rhythms, Lunar and Tidal
Peter P. Fong

Rodentia
Franklin H. Bronson

Rotifera
Robert Lee Wallace

Ruminants
William W. Thatcher

S

Seals
Daniel P. Costa, Daniel E. Crocker

Seasonal Reproduction, Birds
Alistair Dawson

Seasonal Reproduction, Fish
Jon P. Nash

Seasonal Reproduction, Mammals
Robert L. Goodman

Seasonal Reproduction, Marine Invertebrates
John S. Pearse

Sea Urchins
John S. Pearse

Semen
Gail S. Prins

Seminal Vesicles
Lawrence S. Ross

Sertoli Cells, Function
Michael D. Griswold, Lonnie D. Russell

Sertoli Cells, Overview
Lonnie D. Russell

Sex Chromosomes
Baccio Baccetti, Giulia Collodel

Sex Determination, Environmental
Reynaldo Patiño, Carlos A. Strüssman

Sex Determination, Genetic
Józefa Styrna

Sex Differentiation in Amphibians, Reptiles, and Birds, Hormonal Regulation
Tyrone B. Hayes

Sex Differentiation, Psychological
Nancy G. Forger

Sex Ratios
Ronald J. Ericsson, Scott A. Ericsson

Sex Skin
Fred B. Bercovitch

Sexual Attractants
Lee C. Drickamer

Sexual Dysfunction
Steven M. Petak

Sexual Imprinting
David B. Miller

Sexually Transmitted Diseases
Paul Summers

Sexual Selection
Anders P. Møller

SHBG (Sex Hormone-Binding Globulin)
William Rosner

Sheehan's Syndrome
Peter J. Snyder

Sheep and Goats
Duane H. Keisler

Sipuncula
Mary E. Rice

Social Insects, Overview
Wolf Engels, Klaus Hartfelder

Song in Arthropods
Glenn K. Morris

Songbirds and Singing
Eliot A. Brenowitz

Spawning, Marine Invertebrates
John H. Himmelman

Sperm Activation, Arthropods
Julian Shepherd

Spermatogenesis, Overview
Rex A. Hess

Spermatogenesis, Disorders of
Claude Desjardins, Thorsten Diemer

Spermatogenesis, Hormonal Control of
Barry R. Zirkin

Spermatogenesis, in Nonmammals
Gloria V. Callard, Ian P. Callard

Spermatogenetic Cycle in Fish
Takeshi Miura

Spermatophores in the Arthropods
Heather C. Proctor

Spermatozoa
Clarke F. Millette

Sperm Capacitation
J. Michael Bedford, Nicholas L. Cross

Spermiogenesis
Richard Oko, Yves Clermont

Sperm Transport
James W. Overstreet, Mary A. Scott

Sperm Transport, Arthropods
Julian Shepherd

SRY Gene
Grace Lee, Mert Bahtiyar

Sterility
Bradley S. Hurst

Steroid Hormones, Overview
Terry R. Brown

Steroidogenesis, Overview
Margaret M. Hinshelwood

Steroid Hormone Receptors
Nancy H. Ing

Stress and Reproduction
Thomas H. Welsh, Jr.,
Nann Kemper-Green,
Kimberly N. Livingston

Substance Abuse and Pregnancy
Mark A. Morgan, Sara J. Marder

Suckling Behavior
Edward O. Price

Suprachiasmatic Nucleus
Robert Y. Moore

Surfactant
Aron B. Fisher

Symbiosis
Mary Beth Saffo

T

Tardigrada
Roberto Bertolani, Lorena Rebecchi

Teleosts, Viviparity
John P. Wourms

Temperature, Effects on
Testicular Function
Jeffrey B. Kerr

Teratogens
Robert L. Brent, David A. Beckman

Territorial Behavior, Overview
Judy Stamps

Testicular Cancer
Gary D. Steinberg

Testicular Developmental Anomalies
Jay Radhakrishnan

Testis, Overview
Larry Johnson, Tobin A. McGowen,
Genevieve E. Keillor

Testosterone Biosynthesis
Douglas M. Stocco

Theca Cells
Paul F. Terranova, Katherine F. Roby

Theca Cell Tumors
Richard Leach, Nilsa Ramirez

Thyroid Hormones, in Subavian Vertebrates
David O. Norris

Tocolytic Agents
George A. Macones, Martha E. Rode

Transgenic Animals
Jorge A. Piedrahita, Karen Moore

Trophoblast to Human Placenta
Harvey J. Kliman

Tsetse Flies
R. H. Gooding

Tubal Surgery
Alan H. DeCherney, Mikio A. Nihira

Tumors of the Female Reproductive System
Basil C. Tarlatzis, Th. Agorastos

Tunicata (Urochordata)
Andrew Todd Newberry

Turner's Syndrome
David H. Barad

Twinning
Kurt Benirschke

U

Ullrich Syndrome
James Aiman

Ultradian Hormone Rhythms
Johannes D. Veldhuis

Ultrasound
Frank A. Chervenak, Edith D. Gurewitsch

Umbilical Cord
Harvey Kliman

Uterine Anomalies
John A. Rock, Bradley S. Hurst

Uterine Contraction
Robert E. Garfield, Venu Jain, George R. Saade

Uterus, Human
David A. Grainger

Uterus, Nonhuman
Frank F. Bartol

V

Vagina
Raymond E. Papka, Sonya J. Williams

Varicocele
Terry T. Turner

Vasectomy
Sherman J. Silber

Vitellogenins and Vitellogenesis
Gary J. LaFleur, Jr.

Viviparity and Oviparity: Evolution and Reproductive Strategies
Daniel G. Blackburn

Vomeronasal Organ
Michael Meredith

W

Whales and Porpoises
Daniel P. Costa, Daniel E. Crocker

Whitten Effect
John G. Vandenbergh

Wolffian Ducts
Terry W. Hensle, Harry Fisch

Y

Yolk Proteins, Invertebrates
G. R. Wyatt

Yolk Sac
Robert W. McGaughey

Z

Zygotic Genomic Activation
Carol Warner, Ginger Exley

Contributors
Glossary of Key Terms
Subject Index

Contents by Subject Area

GAMETES, FERTILIZATION, AND EARLY EMBRYOGENESIS

Acrosome Reaction
Adrenarche
Allantochorion (Chorioallantois)
Amniotic Fluid
Apoptosis (Cell Death)
Blastocyst
Choriocarcinoma
Chorionic Gonadotropin, Human
Cloning Mammals by Nuclear Transfer
Cotyledonary Placenta
Cryopreservation of Embryos
Cytokines
DHEA (Dehydroepiandrosterone)
Eicosanoids
Embryogenesis, Mammalian
Embryo Transfer
Endogenous Opioids
Endotheliochorial Placentation
Epitheliochorial Placentation
Fallopian Tube
Fertilization
Fetal Adrenals
Follistatin
Gametes, Overview
Germ Layers
Gonadogenesis, Female
Gonadogenesis, Male
Gonadotropes
Granulosa Cells
Hemochorial Placentation
Hormone Receptors, Overview
IGF (Insulin-like Growth Factor)
Implantation
In Vitro Fertilization
Lactotrophs
Leydig Cells
Meiosis
Meiotic Cell Cycle, Oocytes
Neuropeptides
Neurosecretion
Neurotransmitters
Oocyte and Embryo Transport
Oocyte, Mammalian
Oogenesis, in Mammals
Pineal Gland, Regulatory Function
Polyspermy
Primordial Germ Cells
Reproductive Technologies, Overview
Reproductive Toxicology
Sex Chromosomes
Sex Determination, Environmental
Sex Determination, Genetic
Sex Skin
SRY Gene
Testicular Developmental Anomalies
Wolffian Ducts
Yolk Sac
Zygotic Genomic Activation

REPRODUCTION IN HUMANS AND EXPERIMENTAL PRIMATES

Abortion
Adrenal Hyperplasia, Congenital Virilizing
Adrenarche
Altitude, Effects on Humans

Anterior Pituitary
Apgar Score
Artificial Insemination, in Humans
Autonomic Nervous System and Reproduction
Chorionic Gonadotropin, Human
Circumcision
Contraceptive Methods and Devices, Female
Contraceptive Methods and Devices, Male
Cryopreservation of Embryos
Embryo Transfer
Eunuchoidism
Family Planning
Female Reproductive System, Humans
Fetal Alcohol Syndrome
FSH (Follicle-Stimulating Hormone)
Genitalia
Gonadotropin Biosynthesis
Gonadotropin Receptors
Gonadotropin Secretion, Control of
Gonadotropins, Overview
Gynecomastia
Hirsutism
HIV Infection and AIDS
Homosexuality
Hormonal Contraception
Human Placental Lactogen (Human Chorionic Somatomammotropin)
Hyperprolactinemia
Hypogonadism
Hypophysectomy
Hypopituitarism
Hypospadias
Hypothalamic-Hypophysial Complex (Pituitary Portal System)
Hypoxia, Effect on Reproduction
Immunocytochemistry
Immunology of Reproduction
Infertility
In Vitro Fertilization
Kallmann's Syndrome
Klinefelter's Syndrome
Leiomyoma
LH (Luteinizing Hormone)
Lymphokines
Male Reproductive System, Human
Median Eminence
Nervus Terminalis
Neuroendocrine Systems
Orgasm
Osteoporosis
Pituitary Gland, Overview
Prader-Willi Syndrome
Pregnancy in Humans, Overview
Prolactin, Overview
Prolactin, Actions of
Prolactin Inhibitory Factors
Prolactin Secretion, Regulation of
Puberty, in Humans
Puberty, Precocious
Radioimmunoassay
Reproductive Senescence, Human
Reproductive Technologies, Overview
Reproductive Toxicology
Respiratory Distress Syndrome
Sex Differentiation, Psychological
Sex Ratios
Sexual Dysfunction
Sexually Transmitted Diseases
SHBG (Sex Hormone-Binding Globulin)
Sheehan's Syndrome
Sterility
Steroid Hormones, Overview
Steroidogenesis, Overview
Steroid Hormone Receptors
Stress and Reproduction
Suprachiasmatic Nucleus
Turner's Syndrome
Twinning
Ullrich Syndrome

DOMESTIC ANIMALS

Artificial Insemination, in Animals
Breeding Strategies for Domestic Animals
Castration, Effects in Nonhuman Mammals (Female)
Castration, Effects in Nonhuman Mammals (Male)
Cats
Cattle
Chickens, Control of Reproduction in
Cloning Mammals by Nuclear Transfer
Cryopreservation of Sperm
Dogs
Embryo Transfer

Equine Chorionic Gonadotropin
Estrus
Fertility and Fecundity
Freemartin
Guinea Pig, Female
Horses
Pigs
Pregnancy in Dogs and Cats
Pregnancy in Farm Animals
Pregnancy, Maintenance of
Reproductive Technologies, Overview
Ruminants
Sheep and Goats
Transgenic Animals

MAMMALIAN REPRODUCTION

Aggressive Behavior
Allantochorion (Chorioallantois)
Allantois
Androgens, Effects in Mammals
Armadillo
Aromatization
Bruce Effect
Chorionic Gonadotropins, Nonhuman Mammals
Circadian Rhythms
Circannual Rhythms
Copulation, Mammals
Cricetidae (Hamsters and Lemmings)
Deer
Elephants
Embryogenesis, Mammalian
Equine Chorionic Gonadotropin (ECG)
Estrous Cycle
Estrus
Female Reproductive System, Nonhuman Mammals
Freemartin
Global Zones and Reproduction
Hyenas
Internal Fertilization in Birds and Mammals
Intersexuality in Mammals
Lactation, Nonhuman
Magnocellular System
Male Reproductive System, Nonhuman Mammals
Mammary Gland, Overview
Marsupials
Mating Behaviors, Mammals
Microtinae (Voles)
Monotremes
Naked Mole-Rats
Opossums
Ovarian Cycle, Mammals
Parental Behavior, Mammals
Parturition, Nonhuman Mammals
Pheromones, Mammals
Placental Steroidogenesis in Primate Pregnancy
Pregnancy in Other Mammals
Pregnancy, Maintenance of
Pregnancy, Maternal Recognition of
Primates, Nonhuman
Progesterone Actions on Behavior
Protein Hormones of Primate Pregnancy
Puberty Acceleration
Puberty, in Nonhuman Primates
Puberty, in Nonprimate Mammals
Rabbits
Reflex (Induced) Ovulation
Relaxin, Mammalian
Reproductive Senescence, Nonhuman Mammals
Rodentia
Seals
Seasonal Reproduction, Mammals
Sex Ratios
Territorial Behavior, Overview
Uterus, Nonhuman
Whales and Porpoises
Whitten Effect

AVIAN REPRODUCTION

Altricial and Precocial Development in Birds
Altruistic Behavior, Vertebrates
Androgens, Effects in Birds
Avian Reproduction, Overview
Avian Reproductive System, Developmental Endocrinology
Birds, Diversity of
Brood Parasitism in Birds
Chickens, Control of Reproduction in
Corpora Lutea of Nonmammalian Species
Egg, Avian
Energetics of Reproduction
Estrogen, Effects in Birds

Female Reproductive System, Birds
Internal Fertilization, Birds and Mammals
Interrenal Gland, Stress Response and Reproduction
Lactotrophs
Male Reproductive System, Birds
Mate Choice, Overview
Migration, Birds
Molt and Nuptial Color
Nesting, Birds
Ovarian Cycles and Follicle Development in Birds
Parental Behavior, Birds
Photoperiodism, Vertebrates
Pigeons
Pineal Gland, Melatonin Biosynthesis and Secretion
Prolactin, Actions of
Prolactin, in Nonmammalian Vertebrates
Relaxin, Nonmammalian
Seasonal Reproduction, Birds
Sex Differentiation in Amphibians, Reptiles, and Birds, Hormonal Regulation
Sex Ratios
Sexual Imprinting
SHBG (Sex Hormone-Binding Globulin)
Songbirds and Singing
Territorial Behavior, Overview
Viviparity and Oviparity: Evolution and Reproductive Strategies

REPTILES AND AMPHIBIA

Amphibian Ovarian Cycles
Amphibian Reproduction, Overview
Androgens, Subavian Species
Corpora Lutea of Nonmammalian Species
Energetics of Reproduction
Estrogen Effects and Receptors, Subavian Species
Female Reproductive System, Amphibians
Female Reproductive System, Reptiles
Hormonal Control of the Reproductive Tract, Subavian Vertebrates
Interrenal Gland, Stress Response and Reproduction
Leydig and Sertoli Cells, Nonmammalian
Male Reproductive System, Amphibians
Male Reproductive System, Reptiles

Mate Choice, Overview
Migration, Amphibians
Migration, Reptiles
Oogenesis, in Nonmammalian Vertebrates
Photoperiodism, Vertebrates
Placenta and Placental Analogs in Reptiles and Amphibians
Progesterone Effects and Receptors, Subavian Species
Prolactin, in Nonmammalian Vertebrates
Reptilian Reproduction, Overview
Reptilian Reproductive Cycles
Sex Differentiation in Amphibians, Reptiles, and Birds, Hormonal Regulation
Spermatogenesis, in Nonmammals
Territorial Behavior, Overview
Thyroid Hormones, in Subavian Vertebrates
Vitellogenins and Vitellogenesis
Viviparity and Oviparity: Evolution and Reproductive Strategies
Vomeronasal Organ

FISH, ELASMOBRANCHII, AND CYCLOSTOMES

Agnatha
Androgens, Subavian Species
Cephalochordata
Corpora Lutea of Nonmammalian Species
Elasmobranch Reproduction
Energetics of Reproduction
Estrogen Effects and Receptors, Subavian Species
Female Reproductive System, Fish
Fish, Modes of Reproduction
Hormones and Reproductive Behaviors, Fish
Interrenal Gland, Stress Response and Reproduction
Leydig and Sertoli Cells, Nonmammalian
Male Reproductive System, Fish
Mate Choice, Overview
Migration, Fish
Oogenesis, in Nonmammalian Vertebrates
Ovarian Cycle, Teleost Fish
Pheromones, Fish
Photoperiodism, Vertebrates
Pituitary Gland, in Fish
Placenta and Placental Analogs in Elasmobranchs

Progesterone Effects and Receptors, Subavian Species
Prolactin, in Nonmammalian Vertebrates
Relaxin, Nonmammalian
Rhythms, Lunar and Tidal
Seasonal Reproduction, Fish
Sex Determination, Environmental
Spermatogenetic Cycle in Fish
Teleosts, Viviparity
Thyroid Hormones, in Subavian Vertebrates
Vitellogenins and Vitellogenesis
Viviparity and Oviparity: Evolution and Reproductive Strategies

INVERTEBRATES

Acanthocephala
Aedes aegypti
Allatostatins
Altruism in Insect Reproduction
Annelida
Aphids
Aplysia
Asexual Reproduction
Brachiopoda
Bryozoa (Ectoprocta)
Caenorhabditis elegans
Chaetognatha
Chelicerate Arthropods
Cnidaria
Conjugation in Ciliates
Corpus Allatum
Corpus Cardiacum, Insects
Crustacea
Ctenophora
Cycliophora
Diapause
Diploptera punctata
Dorsal Bodies in Mollusca
Drosophila
Ecdysiotropins and Ecdysiostatins
Ecdysteroids
Echinodermata
Egg Coverings, Insects
Female Reproductive System, Insects
Gnathostomulida
Hemichordata
Hemocoelic Insemination
Hermaphroditism
Hybridization
Hydra
Insect Accessory Glands
Insect Reproduction, Overview
Juvenile Hormone
Kamptozoa (Entoprocta)
Kinorhyncha
Locusts
Male Reproductive System, Insects
Marine Invertebrate Larvae
Marine Invertebrates, Modes of Reproduction in
Mating Behaviors, Insects
Mating Behaviors, Invertebrates Other Than Insects
Metamorphosis, Insects
Migration, Insects
Mollusca
Myriapoda
Nematodes and Related Phyla
Nematomorpha
Nemertea
Onychophora
Oocyte Maturation and Spawning in Starfish
Oostatins, Folliculostatins, and Antigonadotropins, Insects
Orthonectida
Oviposition in Molluscs
Ovulation and Oviposition, Insects
Parasites and Reproduction
Parasitoids
Paraspermatozoa
Parental Behavior, Arthropods
Parthenogenesis and Natural Clones
Pheromones, Insects
Phoronida
Placozoa
Platyhelminthes
Poecilogony
Porifera
Priapulida
Protozoa
Rhodnius prolixus
Rhombozoa
Rhythms, Lunar and Tidal
Rotifera
Seasonal Reproduction, Marine Invertebrates

Sea Urchins
Sipuncula
Social Insects, Overview
Song in Arthropods
Spawning, Marine Invertebrates
Sperm Activation, Arthropods
Spermatophores in the Arthropods
Sperm Transport, Arthropods
Symbiosis
Tardigrada
Tsetse Flies
Tunicata (Urochordata)
Yolk Proteins, Invertebrates

REPRODUCTIVE BEHAVIOR

Aggressive Behavior
Altruism in Insect Reproduction
Altruistic Behavior, Vertebrates
Androgens, Effects in Birds
Androgens, Effects in Mammals
Autonomic Nervous System and Reproduction
Brood Parasitism in Birds
Captive Breeding of Wildlife
Circadian Rhythms
Circannual Rhythms
Copulation, Mammals
Critical Period, Estrous Cycle
Endogenous Opioids
Energy Balance, Effects on Reproduction
Estrogen Action, Behavior
Estrogen, Effects in Birds
Estrous Cycle
Estrus
Fertility and Fecundity
Global Zones and Reproduction
Homosexuality
Hormones and Reproductive Behaviors, Fish
Intersexuality in Mammals
Lordosis
Mate Choice, Overview
Mating Behaviors, Insects
Mating Behaviors, Invertebrates Other Than Insects
Mating Behaviors, Mammals
Migration, Amphibians

Migration, Birds
Migration, Fish
Migration, Insects
Migration, Reptiles
Nesting, Birds
Nutritional Factors and Reproduction
Olfaction and Reproduction
Orgasm
Parasitoids
Parental Behavior, Arthropods
Parental Behavior, Birds
Parental Behavior, Mammals
Pheromones, Fish
Pheromones, Insects
Pheromones, Mammals
Photoperiodism, Vertebrates
Pineal Gland, Melatonin Biosynthesis and Secretion
Progesterone Actions on Behavior
Reproductive Senescence, Human
Reproductive Senescence, Nonhuman
Reptilian Reproductive Cycles
Rhythms, Lunar and Tidal
Seasonal Reproduction, Birds
Seasonal Reproduction, Fish
Seasonal Reproduction, Mammals
Seasonal Reproduction, Marine Invertebrates
Sex Differentiation, Psychological
Sexual Attractants
Sexual Imprinting
Sexual Selection
Song in Arthropods
Songbirds and Singing
Spawning, Marine Invertebrates
Stress and Reproduction
Suckling Behavior
Territorial Behavior, Overview
Ultradian Hormone Rhythms
Vomeronasal Organ

FEMALE REPRODUCTIVE SYSTEMS

Activin and Activin Receptors
Amenorrhea
Antiestrogens
Antiprogestins

Breast Cancer
Breast Disorders
Cervical Cancer
Cervix
Circumcision
Clitoris
Contraceptive Methods and Devices, Female
Corpus Luteum (CL)
Corpus Luteum Peptides
Endometriosis
Endometrium
Environmental Estrogens
Estrogen Action, Behavior
Estrogen Action, Bone
Estrogen Action, Breast
Estrogen Action on the Female Reproductive Tract
Estrogen Replacement Therapy
Estrogen Secretion, Regulation of
Estrogens, Overview
Eunuchoidism
Fallopian Tube
Female Reproductive Disorders, Overview
Female Reproductive System, Amphibians
Female Reproductive System, Birds
Female Reproductive System, Fish
Female Reproductive System, Humans
Female Reproductive System, Insects
Female Reproductive System, Nonhuman Mammals
Female Reproductive System, Reptiles
Female Sterilization
Follicular Atresia
Follicular Development
Follicular Steroidogenesis
Genitalia
GnRH (Gonadotropin-Releasing Hormone)
GnRH Pulse Generator
Gonadogenesis, Female
Graafian Follicle
Hormonal Contraception
Hysterectomy
IGF (Insulin-like Growth Factor)
Infertility
Inhibin
Interferons
Leukemia Inhibitory Factor
Local Control Systems in Reproduction

Luteinization
Luteolysis
Luteotropic Hormones
Menarche
Menopause
Menstrual Cycle
Menstrual Disorders
Menstruation
Neurohypophysial Hormones
Osteoporosis
Ovarian Cancer
Ovarian Cycle, Mammals
Ovarian Function in the Perimenopause
Ovarian Hormones, Overview
Ovarian Innervation
Ovary, Overview
Ovulation
Oxytocics
Pelvic Inflammatory Disease (PID)
PMS (Premenstrual Syndrome)
Polycystic Ovary Syndrome
Progesterone Actions on Behavior
Progesterone Actions on the Reproductive Tract
Progestins
Receptors for Hormones, Overview
Relaxin, Mammalian
Theca Cells
Theca Cell Tumors
Trophoblast to Human Placenta
Tubal Surgery
Tumors of the Female Reproductive System
Turner's Syndrome
Ullrich Syndrome
Uterine Anomalies
Uterus, Human
Uterus, Nonhuman
Vagina

MALE REPRODUCTIVE SYSTEMS

Adrenal Androgens
Androgen Inhibitors/Antiandrogens
Androgen Insensitivity Syndromes
Androgens
Andrology: Origins and Scope
Aromatization

Benign Prostatic Hyperplasia (BPH)
Blood-Testis Barrier
Castration, Effects in Humans (Male)
Castration, Effects in Nonhumans (Male)
Circumcision
Contraceptive Methods and Devices, Male
Cryopreservation of Sperm
Cryptorchidism
Dihydrotestosterone
Ejaculation
Epididymis
Erection
Eunuchoidism
Gene Transfer, Sperm-Mediated
Genitalia
Gonadogenesis, Male
Gynecomastia
Hypospadias
Impotence
Infertility
Leydig Cells
Male Reproductive Disorders
Male Reproductive System, Amphibians
Male Reproductive System, Birds
Male Reproductive System, Fish
Male Reproductive System, Human
Male Reproductive System, Insects
Male Reproductive System, Nonhuman Mammals
Male Reproductive System, Reptiles
Orchitis
Pampiniform Plexus
Pelvic Nerve
Penis
Prostate Cancer
Prostate Gland
Prostate-Specific Antigen
Regulation of Sertoli Cells
Semen
Seminal Vesicles
Sertoli Cells, Function
Sertoli Cells, Overview
Spermatogenesis, Overview
Spermatogenesis, Disorders of
Spermatogenesis, Hormonal Control of
Spermatozoa
Sperm Capacitation
Spermiogenesis
Sperm Transport
Temperature, Effects on Testicular Function
Testicular Cancer
Testicular Developmental Anomalies
Testis, Overview
Testosterone Biosynthesis
Varicocele
Vasectomy

PREGNANCY

Abortion
Allantoic Fluid
Amniocentesis
Cardiovascular Adaptation to Pregnancy
CBG (Corticosteroid-Binding Globulin)
Cesarean Delivery
Choriocarcinoma
Corpus Luteum of Pregnancy
Cotyledonary Placenta
Cryopreservation of Embryos
Decidua
Deciduoma
Discoidal Placenta
Ectopic Pregnancy
Endotheliochorial Placentation
Epitheliochorial Placentation
Erythroblastosis Fetalis
Family Planning
Fetal Adrenals
Fetal Alcohol Syndrome
Fetal Anomalies
Fetal Diagnosis, Invasive
Fetal Growth and Development
Fetal Hormones
Fetal Lung Development
Fetal Membranes
Fetal Monitoring and Testing
Fetal–Placental Unit
Fetal Surgery
α-Fetoprotein and Triple Screening
Fetus, Overview
Growth Factors
Hemochorial Placentation
Hormones of Pregnancy
Infections in Pregnancy

Intrauterine Growth Restriction and Mechanisms
  of Fetal Growth
In Vitro Fertilization
Labor and Delivery, Human
Morning Sickness and Hyperemesis Gravidarum
Obstetric Anesthesia
Parturition, Nonhuman Mammals
Pelvimetry
Placenta: Implantation and Development
Placental and Decidual Protein Hormones, Human
Placental Gas Exchange
Placental Lactogens
Placental Nutrient Transport
Placental Steroidogenesis in Primate Pregnancy
Postdate (Postterm) Pregnancy
Postpartum Depression
Preeclampsia/Eclampsia
Pregnancy in Dogs and Cats
Pregnancy in Farm Animals
Pregnancy in Humans, Overview
Pregnancy in Other Mammals
Pregnancy, Maintenance of
Pregnancy, Maternal Recognition of
Pregnancy, Metabolic Changes in
Prenatal Genetic Screening
Preterm Labor and Delivery
Protein Hormones of Primate Pregnancy
Pseudocyesis
Pseudopregnancy
Puerperal Infections
Puerperium
Substance Abuse and Pregnancy
Surfactant
Teratogens
Tocolytic Agents
Twinning
Ultrasound
Umbilical Cord
Uterine Contraction

LACTATION

Breast Disorders
Breastfeeding
Cattle
Galactorrhea
Human Placental Lactogen (Human Chorionic
  Somatomammotropin)
Lactational Amenorrhea
Lactational Anestrus
Lactation, Human
Lactation, Nonhuman
Lactogenesis
Mammary Gland Development
Mammary Gland, Overview
Milk, Composition and Synthesis
Milk Ejection
Nutritional Factors and Lactation
Oxytocin
Placental Lactogens
Pregnancy in Humans, Overview
Prolactin, Overview
Prolactin, Actions of
Prolactin Secretion, Regulation of
Suckling Behavior

Preface

The publication of the *Encyclopedia of Reproduction* comes at a most opportune time. Hardly a day goes by when the news media do not report some new dimension in the treatment of infertility or, conversely, controversies associated with the control of fertility and the ethical issues raised by both. Organismal cloning is a matter of constant debate, and the pharmacological correction of erectile dysfunction has become a preoccupation of international dimensions. Procreation remains a subject of universal interest to every segment of society, from scientists to students, from science reporters to the proverbial person on the street.

The present work should serve as a convenient and comprehensive source of information encompassing all aspects of the subject of reproduction as it relates to the entire animal kingdom. It should be as useful to the expert exploring reproductive phenomena outside his or her own field as it is to students and to the educated public at large. Topics for inclusion were initially generated by forming a matrix of systems (gametes, fertilization, and early embryogenesis; reproductive behavior; female reproductive systems; male reproductive systems; pregnancy; and lactation) and of groups of animals (humans and experimental primates; domestic animals; mammals; birds; reptiles and amphibia; fish, elasmobranchii, and cyclostomes; and invertebrates).

A group of outstanding Associate Editors having expertise in one of more of these areas was then recruited. The preliminary list of entries prepared by the Editors was refined and expanded at a meeting with these Associate Editors, who then identified the appropriate authors. Manuscripts were critically reviewed by the Associate Editors and finally scrutinized by us and the editorial staff at Academic Press.

In a work of this kind, errors of omission and of commission are inevitable and we should appreciate having them called to our attention for correction in possible future editions. The 542 entries constituting the work each contain a glossary of terms, a summary introduction, cross-references to related articles, and a reading list. A standard subject index and an index of reproductive systems and zoological groupings are provided.

Each entry was written to be self-contained, inevitably leading to some overlap of content. We do not view this as a weakness, but instead believe that it will facilitate a reader's search for information by reducing the number of entries that have to be consulted.

The completion of this project demanded the best efforts of a large number of participants. Chief among them are the 700 authors, especially those who wrote articles on short notice so that the publication deadline could be met. The stellar group of 15 Associate Editors, each of whom possesses great breadth of knowledge and who, as a group, span the spectrum of expertise from zoology to animal husbandry to obstetrics and gynecology, also rendered exceptional service.

Finally, we acknowledge the indispensable contributions of the staff at Academic Press: Jasna Markovac, Editor-in-Chief for Biomedical Science, who originally conceived of the Encyclopedia; and Chris Morris, Gail Rice, and Erika Conner, Major Reference Works editors who provided ongoing management of the project.

Ernst Knobil
Jimmy D. Neill

Guide to the Encyclopedia

ORGANIZATION

The *Encyclopedia of Reproduction* is organized to provide the maximum ease of use for its readers. All of the articles are arranged in a single alphabetical sequence by title. Articles whose titles begin with the letters A to En are in Volume 1, articles with titles from Ep through L are in Volume 2, then M through Pri in Volume 3, and Pro to Z in Volume 4.

Volume 4 also includes a complete subject index for the entire work, an alphabetical list of the contributors to the encyclopedia, and a glossary of key terms used in the articles.

Article titles generally begin with the key noun or noun phrase indicating the topic, with any descriptive terms following. For example, "Uterus, Human" is the article title rather than "Human Uterus," and "Migration, Birds" is the title rather than "Bird Migration." This is done so that the same phenomenon or feature can be studied across various groups. For example, all the articles on female reproductive systems in humans, other mammals, birds, etc., appear in one sequence in the Fe- section of the encyclopedia.

INDEX

The Subject Index in Volume 4 contains more than 20,000 entries. The subjects are listed alphabetically and indicate the volume and page number where information on this topic can be found. In addition, the Table of Contents by Subject Area also functions as an index, since it lists all the topics covered in a given area; e.g., the encyclopedia includes 90 different articles dealing with reproduction in invertebrates.

OUTLINE

Each entry in the Encyclopedia begins with a topical outline that indicates the general content of the article. This outline serves two functions. First, it provides a brief preview of the article, so that the reader can get a sense of what is contained there without having to leaf through the pages. Second, it highlights important subtopics that will be discussed within the article. For example, the article "Fallopian Tube" includes subtopics such as "Tubal Disorders," "Tubal Sterilization," and "Assisted Reproductive Techniques Involving the Tube."

The outline is intended as an overview and thus it lists only the major headings of the article. In addition, extensive second-level and third-level headings will be found within the article.

GLOSSARY

The Glossary section contains terms that are important to an understanding of the article and that may be unfamiliar to the reader. Each term is defined in the context of the article in which it is used. Thus the same term may appear as a glossary entry in two or more articles, with the details of the definition varying slightly from one article to another. The encyclopedia has approximately 4,250 glossary entries.

In addition, Volume 4 provides a comprehensive glossary that collects all the core vocabulary of reproductive biology in one A-Z list. This section can be consulted for definitions of terms not found in the individual glossary for a given article.

DEFINING STATEMENT

The text of each article in the encyclopedia begins with an introductory paragraph that defines the topic under discussion and summarizes the content of the article. For example, the article "Energetics in Reproduction" begins with the following statement:

Energetics of reproduction is defined as the amount of energy that an animal expends to reproduce. The energetics of reproduction can include the costs of gamete manufacture, synthesis of secondary sexual characteristics and sex-attractant chemicals (pheromones), and reproductive behavior including territorial defense, nest building, courtship rituals, and parental care. . . .

CROSS-REFERENCES

Almost all articles in the Encyclopedia have cross-references to other articles. These cross-references appear at the conclusion of the article text. They indicate articles that can be consulted for further information on the same topic or for other information on a related topic. For example, the article "Osteoporosis" contains references to the articles "Estrogen Replacement Therapy" and "Menopause."

BIBLIOGRAPHY

The Bibliography section appears as the last element in an article. The reference sources listed there are the authors' recommendations of the most appropriate materials for further research on the given topic. The bibliography entries are for the benefit of the reader and thus they do not represent a complete listing of all the materials consulted by the author in preparing the article.

COMPANION WORKS

The *Encyclopedia of Reproduction* is one of a series of multivolume reference works in the life sciences published by Academic Press. Other such titles include the *Encyclopedia of Human Biology, Encyclopedia of Cancer, Encyclopedia of Toxicology, Encyclopedia of Immunology,* and *Encyclopedia of Microbiology.*

Magnocellular System

R. John Bicknell

The Babraham Institute

I. Neuroanatomy of the Magnocellular System
II. Neurohormone Production and Secretion
III. Magnocellular Oxytocin Neurons
IV. Reproduction and Magnocellular Vasopressin Neurons

GLOSSARY

axon A thin tubular process which is the conducting unit of a neuron and branches to form synapses or secretory nerve endings.

hypothalamus The ventral forebrain region regulating autonomic, endocrine, and visceral integration.

neurohormone A chemical messenger produced in neurons and secreted from nerve endings into the circulation.

neurohypophysis The posterior, neural lobe of the pituitary gland.

neuron A nerve cell.

The magnocellular system, located in the hypothalamic region of the brain, comprises two types of distinctively large neurons producing the peptides oxytocin or vasopressin and secreting these into the systemic circulation from their axon terminals in the neurohypophysis. These oxytocin- and vasopressin-producing populations are independently regulated neuroendocrine systems serving a number of vital reproductive and other physiological functions.

I. NEUROANATOMY OF THE MAGNOCELLULAR SYSTEM

A. Neuronal Cell Bodies

The two magnocellular phenotypes are principally located together in the bilateral supraoptic and paraventricular nuclei of the hypothalamus. The supraoptic nuclei lie at the base of the brain above the lateral edges of the optic chiasm and contain only magnocellular neurons with partial segregation of oxytocin neurons into the more dorsal aspects of the nuclei in rats. The paraventricular nuclei lie on either side of the middorsal third cerebral ventricle and include subdivisions of either intermingled or relatively segregated populations of magnocellular oxytocin and vasopressin neurons. The paraventricular nuclei also contain several other phenotypes including smaller, parvocellular neurons expressing oxytocin, which project to autonomic centers in the brain stem, and others expressing vasopressin which project to the median eminence and, together with corticotrophin releasing-hormone, control adrenocorticotrophin secretion from the anterior pituitary.

A substantial minority of magnocellular neurons are also located in accessory nuclei (e.g., nucleus circularis) outside the paraventricular and supraoptic nuclei. Irrespective of location, the axons of all magnocellular neurons project medially and ventrally through the internal zone of the median eminence and the infundibular stalk to terminate within the neurohypophysis.

B. Neurohypophysis

Each magnocellular axon in the neurohypophysis gives rise to multiple secretory terminals which lie in close apposition to the pericapillary basal lamina where their secreted neurohormones rapidly diffuse into the vasculature. In addition, axons have large swellings, not adjacent to the capillaries, which may represent neurohormone storage or degradation sites. Pituicytes, which are astrocyte-like glial cells, extend ramifying processes which enwrap axons and secretory terminals and can juxtapose between terminals and the capillary basal lamina. The degree of pituicyte enwrapment and basal lamina coverage is reduced in states of high neurohormone output.

II. NEUROHORMONE PRODUCTION AND SECRETION

Both oxytocin and vasopressin are nine-amino acid, cysteine-bridged peptides which are derived by proteolytic cleavage from similar larger precursor polypeptides. The oxytocin and vasopressin genes encoding these precursors are closely linked, being separated by a short intergenic region, and are transcribed in opposite directions from the alternate DNA strands. The active neurohormones together with the other precursor cleavage products are packaged into dense-core vesicles which are actively transported along the axons to the neurohypophysis. Discharge of secretory vesicle contents occurs by exocytosis, triggered primarily by a rise in cytosolic calcium ion concentration within the nerve terminals following invasion of the terminal membrane by depolarizing action potentials generated at the magnocellular cell body.

Patterning of magnocellular neuron electrical activity critically influences the amount of neurohormone secreted per action potential since the release process is subject both to facilitation with increasing action potential frequency and to fatigue during sustained trains of action potentials. Thus, the characteristic 1- to 3-sec high-frequency burst of action potentials generated by oxytocin neurons during suckling greatly facilitates release efficiency and delivers a bolus of oxytocin into the circulation to stimulate milk ejection from the mammary glands.

III. MAGNOCELLULAR OXYTOCIN NEURONS

A. Parturition

The uterus expresses high levels of oxytocin receptor at the very end of pregnancy and oxytocin has long been used clinically to facilitate coordinated uterine contractions at labor. Blood oxytocin measurements and electrophysiological and gene expression studies in the magnocellular oxytocin neurons indicate that these neurons are only strongly activated to secrete in the expulsive phase of parturition. Around this time, antagonists at the oxytocin receptor will disrupt the progress of birth in rats and oxytocin will restore progress when magnocellular neuron activation is prevented by opiate analgesic treatment. Oxytocin secretion from magnocellular neurons is thus not considered to be a major determinant of the onset of parturition. However, in rats oxytocin antagonists can delay onset by several hours and such antagonists may find clinical use in restoring uterine quiescence in potential premature labor. Oxytocin is also expressed in a number of tissues of the reproductive tract, including the uterus, and local paracrine actions of oxytocin may play a role in parturition perhaps before activation of magnocellular secretion.

When the magnocellular oxytocin system does become activated to secrete at parturition, evidence from circulating oxytocin levels in women and in large animal species indicates that output is pulsatile as also suggested by the pattern of electrophysiological activation in rats at this time. The signals to oxytocin neurons that trigger their activation at parturition are ill defined but, in rats, involve spinal or vagal nerve pathways and activation of brain stem noradrenaline and other neurons projecting to the magnocellular system.

B. Lactation

Postpartum, the suckling stimulus of the young activates a reflex pathway to the magnocellular oxytocin neurons, releasing oxytocin, with consequent contraction of the mammary gland myoepithelia and milk ejection. In rats, and similarly in pigs, the continuous suckling stimulus results in a discrete milk ejection event every 5–15 min. Whereas the spinal nerve relay is characterized, the location and nature of the afferent relay pathways directly to the oxytocin neurons are unknown. Characteristically, all the magnocellular oxytocin neurons in the supraoptic and paraventricular nuclei are electrically activated simultaneously during the reflex, accounting for the bolus of oxytocin released into the circulation.

Essential components of this milk-ejection reflex are thought to be common in most mammals but characteristics may vary with suckling behavior. In women, for example, the sight and sound of the young can trigger the reflex without the need for the physical suckling stimulus. The critical role for magnocellular oxytocin secretion in the milk-ejection reflex is demonstrated by its blockade with oxytocin antagonists and the lack of milk transfer to the young in transgenic mice with a disrupted oxytocin gene.

C. Other Functions of the Magnocellular Oxytocin System

In some species oxytocin exerts synergistic actions with vasopressin in the kidney to promote natriuresis. Thus, in the rat, magnocellular oxytocin neurons are activated along with vasopressin neurons by elevations in plasma osmolality. Oxytocin, acting in the brain and released from parvocellular neurons which do not project to the neurohypophysis, is also implicated in a variety of functions in animals, including pair bonding, sexual behavior, maternal behavior, penile erection, and ingestive behavior. It should be emphasized that these non-neuroendocrine oxytocin neural systems are regulated independently, although perhaps coordinately, with the magnocellular oxytocin neurons. It is also evident that the oxytocin peptide is used as a messenger molecule at multiple levels of the reproductive axis.

D. Plasticity of the Magnocellular Oxytocin System during the Reproductive Cycle

Not surprisingly, oxytocin gene expression and biosynthesis are upregulated during the pregnancy–lactation cycle with maximum oxytocin mRNA levels seen just prior to parturition. The content of processed oxytocin peptide in the neurohypophysis of rats is expanded by up to 50% during pregnancy and all of this additional accumulated oxytocin is secreted over the 1 or 2 hr of parturition. The 5′-flanking region of the rat oxytocin gene contains a consensus estrogen response element through which estrogen receptor will induce transcription, suggesting that the elevated estrogen levels in late pregnancy might directly regulate oxytocin biosynthesis. While the magnocellular oxytocin neurons in the rat do not express detectable estrogen receptor-α or progesterone receptor, they may express the more recently identified estrogen receptor-β. The actions of sex steroids on magnocellular oxytocin biosynthesis are thus likely to be mediated directly via estrogen receptor-β, nongenomically, or indirectly via steroid-receptive neurons with afferent inputs to the oxytocin neurons.

In addition, oxytocin neurons co-express opioid peptides which autoinhibit oxytocin secretion via receptors located on the secretory nerve terminals. This restraining system is upregulated through most of pregnancy and may participate in the buildup of neurohypophysial oxytocin content. As parturition approaches, the autoinhibitory restraint desensitizes but premature oxytocin secretion is prevented by induction of inhibitory mechanisms, again involving opioids, acting at the level of the oxytocin cell bodies.

During the pregnancy–lactation cycle in rats, reversible morphological changes also occur in oxytocin neurons, in their synaptic inputs, and in their relationship with glial cells. Thus, magnocellular oxytocin cell bodies hypertrophy and show other indices of enhanced biosynthesis during lactation, their glial coverage is reduced and neuronal apposition increased, and they are contacted by an increased number of synapses. The synaptic remodeling involves both γ-amino butyric acid and glutamate-containing synapses—the major fast inhibitory and excitatory transmitters in the brain. In addition, the numbers of oxytocin neurons contacted by shared synaptic structures (double synapses) are increased. Within the neurohypophysis, glial coverage of the capillary basal lamina is reduced and coverage by secretory nerve terminals is correspondingly increased during lactation. These morphological changes during lactation affect oxytocin but not neighboring vasopressin neurons and all revert to the nonpregnant condition following weaning. Oxytocin itself, secreted from magnocellular dendritic processes or elsewhere, plays a role both in inducing the morphological changes and in enhancing the milk-ejection reflex. It is considered that this remarkable plasticity in the magnocellular system occurs to enhance or enable synchronous activation of oxytocin neurons during the milk-ejection reflex and maximize the efficient secretion of oxytocin into the circulation.

IV. REPRODUCTION AND MAGNOCELLULAR VASOPRESSIN NEURONS

The primary neurohormonal function of vasopressin is to maintain body water balance through its actions in the kidney. Vasopressin also mediates vasoconstriction, in particular vascular beds, in response to acute hypotension. The plasma osmotic threshold for induction of vasopressin (and, in rats, oxytocin) release is reset in pregnancy with the expansion in blood volume and reduced plasma sodium and osmolality. Compared to oxytocin, there may be a relatively minor activation of vasopressin secretion from magnocellular neurons during parturition, although it should be noted that uterine oxytocin receptors in some species are also capable of binding vasopressin.

See Also the Following Articles

MILK-EJECTION; OXYTOCIN; PARTURITION, NONHUMAN MAMMALS

Bibliography

Bicknell, R. J. (1988). Optimising release from peptide hormone secretory nerve terminals. *J. Exp. Biol.* **139**, 51–65.

Fuchs, A., Romero, R., Keefe, D., Parra, M., Oyarzun, E., and Behnke, E. (1991). Oxytocin secretion and human parturition: Pulse frequency and duration increase during spontaneous labour in women. *Am. J. Obstet. Gynecol.* **165**, 1515–1523.

Higuchi, T., Honda, K., Fukuoka, T., Negoro, H., and Wakabayashi, K. (1984). Release of oxytocin during suckling and parturition in the rat. *J. Endocrinol.* **105**, 339–346.

Ivell, R., and Burbach, J. P. (1991). The molecular biology of vasopressin and oxytocin genes. *J. Neuroendocrinol.* **3**, 583–586.

Ivell, R., and Russell, J. A. (Eds.) (1995). *Oxytocin. Cellular and Molecular Approaches in Medicine and Research*. Plenum, New York.

Renaud, L. P., and Bourque, C. W. (1991). Neurophysiology and neuropharmacology of hypothalamic magnocellular neurons secreting vasopressin and oxytocin. *Prog. Neurobiol.* **36**, 131–169.

Russell, J. A., Leng, G., and Bicknell, R. J. (1995). Opioid tolerance and dependence in the magnocellular oxytocin system: A physiological mechanism? *Exp. Physiol.* **80**, 307–340.

Scott Young, W., III, Shepard, E., Amico, J., Hennighausen, L., Wagner, K.-U., Lamarca, M. E., McKinney, C., and Ginns, E. I. (1996). Deficiency in mouse oxytocin prevents milk ejection but not fertility or parturition. *J. Neuroendocrinol.* **8**, 847–853.

Sofroniew, M. V. (1983). Morphology of vasopressin and oxytocin neurones and their central and vascular projections. *Prog. Brain Res.* **60**, 101–114.

Theodosis, D. T., and Poulain, D. A. (1987). Oxytocin-secreting neurones: A physiological model for structural plasticity in the adult mammalian brain. *Trends Neurosci.* **10**, 426–430.

Male Contraceptives
see Contraceptive Methods and Devices, Male

Male Reproductive Disorders

Ivan Damjanov

University of Kansas School of Medicine

I. Classification of Infertility in Males
II. Pathogenesis of Infertility in Males
III. Clinical Evaluation of Infertility in Males

Glossary

androgen insensitivity syndrome A genetic disorder in which the tissues lack androgen receptors and do not respond to testosterone. It is also called *testicular feminization syndrome* because the affected persons have cryptorchid abdominal testes but otherwise appear female. Genetically they are male (46,XY).

azoospermia The absence of spermatozoa in the ejaculate due to faulty spermatogenesis or obstruction of efferent seminal ducts.

cryptorchidism Undescended testes which, instead of being in the scrotum, are retained during development in the inguinal canal or abdominal cavity.

epididymoorchitis Inflammation of the testis and epididymis representing usually a complication of a sexually transmitted disease, such as gonorrhea or syphilis.

hypogonadotropic hypogonadism Hypoplasia or atrophy of the testis due to a lack of gonadotropins, related to hypothalamic or pituitary lesions that decrease gonadotropin production.

impotence Inability to have sexual intercourse because of inadequate erection of the penis.

infertility Failure to conceive after 6–12 months of regular sexual intercourse without contraception.

Klinefelter syndrome A developmental disorder characterized by a trisomy of sex chromosomes (47,XXY), testicular atrophy, and azoospermia.

oligospermia Low sperm count, by convention defined as fewer than 20 million spermatozoa per milliliter of ejaculate.

semen analysis Microscopic and/or biochemical examination of semen ejaculated by masturbation or electrical stimulation.

spermatogenic arrest Incomplete maturation of seminiferous epithelium in the testicular tubules resulting in azoospermia.

testicular biopsy A surgical procedure in which a sample of testis is obtained for histologic examination.

Infertility, defined as an inability to conceive after having tried for 6–12 months of regular sexual intercourse, affects one of every six couples in the United States. In approximately one-third of these cases the cause lies with the male, in one-third of cases with the female, and in the remaining one-third both the male and the female.

I. CLASSIFICATION OF INFERTILITY IN MALES

Male infertility can be classified for clinical purposes as pretesticular, testicular, or posttesticular (Table 1). Pretesticular causes include a variety of endocrine disorders, systemic diseases, and drugs, all of which may adversely affect spermatogenesis. Testicular causes include genetic and developmental disorders, such as Klinefelter syndrome, Noonan syndrome, and cryptorchidism. Infections of the testis, cytotoxic drugs, toxins, and radiation therapy also may damage or destroy the seminiferous epithelium and inhibit sperm formation. Infertility may result from a loss of testes due to tumors, trauma, and surgery. Varicocele, a dilatation of spermatic venous plexus, is associated with infertility and, although not fully understood, it is considered to be the most common cause of testicular infertility. Posttesticular causes of infertility include anatomic obstructions of the seminal excretory ducts or func-

TABLE 1
Clinical Classification of
Infertility in Males

- Pretesticular causes
  - Endocrine disorders
    - Hypothalamic–pituitary disorders
    - Adrenal disorders
    - Thyroid disorders
    - Adrenal disorders
  - Systemic diseases
    - Liver disease
    - Kidney disease
    - Neoplasia
  - Drugs
- Testicular causes
  - Developmental and genetic disorders
    - Klinefelter syndrome
  - Infections
  - Drugs
  - Physical injury
    - Radiation
    - Trauma
    - Surgery
  - Varicocele
  - Idiopathic spermatogenic defects
- Posttesticular causes
  - Congenital anomalies of excretory ducts
  - Infections
  - Surgery
  - Trauma

tional disturbances of ejaculation. Obstructive lesions are either developmental or secondary to infections, surgery, or trauma. Functional disturbances of ejaculation, such as retrograde ejaculation, may be related to a loss of autonomic innervation in diabetics or due to surgery. Paraplegia caused by spinal cord injury is associated with both erectile and ejaculatory disturbances. Antibodies to sperm or seminal plasma may adversely affect the motility and/or the viability of spermatozoa.

II. PATHOGENESIS OF INFERTILITY IN MALES

Male fertility depends critically on the capacity of the male genital system to produce sperm in adequate amounts and to discharge it into the female genital system. If the male cannot perform his normal reproductive function, he is considered to be infertile or sterile. In many instances, infertility is only reduced and such men are considered to be subfertile. With the advances in reproductive technology, many of these concepts have been under revision. Accordingly, some of the prerequisites previously considered to be essential for male fertility have been modified. For example, bilateral agenesis of the epididymis or ductus deferens was an absolute obstacle to fertility in previous years but it can be bypassed today by aspirating the sperm from the distal portion of seminal ducts and using it for *in vitro* fertilization. Likewise, congenitally immotile spermatozoa were considered to be a sign of untreatable infertility until it became apparent that such spermatozoa can be mechanically injected into an ovum to fertilize *in vitro* the female gamete.

These considerations aside, the standard prerequisites for male fertility are still considered to be under the following pathogenetic categories:

Genetic
Developmental
Anatomic
Hormonal
Toxic/medicinal
Immune

These factors can affect spermatogenesis in more than one way and are often interrelated one to another.

A. Developmental/Genetic Causes of Infertility

The development of the male gonads and the remainder of the internal and external genital system are closely interrelated. Most importantly, they depend on the genetic background of the individual and the proper sequence of fetal morphogenetic events that occur during intrauterine life.

Male gonads develop only in individuals who have a normal Y sex chromosome. Typically, normal males have a 46,XY genotype. Those who have an additional X chromosome (47,XXY) develop into males with all the anatomic and clinical features of Klinefelter syndrome. This syndrome is characterized by eunuchoid body proportions, testicular atrophy, lack of secondary male characteristics, gynecomastia, and infertility. Individuals with an abnormal Y chromosome, i.e., Y chromosome missing parts of the sex-determining region, will not develop normally and will have partially or completely feminized external genital organs.

The male gonad develops from the fetal genital ridge that is invaded early in development by primordial germ cells originating in the yolk sac. In genetically male individuals, the fetal testes will form and begin secreting androgens which act on the primordia of the internal and external genital organs, directing their development into the male organs. Individuals who lack androgen receptors (e.g., androgen insensitivity syndrome or testicular feminization syndrome) or are unable to secrete normal androgens (e.g., Leydig cell aplasia, defective androgen synthesis due to 5α-reductase deficiency) do not form male genital organs.

Fetal testes develop inside the abdominal cavity. In the second half of intrauterine life, testes descend through the inguinal canal into the scrotum, and by the end of pregnancy, most males have intrascrotal testes. Approximately 5% of all newborns have one or both testes that are either outside the scrotum or are retractile into the inguinal canal. In most instances, these motile testes become fixed in the scrotum during infancy. By the end of the first year, approximately 0.5–1% of all boys have one or both testes outside their normal anatomic location in the scrotum. Such cryptorchid testes are often abnormal and bilateral cryptorchidism may be accompanied by infertility. Cryptorchid testes can be surgically positioned into the scrotum (orchidopexy) but even so they may be functionally defective.

The normal development of the testis is essential for spermatogenesis. Nevertheless, not all anatomically "normal" testes are able to produce sperm, indicating that spermatogenesis is regulated at several levels. Aspermatogenesis, sperm maturation arrest, and oligospermia may have genetic causes which are poorly understood.

The development of the epididymis, vas deferens, seminal vesicles, prostate, penis, and scrotum is dependent on fetal androgens. Inadequate supply of androgens or lack of androgen receptors will result in feminization and/or abnormal development of these male organs. The development of vas deferens is controlled by other genes as well, as demonstrated by the absence of vas deferens in cystic fibrosis, a common autosomal recessive genetic disorder.

B. Anatomic Causes of Infertility

Anatomic causes of infertility include all the previously mentioned developmental disorders and isolated developmental disorders such as anorchia (lack of testes) or ductular obstruction of rete testis and epididymis which occurs without identifiable causes. Penile disorders which often involve the penile urethra such as epispadia, opening of the urethra on the dorsal side of the penis, or hypospadia, opening of the urethra on the ventral side of the penis, are associated with infertility. Infections of the testis and epididymis (epididymoorchitis) may cause anatomic obstruction preventing the outflow of the sperm from the testis. Disruption of the erection or the ejaculatory reflex due to surgical lesions of the autonomic nervous system could cause impotence, infertility, or both. Diabetes and ischemia due to small blood vessel insufficiency are among the most common causes of reproductive dysfunction in the elderly. Loss of autonomic innervation could also result in reverse ejaculation in which the sperm flow is misdirected into the urinary bladder rather than into the penile urethra.

Varicocele, i.e., dilatation of the veins in the spermatic cord, is an important anatomic cause of infertility. Although it is not known how varicocele inhibits spermatogenesis, it was shown empirically that the surgical correction of the varicosities may improve infertility in the affected person. It was proposed that the abnormal blood flow through the dilated veins may raise the temperature in the testis and thus inhibit spermatogenesis, which normally requires low temperature for completion.

C. Hormonal Causes of Infertility

The maturation of the testes at puberty and their normal reproductive function depend on proper stimulation by the pituitary hormones, follicle-stimulating hormone (FSH) and luteinizing hormone

(LH). These gonadotropins are released from the anterior pituitary under the stimulation of hypothalamic gonadotropin-releasing hormone. Lesions of the hypothalamus or the pituitary (e.g., tumors, surgical injury, trauma, and meningitis) will result in hypogonadotropic hypogonadism and cessation of spermatogenesis. Kallmann syndrome is a rare hereditary X-linked disturbance in the development of hypothalamus characterized by hypogonadism and anosmia.

Other hormonal disturbances known to cause infertility are hypersecretion of prolactin from the pituitary, hypothyroidism or hyperthyroidism, adrenal hyperfunction or hypofunction, and diabetes. Exogenous steroid hormones may also adversely affect fertility.

D. Toxic/Medicinal Causes of Infertility

Spermatogenesis may be adversely affected by a number of environmental toxins, recreational substances, and drugs. It is self-evident that cytotoxic drugs or radiation destroy the rapidly dividing spermatogenic epithelium and thus inhibit spermatogensis. On the other hand, the adverse action of substances such as alcohol, marijuana, or drugs, such as nitrofurantoin, sulfasalazine, or monoamine oxidase inhibitors, is less understood. Withdrawal of these drugs may improve fertility indirectly, implying that they are the cause of infertility in some people. On the other hand, since male fertility may fluctuate seasonally, the alleged infertility effects of the exoge-

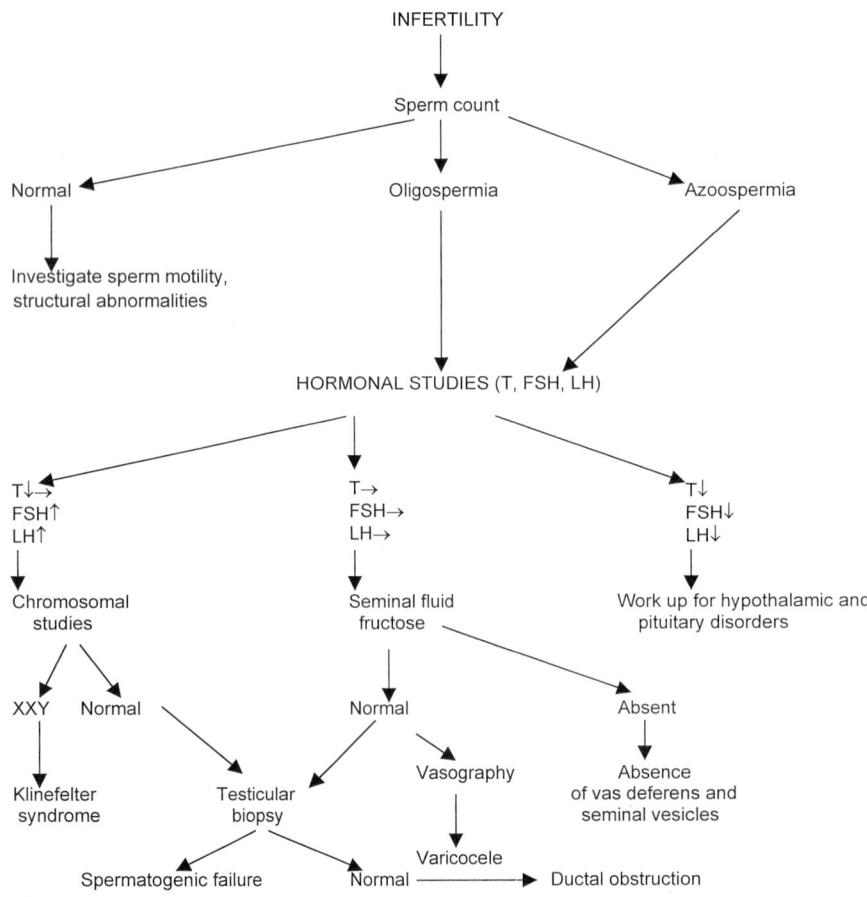

FIGURE 1 Algorithm for diagnosing male infertility. FSH, follicle-stimulating hormone; LH, luteinizing hormone; T, testosterone.

nous substances should be considered not fully documented.

E. Immune Causes of Infertility

A variety of immune mechanisms have been implicated in male infertility but their role remains obscure. Autoantibodies have been demonstrated against spermatozoa and seminal fluid in the serum of some men exhibiting low sperm motility. On the other hand, most vasectomized men have anti-sperm antibodies, which had no adverse effects on the fertility of those who underwent vas reanastomosis. Thus, the role of anti-sperm antibodies remains highly speculative. Even less clinical evidence exists for the role of T cell-mediated immunity, even though there are numerous animal studies suggesting that T cells could adversely affect male fertility.

III. CLINICAL EVALUATION OF INFERTILITY IN MALES

The clinical evaluation of infertility is complex and expensive and requires that both the male and the female partner be examined. These examinations are typically performed according to one of several well-established algorithms (Fig. 1). The evaluation of the male usually begins with a complete history and physical examination and includes basic laboratory tests to rule our an underlying systemic disease. Thereafter, it is customary to perform a sperm count followed by hormonal studies. The outcome of these tests will classify most cases into three categories: those with normal testosterone (T), FSH, and LH; those with low T, FSH, and LH; and those who have low or normal T and high FSH and LH. Testicular biopsy, chromosomal studies, and angiography (vasography) are indicated in cases that cannot be diagnosed during this triage. The correct diagnosis is essential for treatment. Despite major advances in the treatment of infertile couples, many cases of male infertility remain incurable.

See Also the Following Articles

CRYPTORCHIDISM; INFERTILITY; KLINEFELTER'S SYNDROME; VARICOCELE

Bibliography

Damjanov, I. (1993). *Pathology of Infertility*. Mosby, St. Louis, MO..

Dawson, C., and Whitfield, H. (1996). Subfertility and male sexual dysfunction. *Br. Med. J.* **312**, 902–905.

Fisch, H., and Lipshultz, L. I. (1992). Diagnosing male factors of infertility. *Arch. Pathol. Lab. Med.* **116**, 398–401.

Gondos, B., and Wong, T.-W. (1997). Non-neoplastic diseases of the testis and epididymis. In *Urological Pathology* (W. M. Murphy, Ed.), 2nd ed. Saunders, Philadelphia.

Kraus, F. T., Damjanov, I., and Kaufman, N. (Eds.) (1991). *Pathology of Reproductive Failure*. Williams & Wilkins, Baltimore.

Mak, V., and Jarvi, K. A. (1996). The genetics of male infertility. *J. Urol.* **156**, 1245–1257.

Yovich, J. J., and Matson, P. L. (1995). Male subfertility: Concepts in 1995. *Hum. Reprod.* **10**(Suppl. 1), 3–9.

Male Reproductive System, Amphibians

Riccardo Pierantoni

II Università degli Studi di Napoli

I. Introduction
II. Male Determination and Differentiation
III. Hypothalamus–Pituitary–Testicular Axis

GLOSSARY

autocrine communication Regulation of a cell's own functions by one of its secretory products. The secreted substance acts through receptor-mediated mechanisms without passing into the bloodstream.

endocrine communication Regulation of cellular functions elsewhere in the body by the passage of a cell's secretory products into the bloodstream. The secreted substance acts through receptor-mediated mechanisms.

gonadotropin-releasing hormone The hypothalamic decapeptide responsible for the gonadotropin discharge from the pituitary.

gonadotropins Luteinizing hormone and follicle-stimulating hormone; in the males of vertebrates they promote androgen production by Leydig cells and support spermatogenesis acting on Sertoli cells, respectively.

hormone antagonist Molecule which binds to hormone receptors without eliciting cellular response.

hypothalamus The floor of the third brain ventricle; site of production of several substances acting on the adenohypophysis.

Leydig cells Interstitial cells of the testis responsible of the production of androgens.

Müllerian ducts An embryonic genital duct; in mammalian males, it degenerates, whereas in females it is the anlage of the oviducts, uterus, and vagina.

paracrine communication Regulation of the activity of an adjacent, different type cell by the secretory products of a cell without their passing into the bloodstream. The secretory product acts through receptor-mediated mechanisms.

pituitary (hypophysis) A small, rounded endocrine gland attached to the floor of the third brain ventricle. It is divided in adenohypophysis and neurohypophysis.

placode A plate-like epithelial thickening marking in the embryo the anlage of an organ or part.

protooncogenes Cellular genes normally involved in the regulation of cellular functions. Most of them are involved in the regulation of cellular growth and differentiation. Mutations of protooncogenes induce cellular transformation. When they are picked up by certain viruses they generate oncogenic retroviruses.

Sertoli cells Elements of the seminiferous tubules sustaining germ cell progression.

steroid A substance whose molecular structure derives from the cyclopentanoperhydrophenanthrene.

Wolfian ducts The efferent ducts of the mesonephros. This is the middle of the three pairs of embryonic renal structure in vertebrates.

Amphibians are the first tetrapods; therefore, they occupy a position of great interest from the phylogenetic point of view, although it is necessary to remember that modern amphibians are not really primitive tetrapods. As a consequence, speculations about evolution should account for this. As in other vertebrates, the reproductive strategy can be classified as either an associated or dissociated pattern. In the associated reproductive pattern, plasma sex steroids and gametogenic activity rise immediately prior to or during reproduction, whereas in the dissociated reproductive pattern, gonadal activity rises concurrently with plasma sex steroid after mating.

I. INTRODUCTION

The double mode of life, from which the term amphibian is derived, implies that they have acquired features for life on land but retain features of the ancestral fishes; therefore, for reproductive purposes and mode of development they need to return to the aquatic environment. Because little is known about reproduction of apodan amphibians, this article mainly discusses urodeles and anurans.

II. MALE DETERMINATION AND DIFFERENTIATION

In amphibians, heteromorphic sex chromosomes are absent and sex determination appears to be genetic. Male sex is known to be homogametic (ZZ) or heterogametic (XY) depending on the species (Table 1). Sex hormones and temperature influence sex differentiation but a nonsteroidal substance of testicular origin seems to be a major factor. Indeed, ovaries are masculinized after graft or parabiosis experiments, whereas testes are not feminized during gonadal differentiation, which occurs mainly or entirely during larval stages.

Because only following gonadal differentiation are steroid hormones produced by the ovaries or testes, conventional wisdom holds that sex steroid hormones play no role in it. However, sex reversal (genetic males producing eggs or genetic females producing sperm) can be experimentally induced after sex steroid administration in anurans and urodeles. Sex-reversed animals also produce offspring. Anurans can easily be masculinized by testosterone treatment, whereas *Xenopus laevis* and urodeles can easily be feminized by estradiol. Tadpoles of *Rana pipiens* can be reversed to both sexes. At high temperature, females of *R. temporaria* transform into males toward metamorphosis. Generally, sex reversal can be obtained treating animals during a "critical period" (e.g., in *X. laevis* during the third week after hatching).

TABLE 1
Sex Determination in Amphibians

Amphibian	Heterogametic sex
Xenopus laevis	Female (ZW)
Frogs	Male (XY)
Salamanders and newts	Female (ZW)

Wolffian ducts develop as vasa deferentia, are stimulated by androgens, and in adults they function for urine conduction and for transportation of spermatozoa (SPZ) to the cloaca. Müllerian ducts do not degenerate in male urodeles and most anurans. In toads (bufonids), in which the Bidder's organ (a rudimentary mass of ovarian-like tissue available at the anterior end of the adult testis) develops in a functional ovary following testis removal, the rudimentary Müllerian ducts differentiate into oviducts. Besides Wolfian ducts, androgen-dependent characters have been shown to exist in male amphibians and androgen receptors have been characterized, for example, in the thumb pad, skin, and Harderian gland (anurans) or dorsal crest and caudal fin (urodeles). Although fertilization occurs externally in most anurans, *Ascaphus truei* has an everted cloaca for internal fertilization. Among apodan amphibians, the phallodeum is used for internal fertilization, whereas in urodeles the males deposit a coagulated mass of SPZ (spermatophore) which is picked up by the females, following a complicated courtship, by the cloacal lips.

Animals which have been sex reversed have also been sex reversed behaviorally and produce offspring, but male behavior does not appear to be totally dependent on testicular androgens; for example, arginine vasotocin can stimulate amplecting clasping. Mating in anurans is a nonrandom event and larger males are preferred. Often females can choose males of a given size on the basis of call characters. Generally in urodeles male courtship behavior consists of approach (the male sniffs the female head), fan (the male moves the tail near the female head), lashes (the tail of the male hits the female head), and deposition of spermatophore.

III. HYPOTHALAMUS–PITUITARY–TESTICULAR AXIS

As in all vertebrates, testicular activity is mainly under hypothalamic–pituitary control. The hypothalamic gonadotropin-releasing hormone (GnRH) elicits gonadotropin [luteinizing hormone (LH) and follicle-stimulating hormone (FSH)] discharge from the pituitary, and gonadotropins, in turn, act on the

TABLE 2
Mammalian (m)-GnRH and Chicken (c)-GnRH-II Peptides

m-GnRH	pGlu-His-Trp-Ser-Tyr-Gly-Leu-Arg-Pro-Gly-NH$_2$
c-GnRH-II	pGlu-His-Trp-Ser-<u>His</u>-Gly-<u>Trp</u>-<u>Tyr</u>-Pro-Gly-NH$_2$

testis promoting steroidogenesis and spermatogenesis via activation of mechanisms of communication between different cell types within the gonad. Several GnRH molecular forms have been found in the vertebrate brain, and in amphibians the main forms appear to be mammalian and chicken (c)-GnRH-II (Table 2), but a third unidentified form with properties similar to those of salmon GnRH has also been detected.

Recent work has attempted to elucidate the distribution of GnRH molecular forms in the brain. Although controversial results have been obtained, there is a general agreement that hypothalamic GnRH neurons originate from the olfactory placode and migrate centrally during embryogenesis, whereas those located in the mesencephalon may have an intracranial site of origin. Moreover, c-GnRH-II is widely distributed in the amphibian brain suggesting that this peptide, in addition to possessing hypophysiotropic activity, also serves as neuromodulator and/or neurotransmitter. The functional maturation of the brain–pituitary–testicular axis initiates during larval development and SPZ appears few months after metamorphosis. Using iodinated c-GnRH-II, high-affinity, low-capacity binding sites have been detected in frog pituitary and GnRH seasonal cycles have been evidenced in the brain of both urodeles and anurans. The hypothalamic GnRH decreases as a corollary of the increase in plasma LH and androgen which occurs during the reproductive cycle, and pituitary receptors appear during this period. Pituitary GnRH receptors respond to diverse GnRH molecular forms delivering both LH and FSH and, unlike other vertebrate species, they do not become downregulated (loss of cell surface receptors due to a persistent stimulus). Differential response to GnRH molecular forms has been shown to occur during the reproductive cycle. In particular, testicular androgens are strongly stimulated when LH is high in plasma and GnRH receptors are present in the pituitary.

Brain opioid peptides also influence reproductive activity in frogs. Indeed, β-endorphin shows a seasonal profile and injections of naltrexone (an opioid antagonist) increase plasma and testicular androgen levels. It has been suggested that the inhibition of the hypothalamus–pituitary–testicular axis which occurs in postcapture stressed amphibians may be overcome by naltrexone injections.

As in other vertebrates, the structural organization of the testis would predict the presence of internal control systems. Gonadotropins act on the testis via receptors, but how the gonad responds to gonadotropin stimulation depends on local mechanisms. The presence of interstitial and germinal compartments and various cell types within the same compartment demands coordination of the functions of the cells in question to generate efficient mechanisms whereby a primary function, the development of SPZ, results. The existence of cellular communications in amphibian testis has been shown by the use of ethane dimethane sulfonate (EDS), an alkylating agent which provokes degeneration of Leydig cells and disorganization of spermatogenesis in the adjacent zone, thus indicating a local side-by-side "dialogue." EDS and other treatments, which also reduce androgen production, induce mast cell appearance within the testis (Fig. 1) suggesting that these cells are also involved in intragonadal cell-to-cell communication.

The urodele testis is composed of lobes (Fig. 2) which display a zonal distribution of germinal cells. In many species each testis consists of a variable number of lobes and each lobe has zones of germ cells of increasing maturity throughout the cephalocaudal axis. The cephalic part contains early spermatogenic stages and caudally sperm are present followed by the glandular tissue with Sertoli and enlarged Leydig cells often separated by a basement membrane. The anuran testis is composed of convoluted seminiferous tubules (the germinal compartment, in which Sertoli and germ cells are located) and the vascularized interstitial tissue in which Leydig cells (producing androgens), macrophages, and other minor

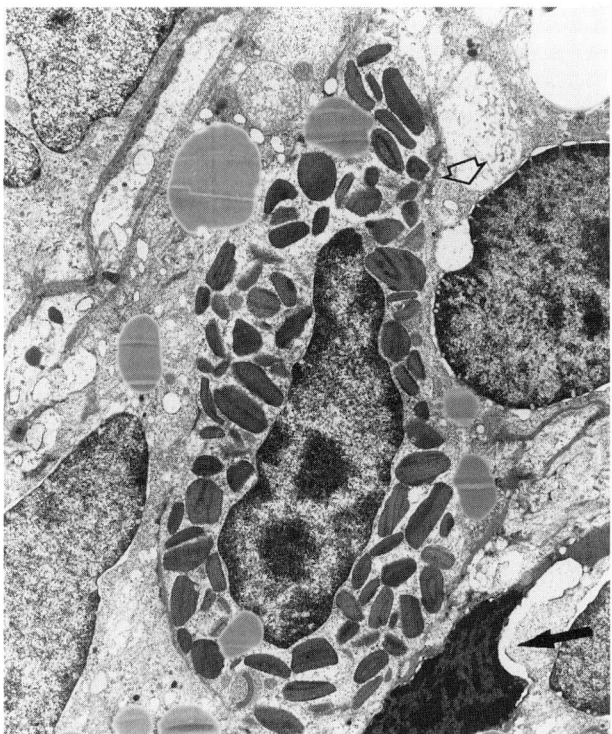

FIGURE 1 Electron micrograph of a mast cell in the interstitial compartment of the frog, *Rana esculenta*, testis after EDS treatment (magnification, ×3800). A degenerating Leydig cell (open arrow) and a Leydig cell characterized by pyknotic nucleus (solid arrow) are shown (photograph kindly supplied by S. Minucci, Dipartimento di Fisiologia Umana e Funzioni Biologiche Integrate "F. Bottazzi," II Università di Napoli, Italy).

antagonist cyproterone acetate. Moreover, SPG multiplication can also be achieved by gonadotropins, primarily FSH, throughout Sertoli cell stimulation. Spermiation is induced by gonadotropins, dopamine, and GnRH-like substances probably of testicular origin; Sertoli cells, after the release of their sperm bundles, become densely lipoidal and cholesterol rich. Then, they eventually become resorbed.

Interstitial (Leydig) cells produce testosterone and 5α-dihydrotestosterone (DHT); DHT appears to be the major metabolite in anurans. Androgen production is stimulated by LH, which acts starting from the 17,20 lyase enzymatic activity. Estradiol produced from androgen by the amphibian testis appears to inhibit androgen biosynthesis via a paracrine and/or autocrine route. The action of estradiol involves inhibition of steroidogenic enzymes starting from 17α-hydroxylase. Contrary to estradiol, testosterone and DHT increase local androgen production—an effect opposite of that seen in mammals. One may speculate that such a mechanism ensures an appropriately high androgen supply within the testis at a

components are present. Anurans have compact ovoid testes.

Spermatogenesis is cystic (Fig. 3), each cyst being composed of cells of the same stage of development. Germ cells within a cyst originate from a single stem cell [spermatogonium (SPG)]. The germinal cyst is formed when the two daughter cells, from a single primary (I) SPG, remain together (secondary; II SPG) and multiply mitotically to give rise to a cluster of cells within a common membranous cyst. During the multiplication period, II SPG transform into I spermatocytes (SPCs) ready to undergo meiosis-1, and then II SPC give rise to spermatids (SPTs; after the meiosis-2) from which SPZs form. SPT experimentally deprived testes show enhanced SPG proliferation. Inhibition of SPT formation can easily be obtained by treating *R. esculenta* with the androgen

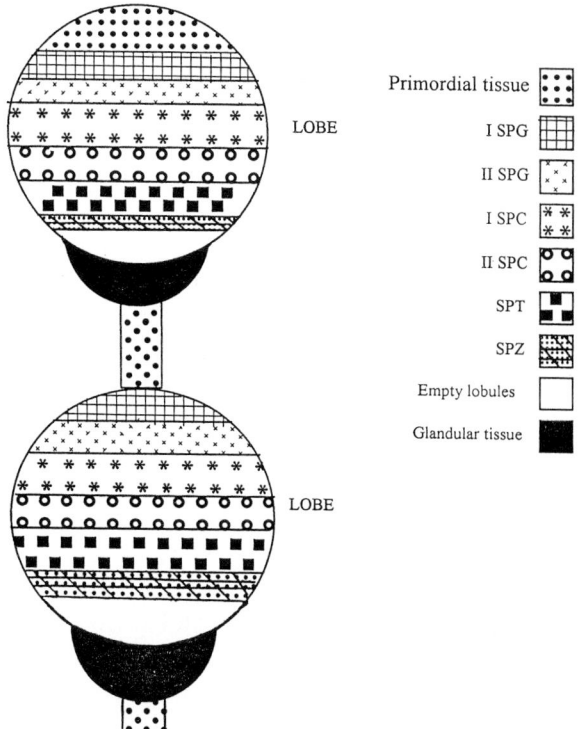

FIGURE 2 Schematic representation of the cephalocaudal zonation in the urodele amphibian testis.

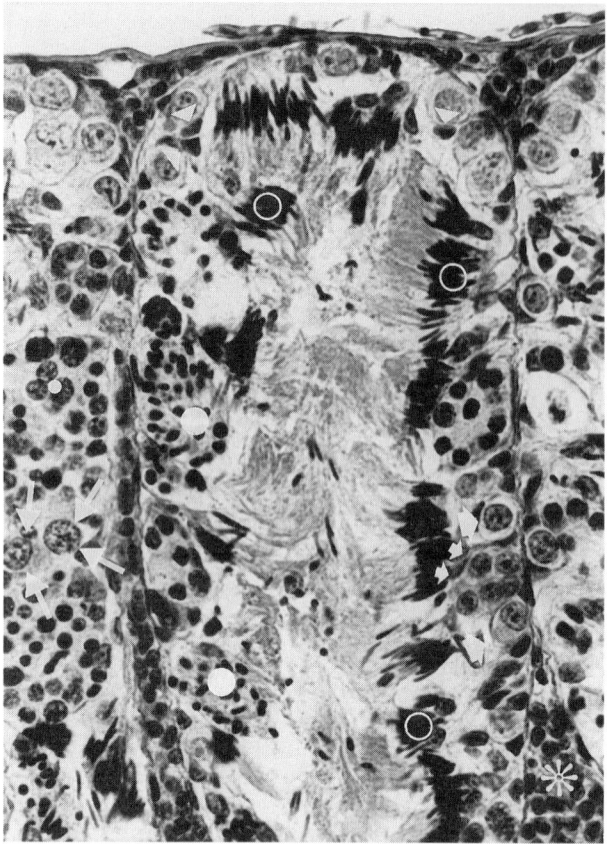

FIGURE 3 Anuran amphibian testis showing the cystic mode of progression of spermatogenesis (magnification, ×300). Interstitial compartment (*); I and II SPG, primary (large arrows) and secondary (small arrows) spermatogonia; I and II SPC, primary (thin arrows) and secondary (small circle) spermatocytes; SPT, spermatids (large circles); SPZ, spermatozoa (O); Sertoli cell nuclei (white triangle) adjacent to I SPG.

low temperature. Androgen and estradiol receptors have been identified in amphibian testis, with the distribution being consistent with a dual localization in Leydig and Sertoli cells. In addition to local (paracrine/autocrine) mechanisms of action that modulate the testicular activity, steroids regulate gonadotropin discharge via endocrine feedback at the brain and pituitary level. Estradiol and DHT are inhibitory at the hypothalamic level, whereas DHT potentiates GnRH activity at the pituitary level. Due to the high level of DHT, anuran males respond better than females to acute GnRH treatment.

Nonsteroidal substances, in concert with steroids, play an important role as local bioregulators. GnRH-like molecules may act on SPG proliferation, spermiation, and androgen production. GnRH binding activity has been found in the testis and GnRH-like material has also been detected by a number of methods. GnRH antagonists decrease GnRH-induced effects in a dose-dependent manner. Metenkephalin and proopiomelanocortin-derived peptides have also been found in testes and, as in mammals, naltrexone (an opioid antagonist) increases androgen production *in vitro* and also provokes the degeneration of germ cells.

Activation of protooncogenes in anuran testis has been observed in both the interstitial and germinal compartments. In *X. laevis*, A-myb expression has been found to decrease dramatically during meiosis, and in *R. esculenta* Myc, Fos, Jun, and Mos proteins show a stage-specific localization in germ cells. Fos and Jun are also present in the interstitial compartment. Increase of *c-fos*-like mRNA appears to correlate with the intratesticular activity of estradiol; therefore, protooncogene activation may be the link between local bioregulators and specific testicular response.

An interesting feature of some amphibian species is the occurrence of seasonal breeding. Therefore, it is often possible to study during a year phenomena which happen in a short time in experimental animal models, such as rats and mice. As a consequence, oncoproteins canonically available in the nuclear compartment, because they are involved in transcriptional activity (Myc, Fos, and Jun), are evidenced in the cytoplasm of frog germ cells for long periods throughout the year corresponding to spermatogenesis shutdown, but they appear in germ cell nuclei when spermatogenesis resumes.

Seasonal reproductive cycles have been described in most amphibians. Thus, testicular activity (steroidogenesis and spermatogenesis) resumes so that reproduction occurs when environmental conditions are the best for the offspring survival. Endogenous rhythms of hypothalamic–pituitary–testicular activity are regulated mainly by temperature, whereas photoperiod is of secondary importance. In urodeles, production of SPZ can be prenuptial (e.g., shortly before breeding) or it can occur postnuptially. In this case, during the breeding period, the gonad is distinguished by a large caudal zone of stored SPZ and a small cephalic zone of SPG and SPC. After breeding, there is intense spermatogenic activity with enlargement of the cephalic region of the testis. In many anuran species living in temperate zones of the Northern Hemisphere, the spermatogenic cycle is interrupted in winter. However, under experimental conditions, spermatogenesis resumes at warm temperatures; these animals are classified as "potentially continuous" breeders. "Discontinuous" cycles are shown by species in which a refractory (insensitive) phase occurs after the breeding period. In this phase, artificial elevation of temperature does not stimulate spermatogenesis until the refractory period is over. Finally, in many tropical and subtropical urodele and anuran species spermatogenesis is "continuous," indicating that mating can occur at any time.

See Also the Following Articles

Amphibian Reproduction, Overview; Female Reproduction System, Amphibians; Male Reproductive System, Fish; Migration, Amphibians; Male Reproductive System, Amphibians

Bibliography

Chieffi, G., Pierantoni, R., and Fasano, S. (1991). Immunoreactive GnRH in hypothalamic and extrahypothalamic areas. *Int. Rev. Cytol.* **127**, 1–55.

Crews, D., and Moore, M. C. (1986). Evolution of mechanisms controlling mating behavior. *Science* **231**, 121–125.

Facchinetti, F., Henderson, I. W., Pierantoni, R., and Polzonetti-Magni, A. (1993). *Cellular Communication in Reproduction.* Journal of Endocrinology, Bristol, UK.

Halliday, T. (1983). Do frogs and toads choose their mates? *Nature (London)* **306**, 226–227.

Norris D. O. (1996). *Proceedings of the International Symposium on Amphibian Endocrinology.* Abstract Book, Boulder. CO.

Norris, D. O., and Jones, R. E. (1987). *Hormones and Reproduction in Fishes, Amphibians, and Reptiles.* Plenum, New York.

Vaudry, H., Roubo, E., and de Loof, A. (1998). Trends in comparative endocrinology. *Ann. N. Y. Acad. Sci.*, in press.

Male Reproductive System, Birds

Tim R. Birkhead
University of Sheffield

I. Introduction
II. Structure
III. Sperm Production, Maturation, and Storage
IV. Sperm Production Rates
V. Sperm Morphology
VI. Ejaculate Size

GLOSSARY

passerines The perching birds, a single order comprising over 5200 species—over half of all bird species. Other birds comprise the nonpasserines (~29 orders).

proctodeum The most external of the three chambers comprising the cloaca; the other chambers are the urodeum and coprodeum.

sperm competition The competition between the sperm from more than one male to fertilize the eggs of a particular female.

The avian reproductive system is characterized by enormous seasonal changes in size and function in both sexes. Outside the breeding season the reproductive system regresses and is minute and nonfunctional, a mass-saving adaptation for flight. During the breeding season the entire system increases in size; the testes, for example, may increase 500-fold in mass.

I. INTRODUCTION

Birds are divided into two groups (nonpasserines and passerines) whose reproductive systems differ in a number of important ways. Most information on avian reproductive systems has been derived from nonpasserines, in particular the domestic fowl *Gallus domesticus* and turkey *Meleagris gallopavo*. However, across different bird species there is considerable variation in the relative size of reproductive structures and in sperm morphology and ejaculate size. Much of this variation is explicable in terms of the intensity of sperm competition, which is widespread in wild birds.

II. STRUCTURE

The male reproductive system comprises the paired testes, epididymis, ductus deferens, areas for the storage of sperm adjacent to the cloaca, and, in a minority of species, a phallus (Fig. 1).

The testes are located within the abdominal cavity. Their seasonal increase in size, triggered by increasing day length under the influence of gonadotrophins, is due mainly to an increase in the size of the seminiferous tubules associated with sperm production. There is also an increase in the number of interstitial cells, responsible for the production of male sex hormones. The left testes is often slightly larger than the right, possibly because the latter has a compensatory role.

The combined mass of the two testes at maximal size across species scales allometrically with body mass with an exponent of approximately two-thirds (log (testes mass [g]) = −1.37 + 0.67 log (body mass

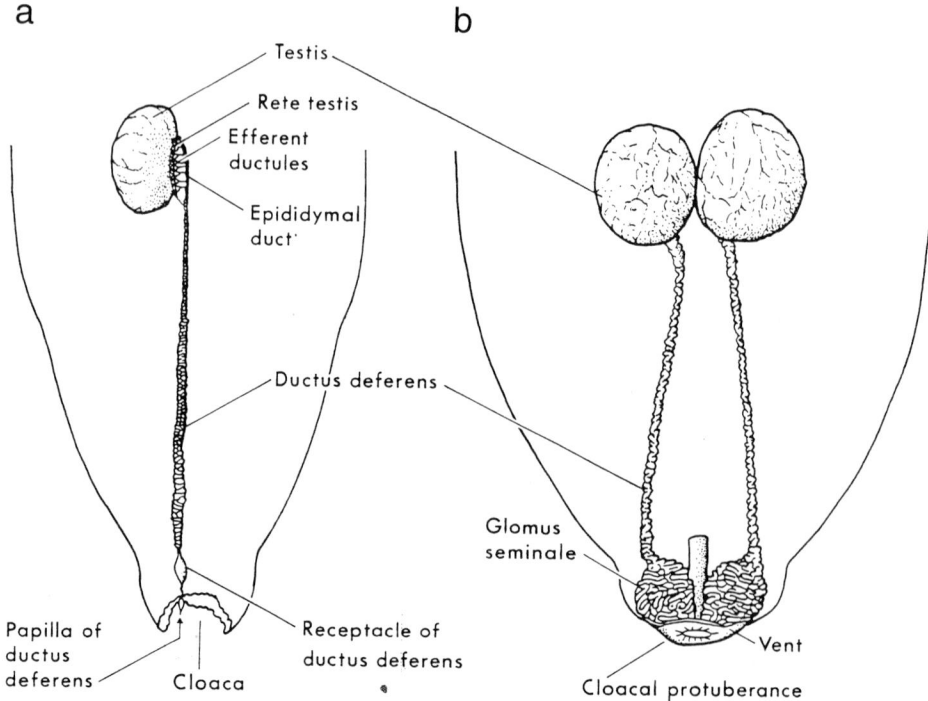

FIGURE 1 The male reproductive system. (a) Ventral view of a nonpasserine (domestic fowl), showing the right testis, ductus deferens, and receptacle. (b) Ventral view of a typical passerine, showing the distinctive coiled seminal glomera forming the cloacal protuberance (reproduced with permission from Birkhead and Møller, 1992).

[g])), presumably since more sperm are required to fertilize the eggs of larger females because of the dilution effects of larger body size. However, relative testes size varies markedly between species. Those species with high copulation rates and with high levels of sperm competition tend to have relatively large testes. For example, in the aquatic warbler *Acrocephalus paludicola*, a species with intense sperm competition, the testes weigh 0.489 g or 4.1% of body mass (~12.0 g), whereas in the sedge warbler *A. schoenbaenus*, a species with less intense sperm competition, the testes weigh 0.161 g or 1.4% of body mass (~11.5 g). Since daily sperm production is positively correlated with testes size, both within and between species, males with larger testes are able to deliver more and/or larger ejaculates. The outcome of sperm competition depends on the relative numbers of sperm in the female tract from each male, and sperm competition has been the main evolutionary force favoring relatively large testes (and sperm storage areas) in some species.

The epididymis is a small, spindle-shaped structure closely associated with the testis and which also undergoes a marked seasonal reconstruction. The ductus deferens is a continuation of the epididymal duct, extending to the urodeum (in the cloaca), through which sperm are transported from the testes to the storage systems adjacent to the cloaca. The sperm storage areas adjacent to the cloaca are referred to as receptacle of the ductus deferens in nonpasserines or the seminal glomera in passerines.

The epididymis, ductus deferens, and receptacle, which together comprise the ducts that convey sperm from the testes, produce secretions which (i) comprise the main component of seminal plasma (the rest originates in the testes), (ii) act as a fluid vehicle for sperm, and (iii) provide a medium in which sperm mature and develop the potential for motility. However, substances in the seminal plasma also suppress sperm motility until after ejaculation. At ejaculation transparent fluid from the lymph folds in the cloaca is added to semen and is thought to trigger sperm motility.

An intromittent phallus exists in 3% of bird species, including waterfowl and ratites. Why some groups of birds should have retained phallus is unclear. The avian phallus differs from those of reptiles and mammals in two important respects: (i) The mechanism of erection is lymphatic rather than blood–vascular as in mammals, and (ii) during ejaculation, semen is transferred along a recess on the outside of the phallus rather than through an internal tube. The male domestic fowl and turkey (and many other species) possess a nonintromittent phallus—a small, heart-shaped erectile structure with a median groove through which semen is discharged. In passerine birds the nonintromittent phallus is either extremely small or absent and semen is transferred via a "cloacal kiss" during which male and female cloacae are juxtaposed, usually for only a few seconds.

III. SPERM PRODUCTION, MATURATION, AND STORAGE

The bean-shaped testes consist of seminiferous tubules and Leydig cells located between the tubules. The seminiferous tubules comprise a series of interconnecting ducts that empty into the rete testis which lies adjacent to the epididymis. The seminiferous tubules are lined with spermatogonia, the first stage of spermatogenesis. The spermatogonia are diploid and divide mitotically to produce spermatocytes which then undergo meiotic divisions to produce haploid spermatozoa. This differentiation from spermatogonia to spermatocytes involves the Sertoli cells (located at the periphery of the tubules) which act as nurse cells to the developing sperm. The duration of spermatogenesis in the domestic fowl is short compared with that of mammals, only 13 days from the start of a spermiogenic activity wave to sperm appearing in the ejaculate (10–12 days from a primary spermatocyte to the completion of spermiogenesis plus an additional 3 or 4 days to pass through the epididymis and ductus deferens). Because of the location of the testes in the abdomen, they function at deep body temperature ($40 \pm 2°C$). However, spermiogenesis shows a marked diurnal, temperature-dependent pattern, being maximum at night when body temperatures are lowest (38 or 39°C), presumably because, as in mammals, high temperatures are deleterious to sperm.

In passerine birds the distal ends of the ductus

deferens are extensively coiled to form the seminal glomera, which lie within the cloacal protuberance. Across species the relative size of the cloacal protuberance covaries with testes mass and reflects the number of sperm stored in the seminal glomera. The seminal glomera thus serves as a site for sperm storage. The maximum number of sperm stored varies from about 10×10^6 in the zebra finch *Taeniopygia guttata* (body mass, ~12 g) to over 3500×10^6 in the white-winged fairy wren *Malurus leucopterus* (body mass, ~9 g). The seminal glomera also serve as a maturation site, possibly analogous to the scrotal epididymis in mammals. Because of their location in the cloacal protuberance the seminal glomera are about 4°C cooler than core body temperature. In the zebra finch, the passerine studied in most detail, sperm taken from the most proximal region of the seminal glomera at the junction with the ductus deferens through to the most distal portion (the ejaculatory duct) show increased motility, proportion of morphologically normal sperm, and swimming velocity. Approximately 84% of the sperm in the seminal glomera is available for ejaculation: Males which copulated to satiation had 16% of the original number of sperm in the seminal glomera. During a normal breeding cycle the male zebra finch relies equally on sperm stored in the seminal glomera and on sperm production. After producing a few relatively large ejaculates, males rely on sperm production and ejaculate relatively immature sperm. It is not known whether immature sperm have reduced ability to fertilize, either on their own or in competition with more mature sperm other males. However, in poultry the degree of sperm motility is the best predictor of fertilizing ability. That passerine birds ejaculate immature sperm differs from the situation in mammals in which immature sperm, located in the lower caudal epididymis, are inaccessible and cannot be ejaculated.

IV. SPERM PRODUCTION RATES

The total number of sperm produced per day by the two testes, the daily sperm production (DSP), is usually expressed as sperm per gram of testes tissue. DSP can be measured using testicular homogenates,

TABLE 1
Sperm Production Rates in Birds

Species	Daily sperm production (DSP) $\times 10^6$	DSP/g testis
Domestic fowl		
Gallus domesticus	2000	80–120
Turkey		
Meleagris gallopavo	520–1120	18–120
Guinea fowl		
Numidia meleagris	70	18.5
Japanese quail		
Coturnix japonica	308	98.7
Zebra finch		
Taeniopygia guttata	1.88	34.91

Note: All species are nonpasserines (order Galliformes) except the zebra finch, which is a passerine.

morphometric analyses of testis tissue sections, or cannulation of the testis; in live birds it can be measured by estimating the daily sperm output from frequent, repeated ejaculations. Table 1 summarizes information on DSP: Most information is available for poultry. In the turkey the rate of sperm production is unaffected by ejaculation frequency.

V. SPERM MORPHOLOGY

Sperm comprise an acrosome, head, midpiece, and tail. The size and form of spermatozoa vary markedly between bird species. There are two basic types: (i) a simple reptilian type, with a smooth, tapering head, typical of many nonpasserines, including the domestic fowl, and (ii) a spiral-shaped type typical of passerine birds—the acrosome is spiral shaped and the tail has a mitochondrial helix running along much of its length. The function of this spiral morphology is unknown, although it is associated with the type of movement which comprises rapid rotation around the longitudinal axis. In contrast, nonpasserine sperm move in an undulating manner. Across species, nonpasserine sperm apparently shows little variation in size, being, like those of the domestic fowl, about 100 μm in total length (data on sperm length are few). Passerine sperm vary from about 50 to 300 μm in length and across species length is positively associated with the intensity of sperm competition, although the advantage conferred by relatively long sperm remains unknown. Avian sperm do not require capacitation within the female tract.

VI. EJACULATE SIZE

The volume of ejaculated semen in birds is small compared with that of mammals of similar size. This is because of the lack of secretory products of the accessory glands (which are absent in birds). Thus, the amount of seminal fluid is low and the concentration of sperm high. At ejaculation, contraction of the muscles in the lower part of the ductus deferens (nonpasserines) or ejaculatory duct (passerines) expels semen through the papillae (which extend into the urodeum of the cloaca) and out through the proctodeal folds (or phallus) and into the female. The numbers of sperm transferred during natural ejaculation is known for very few species. Most researchers working with poultry obtain semen samples manually via the abdominal massage method and these may not be comparable with natural ejaculations. A comparative study showed that for samples obtained manually, ejaculate volume and sperm concentration scaled isometrically with testes mass, but the number of sperm scales positively to testes mass, with a slope >1. Semen samples obtained using a mounted female with a false cloaca can provide better estimates of ejaculate size and have been used successfully in a few passerine species. In the zebra finch, mean number of sperm per ejaculate is about 1×10^6 and in the red-winged blackbird *Agelaius phoeniceus* (body mass, $\sim$74 g) 24×10^6 sperm. In both species sperm numbers per ejaculate vary considerably, even after controlling for time since the previous ejaculation: Similar variation occurs in other taxa. In the zebra finch and in poultry (using manual methods) the number of sperm is reduced in successive ejaculates made over a period of 24 hr or less.

See Also the Following Articles

AVIAN REPRODUCTION, OVERVIEW; FEMALE REPRODUCTIVE SYSTEM, BIRDS; SEASONAL REPRODUCTION, BIRDS

Bibliography

Birkhead, T. R., and Møller, A. P. (1992). *Sperm Competition in Birds: Evolutionary Causes and Consequences.* Academic Press, London.

Birkhead, T. R., Fletcher, F., Pellatt, E. J., and Staples, A. (1995). Ejaculate quality and the success of extra-pair copulations in the zebra finch. *Nature (London)* 377, 422–423.

Etches, R. J. (1996). *Reproduction in Poultry.* CAB International, Oxford, UK.

Howarth, B. (1995). Physiology of reproduction: The male. In *Poultry Production* (P. Hunton, Ed.), pp. 243–270. Elsevier, Amsterdam.

King, A. S. (1981). Phallus. In *Form and Function in Birds* (A. S. King and J. McLelland, Eds.), pp. 107–147. Academic Press, London.

Lake, P. E. (1981). Male genital organs. In *Form and Function in Birds* (A. S. King and J. McLelland, Eds.), pp. 1–61. Academic Press, London.

Møller, A. P. (1991). Sperm competition, sperm depletion, paternal care and relative testis size in birds. *Am. Nat.* 137, 882–906.

Male Reproductive System, Fish

Florence Le Gac and Maurice Loir

Institut National de la Recherche Agronomique, Rennes, France

I. Introduction
II. The Testis: Anatomy, Cellular Organization, and Functional Aspects
III. Excretory Ducts and Accessory Glands
IV. Secondary Reproductive Structures

GLOSSARY

Atherinomorphs A super order of teleostei that comprises numerous fish with internal fertilization, such as Poecilia (guppy), Gambusia (mosquito fish), and Fundulus (killifish).

spermatogenesis The process by which spermatogonial stem cells give rise to spermatozoa.

spermiation In teleosts, the opening of the spermatocysts and release of the spermatozoa into the tubule or sperm duct lumen. In common aquaculture language, this term refers to milt release.

teleostei The division of fish that excludes Agnatha, Elasmobranchs, and Acipenseridae.

The male reproductive system produces the gametes, sex hormones, pheromones, and seminal components necessary for successful breeding. Fish are characterized by the wide variety in the expression of their sexuality and in their reproductive strategies. This is reflected in the considerable diversity in morphology and functional activity of the male reproductive system. Differentiation of synchronous germ cells occurs in spermatogenic cysts, enclosed by Sertoli cells. Production of high levels of 11-oxygenated androgen and androgen glucuronides and high 20-hydroxysteroid dehydrogenase activity of sperm cells are peculiarities of steroidogenesis in these species. Storage, maturation, and excretion of gametes, and possibly pheromones, are often assumed by a simple sperm duct, whereas in other species, various excretory ducts and accessory glands may be involved in these processes. Secondary structures such as gonopodia may also occur.

I. INTRODUCTION

The great systematic diversity of teleostei (more than 20,000 species; four monophyletic groups) and the large variety of their biotopes are mirrored in the morphology of their reproductive systems. They have adopted a considerable diversity of strategies to ensure reproductive success. Reproductive diversity includes unspecialized to highly specialized reproductive modes: gametes released into the aqueous environment (external fertilization, the most common in teleosts) or spermatozoa introduced into the female tract, with internal fertilization being followed by the laying of fertilized eggs or internal gestation of embryos. The age of first sexual maturation, extremely variable between species, can occur earlier in male than in female and also varies between individuals; this is, at least in part, related to body growth in juveniles. Although many tropical species reproduce throughout the year, most of the teleost fish species exhibit an annual rhythm of reproduction related to environmental events (changes in photoperiod, climatic factors, food supplies, etc.). Active spermatogenesis may take place in summer (trout, carp, and pike), in spring (tench, bream, whiting, and sea bream), or begin in autumn and finish in spring (killifish, stickleback, and roach). While teleosts are predominantly gonochoristic, hermaphroditism (simultaneous hermaphroditism, protandry, or protogyny) occurs in a large number of species (Serranidae, Sparidae, and Labridae). Finally, in this poikilothermic vertebrate group, most reproductive events are dependent on water temperature.

II. THE TESTIS: ANATOMY, CELLULAR ORGANIZATION, AND FUNCTIONAL ASPECTS

A: Early Development and General Anatomy

Many uncertainties remain about the origin and migration of primordial germ cells (PGCs) and about differentiation of the gonad in teleosts.

1. Origin

With respect to species such as carp, rosy barb, and medaka, and although experimental proof is lacking, it is likely that the precursor germ cells are not predetermined but will arise epigenetically in extraembryonic layers and segregate from the somatic cells at the onset of gastrulation (50% epiboly). After the completion of epiboly, PGCs can be recognized by their large size, large spherical and pale nucleus, and prominent nucleolus, and they are found in close association with the mesonephric duct and the gut. Within a few days they have been translocated to the gonadal ridges, where they are found surrounded by future cyst cells originating from the coelomic epithelium. PGCs remain in small numbers and retain the potential to give oogonia or prespermatogonia until sexual differentiation of the gonad; even "spermatogonia" could retain bipotentiality until entering meiosis. In protogynous species, the male germline arises from germ cells already present in the previously female gonad.

2. Differentiation

In a few fish species it has been demonstrated that the male sex is determined by heterozygous (XY) or homozygous (WW) sexual chromosomes, but in other cases more complex types of genetic determination have also been proposed. Another important peculiarity of fish (compared to higher vertebrates) is that in many gonochoristic species, sexual differentiation (phenotypic) can naturally diverge from the genotypic sex under the influence of environmental or social factors. Fertile sex-reversed males may also be produced by exogenous administration of androgens to XX fish. In most teleosts, testicular differentiation (from the bipotential primitive gonad) occurs later than ovarian differentiation and is characterized by intense development of the connective stroma and differentiation of the bipotential primordial germ cells into spermatogonia with moderate proliferative activity. The somatic male supporting cells do not differentiate into characteristic Sertoli or peritubular cells until tubule formation and appearance of type B spermatogonia, in the prepubertal period. An interstitial tissue showing detectable steroidogenic activity has also been proposed as an early criteria of male differentiation. The cascade of genetic expression that drives gonadal differentiation in fish is still unknown, but the synthesis of sex steroid hormones and of their cellular receptors should play an important part in this phenomenon. Differentiation occurs during a critical period, specific for each species, that may be from early (embryonic, posthatching, or larval) to very late in the life cycle (from months to several years in freshwater eels and conger eels). This differentiation period (and, in some species, the resulting sex) is temperature dependent.

3. General Anatomy

Developed testes are usually paired organs (they may be partially or totally fused), and most often elongated but may have a lobate or foliaceous shape, and are surrounded by a fibrous tunica. They lie free within the body cavity, attached by a mesorchium (membrane) to the dorsal body wall. Except for some tropical species of teleosts, the morphology, color, and especially the size of the testes undergo cyclic variations throughout the annual reproductive cycle. In salmonids, the testicular weight expressed as a percentage of body weight [gonadosomatic index (GSI)] is below 0.05% in immature males and may reach 10% at the time of spawning, whereas in some tropical species GSI only varies around 0.2%.

The basic organization of the testis is common to all fish and to other vertebrates, consisting of a germinal and an interstitial compartment separated by a basement membrane.

B. The Germinal Compartment

The germinal compartment consists of the acellular basal membrane, often very thin, that surrounds the germinal epithelium consisting of Sertoli cells and germ cells, organized in spermatocysts (or cysts) which represent the spermatogenic functional unit of anamniote testes (Figs. 1 and 2). Fish Sertoli cells surround, and are in contact with, one clone of germ cells at the same stage of differentiation which originate from a single primary spermatogonium and mature synchronously. In contrast, mammalian Sertoli cells are simultaneously in contact with several clones of germ cells at different stages of differentiation.

1. Histological Organization: Spatial and Temporal Changes

Two types of germinal compartment occur in teleosts (Fig. 1). In atherinomorphs (killifish and guppy) and in higher perciform fish (*Oreochromis* sp. and *Sciaenops ocellatus*), the germinal compartment is organized in blind lobules which terminate at the periphery of the testis under the tunica (lobular testis type). In other teleosts, the germinal compartment appears as a network of anastomosing tubules (anastomosing tubular testis type). In both types of testis the tubes converge and are connected by their open ends to the excretory ducts.

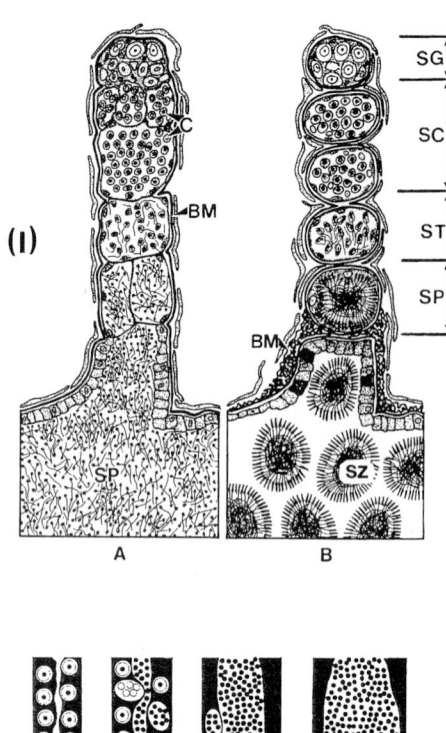

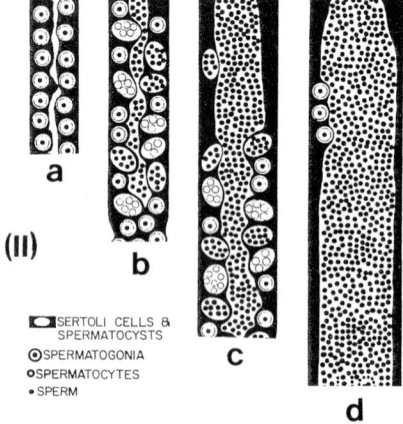

FIGURE 1 Schematic representation of the germinal compartment in teleost fish. (I) Two lobular-type testes with spermatogonia restricted to the distal ends of the lobule. (A) Lobule structure in Cyprinodontidae (external fertilization); (B) lobule structure in Poeciliidae (internal fertilization). Sperm cells are embedded into bundles before release in the efferent ducts. BM, basement membrane; C, cysts; SG, spermatogonia; SC, spermatocytes; ST, spermatids; SP, spermatozoa; SZ, spermatozeugmata. (II) Unrestricted spermatogonial tubular testis type typical of Salmoniformes, Perciformes, and Cypriniformes. The tubules form an anastomosing or branching network. Spermatogonia are present along the entire length of the tubules. (a–d) Evolution of a tubule during the reproductive cycle; see text and Fig. 2. for further details. The unrestricted lobular testis type is more rare; one main difference regarding model II is that basement membrane define and border the blindly ending distal terminus of the lobule (adapted from Grier, 1996).

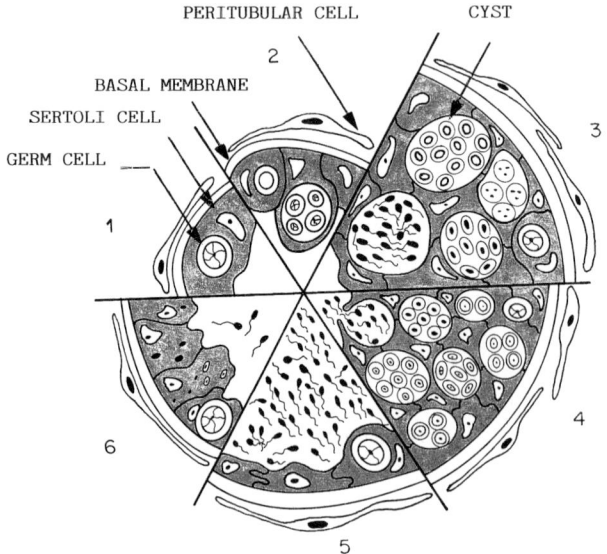

FIGURE 2 Schematic illustration of the spermatogenic cycle in a tubule: 1, Inactive spermatogenesis; 2, spermatogonia multiplication; 3, active spermatogenesis—lobular size increases considerably; 4, Active spermatogenesis—release of early spermatozoa; 5, spermiation; 6, postspawning resorption of spermatozoa (adapted from Billard, 1986).

In the "lobular type" testes (Fig. 1,I), spermatogonia are generally restricted to the distal end of lobules (restricted spermatogonial type). The maturing and growing cysts move down the lobule toward the efferent ducts where they open to release spermatozoa into the ducts. Cyst migration is passive and results mainly from the continuous formation of cysts at the apex of the lobule.

In the "anastomosing tubular type" (Figs. 1,II and 2), the cysts are stationary. Cysts of spermatogonia occur along the entire length of the tubules (unrestricted spermatogonial testis type). In the studied species, primary spermatogonia are permanent cells in the seminiferous epithelium. During the nonbreeding season, Sertoli cells and spermatogonia occur as a solid cord within the tubules. When spermatogenesis develops, the cysts containing successively increasing numbers of secondary spermatogonia, then spermatocytes, spermatids, and spermatozoa, line the tubules, which have by then developed a patent lumen. Spermiation results from the breakdown of the Sertoli cell layer and spermatozoa are released in the lumen. In some species (Blenniidae, *Ophidion* sp), the cysts open earlier and, as a consequence, different postmeiotic germ cell types are mixed in the tubule lumen where spermatogenesis is completed ("semicystic" type of spermatogenesis).

Some tropical species do not display apparent seasonal variations in spermatogenesis, which is qualitatively and quantitatively continuous throughout the year. For species with a clear-cut seasonal reproduction, such as salmonids or pike, at the beginning of testicular recrudescence only cysts of proliferating spermatogonia can be observed, whereas after spermatogenesis is completed tubules are packed with spermatozoa and contain no cysts but some scattered primary spermatogonia. In other cyclic species, the germ cells remaining in the regressed testis may be spermatogonia and spermatocytes (tench) or all germ cell types including spermatozoa (the regression is then essentially quantitative, as in goldfish).

2. Sertoli Cells: Cytology and Functions

Sufficient cytological and functional similarities between the fish "supporting cell" and the higher vertebrate Sertoli cell have now been identified to proclaim an analogy between the two cell types.

Several functions, often based on morphological criteria, have been attributed to fish Sertoli cells. They form the walls of the cysts and therefore provide the structural support for the developing germ cells (Fig. 2). While the cyst increases in size, the "Sertolian" wall becomes thinner, and it is unknown whether in "tubular testis" the Sertoli cells undergo mitosis during this process. In species with "lobular testes" it has been shown that the number of Sertoli cells increases as spermatogenesis progresses (120–136 Sertoli cells per cyst of spermatids in the guppy). Their is no permanent Sertoli cell epithelium in mature fish. It is assumed that in the tubular type of testes, Sertoli cells degenerate at the time of gonadal regression, whereas in the lobular testes they either degenerate or become integrated into the epithelium of the efferent duct.

Occasional adhering junctions have been described between Sertoli cells and mitotic or meiotic germ cells, but no ectoplasmic specializations around spermatids have been observed. The formation of sperm bundles, when present, occurs within the cyst and results from an original structural relationship between spermatids and Sertoli cells: The latter secrete a capsule around the spermatophores, whereas in the case of spermatozeugmata the Sertoli cells enlarge and send projections around the gametes to anchor them at the luminal margin of the cyst (Fig. 1). According to the species, in various teleost orders Sertoli cells may exhibit phagocytic activity at the end of spermatogenesis, engulfing residual bodies cast off by the spermatids and degenerating germ cells and residual sperm cells following the reproductive season. Lysosomes and large digestive vacuoles appear after the cysts open and spermatozoa are released.

Sertoli cells are connected to each other by complex interdigitations, intermediate adhering junctions (desmosome-like junctions), and associated tight junctions localized mainly at the apical side of the cells. At least in some species, these junctions have been shown to restrict the entry of large molecules into the cysts and, in trout, the experimental breaching of this barrier leads to the appearance of anti-sperm antibodies. Unlike in mammals, the fish "Sertoli cell barrier" is established after meiosis and only haploid germ cells are shielded from the vascular compartment and from the immune system.

The total impocketing of the germ cells within the

Sertoli cells implies that the latter are involved in the nutritional support of germ cells (transport, selective filtration, and possible conversion of metabolites or molecules coming from the outside of the cyst). Sertoli cells also take part in the hormonal control of spermatogenesis. They can be involved in steroidogenesis in various species while not in others. This diversity of observations could be due to the fact that Sertoli cells produce steroids only transiently, as is the case in trout. In addition, it is suggested that molecules, such as growth factors and cytokines, are also involved in the Sertoli cell–germ cell dialog. Strong evidence has been obtained for the role of activin (in the Japanese eel) and of the IGF system (in the trout) in proliferation and/or differentiation of premeiotic germ cells.

3. Germ Cells and Male Gametes

i. Spermatogenesis In all vertebrates germ cell development proceeds in a similar way: Spermatogonia divide several times and then develop into spermatocytes which undergo meiosis, the process by which the nucleus of a diploid primary spermatocyte divides to produce four haploid spermatids. These haploid cells then differentiate to form spermatozoa. The fine structure of the germ cell types is similar in all species studied. Cytoplasmic bridges, present between germ cells from the first spermatogonial divisions, connect them in a clonal syncitium inside one cyst. The number of spermatogonial generations, unknown in most species, was found to be 14 in guppy, 10–12 in mosquito fish, and 5 or 6 in brook trout. It is hypothesized that in fish, the duration of the various stages of spermatogenesis is influenced by temperature, with high but physiological temperatures speeding up the events. Data dealing with the timing of spermatogenetic events are scarce. The duration of meiosis plus spermiogenesis varies from 8.5–11 days in two tropical species to 1–3 months in species living in temperate and cold zones. At 25°C, the total duration of spermatogenesis in guppy is 36 days. In atherinomorphs exhibiting inter-

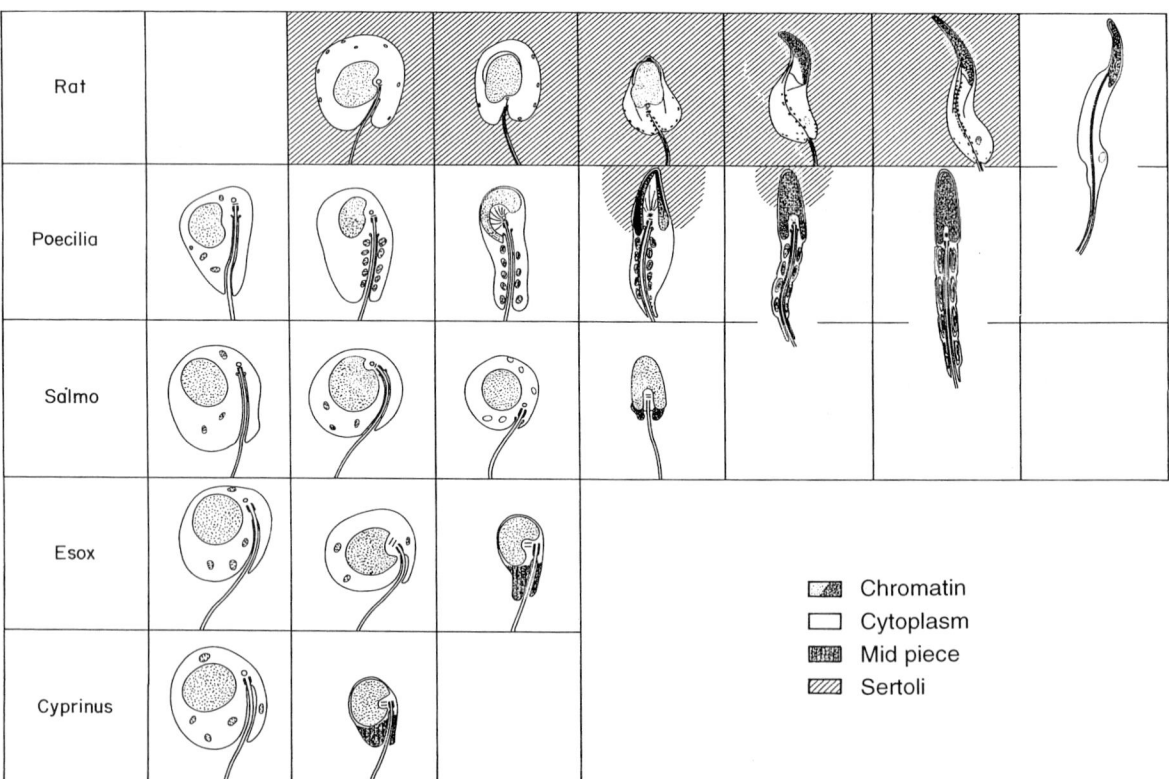

FIGURE 3 Schematic illustration of the progression of spermiogenesis in several teleost genera compared with rat (adapted from Billard, 1986).

nal fertilization, the spermatogenetic process ends by the gathering of spermatozoa into bundles which will be introduced into the female genital tract.

ii. Spermatozoa A great diversity in the ultrastructure of fish spermatozoa is evident for all taxonomic groups (Fig. 3). Although spermatozoa of Elasmobranchs, Chondrostei (sturgeon), Crossopterygians, and Dipnoi possess an acrosome, no true acrosome has been unambiguously identified in teleost spermatozoa. Spermatozoa are most commonly (but not always) uniflagellate, have a classic (9 + 2) or an atypical (9 + 0) axoneme, have from one to a large (undefined) number of mitochondria, and may show cytoplasmic fine-like extensions along the flagellum. The head may range in shape from nearly spherical, or ovoid on an axis perpendicular to the flagellar axis, to markedly elongated and slim. The basic proteins associated with DNA are either histone-like, protamines, or proteins intermediate between the first two types, but never cystine-containing protamines. In species having external fertilization (e.g., *Salmo* sp., *Esox* sp., and *Cyprinus* sp.), the spermatozoon is usually of the primitive type: short head, few mitochondria, and a poorly differentiated midpiece. In most fish employing internal fertilization (Auchenipteridae, Ageneiosidae, and several Atherinomorphs including Poecilia, Cottidae, and Embiotocidae), the head (nucleus) and midpiece are elongated (Fig. 3).

Spermatozoa from several species have the capacity to produce 17α-hydroxy-20α or 20β-dihydroprogesterone (20-HSD activity), which appears to be directly or indirectly involved in the regulation of sperm release or in spawning success (pheromones).

Potential motility is acquired in the tubular lumen or in the efferent ducts and depends on extracellular pH and Ca^{2+} concentration. In the male genital tract, osmotic pressure and, in salmonids, K^+ concentration are the main factors that inhibit sperm movement. At least in external fertilization, hyperpolarization of the membrane is an activating factor initiating motility when the gametes are released into the water. The energy supply depends on endogenous sources and the motility is usually short-lived (0.25–2 min). During this short period, the spermatozoon has to reach a highly specialized channel through the egg walls, the micropyle, to approach the oocyte membrane. In the case of internal fertilization, motility in the female tract lasts for several hours.

C. The Interstitial Compartment and the Steroidogenic Functions

The relative volume of this compartment varies according to the species and throughout the reproductive cycle (from 2 to 14% in the brown trout and from 18 to 46% in a tropical fish, *Myleus ternetzi*).

1. Different Types of Interstitial Cells

Peritubular or -lobular cells form an incomplete layer over the surface of the germinal compartment and are probably not involved in a "blood–testis barrier" but possess characters of contractile myoid cells that could facilitate expulsion of sperm from the tubules. The teleost testis lacks a lymphatic system. In addition to blood capillaries and scattered nerve fibers, typical fibroblasts are present in the interstitial space and macrophages may be observed at the time of testis regression.

Leydig cells are a typical component of the teleost interstitium. They demonstrate the ultrastructural features of steroid-producing cells (smooth endoplasmic reticulum and mitochondria with tubular cristae) and hydroxysteroid dehydrogenase activity (3β-HSD, 11β-HSD, and 17β-HSD). Their abundance, distribution, and cytology vary between species. Ultrastructural changes, sometimes correlated with 3β-HSD activity, suggest maximum steroidogenic differentiation during the periods of full spermatogenesis and of spermiation. Renewal of at least some of them occurs at the beginning of a new spermatogenetic cycle in the rainbow trout. They may originate from interstitial fibroblasts.

2. The Testicular Steroidogenic Tissue in Blennidae and Gobiidae

In several gobiids, an endocrine tissue ("glandular mass") develops separately from the seminiferous area. It is enclosed by an epithelium and composed of steroidogenic cells, whereas the spermatogenic region is devoid of such cells.

In blenniids the male gonad consists of a spermatogenic part with an interstitial tissue containing Leydig cells and a separate accessory testicular gland containing steroidogenic cells. In at least one blenniid species, reductive metabolism of steroids predominates in the spermatogenic part, whereas in the gland a higher production of Δ^4 steroids is noticeable.

3. Steroidogenesis

Teleost testes synthesize mainly androgens (testosterone, androstenedione, and/or their 11β-hydroxy or 11-keto-derived products), progestins [most often 17α-hydroxy-20β-dihydro-progesterone (17,20-P)] and, at least in some species, low levels of estradiol. The biosynthesis of steroids mainly follows the Δ^4 pathway. 5-reduced androgens, 17,20-P sulfates, and different steroid glucuronides may also be synthesized in large amounts. Plasma levels of sex steroids, which partly reflect testicular steroid production, undergo prominent seasonal variations. Androgens and progestins often peak before and/or during the spawning season. Androgens are effective in supporting either the whole process of spermatogenesis or at least some steps such as spermatogonial multiplication and spermatocyte formation (guppy) or maturation (killifish). As in higher vertebrates, male sex steroid hormones play a role in pituitary maturation and regulation of gonadotropin gene expression, in the establishment of male secondary sexual characters and reproductive behavior, and may also participate in the initiation (androgens) or the amplification (17,20-P) of milt production. When released into the water, the steroids or their metabolites can serve as pheromones.

D. Aspects of Testis Regulation by the Pituitary

Duality of gonadotropins (GTH-1 and GTH-2) is well established in many but not all species of fish. GTHs play a prominent role in the regulation of the male steroidogenesis, with GTH-1 and -2 sharing a similar spectrum of steroidogenic activities. Pituitary extracts also have a potent effect on spermatogenesis (maintenance, restoration, and initiation of a precocious cycle), but information is scarce regarding GTH effects on germ cell development that would not be mediated by steroids.

However, testicular development or recrudescence (when spermatogonial proliferation, meiosis, and the beginning of spermiogenesis may take place) is associated, at least in salmonids, with increasing levels of GTH-1. Type I GTH receptors are detected in the germinal epithelium of immature and mature salmon, probably on the Sertoli cells, but there is no evidence of an effect of GTHs on the function of these cells. In salmonids, the end of the reproductive cycle (when spermiogenesis, spermiation, and spawning may take place) is associated with increasing levels of GTH-2 and the appearance of specific GTH-2 receptors in the steroidogenic interstitial tissue. Moreover, GTH-2 has been found more potent than GTH-1 in stimulating steroiogenesis at this stage. Growth hormone may participate in steroid regulation in killifish and during the final stage of the salmonid testicular cycle through interaction with specific binding sites found in testicular membranes at all spermatogenic stages.

E. The Testis in Ambisexual Species

There are three main types of ambisexuality distributed among 13 families belonging to five orders of teleosts: protogyny (functional female first) which is clearly preponderant, protandry (functional male first), and simultaneous hermaphroditism.

In most protogynous species gonads are initially purely ovarian with no apparent testicular tissue. When female and male tissues are present (ovotestes), they are either intermixed or in contact (undelimited types). In all protogynous species the male tissue becomes functional only after the female tissue has ceased to function and a massive degeneration of the former ovarian tissue takes place during sex reversal. Some young oocytes may still be present amid the testicular tissue. The testis retains the ovarian lumen and usually an ovarian lamellar form. It is not well-known how the functional testis develops and from which source the somatic parts of the future testis arise. PGCs, which are undifferentiated bipotential cells, are at the origin of the male germline during sex reversal. In protogynous sparids and maenids, gonads contain male and female tissues separated by a membrane of connective tissue (delimited type), and, after sex reversal, rudiments of the former ovary may sometimes remain adjoined to a fully developed testis.

In all protandric teleosts, previtellogenic oocytes (called "testis–ova" or "oocyte-like cells") are present and newly formed even during the active male phase (they are also observed in testes of some gonochoristic species). Because of the strong structural dimorphism between testis and ovary, sex inversion requires a complete, radical reorganization of the gonad. Macrophages and various immune blood cells can take an active part in the elimination of degenerating testicular tissue. Only small sterile testicular remnants may be attached to one side of the ovary after sex reversal in protandrous species with delimited gonadal tissue (sparids).

In the ovotestes of the simultaneous hermaphrodites (serranidae and *Rivulus marmoratus*) both male and female regions, which are separate but lie in close proximity, mature simultaneously.

Whatever the type of hermaphrodism, in primary as well as in secondary males, the cytological organization of active testes and the functional activity of the various cell types generally do not seem to differ from those of gonochoristic species.

III. EXCRETORY DUCTS AND ACCESSORY GLANDS

A. Efferent and Deferent Ducts

1. Histology and Cytology

The efferent duct system of the male gonad of teleosts is simpler and shorter than in mammals and elasmobranchs: There is no rete testis and no epididymis. It consists of two parts, the efferent ducts (vasa efferentia) and the spermiduct (or sperm duct or deferent duct; vasa deferentia). The section of duct located either along the surface of the testis or within the testicular tissue is sometimes considered to correspond to a third intermediate part. At the time of testicular morphogenesis, teleost vasa deferentia are formed by somatic cells derived from the coelomic wall and are not a part of the nephric duct or Wolffian duct as is the case in mammals. They usually coalesce to an unpaired portion which empties into a genital papilla particularly developed during the breeding season or in a urogenital sinus or a cloaca. In simultaneous hermaphrodites, sperm ducts and oviducts are separate. In protogynous species, the spermiduct arises from a system of interconnected crevices in part of the gonadal wall by splitting of muscle layers of the ovarian capsule; therefore, the form of the spermiduct differs between primary and secondary males. In some protandrous teleosts, little trace of a sperm duct remains in the ovary, whereas in others it is unclear whether it persists in the female.

The morphology of the deferent sperm ducts varies conspicuously between species: They may be twice as long as the testis (trout) or very short (pike), and either narrow or wide. In seasonal breeders, the spermiducts undergo marked morphological changes, especially an increase in volume at the time of spermiation.

In some Atherinomorphs, the epithelial cells of the efferent ducts are considered to be modified pericystic Sertoli cells that have integrated the epithelium. In the pike, the epithelia lining the three consecutive duct parts show only slight differences. The wall of the deferent sperm duct consists of several layers including connective tissue, smooth muscle fibers, and a monolayered epithelium. The epithelial cells may be covered with microvilli and they are interconnected with each other by membrane specializations, including tight junctions. During the breeding season, this epithelium may have a high secretory activity (apocrine secretion and exocytosis). At the same time, the epithelial cells and/or cells in the spermiduct's stroma display 3β-HSD activity and, in some species, ultrastructural features of steroidogenic cells.

2. Functions and Control

Important functions are supported by the excretory ducts. Unlike in mammals and elasmobranchs, in most teleosts, the sperm cells undergo discrete maturational changes along the efferent system: In salmonids they improve their potential for motility as a consequence of the increase in seminal bicarbonate concentration and pH, the latter increase being under the control of 17,20-P. The sperm ducts primarily serve as storage organs and as a transport route for spermatozoa, with the storage capacity greatly differing between species according to the size of the ducts. In trout, the sperm duct is able to contract spontaneously. The epithelial cells form a barrier impermeable to large molecules that is continuous with the Sertoli cell barrier. These cells are involved in ion transport, which controls the ionic composition of seminal fluid. In Salmonids the sperm duct epithelium actively secretes K^+, which maintains the spermatozoa immotile, and absorbs Na^+. Gonadotropins stimulate directly and rapidly this ion transport, and androgens (and possibly 17,20-P) could also participate in this control by maintaining ion transport on a long-term basis. The epithelial cells also participate in the composition of the seminal fluid, which is essential for sperm survival, by secreting proteins, enzymes, lipids, and monosaccharides. In the surfperch, they produce proteins promoting the cohesion of the spermatophores. Especially during the postspawning period, remaining spermatozoa are resorbed by the spermiduct epithelium, which has a high phagocytic activity.

Sperm duct tissues secrete androgens under GTH control; at least in some species, these steroids may control various activities of the cells lining the spermiducts.

3. A Peculiar Situation Encountered in Bleniids

In this family, there is a testicular gland located between the spermatogenetic zone of the testis and the sperm duct (vasa deferentia). This consists of tubules (vasa efferentia?) that germ cells must pass through and it supports the differentiation process of the spermatids, which terminate spermiogenesis there. This gland, which also contains steroidogenic cells, is considered an original structure assuming intermediate functions between those of a testis devoid of stem cells and those of an accessory gland.

B. Seminal Vesicles

1. Anatomy and Histology

In some Siluriformes families and in Gobiidae, seminal vesicles are present in the male reproductive tract, and are connected to the sperm duct. While there is only one pair of vesicles in some catfish species, they can be numerous (up to 44) in others (Fig. 4). Their shape may vary from short and broad lobes to elongated finger-like lobes. In Gobiidae there are two pairs of vesicles. Both testes and seminal vesicles develop from the genital ridge: The vesicles are homologous to testes devoid of spermatogenic cells (and in some species a transition may be observed between successive lobes, with the progressive disappearance of round germ cells). The vesicles, encapsulated in a tunica albuginea, consist of branched tubules, lined by a simple epithelium and surrounded by connective tissue and interstitial cells which presumably are homologous to Leydig cells. These cells are absent in Ictaluridae and in Gobiidae; in some Gobiidae, the epithelium itself has steroidogenic activity. The epithelial cells display features of secretory cells. The seminal vesicles show seasonal changes in size correlated with those of the testes and controlled by androgens. The epithelial and Leydig cells also undergo cyclic seasonal changes in size, morphology, and activity.

2. Functions

When present in the catfish seminal vesicles, the steroidogenic cells synthesize testosterone, 11-oxygenated androgens, and androgen or 5β-reduced steroid glucuronides. In African catfish, the capacity to synthesize steroid glucuronides is more pronounced in the seminal vesicles than in the testis; these water-soluble steroids may serve as sex pheromones. Steroidogenic activity undergoes seasonal variations and is usually strongly enhanced at the time of full spermatogenesis and of breeding. In the seminal vesicles of one Gobiid species, the steroid metabolic patterns closely resemble those in the testes.

In most species, the epithelial cells of the tubules have a secretory activity which is particularly prominent during the breeding period. Although polysaccharides, phospholipids—a possible source of energy for the sperm cells—and proteins are often present, the composition of the secretion varies between the species investigated.

Two main functions have been proposed for the seminal vesicles: temporary storage of sperm and optimization of the fertilizing ability of the gametes. Depending on the species the vesicles may have only one or both of these functions. In the latter case, some vesicles may be devoted to produce seminal fluid while spermatozoa are present in others, mixed with an amorphous material. In African catfish, the

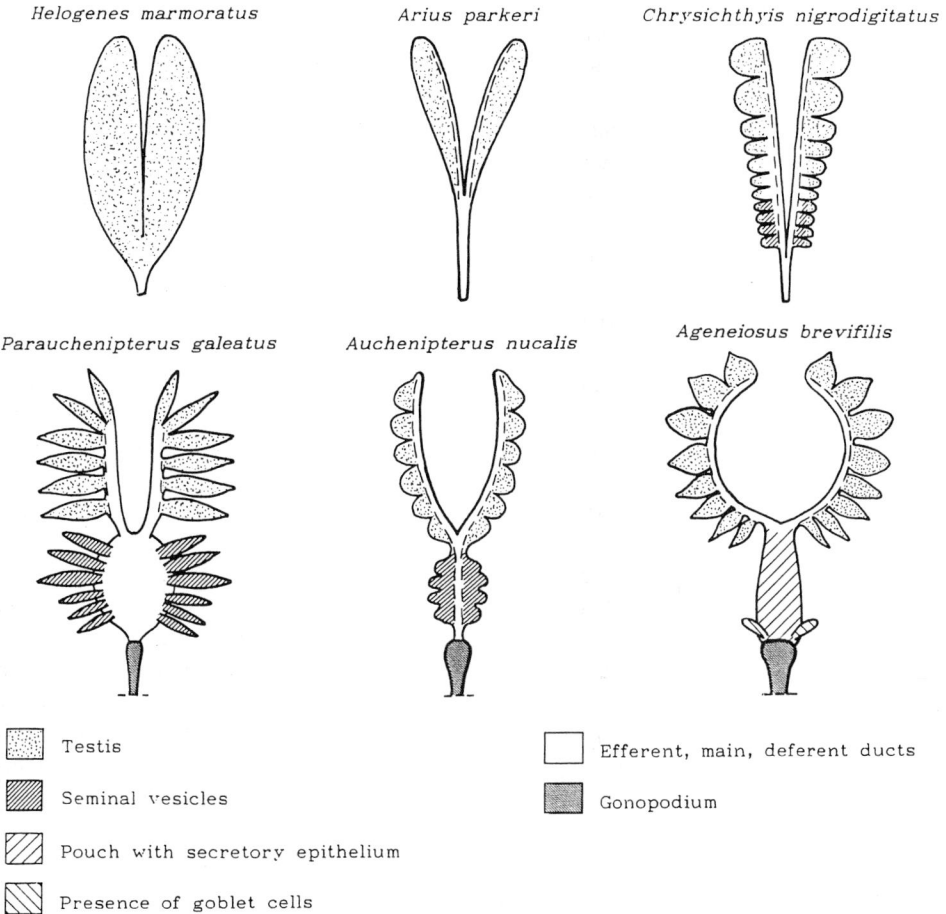

FIGURE 4 Illustration of the male reproductive tract in some Siluriformes (catfishes). Note the diversity in morphology and organization (adapted from Loir *et al.*, Aq. Liv. Res., Gauthier-Villars Editeur, 1989).

seminal vesicle fluid prolongs the period of motility of the sperm cells.

IV. SECONDARY REPRODUCTIVE STRUCTURES

A. Gonopodia

In internally fertilizing teleosts, efficiency of sperm transfer to the female reproductive tract is accomplished by a gonopodium or copulatory organ, either tubular or not, which corresponds to the first ray of the anal fin [in elasmobranchs there are two copulatory organs (claspers) which are derived from the pelvic fins]. The gonopodium is a secondary sexual character developed when the males reach maturity.

B. Priapium

In other species (Phallostetidae), a priapium present under the throat/abdomen of males is a unique copulatory structure in which two elongated parts are used to hold the female during mating; it also comprises a papillary component or seminal papilla used in the transfer of sperm bundles. The priapium is primarily derived from the pelvic skeleton.

C. Anal Glands and Pouches

In male Blenniids, two "anal glands" lie just behind the urogenital papilla; they are well developed during the breeding season. Unpaired or paired "pouches" occur in some Siluridae and Blenniidae. They seem to be outgrowths of the posterior spermiduct. In

Siluridae, they have a conjunctivomuscular wall. Their function and the possible occurrence of changes related to sexual maturation are not known. The presence of cul-de-sacs projecting from the vas deferens, of small diverticulum at the end of the sperm duct, have been mentioned in various teleost species.

Finally, although the structure is not a part of the male reproductive system in the strict sense, we mention the ventral sac-like brood pouch in Syngnathid males. This pouch, in which the eggs are incubated, is controlled by testosterone.

See Also the Following Articles

FEMALE REPRODUCTIVE SYSTEM, FISH; FISH, MODES OF REPRODUCTION IN; MALE REPRODUCTIVE SYSTEM, AMPHIBIANS; MALE REPRODUCTIVE SYSTEM, BIRDS; MALE REPRODUCTIVE SYSTEM, HUMANS; MALE REPRODUCTIVE SYSTEM, INSECTS; MALE REPRODUCTIVE SYSTEM, NONHUMAN MAMMALS; MALE REPRODUCTIVE SYSTEM, REPTILES

Bibliography

Billard, R. (1986). Spermatogenesis and spermatology of some teleost fish species. *Reprod. Nutr. Dev.* **26**, 877–920.

Billard, R., Cosson, J., Crim, L. W., and Suquet, M. (1995) Sperm physiology and quality. In *Broodstock Management and Egg and Larval Quality* (N. R. Bromage and R. J. Roberts, Eds.). Blackwell, Oxford, UK.

Borg, B. (1994). Androgens in teleost fishes. *Comp. Biochem. Physiol. C* **109**, 219–245.

Fostier, A., Jalabert, B., Billard, R., and Breton, B. (1983). The gonadal steroids. *Fish Physiol.* **9**(A), 277–345.

Grier, H. J. (1996). Comparative organization of Sertoli cells including the Sertoli cell barrier. In *The Sertoli Cell* (L. D. Russel and M. D. Griswold, Eds.), pp. 703–739. Cache River Press, Clearwater, FL.

Hamaguchi, S. (1992). Sex differentiation of germ cells and their supporting cells in *Oryzias latipes*. *Fish Biol. J. MEDAKA* **4**, 11–17.

Lahnsteiner, F., Patzner, R. A., and Weismann, T. (1993). The efferent duct system of the male gonads of the European pike (Esox lucius): Testicular efferent ducts, testicular main ducts and spermatic ducts. *J. Submicrosc. Cytol. Pathol.* **25**, 487–498.

Loir, M., Sourdaine, P., Mendis-Handagama, S. M., and Jégou, B. (1995). Cell–cell interactions in the testis of teleosts and elasmobranchs. *Microsc. Res. Technique* **32**, 533–552.

Pudney, J. (1996). Comparative cytology of the Leydig cell. In *The Leydig Cell* (A. Payne, M. P. Hardy, and L. D. Russell, Eds.), pp. 97–142. Cache River Press, Clearwater, FL.

Male Reproductive System, Human

John F. Redman

University of Arkansas for Medical Sciences

I. The Testis
II. The Epididymis, Ductus (Vas) Deferens, and Seminal Vesicles
III. The Scrotum
IV. The Spermatic Cord
V. The Prostate
VI. The Penis

GLOSSARY

corpora cavernosa The primary erectile bodies of the penis.
corpus spongiosum The spongy body which contains the urethral bridge from the prostate to the tip of the glans penis.
ductal (vas) deferens A thin, white muscular tube which contains the duct for sperm transport from the testes to the urethra.

epididymis A crescent-shaped structure covering the posterior aspect of the testis which contains a tightly convoluted epididymal tubule that conducts sperm from the testis to the vas deferens.

Leydig cells Round or ovoid cells which are found in clusters in the interstitium of the testis between the seminiferous tubules and produce the preponderance of testicular steroids.

prostate A complex muscular and glandular organ which surrounds the urethra at its juncture with the bladder neck.

rete testis Sperm transporting ducts which are contained within the mediastinum of the testis and connect the collecting ducts of the testes with the epididymal tubule.

Sertoli cells Highly irregular branched cells which rest on the basement membrane of the seminiferous tubules and support the germ cells during their development.

spermatogonia A first-generation undifferentiated male germ cell.

tunica albuginea The grayish, fibrous covering of both the testes and corpora cavernosa of the penis.

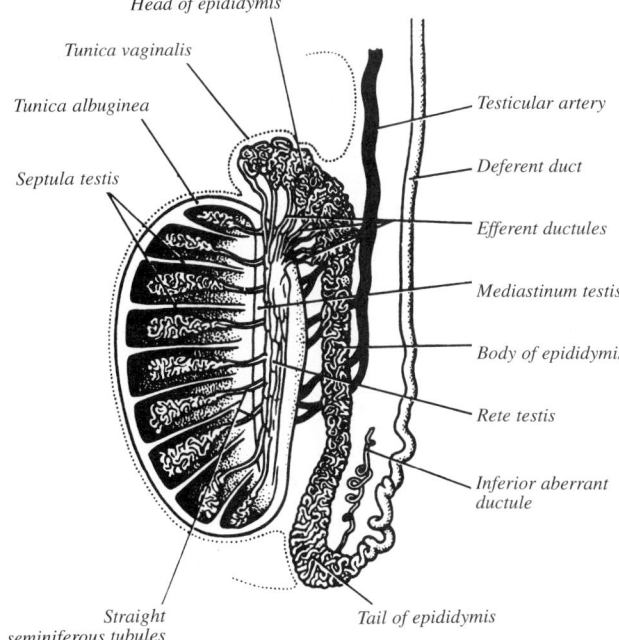

FIGURE 1 Schematic representation of intratesticular, epididymal, and vasal anatomy [reprinted with permission from H. Gray, *Gray's Anatomy*, 36th ed. (British), p. 1411, Saunders, Philadelphia, 1980].

The male reproductive system is composed of the organs of germ cell formation and maturation, the testes; the conduit system for delivery of the germ cells, the epididymis, vas deferens, and the urethra; the accessory organs, the seminal vesicles, Cowper's glands, and prostate gland; and the organ of copulation, the penis.

I. THE TESTIS

A. Gross Anatomy

The testes are paired, elongated, ovoid structures whose function is twofold: production of the male gamete and testosterone. The adult testis ranges from 15 to 30 ml in volume and measures on the average 4 or 5 cm in length and 2 or 3 cm in depth. Normally the testes are positioned so that the long axis is vertical (Fig. 1). Within the scrotum, the testis is free on its lateral aspects, whereas the dorsal aspect is covered by the epididymis. The testis is relatively fixed to the dorsal wall of the scrotum by the investment of serosa, a remnant of the processus vaginalis peritonei, termed the tunica vaginalis. The testis is grayish in coloration, which is the color of the dense capsule of the testicle, the tunica albuginea, which is composed of collagenous tissue interspersed with smooth muscle cells. Just beneath the tunica albuginea is a vascular layer, the tunica vasculosa. Emanating from the tunica albuginea are septa which divide the parenchyma of the testis into 250–400 lobules each of which contain two to four tightly convoluted seminiferous tubules; if stretched to full length each tubule would measure approximately 2 m. Each tubule has a U-shaped configuration with the midportion of the U being located just beneath the tunica albuginea. Each end of the U joins within the lobule to form a straight tube termed the ductuli recti which courses toward the cranial dorsal aspect of the testis to form a series of anastomotic channels termed the rete testis. The rete testis gives rise to 10–15 channels termed the efferent ductuli that pass from the testis into the epididymis, which covers the cranial dorsal aspect of the testis. This area of the testis is called the mediastinum.

B. Vasculature

The primary arterial supply to the testis is by the testicular or gonadal arteries which arise high in the abdomen. The arteries are paired and originate almost from the midline of the ventral aspect of the aorta just caudal to the origin of the renal arteries. Accompanied by the gonadal veins, they course laterally and caudally in the intermediate stratum of retroperitoneal connective tissue and then, joined by the vas deferens at the level of the internal abdominal ring, course sharply medially through the inguinal canal and then caudally into the scrotum to reach the testis. At the level of the testis, or just proximal, the arteries divide to form so-called main branches of the testicular artery and then enter the posterior aspect of the testis through the tunica albuginea but not through the mediastinum of the testis per se. The main branch divides into branches at the level of the rete testis termed centripetal and centrifugal arteries, which further divide into interlobular arterioles. Additional vasculature to the testis derives from vasal, epididymal, and external spermatic (cremasteric) arteries which form anastomotic communications.

The venous drainage of the testis is via two groups of venous channels. Centripetal veins drain toward the rete testis, whereas the centrifugal group drains toward the tunica albuginea, where they join with larger veins which course under the tunica albuginea toward the rete testis. These two groups join to form the pampiniform plexus of veins of the spermatic cord that accompany the testicular artery and the vas passing through the inguinal canal and internal abdominal ring to course medially and cranially through the intermediate stratum of retroperitoneal connective tissue, forming one vein in the vicinity of the internal ring. The left testicular vein drains into the caudal aspect of the left renal vein usually immediately opposite the adrenal vein. On the right side, the gonadal vein joins the vena cava on its lateral aspect just caudal to the juncture of the right renal vein and vena cava. A valve may be present at these junctures, most commonly on the right side. Additional venous drainage of the testis is via communication with the vasal vein and the external spermatic (cremasteric) veins.

C. Innervation

Accompanying its vasculature, the testis is innervated with autonomic sympathetic postganglionic and visceral afferent fibers. These nerves comprise a superior spermatic group deriving from the renal and intermesenteric plexus, a middle spermatic group from the superior hypogastric plexus and cranial aspects of the hypogastric nerves, and an inferior spermatic group from the inferior hypogastric plexus.

D. Lymphatics

The lymphatic drainage of the testis originates from lymphatic capillaries in the septa and ultimately from large lymph channels in the spermatic cord which follow the testicular vasculature. The ultimate drainage is to lymph nodes near the origin of the gonadal vessels. On the right side, these nodes include interaortocaval nodes adjacent to the right renal vein and on the left side preaortic and left periaortic nodes.

E. Microscopic Anatomy

The bulk of the testis is formed of seminiferous tubules with approximately 20–30% of the volume from the interstitial tissue. The interstitium is formed of loose connective tissue through which courses blood vessels, lymphatics, and nerves. Prominent in the interstitium are the Leydig cells, which produce testosterone (Fig. 2). It is estimated that the Leydig cells comprise 5–10% of the testicular volume. Leydig cells usually occur in clusters and are frequently adjacent to capillaries. The seminiferous tubules are surrounded by peritubular tissue, which presents a layered appearance with the outer layer being composed of myofibroblasts and myoid cells and an inner collagenous layer supporting the basement membrane of the tubule. Within the tubules along the basement membrane are two populations of cells: the nonproliferating Sertoli or supporting cells and the proliferating germinal cells (Fig. 3). The Sertoli cells rest on the basement membrane and extend flame-like into the lumen; they have concavities and grooves corresponding to the differentiating germinal cells which they support as they progress from

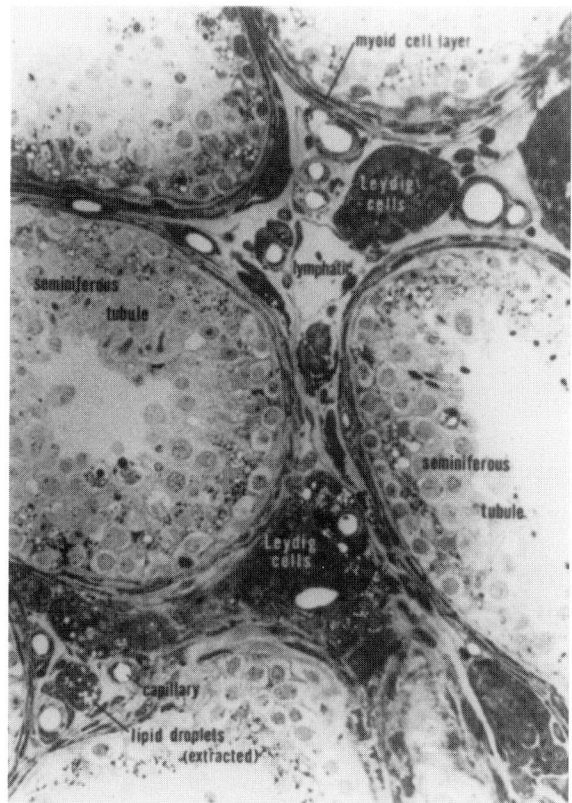

FIGURE 2 Light photomicrograph of testis demonstrating relationships of seminiferous tubules, Leydig cells, and interstitium [reproduced with permission from A. K. Christiansen, Leydig cells, In *Handbook of Physiology* (D. W. Hamilton and R. O. Greep, Eds.), Section 7: Endocrinology, Male Reproductive System, p. 60, American Physiological Society, Bethesda, MD, 1975].

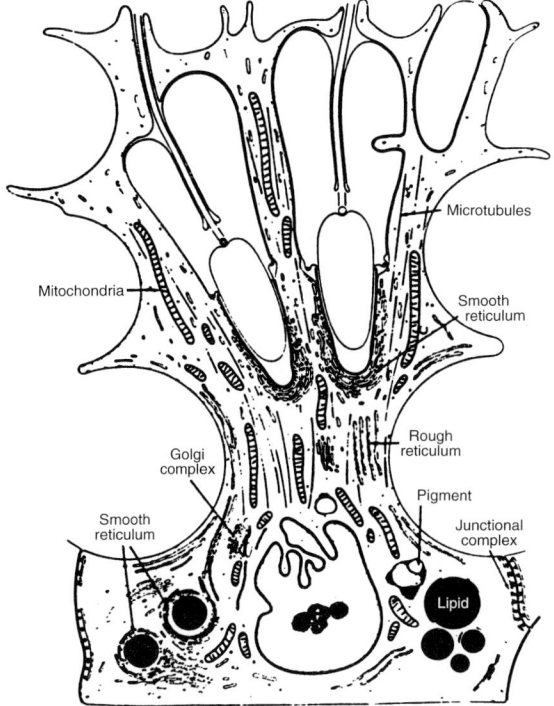

FIGURE 3 Schematic representation of Sertoli cell showing relationships to germ cells [reproduced with permission from A. K. Christiansen, Leydig cells, In *Handbook of Physiology* (D. W. Hamilton and R. O. Greep, Eds.), Section 7: Endocrinology, Male Reproductive System, p. 49, American Physiological Society, Bethesda, MD, 1975].

the least differentiated cells adjacent to the basement membrane to the mature spermatids. The progression from primary germ cell to mature spermatocytes is as follows: The primary germ cell or type A spermatogonia progresses from a pale nucleus stage to a cell with a dark nucleus which progresses to a type B spermatogonia. Type B spermatogonia give rise by a mitotic division to primary spermatocytes termed preleptotene primary spermatocytes. By meiotic division, the following progressing spermatocytes are seen: leptotene, zygotene, and pachytene. The resulting two cells are secondary spermatocytes. Further meiotic division occurs, producing two spermatids that develop without further division into spermatozoa.

II. THE EPIDIDYMIS, DUCTUS (VAS) DEFERENS, AND SEMINAL VESICLES

A. The Epididymis

The epididymis for the most part is a structure consisting of a single, highly convoluted duct (Fig. 1). It is crescentic in form and is held closely to the length of the dorsal aspect of the testis by the visceral tunica vaginalis. The dorsal aspect of the epididymis is covered by the internal spermatic fascia. The epididymis is anatomically described as having three segments or regions: the caput (also termed the globus major or head), the corpus, and the cauda (also termed the globus minor or tail).

The ductal portion of the epididymis is formed as a continuation of the 8–15 ductuli efferente which have passed from the rete testis through the mediastinum of the testis. On entering the head of the epididymis, the ductuli efferente are straight but abruptly become markedly convoluted to form small lobular-like cones. Each duct then drains into the common duct, whose highly convoluted configuration forms the bulk of the epididymis.

The epididymal vasculature courses through connective tissue septa which give the epididymis a lobulated appearance. The head and body of the epididymis are vascularized by a branch of the testicular artery. The tail of the epididymis receives its blood supply from the artery of the vas deferens and from branches of the cremasteric artery.

The epididymis is innervated by intermediate spermatic nerves from the hypogastric plexus and inferior spermatic nerves from the vesical plexus. The lymphatic drainage follows the vasculature and is virtually the same as that of the testis.

The epithelium of the epididymal duct structures transitions from a low cuboidal cell to a high cuboidal epithelium in the ductuli efferente and is characterized by microvilli or stereocilia. The epididymal duct itself is characterized by two distinct cells: a columnar cell with stereocilia, termed principal cells, and round cells arranged along the basement membrane, termed basal cells. The stereocilia progressively become shorter proximally. Proximally the basement membrane is covered with myocytes and fibrocytes in a circular pattern. In the most distal aspects of the epididymal duct a longitudinal muscle coat is acquired.

B. Ductus (Vas) Deferens

The ductus or vas deferens is a white tubular muscular structure 2 or 3 mm in diameter and is found bilaterally (Fig. 3). It originates from the epididymal cauda (tail) and then turns acutely cranially to parallel the epididymis. The caudal portion of the ductus deferens is convoluted, although the convolutions are not nearly as tortuous as that of the epididymis. At approximately the juncture of the corpus and the caput of the epididymis, the vas straightens and courses with the internal spermatic vessels within the spermatic cord, turning laterally at the pubic tubercle to course through the inguinal canal (Fig. 4). The vas then turns acutely medially at the level of the medial aspect of the internal abdominal ring, separating itself from the internal spermatic vessels, and courses cranially turning medially to pass beneath the medial umbilical ligament and then between the terminal portion of the ureter and the base of the bladder. It then turns abruptly caudally and widens in what is known as the ampulla of the vas. As the ampulla of the vas nears the base of the prostate gland, it is joined by the duct to the seminal vesicle on its lateral aspect. The vas then passes through the central zone of the prostate in the ejaculatory duct, which opens into the prostatic urethra as small punctate openings on either side of the prostatic utricle situated on the veru montanum.

The ductus deferens is vascularized by the deferential artery which may arise from either the superior or inferior vesical arteries. Innervation to the ductus deferens is autonomic, with both sympathetic and parasympathetic innervation via the hypogastric and pelvic nerves.

Microscopically, the ductus deferens is noted to have a densely muscular wall which is three layered:

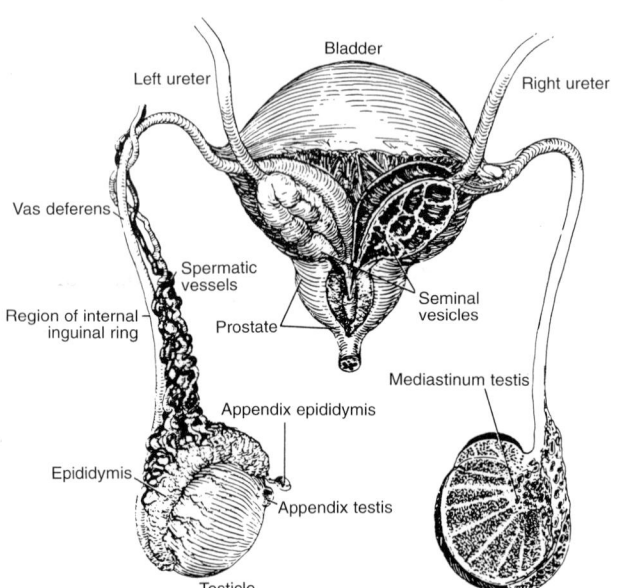

FIGURE 4 Posterior view of bladder showing relationships of testes, epididymis, vas deferens, seminal vesicles, and prostate [reproduced with permission from E. A. Tanagho, Anatomy of the genitourinary tract, In *General Urology* (D. R. Smith, Ed.), 14th ed., p. 8, Lange, Los Altos, CA, 1984].

an outer and inner layer of longitudinal smooth muscle and an inner layer of circularly arranged smooth muscle fibers. Between the inner muscular layer and the epithelium of the lumen is a lamina propria. The epithelium, which is folded, consists of pseudostratified epithelium with stereocilia and basal cells. The ampulla of the vas is similar histologically although it has more elastic fibers.

C. Seminal Vesicles

The seminal vesicles are very grayish, opalescent, cystic-like structures which extend craniolaterally from the midportion of the base of the prostate (Fig. 4). Each connects to the lumen of the ampulla of the vas as it transitions into the ejaculatory duct. They are approximately 3–5 cm in length and 1 or 2 cm in width. The structure has a thin, two-layered muscular wall, an outer longitudinal and an inner circular layer, and is lined with a greatly enfolded epithelium which is composed of cuboidal and columnar cells.

III. THE SCROTUM

A. Gross Anatomy

The scrotum is a complex cutaneous pouch positioned caudal to the base of the penis. It is characterized by corrugated or rugated skin and a raphe or midline seam which extends from the base of the penis to the perineum giving a two-compartment configuration to the scrotum. Closely adherent to the scrotal skin is a layer containing abundant smooth muscle fibers, the Dartos tunic. The involuntary contractions of the Dartos tunic convey alternately a smooth or rugated appearance of the scrotum, a configuration which is influenced by temperature changes. The Dartos tunic, which is superficial membranous fascia, is contiguous with the Colles fascia of the perineum. Beneath the Dartos tunic are three investments, termed fascia, which are contiguous with the coverings of the spermatic cord in the inguinal canal and have their derivation from the abdominal wall. These layers from superficial to deep are the external spermatic fascia, cremasteric muscle and fascia, and the internal spermatic fascia.

The external spermatic fascia is contiguous with the fascia of the external oblique aponeurosis. The underlying cremasteric muscle and fascia layer, which is contiguous with the internal oblique muscle, is characterized by long loops of muscular fascicles interspersed with fascia. The layer is also notable because of the cremasteric vasculature which is distributed throughout the layer. The internal spermatic fascia is contiguous with the transversalis fascia. The tunica vaginalis or the remnant of the processus vaginalis peritonei is considered part of the scrotal wall, particularly the portion which lies loosely adherent to the internal spermatic fascia. The tunica vaginalis actually forms a fluid-filled sac which contains the testis and epididymis, which invaginate the sac so that only the posterior dorsal aspect of the testis and epididymis are not covered. The portion of the sac adjacent to the internal spermatic fascia is termed the parietal layer, whereas the portion that covers the testis and the epididymis is the visceral layer.

B. Vasculature

The scrotal wall is vascularized anteriorly by the superficial external pudendal arteries which arise from the femoral arteries. The posterior aspects of the scrotum are supplied by the scrotal arteries that arise from the perineal vessels, which are branches of the internal pudendal artery.

C. Innervation

Innervation of the anterior scrotum is by the ilioinguinal and genital branch of the genitofemoral nerve. The posterior aspects of the scrotum are innervated by scrotal branches of the perineal nerves which arise from the pudendal nerves.

D. Lymphatics

The lymph drainage of the scrotum follows the vasculature, with a majority of the lymphatic drainage of the scrotum draining to the superficial lymph nodes.

IV. THE SPERMATIC CORD

The spermatic cord is the distinct cord-like aggregate of fascial structures which surrounds the vasculature of the testis and the ductus (vas) deferens. The spermatic cord itself is most precisely the aggregate of fascia and vasculature which extends from the internal inguinal ring through the inguinal canal into the scrotum and terminates at the testis. The essence of the spermatic cord is the internal spermatic vessels and ductus deferens, which join closely as they enter the internal ring from their divergent courses. These structures are encased in the intermediate stratum of retroperitoneal connective tissue which continues with them. The vas is located caudal and medial to the spermatic vessels. The fat-laden intermediate stratum may vary in fat content, giving varying degrees of bulk to the cord. The intermediate stratum has an outer-limiting membrane which some investigators have termed the internal spermatic fascia as opposed to the transversalis fascia.

Overlying the internal spermatic fascia are the cremasteric muscle and fascia which in the inguinal canal form one or more thick fascicles, particularly over the craniolateral aspects of the cord. On the caudal and lateral aspect of the cord, the cremasteric muscle is frequently replaced by fascia through which course the genital branch of the genitofemoral nerve, the external spermatic vessels, which may give off perpendicular branches to the muscle per se within the inguinal canal. Within the scrotum, the cremasteric muscle festoons in wide loops interspersed by cremasteric fascia and branches of the external spermatic vessels. Within the inguinal canal the cremasteric muscle and fascia are overlain by the aponeurosis of the external oblique muscle; however, at the level of the pubic tubercle the aponeurosis ends abruptly, forming the external inguinal ring. The fascia of the external oblique aponeurosis, however, continues caudally closely adherent around the cremasteric muscle and fascia. This continuity is called the external spermatic fascia and is contiguous into the scrotum.

Within the inguinal canal and cranial aspects of the scrotum, the internal spermatic veins are more dilated and numerous and are termed the pampiniform plexus. Also frequently visible are prominent testicular lymphatic vessels.

V. THE PROSTATE

A. Gross Anatomy

The prostate is a complex musculoglandular structure which is described as having a chestnut or blunted cone appearance. The base of the prostate abuts against the base and neck of the bladder, whereas the apex of the structure is in contact with the musculature and fascia bridging the arch of the pubis (Fig. 4). The dorsal aspect of the prostate is covered by the rectum from which it is separated by a derivative of the peritoneum termed Denonvillier's fascia. The ventral aspect of the prostate lies dorsal to the pubic symphysis to which it is attached at its base by condensations of endopelvic fascia termed puboprostatic ligaments. The prostate is covered by a fibrous tissue capsule, which is not a true capsule in that it cannot be separated from the prostatic tissue itself. Extending into the capsule are fibrous septa which separate the parenchyma into approximately 50 lobules which contain the glandular elements of the prostate gland. The glands themselves, which are tubuloalveolar glands, connect by tubules to 20–30 prostatic ducts which open into the prostatic urethra. As observed in the midline coronal plane, the urethra forms a 35° angle in the midportion of the prostate (Fig. 5). The tissue anterior to the urethra is termed the anterior ventral fibromuscular stroma. Dorsal to the urethra, the parenchyma is decidedly glandular and is composed of four regions or zones: peripheral zone, central zone, transition zone, and periurethral gland region. The peripheral zone comprises the bulk (75% of the glandular prostate) while the central zone, which is pyramidal in shape, accounts for 20% of the volume. The wide base of the central zone extends from the base of the prostate and then narrows in the region of the angle of the urethra. The transition zone, which comprises 5% of the glandular prostate, is located on either side of the urethra midway between the base of the prostate and the angle of the urethra. The periurethral gland region which comprises just 1% of the glandular prostate, is located periurethrally just proximal to the transition zone.

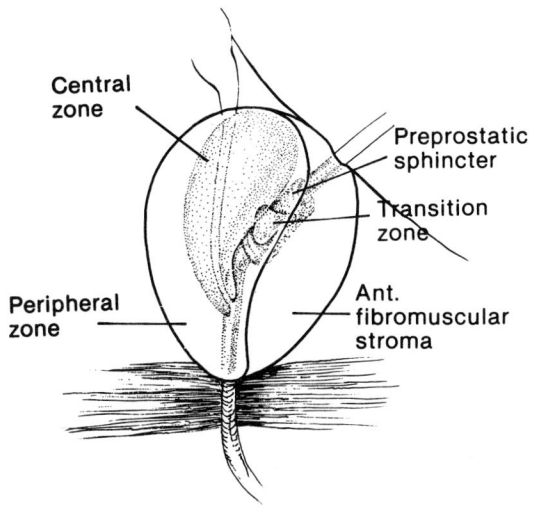

FIGURE 5 Schematic depiction of prostatic zones [reproduced with permission from P. C. Walsh *et al.* (Eds.), *Campbell's Urology*, 5th ed., p. 62, Saunders, Philadelphia, 1986].

The posterior or prostatic urethra is characterized by ridges and a small hillock on its dorsal surface (Fig. 6). In the midline of the dorsum of the urethra is a ridge which extends from the bladder neck to the membranous urethra termed the urethral crest or crista urethralis. At the angle of the urethra in the midportion of the urethra, the crest widens and forms a small hillock, the verumontanum. In the midportion of the verumontanum distally is a short pit, the utricle, which is the caudal remnant of the Müllerian duct. On the dorsal lateral aspects of the verumontanum are the two punctate openings of the ejaculatory ducts. Emanating laterally from the distal aspect of the verumontanum are two folds which course distally and dorsally. Emanating from the verumontanum proximally and extending laterally to the bladder neck are two folds termed the plica urethralis.

B. Vasculature

The arterial supply of the prostate gland is provided by the prostatovesical artery, which is a branch of the inferior vesical artery that arises from the internal iliac (hypogastric) artery. The prostatovesical artery courses along the dorsal lateral aspect of the base of the bladder and then along the dorsal lateral aspect of the prostate, giving off capsular branches which penetrate the prostate at the level of the bladder neck. The prostatovesical artery divides, sending a branch medially which enters the base of the medial aspect of the prostate at the 4- and 8-o'clock positions on the bladder neck. These vessels constitute the urethral group of arteries. Additional vasculature to the apex of the prostate is supplied by branches of the middle hemorrhoidal and internal pudendal arteries.

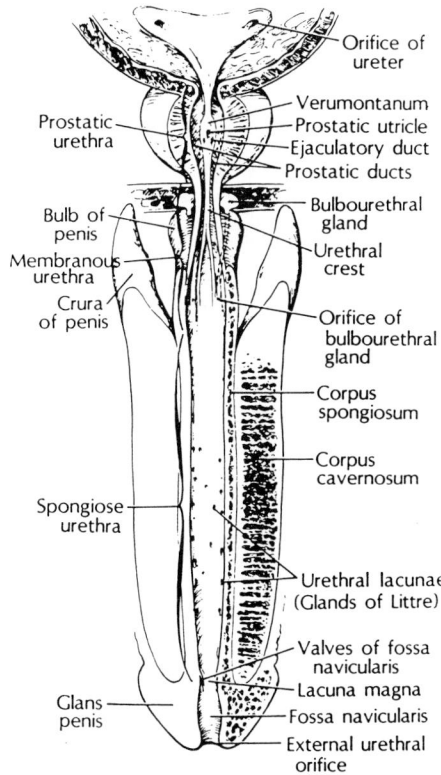

FIGURE 6 Schematic illustration of anterior and posterior male urethra which has been opened to visualize the dorsal aspect of its lumen [reproduced from C. D. Clemente, *Regional Atlas of the Human Body* (C. D. Clemente, Urban, and Schwarzenberg, Eds.), 2nd ed., Fig. 312, 1981].

Venous drainage of the prostate is through the capsular veins which drain into the prostatic venous (Santorini) plexus, which is joined by the dorsal vein of the penis over the ventrum of the prostate. These veins ultimately drain into the hypogastric veins.

C. Innervation

The innervation of the prostate is by parasympathetic and sympathetic nerve contributions, which

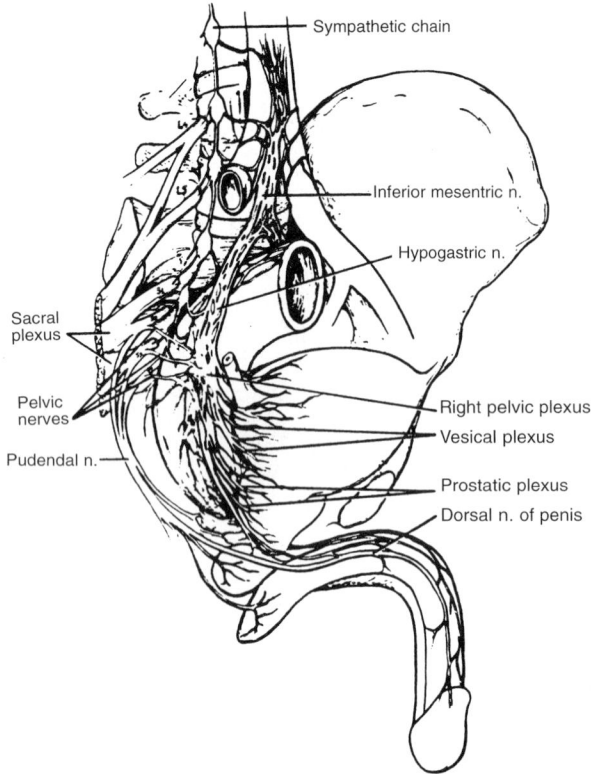

FIGURE 7 Schematic illustration of innervation of pelvic organs [reproduced with permission from J. Y. Gillenwater et al. (Eds.), *Adult and Pediatric Urology*, 2nd ed., Mosby-Year Book, Chicago, 1991].

originate from the rectangular pelvic plexus that lies adjacent to the rectum on either side (Fig. 7). The sympathetic innervation is via the hypogastric nerves, whereas the parasympathetic innervation is by the pelvic splanchnic nerves. The prostatic nerves accompany the prostatovesicular artery forming a neurovascular bundle. The prostatic nerves themselves enter the prostate in the dorsal lateral aspect of the base of the prostate bilaterally from a small prostatic plexus.

D. Lymphatics

Lymphatic drainage of the prostate can be traced from the small lymphatic channels which surround the acini of the prostatic glands. These channels coalesce to form increasingly larger channels which drain peripherally to ultimately form the periprostatic plexus at the level of the surface of the prostate. The lymphatic drainage continues via distinct channels which accompany the vasculature and then to the internal iliac, presacral, and obturator lymph node chains.

E. Microscopic Anatomy

On transverse section, the prostate is noted to have a porous appearance dorsally and laterally because of the glandular composition of these zones of the prostate. The branching glands and ducts arborize in a semicircular pattern from the drainage ducts which empty into the dorsal aspect of the prostatic urethra. The inner mucosal glands drain to the colliculus seminalis.

The acini of the prostate glands are lined by two cell types: a luminal secretory columnar cell and a basal nonsecretory cuboidal cell. Interposed between the secretory cells are neuroendocrine or endocrine–paracrine cells. The ducts are lined by simple or pseudostratified columnar epithelium. The prostatic urethra is lined with columnar epithelium. The prostate's stroma is composed of smooth muscle cells and fibroblasts.

VI. THE PENIS

A. Gross Anatomy

The appendage known as the penis is actually only a portion of the complete structure (Fig. 8). The penile appendage is the pendulous part or corpus, whereas the perineal part is termed the radix or root. The bulk of the penis is formed of three cylindrical erectile structures: two paired corpora cavernosa and a corpus spongiosum. These three structures are surrounded by a fibrous tissue sheath, Buck's fascia.

The corpora cavernosa are filled with a spongy tissue which is in reality a complex system of sinusoids that are large in the central aspects of the corpora and become progressively smaller at the level of the periphery. The spongy tissue of the corpora cavernosa is encased with a dense expansile covering termed the tunica albuginea which is a bilayered structure with multiple sublayers. The corpora cavernosa have their origin along the ventral aspects of the pubic rami on either side and meld together

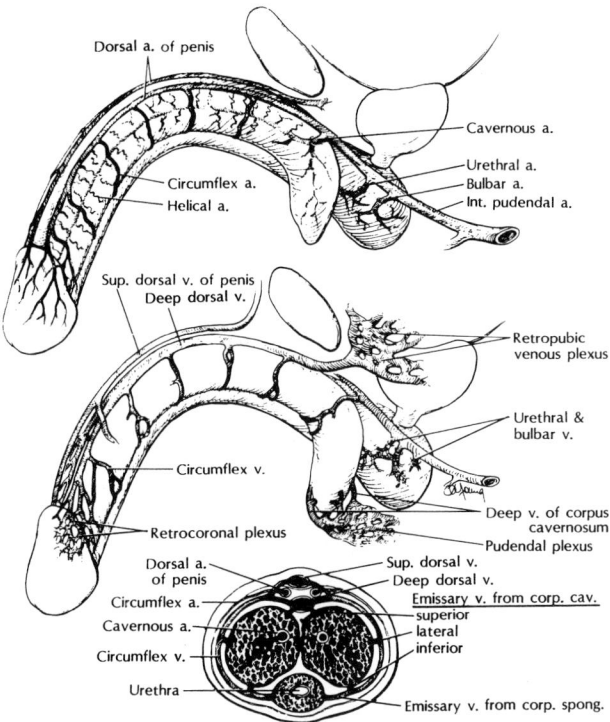

FIGURE 8 Schematic illustration of penis showing vasculature of penis and relationships of the corpora cavernosa to the pubis and cross section of the pendulous portion of the penis [reproduced with permission from J. Y. Gillenwater et al. (Eds.), *Adult and Pediatric Urology*, 2nd ed., Mosby-Year Book, Chicago, 1991].

circular bundles from which radiate intracavernous pillars or struts. In the groove formed on the undersurface of the corpora cavernosa lies the corpus spongiosum, which is also covered by a thin tunic. The sinuses of the spongiosum are large. The terminal portion of the corpus spongiosum fans out distally to form a cap over the tips of the corpora cavernosum, the glans penis. Coursing through the corpus spongiosum is the anterior urethra, which terminates at the ventral tip of the glans as the urethral meatus. The proximal aspect of the corpus spongiosum is bulbous and extends dorsal to the termination of the urethra. On cross section, the corpus spongiosum presents a horseshoe configuration, with the area adjacent to the cavernosa being extremely thin.

The pendulous part of the penis is covered by skin that, over the glans, forms a sheath or prepuce (foreskin), which has an outer or cutaneous leaf and an inner leaf of moist epithelium (Fig. 9). The penile skin, which is thin and hairless, is underlain by a thin layer termed the Dartos tunic which overlies the Buck's fascia. Within the perineum, the bulb of the urethra is covered by a pennate muscle, the bulbospongiosa muscle, whereas laterally the corpora cavernosa are covered by the ischiocavernosus muscle and aponeurosis.

The glans penis is the acorn-shaped terminus of the corpus spongiosum which fits cap-like over the distal ends of the corpora cavernosa. Its flared proximal edges are termed the corona of the glans. Connecting the glans to the prepuce ventrally is a small bridle or frenulum.

beneath the pubic arch. The corpora are separated by a fenestrated septum. The outer layer of the tunica albuginea is composed of longitudinal bundles of collagen and elastin. The inner layer is formed of

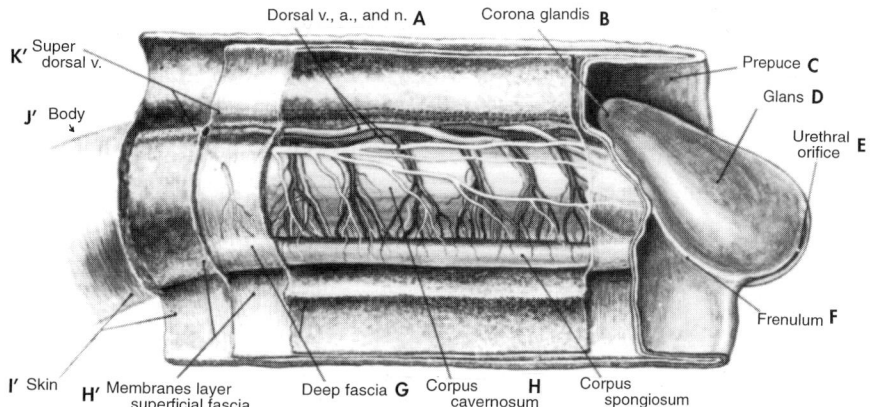

FIGURE 9 Schematic illustration of investments of penis. Note neurovascular bundle and its dispersement (reproduced with permission from R. C. Crafts, *A Textbook of Human Anatomy*, 2nd ed., p. 363, 1979. © 1979 John Wiley and Sons, Inc.).

B. Vasculature

The arterial vasculature of the penis is by three or four pairs of arteries which derive from the internal pudendal arteries: the bulbourethral, which may be represented by separate bulbar and urethral arteries; the profunda or cavernous arteries; and the dorsal penile arteries (Fig. 8). The bulbar arteries vascularize the bulb of the corpus spongiosum, whereas the urethral arteries enter the corpus spongiosum, which supplies the organ throughout its length, even supplying vasculature to the glans. The profunda arteries enter the corpora cavernosa at the level of the divergence of the corpora in the central portion of each corporal body. These vessels give rise to helical vessels, which give rise to short end-arteries which terminate as vascular spaces or lagoons comprising the spongy tissue of the corpora cavernosa. These sinusoids are larger in the central portion of each corporal body and smaller at the periphery. The dorsal penile arteries course in the V-shaped space over the midportion of the dorsum of the corpora cavernosa, giving off circumflex arteries and terminating in the glans penis. The arteries of the penis interconnect via numerous anastomotic vessels.

The venous drainage of the penis is complex and has numerous communications. There are three primary venous drainage systems: the superficial dorsal veins, the deep dorsal veins, and the cavernous veins. The superficial dorsal veins course the length of the penis between the Dartos tunic and Buck's fascia and drain to the saphenous and pudendal veins. The deep dorsal veins accompany the dorsal penile arteries running between Buck's fascia and the tunica albuginea of the corpora cavernosa. These veins arise from a retrocoronal plexus and are joined by circumflex veins which also drain the corpus spongiosum by emissary veins. The superficial and deep dorsal veins communicate at the level of the corona and also just distal to the pubic symphysis. The deep dorsal vein passes beneath the symphysis pubis over the apex of the prostate, where it trifurcates to form Santorini's plexus on the lateral aspects of the prostate with ultimate drainage to the hypogastric veins. The major venous drainage of the corpora cavernosa is via the deep or cavernous veins which drain from the proximal aspects of the corpora into the pudendal venous plexus. The venous drainage of the corpus spongiosa is primarily through the bulbar and urethral veins.

C. Innervation

The somatosensory innervation of the penis is via several nerves: the dorsal penile, the perineal, the pudendal, and the ilioinguinal (Fig. 9). The autonomic innervation of the penis is from the pelvic plexus, which lies retroperitoneally bilaterally alongside the rectum (Fig. 7). The pelvic plexus is formed by visceral efferent parasympathetic preganglionic fibers from S2–4 nerve roots via the pelvic splanchnic nerves (nervi erigentes) and by sympathetic postganglionic fibers from the thoracolumbar ganglia (T12–L2). The innervation from the pelvic plexus to the corpora cavernosa courses through the intermediate stratum of retroperitoneal connective tissue on the dorsolateral aspects of the prostate.

D. Lymphatics

The superficial lymphatic drainage of the penile skin follows the course of the superficial external pudendal arteries which drain to lymph nodes superficial to the fascia lata in the femoral triangle. The lymphatic drainage of the glans and corpora cavernosa follow the dorsal penile vasculature between the corpora cavernosa and Buck's fascia and terminate in the deep inguinal lymph nodes superficial to the femoral vessels in the femoral triangle deep into the fascia lata.

E. Microscopic Anatomy

The tunica albuginea of the corpora cavernosa is composed of collagen bundles which form roughly two layers: an outer longitudinal layer and an inner circular layer. The collagen bundles rest on an irregularly latticed framework of elastic fibers.

See Also the Following Articles

Female Reproductive System, Human; Male Reproductive System, Amphibians; Male Reproductive System, Birds; Male Reproductive System, Fish; Male Reproductive System, Insects; Male Reproductive System, Nonhuman Mammals; Male Reproductive System, Reptiles

Bibliography

Brock, G., Hsu, G., Nunes, L., Von Heyden, B., and Lue, T. F. (1997). The anatomy of the tunica albuginea in the normal penis and Peyronie's disease. **157**, 276–281.

Fawcett, D. W. (1975). Ultrastructure and function of the Sertoli cell. In *Handbook of Physiology. Endocrinology, Vol. V. Male Reproductive System* (R. O. Greep and E. B. Astwood, Eds.), p. 21. American Physiological Society, Washington, DC.

Mawhinney, M. G., and Tarry, W. F. Male accessory sex organs and androgen action. In *Infertility in the Male* (L. I. Lipshultz and S. S. Howards, Eds.), 2nd ed., pp. 124–154. Mosby-Year Book, St. Louis.

McNeal, J. E. (1970). The prostate and prostatic urethra: A morphologic study. *J. Urol.* **104**, 443.

Schlegel, P. N., and Chang, T. Sk. (1992a). The testis, epididymis, and ductus deferens. In *Campbell's Urology* (P. C. Walsh, A. B. Retik, J. A. Stamey, and E. D. Vaughan, Eds.), 6th ed., pp. 190–220. Saunders, Philadelphia.

Schlegel, P. N., and Chang, T. Sk. (1992b). Physiology of erection and pathophysiology of impotence. In *Campbell's Urology* (P. C. Walsh, A. B. Retik, J. A. Stamey, and E. D. Vaughan, Eds.), 6th ed., p. 709. Saunders, Philadelphia.

Yamamoto, M., and Turner, T. T. Epididymis, sperm, maturation, and capacitation. In *Infertility in the Male* (L. I. Lipshultz and S. S. Howards, Eds.), 2nd ed., pp. 103–123. Mosby-Year Book, St. Louis..

Male Reproductive System, Insects

Cedric Gillott
University of Saskatchewan

I. Introduction
II. Structure
III. Development
IV. Functions

GLOSSARY

apyrene sperm A type of sperm unique to Lepidoptera in which there is no nucleus, and the sperm do not fertilize the oocytes.

ectadenia Accessory glands of male insects derived from ectoderm.

eupyrene sperm Nucleate sperm of Lepidoptera, used to fertilize the oocytes.

mesadenia Accessory glands of male insects derived from mesoderm.

The male reproductive system of insects comprises those organs concerned with the production and transfer of sperm to the female. It includes testes and various ductal elements, the latter typically glandular, whose secretions are used both structurally to form the vehicle in which sperm are moved into the female and in a variety of physiological ways, including sperm nourishment and activation, and as pheromones that modify the mated female's fecundity and receptivity.

I. INTRODUCTION

Like other terrestrial animals, insects solved the problem of bringing together the sperm and egg in the absence of an aquatic environment through the evolution of internal fertilization. In most species the sperm are transferred to the female's genital tract within a spermatophore; however, in relatively few species spermatophores are not produced, and sperm transfer is effected by means of an intromittent organ. Though sperm production and transfer are the major functions of the male system, various subsidiary functions have evolved in some species, notably sperm storage, production of pheromones that modify a female's fecundity and/or behavior, and transfer of nutrients to the female.

Reflecting the extreme gross morphological diversity seen within the Insecta, details of the form, function, and biochemical nature of the products of the male system vary widely. This account will attempt to present the generalities that have been established and to point out areas in which knowledge is still lacking.

II. STRUCTURE

A. Internal Organs

The male system includes a pair of testes (in Lepidoptera fused to form a single median organ), paired vasa deferentia and seminal vesicles, a median ejaculatory duct, and in most species accessory glands of varied origin and complexity (Fig. 1A).

In each testis is a varied number of tubular follicles held together by connective tissue. Each follicle connects with a short vas efferens, with the vasa efferentia from each gonad opening into the vas deferens either confluently or in a linear sequence. Typically, within a follicle, the several phases of spermatogenesis can be seen arranged lengthwise along the structure, with the spermatogonia located distally and the spermatozoa adjacent to the vas efferens (Fig. 1B). In follicles of some insect groups, a prominent apical cell is visible and, on the basis of its histochemistry, is believed to provide nutrients for the developing germ cells.

Though principally just tubes for conducting, by peristalsis, sperm from the testes to the seminal vesicles, the vasa deferentia may have important glandular and phagocytic functions in some species. Their

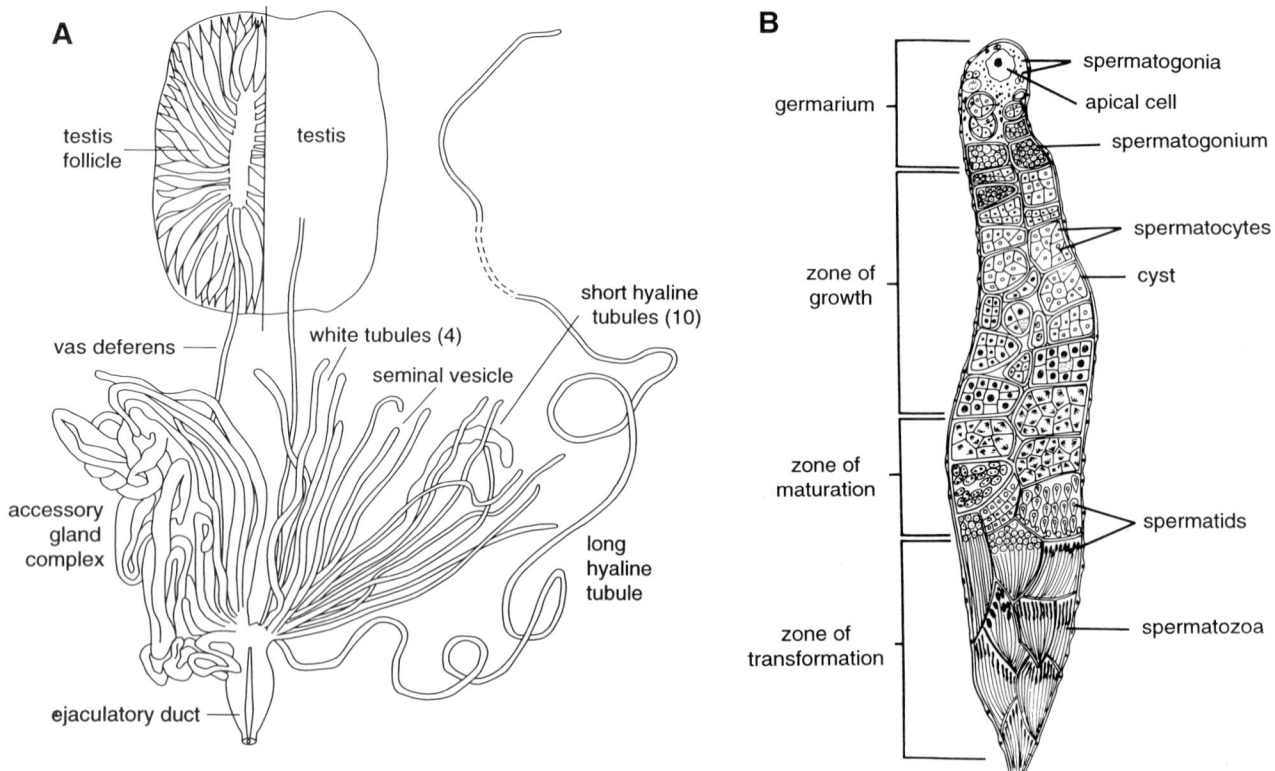

FIGURE 1 Male insect reproductive system. (A) General view of the system in a grasshopper. (B) Sagittal section through testis follicle of grasshopper showing various zones [A, reprinted with permission from *Insect Reproduction* (S. R. Leather and J. Hardie, Eds.). © 1995 by CRC Press, Boca Raton, FL; B, reprinted with permission from *Entomology* (C. Gillott), 2nd ed. © 1995 Plenum, New York).

secretions may nourish sperm or participate in spermatophore formation, whereas their phagocytic cells may get rid of aged or degenerate sperm. The seminal vesicles, in which mature sperm are stored prior to insemination, are for most species simply dilations of the vasa deferentia; however, in Acrididae (short-horned grasshoppers and locusts) the seminal vesicles are a pair of modified tubules of the accessory gland complex.

The vasa deferentia enter the anterior tip of the ejaculatory duct which, in the majority of insects, is an ectodermally derived tube and, as such, has a cuticular lining. In Lepidoptera (butterflies and moths) and a few species in some other orders, however, the anterior end of the ejaculatory duct is formed from mesoderm. The duct is strongly muscular and, in those insects that produce complex spermatophores, it may be divided into a number of distinct regions, each with unique histology and stainability. In addition to producing spermatophore components, the ejaculatory duct may produce seminal fluid, sperm-activating chemicals, and receptivity-inhibiting substances in some species. In Ephemeroptera (mayflies) there is no ejaculatory duct; the paired vasa deferentia open separately to the exterior.

Accessory glands (= collateral glands) occur in most Insecta; however, they are primitively absent in Thysanura (silverfish and firebrats), Ephemeroptera, Plecoptera (stoneflies), Dermaptera (earwigs), and Odonata (dragonflies and damselflies) and have been secondarily lost in some Diptera (true flies). They are typically mesodermal in origin (mesadenia); however, in the few insect groups in which secondary substitution of ectodermal for mesodermal tissues has occurred in the reproductive system, the accessory glands are ectodermal (ectadenia). The accessory glands usually occur as a single pair of tubules whose cytology may show regional or intercellular differences. However, in Coleoptera (beetles) there are two or three pairs of tubules, and in Thysanoptera (thrips) and Acrididae multiple pairs of tubules occur. In this arrangement, each pair of tubules has its characteristic cytology. The secretion of the accessory glands, even of individual tubules, is a complex mixture and, not surprisingly, a variety of functions have been ascribed to these structures, including spermatophore, mating plug and seminal fluid formation; fecundity enhancement and receptivity inhibition; sperm activation; and supply of nutrients to the female.

B. External Genitalia

The external genitalia of male insects are the highly modified remnants of the ancestral paired appendages of the ninth abdominal segment. They serve principally as claspers, ensuring that the male and female genital openings are kept closely apposed during copulation. Secondary functions in some species include molding the spermatophore as it forms and placing this structure in the female genital tract. The male external genitalia take on a huge variety of forms, much to the delight of insect taxonomists, and although determination of homologies is both difficult and controversial, a basic pattern can be identified. The genitalia arise from a pair of primary phallic lobes. In Thysanura the lobes fuse to form a median "penis" (a misnomer, in fact, because it is not an intromittent organ; rather, males deposit spermatophores on the substrate to be picked up later by females). In all other insects, except Ephemeroptera and Odonata (see below), each primary lobe is divided into two secondary lobes—the phallomeres. Between the median pair (mesomeres), the ectoderm invaginates to form the ejaculatory duct. The mesomeres unite to form a tubular intromittent organ, the aedeagus. The outer pair, the parameres, elongate and differentiate into clasping organs.

C. Specialized Systems

In contrast to the general situation outlined previously, the primary phallic lobes of Ephemeroptera remain separate, each serving as a penis with its own gonopore. From each gonopore, an ejaculatory duct invaginates and fuses with the ipsilateral vas deferens.

In male Odonata, the appendages on the ninth segment never develop to a significant degree so that there is no aedeagus and an inconspicuous ejaculatory duct. Instead, a unique method of sperm transfer has evolved using secondary genitalia that develop on the second and third abdominal segments. Prior to mating, but after seizing the female with his legs, the male curves his abdomen forward ventrally and places one or more spermatophores in the secondary genitalia. The female then brings her abdomen forward and upward until its tip contacts the secondary genitalia, allowing sperm transfer to take place.

Sperm transfer in Strepsiptera (stylopoids) and Cimicoidea (bedbugs and relatives) is by means of hemocoelic insemination. In this arrangement, sperm is injected via a rigid penis into the body cavity. From here, the sperm migrate, sometimes along specialized cords of tissue, to the female's sperm-storage organs. En route, a proportion of the sperm is phagocytozed, for example, by blood cells, and it has been proposed that this most unusual method of sperm transfer has evolved as a means of providing nourishment to the female.

III. DEVELOPMENT

A. Embryonic

The origin and development of the testes are varied among Insecta, though two distinct evolutionary trends may be seen—namely, earlier segregation of the primordial germ cells and their restriction to fewer segments. Thus, in Thysanura and Orthoptera (grasshoppers, locusts, crickets, etc.) germ cells arise relatively late in embryonic development and are seen initially in most abdominal somites (though they persist in only two or three segments). In Dermaptera, Psocoptera (booklice), Homoptera (aphids, plant lice, scale insects, etc.), and many endopterygotes (insects with a pupal stage in their life history), the germ cells can be distinguished even during blastoderm formation as roundish cells near the posterior end of the egg. Subsequently, in the exopterygote orders (those without a pupal stage), the germ cells migrate forward to become enclosed by the splanchnic mesoderm of abdominal segments three and four. Later, the germ cells separate into left and right halves, from which the testes form. In endopterygotes, separation of the germ cells into two groups occurs before their anteriorly directed migration.

In exopterygotes and some primitive endopterygotes, the vasa deferentia and, where present, mesadenia develop from one or more pairs of posterior abdominal somites, whereas the ejaculatory duct (and ectadenia) forms by invagination of the ventral ectoderm, typically behind the ninth segment. In these insect groups, all elements of the reproductive system can be identified at hatching. In contrast, the ductal elements of the reproductive system of higher endopterygotes, notably Diptera and Hymenoptera, form from genital imaginal discs that arise late in embryonic development. These groups of cells remain undifferentiated throughout larval life, and the reproductive tract develops from them only at metamorphosis.

B. Postembryonic

Like other systems, the reproductive system of exopterygotes grows throughout the juvenile period, though differentiation of the various components occurs only in the final instar. By contrast, in endopterygotes, there is very little growth in the larval stage; instead, growth and differentiation occur simultaneously and are compressed into the pupal instar. Both differentiation of the tubular elements and spermatogenesis arise as a result of major changes in the levels of two key hormones, juvenile hormone (JH) and β-ecdysone [molting hormone (MH)].

MH alone stimulates growth and differentiation of the reproductive system; however, in all but the final juvenile instar, JH is also present and this hormone inhibits adult differentiation. In the final juvenile instar, the level of JH declines significantly, whereas simultaneously there are one or more surges in the concentration of circulating ecdysteroids. This major change in the ratio of the two hormones allows adult differentiation to occur. Though the nature of the interaction between these two hormones is far from clear, it seems that JH may influence metamorphosis to the adult form in two ways: (i) By inhibiting the release of prothoracicotropic hormone from the brain, it prevents production and/or release of ecdysteroid from the molt glands, and (ii) it directly prevents ecdysteroid from acting at the organ/tissue level. An area requiring much more study is the means by which MH promotes the growth and differentiation of reproductive tissues. At its peak concentration, MH has been shown to stimulate mitosis, specifically to promote the flow of the dividing cells from the G_2 into the G_1 and S phases. Lower MH levels, such as those that occur late in the pupal instar, perhaps induce differentiation.

IV. FUNCTIONS

A. Spermatogenesis and Sperm Storage

1. *Spermatogenesis*

Like that in other animals, insect spermatogenesis includes three phases: (i) the multiplication phase, (ii) the meiotic phase, and (iii) the maturation phase. In many species, this temporal sequence of events is fixed spatially in each testis follicle; that is, the phases occur in the distal, middle, and proximal regions, respectively. In the multiplication phase, each spermatogonium undergoes a species-specific number of mitoses. Generally, these divisions occur within a "cyst," formed by a layer of somatic cells that surround the original spermatogonium. Each of the spermatocytes thus formed undertakes two meiotic divisions, generally resulting in four haploid spermatids; however, especially among Diptera, variants may occur so that only one or two spermatids arise from each spermatocyte. The spermatids then mature (differentiate) into flagellated spermatozoa, by which time the cyst wall has ruptured. However, in many Orthoptera and Coleoptera, the sperm within a bundle (spermatodesm) remain together with their heads enveloped in a mucopolysaccharide "cap." Apparently, this cap, which breaks up after the sperm are transferred to the seminal vesicles, contains enzymes that degrade nutritive molecules secreted by these sperm-storage organs.

In Lepidoptera, two kinds of sperm are formed: eupyrene (nucleate), which fertilize the eggs, and apyrene (anucleate), whose functions have been proposed to include assisting the movement of the eupyrene sperm both from the testes to the vas deferens and within the female reproductive tract; providing nourishment for the eupyrene sperm, the female herself, or the zygote; and playing a role in sperm competition, either by eliminating sperm from earlier matings or by preventing further matings. Eupyrene sperm formation begins earlier than that of the apyrene sperm, and whereas the former type remain associated in bundles until after transfer to the female, apyrene sperm bundles dissociate as they leave the testes.

In most insects, spermatogonia and spermatocyte formation occur in the final nymphal or pupal instar so that the testes of adults contain only spermatids and spermatozoa (only spermatozoa in species with a short adult life). Thus, almost 50 years ago, it was proposed that spermatogenesis was regulated by ecdysone and JH, with the former promoting and the latter inhibiting the process. The discovery that meiotic and even premeiotic phases of spermatogenesis occur in the testes of some adult males led to the proposal that the hormones affected only the rate of spermatogonial mitosis, not differentiation per se; specifically, ecdysone increases the rate of division, whereas JH depresses it, though never below a basal level. Thus, despite the presence of JH in the blood of adult males, spermatogenesis continues, albeit at a very low rate in most species. This proposal was made on the assumption that adult males do not have circulating ecdysteroids. However, several recent reports have demonstrated the presence of ecdysteroids in adult male tissues, including the testes, providing, for a few species at least, a simpler explanation for the occurrence of spermatogenesis in adults.

2. *Sperm Storage*

In most species, bundles of sperm move by peristalsis along the vasa deferentia to the seminal vesicles where they are stored until insemination. However, except for a few species of Lepidoptera, details of the process are lacking. In these Lepidoptera, release of sperm (both eupyrene and apyrene) from the testes exhibits a circadian rhythm. The circadian clock and its associated photoreceptor are in the testis–vas deferens complex. Correlated with the rhythm of sperm movement is cyclic production and release of a carbohydrate-rich secretion by the cells of the upper vas deferens. The function of this secretion is unknown but may be related to the complex changes that the sperm undergo en route to their site of storage. The overall significance of this periodic sperm release and movement lies in the fact that males normally mate only once daily and, at this time, ejaculate all their stored sperm. Thus, the circadian mechanism ensures that sperm will be available for insemination the following day. How sperm release and storage is regulated in species with opportunistically mating males remain unclear.

B. Spermatophore Formation

Essentially, in insects spermatophores are formed from secretions of the accessory glands and occasionally also from the ejaculatory duct. The secretions may mix or remain as separate layers, producing a capsule that encloses the sperm. Though detailed descriptions of the mechanics of spermatophore formation are common, studies of the control of spermatophore formation and of the biochemistry of the process are few. Generally, spermatophores are most complex in primitive groups; in advanced groups, they are relatively simple or have been secondarily lost. Based on their complexity and site of formation, four categories may be recognized. In the most primitive mode of spermatophore formation, the first male-determined method, typical of orthopteroid insects, the complex spermatophore is formed in the ejaculatory duct or copulatory organ of the male. After its transfer to a female, the spermatophore is held between the female's genital plates, and only its anterior tube-like portion enters the vagina or bursa. In the second male-determined method, seen in some Hemiptera (true bugs), Coleoptera, and a few Diptera, the spermatophore again forms in the copulatory sac, but the latter is everted into the bursa copulatrix of the female. After copulation, the sac is withdrawn, leaving the usually less complex spermatophore in the bursa. A relatively simple spermatophore is formed directly in the female tract in Trichoptera (caddis flies), Lepidoptera, some Coleoptera and Diptera, and a few Hymenoptera (bees, wasps, and ants). In this arrangement (first female-determined method), the sperm are still enclosed within the spermatophore. However, in the second female-determined method, the accessory gland secretions do not surround the sperm; rather, they follow the sperm into the female tract and then may harden to form a temporary plug that prevents loss of semen and perhaps further mating.

The spermatophore wall is largely protein. In the mealworm beetle (*Tenebrio molitor*), three structural proteins have been traced from their cells of origin in the accessory glands to their final disposition in the spermatophore wall. Interestingly, in one of these proteins, >25% of the amino acid residues are proline, an amino acid very common in other insect structural proteins (e.g., cuticle, egg shell, and ootheca) and collagen. Undoubtedly, the nature of some of the accessory gland secretions changes as they become part of the spermatophore. Changes in solubility and electrophoretic mobility indicate that some of the proteins may be cleaved into smaller fractions and/or may be combined with other components, though to date there has been only a handful of studies demonstrating the presence of proteolytic enzymes in accessory gland secretions.

Only in crickets (Gryllidae; Orthoptera) is a spermatophore formed before copulation occurs. In these insects, spermatophore formation is under circadian control, though the site of the control center has not been located. In other insects, spermatophore production begins when the male is in the mating position. Mainly tactile, but for some species chemical and visual stimuli trigger the process which is under motor control. An intact central nervous system is critical at the start of the process, perhaps because the brain serves to integrate the initial sensory input. However, subsequent stages are controlled by the terminal abdominal ganglion.

C. Seminal Fluid

In contrast to the mammalian situation in which seminal fluid is a secretion of the seminal vesicles, insect seminal fluid may be derived from any or all the glandular parts of the male system. Furthermore, its nature may change as secretions are added or removed during insemination and when it reaches the female system.

Knowledge of the composition and functions of seminal fluid has been limited by the extremely small quantities of semen available and the high density of sperm that the semen contains. A variety of sugars (e.g., glucose, fructose, and trehalose in the honey bee) have been identified which may serve as energy sources for sperm. Glycogen has been demonstrated histochemically in the seminal fluid of some other species. Lipids, amino acids, and a range of cations have been reported in honey bee semen, though usually it is not clear whether these originate in the sperm or are seminal fluid components.

D. Fecundity Enhancement and Receptivity Inhibition

Though in some insects (e.g., cockroaches) the physical stimulus provided by mating leads to increased fecundity or inhibition of receptivity (willingness of the female to remate), for many species these effects result from chemicals transferred to the female in the seminal fluid. It is appropriate to discuss these fecundity-enhancing substances (FES) and receptivity-inhibiting substances (RIS) in the same context because in some species the same chemical may serve both functions and because the ultimate function of both types of pheromone is to signal to a female that she has been inseminated. Thus, for FES, the significance rests in the fact that insect eggs are fertilized as they are being laid (i.e., passing down the common oviduct). Unfertilized eggs are generally inviable. Hence, it is critical that oviposition does not occur until a supply of sperm is available. The use of RIS ensures that the first (fittest) male's sperm is used to fertilize the eggs. It may also force males to actively search for virgin females and induce changes in the female's behavior, specifically from seeking a mate to finding food and/or an oviposition site.

Generally, the FES and RIS are produced in the accessory glands, though in species which lack these components the source may be the ejaculatory duct (e.g., in the housefly) or testes, as in crickets. The molecules are typically peptides [molecular weight (MW) range, 750–3600 Da] or proteins (MW range, 13–60 kDa). Interestingly, in crickets the FES is a prostaglandin synthetase enzyme complex which promotes prostaglandin production when transferred to the female tract.

The weakest part of our understanding of FES and RIS is their site and mode of action. The most common effect of FES is to stimulate oviposition, though experimentation suggests that the FES does not act directly on the oviduct. Rather, there are intermediate steps, culminating in the release of a myotropic peptide from the brain that enhances peristalsis in the oviducal muscles. For a few species, the FES appears to act at an earlier stage, namely, to induce ovulation or promote vitellogenesis.

Even less evidence is available for how and where RIS act. Most probably, these molecules act (directly or indirectly) on the central nervous system. For some species, the brain seems to be the probable site, though in mosquitoes the terminal abdominal ganglion seems more likely.

E. Other Functions

Several other functions have been proposed for secretory products of the male reproductive tract, mostly relating to sperm transfer to the female system and energy metabolism. In addition, in some species the spermatophore is a source of nourishment for the female.

How sperm transfer from the spermatophore to the spermatheca occurs is largely unknown, though presumably the process involves either the sperm swimming or peristalsis of the muscular wall of the female tract. In the bug *Rhodnius* the male accessory glands produce a factor that induces peristalsis of the female tract. However, the factor may be acting on the peripheral nervous system of the insect rather than directly on the wall musculature per se.

In Saturniidae (giant silk moths) and some other Lepidoptera, the lower part of the mesodermal ejaculatory duct secretes a sperm activator. The activator, released during spermatophore formation, renders the sperm highly motile. The nature of the process has been studied closely in the silk moth *Bombyx mori*, in which the key participant is an endopeptidase "initiatorin" whose roles includes (i) digestion of the coat that surrounds the apyrene sperm (their subsequent activity serves to stir the seminal fluid); (ii) digestion of the intercellular glue that binds the eupyrene sperm, leading to their release and activation; and (iii) initiation of a cascade of metabolic reactions for the provision of substrates for sperm respiration.

An enzyme, esterase 6, produced in the anterior ejaculatory duct of the fruit fly *Drosophila melanogaster*, is transferred in the seminal fluid to the female where it may have two purposes. It has a role in lipid catabolism, thus affecting sperm motility, and it may also be associated with the production of an antiaphrodisiac (*cis*-vaccenol) by the mated female.

In some insect species, notably some crickets and katydids (Tettigoniidae), the male produces a very large spermatophore (up to 40% of the male's body weight), a large part of which protrudes from the female and, after sperm evacuation, is eaten by her. For other species, there are reports that the spermatophore undergoes digestion within the female reproductive tract. A few studies have shown that the male-derived material eventually finds its way into the ovary. In other words, the male makes a nutritive contribution to the female, specifically to egg development.

See Also the Following Articles

Female Reproductive System, Insects; Hemocoelin Insemination; Insect Accessory Glands; Locusts; Molting, Insects; Rhodnius Prolixus; Spermatogenesis

Bibliography

Chen, P. S. (1984). The functional morphology and biochemistry of insect male accessory glands. *Annu. Rev. Entomol.* **29**, 233–255.

Davey, K. G. (1985). The male reproductive tract. In *Comprehensive Insect Physiology, Biochemistry and Pharmacology* (G. A. Kerkut and L. I. Gilbert, Eds.), Vol. 1. Pergamon, Elmsford, NY.

Dumser, J. B. (1980). The regulation of spermatogenesis in insects. *Annu. Rev. Entomol.* **25**, 341–369.

Gillott, C. (1988). Arthropoda–Insecta. In *Reproductive Biology of Invertebrates* (K. G. Adiyodi and R. G. Adiyodi, Eds.), Vol. III, Accessory Sex Glands. Wiley, New York.

Gillott, C. (1995). Insect male mating systems. In *Insect Reproduction* (S. R. Leather and J. Hardie, Eds.). CRC Press, Boca Raton, FL.

Gillott, C. (1996). Male insect accessory glands: Functions and control of secretory activity. *Invertebr. Reprod. Dev.* **30**, 199–205.

Gillott, C., and Gaines, S. B. (1992). Endocrine regulation of male accessory gland development and activity. *Can. Entomol.* **124**, 871–886.

Happ, G. M. (1984). Structure and development of male accessory glands in insects. In *Insect Ultrastructure* (R. C. King and H. Akai, Eds.), Vol. 2. Plenum, New York.

Happ, G. M. (1992). Maturation of the male reproductive system and its endocrine regulation. *Annu. Rev. Entomol.* **37**, 303–320.

Matsuda, R. (1976). *Morphology and Evolution of the Insect Abdomen*. Pergamon, Elmsford, NY.

Male Reproductive System, Nonhuman Mammals

Larry Johnson, Gabrielle U. Falk, and Genevieve E. Spoede

Texas A&M University

I. Introduction
II. Gonads: Testes
III. Excurrent Duct System (Epididymis)
IV. Accessory Sex Glands (Prostate)
V. Copulatory Organ—Penis
VI. Vascular Supply and Innervation
VII. Interactions of Hormones

GLOSSARY

eutherian Referring to placental mammals, excluding the monotremes and marsupials.
perineal Pertaining to the region extending from the external genitalia to the anus.
testicond Having undescended testes.
ungulates Hoofed mammals.

MAMMALIAN ORDERS

Artiodactyla Even-toed ungulates, such as cattle, deer, camels, sheep, goats, and pigs.
Carnivora Meat-eaters such as dogs, cats, and foxes.
Cetacea Mostly marine mammals, such as whales and dolphins.
Chiroptera Bats.
Edentata Mammals having few or no teeth, such as anteaters, sloths, and armadillos.
Hyracoidea Small mammals characterized by short legs, broad nails, and rodent-like incisors, known as hyraxes.
Insectivora Insect-eaters such as shrews, hedgehogs, and moles.
Lagomorpha Rabbits, hares, and pikas.
Marsupialia Mammals which carry and nourish their developing young in a pouch, including koalas, kangaroos, wombats, and oppossums.
Monotremata Egg-laying mammals including the echidna and the platypus.
Perissodactyla Odd-toed ungulates, such as horses, tapirs, zebras, and rhinoceroses.
Proboscidea Elephants.
Rodentia Small mammals with large upper incisors, such as rats, mice, beavers, and squirrels.
Sirenia Aquatic herbivores, such as dugongs and manatees.

The male reproductive system by definition consists of two testes, a system of ducts, a blood and nerve supply, accessory glands, and a penis. It is responsible for producing haploid spermatozoa and delivering them into the female reproductive tract in order for insemination to take place. Before insemination can take place, the spermatozoa must be produced by the hormonally influenced process of spermatogenesis in the testis. Then, the spermatozoa are transferred through an elaborate system of ducts and accessory organs and finally through a copulatory organ in order to internally fertilize the female mammal. Although other organ systems are involved in the process of reproduction, this article describes the basic structure and some species differences of the mammalian male reproductive system. The basic organization and function of the reproductive system is the same for all mammals, but there is much species variation in the morphology of this system.

I. INTRODUCTION

The male reproductive system (Fig. 1) consists of gonads (two testes), their excurrent duct system (efferent ducts, epididymis, ductus deferens, am-

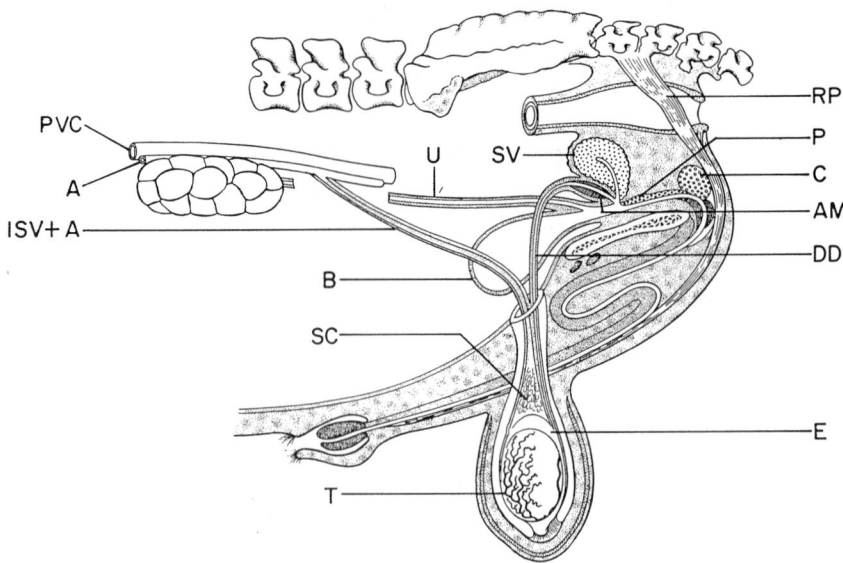

FIGURE 1 Male reproductive organs of the bull. T, testis; E, epididymis; SC, spermatic cord; DD, ductus deferens; ISV & A, internal spermatic vein and artery; A, aorta; PVC, posterior vena cava; AM, ampulla; SV, seminal vesicles; P, prostate; C, bulbourethral glands or Cowper's gland; RP, retractor penis muscle; B, bladder; and U, ureter (reproduced with permission from Setchell, 1991).

pulla, ejaculatory ducts, and, prostatic and penile urethra), and associated accessory sex glands (seminal vesicles, prostate, and bulbourethral glands or Cowper's gland). Testes produce spermatozoa and male sex hormones, testosterone being the most important sex hormone in the male. Spermatozoa in the testis pass from the seminiferous tubules (where they are produced) into the rete testis tubules to exit the testis. Then they pass into the efferent ducts which attach to a single epididymal duct. In the epididymis, spermatozoa acquire the capacity for motility and for fertilizing ova, and matured spermatozoa are stored in the tail of the epididymis. During sexual excitement, the penis becomes erect and fluid from the bulbourethral gland begins to leak out of the end of the penis as it cleans the urine out of the urethra prior to ejaculation. Stored spermatozoa are moved from the tail of the epididymis through the ductus deferens and ampulla into the ejaculatory ducts where they are mixed with fluids from the prostate. The spermatozoa are then washed out through the prostatic urethra and then penile urethra via secretions of the seminal vesicles. Spermatozoa mixed with the secretions of the accessory sex glands constitute the ejaculated semen.

The number of spermatozoa in the ejaculate is only one measure of seminal quality. Certain characteristics of spermatozoa in the ejaculate are necessary for fertilization to take place (Fig. 2). These characteristics include progressive motility, morphology (shape), acrosomal integrity, and the ability of spermatozoa to undergo the acrosomal reaction, to penetrate the zona pellucida, to penetrate the egg, to form a pronucleus, and to produce a live offspring from the fertilized egg. Seminal quality also is influenced by age, season, temperature, and exposure to testicular insults/toxicants.

II. GONADS: TESTES

Testes are the gonads of the male reproductive system. All mammalian species normally have two testes. The basic morphology is generally the same across species, but they can vary in size from up to 1% of the male's body weight to as much as 8% as is seen in the gerbil *Tatera afta*. Size of gonads seems to be related to reproductive strategy. Species with high ejaculatory frequency such as polygamous species tend to have large testes, whereas monogamous

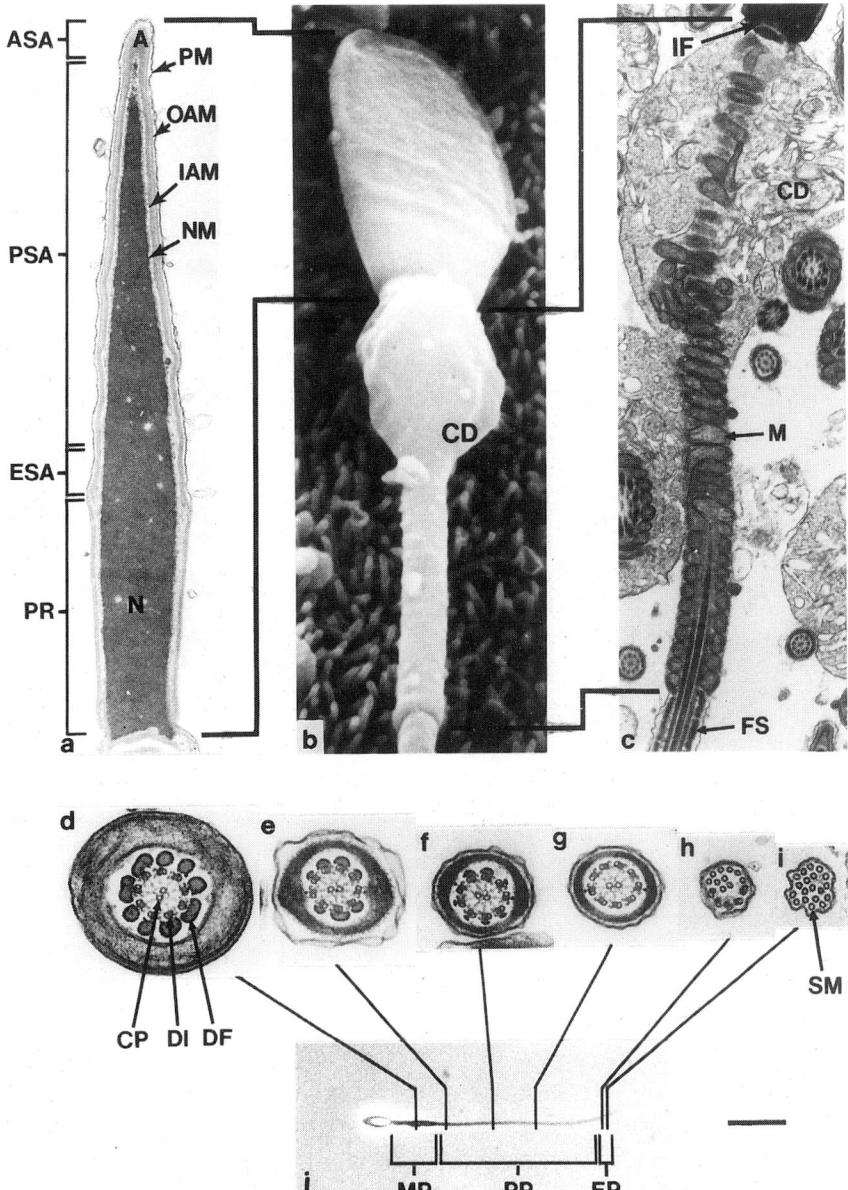

FIGURE 2 Equine spermatozoa of the perissodactyla order as viewed by transmission electron (a, c–i), scanning electron (b), and phase-contrast microscopy (j). Corresponding regions of spermatozoa are indicated by attaching lines between a and b, b and c, or d–i and j. The plasma membrane (PM) or cell membrane encases the entire head and tail of the cell. The head contains the nucleus (N), overlying acrosome (A), and postacrosomal region (PR). Three regions of the acrosome include the apical segment (ASA), principal segment (PSA), and equatorial segment (ESA). The inner acrosomal membrane (IAM) lies over the nuclear membrane (NM), and the outer acrosomal membrane (OAM) lies next to the plasma membrane. The tail is divided into the middle piece (MP), principal piece (PP), and end piece (EP). The tail is attached to the head at the implantation fossa (IF) and contains mitochondria (M), fibrous sheath (FS), the nine doublet microtubules (DI) and central pair microtubules (CP) of the axoneme, and nine dense fibers (DF) that parallel the axoneme and extend to different lengths down the principal piece of the tail. The axoneme gives rise to 20 single microtubules (SM). The cytoplasmic droplet (CD), located at the proximal end of the middle piece, is an indication of immaturity (reproduced with permission from Johnson, 1991).

species with a low ejaculatory frequency in general have smaller testes. In most mammals, the testes are descended through the inguinal canal, outside of the abdominal cavity, and into a species-specific evagination of the body wall known as the scrotum. However, monotremes, edentates, elephants, sirenians, and cetaceans lack scrota and have abdominally placed testes. In bats, most rodents, and some carnivores and ungulates, the testes only descend into the scrotum during breeding season.

The wall of the scrotum consists of the skin (epidermal surface and dermal connective tissue), tunica dartos (smooth muscle), connective tissue, the parietal and visceral layers of the vaginal process, and the tunica albuginea (capsule immediately surrounding the testis). The inner layer of the scrotum is the parietal layer of the tunica vaginalis communis (parietal tunic), which is a continuation of the parietal peritoneum into the scrotum. The testis and epididymis are covered by an inner tube of peritoneum known as the tunica vaginalis propria (vaginal tunic). The space between the two tunics is the vaginal cavity and is continuous with the abdominal cavity at the vaginal ring through which the testes descend from the abdominal wall.

The scrotum encloses two scrotal sacs which are outpocketings of the peritoneal cavity and contain the two testes with their attached epididymides. For optimal spermatogenesis to take place, the internal testicular temperature must be about 5°C lower than internal body temperature. The scrotum provides this temperature regulation function. It holds the testes close to the abdominal cavity in cold conditions or allows the testes to hang further away (down) from the abdomen in hot weather. The skin covering the scrotum is thinner than elsewhere on the body because it has no subcutaneous fat storage and a thin dermis. Also, the skin of the scrotum is less covered with hair. The two peritoneal sacs are enclosed in a smooth muscle coat (tunica dartos). The contraction of the tunica dartos is responsible for the variation in the position of the testes according to temperature. The tunic contracts in cold weather and holds the scrotum closer to the abdominal wall. The cremaster muscle, a thin strip of tissue originating from the internal abdominal oblique muscle, is closely associated with the internal spermatic fascia and also aids in pulling the testes closer to the body, especially in response to noxious stimuli.

The testes themselves are covered by a thick capsule of connective tissue called the tunica albuginea. Connective tissue septa from this capsule penetrate the testes, divide the testes into lobules, and converge in the center of the testes as the central core of connective tissue called the mediastinum testis. The parenchyma of the testes consists of compact seminiferous tubules surrounded by a loose connective tissue that contains blood and lymph vessels, nerves, and the interstitial cells of Leydig that secrete testosterone (Fig. 3a). The percentage of the testicular parenchyma occupied by seminiferous tubules varies: 76% in bulls, 84% in dogs, 72% in horses, 83% in rats, and 62% in humans.

In adults, seminiferous tubules contain two somatic cell types (myoid and Sertoli cells) as well as germ cells (Fig. 3b). The myoid or peritubular cells are the outer boundary cells of the tubule and surround the seminiferous epithelium, composed of Sertoli cells (nurse cells) and germ cells. The number of Sertoli cells varies with species but appears to be very important in regulating the level of spermatogenesis. Germ cells are themselves of three types: spermatogonia, spermatocytes, and spermatids. Spermatogonia (the most immature germ cells) sit on the base of the tubule. Spermatocytes are located in the middle of the seminiferous epithelium (Fig. 3c), and the spermatids (the most mature) are located near the lumen (opening of the tubule) where they are released or spermiated (Fig. 3d) as spermatozoa (Fig. 2).

Spermatogenesis is the sum of events that occur within the seminiferous tubules of the testis that produce spermatozoa. It is a lengthy, chronological process by which stem cell spermatogonia divide by mitosis to maintain their own numbers and to cyclically produce primary spermatocytes that undergo meiosis to produce haploid spermatids which differentiate into spermatozoa. Spermatocytogenesis, meiosis, and spermiogenesis are the three major divisions of spermatogenesis and are characterized by development of spermatogonia, spermatocytes, and spermatids, respectively. Each division takes about one-third the duration of spermatogenesis in any given species. In the bull, these divisions take 21,

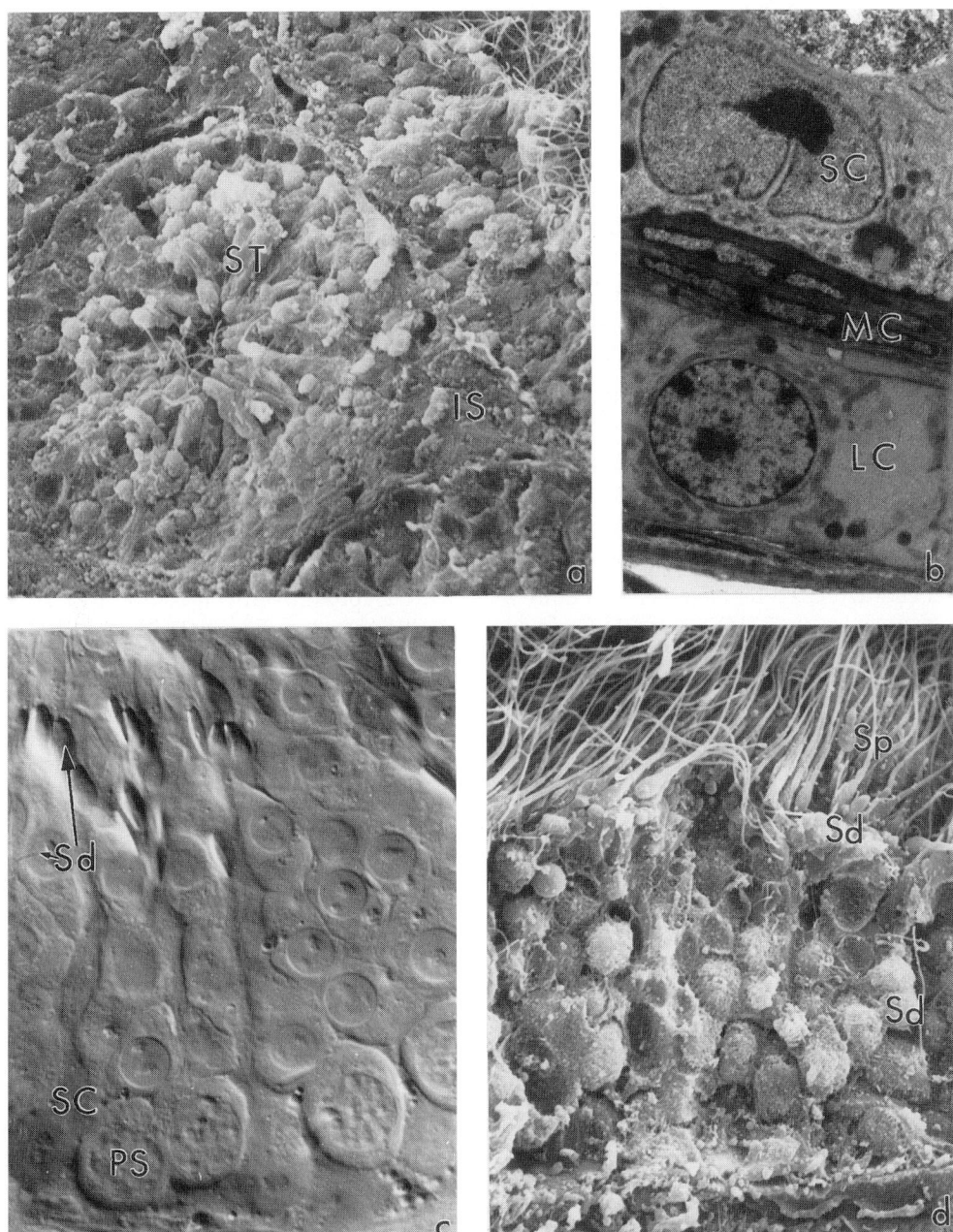

FIGURE 3 The equine testis viewed by scanning electron microscopy (a and d), high-voltage electron microscopy (b), and light microscopy with Nomarski optics (c). (a) The testicular parenchyma is composed of seminiferous tubules (ST) and interstitial space (IS). (b) Leydig cells (LC) that produce testosterone are located in the interstitial space; myoid cells (MC) compose the outer boundary of the seminiferous tubules and lie next to Sertoli cells (SC) or germ cells. (c and d) The seminiferous epithelium is composed of Sertoli cells (SC) and germ cells. The least mature germ cells (spermatogonia) are located at the base of the tubule near the myoid cells, primary spermatocytes (PS) are located in the middle of the epithelium, and spermatids (Sd) are located near the lumen (opening) of the tubule where they are released (spermiated) as spermatozoa (Sp) (reproduced with permission from Johnson, 1997 and Johnson et al., 1978a).

23, and 17 days, respectively for a total duration of spermatogenesis of 61 days. The duration of spermatogenesis is also 61 days in dogs, 60 days in rats, 57 days in stallions, 47 days in rams, 39 days in boars, and 74 days in humans. During spermatocytogenesis, stem cell spermatogonia divide by mitosis to produce other stem cells that continue the lineage throughout the adult life of males. Stem cells give rise to other spermatogonia that cyclically produce committed spermatogonia which proliferate and/or differentiate to produce primary spermatocytes that undergo meiosis. Meiosis allows exchange of genetic material between homologous chromosomes of primary spermatocytes and the production of haploid spermatids. During spermiogenesis, spermatids differentiate from cells with spherical nuclei into mature germ cells shaped like spermatozoa. The species-specific spermatazoon is characterized by flagella and a compressed head containing an acrosome with its enzymes and the male genome in the nucleus. Spermatids are released into the tubular lumen as spermatazoa (Fig. 3d). Spermatozoa travel down the rete testis tubules (ducts) to the excurrent duct system.

III. EXCURRENT DUCT SYSTEM (EPIDIDYMIS)

Once spermatozoa travel through the rete tubules, they exit the testis into the efferent ducts (Fig. 4). The efferent ducts absorb fluid and transport spermatozoa into a single duct of the epididymis (Fig. 4). The epididymis is a tightly coiled, thickly encapsulated single tubule that is present in all mammals and is attached by varying degrees to the surface of one side of the testis and extends from cranial to caudal ends. It consists of a head (caput epididymidis), body (corpus epididymidis), and tail (cauda epididymidis), respectively (Fig. 4). The head and body are areas of spermatozoan maturation (Fig. 5), and the tail is the region in which mature spermatozoa are stored prior to ejaculation. The epididymal tubule is lined with a smooth muscle layer that aids in peristalsis of spermatozoa through the epididymis and out of the epididymis during ejaculation (Fig. 5). This structure also has both absorptive and secretory functions. Upon entering the epididymis, the secretions from the seminiferous tubules are absorbed and the spermatozoa are concentrated up to 20-fold. Concurrently, the secretions of the epididymis are added to the stored spermatozoa and aid in the storage of viable spermatozoa by decreasing their metabolic activity. The epididymis is described as follows in the following mammalian orders:

Rodentia: The rodent epididymis has been the most extensively studied of all species. It has been divided into six to nine areas distinguishable by histological studies. Grossly, the head region is bulbous with a distinct connective tissue segment that is divided into ascending and descending portions. The descending portion narrows into what is referred to as the body, and the tail is also a bulbous region which folds over, tapers, and terminates as the straight ductus deferens.

Lagomorpha: The rabbit epididymis has been used to study spermatozoa maturation. Its head is more spatulate than in rodents, leading into a longer attenuate body and a tail that is also not as bulbous as that seen in rodents.

Artiodactyla and Perissodactyla: In the bull and horse, the epididymis can be divided into 8–10 segments, with maturation of spermatozoa occurring in the head and body of the epididymis. The distal portion of the body is smaller than the other segments but is not as narrow or long in comparison to that of the rat.

Marsupialia: Marsupial epididymides are different from eutherian species in that they are compact, highly lobulated, and not easily divided into the classic three parts.

Proboscidea and Hyracoidea: Both orders have intraabdominal testes, and the epididymal structure is significantly different from that of scrotal mammals. The testes are located just inferior to the kidneys and the body and tail of the epididymis extend caudally into the pelvis.

Edentata: Edentates are also testicond mammals, but their epididymides closely resemble those of scrotal mammals. In the armadillo, the tail of the epididymis is not entirely intraabdominal but appears to occupy a small outpouching in the peritoneum.

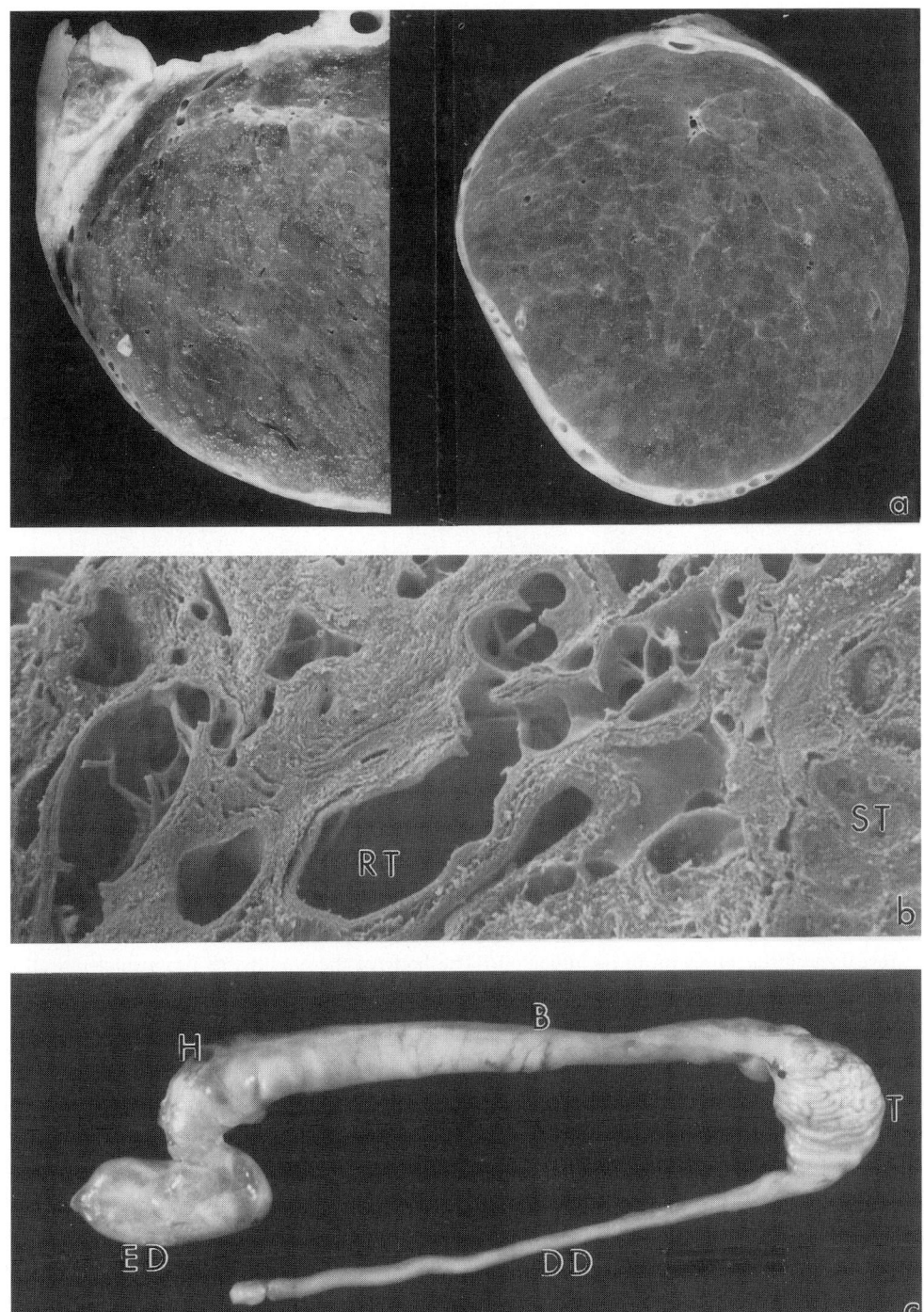

FIGURE 4 Equine testis, efferent ducts, and epididymis. (a) The horse testis viewed in longitudinal and cross sections illustrates dark parenchyma, scattered connective tissue (light regions), and a light connective tissue capsule. (b) Spermatozoa exit the testis by traveling through the lumen of seminiferous tubules (ST) and throughout the rete testis tubules (RT). (c) From the rete testis, spermatozoa travel through the efferent ducts housed in a connective mass (ED) and into the epididymis. The epididymis can be divided into the head (H), body (B), and tail (T) where it attaches to the ductus deferens (DD) (reproduced with permission from Johnson, 1997; Amann *et al.*, 1977; Johnson *et al.*, 1980).

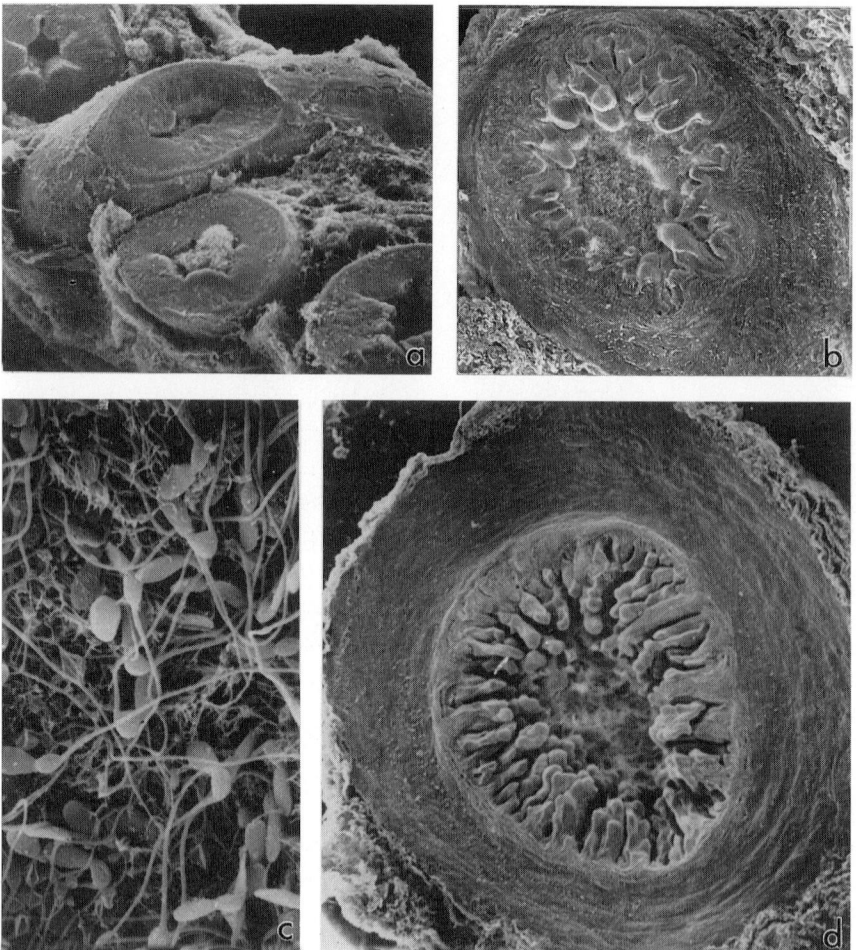

FIGURE 5 Equine epididymis and ductus deferens. a) Four profiles of the epithelium and lumen of the head of the epididymis. One cross section reveals a cluster of sperm in its lumen. (b) The body of the epididymis has a thicker wall than that of the head. (c) The body of the epididymis is also the site of translocation of the cytomplasmic droplet on the spermatozoan tail. Several sperm are seen with the tail bending around the droplet. (d) The tail of epididymis has a large lumen and a thick muscular wall and carries sperm to the ductus deferens (reproduced with permission from Johnson et al., 1978b).

Monotremata: As in other testicond mammals, the distal part of the epididymis is not related directly to the testes and extends into the pelvic region. The typical three-part gross division is also absent, but rather it is divided into two morphologically distinct sections with the distal half serving as a spermatozoa storage area.

Carnivora: The carnivore epididymis has not been extensively studied, but it is grossly divided into three regions as in most scrotal mammals.

During ejaculation, the spermatozoa are transported from the tail of the epididymis, where they are stored, up the tubular ductus deferens by peristaltic contractions of the smooth muscle lining the ductus. Spermatozoa reach the penile urethra after they pass through a short ejaculatory duct. The structure of the ductus deferens is fairly consistent among mammalian species. In all species the terminal portion of the ductus is glandular, but only in some is it thickened into a fusiform structure known as the ampulla (Fig. 1). This structure is particularly well developed in the stallion. It is also consistently present in lagamorphs, proboscides, and perissodactyls and is present only in some species of insectivores, carnivores, primates, chiropterans, arteriodactylids, and rodents. The ampulla is poorly developed in the pig and the cat.

Joining the ejaculatory duct of the ductus deferens is a short duct from the seminal vesicles. These paired structures are not present in all mammals; they are absent in monotremes, marsupials, carnivores, cetaceans, and some insectivores, chiropterans, primates, and lagomorphs. In those in which they are present, they vary primarily in size and degree of lobation. These glands are often referred to as vesicular glands ("seminal vesicles" is a misnomer that was applied earlier when it was thought that these pouches stored spermatozoa).

IV. ACCESSORY SEX GLANDS (PROSTATE)

The male accessory sex glands produce secretions that contribute to the seminal plasma in which spermatozoa are transported (Fig. 1). These glands include the seminal vesicles, prostate, and bulbourethral glands (or Cowper's glands). Not all these organs are present in every species, and each may display a variety of forms and functions among various species. However, all the accessory organs are sensitive to hormonal stimulation which increases the size of the organs. This is why the accessory organs are generally smaller in castrated animals and animals out of season due to lack of hormonal stimulation. The seminal plasma produced collectively by these accessory organs is rich in electrolytes, fructose, ascorbic acid, various enzymes, and vitamins which nourish the spermatozoa, support their metabolism, cleanse the urethra, lubricate the penis and vagina, and neutralize the acidity of the female reproductive tract to make it more conducive to the survival of the spermatozoa.

The prostate is the only unpaired accessory organ and is the only accessory organ that is consistently present in some form in all mammals. The gland can take on either disseminate or discrete forms, both of which drain into the pelvic urethra via several small ducts. The gland is considered disseminate if the glandular acini remain within the lamina propria around the urethra without penetrating the surrounding voluntary muscle. The discrete form is characterized by the gland forming a definite body outside of the urethra and completely encircling it. Most species have either one or the other form, but some may have combinations of the two, as is seen in the boar and bull. The prostate is described as follows in the following mammalian orders:

Carnivora: In carnivores, the canine prostate has been most extensively studied due to its propensity for hyperplasia as is seen in the human prostate. The canine prostate is a discrete organ that has a dorsal depression that grossly divides it into two lobes, but this lobation is not evidenced histologically.

Rodentia: Rodents have discrete prostates, but this order is so varied that no lobes to as many as three lobes may be present, each opening by a duct into the urethra.

Lagomorpha: Lagomorphs have the most grossly different configuration of accessory organs of any group of mammals. The basic structure of the prostate is a bilobed organ fused by a connective tissue capsule. Controversy arises as to the significance of a more cranial organ in close apposition to the prostate and several small peripheral glands to the prostate called paraprostates.

Artiodactyla: As mentioned previously, the bull and boar, as well as the dromedary, have a discrete prostate at the base of the bladder as well as a disseminate prostate that extends the length of the pelvic urethra to the bulbourethral glands. Sheep have an entirely disseminate prostate. Morphology continues to vary among the other species of artiodactyls.

Marsupialia: The marsupial prostate is different in its morphology in that it takes on one of two forms: a cone shape with the tapered end directed along the urethra or a heart shape with the apex directed away from the bladder. The cone shape is most prominent among the marsupials studied.

Chiroptera: Chiropteran prostates display much variability in size, shape, and position.

Hyracoidea: The hyrax appears to have an entirely disseminate prostate.

Proboscidea: Elephants have a compact prostate with two lobes situated on either side of the urethra.

Other groups: Edentates have not been thoroughly studied but appear to be variable across species, and insectivores are subject to much controversy as to the presence of a true prostate gland.

The prostate contributes its secretions to the seminal plasma, and in several species it enlarges considerably during breeding season. In domestic species and in man, this gland is prone to hypertrophy and neoplasia, most often hormonally influenced. The occurrence of this in other species has not been reported.

Another paired set of glands are the bulbourethral glands or Cowper's glands. These glands are located near the bulb of the penis and are connected to the urethra via a short duct. The pressure of the erect penis squeezes the gland. Since the gland has limited luminal storage capacity, its secretions are discharged into the penile urethra prior to ejaculation, rinsing the urine from the urethra and raising the pH of the urethra prior to sperm passage. Most species have only one pair of these glands, but some have as many as three pair (as is seen in some marsupials). These glands are absent in aquatic mammals, mustelids, bears, and dogs.

V. COPULATORY ORGAN: PENIS

The penis is the external copulatory organ of the male (Fig. 1). The basic structure of the penis consists of a central urethra (through which both semen and urine are transported), surrounded by cavernous (erectile) tissue and covered by skin. The penis begins at the caudal border of the ischial arch with two roots (crura). The crura converge and extend cranially as the body of the penis. The two crura converge in the body of the penis as the corpus cavernosum, the main body of erectile tissue. The urethra runs in a ventral groove of the corpus cavernosum surrounded by the smaller mass of erectile tissue called the corpus spongiosum, which begins at the pelvic outlet as an enlargement of spongy tissue referred to as the bulb of the penis. Varying degrees of erectile tissue are seen across species, but there are two basic forms. One type is the musculocavernous form, which has a large amount of erectile tissue with larger blood spaces and more muscle. This type is found in species such as the horse and dog. Artiodactyls tend to have a fibroelastic penis that has a predominance of firm fibroelastic tissue. These species tend to have a sigmoid penis and erectile tissue located at the base of the penis that, when engorged, causes the sigmoid flexure to straighten out and the penis to protrude. The distal end of the penis, the glans penis, is composed of two parts, the bulbus glandis and the more distal pars longa glandis. The bulbus glandis is an expansile structure that is not seen in all species. It is responsible for retaining the penis in the vagina during copulation. The distal pars longa glandis is an extension of the corpus spongiosum. Several species, such as dogs, raccoons, and walruses, have a calcified structure located within the glans penis called the os penis. It originates as the calcified fused ends of the corpora cavernosa. The base and body of the os penis are grooved ventrally to surround the urethra and corpora spongiosum. Still other species (rabbit, ferret, and cats), particularly those in which the female is an induced ovulator, have barbs or spicules on the free end of the penis to enhance mechanical stimulation of the vagina and induce ovulation. As the penis leaves the pelvis, it is covered by a sheath of integument known as the prepuce. At the fornix, the mucosal layer of the prepuce is reflected onto the mucosal layer of the penis. In the erect state, the fornix is eliminated. In the newborn the penis is attached to the prepuce; this attachment breaks down during puberty. The entire unit of the penis is surrounded by a thick fibrous tunic, the tunica albuginea. Muscular attachments include the ischiocavernosus muscle and the retractor penis muscle. The retractor penis muscle originates from the ventral surface of the sacrum and first few caudal vertebrae and is responsible for keeping the nonerect penis within the prepuce and close to the body.

VI. VASCULAR SUPPLY AND INNERVATION

The vascular supply to the penile structures originates primarily from the left and right internal and external pudendal arteries and veins. The internal pudendal artery leads to the artery of the penis which terminates at the level of the ischial arch into three branches. The artery of the bulb of the penis enters the bulb of the penis and supplies the corpus spongiosum and the penile urethra. The deep artery of the penis enters the corpus cavernosum lateral to the bulb at the level of the ischial arch. The third branch, the dorsal artery of the penis, runs on the dorsal surface to the level of the bulbus glandis, where it sends branches to the prepuce and the glans. The external pudendal artery supplies the prepuce, scrotum, and perineal structures. Each of these arteries has an accompanying vein which is important in the mechanism of erection.

The external reproductive organs have somatosensory and somatomotor innervation from the spinal nerves. The penis itself, glans penis, prepuce, and perineal region are supplied by branches of the pudendal nerve. The scrotum, prepuce, and spermatic fascia receive innervation from the genitofemoral branch of the lumbar segments of the spinal nerves. Innervation to the internal reproductive organs is autonomic. The testes are supplied by nerves from the caudal mesenteric ganglion of the sympathetic trunk; the deferent ducts, accessory glands, and urethra are supplied by the pelvic plexus, which receives its primary components from the hypogastric nerves. The erection process is primarily under parasympathetic influence, whereas ejaculation itself is brought about mostly by sympathetic stimulation.

The testes are supplied by the left and right testicular arteries which originate from the abdominal aorta. They elongate and pass through the inguinal canal with the testes in those species that have descended testes. Once through the inguinal canal, the testicular artery becomes tightly coiled. In some species (e.g., marsupials) the artery divides into up to 200 branches only to reunite again prior to supplying the testis. This arrangement eliminates arterial pulses and results in almost no change in mean arterial pressure as arterial blood enters the testes. As the testicular artery reaches the testes, it straightens, runs along the epididymal margin to the distal pole, and branches in a species-specific pattern. As the testicular veins exit the testis, they form a tangled network called the papiniform plexus around the coiled artery. This functions as a method of countercurrent heat exchange, cooling the arterial blood before it enters the testes and warming the venous blood before it returns to the body.

VII. INTERACTIONS OF HORMONES

The hypothalamus of the brain stimulates the pituitary gland to produce hormones [luteinizing hormone (LH) and follicle-stimulating hormone (FSH)]

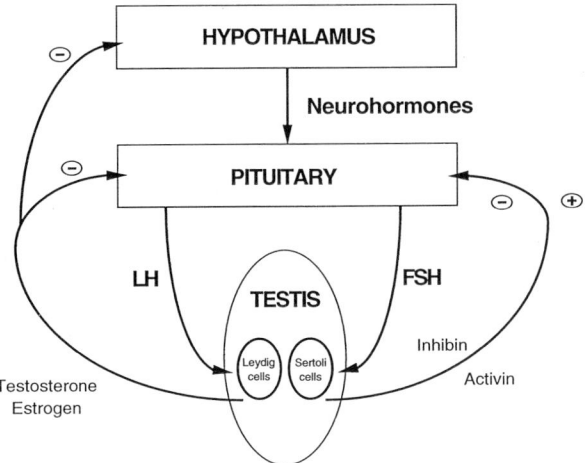

FIGURE 6 A flow chart indicating the interaction of the testis with the hypothalamus and the pituitary of the brain that regulate male reproductive function (reproduced with permission from Johnson 1997).

that are carried by the blood to stimulate the testis (Fig. 6). Under the influence of LH from the pituitary, Leydig cells (located between seminiferous tubules) produce testosterone, the most important male sex hormone. Testosterone is important for the development of the male reproductive tract in the fetus, the maintenance of secondary sex characteristics, spermiogenesis, and male sexual behavior. The Leydig cells also produce estrogen. Both testosterone and estrogen negatively influence both the hypothalamus and the pituitary to decrease stimulation for their production, keeping their levels under control. On the other hand, Sertoli cells (nurse cells) within the seminiferous tubules are stimulated by FSH released from the pituitary to produce both inhibin and activin. Inhibin has a negative feedback loop to the pituitary to decrease FSH secretion. Activin stimulates FSH production to enhance germ cell development.

See Also the Following Articles

EPIDIDYMIS; PAMPINIFORM PLEXUS; PENIS; PROSTATE GLAND; SPERMATOGENESIS

Bibliography

Amann, R. P., Johnson, L., and Pickett, B. W. (1977). Connection between the seminiferous tubules and the efferent ducts in the stallion. *Am. J. Vet. Res.* 38, 1571–1579.

Bone, J. E. (1988). The reproductive system. In *Animal Anatomy and Physiology*, 3rd ed. Prentice Hall, New York.

Evans, H. E., and deLahunta, A (1988). *Guide to the Dissection of the Dog.* Saunders, Philadelphia.

Hamilton, D. W. (1990). Anatomy of mammalian male accessory reproductive organs. In *Marshall's Physiology of Reproduction*, (G. E. Lamming, Ed.), Vol. 2, pp. 691–746. Churchill Livingston, New York.

Johnson, L. (1991). Spermatogenesis. In *Reproduction in Domestic Animals*, pp. 173–219. Academic Press, New York.

Johnson, L. (1995). Efficiency of spermatogenesis. *Microsc. Res. Technol.* 32, 385–422.

Johnson, L. (1997). Efficiency of spermatogenesis in humans and animals. College of Veterinary Medicine, Texas A&M University, College Station, Texas. Copyright © 1997 L. Johnson.

Johnson, L., Amann, R. P., and Pickett, B. W. (1978a). Scanning electron and light microscopy of the equine seminiferous tubule. *Fertil. Steril.* 29, 208–215.

Johnson, L., Amann, R. P., and Pickett, B. W. (1978b). Scanning electron microscopy of the epithelium and spermatozoa in the equine excurrent duct system. *Am. J. Vet. Res.* 39, 1428–1434.

Johnson, L., Amann, R. P., and Pickett, B. W. (1980). Maturation of equine epididymal spermatozoa. *Am. J. Vet. Res.* 41, 1190–1196.

Johnson, L., Welsh, T. H., Jr., and Wilker, C. E. (1997). Anatomy and physiology of the male reproductive system and potential targets of toxicants. In *Comprehensive Toxicology: Volume 10. Reproduction and Endocrine Toxicology* (Sipes, McQueen, and Gandolfi, Eds.). Pergamon, New York.

Meijer, J. C., and Fentener van Vlissingen, J. M. (1993). Gross structure and development of reproductive organs. In *World Animal Science B No. 9* (G. J. King, Ed.), pp. 9–26. Elsevier, New York.

Reece, W. O. (1997). Male reproduction. In *Physiology of Domestic Animals*, 2nd ed. Williams & Wilkins, Baltimore.

Setchell, B. P. (1991). Male reproductive organs and semen. In *Reproduction in Domestic Animals,* 4th ed. (P. T. Cupps, Ed.), pp. 221–249. Academic Press, New York.

Setchell, B. P., and Brooks, D. E. (1988). Anatomy, vasculature, innervation, and fluids of the male reproductive tract. In *The Physiology of Reproduction* (E. Knobil and J. D. Neill, Eds.), pp. 753–819. Raven Press, New York.

Walker, W. F. (1987). *Functional Anatomy of the Vertebrates.* Saunders, Philadelphia.

Varner, D. D., Schumacher, J., Blanchard, T. L., and Johnson, L. (1991). Reproductive anatomy and physiology. In *Diseases and Management of Breeding Stallions.* (P. W. Pratt, Ed.), pp. 1–59. American Veterinary Publications, Goleta, CA.

Male Reproductive System, Reptiles

Daniel H. Gist

University of Cincinnati

I. Anatomy of the Reptilian Male Reproductive System
II. The Testis
III. Control of Testis Function
IV. Actions of Male Sex Hormones
V. Mating Systems
VI. Future Research

GLOSSARY

amniote The amnion of vertebrate eggs, a membrane which encloses a fluid-filled sac in which the embryo develops. Found in reptilian, avian, and mammalian eggs, the amnion is considered a major character in the evolution of vertebrates, contributing to the radiation of vertebrates into the terrestrial environment.

excurrent canals The ducts of the male reproductive tract, derived embryologically from the Wolffian duct, through which sperm are transported from the testis to the intromittant organ.

intromittant organ The male copulatory organ. It is used only for reproduction and possesses erectile tissue to facilitate insertion into the female; sperm are transferred via a groove along its length.

spermatocytogenesis The process by which spermatids elongate, develop acrosomes and flagella, and undergo volume reductions to become morphologically mature spermatozoa; this occurs within Sertoli cell vacuoles.

spermatogenesis The series of meiotic cell divisions occurring in the seminiferous epithelium. It occurs within the vacuoles of Sertoli cells and results in the formation of spermatids containing the haploid number of chromosomes.

The class Reptilia is pivotal in the evolution of vertebrates. Many characters of birds and mammals, including some reproductive characters, made their first appearance in the Reptilia. Turtles (order Testudinata) are the most ancient of the living reptiles, dating to 280 million years before present (BP); morphologically, they have changed little since that time. The alligators and crocodiles (order Crocodilia) arose 240 million years BP; these are the remnants of the archeosaurs, a group which gave rise to the dinosaurs and from which modern-day birds evolved. Lizards and snakes (order Squamata) are of more recent origin, dating to 170 million years BP. Reptiles are the earliest representatives of the amniote vertebrates, and in most species reproduction takes place in the terrestrial environment. This has implications for the male. For example, at mating, sperm are not released to the environment but are transferred directly to the female in a copulatory act. Thus, for reproductive success male reptiles possesses an intromittant organ for sperm transfer, a reservoir for the storage of sperm (epididymis), and behaviors associated with mate selection. Reproduction in most reptiles is seasonal; in males, spermatogenesis occurs for only a portion of the year. However, with epididymal storage of sperm, the time of mating need not correspond to that of spermatogensis. Like other vertebrates, reproduction in reptiles is coordinated by the hypothalamic–pituitary–testicular axis. Most of the advances in our knowledge of reptilian male reproductive biology since the last major review of the subject (1984) have dealt with sexual behavior, mating patterns, and epididymal maturation of sperm.

I. ANATOMY OF THE REPTILIAN MALE REPRODUCTIVE SYSTEM

The male reproductive system consists of the testes, located in the abdominal cavity, and the system of excurrent canals that convey seminal products to the exterior. These canals are derived embryologically from the Wolffian duct. The seminiferous tubules of the testis deliver sperm into one or more efferent ductules (*ductuli efferentes*) which empty into the small ductules (*ductuli epididymis*) of the epididymis. Reptiles lack the rete testis, a series of anastamosing intratesticular canals connecting the seminiferous tubules to the efferent ductules. Rather, the efferent ductules draining the seminiferous tubules lie outside the testis proper and anastamose among themselves and the ductuli epididymis; the number of anastamoses and the histology of these ducts vary among the groups that have been examined. The ductuli epididymis empty at varying levels into the single main duct of the epididymis, the ductus epididymis. The ductus epididymis is highly convoluted, folded over on itself, and forms the epididymis proper. In some species, the epididymis is larger than the testis when full of sperm. The *ductus deferens* conveys sperm from the epididymis to the cloaca, where it is transferred to the intromittant organ. Except for the lizards and snakes, there are no accessory glands which contribute to the seminal fluid.

The intromittant organ of male reptiles is variable in morphology; most squamate males have a pair of hemipenes, each of which has a groove (*sulcus spermaticus*) that serves as a conduit for sperm transfer and an ornate terminus which may facilitate attachment to the female. Each hemipenis is connected to the cloacal wall and, at mating, is everted and extruded from the cloaca. Only one hemipenis is used at mating. Turtles and crocodilians have a single copulatory organ. It contains a sulcus spermaticus and possesses erectile tissue which, when distended, allows insertion into the female.

II. THE TESTIS

Like all vertebrates, the reptilian testis possesses both spermatogenic and endocrine functions. The reptilian testis is a tubular testis, containing seminiferous tubules in which spermatogenesis and spermatocytogenesis occur in intimate association with Sertoli cells. Mature sperm are released into the seminiferous tubule and are carried away from the testis. This contrasts to the testes of nonamniote vertebrates in which sperm are produced in cysts containing germinal elements and one or more Sertoli cells, all enclosed within a connective tissue capsule. In the cystic testis, spermatogenic and spermatocytogenic events occur within the cyst; when sperm are mature the cyst ruptures, discharging them and the accompanying Sertoli cells into the excurrent ducts. There is an additional difference between the reptile testis and that of nonamniotes in the Sertoli cell. Although its size may vary considerably during a reproductive period, the Sertoli cell is a permanent feature of the reptilian testis. By virtue of tight junctions established between adjacent Sertoli cells, a functional blood–testis barrier exists, separating spermatogonia and meiotic cells throughout the year. In contrast, testicular cysts containing Sertoli cells and spermatogonia form annually in nonamniote vertebrates, and a functional blood–testis barrier exists in these vertebrates for only a portion of the annual cycle. The significance of these differences is unknown.

A. Spermatogenesis

Spermatogenesis in most reptiles is seasonal, occurring once a year. Even in tropical species spermatogenesis is discontinuous and is synchronized with environmental events such as rainfall; only a few tropical species produce sperm continually throughout the year. Two basic patterns of spermatogenesis have been described in seasonally reproducing reptiles (Fig. 1). In one, the prenuptial spermatogenic pattern, spermatogenesis and subsequent maturational events (spermatocytogenesis) are timed so that sperm mature just before females are about to ovulate. Typically, spermatogenesis commences at the end of winter and is completed within a period of 2 or 3 months; circulating levels of male sex steroids also follow this same pattern. Matings then coincide with ovulations by females. Postnuptial spermatogenesis, in contrast, commences shortly after females ovulate in the spring and is completed by autumn, approximately 5 or 6 months prior to the next ovulatory period. Matings may occur either in the spring, prior to ovulation, or during the preceding autumn. Utilization of these different reproductive patterns does not follow taxonomic lines; examples of each may be found among turtles, lizards, and snakes.

The cytology of spermatogenesis in reptilian testes resembles that of birds and mammals; spermatogonia proliferate and become incorporated into Sertoli cell vacuoles for spermatogenic and spermatocytogenic events. Cells are displaced toward the lumen as they

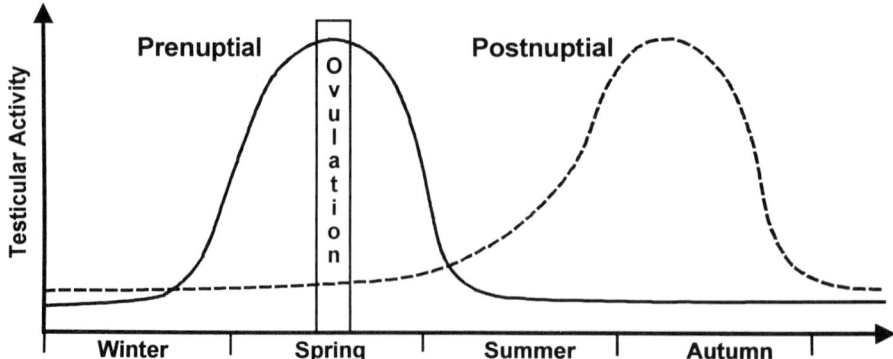

FIGURE 1 Prenuptial and postnuptial spermatogenic cycles of temperate-zone reptiles. In the prenuptial cycle, spermatogenesis, androgen secretion, and mating all occur close to the time when eggs in females mature and are ovulated. In the postnuptial cycle, spermatogenesis does not occur at the same time that eggs are maturing and ovulated in females. In the postnuptial cycle, mating and/or androgen secretion may not occur at the time of spermatogenesis (adapted with permission from Crews, 1984).

develop. Indices of germ cell development based on the presence of easily recognized cell types (e.g., spermatogonia, pachytene spermatocytes, and elongated spermatids) have been developed for several species, but more detailed studies of the spermatogenic and spermatocytogenic events have yet to be performed.

B. Endocrinology

The endocrine portion of the reptilian testis resides in the Leydig (= interstitial) cells, located outside the seminiferous tubules, and the Sertoli cells within the seminiferous tubules. Both structures at some point in the annual cycle possess enzymes essential for sex hormone biosynthesis and cytoplasmic organelles suggestive of steroid production. Teiid lizards have an unusual distribution of Leydig cells, which are almost exclusively located on the periphery of the testis.

A variety of androgens have been identified from testicular tissue, including testosterone (T), androstenedione (A), dihydrotestestosterone (DHT), epitestosterone, and dehydroepiandrosterone. These same steroids may also be found in the blood, though species differences exist. For example, T, DHT, and A have been identified in blood of squamate and crocodilian reptiles as well as in some turtles, but in other turtles epitestosterone and dehydroepiandrosterone are the main testicular and circulating androgens. Varying methodologies have been employed to characterize the androgens isolated from testicular tissues and blood. Thus, the presence or absence of a particular steroid in a given species may reflect the ability or inability of a particular technique to detect that steroid in addition to whatever species differences may exist. Despite the differences noted previously, T, DHT, and A continue to be the androgens most frequently assayed in reptilian blood. Similar to other vertebrates, reptilian sex steroids are transported in blood bound to sex hormone-binding proteins.

Several studies measuring changes in circulating androgens throughout the year or according to the spermatogenic stage of the testis have found a good correlation between the spermatogenic stage and circulating androgen levels in species possessing a prenuptial spermatogenic pattern. Furthermore, histochemical studies of 3-hydroxysteroid dehydrogenase activity, an enzyme involved in the biosynthesis of all major steroid hormones, and of 17-hydroxysteroid dehydrogenase activity, an enzyme more directly involved in the formation androgens, reveal that these enzymes are active in Leydig cells during spermatogenesis. These observations have led many to believe that the Leydig cells are responsible for the synthesis of androgens. However, in reptiles possessing postnuptial spermatogenic patterns, androgen levels and the spermatogenic stage of the testis are not always synchronized with Leydig cell histology and histochemistry. In freshwater turtles, for example, Leydig cells hypertrophy and acquire increased 3-hydroxysteroid dehydrogenase activity in the spring, when the testes are regressed and androgen levels are declining. This has led to the suggestion that both Sertoli and Leydig cells contribute to circulating androgen levels in these species but at different times. Unfortunately, preparations containing only one of these cell types are not yet available for study, and the relative contribution of Leydig cells and Sertoli cells to circulating androgens throughout the year has not been assessed.

Most studies of circulating androgen levels were carried out years ago; new information has since appeared which indicates that capture and maintenance of reptiles constitutes a stress which rapidly and markedly depresses circulating androgens. In the American alligator, for example, plasma testosterone levels were halved within 4 hr following capture and restraint, and by 24 hr they were 10% of their initial value. Thus, both the absolute and relative androgen levels reported in earlier studies, some of which involved purchased animals housed for varying periods of time, may not reflect the feral condition.

Small quantities of estrogen are found in both testicular extracts and blood from male reptiles. In the lacertid lizard *Podarcis sicula*, a species with a prenuptial spermatogenic pattern, measurable quantities (approximately 0.1 that of T) of estradiol are found in blood and vary throughout the year in a pattern different from that of androgens. Blood estrogens begin to rise as testes develop in the spring, but reach their highest levels following testicular regression.

III. CONTROL OF TESTIS FUNCTION

Testicular function in reptiles is controlled by the secretions of the pituitary gland which in turn are influenced by neurohormones released from the hypothalamus of the brain. According to the mammalian paradigm, testicular function is regulated by two pituitary gonadotropins, follicle-stimulating hormone (FSH) and luteinizing hormone (LH), which have separate properties and whose primary actions are upon the seminiferous tubule and the Leydig cell, respectively. Whether the reptilian pituitary possesses the same number of pituitary hormones as that of mammals has been the subject of considerable discussion. A single gonadotropin, chemically resembling mammalian FSH but possessing biological activity of both FSH and LH, is attributed to the pituitary gland of squamate reptiles. Testes of squamate reptiles that possess a prenuptial spermatogenic pattern respond to gonadotropin preparations by initiating or accelerating spermatogenesis and by increasing the secretion of androgen hormones. The steroidogenic response might be indirect because gonadotropin-induced release of prostaglandins has been implicated in the control of steroidogenesis in the testes of *P. sicula* and other species. In other reptilian groups, evidence exists for the existence of two gonadotropic hormones resembling mammalian FSH and LH. Unfortunately, most of these studies have utilized females. Here, FSH will stimulate ovarian development and sex steroid (estrogen) secretion; LH surges at or around the time of ovulation and has been implicated in the control of progesterone secretion. Only in the turtle *Chrysemys picta*, a species with a postnuptial spermatogenic pattern, have gonadotropic hormones been characterized in the male on the basis of their ability to stimulate androgen secretion (*in vivo*) or release from testicular slices (*in vitro*). Clear distinctions between mammalian FSH and LH have not been attained using these assays, possibly because of the heterologous gonadotropin preparations used in these studies or by differential responses of Leydig cells and Sertoli cells.

Gonadotropin secretion from the mammalian pituitary gland is controlled by a neurohormone of hypothalamic origin, gonadotropin-releasing hormone (GnRH), which is released into blood running directly into the anterior pituitary gland. GnRH is a decapeptide and several forms have been isolated from the brains of various reptiles; these differ from one another by only one to three amino acids. Brains of turtles and alligators contain two GnRHs which are identical to the avian GnRH-I and GnRH-2 decapeptides, whereas the squamate brain possesses these and other forms (salmon GnRH and human GnRH) depending on the species. Developmental changes in the composition of GnRH occur in the brains of lizards. Assays based on the ability of GnRH to stimulate FSH or LH release from pituitary fragments incubated *in vitro*, or more indirect *in vivo* assays measuring the ability of GnRH to increase peripheral androgen levels, have demonstrated that turtle and alligator pituitaries respond to GnRHs from several sources. In contrast, in the only study involving squamates, pituitaries from the lizard *P. sicula* responded only to salmon GnRH. The more limited response of the squamate pituitary to GnRHs of various origins may reflect the single gonadotropin thought to be present in this group.

Seasonal changes in the pattern of GnRH release and thus of gonadotropin secretion may occur in response to environmental variables. Of the various environmental factors, attention has focused on temperature and photoperiod as proximate variables affecting reptilian reproductive cycles. However, these variables are intertwined since the range of temperatures is greater at higher latitudes, where photoperiodic changes are also greatest. In general, for species with extensive latitudinal distribution, the reproductive season is extended at lower latitudes. Moreover, experimental studies with a variety of species have indicated that temperature is the more important variable affecting reproduction. In the lizard *Anolis carolinensis*, for example, testes are photorefractory except when maintained close to their preferred temperature. At their preferred temperature, extension of day length will prolong spermatogenesis, and a reduction of day length will accelerate testicular regression, but photoperiodic changes do not affect lizards maintained in the cold. Whether these thermal effects are mediated via the hypothalamic–pituitary–testicular axis or whether they are direct effects on the testis is unclear. Testes of lizards maintained at lower temperatures secrete only small quantities of sex hormones and do not respond to gonadotropin administration. In the freshwater turtle *C. picta*, testes do not respond to photoperiodic changes unless body temperatures are above 17°C. The temperature requirement for optimal testicular development is not absolute; the tuatara lizard (*Sphenodon punctatus*), one of two living representatives of the primitive suborder Sphenodontia, lives on New Zealand islands where the annual temperature ranges from 9 to 16°C. Males have a prenuptial spermatogenic cycle and produce sperm annually, whereas the female cycle takes 4 years to complete. The role of low temperatures in male reproductive physiology has not been examined closely, but exposure to a period of reduced temperatures is essential for normal vitellogenesis in some female lizards and turtles.

IV. ACTIONS OF MALE SEX HORMONES

Until recently, responses to the secretion of male sex hormones in reptiles have received little attention. This is possibly due to the absence of accessory glands in the male reproductive tract or to the absence of strong sexual dimorphisms in some species. Known target organs of the male sex hormones in reptiles undergo changes as secretion of androgens varies. Most organs sensitive to androgen stimulation will respond equivalently to both T and its 5α-reduced metabolite, DHT. The presence of androgen receptors in a target organ is inferred by its response, but there are relatively few studies demonstrating the existence of androgen receptors. The few studies on the identity and distribution of androgen receptors in reptiles indicate a high degree of homology with those of other vertebrates.

A. Excurrent Canals

The excurrent canal system draining the reptilian testis is itself a target of the testicular sex steroids. Portions of this duct system in mammals, such as the epididymis, are key in the maturing of spermatozoa, and acquisition of fertility and motility occur while sperm are stored in the epididymis. Spermatozoa acquire these properties in part by modifications of glycoproteins on their surface resulting from or accompanied by differential secretions of the epididymis along its length. Other portions of the excurrent canal system in mammals are associated with fluid reabsorption and still others receive the secretions of the male accessory glands.

The histology and annual variations of the reptilian excurrent canal are known in only a few species. Unlike the mammal, there are no morphological changes in tubule structure along the length of the reptilian epididymis other than the diameter. The most abundant cell type in the epithelium of the reptilian epididymis is a highly secretory and androgen-sensitive cell resembling the principal cell of the mammalian epididymis. However, histological, cytological, and biochemical studies of the epididymis suggest considerable species differences. The epididymis of *Lacerta vivipara*, a lizard with a prenuptial spermatogenic cycle, has been studied intensively. Annual changes in the epididymal epithelium are highly correlated with the changes in circulating androgens (Fig. 2). When hypertrophied (stage IV), the epithelial cells secrete granules (Fig. 3) into the lumen, where they mix with residing sperm. Production and secretion of the granules is abolished by castration or the administration of androgen inhibitors. The granules contain both insoluble (H) and soluble (L) proteins. The soluble proteins are a mixture of up to nine isoforms which coat the surface of epididymal sperm. Some of these proteins are glycosylated, whereas others are phosphorylated. It is not known whether these secreted proteins affect the motility or fertility of spermatozoa within the epididymis, but at the time when these proteins are secreted maximally, the motility of epididymal sperm is greatest toward the caudal end of the epididymis. Details of the endocrine control of epididymal L proteins have also been studied. While androgen deprivation will prevent L protein synthesis and secretion, early events in the differentiation of the epididymal epithelium remain unaffected. In addition, the stimulatory effects of androgens on L protein secretion are enhanced by corticosterone. Taken together, these findings suggest complex controls for the secretion of L proteins, whose function remains uncertain.

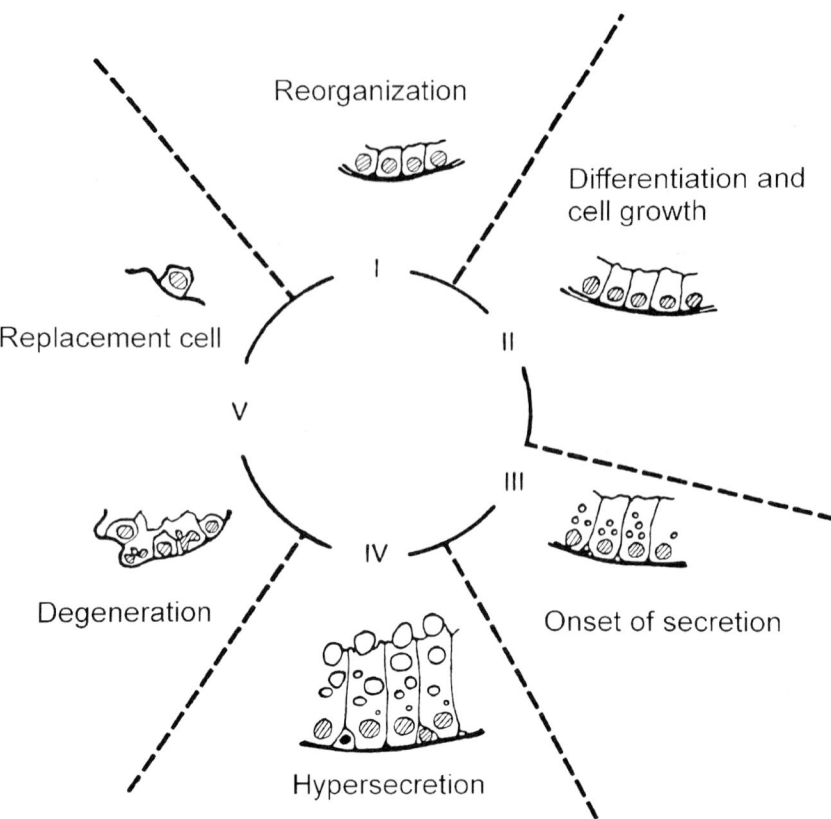

FIGURE 2 Schematic cycle of the lizard (*L. vivipara*) epididymal epithelium. Five stages of epithelial differentiation have been described throughout the annual cycle. Stages II–IV are associated with increased androgen secretion (reproduced with permission from Courty *et al.*, 1987).

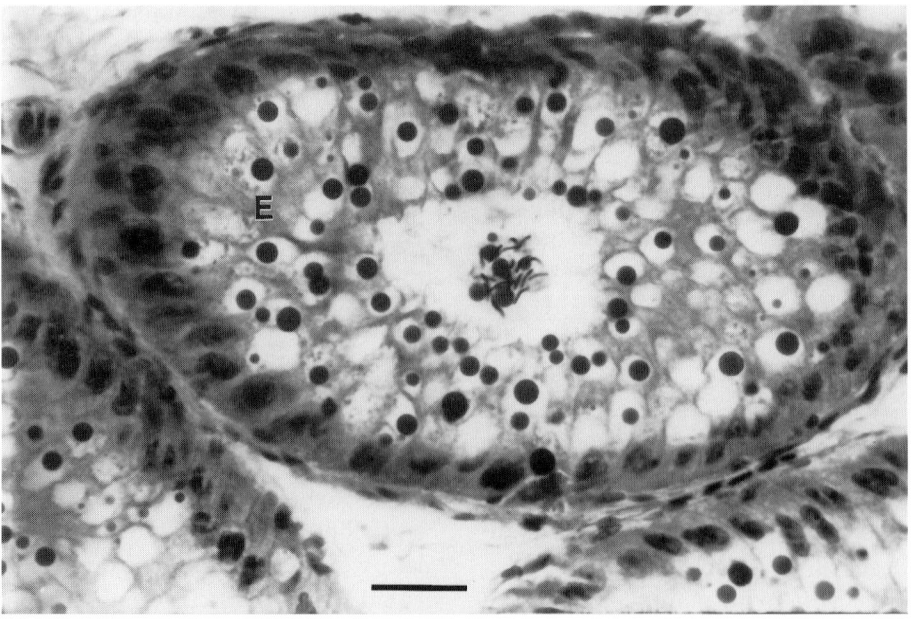

FIGURE 3 Epididymis of the lizard *L. vivipara* during the reproductive period. The epithelium (E) contains numerous secretory granules consisting of a dark central core and a peripheral vacuole. These granules are discharged into the lumen where they mix with sperm. Scale bar = 25 μm (reproduced with permission from Depeiges *et al.*, 1981).

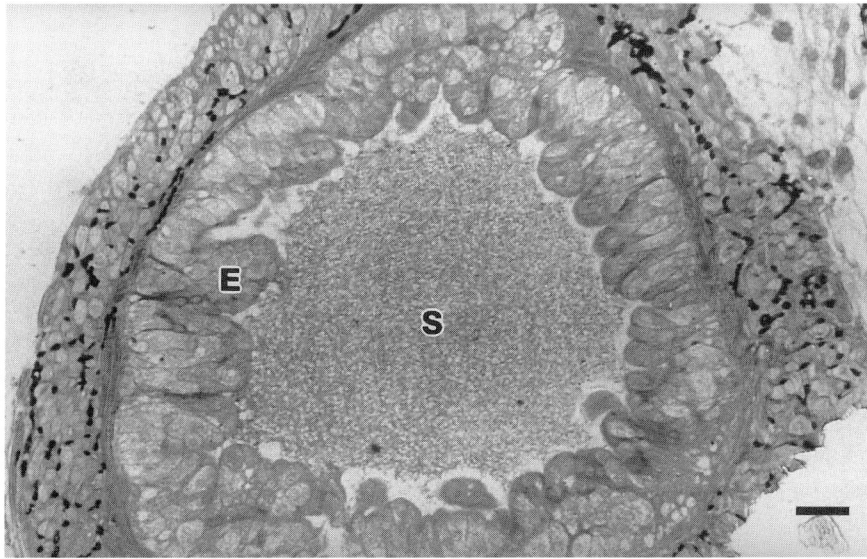

FIGURE 4 Epididymis of the turtle *Chrysemys picta* during the reproductive period (October). The epithelium (E) is hypertrophied but ungranulated. A brush border on the apical surface of the epithelium suggests a secretory or absorptive role. Note sperm (S) in lumen. Scale bar = 25 μm.

In contrast, the variations in the epithelium of the turtle epididymis (Fig. 4) are less conspicuous. Seminal fluids from turtle and snake contain proteins not found in blood, confirming the secretory role for the epididymal epithelium suggested by morphological studies. Some of these secretions are glycosylated, but do not associate with the sperm surface. In the tortoise *Testudo hermanii*, which possesses a postnuptial reproductive pattern, the epithelium lining the ductuli epididymis displays more pronounced changes throughout the year than does that of the epididymis proper. These cells contain N-acetylglucosaminidase, an enzyme associated with altering the glycosylation of membrane proteins. The activity of this enzyme is highest in the ductuli epididymis during the period of androgen synthesis.

The reptilian epididymis possesses receptors to estrogens as well as androgens. In the *P. sicula*, a lizard possessing a prenuptial spermatogenic pattern, circulating levels of estrogen are highest during the regression of the testes and epididymis following mating, and estrogen treatment accelerates regression of the testes and epididymis. In contrast, the epididymis of reptiles possessing the postnuptial spermatogenic pattern remains hypertrophied and contains motile sperm long after regression of the testes.

It is clear that a high degree of variability in epididymal form and function exists among the Reptilia. The epididymis is probably the least studied of the androgen-dependent organs, and consequently it is difficult to address the evolutionary significance of that variability. Because sperm remain within the epididymis longer in reptiles possessing postnuptial spermatogenic patterns, it is possible that sperm maturation within the epididymis occurs more slowly in these species; alternatively, the role of the epididymis in these species may simply be storage.

Reptiles possess no distinct glands along the excurrent canals which contribute to the seminal fluid. The ductus deferens is short in most species and exhibits few specializations. An exception is found in lizards such as *Calotes versicolor* and *Sceloporus*, in which the posterior portion of the ductus deferens, near the cloaca, is swollen into an ampulla lined by a secretory epithelium. This differentiation is androgen dependent, and it is speculated that the ampulla participates in sperm maturation rather than storage. In other lizards, the hemipenis itself is an androgen-dependent organ; castration results in a reduction of

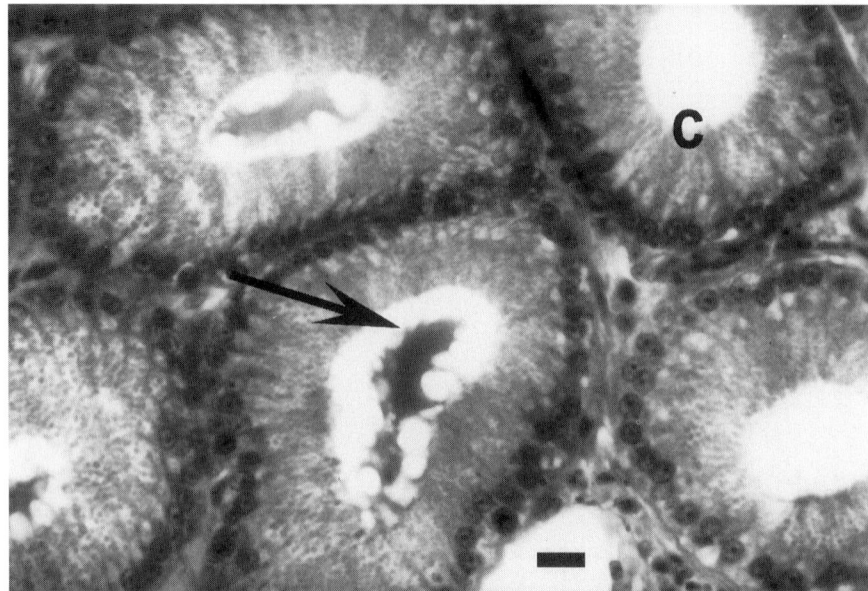

FIGURE 5 Hypertrophied sex segment in the collecting duct (C) of the lizard P. muralis kidney. Note the heightened epithelium containing secretory granules and secretion in the tubule lumen (arrow). In some species, sex segment secretions form a vaginal plug in females. Scale bar = 6 μm.

hemipenis size and an inability to extrude it. These are ameliorated by androgen replacement.

B. Sex Segment

A prominent androgen-dependent character is the sexual segment. Found in both lizards and snakes, the sex segment (Fig. 5) consists of a specialization of the epithelium of collecting ducts in the caudal portion of the kidney. These hypertrophy during the reproductive season and release a granular material into the collecting ducts (Fig. 5). In most species the secretion is merocrine and contains protein, fat, and carbohydrate components. Synthesis of secretory granules by the sexual segment is stimulated by male sex hormones. The function of the sex segment secretions is unknown for most species but it has been speculated that it may serve to separate sperm from urinary products or perhaps to play a pheromonal role. In snakes such as *Thamnophis* and *Natrix*, granules from the sexual segment form a vaginal plug in the cloaca of mated females. In *Thamnophis*, females possessing vaginal plugs are unattractive to males and are neither courted nor mated.

C. Behavior

The central nervous system (CNS) is a target for male sex hormones, and many of the behaviors associated with mate selection and territoriality are initiated, directly or indirectly, as the result of sex hormone secretion. Work with the green anole (*Anolis carolinensis*) clearly established the importance of androgen hormones on male reproductive behavior. Castrated males failed to form territories, court, or mate with females; replacement therapy with either testosterone or DHT, either injected or implanted within the CNS, restored these behaviors. Lizards of other species respond similarly. Functional (hormone-binding) receptors for T and/or DHT and estradiol have been found in the anterior hypothalamus and preoptic areas of the lizard brain, close to the sites of GnRH synthesis and release. However, male sex hormones are not the only factors that affect male reproductive behavior. The plains garter snake *T. sirtalis* possesses a postnuptial spermatogenic pattern and mates in the spring upon emergence from hibernacula. At this time the testes are regressed and androgen levels low. Mating success is achieved using sperm stored over winter in the epididymis.

Neither androgen implants nor injections will induce mating behaviors in this snake, but mating behaviors do appear without additional treatment following exposure to low temperatures for 3 or 4 months at any time of the year.

Progesterone has been implicated in affecting male mating behavior in lizards belonging to the genus *Cnemidophorus*. In *C. ornatus*, which possesses a prenuptial spermatogenic pattern, male behaviors can be induced in castrated males following T, DHT, or progesterone administration. In *C. uniparens*, a parthenogenic hybrid, certain females will exhibit male-type behaviors and will interact with preovulatory females. Those females exhibiting male-type behaviors have little or no circulating androgens but do have elevated progesterone levels. Furthermore, male-type behaviors can be induced in ovariectomized *C. uniparens* females by progesterone administration. Thus, male sex hormones may not be the only inducer of male sexual behavior. The reptilian brain possesses the enzyme aromatase, which will convert aromatizable androgens, notably T and A, into estrogens. Aromatase is concentrated in the hypothalamus and thus the possibility exists that estrogen rather than androgen elicits male-specific behaviors. However, studies utilizing estrogen implants into hypothalamic areas of the lizard (*A. carolinensis*) have shown this not to be the case and the significance of aromatase in the adult male hypothalamus remains unknown.

D. Others

Many reptiles exhibit permanent sex dimorphisms, including size in crocodilians, plastron shape and forelimb modifications in turtles, and coloration in squamates. The mechanisms by which individuals attain these dimorphic characters probably involve exposure to specific hormones during development; these characters have received little attention and will not be considered here. There are, however, several additional male secondary sex characters which are affected by androgens. Most of these are found in squamate reptiles; outside this group, androgen-dependent characters not associated with the male excurrent canals are few.

A large number of male lizards exhibit different or brighter coloration during the reproductive season. In some, the coloration of females also changes during the reproductive period. There are many factors that influence the degree of coloration, including the lighting, background, and stress condition of the animal. Nevertheless, it has been demonstrated in several species of lizards that castration reduces the brightness of sexually dimorphic coloration in males, and that replacement therapy with androgens will restore normal reproductive male coloration. These coloration changes have been demonstrated in male lizards such as *A. carolinensis* and *Urosaurus ornatus* to facilitate conspecific recognition, the establishment of hierarchies, and the maintenance of territories.

Femoral glands are found in a number of lizard species. These exocrine glands, located on the ventral surface of the hindlimbs, are present in both sexes but those of males are more highly developed. In males, these glands enlarge during the reproductive season, regress following castration, and respond to replacement therapy with either T or DHT. The secretions are holocrine, contain proteins and lipids, and form a waxy plug in the duct opening to the exterior; in some species it is possible to obtain samples of the secretion by the application of pressure. In the lizard *Iguana iguana*, the lipid content of femoral gland secretion increases during the reproductive season from 14 to 35% by weight and the absolute amount of secretion is higher in dominant males. A pheromonal function is suggested by the volatile constituents in femoral gland secretion and its ability to elicit conspecific recognition.

V. MATING SYSTEMS

Central nervous system control of male sexual behaviors is but one mechanism contributing to mating success. It is becoming increasing clear that many, if not all, reptiles mate promiscuously, with males copulating with many females and vice versa. Since the interval over which mating occurs may be prolonged, males must aliquot their annual supply of gametes over many females to ensure optimum reproductive success. Storage of sperm in the epididymis allows this to occur, but the mechanism by which sperm are aliquoted is unknown. In reptiles possessing postnuptial spermatogenic cycles, sperm not transferred to females in autumn matings remain over winter within the epididymis and are utilized in matings the following spring, a time when androgen levels are low and the testes incapable of sperm production. Similarly, most female reptiles are capable of storing sperm within their oviducts for extended periods of time, certainly over several egg clutches. This oviductal sperm storage enables females to produce repeated clutches of fertilized eggs without repeated matings. The promiscuity in mating coupled with oviductal retention of sperm suggests that sperm from several matings, possibly with different males, may mix within the oviduct, and that different eggs within a single egg clutch may have different parentage. This is known to occur in some snakes and turtles. Thus, the timing of mating and the subsequent disposition of sperm within the oviduct may play a role equally important as mate selection in ensuring reproductive success.

VI. FUTURE RESEARCH

Our knowledge of the reproductive biology of reptiles is based on intensive studies involving a few species. The Reptilia are a diverse group, and this knowledge base needs to be expanded to additional species, particularly those which have potential economic benefit or are of evolutionary significance. Husbandry of reptiles for food and other purposes requires a detailed knowledge of their reproduction. For males, this includes the duration of the spermatogenic cycle, the number of sperm produced within a cycle, and the number of cycles in a reproductive period. It must also include an understanding of mating systems, including mate selection and sperm storage. The significance of the tubular testis, with its permanent blood–testis barrier, likewise remains to be evaluated. Our knowledge is so fragmentary that information regarding certain facets of reproduction (e.g., epididymal secretions) is based on work done with a single species. There is much to be learned by investigating those features of reptilian male reproductive biology which have evolutionary and functional significance to the amniote mode of reproduction. Acquisition and regulation of male fertility is a poorly known area of mammalian reproductive biology, and sperm acquire these properties in the epididymis and other portions of the excurrent duct system. Studies of these processes in more primitive amniotic vertebrates will yield important information regarding the evolution of mechanisms for the posttesticular processing of sperm.

See Also the Following Articles

REPTILIAN REPRODUCTION, OVERVIEW; REPTILIAN REPRODUCTIVE CYCLES

Bibliography

Callard, I. P., Callard, G. V., Lance, V., Bolaffi, J. L., and Rossett, J. S. (1978). Testicular regulation in nonmammalian vertebrates. *Biol. Reprod.* **18**, 16–43.

Cooper, W. E., and Greenberg, N. (1992). Reptilian coloration and behavior. In *Biology of the Reptilia* (C. Gans and D. Crews, Eds.), Vol. 18E, pp. 298–422. Univ. of Chicago Press, Chicago.

Courty, Y., Morel, F., and Dufaure, J. P. (1987). Characterization and androgenic regulation of major mRMAs coding for epididymal proteins in a lizard (*Lacerta vivipara*). *J. Reprod. Fertil.* **81**, 443–451.

Crews, D. (1984). Gamete production, sex hormone secretion, and mating behavior uncoupled. *Horm. Behav.* **18**, 22–28.

Crews, D. (1992). Behavioral endocrinology and reproduction: An evolutionary perspective. *Oxford Rev. Reprod. Biol.* **14**, 303–370.

Depeiges, A., Betail, G., and Dufaure, J. P. (1981). Time course of appearance *in vivo* and *in vitro* of a specific epididymal protein controlled by testosterone. *Biol. Cell* **42**, 49–56.

Fox, H. (1977). The urogenital system of reptiles. In *Biology of the Reptilia* (C. Gans and T. S. Parsons, Eds.), Vol. 6E, pp. 1–157. Academic Press, New York.

Gist, D. H., and Jones, J. M. (1987). Storage of sperm in the reptilian oviduct. *Scanning Microsc.* **1**, 1839–1849.

Licht, P. (1984). Reptiles. In *Marshall's Physiology of Reproduction* (G. E., Lamming, Ed.), 4th ed., Vol. 1, pp. 206–282. Churchill Livingstone, Edinburgh, UK.

Mammary Gland Development

Jose Russo and Irma H. Russo

Fox Chase Cancer Center

I. Prenatal and Perinatal Development
II. Postnatal Development
III. The Menopausal Breast
IV. Hypothalamic–Pituitary Influences on the Initiation of Ovarian Function
V. Summary and Conclusions

Assessment of mammary development requires the use of morphometric techniques to elucidate the complex process of interaction between two embryologically different, though deeply interconnected tissues—the parenchyma and the stroma. The peculiar tree-like structure of the mammary gland, in which the branches of the tree are the ducts lined by epithelium, as well as the topographic localization of the areas of proliferation require the observation of the organ in three dimensions, such as in whole mount preparations, to establish precisely within the tree the location of every specific structure.

I. PRENATAL AND PERINATAL DEVELOPMENT

The mammary gland parenchyma arises from a single epithelial ectodermal bud. Most authors agree on the successive stages of development of the mammary gland during the embryonic and fetal stages, although there are variations in nomenclature, and the exact time of appearance of each structure varies whether the authors choose to express the age of the embryo based on the estimated time of conception, the last missed menstrual period, or the length of the embryo. Because of difficulties in establishing precisely the day of conception, we personally consider it more accurate to correlate the phases of mammary gland development with embryonal or fetal length. Mammary gland development can be divided into 10 different stages (Table 1). It is important to emphasize the last stage or end vesicle stage in which colostrum reaction is found. This reaction is not a product of lobule-type differentiation because lobule types are never observed in fetal material. In the newborn breast, there are very primitive structures (Fig. 1) composed of ducts ending in short ductules lined by one layer of epithelial and one of myoepithelial cells (Fig. 2). The epithelial cells have an eosinophilic cytoplasm, with typical apocrine secretion. The fine cytoplasmic vacuolization observed in the epithelial cells is due to the presence of lipid droplets, as confirmed electron microscopically. However, secretory activity does not seem to be confined to the primitive alveolar structures since the whole ductal system appears dilated, secretion filled, and lined by a secretory-type epithelium. These observations suggest that secretory activity is a generalized response of all the mammary epithelium to maternal hormonal levels. The secretory activity of the newborn gland subsides within 3 or 4 weeks.

II. POSTNATAL DEVELOPMENT

Mammary gland development during childhood does little more than keep pace with the general growth of the body until the approach of puberty.

A. Adolescence

Although the main changes occurring in the mammary gland are initiated at puberty, ulterior development of the gland varies greatly from woman to woman. Mammary gland development can be defined

TABLE 1
Stages of Prenatal Development
of the Human Breast

Ridge stage (<5-mm embryo)
Milk hill stage (>5.5-mm embryo)
Mammary disc stage (around 10- or 11-mm embryo)
Globule-type stage (11.0- to 25.0-mm embryo)
Cone stage (25- to 30-mm embryo)
Budding stage (30- to 68-mm embryo)
Identation stage (68 mm to 10 cm)
Branching stage (10-cm fetus)
Canalization stage (20 and 32 weeks of gestation)
End vesicle stage, in which the end vesicles are composed of a monolayer of epithelium and contain colostrum (newborn)

from the external appearance of the breast or by determination of mammary gland area, volume, degree of branching, or degree of structures whose appearance indicates the level of differentiation of the gland, such as lobule-type formation.

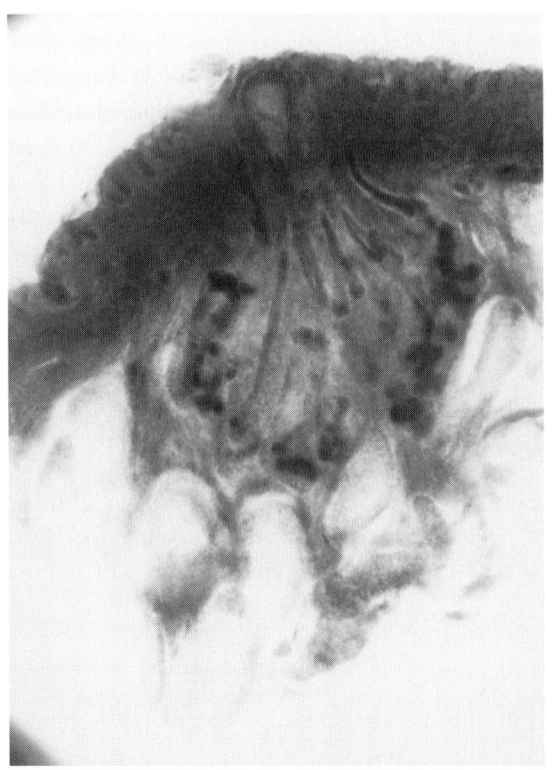

FIGURE 1 Whole mount of a 2-week-old human female breast.

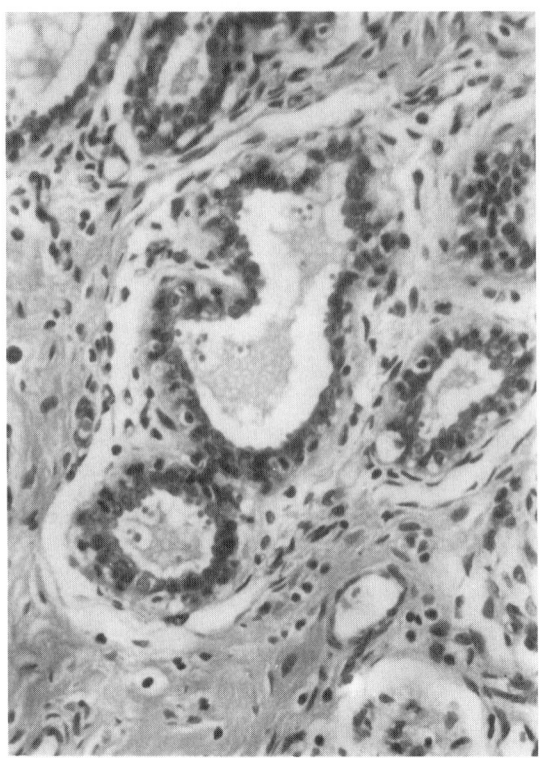

FIGURE 2 Mammary gland of a 2-week-old human female containing ductal structures and primitive ducto lobule-type elements. The luminae of all structures are dilated and filled with proteinaceous fluid (hematoxylin and eosin, ×25).

The adolescent period begins with the first signs of sexual change at puberty and terminates with sexual maturity. Puberty in the female sets in between the ages of 10 and 12 years. With the approach of puberty, the rudimentary mammae begin to show growth activity both in the glandular tissue and in the surrounding stroma. Glandular increase is due to the growth and division of small bundles of primary and secondary ducts (Fig. 3). They grow and divide partly dichotomously (from the Greek word *dichotomos*, or repeated bifurcation) and partly sympodially (from Greek word *syn* + *podion* base, involving the formation of an apparent main axis from successive secondary axes) on a dichotomous basis (Fig. 3). The ducts grow, divide, and form club-shaped terminal end buds. Terminal end buds give origin to new branches, twigs, and small ductules or alveolar buds (Fig. 4). We use the term alveolar bud to identify those structures that are morphologically

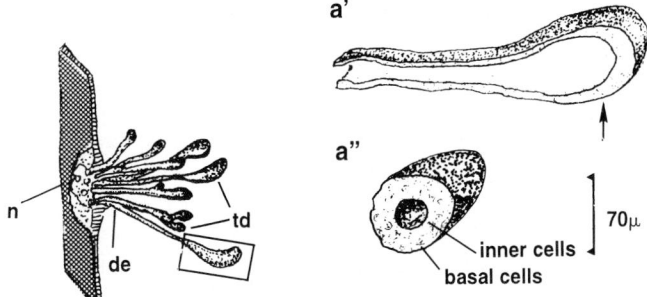

FIGURE 3 Mammary gland of human female at birth formed by several excretory ducts (de), ending in terminal ducts (td). a', detail of the inset showing the club-shaped terminal end bud from which lengthening and further divisions of the virginal ducts originate. a", cross section at the level shown in a'; the duct is lined by the two layers of cells. Proliferation takes place chiefly in the basal cells. The inner cells have secretory properties from which the "witch milk" is formed (n, nipple) (reproduced with permission from Russo et al., 1982).

FIGURE 5 Lobule-type formation occurs after the first menstruation in humans. The number of lobules increases with age. Some portions of the gland do not further develop if pregnancy does not supervene; c', virginal lobule (reproduced with permission from Russo et al., 1982).

more developed than the terminal end bud but yet more primitive than the terminal structure of the mature resting organ, which is called acinus. Alveolar buds cluster around a terminal duct, forming the lobule type 1 or virginal lobule (Figs. 5 and 6) and each cluster is composed of approximately 11 alveolar buds. Terminal ducts and alveolar buds are lined by a two-layered epithelium, whereas terminal end buds in the human fetus are lined by an epithelium made up of up to four layers of cells. Lobule formation in the female breast occurs within 1 or 2 years after onset of the first menstrual period. Full differentiation of the mammary gland is a gradual process taking many years, and in some cases, if pregnancy does not supervene, is never attained.

B. Menstrual Cycle

It is acknowledged that hormonal influences play a significant role in breast development; however, the effect of their fluctuations during the menstrual cycle on parenchymal proliferation has not been definitively elucidated. Normal breast epithelium undergoes cyclic variations of DNA synthesis, as determined in normal breast samples cultured in the presence of [^{3}H]thymidine. Even though cell prolifer-

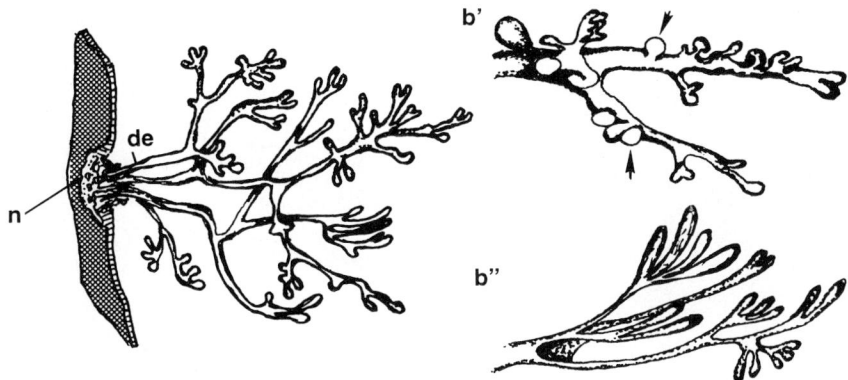

FIGURE 4 Before the onset of puberty in human females, the ducts grow and divide in a dichotomous and sympodial basis. b', a ball-shaped end bud sprouts from the ducts (arrows); b", new branches and twigs develop from the terminal and lateral end buds (n, nipple) (reproduced with permission from Russo et al., 1982).

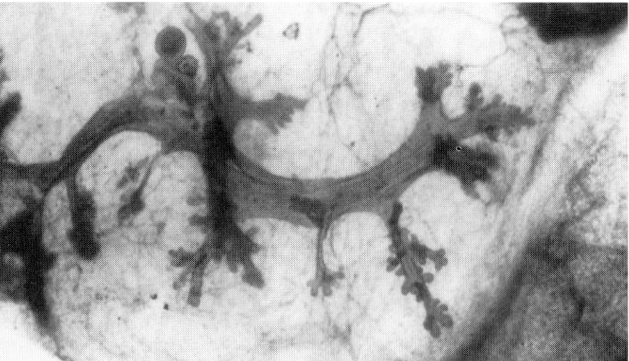

FIGURE 6 Whole mount preparation of breast tissue of an 18-year-old nulliparous woman showing type 1 lobules. Toluidine blue, ×25 (reproduced with permission from Russo et al., 1992).

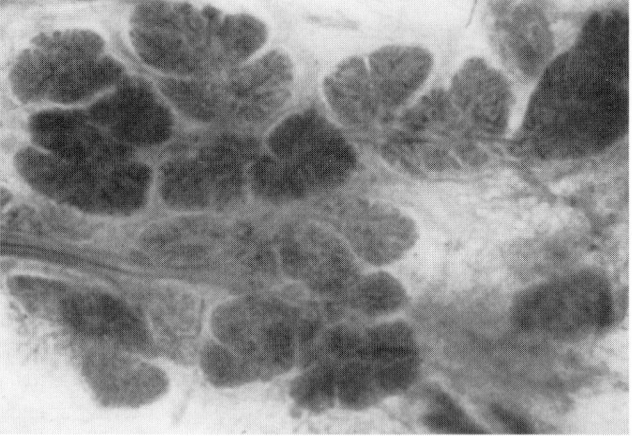

FIGURE 8 Whole mount preparation of human breast tissue of a 35-year-old parous woman containing type 3 lobules. Toluidine blue, ×25 (reproduced with permission from Russo et al., 1992).

ation and cell death seem balanced to maintain the equilibrium of the resting breast, mammary development induced by ovarian hormones during a menstrual cycle never fully returns to the starting point of the preceding cycle. Accordingly each ovulatory cycle fosters slightly more mammary development with new budding of structures that continues until about age 35.

Normal breast tissue of adult women contains two identifiable types of lobules, in addition to the already described type 1: lobule type 2 and type 3 (Figs. 7 and 8). The transition from lobule type 1 to type 2, and of type 2 to type 3, is a gradual process of sprouting of new alveolar buds. In lobule type 2 and type 3, these are now called ductules; they increase in number from approximately 11 in the lobule type 1 to 47 and 80 in lobule type 2 and type 3, respectively (Figs. 9–12 and Table 2). The increase in number results in a concomitant increase in size of the lobules and a reduction in size of each individual structure. The alveolar buds composing a lobule type 1 measure an average of 0.232×10^{-2} mm^2, practically twice the size of the ductules composing lobule type 2, whereas the reduction in size in ductules composing lobules type 3 is less dramatic, although still significant (Fig. 12 and Table 2).

The breast of nulliparous women contains more undifferentiated structures such as terminal ducts and lobule type 1, although occasional lobules type

FIGURE 7 Whole mount preparation of human breast tissue of a 24-year-old nulliparous woman showing type 2 lobules. Toluidine blue, ×25 (reproduced with permission from Russo et al., 1992).

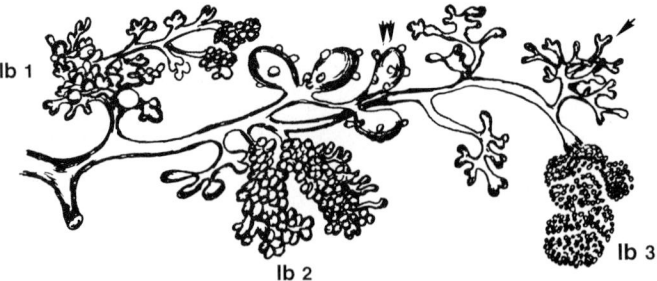

FIGURE 9 Schematic representation of the human breast in the first trimester of pregnancy (reproduced with permission from Russo et al., 1982).

FIGURE 10 Schematic representation of the human breast in the third month of pregnancy (reproduced with permission from Russo *et al.*, 1982).

2 and 3 are seen. In parous women, on the other hand, the predominant structure is the most differentiated lobule type 3. In contrast to the type 1 lobules, which in the nulliparous women remain constant throughout the life span, the type 3 lobules in parous women peak during the early reproductive years, decreasing after the fourth decade of life. In the breast of nulliparous women, type 2 lobules are present in moderate numbers during the early years and sharply decrease after age 23, whereas the number of type 1 lobules remains significantly higher. This observation suggests that a certain percentage of type 1 lobules might have progressed to type 2 lobules, but the number of type 2 lobules progressing to type 3 is significantly lower than that in parous women. In the case of parous women, it is interesting to note that a history of parity between the ages of 14 and 20 years correlates with a significant increase in the number of type 3 lobules that remain present as the predominant structure until the age of 40, the time at which a decrease in the number of type 3 lobules

FIGURE 11 Schematic representation of the human breast in the middle of pregnancy (reproduced with permission from Russo *et al.*, 1982).

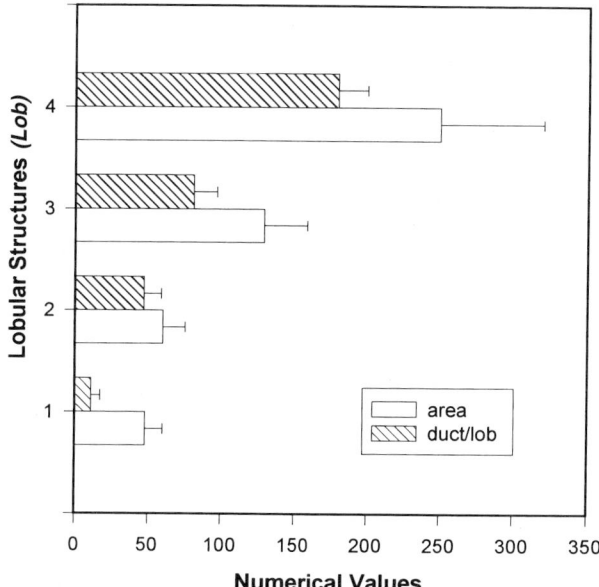

FIGURE 12 Histogram showing the characteristics of the lobule-type structures in the human breast based on lobule area (in μm^2), number of ductules per lobule type, and number of cells per section in lobules type 1 (Lob 1), type 2 (Lob 2), type 3 (Lob 3), and type 4 (Lob 4).

occurs, probably due to their involution to predominantly type 1 lobules (Figs. 13 and 14).

Determination of the proliferative activity DNA-labeling Index (LI) of these structures by measuring the incorporation of [^{3}H]thymidine into the mammary epithelium has shown that the DNA-LI of lob-

TABLE 2
Characteristics of the Lobular Structures
of the Human Breast

Structure	Lobular area[a] (μm^2)	No. of ductules/lobule[b]	No. of cells/ cross section[c]
Lob 1	48 ± 44	11.2 ± 6.3	32.4 ± 14.1
Lob 2	60 ± 26	47.0 ± 11.7	13.1 ± 4.8
Lob 3	129 ± 49	81.0 ± 16.6	11.0 ± 2.0

[a] Student's *t* tests were done for all possible comparisons. Lobular areas showed significant differences between lob 1 vs lob 3 and lob 2 vs lob 3 ($p < 0.005$).

[b] The number of ductules per lobule was different ($p < 0.01$) in all the comparisons.

[c] The number of cells per cross section was significantly different in ductules of lob 1 vs 2 and 3 ($p < 0.01$).

Note. Reprinted with permission from Russo *et al.*, 1992.

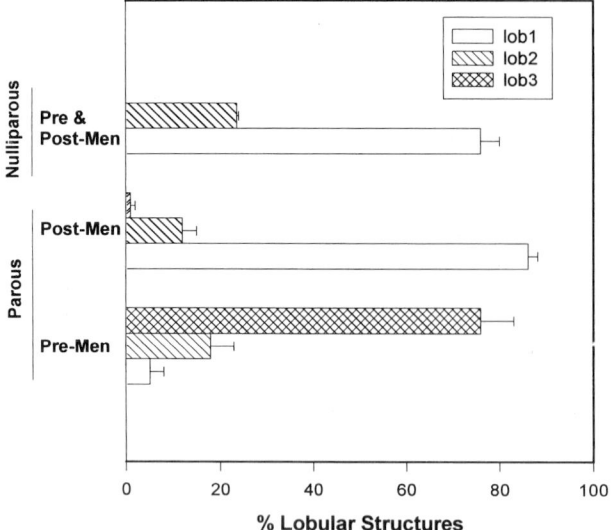

FIGURE 13 Percentage of lobule-type structures in the breast of premenopausal (Pre-Men) and postmenopausal (Post-Men) parous women and pre-and postmenopausal (Pre & Post-Men) nulliparous women. Lobules type 1 (Lob 1), type 2 (Lob 2), and type 3 (Lob. 3).

ule type 2 is around 0.99 and of lobule type 3 is 0.25 (Fig. 15). These values are 5 and 20 times lower than those found in type 1 lobules and up to 60 times lower than in the terminal end bud (Fig. 15). It is important to emphasize that in the study of the proliferative activity of the mammary gland each topographic compartment has to be analyzed individually. There is a gradient in the proliferative activity from the terminal end bud to the type 3 lobules; ductal structures have a proliferative activity intermediate between that of lobules type 1 and type 2 (Fig. 15). This gradient does not seem to be modified with aging, although in older women all proliferative activity is significantly reduced.

C. Pregnancy

During pregnancy, the breast attains its maximum development; it occurs in two distinctly dominant phases characteristic of the early and late states of pregnancy. The early stage is characterized by growth consisting of proliferation of the distal elements of the ductal tree, resulting in the formation of ductules that at this stage can be called acini, thus developing a lobule type 3 (Figs. 9–11) into a lobule type 4 (not shown). The intensity of budding and degree of lobule formation goes beyond what has been observed in the virginal breast. In Fig. 9, a schematic representation of the gland shows abundant budding in various stages from short dichotomous buds up to formed lobules . By the third month of pregnancy, the number of well-formed lobules exceeds the number of primitive budding stages; however, primitive budding stages are still found. In newly formed lob-

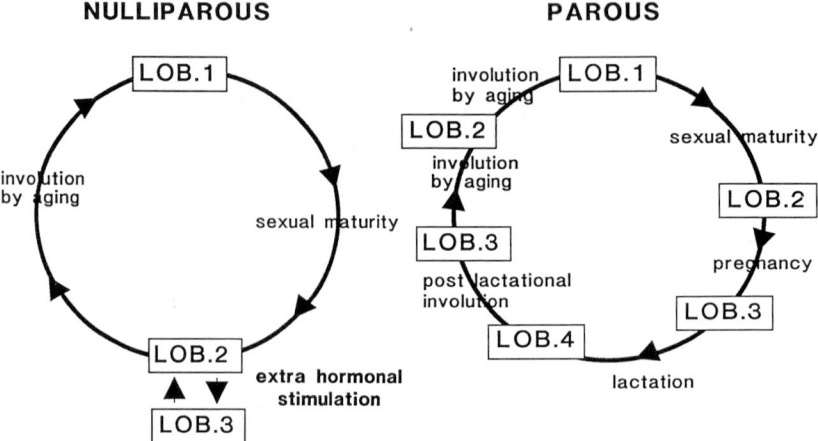

FIGURE 14 Schematic representation of breast development based on relative percentage of lobules present. Nulliparous women's breasts contain primarily lobules type 1 (Lob 1) with some progression to type 2 (Lob 2) and only minimal formation of lobule type 3 (Lob 3). Parous women undergo a complete cycle of development through the formation of type 4 lobules (Lob 4) which later regress (reproduced with permission from Russo et al., 1992).

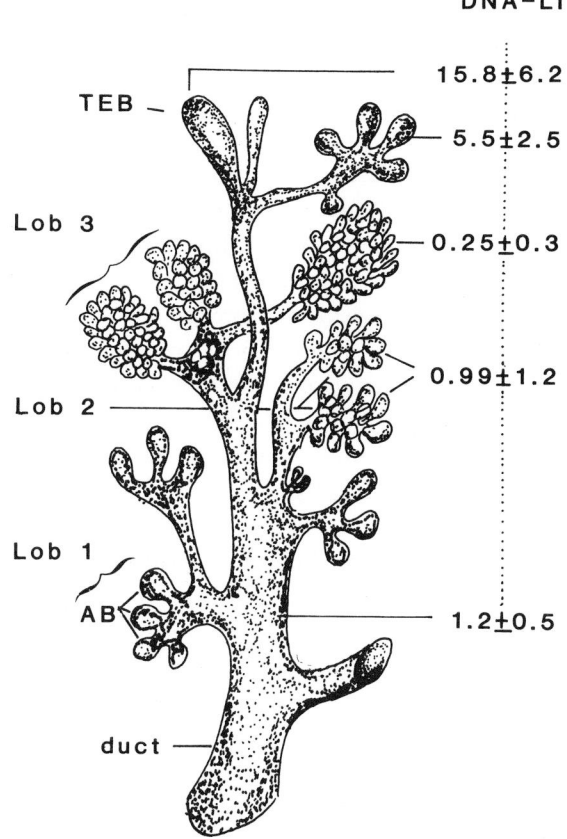

FIGURE 15 Schematic representation depicting the various topographic compartments of the human mammary gland: a terminal end buds (TEB); alveolar buds (AB); lobules types 1, 2, and 3 (Lob 1, Lob 2, and Lob 3); and ducts. On the right-hand column is shown the DNA-labeling index (LI) of each structure (mean ± standard deviation) of nine women ranging in age from 18 to 62 years. A gradient in proliferative activity is observed from the TEB (15.8 ± 5.2), lobules type 1 (5.5 ± 0.5), type 2 (0.9 ± 1.2), type 3 (0.25 ± 0.3), and ducts (1.2 ± 0.5) (reproduced with permission from Russo and Russo, 1987).

ules, the epithelial cells composing each acinus not only increase greatly in number due to active cell division but also increase in size mainly because of cytoplasmic enlargement.

In the middle of pregnancy (Fig. 10), the lobules are further enlarged and increased in number. They surround the duct from which their central branch proceeds so thickly that the chief duct, the terminal or intralobular terminal duct, can no longer be recognized. The transition between the terminal ducts and the budding acini is gradual, making the histological distinction between the two of them difficult since both show evidence of early secretory activity. The definitive structure of the ductal tree is essentially settled by the end of the first half of pregnancy; the mammary changes that characterize the second half of pregnancy are chiefly continuation and accentuation of the secretory activity. Further progressive branching continues with less prominent bud formation (Fig. 11). At this time, the formation of true secreting units or acini, the differentiated structures, becomes more evident. Proliferation of new acini is reduced to a minimum, and the luminae of those already formed become distended by accumulation of secretory material or colostrum. The secretory acinus formed during pregnancy is a terminal outgrowth that marks the end of glandular differentiation. However, just before and during parturition, there is a new wave of mitotic activity with an increase in the total DNA of the gland. During lactation, the process of growth and differentiation may be observed in the same lobule type, side by side with the process of milk secretion.

D. Postlactational Changes

From midpregnancy on, a yellowish fluid containing a high concentration of protein is secreted into the mammary alveoli and may be expelled from the nipple. After postpartum withdrawal of placental lactogen and sex steroids, which appear to prevent the action of prolactin on the mammary epithelium, lactation starts. Colostrum is secreted during the first week postpartum, followed by a 2- or 3-week period of transitional milk secretion, leading to the secretion of mature milk.

No major morphological changes of the mammary gland are observed during lactation. The mammary lobules are enlarged and the acini have a dilated lumen filled with granular, slightly basophilic material admixed with fat. There is a significant variation in lobule size throughout the gland, suggestive of a variation in lactogenic activity from lobule to lobule. Milk is synthesized and released into the mammary acini and ductal system, although it can be stored for up to 48 hr before the rate of milk synthesis and secretion begins to decrease. As long as milk is removed regularly from the mammary gland, the alveolar cells continue to secrete milk almost indefinitely.

The accumulation of milk in the ductoacinar lumina and within the cytoplasm of the lactogenic epithelial cells that occurs after weaning has an inhibitory effect on further milk synthesis. This effect is followed by a series of involutional changes in the mammary gland consisting of a multifocal asynchronous process of reduction in volume of the secretory epithelial cells and further inhibition of their secretory activity.

As a consequence of these changes, mammary lobules undergo atrophy and the stroma shows a marked desmoplastic reaction and fat infiltration. It is considered that postlactational regression is due to two complementary mechanisms: cell autolysis, with collapse of acinar structures and narrowing of the tubules and the appearance of round cell infiltration and phagocytes in and about the disintegrating lobules, and regeneration of the preductal and perilobular connective tissue, with renewed budding and proliferation in the terminal tubules. Until menopausal involution sets in, the parous organ shows more glandular tissue than if pregnancy or pregnancy and lactation had never occurred.

III. THE MENOPAUSAL BREAST

Menopause supervenes as the consequence of the atresia of more than 99% of the 400,000 follicles that are present in the ovaries of a female fetus at a gestational age of 5 months. Gonadotropin-releasing hormone secretion is also implicated in this phenomenon, indicating that a hypothalamic process is involved in the development of menopause. The most characteristic sign of menopause is amenorrhea, which is the result of the almost complete cessation of ovarian estrogen and progesterone production. The years leading up to the final menstrual period, until menopause sets in generally at around age 51 years, constitute the perimenopause. During this time, many women ovulate irregularly, either because the rise in estrogen during the follicular phase is insufficient for triggering a luteinizing hormone (LH) surge or because the remaining follicles are resistant to the ovulatory stimulus. The increase in human longevity occurring in our society has caused a considerable increment in the number of women that will live one-third or more of their lives in the menopausal period, namely without natural estrogen and progesterone. After menopause, the breast undergoes a regressive phenomenon in both nulliparous and parous women. This regression is manifested as an increase in the number of lobule type 1 and a concomitant decline in the number of lobule type 2 and lobule type 3. At the end of the fifth decade of life, the breast of both nulliparous and parous women contains lobule type 1 (Fig. 13). These observations led us to conclude that the understanding of breast development requires a horizontal study in which all the different phases of growth are taken into consideration. For example, the analysis of breast structures at a single given point, i.e., age 50 years, would lead us to conclude that the breast of both nulliparous and parous women appears identical. However, the phenomena occurring in prior years might have imprinted permanent changes in breast biology and affect the potential of the breast for neoplasm but are no longer morphologically observable. Thus, from a quantitative point of view, the regressive phenomenon occurring in the breast at menopause differs in nulliparous and parous women. In the breast of nulliparous women the predominant structure is the lobule type 1, which comprises 65–80% of the total lobule type components and their relative percentage is independent of age. Second in frequency is the lobule type 2, which represents 10–35% of the total. The least frequent is the lobule type 3, which represents only 0–5% of the total lobular population. In the breast of premenopausal parous women, on the other hand, the predominant lobular structure is the lobule type 3, which comprises 70–90% of the total lobule component. Only after menopause do they decline in number, and the relative proportion of the three lobule types present approaches that observed in nulliparous women. These observations led us to conclude that early parous women truly underwent lobule differentiation, which was evident at a younger age, whereas nulliparous women seldom reached the lobule type 3 and never the lobule type 4 stages (Fig. 14).

IV. HYPOTHALAMIC–PITUITARY INFLUENCES ON THE INITIATION OF OVARIAN FUNCTION

In the nonpregnant female, the development of the mammary gland is rigorously controlled by the ovary. Although puberty is often considered to be the point of initiation of ovarian function, the development of the ovary, in fact, is a gradual process. The development and function of the ovary depend on pituitary gonadotropins since receptors for LH and follicle-stimulating hormone (FSH) are present even during the infantile period, when through activating binding to their respective receptors they stimulate the secretion of androgens. The synthesis and release of the pituitary gonadotropins FSH and LH are in turn regulated by gonadotropin-releasing hormone. FSH and LH interact with growth hormone (GH) and prolactin in modulating ovarian steroidogenesis, a function that is also influenced by epinephrine, which is secreted by the adrenal medulla. The ovary also secretes inhibin and activin, glycoprotein hormones that feed back to the pituitary, specifically to modulate the release of FSH. Ductal elongation and branching occurring during puberty are positively regulated by pituitary GH, although its exact mechanism of action is unclear. This hormone directly stimulates ductal growth in hypophysectomized–ovariectomized rats and might also act through its local mediator, insulin-like growth factor-1. Normal duct development, however, requires the presence of estrogen and progesterone, whose respective receptors ER and PR are present in the mammary gland. Estradiol acts locally in the mammary gland, stimulating DNA synthesis and promoting bud formation, probably through an ER-mediated effect. It is also known that the prevailing metabolic condition of an individual animal or human may significantly influence mammary gland response to hormones. The response of the mammary gland to these complex hormonal and metabolic interactions results in developmental changes that permanently modify both the architecture and the biological characteristics of the gland. The mammary gland, in turn, responds selectively to given hormonal stimuli, depending on specific topographic differences in gland development, which modulate the expression of either cell proliferation or differentiation.

V. SUMMARY AND CONCLUSIONS

The human breast undergoes a complete series of changes from intrauterine life to senescence. These changes can be divided into two distinct phases: the developmental phase and the differentiation phase.

The developmental phase includes the early stages of gland morphogenesis, from nipple epithelium to lobule formation. In lobule formation, both processes, development and differentiation, take place almost simultaneously. For example, the progressive transition of lobule type 1 to types 2–4 requires active cell proliferation to acquire the cell mass necessary for the function of milk secretion. This latter process implies differentiation of the mammary epithelium. Therefore, the presence of lobule type 4 is the maximal expression of development and differentiation in the adult gland, whereas the presence of lobule type 3 could indicate that the gland has already been developed, but because the lobule type is not secreting milk, it is not completely differentiated. It is important to point out that the presence of proteins that are indicative of milk secretion, such as α-lactalbumin, casein, or milk fat globule-type membrane protein, also indicates cellular differentiation of breast epithelium. However, only when all the other components of milk (such as lactose, α-lactalbumin, casein, and milk fat) are coordinately synthesized within the appropriate structure can full differentiation of the mammary gland be acknowledged.

Acknowledgment

This work was partially supported by NIEHS Grant SO7280.

See Also the Following Articles

Breast Diseases; Lactation, Human; Mammary Gland, Overview

Bibliography

Anderson, T. J., Ferguson, D. J. P., and Raab, G. M. (1982). Cell turnover in the resting human breast: Influence of parity, contraceptive pill, age and laterality. *Br. J. Cancer* 46, 376–382.

Dabelow, A. (1957). die Milchdruse. In *Handbuch der Mikroskopischen Anatomic des Menchen* (W. Bargmann, Ed.), Vol. 3, Part 3, Haut and Sinnes Organs, pp. 277–485. Springer-Verlag, Berlin.

Edwards, R. G., Howles, C. M., and Macnamee, C. (1990). Clinical endocrinology of reproduction. In *Hormones: From Molecules to Disease* (E.-E Baulieu and P. A. Kelly, Eds.), pp. 457–476. Chapman & Hall, New York.

Ferguson, D. J. P., and Anderson, T. J. (1981). Morphological evaluation of cell turnover in relation to the menstrual cycle in the "resting" human breast. *Br. J. Cancer* 44, 177–181.

Labrie, F. (1990). Glycoprotein hormones: Gonadotropins and thyrotropin. In *Hormones: From Molecules to Disease* (E.-E Baulieu and P. A. Kelly, Eds.), pp. 257–275. Chapman & Hall, New York.

Masters, J. R. W., Drife, J. O., and Scarisbrick, J. J. (1977). Cyclic variations of DNA synthesis in human breast epithelium. *J. Natl. Cancer Inst.* 58, 1263–1265.

Meyer, S. J. (1977). Cell proliferation in normal human breast ducts. Fibroadenomas, and other ductal hyperplasias as measured by tritiated thymidine effects of menstrual phase, age and oral contraceptive hormones. *Hum. Pathol.* 8, 67–81.

Peluso, J. J. (1992). Morphologic and physiologic features of the ovary. In *Pathobiology of the Aging Rat* (U. Mohr, D. L. Dungworth, and C. C. Capens, Eds.), pp. 337–349. ILSI Press, Washington, DC.

Reynolds, E. L., and Wines, J. V. (1948). Individual differences in physical changes associated with adolescence in girls. *Am. J. Dis. Child* 75, 329–350.

Russo, J., and Russo, I. H. (1987). Development of human mammary gland. In *The Mammary Gland Development, Regulation, and Function* (M. C. Neville and C. W. Daniel, Eds.), pp. 67–93. Plenum, New York.

Russo, J., and Russo, I. H. (1994). Toward a physiological approach to breast cancer prevention. *Cancer Epidemiol. Biomarkers Prevention* 3, 353–364.

Russo, J., Tay, L. K., and Russo, I. H. (1982). Differentiation of the mammary gland and susceptibility to carcinogenesis. *Breast Cancer Res. Treatment* 2, 5–73.

Russo, J., Rivera, R., and Russo, I. H. (1992). Influence of age and parity on the development of the human breast. *Breast Cancer Res. Treatment* 23, 211–218.

Russo, J., Ao, X., Grill, C., and Russo, I. H. (1998). Pattern of distribution of cells positive for estrogen receptor α and progesterone receptor in relation to proliferating cells in the mammary gland. *Breast Cancer Res. Treat.*, in press.

Tanner, J. M. (Ed.) (1962). The development of the reproductive system. In: *Growth at Adolescence,* pp. 28–39. Blackwell Scientific, Oxford, UK.

Vorherr, H. (Ed.) (1974). Development of the female breast. In *The Breast*, pp. 1–18. Academic Press, New York.

Mammary Gland, Overview

Isabel Forsyth
The Babraham Institute

I. Introduction
II. Anatomy and Histology of the Mammary Gland
III. Development of the Mammary Gland
IV. Synthesis and Secretion of Milk
V. Milk Removal
VI. Maternal Adaptation to Lactation
VII. Lactation and Fertility

GLOSSARY

alveoli Blind sacs lined by milk-secreting cells which as clusters (lobules) are the basic functional units of the mammary gland.
growth factors Regulatory peptides with effects on survival, growth, differentiation, and function of tissues.
lactogenesis The onset of secretory activity in the mammary gland; stage I is the first appearance of milk-specific products in pregnancy; stage II is the onset of copious milk secretion at parturition.
parenchyma A collective term for the glandular tissue of the mammary gland.
stroma A collective term for the fibrous connective tissue and fat cells that provide support for the parenchyma.
terminal endbuds Specialized transient multilayered epithelial structures at the growing ends of mammary ducts.

The mammary gland is a skin gland and is the organ which defines the mammals. It is a necessary part of their reproductive system. This applies to both subclasses of the class Mammalia: the egg-laying monotremes in the subclass Prototheria and the marsupials and the eutherian or placental mammals in the subclass Theria. The mammary gland synthesizes and secretes milk for the early nutrition of offspring and is essential to postnatal survival and to reproductive success in almost all mammals. The basic microscopic structure of the milk-secreting tissue is remarkably constant among species, but there are wide variations in the gross anatomy of the glands, the composition of milk, the patterns of suckling, and the length of lactation to meet the requirements of each species.

I. INTRODUCTION

About 4000 species of mammals in some 20 orders are known. Detailed study of the mammary gland and lactation has been carried out in remarkably few of them. Indeed, we only know the length of lactation for about 20% of the total. Nevertheless, it is clear that lactation is a most adaptable and efficient method of providing early nutrition for young mammals and has played an important part in the evolutionary success of mammals as a group. Domestication, selective breeding, and improved nutrition have extended the lactations and massively increased the milk yield potential of cows, sheep, and goats for the benefit of human nutrition. Unfortunately, the mammary gland is also the site of a primary cancer affecting about 10% of women. Epidemiological studies show that early pregnancy, and perhaps also lactation, provides a reduced risk of developing breast cancer.

II. ANATOMY AND HISTOLOGY OF THE MAMMARY GLAND

Species vary widely in the number (from 2 to as many as 18), size, and disposition of the mammary glands. They are usually paired structures and are present in both females and males, although male mammary glands are generally poorly developed and in some species lack teats. Teats are also absent in female monotremes—after the young hatch from the eggs, they lick milk from a specialized area of skin. Mammary glands may be found on the thorax (e.g., man, elephants, and bats), inguinally (e.g., cows and other ruminants), on the abdomen (whales), covering the ventral surface of the thorax and abdomen (mice, rats, rabbits, dogs, cats, and pigs), or even along the sides of the body as in the coypu, a relative of the guinea pig. Part of the duct system may be adapted for storage, seen at its most developed in the large cavity or gland cistern in the udder of ruminants (Fig. 1a). In women, terminal ducts leading to the nipple are dilated into storage sinuses.

Microscopic structure is very similar in all species (Figs. 1b and 1d). Milk is made and secreted by epithelial cells that line alveoli in a single layer. In lactation, the secretory cells show the ultrastructural characteristics of active synthesis, having abundant rough endoplamic reticulum and Golgi apparatus for the synthesis and packaging of proteins. The cells are joined just below their luminal surface by tight junctional complexes (Fig. 1d). These cells are continuous with the epithelial cells that line fine ducts, leading to larger ducts with a multilayered epithelium and ultimately to the teat. The secretory cells are covered with a layer of specialized myoepithelial cells (Figs. 1c and 1d). The latter share some of the characteristics of muscle cells and their ability to contract is a vital part of the process of milk ejection (see below). Underlying the epithelial cells is an acellular membrane, the basement membrane, which the

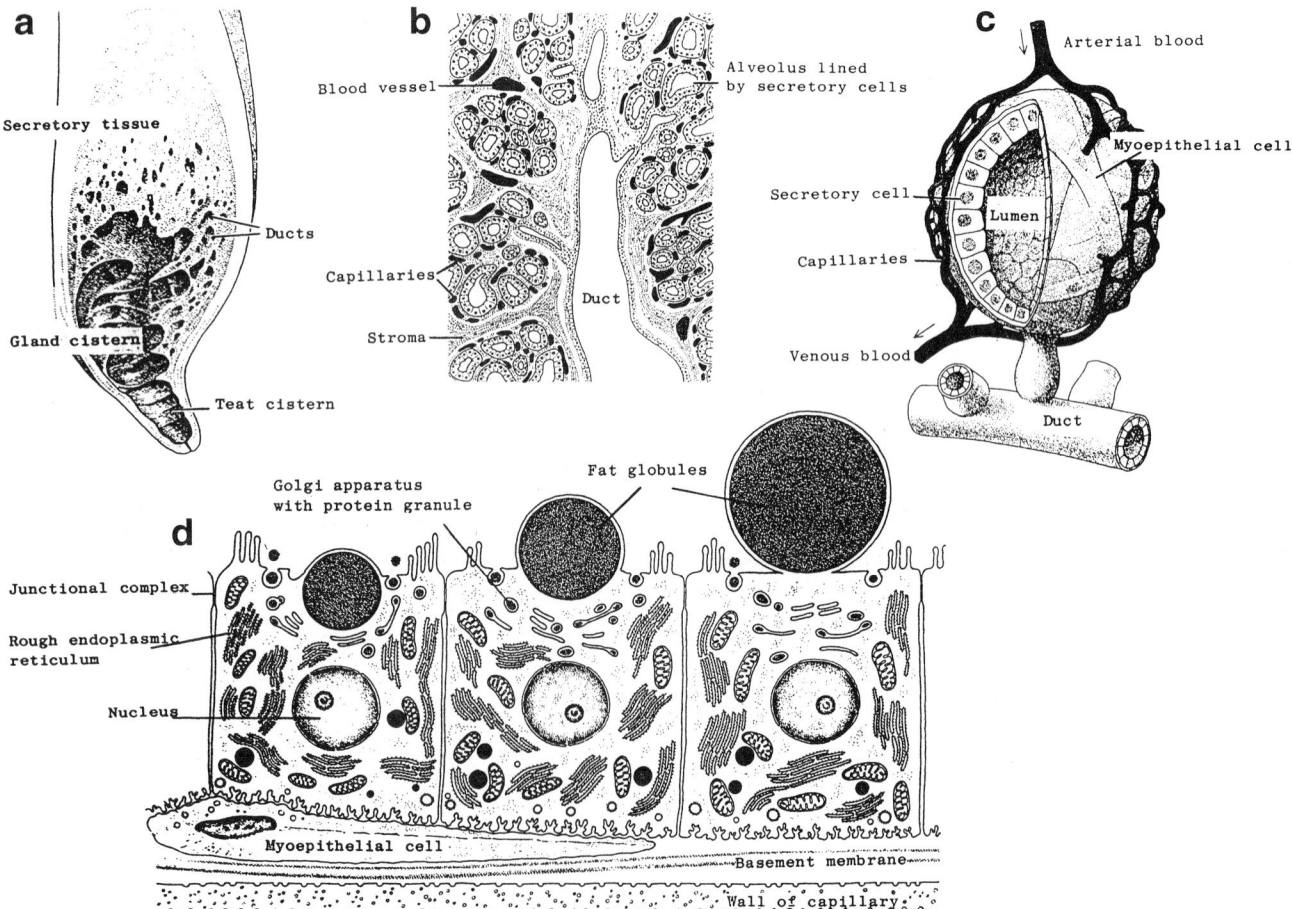

FIGURE 1 The structure of the mammary gland. (a) Section through one gland (quarter) of the udder of a cow. (b) Arrangement of lobules of alveoli and the ducts that drain them. (c) Detail of a single alveolus showing the arrangement of secretory cells, myoepithelial cells, and the blood supply. (d) Ultrastructure of three cells actively secreting milk (a–c, from S. Patton, *Sci. Am.* **221**, 58–68, 1969; d, from Cowie, 1984).

epithelial cells secrete (Fig. 1d). Outside the basement membrane is a rich capillary network (Figs. 1b–1d), bringing the raw materials from which milk is made. The teat receives an abundant nerve supply, but the secretory tissue is not innervated.

Surrounding the glandular tissue, or parenchyma, is a matrix of fat cells and connective tissue (the stroma) known collectively as the mammary fat pad. This functions as supporting tissue, but it has another essential role. By transplantation, it has been shown that mammary epithelial cells will not grow unless placed into an adipose tissue environment.

III. DEVELOPMENT OF THE MAMMARY GLAND

Mammary development begins in the skin of the fetus, by migration of cells to form paired mammary buds in positions corresponding with the mature mammary glands. These cells then begin to divide, leading to the formation of elongated cords of cells that penetrate from the skin into the underlying dermis. The number of these cords determines the number of primary ducts that will open onto the teat: 1 in rodents and ruminants, about 6 in rabbits, and up to 20 in women. Canalization, as a result of death of cells in the core, and some branching then occurs to give a modestly branched tubular gland at birth. Sex differences in mammary gland development are already evident in fetal life. As with other aspects of sexual differentiation, it is the secretion of testosterone from the fetal testes that is responsible for masculinization of the mammary gland. The ovaries play no necessary part.

In the period that precedes puberty, the mammary gland of the female grows for a period at a rate faster than general body growth (allometric growth; Fig. 2). In some species, such as mice, rats, and pigs, this is a period of duct elongation to reach the limits of the mammary fat pad. Elongation occurs by division of cells in the terminal endbud and its subtending duct. The development of alveoli is superimposed on this duct tree. In ruminants, the pattern of mammary growth is somewhat different. There is no unequivocal evidence that endbuds exist. Mammary parenchyma is separated from adipocytes in the mammary fat pad by dense fibroblastic connective tissue and forms a complex tubular structure that does not reach the limits of the fat pad in unmated females. In this respect it shares features with mammary development in humans. The breast development of humans before puberty is largely stromal and duct elongation and increasing complexity of the parenchyma occur only after repeated menstrual cycles. The completion of mammary development to the stage of producing milk for the young varies in different groups. In monotremes, alveoli develop in response to the incubation of eggs. In marsupials, mammary development is essentially the same in pregnant and nonpregnant females, with pregnancy being often shorter than one estrous cycle. The young, born in a very immature state, make their way unassisted to the pouch and remain for a time permanently attached to a teat, a period analogous to the period spent *in utero* by placental mammals. Young can be fostered successfully onto a teat of a virgin female if transferred at a stage of the estrous cycle corresponding with birth, and further development of the gland is dependent on the stimulus of suckling.

In placental mammals, growth of the mammary gland is completed only during pregnancy or even into lactation. The formation of lobules of alveoli reduces the stroma to narrow bands (Fig. 1b). The switch-on of the genes responsible for the synthesis of milk-specific products by the secretory cells also starts from about midpregnancy (lactogenesis stage I). At weaning, the mammary secretory tissue undergoes a phase of rapid involution—cell loss brought about by programmed cell death (apoptosis). Myoepithelial cells are thought largely to survive this process and to play a role in retaining the structural integrity of the tissue.

The growth and differentiation of mammary epithelium are controlled by systemic hormones and locally produced growth factors. Allometric growth before puberty has been shown in several species to be dependent on estrogen and prevented by ovariectomy, although it seems to be independent of the ovaries in lambs. The hormones controlling lobuloalveolar growth are estrogen, progesterone, adrenal corticoids, somatotropin, prolactin, and, in some species, placental lactogen. Insulin and thyroid hormones are also implicated, as are a large number of growth factors produced by the mammary gland

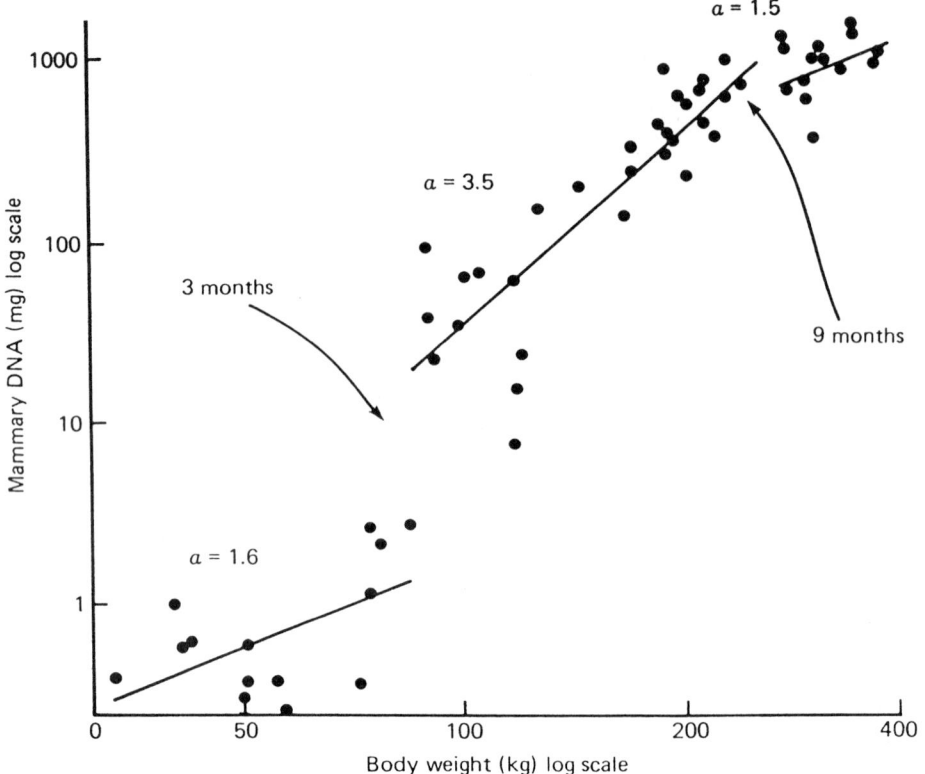

FIGURE 2 Growth of the udder in heifers from birth to 1 year. Allometric growth (constant of allometry $\alpha > 1$) began at 3 months and ceased at 9 months. Estrous cycles began at 6 or 7 months (from Mepham, 1983).

stroma and parenchyma and acting in autocrine, paracrine, and juxtacrine modes to stimulate mitosis. Growth factors known to stimulate mammary development include insulin-like growth factors I and II, members of the epidermal and fibroblast growth factor families, and scatter factor. Inhibitory growth factors are also implicated, notably transforming growth factor-β, influencing patterning of the mammary ductal tree.

IV. SYNTHESIS AND SECRETION OF MILK

Milk provides an almost complete nutrition for young mammals from birth to weaning. There are large variations between species in the composition of milk (Table 1). The major constituents are milk fat, carbohydrates, proteins, ions, and water. In each of these major categories are a large number of individual constituents, which vary between species, between stages of lactation, and even between individuals. Milk also contains other quantitatively minor, but biologically significant components, such as vitamins, hormones, and growth factors. All the components are synthesized by, or transported through, the secretory epithelial cells.

The major carbohydrate in most milks is the dissacharide, lactose, synthesized from glucose. The final step in lactose synthesis is catalyzed by the enzyme, lactose synthetase. This proves to consist of two proteins, the enzyme galactosyl transferase, which occurs in several tissues, and α-lactalbumin, a whey protein unique to the mammary gland. α-Lactalbumin has no enzyme activity but functions to increase the affinity of galactosyl transferase for glucose, allowing the synthesis of lactose. Lactose is also the major osmotically active component of milk and draws water into the secretion. In early marsupial milk, lactose is replaced by complex carbohydrates

TABLE 1
Average Concentrations of Some Constituents in Milk

Species	Total solids (g/liter)	Fat (g/liter)	Casein (g/liter)	Lactose (g/liter)	Calcium (mM)
Cow	127	37	28	48	30
Sheep	193	74	46	48	58
Horse	112	19	13	62	17
Man	124	38	4	70	7
Rat	210	103	64	26	80
Rabbit	328	183	104	21	214
Polar bear	476	331	71	3	72
Blue whale	571	423	72	13	80

that may increase the energy content without increasing the tiny volumes that the pouch young ingest.

The mammary gland synthesizes a variety of proteins from amino acids. The most important, quantitatively and from the point of view of nutrition of the young, are the caseins, a group of phosphoproteins specific to milk. In the milks of species studied in detail, most of the casein is present as spherical bodies known as micelles. Micelles also contain calcium and inorganic phosphate, plus small amounts of citrate and magnesium, thereby supplying components essential for skeletal development.

Immunoglobulins (Igs) form another important protein component of milk and are at particularly high concentrations in the first milk, colostrum. They are selectively transferred from blood or are synthesized locally by cells of the immune system, shown remarkably to migrate from the gut to the mammary gland. Until their own immune systems become mature, young mammals depend on the passive transfer of antibodies from their mothers to protect them from infection. This can occur *in utero* by transfer across the placenta and/or from colostrum, with the gut able to absorb these large proteins intact for some period after birth. Ungulates (cow, sheep, goat, horse, and pig) and some marsupials rely entirely on maternal antibodies, mainly IgG, from colostrum. Rodents, rabbits, and carnivores use both routes and primates (man and monkeys) mainly use the placenta. Even in primates, immunoglobulins, mainly IgA, in colostrum and milk are important in the gut to protect against infection. In addition to antibodies, milk contains other components that can provide protection against pathogens, such as lactoferrin, lysozyme, and the lactoperoxidase system.

Lipids are among the most variable components of milk, differing between species but also with factors such as diet, stage of lactation, and breed (e.g., of cow). Milk fat is secreted from mammary epithelial cells as membrane-bound droplets (Fig. 1d) consisting mainly (>95%) of triacylglycerols, derivatives of glycerol and fatty acids. The latter are made in the mammary gland, from glucose, or in ruminants largely from acetate, or are from plasma lipids coming directly from the diet, or from lipid breakdown in body stores, especially adipose tissue.

Milk synthesis and secretion is under complex systemic and local control. Systemic control is through circulating hormones, of which the most important are insulin, adrenal corticosteroids, progesterone, and the pituitary hormones, prolactin and somatotropin (growth hormone). During pregnancy, progesterone exerts an inhibitory effect on milk synthesis. The differentiation of mammary epithelial cells can begin (lactogenesis stage I), but it is not until progesterone secretion falls at parturition that there is a large increase in the rate of synthesis of milk (lactogenesis stage II; Fig. 3). Adrenal glucocorticoids and prolactin are involved directly in the control of milk protein gene expression. The availability of recombinant bovine somatotropin and its ability to increase milk yield in dairy cows has led to much study of its mechanism of action, but this remains imperfectly understood. It is only quite recently that

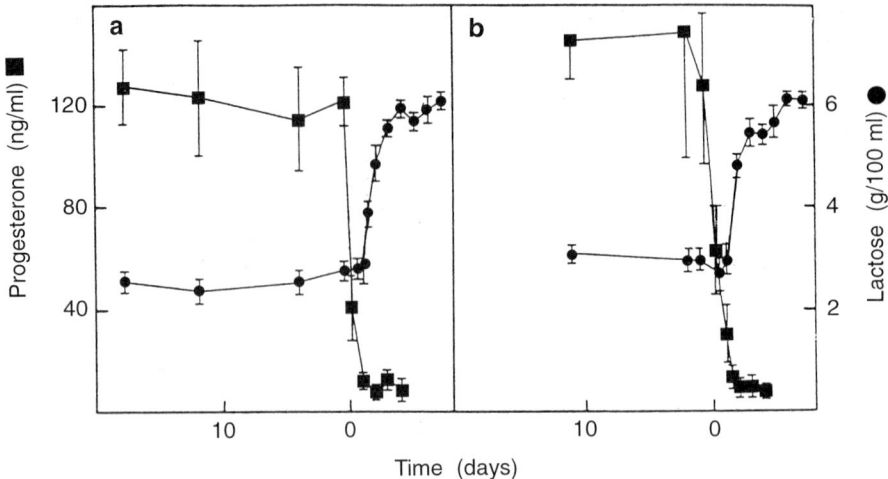

FIGURE 3 Lactogenesis stage II in women: the relationship between falling concentrations of progesterone in blood (mean ± SEM) and rising concentrations of lactose in mammary secretions, a measure of lactogenesis. Zero indicates time of birth. (a) Normal delivery in 12 women. (b) Cesarean section in five women (from J. K. Kulski, M. Smith, and P. E. Hartmann, J. Endocrinol. 74, 509–510, 1977).

receptors for somatotropin have been clearly demonstrated in mammary epithelial cells and, at least in part, somatotropin seems to act as a partitioning agent, diverting metabolites from use by body tissues toward synthesis of milk.

Studies of mammary secretory cells in culture have shown that, in addition to the stimulus of lactogenic hormones, milk production requires cell–cell interactions and also interactions between cells and their basement membrane. Removal of milk from the mammary gland acts as a positive local control mechanism. Thus, increasing the frequency of milking from twice to three or four times daily increases milk yield in dairy cows. Conversely, once the storage capacity of the mammary gland is exceeded, the rate of milk synthesis falls. Suckling patterns appear to be very important in marsupials, such as the wallaby, which practice asynchronous concurrent lactation; that is, the feeding of two offspring of different ages with milk of quite different composition from adjacent mammary glands. One important component of local control is thought to be a glycoprotein made by mammary epithelial cells, termed the feedback inhibitor of lactation. There is evidence that this has a reversible action on the Golgi apparatus of the secretory cells that produce it, blocking constitutive secretion.

V. MILK REMOVAL

Secretion of milk is a continuous process, but milk is obtained intermittently by the young or the milker. Milk stored in large ducts or gland cisterns can be readily removed from the gland. This is not true of the milk in the alveoli and small ducts, in which the major portion of the milk in most species is stored between milkings. Contraction of the myoepithelial cells investing the alveoli leads to a rise in intramammary pressure and ejection of the contained milk. Milk ejection is coordinated with suckling or milking by the neurohormonal milk-ejection reflex. Stimulation of the teat by suckling or milking triggers nerve impulses that pass via the spinal cord to the brain (hypothalamus), where they cause release of the posterior pituitary hormone, oxytocin. This is in turn carried to the mammary gland via the blood and causes the myoepithelial cells to contract.

Patterns of nursing behavior vary enormously between species. In early lactation, rats spend up to 18 hr per day with the litter attached to the teats. Milk ejection recurs regularly every 3–10 min and is sleep related. By contrast, rabbits visit their litters only once every 24 hr and all the milk is delivered in 3 or 4 min.

VI. MATERNAL ADAPTATION TO LACTATION

In lactation, as in pregnancy, maternal metabolism must adapt to meet the nutritional demands of the offspring and also to support the energy costs of developing the mammary gland itself and other organs such as the intestine. The interests of mother and offspring are not equivalent in this. They do not carry identical sets of genes. There is a general relationship (Fig. 4) between milk yield (or energy output in milk) and the metabolic body weight of the mother (i.e., body weight raised to the power of 0.7). Man and other primates fall below this relationship and appear to have evolved a strategy in which maternal investment in lactation is limited in terms of energy per unit time, although the time over which the young remain dependent is long. Human milk is high in lactose but low in fat and especially protein (Table 1). Dairy cows, selected for milk output, fall above the line and may show problems of adaptation to the metabolic demands of lactation. High yields are associated with weight loss in early lactation and the cows may suffer the metabolic diseases of ketosis and milk fever. Extreme examples of mobilization of maternal reserves are provided by mammals that fast during part or all of lactation and may lose up to 40% of their initial body weight. Bears in cold climates give birth while hibernating and suckle their young in dens for 2 or 3 months before emerging. True seals give birth on land and mothers do not eat during relatively short lactations. Baleen whales use blubber accumulated during summer feeding in the highly productive waters of the Artic and Antartic, but then feed little during long lactations of 6 months or more, having migrated to give birth in the subtropics. Milk composition in these species (high fat and low lactose; Table 1) is consistent with the needs of fasting mammals to conserve protein and water. The opposite strategy is seen in rats giving birth to large litters of blind hairless young in which, after an initial decrease, maternal food intake increases by 350%.

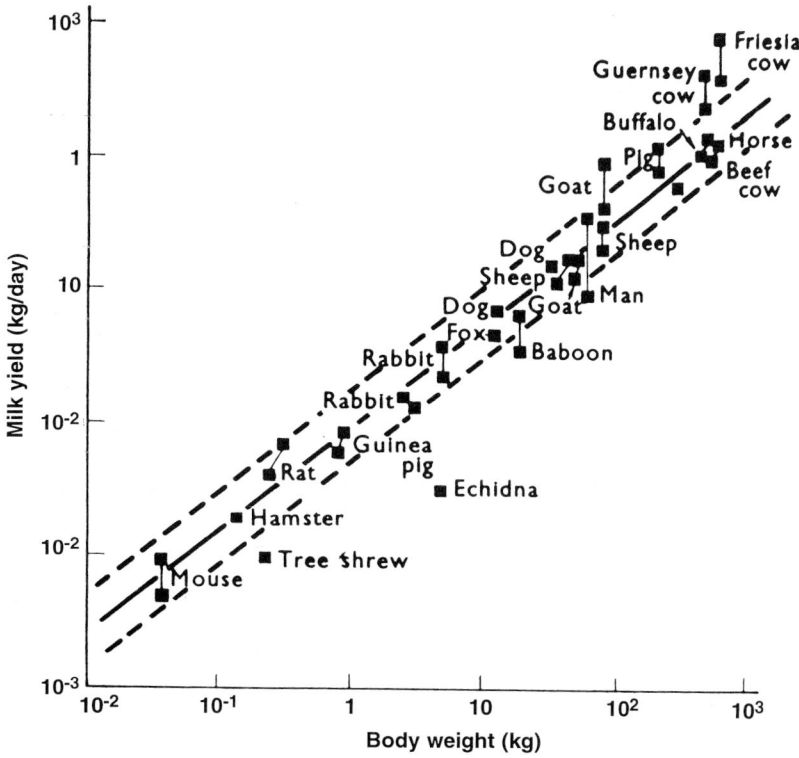

FIGURE 4 Relationship between daily milk yield and maternal body weight (from J. L. Linzell, *Dairy Sci. Abstr.* 34, 351–360, 1972).

VII. LACTATION AND FERTILITY

In addition to the well-known roles of lactation in providing nutrition and immunological protection for young mammals, there is a further important function in spacing births. The stimulus of suckling leads to a period of lactational infertility. Depending on the species, the implantation of an already fertilized egg is delayed or ovulation is suppressed. The mechanisms involved are still imperfectly understood, but it is known that suckling (not milk secretion itself) disrupts the normal pattern of pulsatile release of gonadotropin-releasing hormone in the hypothalamus, resulting in a reduction in the secretion of luteinizing hormone, the hormone responsible for ovulation. As a result, no more young are born until after the weaning of the previous dependent offspring. The effectiveness of lactational amenorrhea as a method in human family planning has been much disputed, but recent studies confirm that women who are fully breast-feeding and do not have menses are at less than 2% risk of becoming pregnant in the 6 months after the birth of a child. Breast-feeding prevents more pregnancies worldwide than all other contraceptive methods.

See Also the Following Articles

Growth Factors; Lactation; Milk, Composition and Synthesis

Bibliography

Cowie, A. T. (1984). Lactation. In *Reproduction in Mammals* (C. R. Austin and R. V. Short, Eds.), 2nd ed., Book 3, pp. 195–231. Cambridge Univ. Press, Cambridge, UK.

Daniel, C. W., Robinson, S., and Silberstein, G. B. (1996). The role of TGF-β in patterning and growth of the mammary ductal tree. *J. Mammary Gland Biol. Neoplasia* 1, 331–341.

Forsyth, I. A. (1991). The mammary gland. *Baillière's Clin. Endocrinol. Metab.* 5, 809–832.

Lincoln, D. W., and Paisley, A. C. (1982). Neuroendocrine control of milk ejection. *J. Reprod. Fertil.* 65, 571–586.

Mepham, T. B. (Ed.) (1983). *Biochemistry of Lactation*. Elsevier, Amsterdam.

Peaker, M., and Wilde, C. J. (1996). Feedback control of milk secretion from milk. *J. Mammary Gland Biol. Neoplasia* 1, 307–315.

Pitelka, D. R., and Hamamoto, S. T. (1977). Form and function in mammary epithelium: The interpretation of ultrastructure. *J. Dairy Sci.* 60, 643–654.

Ronnov-Jenssen, L., Petersen, O. W., and Bissell, M. J. (1996). Cellular changes involved in conversion of normal to malignant breast: Importance of the stromal reaction. *Physiol. Rev.* 76, 69–125.

Short, R. V. (1993). Lactational infertility in family planning. *Ann. Med.* 25, 175–180.

Streuli, C. H. (1995). Basement membrane in the control of mammary gland function. In *Intercellular Signalling in the Mammary Gland* (C. J. Wilde, M. Peaker, and C. H. Knight, Eds.), pp. 141–151. Plenum, New York.

Telemo, E., and Hanson, L. A. (1996). Antibodies in milk. *J. Mammary Gland Biol. Neoplasia* 1, 243–249.

Marine Invertebrate Larvae

Craig M. Young

Harbor Branch Oceanographic Institution

I. What Is a Larva?
II. The Production of Larvae
III. Larval Forms and Diversity
IV. Larval Feeding and Nutrition
V. Larval Orientation, Locomotion, Dispersal, and Mortality
VI. Larval Settlement and Metamorphosis
VII. Ecological and Evolutionary Significance of Larvae
VIII. Economic and Medical Importance of Larvae

GLOSSARY

atrochal larva A uniformly ciliated larva (cilia not arranged in distinct bands).

competent larva A larva that is physiologically and morphologically capable of undergoing metamorphosis.

direct development A life cycle that includes neither a distinct larval form nor a dramatic metamorphosis. In direct development, the embryo develops into a juvenile by a series of gradual changes.

dispersal The process of moving away from the parents and of spreading siblings, either by advection and diffusion of water currents or by active locomotion.

facultative planktotrophy Opportunistic feeding by a lecithotrophic larva that is capable of completing metamorphosis without external food.

indirect development A life cycle which includes a larval stage and metamorphosis.

juvenile In indirect development, a stage of the life cycle following settlement and resembling the adult, yet not reproductively mature. In direct development, any prereproductive stage resembling the adult.

larval ecology The study of factors influencing the distribution and abundance of marine larvae and of processes occurring during the larval stage that influence the distribution and abundance of juveniles and adults.

lecithotrophic larva Nonfeeding larva that receives its nutritional needs entirely from yolk supplies stored in the egg during oogenesis. Lecithotrophic larvae often develop from large, opaque, yolky eggs.

metamorphosis Morphological and physiological changes that occur during the transition from the larval phase to the juvenile phase; often coincides with settlement in benthic species.

mixed development A developmental mode that includes a brooded or encapsulated embryonic stage as well as a free-swimming larval stage.

planktotrophic larva A feeding larva that obtains at least part of its nutritional needs from either particulate or dissolved exogenous sources. Planktotrophic larvae generally hatch from small, transparent eggs.

settlement The permanent transition of a larva from the plankton to the benthos. In sessile organisms, settlement is marked by adhesion to the substratum. It is often closely associated with metamorphosis and may involve habitat selection.

trochal larva A larva with cilia arranged in distinct bands.

The larva of a marine invertebrate is a postembryonic stage of the life cycle which differs from the adult morphologically and which is capable of independent locomotion. Often the most delicate and vulnerable life history stage, it carries out the vital functions of dispersal and habitat selection. The overwhelming majority of marine benthic invertebrates, many pelagic invertebrates, and most marine fishes have complex life cycles that include one or more larval stages. In this respect, marine life cycles superficially resemble those of many terrestrial and aquatic insects. However, the differences are many. In insects, volant adults are generally responsible for dispersal, whereas in marine animals the dispersal stage is most often the larva. Metamorphosis of most insects occurs slowly during the sedentary pupa stage, but in marine animals the pupa is lacking, and the major changes of metamorphosis are very rapid. Metamorphosis marks a transition between habitats in both aquatic insects and benthic marine invertebrates; the former emerge from water to land, whereas the latter settle from plankton to benthos.

I. WHAT IS A LARVA?

The term "larva" has been used in a variety of ways by marine biologists, and there is no firm consensus on the definition. Some definitions, including the one presented in this article, incorporate ecological and behavioral attributes, whereas others use morphological criteria alone. Morphologically, the beginning and ending points of the larval stage may be difficult to define within the developmental continuum and different criteria may be appropriate for different taxa. Some definitions specify that the larval stage ends at metamorphosis, but even this is problematic for some groups (e.g., polychaetes) which have a gradual metamorphosis. Some workers extend the definition of larva to include not only free-swimming stages but also brooded embryos that pass through their entire developmental period within, on, or under the adult yet have some structural characteristics of related larval forms. Even the asexually propagating polyps of benthic hydrozoans and the large physonects of siphonophores are regarded as larval forms by some.

By the definition used in this article, an embryo becomes a larva when it begins to swim or crawl (regardless of whether this occurs at the blastula or gastrula stage) and the larva becomes a juvenile after all exclusively larval structures have been resorbed, transformed, or cast off. Unciliated cleavage stages are not considered larvae by this definition, even though they may drift passively in the plankton. Likewise, brooded or encapsulated developmental stages do not qualify as larvae even though cilia may cause them to rotate rapidly within the confines of their capsules.

II. THE PRODUCTION OF LARVAE

Most, but not all, larvae are products of sexual reproduction. Exceptions include miricidia larvae of trematode flatworms that arise asexually from the redia stage and coronate larvae of some cyclostome bryozoans (order Tubuliporata), which are produced by polyembryony. Although larvae are, by definition, immature stages incapable of sexual reproduction, some oceanic starfish larvae do propagate asexually.

Sexually produced larvae may originate by oviparous, ovoviviparous, or viviparous development. In oviparous development, which is common among marine animals, gametes are shed into seawater, where embryogenesis and larval development occur without any parental protection. The larvae eventually seek suitable adult habitats where they undergo metamorphosis to the juvenile stage. This form of development is found in at least some species of most phyla, with the most notable exception being the phylum Arthropoda. Ovoviviparous development, in which embryos are brooded by the parent without any postovarian nutritive contribution and the larva is the first free-living stage, is common in crustaceans, sponges, and sessile clonal animals such as bryozoans, hydrozoans, and colonial ascidians. Viviparous development, wherein parental nutrients are transferred to the embryo through a direct tissue connection, is relatively uncommon in the marine environment and often is associated with direct development. Nevertheless, some larval forms are produced viviparously among the ascidians and the echinoderms. Many molluscs, flatworms, phoronids, and polychaetes deposit their eggs in capsules or gelatinous egg masses, in which the embryos either develop directly to the juvenile stage or hatch as planktonic larvae. The latter case is known as mixed development.

III. LARVAL FORMS AND DIVERSITY

The shapes and forms of invertebrate larvae are spectacularly diverse (Fig. 1; Table 1). Larvae have been classified by form and ciliary arrangement, nutritional mode, locomotory method, and dispersal potential. Of these, ciliary arrangement appears to be the most useful for elucidating phylogenetic relationships, whereas the other methods of classification find greater application in studies of larval ecology. Yolky atrochal larvae, which have uniform ciliation over all or part of the body, are present in most

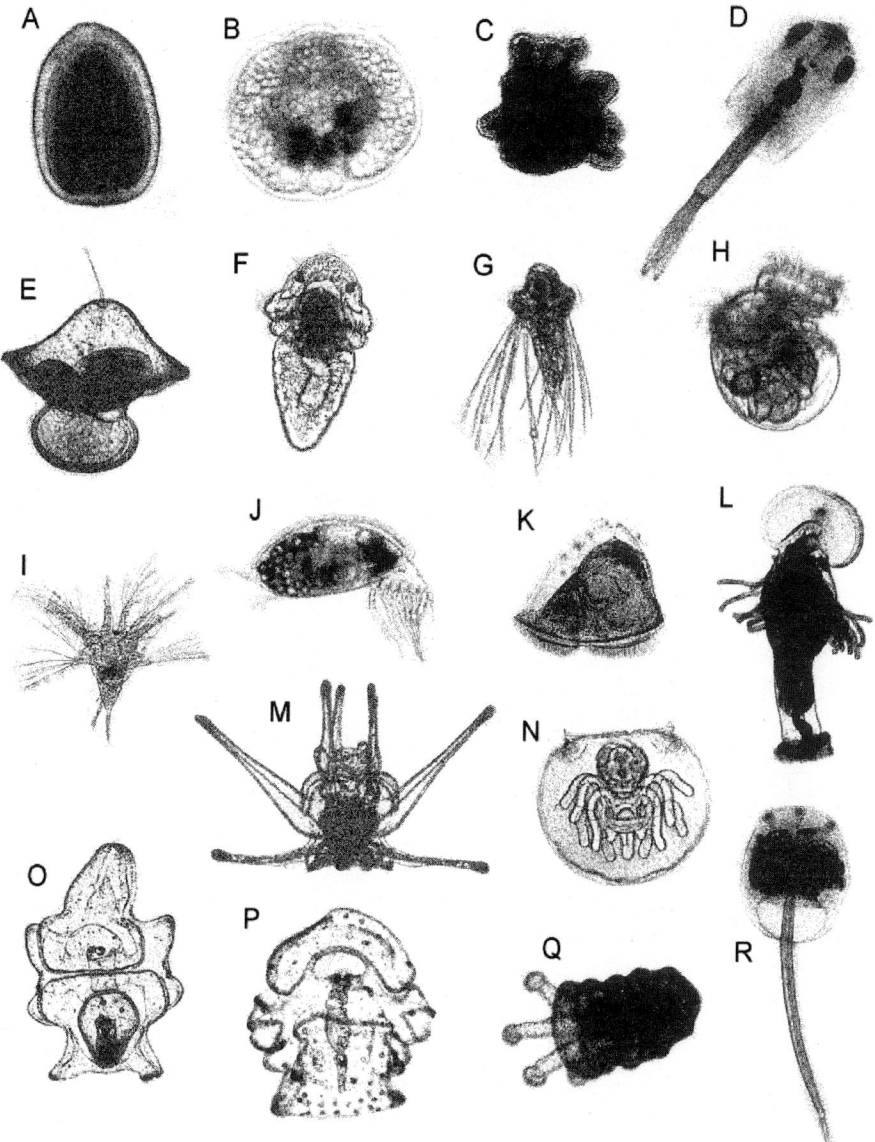

FIGURE 1 Representative larval forms of marine invertebrates. A, Planula larva of a cerianthid anthozoan (sea anemone); B, coral planula; C, Müller's larva of a polyclad flatworm; D, cercaria larva of a marine digenean trematode (fluke); E, pilidium larva of a nemertean; F, late trochophore larva of a serpulid polychaete worm; G, trochophore of a sabellarid polychaete, bearing long protective setae; H, veliger larva of a gastropod mollusc; I, nauplius larva of a cirriped crustacean (barnacle); J,. cyprid larva of a cirripede; K, cyphonautes larva of an anascan bryozoan; L, actinotroch larva of a phoronid; M, echinopluteus larva of an echinoid echinoderm (sea urchin); N, glottidia larva of an inarticulate brachiopod; O, bipinnaria larva of an asteroid echinoderm (starfish); P, auricularia larva of a holothuroid echinoderm (sea cucumber); Q, pentactula larva of a holothuroid echinoderm; R, tadpole larva of a colonial ascidian (sea squirt) [photos provided by W. B. Jaeckle (C, D, L, M, N, P, Q) and C. M. Young (remainder)].

TABLE 1
Known Larval Forms in Major Taxa of Marine Invertebrates

Taxon	Larval forms
Phylum Porifera	Amphiblastula, parenchymula
Phylum Cnidaria	
Hydrozoa	Planula, actinula
Anthozoa	Planula, Edwardsian larva, halcampoides larva, Semper's larva (zoanthina, zoanthella)
Phylum Ctenophora	Cydippid, planuloid larva
Phlum Platyhelminthes	
Turbellaria	Muller's larva, Gotte's larva, Luther's larva
Monogenea	Onchomiricidia
Trematoda	Miricidia, redia, cercaria
Cestoda	Coracidium
Phylum Rhombozoa	Infusoriform
Phylum Nemertea	Pilidium, Desor's larva
Phylum Loricifera	Higgin's larva
Phylum Priapulida	Priapulid larva
Phylum Annelida	Trochophore, nectochaeta, endolarva, exolarva, mitraria, aulophora
Phylum Mollusca	
Polyplacophora	Trochophore
Aplacophora	Trochophore
Bivalvia	Trochophore, bivalved veliger, pediveliger, pericalymma
Gastropoda	Trochophore, veliger, echinospira
Scaphopoda	Trochophore, veliger
Phylum Arthropoda	
Merostomata	Trilobite larva
Pycnogonida	Protonymphon
Crustacea	Nauplius, zoea, megalopa, phyllosoma, puerulus, cyprid, gaucothoe
Phylum Echiura	Trochophore
Phylum Sipuncula	Trochophore, pelagosphera
Phylum Bryozoa	Coronate larva, cyphonautes
Phylum Entoprocta	Trochophore
Phylum Phoronida	Actinotroch
Phylum Brachiopod	Tripartite larva, glottidia (inarticulate) larva
Phylum Echinodermata	
Crinoidea	Doliolaria
Ophiuroidea	Vitellaria, ophiopluteus
Asteroidea	Bipinnaria, brachiolaria, "barrel-shaped larva"
Echinoidea	Echinopluteus
Holothuroidea	Vitellaria, doliolaria, auricularia, pentactula
Phylum hemichordata	Tornaria
Phylum Chordata	
Urochordata	Tadpole
Cephalochordata	Amphioxides

phyla and are typical of sponges, cnidarians, and plathyelminth worms. Although some atrochal larvae, including sponge parenchymulae and cnidarian planulae, are simple ciliated spheres or spheroids, others, including the Müller's larvae of flatworms, have more complicated shapes and may have organized ciliary fields with cilia of different lengths.

Trochal larvae, which have cilia organized into one or more discrete bands, are also found throughout much of the animal kingdom and it has been argued by Nielsen that an organism very like modern trochophore larvae may have been the ancestor of the metazoans. Virtually all trochal larvae may be assigned to one of two larval types, each of which is regarded as characteristic of a major subdivision of the coelomate metazoans. The spiralian eucoelomates or protostomes, including the annelids and molluscs, characteristically have trochophores or trochophore-like larvae. In their basic form, trochophores and veligers have two parallel bands of compound cilia termed the prototroch and mesotroch. Additional ciliary bands (the metatroch, neurotroch, and telotroch) may also be present, particularly in later larval stages. The deuterostome phyla, including the echinoderms and the hemichordates, characteristically have larvae with a single ciliary band, though the shape of this band may be complex and the convolutions vary substantially even within larval forms (Table 1). Larvae in two phyla, Arthropoda and Chordata (subphylum Urochordata) lack external cilia and swim by means of muscular contractions.

Some life cycles include several different larval forms. This is particularly true among parasites, in which there may be definitive and intermediate hosts, each infected initially by a particular larval form. In a typical marine digenean fluke, for example, the ciliated miricidium larva infects a molluscan intermediate host by burrowing into it. Within the intermediate host, the miricidium transforms itself and propagates asexually to form a second larval form, the cercaria, which leaves the host and propels itself by means of a muscular tail to burrow into the definitive host or to locate a settlement site where it is likely to be eaten.

Nonparasitic invertebrates may also have multiple larval forms within a single life cycle. For example, crustaceans typically pass through instar stages which are separated by molting events and designated with separate numbers or names. The early instars of brachyuran crabs are known as zoeal stages and the terminal instar, or megalops, is the settlement stage. Likewise, barnacles pass through numerous feeding naupliar stages then undergo metamorphosis to the nonfeeding cyprid stage, which selects a habitat and undergoes a second metamorphosis to become the juvenile barnacle. In some holothurians, embryos give rise to a yolky, uniformly ciliated vitellaria larva which then organizes its cilia into bands to form a barrel-like doliolaria. The doliolaria becomes a pentactula, the larval stage which settles and grows into a juvenile. Some species of sipunculans pass through both trochophore and pelagosphera larval stages. Similarly, polychaetes begin larval life as trochophores and then become nectochaeta larvae with the onset of segmentation. Molluscan trochophores become veligers as the larval shell develops. Bipinnaria larvae of starfish become brachiolaria larvae with the addition of the attachment organs used at settlement.

IV. LARVAL FEEDING AND NUTRITION

Lecithotrophic larvae rely entirely on maternally provided yolk for their sustenance, whereas planktotrophic larvae acquire nutrients and energy from external sources, either by concentrating and collecting food particles or by absorbing organic molecules from seawater. A few larval forms, including some cnidarian planulae and ascidian tadpoles, carry symbiotic algae from which they may perhaps derive some of their nutrition. However, the lines between planktotrophy and lecithotrophy have been blurred in recent years by the discoveries that some lecithotrophic larvae take up dissolved organic matter and that some phyla have intermediate larval forms that depend to different degrees on maternal provisioning. Facultative planktotrophs, an example of the latter, are fully capable of collecting food particles but have enough yolk to complete metamorphosis without feeding.

Some predatory larvae are found among the Crustacea and the Polychaeta, but most planktotrophic larvae feed on phytoplankton. Ciliated larvae use at least three different mechanisms for collecting food particles. Planula larvae of some cnidarians and a few polychaete trochophores secrete a strand of mucus which is pulled behind the larva like a fishing line. The strand itself is ingested along with any adherent particles. Trochophores, veligers, and related larval forms of protostomous eucoelomates use a dual-band food collection mechanism in which particles are trapped in a food groove that lies between two ciliary bands. The anterior prototroch is composed of large, compound cilia that function simultaneously in locomotion and particle collection. Particles are deflected into the food groove, apparently with the aid of a secondary band, the mesotroch, whose function remains incompletely understood. Echinoderm and hemichordate larvae collect food with convoluted bands of simple cilia. Upon encountering suitable particles, cilia reverse the direction of their beat, deflecting the particles toward the mouth and causing them to concentrate in the circumoral field. They also capture some particles without ciliary reversal, apparently by directing the flow of particles into the oral region. Cyphonautes larvae of anascan bryozoans use stiff cilia of the locomotory band as a sieve to capture particles.

Herbivorous crustacean larvae capture particles by means of fine spines and hairs on their locomotory appendages. Raptorial forms, such as megalopae, feed on individual prey items in much the same way as adult crabs, using the mouthparts to sort food and chelipeds and periopods to hold the prey and disassemble it. The larva of the polychaete *Magelona papillicornis* is a specialist predator that preys on bivalve veligers, which are captured on very long tentacles.

Recent work on nutrition of larval invertebrates indicates that many species have very specialized food requirements. In this respect, polyunsaturated fatty acids have been shown to be especially important. Larvae are most easily cultured on mixed algal diets that ensure a proper balance of essential amino acids and lipids. Many species of both planktotrophic and lecithotrophic larvae are capable of taking up dissolved organic matter from seawater. Although the actual implications of this remain unknown pending a better understanding of the composition and quantity of naturally occurring organic molecules in the sea, it appears that fatty acids are probably more important than dissolved amino acids because of their relatively greater energetic content.

V. LARVAL ORIENTATION, LOCOMOTION, DISPERSAL, AND MORTALITY

Virtually all larvae are capable of locomotion by means of either cilia or muscles. Muscular locomotion is generally faster, at least for short distances, than ciliary locomotion. With the possible exception of late-stage puerulus and megalops larvae of crustaceans, larvae are not capable of swimming faster than the currents they encounter in their environment. Navigation is accomplished by vertical movements that cause larvae to encounter currents moving in different directions and at different speeds. By moving vertically, larvae can apparently control the degree to which they are either exported from or retained in estuaries, enhance offshore dispersal or return to the coast, and ensure retention in the downstream eddies of islands. The behaviors which control these vertical movements may change on diel cycles, or they may shift gradually with ontogeny. Thorson noted long ago that many larvae of benthic invertebrates are initially photopositive and geonegative to maximize dispersal, and they become photonegative and geopositive at the end of larval life to increase chances of encountering the benthos. However, numerous variations on this theme exist, and much of the data that led to this conclusion have recently been challenged as having been generated in unnatural, unidirectional light, which is very different from the scattered light found in the sea. In the most common pattern of diel larval migration, larvae approach the surface at night and then move into deeper water by day.

High fecundities of marine animals attest to the fact that mortality is severe during the embryonic, larval, and juvenile stages. Predation, starvation, and transport to unsuitable habitats are probably the major sources of larval mortality, though fertilization failure may also be a significant source of gamete wastage in free-spawning species. Larvae are preyed upon by a wide assortment of planktonic and benthic predators and filter feeders, but the relative importance of benthic and planktonic predation remains unknown.

The dispersal period varies in marine invertebrates from a few minutes in lecithotrophic larvae of ovoviviparous ascidians and bryozoans to more than 1 year in some planktotrophic species from tropical seas. Larvae in many phyla are capable of riding ocean currents across ocean basins. These far-wandering forms are known as teleplanic larvae. Teleplanic larvae often have adaptations that facilitate locomotion and reduce sinking rates. In a remarkable display of dispersal ability, one lobster larva captured off south Florida was traced to a species that lives only in the Indian Ocean. In some species, the length of the dispersal period is variable. Once larvae become competent to undergo metamorphosis, they can in some cases delay metamorphosis indefinitely until encountering a suitable substratum. The ability of larvae to extend their free-swimming period is dependent on nutritional resources and also on the availability of suitable substrata. In feeding larvae, there appears to be little disadvantage to delaying metamorphosis apart from the risk of planktonic mortality. Indeed, larvae that grow to a larger size in the plankton may achieve a refuge in size that reduces juvenile mortality. In lecithotrophs, however, individuals delaying metamorphosis may often sacrifice some viability, presumably because nutrients needed in the juvenile period are consumed prior to metamorphosis.

At the end of the swimming period, larvae must find the bottom and select an appropriate substratum. Downward movement may be accomplished with a combination of behavioral changes and passive mechanisms that increase sinking rate. In lecithotrophs, specific gravity often increases toward the end of larval life as buoyant lipid yolk is consumed. Some larvae, including those of many echinoderms, grow ever-larger calcite skeletons during the larval stage, causing their sinking rate to increase as a function of age. Many late-stage larvae are less responsive to light and also less active than earlier larvae, so they naturally tend to move toward the sea floor.

VI. LARVAL SETTLEMENT AND METAMORPHOSIS

Settlement and metamorphosis have been studied for many larval groups, though specific inducers of metamorphosis have been isolated and identified for only a few species. Some larvae settle gregariously in response to the presence of other conspecifics, presumably because established individuals indicate that the habitat has recently been suitable or that potential mates are available. Associative settlement, settlement in response to other species, is especially common among predators and herbivores that require specific food items and also among parasites that require a particular host. Many kinds of larvae can be stimulated to settle in the presence of a bacterial biofilm. Some require sediments of a particular grain size. A number of species are known to select habitats in which overgrowth competition or predation is unlikely. Negative phototaxis is common among settling larvae, and this response aids in selecting habitats protected from algal overgrowth, siltation, and large predators. The charge and texture of a surface is also important in determining where a larva will settle.

Metamorphosis can often be stimulated by neuroactive compounds such as γ-aminobutyric acid and by a wide variety of organic and inorganic chemicals, but the specific substances that are involved in metamorphic induction under natural conditions have been largely elusive. A number of molluscs are known to settle in response to compounds found in algae, including but not limited to crustose coralline forms. Gregarious settlement in barnacles and oysters is mediated by proteins found in the adult cuticle or shell, but bacterial metabolites also appear to play a role. The cue for gregarious settlement in sabellariid polychaetes has been more completely characterized than those for other inducers. It consists of free fatty acids found in the matrix that binds together the adult sand tubes.

Most chemical cues that stimulate metamorphosis are bound to surfaces and must be detected by tactile chemoreceptors, so settlement choices tend to occur on tiny spatial scales. However, settlement behavior can also control distribution on larger scales if larvae drifting in the current test the bottom periodically and either accept or reject the site. Larvae of *Phestilla sibbogae*, an opisthobranch that feeds obligately on two species of coral, can detect corals from a distance using waterborne cues. Oyster larvae have also been shown to change their behavior in the presence of waterborne cues associated with adult oysters.

Metamorphosis involves multiple physiological processes and morphogenetic changes, including cell death, resorption of tissues, reorganization of tissues and organs, and activation of organ systems. Some of these processes take place at dramatic speed, whereas others, particularly those that require growth and reorganization of tissues, may take longer to complete. In echinoids (Fig. 2), the juvenile body ("echinus rudiment") forms in an invagination on the left side of the body while the larva is still swimming. Metamorphosis is completed very rapidly by casting off the larval tissues after the juvenile rudiment attaches to the substratum. Ascidian tadpoles provide a contrasting example of rapid metamorphosis. In the initial phase, anterior papillae evert, attaching the animal to the substratum. Immediately after attachment, the muscles and notochord of the tail are rapidly taken into the trunk (Fig. 2). After these initial rapid events, slower processes, including rotation of the body, formation of blood cells, and extension of epidermal ampullae, take place over the next several days. Equally rapid and spectacular metamorphoses have been described in pilidium larvae of nemerteans, coronate and cyphonautes larvae of bryozoans, actinotroch larvae of phoronids, cyprid

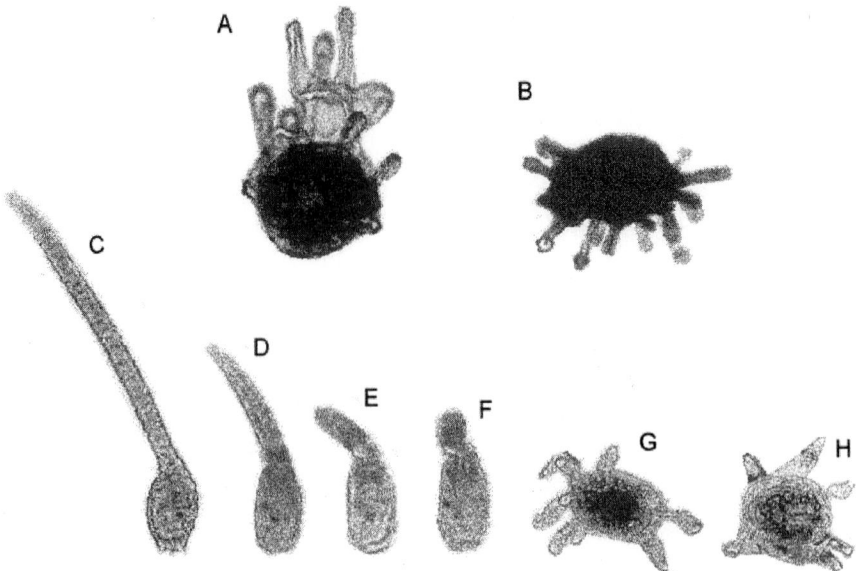

FIGURE 2 Metamorphosis of an echinoid and an ascidian. (A) Late echinopluteus larva of the echinoid *Strongylocentrotus franciscanus* showing the large, opaque echinus rudiment on the left side of the larval body. (B) Juvenile *S. franciscanus* attached by tube feet to the substratum shortly after larval tissues have been cast off. (C) Tadpole larva of the ascidian *Boltenia villosa* just after attachment to the substratum. (D–F) Progressive stages of tail resorption, occurring within minutes of attachment. (G) Juvenile ascidian surrounded by epidermal ampullae 24 hr after settlement. (H) Two-week-old juvenile ascidian with rotated trunk, completed circulatory system, and epidermal ampullae.

larvae of barnacles, veliger larvae of molluscs, and others.

VII. ECOLOGICAL AND EVOLUTIONARY SIGNIFICANCE OF LARVAE

The importance of the larval stage has long been recognized by biologists interested in evolution and ecology. When the major larval forms were identified and described during the latter half of the nineteenth century, many were held forth as examples of the "biogenetic law" (ontogeny recapitulates phylogeny). Walter Garstang refuted the recapitulation argument in the first half of the twentieth century by arguing that larvae are susceptible to mortality and should thus be a focal point for the forces of natural selection. He reasoned, for example, that gastropod torsion, a major developmental event that takes place in the larval stage, may have evolved as a protective mechanism for the larvae themselves and would not, therefore, reflect any particular stage in the phylogeny of the species. The study of selective pressures on early life history stages of marine invertebrates remains an important research focus. A large body of theoretical literature deals with the trade-offs and constraints associated with the evolutionary switch from planktotrophy to lecithotrophy, the control of egg size, the optimal strategy for parental investment and protection, fertilization tactics, and constraints on the evolution of larval body forms.

Because the larval stage is generally responsible for dispersal, larval processes must be taken into account in studies of gene flow, recruitment, and ecology. In recent years, larval supply has received much attention as a potential factor controlling populations of invertebrates and the structure of benthic communities. Marine ecologists now recognize that postsettlement processes, such as predation, competition, and mortality by physical stress, must act on the patterns established initially by larval recruitment. Thus, large research efforts now seek to understand how oceanographic processes influence larval distribution and abundance. Despite a long and rich biological tradition focused on larval evolution and ecology, our understanding remains rudimentary in many respects. Larvae are microscopic, temporary drifters in the plankton and as such are not easily studied in their natural habitats.

VIII. ECONOMIC AND MEDICAL IMPORTANCE OF LARVAE

From a practical standpoint, the study of larvae is particularly important in the control of fouling and in the management of invertebrate fisheries. Fouling studies have been a mainstay of naval research for centuries. Modern work on the fouling of ships and other manmade objects focuses on the study of metamorphic cues and their disruption. Management of fisheries for oysters, clams, prawns, lobsters, abalone, etc. has depended on studies of larval dispersal and abundance since the early twentieth century. In addition, larval invertebrates are often seasonally important foods for the young stages of commercially important finfish.

Cercaria larvae of trematodes and planula larvae of cnidarians may cause minor skin irritations in swimmers. One such larval irritant, known colloquially as "sea lice" or "sea bather's eruption," is the nematocyst-bearing planula of the scyphozoan *Linuche unguiculata*. This larva has caused significant economic losses to coastal communities.

See Also the Following Article

MARINE INVERTEBRATES, MODES OF REPRODUCTION IN

Bibliography

Chia, F. S., and Rice, M. E. (Eds.) (1978). *Settlement and Metamorphosis of Marine Invertebrate Larvae*. Elsevier-North Holland, New York.

Chia, F. S., Buckland Nicks, J., and Young, C. M. (1984). Locomotion of marine invertebrate larvae: A review. *Can. J. Zool.* 62, 1205–1222.

Crisp, D. J. (1974). Factors influencing the settlement of marine invertebrate larvae. In *Chemoreception in Marine Organisms* (P. T. Grant and A. M. Mackie, Eds.), pp. 177–265. Academic Press, New York.

Giese, A. C., Pearse, J. S., and Pearse, V. B. (Eds.) (1987). *Reproduction of Marine Invertebrates, Vol. 9, Seeking Unity in Diversity*. Blackwell/Boxwood Press, Palo Alto, CA/Pacific Grove, CA.

McEdward, L. R. (Ed.) (1995). *Ecology of Marine Invertebrate Larvae*. CRC Press, Boca Raton, FL.

Nielsen, C. (1995). *Animal Evolution. Interrelationships of the Living Phyla*. Oxford Univ. Press, Oxford, UK.

Pawlik, J. R. (1992). Chemical ecology of the settlement of benthic marine invertebrates. *Oceanogr. Mar. Biol. Annu. Rev.* 30, 273–335.

Scheltema, R. S. (1986). On dispersal and planktonic larvae of benthic invertebrates: An eclectic overview and summary of problems. *Bull. Mar. Sci.* 39, 290–322.

Thorson, G. (1950). Reproductive and larval ecology of marine bottom invertebrates. *Biol. Rev.* 25, 1–45.

Young, C. M. (1990). Larval ecology of marine invertebrates: A sesquicentennial history. *Ophelia* 32, 1–48.

Marine Invertebrates, Modes of Reproduction in

Jan A. Pechenik
Tufts University

I. Asexual Reproduction
II. Sexual Development
III. Evolution of Reproductive Pattern

GLOSSARY

gonochoric Exhibiting either male or female sexuality, but never both in the same individual.

hermaphroditic Exhibiting male and female gonadal development within an individual, either simultaneously or in sequence.

larva A free-swimming developmental stage that metamorphoses to adult form and habitat.

lecithotrophic Subsisting on stored energy reserves and incapable of feeding on particulates in the plankton.

planktotrophic Capable of feeding on phytoplankton and other particulates in seawater.

poecilogony Variability of reproductive pattern within a single species.

spermatozeugmata A compound, swimming transport vehicle encountered among gastropods, consisting of numerous normal sperm attached to a single, large abnormal sperm.

Marine invertebrates exhibit a wide range of reproductive patterns. Some patterns are reminiscent of those found among vertebrates and are as sophisticated and complex as any found among terrestrial or freshwater invertebrate or vertebrate species; others are feasible only in aquatic environments and nearly unique to the sea, in large part because seawater is a wet, supportive, nutrient-rich medium whose osmotic concentration closely approximates that of most marine animal tissues. Marine invertebrates exhibit both asexual and sexual reproduction, usually within a single species, and differ considerably among species in how and where eggs are fertilized, where development takes place, and the degree to which development includes a dispersive phase.

I. ASEXUAL REPRODUCTION

Replication that does not involve fertilization of eggs by sperm and the subsequent fusion of genetic material from the two sources is asexual. Typically, the offspring are genetically identical with the parent.

A. Reproduction by Fragmentation and Budding of Adults

In at least some species within the Anthozoa, Ctenophora, Turbellaria, Rhynchocoela (= Nemertea), Polychaeta, Asteroidea, and Ophiuroidea, portions of the original body detach from the parent and differentiate to form a new individual. Some sea anemones, for example, exhibit pedal laceration, in which pieces of the basal disc are routinely left behind as the animal moves across the substrate; each fragment develops into a functional anemone. Similarly, some seastars routinely detach one or more arms together with a portion of the central disc, with each fragment eventually regenerating the missing portions of the seastar body. Among some polychaete species, a new, reproductive individual, complete with separate head and nervous system, buds off from the rear of the original worm, detaches, and swims off to reproduce as either a male or a female. In such cases, the original worm never develops gonads of its own and in this sense never truly reaches adulthood; rather, it produces the adult reproductive form by asexual budding and subsequent differentiation.

Clonal animals, such as hydrozoans, bryozoans, and many anthozoans, replicate by budding off new individuals called zooids, which remain attached to the parental colony.

Some marine sponges (phylum Porifera) exhibit a form of asexual replication in which certain cells within the sponge engulf other cells and then become enclosed in a thick protective covering. Such "gemmules" can withstand levels of dehydration and freezing not tolerated by the parent sponge and can thus persist long after the parental sponge dies. Once environmental conditions improve the cells within each gemmule reactivate and regenerate a number of genetically identical replicas of the original sponge.

B. Asexual Replication of Developmental Stages

A number of species from several invertebrate groups include the asexual replication of developmental stages within what is otherwise a sexual mode of reproduction. Within the Trematoda, for example, a class of parasitic flatworms (phylum Platyhelminthes), adults produce by sexual means, within the body of their vertebrate host, a series of fertilized eggs that give rise to free-swimming ciliated larvae called miracidia. Each miracidium, if it locates the species-specific appropriate intermediate host, produces within that host numerous genetically identical offspring, called cercariae, by asexual replication. An individual miracidium may give rise to thousands, or even tens of thousands, of cercariae, greatly increasing the likelihood that at least one representative of each particular genotype will eventually reach adulthood in the definitive vertebrate host.

A somewhat similar pattern of asexual replication is described for *Symbion pandora*, the sole member of the provisional phylum Cycliophora. The feeding juvenile stage is an external symbiont on the mouthparts of Norwegian lobsters. Within its own body, this juvenile produces one clone of itself at a time in the form of a ciliated "larva," which attaches to mouthparts of the same lobster immediately after emerging from the parent. By continued asexual production of these larvae, a single genotype may progressively dominate the space available on any given lobster.

Some free-living marine invertebrates show a similar system of genotype replication. The sexually produced, ciliated larva of scyphozoan jellyfish, for example, attaches to hard substrate, metamorphoses into an anemone-shaped polyp, and then proceeds to bud off a series of swimming juveniles (called ephyrae) by sequential transverse fission. The swimming, planktonic larvae of some seastar and brittlestar species (phylum Echinodermata) can also bud off new individuals, which swim off and develop into additional larvae that will, in turn, either bud off additional larvae or metamorphose into juvenile seastars or brittlestars.

II. SEXUAL DEVELOPMENT

A. Patterns of Sexuality

Although many marine invertebrates demonstrate separate and permanent male and female sexes (the gonochoric condition), individuals of many species either change sex as they age (sequential hermaphroditism) or produce both male and female gametes simultaneously (simultaneous hermaphroditism). The oyster *Crassostrea virginica* and the sedentary gastropod *Crepidula fornicata* are among the best studied sequential hermaphroditic species. Sequential hermaphrodites generally change sex only once during their lives; typically they are males when young and become females with age (protrandry). Simultaneous hermaphroditism is probably more common, particularly among the tapeworms (phylum Platyhelminthes, class Cestoda), comb jellies (phylum Ctenophora), snails (phylum Mollusca, class Gastropoda), and barnacles (phylum Arthropoda, subphylum Crustacea, class Cirripedia). Such individuals may function simultaneously as male and female so that any two individuals may mate successfully and reciprocally. Self-fertilization occurs routinely among cestodes but is generally rare in other marine invertebrates.

The sexual identity of bivalves, asteroids, echinoids, and many other marine invertebrates is often difficult to discern from external anatomy, although some marine invertebrate species show pronounced sexual dimorphism. In the most extreme cases, males are greatly reduced in size, considerably simplified anatomically and behaviorally, and typically live on or within the female. Such dwarf males are particularly well described for a number of bivalve and cirripede species and in the echiuran *Bonellia viridis* and the recently discovered lobster symbiont *S. pandora*. Among rhizocephalan barnacles, for example, male cyprid larvae enter the female genitalia and subsequently discharge a small mass of tissue that migrates into the distal reaches of the female reproductive system. This amorphous tissue mass initiates gametogenesis and subsequently serves the sole purpose of providing sperm for fertilization.

The case of *B. viridis* is intriguing in that sexual identity is highly labile and influenced by female pheromones. Larvae that metamorphose in isolation become females, reaching lengths of up to 2 m, whereas larvae metamorphosing on a female's proboscis develop into dwarf males, only a few millimeters long; typically the male takes up residence within the female's nephridia.

B. Patterns of Fertilization

Eggs may be fertilized either internally, within the body, or externally, in the surrounding seawater. External fertilization has been especially well studied among anthozoans, hydrozoans, bivalves, polychaete annelids, and echinoderms but occurs in other groups as well. In recent years there has been considerable interest in determining the percentage of eggs that become fertilized among various externally fertilizing marine species and in understanding the morphological and behavioral characteristics of both adults and gametes that affect fertilization success and the role that turbulence and other environmental factors play in controlling fertilization success.

Internal fertilization may be indirect or direct. Indirect internal fertilization involves transfer of sperm to eggs without copulation. Among all bryozoans, and among some echinoderms and bivalves, for example, sperm are shed freely into the sea and must be taken up by neighboring females if eggs are to be fertilized. The males of some noncopulating marine gastropod species apparently use abnormally large and chromosomally deficient sperm to achieve fertilization; in such spermatozeugmata, large numbers of normal, eupyrene sperm are lined up in rows along the tail of a large transport sperm which undulates through the water, apparently facilitating sperm transfer to females.

Many other marine invertebrates transfer sperm in specialized packets called spermatophores. Spermatophore structure and transfer have been particularly well studied in polychaetes, cephalopods, rotifers, chaetognaths, and copepods (phylum Arthropoda, subphylum Crustacea).

Direct sperm transfer involves copulation by means of a specialized penis. Direct sperm transfer is common within some groups, such as the snails (phylum Mollusca, class Gastropoda), but rare or completely absent among many other groups, including the bivalved molluscs and the polychaete and oligochaete worms (phylum Annelida).

C. Protection of Embryos

Eggs that are fertilized externally develop in the surrounding seawater, without protection; females need invest only in gametes, not in protective structures. Fecundities are typically enormous for such externally fertilizing species, with females releasing hundreds of thousands or even millions or tens of millions of eggs per year. In contrast, internal fertilization, whether by direct or indirect means, is usually associated with some degree of embryonic protection and a correspondingly lower fecundity.

Fertilized eggs or embryos may be deposited within gelatinous masses or leathery capsules that are affixed to hard substrates or anchored in soft sediments, as in many polychaete, gastropod, cephalopod, and free-living flatworm species. The functional properties of such encapsulating structures have been studied for few species. They protect embryos against osmotic fluctuation in some gastropod species by reducing the rate of salinity change and provide at least some protection against ultraviolet radiation. Egg masses of only a few species have been shown to be chemically defended against predators, possibly because few workers have examined this issue. However, reports of predation on egg masses and egg capsules by gastropods, flatworms, crustaceans, and fish are common in the literature so that encapsulation provides at best only moderate protection from predators. The ability of encapsulating structures to protect against water-soluble toxicants has been little studied.

Other internally fertilizing species retain their embryos on the body (e.g., copepods) or in specialized brood chambers. Internal retention is the rule among bryozoans, for example, but it also occurs sporadically among polychaetes, gastropods, bivalves, seastars (class Asteroidea), sea cucumbers (class Holothuroidea), and many other groups. Encapsulated or brooded individuals may emerge either as specialized larval stages that are morphologically dissimilar to adults or as miniatures of the adult.

D. Larval Stages

Most marine invertertebrate groups include distinctive, free-living, microscopic larval stages as part of the reproductive pattern. Cephalopods provide a conspicuous exception; all cephalopods emerge from their egg masses as juveniles. Similarly, distinctive larval stages are never found among chaetognaths (arrowworms).

Most marine invertebrate larvae are smaller than 500 μm and few exceed 1000 μm. Larval stages may be feeding (planktotrophic) or nonfeeding (lecithotrophic); that is, they may ingest phytoplankton and other particulates from the surrounding seawater or they may lack the ability to feed on external particulate materials and subsist instead on stored yolk or other nutrients provided by the female parent. Members of a few species produce facultatively planktotrophic larvae: These larvae can complete their development to metamorphosis in the absence of an external food supply but can ingest and assimilate phytoplankton when it is available. Both feeding and nonfeeding larvae are typically able to take up dissolved organic materials (DOM) from the surrounding seawater, although the importance of the nutritional contribution made by such uptake is generally unclear. Only crustacean larvae seem unable to take up DOM.

In some species, larvae can metamorphose to adult form and habitat immediately or within minutes of being released from the parent or encapsulating structure. Larvae of most species, however, must instead develop for a period of time before they become capable of metamorphosing; that is, these larvae must spend hours, days, or sometimes weeks in a precompetent state. It is unknown for all species exactly how the metamorphic pathway develops or what final step makes individual larvae competent to metamorphose. Likely possibilities include the activation of external sensory receptor cells, the completion of particular neural pathways, the activation of neurosecretory capabilities, the activation of receptor systems on target tissues, or the removal or deactivation of inhibitory pathways. This is an active area of inquiry, although the small size of the larvae makes them particularly difficult to study.

During the precompetent period of development larvae may be dispersed great distances from the original parental habitat. In consequence, they are likely to acquire metamorphorphic competence while in habitats inappropriate for juvenile and adult lifestyles. As at least partial compensation for dispersal away from proven parental habitat, the larvae of many—perhaps most—species do not metamorphose immediately upon becoming competent to do so, but rather metamorphose preferentially in response to chemical and physical cues that are typically associated with conditions appropriate for the juvenile. Larvae of some carnivorous species, for example, metamorphose in the laboratory in response to chemical cues produced by their prey. Larvae of sedentary species, including barnacles and tube-dwelling polychaetes, metamorphose in response to chemical cues associated with juveniles or adults of their own species. In the absence of such cues, larvae may delay their metamorphosis. In the laboratory, delay periods range from less than 24 hr in some species to many weeks or even months for other species. In some species, individual larvae become increasingly less selective as they age, eventually metamorphosing in response to conditions that were ineffective earlier in larval life.

E. Metamorphosis

Metamorphosis typically involves a change in habitat and lifestyle. In some species it also involves cataclysmic morphological alterations, with spectacular amounts of tissue degradation and reconstruction taking place within a short time. Such dramatic metamorphoses are particularly common among barnacles, bryozoans, echinoderms, and ascidians. Indeed, among the barnacles there are actually two dramatic metamorphoses: the first from a feeding, triangular "nauplius" to a cylindrical, nonfeeding "cyprid" larva enclosed in a bivalved carapace and the second from the swimming cyprid to the juvenile barnacle. In other groups, metamorphosis involves much simpler morphological changes, as in the loss of the larval swimming organ among prosobranch gastropods.

Metamorphosis to adult form takes place within 5 or 6 hr in some species. In others, 24–48 hr may be required before the major tissue transformations are completed. In most cases metamorphosis to adult form and habitat is irreversible. In some cases, however, including folliculinids (Protozoa) and several coral species (phylum Cnidaria, class Anthozoa), metamorphosed benthic individuals can apparently dedifferentiate to a ciliated form and disperse to new—presumably more favorable—locations before again metamorphosing to benthic form and habitat.

F. Aplanktonic Development

In some species in many marine invertebrate groups individuals emerge from their egg masses, egg capsules, or brood chambers as small versions of the adult; there is no free-living larval stage. Such a pattern is often termed "direct." "Aplanktonic" is a better term, however, since the embryos of many species pass through a clearly recognizable larval morphology before departing the encapsulating structure or brood chamber. Such aplanktonic development is commonly encountered among gastropods and polychaetes and occurs sporadically in many other invertebrate groups, such as the seastars, sea cucumbers, and brittlestars.

Aplanktonic development requires a greater nutritional contribution from the female parent than does the production of feeding larvae. In most species this extra nutritive contribution is seen in the form of larger, yolkier eggs. In other species, females transfer nutrients to eggs while they are being brooded. In both cases, this results in greater egg size and a higher energy content. In other species, particularly among gastropods, additional nutrients are provided in the form of nurse eggs—eggs that are either not capable of being fertilized or, once fertilized, are for unknown reasons incapable of developing. These nurse eggs are ingested by the developing embryos within the encapsulating structures or brood chambers, which require additional nutritive contributions to produce.

G. Mixed Development

As previously implied, many species in many phyla begin their development within brood chambers or encapsulating structures but emerge into the plankton as larvae that are virtually identical to those that develop in the life histories of externally fertilizing species. Such development is said to be "mixed" in that it includes both an encapsulated or brooded phase and a free-living larval phase.

H. Poecilogony

An ability to vary reproductive mode as environmental conditions change, or to reproduce using a combination of reproductive patterns at any given time, would seem adaptively advantageous. Producing both dispersive larvae and nondispersive, aplanktonic embryos, for example, would permit colonization of remote habitats and recolonization of habitat following local extinctions while simultaneously allowing other offspring from the same female to take advantage of favorable parental habitat. It should be similarly advantageous to produce planktotrophic larvae under conditions of high primary productivity and lecithotrophic larvae when external food supplies are less abundant. However, such poecilogenous reproduction, once thought to be widespread among marine invertebrates from many phyla, appears to be rare. In most cases, variable reproductive patterns previously attributed to poecilogony have been shown to involve sibling species that are difficult or impossible to distinguish morphologically. Poecilogony has been convincingly demonstrated for a few spionid polychaetes and a few opisthobranch gastropod species.

III. EVOLUTION OF REPRODUCTIVE PATTERN

Little is known regarding the evolution of reproductive patterns among marine invertebrates, even within particular groups. Much hinges on assumptions about the original reproductive mode among ancestral metazoans or ancestors of particular clades. If external fertilization represents the ancestral mode of fertilization, then free-living larval stages are likely components of the primitive condition. This is the current consensus opinion for most clades, although some workers have suggested that various ancestral metazoans were probably unable to produce large numbers of gametes because of small body size, possibly restricting them to aplanktonic developmental modes; for example, ancestral brooding has been suggested for gastropods and for at least one family of polychaetes.

There is also no agreement about whether larvae within most clades were primitively lecithotrophic or planktotrophic. In many clades there is convincing evidence and argument that many present lecithotrophs are derived from ancestors with planktotrophic development, and that aplanktonic brooded or encapsulated development are similarly derived conditions. Evolutionary change need not be unidirectional, however; some species may have lost feeding larvae from the life cycle only to regain them later. This is apparently less likely to have occurred among echinoderms than among protostomes. In addition, the planktonic, lecithotrophic development of at least one asteroid species has apparently evolved from ancestors that first lost free-living larvae from the life history.

Over evolutionary time scales periodic extinctions seem to have favored the persistence of species with long-lived larvae and extensive dispersive capacity, although in the absence of local extinctions, species with aplanktonic development and correspondingly reduced gene flow may have speciated at faster rates.

Recent studies are revealing much about the genetic mechanisms through which alterations in reproductive pattern can be mediated, although the selective pressures for changes and the conditions under which such changes are more or less likely to take place are still unresolved. The diversity of reproductive patterns seen among marine invertebrates—often within a single genus—implies that, in general, no one pattern is overwhelmingly advantageous, or disadvantageous, for any particular group of marine invertebrates over the long term, and that, on average, increased mortality during development is successfully compensated for by increased fecundity.

See Also the Following Articles

ASEXUAL REPRODUCTION; MARINE INVERTEBRATE LARVAE; POECILOGONY; SPAWNING, MARINE INVERTEBRATES

Bibliography

Adiyodi, K. G., and Adiyodi, R. G. (Eds.) (1983–1994). *Reproductive Biology of Invertebrates*, Vols. 1–6. Wiley, New York.

Giangrande, A., Geraci, S., and Belmonte, G. (1994). Life-cycle and life-history diversity in marine invertebrates and the implications in community dynamics. *Oceanogr. Mar. Biol. Annu. Rev.* **32**, 305–333.

Giese, A. C., Pearse, J. S., and Pearse, V. B. (1987). *Reproduction of Marine Invertebrates*, Vol. IX: *Seeking Unity in Diversity*. Blackwell, Palo Alto, CA.

Jablonski, D., and Lutz, R. A. Larval ecology of marine benthic invertebrates: Paleobiological implications. *Biol. Rev.* **58**, 21–89.

McEdward, L. (Ed.) (1995). *Ecology of Marine Invertebrate Larvae*. CRC Press, New York.

Mileikovsky, S. A. (1971). Types of larval development in marine bottom invertebrates, their distribution and ecological significance: A re-evaluation. *Mar. Biol.* **10**, 193–213.

Pechenik, J. A. (1979). Role of encapsulation in invertebrate life histories. *Am. Nat.* **114**, 859–870.

Raff, R. A. (1996). *The Shape of Life—Genes, Development, and the Evolution of Animal Form*. Univ. of Chicago Press, Chicago.

Strathmann, R. R. (1985). Feeding and nonfeeding larval development and life-history evolution in marine invertebrates. *Annu. Rev. Ecol. Syst.* **16**, 339–361.

Thorson, G. (1950). Reproductive and larval ecology of marine bottom invertebrates. *Biol. Rev.* **25**, 1–45.

Wray, G. A. (1995). Punctuated evolution of embryos. *Science* **267**, 1115–1116.

Marsupials

Marilyn B. Renfree and Geoffrey Shaw

The University of Melbourne

I. Introduction
II. Reproductive Anatomy
III. Patterns of Reproduction
IV. Endocrinology of Pregnancy
V. Embryology
VI. Parturition
VII. Lactation
VIII. Development
IX. Summary

GLOSSARY

altricial young A newborn young that is relatively undeveloped at birth, although particular features may be more advanced.

embryonic diapause A regulated, reversible slowing or halting of embryonic development at the blastocyst stage.

macropodid A member of the kangaroo and wallaby family of marsupials.

pouch An abdominal fold of skin around the mammary glands on a marsupial.

semelparity The state or condition of having a single breeding season in a lifetime.

Marsupials are one of the three groups of living mammals: the familiar eutherian mammals, such as humans, cows, sheep, rats, cats; the marsupials, such as kangaroos and koalas, most of which carry their young in pouches; and the monotremes, which lay eggs. They all have young which are dependent on milk during their early development, the defining feature of all mammals.

I. INTRODUCTION

The marsupial mammals (Fig. 1) diverged from the eutherian line about 100 million years ago, and there are now over 300 living species (Table 1), representing about 6% of the world's mammalian fauna. They are found in Australia and South America, with only one species, the opossum, extending into North America. They are a diverse group occupying almost every ecological niche from tropical rain forest to alpine highlands to desert, with aquatic, arboreal, and terrestrial herbivores, omnivores, and carnivores.

Marsupials are most clearly distinguished from the eutherian mammals by their reproduction (Table 2). The newborn young are altricial, born at a stage when most eutherians are still fetuses in the uterus. Those features essential for postnatal survival, such as a digestive tract, respiratory system, and functional forelimbs needed to climb to the pouch, are disproportionately well developed at birth (Fig. 2). However, by the end of the long lactation, marsupial young are as well developed as equivalent eutherian mammals at weaning.

II. REPRODUCTIVE ANATOMY

The difference between eutherian mammals and marsupials in the development of the urogenital ducts in the embryo has had profound effects on the anatomy of the adult male and female reproductive tracts. In marsupials the ureters enter the bladder medially. In females this keeps the paired female genital ducts apart, whereas in eutherians they enter laterally so that the paired genital ducts can fuse to form a single vagina (Fig. 3). As a result, female marsupials have a complex vaginal structure with paired lateral vaginae and a new connective tissue strand medially which forms the birth canal or median vagina (Figs. 3 and 4a).

FIGURE 1 Tammar wallaby (*Macropus eugenii*) female with a large, nearly weaned pouch young.

The characteristic that gives marsupials their name, the pouch (Latin: marsupium = pouch), is not found in all marsupials. In no species does the male have a pouch, and pouches are also absent in females of some didelphids, caenolestids (South American), and dasyurids (Australian).

Male reproductive anatomy is similar to that of eutherians (Fig. 4b). The testis descends to the scrotum in most species, but the scrotum is always cranial to the penis. There are no seminal vesicles, ampullae, or coagulating glands, but the vasa deferentia enter into the anterior end of a large prostate. The sperm undergo maturation in the epididymis. In American marsupials, with the exception of one species (*Dromiciops*), the spermatozoa form into pairs in the epididymis by intimate association of the acrosomes. However, these sperm pairs separate in the oviduct and fertilization is achieved by a single sperm.

In marsupials there is a single urogenital opening which is identical in appearance in both males and females. Unlike in eutherians, in which the scrotum differentiates through fusion of the labioscrotal folds behind the phallus, the marsupial scrotum forms on the abdomen anterior to the phallus. The pouch and scrotum are probably not homologous but arise from adjacent morphogenetic fields.

III. PATTERNS OF REPRODUCTION

Most marsupials are seasonal breeders, with reproduction timed so that the emergence of the young from the pouch coincides with the most favorable time of year. Photoperiod, nutrition, temperature, and other factors may play a role. Some small dasyurid species are semelparous and the males die off after the breeding season (Fig. 5).

Marsupial reproductive patterns fall into four major groups:

Group 1: Polyovular, polyestrous species in which gestation occupies <60% of the estrous cycle coinciding with the luteal phase. Estrus and ovulation are suppressed during lactation (e.g., possums: Fig. 6, *Trichosurus*).

Group 2: Polyovular, polyestrous species with ultrashort gestation occupying less than the luteal phase which is prolonged into lactation (e.g., bandicoots and opossums: Fig. 6, *Perameles, Didelphis*).

Group 3: Monovular, polyestrous species with gestation extending into the follicular phase, and gestation occupying 94–109% of the estrous cycle length. Postpartum estrous occurs with lactation-controlled diapause in most species (kangaroos, wallabies, and rat kangaroos: Fig. 6, *Macropus*).

Group 4: Polyovular, polyestrous species with a very prolonged preluteal phase and gestation which includes a long period of embryonic diapause after postpartum estrus (e.g., honey possum and some small possums).

TABLE I
Marsupial Orders

Didelphiomorpha: American opossums—75 species
Paucituberculata: shrew opossums—7 species
Microbiotherdae: monito del monte—1 species
Dasyuromorphia: native cats, marsupial mice, numbat—68 species
Peramelomorphia: bandicoots, bilbies—19 species
Diprotodonta: kangaroos, wallabies, possums, gliders, koala, wombats—132 species
Tarsipedidae: honey possum—1 species
Notoryctemorphia: marsupial mole—1 species

TABLE II
Comparison of Reproductive Characters among the Three Groups of Living Marsupials[a]

Character	Monotremes	Marsupials	Eutherians
Bulbourethral glands	Present	Present	Present
Prostate gland	Present, disseminate	Present, disseminate	Present
Seminal vesicles	Absent	Absent	Present
Glans penis	Bifid	Bifid or single	Single
Scrotum	Absent	Prepenial	Pre- or postpenial
Testes	Abdominal	Inguinal or scrotal	Abdominal, inguinal, or scrotal
Testicular blood supply	Simple	Rete mirabile	Pampiniform plexus
Sperm head	Long, fusiform	Short	Short
Ureter's entry	Urogenital sinus, dorsal	Bladder, ventromedial	Bladder, ventromedial
Endometrium	Secretory	Secretory	Secretory
Ovarian follicles	No antrum, liquor folliculi	Antral, liquor folliculi	Antral, liquor folliculi (no antrum in some species)
Corpora lutea	Secretory, autonomous	Secretory, autonomous	Secretory, pituitary or placenta dependent
Ovum	Large, yolk-filled	Small, yolk extrusion	Very small, no yolk
Cleavage	Meroblastic to blastocyst	Holoblastic to blastocyst	Holoblastic to morula (or blastocyst)
Bilaminar blastocyst with trophoblast	Present	Present	Present
Mucoid coat	Present	Present	Present in few, absent in most
Shell membrane	Present	Present	Absent
Shell	Present	Absent	Absent
Embryo formation	From outer layers	From outer layers	Inner cell mass in most
Vascularized yolk sac	Present in all	Present in all	Present in some
Vascularized chorioallantois	Present	Absent (expect Peramelidae)	Present in all
Invasive villous placenta	Absent	Absent (expect Peramelidae)	Present in all
Endocrine function of placenta	Unknown	Slight steroidogenesis; prostaglandin production	Various
Immunoprotection of fetus	Unknown	Uncertain	Present
Delivery of young	Altricial from egg	Altricial from uterus	Altricial to precocial from uterus
Hair and sweat glands	Present	Present	Present
Mammary gland, alveolar, and myoepithelial cells	Present	Present	Present
Mammary hairs	Present	Present	Absent
Mammary anlagen	Areola patches	Areola patches	Mammary lines
Teats	Present	Present	Present
Mammary glands in males	Absent	Absent	Present
Crual spurs and glands	Absent	Absent	Absent
Epipubic bones	Present	Present	Absent
Pouch	Present/absent	Present/absent	Absent
Lactation	Long duration	Long duration	Short duration
Milk composition	Major changes through lactation	Major changes through lactation	Minor changes through lactation

[a] (Adapted from Tyndale-Biscoe and Renfree, 1987).

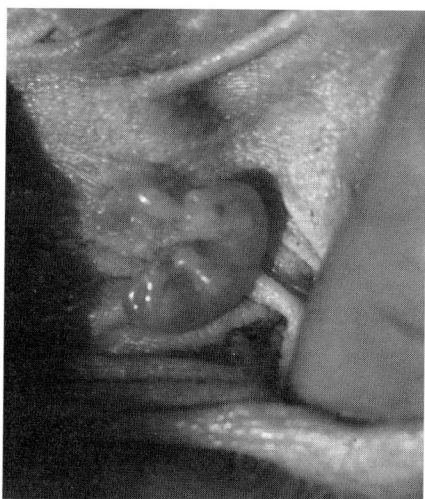

FIGURE 2 Neonatal tammar wallaby attached to a teat in the pouch. Note the size relative to the finger that is holding the pouch open. Many features are poorly developed; note the eye and the hindlimb paddle. Other features are well developed, for example, the forelimbs and mouth.

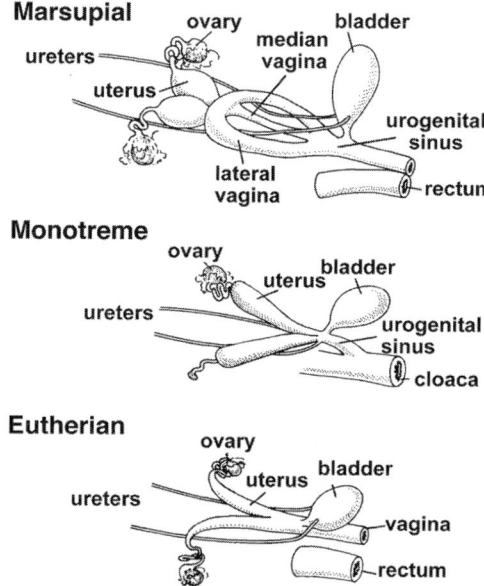

FIGURE 3 Comparison of the urogenital tracts of marsupials, monotremes, and eutherians. Note the relative location of the ureters, which prevents the fusion of the embryonic paired reproductive ducts to form a single vagina as in eutherians. The vaginae fuse cranially to form the anterior vaginal expansion, and a connective tissue strand between this and the urogenital sinus opens to form the median vagina, or birth canal, at first birth (redrawn from Renfree, 1993).

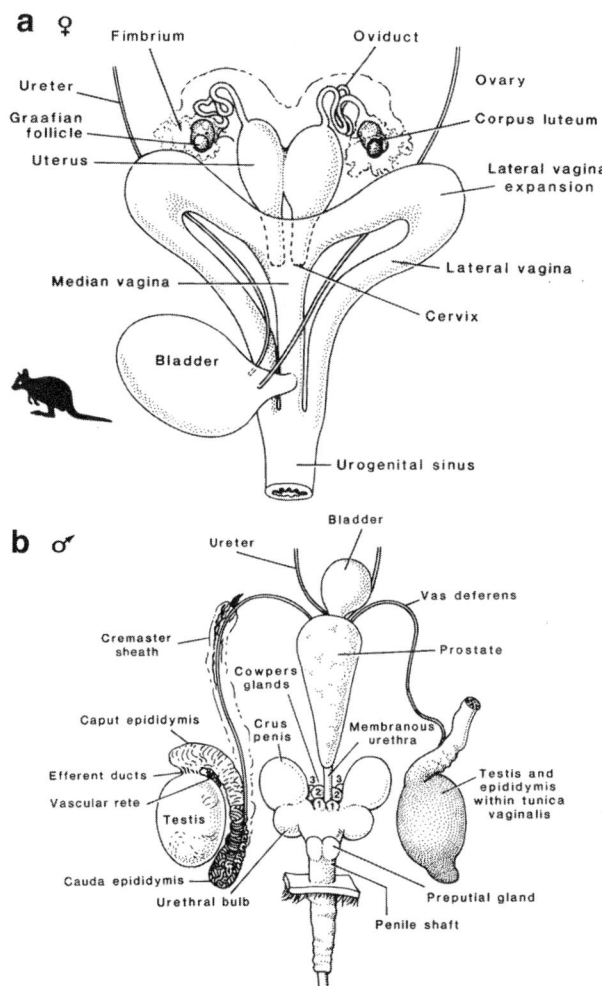

FIGURE 4 Urogenital tracts of typical male and female marsupials. In females (a) the two uteri are completely separate and have separate cervices opening into the anterior vaginal expansion. Males (b) lack seminal vesicles and coagulating glands but have a large prostate and Cowper's glands. In most species the adult testes lie in a pendulous scrotum on the abdomen cranial to the penis, which when erect extends from the common urogenital opening (redrawn from Renfree, 1994).

In all species pregnancy is short relative to the length of lactation. The marsupial estrous cycle varies from 22 to 42 days (mean across known species, 28 days) (Fig. 6). The life of the corpus luteum is not prolonged by the presence of the conceptus, and because the length of gestation is shorter than the estrous cycle in all but one species, anatomical and

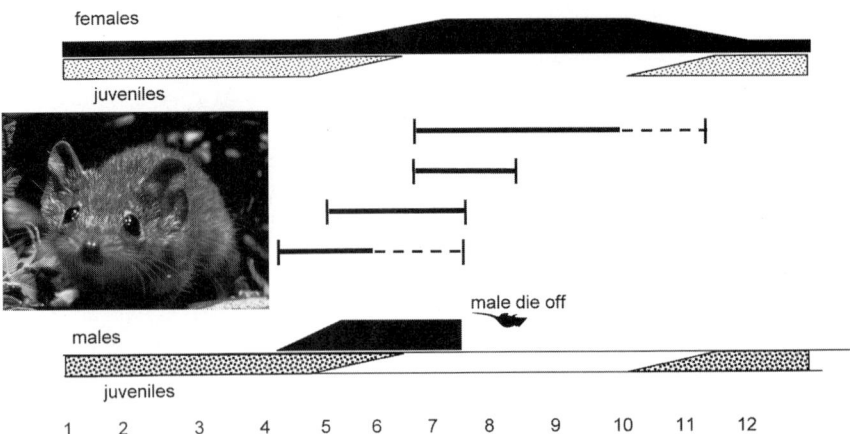

FIGURE 5 Semelparity is a unique marsupial life-history strategy seen in some dasyurid species. The best known example is *Antechinus stuartii*. The young are weaned in summer (the timing varies in different locations in Australia [horizontal bars]), becoming sexually mature in late autumn. Over the next 3 months the males become increasingly territorial, and much fighting and agonistic behavior occurs as females enter their breeding season. By the end of the breeding season, the males almost all die of stress-related diseases so that no males are alive by the time the females are having their litters. For about 2 months the young are permanently attached to the teats and then for a further 2 months are left in a nest while their mother forages. The young are then weaned, and they disperse, becoming sexually mature before the next annual breeding cycle. Few females survive to breed for more than 2 years.

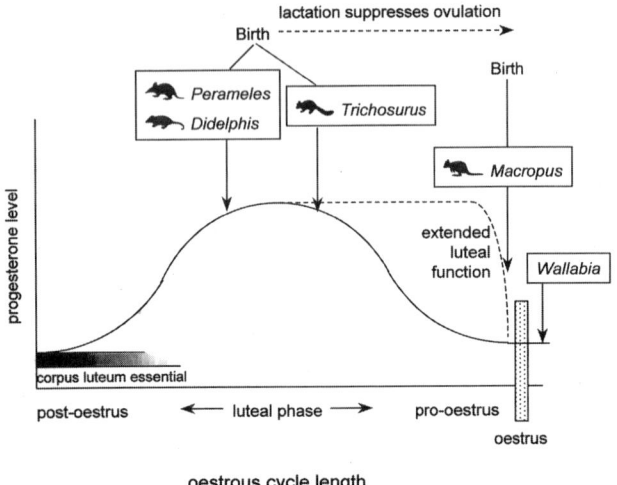

FIGURE 6 Summary of the patterns of estrous cycle and gestation in marsupials. In most species gestation is much shorter than the estrous cycle, and the sucking stimulus from the newborn young in the pouch inhibits ovulation until after weaning. In most macropodids pregnancy is almost as long as the estrous cycle, and the sucking stimulus does not suppress estrus which, can occur a few hours after birth. The resulting conceptus enters diapause, preventing problems of pouch occupancy by two young of different ages.

endocrinological changes of pregnancy and the estrous cycle are similar. This situation in marsupials gave rise to the term pseudopregnancy. This term was later adopted by reproductive biologists in the 1920s to refer to the condition that follows sterile mating in the rabbit and rat. In most species the young are born toward the end of the luteal phase so that the mothers return to estrus is suppressed by the sucking of the neonate. By contrast, in the macropodids (kangaroos and wallabies) the single young is born at the end of the proestrous phase and ovulation and fertilization occur postpartum. The resulting conceptus develops to a 100-cell blastocyst when it enters a dormant state known as embryonic diapause (Fig. 7). Embryonic diapause can also occur in some small possums.

Diapause is maintained by lactation, which stimulates prolactin release from the pituitary, which in turn inhibits the development of the corpus luteum (CL). When the young weans, or if the young is prematurely lost, prolactin falls allowing the CL to reactivate. Luteal progesterone then stimulates the uterus to reactivate the embryo (Fig. 7). The CL, although held in quiescence by prolactin itself, pre-

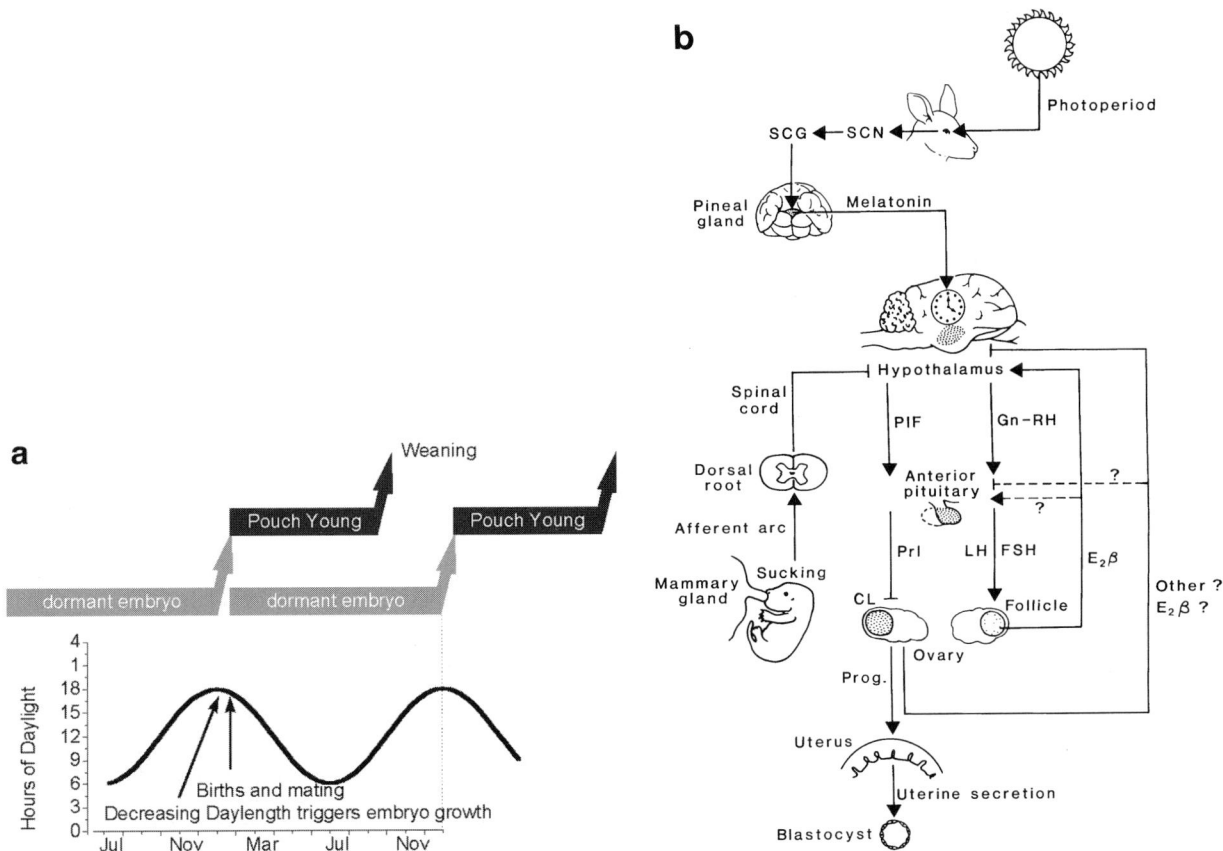

FIGURE 7 Embryonic diapause in the tammar wallaby. (a) Females normally mate after giving birth in January, and hold the embryo in diapause until it reactivates in late December. (b) Between the summer solstice and winter solstice—a sucking stimulus acts via a neural pathway to increase prolactin secretion. In the nonbreeding season photoperiodic signals acting via melatonin produced in the pineal regulate prolactin. High prolactin levels prevent the CL from secreting progesterone. Removal of the sucking or seasonal inhibition or treatment with prolactin-inhibiting drugs allows progesterone to rise, stimulating the uterine secretions which allow the embryo to reactivate (right panel from Renfree, 1994).

vents a new ovulation in the opposite ovary, probably as a result of estrogen secretion.

IV. ENDOCRINOLOGY OF PREGNANCY

The endocrinology of reproduction in marsupials is broadly similar to that of eutherian mammals, with some notable differences. The CL functions autonomously and does not require pituitary support through pregnancy, although the pituitary is needed to supply gonadotrophins for follicular development. The CL secretes progesterone throughout the gestation period, but in the three species in which it has been investigated, progesterone is not needed for the maintenance of pregnancy after blastocyst expansion has been initiated, even though there is little or no placental progesterone secretion. Unlike in many eutherian species, there seems to be no requirement for estrogen during pregnancy. However, it is needed for birth (see Section VI). In the macropodids, reactivation from diapause is associated with a rise in plasma progesterone concentrations with a peak about Day 5 after removal of the pouch young. Progesterone levels then drop, rising again to a plateau between about Days 10 and 26, with a precipitous fall at birth.

V. EMBRYOLOGY

Marsupial oocytes are relatively large compared to those of eutherian mammals, ranging from about 150 to 250 μm diameter. In addition to the primary plasma membrane, the egg has a zona pellucida secreted in the developing follicle, and in tertiary membranes, the mucoid coat and a keratinous shell membrane are secreted by the oviduct and uterus (Fig. 8). Sperm are rapidly transported up the lateral vaginae, not the median vagina, and reach the ampulla within 2–5 hr. Fertilization occurs in the oviduct, and passage to the uterus is rapid (<48 hr).

Cleavage proceeds, without a morula stage, to the formation of a unilaminar blastocyst without an inner cell mass. Formation of the primitive streak occurs on the surface of the embryonic vesicle in a manner akin to the avian embryo. Duration of organogenesis is short and relatively constant (Figs. 9 and 10), even in species of widely differing body masses, ranging from 4 days in the dasyurid *Antechinus* to about 10 days in macropodids.

Marsupials are placental mammals with well-developed placental exchange mechanisms, although these are only established in about the last third

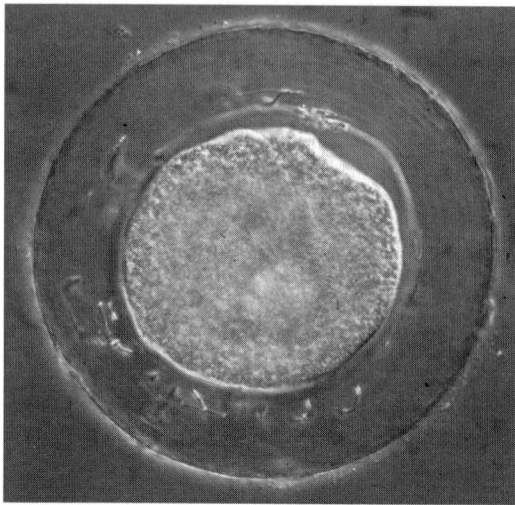

FIGURE 8 Marsupial egg coats. The zona pellucida forms in the follicle. The mucoid coat and shell are laid down in the oviduct and the uterus. Sperm are visible embedded in the muciod coat.

MARSUPIAL RECORD HOLDERS

In the mammalian reproductive Olympiads, marsupials hold many records. The diminutive (8–10 g) honey possum, *Tarsipes rostratus* (left), has the most gold medals. Almost 5% of its body is testes (the human equivalent is a man with two footballs in his scrotum!). Perhaps it needs this size to accommodate the largest of mammalian sperm—360 μm long (right). Despite these two records, females are socially dominant, but they give birth to the smallest mammalian baby, weighing in at <5 mg. They may have the shortest period of organogenesis (conclusive data are not available), but the shortest established period of organogenesis in any mammal is 64 hr in another marsupial species, the Tasmanian devil (*Sarcophilus harrisii*; see Fig. 10). Another marsupial, the bandicoot, may have the shortest recorded gestation length (12.5 days; see Fig. 10). As more is discovered about these remarkable mammals there are bound to be more records broken.

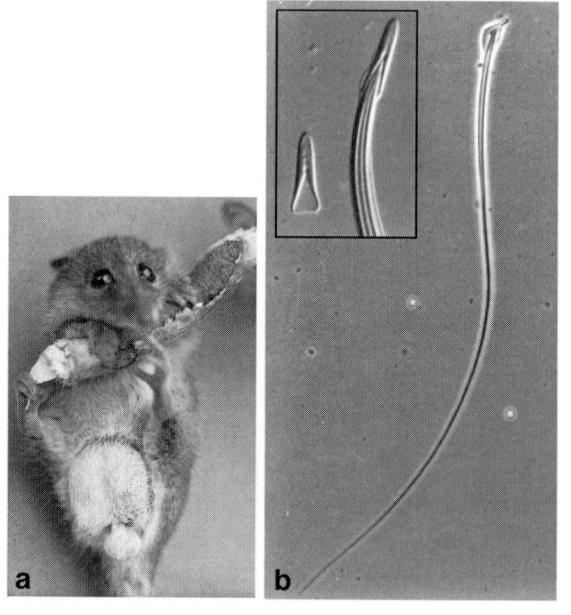

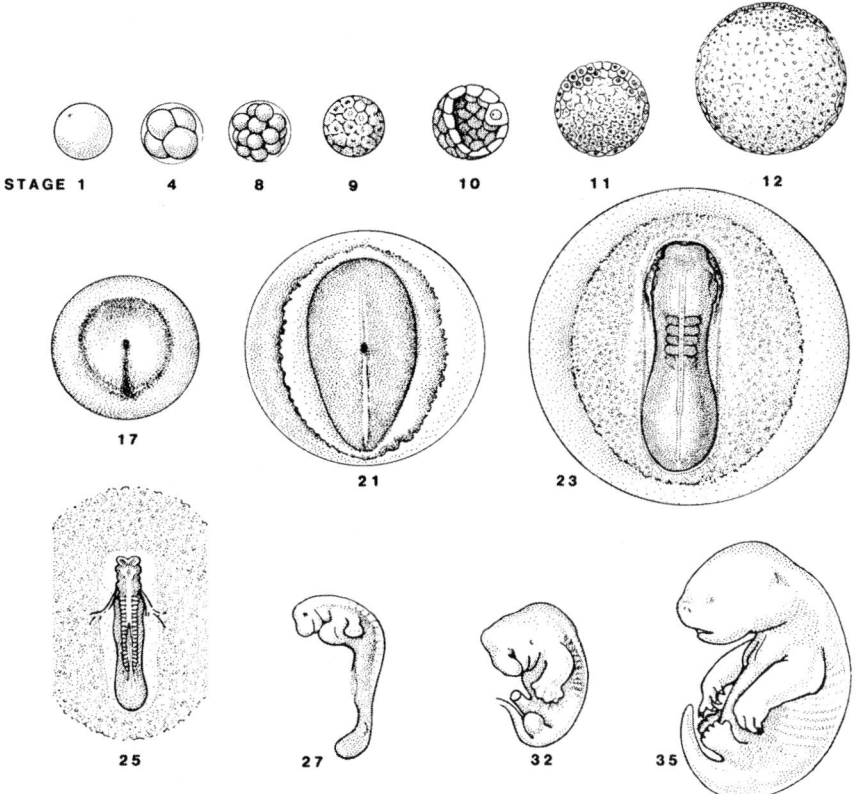

FIGURE 9 Representative stages in the embryology of the opossum. Note the lack of an inner cell mass, superficial development, and the relative development of the forelimbs of the term fetus (stage 35) (from Renfree, 1994).

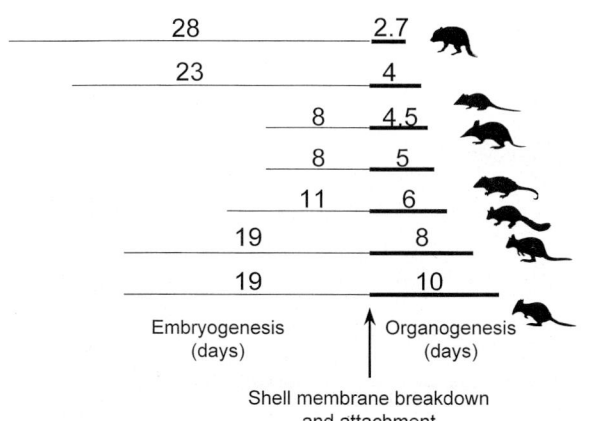

FIGURE 10 Timing of embryogenesis and organogenesis in some marsupials. The species represented are (top to bottom) Tasmanian devil, Antechinus, bandicoot, American opossum, brush tail possum, tammar wallaby, and potoroo.

of gestation. Most marsupials have a choriovitelline (yolk sac) placenta but some have an invasive chorioallantoic placenta as well. Four major placental types have been recognized ranging from those in which the allantois remains closed in the folds of the yolk sac (kangaroos, wallabies, possums, and opossums), to those in which the allantois forms a placental disc with a long umbilical cord in addition to the choriovitelline placenta. All bandicoots have this invasive placentation in which maternal and fetal cells fuse to form a syncytium. The placenta maintains nutritive, excretory, and metabolic functions during gestation. The placenta of the tammar wallaby has minimal steroid synthetic activity but considerable capacity for prostaglandin synthesis. The compositions of the yolk sac and amniotic and allantoic fluids are unique to each compartment, and the yolk sac contains high concentrations of cortisol at term.

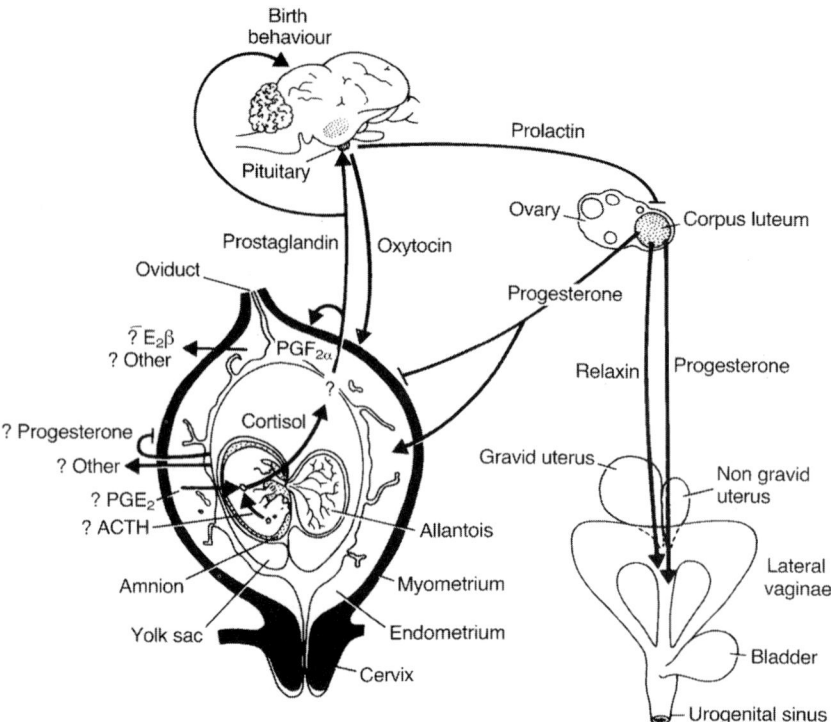

FIGURE 11 Hypothetical model for the endocrine control of parturition in the tammar. Near-term prostaglandin production in the endometrium and placenta increases. There is also a sharp rise in myometrial receptors for oxytocin. Relaxin, together with progesterone and possibly prostaglandins, causes cervical softening and preparation of the birth canal. An increase in cortisol production by the fetal adrenal may stimulate placental PG or uterine PG production may initiate contractions. PG also acts centrally to initiate parturient behavior and induce release of oxytocin (or mesotocin), which further stimulates uterine contractions. The interactions between these hormones sets up a positive feedback cycle that ensures rapid evacuation of the uterus and delivery of the fetus (from Renfree, 1994).

VI. PARTURITION

Despite the small size, a fetal signal is necessary to trigger birth as in many eutherians. The mechanisms involved have been investigated in very few species. The best known is the tammar wallaby (Fig. 11).

Labor is brief and coincides with the onset of a characteristic parturient behavior, which is controlled by prostaglandins (Fig. 12). After the fetuses are expelled from the urogenital sinus they climb rapidly up to the pouch (or abdomen in pouchless forms), where they attach to a teat.

The neonate is altricial. Many organs are poorly developed, but the forelimbs, mouth, digestive tract, and lungs are comparatively well developed and functional to permit the neonate to climb to the teat, attach, and digest milk (Fig. 2).

FIGURE 12 Tammar wallaby in the characteristic birth posture which is adopted minutes before delivery. The young must climb from the urogenital opening to the pouch in order to survive.

VII. LACTATION

Lactation is relatively prolonged due to the immaturity of the neonate. Three phases of lactation are recognized: Stages I and II are the equivalent of lactogenesis stage I of Eutheria. Characteristic changes of milk composition occur that correlate with changes in needs of the young.

In early lactation the milk has a high-protein, low-fat composition, whereas later in lactation there is a high-fat, low-protein milk (Fig. 13). The particular proteins, amino acids, fats, and carbohydrates are all precisely regulated, and in some of the continuously breeding kangaroos adjacent mammary glands supporting a small new young and a large young at foot produce milk of different compositions, a phenomenon known as concurrent asynchronous lactation.

VIII. DEVELOPMENT

The immaturity of the newborn marsupial makes it an excellent model for some developmental studies. Sexual differentiation, which occurs mostly postna-

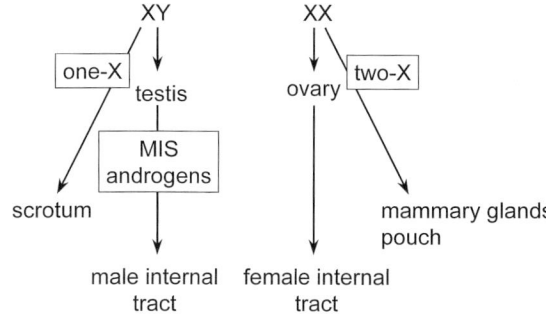

FIGURE 14 Control of sexual differentiation in marsupials. As in eutherian mammals, the Y chromosomal gene *SRY* induces the gonads to form testes and secrete MIS and testosterone, which virilize the internal reproductive tract. In the absence of a Y chromosome, female gonads form into ovaries, which produce neither testosterone nor MIS, allowing a female internal tract to form. However, the sex-specific differentiation of the mammary anlagen and scrotum is not dependent on *SRY* or gonadal hormones. Instead, it is regulated by the number of X chromosomes present (redrawn from Shaw et al., 1990).

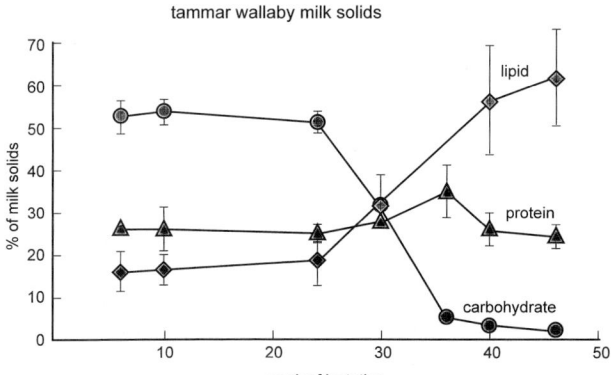

FIGURE 13 Changes in milk composition through lactation in a marsupial. In many species of macropodid the decrease in sucking in late lactation terminates diapause, so a new young may be born and receive milk rich in carbohydrate from one gland, whereas the young at foot is getting milk rich in lipid and low in carbohydrate from an adjacent gland. This is known as concurrent asynchronous lactation (redrawn from Green and Merchant in Tyndale-Biscoe and Janssens, 1988).

tally, has received substantial attention. Most aspects are similar to those of eutherians. They have a homolog of *SRY*, the testis-determining gene on the Y chromosome. The differentiating testes secrete MIS and androgens which control virilization. A notable exception is in the control of the pouch, mammary glands, and scrotum. In the tammar wallaby, for which most information is available, mammary anlagen develop only in females and scrotal bulges form only in males. These sexually dimorphic structures form at least 4 days before gonadal differentiation, independent of gonadal hormones, as a result of unidentified genes on the X chromosome. The presence of one X leads to scrotal differentiation, whereas possession of two X chromosomes leads to mammary gland and pouch formation (Fig. 14). Thus, an XO marsupial intersex has streak gonads, female internal genitalia, but an empty scrotum rather than a pouch and mammary glands.

X chromosome inactivation also differs from eutheria, where X inactivation is random in the fetus but the paternal X is inactivated in the trophoblast. In marsupials the paternal X is inactivated preferentially in both trophoblast and fetus.

IX. SUMMARY

Marsupials are true mammals that have a distinctive reproductive physiology and developmental pathway. Like eutherian mammals, males of most marsupial species have scrotal testes, but the scrotum lies anteroventral to the penis, not posterior. Females have a characteristic paired reproductive system. In most species of macropodids and some possums, there may be a period of embryonic diapause at the blastocyst stage. Organogenesis is compressed into the last few days of gestation. The neonate is altricial and undergoes prolonged development during a complex lactation, in most species held in a pouch. Striking differences in the timing of development have proved especially useful and led to the demonstration that some aspects of somatic sexual differentiation are independent of gonadal hormones and result from a direct effect of the sex chromosomal genes. Because of their unique features, marsupials will continue to make a contribution to the understanding of reproduction in all mammals.

See Also the Following Articles

MONOTREMES; OPOSSUMS

Bibliography

CSIRO (1996). Marsupial gametes and embryos. *Reprod. Fertil. Dev.* 8, 483–834.

Graves, J. A. M., Hope, R. M., and Cooper, D. W. (Eds.) (1990). Mammals from pouches and eggs: Genetics, breeding and evolution of marsupials and monotremes. *Aust. J. Zool.* 37, 143–478.

Hill JP & O'Donoghue CH, 1913, The reproducive cycle in the marsupial *Dasyurus viverrinus*. *Q. J. Microsc. Sci.* 59: 133–174.

Lee, A. K., and Cockburn, A. (1985). *The Evolutionary Ecology of Marsupials*. Cambridge Univ. Press, Cambridge, UK.

Reid, K. C., and Marshall Graves, J. A. M. (Eds.) (1993). *Sex Chromosomes and Sex-Determining Genes*. Harwood Academic, Reading, UK.

Renfree, M. B. (1993). Ontogeny, genetic control and phylogeny of female reproduction in monotreme and therian mammals. In *Mammal Phylogeny* (F. S. Szalay, J. J. Novacek, and M. C. McKenna, Eds.), pp. 4–20. Springer-Verlag, New York.

Renfree, M. B. (1994). Endocrinology of pregnancy, parturition and lactation of marsupials. In *Marshall's Physiology of Reproduction* (G. E. Lamming, Ed.), 4th ed., Vol. 3, Pregnancy and Lactation, pp. 677–766. Churchill Livingstone, Edinburgh, UK..

Renfree, M. B. (1995). Monotreme and marsupial reproduction. *Reprod. Fertil. Dev.* 7, 1003–1020.

Renfree, M. B., and Shaw, G. (1996). Reproduction of a marsupial: From uterus to pouch. *Anim. Reprod. Sci.* 42, 393–404.

Renfree, M. B., Harry, J. L., and Shaw, G. (1995). The marsupial male: A role model for sexual development. *Philos. Trans. R. Soc. London B* 350, 243–251.

Saunders, N. A., and Hinds, L. A. (Eds.) (1997). *Marsupial Biology: Recent Research and New Perspectives*. Univ. of New South Wales Press, Sydney, Australia.

Shaw, G., Renfree, M. B., and Short, R. V. (1990). Primary genetic control of sexual differentiation in marsupials. *Aust. J. Zool.* 37, 443–450.

Tyndale-Biscoe, C. H. (1984). Mammals: Marsupials. In *Marshall's Physiology of Reproduction* (G. E. Lamming, Ed.), 4th ed., Vol. 1, Reproductive Cycles of Vertebrates, pp. 386–454. Churchill Livingstone, Edinburgh, UK.

Tyndale-Biscoe, C. H., and Janssens, P. A. (Eds.) (1988). *The Developing Marsupial: Models for Biomedical Research*. Springer-Verlag, Berlin.

Tyndale-Biscoe, C. H., and Renfree, M. B. (1987). *Reproductive Physiology of Marsupials*. Cambridge Univ. Press, Cambridge, UK.

Tyndale-Biscoe, C. H., Hinds, L. A., and McConnell S. J. (1986). Seasonal breeding in a marsupial: Opportunities of a new species for an old problem. *Rec. Prog. Horm. Res* 42, 471–512.

Walton, D.. and Richardson, B. (1989). *Fauna of Australia*, Vol. 1. Mammalia Australian Government Printer, Canberra, Australia.

Wilson, J. D., George, F. W., and Renfree, M. B. (1995). The endocrine role in mammalian sexual differentiation. *Rec. Prog. Horm. Res.* 50, 349–364.

Mate Choice, Overview

Patricia Adair Gowaty

University of Georgia

I. A Mechanism of Sexual Selection
II. Empirical Observations
III. General Patterns
IV. Functions of Mate Choice
V. Proximate Causation
VI. Relationship to Pair Bonding, Social Systems, and Speciation

GLOSSARY

antagonistic coevolution The process of selection and counterselection between two entities with opposing survival or reproductive interests.

cryptic choice A mechanism of mate choice depending on between-sex interactions that is difficult to observe by investigators, such as postmating or postinsemination mechanisms.

epigamic selection Mechanisms of sexual selection depending on between-sex interactions in which one sex behaviorally or physiologically discriminates among individuals of the opposite sex to bias mating outcomes.

female choice Epigamic selection that occurs when females discriminate among males for gametic sharing.

good genes Two types of mate choice models explaining selective benefit accruing through offspring quality (fecundity or viability) either for specific alleles conferring a particular, beneficial trait to offspring or for alleles complementary to the choosers' alleles.

indicator traits Characteristics in the chosen sex that signal underlying genetic or phenotypic quality of significance to the choosing sex.

intersexual selection A process whereby variation in reproduction of members of one sex occurs because of behavioral, physiological, or morphological interactions with members of the opposite sex.

intrasexual selection A process whereby variation in reproduction of members of one sex and species occurs because of behavioral, physiological, or morphological interactions among individuals of that sex.

natural selection Processes of differential survival and reproduction depending on nonrandom interactions between individuals and their environments, such that environmental factors (pressures) differentially favor the survival and reproduction of some individuals rather than others on the basis of variation among individuals in heritable phenotypic traits.

rare-male mating advantage This occurs when preference is determined by frequency dependence of male phenotypes in which males of rarer phenotypes mate more often than males with more common phenotypes.

runaway selection A selective process in which preferences for traits in choosers coevolve with trait expression in a reinforcing process of positive feedbacks.

sexual conflict Selective processes that occur whenever the fitness interests of mating individuals are not completely congruent.

sexual dialectics Conflict between the sexes distinguished from other modes of between-sex conflict because it is an inevitable consequence of the benefits of mate choice. Because some mates, usually males, will be rejected, they will be under selection to manipulate or control the opposite sexes', usually females', reproductive decisions for their own advantage, whereas the more choosey sex, usually females, will be under selection to resist manipulation or control by others; thus, control and resistance are opposing selective forces hypothetically responsible for evolution of social behavior and social organization.

sexual selection A mechanism of natural selection having solely to do with reproductive competition among members of the same sex and species.

sexually antagonistic coevolution A process of selection and counterselection between the sexes when pairs or potential pairs have opposing survival or reproductive interests; occurs via sexual conflict.

Mate choice is discrimination among potential partners for gametic sharing. Female mate choice theoretically is a selective force responsible for the evolution in males of "bizarre and elaborate" traits such as the display plumage of birds-of-paradise. Male mate choice—in typical sexual species in which female investments in offspring are greater than male investments—theoretically has weaker effects on traits in females; however, male choice in such typical species may provide selection favoring traits in females, including "good mother" traits. Because mate choice should be advantageous whenever perspective mates vary in traits likely to affect offspring fitness, it may be a universal feature of reproductive behavior in sexual species. Despite difficulties in demonstrating freely expressed mate choice in wild-living organisms, few researchers have failed to observe mate choice behavior in controlled laboratory situations. Although the term choice suggests conscious motivation, none is necessary or implied in most discussions of mate choice.

I. A MECHANISM OF SEXUAL SELECTION

Mate choice is a mechanism in sexual selection. Sexual selection "depends on the advantage which certain individuals have over others of the same sex and species solely in respect of reproduction" (Darwin, 1871, p. 568). "Sexual selection is selection that arises from differences in mating success (number of mates that bear or sire progeny over some standardized time interval)" (Arnold, 1994, p. S9). Sexual selection, sometimes simply called differential reproduction, may account for the evolution of traits such as extraordinary ornaments and elaborate behavior that natural selection does not easily explain. Although less often noted, sexual selection may also account for reduction in ornaments and not so elaborate traits. Modern scholars debate differences between natural and sexual selection, with some arguing that all selection reduces to sexual selection; others argue that sexual selection—reproductive selection—is always antagonistic to natural or survival selection.

A. Intra- and Intersexual Selection

Darwin said there were two types of sexual selection and based his categories on interactions either among or between the sexes, intrasexual or intersexual selection, respectively. His discussion focused mostly on two mechanisms of sexual selection, male–male contests and female mate choice. Because Darwin was attempting to explain traits that would seem to have little positive effect, or even negative effects, in the struggle to survive, such as the gaudy colors of mandrills, *Mandrillus sphinx*, or the fanciful horns and antlers of some male ungulates, his discussions seldom included female–female contests or male choice of mates that result in the evolution of traits in females. Mechanisms of sexual selection include both male–male and female–female contests over mates as well as mating discrimination by both females and males.

B. Same-Sex Differential Reproduction

Mechanisms of sexual selection act on various traits related to reproduction possessed by same-sex conspecifics and thereby may cause evolutionary changes. An efficient definition of sexual selection is that it is same-sex differential reproduction. Same-sex differential reproduction occurs because of behavioral, physiological, or morphological interactions between individuals of the same or opposite sex. Therefore, "intersexual selection" or "intersexual mechanisms of sexual selection" cause within-sex variance in reproductive success.

What is considered sexual selection has recently come under reconsideration. Between-sex behavioral or physiological mechanisms accounting for within-sex variation in reproductive success also include mechanisms that coerce choosers' decisions. Therefore, for example, if a female rejects a given male, males may engage antichoice options, such as rape, forced copulation, forced "trades" of copulations for resources, or other manipulative mechanisms, to get rejecting females to mate with them that otherwise would not. Antichoice behavior in one sex creates a selective pressure acting on the opposite sex favoring compensatory resistance to manipulation or coercion.

C. Pre- and Postmating Mechanisms

So-called "cryptic female choice" occurs when females use postmating or postinsemination mechanisms to discriminate among particular males for gametic sharing. Cryptic choice is new jargon, but Darwin's old definition of female choice as a mechanism of discrimination among males readily accommodated the notion that female choice can occur by way of premating, preinsemination, or postmating, postinsemination mechanisms. Most premating mechanisms of mate choice are behavioral; postmating mechanisms include physiological and morphological variations. Mechanisms of postmating discrimination may depend on "double" or "multiple" matings in which the sperm of more than one male occurs simultaneously in a female's reproductive tract. However, even without double or multiple mating, selection also may favor postmating discrimination against the sperm of nonpreferred or nonoptimal males. This would seem to be especially likely in species in which coercion or manipulation of mate choice decisions is characteristic.

D. Mate Choice Questions

Three types of questions characterize most studies of mate choice: (i) Often researchers use mate choice ideas to investigate or explain the maintenance of usually bizarre and/or elaborate traits. Such investigations begin with questions about traits, e.g., Are wing spots of male damselflies indicator traits, honest signals, or sensory traps?; (ii) others focus on benefits of discrimination for choosers, usually with specific reference to particular trait variation, e.g., Do female stalk eyed flies, *Cyrtodiopsis whitei*, more often prefer males with longer than shorter eye-stalks?; and (iii) far less often investigators focus on benefits of discrimination without explicit reference to the phenotypic basis for choice, e.g., Do female *Drosophila pseudoobscura* matings with preferred males produce offspring of higher viability than females mating with less preferred or nonpreferred males?

E. Difficulties in Studies of Mate Choice

Answers to the question, Is mate choice an important factor in the evolution of traits (behavioral, physiological, or morphological) in the opposite sex?, generally remain equivocal. There are three reasons for this. First, in unmanipulated, wild-living populations it is difficult to infer the effects of within-sex contests or between-sex coercion from mate choice; therefore, little certain evidence from natural populations indicates that mate choice actually exists. Second, if choosers have consistent preferences for "good genes," it is difficult to understand why variation in chosen traits remains because the underlying genetic variation in the discriminated trait should go to fixation, making mating discrimination a moot point. Finally, the usual candidate traits that investigators assume choosers use to discriminate, such as elaborately long tails or even colors of novel items such as bower decorations or investigator-applied identification rings, often seem quite irrelevant to the fitness of choosers, i.e., the advantage of female choice for females remains completely opaque in some instances.

Laboratory experiments eliminating or controlling same-sex contests and/or opposite-sexed coercion offer a way to examine the independent effects of mate choice. Resolution of the other difficulties resides in considerations of the ecological problems solved by mate choice for choosers. For example, if pathogens and parasites are ubiquitous ecological problems affecting differential survival, a solution for potential parents is production of offspring with immune systems up to the challenge. Given that parasites and pathogens rapidly mutate, successful immune systems are most likely those capable of mounting responses to a wide variety of potential challenges, i.e., highly polymorphic ones. Thus, selection should favor individuals who choose gametic partners with immune system alleles dissimilar to their own. In this case, good genes are relative to or complementary to the quality of the chooser's genes. This logic provides a solution to the second problem (ii) listed previously because it explains how strong selection favoring mate choice fails to exhaust genetic variation. A theoretical solution to problem iii states that if mating discrimination for complementary alleles is often the case, selection will favor traits indicating immune system characteristics. Thus, bizarre and elaborate signals should be (relatively) honest indicators of immune system functioning (see Section IV,C). At the very least, mating discrimination may ensure that choosers will not expose themselves to pathogens or parasites during mating.

A fourth problem concerns the nature of the information available to discriminating individuals. Can choosers directly assess (unconsciously) immunologically relevant information about the gametes of another or must they rely on correlated phenotypic cues or even signals from the discriminated among individuals? One approach to answering this question begins with experimental comparison of variation in offspring viabilities when choosers mate with preferred versus nonpreferred partners. An unbiased estimate of sources of information available to choosing individuals would come from trials in which those to be discriminated among were offered at random with respect to their obvious phenotypic variations. Given a significant result one could reliably infer selection favoring mate choice and then search for phenotypic correlates including potential sources of direct information, correlated cues, or manipulative signals that differ between preferred and nonpreferred mates.

II. EMPIRICAL OBSERVATIONS

Unless otherwise cited, all references to specific conclusions regarding given species are from Andersson (1994), who provides a definitive, thorough treatment of sexual selection containing a description of studies in over 300 species of animals. A recent review of mate choice in plants is provided by Arnold (1994).

A. Methods of Study

Naturalistic, descriptive studies of mate choice are correlational: Researchers examine trait covariation—preferences and traits—between partners to demonstrate existence of choice, mechanistic bases of choice, and fitness causes and consequences. In most laboratory experimental studies researchers design tests in which choosers discriminate among individuals varying in investigator-determined phenotypic variables (see Section I,D) to examine existence, mechanistic basis, and fitness correlates of choice. Investigators put choosers in arenas with barriers between cells that contain individuals to be discriminated among, and they infer "preference" from behavioral indicators such as how long the choosing individual spends with (remains nearest) one of the to-be-chosen individuals. Conclusions from many of these studies are equivocal because investigators fail to demonstrate if behavioral preferences indicate "mating" preferences, e.g., a female may prefer to share her gametes with an aggressive male but not prefer to stand near him or around him in certain situations. In contrast, unlike most naturalistic, observational studies, arena experiments can eliminate the confounding effects of within-sex contests or between-sex coercion on the expression of mate preferences. Such trials are especially powerful if they evaluate the assumption that "standing near" or "approaching" an opposite-sex individual correlates positively and significantly with likelihood of gametic sharing with that individual.

B. Vertebrates

1. Humans

In contrast to studies of other taxa, most investigations of mate choice in humans depend on what human subjects tell researchers, usually in response to questionnaires about preferences or past behavior. Thus, lies, selective memories, or ignorance—if mating preferences are not available to conscious knowledge—may bias these investigations. Studies based on self-reports make the assumption that subjects have conscious knowledge of their mating preferences. This assumption emphasizes important distinctions of meaning in the word "preference."

Correlational studies of patterns of assortment among humans indicate Kenyan Kipsigis women prefer to marry men with larger, rather than smaller, land holdings. Women, apparently more often than men, all over the world "marry up" socioeconomic classes. In modern Western societies people tend to assort by race, ethnicity, class, socioeconomic status, age, politics, and religion; positive correlation coefficients for these variables are above 0.9. Westerners also tend to assort by personality and intelligence (correlation coefficients around +0.4) and, on average, spouses tend to physically resemble each other (correlation coefficients around +0.2 for morphological variants). Men say they prefer younger women;

women say they prefer men with resources. Men say they prefer women who evidence signs of fertility, and both women and men say they prefer mates who are healthy. A major reason individuals give for marital dissolution is infertility, suggesting that the primary adaptive significance of mate choice has to do with reproduction.

A recent study suggests that individuals assort negatively for some traits, which is in contrast to the usual observation that individuals positively assort. In one of the few studies that experimentally manipulated women's access to sensory cues from men in a manner not dissimilar to some mating preference tests in nonhumans, women preferred the smell of T-shirts previously worn by men whose alleles at loci of the major histocompatibility complex (MHC) were most dissimilar to their own. MHC gene products bind foreign peptide and self-fragments for presentation to T lymphocytes and, thus, are a crucial component of immune recognition in vertebrates; MHC genes also affect individual odors. Wedekind et al. (1995) interpreted their result to mean that women prefer olfactory cues of men with whom they would have a high probability of producing children with highly variable, and presumably highly beneficial, immune system capabilities; mating preferences for MHC dissimilarity also decrease the likelihood of inbreeding, another fitness advantage for females mating with preferred males.

2. Nonhuman Mammals

Experimental and observational studies favor the conclusion that mammal females primarily use olfactory information as the basis for mating choices, though dramatic facial colors and fur of cercopithecine monkeys suggest that visual cues in this group may also be important. Few studies of mating discrimination among male nonhuman mammals exist. Experimental studies in mice, *Mus musculus*, indicate females discriminate among males using odor cues; in trials using mice with *t* locus variability, females preferred males lacking *t* alleles. Wild house mouse females prefer older rather than juvenile males, dominant males rather than subordinates, and familiars (neighbors) rather than strangers. The odors of dominant resident males attract females; females avoid odors of dominant strange males. Females also prefer males whose genes within the MHC are dissimilar to their own.

Arena trials indicate that vole (*Microtus ochrogaster*) females prefer dominant males, as do female rabbits (*Oryctolagus cuniculus*) and lemmings (*Lemmus trimucronatus*). Studies of wild pronghorn (*Antilocapra americana*) indicate that females prefer males with physical vigor, a preference that correlates with past selection against cursorial predators of the North American plains. Female dama gazelle (*Dama dama*) prefer males with longest resident times on leks, and female elephants (*Loxodonta africana*) seem to prefer older, bigger males. There are no published studies of the relative strength or adaptive significance of female versus male choice in nonhuman mammals.

Male harassment of females including forced copulation occurs in a wide variety of nonhuman mammals; coercive, antifemale choice behavior seems to correlate with the possession of intromittant organs and larger size of males in size dimorphic mammals.

3. Birds

There are more studies of mate choice in birds than studies in other vertebrates. Most experiments test for visual or auditory preferences, a bias that arises from efforts to explain elaborate plumage characteristics and rich vocal repertoires characteristic of many birds. Female jungle fowl (*Gallus gallus*) prefer males with brightly colored fleshy head combs; female pigeons (*Columbia livia*) prefer males with variably patterned plumages. In tested passerines with elaborate song types, females prefer males with largest repertoire sizes or particular song types. Female satin bower birds (*Ptilonorhyncus violaceus*) base mating choices on bower quality and decorations, often blue items such as pen caps that surely were not part of traditional or ancestral environments of bower birds.

Among the most important studies of mate choice are experimental and observational studies of zebra finches (*Poephilia guttata*; Burley, 1986). Burley's study was among the first to investigate mate choice based on novel traits. Burley exploited what has since been called "preexisting sensory biases" for color variation in the soft parts and plumages of zebra finches by placing rings of different colors on the legs of females as well as males. Burley measured preferences using a multiarmed choice arena, which required choosing individuals to actively move down a corridor arm that effectively separated the chooser from any further visual stimuli from other to-be-discriminated individuals. Individuals to be discriminated were similar in size and plumage characters. Females preferred males ringed with red color bands, whereas they discriminated against males banded with green rings; males preferred females wearing black bands but discriminated against females wearing orange bands. Importantly, these preferences were associated with variation in fitness measures. In contrast to females mated to green-banded males, females mated to red-banded males had more surviving offspring, produced more sons than daughters, and had sons that, even when unbanded, had higher mating success when they subsequently mated. Furthermore, in experiments with banded males and unbanded females, red-banded males provided less parental care than green-banded males. In contrast to males paired with orange-banded females, males paired with black-banded females had higher offspring production, produced more daughters than sons, and contributed relatively more parental care. Red naturally occurs on the bills and feathers of adult zebra finches, adult legs are black, and beak colors range from orange to red. The importance of these experimental studies goes far beyond the obvious concern that color banding of wild birds may bias fitness variables. They showed that exploitation by conspecifics or by researchers of preexisting sensory biases has a strong, cascading effect on social behavior variation; they showed how microevolutionary process through social behavior variation can work; and they demonstrated that females (and males) respond to coloration and plumage of conspecifics, i.e., that they have "esthetic" sensibilities, a possibility much debated and frequently dismissed by nineteenth century biologists who found Darwin's conclusions wanting because they doubted females had the requisite esthetic capabilities. These extraordinary studies also raise a number of important questions about heritability of fitness and the adaptive significance of sensory exploitation or sensory traps in mate choice.

Long-tailed widowbird (*Euplectes progne*) females prefer to nest on the territories of males with experimentally elongated tails, a result consistent with Fisherian runaway selection and with male exploitation of preexisting sensory biases in females. Peahen (*Pavo cristatus*) females in captivity preferred to mate with males with more eyespots in their trains; most interesting, females in captivity laid more eggs when mated with males with larger trains. Only studies on zebra finch have simultaneously investigated the cues and fitness consequences of mate choice by both females and males in birds.

Most birds lack intromittant organs (prominent exceptions include the ducks and geese); for copulation to occur, females evert their cloacae to collect sperm from males. Thus, although harassment of passerine females by males occurs, forced copulation is unlikely. In contrast, forced copulation is common in ducks and geese. Among the most important studies evaluating fitness consequences of female mate choice is one comparing reproductive success of canvasback duck (*Aythya valisineria*) females in pairs freely chosen by females and in randomly assigned, investigator-determined pairs (Bluhm, 1985). Reproductive success was significantly higher for the freely choosing females than for females mating with randomly assigned males.

4. Reptiles

Mate choice among reptiles is seldom studied. Female anolis lizards (*Anolis carolinensis*) prefer males with brighter and bigger dewlaps and high display rates. There are few or no studies of mate choice in snakes.

5. Amphibians

Ease of experimental manipulation has resulted in multiple studies of mate choice in anurans. Most indicate that females prefer males based on variation in call characteristics. In *Acris crepitans*, *Bufo* species, *Hyla* species, and *Physalaemus pustulosus*, females discriminate male calls; in contrast in many *Rana* species females discriminate on male or territory size.

Researchers investigate male choice as frequently as female choice in salamanders. Both females and males prefer opposite sexes based on size (larger preferred) and pheromonal or odor cues. During "courtship" males deposit spermatophores onto the ground; females walk over the spermatophores and pick them up by their cloacae. Forced copulation would seem to be impossible in salamanders, but females may be pheromonally or hormonally manipulated by males.

6. Fishes

In species of fishes with paternal care, females choose males based on nest characteristics, territory size or quality, or male size or guarding ability. In controlled preference tests and in observational studies of wild fish, males prefer larger females, an observation consistent with male choice for high-fecundity females.

Guppies (*Poecilia reticulata*) are the best studied fish from the perspective of mate choice. They are freshwater fish native to the rivers and streams of Trinidad and northern Venezuela; they are small, internally fertilized, live-bearing (ovoviviparous), and sexually dimorphic with females larger than males. Guppies have highly variable color patterns of melanic, carotinoid, and structural colors. Males display to females. Females discriminate among males' display rates, ornament colors, and the complexity of color, which often correlates negatively with parasite loads. Males possess gonopodia for internal transfer of sperm, males harass females for matings, and forced copulation can occur.

In species that release gametes into the water column for fertilization, forced copulation would seem to be impossible; however, so-called "sneakers" sometimes join spawning pairs. In contrast, in species possessing gonopodia forced copulation attempts seem to be regular features of reported mating behavior.

C. Insects

Female cockroaches (*Nauphoeta cinerea*) exert active mate choice, correlated to variation in male pheromonal cues, and preferred traits in males correlate with developmental rate of offspring (Moore 1988).

Female scorpion flies (*Hylobittacus apicalis*) prefer males offering larger prey items as "nuptual gifts." Females in this species only forage when male densities are low; when male densities are high, females depend on males for food. Duration of copulation positively correlates with the size of male nuptial gifts. Females eat and copulate at the same time. Copulations must last at least 5 min for any sperm to be transferred; copulation durations correlate with number of transferred sperm. In copulations lasting longer than 20 min males transfer an accessory gland substance to the female that inhibits female remating for up to 4 hr.

A comprehensive review of mating system correlates in *Drosophila* is provided by Markow (1996).

D. Role-Reversed Species

Among "role-reversed" vertebrates mate choice is most well-known among pipefishes (*Nerophis ophidion*) (Rosenqvist, 1993). As in all Syngnathidae, only males provide parental care, providing offspring with nutrients and oxygen while brooding them on their ventral surfaces or in modified ventral skin pouches. Females are larger than males and have blue sexual coloring and lateral skin folds that are obvious only during breeding seasons; males lack sexual colors and lateral skin folds. Females place eggs on males' bodies using an ovipositor; males fertilize eggs by "sinking through" their own sperm clouds. Laboratory experiments indicate that energetic expenditures for the production of one offspring are less for males than for females, suggesting that females primarily limit males' reproduction. However, in experimental situations females reproduced more rapidly than males, producing 1.8 clutches during the period of a single male pregnancy; thus, males are a primary limiting factor for females' reproduction. Indeed, females do fight over access to males and demonstrate no preferences for males of different sizes. However, males prefer long females, females with the largest skin folds, and females with the largest blue color patches (Fig. 1).

In katydids (*Requena verticalis*), sex role reversal with greater competition among females for males and increased mate choice by males can be a facultative response to variation in food availability. When food is scarce the value of male food contributions increases so male parental investment increases, and discrimination among males of females also increases. Males rejected females more frequently when food was experimentally limited, whereas females

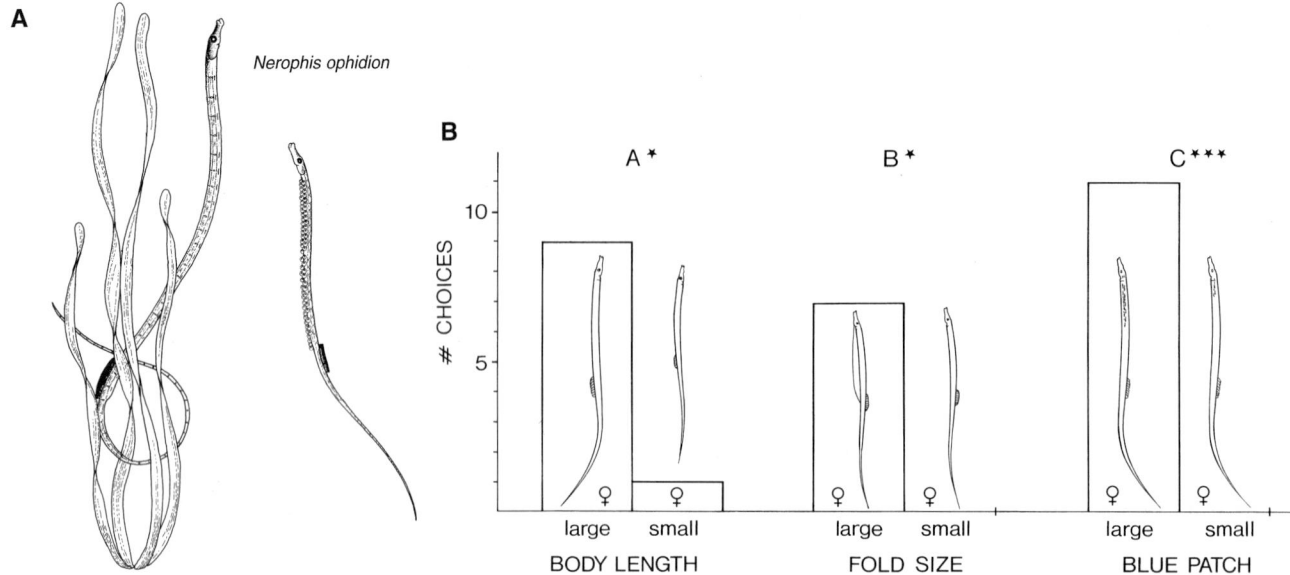

FIGURE 1 A frequency histogram of male choice criteria among pipefish, *Nerophis ophidion*, allowed to discriminate between females experimentally manipulated to differ in (A) length ($n = 10$), (B) size of skin fold ($n = 7$), and (C) size of blue patch ($n = 11$). $*p < 0.05$; $***p < 0.001$ (reproduced with permission from Rosenqvist, 1993; photographic inset of color variation in females courtesy G. Rosenqvist).

rejected males more when food was experimentally supplemented.

III. GENERAL PATTERNS

A. Female Choice or Male Choice

The most frequent examples of within-sex combat in *Descent of Man and Selection in Relation to Sex* (Darwin, 1871) were male–male contests. Furthermore, Darwin's examples of female choice far outnumbered examples of male choice. Investigators explain the usual observation that females are choosier than males using comparisons of relative energetic investments in offspring, relative rates of reproduction, or variation in the operational sex ratio, which is the sex ratio of individuals available for reproduction at a given time. In typical species, females invest more in individual offspring than males, male reproductive rate is greater than that of the female, and more males are generally available for breeding than females. Thus, in typical species intrinsic factors or access to nutritional resources primarily limit female reproductive success, whereas access to females primarily limits male reproductive success. These usual sex differences, with females having more time and energy investment squandered if reproductive attempts fail, result in selection favoring males that have relatively indiscriminate tendencies and females that are relatively more careful about all aspects of reproduction, including being relatively more discriminating about with whom they share their gametes. Thus, in species with highly asymmetric burdens or benefits of maternal and paternal care, researchers expect choosy females and ardent males. Note, however, that the theoretical expectations and empirical observations are about relative, not absolute, choosiness and mating enthusiasm, so that some females are sometimes ardent and some males are sometimes choosy.

B. Mate Choice in Role-Reversed Species

Role-reversed species are those in which the relationship of female limitations on male reproduction is reversed so that access to males limits female reproduction. Some discussions focus on between-sex variation in rates of reproduction: In role-reversed species female reproductive rate exceeds male reproductive rate rather than male rate exceeding female rate. Other discussions focus on whether the opposite sex or extrinsic environmental variation such as food is the "resource" primarily limiting reproduction of females versus males: In role-reversed species females' primary resources for reproduction are males rather than extrinsic resources and intrinsic capabilities. However, other discussions focus on which sex is most modified by mate choice: If females are more modified than males they are called "role reversed," a conclusion often made without simultaneous evaluation of the effects of female choice on males and male choice on females.

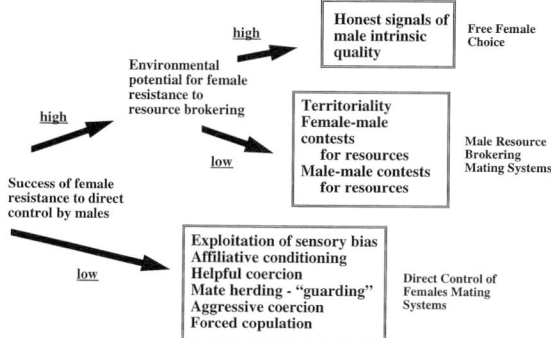

FIGURE 2 Predicted relationship of variation in females' abilities to resist males' manipulations of their reproductive decisions, including mating choices on mechanisms of competition on males and mating outcomes (reproduced with permission from Gowaty, 1997, © Kluwer Academic Publishers).

C. Sexual Dialectics

When one sex of a species is a limiting resource for reproduction in the other sex, selection theoretically favors manipulation of the resource by the resource user and then compensatory selection (counterselection) on the resource user by the resource. Such interactions comprise predicted regularities in the interactions between the sexes. These predictions depend on the notion that the sexes are in ongoing, often antagonistic interactions that create the social environments that exert fundamental selective force on reproduction of opposite-sex individuals (Gowaty, 1997). Predictions include that the mechanisms leading to within-sex variance in reproductive success depend on opposite-sex resistance to attempts to manipulate or coerce. For example, mechanisms of sexual selection on males depend on intrinsic and environmentally determined abilities of females to resist males' efforts to control their reproductive capacities (Fig. 2).

IV. FUNCTIONS OF MATE CHOICE

To understand why individuals discriminate, we need to know what ecological (including social) problems choosy individuals solve by discrimination. Choosy individuals may gain fitness benefits for themselves, for one sex of offspring or the other, or for both offspring sexes simultaneously. For example, females discriminate among males to avoid exposure to pathogens and parasites; to gain immediate access to resources such as food or nesting sites, substrates, or locations; to gain help in raising offspring; to gain protection from predators and other "dangerous solicitors"; and to gain a variety of genetic benefits for their offspring, including the ability of offspring to thrive, survive, and reproduce despite exposure to pathogens and parasites.

A. Components of Sexual Selection via Mate Choice

The ecological problems solved when individuals choose may have different effects on choosers' overall fitnesses. Ranking the fitness effects of selection pressures categorizes the significance of mate choice from the point of view of choosy individuals. This approach, at least initially, specifically ignores assumptions about traits in the opposite sex that mate choice favors or disfavors. If choosing enhances fitness, investigators then may use discriminate function statistical tests to explore the proximate correlates of choice (i.e., the traits choosy individuals prefer) without biasing outcomes with *a priori* investigator expectations. This approach will allow description of the components of selection through mate choice.

B. Bizarre and Elaborate or Run-of-the-Mill and Simple

Run-of-the-mill or average traits are almost never subject to investigation under the trait-based paradigms of normal mate choice studies. Run-of-the-mill traits have seldom, perhaps never, been the subject of models by sexual selection theorists. Nevertheless, although one might guess that run-of-the-mill traits are more likely the result of garden-variety natural selection (differential survival) than sexual selection (differential reproduction), common, even simple, traits are probably also influenced by mate choice. Certainly, it is easy to imagine how male mate choice for high-fecundity females might explain fecundity variation among females.

C. Fisherian Runaway Process, Handicaps, and Indicator Traits

Fitness benefits of mate choice may include those from sons having traits preferred by females. Such female preference can lead to a "runaway" process favoring males with more exaggerated expression of preferred traits and females with more tightly linked and exaggerated preference for the trait. Assuming that sexually selected traits in males indicate the benefit of female choice for discriminating females, females choose males for their handicaps, the possession of which indicates some important quality in males, such as their immune system capabilities. Such traits are indicators of other characteristics. The idea that elaborate or bizarre traits evolve to indicate some underlying superior quality comes from the observation that many bizarre and elaborate traits, especially in males, handicap their possessors, for example, by making it more difficult for them to avoid predation or more energetically costly to feed. Thus, individuals surviving and thriving with elaborate traits indicate underlying superior qualities (usually genes) to choosy individuals. Elaborate or bizarre signals of underlying qualities are more likely to be honest the more it costs the signaler to send them. Courtship signals thus may indicate physiological soundness, good health, and the absence of deleterious mutations; they may indicate social dominance or resource-gathering skills; and they may indicate genetic complementarity with females. Currently, the weight of theory (Kirkpatrick and Ryan, 1991) and data favor the hypothesis that elaborate and bizarre traits honestly signal good genes or, more specifically, predict offspring survival and viability. In keeping with these ideas, plumage ornaments of male great tits, barn swallows, peacocks, and great reed warblers correlate with offspring survival.

D. Courtship as Manipulation, Assessment, and/or Species Recognition

Animal communication theories argue that signs or signals manipulate the behavior of receivers so that signals are often deceptive. Thus, "courtship" signaling may have originated as attempts, usually by males, to manipulate the reproductive decisions of females. Alternatively, courtship may provide choosers the opportunity to assess important behavioral characteristics such as the quality of parental care a mate is likely to provide. Courtship rituals may be a way of coordinating the behavior of individuals when high levels of cooperation in some aspect of reproduction enhance the fitness of each member of a pair. In addition, some courtship signals may specifically provide information on species identification.

V. PROXIMATE CAUSATION

Most published studies of mate choice have been attempts to explain the existence of or variation in a given trait. There are few published studies of the functions of mate choice independent of some *a priori* unexplained trait, i.e., courtship and discrimination itself. Ongoing investigations are turning the process around by asking whether individuals discriminate even in tests not based on variation in any specific traits. Often, investigators offer to choosers individuals to be discriminated at random with respect to obvious phenotypes. In these tests the dependent variable is a component of fitness in the choosy individual or the offspring. Such experiments allow strongly inferential conclusions about the adaptive significance of mate choice for choosy individuals.

A. Mate Choice without Signals

If courtship signals are manipulative means for influencing reproductive decisions, mating choices might occur without courtship. This logic predicts that mate choice without courtship will occur, e.g., when the probability of producing offspring of high relative viabilities is equal for each partner with the other. The logic further predicts that, even in species with sexually reluctant, highly discriminating females, some females in the presence of certain males will be facultatively ardent or sexually enthusiastic. In contradiction to dogma, it suggests that males, even in species in which males seem universally ardent and sexually enthusiastic, under some ecological and social situations will be "coy" and sexually reluctant. An empirical implication is that sexual enthusiasm in females and male mating reluctance may be significant evolutionary forces even in non-role-reversed species.

B. Rare Male Advantage

Experiments using *Drosophila melanogaster*, *D. pseudoobscura*, *D. persimilis*, *D. willistoni*, *D. equinoxialis*, *D. tropicalis*, and *D. immigrans* indicate that rare males have an advantage in mating over common males. Rare male mating advantage is frequency dependent on rare males having a mating advantage over more common males. Rare male mating advantage exists when strains differ in chromosomal arrangements, geographic origins, and between strains with different alleles at a single locus. Females typically use odor cues to discriminate rarer males among a sample of males, but the influence of tactile cues remains a possibility. Although it is possible that some of the many demonstrations of rare male advantage are due to unconscious bias in the use of more active males that congregate at the top of collection vials rather than less active males that congregate toward the bottom of collection vials, it seems unlikely that experimental bias explains all observations of rare male advantage. What seems missing from most discussions of the phenomenon is a convincing argument about the selective advantage of rare male advantage. Female preference for rare males would favor offspring heterozygosity in much the same way that preferences based on major histocompatibility loci dissimilarities do in mice and humans. The connections of rare male mating advantage to questions of immune system function have yet to be systematically explored in any taxa. New experiments evaluating simultaneously preferences of females and males from rare and common strains exposed to parasitic challenges may test the hypothesis that mate choice for rare phenotypes occurs because selection favors complementarity of partners' immune system alleles for viability benefits in progeny.

C. Exploitation of Preexisting Sensory Bias

A new trait may originate in one sex because it stimulates existing sensory machinery in the discriminating sex. This is a hypothesis for the origin of traits used in mate choice known as "exploitation of preexisting sensory bias." Chooser responses used in other contexts (e.g., avoiding predation, care of offspring, predation, or foraging) may constitute a "sensory trap" that courting individuals may use to increase the probability that choosing individuals will discriminate in their favor. The most frequently cited demonstration of female choice for a trait based on preexisting sensory bias in females is platyfish (*Xiphophorus maculatus*). Platyfish lack tails, but females prefer congeneric swordtails, *X. helleri*, which have long caudal fins known as swords, and platyfish males with experimentally provided swords. Likewise, female swordtails prefer males with swords. This result suggests that female swordtails had the preference for males with long swords before long caudal fins evolved. Preference among both female and male birds for color variations in novel leg bands is also, no doubt, based on preexisting sensory biases. Exploitation of preexisting sensory biases can explain Fisherian runaway coevolution between the preference in one sex and the trait in the other. Exploitation of preexisting sensory biases may be a common way that individuals of the chosen sex manipulate choosers into mating with them that otherwise would not. Manipulations of reproductive decisions via sensory traps are unlikely to remain stable unless coupled to an underlying trait(s) advantageous to the chooser, i.e., unless the sensory trap becomes an indicator trait.

D. Preinsemination Courtship Signals

Mate choice that discriminates variation in colors, sizes, odors, and complex motor patterns functions when opposite sexes are within eyesight of each other. In contrast, mate choice that discriminates variation in acoustic characteristics can occur over larger distances. Senders can delay olfactory signals temporally and spatially. Thus, theories of courtship often focus on the ecological contexts of males and females in time and space, e.g., guppies do not use courtship signals based on color variation in streams with predators that could also be attracted to their bright coloration.

E. Postinsemination Courtship Signals

Postmating or postinsemination mate choice depends on tactile, chemical (physiological), and morphological variation in the discriminated sex (Eberhardt, 1996). Observation of exploitation of preexisting biases and traps in the machinery of touch and pressure sensation, physiology, and morphology of discriminating individuals is undocumented.

VI. RELATIONSHIP TO PAIR BONDING, SOCIAL SYSTEMS, AND SPECIATION

Female mating discrimination and resistance to male attempts to coerce or manipulate female mating choices may be the basis for variation in social organization known as mating systems as well as an evolutionary force in speciation.

A. Pair Bonding and Gametic Sharing

Recent molecular marker studies of genetic parentage in "monogamous" birds indicate pair bonding and gametic sharing seldom overlap completely. In more than 75% of tested socially monogamous bird species, females produce broods sired wholly or in part by "extra-pair males." Extra-pair males sire between 5 and 95% of offspring. Implications of these findings include the following: Some classic studies of mate choice in wild-living birds suffer from criticisms leveled at experimental lab arena studies. Namely, social associations may be inaccurate indicators of likelihood of gametic sharing. In addition, mate choice may have several functions because the fitness benefits of social pairing may be different from those for gametic sharing. Finally, the proximate causation of social pairing may not be the same as that for gametic sharing.

B. Female Choice and the Polygyny Threshold

In species in which males find it impossible or difficult to coerce females' reproductive decisions, female choice may be the basis for social polygyny (of red-winged blackbird, *Agelaius phoenicius* type). Theoretically, females choose to mate with already mated males rather than unmated males whenever already mated males defend superior territories with resources capable of making up fitness deficits associated with decreases in paternal care that would be shared among offspring of several females. This idea, the polygyny threshold model, motivated much research during the past 25 years on social behavior evolution in birds and other taxa. However, because few species conform to its assumptions, there are few empirical supporting examples. Nevertheless, rejection of the idea that female choice operates in the evolution of social organization is unlikely.

C. Sexual Conflict and the Evolution of Female-Resistance Traits

Whenever variation in offspring viability favors female discrimination among males, the conditions occur for one type of intersexual conflict. This is because whenever females reject males, selection will favor those males that attempt to influence females' reproductive decisions (via coercion, force, brokering of resources, etc.). Whenever such manipulative males are successful, selection will favor females that resist males' efforts to modify females' decisions. Therefore, sexual conflict in the form of male attempts to manipulate female reproduction and female attempts to resist manipulation may be an almost inevitable outcome of original selection favoring female discrimination among males. Elaborate and bizarre male traits may originate as signals serving males' attempts to manipulate and control females' reproductive decisions. Likewise, bizarre traits in females may be resistance morphologies that decrease the likelihood that nonpreferred males will be able to mate with them. Examples of resistance morphologies include the midleg spines of female houseflies used to decrease the likelihood of successful copulation by nonpreferred males. Experimental elongation of abdominal spines in water striders (*Gerris incognitus*) demonstrated that females use the spines to dislodge harassing, nonpreferred males in their attempts at forced copulations (Fig. 3). Females with experimentally elongated spines had fewer copulations with shorter durations than females with experimentally shortened spines, which experienced more copulations of longer duration (Arnqvist and Rowe, 1995). Female butterflies (*Helioconius erato*) have modified storage chambers for a pheromonal substance repellent to other males transferred to them during copulation by males. Females control release of this substance; thus, they are likely to use it only when it is to their advantage to do so, perhaps when nonpreferred males attempt to copulate.

D. Female Constraint and the Evolution of Monogamy

Males in typical species should exploit variations among females that increase the likelihood that females will mate with them rather than some other male. In socially monogamous cavity-nesting birds, males often broker females' access to nesting sites so that females may trade copulations and fertilizations for access to critically limiting resources. If females vary in their abilities to feed their offspring, males may adjust paternal provisioning for trades with females of copulations and inseminations. This charac-

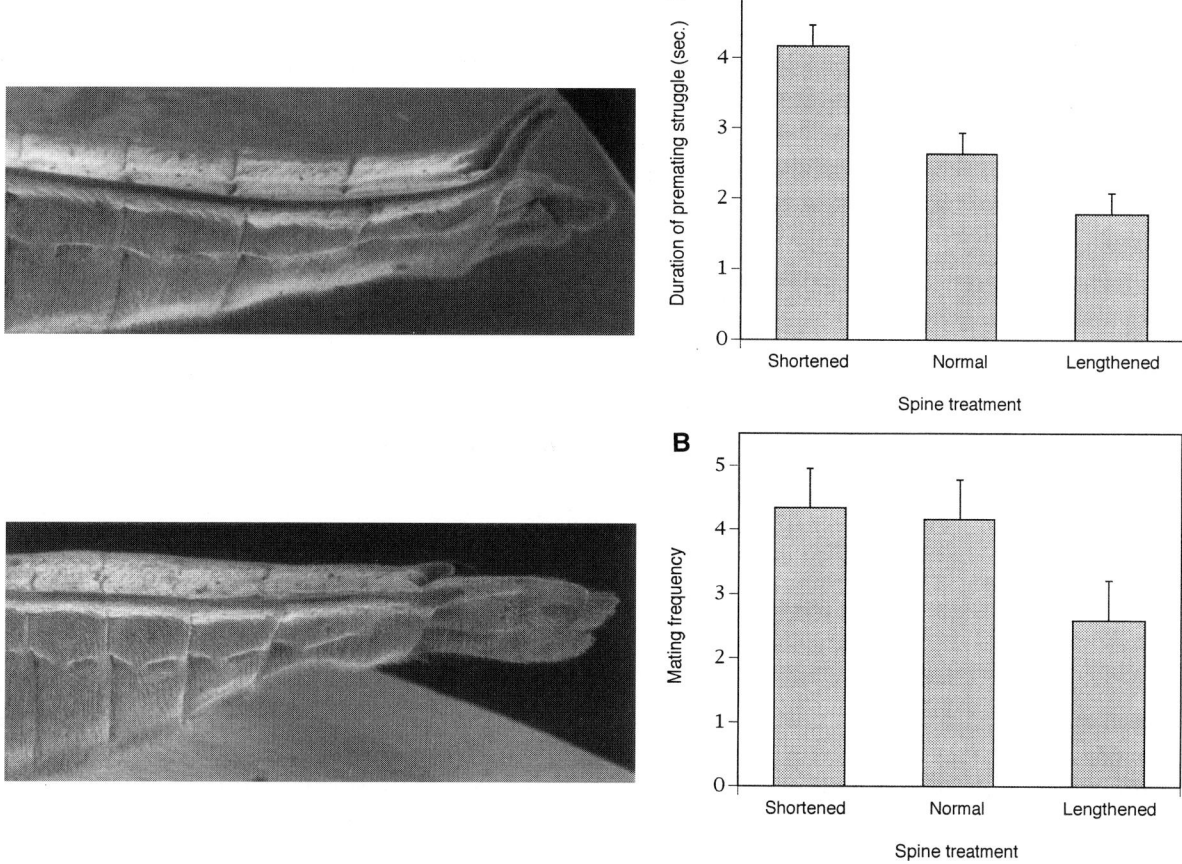

FIGURE 3 Frequency histogram of duration of "premating struggle" (A) and mating frequency (B) as a function of experimental manipulation of lateral spine length of females. (Insert) Scanning electron micrograph of female (top) and male (bottom) *Gerris incognitus*.

terization of interactions between the sexes in socially monogamous birds predicts the occurrence and distribution of extra-pair paternity among females as a function of their abilities to provision their offspring without male help. If females behaviorally control copulations in birds, it predicts that highly skilled females or females in resource-rich circumstances will be more likely than unskilled or unlucky females to share their gametes with additional social partners; it also predicts that these highly skilled or lucky females will be less likely to share any gametes with nonoptimal males, whether or not these are her social partners. Thus, male attempts to manipulate females' reproductive decisions may be pivotally important determinants of the degree of match between social pairing and gametic sharing. This idea, like the polygyny threshold model, depends on fitness advantages for females when they freely choose among available males for gametic sharing.

E. Speciation by Sexual Dialectics

Rice (1996) experimentally arrested female counterresponse to male adaptations that manipulate female reproduction for male advantage using captive populations of *D. melanogaster*. Using clever genomic "tricks" and selective breeding, Rice created males in which most of their genome behaved like a Y chromosome so that the sons' genes were inherited through fathers only. In this way he removed mother's genetic contributions to succeeding generations and thereby eliminated from the evolving population genes coding female resistance to male manipulations: He arrested female evolution. Rice showed that without coevolving female resistance mechanisms, experimental males were able to more quickly remate with females previously mated to nonexperimental competitors, whereas competitor males had decreased capacity to remate with females mated to experimental males. He showed that female mortality was positively correlated to female remating frequency. Rice's experiment provides the strongest experimental evidence that coevolutionary responses of the sexes to each other are evolutionary forces capable of driving sympatric speciation. If freely expressed female choice benefits female fitness, the mechanisms males use to manipulate female reproduction for their own advantages may be detrimental to females, in which case selection may favor female counteradaptations. In *D. melanogaster* males' accessory gland products cause females to delay remating, lay more eggs, and die sooner.

Male manipulation of female remating interval and fecundity easily fits criteria for antifemale choice mechanisms. Resistance mechanisms in females should operate to ameliorate the effects of delayed remating, higher fecundity, and higher mortality. Such sexual dialectics could lead to sympatric speciation whenever male manipulation or female resistance exhausts variation on a given trait (Gowaty, 1997). Within such lineages, the traits subject to sexually antagonistic coevolution may shift so that coevolutionary responses would continue operating but on other traits, thus facilitating male manipulation or female resistance. That is, sexual dialectics may use up available variation in given traits quickly, moving manipulation and resistance to another stage. Individuals in these lineages would be acting out the evolutionary play on different ecological stages. Fixation in one lineage and subsequent shift of selection via manipulation and resistance onto other traits in alternate lineages would amount to sympatric speciation. Sexual dialectics theory predicts and explains species-specific reproductive morphologies and physiologies among sympatric congeners, even without historical geographic isolation.

Acknowledgments

I thank J. Byers, L. Ehrman, A. Moore, G. Rosenqvist, and J. Wingfield for useful comments on a previous draft and G. Rosenqvist, G. Arnqvist, and L. Rowe for permission to use previously published figures, drawings, and photographs. Grants from the NSF and NIMH supported the preparation of the manuscript.

See Also the Following Articles

Mating Behaviors; Sexual Attractants; Sexual Selection

Bibliography

Andersson, M. (1994). *Sexual Selection*. Princeton Univ. Press, Princeton, NJ.

Arnold, S. (1994). Sexual selection in plants and animals: A symposium. *Am. Nat.* **144**(Suppl.).

Arnqvist, G., and Rowe, L. (1995). Sexual conflict and arms

races between the sexes: A morphological adaptation for control of mating in a female insect. *Proc. R. Soc. London B* **261**, 123–127.

Baker, R., and Bellis, M. (1995). *Human Sperm Competition: Copulation, Masturbation, and Infidelity.* Chapman & Hall, London.

Bluhm, C. K. (1985). Mate preferences and mating patterns in Canvasbacks (*Aythya valisineria*). In: *Avian Monogamy* (Gowaty, P. A., and D. W. Mock, Eds.), pp. 45–56. Ornithological Monographs 37: American Ornithologists Union, Washington, DC.

Burley, N. (1986). Sex ratio manipulation in color-banded populations of zebra finches. *Evolution* **40**, 1191–1206.

Darwin, C. (1871). *The Descent of Man, and Selection in Relation to Sex.* Murray, London.

Eberhardt, W. G. (1996). *Female Control: Sexual Selection by Cryptic Female Choice.* Princeton Univ. Press, Princeton, NJ.

Gowaty, P. A. (1997). Sexual dialectics, sexual selection, and variation in mating behavior. In *Feminism and Evolutionary Biology: Boundaries, Intersections, and Frontiers* (P. A. Gowaty, Ed.), pp. 351–384. Chapman & Hall, New York.

Hrdy, S. (1997). Raising Darwin's consciousness: Female sexuality and the prehomind origins of patriarchy. *Hum. Nature* **8**, 1–50.

Kirkpatrick, M., and Ryan, M. J. (1991). The evolution of mating preferences and the paradox of the lek. *Nature* **350**, 33–38.

Markow, T. (1996). Evolution of *Drosophila* mating systems. In *Evolutionary Biology* (M. K. Hecht *et al.*, Eds.), Vol. 29, pp. 73–106. Plenum, New York.

Moore, A. J. (1988). Female preferences, male social status, and sexual selection in *Nauphoeta cinerea*. *Anim. Behav.* **36**, 303–305.

Rice, W. (1996). Sexually antagonistic male adaptation triggered by experimental arrest of female evolution. *Nature* **381**, 232–234.

Rosenqvist, G. (1993). Sex role reversal in a pipefish. *Mar. Behav. Physiol.* **23**, 219–230.

Trivers, R. L. (1972). Parental investment and sexual selection. In *Sexual Selection and the Descent of Man, 1871–1971* (B. Campbell, Ed.), pp. 136–139. Heinemann, London.

Wedekind, C., Seebeck, T., Bettens, F., and Paepke, A. J. (1995). MHC-dependent mate preferences in humans. *Proc. R. Soc. London B* **260**, 245–249.

Mating Behaviors, Insects

William H. Cade
Brock University

I. Introduction
II. Long-Range Signaling and Female Attraction
III. Female Choice
IV. Male–Male Competition
V. Courtship
VI. Copulation, Sperm Transfer, and Sperm Competition
VII. The Costs of Mating

GLOSSARY

costs of mating The negative consequences of mating in terms of energy and time expenditure and increased chances of being located by predators and parasites.

female choice The selection of a male by a female based on different levels of criteria, such as the male belonging to the same species, being sexually mature and competent, or that he is the best male available when compared with other males.

mate attraction The use of specialized acoustical, visual, or chemical signals (pheromones) to attract mates from a distance, sometimes over several hundred meters or more.

mate guarding Aggressive behavior by a male who has just mated toward other nearby males that prevents others from mating with the same female.

reproductive competition The various behavior patterns, such as territoriality and aggression, that males use to compete for access to females and the opportunity to mate and to prevent other males from mating.

sperm competition The competition within a female's reproductive tract between sperm of different mates for fertilization of ova often resulting in the last male to mate fertilizing most of the eggs.

M ating behavior includes a wide range of activities that male and female animals engage in resulting ultimately in the fertilization of eggs. Mating behavior includes signaling for and finding a mate, competition with other individuals for access to mates, evaluating and choosing mates, and the coupling of male and female genitalia, transfer of sperm to a female, and guarding of a mate after mating. Each one of these components involves very detailed patterns of behavior by both males and females. Since the currency of natural selection is ultimately the reproduction of offspring in future generations, it is not surprising that some of the most complex and fascinating activities of animals characterize their mating behavior, a generalization that is especially true in insects.

I. INTRODUCTION

Insects are the most successful animal group by various measures, the most important being the number of species. There are approximately 1 million known species and certainly many times that number that are still to be discovered. With so many species, it follows that there is tremendous variation in mating behavior and other life history traits from group to group. It is usually possible, however, to identify the same basic components of mate attraction, reproductive competition, mate choice, fertilization, and postmating behavior in a given species or group.

There is a vast amount of information available on insect mating behavior, although some groups are much better understood than others. One family in which mating behavior has been studied extensively is the crickets, a very diverse group with many species. In many cricket species males use specialized acoustical signals to compete with other males and to attract females. Crickets also show all of the basic components of insect mating behavior. This article will describe the basic components of insect mating behavior by focusing on cricket mating behavior, especially that of field crickets in the genus *Gryllus*. Special cases of mating behavior in other insect groups will also be described.

II. LONG-RANGE SIGNALING AND FEMALE ATTRACTION

In many animals, including insects, mating behavior begins with the attraction of mates often from distances of several hundred meters or more. The means of attracting mates can take various forms depending on the insect group. Firefly beetle males, for example, generate their own light and flash the light while they fly at night. The flashing and flight patterns are specific to the species. Females resting on the substrate answer with a flash and males land nearby. In some species female insects do the attracting. Moths, for example, release very small amounts of chemicals or pheromones which are carried by air currents (Fig. 1). Males in the same species detect the pheromone by fine hairs on their antennae. Males fly upwind until they find the female. In some cases males and females are both attracted to each other as in butterflies with brilliant coloration. Members of the opposite sex see the species-specific color patterns and fly to meet a potential mate.

At night one of the best ways to communicate is by production of acoustical signals and crickets are well-know for their rhythmic songs on warm summer evenings. These "calling songs" are specific to each species and are produced by males rapidly rubbing

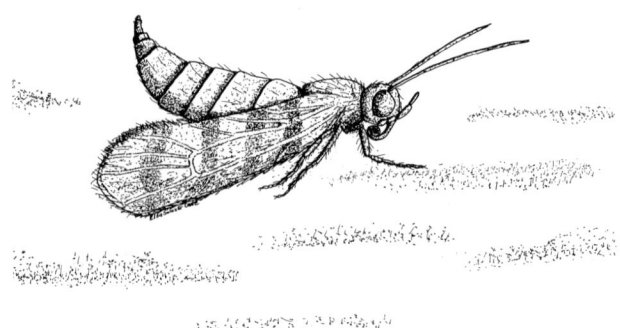

FIGURE 1 Female moths of various species arch their abdomens in the air, so-called calling behavior, while releasing male-attracting pheromones from glands in the abdomen (drawing by E. Salazar-Cade).

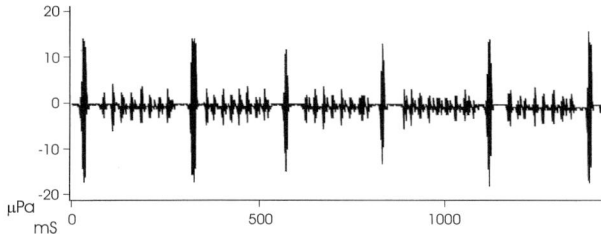

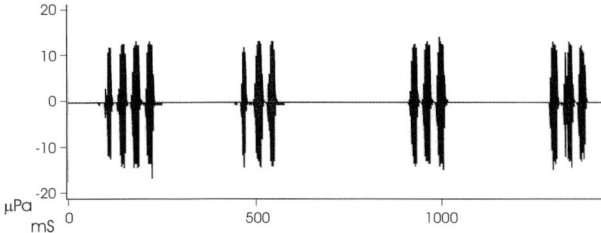

FIGURE 2 Spectrographs of male *Gryllus bimaculatus* songs. The courtship song (top) has a series of loud ticks shown by long spikes separated by fainter sounds resembling a rustling of leaves. By contrast, the calling song (bottom) consists of three or four pulses of sound grouped into distinct chirps.

their wings together. Each closure of the wings produces a pulse of sound and these pulses are produced in groups of chirps or trills (Fig. 2). In the genus *Gryllus*, male crickets call for as long as several hours each night. Phonotaxis or the orientation to or away from a sound source has been studied in many cricket species. A general finding is that female crickets are attracted to the male calling song of their species and not to the songs of other species. One of the most striking examples of female cricket phonotaxis involves flying mole crickets in Florida. Male and female *Scapteriscus acletus* and *S. vicinus* fly at night in large numbers and land near calling males. Ulagaraj and Walker used tape-recorded cricket songs broadcast outdoors to attract hundreds of crickets each night to special traps fitted to the loudspeakers.

The cricket calling song provides information on the location of potential mates and is very important to females in choosing males. Several related findings from different experiments demonstrate the important role of phonotaxis to mating behavior. Female crickets that have recently mated are not very phonotactic. The level of phonotaxis increases over time, however, when females are isolated from males. Virgin female crickets are usually more phonotactic than previously mated females and the intensity of phonotaxis to male calling song increases with age. Old virgin females (20+ days of adult age) are the most phonotactic since they have less time remaining than younger females in which to mate and lay eggs.

III. FEMALE CHOICE

Parental investment is the expenditure of time and energy by parents in their offspring. It involves the nutrients invested in eggs and feeding and protection of young. Female animals usually have a much higher level of parental investment than males and this is thought to lead to a situation in which females discriminate in their choice of mates. Males, by contrast, are usually less discriminating and show a higher level of competition for mates than females. Male beetles, for example, fight for females using horns and other protrusions from their bodies. Females wait nearby and mate with the victorious male. The general trend of high female parental investment compared to males is especially true in the insects, but there are exceptions. Giant water bug males, for example, allow females to lay eggs on their backs. Males protect the eggs until they hatch. Male giant water bugs are very discriminating in their choice of mates as a consequence of this high investment and only allow females to lay eggs on their backs after a long period of mating.

In crickets female choice involves evaluation of males using information in their calling songs. The relative attractiveness of male cricket song to females depends on a variety of factors. A common factor across many species involves the relative loudness or intensity of the song. In Ulagaraj and Walker's study on mole crickets, for example, the number of crickets attracted to broadcast song increased several-fold when the loudness of the signal was increased. It was later discovered that large mole cricket males produce louder songs than small males. Large size may confer advantages on males in terms of male–male competition. In mole crickets the loudness of a male's song provides information to females on the size of the caller.

The amount of time that male crickets call is very important in terms of attracting females. All night studies on individually marked field crickets (*Gryllus* species) in a natural setting showed that the amount of calling varies greatly between individual males. Some males call several hours each night, whereas others call only a short time or not at all. The amount of time that males call is termed the duty cycle. The higher the duty cycle the more likely a male is to attract females, although other variables are known to be important.

The time at which males call is another important variable influencing female phonotaxis and choice of mates. Cricket calling behavior is generally restricted to the night, but there are many different patterns of calling depending on the species. In some species males call all night with little change in the number of callers with time. In others, males call for a short time immediately after sunset. Still another pattern involves a dawn chorus where the number of callers increases greatly at sunrise. In the field cricket *G. integer*, males start calling at sunset but show a large increase in the number of callers before sunrise. This dawn chorus corresponds to the time that most matings occur in *G. integer*. There may be many factors influencing females postponing mating until dawn, but one possibility is that dawn matings occur after females have observed male behavior for several hours and thus have information on which to base mate preferences.

The fine structure of a male's crickets song is another factor determining how attractive the signal is to females. Calling songs are species specific, but there is much variation between individual males. Males may differ, for example, in the number of sound pulses in a trill or chirp, the rate at which the pulses are given, the timing of silent intervals between parts of the song, and other features. A variety of different experimental approaches to female phonotaxis have been used on many cricket species in the laboratory. A common approach is to broadcast recordings of different male songs through different loudspeakers in an enclosure. Female crickets walk in the enclosure and approach the loudspeaker broadcasting the most attractive song. Another series of experiments have been performed in which females run on a sphere and songs are broadcast through different loudspeakers around the sphere. A computer records the direction and speed the female runs on the sphere (Fig. 3). These experiments have shown that in some crickets male songs

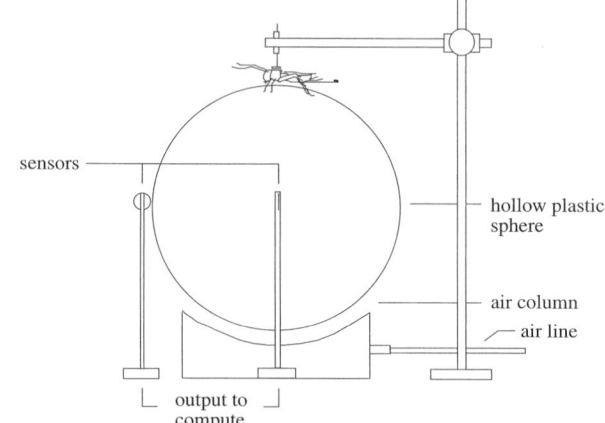

FIGURE 3 A female cricket runs on a sphere suspended on a column of air. Computer-synthesized cricket songs are broadcast from around the sphere. The Kugel device is essentially an inverted computer mouse and the computer records the position of the female with respect to the songs (drawing by W. Wagner).

with the greatest number of pulses per trill are the most successful at female attraction. More information is needed on cricket song structure and female phonotaxis. Most important is the need to understand the relationship between song structure and the quality of a male as a mate. In this connection, the structure of a male's songs may indicate his ability to withstand parasite infections, an important trait to females in evaluating males.

IV. MALE–MALE COMPETITION

Male insects, in general, show very high levels of reproductive competition given their low parental investment. This competition takes many forms and may involve defense of territories, use of aggression and dominance over other males, and the ability to search for and locate females. Dragonfly males, for example, spend much time flying along stream banks and other bodies of water where they defend territories. Other males which enter the territory are quickly challenged and usually expelled.

In some species of field crickets male calling songs attract not only females but also other conspecific males. Attracted males take up positions near other callers and begin singing. In this way, phonotaxis by males leads to the formation of mating aggregations similar to the leks or mating territories of some birds and mammals. Within aggregations field cricket males are separated by a minimum distance of approximately 1 m. Playback experiments involve broadcasting song to calling males at different distances between the male and loudspeaker. At distances <1 m separating a calling male and what sounds like a nearby caller, male crickets are likely to show aggressive behavior. That is, they leave their singing station, run toward the loudspeaker, and attack conspecific males tethered to the loudspeaker. In natural populations a calling male attacks any male that sings at a distance of <1 m from his calling station. The calling song therefore attracts females for mating and serves to maintain distances between nearby callers.

Specialized chorusing behavior within aggregations of crickets is another form of reproductive competition. Field cricket males often sing simultaneously from burrows, under rocks, or other protected locations. Some tree crickets chew a hole in a leaf and position their body so that the leaf reflects the song, thus resulting in a more directional signal (Fig. 4). Previously silent males often start calling when a neighbor starts to call. Field cricket songs are high intensity with signals in the 80- to 90-dB range. This is one of the loudest female attracting signal in the animal kingdom. In some cases males will increase the intensity of their song when a nearby male starts to call.

Synchronous calling is another aspect of specialized chorusing behavior in some cricket species. In species with distinct chirps of song, nearby males may produce the chirps at the same time so that it sounds like a single loud chirp from an area occupied by several males. Alternation calling, the opposite of simultaneous calling, also occurs in some species. Here a male produces his chirps during the silent intervals of adjacent males. This leads to a "back and forth" type of calling. Similar patterns of specialized chorusing behavior occur in other insects and in frogs and other vertebrates that use acoustical communication in their sexual behavior.

Reproductive competition in many species of insects often involves fighting and field crickets are some of the most aggressive insects. Male *Gryllus* also have a specialized aggressive song they use when fighting. The aggressive song is similar in structure to the female-attracting calling song, but it is produced in bursts of a few seconds rather than continuously over long periods. Males bite, kick, and head butt each other while singing aggressively in what are often prolonged fights (Fig. 5). The winner of a fight is always obvious because the defeated male turns and runs. Fights do not appear to cause damage to combatants, but defeated males are prevented from calling for females in the vicinity of the winner. Large cricket males readily defeat smaller males and females that choose large males based on their songs may pass on genes for large body size to their sons. Most important, the most aggressive males mate more often than males who are less successful in fighting, at least in laboratory experiments.

The males of many species of *Gryllus* also show satellite behavior. That is, they walk and remain motionless near calling males but do not call. Satellite or similar behavior is common in many insects and other animals, and silent males intercept females attracted by the other male's song without the energetic costs of calling. The tendency of individual males to call or to perform satellite behavior varies within species. In *G. integer* some males call very little, whereas others show both calling and satellite behav-

FIGURE 4 Tree crickets use a leaf as a reflector (drawing by E. Salazar-Cade).

FIGURE 5 Male *Gryllus bimaculatus* fight (photographs by G. Tachon).

ior. The tendency for individuals to perform calling or satellite behavior depends on several variables, and population density is especially important. At high population density silent males are just as likely to find females and mate as males that call frequently. Satellite behavior is common at high densities. At lower densities calling behavior is the most successful reproductive tactic and satellite behavior is less common.

V. COURTSHIP

Once in close proximity male and female insects show very intricate patterns of behavior that, if successful, lead to copulation and the transfer of sperm. Male fruit flies, for example, extend and vibrate their wing that is closest to the female. Movement of the wing causes a sound while the male walks around the female. Cockroach males use both wings and raise them in front of a potential mate. The fanning of the wings may also help disperse mating pheromones which cockroaches release at close range. Grasshoppers are probably one of the most complex insects in terms of close-range courtship. Here the males of various species move antennae, mouthparts, legs, and abdomen in a concert of visual displays.

In crickets, close-range courtship behavior involves another specialized song. The courtship song is less species specific and is usually a soft shuffling-like sound. Male *Gryllus* dance and sing in front of females. Dancing males rock back and forth and shake their bodies. This courtship behavior can last for several minutes and functions to increase female receptivity.

Females may further evaluate males by their courtship display. The song and some elements of the male dance appear to be especially important if a mating is to take place. In a set of experiments in house crickets, *Acheta domesticus*, males that had their wings removed danced in front of females and moved the remainder of their wings. There was, of course, no song and no matings took place. Taped courtship song was then broadcast from loudspeakers, and female *A. domesticus* then mated with the silenced males.

VI. COPULATION, SPERM TRANSFER, AND SPERM COMPETITION

Insects assume various positions when copulating. In cockroaches, water bugs, and many groups the female climbs on top of the male such that their genitalia at the tips of their abdomens meet. In some cases the male is in the "superior position" as in lovebugs, in which the male climbs on the back of the female. In some insects sperm swim directly from the male into the female, and in others the males produce a capsule of varying sizes, termed a spermatophore, which contains many sperm cells. The spermatophore is attached to the female and sperm migrate into the female. An unusual variation on the use of a spermatophore is the spermatostyle in whirligig beetles. Here males produce a rod-like structure, the spermatostyle, and many sperm are embedded head first into the rod. The entire spermatostyle is then transferred to the female beetle.

Many insect females mate repeatedly, often with different males, and in some species they store enough sperm from a single mating to fertilize most or all of their eggs. Sperm storage and multiple mating lead to sperm competition or the competition between the sperm of different males within the female for access to ova. In many species of insects, sperm competition results in the last male to mate with a female fertilizing most of her eggs. This pattern of second male precedence may result from some ordering of sperm within the female's spermatheca. That is, the last sperm deposited in the female are the first to be used in fertilization. Sperm competition results in several adaptations by males, including mate guarding and other mechanisms which ensure that sufficiently large amounts of sperm are delivered to females. Hangingflies and scorpionflies are especially interesting with respect to the amount of sperm transferred. In these species males hunt for prey, usually smaller insects. Once males have their prey item they produce a pheromone that attracts females. Males present the prey to mates while mating, a form of nuptial feeding. Large prey items take longer for females to eat and, as a consequence, they remain coupled with the male for a longer time. The longer the copulation, the more sperm the male transfers.

In field crickets a mating lasts only 10–20 sec. A female *Gryllus* climbs on the back of a male, and the male thrusts the tip of his abdomen toward her abdomen and then produces a spermatophore. Using special genitalia, the male guides the neck of the spermatophore into the opening of the female's reproductive tract, the bursa copulatrix, and sperm enter the female.

The time that the spermatophore remains on the female correlates with increased numbers of sperm deposited into the female. In *Gryllus* it takes 30 min to 1 hr for the contents of the spermatophore to empty. Although it is in a male's interest for the spermatophore to remain attached for some time, females often remove and eat the spermatophore after only a few minutes. Females also rub their abdomens against the substrate and remove spermatophores. It has been suggested that premature removal of the spermatophore is one way that females can exercise mate choice by controlling the amount of sperm entering their bodies from individual males. Some crickets have evolved a form of nuptial feeding in which males have special glands and females feed

FIGURE 6 Female Mormon cricket feeds on attached spermatophylax (photograph by D. T. Gwynne).

on the secretions from these glands while the sperm migrate from the spermatophore. Mormon crickets, *Arabrus simplex*, produce a large spermatophore termed a spermatophylax. Here the ampul containing sperm is accompanied by a large mass of proteins and other nutrients that a female feeds on while the sperm enter her body (Fig. 6). The spermatophylax can represent up to 25% of the male's weight.

In addition to remaining with a recent mate, male crickets actively guard females and prevent other males from mating. Mate guarding behavior also keeps the female nearby and leads to additional matings between the same pair.

VII. THE COSTS OF MATING

The benefit of mating behavior is obvious: It leads to the production of offspring. Mating behavior is also costly in terms of the time and energy involved in mate attraction and location, fighting, and other activities. Another cost of mating in some insects is an increased likelihood of encountering predators or parasites. In hangingflies and scorpionflies, for example, males hunting for nuptial prey have an increased likelihood of flying into spiderwebs. The visual, chemical, and acoustical displays used to attract mates should also make the signaler more obvious to other species. A special case of predators using the mating signal of their prey occurs in firefly beetles. Male fireflies in the genus *Photinus* flash for females as they fly above an area. Female fireflies in another genus, *Photuris*, flash back to the male as if they were potential mates. When the male lands near the female she eats him.

Bats, lizards, and domestic cats have all been shown to orient to the calling songs of crickets and their relatives. A special type of relationship exists between fly parasites in the tribe ormini and different species of field crickets and bush crickets around the world. In *Gryllus* and their relatives, female ormine flies hear male calling songs and orient to the source of the signal. Flies deposit living larvae which burrow into and consume the cricket. After 7 days fly larvae pupate and the cricket always dies (Fig. 7).

Ormines are known to attack singing crickets and katydids or bush crickets in North America, the Hawaiian Islands, Australia, and Greece. The effect of flies on host behavior is different depending on the species. In Texas populations of *Gryllus* acoustically orienting flies do not attack satellite male crickets as often as they do calling males. Flies have probably selected for reduced levels of calling in this cricket species. In the Hawaiian Islands, by contrast, flies may have selected for fine differences in the structure of song such that males singing particular songs are not parasitized as often. In Greece, flies attack several species of bush crickets but are not as successful at parasitizing species in which males call very little.

There is little information on the female fly phonotaxis to the fine structure of cricket song, but some data suggest that flies have the same song preferences for male cricket song as do female crickets. If so, then male crickets in a population with songs that are most attractive to females are also the most likely to be attacked by flies. Mating behavior in crickets and other insects should generally reflect a trade-off between the costs and benefits.

See Also the Following Article

INSECT REPRODUCTION, OVERVIEW

Bibliography

Alexander, R. D. (1961). Aggressiveness, territoriality and sexual behavior in field crickets (Orthoptera: Gryllidae). *Behaviour* **17**, 130–223.

Alexander, R. D. (1975). Natural selection and specialized chorusing behavior in acoustical insects. In *Insects Science and Society* (D. Pimental, Ed.), pp. 35–77. Academic Press, New York.

Bailey, W. J. (1991). *Acoustic Behavior of Insects.* Chapman & Hall, New York.

Cade, W. H., and Cade, E. S. (192). Male mating success, calling and searching behaviour at high and low densities in the field cricket, *Gryllus integer. Anim. Behav.* **43**, 49–56.

Campbell, D. J., and Shipp, E. (1979). Regulation of spatial pattern in populations of the field cricket *Teleogryllus commodus* (Walker). *Z. Tierpsychol.* **51**, 260–268.

Doherty, J. A., and Storz, M. M. (1992). Calling song and selective phonotaxis in the field crickets *Gryllus firmus* and *G. pennsylvanicus* (Orthoptera: Gryllidae). *J. Insect Behav.* **5**, 555–569.

French, B. W., and Cade, W. H. (1987). The timing of mating, calling, and locomotion in the field crickets, *Gryllus veletis, G. pennsylvanicus,* and *G. integer. Behav. Ecol. Sociobiol.* **21**, 157–162.

Matthews, R. W., and Matthews, J. R. (1978). *Insect Behavior.* Wiley, New York.

Simmons, L. W., and Zuk, M. (1992). Variability in call structure and pairing success of male field crickets *Gryllus bimaculatus*: The effect of age, size and parasite load. *Anim. Behav.* **44**, 1145–1152.

Thornhill, R., and Alcock, J. (1983). *The Evolution of Insect Mating Systems.* Harvard Univ. Press, Cambridge, MA.

Ulagaraj, S. M., and Walker, T. J. (1973). Phonotaxis of crickets in flight: Attraction of male and female crickets to male calling songs. *Science* **182**, 1278–1279.

Walker, T. J. (1983). Diel patterns of calling in nocturnal Orthoptera. In *Orthopteran Mating Systems: Sexual Competition in a Diverse Group of Insects* (D. T. Gwynne and G. K. Morris, Eds.), pp. 45–72. Westview, Boulder, CO.

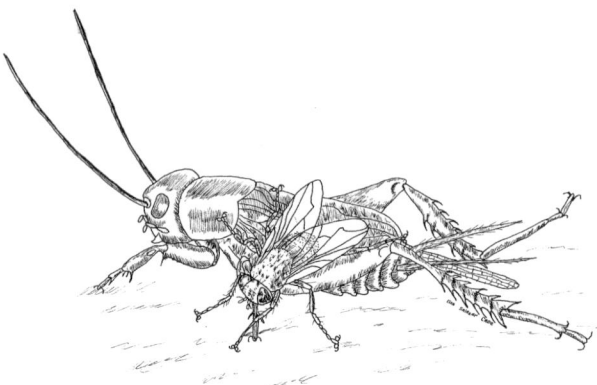

FIGURE 7 Female flies acoustically orient to cricket songs and parasitize males (drawing by E. Salazar-Cade).

Mating Behaviors, Mammals

Michael J. Baum

Boston University

I. Neuroendocrine Control of Feminine Sexual Behavior
II. Neuroendocrine Control of Masculine Sexual Behavior
III. Sexual Differentiation of Mating Behavior

GLOSSARY

aromatase A microsomal enzyme that converts androgenic precursors to estrogens.

mounting behavior A component of masculine sexual behavior which brings the male's body into contact with the female in an orientation that facilitates penile intromission into the female's vagina.

proceptive behavior The suite of behaviors which females display in order to solicit sexual behavior from male conspecifics. (A term coined by Frank Beach, the father of behavioral endocrinology.)

receptive behavior A component of feminine sexual behavior, usually involving immobility and orientation of the perineal region towards a male conspecific in a posture which facilitates penile intromission into the female's vagina.

steroid receptors A superfamily of transcription factors whose activity depends on the binding of specific hormonal ligands, such as estradiol, progesterone, or an androgen (i.e., testosterone or 5α-dihydrotestosterone).

Mating behavior in mammalian males and females is composed of species- and sex-specific sequences of motivated behavior. Males typically pursue female conspecifics for the purpose of mounting, intromitting the penis, ejaculating, and passing the sperm which will fertilize ovulated gametes. Females concurrently display patterns of mating behavior which include approaching a male conspecific, soliciting mounting, and displaying receptive behaviors to males' mounts which facilitate penile intromission and the ejaculation of sperm. Among mammalian species, these respective patterns of masculine and feminine sexual behavior are subject to variable degrees of neuroendocrine and social and environmental control. Mating behavior has evolved in animals to facilitate sexual reproduction; however, males and females of several primate species, including man, engage in this behavior at times when reproduction per se is not possible (e.g., at infertile times during the female's ovarian cycle). Engaging in sexual behavior at such times may reflect the intrinsic rewarding nature of mating to individuals. Such behavior may also facilitate social bonding or stabilize familial relationships.

I. NEUROENDOCRINE CONTROL OF FEMININE SEXUAL BEHAVIOR

Mating behavior has evolved in females as a means of obtaining sperm, ejaculated by the male into the female's vagina, which are needed to fertilize recently ovulated ova. In some species, females' mating behavior also facilitates the ability of males to deliver vaginal–cervical stimulation (via penile intromission) which is required by females in order to induce (i) a surge in the secretion of luteinizing hormone (LH) by the pituitary gland needed for ovulation and/or (ii) daily peaks in the secretion of prolactin by the pituitary gland and the resultant production of progesterone by the ovarian corpora lutea needed to initiate and sustain pregnancy. In the vast majority of mammalian species (with the exception of some primates, including man) feminine sexual behavior is exhibited only at very discrete times during the female's ovarian cycle, so as to coincide with ovulation. The expression of feminine sexual behavior during these periods of estrus (which last from 12 to 72 hr) depends critically on the action in the central nervous system of the steroid hormone, estradiol, which is secreted by the ovaries prior to ovulation. In certain species (e.g., the ferret, rabbit, and cat) in which ovulation is induced by the receipt of intromissive stimulation from a male conspecific, estradiol alone suffices to facilitate proceptive and receptive components of females' sexual behavior. In other species (e.g., rat, guinea pig, and hamster) the preovulatory secretion of LH and resulting ovulation occur "spontaneously" in response to the neural actions of estradiol. In these species feminine sexual behavior is activated by the sequential actions of estradiol and progesterone. In rodents, the female's receptive response, shown after priming with estradiol and progesterone, includes an immobile posture and a reflexive, concave arching of the back (lordosis), coupled with an elevation of the perineum which facilitates penile intromission by the male. In other species the estrous female's receptive posture ranges from a passive acceptance of a male's neck grip (e.g., ferret) to a rigid, immobile stance which is assumed whenever a male is in close proximity (e.g., pig).

Studies carried out using female rats, guinea pigs, and hamsters have shown that the ovarian steroids, estradiol and progesterone, facilitate females' display of sexual behavior (shown in response to the presence of a male) via genomic actions, mediated by specific intracellular receptor proteins, in neurons located in the ventromedial nucleus of the hypothalamus. Estradiol, secreted in increasing amounts by the ovaries during the preovulatory phase of the estrous cycle, acts on estradiol receptors located in ventromedial hypothalamic neurons to promote the synthesis of progestin receptors. Progesterone, secreted by the ovaries, then binds to these progestin receptors and together this steroid–receptor complex promotes further neuronal gene transcription which leads to the expression of feminine sexual behavior. A wide variety of experimental methods have provided incontravertible evidence of the obligatory role of estradiol and progesterone in the control of feminine sexual behavior among female rodents. These include the experimental control of lordosis behavior following ovariectomy and administration of estradiol and progesterone (either systemically or intracerebrally), as well as the systematic manipulation of sexual behavior in female rodents by administration of antiestrogenic and antiprogestational drugs or by intracerebral administration of antisense oligonucleotides which hybridize with estradiol or progestin receptor mRNA, thereby preventing the synthesis of these receptor respective receptor proteins. Recently, several groups have shown that the display of receptive sexual behavior in response to exogenous ovarian steroids is disrupted in transgenic female mice which lack the capacity to express either estradiol or progestin receptor genes. Considerably less is known about the gene products (other than progestin receptor) which mediate the behavioral actions of estradiol and progesterone in the female. Candidates include receptors for several different neuropeptides and/or neurotransmitters including oxytocin, GABA, serotonin, and acetylcholine.

The requirement of estradiol (in some species working synergistically with progesterone) for the activation of females' sexual behavior is absolute in all infraprimate species. In various primate species which have been systematically studied, including macaque, chimpanzee, and human, the expression of feminine sexual behavior, especially the proceptive or appetitive aspects of this behavior, varies over the stages of the ovarian cycle. Females typically are maximally proceptive (i.e., they most frequently initiate sexual activity with a male conspecific) around the time of ovulation. Although social relationships can strongly modulate the ability of ovarian steroids to activate sexual behavior in females of several infraprimate species, such effects are most dramatically shown in social primates. Thus, dominant females in established macaque groups will have preferential sexual access to males in the group. Under such conditions estradiol has been shown to facilitate females' proceptive sexual behavior, with the effects being most evident in more subordinate individuals within the group hierarchy. Progesterone acts in female primates to reduce their attractiveness to males, thereby reducing the incidence of mating. In contrast to infraprimate species, however, none of these effects of ovarian steroids on females' sexual behavior are absolute. Thus, after ovariectomy and in the absence of hormone replacement, female stump-tailed macaques and women readily display all aspects of feminine sexual behavior with a male conspecific, especially in the context of established heterosexual relationships. There are numerous clinical reports that postmenopausal women, or women who have been ovohysterectomized, show increased levels of sexual behavior in response to treatment with testosterone. The available evidence suggests, however, that endogenous androgens of ovarian or adrenal origin, at best, play only a minor role in the activation of sexual behavior in female primates, including women.

II. NEUROENDOCRINE CONTROL OF MASCULINE SEXUAL BEHAVIOR

Males of different mammalian species show a variable sequence of pursuit and mounting patterns which culminate in penile intromission and the ejaculation of sperm. In each instance the male is attracted to the female as a result of behavioral (proceptive), chemosensory, and/or auditory cues emanating from the female. The male rat (and rhesus monkey) displays a series of mounts with pelvic thrusting and discrete penile intromissions which (after 5–15 such intromissions) culminate in an intromission with an accompanying ejaculation. In other species (hamster, mouse, and man) the male exhibits a single mount with intromission and prolonged pelvic thrusting leading to an ejaculation. In all species there is a postejaculatory period during which the male's interest in the female is diminished. There is considerable species variation in the duration of this postejaculatory interval, ranging from a few minutes to several days. The particular pattern of mounting (multiple, spaced mounts and intromission as opposed to a single mount/intromission with intravaginal thrusting) displayed by males of different species has evolved to maximize the type of vaginal–cervical stimulation needed by female conspecifics to induce a preovulatory LH surge and/or to augment prolactin secretion needed to establish luteal function. Finally, the receipt of vaginal–cervical stimulation by estrous female rats has also been shown to facilitate sperm transport into the uterus; the occurrence of orgasm during mating in women reportedly also facilitates sperm transport.

One of the first endocrine experiments was the demonstration by Berthold in 1837 that castration diminished the sexual behavior of roosters. When synthetic sex steroids became available in the 1930s it was shown that castration abolished mating behavior in the rat and that subcutaneous injections of testosterone restored this behavior. Testosterone is normally secreted by the Leydig cells of the testes, and this steroid has now been implicated in the regulation of sexual arousal and masculine sexual behavior in species representing all mammalian orders, including primates such as rhesus monkey and man. Studies suggest that testosterone activates sexual behavior as a result of its intracellular actions at several sites in the forebrain and spinal cord as well as in the penis. Some of these actions are mediated via intracellular androgen receptors, which bind either testosterone itself or a potent, androgenic metabolite, 5α-dihydrotestosterone (DHT). Androgens likely act in the spinal cord and penis to augment erectile function, although several studies with rats suggest that androgens may also contribute to masculine sexual arousal and mating performance by acting in forebrain sites including the medial amygdala and the lateral septal region. In addition to these androgen receptor-mediated actions, circulating testosterone also promotes sexual behavior in the male after its conversion to estradiol by a brain enzyme, aromatase. Aromatase is found in several forebrain regions, including the medial preoptic/anterior hypothalamic area (mPOA/AH) and the medial amygdala. Estradiol receptors are also present in these regions in the male (as in the female), and several studies suggest that they mediate the behavioral actions of estradiol, formed in adult male mammals via local aromatization of circulating testosterone. Supportive evidence includes the demonstration in castrated male rats that combined treatment with estradiol and DHT restores mating behavior as readily as testosterone. In addition, in different studies localized implantation of either testosterone or estradiol into the mPOA/AH activated mating in long-term castrate male rats. The role of estradiol in the activation of behavior has been established by several studies showing that intra-mPOA/AH infusions of drugs which block aromatase activity (thereby inhibiting estradiol synthesis) also inhibited the ability of systemic testosterone treatment to restore mating behavior in castrated males. This drug effect was reversed by concurrent administration of estradiol. However, estradiol given by itself to male castrates only partly restored masculine sexual behavior, suggesting that in gonadally intact rats both androgen (either testosterone or DHT) and estradiol, formed from circulating testosterone, are needed for the activation of masculine sexual behavior. Evidence of estrogen's involvement in the activation of sexual behavior in males exists for a wide variety of mammalian orders, including carnivores (ferrets), ungulates (red deer), and a nonhuman primate (cynomolgus monkey).

Numerous studies, in addition to the previously mentioned experiments in which steroids were manipulated, have implicated mPOA/AH neurons in the control of masculine sexual behavior and in the male-typical profile of heterosexual sexual partner preference. Bilateral destruction of the mPOA/AH has been shown to reduce masculine mating performance in tests with estrous female conspecifics of a wide variety of vertebrate species. In rat and ferret such lesions also caused males to display a female-typical profile of sexual partner preference (lesioned males, like females, preferred to approach and interact sexually with sexually active males). Neurons in the mPOA/AH express estradiol and androgen receptor. They also receive inputs via polysynaptic pathways which convey reproductively relevant chemosensory information from the primary (main olfactory epithelium) and accessory (vomeronasal organ) olfactory bulbs. Although the details vary considerably among males of different mammalian species, much evidence suggests that chemosensory cues (vaginal secretions, urine, and skin gland secretions) derived from estrous females and steroidal signals (testosterone acting via androgen receptors and/or estradiol acting via estradiol receptors) converge to activate mPOA/AH neurons, thereby promoting male-typical partner selection and masculine sexual behavior.

Several clinical experiments point to an activational role of testosterone in promoting sexual motivation among men, whereas erectile function in man seems less dependent on the action of androgen than it does in animals such as the male rat. Supportive data include the observation that prepubertal boys and adult, hypogonadal men are not motivated to establish romantic (either hetero- or homosexual) relationships, although penile erections do occur routinely in such individuals. Administration of testosterone to hypogonadal men stimulated heterosexual interest and (when sexual partners were available) increased the incidence of sexual intercourse. Further evidence of the activational role of testosterone in the human male is derived from studies showing that the administration of an LHRH antagonist (which blocks pituitary LH and thus testicular testosterone secretion) caused significant reductions in heterosexual interest and the incidence of sexual intercourse in normal, adult men. This reduction in masculine sexual interest and behavior was reversed by concurrent treatment with testosterone. Inhibition of peripheral (nonneural) aromatase activity in such men failed to affect their sexual arousal or mating performance, suggesting that circulating estradiol is not essential for the activation of sexual behavior in man. This result does not, however, rule out a possible role of neural aromatization of testosterone in the regulation of sexual arousal and mating in men such as the one known to exist in nonhuman mammals.

III. SEXUAL DIFFERENTIATION OF MATING BEHAVIOR

Experiments on species representing essentially all mammalian orders have shown that sex differences in sociosexual partner preference and in mating behavior per se result from the perinatal action of testosterone and/or of estrogenic metabolites of testosterone in the developing male's nervous system. Supportive evidence includes the demonstration that administering testosterone to females during a critical, species-specific perinatal period organizes male-typical mating capacity and a preference to seek out and attempt to mount normal, estrous females in adult tests given while subjects were again treated with testosterone. Perinatal administration of testosterone to females also reduces their later capacity to display female-typical patterns of mating behavior. Evidence that perinatal exposure of males to endogenous testosterone, secreted perinatally by the testes, plays a critical role in masculine psychosexual development is provided by a variety of studies. In species such as rat, hamster, mouse, and ferret, in which some of the differentiating actions of testosterone occur neonatally, studies have shown that castration shortly after birth attenuated subjects' later ability to display masculine sexual behavior and (in some species) enhanced subjects' lordotic (feminine) responses to ovarian hormones. A variety of experimental methods, including the perinatal administration of estrogen to female subjects as well as the perinatal administration of aromatase-inhibiting drugs to male subjects, have implicated estrogens, formed perinatally via neural aromatization of circulating testosterone, in both the masculinization and the defeminization of mating capacity in male mammals. Intracerebral administration of antisense oligonucleotides to estradiol receptor mRNA attenuated the defeminizing effect of neonatally administered testosterone on female rats' sexual behavior, and male mice with a null mutation of the estradiol receptor gene showed less masculine sexual behavior than wild-type controls. These findings corroborate earlier studies which implicated estrogens in the differentiation of male-typical patterns of mating behavior.

See Also the Following Articles

NEUROENDOCRINE SYSTEMS; PROGESTERONE ACTIONS ON BEHAVIOR; TESTOSTERONE BIOSYNTHESIS

Bibliography

McCarthy, M. M., and Albrecht, E. D. (1996). Steroid regulation of sexual behavior. *Trends Endocrinol. Metab.* 7, 324–327.

Meisel, R. L., and Sachs, B. D. (1994). Male sexual behavior. In *The Physiology of Reproduction* (E. Knobil and J. D. Neill, Eds.), 2nd ed., Vol. 2, pp. 3–106. Raven Press, New York.

Pfaff, D. W., Schwartz-Giblin, S., McCarthy, M. M., and Kow, L.-M. (1994). Cellular mechanisms of female reproductive behaviors. In *The Physiology of Reproduction* (E. Knobil and J. D. Neill, Eds.), 2nd ed., Vol. 2, pp. 107–220. Raven Press, New York.

Robbins, A. (1996). Androgens and male sexual behavior; from mice to men. *Trends Endocrinol. Metab.* 9, 345–350.

Special issue (1997). Single gene mutations, gene knockouts, and behavioral neuroendocrinology. *Horm. Behav.* 31, 186–255.

Vagell, M. E., and McGinnis, M. Y. (1997). The role of aromatization in the restoration of male rat reproductive behavior. *J. Neuroendocrinol.* 9, 415–421.

Mating Behaviors, Invertebrates Other Than Insects

Janet L. Leonard

Oregon State University and University of California, Santa Cruz

I. Introduction
II. Mating Styles
III. Exotic Sexual Phenomena

GLOSSARY

anisogamy vs isogamy In anisogamy the two haploid gametes that unite in fertilization are of very unequal size (egg and sperm); in isogamy they are of equal size (isogametes).

cryptic female choice Occurs when a female's reproductive tract is able to accept or reject (or bias) the use of sperm from one male vs another.

evolutionary arms race A process whereby two sets of elaborate adaptations evolve as counters to each other, in a moving stalemate.

genitalia Here used to mean all reproductive structures other than the gonads and structures involved in care of zygotes.

hypodermic insemination Introduction of sperm into the body of the female not through a specialized reproductive opening but by forcing their way through the body wall, either mechanically or chemically. In some species they are met by specialized cellular adaptations of the female.

indirect sperm transfer A case in which sperm (or spermatophores) do not move directly from the male's body into the body of the female but come in contact with a third structure, usually either the substrate or an appendage of the male's body; the rule in arachnids.

internal vs external fertilization Internal fertilization occurs within the body of the female; in external fertilization the eggs are outside the body of the female, although perhaps still attached to or sheltered by the female's body. The distinction is sometimes unclear in invertebrates due to our ignorance and/or strange and wonderful invertebrate adaptations.

mating system The species-typical pattern of reproductive behavior and parental care.

polymorphism (i) Sperm polymorphism: in some species males produce more than one type of sperm, the functions of which are often unclear; (ii) sexual polymorphism, most often in males, where different body forms are associated with different mating strategies.

role reversal Males are normally considered to compete for mates, whereas females are choosy about the male with whom they mate; in role reversal the opposite occurs to a greater or lesser extent.

secondary sexual characteristics Darwin used this term to refer to sexually dimorphic traits that appeared in mature individuals and did not function directly in improving fertilization efficiency but rather served to attract mates or repel rivals. Eberhard extended the term to include all genitalia.

sex allocation Charnov's term for the process (and his theory) of partitioning available resources between reproduction through eggs vs sperm; usually refers to hermaphrodites.

sexual conflict A situation that occurs when the members of a mating pair have incongruent interests due to differing selective pressures; usually involves a male wanting to mate with a reluctant female.

sexual selection A distinction is made between sexual selection *in sensu latu* and *in sensu strictu*: The former refers to selection acting on sexual performance and mating success; the latter is how Darwin used the term—to refer to an effect of mating success that runs counter to natural selection.

sperm competition Competition between sperm or ejaculates from more than one male to fertilize the eggs of a single female.

Mating is the series of events by which the sperm of one individual and the eggs of another are brought into proximity so that fertilization may occur. The invertebrate animals have evolved a wealth of ways of accomplishing this task. Asexual reproduction, parthenogenesis, and self-fertilization occur in many invertebrate groups but will not be considered here. Adaptations for mating may include complex and elaborate behaviors, specialized sensory systems, pheromones, developmental processes, biochemical signaling, physiological manipulation, and elaborate genitalia. The complexity of mating adaptations is thought to be to a large extent the product of an evolutionary arms race between the sexes. The following are the three best generalizations about invertebrate mating: (i) The mating systems are poorly understood; (ii) they prove that the complexity of mating behavior and mating systems is not dependent on taxon or behavioral sophistication; and (iii) these systems offer a wealth of exciting opportunities for testing current theory in sexual selection and evolutionary ecology and will be important in developing new theory.

I. INTRODUCTION

The multicellular invertebrates (excluding insects) represent the bulk of the diversity of the animal kingdom, not just taxonomically and morphologically but also in terms of modes of reproduction and mating systems. Among the invertebrates, one finds all possible variations on the theme of anisogamy. There are not just species but major taxa (phylum, class, etc.) that are simultaneous hermaphrodites, sequential hermaphrodites, and/or have separate sexes (Table 1). There is also a wide range of modes of fertilization and/or mating between and within taxa, varying from simple broadcast spawning in which gametes are shed into the medium, usually seawater, and meet there for fertilization to species with very prolonged and elaborate courtship and mating behavior, with or without internal fertilization. Among the invertebrates particular sexual phenomena have evolved independently in many taxa; i.e., brooding, internal fertilization, dwarf males, hypodermic insemination, haplodiploidy, sexual size dimorphism, and male–male combat (Table 1).

Modern biology views a species-typical mating system, not only as an adaptation to an environment but

TABLE 1
Taxonomic Survey of Metazoan Mating Behavior[a]

Porifera: Usually hermaphroditic but not simultaneously, with broadcast spawning of sperm ("smoking sponges") and internal fertilization; amebocytes transport sperm to egg. Spawning may or may not be synchronous in a population.

Placozoa: One species, *Trichoplax adhaerens*, little known about sexual reproduction.

Mesozoa: Poorly known parasitic forms; orthonectids have separate sexes with dwarf males and internal fertilization; dicyemids are hermaphroditic.

Cnidaria: A large and diverse phylum; both hermaphroditism and dioecy occur. Usually broadcast spawners with either internal or external fertilization; pair mating with transfer of a spermatophore described in cubozoan medusae.

Ctenophora: Mostly simultaneous hermaphrodites with broadcast spawning and external fertilization, some brood eggs.

Platyhelminthes: Mostly simultaneous hermaphrodites; a few with separate sexes, examples of dwarf males. Most have internal fertilization with copulation and often very elaborate, sometimes species-specific, genitalia and/or courtship behavior. Hypodermic insemination; sperm storage. Sexual selection probably important at least in turbellarians; little studied (see text).

Gnathostomulida: Largely simultaneous hermaphrodites; polymorphic sperm; mostly if not exclusively pair mating with internal fertilization.

Nemertea: Largely dioecious, some simultaneous or protandrous hermaphrodites; pair mating or group mating ("mating balls") with either external or internal fertilization. Simple genitalia, no copulation.

Nematoda: Famous for ameboid sperm; functionally dioecious with copulation; dwarf males occur; female pheromones attract males; sperm of hermaphrodites apparently used only for self-fertilization; genitalia may be complex and species specific.

Nematomorpha: Dioecious with pair mating and internal fertilization; some sexual dimorphism, males are more active; receptive females produce a pheromone to attract males.

Acanthocephala: Dioecious with copulation and some sexual dimorphism; mating plug produced by male; famous for homosexual rape in which males prevent rivals from mating by cementing genitalia with mating plug (see text).

Rotifera: Mostly females, males are produced occasionally and may be diploid or haploid (dwarf) depending on taxon; males unknown in some taxa. Fertilization occurs by copulation or hypodermic insemination.

Gastrotricha: Hermaphroditic when sexual; only parthenogenetic females known in some taxa; internal fertilization; mutual cross-fertilization; pair mating with copulation in some forms; hypodermic insemination described.

Kinorhyncha: Dioecious with internal fertilization; simple genitalia; males have spicules presumably used in copulation (never described); some forms deposit spermatophore on body of female.

Loricifera: Newly discovered and poorly known; dioecious with some sexual dimorphism; internal fertilization considered probable.

Tardigrada: Largely dioecious; males unknown in some taxa; dwarf males occur in others. Pair mating; fertilization largely internal through copulation or hypodermic insemination. External fertilization occurs when, after courtship by male, females deposit eggs on substrate where they are covered by sperm, or sperm are deposited under the cuticle of the female fertilize eggs as they are deposited into the shed cuticle during moulting.

Priapula: Dioecious, typically broadcast gametes with external fertilization; males unknown in one species.

Mollusca: Large phylum with tremendous reproductive diversity.

    Class Monoplacophora: Dioecious broadcast spawners with external fertilization.

    Class Polyplacophora: Mostly dioecious, broadcast spawners, external fertilization usual; brooding of eggs in mantle cavity or internal fertilization are found.

    Class Aplacophora: Dioecious or hermaphroditic broadcast spawners with external fertilization.

    Class Bivalvia: Dioecious or hermaphroditic; usually broadcast spawners with external fertilization; dwarf and/or parasitic males found attached to female in some species, often two males per female; sexual dimorphism also found where eggs/larvae are brooded. Hermaphrodites may be simultaneous, sequential (usually protandric but sometimes protogynous), or "rhythmical" in which individuals repeatedly change sex over their lifetime, i.e., *Ostrea* oysters produce sperm while brooding larvae but change to produce only eggs after larvae are discharged, then again produce sperm while brooding the new clutch. "Erratic" or irregular sex change is known in *Crassostrea* and a few other oysters. Many populations contain individuals

continued

Continued

that deviate from the usual sexual pattern, i.e., "pure" males or females in hermaphroditic taxa or the reverse. Fascinating material for study of sex allocation.

Class Scaphopoda: Dioecious broadcast spawners with internal fertilization; eggs released either singly or in a cloud.

Class Gastropoda: A large and diverse group in its own right.

    Subclass Prosobranchia: Typically dioecious with copulation but includes forms that have external fertilization with broadcast gametes; sequential hermaphrodites are not uncommon; a few simultaneous hermaphrodites; very elaborate genitalia found in many groups; elaborate courtship behavior with male–male combat and apparently female choice described for some species; pairing may occur even in broadcast spawners; spermatophores, sperm storage; digestion of unwanted sperm; female pheromones; polymorphic sperm. Sexual selection probably important but little studied.

    Subclass Opisthobranchia: Probably all functionally simultaneous hermaphrodites; gametes often mature protandrously; internal fertilization with sperm storage and digestion of excess sperm in many but not all taxa; genitalia have taxonomic value in some groups; pair mating usual, either simultaneously or sequentially reciprocal insemination found in some groups; group sex, often involving chain copulation (see text) is not uncommon; female pheromones and location of mates by trail following described; monogamy described; reciprocal or unilateral spermatophore transfer found in aphallic forms; precopulatory courtship common and may be elaborate; polymorphic sperm. Sexual selection probably important but as yet little studied (see text).

    Subclass Pulmonata: All simultaneous hermaphrodites with internal fertilization; genitalia, particularly penis, often species specific in morphology; genitalia always complex; genitalic polymorphism in some species; aphallic forms may have eversible vagina that picks up spermatophore from partner; penial stylets and/or darts in some taxa; other morphological specializations for courtship found; sperm storage; digestion of unwanted sperm; sperm competition and evidence of cryptic female choice in some species; multiple paternity of egg clutches demonstrated in a few species, probably very common; spermatophores, often species specific, in some groups; aphallic taxa; courtship fairly to very elaborate; copulation and/or spermatophore transfer may be simultaneously reciprocal, sequentially reciprocal, or unilateral, or groups may form copulatory chains (see text). In terrestrial forms genitalia regress during nonbreeding season or under food stress; consistent with secondary sexual characteristic. Sexual selection probably very important; only a few species studied.

Class Cephalopoda: All dioecious, almost all sexually dimorphic, dwarf males in some taxa; pair mating; copulation; taxonomically significant spermatophores; a specialized arm (hectocotylus) holding spermatophores is detached and left in the mantle of the female in some groups; hypodermic insemination from spermatophores attached to the skin in some groups; sperm storage; many die after spawning and/or period of egg guarding; courtship well developed, often very elaborate; visual stimuli important, in some species mature females develop sex-specific photophores; males aggressive and initiate encounter; male–male combat described in several; consortships established after female choice; several ethological studies; little attention to sexual selection *per se*.

Kamptozoa (Entoprocta): Mostly if not exclusively hermaphroditic, may be protandric; broadcast spawning with internal fertilization.

Pogonophora: Dioecious; sperm in spermatophores, apparently released singly into water; fertilization apparently takes place on body or within tube of female, not known whether internal or external; poorly known.

Sipuncula: Dioecious except for one species; broadcast spawning with external fertilization; males spawn first; females spawn in response to the presence of sperm.

Echiura: Dioecious, broadcast spawners, usually epidemic, with external fertilization; in one family, Bonellidae, world record sexual dimorphism; dwarf males 1 or 2 mm long live on proboscis of female up to 2 m long.

Annelida:

    Class Polychaeta: Mostly dioecious, some sequential (either protogynous or protandric) and simultaneous hermaphrodites; simple reproductive structures, lack even permanent gonads; largely broadcast spawners with external fertilization, some internal fertilization with brooding; some taxa form specialized reproductive individual (or transform body form) which swims to surface in response to an environmental cue for mass broadcast spawning (epitoky). Sex allocation and sexual selection, including examples of monogamy and egg trading, have been studied in some sequential and simultaneous hermaphrodites.

continued

Continued

Class Clitellata: Simultaneous hermaphrodites with internal fertilization, a specialized structure (clitellum) forms over reproductive segments on mature worms; relatively elaborate genitalia and sexual behavior. Sexual selection probably important but very little studied.

Subclass Oligochaeta: Simultaneous hermaphrodites, relatively complex genitalia with permanent gonads; reciprocal mating with some preliminary courtship occurs; either internal fertilization with copulation or sperm transport along an exterior ciliated groove within a mucous sheath formed by the clitellum, to the sperm storage organs of the sperm recipient.

Subclass Hirudinida: Simultaneous hermaphrodites with internal fertilization and complex genitalia, often species-specific copulatory apparatus; usually (not always) reciprocal mating, either copulation or hypodermic insemination with spermatophore.

Arthropoda: The largest and most complex phylum (or superphylum); largely dioecious.

Subphylum Crustacea: Most diverse group of arthropods sexually.

Class Remipedia: Poorly known simultaneous hermaphrodites; apparently pair mating with spermatophore transfer; no penis; copulation not observed.

Class Cephalocarida: A small group of poorly known hermaphrodites.

Class Branchiopoda: Dioecious; females usually more common than males; males unknown in some species; dwarf males in some groups; pair mating, penis found in some groups; copulation or transfer of sperm to external uterine chamber; sexual selection by scramble-competition polygyny studied in anostracans (fairy shrimp).

Class Malacostraca: A large and relatively well-studied group, includes stomatopods, crabs, lobsters, shrimps, amphipods, and isopods; Darwin cited examples of sexual dimorphism and secondary sexual characteristics and predicted male–male combat over females; some hermaphroditism; mostly dioecious with pair mating and copulation; external fertilization via a spermatophore in some groups; genitalia complex and often species specific; egg laying and/or insemination often associated with molting of the cuticle; mate guarding by males and male–male competition over large or about-to-molt females described from several groups; very sophisticated behavior and sensory capabilities involved in courtship and mating; wide range of mating systems described; monogamy; polymorphisms including harem polygyny with male polymorphism; resource defense polygyny in fiddler crabs (see text); sperm competition and/or cryptic female choice. Sexual selection has been studied in many groups (see text) and is probably much more widespread.

Class Maxillopoda: A diverse and taxonomically controversial group; here including copepods, ostracods, and cirripedes (barnacles) along with less familiar forms; includes both dioecious and simultaneously hermaphroditic forms; dwarf or complemental males among the barnacles, males unknown in some ostracods, smaller than females in copepods; largely pair mating with copulation in most groups; transfer of spermatophores occurs in copepods; in barnacles one sees pseudocopulation where sperm are deposited in the mantle cavity of the female-acting hermaphrodite, often by more than one adjacent hermaphrodite; barnacle mating systems have been reviewed from the standpoint of sex allocation and sexual selection by Charnov. Morin and Cohen have reviewed ostracod courtship and mating behavior, and Blades-Eckelbarger has reviewed mating, spermatophores, and genitalia for some copepods (both in Bauer and Martin, 1991).

Subphylum Cheliceriforms: Dioecious

Class Chelicerata: Living members include horseshoe crabs which have been studied from the standpoint of sexual selection (see text); external fertilization with simple genitalia; male–male competition for mates causes mating chains; larger females preferred; external fertilization; sperm competition including unattached males.

Class Arachnida: Large, diverse, and comparatively well-studied group including spiders, scorpions, ticks, mites, and less familiar forms; mostly terrestrial; Darwin considered sexual selection likely in spiders at least, noted male stridulatory organs and predicted (correctly) that they would be used in courtship; sexual dimorphism not uncommon; copulation unusual; usual pattern involves indirect sperm transfer either by spermatophores picked up by female or by the use of specialized male appendages to transfer sperm from the male gonopore to the body of the female; pair mating usual but not found in some groups. Elaborate courtship and mating behavior well described in many groups, many species specific; female pheromones known from some groups; sexual cannibalism in some groups;

continued

polymorphisms in males, females, or both sexes known in scorpions but little understood; male–male combat over access to females known in some groups; male polymorphism and sneaky male strategies known in harvestmen; dwarf males, nuptial gifts known in spiders; species-specific genitalia and/or spermatophores; sperm competition and/or cryptic female choice studied in some taxa.

Class Pycnogonida: External fertilization with pair mating; simple genitalia; sexual dimorphism common; male parental care predicts role reversal in courtship; sexual selection should be important but little studied.

Subphylum Uniramia: Dioecious, largely terrestrial.

Superclass Myriapoda: Centipedes, millipedes, and some less familiar forms; dioecious; indirect sperm transfer, in one group involving female ingestion of spermatophores; pair mating usual, courtship often complex and species specific, female pheromones found in some; spermatophores in many groups, often species specific, empty spermatophore may serve as nuptial gift in some centipedes; species-specific genitalia and potential for cryptic female choice; role reversal involving a clasping organ found on females of a centipede known to Darwin.

Superclass Insecta[b]

Onychophora: Dioecious, terrestrial; sexual behavior poorly known; spermatophores carry sperm; copulation seldom observed; hypodermic insemination from spermatophore known; sperm storage by the female.

Chaetognatha: Simultaneous hermaphrodites with pair mating in a head-to-tail orientation and reciprocal spermatophore transfer; poorly known.

Phoronida: Both dioecious and simultaneously hermaphroditic species; broadcast spawners with either internal or external fertilization; fertilization is poorly understood but involves lysis of female body wall by sperm proper in one species.

Brachiopoda: Mostly dioecious with broadcast spawning and external fertilization; some species retain eggs and have internal fertilization followed by brooding.

Bryozoa (Ectoprocta): Mostly simultaneous hermaphrodites; so-called dioecious species usually (but not always) involve colonies in which individual zooids are either male or female but the colony as a whole is hermaphroditic; broadcast spawners with either external or internal fertilization (brooding).

Echinodermata: A largely dioecious group; simultaneous hermaphrodites known in some groups, often associated with brooding; broadcast spawners with external fertilization except for brooding forms; gonads lacking in crinoids; genitalia simple in all forms.

Hemichordata: Largely, if not exclusively, dioecious, broadcast spawners with simple genitalia, external fertilization; in enteropneusts females spawn mucoid egg masses followed by spawning of sperm by neighboring males; in the sessile colonial pterobranchs gametes are apparently released into the tube of the colony where fertilization occurs and the larvae are brooded.

Chordata: Largely dioecious; some hermaphroditic groups; diverse mating systems.

Subphylum Urochordata: Largely simultaneous hermaphrodites with simple reproductive systems; broadcast spawners often with external fertilization; internal fertilization associated with brooding in some forms; incompatibility mechanisms which prevent inbreeding known in some forms.

Subphylum Cephalochordata: Dioecious, serially arranged gonads in both sexes; broadcast spawners, usually at dusk; external fertilization.

Subphylum Vertebrata[b]

---

[a] Taxon listed is a phylum unless otherwise indicated.
[b] Mating behavior discussed in separate article(s).

---

also as a compromise between what natural selection favors for females and what it favors for males. That is, rather than view a mating system as operating for "the good of the species," one starts from the assumption that individuals, male and/or female, are acting selfishly to increase their own individual fitness. This creates a potential for sexual conflict and/or sexual selection. Darwin developed his theory of sexual selection to account for the evolution of traits that seemed detrimental to the survival of the individual that bore them but served to further mating success through either male–male competition or female choice of ornamented males. He considered that the evidence for sexual selection was found in the existence of dimorphism in secondary sexual characteristics and remarked on their absence even in the "higher" invertebrates, aside from some arthropods—even in those such as cephalopods and some snails with well-developed sensory systems and elaborate behavior. He did believe that there was evidence, based on morphological sexual dimorphisms, for sexual selection in many of the arthropods, including many insects and crustaceans, some spiders, and some centipedes. In fact, he predicted, in *Descent of Man*, from instances of sexual size dimorphism and morphological evidence of weapons and clasping organs that male–male combat for access to females would be found in crustaceans, as has been the case. Darwin very explicitly excluded the "organs of reproduction" from his definition of secondary sexual characteristics and expected that the various bizarre sexual phenomena of invertebrates such as dwarf males, of which he was well aware, would be due to differing natural selection pressures associated with differing habits of life.

Modern mating systems theory, on the other hand, has tended to view all very elaborate, expensive, and/or bizarre reproductive structures or behaviors as evidence for sexual conflict and/or sexual selection *in sensu latu* until proven otherwise. A distinction is made between Darwinian sexual selection or sexual selection *in sensu strictu*, which involves a force of selection explicitly and directly opposed to that of natural selection, and the more general idea of selection acting through the process of mating success—sexual selection *in sensu latu*. Recently, Eberhard argued persuasively that the genitalia and most of the reproductive system of both males and females should be regarded as the result of evolutionary arms races in which females attempt to choose the father of their offspring and males compete both to exclude each other and to circumvent the mechanisms of female choice. That is, both male–male competition and female choice continue within the female reproductive tract and only end with fertilization. In this view the only primary sexual characters are the gonads; all other reproductive characters which are involved in fertilization have evolved through sexual selection *in sensu latu* and/or sexual conflict. The logical extension of this is to extend the concept of sexual selection and sexual conflict to all sexual interactions, including those between plants and single-celled organisms. The concept has even been extended to explain pheromonal signaling between isogametes of different mating types. While the exact nature and degree of selective forces in any particular case are difficult to characterize, it is clear that sexual selection *in sensu strictu*, if it exists at all, is like natural selection, a law of nature that will crop up wherever sex exists. The view that reproductive adaptations should be viewed not as adaptations for the good of the species but rather as the result of selfish individuals attempting to increase their reproductive success dates back to the integration of genetics into evolutionary theory in the 1930s but was developed explicitly for the whole range of organisms with special attention to the invertebrates by Michael T. Ghiselin. His review of the sexual diversity of the living world in light of individual selection remains a fruitful source of research topics and an authoritative review of invertebrate sexuality.

II. MATING STYLES

A. The Broadcast Spawners

Perhaps the most common form of mating behavior among marine organisms, including many groups of invertebrates (Table 1), is to release gametes directly into the water column. This may involve release of both types of gametes with external fertilization in the water column or, as is characteristic of sponges, only sperm may be released into the water column which then make their way to the body of the female where the eggs have been retained. In sponges the sperm are ingested by the cells of the female and then transported to the eggs for fertilization. Where sperm are transported by the body of the female to the eggs, there is the potential for female choice. Therefore, brooding of eggs and/or larvae is expected to be associated with increased male–male competition. Although little studied, there are other variables of reproductive behavior in broadcast spawners which may be subject to sexual selection. These include the chemoattractants found in eggs, chemotactic behavior of sperm, timing of gamete release relative to environmental cues (light, temperature, tidal cycle, salinity, the presence of other gametes in the water, etc.), and the degree of synchronicity within an individual and with the other individuals in a population. Sessile invertebrates, such as corals and tunicates, may face a trade-off between natural and sexual selection if the location and/or morphology of the individual or colony which is optimal for feeding and/or growth (including asexual reproduction) is not necessarily optimal for fertilization. In hermaphroditic animals, sex allocation, the relative timing and energy allocation to the production and timing of eggs versus sperm, is an aspect of sexual selection (*in sensu latu*) that has been treated extensively by Eric Charnov. In broadcast spawners male–male competition in general would be expected to be of the scramble rather than the contest type, although in those motile species with broadcast spawning, such as chitons, some of the marine gastropods, and echininoderms, males (or hermaphrodites when acting as males) might compete for positions which improve the probability of fertilizing eggs.

B. Species with Pair Mating

The invertebrates offer a wealth of behaviors in which the direct interaction of specific individuals is a normal preliminary to fertilization. This type of behavior offers more opportunity for observation of the classic forms of sexual selection, i.e., male–male competition and female choice of the father of her offspring, than does broadcast spawning and has consequently received more attention. Here the broad spectrum of reproductive behavior and mating systems has been broken down into a few general categories.

1. Classical Dioecious Mating Systems

It is now clear that Darwin was incorrect in thinking that most of the invertebrates lacked the sensory and behavioral sophistication that would permit sexual selection to operate; there is good evidence for sexual selection in the familiar form of male–male competition for access to females and female choice of mates, both in the higher crustaceans and spiders in which Darwin anticipated that it would occur, and in a variety of other taxa. Courtship behavior and displays have been described in many invertebrate taxa (Table 1), including copepods, ostracods, tardigrades, fairy shrimps, many gastropods, most arachnids including mites and ticks, millipedes, centipedes, and many hermaphroditic taxa, including flatworms and earthworms. Fiddler crabs have a mating system based on male–male competition for a resource (burrow sites) and female choice of males based on burrow quality; fairy shrimps have scramble-competition polygyny; fighting conchs (a gastropod), mud snails, cephalopods, and horseshoe crabs (a chelicerate) have male–male combat over access to females; size-assortative mating is found in flatworms, sea slugs, amphipods, and stomatopods. Male–male interference competition in the form of homosexual rape has been described in acanthocephalan worms. Harem polygyny with a balanced polymorphism of three male morphs (large dominant harem-holders and two types of "sneaky" males) has been described for a marine isopod. Eberhard has argued convincingly that the secondary sexual characteristics that are indicative of sexual selection include most of the elaborate genitalia and accessory sexual structures found in the animal kingdom. There has been increased emphasis on sperm competition as a form of male–male competition and on cryptic female choice, i.e., female acceptance or rejection of sperm or ejaculates after mating. A simple form of pair courtship and mating with spermatophore transfer has been described for a cubozoan jellyfish. The occurrence of sophisticated mating systems and sexual selection seem to depend less on taxon than on ecological factors (not entirely understood) and the detail and sophistication with which they have been studied.

2. Hermaphrodite Mating Systems

One aspect of reproductive biology that does seem to be determined to a large extent by taxonomy and to be resistant to environmental selective pressures is the mode of sexuality. As G. C. Williams pointed out, the occurrence of hermaphroditism versus dioecy in the animal kingdom is relatively invariant within major taxa (phyla, class, etc.) (Table 1) and explaining the current distribution of hermaphroditism remains one of the most difficult problems of reproductive biology.

Although Darwin was familiar with the elaborate mating behavior and genitalia of many hermaphrodites, including the pulmonate snails and slugs, he did not consider it possible that sexual selection would be found in this group. Discussion of mating behavior in invertebrate hermaphrodites in terms of sexual selection was largely the work of Charnov. He predicted that hermaphrodites should compete for opportunities to mate in the male role and be coy in the female role. These suggestions have stimulated interest in the mating behavior of hermaphrodites with the result that an exciting variety of new mating systems are being identified. Some of the mating systems are based on sequential reciprocation between a pair of hermaphrodites. In an opisthobranch, *Navanax inermis,* normal sexual behavior involves bouts of unilateral copulations over the course of a few hours, with the members of a pair alternating sexual roles. In a hermaphroditic polychaete members of a monogamous pair alternate sexual roles in spawning bouts. It has been suggested that such serial reciprocal mating systems represent "gamete-trading," which may resolve sexual conflict due to a preferred sexual role in hermaphrodites. Where the female role is preferred, the system is sperm trading; where, as in the polychaetes, the male role is preferred, consistent with Charnov's prediction, the system is egg trading. Existing data suggest that internally fertilizing hermaphrodites such as gastropods are sperm trading.

Charnov also predicted that elaborate courtship behaviors and/or structures such as the love dart of *Helix* snails, a large calcareous dart shot into the partner prior to copulation (the function of which has been a zoological mystery since Aristotle's day), should serve to induce the recipient to use the shooter's sperm to fertilize eggs and/or as a gift of calcium to aid in egg laying. Recent evidence suggests that the gift of calcium hypothesis is inviable (Koene and Chase, 1998). An alternative suggestion, that *Helix* have a sperm-trading mating system and that the dart serves as an expensive and therefore reliable signal of an individual's intention to donate sperm, thus inducing the partner to reciprocate, remains untested. Charnov's sperm utilization hypothesis also remains viable. Further understanding of mating systems in hermaphrodites with singly successively reciprocal mating, nonreciprocal mating, or group (chain copulation) mating should shed light on the fundamental causes of sexual selection and sexual conflict.

III. EXOTIC SEXUAL PHENOMENA

A. Dwarf and/or Parasitic Males

Several phyla of invertebrates (Table 1), along with some plants and a few insects and vertebrates, have a remarkable sexual system in which the males are very much smaller than the females and/or hermaphrodites that are their mates. Although this phenomenon was well-known to Darwin, its explanation in terms of sexual selection was left to Ghiselin, who pointed out that dwarf males are associated with populations with low encounter probabilities, due to lack of motility and/or low population density. Dwarf males then have evolved as a result of male–male scramble competition to locate and fertilize mates. Sexual selection has sacrificed growth in favor of early maturity and high sperm production. Familiar examples of dwarf males are found in some of the common orb-weaving spiders and the nematode *Caenorhabditis elegans* in which populations consist of hermaphrodites and much smaller males. Burrowing barnacles have separate sexes with dwarf males, whereas some of the familiar intertidal species are hermaphrodites with small dwarf (complemental) males attached. Contest competition by male–male combat is apparently found in some of the spiders with dwarf males, with the larger male usually the victor.

B. Monogamy

A lifelong (or prolonged) association between a male and female (or two hermaphrodites) has been reported in a variety of invertebrate groups. It often seems associated with low encounter probabilities and may take the form of a dwarf or parasitic male attached to a very large female or hermaphrodite. A well-known example is the parasitic flatworm *Schistosoma*, a rare dioecious genus in a predominately hermaphroditic group. In this genus there is strong sexual dimorphism, the body of the male being folded around that of the female, and females may not mature sexually until they acquire a male. Long-term monogamy without dwarf males has also been reported in a variety of other taxa, including some stomatopods and other crustacea, some simultaneous hermaphrodites including sea slugs found on coral heads, and tube-dwelling polychaetes. These examples may again be related to difficulties in dispersal or searching for mates.

C. Group Sex

There are a variety of examples of species in which normal mating behavior involves mating interactions among a group of individuals. One obvious case is the many broadcast spawners, including the famous swarming of the palolo worm, a polychaete, in which an entire local population may release gametes simultaneously in response to an environmental cue. This epidemic breeding is usually explained as a selfish herd phenomenon in which each individual can reduce the chance that all of its own gametes and/or the resulting larval offspring will fall victim to predation by spawning when other gametes and/or larvae are most abundant. The selection pressure in this case would favor ever tighter synchronization of the population. Sexual selection theory would predict evolution of stronger chemoattractants in eggs, better chemotaxis behavior in sperm, and larger gonads and greater gamete production. Comparison of epidemic vs isolated broadcast spawners would offer opportunities to test sexual selection theory.

Other examples of group mating seem to be fundamentally different. In dioecious species male–male competition may lead to clusters of males around a female but where copulation occurs (i.e., a penis or spermatophore is introduced into a specialized female vagina or reproductive opening) only one male can mate at a time. Where fertilization is external or hypodermic insemination can introduce sperm directly into the female's body, it becomes possible for multiple males to inseminate a female simultaneously. The pairs of mating horseshoe crabs that are so familiar on beaches of the east coast of North America are often accompanied by satellite males that crowd in at the time of spawning and achieve some reproductive success through sperm competition, although less than that of the paired male, resulting in a mildly polyandrous mating system. Somewhat similar behavior occurs in hermaphrodites; once a barnacle has deposited eggs in its mantle cavity, several adjacent individuals may insert their pseudopenes into the mantle cavity and deposit sperm on the eggs. Charnov discussed the theoretical ramifications of this type of local mate competition. Group hypodermic insemination has been described in *Palio* nudibranchs.

Even more mysterious and fascinating is another form of group mating found in hermaphroditic gastropods: chain copulation. In chain copulation, the lead individual acts as a female to the second individual; the second is both male to the first animal and female to the third, and so on, with the rear individual acting solely as male to the next to last. This type of behavior has been described in the field for a variety of opisthobranchs and pulmonates. The two basic prerequisites seem to be a long separation between the penis and common genital aperture (female opening) and high population density. The fact that chain copulation is a normal part of the mating system in these species is reflected in an anatomical specialization, a partition of the large hermaphroditic duct found in such genera as *Aplysia* and *Bulla* that typically show chain copulation. This type of mating system represents an opportunity to test current theories of sexual selection in hermaphrodites.

The biology of most invertebrate groups is very poorly known; in many cases most of the information has come from taxonomic studies. The invertebrates have already demonstrated the generality of the processes of sexual selection and comparative studies of the many instances in which particular phenomena have evolved independently should make it possible to test alternative models and identify the environmental circumstances that favor particular forms of sexual selection.

See Also the Following Articles

Caenorhabditis elegans; Hermaphroditism; Mate Choice, Overview; Mating Behaviors, Insects; Mating Behaviors, Mammals; Sexual Selection

Bibliography

Andersson, M. (1994). *Sexual Selection*. Princeton Univ. Press, Princeton, NY.

Bauer, R. T., and Martin, J. W. (Eds.) (1991). *Crustacean Sexual Biology*. Columbia Univ. Press, New York.

Brusca, R. C., and Brusca, G. J. (1990). *Invertebrates*. Sinauer, Sunderland, MA.

Charnov, E. L. (1979). Simultaneous hermaphroditism and sexual selection. *Proc. Natl. Acad. Sci. USA* 76, 2480–2484.

Charnov, E. L. (1982). *The Theory of Sex Allocation*. Princeton Univ. Press, Princeton, NJ.

Darwin, C. (1871). *The Descent of Man and Selection in Relation to Sex*. J. Murray, London.

Eberhard, W. G. (1985). *Sexual Selection and Animal Genitalia*. Harvard Univ. Press, Cambridge, MA.

Eberhard, W. G. (1996). *Female Control: Sexual Selection by Cryptic Female Choice*. Princeton Univ. Press, Princeton, NJ.

Ghiselin, M. T. (1974). *The Economy of Nature and the Evolution of Sex*. Univ. of California Press, Berkeley.

Koene, J. M., and Chase, R. (1998). *J. Moll. Stud.* **64**, 75–80.

Leonard, J. L. (1990). The Hermaphrodite's Dilemma. *J. Theor. Biol.* **147**, 361–372.

Leonard, J. L. (1991). Sexual conflict and the mating systems of simultaneously hermaphroditic gastropods. *Am. Malac. Bull.* **9**, 45–58.

Leonard, J. L. (1992). The "love-dart" in helicid snails: A gift of calcium or a firm commitment. *J. Theor. Biol.* **159**, 513–521.

Meglitsch, P. A., and Schram, F. R. (1991). *Invertebrate Zoology*, 3rd ed. Oxford Univ. Press, New York.

Premoli, M. C., and Sella, G. (1995). Sex economy in benthic polychaetes. *Ethol. Ecol. Evol.* **7**, 27–48.

Preston-Mafham, R., and Preston-Mafham, K. (1991). *The Encyclopedia of Land Invertebrate Behaviour*. MIT Press, Cambridge.

Smith, R. L. (Ed.) (1984). *Sperm Competition and the Evolution of Animal Mating Systems*. Academic Press, Orlando, FL.

Tompa, A. S., Verdonk, N. H., and van den Biggelaar, J. A. M. (Eds.) (1984). *The Mollusca, Vol. 7 Reproduction* (K. M. Wilbur, Editor-in-Chief). Academic Press, Orlando, FL.

Wickler, W., and Seibt, U. (1981). Monogamy in Crustacea and man. *Z. Tierpsychol.* **57**, 215–234.

Williams, G. C. (1975). *Sex and Evolution*. Princeton Univ. Press, Princeton, NJ.

Zahavi, A., and Zahavi, A. (1997). *The Handicap Principle*. Oxford Univ. Press, New York.

Median Eminence

Ann-Judith Silverman
Columbia University

I. Introduction
II. Zonation within the Median Eminence
III. The Essential Vascular Arrangement: A Venous Portal System
IV. Capillary Structure
V. Specialized Glial Cells
VI. The GnRH Axons
VII. Dynamic Interactions in the Median Eminence

GLOSSARY

magnocellular neurosecretory system Neurons whose cell bodies (large, hence *magno*) are located primarily in the paraventricular and supraoptic nuclei of the hypothalamus. These cells make oxytocin and vasopressin and send axons to the posterior pituitary.

neural–hemal tissue/organ Denoting a zone where neurosecretory nerve terminals end on or near blood vessels which will carry the neurohormone into the circulatory system. The median eminence is one of these.

neurohormone The secretory product of a neuron which is released into the vascular supply and interacts with a target cell at some distance from the site of release via specific receptors.

neurosecretion The release of the products of a neuron into the vascular supply.

parvicellular neurosecretory system An umbrella term to include all neurosecretory neurons that project to the median eminence and which regulate anterior pituitary function. One such parvicellular subgroup synthesizes and releases gonadotropin releasing hormone which stimulates gonadotropin release, and another releases dopamine, which inhibits prolactin (*parvi* = small).

releasing hormone/release-inhibiting hormone The products of the parvicellular neurosecretory system which stimulate secretion from the anterior pituitary or inhibit such release, respectively.

The median eminence is one of the seven circumventricular organs or "windows on the brain" and forms the final common pathway as a neural–hemal connection between the central nervous system and the adenohypophysis. The overriding function of the median eminence is to serve as an efficient gateway for neurosecretory signals to reach the periphery. To accomplish that goal it is characterized by (i) its "leaky" capillaries with fenestrated endothelia, (ii) the venous portal arrangement of the blood supply and drainage to the anterior pituitary which facilitates delivery of secreted molecules, (iii) the specialized glial cells or tanycytes that span from the third ventricular infundibular recess to the ventral surface of the brain, and (iv) most important, the numerous axonal terminations of the multitudinous neurosecretory systems that regulate the function of the anterior pituitary. It also serves as a highway for the axons originating in the supraoptic and paraventricular nuclei that terminate in the neurohypophysis.

I. INTRODUCTION

The original concept of vertebrate neurosecretion stems from the works of Ernest Scharrer and Green and Harris (see Knigge and Silverman, 1974; Porter *et al.*, 1974). Scharrer demonstrated using cytological measures that specific cell groups in the midbrain changed their cytological properties in response to specific physiological challenges. Later, Bargmann and Scharrer provided the basic evidence for the existence of the projection of the supraoptic and paraventricular nuclei to the capillaries of the posterior pituitary. They proposed that these cells released their product into the blood supply. It took many decades before this concept was accepted. The second major step occurred when Green and Harris (1947) proposed that the central nervous system (CNS), and the hypothalamus in particular, sent signals to the anterior pituitary via the specialized vascular arrangement and blood flow of the median eminence. They demonstrated that the capillary plexus in the median eminence (now known as the primary portal plexus) drains into portal veins which break up into a second capillary system in the anterior pituitary. Blood flow was in the direction described. Finally, the work of Halasz and the Hungarian school used pituitary transplants under the kidney capsule or into the hypothalamus to demonstrate that the secretions of the anterior pituitary were dependent on signals arising from the CNS. As the interface between the brain and the adenohypophysis, the median eminence then became a subject of study.

II. ZONATION WITHIN THE MEDIAN EMINENCE

The median eminence has traditionally been divided into two zones based on the types of axons within the zone. It might be easier for the reader to view this tissue as divided into four regions. From dorsal to ventral, the first would be the zone comprising the cell bodies of the specialized glial cells, tanycytes, lining the ventricular wall (see Section V). The next is the internal zone, which contains axons of passage. These axons are derived from the magnocellular neurosecretory system and contain the neuropeptides oxytocin or vasopressin. These fibers will extend through the median eminence, down the pituitary stalk, and into the posterior pituitary. This zone will not be discussed further in this chapter. The next territory, the zona externa, is occupied by the arriving axons of the parvicellular neurosecretory system and their nerve terminals. This includes axons which carry hormones that regulate reproductive function: gonadotropin-releasing hormone (GnRH) (stimulates gonadotropins) and dopamine (inhibits prolactin release). These nerve terminals then abut on the perivascular space of the fenestrated capillaries (see Section IV). They are frequently separated from this extracellular space by the tanycytic end feet.

III. THE ESSENTIAL VASCULAR ARRANGEMENT: A VENOUS PORTAL SYSTEM

A. Definition

Anatomically a portal system is a vascular arrangement in which two capillary beds are joined by a blood vessel which has drained the first and breaks up into the second. Examples include the venous portal system between the small intestine and the liver and the arteriolar portal system between the afferent and efferent arterioles of the kidney glomerulus tuft.

B. The Hypophysial Portal System

In 1930 and 1933, Popa and Fielding (see Knigge and Silverman, 1974) described the hypophysial portal vessels connecting the median eminence with the pars distalis of the anterior pituitary (Fig. 1). The median eminence is supplied by the superior hypophysial artery. This then gives rise to the capillaries within the median eminence proper which elaborate intricate loops within the tissue. It is onto these capillaries that the neurosecretory terminals will abut (see Section IV). This plexus is then drained by the portal veins which go directly to the pars distalis. This gland receives 85% or more of its blood supply via this route. Since the volume of blood within this system is very small, the neurohormones are delivered with very little dilution. The second consequence of this vascular arrangement as a delivery system is that materials released into the primary portal plexus reach their target very rapidly. This rapidity is an essential feature for the efficacy of this neurosecretory system because neurohormones are degraded probably beginning at the site of release and certainly within the blood.

It should be noted that this neural–hemal arrangement, though present in most vertebrates, is absent in some species of teleosts in which there is a direct innervation from the CNS onto the secretory cells of the anterior pituitary. A direct innervation can occur mammals but is very sparse. For example, axons positive for pituitary adenylate cyclase activating peptide 38 and its derivatives are seen in the

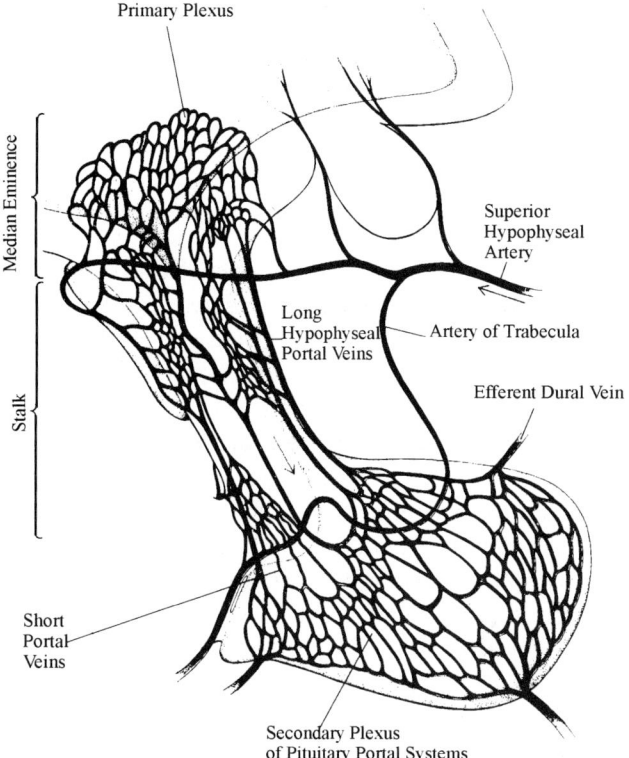

FIGURE 1 Schematic of the median eminence region as viewed in sagittal section. The median eminence is supplied by the superior hypophyseal artery. This artery breaks up into a complex capillary bed, known as the primary portal plexus. As explained in the text, these capillaries are fenestrated and collect the secretion of neurosecretory terminals that end here. The capillaries collect into long and short portal veins that carry these secretions directly to the anterior pituitary.

median eminence but are also abundant in the anterior pituitary proper (Miccelsen et al., 1995).

IV. CAPILLARY STRUCTURE

A. The Blood–Brain Barrier

In most regions of the CNS the capillaries are specially modified to provide a tight seal, termed the blood–brain barrier, between the CNS and the blood. This seal is composed of the nonfenestrated endothelia which have many tight junctions among them. Additionally, these vessels are invested with astrocytic processes and pericytes, which are thought to participate in further elaborations of this barrier. The barrier does not permit the flow of ions and essential substrates such as glucose and amino acids are transported across the endothelium.

B. Fenestrated Capillaries and the Lack of the Blood–Brain Barrier

In the median eminence (and in other circumventricular organs: choroid plexus, subfornical organ, OVLT, area postrema, and posterior pituitary), the arrangement described previously is abrogated. This can be demonstrated in many ways. At the ultrastructural level (Fig. 2) it is clear that the endothelial cells of the capillaries within the median eminence are fenestrated. This morphological feature is equivalent to the distribution of endothelial cell tight junctions: continuous in the brain and discontinuous or absent in the circumventricular organs, including the median eminence (Petrov et al., 1994). Tracers injected into the general circulation, such as horseradish peroxidase, which cannot traverse brain capillaries readily enter the median eminence and diffuse from its midline location to the adjacent brain neuropil (the arcuate or infundibular nucleus). Func-

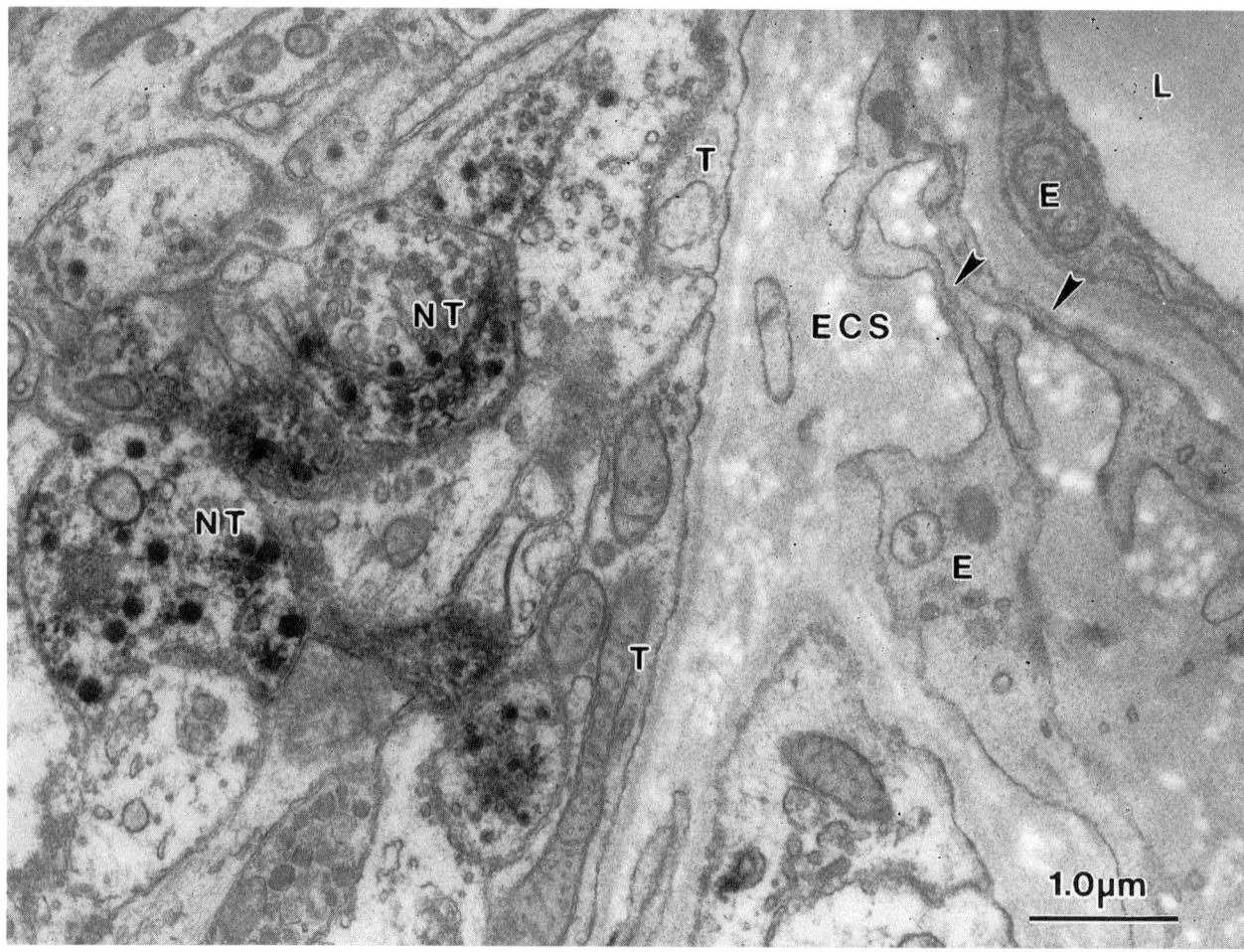

FIGURE 2 Electron micrograph showing the relationship between GnRH-containing nerve terminals (NT), as defined by the deposition of immunocytochemical reaction product, and the primary portal capillaries. The lumen (L) of the capillary is in the upper right. The walls of the endothelial cell (E) of this and other capillaries are very thin (arrows). The secretory materials are released into the extracellular space (ECS). Note the feet of the tanycytes between the GnRH terminals and the ECS; this is described in Section V.

tionally the most important feature of these leaky capillaries is that neurohormones which are released by neurosecretory axons gain access to the capillaries and can be measured in much higher concentrations in the draining portal veins.

V. SPECIALIZED GLIAL CELLS

The ventricular space is lined everywhere by ependymal cells derived from the original neuroepithelium. Within the median eminence these ependymal cells have a unique phenotype which can be seen by their retention of an embryologically early intermediate filament (vimentin). There are also classical astrocytes in this tissue as defined by the presence of the more mature intermediate filament, glial fibrillary acidic protein (Chauvet et al., 1995). Along most of the wall of the ventricles, ependymal cells form a simple cuboidal epithelium. The tanycytes have their cell body region (with their nucleus) at the ventricular wall but then send long processes which span from the ventricular to the pial surface. These tanycytes are found along the lateral walls adjacent to the arcuate nucleus and along the floor of the infundibular recess which forms the dorsal aspect of the median eminence.

The tanycytes clearly form a structural scaffold for the numerous axons which pass through the median eminence (Fig. 3). Other roles for these cells have been postulated and include (i) the active transport of materials from the cerebrospinal fluid into the median eminence (Knigge and Silverman, 1974), (ii) channels which guide axons to the median eminence during development (Kozlowski and Coates, 1985), (iii) regulation of the efficacy of neurosecretion (King and Rubsin, 1994; see Section VII), and (iv) recently, the production of both growth factors and their receptors (Gonzalez et al., 1994; Garcia-Segura et al., 1996).

Many of the tanycytes which lie on the floor of the infundibular recess have a rich microvillous border (see Knigge and Silverman, 1974). Such an anatomical modification is frequently associated with an absorptive and/or transport function (e.g., cells in the proximal convoluted tubule of the kidney). Such transport was strongly suggested by the early work of Silverman and Knigge (see review) and has been substantiated recently by tracking specific molecules into the tanycyteic compartment (Bjelke and Fuxe, 1993).

VI. THE GnRH AXONS

All the axons containing releasing hormones and release-inhibiting hormones enter the median eminence and terminate on or near the portal capillaries. For the purposes of this article, I discuss those containing GnRH since it is the most immediately relevant to reproduction.

The GnRH neurons are widely scattered in the ventral forebrain (Silverman et al., 1994). One question that arises is which among these are neurosecretory. This question has been addressed in two ways. The first is by the placement of a retrograde tracer into the median eminence *in vivo* which will label the cell bodies of origin. Such experiments have been performed in rats and in monkeys using different tracers and different survival times. Hence, the percentage of GnRH cells backfilled has varied from 50 to 75% of the total. However, all agree that the cells innervating the median eminence, and therefore participating directly in the regulation of gonadotropin secretion, are not confined to a single area but distributed throughout the entire population with the exception, perhaps, of those in the main and accessory olfactory bulbs.

A second approach has been the intraperitoneal injection of a different tracer molecule (Fluorogold) from which the tracer gains access to the general circulation. The advantage of this method is that it does not depend on the survival of an animal following a traumatic surgery. The disadvantage is that it will label all neurons that project to any circumventricular organ. GnRH neurons project robustly to the OVLT. In rats this approach has labeled all GnRH neurons, regardless of physiological condition, whereas in the mouse the capture of the tracer appeared to be related to the activity of the GnRH system (Silverman et al., 1994).

A brain transplantation paradigm has shown how essential is the contact between the GnRH axon terminal and the hypophysial portal vasculature. The

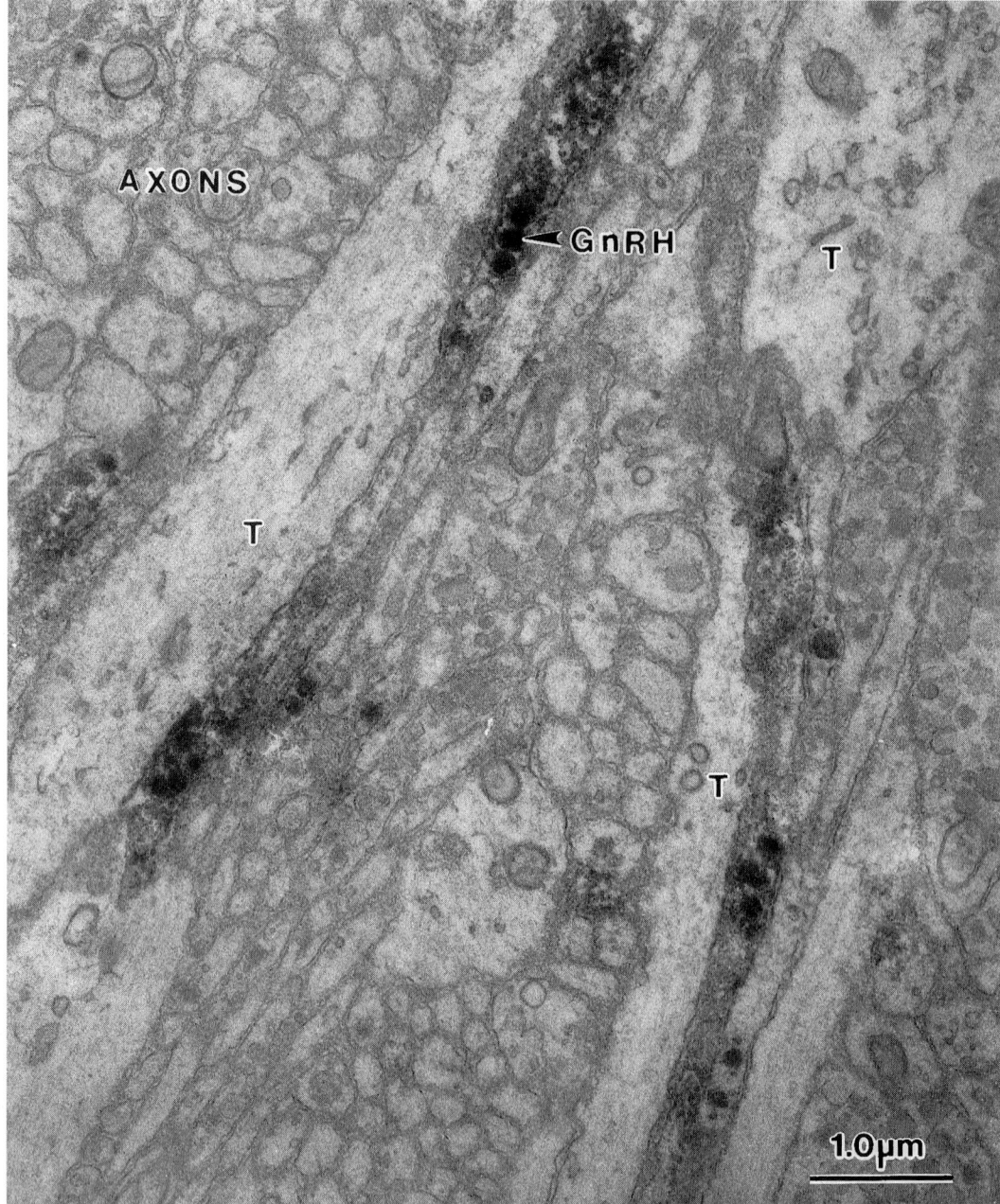

FIGURE 3 This electron micrograph highlights the relationship between the tanycyte processes (T), GnRH axons (arrowhead), and other axon bundles (axons).

hpg mutant cannot synthesize GnRH. If normal brain tissue containing GnRH neurons is placed in the third ventricle, then GnRH axons will exit from the graft into the host (Fig. 4). One does not need many cells for a functional hypothalamic–pituitary–gonadal axis; females with a dozen or fewer GnRH neurons can get pregnant, but only if the axons of such cells reach the primary portal plexus. Survival of GnRH cells in the lateral ventricle without innervation of the median eminence does not result in stimulation of gonadotropin secretion (Silverman and Gibson, 1990).

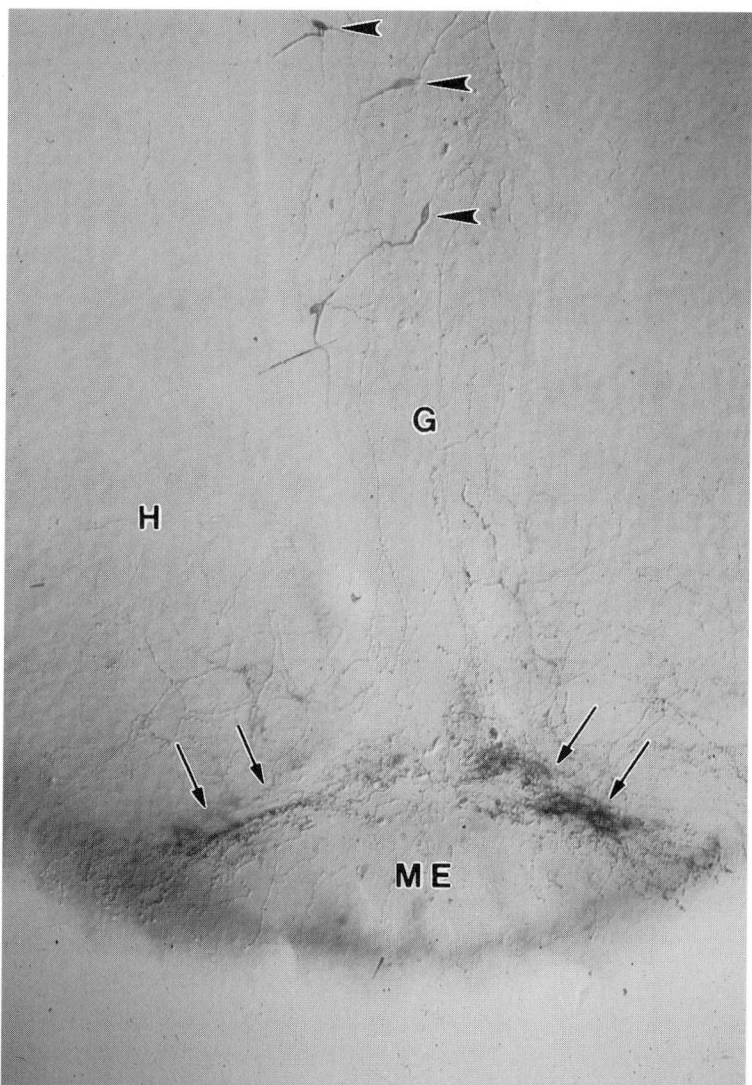

FIGURE 4 This figure shows coronal section though the median eminence (ME) of a mouse and illustrates two points. First is the relationship of the GnRH axon terminals to the lateral aspects (arrows) of the ME. Such a distribution would be seen in a normal animal. This particular animal is a host (H) hypogonadal mouse and the GnRH axons are derived from GnRH neurons (arrowheads) in a third ventricular graft (G).

The distribution of fibers within the median eminence shows some species variations. In most animals GnRH axons are concentrated at the lateral corners of the median eminence (viewed in cross section). The distance down the infundibular stalk that such axons travel and whether or not they reach into the posterior pituitary varies much more. It is unknown whether there is any functional significance to the broader distribution.

VII. DYNAMIC INTERACTIONS IN THE MEDIAN EMINENCE

A. Axon to Axon

The possibility that the rate/timing/quantity of release of GnRH from nerve terminals in the median eminence might be regulated by other nerve terminals therein has been considered for at least two

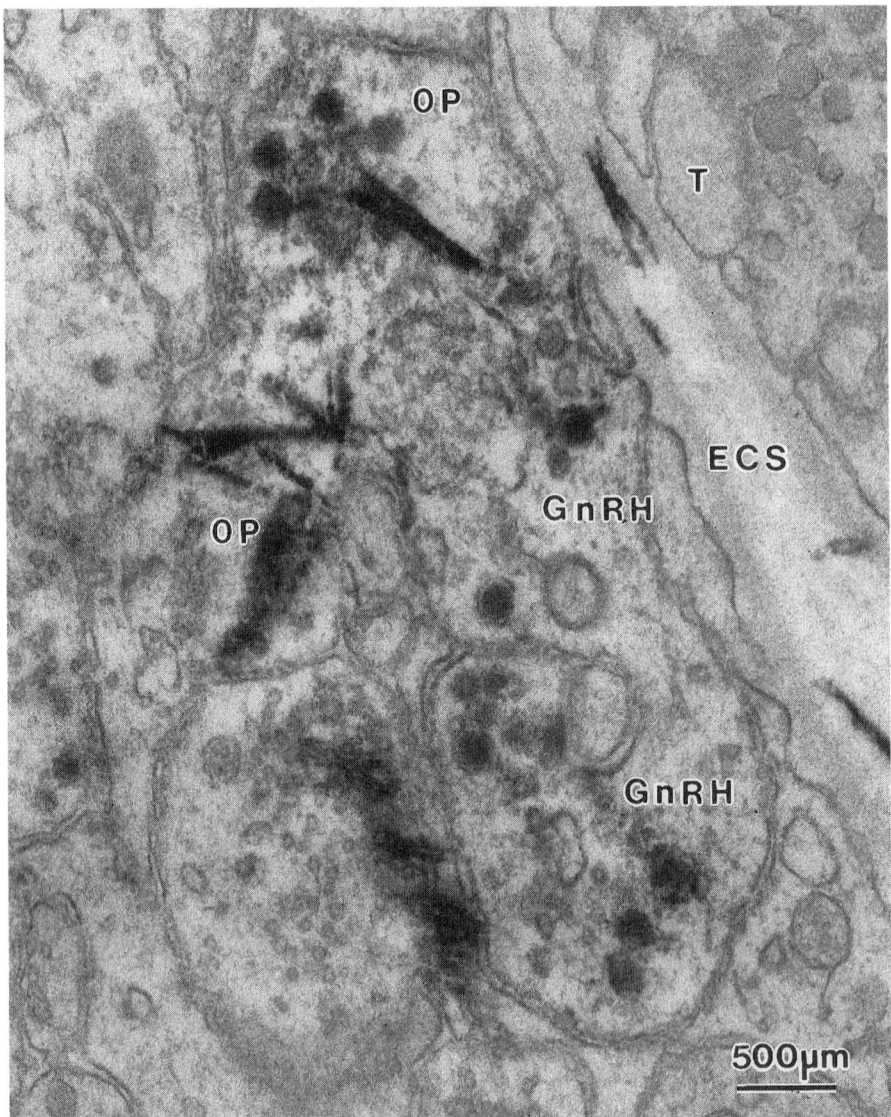

FIGURE 5 Neuron terminals in the ring dove median eminence. GnRH axons are immunostained with diaminobenzidine (diffuse) and opsin axons with tetramethylbenzidine (crystals) abutting the extraceullular space (ECS) of a portal capillary (some of the crystals stick nonspecifically to the matrix materials in the ECS). Such intimate relationships suggest that the products released from either could influence the secretion pattern of its neighbor.

decades. Although "synaptoid" interactions have been reported among elements in the median eminence, true synapses have not been documented. However, in the peripheral nervous system, sympathetic terminals do not make formal synapses with their targets but instead release their transmitter into a space from which it diffuses to influence many cells. The concept of "volume transmission" in the CNS is being discussed more frequently (Agnati et al., 1995). A similar arrangement is certainly possible in the median eminence and has been the basis for pharmacological studies examining changes in release of GnRH from median eminence fragments *in vitro*. Similarly, this is the rationale for using push–pull cannulation to collect secreted molecules at this level. This is a vast literature and will be covered experimentally in other chapters. A few current examples are presented to illustrate the point.

Axonal terminals that synthesize the gaseous neurotransmitter nitric oxide (NO) are found within the median eminence and show considerable overlap with GnRH axons in the lateral aspect of the median eminence (Bhat *et al.*, 1995). NO is effective in stimulating the release of GnRH *in vivo* and *in vitro* from arcuate nucleus/median eminence preparations (Rettori *et al.*, 1993). Another intriguing observation is that intravenous administration of GnRH (which would reach the median eminence) or local injection of GnRH into the median eminence result in a burst of multiunit electrical activity correlated with a luteinizing hormone pulse. There was no effect of administering GnRH into the preoptic area where GnRH cell bodies reside (Hiruma and Kimura, 1995). In birds, photoperiodic signals that regulate reproductive activity are located within the brain itself. Recent work using double-label light and electron microscopy has shown a multitude of close associations between nerve terminals containing an opsin protein (and presumably from these encephalic photoreceptors) and those containing GnRH within the ring dove median eminence (Fig. 5). The abundance and closeness of these axonal types suggest a functional connection which can now be tested.

B. Axon to Glia

Another area of investigation concerns the dynamic changes that occur between neurosecretory terminals and tanycytic or astrocytic processes. It has been demonstrated quantitatively in the CNS that activity states alter the percentage of a neuron's plasma membrane in contact with an astrocytic process. The association between astrocytes and GnRH neurons was altered by exogenous steroids.

From the earliest ultrastructural studies on GnRH axons and terminals in the median eminence, the observation was made that such terminals rarely (if ever) touched directly on the perivascular space but were separated from it by thin glial processes. Even prior to the use of immunocytochemistry, Kumar and colleagues (among others) postulated alterations in the tanycytes during the reproductive cycle (see Knigge and Silverman, 1974). This finding has been extended recently by the confocal double-label microscopy of King and Rubsin (1994). Confocal microscopy permits a wider survey of tissue and the results suggest a dynamic interaction between GnRH terminals and tanycytic end feet.

The studies of the median eminence have now clearly moved into a dynamic phase in which the interfaces of cell biology and physiology shall intersect.

See Also the Following Articles

GnRH (Gonadotropin-Releasing Hormone); Hypothalamic–Hypophysial Complex; Magnocellular System

Bibliography

Agnati, L. F., Zoli, M., Stromberg, I., and Fuxe, K. (1995). Intercellular communication in the brain: Wiring versus volume transmission. *Neuroscience* 69, 711–726.

Bhat, G. K., Mahesh, V. B., Amar, C. A., Ping, L., Aguan, K., and Brann, D. W. (1995). Histochemical localization of nitric oxide in the hypothalamus: Association with GnRH neurons and colocalization with NMDA receptors. *Neuroendocrinology* 62, 187–197.

Bjelke, B., and Fuxe, K. (1993). Intraventricular β-endorphin accumulates in DARPP-32 immunoreactive tanycytes. *Neuroreport* 5, 265–268.

Chauvet, N., Parmentier, M. L., and Alonso, G. (1995). Transected axons of the adult hypothalamo-neurohypophysial neurons regenerate along tanycytic processes. *J. Neurosci. Res.* 41, 129–144.

Garcia-Segura *et al.* (1996). *Frontiers Neuroendocrinol.* 17, 180–211.

Gonzalez, A. M., Logan, A., Ying, W., Berry, M., and Baird, A. (1994). Fibroblast growth factors in the hypothlamic–pituitary axis: Differential expression of FGF-2 and a high affinity receptor. *Endocrinology* 5, 2289–2297.

Hiruma, H., and Kimura, F. (1995). LHRH is a putative factor that causes LHRH neurons to fire synchronously in ovariectomized rats. *Neuroendocrinology* 61, 509–516.

King, J. C., and Rubsin, B. S. (1994). Dynamic changes in LHRH neurovascular terminals with various endocrine conditions. *Horm. Behav.* 28, 349–356.

Knigge, K. M., and Silverman, A. J. (1974). Anatomy of the endocrine hypothalamus. In *Handbook of Physiology*, Section 7: Endocrinology, Vol. IV, pp. 1–32.

Kozlowski, G. P., and Coates, P. (1985). Ependymo-neuronal specializations between LHRH fibers and cells of the cerebrospinal fluid. *Cell Tissue Res.* 242, 301–311.

Petrov, T., Howarth, T. L., and Stevenson, B. R. (1994). Distribution of the tight junction associated protein in the circumventricular organ of the CNS. *Mol. Brain Res.* **21**, 235–246.

Porter, J. C., Ondo, J. G., and Cramer, O. M. (1974). Anatomy of the hypothalamus. In *Handbook of Physiology*, Section 7: Endocrinology, Vol. IV, pp. 33–43.

Rettori, V., Belova, N., Dees, W. L., Nyberg, C. L., Gimeno, M., and McCann, S. M. (1993). *Proc. Natl. Acad. Med.* **90**, 10130–10134.

Silverman, A. J., and Gibson, M. J. (1990). Hypothalamic transplantation: Repair of reproductive defects in hypogonadal mice. *Trends Endocrinol. Metab.* **1**, 403–405.

Silverman, A. J., Livne, I., and Witkin, J. W. (1994). The GnRH neuronal systems: Immunocytochemistry and in situ hybridization. In *The Physiology of Reproduction* (E. Knobil and J. Neill, Eds.), 2nd ed., Vol. 1, pp. 1683–1710. Raven Press, New York.

Meiosis

Zbigniew Polański
Jagiellonian University
Poland

Jacek Z. Kubiak
Institut Jacques Monod
France

I. Features of Meiosis
II. Physiology of Meiosis
III. Genetics of Meiosis
IV. Meiosis in Males and Females
V. Cell Cycle Control during Meiosis

GLOSSARY

bivalent A functional unit formed by two homologous chromosomes during prophase of the first meiotic division.

crossing-over Exchange of a part of genetic material between homologous chromosomes resulting from symmetrical breaks of chromatids and their reciprocal reannealing.

cytostatic factor (CSF) Activity responsible for the arrest at the second meiotic metaphase in amphibian and mammalian oocytes.

diploid Having two copies of each chromosome; physiologically one comes from the mother and the other from the father.

DNA repair A process in which erroneous sequences of DNA are removed and substituted by correct ones; involves the cutting and reannealing of DNA strands.

genetic recombination A sum of processes resulting in the appearance of new combinations of genes in the progeny.

haploid Having one copy of each chromosome.

homologous chromosomes Chromosomes carrying the same genetic loci. Diploid organisms contain pairs of homologous chromosomes, one deriving from each parent.

maturation-promoting factor (MPF) The major kinase controlling the entry into M phase of the cell cycle in mitotic and meiotic cells.

sister chromatids Identical copies of a chromosome, derived by DNA replication.

univalent A single chromosome present during the first meiotic division (when homologous chromosomes are arranged in bivalents).

Meiosis, first described by Edouard Van Beneden in 1883, is a process associated with sexual reproduction which precedes the formation of gametes. It is accomplished as two successive cell cycles during which the number of chromosomes is halved, giving rise to haploid gametes. The diploid amount of genetic material is restored following the union of male and female gametes at fertilization. Thus, meiosis ensures the stable maintenance of DNA content through generations. In addition, meiosis allows genetic recombination, a key requisite for biological evolution.

I. FEATURES OF MEIOSIS

The key events contributing to the unique features of meiosis are (i) pairing of homologous chromosomes, (ii) crossing-over, and (iii) random segregation of homologous chromosomes at the first meiotic division. The first two events take place during the first meiotic prophase, while the third, although executed at first meiotic anaphase, is a consequence of processes taking place at this time. Thus, the period of first meiotic prophase, which may take more than 90% of the overall time of meiosis, is of great importance for the successful progression of meiosis.

A. Details of First Meiotic Prophase

The first meiotic prophase is traditionally divided into five stages (Fig. 1). The first is leptotene, which begins the first meiotic prophase after a series of mitotic divisions in oogonia or spermatogonia. The sister chromatids of each chromosome condense slowly and become stably attached to each other by a newly appearing proteinaceous structure, termed the "axial core." Some parts of chromatids (probably specialized DNA sequences) bind directly to axial core, whereas other segments of chromatids stick out loosely as chromatin loops. During the next stage, zygotene, the homologous chromosomes approach each other at certain homologous sites. Starting from these sites a ribbon-like, proteinaceous structure develops between each pair of homologous chromosomes bringing them into close apposition along their entire lengths. This structure, called the synaptonemal complex (SC), consists of a "central element" assembled between two "lateral elements" (developed from the axial cores). Completion of the binding of the homologous chromosomes (synapsis) terminates the formation of bivalents. The SC persists through pachytene. Other specific structures called recombination nodules (RN) may also be seen along the SC between its lateral elements at this time. At the beginning of the next stage, diplotene, the SC disappears gradually, enabling the homologous chromosomes of each bivalent to separate to some extent. They remain, however, bound to each other at the chiasmata (singular chiasma), which mark the sites where crossing-over has occurred. Finally, during diakinesis, the bivalents undergo extensive condensation. The process of chiasma shifting toward the bivalent ends (chiasma terminalization) observed during this stage in the light microscope seems to be an artifact linked to the chromosome condensation. Chiasmata are conserved in their position from the time of first appearance in diplotene.

B. Further Stages and the Outcome of Meiosis

The first meiotic prophase terminates with nuclear membrane breakdown. The condensed bivalents now align on the metaphase plate of the first meiotic division spindle. With appropriate staining, their characteristic morphology is easily distinguishable within the resolution of the light microscope (Figs. 2A and 2B). At the first meiotic anaphase, the homologous chromosomes (each composed of two sister chromatids) migrate to the opposite poles of the meiotic spindle. The chromatid segments of the homologous chromosomes distal to the chiasmata have changed associations and migrate accordingly. Thus, the altered composition of the homologous chromosomes is finalized at that time. Cytokinesis concludes the first meiotic cycle by the separation of the two sets of homologous chromosomes into the first two daughter cells.

In the second meiotic cycle the daughter cells enter the second meiotic metaphase without replication of DNA and in many cases without the reformation of the nuclear envelope. At the second meiotic anaphase and subsequent cytokinesis the sister chromatids of each chromosome are separated into the daughter cells, as occurs during mitotic divisions. Suppression of DNA replication between the two meiotic divisions is a key phenomenon, ensuring that the resulting gametes are haploid.

The process of meiosis creates four genetically different haploid cells from a single precursor. The new combination of genes present in the gametes depends on the random segregation of the homologous chromosomes during first meiotic anaphase, anticipated by Gregor Mendel in his law of independent assortment, and the reciprocal exchange of genetic material between homologous chromosomes. The extent of recombination is considerable. In the human, bearing

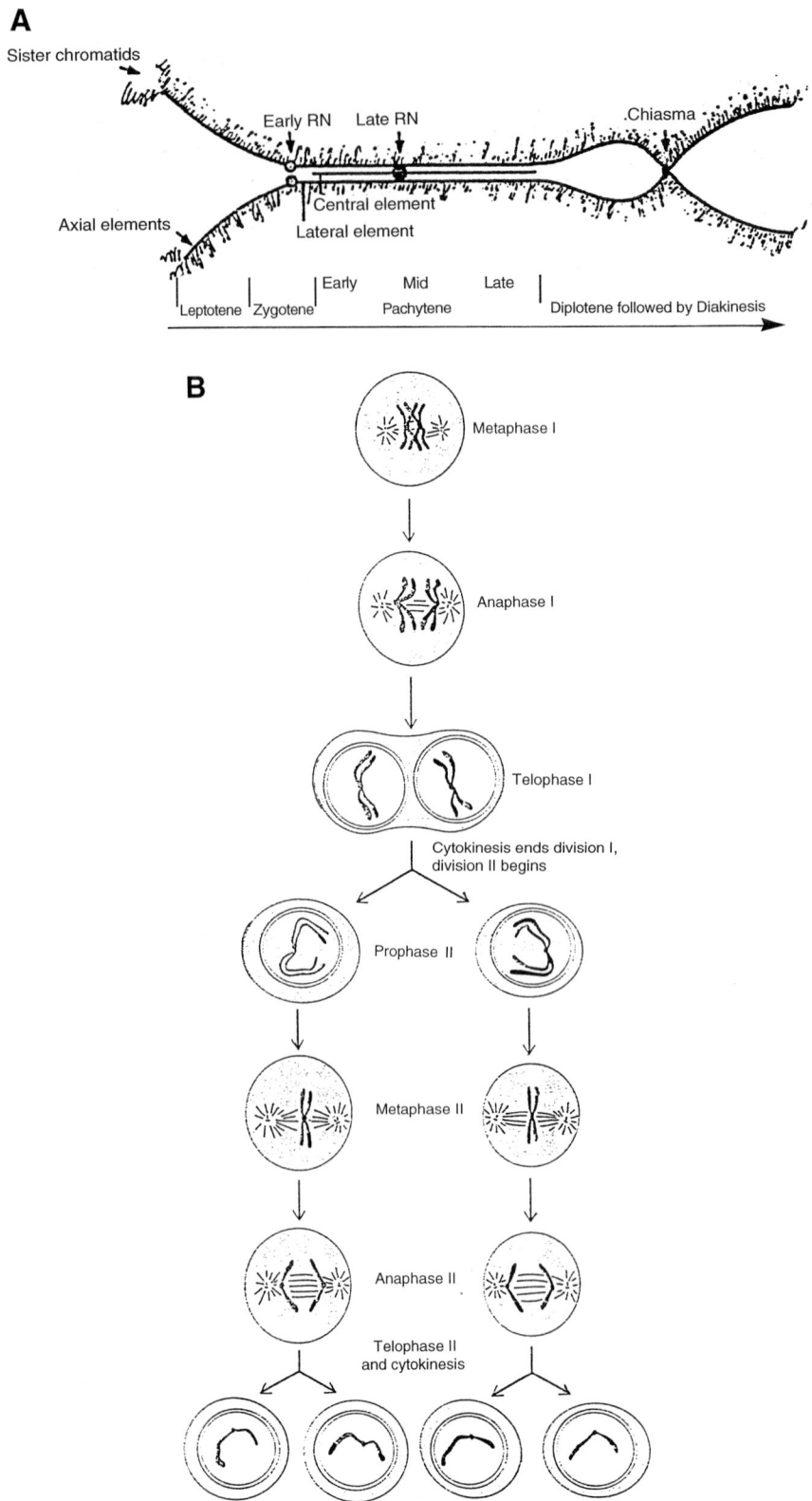

FIGURE 1 Schematic representation of meiosis. (A) First meiotic prophase; (B) further stages of meiosis (reproduced from (A) S. M. Baker *et al.*, *Nature Genet.* **13**, 336–342, 1996; (B) Alberts *et al.*, 1989).

stood. It is, for example, the issue of the molecular basis for the precise pairing of homologous chromosomes for which some alternative concepts are proposed (most of them coming from the studies of the same organism, the yeast).

Concerning further events, the SC appears to play a central role in holding the homologous chromosomes in close apposition; however, the details are still unclear. In oocytes of *Lepidoptera* or *Trichoptera*, chromosome pairing occurs in the presence of SC but no recombination is observed, whereas during male meiosis in many *Diptera*, no SC is found between the homologs and no crossing-over occurs. These observations suggest that the SC is necessary but not sufficient for genetic exchange to occur.

Crossing-over probably takes place between the early zygotene and the end of pachytene since close association of recombining elements seems to be necessary. However, this phenomenon has never been observed at the electron microscopy level and its precise timing is unknown. It has been hypothesized that recombination nodules may represent multienzyme complexes catalyzing reciprocal exchange between two molecules of DNA since their numbers and distribution correlate well with the sites at which chiasmata are later distinguished. The distribution of the crossover sites deviates from random in two ways: (i) Two close neighbor crossover events are rarely detected, this so-called "crossover interference" probably being provoked by transmission of an inhibitory signal along the chromosome from an established crossover site; and (ii) in some regions of chromosomes, termed recombination hot spots, the meiotic exchange occurs much more frequently than in others. These may cover both large, cytologically recognizable parts of chromosomes or be limited to segments of one locus resolution. In mice and humans the frequency of the crossover events between different regions of the same chromosome may differ by 800-fold.

The proper segregation of the chromosomes or chromatids at anaphase requires them to be stably anchored in the equatorial plane of the metaphase spindle. This may be achieved only after they come under the balanced tensions directed toward opposite spindle poles, which in turn requires that they must be bound to each other before they start to separate. During the first meiotic division the mainte-

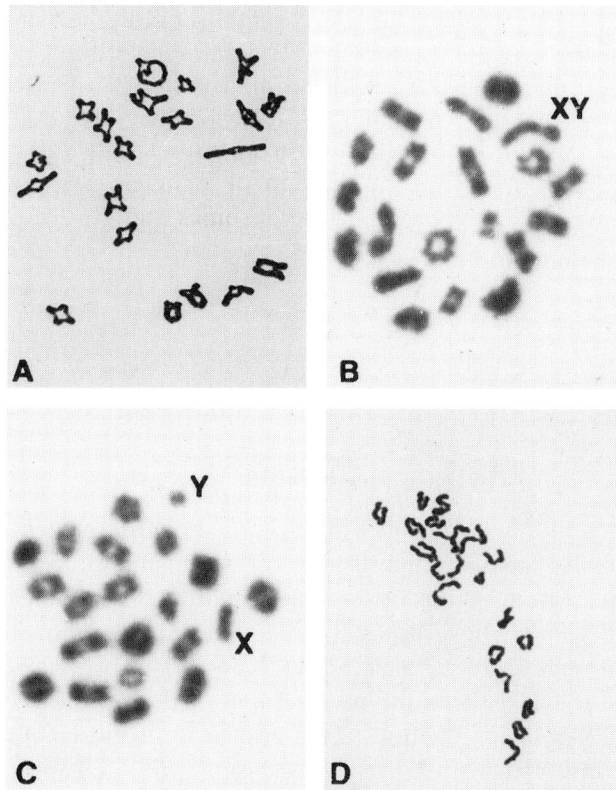

FIGURE 2 Morphology of meiotic chromosomes. First meiotic metaphase in mouse (A) oocyte and (B) spermatocyte (XY bivalent is marked). (C) Precocious separation of sex chromosome bivalent at first meiotic metaphase in mouse spermatocyte (X and Y univalents are marked). (D) Second meiotic metaphase in mouse oocyte (reproduced from (A) Z. Polanski, *Int. J. Dev. Biol.* **39**, 1015–1020, 1995; (B, C) H. Krzanowska and B. Wabik-Sliz, *Mol. Reprod. Dev.* **39**, 347–354, 1994).

23 pairs of homologous chromosomes, there are 2^{23} possible haploid sets, and since at least one chiasma is formed per bivalent in most organisms, the actual level of recombination is much higher.

Each gamete resulting from meiosis contains only one haploid set of chromosomes. In the animal kingdom, such reduction in genetic material is unique to the meiotic process.

II. PHYSIOLOGY OF MEIOSIS

Although the particularities of chromosome behavior during meiosis have been well described, the underlying molecular mechanisms are poorly under-

associations have not been established at the end of the synaptic period the bivalents separate to the univalents (Fig. 2C) which segregate independently of each other or split precociously into individual chromatids at the first meiotic anaphase. Both cases may result in the production of gametes with an unbalanced number of chromosomes and give rise to aneuploid embryos.

III. GENETICS OF MEIOSIS

A. Genes Affecting Meiosis

A number of avenues of research, including screening for meiotic mutants in lower organisms and gene disruption in the mouse, have identified a number of genes whose products affect the progression of meiosis (Table 1). A common characteristic of certain of these genes is the involvement of their products in DNA repair. These genes appear to be extremely conserved evolutionarily; for example, *Mlh1* and *Pms2* are homologs of the bacterial DNA repair gene *Mut1*. This homology supports the "double strand break repair" model to explain the origin of reciprocal recombination during meiosis. This model postulates that DNA repair mechanisms, which find and use the undamaged homolog for repair, have evolved into meiotic mechanisms for recognition, pairing, and exchange of homologous sequences.

B. Meiosis and Aneuploidy

The erroneous allocation of chromosomes into gametes may give rise to aneuploid embryos. In humans aneuploidy is the most frequent case of pregnancy loss (about 35% of spontaneous abortions) and when resulting in live birth it is predominantly associated with mental retardation.

There are several possible causes of erroneous segregation of chromosomes at meiosis. Molecular studies on the origin of an extra chromosome in trisomics have revealed that the main contributions to the aneuploidy in humans are errors of maternal meiosis occurring during the first meiotic division (Table 2).

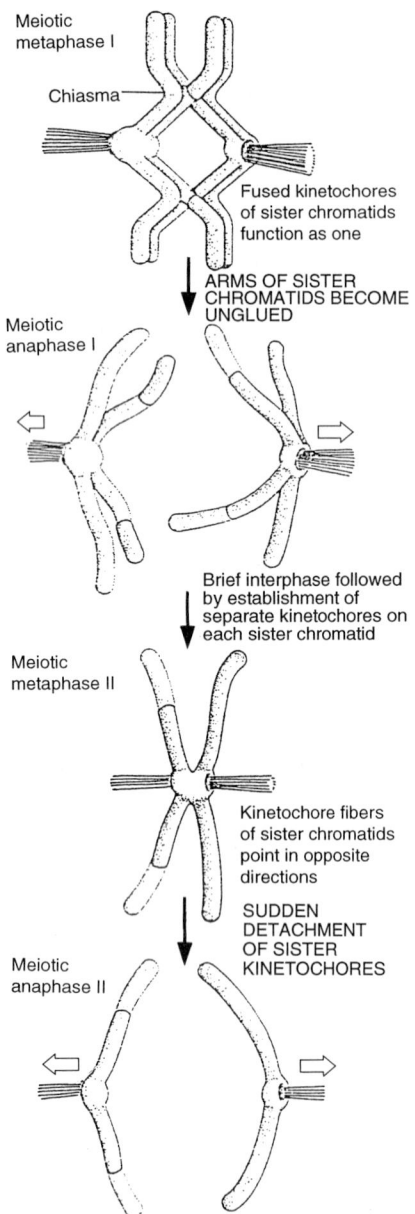

FIGURE 3 Comparison of the mechanisms of chromosome alignment (at metaphase) and separation (at anaphase) in meiotic division I and meiotic division II. The mechanisms used in meiotic division II are the same as those in normal mitosis (reproduced from Alberts et al., 1989).

nance of this binding relies solely on chiasmata (Fig. 3). Thus, these structures are not only the cytological manifestation of the crossing-over but also play a central role in ensuring the proper segregation of homologous chromosomes. Conversely, if chiasmata

TABLE 1
Examples of Genes Affecting Meiosis

Organism	Gene	Effect of mutation on		Other effects/comments
		Crossing-over	Chromosome segregation	
Caenorhabditis elegans	bim-6	Decreased frequency	Increased number of errors	
	rec-1	Altered distribution	Not reported	
	mei-1	Probably not affected	Increased number of errors	Acts via meiotic spindle function
	rad-4	Not reported	Reduced number of errors	Radiation sensitive
Drosophila melanogaster	ald	Not affected	Increased number of errors	May act via mechanism of chiasma maintenance
	mei-9	Decreased frequency	Increased number of errors	Element of DNA repair system
	mei-41	Decreased frequency	Increased number of errors	Abnormal morphology and reduced number of RN; mutants defective in DNA repair
	ncd	Not affected	Increased number of errors	Acts via meiotic spindle function
	grau, cort	Not reported	Increased number of errors during female meiosis I	Oocytes arrested at the metaphase II
Mus musculus	Mlh1	Decreased frequency	Male meiosis arrested at the metaphase I	Element of DNA repair system
	Pms2	Not reported	Not reported	Abnormalities in synapsis of homologs; element of DNA repair system
	Mos	Probably not affected	Probably not affected	Lack of the metaphase II arrest in oocytes

The limited paternal contribution to the origin of aneuploid embryos does not necessarily indicate that in male meiosis fewer errors occur. There is some evidence from studies on mice that spermatocytes abnormal in respect to chromosome number or configuration (e.g., those from *Mlh1* mutants; Table 1) are efficiently eliminated during spermatogenesis.

There exists an apparent correlation between the advanced maternal age and virtually all kinds of trisomy. The incidence of clinically recognized trisomic pregnancies increases more than 10-fold in women over 40 in comparison to those aged 20–25. Interestingly, there is a general tendency, seen in many organisms studied, in which the number of chiasmata decreases in older individuals. As mentioned earlier, the chiasmata are believed to ensure the proper segregation of homologs. This is in agreement with observations that in mutants with reduced frequency of reciprocal recombination the proportion of univalents drastically increases at the first meiotic metaphase (see Table 1). Whereas several models have already been proposed to explain the maternal age effect, the data on the potential effect of the paternal age are contradictory.

Recent studies in human and *Drosophila* demonstrate a strong correlation between the positioning of the sites of recombination and the susceptibility to meiotic error at the first or the second meiotic division. Distal exchanges result in the increased frequency of missegregation during the first meiosis, whereas exchanges in the centromeric region predispose to the failures in the second one. These observations demonstrate that processes occurring at the first meiotic prophase affect segregation of chromosomes

TABLE 2
Contribution of Parental Meiotic Errors
to the Origin of Trisomy in Humans

Parental origin	Stage of error	Number of informative cases (%)
Paternal	Total M I and M II	103 (12.8)
Maternal	M I	422 (52.5)
	M II	125 (15.5)
	M I or M II	154 (19.2)
Total		804 (100)

Note. Adapted from T. Hassold *et al.*, *Curr. Opin. Genet. Dev.* 3, 398–403, 1993.

not only during the first but also during the second meiotic division.

IV. MEIOSIS IN MALES AND FEMALES

The production of the four functional gametes from one cell entering meiosis occurs only in males. In female meiosis, cytokinesis at both the first and second meiotic division is unequal, in each case forming one large and one much smaller cell (Fig. 4). The small cells, termed polar bodies (the first or the second, depending on the meiotic cycle in which they are produced), are discarded and exist only as a means to remove excess genetic material from the cell which will become the functional ovum. Such a mode of meiosis minimalizes the loss of products stored in the oocyte necessary for the early embryonic development.

The male meiosis is characterized by an uninterrupted sequence of events. Populations of spermatogonia enter meiosis and progress smoothly to its end. In females, meiotic 'blocks' generally arrest the process at specific stages (Fig. 4). Most commonly, two meiotic blocks occur: the primary at the diplotene stage (the 'germinal vesicle' stage), and the secondary at metaphase (the first or second depending on species). It is a common feature of female meiosis that the fertilizing spermatozoon releases the oocyte from the secondary meiotic block (or from the primary one in the few species which exhibit a single meiotic arrest; Fig. 4). Thus, female meiosis is in most cases completed following fertilization, although the maternal- and the paternal-derived chromatin do not mix until the oocyte meiosis has finished.

The occurrence of meiotic blocks may greatly lengthen the period of female meiosis. In eutherian mammals, the whole population of oocytes enters meiosis synchronously in fetal ovaries, and around the time of birth all oocytes are already arrested at diplotene. The release from this block occurs in sexually mature females throughout the whole reproductive life. Within each estrous cycle, oocytes com-

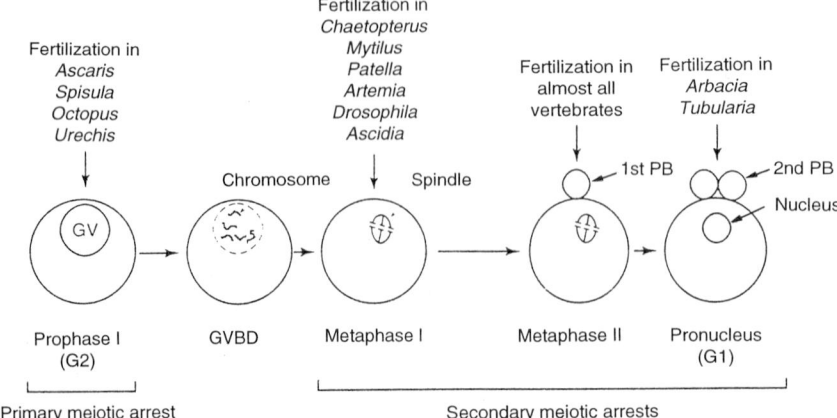

FIGURE 4 Stages of meiotic arrest in oocytes at which sperm penetration occurs in different animals. GV, germinal vesicle; GVBD, germinal vesicle breakdown; 1st PB, first polar body; 2nd PB, second polar body (reproduced from Sagata, 1996).

mitted to be ovulated resume meiosis. Thus, in humans the overall length of time of meiosis is about 6 weeks in males but 12–50 years in females.

V. CELL CYCLE CONTROL DURING MEIOSIS

The appearance of meiotic divisions during evolution requires modifications of cell cycle control. Cell cycle arrests during the meiotic prophase or metaphase of oocytes depend on specific molecular mechanisms. The first requires a mechanism preventing activation of the major M-phase or maturation-promoting factor (MPF), whereas the second requires the stabilization of this MPF. It is widely accepted that arrest of oocyte meiosis at meiotic prophase is achieved by the stable activity of protein kinase A. The later metaphase arrest may be maintained either physically by mechanical forces generated by the chiasmata (M I arrest in *Drosophila* oocytes) or by a cytoplasmic factor CSF (cytostatic factor) in unfertilized amphibian and mammalian eggs.

Another key event of meiosis depending on modified cell cycle regulation is the suppression of DNA replication during the second meiotic cycle. The regulation of this event, which occurs in all meiotic cells, is poorly understood. A possible mechanism could involve a protein kinase termed Mos (product of the c-*mos* protooncogene) which is expressed preferentially in all cells undergoing meiosis.

Although the function of Mos during male meiosis remains unknown, in *Xenopus laevis* oocytes it seems to be involved in the release from the prophase arrest and the repression of DNA replication. Mos is also implicated in the CSF activity during M II arrest in amphibian (*X. laevis*) and mammalian (mouse) oocytes. These divergent functions of a single protein in *Xenopus* might be determined by the presence of different substrates in different cells (oocytes vs spermatocytes) or during different periods of meiosis (primary vs secondary meiotic arrest in oocytes). One must stress, however, that these roles of Mos do not seem to be fulfilled in oocytes of some other species and phyla. It appears that different modifications of the cell cycle have evolved during evolution, as illustrated by the mechanisms of metaphase arrest in *Drosophila* and amphibians.

Bibliography

Alberts, B., Bray, D., Lewis, J., Raff, M., Roberts, K., and Watson, J. D. (1989). *Molecular Biology of the Cell*. Garland, New York.

Baker, S. M., Plug, A. W., Prolla, T. A., Bronner, C. E., Harris, A. C., Yao, X., Christie, D.-M., Monell, C., Arnheim, N., Bradley, Ashley, T., and Liskay, R. M. (1996). Involvement of mouse *Mlh1* in DNA mismatch repair and meiotic crossing over. *Nature Genet.* **29**, 336–342.

Eichenlaub-Ritter, U. (1996). Parental age-related aneuploidy in human germ cells and offspring: A story of past and present. *Environ. Mol. Mutagen.* **28**, 211–236.

Ferguson, L. R., Allen, J. W., and Mason, J. M. (1996). Meiotic recombination and germ cell aneuploidy. *Environ. Mol. Mutagen.* **28**, 192–210.

Griffin, D. K. (1996). The incidence, origin, and etiology of aneuploidy. *Int. Rev. Cytol.* **167**, 263–296.

Hassold, T. J. (1996). Mismatch repair goes meiotic. *Nature Genet.* **13**, 261–262.

Hawley, S. R., McKim, K. S., and Arbel, T. (1993). Meiotic segregation in *Drosophila melanogaster* females: Molecules, mechanisms, and myths. *Annu. Rev. Genet.* **27**, 281–317.

Lichten, M., and Goldman, A. S. H. (1995). Meiotic recombination hotspots. *Annu. Rev. Genet.* **29**, 423–444.

Maro, B., Kubiak, J. Z., Verlhac, M.-H., and Winston, N. J. (1994). Interplay between the cell cycle control machinery and the microtubule network in mouse oocytes. *Sem. Dev. Biol.* **5**, 191–198.

Moens, P. B. (Ed.) (1987). *Meiosis*. Academic Press, New York.

Moens, P. B. (1994). Molecular perspectives of chromosome pairing at meiosis. *BioEssays* **16**, 101–106.

Orr-Weaver, T. (1996). Meiotic nondisjunction does the two-step. *Nature Genet.* **14**, 374–376.

Sagata, N. (1996). Meiotic metaphase arrest in animal oocytes: Its mechanisms and biological significance. *Trends Cell Biol.* **6**, 22–28.

Wagner, R. P., Maguire, M. P., and Stallings, R. L. (1993). *Chromosomes: A Synthesis*. Wiley-Liss, New York.

Zetka, M., and Rose, A. (1995). The genetics of meiosis in *Caenorhabditis elegans*. *Trends Genet.* **11**, 27–31.

Meiotic Cell Cycle, Oocytes

Nava Dekel
The Weizmann Institute of Science

I. Introduction
II. Maintenance of Meiotic Arrest
III. Resumption of Meiosis
IV. The Meiotic Cell Cycle
V. Epilogue

GLOSSARY

gap junctions Specialized regions in adjacent membranes of neighboring cells that allow transcellular flow of ions and molecules that are smaller than 1 kDa.

maturation-promoting factor A key regulator of oocyte maturation composed of the regulatory cyclin B and the catalytic p34^{cdc2} kinase.

oocyte maturation Transformation of an oocyte into a fertilizable egg occurring upon release of the first prophase arrest and progression through metaphase I to the second metaphase of meiosis.

Meiosis is a particular example of cell division occurring in germ cells. This specialized cell cycle consists of two successive rounds of chromosome segregation that follow one round of DNA replication. Meiosis produces progeny cells with half as many chromosomes as their parents, thus making sexual reproduction possible. This article is concerned with the events that have been implicated in the control of meiosis in mammalian oocytes. Research in progress may reveal additional regulatory processes that govern the meiotic cell cycle.

I. INTRODUCTION

The entry of the oocyte into meiosis takes place in the fetal ovary and is accompanied by DNA synthesis analogous to S phase in mitotic cells. Meiosis proceeds up to the diplotene stage of the first prophase and is arrested at diakinesis, just prior to or shortly after birth. This stage corresponds to the G_2 phase of the cell cycle and is characterized by diffused chromosomes surrounded by an intact nuclear membrane termed "germinal vesicle" (GV). Reinitiation of meiosis, which occurs in fully grown oocytes after puberty, represents transition from G_2 to M phase and involves condensation of the interphase chromatin, dissolution of the nuclear membrane referred to as GV breakdown (GVB), spindle formation, and chromosome segregation. These oocytes complete the first meiotic division by the formation of the first polar body that is immediately followed by their maturation into unfertilized eggs arrested at the second metaphase of meiosis. Completion of the second meiotic division is triggered at fertilization, upon sperm penetration.

The protracted nature of the meiotic division, which is unique to the female gamete, is apparently subjected to a complex mode of regulation. The following three aspects related to the control of meiosis in mammalian oocytes will be discussed in detail: (i) the role of the ovarian follicle in maintenance of meiotic arrest, (ii) the mode of luteinizing hormone (LH) action that stimulates meiosis resumption, and (iii) biochemical and molecular events that govern the meiotic cell cycle.

II. MAINTENANCE OF MEIOTIC ARREST

Oocyte maturation *in vivo* is stimulated by the preovulatory surge of the pituitary LH. However, when meiotically arrested oocytes are removed from antral follicles, they resume meiosis spontaneously. This observation, which was initially reported for the rabbit and later extended to include other mammalian species, led to the hypothesis that the follicle provides an inhibitory factor that maintains the oocyte in meiotic arrest.

A. Oocyte Maturation Inhibitor

The demonstration that spontaneous maturation is inhibited when isolated oocytes are incubated in the presence of follicular fluid indeed supported the idea that a meiosis arrestor is produced by the ovarian follicle, further suggesting that this factor is secreted into the follicular antrum. The most extensive attempts to characterize the oocyte maturation inhibitor (OMI) present in the follicular fluid were conducted in the pig. These studies attributed the OMI activity to a small (<2 kDa) peptide apparently generated by the granulosa cells. They further revealed that the inhibitory activity in porcine follicular fluid declines in the course of follicular growth and could be reversed by LH. Follicular fluids from ovaries of rabbit, ovine, bovine, porcine, hamster, and human origin have been shown by several laboratories to exert an inhibitory effect on spontaneous maturation of isolated oocytes, whereas other investigators were unable to demonstrate OMI activity in preparations of some of the previously mentioned species.

B. Hypoxanthine

Problems in reproducing the inhibitory effect of follicular fluid on oocyte maturation, as well as the failure to purify the OMI to homogeneity, raised some doubts regarding the relevance of this factor. However, a later report demonstrated that, even though porcine follicular fluid exhibits a low inhibitory activity by itself, it synergizes dramatically with cAMP to suppress murine oocyte maturation. These studies claimed that hypoxanthine is the principal low-molecular-weight component of follicular fluid that inhibits oocyte maturation. Further studies by these investigators demonstrated that, in addition to its inhibitory action on GVB, hypoxanthine also sustained relatively high intraoocyte concentrations of cAMP. The conclusion of these studies, therefore, was that suppression of the oocyte cAMP phosphodiesterases activity appears to be one mechanism by which hypoxanthine maintains the oocyte in meiotic arrest.

C. Cyclic AMP

The aforementioned suggested cooperation between cAMP and hypoxanthine took into account previously reported observations that a membrane permeant derivative of cAMP or a phosphodiesterase inhibitor reversibly block the spontaneous maturation *in vitro* of isolated mouse oocytes. The negative regulation of meiosis by cAMP, which was further demonstrated for oocytes of several mammalian species, led to the hypothesis that cAMP could possibly serve as the follicular arrestor of meiosis. This intriguing possibility was later supported by a large body of accumulated evidence, including the following: (i) Spontaneous maturation *in vitro* is preceded by a sharp drop in intraoocyte cAMP, whereas no decrease in cAMP concentrations is observed in oocytes maintained meiotically arrested by a phosphodiesterase inhibitors; (ii) experimental elevation of intraoocyte cAMP concentrations is always associated with maintenance of meiotic arrest, whereas a drop in cAMP levels is followed by resumption of meiosis; (iii) postovulatory mature oocytes contain lower concentrations of cAMP than follicular immature oocytes; and (iv) the presence of cAMP derivatives or inhibitors of cAMP phosphodiesterase prevent the LH-stimulated maturation of follicle-enclosed oocytes *in vitro*. These observations laid the groundwork for the currently accepted idea that cAMP concentrations within the oocyte regulate its meitotic status.

Intraoocyte concentrations of cAMP that drop sharply upon separation of the oocyte from the ovarian follicle suggest that the oocyte lacks the ability to produce this inhibitor on its own. Supporting this idea, it has been demonstrated that in the rat neither cholera toxin nor forskolin, which interact with the regulatory GTP-binding component and the catalytic subunit of adenylate cyclase, respectively, were able to induce a rise in oocyte concentrations of cAMP. Mouse oocytes also failed to respond to cholera toxin; however, when exposed to forskolin, they did generate sufficient cAMP levels to inhibit their spontaneous maturation. Nevertheless, their response to forskolin is apparently of no physiological relevance since, similar to the rat, a rapid decrease in cAMP concentrations followed by spontaneous maturation is observed in mouse oocytes immediately after their isolation from the ovarian follicle. These latter findings point toward a limited ability of the oocyte, of both these species, to produce cAMP under physiological conditions at levels that are sufficient to maintain meiotic arrest.

Despite their limited potential to generate cAMP, the follicular oocytes do contain inhibitory levels of this nucleotide. This cAMP is apparently transmitted to the oocyte from the follicle cells via gap junctions present at the regions of contact between the projections of the follicular cumulus cells and the oolema. Transfer of cAMP from the cumulus cells to the oocytes is suggested by the significantly higher concentrations of the nucleotide present in oocytes derived from forskolin-stimulated cumulus–oocyte complexes compared to similarly stimulated cumulus-free oocytes. Transfer of cAMP to the oocyte can also explain earlier findings that activators of adenylate cyclase could delay the spontaneous maturation of oocytes cultured within their cumuli but failed to affect the oocytes in the absence of the attached cumulus cells. Taken together, the results of these experiments suggest that cAMP, generated by the cumulus cells, is transferred to the oocyte to maintain it in meiotic arrest.

III. RESUMPTION OF MEIOSIS

A. Oocyte Maturation-Inducing Agents

As mentioned earlier, oocyte maturation is triggered by the pituitary LH. Thus, blocking the preovulatory surge of gonadotropins in the rat by nembutal, hypophysectomy, or administration of an antiserum to the β subunit of LH all prevent oocyte maturation. Conversely, administration of exogenous LH to nembutal-treated rats results in reinitiation of meiosis. LH-induced resumption of meiosis in oocytes has also been demonstrated *in vitro* in explanted ovarian follicles. When incubated in hormone-free medium, the oocytes that reside in these follicles remain meiotically arrested; upon addition of LH, these oocytes resume meiosis in culture.

The stimulatory action of LH on oocyte maturation is limited to the large Graafian follicles. In small antral follicles in which the granulosa cells do not possess LH receptors, resumption of meiosis cannot be induced by this gonadotropin. Thus, the selection of the cohort of ovarian follicles that will produce mature oocyte ready to ovulate at a specific sexual cycle is determined by the expression of the relevant receptors by the granulosa cells. The immediate cellular response to the interaction of LH with its receptor is activation of the adenylate cyclase and generation of cAMP. Forskolin that bypasses this hormone–receptor interaction was found to be a potent inducer of maturation in follicle-enclosed oocytes. Induction of oocyte maturation by forskolin was associated with elevation of cAMP in the follicle and was potentiated by a transient exposure to a phosphodiesterase inhibitor, suggesting that LH-induced oocyte maturation is a cAMP-mediated response. This last conclusion seems to present an apparent paradox. If cAMP mediates LH action to induce oocyte maturation, what mechanism allows the oocyte to mature under the influence of LH since cAMP inhibits oocyte maturation?

The contradiction could possibly be explained by assuming that these two responses represent different sensitivities to cAMP. This assumption was tested by the analysis of the dose response of cAMP-mediated induction compared to inhibition of oocyte maturation. Maturation was induced in follicle-enclosed oocytes by transiently exposing them to either the permeant derivative of cAMP, dibutyryl cAMP (dbcAMP) or the phosphodiesterase inhibitor methylisobutylxanthine (MIX). Inhibition of maturation was obtained by addition of the previously mentioned agents either to follicle-enclosed oocytes incubated in the presence of LH or to isolated oocytes that mature spontaneously *in vitro*. These experiments revealed that the concentrations of either dbcAMP or MIX, needed for induction of meiosis resumption, were fourfold higher than those required for inhibition of oocyte maturation suggesting that cAMP plays a dual role in regulation of meiosis. Basal levels of cAMP, transferred to the oocyte, result in maintenance of meiotic arrest, whereas elevated, LH-stimulated levels of the cyclic nucleotide mediate the induction of oocyte maturation.

The apparent opposite responses of the oocyte to cAMP are not only due to sensitivity variations but also to the different specific target cells for the nucleotide action. The oocyte is the target for the inhibitory action of cAMP. On the other hand, it is the response of the cumulus cells to this cyclic nucleotide that leads to oocyte maturation. The mechanism by which the follicular somatic cells mediate LH-induced oocyte maturation will be discussed later.

Meiosis can also be induced in follicle-enclosed rat oocytes by factors which do not enhance follicular cAMP production. Thus, the hypothalamic gonadotrophin-releasing hormone (GnRH) or its agonistic analogs successfully promote resumption of meiosis in rat oocytes both *in vivo* and *in vitro*. The demonstration that activators of protein kinase C (PKC), such as phorbol ester, a membrane permeate derivative of diacylglycerol and phospholipase C, also induce meiosis in follicle-enclosed oocytes suggests that PKC activation may mediate the effect of GnRH on the oocyte. Induction of oocyte maturation by epidermal growth factor (EGF), transforming growth factor-α, and basic fibroblast growth factor, suggests that ligand-activated tyrosine kinases may represent an additional signaling cascade capable of stimulating resumption of meiosis. The mechanism of tyrosine kinase receptor-induced resumption of meiosis has not been studied. The physiological relevance of either PKC or tyrosine kinase-mediated induction of oocyte maturation is largely unknown.

B. The Mechanism of LH Action

Evidence that elevated levels of cAMP in the oocyte are responsible for arrest of meiosis indicated that any mechanism involved in LH-induced meiosis reinitiation should lead to a drop in intraoocyte cAMP concentrations. It was predicted that if maintenance of meiotic arrest is dependent on communication of cAMP, then uncoupling of the oocyte from the cumulus cells will terminate the supply of this inhibitor, providing the appropriate conditions for resumption of meiosis. This assumption was supported by early demonstrations that LH as well as human chorionic gonadotropin (hCG), which binds to the ovarian follicle LH receptors, uncouple the oocyte from the cumulus cells. Time analysis performed later revealed that coupling in the rat cumulus–oocyte complexes is sharply decreased at 1 hr of incubation with LH, a time point at which almost all the oocytes are meiotically dormant. Similar chronological relationships between uncoupling and reinitiation of meiosis were found *in vivo* in cumulus–oocyte complexes isolated from hCG-treated rats. Uncoupling in the cumulus–oocyte complex was also correlated to meiosis resumption when other inducing agents, such as GnRH or EGF, were employed. A recent study demonstrated that heptanol, a seven-carbon alcohol that blocks cell-to-cell communication in various experimental systems including the cumulus–oocyte complex, reduces intraoocyte concentrations of cAMP and promotes maturation of rat follicle-enclosed oocytes. These findings suggest that breakdown of communication can serve as a sufficient signal for the induction of oocyte maturation.

A decrease in communication in the cumulus–oocyte complex during early stages of meiosis was also described in the hamster. However, the results reported in the sheep and mice seem to suggest that reinitiation of maturation precedes the reduction in cumulus cell–oocyte coupling. Later studies performed in these and other animal species suggest that the granulosa cells serve as the major source for cAMP, whereas the cumulus cells provide mainly the channels for transmission of this inhibitory signal. Therefore, uncoupling of the oocyte from the cumulus cells creates conditions that are sufficient, but not necessary, for resumption of meiosis. Alternatively, in those species in which a decrease in coupling in the cumulus–oocyte complex could not be detected prior to onset of meiosis, downregulation of gap junctions at the interface between the cumulus stalk and moral granulosa cells could possibly lead to a substantial decrease in transfer of cAMP to the oocyte, resulting in oocyte maturation.

The response of the junctional communication network in rat ovarian follicles to LH has recently been analyzed at the biochemical and molecular levels. It was revealed that the ovarian gap junction protein, connexin 43 (Cx43), undergoes phosphorylation within 10 min of exposure to this gonadotropin which is immediately followed by dephosphorylation of the protein. At later time points of exposure to LH a progressive decrease in the amount of the protein followed by its total elimination could be detected. Disappearance of Cx43 was associated with a reduction in the level of its mRNA. This study suggests that the immediate LH-induced phosphorylation/dephosphorylation of Cx43 may result in conformational changes of the protein that can possibly lead to a decrease in coupling in the ovarian follicle, observed in previous studies shortly after exposure to this gonadotropin. The later effect of LH on gene expression, which results in a reduced synthesis of the protein, apparently accounts for the disappearance of the gap junctions observed by morphological studies at later time points of exposure to the gonadotropin.

IV. THE MEIOTIC CELL CYCLE

Regulation of oocyte maturation by the somatic components of the ovarian follicle has been discussed in detail previously. This section will focus on the intraoocyte biochemical mechanisms involved in the regulation of the meiotic division of the female gamete.

A. Cell Cycle-Regulated Kinases

Until quite recently, very little was known about the control of cell division. However, extensive studies during the past 10 years have substantially advanced our understanding of the biochemical and molecular basis of the cell cycle machinery, with most impressive progress made toward description of the events leading to entry into M phase. The initial landmark in this field was established by the discovery that maturing *Rana pipiens* oocytes produce a factor that causes them to resume meiosis. This factor, which was termed "maturation-promoting factor" (MPF), has been demonstrated in other amphibian as well as starfish oocytes. In addition to maturing oocytes, MPF activity has been found in mitotically dividing cells. Because MPF appears to control the transition from G_2 to M phase in both meiosis and mitosis, the acronym MPF was later proposed to mean "M phase-promoting factor."

Purification of *Xenopus* oocyte MPF allowed its identification as a hetrodimer, one component of which is a 34-kDa protein homologous to the product of the *cdc2* gene of the fission yeast, which is termed $p34^{cdc2}$. This protein, which is highly conserved among eukaryotes, is a serine/threonine kinase that exhibits a strong preference for histone H1 as a substrate. The second component of MPF is a 45-kDa B-type cyclin. Cyclin is subjected to periodic synthesis and degradation, whereas the levels of $p34^{cdc2}$ remain constant throughout the cell cycle. Association with cyclin is regarded as essential for $p34^{cdc2}$ kinase activity. Another obligatory step for activation of the cyclin/$p34^{cdc2}$ complex is the removal of the phosphates from threonine 14 and tyrosine 15 of $p34^{cdc2}$. This step is regulated by specific protein tyrosine phosphatases and kinases homologous to the protein products of the *cdc25* and *Wee1* genes of fission yeast, respectively.

The central regulatory role of MPF at the onset of M phase has been initially established, mostly by studies conducted on either invertebrate oocytes that resume the meiotic division or other cells that undergo mitosis. Later, experiments demonstrated that MPF is also activated in maturing oocytes of various mammalian species. These experiments revealed that MPF activity oscilates in exact correspondence with the meiotic cell cycle; it appears initially at GVB, reaches a high level at metaphase I, and then disappears transiently at the time of first polar body emission. Thereafter, MPF activity reappears at metaphase II and remains at an elevated level until fertilization.

Reinitiation of meiosis in rat and mouse oocytes is associated with tyrosine dephosphorylation of $p34^{cdc2}$. Removal of the phosphates from $p34^{cdc2}$ kinase is essential for meiosis resumption since addition of vanadate, an inhibitor of tyrosine phosphatases that prevents tyrosine dephosphorylation of $p34^{cdc2}$ in these oocytes, blocks their spontaneous maturation.

The phosphorylated form of $p34^{cdc2}$ present in G_2-arrested oocytes disappears soon after the onset of meiosis, with no further change until metaphase II. On the other hand, histone H1 kinase activity, which was high during metaphases I and II, did decrease at polar body extrusion. These results suggest that, in maturing mammalian oocytes, dephosphorylation of $p34^{cdc2}$ triggers its activation upon entry into the first metaphase. However, the previously mentioned cell cycle-dependent oscillatory pattern of MPF activity, observed upon exit from the first and entry into the second meiotic metaphase, is not associated with another round of phosphorylation/dephosphorylation of $p34^{cdc2}$.

Information related to cyclin in mammalian oocytes is scarce. Cyclin synthesis, in the first but not the second meiotic cell cycle, is denied indirectly by early reports that in mouse oocytes treated with inhibitors of protein synthesis, meiotic maturation is blocked before the first polar body emission, but GVB and chromosome condensation proceed. Accordingly, a later study demonstrated that in maturing mouse oocytes, the formation of MPF requires protein synthesis during the second but not the first metaphase. Taken together, these findings suggest that, in rodent oocytes, cyclin synthesis requirement is completed well before transition into M phase and a pool of inactive $p34^{cdc2}$/cyclin heterodimers or "pre-MPF" is held in an inactive state by a posttranslational block.

On the other hand, in species with a longer interval between the stimulus and GVB, such as pig, cow, and sheep, *de novo* protein synthesis is required for GVB. These observations possibly suggest that in larger mammals, accumulation of cyclin in G_2-arrested oocytes did not reach the particular threshold level sufficient to support reinitiation of meiosis, requiring further *de novo* synthesis of the protein. However, this assumption is not supported by a recent report of the presence of B-type cyclin in prophase-arrested pig oocytes. Nevertheless, it is still possible that other, as yet unknown, regulatory proteins, localized upstream to MPF, are absent in G_2-arrested porcine, bovine, and ovine but not rodent oocytes.

One such protein candidate could possibly be Mos, a serine/threonine kinase that, at least in *Xenopus* oocytes, has been identified as a regulator of MPF activity throughout meiosis. The c-*mos* protooncogene is expressed exclusively in male and female germ cells, suggesting a specific function of its protein product in the meiotic cell cycle. Recent experiments used targeted disruption of the c-*mos* gene in embryonic stem cells to generate a *mos*-deficient mouse model. Analysis of the oocytes of these mutant mice revealed that their capacity to reinitiate meiotic maturation and their potential to progress through metaphase I to metaphase II were not affected. However, their ability to arrest at metaphase II was severely impaired resulting in spontaneous parthenogenic activation. Recent studies demonstrated that oocytes of these mutant mice kept their capacity to activate histone H1 kinase upon entry to the first as well as the second metaphase. However, they were unable to maintain the elevated level of the active MPF in metaphase II oocytes. These observations seem to deny a major role for Mos in MPF activation at the onset of meiosis, at least in mouse oocytes (Fig. 1).

Mos is a potent activator of mitogen-activated protein (MAP) kinases. Members of the MAP kinase family are serine/threonine protein kinases that are activated upon their phosphorylation on specific tyrosine and threonine residues. The presence of two

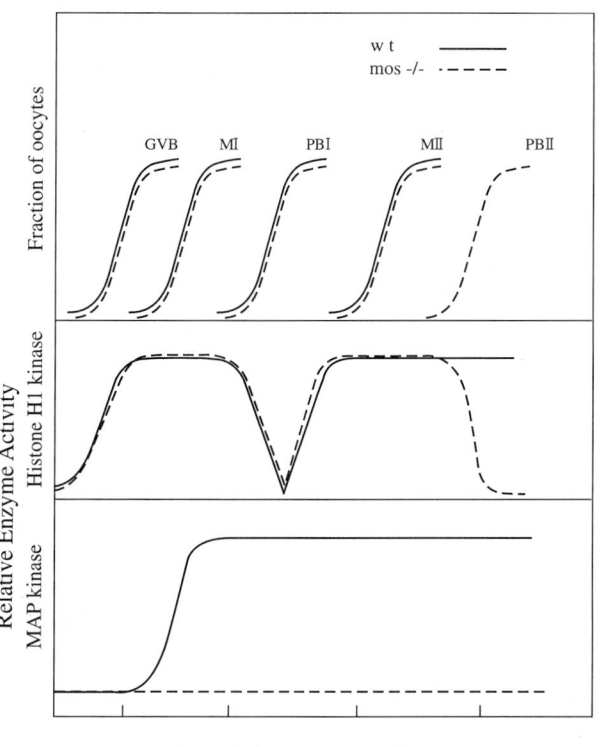

FIGURE 1 Cellular and biochemical events throughout resumption of meiosis in wild-type (wt) and mutant (mos−/−) mouse oocytes.

isoforms, 42 and 44 kDa of MAP kinase, has been identified in oocytes of various mammalian species and their activity has been shown to increase after reinitiation of meiosis. The elevation of MAP kinase activity is somewhat delayed when compared to histone H1 activation, correlating better temporally with spindle formation rather than GVB. MAP kinase-independent GVB is also demonstrated in the previously mentioned *mos*-deficient mouse model in which dissolution of the GV occurred normally in the absence of activation of MAP kinase. The demonstrated capacity of oocytes of this mutant mouse to progress through metaphases I and II of meiosis seems to suggest that, similar to Mos, MAP kinase in mammals is only responsible for maintenance of metaphase II arrest (Fig. 1). Other studies, however, suggest that MAP kinase is involved in reorganization of the microtubules and appropriate orientation of the spindle at metaphase I in addition to its contribution to the stabilization of the spindle at metaphase II.

B. Inhibition by cAMP

The information summarized so far suggests that GVB in mammalian oocytes is subjected to regulation by $p34^{cdc2}$ kinase, whereas MAP kinase involvement in oocyte maturation is mainly at the level of spindle microtubule organization and metaphase II arrest. The regulatory relationships between these two cell cycle kinases are largely unknown but available experimental evidence implies that both of them are subjected to negative regulation by cAMP.

Inhibition of MPF activation by cAMP was suggested by early experiments demonstrating that *Xenopus* oocytes are induced to resume meiosis following fusion with maturing but not with dbcAMP-arrested mouse oocyte. On the other hand, fusion of metaphase II mouse oocytes rapidly induced GVB in dbcAMP-arrested homologous oocytes, suggesting that once cytoplasmic MPF has been activated, its ability to induce the transition of nuclei to metaphase is no longer sensitive to cAMP. Recent studies suggest that it is not the synthesis of $p34^{cdc2}$ but rather post-translational modification of this protein that is regulated by cAMP in the meiotic prophase. Specifically, it has been demonstrated that dephosphorylation of $p34^{cdc2}$, which is associated with reinitiation of meiosis of mouse and rat oocytes, is reversibly inhibited by the phosphodiesterase inhibitor MIX that maintains a relatively high intraoocyte concentration of cAMP, inhibiting the resumption of meiosis. These findings offer evidence for a possible negative regulation of $p34^{cdc2}$ kinase activity by cAMP. Further demonstrations of vanadate-inhibited meiosis resumption that is independent of intraoocyte concentrations of cAMP indicate that the maturation-associated decrease in intracellular concentrations of cAMP is essential, but not sufficient, for reinitiation of meiosis. They may also suggest that $p34^{cdc2}$ dephosphorylation is distal to the decrease in cAMP.

The elevation of MAP kinase activity associated with resumption of meiosis is effectively prevented by relatively high intraoocyte concentrations of cAMP. As mentioned previously, these conditions also block the dissolution of the GV membrane. Recent unpublished results in our laboratory demonstrate that selective inhibition of MAP kinase activation in rat oocytes does not interfere with GVB, despite the fact that both these events are subjected to regulation by cAMP. Dissolution of the GV that is independent of MAP kinase has actually been demonstrated in the *mos*-deficient mouse model. Taken together, these findings suggest that the MPF-mediated GVB and the MAP kinase-mediated spindle organization may share a common upstream regulator that is controlled by cAMP.

It has been reported that an inhibitor of cAMP-dependent protein kinase (PKA) can induce maturation in MIX-arrested mouse oocytes. This observation suggests that an activated PKA mediates the negative action of cAMP on resumption of meiosis. Further experimental evidence related to the biochemical nature of the phosphoprotein that links the cAMP-activated PKA to the downstream regulatory kinases of the cell cycle in the mammalian oocyte is not available.

V. EPILOGUE

The studies discussed herein suggest that the follicle acts as one physiological unit in which the functional syncytium generated by gap junctions between granulosa cells, cumulus cells, and the oocyte plays

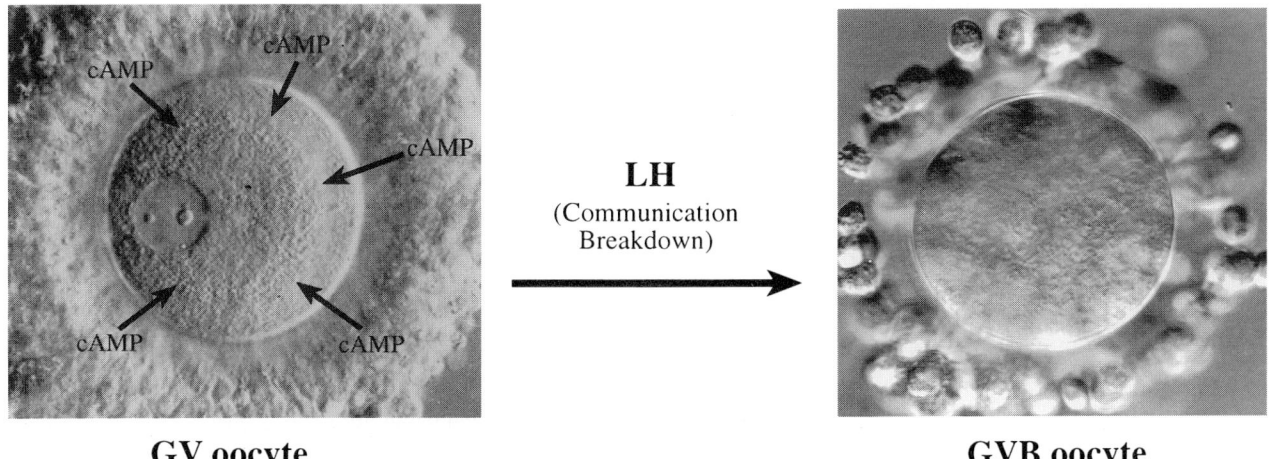

FIGURE 2 LH-induced oocyte maturation mechanism of action.

a crucial role in regulation of oocyte maturation. The basal levels of cAMP generated by the somatic cells in small antral follicles are continuously transferred to the oocyte to maintain it in meiotic arrest. Upon formation of LH receptors in the cumulus–granulosa cells of a selected group of maturing follicles, they acquire the ability to respond to the preovulatory surge of LH. This stimulated follicle generates increasing amounts of cAMP. The elevated concentrations of cAMP mediate the action of LH on the ovarian gap junction protein Cx43. Phosphorylation/dephosphorylation reactions probably result in conformational changes of Cx43, inducing an immediate reduction of cell-to-cell communication within the

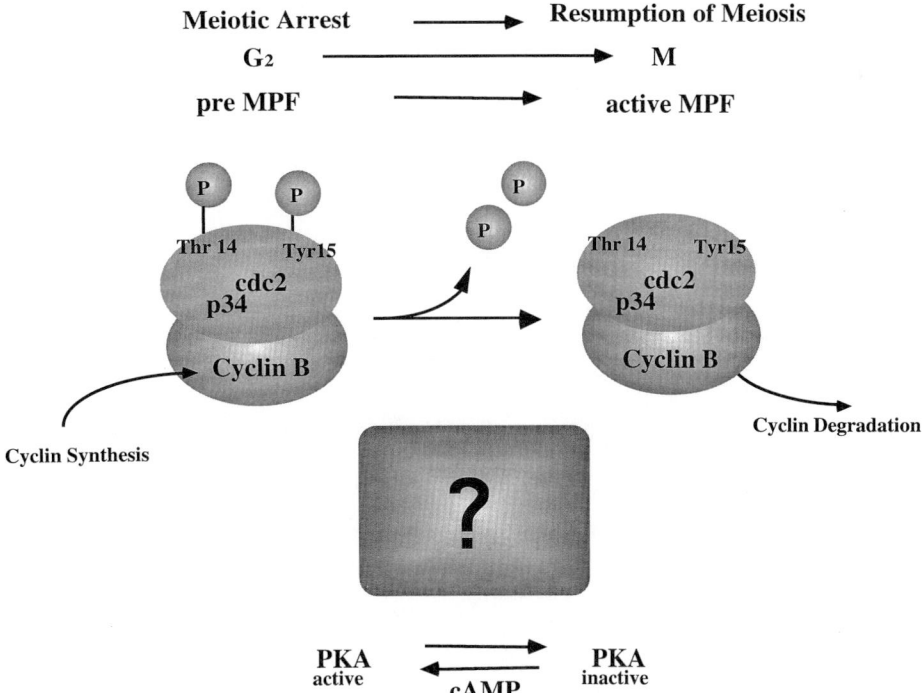

FIGURE 3 The meiotic cell cycle in mammalian oocytes.

ovarian follicle. Under these conditions, the flow of cAMP from the follicle cells to the oocyte decreases below the threshold level required to maintain meiotic arrest and the oocyte resumes meiotic maturation (Fig. 2). The current knowledge regarding the mechanism by which cAMP elicits its negative regulation on the meiotic cell cycle is limited. Nevertheless, there is evidence that cAMP action in the oocyte is mediated by PKA that, by an unknown cascade of biochemical reactions, prevents the dephosphorylation of p34^{cdc2}, maintaining MPF in an inactive state (Fig. 3).

See Also the Following Article

MEIOSIS

Bibliography

Araki, K., Naito, K., Haraguchi, S., Suzuki, R., Yokoyama, M., Inoue, M., Aizawa, S., Toyoda, Y., and Sato, E. (1996). Meiotic abnormalities of c-*mos* knockout mouse oocytes: Activation after first meiosis or entrance into third meiotic metaphase. *Biol. Reprod.* 55, 1315–1324.

Choi, T., Fukasawa, K., Zhou, R., Tesssarollo, L., Borror, K., Resau, J., and Vande Woude, G. F. (1996). The Mos/mitogen-activated protein kinase (MAPK) pathway regulated the size and degradation of the first polar body in maturing mouse oocytes. *Proc. Natl. Acad. Sci. USA* 93, 7032–7035.

Dekel, N. (1996). Protein phosphorylation/dephosphorylation in the meiotic cell cycle of mammalian oocytes. *Rev. Reprod.* 1, 82–88.

Downs, S. M. (1993). Purine control of mouse oocyte maturation: Evidence that nonmetabolized hypoxanthine maintains meiotic arrest. *Mol. Reprod. Dev.* 35, 82–94.

Goren, S., and Dekel, N. (1994). Maintenance of meiotic arrest by a phosphorylated p34^{cdc2} is independent of cyclic adenosine 3′,5′-monophosphate. *Biol. Reprod.* 51, 956–962.

Granot, I., and Dekel, N. (1994). Phosphorylation and expression of connexin-43 ovarian gap junction protein are regulated by luteinizing hormone. *J. Biol. Chem.* 269, 30502–30509.

Granot, I., and Dekel, N. (1997). Development expression and regulation of the gap junction protein and transcript in rat ovaries. *Mol. Reprod. Dev.* 47, 231–239.

Tsafriri, A., and Dekel, N. (1994). Molecular mechanisms in ovulation. In *Molecular Biology of the Female Reproductive System* (J. K. Findlay, Ed.), pp. 207–258. Academic Press, San Diego.

Verlhac, M.-H., Kubiak, J. Z., Weber, M., Geraud, G., Colledge, W. H., Evans, M. J., and Maro, B. (1996). Mos is required for MAP kinase activation and is involved in microtubule organization during meiotic maturation in the mouse. *Development* 122, 815–822.

Zernicka-Goetz, M., Verlhac, M.-H., Geraud, G., and Kubiak, J. Z. (1997). Protein phosphatases control MAP kinase activation and microtubule organization during rat oocyte maturation. *Eur. J. Cell Biol.* 72, 30–38.

Melatonin

see Pineal Gland, Melatonin Biosynthesis and Secretion

Menarche

Harry Hatasaka

University of Utah School of Medicine

I. Demographics
II. Physiology
III. Factors Influencing the Onset of Menarche
IV. Clinical Evaluation

GLOSSARY

adolescence The period of time from the appearance of secondary sex attributes until the completion of physical maturation.

adrenarche A physiologic change resulting in augmentation of adrenal cortex function marked especially by androgen production and release. It generally occurs at approximately 8 years of age and is manifest by the appearance of pubic and axillary hair.

gonadarche The reactivation of the quiescent but competent pituitary–ovarian axis by maturation of hypothalamic pulsatile gonadotropin-releasing hormone release ultimately resulting in the ovarian production of estrogen.

menstruation The physiologic sign marked by the cyclic passage of endometrial effluvium during an ovulatory cycle when pregnancy does not occur. Menstruation is distinguished from withdrawal bleeding, which is characterized by the lack of ovulation, such as uterine bleeding following a course of oral contraceptives.

pubarche The initiation of growth of pubic hair

puberty The period of time analogous to adolescence when physiologic maturation culminates in the capability of reproduction.

thelarche The pubertal milestone representing the start of breast development.

Menarche is defined as the occurrence of a woman's first menstrual period. It is therefore a discrete event as opposed to puberty, which refers to a progressive series of events leading up to reproductive competence.

I. DEMOGRAPHICS

The appearance of menarche is a conspicuous personal event that few women forget. To the unprepared girl, the appearance of the first bleeding can be frightening. In the Arab culture, both genders speak freely of the event, so to Arab girls menarche is anticipated. For Arabs, special meaning is attached to the menarche because it represents the all-important transition from girlhood to womanhood. The first menses signifies the time for Arab women to don veils and also signifies that the woman has become eligible for marriage.

When it occurs at an expected age, menarche is a reassuring sign that the maturation of the reproductive system is progressing normally. The reported normal age range for menarche in the United States is from 9.1 to 17.7 years. The mean chronological age of menarche in North American girls is 12.7 ± 1.0 (mean ± SD).

It appears that the age of menarche has fallen substantially in the past century in developed countries. In Europe during this time there has been a 3- or 4-month earlier onset of menarcheal age on average for every decade. This is likely due to better nutrition and the higher recorded average body weights of girls. The trend appears to have leveled in the United States in the past generation.

II. PHYSIOLOGY

Menarche does not necessarily represent reproductive competence. The event generally occurs in the middle of the pubertal process and signifies advancing maturation of the hypothalamic–pituitary–ovarian–uterine axis. The first sign of puberty tends to be the start of the growth spurt first noted at the approximate age of 8, followed by thelarche (mean age, 9.8), pubarche (mean age, 13.7), and then menarche at a median age of 12.8 (range, 10–16 years). Menarche tends to occur approximately 1 year after a girl's growth rate has peaked and within 2 years of the start of thelarche. In fact, there is a rather constant interval of 2.3 ± 1.0 (SD) years between thelarche and menarche regardless of the chronologic age of thelarche. For a human female, by the second trimester, the hypothalamic–pituitary axis is capable of gonadotropin secretion, which is subject to negative feedback control from sex steroids by the late third trimester. Likewise, the uterus is capable of physical menstruation by birth. This is evidenced by the occasional postpartum withdrawal bleed noted in some female infants primed by the gestational estrogens. Moreover, premenarchal ovaries will respond to follicle-stimulating hormone (FSH) with follicular development and atresia and estradiol production. Despite the functional nature of the hypothalamic–pituitary–ovarian–uterine axis, only very low levels of gonadotropin secretion are noted throughout infancy until approximately 9 years of age. Two mechanisms appear to subdue the axis during these years. First, the negative feedback mechanism of sex steroids on gonadotropin release is quite sensitive. There is also a centrally mediated repression of gonadotropin-releasing hormone (GnRH) release from the hypothalamus. Eventually, de-repression of the central mechanism allows progressive increases in FSH which correlate with increasing ovarian follicular diameters. Menarche occurs when adequate endometrial development induced by rising estrogen exposure allows a first withdrawal bleed. Ovulation typically does not occur for 12–24 months after menarche.

The molecular and even the cell signaling mechanisms involved in the control of puberty and menarche initiation are not yet elucidated. However, it is instructive to try to determine the typical hormonal milieu at the time of menarche. The main tools used in studying the pubertal sequence have been the observation of hormonal expression, including GnRH, gonadotropins, sex steroids, growth hormone, and insulin-like growth factor-1, in correlation with physical expressions of secondary sexual characteristics. The recent availability of highly sensitive assay techniques for gonadotropins and sex steroids has further elucidated the dynamic sequence of these hormones leading up to menarche. Indeed, it is not possible to measure the direct hypothalamic production of GnRH *in vivo* so that its release must be inferred from peripheral measurements of gonadotropins. Using frequent sampling, algorithms such as Cluster analysis have been developed to decipher the pulsatility of GnRH release. It is the "pulse generator" located within the arcuate nucleus of the hypothalamus that ultimately directs the progression of pubertal events including menarche. The seminal work of Knobil and colleagues led to the understanding that pulsatile release of GnRH is imperative to effect pituitary gonadotropin release. GnRH given in a continuous fashion results in the "downregulation" of GnRH receptors on pituitary gonadotropes and the consequent absence of gonadotropin production and secretion. This is a property that translates into important clinical utility for the treatment of centrally mediated precocious puberty.

GnRH activity manifests as irregular, low-amplitude pituitary luteinizing hormone (LH) and FSH pulses in the midchildhood ages of approximately 1 to 9. The presence of these pulses and their eradication with the use of GnRH antagonists demonstrates the presence and functional activity of a GnRH pulse generator in prepubertal girls.

Notably, as maturation of the GnRH center gradually occurs with time, these pulses are amplified within an hour of the onset of sleep in prepubertal and pubertal girls. Gradually, the pulses tend to become more regular and of higher amplitude.

Even prepubertally, a difference in estradiol serum concentrations is apparent because there is an incomplete central blockade (tonic central nervous system inhibition) of GnRH production. Using a highly sensitive recombinant cell bioassay, prepubertal girls were found to have eightfold higher concentrations of estrogen levels than prepubertal boys. Earlier female gonadal differentiation in the presence of tonic, ambient gonadotropins may explain the difference between the genders. Thus, "ovarian activation" may take place much sooner than previously presumed, indicating that the endocrinologic basis for puberty begins well before the manifestation of secondary sexual characteristics.

Not only do gonadotropin concentrations begin to rise prepubertally due to increasing GnRH exposure but also the response is amplified by a remarkable increase in sensitivity of the pituitary to GnRH. A centrally mediated alteration in the pattern of GnRH release (an infrequent, low-amplitude release) may serve to "prime" the gonadotropes and may account for the sensitivity of the pituitary to higher GnRH stimulation. Only with this reintroduction of higher concentrations of gondadotropins are the full effects of ovarian function manifested.

In a cross-sectional study of prepubertal girls, those who underwent the onset of puberty (as detected by the onset of breast development) within 8 months of hormonal measurement had significantly higher concentrations of LH, FSH, and estradiol than their counterparts who did not begin puberty within 8 months.

By the onset of puberty, LH concentrations increase markedly and are associated with measurable increases in sex steroid production. As puberty progresses, daytime pulsatility (amplitude > frequency) increases while the sleep-entrained LH amplification diminishes until an adult pattern characterized by sleep LH concentrations being less than awake concentrations is attained at a mean age of about 15.7.

The increase in FSH is less marked than that for LH and the differential regulation of the two gonadotropins may be due to the increasing inhibin production from granulosa cells which imparts a selective negative feedback effect in conjunction with estradiol at the pituitary level. After menarche, the ratio of LH to FSH has increased and approaches unity (Fig. 1).

To summarize, the physical components are all competent by birth for menarche to occur but the system is repressed at the level of the central nervous system. The progression of function up until menarche finds the prepubertal ovary responding to circulating tonic FSH (mean 1 IU/liter compared to an LH mean of 0.4 IU/liter) with follicular growth and atresia. A relatively high female estradiol concentration in the 2.2 pmol/liter range compared to 0.3 pmol/liter for males of the same age results. However, the ability to initiate menarche requires "reactivation" of the GnRH pulse generator, resulting in a sharp increase in pituitary LH (pulse amplitude > pulse frequency). Overall increases in pituitary gonadotropins ensue (approximately 2.5-fold for FSH and about 4.5-fold for LH) due both to more exposure to pulsatile GnRH and to an increasing sensitivity of the pituitary to GnRH. Consequent ovarian stimulation results in steroidogenesis initiating endometrial growth. Generally, postmenarchal circulating estradiol concentrations exceed 25 pg/ml. In premenarchal girls, strong positive correlations have been observed between their ages, ovarian and uterine volumes, and serum gonadotropin and estradiol concentrations. Even before the onset of secondary sexual characteristics, as FSH increases, the ovarian volume and the number of small ovarian follicles

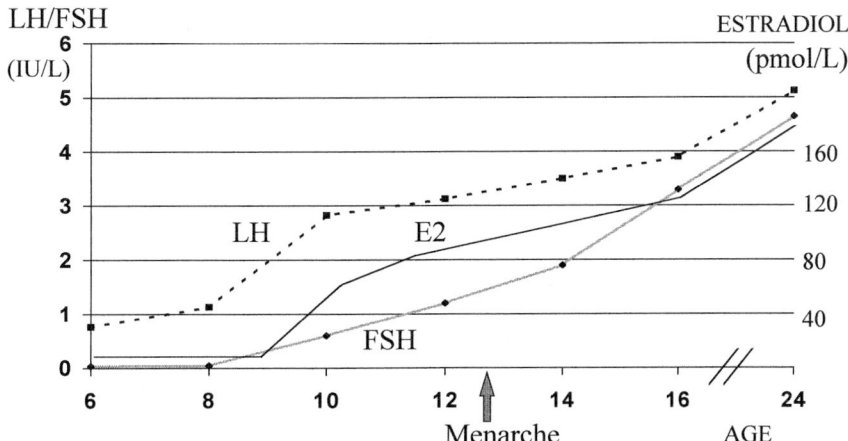

FIGURE 1 Mean female hormonal concentrations across puberty (extrapolated from Apter et al., 1989).

increase. Between Tanner breast stages I and II, significant mean increases in estradiol and gonadotropin concentrations are noted. At this phase (approximately 10 years of age) there is also a dramatic increase in uterine volume, especially of the corpus compared to the cervical tract. Even after menarche the uterine volume continues to increase under the influence of increasing sex steroid exposure.

In essence, menarche can be viewed as another secondary sexual characteristic such as thelarche. Whereas menarche does not reflect an acute dramatic change in the hormonal control of the reproductive system, it does represent maturation and intact function of the components of the hypothalamic–pituitary–ovarian–uterine axis. An individual threshold of circulating estradiol and the normal activation of endometrial growth and differentiation are prerequisites for menarche. In general, by the time menarche occurs, FSH and LH are in the normal adult ranges and resultant estradiol levels tend to achieve adult midfollicular values. Other classical hormones are not needed to explain the genesis of menarche. The physical event of menarche itself does not appear to impart any required property for successful reproduction because there are case histories of girls delivering their first children before menarche ensues.

III. FACTORS INFLUENCING THE ONSET OF MENARCHE

A. Exercise and Menarche

The age of menarche of a group of ballet dancers was noted to lag 3 years behind that of an age-matched control population. Frisch found that each year of training among competitive swimmers and runners delayed menarche by 5 months when rigorous training began before menarche.

Some effects of strenuous exercise physiologically include a decrease in the metabolic clearance rate of estradiol and an increased production of corticotropin-releasing hormone which can lead to inhibition of GnRH release with consequent alteration of menstrual cyclicity. Furthermore, levels of the so-called "antireproductive hormones," including prolactin, cortisol, melatonin, androgens, and β-endorphin and β-lipotropin, have been noted to rise with exercise. Ultimately, the hormonal consequences for some active athletes result in perturbation of the pulsatile secretion of LH which may lead to menstrual disturbances including the inhibition of menarche. Such women are at risk for osteoporosis due to hypoestrogenism, so a reduction in exercise or dieting and/or the addition of hormone replacement is indicated.

B. Stress and Menarche

Stress has been proposed as a factor in delaying menarche. Intensive athletics of various disciplines, such as ballet dancing and swimming, are associated with delayed menarche. Interestingly, some amenorrheic ballet dancers have been observed to regain their menses during holidays when not under the psychological stress of having to perform yet without having gained measurable weight. An hypothesized mechanism is an enhanced secretion of adrenal corticosteroids and catecholamines impacting gonadotropin secretion. However, the relative contributions of exercise, stress, metabolic energy redistribution, weight, and nutritional deprivation leading to delayed menarche or amenorrhea have been difficult to assess. Young musicians who are under heavy performance stress at early ages have not been found to have delayed menarche as have their ballet dancer peers.

C. Body Weight and Lean/Fat Content

That weight may be a determinant of the onset of menarche was suggested by the fact that the frequency distribution of age of menarche and weight is skewed toward earlier menarche at higher weights. However, inactivity, with its decreased metabolic drain, rather than the obesity could contribute as well. Frisch and McArthur have suggested that a critical weight for height, which represents a minimum fat to lean body weight ratio, is necessary for the onset of menarche, possibly because adipose tissue helps to convert precursor androgens to bioactive estrogens. This peripheral conversion accounts for approximately one-third of the circulating estrogen concentration in premenopausal women. Assuming that a critical level of estrogen is required to activate the pre-prepared reproductive axis to initiate the first cycle, several other effects of adipose tissue may be relevant. Adipose tissue can serve as a storage source of estrogen steroids. Moreover, obesity has been associated with a relative diminution in sex hormone-binding globulin which can contribute to a higher concentration of free and therefore active sex steroid fractions. Finally, body fat content helps to control the potency of estrogen activity within the body by promoting the metabolism toward the more potent 16-hydroxylated form rather than to the less potent 2-hydroxylated form.

In support of a role for body fat in the genesis of menarche, there is a large increase in the proportion of body fat (and a concomitant decrease of body water) in girls preceding menarche. Both critical absolute and relative amounts of fat have been proposed as being necessary for reproduction. At menarche, the average weight for girls in the United States is 47.8 (SE 0.5) kg and the average proportion of fat content to total body weight is 22%. Frisch has proposed that for Caucasian U.S. girls, a clinically helpful assumption is that girls with amenorrhea due to lack of adipose tissue need to attain a minimum body fat content of about 17% (which corresponds to the 10th percentile) for menarche to occur. The minimum absolute amount of body fat required for menarche is dependent on a girl's age and height. In support of a critical weight hypothesis is the observation that when menarcheal women lose approximately 10–15% of their body weight (about one-third of their body fat content), oligomenorrhea or amenorrhea usually ensues yet is reversible with weight gain. Also, after menarche, girls tend to gain another 4.5 kg of fat by 18 years of age. During this time period, cycles usually become ovulatory and total body fat increases to adult levels in the range of 26–28% of total body weight.

However, controversy exists because a number of researchers have been unable to identify a critical weight for the inducement of menarche. In fact, Zacharias and colleagues noted that the correlation between menarcheal age and weight is so weak with respect to a girl's age that the relationship between her weight and age of menarche is random. Thus, they regard a girl's size as a correlate rather than a determinant of menarche. They also noted that prematurely born girls tend to be smaller at the time of menarche, but their ages at the onset of menarche are identical to their larger counterparts born at full term. Likewise, a study of healthy U.S. girls raised in Rio de Janeiro noted that they were thinner, shorter, and lighter than a group of U.S. girls matched for socioeconomic background and ancestry. However, the mean age of menarche between the two groups remained the same.

D. Other Variables

Beside the factors already discussed, a number of other interesting environmental factors have been reported to affect the average ages of menarche among populations of girls. Concrete explanations for these influences have not been established (Table 1).

Put into perspective, when the environment is optimal, twin studies indicate that the next most influential factor on the age of menarche is genetic. Some apparent genetic predispositions have been reported. A significant correlation has been noted between the menarcheal ages of mothers and their daughters in

TABLE 1
Environmental Factors Influencing the Age of Menarche

The more severe the visual loss suffered by blind girls, the earlier the onset of menarche
Living at higher altitudes is associated with later onset of menarche
Living closer to the equator is associated with earlier menarcheal ages
Girls living in urban rather than rural areas tend to experience menarche at earlier ages
Socioeconomic deprivation has been associated with delayed onset of menarche
Twins tend to have later menarche than singletons studied in the same population

general. Also, African American girls begin menses 0.1 year earlier than the 12.7 U.S. average recorded in one study, whereas Caucasian American girls began 0.1 year later than the average.

IV. CLINICAL EVALUATION

A. Premature Menarche

Because adrenarche and puberty are dissociated physiologically, variation of their manifestations exist. However, the usual sequence for puberty is for the growth spurt to begin first, followed by thelarche. Physiologic menarche almost always follows thelarche. Thus, onset of puberty is rarely heralded by menarche first. When it is, although it is most likely to be a normal variation of the pubertal sequence, it is best to evaluate the situation by considering a state of hyperestrogenism, either endogenous or exogenous, as well as evaluating local causes of bleeding (foreign objects, trauma, neoplasms, infections, and sexual abuse).

The onset of puberty prior to the age of 8 (2.5 SD from the mean) is generally considered precocious and merits evaluation. Precocious puberty is classified as either premature activation of hypothalamic GnRH pulsatile release (central, complete, isosexual, or true precocity) or a GnRH-independent mechanism (precocious pseudopuberty) of sex steroid secretion.

The development of the secondary characteristics suggests the establishment of adequate function of the hypothalamic–pituitary–ovarian axis to produce the necessary sex steroids to initiate pubertal changes. Although quite variable, pubertal changes are apparent in most girls by age 13. By the time menarche appears, breast and pubic hair development is usually (62%) at Tanner stage 4. An exception to the normal order of pubertal development was noted among ballet dancers. Although they had a normal age of pubarche, their growth progression and menarcheal ages lagged significantly and thelarche tended to follow rather than precede menarche. In attempting to predict the age of menarche in a particular girl, bone age, height, and ponderal index (height/cube root of weight) tend to be the most predictive variables, whereas weight is the least predictive measure and age is intermediate. A useful generality is that menarche usually occurs by the time a girl has a bone age of 13 years.

B. Delayed Menarche

The absence of menarche can be associated with either no evidence of secondary sexual characteristics or their presence. When no secondary sexual characteristics are identified by age 13, an initial evaluation for delayed puberty is appropriate.

When there are signs of secondary sexual characteristics but menarche does not occur within 2 or 3 years of the onset of growth acceleration, then evaluation of primary amenorrhea is warranted. The possibility of pregnancy should always be considered. Anatomical malformation or the absence of the genital structures should also be evaluated, including imperforate hymen, transverse vaginal septum, congenital absence of the cervix, Mayer–Rokitansky–Kuster–Hauser syndrome, Asherman's syndrome, and Müllerian malformations.

Occasionally, despite normal initial physiologic development and normal reproductive anatomy, secondary abnormalities causing compromise of the hypothalamic–pituitary–ovarian axis could arise after initiation of the secondary sexual characteristics but prior to menarche. Examples include excessive exercise, eating disorders, and constitutional delay. This situation requires careful evaluation for primary amenorrhea.

C. Ovulation in Relation to Menarche

Even though the occasional girl conceives before she has her first menses, most cycles in menarcheal girls generally remain anovulatory for a year or more. Among healthy 12- to 14-year-old girls, 90% are anovulatory and up to half of these anovulatory girls are still anovulatory by 4 years after menarche. However, investigation should be initiated when ovulation is not detected within 2 years of menarche because half of the girls in this category will go on to have cycle abnormalities.

See Also the Following Articles

ADRENARCHE; AMENORRHEA; MENSTRUATION; PUBERTY, PRECOCIOUS; PUBERTY, IN HUMANS

Bibliography

Apter, D., Cacciatore, B., Alfthan, H., and Stenman, U.-H. (1989). Serum luteinizing hormone concentrations increase 100-fold in females from 7 years of age to adulthood, as measured by time-resolved immunofluorometric assay. *J. Clin. Endocrinol. Metab.* **68**, 53–57.

Apter, D., Butzow, T. L., Laughlin, G. A., and Yen, S. S. C. (1993). GnRH pulse generator activity during pubertal transition in girls: Pulsatile and diurnal patterns of circulating gonadotropins. *J. Clin. Endocrinol. Metab.* **76**, 940–949.

Arena, B., Maffulli, N., Maffulli, F., and Morleo, M. A. (1995). Reproductive hormones and menstrual changes with exercise in female athletes. *Sports Med.* **19**, 278–287.

Cacciatore, B., Apter, D., Alfthan, H., and Stenman, U.-H. (1991). Ultrasonic characteristics of the uterus and ovaries in relation to pubertal development and serum LH, FSH and estradiol concentrations. *Adolescent Pediatr. Gynecol.* **4**, 15–20.

Frisch, R. E. (1994). The right weight: body fat, menarche and fertility. *Proc. Nutr. Soc.* **53**, 113–129.

Harlan, W. R., Harlan E. A., and Grill, G. P. (1980). Secondary sex characteristics of girls 12 to 17 years of age: The U. S. Health Examination Survey. *J. Pediatr.* **96**, 1074–1078.

Knobil, E. (1980). The neuroendocrine control of the menstrual cycle. *Recent Prog. Horm. Res.* **16**, 53–88.

Plant, T. M. Fourth international conference on the control of the onset of puberty. In *The Neurobiology of Puberty* (T. M. Plant and P. A. Lee, Eds.), pp. 337–342. Journal of Endocrinology, Bristol, UK.

Rosenfield, R. L. (1991). Puberty and its disorders in girls. *Endocrinol. Metab. Clin. North Am.* **20**, 15–41.

Veldhuis, J. D., and Johnson, M. (1986). Cluster analysis: A simple, versatile and robust algorithm for endocrine pulse detection. *Am. J. Physiol.* **250**, E486–E490.

Warren, M. (1980). The effects of exercise on pubertal progression and reproductive function in girls. *J. Clin. Endocrinol. Metab.* **51**, 1150–1157.

Wheeler, M. D. (1991). Physical changes of puberty. *Endocrinol. Metab. Clin. North Am.* **20**, 1–14.

Zacharias, L., Rand, W. M., and Wurtman, R. J. (1976). A prospective study of sexual development and growth in American girls: The statistics of menarche. *Obstet. Gynecol. Surv.* **31**, 325–337.

Menopause

Laurence C. Udoff and Eli Y. Adashi
University of Utah

I. Changes in the Hormonal Milieu
II. Consequences of the Menopause
III. Hormone Replacement Therapy
IV. Summary

GLOSSARY

follicles An intraovarian structure composed of an oocyte and specialized cells such as granulosa and theca cells that work in concert to synthesize estrogen, progesterone, and androgens.

hormone replacement therapy The use of estrogens, progestins, and/or androgens in the treatment of the consequences of a hormone-deficiency state such as menopause.

hot flash The subjective sensation of intense heat mainly of the upper body that lasts a few minutes and can be due to a decline in circulating estrogen levels.

menopause A point in time when the loss of normal ovarian function occurs resulting in cessation of menses and a change in circulating hormone levels characterized mainly by a sharp decline in circulating estrogen levels.

menopause transition The point of time encompassing the beginning signs of ovarian failure, such as shortened or irregular menstrual cycles and fluctuations in hormone levels, up to the onset of menopause.

osteoporosis A condition characterized by the loss of bone tissue leading to bone fragility and an increased risk of fracture.

selective estrogen receptor modulators A new class of drugs that can selectively act as an agonist or antagonist of estrogen receptors in various tissues (e.g., raloxifene is an estrogen receptor agonist in the bone and an antagonist in breast and endometrial tissue).

The menopause defines a point in time when menstruation ceases due to the loss of normal ovarian function. In women with regular menstrual cycles, the ovary is active in synthesizing sex steroids, such as estrogens, progesterone, and androgens, in a very precise manner regulated by the hypothalamus and pituitary and a host of intermediary compounds. As a women ages, the number of follicles that respond to stimulation decreases. Eventually, the natural course of events dictates that the ovary will become unresponsive, despite significant stimulation from the pituitary gonadotrophs, and the process of follicle recruitment, development, maturation, ovulation, and corpus luteum formulation will no longer occur. The result is that menstruation ceases and the hormonal environment changes. The average age for menopause in women in the United States is 50 or 51 years old. This has been remarkably constant over time. When occurring naturally, this is a gradual process. The menopause transition defines the period of time just before the menopause, when ovarian function is waning. This is characterized mainly by irregular menses in women who previously experienced regular cycles. The menopause transition is thought to be on average 4 years in duration. The term menopause is also used to describe the hormonal state of women who, due to some "unnatural" event, have lost ovarian function. This includes patients that have undergone surgical castration, radiation or chemotherapy that has resulted in the loss of ovarian function, or the presence of genetic or autoimmune factors leading to a premature loss of ovarian function (premature ovarian failure). In the latter examples, there is near unanimous agreement within the medical community to consider these entities as disease states requiring prompt diagnosis and treatment, especially in young patients. However, in patients that undergo natural menopause, there still remains a significant debate as to whether this is a condition that requires treatment, especially in the asymptomatic patient. This article will focus on the changes that are associated with the menopause and the data that link these changes to potential health risks as well as to various physical and psychological complaints. The risks and benefits of hormone replacement will also be reviewed to give the reader a comprehensive overview of this controversial subject.

I. CHANGES IN THE HORMONAL MILIEU

A. Changes in Estrogen Production

The most dramatic endocrinologic alteration of the menopause is the decline in circulating levels and excretion rates of estrogens. Production rates for two of the major estrogens, estradiol and estrone, decline to 6 and 24 mg per 24 hours, respectively. Circulating levels of estradiol drop from an average of 200 pg/ml in young women with normal menstrual cycles to <30 pg/ml in naturally menopausal women. In surgically menopausal women, circulating estradiol levels drop to <10 pg/ml within 48 hr of surgery. Also, in menopausal women there is a reversal in the ratio of estradiol to estrone. During the reproductive years, the primary circulating estrogen is estradiol, synthesized mainly by the ovary. In the menopause, the primary circulating estrogen is estrone, synthesized mainly in peripheral tissues by the enzyme aromatase which converts androstenedione (mostly of adrenal origin) into estrone. The production of estrogen through this pathway is related, at least in part, to body weight (i.e., adipose-based aromatase) and perhaps also to age. Though the data overwhelmingly suggest that estrogen production in menopausal women is mainly via this route of extraglandular conversion of androgens to estrogens, studies have found that the ovarian stroma of menopausal women may continue to have aromatase activity suggesting that the climacteric ovary may, in a small way, contribute to the circulating estrogen pool.

B. Changes in Androgen Production

The major androgens found in the circulation of normally menstruating women include (in order of highest to lowest concentrations) dehydroepiandrosterone sulfate (DHEAS), dehydroepiandrosterone (DHEA), androstenedione (A), and testosterone (T). Of the major androgens, A appears to be the most affected by the transition to the menopause. The production rate of A decreases by 50% to 1.5 mg/24 hr, resulting in serum levels of 75 ng/dl. This decrease is mainly due to the decline in A production by the menopausal ovary. The rate of testosterone production also declines in the menopause. A decline of 28% has been reported, lowering the production rate to 120 mg/24 hr and the serum level to approximately 25 ng/dl. This decline is mainly due to the decrease in peripheral conversion of A to T, with the ovary maintaining a relatively constant level of production. Circulating levels of DHEA and DHEAS are also thought to decline with age but may be part of a process primarily related to changing adrenal gland function rather than altered ovarian function which characterizes the menopause.

II. CONSEQUENCES OF THE MENOPAUSE

A. Menopause-Related Symptoms

Estrogens and androgens may affect many different tissues and organ systems throughout the body. A partial listing of sites known to be receptive to sex hormone stimulation includes the brain, skeleton, bone marrow, heart and arteries, external genitalia, ovaries, breasts, muscle, skin, hair, liver, and kidneys. With this in mind, a change in the hormonal milieu, such as described with the menopause, could have far-reaching consequences. In some women, this change in hormonal environment leads to the development of characteristic symptoms. One of the most notable of these is the hot flash or vasomotor instability. The hot flash is the subjective sensation of intense warmth of the upper body, typically lasting for a few minutes. It may be preceded by a prodrome of palpitations or headache and is often associated with weakness, faintness, or vertigo. The episode culminates in profuse sweating and a cold sensation. The objective component of this phenomena is a visible ascending flush of the thorax, neck, and face. Hot flashes are thought to be due to the withdraw of estrogen from the circulation. Therefore, they are associated with menopause of natural, surgical, or medical origins (e.g., medications that result in a decline in estrogen levels or effect). It has been noted that approximately 70% of women will experience hot flashes within 3 months of a natural or surgical menopause. It is also common for women to note hot flashes during the menopause transition phase since this is a time characterized by relatively sudden fluctuations in estrogen levels. Studies have suggested that 50% of women will have persistent symptoms 5 years after the menopause, and in 10 years, 25% of women will still be effected. Other common symptoms noted in menopausal women include sleep disturbances (may be due to hot flashes causing awakening), sexual dysfunction (mainly a decrease in libido thought to be due to the decline in androgen levels), urogenital symptoms (e.g., vaginal dryness, irritation, difficulty urinating, or incontinence), and psychological/somatic complaints such as depression, mood swings, irritability, and an overall diminished sense of well-being. It should be noted that not all women experience adverse symptoms with the onset of menopause. The transition may be relatively asymptomatic, discounting the cessation of menses.

B. Long-Term Health Consequences of the Menopause

Epidemiologic data suggest that the risk for several serious health disorders increases once women enter menopause. The cause of this risk is thought to be due to the change in the hormonal environment that occurs with menopause, in particular the decline in circulating estrogen levels. For example, it has been shown that changes in bone metabolism that are linked to an increased risk for osteoporosis and bone fracture are closely related to the loss of ovarian function. When estradiol levels begin to decline, even before the onset of menopause, bone mineral density decreases. With the beginning of menopause, women begin to loose bone at an accelerated rate, losing approximately 10–20% of bone mineral density within the first 10 years after menopause. Similar or accelerated effects may be seen in younger women who are menopausal due to surgery, premature menopause, or as an iatrogenic response to medical treatment. According to data from numerous clinical trials, estrogen replacement therapy can reverse this process and prevent osteoporotic bone fractures and inhibit further bone loss. Current hypotheses suggest that estrogen may regulate cytokine secretion in bone that balances bone resorption and formation in favor of a net status quo or even an increase in bone mineral density. Though osteoporosis is certainly a multifactorial disease (other risk factors include a history of cigarette smoking, alcohol abuse, sedentary lifestyle, thinness, and steroid use), entrance into menopause significantly elevates the relative risk.

Menopause has also been linked to an increase in the risk of cardiovascular disease (CVD), the major cause of death in women. This is based mainly on the findings of several epidemiologic studies that have reported an increase in CVD risk in association with either natural or surgical menopause. Regarding natural menopause, the data have been criticized for a lack of adequate control for confounding variables such as smoking history, weight, and age, all known to greatly impact the risk of CVD. With closer scrutiny, it has been suggested that natural menopause is not associated with a sudden, sharp increase in CVD risk. However, there is strong evidence to associate estrogen replacement therapy with a reduction in CVD risk. Data examining the impact of estrogen replacement in menopausal women have clearly shown a benefit with as much as a 50% decline in the relative risk of cardiovascular disease and mortality in menopausal women receiving estrogen replacement therapy.

Menopause has also been associated with other long-term complications. The negative effects of hypoestrogenism on urogenital tissue are well documented as are the beneficial responses to estrogen replacement. Data regarding other chronic diseases, such as Alzheimer's disease, macular degeneration, and colon cancer, have recently been presented that suggest the risk for these diseases may be related to hypoestrogenism and that estrogen replacement may decrease this risk.

III. HORMONE REPLACEMENT THERAPY

A. The Role of Hormone Replacement Therapy

Hormone replacement therapy (HRT) is the mainstay of treatment for menopause-related symptoms. The efficacy of HRT is well documented in relieving hot flashes, urogenital symptoms (e.g., atrophic vaginitis and urinary tract complaints), and the psychological complaints related to the menopause such as mood swings, irritability, depression, insomnia, and loss of libido. Regarding the latter set of symptoms, controversy exists as to whether these symptoms are due to a decline in estrogen levels or whether they are mainly due to other psycho/social environmental factors that commonly occur in women in the menopausal age group (e.g., death of spouse, lifestyle changes due to maturation of children, and decline in overall health). Psychological counseling and/or antidepressant medication may be more appropriate for some of these instances. This highlights the point that menopause itself is not a disease state requiring treatment. Rather, the role of HRT may best be characterized as treatment for the short-term complaints related to hormone deficiency and as preventive therapy for serious health problems such as osteoporosis and CVD in those who the long-term benefits of treatment outweigh the risks.

B. Treatment Regimens

Hormone replacement therapy in menopausal women entails the use of estrogens, progestins, and androgens. Various hormonal formulations, delivery systems, and dosing schedules may be used. Regarding estrogens, many compounds are available, though conjugated equine estrogen in the oral form is the most commonly used preparation in the United States. Other estrogen choices include oral esterified estrogens, estrone, and estradiol. All of these compounds are extensively metabolized after oral ingestion. Transdermal patches are also available that release estradiol directly into the circulation, bypassing the "first pass" metabolism in the liver. Though not popular in the United States, estrogen implants and vaginal rings that release estrogen are available.

In patients that have a uterus *in situ* it is recommended that a progestin hormone be added. This is to counteract the stimulatory effect of unopposed estrogen therapy on the endometrium (i.e., uterine lining) which can lead to the development of endometrial hyperplasia in a significant number of patients (as much as 60% in some studies). Hyperplasia of the endometrium has the potential to progress further to endometrial cancer. Data examining the use of progestins to protect the endometrium from hyperplasia and cancer have suggested that it should be used for 12–14 days out of the month. This has led to the use of sequential hormone replacement regimens in which the estrogen is given from 21 to 30 days each month with the placement of a course of progestin usually at the beginning or end of the month. Uterine bleeding commonly occurs after the progestin course. Another regimen that has gained popularity, due in part to its ease of use, is the combined continuous administration of an estrogen and a progestin. In many women, this regimen eventually leads to the absence of any uterine bleeding and is thought to help improve patient compliance by also reducing progestin-related side effects (due to the lower continuous dose of progestin that may be used to protect the endometrium). In women who still remain intolerant to progestin-related side effects, data have shown that less then monthly progestin courses (e.g., quarterly) may also protect the endometrium from the development of hyperplasia and cancer. In the United States, the most commonly used progestin is the oral form of medroxyprogesterone acetate. Other available preparations included oral micronized "natural" progesterone and norethindrone acetate. Progesterone implants and progesterone-releasing intrauterine devices are also available but are not commonly used.

The use of androgens for the treatment of menopausal symptoms has remained controversial even though the preparations have been in clinical use for many years. As previously noted, the hormonal environment of menopausal women is characterized by a decline in circulating androgens. However, the impact of this is still not clear, nor are data available to outline the optimal replacement regimen. Androgen replacement therapy is mainly used in patients who complain of loss of libido after menopause or in patients with persistent hot flashes despite estrogen replacement, though data in support of the latter are not unequivocal. Androgen replacement in the menopause may have other benefits such as improving mood and sense of well-being, maintaining bone density, and increasing muscle strength.

C. Side Effects and Risks Associated with Hormone Replacement Therapy

There are numerous side effects that may be associated with the use of HRT in menopausal women, many of which are thought to be linked to the overall poor long-term compliance rates with this treatment (often reported as <25%). Women may experience bloating, breast tenderness, weight gain, headaches, nausea, irritability, and premenstrual-like symptoms. Additionally, most women can expect uterine bleeding. Women receiving a sequential regimen have a 80–90% chance of continued uterine bleeding, though it is usually a predictable light flow. Women who are utilizing a continuous daily estrogen/progestin regimen are much more likely to experience amenorrhea, though in the first 6 months of treatment irregular and unpredictable bleeding episodes are common.

One of the most publicly debated issues surrounding HRT is whether or not treatment is linked to an increase in the risk of breast cancer. Over 40 epidemiologic studies have been performed and no clear answer has emerged. It has been suggested that if estrogen replacement had a large impact on the risk of breast cancer, the results of most of these studies would be consistent and uniform. The finding of conflicting data from large, well-designed studies suggests that the epidemiological methodology employed is unable to overcome whatever recognized or unrecognized biases are present within these studies. This implies that at most the effect would be small. In addition, several analyses have shown that even if one assumes an increase in risk of breast cancer with the use of HRT the benefits of treatment (e.g., lower CVD risk) outweigh the risks. Despite this line of reasoning, the fear of breast cancer remains one of the primary reasons women do not use HRT and now is part of the driving force behind a new generation of medications that selectively bind to estrogen receptors in the bone but not the breast (selective estrogen receptor modulators).

There are other well-documented complications from the use of HRT. Several of these are likely due to the effects of estrogen on the liver. This includes an increase, albeit small, in the incidence of venous thromboembolism, an increase in serum triglyceride levels, and an increase in gallbladder disease. The effect of estrogen replacement on carbohydrate economy is unclear, but in general HRT it is not contraindicated in diabetics. HRT is not associated with detrimental effects in patients with CVD, hypertension, or a history of cancer of the cervix, ovary, or vulva. If a progestin is used, there is no increased risk in endometrial cancer expected, though the use of HRT in patients with a history of endometrial cancer is somewhat controversial, as it is with patients with a history of breast cancer or a strong family history of breast cancer.

IV. SUMMARY

The menopause is clearly a time of change in a women's life from the perspective of the hormonal environment as well as life circumstances. It may also be considered as an opportunity to make adjustments for the future to help ensure good health and well-being. In many women, the use of hormone replacement therapy may be appropriate to help reach this goal.

See Also the Following Articles

AMENORRHEA; ESTROGEN REPLACEMENT THERAPY; MENSTRUATION; OSTEOPOROSIS

Bibliography

Adashi, E. Y., Rock, J. A., and Rosenwaks, Z. (Eds.) (1996). The climacteric consequences of follicular depletion. In *Reproductive Endocrinology, Surgery and Technology*. Lippincott-Raven, New York.

Grodstein, F., Stampfer, M. J., Colditz, G. A., *et al.* (1997). Postmenopausal hormone therapy and mortality. *N. Engl. J. Med.* 336, 1769–1775.

Speroff, L., Glass, R. K., and Kase, N. G. (1994). Menopause and postmenopausal hormone therapy. In *Clinical Gynecological Endocrinology and Infertility*. Williams & Wilkins, Baltimore.

Udoff, L. C., and Adashi, E. Y. (1997). Combined continuous hormone replacement therapy: An update. *Reprod. Med. Rev.* 6, 11–22.

Menstrual Cycle

Sarah L. Berga
The University of Pittsburgh School of Medicine

I. The Role of the Hypothalamus
II. Ovarian Events
III. Clinical Consequences

GLOSSARY

corpus luteum An ovarian apparatus formed after the oocyte is released and composed of transformed granulosa cells that secrete progesterone; literally means yellow body.

endometrium The glandular epithelium that lines the uterine cavity.

estradiol An 18-carbon steroid synthesized by the granulosa cells of the ovarian follicle from androgen precursors, particularly androstenedione, which are supplied by surrounding theca cells.

follicle A spherical ovarian apparatus composed of an oocyte encircled by a single or multiple granulosa cells, which in turn are surrounded by androgen-secreting theca and stroma cells; when the antral stage is formed, fluid accumulates in the center of the follicle, making it possible to image it by ultrasound.

follicle-stimulating hormone (FSH) A hormone that initiates and sustains follicular development, in part by inducing the enzyme aromatase, which converts androstenedione into estradiol.

follicular phase Ovarian events that occur during the first half of the menstrual cycle, namely, the growth of a primary follicle into a preovulatory or Graafian follicle; characterized by ever-increasing levels of circulating estradiol.

gonadotropins Large dimeric, glycosylated polypeptide hormones, luteinizing hormone (LH) and FSH, each of which is composed of a common α and a unique β subunit; made in and secreted by the pituitary gland when stimulated by pulsatile gonadotropin-releasing hormone (GnRH). Because of the difference in the β subunit composition, the half-life of LH is about 20 min, whereas that of FSH is approximately 4 hr.

GnRH pulse generator A group of endogenously pulsatile, interconnected neurons located in the mediobasal hypothalamus that synthesize and release GnRH into the portal vasculature to thereby stimulate pituitary gonadotropin synthesis and secretion.

hypothalamus Part of the brain located around the inferior portion of the third ventricle; receives ascending neural signals from the brain stem and descending neuronal input from the limbic lobe and frontal cortex and projects primarily to the median eminence and posterior pituitary.

LH surge A large release of LH that is triggered by exponentially rising estradiol concentrations in the circulation and causes follicular rupture and release of the oocyte.

luteal phase Ovarian events during the second half of the menstrual cycle after the LH surge and ovulation, namely, the transformation of the follicle into the progesterone-secreting corpus luteum.

luteinizing hormone (LH) A hormone that is released by the pituitary and stimulates theca and stroma cells to secrete the androgen, androstenedione.

ovary Gonadal tissue composed of the cortex of the undifferentiated fetal gonad into which germ cells that become oocytes migrate prior to 20 weeks of fetal life.

pituitary Epithelium composed of the anterior and intermediate glandular lobes and the neurohypophysis or neural lobe which is an extension of the hypothalamus.

progesterone A 21-carbon steroid made and secreted by the corpus luteum in high concentrations after ovulation for about 14 days.

proliferative phase Endometrial events that occur during the first half of the menstrual cycle, namely, the growth of the endometrium in response to sustained exposure to estradiol.

secretory phase Endometrial events that occur during the second half of the menstrual cycle, namely, the transformation of proliferative endometrium into secretory endometrium.

The menstrual cycle refers to an orderly progression of events that produces a mature ovum ready for fertilization and an endometrium primed for implantation. Paradoxically, the term menstrual cycle refers to the periodic shedding of the endometrium that occurs when fertilization and implantation do not occur. The tightly orchestrated sequence of events depends directly on appropriate input or drive from the hypothalamic gonadotropin-releasing hormone (GnRH) pulse generator and also requires intact pituitary function, responsive oocytes, and endometrium capable of responding to sex steroids and other mediators of cellular differentiation. The systemic effects and health consequences of the menstrual cycle are not confined to reproductive organs alone, however. Sex steroid excursions characteristic of the menstrual cycle affect every tissue in the body, including bones, brain, breasts, the cardiovascular tree, skin, hair, and even the pattern of fat deposition. The menstrual cycle is generally viewed as being composed of two halves. The first half is termed the follicular phase when referring to ovarian events and the proliferative phase when referring to endometrial events. During the follicular phase, the oocyte grows from an antral follicle to a preovulatory follicle and this progression is accompanied by a progressive increase in estradiol. The increase in estradiol causes the endometrium to develop or proliferate. Likewise, the second half of the menstrual cycle is termed luteal when ovarian events are the focus and secretory when the endometrium in being considered. In the luteal phase, the corpus luteum secretes sufficient progesterone to cause the proliferative endometrium to transform into a secretory pattern. By convention, the first day of menstrual shedding, termed menses, is Day 1. However, if the day of the luteinizing hormone (LH) surge is designated Day 0, then the follicular phase is numbered as -1, -2, and so on and the luteal phase is numbered as 1, 2, and so on. A typical menstrual interval is 28 or 29 days, but the interval is erratic for the first and last 5 years of menstrual life, termed gynecologic age. The menstrual interval is approximately 30 or 31 days in length once regular cyclicity is established after puberty, and it becomes progressively shorter. The interval is about 25 or 26 days by gynecologic age 25.

I. THE ROLE OF THE HYPOTHALAMUS

Menstrual cyclicity is a centrally initiated process that depends on reactivation of the pulsatile release of hypothalamic GnRH. GnRH is a decapeptide with a very short half-life that is secreted into the portal vasculature to drive the pituitary synthesis and release of gonadotropins. GnRH neurons originate in the olfactory placode, migrate into the hypothalamus during fetal development, and display synchronized GnRH pulsatility sufficient to drive the fetal gonad by 19–21 weeks of fetal age. Since GnRH neurons are endogenously pulsatile, the childhood quiescence of the GnRH pulse generator likely involves an alteration in GnRH-to-GnRH networking so that the synchrony is disrupted. The expression of GnRH pulses sufficient to drive pituitary and gonadal function reflects the episodic but coordinated firing of a group of GnRH neurons such that a bolus of GnRH is released into the portal vasculature. It has been suggested that one mechanism for disrupting vital GnRH-to-GnRH connections involves glial interposition, which prevents synchronous firing of the GnRH neurons. Other cellular mechanisms that may mediate this type of neuronal plasticity remain a mystery, but it has been shown that the GnRH pulse generator can be prematurely reactivated by excitatory amino acids. The hypothalamus is located around the inferior portion of the third ventricle and receives ascending input from the brain stem and descending signals from the limbic lobe and frontal cortex. It projects primarily to the posterior pituitary and median eminence. Thus, neuronal systems implicated in the regulation of GnRH include corticotropin-releasing hormone, endogenous opioid peptides, neuropeptide Y, norepinephrine, excitatory amino acids, serotonin, γ-aminobenzoic acid, and dopamine. Because of its short half-life, GnRH is not detectable in the peripheral circulation. In humans, GnRH pulse patterns are inferred by tracking the pulsatile release of LH in the peripheral circulation. Frequencies that are too slow or too rapid result in anovulation.

II. OVARIAN EVENTS

A menstrual cycle starts with the development of a cohort of follicles, one of which will become dominant. Follicles are composed of an oocyte encircled by granulosa cells, which in turn are surrounded by theca cells. Follicular development requires the sustained release of GnRH at a pulse frequency of once every 90 min for approximately 14 days. GnRH stimulates the release of the pituitary gonadotropins, LH and follicle-stimulating hormone (FSH). LH stimulates ovarian theca cells to synthesize and release androgens, whereas FSH induces granulosa cell development, including the enzyme aromatase. Aromatase converts thecally produced androstenedione into estradiol. In the presence of sustained GnRH pulsatility, the secretion of FSH will be regulated primarily by estradiol feedback at the level of the pituitary. A variety of polypeptides have been implicated as modulators of FSH action at the ovarian level, including activin, inhibin, cytokines, and growth factors. Many of these factors serve as autocrine and paracrine regulators of ovarian and pituitary function. In the midfollicular phase, progressive increments in estradiol released by the developing follicle suppress FSH below the threshold needed to sustain folliculogenesis, thereby preventing further multifollicular development. The dominant follicle escapes demise by becoming FSH independent. Selection of the dominant follicle occurs about day 8 or 9 of the follicular phase. The dominant follicle continues to mature because FSH induces LH receptors on granulosa cells. Although FSH levels decline as estradiol rises across the follicular phase, LH levels may increase somewhat across the follicular phase.

In the late follicular phase, the exponential rise in estradiol secreted by the nearly mature dominant follicle triggers a LH surge. The LH surge lasts about 36 hr and depletes the pituitary of LH. Ovulation ensues approximately 36 hr following the onset of the LH surge. Thereafter, granulosa cells transform into progesterone-secreting luteal cells and the ovulated follicle sac becomes the corpus luteum. The progesterone secreted by the corpus luteum causes the endometrium to transform from a proliferative state into a secretory state capable of sustaining implantation. Progesterone also has other actions. It causes myometrial smooth muscle relaxation to facilitate implantation should the ovum be fertilized. The combined exposure to estradiol and progesterone induces central β-endorphin tone that slows GnRH and LH pulse frequency to about one pulse every 4 hr. This slowing of GnRH pulse frequency further ensures that FSH levels do not rise and follicular recruitment is constrained until the demise of the corpus luteum. Figure 1 provides a schematic overview of the anatomic and hormonal relationships characteristic of the menstrual cycle.

The process by which a primordial follicle leaves the resting pool and starts to grow is termed recruitment. Follicles that are recruited to grow beyond the antral stage need sustained FSH input. Prior to recruitment, however, follicles must undergo a maturational process that is largely gonadotropin independent. Germ cells arrive in the gonadal ridge from the yolk sac endoderm by the seventh gestational week. Then the oogonia transform into oocytes and enter the first stage of meiosis until about the 20th week of gestation. Thereafter, oogonia form primordial follicles, a process that lasts until shortly after birth. Primordial follicles are defined by their simple structure, consisting of an ovum surrounded by a single layer of flattened epithelial cells. The store of primordial follicles is fixed *in utero* and after birth the number of primordial follicles declines. Across each interval of time, a certain percentage of these resting follicles die or become atretic, regardless of whether FSH is available. Follicles that remain in the resting state do so because of the process of active inhibition that requires oocyte–granulosa cell communication. It is estimated that the number of primordial follicles at birth is about 500,000–2 million and that the rate of atresia is relatively constant until the number of remaining follicles is 25,000, at which point the rate of atresia accelerates. Very high levels of FSH may accelerate the rate of atresia. Menopause is estimated to occur when the follicle number approximates 1000. Before puberty, primordial follicles initiate growth up to the FSH-independent stage and then they undergo atresia because there is insufficient FSH to support recruitment. It takes 120 days for a primordial follicle to grow to the secondary stage and

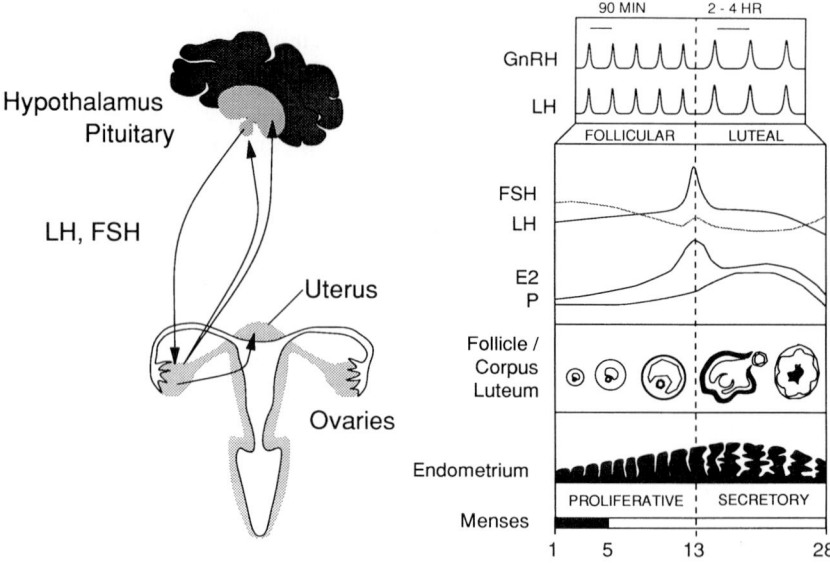

FIGURE 1 The anatomic (left) and physiological (right) relationship characteristic of the human menstrual cycle. Secretory hormones connect the hypothalamic–pituitary–ovarian–endometrial axis. GnRH is secreted into the portal vasculature in pulses to stimulate pituitary secretion of LH and FSH. The coordinated effects of FSH and LH cause recruitment of antral follicles, selection of the dominant follicle, and ovulation. Rising estradiol levels of follicular origin are secreted into the circulation and cause the endometrium to proliferate. After ovulation is triggered by the LH surge, the follicle sac is transformed into a corpus luteum that secretes large amount of progesterone. Progesterone transforms the proliferative endometrium into a secretory state capable of permitting implantation.

85 days for it to grow from a secondary follicle to an antral follicle that can be recruited by FSH. If FSH is available, then follicles that have become recruitable will develop until FSH drops below the threshold needed to sustain further maturation. It takes about 15 days for a follicle to grow from an antral to a preovulatory follicle. Oocytes in these follicles have acquired a zona pellucida and are competent to undergo germinal vesicle breakdown to complete meiosis. As noted previously, selection of the dominant follicle occurs when the lead follicle secretes enough estradiol to suppress FSH below the threshold needed for follicular development, but the dominant follicle escapes demise by developing LH receptors and becoming FSH independent.

The process of ovulation involves a series of tightly orchestrated events. Meiosis is arrested until just before ovulation occurs, when oocytes undergo nuclear progression from dictyate of the first meiotic prophase (four times the haploid DNA complement) to metaphase II (two times the haploid DNA complement). Germinal vesicle breakdown and extrusion of the first polar body accompany the first meiotic division. If fertilization occurs, meiosis is completed with reduction to a single DNA haploid complement and extrusion of the second polar body. The release of the ovum from the follicle requires the degeneration of the follicle wall by proteolytic enzymes, including plasminogen activator. Prostaglandins are formed in follicular fluid and the concentrations peak at the time of ovulation and participate in the proteolysis of the follicle wall as well as extrusion of the oocyte–cumulus cell mass by smooth muscles in the ovary.

III. CLINICAL CONSEQUENCES

Understanding the fundamental principles that underlie menstrual cyclicity permits the clinician to make more accurate diagnoses or to devise more rational interventions. For instance, it has been argued that the luteal phase invariably lasts 14 days. While it is true that the corpus luteum has the capacity to sustain progesterone secretion for 14 days if there is appropriate LH stimulation, if LH secretion is insufficient, progesterone secretion may decline to such an extent that the endometrium is destabilized and withdrawal bleeding may ensue. In this situation, the corpus luteum may be rescued if LH levels recover or human chorionic gonadotropin is available, but the endometrium may not be able to support implantation if there has been a sufficient decline in progesterone. Thus, luteal length is not automatically 14 days once ovulation has occurred. Women whose luteal phase length is short or those with luteal lengths of normal duration but with reduced progesterone secretion have what is commonly termed luteal phase deficiency. Based on the foregoing considerations, it stands to reason that luteal lengths would be 14 days in conception cycles, so the obstetrician's assumption of such when estimating the day of confinement is reasonable.

Another common assumption is that withdrawal bleeding signals that the preceding ovarian cycle was ovulatory. Gonadotropin stimulation can be inconsistent or inadequate. The aging follicle may fail to respond to appropriate gonadotropin stimulation. In these instances, the follicle may start to develop but become atretic before a preovulatory follicle forms or ovulation occurs. A significant decline in estrogen levels may destabilize the endometrium and lead to withdrawal bleeding. This type of bleeding is often termed anovulatory cycling. Based on what is known about the need for sustained GnRH drive, it should be obvious that stress-induced disruption of GnRH has the potential to produce luteal insufficiency or anovulation. The same sequela also can result from unresponsive or poorly responsive oocytes. Thus, menstrual pattern alone is a poor indicator of underlying hormonal dynamics.

Follicular number is endowed *in utero* and menopause occurs when this cohort is exhausted. Early puberty is unlikely to significantly alter the timing of menopause because the pubertal rise in FSH only rescues follicles from atresia. However, unilateral oophorectomy will advance the age of menopause. Furthermore, the later in life that a unilateral oophorectomy is performed, the more the advance in the age of menopause, suggesting that there are compensatory mechanisms for early declines in the pool. On the other hand, exogenous gonadotropin administration for the purpose of superovulation also only rescues follicles that would otherwise become atretic and thus this form of infertility treatment is unlikely to advance the age of menopause.

Although estradiol and progesterone levels change dynamically across a menstrual interval, the range of these excursions is minimal from individual to individual during the midreproductive years. When considering fertility, the focus is on the effects of these steroid excursions on the endometrium. However, sex steroid excursions also have important effects on other target tissues. Bone density and linear growth are gated by exposure to estradiol in particular. Breast tissue undergoes cyclic changes. Glandular development and mitotic index are maximal during the luteal phase. The brain is highly responsive to hormonal milieu, particularly those regions in the limbic lobe that subserve emotional states. The high levels of estradiol secreted by the preovulatory follicle enhance sexual drive. Progesterone has profound behavioral effects and can augment depressive symptoms in predisposed women. In addition, estrogens may maintain cognitive faculties or facilitate memory. Epithelial and glandular tissues and smooth muscle are highly sensitive to sex steroid exposure. Thus, habitus is greatly affected by sex steroids. Pigmentation and hair shaft diameter and rate of hair growth reflect the ambient hormonal background. Estrogens mediate the deposition of fat in a gynecoid pattern. Metabolism, appetite, and sensitivity to metabolic and psychogenic stressors are modulated by sex steroids. It is these important systemic effects of the cyclic sex steroid exposure that provide the rationale for intervening when menstrual cyclicity is disturbed.

In devising rational treatment strategies for women seeking to maintain health after menopause with hormone replacement therapy, understanding the physiological ranges achieved in the normal menstrual cycle can help to guide therapy. By way of analogy, if there is a concern that pharmacological levels will aggravate an underlying medical condition, but this condition was not aggravated by normal ovarian function, then it should be safe to administer sex steroids in levels that approximate those observed during the menstrual cycle. For estradiol, the mean level is approximately 100 pg/ml, with a low value of 20–30 pg/ml on Day 2 or 3 and a peak of 300–400 pg/ml at midcycle. Mean progesterone levels in a luteal phase approximate 5–10 ng/ml. It is possible to replace these hormones in a physiological fashion in any functionally agonadal woman of any age.

In summary, understanding the menstrual cycle provides a necessary cognitive foundation for devising rational clinical approaches to many gynecological and medical conditions, including menstrual disturbances, anovulation, delayed and early puberty, hormonal contraception, infertility, miscarriage, osteoporosis, and endometriosis.

See Also the Following Articles

Corpus Luteum; Endometrium; FSH (Follicle-Stimulating Hormone); GnRH Pulse Generator; LH (Luteinizing Hormone); Menstruation; Progesterone Actions on Reproductive Tract

Bibliography

Caviness, V. S., Jr. (1992). Kallmann's syndrome—Beyond migration. *N. Engl. J. Med.* **326**, 1775–1777.

Conte, F. A., Grumbach, M. M., and Kaplan, S. L. (1975). A diphasic pattern of gonadotropin secretion in patients with the syndrome of gonadal dysgenesis. *J. Clin. Endocrinol. Metab.* **40**, 670–674.

Erickson, G. F. (1986). An analysis of follicle development and ovum maturation. *Sem. Reprod. Endocrinol.* **4**, 233–254.

Faddy, M. J., Gosden, R. G., Gougeon, A., Richardson, S. J., and Nelson, J. F. (1992). Accelerated disappearance of ovarian follicles in mid-life: Implications for forecasting menopause. *Hum. Reprod.* **7**, 1342–1346.

Filicori, M., Santoro, N., Merriam, G. R., and Crowley, W. F. (1986), Characterization of the physiological pattern of episodic gonadotropin secretion throughout the human menstrual cycle. *J. Clin. Endocrinol. Metab.* **62**, 1136–1143.

Gay, V. L., and Plant, T. M. (1988). Sustained intermittent release of gonadotropin-releasing hormone in the prepubertal male rhesus monkey induced by N-methyl-D,L-aspartic acid. *Neuroendocrinology* **48**, 147–152.

Hurley, D. M., Brian, R., Outch, K., Stockdale, J., Fry, A., Hackman, C., Clarke, I., and Burger, H. G. (1984). Induction of ovulation and fertility in amenorrheic women by pulsatile low-dose gonadotropin-releasing hormone. *N. Engl. J. Med.* **310**, 1069–1074.

Miller, D. S., Reid, R. R., Cetel, N. S., Rebar, R. W., and Yen, S. S. C. (1983). Pulsatile administration of low-dose gonadotropin-releasing hormone in women with hypothalamic amenorrhea. *J. Am. Med. Assoc.* **250**, 2937–2941.

Pan, J. T., Kow, L. M., and Pfaff, D. W. (1986). Single-unit activity of hypothalamic arcuate neurons in brain tissue slices. *Neuroendocrinology* **43**, 189–196.

Rasmussen, D. D., Liu, J. H., Wolf, P. L., and Yen, S. S. C. (1983). Endogenous opioid regulation of gonadotropin-releasing hormone release from the human fetal hypothalamus in vitro. *J. Clin. Endocrinol. Metab.* **57**, 881–884.

Treloar, A. E., Boynton, R. E., Bohn, B. G., and Brown, B. W. (1967). Variation of the human menstrual cycle through reproductive life. *Int. J. Fertil.* **12**, 77–126.

Watanabe, G., and Terasawa, E. (1989). *In vivo* release of luteinizing hormone releasing hormone increases with puberty in the female rhesus monkey. *Endocrinology* **125**, 92–99.

Weiner, R. I., and de la Escalera, G. M. (1993). Pulsatile release of gonadotropin releasing hormone (GnRH) is an intrinsic property of GT1 GnRH neuronal cell lines. *Hum. Reprod.* **8**, 13–17.

Wildt, L., Marshall, G., and Knobil, E. (1980). Experimental induction of puberty in the infantile female rhesus monkey. *Science* **207**, 1373–1375.

Zeleznik, A. J. (1981). Premature elevation of systemic estradiol reduces serum levels of follicle-stimulating hormone and lengthens the follicular phase of the menstrual cycle in rhesus monkeys. *Endocrinology* **109**, 352–355.

Menstrual Disorders

Shimon Segal and Isaac Schiff

Harvard Medical School

I. Normal Menstrual Cycle
II. Diagnosis of Menstrual Disorders
III. Treatment of Menstrual Disorders

The causes of menstrual disorders include abnormal function of the hypothalamic–pituitary axis, hormonal disorders, a wide spectrum of diseases of the reproductive system, nongynecological causes, pelvic infection, systemic diseases, iatrogenic causes, and coagulation disorders. In cases in which an organic cause of menstrual disorder is not identified, the abnormal menstrual bleeding is considered to be dysfunctional uterine bleeding. In order to understand the pathophysiologic mechanisms of menstrual disorders, we review the normal menstrual cycle, and then we review the different causes of menstrual disorders and their pathophysiologic mechanism in relation to the age of women.

I. NORMAL MENSTRUAL CYCLE

Normal puberty and adult reproductive function are established during fetal life with the development of the hypothalamus, pituitary gland, ovaries, uterus, cervix, and vagina. The hypothalamus secretes gonadotropin-releasing hormone (GnRH) in a pulsatile fashion and stimulates the fetal pituitary to secrete luteinizing hormone (LH) and follicle-stimulating hormone (FSH) as early as 5 weeks of gestation. LH and FSH are detectable in fetal plasma by 10 weeks of pregnancy.

Germ cells increase in number in the ovaries by meiotic division from 600,000 in the second month to 7 million in the seventh month of fetal life. The primary oocyte number then decreases to 2 million at birth and to 300,000 at puberty. The oocytes established within primordial follicles are arrested in the diplotene stage of the meiotic prophase and each oocyte is surrounded by a single layer of granulosa cells.

The menstrual cycle can be divided into three stages: (i) The follicular phase, at which time a new cohort of follicles is recruited and the dominant follicle is selected; (ii) the ovulatory period, during which final maturation of the oocyte occurs; and (iii) the luteal phase, when the corpus luteum is formed and hormones are secreted in preparation for embryo implantation. The menstrual cycle, the interval between periods, lasts between 25 and 30 days in the majority of women.

A. The Follicular Phase

During the follicular phase several follicles are recruited. However, only one follicle will grow to maturity. The selection process of the dominant follicle is not completely understood but occurs early in the follicular phase and is completed by Day 6 of the cycle.

The dominance of one follicle is achieved through its ability to increase peripheral estradiol concentrations which in turn inhibit FSH secretion. FSH secretion is suppressed to a concentration insufficient to support the growth of the other follicles of the cohort.

The developing follicle produces its own estradiol microenvironment. Estradiol is the main factor that promotes granulosa cell growth, either directly or indirectly, through the promotion of local growth factors. These factors may include epidermal growth factor, insulin-like growth factors (IGFs), and interleukins.Optimal follicular growth requires appropriate pituitary hormone (FSH) pulse secretion. Estrogen causes LH pulses to be secreted in small amplitude but high frequency. With increased estrogen secretion throughout the follicular phase, the endometrium proliferates, and the endometrial glands undergo mitoses and becomes thicker. Mucus secretion by the endocervical glands increases and the mucus becomes more alkaline with increased elasticity.

B. The Midcycle Period

Maturity of the follicle is marked by high circulatory concentrations of estrogen. When a threshold is reached (approximately 200 pg/ml of estradiol for at least 36 hr), estradiol activates the positive feedback loop, thereby signaling to the hypothalamus and the pituitary gland that the follicle is ready for the ovulatory surge of LH. In response, there is a release of GnRH and gonadotropins.

The ovulatory gonadotropin surge induces changes in the graafian follicle, resulting in a change in ovarian steroid hormones secretion. The LH surge arrests granulosa cell proliferation thus decreasing the estrogen production. Androgen production also decreases because the high LH surge inhibits 17α-hydroxylase, further depressing estradiol secretion. The process of luteinization is initiated, and a preovulatory rise in progesterone occurs. The LH surge initiates the resumption of meiosis and the fully grown oocyte progresses from the diplotene stage to methaphase II of the second meiotic division. At ovulation, meiosis is arrested again. The second meiotic division will be completed at the time of fertilization. The release of the oocyte occurs about 18 hr after the gonadotropin surge.

C. The Luteal Phase

After ovulation, the corpus luteum is formed. The corpus luteum is derived from the graafian follicle after changes induced by the gonadotropin surge, including the vascularization of the previously avascular granulosa cells and the production of estradiol and progesterone. With increased progesterone secretion, endometrial glandular proliferation is arrested and the glands undergo a secretory differentiation. Cervical mucus becomes acidic and viscous, and it is secreted in decreased amounts. Progesterone alters the central nervous system GnRH and gonadotropin pulsatile patterns with a drastic decrease in pulse frequency.

During the final days of the luteal phase, and in the absence of implantation of a fertilized egg, ovarian steroid secretion decreases. The withdrawal of the hormonal support causes the endometrium to undergo necrotic changes, and menstrual bleeding results. Normal menses occurs at an interval of 28 ± 4 days, with a duration of 4 ± 2 days, and 40 ± 20 ml of blood is generally lost with each menses.

II. DIAGNOSIS OF MENSTRUAL DISORDERS

We will discuss the common causes of menstrual disorders in relation to their pattern and to age of women (Table 1). The following are abnormal uterine bleeding patterns:

Oligomenorrhea: Intervals >35 days
Polymenorrhea: Intervals <21 days
Menorrhagia: Regular intervals, excessive flow and duration
Metrorrhagia: Irregular intervals, excessive flow and duration
Intermenstrual bleeding: Bleeding between otherwise regular menstrual cycles

A. Newborn

A newborn may experience uterine bleeding caused by withdrawal of the maternal estrogen transferred from the placenta.

B. Childhood

Hypothalamic disorders are those involving premature maturation of the hypothalamic–pituitary–ovarian axis. Sex steroids are increased as a result of a premature increase in pulsatile GnRH affecting the pituitary and ovaries, resulting in premature menses. Additionally, a number of central nervous system (CNS)-related problems, such as hematomas in the hypothalamus, encephalitis, meningitis, and hydro-

TABLE 1
Menstrual Disorders[a]

Reproductive life	
Anovulatory	Ovulatory
Central	Infections
Neurogenic	Uterine myomas
Tumors	Von Willebrand
Psychogenic	Liver disease
Stress	Iatrogenic
Nutritional	Anticoagulants
Endocrine	IUD
Adrenal	
Thyroid	
PCO	
Gonadal	
Tumor	
Adolescence	Perimenopause
Hypothalamic immaturity	Endometrial hyperplasia
Psychogenic	Tumor
Nutritional	
Delayed puberty	
Childhood	
Hypothalamic immaturity	
Precocious puberty	
Ovarian tumors	
Newborn	
Maternal estrogen	

[a] Courtesy of Dr. Jeffrey M. Goldberg.

cephalus, may similarly stimulate reproductive organs development, although the mechanism of action of these conditions is unknown.

An ovarian estrogen-producing cyst is the cause of precocious puberty and irregular menses in 11% of girls. Of these, 5% have a granulosa cell tumor and 1% have a theca cell tumor.

McCune–Albright syndrome, manifested by the autonomous early production of estrogen by the ovaries, is caused by a defect in cellular regulation of the G protein cAMP–kinase function. The syndrome presents with sexual precocity and premature menstruation.

Delayed puberty (primary amenorrhea in 16-year-old women or older) may be caused by delayed maturation of the hypothalamic–pituitary–ovarian axis. This condition may be either hypothalamic or pituitary in origin due to a defect in gonadotropin release.

C. Adolescence

Hypothalamic–pituitary–ovarian axis immaturity may persist during adolescence and may cause anovulatory or irregular menstrual cycles. Hypothalamic amenorrhea can be due to excess dopamine secretion, which inhibits GnRH pulse frequency and this reduces secretion of FSH, LH, and prolactin.

Anorexia nervosa, a psychiatric disease associated with food aversion, fear of weight gain, and distorted body image, results in a very limited caloric intake that causes severe weight loss. A loss of body weight in the range of 10–15% of normal weight for height represents a loss of about one-third of total body fat. The decreased body fat component will result in decreased gonadotropin secretion, which often produces amenorrhea.

D. Exercise

Among runners, an increase in endogenous opioids (endorphins) after exercise suppresses GnRH secretion, thus reducing pituitary LH pulse frequency and amplitude, leading to menstrual irregularity.

E. Reproductive Life

Menstrual disorders during reproductive life are differentiated between ovulatory and anovulatory cycles.

1. Menstrual Disorders Associated with Anovulatory Cycles

Pituitary prolactinoma: A pituitary tumor that secretes excessive prolactin can inhibit the pulsatile secretions of GnRH and can cause amenorrhea.

Hyperthyroidism or hypothyroidism: Thyroid hormone secretion is controlled by hypothalamic thyrotropin-releasing hormone (TRH), and TRH also stimulates prolactin secretion by the pituitary. Thus, abnormal TRH secretion can inhibit the pulsatile secretion of GnRH and cause abnormal uterine bleeding.

Adrenal hyperplasia: This is manifested by an increase in androgen levels which either by themselves can cause amenorrhea or are aromatized to estrogen creating an environment which prevents the LH surge and ovulation.

Ovarian–granulosa cell tumors: These tumors continuously secrete high levels of estrogen. The high estrogen level inhibits the LH surge and ovulation. There is no progesterone secretion and the unopposed estrogen causes endometrial hyperplasia and anovulatory irregular menstrual bleeding.

Polycystic ovarian cyst: Polycystic ovarian cyst (PCO) is an endocrine disorder characterized by excessive androgen production, abnormal gonadotropin secretion, and chronic anovulation, and it can present itself with menstrual irregularity.

Dysfunctional uterine bleeding: A diagnosis of exclusion after organic causes have been ruled out. It is usually due to anovulation with breakthrough bleeding resulting from chronic and unopposed estrogen stimulation. It is most commonly seen at menarche or perimenopause. It is usually cyclic and painless and can vary greatly in interval, duration, and amount of flow.

2. Menstrual Disorders Associated with Ovulatory Cycles

Pelvic inflammatory disease: It is possible that the severe endometrial inflammation may cause capillary oozing with spotting or menorrhagia in 40% of patients with the disease.

Uterine fibroids: Particularly for submucosal fibroids there are some theories that the excessive menstrual bleeding is caused by compression of the overlying endometrium or by increased endometrial surface.

Von Willebrand disease: A coagulation disorder which may cause menorrhagia and represents up to 20% of menstrual disorders occurring during adolescence.

Iatrogenic causes: The causes of abnormal uterine bleeding include sex steroids, anticoagulants, and use of intrauterine contraceptive devices.

Systemic diseases: Diseases such as hypothyroidism or liver disease alter estrogen metabolism, leading to menstrual irregularity.

F. Perimenopause

Menstrual disorders that occur during perimenopause are caused by anovulatory cycles when ovarian follicles no longer respond to increasingly elevated LH and FSH levels. Bleeding from malignant endometrial neoplasms is also of great concern in this age group.

G. Evaluation

In most cases, the history and physical exam will exclude extragential sources of bleeding and suggest the differential diagnosis and the appropriate laboratory investigations. The following are diagnostic laboratory tests for evaluation of menstrual disorders:

Blood pregnancy test: human chorionic gonadotrophin level
Hematocrit, platelet count, and coagulation profile
Serum LH and FSH
Serum prolactin
Serum thyroid hormones and thyroid-stimulating hormones
Pelvic ultrasonography: useful to rule out myomas and adnexal masses
Endometrial biopsy: indicated for all patients over 35 years of age with irregular bleeding (e.g., women at greatest risk for endometrial hyperplasia and carcinoma); of very limited use in the perimenarcheal group; may reveal proliferative endometrium, endometrial hyperplasia, carcinoma, polyps, endometritis, or retained products of conception
Office-based hysteroscopy: indicated for patients with recurrent bleeding unresponsive to medical therapy or dilatation and curettage

III. TREATMENT OF MENSTRUAL DISORDERS

Treatment is specific for each patient and is based on the patient's age, desire for contraception or fertility, the severity or chronicity of the bleeding, and any other associated conditions.

A. Conservative Medical Management

If bleeding is not severe and the patient is not concerned about contraception or fertility, observation may be appropriate. Most adolescent girls with abnormal bleeding will establish normal ovulatory cycles within 2 years. Medical conditions, such as hypothyroidism or hyperprolactinemia, once evaluated and treated will no longer affect the menstrual cycle. Oral contraceptives will stop the bleeding in most cases and regulate the menstrual cycle and reduce menstrual blood loss by 60%. Additional benefits include contraception and a reduction of androgens in PCO. For the acute episode, the administration of one pill twice a day for 1 week will usually stop the bleeding. The mechanism of action of estrogen is believed to be a stimulus to clotting at the capillary level.

Estrogens alone will temporarily stop acute, heavy bleeding but will not correct the underlying defect. Conjugated equine estrogen is administered orally (2.5 mg) four times per day or intravenously (iv; 25 mg) every 4 hr up to three doses. No study has shown iv therapy to be more effective than oral treatment. If bleeding stops, the patient should be maintained on oral estrogen for 21 days with 10 mg medroxyprogesterone acetate per day added for the last 5 days.

Progesterone alone is used for annovulatory bleeding. Progesterone does not interfere with the normal hypothalamic maturation and may be used for adolescents and women of reproductive age who do not require contraception and do not desire fertility. Progesterone may also be used for the perimenopausal woman for whom oral contraceptives may be contraindicated. A dose of 10 mg medroxyprogesterone acetate is administered orally once a day for 10 days every 1 or 2 months.

Antiprostaglandins or the nonsteroidal antiinflammatory agents, naproxen sodium and mefanamic acid, decrease menstrual blood loss by altering the balance between the platelet aggregating vasoconstrictor, thromboxane, and the antiaggregating vasodilator prostacyclin (PGI_2). The beneficial effect is observed only in ovulatory cycles with heavy menses.

GnRH agonists are effective in suppressing the pituitary gonadal axis by downregulating the gonadotrophs of the anterior pituitary, thereby inhibiting gonadotropin release and reducing ovarian hormone secretion. GnRH may be used in precocious puberty and in cases in which androgens, estrogen, and progesterone are contraindicated.

Ovulation induction is indicated only for patients attempting conception. Clomiphene citrate is the first line of therapy. Human menopausal gonadotropins and/or GnRH are reserved for clomiphene-resistant patients.

The treatment for Von Willebrand disease is desmpressin acetate, a synthetic analog of vasopressin that promotes the release of normal-functioning Von Willebrand factor from storage sites.

B. Surgical Management

Dilatation and curettage (D&C) is indicated for profuse bleeding, recurrent bleeding unresponsive to medical treatment, or when endometrial biopsy reveals premalignant neoplasia. Hysteroscopy at the time of D&C may reveal a polyp or submucosal myoma which a D&C might not diagnose.

Hysterectomy is the last resort when other measures fail to control bleeding. It can also be considered in patients with other coexisting indications, especially in the perimenopausal patient with myomas, endometrial hyperplasia, or pelvic relaxation.

Endometrial ablation is an alternative to hysterectomy in the patient who is a poor surgical candidate or who desires to retain her uterus. It is performed under hysteroscopic control using coagulation via a resectoscope.

Acknowledgment

The authors thank Jonathan Tilly for his help in preparing the manuscript.

See Also the Following Articles

Corpus Luteum; FSH (Follicle-Stimulating Hormone); GnRH (Gonadotropin-Releasing Hormone); LH (Luteinizing Hormone); Menstruation

Bibliography

Berga, S. L., Mortola, J. F., Suh, G. B., Laughlin, G. A., Pham, P., and Yen, S. S. C. (1989). Neuroendocrine aberrations in women with functional hypothalamic amenorrhea. *J. Clin. Endocrinol.* 68, 301.

Brenner, P. F. (1996). Differential diagnosis of abnormal uterine bleeding. *Am. J. Obstet. Gynecol.* 175, 766.

Bullen, B. A., Skrinar, G. S., Beitins, I. Z., von Mering, G., Turnbull, B. A., and McArthur, J. W. (1985). Induction of menstrual disorders by strenuous exercise in untrained women. *N. Engl. J. Med.* 312, 1349.

Chuong, C. J., and Brenner, P. F. (1996). Management of abnormal uterine bleeding. *Am. J. Obstet. Gynecol.* 175, 787.

Cook, C. B., Nippoldt, T. B., Kletter, G. B., Kelch, R. P., and Marshall, J. C. (1991). Naloxone increases the frequency of pulsatile luteinizing hormone secretion in women with hyperprolactinemia. *J. Clin. Endocrinol. Meteb.* 73, 1099.

Falsetti, L., Pasinetti, E., Mazzani, M. D., and Gastaldi, A. (1992). Weight loss and menstrual cycle: Clinical and endocrinological evaluation. *Gynecol. Endocrinol.* 6, 49.

Giuldice, L. C. (1992). Insulin-like growth factors and ovarian follicular development. *Endocr. Rev.* 13, 641–669.

Goodman, A. L., and Hodgen, G. D. (1983). The ovarian triad of the primate menstrual cycle. *Recent Prog. Horm. Res.* 39, 1–67.

Hillier, S. G., Reichart, L. E. J., and Van Hall, E. V. (1981). Control of preovulatory follicular estrogen biosynthesis in the human ovary. *J. Clin. Endocrinol. Metab.* 52, 847.

Hodgen, G. D. (1989). Biological basis of follicle growth. *Hum. Reprod.* 4, 37.

Howlettt, T. A., Tomlin, S., Hgahfoong, L., Rees, L. H., Bullen, B. A., Skrinar, G. S., McArthur, J, W (1984). Release of beta-endorphin and met-enkefalin during exercise in normal women: Response to training. *Br. Med. J.* 288, 1950.

Knobil, E. (1980). The neuroendocrine control of the menstrual cycle. *Recent Prog. Horm. Res.* 36, 53.

Lee, P. A., Van Dop, C., and Migeon, C. J. (1986). McCun–Albright syndrome: Long term follow-up. *J. Am. Med. Assoc.* 256, 290.

Monroe, S. E., Levine, L., Chang, R. J., Keye, W. R., Jr., Yamamoto, M., and Jaffe, R. B. (1981). Prolactin-secreting pituitary adenoma: Increased gonadotropin responsivity in hyperprolactinemic women with pituitary adenomas. *J. Clin. Endocrinol. Metab.* 52, 1171.

Rabinovici, J. (1993). The differential effects of FSH and LH on the human ovary. *Bailieres Clin. Obstet. Gynecol.* 7, 263.

Waren, M. P. (1985). Effect of exercise and physical training on menarche. *Sem. Reprod. Endocrinol.* 3, 17.

Menstruation

Linda R. Nelson
University of Illinois at Chicago

I. Normal Physiology
II. Endometrial Changes across the Menstrual Cycle
III. Pathophysiology
IV. Therapeutic Options

GLOSSARY

anovulatory uterine bleeding Shedding of endometrial tissue that occurs on a sporadic basis in the absence of ovulatory cycles.

dysfunctional uterine bleeding Shedding of endometrial tissue that occurs on an irregular basis that cannot be attributed to pelvic or systemic pathology.

hypermenorrhea (menorrhagia) Increased menstrual flow and/or duration.

hypomenorrhea (cryptomenorrhea) Decreased menstrual flow and/or duration.

menstruation The periodic shedding of endometrial tissue that occurs when an ovulatory cycle has not resulted in a pregnancy.

metorrhagia Bleeding between menstrual periods.

oligomenorrhea Menstrual cycles that occur at intervals more that 35 days apart.
proliferative phase The descriptive term for the alterations in the endometrium that occur during the preovulatory or follicular phase of the menstrual cycle.
secretory phase The descriptive term for the alterations in the endometrium that occur during the postovulatory or luteal phase of the menstrual cycle.

Menstruation refers to the regular shedding of endometrial tissue that occurs under the influence of the hypothalamic–pituitary–ovarian axis.

I. NORMAL PHYSIOLOGY

The menstrual cycle is a characteristic of all higher primates, including the Old World monkeys, great apes, gibbons, and humans. The menstrual cycle is dependent on the appropriate functioning of all components of the hypothalamic–pituitary–ovarian (HPO) axis. The details of the endocrine, paracrine, and autocrine signals involved in each level of this axis are described in detail in other articles in this encyclopedia. In addition, many other organs and signals can impact upon the functioning of the HPO axis. For example, neurotransmitters may relay information from higher cortical areas or from areas of the limbic system that can alter the gonadotropin-releasing hormone (GnRH) neuronal output from the arcuate nucleus. Disruption of the normal functioning of the thyroid or adrenal gland may also impact the HPO axis and lead to menstrual dysfunction. This article will focus on menstruation itself and some of the most common abnormalities in the menstrual cycle.

Menstruation begins at menarche (the first menstrual bleeding) and continues until menopause (the last menstrual cycle). During the reproductive years the average women will have approximately 400–500 menstrual cycles. The menstrual cycles that occur in the first year or more following menarche are likely to be irregular, as are those in the perimenopausal years. However, once ovulatory cycle are established, there are "normal" indices for menstruation. The onset of bleeding is always noted as cycle Day 1. The normal interval between menses is 28 ± 7 days. The normal duration of bleeding is 4–6 days, with an acceptable range of 2–8 days. The average volume of blood loss is 30 ml but the upper limit of normal is 80 ml.

II. ENDOMETRIAL CHANGES ACROSS THE MENSTRUAL CYCLE

Menstruation refers to the periodic shedding of endometrial tissue when an ovulatory cycle has not resulted in a pregnancy. The human endometrium can be morphologically divided into two "layers." The upper two-thirds comprises the functionalis layer and the inner one-third is the basalis layer. In each menstrual cycle, the functionalis layer undergoes proliferation, differentiation, and degeneration. The basalis layer remains intact and underlies the regeneration of the endometrium following menstruation. The endometrial glands, stroma, and vasculature respond to a variety of signals including endocrine signals, such as steroid hormones, and paracrine or autocrine signals, such as growth factors. The preovulatory phase of the menstrual cycle is often referred to as the follicular phase since the ovary is developing the dominant follicle destined to ovulate. During this phase, circulating estrogen levels rise and effect alterations in the endometrium. Since the primary response of the endometrium during this phase is proliferation, this is also referred to as the proliferative phase of the menstrual cycle. Following ovulation, the corpus luteum produces progesterone and estradiol and the predominant response of the endometrium is the glandular differentiation and production of secretory products to aid in the implantation and maintenance of an embryo. This phase of the menstrual cycle is thus referred to as either the luteal or secretory phase based on the activity of the ovary or endometrium, respectively. Menstrual endometrium refers to the endometrial changes that occur during the time of menstrual bleeding.

The histological alterations that occur in the endometrium can be noted in the glands, stroma, and vasculature. Classic studies of Noyes detail these phenomena and provide a standardized system for "dating" the endometrium by normalizing it to a 28-day cycle. During the follicular phase of the menstrual cycle the increasing estrogen levels act via estrogen receptors which are expressed at their highest level during this phase of the cycle. Proliferation occurs in all cellular constituents of the endometrium as evidenced by DNA and RNA synthesis. The thickness of the endometrium increases to approximately 5 or 6 mm in height and this growth can be appreciated by serial measurements made with transvaginal sonography. Within the functionalis layer, the glands become larger and shift from a narrow, tubular configuration to become more tortuous and contiguous with each other. The spiral arteries also undergo proliferation and extend to traverse the entire thickness of the growing functionalis layer. The stomal fibroblasts proliferate and form a less dense supporting layer. Recent research has demonstrated that there are also changes in the levels of many growth factors, such as insulin-like growth factors-I and -II, epidermal growth factor, platelet-derived growth factor, and transforming growth factor-β, as well as their receptors.

Following ovulation, estrogen continues to circulate, and progesterone levels increase to their midluteal peak. Progesterone leads to a decreased proliferation, partially by its inhibition of estrogen receptors and partially by its stimulation of enzymes such as sulfotransferase and 17-OH dehydrogenase, leading to estrogen degradation. Endometrial glands and spiral vessels become increasingly tortuous and the surrounding stroma becomes more edematous. Secretory activity by the glandular epithelium is stimulated early in this phase. Subnuclear intracytoplasmic glycogen vacuoles form and there is an increased secretion of peptides and glycoproteins into the glandular lumen. In the second half of the secretory phase, the endometrium appears to have three layers: the inner basalis, the midlevel stratum spongiosum, and a superficial layer of larger stromal cells called the stratum compactum. Elevated progesterone levels lead to further changes in the histology and function of the stomal cells called predecidualization. Decidual and predecidual changes include increased mitotic activity, cellular enlargement, and increased protein secretion.

If a viable embryo does not implant in the endometrium during the implantation window the endometrium will begin to degenerate partially due to the falling output of progesterone and estradiol from the corpus luteum. An early pregnancy will prevent these events through the stimulation of the corpus luteum by human chorionic gonadotropin production. If a pregnancy has not occurred, a series of intercurrent events that lead to menstruation follows. Lysosomal enzymes including acid phosphatase are released within the endometrium, leading to tissue necrosis, vascular thrombosis, and extravasation of red and white blood cells. In addition, in this phase of endometrial breakdown, enzymes such as matrix metalloproteinases degrade the stroma and basement membrane. Prostaglandins ($PGF_{2\alpha}$ and PGE_2) increase in late secretory endometrium and achieve maximal concentrations during menstruation. $PGF_{2\alpha}$ may stimulate the degeneration of the endometrium and enhance myometrial contractions to empty the uterus of menstrual efflux.

The vascular events that precede menstruation occur mainly within the spiral arterioles. Initially, blood flow decreases and this is followed by vasodilation. The spiral arterioles then undergo vasoconstriction and relaxation in a spasmodic manner. These alterations in blood flow lead to endometrial ischemia. Vascular permeability increases and leakage into the endometrium occurs. With increasing tissue ischemia and necrosis, the functionalis layer of the endometrium separates from the basalis layer and is shed into the endometrial cavity. The basalis layer proceeds to repair the endometrium and regenerate the functionalis layer. Rising estrogen from the new follicular phase aids in the stabilization of the extracellular matrix and the clot formation in the spiral vessels.

III. PATHOPHYSIOLOGY

Menstrual disorders are very common and account for approximately one-third of office visits to a gynecologist. Hypermenorrhea is increased menstrual flow and/or duration. In addition, there may be passage of large blood clots which are normally not formed due to the fibrinolytic activity of plasmin within the menstrual tissue. The initial approach to hypermenorrhea is to rule out etiologies, such as pregnancy or iatrogneic drug use. In addition, local pelvic pathology needs to be ruled out. Benign lesions, such as endometrial polyps or submucosal/intramural fibroids, need to be evaluated with imaging studies, such as transvaginal sonography, hysterosalpingography, or saline infusion sonography. Endometrial lesions, such as hyperplasia or carcinoma, need to be evaluated with an endometrial biopsy. Systemic pathology such as other endocrinopathies (e.g., thyroid), as well as platelet or other hematologic dysfunction, need to be ruled out with the appropriate lab tests.

The term dysfunctional uterine bleeding refers to abnormal menstrual bleeding that does not have a structural basis. Some authors include anovulatory bleeding in this category and others prefer to use the term anovulatory uterine bleeding to define this group and reserve dysfunctional uterine bleeding for patients without a structural or endocrinological basis for their bleeding. Anovulatory bleeding is generally not accompanied by molimina (premenstrual signs such as breast tenderness) and is variable in quantity and duration.

Hypomenorrhea refers to decreased menstrual flow and/or duration. The workup of this disorder includes ensuring that there are no structural lesions of the uterus, cervix, and vagina that are inhibiting menstrual flow. In addition, adhesions within the endometrial cavity (Asherman's syndrome) may lead to hypomenorrhea. As with patients presented with hypermenorrhea, other endocrinopathies, such as thyroid, adrenal, or pituitary disorders, need to be ruled out. Systemic pathology involving eating disorders or hepatic dysfunction should also be addressed. When menstrual cycles are missed for a significant length of time (e.g., 6 months), the original diagnosis of anovuation changes to amenorrhea.

IV. THERAPEUTIC OPTIONS

As previously indicated, patients with alterations in menstrual flow need to have a thorough evaluation to rule out the presence of structural lesions and systemic disorders. However, once this has been done, and the diagnosis of dysfunctional bleeding has been made, there are several therapeutic options available. One of the most common abnormalities in patients with irregular menstrual bleeding is oligo- or anovulation. This anovulatory or oligoovulatory bleeding that occurs after a prolonged interval of estrogenic stimulation of the endometrium will often be heavy and prolonged. Acute, heavy bleeding may require emergent intervention due to a rapid decrease in the hemoglobin concentration and intravascular loss contributing to hypotension and orthostatic changes. In this acute setting two approaches can be taken: surgical therapy with a dilation and curettage and medical therapy with estrogens or progestins. In the setting of prolonged, heavy bleeding, estrogen treatment provides the most rapid stabilization of the endometrium and cessation of bleeding. Estrogen can be administered intravenously (iv) (conjugated estrogens 25 mg iv QD 2-4 hr for a total of four injections). Following this treatment, oral contraceptives need to be given in a tapering dose regimen in order to further stabilize the endometrium. Withdrawal bleeding can then be delayed for at least 2 weeks, during which time the hemoglobin and hematocrit will increase toward the baseline level. If bleeding is less acute, oral contraceptives alone can be used in the tapering method. Oral contraceptives (monophasic, 30–35 mg ethinyl estradiol) are administered in a gradually taping dose with four pills QD for 5 days, then three pills QD for 4 or 5 days, then two pills QD for 4 or 5 days, and finally one pill QD for 5 days. In the following week withdrawal bleeding will occur and this will generally be a heavy menses but should not require surgical intervention. Patients with oligo- or anovulation need to have chronic therapy to avoid the episodes of acute, heavy bleeding. As long as these patients have adequate estrogenic stimulation (World Health Organization group II anovulation), progestins can be added to substitute for the usual luteal phase and effect controlled withdrawal bleeding. Cyclic progestin therapy with norethindrone or norethindrone acetate (5 mg postoperative QD) or medroxyprogesterone acetate (10–30 mg postoperative QD) can be given for 10–14 days each month. Additional formulations of progesterone such as vaginal gel or suppositories and oral formulations are currently under active investigation. The major disadvantage of this regimen is that there is no contraception provided and synthetic progestins are contraindicated in pregnancy. The other option for oligo- or anovulatory patients not seeking pregnancy is that of oral contraceptives. Withdrawal bleeding is generally light in quantity and is well tolerated by patients.

Other medical therapies also exist for menorrhagia and dysfunctional uterine bleeding. Danazol is particularly useful even in patients whose dysfunctional uterine bleeding (DUB) occurs in ovulatory cycles. Women with hypermenorrhea may also be aided by the administration of nonsteroidal antiinflammatory agents. These agents lead to a decrease in menstrual flow by interfering with the prostaglandin milieu within the endometrium. Finally, the entire HPO axis can be "shut down" in women by the continuous administration of gonadotropin hormone-releasing hormone. The charateristics of the pituitary GnRH receptor that lead to desensitization and downregulation are described elsewhere in this encyclopedia. Clinically, continuous delivery of synthetic GnRH can be accomplished by subcutaneous, intramuscular, or IN delivery of a GnRH analog. This therapy results in abolition of menstrual bleeding due to the pronounced decrease in follicle-stimulating hormone and luteinizing hormone which leads to suppression of folliculogenesis and menopausal levels of estradiol. This therapy may be useful for short-term use, but the menopausal levels of estradiol lead to bone loss and signs of estrogen deficiency in other tissues as well, which limit the use to 6 months.

Surgical treatments for dysfunctional uterine bleeding and menorrhagia are also an option. As mentioned previously, the diagnostic workup of abnormal menstrual bleeding (either menorrhagia or metorrhagia) may include investigation of the uterine cavity to rule out structural lesions and hyperplasia or carcinoma of the endometrium. Hysteroscopy and dilation and curettage (D&C) may be utilized as both diagnostic and therapeutic procedures. However, some patients will have no structural lesions present and may have the recurrence or persistence of DUB following a D&C, or they may not be able to tolerate medical management. Hysterectomy has historically been utilized as the definitive therapy for abnormal bleeding in women who have completed childbearing. Up to 50% of the hysterectomies performed in the United States have abnormal bleeding as an indication. Recently, however, resection of the endometrium has been popularized as a more conservative treatment. Hysteroscopic resection can be achieved with laser ablation or cautery and mechanical resection. Within the past year a simplified technology has been introduced that allows the endometrium to be ablated via thermal energy delivered by a balloon that conforms to the uterine cavity. Any of the ablation procedures may lead to compete cessation of menstruation in some patients but the majority of patients will have lighter menses and thus have improvement in their symptoms. It is likely that the technology will continue to evolve for ablation of the endometrium in patients that might otherwise require a hysterectomy for abnormal bleeding.

See Also the Following Articles

ENDOMETRIUM; MENARCHE; MENOPAUSE; MENSTRUAL CYCLE; MENSTRUAL DISORDERS

Bibliography

Giudice, L. C., and Ferenczy, A. (1996). The endometrial cycle. In *Reproductive Endocrinology, Surgery, and Technology* (E. Y. Adashi, J. A. Rock, and Z. Rosenwaks, Eds.), pp. 271–300. Lippincott-Raven, Philadelphia.

Hulboy, D. L., Rudolph, L. A., and Matrisian, L. M. (1997). Matrix metalloproteinases as mediators of reproductive function. *Mol. Hum. Reprod.* 3, 27–45.

Noyes, R. W., Hertig, A. W., and Rock, J. (1950). Dating the endometrial biopsy. *Fertil. Steril.* 1, 3.

O'Conner, H., and Magos, A. (1996). Endometrial resection for the treatment of menorrhagia. *N. Engl. J. Med.* 335, 151–156.

Mesoderm

see Germ Layers

Metamorphosis, Insects

Fred Nijhout

Duke University

I. The Evolution of Metamorphosis and Reproductive Periods
II. The Hormonal Control of Metamorphosis
III. Developmental Hormones Change Function after Metamorphosis
IV. Polyphenism: The Development of Alternative Reproductive Forms at Metamorphosis
V. Maternal Effects

GLOSSARY

Apterygota Wingless insects without metamorphosis. This group consists of four small orders: Protura, Diplura, Microcorrhyphia, and Thysanura. Apterygotes branch off the insect phylogenetic tree near its base and posses many characters that are believed to be primitive for the insects. Also called **Ametabola**.

commitment In development, the point at which the developmental fate of a tissue has become fixed to follow a particular developmental pathway. Also called **determination**.

ecdysteroids A generic term referring to an assortment of steroid hormones used in the regulation of insect development and reproduction. Most insects secrete a prohormone, ecdysone, which is transformed in the epidermis and fat body to the active hormone, 20-hydroxyecdysone.

Hemimetabola Insects with incomplete metamorphosis. These insects molt to the adult stage directly from the larval stage.

Holometabola Insects with complete metamorphosis. These insects molt from a larval to a pupal stage before molting to an adult.

imaginal disc The tissues in holometabolous insect larvae that will give rise to adult appendages and other structures. Imaginal discs can exist as slightly thickened epidermal placodes or as relatively complex invaginated pouches of epidermal cells.

juvenile hormone A hormone that is used in larval insects to regulate the progression to successive developmental stages and by adult insects as a gonadotropic hormone and a regulator of vitellogenin synthesis.

maternal effect A mechanism by which a female affects the characteristics of her offspring that does not involve the transmission of nuclear genes. Typically this is achieved through the transmission of chemicals to the egg.

polyphenism The ability of individuals with identical genotypes to develop into discrete alternative phenotypes, usually in response to an environmental stimulus.

Metamorphosis refers to the morphological and physiological transformation of a nonreproductive larval insect to a reproductive adult. In many insects the physiological control of metamorphosis is intimately related to the physiological control of reproduction.

I. THE EVOLUTION OF METAMORPHOSIS AND REPRODUCTIVE PERIODS

The relationship between reproduction and metamorphosis has undergone great changes in the course of insect evolution. The primitive or ancestral relationship is believed to have consisted of an alternation of reproductive cycles and molting cycles, preserved today in the apterygote insects. The best studied example is that of the firebrat, *Thermobia domestica* (Thysanura). In this species larvae grow by molting until they reach a size characteristic of the adult, at which point external genitalia appear and gonads develop internally. Except for the possession of external genitalia, the external morphology of adults is like that of large larvae. The larvae themselves undergo a kind of simple metamorphosis in the middle of larval life when they develop a body covering of scales, typically after the third larval molt. Once it has reached the adult stage, *Thermobia* continues to undergo stationary molts, without growing. Each molting cycle is divided into two distinct phases: a reproductive phase, during which the animal mates and a new batch of eggs develops and is deposited, and a molting phase, during which the animal sheds its cuticular exoskeleton and develops a new one. When the female molts, stored sperm are lost from the spermatheca so she must mate again in order to reproduce after molting. The reproductive and molting phases are both under hormonal control. It appears that mating is the stimulus that initiates each gonadotropic cycle. Mating stimulates the secretion of juvenile hormone (JH) which provokes vitellogenesis of the terminal follicle of each ovariole. Oviposition of those eggs appears to be the stimulus that triggers the secretion of the molting hormone, ecdysone, and the cessation of JH secretion. Ecdysone then induces the next molt, which occurs in the absence of JH. The reproductive phase of each molting cycle is therefore characterized by high titers of JH and low titers of ecdysone, whereas the converse is true during the subsequent molting phase.

The evolution of winged insects has been marked by the progressive divergence of adult and larval morphology and specialization. With the evolution of wings, the adult stage became specialized for reproduction and dispersal, whereas the wingless larval stage became specialized for growth. There are two major lineages of winged insects, the Hemimetabola (insects with incomplete metamorphosis) and the Holometabola (insects with complete metamorphosis, comprising the great majority of insect species). In the Hemimetabola, larval and adult morphologies have diverged to varying degrees. In cockroaches and crickets, for instance, adults differ from larvae mainly in the development of internal reproductive organs, external genitalia, and wings. By contrast, in groups with aquatic larvae, such as the dragonflies and mayflies, adults and larvae have evolved very different body forms and physiology, each highly specialized in its own way. The most dramatic divergence, however, has occurred in the Holometabola, in which larval and adult differences are of a degree that typically characterizes different orders of insects: Other than sharing the general insect groundplan, the morphology of a caterpillar has nothing in common with that of a butterfly, nor the morphology of a maggot with that of a fly.

II. THE HORMONAL CONTROL OF METAMORPHOSIS

The transformation of a larva to a dramatically different adult is orchestrated by the same pair of hormones that in the Apterygota control the alternation of reproduction and molting. The molting hormone, ecdysone, stimulates not only the molt from larva to adult (or, in the case of Holometabola, the sequence of molts from a larva to a pupa and then to an adult) but also the breakdown of uniquely larval tissues and the growth and differentiation of adult structures from their larval counterparts, or *de novo* from nests of specialized cells that are kept in reserve for this function (such as the imaginal discs, histoblasts, and neuroblasts). In holometabolous insects the external genitalia and accessory glands develop from imaginal discs that are associated with the ventral epidermis of the eighth and/or ninth abdominal segments, whereas the gonads develop from primordia lying dorsally in the abdomen. The juvenile hormone acts as a status quo hormone during larval life. While it is present, tissues retain their current state of commitment, and periodic secretion of ecdysone results in molts to ever larger larval instars. Juvenile hormone secretion ceases at the end of larval life, and after the JH level declines, ecdysteroid secretion stimulates a switch in commitment so that when the animal next molts it develops the characteristics of the following developmental stage.

The switch in the developmental program from that of a larva to that of a pupa has been the best studied process in metamorphosis. It appears that pupal commitment can only happen after the level of JH has fallen below a tissue-specific threshold. Different tissues switch from larval to pupal commitment at different times. The imaginal discs switch long before the general epidermis, and imaginal discs each have a characteristic time, usually early in the last larval instar, when they switch from larval to pupal commitment. Tissues, then, differ in their sensitivity to JH and in the time at which they can become committed to the next developmental stage. Fluctuations in JH titers during larval life, and particularly during the last larval instar, appear to be involved in orchestrating the sequential commitment of different tissues.

III. DEVELOPMENTAL HORMONES CHANGE FUNCTION AFTER METAMORPHOSIS

After metamorphosis, the maturation of ovaries and testes, the synthesis of vitellogenin by the fat body, and vitellogenesis of the eggs are all controlled by hormones. Interestingly, the hormones that earlier controlled molting and metamorphosis—ecdysteroids and JH—are also the ones that control the various processes in the reproductive cycle of insects. The fact that these hormones are used for completely different functions in the adult means that during metamorphosis there is a significant reprogramming of tissue responsiveness to these hormones, such as the elimination of the epidermal response to ecdysone (so the animal no longer molts in response to ecdysteroid secretion) and the new acquisition of vitellogenin synthesis capability by the female fat body in response to hormones (JH or ecdysteroids, depending on the species). In the Hemimetabola and some Holometabola the fat body persists from the larva to the adult and must therefore be reprogrammed at metamorphosis to produce vitellogenin in response to JH or ecdysteroids. In some Diptera and Hymenoptera, by contrast, the larval fat body breaks down at metamorphosis and the adult fat body develops from undifferentiated cells. These various changes in tissue responsiveness to hormones after metamorphosis are presumably mediated by changes in the types and distribution of hormone receptors among tissues.

This dual role for JH and ecdysteroids at different stages in the life cycle has raised the question of whether the primitive function of these hormones was developmental (molting and metamorphosis) or gonadotropic. In the case of ecdysteroids, the primitive function in insects is certainly developmental because all arthropods use ecdysteroids as molting hormones. The JH is clearly a gonadotropic hormone in the most primitive insects (such as *Thermobia*) and probably also plays the role of a developmental status quo hormone. The dual function of JH, therefore, probably arose very early in insect evolution, possibly before the evolution of true metamorphosis in the Apterygota.

IV. POLYPHENISM: THE DEVELOPMENT OF ALTERNATIVE REPRODUCTIVE FORMS AT METAMORPHOSIS

In regard to reproduction, probably the most interesting evolutionary innovation of insect metamorphosis is the ability of many species to metamorphose into one of several alternative adult forms, often with very different reproductive habits and potentials. In the ants and termites, for instance, a larva can metamorphose into a sterile worker, a sterile soldier, or a reproductive queen (and, in the case of termites, in which castes can often be of both sexes, a king). All these castes are genetically identical, and a young larva is totipotent and has the capacity to develop into any of them. The choice to metamorphose into one or the other is made on the basis of nutritional and pheromonal signals the larva is exposed to during critical periods in its development. The social bees and wasps also have sterile and reproductive castes as alternative adults forms. This ability of genetically identical animals to develop into very different adult phenotypes is called polyphenism.

Polyphenisms are widespread in insects and typically evolve as flexible developmental adaptations to changing environments so that adult phenotypes can be produced that have unique adaptations to an environment that is not encountered in every generation or that is not encountered by all members of a population. Particularly widespread are the so-called dispersal polyphenisms, in which one of the alternative forms is specialized for migration, with long wings and relatively small ovaries that produce few eggs, and the other is more sedentary, with small wings (in some species this form is actually wingless) and large ovaries that produce many more eggs. The migratory form typically develops when the environment deteriorates. Depending on the species, a decline in food quality or quantity, or changes in temperature, humidity, or photoperiod experienced during a critical period in development provides the signal for a developmental switch that produces the migratory instead of the sedentary form.

The environmental signals alter development by altering the temporal pattern of hormone secretion. All polyphenisms studied so far are the result of changes in the secretion of either JH (mostly) or ecdysteroids (less frequently). In each case there are well-defined windows of hormone sensitivity during the embryonic, larval, or pupal stage, when the future developmental pathway can be altered. In the case of the ant *Pheidole bicarinata*, an elevated level of JH in the embryo causes the animal to develop into a queen, whereas elevated JH during a critical window in the last larval instar reprograms the growth pattern of the larva so that it develops into a soldier instead of a worker. In honey bees (*Apis mellifera*), there is a JH-sensitive period at the beginning of the last larval instar during which elevated levels of JH can induce queen development. In dispersal polyphenism, the more highly reproductive sedentary phase appears to develop if JH is elevated during a particular critical period in larval development.

V. MATERNAL EFFECTS

The hormonal control of development in insects can extend across generations. The eggs of many insects enter diapause at some point in embryonic development. In the control of dispersal polyphenism in migratory locusts, the elevated level of JH in females of the sedentary form biases her offspring toward the development of sedentary form morphology. Also, if the JH level of such a female is artificially reduced, her offspring have a greater propensity to develop into the migratory form. It appears that the level of JH that a female's eggs are exposed to can alter the developmental fate of her offspring, presumably by altering their pattern of JH secretion so that it favors development of one or the other alternative polyphenic forms.

See Also the Following Articles

Ecdysteroids; Juvenile Hormone; Mating Behaviors, Insects

Bibliography

Hagedorn, H. H. (1985). The role of ecdysteroids in reproduction. In *Comprehensive Insect Physiology, Biochemistry, and Pharmacology* (G. A. Kerkut and L. I. Gilbert, Eds.), Vol. 8, pp. 205–262. Pergamon, New York.

Happ, G. M. (1992). Maturation of the male reproductive system and its endocrine regulation. *Annu. Rev. Entomol.* 37, 303–320.

Koeppe, J. K., Fuchs, M., Chen, T. T., Hunt, L.-M., Kovalek, G. E., and Briers, T. (1985). The role of juvenile hormone in reproduction. In *Comprehensive Insect Physiology, Biochemistry, and Pharmacology* (G. A. Kerkut and L. I. Gilbert, Eds.), Vol. 8, pp. 165–203. Pergamon, New York.

Mousseau, T. A., and Dingle, H. (1991). Maternal effects in insect life histories. *Annu. Rev. Entomol.* 36, 511–534.

Nijhout, H. F. (1994). *Insect Hormones*. Princeton Univ. Press, Princeton, NJ.

Riddiford, L. M. (1985). Hormone action at the cellular level. In *Comprehensive Insect Physiology, Biochemistry, and Pharmacology* (G. A. Kerkut and L. I. Gilbert, Eds.), Vol. 8, pp. 37–84. Pergamon, New York.

Wheeler, D. E. (1986). Developmental and physiological determinants of caste in social Hymenoptera: Evolutionary implication. *Am. Nat.* **128**, 13–34.

Wyatt, G. R., and Davey, K. G. (1996). Cellular and molecular actions of juvenile hormone. II. Roles of juvenile hormone in adult insects. *Adv. Insect Physiol.* **26**, 1–155.

Zera, A. J., and Denno, R. F. (1997). Physiology and ecology of dispersal polymorphism in insects. *Annu. Rev. Entomol.* **42**, 207–231.

Mice

see Rodentia

Microtinae (Voles)

Theresa M. Lee

University of Michigan

I. Introduction to Taxonomy, Evolution, and Ecology of Microtine Rodents
II. Reproductive Traits of Microtine Rodents and Life Histories
III. Physiological and Behavioral Traits Influenced by Gonadal Hormones

GLOSSARY

altricial Referring to an immature state at birth which requires a period of neonatal dependency on adult care.

Bruce effect Termination of pregnancy prior to implantation when a female is exposed to the odors of an unfamiliar, reproductively active male.

corpus luteum The yellow endocrine body formed on the ovary at the site of a ruptured follicle, immediately following ovulation; the organ secretes estrogen and progesterone and is necessary for maintenance of pregnancy.

facultative Capable of living under conditions other than those which are typical; adaptable to changes in the environment.

gonadal involution Degeneration of testes and ovaries to a state containing primarily primordial germ cells, leaving the animal infertile with very low concentrations of gonadal steroids.

irruption A sudden rapid increase in local population density, well above the typical numbers found during peaks in population cycles; population explosion.

6-methoxybenzoxazolinone A nonsteroidal chemical produced in grass plants when the growth portion of the plant is damaged.

monogamy A mating system in which a single male and female mated pair cohabit, rear young together, and together exclude strange males and females from the territory.

pheromones Odors produced in the urine, feces, or scent glands of animals that can trigger physiological or behavioral changes in other individuals, usually through the vomeronasal organ.

photoperiod The ratio of hours of light and dark within 24 hr. Changes in photoperiod trigger many physiological changes to prepare organisms for changes in season.

polygyny A mating system in which one male mates with several females such that each female will likely be impregnated by a single male.

promiscuity A mating system in which multiple males mate with the same female such that the female will likely be impregnated by more than one male.

Microtinae (voles) are herbivorous, short-tailed rodents with stocky bodies (species range in weight from 20 to 170 g and from 83 to 250 mm in length), small ears and eyes, and short legs adapted for life in areas with snowy, cold winters. Despite the common name of meadow mice, voles are not mice and they are not found exclusively in meadows. The biology and behavior of these rodents has been studied extensively, initially because of the heavy damage inflicted by some species on crops and trees during the peaks of population cycles, and because of their importance as a major prey item for avian and mammalian carnivores. Recently, research has focused on the facultative response by many species to a variety of environmental and social cues for timing reproduction, puberty, and altering social interactions. Additionally, the variety of species demonstrating different mating strategies has allowed for extensive comparative behavioral and physiological studies to examine both distal (evolutionary) and proximate (physiological and behavioral) mechanisms responsible for interspecies variation.

I. INTRODUCTION TO TAXONOMY, EVOLUTION, AND ECOLOGY OF MICROTINE RODENTS

Voles are the most abundant group of rodents on earth (Fig. 1). While their shared ancestry can be recognized in the common body form and some similar habits, the species have adapted to a much broader array of habitats than the original ancestral species.

A. Classification of Voles and Closely Related Species

The taxonomic classification of vole species is complex and there are many reasonable disagreements. There are approximately 45 species in the genus *Microtus* divided into six to eight subgenera which combine the various species in different ways (including subgenera *Microtus*, *Neodon*, *Pitymys*, *Stenocranius*, *Proedromys*, and sometimes *Lasiopodomys*, *Aulacomys*, *Chilotus*, *Herpetomys*, and *Orthriomys*). The genus *Microtus* is just one of several in the larger family group Arvicolidae, which includes other non-*Microtus* species which are called voles, such as *Clethrionomys*, *Arborimus*, *Arvicola*, and *Lagurus*, and species which are commonly called lemmings (*Dicrostonyx*, *Lemmus*, *Synaptomys*, and *Mictomys*). The family Arvicolidae is sometimes included in the larger family Cricetidae (order Rodentia). For the purpose of this article the terms vole and microtine rodents will refer only to the genus *Microtus*.

B. Evolution

Arvicolidae appeared in the late Miocene period (approximately 12–14 million years ago) in North America but not until the Pliocene (approximately 10 million years ago) in Eurasia. It is unclear where the first species appeared, but it is likely to have been in the boreal grasslands and forests of North America or Eurasia. Modern forms of the genus *Microtus* ap-

Figure 1 *Microtus pennsylvanicus* is the most common microtine rodent in North America (photo by Tammy Jo Jechura).

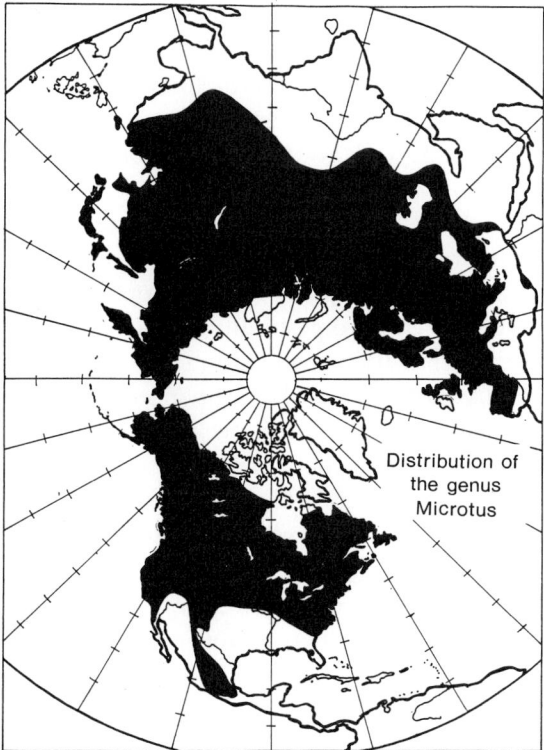

Figure 2 World distribution of predominantly boreal-adapted genus *Microtus* (reproduced from Merritt, 1984, p. 19).

peared in the northern regions of North America and Eurasia about 2 million years ago. *Microtus* species are most numerous in northern, boreal habitats, but extend as far south as southern China, Greece, and Guatemala at high altitudes that provide a boreal habitat (Fig. 2). The most common species in North America is *M. pennsylvanicus*, with a range that extends from the Ohio River valley north to the arctic and from coast to coast above 40°N latitude. The closely related *M. agrestis* is found throughout Europe and large parts of Asia.

C. Ecology

Modern *Microtus* inhabit a variety of common temperate and arctic habitats: dry grasslands, grassy areas in deciduous and pine forests, wet meadows, tundra, and salt marshes. Less common sites include talus slopes, rocky outcrops, and forests with little grass. The majority of species burrow 100 mm or less beneath the surface of grass thatch, leaf mold, or loose soil. It is common to find runways through the tall grasses or leaf litter produced by the travel of voles. Just off the runway animals build nests under protective logs and rocks or burrow beneath the surface of the ground. In contrast, some species (e.g., *M. ochrogaster*) excavate much more elaborate burrow systems which may contain a nest site as well as chambers with stored food. Voles are terrestrial, but species living in wet environments often swim and dive well.

The majority of microtine rodents have well-defined home ranges in which individuals have nests, food, and water. Individuals from some species aggressively exclude most other individuals from their home range, thereby forming well-guarded territories. However, among species with cycles of population growth and decline, the home ranges or territories at various population sizes can be dramatically different in size, and intensity of defensive behaviors can be quite altered. For example, *M. ochrogaster* in eastern Kansas may fluctuate between 4 and 115 animals per hectare every 2–4 years and peak at 350 individuals per hectare during a population irruption. Several populations of North American species (e.g., *M. pennsylvanicus* and *M. montanus*) have been reported at densities of over 1000 animals per hectare during irruptions. At high densities, social and territory structures typically break down despite high levels of interanimal aggression. The stressful social interactions at high densities, which lead to injury and disease, along with eventual reduction of favored food supplies and increased predator load are hypothesized to play roles in the rapid population decline after an irruption in an area. At high densities, species can become serious agricultural pests, destroying growing grain and hay crops, forest plantings, and fruit orchards.

Voles are generally considered to be herbivores (eating only grasses, leaves, and twigs), but many species have expanded their diets to include seeds, nuts, bulbs, and tubers. Some species store food for the winter, but, most likely, the majority consume dried grasses, roots, and tubers through the winter months. The lower calorie content of winter food materials plays a role in the reduced winter reproductive output.

Voles are active throughout the year. They do not hibernate in winter nor is there evidence that any species enter torpor, and of course they do not migrate to better winter climates. Rather, they maintain their body temperatures within a well-defined range and survive winter by making both physiological changes (increased size and activity of brown adipose tissue for heat production; decreased body mass which decreases needed food consumption; increased pelage length, depth, and density; and reduced reproduction) and behavioral changes (aggression levels decline and territorial animals live in communal winter nests increasing huddling heat conservation, food choices may be altered, and some species alter their daily activity patterns). Photoperiod (day length) is the key environmental variable which drives the physiological and behavioral changes for most microtine rodents to allow adaptation to seasonal changes in temperature, food, and population density. However, voles are noted for their facultative seasonal strategies. Thus, year-to-year variation in food quality, rate of temperature decline, and population structure can influence the type and timing of winter adaptations. For example, *M. montanus* becomes reproductively inactive in response to short photoperiods, but the presence of green-growing grasses that contain the plant derivative 6-methoxybenzoxazolinone (6-MBOA) can stimulate reproductive development even in midwinter. Some species have adapted to quite atypical vole environments by evolving local adaptations. *Microtus californicus*, for example, breeds under long summer day lengths in the laboratory but in the field only during the late autumn and winter. The extremely dry summer climate found in most of California prohibits reproduction in the summer, and the species breeds during the rainy winter months because the presence of water, rather than photoperiod, has become the more critical reproductive signal.

The daily activity patterns of microtine rodents are described as ultradian (activity/rest cycles at intervals of less than 12 hr, usually 2–4 hr) but are influenced by the animal's size, diet, season of the year, and local predators. Because microtine rodents are herbivores, most of the calories in their diet are derived from fibrous plant walls which are broken down by bacterial fermentation in the hindgut. The calories produced by fermentation in voles are relatively low compared to those of other herbivores (cattle and camels), and the capacity of the hindgut is relatively small. As a result, voles must eat often to gain enough calories for survival. Younger or smaller animals may become active and eat as often as every 2 hr, whereas larger animals (with larger guts) can eat at less frequent intervals. Voles demonstrate a high preference for young, growing parts of plants presumably because of the easier digestibility, thereby decreasing the duration of time spent eating.

Activity patterns are also influenced by circadian rhythms (24-hr activity/rest cycles), such that more activity (particularly noneating activity) occurs during the night and dawn/dusk transitions. However, during the winter several species alter the activity pattern to a predominantly diurnal (day active) pattern. These seasonal changes in activity are driven by changes in photoperiod and are thought to be adaptive by keeping animals in warm, communal nests during the coldest time of a winter day. The exact timing of activity periods appears to be synchronized by sunrise for *M. arvalis*, thereby synchronizing the population in a given area. In a typical microtine facultative adaptation, the daily patterns of activity are altered when predators with regular behavior patterns appear. For example, avian predators will adapt the time they arrive at a field to feed to the timing of activity of the local voles, but within a few days the voles alter their daily pattern of activity in response to the pattern of predator arrival. The ensuing game of "cat-and-mouse" (or bird-and-vole) is an excellent example of the adaptations of predator and prey to each other's behavioral patterns.

II. REPRODUCTIVE TRAITS OF MICROTINE RODENTS AND LIFE HISTORIES

Microtine reproductive traits, as with their ecology, have certain consistencies across the many species but also demonstrate the ability of voles to adapt to a wide variety of situations. Many of the variations in reproductive traits between species are likely related to variations in mating strategies. Among the well-studied *Microtus*, the majority have polygynous (one male with several female mates such that each female will likely be impregnated by a single male) or promiscuous (multiple males mating sequentially with the same female such that the female will likely be impregnated by more than one male). Polygynous mating systems are associated with large male ranges or territories which contain multiple adult females often defending smaller individual territories, and matings likely occur between unrelated, but familiar, individuals (e.g., *M. pennsylvanicus* and *M. agrestis*). Promiscuous mating systems are described for species that have males with overlapping ranges that include several females, but the exclusion of males by other males so as to form true territories is rare, and both sexes actively seek to mate with strangers (e.g., *M. montanus*). In contrast, two well-studied species (*M. ochrogaster* and *M. pinetorum*) are described as monogamous (single male and female mated pair cohabit, rear young together, and together exclude strange males and females from the territory). Many of the species variations in reproduction and related behaviors and physiology are hypothesized to be the result of adaptations to a monogamous mating system from the more common microtine nonmonogamous systems. The facultative nature of voles is evident even in the study of reproduction. For example, at high population densities the monogamous *M. ochrogaster* male is very likely to mate with multiple females, whereas *M. pennsylvanicus* males exhibit many behaviors in common with monogamous males (paternal behavior and long-term cohabitation with a female) at very low population densities and in winter communal nests.

A. Timing of Reproductive Function

1. Pubertal Development

Under the best of conditions (long day lengths, good nutrition, and low to moderate population density), for most *Microtus*, females reach maturity at approximately 30 days of age and males at 6–8 weeks of age. This difference assists in preventing inbreeding among most species and is associated with the greater dispersal of male siblings. However, it is well established that a variety of environmental variables influence the timing of puberty and the timing of the primary breeding season for overwintering adults. These factors include photoperiod, nutrition, and social interactions.

i. Photoperiod As with most other mammals living in the temperate and more northerly climates, photoperiod is the primary environmental cue for timing the yearly reproductive effort. Field data from the majority of microtine rodents indicate that young born late in the summer or in early autumn overwinter in a prepubertal state. In contrast, those born in the spring or early summer rapidly reach puberty and enter the breeding population. Photoperiod is the major determinant in whether young delay puberty. The photoperiod experienced by the pregnant female (*M. montanus* and *M. pennsylvanicus*) controls the production of melatonin by the pineal gland. The nightly elevation in melatonin reflects the changing length of the night, such that more melatonin is produced during the longer nights of August–December than March–July. Melatonin is able to cross the placenta to the fetuses and to enter the milk during lactation. Work with *M. pennsylvanicus* has demonstrated that maternal melatonin during both gestation and lactation interact with the photoperiod being experienced by the young postnatally to determine the rate of growth, development of pelage, and timing of puberty. Animals born in decreasing or short day lengths, all other things being equal, remain prepubertal for 18–26 weeks. In the field, young born in late summer and autumn reach maturity the following early spring. Remaining small in size and prepubertal through the winter months reduces the energy needs of the animals until reproductive activity is likely to be successful.

Adult microtine rodents are also sensitive to photoperiod. As day lengths decrease in late summer, body weight declines and a winter pelage develops in response to changing melatonin, which alters pituitary release of prolactin, luteinizing hormone (LH), follicle-stimulating hormone (FSH), and thyroid-stimulating hormone. These pituitary hormones in turn control gonadal production of testosterone or estrogen and thyroxine. Thus, melatonin controls the hormonal changes necessary to prepare the adult animals that survive at summer's end for the winter. Gonadal involution reliably occurs in response to decreasing fall day lengths in the more northern ranges of voles, but in the more temperate or southern areas, gonadal involution is often more sensitive to other factors. Thus, all microtines likely respond to photoperiod to prepare for the cooler winter temperatures, but in the more southern ranges it is not unusual for breeding to continue in the winter if temperatures are relatively mild, cover is good, nutrition is high, or population density is accommodating.

ii. Nutrition It has long been recognized that voles begin producing young when there is plenty of newly growing green plants. It is quite clear that the ease of digestion and increased nutritional content of newly growing grasses and other plants is important for supporting pregnancy, lactation, and early development. For example, *M. pennsylvanicus* born in February/March reach puberty more quickly if the spring growth of alfalfa, sprouted wheats, and wild grasses has already begun. Similarly, *M. arvalis* develops more rapidly when fed the first, spring cutting of alfalfa than if they are fed the third, late-summer cutting of the same field. The protein content is higher in the first than the third cutting, and dietary protein is a growth-limiting factor in the diets of many voles. For those species that live in relatively dry climates, such as *M. californicus*, the new growth of green plants in autumn is in response to the annual rainy season. The green plants are not as critical as the water content they provide the animals.

In the context of timing of reproduction, however, plant compounds have been demonstrated as important signals for the hypothalamic–pituitary system to undergo changes appropriate to the season of the year. The same species are often also sensitive to photoperiod, and one can think of the chemical cues in the food as signals that fine-tune the responses to annual variations in the exact timing of nutritional changes. Both *M. montanus* and *M. ochrogaster* are sensitive to the presence of 6-MBOA in newly growing grasses (such as wheat). This nonsteroidal plant compound stimulates more rapid growth, the earlier onset of puberty in young animals, more successful postpartum matings, and increased litter survival for adults. The effect is greatest on females with diets containing 6-MBOA, resulting in an increase in uterine weight, ovarian weight, and follicle number.

When oats treated with or without 6-MBOA were added to the environment of two adjacent *M. montanus* populations during the winter, those receiving oats with 6-MBOA underwent gonadal development, whereas those without the chemical additive did not.

At the end of the growing season, the nutritional value of plants declines. In addition, many plants after flowering begin to produce phenolic compounds. When these compounds are artificially added to the diet of *M. montanus* they cause a decrease in uterine weight, inhibition of follicular development, and diminished reproductive activity. Whether these reproductive inhibitors play a major role in the timing of field animals also exposed to decreasing day length is unclear. However, one can hypothesize that as with 6-MBOA, they may be important for fine-tuning the end of the reproductive season in a part of the country where the end of the growing season may arrive earlier than the decreasing day lengths would predict.

iii. Social Cues Two types of social interactions have been identified to influence reproduction: population density and specific olfactory odors (pheromones) that can stimulate or inhibit the onset of puberty. High population densities have been demonstrated to delay the onset of puberty in *M. pennsylvanicus*, *M. pinetorum*, and *M. agrestis*, whereas moderately high population densities stimulated earlier onset of puberty in *M. ochrogaster*. The differences between species are related to differences in social structures and the population densities observed. It is likely that under very high densities associated with irruptions in the field, all microtines will demonstrate decreased reproductive success and delayed puberty. A high degree of aggression is associated with such crowding and results in high natality, mortality, and stress-induced disease. Additionally, the increased production of corticosterone under these circumstances has a direct negative feedback effect on the reproductive system. Under moderately crowded conditions the monogamous mating system of *M. ochrogaster* breaks down. Under lower densities, the offspring remain with the parents, often well past weaning, and puberty is inhibited in the females by the presence of familiar male siblings and the father. However, with higher densities the young females are more likely to have contact with a unrelated male and puberty will ensue.

Female *M. ochrogaster* and *M. californicus* delay pubertal development so long as they are exposed to the odors of related males (father or male siblings). A single drop of urine from an unfamiliar (or "strange") male is sufficient to bring about pubertal maturation in these females by rapidly increasing gonadotropin-releasing hormone (GnRH) from the hypothalamus to stimulate pituitary release of LH and FSH and the development of the ovary. The pheromone (odor signal) in the male's urine has been identified as an androgen derivative. This is consistent with data indicating that urine of males with low testosterone levels (due to short day lengths, poor diet, or castration) will not bring about puberty in these young females. Studies examining whether the timing of puberty for male *M. ochrogaster* is similarly influenced by adult female pheromones demonstrated that males reached puberty at the same time regardless of social condition. Photoperiod, however, does influence the rate of development for young males. Interestingly, under low population densities, males reach puberty but often do not leave the parent's home range. The reason they do not leave is not understood. Females also remain on the parent's home range so long as they do not contact an unfamiliar male and enter puberty. If this occurs, then the new pair establishes and defends their own nest and territory. As a result, the adult parents may rear multiple litters with the assistance of older siblings in the nest.

Nonmonogamous species, such as *M. pennsylvanicus*, are also responsive to the odors of adults, although timing of puberty is strongly influenced by photoperiod for both males and females. Females exposed to the odors or urine of reproductively active males are stimulated to enter puberty as early as 20 days of age. Young males exposed to the same odors delay puberty. Unlike Mus, however, the odors of adult females apparently have no influence on development of young males or females.

Thus, the timing of reproduction in microtine rodents can be influenced by a suite of environmental cues interacting with the animals' reproductive endocrine system differently at different ages. Puberty (GnRH stimulation of LH and FSH, as well as increased prolactin and thyroxin needed for growth) is carefully timed by a combination of photoperiod (melatonin), nutrition (calories and protein), dietary chemical cues (6-MBOA and phenols), population density, and pheromonal social cues (androgen derivatives). Adult reproduction is apparently timed by photoperiod, nutrition, and dietary cues, but once puberty has occurred, social cues become important primarily for making choices among appropriate mates. *Microtus pennsylvanicus*' aggressive and affiliative (and mating) interactions are strongly influenced by the odors produced by adult and juvenile animals of both sexes. The olfactory signals and responses to signals are controlled by androgen and estrogen levels, which are in turn controlled by photoperiod and age. The responses to odor cues guide adult male and female choices of mates and influence aggressive interactions that establish and maintain home ranges and territories.

B. Estrus, Ovulation, and Gestation

1. Estrus

The male urinary pheromonal cues described previously have been described as critical to bring the females into a state of estrus. Studies in the 1950s and 1960s, in which males and females were housed continuously, seemed to indicate that estrus was spontaneous in most, if not all, microtine females. However, it has since become clear that most of these estrous events were the result of spontaneous postpartum estrus (estrus 12–48 hr after parturition). In the 1970s and 1980s many of the studies examining estrous occurrences that were not associated with parturition were carried out with monogamous *M. ochrogaster* and *M. californicus*. The females of these two species do not typically enter estrus spontaneously but rather require contact with male urine. However, as described previously, these females have likely not yet reached puberty, and male pheromones stimulate both the onset of final pubertal development and estrus after several days of exposure.

In contrast, the data from nonmonogamous *M.*

montanus and *M. pennsylvanicus* indicate that the females become pubertal without any contact with males. By the late 1980s it was commonly thought that all microtines required male urinary contact to induce a state of estrus; however, this has been proven untrue for *M. pennsylvanicus*. More than half of all adult virgin females housed in long day lengths will mate with a male within 9 hr of pairing, with many mating within minutes of pairing. After removal of the vomeronasal organ (necessary for "smelling" pheromonal odors) even more females mated within the first 9 hr of pairing with a male. Thus, it appears possible that monogamous species require contact with male pheromones to induce the onset of estrus, whereas nonmonogamous species spontaneously enter estrus in response to the environmental cues described previously.

2. Ovulation

A consistent feature of microtine reproductive biology is that all females require copulatory stimulation to induce ovulation (reflex ovulation) once receptivity (estrus) has been established. The copulatory pattern [timing of mounts, intromissions (penial insertion), and ejaculations] is critical for producing the species-appropriate cervical stimulation required to activate the firing of GnRH neurons and ultimately the LH surge and ovulation. The length of time between the onset of coitus and ovulation also varies between species, with a range of 6–18 hr.

The pattern of mating can also vary within a species. *Microtus pennsylvanicus* mate in long day lengths for approximately 60 min with an average of nine evenly spaced ejaculations, which occur at the end of each mount/intromission, resulting in an LH surge within 30 min of the onset of copulation and lasting approximately 180 min. When females are housed in short day lengths, a pair will mate for 10 hr or more, with many mounts and intromissions preceding each ejaculation. Even after such extensive copulatory activity, only half of the females produce a sufficient LH surge to allow ovulation.

Corpora lutea (CL) form between 15 and 18 hr following ovulation and are fully functional at 48–72 hr. The full development of CL is apparently dependent on more copulatory stimulation than is necessary to cause ovulation. A single ejaculatory series was sufficient to cause ovulation in *M. montanus*, but the CL did not last until Day 8 of gestation and the embryos failed to implant. Similar findings have been reported for *M. agrestis* and *M. pennsylvanicus*. Prolactin is also stimulated by coitus and in many species is important for maintaining the CL and allowing implantation. Half of the *M. pennsylvanicus* mated in a short photoperiod did not maintain elevated postcoital prolactin concentrations, even when the LH surge and ovulation had occurred, and pregnancies failed. The additional copulatory stimulation required by *M. montanus* and *M. agrestis* to maintain the CL and implantation may be necessary to stimulate the high gestational prolactin levels also found in rats and mice.

3. Gestation

The duration of gestation for most microtines is 20–25 days. While some related genera are reported to have longer gestations (lactational delays) if females have been impregnated during a postpartum estrus and are therefore nursing at the same time, this appears not to be the case for any *Microtus*.

The continued presence of the stud male for 1–4 days following mating has been demonstrated to increase the success rate of ensuing pregnancies for several *Microtus* species. Because *M. ochrogaster* forms permanent pair bonds as a function of 24 hr of mating and cohabitation, it is not surprising that removal of the male within the first 4 days can cause the failure of the majority of pregnancies. More surprising, however, are the reports of similar positive effects of the presence of the stud male on pregnancy success for *M. montanus*, *M. agrestis*, and *M. pennsylvanicus*. For each of these species, the presence of the stud male during the first few days of pregnancy was associated with a significant increase in the number of successful outcomes. When males were removed early, the CL degenerated rapidly in some females, resulting in implantation failure. Exactly how the males' presence influences the maintenance of CL and enhances implantation is not known.

There have been few studies on the partial loss of embryos or fetuses during pregnancy (as opposed to total termination of pregnancy). However, pregnancies that occur while *M. pennsylvanicus* is lactating produce more pups than for equivalently aged, previously pregnant but nonlactating females. The increase in pups is male biased. Recent evidence for the same species indicates that short day lengths or increased melatonin selectively decreases the number of male fetuses. Both effects are apparently not on the number of embryos produced but rather the number that implant and survive until birth.

Termination of preimplantation pregnancy after exposure to a strange male (or only the odors/pheromones of a strange male; Bruce effect) has been demonstrated in European and North American microtine rodents. As with mice, this mechanism hastens the onset of a new estrus and the opportunity for a new male to mate with the female. *Microtus pennsylvanicus* exposed to strange males every 2–5 days underwent repeated pregnancy terminations and new matings. The presence of the stud male defending his range/territory would prevent such interactions except under high population densities when such encounters may be frequent. The Bruce effect is the result of a strange male's odor blocking the release of prolactin necessary for maintenance of the CL until implantation. When *M. pennsylvanicus* (and probably other microtines as well) are nursing, and thereby maintaining high prolactin concentrations, the presence of a strange male does not cause termination of the new pregnancy.

Male-induced postimplantation pregnancy block has also been demonstrated in several New World *Microtus* through Days 14–16 of pregnancy. As with the Bruce effect, the odors in male bedding are sufficient to cause pregnancy loss. The mechanism underlying this late pregnancy blockage is uncertain; however, lactating females are far less susceptible than those that are not lactating. This would suggest that prolactin might again be involved. However, removal of the pituitary gland after implantation and establishment of the CL does not result in pregnancy termination, indicating that prolactin levels are not a critical factor in pregnancy maintenance after implantation.

C. Postpartum Reproductive Events

1. Postpartum Estrus

In all *Microtus* studied (10 species), females have a spontaneous postpartum estrus beginning during parturition or within the next 48 hr. Ovulation still requires the copulatory stimulus, but the success rate of postpartum matings appears to be considerably higher than those of nonlactating females. There is no lactation-induced delay of gestation.

2. Lactation and Maternal Behavior

Microtine pups are altricial—born without fur, eyes and ears closed, capable only of nursing and huddling, and weigh 2 or 3 g at birth. Litter sizes range from 1 to 12 pups, with larger litters typical of more northern species and multiparous females (having produced previous litters). Eyes open between 8 and 14 days, and weaning occurs between 12 and 20 days (species dependent). Mothers maintain nearly constant contact (18 hr or more per day) for the first few days of life. The control of lactation and onset of maternal behavior have not been much studied in *Microtus*, but to the extent that they have, they appear to be essentially identical to those of rats and mice. That is, estrogen, progesterone, and prolactin during pregnancy prepare the mammary gland and the neural centers for lactation and other maternal behaviors. At parturition, the rapid decline in progesterone, rise in estrogen, and appearance of oxytocin trigger milk production and maternal care. Suckling stimulation, the odors of the pups, and physical contact with the pups maintain the maternal behavior until the pups reach weaning. After the first few days, species variation, influence of ambient temperature, and photoperiod influence maternal behaviors. For example, *M. ochrogaster* females spend more time in contact with their pups than do *M. pennsylvanicus*, but the latter grow faster and are weaned earlier. It is likely that the mechanism underlying this difference is that *M. pennsylvanicus* produce more milk during each nursing bout, thereby decreasing the time necessary on the nest and allowing more time to forage. *Microtus ochrogaster* pups usually remain in the home range in a nonreproductive state for some time after weaning; therefore, a rapid growth rate is less essential (as long as they are weaned prior to the arrival of the next litter). Similarly, the slower growth rate of *M. pennsylvanicus* young in short autumn-like day lengths is enhanced by the female spending less time with the pups than similar females will when the pups are growing more rapidly. Of course, all rodent mothers increase their contact with pups when temperatures are lower.

D. Dispersal of Pups

Dispersal of the pups is influenced by species variations, maternal behavior, photoperiod, and social cues (see the previous description of puberty delay and induction). Monogamous species, such as *M. ochrogaster* and *M. californicus*, allow the weaned young to remain in the home nest and range. Under short day lengths, the pups of nonmonogamous species which gather into communal winter nests (*M. montanus* and *M. pennsylvanicus*) also remain in the home nest and range in a prepubertal state. Thus, one might think that pubertal development is critical to determining the timing of dispersal. However, *M. ochrogaster* males stay in the home nest well after attaining postpubertal status, and *M. pennsylvanicus* leave the maternal nest at 16 days of age, well before puberty and the arrival of the next litter. For nonmonogamous species, dispersal during the summer is likely related to maternal behaviors (excluding pups from the nest or the female's establishment of a new nest). Apparently in all microtine species, the females will establish their own nests once they have attained puberty and mated, if not before.

E. Summary of Life Histories

Because of the variation in timing of development and onset of puberty as a function of photoperiod, nutrition, and social cues, the life history of voles varies between species and seasons of the year. *Microtus pennsylvanicus* born in the spring have an average life span of 4 weeks, with males disappearing faster than females (presumably due to greater predator exposure during dispersal). On average, half of spring-born animals will produce one litter before disappearing. As few as 1 or 2% (and none in many years) will survive to overwinter. Under laboratory conditions, multiparous females do not cease mating and producing litters in decreased day lengths (although they undergo other short photoperiod-induced changes in preparation for winter) as long as there are fertile males. Thus, overwintering *M. pennsylvanicus* females with litters in communal nests may be primarily multiparous females.

Young born in late summer/autumn that delay puberty have an average life span of 6 months. However, only half of this cohort will also produce one litter. Thus, the determination of longevity appears to be increased exposure to predation as activity outside the protected nest environment occurs when dispersing, establishing territories, and looking for mates. Therefore, male mortality is greater than that of females. In species that delay dispersal, such as *M. ochrogaster*, individual longevity may be greater, but because of the great social control of puberty, many animals never leave the home nest and establish their own territories and litters.

When housed in a laboratory environment microtine rodents can live well beyond a year. Typically, *M. pennsylvanicus* females quit producing litters at 11–13 months of age, whereas stud males are quite capable of fertile matings through 18–24 months of age. Therefore, microtine life history and longevity are strongly influenced by exposure to predation.

III. PHYSIOLOGICAL AND BEHAVIORAL TRAITS INFLUENCED BY GONADAL HORMONES

A number of physiological and behavioral traits of *Microtus* are directly influenced by reproductive hormones (androgens and estrogens) and are related to mating systems or reproductive behavior.

A. Body and Brain Size

Males of nonmonogamous *Microtus* are substantially larger than females. This is the direct result of testosterone. In the absence of testosterone, males do not acquire as large a brain or body size as when the hormone is present. *Microtus pennsylvanicus* born in short day lengths maintain body weights and lengths marginally greater than those of females and maintain equivalent brain weights. When the number of hours of light/day are increased or the males become insensitive to the short photoperiod signal after 4–6 months of exposure, the gonads develop, and body weight and brain weight increase. Adult males weigh 50–100% more than females. Monogamous males (e.g., *M. ochrogaster*) are not larger than females and do not have larger brains.

B. Learning

Testosterone also drives a sex difference in territory/range size in nonmonogamous species which is related to the larger male body and brain size. Male *M. montanus* and *M. pennsylvanicus* maintain larger ranges than females of the same species, but only when they are reproductively active. In contrast, monogamous *M. ochrogaster* and *M. pinetorum* demonstrate no sex difference in range size. Researchers hypothesized that the reported sex difference in learning spatially related tasks (various types of symmetrical mazes or Morris water maze) in rats and mice might be related to the larger range size of males compared with females and perhaps an evolved adaptation by males to learn to maneuver over large areas rapidly. When reproductively active nonmonogamous male and female voles were compared, the results were like those of rats and mice: Males learned more quickly than females. However, as predicted, the monogamous males and females, which do not have different range sizes, did not differ in their rates of learning. Research further demonstrated that the sex difference in nonmonogamous species disappeared under short photoperiods when male gonads were regressed. In addition, the sex difference in learning, controlled by testosterone and photoperiod, is correlated with the size of portions of the hippocampus involved in learning spatial tasks. It is unknown whether this area shrinks when testosterone and rate of learning are reduced.

Recently, an examination of the potential role of female hormones on the learning of spatial tasks has been performed for *M. pennsylvanicus*. It appears that high estrogen levels reduce the rate of learning, and lower estrogen levels improve female performance. An analysis of the underlying neural structures has not yet been reported.

C. Affiliative/Aggressive Interactions

It was previously noted that photoperiod influenced levels of male–male and female–female aggression, same-sex and opposite-sex affiliative behaviors, and response to olfactory odors of adults and juveniles. The differential responses to individuals or their odors are greatly influenced by gonadal hormones in the individual smelling the odors as well as by steroid-independent effects of photoperiod. The production of seasonally specific odors is controlled by gonadal steroids and prolactin. As previously noted, these odors and responses to them likely play an important role in defending home ranges/territories and choosing mates.

D. Pair Bonding and Paternal Behavior

A great deal of work has been performed to examine the neuroendocrinological underpinnings of pair bond formation and paternal behavior in the monogamous *M. ochrogaster*. While male androgens are important for inducing a state of estrus (high estrogen) to allow copulation of a pair, the formation of the pair bond which follows approximately 24 hr of mating is influenced by changing concentrations of oxytocin, vasopressin, and corticosterone. Copulation results in increased vasopressin (neuropeptide) in males, which correlates with increased aggression toward strange males and to a lesser extent strange females. The combination of increased vasopressin and corticosterone or stress-induced release of vasopressin (or perhaps direct effects of corticosterone) facilitates male–male aggression and pair bonding (determined by how much time the male chooses to spend with his female mate versus a strange female) with a female even in the absence of mating. In contrast, stress or high levels of corticosterone in females inhibit the formation of strong partner preferences, whereas elevations in oxytocin facilitate the formation of such bonds. Thus, the male preventing access of strangers to the female partner likely reduces the stress (and corticosterone levels) of the female, enhancing the probability of a successful pregnancy as well as pair bonding following the elevation of oxytocin levels during copulation.

Paternal behavior is common among monogamous male mammals, but the hormonal determinants have been little studied. Paternal behavior (retrieval, cleaning, huddling, and guarding from intruders) can be induced in *M. ochrogaster* by injection of vasopressin into the brain. The amount of vasopressin present in areas of the brain involved in producing parental behaviors is less in nonmonogamous than monogamous males, suggesting that monogamous males have neural adaptations which cause them to easily exhibit paternal behaviors. Interestingly, *M. ochrogaster* display these behaviors without ever fathering a litter, and this is thought to be related to the filial care of siblings demonstrated by the species. No data have been reported on whether other nonmonogamous species which become paternal after exposure to pups show increased sensitivity to vasopressin as nonmonogamous females show increased sensitivity to oxytocin near the time of birth.

There is still much to learn about the extraordinarily successful genus *Microtus*. The taxonomy is changing as new biochemical and genetic methods are applied to try to disentangle the history of Arvicolidae. The relationship of a variety of physiological and behavioral differences between *Microtus* species with mating systems and ecology remains to be completely elucidated. Lastly, we have only begun to examine the neural and endocrinological control mechanisms that vary between species and why those variations exist.

See Also the Following Articles

Bruce Effect; Cricetidae (Hamsters and Lemmings); Pheromones; Photoperiodism; Rodentia (Rats, Mice, Etc.); Seasonal Reproduction

Bibliography

Carter, C. S., DeVries, A. C., Taymans, S. E., Roberts, R. L., Williams, J. R., and Getz, L. L. (1997). Peptides, steroids, and pair bonding. In *The Integrative Neurobiology of Affiliation* (C. S. Carter, I. I. Lederhendler, and B. Kirkpatrick, Eds.), Vol. 807, pp. 260–272. New York Academy of Science, New York.

Chepko-Sade, B. D., and Halpin, Z. T. (Ed.) (1987). *Mammalian Dispersal Patterns: The Effects of Social Structure on Population Genetics*. Univ. of Chicago Press, Chicago.

Daan, S. (1981). Adaptive daily strategies in behavior. In *Handbook of Behavioral Neurobiology* (J. Aschoff, Ed.), Vol. 4, Biological Rhythms, pp. 275–298. Plenum, New York.

Drickamer, L. C., Davis, D. E., Vandenbergh, J. G., Madison, D. M., and McShea, W. J. (1987). Symposium on behavior as a factor in the population dynamics of rodents: New concepts and approaches. *Am. Zool.* 27, 821–969.

Ferkin, M. H., Sorokin, E. S., and Johnston, R. E. (1995). Seasonal changes in scents and responses to them in meadow voles: Evidence for the co-evolution of signals and response mechanisms. *Ethology* 100, 89–98.

Gaulin, S. J. C., FitzGerald, R. W., and Wartell, M. S. (1990). Sex differences in spatial ability and activity in two vole species (*Microtus ochrogaster* and *M. pennsylvanicus*). *J. Comp. Psychol.* 104, 88–93.

Jannett, F. J., Madison, D. M., Wolff, J. O., Lidicker, W. Z., Jr., Getz, L. L., and Carter C. S. (1980). Symposium on social systems of microtine rodents. *Biologist* 62, 3–69.

Johnson, M. L., and Johnson, S. (1982). Voles. In *Wild Mammals of America* (J. A. Chapman, and G. A. Feldhamer, Eds.), pp. 326–354. Johns Hopkins Univ. Press, Baltimore.

Lee, T. M. (1993). Development of meadow voles is influenced postnatally by maternal photoperiodic history. *Am. J. Physiol.* 265, R749–R755.

Meek, L. R., and Lee, T. M. (1993). Prediction of fertility by mating latency and photoperiod in nulliparous and primiparous meadow voles (*Microtus pennsylvanicus*). *J. Reprod. Fertil.* 97, 353–357.

Merritt, J. F. (1984). Winter Ecology in Small Mammals, Spec. Publ. No. 10. Carnegie Museum of Natural History, Pittsburgh.

Rowsemitt, C. N. (1986). Seasonal variations in activity rhythms of male voles: Mediation by gonadal hormones. *Physiol. Behav.* 37, 797–803.

Sanders, E. H., Gardner, P. D., Berger, P. J., and Negus, N. C. (1981). 6-Methoxybenzoxazolinone: A plant derivative that stimulates reproduction in *Microtus montanus*. *Science* 214, 67–69.

Tamarin, R. H. (1985). Biology of New World Microtus, Spec. Publ. No. 8. American Society of Mammalogists, Shippensburg, PA.

Zucker, I., Lee, T. M., and Dark, J. (1991). The suprachiasmatic nucleus and annual rhythms of mammals. In *Suprachiasmatic Nucleus: The Mind's Clock* (D. Klein, S. Reppert, and R. Moore, Eds.), pp. 246–259. Oxford Univ. Press, New York.

Migration, Amphibians

Raymond D. Semlitsch and Travis J. Ryan

University of Missouri

I. Ultimate Mechanisms
II. Proximate Mechanisms
III. Case Study: *Ambystoma talpoideum*

GLOSSSARY

complex life cycle A life cycle with at least two distinct postembryonic morphologies, each of which is adapted for a fundamentally different way of life or niche (e.g., aquatic larvae and terrestrial adults).

emigration Movement of individuals away from an aquatic breeding site once reproduction and/or metamorphosis has concluded.

immigration Movement of individuals toward an aquatic breeding site in advance of reproductive activity.

natural selection Differential survival and reproduction of individuals.

philopatry Faithful use of natal sites during an individual's lifetime.

Amphibian migrations consist mainly of seasonal movements toward or away from aquatic breeding sites. During the breeding season, migrations include breeding adults moving toward the site (immigration), whereas once breeding has concluded, migrations away from the aquatic site include spent males and females and a short time later include recently metamorphosed juveniles (emigration). Migrations may be highly synchronized, where most of the population move simultaneously en masse toward the breeding site over one or a few nights, or markedly individualistic, with migrants arriving over a longer time period. In addition to the conventional sense, migrations may also include movements among terrestrial sites (e.g., feeding habitats and aestivation/hibernation sites) or daily vertical migrations of larval amphibians within the water column.

I. ULTIMATE MECHANISMS

The evolution of migratory behavior in most animals likely results from the need to follow transient resources through time and space. Amphibians with complex life cycles frequently leave the aquatic habitats in which they have completed embryonic, larval, and metamorphic development to live in terrestrial habitats. Most of these predominantly terrestrial amphibians return to aquatic sites for reproduction and the "completion" of the life cycle. Amphibian taxa which have abandoned the complex life cycle are either direct developing (i.e., larval development and metamorphosis is completed prior to hatching) or perennibranchiate (i.e., metamorphosis is incomplete and adults retain many aquatic larval characteristics, including external gills) and are specialized for a single habitat type. These specialized types do not exhibit migratory behavior in the conventional sense, having lost the ability to move between aquatic and terrestrial habitats. Species of amphibians, such as frogs of the genus *Rana* and salamanders of the genus *Ambystoma*, exhibit complex life cycles, with aquatic larvae metamorphosing into terrestrial juveniles, and therefore must use aquatic habitats such as ponds for reproduction and larval growth. Most current knowledge of amphibian migrations comes from studies conducted with pond-breeding amphibians because ponds and similar lentic habitats, though often temporally variable, offer extremely high primary productivity for growing larvae and thus serve as the nuptial/natal site for many amphibians. Additionally, because they are spatially fixed, ponds are easily monitored. In contrast, lotic systems, such as streams, are often much larger bodies of water with regard to both volume and surface area, have much lower primary productivity, and are logistically more difficult to monitor. Because immigration is more tightly linked to reproductive success than emigration, immigration has been the more actively investigated aspect of migration for most amphibians.

A. Timing of Reproduction

Migratory behavior and the use of ponds as breeding sites likely evolved simultaneously with the need to find high-quality reproduction sites and high-quality hibernation sites each year. Among migrating species, natural selection presumably favored individuals with the ability to find ponds and hibernation sites at the appropriate time of the year and has acted against those individuals arriving at the wrong time or place. Although ponds used for breeding often fill and dry at varying times, there is usually one season of the year when the probability that a pond contains water is the highest. The seasonal pattern of a species' reproductive cycle (i.e., egg maturation and sperm production) must be synchronized with this seasonal pattern of pond filling. Environmental conditions often fix the breeding season, for example, into categories of spring, summer, or autumn breeding species. This means that breeding adults are typically not found outside a particular season and an individual's physiological machinery is at maximum performance within this season. Once the season of reproduction has been fixed, the precise initiation of breeding migrations is generally dependent on associating reliable proximate environmental cues with terrestrial conditions for migration and pond quality. Furthermore, natural selection may fine-tune the annual timing of reproductive and migratory events to very narrow windows of time. Conversely, in the absence of reliable environmental cues for migration and reproduction, some species found in desert regions (e.g., *Scaphiopus*) have evolved flexible behaviors that are triggered anytime of the year by stochastic but highly efficient cues such as heavy rains (typically >7–10 mm in 1 day).

B. Philopatry

Each year an individual is faced with the question of whether to breed in the same pond and to overwinter at the same hibernation site. For adults with previous experience and knowledge of a particular site, the decision may be based on past success or failure. However, because newly matured adults breeding for the first time have no experience finding or selecting ponds and because both breeding and hibernation sites can change with time, and hence the probability of success may change, natural selection is thought to favor philopatry. That is, without absolute knowledge of the condition of the natal pond and all alternative sites, individuals will be favored over time (i.e., accumulate higher reproduction success and exhibit higher survival) if they simply return to their natal pond for breeding and then return to their original hibernation sites for overwintering. The logic is that an individual is a product of a site's quality (i.e., the pond as a larva and the terrestrial hibernation site as a juvenile), and without knowledge of other sites an individual will, on average, be more successful at its natal site. The primary constraint on obtaining knowledge of alternative sites is that overland "exploratory" activity is energetically costly for amphibians, which are vulnerable to desiccation and terrestrial predators. In addition, once arriving at an alternative site, its quality may be the same or lower than that of the abandoned natal site. Thus, on average, natural selection likely acts against costly exploratory behavior and the abandonment of natal sites.

Philopatry is widespread among all species of pond-breeding amphibians. However, newly created ponds are often colonized, suggesting that philopatry is not absolute and colonization of high-quality empty ponds is a boom for populations founded by some individuals. Additional evidence suggests that newly created ponds within the migratory route of adults can actually attract individuals to use these new sites rather than old natal sites. Philopatry is likely stronger in species preferring ponds in habitats in later stages of succession (e.g., *Ambystoma* using woodland ponds) because of their stability and predictability. Philopatry is likely weaker in species that prefer or require early successional ponds or some level of disturbance (e.g., *Pseudacris*) because of their rapid change in physical features, and hence quality, over time.

II. PROXIMATE MECHANISMS

Although the evolutionary pressures that have likely resulted in breeding migrations, whether highly synchronized or individualistic, may be understood both empirically and theoretically, the ultimate mechanisms alone do not explain how amphibians know precisely when and where to migrate. Many endogenous (physiological) signals, such as hormones, prepare amphibians for migrations, but a different set of mechanisms must be invoked to explain the initiation of migration and the means by which amphibians locate a specific breeding site from a few hundred to several thousand meters away. The initiation of migrations depends on the interaction of exogenous (environmental) with endogenous (physiological) signals, whereas the homing response is likely due to a combination of several different cues.

A. Environmental Cues

The breeding migrations of most amphibians are highly seasonal as well as taxon and population specific. Even species using the same breeding site may have different patterns of migration and reproductive activity and thus initiate migrations in response to different exogenous factors because endogenous factors are likely to be similar if not identical. For instance, *Ambystoma opacum*, the marbled salamander, breeds in ephemeral ponds in southeastern North America and completes migration, courtship, and oviposition in the autumn, prior to the pond basin filling with rainwater. *Ambystoma talpoideum*, the mole salamander, is a closely related species that utilizes the same ephemeral ponds as *A. opacum* but usually does not migrate until the pond basin has flooded. Despite species differences in the timing of migrations, nearly all amphibians rely on some suite of exogenous or environmental cues indicating the quality of terrestrial conditions for migration and aquatic conditions for courtship and oviposition. The particular climatic regimes of different locales and the breeding ecology of different taxa will determine the precise combination of environmental factors that will result in the initiation of a breeding migration. Regardless of regional differences in climatic patterns and taxon-specific requirements, there are at least two environmental factors in temperate regions that are critical to most migratory amphibians: temperature and rainfall.

1. Temperature

Temperature often plays a critical role in the timing of migrations. Amphibians are, of course, ectothermic organisms; they are dependent on external environmental temperature to drive physiological processes and, in particular, to provide the means for locomotion. Although many amphibians are cold-hardy (if not freeze resistant in some cases), few are capable of sustained locomotor activity in subfreezing temperatures. At these temperatures, standing water in breeding ponds is most often frozen or near freezing, limiting the utility of migratory activity. Therefore, most migrations occur when air temperatures are sufficiently warm to allow for locomotor activity and access to aquatic breeding sites. Temperature is more of a limiting factor for immigrations in regions where extremely low temperatures are likely (e.g., the northern latitudes in the Northern Hemisphere or high altitudes). Extremely high temperatures are also likely to limit emigrations and other movements in the summer in regions where the temperatures may rise to such a point that desiccation becomes an additional concern (e.g., the southern latitudes in the Northern Hemisphere). In general, at colder temperatures overland movements may be restricted because of the ectothermic nature of amphibians, and at warmer temperature overland movements may be restricted because of increasing water loss due to the semipermeable nature of amphibian skin and the need for water conservation in terrestrial situations.

Most migratory amphibians are semiaquatic pond breeders that overwinter in terrestrial refugia, often abandoned mammal burrows or other small subterranean hibernicula. Because of this, the soil temperature may be more important than air temperature in determining the onset of migratory behavior. Because of the thermal inertia of soil, prolonged changes in surface air temperatures are required to elicit a similar change in soil temperature. Only when this change in soil temperature arrives will individuals break hibernation in advance of the breeding migrations. Thermal inertia of soil thus protects breeding amphibians from misinterpreting rapid, localized changes in temperature that might result in freezing in the case of reversal in short-term trends.

2. Rainfall

Rainfall also serves as an environmental cue for breeding amphibians in at least two distinct ways. First, rainfall is required to fill the dry basins of ephemeral (temporary) ponds, rendering them suitable for reproductive activities. Even in more or less permanent ponds, significant rains may be required to raise the water level appreciably. In regions that experience heavy winter snowfall, rains hasten the spring thaw both above and below the soil surface. Rainfall may also function in consort with rising temperatures to activate other organisms of pond communities (e.g., bringing dormant phytoplankton and zooplankton out of diapause) on which most hatchling amphibian larvae depend. All these functions of rainfall implicate long-term or cumulative rainfall as an important abiotic factor. One of the earliest detailed ecological studies of the environmental components of amphibian breeding cycles implicated cumulative rainfall as the most important factor (compared to temperature, humidity, wind, etc.) in determining the onset of migratory activity for *Rana temporaria* in England. A similar conclusion is reached when considering data collected during a 20-year study of the breeding dynamics of *Ambystoma talpoideum* in South Carolina. This latter study has also indicated that the absence of significant rainfall (i.e., enough to fill the pond basin) can result in decreased reproductive activity in an amphibian population or community. Data from several other investigations covering a diverse group of amphibians over a broad geographic range corroborate these general trends.

Short-term, local, or daily rainfall is also critical for successful amphibian migrations, and of course short-term rainfall is inextricably connected to long-term patterns of precipitation. In extreme habitats (e.g., deserts of southwestern United States) the short-term rain may be sufficient to fill extremely ephemeral habitats. In these areas amphibians may be active above ground long enough only to migrate, breed, feed, and return to their subterranean refuges (e.g., *Rhinophrinus dorsalis*). Generally, local rainfall indicates that the terrestrial conditions are favorable for movements rather than marking the gradual change of season. A saturated substrate provides protection from desiccation, one of the greatest risks of overland travel for many amphibians. This is a special concern for groups reliant on cutaneous respiration for all or a considerable portion of gas exchange (e.g., lungless salamanders). For species adapted to relatively arid regions this is less of a concern.

3. Limiting Factors

Migratory behavior is a complex phenotype, underpinned by genetic and physiological mechanisms but still strongly impacted by the current environmental conditions. An emergent pattern is that, for most species studied, there appears to be one critical climatic factor, and breaching a threshold value for this factor is likely to initiate migrations. For several species, either temperature or rainfall is often the critical or limiting factor. In regions where winter temperatures are benign (e.g., southern latitudes in the Northern Hemisphere), rainfall is often the limiting factor, whereas in regions where winter temperatures become severe, migrations become increasingly dependent on temperature thresholds, and rainfall becomes less of a pivotal issue.

B. Homing Ability

There are at least four physiological tools that allow an individual amphibian to find its natal breeding site, and in several aspects, they constitute one of the most poorly studied aspect of amphibian migratory behavior. Although several model systems have been vigorously studied, the means of orientation and homing in the majority of amphibian taxa remain unexplored. Several examples from the literature do exist, however, indicating the types of signals that amphibians use to locate breeding sites.

1. Acoustic Cues

The auditory system is poorly developed in most caudate amphibians and is not likely to play a substantial role in the location of breeding sites. On the other hand, the ability of anuran amphibians to broadcast and receive acoustic signals is highly developed and has been suggested to function in orientation. Several aspects of the anuran acoustic system, however, suggest that airborne sounds are not the primary mechanisms for site location. Furthermore, their role is probably limited to location of favorable sites within the pond. With the exception of a few notable taxa, the sound pressure levels achieved by most individuals or choruses are probably too low to permit any reliable long-range signaling. Studies have shown that females, and perhaps even males, of many species respond phonotactically over distances <50 m. Thus, acoustic cues may prove reliable only after the breeding site has been located. Few anurans (e.g., *Pseudacris triseriata* and *Bufo calamita*) can broadcast signals approaching 1000 m, a distance anurans may travel during immigration. Although location of and orientation to acoustic signals may be functional over an extremely short distance, the costs to the callers (in terms of energy expenditure, increased vulnerability to predation, and increased competitor density) eliminate calling as an efficient mechanism of signaling breeding site location among anurans. Of course, it also raises the question of how the first calling male or males find the breeding site.

2. Magnetic Cues

An internal compass, calibrated to the earth's magnetic field and based on changes of inclination of the same, is known to be an efficient mechanism for homing and navigation in many animals. This compass is most reliable for animals migrating large distances, however, because the inclination is likely to change little over 50–5000 m, a distance typical of amphibian migrations. Magnetic cues may therefore appear to be of little use for amphibian homing. Nonetheless, several frogs and salamanders have been found to be able to recognize small changes in magnetic signals; at least one salamander, *Eurycea lucifuga*, can detect the changes likely to be experienced within many amphibian migratory distances. Although magnetic cues may be detected, there is little evidence to show that these cues are being generally employed to locate the breeding sites. It remains to be demonstrated that amphibians rely on an internal compass based on sensitivity to the earth's magnetic field to locate breeding sites.

3. Visual Cues

Any number of visual cues are used by animals for orienting during migrations. Most common among these is sun orientation. This particular cue is probably quite ineffective for most amphibians since the majority migrate at night. Nonetheless, some anurans and salamanders [e.g., toxic species such as the California newt (*Taricha granulosa*) and the American toad (*Bufo americanus*)] are known to exhibit some level of diurnal migration. Dependence on sun orientation carries the risk of a loss of primary cues on overcast days; experiments have demonstrated that *T. granulosa* orients toward home sites during both clear days and clear nights but randomly under overcast conditions. Though *T. granulosa* was found to exhibit some degree of lunar orientation, no amphibians are known to exhibit stellar orientation. The sun itself may not be required for diurnal migration, however. *Ambystoma tigrinum* (tiger salamander) and *Rana catesbeiana* (bullfrog) have both been shown to be able to orient toward polarized light, which might serve as a surrogate for the actual position of the sun. The demonstrated ability to use polarized light to orient toward a natal or experimental site does not necessarily indicate that these cues are actually employed for homing, and the evidence that polarized light is actively used by migrating amphibians is particularly weak. Nonetheless, the sun, the moon, and polarized light serve to indicate the general direction of the natal site; thus, their use will decrease as an individual approaches its goal. Fine-scale visual orientation is more likely determined by the use of fixed landmarks, familiar objects, or indicators near the pond. However, there are few data to suggest that migrating amphibians use landmarks to any appreciable extent.

4. Olfactory Cues

Olfaction in amphibians has been well studied and implicated in the detection of food, predators, conspecifics, sex identification, kin recognition, and perhaps even neighbor recognition. The apparent acuity of olfaction is in part attributable to the sensitivity of Jacobson's (or vomeronasal) organ, an accessory olfaction system that operates independently of, but in consort with, the main olfactory system.

Because most pond-breeding amphibians rarely migrate more than a couple hundred meters from their natal ponds, there is a premium on orientation cues which are temporally stable and particularly strong over short distances. Chemical cues, or odors, and more likely gradients of the same, are critical in nonmigratory territorial salamanders (e.g., *Plethodon* sp.) as well as in a diverse array of semiaquatic caudate and anuran amphibian species. The evidence for the primacy of olfaction in homing behavior comes from numerous experiments conducted on many temperate North American amphibian species. For instance, *Ambystoma maculatum* orient more toward "mud-water" from their natal pond than "mud-water" from a foreign pond. The most famous experiments were conducted by Victor Twitty in the late 1950s and early 1960s on the red-bellied newt, *Taricha rivularis*. Twitty found that red-bellied newts, collected at a breeding site at a specific section of a stream, would return to the same breeding site when displaced several hundred meters either up- or downstream. Furthermore, the newts would return to the same breeding site when displaced up to 8 km and released in different, presumably novel, drainages. Though this homing behavior is noteworthy, it is remarkable that many newts were able to navigate these distances over severe terrain when blinded. In later experiments, Twitty and colleagues rendered olfactory nerves of some newts useless, and the resultant anosmic individuals were unable to return to their origin in the stream with the accuracy of either blinded or control animals. Adult *Plethodon jordani*, although nonmigratory, demonstrate a similar response: Individuals with severed olfactory nerves were unable to return to their home ranges when displaced, whereas blinded and untreated individuals returned to their home ranges with great regularity.

III. CASE STUDY: *AMBYSTOMA TALPOIDEUM*

Ambystoma talpoideum is a small, pond-breeding species of mole salamander from the southeastern United States. Its typical life cycle includes annual immigration to temporary ponds where mating and oviposition occur in winter and early spring, followed by emigration from the pond to terrestrial woodland habitats. Extensive studies on *A. talpoideum* have been conducted over the past 20 years at a site in South Carolina called Rainbow Bay. This site also serves as the breeding site for up to 27 species of amphibians indigenous to the southeastern Atlantic Coastal Plain of the United States. Rainbow Bay is a temporary pond with a surface area of approximately 1 ha and a maximum water depth of 1 m. The terrestrial area surrounding Rainbow Bay is dominated by mixed pine and hardwood forests.

The migrations of *A. talpoideum* have been studied at Rainbow Bay using a terrestrial drift fence with pitfall traps. The pond was encircled by a drift fence of aluminum flashing (440 m) with pitfall traps (40 liters) buried inside and outside the fence flush to the ground and next to the fence at 10-m intervals. These traps have been checked daily for the past 18 years; all individuals captured were recorded (some were measured and marked for future identification) and immediately released on the opposite side of the fence. For species such as *A. talpoideum*, this sampling technique provided a nearly complete annual census of the number of breeding adults and metamorphosing juveniles as well as the timing and magnitude of migrations. Adults of species that are proficient climbers (e.g., treefrogs, *Hyla* sp.) or jumpers (e.g., bullfrogs, *R. catesbeiana*) can trespass the fence without being captured and provide only qualitative data on migratory activity.

Over the period of study, Rainbow Bay has filled as early as September 6 and as late as April 24; however, in most years it filled in December, January, or February. The average date of arrival to the pond for males and females is significantly correlated with the date of pond filling. Males arrived, on average, 18 days earlier than females. It has also been found that most adults breed at the same site repeatedly, with many adults breeding four to six times over their life time but not necessarily in consecutive years. These results suggest that breeding adults synchronize their breeding migrations with pond filling, and that males derive more benefit than females from arriving at the site early in the season, presumably to maximize opportunities for mating. Females derive little benefit from arriving before males and may actually benefit from waiting until a larger selection of males is present. Finally, these results indicate that adults exhibit strong reproductive philopatry to natal breeding sites.

Studies of environmental cues that influence the migrations of *A. talpoideum* have also been conducted at Rainbow Bay. These studies indicate that the initiation of migrations is triggered by declining air temperatures in autumn coupled with rainfall. Apparently, adequate amounts of rainfall earlier in the year consistently fail to trigger migrations. Once breeding migrations have initiated, some adults are captured nearly every night there is a measurable amount of rainfall and the air temperature is above freezing. Thus, rainfall becomes more important than temperature once migrations have begun. Very large mass migrations, and ultimately the cumulative number of breeding adults, are determined by cumulative rainfall during the migration season. Additionally, rainfall occurring at night results in more migrating adults than rainfall during the day because *A. talpoideum* is strictly nocturnal. In drought or dry years, few if any adults migrate to breeding sites and many individuals appear to "skip" reproduction all together. The data indicate that these adults then appear in subsequent years and likely trade off reproduction in any one year for future survival.

Emigration of adults after the breeding season begins in March and is highly consistent among years. Adults typically emigrated out of the pond at the same place they had entered, with the population using some terrestrial corridors (i.e., wooded areas continuous outward from the shoreline of the pond) more often than others (i.e., open fields). When adults leave the pond, the majority make several long movements (81–261 m) into adjacent forested areas. Once a suitable habitat or summer home range is found, movements are then restricted to that prescribed area until migratory movements are initiated back to the breeding site the following autumn or winter. The maximum cumulative distance moved by an individual was 287 m away from the pond. Once migrations cease, individuals spend the duration of the summer and autumn underground in small activity centers (0.02–0.21 m^2), only occasionally moving among other such sites.

Metamorphosed juveniles emigrate from Rainbow Bay as early as May, continuing through September or until the pond dries. Rainfall is the primary determinant of emigration in juveniles but it appears that maximum summer temperatures may also restrict terrestrial movements. Upon leaving the pond, juveniles also make several long-distance movements (14–204 m) but generally remain closer to the pond than adults. The case study of *A. talpoideum* is typical for a pond-breeding salamander and illustrates how migration behavior has evolved to be tightly coupled with environmental variation, especially that associated with important but transient resources such as ponds.

See Also the Following Articles

AMPHIBIAN REPRODUCTION, OVERVIEW; MIGRATION, FISH

Bibliography

Gauthereaux, S. A., Jr. (1980). *Animal Migration, Orientation, and Navigation*. Academic Press, New York.

Phillips, J. B. (1977). Use of the earth's magnetic field by orienting cave salamanders (*Eurycea lucifuga*). *J. Comp. Physiol.* 121, 273–288.

Savage, R. M. (1935). The influence of external factors on the spawning date and migration of the common frog, *Rana temporia temporia* Linn. *Proc. Zool. Soc. London* 1935, 49–98.

Schmidt-Koenig, K., and Keeton, W. K. (1978). *Animal Migration, Navigation, and Homing*. Springer-Verlag, Berlin.

Semlitsch, R. D., Scott, D. E., Pechmann, J. H. K., and Gibbons, J. W. (1996). Structure and dynamics of an amphibian community: Evidence from a 16-year study of a natural pond. In *Long-Term Studies of Vertebrate Communities* (M. L. Cody and J. Smallwood, Eds.). Academic Press, New York.

Sinsch, U. (1990). Migration and orientation in anuran amphibians. *Ethol. Ecol. Evol.* 2, 65–79.

Twitty, V. C. (1959). Migration and speciation in newts. *Science* 130, 1735–1742.

Migration, Birds

Peter Berthold

Research Unit for Ornithology of the Max Planck Society

I. Introduction
II. Prevalence, Ultimate Factors, the Network of Migration Routes, and Records
III. Migratory Drive and Control of the Process of Migration
IV. Migratory Disposition, "Fuel," and Metabolic Adaptations
V. Ecophysiological Aspects
VI. Navigation
VII. Capacity to Adapt by Microevolution

GLOSSARY

endogenous annual rhythms The expression of rhythmic physiological processes with a period of approximately a year; biological "calendars"; also called *circannual rhythms*.

migration drive An innate behavior that causes birds to begin migration; also called *urge to migrate*.

migratory disposition A state of readiness for migration which comprises many morphological, physiological, and behavioral adaptations.

migratory restlessness The migratory activity of captive migrants; an expression of migration drive.

vector navigation The only form of navigation that has been explained; migratory birds fly according to a vector that consists of genetically determined migration directions and programs governing the duration of migration.

Bird migration as it is most broadly defined—the act of moving from one spatial unit to another or the act of moving from one place of abode to another—is hardly distinguishable from the everyday activities in a given home range. More classical definitions refer chiefly to the regular annual (biannual or seasonal) migration in which a bird commutes between breeding grounds and wintering areas. However, they normally also include more irregular movements, such as postjuvenile dispersal, irruptive movements, or invasions of new ranges.

I. INTRODUCTION

Bird migration has been a subject of research at least since the time of Aristotle. The two hypotheses that are his legacy now seem rather ridiculous: (i) That birds might not go away at all but rather spend the winter in the mud under bodies of water as amphibians do, or, alternatively, (ii) that the species present in summer are transformed into the species that spend the winter in the same region (transmutation theory). As happened in many other areas of the natural sciences, bird migration began to be studied systematically in the past century, at first by observation and description alone. A milestone in this research was the introduction of scientific bird banding around the turn of the century; that is, the attachment of individually labeled metal bands to the birds' legs. To date, about 200 million birds have been banded, and the data on recovery sites in combination with direct observations have provided at least a rough idea of the migration routes followed by many species. The recent development of satellite tracking has made it possible to monitor migration almost continuously, even over transcontinental distances (Fig. 1), opening up a new dimension for bird migration research. The experimental study of migratory birds began in the 1920s, and especially during the past 25 years, thousands of experiments have clarified the most important control mechanisms, metabolic adaptations, and means of navigation. Currently, there is considerable emphasis on research into the genetic bases of migration and its evolutionary origins.

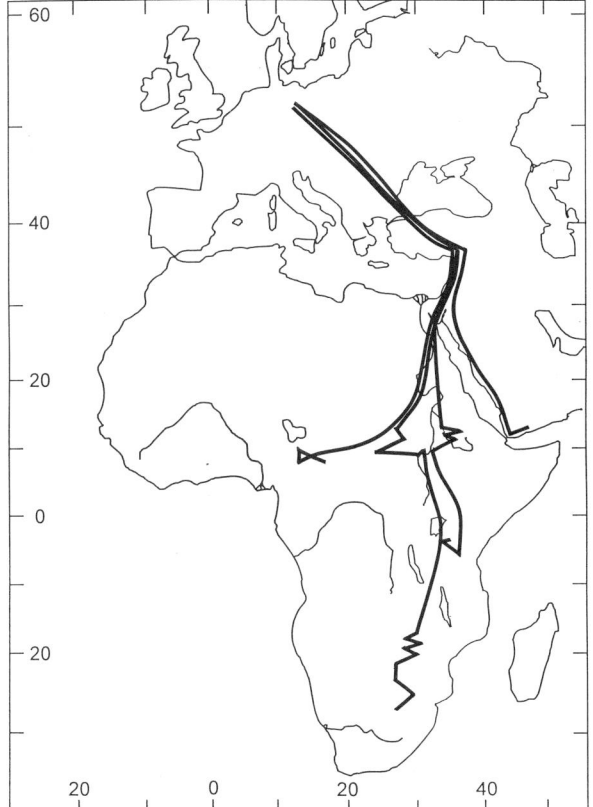

FIGURE 1 Migration routes of white storks (*Ciconia ciconia*) from Central European breeding grounds into diverse African winter quarters by way of the eastern detour around the Mediterranean Sea. The routes were monitored by satellite tracking, which specified as many as 15 locations per day. The data show the extraordinary directedness of the migration routes as far as the Mediterranean, before a sharp inflection takes the birds in a more southerly direction toward Africa (reproduced with permission from P. Berthold et al., *Falke* 44, 134, 1997).

II. PREVALENCE, ULTIMATE FACTORS, THE NETWORK OF MIGRATION ROUTES, AND RECORDS

A. Number and Distribution of Migratory Bird Species

Bird migration used to be considered almost entirely a phenomenon of higher latitudes, but recently migration has been documented for many tropical species once thought to live permanently in their breeding grounds. Of the approximately 10,000 bird species now extant, probably over half engage in more or less regular migrations. The number of individuals that migrate annually has been estimated to be on the order of 50 billion.

B. Factors Giving Rise to Migration

In view of the fact that birds throughout the world migrate, it is clear that migration cannot have come about as a result of geological events such as continental drift or ice ages. There are more than 10 different hypotheses and theories about the origin of bird migration, but only 1 factor dominates: the availability of food. Food supply evidently provides two kinds of incentives to migrate: either to escape from regions where there is a seasonal shortage (e.g., the Arctic in winter) or to seek out regions with a seasonal surplus of food (such as the Arctic during its brief summer, with long days, little competition, low predator pressure, and other advantages that make it practically a paradise on earth).

C. Migration Routes and Records

The world's migratory birds have established so many routes that altogether they form a network enclosing the earth. If they were all to be plotted on a map, large regions would be completely covered with the resulting lines. This is because many species are "broad-front migrants": The various populations travel on parallel routes, moving like a wave across whole continents. Such species know no ecological barriers and can cross deserts as vast as the Sahara, oceans, or the highest mountains, including the Himalayas. In contrast, many species—especially large birds that fly mainly by gliding and waterfowl—tend to be concentrated within narrow migration corridors. Through the most important corridor, in Israel at the eastern edge of the Mediterranean, 500 million birds migrate each year.

The record holder among the long-distance migrants is the Arctic tern (*Sterna paradisaea*), which flies annually from Arctic breeding grounds to 50,000 km away, and the distance traveled in a lifetime may amount to 1 million kilometers. Small birds are hardly less remarkable: The wheatear (*Oenanthe*

oenanthe) breeds in Eurasia and as far as Alaska and overwinters in southeast Africa, flying up to 30,000 km per year. Also, some swallows travel as far as 15,000 km. The maximal nonstop flight distance is on the order of 5000–7000 km, a distance that many shorebirds can cover in flights lasting as long as 100 hr, e.g., from Alaska to Hawaii or from northern Siberia to Australia or Tasmania. The flight altitude over most areas of land or water is moderate, up to about 2000 m, but over ecological barriers, such as mountains, deserts, or oceans, altitudes of 6000–10,000 m are attained, and the absolute record measured so far (for a vulture) was 11,300 m.

III. MIGRATORY DRIVE AND CONTROL OF THE PROCESS OF MIGRATION

A. Initiation of Migration

A considerable number of bird species do not leave their breeding grounds until the snow falls or, in the case of waterfowl, the water freezes; indeed, some species migrate only in years when such events actually occur. It is evident that the birds' flight from the snow or ice is triggered directly by these exogenous factors. However, most migratory birds migrate every year, often leaving in midsummer when the conditions in the breeding grounds still appear to be optimal. What induces them to depart? As early as the beginning of the eighteenth century the existence of some endogenous factors (i.e., within the birds themselves) was postulated, and much later it was confirmed. Since the end of the 1960s a number of species have been found to have an endogenous annual rhythmicity, a kind of built-in calendar that is based on physiological rhythms in the body, called "biological clocks," and controls all the major processes that depend on the time of year, including migration (Fig. 2). Furthermore, recent breeding experiments in which individuals of migratory warbler populations were crossed with nonmigratory ones and a redstart species that begins migration early was crossed with another that leaves later have demonstrated that the migration drive and the exact time of departure are both genetically determined. This finding explains the amazing seasonal precision of

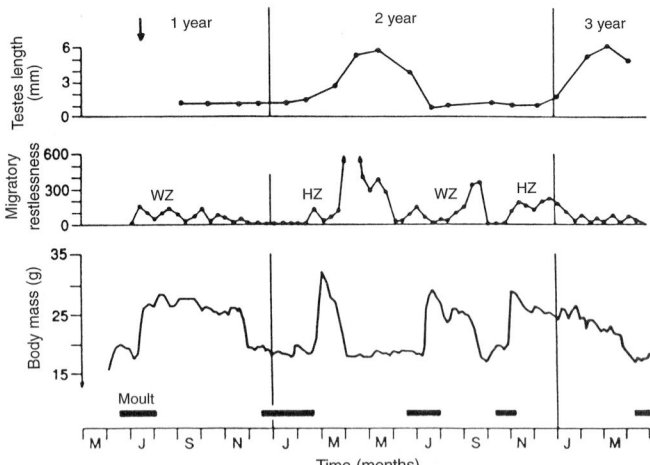

FIGURE 2 Endogenous annual periodicity of (from bottom to top) molt, body mass (showing periods of migratory fat deposition), migratory activity (during WZ, outward migration, and HZ, return migration), and length of testes in a hand-raised garden warbler (*Sylvia borin*) hatched at the end of May and then kept for years in constant conditions (reproduced with permission from Berthold, 1996).

the arrival of many migratory birds at their destinations. These "calendar birds," so called because one could set the calendar by them, appear (e.g., in Arctic breeding grounds) on almost the same day every year; they can do so because their internal, genetically controlled calendar tells them what time of year it is, and this mechanism is fine-tuned by external factors such as photoperiodicity, the annual cycle of day length, so that it is highly precise.

B. The Process of Migration and Its Termination

Most migratory birds travel to "goal areas" that are geographically well-defined so that they must have reliable means of ending the journey. We now know that in many species the distance the bird flies and the details of that flight are controlled by genetically determined migration timing programs, operating in combination with inherited programs governing the direction of migration (see Section VI). That is, the birds migrate according to a vector that specifies, in addition to a particular direction, a particular period of time during which a long, medium, or short distance may be covered, depending on the species and population concerned. When the time comes for the preprogrammed migratory activity to stop, in normal circumstances the goal area typical of the species will have been reached. This so-called vector navigation allows even young birds that have never migrated before to find their goal area, even if they are migrating alone with no guidance from experienced conspecifics.

IV. MIGRATORY DISPOSITION, "FUEL," AND METABOLIC ADAPTATIONS

A. Migratory Disposition

Long-distance migrants, especially species that must fly for long distances nonstop and cross ecological barriers such as deserts or oceans, prepare for migration by adopting a particular migratory disposition—a state of readiness for migration. As a rule, this state is characterized by hypertrophy of the flight muscles, deposition of fat as a fuel and of protein for metabolic processes, metabolic adaptations including integration of enzyme systems, the development of migratory behavior, and perhaps certain hormonal adjustments. Behavioral adaptations are most necessary for species that migrate at night that normally, when not migrating, are active by day. It is thought that the endogenous clock responsible for diurnal activity, the circadian rhythmicity, splits to produce two locomotor components—one for non-migratory diurnal and one for migratory nocturnal activity. The role of the hormones in migratory disposition remains relatively unclear. There is evidently no specific migration-inducing constellation of hormones, but it appears that testosterone, prolactin, and thyroid hormones play an important role in certain migratory events.

B. Fuel

The fuel for all long-distance migrations is exclusively fat. Fat has at least four advantages for this purpose: (i) It provides the highest concentration of metabolic energy of all body chemicals; (ii) it can be stored without water, (iii) it can be handled efficiently in intermediary metabolism and completely oxidized to give energy, water, and carbon dioxide; and (iv) muscle fibers with fat metabolism become fatigued relatively slowly. Many migratory birds begin to lay down fat depots before beginning their journey, and the smaller species that travel very long distances often accumulate so much fat that they double their body weight, especially before they cross barriers that make feeding impossible.

C. Metabolic Adaptations and Body Composition

In order to build up fuel reserves, the whole body of a bird converts from carbohydrate to fat metabolism during the migration period. In addition, stored carbohydrate is broken down and in some cases the water content of the body is reduced to decrease ballast. Not only fat but also, to a certain extent, protein reserves are accumulated, to be used for enlarging the flight muscles, maintaining bodily functions, and, in some species, for developing eggs after arrival in the breeding grounds. The large fat depots are acquired almost entirely by hyperphagia (eating more than normal), though the consumption of special kinds of food can also contribute.

V. ECOPHYSIOLOGICAL ASPECTS

During migration a bird is exposed to many ecological conditions that change continually, sometimes from one extreme to its opposite. Waders, for instance, after leaving the warm waters in their near-equatorial winter home often find themselves standing in a snowstorm on the permafrost of the Arctic tundra only a few weeks later. Migratory birds must therefore develop a great many special adaptations and strategies that allow them to cope with these diverse ecological conditions. Here, examples are given of two aspects: ecological/economic adaptations in feeding at the time of migration and strategies for crossing large deserts.

A. Feeding Strategies

Many insectivorous songbirds are faced with a dilemma in the migration period. The development of hyperphagia, a requirement for fat deposition, often occurs at a time when the supply of insects is declining, and an increase in locomotor activity in order to catch more of the progressively scarcer insects would make it still harder to accumulate fat efficiently. Many species have solved the problem by switching to a different kind of food: In the migration season, they become partially frugivorous. Little effort is needed to eat juicy fruits in quantities as large as can be metabolized, and in so doing they take in many carbohydrates, which promote the deposition of fat. Proteins and essential amino acids are acquired in the smaller amounts of animal food with which the birds supplement their diet by expending only moderate effort.

B. Strategies for Crossing Deserts

Large deserts represent enormous geographical barriers to migrating birds—in particular the Sahara, which a bird traveling between Eurasia and Africa encounters immediately before or after the Mediterranean Sea. Deserts are conquered with three strategies. First, a bird can fly around them; many species detour around the Sahara by way of migration divides that channel the individuals into eastern and western routes as soon as they leave their breeding areas. The alternatives are to cross the desert in a nonstop flight (which for small birds can last up to 60 hr and probably requires a tailwind) or to use an intermittent strategy. In the latter, small birds that fly at night make one or two stopover landings in the desert during the day, spending the daytime in the shade behind stones or the like if possible and taking off in the evening for the next stage of nocturnal migration. The stopover probably helps the bird avoid flying when the wind conditions are unfavorable and the temperatures so high as to possibly cause dehydration.

VI. NAVIGATION

The precision with which migratory birds find their way borders on the fabulous. During their few years of life many small birds commute between a nesting site no larger than a human hand, in their breeding grounds, and a preferred sleeping place of the same size in their winter quarters. The ability to target these destinations so reliably, even over distances of 10,000 km or more, merits the term "pinpoint navigation." Furthermore, many birds migrate in amazingly straight lines; for instance, when white storks fly from Central Europe to the eastern Mediterranean (Fig. 1), their flight path often departs by only a few percentage points from a ruler line drawn on a map. Clearly, migratory birds must possess excellent orientation mechanisms and have an equally remarkable ability to exploit these mechanisms in directing their flights through the natural surroundings. The mechanisms underlying compass orientation (flying "innate" courses) are well-known, but apart from the vector navigation mentioned previously, hardly anything is known about goal-oriented homing (or true navigation).

A. Compass Orientation

In order actually to fly in the migration directions that are part of their genetic inheritance, birds–not only migratory birds–have available a number of biological compasses. The sun compass makes use of the azimuth of the sun, correcting for the sun's movement during the day by reference to the bird's internal 24-hr clock. The magnetic compass operates by measuring the angle of incidence of the field lines of the earth's magnetic field at the surface of the earth ("inclination compass"). Directional orientation by means of the star compass is accomplished by way of the apparent rotation of the sky about the polar star. In addition to these three compasses, there may be other guides, such as sunset cues and infrasound. The magnetic compass evidently functions as the primary compass, whereas the others tend to be secondary compasses that can be employed only after the positions and movements of the features they use for reference have been learned by observation.

B. True Navigation

The question of how birds (not only migrants but also, for example, pigeons that have been experimentally displaced from their home area) can so accurately find the way back to a familiar destination (breeding grounds or pigeon loft), even over vast distances, is the last great puzzle in the study of migration. It is certain that as soon as they take off from the site where they begin the return flight (or are released in displacement experiments), birds use a site-based orientation mechanism to determine the direction of their target destination. This ability has been demonstrated by experiments in which many birds fly toward the target immediately upon being released. It is thought to be a case of true navigation by a map-and-compass system. Here, the "map" is based on gradients in at least two geophysical features—for example, in properties of the earth's magnetic field and in odor fields in the atmosphere—that specifically characterize each point on the earth's surface. As the bird nears the target, it can additionally make use of landmarks learned when it was previously in the area. Unfortunately, the factors that could underlie a postulated map-and-compass system are completely unknown.

VII. CAPACITY TO ADAPT BY MICROEVOLUTION

Bird migration is not a rigid behavioral complex maintained over the long term as a species-specific character; on the contrary, in changing environmental conditions it can be adjusted with astonishing speed to suit the new situation. Currently, in the course of "global warming" we observe many species leaving the breeding grounds later and returning sooner, breeding earlier in the year, shortening the distance over which they migrate, and establishing new winter quarters closer to home. These changes are accompanied by changes in migration direction. Furthermore, in partially migrant populations (those in which some individuals stay in the breeding grounds while others regularly migrate), there has been an increase in the nonmigratory fractions, i.e., a shift toward being completely sedentary. Experimental studies of the genetic bases of bird migration, selection experiments, and experiments to elucidate the changes in migratory behavior observed in the field indicate that the alterations in schedule and distance have probably arisen mainly by processes of microevolution and selection. The phenomenon of partial migration seems to play a key role. It has been found among birds throughout the world and may well serve as a sort of turntable from which evolution can quickly take off in the direction from migratory to permanently resident behavior or the opposite direction, depending on what the environmental conditions require at the time. That is, partial migration probably makes the migration/nonmigration behavioral complex extremely adaptable.

See Also the Following Articles

CIRCANNUAL RHYTHMS; GLOBAL ZONES AND REPRODUCTION; MIGRATION, AMPHIBIANS; MIGRATION, FISH; MIGRATION, REPTILES; NUTRITIONAL FACTORS AND REPRODUCTION; PHOTOPERIODISM, VERTEBRATES; SEASONAL REPRODUCTION, BIRDS

Bibliography

Berthold, P. (1990). Patterns of avian migration in the age of "greenhouse" effects: A central European perspective. *Acta XX Congr. Int. Ornithol. Christchurch*, 267–268.

Berthold, P. (Ed.) (1991). *Orientation in Birds*. Birkhäuser, Basel.

Berthold, P. (1994). *Bird Migration: A General Survey*. Oxford Univ. Press, New York.

Berthold, P. (1996). *Control of Bird Migration*. Chapman & Hall, London.

Berthold, P., and Pulido, F. (1994). Heritability of migratory activity in a natural bird population. *Proc. R. Soc. London B* **257**, 311–315.

Berthold, P., and Querner, U. (1995). Microevolutionary aspects of bird migration based on experimental results. *Israel J. Zool.* **41**, 377–385.

Dingle, H. (1996). *Migration. The Biology of Life on the Move*. Oxford Univ. Press, New York.

Gwinner, E. (1986). *Circannual Rhythms*. Springer, Berlin.

Pulido, F., Berthold, P., and van Noordwijk, A. J. (1996). Frequency of migrants and migratory activity are genetically correlated in a bird population: Evolutionary implications. *Proc. Natl. Acad. Sci. USA* **93**, 14642–14647.

Wiltschko, R., and Wiltschko, W. (1995). *Magnetic Orientation in Animals*. Springer, Berlin.

Wingfield, J. C., Schwabl, H., and Mattocks, P. W. (1990). Endocrine mechanisms of migration. In *Bird Migration: Physiology and Ecophysiology* (E. Gwinner, Ed.), pp. 232–256. Springer, Berlin.

Migration, Fish

Graham Young

University of Otago

I. Introduction
II. Reproductive Migration of Anadromous Salmon
III. Imprinting and Homing during Reproductive Migration of Other Fishes
IV. Freshwater Eel Migration
V. Osmoregulatory Adjustments during Spawning Migration of Diadromous Fishes
VI. Semelparity in Migratory Fish and the Role of the Interrenal Gland

GLOSSARY

17,20β-dihydroxy-4-pregnen-3-one A progesterone derivative responsible for inducing final maturation of oocytes in females and spermiation in males of a number of teleost fish species.

gonadotropin-releasing hormone A peptide hormone synthesized in the brain which stimulates secretion of gonadotropins from cells in the pituitary; also involved in neurotransmission.

interrenal gland Found in the anterior kidney of fishes, this diffuse gland produces corticosteroids and is the homolog of the adrenal cortex.

11-ketotestosterone A potent androgen of teleost fishes which promotes spermatogenesis. Once thought to be a male-specific androgen, it has been found in substantial quantities in females of certain species.

thyroid hormones The thyroid gland mainly produces thyroxine. Thyroxine is converted to a biologically active form, triiodothyronine, in peripheral tissues.

thyrotropin-releasing hormone A peptide hormone synthesized in the brain which stimulates secretion of thyroid-stimulating hormone from cells in the pituitary; also involved in neurotransmission.

Reproductive migration, the movement of a species from a feeding area to a place suitable for spawning, is a widespread phenomenon, occurring in all major classes of fish from the primitive jawless fishes, the Agnatha (e.g., lampreys), to the advanced bony fishes, the Teleostei (e.g., salmon). Many fish species move relatively modest distances from feeding areas to spawning sites; for example, migrations which may involve no more than a displacement from the sea floor to the sea surface. At the other extreme, exemplified by the spectacular journeys of species such as salmon and freshwater eels, reproductive migration may be over long distances, precisely timed, of high geographical precision, and may involve transitions between fresh water and seawater. The long-distance migrators have evolved an array of specialized physiological, biochemical, morphological, and behavioral processes which provide the resources that fuel their journeys and enable them to navigate to the spawning area at an appropriate time, to adjust osmoregulatory mechanisms if necessary, and to spawn successfully. The study of reproductive migration of fishes thus involves research at all levels of organization, from molecular to ecosystem, in order to understand the evolution and ecology of migratory strategies, and the coordination of the biochemical, physiological, and behavioral processes which permit fishes to undergo a reproductive migration and spawn. Knowledge of many of these processes is incomplete, even in salmon and eels, which are commercially important species to which considerable research has been devoted.

I. INTRODUCTION

Fundamentally, fishes undertake migrations to maximize the chances of reproductive success. A number of variations exist on a common theme of spatial separation of growth areas from spawning areas, and while categorizing migrations may risk

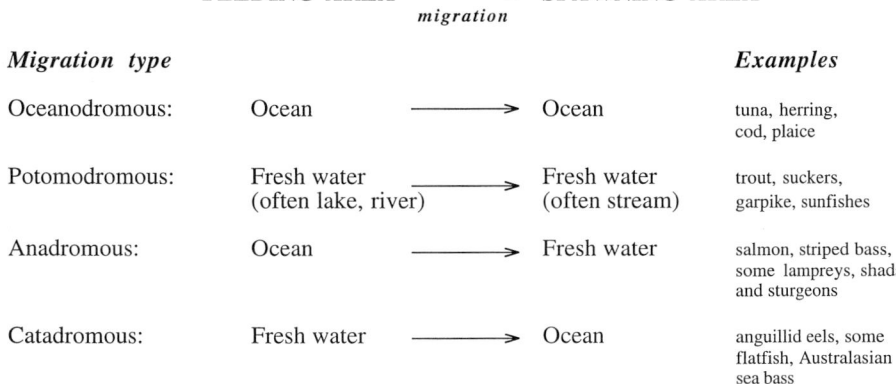

FIGURE 1 Types of reproductive migrations of fishes. Not included are littoral migrations, which generally involve short movements in or between fresh water and seawater to the shallow littoral zone for spawning.

oversimplification, four major types can be recognized.

A. Types of Reproductive Migrations

Littoral reproductive migrations (Fig. 1) are generally over short distances in or between freshwater and marine environments and usually involve movement to the shallow littoral zone for spawning. Some species, such as the Californian grunion, *Lueresthes tenuis*, and the New Zealand inanga, *Galaxias maculatus*, show an extreme littoral migration: They migrate on the highest tides of the month to deposit their eggs in terrestrial predator-free environments, either well up on sandy shores (grunion) or in the fringing vegetation of estuaries (inanga), where they will develop until hatching is triggered by immersion on the next highest tide.

Oceanodromous migrations of marine fishes range from modest local migrations to the extensive regular circuits of species such as cod, tuna, and herring. These latter journeys may be up to 10,000 km, generally following oceanic currents, and encompass areas specific for each part of the life cycle (e.g., nursery, feeding, winter and summer, and spawning) which must be entered at the appropriate time to be of benefit to the animal. How these species recognize specific sites and navigate and time journeys between sites has been the subject of much speculation but few data exist. They likely share some of the mechanisms (such as olfactory memory and compass sense) identified in salmon.

Potomodromous reproductive migrations are restricted to fresh water, are common in all the major fish classes, and consist of seasonal return movements of fish to spawning areas, which are often upstream of feeding areas. Spawning areas provide suitable environmental conditions for egg development and may also have a higher production of food suitable for the young.

Diadromous reproductive migrations represent a further general elaboration of the potomodromy pattern in that adult fish move from a freshwater (catadromous) or marine (anadromous) environment, where they have spent most of their life, to the opposite environment for spawning. Thus, migration in these species also involves mechanisms that switch osmoregulatory processes between salt conservation in fresh water to salt excretion and water conservation in seawater. These species vary in their fidelity to their natal areas, ranging from extreme homing precision to movement into a larger area. Diadromy is not geographically uniform. As a broad generalization, catadromy predominates in warmer regions where freshwater environments have high productivity, whereas anadromy is common in the highly productive marine environments of colder regions. Diadromy can therefore be viewed as the exploitation of two contrasting environments: One optimal for reproduction and the other optimal for growth and the laying down of energy reserves for the spawning migration and gamete production.

B. Migration versus Residence

Many populations of fish species which undertake migrations contain both migratory and resident individuals. These include the well-known anadromous (e.g., Pacific and Atlantic salmon) and potomodromous (e.g., the brown trout) salmonids and also species such as suckers and anadromous sticklebacks, marine Atlantic cod, herring, and smelt. The fecundity and fertility of fishes increases with body size, and body size is frequently limited by food supply. Fish often commence sexual maturation once their growth rate starts leveling off. By migrating to areas of richer food resources before food becomes limiting to growth, migrants may delay maturity in comparison to residents by 1 or more years and increase their reproductive output as a consequence of this shift in environment, as long as the costs (increased mortality rates, delayed maturity, and the energetic demands of migration) do not outweigh the benefits of increased reproductive potential. Long migrations, such as those undertaken by salmon, are extremely energy demanding: For example, female sockeye salmon may expend 50% of their total energy content during a 500-km river migration. Mortality of migrants is probably also higher compared to that of resident individuals. The benefits of migration lie in access to improved food resources, often coupled with the avoidance of adverse conditions such as low temperatures or water flow. Females, whose reproductive success is more or less directly related to body size, often form a disproportionately large percentage of migrants, whereas males often dominate among residents. Resident males may offset the disadvantage of smaller size by employing alternative tactics such as sneaker spawning—darting in to fertilize some of the eggs of females that are spawning with large migratory males.

C. Imprinting and Homing

The return to a natal site for reproduction (homing) is known to occur in a large number of marine and freshwater species. The environmental cues used by migrating fish have been studied in a restricted number of mainly teleostean species, and even less information is available on the physiological mechanisms underlying this behavior. However, many of the species studied appear to employ a compass sense based on magnetic and celestial information for open-water navigation, whereas in at least some species olfactory cues guide the final stages of the homing migration. The idea that each body of water possesses characteristic odors, and that the basis of the homing ability of species such as salmon is the ability to detect these odors, was proposed over 100 years ago. Fish have an acute olfactory sense which can discriminate between home and nonhome waters and between different nonhome waters, and they can detect odors from prey, predators, and the same species (conspecifics). Detection thresholds for a range of molecules, such as amino acids, steroids, prostaglandins, and bile salts, are in the range of 10^{-7}–10^{-14} M. The most intensively studied migratory fish species are the anadromous salmon and catadromous freshwater eels. Recent advances in telemetry, electrophysiology, molecular biology, and endocrinology have led to an improved understanding of the physiological basis for homing and reproduction in migrating salmon, and it is likely that some of the findings are applicable in broad terms to other migratory fish species.

II. REPRODUCTIVE MIGRATION OF ANADROMOUS SALMON

A. Imprinting and Homing

Salmon spawn in fresh water (Fig. 2). While juveniles of some species (chum and pink salmon) are able to adapt to seawater at a very early age and migrate to the ocean within a few months of hatching, those of other species remain in fresh water for a variable period of time, often 1 or more years. These young salmon (parr) then undergo a developmental process, the parr–smolt transformation, driven by the endocrine system and consisting of biochemical, physiological, morphological, and behavioral changes that preadapt them to oceanic life. The resulting fish, smolts, are able to migrate to the ocean and osmoregulate and grow there before returning to fresh water to spawn. Experimental studies indi-

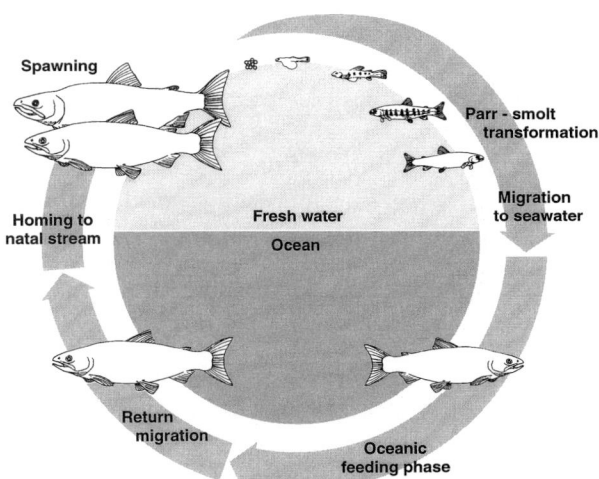

FIGURE 2 Life cycle of anadromous salmon (genus *Salmo* and *Oncorhynchus*). Species such as chum and pink salmon do not have a substantial residence period in fresh water but migrate to the ocean within a few months of hatching. Freshwater residence time of other species is variable but may be 1 or more years. Not all individuals in a population migrate, and males especially may become sexually mature at the parr stage.

cate that salmon undergoing the parr–smolt transformation imprint strongly to site-specific odors in their home stream and adults use odorant memory to guide them on their return journey. Evidence exists, however, for imprinting at other times during freshwater residence. Indeed, some wild juvenile salmon display seasonal movements away from natal areas prior to the parr–smolt transformation, suggesting that some imprinting must occur much earlier than the time of seaward migration. Olfactory memory may therefore arise from a temporally and physiologically complex process resulting in a suite of site-specific odor memories.

The neurophysiological basis for long-term olfactory memory has not been firmly established, but both central (olfactory bulb and olfactory lobe of the telencephalon) and peripheral (olfactory epithelium) olfactory systems are implicated.

Homeward migration of maturing salmon from the open ocean to natal streams involves compass orientation based on celestial and magnetic information as well as other factors such as temperature and salinity profiles. Environmental cues such as photoperiod and temperature prompt salmon to commence their return journey and to begin gametogenesis. The marked variation in the time spent in the oceanic feeding phase displayed within and between species suggests that both genetics and metabolic status influence the "decision" to respond to environmental cues and undertake a spawning migration in a given year. Once in coastal waters, olfactory stimuli and rheotactic responses appear to dominate movements, although vision appears to be important when migration involves journeys across large lakes.

B. Regulation of Imprinting and Homing

The factors responsible for heightened imprinting during the parr–smolt transformation remain to be clarified. Increased activity of several endocrine glands occurs during this developmental process, including that of the thyroid. Thyroid hormones are known to regulate neurogenesis in pre- and neonatal vertebrates and maintenance of peripheral olfactory systems. Artificial elevation of the thyroid hormone thyroxine in earlier life stages stimulates imprinting, and imprinting appears strongest at times of elevated endogenous thyroxine levels. Receptors for the biologically active thyroid hormone, triiodothyronine, have been detected in all regions of the masu salmon brain except the olfactory bulb, with levels being highest in the olfactory epithelium. Specific binding of triiodothyronine to the olfactory epithelium was elevated twofold in smolts compared to earlier developmental stages. Thyroid hormones (and other hormones) may promote differentiation of the olfactory system and proliferation of olfactory receptor neurons to facilitate imprinting, a potentially unique action of thyroid hormones at such a late stage of development. Other hormones and neurotransmitters are likely to be involved in the mechanisms underlying imprinting. Surges in brain content of the neurotransmitters serotonin, dopamine, norepinephrine, and glutamine coincide with elevated plasma thyroxine levels during the parr–smolt transformation in coho salmon. Furthermore, a variety of stimuli, including "novel" water, temperature changes, and water flow cause short-term increases in thyroxine levels, and Dittman and Quinn have hypothesized that changing environmental conditions before and during smolt migration may also facilitate imprinting by stimulating thyroid hormone production and thus allow migrating smolts to learn olfactory "waypoints" during their journey downstream. The importance of sequential imprinting may be reflected in the impaired homing ability of coho salmon smolts which were not allowed to migrate naturally. However, the depression in olfactory learning and olfactory system monoamine levels resulting from treatment of Atlantic salmon smolts with thyroxine has led Morin and colleagues to propose that a final surge in thyroxine toward the end of the parr–smolt transformation acts to terminate the period of enhanced olfactory learning and preserve home stream odor memory as fish migrate downstream.

Olfactory system salmon gonadotropin-releasing hormone (sGnRH) may be involved in migratory behavior. Brain sGnRH content increases, coincident with elevations in plasma thyroxine levels, during the parr–smolt transformation of chinook salmon, and increased expression of sGnRH in the olfactory nerve occurs during downstream migration of chum salmon, along with the proliferation of GnRH-containing fibers in the forebrain and midbrain.

Although plasma sex steroids may increase moderately in salmon smolts, the parr–smolt transformation and maturation are conflicting physiological processes. Parr which mature in fresh water do not undergo the parr–smolt transformation, and administration of sex steroids to juvenile salmon prevents them from becoming smolts. Recently, it has been shown that androgens inhibit downstream migratory behavior in smolts.

The role of gene expression and protein synthesis in imprinting/homing has been little explored. One early study on adult salmon showed that discrimination of the odor of natal streams by their olfactory bulb was reduced temporarily by inhibitors of RNA and protein synthesis, suggesting that continuous protein synthesis in the olfactory system is required during homing migration. Recent findings have implicated several proteins and hormones in the olfactory system in imprinting and homing. A cytoplasmic protein (N24) specific to the olfactory system has been identified in sockeye salmon and similar proteins exist in the olfactory systems of two nonsalmonid migratory species but not in the olfactory systems of nonmigratory species. N24 is confined to peripheral olfactory receptor cells and N24-containing axons terminate in the olfactory bulb near the mitral cells, which are important in the processing of peripheral olfactory information. N24 content appeared stronger at the time of imprinting and in adults at the time of homing to the natal stream, suggesting that this protein is involved in imprinting and homing in salmonids and possibly other migratory species.

GnRH and thyrotropin-releasing hormone (TRH) neurons in the olfactory system appear to be involved in the control of the homeward migration. Compared with chum salmon caught on river spawning sites, those caught in the coastal sea showed higher expression of the gene for the sGnRH precursor, pro-sGnRH, and higher GnRH content and neuron numbers in the olfactory nerve and the transitional area between the olfactory nerve and olfactory bulb. In contrast, pro-sGnRH expression and sGnRH content in the telencephalon and the preoptic area (controlling gonad maturation via stimulating secretion of pituitary gonadotropins) were higher in samples taken from salmon on the spawning ground (Fig. 3).

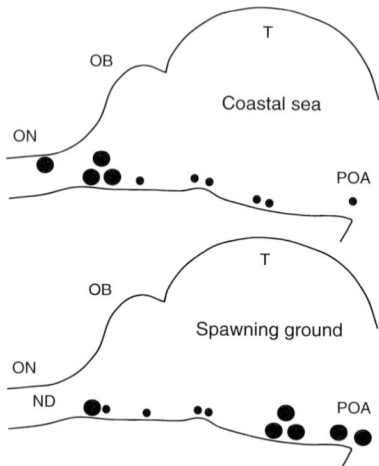

FIGURE 3 Expression of the gene for the precursor of salmon gonadotropin-releasing hormone, pro-GnRH, in brains of chum salmon caught in the coastal sea and on river spawning grounds. The size of each circle represents the abundance of pro-GnRH messenger RNA. ND, nondetectable; ON, olfactory nerve; OB, olfactory bulb; T, telencephalon; POA, preoptic area (reproduced with permission from Ueda, 1995).

TRH levels in the hypothalamus and pituitary, along with plasma thyroxine levels, gradually declined during the homing journey of chum salmon, whereas TRH concentrations and the numbers of cells expressing pro-TRH in the olfactory bulb increased just before entry into the home stream. These results are consistent with the involvement of GnRH and TRH neurons in the olfactory system in neuromodulation or neurotransmission of olfactory signals in homing salmon and those of the preoptic area and hypothalamus in controlling pituitary hormone secretion. Increased production of sex steroids during spawning migration could also influence olfactory discrimination since it has been shown that they augment electrical potentials in the olfactory bulb of goldfish.

C. Reproductive Hormones in Migratory Salmonids: Variations in Life History Strategy

Although only a few studies have been carried out on wild salmon during the spawning migration, the temporal patterns of reproductive hormones generally comply with those reported during reproduction of captive freshwater salmon and nonmigratory salmonids. There seems to be no fixed pattern as to the stage of gametogenesis achieved by the time salmon enter fresh water. Female chum salmon migrating to spawning grounds in northern Japan appear to initiate oocyte development in the North Pacific Ocean, and blood estradiol-17β (controlling yolk protein production) levels peak in the coastal sea but drop sharply during upstream migration. Levels of oocyte maturation-inducing hormone, 17,20β-dihydroxy-4-pregnen-3-one, rapidly increase during the spawning period (Fig. 4). In males, the androgens testosterone and 11-ketotestosterone (controlling spermatogenesis) gradually increase during the homeward migration to peak at the prespawning riverine period and rapidly decline during spawning with increasing levels of 17,20β-dihydroxy-4-pregnen-3-one, which controls spermiation. However, marked variations occur within species. For example, the chinook salmon exhibits wide variations in life history strategy, with two racial forms, named "spring" and "fall" because of the time of the reproductive migration into fresh water. Spring chinook return in a relatively

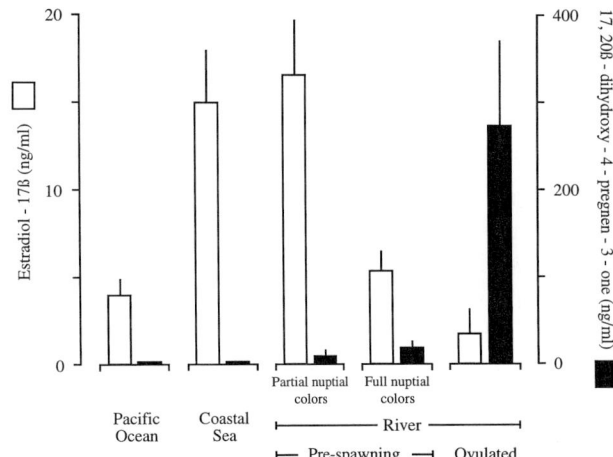

FIGURE 4 Sex steroid levels in female chum salmon during reproductive migration to natal streams. Chum salmon assume a prespawning nuptial coloration during riverine migration. Those with partial nuptial coloration are less reproductively advanced. Each bar represents mean + standard error. Graph drawn from data from Ueda et al. (*Gen. Comp. Endocrinol.* 53, 203–211, 1994, and *Nippon Suisan Gakkaishi* 57, 881–884, 1991).

immature stage to their natal rivers 6–9 months before spawning and the greater part of gonadal growth is accomplished in fresh water, a time when the animals do not feed. Consequently, oocyte growth is dependent on the metabolic stores laid down before entry into fresh water. Fall chinook remain in the ocean, feeding during most of the vitellogenic phase of oocyte development. This plasticity in gonadal development in relation to the timing of migration into fresh water is reflected in levels of pituitary gonadotropins and sex steroids and suggests that reproductive hormones per se are not significant triggers of the spawning migration.

III. IMPRINTING AND HOMING DURING REPRODUCTIVE MIGRATION OF OTHER FISHES

Olfactory imprinting as the basis for homing is not restricted to salmonid species, although few detailed studies of other fishes are available. The white sucker, for example, shows extreme fidelity to natal streams and depends on olfaction to locate the home stream. However, unlike salmon, their larvae migrate from the stream before full development of the peripheral olfactory system, so it is uncertain whether imprinting occurs prior to migration downstream. Possibly, larvae remain near the mouth of the home stream until imprinting occurs later. An alternative or additional mechanism to juvenile imprinting on site-specific odors may be that fish on their spawning migration are attracted to pheromones from resident conspecific juveniles living in natal streams. The importance of pheromones in homing of salmon remains uncertain, although it has clearly been shown that salmon and other fishes can detect the odor of conspecifics.

Guidance of prespawning adults to spawning areas by conspecific odors seems to have evolved in a very specialized fashion in sea lamprey. Behavioral experiments show that migratory adult sea lamprey are attracted to larval odors, and removal of larvae from streams results in a reduction in adults migrating into these streams. Homing of adult sea lamprey to freshwater streams (not necessarily their natal streams) has been experimentally linked to the release of two unique bile acids, petromyzonal sulfate and allocholic acid, from conspecific larvae. The olfactory system of adult sea lamprey has a detection threshold of 10^{-12} M for these compounds. Intriguingly, during the radical metamorphosis of freshwater larvae into the migratory parasitic phase of the life cycle (involving a migration to lakes or oceans), the gall bladder is lost and larval bile acids are not found in the adult.

Prespawning and spawning eels are characterized by a pungent "odor of ripeness," detectable by humans, which may serve a pheromonal function in encouraging congregations of individuals in the ocean.

IV. FRESHWATER EEL MIGRATION

Freshwater eels (genus *Anguilla*) reproduce in oceanic spawning areas, presumably at great depths, although this has never been observed (Fig. 5). Larval stages are thought to span about 15 months, during which time they are carried by ocean currents. In

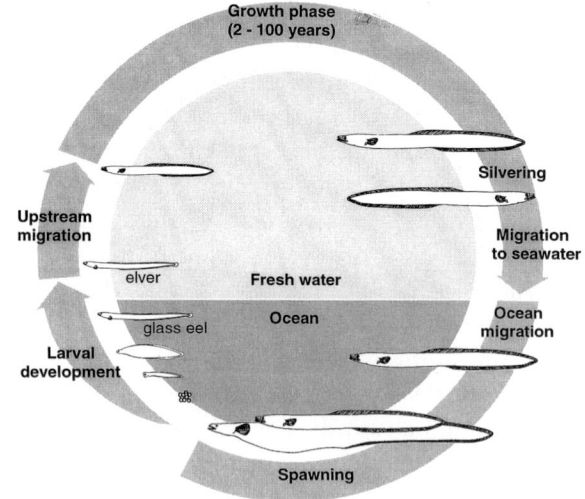

FIGURE 5 Life cycle of catadromous freshwater eels (genus *Anguilla*). Larval stages are thought to span about 15 months. In coastal areas, larvae metamorphose into "glass" eels and become pigmented ("elvers") during migration upstream. Residence period in fresh water may last for 2–100 or more years, during which time eels remain sexually immature. "Silvering" (deposition of guanine in the skin) is characteristic of eels about to undertake a reproductive migration to oceanic spawning grounds.

coastal areas, the larvae metamorphose into "glass" eels and actively migrate up freshwater streams and rivers, become pigmented ("elvers"), and reside in fresh water for 2–100 or more years, depending on sex and species. The gonads remain immature throughout this freshwater residency period. In response to an inconclusively identified trigger, the adult eel (often in the fall) undergoes another transformation, know as "silvering," preparatory for the long journey back to the oceanic spawning ground. The skin takes on a silver color due to deposition of the pigment guanine, the eyes increase in size, whereas the eels cease feeding, their gastrointestinal tracts atrophy, gonadal development begins, and they acquire the mechanisms to osmoregulate in seawater. With the exception of a few eels caught in coastal areas, migrants have never been caught in the open ocean, and the details of their navigation, homing, reproduction, and spawning are a matter of speculation. The identification of spawning grounds of the European and American eel as lying in the Sargasso

Sea (a journey of up to 5000 km for European eels) was based on surveys of catches of eel larvae. Recently, this approach has identified the spawning grounds of the Japanese eels as being in the Pacific Ocean east of the Philippines. Major oceanic currents appear to be the key determinant of spawning area location and larval distribution. On this basis, two areas in the southwest Pacific Ocean have been proposed as the spawning grounds of two species endemic to parts of Australia and New Zealand.

Even eels of the same species show huge variation in the time they reside in fresh water. What triggers the onset of puberty and the spawning migration of eels? Although several theories have been proposed, all are related to growth rate and/or energy reserves. The most plausible seems to be the achievement of a critical level of fat stores, especially because fat content is known to influence age at puberty in higher vertebrates.

Although GnRH treatment has only relatively minor effects on gonadal growth, treated eels attempted to escape from holding tanks more frequently, suggesting that GnRH may stimulate brain centers controlling migratory behavior. Changes in the GnRH systems in the eel brain in relation to downstream migration have not yet been documented. Giorgi and colleagues reported high levels of dopamine and its major metabolite dihydroxyphenylacetic acid in the olfactory bulb in nonmigratory European eels, and both increased substantially at the time of silvering, whereas levels in other brain areas did not vary with developmental stage. These observations are somewhat similar to the changes in neurotransmitters reported during the salmon parr–smolt transformation.

The reproductive physiology of wild eels is not well understood for several reasons: The gonads of most species are relatively immature, generally in the very early stages of gametogenesis at the time of entry into the ocean; migrants cannot be caught in the open ocean; and gonad development is arrested in migrants held captive due to inhibition of hypothalamic GnRH and pituitary gonadotropin secretion by dopamine. Many attempts have been made with captive eels to overcome the arrest of gametogenesis by manipulating environmental factors such as water salinity, wavelength and quantity of light, and temperature. The only treatment with a discernible effect was the submergence of silver European eels at 450 m for 3 months, which induced a slight increase in relative gonad weight. Most information on eel reproductive physiology, therefore, has come from inducing gonad maturation artificially with gonadotropic preparations (usually salmon pituitary homogenate for females and human chorionic gonadotropin for males). Generally, patterns of sex steroids conform to the general teleost pattern, with a few notable exceptions: Estradiol-17β does not increase substantially in plasma and aromatase activity does not increase in the ovarian follicle until very late vitellogenic or postvitellogenic stages (possibly associated with the vitellogenic development of the next crop of oocytes). In most cases salmon pituitary homogenate treatment alone does not induce final oocyte maturation, which requires injection of 17,20β-dihydroxy-4-pregnen-3-one or its immediate precursor, 17α-hydroxyprogesterone. High activity of the enzyme 20β-hydroxysteroid dehydrogenase, which catalyzes the final step of the synthesis of 17,20β-dihydroxy-4-pregnen-3-one, is found even in vitellogenic eel ovarian follicles. Control of this enzyme appears to differ markedly between salmon and eels: A preovulatory surge of gonadotropin increases 20β-hydroxysteroid dehydrogenase activity in salmon ovarian follicles. The failure of pituitary homogenate-treated eels to undergo final oocyte maturation therefore appears to be due to lack of substrate.

The only information on sex steroid levels in wild eels at later stages of gametogenesis comes from studies on the New Zealand long-finned eel, *Anguilla dieffenbachii*, a species which shows relatively advanced gonadal development at the time of migration into seawater. Males of this species are in late spermatogenesis with some mature spermatozoa in the testes, whereas the most advanced oocytes in females are midvitellogenic. In general, the pattern of sex steroids in migratory long-finned eels is similar to that seen for equivalent stages during artificial maturation of eels. However, surprisingly high levels of 11-ketotestosterone, generally regarded as a male-specific androgen, are found in migratory females of this species and those of the short-finned eel (*A. australis*) (Fig. 6). Recent reports for migratory female salmonids and sturgeon, species which under-

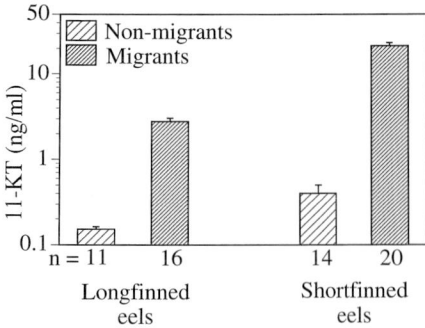

FIGURE 6 11-Ketotestosterone (11-KT) levels in nonmigrant and migrant female long-finned and short-finned eels. Note that the vertical axis is logarithmic. Number of eels sampled is indicated under the columns (unpublished data from P. M. Lokman and G. Young).

take long spawning migrations, and of the stimulatory effects of this androgen on heart size and red muscle mass in rainbow trout suggest that 11-ketotestosterone may play a role in preparing both males and females for the physiological demands of long spawning migrations.

V. OSMOREGULATORY ADJUSTMENT DURING SPAWNING MIGRATION OF DIADROMOUS FISHES

Species whose spawning migration involves movement from fresh to seawater or vice versa must coordinate reproductive processes with a change in osmoregulatory physiology. Teleost species such as salmon, which move into fresh water from seawater, must switch from a salt-excreting, water-conserving mode (hypoosmoregulation) to hyperosmoregulation, characterized by salt retention and water excretion. Thus, seawater teleosts drink seawater, actively transport sodium, which is followed by water across the intestine, and excrete the resulting salt load by specialized gill "chloride" cells. In contrast, teleosts in fresh water do not drink, and they combat hydration by excreting copious dilute urine, while the gills absorb ions. The interrenal hormone, cortisol, and, in some species, growth hormone from the pituitary are implicated in the control of many aspects of osmoregulation in seawater, whereas the pituitary hormone prolactin acts as a salt-conserving hormone in freshwater teleosts.

Although some salmonids can complete gametogenesis in seawater, other salmonid species show a progressive loss of ability to osmoregulate in seawater as gonadal development advances, accompanied by elevated (possibly compensatory) plasma levels of cortisol and growth hormone but with no change in the low levels of prolactin that are characteristic of teleosts in seawater. The loss of ability to osmoregulate in seawater may be linked to a negative effect of sex steroids on hypoosmoregulatory processes, similar to their negative effect on the parr–smolt transformation.

As in juvenile salmon smolts, the thyroid gland and the interrenal gland of freshwater eels are activated in migrants, and eels undergo a preadaptive adjustment in osmoregulatory processes prior to entry into seawater, including enlargement and proliferation of salt-secreting chloride cells in the gill and increased salt and water uptake by the intestine. Eels in seawater have lower prolactin and higher cortisol levels than their freshwater counterparts, and their gill chloride cells increase in number and size. Feeding ceases and the gastrointestinal tract shows atrophy, losing its nutrient absorption function but increasing solute-linked absorption of water that the eel now actively drinks. Experimentally, cortisol treatment, in addition to stimulating various hypoosmoregulatory mechanisms, also promotes silvering, the deposition of guanine. Thyroid hormones, perhaps through stimulatory effects on the interrenal gland, may also promote the development of hypoosmoregulatory mechanisms.

Little is known about the hormonal control of hyperosmoregulation in the adult sea lamprey (or agnathans in general) during their migration into fresh water.

VI. SEMELPARITY IN MIGRATORY FISH AND THE ROLE OF THE INTERRENAL GLAND

Some fish species which undertake spawning migrations are semelparous, that is, they spawn once and then inevitably die ("programmed death"). The most famous examples are Pacific salmon. Pacific salmon display greatly elevated cortisol levels and interrenal hyperplasia in the spawning and postspawning period. This apparently is a result of gonadal steroids stimulating the hypothalamus–pituitary–interrenal axis since gonadectomy results in a rapid involution of the interrenal tissue, a reduction in cortisol levels, and prolonged survival. Androgens and estrogens induce interrenal hyperplasia and 17,20β-dihydroxy-4-pregnen-3-one may serve as a substrate for cortisol synthesis by the interrenal during the final stages of spawning. The anadromous Atlantic salmon is iteroparous, i.e., may spawn several times between return trips back to the ocean. Sex steroid control of the hypothalamus–pituitary–interrenal axis does appear to vary between semelparous and iteroparous species: Both androgens and estrogens activate the axis in semelparous species, whereas marked gender differences in the response of this axis to stress and to sex steroids occur in the iteroparous rainbow trout.

Lampreys also inevitably die after spawning. Although evidence is circumstantial, interrenal steroids have also been implicated in programmed death in these species. Surgical removal of the pituitary resulted in prolonged survival. Although gonad removal delays death, sex steroid treatment of intact animals also prolongs survival, suggesting that the model for programmed death in salmon, sex steroid activation of the interrenal, does not apply. Pituitary removal may promote survival through the removal of corticotropic pituitary hormones.

Eels have long been considered to be semelparous, but evidence from induced maturation experiments in which successive crops of eggs are spawned suggests that females at least have the potential to spawn several times in nature. Nonetheless, there is little evidence to suggest that they survive for prolonged periods after spawning or undergo prolonged cycles of gonadal recrudescence in the ocean. Though cycles of spawning and feeding can occur in captivity, the ovary is devoid of developing oocytes after releasing three or four crops of eggs. Whether death comes simply from starvation or is promoted through interrenal corticosteroids has not been addressed.

See Also the Following Articles

Fish, Modes of Reproduction in; GnRH (Gonadotropin-Releasing Hormone); Migration, Amphibians; Migration, Birds; Migration, Insects; Migration, Reptiles; Photoperiodism, Vertebrates; Seasonal Reproduction, Fish

Bibliography

Dickhoff, W. W. (1989). Salmonids and annual fishes: Death after sex. In *Development, Maturation, and Senescence of Neuroendocrine Systems: A Comparative Approach* (M. P. Schreibman and C. G. Scanes, Eds..), pp. 253–266. Academic Press, New York.

Dittman, A. H., and Quinn, T. P. (1996). Homing in Pacific salmon: Mechanisms and ecological basis. *J. Exp. Biol.* **199**, 83–91.

Ebbesson, S. O. E., Smith, J., Conchita, C., and Ebbesson, L. O. E. (1996). Transient alterations in neurotransmitter levels during a critical period of neural development in coho salmon (*Oncorhynchus kisutch*). *Brain Res.* **742**, 339–342.

Giorgi, O., Deiana, A. M., Salvadori, S., Lecca, D., and Corda, M. G. (1994). Developmental changes in the content of dopamine in the olfactory bulb of the European eel (*Anguilla anguilla*). *Neurosci. Lett.* **172**, 35–38.

Hasler, A. D., and Scholz, A. T. (1983). *Olfactory Imprinting and Homing in Salmon.* Springer-Verlag, Berlin.

Larsen, O. L., and Dufour, S. (1993). Growth, reproduction and death in lamprey and eels. In *Fish Ecophysiology* (J. C. Rankin and F. B. Jensen, Eds.), pp. 72–104. Chapman & Hall, London.

Li, W. M., Sorensen, P. W., and Gallaher, D. D. (1995). The olfactory system of migratory adult sea lamprey (*Petromyzon marinus*) is specifically and acutely sensitive to unique bile acids released by conspecific larvae. *J. Gen. Physiol.* **105**, 569–587.

McKeown, B. A. (1984). *Fish Migration.* Timber Press, Beaverton, OR.

Morin, P.-P., Hara, T. J., and Eales, J. G. (1995). T4 depresses olfactory responses to L-alanine and plasma T3 and T3 production in smoltifying Atlantic salmon. *Am. J. Physiol.* **269**, R1434–R1440.

Morin, P.-P., Winberg, S., Nilsson, G. E., Hara, T. J., and Eales, J. G. (1997). Effects of L-thyroxine on brain monoamines during parr–smolt transformation of Atlantic salmon (*Salmo salar* L.). *Neurosci. Lett.* **224**, 216–218.

Parhar, I. S., and Iwata, M. (1996). Intracerebral expression of gonadotropin-releasing hormone and growth hormone-releasing hormone is delayed until smoltification in salmon. *Neurosci. Res.* **26**, 299–308.

Smith, R. J. F. (1985). *The Control of Fish Migration*. Springer-Verlag, Berlin.

Ueda, H. (1995). Homing mechanisms in salmon: Roles of vision and olfaction. In *Proceedings of the Fifth International Symposium on the Reproductive Physiology of Fish* (F. W. Goetz and P. Thomas, Eds.), pp. 218–220. Fish Symposium 95, Austin, TX.

Ueda, H., and Yamauchi, K. (1995). Biochemistry of fish migration. In *Biochemistry and Molecular Biology of Fishes, Volume 5, Environmental and Ecological Biochemistry* (P. W. Hochachka and T. P. Mommsen, Eds.), pp. 265–279. Elsevier, Amsterdam.

Werner, R. G., and Lannoo, M. J. (1994). Development of the olfactory system of the white sucker, *Catostomus commersoni*, in relation to imprinting and homing: A comparison to the salmonid model. *Environ. Biol. Fish* 40, 125–140.

Migration, Insects

Kenneth Wilson
University of Stirling

I. Background
II. The Oogenesis–Flight Syndrome
III. Migration and Reproduction in Males
IV. The Trade-Off between Migration and Reproduction
V. Genetic and Environmental Regulation
VI. Hormonal Control

GLOSSARY

juvenile hormone A hormone produced in the corpora allata (paired neurosecretory organs located just behind the insect brain) and, in adult insects, associated with the promotion of both flight and oogenesis.

migration As defined by Kennedy (1985), a persistent and straightened-out movement effected by the animal's own locomotory exertions or by its active embarkation on a vehicle. It depends on some temporary inhibition of station-keeping responses but promotes their eventual disinhibition and recurrence.

migration syndrome A suite of interacting physiological, morphological, life-history, and behavioral adaptations that have coevolved to maximize the benefits of the migratory lifestyle.

oogenesis–flight syndrome A term coined by Johnson (1969) to describe the temporal and physiological relationship between migration and reproductive development in female insects. It proposes that migration occurs in postteneral females before reproduction or between reproductive cycles.

prereproductive period The period between adult eclosion and attainment of sexual maturity. In females it is often termed the precalling period and is measured by the interval between eclosion and the initiation of pheromone release ("calling"). In males it is the interval between eclosion and the onset of responsiveness to calling females.

wing polymorphism A polymorphism or polyphenism in which wing length varies in response to genetic or environmental factors, respectively. In wing dimorphic species, two extreme forms are recognized: Macropterous individuals are long-winged and generally capable of sustained flight, whereas brachypterous (or apterous) individuals have small wings (or no wings at all) and are incapable of migratory flight.

Migration and reproduction are intimately linked at both the functional and mechanistic levels. Functionally, strategies for migration and reproduction have coevolved to allow insects to escape deteriorating and potentially hostile habitats and to colonize and rapidly exploit new favorable ones. Mechanistically, an antagonistic relationship between flight and oogenesis ensures that competition between the two activities for valuable resources is minimized. The confusion that dominated the study of insect migration in previous decades has been replaced by a clearer understanding of the adaptive significance of migration and the hormonal and genetic mechanisms involved.

I. BACKGROUND

The study of insect migration has a long and checkered history and, in 1966, C. G. Johnson observed that the subject was "confused" and lacked "system." Much of the confusion and controversy associated with the study of migration was due to the multitude of definitions used by various workers in the field. The definition used here, which was initially proposed by J. S. Kennedy to explain the behavior of migratory insects, is now generally accepted as a good description of migration in most animal taxa. Central to Kennedy's definition is an emphasis on the behavior and physiology of the individual rather than on the consequences of migration for the spatial distribution of the population as a whole. This point is crystallized in the observation that migration is a behavioral process with ecological consequences.

Another key element of Kennedy's definition is the temporal partitioning of migration and so-called station-keeping responses (feeding, mating, oviposition, etc.). These responses are temporarily inhibited during migration, but migration ultimately stimulates their expression. The temporal and physiological relationship between migration and reproduction was first discussed by C. G. Johnson in a concept known as the oogenesis–flight syndrome.

II. THE OOGENESIS–FLIGHT SYNDROME

Johnson saw the ontogenetic relationship between oogenesis and flight as central to the understanding of insect migration. In his original description of the oogenesis–flight syndrome, Johnson viewed migration and reproduction as essentially mutually exclusive activities; migration being undertaken by individuals that were sexually immature or between reproductive cycles. In order to assess the evidence pertaining to the oogenesis–flight syndrome it is useful to consider separately the reproductive status of individuals when they initiate their first migratory flight, when they undertake interreproductive flights, and when they terminate migration.

For the vast majority of insects, their first migratory flights appear to be undertaken while the female is sexually immature. Moreover, seasonal cues or cues associated with the imminent deterioration of the current habitat often lead to an extension of the prereproductive period (PRP) and hence the window of opportunity available for migration. These observations have led some authors to use the duration of the female PRP as a tool for identifying migratory species or individuals. However, this technique, and the assumptions underlying it, have been criticized by others for their lack of generality and failure to account for migratory flights that occur between reproductive episodes. A recent review of interreproductive migration concluded that while there is evidence that some species migrate between egg-laying sites with undeveloped eggs in their ovaries, few species appear to do so when their oocytes are fully mature. Regarding the relationship between reproduction and the termination of migration, the observation that usually <10% of females captured during migratory flights are carrying spermatophores or fully developed oocytes indicates that although migration is not terminated immediately upon maturation, it probably does not persist for long after oogenesis has begun.

Tethered-flight and wind-tunnel studies also suggest that migratory flights may be largely restricted to sexually immature females: In 90% of the species in which it has been examined, there was a significant decline in tethered-flight duration with the onset of reproductive maturity. Tethered-flight studies of the lesser migratory grasshopper (*Melanoplus sanguinipes*) suggest that migration continues as ovarian development is progressing and stops only when oocyte maturation is complete. Indeed, in this insect (and several others including milkweed bugs, aphids, fruit flies, and locusts), laboratory studies suggest that flight activity actually accelerates oogenesis. Caution should be exercised in interpreting the results of tethered-flight studies, however, because they are generally incapable of distinguishing between truly migratory and nonmigratory flights (Kennedy, 1985).

III. MIGRATION AND REPRODUCTION IN MALES

One of the early criticisms of the oogenesis–flight syndrome was that it failed to encompass male migrants. However, as Johnson himself noted, the selection pressures influencing the migratory and reproductive strategies of males may be very different from those of females. Whereas natural selection favors females that maximize their chances of reaching new habitats suitable for oviposition, it favors males that maximize the number of these females they inseminate, and insemination can occur anywhere, not just at the oviposition site.

We still know relatively little about the integration of migration and reproduction in males. For some species, there is good evidence that both sexes undertake their first migratory flights as prereproductive adults. This evidence includes the failure to catch males in pheromone traps at emergence sites, the observed decline in tethered-flight activity with the onset of sexual maturation, and the observation that prereproductive periods in males and females are often of comparable length and regulated by the same genetic and/or environmental factors. However, in other species, including the monarch butterfly (*Danaus plexippus*), males are known to mate with females at emergence sites before undertaking their first migratory flights. In some other insects, the males fly only very short distances following mating, leaving immature females to migrate considerable distances alone. Even for species in which the males are believed to commence migration prereproductively, there is evidence to suggest that migration may continue after sexual maturity is attained. In these species, males often fail to exhibit a decline in tethered-flight activity after the onset of sexual maturation. Indeed, in male *D. plexippus* tethered-flight capacity is positively correlated with reproductive development (the reverse is observed in females).

IV. THE TRADE-OFF BETWEEN MIGRATION AND REPRODUCTION

Insects migrate in order to reach habitats that will yield greater fitness than can be gained by staying in the present ones. Although the benefits of migrating out of a poor habitat may be substantial, the act of migration may be costly. A recent review of the subject identified four potential costs: (i) the actual metabolic cost; (ii) the risk of increased predation or failure to reach a suitable habitat; (iii) energetic and developmental costs of constructing and maintaining the flight apparatus; and (iv) potential reproductive costs due to increased time to first oviposition, decreased energy reserves available for reproduction, shortened life span, and/or a decrease in overall fecundity.

Because of the apparent ease of distinguishing between migratory and nonmigratory individuals, much attention has focused on examining the reproductive costs of the capacity for migration in wing polymorphic species. These studies generally identify two penalties of possessing flight competence: an extended prereproductive period and a reduced level of fecundity (little difference is usually observed in the development times or longevities of winged and wingless morphs). Of course, these costs are not necessarily incurred by individuals outside the confines of the laboratory, where winged insects may be able to recoup any deficit in resources by flying to neighboring plants in which those resources are available. Evidence in support of this assertion is provided by studies of coroxid beetles and planthoppers.

Although most of the research on the costs of the capacity for migration has concentrated on wing polymorphic insects, the majority of studies examining the actual costs of migratory flight have focused on species that are monomorphically winged. These studies indicate that although the metabolic costs of flight are high (because both flight and reproduction draw heavily on lipid reserves), under conditions of adequate nutrition many species fail to exhibit any reproductive cost of flight. For example, in the milkweed bug (*Oncopeltus fasciatus*) flights of several hours over 6 days affected fecundity and longevity only in females that were starved. Even more remarkable, in the lesser migratory grasshopper (*M. sanguinipes*), migratory individuals flown to exhaustion produced more offspring than nonmigratory conspecifics and, in both species, insects that were flown on flight mills matured significantly earlier than those that were not, thus reducing the real cost of flight substantially.

An explanation for the apparent lack of a cost to flight activity lies in the observation that traits do not evolve in isolation; the propensity to undertake migratory flight is associated with the evolution of a whole suite of physiological, hormonal, and life-history adaptations geared to minimize the costs of migrating. This assemblage of adaptations is often referred to as the migration syndrome. Genetic studies have shown that when artificial selection is applied to one migratory character, such as wing length, correlated responses are observed in a number of other traits that facilitate migration, such as lipid reserves, flight capacity, and early fecundity. For example, when milkweed bugs from a migratory population were selected on the basis of tethered-flight activity, a significant response was observed in both sexes after just two generations. Moreover, the line selected for long flights showed a correlated increase in both fecundity and wing length. Thus, although all individuals may incur an energetic cost of flying (which can often be ameliorated by dietary intake of carbohydrate), the migration syndrome ensures that migratory individuals express the physiological and behavioral traits required to minimize those costs. It is interesting to note that genetic studies have so far failed to show a positive genetic correlation between flight capacity and prereproductive period. It seems likely that this genetic uncoupling of PRP from other migratory traits releases the insects from genetic constraints on the relative timing of reproduction and migration, thus allowing them to respond to environmental uncertainty.

V. GENETIC AND ENVIRONMENTAL REGULATION

The vast majority of migrating insects do so facultatively in response to environmental cues coinciding with or anticipating deterioration of the current habitat. Because new habitats are frequently created and ultimately destroyed by seasonal developments in the weather and climate, environmental cues associated with seasonal change are frequently used by insects. Outside the tropics, the most reliable seasonal predictor is the change in day length, and a number of insects use photoperiod as a cue to synchronize their migratory movements and reproductive activities. In adults of both the large milkweed bug and the monarch butterfly, short days in the autumn result in a shutdown of all reproductive activities which facilitates their long migratory flights from temperate breeding areas in the north to subtropical overwintering sites further south. Similar responses are observed in other temperate migrants, including a number of noctuid moths. In all these species, the photoperiodic response may be modified by responses to temperature and other seasonal predictors including rainfall and host–plant quality. Larval crowding is often indicative of a deteriorating habitat and is frequently used by both temperate and tropical migrants as a cue to delay oocyte development and/or to migrate. There is some evidence from the African armyworm moth and the lesser migratory grasshopper that tethered-flight activity is also enhanced in adults experiencing high densities as larvae.

In some instances, environmental cues perceived in one generation may lead to responses in the following one. These responses are termed maternal effects and are most likely to evolve in species for which ontogenetic constraints require that migratory phenotype be determined early in development and/or for which cues forecast conditions well in advance. For example, in many species of aphids, the environmental conditions experienced by virginoparous females are reflected in the proportion of winged, and thus potentially migratory, offspring they produce.

In the tropics, there are few reliable cues to the impending state of the environment, particularly for phytophagous insects which depend on the temporal and spatial distribution of rainfall to generate patches of host plants. In these circumstances it can be expected that variation in the capacity for migration will be primarily under genetic control and the proportion of migrants in any particular population will be determined by the genotypes of the parent moths

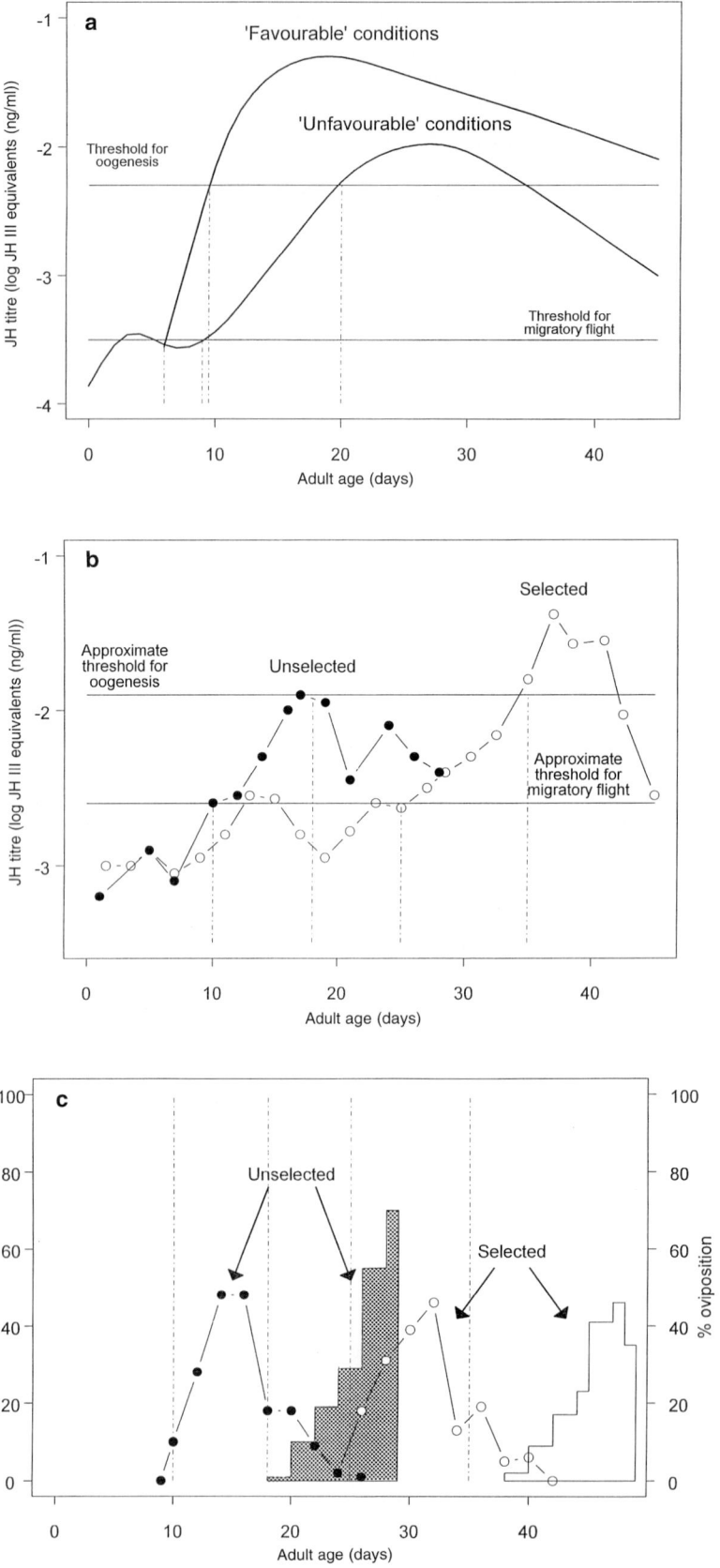

colonizing the habitat. There is now good evidence from a number of tropical migrants, including the African armyworm moth, that this is the case.

Many other studies have identified a significant additive genetic component to PRP and/or tethered-flight activity, even in species that also respond to environmental cues. Thus, the regulation of migration is best viewed in terms of an interaction between genetic and environmental components. For any particular species or population, the dominant component is likely to be determined by the relative stability of the habitats occupied by the insect and the reliability and utility of the environmental cues associated with habitat persistence.

VI. HORMONAL CONTROL

Johnson's proposal of an oogenesis–flight syndrome was influenced by his belief that migratory flight was a symptom of a hormonal deficiency which was obviated at the onset of ovarial development. Thus, he assumed that a hormone (originating from the corpora allata) had a positive influence on oogenesis but a negative effect on flight. In fact, it is now apparent from studies on a number of migratory insects (including ladybird beetles, milkweed bugs, and some lepidoptera) that juvenile hormone (JH) has a positive influence not only on oogenesis in females but also on flight activity in both sexes. An explanation for this apparent paradox is that the effect of JH on migration and oogenesis depends critically on the level of the hormone in the hemolymph (Fig. 1). When JH titers are low, neither activity is stimulated; at intermediate titers, migratory flights are initiated; and when titers are high, JH promotes oogenesis and suppresses migration. Experimental studies suggest that environmental factors that influence oogenesis and migratory activity (e.g., day length and ambient temperature) also influence the JH trajectory, with short days and cool temperatures producing slower increases in JH titers, longer PRPs, and earlier flight activity. In the true armyworm, *Pseudaletia unipuncta*, environmental conditions that induce delayed ovarian development and slow rates of JH production in females produce delayed sexual responsiveness and low rates of JH acid release in males.

Because wingless adults of wing polymorphic species often resemble juvenile stages, early workers in the field suggested that JH may be involved in the production of apterous individuals. However, only recently have studies of aphids, planthoppers, and orthopterans confirmed the role of JH in wing form determination. Studies on the ground cricket *Gryllus rubens* showed that application of JH late in nymphal development induced short-winged adults, even in insects that were genetically or environmentally "programmed" to become long-winged adults. Interestingly, although nymphs destined to develop into

FIGURE 1 The relationship between juvenile hormone (JH) titer and initiation of migration and oogenesis in the large milkweed bug (*Oncopeltus fasciatus*). (a) A schematic diagram illustrating the proposed thresholds of JH levels required to intiate flight and oogenesis. Under "favorable" conditions (e.g., high temperatures and long photoperiods), JH titers rise and rapidly exceed the threshold for oocyte development, leading to the onset of reproduction. When environmental conditions are unfavorable (e.g., low temperatures and short photoperiods), migration will occur when JH titers exceed the threshold for migratory flight but remain below that for ovarian development (redrawn from Rankin and Riddiford, 1978). (b) Actual data for JH titers are presented for an unselected line of milkweed bugs held under favorable conditions (●) and a line selected for delayed onset of flight (○). In the selected line, the trajectory of JH increase is approximately 15 days later than that of the unselected line. (c) Data from the same experiment illustrating the timing of tethered flight activity in the unselected and selected lines (● and ○ symbols, respectively), as well as oviposition behavior (shaded and open bars, respectively). This figure illustrates that flight occurs at intermediate JH titers, whereas oviposition follows titers at maximal levels; Figs. 1b and 1c redrawn from Dingle (1985) after Rankin (1978).]

long- or short-winged adults differed little in their titers of JH, there were very marked differences in the activity levels of JH esterase, JHE (an enzyme involved in the breakdown of JH). Midway through the final instar, nymphs of long-winged crickets exhibit considerably higher JHE activity than nymphs of short-winged individuals at a similar age, suggesting that it is the removal of JH by JHE that permits the production of wings in the adult.

Wing muscle histolysis is a process that occurs after a period of intense flight activity or prior to reproduction. In species that undergo wing muscle degeneration at the onset of reproduction, histolysis appears to free up resources for reproduction that had been previously allocated to flight-related activities. In these insects, elevated JH titers are associated with wing muscle degeneration. However, in species that histolyze their wing muscles on arrival at diapause sites, the opposite effect is observed, with increased JH titers being associated with flight muscle regeneration prior to emigration from the diapause site. Thus, the same hormone appears to have opposite effects in different species.

A range of other hormones also affect the relationship between migration and reproduction. These include adipokinetic hormone (secreted by the glandular lobes of the corpora cardiaca), which affects the mobilization and use of flight fuels. In monarch butterflies, injections of both this hormone and JH (either alone or in combination) increased the propensity for flight. A number of neuropeptides also appear to affect flight activity, though their effects are still poorly understood. Octopamine (a biogenic amine) not only influences flight metabolism but also appears to have a direct stimulatory effect on the flight motor patterns in the thoracic ganglia.

See Also the Following Articles

JUVENILE HORMONE; LOCUSTS; MIGRATION, BIRDS

Bibliography

Dingle, H. (1996). *Migration: The Biology of Life on the Move.* Oxford Univ. Press, Oxford, UK.

Drake, V. A., Gatehouse, A. G., and Farrow, R. A. (1995). Insect migration: A holistic conceptual model. In *Insect Migration: Tracking Resources through Space and Time* (V. A. Drake and A. G. Gatehouse, Eds.), pp. 427–457. Cambridge Univ. Press, Cambridge, UK.

Gatehouse, A. G., and Zhang, X.-X. (1995). Migratory potential of insects: Variation in an uncertain environment. In *Insect Migration: Tracking Resources through Space and Time* (V. A. Drake and A. G. Gatehouse, Eds.), pp. 193–242. Cambridge Univ. Press, Cambridge, UK.

McNeil, J. N., Cusson, M., Delisle, J., Orchard, I., and Tobe, S. S. (1995). Physiological integration of migration in Lepidoptera. In *Insect Migration: Tracking Resources through Space and Time* (V. A. Drake and A. G. Gatehouse, Eds.), pp. 279–302. Cambridge Univ. Press, Cambridge, UK..

Rankin, M. A. (1989). Hormonal control of flight. In *Insect Flight* (G. J. Goldsworthy and C. H. Wheeler, Eds.), pp. 139–163. CRC Press, Boca Raton, FL.

Rankin, M. A., and Burchsted, J. C. A. (1992). The cost of migration in insects. *Annu. Rev. Entomol.* 37, 533–559.

Rankin, M. A., McAnelly, M. L., and Bodenhamer, J. E. (1986). The oogenesis–flight syndrome revisited. In *Insect Flight: Dispersal and Migration* (W. Danthanarayama, Ed.), pp. 27–48. Springer-Verlag, Berlin.

Roff, D. A., and Fairbairn, D. J. (1991). Wing dimorphisms and the evolution of migratory polymorphisms among Insecta. *Am. Zool.* 31, 243–251.

Wilson, K. (1995). Insect migration in heterogeneous environments. In *Insect Migration: Tracking Resources through Space and Time* (V. A. Drake and A. G. Gatehouse, Eds.), pp. 279–302. Cambridge Univ. Press, Cambridge, UK.

Migration, Reptiles

David W. Owens

Texas A&M University

I. Introduction
II. Nonreproductive Forms of Migration in Reptiles
III. Reproductive Forms of Migration in Reptiles
IV. Physiological Control

GLOSSARY

amniotic vertebrate The Amniota, which are higher vertebrates in which the embryo is enclosed in a fluid-filled extraembryonic membrane that forms a sac and facilitates terrestrial development.

developmental migration The ontogenetic movement of animals from one foraging area to another as animal size and feeding preferences change. This is often a one-way migration.

hibernaculum The winter dwelling place of an animal during hibernation.

migration Regular movement of animals to and from feeding, basking, or reproductive and nesting areas.

poikilothermous Describing animals that have a body temperature which varies with the environment, such as the "cold-blooded" reptiles and amphibians.

Reptilian migration adaptations include all of the variations seen in other vertebrate groups. In this review a brief discussion of several of the common "regular movement" patterns seen in Reptilia is included prior to a more complete treatment of some dramatic examples. Migration can, in the most common cases, involve a predictable daily movement of just a few centimeters or, in the greatest extreme, a multiannual trek of thousands of kilometers.

I. INTRODUCTION

There are at least five key selection factors in the life history of reptiles favoring the evolution of migratory patterns to varying degrees in many different taxa. The first, and probably the one best developed in the poikilothermic reptiles, is the need to use behavioral thermoregulation to improve (optimize?) body temperature for daily or annual activities. The second is the regular and often distinctive adaptations which each species uses to find food. The third is the necessity of most reptiles to seek sheltered/protected areas from predators at predictable times of the day or year. The fourth form seen in many reptiles occurs as an animal ages and grows to a much larger size, necessitating a move to a completely different place to find food and shelter. These special migrations have been termed "developmental migrations." The fifth and best known form of reptilian migrations are adaptations related to reproductive processes such as mating and nesting.

II. NONREPRODUCTIVE FORMS OF MIGRATIONS IN REPTILES

A. Thermoregulation

The need to regulate body temperature (thermoregulate) by moving from one microhabitat to another virtually defines the daily activity patterns of many snakes, lizards, crocodilians, and turtles. These may have been the first migrations to have evolved in terrestrial animals. What may not be obvious to most is that, for temperate reptiles in particular, cooling off at night or in the winter months amounts to a tremendous conservation of energy compared to homeotherms that require metabolic heat for enzyme function and movement. Forest and rock-dwelling lizards often shift only a few centimeters back and forth to maintain their preferred body temperature. Many basking reptiles even change their orientation to the sun to make fine scale adjustments in the solar radiation they accumulate (facing the sun reduces input, whereas a perpendicular angle increases solar heating). Without these predictable movements (minimigrations) reptiles would either become too cool for movement, digestion, and reproduction or become too hot, with resultant problems such as protein degeneration. Indeed, these relatively small movements to accumulate the correct amount of thermal input are summed in the animal over the year and must be optimized for each species to permit proper development of gonads for reproduction.

A larger scale thermoregulatory movement pattern is thought to develop in the marine turtles, which can move many hundreds of kilometers each year either north and south or from inshore to offshore in order to locate optimal temperatures. Some crocodilians, which are the most socially oriented extant reptiles, actually compete for basking sites, with the low-ranking individuals prevented from the daily thermoregulatory bouts and thus realizing a slower growth rate.

B. Foraging and Sheltering

Finding food (foraging) and seeking shelter or protection are the dual goals which drive movement of many animals each day. Reptiles exhibit excellent examples of these simple forms of migration. Reptiles may live very close to where they eat, as do many lizards, or make regular forays into a home range to forage. Green sea turtles regularly travel to and repeatedly crop the same small patches of sea grass because young shoots are more digestible and nutritious. The important consideration seems to be how far food is from the protected resting area. Rock lizards and crocodilians often wait near their resting area and eat what happens along. On the other hand, there are numerous examples in which reptiles forage in suboptimal areas and retreat back to a resting (digesting) area. The American crocodile will forage in the coastal ocean for extended periods of up to days and retreat to an estuary to rest, finish a meal, or bask. Loggerhead sea turtles rest at night on shallow reefs (10–50 m) or even oil platforms but move off for foraging in much deeper water (100+ m) during the day. For example, a 150-kg+ subadult loggerhead male with a satellite transmitter on its back has been doing this continuously for 2 years using the west bank of the Flower Gardens National Marine Sanctuary as home (Fig. 1).

One of the most dramatic seasonal migrations in terms of numbers of individuals occurs in some lizards and snakes which come together in the fall as temperatures drop to form hibernating aggregations for the winter. In the case of the Canadian red-sided garter snake (*Thamnophis sirtalis parietalis*) in Manitoba, there are reported to be as many as 10,000 individuals in one overwintering den or hibernaculum. A reproductive purpose for the aggregation has also evolved since mating is often seen in the spring as the snakes emerge from the hibernaculum. Rattlesnakes are also well-known for their aggregating use of hibernacula in winter. It is commonly believed that lizards and snakes use chemical trails deposited on the ground as pheromones when they move along to the dens. Individuals then follow each other to regularly used sites in this unusual example of social behavior. Unfortunately for the snakes, these sites have sometimes been exploited by the notorious "rattlesnake roundups" in which hundreds of individuals may be collected and killed as a form of sport.

C. Developmental Migrations

Developmental migrations appear to be important in animals which are long-lived and precocious as young. Turtles and alligators, among others, show this life history characteristic. Green sea turtle hatchlings leave the natal beach and head to the open sea as rapidly as possible to begin what is often called the "lost year." During this first of a series of developmental migrations the small turtle rests and hides in flotsam and *Sargassum* mats in the open sea, foraging on animal and plant matter until it is about 20–30 cm or more in carapace length (the age is still not known). The next developmental migration occurs when the turtles become completely herbivorous and seek a shallow coral reef or rocky substrate of some form where there are plants. The third developmental movement for the green sea turtle occurs after many years, when the turtle reaches 35–50 cm carapace length and moves into a less exposed shallow habitat with large volumes of algae or marine flowering plants such as turtle grass. Some individuals seem to make a fourth developmental migration to a final foraging ground which they use as a base for the rest of their lives. Details for the various developmental migrations for sea turtles are not very well-known and this green turtle example may be quite variable. For example, Japanese investigators have found that the loggerhead *Caretta caretta* may use the entire oceanic Pacific in their pelagic developmental habitat, with juveniles spread out all the way from Japan to the west coast of the Americas, eventually returning as adults to southern Japan to nest. Although the track of this migration is not known, it would have to be at least 6000 km.

III. REPRODUCTIVE FORMS OF MIGRATION IN REPTILES

A. The Sea Turtles

Although fishermen and human beach dwellers have long told stories of sea turtle migrations, it was not until Dr. Archie Carr and several students began their pioneering work on the green sea turtles of the Caribbean and Ascension Island that the magnitude of these migrations became obvious to the scientific community. What the Carr group and others found was that while there is some variation among the species, all sea turtles are migratory as the females and the males move away from established resident feeding grounds and routinely travel up to 1000 km, and some travel up to 2600 km, to locate specific beaches. It was also found that sea turtles are very site specific and have high site fidelity, returning to the same beach migration after migration. The females of most species and individuals migrate to their nesting beaches on multiannual intervals, with 3–5 nonreproductive years in between nesting bouts being common. The genus *Lepidochelys*, the ridleys, has proven to be the exception, however, in that in both species there are some individuals which migrate annually.

By using molecular markers, several labs have recently shown that at the time of reproduction sea turtles actually return to the same beach where they initially developed and hatched. This strong evidence of natal beach homing supports the imprinting hypothesis as suggested by Carr. One reason it has been so difficult to document imprinting is that it is now know that it takes from 15 to 50 years for a sea turtle to reach sexual maturity and thus make the return migration. This may be the most extreme example of long-term memory in nature. Turtles have a very small brain, but their memory is quite good! How do they find that original beach?

Dr. Carr thought that sea turtles might use a keen olfactory system and key on the distinctive smells of each beach to make their way "home" as has been shown in salmon. Limited laboratory experimentation supports this as a possible component of a multiple cueing system. Recently, Lohmann and colleagues conducted several clever experiments supporting the possibility that sea turtles may have the capability to use the earth's magnetic fields and magnetic inclination angles to develop an internal "magnetic map" as a way of orienting on a global scale. Even though sea turtles are not thought to have the visual acuity appropriate for celestial (star) navigation, they could use a sun compass system to assist with general headings and may even use their low-frequency hearing to identify distinctive beach or wave sounds. Taken together, sea turtles, like birds, appear to have a complex and overlapping multiple navigation cueing system that is used to find their natal beach.

Work with radio/satellite tracking systems has provided a better appreciation for the migration pathways of sea turtles in recent years (Figs. 1 and 2). Some species, such as the critically endangered Kemp's ridley (*Lepidochelys kempi*), migrate fairly close to the coast as they make their way from feeding grounds along Louisiana and Texas or Yucatan to the far western Gulf of Mexico, where they nest in Tamaulipas, Mexico. The closely related olive ridley (*L. olivacea*), however, migrates out across the open ocean in the eastern tropical Pacific and is highly pelagic in its movements and living style (Fig. 3), as

FIGURE 1 A satellite transmitter being attached to a large subadult loggerhead male on board a ship. The hard shells of sea turtles have proven ideal as a substrate for attaching transmitters. Sea turtles, in turn, have proven ideal platforms for the study of oceanic migration in marine animals.

FIGURE 2 An adult female that has just completed nesting and is preparing to return to the ocean. She has been fitted with a satellite transmitter. Behind her, several other females are nesting in the mass behavior known as an "arribada."

is also the case for the giant leatherback. These ridley species are also unique among the sea turtles in that they nest in mass aggregations known as arribadas in which thousands of individual females ascend the beach to nest in an apparently synchronized manner. Green sea turtles always forage in nearshore areas as adults but do migrate across open sea areas, as demonstrated by their well-documented migration from Brazil to Ascension Island in the mid-Atlantic.

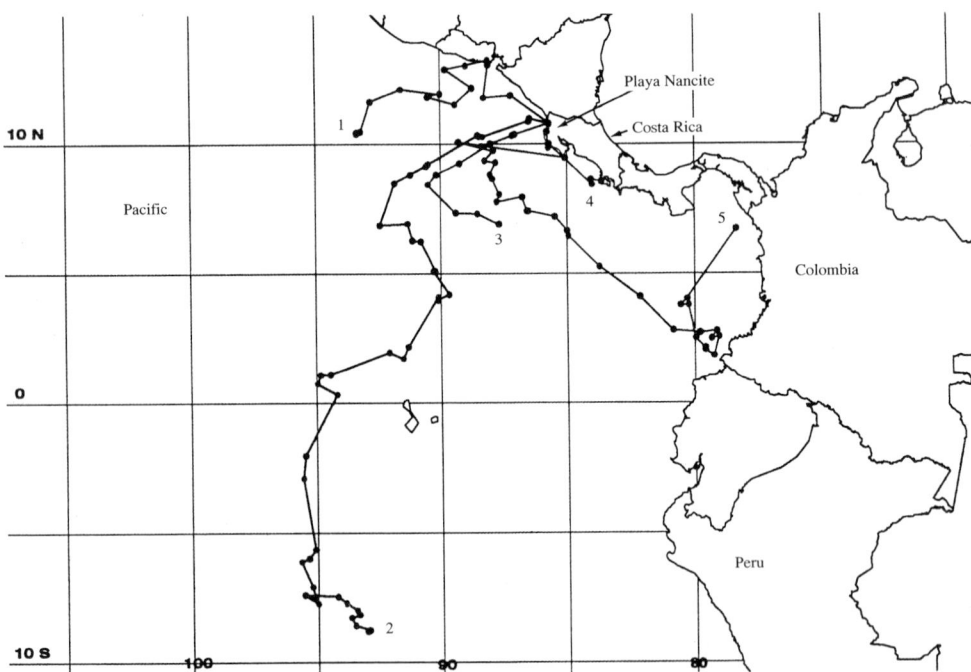

FIGURE 3 Satellite transmitters were attached to the shells of five nesting females at Playa Nancite nesting beach in northwestern Costa Rica. These postnesting migratory tracks show how the olive ridleys typically move into the open ocean to forage. Note also that the turtles show no inclination toward any form of social behavior in their migrations (reproduced with permission from Plotkin et al. (1995) Mar. Biol. 122, 137–143).

B. Other Reproductive Migrants

A few terrestrial reptiles are also known to make reproductive migrations. In each case it appears that the females are moving to an area which will improve developmental and hatching prospects for the offspring. In the Galapagos Islands, for example, some of the subspecies of the giant tortoise (*Geochelone nigra*) are known to migrate from foraging areas in the highlands down to the coastal lowlands to nest. The females travel well-worn tortoise trails to nest in sunny areas known as "campos" which have little vegetation and a silty soil. There are also two species of lizards which show short but interesting migrations. The young adults of the Australian ornate crevice dragon (*Ctenophorus ornatus*) migrate to larger granite outcrops for breeding purposes where they displace the young from the previous year in a repeating cycle of displacement and return migration. There is also a well-documented Panamanian population of *Iguana iguana* in which the adults undertake a short swim to a small island where a more protected breeding site can be found.

Among the Crocodilians the American alligator shows an interesting nesting behavior in which the female, after mating, often moves to a smaller pond away from the main population of adults. She builds a large nesting structure on land, deposits the eggs in the nest, and protects the area until the hatchlings emerge. It is thought that she does this to keep the hatchlings away from other adults (males?) that are known to eat the young alligators.

IV. PHYSIOLOGICAL CONTROL

Very little is known about the physiological control of migration in reptiles. We have hypothesized that testosterone in both the male and female may drive the initiation of the reproductive migration in sea turtles. Female loggerheads (*C. caretta*) monitored with radio tags and repeatedly recaptured over several months at Heron Island on the Great Barrier Reef of Australia showed a rise in testosterone as they left the foraging grounds for their nesting beaches. Olive ridleys, arriving from an open-ocean migration, similarly show elevated testosterone levels in the blood (Table 1) which drop to very low levels by the time of the final nesting for the season. In the several weeks prior to the migration, estrogen is elevated as the ovaries develop and testosterone is relatively low. Thus, the dramatic shift in these two steroids is most interesting and implicated in the behavioral changes required for migration. Males show an even earlier rise in testosterone and are well-known for departing to the mating areas prior to the females. In both sexes the testosterone levels in the blood drop gradually over the reproductive season until they are at much lower relative levels during the return migrations. Even though females often lay 2–10 clutches at about 2-week intervals, it appears that mating only occurs during a discrete receptive period prior to the onset of nesting. Once again, the males appear to begin the return trip first since males are rarely seen after the middle of the nesting season.

TABLE 1

Testosterone Levels in Sea Turtles in the Middle and at the End of Migration Season[a]

Turtle	Capture at First Nesting		Recapture at Last Nesting	
	Date	Testosterone (pg/ml)	Date	Testosterone (pg/ml)
1	Sept. 18/19	192.8	Nov. 24	16.7
2	Sept. 18/19	102.9	Not recaptured	
3	Sept. 18/19	187.6	Nov. 24	3.2
4	Sept. 18/19	205.8	Nov. 24	5.2
5	Sept. 18/19	224.3	Nov. 24	6.1
6	Sept. 18/19	299.6	Nov. 24	11.7

[a] Female olive ridley sea turtles from Costa Rica as shown in Fig. 3. A striking drop in testosterone levels is seen at the end of the nesting/migration season. Turtles returned to the deep ocean pelagic feeding grounds soon after the last nest (reproduced with permission from Plotkin *et al.* (1995) *Mar. Biol.* **122**, 137–143.

Acknowledgments

I thank Heather Kalb and Patricia Vargas for commenting on a draft of this article.

See Also the Following Articles

MIGRATION, AMPHIBIANS; MIGRATION, FISH; REPTILIAN REPRODUCTION, OVERVIEW

Bibliography

Bjorndal, K. (1995). *Biology and Conservation of Sea Turtles*, Rev. ed. Smithsonian Institution Press, Washington, DC.

Bowen, B., and Karl, S. (1997). Population genetics, phylogenetics, and molecular evolution. In *The Biology of Sea Turtles* (P. Lutz and J. Musick, Eds.), pp. 29–50. CRC Press, Boca Raton, FL.

Carr, A. (1967). *So Excellent a Fishe: A Natural History of Sea Turtles*. Schribner, New York.

Gans, C., and Crews, D. (1992). *Biology of the Reptilia: Volume 18, Physiology E : Hormones, Brain and Behavior*. Univ. of Chicago Press, Chicago.

Gauthreaux, S. A., Jr. (Ed.) (1980). *Animal Migration, Orientation, and Navigation*. Academic Press, New York.

Lohmann, K. J., Witherington, B., Lohmann, C., and Salmon, M. (1997). Orientation, navigation, and natal beach homing in sea turtles. In *The Biology of Sea Turtles* (P. Lutz and J. Musick, Eds.), pp. 107–136. CRC Press, Boca Raton, FL.

Miller, J. D. (1997). Reproduction in sea turtles. In *The Biology of Sea Turtles* (P. Lutz and J. Musick, Eds.), pp. 51–82. CRC Press, Boca Raton, FL.

Musick, J. A., and Limpus, C. J. (1997). Habitat utilization and migration in juvenile sea turtles. In *The Biology of Sea Turtles* (P. Lutz and J. Musick, Eds.), pp. 137–164. CRC Press, Boca Raton, FL.

Orr, R. T. (1970). *Animals in Migration*. Macmillan, London.

Owens, D. W. (1997). Hormones in the life history of sea turtles. In *The Biology of Sea Turtles* (P. Lutz and J. Musick, Eds.), pp. 315–341. CRC Press, Boca Raton, FL.

Pritchard, P. C. H. (1979). *Encyclopedia of Turtles*. T. F. H. Publications, Neptune, NJ.

Spotilla, J. R., O'Connor, M. P., and Paladino, F. V. (1997). Thermal biology. In *The Biology of Sea Turtles* (P. Lutz and J. Musick, Eds.), pp. 297–314. CRC Press, Boca Raton, FL.

Milk, Composition and Synthesis

Harold M. Farrell, Jr.
U.S. Department of Agriculture

I. Introduction
II. Synthesis and Secretion of Milk
III. Composition

GLOSSARY

casein micelle A white, colloidal aggregate that is found in milk and that is composed of several proteins together with calcium and phosphorus.

fat globule membrane The membrane that surrounds fat droplets of milk as they approach the apex of the plasma membrane of a milk-secreting cell.

Golgi apparatus A complex cytoplasmic organelle consisting of a series of layered fluid-containing sacs and associated small vesicles; involved in the delivery of cellular products to the cell surface or an intercellular destination.

secretory epithelial cell One of the system of mammary cells involved in the secretion of milk.

The common attribute that defines the vertebrate class mammalia is the production of milk, as the primary nutrient for the neonate, by mammary tissue. While the morphology and physiology of the mammary gland varies considerably from species to species, at the ultrastructural level the mammary epithelial cells have a common cellular motif. This cellular motif is quite adaptable, and by the regulation of its elements, each species can respond to a variety of nutritional circumstances and efficiently produce a milk with a composition suited to the requirements of its neonate.

I. INTRODUCTION

The virtual image of milk which would be constructed by most people is that of a creamy white fluid. The lubricity and taste of milk are related to this perception and are based on three unique biological structures: the colloidal calcium–protein complexes (the casein micelles), the milk fat globules with their limiting membrane, and the milk sugar, lactose. The complexity of these structures is necessitated by the fact that milk is in essence predominantly water. It is the accommodation of these ingredients to an aqueous environment that forms the basis for the structure of milk at the molecular level and calls for the unique secretory process of milk synthesis.

II. SYNTHESIS AND SECRETION OF MILK

A. Cell Physiology

The evolution of the mammary gland, presumably from external sweat glands, has yielded a great variety of exterior appearances in many species, but at the tissue level there is a common organizational theme as shown in Fig. 1a. Mammary secretory cells are epithelial in nature and are arranged in alveoli which are connected to ductal tissue. The secretory epithelial cells (SECs) are surrounded by a layer of myoepithelial cells, which are able to contract and expel milk into the ducts in response to the hormone oxytocin. The alveoli are highly vascularized to ensure a constant flow of the metabolic precursors needed for milk synthesis and secretion. Finally, the vascularized alveoli are embedded in an extracellular matrix. This matrix not only supports the cells but also through cell–cell interactions is responsible for the full expression of the genes that control milk synthesis.

B. Protein Synthesis and Secretion

Adaptation of milk components to their ultimate aqueous environment begins during secretion. Lipid and protein synthesis are partitioned from the start. Amino acids and their metabolic precursors are actively transported into the SECs and assembled into proteins on the ribosomes of the highly developed

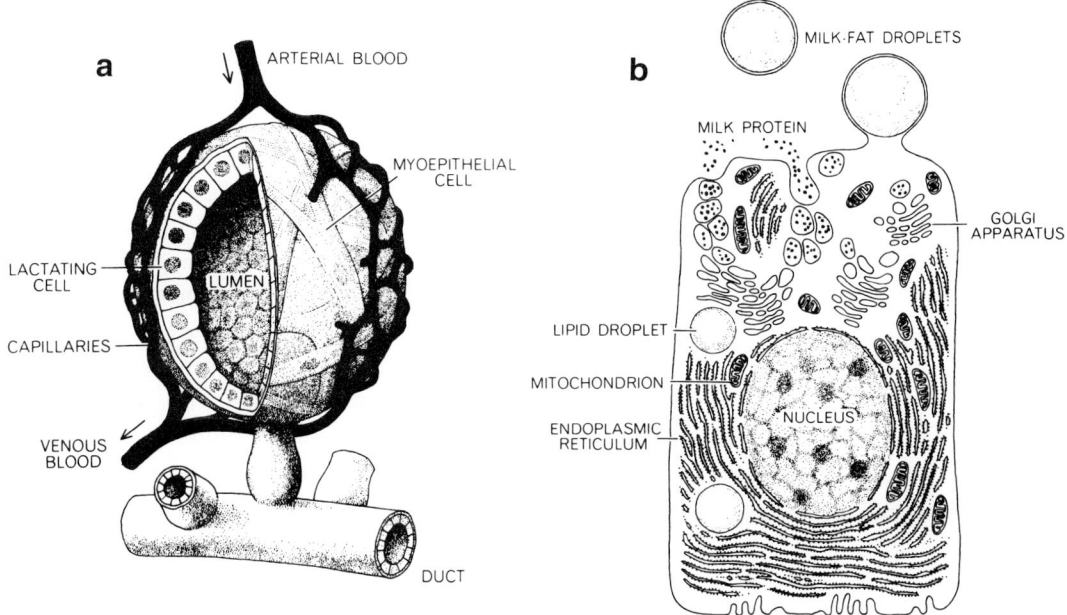

FIGURE 1 Cell physiology of lactating mammary gland. (a) A single alveolus consisting of lactating epithelial cells (SECs) surrounding the lumen. (b) A typical lactating cell indicating active secretion of protein and lipid by distinct mechanisms (reprinted with permission from S. Patton, Sci. Am., July 1969).

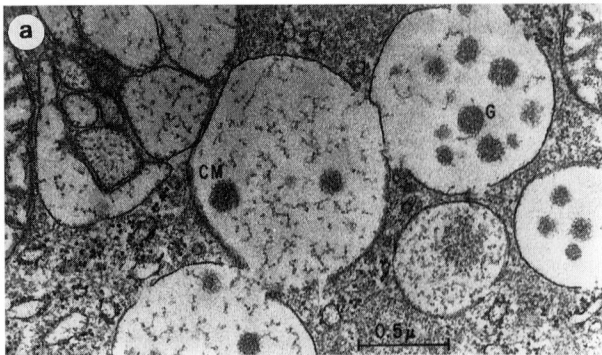

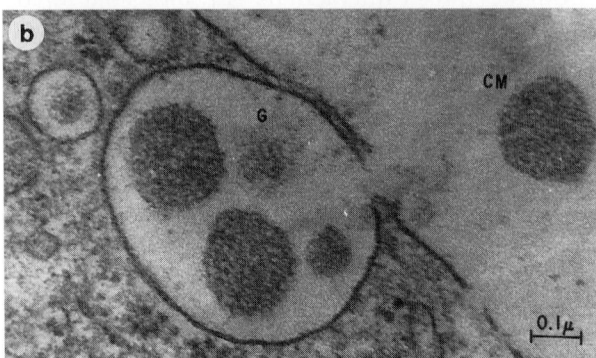

FIGURE 2 (a) Formation of casein micelles (CM) within Golgi vacuoles (G) and the aggregation of small submicellar particles into larger micelles. (b) A Golgi vacuole about to discharge its contents into the alveolar lumen; a CM is already present in the lumen (reproduced with permission from Farrell, 1988).

rough endoplasmic reticulum. All proteins of mammary origin have conserved leader sequences which cause insertion of the nascent proteins into the lumen of the endoplasmic reticulum (ER) shown in Fig. 1b. The proteins are then transported through the Golgi apparatus as shown in Fig. 1b; presumably the globular proteins of milk are folded during this period. In the Golgi apparatus, the caseins, which are the major milk proteins in most species, are phosphorylated to begin the process of calcium transport. In general, when milks that contain >2% protein are analyzed, the accompanying inorganic phosphate and calcium levels found yield insoluble precipitates (apatite or brushite). Conversely, in the absence of these salts, the casein components, as a result of their open structures, have a high viscosity. Thus, the gradual intercalation of calcium, casein, and phosphate into colloidal casein micelles ensures the effective transport of these vital minerals. This process can be visualized in Fig. 2a, in which small submicellar particles are seen in the secretory vacuoles nearest the *trans*-Golgi. Through the binding of calcium, which is actively transported by a Ca^{2+}-ATPase, and the accretion of phosphate, which is a hydrolysis product of nucleotide diphosphates, the colloidal casein micelles are formed and finally secreted by reverse pinocytosis (Fig. 2b).

C. Lactose Synthesis and Secretion

Another important biochemical event occurs in the Golgi apparatus: the synthesis of lactose from galactose and glucose. The Golgi membranes allow free transport of glucose but not disaccharides. The accumulation of lactose in the Golgi vesicles ensures the delivery of concentrated carbohydrate to the neonate. Interestingly, the water solubility of α-lactose at 30°C is 9.8%; primate milks at 7% lactose approach this limit. Lactose is found only in milks. This singularity is derived from the fact that the Golgi apparatus universally glycosylates proteins via the enzyme galactosyl transferase, but in the SEC the globular protein α-lactalbumin is inserted into the lumen of the ER and transported to the secretory vacuoles. In the secretory vesicles, the protein binds to galactosyltransferase and induces a modification of the active site to allow glucose to become the acceptor for galactose. Here a constitutive enzyme of the membranes has been subverted to carbohydrate synthesis. Lactose, α-lactalbumin, and the casein micelles are cosecreted in the Golgi vacuoles. It has been postulated that lactose acts as the primary osmoregulator in the secretion of the aqueous phase of milk (skim milk). Indeed, a "knockout" gene experiment on α-lactalbumin in mice virtually stopped milk synthesis. In marine mammals the lactose content is characteristically low, but the total protein contents are high, approaching 10% in some species. In many species (but not rodents and humans) β-lactoglobulin is the major globular protein cosecreted with lactose and casein. Currently, no incontrovertible evidence has been presented for a biological role for this globular protein; however, it has been postulated to be related to the calcium–phosphate balance necessary for casein micelle formation.

D. Lipid Synthesis and Secretion

At the onset of lactation there is an apparent increase in the lipoprotein lipase activity in the capillaries surrounding the SEC and a concomitant decrease of the enzyme in adipose tissue. The active lipoprotein lipase assures the SEC of a steady supply of dietary lipid for milk synthesis. Triacylglycerols constitute >98% of the milk fat of all species studied to date. Fatty acids with carbon numbers ≥16 are derived from the diet, whereas fatty acids with carbon numbers ≤16 are synthesized by the SEC. Dietary constraints and synthetic capability of the SEC determine the ratio of *de novo* synthesized to dietary fatty acid. Variations in the amounts and types of thioesterases present are responsible for the flavor notes created by the mixture of short-chain fatty acids common to some species. The hydrophobic lipid droplets form near the ER and are transported to the apical plasma membrane for secretion. As they traverse the cytosol, the growing lipid droplets, in accommodation to their aqueous environment, are encapsulated by a lipid monolayer that stains with osmium and contains enzymes and proteins that are cosecreted with milk lipids.

The secretion of protein, lactose, and salts in water (skim milk) appears to generate a surplus of apical plasma membrane. This surplus membrane either directly or indirectly gives rise to the milk fat globule membrane (FGM) that surrounds the fat droplets as they approach the apical plasma membrane as shown in Fig. 1b. Since the FGM is derived from SEC membrane, it shares many common properties with such membranes in that it contains phospholipids, cholesterol, and membrane-like proteins comprising a typical "unit membrane." However, the FGM has attached to it an inner coat that is thought to be proteinaceous and to bridge the gap between the inner phospholipids and the core triglycerides of the fat globules. The origin of this layer may be twofold: a portion arising from an inner coat of the plasma membrane and a portion arising from the interfacial layer that surrounds the nascent lipid globules. Since

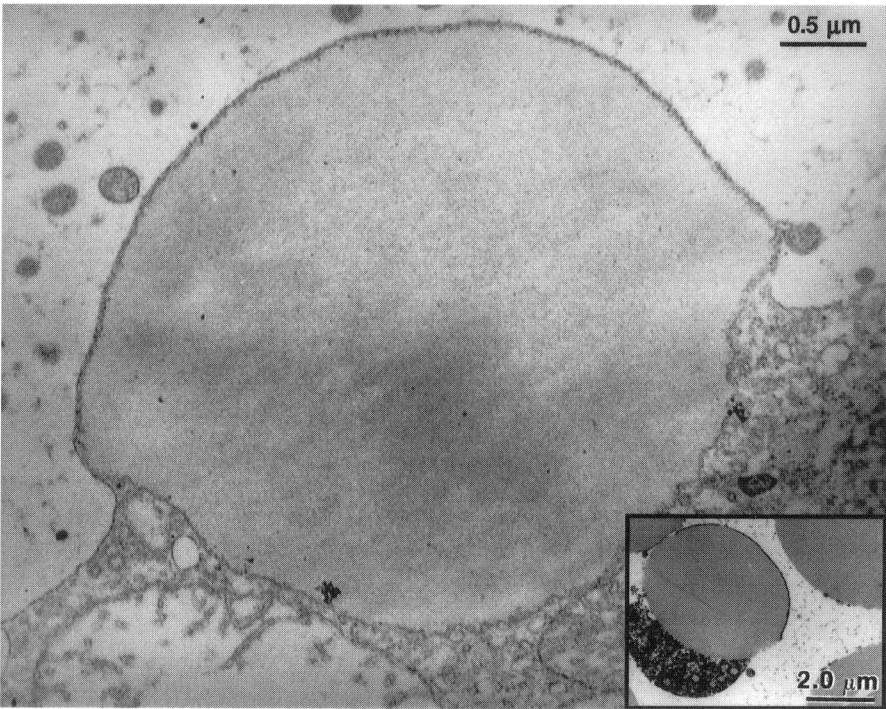

FIGURE 3 A fat globule passing through the plasma membrane of a secretory epithelial cell. The edge of the membrane shows the origin of the fat globule membrane (reprinted with permission from H. M. Farrell and M. P. Thompson, *J. Dairy Sci.* 54, 1219–1228, 1971). (Inset) Mature fat globule of human milk demonstrating the occurrence of "signets" of cytoplasm (reprinted with permission from R. J. Carroll *et al.*, *Food Microstruct.* 4, 323–331, 1985).

the lactating mammary gland contains few active lysozomes, it has been theorized that there is little membrane recycling and that the surplus membrane generated by skim milk secretion is removed as FGM.

In the strictest sense, the secretion of milk can be defined as eccrine, in that after a cycle of protein and fat secretion the cells remain unchanged. In addition, well-developed tight junctions in the alveoli prevent direct leakage of serum components. A type of limited apocrine secretion occurs as well in that parts of mammary cells are often attached to the FGM. When fat globules are pinched off as shown in Fig. 3 where the plasma membrane clearly envelops the globule, pieces of cellular material can become entrapped with the globule. An example of such a cytoplasmic "signet" is shown for human milk in Fig. 3 (inset). The degree of signet formation varies considerably within and across species, with marsupial milks being highest.

III. COMPOSITION

A. Gross Composition

The composition of milk varies widely across species, with stage of lactation, and in response to diet. For the sake of comparison, the compositions of goat, cow, and human milk will be presented since they are well studied and of nutritional and commercial importance to most readers. The total solids contents of the three milks are approximately the same (Table 1). The major differences between the two ruminant milks and human milk arise from changes in the protein and lactose contents, which generally are negatively correlated. The major difference in protein content arises from the decidedly lower casein content of human milk. Indeed, the caseins are not the major proteins of human milk. The lower casein content is also reflected in the lower ash content because casein and calcium + phosphorus are positively correlated.

The total caloric values of the three are approximately equivalent (Table 2). Thus, for an adult, about 1 liter of unprocessed milk would supply much of a person's protein requirement and a sufficiency of calcium (depending on the excess chosen with respect to gender). Commercial milks are always fat adjusted to 3–3.5% fat depending on state requirements, and reduced fat products are also available. Thus, the caloric content of milk can be reduced while maintaining the calcium level; the majority of this mineral is associated with the caseins in the skim milk phase (Table 2). For an infant, about 1 liter of human milk (which is about the average daily yield of a nursing mother) supplies a good balance of protein, calcium, and calories. Both cow and goat milks are superabundant in nutritional content for infants and, accordingly, properly balanced infant formulas are constructed to take these concepts into account. It is important to note that the data given in Tables 1 and 2 are composite averages, and that individuals will differ from day to day.

B. Protein Composition

The dominant feature of skim milk is the casein micelle (Fig. 4a). This unique supramolecular aggregate imparts the opalescence characteristic of skim milk. As noted previously, the chief function of the

TABLE 1
Weight Percentage Composition of Goats, Cow, and Human Milk[a]

	Total solids	Fat	Casein	Whey protein	Lactose	Ash	N
Goat	13.2	4.5	2.5	0.4	4.1	0.8	2662
Cow	12.7	3.7	2.8	0.6	4.8	0.7	comp.
Human	12.4	3.8	0.4	0.6	7.0	0.2	comp.

Note. N, number or compilation (comp.).
[a] From Jenness and Sloan, *Dairy Sci. Abstr.* 32, 599 (1970).

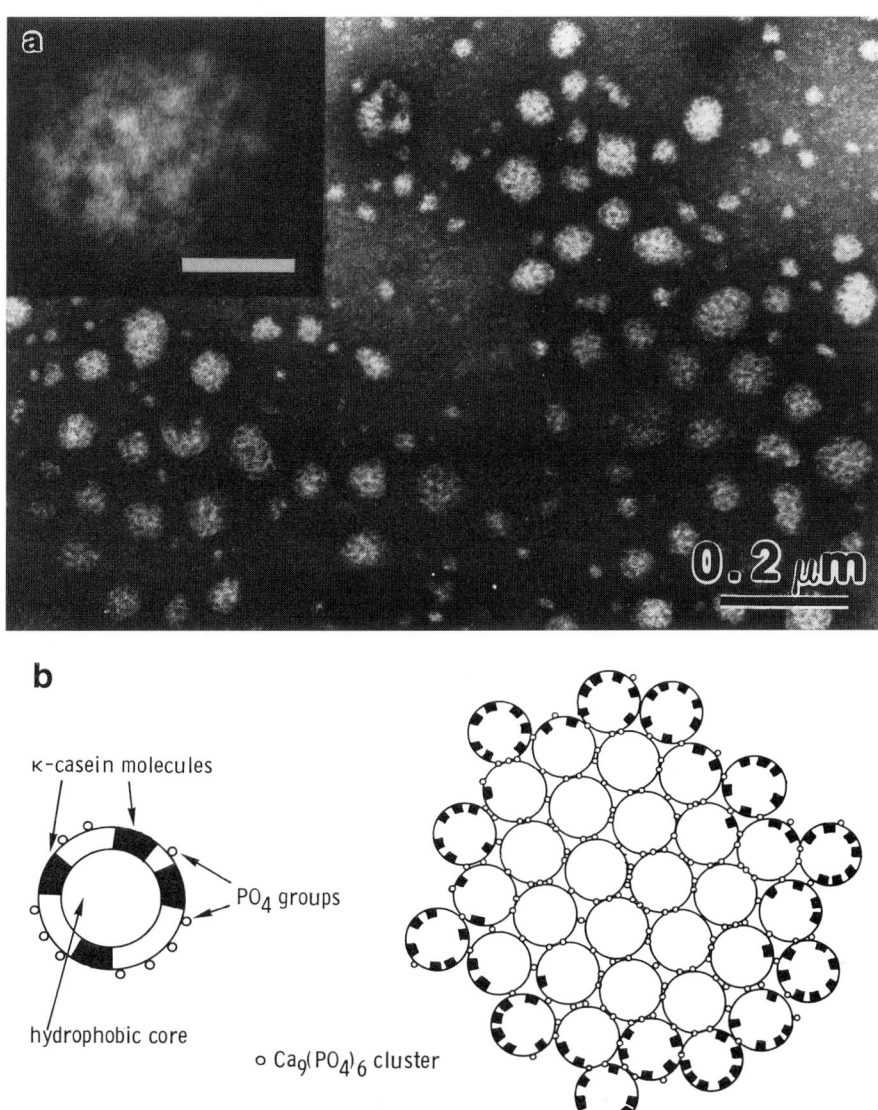

FIGURE 4 (a) Human casein micelles showing a wide range of sizes. (Inset) An enlarged micelle with well-defined submicellar structures (white particles). Scale bar = 30 nm (reprinted with permission from R. J. Carroll *et al.*, *Food Microstruct.* 4, 323–331, 1985). (b) Model for casein micelle structure (bovine) showing submicellar structure and surface arrangement of κ-casein [reprinted with permission from D. G. Schmidt, In *Developments in Dairy Chemistry-1*, *Proteins* (P. F. Fox, Ed.), Applied Science, London, 1982].

micelle is to fluidize the protein and solubilize the calcium and phosphate. From research on the characterization of the caseins of cows' milk, four major casein components are recognized: α_{s1}-, α_{s2}-, β-, and κ-casein. Caseins studied by protein or gene sequencing have been found to be homologous to these proteins in all species examined to date. The α_{s1}-, α_{s2}-, and β-casein are precipitated by calcium at the concentrations found in most milks. However, κ-casein is not only soluble in calcium but also interacts with and stabilizes the other calcium caseinates to initiate formation of the stable colloidal state. The casein micelle is thought to be composed of spherical aggregates of the individual caseins (submicelles) that are

TABLE 2
Nutritional Values of Milk Compared to Recommended Dietary Allowances

Dietary component	RDA		Supplied by 1 liter of milk		
	Adult[a]	Infant[b]	Goat	Cow	Human
Energy (calories)	2700 (M)	650	678	650	648
	1925 (F)				
Protein (g)	65 (M)	13	29	34	10
	55 (F)				
Calcium (g)	0.8	0.4	1.38	1.44	0.32
Phosphorus (g)	0.8	0.3	.72	.72	0.16

[a] 80-kg male >25 years; 60-kg female >25 years.
[b] 6-kg infant at 6 months.

held together by calcium–phosphate linkages. κ-Casein is thought to predominate on the micellar surface (Fig. 4b). In milk clotting in the stomach, the enzyme chymosin (rennin) specifically cleaves one bond in κ-casein to initiate aggregation of the micelles. At the ultrastructural level, the casein micelles of most species appear similar; however, the proportions of the various caseins vary widely. In goats, there is a high degree of variance in casein proportions among animals, which appears to be genetically controlled (Table 3).

In analogy with blood, the clotting of milk *in vitro* by chymosin leads to the generation of a milk serum. Milk serum can also be generated by ultracentrifugation or dialysis. All three methods yield similar products. Acid precipitation during cheese production yields a different product, richer in calcium, called whey. Milk serum contains some caseins and salts not associated with the micelles as well as lactose and the whey (globular) proteins of milk. The major whey proteins of mammary origin are α-lactalbumin, β-lactoglobulin, and lactoferrin; serum albumin and immunoglobulins are derived from blood by passive and active transport, respectively. In addition, there are a wide variety of milk components related to the biological origins of milk. These components arise from SECs, white blood cells (leukocytes) that move to the mammary gland during infection, and blood components that diffuse passively into milk through damaged tight junctions in the secretory alveoli.

C. Lipid Composition

When whole milk is either centrifuged or allowed to stand quiescently, a cream layer separates from the skim milk. The cream phase contains the fat globules, surrounded by FGM, and an aqueous phase similar to skim milk. The diameters of the fat globules range from 1 to 15 μm, but there is considerable variation across species. In freshly secreted milks, the FGM contains virtually all the polar lipids and cholesterol found in milk ($\sim$1% of total lipid) as well as true membrane-associated proteins. Pooling and processing of milk can lead to changes in lipid distributions. The major mammary-derived membrane protein has been called butyophilin; this protein may be a transmembrane protein and thus is

TABLE 3
Percentage of Various Caseins in Milk

Milk	α_{s1}	α_{s2}	β	κ
Goat[a]	5–17	6–20	50	15
Cow[a]	38	10	40	12
Human[b]	Trace	n	70	27

Note. n, not reported.
[a] From Mora-Gutierrez *et al.*, *J. Dairy Sci.* 76, 3689–3710, 1993.
[b] From Carroll *et al.*, *Food Microstruct.* 4, 323–331, 1985.

TABLE 4
Quantitative Analysis of Phospholipids of Goat, Cow, and Human Milk[a]

Phospholipid	% Phospholipid[a]		
	Goat	Cow	Human
Phosphatidyl ethanolamine	33.3	25.5	31.8
Phosphatidyl serine	6.9	5.8	3.1
Phosphatidyl inositol	5.6	4.4	4.7
Phosphatidyl choline	25.7	27.6	34.5
Sphingomyelin	27.9	31.9	25.2

[a] From W. W. Christie, *Developments in Dairy Chemistry—2 Lipids* (P. F. Fox, Ed.), Applied Science, London, 1983.

found on the luminal side of the FGM as well as in the inner coat. The relative compositions of polar lipids are given in Table 4.

The triacyl glycerides, at >98% of the total, dominate the lipid composition of all species. Differences arise in the individual fatty acids of the triglycerides. These differences are due to dietary, seasonal, and species influences. Typical distributions of the major fatty acids found in goat, cow, and human milk are given in Table 5. For the ruminants, the distribution of fatty acids is highly influenced by the biohydrogenation processes that occur in the rumen and lead to saturation of feed lipids. For monogastric animals such as humans, milk fatty acids are more directly influenced by changes in diet on a day to day basis. The appearance of $C_{18:1}$ in ruminant milk is due to the occurrence of an active Δ-9 desaturase in mammary tissues.

TABLE 5
Comparison of the Fatty Acid Distribution of Milk Trigylcerides[a]

Me ester	Weight (%)		
	Goat	Cow	Human
C_4	2.0	3.3	Trace
C_6	2.5	2.1	Trace
C_8	3.4	1.1	Trace
C_{10}	11.3	3.0	1.6
C_{12}	5.0	2.9	6.9
C_{14}	12.1	9.0	8.5
C_{16}	27.8	24.0	20.9
$C_{18:0}$	7.4	13.2	7.3
$C_{18:1}$	27.6	33.3	39.5
$C_{18:2}$	2.7	3.8	10.1
$C_{18:3}$	Trace	Trace	1.3

[a] From W. W. Christie, *Development in Dairy Chemistry—2 Lipids* (P. F. Fox, Ed.), Applied Science, London, 1983.

See Also the Following Articles

LACTATION, HUMAN; LACTATION, NONHUMAN; LACTOGENESIS; MAMMARY GLAND DEVELOPMENT; MAMMARY GLAND, OVERVIEW; MILK EJECTION

Bibliography

Farrell, H. M., Jr. (1988). Physical equilibria: Proteins. In *Fundamentals in Dairy Chemistry* (N. B. Wong, Ed.), 3rd ed. Van Nostrand–Reinhold, New York.

Farrell, H. M., Jr., and Thompson, M. P. (1988). Caseins as calcium binding proteins. In *Calcium Binding Proteins* (M. P. Thompson, Ed.), Vol 2, CRC Press, Boca Raton, FL.

Fox, P. F. (Ed.) (1983). *Developments in Dairy Chemistry—2 Lipids*. Applied Science, London.

Holsinger, V. H. (1988). Lactose. In *Fundamentals in Dairy Chemistry* (N. B. Wong, Ed.), 3rd ed. Van Nostrand–Reinhold, New York.

Holt, C. (1992). *Adv. Protein Chem.* **43**, 63–151.

Jensen, R. G. (Ed.) (1995). *Handbook of Milk Composition*. Academic Press, New York.

Protoplasma **159**, 75–208 (1990). [This two-volume issue dedicated to Dr. S. Patton contains an excellent summary of novel concepts regarding milk synthesis and secretion.]

Whitney, R. McL. (1988). Milk proteins: Composition. In *Fundamentals of Dairy Chemistry* (N. B. Wong, Ed.), 3rd ed. Van Nostrand–Reinhold, New York.

Milk Ejection

Jonathan B. Wakerley

University of Bristol

I. Introduction
II. Characteristics of Milk Ejection in Different Species
III. Hypothalamic Regulation of Milk Ejection
IV. Transmission of Afferent Stimuli for Triggering Milk Ejection
V. Role of Neurotransmitters and Neuromodulators in the Milk-Ejection Reflex
VI. Disorders of Milk Ejection

GLOSSARY

afferent pathway A neural pathway carrying impulses (usually derived from the nipples) toward the oxytocin-releasing cells of the hypothalamus.

bursting activity A brief episode of high-frequency firing characteristically displayed by oxytocin neurons prior to milk ejection.

hypothalamic oxytocin pulse generator The hypothalamic circuitry (including the oxytocin neurons themselves) responsible for the generation of oxytocin pulses during suckling.

milk transfer The process by which milk is moved from the mammary gland to the buccal cavity of the offspring.

neuromodulators Substances other than conventional fast-acting neurotransmitters that alter the electrical activity of neurons.

nursing The act of feeding the offspring by enabling them to suck the nipples.

sucking The action of holding the nipple in the mouth and applying rhythmical negative pressure.

suckling Strictly, a synonym for nursing. However, it is frequently used as an interchangeable term for sucking.

Milk ejection refers to the active process by which recently synthesized milk held within the mammary gland is actively ejected and thereby made available to the sucking young. This process is also referred to as "milk letdown" or "the draft." It is important to appreciate that milk secretion within the lactating mammary gland occurs as a continuous process, and the newly secreted milk is stored within the mammary alveoli or in specially modified parts of the duct system (Fig. 1). Because of the effects of surface tension, milk within the alveoli cannot be removed solely by sucking and has to be actively ejected by the action of special contractile cells called myoepithelial cells (Fig. 2). The process of milk ejection (galactokinesis) is therefore quite separate from milk secretion (galactopoiesis) and is under separate physiological regulation. Active milk ejection is needed for successful nursing in all mammalian groups, although in ruminants a portion of the milk is stored in a large cistern within the udder, from where it can be passively removed. Milk ejection is controlled by a neuroendocrine mechanism referred to as "the milk-ejection reflex," which is activated in response to suckling. When the offspring suck the nipples, the excitatory stimulus is transmitted to the hypothalamus to cause release of the hormone oxytocin from the posterior lobe of the pituitary. Oxytocin circulates to the mammary gland where it causes contraction of the myoepithelial cells and ejection of the milk. The milk-ejection reflex is often regarded as a "classical" example of a neuroendocrine reflex and continues to provide an invaluable model for understanding the organization of other less accessible neuroendocrine systems. This article covers all aspects of milk ejection but concentrates especially on the afferent control and organization of the hypothalamic oxytocin neurons. These are key areas in understanding operation of the milk-ejection reflex, and still are a focus of much research.

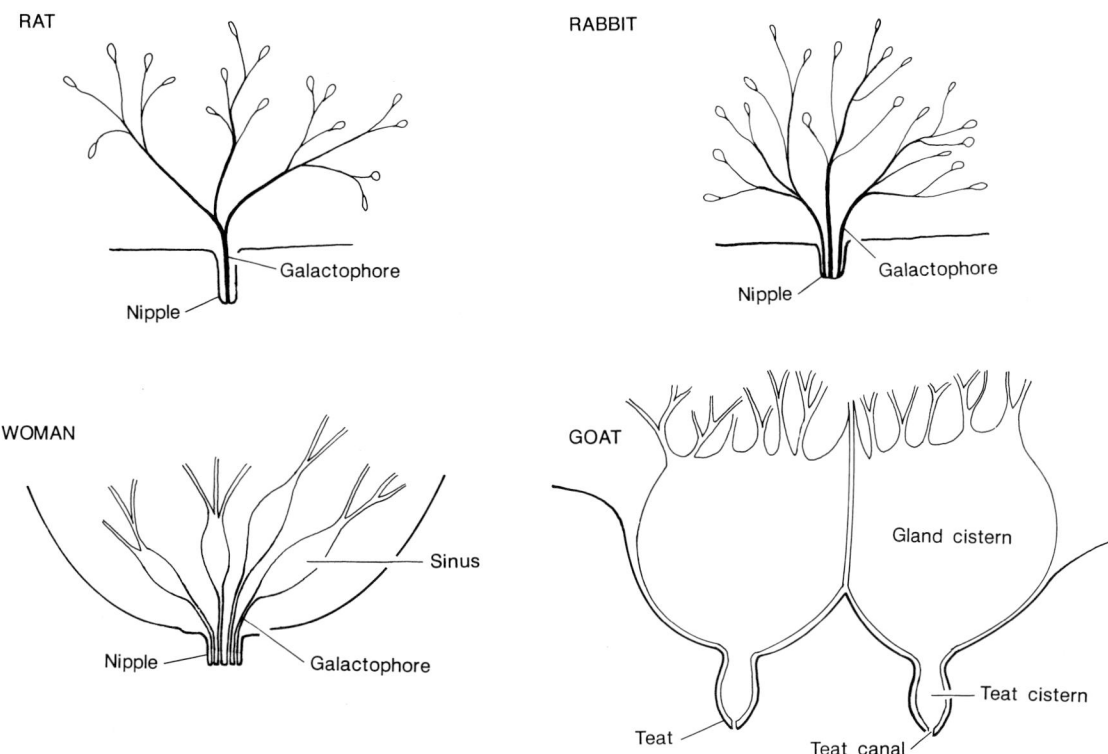

FIGURE 1 Arrangement of milk ducts and specializations for milk storage in various species. In the rat and rabbit there is no special storage space, and the alveoli feed into one or several narrow ducts that open onto the surface of the nipple. In women the ducts draining the alveoli have special dilated regions (sinuses) in which some milk can be stored. The mammary gland of the goat (like other ruminants) has a large gland cistern, connecting with the teat cistern, allowing storage of a considerable volume of milk (reproduced with permission from Wakerley et al., 1994).

I. INTRODUCTION

The phenomenon of milk ejection has been recognized for centuries, and there are many references to this process in art and literature. One well-known example is the painting by Tintorreto titled "The Origin of the Milky Way," which shows milk spurting from the nipples of the Goddess Juno after the infant Hercules is plucked from her breast. However, our understanding of this process really began when Gaines in 1915 reported the effects of posterior pituitary extracts on intramammary pressure in the goat. The concept of the milk-ejection reflex was first put forward by Ely and Petersen in the 1940s, but the neuroendocrine basis and the important role of the hypothalamus was not established until the 1950s with the work of Cross and Harris. At around the same time, organization of the myoepithelial cells surrounding the mammary alveoli was described, and oxytocin was artificially synthesized. Another important landmark was the advent of the technique of radioimmunoassay in the 1970s, which allowed determination of oxytocin in serial blood samples during nursing. This unequivocally established the pulsatile nature of oxytocin release during activation of the milk-ejection reflex and paved the way for studies to elucidate the neural mechanisms underlying generation of oxytocin pulses, work which continues today. Exploration of the milk-ejection reflex has ranged across a number of domestic and laboratory species, as well as the human, but there are remain many mammalian groups in which details of this fascinating mechanism remain completely unknown.

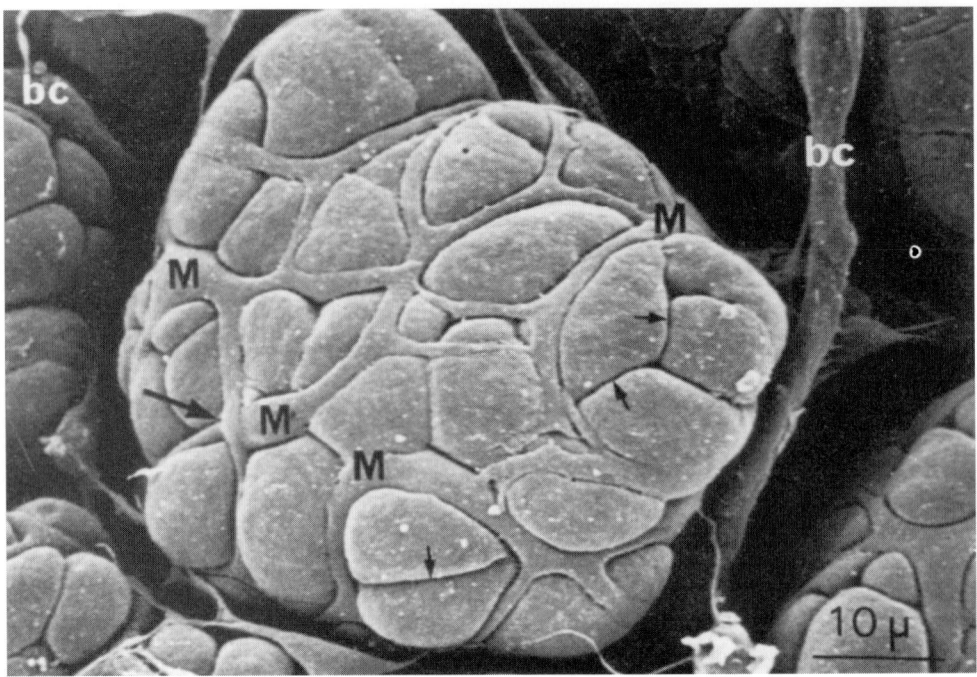

FIGURE 2 Scanning electron micrograph of an alveolus of the rat mammary gland to show the network of myopeithelial cells (each separate cell is marked by M). The large arrow indicates overlap of adjacent myoepithelial cells, and the small arrows indicate the borders of adjacent secretory cells. The arrangement of the myoepithelial cells enables them to squeeze the alveolus and eject the milk (bc, blood capillary) (reproduced with permission from Wakerley et al., 1994).

II. CHARACTERISTICS OF MILK EJECTION IN DIFFERENT SPECIES

A. Variations in Nursing Behavior

The milk-ejection reflex operates within the context of nursing behavior, and this varies considerably according to habitat and lifestyle. Nursing may take place with the mother standing in open grassland, crouching in a nest, perched in a tree, or even swimming in water. Frequency of nursing may be two or three times per hour (e.g., sheep and horses), once per day (rabbits), or once per week (seals). In marsupials, the young make their way to the pouch and become firmly attached to a nipple shortly after birth and remain in this position for several weeks.

Nursing often follows a sequence of stereotyped behavior starting with cleaning the young, which helps to reinforce mutual recognition and stimulates the young to seek the nipples. After attachment to the nipples, there may be a quiescent or latent phase (expressed in the behavior of both mother and offspring) before a period during which one or several milk-ejection responses take place. In some species (rat) milk ejection only occurs when the mother is in a somnolent state, whereas in other species (rabbit) the mother is alert during milk ejection. Certain species (pig) may show characteristic vocalization leading up to milk ejection.

B. Patterns of Milk Ejection and Plasma Oxytocin Profiles

Patterns of milk ejection can be studied by direct recording of intramammary pressure through a cannula inserted into a milk duct. Such recordings have revealed that milk ejection occurs as one or more discrete intrammary pressure responses, which usually last between 0.5 and 2 min. Each response is characterized by an abrupt rise in pressure (10–20 mm Hg) followed by a slower decay phase, perhaps involving one or more secondary oscillations. This basic pattern seems common to most species, although there is considerable variation in the number and frequency of responses that may be observed during a given nursing episode. Pigs and rabbits, for example, show just a single response, whereas rats show a whole series of regularly spaced milk ejections (Fig. 3A).

Measurements of plasma oxytocin, combined with intramammary pressure recordings, show that this discontinuous or episodic pattern of milk ejection arises because of the highly pulsatile nature of oxytocin release (Fig. 3B), coupled with the short half-life (<2 min) of this hormone in blood. Women show a similar pulsatile pattern so that suckling is associated with highly fluctuating levels of oxytocin (Fig. 4). Pulsatile oxytocin release ensures that oxytocin reaches the periphery in a highly concentrated bolus of hormone, providing the most effective signal for evoking a mammary response. It also ensures that contractions occur synchronously in all the mammary glands, and in polytocous species this enables milk-ejection-related behavior to be coordinated within the whole litter. An exception to these general principles is provided by marsupials, such as the Wallaby, in which there is concurrent asynchronous lactation of adjacent mammary glands. The gland to which the youngest offspring is attached is unusually responsive to slowly rising background levels of oxytocin and can show rhythmical contractions quite independently of the less sensitive gland being suckled by the older offspring.

C. Role of Sucking Behavior of the Young in Milk Transfer

Detailed studies of how sucking contributes to milk removal suggest that it may be oversimplistic to ascribe creation of negative pressure within the buccal cavity as being the major factor, although this is certainly essential for maintaining a grasp of the nipple. Actual removal of milk may rely more on rhythmical compression of the milk duct(s) by squeezing the nipple between the tongue and hard palate ("stripping"), somewhat akin to manual milking. As well as sucking, the offspring may provide other stimuli that contribute to milk removal, such as butting (calves) and treading (kittens). Such activity may help to create positive pressure within the mammary gland and may evoke local mechanically induced contractions of the myoepithelium.

In those species in which the milk is stored solely within the alveoli, milk transfer is mainly restricted to periods of active milk ejection. However, in certain species of ruminants, most notably the goat, much of the milk is stored in an udder cistern from which it can be easily removed so that milk transfer during suckling is less dependent on raised intramammary pressure. The human mammary gland has a small

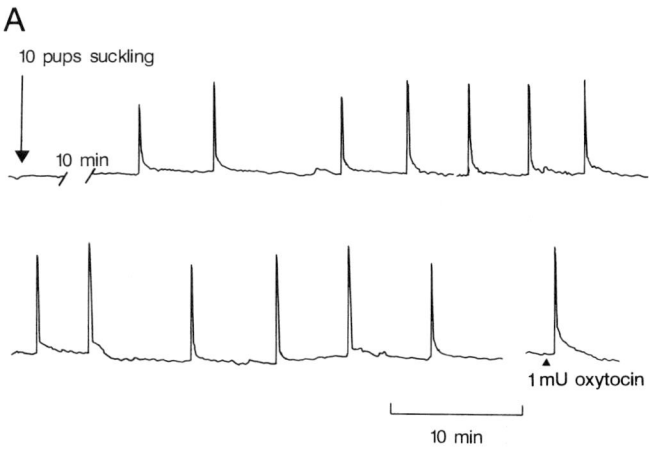

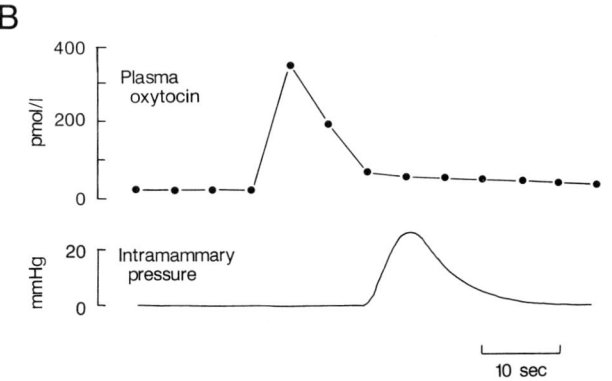

FIGURE 3 (A) Intramammary pressure recordings taken from an anesthetized rat during suckling, showing the intermittent milk-ejection responses. Each reflex response involved a transient pressure rise (approximately 15 mm Hg) and a similar pressure response was seen after the rapid iv injection of 1 mU (2.2 ng) oxytocin. It should be noted that, somewhat unusually, the milk-ejection reflex of the rat is not abolished by anesthesia. (B) Plasma oxytocin (measured in jugular venous blood) shows a transient rise before a milk-ejection response (note the more expanded time scale in this record) confirming release of a pulse of hormone (reproduced with permission from Wakerley *et al.*, 1994).

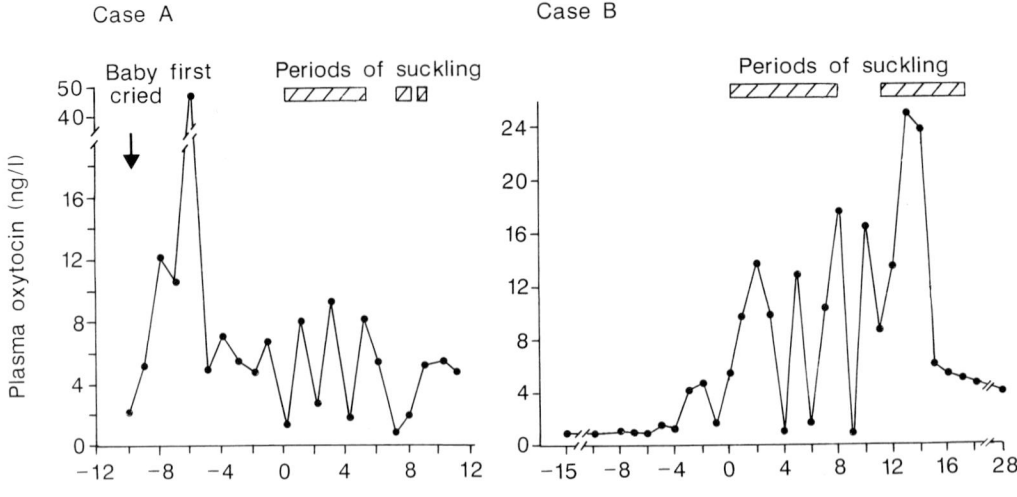

FIGURE 4 Elevated plasma levels of oxytocin in the woman in response to suckling. Note in both subjects hormone levels fluctuated wildly, indicating pulsatile release. In case A, there was a large rise of oxytocin in response to the baby crying prior to being placed on the nipple. Such anticipatory release (a "conditioned" response) is common in women. In case B, oxytocin levels only rose significantly after suckling had commenced (reproduced with permission from Wakerley et al., 1994).

milk sinus at the base of each lactiferous duct, and this may explain why the baby is able to obtain milk as soon as it attaches to the nipple without necessarily requiring a definable milk-ejection response. Women often show release of oxytocin in anticipation of suckling (Fig. 4) and this may help move milk toward the duct sinuses.

Milk ejection may trigger characteristic behavior from the young. Rat pups, for example, show a stretch reaction involving extension of the forelimbs, and this may help milk transfer by pressing on the mammary gland. This reaction is quickly followed by nipple shifting, whereby the pups search for an unoccupied nipple in anticipation that the gland will still be replete.

III. HYPOTHALAMIC REGULATION OF MILK EJECTION

Release of oxytocin from the neurohypophysis is regulated by neuroendocrine oxytocinergic neurons of the hypothalamus and these cells may be considered as the "efferent path" of the milk-ejection reflex.

As noted previously, an intermittent pattern of milk ejection is a feature common to virtually all mammalian species, reflecting the fact that release of oxytocin during suckling is highly pulsatile. This pulsatile pattern is achieved because the oxytocin neurons are able to function as a coordinated unit, sometimes referred to as the "hypothalamic oxytocin pulse generator."

A. Organization of Hypothalamic Oxytocin Neurons

The oxytocinergic neurosecretory neurons responsible for controlling milk ejection are located in two pairs of magnocellular hypothalamic nuclei: (i) the paraventricular nuclei, which lie on either side of the third ventricle, and (ii) the supraoptic nuclei, which are located on the lateral border of the optic chiasm (Fig. 5A). The axons of these neurons project via the hypophysial stalk to terminate in the posterior lobe of the pituitary gland. Within these nuclei, the oxytocin cells are intermingled with vasopressin neurons which, of course, subserve quite separate physiological roles, mainly involving water balance. Mor-

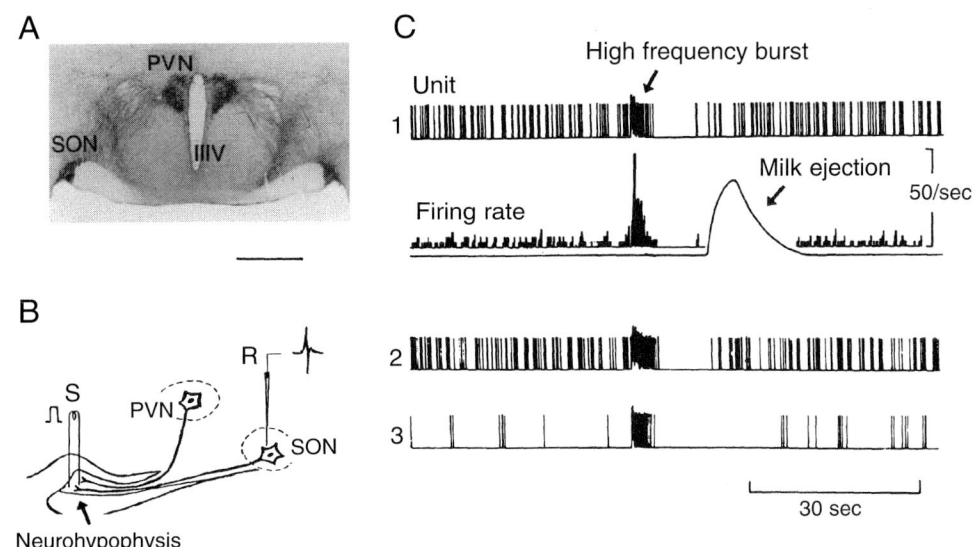

FIGURE 5 (A) Location of oxytocin-immunoreactive neurons in supraoptic nuclei (SON) and paraventricular nuclei (PVN) of lactating rat hypothalamus. Scale bar = 1 mm. (B) Electrophysiological recording arrangement for antidromic identification of magnocellular neurons in SON and PVN that project to the neurohypophysis. Electric shocks are applied to the neurohypophysis through a stimulating electrode (S), and a recording electrode (R) is positioned in one or another of the magnocellular nuclei to record the antidromic action potential, confirming that the neuron under study projects to the neurohypophysis. (C) Pattern of firing of oxytocin neurons in the PVN and SON during activation of the milk-ejection reflex. The traces shown were obtained from three individual oxytocin neurons (numbered 1–3) recorded from different rats during suckling under anesthesia. For neuron 1 is shown the unit trace (corresponding to individual action potentials), a trace of the second-by-second firing rate (see scale given on right), and the intramammary pressure trace (the abrupt rise signals the occurrence of milk ejection). Note the characteristic high-frequency burst of firing in this neuron approx 12 sec before milk ejection. The delay to milk ejection reflects the time required for oxytocin to reach the mammary gland following its release from the neurohypophysis. For neurons 2 and 3 only the unit traces are shown. Note the remarkable similarity in the pattern of firing of these cells over the same corresponding period of a milk-ejection response. This highly stereotypic response occurs synchronously within the whole population of oxytocin neurons and is repeated every 5–10 min during suckling, thereby giving rise to an intermittent pattern of milk ejection (see Fig. 3A) (reproduced with permission from Wakerley and Ingram, 1993).

phologically, oxytocin neurons consist of a soma with one or two simple unbranched dendrites, from which arises a single nonmyelinated axon. They receive numerous axodendritic and axosomatic connections, giving rise to frequent excitatory and inhibitory synaptic potentials. It should be remembered that oxytocin neurons are involved in several other functions so that only some of their synaptic inputs will be directly concerned with controlling milk ejection.

The gene directing synthesis of the oxytocin prohormone is located on chromosome 20 in the human and comprises three exons (regions which encode the amino acid in the precursor) separated by two introns, an arrangement which seems common to other mammalian species. Oxytocin, together with its associated neurophysin molecule, is packaged into granules within the Golgi apparatus and transported to the neurosecretory terminals of the neurohypophysis. The neurohypophysis normally contains enormous reserves of the hormone, but these stores are somewhat depleted during lactation. Oxytocin release involves exocytosis of packaged oxytocin granules, a calcium-dependent process, which is triggered by the rise in calcium following arrival of impulses at the neurosecretory terminals. Although impulse traffic descending from the hypothalamus is the primary determinant of oxytocin release, additional "presynaptic-like" control mechanisms operate within the neurohypophysis.

B. Adaptations of Oxytocin Neurons during Lactation

Toward the end of pregnancy, oxytocin neurons undergo a number of morphological and biochemical adaptations in preparation for the increased functional demands associated with lactation. The cells increase their diameter, there is hypertrophy of the cellular organelles involved in hormone synthesis, and there is an increase in oxytocin mRNA. There is also reorganization of the synaptic input to the oxytocin neurons, with an increase in double synapses, and a retraction of the glia surrounding their soma and dendrites, such that the degree of direct membrane apposition between adjacent oxytocin neurons is increased. The increased number of double synapses and membrane apposition may favor the occurrence of synchronous bursting activity. Changes also take place in the neurohypophysis, where there is an unwrapping of the oxytocin-containing neurosecretory terminals from their glial ensheathments. These peripartum adapatations appear to be triggered by the combined influence of ovarian steroids as well as by oxytocin itself.

C. Electrical Activity of Oxytocin Neurons during Milk Ejection

Most of the information concerning the electrical behavior of oxytocin neurons during suckling is derived from the rat since in this species it is possible to obtain combined electrophysiological and intramammary pressure recordings during reflex milk ejection using an anesthetized preparation. Approximately 50% of paraventricular and supraoptic neurosecretory cells projecting to the neurohypophysis (i.e., cells identified by antidromic identification; Fig. 5B) show activity that is highly correlated with milk ejection and these are presumed to be oxytocin-releasing neurons. Such neurons show no change in firing at the start of suckling, but following a latency of 5–10 min they show intermittent bursts of firing, with each burst comprising a 2- to 4-sec period of high-frequency firing followed by inhibition (Fig. 5C). These bursts occur at intervals of approximately 5 min and are followed within 12–15 sec by a milk-ejection response. The precise timing of the bursts of activity on the oxytocin neurons does not appear to be related to fluctuations in the intensity of the suckling stimulus. Instead, the pattern of intermittent responses depends on the complex nature of the afferent circuitry carrying the input from the nipples, which seems to involve stimulus "gating" as well as a timing mechanism for determining the interval between each burst. More or less identical high-frequency bursting responses can be recorded from both the paraventricular and supraoptic nuclei, and simultaneous double recordings indicate a high degree of synchrony within the whole population of oxytocin neurons.

Milk-ejection responses can be evoked by electrical stimulation of the neurohypophysis using parameters based on the high-frequency bursting responses shown by the oxytocin neurons. However, if these parameters are changed, for example, by using longer trains of stimuli at lower frequency, the stimulation is far less effective at evoking milk-ejection responses. Thus, the pattern of firing shown by the oxytocin neurons is optimally suited to the needs of the system, maximizing frequency facilitation of oxytocin release at the level of the neurohypophysial terminals.

IV. TRANSMISSION OF AFFERENT STIMULI FOR TRIGGERING MILK EJECTION

The principal stimulus for activation of the milk-ejection reflex is derived from the offspring suckling of nipples. Investigation of the route by which the excitatory input reaches the hypothalamus has mainly utilized electrical stimulation and recording, or lesioning techniques. Studies using markers of neuronal activation (e.g., Fos protein induction) have met with limited success.

A. Peripheral Receptors and Spinal Pathways

There are dense intradermal nerve plexuses at both the base and the tip of the nipples so that they constitute one of the most densely innervated regions of the body. Nerve endings in the nipple are mostly of the nonencapsulated type and are responsive to mechanical stimulation and to temperature. There are also a few nerve endings within the mammary gland itself, but these are mostly derived from sympathetic fibers innervating arterioles and have no role in activating the milk-ejection reflex.

The unmyelinated afferent fibers from each nipple pass through the dorsal roots of the appropriate spinal nerves and synapse within the substantia gelatinosa of the dorsal horn over two or three adjacent segmental levels of the spinal cord. Neurons in the dorsal horn respond to pulling and stretching of the nipples and fire rhythmically in a pattern which mirrors the sucking activity of the offspring. In species having more than one pair of mammary glands individual dorsal horn neurons receive inputs from several adjacent nipples, giving summation of the sucking stimulus.

Within the spinal cord, the input ascends within the lateral funiculi without decussating (crossing over), probably by passing up the spinocervical tract to relay in the lateral cervical nucleus, although other sensory pathways may be important. The dorsal vagal complex of the medulla oblongata may function in relaying oxytocin-releasing stimuli and may receive somatosensory inputs via the spinosolitary tract.

B. Pathways from Brain Stem to Hypothalamus

Ascending fibers from the lateral cervical nucleus decussate beneath the central canal and ascend in the ventrolateral medulla, joining the lateral aspect of the medial leminiscus. The noradrenergic A_2 cell group in the medulla may also contribute to excitation of oxytocin neurons during suckling. After traversing the pons, the main afferent pathway passes dorsally to enter the lateral tegmentum of the midbrain via the spinal leminiscus, possibly relaying within the external nucleus of the inferior colliculus. Importance of the lateral tegmentum in the milk-ejection reflex is well supported by lesioning and electrical stimulation studies. Above the midbrain, the sucking stimulus is conducted through the subthalamic peripeduncular region (Fields of Forel and zona incerta) and posterior lateral hypothalamus, with further multiple synaptic relays before reaching the paraventricular and supraoptic nuclei. A summary diagram of the afferent pathway of the milk-ejection reflex is provided in Fig. 6.

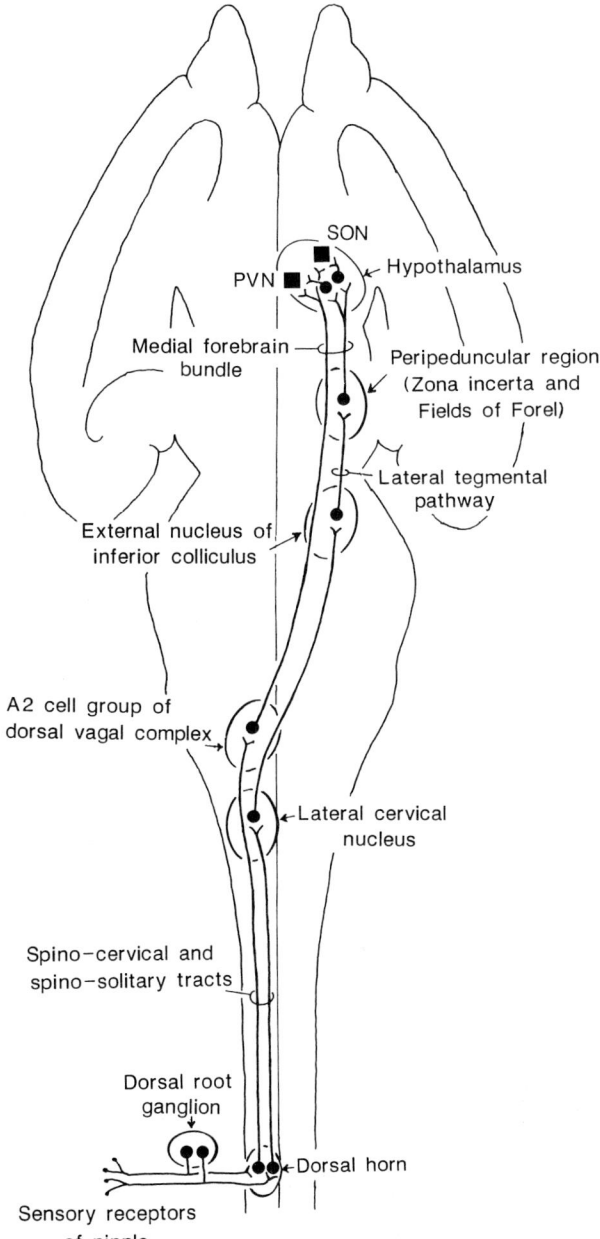

FIGURE 6 Route of the afferent pathway of the milk-ejection reflex based on evidence obtained in the rat. This complex pathway involves spinal tracts that ascend ipsilaterally to the lateral cervical nucleus and to the A_2 noradrenergic cell group of the dorsal vagal complex in the medulla. The A_2 cell group projects directly to the oxytocin neurons of the paraventricular (PVN) and supraoptic nuclei (SON), whereas the pathway through the lateral cervical nucleus involves further synaptic relays within the external nucleus of the inferior colliculus, the peripeduncular region, and the hypothalamus (reproduced with permission Wakerley et al., 1994).

The means by which activation of the two pairs of magnocellular nuclei is synchronized is not yet understood. It is known that after unilateral lesioning of the midbrain lateral tegmentum, bursting responses can still be recorded on both sides of the brain, suggesting that there is a coordinating center within, or near, the hypothalamus. Neurons of the dorsomedial nucleus in the posterior hypothalamus show brief increases in firing in association with suckling-evoked bursting activity of oxytocin neurons, and this structure could have a role in burst coordination. The close membrane apposition and shared synaptic inputs of the oxytocin neurons (see Section III,B) are features which might be expected to further enhance their synchronicity.

C. Descending Modulatory Pathways

The milk-ejection reflex can be modulated by higher cognitive centers in the forebrain through so-called "descending pathways" that mediate conditioning of the reflex. The phenomenon of conditioning, whereby other stimuli normally associated with nursing (or milking in the case of domesticated species) can activate the milk-ejection reflex, is well-known and is particularly evident in the human, in which the sound of the crying baby can cause copious discharge of milk. Higher centers can also bring about suppression of the milk-ejection reflex, for example, during fear or anxiety, so that some descending pathways are inhibitory to oxytocin release.

Many regions of the forebrain are connected to the magnocellular nuclei and may be involved in cognitive modulation of the milk-ejection reflex. Regions in which electrical stimulation evokes milk-ejection responses include the cingulate and pyriform cortex, nucleus accumbens, diagonal band of Broca, bed nucleus of stria terminalis, medial septum, hippocampus, and fimbria of fornix. These structures might be expected to have a facilitatory role in the milk-ejection reflex, and in this regard the bed nucleus is of particular interest because it contains neurons that show cyclical firing correlated with the ongoing pattern of milk ejections. The bed nucleus is also strongly implicated in facilitation of the milk-ejection reflex by central oxytocin (see Section V,A,1). Other forebrain structures have an inhibitory role; for example, stimulation of the lateral septum suppresses the firing of oxytocin neurons and interrupts milk ejection.

D. How Is the Bursting Pattern of Electrical Activity in Oxytocin Neurons Generated?

There is no evidence that the intermittent bursting of oxytocin neurons is regulated by fluctuations in the level of the sucking stimulus, and indeed a similar pattern of firing occurs during unpatterned electrical stimulation of the nipples. Thus, this episodic excitation is fashioned centrally, perhaps owing to special properties of the oxytocin neurons. These neurons do indeed possess certain unusual electrophysiological properties, one example being a noninactivating outward potassium current. Blockade of this current by transient inhibitory synaptic input, for example, by release of γ-aminobutyric acid (GABA), could create rebound depolarization sufficient to generate a burst of firing. However, an intrinsic mechanism is unlikely to be the sole explanation for the generation of bursting in oxytocin neurons because bursting is so specific to suckling and is never observed in response to other stimuli even though these may produce a change in afferent excitation or inhibition.

From studies of the effects of electrical stimulation in brain and spinal cord on milk ejection, it has been proposed that there exists some form of gating mechanism whereby the flow of afferent stimuli from the nipples can only reach the oxytocin neurons and trigger a burst of firing at defined intervals. This mechanism could be based on blockade of transmission across excitatory synapses along the pathway that is removed periodically or might take the form of a complex neural network in which reverberating excitation induced by suckling leads to a periodic output to the oxytocin neurons. The location of the gating mechanism is also uncertain, although it is tempting to propose that it lies within the posterior hypothalamus since this region appears to have a special role in coordinating bursting between the different magnocellular nuclei (see Section IV,B). Finally, it remains to be determined whether the oxytocin neurons themselves directly contribute to the circuitry that times their own intermittent responses or whether they only act as "follower" cells. The dramatic increase in burst frequency when oxytocin is injected into the magnocellular nuclei to mimic local release of the peptide suggests that they could certainly exert some influence on the timing of their own activation.

V. ROLE OF NEUROTRANSMITTERS AND NEUROMODULATORS IN THE MILK-EJECTION REFLEX

There has been extensive work on the roles of different neurotransmitters and neuromodulators in the milk-ejection reflex, mainly involving investigation of the effects of agonists and antagonists on reflex oxytocin release or electrical activity of oxytocin neurons. Interpreting the results of such studies, particularly those in which drugs are given systemically or by intracerebroventricular injection, is not always easy because of the difficulty of defining the site of action. An ever-increasing number of different neurochemical agents have been implicated in the milk-ejection reflex, and this section provides only a brief synopsis.

A. Neuropeptides

1. Oxytocin

Oxytocin has a major role as a central neuromodulator of the milk-ejection reflex. Intracerebroventricular injection of 1 or 2 ng oxytocin during suckling is followed by a profound increase in the frequency and amplitude of bursting responses. This so-called "facilitatory" effect is specific for the milk-ejection reflex and does not result from a generalized increase in the excitability of oxytocin neurons. Oxytocin receptors have now been detected within the magnocellular nuclei, and facilitation may result from a local action on the oxytocin neurons. It is known that suckling causes local release of oxytocin in the magnocellular nuclei, so this may represent a physiologically important mechanism. A high density of oxytocin receptors are also found in the limbic forebrain, most notably the bed nucleus of stria terminalis and, from microinjection and lesion studies, this represents another target site for the central action of oxytocin on the milk-ejection reflex. The bed nucleus, which receives an oxytocinergic innervation from the rostral hypothalamus, is also strongly implicated in oxytocin-induced maternal behavior. Oxytocin receptors within the bed nucleus are upregulated in the peripartum period probably under the influence of ovarian and placental steroids.

2. Opioids

The oxytocin-rich regions of magnocellular nuclei are innervated by enkephalin and endorphin-containing fibers, and enkephalin is colocalized within oxytocin neurons. Opioids active at μ receptors powerfully inhibit the firing of oxytocin neurons at the level of the hypothalamus, and there is good evidence for kappa-receptor-mediated opioid inhibition of oxytocin release at the level of the neurohypophysis. The general opioid antagonist, naloxone, enhances oxytocin release in response to suckling, suggesting that opioids may provide a tonic restraint of oxytocin release. Physiologically, the influence of opioids may become especially significant under conditions of stress-induced inhibition of the milk-ejection reflex.

B. Excitatory and Inhibitory Amino Acids

1. Glutamate

Glutamate has been identified as a major excitatory neurotransmitter within the hypothalamus and acts through two glutamate receptor subtypes, one being responsive to N-methyl-D-aspartic acid (NMDA receptors) and the other being responsive to α-amino-3-hydroxy-5-methyl-4-isoxazolepropionic acid (AMPA receptors). Oxytocin neurons appear to be predominantly controlled by the AMPA rather than the NMDA glutamate receptor subtype, and microinjection of AMPA into the supraoptic nuclei is much more effective than NMDA at evoking oxytocin release. Locally applied NMDA antagonists do not prevent suckling-evoked oxytocin release, whereas release is almost completely abolished by antagonism of AMPA receptors, indicating the important role of glutamatergic transmission operating through these receptors.

2. GABA

Oxytocinergic cells receive a rich input from local GABAergic neurons, many of which form double synapses in the lactating animal, and GABA exerts a powerful inhibition of their background firing. In view of this powerful inhibitory effect, the reported action of GABA on bursting activity is seemingly paradoxical since intrahypothalamic injections of GABA facilitate rather than inhibit bursting. Conversely, GABA antagonists have a suppressive effect on bursting responses. One explanation as to why injections of GABA increase burst generation is that a certain level of GABA inhibition is required to bring the membrane potential of the oxytocin neurons within a critical range compatible with bursting activity, although this requires further investigation.

C. Catecholamines (Monamines)

Oxytocin neurons are innervated by noradrenergic fibers arising mainly from the A_2 noradrenergic cell group of the brain stem. Destruction of noradrenergic terminals within the magnocellular nuclei prevents oxytocin release during suckling, and suckling is associated with an increase in noradrenaline turnover, strongly implicating this neurotransmitter in the milk-ejection reflex. Numerous studies have examined the effects of adrenoceptor agonists and antagonists on reflex oxytocin release or electrical activity of oxytocin neurons. The consensus is that noradrenaline contributes to the suckling-induced activation of oxytocin neurons through α_1 adrenoceptors located in, or close to, the magnocellular nuclei. Noradrenaline may also be involved in β receptor-mediated inhibition of the milk-ejection reflex, this action being exerted at a more distant site, possibly outside the hypothalamus. Studies have also been undertaken with α_2 adrenoceptor agonists, revealing complex dose-related facilitatory and inhibitory effects, which probably indicate the additional widespread involvement of noradrenaline in the milk-ejection reflex.

Another catecholamine that may be important is dopamine, which is also found in synapses contacting oxytocin neurons. Intracerebroventricular dopamine increases the frequency of milk-ejection responses, an effect that may be mediated through D_1 receptors located on the oxytocin neurons.

VI. DISORDERS OF MILK EJECTION

In view of the essential need for oxytocin in female reproduction, and hence species survival, it is hardly surprising that there are no recognized inherited conditions associated with failure of oxytocin secretion in which the milk-ejection reflex would be expected to be impaired. There are, however, reports of transgenic mice in which a functional oxytocin gene has been eliminated, and such animals are unable to lactate successfully unless they are given exogenous oxytocin. It is recognized that neurohypophysial hormone secretion may be permanently prevented by damage arising from cranial trauma, or neoplasia, and this would disrupt the milk-ejection reflex; however, this is a rare occurrence and therefore of little clinical importance in either human or veterinary medicine.

The milk-ejection reflex can be commonly prevented from functioning on a temporary basis owing to its susceptibility to blockade by fear, emotional disturbance, or stress. Under these conditions, activation of inhibitory pathways (see Section IV,C) will stop release of oxytocin even in the presence of an adequate suckling stimulus. Under very extreme conditions peripheral vasoconstriction can prevent oxytocin from reaching the mammary myoepithelium. The milk-ejection reflex will also fail if it is not possible for the mother to display normal nursing behavior, owing to inappropriate environmental conditions or disturbance. For species subjected to unaccustomed captivity or suffering loss of habitat owing to human activity one may envisage that such problems could significantly contribute to loss of reproductive potential.

Some sources report that up to 40% of women in Western societies who decide to breast-feed their babies are unable to establish lactation or give up prematurely. It is difficult to be sure what proportion of breast-feeding failures arise through problems associated with milk ejection, but most authorities providing advice on breast-feeding emphasize the need for the mother to be in a calm state for successful feeding to take place, and it is recognized that mothers who are emotionally "highly strung" tend to be prone to breast-feeding difficulties. It has also been suggested that failure of milk transfer during breast-feeding can arise as a result of inadequate conditioned release of oxytocin that women commonly experience prior to the onset of nursing. Successful attempts have been made to improve milk transfer using oxytocin administered by nasal sprays, but this has not gained acceptance, probably because of reluctance to sanction routine use outside of the hospital environment. Theoretically, it might be possible to overcome emotional inhibition of milk ejection using pharmacological agents but again this would be unlikely to be sanctioned on the grounds of possible harmful side effects. Reassurance and emotional support remain the most appropriate means of overcoming breast-feeding problems.

See Also the Following Articles

BREAST FEEDING; LACTATION, HUMAN; LACTOGENESIS; MAMMARY GLAND DEVELOPMENT; MAMMARY GLAND, OVERVIEW; OXYTOCIN; SUCKLING BEHAVIOR

Bibliography

Cross, B. A. (1977). Comparative physiology of milk removal. *Symp. Zool. Soc. London* **41**, 193–210.

Crowley, W. R., and Armstrong, W. E. (1992). Neurochemical regulation of oxytocin secretion in lactation. *Endocr. Rev.* **13**, 33–65.

Hatton, G. I. (1990). Emerging concepts of structure–function dynamics in adult brain: The hypothalamo-neurohypophysial system. *Prog. Neurobiol.* **34**, 437–504.

Ingram, C. D., Adams, T. S. T., Jiang, Q. B., Terenzi, M. G., Lambert, R. C., Wakerley, J. B., and Moos, F. (1995). Limbic regions mediating the central actions of oxytocin on the milk-ejection reflex in the rat. *J. Neuroendocrinol.* **7**, 1–13.

Lambert, R. C., Moos, F. C., and Richard, Ph. (1993). Action of endogenous oxytocin within the paraventricular or supraoptic nuclei: A powerful link in the regulation of the bursting pattern of oxytocin neurons during the milk-ejection reflex in rats. *Neuroscience* **57**, 1027–1038.

Lincoln, D. W., and Paisley, A. C. (1982). Neuroendocrine control of milk ejection. *J. Reprod. Fertil.* **65**, 571–586.

Theodosis, D. T., and Poulain, D. A. (1992). Neuronal-glial and synaptic plasticity of the adult oxytocinergic system: Factors and consequences. *Ann. N. Y. Acad. Sci.* **652**, 303–325.

Wakerley, J. B., and Ingram, C. D. (1993). Synchronisation of bursting in hypothalamic oxytocin neurons: Possible coordinating mechanisms. *News Physiol. Sci.* **8**, 129–133.

Wakerley, J. B., Clarke, G., and Summerlee, A. J. S. (1994). Milk ejection and its control. In *The Physiology of Reproduction* (E. Knobil and J. D. Neill, Eds.), 2nd ed., pp. 1131–1177. Raven Press, New York.

Mites

see Chelicerate Arthropods

Mollusca

A. Saber M. Saleuddin

York University

I. Introduction
II. Reproduction in Minor Groups
III. Reproduction in Major Groups
IV. Gametes
V. Sexuality
VI. Control of Reproduction
VII. Developmental Strategy
VIII. Larval Forms

GLOSSARY

γ-aminobutyric acid A biogenic neurotransmitter.
bursa copulatrix The female reproductive tract organs responsible for receiving excess spermatozoa during copulation.
dorsal bodies Putative endocrine glands in pulmonate molluscs.
ferritin An iron-containing protein.
gonochoristic Possessing either female or male gonads.
hectocotylus Structurally modified arms of certain male cephalopods involved in spermatophore transfer to females.
parthenogenesis Reproduction without fertilization of eggs.
serotonin A biogenic neurotransmitter.
spadix A male reproductive organ used for sexual arousal in females.
spermatheca A female reproductive organ for storing spermatozoa.
spermatophores Packets of spermatozoa transferred to females during copulation.
vitellogenic Describing mature eggs containing egg-specific protein(s).

I. INTRODUCTION

The Mollusca constitutes the second largest phylum in the animal kingdom and it includes animals such as squids, octopuses, oysters, scallops, escargot, limpets, mussels, and cowries. The phylum consists of seven classes: Monoplacophora (single external shell), Aplacophora (no shell), Polyplacophora (jointed external shell), Scaphopoda (tusk-like shell), Gastropoda (mostly coiled shell), Bivalvia (two jointed shell pieces), and Cephalopoda (arms/tentacles in the head region). The last three classes show immense diversity in number of organisms, distribution, and habitat. Molluscs are found on land but the majority are aquatic, existing as marine as well as freshwater species. Because of the presence of their external shell, molluscs have left behind a rich fossil record dating back to the Cambrian period. Molluscs are recognized by the presence of several or all of the following features: external/internal shell, mantle cavity, prominent foot, gills, and radula. Many species of molluscs are good to eat and thus shell fisheries in many developed countries make major economic contributions. Several molluscs produce pearls and this function has been best exploited by Japan and to certain extent by China, where the pearl industry is a major economic force. A number of molluscs are intermediate hosts of dreadful parasites which cause debilitating conditions in humans. Crop/vegetation damage caused by a number of molluscs is well documented. Several species, such as the shipworm, cause extensive destruction to wooden structures. One of the primary reasons for the success of molluscs as a dominant group in the animal kingdom is the diversification of their reproductive systems. Although reproduction has been dealt with in several special books on molluscs, the most comprehensive reviews on molluscan reproduction appear in Tompa *et al.* (1984). The reproductive strategy used by molluscs is indeed variable and to write an overview on reproduction on such a diverse groups of animals within the size constraint imposed is a challenge I accept with humility.

II. REPRODUCTION IN MINOR GROUPS

Primitively molluscs were gonochoristic and the gametes were discharged to the outside through the nephridial pore, where fertilization occurred. This situation persists in some members of the Aplacophora, Polyplacophora, and Scaphopoda and in a very few members of Bivalvia and Gastropoda. A majority of aplacophorans, however, are hermaphrodites. Fertilization is external and the trochophore larval stage is present in some species, whereas in others direct development occurs. In chitons (Polyplacophora), the sexes are separate and the gametes pass from the single gonad to the outside through the genital pores. The eggs typically have little yolk, development is external, and short, modified nonfeeding trochophore and/or veliger larval stages are present in several species. Sexes are separate in *Neopilina* (Monoplacophora), which have paired gonads, and the sex cells are liberated to the outside through the two nephridiopores.

III. REPRODUCTION IN MAJOR GROUPS

Among the major groups, gastropods exhibit considerable variations not only in sexuality and structure of the reproductive tract but also in reproductive physiology, including hormonal control. All gastropods have a single gonad but in a few primitive paraphyletic prosobranchs (archaeogastropods), the right kidney is used for the passage of gametes to the outside. In these species, there is no copulation, the eggs have little yolk, fertilization is external, and nonfeeding larval stages (trochophore/veliger) are found in most cases. Sexes are separate among most members of mesogastropods and neogastropods (prosobranchs) but the majority of higher gastropods (opisthobranchs and pulmonates) are hermaphrodites. In these molluscs, the reproductive tract is complex and although wide variations are found among members, a common structural plan emerges. In males, the tract has evolved to allow sperm transfer through copulation (use of penis), and the appearance of accessory glands (e.g., prostate) perhaps to provide nutrients for spermatozoa. In females, the tract consists of a temporary sperm storage structure (bursa copulatrix), accessory sex glands for the production of secretion(s) to coat eggs with perivitelline fluid (albumen gland or equivalent), and accessory sexual structures (uterus/oviduct) to envelope eggs in jelly and/or to secrete egg capsules. Among higher prosobranchs, eggs are encapsulated and fertilization is internal. In simultaneous hermaphroditic species of gastropods, which include the majority of opisthobranchs and pulmonates, normally there is a single ovotestis and the male and female components of the reproductive tract are combined with varying degrees of fusion in the distal part of the tract. In these species, both male and female gametes are produced in a single ovotestis and these are carried by a hermaphroditic duct to the carrefour/fertilization pouch where, following fertilization (self or cross), the eggs are coated by the perivitelline fluid secreted by the albumen gland or its equivalent. They then pass through the female tract where the eggs are coated in a jelly mass and/or put in capsules before being extruded through the vagina (Figs. 1 and 2). Most of the hermaphroditic species do reproduce by cross-fertilization although selfing, exceptional or usually, has been reported for some. During copulation reciprocal exchange of gametes may occur or one individual acts as a male donor and the other as the female recipient (unilateral transfer). The foreign spermatozoa introduced by the penis of the donor into the vagina of the recipient move up to the carrefour via the female/spermioviduct tract (Fig. 1). The excess spermatozoa are stored temporarily in the bursa copulatrix, where they eventually undergo enzymatic degradation. In many pulmonates, such as *Helisoma, Helix,* and the majority of opisthobranchs, copulation is mandatory for the production of viable eggs, whereas in several others selfing is the primary mode of reproduction. Many opisthobranchs and pulmonates exhibit a plethora of courtship behaviors prior to copulation. In members of the Stylommatophora, spermatozoa are packaged into spermatophores (bags of spermatozoa) secreted by the epiphallum and the flagellum of the male tract and during copulation the spermatophores are transferred into the vagina of the partner acting as a female recipient. In certain prosobranchs and opistho-

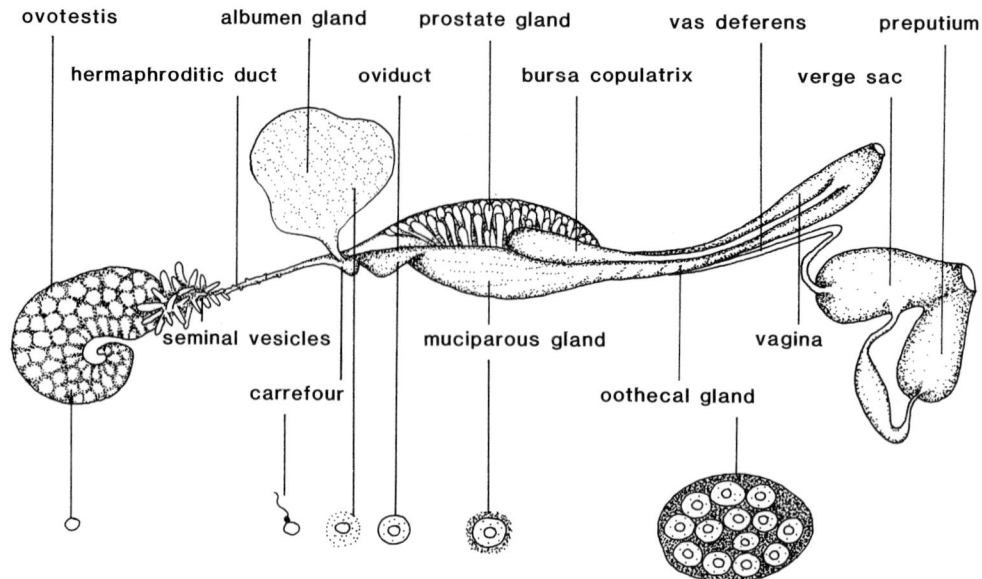

FIGURE 1 Diagrammatic representation of the reproductive system in a simultaneously hermaphroditic pulmonate (e.g., *Helisoma*). Note the female (upper) and the male (lower) parts of the tract. Following fertilization in the carrefour, the eggs are coated by albumen (perivitelline coat) secreted by the albumen gland and as the fertilized eggs travel down the female tract where mucous and egg jelly matrix are secreted by the oviduct–muciparous and the oothecal glands, respectively. Not to scale (courtesy of Sharon Miksup).

branchs, spermatophores are also used as a vehicle of spermatozoa transfer during copulation. In many members of Helicidae (*Helix* and *Philomycus*), a distal part of the female tract is modified into a dart sac which produces calcareous darts. The darts are fired reciprocally by each mating partner of *Helix aspersa* during courtship and thus become lost to the individuals. New darts are secreted by the sac. In *Philomycus*, the fired darts are retracted into the sac. The number of darts in a sac varies from one to many and the dart sizes are also variable. In certain slugs (*Agriolimax* and *Limax*) the dart sac is replaced by a long extensible structure called the sarcobelum and during courtship it, along with the penis, is everted and entwines the partners. The sarcobela of the mating partners are used to rub each other's body to produce sexual arousal. In certain members of the Stylommatophora (*Helix* and *Agriolimax*), the uterus (part of the spermioviduct) is capable of coating each egg with a calcified ($CaCO_3$) shell. Developing embryos utilize the eggshell as a major source of calcium.

The majority of bivalves (mussels, oysters, and scallops) are gonochoristic but several hermaphroditic species are known (e.g., *Pisidium*). Among members of the former category, paired gonads are located near the digestive gland and the gametes are extruded from each gonad via a gonoduct which can be fused to the nephridiopore or can be separate. In rare cases, a short gonoduct can open into the kidney from whence the gametes are discharged into the suprabranchial chamber via the nephridiopore. In hermaphroditic species the paired gonads (male and female), including the gonoducts, can be joined (*Cardium serratum*) or each is spatially distinct (*Entodesma saxicola*). Each gonoduct may discharge gametes through the kidney or directly into the suprabranchial chamber. The extrusion of gametes (spawning) is synchronous and can be seasonal, and fertilization is external in the majority of bivalves.

Sexes are separate in cephalopods (*Nautilus*, squids, and octopuses) and sexual dimorphism is evident in almost all species. Sexual dimorphism is exhibited by the presence of specialized structures such as the spadix in male *Nautilus*. Males are smaller

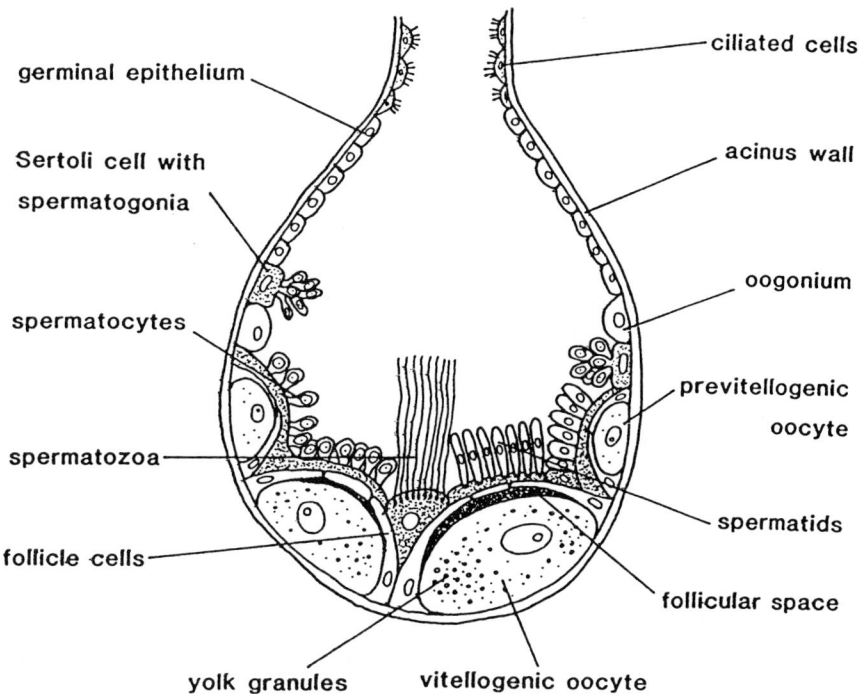

FIGURE 2 Detailed structures of an acinus of the ovotestis of a typical pulmonate, e.g., *Helisoma*, showing both spermatozoa, a Sertoli cell, and the developing oocytes (eggs). Not to scale (courtesy of Sharon Miksys).

in a few species (e.g., *Argonauta*), possess hectocotylus arms (squids and octopuses), and in some squids show highly aggressive behavior during courtship. In both sexes there is a single median gonad. The reproductive tract consists of specialized areas and glands for the transport of gametes and to provide nutrients to coat eggs with extravitelline materials (protective jelly). In *Nautilus*, it is not known what roles, if any, the penis and spadix play in the transfer of spermatophores. In coleoids (squids and octopuses), the hectocotylus arm(s) is used for the transfer of spermatophores. The male tract consists of structures such as spermatophoric organ and vas deferens, which are involved in the formation of spermatophores. They are then stored temporarily in a "Needham's sac." The female tract includes nidamental glands (secretes the chorion), one or paired oviducts, and spermatheca. Many cephalopod males exhibit an array of courtship behaviors during mating and mate selection, such as aggressiveness, intensified display of body parts, changes in coloration, and posture. In *Loligo paelei*, pairing seems to be permanent and a hierarchy of dominance among males is established which can change depending on the availability of a new mature female. Copulation in cephalopods may last from several seconds to nearly an hour, during which the spermatophores are transferred by the hectocotylus arm(s) of a male onto the tip of the oviduct which is modified into a seminal receptacle. A modified buccal seminal receptacle to receive spermatophores is present in *Loligo*. In some species, spermatophore transfer can occur directly from the penis to the mantle cavity of a female. Common copulation positions between partners are either head to head or side to side.

IV. GAMETES

A. Spermatozoa

The basic structure of molluscan spermatozoa has been described. The length of spermatozoa varies from 50 to 900 μm. In prosobranchs, functional haploid eupyrene as well as nonfunctional oligopyrene, hyperpyrene, and apyrene paraspermatozoa are found. Nonfunctional spermatozoa are longer (may reach 200 μm) than their eupyrene counterparts and their functional roles have not been ascertained. A complex structure, spermatozeugma, has been described which consists of a large number of eupyrene spermatozoa attached to a nonfunctional paraspermatozoon. It has been proposed that spermatozeugma are vehicles for eupyrene spermatozoa and they are able to enter into the oviduct easily. The existence of atypical or nonfunctional spermatozoa has also been reported in several opisthobranchs. Among the members of Unionacea (Bivalvia) and in a few marine bivalves (*Ostrea*) spermatozoa occur as sperm morulae (sperm balls). In many species both functional and nonfunctional spermatozoa are found. The morulae, like spermazeugma in some prosobranchs, may have transport functions to ensure fertilization. It was mentioned earlier that in several groups of molluscs spermatozoa are packaged into specialized and complex structures called spermatophores. The size of a spermatophore varies from 8 mm in some land pulmonates to 11 cm in *Nautilus*. Commonly, one spermatophore is formed at each copulation in cephalopods. The process of spermatogenesis is basically similar to that in other animals.

B. Eggs

The size and the number of eggs produced by molluscs, as in other animals, depends partly whether or not the fertilization is external or internal and is partly related to the modes of development (direct or indirect). Generally, molluscs which spawn (external fertilization) produce a large number of small eggs (many bivalves and several prosobranchs) with no specific yolk protein and many of these species have indirect development. Many prosobranchs produce a variable number of eggs per capsule or gelatinous strings. Direct development in some prosobranchs is dependent not only on large egg size but also on the presence of nurse eggs. An interesting situation exists in certain opisthobranchs in which eggstrings contain extracapsular yolk material which provides nutrition to the developing larvae. Compared to freshwater pulmonates, terrestrial pulmonates produce large (approximately 5 mm or larger) eggs, and in several cases the eggs are calcified. Even in large eggs the actual size of the ovum itself is approximately 100 μm in diameter and the rest is composed of extravitelline jelly and the calcified layers. The presence of well-characterized vitellogenic protein(s) in molluscan oocytes, either gonadal or extragonadal, has not been shown conclusively. Membrane-bound granules containing ferritin aggregates have been reported in mature oocytes of several pulmonates. It has been suggested that developing oocytes take up ferritin synthesized extragonadally and that the uptake of ferritin and not synthesis is controlled by the endocrine dorsal bodies. Cephalopods eggs are large (1 or 2 cm) and they have several egg coats secreted primarily by the uterus and the nidamental gland. The eggs are laid individually or in clusters cemented to a substratum. Oogenesis in molluscs is essentially similar to that in other animals. Structural features of an ovotestis are depicted in Fig. 2.

V. SEXUALITY

As mentioned earlier, the majority of molluscs are unisexual, except those in the classes Monoplacophora, Scaphopoda, Cephalopoda, which are hermaphrodites, either simultaneous or consecutive. Moreover, protandry (initial male phase) is commonly encountered in bivalves and gastropods. Within predominantly unisexual genera, one or two species are often hermaphrodites, whereas among species in which hermaphoditism predominates, unisexuality may appear. Although hermaphroditism occurs in most major groups of molluscs, mention must be made about this condition in *Crepidula fornicata*. The members of this species live in a chain in which the larger individuals at the bottom of the chain are females and the members on top of the chain are small males. The members in the middle are simultaneous hermaphrodites. Young *Crepidula* start life as males (protandric hermaphrodites) followed by a hermaphrodite phase and finally become females. A similar situation exists in other prosobranchs, such as *Patella vulgata*, *Lottia gigantic*, and *Calyptraea sinensis*. In several bivalves, including *Ostrea* (oyster) and *Teredo* (shipworm), more than one sex change in the life of an animal can occur. This rhythmic sex change from male to female either annually or seasonally is termed rhythmic consecutive sexuality. For prosobranch species, factors from the central nervous system (CNS) and the tentacles play roles in sex change, whereas in *Ostrea* environmental conditions seem to influence the sex change. When young *Crepidula* (male phase) were raised in the company of female members, 90% continued as males, but when raised in isolation about 68% continued as males, whereas in others (32%) no male phase could be detected. In other species of *Crepidula*, sex change is size specific and apparently not under environmental control. Finally, parthenogenesis has been reported in several molluscs including a prosobranch, *Paludestrina jenkinsi*, which lives in fresh and brackish waters. No male phase has been found. All individuals are females, judged by the presence of oocytes in the gonad. Induced parthenogenesis has been obtained in *Crassostrea virginica*.

VI. CONTROL OF REPRODUCTION

A. External Inputs

For the majority of molluscs, several external factors, such as higher temperatures, longer photoperiods, increased food availability, and high humidity, seem to affect reproduction. Additionally, lunar cues, tides, crowding, parasites, and salinity may also influence reproductive capacity of many molluscs. It is well established that reproduction and its strategies are related to the geographic locations of molluscs. Molluscs in low latitudes have an extended reproductive season, whereas those living in higher latitudes have a limited season. This restriction is primarily due to food availability and temperature. Finally, organic and inorganic pollutants affect reproduction.

B. Internal Inputs

Factors from the CNS, the gonads, the optic tentacles, and the putative endocrine structures (dorsal bodies in pulmonates and optic glands in cephalopods) have been implicated in reproduction of all major groups (classes) of molluscs. The role of the CNS in sex reversal in prosobranchs has already been mentioned. In all pulmonates thus far studied, the presence of dorsal bodies (DBs) has been described and their roles in female reproduction have been established. The DB hormone (DBH) is involved in oocyte maturation, ovulation, differentiation of female accessory organs/glands, and the secretion of albumen by the albumen gland. A pair of optic glands present on the optic tracts in cephalopods is involved in reproduction. However, the chemical nature of DBH and optic gland hormone is not known. The endocrine control of egg laying is well documented in a number of molluscs. In *Aplysia californica* (an opisthobranch), neurosecretory bag cells in the CNS produce the peptidergic egg-laying hormone (ELH). ELH also induces egg-laying behavior in this mollusc. In *Lymnaea stagnalis* (a basommatophoran pulmonate), neurosecretory caudodorsal cells in the CNS produce the peptidergic caudodorsal hormone, which has been sequenced. Like ELH in *Aplysia*, this hormone controls egg laying and egg-laying behavior. In general, egg laying itself and the repertoire of egg-laying behaviors in pulmonates and opisthobranchs are certainly under the control of hormones produced by the CNS. In some prosobranchs, the juxtaganglionar organ, an analogous structure to the DBs in pulmonates, is involved in reproduction.

Both male and female bivalves extrude a large number of sex cells in a synchronized fashion (spawning) to ensure fertilization. However, in rare cases fertilization in the mantle cavity does occur. The involvement of the CNS in the sex cell maturation and spawning in bivalves was suggested as early as the 1950s, but only recently has the role of serotonin in bivalve reproduction been established. Micromolar concentrations of serotonin cause oocyte maturation through germinal vesicle breakdown *in vitro*, and injection of serotonin induces spawning in male and female *Spisula solidissima*. Dopamine also causes spawning in *Spisula* but only at a high nonphysiological concentration. Serotonin injection induces spawning in several other bivalves, *Crassostrea*, and *Pecten* and both oocyte maturation and spawning in *Dreissenia polymorpha*. Serotonin has been detected in the hemolymph and serotonin-containing neurons are common in the CNS of bivalves.

The role of the optic glands in cephalopod reproduction has been reviewed. A factor(s) from the glands regulates the production of spermatozoa and eggs and the differentiation of the reproductive tracts in both sexes. The majority of the cephalopods are semelparous, in which death follows after the reproductive period in both sexes. However, if the optic glands are removed from brooding females, the animals abandon their eggs, resume feeding, and the programmed death is delayed for a considerable period.

VII. DEVELOPMENTAL STRATEGY

Except in basommatophoran pulmonates, oviparity is the most common reproductive strategy among molluscs, in which eggs and spermatozoa are released to the outside and fertilization is external. The majority of molluscs are iteroparous (breeding in several seasons). In some cases (some bivalves and polyplacophorans), the spermatozoa enter into the mantle cavity of the female, where waiting eggs are fertilized before being released outside. In most groups, ovoviviparity is encountered where the eggs are retained in the brood spaces (gills and mantle) of the body. Several prosobranchs will protect egg capsules with their foot and some will carry them attached to the shell. Ovoviviparity is commonly seen in some land snails in which the calcified eggs are retained in the oviduct during adverse environmental conditions. During this retention period the eggs will develop into embryos which hatch with the return of favorable conditions. Parental care of eggs in cephalopods consists of guarding the eggs from predators and flushing them with water jets for cleaning purposes. Only female *Argonauta* have a shell onto which eggs are attached and this provides protection for the eggs. Viviparity is well documented in a few pulmonates (*Stylodon* and *Tekoulina*) in which a placental connection has been proposed. In several bivalves (*Ostrea, Teredo,* and *Sphaerium*), the embryos develop either within the gills (*Sphaerium*) or in the mantle cavity (*Ostrea*) and no placental connections have been observed. In *Ostrea* (oysters), larvae are released outside, whereas in *Sphaerium* juvenile bivalves are liberated from the mantle cavity. In *Anodonta* (clam), numerous larvae called glochidia are released which must attach themselves to a freshwater fish to complete the life cycle. Eventually, the larvae metamorphose into miniature adults and fall off the host.

VIII. LARVAL FORMS

Larval forms are present in most classes of the phylum. Exceptions are the cephalopods, in which larval stage is not present, and the monoplacophorans, in which no information on larval form is available. In aplacophorans, polyplacophorans, scaphopoda, many marine bivalves, and primitive gastropods (paraphyletic archaeogastrpods), a trochophore larval stage is found. In the last two groups, the trochophore larval stage may change to a veliger, which is usually nonfeeding. In most prosobranchs and opisthobranchs, a planktotrophic veliger is the larval form. Larval forms are absent in most pulmonates and freshwater prosobranchs. A few members of aquatic pulmonates (*Melampus* and *Siphonaria*) have feeding veliger larvae. Among freshwater bivalves, a larval stage is absent in Sphaeridae, but in *Dreissena* and *Anodonta* feeding veliger and nonfeeding glochidia larval stages are found, respectively. From an ecological perspective three types of larva are represented in molluscs:(I) the planktotrophic type, found in most bivalves and prosobranchs, has a long larval stage (up to 3 months);(ii) the planktotrophic type, with a very short larval life (about a week), found among some opisthobranchs; and (iii) the lecithotrophic type, which is derived from yolky eggs. Lecithotrophic larvae are found in deep sea bivalves, scaphopods, polyplacophorans, and most archaeogastropods. It is worth noting that larval forms found in various groups are important for dispersion of the species, particularly for those which are less mobile. Larvae eventually settle to the substrate and undergo metamorphosis. The stimuli for metamosphosis can depend on food availability, types of substrata, and chemicals such as choline or γ-aminobutyric acid.

In conclusion, primitively molluscs are gonochoristic but hermaphroditism is found in several classes of the phylum, especially gastropods. In higher classes, the reproductive tracts in both sexes are complex and modified to allow copulation between sexes (internal fertilization), to store spermatozoa, to provide nutrients to the gametes, to encapsulate eggs, and, in some cases, to provide the space for the development of embryos. External fertilization is common in those species which spawn. Both external and internal inputs influence egg maturation. Various reproductive strategies have been adopted to ensure high fecundity in many molluscs.

Bibliography

Arnold, J. M. (1984). Cephalopods. In *The Mollusca* (A. S. Tompa, N. H. Verdonk, and J. A. M. van den Biggelaar, Eds.), Vol. 7 (Reproduction), pp. 419–454. Academic Press, New York.

Fretter, V. (1984). Prosobranchs. In *The Mollusca* (A. S. Tompa, N. H. Verdonk, and J. A. M. van den Biggelaar, Eds.), Vol. 7 (Reproduction), pp. 1–45. Academic Press, New York.

Hadfield, M. G., and Switzer-Dunlap, M. (1984). Opisthobranchs. In *The Mollusca* (A. S. Tompa, N. H. Verdonk, and J. A. M. van den Biggelaar, Eds.), Vol. 7 (Reproduction), pp. 209–350. Academic Press, New York.

Joosse, J. (1988). The hormones of molluscs. In *Endocrinology of Selected Invertebrate Types* (H. Laufer and R. G. H. Downer, Eds.), pp. 89–140. A. R. Liss, New York..

Juneja, R., and Koide, S. S. (1996). Biochemical pathways involved in serotonin-regulated *Spisula* oocyte maturation and fertilization. *Invertebr. Reprod. Dev.* 30, 47–53.

Mackie, G. L. (1984). Bivalves. In *The Mollusca* (A. S. Tompa, N. H. Verdonk, and J. A. M. van den Biggelaar, Eds.), Vol. 7 (Reproduction), pp. 351–418. Academic Press, New York.

Morton, J. E. (1979). *Molluscs*. Hutchinson, London.

Purchon, R. D. (1977). *The Biology of the Mollusca*. Pergamon, Oxford, UK.

Saleuddin, A. S. M., Mukai, S. T., and Khan, H. R. (1994). Molluscan endocrine structures associated with the central nervous system. In *Perspectives in Comparative Endocrinology* (K. G. Davey, R. E. Peter, and S. S. Tobe, Eds.), pp. 257–263. National Research Council of Canada, Ottawa.

South, A. (1992). *Terrestrial Slugs*. Chapman & Hall, London.

Tompa, A. S. (1984). Land snails (Stylommatophora). In *The Mollusca* (A. S. Tompa, N. H. Verdonk, and J. A. M. van den Biggelaar, Eds.), Vol. 7 (Reproduction), pp. 47–140. Academic Press, New York.

Molt and Nuptial Color

Christopher W. Thompson

Washington Department of Fish and Wildlife and University of Washington

I. Molt
II. Nuptial Coloration

GLOSSARY

activation Target tissues that exhibit a response to a hormone only when the hormone is present (e.g., circulating plasma levels are elevated) and whose response disappears when the hormone is absent (or at basal levels) are said to be activated by the hormone. Activational effects typically are exhibited during adulthood.

body plumage All feathers on a bird except for the flight feathers of their wing (primaries and secondaries, collectively called remiges) and tail (rectrices).

generation of feathers All feathers grown during a single molt comprise a feather generation, even if the molt is temporarily stopped and subsequently reinitiated and completed prior to the onset of the next molt, i.e., even if the molt is divided into two discrete time periods.

organization Target tissues that are exposed to elevated levels of a hormone and exhibit a long-term (typically about a year) or permanent response that does not disappear when the hormone level returns to its basal level are said to be organized by the hormone. Organizational effects typically occur in early life prior to adulthood.

subadult A bird wearing a plumage more advanced than its first (juvenal) plumage, but that has not yet attained its adult (definitive) plumage. The extent to which subadults are physiologically and behaviorally capable of breeding differs dramatically among species.

M olt is the normal shedding of feathers and the replacement of most or all of these feathers by a new generation of feathers. In species in which reproductively mature (hereafter adult) birds replace their body plumage only once per year during a single molt, this molt results in a plumage that is worn for an entire year, including the subsequent breeding season, until their next molt. In the Northern Hemisphere this molt typically occurs in late summer, fall, or early winter at the end of the breeding season and/or after fall migration (in migratory species). In the Southern Hemisphere the seasonal timing of molt is the opposite. In equatorial and tropical species, molt overlaps with breeding more so than in other species and, both within and among species, occurs during more months of the year.

In species in which adult birds replace their body plumage twice per year, the molt preceding breeding results in a plumage that is worn for most or all of the breeding season and then is replaced by another plumage for the nonbreeding season. Regardless of whether birds have one or two body molts per year, the nuptial (breeding) color of plumage during the breeding season varies among species as well as between sexes and ages within species. This variation spans a spectrum from being dull and cryptic to bright and conspicuous, often including elaborate sexually selected plumage characters, such as long tail plumes and head crests (Fig. 1). Changes in plumage color, usually from dull to bright, between the nonbreeding and breeding season may not always result from molt. They also may result from (i) feather wear, especially in birds that molt only once a year shortly after breeding; (ii) a combination of both molt and feather wear on different parts of their body; and/or (iii) waxes or oils that birds apply to their plumage from their uropygial gland (the preen or oil gland at the base of the spine in birds). One species, the Toki or Japanese crested ibis (*Nipponia nippon*), secretes a black tar-like substance from a well-defined patch of skin in the throat and neck

FIGURE 1 (A) Highly modified and exaggerated tail feathers of a superb lyrebird (*Menura novaehollandiae*). (B) Highly modified and exaggerated crown feathers of a victoria crowned pigeon (*Goura victoria*).

region that it daubs on its plumage to achieve nuptial color. Otherwise, plumage color does not change between molts except for (i) slight fading (oxidation) of carotenoid pigments in some species due to prolonged exposure to sunlight and (ii) occasional staining of plumage by environmental minerals (e.g., in some waterbirds that live on bodies of water with high iron content, their ventral plumage often becomes stained a rusty color).

I. MOLT

A. Why Do Birds Molt?

Birds molt for two reasons: (i) to replace worn feathers (Fig. 2) and (ii) to change their appearance, typically from cryptic to conspicuous plumage or vice versa.

B. Number of Molts per Molt Cycle

Molt typically occurs before or after breeding and not during migration in migratory species. Since most birds breed only once a year, their molt cycles are 12 months long. However, in many tropical species, birds breed in cycles less than or more than a year in length, resulting in molt cycles that also are less than or more than a year. In the following discussion, I use the term "year" synonomously with "molt cycle." Most birds molt once, twice, or rarely three (and possibly four) times during a molt cycle (e.g., various ptarmigans, *Lagopus* spp., Ruffs, *Philomachus pugnax*, and Oldsquaws, *Clangula hyemalis*).

C. Control of Molt

1. Natural Selection for Molt

Whether a given bird species has one, two, or three molts per year is a function of four natural selection pressures: (i) its rate of feather wear, (ii) the extent to which changing plumage color between breeding and nonbreeding season is advantageous, (iii) the extent to which risk of predation may be greater during the molting versus nonmolting period, and (iv) the energetic cost of molting including increases in (a) basal, bone, and protein metabolism, (b) requirements for specific vitamins, minerals, and fatty acids, (c) erythropoeisis, and (d) thermoregulatory expenses due to decreased plumage insulation.

To replace worn plumage, birds must replace, at a minimum, most to all of their body plumage and some to all of their wing and tail feathers (hereafter flight feathers) once per year. Many large birds (e.g., albatrosses and eagles) require at least 2 or 3 years to replace all their flight feathers, and at least some of these species may also take as long to replace their body plumage. Conversely, some birds [e.g., certain Emberizid sparrows (*Ammodramus* spp.) and diving ducks (*Oxyura* spp.)], especially those that live in habitats that are especially abrasive (e.g., saltmarshes and sandy deserts), may replace their body plumage and, in some cases, some to all flight feathers twice per year.

In addition to the need to replace plumage because of feather wear, differences in natural selection pressures on birds during the breeding versus nonbreeding season may favor different species, or ages or sexes within species, to wear plumages that differ in conspicuousness during these two seasons, i.e., to have two molts per year.

In the life of birds, the three most energetically expensive activities are reproduction, molt, and, in migratory species, migration. Many species have only one complete body molt per year (typically after breeding) or, if they have a second body molt (usually

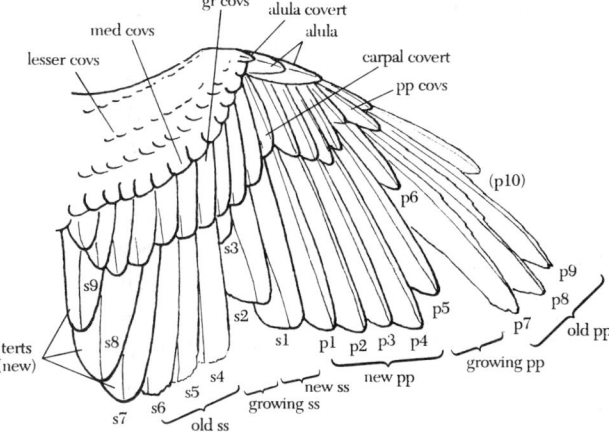

FIGURE 2 A partially extended wing of a bird during molt indicating new, old, and growing primary (p) and secondary (s) flight feathers.

shortly before breeding), it is incomplete rather than complete. In these species, the high energetic cost of molt may preclude certain species, or ages or sexes within species, from undergoing a complete second body molt.

2. Physiological Regulation of Molt

i. Phylogenetic Constraints The aspects of the molting process that can vary are its (i) timing in relation to annual cycle events (e.g., reproduction and migration) or calendar date, (ii) sequence, (iii) extent, and/or (iv) rate (duration). In addition, these aspects may vary between populations, sexes, age classes, or even individuals within age classes (e.g., successful versus unsuccessful breeders and young fledged from early versus later clutches in species that lay multiple clutches per breeding season). Determining the extent to which these aspects of molt may be constrained by phylogenetic relationships is complicated by two factors. First, to address this issue, there must be variation in molt strategies within and among different taxonomic levels of taxa. Unfortunately, most of what we know about molt is confined to passerine birds (order Passeriformes) or so-called "songbirds" which exhibit little variation in these aspects of molt relative to the 28 orders of nonpasserine birds. Second, there are very few groups of birds for which we know both their phylogenetic relationships and molt strategies. However, for those few groups for which we do have these data (e.g., Emberizidae: *Passerina* spp.), it appears that molt sequence is evolutionarily the most conservative aspect of molt; all other aspects appear to be capable of evolving changes relatively rapidly and extensively in response to natural selection pressures.

ii. Genetic Control Laboratory studies, including various breeding experiments, have conclusively demonstrated that, under constant photoperiodic conditions (12 hr light:12 hr dark) and in the absence of unusual external environmental stimuli (e.g., low temperature or reduced food supply) that might otherwise influence the timing and rate of molt, the onset and duration of molt in some migratory species, and possibly in most or all species of birds, are regulated by endogenous circannual rhythms that are genetically based. However, as discussed later, the timing and rate of molt can be altered by various external environmental stimuli.

iii. Environmental Control Mediated by the Neuroendocrine System The timing and rate of molt are dictated by an endogenous circannual cycle in many, and possibly most or all, species of birds. However, in all species from highly seasonal temperate and boreal breeders to tropical and opportunistic breeders, the onset and rate of molt can be strongly influenced by environmental stimuli. These stimuli are transduced by the senses, including extraocular light receptors in the hypothalamus, into neuroendocrine signals; these may be classified into four general categories:

1. Initial predictive information: This refers to environmental cues (usually photoperiod) that determine the window of time during the year when prenuptial and postnuptial molt may occur well in advance of their onset. This information is predictive in that it allows behavioral, physiological, and hormonal preparation for molt in advance and ensures that molting occurs when conditions are suitable (e.g., relatively abundant food supply and moderate temperature). As discussed previously, in many species the onset and duration of molt are determined by endogenous circannual rhythms; this can be thought of as the default molt schedule. However, because the internal clocks of birds are not exactly 365 days, they are not synchronized with the external environment. Initial predictive cues serve to entrain the default molt schedule appropriately from year to year, i.e., to advance or delay the onset and duration of molt. Additional environmental cues may further modify the timing and rate of molt as the onset of molt approaches and during molt itself.

2. Supplementary information: This provides short-term predictive cues to fine-tune the onset and rate of molt to local conditions. Examples include food availability and temperature which could advance or delay onset of molt, as well as increase or decrease rate of molt, by direct effects on neuroendocrine mechanisms or by indirectly affecting energy and nutrient balance, which in turn cause neuroendocrine changes. Supplementary cues do not appear

to influence the onset of molt as much as the rate of molt.

3. Synchronizing and integrating information: This includes all intraspecific behavioral interactions within and between sexes in adults as well as between parents and their young. Such cues are critical for coordinating reproductive activities. Coordination of molt between parents and their young is not necessary for as many species of birds and is virtually undocumented. However, such coordination must be essential in certain taxa. One well-known example is that failed breeders molt earlier than successful breeders in most species. Similarly, adults that renest after a failed breeding attempt finish breeding later, and therefore molt later, than adults that successfully rear their first clutch. Another example is that the timing of molt within mated pairs of many species of geese and swans is always closely coordinated, the male preceding the female or vice versa, depending on the species or population involved. It also has been demonstrated experimentally in song sparrows, *Melospiza melodia*, that males will delay molt for months after the usual date of onset if the females to which they are paired continue to display sexually receptive behavior (artificially induced by giving them estradiol implants). Perhaps one of the most fascinating examples regards cavity-nesting hornbills (Bucerotidae). During nesting, female hornbills are enclosed within their nesting cavity except for a thin slit through which their mate feeds them and their young. During this time, the female molts by losing all of her flight feathers synchronously or nearly so and rapidly regrowing them, whereas breeding males replace their flight feathers more slowly and sequentially. However, females that fail to successfully hatch eggs and/or rear young will break out of their nest cavity and follow a molt sequence similar to that of males. Thus, the effect of synchronizing and integrating information on molt may be more widespread than is generally recognized.

4. Modifying information: This includes factors that, once molt has begun, can alter its rate or stop it temporarily (suspended molt) or permanently (arrested molt). For most species, the factors most likely to decrease the rate of molt or stop molt are low food supply and/or low temperature, e.g., a sudden winter storm in the arctic. However, in opportunistic breeders such as some desert birds, especially favorable food supply may cause them to interrupt their molt in order to breed.

This raises the question: How do all these environmental stimuli affect the neuroendocrine system so that birds molt in an adaptive way. Unfortunately, despite considerable research on neuroendocrine control of molt spanning nearly a century, the answer is that we know very little about neuroendocrine control of molt, especially compared to our knowledge of the neuroendocrine regulation of reproduction and migration. Because birds vary tremendously in their life histories, one should expect the neuroendocrine regulation of major events in their annual cycles, including molt, to be equally diverse. In fact, this seems to be the case.

One must first recognize that postjuvenal molt (first prebasic molt of Humphrey and Parkes), prenuptial molt (prealternate molt of Humphrey and Parkes), and postnuptial molt (definitive prebasic molt of Humphrey and Parkes) may be hormonally regulated in different ways from one another. In addition, although little is known about neuroendocrine regulation of postnuptial molt, far less is known about neuroendocrine regulation of prenuptial molt, and almost nothing is known about neuroendocrine regulation of postjuvenal molt, which is not discussed further. Indeed, so few studies exist regarding neuroendocrine control of prenuptial molt, and the results of these studies are so variable, that no generalizations can be drawn from the data. However, the data are generally consistent with three hypotheses. First, an increase in thyroxine (T_4) immediately prior to and/or at the onset of and/or during part to all of prenuptial molt is essential in most species for molt to occur and for feathers to grow normally, as is true for postnuptial molt. However, at least a few species exhibit no apparent increase in T_4 at any time of year, suggesting that T_4 may not play a role in molt in these species. Second, testosterone (T) and estradiol (E_2) in adult males and females, respectively, may have a less negative effect, or possibly even a stimulatory effect, on feather growth during prenuptial molt in contrast to their inhibitory effects during postnuptial molt. Prenuptial molt typically occurs for 2 or 3 months immediately preceding breeding and during gonadal recrudescence, i.e., when T and E_2 levels are increasing and reaching maximal annual levels. Thus, if T and E_2 had suppressive effects on molt at this time, prenuptial molt could not occur. Third, since molt can be suspended or arrested for events other than breeding and that are not associated with elevated sex steroid levels (e.g., migration, lack of food, and inclement weather), such suppression of molt must be caused by neuroendocrine mechanisms independent of sex steroids or other hormones upstream (e.g., pituitary gonadotropins).

Regarding postnuptial molt, at least three generalizations appear to be true. First, as in prenuptial molt, elevated levels of T_4 usually are necessary for normal molting and feather growth. In most species, T_4 increases dramatically at the onset of feather loss which is caused by active growth of new feathers within feather follicles. Severe reduction or elimination of T_4 by surgical thyroidectomy or administration of a T_4 antagonist during the breeding season or during molt prevents most species from molting at all or at least for many months after molt normally would occur; in addition, thyroidectomized birds that do eventually undergo molt often grow feathers that are abnormal in structure and often in pigmentation as well. However, some species will molt normally even if thyroidectomized as long as they are thyroidectomized after the beginning of the breeding season (e.g., typically 13 hr or more of day light in temperate species) and experience elevated T_4 levels for as little as 1 or a few days prior to thyroidectomy. This implies that, in these latter species, T_4 has an organizational effect on the neuroendocrine mechanisms that regulate molt, whereas in the former species T_4 appears to have an activational effect on the these same mechanisms. In addition, as mentioned previously, at least a few species exhibit no apparent increase in T_4 throughout their annual cycle, including their postnuptial molt, suggesting that molt in these species may be regulated by hormones other than T_4. Another thyroid hormone, triiodothyronine (T_3), also increases dramatically when molting is most intense, i.e., when feather insulation is reduced to its lowest point during the molt. However, T_3 probably serves no function in the molting process per se, but rather increases metabolic rate to compensate for the increased thermoregulatory cost resulting from decreased plumage insulation during molt.

Second, in most species, sex steroid hormones (T in males and E_2 in females) and T_4 reciprocally antagonize one another so that high T and E_2 levels can inhibit postnuptial molt, even in the presence of high T_4 levels, and vice versa. Because both reproduction and molting are energetically expensive processes, it is adaptive for most birds to separate the events temporally. As a result, natural selection presumably has favored the evolution of as yet unknown physiological mechanisms whereby T and E_2 inhibit the stimulatory effects of T_4 on the postnuptial molting process. However, there are at least three situations in which selection may have favored T and E_2 to have less suppressive effects on postnuptial molt. First, T and E_2 may have less effect on molt in species that both breed and molt opportunistically, such as many desert species in the interior of Australia or certain nomadic species such as crossbills (*Loxia* spp.); some opportunistically breeding species molt at a specific time of year no matter what, even if conditions are suitable for breeding; it is likely that T_4 suppresses sex steroid levels and breeding in these species. Conversely, other opportunistic breeding species breed whenever conditions are suitable and molt at variable times of year between breeding attempts; in these species it seems likely that T and E_2 effectively shut down molt. Second, T and E_2 may have less suppressive effect on molt in species in which molt temporally overlaps to a substantial extent with breeding; such species presumably are forced to molt during the breeding season because time and/or energetic constraints following the breeding season are greater than those that exist during that portion of the breeding season when they molt. Examples include boreal breeding species, such as the Dunlin, *Calidris alpina*, and two crows (*Corvus corax* and *Nucifraga columbiana*), and many equatorial and tropical breeding species such as hornbills, in which molt and breeding overlap completely from egg laying to fledging of chicks.

Third, as with prenuptial molt, postnuptial molt can be suspended or arrested due to various factors unrelated to breeding and elevated sex steroid levels (e.g., migration, food shortage, short days, low temperature, and inclement weather). As a result, suppression of this molt must be mediated by hormones other than sex steroids; a likely candidate may be corticosterone, although elevated corticosterone levels have not been observed in some species during molt.

Many studies have suggested that various hormones serve various functions in the molting process of various species. For example, female white-crowned sparrows, Zonotrichia leucophrys, must be exposed to elevated levels of E_2 during early winter (e.g., November) or they will not undergo postnuptial molt the following year; prolactin levels during molt may be higher, lower, or the same as those during breeding suggesting that prolactin may stimulate, inhibit, or have no effect on molt in various species. Similarly, progesterone, growth hormone, and corticosterone levels are higher during molt in some species but not others, suggesting that they may stimulate molt in some of these species. However, the extent to which these hormones may serve a permissive, stimulatory, or suppressive role in the molting process is unclear. In addition, because hormones may have organizational effects on target tissues involved in the molting process months before molt occurs, and not solely during the molting process itself, it is exceptionally difficult to experimentally determine the function that any hormone plays in molt.

II. NUPTIAL COLORATION

A. Colors of Feathers and Soft Parts

Colors in feathers and soft parts [beaks, ceres (soft and thickened portion of the base of the bill), feet and/or legs, irides of eyes, wattles, combs, or caruncles] may be caused by pigments deposited in them or by their microscopic structure. Four classes of pigments occur in birds: melanins, carotenoids, porphyrins, and pigments that are chemically characterized but unidentified. Melanins are polymers of indole-5,6 quinone bound to a protein matrix; they are derived from the amino acid tyrosine, synthesized in melanocytes, and exist in all orders of birds. Melanin occurs in two chemically distinct forms: (i) eumelanin, which produces dark browns, grays, and blackish colors, and (ii) phaeomelanin, which produces dull yellows, oranges, and reddish colors.

In contrast to melanin, no animals, including birds, can synthesize carotenoids. Carotenoids are hydrocarbon compounds that contain cyclic endgroups which, in turn, often contain hydroxy or keto groups. Carotenoids are synthesized only by photosynthetic organisms and plants. Thus, birds must ingest them in their diet, either directly from plants (e.g., fruit and seeds) or indirectly from prey that have ingested dietary carotenoids. However, to varying degrees, birds may make minor chemical alterations to them which often change their colors prior to being deposited in tissue (including feathers). Carotenoids occur in about 50% of all orders of birds and are responsible for bright yellows, oranges, and reddish colors (except as discussed later). Carotenoids also frequently overlay structural colors (discussed later) to make a wide variety of brilliant colors, including nearly all greenish colors. No green carotenoids are known, with the possible exception of the green facial feathers of eiders, Somateria spp.

Porphyrins (tetrapyrroles), other than hemoglobin itself, are synthesized from catabolism of hemoglobin or are synthesized from the amino acid glycine. They occur in about 25% of all orders of birds, especially in owls (Strigiformes), bustards (Otididae), and turacos (Musophagidae), and they produce mainly dull pinkish, brownish, or rusty colors. However, turacos contain two copper-containing porphyrins: turacin, which is blue, and turacoverdin, which is green and is thought to be the oxidized form of turacin.

There are two classes of structural colors: structural blue caused by Tyndall or Raleigh scattering (Fig. 3) and iridescence caused by interference of light waves reflected from the surfaces of very thin multiple laminations separated by equally thin or thinner layers of material possessing a contrasting refractive index. Other than blue in turacos, all blue plumage and soft part color is structural. Iridescent colors span the visible spectrum from blue to red and are best exemplified by New World hummingbirds (Trochilidae) (Fig. 4) and their Old World ecological counterparts, sunbirds (Nectariniidae).

Recent advances in technology have improved our ability to characterize and correctly identify various chemically labile pigments. However, researchers have not yet been able to identify many pigments in the plumage of some birds, e.g., many of the yellow and greenish colors in parrots (Psittaciformes), Rollulus spp. (Phasianidae), and Corythaeola spp. (Musophagidae).

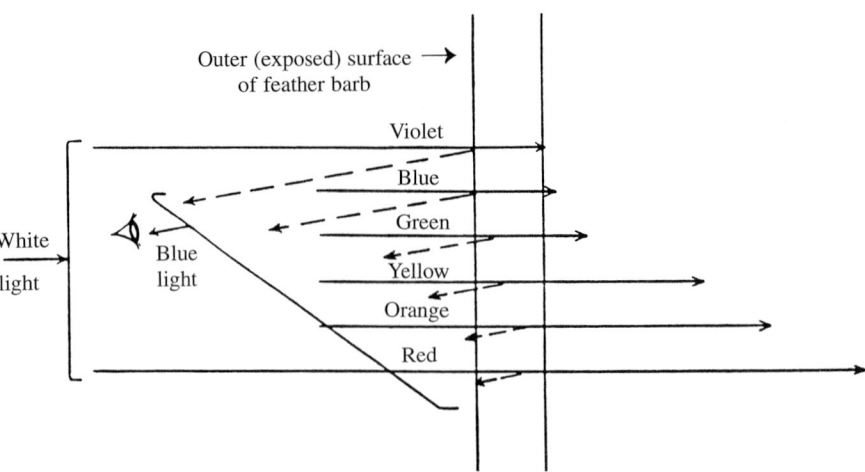

FIGURE 3 Tyndall or Raleigh effect. Within feather barbs (the structures attached to each side of the shaft of each feather) tiny particles or air spaces that are about the same size as blue wavelengths (450 nm or 0.45 μm) reflect the blue wavelengths of incident white light back to the observer, whereas longer wavelengths either pass through the tissue or more commonly are absorbed by a dense underlying layer of melanin.

B. Cryptic versus Conspicuous Plumage Color

The terms cryptic and conspicuous refer to the coloration of an object in relation to its background, i.e., an object cannot be inherently cryptic or conspicuous. For example, parrots, although generally quite colorful, are also quite cryptic in the tropical forests in which they live. Thus, birds that are highly visible against their background are considered conspicuous, whereas birds that are very difficult to discern against their background are considered cryptic. Crypsis may be achieved by mimicking background coloration [e.g., nightjars (Caprimulgiformes)], countershading (most seabirds), or disruptive coloration (many shorebirds). This raises the question: Why are birds conspicuous versus cryptic in a given season? In a general sense, the answer is simple: Natural selection favors birds that maximize their reproductive success and survival. Thus, in species that have two complete body molts per year, the advantages of wearing a conspicuous plumage in the breeding season (and possibly during the nonbreeding season as well) must outweigh the costs of doing so. However, identifying the specific ecological advantages and disadvantages of cryptic versus conspicuous plumage in the breeding or nonbreeding season is difficult.

C. Seasonal Monochromatism

Adult birds that do not change plumage color throughout the year are seasonally monochromatic. These species may be sexually monochromatic as well, such as crows (Corvidae), titmice (Paridae), swallows (Hirundinidae), waxwings (Bombycillidae), and most parrots (Psittaciformes), or sexually dichromatic such as woodpeckers (Picidae) and kingfishers (Alcedinidae). The plumages of these species may be either cryptic [e.g., many sparrows (Emberizidae)] or conspicuous [e.g., certain kingbirds (Tyrannidae) and waxwings (Bombycillidae)].

FIGURE 4 Photograph of a hummingbird with an iridescent forehead, chin, and throat.

In species that are sexually dichromatic, males typically are brighter and more conspicuous than females, e.g., northern cardinals (*Cardinalis cardinalis*), painted buntings (*Passerina ciris*), and phainopeplas (*Phainopepla nitens*).

D. Seasonal Dichromatism

Adult birds whose coloration in the breeding season differs significantly from that in the nonbreeding season are seasonally dichromatic. These color changes usually result exclusively from changes in plumage color, but in many species they can also result from changes in the colors of soft parts. Most soft parts are permanent structures that are not molted. However, the outer covering of the beak, known as the rhamphotheca, may become exaggerated in shape and color during the breeding season by growing extra plates or projections on its existing structure. This occurs, for example, in male white pelicans, *Pelicanus erythorhynchus*, many auks (Alcidae) such as puffins (*Fratercula* spp.), and some penguins (*Aptenodytes* spp.). These extra plates and projections are molted at the end of the breeding season and are regrown many months later during prenuptial molt, i.e., shortly before the next breeding season. Species that change color seasonally may be sexually monochromatic, such as starlings (Sturnidae), most seabirds [e.g., auks, albatrosses (Diomedidae), shearwaters (Procellariidae)], and gulls and terns (Laridae), or sexually dichromatic, such as wood warblers (Parulidae), tanagers (Thraupidae), and grosbeaks (Emberizidae). Breeding plumage in most of these species is brighter and more conspicuous than their nonbreeding plumage, although rare exceptions exist, e.g., marbled murrelets (*Brachyramphus marmoratus*). Similarly, in sexually dichromatic species, males usually are brighter and more conspicuous than females, although rare exceptions also exist, e.g., phalaropes (Scolopacidae) and some kingfishers. As mentioned previously, birds may change plumage color between nonbreeding and breeding seasons by molt (most species), feather wear [e.g., European starlings (*Sturnus vulgaris*) and many blackbirds (Icteridae)], or a combination of both [e.g., varied buntings (*Passerina versicolor*)].

E. Age Monochromatism versus Dichromatism

In a few very reproductively prolific species [e.g., mourning doves (*Zenaida macroura*)], birds may become reproductively mature and breed successfully while still in their first (juvenal) or a subsequent subadult plumage. However, most species acquire adult plumage at the same time or before they acquire reproductive maturity. Indeed, in long-lived species, birds may acquire this plumage long before sexual maturity, e.g., many seabirds such as auks and albatrosses. The color pattern of this plumage typically does not change with increasing age, although some characters may become slightly larger or brighter, and females may acquire some subtle male-like coloration. However, in a significant percentage of taxonomically, ecologically, and geographically diverse sexually dichromatic species, males exhibit delayed plumage maturation (DPM). Typically restricted to their first potential breeding season [but potentially for many years in some tropical manakins (Pipridae) and birds-of-paradise (Paradisaeidae)], reproductively mature males wear subadult plumage that in most species is female-like or intermediate between males and females in coloration. DPM usually occurs in species in which adult males significantly outnumber adult females; this increases competition among males for females and allows females to be more selective about males with whom they choose to pair and/or mate. Adult male plumage in these species is typically very bright and conspicuous, and the cost to males of wearing this plumage is high. These males typically suffer considerable mortality from predators because of their conspicuous plumage color; they also expend substantial time and energy obtaining and defending a territory as well as attracting and defending a mate. In addition, subadult males are less experienced and thus less able to compete successfully with adult males and/or to attract adult females. Thus, their chances of gaining any reproductive success during their first potential breeding season are low, whereas their chances of dying due to predation and/or interspecific competition are significant. As a result, natural selection presumably has favored subadult males to remain relatively cryptic during their first potential breeding season by delaying acquisition of adult plumage until their second or subsequent potential breeding season. Much more rarely, DPM is known to occur in females but not males [e.g., tree swallows (*Tachycineta bicolor*)]; rarer still, DPM may occur in both males and females, e.g., waxwings (*bombycilla* spp.) and orange-breasted or Leclancher's buntings (*Passerina leclancherii*). The reasons for DPM in females of these species are not clear but presumably relate to competition, e.g., competition among females for limited nest holes in tree swallows.

F. Polymorphism

Many geographically, ecologically, and taxonomically diverse species exhibit polymorphisms in plumage color that are not related to age or sex. Some polymorphisms are correlated with gradients or clines in environmental conditions such as humidity, temperature, and soil coloration [e.g. various white-eyes (*Zosterops* spp.) and sooty-capped tanager (*Chlorospingus pileatus*)]. Rarely, polymorphisms may play a direct function in breeding or mate attraction. For example, different color morphs of white-throated sparrows (*Zonotrichia albicollis*) and snow/blue geese (*Chen caerulescens*) mate assortatively. However, empirical evidence indicating the adaptive significance of most polymorphisms is weak or nonexistent. Despite this, most polymorphisms are thought to serve functions related to foraging [e.g., in herons (Ardeidae)], thermoregulation [e.g., in screech owls (*Otus asio*)], avoiding detection by prey (e.g., in *Buteo* hawks), or individual recognition [e.g., ruffs and ruddy turnstones (*Arenaria interpres*)].

G. Genetic Control

Ultimately, of course, all phenotypic expression is controlled by the genome. The issue is the degree to which this expression may be influenced by environmental cues mediated by neuroendocrine mechanisms in various species of birds. Not surprisingly, the answer appears to be that species span the continuum from those that are "hard-wired" to develop certain plumage color patterns [e.g., mute swan (*Cygnus olor*), Ross' goose (*Anser rossii*), and color morphs of various domesticated parrots such as budgerigars (*Melopsittacus undulatus*)] regardless of external environmental conditions, to those that are very sensitive and responsive to various external stimuli.

H. Phylogenetic Constraints

Color patterns of birds are obviously subject to very strong and varied selection pressures. There is no evidence that birds are phylogenetically constrained from changing their color patterns in response to selection pressures, or that seasonal color change is constrained to occur by molt versus feather wear or vice versa. For example, the genus *Passerina* is composed of six closely related species of North and Central American buntings that differ tremendously in their color patterns both intraspecifically between ages and sexes and interspecifically. All six species are sexually dichromatic, but the extent of sexual dichromatism differs from slight to extreme. Male DPM exists in all six species, but female DPM exists in only two [orange-breasted and rose-bellied buntings (*P. rositae*)]. Subadult male plumages vary from female-like (e.g., painted and varied buntings) to male-like or intermediate between males and females [e.g., indigo buntings (*P. cyanea*)]. Also, although all species have at least a partial prenuptial molt and a complete postnuptial molt, two species (painted and rose-bellied bunting) are seasonally monochromatic despite having a prenuptial molt, whereas the remaining four species are seasonally dichromatic. Last, seasonal color change in the latter four species occurs exclusively by molt in one species (indigo bunting), almost exclusively by wear in two species [varied and lazuli buntings (*P. amoena*)], and by a combination of molt and wear in one species (orange-breasted bunting).

I. Neuroendocrine Control

Because the physiological mechanisms for (i) synthesizing melanins, (ii) synthesizing porphyrins, and (iii) regulating carotenoid uptake, transport, and metabolism differ dramatically from one another, one might expect that neuroendocrine mechanisms that control deposition of these pigments into plumage and soft parts might also differ substantially. However, this does not appear to be the case. Since sex steroids, and the hypothalamic and pituitary hormones that regulate them, typically increase and decrease at the onset and termination, respectively, of the breeding season, one might expect these hormones to regulate plumage color in seasonally dichromatic species but not in seasonally monochromatic species. Available data generally support this expectation. In all birds that are seasonally and sexually monochromatic [e.g., blue jays (*Cyanocitta cristata*)] and most birds that are seasonally monochromatic but sexually dichromatic [e.g., bullfinches (*Pyrrhula pyrrhula*)], their plumage color appears to be under direct genetic control and is not affected by sex steroids or gonadotropins. However, some species that are seasonally monochromatic but sexually dichromatic are affected by both steroids and gonadotropins; for example, in chickens (*Gallus gallus*) high luteinizing hormone (LH) levels cause male-like plumage to grow during molt, whereas high E_2 levels cause female-like plumage to grow during molt by depressing LH levels. As expected, in sexually monochromatic and sexually dichromatic species that are seasonally dichromatic, male-like plumage and soft part coloration is promoted by sex steroids and gonadotropins, usually in one of two ways—either by testosterone [e.g., various *Larus* gulls, ruffs, house sparrows (*Passer domesticus*), and European starlings] or by LH which is suppressed by E_2 but is unaffected by T [e.g., ducks (Anatidae), *Euplectes* weaver finches, estrildid finches, indigo bunting, and paradise Whydah (*Vidua paradisaea*)]. However, there are many exceptions and variations on this general scheme. For example, in masked weavers (*Quelea quelea*), bill color changes in females, but not males, in response to E_2 but does not change in either sex in response to T. Also, in species with so-called reverse sexual dichromatism in which females have more conspicuous and colorful plumage than males [e.g., phalaropes, painted snipes (*Rostratula* spp.), and button quail (*Turnix* spp.)], the brighter plumage of females results from higher circulating plasma levels of T in females (from their ovaries) than in males.

J. Effect of Diet, Body Condition, and Stress on Plumage Color

For more than a century, aviculturalists have known that both diet and stress [e.g., cage conditions reflected by cage size, cage content (e.g., the presence or absence of plants and shelter), and bird density] can influence plumage color in some bird species but not in others. Unfortunately, we know little more today than we did a century ago about the underlying physiological and neuroendocrine mechanisms that regulate variation in plumage color in those species that are affected by conditions in captivity. In short, we know that plumage and soft part color patterns resulting solely from melanins, structural colors, or iridescent colors are unaffected by stress and diet. However, birds whose plumage and soft part colors result, at least in part, from carotenoids are extremely susceptible to such conditions; this includes most species with bright yellow, orange, or red plumage color—well-known examples include house finches (*Carpodacus mexicanus*) and painted buntings. When such species are exposed to limited supplies of certain carotenoids or become physically debilitated during the time of year that they molt (e.g., due to stress caused by high cage density or poor health), they grow plumage that is less brightly colored than normal, e.g., yellowish or orangish plumage instead of reddish plumage. In addition, certain frugivorous species, such as cedar waxwings (*Bombycilla cedrorum*), occasionally eat fruits from nonnative plants that contain red carotenoids that they have not been exposed to evolutionarily. These carotenoids apparently circumvent the normal carotenoid metabolism of these birds and are deposited into their plumage so that when they eat these fruits, they grow red instead of yellow tips on the ends of their tail feathers.

K. Ecological Functions of Plumage Color during the Breeding Season

Most birds wear one or two plumages during each year. In birds that change plumage seasonally by undergoing complete body molts both before and after breeding, it is reasonable to assume that the plumages grown during prenuptial and postnuptial molt are adapted to the breeding season and nonbreeding season, respectively. However, in species that wear only one plumage throughout the year or that have an incomplete prenuptial molt, two explanations are possible. First, selection pressures during winter and summer may be sufficiently similar that it is not advantageous to undergo a prenuptial molt. Alternatively, some of these species may be constrained (e.g., energetically) from undergoing two complete body molts per year. As a result, plumages of these species may not be optimally adapted to either the breeding or nonbreeding season. Given this caveat, adult plumages may serve the following functions during the breeding season:

1. Individual recognition: Intraspecific variation in color patterns (within or between age or sex classes) may facilitate individual recognition. However, this function would not favor evolution of conspicuous plumage more than cryptic plumage or vice versa.
2. Male quality: In male–male interactions, bright coloration may indicate male status (quality) to potential aggressors, thereby reducing the number and intensity of male–male aggressive interactions. Similarly, bright color may help attract conspecific females.
3. Parasite resistance: As a specific example of the "male quality" argument discussed previously, bright males (or females in species with reverse sexual dichromatism) may indicate to potential mates (usually females) that they are healthy and free of disease and parasites. Such males may provide direct benefits by reducing the chance of transmitting disease to their mate and/or offspring and by providing better care for them. If plumage brightness is correlated with genetically based resistance to disease, then males also may provide indirect benefits to females by contributing better genes to their offspring than duller males could contribute.
4. Unprofitable prey: Although lacking much empirical support, bright coloration of birds may indicate to potential predators that they are sufficiently adept at avoiding predation that it is not worth the predator's energy to try to predate them.
5. Aposomatism: Also lacking much empirical support, bright plumage and soft part coloration may indicate that birds are unpalatable to eat and thus should not be preyed upon.
6. Reproductive isolation: In the same way that vocalizations and behavioral displays facilitate reproductive isolation, so may color patterns serve this function in whole or part. However, there are precious few data to support this idea since many taxa that are visually quite distinct from one another often interbreed extensively [e.g., red- and yellow-shafted flickers (*Colaptes auratus auratus* and *C. a. cafer*, respectively) and Myrtle and Audubon's Warblers (*Dendroica coronata coronata* and *D. c. auduboni*, respectively)] and are considered single species, whereas other pairs or complexes of species are composed of species that are visually nearly identical and interbreed rarely if at all [e.g., oak titmouse (*Baelophus inornatus*) and juniper titmouse (*B. ridgwayi*), Pacific slope flycatcher (*Empidonax difficilis*) and cordilleran flycatcher (*E. oberholseri*), and black-tailed gnatcatcher (*Polioptila melanura*) and California gnatcatcher (*P. californica*)].

See Also the Following Articles

ANDROGENS, EFFECTS IN BIRDS; ESTROGEN, EFFECTS IN BIRDS; PHOTOPERIODISM, VERTEBRATES; SEASONAL REPRODUCTION, BIRDS; SEX DIFFERENTIATION IN AMPHIBIANS, REPTILES, AND BIRDS, HORMONAL REGULATION OF

Bibliography

Brush, A. H. (1978). Avian pigmentation. In *Chemical Zoology* (A. H. Brush, Ed.), Vol. X, pp. 141–164. Academic Press, New York.

Brush, A. H. (1990). Metabolism of carotenoid pigments in birds. *Fed. Am. Soc. Exp. Biol. J.* **4**, 2969–2977.

Burtt, E. H., Jr. (1986). An analysis of physical, physiological, and optical aspects of avian coloration with emphasis on wood-warblers, Ornithological Monographs No. 38. American Ornithologists' Union, Washington, DC.

Butcher, G. S., and Rohwer, S. (1989). The evolution of conspicuous and distinctive coloration for communication in birds. In *Current Ornithology.* Vol. 6 (D. M. Power, Ed.), pp. 51–108. Plenum Press, New York

Fox, D. L. (1979). *Animal Biochromes and Structural Colors*, 2nd ed. Univ. of California Press, Berkeley.

Hahn, T. P., Swingle, J., Wingfield, J. C., and Ramenofsky, M. (1992). Adjustments of the prebasic molt schedule in birds. *Ornis Scand.* **23**, 314–321.

Hailman, J. P. (1977). *Optical Signals: Animal Communication and Light*. Indiana Univ. Press, Bloomington.

Humphrey, P. S., and Parkes, K. C. (1959). An approach to the study of molts and plumages. *Auk* 76, 1–31.

Jenni, L., and Winkler, R. (1994). *Moult and Aging of European Passerines*. Academic Press, New York.

Murphy, M. E. (1996). Energetics and nutrition of molt. In *Avian Energetics and Nutritional Ecology* (C. Carey, Ed.), pp. 158–198. Chapman & Hall, New York.

Palmer, R. S. (1972). Patterns of molting. In *Avian Biology* (D. S. Farner, J. R. King, and K. C. Parkes, Eds.), Vol. 2, pp. 65–102. Academic Press, New York.

Payne, R. B. (1972). Mechanisms and control of molt. In *Avian Biology* (D. S. Farner, J. R. King, and K. C. Parkes, Eds.), Vol. 2, pp. 104–155. Academic Press, New York.

Pyle, P., Howell, S. N. G., DeSante, D. F., Yunick, R. P., and Gustafson, M. (1997). *Identification Guide to North American Birds. Part 1. Columbidae to Ploceidae*. Slate Creek Press, Bolinas, CA.

Voitkevich, A. A. (1966). *Feathers and Plumage of Birds*. Sidgwick & Jackson, London.

Wingfield, J. C. (1983). Environmental and endocrine control of reproduction: An ecological approach. In *Avian Endocrinology: Environmental and Ecological Aspects* (S.-I Mikami, S. Ishii, and M. Wada, Eds.), pp. 121–148. Japanese Scientific Societies/Springer-Verlag, Tokyo/Berlin.

Wingfield, J. C., Ishii, S., Kikuchi, M., Wakabayashi, S., Sakai, H., Yamaguchi, N., Wada, M., and Chikatsuji, K. (1998). Biology of a critically endangered species, the "Toki" (Japanese Crested Ibis), *Nipponia nippon*. *Ibis*, in press.

Witschi, E. (1961). Sex and secondary sexual characters. In *Biology and Comparative Physiology of Birds* (A. J. Marshall, Ed.), Vol. 2, pp. 115–168. Academic Press, New York.

Monotremes

Mervyn Griffiths
CSIRO Division of Wildlife and Ecology

I. *Tachyglossus*
II. *Ornithorhynchus*
III. *Zaglossus*
IV. Conclusions

GLOSSARY

anlage During embryonic development, the earliest discernible development of an organ or part.

bifid Divided into two distinct parts.

cloacate Having a cloaca (a chamber for the pasage of reproductive and excretory products); involving or transported by the cloaca.

intussusception Growth of an organ or part by means of the reception of new matter from an external source.

sauropsid Relating to or resembling modern reptiles and birds.

testicond Having the testes normally retained within the abdominal cavity.

The Monotremata, the egg-laying mammals, comprise an order of the infraclass Prototheria of the class Mammalia. There are three living representatives: the short-beaked echidna (spiny anteater) *Tachyglossus aculeatus,* occurring in Australia and New Guinea; the long-beaked echidna *Zaglossus bruijnii,* found in the Central Cordillera of New Guinea; and the platypus *Ornithorhynchus anatinus,* an amphibious mammal found in the freshwater rivers and lakes of eastern Australia from northern Queensland to southern Tasmania. Nothing is known of breeding in *Tachyglossus* in New Guinea but some data are available for *Zaglossus*; the only detailed information on breeding in monotremes has come from studies of *Tachyglossus* and *Ornithorhynchus* in southeastern Australia. Since the anatomical layout of the reproductive organs is similar and the cytology of the gonads is identical in the three genera, a procedure has been adopted for the descriptions in this article that uses an illustration of a condition in one genus to serve for those in the other two.

I. TACHYGLOSSUS

A. Male

Echidnas and platypuses are testicond and cloacate (Fig. 1). The seminiferous tubules of each testis communicate with an epididymus, differentiated into a caput and cauda, which leads into a vas deferens (Djakiew and Jones, 1981). Sperm pass through this to a median urogenital sinus from which at its posterior ventral surface there is a muscular penis with an incipiently bifid glans, the rounded surface of which bears two soft epidermal rosettes giving, upon erection, a quadripartite anemone-like appearance.

The penis is housed in a preputial sac and upon erection it passes through an opening into the cloaca and through this to the exterior. The penis conveys sperm only.

Echidnas breed only in spring. The testes show marked seasonal variations: during October–March (summer in Australia) they are quiescent with

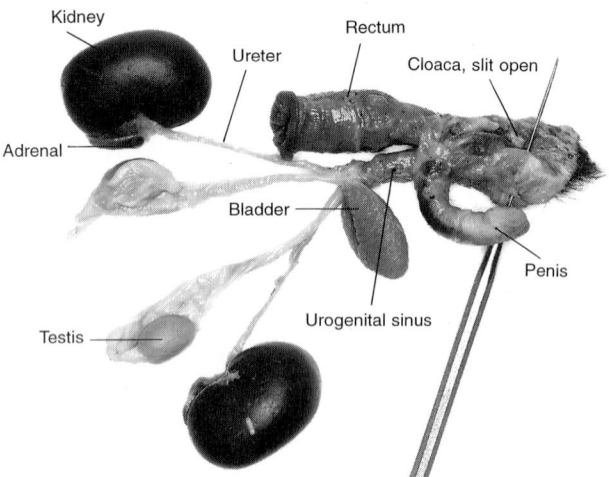

FIGURE 1 *Tachyglossus* male reproductive organs. Magnification, ×0.5. The preputial sac has been removed to show the penis; the testes are in the regressed state. The probe indicates the passage from preputial sac to cloaca (from Griffiths, 1968).

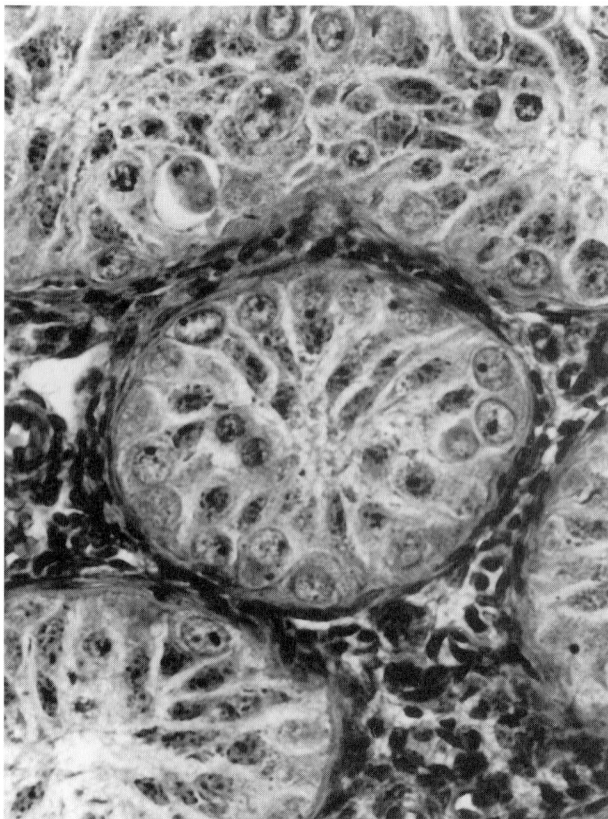

FIGURE 2 *Tachyglossus* testis. Transverse sections of resting-phase tubules showing spermatogonia at the periphery and Sertoli cells with elongated nuclei. Magnification, ×346 (from Griffiths, 1978, with permission of Academic Press).

weights ranging from 1 to 3 g/kg body weight and with tubule diameters measuring 70–150 μm. After March, they increase gradually to a peak in late August (spring) of 17 g/kg, the set with tubules exhibiting a diameter of 370 μm. At the end of September a sudden regression takes place and the testes resume the condition they had the previous October (Griffiths, 1978). Along with these events the cytology of the testes changes. In summer and autumn the seminiferous tubules have no lumina and consist of a peripheral ring of spermatogonia (Fig. 2) interspersed with elongated Sertoli cells. By June the tubules exhibit primary and secondary spermatocytes, spermatids, and the first signs of a lumen (Fig. 3). In July the tubules are ripe, with fully developed lumina containing spermatozoa and exhibiting an epithelium consisting of spermatogonia, primary and

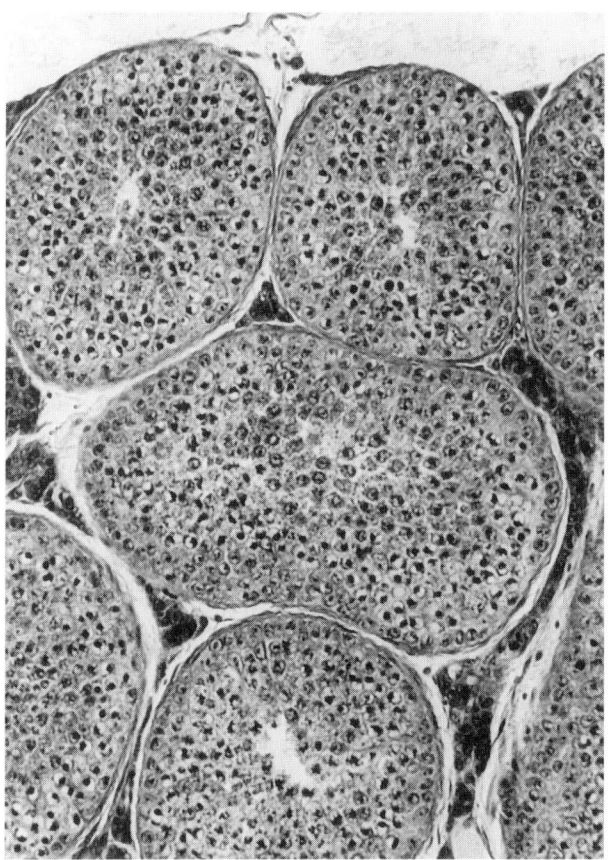

FIGURE 3 *Ornithorhynchus* testis. Sections of tubules showing advanced stage of spermatogenesis. Lumina appear with spermatids at the border of lumen. Magnification, ×170 (from Griffiths, 1978, with permission of Academic Press).

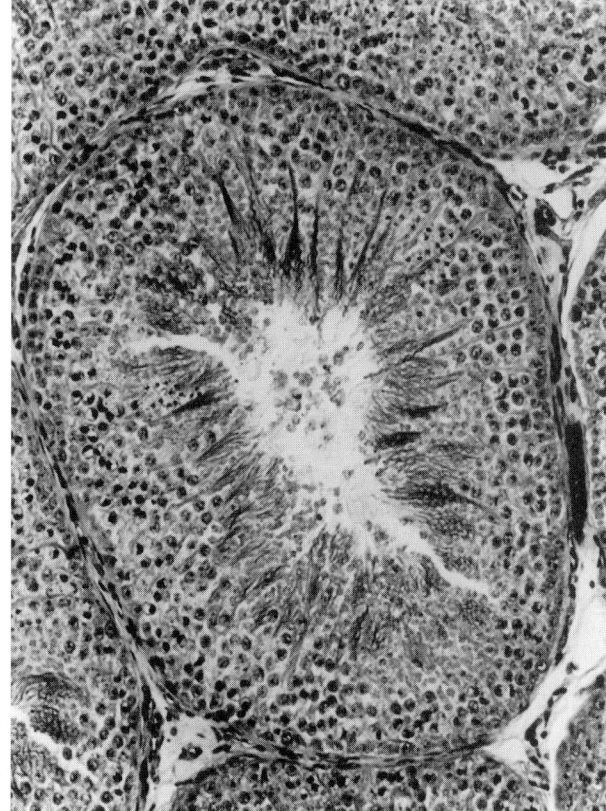

FIGURE 4 *Zaglossus* testis. Section of ripe tubule showing filiform spermatozoa with their coiled heads embedded in the elongated Sertoli cells. Magnification, ×200 (from Griffiths, 1978, with permission of Academic Press).

secondary spermatocytes, spermatids, and spermatozoa embedded in Sertoli cells (Fig. 4). Benda (1906) and Carrick and Hughes (1982) have given details of spermiogenesis; the end result is a filiform spermatozoan, with a corkscrew-shaped head, that very much resembles those of sauropsidan reptiles. The spermatozoa shed into the lumen are immobile but on passing to the epididymis they undergo a maturation process and become motile (Bedford and Rifkin, 1979; Jones *et al.*, 1992). In the Monotreme the males are the heterogamatic sex and the sex-determining mechanism consists of a multivalent system of X and Y chromosomes (Bick and Sharman, 1975; Watson *et al.*, 1992; Bick, 1992). According to the latter, the multivalent takes the form $X_1Y_1X_2Y_2X_3Y_3X_4Y_4X_5$ in *Tachyglossus* and *Zaglossus* and $X_1Y_1X_2Y_2X_3Y_3X_4Y_4$ in the platypus, with the diploid number in the echidnas being 63♂ 64♀ and 52 52 in the platypus.

B. Female

The female tract consists of paired ovaries exhibiting ripe follicles up to 5 mm in diameter. Each ovary is enclosed by an infundibular funnel leading to a convoluted Fallopian tube which in turn leads to a muscular uterus, the mucosa of which is packed with convoluted glands opening into the lumen of the uterus. The paired uteri open separately into the anterior end of a urogenital sinus opening to the cloaca (Fig. 5; although this figure illustrates the platypus tract it serves for that of the echidnas, except that they have two functional ovaries).

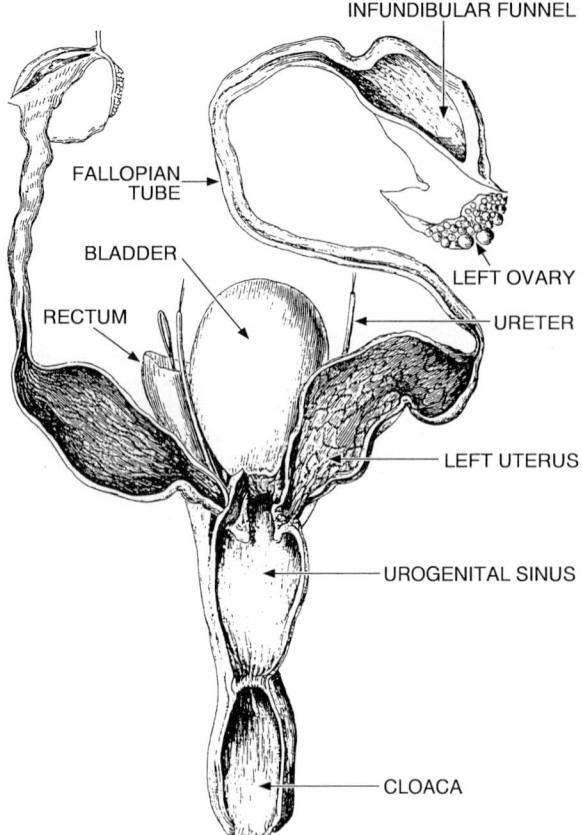

FIGURE 5 *Ornithorhynchus* female reproductive system. Left ovary only is functional. Magnification, ×1.2.

The only data on seasonal changes in the ovaries indicate that the ovaries are very small in the summer months; the uteri are thin and strap-like at that time. It is known that laid eggs can be obtained only from late June to late October (Griffiths, 1984). *Tachyglossus* does not breed every year—one egg, rarely two, is laid.

Flynn and Hill (1939) provided the definitive description of oogenesis. Preparations from *Tachyglossus* and *Ornithorhynchus* were used to give a sequential account of "monotreme" oogenesis. They distinguished three phases of development of the egg within a follicle from a tiny oocyte ≈0.05 mm in diameter and a relatively huge yolk-laden egg 4 mm in diameter housed in a 5-mm follicle. The egg exhibits an organelle found in the eggs of reptiles: a flask-shaped area of yolk-free cytoplasm known as a latebra, with the neck of the flask contacting the eccentrically placed nucleus as in the eggs of reptiles.

Following eruption from the follicle a corpus luteum forms therein and the ripe egg is passed to the Fallopian tube to await fertilization following copulation. We are indebted to Dr. Peggy Rismiller's devoted and patient fieldwork for description of copulation in wild echidnas (Rismiller and Seymour, 1991): At the start of the breeding season "trains" and groups of echidnas come together that consist of one female and two to seven males. These follow the female nose-to-tail, sometimes for days. Finally, one male takes over, digs a doughnut-shaped trench around the posterior end of the female, squirms down beside her, and manages to place his cloacal aperture on hers and to insert his penis. They may remain in this position for as long as an hour.

Following fertilization the anlagen of the first two layers of the three-layered keratinous eggshell are laid down in the infundibulum at its junction with the uterus. The ovum then passes to the uterus, where it undergoes cleavage which is meroblastic as in the Sauropsida (Caldwell, 1887; Flynn and Hill, 1947). The gestation period in the uterus is 21–23 days (P. D. Rismiller, personal communication), during which the eggshell grows by intussusception and the embryo grows and differentiates. The nourishment for this comes overwhelmingly from protein-rich secretions from the uterine glands (C. J. Hill, 1933, 1941) as happens in the gestation of marsupials; the contribution of the yolk in the 4-mm egg to development is almost negligible.

When the egg attains a size of 12 mm diameter and contains a 10-somite embryo, the third, protective, layer of the shell is laid down and it grows to a diameter of 15–16.5 mm (J. P. Hill, 1933); the embryo is now at a 40-somite stage. The egg is then laid into a ventrally located pouch, probably by apposition of the cloaca which is eversible at this stage (Fig. 6). Here it is incubated at the mother's body temperature of 32°C for 10 or 11 days (Griffiths, 1978); this is the period of organogenesis, which takes place at the expense of uterine secretions stored in the egg before laying. At the end of this period the egg hatches (Fig. 7). The first ever observation of this process took place at September 17, 1967 (Griffiths *et al.*, 1969) and was filmed in slow motion [see CSIRO Film and Video Centre, *The Comparative Biology of Lactation*(film), 1974].

The hatchlings have an eggtooth and a caruncle

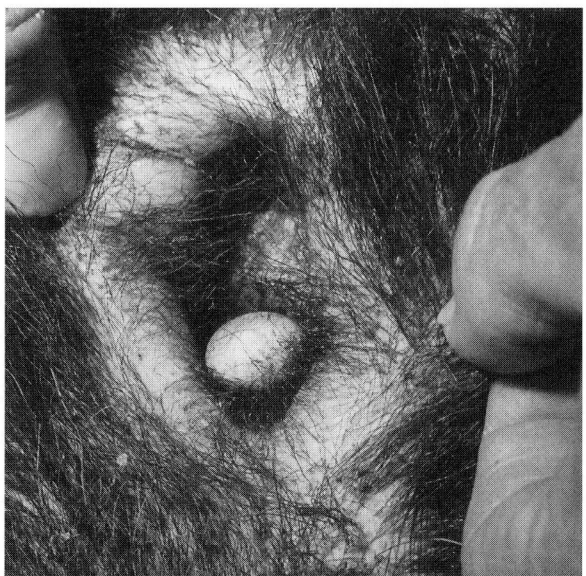

FIGURE 6 *Tachyglossus* egg in the pouch on day 8 of incubation (from Griffiths *et al.*, 1969, with permission of the Zoological Society of London).

on the premaxilla bone for tearing open the shell and weigh about 380 mg (Griffiths *et al.*, 1969; P. D. Rismiller, personal communication). Figure 8

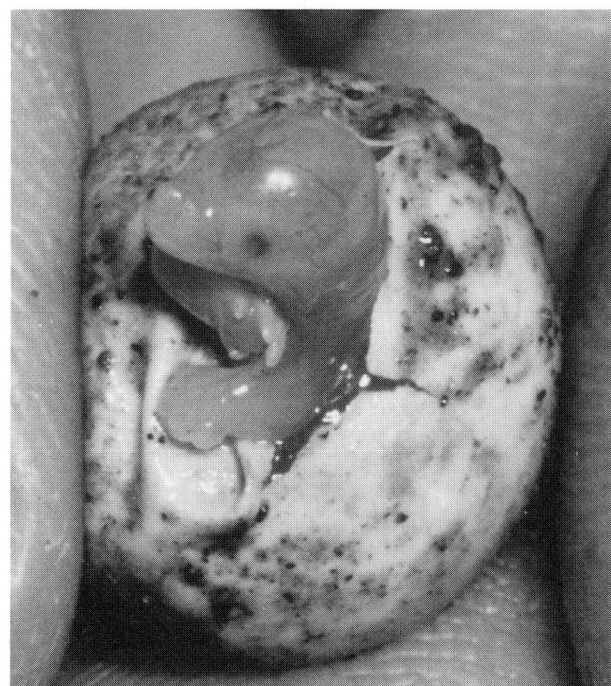

FIGURE 7 *Tachyglossus* young emerging from the egg after ≈10 days and 6 hr incubation (CSIRO photograph by Ederic Slater).

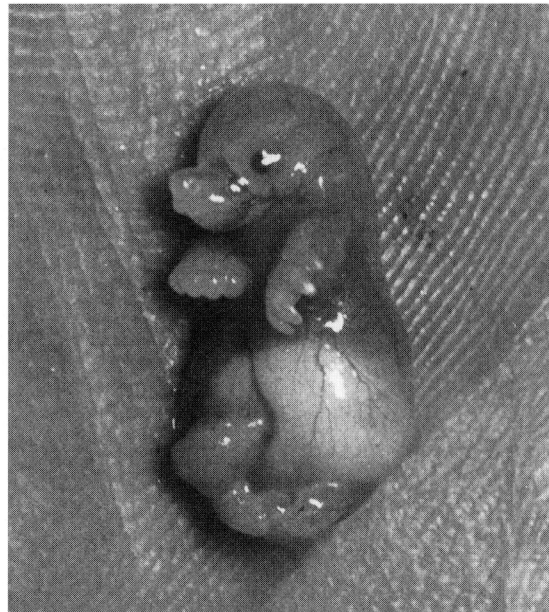

FIGURE 8 *Tachyglossus* newly hatched showing relatively great development of the forelimbs. Magnification, ×4.6 (from Griffiths *et al.*, 1969, with permission of the Zoological Society of London).

shows a recently hatched young. It had already imbibed milk from one of the two mammary glands opening into the pouch. The young one stays here for an average of 53 days (Griffiths, 1989) until it starts to grow spines, at which point it is ejected and placed in a burrow where it is suckled from time to time. It is weaned after a total lactation period of 200–210 days (Griffiths *et al.*, 1988; P. D. Rismiller, personal communication).

II. ORNITHORHYNCHUS

A. Male

Apart from the fact that the penis is quite bifid, the left glans being longer than the right, the reproductive organs are similar to those of *Tachyglossus* and show the same changes associated with a distinct breeding season: peak testes weights and tubule diameters (14.5 g/kg and 320 μm, respectively) being attained in August, with spermiogenesis in full swing (Temple-Smith, 1973; Griffiths, 1978). Along with those peaks, spermatic vein plasma testosterone increases from 10 ng/ml in summer to 400 ng/ml in August (Carrick and Cox, 1977).

Regression is not as precipitous as in *Tachyglossus*; spermiogenesis may be seen in some tubules in December. The spermatozoa are filiform as in *Tachyglossus*.

B. Female

Only the left ovary is functional (Fig. 5). Ovaries are small in summer, averaging 0.2 g/kg in weight and attaining a peak weight of 0.6 g/kg in August (Temple-Smith, 1973). Along with that growth plasma progesterone levels increase from a summer low of 2.6n mol/l to 27 n mol/l in August (Handasyde *et al.*, 1992). As in *Tachyglossus* mating occurs in the spring but with the platypus courtship and copulation take place in the water (Strahan and Thomas, 1975).

The cytology of the secretion of nutritive fluids and of the precursors of the shell layers has been described by C. J. Hill (1933, 1941). The shell is three layered as in *Tachyglossus*. The gestation period is unknown but is at least 9 days (F. N. Carrick, personal communication); the end result of gestation is an egg slightly larger, up to 17.5 mm diameter, than that of *Tachyglossus*. Despite having only one functional ovary, platypuses can lay 1–3 eggs at a time (Burrell, 1927). The platypus has no pouch but it is surmised that, in the privacy of a nesting burrow, she places her eggs via an eversible cloaca between her broad curled-over tail and her abdomen up against the areolae (milk patches) of her two mammary glands (Burrell, 1927). The fact that the eggs are sticky reinforces the notion that they are retained there for incubation.

Incubation time is unknown but it is thought to be ≈11 days as in *Tachyglossus* since the temperature between curled-over tail and abdomen is 31.8°C at an ambient temperature of 20°C (Grant, 1976), the eggs are practically the same size as those of *Tachyglossus*, and the embryo within at laying is at the same stage, i.e., it has 40 somites (Hughes and Carrick, 1978).

The female suckles her young for at least 4 months (Grant and Griffiths, 1992) and the nestlings leave the burrow and take to the water in late January to early March.

III. *ZAGLOSSUS*

The male and female reproductive systems are the same as those in *Tachyglossus* (Kolmer, 1925; Griffiths, 1984). There are no data available on gestation period, size of eggs, degree of development of embryos at time of laying, incubation period, or on how long the young is carried in the pouch.

Whether *Zaglossus* has a breeding season is debatable. The few facts regarding this issue are as follow: Five adult males taken in the Wharton Ranges of Papua New Guinea in the month of July had testes at varying degrees of development (Table 1). One with a tubule diameter of 84 μm had spermatogonia that divided and gave rise to primary and secondary spermatocytes only; the others showed advanced spermatogenesis ranging to full spermiogenesis (Fig. 4). The testis of the animal taken in late October showed a mixture of spermiogenesis, partial regression, and masses of trapped degenerating spermatozoa. These data suggest that in this part of New Guinea *Zaglossus* starts to come into breeding condition in late June; the breeding season reaches a peak in July and regresses in October. Two lone observations of females tend to support this notion: One animal taken in July had oocytes up to 2.1 mm in diameter, uteri with well-developed glands, and mammary glands at a stage comparable to those of *Tachyglossus* at the start of its breeding season (Griffiths, 1984). The ovaries of the other female, which was taken October 4 in another part of the Central Cordillera, had large follicles up to 4.0 mm in diameter (M. Griffiths, unpublished data).

IV. CONCLUSIONS

Reproduction in monotremes is a mosaic of that found in marsupials and in reptiles: internal testes, filiform spermatozoa, cloacas for passage of reproductive and excretory products, oviparity showing large yolk-laden eggs with latebras, and meroblastic cleavage are characteristics of sauropsid reproduction. On the other hand, a prolonged gestation period during which protein-rich nutrients are taken up through the eggshell followed by emergence of a hatchling resembling very much a marsupial neonate are characteristic of marsupial reproduction (Grif-

TABLE 1
Weight and Mean Tubule Diameter of Testes of Adult *Zaglossus brujnii* from Central Papua[a]

Date sampled	Body wt (kg)	Testis wt (2)(g)	Testis wt/body wt (g/kg)	Mean tubule diameter (μm)
July 5, 1972	8.0	100	15.0	350
July 7, 1972	6.0	86	14.3	300
July 7, 1972	6.6	66	10.0	240
July 10, 1972	7.6	4	0.5	84
July 14, 1972	7.0	88	12.6	300
October 27, 1973	7.0	65 (32.5 × 2, only one weighed)	9.3	240

[a] From Griffiths (1978) with permission of Academic Press.

fiths, 1984), especially because marsupial blastocysts are enclosed in a keratinous eggshell for a major part of gestation (Renfree, 1982).

See Also the Following Articles

Marsupials; Oogenesis, in Mammals; Seasonal Reproduction, Mammals

Bibliography

Bedford, J. M., and Rifkin, J. M. (1979). An evolutionary view of the male reproductive tract and sperm maturation in a monotreme mammal—The echidna, *Tachyglossus aculeatus*. *Am. J. Anat.* **156**, 207–215.

Benda, C. (1906). Die Spermiogenese der Monotremen. *Denkschr. Med.-Naturwiss. Ges. zu Jena* **6**, 413–438.

Bick, Y. A. E. (1992). The meiotic chain of chromosomes of Monotremata. In *Platypus and Echidnas* (M. L. Augee, Ed.). Royal Zoological Society of New South Wales, Sydney.

Bick, Y. A. E., and Sharman, G. B. (1975). The chromosomes of the platypus (Ornithorhynchus; Monotremata). *Cytobios* **14**, 17–28.

Burrell, H. (1927). *The Platypus*. Angus & Robertson, Sydney.

Caldwell, W. H. (1887). The embryology of the Monotremata and Marsupialia. Part 1. *Philos. Trans. R. Soc. London. B* **178**, 463–486.

Carrick, F. N., and Cox, R. I. (1977). Testicular endocrinology of marsupials and monotremes. In *Reproduction and Evolution* (J. H. Calaby and C. H. Tyndale-Biscoe, Eds.). Australian Academy of Science, Canberra.

Carrick, F. N., and Hughes, R. L. (1978). Reproduction in female monotremes. *Aust. Zool.* **20**, 233–253.

Carrick, F. N., and Hughes, R. L. (1982). Aspects of the structure and development of monotreme spermatozoa and their relevance to the evolution of mammalian sperm morphology. *Cell Tissue Res.* **222**, 127–141.

Djakiew, D., and Jones, R. C. (1981). Structural differentiation of the male genital ducts of the echidna (*Tachyglossus aculeatus*). *J. Anat.* **132**, 187–202.

Djakiew, D., and Jones, R. C. (1983). Sperm maturation, fluid transport and secretion and absorption of protein in the epididymis of the echidna, *Tachyglossus aculeatus*. *J. Reprod. Fertil.* **68**, 445–456.

Flynn, T. T., and Hill, J. P. (1939). The development of the Monotremata. IV. Growth of the ovarian ovum, maturation, fertilization and early cleavage. *Trans. Zool. Soc. London* **24**, 445–622.

Flynn, T. T., and Hill, J. P. (1947). The development of the Monotremata, Part 6. The later stages of cleavage and the formation of the primary germ layers. *Trans. Zool. Soc. London* **26**, 1–151.

Grant, T. R. (1976). Thermoregulation of the Platypus, *Ornithorhychus anatinus*. PhD thesis, University of New South Wales, Sydney.

Grant, T. R., and Griffiths, M. (1992). Aspects of lactation and determination of sex ratios and longevity in a free-ranging population of platypuses, *Ornithorhynchus anatinus* in the Shoalhaven River, NSW. In *Platypus and Echidnas* (M. L. Augee, Ed.). Royal Zoological Society of New South Wales, Sydney.

Griffiths, M. (1968). *Echidnas*. Pergamon Press, Oxford, UK.

Griffiths, M. (1978). *The Biology of the Monotremes*. Academic Press, New York.

Griffiths, M. (1984). Mammals: Monotremes. In *Marshall's Physiology of Reproduction* (G. E. Lamming, Ed.). Churchill Livingstone, Edinburgh, UK.

Griffiths, M. (1989). Tachyglossidae. In *Fauna of Australia, Vol. 1B, Mammalia* (D. W. Walton and B. Richardson, Eds.). Australian Government Publishing Service, Canberra.

Griffiths, M., McIntosh, D. L., and Coles, R. E. A. (1969). The mammary glands of the echidna *Tachyglossus aculeatus*, with observations on the incubation of the egg and on the newly-hatched young. *J. Zool. London* 158, 371–386.

Griffiths, M., Kristo, F., Green, B., Fogerty, A. C., and Newgrain, K. (1988). Observations on free-living lactating echidnas. *Tachyglossus aculeatus* (Monotremata: Tachyglossidae) and sucklings. *Aust. Mammal.* 11, 135–143.

Handasyde, K. A., McDonald, I. R., and Evans, B. K. (1992). Seasonal changes in plasma concentrations of progesterone in free-ranging platypus (*Ornithorhynchus anatinus*). In *Platypus and Echidnas* (M. L. Augee, Ed.). Royal Zoological Society of New South Wales, Sydney.

Hill, C. J. (1933). The development of the Monotremata. 1. The histology of the oviduct during gestation. *Trans. Zool. Soc. London* 21, 413–443.

Hill, C. J. (1941). The development of the Monotremata. V. Further observations on the histology and secretory activities of the oviduct prior to and during gestation. *Trans. Zool. Soc. London* 25, 1–31.

Hill, J. P. (1933). The development of the Monotremata. II. The structure of the egg-shell. *Trans. Zool. Soc. London* 21, 443–476.

Jones, R. C., Stone, G. M., and Zupp, J. (1992). Reproduction in the male echidna. In *Platypus and Echidnas* (M. L. Augee, Ed.). Royal Zoological Society of New South Wales, Sydney.

Kolmer, W. (1925). Zur Organologie und microscopische Anatomie von Proechidna (Zaglossus) bruynii. *Z. wiss. Zool.* 125, 448–482.

Renfree, M. B. (1982). Implantation and placentation. In *Reproduction in Mammals* (C. R. Austin and R. V. Short, Eds.) 2nd ed. Cambridge Univ. Press, Cambridge, UK.

Rismiller, P. D., and Seymour, R. S. (1991, February). The echidna. *Sci. Am.*, 80–86.

Strahan, R., and Thomas, D. E. (1975). Courtship of the platypus, *Ornithorhynchus anatinus. Aust. Zoologist* 18, 165–178.

Temple-Smith, P. D. (1973). Seasonal breeding biology of the Platypus, *Ornithorhynchus anatinus* (Shaw, 1799) with special reference to the male. PhD thesis, Australian National University, Canberra.

Watson, J. M., Meyne, J., and Graves, J. A. M. (1992). Chromosome composition and position in the echidna meiotic translocation chain. In *Platypus and Echidnas* (M. L. Augee, Ed.). Royal Zoological Society of New South Wales, Sydney.

Morning Sickness and Hyperemesis Gravidarum

Marcelo F. Noguera and Marc Jackson

University of Pennsylvania School of Medicine

I. Definition and Prevalence
II. Physiology of Vomiting
III. Etiology
IV. Treatment

GLOSSARY

hyperemesis gravidarum A syndrome of nausea and vomiting of great intensity requiring hospitalization for intensive treatment and prompt correction of fluid and electrolyte imbalance.

morning sickness Nausea and vomiting during the first half of pregnancy, typically commencing between the first and second missed period.

I. DEFINITION AND PREVALENCE

Nausea and vomiting are an almost invariable sign of pregnancy, estimated to occur in up to 90% of gestations. Although distressing for many patients, morning sickness can be considered a favorable sign since women who have nausea and vomiting during pregnancy have a lower rate of miscarriage in the first 20 weeks of gestation than those who do not (Weigel and Weigel, 1989).

In the first trimester, these symptoms are looked upon as one of the major discomforts accompanying pregnancy. However, in a small proportion of pregnancies these conditions persist and worsen, eventually interfering with fluid intake and nutrition and subsequently causing dehydration and electrolyte disturbances.

Nausea and vomiting during pregnancy is almost exclusively a disorder of the first half of gestation. These symptoms typically appear within 8 weeks of the last normal menstrual period and continue to about 16–20 weeks gestational age. Nausea and vomiting are usually worse in the morning hours, but they may continue throughout the day. In nearly every case, the nausea and vomiting are self-limited. Fortunately, the symptoms nearly always abate by the second half of pregnancy.

The syndrome of nausea and vomiting that becomes so severe that it requires hospitalization for management is known as hyperemesis gravidarum (HG). More specifically, Fairweather (1968) defined hyperemesis gravidarum as a condition characterized by vomiting that occurs for the first time before the 20th week of gestation and is sufficiently pernicious to produce weight loss and fluid, electrolyte, and acid-base imbalances. This entity is not well understood and is one for which many practitioners have little sympathy, possibly because we find it difficult to understand its pathogenesis and how to treat it. A number of etiologies have been proposed to explain the syndrome, ranging from impaired psychological behavioral to a wide spectrum of endocrine derangements. However, no single theory seems to provide an adequate explanation for HG.

II. PHYSIOLOGY OF VOMITING

Nausea describes the feeling of an imminent desire to vomit, usually referred to the throat or epigastrium. Vomiting or emesis refers to the forceful oral expulsion of the gastric contents. Vomiting is under control of two distinct medullary centers, the vomiting center and the chemoreceptor trigger zone. The vomiting center controls and integrates the actual act of emesis. It receives afferent stimuli from the gastrointestinal tract and other parts of the body and from higher cortical and specialized brain stem centers, especially from the labyrinthine apparatus and from the chemoreceptor trigger zone. Because the vomiting center is located near other medullary centers regulating respiratory, vasomotor, and autonomic functions, nausea may be incited by illness involving either the gastrointestinal system or the neurological pathways controlling the vomiting reflexes or by other cerebral impulses. Also, certain chemical or pharmacological agents transported by the blood can cause vomiting. The important efferent pathways in the act of vomiting are the phrenic nerves (innervating the diaphragm) and the spinal nerves (innervating the intercostal musculature).

III. ETIOLOGY

The specific cause of morning sickness and hyperemesis gravidarum is unknown. A variety of hypotheses have been proposed, but none has been proven satisfactorily. In fact, the complex physiology of pregnancy and vomiting suggest that several mechanisms may interact to produce the clinical syndrome. Hormonal factors during pregnancy, such as the rise of chorionic gonadotropin, have been implicated, and emotional and psychological factors undoubtedly contribute to the severity of the syndrome. It seems likely that the cause of nausea and vomiting in pregnancy may differ between patients, or at least that the relative contribution of various factors differs.

A. Hormonal Factors

Human chorionic gonadotropin (hCG) is thought to be important in causing nausea and vomiting of pregnancy. The time of peak secretion of hCG coincides with the period of highest incidence of morning sickness. Also, the incidence of HG is higher in twin pregnancy, molar pregnancy, and Down's syndrome, conditions known to have increased maternal hCG levels. The exact role of hCG in causing morning sickness is unclear, though. Measurements of serum hCG, the β fraction of hCG, and urine hCG in women with hyperemesis gravidarum have yielded conflicting results, with higher, lower, and unchanged values reported, respectively, compared to uncomplicated pregnancy (Goodwin et al., 1994)

The steroid hormone 17-hydroxyprogesterone, which is produced by the corpus luteum in early pregnancy, has also been suggested as a contributor to morning sickness and hyperemesis gravidarum. Serum levels of β-hCG and 17-hydroxyprogesterone have been correlated in pregnant patients, but a convincing relation between either of these hormones and a higher incidence or increased severity of nausea and vomiting has not been established.(Soules et al., 1980).

Similarly, abnormal serum concentrations of maternal adrenocorticotropin, cortisol, estriol, estrone, estradiol, follicle-stimulating hormone, thyroid-stimulating hormone (TSH), growth hormone, and prolactin have all been associated with morning sickness, but none has been shown to have a clear causative relationship with the development of HG (Kauppila et al., 1979).

B. Psychological Factors

Among the oldest theories of the pathogenesis of morning sickness and HG are those that attribute the disorder to psychological and behavioral factors. It is noteworthy that in primitive civilizations, excessive morning sickness is essentially unknown; only after the concept of cultural status is achieved do pregnant women suffer from this complaint in significant numbers. (Also, the importance of psychological factors has been demonstrated by cases with disappearance of symptoms when a patient is separated from her family and the relapse that can occur upon returning to the family environment; Caruso et al., 1990). The fact that HG is only seen in human primates, that it is treatable with hypnosis and other forms of suggestion, and that its incidence markedly decreases during wartime are further supporting evidence of at least a partial psychosomatic origin (Katon et al., 1980).

In the early period of psychotherapy morning sickness was referred to as one of the hysterias or neurosis syndromes. Later, it was thought that the vomiting represented an unconscious rejection of the pregnancy. Recent theories have suggested that HG is caused by a mixture of feelings of ambivalence regarding pregnancy and femininity rather than outright rejection of pregnancy. Some authors have reported HG to be more prevalent in women exhibiting particular emotional and personality problems, such as immaturity, dependency, depression, and anxiety. Others have stated that HG is a protest reaction against pregnancy, the result of psychic conflicts deriving from family and marital problems. Katon et al. (1980) suggested a biopsychosocial perspective for HG. In their opinion, HG is a special type of somatization—a pathological behavior in which a dysphoric effect is expressed as physical symptoms. These physical symptoms are expressed in order to elicit care by physicians and family and to gain relief from stressful home situations (Katon et al., 1980).

Currently, though, psychological factors are not thought to be the primary cause of HG. Psychological issues are probably important in only a small minority of cases, and much of the psychological dysfunction seen with HG is likely a result, and not a cause, of the physical ill of the disorder (Macy, 1986).

C. Transient Hyperthyroidism during Pregnancy

Normal changes in thyroid function during pregnancy are well described, and a relation between abnormal thyroid function and HG has been postulated. It has been suggested that a pathologic activation of the thyroid gland is induced by circulating hCG in early pregnancy, and that chemical thyrotoxicosis with increased T4 and suppressed TSH is related to the severity of HG (Kimura et al., 1993). Also, a high percentage of patients with HG have increased serum thyroid hormones. The significance of this phenomenon of transient thyrotoxicosis in relation to the occurrence of HG remains unclear.

Thyrotoxicosis in early pregnancy, hyperemesis gravidarum, and molar pregnancy are all similar in presentation, and all are in the differential diagnosis for a gravida with intractable nausea and vomiting. Because of the molecular similarity between the α subunit of hCG and TSH, patients with significantly elevated hCG levels often manifest thyroid dysfunction and abnormal thyroid function testing. The normal changes in thyroid function in early pregnancy should be recognized before thyrotoxicosis is diagnosed and treatment is started in a patient with HG.

D. Nutritional and Dietary Factors

Perhaps because of the involvement of the gastrointestinal system in morning sickness and hyperemesis gravidarum, much attention has been directed at nutritional and dietary factors as possible causes. Reinken and Gant (1974) suggested that vitamin B_6 deficiency may be important in the pathophysiology of morning sickness and HG. The proposed mechanism involves an altered protein metabolism with increased need for the coenzyme pyridoxal phosphate.

Other investigators have noted alterations in serum lipids and lipoproteins in pregnant women with HG. One study compared women with HG, another group of patients who were pregnant but not nauseated, and a control group of nonpregnant, otherwise healthy subjects. This study demonstrated an increase in free and total cholesterol levels and in phospholipids in patients with vomiting compared to the nonvomiting patients (Jarnfelt-Samsioe et al., 1987).

Another theory of HG is that nutritional deficiencies, such as abnormalities of trace elements, might be associated with HG. Lao et al. (1988) found a correlation between plasma zinc concentration and thyroxine levels, correlating plasma zinc levels, thyroid function in pregnancy, and nausea and vomiting.

Endorphin levels may also be involved in morning sickness. In this model, a defect in endorphin production, abnormal binding to placental receptors, or an altered central or peripheral response to endorphins may be responsible for vomiting during pregnancy (Stark, 1984).

Although some of these hypotheses are attractive and the data persuasive, the cause of morning sickness and hyperemesis gravidarum is still uncertain. It seems likely that a number of mechanisms exist with the common endpoint of nausea and vomiting in pregnancy, and that different patients may have similar symptoms caused by remarkably different physiological derangements.

IV. TREATMENT

Mild gestational nausea with occasional vomiting is usually treated with dietary modification and reassurance. Although most practitioners recommend a bland diet, each patient has different triggers for her nausea. Patients should be encouraged to keep track of the things that aggravate their nausea and vomiting and to avoid them. Among the most common triggers for nausea and vomiting in early pregnancy are coffee, cigarette smoke, strong odors such as perfumes, and greasy foods. Fatty foods prolong the already lengthened emptying time of the stomach, and many patients do not tolerate large meals. Thus, a regimen of frequent, low-fat snacking often provides relief while ensuring adequate fluid volume and caloric intake.

Patients also frequently find that their prenatal vitamins provoke nausea, presumably because of the iron content. Discontinuing iron supplementation until the end of the first trimester can help reduce nausea and vomiting. Finally, a variety of traditional remedies have been used with varying success for treatment of gestational nausea, including mint teas, ginger capsules, hypnosis, acupuncture, and acupressure.

When conservative measures fail, medical therapy is instituted, generally beginning with medications with a lower risk and side effect profile, reserving drugs with more potential for harm for intractable cases.

Pyridoxine (vitamin B_6) and the antihistamine doxylamine have been used to treat nausea and vomiting of pregnancy. Formerly combined in a single pill under the brand name Benedectin, the product was removed from the market more than 10 years ago despite the fact that controlled trials failed to demonstrate teratogenicity. Currently, pyridoxine and doxylamine are prescribed separately. Pyridoxine alone has been shown in blinded, placebo-controlled trials to reduce the severity of gestational nausea (Sahakian et al., 1991; Vutyavanich et al., 1995).

Although they can be very useful in prevention, antihistamines alone appear to be less effective in controlling nausea and vomiting. As mentioned previously, they are probably most helpful in combination with other medications. Scopolamine, diphenhydramine, meclizine, and hydroxazine have all been used with some success. Drowsiness is a frequent but variable side effect of these medications.

Phenothiazines are also commonly used to treat nausea and vomiting of pregnancy. Prochlorperazine and promethazine are effective in most cases, and each is available as a rectal suppository. Drowsiness is their most common side effect, and extrapyramidal symptoms are infrequent. Although phenothiazines are not thought to be teratogenic, neonatal platelet aggregation may be impaired if the drugs are used near delivery. Well-controlled data on the use of phenothiazines in human pregnancy are lacking, and although they are considered safe, they should be given only if the potential benefit is thought to outweigh the possible risk.

Metoclopromide is sometimes prescribed, especially for patients with vomiting. It increases the rate of gastric emptying, relaxes the pyloric sphincter, and enhances the motility of the duodenum and jejunum. Like the phenothiazines, drowsiness is the most common side effect, and extrapyramidal symptoms can occur. While animal studies have not shown any adverse fetal effects, metoclopromide has not been widely studied in human pregnancy. Its usefulness is not clear, and its use is not advocated unless strongly indicated.

Patients with nausea and vomiting so severe as to cause dehydration, persistent ketonuria, or abnormalities in serum electrolytes are admitted to the hospital and treated with a more intense regimen (Briggs, 1997). Because hyperemesis gravidarum is essentially a diagnosis of exclusion, it is important to consider and, where necessary, rule out other serious disorders in the differential diagnosis. These include gastrointestinal diseases (gallbladder disease, hepatitis, pancreatitis, and inflammatory bowel disease), endocrine disorders (especially thyrotoxicosis), obstetrical abnormalities (multiple gestation and molar pregnancy), psychological dysfunction, and miscellaneous etiologies such as atypical migraine headache, alcoholism, and brain tumor. At the time of hospitalization and initiation of intensive treatment for HG, a limited laboratory and diagnostic evaluation can effectively exclude the huge majority of medical/obstetric disorders which manifest with intractable nausea and vomiting.

After admission, oral intake is restricted while intravenous hydration is given. Intravenous fluids with dextrose are given in an amount sufficient to correct any volume deficit and electrolyte derangement. Also, multivitamins and thiamine are often added to the fluids. An antiemetic drug regimen is begun on admission, and fluids by mouth are not begun until nausea is well controlled, typically on the second or third hospital day. While the drug regimen continues, patients are first given clear liquids, with hot, cold, and excessively salty or sweet drinks avoided. Later, bland crackers or bread are offered. If these are tolerated for a day, diet is advanced to a bland low-fat diet, best given in small amounts as snacks throughout the day.

Perhaps because treatment of HG has not been well studied, there are likely as many drug regimens in use as there are physicians treating HG. Most of these regimens involve some combination of phenothiazines, anatihistamines, pyridoxine, and sedatives. One drug regimen recently shown to be effective in controlling HG includes intravenous droperidol and diphenhydramine, in combination with a standardized fluid and diet regimen much like that outlined previously (Nageotte et al., 1996; Briggs, 1997). Droperidol and diphenhydramine are given on a regular schedule and not just for symptoms. Once patients are tolerating a bland diet, the intravenous antiemetic regimen is discontinued and patients are begun on oral metaclopromide and the antihistamine hydroxyzine. Metoclopromide and hydroxazine are given before meals for about a week and are then discontinued. This standardized protocol produced small reductions in length of hospitalization compared to that of patients treated with various regimens; significantly, readmissions for hyperemesis were reduced by 50% with this organized protocol compared to historical controls.

See Also the Following Articles

HORMONES OF PREGNANCY; NUTRITIONAL FACTORS AND REPRODUCTION; PREGNANCY IN HUMANS

Bibliography

Briggs, G. G. (1997). A guideline for treating hyperemesis gravidarum. *Contemp. Ob/Gyn.* **42**, 70–79.

Caruso, S. M., El-Mallakh, R. S., and Mahlan, H. S. (1990). System dynamics in hyperemesis gravidarum. *Family Syst. Med.* **8**, 91–95.

Fairweather, D. V. (1968). Hyperemesis gravidarum. *Am. J. Obstet. Gynecol.* **102**, 135–175.

Goodwin, M. T., Hershman, J. M., and Cole, L. (1994). Increased concentration of the free beta-subunit of human chorionic gonadotropin in hyperemesis gravidarum. *Acta Obstet. Gynecol. Scand.* **73**, 770–772.

Jarnfelt-Samsioe, A., Eriksson, B., and Mattsson, L. A. (1987). Serum lipids and lipoproteins in pregnancies associated with hyperemesis gravidarum. *Gynecol. Endocrinol.* **1**, 51–60.

Katon, W. J., Ries, R. K., Bokan, J. A., and Kleinman, A. (1980). Hyperemesis gravidarum: A biopsychosocial perspective. *Int. J. Psych. Med.* **10**, 151–162.

Kauppila, A., Ylikorkala, O., and Jarvinen, P. A. (1979). The function of the anterior pituitary–adrenal cortex axis in hyperemesis gravidarum. *Br. Med. J.* **1**, 1670.

Kimura, M., Amino, N., Tamaki, H., Ito, E., Mitsuda, N., Miyai, K., and Tanizawa, O. (1993). Gestational thyrotoxicosis and hyperemesis gravidarum: Possible role of HCG with higher stimulating activity. *Clin. Endocrinol.* **38**, 345–350.

Lao, T. T. H., Chin, R. K. H., and Mak, Y. T. (1988). Plasma zinc concentration and thyroid function in hyperemetic pregnancies. *Acta Obstet. Gynecol. Scand.* **67**, 599–604.

Macy, C. (1986). Psychological factors in nausea and vomiting in pregnancy: A review. *J. Reprod. Infant Psych.* **4**, 23–55.

Nageotte, M. P., Briggs, G. G., Towers, C. V., and Asrat, T. (1996). Droperidol and diphenhydramine in the management of hyperemesis gravidarum. *Am. J. Obstet. Gynecol.* **174**, 1801–1806.

Reinken, L., and Gant, H. (1974). Vitamin B_6 nutrition in women with hyperemesis gravidarum during the first trimester of pregnancy. *Clin. Chim. Acta* **55**, 101–102.

Sahakian, V., Rouse, D., Sipes, S., *et al.* (1991). Vitamin B_6 is effective therapy for nausea and vomiting of pregnancy: A randomized double-blind placebo-controlled study. *Obstet. Gynecol.* **78**, 33–36.

Soules, M. R., Hughes, C. L., Garcia, J. A., Livengood, C. H., Prystowsky, M. R., and Alexander, E. (1980). Nausea and vomiting of pregnancy: Role of human chorionic gonadotropin and 17-hydroxyprogesterone. *Obstet. Gynecol.* **55**, 696–700.

Stark, G. C. (1984). Pregnancy induced hyperemesis: A reassessment of therapy and proposal of a new etiology. *Mis. Med.* **81**, 253–256.

Vutyavanich, T., Wongtra-ngan, S., and Ruangsri, R. (1995). Pyridoxine for nausea and vomiting of pregnancy. A randomized double-blind placebo-controlled trial. *Am. J. Obstet. Gynecol.* **173**, 881–884.

Weigel, M. M., and Weigel, R. M. (1989). Nausea and vomiting of early pregnancy and pregnancy outcome: An epidemiological study. *Br. J. Obstet. Gynaecol.* **96**, 1312–1318.

Mosquitoes

see Aedes aegypti

Müllerian Duct

see Vagina

Multiple Birth

see Twinning

Myoma

see Leiomyoma

Myriapoda

Richard L. Hoffman
Virginia Museum of Natural History

I. Class Chilopoda (Centipeds)
II. Class Diplopoda (Millipeds)
III. Class Pauropoda (Pauropods)
IV. Class Symphyla (Symphylids)

GLOSSARY

anamorphosis A developmental pattern in which body segmentation is completed after escape of the embryo from the egg capsule by successive addition of somites and appendages from a growth zone at the posterior end of the body.

coxa The basalmost unit (podomere) of a walking leg of an insect or myriapod, pivoted against the thoracic region by usually a dorsal and ventral condyle.

cyphopods Sclerotized receptacular structures at the external end of the oviducts of millipeds that grasp the gonopods during mating and contain seminal receptacles for temporary sperm storage; anatomically may be derived from basal elements of an otherwise disappeared pair of legs.

diplosegments The body unit in Diplopoda, composed of two adjoining embryonic somites fused with the dislocation of legs and ganglia into the posterior subunit of each.

epimorphosis A developmental pattern in which segmentation of the body is achieved prior to eclosion of the egg; subsequent development involves primarily an increase in body size and the maturation of adult structures.

gonopods The sperm transfer organs of most diplopods, composed of appendages of the seventh body segment modified during the final one or several immature stadia; highly specific in structure and mandatory for classification and identification.

homononymous The condition of an animal's body being composed of a series of similar structural units without modification into regional functional units (e.g., as in earthworms).

labrum The anteriormost edge (or separate sclerite) of an arthropod head, functionally a kind of preoral "upper lip."

opisthogoneate The condition in which the reproductive systems debouch to the exterior at the posterior end of the body.

podomere Any one of the several structural units composing the leg of an arthropod.

progoneate The condition in which the reproductive systems open to the exterior near the anterior end of the body, for example, in Diplopoda, through or behind the coxae of the second pair of legs.

periodomorphosis A phenomenon occurring in some families of juliform Diplopoda in which males alternate reproductive and nonreproductive forms after becoming sexually mature (analogous to a similar pattern in many decapod crustaceans).

somite An embryonic structural unit; in arthropods, it typically contains a ganglion of the ventral nerve cord and is capable of producing one pair of appendages during later development.

spermatophore A droplet of reproductive fluid, varying in consistency from liquid to gelatinous, used in most myriapod groups to achieve fertilization in lieu of a copulatory organ modified from the vasa deferentia.

teloblastic A postembryonic developmental pattern in which body units are produced by an embryonic growth zone located at the rear of the body, immediately in front of the telson segment.

The term "Myriapoda" was at one time a unit of classification equivalent to such taxa as Insecta, Crustacea, and Arachnida, but is now used in an informal collective sense. Initially proposed as a group containing any terrestrial arthropods having more than three pairs of legs, Myriapoda was abandoned as a taxonomic category with the realization that some of its components are more closely related to insects, for instance, than to each other.

With this qualification, the concept of myriapods includes animals now referred to the four classes Chilopoda, Symphyla, Pauropoda, and Diplopoda, collectively embracing probably as many as 100,000 species worldwide. In Chilopoda and Insecta the reproductive systems debouch at the posterior end of the body; this organization is referred to as opisthogoneate. In Pauropoda, Symphyla, and Diplopoda, the gonoducts emerge immediately behind the head, in association with the second pair of legs; thus, they are progoneate. It is not established whether this unusual condition was derived from a common ancestor or originated by random parallel evolution among three unrelated lineages.

Aside from the polypodous, multisegmented body, these four kinds of animals have little in common either biologically or anatomically. Chilopods ("centipeds") are invariably carnivores whose structure reflects a predatory lifestyle. Diplopods ("millipeds"), pauropods ("pauropods"), and symphylids ("symphylids") are predominantly detritivores, feeding largely on decomposing plant material, although a few kinds of millipeds are opportunistical scavengers on animal tissue. These diverse adaptations in turn are reflected in all aspects of reproductive biology. The most pragmatic approach to a brief review of an extensive and complex subject may be that of treating each of the four groups under the following categories: (i) gross anatomy of the reproductive system, (ii) courtship strategies and mechanisms of sperm transfer, and (iii) embryonic and postembryonic development. Of the several kinds of myriapods, centipeds are arguably the most generalized in structure and may be placed first in the sequence.

I. CLASS CHILOPODA (CENTIPEDS)

A. General

Centipeds occur worldwide except in polar regions. There are currently approximately 3000 species recognized, dispersed among five orders and about 22 families. So far as known, all are carnivores, and the group is unique in that the first pair of legs has been modified into a pair of large poison fangs (prehensors) turned forward beneath the head. The body segments in the adult correspond to embryonic somites, and each except the last two bears one pair of legs. The segmentation is homonymous, with no trace of tagmosis or distinction into thorax and abdomen. Pedal segments vary between 15 and 150. The smallest species are <10 mm long and resemble tiny bits of thread, and the largest attain lengths of 200 mm. Such giants can inflict serious "bites" requiring medical treatment; smaller species often invade human habitations and visit unprovoked attacks on the occupants.

Virtually all centipeds are bisexual (one or two are considered to be parthenogenic). External sexual dimorphism is uncommon and is usually expressed only in modifications of the last pair of legs in males; females are frequently the larger sex (on the other hand, young individuals often differ considerably from adults and have often been described as different species). Fertilization is always internal, following transfer of a spermatophore, and shelled eggs are released from the mother to develop with or without parental care. The Chilopoda is currently divided into two subclasses and five orders, dispersed as follows:

Subclass Anamorpha
Order Lithobiomorpha
Order Scutigeromorpha
Order Craterostigmomorpha
Subclass Epimorpha
Order Scolopendromorpha
Order Geophilomorpha

There is no unanimity among specialists regarding details of classification. A review of the various arrangements is given by Lewis; the only complete review of the higher categories in English is that of Hoffman, which requires substantial emendation. No centipeds have acquired vernacular names, and reference must be made in terms of the anglicized scientific names (e.g., "lithobiomorphs," and "lithobiids,") in the following treatment.

B. Reproductive Anatomy

In general there is little external differentiation of the sexes, and in many species of Scolopendromorpha gender can be determined only by dissection. Sexually related modifications occur in males and usually relate to the last pair (or two pairs) of legs. In some species of scolopendromorphs, the last pair may be thickened or some podomeres may have enlarged ridges or crests; in others, the basal segment is produced into an elongated cylindrical process, the appearance of which suggests a possible clasping function. In many species of Geophilomorpha the last pair of legs is prominently thickened and densely pubescent in males; a secretory function has been implied but not established.

In Lithobiomorpha the first postpedal segment of females bears two small jointed appendages ("gonopods") adjacent to the gonopore that are used in holding and manipulating both newly extruded eggs and spermatophores; these structures vary specifically and have been extensively used in classification of the group. In female scutigeromorphs the basal podomeres of the gonopods are medially fused; only the small cylindrical telopodites are moveable. Male lithobiomorphs frequently modify either the last (15th) or last two (14th and 15th) segments of legs in curious ways: generally incrassate, with several podomeres deeply grooved, crested, or provided with tufts of setae. The relevance of these elaborate developments is unknown; possible species recognition by females and/or dispersal of pheromones have been suggested in the absence of direct observation. The normal distribution of stout setae (spurs) on the podomeres is typically much reduced on these terminal modified legs. In one Asiatic genus of Lithobiidae several of the midbody terga are broadly expanded (wing-like) in males only, to what end cannot be guessed; in one American genus the prehensors of males are greatly enlarged in one species only.

In Chilopoda there are three small postpedal segments at the posterior end: The first (intermediate) is unmodified; the second (first genital) carries the much reduced, biarticulate gonopods in both sexes; and the penis and vagina are located on the third (second genital). The penis is not intromittent, very weakly sclerotized, and in smaller species essentially invisible except in prepared slides. The primary sexual organs (testes and ovaries) are located in the hemocoel, dorsal to the enteron in all centipeds. In geophilomorph species there are two fusiform testes drained by a median dorsal vas deferens varying in thickness; distally this tube is joined at the genital atrium by paired dorsal and ventral "accessory glands." A similar pattern occurs in scolopendromorphs, but the testes are shorter and more numerous (up to 13 pairs); they may extend nearly to the anteriormost part of the body cavity. In lithobiomorphs there is a long, single testis, which is relatively small compared to the enlarged seminal vesicles and accessory glands. Male scutigeromorphs have a single pair of testes, drained by a very long, highly coiled vas deferens; the accessory glands are absent, replaced by enlarged seminal vesicles, and the vasa deferentially are modified into ejaculatory ducts.

Spermatozoa are exceptionally variable through the class. Spermatids are produced by two mitotic divisions followed by two meiotic divisions, and in turn they give rise to very long filiform spermatozoa (up to 4 mm long in some species) which are tightly enrolled until placement in the female, when they become motile. In some centipeds two sizes of sperm are produced, whereas in others only one size is produced. In scutigeromorphs the two size classes are produced in two separate regions of the testes.

The female system is basically similar in geophilomorphs and scolopendromorphs: A large, single median dorsal ovary merges into a single long oviduct, and this is joined at the genital atrium by a pair of seminal vesicles and one pair of accessory glands. In females of lithobiomorphs and scutigeromorphs there are two pairs of the latter, and the oviduct is relatively very short. Oogenesis occurs in four stages, culminating in vitellogenesis in which yolk reserves accumulate and the oocyte migrates toward the posterior end of the ovary where the chorion is deposited from the endothelium. Like spermatogenesis, oogenesis is controlled by hormones produced by protocerebral neurosecretory cells.

C. Courtship Strategies and Sperm Transfer

Copulatory sperm transfer does not occur in Chilopoda, in contrast to the often prolonged intromittent encounters common among insects. Typically, the process involves deposition of spermatophores on a silken web constructed by the male, from which they are taken up by the female gonopore. This indirect technique is also practiced by unrelated arthropods, such as symphylids and polyxenoid millipeds, and requires notification of the receptive female as to the location of the spermatophore. This is done in part by antennal contact by both sexes and in part by the use of thread guidelines or a guide pathway.

In the several scolopendrids which have been observed, males initiate courtship; the partners first seize each other with the prehensors and then with the legs, and they assume close body contact, often head to tail, in which the antennae of the male are used to tap the posterior part of the female's body (usually her legs). This phase may last many hours. The male then spins a silken web (source of silk not known, but probably is produced from the accessory glands) in a tunnel or similar crevice and deposits a single spermatophore, which is then picked up by the genital aperture of the female. After a few minutes the spermatophore ruptures and the sperm enter the genital atrium and the seminal receptacles. The female usually eats the vehicular part of the spermatophore. A single account has been published of direct transfer of the spermatophore from male to female, without courtship or web, by an Indian species; this is so unusual that confirmatory evidence is desirable. In some geophilomorphs, the "tapping" form of courtship is followed by departure of the female while the web is constructed; she may not return for several hours, and when she does return she locates the spermatophore (chemotactically?) with her antennae. Lithobiomorphs likewise utilize the web technique but with more sophisticated courtship that involves mutual stimulation with the antennae prior to web construction. In this group the spermatophore is produced externally and consists of an initial droplet of clear fluid onto which a drop of spermatozoa is subsequently placed. The male then signals the female with his antennae, and then she moves over the web, picks up the spermatophore with her gonopods, and leaves. The male eats the web, and the female eats the spermatophore after the sperm have transferred into her seminal receptacles. During the courtship phase, the female antennae often contact the terminal legs of the male, suggesting a possible species-recognition role for modifications of those appendages which is common among lithobiids.

Scutigeromorphs practice somewhat similar courtships but no web is produced. In one species, the male conducts the female over the spermatophore, which she perceives tactically and retrieves with her gonopods. In another species, the female is positioned by her partner closer to the spermatophore, which he then picks up with his prehensors and inserts into the female genital atrium. This is apparently a derived form of the lithobiid pattern.

The place at which fertilization occurs in the female system appears not to have been determined. Presumably the spermatozoa become motile within the seminal receptacles and migrate up the oviduct to intercept the oocyte, perhaps before the chorion has been applied.

D. Development

Chilopods utilize two basically different modes of embryonic development, reflected also in parental behavior. In the orders Scolopendromorpha and Geophilomorpha, epimorphic development occurs within the eggs prior to eclosion; juveniles emerge with the adult complement of body segments and legs and subsequent stadia and molts merely increase body size. In the Lithobiomorpha and Scutigeromorpha development is anamorphic: The first instar has far less than the adult number of segments, which are (as in millipeds) added at each molt from a teloblastic growth zone.

Early cleavage stages produce a blastula stage in which a single layer of ectodermal cells surround a solid core of enucleated macromeres; in gastrulation incipient endoderm cells are split off from the ectoderm and migrate inward to envelope the diminishing yolk material. Mesoderm is ectodermally derived, at first organized into somites, forming a germ band with incipient segmentation and appendage anlagen on the surface of the yolky gastrula. Fairly copious yolk persists even into eclosion. The body cavity, initially a sequence of coelomic sacs, is derived by schizocoely from the mesodermal layer.

In the two epimorph orders, the eggs are typically gathered into a cluster by the mother, who coils around and cares for them, frequently manipulating them with the mouthparts (cleaning?). After hatching, the young remain close to their parent for some time, during several stadia, before becoming motile enough to disperse.

In geophilomorphs, hatching involves the last embryonic stadium; this is succeeded by the "peripatoid" stadium in which antennae are differentiated but approximate body segmentation and appendages are only indistinctly indicated; the next stadium is the "fetus" in which organization is more defined, segmental sclerites appear, and the centipede is capable of motion. Subsequent postlarval stadia are distinguished as Adolescens I, Adolescens II, Adolescens III, Maturus junior, and Maturus senior: These stages are identifiable by various external characters which vary from one taxon to another. Sexual maturity appears to be reached during the third year; adults continue to molt at yearly intervals (in temperature regions at least) and may live for 5 or 6 years.

The development in Scolopendromorpha is generally similar to that just outlined, but with differing numbers of postembryonic stadia. Development studies of a number of scolopendromorphs suggest that generalizations are difficult to make across higher taxa because related species appear to differ in the number of stadia, degree of maturity at each stadia, and so on. The young begin to leave the mother (or brood chamber) in the third adolescens stadium. Individuals are thought to live for at least 6 years; the very large tropical species probably live much longer. In arid West Africa, two broods are produced each year, corresponding to climatic periods. In eastern North America, the common *Scolopocrytops sexspinosus* is frequently found brooding eggs or young inside rotting logs during midsummer, and there is apparently only one brood. A very closely related, syntopic species, *S. nigridius,* is equally abundant but brooding females have never been found and apparently resort to a brood chamber far underground.

There is no parental care provided in the three anamorphic orders, the eggs being laid singly and abandoned after being coated with soil particles. Young emerge from the egg resembling the adult, but with incomplete numbers of body segments and legs. Development is through a number of anamorphic stadia in which legs, antennomeres, ocelli, and other quantitative characters are increased at each molt; when the normal complement of legs is reached there follows an epimorphic series of stadia resulting in sexually mature animals. Collectively as many as 12 posteclosion stadia have been distinguished, although a fair amount of variation seems to occur not only between species but also within species. The final epimorphic phase is characterized by the definitive appearance of secondary sexual characters, such as gonoforceps in females and modification of the terminal legs in males. Postmaturational molting has been observed in female lithobiids.

II. CLASS DIPLOPODA (MILLIPEDS)

A. General

The Diplopoda is the fourth largest class of arthropods (following Insecta, Arachnida, and Crustacea) with a currently known total of about 8000 described species dispersed among 115 families and 15 orders. There is reason to believe that the number of recent species is as high as 80,000, but destruction of tropical forests ensures that most of these will never be collected and classified prior to their extinction.

Millipeds (millipedes in England) are exceptionally interesting animals with diverse structure, lifestyles, and behaviors, and the group is probably unsurpassed for its potential in biogeographic and evolutionary studies.

Essentially all species are detrivores, and in many parts of the world they are major factors in the conversion of leaf litter into soil by both mechanical and chemical breakdown. The majority of species have paired segmental glands (ozadenes) capable of producing a wide array of volatile secretions (allomones) having both antipredator and fungicidal properties. The potential for significant pharmaceutical application is considerable but has scarcely been investigated.

All diplopods are bisexual (except for a very few, mostly very small parthenogenic species); all practice internal fertilization followed by external development of the eggs, which hatch into larvae with three pairs of legs. Subsequent development is by several varieties of anamorphosis through successive stadia in which body segments are added from a teloblastic growth zone at the posterior end of the body. The body units, superficially appearing to be discrete segments, are in fact diplosegments consisting of two coalesced embryonic somites in which the ganglion and leg pair of the anterior somite have migrated posteriad to form two pairs of each in the posterior. The class name Diplopoda is thus derived from the apparent double pair of legs on each body segment (or "ring" in some orders). With only a few exceptions, millipeds accomplish sperm transfer either by oral transport of a spermatophore by the male or by the use of appendages of the seventh body segment modified as sperm transfer organs.

Owing to their superficial similarity and generally retiring lifestyle, millipeds attract little popular attention and few have acquired vernacular names. As in the case of centipeds, it is necessary to adapt the existing scientific group names by use of an appropriate suffix ("polydesmids," "glomerids," etc.). An outline of one currently available classification is provided here for orientation of the information discussed in the subsequent sections. To a major extent, the higher categories are distinguished by modifications of reproductive structures, such as the male gonopore and genitalia of the seventh body segment. The latter are the primary criteria for distinguishing species and generic groups.

Subclass Penicillata ("bristly millipeds" with indirect sperm transfer by male web)
Order Polyxenida
Subclass Pentazonia ("pill millipeds" with oral sperm transfer)
Order Glomeridesmida
Order Glomerida
Order Sphaerotheriida
Subclass Helminthomorpha ("typical millipeds" sperm transfer by way of male gonopods)
Order Polyzoniida
Order Stemmiulida
Order Spirobolida
Order Spirostreptida
Order Julida
Order Siphoniulida
Order Siphonophorida
Order Platydesmida
Order Callipodida
Order Chordeumatida
Order Polydesmida

B. Reproductive Anatomy

As in chilopods, there is little significant external sexual dimorphism in the majority of Diplopoda. In many orders, females tend to be slightly larger in body mass, with smaller and shorter legs and antennae than in males. In those species with elongate, cylindrical bodies, the sixth and seventh segments are often somewhat enlarged to accommodate the male genitalia (which are carried inside the seventh segment); in these forms the legs of males usually have membranous pads on the ventral surface of the postfemora, tibiae, and/or tarsi. In several orders, there is no sexual difference in structure aside from the external genitalia.

An important feature of the primary reproductive organs is their location ventral to the enteron. The male gonads are located in the posterior two-thirds (or more) of the body and typically show a "ladder" pattern of cross-connectives between the two tubular testes. Anteriad, each testis merges into a vas deferens, which unites with the opposite for a short distance, then separates again prior to the male gonopore, located on the ventral side of the third body segment, between the second and third pairs of legs. There are two basic modes of termination of the vasa deferentia; the anatomical homologies are not established. In one, perhaps the more generalized, the coxae of the second pair of legs are perforated on the posterior or ventral surface. The gonopore may open flush on the surface or through an elongate gonapophysis. In the other type, the vasa deferentia open through membranous tubular structures (traditionally called "penes") surrounded by small sclerites, located on the intersegmental membrane behind the second legs. The type of gonopore accommodation is to some extent correlated with taxonomic groupings. It has been suggested that the sclerite complex represents the remnants of a largely dismantled third leg pair; this is an interesting idea but one which generates problems with serial homology of the appendages involved.

A characteristic of virtually all known species of Helminthomorpha is the modification of one or both pairs of legs of the seventh body segment in males into sperm transfer structures called gonopods which are functionally analogous to the first abdominal appendages of decapods or the first legs (pedipalps) of spiders. Although some variation (geographic and individual) does occur, in general gonopods are remarkably constant and afford the best characters for the definition of species and genera.

The simplest, most generalized form of gonopods occurs in the order Platydesmida, in which they are merely strongly reduced versions of the adjacent walking legs. Sperm transfer in this group has not been documented. In the related orders Polyzoniida and Siphonophorida, the gonopods are still leg-like, but the two pairs are dissimilar (anterior pair is shorter and thicker, and the posterior pair is more slender and attenuated), indicating a step toward specialization.

In Polydesmida, the legs of the seventh segment of males are unmodified through the first several juvenile stadia. In the sixth stadium they appear only as small rounded knobs in front of the unmodified posterior pair, whereas in the penultimate stadium they are larger and often divided into coxa and telopodite. During the final molt the gonopods assume their characteristic form in a metamorphosis as dramatic as that of lepidopteran pupae.

In the more generalized species, the gonopods retain the original sternal element, which separates the coxae and to which they are often fused. The coxal musculature is retained, and a pair of small muscles motivates a special small tubular structure, the cannula, which originates at the distal end of the coxae and inserts into a cavity at the base of the telopodite groove (contraction of one of the muscles withdraws the tip of the cannula from its insertion and presumably allows exit of the secretion of the coxal gland). The cannula is conceivably homologous with the flagellum of the coxae in Julida, and both may represent atavistic homologs of the eversible (and retractable by muscle) coxal sacs present in some other taxa.

In one species of polydesmid, *Aenigmopus alatus*, endemic in Guatemala, no gonopods are developed and it is not known how sperm transfer is effected. The second pair of legs are enlarged and incrassate and presumably have become involved in the process. A closely related genus, *Tridontomus*, has normal gonopods. Since both of these species are known only from preserved material, a study of living specimens would be of exceptional interest.

In other orders the gonopods may develop more or less incrementally, becoming larger and more sclerotized in the final juvenile stadia but not functional until a molt into maturation occurs and the definitive species shape is assumed. There is extensive variation in pattern through the various taxa. In some (particularly the spirostreptoids) the posterior pair becomes progressively suppressed and may be represented in the adult by the original sternum, by reduced musculature, or by nothing. Correspondingly, the anterior pair is complexly modified; its original telopodite is in the form of a separate monomeric shaft largely enclosed within the coxal region. Frequently, the coxal elements are enlarged and modified, with occasional retention of the telopodite as a few small podomeres. In the Julida, the posterior gonopods are large and function in spermatophore transfer. In the orders Spirostreptida, Chordeumatida, and Polydesmida, the distal element (telopodite) of the gonopod often becomes remarkably elaborated with branches, flaps, frills, lobes, projections, and membranes: It is difficult to visualize how some of these appendages are able to function.

In the orders that have been examined for the character, there is an internal coxal (or coleomic) gland which provides a secretion conducted along the length of the telopodite by a deep groove; the chemical nature of this secretion is uncertain. In Chordeumatida, the groove is present only marginally in a few families, and in this order it appears that the coxal region of several legs immediately behind the gonopods is involved in sperm transfer.

The ovaries are paired and (enclosed in a common ovarial sac) extend throughout much of the body cavity, ventral to the enteron, terminating in oviducts which debouch through sclerotized, bivalvular structures (cyphopods) located between the second and third leg pairs or, in pentazonians, on the coxae of the second legs. The shape and structure of cyphopods varies with each order. In the juliform taxa (Spirobolida and Spirostreptida) they are retracted deeply into eversible membranous pockets, whereas in polydesmids they tend to be placed behind the second leg pair and are capable of eversion to engage the gonopods during mating; the two valves contain a variable number of seminal receptacles and typically are at least partially enclosed by a curved sclerite. In callipodids, the cyphopods are relatively small but placed at the ends of inverted tubes capable of great protrusion, suggesting use in placing eggs into deep crevices. Similar but even longer telescoping cyphopods occur in glomeridesmids, in which they were originally mistaken as penes. In several taxa the shape of the cyphopods is specifically distinct and constant, making identification of females possible. Generally, their structure is basically simple and conservative and in no case do they match the often baroque complexity of male genitalia of the same species.

C. Courtship Strategies and Sperm Transfer

Each of the three subclasses of Diplopoda is characterized by major differences in both overall structure and sperm transfer techniques. In the tiny species of Penicillata ("polyxenids"), no appendages are modified, and "courtship" involves production of a fertilization web by the male, generally similar in pattern to that employed by some chilopods. With silk produced by penial glands the male spins a simple web of several mostly parallel strands, on one of which he deposits two droplets of spermatic fluid. Then, using silk from glands on the coxae of the eighth and ninth legs, he produces a "guidance" track composed of two parallel bands of threads; these fibers are thought to contain olfactory chemicals recognizable by females. A female encountering such a trackway follows it to the web and takes up both droplets with her gonopores. The spermatids become mature (motile) once inside the seminal receptacles. In this very simple pattern the two participants are never in contact or necessarily even near each other.

Pentazonoid millipeds have modified the last two or three pairs of legs of adult males for use in sperm transfer. The last pair is by far the largest and projections from one or two podomeres produce a forciculate device (telopod). In the order Glomerida of the Northern Hemisphere, males, to initiate mating, will approach (backing up to) a receptive female and "display" by extending these large appendages outward toward her anterior end, usually seizing one antenna and a cyphopod. A gland at the base of the telopods then secretes a fluid, thought to be a possible arrestant or tranquilizer, which is ingested by the female. At this point, the female's antenna is released and the telopod seizes the opposite cyphopod. The male's body is curved around in a "C" shape, so his head is near the site of action; he secrets droplets of spermatic fluid from his penes (gonophores) which he takes up in his mouth and transfers to the nearby cyphopods. In the usually much larger species of Sphaerotheriida (South Africa and Southeast Asia), the telopodites are generally similar to those of glomerids, but in many species one of the basal podomeres may be provided with prominent ribs which are rubbed against knobs on the inside of the terminal tergite, having thus a stridulatory capability. The few observations available suggest that the sound produced may function both as "advertisement" of the male's intentions and as species-recognition signals.

The subclass Helminthomorpha contains by far the greatest number of both species and higher taxa, with a corresponding diversity of body forms and courtship protocols. As previously noted, the common structural denominator is the modification of one or both pairs of legs of the seventh body segment into sperm transfer devices (gonopods). Because of the different location of the cyphopods (second segment of female) and gonopods (seventh segment of male), the male body must be considerably offset anteriad to achieve contact of the two structures. Within this subclass there is a complete spectrum of modification, from simply reduced walking legs to structures of byzantine complexity. Courtship likewise varies from none (physical domination by the male) to the purchase of female complicity by provision of an edible secretion. No cases are known in which the female initiates courtship. A widespread behavioral trait is tapping of the body (and/or legs) with the antennae, perhaps as notification of intentions or to establish receptivity.

In many of the larger, cylindrical species (Spirobolida and Spirostreptida), males are provided with membranous pads on one or more of the podomeres which are useful for maintaining position on the female's smooth body. Aggressive males crawl along the female's back and down over the front of her head, stopping her forward motion and often throwing her to one side to achieve venter-to-venter con-

FIGURE 1 Mating pair of giant millipeds (Spicostreptidae). Photo courtesy of Dr. Stephen W. Bullington.

tact; the legs of the partners often "zip" together for much of the body length (Fig. 1). Nonreceptive females endeavor to escape, most effectively by releasing allomones from those segments closest to the male's head. Receptive females may arch the anterior segments to facilitate insertion of the gonopods into the cyphopod space between the second and third pairs of legs. Contact is often maintained for several hours. Females are able to mate repeatedly, and recent observations suggest that some gonopods are modified to scoop out or dislodge spermatophores placed by a previous partner.

Several mating patterns are recorded for the large order Julida, which contains some of the most common milliped species. In the more evolutionarily generalized species (parajulids and blaniulids) the first pair of legs of males may be greatly enlarged (functioning as clasping organs to capture the female and then stabilize her head position during mating) or greatly reduced and hook-like (achieving the same effect by a more local engagement of her labrum); in either case the male captures the female without any evident courtship and immobilizes her with the mechanism just described as well as a loop of his body around hers, near the anterior end. In this pattern, the gonopods are engaged first, then the male provides them with a spermatophore that is transferred with his mouthparts. The apparently plastic material molds itself to the surface of the cyphopods and spermatozoa then move into the seminal receptacles. In at least one European species of Julidae (a more derived group), the coxae of the first legs of the male secrete an edible fluid, which he presents to the female in a frontal approach by raising the anterior third of his body. If receptive, the female adopts the same posture, the bodies meet, and while she licks at the coxal secretions he seizes the opportunity to insert the gonopods. This form of compliance by prandial distraction seems to have originated many times throughout the animal kingdom.

Males of Polydesmida simply overpower the female, and the partners assume a venter-to-venter position, with the legs interlocked, sometimes maintaining the stance for several hours. Species recognition is presumably chemosensory since all species are eyeless, and males will not mate with females of a congeneric species even if of the same body size and virtually identical structure.

In the orders Julida, Spirobolida, and Spirostreptida the spermatophore is placed on the gonopods after their engagement in the female's body, and in Polydesmida it is placed prior to the search for a female. In all these orders, the gonopod is provided with a deep groove extending from base to apex and connected to a gland in the coxa or even within the body cavity. The function of the secretion transmitted by capillary action along the groove is unknown; both a prostatic (nutritive/buffering) and enzymatic (dissolving the spermatophore) function have been suggested.

Oogenesis involves the production by oocytes of a yolk nucleus which disappears following the arrival of protein yolk granules in the cytoplasm. Coincident with the initiation of vitellogenesis, a class of cytoplasmic inclusions rich in calcium appears; presumably these function as calcium reserves used in exoskeleton formation by the embryo.

The number of eggs produced varies remarkably, to some extent as a function of body size. A large female spirostreptid, *Archispirostreptus gigas* (body diameter ≈20 mm), was found to contain 230 eggs, with an average diameter of 2.2 mm. A large polydesmoid, *Pachydesmus crassicutis* (body diameter 8 mm), contained 2500 shelled eggs of 0.66 mm average diameter. A slightly smaller species, *Rhododesmus mastophorus* (diameter 6 mm), contained 126 shelled eggs with a diameter of 0.4 mm plus an equal number of similar size eggs but with no chorion; this specimen also had about 450 smaller, shellless ova intermixed randomly throughout the reproductive tract with the mature eggs. Smaller species reduce the number of eggs produced, but not the diameter, so that a female pyrgodesmid 1.0 mm in diameter will produce only four to seven eggs, each of which is as wide as the body cavity: It is difficult to imagine how they are expelled through the oviduct and cyphopod aperture.

D. Development

Early embryonic processes (histogenesis and organogenesis) are basically similar to those described for Chilopoda. However, mesoderm originates from a plug of cells at the posterior end of the endodermal germ band that pushes into the yolk mass. In later stages, somites destined to form the body group in pairs, with the legs, ganglia, and associated musculature of the anterior member migrating into the posterior, resulting in the formation of diplosegments characteristic of this class. However, there is reason to suspect that the segmentation observable in the adult or late juvenile may not correspond exactly to the early embryonic segmentation; considerable investigation is required to resolve some of the current uncertainties. Diplosegments appear first in the second postembryonic stadium, when the animal has seven pairs of legs.

Egg care is accomplished in a variety of ways. Female polyxenids lay their eggs in a moniliform string, which is then arranged in a flat spiral which is coated with loose setae derived from the caudal brush of the parent. In some spiroboloid species, each is taken into the mouth of the female to be coated with a layer of chewed humus, then passed back along the legs and taken into the rectum for removal of excess moisture. The resultant coated eggs very much resemble the fecal pellets produced by adults and are scattered randomly among the latter. Possibly such eggs are also impregnated with some of the mother's defensive allomones during passage along the legs; the newly hatched young feed on the egg coating for their initial nourishment. In glomerids, each egg is encapsulated in a small pellet of mud formed by the mother's mouth and sometimes the pellets are joined into pairs; the construction details are precise enough to often permit identification of the species that produced them.

Juloid and spirostreptoid millipeds frequently construct a subterranean chamber for oviposition, which when sealed by the female provides protection against predators as well as some climate control. In most polydesmid taxa, the female produces an "igloo" or spherical chamber composed of mud, chewed humus, and/or fecal material; eggs are introduced just before the structure is closed. Females of chordeumatoids produce a "nest" composed of silk from the spinnerets on their telson. In one genus of platydesmids (*Brachycybe*, native to North America and Japan), a male gathers up the eggs as they are released from the female and coils around (broods) them until hatching occurs: This is the only instance of male care known for any myriapod species.

Embryonic development requires from 2 to 4 weeks, varying greatly from group to group. Eggs hatch into a pupoid stage usually having three pairs of leg primordia. This is followed by a variable number of stadia, of which the first and second generally remain within the split eggshell, and active feeding begins during the third stadium. The first stadium generally has three pairs of legs, but there are exceptionally greater numbers (in one rare case, as many as 41).

The postembryonic development of all diplopods is "anamorphic" in the broad sense of that term, e.g., gradual increase in number of body segments through a number of stadia until the definitive number has been reached. During the process, the body has a number of legless segments in front of the growth zone, and at each molt it adds new segments and simultaneously legs for those previously legless. Sexual maturity is generally reached with the conclusion of the final molt, when both male and female genitalia assume the form characteristic for the species; however, in some forms (e.g., pentazonoids and platydesmoids) the change from penultimate to final condition may be very slight in these appendages and some "new species" have been described on the basis of (unsuspected) immature males with "simple" gonopods. Spermatogensis in some forms commences prior to the final molt. In a group as diverse as the Diplopoda, there are several varieties of anamorphism, recently distinguished as follows:

Euanamorphosis, in which every molt results in the addition of new segments, even after the attainment of sexual maturity.
Hemianamorphosis, in which molting continues after maturity is reached but no further segments are added.
Teloanamorphosis, in which no further segments are added beyond the definitive number, and no further molting occurs.

The entire subject of milliped development is reviewed in detail by Enghoff *et al.*, whose paper must be consulted for information on the known diversity and variability in the process. Some examples will provide an idea of the different adaptations.

Members of some orders (Polyzoniida, Siphonophorida, Platydesmida, and Julida in part) become sexually mature despite having a number of legless segments, and in these groups the number of adult segments may be extremely variable. During anamorphosis, the male genitalia change gradually into the adult condition, and it may not always be possible to differentiate at which stadium this occurs. The number of segments added at a given molt is not always the same within a particular species and may vary from 4 or 5 to as many as 15. In spiroboloids and perhaps other orders, molting occurs after sexual maturity is reached, especially in females, which may live for a decade of more and continue reproducing, although no new segments are added and only the body size increases. In polydesmids, there is a fixed number of molts (usually seven), the gonopods are produced during the final molt, and adults live only a short time, mating only once. In some spirostreptoids and callipodoids, females attain the complete number of segments and legs with still unformed cyphopods; presumably a postmaturity molt is required to complete these structures.

In many species of the order Julida, males undergo

a curious process called peridomorphosis in which, after they become mature sexually, stadia with complete genitalia capable of sperm transfer molt into alternate stadia with nonfunctional gonopods and back again. In general, this sequencing is reminiscent of that found in freshwater cambarid decapods. In julids, the process occurs in a number of permutations, and a number of theories have been advanced to account for it. In the majority of millipeds, diapause is spent within the protection of a "molting chamber" constructed from soil or mud.

III. CLASS PAUROPODA (PAUROPODS)

Pauropods are minute progoneate, eyeless arthropods, with a maximum size of about 1.5 mm. They are characterized by the remarkably formed antennae which consist of a basal cylindrical stalk and three long moniliform distal flagellae, and usually a spherical sensory organ (globulus) is also present at the end of a short stalk. Mouthparts consist of mandibles and one pair of maxillae. Although the 9–11 leg-bearing trunk segments are not diplosegments, and the mouthparts do not have a gnathochilarium, pauropods are considered by some specialists to be the sister group to Diplopoda.

There are about 500 species of pauropods, dispersed among five families. There is only one order. A large number of the known species are entirely cosmopolitan or are known from several continents, suggesting the distribution of other minute animals such as rotifers and protistans. Apparently all feed on fungi or decomposing litter.

Because of their small size and the lack of interest from specialists, little is known of the reproductive biology of pauropods. The sexes are separate, and there is no obvious sexual dimorphism. A single ovary lying ventral to the enteron merges anteriad into an oviduct and then enters a larger vagina which opens to the surface behind the second pair of legs. There are four thread-like testes placed dorsal to the enteron; they merge into an elongate vas deferens which distally divides into two "ejaculatory ducts," each opening into a membranous penis located behind the coxae of the second legs. Courtship and sperm transfer appear to be undocumented, although the use of webs, or at least spermatophores, seems probable because there are no external male or female genitalia. The embryonic development of one species has been studied. Eggs are laid separately without parental care provided. Postembryonic development entails up to six stadia (5, 6, and 8–11 pairs of legs).

IV. CLASS SYMPHYLA (SYMPHYLIDS)

Symphylids comprise a small class of eyeless progoneate arthropods, characterized by long, moniliform antennae, a pair of postantennal organs, mandibles, two pairs of maxillae, a 14-segmented trunk with 12 pairs of legs in adults, and two conical cerci on the last tergite. The body is small (up to 8 mm long), thin, and delicate, with reduced pigmentation. All species live in moist woodland and arable soil, and some may become very numerous. About 170 species are known, dispersed throughout two families and a single order.

Sexes are separate, without no external differentiation. Both the ovaries and the testes are paired and located ventral to the enteron; the gonoducts in each sex unite prior to debouching at a median gonopore behind the second pair of legs.

Sperm transfer and fertilization have been studied in one species (*Scutigerella immaculata*). The male produces a long, thin stalk from a secretion from the gonopore and attaches a droplet of semen (or spermatophore) to its apex. When a female happens to find such a device, she takes the droplet into her mouth, and then the sperm move into a special internal cheek pouch (an oral seminal receptacle). She then removes an egg from the female gonopore with her mouth and attaches it to a suitable substrate, such as a moss plant. That done, she anoints the egg with sperm discharged from the cheek pouch and fertilization ensues. The eggs are laid in clusters of 9–25 and guarded by the female. Hatching requires a week to a month, depending on temperature. There are seven juvenile stadia, during which the pairs of legs are increased from 6 to 12. Postmaturational molting continues to a recorded maximum of 50; a life span of 7 years has been established.

Bibliography

Attems, C. (1926-. Progoneata. In *Handbuch der Zoologie*, Vol. 4, pp. 1–402. de Gruyter, Berlin.

Enghoff, H., Dohle, W., and Blower, J. G. (1993). Anamorphosis in millipedes (Diplopoda)—The present state of knowledge with some developmental and phylogenetic considerations. *Zool. J. Linnean Soc. London* 109, 103–234.

Hoffman, R. L. (1982). Classes Chilopoda and Diplopoda. In *Synopsis and Classification of Living Organisms* (S. P. Parker, Ed.), Vol. 2, pp. 681–724. McGraw-Hill, New York.

Hopkin, S. P., and Read, H. J. (1992). *The Biology of Millipedes*. Oxford Univ. Press, Oxford, UK.

Johannsen, O. A., and Butt, F. H. (1941). *Embryology of Insects and Myriapods*. McGraw-Hill, New York.

Lewis, J. G. E. (1981). *The Biology of Centipedes*. Cambridge Univ. Press, Cambridge, UK.

Scheller, U. (1982). Classes Pauropoda and Symphyla. In *Synopsis and Classification of Living Organisms* (S. P. Parker, Ed.), Vol. 2, pp. 688–689, 724–726. McGraw-Hill, New York.

Naked Mole-Rats

Christopher G. Faulkes
Zoological Society of London

I. Natural History and Social Structure of Mole-Rat Colonies
II. The Physiology of Reproductive Suppression
III. Cues Regulating Male and Female Reproduction
IV. Ecological Constraints and the Evolution of Cooperative Breeding

GLOSSARY

cooperative breeding A social system displayed by some species in which members of the social group assist in rearing young that are not their own offspring.

eusocial A social strategy arising from the study of certain colonial insects, such as bees and termites, that have overlapping generations, cooperative care of young, and a reproductive division of labor.

inclusive fitness The concept that an animal's fitness (the ability of genetic material to perpetuate itself in the course of evolution) depends not only on its own individual reproductive success but also on that of its close relatives who share many genes in common with the individual.

reproductive skew The partitioning of direct reproduction among individuals in societies. Where all individuals have equal chances of reproduction, skew is said to be low, whereas in societies in which reproduction in restricted to one or a small number of individuals, reproductive skew is high.

reproductive suppression A failure or delay in the ability of an animal to reproduce, resulting from a disruption of normal behavioral and/or physiological processes brought about by environmental/social cues.

Naked mole-rats are rodents that provide an extreme example of how social cues can regulate and inhibit reproduction. In colonies that contain on average approximately 100 individuals, reproduction is normally monopolized by a single breeding "queen," who mates with one to three breeding males and suppresses reproductive function in subordinate group members of both sexes.

I. NATURAL HISTORY AND SOCIAL STRUCTURE OF MOLE-RAT COLONIES

The naked mole-rat is one member of the family Bathyergidae (the African mole-rats), a group of Hystricomorph rodents endemic to sub-Saharan Africa containing at least 18 species. Within the family, social strategies include solitary, social/colonial dwelling, and, at the extreme in the naked mole-rat *Heterocephalus glaber*, eusocial. A second species, the Damaraland mole-rat, *Cryptomys damarensis*, is also known to be eusocial, although the maximum group sizes attained are less than those of the naked mole-rat. Thus, within a single taxonomic group the entire spectrum of sociality is represented, making the Bathyergidae a unique model with which to investigate the evolution of social behavior and the mechanisms underlying the social control of fertility.

Eusocial animals characteristically live in groups which contain overlapping generations that cooperate in the rearing of offspring, with reproduction restricted to one or a small number of specific breeders of each sex. The nonbreeding group members are reproductively inactive and in mammalian examples of eusociality are quiescent rather than irreversibly sterile (as in some eusocial invertebrates). Eusocial species fall at the extreme end of a continuum of sociality and cooperative breeding strategies, of which many examples can be found among mammals and birds. In these "high reproductive skew" societies, social cues lead to suppression of reproduction in subordinate individuals, whereas reproduction is monopolized by a small number of dominant group members.

Naked mole-rats are found in the arid regions of Kenya, Ethiopia, and Somalia—areas that typically have low, unpredictable rainfall. Living totally underground, the burrow systems that they inhabit are extensive, commonly total 3 or 4 km in tunnel length, and are mostly composed of foraging tunnels built in search of the underground roots and tubers that form their staple diet. Interspersed among the tunnels of the central part of the burrow are communal nest and toilet chambers, which are important focal points where mole-rats can interact socially. Highly adapted for life underground, naked mole-rats have evolved many unusual features in addition to their social system. Apart from their naked skin and extrabuccal incisors which are used to excavate the burrows, they have a low body temperature and metabolic rate and are effectively cold-blooded (ectothermic) mammals. They are also extremely long-lived for a small rodent (average body size ≈34 g); longevity in captivity can be in excess of 21 years.

Colonies of this mammalian equivalent of a social insect commonly contain approximately 100 but sometimes in excess of 295 individuals; even in such large groups reproduction is normally restricted to a single dominant "queen," who selects and mates with one to three specific breeding males. Reproductive output from this one individual can be high: Litter sizes can number up to 27, and one queen in a captive colony is known to have produced over 900 offspring in her lifetime. The rest of the colony of both sexes does not breed because the members are reproductively suppressed by the queen. Instead, they act as "workers," maintaining the burrow system, or "soldiers," defending the colony against foreign mole-rats or several species of snake which are the main predators of the mole-rat. Despite their longevity, more than 99% or naked mole-rats are destined to a life of chronic infertility and never reproduce.

II. THE PHYSIOLOGY OF REPRODUCTIVE SUPPRESSION

Naked mole-rats are unusual among cooperative breeders in that a clear physiological block to reproduction is apparent in nonbreeders of both sexes. While suppression of reproductive physiology has been reported in nonbreeding female marmosets and other *Callitrichid* primates, dwarf mongooses, and Damaraland mole-rats, in most species reproductive suppression among males appears to be predominantly behavioral. Endocrine deficiencies are not usually observed among nonbreeding males, and suppression is thought to be due to exclusion from mating as a result of interactions with more dominant individuals.

Research suggests that in colonies of naked mole-rats, social cues acting between the dominant queen and nonbreeders are translated into a disruption of the normal patterns of secretion of luteinizing hormone (LH) and follicle-stimulating hormone from the anterior pituitary gland, resulting in inadequate hormonal stimulation of the gonads by these pituitary gonadotrophins and ultimately in a state of infertility. When compared with breeders, both male and female nonbreeding naked mole-rats have reduced concentrations of plasma LH, and their pituitaries are less responsive to exogenous gonadotrophin-releasing hormone (GnRH), suggesting a lack of priming of the pituitary gland as a consequence of altered secretion of endogenous GnRH. In females, the disruption of gonadotrophin secretion ultimately results in anovulation, and the reproductive tract and ovaries of suppressed nonbreeders remain in a prepubescent state. Histological examination of ovaries removed from nonbreeding females has shown an absence of corpora lutea or corpora albicantia and only the very occasional presence of a preovulatory follicle, in contrast to breeding females. In addition, the mass of the ovary and the whole reproductive tract was vastly greater in breeders than in nonbreeders [ovary: (breeder; $n = 8$) 0.79 ± 0.24 mg/g body weight (BW) versus (nonbreeder; $n = 14$) 0.25 ± 0.03 mg/g BW, $P < 0.05$; reproductive tract: (breeder; $n = 7$) 27.1 ± 8.9 mg/g BW versus (nonbreeder; $n = 59$) 1.5 ± 0.1 mg/g BW, $P < 0.01$). Nonbreeding female naked mole-rats captured from colonies in the wild had similarly diminutive ovarian and reproductive tract measurements to those taken from nonbreeding females in captive colonies.

Among male naked mole-rats, altered secretion of hypothalamic GnRH and pituitary LH secretion gives rise to clear physiological differences between breeders and nonbreeders, with the latter having lower concentrations of urinary testosterone compared with the former [mean ± SEM: 23.8 ± 2.3 vs 5.2 ± 1.4 ng/mg creatinine (Cr), respectively; $P < 0.001$]. Histological examination of naked mole-rat testes showed that nonbreeding males (captive and wild-caught) had fewer Leydig cells (testosterone secreting) than breeding males. While these observations suggest that suppression of reproductive physiology occurs in nonbreeding male naked mole-rats, levels of reproductive hormones are sufficient to support some spermatogenesis, and these males apparently produce some mature spermatozoa. However, closer investigation of spermatozoa in nonbreeders has shown that these males produce significantly lower numbers of sperm compared with breeders, and in most nonbreeding males these sperm are not motile.

The neuronal mechanisms that bring about socially induced suppression reproduction in nonbreeding naked mole-rats are unknown. In socially suppressed female common marmoset monkeys, GnRH attenuation is mediated by a mechanism involving both opioid peptides and an increased sensitivity to the negative feedback effects of estradiol. Our studies on the naked mole-rat also suggest a specific neuroendocrine regulatory mechanism controlling reproduction and, so far, a causal link between elevated levels of the stress hormone cortisol and reproductive suppression has been ruled out. Although nonpregnant breeding queens had significantly lower concentrations of urinary cortisol than nonbreeding females and pregnant queens, when nonbreeding females were separated from their colonies, their concentrations of urinary cortisol remained high, but they still ovulated and became reproductively active.

In males, breeders had significantly lower urinary cortisol concentrations than nonbreeders (73.8 ± 7.4 vs 164.9 ± 19.7 ng/mg Cr, respectively), but when nonbreeders were separated from their colonies, they became reproductively active even though urinary cortisol concentrations increased from 165.7 ± 20.6 to 234.1 ± 46.3 ng/mg Cr. These observations are similar to findings in nonhuman primates, in which differences in glucocorticoids between individuals of different social rank do not seem to be directly involved in the physiological suppression of reproduction.

In female naked mole-rats, a possible role for prolactin, another hormone implicated in stress responses, in the mechanism of suppression has proved difficult to ascertain. However, circumstantial evidence suggests that this hormone may not be directly implicated: The breeding queen has a postpartum estrus, which may often occur when she is still lactating (8–11 days postpartum) and therefore at a time when prolactin levels would be expected to be elevated. Again, this is similar to the situation suggested for another aseasonal, cooperatively breeding mammal, the common marmoset, in which prolactin has not been functionally linked to anovulation in female subordinates. A simple, direct relationship between chronic physiological stress and reproductive inhibition therefore does not seem to be operative in the naked mole-rat, but more definitive data are still required.

In both male and female nonbreeding naked mole-rats, reproductive suppression is readily reversible if the social cues maintaining reproductive suppression are removed. For example, if nonbreeding females are removed from their colonies and either paired with a male or housed singly and then paired with a male, urinary progesterone concentrations rise for the first time to levels indicative of a luteal phase of an ovarian cycle after approximately 8 days. Likewise, if males are similarly separated from their colonies, concentrations of plasma LH and urinary testosterone increase significantly, with urinary testosterone reaching levels comparable to those of breeding males after approximately 5 days.

Separation experiments such as these, together with studies following the removal or death of a queen, prove conclusively that the breeding queen plays the central role in bringing about suppression of reproduction in both male and female nonbreeding naked mole-rats. The suppression among females by the queen is readily apparent because when a queen dies she is replaced by a former nonbreeder, usually a high ranking individual who may "fight it out" with a rival contender.

The queen not only suppress reproduction in nonbreeding males but also appears to modulate reproductive function in breeding males. This phenomenon can be seen both in a colony situation and in male–female pairs. In the latter case, when urinary testosterone profiles of reproductively active males paired with females were plotted relative to the mate's ovarian cycle, testosterone concentrations in the male peaked during the follicular phase of the cycle, just prior to estrus.

The central role of the breeding queen in imposing reproductive suppression on both breeding and nonbreeding males within a colony was seen in captivity when measurements of male urinary testosterone were made before and after the removal of the breeding queen. Upon her removal, urinary testosterone concentrations rose significantly in both breeders and nonbreeders, indicating that reproductive suppression was released and the hypothalamic–pituitary–gonadal axis had become active in the nonbreeder males and was no longer being modulated by the queen in breeding males.

III. CUES REGULATING MALE AND FEMALE REPRODUCTION

The socially induced suppression of reproduction in male and female naked mole-rats appears to be mediated by a mechanism involving direct contact with the breeding queen rather than by primer pheromones (chemical signals acting between individuals of the same species that may elicit a physiological response in the recipient). The lack of an obvious pheromonal effect in the inhibition of naked mole-rat reproduction is surprising given their communal toilet habits. Each colony has one or more specific toilet chambers and following urination or defecation, naked mole-rats roll around in the soiled litter, thereby facilitating transfer of any potential urinary or secreted pheromones. However, reproductive activation could not be prevented or delayed in nonbreeding males and females separated from but maintained in olfactory contact with their colonies by daily transfer of bedding and litter from the nest, food, and toilet chambers. Females undergoing bedding transfer and control procedures showed no significant difference in time to the first sustained elevation of urinary progesterone, indicative of a luteal phase of the ovarian cycle. Similarly, in males, there was no difference between control and bedding transfer groups in the time from separation to elevations of urinary testosterone to concentrations comparable with breeding males. Evidence from these and other experiments points to a behavioral mechanism, initiated by the queen, as the primary cue in suppression of reproduction in nonbreeding male and female naked mole-rats. A current hypothesis is that overt and subtle agonistic interactions between the queen and the nonbreeders cause a neuroendocrine response in the latter, the physiological result of which is inhibition of gonadal function and infertility. Overt aggression in naked mole-rat colonies is rare except during queen succession. It seems likely that such events occur during critical periods when the social hierarchy is rearranged, and the new queen enforces her dominance over the other colony members and reestablishes reproductive inhibition. After succession, only subtle cues from the new queen may be required to maintain suppression in most colonies.

Naked mole-rats lie at one extreme of the continuum of mammalian cooperative breeding strategies. However, they share many similarities with other singular cooperative breeders with respect to the ultimate and proximate factors involved in the evolution and maintenance of their social system. The main difference that sets naked mole-rats apart from other African mole-rats and all other mammals is the intense degree of inbreeding, giving rise to very high levels of intracolony genetic relatedness. Nucleotide sequence variation in the mitochondrial DNA D-loop region, which is normally highly variable, was absent within and between adjacent colonies in Mtito Andei, Kenya. Multilocus minisatellite DNA fingerprints have produced band sharing coefficients as high as 1.0 within some wild colonies from the same region of Kenya, with average relatedness within colonies being very high ($r = 0.81$ compared with 0.5 in siblings of an outbred species). The high level of inbreeding results from, and is enforced by, the high cost of dispersal for nonbreeders. In other singular cooperative breeders, such as the Damaraland mole-rat and the dwarf mongoose, intragroup relatedness is less than that found in the naked mole-rat. However, the opportunity for dispersal is greater among the former two species than the latter. In eusocial Damaraland mole-rats, inbreeding is avoided and dispersal occurs after periods of heavy rainfall. In 13 of 14 newly formed colonies studied in Namibia, the breeding male and female were from different colonies. Nevertheless, even with such a dispersal phase, 90% of Damaraland mole-rat nonbreeders do not get an opportunity to reproduce.

IV. ECOLOGICAL CONSTRAINTS AND THE EVOLUTION OF COOPERATIVE BREEDING

Comparative studies of the ecology and behavior of African mole-rats have led to the development of a "food-aridity" hypothesis to explain the spectrum of sociality seen in the family. It is currently thought that eusociality has evolved in mole-rats through kin selection and in response to constraints imposed by ecological factors, in particular the rainfall pattern of their environment and the distribution of food in the form of underground roots and tubers. Both eusocial species of mole-rats are found in arid regions with unpredictable rainfall. Because the underground roots and tubers on which they feed are more widely distributed in arid regions, and in general the soil hardness is greater, the energetic costs of burrowing and the risk of unsuccessful foraging by digging tunnels at random are larger than those in more mesic habitats, and consequently the costs of dispersal and individual reproduction are high. Thus, by living in large social groups with cooperation and reproductive suppression, eusocial mole-rats are able to exploit an ecological niche where solitary animals or small groups would be unlikely to survive. However, balanced against the benefits of sociality is the reproductive sacrifice made by most of the colony members. This is offset by the close genetic relatedness of individuals within colonies because nonbreeders benefit indirectly by increases in their inclusive fitness when close kin successfully reproduce.

See Also the Following Article

RODENTIA

Bibliography

Brett, R. A. (1991). The population structure of naked mole-rat colonies. In *The Biology of the Naked Mole-Rat* (P. W. Sherman, J. U. M. Jarvis, and R. D. Alexander, Eds.), pp. 97–136. Princeton Univ. Press, Princeton, NJ.

Creel, S. R., and Waser, P. M. (1997). Variation in reproductive suppression among dwarf mongooses: Interplay between mechanisms and evolution. In *Cooperative Breeding in Mammals* (N. G. Solomon and J. A. French, Eds.), pp. 150–170. Cambridge Univ. Press, New York.

Faulkes, C. G., and Abbott, D. H. (1997). Life in a reproductive dictatorship: Regulation of male and female reproduction by a single breeding female in colonies of naked mole-rats. In *Cooperative Breeding in Mammals* (N. G. Solomon and J. A. French, Eds.), pp. 302–334. Cambridge Univ. Press, New York.

Faulkes, C. G., Abbott, D. H., O'Brien, H. P., Lau, L., Roy, M., Wayne, R. K., and Bruford, M. W. (1997). Micro- and macro-geographic genetic structure of colonies of naked mole-rats, *Heterocephalus glaber*. *Mol. Ecol.* 6(7).

French, J. A. (1997). Proximate regulation of singular breeding in Callitrichid primates. In *Cooperative Breeding in Mammals* (N. G. Solomon and J. A. French, Eds.), pp. 34–75. Cambridge Univ. Press, New York.

Jarvis, J. U. M. (1981). Eusociality in a mammal—Cooperative breeding in naked mole-rat Heterocephalus glaber colonies. *Science* 212, 571–573.

Jarvis J. U. M. (1991). Reproduction of naked mole-rats. In *The Biology of the Naked Mole-Rat* (P. W. Sherman, J. U. M. Jarvis, and R. D. Alexander, Eds.), pp. 384–425. Princeton Univ. Press, Princeton, NJ.

Jarvis, J. U. M., O'Riain, M. J., Bennett, N. C., and Sherman, P. W. (1994). Mammalian eusociality: A family affair. *TREE* 9, 47–51.

Lacey, E. A., and Sherman, P. W. (1991). Social organization of naked mole-rat colonies: Evidence for a division of labor. In *The Biology of the Naked Mole-Rat* (P. W. Sherman, J. U. M. Jarvis, and R. D. Alexander, Eds.), pp. 275–336. Princeton Univ. Press, Princeton, NJ.

Lacey, E. A., and Sherman, P. W. (1997). Cooperative breeding in naked mole-rats: Implications for vertebrate and invertebrate sociality. In *Cooperative Breeding in Mammals* (N. G. Solomon and J. A. French, Eds.), pp. 267–301. Cambridge Univ. Press, New York.

Lovegrove, B. G., and Wissel, C. (1988). Sociality in mole-rats: Metabolic scaling and the role of risk sensitivity. *Oecologia* 74, 600–606.

O' Riain, M. J., Jarvis, J. U. M., and Faulkes, C. G. (1996). A dispersive morph in the naked-mole rat. *Nature* 380, 619–621.

Reeve, H. K., Westneat, D. F., Noon, W. A., Sherman, P. W., and Aquadro, C. F. (1990). DNA "fingerprinting" reveals high levels of inbreeding in colonies of the eusocial naked mole-rat. *Proc. Natl. Acad. Sci. USA* 87, 2496–2500.

Sherman, P. W., Jarvis, J. U. M., and Braude, S. H. (1992). Naked mole-rats. *Sci. Am.* 267(2), 42–48.

Nematodes and Related Phyla

Denis J. Wright
Imperial College of Science, Technology, and Medicine

I. Introduction
II. Mechanisms of Reproduction in Nematodes
III. Structure of the Nematode Reproductive System
IV. Other Phyla

GLOSSARY

Adenophora One of the two classes into which the Nematoda are still commonly divided; they lack phasmids (posterior sense organs), have a single-celled "secretory–excretory system" and are usually free-living, mostly in an aquatic environment.

cuticle A collagenous, multilayered, elastic outer region of the nematode body wall which acts antagonistically to the longitudinal body wall muscles in locomotion; also invaginates the buccal cavity and various other openings and forms structures such as the male (copulatory) spicules and bursa.

entomopathogenic Rhabditid nematodes (*Steinernema* and *Heterorhabditis* spp.) infecting insects which release symbiotic bacteria killing the host. The nematodes undergo several life cycles within the cadaver, emerging eventually as infective nonfeeding juveniles.

intersexes Males or females possessing secondary female or male sexual characteristics.

pseudocoelom The space between body wall and alimentary canal which is surrounded by tissues of mixed lineage (e.g., a coleom, which is mesodermal in origin).

rhabditid nematodes (Class Secernentea: Order Rhabitida); a major group of free-living and animal parasitic nematodes.

Secernentea One of two nematode classes; they possess phasmids and a tubular secretory–excretory system and are free-living in mostly terrestrial environments; they also contain the majority of the animal and plant parasitic nematode species described.

tylenchid nematodes (Class Secernentea: Order Tylenchida); a major group of plant parasitic species which possess a hollow collagenous "stylet" for feeding.

The nematodes or "roundworms" are bilaterally symmetrical, elongate cylindrical invertebrates which lack appendages. They have a uniform and simple life cycle of an egg, four vermiform larval (juvenile) stages, and, in most species, a vermiform adult. Nematodes are placed traditionally in a heterogeneous group of "pseudocoelomate" phyla, the "Aschelminthes," which also includes the Kinorhyncha, Loricifera, Nematomorpha, Rotifera, and Acanthocephala, and, in some classifications, the Gatrotricha, Tardigrada, and Priapulida. Recent phylogenetic analysis of 18S ribosomal DNA sequences has provided evidence for a clade (termed the Ecdysozoa) containing the arthropods and all other molting invertebrate phyla including the nematodes, nematomorphs, kinorhynchs, tardigrades, and priapulids. Nematodes are by far the largest group of pseudocoelomates and are thought to be the most species-rich group of organisms after the arthropods. Almost 50% of nematode species described are parasites of animals and plants; the remainder are free-living in the soil and in freshwater and marine sands and muds. The sheer abundance of free-living nematodes indicates their ecological importance although they have been relatively poorly studied and it is clear that the vast majority of such species are as yet undescribed. Nematodes exhibit a wide range of reproductive strategies and this has been suggested to be one of the most important factors in their enormous and diverse ecological success as both free-living and parasitic organisms.

I. INTRODUCTION

There are ≈15,000 described nematode species but current estimates of undescribed species are as high as several million. All nematodes, and other "pseudocoelomates," reproduce via gametes produced from germ cell lines. Asexual reproduction, where it occurs, is always by the production of "diploid" eggs by various forms of parthenogenesis and not from the subdivision of somatic cell lines by budding or fission as can occur in some other groups of invertebrates, such as Cnidaria, Platyhelminthes, and Annelida, or by polyembrony as is the case in some insects and bryozoans.

In general, egg production in nematodes is correlated with body size, with the larger, animal parasitic species having the greatest fecundity. The cosmopolitan human parasite *Ascaris suum* (length, ∼30 cm) can lay 200,000 eggs per day. However, the rate of development is also critical. For example, the widely studied soil rhabditid species, *Caenorhabditis elegans* (length, ∼1 mm), which produces ∼300 eggs in a life cycle of less than 3 days under optimal conditions, also has an enormous reproductive potential.

II. MECHANISMS OF REPRODUCTION IN NEMATODES

It is generally accepted that most nematode species are dioecious (gonochoristic) and reproduce by amphimixis (e.g., *A. suum*; the free-living soil rhabditid species, *Panagrellus redivivus*; and the tylenchid plant parasites, *Globodera* and *Heterodera* spp.). However, there are frequent exceptions to this, particularly among rhabditid and tylenchid species, in which hermophroditism and parthenogenesis respectively are particularly common.

Most species have a sex ratio of 1:1, although the proportion of males may increase as a result of differential survival. Environmentally influenced sex determination also occurs in some species. For example, in the adenophoran, mermithid parasites of insects, the presence of existing juveniles appears to increase the proportion of invading juveniles which develop into adult males. Similarly, in the tylenchid, root-knot nematodes (*Meloidogyne* spp.) overcrowding within the roots in at least some species can lead to a change in the development of the genital primordium at the second juvenile stage resulting in a much greater proportion of "males" (intersexes with two rather than one testis).

In the Adenophora, the few amphimictic species examined appear to have an XX (female), XY (male) type of mechanism with a typical haploid chromosome number of 6, although some parthenogenetic plant parasitic species are polyploid. In the more widely studied Secernentea, many free-living species also have a haploid number of 6 and an XX (female), XO (male) mechanism appears to be the most widespread although an XX–XY type occurs in some groups. Plant parasitic nematodes do not appear to possess distinct sex chromosomes. Greater cytogenetic variation is found within the Secernentea, and many parasitic forms are polyploid and are hermaphroditic, parthenogenetic, or show heterogony (see Section II,D). Sex determination has been best studied in *C. elegans*, in which there is an XX (hermaphroditic "female"), XO (male) sex chromosome system with dosage compensation to equalize the amount of X-linked gene products. In this species, both the hermaphrodite and the male are diploid with five pairs of autosomes. Males are produced from nondisjunction events on the X chromosome and as half the progeny from a mating between a hermaphrodite and a male. Intersexes occur across a range of free-living and parasitic genera, usually female with male accessory sex organs (see Section III,A) but in some *Meloidogyne* species they can be male.

A. Amphimixis

Amphimictic species and races of nematodes possess various physiological and structural adaptations to assist in mate location, sperm transfer, and utilization. The presence of sex pheromones has been indicated in a number of parasitic and free-living species. For example, the swollen sedentary females of cyst nematodes, *Globodera* and *Heterodera* spp., are known to attract the mobile, vermiform male. In *C. elegans*, males are attracted to hermaphrodites but the latter are not attracted to each other. In the amphimictic rhabditid species, *P. redivivus*, both adult males and females are mutually attracted to each other but males are much less attracted to gravid females compared with unmated females.

Male nematodes possess a variety of accessory reproductive structures (see Section III,A) that are involved in copulation. In some species, the male coils its tail end around the vulval region of the female, whereas in other species (many secernentean and a few adenophoran species) lateral extensions of the cuticle (caudal alae) in the male tail form a "copulatory bursa" which is used to grip the female. Males of most species have spicules which are extended to enter and open the vulva and vagina. In species with paired spicules, these may form a channel to aid sperm transfer to the female (Fig. 1). The spicules possess one or more nerves which end at pores in the tip and in at least some species they are known to have a sensory function before or during copulation. Other caudal sensory structures (sensillae) are also implicated in copulation (Fig. 1). In *C. elegans*, there are nine pairs of caudal papillae (sensory rays) embedded in the fan-shaped bursa which are thought to be chemosensory. The sensory rays are involved in the orientation of the male tail in relation to the hermaphroditic female (see Section II,B). Vulval location is then mediated first by the hook sensillum

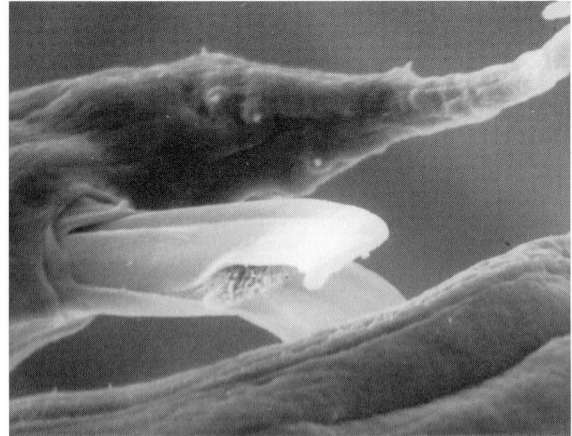

FIGURE 1 Scanning electron micrograph (×2000) of the tail region of an adult male *Panagrellus redivivus* (Order Rhabditida), showing paired curved spicules extended through the gubernacular sheath. The membranous velum (attached to the curved backbone of the spicule) on each spicule combine to form a channel through which sperm (shown) pass during copulation. This species does not possess a copulatory bursa and instead the male coils around the female (see Section II,A). Several caudal papillae (sensillae) can be seen on the dorsal (two) and ventromedian (three) postcloacal region of the tail.

(adjacent to the male cloaca) and then by spicule and postcloacal sensillae.

In most species, the female stores sperm in the region between the uterus and the ovary (modified to form a seminal receptacle in some species). Fertilization usually occurs in this region and eggshell formation commences (see Section III,B). In one animal parasitic species, *Syngamus tracheae*, the male is permanently attached to the female. In a few other animal parasites, the spicules are absent. In one case, the male lives within the uterus of the female. In another, the male inserts the posterior end of its body into the uterus of the female.

B. Hermaphroditism

Automixis (self-fertilization) is found among free-living (e.g., *C. elegans*) and entomopathogenic (*Heterorhabditis* spp.) rhabditid species and in diplogastrids and aphelenchids. Some adenophoran species are also known to be hermaphorodites. Cross-fertilization, the usual mechanism in hermaphroditic animals, has not been found within the Nematoda. Hermaphroditic nematodes are all reported to be protandrous, with the ovotestis first producing sperm then eggs. The presence of an ovotestis may avoid most of the energy costs incurred by the majority of other groups of hermaphroditic animals which have separate male and female reproductive systems.

Some nematode species have automictic and amphimictic races (e.g., *Rhabditis marionis*) and in *C. elegans* hermaphroditism is facultative. Males of the latter species, which are normally very rare, can occur at higher frequencies at elevated temperatures and amphimixis with the "hermaphroditic" females displaces automixis.

C. Parthenogenesis

Parthenogenesis is a common form of reproduction among free-living nematodes and rivals hermaphroditism as an alternative to amphimixis, but it is particularly common in plant parasitic nematodes. It also occurs in some animal parasitic species, e.g., *Strongyloides stercoralis*. Some species are also reported with parthenogenetic and amphimictic races, e.g., the mycophagous species, *Aphelenchus avenae*. About 20% of the plant parasitic species in which the mechanism of reproduction has been examined are parthenogenetic (facultative or obligate) and a further 20% are probably parthenogenetic. This compares with the animal kingdom as a whole in which approximately 0.1% of species are thought to be parthenogenetic.

Parthenogenesis may be mitotic or meiotic. In *Meloidogyne* spp., facultative meiotic parthenogenesis is thought to prevail under conditions favorable for rapid development and reproduction and here males are absent or rare, whereas amphimixis is common under more adverse conditions in which males are more abundant (e.g., *Meloidogyne graminicola, M. naasi,* and *M. graminis*). In some cases, the same female may have oocytes developing amphimictically and other oocytes developing by meiotic parthenogenesis. Other genera of plant nematodes show obligatory mitotic parthenogenesis (e.g., *M. incognita, M. javanica,* and *M. arenaria*). In *M. hapla*, different races are facultatively (meiotic) or obligatorily (mitotic) pathenogenetic.

Mitotic parthenogenesis (ameiotic thelytoky) is the most usual type of parthenogenesis in the plant kingdom but also occurs among various animal groups. It is probably the more common form of parthenogenesis in plant parasitic nematodes. Here, a mitotic division results in the formation of a diploid egg pronucleus: for example, *M. javanica*, which also has polyploid or aneuploid derivatives.

In meiotic parthenogenesis, diploidy is reestablished by (i) the second polar nucleus formed during meiosis fusing with the oocyte pronucleus instead of being extruded (e.g., *M. hapla*), (ii) chromosomes regrouping at telophase II to form a single pronucleus (most populations of *A. avenae*), or (iii) an endomitotic division which occurs during prophase of the first cleavage division (doubling of chromosome number without division of the nucleus), e.g., the cyst species *Heterodera betulae*.

Pseudogamy (gynogenesis or apomixis) can be regarded as an unusual form of parthenogenesis which has been reported to be found in some apparently hermophroditic (e.g., *R. anomala*) and dioecious species (e.g., *R. maupasi*) in which the egg must be activated by penetration of the sperm but without the subsequent fusion of the sperm with the egg pronucleus. Diploidy has been reported to be either maintained by mitotic parthenogenesis or restored by an endomitotic division of the oocyte pronucleus. However, the original studies on pseudogamy in nematodes were cytological and open to doubt, and recent microsatellite DNA fingerprinting studies have shown that in at least one animal parasitic species, *Strongyloides ratti*, that reproduction is amphimictic rather than pseudogamous.

D. Heterogony

Heterogony (variation in breeding form) has only been conclusively demonstrated among parasitic species of the Rhabitida and Strongylida, although it may also occur in other parasitic and free-living groups. In most parasitic species which show heterogony, there is a parthenogenetic life cycle in the host and a free-living cycle in which the nematodes reproduce by amphimixis for one or more generations (e.g., *S. ratti*). In entomopathogenic *Heterorhabditis* species, the first generation following invasion of the host is hermaphroditic, with subsequent generations being amphimictic.

E. Evolution of Reproduction in Nematodes

In contrast with most other animal groups, hermaphroditism is not generally considered to be primitive in nematodes and is thought (as in the case of parthenogenesis) to have arisen on a number of occasions from amphimictic ancestors. Pseudogamy in nematodes has sometimes been considered to be a transitional stage in the development of parthenogenetic systems in which the presence of males is not required, although in the absence of any molecular genetic evidence this suggestion is purely speculative.

The dominance of amphimixis among animal parasitic nematodes has been associated with the need to maintain maximum genetic variability in relation to the continual evolutionary "arms race" between the host immune response and the parasite. The widespread occurrence of hermaphroditism and parthenogenesis in free-living and plant parasitic nematodes has also been related to some extent with the degree of habitat heterogeneity and stability. Obligatory (mitotic) parthenogenesis might be expected to be particularly favored in environments that are relatively stable and *Meloidogyne* species with this mode of reproduction are largely tropical in distribution with wide host plant ranges. In contrast, facultatively parthenogenetic or amphimictic *Meloidogyne* species tend to have a more temperate, restricted distribution with a relatively narrow host range. Likewise, the largely amphimictic, cyst nematodes predominate in temperate habitats and have relatively specific host–plant relationships.

III. STRUCTURE OF THE NEMATODE REPRODUCTIVE SYSTEM

The reproductive system is formed from a single or a double genital tube. In both females and males, the genital primordia differentiate as early as the second-stage juvenile. Adult females are usually larger than males.

A. Male Reproductive System

The male reproductive system in most secernentean nematode species consists of a single genital tube (termed monorchic), whereas the adenophoreans often have two (diorchic). The genital tube consists of a distal testis leading in turn to the seminal vesicle and vas deferens, which opens into the cloaca (Fig. 2a). Some species also have a vas eferens between the testis and the seminal receptacle.

Spermatogenesis in the majority of nematodes is telogonic, in which germ cell formation is limited to a relatively short region, the germinal zone, at the distal end of the testis. In these nematodes, the remainder of the testis is termed the growth zone. In some adenophorans germ cell formation is hologonic and occurs along the length of the testis. Mitotic divisions lead to the production of spermatogonia which change shape and enlarge in the growth zone to form spermatocytes. In some species, the cells in the growth zone are arranged radially around a central core, the rachis. The spermatocytes undergo meiotic divisions to form haploid spermatids which either develop into mature spermatozoa in the male or complete their maturation in the uterus of the female following copulation. Nematode sperm lack an acrosome or a nuclear membrane (which is lost during spermatogenesis) and, with a few possible exceptions, appear to be nonflagellate and move by the production of pseudopodia (ameboid movement).

Male nematodes possess a range of accessory sex structures which are involved in copulation (see Section II,A). The spicules are elongate, partly sclerotized, cuticular structures in invaginations of the cloaca which are extended and retracted by associated muscles. A cuticularized part of the spicule pouch, the gubernaculum, can help guide the spicules. Nearly all male nematodes usually have a pair of spicules, although in some species a fused or single spicule is present. The spicules vary in size and shape between species and a number of species have asymmetric spicules. In some species, the spicules are

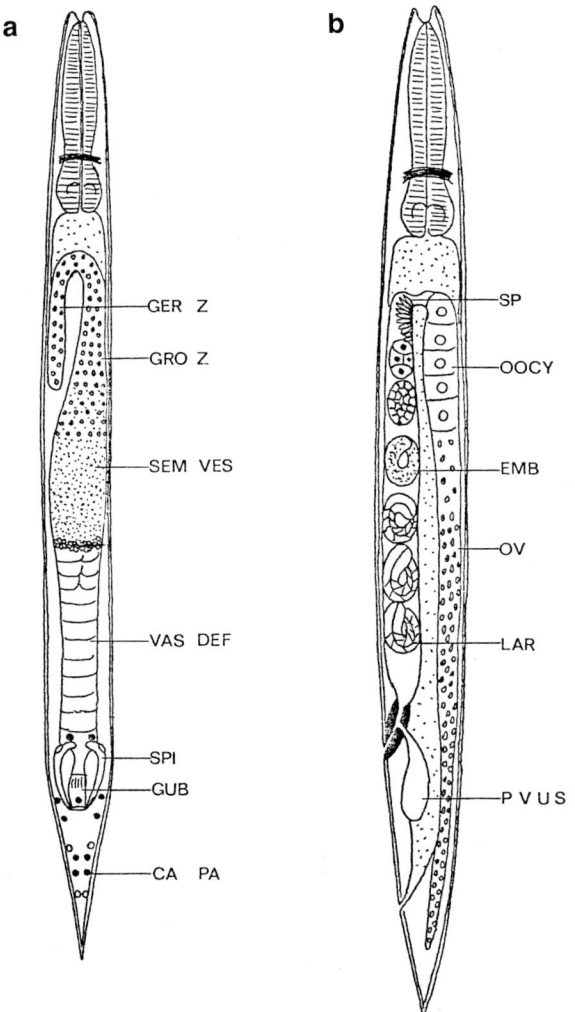

FIGURE 2 Diagrams of the reproductive system in adult male (a) and female (b) *Panagrellus redivivus*. In both cases, the reproductive system can be seen to extend from the anterior intestinal region (behind muscular pharynx and associated nerve ring) to the anal region in the tail. Both sexes have a single genital tube (monorchic/monodelphic for male/female); the postvulvar uterine sac (PVUS) in the female appears to be a vestigial, second genital tube. GER Z, germinal zone of testis; GRO Z, growth zone of testis; SEM VES, seminal vesicle; VAS DEF, vas deferens; SPI, spicule; GUB, gubernaculum; CA PA, caudal papilla (sensilla); OV, ovary; OOCY, oocytes in oviduct; SP, sperm in seminal receptacle; EMB, fertilized egg with developing embryo; LAR, fully developed first larval (= juvenile) stage (*P. redivivus*, unlike most nematodes, is ovoviparous).

known to interlock, forming a channel which appears to aid transfer of sperm; in *P. redivivus*, membranous extensions of each spicule (the velum) overlap to form a partial tube (Fig. 1). In a few adenophoran and many secernentean species, lateral extensions of the cuticle (alae) in the tail region form a copulatory bursa which is used to grasp the female prior to copulation (see Section II,A). Sensory structures located in the cloacal region (a hook sensillum and two postcloacal sensillae in *C. elegans*), in the spicules, and in caudal papillae (often associated with the bursa when present) assist orientation of the male for copulation (see Section II,A). Among the Adenophora, various copulatory "supplements" are found comprising ventromedian and subventral papillae, setae (sensory appendages), and cuticular depressions in the pre- and postcloacal regions, although their functions are not fully understood.

B. Female Reproductive System

In the majority of nematode species examined, the reproductive system consists of two genital tubes (termed didelphic) leading from a common, transverse vagina. The genital tube shows regional differentiation into a distal ovary, an oviduct, and a proximal uterus. Didelphic species are termed amphidelphic if their uteri are apposed at their origin; they are termed prodelphic or opisthodelphic if they are parallel and run anteriorily or posteriorily, respectively (terms also used to describe the orientation of the uterus in single-tubed or monodelphic species; see Fig. 2b).

The genital aperture, the vulva, is muscular and lined with cuticle and is situated ventrally although its position along the body varies between species. The vagina is also lined by cuticle and in species which produce large numbers of eggs (some animal parasitic species) it has a muscularized region—the ovijector. The uterus consists of squamous polyhedral epithelial cells and may be muscularized. The distal end of the uterus forms a distinct sperm storage chamber, the seminal receptacle (spermatheca) in some nematodes. The ovary in most species is telogonic, with the remainder of the ovary forming the growth and ripening zones.

The germ cells differentiate into oogonia, which

in secernenteans (e.g., adenophorans) are arranged radially around a central rachis to which they are linked by cytoplasmic bridges. In the secernentean telogonic ovary, oogonia are formed by mitosis in the germinal zone, develop into oocytes which undergo a meiotic division, detach from the rachis, increase in size, and pass down the oviduct into the spermathecal region where fertilization generally occurs. Eggshell formation also commences and is completed further down the uterus.

Fertilization in at least some secernentean nematodes (e.g., *C. elegans* and *P. redivivus*) is extraordinarily efficient and nearly every sperm may fertilize an egg. In such cases, it is the number of sperm which is limiting rather than the number of oocytes (unlike most adenophorans and other animals).

The basic structure of the eggshell appears to be similar in all nematodes, consisting of an outer vitelline layer, a middle chitinous layer (the only chitinous structure found in nematodes), and an inner lipid layer. It is a very impermeable structure, apart from allowing gas exchange, and in many species it is a key to survival under unfavorable environments. A few nematodes are known to be ovoviparous (e.g., the adenophoran animal parasite, *Trichinella spiralis*, and the rhabditid, *P. redivivus*; Fig. 2b); the juveniles hatch from the egg within the uterus.

IV. OTHER PHYLA

In comparison with the nematodes, the other pseudocoelomate phyla have been poorly studied and contain few species.

The Nematomorpha or "hairworms" (~320 species) include marine, freshwater, and semiterrestrial groups, all of which are parasites of invertebrates during their larval stages with free-living adults. All nematomorphs are dioecious, with two tubular gonads extending along most of the elongated, threadlike, cylindrical body (typically 5–10 cm in length). Reproduction appears to be solely by amphimixis but the possibility of parthenogenesis cannot be completely excluded because the reproductive mechanism has only been studied in a few species. Little is known of the sexual behavior of nematomorphs, although there is evidence that males are attracted to females. During copulation the male forms tight coils around the female and sperm are deposited externally near the female cloacal aperture. Males do not possess copulatory spicules (cf. nematodes). The sperm eventually reaches the female seminal receptacle and strings of fertilized eggs are released into the environment.

The Priapulida (~16 species) are cylindrical, worm-like or cucumber-shaped benthic animals (0.5 mm to 30 cm in length) found in marine muds and sands. They are described in some textbooks as gonochoristic, but although this appears to be true in two families (Priapulidae and Tubiluchidae), in the third family (Maccabeidae) only females have been found, suggesting that parthenogenesis, although protogynous (ova produced before sperm), hermaphroditism remains a possibility. In the Priapulidae, the sexes are similar and fertilization is external. In the Tubiluchidae, there is marked external sexual dimorphism and fertilization is internal.

The Tardigrada or "water bears" (~600 species) are microscopic (~0.5 mm) with a short, cylindrical body with four pairs of ventral legs. The relatively few marine interstitial species are almost all gonochoristic with only one hermaphroditic species described. In freshwater forms and in the most numerous group which live in water films on terrestrial mosses and lichens, although most species appear to be gonochoristic a number of cases of hermaphroditism have been reported. In most species there is relatively little sexual dimorphism. The male has two vasa deferentia and the female one oviduct, but these structures are often difficult to distinguish and confusion regarding identification of the testis has also contributed to confusion of sexuality in the earlier literature. Hermaphroditic tardigrades have one gonoduct which appears to be a typical ovotestis, and hermaphroditism is possibly simultaneous rather than protandrous or protogynous.

The Kinorhyncha (~150 species) are also microscopic (typically <1 mm) and are found in marine muds and sands. They have a short grub-like body, flattened dorsoventrally. They are dioecious (gonochoristic), having a pair of gonads. Copulation has not been observed but males in two genera extrude spermatophores which appear to be directed toward the female by spines or spicules. The sperm is stored within the seminal receptacle of the female and fertilization is thought to be internal.

Of the other, possibly more distantly related pseudocoelomates, the Loricifera (9 microscopic marine species) and Acanthocephala (~1200 animal parasitic species) are reported to be dioecious (gonochoristic). The Rotifera (~1500 microscopic, predominantly freshwater species) are dioecious but males are often rare and they are parthenogenetic or cyclically parthenogenetic–gonochoristic. The Gastrotricha (~450 mostly microscopic, marine and freshwater species) are reported to be mainly monoecious (hermaphroditic), with some parthenogenetic species.

See Also the Following Articles

CAENORHABDITIS ELEGANS; HERMAPHRODITISM; NEMATOMORPHA; PARTHENOGENESIS AND NATURAL CLONES; ROTIFERA; TARDIGRADA

Bibliography

Adiyodi, K. G., and Adeyodi, R. G. (1983-1992). *Reproductive Biology of Invertebrates*, Vols. 1–4A and 5. Wiley, Chichester, UK.

Aguinaldo, A. M. A., Turbeville, J. M., Linford, L. S., Rivera, M. C., Garey, J. R., Raff, R. A., and Lake, J. A. (1997). Evidence for a clade of nematodes, arthropods and other moulting animals. *Nature* 387, 489–493.

Bird, A. F., and Bird, J. (1991). *The Structure of Nematodes*, 2nd ed. Academic Press, San Diego.

Riddle, D. L., Blumenthal, T., Meyer, B. J., and Priess, J. R. (1997). *C. elegans II*. Cold Spring Harbor Laboratory Press, Cold Spring Harbor, NY.

Ruppert, E. E., and Barnes, R. D. (1991). *Invertebrate Zoology*, 6th ed. Saunders, Fort Worth, TX.

Viney, M. E., Matthews, B. E., and Walliker, D. Mating in the nematode parasite *Strongyloides ratti*—Proof of genetic exchange. *Proc. R. Soc. London B* 254, 213–219.

Wharton, D. A. (1986). *A Functional Biology of Nematodes*. Johns Hopkins Univ. Press, Baltimore.

Nematomorpha

Andreas Schmidt-Rhaesa
University of South Florida in Tampa

I. Introduction
II. Morphology of the Reproductive Systems
III. Morphology of the Gametes
IV. Reproductive Behavior
V. Early Development
VI. Larva and Parasitic Development

GLOSSARY

cloaca A structure in *Gordiida* formed by fusion of the reproductive and intestinal systems.

extracellular matrix A matrix secreted from cells that serves mainly as an attachment for cells and surrounds primary body cavities.

gonoparenchyma A tissue rich in vesicles and endoplasmatic reticulum; found in young females of *Nectonema*. It has a presumed function in oogenesis.

Gordiida A monophyletic subtaxon of the Nematomorpha with all freshwater species (well-known genera: *Gordius*, *Paragordius*, and *Chordodes*).

Nectonema A monophyletic genus and subtaxon of the Nematomorpha, sister group of the Gordiida. *Nectonema* contains four marine species.

parenchyma A tissue with cells rich in vesicles; absent in *Nectonema*. It probably functions as nutrient storage.

postseptum The posterior part of the larva.

preseptum The anterior part of the larva.

testis tubes Paired longitudinal tubes lined with epithelium in early stages; this epithelium probably gives rise to the spermatozoa.

Nematomorpha is a small group of about 300 species of long and slender worms. They are closely related to nematodes within a larger group to which the names Aschelminthes, Nemathelminthes, or pseudocoelomates are applied. All Nematomorpha reproduce in water; four species of the genus *Nectonema* are marine and the majority of species, called Gordiida, are freshwater species. The main part of the life cycle is parasitic, in which larval nematomorphs infect hosts that are usually arthropods. Knowledge about nematomorph reproduction and development is incomplete and many aspects need further investigation.

I. INTRODUCTION

All nematomorphan species are gonochoric with dimorphic males and females which are easily distinguishable by their external appearance. Asexual reproduction does not occur in nematomorphs. The taxa *Nectonema* and Gordiida differ significantly in their reproductive systems and in their mode of reproduction. The life cycle of both taxa starts with a unique type of larva that infects hosts. Within the host, the larva grows to about adult size, molts, and emerges from the host to reproduce in the aquatic environment.

II. MORPHOLOGY OF THE REPRODUCTIVE SYSTEMS

A. *Nectonema*, Females

Females of *Nectonema* are usually larger than males (maximum lengths up to 960 mm have been reported) and have a terminal genital opening. In contrast to Gordiida, the posterior opening in both sexes of *Nectonema* is not a cloaca because the intestine ends blindly and does not communicate with the reproductive system. Mature females of *Netonema* have a large primary body cavity which is filled with large amounts of oocytes (Fig. 1A). There are no defined reproductive organs and the oocytes are not surrounded by an epithelial layer. The body cavity is lined by the extracellular matrix (ECM) of the

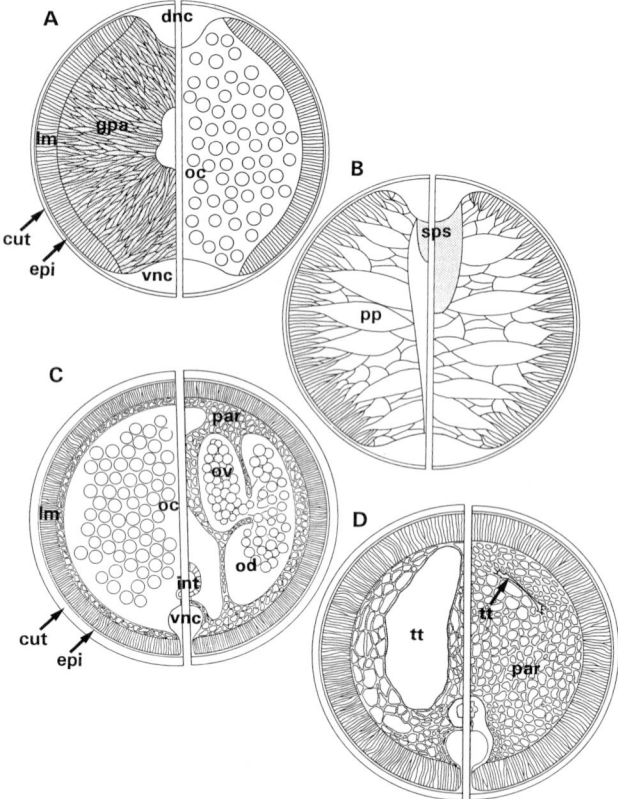

FIGURE 1 Cross section through different stages of male and female nematomorphs. The younger stages are on the left of in A and B and on the right in C and D. (A) *Nectonema* female with gonoparenchyma (gpa) or large body cavity filled with oocytes (oc). (B) *Nectonema* male with large protoplasmatic extension of the muscle cells (pp) and different sizes of the sperm sac (sps). (C) Gordiida female with ovarium (ov) and oviduct (od) embedded in parenchyma (par) or with large body cavity filled with oocytes (oc). (D) Gordiida male with different extensions of the testis tubes (tt). cut, cuticle; dnc, dorsal nerve cord; epi, epidermis; int, intestine lm, longitudinal musculature; vnc, ventral nerve cord (C and D reproduced from Schmidt-Rhaesa, 1997a, with kind permission of Fischer-Verlag, Stuttgart).

surrounding longitudinal muscle cells. In earlier stages of female *Nectonema* the body cavity is preceded by a solid tissue with cells containing numerous vesicles and endoplasmatic reticulum (Figs. 1A and 2A). This gonoparenchyma either completely fills the body or surrounds a central lumen of varying size. Into this lumen round compartments are budded off by the gonoparenchyma cells (Fig. 3B). Al-

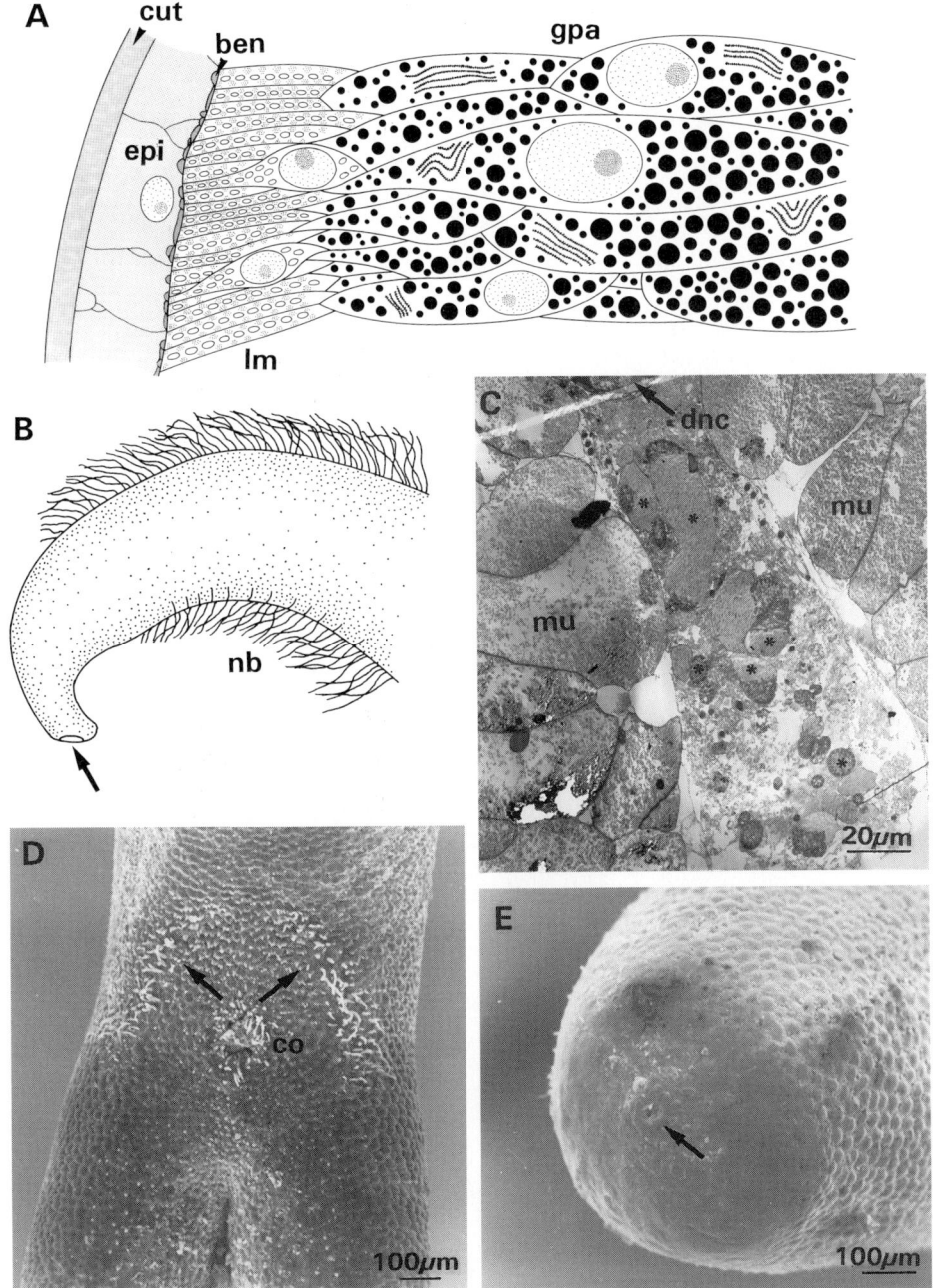

FIGURE 2 Reproductive organs in Nematomorpha. (A) Gonoparenchyma (gpa) in an early stage of female *Nectonema*. ben, basiepidermal nervous system; cut, cuticle; epi, epidermis; lm, longitudinal musculature. (B) Ventrally curved posterior end with terminal genital opening (arrow) in *Nectonema* males. nb, natatory bristles. (C) Sperm sac of male *Nectonema munidae*, attached to dorsal nerve cord (dnc) and surrounded by protoplasmatic extensions of the longitudinal muscle cells (mu). Note muscle strands (*) running through the sperm sac. (D) Ventral view of *Gordionus violaceus* (Gordiida) with cloacal opening (co), the base of two tail lobes posterior of the cloacal opening, and cuticular structures anterior to it (arrows). (E) Terminal cloacal opening (arrow) in a female *G. violaceus* (A, reproduced from Schmidt-Rhaesa, 1997b, with kind permission of Springer-Verlag, Berlin; B, modified from Beattie, 1987; D and E, reproduced from Schmidt-Rhaesa, 1997a, with kind permission of Fischer-Verlag, Stuttgart).

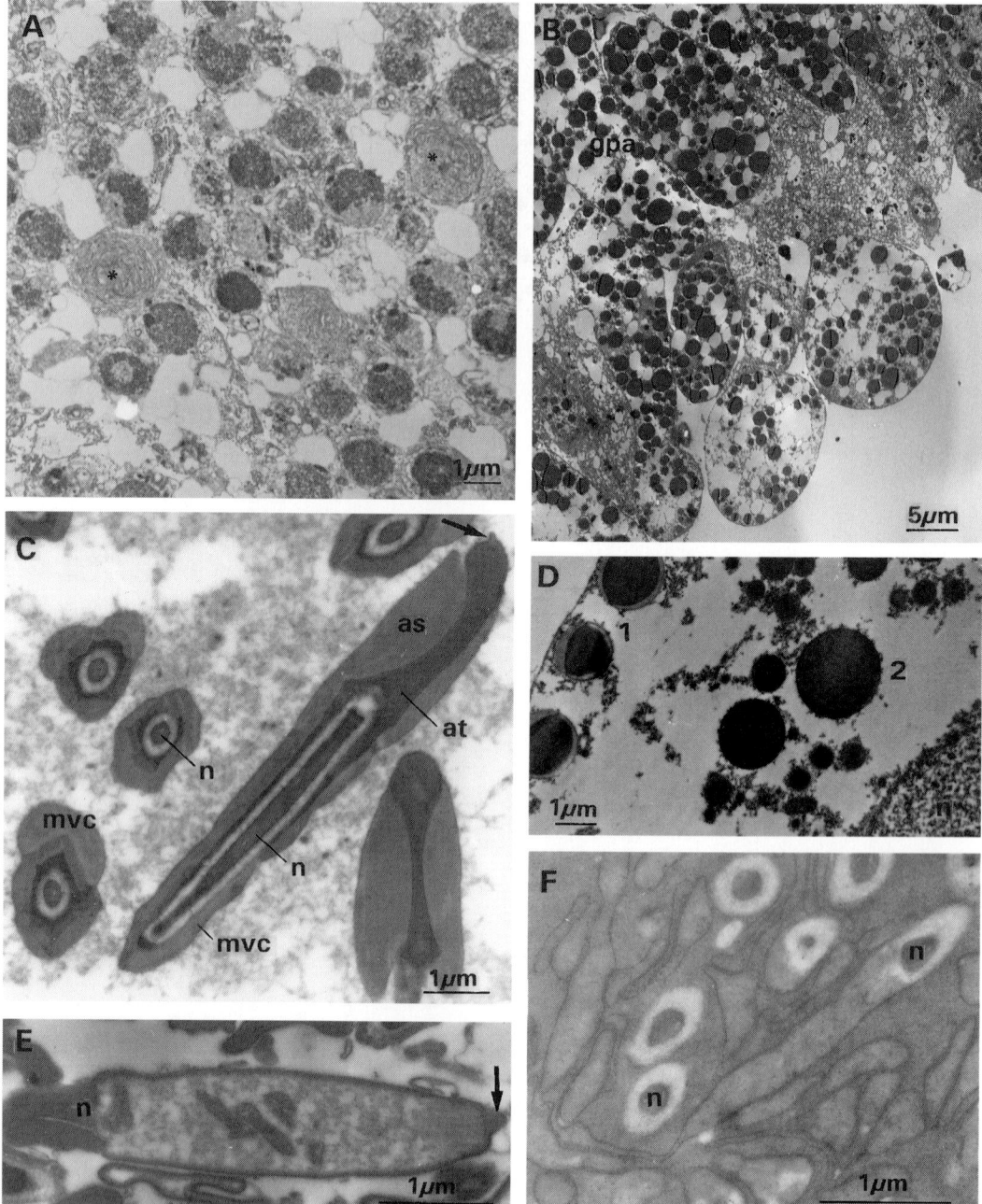

FIGURE 3 Gametes in Nematomorpha. (A) Gametes in the sperm sac of male *Nectonema munidae*. Note the arrangement of some gametes around a central structure (*). (B) Female of *N. munidae*: round compartments are budded from the gonoparenchyma (gpa) into the body cavity. (C) Spermatozoa of *Gordius aquaticus* inside the male testis tubes with nucleus (n), multivesicular complex (mvc), acrosomal tube (at), acrosomal sheath (as), and acrosome (arrow). (D) Oocyte of *N. munidae* with peripheral (1) and central (2) types of vesicles. (E) Anterior end of spermatozoon of *G. aquaticus* deposited in the sperm drop on the posterior end of the female. n, nucleus; arrow, acrosome. (F) Tightly packed spermatozoa of *G. aquaticus* in the female seminal receptacle. Note the variable shape of the anterior end (B, D, reproduced from Schmidt-Rhaesa, 1997b, with kind permission of Springer-Verlag, Berlin; C, E, F, reproduced from Schmidt-Rhaesa, 1997c, with kind permission of Balaban Publ., Rehovot).

though nuclei are rarely found in these compartments, it is presumed that this is the mode of oogenesis in Nectonema.

B. Nectonema, Males

The posterior end of male Nectonema is ventrally curved and tapers toward the end, forming a short, tube-like structure (Fig. 2B). The genital opening is situated terminally. Internally, a unique and incompletely investigated reproductive organ is found: the sperm sac. This structure is attached to the dorsal epidermal cord (Figs. 1B and 2C). Several muscle strands pass through the sperm sac in the posterior end of the body (Fig. 2C), and presumably participate in sperm ejection.

C. Gordiida, Females

The female cloacal opening is always terminal on the posterior end (Fig. 2E). The earliest structures of the reproductive system are iterative accumulations of cells that develop into ovaries. In later stages, these ovaries are iterative sac-like extensions of paired longitudinal oviducts (Fig. 1C). The oviducts are lined with an epithelium which rests on the ECM of the surrounding parenchyma cells. In the posterior region of the body, the oviducts lead into an atrium into which the intestine joins slightly posterior to form a cloaca. An anterior extension of the atrium serves as a seminal receptacle. In more mature stages, a large oocyte-filled primary body cavity is present (Fig. 1C). The transition between the stage with ovarial ducts embedded in parenchyma and the stage with the large body cavity filled with oocytes is not known. It should be mentioned that the parenchyma of Gordiida differs structurally from the gonoparenchyma of Nectonema and the two tissues are not homologous.

D. Gordiida, Males

In males of all species of Gordiida the cloacal opening is on the ventral side (Fig. 2D). In many species, two lobes are present posterior of the cloacal opening which seem to play a role in copulation. In other species, there is only a longitudinal furrow or no division of the posterior end at all. Around or in the vicinity of the cloacal opening, different cuticular structures (spines, bristles, warts, and cuticular folds) can be present which may be important during copulation as sensory structures or for optimization of the exact copulatory position. Internally, the reproductive system consists of two longitudinal testis tubes (Fig. 1D). These tubes can be narrow and inconspicuous or the dominating structure in a cross section of the animal, depending on their sperm content (Fig. 1D). The extension of the testis tubes corresponds with the surrounding parenchyma, which probably has nutritive function in gametogenesis. The fine structure of the testis tubes consists of a broad ECM, surrounding an endothelium. This endothelium disintegrates during further development and is not present (or present only as a few scattered cells) in mature males. It is presumed that gametes originate from this epithelium. In the posterior part of the body, the testis tubes either run separately into the intestine or join ventrally to the intestine and lead into it with a single duct.

III. MORPHOLOGY OF THE GAMETES

A. Nectonema, Females

Oocytes of Nectonema munidae are about 40 mm in diameter. The nucleus is situated centrally and two types of vesicles occur (Fig. 3D). One type is restricted to the periphery and may be responsible for spines that are formed by the egg after contact with seawater. Nurse cells are seen as clusters of vesicle-containing cells between the oocytes in mature females. The budded compartments of the gonoparenchyma contain only one type of vesicle similar to those in the interior of mature oocytes (Fig. 3B).

B. Nectonema, Males

Only few observations have been made of male Nectonema gametes. They are round structures in which only a large nucleus can be found (Fig. 3A). Some of these gametes are clustered around a central structure which probably represents a cytophor (Fig. 3A), suggesting that all male gametes investigated so far are in the early stages of spermatogenesis and that the mature sperm has not yet been investigated.

C. Gordiida, Female

The oocytes have an average diameter of about 20 mm. Comparable to *Nectonema* oocytes, the nucleus is central and many vesicles are present. The peripheral type of vesicles seen in *Nectonema* are not found in Gordiida and eggs of gordiid species do not form spines when they come into contact with water. Nurse cells are present between the oocytes.

D. Gordiida, Males

The spermatozoa of two *Gordius* species are well investigated ultrastructurally. The mature spermatozoon is unique among all animal sperm cells described to date. In the spermatozoa found in the testis tubes the nucleus is rodlike and positioned in the posterior half of the sperm (Fig. 3C). It is surrounded by an electron-lucent perinuclear cisterna and a multivesicular complex with up to three layers. The anterior end is composed of an acrosomal tube surrounded by an acrosomal sheath. The acrosome is the apical structure. The spermatozoa deposited on the posterior end of the female and those in the seminal receptacle differ morphologically from the spermatozoa described previously (Figs. 3E and 3F). Either the acrosomal tube or the sheath disintegrates, leaving one uniform compartment with the acrosome in the anterior end. This compartment becomes slender and variable in shape within the seminal receptacle (Fig. 3F). In the posterior region, the outer layers of the multivesicular complex disappear, whereas the inner vesicles fuse to form one narrow sheath around the nucleus. Flagella or centrioles are not present.

IV. REPRODUCTIVE BEHAVIOR

Current knowledge of the reproductive behavior of Gordiida is summarized by Schmidt-Rhaesa (1997a), and a main contribution is provided by Dorier (1930). All knowledge about *Nectonema* is found in Huus (1932) and an unpublished investigation by Beattie (1987).

In *Nectonema*, the male coils its posterior end around the female and injects its slender posterior end into the female genital opening. Two or three hours later the female releases the eggs separately into the water.

In Gordiida, reproduction has been observed only in *Gordius aquaticus*. The male coils its posterior end around the female and glides with the aid of the bifurcated "tail" along the female until the cloacal openings are in close vicinity (Fig. 4A). Then the male ejects a mass of spermatozoa which are deposited in a "drop" surrounding the entire posterior end of the female. A large number of rods run from the female cuticle through this sperm drop (Fig. 4B), presumably maintaining the structure. The origin and formation of these rods are unknown. It is not known how copulation is performed in species without a bifurcated posterior end, but it would be of interest to determine how structures such as circumcloacal spines or adhesive warts participate. After deposition of the sperm drop on the female, at least part of the spermatozoa enter the cloacal opening and are stored in the seminal receptacle. Fertilization occurs internally in the atrium or in the posterior region of the oviducts. Eggs are laid in strings into the water and both sexes die soon after discharging gametes.

V. EARLY DEVELOPMENT

The early development is still incompletely known. Descriptions by different authors are often contradictory, which could be due to insufficient observations or to differences between species.

The first cleavage is total and either equal or slightly unequal. At the four-cell stage, a tetrahedral figure (with two blastomeres lying in one plane and two in a perpendicular plane) was observed by some authors. This particular figure is also known from nematodes and gastrotrichs. After this step the cleavages are not simultaneous. Observations vary concerning the time and type of the blastula (sterro- or coeloblastula) and the mode of gastrulation (invagination or unipolar ingression). The blastopore closes

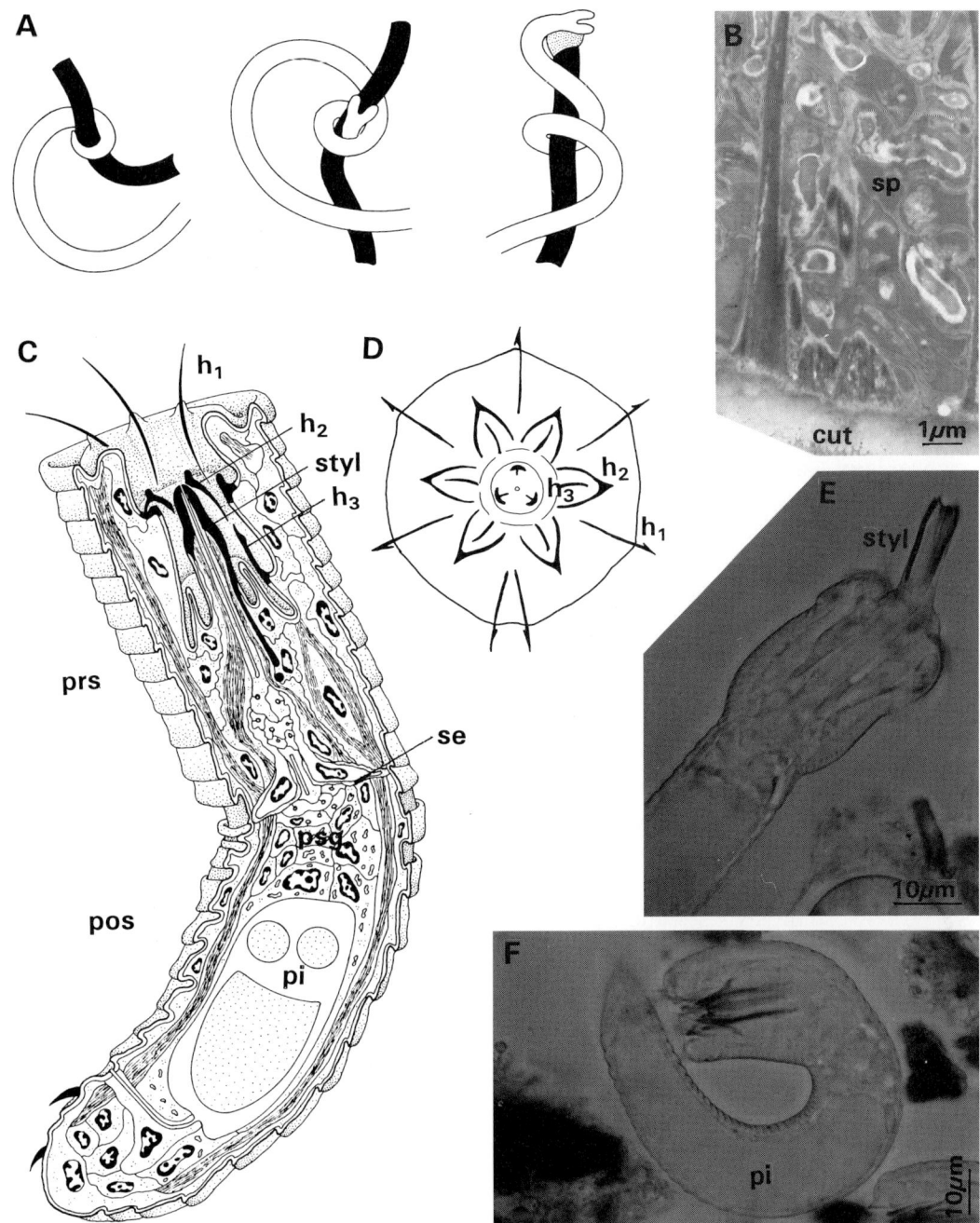

FIGURE 4 Copulation in *Gordius aquaticus*. (A) The male (white) coils around the female (black), glides with its bifurcated tail along it, and finally deposits a drop of sperm on the posterior end. (B) Rods in the sperm drop, originating on the female cuticle (cut). sp, spermatozoa. (C) Section of a larva of *Paragordius varius* with preseptum (prs) and postseptum (pos). h_1–h_3, circlets of hooks; pi, pseudointestine; psg, postseptal gland; se, septum; styl, stylet. (D) Arrangement of hooks and stylets on the preseptum; note ventral double hooks in outermost circlet. (E) Larva of *G. aquaticus* with everted hooks and stylet. (F) Larva of *G. aquaticus* with inverted hooks (A, C, D, reproduced from Schmidt-Rhaesa, 1997a, with kind permission of Fischer-Verlag, Stuttgart).

early. The pore on the posterior end of the larva opens close to the region of the blastopore. This "deuterostomy" is also present in some other "protostome" species and does not indicate a relationship with the Deuterostomia. A second invagination forms the anterior end of the larva.

VI. LARVA AND PARASITIC DEVELOPMENT

The nematomorph larva, which emerges from the eggs, is completely different from the worm-like adults (Figs. 4C–4F). It is only about 100 mm long and has an array of hooks and stylets adapted to start the parasitic phase of the life cycle.

All gordiid species observed so far have a uniform larva. The body consists of an anterior (preseptal) and posterior (postseptal) end; both are divided by a septum that is made up of a layer of cells and surrounding ECM on which muscles attach (Fig. 4C). The preseptum is superficially annulated and bears three circlets of hooks, with six hooks in the two inner circlets and seven hooks in the outer circlet. Two hooks of the outer circlet are close together on the ventral side, thus forming a hexaradial symmetry (Fig. 4D). The region of the three circlets is in- and evaginable and three central stylets can be everted (Fig. 4E). Between the stylets, a duct opening which leads to a postseptal gland can be observed. A second postseptal structure is the pseudointestine, a large and probably glandular organ (Fig. 4F) which opens subterminally with a pore. Some authors observed the secretion of a substance from the pseudointestine which hardens in water and may be responsible for cyst formation. All larvae encyst in their hosts. In some cases it was observed that larvae can form a cyst outside the host, e.g., on vegetation. The heavy armature of the gordiid larva obviously helps to infect a host. While it was once believed that infection takes place through the external surface of the host, it is likely that larvae are swallowed, bore through the intestinal epithelium, and then encyst in the host tissue. The larvae themselves can be found in almost any aquatic taxon, from platyhelminths to vertebrates, but successful development takes place only in arthropods (with the exception of two reports of leeches as hosts). Infection can be direct, as is the case in the North American species *Gordius robustus*, or indirect, as in the Japanese species *Chordodes japonensis*, but the mode is unknown for the majority of species.

Thorough investigations of the parasitic phase are lacking, but there is support for the following data: It seems that growth takes place only in the postseptal region of the larva. The preseptum remains on the growing body as a tiny cone on the anterior end. Once the nematomorph has grown to adult length, it molts once, shedding the larval cuticle together with the rest of the larval preseptum, and is then ready to leave the host.

In *Nectonema*, the data are much weaker. The larva resembles that of gordiid species, but it has only two circlets with six hooks each. A septum is probably not present. All known hosts are decapod crustaceans. Nothing is known about the mode of infection, i.e., if planctonic crustacean larvae or benthic adults are infected. Growth and molting are similar to those described for gordiids.

Bibliography

Beattie, D. A. (1987). *Nectonema agile* (Nematomorpha) from Passamaquoddy Bay, New Brunswick. Master's thesis, University of New Brunswick, Frederickton Canada.

Bresciani, J. (1991). Nematomorpha. In *Microscopic Anatomy of Invertebrates* (F. W. Harrison and E. E. Ruppert, Eds.), Vol. 4, pp. 197–218. Wiley-Liss, New York.

Dorier, A. (1930). Recherches biologiques et systematiques sur les Gordiacs. *Trav. Lab. Hydrobiol. Piscicult. Univ. Grenoble* **22 anne**, 1–183.

Feyel, T. (1936). Recherches histologiques sur *Nectonema agile* Verr. Etude de la forme parasitaire. *Arch. Anat. Microsc.* **32**, 197–234.

Huus, J. (1932). Über die Begattung bei *Nectonema munidae* Br. und über den Fund einer Larve dieser Art. *Zool. Anz.* **97**, 33–37.

Lanzavecchia, G., Eguileor, M. de, Valvassori, R., and Scari, G. (1995). Body cavities of Nematomorpha. In *Body Cavities: Function and Phylogeny* (G. Lanzavecchia, R. Valvassori, and M. D. Candia Carnevali, Eds.), Selected Symposia and Monographs U.Z.I. 8, pp. 45–60. Muchi, Modena.

Lora Lamia Donin, C., and Cotelli, F. (1977). The rod-shaped sperm of Gordioidea (Aschelminthes, Nematomorpha). *J. Ultrastruct. Res.* **61**, 193–200.

Malakhov, V. V., and Spiridonov, S. E. (1984). The embryogenesis of *Gordius* sp. from Turkmenia, with special reference to the position of Nematomorpha in the animal kingdom. *Zool. Zh.* 63, 1285–1296.

Schmidt-Rhaesa, A. (1996). Ultrastructure of the anterior end of three ontogenetic stages of *Nectonema munidae* (Nematomorpha). *Acta Zool.* 77, 267–278.

Schmidt-Rhaesa, A. (1997a). Nematomorpha. In *Süßwasserfauna von Mitteleuropa* (Freshwater Fauna of Middle Europe) (J. Schwoerbel and P. Zwick, Eds.), Vol. 4 (4). Fischer-Verlag, Stuttgart.

Schmidt-Rhaesa, A. (1997b). Ultrastructural features of the female reproductive system and female gametes of *Nectonema munidae* Brinkmann 1930 (Nematomorpha). *Parasitol. Res.* 83, 77–81.

Schmidt-Rhaesa, A. (1997c). Ultrastructural observations of the male reproductive system and spermatozoa of *Gordius aquaticus* L. 1758. *Invertebr. Reprod. Dev.* 32, 31–40.

Zapotosky, J. E. (1974). Fine structure of the larval stage of *Paragordius varius* (Leidy, 1851) (Gordioidea, Paragordiidae). I. The preseptum. *Proc. Helminthol. Soc. Wash.* 41, 209–221.

Zapotosky, J. E. (1975). Fine structure of the larval stage of *Paragordius varius* (Leidy, 1851) (Gordioidea, Paragordiidae). II. The postseptum. *Proc. Helminthol. Soc. Wash.* 42, 103–111.

Nemertea

J. M. Turbeville
University of Arkansas

I. Introduction
II. Asexual Reproduction
III. Sexual Reproduction
IV. Embryology
V. Larval Development

GLOSSARY

direct development Development in which embryos develop directly into juveniles without an intervening larval stage.

imaginal disc A discrete plate of ectodermal cells formed by invagination or delamination of the larval ectoderm that differentiates into definitive ectodermal structures in indirect developing nemerteans.

indirect development Development in which embryos pass through a stage that is morphologcally distinct from the juvenile.

lecithotrophic larva A nonfeeding larva that obtains nutrition from yolk stores.

planktotrophic larva A larva that obtains nutrition by feeding on planktonic organisms.

rhynchocoel A fluid-filled coelomic cavity enclosing the nemertean proboscis apparatus.

Nemertean worms are bilateral metazoans found in marine, terrestrial, and freshwater habitats. The phylum contains both dioecious and hermaphroditic species. The gonads are saccular, and accessory reproductive organs are typically absent. Fertilization is either external or internal. Cleavage is spiral and gastrulation occurs primarily by invagination. The mouth forms at or near the site of the blastopore. Endomesoderm is derived from the 4d cell. Embryos of most species develop directly into juvenile worms without an intervening larval stage, but in one group embryos develop into larvae that subsequently undergo a unique, radical metamorphosis.

I. INTRODUCTION

The phylum Nemertea comprises about 1000 species of unsegmented worms that possess an eversible proboscis apparatus contained in a fluid-filled coelomic cavity or rhynchocoel. The phylum has a worldwide distribution and includes marine, fresh-

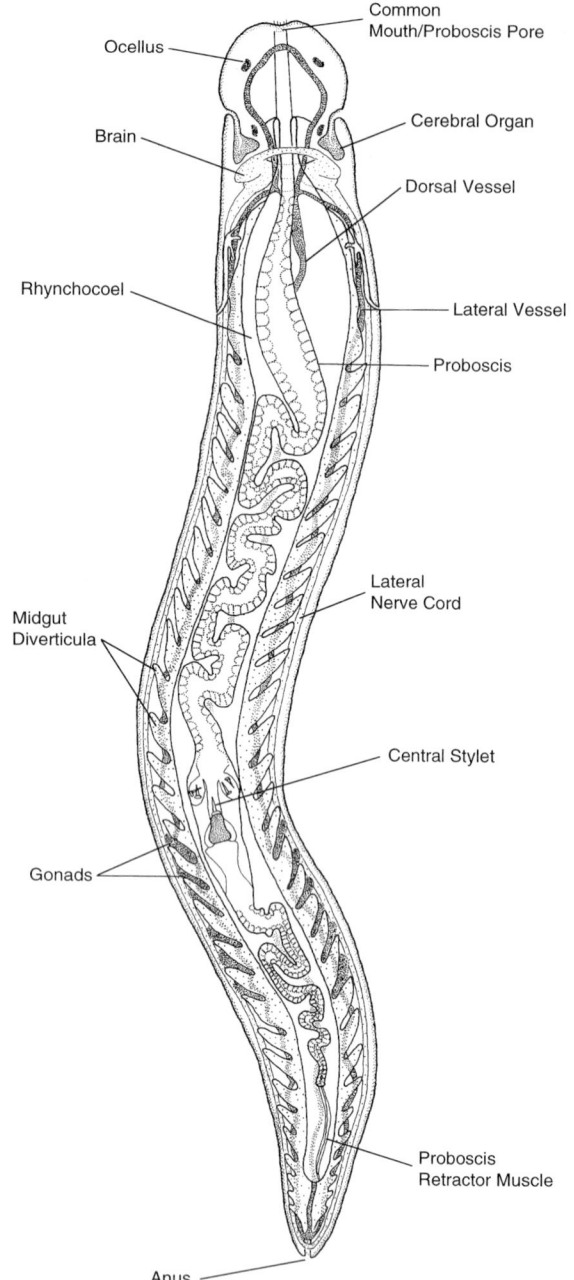

FIGURE 1 Schematic diagram of a hoplonemertean: dorsal view (from Turbeville, 1996, after Bürger, 1895).

some species. External appendages are present in a few pelagic species.

The eversible proboscis apparatus is a unique feature of the phylum and is situated in the rhynchocoel when uneverted. The rhynchocoel–proboscis complex is independent of the digestive system and is situated dorsal to it (Figs. 1 and 2). Contraction of muscles associated with the wall of the rhynchocoel raises the fluid pressure in this cavity and causes the proboscis to evert. The proboscis is drawn into the rhynchocoel by contraction of the highly extensible proboscis retractor muscle (Fig. 1). The proboscis is used principally in prey capture, and in hoplonemerteans it contains a nail-shaped calcareous stylet that is used to stab prey (Fig. 1). The nemertean nervous system is composed of a four-lobed brain, two major lateral nerve cords joined by transverse connectives and peripheral nerve networks. The anterior end of nemerteans may contain ocelli and cephalic slits or grooves that are presumed sensory in function (Fig. 1). Some species posses a pair of chemosensory–neuroendocrine structure near the brain called a cerebral organ. The digestive tract is complete, possessing both a mouth and an anus. The circulatory system consists of a continuous system of cell-lined channels and, like the rhynchocoel, is a probable coelom homolog. The worms have a protonephridial excretory system. Major features of nemertean anatomy are summarized in Figs. 1 and 2.

Traditionally, nemerteans have been considered acoelomate animals closely related to flatworms, but recent morphological and molecular analyses suggest that they are most closely related to coelomate spiralians, such as molluscs and annelids.

II. ASEXUAL REPRODUCTION

Certain species of the heteronemertean genus *Lineus* reproduce asexually. Sexually immature adults undergo spontaneous transverse fragmentation with subsequent regeneration. Fragments containing a piece of the brain or lateral nerve cord are capable of regenerating an entire worm. Small fragments typically encyst and undergo differentiation within the protective coat. In these species the asexual reproductive phase occurs during warm months of the

water, and terrestrial species. These animals range in length from a few millimeters to several meters; a length of 30 m has been estimated for *Lineus longissimus*. Nemerteans are unsegmented, although regularly spaced transverse constrictions are present in

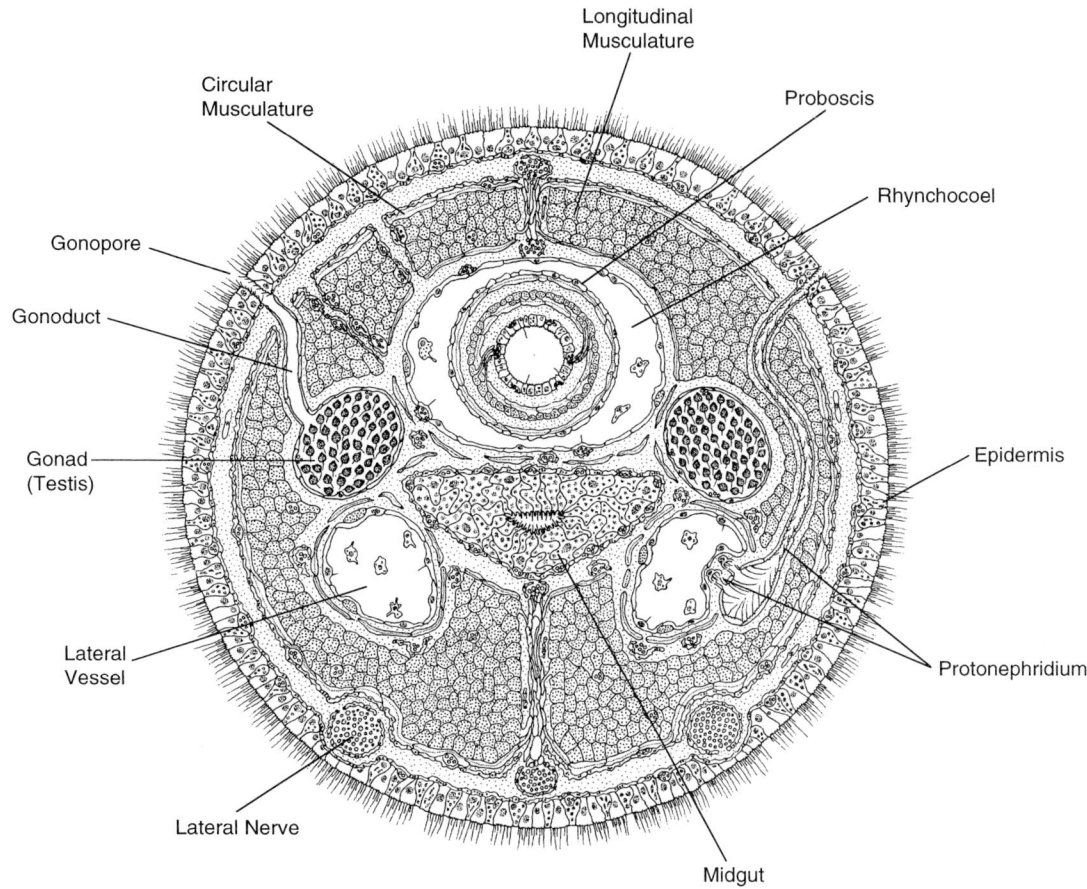

FIGURE 2 Composite diagrammatic transverse section of a palaeonemertean (from Turbeville and Ruppert, 1985).

year, alternating with a sexual reproductive phase in winter.

III. SEXUAL REPRODUCTION

Nemerteans are predominantly gonochoristic, but a number of hermaphroditic species occur. Terrestrial and freshwater nemerteans account for many of the hermaphroditic species. Most hermaphroditic species are protandric, but a few are simultaneous hermaphrodites (e.g., *Prosorhochmus claparedii*). Viviparity is common among hermaphroditic species.

A. Gonads

Nemerteans possess discrete, saccular gonads that are typically arranged in longitudinal rows on each side of the midgut (Figs. 1 and 2). They are situated between the intestinal diverticula in those species that exhibit such gut outpockets. In males of some pelagic species the testes are situated anteriorly near the brain.

The organization of the testes and ovaries is similar. The gonadal epithelium is composed of somatic cells, germinal cells, and, at least within the Hoplonemertea, muscle cells (Fig. 3). Germinal cells develop into mature gametes. The developing gametes eventually lose contact with the extracellular matrix wall of the gonad and come to reside in the gonad lumen. At maturity, an evagination of the gonad meets an epidermal invagination to form a gonoduct in most species (Fig. 2). In *Carcinonemertes*, ducts from individual testes join a common dorsal vas deferens that opens into the hindgut. In the interstitial *Cephalothrix pacifica* oviducts open into the gut.

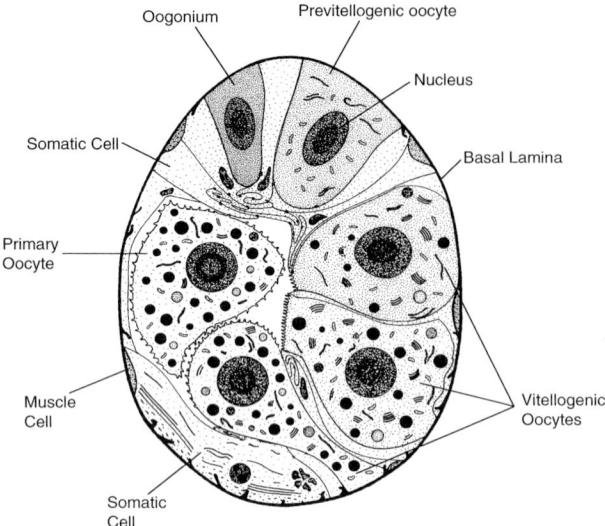

FIGURE 3 Cross section of the ovary of the hoplonemertean *Carcinonemertes epialti* (from Stricker, 1986).

B. Gametogenesis

1. Oogenesis

During oogenesis germinal cells of the ovarian epithelium termed previtellogenic oocytes increase in size and exhibit mitochondria and rough endoplasmic reticulum. Vitellogenic oocytes are typically characterized by Golgi complexes and lipid inclusions (Fig. 3). Vitellogenic oocytes possess apical microvilli and may be surrounded by a vitelline envelope. Fine structural details suggest that both autosynthetic and heterosynthetic pathways may contribute to yolk formation in nemerteans, although experimental evidence is lacking. Three glycoproteins that are likely vitellin subunits have been identified for *Lineus lacteus*. Vitellogenic oocytes eventually lose their investment of somatic cells and lose contact with the basal lamina of the gonadal epithelium (Fig. 3). Nemertean eggs are typically isolecithal.

The oocytes of many nemertean species exhibit a small stalk at the vegetal pole that may be the region of attachment of the oocyte to the ovarian basal lamina. Most nemertean ovaries contain 4–10 ripe eggs at maturity, but ovaries of several species contain only a single oocyte and a few species produce 20–50 eggs per ovary. Eggs range in size from 50 μm to 2.5 mm in diameter and are typically surrounded by a chorion and an outer jelly layer.

2. Spermatogenesis

Spermatogonia undergo mitosis that results in groups of spermatocytes interconnected by cytoplasmic bridges. Meiosis of spermatocytes subsequently occurs, forming spermatids. The syncytial groups of spermatids differentiate into spermatozoa.

Spermatozoa consist of a nucleus, an apical acrosome, a middle piece containing mitochondria (or a mitochondrion), and a single flagellum with an axoneme possessing the typical 9+2 arrangement of microtubules. Both "primitive" and modified spermatozoa occur within the phylum, and sperm morphology appears to be correlated with mode of fertilization (Fig. 4).

C. Fertilization

Observations of fertilization are limited but suggest that external fertilization predominates. Gametes are either spawned into seawater or released into a mucus sheath secreted around a pair of sexually mature worms (pseudocopulation). In some species (e.g., *Carcinonemertes epialti* and *Lineus viridis*) fertilization is internal; sperm are spawned into the mucus sheath surrounding a mating pair and enter the ovaries of the female where fertilization takes place. Copulatory organs are present in a few bathypelagic species. For example, the pelagic *Phallonemertes* possess tubular penes that may function in direct transfer of sperm or in spermatophore transfer.

Germinal vesicle breakdown occurs in *Cerebratulus lacteus* oocytes after eggs are spawned into seawater, and meiotic maturation divisions then commence. The eggs become arrested in metaphase of the first meiotic division, but following fertilization, meiotic divisions are completed and polar bodies are formed at the animal pole (Fig. 5). Experiments suggest that a soluble protein within the sperm is responsible for initiating meiotic maturation and repetitive calcium waves.

In the nemertean *C. lacteus* the oocyte membrane becomes polarized at fertilization, shifting from a negative charge to a positive charge. Polarization

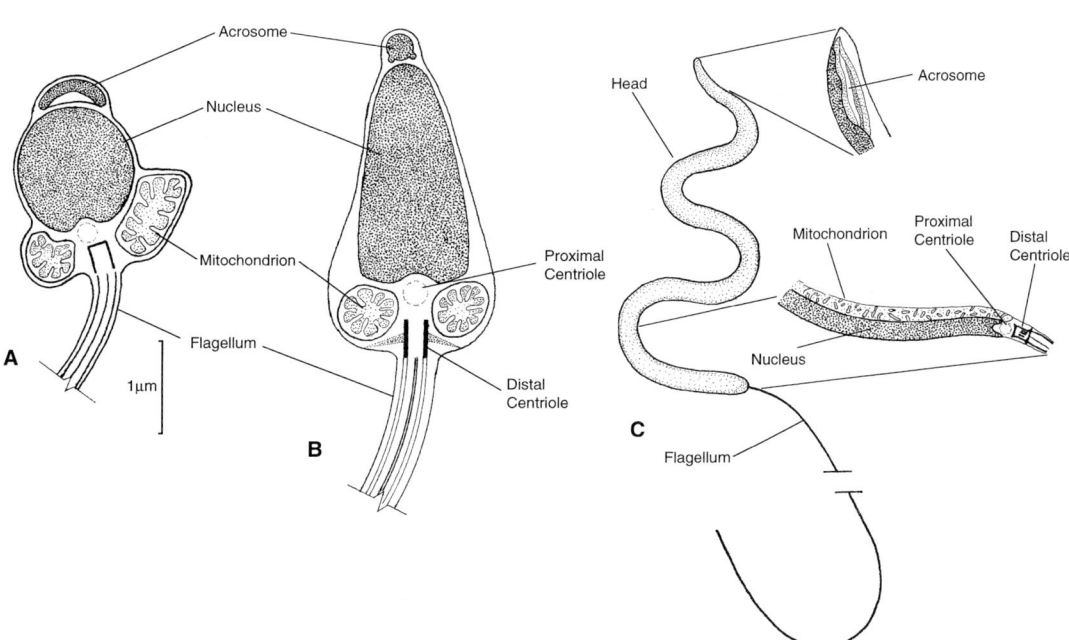

FIGURE 4 Nemertean spermatozoa. (A) Primitive sperm of the palaeonemertean *Procephalothrix spiralis* (after Turbeville and Ruppert, 1985, and J. Turbeville, unpublished observations). (B) Primitive spermatozoan of the heteronemertean *Micrura fasciolata* (redrawn from Franzén, 1983). (C) Modified sperm of the hoplonemertean *Tetrastemma phyllospadicola* (redrawn from Stricker and Cavey, 1986).

initially requires and influx of extracellular calcium, but the maintenance of polarization may involve the release of intracellular calcium. Polarization lasts for approximately 1 hr and acts as a temporary block to polyspermy (fast block to polyspermy). A cortical reaction typical of other animals has not been described for nemerteans, but a putative fertilization membrane was reported for *Parborlasia corrugatus*.

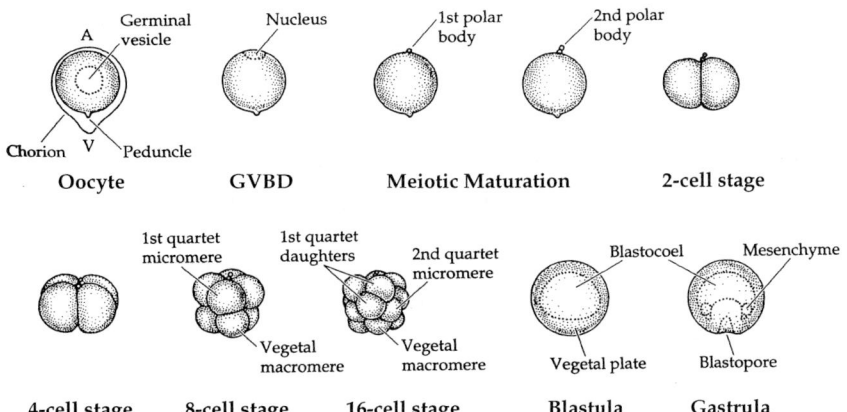

FIGURE 5 Summary diagram of nemertean development. Animal pole directed upward in all embryos. A, animal pole; B, vegetal pole; GVBD, germinal vesicle breakdown; mesenchyme = embryonic nonepithelial mesoderm (modified from Henry and Martindale, 1997).

Fertilized eggs of some nemertean species are deposited on substrates in mucus strings or masses.

D. Control of Reproduction

The brain of the heteronemerteans *Lineus ruber* and *L. lacteus* and the hoplonemertean *Amphiporus lactifloreus* produce a neurohormone during the quiescent phase of the reproductive cycle that has been implicated in the control of reproduction. This hormone inhibits the precocious formation of gonads and is referred to as gonad-inhibiting hormone (GIH). The hormone apparently exerts its effect by inhibiting RNA synthesis. Ecdysteroids have also been isolated from certain species (e.g., *Paranemertes perigrina*) and concentrations are typically higher in females during the reproductive season, suggesting a possible role in reproduction.

Results of experiments designed to test the effect of temperature and light on gonadogenesis in the heteronemertean *L. ruber* suggest that temperature rather than photoperiod is the main environmental factor influencing the onset of gonadogenesis and thus reproduction in this species.

IV. EMBRYOLOGY

Zygotes undergo spiral quartet cleavage. The first two cleavage planes are meridional with respect to the animal–vegetal axis of the zygote. The second cleavage plane is perpendicular to the first. The first two cleavages are equal and result in four vegetal macromeres of equivalent size that lie in a single plane. The third and subsequent cleavages are oblique to the animal–vegetal axis and produce animal micromere quartets. These cleavages are alternately dextral (clockwise) and sinistral (counterclockwise) relative to the vegetal macromeres. Micromeres of the first quartet are larger than the parent macromeres in most nemerteans as a result of the subequatorial third cleavage plane. This phenomenon is not unique to nemerteans. Cell cross-furrows as occur in annelid and mollusc embryos are not common in nemerteans. Subsequent cleavages give rise to a coeloblastula in most species. Gastrulation generally occurs by invagination, but both epiboly and ingression apparently participate in some species. The blastopore generally closes and the mouth forms secondarily at or near the site of the closed blastopore. Early development is summarized in Fig. 5.

As in other spiralian taxa, mesoderm is derived from both ectodermal micromeres and the 4d cell of the fourth micromere quartet. Thus, both ecto- and endomesoderm are present as in other spiralians. A study using fluorescent tracers in embryos of *C. lacteus* confirms the dual origin of mesoderm. For example, larval muscles (ectomesoderm) differentiate from derivatives of the 3a and 3b cells of the third quartet. Mesodermal bandlets (endomesoderm) are derived from the proliferation of the 4d cell and are present at the pilidium larval stage.

Classic experiments with *Cerebratulus* embryos demonstrated that isolated animal and vegetal half-embryos at the 8- or 16-cell stage exhibit mosaic development (lack regulative capacity), producing incomplete larvae. Cells of the animal and vegetal halves differentiate structures typical of their normal fates. Further experiments indicated that cytoplasmic factors specifying cell fates in this nemertean are segregated along the animal–vegetal axis sometime between fertilization and the 8-cell stage.

Blastomere isolation studies at the two- and four-cell cleavage stages reveal that embryos of the indirect developer *C. lacteus* exhibit some regulation. Blastomeres separated at the two-cell stage are capable of developing into normal larvae, but those separated at the four-cell stage give rise to incomplete larvae. However, some regulation is also observed at this stage. For example, isolated four-cell stage blastomeres are all capable of forming muscles. During normal development, the blastomeres of the D and C quadrants do not give rise to muscle cells, suggesting that the cells of the A and B quadrants prevent progeny of the D and C quadrants from forming muscle cells.

In contrast to isolated two-cell stage blastomeres of *C. lacteus*, those of the direct developer *Nemertopsis bivittata* exhibit a lower degree of regulation. In *Nemertopsis* each of the isolated blastomeres form half "larvae" that possess only a single pigmented ocellus rather than the usual two.

V. LARVAL DEVELOPMENT

A. Direct Development

Most nemerteans (Palaeonemerteans and Hoplonemerteans) lack a distinct larval stage and development is direct (Fig. 6). The embryo develops into a juvenile without radical metamorphosis. The vermiform, ciliated larvae of some species are planktonic and feed (planktotrophic), whereas others are planktonic lecithotrophs (Fig. 6). These larvae eventually settle and take up a benthic existence. Some direct-developing larvae do not exhibit a planktonic phase.

B. Indirect Development

A distinct larval stage is present in the heteronemerteans. Many species possess a planktotrophic larva that resembles a helmet or hat called the pilidium larva (Fig. 6). The pilidium is characterized by ciliated lobes or lappets that are involved in locomotion and feeding. The blastocoel of the larva contains mesodermal cells and larval muscle bands. A pelagic lecithotrophic larva called Iwata's larva occurs in *Micrura akkeshiensis*. An apical ciliary tuft is present in pilidium larvae and Iwata's larva, but the Iwata's larva lacks lappets. The apical tuft may have a sensory

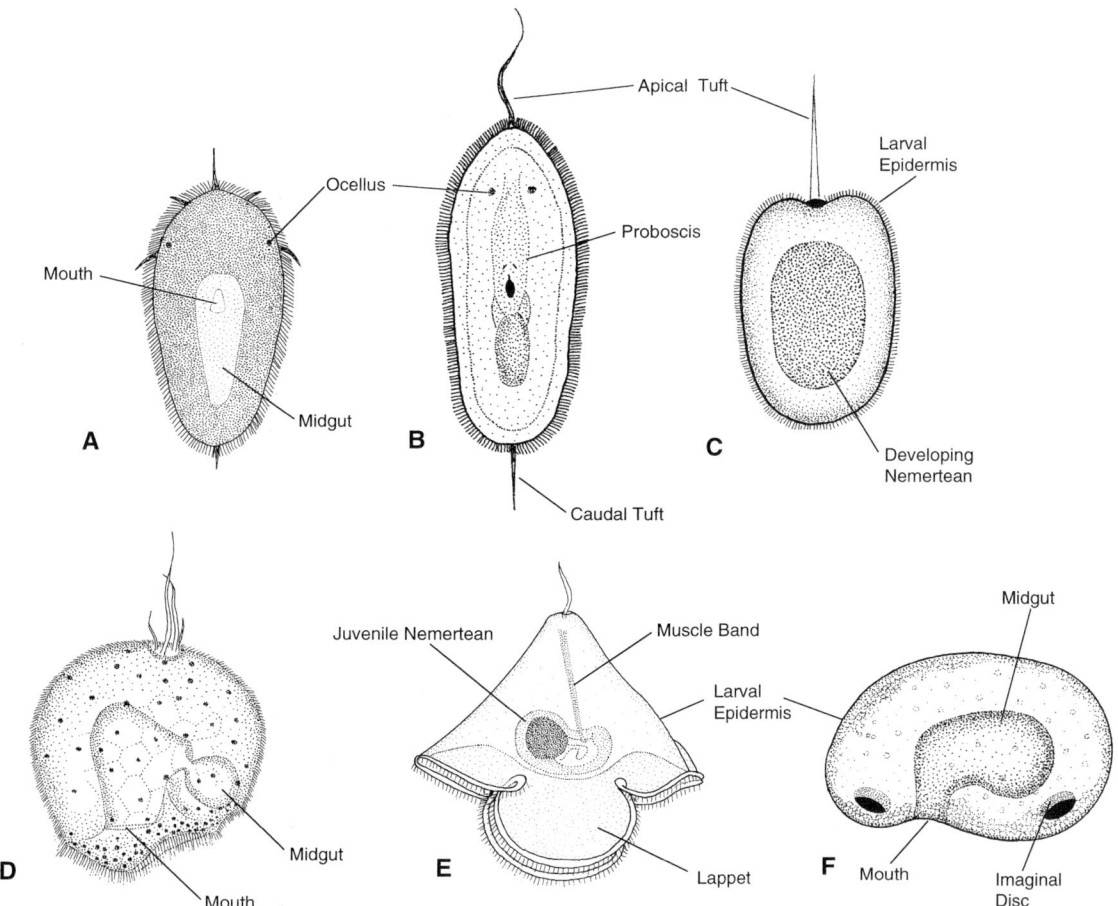

FIGURE 6 Nemertean larvae. (A) Direct-developing "larva" of the palaeonemertean *Proephalothrix simulus* [redrawn from Iwata (1960) and Friedrich, 1979; from Turbeville, 1996]. (B) Direct-developing larva of the hoplonemertean *Emplectonema gracile* (redrawn from Iwata, 1960). (C) Larva (Iwata's larva) of the heteronemertean *Micrura akkeshiensis* (redrawn from Iwata, 1958). (D) Pilidium larva (premetamorphosis) of *Micrura caeca* (after Coe, 1899; from Turbeville, 1996). (E) Unidentified pilidium larva (postmetamorphosis; after Bürger, 1895; from Turbeville, 1996). (F) Desor's larva (redrawn from Barrois, 1877).

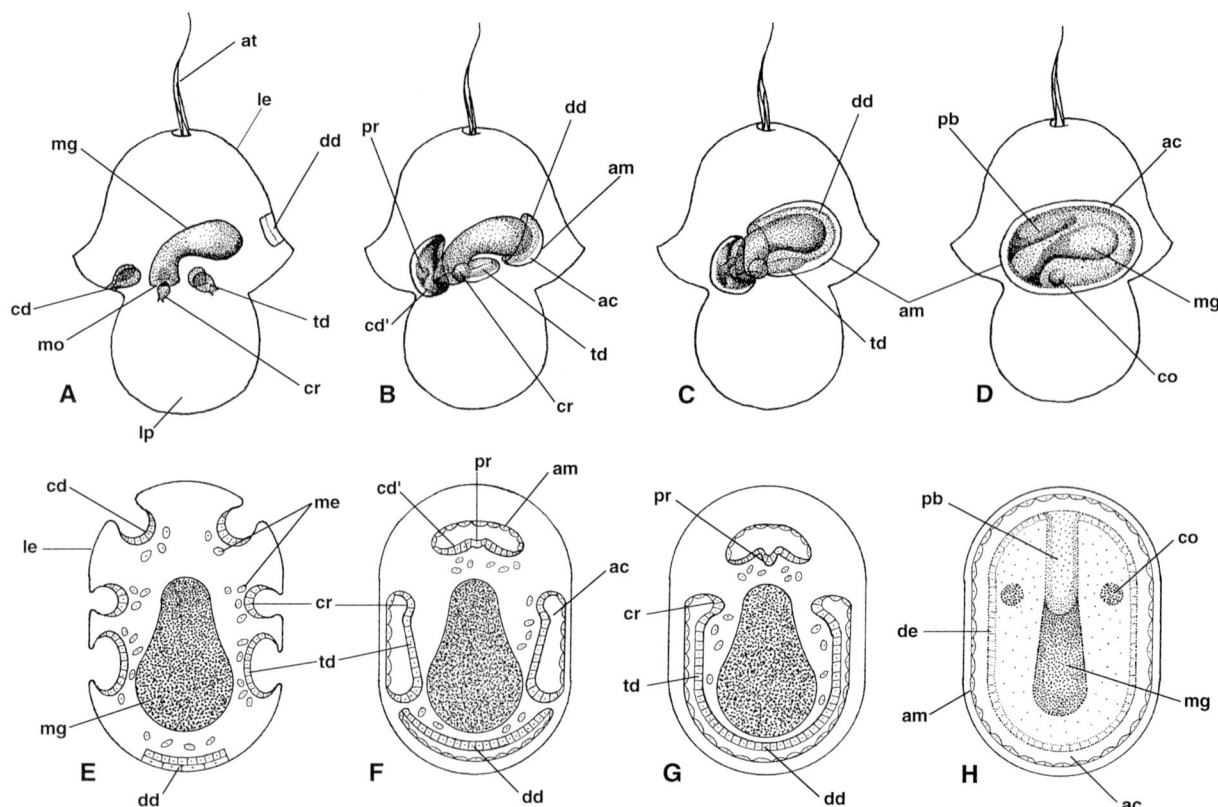

FIGURE 7 Metamorphosis of the pilidium larva. A–D, lateral views; E–H, frontal sections corresponding to stages a–d (A–D after Salensky, 1912; E–H after descriptions in Salensky, 1912). (A, E) Imaginal disc formation. Three pairs are formed by invagination of the larval epidermis (ectoderm) and one, the dorsal disc (dd), is formed by delamination of the epidermis. (B, F) Later stage showing fusion of cephalic discs (cd′) and their amnions (am), fusion of cerebral discs (cr) with trunk discs (td), partial fusion of trunk discs, and initial expansion of the dorsal disc (dd). The proboscis rudiment (pr) forms at the junction of the two cephalic discs. (C, G) Subsequent stage revealing fusion of dorsal disc (dd) with trunk discs (td). Trunk disc fusion is subsequently completed. (D, H) Juvenile worm surrounded by continuous amniotic cavity (ac), following fusion of dorsal and trunk discs with cephalic discs and subsequent cell differentiation. ac, amniotic cavity; am, amnion; at, apical tuft; cd, cephalic disc; cd′, fused cephalic discs; co, cerebral organ; cr, cerebral disc; dd, dorsal disc; de, definitive epidermis; le, larval epidermis; lp, lappet; me, mesoderm; mg, midgut; mo, mouth; pb, proboscis; pr, proboscis rudiment; td, trunk disc.

and locomotory function. A lecithotrophic benthic larva that develops in an egg capsule, called Desor's larva, is characteristic of *L. viridis*. *Lineus ruber* possesses a benthic larva, Schmidt's larva, which feeds on abortive oocytes. Desor's and Schmidt's larvae lack lappets and an apical tuft (Fig. 6). All nemertean larvae possess an incomplete gut (anus is absent) and lack nephridia. The nephridia and anus develop later.

Metamorphosis of nemertean larvae is complex and unique. It involves the formation of imaginal discs from the larval epidermis (ectoderm; Fig. 7). Seven discs develop in the pilidium larva: paired cephalic discs, cerebral discs, and trunk discs and an unpaired dorsal disc. The paired discs develop from the invagination and pinching off of the larval epidermis, whereas the unpaired dorsal disc forms by delamination of the epidermis (Fig. 7). The inner surface of cuboidal to columnar cells becomes the imaginal disc (secondary ectoderm) and an outer squamous cell layer forms an amnion. These two cell layers enclose an amniotic cavity. These imaginal discs and their respective amnions gradually expand and fuse as follows (Fig. 7): The cephalic discs fuse together first and the proboscis rudiment forms near the center of these fused discs. The cerebral discs then fuse with the anterior end of trunk discs. The dorsal disc fuses with the posterior end of the trunk discs as the latter complete fusion anteriorly along the ventral side of the developing worm. Finally, the cephalic disc fuses with the trunk and dorsal discs. Amnion fusion during this phase forms a continuous amniotic cavity around the developing worm. As the imaginal discs expand and fuse they surround the larval gut and mesodermal cells that are situated at the base of the discs, along the outer wall of the larval gut, and scattered within the larval blastocoel. The fused imaginal discs form the definitive epidermis and other ectodermal derivatives. The larval gut differentiates into the definitive gut, with an anus breaking through as an ectodermal invagination (proctodaeum). The mesoderm differentiates into muscles and other mesodermal derivatives. The fully differentiated worm eventually breaks out of the larval ectoderm ("envelope"). In some species the juvenile worm consumes the larval tissue when "hatching." Metamorphosis of Iwata's, Desor's, and Schmidt's larvae is similar to that of the pilidium, differing primarily in the number of imaginal discs and order of their fusion. In the Desor's larva a total of eight discs develop, including a distinct proboscis disc, formed as an anterior ectodermal invagination. In Iwata's larva only five imaginal discs are formed and, like the pilidium larva, a proboscis disc is absent. Although unconfirmed, some classic observations suggest that a proboscis imaginal disc may form in the pilidium larvae of certain species.

See Also the Following Articles

ASEXUAL REPRODUCTION; HERMAPHRODITISM

Bibliography

Bierne, J. (1983). Nemertina. In *Reproductive Biology of Invertebrates: Oogenesis, Oviposition and Oosorption* (K. G. Adiyodi and R. G. Adiyodi, Eds.), Vol. 1, pp. 147–167. Wiley, Chichester, UK.

Cantell, C.-E. (1989). Nemertina. In *Reproductive Biology of Invertebrates: Ferilization, Development and Parental Care* (K. G. Adiyodi and R. G. Adiyodi, Eds.), Vol. 4, Part A, pp. 147–165. Wiley, Chichester, UK.

Franzén, Å. (1983). Nemertina. In *Reproductive Biology of Invertebrates: Spermiogenesis and Sperm Function* (K. G. Adiyodi and R. G. Adiyodi, Eds.), Vol. 2, pp. 159–170. Wiley, Chichester, UK.

Friedrich, H. (1979). Nemertini. In *Morphogenese der Tiere, Lieferung 3: D5-I* (F. Seidel, Ed.). Fisher-Verlag, Stuttgart.

Henry, J. J., and Martindale, M. Q. (1996). The origins of mesoderm in the equal-cleaving Nemertean worm *Cerebratulus lacteus. Biol. Bull.* **191**, 286–288.

Henry, J. J., and Martindale, M. Q. (1997). The Nemertea. In *Embryology: Constructing the organism.* (S. F. Gilbert and A. M. Raunio, Eds.). Sinauer, Sunderland, MA.

Salensky, W. (1912). Uber die Morphogenese der Nemertinen. I. Entwicklungsgeschichte der Nemertinen im Inneren des Pilidiums. *Mäm. Acad. Imp. Sci. St. Petersburg* **30**, 1–74.

Stricker, S. A. (1986). An ultrastructural study of oogenesis fertilization and egg laying in a nemertean ectosymbiont of crabs, *Carcinonemertes epialti* (Nemertea, Hoplonemertea). *Can J. Zool.* **64**, 1256–1269.

Stricker, S. A. (1987). Phylum Nemertea. In *Reproduction and Development of Marine Invertebrates of the Northern Pacific Coast* (M. F. Strathmann, Ed.), pp. 129–137. Univ. of Washington Press, Seattle.

Stricker, S. A. (1997). Intracellular injections of a soluble sperm factor trigger calcium oscillations and meiotic maturation in unfertilized oocytes of a marine worm. *Dev. Biol.* **186**, 185–201.

Stricker, S. A. and Cavey, M. J. (1986). An ultrastructural study of spermatogenesis and the morphology of the testis in the nemertean worm *Tetrastemma phyllospadicola*. (Nemertea, Hoplonemertea). *Can. J. Zool.* **64**, 2187–2202.

Turbeville, J. M. (1991). Nemertinea. In *Microscopic Anatomy of Invertebrates* (F. W. Harrison and B. Bogitsch, Eds.), Vol. 3, pp. 285–328. Wiley-Liss, New York.

Turbeville, J. M. (1996). Nemertini, Schnurwürmer. In *Spezielle Zoologie* (W. Westheide and R. M. Rieger, Eds.), Vol. 1, pp. 265–275. Fischer, Stuttgart.

Turbeville, J. M. and Ruppert, E. E. (1985). Comparative ultrastructure and the evolution of nemertines. *Am. Zool.* **25**, 53–71.

Nervus Terminalis

Marlene Schwanzel-Fukuda
The State University of New York at Brooklyn

I. Brief Description and History of the Nervus Terminalis
II. Early Speculations on the Function of the Nervus Terminalis
III. Immunocytochemical Techniques Provide a Marker for the Nervus Terminalis and a Renewal of Interest in Its Study
IV. Recent Studies of the Nervus Terminalis
V. The Role of the Nervus Terminalis in the Migration of LHRH Neurons
VI. Kallmann's Syndrome: A Failure of LHRH Cell Migration
VII. Summary
VIII. Future Directions

GLOSSARY

accessory olfactory system A system composed of the vomeronasal organ, the vomeronasal nerves, and the accessory olfactory bulb.

olfactory pit One of a pair of structures formed during early development when the olfactory placodes sink below the surface. Either olfactory pit then forms two recesses, one medial and one lateral. The lateral recess of the olfactory pit gives rise to the olfactory nerves, and the medial recess gives rise to the vomeronasal and terminal nerves.

olfactory placode A thickening of the ectoderm on the lateral sides of the head of vertebrate embryos that gives rise to the nasal cavity.

olfactory system A specialized system composed of the olfactory receptor cells in the olfactory epithelium of the nose, their axons which form the olfactory nerves, and the main olfactory bulb.

pheromone A species-specific chemical attractant found in the glandular secretions and/or the urine of many animals. The action of a pheromone, perceived by an individual of the same species, is to produce a change in the sexual or social behavior of that individual.

vomeronasal organ A special chemosensory receptor found in the nose of many vertebrates which responds to pheromones; formed by the epithelium of the medial recess of the olfactory pit.

I. BRIEF DESCRIPTION AND HISTORY OF THE NERVUS TERMINALIS

The nervus terminalis, or terminal nerve, is a plexiform, ganglionated cranial nerve (or complex of nerves) of unknown function found in the nose of most vertebrates, including humans. It is located rostral to the olfactory and optic nerves, and since it was discovered after the 12 cranial nerves had been named, it is also called the "zeroeth cranial nerve."

Similar to the olfactory nerves, the nervus terminalis is derived from cells which originate in the olfactory placodes, thickenings of the ectoderm on the lateral sides of the head. Comparatively early in gestation, in mammals, the olfactory placodes invaginate to form simple olfactory pits, each of which soon develops a secondary recess in its medial wall on either side of the developing nasal septum. The axons of the olfactory nerves originate from receptor cells in the epithelium of the lateral part of the olfactory pit, whereas the vomeronasal and terminal nerves originate from the epithelium of the medial wall of the olfactory pit. The nervus terminalis crosses the nasal mucosa in the form of a loose plexus and traverses the cribriform plate of the ethmoid bone medial to the olfactory and vomeronasal nerves. The axons of the vomeronasal nerve end in the accessory olfactory bulb, whereas the axons of the nervus terminalis, forming three or four short rootlets, enter the medial forebrain, medial and caudal to the olfactory bulbs. The nervus terminalis can be distin-

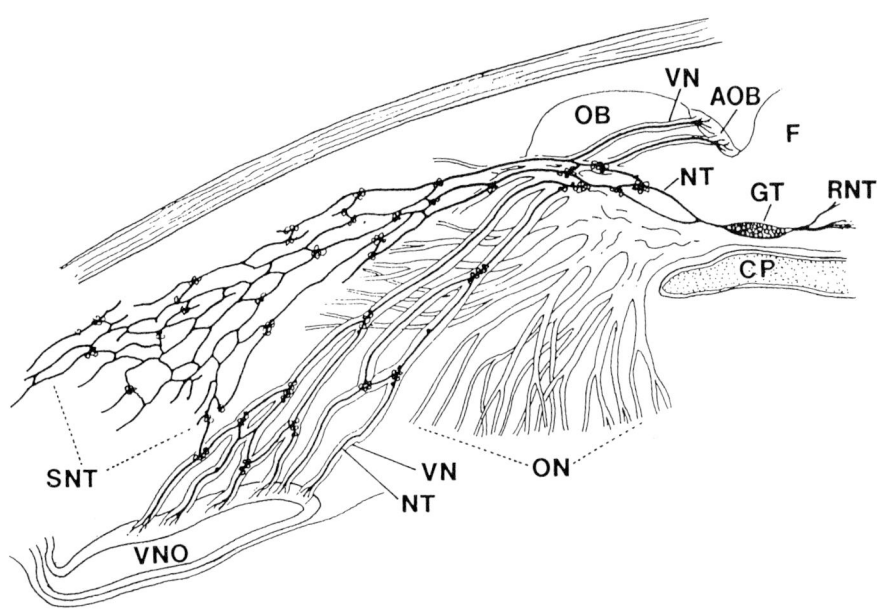

FIGURE 1 Graphic reconstruction of the right side of the olfactory bulb and stalk, olfactory and vomeronasal nerves, and the nervus terminalis based on a series of sagittal sections of a head of a 1-day-old rabbit. The nervus terminalis (NT) is shown in black, and the ganglion cells along its course are indicated as small open circles. The olfactory (ON) and vomeronasal (VN) nerves are given in outline. F, forebrain; OB, olfactory bulb; AOB, accessory olfactory bulb; CP, cribriform plate; GT, ganglion terminale; RNT, roots of the nervus terminalis; SNT, septal distribution of the nervus terminalis; VNO, vomeronasal organ (adapted from G. C. Huber and R. S. Guild, p. 259, 1913; reproduced with permission from Schwanzel-Fukuda and Pfaff, 1995).

guished from the vomeronasal and olfactory nerves by the presence of ganglia at nodal points along the plexus and ganglion cells intermingled with axon fascicles in the nasal mucosa. The largest ganglion of the terminal nerve, the ganglion terminale, is commonly found just above the cribriform plate, medial and caudal to the olfactory bulbs (Fig. 1).

The nervus terminalis was first reported as a "supernumary cranial nerve" in the brain of a shark (*Galeus canis*) by Gustav Fritsch in 1878. The name "nervus terminalis" was given to this nerve by William Locy, who, in 1905, was the first to carry out a detailed study of its embryology in *Squalis acanthus* embryos. Locy observed that the central roots of the nervus terminalis almost always could be traced into the lamina terminalis, the site of closure of the anterior neuropore. The first and most comprehensive description of the nervus terminalis in a mammal is attributed to Huber and Guild. In rabbits, they described and named the large ganglion associated with this nerve, the "ganglion terminale" (Fig. 1). The nervus terminalis was even found to be present in animals such as the porpoise, which lack both an olfactory and a vomeronasal system, and in bats and chickens, which lack only the vomeronasal system. Comparative anatomists were fascinated with the nervus terminalis. They recognized in it the features of a phylogenetically ancient nerve that has retained its position in regions of the head that have undergone considerable change throughout evolution. Using specialized silver stains (including a modified pyridine silver method) they were able to differentiate the nervus terminalis from the olfactory and vomeronasal nerves, and it was studied intensively in a wide variety of vertebrates from the late nineteenth century through the early part of this century.

II. EARLY SPECULATIONS ON THE FUNCTION OF THE NERVUS TERMINALIS

Many elegant descriptive studies of the nervus terminalis in a variety of species were developed at the beginning of this century, including those in humans. However, its function remained obscure. The major reason for this is that it is very difficult to isolate the nervus terminalis from the vomeronasal and trigeminal and the olfactory nerves in order to test its function. This problem is compounded by the fact that the nervus terminalis is located deep within the head, just lateral to midline in both the brain and nasal regions. Most of the postulations about its functions were therefore based on morphological observations rather than on experimental data. Both sensory and autonomic functions were proposed for the terminal nerve, based on its origin from the olfactory placode and the resemblance of the ganglion cells to sympathetic neurons. Autonomic (vasomotor) functions were also considered for the ganglion cells on the basis of the proximity of the nervus terminalis plexus to Bowman's glands and the blood vessels of the nasal cavity. However, since the nervus terminalis could not readily be tested experimentally, interest in its study faded. It came to be considered an "anatomical curiosity," the vestige of an ancient nerve whose function was lost or superseded by other parts of the nervous system over the long course of vertebrate evolution. After about 1950, reference to the nervus terminalis was dropped from many embryology and neuroanatomy textbooks.

III. IMMUNOCYTOCHEMICAL TECHNIQUES PROVIDE A MARKER FOR THE NERVUS TERMINALIS AND A RENEWAL OF INTEREST IN ITS STUDY

The production of antibodies to specific peptides and the development of immunocytochemical procedures during the 1970s made possible new approaches to the study of the nervus terminalis. The

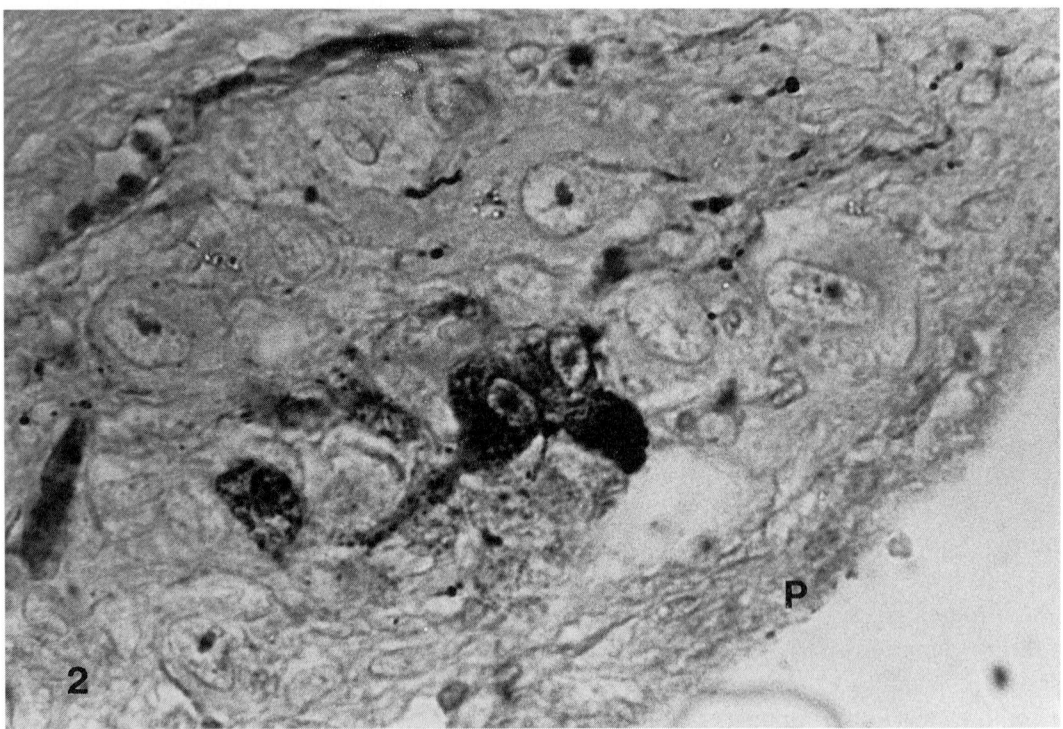

FIGURE 2 LHRH-immunoreactive (dark gray cells) and nonimmunoreactive neurons are seen within the ganglion terminale of an adult guinea pig, found along the ventromedial forebrain. This large ganglion lies deep to the epipial layer of the pia mater (P) (sagittal 8-μm section) (reproduced with permission from M. Schwanzel-Fukuda and A.-J. Silverman, *J. Comp. Neurol.* **191**, 213–225, 1980).

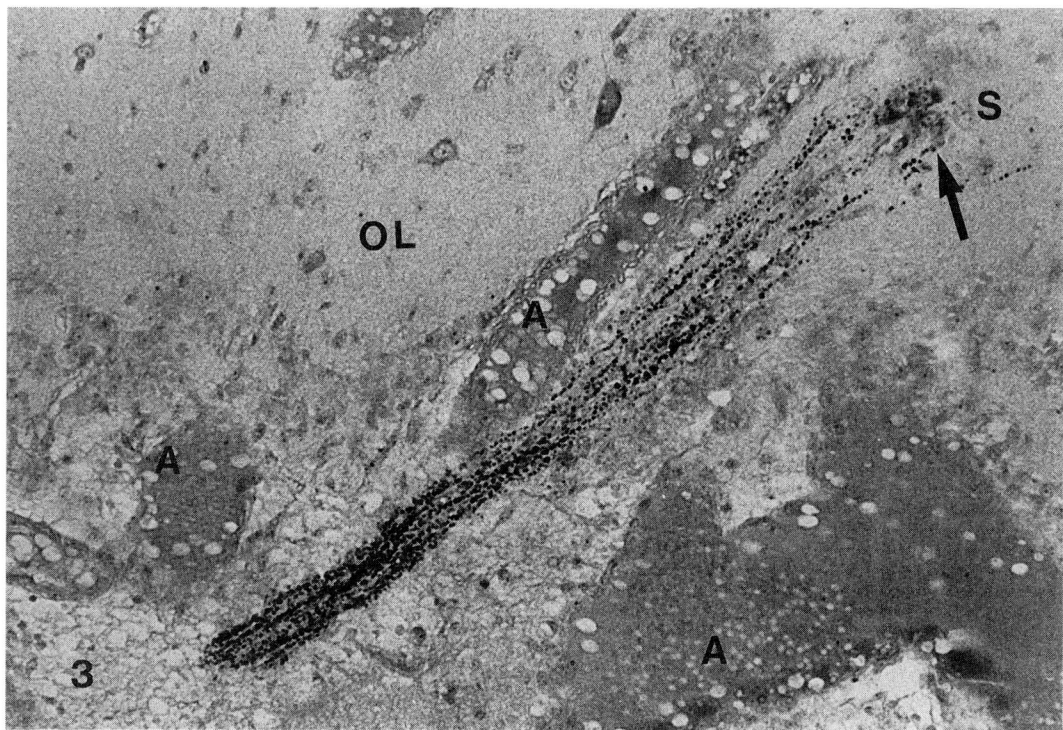

FIGURE 3 LHRH-immunoreactive axons of the terminal nerve crossing branches of the anterior cerebral artery (A) at the junction of the septum (S) and the olfactory lobe (OL). A small cluster of LHRH-immunoreactive cells (arrow) is seen in the ventromedial septum (adult guinea pig; sagittal 8-μm section) (reproduced with permission from M. Schwanzel-Fukuda and A.-J. Silverman, *J. Comp. Neurol.* **191**, 213–225, 1980).

discovery, in guinea pigs, that a population of ganglion cell bodies and axons of the nervus terminalis contain luteinizing hormone-releasing hormone (LHRH, a crucial hormone for normal reproductive function) provided a marker for this nerve (Figs. 2 and 3) and brought about a renewal of interest in its study. LHRH immunoreactivity was detected by immunocytochemical procedures in the nervus terminalis of almost all vertebrates examined and served to distinguish this nerve from the olfactory, vomeronasal, and trigeminal nerves, which are also present in the nasal cavity.

IV. RECENT STUDIES OF THE NERVUS TERMINALIS

Demski and Northcutt, in the early 1980s, studied the terminal nerve of the goldfish by injecting horseradish peroxidase (HRP) into the nasal sac. They discovered that the terminal nerve has complex associations with the olfactory and visual systems and is in a position to modulate both the physiological and the behavioral aspects of reproduction. They found, as expected, HRP-filled axons in the olfactory bulb. They also found labeled axons extending via the medial olfactory tract into the basal forebrain, an area homologous to the septal region and a known projection site for terminal nerve axons involved in reproductive functions. In addition, injection of HRP into the retina labeled axons of the terminal nerve. They found that electrical stimulation of the medial olfactory tract or the optic nerve caused sperm release. Thus, LHRH-immunoreactive cells in the nervus terminalis might be involved in the control of reproductive physiology and behavior. White and Meredith carried out electrophysiological studies of the nervus terminalis in Elasmobranchs and demonstrated that the ganglion terminale is tonically active and this activity is suppressed by efferent impulses from the brain. In hamsters, Wirsig and Leonard showed that severance of the nervus terminalis in neonatal male hamsters caused deficits in mating behavior. A common factor in these studies, in all species, is the implication of axons of the nervus terminalis, with their content of LHRH, in areas of the brain which are known to mediate the physiological and behavioral aspects of reproduction.

V. THE ROLE OF THE NERVUS TERMINALIS IN THE MIGRATION OF LHRH NEURONS

Tracing the embryological history of the LHRH cells, in mice, with antibodies to LHRH and tritiated thymidine led to the discovery that the LHRH cells originate in the same part of the epithelium of the medial olfactory pit as that which gives rise to the vomeronasal and terminal nerves. In serial, sagittal sections through the whole heads of embryonic mice from 11 to 18 days of gestation, the LHRH-immunoreactive cells were traced across the nasal mucosa migrating on or encompassed by axons of the nervus terminalis (Fig. 4). The LHRH cells are accompanied a short distance into the forebrain by the central processes of the nervus terminalis. The actual migration of LHRH cells into the brain is largely over by about 16 days of gestation, although a few LHRH-immunoreactive cells are detected in ganglia of the nervus terminalis into old age. The phenomenon of LHRH cell migration along axons of the nervus terminalis has been generalized to other species, including humans.

VI. KALLMANN'S SYNDROME: A FAILURE OF LHRH CELL MIGRATION

Support for the hypothesis of LHRH cell migration on axons of the nervus terminalis in humans comes

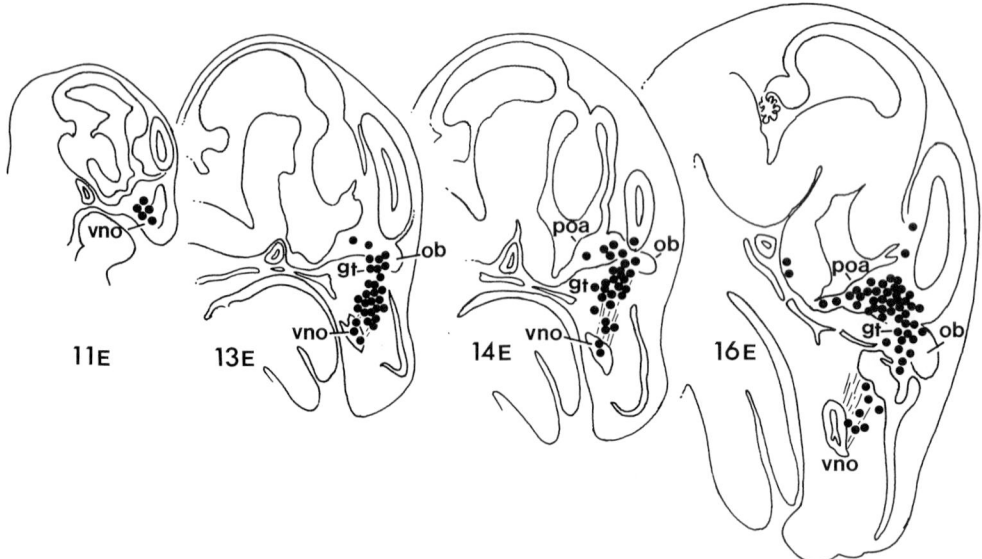

FIGURE 4 The migration route of LHRH-immunoreactive neurons from the medial olfactory placode to the forebrain is shown in microprojection drawings in the sagittal plane of 6-μm sections through the whole heads of embryonic mice on Days 11, 13, 14, and 16. The black dots represent LHRH-immunoreactive neurons. The solid circles represent LHRH-immunoreactive neurons. gt, ganglion terminale; ob, olfactory bulb; vno, vomeronasal organ; poa, preoptic area. On Day 11, LHRH neurons are seen in the anlage of the vomeronasal organ and medial wall of the olfactory placode. On Day 13, most of the LHRH-immunoreactive cells are seen on the nasal septum with axons of the vomeronasal and terminal nerves. On Day 14, most of the LHRH cells are in the ganglion terminale and in the central roots of the nervus terminalis. The 16-day-old embryonic brain showed most of the LHRH neurons arching through the forebrain into the hypothalamus and preoptic area. The migration is largely over at this age (reproduced with permission from M. Schwanzel-Fukuda and D. W. Pfaff, *Nature* 338, 161–164, 1989).

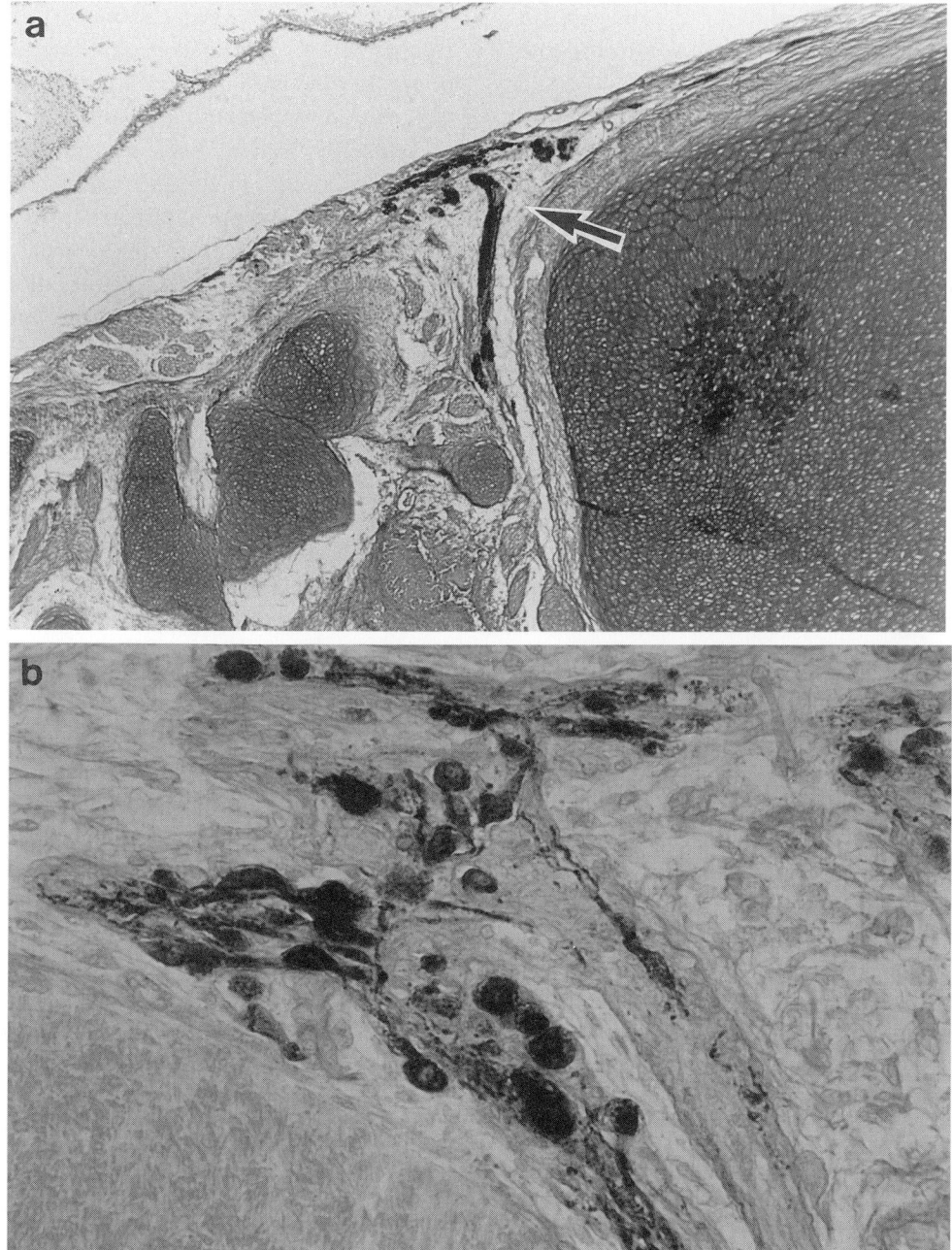

FIGURE 5 (a) Sagittal 10-μm section through the brain and nasal regions of a 19-week-old male fetus with Kallmann's syndrome, immunoreacted with antiserum to LHRH and lightly counterstained with cresyl violet. Clusters of LHRH-immunoreactive cell bodies and axons (arrow) in the nervus terminalis, just lateral to midline, are seen emerging from the nasal cavity through a perforation in the cribriform plate and extending along the dorsal surface of the cribriform plate deep to the meninges. (b) The same section, at higher magnification, shows LHRH-immunoreactive neurons in ganglia on the dorsal surface of the cribriform plate (reproduced with permission from Schwanzel-Fukuda, 1997).

from a study of the brain and nasal regions of a 19-week-old fetus who had Kallmann's syndrome and those of three normal fetuses of the same age and sex. Immunocytochemical localization of LHRH immunoreactivity in the Kallmann's fetus showed an absence of the olfactory bulbs and a complete absence of LHRH in the brain. In contrast, the nasal regions showed thick fascicles of LHRH-immunoreactive axons and ganglion cells in a distribution corresponding to that of the nervus terminalis. On the dorsal surface of the cribriform plate the LHRH cell bodies and axons of the nervus terminalis were observed between the meninges and the cribriform plate and no axons extended into the brain. Clusters of ganglion cells of the nervus terminalis were seen immediately adjacent to the unpaired crista galli, just lateral to midline and medial to the vomeronasal and olfactory nerves. Thus, in Kallmann's syndrome, the axons of the nervus terminalis do not enter the brain but remain "trapped" on the cribriform plate deep to the meninges (Fig. 5). As a result, the LHRH cells were also trapped and could not migrate into the brain. In this Kallmann fetus, a failure of the olfactory nerves to make contact with the anlage of the forebrain resulted in the tangles of olfactory nerve axons, or neuromas, on the surface of the cribriform plate either side of midline. These observations support the hypothesis that all LHRH cells originate from the olfactory placode and suggest that at least one function of the nervus terminalis may be that of a guide for the migration of LHRH cells from the olfactory placode into the brain.

VII. SUMMARY

1. The nervus terminalis is not a vestigial nerve. It is present in adult animals of all species, including humans.
2. A population of ganglion cells of the nervus terminalis contain LHRH, a brain hormone essential for normal reproductive function. The LHRH cells share a common origin with the nervus terminalis and vomeronasal nerves in the epithelium of the medial olfactory pit.
3. The nervus terminalis is very important during early development since this nerve forms the migration route for LHRH cells from the epithelium of the olfactory placode into the forebrain.
4. The nervus terminalis, unlike the vomeronasal and olfactory nerves, is the only cranial nerve which projects directly from the nose to the septal and preoptic areas of the brain. It is in a position to receive information from the external environment, to relay and integrate this information through its ganglia, and to convey it to the central nervous system.
5. The function of the nervus terminalis, other than its role in the migration of LHRH neurons, is still the subject of speculation much as it was more than 100 years ago. The interesting data collected in fishes suggest that the nervus terminalis may be involved a complex integration of sensory and autonomic signals geared to the regulation of reproductive behavior and functions.

VIII. FUTURE DIRECTIONS

Major questions still remain unanswered about the structure and function of the nervus terminalis. Does the nervus terminalis act in concert with the vomeronasal system and is there a possible association between the vomeronasal nerve and nervus terminalis in chemosensory processes? Do signaling molecules within or on axons of the nervus terminalis plexus play a role in LHRH cell migration and guidance? Do early arriving axons the nervus terminalis influence development of the forebrain? The nervus terminalis provides a fine model system for studying cell migration. Its presence in a wide range of vertebrates from zebrafish to primates, including humans, should be useful to many researchers in the biological and medical sciences. The ongoing advances in immunocytochemistry and molecular biology provide new and powerful tools to unravel the complex functions of this interesting cranial nerve.

See Also the Following Article

KALLMANN'S SYNDROME

Bibliography

Demski, L. S., and Schwanzel-Fukuda, M. (1987). The terminal nerve (nervus terminalis): Structure, function and evolution. *Ann. N. Y. Acad. Sci.* **519**.

Demski, L. S. (1993). Terminal nerve complex. *Acta Anat.* **148**, 81–95.

Muske, L. E. (1993). Evolution of gonadotropin-releasing hormone (GnRH) neuronal systems. *Brain Behav. Evol.* **42**, 215–230.

Schwanzel-Fukuda, M. S., and Pfaff, D. W. (1995). Structure and function of the nervus terminalis. In *Handbook of Olfaction and Gustation* (R. L. Doty, Ed.). Dekker, New York.

Schwanzel-Fukuda, M. S. (1997). The origin and migration of LHRH neurons in mammals: A comparison between species including humans. In *GnRH Neurons: Gene to Behavior* (I. S. Pahar and Y. Sakuma, Eds.). Brain Shuppan, Tokyo.

Nesting, Birds

Joanna Burger
Rutgers—The State University of New Jersey

I. Introduction
II. Nesting Habitat
III. Types of Nests
IV. Spatial Pattern of Nesting
V. Temporal Patterns of Nesting
VI. Predation, Competition, and Nesting Behavior
VII. Special Cases

GLOSSARY

cannibalism Eating of members of the same species, usually an adult bird killing and eating eggs or young birds that are not yet fully grown.

coloniality The behavior of birds (or other animals) whereby many breed in close proximity, interact, and usually join in group defense against predators.

competition When two or more individuals use the same resource that is in short supply, such as nesting sites, nest material, mates, or food.

dispersion pattern The spatial distribution of nests, often measured as nearest neighbor distance; refers to whether nests are clumped or randomly distributed in space.

piracy When an animal steals resources, such as food or nest material, from another animal.

predation When an egg, young, or adult bird is killed and eaten by another animal.

social facilitation When the presence of many birds is so stimulating that there are more behaviors (or more intense behaviors) than might occur otherwise. Such stimulation presumably leads to earlier egg laying and, in general, early nesting pairs are more successful than later nesting pairs. Social facilitation occurs in courtship, feeding, and preening.

solitary When a pair of birds nests alone in the habitat, separated from other members of its species; territories may abut, but nests are far apart.

substrate The surface on which a bird builds its nest.

territory A space within which individuals are aggressive toward, and usually dominant over, other members of the same species; territories are usually used for nesting (courtship, incubation, and chick care).

Nesting in birds is the process by which birds breed and produce offspring. It usually involves a bird, or a pair of birds, selecting a nest site, building a nest or nest scrape (depression in the sand or ground), laying eggs (only the female), incubating (sitting on the eggs to keep them at a constant, warm temperature) or otherwise keeping the temperature constant, and caring for the young chicks (baby birds) until they are able to take care of themselves (independence or fledging). The study of nesting in birds involves observing their behavior during all of these phases, recording how many eggs are laid and young are fledged (measures of reproductive success), and recording the behavioral interactions between a bird, or pair of birds, and neighbors, intruders, and predators. The study of nesting is intimately involved with the study of mating systems (who mates with whom and for how long), nest building and nest sites, clutch size (the number of eggs laid in a nest), and reproductive success (how many young a female or pair produces per year).

I. INTRODUCTION

The advances in behavioral ecology and evolutionary theory during the past three decades have facilitated insight into the short-term and evolutionary mechanisms of nesting behavior in birds. For many decades naturalists merely described various aspects of nesting in birds without benefit of understanding these behaviors in light of ecological and evolutionary processes. In the 1940s and 1950s Konrad Lorenz revolutionized the understanding of behavior, particularly breeding and nesting behavior, by defining four ways that behavior can be examined: (i) biological function (What does the nesting behavior do for the animals?), (ii) causation (What stimuli result in a particular aspect of nesting behavior?), (iii) biological significance (How does nesting behavior contribute to survival of the bird?), and (iv) evolution (How did various aspects of nesting behavior evolve?). With this framework, people in different disciplines have examined the nesting behavior of birds. Physiologists examine the hormonal basis of nesting behavior, neurobiologists examine the interaction of the brain with nesting behavior, and ethologists examine the function and biological significance of particular aspects of nesting behavior in wild birds. Niko Tinbergen, a student of Konrad Lorenz and cowinner with him and Karl von Frisch of the Nobel prize for physiology and medicine in 1973, studied nesting behavior of gulls, thus transforming the field from merely observational to that involving field testing of hypotheses about the causes and adaptiveness of nesting behavior. Adaptiveness refers to the ability of the behavior pattern to result in high levels of survival and reproductive success.

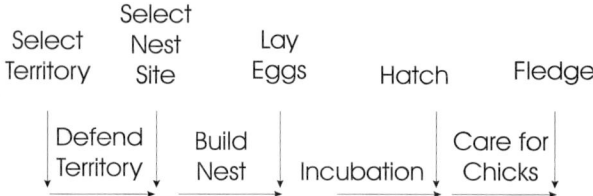

FIGURE 1 Nesting birds have a breeding cycle that starts with selection of a territory and ends when the chicks fledge and are able to care for themselves.

Nesting behavior (Fig. 1) in birds can be examined from the standpoint of nesting habitat (Where do birds nest?), spatial patterns of nesting, temporal patterns of nesting, the role of predators and competitors, and the role of parental care in nesting. Nesting in birds is greatly influenced by mating systems, habitat, predators and competitors, and timing aspects.

II. NESTING HABITAT

All birds reproduce by laying eggs and keeping the eggs at a constant, high (about 37–39oC) temperature while they are developing. In most cases, birds maintain a constant temperature by incubating (sitting on) the eggs until the chicks hatch. All birds are limited spatially during the incubation phase in that they are restricted to the site where the eggs are laid, and they must select an appropriate nesting habitat and substrate for their eggs.

Upon hatching, the chicks show one of the following three patterns: They are able to walk and take care of themselves (precocial), they are able to walk and partially take care of themselves (semiprecocial), or they are unable to walk, find food, protect themselves, or maintain a constant body temperature (altricial). Except for precocial species, birds continue to be tied to their nest sites until the young are able to move about easily. For many species, parents remain at the nest site until the chicks fledge (able to fly and leave the nest). Some parental care usually continues after fledging, and the amount of postfledging care depends on the species of bird.

A. Selection of Nesting Habitat

The selection of nesting habitat is very important to all species of birds because the parent or parents will be tied to that place until the young fledge. Thus, its own safety and that of its eggs or offspring depend on selecting a site that is free from humans, other predators, or inclement weather. Selection of habitat involves choosing a geographical area (e.g., New Jersey or Texas), a general habitat (e.g., field, marsh, or woods), an elevation (e.g., ground or above ground), and a specific habitat (e.g., a tree, under grass, or in a dense shrub).

Most birds have preferred nesting habitat which includes both general habitat features and specific habitat substrates. For example, some songbirds nest in forests (e.g., ovenbird *Seiurus aurocapillius* and rufous-sided towhee *Pipilio erythrophthalmus*), whereas others nest on the edge (e.g., catbird *Dumatella carolinensis*, indigo bunting *Passerina cyanea*, and robin *Turdus migratorius*), and still others nest in the open (e.g., song sparrow *Melospiza melodia*, killdeer *Charadrius vociferus*, and meadowlark *Sturnella magna*). Within this general habitat, birds nest in particular places: Ovenbirds, towhees, and meadowlarks nest on the ground, song sparrows nest in hedges, catbirds nest in dense shrubs, and robins nest in shrubs or trees. The ways that birds within a given habitat (such as a forest or backyard) divide up the nesting space is called habitat partitioning.

B. Habitat Partitioning

Ornithologists who study nesting behavior have devoted considerable attention to figuring out how birds partition space, particularly among members of the same species, and between closely related species. Since the nesting requirements of members of the same species are similar, they share the greatest overlap. Partitioning of nesting habitat is particularly interesting in the case of different species of birds that nest in close proximity (such as a nesting colony).

In addition to selecting a nesting habitat, birds select a nest substrate, such as the ground (dirt, sand, and flat rocks), vegetation (shrubs and trees), rocky cliffs, or other structures. Nesting on the ground has the advantage that eggs and young will not fall out of the nest, but it has the disadvantage of exposing the offspring to increased human disturbance and predation. Nesting on trees and rocky cliffs has the advantage of removing the offspring from ground predators [such as cats and foxes (*Vulpes vulpes*)], but the young can fall out of nests.

C. Competition for Nesting Habitat

There is often intense competition for nest sites, and birds may devote many days or weeks to defending their nest site at the beginning of the nesting season. Some species, particularly those in crowded conditions, must defend their nest site throughout the nesting season or they will lose it (and their offspring) to neighbors (of the same or a different species). Understanding how birds compete for nesting habitat has been an important aspect of the study of nesting behavior, particularly regarding which sites are preferred, why they are preferred, why some individuals get the preferred sites, and how the sites are protected throughout the nesting period.

III. TYPES OF NESTS

Once a pair of birds has selected a nest site, they have to construct the nest. Both members of a pair construct the nest in many species of birds, but in some only the female makes the nest. Both members of a pair usually construct the nest in species in which both the male and female incubate the eggs and take care of the young.

The type of nest that birds build varies by both where they build it and the species (Fig. 2). Some birds that nest on the ground, such as piping plovers (*Charadrius melodus*), hardly build a nest at all but rather merely make a scrape or small depression in the ground and lay their eggs on the sand. Sometimes they add a few pebbles or shells, but otherwise there is no nest. Other species that nest on the ground construct a nest of grass or twigs that may be from 1 or 2 in. to 1 ft high. Species that construct larger nests on the ground usually do so because their eggs and chicks can be exposed to floods.

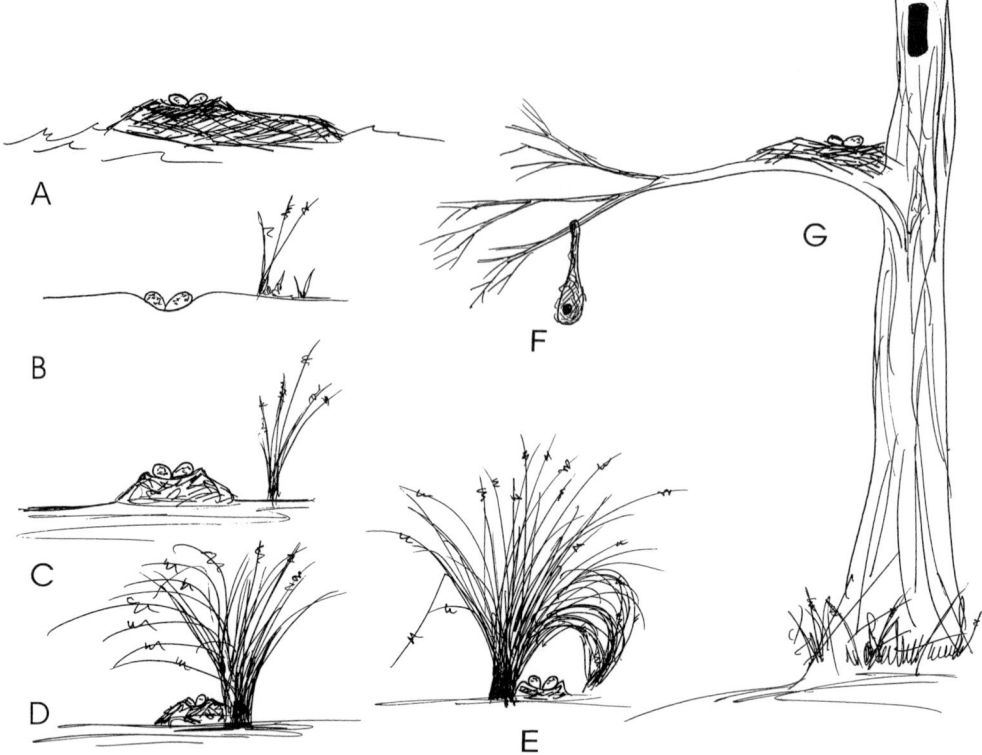

FIGURE 2 Most birds construct nests, ranging from elaborate nests built over water to bulky nests that sit on the crotch of a tree or hang down from the branches. Some build nests over water (A; grebes and gulls), make depressions in the sand (B; shorebirds), build nests on the ground (C and D; shorebirds and gulls), hide them under vegetation (E; sparrows and rails), hang them on branches (F; orioles), or place them on the tops of branches (G, hawks and most passerines). Black hole in tree above G used by woodpeckers, parrots, and other hole nesters.

Species that nest solitarily on the ground often build their nests in vegetation and some try to hide them from predators by pulling vegetation over the top to make a dome. Others actually build an elaborate cover or roof so they are hidden. Species that nest on the water, such as grebes and some gulls, also construct an elaborate nest from vegetation, and these nests are usually 1 or 2 ft across so they provide not only a place for the eggs but also a place for both members of the pair to stand on the nest when they feed the chicks. Species that nest on rocks on cliffs usually construct a nest that will keep the eggs from falling off the cliff.

Most birds that place their nests in bushes or trees build elaborate nests so that their eggs and chicks do not fall out of the tree. However, there are exceptions, such as the white tern (*Gygis alba*) that nests on tropical oceanic islands. White terns lay their egg on a small depression in a branch or the crotch of a tree, and the young hatch with longer claws than most chicks so that they can cling to the branches.

There is great variation in how elaborate birds that nest in shrubs or trees build there nests. Some species build a nest that is situated on the crotch of the tree and that is merely a platform. Others birds, such as some hawks and herons, build a more elaborate platform nest made of sticks that they may use for many years. Each year, they just add a few more sticks to repair any damage caused by winter storms.

Other species build elaborate nests that hang from bushes or trees. In North America, the Baltimore oriole (*Icterus galbula*) builds a nest that hangs 5 or 6 in. from a tree. In South America, some oropendolas (*Psarcolius* spp.) build elaborate nests that may hang more than 3 ft, and they often nest in trees with social wasps to further deter mammalian predators. In Africa and Asia, some weaverbirds build nests that hang from trees, and they nest in colonies in which up to 100 birds may occupy one large tree. The advantage of building a nest that hangs from the tree is that it is more difficult for predators to reach, particularly tree-climbing mammals and snakes.

IV. SPATIAL PATTERN OF NESTING

One important aspect of nesting is how birds use space. Spacing patterns range along a continuum from birds that nest solitarily by themselves to those that nest very close together in dense colonies. Some species always nest at a particular nesting density, whereas others may exhibit nearly the full range of nesting patterns.

A. Solitary Nesting

The vast majority of bird species nest solitarily; although their territory boundaries may cross those of another of the same species, their nests are far from each other. For example, about 85% of songbirds nest solitarily. This means that there are no other birds of the same species nesting close by. Species that nest solitarily often have their nesting and feeding territories in the same space. Many songbirds nest solitarily, and they acquire all their resources from the area around their nest. Species that nest solitarily may spend a great deal of time defending their space against intruders of the same species that are looking for territories, although territory defense usually occurs at the beginning of the nesting cycle.

The main advantage of a solitary nesting pattern is reduction of the conspicuousness of the nest to predators, thus reducing predation on the eggs, chicks, and incubating birds. Most solitary birds accomplish this by being difficult to see or find, thus avoiding most predators. Species that nest solitarily nest in places where they can avoid mammalian predators (e.g., hawks that nest in trees or on cliffs), nest in the open where they have a good view of any approaching predator and can fly away (e.g., ground-nesting species), or nest so that they are mostly hidden. Some species nest in holes in trees, where their nests are difficult for predators to reach. Ducks often nest in thick vegetation, and their nests are difficult to find.

FIGURE 3 Some species of birds that nest solitarily have distraction displays to lure predators away from their eggs of chicks.

The eggs of species that nest on the ground usually match the color of the substrate and are speckled to further avoid detection. People and predators often walk right past ground-nesting birds without detecting them. Once discovered, the birds sneak from the nest, hide in the grass, or burst forth so explosively that the predator is disoriented and follows the adult (soon forgetting the location of the nest). Some ground-nesting species have a distraction display. They feign injury and lead the predator away from the nest. Once the predator has been drawn far away, the bird flies up and away. Birds that give distraction displays include shorebirds such as the killdeer and piping plover (Fig. 3). The disadvantage of nesting solitarily is that once a nest is discovered, it is highly likely to be preyed on. Furthermore, usually there are no other birds to provide early warning of approaching predators or to help with defense.

B. Colonial Nesting

Nearly all species of seabirds (birds that live primarily along the coasts or on the open oceans) nest in colonies. Seabirds include albatrosses, auks, frigate birds, gannets and boobies, gulls and terns, pelicans, penguins, petrels and shearwaters, skimmers, and tropic birds. Other species, however, also nest in colonies, including blackbirds, egrets, herons, ibises, and swallows. Birds sometimes nest in colonies of

FIGURE 4 Cape gannets (*Sula capensis*) nest in very dense colonies in which there is barely enough room for them to walk between nests.

only one species, but more often they nest in colonies with several different species (mixed-species colonies). Some species, such as herring gulls (*Larus argentatus*) and common terns (*Sterna hirundo*), may nest solitarily as well as in dense colonies. Other species, such as gannets, boobies, and Adelie penguins, almost always nest in colonies (Fig. 4).

C. Why Do Birds Nest in Colonies?

For nearly three decades the relative advantages and disadvantages of coloniality have been discussed. The possible advantages of nesting in a colony are increased early warning of predators (more eyes to watch for predators), increased predator protection (more birds to chase away predators), increased social facilitation (Fig. 5), and decreased individual chance of predation. The decreased chance of individual predation is due to predator swamping; the predator can only eat so many eggs or chicks (or adults), and if there are tens or hundreds of birds nesting together, an individual's chance of being eaten is less than that of a solitary-nesting bird discovered by a predator. Furthermore, the chance of being eaten in a colony of 150 birds is less than that in a colony of 50, assuming there is no increase in the number of predators. Many colonial-nesting species engage in group defense, in which they fly overhead and dive at intruders that enter the colony (mobbing). Colonially nesting species that do not have such defenses either nest on remote oceanic islands far from predators (e.g., albatrosses and penguins) or have underground burrows to which they can escape (e.g., auks, puffins, and petrels).

Another advantage that is still being debated was proposed by Zahavi and others, who suggested that birds nest in colonies so that they can exchange information about the location of foods. This would be particularly adaptive for species that have an ephemeral (unpredictable) food supply, such as most seabirds. There are two kinds of information transfer: active and passive. In passive information transfer, one bird that is having trouble finding food watches

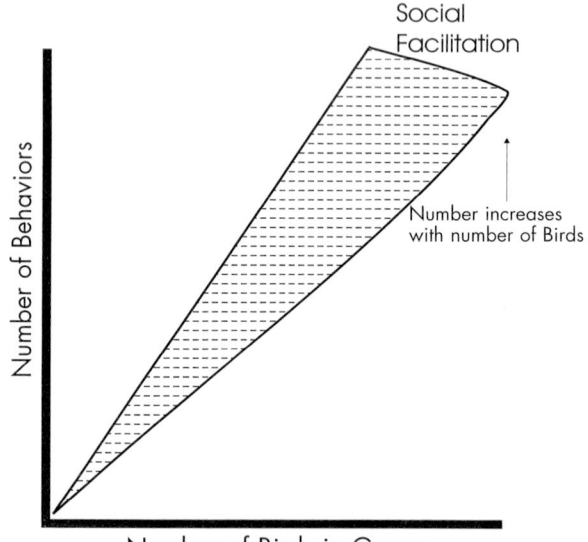

FIGURE 5 As the number of birds in a colony increases, the number of displays or other behaviors can increase simply as a result of the number of birds. However, it can also increase at a greater rate than the number of birds (shown as the dashed area), and this is called social facilitation.

its neighbors, and when one brings back a large prey item, it follows that neighbor out to its foraging ground. Active transfer of information is much harder to show, and it has not been demonstrated convincingly for any species.

The disadvantages of nesting in colonies are increased visibility (a large colony is more visible than a small one), increased competition for nesting space and nesting materials, increased rate of attempted rapes or interference with mating, increased rate of parasitism, and increased rates of cannibalism and piracy. There is also a possibility of increased competition for food because there are so many birds in the same area.

It has been suggested that coloniality, in some species, may result from the passive selection of nest sites when nest sites are limited rather than an active seeking of other pairs for nesting companions. For example, if 500 terns prefer to nest on a salt marsh island and there is only one small salt marsh island that is suitable, then all 500 terns may nest on one island because it is the only space and not because they "want" to nest close together. The examination of the advantages and disadvantages of coloniality is and will continue to be a fertile area of research as more sophisticated means of following the nesting behavior and reproductive success of individuals and groups become available.

D. Why Nest in Mixed-Species Colonies?

Most species that nest in colonies do so in mixed species colonies of two to several species. Nesting in mixed-species colonies may partially reduce competition for space (since each species may have a slightly different habitat preference) as well as provide protection (bigger species may defend the colony against predators). Some species that do not have group defense, such as black skimmers or grebes, often nest in colonies with species that have active defense (such as terns or gulls). They thus derive benefits from early warning as well as from predator defense (Fig. 6). Neucterlein coined the term "social parasitism" for species that nest within the colonies of other species for protection. Nesting in mixed-species colonies has costs: Larger species may claim

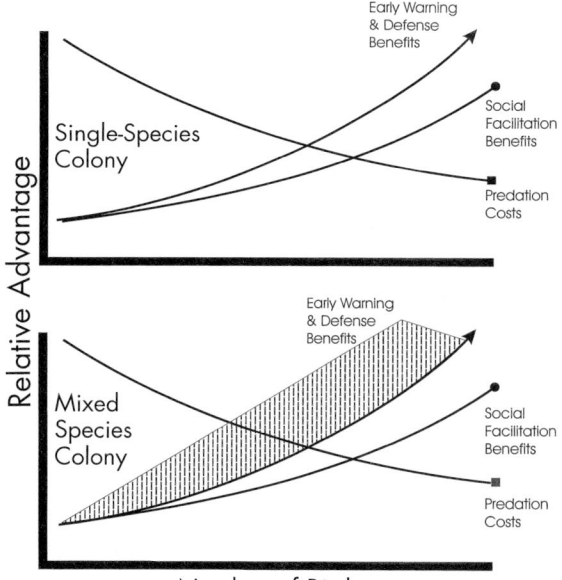

FIGURE 6 Birds that nest in colonies usually derive many benefits, such as increased warning and defense from predators, increased social facilitation, and lowered losses due to predators. Birds that nest in colonies with other species can increase the benefits while further lowering the costs (such as predation).

the best territories or prey on the eggs and young of smaller species.

V. TEMPORAL PATTERNS OF NESTING

There are three interesting temporal aspects of nesting in birds: (i) when birds initiate nesting, (ii) how long each phase of the nesting cycle takes, and (iii) how synchronous breeding is for any given species.

A. Timing of Breeding

There is much debate about the factors that are responsible for the timing of breeding in birds. Even in tropical regions birds do not normally nest year-round; instead, there is usually a discrete nesting season. In temperate regions, most birds breed in the spring. Nesting occurs progressively later from south to north. For example, robins begin nesting in late March in the south, in April in the midregions of the United States, and in May in the far north. Species that breed only in the far north, such as many species of shorebirds, begin breeding in mid-June. In the case of temperate and arctic breeding, nesting is usually influenced by temperature and food supply. Birds cannot begin to initiate nesting and lay eggs until there are sufficient food resources to do so (unless they arrive from migration with sufficient food reserves to lay eggs). During some years in the Canadian tundra, shorebirds, gulls, and snow geese do not breed if the snows linger into early July or food is scarce. These two proximate factors (weather and food supply) thus have the ability to delay breeding in any given year.

Birds that breed in tropical regions also have a discrete breeding cycle, although the total breeding period may be longer. Birds nesting in tropical regions may be influenced more by rainfall than by temperature, but they are also influenced by food supply.

B. Length of the Nesting Cycle

The length of the individual phases of the reproductive cycle is fairly constant from year to year within a given population or species. That is, the period of territory acquisition usually takes the same length of time within a given population (although it may take from a few days to several months in different species). Egg laying takes 1–12+ days depending on the species of bird and the number of eggs laid. The incubation period is fairly constant within a species but varies from 10 (brood parasites) to 78 (wandering albatross *Diomedea exulans*) days, and the fledging period may take up to 278 days (wandering albatross). Some species, notably seabirds, have extended postfledging parental care for up to 40 weeks (black-footed albatross *Diomedea nigripes*). Thus, for most individual birds, the total nesting time is fairly fixed, with longer periods when food is scarce. The most variable phase is the pre-egg-laying period when birds are finding mates and defending territories. In general, however, it is advantageous for the parents to incubate for as short a time as possible and for the young to grow as quickly as possible, thus decreasing the time the young are vulnerable to predators and inclement weather.

Songbirds (passerines) usually have short nesting cycles of only a few weeks since the incubation period is about 2 weeks and the chick care phase may be as short. This allows some species to raise two (or more) broods in a year, such as robins and house wrens (*Troglodites aedon*). This increases their chances of raising young in any given breeding season.

C. Synchrony of Nesting Cycles

Although the time for nesting by any individual bird is relatively fixed, the total nesting period for the species may be short (all birds initiate egg laying at the same time), or it may be extended (birds initiate egg laying over several weeks or months). For example, a colony of thousands of pairs of red-billed queleas (*Quelea quelea*) in Africa may initiate nest building and egg laying in the same 3- or 4-day period, whereas white terns (*Gygis alba*) in the tropics may initiate egg laying over a period of 3 months or more.

The factors that cause synchrony, or lack thereof, are of considerable interest. A lack of synchrony may serve to increase partitioning of food resources in a region with low prey availability. Extreme synchrony, as in the case of the queleas, may be a result of traveling in very large flocks and increased social facilitation.

One advantage of synchronous nesting is that it reduces the time that eggs and chicks are available to predators. If the number of predators in a region is constant, then a highly synchronous nesting pattern will result in fewer young taken overall. If nesting is spread over several months, compared to only 2 or 3 months, then predators have a longer period to forage and can take a higher percentage of the eggs and young.

VI. PREDATION, COMPETITION, AND NESTING BEHAVIOR

The phenology (seasonal timing of events), habitat and nest site selection, and nesting behavior are all influenced, both proximally and ultimately (in evolutionary terms), by predation and competition.

FIGURE 7 Herring gull eating egg of conspecific.

A. Predation

Predation has shaped all aspects of habitat selection, the types of nest site selected, the dispersion pattern of nesting, the spatial pattern of nesting (solitary or colonial), and nearly all aspects of nesting behavior. Predators can take eggs, chicks, and, in some cases, adults (Fig. 7). For most nesting birds, the primary predators are either mammalian or avian, although some snakes can be devastating, as the brown tree snake (*Boiga irreguluaris*) has been on nesting birds in Guam. In only a few years the nonnative tree snake wiped out most of the native birds; there were no snakes on the island before the brown tree snake invaded.

The type of predators to which nesting birds have been exposed influences their defensive behavior. If the threat is only to the eggs, then incubating adults may sit tight, preventing the predator from having access. When the threat is to both eggs and young, adults usually remain and defend the nest against predators. When the threat includes adults, however, birds usually remain at a respectable distance from the predator. For example, nesting common terns will remain on their nest when a turnstone (*Arenaria interpres*) attempts predation on eggs, they will circle overhead and dive-bomb (sometimes actually striking its head) a herring gull that tries to prey on eggs or chicks, but they usually fly well above a fox that is capable of eating them.

Predation is theorized to be one of the primary causes of the evolution of coloniality since in a colony the nesting birds have early warning and increased predator defense. However, colonies are usually located either on cliffs or high in trees or on remote islands where mammalian predators are not found.

B. Competition

Competition is another primary force in the evolution of nesting patterns and nesting behavior in birds. Birds compete for access to mates, territories, nest material, nest sites, and food. Competition for nesting territories is strong in most birds, whether they nest solitarily or in colonies. In colonial species, competition for territories and nest sites can be somewhat reduced by nesting in mixed-species colonies in which the nesting requirements of each species differ somewhat, and competition in solitary species can vary markedly depending on the breeding density of birds. Species such as piping plovers that are endangered and are at a relatively low population level have little nest competition, except in places where the competition is artificially high because of human encroachment onto beach nesting habitats.

VII. SPECIAL CASES

There are a number of interesting ways that birds deal with nesting, including brood parasitism and incubator birds. Brood parasites are species that lay their eggs in the nests of other species, such as brown-headed cowbirds (*Molothrus ater*). Cowbird females watch for females of other passerines that are going to their nests to lay their own eggs. When they find one, they lay their egg in the nest (called a host) and sometimes remove one of the host's eggs. Cowbird young usually hatch before the host young hatch, and because of their rapid growth they grow faster than the host's young. The host's young often die because they do not obtain enough food. Favorite hosts of the brown-headed cowbird are warblers, sparrows, thrushes, and vireos.

Other brood parasites include cuckoos. Cuckoo young are far more aggressive than brown-headed cowbirds, and they actively kick the host's young out of the nest, thereby removing any competition for food that the parents may bring back.

Incubator birds, in the family Megapodiidae, live in Australia and the Pacific Islands. They lay their eggs in mounds of rotting vegetation instead of building a nest and sitting on the eggs to incubate them. They visit the mounds daily to adjust the temperature by removing or adding more vegetation to the mound. When they hear the calls of their hatching young, they may remove some of the vegetation to help the young get out of the mounds. One species of incubator bird even lays its eggs in soil heated by volcanic activity.

See Also the Following Articles

AGGRESSIVE BEHAVIOR; AVIAN REPRODUCTIVE, OVERVIEW; BROOD PARASITISM, BIRDS; EGG, AVIAN; PARENTAL BEHAVIOR, BIRDS; SEASONAL REPRODUCTION, BIRDS

Bibliography

Barnard, C. J., and Behnke, J. M. (Eds.) (1990). *Parasitism and Host Behavior.* Taylor & Francis, London.

Brown, C. R., and Brown, M. B. (1996). *Coloniality in the Cliff Swallow.* Univ. of Chicago Press, Chicago.

Burger, J. (1991). Coastal landscapes, coastal colonies and seabirds. *Aquat. Rev.* **4**, 23–43.

Cody, M. L. (1985). *Habitat Selection in Birds.* Academic Press, Orlando, FL.

Lack, D. (1968). *Ecological Adaptations for Breeding in Birds.* Methuen, London.

Moller, A. P. (1994). Parasites as an environmental component of reproduction in birds as exemplified by the swallow *Hirundo rustica. Ardea* **83**, 161–172.

Newton, I. (Ed.) (1989). *Lifetime Reproduction in Birds.* Academic Press, London.

Skutch, A. F. (1976). *Parent Birds and Their Young.* Univ. of Texas Press, Austin.

Wittenberger, J. F., and Hunt, G. L., Jr. (1985). The adaptive significance of coloniality in birds. In *Avian Biology* (D. S. Farner, J. R. King, and K. C. Parkes, Eds.), Vol. 3, pp. 1–79 Academic Press, New York.

Neuroendocrine Systems

George Fink
Medical Research Council Brain Metabolism Unit

I. Central Neuroendocrine Systems: Hypothalamic–Pituitary Axis
II. The Hypothalamo–Adenohypophysial System
III. Pineal Gland and Photoperiodic Control of Reproduction
IV. The Hypothalamo–Neurohypophysial Systems
V. The Intermediate Lobe of the Pituitary Gland
VI. Hormonal Effects on the Nervous System

GLOSSARY

action potentials Electrical signals which propagate messages along nerve fibers. The potentials are produced by the flux of ions, in particular sodium and potassium ions, through ion channels along the nerve cell membrane.

limbic system A system whose name derives from the fact that it forms a border between the new and the old brain. It is composed of a number of structures, all interconnected to form a circuit which is concerned primarily with emotions, memory, olfaction, and neuroendocrine control.

neurohemal junction A specialized junction between nerve terminals and a plexus of capillaries which facilitates the transmission of neurohormones from the nervous system into the bloodstream or conversely the transfer of hormones from the bloodstream into the nervous system.

neurohormone A chemical substance which is released from nerve terminals into the bloodstream which transports the substance to its target cell. Neurohormones are often the same chemical substance as neurotransmitters.

neurosecretion The secretion of hormones by nerve cells.

neurotransmitter A chemical compound released by nerve

cells which either excites or inhibits a target cell. The target cell could be another nerve cell, a glandular cell, or a muscle fiber. Chemical neurotransmission is widely used in the nervous system and since it is capable of analog functions it has a considerable advantage over electronic transmission, which is necessarily digital.

portal vessels Vessels that transport blood from one capillary bed to a second capillary bed before joining the systemic circulation.

steroid hormones Composed of a basic unit of three six-membered and one five-membered carbon rings.

synapse A specialized junction between two nerve cells where signaling occurs by way of the release of neurotransmitters.

Neuroendocrinology is the study of the interactions between the nervous and endocrine systems. The discipline is divided into central and peripheral neuroendocrinology. Central neuroendocrinology deals with how the brain controls pituitary hormone secretion and in turn how hormones control the brain. The brain–pituitary system is the interface between the central nervous and the endocrine systems by which external factors, such as day length and stress, and internal factors, such as emotion, trigger endocrine responses. Peripheral neuroendocrinology encompasses the interactions between nerves and endocrine cells and organs in the periphery. Here, attention will be focused on central neuroendocrinology, which is more important for reproduction.

I. CENTRAL NEUROENDOCRINE SYSTEMS: HYPOTHALAMIC–PITUITARY AXIS

A. Brief History

The central importance of the anterior pituitary gland as "conductor of the endocrine orchestra" was not understood until the early 1930s when P. E. Smith published his parapharyngeal method for removing the gland ("hypophysectomy"). The effects of hypophysectomy proved to be so dramatic that for a short period most scientists in the field, led by the distinguished Harvey Cushing, thought that the pituitary gland was autonomous. However, around the same time, William Rowan, working in Alberta, Canada, on the annual migration of birds, showed that day length had a potent effect on the growth of the gonads. Rowan's experiments, together with those on seasonal breeding in animals and Hans Selye's observations on the effect of stressful stimuli on endocrine organs and especially the adrenal gland, led to the concept that the anterior pituitary gland must be under central nervous control, a view which Cushing soon adopted. The observational and experimental evidence which supported this concept was summarized by Marshall in his 1936 Croonian lecture. It had long been known that the pituitary gland and brain were connected by the pituitary stalk, but several lines of evidence suggested that neural control of the anterior pituitary gland was mediated not by nerve fibers in the pituitary stalk but by chemical substances released into the hypophysial portal vessels. These vessels, first described by Popa and Fielding in 1930, surround the pituitary stalk linking a plexus of capillaries at the base of the hypothalamus with a second capillary plexus in the anterior pituitary gland. Throughout the 1930s and 1940s a debate raged about the direction of blood flow in these portal vessels. This debate, based on histological evidence, could have been avoided had someone read the 1935 report by Bernado Houssay and associates that showed that in the living toad blood flowed from the hypothalamus down to the pituitary gland. Alas, Houssay's paper was published in French.

The neurohumoral hypothesis of the control of the anterior pituitary gland was first formally advanced by Friedgood in 1936 and Hinsey in 1937. However, it was the elegant pituitary graft experiments of Harris and Jacobsohn which showed beyond doubt that the anterior pituitary gland was controlled by substances released by nerve terminals in the median eminence at the base of the hypothalamus and transported to the pituitary gland by the hypophysial portal vessels. The characterization of the first three of these substances, thyrotropin-releasing factor, luteinizing hormone-releasing factor, and somatostatin, was to take a further 18–21 years of hard work in the laboratories of Guillemin and Schally for which they were awarded the 1977 Nobel Prize for Physiology and Medicine. Soon after its characterization as a decapeptide in 1971, luteinizing hormone-releasing factor (later termed gonadotropin-releasing hormone) was measured by radioimmunoassay in hypophysial portal blood of the anesthetized rat and shown to be increased nearly fivefold by a small electrical stimulus applied to the medial preoptic area of the hypothalamus.

B. General Anatomy and Development

The hypothalamic–pituitary axis is the functional unit of the central neuroendocrine system. Located at the base of the brain, the hypothalamus is linked to the pituitary gland by the pituitary stalk (Fig. 1). The hypothalamus is composed of a medial portion adjacent to the third cerebral ventricle and in which are located the major hypothalamic nuclei and a lateral portion. The latter is composed mainly of a large cable of nerve fibers which runs from the midbrain to the forebrain, the medial forebrain bundle (Fig. 2), together with relatively few aggregations of nerve cell bodies. Axons from nerve cells located in the hypothalamic nuclei project to the median eminence, where they either terminate on the loops of primary capillaries of the hypophysial portal vessels in the external layer of the median eminence or form a cable which passes through the internal layer of the median eminence to form the bulk of the pituitary stalk (Fig. 3). The median eminence, so called because it protrudes as a small dome in the midline from the base of the hypothalamus, forms the floor of the third ventricle and is delineated by the optic chiasm in front, the mammillary bodies behind, and a depression (hypothalamic sulcus) on either side. Arising from the median eminence is the neural stalk, which links the pituitary gland to the brain.

The pituitary gland weighs about 1 g in the human and is located in a fossa in the basisphenoid bone at the base of the skull, the "sella turcica," so called

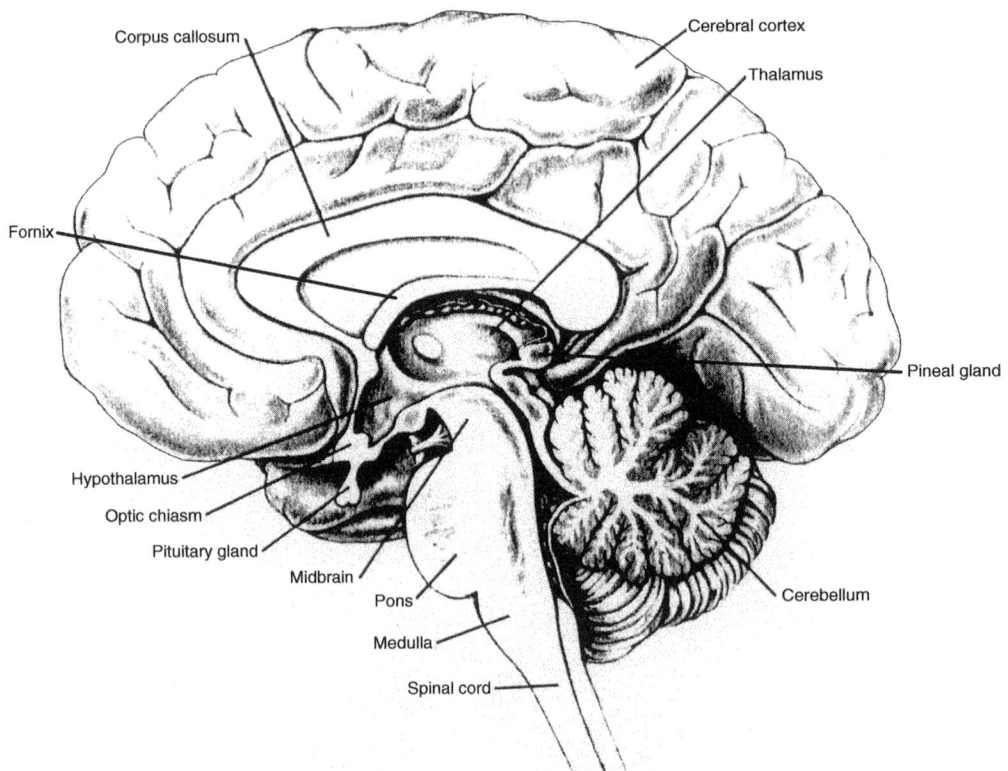

FIGURE 1 A midsagittal section of the human brain showing the inside surface. Note the pituitary gland attached by way of the pituitary stalk to the floor of the hypothalamus. The hypothalamus and the thalamus, which lies above it, form the wall of the third cerebral ventricle at the posterior end of which is the pineal gland.

hypothalamo–neurohypophysial axis. The neural stalk is made up of numerous nerve fibers which project mainly from the paraventricular and supraoptic nuclei (PVN and SON) of the hypothalamus to terminate on a capillary bed derived from the inferior hypophysial artery and located in the neural lobe of the pituitary gland. The three pituitary lobes are also termed the pars distalis, pars intermedia, and pars nervosa, respectively (Fig. 3). The pituitary stalk and median eminence are covered by a single layer of cells termed the pars tuberalis which is continuous with the pars distalis.

The adenohypophysis develops from an outgrowth of the ectodermal placode which forms the roof of the embryonic mouth (or "stomadeum"). This ectodermal outgrowth forms Rathke's pouch and meets the neurohypophysis, which grows down from the floor of the embryonic third ventricle. Rathke's pouch closes and separates from the roof of the mouth. The caudal (rear) part of the pouch remains thin to form the pars intermedia, which becomes tightly juxtaposed to the rostral surface of the neurohypophysis (Fig. 4). The rostral part of the pouch develops into the pars distalis (Fig. 4). Vasculariza-

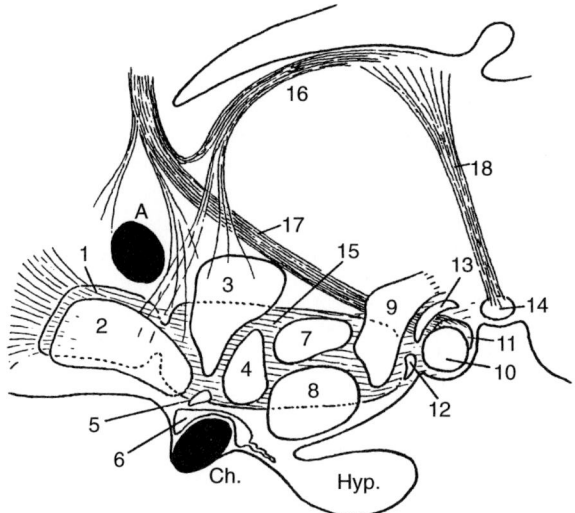

FIGURE 2 Diagram showing the relative positions in a sagittal plane of the hypothalamic nuclei in a typical mammalian brain and their relation to the fornix, stria habenularis, and fasciculus retroflexus. A, anterior commissure; Ch, optic chiasma; Hyp, hypophysis (pituitary gland); 1, lateral preoptic nucleus (permeated by the medial forebrain bundle); 2, medial preoptic nucleus; 3, paraventricular nucleus; 4, anterior hypothalamic area; 5, suprachiasmatic nucleus; 6, supraoptic nucleus; 7, dorsomedial hypothalamic nucleus; 8, ventromedial hypothalamic nucleus; 9, posterior hypothalamic nucleus; 10, medial mamillary nucleus; 11, lateral mamillary nucleus; 12, premamillary nucleus; 13, supramamillary nucleus; 14, interpeduncular nucleus (a mesencephalic element in which the fasciculus retroflexus terminates); 15, lateral hypothalamic nucleus (permeated by the medial forebrain bundle); 16, stria habenularis; 17, fornix; 18, fasciculus retroflexus of Meynert (habenulopeduncular tract) [modified from W. E. Le Gros Clark, in *The Hypothalamus* (W. E. Le Gros Clark *et al.*, Eds.), Oliver & Boyd, Edinburgh, UK, 1938].

because its shape resembles a Turkish saddle. The close proximity of the hypothalamus and pituitary gland to the optic chiasm means that tumors either in the hypothalamus or the pituitary gland may lead to visual symptoms and signs.

The hypothalamic–pituitary axis is divided functionally into two systems. The hypothalamus, hypophysial portal vessels, and the anterior and intermediate lobes of the pituitary gland constitute the hypothalamo–adenohypophysial axis. The hypothalamus, neural stalk, and neural lobe constitute the

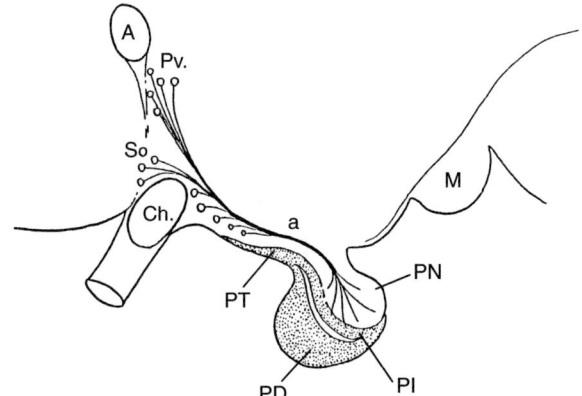

FIGURE 3 Schematic section of the mammalian hypothalamus and pituitary gland showing the neurohypophysial tract (a) which is composed mainly of fibers derived from the paraventricular (Pv) and supraoptic (So) nuclei. A, anterior commissure; Ch., optic chiasma; M, mamillary bodies; PT, pars tuberalis; PD, pars distalis; PI, pars intermedia; PN, pars nervosa [modified from W. E. Le Gros Clark, in *The Hypothalamus* (W. E. Le Gros Clark *et al.*, Eds.), Oliver & Boyd, Edinburgh, UK, 1938].

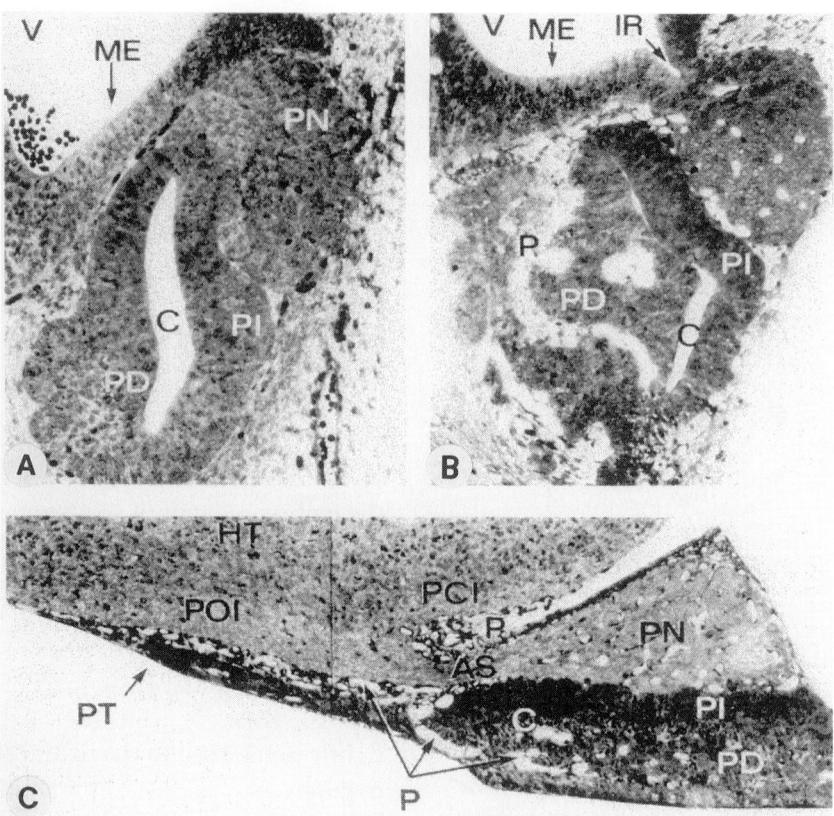

FIGURE 4 Development of the pituitary gland in the rat. Photomicrographs of midline sagittal sections through the hypothalamic–pituitary complex of rats at Embryonic Days 15 (A), 17 (B), and 20 (C). (A) The pituitary anlage shortly after closure of Rathke's pouch which migrates dorsally to meet the neurohypophysial downgrowth from the floor of the hypothalamus. Rotation of the pituitary gland caudally through 135° with respect to the base of the diencephalon (hypothalamus) is seen, as is the invasion of the pars distalis (PD) by the leash of portal vessels (P) at E 17. AS, anatomical stem; C, hypophysial cleft; HT, hypothalamus; IR, infundibular recess; ME, median eminence; PCI, pars caudalis infundibuli; PI, pars intermedia; PN, pars nervosa; POI, pars oralis infundibuli; V, third ventricle. 1µ Araldite sections, toluidine blue stain. Magnification, ×80 (reproduced with permission from G. Fink and G. C. Smith, Z. Zellforsch. 119, 208–226, 1971).

tion of the median eminence and the pituitary gland begins at about Day 15 of embryonic (E 15) life in the rat, and the hypophysial portal vessels become defined by E 18. Nerve terminals in the median eminence with granular vesicles (presumably neurohormones) are first evident on E 16 and the first secretory granules appear in pars distalis cells on E 17. This sequence of embryonic development suggests that the development of secretory cells in the pars distalis depends on the development of nerve terminals in the median eminence and the anlage of the hypophysial portal vessels.

II. THE HYPOTHALAMO–ADENOHYPOPHYSIAL SYSTEM

A. Neurohormonal Control of Anterior Pituitary Hormone Secretion

The hypothalamo–adenohypophysial system is composed of the hypothalamus, pituitary stalk, and anterior pituitary gland (Figs. 3–5). Transmission of signals between the brain and anterior pituitary gland occurs not through direct innervation of the anterior pituitary cells but rather through chemical messengers (neurohormones) which are transported by the

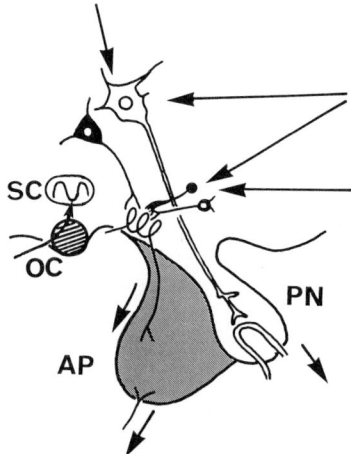

FIGURE 5 Schematic diagram of the hypothalamic–pituitary system showing the magnocellular (white) projections directly to the systemic capillaries of the pars nervosa (PN) and the parvocellular (black) projections to the primary plexus of the hypophysial portal vessels which convey neurohormones to the pars distalis of the anterior pituitary gland (AP). Dorsal to the optic chiasm (OC) are the suprachiasmatic nuclei (SC), which receive direct projections from the retina and play a key role in the control of circadian rhythms (indicated by the sinusoidal curve). The activity of the intrinsic neurons of the hypothalamus is greatly influenced by projections (arrows) from numerous areas of the forebrain, midbrain, and hindbrain, particularly the limbic system, as well as by hormones, mainly estrogen and progesterone in the case of the hypothalamic–pituitary–gonadal system.

B. Hypophysial Portal Vessels

The portal vessels are so called because they transport the chemical messengers from one capillary bed (primary capillaries) to a second capillary bed before entering the general circulation. In principle, this is identical to the hepatic portal system, which transports substances from the primary bed of capillaries in the intestine and its appendages (e.g., pancreas) to a second bed of capillaries or sinusoids in the liver. Both the primary and secondary (sinusoids) plexus of capillaries are fenestrated (Fig. 6), which presumably facilitates the transport of substances across the capillary wall. The hormones released from anterior pituitary cells are transported by pituitary veins into the systemic circulation by which they are transported to their major target organs, the gonads and the adrenal and thyroid glands.

C. Neurohemal Junctions and Circumventricular Organs

The close juxtaposition of nerve terminals and capillaries which facilitates the release of chemical messengers from nerve terminals into the bloodstream (Fig. 6) is termed a neurohemal junction, the fundamental functional module of a neuroendocrine system. Neurohemal junctions are also the fundamental functional units of the neurohypophysis and of the circumventricular organs such as the organum vasculosum of the lamina terminalis, the subfornical organ, and the pineal gland, which are located at various sites around the third cerebral ventricle, and the area postrema adjacent to the fourth cerebral ventricle. The median eminence is also a circumventricular organ. All circumventricular organs are characterized by a dense plexus of fenestrated vessels and the fact that the blood–brain barrier is inoperative at these sites. The major difference between the median eminence, neurohypophysis, and pineal gland and the other circumventricular organs is that the neurohemal junctions in the median eminence, neurohypophysis, and pineal gland facilitate the transport of neurohormones from the nerve terminals or nerve cell derivatives (pineal) into the bloodstream, whereas at the other circumventricular or-

hypophysial portal vessels from the hypothalamus to the anterior pituitary gland, where they either stimulate or inhibit the release of anterior pituitary hormones. Synthesized in nerve cells of the hypothalamic nuclei, the neurohormones are released from nerve terminals into the plexus of primary capillaries of the hypophysial portal vessel system. These capillaries are derived from the superior hypophysial arteries and coalesce to form the hypophysial portal veins which run on the surface or through the pituitary stalk to the anterior pituitary gland, where they form a secondary plexus of vessels called the pituitary sinusoids. The vessels on the surface of the stalk are the long portal vessels, whereas those within the substance of the stalk are the short portal vessels.

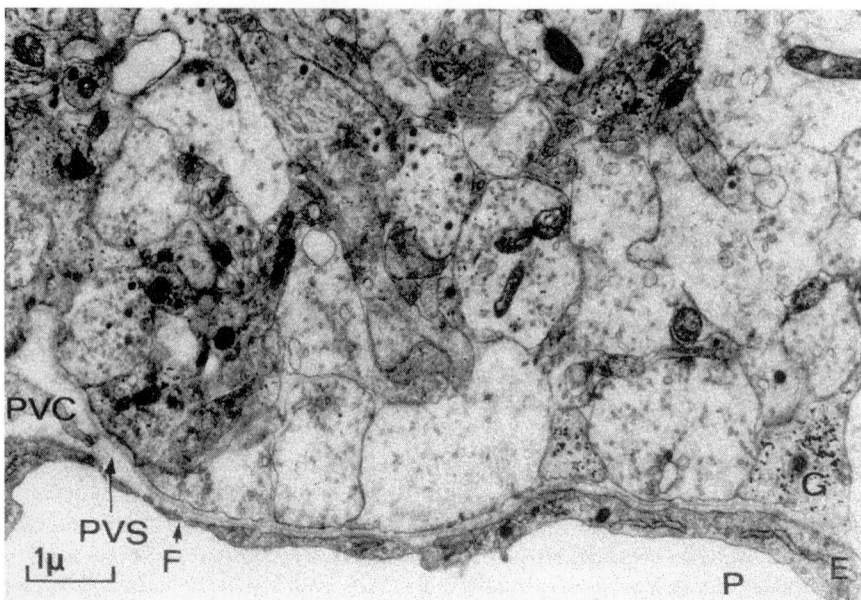

FIGURE 6 Electromicrograph of the external layer of the median eminence of a rat at the first postnatal day. Note the high density of nerve terminals around part of a primary portal capillary vessel (P), which is fenestrated (F). Note also the large number of agranular and granular vesicles in the nerve terminals. These vesicles contain the packets (quanta) of neurohormone or neurotransmitter which are released on nerve depolarization as a consequence of nerve action potentials. The neurohormones are released into the perivascular space and from there move rapidly into portal vessel blood for transport to the pituitary gland. This arrangement is typical of neurohemal junctions found in the several circumventricular organs of the brain. E, endothelial cell; F, fenestration; G, glial process; P, portal vessel; PVC, perivascular cell; PVS, perivascular space. Magnification, ×13,200 (reproduced with permission from G. Fink and G. C. Smith, Z. Zellforsch. 119, 208–226, 1971).

gans, the neurohemal junctions facilitate the transport of neurohormones from the blood to nerve cells. The latter mechanism has been implicated in the "cross talk" between peripheral organs and the brain so that the peptide angiotensin, for example, increases blood pressure by activating neurons of the subfornical organ which have a high density of angiotensin receptors. In addition to their importance as sites for the transfer of chemical messengers from nerve terminals into blood vessels or vice versa, the neurohemal junctions of circumventricular organs are also sites at which drugs or toxins, which are normally excluded by the blood–brain barrier, can penetrate the brain and affect brain function.

D. Hypothalamic Neurohormones

1. General

Some of the hypothalamic neurohormones released into hypophysial portal blood are also released at synapses in brain and therefore can also serve as "neurotransmitters." Most of the neurohormones are peptides, synthesized in discrete hypothalamic nuclei, which mediate the neural control of anterior pituitary hormones. The neurohormone-secreting neurons are "the final common pathway" neurons for neural control of the anterior pituitary gland, a term borrowed by Harris from Sherrington's description of the α motor neurons of the spinal cord which innervate and control the contraction of skeletal muscles. Like the α motor neurons, the hypothalamic neurons are controlled by inputs to the hypothalamus from the brain stem and midbrain as well as from higher brain centers. The hypothalamic neuroendocrine neurons are therefore connected with many other regions of the nervous system and in particular the components of the limbic system, which is involved in several important higher brain functions including emotion, olfaction, and memory.

The neural control of all established anterior pituitary hormones is mediated by at least one or more neurohormones. In some cases the neurohormones may act synergistically, as is the case for adrenocorticotropin (ACTH), the release of which is stimulated by both the 41-amino acid (aa) residue peptide, corticotropin-releasing hormone (CRH), and the nonapeptide, arginine vasopressin (AVP). In other cases the neurohormones act antagonistically, as is the case for growth hormone (GH), the release of which is stimulated by the 44-aa peptide, GH-releasing hormone (GHRH), and inhibited by the 14-aa peptide, somatostatin.

2. Gonadotropin-Releasing Hormone

The anterior pituitary hormones which are of special relevance to reproduction are follicle-stimulating hormone (FSH), luteinizing hormone (LH), and prolactin. As discussed in greater detail elsewhere in this encyclopedia, the primary function of these hormones in the female is to stimulate the development of ovarian follicles and estrogen secretion (FSH); ovulation, corpus luteum formation, and progesterone secretion (LH); and breast development and lactation (prolactin). In the male, FSH stimulates spermatogenesis, LH stimulates testosterone secretion, and the function of prolactin is unknown. The release of LH and FSH is stimulated by one and the same decapeptide—gonadotropin-releasing hormone (GnRH). Despite intensive investigations carried out over more than 30 years, no specific FSH-releasing hormone has been discovered. Nonetheless, because it is extremely difficult to prove the nonexistence of a factor, some journals and authors prefer the term LH-releasing hormone (LHRH), which leaves open the possibility of the existence of a separate FSH-releasing hormone. The release of FSH is inhibited by a large peptide (inhibin) which is synthesized by the gonads. FSH secretion is also influenced by other peptides, such as activin and follistatin. In addition to stimulating the release of LH and FSH, GnRH also induces their synthesis, maintains the structure and function of the gonadotropes, and has the apparent unique property among the neurohormones of being able to increase the responsiveness of the anterior pituitary gland to itself—the "self-priming" effect of GnRH. In women, at the time of the midcycle, when the pituitary gland has already been primed by the endogenous surge of estradiol-17β, one pulse of GnRH can increase the LH response to a second pulse administered 2 hr later by more than twofold. In the proestrous rat, the GnRH-induced increase in pituitary responsiveness to a second pulse of GnRH administered 1 hr after the first is about sevenfold. The self-priming effect of GnRH also enables the ovulatory surge of LH to be triggered by a series of GnRH pulses as well as the GnRH surge.

3. Prolactin Control: Dopamine

In contrast to other pituitary hormones, the brain predominantly inhibits prolactin secretion. Thus, pituitary grafts under the kidney capsule, where they are removed from the influence of the high concentrations of hypothalamic neurohormones present in the hypophysial portal vessels, undergo atrophy and stop secreting all known pituitary hormones with the exception of prolactin, the secretion of which is increased. However, despite the fact that this discovery was made 40 years ago by Everett and Nikitovitch-Winer, the neurohormone(s) that regulates prolactin release under physiological conditions remains to be established. Pharmacologically, dopamine and its agonists are powerful inhibitors of prolactin release, and the potent dopamine agonist, bromocriptine, and its modern derivates are highly effective in reducing high plasma concentrations of prolactin (hyperprolactinemia) which are relatively common in women, often produced by benign prolactin-secreting tumors of the pituitary gland, and are associated with infertility. These and some experimental observations have led to the view that dopamine is the prolactin-inhibitory neurohormone. If this is eventually proven, then dopamine will be the only nonpeptide hypothalamic–pituitary regulatory neurohormone. Pharmacological studies have shown that prolactin release can be stimulated by the tripeptide, thyrotropin-releasing hormone (TRH), and also by vasoactive intestinal peptide, but again there are no robust data which show that these neurohormones are effective under physiological conditions.

4. Site of Neurohormone Synthesis

The neurohormones are synthesized in different hypothalamic nuclei. Thus, GnRH neurons are located mainly in the medial preoptic area, from where they project to the median eminence as well as to the organum vasculosum of the lamina terminalis. The function of the latter remains unresolved. The GnRH neurons migrate to the medial preoptic area from the olfactory placode: Failure of this migration, caused by a genetic defect, underlies Kallman's syndrome, which results in hypogonadism and infertility. Dopamine, the putative prolactin-inhibiting neurohormone, is synthesized in the arcuate nucleus of the hypothalamus, which sends a dense projection of dopamine-containing axons to the median eminence as well as down the neural stalk to the intermediate lobe of the pituitary gland. In addition to dopamine's putative role in inhibiting the release of prolactin, the dopamine projections may also play a role in modulating the release of other neurohormones, GnRH in particular, by a "presynaptic" action at the nerve terminals in the median eminence. Dopamine also mediates the neural control of melanocyte-stimulating hormone (MSH).

5. Teleological Advantages of Neurohormonal Control

The hypothalamic–pituitary axis is another illustration of the remarkable economy of physiological systems. First, and perhaps most impressive, are the exquisite hypophysial portal vessels which, by transporting the neurohormones from the hypothalamus to the pituitary gland (undiluted by mixture in the systemic circulation), ensure that the release of small amounts of hypothalamic neurohormones will reach the pituitary gland at concentrations sufficient to exert their effects. The importance of this fact is demonstrated by the fact that hypothalamic concentrations of the hypothalamic–anterior pituitary neurohormones, such as GnRH, are about three orders of magnitude lower than those of the neurohypophysial nonapeptides, vasopressin and oxytocin, which reach their peripheral targets by the systemic circulation. Second, the transport of neurohormones at effective concentrations by the hypophysial portal vessels also protects the body from potential adverse effects of the high concentrations of the neurohormones necessary to affect pituitary hormone secretion. Thus, for example, the high portal blood concentrations of somatostatin may, in the systemic circulation, have adverse effects on the gut and insulin secretion by the β cells of the pancreas. Similarly, the portal plasma concentrations of atrial natriuretic peptide which inhibit ACTH secretion would, in the systemic circulation, cause a lethal drop in blood pressure. Third, the neurohormones in the hypothalamo–adenohypophysial system are chemical messengers that are also utilized in other systems. Thus, all the neuropeptides of the hypothalamic–pituitary system have been implicated as neurotransmitters, neuromodulators, or neurotropins elsewhere in the nervous system, although robust evidence for their precise function needs to be established. Somatostatin stands out in this respect in that it is secreted by cells of the pancreatic islets and inhibits insulin secretion. GnRH is also present in the placenta, and at high concentrations it affects the gonads and has also been implicated in mating behavior.

E. The Hypothalamic–Pituitary–Gonadal System

A brief account of the functioning of the hypothalamic–pituitary–gonadal axis is given here to consolidate the several points made previously on the neurohumoral control of the pituitary gland and to introduce some aspects of steroid hormone action on the brain. For most of reproductive life, the secretion of LH and FSH is modulated by a negative feedback action of gonadal steroids—estrogen in the female and testosterone in the male. The secretion of FSH is further modulated by the peptides, inhibin, activin, and follistatin. The physiological power of steroid hormone negative feedback is shown by the marked increase in the plasma concentrations of LH and FSH which occurs after menopause, when the ovary stops secreting estrogen.

The female menstrual cycle in the human and estrous cycle in the rat are punctuated by a massive surge of LH which triggers ovulation (Fig. 7). The LH surge, in turn, is triggered by the spontaneous surge of estradiol-17β which occurs during the late follicular phase in the human and reaches a peak at noon of proestrus in the rat. This positive feedback action of estrogen involves (i) a direct action of estradiol-17β on the brain to stimulate the surge release

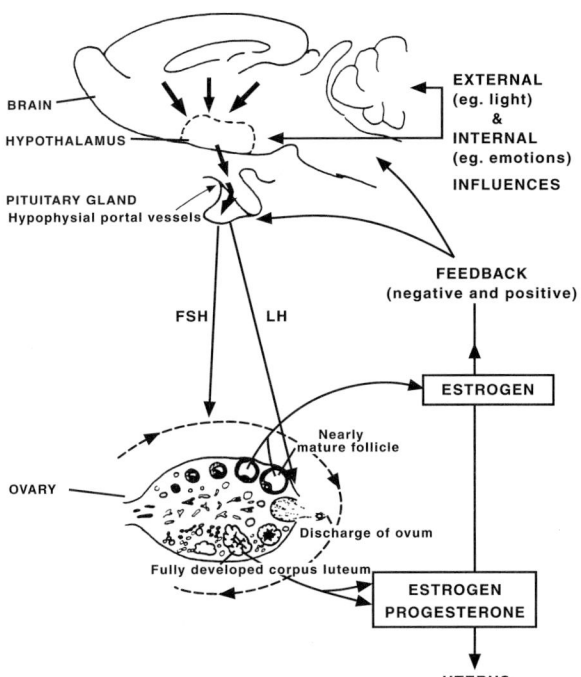

FIGURE 7 Schematic diagram of the control of the ovarian cycle by follicle-stimulating hormone (FSH) and luteinizing hormone (LH) released from the anterior pituitary gland. The secretion of LH and FSH is controlled by the brain by way of GnRH, a decapeptide that is released from hypothalamic neurons into the hypophysial portal vessels. The release of GnRH from hypothalamic neurons is influenced by external and internal factors acting by way of central nervous pathways, and the system is regulated by positive- and negative-feedback control involving estrogen and progesterone secreted by the ovary. Estrogen and progesterone act on the uterus to prepare the endometrium for implantation of the zygote should fertilization occur. Not shown for the sake of clarity is the peptide inhibin, which is secreted by ovarian follicles and inhibits FSH release. FSH secretion is also influenced by activin and follistatin (reproduced with permission from G. Fink, 1988b).

of GnRH and/or to increase the pulse frequency of GnRH release, and (ii) an increase in pituitary responsiveness to GnRH. The latter, which in the human and the rat is on the order of 20- to 50-fold, is pivotal for the occurrence of the ovulatory surge since the amount of GnRH released during the surge or pulses is far too small to release an ovulatory surge of LH were it not for the enormous amplification of the system produced by estrogen and the self-priming effect of GnRH. GnRH self-priming serves also to coordinate the release of GnRH with the increase in pituitary responsiveness to GnRH so that both events reach a peak at the same time and thereby ensure the occurrence of the ovulatory LH surge (Fig. 8).

F. GnRH Surge and Pulse Generators

Numerous studies in the rat have shown that the integrity of the medial preoptic area and the nearby suprachiasmatic nuclei is essential for the occurrence of regular estrous cycles and the spontaneous surge of GnRH and LH. This has led to the concept that this most rostral area of the hypothalamus (Fig. 2) is the site for the GnRH surge generator. Pulsatile GnRH release is thought to be generated by the medial basal hypothalamus, the pulse generator. The rhesus monkey differs from the rat in that the LH surge and ovulation can occur even after total surgical isolation of the medial basal hypothalamus. Experimental studies suggest that in the rhesus monkey, the LH surge is triggered by pulses of GnRH with a frequency of about one per hour, which is the optimal frequency for GnRH self-priming in an estrogen-primed rat and similar to that in the human female toward midcycle. This pulse frequency is also optimal for generation of a LH surge in the estrogen-primed human female. Thus, while the occurrence and importance of the spontaneous GnRH surge cannot be discounted in the human, these observations indicate that in the monkey and the human the GnRH pulse generator may play a pivotal role in the control of ovulation. The GnRH pulse generator also plays a crucial role in the onset of puberty and in the control of seasonal reproduction.

The neural control of both the pulse and the surge generators is far from understood, but evidence suggests that several stimulatory (e.g., serotonergic, noradrenergic, and peptidergic) and inhibitory (e.g., GABA and endogenous opioids) mechanisms control the GnRH surge and GnRH pulses. The existence of several primary and/or backup systems for GnRH control might be expected since the control of GnRH release is central for the reproduction of mammalian species.

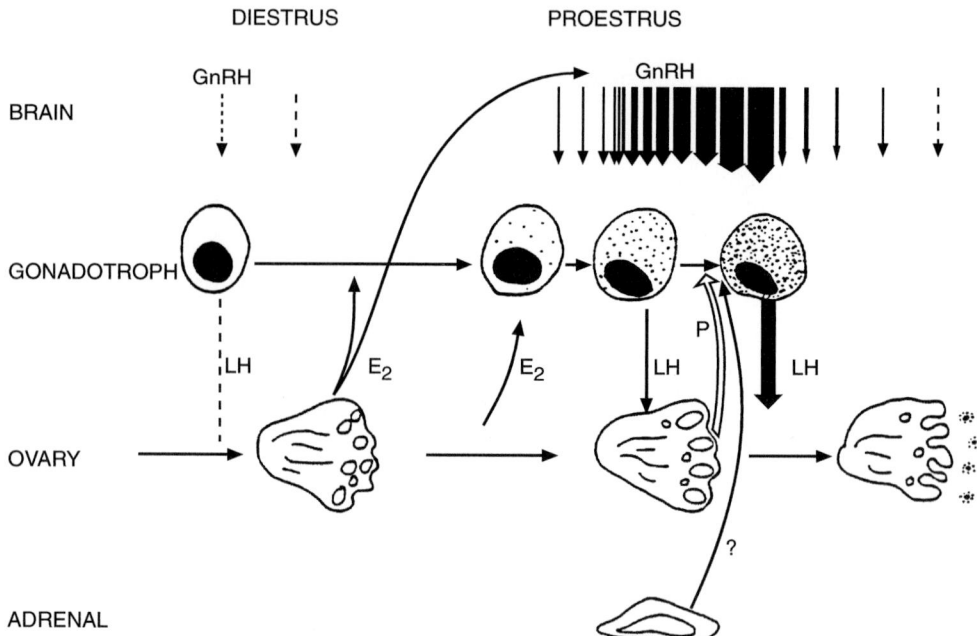

FIGURE 8 Schematic diagram which shows the cascade of events which generate the spontaneous ovulatory LH surge in the rat. The increase in plasma concentrations of estradiol-17β (E2) increases the responsiveness of the pituitary gonadotropes (increased stippling) to GnRH and also triggers the surge of GnRH. Pituitary responsiveness to GnRH is further augmented by the priming effect of GnRH, the unique capacity of the decapeptide to increase pituitary responsiveness to itself. Progesterone (P) secreted by the ovary in response to the LH released during the early part of the LH surge may also enhance pituitary responsiveness to GnRH. The priming effect of GnRH coordinates the surge of GnRH with increasing pituitary responsiveness so that the two events reach a peak at the same time. The conditions are thereby made optimal for a massive surge of LH. This cascade, which represents a form of positive feedback, is terminated by destruction of a major component of the system in the form of the rupture of the ovarian follicles (ovulation). The human female shows a similar increase in pituitary responsiveness and GnRH self-priming just before the midcycle LH surge.

III. PINEAL GLAND AND PHOTOPERIODIC CONTROL OF REPRODUCTION

The pineal gland is a circumventricular organ (Fig. 1) which secretes melatonin into the circulation. The gland has held the fascination of scientists for many years because the secretion of melatonin is exquisitely sensitive to light, pinealocytes in submammalian species are photoreceptors, and the gland offers an excellent experimental model for studies of the transduction of light into nerve impulses and neurohormone secretion. The secretion of melatonin is thought to play a key role in the photoperiodic control of reproduction in some species, but the precise mechanism of action of this hormone remains unknown.

The outer segment (sensory pole) of the pinealocyte in fish, amphibia, and reptiles has all the ultrastructural characteristics of a true photoreceptor, but these features are only vestigal in mammals and intermediate forms exist in birds. In fish, amphibia, and reptiles, the effector pole of the pinealocyte "synapses" with secondary pineal neurons that give rise to the pineal tract, which propagates signals to the central nervous system. In birds and mammals, however, the pinealocytes secrete melatonin directly into the circulation (or cerebrospinal fluid) in a neuroendocrine manner.

Melatonin, a derivative of serotonin, is synthesized within the pinealocytes in two steps. First, serotonin is converted by the rate-limiting enzyme, N-acetyltransferase (NAT), to N-acetyl serotonin, which is then converted to melatonin by hydroxyindole-O-methyltransferase. Since little if any melatonin is stored, the rate of melatonin secretion is tightly linked to its synthesis, which depends on NAT action which in turn depends on noradrenaline release from the dense sympathetic innervation of the gland. In mammals, the control of melatonin secretion by light is mediated by a multisynaptic pathway which starts at the retina of the eye and successively involves synapses in the suprachiasmatic nuclei, the PVN, the intermediolateral column of the spinal cord, and the neurons of the superior cervical ganglion of the sympathetic nervous system. Sympathetic terminals in the pineal gland release noradrenaline, which stimulates melatonin secretion by an action on adrenoreceptors on pinealocytes. Cyclic AMP seems to be the main intracellular second messenger which mediates the action of noradrenaline in this system, and recent studies suggest that the main point of regulation is NAT, which, depending on the species, can be affected at the level of NAT gene expression as well as by posttranslational modifications which alter the activity of the enzyme.

The secretion of melatonin starts with the onset of the dark period (night) and stops with the onset of the light period (day). The secretion of melatonin during the dark period is stopped abruptly by exposure to light. In blind persons the secretion of melatonin takes on the typical 25-hr free-running period. These and other data suggest that the secretion of melatonin is predominantly controlled by light exposure superimposed on the intrinsic rhythm of the major neural clock, the suprachiasmatic nuclei.

The action of melatonin is predominantly inhibitory with respect to reproduction, and the effects of pinealectomy and manipulation of melatonin levels are most pronounced in seasonal breeding animals such as the wallaby, hamster, vole, and sheep. However, the precise role of melatonin in reproduction has yet to be established.

It is well established that the suprachiasmatic nuclei constitute the central generator of circadian rhythms of the body and that the functional integrity of these nuclei are indispensable for normal reproductive rhythms. The relative importance of the pineal gland and its precise role in reproductive control in relation to the suprachiasmatic nuclei await to be determined. With the discovery of genes that regulate circadian rhythms and melatonin synthesis the prospect of defining what is likely to be a remarkably elegant and robust molecular control system seems excellent.

IV. THE HYPOTHALAMO–NEUROHYPOPHYSIAL SYSTEMS

A. Overview

The neural lobe is the site of release of the nonapeptides, oxytocin and vasopressin, into the systemic circulation. Presumably because a considerable distance removes the target cells of these peptides from their site of release in the neural lobe, and because there is massive dilution in the systemic circulation, the rate and amount of synthesis of oxytocin and vasopressin and their content in the hypothalamus is about three orders of magnitude greater than that of the neurohormones in anterior pituitary control. This fact, together with their synthesis in discrete hypothalamic nuclei, their ease of assay, and the fact that both nonapeptides contain disulfide bridges which allow incorporation of the radioactive tracer, ^{35}S-cystine, contributed to four important landmarks of neuroendocrinology. First, oxytocin and vasopressin were the first of the hypothalamic neurohormones to be sequenced (by Du Vigneaud and colleagues). Second, the glycoprotein components of their precursor proteins led to their conspicuous staining and thereby to the concept of neurosecretion, the term applied to neurons whose main purpose is to secrete neurohormones rather than transmit signals by propagated action potentials. Third, the fact that both peptides have disulfide bridges and their synthesis involves the incorporation of cystine, which can be labeled with the relatively high-energy radionuclide ^{35}S, made them excellent models for the first studies of neuropeptide synthesis and transport (by Sachs and associates). Fourth, oxytocin and vasopressin-containing neurons were the first neuroendocrine neurons from which electrophysiological recordings were made (by Cross and Green).

Oxytocin and vasopressin-containing neurons facilitate electrophysiological recording because they are large and because their axons terminate in the surgically accessible neural lobe. This makes it relatively easy to locate and verify by antidromic stimulation the identity of the neurons and record from them. Furthermore, the electrophysiological activity of the hypothalamic magnocellular neurons can be correlated with secretion of vasopressin and oxytocin. Finally, the neural lobe is a "bag" of nerve terminals and, therefore, proved to be a useful model for the study of excocytosis and stimulus-secretion coupling—that is, the coupling between the cascade of ion fluxes through membrane channels triggered by action potentials and the calcium-dependent release of neurohormones or neurotransmitters. The neurohormones, like all other known neurotransmitters and peptide and protein hormones, are packaged in secretory vesicles (Fig. 6) which are released in packets or "quanta."

B. Oxytocin: Lactation and Parturition

Oxytocin, concerned mainly with milk ejection during lactation and parturition (the birth process), provides the perfect example of a neuroendocrine reflex. Oxytocin is the neurohormonal component of the milk ejection reflex whereby suckling at the nipple of lactating mothers triggers volleys of impulses which travel through the mammary nerves to the spinal cord and by way of a multisynaptic pathway reach the hypothalamus, where they trigger the release of oxytocin. Oxytocin transported by the systemic circulation stimulates the contraction of the myoepithelial cells of the breast acinar resulting in milk ejection. During parturition, oxytocin coordinates and reinforces uterine contractions. Here too a reflex is involved in that as uterine contractions force the head of the fetus against the cervix, volleys of impulses are triggered which ascend through multisynaptic pathways involving the pelvic nerves and the spinal cord to the hypothalamus to trigger the release of oxytocin, which acts on the smooth muscle cells of the uterus. This is a classical positive feedback system.

Vasopressin (or the antidiuretic hormone) is concerned mainly with the control of body water, although, as its name implies, it also induces vasoconstriction and thereby can increase blood pressure. The vasopressin cells of the supraoptic and paraventricular nucleus respond to osmotic stimuli—an increase in plasma osmolality triggers the release of vasopressin which increases water reuptake in the nephron, as a consequence of which there is an overall increase in body water with a fall in plasma osmolality. This is a perfect example of a homeostatic regulatory mechanism. In addition to its role in osmoregulation, vasopressin synthesized in the smaller (parvicellular) neurons of the PVN acts synergistically with CRH to release ACTH.

V. THE INTERMEDIATE LOBE OF THE PITUITARY GLAND

The major secretion of the intermediate lobe of the pituitary gland is MSH, a 13-amino acid residue peptide which together with ACTH and β-endorphin is derived from the precursor proopiomelanocortin (POMC). In addition to αMSH, the pars intermedia also contains other derivatives of POMC: βMSH, γMSH, CLIP, and β-endorphin. The fact that in the pars distalis ACTH is the major hormone derived from posttranslational processing of POMC, whereas in the pars intermedia MSH is the major active hormonal product of POMC processing, reflects the presence of different enzymatic processing pathways in the two parts of the gland. The release of MSH is inhibited by dopaminergic neurons that originate in the arcuate nucleus and reach the intermediate lobe by way of the neural stalk. Man is conspicuous among mammals in that the pars intermedia is not defined as a separate lobe of the human pituitary gland.

VI. HORMONAL EFFECTS ON THE NERVOUS SYSTEM

While neuropeptides such as angiotensin have potent effects on brain function, steroid and thyroid hormones, which are secreted by the three major pituitary target organs, have by far the most prominent effects on the brain. The effects of the steroid and thyroid hormones may be classified in terms of (i) feedback actions, (ii) brain differentiation and neural plasticity, (iii) neurotransmission, and (iv) membranes and ion channels.

A. Feedback Actions

The feedback actions of steroid and thyroid hormones have been known since the 1930s. There are two types of feedback: negative and positive.

1. Negative Feedback

Negative feedback is deployed in most systems of the body, and its "purpose" is homeostasis (a term introduced by Walter Cannon); that is, to maintain the functioning of a system at a constant predetermined level. Man, in the design of machines and control systems, has borrowed the concept of negative feedback from biology. The analogy most often used to explain negative feedback is central heating of a house. Central heating systems are composed of a heater which is controlled by a thermosensitive device that can be set to maintain the house at a certain temperature. The thermosensor is composed of a detector, comparator, and a drive. If the temperature of the house drops below the preset temperature, the thermosensor switches on the heater. Once the preset temperature has been reached or exceeded (e.g., due to high ambient temperature) the thermosensor switches the heater off. Similarly, the range of the blood concentrations of gonadal steroid hormones (especially estrogen, testosterone, and progesterone), adrenal corticosteroids (especially cortisol in man), and thyroid hormones (thyroxine and triiodothyronine) have been preset at levels consistent with requirements for normal body function. The sensor-comparator which "measures" and compares the hormonal level with the preset level and the drive or regulator are probably located within the hypothalamic–pituitary unit, and the setting is probably determined genetically. Decrease of the gonadal, adrenal, or thyroid hormones below the preset level results in the increased secretion of the pituitary gonadotropins, ACTH or TSH, which increase the synthesis and release of gonadal steroids, adrenal steroids or thyroid hormone, respectively. Increased levels of the target hormones above the preset level reduces or inhibits the secretion of the corresponding pituitary "tropic" hormone—hence "negative feedback." Interruption of negative feedback results in oversecretion of the pituitary hormones as occurs, for example, at menopause, when gonadal steroid concentrations in blood are low or absent and pituitary gonadotropin concentrations reach very high levels.

The negative feedback actions of the pituitary target hormones have been used extensively in the clinic. The negative feedback of estrogen and progesterone, for example, is the basis of the "contraceptive pill," different formulations of which are used widely to block the secretion of FSH and LH and thereby prevent the development of ovarian follicles and ovulation.

2. Positive Feedback

Positive feedback, in which increased output of the system increases the drive of the regulator, is far less common than negative feedback. The likely reason for this is that positive feedback can only be terminated by destruction of a component of the feedback loop. The replacement of negative by positive feedback in the hormonal systems mentioned previously or in the blood glucose or blood pressure control systems would be deleterious and possibly lethal. Nevertheless, positive feedback is essential for two crucial events in reproduction: ovulation and parturition. In both systems, termination of the positive feedback cascade is in fact associated with the "destruction" of the output component of the system—rupture of the ovarian follicle and ovulation in the case of estrogen positive feedback and expulsion of the fetus in the case of the parturition–oxytocin reflex. However, while in the latter the cause of feedback termination is clear (i.e., expulsion of the fetus prevents further ascending nerve volleys from the uterine cervix which are necessary for triggering the reflex release of oxytocin), the factors which determine the cessation of the ovulatory LH surge have yet to be determined.

B. Brain Differentiation and Plasticity

Thyroid and gonadal steroid hormones exert both reversible and irreversible effects on brain structure, connectivity, and synapses. Thus, if left untreated, congenital lack of thyroid hormone results in irreversible cretinism due to serious defects in brain development. Reduced secretion of thyroid hormones in adult life (myxedema) also results in cognitive and other neurological deficits which can be reversed by thyroid hormone administration. Here, attention will be focused on gonadal steroids and their irreversible effects on sexual differentiation of the brain as well as reversible effects on certain neuronal systems.

1. Sexual Differential of the Brain

The early studies of Steinach and Pfeiffer showed that in the rodent the differentiation of neural control of reproductive function (cyclical in female and acyclical in the male) is determined by exposure to sex steroids rather than the genetic sex of the individual. Thus, transplantation of testes to genetic female rodents before a critical period of brain development permanently abolishes estrous cycles and ovulation and induces male behavior. The classical studies with pituitary grafts carried out by Harris and Jacobsohn and confirmed by Adams, Smith, and Peng showed that this action of the gonads was on the brain. Androgens were found to be as effective as the testis in producing masculinization of the brain. By the early 1960s it was clear that, irrespective of the genetic sex, in the rodent as well as in several other mammals the brain is at first neuter or feminine, and in the male it is converted to the masculine form by exposure to androgens either *in utero* or during the early neonatal period.

The effect of testosterone is mediated by its enzymatic conversion by aromatase to estradiol-17β. The female brain is thought to be protected from the masculinizing effects of circulating estradiol by the presence of α-fetoprotein, which has a very high affinity for estrogen. Male animals in which the brain is protected from the masculinizing actions of androgens by castration shortly after birth have the capacity to show estrous cyclicity, as assessed by the formation of corpora lutea in grafted ovaries, as well as female mating behavior.

The mechanism of action of androgens and estrogens in masculinizing the brain has not been established. Estrogen has been shown to stimulate neurite outgrowth from hypothalamic explants in culture, and ultrastructural studies demonstrated sex differences in synapse formation in hypothalamus. The most striking morphological sex difference in the mammalian brain is the sexually dimorphic nucleus of the preoptic area. In the male rat this nucleus is about three to five times the size of that in the female, and the size of the nucleus in the female can be converted to that in the male by the administration of androgens before the critical period of brain development (Postnatal Day 5 in the rat). However, despite its anatomical prominence, the function of the sexually dimorphic nucleus remains unknown. More meaningful effects of sex steroids on sexual differentiation of the brain come from studies of the sex differences in the telencephalic nuclei that control song in some songbirds because the functional anatomy of these nuclei is reasonably well understood. Androgens switch the song system on, but it seems that the central nuclei which control song are only responsive to androgen in adults if the birds have been exposed to a surge of androgen or estrogen at hatching. The female brain in birds may be protected from the masculinizing effects of androgens by enzymatic inactivation.

What significance do these findings have for man? In man sexual outlook and behavior is governed more by gender assignment and social factors and, therefore, the precise role and importance of sexual differentiation of the brain by sex steroids is more difficult to identify. However, there are three genetic deficiencies which suggest that androgen exposure *in utero* and possibly during the early neonatal period does influence sexual differentiation of the brain in man. In the adrenogenital syndrome, the adrenal gland secretes large quantities of androgen during fetal and postnatal life due to a mutation in one of several enzymes. A cohort of women with the adrenogenital syndrome followed at Johns Hopkins hospital for about 20 years showed that, although most have normal menstrual cycles, they also have a greater degree of male characteristics on psychosocial tests compared with women with androgen insensitivity or Müllerian duct aplasia. The testicular feminization or androgen-insensitivity syndrome, which occurs also in cattle, rats, and mice, is the second genetic deficiency. Individuals with the disorder are genotypically male but phenotypically female as a consequence of androgen receptors being absent or defective. Thus, although testes are present and secrete androgens, the tissues cannot respond to testosterone or 5α-dihydrotestosterone. As a consequence of normal plasma concentrations of testicular estrogens, the individual undergoes a "feminizing" puberty which results in the development of breasts and female appearance but no pubic or axillary hair. Individuals with testicular feminizing syndrome show feminine behavior, as would be predicted from androgen-deprived rodents. Third, testosterone is normally converted in the periphery to 5α-dihydrotestosterone, a powerful androgen. This steroid cannot masculinize the brain but is important for masculinizing the genitalia. Imperato-McGinley *et al.* carried out an important study, in the Dominican Republic, on pseudohermaphrodites with a deficiency of 5α-reductase, which converts testosterone to 5α-dihydrotestosterone. At birth, the external genitalia of these individuals are female in appearance and consequently the affected individuals are raised as girls. Under the influence of testosterone secreted in large amounts at puberty, the external genitalia become male in appearance, and despite the fact that they have been brought up as girls, the individuals take on a male gender role. This led Imperato-McGinley *et al.* to conclude that exposure of the brain to testosterone *in utero*, neonatally, and at puberty appears to contribute substantially to the formation of male gender identity. The syndrome of 5α-reductase deficiency is in keeping with the rodent model of sexual differentiation of the brain which predicts that 5α-dihydrotestosterone plays no significant role in masculinizing the brain; the brain in this syndrome will have been masculinized by testosterone or estrogen. These three genetic deficiencies, together with other data, suggest that although the effect of testosterone in man is not as dramatic in sexual differentiation of the brain as it is in the rodent, behavior and psychosocial attitudes may be permanently influenced by the nature of steroid exposure of the brain during early development.

2. Neural Plasticity: Long-Term Reversible Effects of Steroids

The bed nucleus of the stria terminalis (BNST) provides a dramatic example of the effects of sex steroids on neural plasticity. The stria terminalis is the main nerve fiber tract that connects the hypothalamus with the amygdala, an important component of the limbic system concerned with emotion, olfaction,

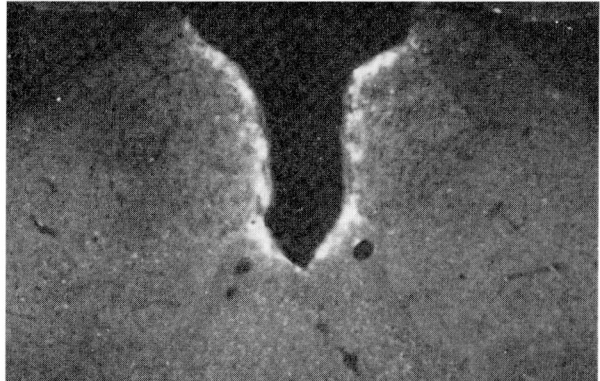
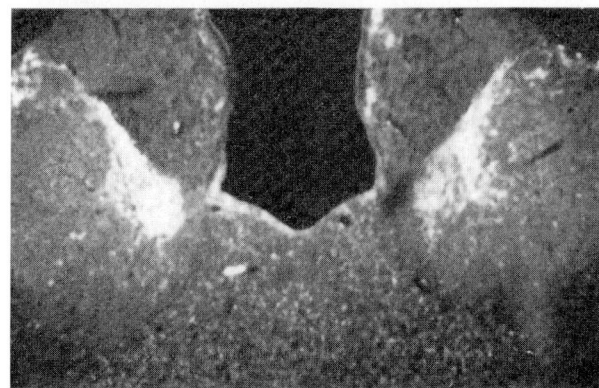

FIGURE 9 Photomicrographs of coronal sections taken through the habenula of the hypogonadal (*hpg*) mouse, a mutant deficient in GnRH and therefore in estrogen or testosterone. (Bottom) Section taken from an *hpg* mouse treated with testosterone. (Top) Section from an untreated *hpg* mouse. Note the high density of AVP-containing fibers in the lateral habenula of the mouse treated with testosterone. Similar results were obtained by treatment with estrogen and by the transplantation of a hypothalamic graft from a normal mouse into the third ventricle of an *hpg* mouse. The development of the dense plexus of AVP terminals in the lateral habenula of the *hpg* mouse treated with testosterone is due to stimulation of AVP gene transcription shown in Fig 10 (data modified with permission from C. Mayes *et al.*, *Neuroscience* 25, 1013–1022, 1988).

aggression, and the control of related behaviors and central neuroendocrine systems. Equivalent to a junction box within the stria terminalis, the BNST contains neurons, concerned with olfactory or "social" memory, that project to the lateral habenula and lateral septum. These neurons utilize arginine vasopressin (AVP) as a neurotransmitter. The expression of AVP in the BNST, but not in the PVN or

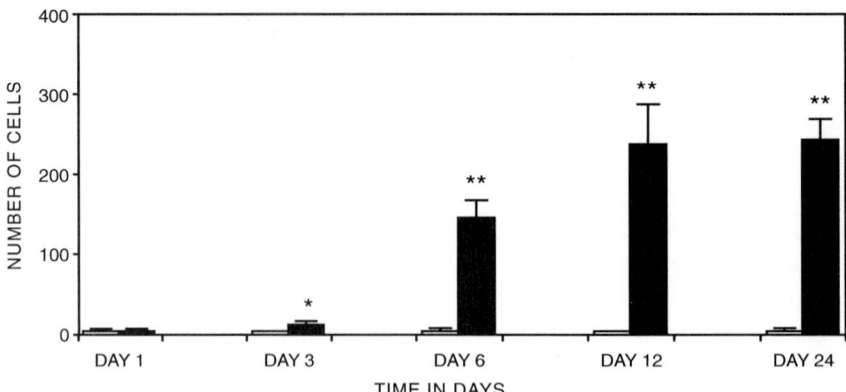

FIGURE 10 The dramatic stimulatory effect of testosterone on expression of the AVP gene in the bed nucleus of the stria terminalis (BNST). Mean (± SEM) number of cells expressing AVP mRNA in the BNST of hypogonadal mice at different times after implanting either empty (open bars) or testosterone propionate-containing silicone elastomer capsules (solid bars). Significance of differences (Mann–Whitney U test): $*p < 0.05$; $**p < 0.01$ (reproduced with permission from R. Rosie et al., Mol. Cell. Neurosci. 4, 121–126, 1993).

SON of the hypothalamo–neurohypophysial system, is dependent on normal levels of testosterone or estrogen. Thus, the AVP concentrations in the BNST neurons fall to undetectable levels after castration or ovariectomy and can be restored to normal by the administration of either of these two sex steroids. As in the case of sexual differentiation of the brain, the action of testosterone is dependent on its conversion to estrogen, and the action of estrogen or testosterone is mediated by AVP gene transcription (Figs. 9 and 10). The reason for the exquisite sensitivity to sex steroids of the BNST, but not the PVN or SON, remains to be established, but perhaps one relevant difference is that the BNST contains high concentrations of both α and β estrogen receptor and aromatase, whereas the AVP and SON contain only relatively low concentrations of the β estrogen receptor and little if any aromatase.

C. Effects of Sex Steroids on Neurotransmission

Many studies have demonstrated that adrenal and gonadal steroids affect the turnover of monoamine neurotransmitters by actions on the enzymes that synthesize and metabolize monoamines. Recent studies in rodents have shown that estrogen increases the expression of the genes for the serotonin 2A receptor and serotonin transporter in the dorsal raphe nucleus of the midbrain, with a concomitant increase in the density of serotonin 2A receptors and serotonin transporter sites in higher forebrain centers which in man are concerned with the control of mood, mental state, emotion, and memory. Estrogen also affects the density of dopamine 2 receptors in the striatum of the brain. These findings provide a rational biological basis for the depressive symptoms which occur in some women premenstrually, around the time of the menopause, or after childbirth, times at which estrogen levels fall precipitously. The effect of estrogen on serotonin 2A receptors may also help to explain the sex differences in schizophrenia.

D. Effects on Cell Membranes and Ion Channels

The classical mechanism of steroid action involves transport of the steroid into the cell cytoplasm, where it binds to and activates steroid receptors which in turn bind to a specific binding site (steroid response element) of the promoter region of a gene. The activated steroid receptor thereby stimulates or suppresses gene expression. In addition to this classical mechanism of steroid action, some steroid effects are likely to involve rapid effects on membranes and membrane receptors. Progesterone and its cogeners, for example, activate the GABAA receptor and thereby potentiate the influx of chloride ions, which results in marked hyperpolarization of the cell. This action of progesterone and its derivatives explains its anesthetic/sedative effect which, first noted by Hans Selye, led to the development of alpha-xalone (3α-hydroxy-5α pregnane-11,20 dione), a progesterone derivative which is a potent anesthetic. The apparent sedative effect of progesterone may explain why some women feel remarkably calm during the second and third trimesters of pregnancy when progesterone levels are elevated. Similarly, progesterone may also exert an anxiolytic-like effect during the luteal phase of the menstrual cycle. Electrophysiological and other studies show that some of the effects of estrogen may also be mediated by actions at the cell membrane as well as through its cytoplasmic receptors.

See Also the Following Articles

GnRH Pulse Generator; Neurohypophysial Hormones; Neurotransmitters; Oxytocin; Photoperiodism, Vertebrates; Prolactin Secretion, Regulation of

Bibliography

Baulieu, E. E. (1997). Neurosteroids: of the nervous system, by the nervous system, for the nervous system. *Recent Prog. Horm. Res.* 52, 1–32.

Cross, B. A., Dyball, R. E. J., Dyer, R. G., Jones, C. W., Lincoln, D. W., Morris, J. F., and Pickering, B. T. (1975). Endocrine neurons. *Recent Prog. Horm. Res.* 31, 243–294.

Crowley, W. F., Jr., Filicori, M., Spratt, D. I., and Santoro, N. F. (1985). The physiology of gonadotropin-releasing hormone (GnRH) secretion in men and women. *Recent Prog. Horm. Res.* 41, 473–531.

Cyr, M., Bosse, R., and Di Paolo, T. (1998). Gonadal hormones modulate 5-hydroxytryptamine$_{2A}$ receptors: emphasis on the rat frontal cortex. *Neuroscience* 83, 829–836.

Fink, G. (1979). Feedback actions of target hormones on hypothalamus and pituitary with special reference to gonadal steroids. *Annu. Rev. Physiol.* 41, 571–585.

Fink, G. (1988a). The G. W. Harris lecture. Steroid control of brain and pituitary function. *Q. J. Exp. Physiol.* 73, 257–293.

Fink, G. (1988b). Gonadotropin secretion and its control. In *The Physiology of Reproduction* (E. Knobil and J. D. Neill, Eds.), pp. 1349–1377. Raven Press, New York.

Fink, G. (1995). The self-priming effect of LHRH; A unique servomechanism and possible cellular model for memory. *Front. Neuroendocrinol.* 16, 183–190.

Fink, G. (1996). The psychoprotective action of oestrogen is mediated by central 5-hydroxytryptamine as well as dopamine receptors. In *Molecular Mechanisms of Neuronal Communication* (K. Fuxe, T. Hŭkfelt, L. Olson, D. Ottoson, and A. Bjorklund, Eds.), pp. 177–204. Pergamon, Oxford, UK.

Fink, G. (1997). Mechanisms of negative and positive feedback of steroids in the hypothalamic–pituitary system. In *Principles of Medical Biology* (E. Bittar and N. Bittar, Eds.), pp. 29–100. JAI, Greenwich, CT.

Fink, G., and Sumner, B. E. H. (1996). Oestrogen and mental state. *Nature* 383, 306.

Guillemin, R. (1978). Control of adenohypophysial functions by peptides of the central nervous system. *Harvey Lect.* 71, 71–131.

Harris, G. W. (1955). *Neural Control of the Pituitary Gland.* E. Arnold, London.

King, D. P., Zhao Y., Sangoram, A. M., Wilsbacher, L. D., Tanaka, M., Antoch, M. P., Steeves, T. D., Vitaterna, M. H., Kornhauser, J. M., Lowrey, P. L., Turek, F. W., and Takahashi, J. S. (1997). Positional cloning of the mouse circadian *clock* gene. *Cell* 89, 641–653.

Klein D. C., Coon, S. L., Roseboom, P. H., Weller, J. L., Bernard, M., Gastel, J. A., Zatz, M., Iuvone, P. M., Rodriguez, I. R., Begay, V., Falcon, J., Cahill, G. M., Cassone, V. M., and Baler, R. (1997). The melatonin rhythm-generating enzyme: molecular regulation of serotonin N-acetyltransferase in the pineal gland. *Recent Prog. Horm. Res.* 52, 307–357.

Knobil, E. (1980). The neuroendocrine control of the menstrual cycle. *Recent Prog. Horm. Res.* 36, 53–58.

Lincoln, G. A., and Short, R. V. (1980). Seasonal breeding: nature's contraceptive. *Recent Prog. Horm. Res.* 36, 1–52.

McQueen, J. K., Wilson, H., and Fink, G. (1997). Estradiol-17β increases serotonin transporter (SERT) mRNA levels and the density of SERT binding sites in female rat brain. *Mol. Brain Res.* 45, 13–23.

Moss, R. L., Gu, Q., and Wong, M. (1997). Estrogen: Nontranscriptional signaling pathway. *Recent Prog. Horm. Res.* 52, 33–69.

Reiter, R. J. (1997). The pineal gland. In *Principles of Medical Biology* (E. E. Bittar and N. Bittar, Eds.), pp. 145–164. JAI, Greenwich, CT.

Schally, A. V., Arimura, A., and Kastin, A. J. (1973). Hypothalamic regulatory hormones. *Science* 179, 341–358.

Shughrue, P. J., Lane, M. V., Merchenthaler, I. (1997). Comparative distribution of estrogen receptor-alpha and -beta mRNA in the rat central nervous system. *J. Comp. Neurol.* 388, 507–525.

Sumner, B. E. H., and Fink, G. (1995). Estrogen increases the density of 5-hydroxytryptamine$_{2A}$ receptors in cerebral cortex and nucleus accumbens in the female rat. *J. Steroid. Biochem. Mol. Biol.* **54,** 15–20.

Sumner, B. E. H., and Fink, G. (1997). The density of 5-hydroxytryptamine$_{2A}$ receptors in forebrain is increased at pro-estrus in intact female rats. *Neurosci. Lett.* **234,** 7–10.

Takahashi (1996). The biological clock: its all in the genes. *Prog. Brain Res.* **111,** 5–9.

Wagner, C. K., and Morrell, J. I. (1996). Distribution and steroid hormone regulation of aromatase mRNA expression in the forebrain of adult male and female rats: a cellular-level analysis using *in situ* hybridization. *J. Comp. Neurol.* **370,** 71–84.

Neurohypophysial Hormones

Hans H. Zingg
McGill University

I. Overview
II. The Hypothalamo–Neurohypophysial System
III. Biosynthesis of Neurohypophysial Hormones
IV. Mechanism of Release
V. Hormone Structures
VI. Physiology
VII. Mechanisms of Action
VIII. Pathophysiology

GLOSSARY

action potential A short-lasting but intense depolarization of the neuronal cell membrane that is propagated along the neuronal processes.

axon The major process of a neuron along which action potentials as well as secretory materials are propagated.

depolarization A reversal of the electric potential present across the cell membrane.

exocytosis The process by which a secretory granule fuses with the cell membrane and releases its content into the extracellular space.

G-protein-linked receptors A large class of cell surface proteins that mediate the action of diverse extracellular mediators, including many peptide hormones. These molecules span the cell membrane seven times and are linked to guanine nucleotide-binding proteins (G-proteins) which mediate the generation of the intracellular signal.

hypothalamus A small brain region close to the ventral brain surface, immediately caudal to the optic chiasm and above the pituitary gland to which it is connected via the median eminence and the pituitary stalk.

neurosecretion The process by which a neuron is able to release soluble messengers from its nerve endings. A more restricted definition refers to the process by which specific neuroendocrine cells release hormones directly into the bloodstream. Neurons of the hypothalamo–neurohypophysial system have served as an important model system for the development of the concept of neurosecretion.

secretory granules Specific intracellular particles surrounded by a bilayer membrane containing secretory peptides.

In most mammalian species, the neurohypophysis releases two important peptide hormones: arginine vasopressin (AVP) and oxytocin (OT). These two small neuropeptides exert a wide spectrum of biological functions including functions related to fluid homeostasis (AVP) and reproduction (OT). Although these peptides are released at the level of the neurohypophysis, their biosynthesis occurs in the cell bodies of specialized neurosecretory neurons located in the hypothalamus. Therefore, the hypothalamo–neurohypophysial system has long served as a classical model system for the study of neurosecretion and for the mechanism that couples nerve cell excitation to secretion. With the cloning of the genes encoding AVP and OT and, recently, with the cloning of the family of the receptors that mediate the actions of AVP and OT, new insights into the mechanisms of biosynthesis and action of these peptides have been gained. Agonists as well as novel specific antagonists of AVP and OT are of considerable pharmacotherapeutic importance.

I. OVERVIEW

The main action of arginine vasopressin (AVP) is stimulation of renal water reabsorption at the level of the distal convoluted tubules and the medullary collecting ducts of the kidney. Thus, AVP is also referred to as "antidiuretic hormone" (ADH). Other activities of AVP include stimulation of vasoconstriction, pituitary corticotropin release, adrenal steroid secretion, platelet aggregation, and hepatic glycogenolysis. The main actions of oxytocin (OT) include milk ejection from the mammary gland during lactation and uterine contractions during parturition. Additional functions of OT include stimulation of endometrial prostaglandin $F2_\alpha$ production, pituitary prolactin and luteinizing hormone secretion, luteolysis, sperm transport, and natriuresis. Both AVP and OT act as neurotransmitters within the brain. AVP is involved in temperature regulation and learning and memory processes, whereas OT mediates sexual, affiliative, and maternal behaviors. AVP deficiency or resistance to AVP action leads to diabetes insipidus, a condition accompanied by major renal fluid losses. OT is absolutely required for milk ejection during lactation. Although OT represents the most potent uterotonic agent know and is commonly used in obstetrical practice, its physiological role in the process of human parturition is not clearly defined. Recent clinical studies indicate, however, that OT antagonists represent promising pharmacotherapeutic tools for the treatment of premature labor.

II. THE HYPOTHALAMO–NEUROHYPOPHYSIAL SYSTEM

A. The Neurohypophysis

The neurohypophysis constitutes the posterior lobe of the pituitary gland. It is derived from the neural ectoderm of the floor of the forebrain and represents a ventral extension of the hypothalamus. The neurohypophysis contains nerve endings (i.e., axon terminals) of neurons whose cell bodies reside within the hypothalamus, specifically within the supraoptic nucleus and the paraventricular nucleus of the hypothalamus (Fig. 1).

B. The Magnocellular System

Due to their relatively large size, the hypothalamic neurons projecting to the neurohypophysis have been termed magnocellular neurons. Together with their axonal extensions, these neurons form the hypothalamo–neurohypophysial tract. On their way to the neural lobe of the pituitary, the axons of the magnocellular neurons traverse the internal zone of the median eminence, a structure representing the expanded upper end of the pituitary stalk. The nerve fibers then continue down the nervous part of the pituitary stalk (also known as the infundibular stem) to end on basement membranes of capillaries in the posterior pituitary, where the neurohormones are released into the general circulation.

C. The Parvicellular System

AVP and OT are also produced in smaller neurons, so-called parvicellular neurons, in the paraventricular nucleus. Axons emanating from these neurons terminate on capillaries at the level of the external

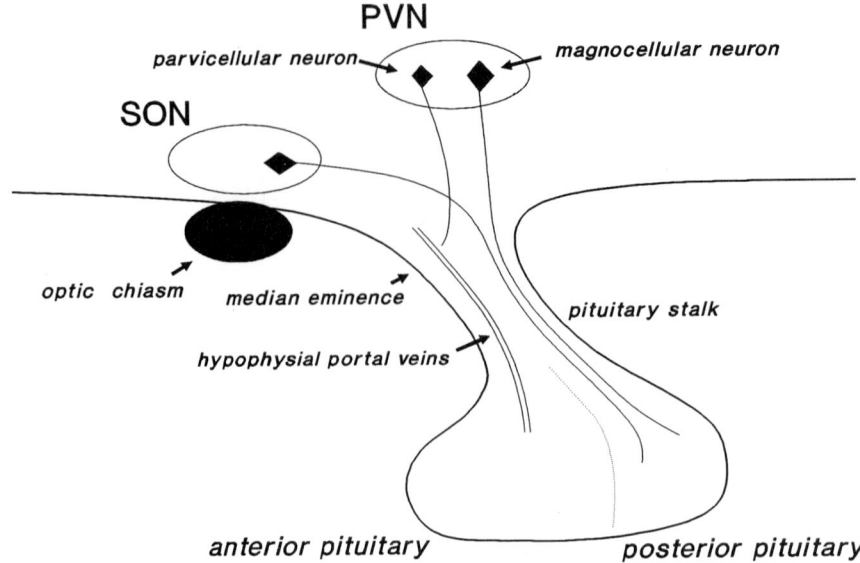

FIGURE 1 Schematic diagram of the hypothalamo–neurohypophysial system. SON, supraoptic nucleus of the hypothalamus; PVN, paraventricular nucleus of the hypothalamus.

zone of the median eminence, where the secretory products are released into the pituitary portal system, a capillary system that allows further transport of the hormones to the anterior pituitary gland. Some axons of parvicellular neurons also project to different areas in the brain where AVP and OT are known to act as neurotransmitters.

III. BIOSYNTHESIS OF NEUROHYPOPHYSIAL HORMONES

In each species studied, the genes encoding AVP and OT are located in close linkage on the same chromosome. They are positioned tail-to-tail in opposite orientation and are, therefore, transcribed from opposite strands. OT and AVP are synthesized as larger precursor molecules which contain, in addition to the biologically active nonapeptide, a 10-kDa peptide called neurophysin (NpI and NpII for OT and AVP, respectively) (Fig. 2). In addition, the AVP precursor molecule also contains at its C terminus a glycoprotein whose function is unknown. The genes encoding AVP and OT are expressed in distinct, mutually exclusive sets of hypothalamic neurons. Following synthesis at the level of the neuronal cell bodies, the precursor peptides are packaged into secretory granules and transported along the axons to the axon terminals where they are released. Within the secretory granules, the precursor peptides are processed to their bioactive forms. Even after AVP and OT have been separated from their respective neurophysins by proteolytic processing, AVP and OT are still able to bind noncovalently to their respective neurophysins within the secretory granule. However, once released, AVP and OT circulate free in plasma. Thus, the biological significance of the neurophysins remains unclear.

OT is also synthesized peripherally in nonneural cells. This includes the uterus, where OT is synthesized specifically prior to parturition, as well as the ovarian follicles of ruminants. Thus, in addition to

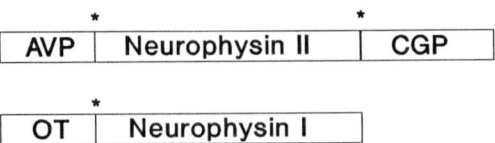

FIGURE 2 Schematic structure of the AVP precursor molecule (pro-vasopressin) and of the OT precursor molecule (pro-oxytocin). AVP, arginine vasopressin; OT, oxytocin; CGP, C-terminal glycopeptide; *, sites of proteolytic cleavage during precursor processing.

its roles as circulating hormone and neurotransmitter, OT may also act as a local paracrine mediator.

IV. MECHANISM OF RELEASE

AVP and OT are classical neurosecretory products and are released from nerve endings according to the principles of neurosecretion. Like any other neuron, the neuroendocrine AVP- and OT-producing neurons are capable of generating action potentials which are propagated from the cell body along the axon to the nerve endings. Each action potential represents a short-lasting membrane depolarization. Specific activation of a neuron results in an increased frequency of action potentials. At the level of the nerve endings, the arrival of action potentials leads to membrane depolarization which, in turn, triggers the opening of voltage-sensitive calcium channels and, as a result, an increase in the intracellular calcium concentration. The increase in intracellular calcium stimulates, in turn, a process termed exocytosis, by which peptide-containing secretory granules fuse with the cell membrane and release their content into the extracellular space. Through this mechanism which couples neuronal excitation to secretion, an increase in the frequency of action potentials arriving at the nerve terminus results in a proportional increase in neuropeptide secretion. In summary, the axon of a neurosecretory cell fulfills two essential functions. First, through axonal transport, it delivers the secretory granules, which are formed in the nerve cell body, to the nerve ending. Second, the axonal membrane propagates action potentials to the nerve ending where they trigger peptide release.

V. HORMONE STRUCTURES

The primary structures of the nonapeptides OT and AVP are shown in Table 1. The two peptides are highly related and differ from one another in positions 3 and 8. They both possess a disulfate-linked ring structure, involving the cysteine residues in positions 1 and 6. In most mammalian species, vasopressin contains an arginine residue in position 8 and is therefore called arginine vasopressin. In some mammalian species, however, including the pig, vasopressin contains a lysine residue in position 8 and is thus referred to as lysine vasopressin. Only the C-terminally amidated forms are biologically active.

AVP and OT are the latest offsprings of a long phylogenetic evolution that can be traced back to mollusks and insects. Vasotocin (VT; Table 1) is an example of an ancestral molecule that is widely distributed among nonmammalian vertebrates. Throughout species evolution, vasopressin-like mol-

TABLE 1
Sequences of Vasopressin, Oxytocin, and Related Molecules

Molecule	Position								
	1	2	3	4	5	6	7	8	9
AVP	Cys–	Tyr–	Phe–	Gln–	Asn–	Cys–	Pro–	Arg–	Gly–NH$_2$
LVP	=	=	Phe	=	=	=	=	Lys	=
VT	=	=	Ile	=	=	=	=	Arg	=
OT	=	=	Ile	=	=	=	=	Leu	=
dDVAP	dCys	=	Phe	=	=	=	=	D-Arg	=
Atosiban	dCys–	D-Tyr(OEt)–	Ile	Thr	=	=	=	Orn	=

(S–S disulfide bond between positions 1 and 6)

Note. Abbreviations used: AVP, arginine vasopressin; LVP, lysine vasopressin; VT, vasotocin; OT, oxytocin; dDAVP, 1-desamino-8-D-arginine vasopressin (desmopressin); Atsobian, a clinically tested OT antagonist; S–S, disulfide bond; =, same residue as AVP; dCys, desamino cysteine; Orn, ornithine.

ecules played an important role in fluid homeostasis. In mammals, AVP has maintained this role through its specific actions at the level of the kidney. The related OT molecule, however, has assumed very different functions in mammals—functions that are essential for mammalian reproduction, such as milk ejection and uterine contractions.

VI. PHYSIOLOGY

A. Vasopressin: Main Action

The main action of AVP is the enhancement of water reabsorption at the level of the distal convoluted tubule and medullary collecting duct of the kidney. The level of water reabsorption has a direct effect on blood osmolality. Therefore, AVP plays a crucial role in the regulation of plasma osmolality. Several specialized brain cells, called osmoreceptors, are capable of sensing very small changes in blood osmolality. In addition, AVP neurons themselves are thought to be osmosensitive. Stimulation of osmoreceptors by increased blood osmolality leads to specific activation of AVP neurons, i.e., an increase in their rate of action potential generation. This, in turn, stimulates the process of neurosecretion and leads to enhanced AVP secretion and the resulting increase in renal water reabsorption leads to a decrease in blood osmolality. This system represents a classical negative feedback mechanism that is capable of maintaining blood osmolality within narrow limits. The system is exquisitely sensitive and responds to osmolality changes as small as 1%.

Additional factors that stimulate AVP secretion include orthostatic hypotension and a fall in blood volume (hemorrhage). These stimuli are mediated via left atrial, carotid and aortic baroreceptors. Additional stimuli that affect AVP secretion include pain, stress, increased temperature, and nicotine. On the other hand, low temperatures and ethanol exert inhibitory actions on AVP release.

B. Additional Vasopressin Actions

AVP-containing fibers projecting to the median eminence deliver AVP into the pituitary portal system and thus to the anterior pituitary gland, where AVP acts in synergism with corticotrophin-releasing factor to stimulate the release of adrenocorticotropic hormone from pituitary corticotrophs.

AVP also acts as a contractor of vascular smooth muscle cells and is able to increase total peripheral resistance. Its effect on platelets leads to increased platelet aggregation. At the level of the adrenal cortex, AVP stimulates a secretion of corticosteroids. Finally, in the rat, AVP stimulates hepatic glycogenolysis. However, in the human, this effect seems to be negligible. As a neurotransmitter, AVP has been reported to stimulate learning and memory processes. Moreover, AVP acts as an antipyretic and decreases body temperature.

C. Oxytocin

The two main physiological actions of OT include milk ejection and uterine contraction. OT secretion (i.e., the activity of hypothalamic OT neurons) is specifically stimulated by activation of touch receptors in the nipples of the mammary gland and by vaginal stimulation during intercourse or delivery. In addition, estrogen, hemorrhage, plasma hyperosmolality, and mild stress also act as stimulants.

The role of OT in lactation is firmly established. In the rat, in which lactation occurs in a phasic pattern, each period of milk ejection is preceded by a dramatic increase in the activity of OT neurons, which is immediately followed by a rise in circulating OT levels. In women, lactation is also accompanied by increased OT plasma levels. In rats, application of OT antibodies or OT antagonists blocks milk ejection. Moreover, mice, in which the OT gene has been inactivated by homozygous recombination ("OT knockout mice"), are unable to lactate.

The role of OT in parturition is less clear. In women, OT secretion is pulsatile at term and frequency and duration of pulses increase with advancing labor. However, in OT-deficient OT knockout mice, the timing and progress of parturition is apparently normal. On the other hand, in humans, a recent international study showed that the application of an OT antagonist (Atosiban; Table 1) represents an efficient strategy to delay preterm labor. It thus appears that, although OT is not the only system involved in inducing uterine contractions at parturition, premature activation of the OT system may be an underlying cause of spontaneous premature labor.

VII. MECHANISMS OF ACTION

The specific actions of AVP and OT are mediated by a family of specific cell membrane receptors that belong to the large family of G-protein-coupled receptors.

A. Vasopressin Receptors

Based on the second messenger systems to which they are coupled, AVP receptors have been subdivided into two main subclasses. Activation of V1-type receptors results in an increase in intracellular calcium via stimulation of phosphoinositol turnover. In contrast, activation of V2-type receptors (V2R) leads to stimulation of adenylate cyclase and results in an increase in intracellular cyclic adenosine monophosphate. V2Rs are mainly present in the distal tubules and collecting ducts of the kidney where they mediate the most essential function of AVP, namely, renal water reabsorption. Two V1 receptor subtypes have been identified: V1b receptors are present in the anterior pituitary, where they mediate corticotrophin release, and V1a receptors are present in liver, vascular smooth muscle cells, adrenal cortex, uterus, platelets, and central nervous system.

B. OT Receptors

To date, only one type of OT receptor (OTR) has been identified. Within the uterus, OTRs are present on myometrial cells as well as on endometrial epithelial cells where they mediate uterine contraction and prostaglandin $F_{2\alpha}$ release, respectively. OTRs are also present in all the other OT target tissues, including mammary gland, pituitary gland, brain, kidney, thymus, and testes.

An outstanding feature of the OTR is that it undergoes dramatic up- and downregulation in a stage- and tissue-specific fashion. Whereas uterine OTRs are highly upregulated prior to term and are acutely downregulated after delivery, mammary gland OTRs slowly rise during pregnancy and remain highly expressed throughout lactation. This tissue- and stage-specific regulation of OTRs enables switching of the targets for OT action and allows circulating OT to play a dual role: to promote uterine contractions specifically during parturition and to mediate milk ejection during lactation.

Estrogens induce OTR expression in several target tissues, including the hypothalamus, where OT mediates behavioral functions. Thus, estrogen priming is a prerequisite for the central behavioral effects of OT. Molecular studies have demonstrated that OTR upregulation occurs at the level of OTR gene transcription and the underlying molecular mechanisms are currently under investigation. Specifically, in the uterus, there is a dramatic upregulation of OTRs prior to parturition. This receptor upregulation leads to a strong increase in uterine OT sensitivity. Since this upregulation occurs in all mammalian species studied, it is thought that this mechanism is involved in triggering parturition and premature upregulation of uterine OTRs may be a contributing cause to spontaneous preterm labor. The precise role that uterine OTR upregulation plays in the human remains still uncertain, however.

VIII. PATHOPHYSIOLOGY

A. Abnormalities of Vasopressin Secretion or Action

Under certain conditions, AVP can be overproduced, leading to what is referred to as syndrome of inappropriate ADH production (SIADH). The hallmark of this condition is an AVP-induced increased in water retention, leading to a lowering of plasma sodium (hyponatremia). The most common cause for SIADH is ectopic AVP production by a tumor, specifically by small cell lung carcinomas. In addition, pulmonary and cerebral disorders can also give rise to SIADH, especially in children.

Dysfunction or destruction of the hypothalamo–neurohypophysial complex by disease, accident, or surgical intervention can result in insufficient AVP secretion. This condition is called central diabetes insipidus and is easily treated by AVP administration. A synthetic analog with a longer half-life and a specificity for the renal V2 receptor is used for therapy (1-desamino-8-D-arginine vasopressin; Table 1). Autosomal-dominant, inherited forms of central diabetes insipidus also exist. In all cases analyzed, a mutation in the AVP gene has been discovered. It is not clear, however, how these mutations that involve, in each case, only one allele lead to the observed dominant phenotype.

Diabetes insipidus can also result from a failure of the kidney to respond to AVP (nephrogenic diabetes insipidus). The inherited form (familial nephrogenic diabetes insipidus) is due to a mutation in the gene encoding the V2 receptor. Since the V2R gene is on the X chromosome, this condition is X-linked. Over 80 different mutations have been identified. Since, in this case, the kidneys are unresponsive to AVP, this condition cannot be treated by AVP administration. Patients are required to drink large amounts of water to maintain water balance and any fluid restriction represents an immediate danger. If the condition is not recognized at birth, infants with familial nephrogenic diabetes insipidus are likely to be severely dehydrated. In many cases, this has led to permanent brain damage. Early recognition at birth is now feasible through genetic testing.

B. Oxytocin Deficiency

No condition of specific OT deficiency or overproduction has been described in the human. It remains to be determined whether premature activation of uterine OT receptor gene expression or premature OT gene expression or secretion may be a factor contributing, in some cases, to premature onset of parturition. Clinical studies using the OT antagonist Atosiban lend some support to this hypothesis. A recent report demonstrates that homozygous inactivation of the gene encoding the receptor for prostaglandin $F_{2\alpha}$ in mice leads to complete suppression of parturition despite normal fetal development. In addition, in these mice, there is complete suppression of uterine OT receptor expression. These recent data indicate that prostaglandin $F_{2\alpha}$ action is necessary for uterine OTR induction and that this mechanism may represent an essential link in the induction of parturition.

See Also the Following Articles

Neurosecretion; Oxytocin

Bibliography

Gainer, H., and Wray, S. (1994). Cellular and molecular biology of oxytocin and vasopressin. In *The Physiology of Reproduction*, pp. 1099–1129. Raven Press, New York.

Ivell, R., and Russell, J. A. (1995). *Oxytocin: Cellular and Molecular Approaches in Medicine and Research*. Plenum, New York.

Zingg, H. H. (1996). Vasopressin and oxytocin receptors. In *Membrane Surface Receptors* (M. C. Sheppard and J. A. Franklyn, Eds.), Bailliere's Clinical Endocrinology and Metabolism, Vol. 10, No. 1, pp. 75–96. Bailliere Tindall, London.

Zingg, H. H., Bichet, G., and Bourque, C. (1998). *Vasopressin and Oxytocin: Synopsis of Recent Advances in Molecular, Cellular and Clinical Research*. Plenum, New York.

Neuropeptides

Brian J. Arey
Wyeth-Ayerst Research

Francisco J. López and Andrés Negro-Vilar
Ligand Pharmaceuticals Inc.

I. Introduction
II. The Role of Neuropeptide Y
III. The Role of Endothelins
IV. The Role of Galanin
V. The Role of Opioids and Integration of Neuropeptide Action in the Control of Reproduction

GLOSSARY

estrous cycle The 4-day reproductive cycle of the female rat.
gonadotropins Any of a group of hormones secreted by the anterior pituitary gland that regulate gonadal function.
hypothalamus An area lying along the ventral border of the brain that is intimately involved in the regulation of pituitary hormone secretion, autonomic functions and reproductive behaviors.
pituitary A gland composed of a heterogeneous population of cells lying at the base of the skull just below the brain that is a major source of endocrine factors that regulate most physiological functions within the body.
receptor A cellular protein that specifically binds endocrine signals. Binding of the hormone to its specific receptor leads to a biological response within the cell.

Neuropeptides are crucial to the proper timing and coordination of reproductive events in both males and females. These neurotransmitters have been found to be involved in the regulation of reproduction at all levels of the hypothalamo–pituitary–gonadal axis. Many of these functions have been conserved across evolution. The study of neuropeptides includes the analysis of the physiology of neuropeptides, discovery of novel peptidergic transmitters, their receptors, signaling mechanisms, and interactions. Thus, this broad area of research involves the integration of many different fields, including physiology, pharmacology, and cellular and molecular biology. Study of the anatomy, physiology, and expression of these factors has been important in understanding how these proteins interact with target cells and each other to lend exquisite control to propagation of species.

I. INTRODUCTION

Since the initial description of the relationship between the hypothalamus and the pituitary, there have been significant advances in our understanding of the regulation of the hypothalamo–pituitary–gonadal axis. Despite these large strides forward, data presented in the literature provide a far more complex picture than that originally envisioned by the pioneers of neuroendocrinology. Rather than one releasing or inhibiting hormone functioning to regulate the secretion of one specific pituitary hormone, data from the past 20 years of research have clearly demonstrated a great deal of complexity and redundancy for the regulation of pituitary function. Thus, we can now imagine a system in which several neuroendocrine signals intermingle as modulators of responses rather than regulators. From an evolutionary perspective, one can envision that such depth of regulation would be advantageous since the integrated unit of the hypothalamus and the master gland known as the pituitary controls in some way every metabolic and endocrine function in the body. Therefore, this ensures not only a balance of physiological processes

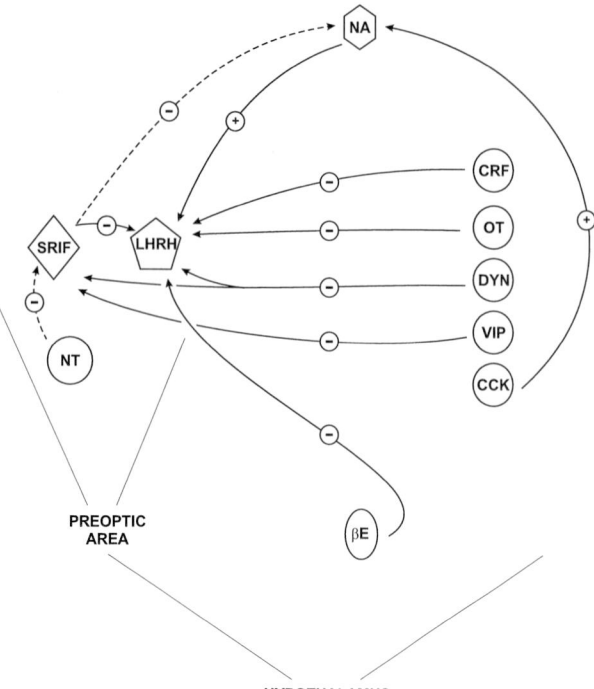

FIGURE 1 Schematic representation of the complex nature of the interaction between neuropeptides to regulate reproduction. The interaction of neuropeptides in the control of LHRH secretion is shown as an example. Note that numerous neuropeptide systems integrate not only with LHRH neurons but also with other neuropeptide-containing cells to regulate LHRH secretion. These interactions include both negative and positive synaptic inputs arising from numerous areas of the hypothalamus and extra-hypothalamic regions of the CNS. NA, noradrenaline; CRF, corticotropin-releasing factor; OT, oxytocin; DYN, dinorphin; VIP, vasoactive intestinal polypeptide; CCK, cholecystokinin; SRIF, somatostatin; NT, neurotensin; LHRH, luteinizing hormone-releasing hormone; βE, β-endorphin (adapted with permission from Kordon et al., 1994).

but also an integrative function of the different physiological modulators produced by the pituitary gland. One area of vigorous research in the regulation of reproductive processes has focused on the role of small peptidergic modulatory factors at all three levels of the hypothalamic–hypophyseal–gonadal axis.

Several key peptidergic systems have emerged as major players in the integration of signals that control reproductive function. Indeed, neuropeptides are found to exert their effects along the entire pathway that connects the brain to the gonad. The primary center of the regulatory loop controlling reproduction is found within the hypothalamus. This center is thought to be composed of a number of both peptidergic and nonpeptidergic neuronal systems which are integrated in a highly complex system (Fig. 1). The end result of these complex interactions is the release of luteinizing hormone-releasing hormone (LHRH) from nerve terminals in the median eminence. Secretion of LHRH into the portal vasculature is the key regulatory event in the secretion of the gonadotropins LH and follicle-stimulating hormone (FSH) from the pituitary gland which prepare gametes in the gonad for reproduction. LHRH is a neuropeptide composed of 10 amino acids which is secreted in a pulsatile fashion. Many studies have been performed to dissect the nature and control of pulsatile LHRH release. From these studies neuropeptidergic systems have been found to be crucial to the proper timing and generation of pulses of LHRH. Among these systems are three neuropeptides that have been studied in detail only recently. In this article, we summarize data on these three neuropeptide modulators in the control of reproduction. We exclude from this compilation a detailed analysis of the endogenous opioid peptides (EOPs) since several comprehensive reviews have covered this topic. However, we provide an integrative perspective involving the interactions of these systems with the EOPs within the obvious limitations of our current knowledge.

II. THE ROLE OF NEUROPEPTIDE Y

Neuropeptide Y (NPY) is a 36-amino acid peptide neurotransmitter found ubiquitously in the central and peripheral nervous systems. It shares significant homology with other members of the pancreatic peptide family, including peptide YY and pancreatic peptide, all of which are derived from the common precursor peptide, pre-proNPY. Within the hypothalamus, NPY-positive neurons are organized in a centralized fashion with cell bodies primarily located within the arcuate nucleus. This nucleus of NPY-positive perikarya radiates axons to a number of the other hypothalamic nuclei. However, it is the projections to the median eminence (ME) and medial preoptic area (MPOA) which are of paramount importance to the regulation of reproductive events. Neuropeptide Y-positive terminals have been found on LHRH axons in the ME as well as impinging on LHRH cell bodies in the MPOA. Interestingly, NPY terminals and axons have also been identified projecting to the posterior pituitary from both central and peripheral pathways and impinging on perivascular structures. These data suggest that NPY may alter pituitary function through alteration of hypophyseal blood flow. Perhaps more intriguing, though, is the recent identification of NPY mRNA expression in the Leydig and Sertoli cells of the rat since this suggests an autocrine or paracrine role for NPY in the regulation of gonadal function. Thus, at key points within the entire hypothalamo–pituitary–gonadal axis, NPY neurons are poised to significantly impact reproductive processes.

Endogenous NPY is known to play a physiological role in the regulation of both gonadotropin and prolactin (PRL) secretion in the rat. This is clear from studies using NPY receptor antagonists and passive immunization paradigms with blocking antisera. In this regard, administration of the NPY receptor antagonist BIBP3226 attenuates the surge of LH on proestrus, indicating that the endogenous peptide contributes to the full expression of the preovulatory surge of LH. In addition, intracerebroventricular administration of NPY antisera led to a diminution in pulse frequency and pulse amplitude of episodic LH secretion in female rats, again providing evidence for a physiological role of NPY in controlling basal LH levels. The action of NPY to control gonadotropin secretion is multifaceted through its ability to exert control at multiple sites. In this regard, the peptide modulates the activity of LHRH neurons directly since challenge of immortalized LHRH-secreting neurons (GT1) with NPY leads to a rapid increase in the release of LHRH. Similarly, intracerebroventricular administration of NPY significantly increases the levels of LHRH mRNA in the preoptic area of the rat. These data suggest a role of NPY to regulate gonadotropin secretion via modulation of the major stimulatory input to LH and FSH secretion (i.e., LHRH neurons). In addition, the peptide also appears to modulate gonadotropin secretion directly at the pituitary level. Cultured anterior pituitary cells respond to NPY with an increase in both LH and FSH secretion. Moreover, it has been shown that NPY is involved in the priming of pituitary gonadotropes to the arrival of LHRH prior to the ovulatory surge of LHRH on proestrus. This effect may be due to the activation of spare LHRH receptors on the gonadotrope cell surface. Thus, NPY appears to be critical to the proper timing and secretion of pituitary gonadotropins during the preovulatory surge of LH as well as the control of pulsatile LH secretion using a hypothalamic–pituitary site of action.

One feature of the modulatory actions of NPY on gonadotropin secretion is its dependence on gonadal steroids. In the absence of the gonad, administration of NPY is inhibitory to LH secretion in both male and female rats. However, treatment of ovariectomized animals with estrogen changes the phenotypic inhibitory action to a stimulatory effect of NPY on LH secretion. The effect of estrogen on the activity of NPY expressing neurons is also apparent prior to the proestrus surges of LH and PRL since levels of NPY mRNA increase during the morning of proestrus. Furthermore, treatment of immature rats with progesterone elevates the hypothalamic concentrations of LHRH and NPY in parallel. Therefore, depending on the prevailing physiological state of the ovary, NPY can elicit different effects on gonadotropin secretion not only by increasing the responsiveness of the system to the peptide but also by elevating NPY input to the axis. Overall, these data emphasize the importance of NPY in determining proper timing of the reproductive cyclicity.

Prolactin is an important reproductive hormone in its own right since it plays a requisite role in the initiation and maintenance of pregnancy in a wide range of species. Hypothalamic control of PRL secretion is inhibitory in nature via a PRL-inhibiting factor, i.e., dopamine. Release of dopamine from nerve terminals in the mediobasal hypothalamus exerts tonic inhibitory control of PRL secretion, clamping release of the hormone for the majority of the reproductive cycle in female rats. This tonic inhibition masks rhythmic stimulatory inputs from being expressed as hormone secretion unless the appropriate physiological conditions are present. Studies using primary cultures of pituitary cells have demonstrated a direct effect of NPY on the release of PRL. In contrast to its effects on gonadotropin secretion, NPY is inhibitory to PRL secretion and also reduces the expression of PRL mRNA in pituitary lactotrophs.

This effect is additive with that of dopamine not only because of the negative input but also, and perhaps more important, because of the anatomical localization of both neurotransmitters. In this respect, NPY has been found to be colocalized with dopamine in tuberoinfundibular neurons during lactation. Therefore, the direct inhibitory action of NPY at the pituitary level is partially due to its cosecretion with the dominant physiological regulator of PRL secretion, dopamine. The result of this cosecretion is a modulation of the dopaminergic input to the pituitary gland such that dopamine and NPY inhibit PRL secretion most likely through a negative coupling to the flux of Ca^{2+} into the cell.

III. THE ROLE OF ENDOTHELINS

The endothelins represent an interesting family of neuropeptides that are still in the early stages of investigation regarding their role in the regulation of reproductive hormone secretion. Endothelins are 21-amino acid peptides that were originally isolated and identified as potent vasoconstrictor agents released from vascular endothelial cells. Recent evidence from a number of laboratories indicates that endothelin is an important modulator of hormone secretion within the hypothalamo–pituitary–gonadal axis. The endothelins represent a family of three peptides of close homology (50%; ET1–ET3). Ten amino acids, including four cysteine residues, are strictly conserved throughout the family. In terms of signaling, three separate endothelin receptor proteins have been identified, designated A–C (ET_A-R, ET_B-R, and ET_C-R, respectively). They differ both in their anatomical distribution and in their pharmacological profiles. Based on the distribution of the three subtypes of endothelin receptors, ET_A-R and ET_B-R may have a physiological role in reproduction. ET_A-R is expressed mainly in the endothelial cells of the vasculature as well as in the anterior pituitary gland, whereas the expression of ET_B-R is mainly found in the brain. In addition, recent pharmacological data (Schild analysis) suggest that more than one type of ET_A-R may be present in the pituitary. In the ovary, both ET_A-R and ET_B-R have been localized but the expression of the two receptors is strikingly different. The ET_A-R subtype is found in abundance in the Fallopian tube, whereas very little is found in the ovarian follicles. In contrast, the ET_B-R subtype is robustly expressed in the granulosa cell layer of developing granulosa cells and to a lesser degree in atretic follicles. In the rat testis, ET_A receptors are found abundantly in both the Leydig and Sertoli cells, but unlike the ovary, only one form of the receptor has been located in the testis.

In terms of the expression of endothelin peptides in the hypothalamus, the endothelin-1 peptide has been localized to the magnocellular subdivision of the paraventricular nucleus. However, all three endothelin peptides have been found to be expressed in the pituitary gland of the rat. As a result of this expression pattern of both endothelins and their receptors, a great deal of literature has been devoted to the autocrine and paracrine effects of endothelins to regulate pituitary hormone secretion. Indeed, endothelin-3 stimulates LHRH release from both ME fragments and the LHRH neuronal cell line GT1 *in vitro*.

Functionally, a number of laboratories have demonstrated a direct effect of endothelin on various cell types within the anterior pituitary gland. Like NPY, endothelins have a direct stimulatory effect on gonadotropin secretion while inhibiting PRL secretion. The differential effects of endothelins on gonadotropin and prolactin secretion appear to be very consistent with the opposing effects of other neuropeptides on the secretion of these two hormones.

The physiological relevance of endothelin in the regulation of pituitary hormone secretion is supported by the observations that the expression of both endothelins and endothelin receptor subtypes is regulated by ovarian steroids. The responsiveness of pituitary cells to endothelins is enhanced in the presence of estrogen and inhibited by progesterone. Therefore, the ability of endothelins to regulate anterior pituitary function and, more important, reproduction is tightly linked to the stage of the estrous cycle in female rats. Although it has not yet been demonstrated that endothelin is absolutely required for the surges of the gonadotropins or prolactin on proestrus, a recent report found that infusion of endothelin-3 in pentobarbital-blocked rats can restore ovulation. However, it is more likely that endothelin activity, at least in the pituitary, represents a modulator for input from other neurohormone systems. This idea is supported by data showing that endothelin and dopamine work in concert to both stimulate and inhibit prolactin secretion. These effects are mediated via different signaling mechanisms that are activated depending on the degree of activation of the receptor-signaling mechanism.

The site for endothelin regulation of reproduction is not restricted to the pituitary gland since endothelins directly affect gonadal function. Endothelin-1 was found to inhibit the FSH-induced accumulation of progesterone in primary cultures of rat granulosa cells. Furthermore, endothelin-like immunoreactivity has been described in human seminal fluid. These data suggest that endothelin represents an important modulator of fertility at the gonadal level. However, given the scarcity of data concerning the effects of endothelins in the gonad, the overall physiological significance or complexity of endothelin activities in the ovary, testis, and uterus is not fully appreciated.

IV. THE ROLE OF GALANIN

Galanin (GAL) is a 29-amino acid neuropeptide that is found ubiquitously in both the peripheral and the central nervous systems (CNS). In the CNS, detailed analysis of GAL binding sites has revealed widespread distribution of potential targets for GAL action. Furthermore, mapping of the expression of GAL in the hypothalamus using immunocytochemistry and *in situ* hybridization techniques has demonstrated a high degree of expression of GAL in the ME as well as in the paraventricular, supraoptic, arcuate, and dorsomedial nuclei. More importantly, GAL expression has been found in LHRH neurons. Therefore, GAL is found in numerous hypothalamic nuclei known to be involved in regulating reproductive hormone secretion and behavior. Within the rat pituitary gland, GAL is also expressed within gonadotropes and lactotrophs, in which it is thought to play an autocrine/paracrine role in the regulation of pituitary hormone secretion and/or responsiveness to hypothalamic neurohormones.

From a functional perspective, investigators from a number of laboratories have studied the effect of GAL on pituitary hormone secretion. In terms of its expression, the GAL peptide is found abundantly in the external layer of the ME. Furthermore, GAL is found in hypophysial portal blood in higher levels than those observed in peripheral blood, consistent with its hypothesized role as a hypothalamic hypophysiotropic hormone. These studies have provided a solid basis for a direct role for GAL in regulating pituitary hormone secretion. Indeed, studies using cultured pituitary cells have found that GAL stimulates both basal and LHRH-induced LH release. Similarly, GAL was also found to stimulate PRL secretion *in vitro*. Therefore, these data indicate that GAL may be a primary regulator of pituitary hormone secretion at the anterior pituitary level.

In addition to the data demonstrating a direct effect of GAL on reproductive hormone secretion from the pituitary, a large volume of data has been published verifying the ability of GAL to regulate anterior pituitary hormone secretion indirectly via action on the physiological mediators of LH and PRL secretion. It has been shown that GAL is a potent stimulator of LHRH release from the ME, although this effect appears to be indirect because it is blocked with α_2-adrenergic receptor antagonists. Intracerebroventricular administration of GAL stimulates PRL secretion, suggesting a central site of action of the peptide in regulating PRL secretion. Perhaps the most convincing argument for a physiological role for GAL, however, stems from *in vivo* data that have utilized GAL antisera. Using passive immunization procedures GAL was clearly shown to be responsible for the modulation of both LH and PRL surges on proestrus. Interestingly, administration of the GAL antiserum had little effect on FSH secretion, suggesting an independent regulation of FSH and LH secretion on proestrus. These data have been confirmed and expanded by showing that centrally administered GAL receptor blockers abolish the preovulatory surge of LH.

In addition to providing hormonal regulation for reproductive processes, GAL also has been shown to play a role in regulating mating behavior. These effects have been demonstrated for both male and females. In females, administration of GAL directly into the MPOA increases lordosis behavior in an estrogen-dependent manner. Interestingly, testosterone priming of female rats led to the induction of male-typical sexual behavior in female rats upon GAL administration into the hypothalamus, suggesting a sex hormone-specific effect of galaninergic control

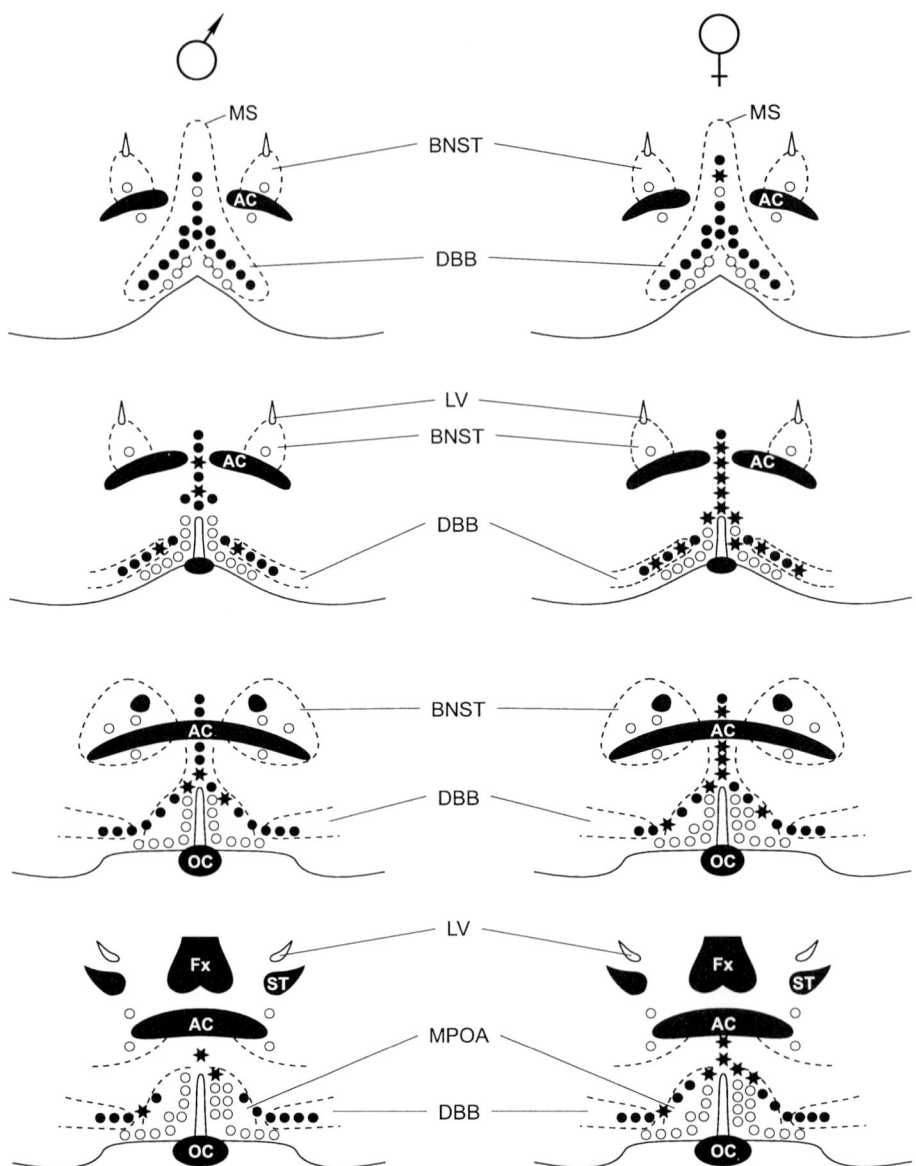

FIGURE 2 Schematic representation of GAL (○)-containing and LHRH (●)-containing cell bodies in the preoptic area of the hypothalamus in the male (left) and female (right) rat. Stars depict cells expressing both neuropeptides. Both sexes contain approximately equal numbers of GAL- and LHRH-containing cell bodies. However, the female rat hypothalamus has an increased incidence of neurons expressing both peptides. AC, anterior commissure; BNST, bed nucleus of the stria terminalis; DBB, diagonal band of Broca; Fx, fornix; LV, lateral ventricle; MPOA, medial preoptic area; MS, medial septum; OC, optic chiasm; ST, stria terminalis (adapted with permission from Merchenthaler et al., 1993).

of reproductive behavior. A role for GAL in regulating male copulatory behavior has also been described, although experiments from different laboratories have led to opposing results. Intracerebroventricular administration of GAL stimulates male copulatory behavior, whereas injection of the peptide into the MPOA leads to a decrease in male reproductive behavioral indices. In toto, these data illustrate that GAL may act at multiple sites within the hypothalamus to coordinate mating behaviors.

One interesting feature of the GAL regulatory input to the pituitary gland is its strong dependence on gonadal steroids. Initially, it was found that the expression of GAL in the pituitary was exquisitely

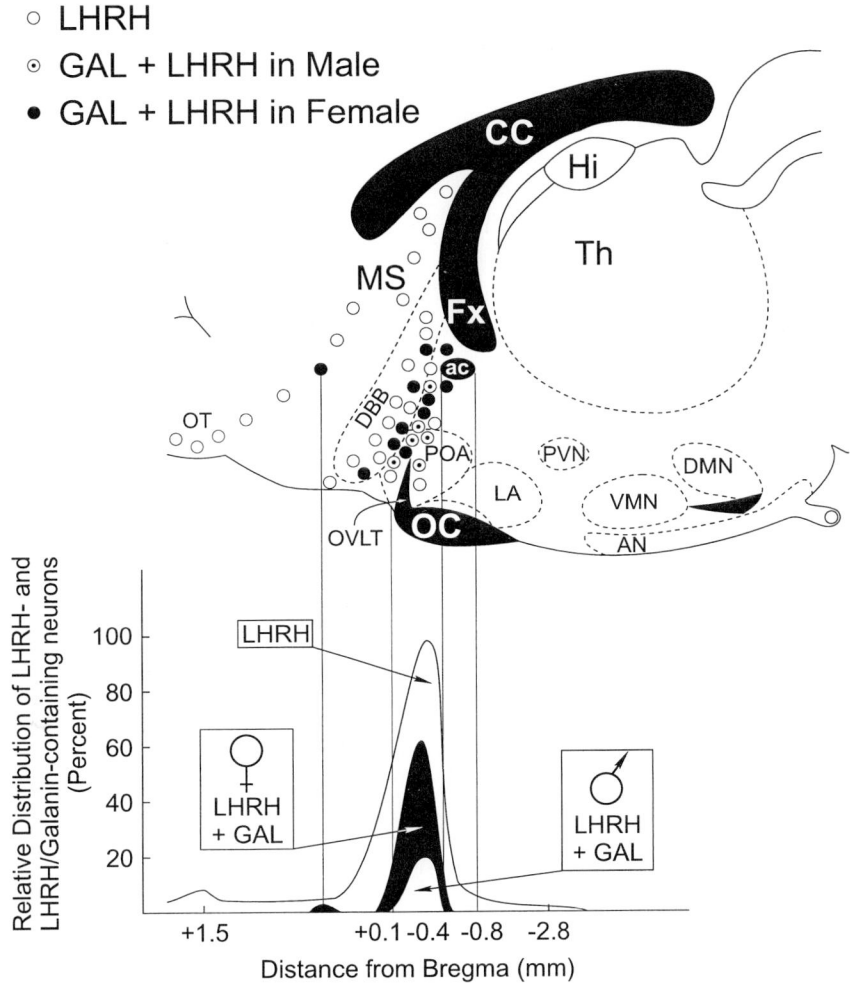

FIGURE 3 Schematic representation of the sagital distribution of neurons coexpressing GAL and LHRH in the brain of the female and male rat. (Top) The relative locations of GAL and LHRH-expressing neurons. Note that in both sexes the majority of double-labeled perikarya are found along the border of the DBB and POA. In addition, a few neurons expressing both peptides are found in the anterior paraolfactory area of the female rat but not in the male. (Bottom) Relative distribution of LHRH-expressing perikarya and GAL and LHRH-expressing cells in the male and female rat brain. In the female rat, 65% of LHRH neurons coexpress GAL and LHRH, whereas in the male only 20% of LHRH neurons contain GAL. ac, anterior commissure; AN, arcuate nucleus; CC, corpus callosum; DBB, diagonal band of Broca; DMN, dorsomedial nucleus; Fx, fornix; Hi, hippocampus; LA, lateral hypothalamus; MS, medial septum; OC, optic chiasm; OT, olfactory tubercle; OVLT, organum vasculosum lamina terminalis; POA, preoptic area; PVN, paraventricular nucleus; Th, thalamus; VMN, ventromedial nucleus (adapted with permission from Merchenthaler et al., 1993).

sensitive to the presence of estrogen. Further studies suggested that this was also the case for GAL expression in the hypothalamus. In male rats, approximately 10–15% of LHRH neurons coexpress GAL. The LHRH neuronal system of female rats, however, consists of a population of neurons of which the vast majority of cells coexpress these neuropeptides. The expression of GAL in LHRH neurons, therefore, is sexually dimorphic (Fig. 2). Note that both the male and the female rat hypothalamus contain approximately equal numbers of LHRH- and GAL-expressing neurons. However, examination of LHRH neurons that express GAL reveals that the female rat contains more than twofold more GAL-expressing LHRH neurons than does the male rat. Interestingly, the distribution of dual-labeled cells in the hypothalamus is strikingly similar in both sexes (Fig. 3), with the major focus of GAL-expressing LHRH neurons present in the transitional area between the diagonal band of Broca and the preoptic region. Subsequent studies using gonadectomized rats treated with estrogen have shown that the sexually dimorphic coexpression of LHRH and GAL is heavily dependent on the presence of estrogen. However, in male rats, treatment with either testosterone or estrogen following gonadectomy does not alter the expression of GAL in LHRH neurons. Thus, estrogen-dependent expression of GAL in LHRH neurons must be neonatally determined during the critical period of sexual differentiation. Sex-reversal studies have demonstrated that one can induce estrogen responsiveness to GAL expression in LHRH neurons of male rats if the animals are neonatally orchidectomized. Recent studies using the GT1 cell line have also demonstrated regulation of GAL expression *in vitro*. Immortalized LHRH neurons were found to contain the α subtype of the estrogen receptor by RT-PCR and sequencing of the amplification product. More important, estrogen receptors were functional since treatment of these cells with physiological concentrations of various estrogens led to stimulation of a heterologous gene in transfection studies. In addition, GT1–7 cells express GAL mRNA as revealed by RT-PCR. Using a quantitative RT-PCR paradigm, we were able to demonstrate that GAL mRNA expression in immortalized LHRH neurons is estrogen dependent. As a whole these data provide a strong case for GAL as a major factor in the regulation of reproductive processes at both the hypothalamus and the pituitary levels and reveal a physiological role for endogenous GAL in reproductive functions.

V. THE ROLE OF OPIOIDS AND INTEGRATION OF NEUROPEPTIDE ACTION IN THE CONTROL OF REPRODUCTION

From a review of the literature, it is evident that not only do neuropeptides act at multiple sites to modulate reproduction but also they interact with each other to fine-tune the reproductive signal. Such cross talk between modulators has been illustrated for many of the neuroendocrine factors known to play a role in the control of anterior pituitary function. Numerous peptide and nonpeptide factors interact at the hypothalamic level to strictly regulate the secretion of pituitary hormones (Fig. 1). Both the opioid peptides and the vasoactive intestinal peptide are classically known for their many interactions with other neuropeptides in the regulation of hypothalamic–pituitary function. Indeed, one of the best described examples in which multiple neuropeptides interact to regulate a physiological response is the hypothalamic control of LHRH release. Figure 4 summarizes some of the interactions of these neuropeptides as they occur at different levels within the hypothalamic–pituitary axis. LHRH itself is a decapeptide that interacts with numerous neuropeptides to coordinate its pulsatile release which subsequently leads to the secretion of LH from the pituitary gland. Opioid peptides exert a chronic inhibitory tone on LHRH neurons, thus suppressing LHRH secretion into the portal vasculature. Similar to other neuropeptides discussed here, the actions of endogenous opioids are exerted at multiple sites within the hypothalamus. In addition, at the pituitary level locally produced peptides significantly contribute to fine-tune the signal arriving at the gonadotrope.

As mentioned previously, NPY has been shown not only to regulate LHRH secretion but also to act in concert with GAL to regulate the release of LHRH. Using immunocytochemical methods, it has been found that GAL neurons branch out to impinge on other neuronal types within the hypothalamus to regulate LHRH release. For example, GAL-positive neurons are in close proximity to NPY neurons in the arcuate nucleus. These data suggest that GAL and NPY may regulate the release of each other, adding another level of complexity to an already crowded network of factors and feedback loops. Indeed, there is evidence suggesting that GAL and NPY act in concert to regulate the LHRH surge on proestrus. Given the data concerning the interactions between GAL and NPY, one wonders what role opioids play in this system. Recently, it has been shown that GAL not only interacts with NPY but also may regulate the activity of the opioid system in the hypothalamus. Immunocytochemical and electron microscopy studies have found that NPY-positive axons establish what appear to be synaptic contacts in the arcuate nucleus with opioid neurons. These observations and the fact that NPY treatment into the third ventricle leads to increased release of β-endorphin

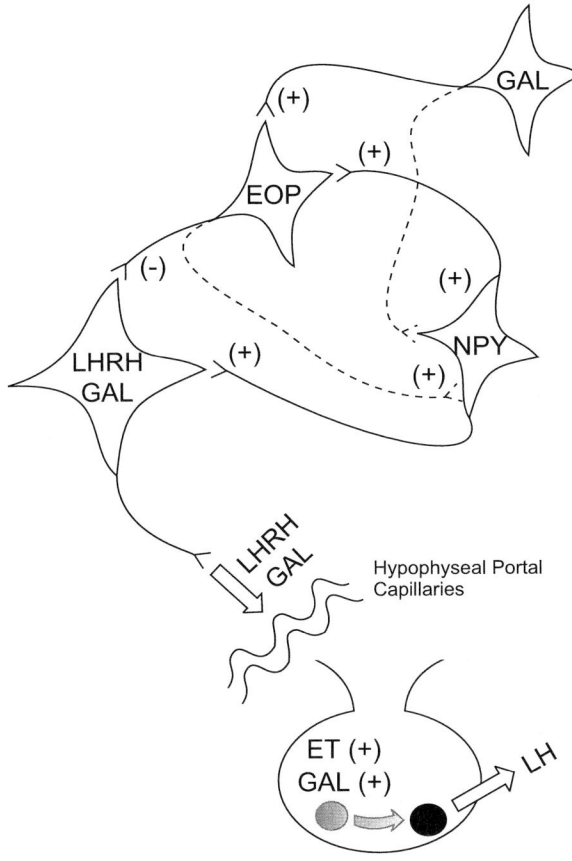

FIGURE 4 Schematic representation of the regulatory interactions among luteinizing hormone-releasing hormone-, galanin-, neuropeptide Y-, and endogenous opiod-cointaining neurons. The drawing is intended to illustrate the actual possible connections; however, this connectivity may not represent anatomical connections. In addition, the actions of the different neuropeptides are conditioned in many cases to the presence or absence of steroid hormones. Hypothalamic, neuroendocrine, and paracrine interactions are depicted in the scheme at the different levels in the hypothalamic–pituitary axis. Many other regulatory systems are known to impinge on the gonadal system and are not represented for the sake of simplicity. EOP, endogenous opiods; ET, endothelin; GAL, galanin; LHHR, luteinizing hormone-releasing hormone; NPY, neuropeptide Y.

suggest a role for NPY in the regulation of opioid release within the hypothalamus. Conversely, axons from enkephalin-positive neurons also impinge on NPY-positive cells in the arcuate nucleus, suggesting that opioids may also influence NPY neurons. Thus, these morphological and physiological studies provide the basis for regulatory loops between neuropeptides to control reproduction.

In conclusion, data are abundant demonstrating the interaction of multiple neuropeptides to regulate hormone secretion at both the hypothalamus and the pituitary. Moreover, it is evident that neuropeptides also regulate reproductive events at sites distant from the hypothalamic–pituitary axis including the gonad itself. It seems likely that these factors may also interact with each other at these "peripheral" sites to provide additional redundancy and complexity to the miraculous physiological phenomenon that reproduction of the species represents.

See Also the Following Articles

LH (Luteinizing Hormone); Neurotransmitters; Prolactin Secretion, Regulation of

Bibliography

Freeman, M. E. (1994). The neuroendocrine control of the ovarian cycle of the rat. In *The Physiology of Reproduction* (E. Knobil and J. D. Neill, Eds.), pp. 613–658. Raven Press, New York.

Houben, H., and Denef, C. (1994). Bioactive peptides in anterior pituitary cells. *Peptides* 15, 547–582.

Kalra, S. P. (1993). Mandatory neuropeptide-steroid signaling for the preovulatory luteinizing hormone-releasing hormone discharge. *Endocr. Rev.* 14, 507–538.

Kalra, S. P., and Kalra, P. S. (1996). Nutritional infertility: The role of the interconnected hypothalamic neuropeptide Y-galanin-opioid network. *Front. Neuroendocrinol.* 17, 371–401.

Kordon, C., Drouva, S. V., Martínez de la Escalera, G., and Weiner, R. I. (1994). Role of classic and peptide neuromediators in the neuroendocrine regulation of luteinizing hormone and prolactin. In *The Physiology of Reproduction* (E. Knobil and J. D. Neill, Eds.), pp. 1621–1681. Raven Press, New York.

Merchenthaler, I., López, F. J., Lennard, D. E., and Negro-Vilar, A. (1991). Sexual differences in the distribution of neurons coexpressing galanin and luteinizing hormone-releasing hormone in the rat brain. *Endocrinology* 129(4), 1977–1986.

Merchenthaler, I., López, F. J., and Negro-Vilar, A. (1993). Anatomy and physiology of central galanin-containing pathways. *Prog. Neurobiol.* 40, 711–769.

Sakurai, T., Yanagisawa, M., and Masaki, T. (1992). Molecular characterization of endothelin receptors. *Trends Pharmacol. Sci.* 13, 103–108.

Neurosecretion

Harold Gainer
National Institutes of Health

I. Introduction
II. Cell Biology of the Neurosecretory Cell
III. Biosynthesis of the Neurosecretory Product
IV. The Neurosecretory Vesicle
V. Mechanisms of Secretion
VI. Concluding Remarks

GLOSSARY

autocrine The release of a chemical messenger by a cell into the extracellular space, that has as its target the secretory cell itself (the suffix *crine* denotes secretion and derives from the Greek word *krinein,* meaning to separate). In contrast, in *paracrine* secretion the target is the adjacent cell population, and *endocrine* excretion denotes secretion directly into the bloodstream to act on distant target organs.

axonal transport The mechanism by which proteins made in the cell body can be delivered to the axon and nerve terminals. Membrane proteins are transported as membrane-bounded vesicles.

endocytosis The process by which the plasma membrane buds off into the cell as a vesicle. This serves to reduce the membrane surface area after exocytosis and also allows the cell to sample the extracellular space which is trapped in the vesicle interior.

exocytosis The process by which chemical substances within membrane-bounded vesicles in the cell can be secreted into the extracellular space. This involves fusion of the spherical vesicle membrane with the planar plasma membrane.

neuropeptides Small molecules in neurons made from amino acids that are connected by peptide bonds and are secreted as intercellular messengers, i.e., as hormones, neurotransmitters, etc.

neurosecretory cell Any cell that secretes neuropeptides or proteins from neurosecretory vesicles.

neurosecretory vesicles Large vesicles (100–350 nm in diameter), usually with electron-dense cores that contain peptides and/or proteins as secretory material (also called *neurosecretory granules* and *large dense core vesicles*).

plasmalemma The name given to the cell membrane that separates the extracellular space and the cytoplasmic space of the cell. Specialized regions of the plasma membrane participate in exocytosis.

posttranslational processing Enzymatic steps that serve to modify newly synthesized proteins so as to influence their functions. Peptides are made from precursor proteins that are proteolytically cleaved, amidated, sulfated, etc. in the cell to produce a biologically active secreted peptide.

vesicle A membrane-bounded organelle, usually spherical, in cells that allows for the intracellular transport and membrane trafficking of proteins and which participates in exocytosis.

Neurosecretion refers to the process of secretion (or exocytosis) of chemical messengers by nerve cells, usually from their nerve terminals. In the classical definition of neurosecretion, the secreted molecule was conceived as peptidic or proteinaceous in nature and directly secreted into the blood in order to act at distant target organs. Consequently, neurons exhibiting this function were termed neurosecretory cells or neuroendocrine cells since they had the ontogenetic history and morphology of neurons but also the intracellular organization and functions of endocrine cells. Recently, the term neurosecretion has been extended to include the secretion of peptides from neurons irrespective of the sites of their receptor-bearing targets. Neurons secreting neuropeptides close to their targets in an autocrine, paracrine, or even a synaptic configuration in central as well as peripheral nervous systems are also said to be participating in neurosecretion.

I. INTRODUCTION

The original concept of neurosecretion was first proposed in a paper by Ernst Scharrer in 1928, who described "gland-like" nerve cells in fish brain. Subsequent work by his wife, Berta Scharrer, on invertebrate species and their collaborations with W. Bargmann gave rise to the revolutionary idea that neurons can also behave as endocrine cells. This idea was not easily accepted by the field, and it was not until the 1940s and 1950s, when the chemical natures of the neurosecretory products in the posterior pituitary were definitively identified as the neuropeptides (oxytocin and vasopressin) by the future Nobel laureate, V. du Vigneaud and colleagues, that the concept of neurosecretion entered the mainstream of endocrinological thinking. Shortly thereafter, the bold hypothesis of Geoffrey Harris, that other neurosecretory cells in the hypothalamus secrete specific "factors" into the portal vascular system to directly regulate various hormone secretions from the anterior pituitary gland, was proposed. The latter view was validated by the subsequent identifications and characterizations of some of these factors by the Nobel laureates, R. Guillemin and A. Schally, as neuropeptides. To date, there are nine such hypothalamic hypophysiotrophic hormones which have been definitively identified in the regulation of the adenohypophysial hormone. These are alternately referred to as hypothalamic-releasing factors (or hormones) or hypothalamic regulatory hormones. Taken together, these hypotheses and findings reinforced the original concept of neurosecretion and established the field of neuroendocrinology.

II. CELL BIOLOGY OF THE NEUROSECRETORY CELL

The cellular component of neurosecretion is the "neurosecretory cell." This cell is neuron-like in its morphology in that it has a cell body region responsible for gene expression and protein synthesis and neuritic processes which extend from the cell body to receive information (the dendrites) or to convey it to the next cell via the most unique process in neurons, the axon. The axon is responsible for transporting the secretory material to the nerve terminal where it is secreted in response to an electrical nerve impulse conducted by the axon coupled to the increase of calcium ion concentrations in the nerve terminal. A schematic illustration of a neurosecretory cell and the physiological and molecular processes within it is shown in Fig. 1.

Unlike "conventional" neurons that secrete small neurotransmitters, such as acetylcholine, excitatory and inhibitory amino acids, and various monamines that can be synthesized by enzymes in the nerve terminal, the peptidic secretory products in "neurosecretory cells" require *de novo* protein synthesis on the rough endoplasmic reticulum as well as packaging by the Golgi apparatus in the cell body. Consequently, the regulation of secretion in the neurosecretory cell requires an intimate association with gene expression and protein synthesis in the cell body and continual supply to the nerve terminal by axonal transport, in contrast to the conventional neuron in which much of this regulation occurs locally in the nerve terminal itself. In the conventional neuron the small secretory vesicles (about 30–50 nm in diameter) that fuse with the plasma membrane during secretion (or exocytosis) are reformed by endocytosis, recycled, and refilled with neurotransmitter in the nerve terminal for repeated rounds of secretion. Thus, while the nerve terminal in conventional neurons is relatively autonomous in neurotransmitter secretion, neurosecretion of peptides from all neurons requires the "on-line" participation of the cell body and axon.

III. BIOSYNTHESIS OF THE NEUROSECRETORY PRODUCT

As noted previously, all peptide intercellular messengers secreted from nerve terminals are first synthesized as protein precursors and then processed by enzymatic mechanisms into biologically active peptides before secretion. The first step in the biosynthesis of a peptide is the expression of the gene that encodes the protein precursor. In the cases of the classical neurosecretory peptides, oxytocin and vasopressin, these 9-amino acid-long peptides are first made as part of two separate precursor proteins con-

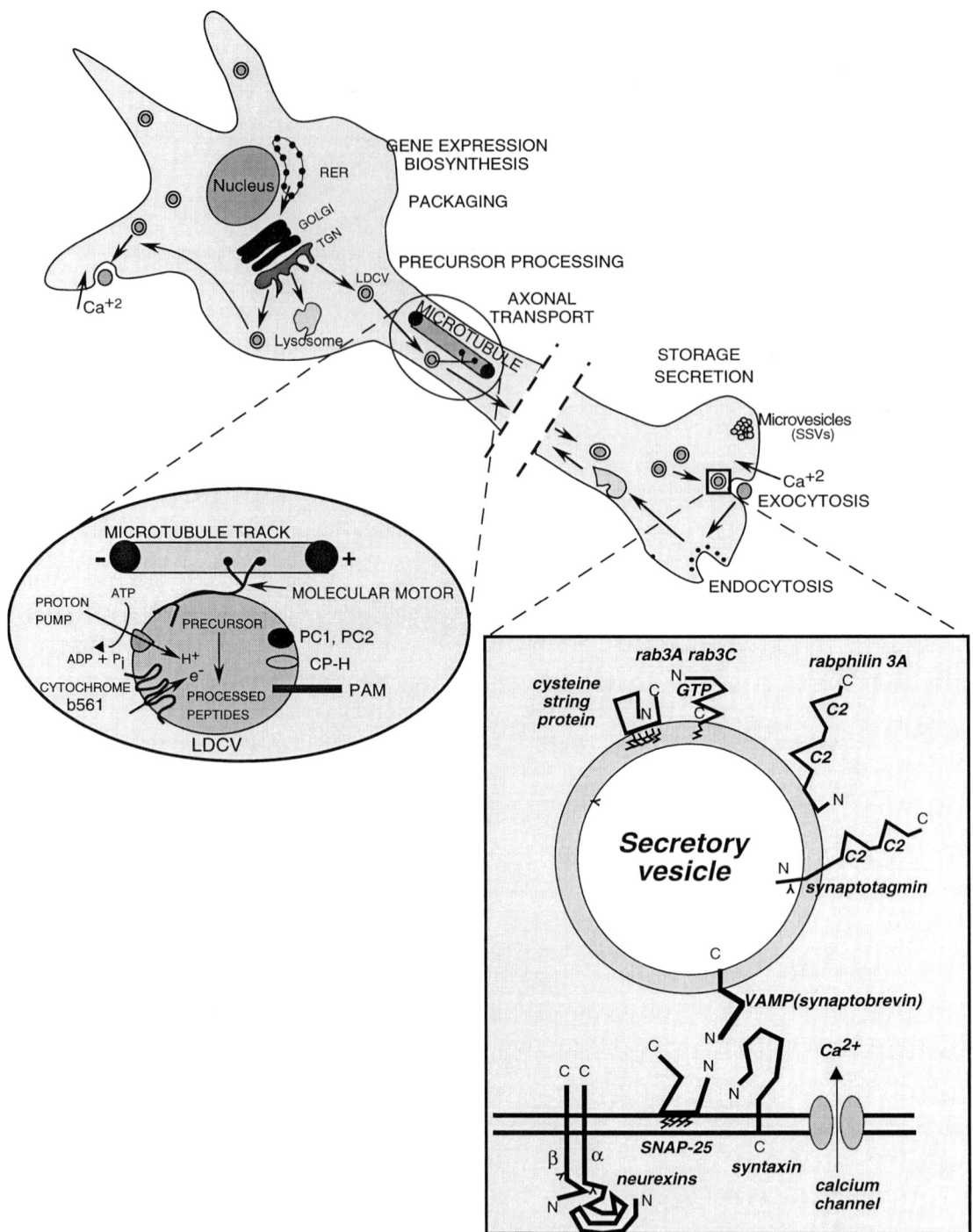

FIGURE 1 Cellular and molecular properties of a neurosecretory cell. The structure of the neurosecretory cell is depicted schematically with notations of the various cell biological processes that occur in each topographic domain. Gene expression, protein biosynthesis, and packaging of the protein into LDCVs occur in the cell body, where the nucleus, RER, and Golgi apparatus are located. Enzymatic processing of the precursor proteins into the biologically active peptides occurs primarily in the LDCVs (see inset), often during the process of anterograde axonal transport of the LDCVs to the nerve terminals on microtubule tracks in the axon. Upon reaching the nerve terminal, the LDCVs are usually stored in preparation for secretion. Conduction of a nerve impulse (action potential) down the axon and its arrival in the nerve terminal causes an influx of calcium ion through calcium channels The increased calcium ion concentration causes a cascade of molecular events (see inset and text) that leads to neurosecretion (exocytosis). Recovery of the excess LDCV membrane after exocytosis is performed by endocytosis, but this membrane is not recycled locally; instead, it is retrogradely transported to the cell body for reuse or degradation in lysosomes. Abbreviations used: RER, rough endoplasmic reticulum; TGN, trans-Golgi network; LDCV, large dense core vesicle, SSV, small secretory vesicles; PC1 or -2, prohormone convertase-1 or -2; CP-H, carboxypeptidase H; PAM, peptiylglycine α-amidating monooxygenase (adapted from H. Gainer and H. Chin, *Cell. Mol. Neurobiol.* **18**(2), 211–230, 1998).

taining about 106 and 145 amino acids, respectively. Their mRNAs are translated on ribosomes attached to the rough endoplasmic reticulum (RER) in the cell body of the neurons, and their resultant precursor proteins undergo cotranslational processing steps in the RER which includes removal of the signal sequence amino acids from their N terminus of the precursor, the formation of disulfide bonds, and the initial stages of glycosylation of the precursor. There are known to be about 100 neuropeptides that are secreted in the nervous system. These include various endogenous opiate peptides, families of gastrointestinal-related peptides, peptide hormones, etc. All these undergo similar biosynthesis mechanisms, from initial precursor synthesis to posttranslational processing, yielding the final peptide products to be secreted.

The posttranslational processing steps involve a variety of proteolytic (endo- and exopeptidase) as well as nonproteolytic (e.g., amidation, sulfation, etc.) mechanisms. These steps occur within membrane-bounded compartments in the cell, starting within the trans-Golgi network and then moving into the secretory vesicle [large dense core vesicle (LDCV)] itself (Fig. 1). The first processing step is to proteolytically cleave the intact protein precursor into its correct peptide fragments. It should be noted that identical precursor proteins can be converted into different peptide fragments by different cell types, and that these different posttranslational processing programs contribute to the diversity of peptides secreted by specific neurons. Thus, the diversity of neuronal phenotypes is, in part, a result of variations in processing mechanisms from cell to cell. The decisive factor is the particular combination of endoproteases that are active in a given cell. There are at least eight species of prohormone (precursor) convertases (PCs) found in mammalian cells, and the presence or absence of any one of these can determine whether a precursor protein will generate a particular peptide (e.g., the presence of PC1 in corticotropes causes the production of ACTH and β-lipotropin from their precursor, proopiomelanocortin, but further cleavage of these peptides to α-MSH and β-endorphin occurs only when PC2 is present, as is the case in the melanotropes). Since the PCs usually cleave at paired basic amino acid motifs, an exopeptidase (carboxypeptidase H) that trims the remaining C-terminal basic amino acid residue is required in most cells. Another critical step in fashioning many biologically active peptides is the conversion of their Cterminal glycines into amides by an enzyme system referred to as peptidylglycine α-amidating monooxygenase. All these enzymes are found in the secretory vesicles (LDCVs) and operate optimally at the mildly acidic conditions (pH 5 or 6) found in these organelles (Fig. 1).

IV. THE NEUROSECRETORY VESICLE

In addition to being the primary site of posttranslational processing of precursors to peptides, neurosecretory vesicles (NSVs; also called neurosecretory granules, or LDCVs) serve as the vehicles for the axonal transport, storage, and secretion of the peptides (Fig. 1). The NSVs are large membrane-bounded organelles, usually between 100 and 350 nm, that have a high protein and peptide content that cause them to have electron-dense cores (hence the name LDCV) when viewed by electron microscopy. Indeed, the large sizes of the NSVs and their intracellular accumulations at high concentrations often cause neurosecretory cells and terminals in invertebrates to have a blue-white reflection visible through the light microscope upon epiillumination. This is particularly notable in the bag cells and R15 in Aplysia and in the X-organ–sinus gland complex in crustacea.

NSVs are formed in the Golgi apparatus and bud off as immature granules from the trans-Golgi network (Fig. 1). The NSV membranes contain proton and electron transport systems which are involved in precursor processing as well as multiple proteins that are involved in the secretory process (see insets in Fig. 1). The anterograde (from cell body to terminal) transport of the NSVs occurs on microtubular tracks and, hence, probably involves members of the kinesin family of molecules as the molecular motors. This kinesin gene family has many members and it is not clear which of the specific molecules in this family are associated with NSV axonal transport. The KIF2 and -3 kinesin proteins are reputed to transport vesicles in the 90- to 160-nm diameter range and thus may be candidates for this function.

In neurosecretory cells, most of the NSVs are stored in the nerve terminals where they are mobilized for secretion by electrical activity. However, some NSVs can also be stored in cell bodies and dendrites and calcium-dependent secretion can also occur in the dendrites. An example of this dual mode of secretion occurs in the magnocellular oxytocin neurons in the hypothalamus of lactating animals. As part of the suckling reflex, these neurons secrete large boluses of oxytocin from nerve terminals in the pituitary into the bloodstream to act on their distant targets, the mammary glands. They also secrete oxytocin from their dendrites into the hypothalamus itself for a paracrine function, i.e., to modulate the synchrony of the oxytocin neuron population's electrical response to suckling.

V. MECHANISMS OF SECRETION

The molecular mechanisms that subserve the secretion of peptides from LDCVs in neurosecretory terminals appear to be very similar to those which underlie secretion from small synaptic vesicles at synapses. In both cases, the secretory event is preceded by an influx of calcium ions through a voltage-gated calcium channel located near a secretory vesicle (see lower inset in Fig. 1). This close apposition of the secretory vesicle to the calcium channel is very important since the secretory event requires a relatively high concentration of calcium ions (10–100 mM) which occurs only immediately adjacent to open calcium channels in the plasma membrane. Secretory vesicles in such locations are said to be in "readily releasable" pools and at active zones of synapses they are referred to as "docked." Usually about 1–5% of NSVs are docked; hence, only these are subject to secretion upon excitation. An intensive research effort is under way to identify the molecules that are involved in the cascade of events that cause docked secretory vesicles to fuse with plasma membranes and thereby release their intravesicular contents into the extracellular space. The latter event is termed "exocytosis" (Fig. 1) and this fusion event is the final step in secretion. Following exocytosis, the nerve terminal membrane is increased in surface area, and this additional membrane is subsequently returned to the cell through a budding process known as "endocytosis" (Fig. 1). Extensive cloning studies have uncovered a very large number of protein families that are associated with either the secretory vesicles membranes or the active zones of secretion on the plasma membrane (Fig. 1, lower inset). Although the list of membrane and soluble proteins that are associated with secretion continues to grow at a rapid pace, several specific proteins are already known to be important in the secretory process. These are the so-called v-SNARES, e.g., synaptobrevin [also called vesicle-associated membrane protein (VAMP)], on the vesicle membrane, and the t-SNARES (t stands for "target"), e.g.. syntaxin and SNAP-25, on the plasma membrane (Fig. 1, lower inset).

Currently, the most coherent theoretical view of secretory mechanisms is represented by the "SNARE" hypothesis, first formulated by J. Rothman and colleagues in 1993. The term SNARE originally referred to the soluble N-ethylmaleimide-sensitive (fusion) attachment (NSF) protein, which is an ATPase that complexes with various other soluble proteins (which have been given the unfortunate names of α-, β-, and γ-SNAP) and the v- and t-SNARES described previously. One of the t-SNARES, syntaxin, is able to bind to the N-type calcium channel on the plasma membrane as well as a calcium-binding protein on the vesicle membrane called synaptotagmin. In addition, the v-SNARE, VAMP, binds to both t-SNARES, syntaxin and SNAP-25, and this "7S complex" is believed to represent the molecular basis of the docking mechanism. Addition of the NSF, soluble SNAPS, and ATP causes a larger 20S complex and this is believed to lead to "primed" (for release) vesicles. These primed vesicles presumably represent the so-called readily releasable pool described earlier and would undergo fusion with the plasma membrane (exocytosis) upon presentation of the appropriate calcium signal. Synaptotagmin is currently the leading molecular candidate for being the calcium sensor. The SNARE hypothesis has proven to be highly heuristic; however, much more work is still needed to fully comprehend the dynamics and mechanisms of secretion.

VI. CONCLUDING REMARKS

Since the initial proposal of the concept of neurosecretion by Ernst Scharrer approximately 70 years ago, there has been an impressive affirmation and expansion of this idea. We now know that all endocrine functions in mammals are orchestrated by the secretion of regulatory neuropeptides into the portal system by hypothalamic neuroendocrine cells. The emergence of neuropeptide secretion as a major modulatory influence within the central nervous system itself has further enhanced the significance of neurosecretion. It is now clear that all neurons secreting biologically active peptides, independent of function (endocrine versus synaptic), have common cell biological properties and are participating in the process of neurosecretion.

Bibliography

Arch, S., and Gainer, H. (1985). Neurosecretion: *Handbook Neurochem.* **8**, 281–307.

Bauerfeind, R., and Huttner, W. B. (1993). Biogenesis of constitutive secretory vesicles, secretory granules and synaptic vesicles. *Curr. Opin. Cell Biol.* **5**, 628–635.

Bennett, M. K., and Scheller, R. H. (1993). The molecular machinery for secretion is conserved from yeast to neurons. *Proc. Natl. Acad. Sci. USA* **90**, 2559–2563.

Castel, M., Gainer, H., and Dellmann, H. D. (1984). Neuronal secretory systems. *Int. Rev. Cytol.* **88**, 303–459.

De Camilli, P., and Jahn, R. (1990). Pathways to regulated exocytosis in neurons. *Annu. Rev. Physiol.* **52**, 625–645.

Gainer, H. (1992). Intracellular protein trafficking and proprotein processing: An overview. In *Mechanisms of Intracellular Trafficking and Processing of Proproteins* (Y. P. Loh, Ed.), pp. 1–17. CRC Press, Boca Raton, FL.

Hokfelt, T. (1991). Neuropeptides in perspective: The last ten years. *Neuron* **7**, 867–879.

Jahn, R., and Sudhof, T. C. (1994). Synaptic vesicles and exocytosis. *Annu. Rev. Neurosci.* **17**, 219–246.

Joose, J., Buijs, R. M., and Tilders, F. J. H. (Eds.) (1992). The peptidergic neuron. *Prog. Brain Res.* **92**, 1–408.

Maddrell, S. H. P., and Nordmann, J. J. (1979). *Neurosecretion.* Blackie & Son, Glasgow.

Nassel, D. R. (1996). Peptidergic neurohormonal control systems in invertebrates. *Curr. Opin. Neurobiol.* **6**, 842–850.

Posiner, A. M., and Trifaro, J. M. (Eds.) (1982). *The Secretory Granule.* Elsevier, Amsterdam.

Scharrer, B. (1987). Neurosecretion: Beginnings and new directions in neuropeptide research. *Annu. Rev. Neurosci.* **10**, 1–17.

Zupane, G. K. H. (1996). Peptidergic transmission: From morphological correlates to functional implications. *Micron* **27**, 35–91.

Neurotransmitters

William F. Ganong
University of California, San Francisco

I. Chemical Messengers
II. Neurons
III. Synapses
IV. Extrasynaptic Effects of Neurotransmitters
V. Metabolism and Reuptake
VI. Chemistry
VII. Multiple Roles
VIII. Cotransmitters
IX. Multiple Receptors
X. Synaptic Plasticity

GLOSSARY

cotransmitter A neurotransmitter secreted by a neuron that secretes two or more neurotransmitters.

neural hormone A chemical messenger secreted by a neuron into the circulating blood.

neuromodulator A substance secreted by a neuron that does not change the membrane potential of other neurons but alters their excitability.

neurotransmitter A chemical messenger secreted by neuron at a synapse.

synapse A specialized junction between the ending of a presynaptic neuron and the dendrite, cell body, or axon of a postsynaptic neuron.

synaptic plasticity Modification of the strength of synaptic transmission by past experience.

I. CHEMICAL MESSENGERS

Neurotransmitters are a special form of chemical messenger by which neurons communicate with other neurons, muscle cells, and gland cells (Fig. 1). Almost all cells communicate with other cells by way of chemical messengers. Some of these substances pass directly from cell to cell via gap junctions (Fig. 2). Others are secreted by one cell, pass in the extracellular fluid to other cells, and produce physiologic responses in these target cells by acting on receptors in the cell membrane or inside the cell. If the secreted messenger enters the circulating body fluids, it is known as a hormone (Fig. 2). If it simply crosses from one nerve cell to another a few nanometers away at synapses, the special junctions between the nerve cells, it is known as a neurotransmitter. Between these extremes, some substances diffuse in extracellular fluid without entering the bloodstream and act on receptors in cells some distance away. These substances are known as paracrine mediators. Both hormones and neurotransmitters can act in this fashion. Others can act back on the cells that secreted them (autocrine mediators).

II. NEURONS

Neurons generally have a cell body that contains the nucleus and the protein synthesizing machinery of the cell, many branched dentrites, and a long axon that may also be branched and can extend for relatively great distances in the body (Fig. 1). The endings of most axons in the central nervous system terminate in synapses on other neurons.

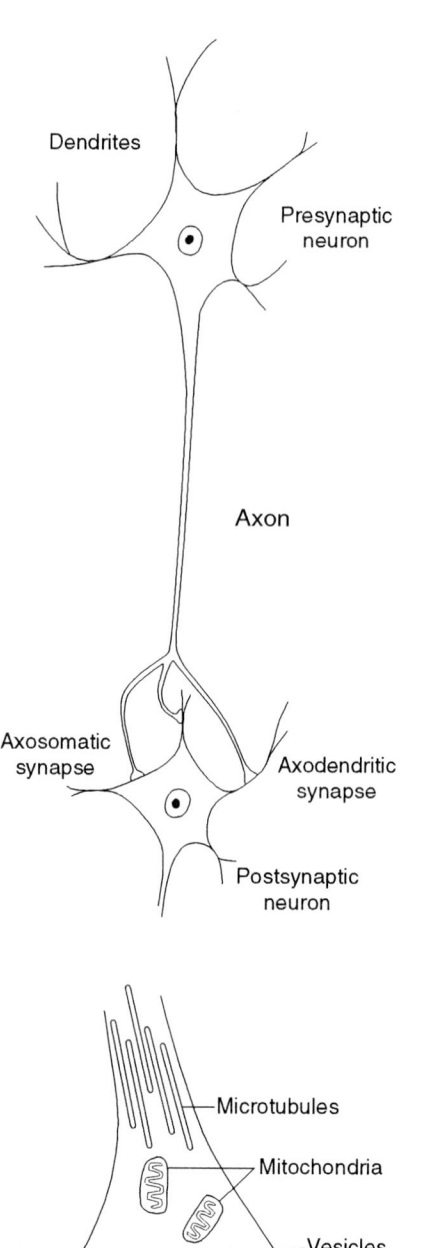

FIGURE 1 (Top) Relation between presynaptic and postsynaptic neurons, showing two axodendritic synapses and one axosomatic synapse. (Bottom). Enlargement of synaptic junction, showing the vesicles that contain neurotransmitters in the ending of the presynaptic neuron and the specialization of the postsynaptic membrane.

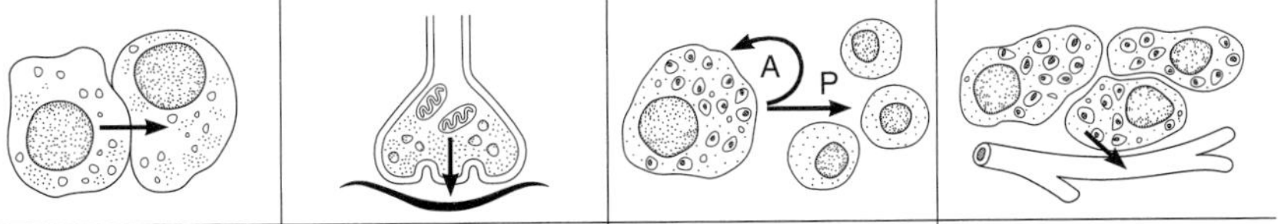

	GAP JUNCTIONS	SYNAPTIC	PARACRINE AND AUTOCRINE	ENDOCRINE
Message transmission	Directly from cell to cell	Across synaptic cleft	By diffusion in interstitial fluid	By circulating body fluids
Local or general	Local	Local	Locally diffuse	General
Specificity depends on	Anatomic location	Anatomic location and receptors	Receptors	Receptors

FIGURE 2 Intercellular communication by chemical messengers. A, autocrine; P, paracrine (reproduced with permission from W. F. Ganong, *Review of Medical Physiology*, 18th ed., Appleton & Lange, Stamford, CT, 1997).

III. SYNAPSES

A. Morphology

At synaptic junctions there are many mitochondria and vesicles containing neurotransmitters in the presynaptic nerve ending. In general, the synaptic vesicles containing acetylcholine or amino acid transmitters are small and clear and those containing amine transmitters are small and granulated, i.e., they have a dense central core. Polypeptides are found in large granulated vesicles, and these may also contain amines. Between the presynaptic and postsynaptic neurons there is a narrow synaptic cleft that is 20–40 nm wide. Neurotransmitters are released into this cleft and act on the membrane of the postsynaptic cell. This membrane is thickened and specialized, and it contains many receptors. Somewhat similar junctions occur at the locations where neurons innervate muscle cells or gland cells.

B. Electrical and Chemical Events

The language of neurons is the action potential, the all-or-none electrical disturbance that is propagated from the cell body to the axon terminals. At the axon terminals, the action potential increases calcium permeability, and the calcium that enters the endings brings about exocytosis of the synaptic vesicles, releasing their contents into the synpatic cleft. The neurotransmitters generally open ion channels in the postsynaptic neuron. When this results in a net influx of positive charge, the neuron becomes more excitable. When this results in a net efflux of positive charge, the neuron becomes hyperpolarized and less excitable. These local, nonpropagated depolarizing and hyperpolarizing potentials sum temporally and spatially. If the net potential change is depolarizing enough to reach the firing level of the postsynaptic neuron, an axon potential is produced.

In some instances, chemical messengers produced by neurons do not produce changes in membrane potential but instead by chemical mechanisms alter the sensitivity of neurons to neurotransmitters. These substances are neuromodulators, and their effects can be widespread. Many are polypeptides, but some may be neurosteroids, steroids produced by the brain that act locally in this fashion.

IV. EXTRASYNAPTIC EFFECTS OF NEUROTRANSMITTERS

In addition to acting at synaptic junctions, neurotransmitters can diffuse away from the synapse and act on nearby receptors in a paracrine fashion. Indeed, the correspondence between presynaptic and postsynaptic specializations is rarely one to one, and paracrine action combined with synaptic action may be relatively common.

It should also be noted that there are neurotransmitter receptors in the cell membranes of presynaptic as well as postsynaptic neurons. When neurotransmitters occupy these presynaptic receptors, they facilitate or inhibit neurotransmitter release at the synaptic junction. Presynaptic inhibition is more common than presynaptic facilitation and provides a feedback mechanism to prevent excessive or prolonged neurotransmitter release.

V. METABOLISM AND REUPTAKE

The action of neurotransmitters is generally rapid in onset and short in duration. Extracellular transmitter is removed from the synaptic cleft and its environs by metabolism and reuptake. Enzymes specific for the secreted transmitter are concentrated at synaptic junctions, although they are also found in other locations. Quantitatively more important in terms of terminating the action of a neurotransmitter and conserving the transmitter is reuptake into the presynaptic terminal. Transporter proteins specialized for this purpose have been cloned for norepinephrine, dopamine, serotonin, histamine, glycine, and γ-aminobutyric acid (GABA), and there are several transporters that are responsible for the reuptake of glutamate. Acetylcholine is not taken up directly, but there is an active uptake of its precursor choline. Once taken into the cytoplasm of the presynaptic neuron, other transporters concentrate the neurotransmitter in the secretory vesicles. The importance of reuptake is demonstrated by the fact that drugs which inhibit reuptake of specific neurotransmitters produce significant increases in the concentration of the neurotransmitter in the synaptic cleft. This augments and prolongs the action of the neurotransmitter, producing marked clinical effects. For example, serotonin deficiency appears to be involved in the pathogenesis of depression, and the widely used antidepressant Prozac acts by inhibiting serotonin reuptake.

VI. CHEMISTRY

A somewhat arbitrary and probably incomplete list of established neurotransmitters and neuromodulators is presented in Table 1. For the most part, the transmitters are simple molecules. Acetylcholine, the acetyl ester of the quaternary ammonium compound choline, is ubiquitous and in a chemical class by

TABLE 1
Neurotransmitters and Neuromodulators in the Nervous System of Mammals[a]

Substance	Location
Acetylcholine	Myoneural junction; preganglionic autonomic endings, postganglionic sympathetic sweat gland, and muscle vasodilator endings; many parts of brain; endings of some amacrine cells in retina
Amines	
Dopamine	SIF cells in sympathetic ganglia; striatum, median eminence, and other parts of hypothalamus; limbic system; parts of neocortex; endings of some interneurons in retina
Norepinephrine	Most postganglionic sympathetic endings; cerebral cortex, hypothalamus, brain stem, cerebellum, spinal cord
Epinephrine	Hypothalamus, thalamus, periaqueductal gray, spinal cord
Serotonin	Hypothalamus, limbic system, cerebellum, spinal cord; retina
Histamine	Hypothalamus, other parts of brain
Excitatory amino acids	
Glutamate	Cerebral cortex, brain stem
Aspartate	Spinal cord, other parts of CNS?
Inhibitory amino acids	
Glycine	Neurons mediating direct inhibition in spinal cord, brain stem, forebrain; retina
γ-Aminobutyrate	Cerebellum; cerebral cortex; neurons mediating presynaptic inhibition; retina

continues

Continued

Polypeptides	
Substance P, other tachykinins	Endings of primary afferent neurons mediating nociception; many parts of brain; retina
Vasopressin	Posterior pituitary; medulla; spinal cord
Oxytocin	Posterior pituitary; medulla; spinal cord
CRH	Median eminence of hypothalamus; other parts of brain
TRH	Median eminence of hypothalamus; other parts of brain; retina
GRH	Median eminence of hypothalamus
Somatostatin	Median eminence of hypothalamus; other parts of brain; substantia gelatinosa; retina
GnRH	Median eminence of hypothalamus; circumventricular organs; preganglionic autonomic endings; retina
Endothelins	Posterior pituitary, brain stem
Enkephalins	Substantia gelatinosa, many other parts of CNS; retina
β-Endorphin, other derivatives of proopiomelanocortin	Hypothalamus, thalamus, brain stem; retina
Endomorphins	Thalamus, hypothalamus, striatum
Dynorphins	Periaqueductal gray, rostroventral medulla, substantia gelatinosa
Cholecystokinin (CCK-4 and CCK-8)	Cerebral cortex; hypothalamus; retina
Vasoactive intestinal polypeptide	Postganglionic cholinergic neurons; some sensory neurons; hypothalamus; cerebral cortex; retina
Neurotensin	Hypothalamus; retina
Gastrin-releasing peptide	Hypothalamus
Gastrin	Hypothalamus; medulla oblongata
Glucagon	Hypothalamus; retina
Motilin	Neurohypophysis; cerebral cortex, cerebellum
Secretin	Hypothalamus, thalamus, olfactory bulb, brain stem, cerebral cortex, septum, hippocampus, striatum
Calcitonin gene-related peptide-α	Endings of primary afferent neurons; taste pathways; sensory nerves; medial forebrain bundle
Neuropeptide Y	Noradrenergic, adrenergic, and other neurons in medulla, periaqueductal gray, hypothalamus; autonomic nervous system
Activins	Brain stem
Inhibins	Brain stem
Angiotensin II	Hypothalamus, amygdala, brain stem, spinal cord
FMRF amide	Hypothalamus, brain stem
Galanin	Hypothalamus, hippocampus, midbrain, spinal cord
Atrial natriuretic peptide	Hypothalamus, brain stem
Brain natriuretic peptide	Hypothalamus, brain stem
Purines	
Adenosine	Neocortex, olfactory cortex, hippocampus, cerebellum
ATP	Autonomic ganglia, habenula
Gases	
NO, CO	CNS
Lipids	
Anandamide	Hippocampus, basal ganglia, cerebellum

[a] Transmitter functions have not been proved for some of the polypeptides (modified and reproduced with permission from W. F. Ganong, *Review of Medical Physiology,* 18th ed., Appleton & Lange, Stamford, CT, 1997).

itself. Other neurotransmitters are amines. Three of these—dopamine, norepinephrine, and epinephrine—are derivatives of catechol and are known as catecholamines. Two other important amine transmitters are serotonin (5-hydroxytryptamine) and histamine. Amino acids comprise a third group of neurotransmitters. Glutamate is the main excitatory neurotransmitter in the brain, and glycine and GABA are the main inhibitory transmitters—GABA in the brain and glycine in the spinal cord. Purines such as adenosine and ATP function as neurotransmitters. Apparently, so do two compounds that are gases: nitric oxide and carbon monoxide. Recent research has established that the lipid anandamide is the endogenous ligand for the receptor that binds tetrahydrocannabinol, the psychoactive ingredient in marijuana. Finally, a large number of neurotransmitters and presumed neuromodulators are polypeptides. These include substance P, which is a major transmitter at the first synapse in pathways mediating pain; opioid peptides, the endogenously produced compounds that bind to morphine receptors ("the body's own morphine"); and polypeptides related to inhibins in the gonads. They also include gastrointestinal hormones such as vasoactive intestinal polypeptide and various forms of cholecystokine, angiotensin II, and at least seven hypothalamic hormones.

VII. MULTIPLE ROLES

It is worth noting that neurons can secrete hormones as well as neurotransmitters; for example, vasopressin and corticotropin-releasing hormone. Furthermore, the same chemical messenger can serve as a neurotransmitter in one location, a hormone secreted by neurons in another, and a hormone secreted by gland cells in a third. Somatostatin is an example: It is secreted as a neurotransmitter or a neuromodulator in various parts of the brain; as a growth hormone-inhibiting hormone by neurons in the hypothalamus; and as a paracrine and hormonal product by the endocrine D cells in the pancreatic islets and gastrointestinal mucosa. The same is also true for other polypeptides and for the amine transmitters. Thus, for example, dopamine, norepinephrine, and epinephrine are secreted as hormones by the adrenal medulla and serotonin is probably secreted as a hormone by cells in the gastrointestinal mucosa. The important point is that at least for amines and polypeptides, the same substances are secreted by very different cells and are chemical messengers in both the endocrine and the nervous systems.

VIII. COTRANSMITTERS

A unique feature of some neurons is the presence of two or more neurotransmitters in them. This is especially true in hypothalamic paraventricular neurons, which in certain circumstances contain four or five neurotransmitters. A common combination is a catecholamine or serotonin and a polypeptide, although various combinations of polypeptides can also occur. Of course, the presence of neurotransmitters does not prove secretion, but in some instances release of cotransmitters has been demonstrated experimentally. There is evidence that one neurotransmitter sometimes potentiates the action of the other, but the exact significance of the cotransmitter arrangement is still unknown.

IX. MULTIPLE RECEPTORS

The receptors on which neurotransmitters act trigger a variety of intracellular responses. Many act directly on ion channels, and many others act via heterotrimeric G proteins to alter intracellular cyclic AMP, Ca^{2+}, and protein kinases. The molecular biology of receptor action is beyond the scope of this article. However, it is worth noting that not only are the neurotransmitters numerous and varied but also the receptors for each neurotransmitter that have been studied in detail are numerous and varied. Thus, for example, there are at least 5 different dopamine receptors, 9 different norepinephrine and epinephrine receptors, and 13 different serotonin receptors. In each case, all are related but have slightly different biochemical structures and effects as well as different localizations in the brain and other parts of the nervous system. In the case of the amino acid receptors, the number is even greater because the receptors are constructed of various combinations of many different subunits. This permits much greater specialization and diversification in the physiological effects of a given neurotransmitter than would be possible with neurotransmitter diversity alone.

X. SYNAPTIC PLASTICITY

Not only are there multiple neurotransmitters and multiple receptors for many and perhaps all of the neurotransmitters but also the strength of synaptic transmission in a given synapse can be altered by experience. For example, repeated rapid (tetanizing) stimulation of many presynaptic neurons leads to a subsequent prolonged increase in the postsynaptic depolarization produced by a subsequent test stimulus (long-term potentiation). Conversely, stimulation of other presynaptic neurons or stimulation at different parameters leads to a subsequent prolonged decrease in the postsynaptic response (long-term depression). Both long-term potentiation and long-term depression appear to be widespread phenomena at synaptic junctions in the nervous system, and they are of particular interest because of their implications regarding the physiology of learning and memory.

See Also the Following Articles

NEUROENDOCRINE SYSTEM; NEUROPEPTIDES

Bibliography

Baulieu, E.-E. (1997). Neurosteroids: Of the nervous system, by the nervous system, for the nervous system. *Recent Prog. Horm. Res.* **52**, 1–30.

Brann, D. W., and Mahesh, V. B. (1994). Excitatory amino acids: Function and significance in reproduction and neuroendocrine regulation. *Front. Neuroendocrinol.* **15**, 3–50.

Hoffman, B., *et al.* (1998). Distribution of monoamine neurotransmitter transporters in the rat brain. *Front. Neuroendocrinol.* **19**, in press.

Lenkei, Z., Palkovits, M., Corvol, P., and Llorens-Cortès, C. (1997). Expression of angiotensin type-1 (AT1) and type-2 (AT2) receptor mRNA in the adult rat brain: A functional neuroanatomical review. *Front. Neuroendocrinol.* **18**, 383–439.

Rouillè, Y., *et al.* (1995). Proteolytic processing mechanisms in the biosynthesis of neuroendocrine peptides: The subtilisin-like protein convertases. *Front. Neuroendocrinol.* **16**, 322–361.

Vidal, C., and Changeux, J.-P. (1996). Neuronal nicotinic acetylcholine receptors in the brain. *Newslett. Physiol. Sci.* **11**, 202–205.

Nursing Behavior
see Suckling Behavior

Nutritional Factors and Reproduction

Gary L. Williams

Texas A&M University

I. Energy Nutrition and the Central Control of Reproduction
II. Metabolic Signals to the Gonads
III. Reproductive Sequelae to Altered Energy Balance
IV. Nitrogen Balance and Reproduction
V. Vitamins, Cofactors, Minerals, and Essential Fatty Acids in Reproduction

GLOSSARY

adipocyte A fat cell.

aspartate An acidic amino acid with "excitatory" or neurotransmitter activity that is released from presynaptic nerve terminals within the brain.

body condition score A score, usually determined visually and/or by external palpation, used to subjectively estimate the degree of fatness of an animal.

glutamate An acidic amino acid with "excitatory" or neurotransmitter activity that is released from presynaptic nerve terminals within the brain.

gonadostat hypothesis A hypothesis that puberty occurs as a function of decreased negative feedback sensitivity to estradiol-17β by brain/hypothalamic centers controlling gonadotropin secretion.

insulin-like growth factor-binding proteins A class of proteins that bind and carry insulin-like growth factors in the circulation.

insulin-like growth factors I and II Peptide growth factors (somatomedins), produced primarily in the liver but also throughout other tissues and organs, that mediate the local effects of growth hormone (somatotropin).

leptin A newly discovered hormonal product of the obese (*ob*) gene and expressed by adipocytes; believed to be involved in regulating food intake, metabolism, and reproductive function.

neuropeptide Y An orexigenic agent linking nutrient status to hypothalamic regulation of feeding behavior and function of the hypothalmic–hypophyseal axis; an intermediary of leptin activity.

volatile fatty acids Short-chain fatty acids (e.g., acetic, propionic, and butyric acids) that are products of rumen fermentation, are absorbed directly into the bloodstream, and serve as the primary energy sources in ruminant animals.

The most limiting factor for reproductive success in wild and domesticated mammals is nutrition. Nutritional status regulates age at puberty, physiological events that control the estrous cycle, the efficiency of fertilization, implantation, pregnancy maintenance, and length of the anovulatory interval postpartum. In nature, the ability of the reproductive system to respond to changes in nutrition is a critically important adaptation for survival under changing environmental conditions. The study of how nutrients and their metabolism affect reproductive processes has gained great momentum in recent years. The purpose of this chapter is to provide a succinct overview of the current status of our understanding of nutrition–reproduction interactions in domestic mammals.

I. ENERGY NUTRITION AND THE CENTRAL CONTROL OF REPRODUCTION

The master control center for reproduction in both males and females resides within the hypothalamus. As described in detail elsewhere in this volume, the hypothalamus is characterized by a neural oscillator that regulates the timing and pattern of gonadotropin-releasing hormone (GnRH) discharge from widely dispersed secretory neurons. These cells have terminals that end within the arcuate region of the medial basal hypothalamus and discharge GnRH into the hypophyseal portal vessels of the anterior pituitary. The cellular origins of the electrical pacemaker underlying the integration of these widely dispersed neurons have not been clearly elucidated; however, electrodes placed in the arcuate region of the medial basal hypothalamus in monkeys and goats readily record its multiunit electrical activity. Gonadotropin-releasing hormone in turn binds to specific, high-affinity, membrane-bound receptors on gonadotrophs, resulting in the synthesis, storage, packaging, and release of luteinizing hormone (LH) and follicle-stimulating hormone (FSH) into the peripheral circulation.

Dietary energy intake and metabolism exert profound effects on this neurohumoral communication system, both before and after sexual maturation. Indeed, studies to determine how energy intake influences reproductive processes have focused to a large degree on how the brain receives, interprets, and responds to changes in energy substrates.

A. Perception of Metabolic State by the Brain

It is clear that the brain is able to detect even subtle changes in metabolic fuel availability. As a result, the pattern of GnRH, and thus the gonadotropins, is either diminished or enhanced, depending on the direction of change of metabolic fuels. Efforts to identify and characterize the chemical signals that communicate somatic metabolism to the brain, including those areas involved in regulation of the pulse generator, are ongoing. It is likely that there are a variety of metabolic cues that impose a degree of redundancy, and that metabolic cues may affect the control of GnRH and gonadotropin secretion at multiple loci.

In the female, in which most of the research has been conducted, acute and severe undernutrition almost immediately decreases GnRH/LH secretory activity in monogastric species previously maintained on normal planes of nutrition. However, fasting for short periods is less disruptive to gonadotropin secretion in ruminants (cattle, sheep, and goats) because of the fermentation processes within the forestomach (rumen). Therefore, changes in the production and absorption of energy-rich volatile fatty acids from the rumen occur much more slowly in response to dietary modifications. Several days are required for fasting to suppress basal LH secretion in ruminants as opposed to only hours in monogastrics. Nevertheless, once caloric restriction suppresses peripheral LH levels in previously well-nourished animals, refeeding in both monogastrics and ruminants results in a rapid restoration of LH secretion. However, in chronically undernourished animals, in which most of the body fat has been metabolized, refeeding does not result in an immediate restoration of normal gonadotropin secretion or reproductive activity. Instead, a critical level of "fatness" is required in both animals and man before reproductive cyclicity, at least in the female, resumes. In animal agriculture, this degree of fatness is referred to as "body condition," and numerical scoring systems have been developed, particularly for cattle and horses, to subjectively quantify body condition. Body condition scores are directly correlated with reproductive potential.

B. Metabolic Signals to the Brain and Hypothalamus

The recent scientific literature is rich in information regarding putative mechanisms for chemically signaling metabolic state to the brain. Based on these studies, it has been possible to form a hypothetical working model that includes an array of metabolic fuels, hormones, growth factors, and other elements as part of a dynamic communication network.

1. Metabolic Fuels

Glucose has considerable appeal as a metabolic signal for neuroendocrine tissue because it is the primary energy fuel for the brain, and changes in serum glucose often parallel modifications in LH secretion, particularly in monogastric species. In pathological states such as diabetes and following insulin-induced hypoglycemia, the frequency of LH pulses is diminished and reproduction is compromised. A variety of experiments have been conducted to understand the relationship between glucose and insulin in signaling nutritional status and modulating gonadotropin release. The relationship is not a simple one. Although drugs that block glycolysis reduce serum glucose and suppress the pulse generator, intracerebroventricular administration of insulin decreased, rather than increased, LH secretion. However, when insulin and glucose were administered together, LH secretion did increase. Sensors of glucose tone are scattered throughout the brain, and a number of experiments show that the area postrema, a caudal region of the brain adjacent to the ventricular system, seems to have special importance for detecting glucose.

The roles of the volatile fatty acids (VFAs) and the nonesterified fatty acids (NEFAs) have also been of interest, particularly in ruminants. The VFAs serve as primary sources of energy for ruminant animals, and NEFAs increase in the peripheral circulation of all animals in negative energy balance. While direct infusion of NEFA is without effects on LH secretion, infusion of propionate or feeding diets that altered propionate production in the rumen were gluconeogenic. Thus, both peripheral insulin and glucose increased, which in turn stimulated increases in both basal and GnRH-mediated LH release. Interestingly, in fasted monkeys, an increase in serum glucose was not required for feeding-induced increases in LH secretion. Therefore, it appears that changes in metabolic state are more dependent on the general availability of metabolic fuel than on glucose signaling per se.

2. Leptin, Growth Hormone, and NPY

Leptin, a hormonal peptide produced by adipocytes, was first discovered in genetically obese (ob/ob), infertile mice in 1994. Injection of leptin into these mice leads to a return to normal weight and induces reproductive development and normal gonadal function in both male and female ob/ob mice. Moreover, leptin administration hastens puberty in normal mice. Leptin decreases appetite and energy expenditure, and the amount of leptin in normal animals is determined by the amount of fat in the body. Additional work has now resulted in the working hypothesis that leptin may serve as the long sought after link between metabolic state and the reproductive system. Leptin receptors are located throughout the body, but importantly are located in areas of the brain in which the blood–brain barrier is weak.

A central mediator of leptin action appears to be the neurotransmitter, NPY. Leptin treatment reduces NPY mRNA in the hypothalamus and may mediate the effects of leptin on GnRH secretion. NPY induces feeding behavior and inhibits gonadotropin secretion when infused into the third ventricle. Some controversy exists about the nature of NPY action on the hypothalamic–hypophyseal axis. In normal fed rats, there is overwhelming evidence that intermittent injections of NPY actually increase LH secretion in the presence of the gonadal hormones and amplify the action of other interacting stimulatory signals. To the contrary, animals subjected to dietary energy restrictions exhibit increased levels of NPY and NPY mRNA within the hypothalamus and decreased release of LH. Infusion of NPY into the third ventricle results in an immediate inhibition of LH release. Hence, it is conceivable that the role of NPY is to serve as a positive, intermittent signal during states of normal nutrition, but during chronic undernutrition it may act as a central mediator of undernutrition to suppress reproduction. Interestingly, mouse mutants lacking NPY appear normal, leading to the conclusion that there must be alternative neuromodulators of leptin action within the brain.

The actions of NPY are also coupled to growth hormone (GH) secretion. One of the most conspicuous features of dietary energy restriction and negative energy balance in most mammals, including man and domestic farm species, is a dramatic increase in peripheral levels of GH and a fall in serum insulin-like growth factor-I (IGF-I). The rise in GH has been attributed to a fall in somatostatin in the hypophyseal portal system, with no change in GH-releasing hormone release. To date, no relationship has been found directly linking elevated GH secretion to LH release; however, as discussed later, GH plays an important role in mediating the effects of gonadotropins at the gonadal level. Under normal dietary conditions, GH secretion is tightly coupled to the hormone, IGF-I, and its binding proteins which mediate GH effects on target tissues. However, the GH/IGF-I axis becomes uncoupled during states of severe undernutrition, IGF-I is suppressed, and shifts occur in the proportions of IGF-binding proteins (BPs). During periods of recovery from anovulatory states, such as lactational anestrus, GH declines and IGF-I increases in the peripheral circulation. This has been taken to indicate that IGF-I and its bioavailability may be linked with changes within the central nervous system that regulate gonadotropin secretion or

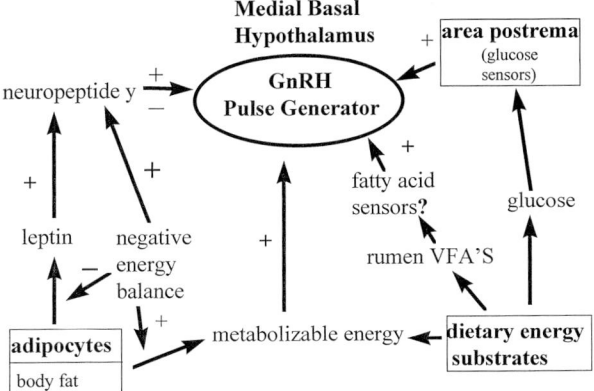

FIGURE 1 Hypothetical Illustration of how the central nervous system perceives and responds to changes in nutrient availability. Diagram shows the positive influence of dietary energy substrates on the hypothalamic pulse generator via increased availability of metabolizable energy and through putative direct effects of glucose and volatile fatty acids (ruminants) on sensors within the brain. During normal states of nutrition, leptin, a hormonal product of adipocytes, positively modulates the GnRH pulse generator through neuropeptide Y. When dietary energy intake is restricted, negative energy balance results in the net metabolism of body fat reserves, and leptin production declines. Ironically, during states of severe undernutrition, the positive link between leptin and neuropeptide Y is disturbed. While leptin synthesis declines, neuropeptide Y synthesis increases dramatically, and LH secretion is suppressed. It has not been determined conclusively whether this effect of neuropeptide Y occurs at the hypothalamic level, pituitary level, or both.

other functions at specific gonadal target sites. The rat and mouse seem to represent exceptions to the relationships described previously because the GH/IGF-I axis does not appear to be responsive to hypoglycemia.

3. Excitatory Amino Acids and Gastrointestinal Signals

Glutamate and aspartate, two acidic amino acids, are also known as excitatory amino acids and have neurotransmitter activity within the brain. Glutamate, N-methyl-D-aspartate, and other related agonists stimulate LH secretion when infused directly into the bloodstream or brain ventricular system. Both aspartate and glutamate are found in large concentrations in presynaptic areas of several hypothalamic nuclei and during nutritional restriction may serve as another of the redundant mechanisms through which metabolic state is interpreted by the brain. However, the precise role of excitatory amino acids as metabolic signals for reproduction remains to be determined. Both the nervous system and the gastrointestinal peptide, cholecystokinin (CCK), have been implicated in the signaling mechanisms linking nutrition to reproduction. In studies with fasted rats, transection of the subdiaphragmatic and gastric vagal nerve branches quickly restored the pulsatile pattern of LH release. Transection of these nerve tracts was shown to cause the loss of opioid receptors, implicating the opioid peptides as potential factors in the regulatory process. Evidence for a role for CCK is less convincing, and current evidence does not justify emphasis on this peptide as a viable metabolic signal to the brain. Figure 1 is a hypothetical illustration that summarizes how the central nervous system perceives and responds to changes in metabolizable energy.

II. METABOLIC SIGNALS TO THE GONADS

Traditionally, the effects of undernutrition or malnutrition on reproduction have been considered from two perspectives: (i) their effects on the brain and thus on the central control of reproduction, or (ii) their influence on overall well-being, which in turn would be assumed to negatively impact reproductive potential. Recently, the roles that specific hormones, metabolites, and growth factors have in mediating the effects of nutrition on peripheral reproductive tissues have been considered.

A. The GH/IGF-I Axis and Insulin

There are a host of growth factors and hormones, acting as autocrine or paracrine agents, that influence the function and metabolism of both male and female gonadal cells. The precise mechanisms through which any of these agents are affected by nutrition have not been deduced. The most studied factors related to nutrition are IGF-I and IGF-BPs. IGF-I stimulates granulosa cell proliferation, aromatase activity, and progesterone biosynthesis. Ovarian cells, particularly luteal cells, contain GH receptors, and GH sensitizes the ovary to gonadotropin stimulation. In cattle, exogenous GH (bovine somatotropin) increases numbers of ovarian follicles and increases the weight and progesterone output of corpora lutea. GH stimulates the production of IGFs throughout the body, particularly in the liver, which increases circulating IGFs under adequate nutritional conditions. During follicular development, IGF-I concentrations increase in follicular fluid as a function of follicle size. Similarly, mRNA concentrations for IGF-I increase in ovarian cells during follicular growth (either granulosa or theca, depending on species). Therefore, GH can potentially affect ovarian function through liver-derived IGF-I, directly at the level of the ovary, or both. Insulin also interacts with ovarian granulosa cells to increase IGF-I binding *in vitro* and enhances granulosa cell steroidogenic potential; hence, insulin probably modulates the function of ovarian granulosa cells either directly or through IGF-I under changing nutritional conditions that influence the secretion of both hormones. There are a host of other growth factors involved in the neuroendocrine–gonadal axis of reproduction; however, how nutritional status precisely regulates their function has not been elucidated.

B. Metabolic Targeting to Enhance Reproduction

Defining dietary energy requirements for reproduction, optimizing energy intake, and exploiting specific metabolic pathways to maximize reproductive efficiency are both a science and an art. For many years, animal husbandmen have manipulated the diets of various classes of livestock to enhance one or more aspects of reproductive performance.

1. Flushing

The term "flushing" is commonly used to describe the process of dramatically increasing energy intake in ewes, sows, and mares for a few weeks before breeding. In polytocous species, such practices are known to increase ovulation rate. However, this practice is not effective if the animals are already in optimal or fat condition. Hence, the effect of flushing, from a scientific perspective, is simply the result of increased energy availability in animals previously maintained in less than optimal body condition. However, because of unique features of the digestive systems of different animal species, there have been a number of other creative methods developed to enhance reproductive performance through targeted manipulation of energy metabolism. These practices are particularly relevant to grazing ruminants in which body condition is often compromised due to harsh environmental conditions that reduce feed availability.

2. Sodium Ionophores

A class of compounds known as sodium ionophore antibiotics have been used extensively in ruminants, particularly cattle, to improve feed efficiency by modifying rumen fermentation processes. A common sodium ionophore, monensin, is produced by *Streptomyces cinnamonensis* and, when fed to ruminants, improves energy efficiency by shifting rumen microbial fermentation toward the production of propionate and away from butyrate and acetate. Propionic acid is gluconeogenic, thus increasing basal insulin secretion. A number of studies have indicated that such changes in cattle have the potential to positively modify basal gonadotropin secretion and pituitary responsiveness to GnRH. In some cases, reproductive performance in the field was improved, particularly in cattle maintained under less than optimal dietary conditions.

3. Dietary Fat

With the exception of the essential fatty acids, which are discussed later, fat serves primarily as a concentrated source of calories, particularly in monogastric species. Therefore, efforts to quickly increase energy density in the diet are benefited by the addition of fat. In nature, ruminant species do not consume significant quantities of fat, but can tolerate amounts of up to 5% of total dry matter intake without detrimental effects on digestion. If the fat is encumbered within whole oilseeds, this level can be exceeded because it is not spread throughout diet dry matter. Instead, it is slowly released in the rumen and much of it actually passes through the rumen undigested. When fat is fed to ruminants, rumen propionate production increases markedly, similar to that described as occurring in response to sodium ionophores. However, when cattle are fed isoenergetic diets containing either 3–8% fat or no fat, those fed fat exhibit an array of metabolic changes that can positively influence ovarian physiology. Feeding fat, particularly polyunsaturated plant oils to cattle, increases the basal secretion of insulin and GH and markedly increases medium-sized follicle populations. Saturated and highly polyunsaturated fats are less effective in creating these changes. Interestingly, all fats, including saturated, polyunsaturated, and highly polyunsaturated, are cholesterologenic in ruminants. Marked differences in serum cholesterol concentrations are not observed in cattle fed these different types of fats. Collectively, the metabolic effects of fat supplementation in cattle are to enhance ovarian cellular proliferation, follicular fluid IGF-I, and, in field studies, rebreeding performance of postpartum females without effects on net energy balance or body weight per se.

III. REPRODUCTIVE SEQUELAE TO ALTERED ENERGY BALANCE

A. Puberty

The majority of work describing the role of nutrition in attainment of puberty has been conducted in the female. Early demographic studies in human females, and extensive work in agricultural species, have demonstrated a close relationship between growth, metabolic rate, and age at onset of puberty. However, until recently, no metabolic, hormonal, or biochemical signal had been identified that could reasonably explain how the brain perceived the critical body mass or basal metabolic rate appropriate for the onset of puberty. As explained earlier, a peptide (leptin) produced and released from adipocytes has recently been identified as a leading candidate for this signal.

During the prepubertal period, the hypothalamus exhibits a heightened sensitivity to the negative feedback effects of estrogens, principally estradiol-17β, relative to that which exists after sexual maturation. This so-called "gonadostat" hypothesis was first proposed over 30 years ago and has been confirmed over time in a variety of species. In the prepubertal heifer, this status is characterized by an increased number of hypothalamic receptors for estradiol compared to postpuberty. Sensitivity to estradiol appears to also play an important role in mediating the effects of other modulators of reproductive function, including photoperiod, suckling, and undernutrition in the adult. Dietary energy restriction retards growth and delays the onset of pulsatile GnRH and LH release and puberty until an adequate nutritional regimen is reinstated to allow the eventual establishment of a "proper" body weight index (Fig. 2). Hence, age at puberty in the female can be hastened or markedly delayed by modifications in diet. Unfortunately, there are complications associated with the strict interpretation of these relationships, and what constitutes the proper body mass or body weight index for puberty has remained undefined. Extreme under- or overfeeding can have additional long-term effects on reproductive development and lactational performance. For example, chronic malnutrition can result in stunting, with attendant failure of normal skeletal development. This has ramifications beyond reproduction alone. Overfeeding of prepubertal heifers and ewe lambs, on the other hand, decreases lactational performance because the amount of mammary

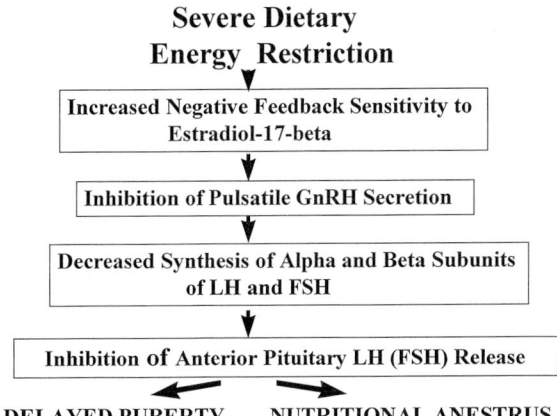

FIGURE 2 Effects of severe dietary energy restriction on the hypothalamic–pituitary axis of pubertal and postpubertal female mammals. In the prepubertal female, dietary energy restriction delays the onset of puberty. In sexually mature females, severe energy deprivation causes nutritional anestrus.

glandular parenchyma for milk production is replaced by fat.

B. Spermatogenesis in the Male

Gross dietary energy restriction delays puberty in the male as it does in the female, through neuroendocrine mechanisms previously described. However, in mature male sheep and goats, there is evidence for a dissociation of the nutrition–gonadotropin connection. Dietary energy restriction results in a temporary (few weeks) suppression of pulsatile LH secretion coincident with the initial loss in body mass. The secretion of LH then returns to normal. To the contrary, maintenance of the nutritional restriction causes a continued divergence in testicular function, with continued negative effects on testicular mass and sperm output for several months. These effects in the male have been termed "GnRH-dependent" and "GnRH-independent" pathways. Overfeeding of young, growing bulls also negatively impacts future reproductive performance by reducing testicular size, daily sperm production, and epididymal reserves of sperm. This effect is the result of an infiltration of testicular parenchyma with fat, similar to the effect of overfeeding on mammary development in the female.

C. Estrous Cycles

Once estrous cycles have been established, it is much more difficult for nutritional restrictions to cause their cessation than for them to be initiated in anestrous, undernourished animals. However, chronic restriction of dietary energy in sexually mature heifers results in decreased secretion of progesterone by the corpus luteum and lowered responsiveness of luteal tissue to gonadotropic stimulation *in vitro*. Young female swine (gilts) and cattle (heifers) fed 40–60% of their dietary energy requirement ceased ovulation within 2 (gilts) to 5 (heifers) months. Cessation of ovulation was not preceded by alterations in estrous cycle lengths. Pulsatile administration of GnRH restored a normal pattern of LH secretion, ovulation, and formation of corpora lutea. In ovariectomized ewes fed 60% of their energy requirements, anterior pituitary contents of FSH and LH declined by 60–70%. This was associated with concomitant declines in mRNAs for the α, LHβ, and FSHβ subunits. Frequent injections of GnRH completely restored gonadotropin contents of the pituitary and coincident mRNAs for the respective subunits.

D. Pregnancy, Parturition, and Postpartum Reproduction

The role of energy nutrition in regulating reproduction during gestation, parturition, and the postpartum period has been studied in a number of species. Both prepartum and postpartum nutrition affect postpartum reproductive performance. A significant depletion of body energy reserves before parturition (e.g., low body condition) lowers the number of females that carry fetuses to term. If energy restriction is severe and occurs early in pregnancy, embryonic mortality is increased relatively soon after fertilization. In cattle, dietary energy intake interacts with suckling to markedly delay the resumption of ovulatory cycles postpartum. This occurs as a consequence of extended periods of low gonadotropin secretion. In cattle and sheep, high levels of circulating placental-derived estradiol characteristically deplete anterior pituitary stores of LH during late gestation, and these stores must be repleted before responsiveness to hypothalamic GnRH stimulation resumes. Low-energy intake and loss of body energy reserves delays the postpartum repletion process and interacts with suckling to inhibit the secretion of GnRH (Fig. 2). Even when estrous cycles resume, first-service conception rates suffer if dietary energy intake remains suboptimal. In species that do not exhibit gestational depletion of anterior pituitary LH, suboptimal energy intake also delays rebreeding. Restoration of a normal dietary intake in animals undernourished during gestation can reverse most of the effects of undernutrition given time, but rebreeding intervals remain extended compared to those receiving adequate nutrition during gestation. Undernutrition during gestation with attending reproductive sequelae have large negative impacts on agricultural production systems. This effect is greater in young (primiparous) females in which growth, lactation, and reproduction must occur simultaneously. Moreover, the effects of extreme undernutrition during gestation extend to lactation, the maternal immune system, and the fetus, resulting in weak, lightweight offspring whose survival are further compromised by the lack of adequate passive transfer of immunoglobins from the mother.

IV. NITROGEN BALANCE AND REPRODUCTION

A. Role of Protein

1. Protein Deficiency

In monogastric species, such as the laboratory rat, very low protein diets result in cessation of estrous cycles. If fertilization occurs, there is a tendency for high embryonic or fetal death and birth of premature and weak offspring. The male also experiences reduced fertility. In other animals, protein deficiency delays puberty and lowers pregnancy rates. It is not clear whether inadequate dietary protein delays puberty as a specific effect of protein itself or simply as the result of hindered growth and maturity. In monogastrics, the 10 essential amino acids are critical elements of protein nutrition. In swine, the amino acids lysine and methionine are of particular importance to reproduction because of the extensive use of corn–soybean-based diets that are low in these amino acids. This is less of a concern in ruminants because the rumen microflora are capable of providing adequate quantities of the essential amino acids not synthesized by the animal's own tissues. Protein requirements increase during the late stages of gestation and level of protein is an important regulator of the level of milk production.

2. Protein Excess

Excess protein has been implicated in reproductive failure (e.g., failure to conceive and early embryonic death) as well, but effects can be subtle and research findings are controversial. Based on studies in dairy cattle, it is hypothesized that excess protein results in greater concentrations of urea nitrogen within the uterine environment, thus changing the pH and reducing conception rates. Since the first report of these findings, there has been no further convincing evidence to support the hypothesis. In general, most research indicates that very high protein levels, as well as high intake of nonprotein nitrogen (NPN) by ruminants adapted to dietary NPN, while not beneficial, do not have negative effects on reproduction.

B. NPN in Ruminants

Because of the unique features of their forestomach, ruminants can utilize NPN sources, such as urea, to meet protein demands. Rumen microflora directly synthesize amino acids from NPN when adequate energy is provided. Hence, even low-protein, high-fiber feed sources are utilized effectively by cattle, sheep, and other ruminant species if animals are adapted to supplemental NPN and provided digestible energy of the diet is adequate to accommodate the NPN load. In unadapted animals, as little as 1% NPN of total dry matter intake can be fatal. Therefore, caution must be employed when using NPN. In adapted ruminants, 25% or more of total dietary nitrogen can be provided by NPN without detrimental effects on reproduction or other variables.

V. VITAMINS, COFACTORS, MINERALS, AND ESSENTIAL FATTY ACIDS IN REPRODUCTION

A. Vitamins and Cofactors

Technically, vitamins are defined as essential organic micronutrients that must be supplied in the diet. What is considered as a vitamin for some species may only be a necessary metabolite for others because it is synthesized within the body (e.g., vitamin C). Observations regarding vitamin deficiency in man and animals appear in recorded history as early as 2600 BC (e.g., beriberi). Although some redefinition and extension of vitamin terminology, nutrition, and biochemistry has occurred in recent years, most of the research regarding the effects of gross vitamin deficiency, including their roles in reproduction, occurred early in this century. Table 1 provides a brief overview of the biochemical actions of the common vitamins and cofactors and some of their general and reproduction-related deficiency symptoms. For a detailed treatise on the role of vitamins in reproduc-

TABLE 1
Overview of Vitamins and Cofactors That Influence Reproduction: Biochemical Actions and General and Reproduction-Related Symptoms Associated with Their Deficiency

Vitamin or cofactor	Biochemical actions	General deficiency symptoms	Reproductive deficiency symptoms
Vitamin A	Multiple actions; not well defined	Blindness; xerophthalmia; poor growth	Failure of spermatogenesis; fetal death
Vitamin D	Absorption/mobilization of Ca^{2+}	Rickets	Congenital malformations
Vitamin E	Biological antioxidant	Multiple system defects	Testicular degeneration; fetal death
Thiamine	Coenzyme; neural factor	Nervous disorders; sudden death	General reproductive failure in horses
Riboflavin	Coenzyme (flavoproteins)	Multiple system defects	Abortion; fetal resorption; weak litters
Niacin	Coenzyme (NAD; NADP)	Multiple system defects	Weak/dead fetuses; involution of gonads
Pyridoxine (B_6)	Coenzyme (protein metabolism)	Multiple system defects	Testicular degeneration
Pantothenic acid	Coenzyme constituent; acyl carrier	Multiple system defects	Reproductive failure; embryonic/fetal death
Biotin	Coenzyme (CHO/fat/protein metabolism)	Multiple system defects	Reproductive failure; embryonic/fetal death
Folic acid	Coenzyme (single carbon transfer)	Megaloblasic anemia; leukopenia	Poor hatchability (poultry)
B_{12}	Coenzyme (cobalamine; methyltransferase)	Anemia; skin/digestive disturbances	Decreased litter size/survivability
Choline	Lipotropic factor; acetylcholine synthesis	Decreased growth; fatty liver; perosis	Decreased conception and litter size
Vitamin C	Antioxidant; cofactor	Scurvy	Decreased fertility; fetal/neonatal death

tion, with specific reference to mechanisms of action and species variability, the reader is referred to McDowell (1989).

B. Minerals

Minerals are an essential component of normal animal diets, and requirements for them vary according to genotype, physiological status, and environmental conditions. The primary macrominerals that play important roles in reproduction are calcium and phosphorous. Several microminerals also play critical roles in reproduction, including cobalt, copper, iodine, manganese, selenium, and zinc. The prevalence of most mineral deficiencies varies by region. Both mineral deficiencies and excesses are often a function of regional soil variations, and these variations in one mineral can interact to produce either deficiencies or excesses of others within the animal. The main minerals that create toxic effects due to soil content are selenium, molybdenum, and fluorine. Table 2 summarizes the primary biochemical functions and general and reproductive symptoms of macro- and micromineral deficiencies that are related directly or indirectly to reproduction.

C. Essential Fatty Acids

There are three fatty acids that animals do not synthesize in adequate amounts and which are necessary for maintaining normal cellular integrity and physiological function. They are linoleic, linolenic, and arachidonic acids. These polyunsaturated fatty acids occur primarily in plant oils, and the latter

TABLE 2
Dietary Minerals That Influence Reproduction: Biological Functions and General and Reproduction-Related Deficiency Symptoms

Mineral	Function	Primary deficiency symptoms	
		General	Reproductive
Macrominerals			
Phosphorous	Bone mineralization	Osteoporosis; osteomalacia	Reduced conception; embryonic/fetal mortality
Calcium	Milk production; bone mineralization	Osteoporosis; osteomalacia	Fetal skeleton deformation
Microminerals			
Cobalt	B_{12} synthesis	Anemia	Decreased litter size/survivability; decreased egg size/hatchability; neonatal mortality
Copper	Enzyme cofactor	Hair bleaching; dermatosis; bone/joint pathology	Delayed/suppressed estrus; neonatal ataxia
Iodine	Component of thyroid hormones	Goiter	Neonatal goiter; irregular estrous cycles; low conception; decreased libido/semen quality
Manganese	Enzyme cofactor	Bone malformations	Low conception; abortion; stillbirths
Selenium	Component of metalloenzymes	Muscle degeneration	White muscle disease; neonatal mortality
Zinc	Enzyme cofactor	Impaired immune function; reduced growth; skin lesions	Testicular dysgenesis

serve as excellent dietary sources. Symptoms of essential fatty acid deficiency include disturbed water balance, dermatosis, poor growth, and marked reductions in reproductive and lactational performance. Linoleic acid can meet most, if not all, of the essential fatty acid requirements because it is metabolized to both linolenic and arachidonic. Linoleic acid is essential for the arachidonic acid cascade that results in the synthesis of prostaglandins, important hormones associated with normal uterine and luteal function. Essential fatty acid deficiencies impair reproduction in multiple ways, including effects on libido and spermatogenesis in males and embryo survival in females.

See Also the Following Articles

ESTROUS CYCLE; GROWTH FACTORS; NEUROTRANSMITTERS; RUMINANTS

Bibliography

Armstrong, J. D., and Benoit, A. M. (1996). Paracrine, autocrine, and endocrine factors that mediate the influence of nutrition on reproduction in cattle and swine: An in vivo IGF-I perspective. *J. Anim. Sci.* 74(Suppl).

Barash, I. A., Cheung, C. C., Weigle, D. S., Ren, H., Kabigting, E. B., Kuijper, J. L., Clifton, D. K., and Steiner, R. A. (1996). Leptin is a metabolic signal to the reproductive system. *Endocrinology* 137(7), 3144–3147.

Ferrell, C. L. (1991). Nutritional influences on reproduction. In *Reproduction in Domestic Animals* (P. T. Cupps, Ed.). Academic Press, San Diego.

Foster, D. L., Nagatani, S., Bucholtz, D. C., Tsukamura, H., Tanaka, T., and Maeda, K. (1997). Metabolic links between nutrition and reproduction: Signals, sensors, and pathways controlling GnRH secretion. In *Nutrition and Reproduction* (W. Hansel and G. Bray, Eds.). in press. Louisiana State Univ. Press, Baton Rouge.

Keisler, D. H., and Lucy, M. C. (1996). Perception and interpretation of the effects of undernutrition on reproduction. *J. Anim. Sci.* 74(Suppl.).

Kinder, J. E., Bergfeld, E. G. M., Wehrman, M. E., Peters, K. E., and Kojima, F. M. (1995). Endocrine basis for puberty in heifers and ewes. *J. Reprod. Fertil.* 49(Suppl.).

Martin, G. B., and Walkden-Brown, S. W. (1995). Nutritional influences on reproduction in mature male sheep and goats. *J. Reprod. Fertil.* 49(Suppl.).

McDowell, L. R. (1989). *Vitamins in Animal Nutrition* (T. J. Cunha, Ed.). Academic Press, San Diego.

Nutrient Requirements of Domestic Animals: A Series (1996). National Academy Press, Washington, DC.

Williams, G. L. (1997). Fat, follicles and fecundity: The ruminant paradigm. In *Nutrition and Reproduction* (W. Hansel and G. Bray, Eds.), in press. Louisiana State Univ. Press, Baton Rouge.

Nutritional Factors and Lactation

Michael J. VandeHaar
Michigan State University

I. Nutrition and Lactation in the Dairy Cow
II. Metabolic Needs of the Mammary Gland
III. Digestion
IV. Metabolism in Support of Lactation
V. Nutrient Partitioning
VI. Practical Nutrition during the Lactation Cycle
VII. Nutrition and Lactation in Nonruminants and Humans

GLOSSARY

digestion The processes by which foodstuffs are broken down into absorbable units.

gluconeogenesis The formation of glucose from noncarbohydrate compounds.

homeorhesis Coordination of metabolism in support of a dominant physiological process.

lipogenesis The formation of fatty acids from acetyl-CoA.

lipolysis The breakdown of triglycerides to release fatty acids.

metabolism The complex physical and chemical processes involved in the maintenance of life.

Nutrition is the series of processes by which an organism takes in and assimilates food for promoting growth, synthesizing secreted products, replacing worn and injured tissue, and doing metabolic work. One distinguishing characteristic of reproduction in mammals is that it is accompanied by lactation. Lactation enables the neonate to use the maternal digestive system to process available foods into the proper balance of required nutrients for growth and survival. Milk composition varies considerably in the different mammalian species, but in most situations the maternal system produces milk in the correct quantity and composition for proper growth of offspring. In addition, humans have used the lactation of other species, especially ruminants, to enhance their own food supply. Thus, humans have used the capability of the ruminant digestive system to process fibrous plant material with little nutritional value for humans into the high-quality nutrients of milk. To produce milk, the lactating mammary gland must be provided with specific metabolites from blood. These metabolites serve as the building blocks for synthesis of milk components and as the fuel to power the synthetic reactions. To enable the gland to produce milk in the desired quantity, all the necessary metabolites must be in adequate supply. A shortage of nutrients in the maternal diet can be overcome, at least in the short run, by mobilization of nutrients from maternal body stores, but eventually milk production will be impaired as maternal supplies are depleted. The mechanisms by which inadequate maternal nutrition impairs lactation are likely twofold: direct effects of nutrient shortage on the mammary gland and indirect effects on the mammary gland through effects of nutrition on the endocrine system.

I. NUTRITION AND LACTATION IN THE DAIRY COW

The effects of nutrition on lactation are perhaps most pronounced and best known in the dairy cow because of her tremendous ability to produce large quantities of milk. During peak lactation, a modern high-producing dairy cow typically produces more than 50 kg of milk per day. Her daily intake of nutrients is at least four times that of a cow at maintenance (not lactating), and more than 75% of her net energy intake is captured as milk.

The quantity of milk produced by a cow is a function of the ability of the mammary gland to produce milk and the ability of the cow to provide the necessary nutrients to the mammary gland. The ability of the gland to produce milk is largely dependent on the number of epithelial cells present in the mammary parenchymal tissue. The number of mammary epithelial cells is partly dependent on mammogenesis before puberty, during pregnancy, and around parturition and on involution which occurs gradually as lactation progresses. Animals that gain more than ~1 kg of body weight per day before puberty often have reduced mammogenesis and decreased subsequent milk production. The reason for this impaired mammogenesis is not clear but the greatest impairment of mammogenesis has occurred in studies in which diets were high in energy but low in protein. Inadequate energy or protein during mid- or late lactation (100 days after parturition or later) can advance the involution process and decrease milk production for the duration of the lactation.

The principal organic components of milk are lactose, triglycerides, and proteins. To make these components, the ruminant mammary gland uses glucose, acetate, ketones, fatty acids, and amino acids. Milk production will be greatest and most efficient when the optimal amounts of each metabolite are supplied to the mammary gland.

II. METABOLIC NEEDS OF THE MAMMARY GLAND

Lactose is the predominant carbohydrate of milk, and lactose synthesis is the major osmotic regulator of total milk production. Bovine milk is ~5% lactose, so a cow producing 50 kg of milk per day synthesizes 2500 g of lactose daily. Of total glucose used by the gland, 70% is for lactose production, and the ruminant mammary gland is adapted to preserve glucose for this function. In nonruminants, mammary tissue converts glucose to lactose and fatty acids and oxidizes glucose in the pentose phosphate pathway and citric acid cycle. The pentose phosphate pathway produces the reduced form of nicotinamide adenine dinucleotide phosphate (NADPH), which is needed for *de novo* fatty acid synthesis in the gland. The citric acid cycle in conjunction with oxidative phosphorylation produces adenosine triphosphate (ATP), which is used to drive the anabolic reactions of milk synthesis. However, in the mammary gland of ruminants, glucose is not converted to fatty acids, almost no glucose is oxidized in the citric acid cycle, and only half of the NADPH is produced by oxidation of glucose. Instead, acetate is the building block for fatty acids, the major fuel for ATP production, and supplies half the NADPH via the isocitrate pathway. Some glucose is converted to glycerol for fatty acid esterification.

The fatty acids of milk fat are derived from two sources: preformed lipids from blood and *de novo* synthesis of fatty acids within the gland, each accounting for half of the milk fatty acids. In ruminants, the primary substrates for *de novo* synthesis of fatty acids are acetate and β-hydroxybutyrate; of these, acetate accounts for ~80% of *de novo* fatty acid–carbon. Nearly all the fatty acids with 4–14 carbons (C4–C14 fatty acids) and 60% of the palmitic acid (C16) are synthesized *de novo* in the gland. Palmitic acid and the C18 fatty acids, stearic and oleic, are the principal fatty acids in plasma lipoproteins of ruminants. Lipoprotein lipase within the capillary wall hydrolyzes the triglycerides of plasma lipoproteins and thus enables transport of these fatty acids into mammary cells, whereupon some of the stearate is desaturated to oleate. Plasma fatty acids are used almost exclusively for milk fat synthesis and account for all of the C18 and nearly half of the C16 milk fatty acids. Mammary tissue may also use nonesterified fatty acids (NEFA) from blood, and NEFA uptake by mammary tissue is directly related to the NEFA concentration of blood. Net mammary uptake of NEFA occurs when cows are in negative energy balance and NEFA release from adipose tissue is high; such is the case during the first month or two of lactation when feed intake is not adequate to support the large amount of milk produced. High concentrations of long-chain fatty acyl-CoA seem to inhibit *de novo* fatty acid synthesis within the mammary gland. Therefore, in early lactation, when much body lipid is being mobilized and is available to the mammary gland, milk fat will contain less short-chain and more long-chain fatty acids.

The major milk proteins of ruminants are casein, β-lactoglobulin, and α-lactalbumin. Excess essential amino acids and insufficient nonessential amino acids are taken up by the mammary gland, and deamination of essential amino acids and synthesis of nonessential amino acids occurs before amino acids are activated for protein synthesis. Although the branched chain amino acids may be used as energy sources for mammary tissue, in general amino acids are not major fuels for lactating mammary gland. Protein synthesis in mammary tissue is similar to that of other tissues. The amino acid profile of individual milk proteins is predetermined by the genetic code, not by dietary constraints. Thus, a deficiency of a single amino acid relative to its requirement by the gland will decrease total synthesis of protein. In ruminants, the amino acid that seems most limiting for milk synthesis is methionine. Lysine may be limiting in some situations, but protein produced by ruminal microbes during fermentation is relatively high in lysine, and thus the gland typically receives enough lysine if total protein nutrition is adequate.

Nutrient supply to the mammary gland directly influences milk synthesis, but it seems quantitatively important only if demands of the mammary gland are not already met. In fasted animals, when glucose was in short supply, milk secretion rate was decreased, and, if glucose then was infused into the mammary blood supply, milk yield could be improved dramatically. However, providing extra glucose, amino acids, or acetate to an already well-supplied mammary gland generally does not increase milk synthesis. Providing extra long-chain fatty acids to the gland seldom alters milk synthesis but may increase milk triglyceride output and decrease the proportion of fatty acids from *de novo* fatty acid synthesis.

III. DIGESTION

The processes involved in converting dietary carbohydrates, fats, and proteins to the metabolites taken up by the mammary gland are shown in Fig. 1. The diet of a cow contains carbohydrates from the cell wall of plants as well as from the contents of plant cells. Cell-wall carbohydrates and related compounds include cellulose (the most abundant carbohydrate on earth), hemicellulose, lignin, and pectins. Of these, lignin is not digestible, pectin is very digestible, and the cellulose and hemicellulose are partly digested. Cell-wall carbohydrate can only be digested by microbial fermentation because the digestive enzymes of mammals are unable to hydrolyze most β-glycosidic bonds. Thus, rumen fermentation enables the cow to utilize fibrous feeds that humans cannot effectively use as a food. Physical dynamics of the rumen enhance the efficiency with which the ruminant digests fiber. Rumination or cud-chewing aids in breaking down fibrous particles for access by rumen microbes. During rumination, saliva is produced that contains bicarbonate and phosphates to help keep the rumen at a pH of 6 or 7 for an optimal environment for rumen microbes. To achieve adequate salivation and an optimal rumen pH, diets for cows should contain at least 25% cell-wall carbohydrate. Furthermore, particles in the rumen are selectively retained—digesta particles in a cow generally do not leave the rumen and pass on to the omasum until they are small enough (~2–4 mm) and dense enough (1.1–1.3 g/ml). Once a fiber particle leaves the rumen, it may be fermented in the hindgut, and the hindgut accounts for ~10% of total production of fermentation acids. Most fermentation acids are absorbed directly through the epithelium of the rumen into the portal bloodstream.

Carbohydrates of plant cell contents include starches and sugars. Although some starch will es-

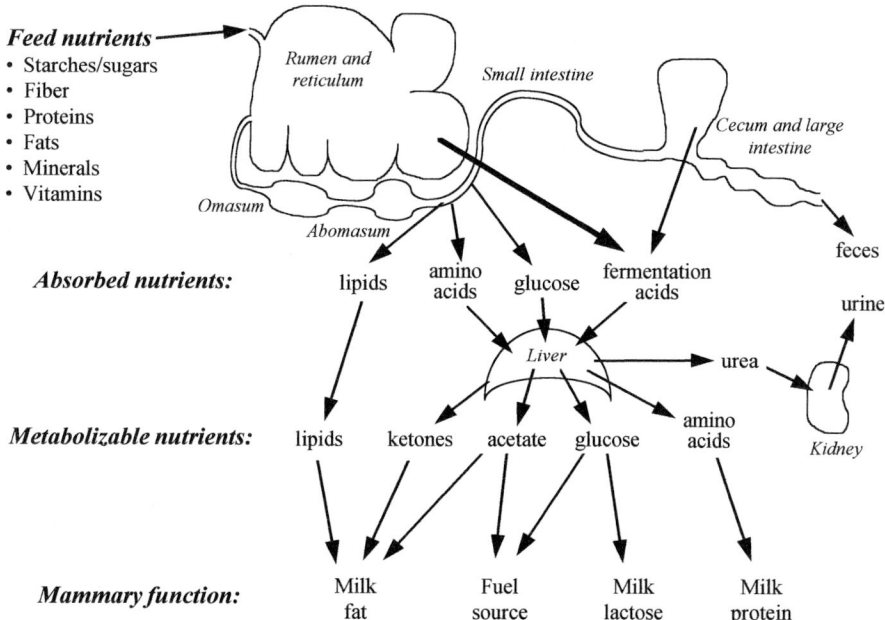

FIGURE 1 Conversion of feed nutrients to milk.

cape rumen fermentation, most nonfiber carbohydrate is fermented. Fermentation of carbohydrates produces acetic, propionic, and butyric acids. As the amount of starch in a diet increases, production of propionate increases relative to production of acetate and in very high starch diets, significant amounts of lactate may also be produced. Starch that escapes rumen fermentation can be digested in the small intestine via amylase and maltase, and the resulting glucose that is absorbed accounts for ~10% of total glucose needed by the cow.

Feed triglycerides are partly hydrolyzed in the rumen, but fatty acids are not hydrolyzed in the anaerobic rumen environment. Unsaturated fatty acids are partly hydrogenated. Fat is digested in the small intestine via the action of lipase and bile salts to form lipid micelles, and fatty acids are absorbed across intestinal cells where they are reesterified to form triglycerides.

Most protein (60–70%) also will be broken down in the rumen. In the process of fermentation, bacteria in the rumen use the resulting ammonia as their primary source of N to make proteins for growth and reproduction. Nonprotein N in the diet can be used as a source of some of this rumen ammonia, thus enhancing overall efficiency of dietary protein use. However, for maximizing milk output, a lactating cow must be fed some protein that will escape rumen fermentation; feeds such as blood meal and heat-treated soybean meal are high in rumen-escape protein. Many rumen bacteria are washed down the tract. Along with the protein of escaped feed particles, the protein of these bacteria is digested in the abomasum (true stomach) and small intestine via pepsin, trypsin, chymotrypsin, aminopeptidase, and carboxypeptidase enzymes. Rumen microbes supply about half of the amino acids absorbed by the cow.

IV. METABOLISM IN SUPPORT OF LACTATION

Metabolism is the process of converting digested nutrients into their final products. Metabolism in support of lactation is the process by which the body supplies the mammary gland with the metabolites needed for milk synthesis.

A. Lipid Metabolism

Ruminant lipid metabolism is different from that of nonruminants in that the microbial population of the rumen hydrogenates most dietary fatty acids. The major fatty acids of forages and cereal grains are linolenic acid and linoleic acid, respectively. Most of these unsaturated acids are converted to stearic acid in the rumen so that unsaturated fatty acids make up <20% of absorbed fatty acids. After absorption, fatty acids are incorporated into intestinal lipoproteins, which have less triglyceride and higher density than the typical chylomicrons of nonruminants. These lipoproteins travel through the lymphatic system to bypass the liver and enter directly into circulating blood. Ruminant liver does not secrete appreciable amounts of lipoprotein triglyceride and lacks enzymes necessary to use dietary triglyceride from intestinal lipoproteins. Thus, dietary fat is directed to extrahepatic tissues. Throughout most of lactation in a high-producing dairy cow, daily turnover of triglyceride fatty acids is 1.0–1.5 kg and more than half of these fatty acids are used by the mammary gland.

As in other vertebrates, lipid is the major form of stored energy in ruminants, presumably because it is energy dense and so can be stored in fat depots with minimal weight. Most of the stored lipid is in the form of triglycerides, with the predominant fatty acids being palmitic, stearic, and oleic acid. The diet of most ruminants is generally low in lipid (3–8% of diet dry matter), so fatty acids for storage are also synthesized. In the nonlactating ruminant, most fatty acid synthesis occurs in adipose tissue with acetate, ketones, and some lipogenic amino acids as the major precursors. These synthesized fatty acids, along with preformed fatty acids from blood, are used to form triglycerides. Unlike the very young ruminant, the mature ruminant has a relatively inactive citrate cleavage pathway and uses very little glucose carbon for fatty acid synthesis. Glucose, however, does serve as a precursor for the glycerol moiety of triglyceride and is a major source of NADPH to drive fatty acid synthesis. Adipose tissue synthesizes mostly C16 and C18 acids and has a desaturase enzyme that converts stearic to oleic acid. A comparison of the fatty acid composition of lipids in the diet, milk, and intermediate points is presented in Table 1.

Changes in adipose tissue lipid metabolism during lactation are summarized in Table 2. Early in lactation, the energy demands of milk production are

TABLE 1

Fatty Acid Composition (Percentage by Weight) of Lipid from Alfalfa Hay, Rumen Contents after Feeding, Bovine Plasma in Midlactation, Ovine Fat Depots, and Bovine Milk in Midlactation

Fatty acid	Alfalfa hay	Rumen contents	Blood triglycerides	Subcutaneous adipose	Milk fat
4:0					3
6:0					2
8:0					1
10:0					4
12:0					4
14:0	1	2	5	3	13
16:0	35	33	29	20	30
16:1	1	0	4	4	3
18:0	4	45	27	19	11
18:1	3	8	29	49	24
18:2	24	4	5	3	4
18:3	32	8	1	2	1

greater than net energy intake. During the first month of lactation, as much as one-third of the metabolites for milk production may be from body reserves, and during the first 5 or 6 weeks, the total amount of body fat mobilized is commonly 40–60 kg. After peak lactation, however, energy intake matches energy needs so body fat is no longer lost; body stores are then replenished as lactation progresses further.

To mobilize fatty acids from adipose tissue, hormone-sensitive lipase catalyzes their hydrolysis from stored triglycerides, and they are released into blood as NEFA. During early lactation, this enzyme is active and the rate of lipolysis is high; in addition, lipogenesis and fatty acid esterification are suppressed. Thus, during early lactation, mammary gland is favored at the expense of adipose. Ruminant liver takes up NEFAs proportional to their concentration in blood; generally, this amounts to ~25% of whole body NEFA flux during lactation. Most NEFAs cleared by the liver are oxidized and released as ketones, with the rest oxidized and released as acetate, completely oxidized to CO_2, or reesterified. The rate-limiting step in the oxidation of NEFAs in liver cells is likely entrance into the mitochondria; once in the mitochondria, fatty acids undergo β-oxidation to acetyl-CoA. Acetyl-CoA may enter the citric acid cycle;

TABLE 2
Changes in Adipose Tissue Lipid Metabolism during the Lactation Cycle

Variable	Late gestation	Early lactation	Mid lactation	Late lactation
Energy balance	+	−	0	+
Body fat stores	High	Medium	Low	Medium
Plasma NEFA concentration	Low	High	Medium	Low
Rate of lipolysis	Low	High	Medium	Low
Activity of lipoprotein lipase	Medium	Low	Medium	High
Fatty acid synthesis	Medium	Low	Medium	High
Fatty acid esterification	Medium	Low	Medium	High

however, ruminant liver is focused on gluconeogenesis, so oxaloacetic acid may be limiting for complete oxidation of fatty acids. Consequently, liver has elevated mitochondrial acetyl-CoA concentrations, which promote ketogenesis. Liver NEFAs that are not oxidized will be esterified, and because lipoprotein synthesis in ruminant liver is very low, the reformed triglycerides are not quickly exported. Thus, in cases of high lipid mobilization, especially in the last week before parturition when cows often have poor appetites, a significant amount of triglyceride may accumulate in the liver, causing hepatic lipidosis. During early lactation, mammary tissue removes some of the NEFAs from blood and thus helps prevent hepatic lipidosis despite elevated lipid mobilization. Blood NEFAs also are used by several other tissues, including muscle, as a major metabolic fuel.

As lactation progresses, the cow eats enough to meet her energy requirements, net lipid mobilization ceases, and repletion of adipose tissue begins to occur. In mid- and late lactation, almost all net fatty acid uptake by mammary gland is from the triglyceride–fatty acids associated with intestinal lipoproteins.

B. Ketone Metabolism

During early lactation when cows are in negative energy balance, NEFA oxidation may account for half of ketogenesis. Usually, however, most ketones are synthesized from ruminally produced butyrate in the epithelium of the rumen and in the liver. Ketones serve as an alternative fuel source for body tissues, including heart, kidney, and muscle, when carbohydrates are in short supply, such as in early lactation. Ketones have several effects on metabolism; they may spare glucose from oxidation, but they also inhibit muscle proteolysis and thereby decrease availability of glucogenic substrates and decrease adipose tissue lipolysis. Decreased lipolysis in response to high ketones serves as a protective measure to help prevent excess lipolysis and subsequent excess ketogenesis. If this feedback loop is not sufficient to counter the mechanisms that stimulate lipolysis in early lactation, a condition known as ketosis may occur in which ketone concentrations in blood reach toxic levels. The major factor influencing the utilization rate of blood ketones is concentration. Bovine mammary gland extracts about 20% of acetoacetate and 45% of β-hydroxybutyrate from arterial blood and uses ~30% of daily ketone flux in a cow producing 50 kg of milk per day.

C. Acetate Metabolism

Blood acetate in ruminants is derived from two sources: microbial fermentation and endogenous production. In a fed lactating cow in early lactation, the amount of acetate entering blood may exceed 5000 g/day, with >60% from ruminal fermentation, 8% from intestinal fermentation, and 30% from partial catabolism of lipids and amino acids. Liver uses <10% of absorbed acetate from the gut and, in many cases, liver is a net producer of acetate. Gut tissues, however, may use 30% of absorbed acetate, and most of this acetate is oxidized. Acetate also is a major metabolic fuel for muscle, and acetate uptake is directly proportional to its concentration in blood. In late lactation, acetate serves as the major source of carbon for fatty acid synthesis in adipose tissue as lipid stores are repleted. The mammary gland extracts ~60% of arterial acetate and accounts for 20–40% of whole body acetate use during lactation.

D. Glucose Metabolism

During lactation, the mammary gland of ruminants extracts ~30% of arterial glucose. A lactating cow producing 50 kg of milk per day has a whole body glucose flux of 4000–5000 g/day, and her mammary gland takes up ~3500 g of glucose daily. Because <10% of metabolizable energy intake is absorbed glucose in ruminants, ruminants must rely on gluconeogenesis to meet almost all their glucose needs. This major and continuous function of the liver is evidenced by the constantly high activity of liver gluconeogenic enzymes. Thus, in contrast to nonruminants, gluconeogenesis in ruminants is greatest immediately postprandially, and it is directly proportional to intake of glucogenic compounds. The major precursors for gluconeogenesis are propionate, lactate, glycerol, and amino acids with estimates for gluconeogenic carbon at 50–60% propionate, ~10% gut lactate, ~10% endogenous lactate, 5–10% glycerol, and 5–20% amino acids. The contribution of amino acids to gluconeogenesis is not clear and some studies have found as much as 40% of glucose comes from amino acids.

Because of the high demand of the mammary gland for glucose, ruminant metabolism is altered during lactation so that glucose is preserved for lactose synthesis as much as possible. Hence, glucose is not used for fatty acid synthesis in lactating ruminants, and in its place acetate and ketones are used. Furthermore, oxidation of glucose in muscle and adipose tissue is decreased during early lactation.

E. Protein Metabolism

A lactating cow producing 50 kg milk secretes 1600 g of milk protein per day; this is equivalent to 8 kg of muscle tissue accretion per day. To produce this much milk, the cow must absorb ~3200 g of amino acids and dipeptides daily. Thus, the absorbed amino acids on average are captured in milk with 50% efficiency. During most of lactation, the cow is in body protein equilibrium, and those amino acids that are not captured in milk are catabolized. Catabolism of amino acids is the result of (i) absorption of amino acids with a profile that does not match requirements, (ii) the need for glucogenic compounds and metabolic fuels, and (iii) the extent to which proteins are "turned over." Body protein turnover is the daily degradation and resynthesis of body proteins. For a cow producing 50 kg of milk per day, the amount of body protein that is broken down and resynthesized each day is likely 2–4 kg. The gastrointestinal tract, liver, heart, and skeletal muscle account for almost all of this turnover. Whereas some protein turnover is required to maintain a dynamic system, many of the amino acids from endogenous proteolysis are not recaptured in proteins but instead are deaminated and oxidized.

During the first month of lactation, body protein catabolism is greater than resynthesis, and the cow typically loses 5–10 kg of body protein, which is equivalent to 25–50 kg of muscle because muscle is ~20% protein and ~80% water. If half of the protein lost were captured by milk protein synthesis, these endogenous amino acids would account for 5–10% of milk protein secretion in the first month of lactation. However, much of these amino acids are deaminated and used for gluconeogenesis, which although important, is a very inefficient use of body protein. Furthermore, some amino acids may be oxidized in other tissues as an energy source because of decreased glucose use by nonmammary tissues. Deamination of absorbed and endogenous amino acids largely occurs in the liver, and the resulting amino group can either be used to make other amino acids or is converted to urea and excreted. When cows are fed protein-deficient diets, up to 20 kg of body protein can be lost during early lactation. Any body protein lost during early lactation must be replenished during mid- and late lactation.

F. Maintenance Functions

Energy expenditure for maintenance of an animal consists of three major parts: 40–50% is for work functions (liver, heart, kidney, nerve, and lung work), 15–25% is for replacement of proteins, membrane lipids, and other cell components, and 25–35% is associated with transport of compounds across membranes. Ruminant splanchnic tissues account for half of total heat production. The weights of the digestive tract, liver, and heart are all increased in lactating compared to nonlactating cows. These larger organs are metabolically more active and thus increase the basal heat production of a lactating cow. The greater energy expenditure associated with these larger organs during lactation is part of the energetic expense of digesting and metabolizing more nutrients to support lactation and should not be considered part of the animal's basal maintenance requirement.

V. NUTRIENT PARTITIONING

Possible uses of metabolizable nutrients are shown in Fig. 2. Upon entering the general circulation, absorbed or endogenous nutrients can be used for one of several functions. The most important priority for nutrient flow is maintenance of the body. Even if an animal is not fed, nutrients must be available to support maintenance functions or death will occur.

The next most important metabolic function for a lactating animal, especially early in lactation, is milk synthesis. In the first month of lactation, feed intake is insufficient to meet the cow's energy needs. Thus, the rate of fatty acid esterification falls below the rate of lipolysis in adipose tissue. In addition, the rate of protein synthesis falls below the rate of proteolysis in muscle to release amino acids. Nonmammary peripheral tissues use less glucose and more NEFA, acetate, and ketones as energy sources, and mammary gland rapidly draws nutrients from blood. Thus, the priority of the mammary gland early in lactation is so high in dairy cows that nutrients are mobilized from other tissues. This repartitioning of body nutrients is one of the best examples of the concept of homeorhesis, in which all metabolism is focused on the support of a dominant physiological process.

Once the animal is pregnant, uterine functions also

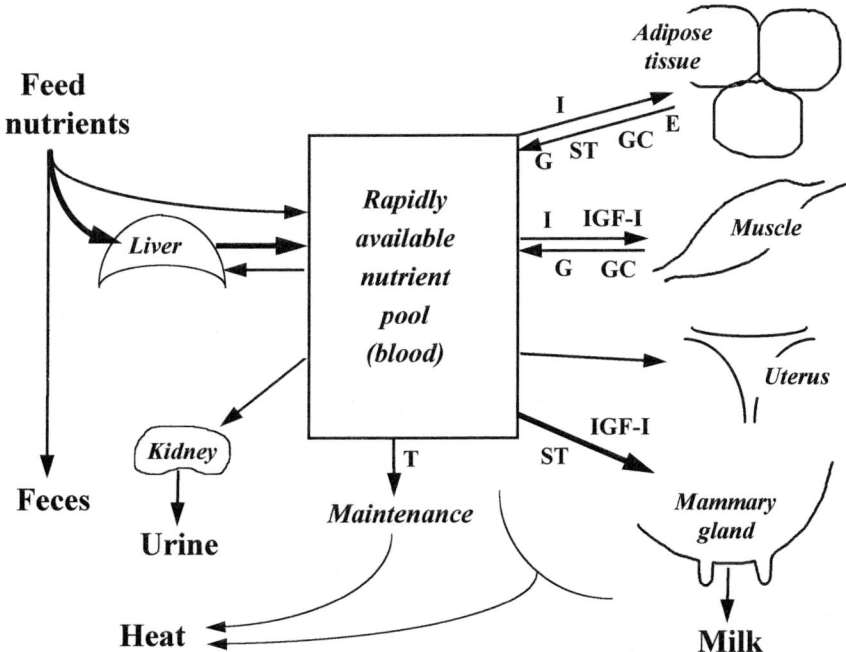

FIGURE 2 Partitioning of feed nutrients to the various body functions and the involvement of the major metabolic hormones. Somatotropin (ST), insulin (I), glucagon (G), gluococorticoids (GC), epinephrine (E), insulin-like growth factor-I (IGF-I), and thyroid hormones (T).

take priority and will become more important as gestation progresses to the detriment of concurrent lactation. In later lactation, as the mammary gland gradually involutes and milk yield decreases, replacement of muscle and adipose tissues also begin to take greater priority. Muscle generally has greater priority than adipose, and net accretion of adipose tissue only occurs when the nutrient pool is filled because the needs of all other tissues have been met.

The partitioning of nutrients to the various functions is regulated by the animal's endocrine system. The major lactogenic hormone in cattle is somatotropin; in nonruminants, somatotropin is also important, but prolactin also directs metabolism toward the mammary gland. Long-term effects of somatotropin may be partly mediated by insulin-like growth factors (IGF). Insulin-like growth factor-I is clearly mitogenic for mammary cells, and undernutrition decreases IGF-I concentrations. Currently, the only known function of IGF-I with regard to the mammary gland is mammogenesis. Although IGF-I certainly mediates some of somatotropin's action on skeletal growth, the mechanism by which somatotropin directs metabolism toward support of milk synthesis is not clear. Insulin is not needed for glucose uptake by ruminant mammary gland but it is needed for glucose uptake by other tissues; hence, low insulin concentrations in early lactation favor use of glucose for milk synthesis. In later lactation, insulin concentration is higher, thus increasing the priority of other tissues.

VI. PRACTICAL NUTRITION DURING THE LACTATION CYCLE

Maximization of milk yield generally maximizes efficiency and profitability. To maximize yield of milk during a lactation cycle, special attention must be given to the cow during the first month of lactation. Increases in milk production early in lactation are associated with increased milk production throughout the remainder of the lactation. To maximize milk yield, cows should be encouraged to consume maximum energy intake. This requires adequate fiber to promote rumen function, high-quality feeds, balanced amounts of protein, minerals, and vitamins, and good feeding management. Energy is the most limiting nutrient for cows in early lactation; however, inadequate protein also impairs milk production. Moreover, additional protein should be fed during the first month of lactation to balance the energy released from adipose tissue. Excessive loss of body fat can cause lactational anestrus, thus added dietary fat is often beneficial in early lactation. Added dietary fat generally increases the proportion of C18 fatty acids in milk fat and often decreases milk protein content. Diets high in grain are necessary to provide ample glucogenic compounds, but too much fermentable grain and too little fiber or a large meal of grain causes rumen acidosis. During rumen acidosis or when high levels of unsaturated fats are fed, the trans fatty acid vaccenic acid is produced in the rumen. Vaccenic acid inhibits *de novo* fatty acid synthesis in the mammary gland, and consequently low fiber or high oil diets can decrease milk fat content. On the other hand, diets with excessive fiber decrease intake of energy and glucogenic precursors; thus, excessive fiber decreases milk production and increases mobilization of lipid from body stores.

In later lactation, the animal can be fed a diet with poorer quality feeds and less protein. The goal of nutrition planning in late lactation is to achieve the desirable amount of body fat stores for the next lactation cycle. Throughout the lactation cycle, rumen microbes synthesize all water-soluble vitamins needed by the cow, but occasionally supplementation with some B vitamins, such as niacin, may be helpful. The fat-soluble vitamins A, D, and E should be supplemented. In most situations, lactating cows also require supplementation of calcium, phosphorus, sodium, selenium, iodine, zinc, copper, and cobalt.

During the "dry" period after cessation of milk production and before the next parturition, the cow has very low nutrient requirements. However, in the last 3 weeks of gestation, nutrient requirements increase and feed intake typically decreases. During this time, cows should be fed high-quality diets similar to those of lactation to promote positive nutrient balance. Mobilization of body reserves before calving increases the risk of the cow for postpartum metabolic disorders and for infectious diseases. Agents, such as calcium chloride, that acidify the urine also should be fed to increase the amount of rapidly exchangeable bone and prepare the calcium regulatory system for the calcium stress of parturition.

VII. NUTRITION AND LACTATION IN NONRUMINANTS AND HUMANS

For nonruminants with large fermentation capacities in the hindgut, such as a horse, the absorbed nutrients available for milk synthesis are similar to those of ruminants. One major difference between hindgut and foregut fermentation is that the hindgut fermenters will digest starches, sugars, and protein in the small intestine rather than the rumen. Thus, the hindgut fermenter depends less on gluconeogenesis for glucose needs. In addition, the hindgut fermenter does not use the microbial protein produced during fermentation, so the amino acid profile of feed proteins is an important nutritional concern. Mammals consuming diets high in protein and fat but low in carbohydrate also must depend on gluconeogenesis for meeting glucose needs of the mammary gland. This may be the reason that many carnivores and piscivores produce milk with a low content of lactose relative to fat and protein.

For omnivores such as humans, milk synthesis requires glucose, amino acids, and fatty acids. Glucose is converted not only to lactose and glycerol but also to fatty acids, and glucose is used to generate the ATP and NADPH to drive the milk synthetic reactions. Amino acid interconversions occur as in the bovine gland.

In contrast to animals that are milked, the volume of milk produced in humans, as in most nursing mammals, is determined largely by the demand of the infant, not maternal lactational capacity or maternal nutrition. Thus, actual milk yield is usually less than potential yield, especially early in lactation. Furthermore, maternal energy stores can be drawn on if dietary energy is inadequate; consequently, milk yield is not closely related to maternal energy intake during lactation. However, malnutrition can decrease or even stop the flow of milk, especially if maternal body stores are marginal. In such cases, decreases in food intake are paralleled by decreases in milk output. Malnutrition during pregnancy also can decrease lactational performance, perhaps because infants are born smaller and with less ability to provide a strong sucking stimuli. Given good nutrition after parturition, women with marginal body stores can produce adequate milk for healthy infant growth. However, providing dietary supplements to chronically undernourished women once lactation has been established may do little to improve milk output because the extra nutrients may be partitioned to maternal body stores rather than to milk.

Protein deficiency may be at least as important as energy deficiency in reducing milk secretion, and protein quality is just as important as protein quantity. For lactation, high-quality dietary protein is that which results in a profile of absorbed amino acids similar to the profile found in milk. The highest quality proteins are those of animal sources, followed by proteins from legume leaves, legume seeds, grains, roots, and tubers. In most cases, a proper blend of plant proteins is completely adequate to meet the needs of the lactating vegetarian.

The intake of most minerals and vitamins is generally adequate if food intake is kept proportional with energy needs. Very little change in dietary habits is needed if total food intake is increased. Calcium content, however, should be increased to prevent bone demineralization. Other minerals and vitamins for which supplementation may be necessary include magnesium, zinc, vitamin B_6, and folate. Extra iron is not needed for lactation but should be consumed to replenish body stores lost during pregnancy. For most minerals, bioavailability of dietary sources should be considered in meeting requirements, and the bioavailability of minerals from animal products (such as meat or dairy products) generally is greater than that of plant foodstuffs.

See Also the Following Articles

Cattle (Bovidae); Lactation, Human; Lactogenesis; Mammary Gland Development; Mammary Gland, Over-view; Milk, Composition and Synthesis; Nutritional Factors and Reproduction

Bibliography

Allen, L., King, J., and Lonnerdal, B. (1994). *Nutrient Regulation during Pregnancy, Lactation, and Infant Growth.* Plenum, New York.

Berger, H. (1988). *Vitamins and Minerals in Pregnancy and Lactation.* Raven Press, New York.

Garnsworthy, P. C. (1988). *Nutrition and Lactation in the Dairy Cow.* Buttersworth, London.

Institute of Medicine (1991). *Nutrition during Lactation.* National Academy Press, Washington, DC.

Rasmussen, K. M. (1992). The influence of maternal nutrition on lactation. *Annu. Rev. Nutr.* **12**, 103–117.

Sniffen, C. J., and Herdt, T. H. (1991), Dairy nutrition management. *Vet. Clin. North. Am. Food Anim. Practice* 7, 311–632.

Worthington-Roberts, B. S., and Williams, S. R. (1997). *Nutrition in Pregnancy and Lactation,* 6th ed. Brown and Benchmark, Madison, Wisconsin.

Obstetric Anesthesia

David L. Hepner and Brett B. Gutsche

University of Pennsylvania Medical Center

I. Introduction
II. Physiological Changes of Pregnancy
III. Systemic Medications during Labor and Delivery
IV. Inhalation Analgesia
V. Pain Pathways during Labor and Delivery
VI. Paracervical and Pudendal Block
VII. Continuous Epidural Analgesia during Labor and Delivery
VIII. Anesthetic Techniques for Cesarean Section
IX. Conclusion

GLOSSARY

analgesia The act of ablating or attenuating pain without loss of consciousness.
anesthesia Loss of sensation, usually produced by loss of consciousness, induced for the performance of surgery.
epidural space A space between the dural membrane and vertebral column that contains nerve roots that are the sites of action for local anesthetics.
general anesthesia Pain relief over the entire body induced by drugs that produce unconsciousness.
local anesthetics Drugs that block the neural impulses, thereby blocking nerve conduction and preventing pain in a particular area of the body.
obstetric Referring to obstetrics as it deals with the management of pregnancy, labor and delivery, and postpartum care.
pain An unpleasant sensory and emotional experience associated with a stimulus.
regional anesthesia Blockade of nerves that provide pain relief or prevent pain in a particular area of the body while maintaining consciousness.

Obstetric anesthesia encompasses all forms of pain control used during labor and vaginal delivery and for the performance of a cesarean section. This broadly includes an expertise in systemic intravenous (iv) medications, all forms of local and regional analgesia, and general inhalation anesthesia and analgesia. The ultimate goals during the administration of anesthesia are to use the smallest amount of drug that will provide adequate analgesia to the mother while minimizing drug diffusion through the placenta into the fetus and to maintain maternal consciousness. Amnesia is usually not desired during the delivery of a newborn because it will delay the bonding between the mother and the newborn. Since vaginal delivery of a baby is commonly thought of as a natural and participatory process, maternal awareness is highly desirable during the birth process.

I. INTRODUCTION

Sir James Young Simpson, a professor of midwifery at the University of Edinburgh and a prominent British obstetrician, was the first physician to use anesthesia for obstetrics. On January 19, 1847, he used inhaled ether to allow internal version extraction of a dead baby in a mother with a severely contracted pelvis. Although very intrigued and pleased with the results of the newly discovered ether, he was not happy with its lingering odor and with the bronchial irritation it caused. After Simpson and two of his assistants became unconscious while inhaling chloroform, suggested by a Liverpool chemist, he began to use it for analgesia during labor and delivery.

Despite this new discovery, many religious and obstetric authorities objected to the use of anesthesia during labor and delivery. The religious objection dealt with Genesis Chapter 3, Verse 16, which states, "I will greatly multiply thy sorrow and thy conception; in sorrow thou shalt bring forth children." Many religious authorities interpreted this as implying that childbirth was ordained by God to be painful. The second reason was a medical one and was expressed by many medical authorities including Dr. Charles Meigs, a professor of obstetrics in Philadelphia. He stated that the pain of childbirth was physiologic, that pain was the best guide for forceps placement, and that anesthesia would slow down the uterine contractions, prolonging labor and delivery, and in addition it would increase maternal risks. The debate over the use of obstetric anesthesia continued until Dr. John Snow performed the successful administration of chloroform analgesia to Queen Victoria on April 7, 1853, for the birth of her eighth child, Prince Leopold.

Patient demand became very important in the continued use of obstetric anesthesia. Indeed, Simpson once said, "Obstetricians may oppose it but I believe our patients themselves will force the use of it upon the profession." The use of anesthesia for obstetrics was taking a very prominent role by the turn of the century following the introduction of the hypodermic syringe and needle. Von Steinbuchel, a German physician, introduced the concept of "Dammerschlaf" in 1902. Amnesia and analgesia were produced by injection of scopolamine plus morphine. By 1914, this concept was introduced as twilight sleep into the United States, but by 1917 cases of neonatal depression and maternal deaths sharply curtailed this approach. In the 1930s Grantly Dick-Read advanced the concept of natural childbirth. He believed that since childbirth was a normal physiologic process it should not be painful. His goal was to teach parturients relaxation techniques by specific types of breathing and muscular exercises. In the 1930s, the "Soviet method" was introduced by Velvosky and Nicolaiev and included psychological preparation, education, and reconditioning. This method was introduced by Dr. Fernand Lamaze to France in the 1950s and then by one of his patients into the United States. This led to the introduction of the psychoprophylaxis technique into the United States which included education about pregnancy, labor and delivery, relaxation techniques, and breathing exercises. Despite decreasing the use of systemic analgesics and major conduction block, many patients using psychoprophylaxis in labor continued to require medication. Indeed, Melzack et al. (1981) demonstrated that labor is still very painful for most women even after prepared childbirth training. The primiparas who had received prepared childbirth training had only minimally lower pain scores than those who had received no such training, and most patients (81%) who received it still requested epidural analgesia. Melzack and colleagues concluded that prepared childbirth training and epidural analgesia should be regarded as compatible, complementary procedures aimed at assisting women in childbirth to suffer less fear, anxiety, and pain. Melzack and colleagues (1984) also concluded that childbirth is more painful than most clinical pain syndromes and pain after an accident (Fig. 1).

Physical variables are an important determinant in the severity of labor pain. The frequency and intensity of the contractions, the amount of cervical dilatation, the weight of the parturient and of the infant, and the shape of the maternal pelvis are physical factors that may account for the variability in the perception of pain by a parturient.

In summary, both physical and psychological factors are involved in the amount of pain experienced during parturition. Epidural analgesia has been proven to be effective and beneficial (see Section II) during labor and delivery. Its effectiveness in reducing the pain of childbirth and improving the maternal experience will be complemented by childbirth classes. A discussion on obstetric anesthesia techniques should be included in childbirth classes since many anesthetic options, including epidural and/or subarachnoid block, are readily available and safe when administered by a competent anesthesiologist.

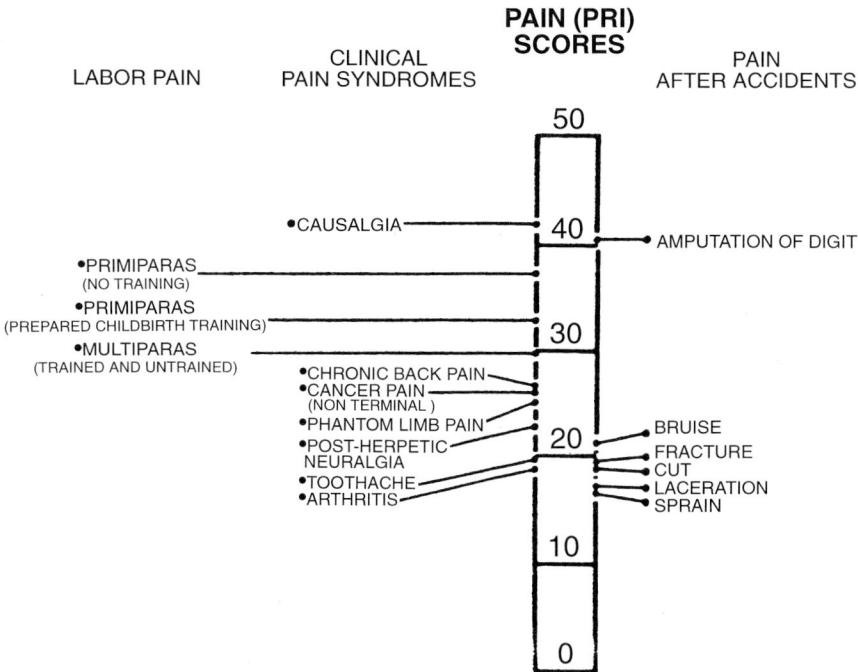

FIGURE 1 Only amputation of a digit and causalgia (burning pain after partial injury of a nerve or one of its major branches) are more painful than labor (reprinted from Melzack, 1984, with kind permission of Elsevier Science-NL, Sara Burgerhartstraat 25, 1055 KV Amsterdam, The Netherlands).

II. PHYSIOLOGICAL CHANGES OF PREGNANCY

Maternal physiologic changes associated with pregnancy have significant effects on the anesthetic management of the gravid patient. There is no evidence in humans that either general anesthesia or regional anesthesia is safer in the gravid patient from the standpoint of teratogenecity, premature labor, or final outcome. However, a regional technique, by allowing the parturient to remain awake and to protect her airway, is safer than a general anesthetic because it significantly reduces the risks of maternal pulmonary aspiration and loss of the maternal airway from a failed intubation. In the obstetric population the risk of failed intubation has been reported to be as great as 1 in 300 undergoing cesarean section, which is about eight times the rate in the general surgical patient population.

Physiological changes of pregnancy begin as early as the first trimester. The parturient's airway is characterized by edema and vascular engorgement of respiratory mucus membranes. This, together with a 20% decreased functional residual capacity (FRC) and an increased oxygen consumption (15–20%), predisposes to the development of rapid hypoxia during induction and emergence of general anesthesia. Oxygen consumption is further increased with painful uterine contractions. Epidural analgesia during the first and second stages of labor decreases this additional increase in oxygen consumption.

The decreased FRC along with an increased minute ventilation (50%) and a 25–40% decreased minimum alveolar concentration (MAC) for the inhalation agents result in a faster anesthetic induction with the inhalation anesthetics. The maternal hyperventilation in pregnancy is secondary to increased carbon dioxide production and to the respiratory stimulant effects of progesterone. The decrease in $PaCO_2$ is compensated by a renal excretion of bicarbonate resulting in a decreased buffering capacity of maternal blood. Marked passive maternal hyperventilation by the anesthetist during general anesthesia may be associated with fetal hypoxia and metabolic acidosis. Huch and associates noted that uterine contractions led to pain and marked maternal hyperventilation. There was hypoventilation between contractions because pain no longer stimulated respiration and maternal hypocapnia decreased respiratory drive. This led to a decrease in the maternal PaO_2 by 10–50%, which led to a decrease in fetal PaO_2 and late fetal heart rate decelerations. These changes were prevented by epidural analgesia but not by intravenous narcotics.

Parturients have an increased heart rate and stroke volume that lead to an increased cardiac output and left ventricular work. This is usually well tolerated by healthy parturients but may not be in parturients with heart disease, preeclampsia, pulmonary hypertension, or severe anemia. Effective regional analgesia attenuates the increase in blood pressure and cardiac output during labor (Fig. 2). As early as the end of the first trimester aortocaval compression may occur which can compromise both the mother and the fetus. The mother may experience supine hypotension. The fetus may be compromised as a result of the decreased intervillous perfusion from both compression of the aorta and a decreased maternal cardiac output as a result of vena caval compression.

In the second half of pregnancy serum gastrin levels are elevated. Before 20 weeks of gestation one-third of parturients have gastric volumes >25 ml with a gastric pH <2.5. As early as 15 weeks of gestation many parturients develop signs and symptoms of gastric incompetence with reflux. Gastric emptying is delayed as early as 8–11 weeks' gestation. Thus, even before the end of the first trimester of pregnancy the gravid patient is at high risk for pulmonary aspiration.

III. SYSTEMIC MEDICATIONS DURING LABOR AND DELIVERY

Systemic medications are widely used during labor and delivery often in early labor before central neuraxis block is initiated, when a major conduction block is contraindicated, or when the patient does not desire a major conduction block. Commonly used systemic medications consist of opioids and sedative tranquilizers.

Opioids are the most effective of the systemic medications because their action at the μ and κ receptors provides analgesia. Morphine is the prototype of the opioids but has a slow onset and a long duration. Meperidine (Demerol) is probably still the most widely used opioid in obstetrics today. It achieves a rapid placental transfer and can cause a decrease of the beat-to-beat variability in the fetal heart rate tracing. Neonatal depression depends on the relationship between the time of administration of meperidine and the delivery of the neonate. Neonates born to mothers who receive the drug 2 or 3 hr prior

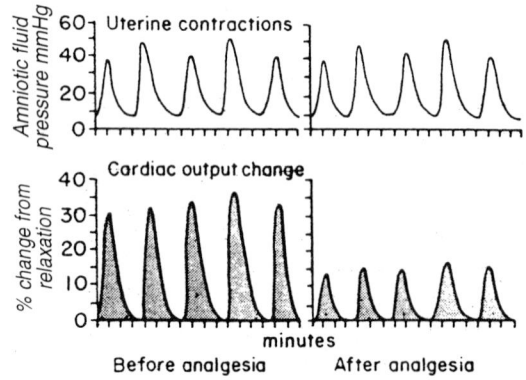
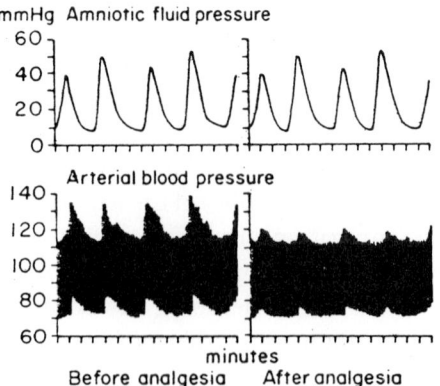

FIGURE 2 Attenuation of the increases in cardiac output and blood pressure, secondary to labor pain, with continuous epidural analgesia in a primipara (reprinted with permission from Bonica, 1994, p. 628).

to delivery exhibit the greatest amount of neonatal depression, whereas those receiving it an hour or less prior to delivery did not show such depression.

The mixed agonists–antagonists opioids given during labor and delivery include butorphanol (Stadol) and nalbuphine (Nubain). They provide analgesia when administered alone but antagonize the analgesia of μ opioids (morphine) when given with them. The reported benefit of these drugs is limited respiratory depression with increasing doses. However, there is a rapid placental transfer resulting in no difference in Apgar or neonatal neurobehavioral scores between this group of drugs and meperidine.

Overall, the main concern with the opioids is ventilatory depression secondary to a shift to the right of the carbon dioxide response curve in both the mother and the neonate. These analgesics also decrease maternal gastric emptying time and are associated with maternal sedation, occasional orthostatic hypotension, increased nausea and vomiting, and urinary retention. Although some argue that the partial agonist–antagonists have a decrease in the respiratory depressant effects and other side effects ("ceiling effect"), there is also a concern regarding the limited amount of analgesia available.

Commonly used sedative tranquilizers are the phenothiazine derivatives (e.g., promethazine and pipermethazine) and hydroxyzine (an antihistamine). Promethazine (phenergan) and hydroxyzine (vistaril) are the most common drugs in this group. These drugs, usually administered with opioids, decrease anxiety, nausea, and vomiting. Contrary to popular belief, they do not increase the analgesia of opioids. They rapidly cross the placenta, resulting in a decreased beat-to-beat variability of the fetal heart rate tracing. Intravenous hydroxyzine is contraindicated because it is associated with thrombophlebitis.

Ketamine is a potent agent that could be used as an amnesic, analgesic, or anesthetic. Initial sedating doses provide a short period of amnesia and analgesia with minimal or no side effects. Side effects include hallucinations, unpleasant dreams, nystagmus, increased uterine tone, and loss of consciousness. Ketamine crosses the placenta, and doses in excess of 1 mg/kg are associated with neonatal respiratory depression and muscle rigidity. It is mainly used during the second stage of labor to provide or supplement analgesia during a spontaneous or instrumental vaginal delivery.

IV. INHALATION ANALGESIA

Inhalation analgesia has largely been abandoned because of concerns of maternal airway compromise in a parturient with a full stomach. However, it may be beneficial as a supplement either to a partially working regional technique or to an intravenous analgesic in cases of a difficult breech extraction or of a midforceps delivery. Nitrous oxide, in concentrations of 50% or less in O_2, provides sedation and analgesia without depressing the maternal airway reflexes. This can be supplemented with intravenous ketamine. If this is not sufficient, very small doses of isoflurane or enflurane are added to the nitrous oxide. The goal is to produce analgesia, with maternal consciousness and airway reflexes maintained, and not anesthesia. It is very important to keep verbal contact with the patient and to avoid inhalation induction during attempts at inhalation analgesia.

V. PAIN PATHWAYS DURING LABOR AND DELIVERY

There are two principal pain pathways associated with labor and vaginal delivery. Pain in labor initially is visceral in nature and results from stretching and dilatation of the cervix and lower uterine segment. It is transmitted by the $A\delta$ and C primary afferent fibers. These visceral sensory fibers traverse through the Frankenhauser ganglion (paracervical plexus) and then sequentially through the inferior, middle, and superior hypogastric plexi of the pelvis. From the superior hypogastric plexus these fibers pass to the lumbar sympathetic chain and through lumbar splanchnic nerves, where they proceed cephalad through the lower thoracic chain and then leave it by coursing through the white rami communicantes connected with T10, T11, T12, and L1 spinal nerves. Finally, the fibers pass through the posterior roots of these spinal nerves to enter the spinal cord and

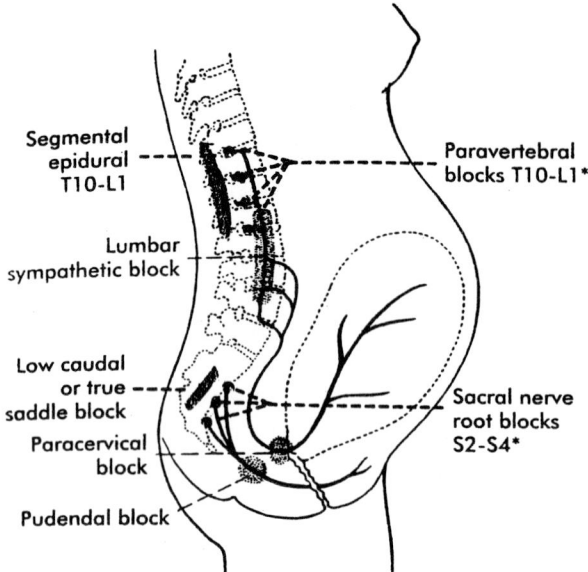

FIGURE 3 Peripheral nociceptive pathways involved in labor and the various nerve blocks used to alleviate labor pain (reprinted with permission from Chestnut, 1994, p. 317).

make contact with dorsal horn neurons (Fig. 3). The pain of the first stage of labor has a visceral component arising from the cervix and lower uterine segment referred to the dermatomes supplied by the T10–L1 spinal nerves. During the early first stage of labor the pain may only be referred to the T11 and T12 dermatomes, but as labor progresses the referred pain will spread to the T10 and L1 dermatomes.

As labor continues, the presenting part begins to transverse the birth canal which leads to stretching of the canal and often tearing of the perineum, causing somatic pain. This pain results in the mother having the uncontrollable urge to bear down and push, bringing into play the secondary forces of labor, which help expel the fetus. The peripheral somatic nerve pathways are transmitted primarily via the pudendal nerve that is derived from the posterior roots of S2–S4 spinal nerves. In addition, the iliohypogastric (T12 and L1) and genitofemoral nerves (L1 and L2) may convey these somatic sensations.

While cervical and uterine discomfort is usually associated with the first stage of labor and perineal pain with the second stage, these sensations may overlap, particularly in the primagravida. When she reaches 7 or 8 cm cervical dilation, the presenting part usually begins to descend, causing the mother both visceral and somatic pain. It results in the mother having the urge to bear down, which, at this point of time, will be useless.

VI. PARACERVICAL AND PUDENDAL BLOCK

Regional techniques are widely employed for the relief of labor pain. Pudendal block has long been used by obstetricians for perineal analgesia at delivery. Paracervical block (PCB) is rarely used during labor with a viable fetus. PCB interrupts the pain pathways to the cervix and lower uterine segment, thus suppressing the pain of the first stage of labor (Fig. 4). Unfortunately, PCB is associated with fetal bradycardia resulting in fetal acidosis and even death. Explanations for the fetal bradycardia seen with PCB include a reflex decrease in heart rate secondary to manipulation of the fetal head, increased uterine arterial and myometrial tone, and umbilical vasoconstriction secondary to a direct diffusion of the local anesthetic into the fetal circulation. PCB is most useful to provide analgesia when the fetus is not a consideration as for dilatation and curettage, repair of cervical laceration, and for any postpartum intrauterine manipulation including removal of a retained placenta. From a maternal standpoint it is safe because it is not associated with hypotension and other serious side effects.

Bilateral pudendal nerve block is indicated to alleviate vaginal and perineal pain. It is performed bilaterally by a transvaginal approach through the sacrospinous ligament to reach the pudendal nerve and provides analgesia of the vagina, vulva, and perineum (Fig. 4). Initiation after full cervical dilatation provides good analgesia for the second stage of labor. It is usually performed by the obstetrician prior to delivery of the neonate in a parturient without any other form of nerve block or in a parturient with an epidural with "sacral sparing." It will provide complete analgesia for episiotomy with repair and with opioid or other supplementation will usually allow forceps and vacuum extraction. It will not provide analgesia for any intrauterine or cervical procedure. Complications are very rare and include direct injec-

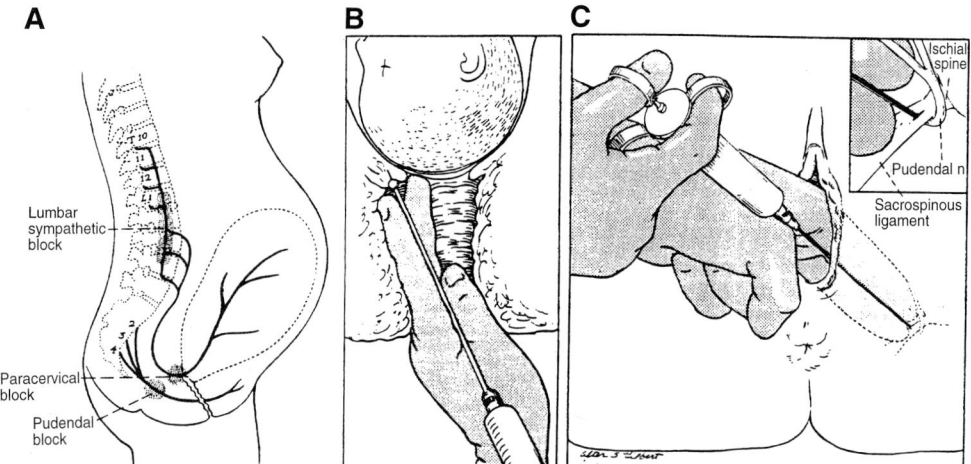

FIGURE 4 (A) The paracervical block interrupts the Frankenhauser (paracervical) ganglion lateral to the cervix. (B) Paracervical block technique: A needle guide is inserted between the index and middle fingers and into the lateral fornix of the vagina at the 4- and 8-o'clock position close to the cervix. The needle protrudes less than 0.5 cm from the guide and after negative aspiration, 5–7 cc of a dilute local anesthetic (1% lidocaine or 2% chloroprocaine) is injected. (C) Transvaginal approach to the pudendal nerve. The needle guide is introduced through the vaginal mucosa and then the needle protrudes about 1 cm through the sacrospinous ligament (located 1 cm medial and posterior to the ischial spine) (reprinted with permission from Bonica, 1990, p.1336).

tion into the fetus, direct intravascular injection, and systemic toxicity. Properly performed, it is not associated with maternal hypotension or ill effects on the fetus and newborn.

VII. CONTINUOUS EPIDURAL ANALGESIA DURING LABOR AND DELIVERY

Epidural analgesia for labor and delivery remains the standard against which all other methods are measured because it provides complete pain relief without obtunding maternal awareness while protecting the mother and fetus from the stresses of labor (Fig. 5). It provides rapid analgesia (3–5 min), and it allows for continuous analgesia via an epidural catheter infusion and for supplementation of the analgesia via patient-controlled epidural analgesia pump. It can provide excellent sensory analgesia with minimal or no motor block, especially when using the local anesthetics bupivacaine or ropivacaine plus opioids (fentanyl or sufentanil). If the need for a cesarean section arises, the epidural block can also be rapidly intensified to provide satisfactory analgesia.

Continuous epidural analgesia (CEA) is not associated with, nor does it interfere with, the maternal protective reflexes provided maternal hypotension and local anesthetic toxicity are avoided.

The epidural space contains nerve roots, fibrous tissue, fat, blood vessels, and lymphatics. The epidural venous plexus is part of the vertebral plexus and is likely to be engorged in pregnancy because it forms important collaterals when the inferior vena cava is obstructed by the gravid uterus. Because of this venous engorgement, the epidural needle or catheter more easily penetrates these vessels.

The L2–3 and L3–4 interspaces (lumbar epidural technique) are the most common interspaces used for a lumbar epidural because they are below the terminus of the spinal cord. A Hustead, Touhy, or Crawford needle is advanced through skin, subcutaneous tissue, supraspinous and interspinous ligaments, and ligamentum flavum. Entry of the needle tip into the epidural space is signaled by an abrupt loss of resistance to either air or fluid. A plastic epidural catheter is passed 3–5 cm into the epidural space through the needle, which is then removed over the catheter.

The epidural technique has specific contraindica-

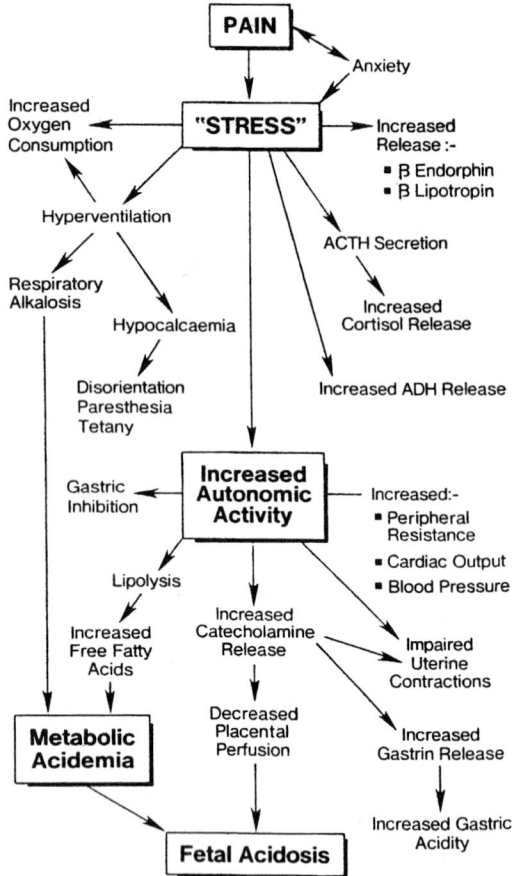

FIGURE 5 Physiologic and biochemical changes secondary to labor pain (reprinted with permission from Brownridge and Cohen, 1988, p. 600).

tions. CEA is absolutely contraindicated if the patient refuses or if there is an infection at the puncture site. Relative contraindications include coagulopathy (risk of epidural hematoma), hypovolemia (risk of hypotension and decreased uterine blood flow), frank sepsis (risk of epidural abscess), and preexisting neurologic disease (risk of worsening neurologic disease). Drawbacks include hypotension, motor blockade, and a potential for prolongation of labor, which can be avoided by minimizing the density of the motor block.

A decrease in sympathetic tone caused by epidural or spinal block will result in a greater significant decrease in blood pressure in a parturient compared to a nongravid individual due to aortocaval compression. Because uterine blood flow is poorly autoregulated, it decreases proportionally to the decrease in maternal blood pressure. Hypotension can be minimized by adequate prehydration and by left uterine displacement. If hypotension (systolic blood pressure <100 mmHg or >30% decrease from baseline) does occur, immediate treatment is necessary. The most effective treatment is an intravenous bolus injection of 5–10 mg of ephedrine with repeated treatments as necessary to keep the systolic blood pressure >100 mmHg. As long as hypotension is corrected quickly, it has little demonstrable clinical effect on the fetus or newborn. While previous animal studies have shown decreased uterine blood flow and fetal asphyxia associated with a pure α agonist such as phenylephrine or methoxamine, recent clinical studies comparing ephedrine to phenylephrine for the treatment of regional anesthesia-induced hypotension have not shown any clinical difference in the neonate as assessed by umbilical blood gases and Apgar scores. While ephedrine remains the vasopressor of choice, phenylephrine does offer an alternative in cases in which ephedrine is ineffective or is relatively contraindicated (such as increased maternal heart rate).

Many obstetricians argue that motor blockade may lead to ineffective pushing during the second stage of labor with prolongation of labor and increased usage of instrumental deliveries. However, the mechanics involved in pushing include the usage of the valsalva maneuver, which involves the diaphragm, and pelvic musculature relaxation. To involve the diaphragm would only result in an inordinately high motor block extending above C5. Pelvic musculature relaxation is obtained with a dense epidural motor block. A parturient should be able to adequately push even with a high degree of motor blockade as long as she is coached to do so with her contraction. However, some parturients feel more comfortable pushing when they have some sensation of uterine contractions. Dilute solutions of local anesthetics combined with fentanyl or sufentanil achieve a very good sensory level of analgesia with minimal if any motor blockade and will allow the mother to sense perineal pressure associated with a contraction in the second stage of labor.

The concept of the "walking epidural," introduced in Great Britain by Barbara Morgan, has gained increasing popularity in the United States. It often involves the use of the combined spinal–epidural (CSE) technique. Through an epidural needle sited in the epidural space, a longer pencil point spinal needle is advanced through the epidural needle and into the subarachnoid space. Fentanyl or sufentanil produce immediate analgesia without a sympathectomy or a motor block. The addition of small doses of bupivacaine or ropivacaine prolongs the duration of analgesia without causing significant sympathectomy or motor blockade. The spinal needle is then withdrawn and an epidural catheter is inserted. The catheter is tested and analgesia is initiated only after the parturient starts to feel discomfort with the dissipation of the subarachnoid block. A drawback to this technique is the absence of a tested functioning epidural catheter in place in the event of the need for a cesarean section.

Great controversy exists over the effect of CEA on the progress of labor. Some retrospective and less controlled studies indicate CEA is associated with a prolongation of labor, increased usage of oxytocin, increased frequency of instrumental deliveries, and increased incidence of cesarean section. Thorp *et al.* conducted a prospective randomized trial in nulliparous women comparing epidural analgesia and parenteral meperidine. They found longer first and second stages of labor, an increased frequency of instrumental vaginal deliveries, and an increased rate of cesarean sections in the CEA group. They further recommended that an epidural catheter should not be placed prior to 5-cm cervical dilatation because all cesarean sections in the epidural group occurred when the epidural catheter was placed prior to this. However, the study was flawed by early termination after 93 patients, even when the initial power analysis suggested 200 patients were needed, and by the senior investigator being the one making the decisions for instrumental vaginal deliveries and cesarean sections. Chestnut *et al.* conducted two studies comparing early (3- or 4-cm cervical dilatation) to late (≥5-cm cervical dilatation) administration of epidural analgesia. One study included nulliparous women in spontaneous labor (Chestnut *et al.*, 1994a) and the other one nulliparous women receiving intravenous oxytocin (Chestnut *et al.*, 1994b). In neither study was the obstetric outcome altered if the epidural analgesic was started prior to or after 5-cm cervical dilation. Furthermore, in both studies the cesarean section rate was considerably lower than that of the epidural group in the Thorp study. The latest prospective study (Sharma *et al.*, 1997) was a randomized trial of epidural versus patient-controlled intravenous analgesia with meperidine. This study was designed to minimize the crossover rate, which it did successfully. Only five women in the meperidine group crossed over to the epidural group. Cesarean deliveries were not increased in the epidural group.

It is very difficult to compare epidural analgesia to no analgesia or intravenous medications. It is not ethical to withhold epidural analgesia since it is the most effective way to prevent labor pain. The patients that request epidural analgesia may be having more pain associated with abnormal labors; therefore, the association between longer, more painful labors, dystocia, and epidural analgesia may be coincidental and not causal. Also, some of the earlier studies used a higher concentration of local anesthetics (0.25% bupivacaine in the Thorp study) than is currently used. Chestnut *et al.* (1990) demonstrated that using lower doses of local anesthetics (0.0625% bupivacaine along with 2 μg/cc fentanyl) in the epidural infusion decreases the frequency of motor block and instrumental deliveries. Therefore, it is very hard to compare some of the earlier studies to today's practice. As stated jointly by the American College of Obstetricians and Gynecologists and the American Society of Anesthesiologists (1992),

> Labor results in severe pain for many women. There is no other circumstance where it is considered acceptable for a person to experience severe pain amenable to safe intervention while under a physician's care. Maternal request is sufficient justification for pain relief during labor.

We do not believe that any cervical dilatation has to be achieved prior to the beginning of the epidural technique, as long as intrathecal narcotics or low doses of local anesthetics with narcotics are used in the epidural space and the patient is committed to deliver.

VIII. ANESTHETIC TECHNIQUES FOR CESAREAN SECTION

Regional anesthesia (spinal and CEA) has largely replaced general anesthesia for cesarean section. Data from both the United Kingdom and the United States outline the safety of regional anesthesia compared to general anesthesia (GA). Over two-thirds of maternal anesthetic deaths in England and Wales (1970–1987) and in the United Kingdom (1988–1993) were the result of either pulmonary aspiration or failure to maintain the airway. The vast majority of deaths occurred during anesthesia for emergency cesarean section. In recent years regional anesthesia has largely replaced GA in the United Kingdom and in the United States. Data from the United States show that the anesthesia-related mortality rate declined from 4.3 (1979–1981) to 1.7 (1988–1990) per million live births. However, the number of deaths due to complications of GA remained stable over that time period. In contrast, the number of deaths due to complications of regional anesthesia decreased over that time period and this accounts for the decrease in anesthesia-related mortality. It is best to avoid a general anesthetic unless the situation warrants an immediate onset of analgesia for an urgent cesarean section, the patient refuses regional anesthesia, or in a relatively rare situation in which it is contraindicated. Examples of specific situations indicating GA include continuing maternal hemorrhage, an imminent footling breech vaginal delivery, a prolapsed umbilical cord, and evidence of extremely severe fetal compromise. If the anesthesiologist is not comfortable with the maternal airway, a decision may be made to proceed with a subarachnoid block, even though this may possibly delay birth. The obstetrician should be able to identify the patient with a potentially difficult airway and seek early anesthesia consultation. The maternal airway is assessed by the following criteria: mouth opening (ideally >4 cm), distance from the mandibular mentum to the superior thyroid notch during neck extension (ideally >6.5 cm), and ability to extend the neck at least 35° from the neutral position (cervicooccipital extension). Visualization of the faucial pillars, soft palate, and uvula during mouth opening correlates with the success of intubation. If all of these structures are visualized during mouth opening, the likelihood of being able to intubate a patient is much greater than if these structures were partially visualized or not visualized. Other physical characteristics are also important, such as the presence of prominent maxillary incisors, the maternal body weight and habitus, and a history of diabetes mellitus, which is associated with limitation of joint movement.

Preceding induction the parturient should receive a clear antacid, a gastric emptier, and an H2 blocker to neutralize the gastric acidity and volume. Preoxygenation, which can rapidly be accomplished during surgical preparation, is essential to avoid maternal hypoxia on induction. A rapid sequence induction is performed to be completed when the surgeon is ready to make the incision, but only after laryngoscopy and confirmed endotracheal intubation does surgery commence. Maintenance of GA before delivery usually utilizes a low concentration of a potent inhalation agent in 50–100% oxygen with nitrous oxide to ensure both maternal anesthesia and satisfactory fetal oxygenation. Following delivery, the nitrous oxide can be increased to 70% and the inhalation agent is continued at only two-thirds MAC or less because higher concentrations can prevent uterine contraction and result in increased blood loss. Opioids and sedatives may also be used after delivery to further supplement the low level of inhalation analgesia.

A subarachnoid block (SAB) is often the preferred regional technique for an elective cesarean section or in cases in which an epidural catheter is not in place because it has a faster onset of a denser block than an epidural. Conical or pencil-point needles of a 22 or higher gauge are used to minimize the incidence of postdural puncture headaches. Sufficient hyperbaric local anesthetic (e.g., bupivacaine, lidocaine, or tetracaine) is given to produce a minimum T4 sensory level required to avoid marked maternal visceral discomfort experienced with uterine exteriorization and peritoneal retraction. Addition of small amounts of fentanyl or sufentanil will further attenuate visceral pain. Finally, addition of morphine to the injectate will provide 18–24 hr of maternal analgesia. Maternal hypotension is more frequent in occurrence, rapid in onset, and severe than seen with an epidural block. Prevention and treatment include uterine displacement, vasopressors (ephedrine), and rapid infusion of full-strength balanced salt solutions. High or total spinal blockade, a rare occurrence, associated with hypotension and respiratory insufficiency requires both cardiovascular and respiratory support until the block recedes. Although some advocate immediate tracheal intubation to provide respiratory support and to protect the maternal airway from aspiration of gastric contents until the block recedes, adequate mask ventilation with cricoid pressure may be all that is needed. "Spinal headache," a complication of dural puncture, can result after either SAB or accidental dural puncture with the epidural needle. In general, the larger the needle puncturing the dura, the greater the incidence and severity of the headache. It has a marked decreased incidence using pencil-point needles compared to the same gauged classical sharp-ended, Quinke-type needles. As cerebrospinal fluid (CSF) leaks, traction is applied to neural structures and a headache may develop. The headache is positional because more CSF leaks with the patient standing or sitting. By lying supine the parturient will not feel the headache but the time for the resolution of the headache will not be altered. Conservative treatment includes analgesics, caffeine, and abdominal pressure with a tight binder. If conservative therapy fails, an epidural blood patch is performed in which nonanticoagulated blood drawn aseptically from the patient is injected into the epidural space, which seals the hole in the dura.

IX. CONCLUSION

Physical and psychological factors are involved in the pain of childbirth and should be included in the decision to provide analgesia during labor and delivery. Childbirth classes that teach breathing techniques help to decrease somewhat labor pain, are very helpful during the early stages of labor, and may lead to a delivery without pharmacological analgesia. However, a majority of parturients will request some form of analgesia despite childbirth classes. Analgesic techniques for labor pain include opioids, sedative tranquilizers, ketamine, inhalation analgesia, PCB, pudendal block, subarachnoid block, and epidural analgesia or a combination of these. Opioids, the most common systemic medication, are usually given early in labor. Some obstetricians are hesitant to use major conduction analgesia early in labor because of a few poor or uncontrolled studies that show a longer duration of labor and a higher incidence of cesarean sections with early initiation of epidural analgesia. Despite this, epidural analgesia remains the gold standard against which other anesthetic techniques for labor are compared.

There are many maternal physiological changes in pregnancy secondary to labor pain that lead to increased autonomic activity, maternal metabolic acidosis, and fetal acidosis. Epidural analgesia has been proven to attenuate these physiologic changes and to protect the mother and fetus from the stresses resulting from labor pain.

Parturients are at a higher risk for general anesthesia because of the increased risk for a difficult intubation and for aspiration of gastric contents compared to the general population. General anesthesia in obstetrics has many problems and shortcomings, including loss of the maternal airway, maternal aspiration of gastric contents, neonatal depression, uterine relaxation predisposing to postpartum hemorrhage, and delayed maternal infant bonding. Regional anesthesia properly administered is safe and effective and is the preferred technique for labor analgesia and in general for cesarean section.

See Also the Following Articles

Cesarean Delivery; Labor and Delivery, Human; PreTerm Labor and Delivery

Bibliography

American Society of Anesthesiologists (ASA)and American College of Obstetricians and Gynecologists (ACOG) (1992). *Pain Relief during Labor*. ASA/ACOG.

Bonica, J. J. (1967/1969). *Principles and Practice of Obstetric Analgesia & Anesthesia*, Vols. 1 and 2. Davis, Philadelphia.

Bonica, J. J. (1990). The pain of childbirth. In *The Management of Pain* (J. J. Bonica, Ed.), 2nd ed. Lea & Febiger, Malvern, UK.

Bonica, J. J. (1994). Labour pain. In *Textbook of Pain* (P. D. Wall and R. Melzack, Eds.), 3rd ed. Churchill Livingstone, Edinburgh, UK.

Bromage, P. R. (1978). *Epidural Analgesia*. Saunders, Philadelphia.

Brownridge, P., and Cohen, S. E. (1988). Neural blockade for obstetrics and gynecologic surgery. In *Neural Blockade in Clinical Anesthesia and Management of Pain* (M. J. Cousins and P. O. Bridenbaugh, Eds.), 2nd ed. Lippincott, Philadelphia.

Chestnut, D. H. (1994). *Obstetric Anesthesia*. Mosby-Year Book, St. Louis.

Chestnut, D. H., Laszewski, L. J., Pollack, K. L., *et al.* (1990). Continuous epidural infusion of 0.0625% bupivacaine–0.0002% fentanyl during the second stage of labor. *Anesthesiology* 72, 613–618.

Chestnut, D. H., McGrath, J. M., Vincent, R. D., *et al.* (1994a). Does early administration of epidural analgesia affect obstetric outcome in nulliparous women who are in spontaneous labor? *Anesthesiology* 80, 1201–1208.

Chestnut, D. H., Vincent, R. D., McGrath, J. M., *et al.* (1994b). Does early administration of epidural analgesia affect obstetric outcome in nulliparous women who are receiving intravenous oxytocin? *Anesthesiology* 80, 1193–1200.

Dewan, D. M., and Hood, D. D. (1997). *Practical Obstetrical Anesthesia*. Saunders, Philadelphia.

Greiss, F. C., Jr., and Gobble, F. L., Jr. (1967). Effect of sympathetic nerve stimulation on the uterine vascular bed. *Am. J. Obstet. Gynecol.* 97, 962–967.

Hawkins, J. L., Koonin, L. M., Palmer, S. K., *et al.* (1997). Anesthesia related deaths during obstetric delivery in the United States, 1979–1990. *Anesthesiology* 86, 277–284.

Her Majesty's Stationery Office. Report on confidential enquiries into maternal deaths in England and Wales, 1970–72 through 1985–87. Her Majesty's Stationery Office, London.

Her Majesty's Stationery Office. Report on confidential enquiries into maternal deaths in the United Kingdom 1988–90 through 1991–1993. Her Majesty's Stationery Office, London.

Huch, A., Huch, R., Schneider, A., *et al.* (1977). Continuous transcutaneous monitoring of fetal oxygen tension during labour. *Br. J. Obstet. Gynaecol.* 84(Suppl.), 1–39.

Melzack, R. (1984). The myth of painless childbirth (The John J. Bonica Lecture). *Pain* 19, 321–337.

Melzack, R., Taenzer, P., Feldman, P., *et al.* (1981). Labour is still painful after prepared childbirth training. *Can. Med. Assoc. J.* 125, 357–363.

Ramanathan, S., and Grant, J. C. (1988). Vasopressor therapy for hypotension due to epidural anesthesia for cesarean section. *Acta Anaesthesiol. Scand.* 32, 559–565.

Sharma, S. K., Sidawi, J. E., Ramin, S. M., *et al.* (1997). Cesarean delivery: A randomized trial of epidural versus patient-controlled meperidine analgesia during labor. *Anesthesiology* 87, 487–494.

Shnider, S. M., and Levinson, G. (1993). *Anesthesia for Obstetrics*. Williams & Wilkins, Baltimore.

Thorp, J. A., Hu, D. H., Albin, R. M., *et al.* (1993). The effect of intrapartum epidural analgesia on nulliparous labor: A randomized, controlled, prospective trial. *Am. J. Obstet. Gynecol.* 169, 851–858.

Olfaction and Reproduction

Dietrich L. Meyer
University of Goettingen School of Medicine

Rakesh K. Rastogi
University of Naples

I. Introduction
II. Pheromone Information in the Vertebrate Brain
III. Where and with Whom to Mate
IV. How Sex Partners Are Synchronized
V. Child Behavior and Maternal Care
VI. Molecules That Encode Messages to Conspecifics
VII. Hormones Alter Olfaction and Pheromone Release

GLOSSARY

conspecifics Members of the same species.
hypothalamus A brain area that is intimately involved in pituitary gland control.
limbic system An ancestral part of the vertebrate brain involved in smell, emotion, and memory.
neuronal Relating to nerve cells.
olfactory bulb A brain structure in which most primary olfactory nerve fibers are first relayed.
pheromones Chemical signals between conspecifics.
terminal nerve A cranial nerve that may mediate information on pheromones.
vomeronasal organ An isolated island of olfactory tissue; involved in pheromone perception.

Accurate timing of a sequence of events is a prerequisite for (successful) mating. Usually, communication is involved and, often, the signaling is mediated by molecules. Furthermore, in many species offspring survival depends on chemical cues which play a crucial role in the control of parent and/or child behavior. The spectrum reaches from the prevention of cannibalism by maternal child recognition to the newborn finding a nipple. In addition to molecules released by conspecifics, odorants from other animals, plants, or mineral sources may have a decisive role for elicitation/timing of behavioral patterns that are essential for the reproduction of a species. Chemoinformation related to reproduction appears to be as important for ancestral life forms (e.g., bacteria and algae) as it is for animals: at the lower end of the phyletic hierarchy, female gametes of brown algae use mixtures of unsaturated cyclic hydrocarbons to trigger the release and attraction of male gametes and, at the upper end, the perfume industry is proof for the fact that chemoattraction of the opposite gender has not been discarded during the evolutionary development of species.

I. INTRODUCTION

It was long known that animals communicated with conspecifics by chemical signals when these were termed pheromones by Karlson and Luscher. Pheromones have since been demonstrated to play a key role for a variety of behaviors. Because reproductive behavior mainly depends on the signaling between conspecifics, this article is primarily concerned with pheromonal chemical cues. Where relevant, other types of molecules that supply biologically relevant information (i.e., molecules from sources other than members of the same species) are briefly mentioned.

Neither in invertebrates nor in vertebrates does a singular chemoreceptive sensory system serve to survey the entire chemical environment. Several receptive structures and their projection sites in the brain are usually involved. The primary distinction is between the sense of smell and the sense of taste. It is still debated what the true differences are and, as frequently argued, whether threshold levels are the main factor. Laverack proposed the following definitions (note: definitions do not bear on system sensitivity):

Olfaction: A sense in which chemical signals alone are received by a population of receptors. Simultaneous contact is not essential.

Taste: A combined sense, with both chemical and mechanical sensors active in the same end organ. Contact is essential.

For several of the subsequently described biological phenomena, all of which are triggered and/or modulated by the detection of chemical cues, it is undetermined which sensory modality is involved. The matter is confused by the existence of additional sensory structures and neuronal pathways that cannot easily be incorporated into either the smell or the taste category of chemosensitivity (e.g., vomeronasal, terminal nerve, and extrabulbar olfactory systems of the vertebrates).

II. PHEROMONE INFORMATION IN THE VERTEBRATE BRAIN

Pheromones have distinct patterns of release, are innately recognized by specific neural systems, and elicit immediate responses of adaptive significance. It appears that several chemoreceptive organs and pathways are involved in vertebrate pheromone detection. Chemoreceptors may be located anywhere on the animal body, not just on the head. While aggregations of chemosensors may form organs, in many cases (particularly in aquatic vertebrates) the receptors are scattered all over the body surface.

A. The Receptors

There is no evidence that specialized pheromone receptor cells exist in any vertebrate. Nevertheless, one may speculate that they will be found in the future.

A description of receptor anatomy, turnover, biochemistry, and transducer mechanisms is beyond the scope of this article. The interested reader is referred to articles by Duvall *et al.* (1986) and Finger and Silver (1991).

B. Different Sensory Channels

Each of the various vertebrate chemosensory systems has, at one time or another, been claimed to mediate some kind of pheromone perception. It may turn out that, indeed, several sensory channels are involved, possibly even in one species. However, many of the reports that are based on behavioral studies of lesioned animals are not conclusive since surgically or chemically inflicted destruction of nervous tissues is not sufficiently selective to determine unequivocally which sensory structure mediates a certain pheromone response.

For instance, several of the studies described later used a zinc compound to destroy the olfactory mucosa. This procedure certainly interfered with terminal nerve functions and may also have irritated accessory olfactory system sensors. Similarly, sperm release induced by electrical excitation of the teleost medial olfactory tract, probably by activation of the brain's preoptic area that controls sperm duct contractions, is no proof for a pheromone-related function of the olfactory system: The terminal nerve runs in close association with the medial olfactory tract and selective electrical stimulation is impossible. Consequently, results of lesion studies and stimulation experiments, designed to identify pheromone receptive sensory channels, are to be interpreted with care.

1. The Olfactory System Proper

In all nonteleost vertebrates, the principal olfactory system (or the olfactory system proper) originates from ciliated nerve cells, each of which projects a fiber to the olfactory bulb. Here, inputs from the receptors are relayed and, via the medial and the lateral olfactory tracts, processed information is distributed to many parts of the vertebrate brain. Most of the recipient regions are part of the limbic system, which is not only related to olfaction but also influences emotions, plays a role in memory functions, and mediates aspects of the making of decisions. Furthermore, olfactory projections reach the reticular formation from where arousal and the degree of an organism's wakefulness are regulated.

In addition to afferents from sensory receptors, the olfactory bulbs receive centrifugal nerve fibers from several parts of the brain. It is unclear which functions they serve and how these modulate the perception of pheromones and other olfactory cues. One hypothesis is particularly fascinating as it relates to major unsolved problems of olfactory system research. In essence, it is hypothesized that the brain tells the olfactory bulbs which odorants to check for, in dependence of environmental circumstances (i.e., other sensory inputs).

According to this notion, the olfactory system does not produce a unique nervous signal in response to each substance that can be discriminated on the behavioral level, but the learning of which molecules may be encountered in a particular environmental/social situation is a major factor in olfactory sensing. Even if this hypothesis is substantiated with regard to the perception of nonpheromonal odorants, it is not automatically clear that the detection of pheromones works according to a similar biological principle.

2. The Accessory Olfactory System

The vomeronasal system plays a major role in the perception of chemical signals of a social nature and modulates reproductive behavior in many vertebrates. Unlike receptors in the mucosa of the olfactory system proper, vomeronasal receptors lack cilia and possess microvilli. The vomeronasal organ (often termed Jacobson's organ) is a highly developed diverticulum in anuran amphibians. In snakes and lizards, it forms a blind sac on each side, opening through the hard palate into the mouth, completely separated from the nasal cavity. The prongs of the forked tongue of some reptilians insert into this pair of palatal openings, delivering odorous particles collected by the tips of the tongue to the chemoreceptors. Most mammals have vomeronasal organs opening into the nasal cavity or into the nasopalatine canal.

Typically, receptors of the accessory olfactory system project their fibers to the accessory olfactory bulbs. These are small but distinct regions adjacent to the principal olfactory bulbs. The central connections of the accessory olfactory system are similar but not identical to those of the olfactory system (proper). In some mammals, fibers originating from the accessory olfactory bulbs have been found to terminate in a portion of the amygdala not receiving afferents from the main olfactory bulb. The amygdala appears to relay afferent input to the hypothalamus where modulation of pituitary functions is possible. Thus, this pathway may serve as the neuronal substrate to turn incoming chemical signals into hormonal responses of the sensing animal.

Of all vertebrates, snakes appear to be most dependent on the vomeronasal system. In fact, none of the known behavioral responses to chemical stimuli shown by snakes has been unequivocally attributed to the olfactory system proper. In addition using vomeronasal input to select and to trail prey, snakes use it to aggregate at certain nesting sites, to court, and to identify their mates.

Various families of olfactory receptor genes from vomeronasal epithelia have been cloned and it is often taken for granted that these receptors respond to pheromones. While this may be true in many cases, in some species (e.g., snakes), the vomeronasal organ is involved in other tasks such as foraging. Hence, an unequivocal difference between a pheromone-sensing system and a system that senses other odors cannot be defined.

i. Dual Nature of the Teleost "Olfactory" System? Teleost fish do not have Jacobson's organ. The sensory epithelium of their olfactory organ contains both ciliated and microvillous receptor cells and it has been proposed that the olfactory system proper and the accessory system are combined into one organ in the nose of teleosts.

3. The Terminal Nerve and the Preoptic Nerve

Around the turn of the century, a novel cranial nerve was discovered and, subsequently, described in almost all vertebrate classes, including Man. It is located medial to the olfactory nerve/tract and, hence, may be referred to as cranial nerve No. zero or terminal nerve. Soon it was postulated that this nerve (or components thereof) was chemosensory. Species differences and unclear definitions led to confusion regarding which neuronal structures belong to the terminal nerve and which are not a part of it. The confusion still persists in the literature. It is augmented by the fact that, actually, two new nerves were found. The parallel discovery of the so-called preoptic nerve remains unknown to many scientists in the field. This nerve runs below the anterior portion of the brain and connects chemosensory structures with diencephalic regions that are known to control the pituitary gland. The preoptic nerve is only found in ancestral fishes (e.g., lungfish) which evolved long before the recent teleosts. Its fibers have a unique chemical signature. In vertebrates lacking a preoptic nerve, the same signature is displayed by a subpopulation of terminal nerve fibers. These projection have been the subject of most of the recent investigations carried out on the cranial nerve No. zero.

We have proposed that the preoptic nerve has been fused with the terminal nerve during evolutionary history and now courses toward its diencephalic targets via the telencephalon instead of below it. The chemical feature that characterizes the preoptic nerve in the fish that have it or, in other vertebrates, one of the terminal nerve components, is the gonadotropin-releasing hormone immunoreactivity (GnRH-ir).

GnRH-ir nerve fibers may innervate the olfactory mucosa but their main input appears to be from the olfactory bulb. Because of the GnRH expression, they have often been implicated in pheromone perception related to reproductive behavior. However, all attempts to gain direct evidence by recording neuronal activity in response to pheromone application to the nose have failed. One impediment to the testing of the hypothesis has been insufficient information concerning the chemical nature as well as the physiological and behavioral effects of vertebrate sex pheromones. Hence, it remains undetermined whether the GnRH subsystem of the terminal nerve is a specialized pheromone detector system.

4. The Olfactoretinalis Projection

This pathway appears to exist only in teleost fishes. Often, it is considered to be part of the GnRH-ir component of the terminal nerve mainly because it displays the same immunoresponse. Nerve cells that are either scattered along the olfactory tract or, in other species, are aggregated in a nucleus of the rostral forebrain receive input from olfactory structures (e.g., olfactory bulb) which is relayed directly to the retina by centrifugal fibers in the optic nerve. Release of GnRH modulates neural activity in the retina. Consequently, the olfactoretinalis projection is suspected to be the link between the perception of sex pheromones and intraretinal changes of neuronal filter properties. Many fishes change colors and/or perform particular patterns of locomotion when ready to mate. Recognition of such visual signals may be enhanced by pheromone-dependent retinal feature detection.

Considerable efforts have been made to verify the previous hypothesis. Future studies will have to determine whether the olfactoretinalis pathway is a part of the GnRH-ir system of the terminal nerve system or has evolved independently in teleosts. The latter is to be presumed if, indeed, the GnRH component of the terminal nerve represents the preoptic nerve of earlier evolutionary stages since in the ancestral nonteleost bony fishes (sturgeons, lungfish, etc.) in which the preoptic nerve is found, no GnRH-ir fibers innervate the retina.

5. The Extrabulbar Olfactory System

In recent years, evidence suggesting that another component of the terminal nerve may actually belong to the olfactory system has mounted. A small portion of olfactory receptors, which cannot be distinguished from other such cells, gives rise to nerve fibers which bypass the olfactory bulb to innervate targets in the forebrain and in the diencephalon directly (i.e., without being relayed in the olfactory bulb). These fibers resemble their companions of the olfactory system proper in a distinct carbohydrate signature of their membranes. It has been speculated that these extrabulbar olfactory projections mediate information on molecules that are easily discriminated from others and that they are specialized for pheromone perception.

Because extrabulbar olfactory fibers have only been reported in aquatic vertebrates (fishes and amphibians), this system would appear to play a role only for the perception of water-dissolved pheromones, whereas volatile substances may require other sensory structures. The only terrestrial vertebrate that has been found to have extrabulbar olfactory fibers is the embryonic/newborn rat. Mammalian embryos, however, are not yet terrestrial but are surrounded by an aquatic environment. The amniotic fluid is loaded with maternal chemical stimuli and neurobiology is far from appreciating molecular communication during this early phase of life.

6. The Gustatory (Taste) System

Taste is perceived by specialized receptors (taste buds) and is mediated by special visceral fibers of the facial (glossopharyngeal) and vagal nerves. However, many sensations that humans refer to as taste are truly olfactory. Correlation of taste and smell is important to identify food in primates and probably also in other vertebrates: Most persons experiencing temporary impairment of olfaction due to nasal congestion testify to the bland taste of food.

To our knowledge, the taste system has not been demonstrated to play a role in premating or mating behavior. Nevertheless, it has been proposed that sex pheromones may be exchanged when humans kiss each other. The gustatory system is the prime candidate for the detection of such molecules.

It also is unclear whether newborns use the taste system for the sensing of pheromones (e.g., in the context of nipple search behavior), but it is undisputed that taste is utilized to identify possible food sources and, thus, is crucial for offspring survival: Human newborns often reject water or sugar–water feedings but avidly take milk. Young infants tend to prefer sweet fruits to less sweet vegetables. These food preferences appear to be innate; they may have developed to ensure an adequate supply of certain vitamins and other essential nutritional compounds that have no taste of their own.

7. The Trigeminal System

The trigeminal nerve became specialized early in phylogeny as the principal cutaneous sensory nerve of face and head. It combines two branchial nerves of different origin, an ophthalmic and a maxillomandibular branch.

The trigeminus is a large nerve, even in the most primitive recent vertebrates. In phylogeny, the distribution of the trigeminal nerve increases while the cutaneous distribution of the facial, glossopharyngeal, and vagal nerves diminishes. With regard to chemoperception, the trigeminus is usually associated with the initiation of reflexes of avoidance but not of attraction.

III. WHERE AND WITH WHOM TO MATE

Most behavioral responses are not triggered by a singular sensory cue. This also holds true for the elicitation of the various components of reproductive behavior which occur in temporal succession. Chemical signals may be crucial, but often simultaneous visual or somatosensory stimuli are required to make an organism respond.

A. En Route to the Mating Site

Species that are thinly spread in large habitats tend to congregate at certain locations to reproduce. Long-range guidance of their migrations often depends on olfactory cues that are not signals from their (possible) mates.

Within the vertebrates, salmon are probably the most thoroughly studied model of investigations into olfactory guided attraction to the mating site. For juvenile silver salmon (a Pacific salmon), it has been demonstrated that imprinting to the natal tributary occurs during a metamorphosis that is commonly referred to as smolt transformation. After oceanic migration, salmon return to the site of birth to spawn. Unidentified olfactory cues are essential for salmon homing. Two hypotheses exist: On the one hand, it has been suggested that the unique mix of dissolved minerals and organic substances from each location allows a fish to detect its home from some distance downstream; on the other hand, juvenile conspecifics may be the source of relevant odorants in a stream (pheromone hypothesis of salmon homing). Analysis of the data currently available suggests that, at least for Pacific (rather than Atlantic) salmon, pheromone guidance is unlikely or plays a secondary role.

B. Attraction and Selection of a Mate

1. Chemical Lighthouses

One sex attracts the other. However, in many species this only holds true for the mating season. Often pheromones are employed to temporarily unite males and females that are (more or less) separated from each other for most of the year (e.g., bears). Typically, these substances are of female origin, designated to guide the males.

Many invertebrate sex attractants have been chemically analyzed and are now synthesized for research purposes as well as for pest control (pheromone traps). Despite the fact that, when in heat, dogs, cats, and livestock are impressive examples of the efficiency of female pheromone signaling to males, not much is known about the chemical nature of vertebrate long-range sex attractants.

2. Olfaction and the Problem of Inbreeding

Body scents reflect an animal's genetic constitution at the extremely polymorphic major histocompatibility complex of genes. Behavioral manifestations of communication systems that depend on odors which are specific to individuals include mating preferences and neuroendocrine responses.

To the surprise of many, recent research has demonstrated that some vertebrates prefer to mate with partners that differ, rather than resemble, their own chemical signature. Because close relatives of one species have similar genes and, therefore, similar smells, the preference for mates with different scents may be an effective way to prevent inbreeding in animals that form stable groups or herds.

IV. HOW SEX PARTNERS ARE SYNCHRONIZED

Reproductive success of a species depends on the precision of the timing of partner behavior. Coordination can either depend on conspecific signaling or on information provided by other sources (e.g., length of day, moonlight intensity, rainfall, tidal cycle, and temperature). For a high precision of the timing, exchange of information between partners is compulsory.

A. Puberty and Gonadal Maturation

If housed together, female mice mature more slowly after exposure to the odor of adult females and pheromones can delay puberty in voles and tamarins. Whether the olfactory system or the vomeronasal system mediates the neuroendocrine response resulting in puberty delay is unclear. However, the vomeronasal system has been implicated in the release of the maturation delay substance(s) from group-housed female mice. The delay substance is present in bladder urine of all adult females, but in individually housed females, the substance is deactivated in the urethra. Group housing prevents the deactivation of the delay substance and, consequently, the urine becomes effective in delaying the onset of female puberty. The chemosensory stimuli indicative of grouping are sensed by the vomeronasal organ. After removal of the organ, females continue to deactivate the delay substance when group housed.

Exposure to chemical signals from one sex can also advance the onset of puberty in the opposite gender. This was revealed by studies on rodents and, additionally, it was demonstrated that destruction of the vomeronasal system eliminated this effect.

In goldfish, preovulatory pheromones (pheromones that are released at peak levels during oocyte final maturation, e.g., 17,20 β-P) stimulate rapid reproductive responses in males, increasing serum gonadotropin levels as well as the volume of milt, and postovulatory certain prostaglandins are responsible for the elicitation of male sex behavior. Most or all substances serving as pheromones in goldfish reproductive behavior are also present and biologically active in the serum of fish that are not breeding.

B. Female Cycles

Pheromones regulate estrous cycles in many species. When subjected to constant (bright) illumination, female rats do not ovulate if they are housed individually or only with other females. Upon contact with the soiled bedding from male rats, many females ovulate despite permanent light exposure. Lesions of the vomeronasal system virtually eliminate this response.

Under normal (cyclic) light conditions, exposure of female rats to males or their chemical signals shortens the 5-day cycle of the females by 1 day. Surgical or chemical destruction of either the olfactory system proper or the vomeronasal system eliminates cycle shortening by male urine. Additional work is necessary to comprehend the individual roles of olfactory and vomeronasal systems in the mediation of pheromone-induced influences on the hormones that control reproductive cycles in female rats.

Even less is known about the neurobiology of the college phenomenon: In 1971 the scientific community first became aware of a phenomenon that must have been known for centuries but remained without public attention. Women who live together in college dormitories or nunneries tend to synchronize their menstruation cycles. This phenomenon is also observed in ordinary families in which sisters or mother and daughter(s) ovulate in synchrony, more often than statistically expected. It is now clear that female underarm perspiration contains volatile chemical signals which can shift the onset of menstrual cycles of other women and that the effect mainly depends on the time females spend in close proximity.

In the course of studies on the college phenomenon another pheromonal influence on human menstruation cycles was discovered: The glands of male armpits emit an unidentified chemical signal that stabilizes, as well as shortens, "monthly" cycles of women. Consequently, its presence and biological action is most evident in females who have long spontaneous menstruation periods and/or frequent interactions with men.

C. "Pregnancy Block"

Female mice which have mated and are subsequently exposed to the odor of a strange male undergo hormonal changes resulting in termination of the pregnancy (pregnancy block). This implies the formation of a memory or some form of recognition process by the female for its mate's pheromones at the time of insemination. As a result of this memory, males made familiar by mating are recognized by the females, thereby mitigating pregnancy block. Such a memory function is biologically important to the female because it is required to sustain pregnancy when the mate remains in the vicinity. The active odorant is contained in the urine. Formation of the imprint is initiated by vaginocervical stimulation during a single intercourse. The minimum time of postcoitus exposure to the male pheromone that is required for establishment of the olfactory memory trace is between 3 and 4.5 hr. If male and female are separated before his chemoidentity has been memorized by the female, her mate's odor can induce pregnancy failure when he returns. A stabilized imprint is long-lasting unless pregnancy ensues. In the event of pregnancy the olfactory memory fades fast, an effect which can be replicated by implants of estradiol in nonpregnant females. Pregnancy-induced fading of the female's memory of her former mate's chemoidentity allows successful reproduction with other males.

The olfactory block to pregnancy is caused by pheromonal action which is probably mediated via the accessory olfactory system. The olfactory bulb proper and the accessory olfactory bulb receive noradrenergic innervation from the brain stem via the olfactory striae. Destruction of these centrifugal neuronal connections, 6 days before mating, depletes noradrenaline in both bulbs and results in a failure by the female to recognize her mate. Hence, his odor now blocks his own pregnancy. However, noradrenaline depletion carried out after imprint formation at mating does not prevent recognition of the mate, implying that noradrenaline is required for formation, but not for recall, of male pheromone(s).

It is now undisputed that two types of synaptic plasticity can occur in the mouse accessory olfactory bulb (probably influencing dendrodendritic synapses between mitral and granule cells): The association of mating and pheromone exposure induces memory formation by increasing the inhibition of the pheromonal signal, thereby preventing activation of the neuroendocrine block to pregnancy. Male odors perceived independent of mating have the opposite effect—decreasing the inhibition of the pheromonal signal and, consequently, promoting estrous.

V. Child Behavior and Maternal Care

Pheromone communication between mother and child increases the survival chances of newborns. In several of the species of mammals investigated, the mother needs several hours to learn recognition of her newborn's chemoidentity. During this period adoption of alien babies is possible. Because of this delay, mothers with stillborn offspring can contribute to species survival by substituting for a female that has died or was disabled during parturition.

A. Mother to Child Signaling

Numerous cases have been reported in which maternal pheromone signals are utilized by newborns: In many (all?) marsupials the (very immature) babies have to find their way to the pouch without maternal support. Apparently, the female is not concerned when her (still blind) baby fights for its life while climbing from the vagina. This behavioral negligence is compensated for by pheromone signals: Abdominal skin glands lay a chemical trail to guide the baby from the vagina to the pouch.

Also, rabbits give their young little care. The doe does not even retrieve pups which stray from the burrow. When entering the nest, the mother simply positions herself over the litter. She does not provide the young with any direct behavioral assistance to suckle. Furthermore, to reduce the risk of predators trapping her (and the pups), it is important that the time spent with the offspring be kept to a minimum. These strategies place great demands on the pups. They have to find the nipples quickly. A pheromone that is contained in the milk provides the pups with information on the nipple location. Interestingly, production of the pheromone is not confined to lactating does. By using the response of newborns to test for the presence of the pheromone, it can be shown that all mature does produce it. When rendered anosmic by removal of the olfactory bulbs or irrigation of the nasal mucosa with zinc sulfate, the newborns are completely unable to suckle from the mother. Probably, the olfactory system proper is responsible for detection of the nipple-search pheromone.

B. Child to Mother Signaling

Herds of grazing mammals often give birth in synchrony. It is therefore important for the mother to form a rapid recognition of her own offspring to distinguish them from others. In sheep, the ewe develops a selective bond with her lamb within a few hours after parturition based on chemosensory signals from the newborn. Recognition requires an intact olfactory bulb (proper). To allow the imprinting, processing of olfactory signals is altered by vaginocervical feedback to the brain, stimulating interest in lamb odors.

Lamb odors have little effect on either transmitter release or neuronal activity in the olfactory bulb (proper) prior to birth. After birth, there is an increase in the number of mitral cells, and a subset of these neurons has a distinct electrical responsiveness to the odor of the mother's child. Furthermore, the release of glutamate and γ-aminobutyric acid from the dendrodendritic synapses between mitral and granule cells is increased. Preparturition lesions of the olfactory bulbs (proper) or of their noradrenergic centrifugal innervation eliminate these effects as well as child recognition, enabling the ewe to adopt alien lambs even many hours postpartum.

Both olfactory recognition of the juvenile and maternal experience are important determinants of successful maternal care. In primiparous (i.e., inexperienced) mice olfactory cues play the dominant role: Noradrenaline depletion of the olfactory bulbs, induced by lesions of the centrifugal noradrenergic pathways, results in cannibalism at birth without producing general anosmia. Similar lesions made after parturition and maternal experience are completely without effect. Noradrenaline depletion of the olfactory bulbs in multiparous mice does not result in cannibalism. There is good evidence that the olfactory bulbs proper rather than the accessory olfactory bulbs mediate information on child-emitted pheromones in mice.

The current working hypothesis of many scientists is that the centrally controlled release of noradrenaline exerts morphogenetic effects on the intrinsic circuits of the olfactory bulbs (proper and accessory) in dependence of olfactory experience. These changes are supposed to result in a sharpening of the odor-induced pattern of activity, due to increases in neuronal lateral inhibition, in the olfactory bulb proper. This is in contrast to the prevailing assumption regarding mechanisms involved in accessory olfactory bulb learning in which increased self-inhibition of mitral cells is invoked and it is postulated that experience-dependent changes in responsiveness are the result of a disruption of pheromone information transmission.

VI. MOLECULES THAT ENCODE MESSAGES TO CONSPECIFICS

Pheromones may have evolved from hormones. One hypothesis on the evolutionary development of pheromones implies that substances (or their metabolites) that had been used for signaling within organisms were subsequently employed for communication between individuals. An opposing concept of pheromone evolution suggests that external chemical signals that were used by single-cell organisms (e.g., bacteria) became hormones in species with multiple cells. Regardless of which one of the two hypotheses correctly describes the phylogenetic events, the problem of "undesired" interspecies cross talk is imminent.

A pheromone is a molecule that is secreted by an organism and emitted into the environment; when binding to pheromone receptors of a conspecific organism, it triggers biochemical and/or neuronal cascades in the target organism. By itself or in concert with other pheromones, it ultimately mediates social and/or reproduction-associated behavior. This definition does not exclude effects on nonconspecific organisms; it parallels the one of hormones—hormones propagate signals within an organism, whereas pheromones take an extracorporal path to reach other individuals of the same species.

A. Specificity of Pheromones

When Karlson and Luscher coined the term pheromone, they did not require species specificity. Many subsequent authors, however, suggested that a species-dependent molecular specificity existed when they applied the term pheromone to a chemical signal of social significance. A need to readopt the original definition is suggested by some experimental evidence as well as by appreciation of the fact that, most probably, more species of teleost fish exist than possible pheromones that they can rely on to coordinate their sex lives. Hence, it was not surprising when research demonstrated that mixes of molecules are employed for signaling. Each mixture can be species specific, but molecules in the mix may be "shared" by several species.

Furthermore, experimental evidence shows that some species emit pheromone messages to conspecifics which are "understood" across species boundaries. The expectation of true species specificity appears unwarranted and it has become undisputed that species-specific and non-species-specific pheromones are to be distinguished. The prevailing assumption is that taxonomically related species may not have come to evolve species-specific pheromones if there was no advantage that enforced their development.

The effects of individual pheromones as well as of blends of several pheromones are most thoroughly studied in insects. About 4500 lepidopteran species communicate by only $\approx$150 such molecules. In the moth *Bombyx mori*, just one chemical compound [bombycol (E,Z-10,12-hexadecadien-1-ol)], secreted from a female gland, is sufficient to induce the male sex behavior.

Most insect species use pheromone blends with particular ratios, e.g., pheromones A and B [A =(Z)-9-tetradecenyl acetate, B =(Z)-11-tetradecenyl acetate], serve as attractants for one kind of moth when present in a concentration ratio 1A:1B, whereas they are an attractant for another moth when blended in a ratio of 1A:10B. In summary, utilization of pheromone mixtures enhances the (chemical) information transfer capacity.

Because pheromone mixtures are used for communication, it is feasible that some of the molecules involved, which in concert with others influence the behavior of conspecifics, also act on nonconspecifics (mostly as deterrents). This is important in sympatric species which share the same habitat and was first demonstrated in bark beetles. Hence, pheromones are species-specific signals as well as messages from one species to another, at least in insects.

B. The Chemical Nature of Pheromones

All major classes of molecules appear to play a role in pheromone signaling. The insect pheromones mentioned previously are carbonic acids or derivatives thereof. This class of pheromones is, however, not just found in insects. For instance, a few weeks prior to ovulation, female elephants release (Z)-7-dodecen-1-yl acetate (Z7–12:Ac) and the urine concentration of Z7–12:Ac increases from undetectable levels (luteal phase) to 2 μM (early follicular phase) and to about 150 μM just prior to ovulation. This compound elicits premating and mating-associated behavior in adult male elephants.

On a much lower level of phylogeny, pheromone signaling has evolved: Yeast cells and bacteria use peptides for specific communication. Haploid *Saccharomyces cervisiae* (beer yeast) approach each other to conjugate by orienting growth along pheromone gradients, a mechanism referred to as chemotropism. The mediating molecule is a dodecapeptide pheromone [YIIKGVFWDPAC(farnesyl)-OCH$_3$]. Some bacteria can control their growth by a peptide pheromone that is released into the environment. This class of pheromones (peptides) also exists in vertebrates: Sodefrin is a decapeptide pheromone of red-bellied newts (*Cynops pyrrhogaster*), and aphrodisin, a glycoprotein which was isolated from hamster vaginal discharge, acts as a mounting-inducing pheromone in male hamsters. Aphrodisin is also expressed in hamster ovaries and in the uterus where it may affect egg maturation and implantation, an example of the plurifunctional potency of some "pheromones." Another case of mediators/hormones which act as pheromones is provided by prostaglandins. First isolated from sperm fluid, they are now known to regulate functions of several organs. Furthermore, F-type prostaglandins are pheromones in goldfish and presumably also in other vertebrates.

One of the most important classes of pheromones are steroids. Preovulatory female goldfish secrete 20β-dihydroxy-4-pregnen-3-one and F-prostaglandins into the urine and, when exposed to the urine, male goldfish respond with increased plasma levels of gonadotropin-II, 17,20β-dihydroxy-4-pregnen-3-one, testosterone, and 11-ketotestosterone as well as an increase of milt volume. The water-borne prostaglandins (PGF$_{2\alpha}$ and 15-keto-prostaglandin F$_2$) are detected by the goldfish olfactory mucosa at low nanomolar concentrations.

Steroids can also be powerful pheromones in other vertebrates. In boars, androsterones elicit mounting behavior and the perspiration secreted from human apocrine glands of axillary and pubic hair regions appears to contain steroid hormones. It has been proposed that these molecules and/or their metabolites act as human pheromones. A study sponsored by Erox Corporation has revealed that some steroids elicit stronger responses in the human male than in the female vomeronasal epithelium, whereas others induce more pronounced reactions in female than in male Jacobson's organs. Nevertheless, the prevailing (but unproven) assumption is that the adult human vomeronasal organ is not functional. This does not exclude a role of pheromones in human communication: As in fish, a receptor cell population of the (principal) olfactory epithelium may be responsive to pheromones.

C. "Aquatic" (Water-Dissolved) and "Aerial" (Volatile) Pheromones

It is often implied or claimed that pheromones can be subdivided in two classes defined as aquatic and aerial. This distinction does not appear to be appropriate because most if not all classes of pheromones can be found in air and in water. Obviously, molecules of low volatility (steroids, proteins, and large peptides) must be highly concentrated in the secreted fluid and, since the range of diffusion is limited, sniffing behavior is required for detection.

A particularly interesting problem arises when volatile pheromones enter the water/mucous interface of a sensory epithelium. In this case, only negligible portions of the pheromones would be dissolved. Nature has solved this problem by pheromone-binding proteins being present at high concentrations in the mucous of the chemosensory epithelia of many species. Potentially, pheromone-binding proteins can facilitate the diffusion of such molecules and, at the same time, exclude other molecules from passing to the receptors. Consequently, it is to be presumed that pheromone-binding proteins act as chemical filters.

D. Volatile Pheromones in Aquatic Environments

Despite the fact that many pheromones do not dissolve in water, they may nevertheless be used for chemocommunication in aquatic environments: (i) Volatile pheromones are dissolved in water if concentrations are sufficiently low. Since pheromones act at extremely low concentrations, this requirement may be fulfilled; (ii) after drifting to the water surface, such substances form high concentration layers, which can be sensed by animals that skim the surface such as the teleost fish *Anableps*; (iii) pheromones may become attached to bacteria and other organic matter present in a body of water and thereby be distributed; and (iv) little is known regarding why "volatile" substances may serve pheromonal functions in marine habitats. Many of the molecules involved are easily dissolved under the high-pressure conditions present at sufficient oceanic depth. Hence, deep-sea fish may utilize pheromones that are usually referred to as volatile.

VII. HORMONES ALTER OLFACTION AND PHEROMONE RELEASE

Pheromone perception and emission is modulated by serum levels of hormones. Evidence for hormonal influences on the sensitivity of chemosensory systems exists for many vertebrate classes.

Prior to scientific evaluation, it was common knowledge that women are more sensitive than men to many odorants. Research has since confirmed this belief and demonstrated that adult males and prepubital females or males are less sensitive than adult females. Furthermore, it was noted that female sensitivity varies during the menstrual cycle. It peaks during the second half of menses, midcycle, and midluteally. It is unclear whether the ovaries modulate olfactory sensitivity directly or by cyclic changes in the mucose layer covering the olfactory receptors. Only in pregnant mice has enhanced neurogenesis of (vomeronasal) receptors been demonstrated unequivocally.

Also, clinical evidence provides indications for interrelationships between the olfactory and reproductive systems. Patients suffering from adrenal cortical failure (Addison's disease) have significantly increased olfactory acuity, whereas persons with chromatin-negative gonadal dysgenesis (Turner's syndrome) display massive deficits. For the neurobiology of smell, in general, and pheromone perception, in particular, the pathology of Kallmann's syndrome is of prime interest. Kallmann's syndrome patients display hypogonadotropic hypogonadism in combination with anosmia. Histology has revealed that these symptoms are associated with a lack of migration of GnRH-ir neurons from the embryonic olfactory placode (where they originate) into the brain. This finding emphasizes the functional significance of the terminal nerve/preoptic nerve for (human) olfaction and gonadal development.

In addition to their influence on sensory systems, hormones also modulate pheromone release: Attraction of conspecific males during the time a female is in estrus is undesired to many owners of dogs and cats. Further evidence for the involvement of hormones in the control of pheromone emissions came from ovariectomized rabbits. The nipple-search behavior of pups was used as a bioassay and the experiments revealed that nipple pheromone release is induced during parturition by the combined action of estrogen and progesterone and maintained during lactation by the dual action of estrogen and prolactin.

See Also the Following Articles

Kallmann's Syndrome; Nervus Terminalis; Vomeronasal Gland.

Bibliography

Ackerman, D. (1990). *A Natural History of the Senses*. Random House, New York.
Burton, R. (1976). *The Language of Smell*. Routledge Kegan Paul, Boston.
Doty, R. L. (Ed.) (1976). *Mammalian Olfaction, Reproductive Processes, and Behavior*. Academic Press, New York.

Duvall, D., Mÿller-Schwarze, D., and Silverstein, R. M. (Eds.) (1986). *Chemical Signals in Vertebrates*. Plenum, New York.

Finger, T. E., and Silver, W. L. (Eds.) (1991). *Neurobiology of Taste and Smell*. Krieger, Malabar.

Karlson, P., and Luscher, M. (1959) "Pheromones": A new term for a class of biologically active substances. *Nature (London)* **183**, 55–56.

Keverne, E. B. (1983). Pheromonal influences on the endocrine regulation of reproduction. *TINS* **6**, 381–384.

Laverack, M. S. (1988). The diversity of chemoreceptors. In *Sensory Biology of Aquatic Animals* (J. Atema, R. R. Fay, A. N. Popper, and W. N. Tavolga Eds.). Springer, New York.

Leach, M. (Ed.) (1950). *Standard Dictionary of Folklore, Mythology, and Legends*. Funk & Wagnalls, New York.

Liley, N. R., and Stacey, N. E. (1983). Hormones, pheromones, and reproductive behavior in fish. In *Fish Physiology* (W. S. Hoar, D. J. Randall, and E. M. Donaldson, Eds.), Vol. 9. Academic Press, New York.

Novotny, M. (1987). The importance of chemical messengers in mammalian reproduction. In *Masculinity/Femininity: Basic Perspectives* (J. M. Reinisch, L. A. Rosenblum, and S. A. Sanders, Eds.). Oxford Univ. Press, New York.

Pfaff, D. W. (Ed.) (1975). *Hormonal Factors in Brain Function*. MIT Press, Cambridge.

Stoddart, D. M. (1990). *The Scented Ape*. Cambridge Univ. Press, Cambridge, MA.

Theimer, E. D. (Ed.) (1982). *Fragrance Chemistry*. Academic Press, New York.

Vandenbergh, J. G. (Ed.) (1983). *Pheromones and Reproduction in Mammals*. Academic Press, New York.

van Toller, S., and Dodd, G. H. (1988). *Perfumery*. Chapman & Hall, London.

Weaver, N. (1983). Pheromones and behavior. In *Endocrinology of Insects* (R. G. H. Downer and H. Laufer, Eds.). A. R. Liss, New York.

Onychophora

Michael T. Ghiselin
California Academy of Sciences

I. Evolutionary Background
II. Modes of Reproduction
III. Evolutionary Ecology

GLOSSARY

cryptic female choice A mode of sexual selection in which the females affect male reproductive success after copulation.

gonochoric Separate-sexed (nonhermaphrodite).

oviparity A condition in which the egg develops outside of the body of the parent.

ovoviviparity A condition in which the fertilized egg is retained within the parental body and develops there but does not receive nutriment.

parthenogenesis Uniparental reproduction with development from an egg; if genetic recombination does not occur it is a form of asexual reproduction.

pseudospecies A group of related organisms produced strictly by asexual reproduction but otherwise resembling a (biological) species.

spermatophore A packet of sperm that is transferred from the male to the female.

viviparity A condition in which the fertilized egg is retained within the parental body, and both develops and receives nutriment there.

Onychophorans are a small group of animals that are sometimes treated as a separate phylum. The best known genus, *Peripatus*, often serves as a vernacular name. They are sometimes called "velvet worms" but they crawl about on claw-bearing legs and look similar to a caterpillar or a millipede. They are reasonably treated as primitive Arthropoda, even though the soft body and unjointed limbs make "arthropod" somewhat of a misnomer. They do have a cuticle and other "typical" arthropodan traits in addition to some that point to the derivation of arthropods from something such as annelids.

I. EVOLUTIONARY BACKGROUND

Onychophorans are often considered "living fossils." A variety of fossils found in Paleozoic deposits resemble the modern forms in general morphology, but they were considerably more diverse and were marine. All extant onychophorans are terrestrial. Their soft bodies allow them to squeeze through remarkably narrow openings, but their permeable integument restricts them to humid situations such as inside rotting logs, and they are mainly active at night. They feed on small animals such as insects.

The Onychophora evidently invaded the land at a very early period and have evolved remarkably slowly ever since their initial adaptation to terrestrial conditions. Their biogeography indicates that the basic diversification occurred prior to the breakup of the Gondwanaland supercontinent about 116 million years B.P. Two families are recognized. The first of these, Peripatidae, occurs in three tropical areas. One lineage of Peripatidae, which includes *Peripatus* proper, is widely distributed in South America, Central America, and Mexico and the Caribbean. Probably its closest relatives, *Mesoperipatus*, occur in a small area of tropical East Africa. A third group occurs in Southeast Asia but does not penetrate across Wallace's line. The other family, Peripatopsidae, has a more southerly range. Within it there is a South African assemblage that includes the type genus *Peripatopsis*. It seems to have branched off before two other lineages. One of these occurs in southern Chile and the second occurs in New Zealand, Tasmania, and Australia and New Guinea, again not crossing Wallace's line. Although until recently only about 100 species of extant onychophorans were recognized, biochemical work indicates that there are many "cryptic species" that are otherwise hard distinguish. This indicates that the group has been slowly diversifying physiologically while changing little in structure over a long period of time.

II. MODES OF REPRODUCTION

Reproductively, the onychophorans are some of the most remarkable creatures in the entire animal kingdom. Although in general a homogeneous group, they span the gamut from oviparity, through ovoviviparity, to viviparity sometimes with a placenta. In correlation, the males have also become much modified, and an impressive pattern of sexual competition and sexual dimorphism has evolved. Onychophorans therefore provide superb, but still underutilized, materials for the study of reproductive strategies.

Onychophorans are strictly gonochoric (old records of hermaphroditism are mistaken). Parthenogenesis has been reported in one species or pseudospecies, but otherwise reproduction is strictly sexual. Restricted motility implies a certain amount of inbreeding, and the scarcity of males in collections of some species has led some authors to suspect that there is local mate competition. So far as is known, however, the real sex ratio is 1 : 1, with the apparent sex ratio being due to the males having a shorter life span.

The reproductive tract in both males and females opens ventrally near the posterior end of the body. With few exceptions the sperm are packaged as spermatophores. In the Peripatidae, they are inserted into the genital opening. In the Peripatopsidae, they are attached to the female's body and hypodermic impregnation occurs, after which the sperm make their way either to the ovaries or, in some cases, to seminal receptacles. In some Australian onychophorans specialized structures on the head aid in the transfer of spermatophores. The ability to store sperm, together with other features of reproductive biology such as a long gestation period, places a premium on the males sometimes inseminating the females when both are still quite young.

In what would seem to have been the ancestral condition, oviparity, the eggs are yolky and covered with a protective shell. They develop externally. Ovoviviparity evolved by a straightforward retention of the eggs within the body. The possibility that oviparity is secondary has been repeatedly suggested by various authors, but if this were so an eggshell such as is present in ovoviviparous forms would have had to evolve before its protective function. There has followed reduction in the amount of yolk and transfer of nutrients from the mother to the developing embryos. In the viviparous Peripatidae there is a placenta with a stalk that attaches the embryo to the wall of the uterus, but nutrient uptake is via the surface of the embryo and not via the placenta itself. Viviparous Peripatopside lack a placenta, but some have a specialized area, the trophic vesicle, through which the developing embryo takes up nutrients. Retention of the developing young within the mother's body gives protection and also provides for the progressive provisioning of the young. In many cases the young are the same age; however, they may be produced on a type of assembly line basis, with a graded series of offspring of different ages occurring within the uterus of the mother.

The young of viviparous species can be quite large. They do not undergo a metamorphosis and are born with the same number of legs as the adult. The young may stay with the mother for quite some time, and there have been reports of maternal care for the young.

III. EVOLUTIONARY ECOLOGY

Judging from the patterns of geographical distribution, at least some of the extreme modifications in reproductive physiology must have occurred many millions of years ago, and the group remains in a condition of relative evolutionary stasis.

Within the Peripatidae, the Oriental *Eoperipatus* is more primitive insofar as there is a large egg and no placenta. The common possession of a placenta by the African *Mesoperipatus* and the tropical American forms is one indication of their closer relationship. Within the Peripatopsidae there are oviparous and ovoviviparous forms, as well as viviparous ones, but none of these have a placenta.

Oviparity correlates with relatively cool and harsh climates, relatively small body size, very little sexual dimorphism, and no deviation from the usual 1:1 apparent sex ratio. The viviparous species are characteristic of more tropical climates, are of relatively larger body size, have smaller males with fewer legs than the females, and have a short life span. It appears that larger female body size evolved because of the advantage, to an organism living in a stable climate, of investing resources in the developing young. The only known parthenogenetic assemblage also fits this pattern quite well, despite being viviparous. Typical of animals that have lost sexuality, they have evolved more rapid reproduction and live in relatively unstable habitats.

The ability of the females to store sperm may also have placed a premium on early reproduction in the males, which means less resources invested in growth or survival. The sexual dimorphism in size is accompanied by a dimorphism in the number of legs, which are fewer in the males. That this dimorphism is adaptive is indicated by the fact that the number of legs is congenital. As one might expect, the mother invests equal amounts of resources in sons and daughters. At birth, the sex ratio is 1:1 and the males and females are the same size. Much of the onychophoran reproductive behavior can be explained in terms of sexual selection. Male–male competition in onychophorans is a kind of scramble in which male reproductive success depends on getting to the females first (so-called "male dispersal"). Some aggression between males suggests that there is also sexual selection by male combat, but very little information is available. Spermatophores and hypodermic impregnation suggest cryptic female choice, but this possibility has not yet been investigated.

See Also the Following Article

Parthenogenesis and Natural Clones

Bibliography

Briscoe, D. A., and Tait, N. N. (1995). Allozyme evidence for extensive and ancient radiations in Australian Onychophora. *Zool. J. Linnean Soc.* 114, 91–102.

Ghiselin, M. T. (1974). *The Economy of Nature and the Evolution of Sex.* Univ. of California Press, Berkeley.

Ghiselin, M. T. (1984). *Peripatus* as a living fossil. In *Living Fossils* (N. Eldredge and S. M. Stanley, Eds.), pp. 214–217. Springer-Verlag, New York.

Ghiselin, M. T. (1985). A movable feaster. *Nat. History* 94, 54–60.

Havel, J. E., Wilson, C. C., and Hebert, P. D. N. (1989). Parental investment and sex allocation in a viviparous onychophoran. *Oikos* 56, 224–232.

Monge-Nájera, J. (1995). Phylogeny, biogeography and reproductive trends in the Onychophora. *Zool. J. Linnean Soc.* **114**, 21–60.

Ruhberg, H., and St. J. Read, V. M. (1992). Onychophora. In *Reproduction of Invertebrates, Volume 5: Sexual Differentiation and Behavior* (K. G. Adiyodi and R. G. Adiyodi, Eds.), pp. 267–280. Wiley, New York.

Storch, V., and Ruhberg, H. (1993). Onychophora. In *Microscopic Anatomy of Invertebrates, Volume 12: Onychophora, Chilopoda and Lesser Protostomia* (F. W. Harrison and M. E. Rice, Eds.), pp. 11–56. Wiley-Liss, New York.

Tait, N. N., and Briscoe, D. A. (1990). Sexual head structures in the Onychophora: Unique modifications for sperm transfer. *J. Nat. History* **24**, 1517–1527.

Oocyte and Embryo Transport

Manuel Villalón, Luis Velasquez, and Horacio Croxatto

Pontificia Universidad Católica de Chile

I. Introduction
II. The Pathway
III. The Time Course
IV. Mechanics
V. Regulation

GLOSSARY

ampullary–isthmic junction The short tubal segment in which the structural and functional transition from ampullary to isthmic segment takes place.

ampullary segment The distal oviductal segment which opens through the infundibulum and fimbria into the peritoneal cavity or ovarian bursa.

embryo transport Passage of the embryo from the site of fertilization, which is the ampullary segment of the oviduct, to the site of implantation in the uterus.

endosalpinx The inner layer of mucous membrane lining the oviductal lumen, composed of epithelium, basal membrane, and subepithelial connective tissue, blood and lymphatic vessels, and nerve fibers.

Fallopian tube The name given to the oviduct in the human female.

fimbria A complex expansion of the mucous membrane of the oviduct outside the lumen at the ovarian end.

isthmic segment The proximal oviductal segment connected with the uterine cavity.

mesosalpinx The external layer of mesothelium and loose connective tissue covering the oviduct.

myosalpinx The muscle layer of the oviduct.

oocyte transport Passage of the oocyte from the ovarian surface into and through the oviduct, with further passage into uterus and expulsion per vaginam. In some species, namely, equids and bats, oocytes are retained in the oviduct where they disintegrate.

oviduct The tubular organ connecting the periovarian space with the uterine cavity.

oviductal transport The stage of oocyte or embryo transport that takes place in the oviduct.

uterotubal junction The site at which the isthmic segment ends in the uterine cavity.

The oocyte is the only single-cell entity from which new individuals are generated in natural circumstances. In mammals, the oocyte needs to be transferred from the ovarian follicle to appropriate sites for fertilization and development up to a viable fetus. Egg transport in mammals refers to the passage of the oocyte or embryo from the ovarian surface toward the site of implantation in the uterus. It encompasses cumulus–oocyte pickup by the fimbria and transport through the oviduct and within the uterus until attachment of the blastocyst at the implantation site takes place.

I. INTRODUCTION

The oviduct and uterus (Fig. 1) are connected in line, offering a great diversity of structural and functional features across species in accordance with their reproductive strategies and evolutionary adaptations, to meet functions as distinct as conveying spermatozoa to the egg on time, in quantities and qualities that ensure safe fertilization; transporting and nurturing the developing egg; sustaining the implanted embryo; delivering the fetus; and limiting the access of invading microorganisms that penetrate through the external opening of the genital tract.

Unfertilized oocytes may or may not enter the uterus, or they may disintegrate within the tract or be expelled to the vagina depending on the species. When fertilization takes place, usually in the ampullary segment of the oviduct, the resulting zygotes may develop up to morula or blastocyst stage within the oviduct. After entering the uterus, they continue to develop and display great mobility within the endometrial cavity before the onset of implantation. In some species implantation is preceded by a state of embryonic dormancy or diapause in which development is temporarily halted.

The oviduct or Fallopian tube accomplishes several interrelated functions. It captures the oocyte once it has been released from the follicle at the completion of the ovulatory process, it controls the ascent of spermatozoa to the site of fertilization, it provides the space and biological milieu where fertilization takes place, it contributes to remove or add coatings to the egg and to its nurture as it is being transported toward the uterus and it delivers the developing embryo at an appropriate developmental stage and time to the endometrial cavity where development continues. In polytocous species, when several embryos enter the uterus, further transport takes places for the purpose of spacing them along the uterine horn. In some monotocous species the single embryo that enters the uterus is transported up and down and from one to another uterine horn until attachment takes place. Such traveling allows the embryo to signal to the entire endometrium its presence so that the luteolytic mechanism that normally originates in the epithelial lining of the uterus is halted or prevented. This allows continued secretion of progesterone by the corpus luteum, a hormone essential for the establishment and continuation of pregnancy.

The mechanics of oviductal egg transport encompass continuous ciliary beat toward the uterus, intermittent muscle contractions which generate transient pressure gradients and mobilize the luminal contents, sphincteral tonic muscle contractions that temporally close the junction between segments, changes in the quality of the secretions that plug and unplug

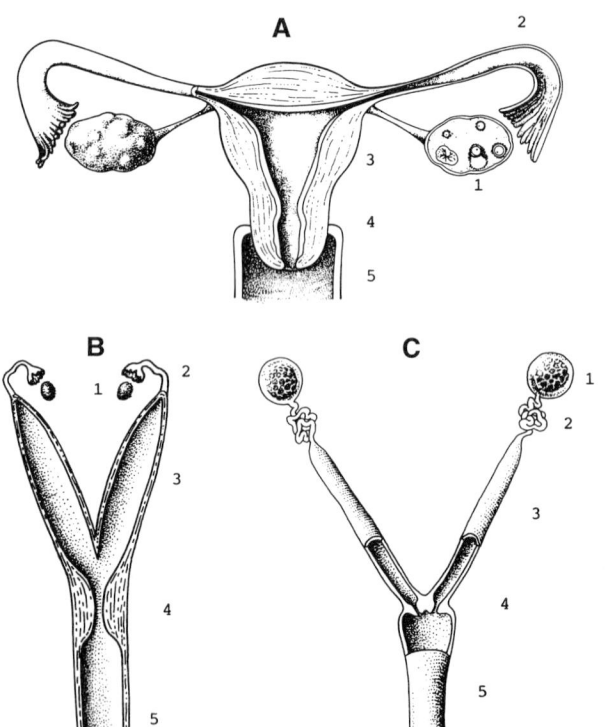

FIGURE 1 Three different types of female genital tracts encountered among mammals. A, human; B, pig; C, rat. 1, ovary; 2, oviduct; 3, body of the uterus or uterine horn; 4, cervix; 5, vagina. Note contrast between the simplex uterus in A and the uterine horns in B and C; communicating uterine horns and single cervix in B and independent uterine horns and double cervix in C; and highly coiled oviduct opening into the periovarian sac in C and more straight tubes opening into the peritoneal cavity in A and B. Drawings are not to scale.

the lumen, and changes in vascular engorgement and interstitial fluid dynamics that affect tissue elasticity. The interaction between these factors determines the rate and pattern of oviductal egg transport.

Egg transport is said to be a regulated process because the passage of the egg to the site of implantation does not follow a fixed invariable time course but one that changes with the physiologic environment external to the genital tract and with the condition of the eggs being transported. The terms endocrine and embryonic regulation of egg transport allude to the source of signals that act on oviductal or uterine cells to control the rate or direction of movement of eggs within the lumen or to time their passage from one segment to the next. These signals are believed to trigger paracrine and autocrine signaling within the oviduct and uterus.

The following sections describe the structure of the oviduct, the pattern and time course of egg transport, the physiology of the cells involved in the mechanics of transport, and how the process is controlled by ovarian hormones and signals arising from the egg.

II. THE PATHWAY

Gabriele Fallopius of Padua provided, in the sixteenth century, a correct anatomical description of the oviduct, or tube as he called it, as the tubular organ connecting the ovary with the uterus. It was not until two centuries later that William Cruikshank demonstrated that eggs pass from the ovary to the uterus through the oviduct.

Depending on the species, the distal end of the tube may open into the abdominal cavity or into a bursa that surrounds the ovary. The proximal end opens into the uterine cavity. In the human, the length of the extrauterine portion of the oviduct measures from 6 to 15 cm and is fairly straight, whereas in small rodents it measures a few millimeters and is highly coiled.

Four segments constitute the oviduct. From distal to proximal end they are the fimbria and infundibulum, the ampulla, the isthmus, and the intramural segment. Two transitions from one segment to the next are distinguished and are designated as ampullary–isthmic junction (AIJ) and uterotubal junction (UTJ).

The wall of the oviduct consists of an outer connective tissue layer (the tunica serosa), a middle muscular layer (the myosalpinx), and an inner, highly folded mucosa (the endosalpinx) (Fig. 2). The serosa encircles the oviduct forming the mesosalpinx, which loosely connects the oviduct to the posterior peritoneum. The two serosal sheets of the mesosalpinx enclose a thin smooth muscle layer as well as supporting ligaments, nerves, lymphatics, and blood vessels. The mucosa is lined by a columnar monostratified epithelium containing mainly ciliated and secretory cells. Ciliated cells predominate in the ampulla and secretory cells in the isthmus. Morphological and functional changes of the tubal epithelium occur in association with the hormonal fluctuations of the ovarian cycle.

The fimbria and infundibulum are an expansion of the mucous membrane that adopts diverse forms across the species (funnel shaped, flower shaped, or simple eversion) having in common the highest density of ciliated cells and the thinnest layer of muscle fibers. The active stroke of the cilia beats in a centripetal direction, converging toward the entrance to the ampulla. The ampulla is the longest segment in the human and the shortest in rodents. Abundant and complex mucosal folds fill the lumen, leaving only a virtual maze-like space. The surrounding muscle layer is very thin.

Throughout the isthmus, the mucosa is reduced to a few primary folds, whereas the myosalpinx is thick and organized in layers: a thin inner longitudinal layer that runs at the foot of the mucosal folds, an intermediate circular layer that occupies the largest portion of the wall thickness, and an outer helicoidal layer which forms a continuum with the interserosal muscle of the mesosalpinx. In the human, the two outer layers are replaced by muscle bundles interlaced in all directions.

At their caudal end, the oviducts penetrate the uterine wall to form the intramural segment. The narrow lumen and the drastic increase in the thickness of the surrounding smooth muscle of this segment are probably responsible for the high resistance

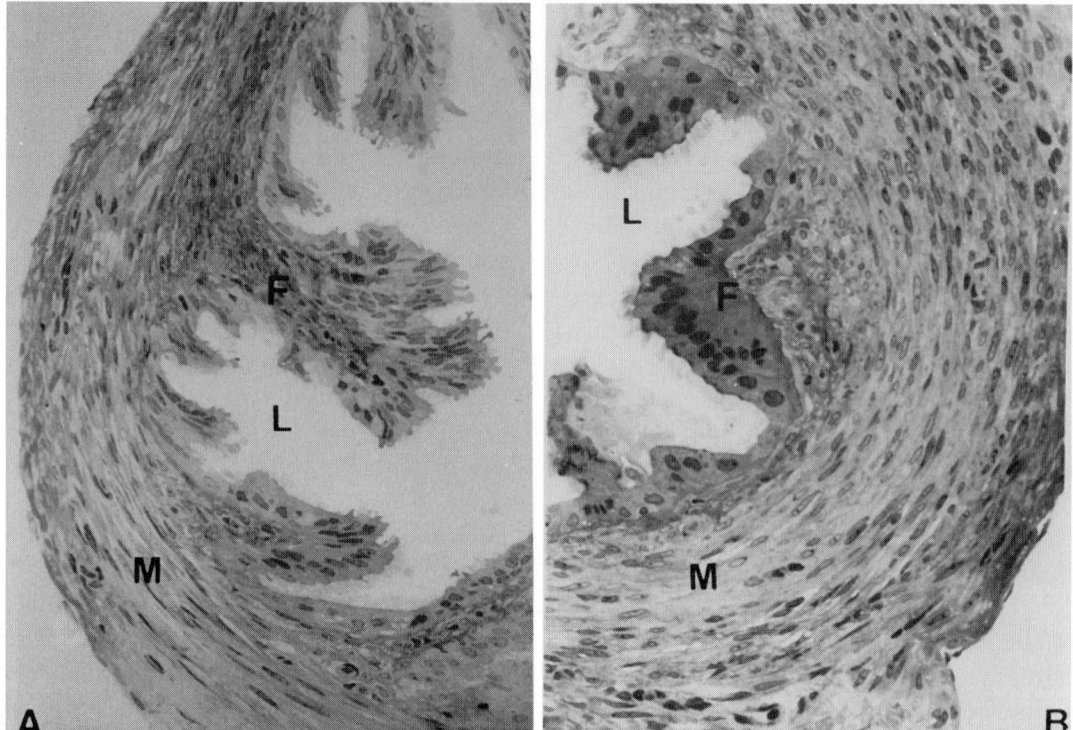

FIGURE 2 Cross sections of the wall and lumen (L) of the rat oviduct. (A) Ampullary segment: note the thin muscle layer (M) and complex mucosal folds (F) protruding into the lumen. (B) Isthmic segment: note thicker muscle layer and more simple and smaller mucosal folds.

to the passage of fluids found at the UTJ in some species.

The oviduct receives its blood supply from the uterine and the ovarian arteries. A close relationship between arteries and veins in the ovarian pedicle provides the anatomical basis for a local countercurrent transfer system of hormonal signals between the ovary and the tube. Lymphatic vessels are most developed in the isthmic segment and drain the mucosa, smooth muscle layer, and the serosa to the paraaortic nodes.

The tube is innervated by short and long postganglionic sympathetic fibers and by parasympathetic and afferent visceral nerves. Cholinergic and short adrenergic fibers supply the mucosa and muscle, respectively. The sympathetic innervation of the ampulla is sparse and reduced to vascular terminals, whereas that of the isthmus is abundant and also supplies the myosalpinx. There is also abundance of peptidergic terminals.

III. THE TIME COURSE

Embryo transport up to the site of implantation in several groups of mammals is shown in Table 1. It comprises three main phases.

TABLE 1
Duration of Embryo Transport Across Groups of Mammals

Species	Interval ovulation–implantation (days)
Rodents	4–7
Primates	8–11
Carnivores	11–14
Bats	10–16
Ungulates	15–60

A. Oocyte Release from the Follicle and Uptake by the Fimbria

The transit of the oocyte from the ovulating follicle into the tubal lumen lasts from a few minutes to a few hours. This process requires follicular rupture with extrusion of the cumulus oophorus, contact of the latter with the fimbria, and the action of a mucociliary current that drives it into the infundibulum. In small rodents, the follicular content is emptied into the periovarian bursa, into which the fimbria protrudes, and this ensures the cumulus–fimbria encounter. In the human, the ruptured follicle voids its contents directly to the peritoneal cavity, unless the fimbria is covering it. During ovulation, the distal end of the tube and the ovary move actively using the tuboovarian ligaments as a bascule, allowing the fimbria to sweep the surface of the ovary, thus enhancing the probability of capturing the cumulus oophorus. Cilia pull very effectively the mucous matrix of the cumulus, detaching it from the ruptured follicle and carrying it into the infundibulum. However, normal or near normal fertility has been observed in animals following microsurgical disruption of the tuboovarian anatomic relationships, including fimbriectomy, thus challenging the functional importance of this arrangement.

B. Oviductal Transport

This phase extends until the egg is delivered to the uterine cavity. Its duration varies among the species, ranging from 1 to 12 days (Table 2) and it is unrelated to oviductal length. It is fairly constant among conspecific individuals observed under the same physiologic condition—an indication that it is a highly programmed and controlled process. On the other hand, it varies under different conditions, e.g., mated versus nonmated. This suggests that the program is not fixed but subject to physiologic regulation.

Transport through the ampullary segment takes minutes, whereas transport through the isthmus takes from several hours to several days. When eggs reach the AIJ, their transport stops for several hours or days. Most stages of fertilization and the loss of cumulus cells occur while eggs are arrested at the AIJ. Once eggs pass the AIJ they tend to travel close together. Rabbit eggs acquire a thick mucin coat as they traverse the isthmus. In some species eggs are also retained at the UTJ for a while before entering the uterus. The proportion of time spent by the ova in each segment varies among species. Using 100% as the total time of oviductal transport, the egg spends 90% of the time in the ampulla in the human, whereas in rodents the eggs spend only 25% of the time in this segment.

During transport through the ampulla eggs move back and forth for short distances at a speed that exceeds by far the average speed of their net transport toward the uterus. In the isthmus, eggs are transported within a bolus of fluid which also is displaced back and forth at high speed for variable distances by changes in intraluminal pressure gradients caused by simultaneous and successive contractions and relaxations of the myosalpinx at various points. Passage of embryos to the uterus at the normal time is important since early arrival is often followed by their expulsion to the vagina and undue retention in the oviduct causes loss of viability in most mammals.

C. Uterine Transport

Once the eggs enter the uterus they can be retained or expelled. At the normal time they are due to enter, the uterus shifts from an expulsive to a retentive state. During retention embryos are not still. In animals which have uterine horns, embryos are transported throughout the full length of the uterus and, when there is luminal continuity between the two horns, they can also migrate from one to the other.

TABLE 2
Approximate Duration (Hours) of Oviductal Egg Transport

Opposum	24	Rhesus monkey	72
Pig	48	Human	80
Guinea pig	48–72	Musk Shrew	85
Rabbit	55	Rat	88
Hamster	60	Goat	98
Baboon	71	Horse	144
Ewe	66	Cat	144–168
Mouse	72	Bat	188
Cow	72	Dog	192–240

In ungulates and other groups, the whole length of the uterus needs the proximity of the embryos or their signals to deactivate the luteolytic mechanism that normally originates in the endometrium when embryos are not present. If embryos are prevented from entering a significant section of the uterus during the period of mobility, pregnancy will not continue. Embryos are transported to their site of implantation as a result of myometrial activity. In polytocous species, they are transported along the uterine horns in a way that results in implantations much more regularly spaced than would be expected if spacing were to occur at random. Chemical signals produced by the embryo or distension of the uterine wall caused by blastocyst expansion initiate waves of myometrial contractions that lead to embryo spacing. In the simplex uterus of the human the blastocyst is unlikely to move much and most often implants medially in the posterior wall.

IV. MECHANICS

Smooth muscle, ciliated, and secretory cells are recognized as the oviductal mechanical effector cells that generate the forces that move or prevent the eggs from moving or progressing within the lumen. A description of the most relevant aspects of the physiology of these cells is provided in the following sections.

A. Smooth Muscle Cells

The smooth muscle of the oviduct exhibits both phasic and tonic activity. Detailed measurements of transport of microspheres or supravitally stained eggs in cumulus inside the oviduct by cinematographic recording and computer image analysis have revealed that phasic muscle contractions of the tube are responsible for the typical random pendular movements of the egg observed in a variety of species, including primates.

Along the wall of excised oviducts there are multiple pacemaker sites which change location, alternating periods of short-ranged directed (biased) propagation of the myoelectrical activity with periods of complete randomness of propagation. Directional propagation of myoelectric activity and the corresponding movement of material within the lumen take place through very short tubal lengths at speeds that can reach up to 1 or 2 mm/sec. Contractions last for only a few seconds before reversing direction, resulting in a characteristic to-and-fro pattern of intraluminal motion and very slow rates of net egg transport. However, oviductal motility can be partially derandomized resulting in biased contraction-driven egg movements. There is little information to explain how muscle cells integrate their action in time and space to account for the pattern of egg transport. Gap junctions are present in epithelial and smooth muscle cells of the oviduct and their number and functional state are regulated by sex hormones. Thus, specific molecules such as connexins might be involved in integrating the action of mechanical effector cells in the oviduct.

In summary, the movement of material within the lumen is determined by the frequency and distance that phasic contractions of the circular muscle propagate along the oviduct. On the other hand, pauses of net tubal transport at the junctions are probably due to the tonic activity of the myosalpinx and to the thickening of the smooth muscle layer in the next segment.

The phasic and/or tonic behavior of the myosalpinx can be independently affected by hormones or neurotransmitters, and longitudinal and circular muscle fibers can respond differently to the same substance. Estradiol (E_2) and progesterone (P) receptors and uptake of these steroids occur in the oviducts of all species examined. These receptors are localized in secretory, stromal, and smooth muscle cells and their levels are subject to differential regulation. Within physiologic concentrations, steroids do not cause immediate contraction or relaxation of smooth muscle, but they change the cell membrane potential and the responsiveness of the muscle cell to inotropic substances. Sex steroids also concentrate in sympathetic ganglia innervating the genital tract and part of their action on oviductal contractility may be mediated through the extrinsic innervation. In general, estrogen enhances and progesterone inhibits contractility of oviductal smooth muscle, probably by regulating the synthesis and/or release of sympathetic agonists and prostaglandins and the turnover of their corresponding receptors. This affords a complex control system in which the same agent has different effects. For instance, norepinephrine, a dual α- and β-receptor agonist with preferential α-receptor stimulation, causes contraction or relaxation of the myosalpinx depending on the tubal segment, the muscle layer, and the type of steroidal domination.

Several neuropeptides occur in the oviduct of the human and other species. They include neuropeptide Y, vasoactive intestinal peptide, and substance P. γ-Aminobutyric acid is highly concentrated in the rat oviduct but it is localized in the epithelium rather than the nerve terminals. Other substances which also affect the contractile activity of oviductal muscle are the prostaglandins (PGs). Their content, distribution, and effect on the oviduct depend on the species, the segment of the tube, type of prostaglandin, and the preexisting steroid levels. In the rabbit, for example, the number of binding sites of $PGF_{2\alpha}$ and PGE_2 in the myosalpinx changes in the various segments of the tube and at different times after ovulation. Estrogen enhances the synthesis of $PGF_{2\alpha}$ and the stimulation of tubal contractions produced by $PGF_{2\alpha}$. Conversely, progesterone inhibits the local synthesis and the response of tubal smooth muscle to $PGF_{2\alpha}$ and it potentiates the inotropic effect of PGE_1 on the proximal isthmus.

In the monkey, PGE_1 and PGE_2 have no effect on spontaneous tubal contractions during the preovulatory phase, but during the luteal phase they suppress spontaneous tubal contractions. On the other hand, $PGF_{2\alpha}$ has no effect on contractility during the follicular phase, but it stimulates contractions immediately before and during ovulation.

Human oviducts contain PGs, and, with some exceptions, $PGF_{2\alpha}$ stimulates and PGE_2 inhibits contractions in the different tubal segments at various times during the menstrual cycle. Clearly, PGs are involved in the control of tubal contractility in some species and their synthesis and effects are modulated by ovarian steroids.

B. Ciliated Cells

In several species, estrogen stimulates the development, persistence, and density of cilia in the mucosa of the oviduct. In primates the density of oviductal cilia and the height of the ciliated epithelium undergo cyclic variations in relation to the menstrual cycle. The oviductal epithelium is low (10 μm) in periods of progesterone dominance, its height doubles in periods of estrogen dominance, and it shows signs of atrophy in the late menopausal state.

Oviductal cilia are involved in oocyte transport, and in the absence of muscle contractions, ciliary activity is sufficient to drive the cumulus through the ampulla in the rabbit and the rat. Although ciliary action is sufficient to power tubal transport, it may not be essential. Some women suffering from immotile cilia syndrome, who are assumed to have no ciliary activity in their tubes, have intrauterine pregnancies. If the role of muscle and cilia in the oviduct is considered mutually exclusive, this appears paradoxical. However, it can be explained on the basis of redundancy in oviductal function, i.e., both muscle contractions and ciliary movement might not be necessary, yet each could be sufficient to ensure the transport of gametes.

Cilia beat continuously in a prouterine direction, but the frequency of ciliary beat changes throughout the ovarian cycle. Oviductal ciliated cells *in vitro* are responsive to neurotransmitters and hormones. For instance, PGs $F_{2\alpha}$, E_1 and E_2, and ATP in concentrations equivalent to those found *in vivo* increase the frequency of ciliary activity. β-Adrenergic agonists also stimulate ciliary motion and, although ovarian steroids do not cause immediate changes in ciliary activity, either estrogen or progesterone alone strongly potentiates and both combined inhibit β-adrenergic stimulation. Embryo secreted factors such as platelet-activating factor (PAF) and PGE_2 also increase the frequency of ciliary beat at concentrations in the nanomolar range. Oviductal ciliary activity can be modified by locally produced chemical signals, including embryo-secreted factors, and sex steroids can modulate the response to these signals.

C. Secretory Cells

Oviductal fluid is formed mainly by transudation from the blood and active secretion from the endosalpinx and varies in composition, rheological properties, and volume depending on the stage of the sex cycle. Water channels present in rat oviductal epithelium indicate that they might play a role in fluid formation, especially since their number varies during the estrous cycle. The fluid volume is greatest near ovulation and minimal during the luteal phase. Because muscle contractions cause displacement of fluid within the lumen, which results in the characteristic pendular movement of eggs in the isthmus, the amount and rheologic properties of the fluid produced might be important for egg transport. In humans and rabbits tenacious mucus that fills the isthmus near the time of ovulation and that disappears several days later has been suggested to play a role in timing the entrance of the egg into this segment.

V. REGULATION

The transport of oocytes and embryos is a regulated process. Premature arrival or surgical transfer of the embryos to the uterus is followed by decreased rate of implantation at least in rodents and rabbits. On the other hand, embryos retained in the oviduct by mechanical or hormonal manipulations stop developing and lose viability. The human oviduct is exceptional in that it allows full development and even implantation of the embryo, leading to tubal pregnancy, a life-threatening condition representing nearly 1% of all pregnancies. The idea of an optimal time for the embryo to move from the oviduct to the uterine milieu is widely held as a teleological argument in favor of the existence of control mechanisms that time its passage into the uterus. Three regulatory pathways have been investigated: (i) endocrine regulation exerted by the ovarian steroid hormones, (ii) nervous regulation through the autonomic innervation of the oviduct, and (iii) paracrine regulation exerted by the embryo. The relative importance of these regulatory pathways appears to differ widely between species.

A. Endocrine Regulation

The concept that ovarian steroids control oviductal transport arose from experiments designed to assess the effects of ovariectomy or exogenous hormones on this process. Diverse species react differently to these manipulations, and within a species the same hormone exerts qualitatively different effects on oviductal transport depending on the dose and time of administration. Selective hormone deficits produced by inhibiting synthesis and secretion or neutralizing the hormone in the circulation by means of antibodies or by intercepting the hormone at the receptor level by antihormones have confirmed that estrogens and progestins have opposite effects on egg transport and act disparately in different species. Endogenous estradiol speeds oviductal transport in the rat and has the opposite effect in rabbits. Endogenous progesterone delays oviductal transport in the rat and has the opposite effects in rabbits. Acute administration of fairly high doses of estradiol or progesterone to women in the immediate postovulatory period does not alter the recovery of the oocytes from the tubes up to the time they would normally pass into the uterus. Thus, these hormones do not have an important role in the control of oviductal transport in the human.

B. Nervous Regulation

Despite the profuse autonomic innervation of the oviduct, the high responsiveness of the myosalpinx to nerve stimulation, and the assortment of neurotransmitters found in the nerve terminals, there has been little or no evidence that any of these play a significant role in the regulation of egg transport. The majority of the studies on this topic were performed in rabbits and the conclusions may not apply to other species. Various procedures to achieve denervation of the oviduct do not affect the pattern of egg transport in the rabbit and systemic administration of adrenergic agonists or antagonists causes minimal alteration of transport. Various forms of stress have little if any consequence on embryo transport in rats and denervation of the oviduct has no deleterious effect on fertility in these rodents.

C. Paracrine Regulation

This type of regulation refers to signals arising from cells located in the vicinity of the cells they control and which diffuse through the fluids that bathe them. Signals produced by the embryo within the oviductal lumen are targeted to neighbor oviductal cells whose response directly or indirectly affects embryo transport. On the other hand, some oviductal cells may function as a relay station between endocrine and embryonic factors and the mechanical effector cells on which they act via paracrine mechanisms. The following sections provide examples of this recently emerging concept.

1. Embryo-Derived Factors

The embryo has been proposed as a source of signals that regulate its transport because marked differences in oviductal transport of fertilized and unfertilized ova have been found in the mare, the bat, the hamster, and the mouse. Oocytes are retained in the oviduct in equids and bats and they enter one day later than embryos in hamsters. Equine embryos were found to produce PGE_2 in increasing amounts from Day 5 after ovulation—that is, shortly before entering the uterus—and intraoviductal infusion of PGE_2 induced passage of embryos and unfertilized oocytes to the uterus by Day 4. Thus, PGE_2 secreted by the embryo may account for the differential transport of oocytes and embryos to the uterus of the mare. Hamster embryos are also believed to secrete a factor that acts on the oviduct to hasten their transport to the uterus. Since treatment with prostaglandin synthesis inhibitors does not alter oviductal embryo transport, prostaglandins are unlikely to play such a role in this species. Platelet-activating factor is another widely distributed cell to cell signaling substance which is secreted by oviductal-stage embryos of the mouse, human, sheep, and rabbit. Several facts suggest it may be involved in the control of hamster embryo transport. Antagonists to PAF delay the transport of embryos, but not oocytes, to the uterus in the hamster; local administration of PAF accelerates oviductal transport of oocytes; and PAF-like activity is found in spent media of two-cell through morula-stage hamster embryos. PAF receptor mRNA is expressed in the endosalpinx in the hamster and is most prominent in the subepithelial cells located in the mucosal folds that protrude in the lumen. This localization is most favorable for a paracrine function of embryo-derived PAF during oviductal embryo transport. Therefore, PAF is believed to be the embryonic signal that acts on the oviduct to hasten embryo transport to the uterus in the hamster. PAF may also play a role in the control of embryo transport in other species whose embryos secrete this substance during tubal transport. If PAF is the embryonic signal that hastens embryo transport to the uterus, and this is achieved, as in the rat, by increased frequency of myosalpinx contractions, the endosalpinx may function as a relay station between the embryo and the smooth muscle cells.

2. Endosalpinx-Derived Factors

The endothelins are a family of peptides produced by various cell types, including the endothelial cells, with prominent action on smooth muscle cells of blood vessels. Oviductal epithelial cells in culture produce endothelin and the endothelin receptor is expressed in the myosalpinx. Thus, endothelin produced by the oviductal epithelial cells could act as a paracrine factor on smooth muscle cells. Endothelin-1 (ET-1) contracts bovine oviduct rings but also induces transient relaxations by releasing nitric oxide (NO). Hence, a balanced synthesis of endogenous ET and NO may contribute to the physiologic contraction and relaxation of the oviduct in an autocrine/paracrine fashion. NO is generated from L-arginine by a group of enzymes termed NO synthase. NO synthase expression and activity in reproductive tissues is influenced by the levels of sex hormones and in the oviduct it changes during the estrous cycle.

See Also the Following Articles

FALLOPIAN TUBE; OOCYTE, OVERVIEW

Bibliography

Croxatto, H. B. (1996). Gamete transport. In *Reproductive Endocrinology, Surgery and Technology*. (E. Y. Adashi, J. A. Rock, and Z. Rozenwaks, Eds.), pp. 385–402. Lippincott-Raven, Philadelphia.

Croxatto, H. B., and Villalón, M. (1991). International symposium in the biology of the oviduct. *Arch. Biol. Med. Exp.* 24, 213–422.

Croxatto, H. B., and Villalón, M. (1995). Oocyte transport. In *Gametes—The Oocyte* (J. G. Grudzinskas and J. L. Yovich,

Eds.), pp. 253–276. Cambridge Univ. Press, Cambridge, UK.
Croxatto, H. B., Ortiz, M. E., Villalon, M., Cardenas, H., Imarai, M., Hermoso, M., Velasquez, L. A., and Orihuela, P. (1997). Basic aspects of oviduct function. In *New Horizons in Reproductive Medicine* (C. Coutifaris and L. Mastroianni, Eds.), pp. 233–239. Parthenon, Casterton, UK.
Hafez, E. S. E., and Blandau, R. J. (1969). *The Mammalian Oviduct*. Univ. of Chicago Press, Chicago.
Harper, M. J. K. (1994). Gamete and zygote transport. In *The Physiology of Reproduction* (E. Knobil and J. D. Neill, Eds.), pp. 123–187. Raven Press, New York.
Harper, M. J. K., Pauerstein, C., Adams, C. E., Coutinho, E. M., Croxatto, H. B., and Paton, D. M. (1976). *Ovum Transport and Fertility Regulation*. Scriptor, Copenhagen.
Hogdson, B. J., and Eddy, C. A. (1975). The autonomic nervous system and its relationship to tubal ovum transport—A reappraisal. *Gynecol. Invest.* **6**, 162–185.
Hunter, R. H. F. (1988). *The Fallopian Tubes*. Springer-Verlag, New York.
Jansen, R. P. S. (1984). Endocrine response in the Fallopian tube. *Endocr. Rev.* **5**, 525–551.
Johnson, A. D., and Foley, C. W. (1974). *The Oviduct and Its Functions*. Academic Press, New York.
Siegler, A. M. (1986). *The Fallopian Tube*. Futura, New York.
Verdugo, P., and Villalón, M. (1993). Functional anatomy of the Fallopian tube. In *Infertility Male and Female* (V. Insler and B. Lunenfeld, Eds.), pp. 53–84. Churchill Livingstone, London.

Oocyte, Mammalian

Maria Strömstedt and Anne Grete Byskov

Copenhagen University Hospital

I. Introduction
II. The Developing Ovary and Initiation of Meiosis
III. Meiotic Arrest and Follicular Formation
IV. Oocyte and Follicular Growth
V. Resumption of Meiosis
VI. Ovulation
VII. Fertilization and Second Meiotic Division

GLOSSARY

cumulus cells The granulosa cells attached to the oocyte with gap junctions.
egg cell The female germ cell arrested in metaphase of the second meiotic division.
follicle The oocyte-granulosa cell compartment delineated with a basement membrane.
follicular fluid The fluid-filled cavity among the granulosa cells.
germinal vesicle The nucleus of the oocyte.
germinal vesicle breakdown The disappearance of the nuclear membrane as one of the first signs of resumption of oocyte meiosis.
granulosa cells Somatic cells enclosed with the oocyte in the follicle.
meiosis The two consecutive divisions unique for germ cells, which result in four haploid gametes, all with different genetic constitutions.
oocyte The female germ cell arrested in the first meiotic prophase.
oogonium The mitotically dividing female germ cell.
ovary The primary reproductive organ of the female which contains the female germ cells and produces egg cells.
polar body The abortive cell division products of the first and second meiotic divisions of the female germ cell line.
pronucleus The haploid nucleus of a male (spermatozoan) or a female (egg) germ cell present in the fertilized egg before the male and female pronuclei fuse.
zona pellucida The glycoprotein coat surrounding the oocyte, produced by the oocyte.

The female germ cell is termed "oogonium" when it is dividing mitotically in the ovarian anlage. It becomes an "oocyte" (Greek: *oos* = egg and *kytos* = empty, small cavity) when it enters the first meiotic prophase, where it remains for an extended period. Thus, the oocyte is a diploid cell (2n) with 4c DNA which has not yet finished the first meiotic division.

I. INTRODUCTION

When the oocyte has grown to full size it can respond to hormone stimulation by resuming the first meiotic division and beginning the second meiotic division. It is now called an "egg cell." The egg cell can be fertilized and begin preembryogenesis. Oogenesis is the process of oocyte formation from initiation of meiosis to fertilization.

The oocyte must be accompanied by some epithelial cells, the granulosa cells, and surrounded by a basement membrane. This unit is termed a follicle. It makes no sense to describe the oocyte without its follicle. The oocyte cannot exist alone and the follicle cannot exist without an oocyte. In addition to the oocyte and the granulosa cells, a third cell type, the theca cells, becomes essential for the growing follicle. The theca cells are hormone-producing cells that furnish the granulosa cells with hormone precursors crucial for follicular and oocyte growth.

II. THE DEVELOPING OVARY AND INITIATION OF MEIOSIS

A. Ovarian Formation

The gonads are formed on the ventral side of the mesonephros, the second embryonic kidney anlage. During embryogenesis and early fetal life the primordial germ cells migrate from the extraembryonic yolk sac endoderm to this area simultaneously with cells of mesonephric origin. Together with mesenchymal cells and cells of the peritoneum, they form the future gonad.

When the germ cells arrive at the gonadal anlage the male germ cells are termed spermatogonia and female germ cells oogonia. Both continue to multiply by mitosis. When gonadal sex differentiation begins in fetal life, the male germ cells become enclosed in testicular cords. In the female fetus ovarian differentiation follows shortly hereafter. The transformation of the oogonia to an oocyte begins when meiosis is initiated some time after gonadal sex differentiation. Initiation of meiosis signifies a one-directional differentiation pattern toward the formation of an egg cell. In contrast to oogonia, the spermatogonia do not initiate meiosis until after puberty.

Meiosis in the developing ovary is always first seen in the innermost placed oogonia, in contact with the invading mesonephric cells. Gradually, meiosis spreads toward the periphery of the ovary. During early ovarian formation a peripheral cortex and a basal/central medulla form. The cortex is rich in oocytes, whereas the medulla becomes packed with somatic cells, mainly of mesonephric origin, and few oocytes.

The timing of meiotic initiation in the ovary varies between species. In some species meiosis is delayed in respect to gonadal sex differentiation (delayed meiosis), as in the human and pig, whereas in others, such as the mouse, meiosis begins immediately after sex differentiation (immediate meiosis).

B. Meiosis

Meiosis is the reduction division unique for germ cells. It consists of two divisions which result in the production of the haploid gametes. The premeiotic DNA synthesis introduces the first meiotic division. The germ cell then contains 4c DNA and a chromosome number of 2n as in normal diploid cells.

The first meiotic division differs from a mitotic division in two major respects: (i) In the first meiotic prophase a unique exchange between the maternal and paternal genes takes place, and (ii) during the first meiotic division the homologous chromosomes are separated into the daughter cells because the centromere, which binds the chromosomes together, does not divide as in mitotic divisions. The two resulting daughter cells thus contain 1n chromosomes and 2c DNA and are equipped with different genetic constitutions.

The second meiotic division resembles a normal

mitotic division except that it is not preceded by DNA synthesis. The sister chromatids separate and segregate, resulting in the haploid gametes with 1n chromosomes and 1c DNA. All four daughter cells are now haploid and genetically different (Fig. 1).

As mentioned previously, the oocyte embarks on meiosis early in life, often during fetal life. However, it is arrested in the last part of the first meiotic prophase, the diplotene stage.

The first meiotic prophase consists of four consecutive stages: leptotene, zygotene, pachytene, and diplotene stages. During the first three stages the homologous chromosomes line up and condense increasingly. In the pachytene stage the synaptonemal complexes, consisting of paired elements from the condensed sister chromatids, assemble. When the synapsis between the chromosomes is complete, the genetic exchange between the chromatides, the crossing over, and the formation of chiasmata take place. It seems that the formation of the synaptonemal complex is crucial for normal segregation of the chromosomes in the first meiotic division, which takes place much later in the fully grown oocyte at the time of ovulation. In diplotene stage the chromosomes decondense. Interestingly, the nuclear membrane is present throughout the first meiotic prophase.

When the oocyte reaches the last part of the first meiotic prophase (the diplotene stage), the meiotic process stops. The oocyte remains in the diplotene stage of the first meiotic prophase until it is either eliminated by atresia or apoptosis or succeeds in reaching the maturation stage and resumption of meiosis at ovulation time.

The two meiotic divisions of the female germ line are asymmetric divisions, each with one daughter cell (the polar body) being very small and abortive. Therefore, only one egg cell results from the two meiotic divisions in the female germ line.

C. The Number of Oocytes Throughout Life

All germ cells in the developing ovary will enter meiosis during a rather short period, often before birth. A meiotic germ cell cannot return to mitosis. When all oogonia have entered meiosis the total number can therefore only diminish with time. By the time of menopause none or very few oocytes remain in the woman's ovary, whereas other mammals may be fertile for the entire life span (Fig. 2).

The number of oocytes declines steadily throughout life, but the most serious oocyte death occurs in the transitory stages of the meiotic prophase. The largest number of germ cells in the mammalian ovary is found early in life when some of the oogonia are still dividing mitotically and others have entered meiosis. In one human ovary the maximal number of oocytes/oogonia is around 7 million, which is reached in the fifth month of fetal life. An abrupt

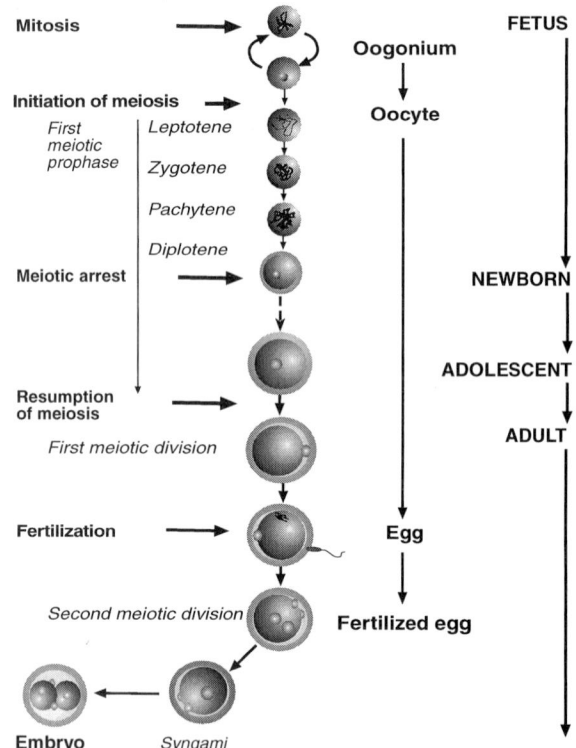

FIGURE 1 Oogenesis: Female germ line from oogonium to fertilized egg. In most female mammalian species meiosis is initiated in the oogonia, now termed oocytes, in fetal life. The oocyte is arrested in the diplotene stage of the first meiotic prophase until the it reaches ovulation time and full size when meiosis finally is resumed. Soon after, the first meiotic division takes place. The second meiotic cycle begins shortly hereafter but is arrested in the second meiotic metaphase. Meiosis is resumed for the second time when the spermatozoan enter the oocyte. When the two pronuclei merge together (syngamy), fertilization is fulfilled and embryogenesis can begin.

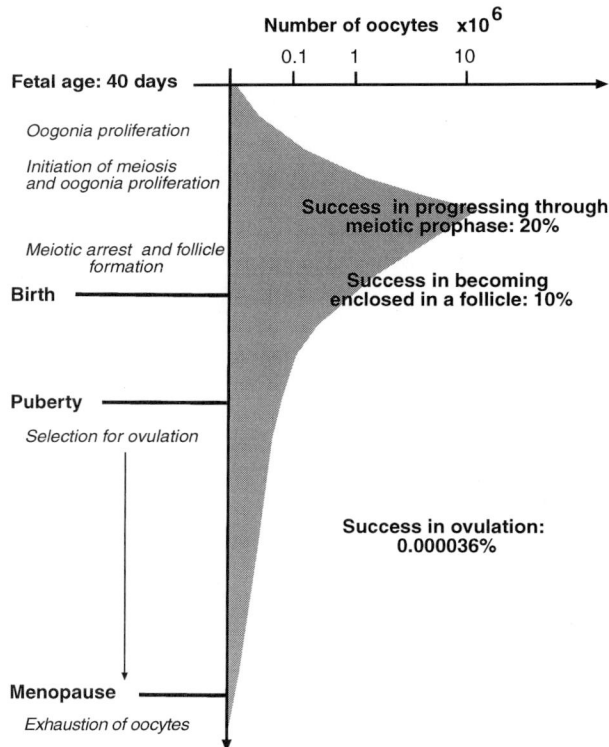

FIGURE 2 Number of oocytes in the human throughout life. From the time when the first oogonia reach the ovarian anlage, around Day 35 of fetal life, they proliferate mitotically until entering meiosis. The first meiotic germ cells are seen around the 11th week of fetal life. Successively all germ cells enter meiosis and by the time of birth they are all oocytes. The largest number of germ cells, around 14 millions, are found in the female human during the fifth month of fetal life. The germ cell number decreases dramatically thereafter, probably the result of apoptosis. Follicle formation occurs concomitantly with the oocytes reaching the diplotene stage. Only approximately 20% of the 14 million oocytes reach diplotene stage and probably only 10% become enclosed in a follicle. At the time of birth around 1.5 millions oocytes are left. Of these, at the most 500 will succeed in ovulation, i.e., 0.000036% of the 14 million oocytes found in the fifth month of fetal life. Menopause, which usually occurs between 45 and 55 years of age, is most likely reached slightly before the ovary is exhausted of oocytes.

decrease in the oocyte number during the last month of fetal life leaves each of the ovaries of the newborn girl with a total of only approximately 1.5 million oocytes. Thus, in the human female fetus only about 1 out of 10 oocytes which begin meiosis will survive. Moreover, during childhood many more oocytes disappear, leaving the oocyte population at 200,000 at the age of 20 years. During fertile life, at most 500 oocytes of the 14 million that entered meiosis will ovulate. All others disappear. Perhaps the low success rate is the result of an important selection for good quality!

III. MEIOTIC ARREST AND FOLLICULAR FORMATION

The mechanisms which control meiotic arrest of the oocyte in the diplotene stage are not known. It is likely that the meiotic arrest is needed as an important checkpoint to ensure that the oocyte has time to grow big enough before fertilization in order to sustain the following embryogenesis.

As soon as the oocyte reaches diplotene stage it must become enclosed by the granulosa cells to form a follicle. Oocytes without granulosa cells ("naked oocytes") have not been found in the mammalian ovary. Enclosure within a follicle seems to be mandatory for the survival of oocytes. When the granulosa cells surround the oocyte they simultaneously form a basement membrane around them and the follicular compartment is formed. Perhaps the enclosure of the germ cell compartment creates an important immunological barrier around the oocyte.

In the developing ovary of many mammalian species, the mesonephric connections to the ovary are clearly in contact with the centrally placed oocytes, indicating that the granulosa cells are of mesonephric origin. Growing oocytes can be seen in such intraovarian cell cords. Individual follicles are only created when the granulosa cells/oocyte enclose themselves by a basement membrane and cut off the mesonephric connections. Many of these early growing oocytes do not succeed in this process and will not survive for a long time. Oocytes closer to the periphery in the ovarian cortex will be surrounded by a few granulosa cells and are immediately enclosed by a basement membrane to form the primordial follicles (Fig. 3). Oocytes of the primordial follicles will remain small and nongrowing for different periods of time. They form the pool which will be drawn on until exhausted at menopause.

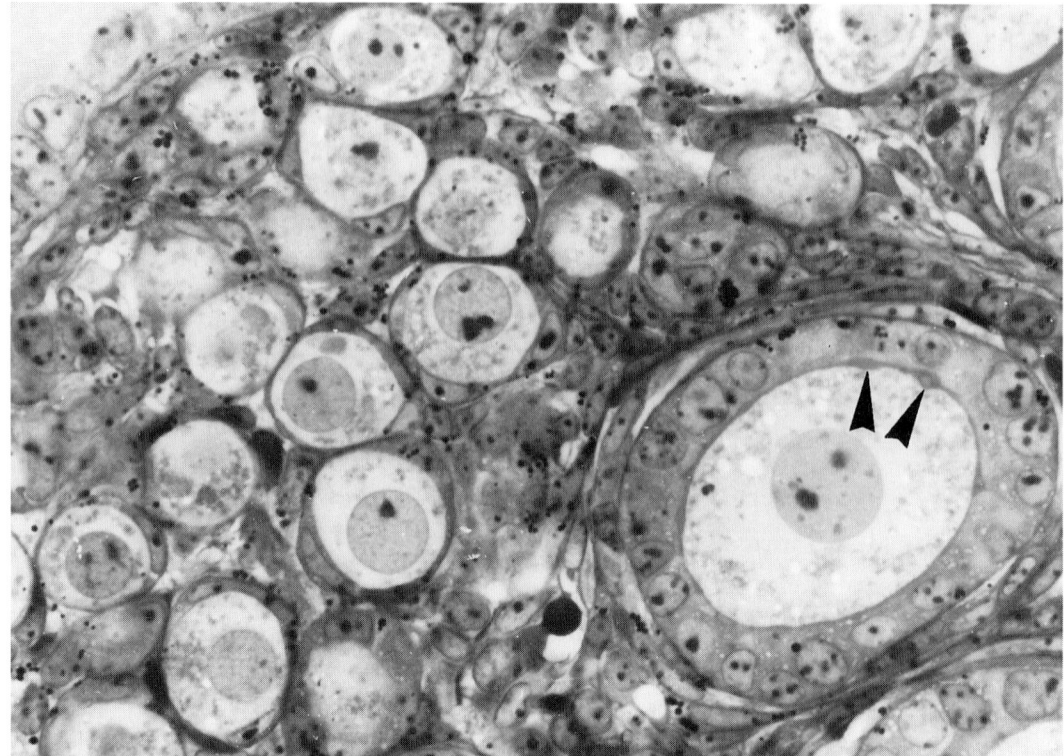

FIGURE 3 A histological section of the ovarian cortex of a 7-day-old mouse ovary. Small oocytes of the primordial follicles are surrounded with a few flattened granulosa cells and crowded in the periphery of this 7-day-old mouse ovary. A growing oocyte of a primordial follicle is seen surrounded by a layer of cuboid granulosa cells. The zona pellucida is beginning to form (arrowheads).

What keeps the primordial follicle and its oocyte from assuming growth is a mystery. Perhaps the small follicles/oocytes exert growth inhibition on each other. This would explain why recruitment of growing follicles takes place at the border between the oocyte-dense cortex and the medulla with relatively many more somatic cells and fewer oocytes.

When the first wave of abnormally growing follicles of the developing ovary has gone, a new follicular growth pattern becomes established.

IV. OOCYTE AND FOLLICULAR GROWTH

A. Follicular and Oocyte Growth

It should be emphasized that a follicle cannot exist without its oocyte. Removal of the oocyte from any follicle will immediately lead to granulosa cell apoptosis (programmed cell death).

Early follicular growth is recognized by multiplication of the granulosa cells and an almost simultaneous enlargement of the oocyte. It appears that the granulosa cells begin to multiply before the oocyte starts to enlarge. During follicular growth, the oocyte volume increases around 250 (mouse) to 300 (human) times and the granulosa cells divide 18–25 times.

From early stages of follicular development the oocyte and the granulosa cells are connected via gap junctions. Gap junctions also interconnect the granulosa cells. When the oocyte grows and the granulosa cells multiply, the network of gap junctions facilitates transport of nutrients from the blood vessels surrounding the follicle through the avascular granulosa layer to the oocyte.

Generally, follicular growth always begins at the

inner part of the ovarian cortex and growing and larger follicles are situated in the well-vascularized medulla. Follicular growth is also accompanied by increasing vascularization of the theca. Very few blood vessels are present in the cortex and the outermost primordial follicles may not have sufficient access to nutrients to support growth initiation.

The factors which initiate follicular growth have been obscure until recently when a growth differentiation factor, GDF-9, locally produced within the follicle, was found to be crucial for granulosa cell multiplication. Follicles of GDF-9-deficient mice never develop beyond the primary stage, although the oocyte can grow to almost full size in such follicles. However, an interaction between the oocyte and the granulosa cells is considered crucial for the normal follicular and oocyte growth pattern.

1. Cytoplasm of the Growing Oocyte

Oocyte growth is closely correlated to a substantial increase in number of organelles and biogenesis of new organelles/substances. The numbers of mitochondria and Golgi apparatus increase dramatically and change in structure. The number of ribosomes also increases, but relative to the enlargement of the cytoplasm their density decreases 10-fold. Most of the RNA synthesis and turnover of the growing oocyte is relatively slow and the RNA itself is unusually stable. Structural proteins and enzymes are synthesized and stored throughout oocyte growth and exhibit a slow turnover like that of the RNA. During later stages of growth a network of microtubules and filaments, e.g., tubulin and actin, develops. Cortical granules, $0.1–0.5$ μm in diameter, originating in the Golgi complex during late preantral and antral stages develop. They contain mucopolysaccharides, proteases, tissue-type plasminogen activator with serine protease activity, acid phosphatase, and peroxidase. In the fully grown oocyte the cortical granules move to the periphery of the egg. They play an important role in preventing polyspermia.

2. Nucleus of the Growing Oocyte

The size of the nucleus increases in parallel to growth, from 8 μm in primordial mouse oocytes to 22 μm in fully grown ones. In small oocytes the nucleus is smooth and spherical but during growth it shows irregularities and undulations. Also, the nucleolus enlarges from 2 to 10 μm in primordial and tertiary mouse follicles, respectively. Progressively, its structure changes from a loose fibrillogranular structure to a dense fibrillar body, indicating changes to an intense ribosomal RNA synthesis.

3. Follicle Types

The terminology for different follicle types and their oocytes is mainly based on the size of the oocyte and the number of granulosa cells or the number of granulosa cell layers. Other important parameters are the presence of a theca layer of hormone-producing, spindle-formed cells surrounding the follicular basement membrane, the presence of a liquid-filled antrum in the granulosa layer, the ability of the granulosa cells to respond to hormone stimulation, and finally the maturation stage of the oocyte itself. The basement membrane of a follicle is usually depicted as the border of the follicle (Fig. 4).

Primordial follicle: A nongrowing unit of squamous granulosa cells and a small oocyte with a diameter of 18 μm in the mouse and 30 μm in the bovine oocytes. The nucleus of the oocyte is around 9 μm. The number of granulosa cells ranges from 7 in the mouse and 15 in the human to 24 in the cow. The granulosa cells are attached to the oocyte plasma membrane (the oolemma) by gap junctions (membrane specializations) which permit exchange of small molecules. Small gap junctions also interconnect the granulosa cells. Follicle diameter ranges from 25 μm in the mouse to 35 μm in the cow.

Primary follicle: A single layer of cuboidal granulosa cells and the oocyte (and its nucleus) begin to enlarge and are around 30 μm in the mouse and 50 μm in the human. This follicle type grows slowly. An amorph glycoprotein cell coat, the zona pellucida, gradually surrounds the oocyte, leaving cellular protrusions from the granulosa cells to connect with gap junctions on the oolemma. The zona is produced by the oocyte alone. Gap junctions between the granulosa cells and the oocyte increase in number and size.

Secondary follicle: Two or three layers of granulosa cells. The granulosa cells surrounding the oocyte remain attached to the oocyte with gap junctions.

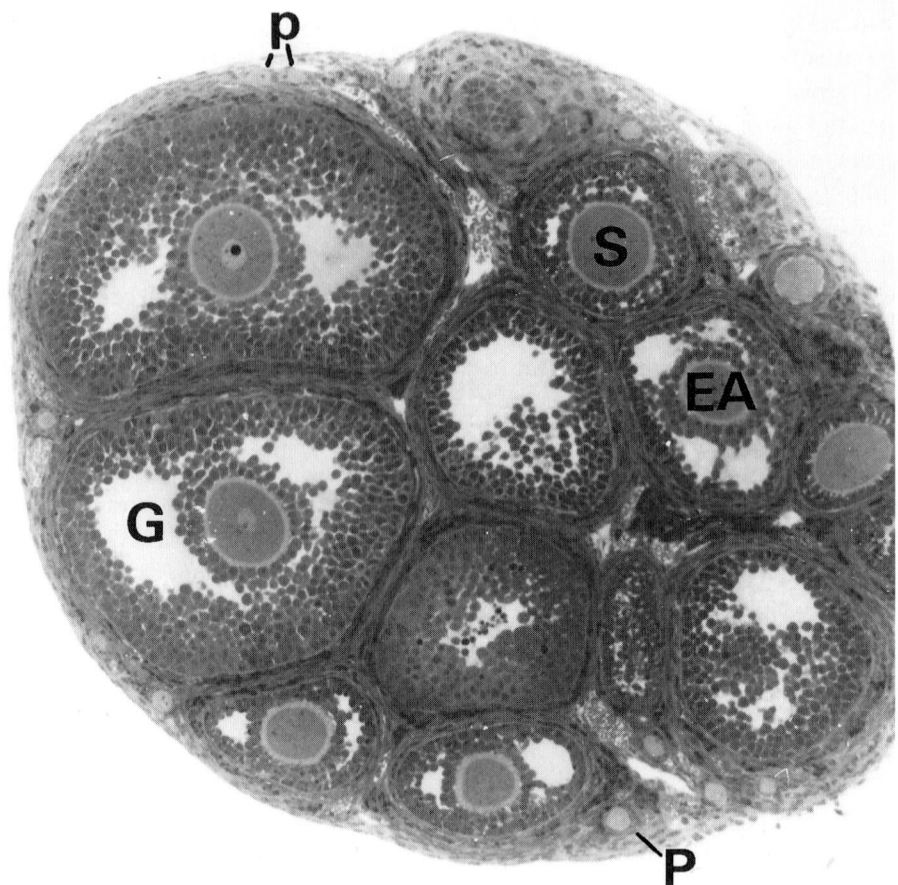

FIGURE 4 A histological section of a pubertal, 21-day-old mouse ovary. At puberty the mouse ovary is full of all types of follicles, except preovulatory ones: primordial (p), primary (P), secondary (S), early antral (EA), and Graafian follicles (G).

These cells are termed cumulus cells. The oocyte enlarges considerably, to around 50 μm in the mouse and 70 μm in the human. The nucleus of the oocyte is around 18 μm (mouse). Zona pellucida gradually increases in thickness. The follicle begins to grow a little faster and a vascularized theca layer is created around the basement membrane.

Preantral follicle: The oocyte has almost reached its final size—in the mouse around 80 μm and in the human 110 μm in diameter, and the nucleus is almost 20 μm. The thickness of the zona pellucida is now around 6 μm in the mouse and contains three different glycoproteins: ZP1, ZP2, and ZP3. Several layers of granulosa cells are present. The growth rate has increased and the theca layer, including the blood vessels, is growing as well.

Early antral follicle: The oocyte is fully grown: 85 and 120 μm in the mouse and human, respectively. The mouse nucleus is around 22 μm. Zona pellucida has reached its final thickness, e.g., 7 μm in the mouse. Between the fast-growing granulosa cells small, cell-free lakes of follicular fluid, the beginning "antrum" (Greek word for cave) accumulates. The growth of the granulosa cells and the antrum formation is dependent on the pituitary gonadotropin stimulation, in particular follicle-stimulating hormone (FSH). The granulosa cells of antral follicles develop an increasing number of FSH receptors. No other cell in the female body is known to possess FSH receptors. Binding of FSH to its receptor results in expression of a battery of genes coding for, e.g., enzymes and other proteins involved in steroidogenesis, growth factors, and peptides regulating gonadotropin release such as activin, inhibin, and follistatin. The production of androgens from the theca cells by stimulation by luteinizing hormone (LH) is also crucial for the function of the granulosa cells which convert these hormones further, in particular to estradiol.

Graafian follicle: One big antrum is formed by coalescence of the smaller ones and is enlarged concomitantly with the enlargement of the follicle itself. In the follicular fluid, steroids, growth factors, and peptide hormones accumulate. The theca/granulosa cells are fully dependent on stimulation by gonadotropins. The "dominant" follicle(s) will be selected for ovulation among the Graafian follicles.

Preovulatory follicle(s): At a certain time before the preovulatory peak of gonadotropins, a species-related number of the antral follicles will be "selected" to continue their growth and reach the preovulatory stage. These are the follicles with the largest number of FSH receptors, which sometimes are also the largest ones, having access to sufficient amount of gonadotropins to support their growth.

Shortly after the preovulatory peak of gonadotropins, which eventually result in ovulation, the selected follicle begins a rapid expansion as the result of accumulation of follicular fluid which expands the antrum. This expansion is not caused by proliferation of the granulosa cells. In contrast, granulosa cell divisions decrease abruptly. The high concentrations of gonadotropins within the follicle change the steroid synthesis pattern of the granulosa and theca cells and prepare the oocyte for maturation and resumption of meiosis.

B. Oocyte Depletion and Follicular Atresia

As mentioned previously, a high percentage of those oocytes which enter meiosis will die. In fact, in all mammals the majority of the follicles are destined to become atretic and die, leaving only a few percent for ovulation. In the human less than 1% of the newborn girl's oocytes will reach ovulation. Oocyte attrition may take place at any time of follicular development. It is usually the result of atretic degeneration of the granulosa cells by apoptosis (programmed cell death). Early stages of apoptosis are characterized by the generation of "DNA ladders—internucleosomal cleavage of DNA in multiples of 185 base pairs caused by nucleases. It seems that the highest rate of follicular atresia occurs among preantral and antral follicles. The theory is that some of these follicles, which are developing a dependence on gonadotropins, undergo granulosa cell apoptosis when the amount of gonadotropins is insufficient. When the degenerating granulosa cells no longer are able to support the oocyte, it will also die.

The oocyte may rest in the meiotic prophase for a long time, e.g., in the human for 55 years and even longer. During all these years the oocyte risks exposure to harmful influences, such as irradiation, compounds with hormone activity, and chemicals affecting metabolism and growth. Since the oocyte is equipped with DNA, which is synthesized in fetal life, it is likely that the risk for DNA damage increases with age, although it is known that some DNA repair occurs.

V. RESUMPTION OF MEIOSIS

During the growth phase, the mammalian oocyte is arrested at the diplotene stage of prophase I of the first meiotic division. It does not resume meiosis until puberty, when it is stimulated by the preovulatory LH surge. At this time the nuclear membrane, i.e., germinal vesicle (GV) (Figs. 5 and 6), breaks down and the chromosomes condense into bivalents which align on the meiotic spindle at metaphase I. The homologous chromosomes segregate until one set is extruded in the first polar body, whereas the other remains within the oocyte. Meiosis is then arrested again and does not resume until after the oocyte has been fertilized. The purpose of meiosis is twofold: a reduction to the haploid number of chromosomes and recombination of genetic information. Only oocytes that have undergone this maturation process and have been arrested at the metaphase II are capable of being fertilized and developing normally.

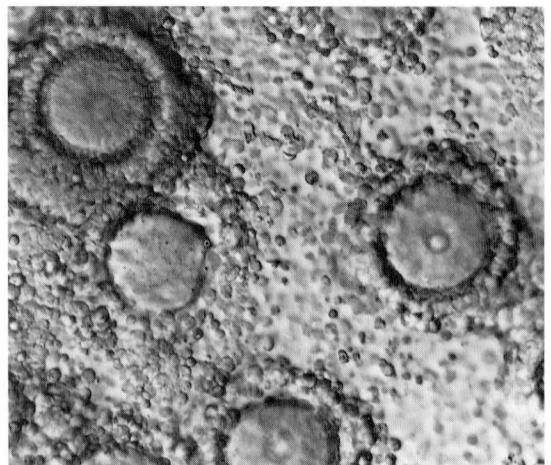

FIGURE 5 Four cumulus-enclosed oocytes isolated from a gonadotropin-stimulated mouse ovary. The oocytes have been cultured for 20 hr. Three of the oocytes have a nucleus, whereas one is in the GVBD stage.

A. Maturation-Promoting Factor

Maturation-promoting factor (MPF) was initially described in amphibian oocytes as an activity which appears in the cytoplasm prior to GV breakdown (GVBD) and which induces meiosis. MPF was later found to be a universal regulator of the G_2 to M transition of both mitosis and meiosis from yeast to man. MPF consists of two proteins: a protein kinase and a cyclin. The activity of MPF depends on the phosphorylation state of the protein kinase. Active MPF is required for chromosome condensation and cytoplasmic reorganization. MPF activity appears during meiotic maturation, peaks at the metaphase I and II of meiosis, and decreases at anaphase I and II.

B. c-mos

The oncogene c-*mos* has been shown to play a role in meiosis in vertebrates. In *Xenopus*, c-*mos* functions at several steps during meiosis, whereas in murine oocytes its function is restricted to the second meiotic arrest. Mos is a serine-threonine protein kinase and is believed to be a component of cytostatic factor (CSF), which was first described in the cytoplasm of unfertilized *Xenopus* eggs. In mice, c-*mos* mRNA is expressed in growing and fully grown oocytes, but the Mos protein is only present in fully grown oocytes during maturation and in unfertilized eggs. Mos protein is not present in fertilized eggs and is also absent from most adult tissues. The female c-*mos*-deficient mice have substantially reduced fertility and fail to arrest at metaphase II. Oocytes from these mice frequently undergo parthenogenesis, division without preceding fertilization, and their ovaries often contain cysts and teratomas. c-*mos* seems to be crucial for the checkpoint at the arrest in metaphase II: The egg degenerates if not fertilized at this stage.

C. Cyclic Adenosine 3′,5′-Monophosphate

Removal of the oocyte from the follicle induces meiosis, suggesting that a substance present in follicular fluid is the inhibiting factor. Several lines of evidence show that cyclic adenosine 3′,5′-monophosphate (cAMP) maintains the oocyte in meiotic arrest. Membrane-permeable analogs of cAMP as well as inhibitors of phosphodiesterase (PDE), an enzyme that hydrolyzes cAMP, block spontaneous meiosis in isolated oocytes as well as the gonadotropin-induced

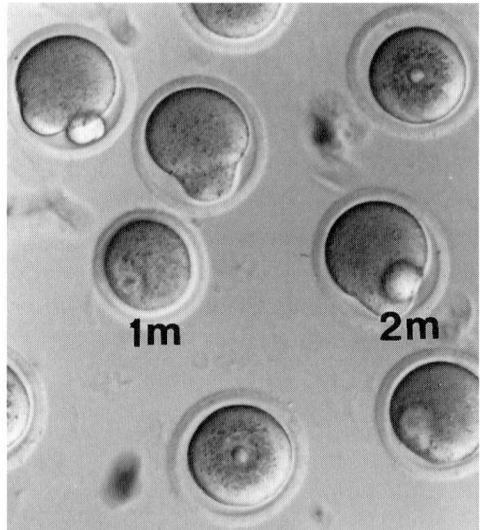

FIGURE 6 Mouse oocytes, mechanically deprived of their cumulus cells, isolated from a gonadotropin-stimulated mouse ovary. The oocytes have been cultured for 20 hr. The oocytes are in GV stage and GVBD stage, and they are arrested in the first meiotic metaphase (1m) and in the second meiotic metaphase (2m). A zona pellucida is surrounding each of the oocytes.

meiosis in follicle-enclosed oocytes. Lower concentrations of cAMP maintain meiotic arrest, whereas higher levels mediate gonadotropin action and induce resumption of meiosis. The effect of gonadotropins on oocyte resumption of meiosis is indirect since no receptors appear to be present on the oocyte.

Protein kinase A (PKA) mediates the action of cAMP in cells by phosphorylating target proteins. The concentration sensitivity to cAMP may be explained by the fact that only type I PKA is present in the oocyte, whereas both type I and type II are present in the cumulus cells. Since the regulatory subunit of PKA type I has a higher affinity for cAMP than that of type II, a low level of cAMP will stimulate PKA type I in the oocyte, maintaining the oocyte in meiotic arrest. The elevation in cAMP levels brought about by gonadotropins at the time of ovulation will stimulate the activation of the type II PKA in the cumulus cells. This leads to the production of a signal which overcomes the inhibition and leads to reinitiation of meiosis.

Although LH is considered to be the physiological stimulator of meiotic resumption, both LH and FSH stimulate the resumption of meiosis in follicle-enclosed oocytes *in vitro*. When isolated cumulus-enclosed oocytes are used, only FSH is active, which is explained by the fact that the cumulus cells have receptors for FSH but not for LH. The mural granulosa cells contain receptors for both LH and FSH. Both gonadotropins act by increasing the intracellular levels of cAMP. When the levels of LH in the circulation increase, cAMP levels in granulosa cells also increase as a result of stimulation of adenylate cyclase. Somehow this also leads to an increase in cAMP concentration in the cumulus cells, perhaps by a direct action of FSH on these cells or perhaps by transport of cAMP from mural granulosa cells to the cumulus cells through gap junctions.

D. Purines

Several studies suggest that hypoxanthine, as well as other purines or pyrimidines, has a function in the maintenance of meiotic arrest. Hypoxanthine is present in follicular fluid of several mammals in concentrations high enough to maintain meiotic arrest. The major pathway of hypoxanthine metabolism in cumulus-enclosed oocyte complexes (CEOs) is the salvage of hypoxanthine to inositol monophosphate (IMP) and further metabolism to adenyl nucleotides. The adenyl nucleotides have little effect on meiotic arrest and several lines of evidence suggest that it is hypoxanthine itself which inhibits meiotic maturation. Hypoxanthine is believed to act through the inhibition of PDE, leading to an increase in cAMP. The order of potency for the inhibition of meiotic resumption measured as GVBD between different purines is the same as that for inhibiting PDE.

E. Steroids

Although there is no evidence for any effect of steroid hormones on the resumption of meiosis in mammalian oocytes, different steroids have been shown to induce meiotic resumption in several non-mammalian species. In *Xenopus*, the meiosis-inducing substance is progesterone, which acts on a receptor in the oocyte plasma membrane, leading to the inhibition of oocyte adenylate cyclase activity. This is in contrast to the classical steroid hormone receptors which are all intracellular proteins. In fish, two other steroids have similar effects, also acting on oocyte membrane receptors.

F. Meiosis-Inducing Sterols

Mouse cumulus cells secrete a meiosis-activating substance in response to FSH, forskolin, or dibutyryl cAMP. Although intact cumulus–oocyte connections are crucial for initiating the response, the substance produced is diffusible and its transfer from cumulus cells to the oocyte does not require the presence of gap junctions. Meiosis-activating substances have been purified from human follicular fluid (FF-MAS) as well as from bull testis and have been identified as two different but closely related sterols, both intermediates in the cholesterol biosynthetic pathway. Gonadotropins stimulate the activity of the enzyme producing FF-MAS, suggesting a mechanism for the effect of FSH on the production of a meiosis-activating substance. These two sterols induce meiosis in hypoxanthine-arrested naked oocytes as well as in cumulus-enclosed oocytes. The mechanism of action of the sterols is not yet known.

G. Role of Calcium in Meiosis

A role for calcium in mammalian meiosis has been proposed. In studies on invertebrates or lower vertebrates, control points of mitosis as well as of meiosis are triggered by increases in intracellular calcium and treatment of cumulus-enclosed oocytes from cow with LH induces calcium oscillations in the oocyte. The evidence for the involvement of calcium in the initiation of meiotic resumption is somewhat contradictory and there may be differences between mammalian species. There is better evidence that calcium is involved in the progression of meiosis past metaphase I to polar body formation and during fertilization.

H. Resumption of Meiosis: Summary

Based on the available data on the role of cAMP and MAS in resumption of meiosis, a hypothesis may be proposed for the mechanism by which the preovulatory gonadotropin surge reinitiates oocyte meiosis. Before the gonadotropin surge, intracellular levels of cAMP in the cumulus–granulosa cells are low. In the oocyte a certain level of cAMP is obtained due to the inhibition of PDE by hypoxanthine or other purines present in the follicular fluid. This level of cAMP activates PKA and somehow prevents GVBD. The preovulatory gonadotropin surge increases transiently the intracellular concentration of cAMP in the cumulus cells, leading to an increased production of MAS. These sterols will then diffuse into the oocyte and somehow overcome the meiotic arrest, leading to induction of GVBD.

VI. OVULATION

Ovulation is triggered by the midcycle LH surge, which in turn is triggered by estradiol produced by the preovulatory follicle. The LH surge leads not only to reinitiation of meiosis and nuclear maturation but also to cytoplasmic maturation and maturation of the zona pellucida. The cytoplasmic maturation includes the migration of cortical granules to the outer cortex of the oocyte. LH also stimulates the transformation of the follicular mural cells from estrogen production to progesterone production (luteinization), the latter hormone being the major steroid hormone produced by the corpus luteum after ovulation. LH stimulates local prostaglandin production and enzymatic digestion of the follicular wall, leading to its rupture approximately 36 hr later. Also, the production of leukotrienes, cytokines, release of plasminogen activator, and the expansion of the cumulus due to release of glucosaminoglycans are stimulated. Just before ovulation, the gap junctions between the oocyte and the cumulus cells disappear and the cumulus–oocyte complex is detached from the follicular wall. When released from the follicle, the cumulus–oocyte complex stays near the ovarian surface and is picked up by the fimbriae of the fallopian tube. The presence of the cumulus cells helps in the adhesion to the cilia on the fimbriae. It is believed that the fertilizable life span of a human oocyte is 12–24 hr after ovulation.

VII. FERTILIZATION AND SECOND MEIOTIC DIVISION

Before ejaculated spermatozoa can fertilize an egg, it must spend a period of time in the female genital tract. This process is known as capacitation and is not fully understood, but it includes changes in surface characteristics. The removal of seminal plasma factors has been suggested to play a role and one such factor may be cholesterol. The generation of hydrogen peroxide in the spermatozoa increases tyrosine phosphorylation and is essential for capacitation and the influx of extracellular Ca also seems to be involved. Capacitated sperm can undergo the acrosome reaction, bind to the zona pellucida, and acquire hypermotility. The acrosome is a membrane-enclosed, flat, lysosome-like organelle loaded with hydrolyzing enzymes. It covers the anterior part of the sperm nucleus. During the acrosome reaction, the acrosomal membrane fuses with the sperm plasma membrane and the acrosomal content is released. The timing of the acquisition of hypermotility appears to be important since infertile men have higher levels of prematurely activated spermatozoa than normally fertile men and since hyperactivation impairs transport through the female reproductive tract.

Spermatozoa can also be capacitated *in vitro* by different washing procedures and thus the female reproductive tract is not an absolute requirement.

A. Cumulus–Oophorus

The first obstacle encountered by the spermatozoa as it approaches the unfertilized egg is the cumulus. This consists not only of the cumulus cells but also of the fibrous matrix they secrete, which to a large extent is made up of hyaluronic acid. The mechanism by which the sperm penetrates the cumulus is not known, but it has been speculated that enzymes on the surface of the sperm membrane aid in this process. PH-20 may be such a protein since it is present on the plasma membrane of sperm from all species examined and contains hyaluronidase activity. Antibodies against this protein which block its hyaluronidase activity also block sperm penetration through the cumulus. Although oocytes without their cumulus–granulosa cells can be fertilized *in vitro*, the cumulus appears to be beneficial for fertilization and its presence reduces differences in male fertility.

B. Zona Pellucida

After penetrating the cumulus, the sperm also has to pass through the zona pellucida, which in humans is a 15- to 18-μm-thick coat made up of three glycoproteins: ZP1, ZP2, and ZP3. Capacitated spermatozoa bind tightly to the surface of the zona before penetrating it. ZP3 is the receptor for the spermatozoa in several mammalian species. In addition to binding sperm, ZP3 also triggers the acrosome reaction. In most species, the sperm pass through the zona within a few minutes. The exact mechanism by which this occurs is not known but most likely includes enzymatic digestion of the zona as well as mechanical disruption. The oocytes of mice lacking the ZP3 gene grow normally and secrete ZP1 and ZP2, but no zona is formed. The cumulus–granulosa cells closest to the oocyte are disorganized and the females are sterile. An intact zona pellucida is, however, not necessary for successful fertilization *in vitro*. Thus, the function of the zona is to protect the eggs and early embryos, to prevent cross-species fertilization, and to prevent polyspermia.

C. Fusion of the Spermatozoan with the Oocyte

After passing through the zona, the sperm head binds to the plasma membrane of the oocyte (the oolemma) and the whole spermatozoa is subsequently taken up into the ooplasm. The mitochondria present in the tail of the spermatozoa becomes degraded and only maternal mitochondria will be passed on to the next generation. Only spermatozoa that have undergone the acrosome reaction are able to fuse with the oolemma. There is some species specificity in the interaction between spermatozoa and the oolemma, but it is much less strict than the interaction with the zona. The fusion of the membranes most likely involves more than one factor in the sperm as well as in the oocyte plasma membrane. Immediately after the fusion of the sperm with the oocyte, the latter displays a series of membrane hyperpolarizations and Ca oscillations. The first rise in intracellular Ca levels occur 10–30 sec after the attachment of the spermatozoa and the oscillations continue for another hour. A protein present in cytoplasm of spermatozoa which gives rise to Ca oscillations in oocytes has recently been identified. An early sign that fertilization has occurred is the exocytosis of the cortical granules. This begins within a few minutes of sperm–egg binding and most of the granules are released a few minutes later. Ca is necessary for the release of cortical granules.

D. Blocking of Polyspermia

Oocytes of all species have some way of preventing more than one sperm from fusing with the oolemma; that is, a block to polyspermia. In mammals different strategies are employed; in some species, including the human, the major mechanism is the zona reaction caused by the released cortical granules. In other species a plasma membrane block is most important, whereas in yet other species both these mechanisms are employed. The zona reaction involves the proteolytic modification of ZP2 and also some modification of ZP3, inactivating its sperm receptor function as well as its ability to induce the acrosome reaction. The plasma membrane polyspermia block is also very rapid and is completed minutes after sperm–egg binding, but the mechanism behind it is not well understood. In the sea urchin oocyte, the first block to polyspermia is depolarization of the oolemma, but this appears not to be the case in mammalian oocytes.

Immediately following fertilization, the oocyte completes meiosis, extrudes the second polar body, and the remaining nucleus forms the egg pronucleus. Once inside the oocyte, the nuclear membrane of the sperm breaks down and the nucleus decondensates. A new nuclear membrane is formed around the sperm chromatin to give rise to the sperm pronucleus. DNA synthesis begins simultaneously in the egg and sperm pronuclei some hours after fertilization. Finally, the two pronuclei fuse and their chromosomes mix (syngamy) before the first mitotic division takes place, marking the beginning of embryonic development.

See Also the Following Articles

Follicular Development, Control of; Granulosa Cells; Meiosis

Bibliography

Albertini, D. F. (1992). Regulation of meiotic maturation in the mammalian oocyte: Interplay between exogenous cues and the microtubule cytoskeleton. *BioEssays* **14**, 97–103.

Byskov, A. G., Yding Andersen, C., Nordholm, L., Thøgersen, H., Guoliang, X., Wassmann, O., Andersen, J. V., Guddal, E., and Roed, T. (1995). Chemical structure of sterols that activate oocyte maturation. *Nature* **374**, 559–562.

Downs, S. M., and Hunzicker-Dunn, M. (1995). Differential regulation of oocyte maturation and cumulus expansion in the mouse oocyte–cumulus cell complex by site-selective analogs of cyclic adenosine monophosphate. *Dev. Biol.* **172**, 72–85.

Eppig, J. J. (1996). The ovary: Oogenesis. In *Scientific Essentials of Reproductive Medicine* (S. C. Hillier, H. C. Kitchener, and J. P. Neilson, Eds.), pp. 147–159 Saunders, London.

Gougeon, A. (1996). Regulation of ovarian follicular development in primates: Facts and hypotheses. *Endocr. Rev.* **17**, 121–155.

Hirshfield, A. (1991). Development of follicles in the mammalian ovary. *Int. Rev. Cytol.* **124**, 43–101.

Homa, S. T. (1995). Calcium and meiotic maturation of the mammalian oocyte. *Mol. Reprod. Dev.* **40**, 122–134.

Snell, W. J., and White, J. M. (1996). The molecules of mammalian fertilization. *Cell* **85**, 629–637.

Speroff, L., Glass, R. H., and Kase, N. G. (1994). *Clinical Gynecologic Endocrinology and Infertility*, 5th ed. Williams & Wilkins, Baltimore.

Wassarman, P. M., and Albertini, D. F. (1994). The mammalian ovum. In *The Physiology of Reproduction* (E. Knobil and J. D. Neill, Eds.), pp. 79–122. Raven Press, New York.

Yanagimachi, R. (1994). Mammalian fertilization. In *The Physiology of Reproduction* (E. Knobil and J. D. Neill, Eds.), pp. 189–317. Raven Press, New York.

Yding Andersen, C., and Byskov, A. G. (1996). Gonadal differentiation. In *Scientific Essentials of Reproductive Medicine* (S. C. Hillier, H. C. Kitchener, and J. P. Neilson, Eds.), pp. 105–119. Saunders, London.

Oocyte Maturation and Spawning in Starfish

Takeo Kishimoto

Tokyo Institute of Technology

I. Starfish Reproductive Phenomena
II. History and Gonad-Stimulating Substance
III. Maturation-Inducing Substance and 1-Methyladenine
IV. Maturation-Promoting Factor and Cdc2 Kinase
V. Meiotic Maturation and Fertilization
VI. Concluding Remarks

GLOSSARY

cdc2 kinase A Ser/Thr protein kinase which is the complex of cyclin B with a homolog of fission yeast cdc2 gene product. Cdc2 kinase is not only the molecular entity of maturation-promoting factor but also governs entry into and exit from M phase in all eukaryotic cells.

gonad-stimulating substance (GSS) A primary hormone which is released from radial nerves and causes spawning and oocyte maturation in starfish. GSS acts directly on gonadal follicle cells to produce a secondary hormone, a maturation-inducing substance which is identified as 1-methyladenine.

MAP kinase Originally the abbreviation of mitogen-activated kinase, but it is also activated during oocyte maturation. MAP kinase activity is necessary and sufficient to prevent entry into S phase or parthenogenesis in mature starfish eggs at female pronucleus stage.

maturation-inducing substance (MIS) A secondary hormone which is produced in gonadal follicle cells under the influence of a primary hormone, GSS, released from radial nerves in starfish. MIS acts directly on oocyte surface to activate a maturation-promoting factor, which in turn causes both spawning and oocyte maturation. Molecular entity of starfish MIS is identified as 1-methyladenine.

maturation-promoting factor (MPF) Originally identified as an activity which mediates in oocyte cytoplasm the maturation-inducing effect of 1-methyladenine on oocyte surface. Subsequently, MPF was found to be a general inducer of M phase common to all eukaryotic cells and hence was renamed M-phase-promoting factor. The molecular identity of M-phase-inducing activity contained in MPF is cdc2 kinase, the cyclin B/Cdc2 complex.

The molecular mechanism of starfish oocyte maturation and spawning is described from the aspect of three mediators involved in these processes. The primary mediator is the neurosecretory hormonal peptide, a gonad-stimulating substance (GSS), which acts on the gonadal follicle cells; the secondary mediator is maturation-inducing hormone, which is identified as 1-methyladenine (1-MeAde), produced in the follicle cells under the influence of GSS and acting on oocyte surface; and the tertiary mediator is maturation-promoting factor (MPF), whose activity is generated by cdc2 kinase. It is activated in 1-MeAde-stimulated oocytes, and is directly responsible for oocyte maturation. Although the article begins from a historical perspective on the birth of the endocrinology of starfish reproduction, emphasis will be focused on current cell biology of the cell cycle control during starfish oocyte maturation and fertilization.

I. STARFISH REPRODUCTIVE PHENOMENA

Starfish has a diffuse nervous system consisting of five major radial nerves and a circumoral nerve ring (Fig. 1). There are 10 gonads consisting of a cluster of elongate, tubular lobes, and 2 in each arm that are suspended freely in the celomic cavity are connected with gonoducts which open to the body surface. Starfish is dioecious, though the sexes cannot be distinguished externally. In general, starfish goes

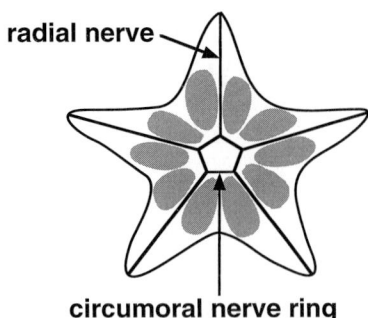

FIGURE 1 Major nervous system in the starfish consisting of five radial nerves and a circumoral nerve ring. During the breeding season, the body cavity is filled with 10 ripe gonads (gray area).

through an annual reproductive cycle, and at the breeding season the body cavity of female starfish is filled with ripe ovaries which appear orange due to yolk, whereas the testes appear pale yellow or white due to sperm. Ripe ovaries are filled with millions of fully grown immature oocytes which are characterized by the presence of a single large nucleus called germinal vesicle with 4n ploidy (Fig. 2). These immature oocytes are surrounded by a follicle cell layer in the ovary and are arrested in the prophase of the first meiotic cycle.

At the time of spawning, the prophase arrest is released and immature oocytes resume meiosis, as indicated by the breakdown of germinal vesicle (GVBD), to undergo the subsequent sequential processes of oocyte maturation until the completion of the second meiosis in the absence of fertilization (Fig. 2). In *Asterina pectinifera*, at room temperature GVBD generally occurs at ~20 min, the first meiotic metaphase at ~45 min, and the first polar body, which is a hallmark of the completion of the first meiotic cycle, is emitted at 60–70 min. Thereafter, the second polar body is discharged at 100–110 min, resulting in the arrest at the female pronucleus stage with a haploid set of genomes. Immature oocytes fail to respond normally to the fertilization stimulus and fertilizability is acquired during oocyte maturation. Naturally, fertilization occurs near the end of first meiosis, although mature eggs with a female pronucleus are still fertilizable; fertilization is followed by cleavage cycles (Fig. 2).

For *in vitro* experiments, (i) hundreds of small fragments of ovary are easily taken without injury from an individual, and isolated ovarian fragments can simply be kept in seawater for hours without losing their physiological activity; (ii) millions of fully grown oocytes, which are almost equal in size

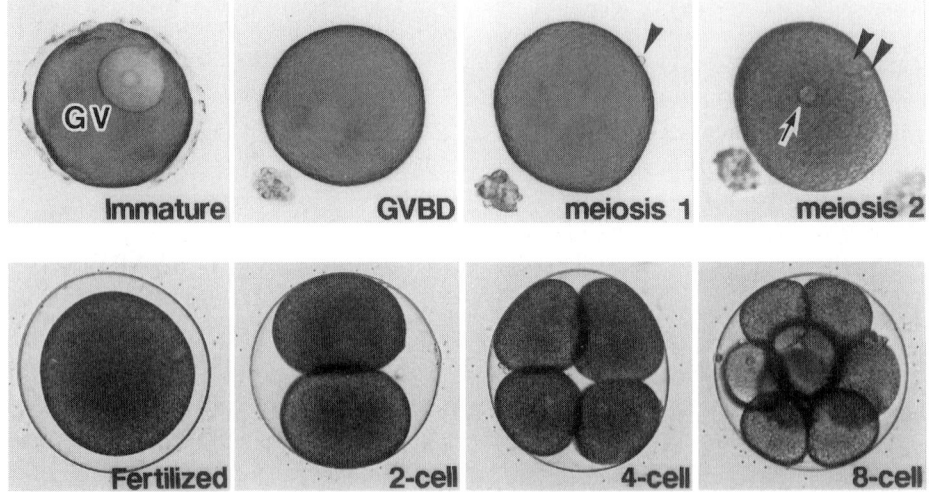

FIGURE 2 Oocyte maturation and early cleavage in the starfish, *Asterina pectinifera*. A fully grown immature oocyte arrested at the prophase of the first meiotic cycle contains a germinal vesicle (GV) and is surrounded by a follicular layer. Breakdown of germinal vesicle (GVBD) and follicular layer is the first indication of meiosis reinitiation. Thereafter, meiotic cycles complete with the extrusion of two polar bodies (arrowheads) in the absence of fertilization, resulting in the formation of a female pronucleus (arrow) with haploid genomes. Upon fertilization, fertilization envelope elevates, followed by early cleavage cycles.

and arrest at the same prophase stage of the first maturation division, are easily available and kept in seawater; (iii) the process for spawning and oocyte maturation requires a relatively short time and is easily observable under a light microscope. Thus, starfish provides suitable material for the biochemical and physiological study of the mechanisms of oocyte maturation and spawning.

II. HISTORY AND GONAD-STIMULATING SUBSTANCE

In related marine invertebrates, sea urchin eggs and spermatozoa have been extensively utilized for the study of fertilization and early development since the nineteenth century. In contrast, significant investigations in starfish of the mechanism of spawning and oocyte maturation started only at the end of 1950s despite the advantages mentioned previously for using the starfish in experiments. This may be ascribed to a lack of information on the starfish reproductive endocrinology involved in oocyte maturation and spawning, whereas sea urchin oocytes accomplish meiotic maturation within the ovary before spawning and hence are ready for use in fertilization experiments.

The study of the endocrinology of starfish reproduction was aided by a finding by Chaet and McConnaughy in 1959 that a hot-water extract of radial nerve of *Asterias forbesi* can induce the shedding of gametes upon injection into the celomic cavity of ripe animals. The active substance contained in the radial nerve was called the gamete-shedding substance (GSS). GSS induces spawning not only in an individual but also in an isolated ovarian fragment. GSS is detectable in the celomic fluid only when starfishes are undergoing natural spawning, indicating that GSS is a hormone acting directly on the gonads. GSS is present in the radial nerve throughout the year and its contents are nearly the same irrespective of the breeding season. Cross experiments among different starfishes indicate that GSS acts, with some exceptions, non-species-specifically.

GSS induces oocyte maturation as well as spawning upon addition to isolated ovarian fragments. Although GSS was thought to be a direct inducer of spawning in the early studies, later experiments on the induction of oocyte maturation revealed that the action of GSS is indirect. This hormone acts directly on the ovarian follicle cells around oocytes to produce the second hormone, a maturation-inducing substance (MIS) which acts directly on oocytes to induce maturation and spawning. As a result, the GSS was renamed the gonad-stimulating substance (GSS) in 1969 by Kanatani.

GSS has been purified from the radial nerves of *Asterias amurensis* and identified to be a single peptide consisting of 22 amino acids with a molecular weight of ~2.1 kDa. However, the amino acid sequence of GSS has not been yet determined in any starfish species.

Microsurgical experiments indicate that GSS-containing granules are localized in the supporting cells located just beneath the outer sheath of the radial nerves. However, the mechanism of release of GSS is not known, although a releasing factor may be present in the ripe ovary.

III. MATURATION-INDUCING SUBSTANCE AND 1-METHYLADENINE

Oocyte maturation occurs when GSS is added to isolated oocytes surrounded by follicles, whereas it fails in follicle-free oocytes. Furthermore, the supernatant obtained from the incubation mixture of isolated follicle cells with GSS can induce maturation in follicle-free oocytes. Thus, Kanatani and Shirai demonstrated in 1967 the presence of the secondary mediator, MIS, which is synthesized and released in follicle cells under the influence of GSS and acts directly on oocytes to induce maturation. In 1969, Kanatani and colleagues purified and isolated this secondary hormone from the incubation mixture of GSS and ovarian fragments of *A. amurensis* and identified the hormone as 1-methyladenine (1-MeAde). 1-MeAde is the first substance chemically identified as MIS in all organisms. 1-MeAde acts directly on isolated follicle-free oocytes, and its effective dose for oocyte maturation is approximately $0.1-1.0\ \mu M$.

1-MeAde is produced specifically under the influence of GSS. The only other substance known to act on follicle cells to produce 1-MeAde is concanavalin A. The biochemical role of GSS is ascribed to induce transfer of a methyl group from methionine, possibly through S-adenosylmethionine, to the N1 site of the purine nucleus of a precursor of 1-MeAde, though neither the enzyme for the methyl transfer nor its acceptor have been identified. Thereafter, 1-MeAde is biosynthesized via 1-methyladenosine monophosphate and finally 1-methyladenosine.

1-MeAde also causes spawning when injected into the celomic cavity of ripe starfishes or when applied to isolated ovarian fragments, though this effect is indirect. Once the follicular envelopes around the oocytes break down and are removed, the denuded oocytes become freely movable within the ovary and hence are forced out by the contraction of the ovarial wall. Follicular envelope breakdown is triggered by the disruption of desmosomal contacts between the processes of the follicle cell and the oocyte surface. The disruption is caused inside the oocytes by the action of maturation-promoting factor (MPF), which is activated in oocyte cytoplasm under the influence of 1-MeAde. Then ovarian contraction is induced by the jelly substance which is allowed to come into contact with the ovarian wall following follicular envelope breakdown.

In some starfishes, such as *A. amurensis*, *A. forbesi*, and *Marthasterias glacialis*, immature oocytes undergo so-called "spontaneous" maturation when they are isolated in seawater. This is ascribed to the production of 1-MeAde in follicle cells in response to Ca^{2+} contained in seawater, based on the fact that spontaneous maturation does not occur in Ca^{2+}-free seawater. Naturally, asterosaponin prevents the production of 1-MeAde even in the presence of Ca^{2+}, whereas despite the presence of asterosaponin, GSS can produce 1-MeAde in follicle cells to induce spawning and oocyte maturation.

When applied externally to isolated immature oocytes, 1-MeAde induces maturation invariably, whereas it fails upon injection into the inside of immature oocytes, indicating that 1-MeAde acts on oocyte surface externally to cause maturation. Thus, it is anticipated that 1-MeAde receptor might be present in the plasma membrane of the oocyte and that the presumable binding site of 1-MeAde with its receptor is in the N9 or N7–N9 region, though the receptor has not been isolated. The putative 1-MeAde receptor is coupled with heterotrimeric G protein, which is sensitive to pertussis toxin. While the injection of $G_{i\alpha}$ into immature oocytes prevents 1-MeAde-induced maturation, injection of $G_{\beta\gamma}$ induces germinal vesicle breakdown in the absence of 1-MeAde stimulus, indicating that the release of $G_{\beta\gamma}$ in response to 1-MeAde is sufficient to induce oocyte maturation (Fig. 4).

IV. MATURATION-PROMOTING FACTOR AND Cdc2 KINASE

In a hormonal cascade leading to oocyte maturation and spawning in starfish, GSS is regarded as the primary mediator and 1-MeAde as secondary (Fig. 3). Accordingly, it is anticipated that the tertiary mediator must be present within oocyte cytoplasm to execute maturation in response to 1-MeAde signal at the oocyte surface. In fact, Kishimoto and Kanatani demonstrated in 1976 that injection of the cytoplasm taken from 1-MeAde-treated maturing oocytes, but not from intact immature oocytes, induces matura-

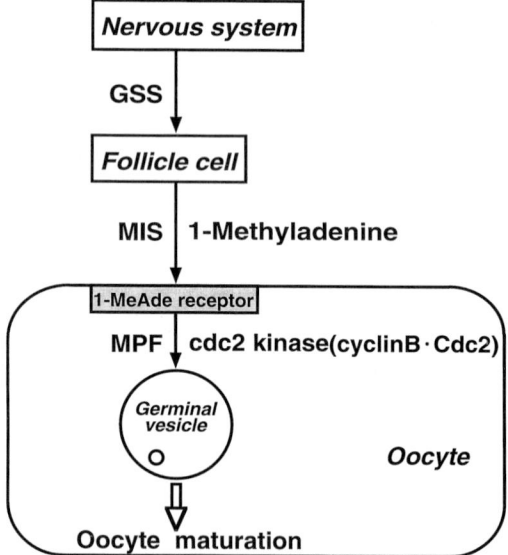

FIGURE 3 Three mediators are involved in starfish oocyte maturation and spawning. For the hormonal control of starfish oocyte maturation and spawning, the primary mediator is GSS (gonad-stimulating substance); the secondary is MIS (maturation-inducing substance), which is identified as 1-methyladenine (1-MeAde); and the third is MPF (maturation-promoting factor), whose activity is generated by cdc2 kinase, the complex of cyclin B with Cdc2 protein.

tion in recipient intact immature oocytes. Such a cytoplasmic tertiary mediator in inducing oocyte maturation has been called MPF after the mode of amphibian hormonal cascade in which the equivalent tertiary mediator was designated to be MPF by Masui and Markert.

Originally, starfish MPF had been identified in oocytes as mediating the GVBD-inducing action of 1-MeAde at the resumption of the first meiotic cycle. However, later studies revealed that similar activity is detectable in the second meiotic cycle and even in each M phase of early cleavage cycles after fertilization in starfish. Furthermore, MPF activity is detectable in M-phase extracts from yeast to mammalian cultured cells and lacks species specificity. Thus, around 1980 MPF was much more generalized to be an "M-phase-promoting factor" that is common to the M-phase control in all eukaryotic cells. As a result, the focus of MPF studies changed from the endocrinology of starfish reproduction to the cell biology of M-phase control common to all eukaryotic cells.

What is the molecular entity of MPF? Cell cycle research revealed at the end of the 1980s that an active component of MPF is cdc2 kinase, which is the complex of cyclin B protein with the homolog of fission yeast cdc2 gene product. In fact, inhibition of cdc2 kinase activation prevents 1-MeAde-induced starfish oocyte maturation, whereas sole injection of cdc2 kinase into immature starfish oocyte induces the whole process of oocyte maturation, including the resumption of meiosis and the completion of the subsequent meiotic cycles, follicular envelope breakdown, and the acquisition of fertilizability.

In immature oocytes, the inactive form of the cyclin B/Cdc2 complex, its direct activator, the Cdc25 phosphatase which dephosphorylates Thr14 and Tyr15 residues in cyclin B-associated Cdc2, and its direct inactivator, Wee1 family kinase which phosphorylates both residues, are present, indicating that the activation/inactivation balance is inclined to the inactivation in G_2-phase-arrested oocytes. Accordingly, the major issue at meiosis reinitiation is how this balance is reversed toward the activation of cdc2 kinase; currently there are two pathways proposed for this tipping (Fig. 4). In one pathway, the activity of the Wee1 family inactivator is downregulated by its putative suppressor prior to the cdc2 kinase activation. In the other pathway, Cdc25 undergoes an initial cdc2 kinase-independent phos-

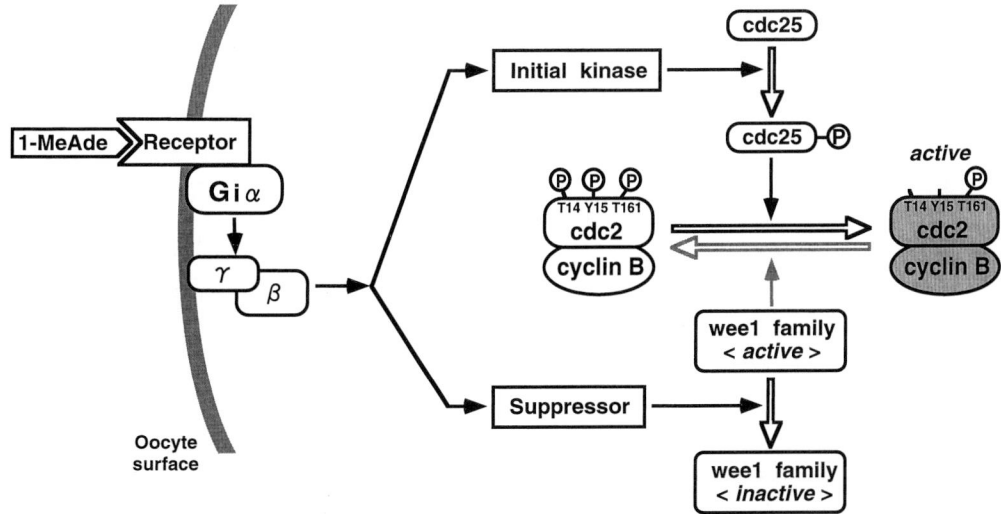

FIGURE 4 A model leading to the initial activation of cdc2 kinase at meiosis reinitiation in starfish oocytes. A maturation-inducing hormone, 1-methyladenine (1-MeAde), interacts with the putative oocyte surface receptor which is coupled with heterotrimeric G protein. The released $G_{\beta\gamma}$ causes the activation of two pathways that are independent of cdc2 kinase activity. In one pathway, the putative initial kinase upregulates Cdc25 phosphatase, whereas in the other pathway the putative suppressor downregulates Wee1 family kinase. These two pathways cooperatively initiate the activation of cdc2 kinase, the cyclin B/Cdc2 complex.

phorylation and activation by the putative initial kinase. These two pathways that are activated in response to 1-MeAde but independently of the cdc2 kinase activity result in the initial increase in cdc2 kinase activity by simultaneously removing the inhibitor and stimulating the activator. Once initially activated, further increase in cdc2 kinase activity is brought about by the cdc2 kinase-dependent feedback loops that upregulate Cdc25 phosphatase and downregulate Wee1 family kinase. After 1-MeAde stimulation, however, the pathway that links $G_{\beta\gamma}$ to the initial kinase and the suppressor is still missing.

Due to its own M-phase-inducing activity, the active cdc2 kinase is no doubt a minimally essential component of MPF. In addition, considering that MPF was originally identified as a cytoplasmic activity, the model proposed in Fig. 4 supports the idea that the suppressor, the initial kinase, and Cdc25 phosphatase would also be included in responsible components for the original cytoplasmic MPF activity.

V. MEIOTIC MATURATION AND FERTILIZATION

Once the G_2-phase arrest in immature starfish oocytes is released by cdc2 kinase, meiosis, which is composed of two successive M phases, is completed in the absence of fertilization, resulting in the formation of mature eggs with a female pronucleus. The activity of the cyclin B/Cdc2 complex (cdc2 kinase) oscillates along with the progression of meiosis I and II, with a peak at each metaphase (Fig. 5). In contrast, the other M-phase cyclin, cyclin A, is undetectable during meiosis I, and thereafter the activity of the cyclin A/Cdc2 complex peaks at meiotic metaphase II as does that of the cyclin B/Cdc2 complex. Although the molecular mechanism which ensures the lack of S phase between meiosis I and II is unknown, it is supposed that the ability to replicate DNA is acquired by the interkinesis period but is suppressed by an immediate reactivation of the cyclin B/Cdc2 complex at the entry into meiosis II; this is supported by the lack of the Tyr15 phosphorylation/dephosphorylation step in Cdc2 subunit.

In addition to M-phase cyclins, another feature during starfish meiotic cycles is the dynamics of the MAP kinase activity (Fig. 5). Following cdc2 kinase activation at meiosis reinitiation, MAP kinase is activated at germinal vesicle breakdown, and thereafter it is maintained at elevated levels of activity even after the completion of meiosis II. When mature eggs are inseminated, MAP kinase activity decreases immediately and then initiation of S phase in the first cleavage cycle follows. At least in *A. pectinifera* and *A. forbesi*, MAP kinase activity is necessary and sufficient to support the G_1-phase arrest in unfertilized mature eggs. In fact, sole inactivation of MAP kinase in the absence fertilization causes S phase in

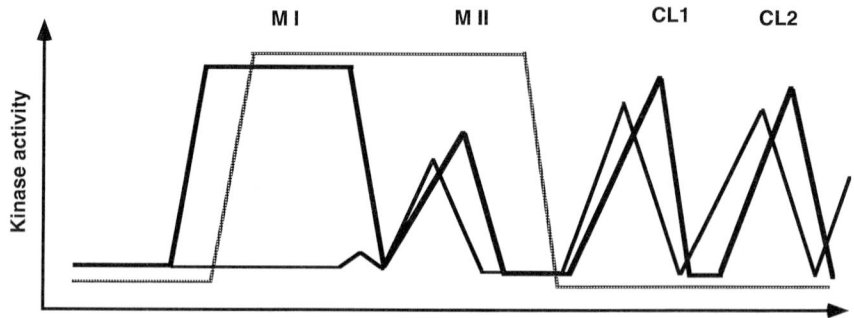

FIGURE 5 Dynamics in the activity of the cyclin B/Cdc2 complex (cdc2 kinase), the cyclin A/Cdc2 complex, and MAP kinase through starfish oocyte maturation and early cleavages. The activity of the cyclin B/Cdc2 complex (cdc2 kinase) (bold line) cycles along with each meiotic (MI and MII) and early cleavage cycles (CL1 and CL2). In contrast, the cyclin A/Cdc2 complex (thin line) is almost undetectable during meiosis I (MI), and thereafter its activity cycles along with the cell cycle. After the cdc2 kinase activation, MAP kinase (dotted line) is activated at germinal vesicle breakdown and is maintained at elevated levels of activity until the completion of meiosis. Upon fertilization it is inactivated, and thereafter its activity is undetectable during early cleavage cycles.

mature eggs, whereas artificial maintenance of MAP kinase activity prevents S phase in fertilized mature eggs. Thus, the transition to S phase following fertilization is executed through a dual repression; that is, the repression of the activity that represses DNA replication.

In contrast, female pronuclei of *M. glacialis* undergo S phase in the absence of fertilization under the elevated levels of MAP kinase activity and then arrest at G_2 phase of the first cleavage cycle. It is difficult to explain the reason for the discrepancy between these different types of cell cycle arrest, though both prevent unfertilized mature eggs from the parthenogenetic activation.

Although immature oocytes do not respond to insemination, on and after GVBD maturing oocytes elevate fertilization envelope immediately in response to insemination, indicating the acquisition of fertilizability. This response is mediated by phospholipase Cγ, which stimulates the production of inositol triphosphate resulting in intracellular Ca^{2+} release. When insemination is performed during meiosis I, however, the decrease in MAP kinase activity occurs with a significant lag, that is, only after the completion of meiosis I, whereas it occurs immediately at the insemination during meiosis II. Thus, fertilization stimulus for the MAP kinase inactivation appears to be retained during meiosis I, possibly ensuring the successful repression of S phase during the transition from meiosis I to meiosis II, that is, the formation of haploid female genomes.

VI. CONCLUDING REMARKS

Almost 40 years have passed since the preliminary studies of the reproductive biology in starfish. Currently, a major interest in starfish oocytes concerns the cell and developmental biology of the cell cycle control. Indeed, it is remarkable how the study of starfish oocyte maturation has contributed to the study of the cell cycle control common to all eukaryotic cells. Once again, however, it should be remembered that all starfish studies originated from the endocrinology of reproduction, despite that even the amino acid sequence of GSS is not determined. Major problems to be resolved are the pathway leading to GSS production and its release; the GSS receptor in follicle cells and the pathway producing 1-MeAde; the 1-MeAde receptor in oocyte plasma membrane and its linkage to cdc2 kinase activation; and the pathway leading to MAP kinase inactivation upon fertilization. Although most of these problems are apparently peculiar to starfish system, it is expected that their elucidation would reveal a new general concept in biology.

Acknowledgments

I thank Dr. John Pearse for reading the manuscript. This work was supported by grants from the Ministry of Education, Science and Culture and the CREST. The author is an investigator of the CREST of Science and Technology Corporation, Japan.

See Also the Following Article

MEIOTIC CELL CYCLE, OOCYTES

Bibliography

Carroll, D. J., Ramarao, C. S., Mehlmann, L. M., Roche, S., Terasaki, M., and Jaffe, L. A. (1997). Calcium release at fertilization in starfish eggs is mediated by phospholipase Cγ. *J. Cell Biol.* **138**, 1303–1311.

Chaet, A. B., and McConnnaughy, R. A. (1959). Physiologic activity of nerve extracts. *Biol. Bull.* **117**, 407–408.

Coleman, T. R., and Dunphy, W. G. (1994). Cdc2 regulatory factors. *Curr. Opin. Cell Biol.* **6**, 877–882.

Jaffe, L. A., Gallo, C. J., Lee, R. H., Ho, Y.-K., and Jones, T. L. Z. (1993). Oocyte maturation in starfish is mediated by the $\beta\gamma$-subunit complex of a G-protein. *J. Cell Biol.* **121**, 775–783.

Kanatani, H. (1973). Maturation-inducing substance in starfishes. *Int. Rev. Cytol.* **35**, 253–298.

Kanatani, H. (1985). Oocyte growth and maturation in starfish. In *Biology of Fertilization*, Vol. 1 (C. B. Metz and A. Monroy, Eds.), pp. 119–140. Academic Press, San Diego.

Kishimoto, T. (1986). Microinjection and cytoplasmic transfer in starfish oocytes. *Methods Cell Biol.* **27**, 379–394.

Kishimoto, T. (1988). Regulation of metaphase by a maturation-promoting factor. *Develop. Growth Differ.* **30**, 105–115.

Kishimoto, T. (1996). Starfish maturation-promoting factor. *Trends Biochem. Sci.* **21**, 35–37.

Kishimoto, T. (1998). Cell cycle arrest and release in starfish oocytes and eggs. *Seminars Cell Dev. Biol.* **9**(4), in press.

Meijer, L., and Guerrier, P. (1984). Maturation and fertilization in starfish oocytes. *Int. Rev. Cytol.* **86,** 129–196.

Nurse, P. (1990). Universal control mechanism regulating onset of M-phase. *Nature* **344,** 503–508.

Okano-Uchida, T., Sekiai, T., Lee, K., Okumura, E., Tachibana, K., and Kishimoto, T. (1998). *In vivo* regulation of cyclin A/Cdc2 and cyclin B/Cdc2 through meiotic and early cleavage cycles in starfish. *Dev. Biol.* **197,** 39–53.

Picard, A., Galas, S., Peaucellier, G., and Doree, M. (1996). Newly assembled cyclin B–cdc2 kinase is required to suppress DNA replication between meiosis I and meiosis II in starfish oocytes. *EMBO J.* **15,** 3590–3598.

Tachibana, K., Machida, T., Nomura, Y., and Kishimoto, T. (1997). MAP kinase links the fertilization signal transduction pathway to the G1/S-phase transition in starfish eggs. *EMBO J.* **16,** 4333–4339.

Oogenesis, in Mammals

Helen M. Picton and Roger G. Gosden

University of Leeds

I. Introduction
II. From Primordial Germ Cells to Oocytes
III. Oocyte Growth
IV. Oocyte-Specific Gene Expression
V. Cell Interactions
VI. Oocyte Meiotic Maturation

GLOSSARY

cytoplasmic maturation The process which accompanies oocyte nuclear maturation to prepare the cytoplasm for fertilization and embryo development.

meiotic maturation The final step of oogenesis which includes condensation of chromatin, nuclear envelope breakdown, and separation of homologous chromosomes; it results in the formation of female gametes capable of being fertilized by sperm.

oocytes Large, active secretory meiotic cells derived from oogonia which, after meiotic division, give rise to the unfertilized female gamete.

oogonia The mitotic female stem cells derived from primordial germ cells and from which all oocytes are derived.

primordial germ cell The migratory, undifferentiated cell which populates the primitive fetal ovary early during embryonic development and divides to form ovarian stem cells—the oogonia.

zona pellucida A glycoprotein coat, secreted by the oocyte, which is responsible for sperm binding during fertilization and the prevention of polyspermy after fertilization.

In mammals oogenesis is initiated early in fetal development and ends months to years later in the sexually mature adult. Oogenesis begins with primordial germ cell (PGC) formation and encompasses a series of developmental milestones and cellular transformations, from PGCs to oogonia and oocytes (occurring in the fetus), up to the production of a highly specialized gamete (occurring in the adult), which is capable of transmitting genetic information to the developing preimplantation embryo.

I. INTRODUCTION

The oocyte is not only the rarest and the largest cell in the body but also has one of the most remarkable life histories. Considering the complexity of the oocyte, experimental science has only just touched the surface of the topic. In this article, we describe oocyte growth and differentiation and the cytoplasmic and nuclear alterations which occur during development from primordial stages to the emergence of a mature oocyte in a Graafian follicle.

II. FROM PRIMORDIAL GERM CELLS TO OOCYTES

Oogenesis begins early in fetal development with the formation of primordial germ cells (PGCs) which populate the primitive fetal ovary. PGCs are characteristically large, possessing a round nucleus with one or more conspicuous nucleoli and, in their cytoplasm, many glycogen granules, ribosomes, mitochondria, and variable numbers of lipid droplets. The transport of PGCs toward the presumptive ovary depends initially on passive transfer and mass movement of the surrounding tissues and, subsequently, on both the inherent capacity of PGCs for independent ameboid-like movement and their response to chemotactic substances such as transforming growth factor β_1. During migration along the hind gut to the primitive gonad, PGCs undergo a species-specific number of mitoses. The survival of PGCs is promoted by kit-ligand produced by the somatic cells. PGCs are the sole source of adult germ cells and their early history is identical between male and female gonads (Fig. 1).

Once established in the developing ovary, PGCs lose their motility and become more spherical with fewer cytoplasmic organelles (Fig. 2). These cells, often referred to at this stage as oogonia, are stem cells that expand the definitive population of germ cells. Oogonia are connected by intercellular bridges and have a high frequency of mitotic division. Germ cells in the mouse ovary undergo approximately four mitotic cycles before entering meiosis between Days 14 and 16 of a 20-day gestation period, whereas in humans, as in other large mammalian species, they undergo many more rounds of division over a period of several months until shortly before birth. Thus, in these species by mid-to-late gestation many stages of germ cell development are present in the ovary simultaneously (Fig. 3).

After a number of rounds of mitotic divisions, meiosis is initiated in the oogonia. The cells, now called primary oocytes, progress through prophase I of the first meiotic division before becoming ar-

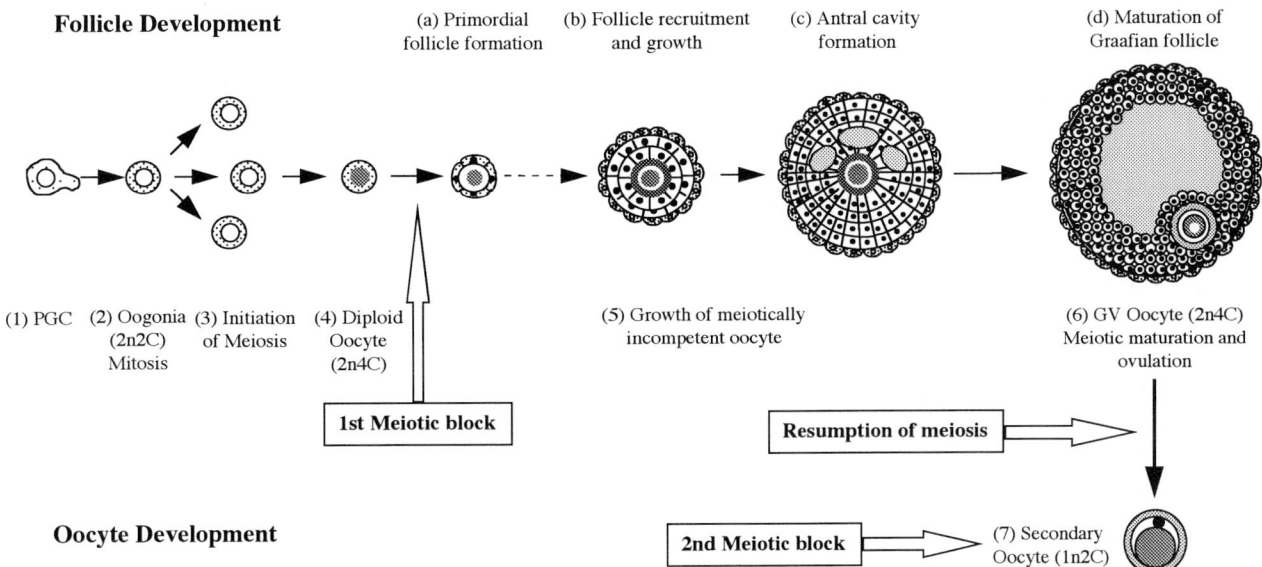

FIGURE 1 The checkpoints of oogenesis from primordial germ cells (PGC) to secondary oocytes. The chromosomal (n) and DNA content (C) are illustrated.

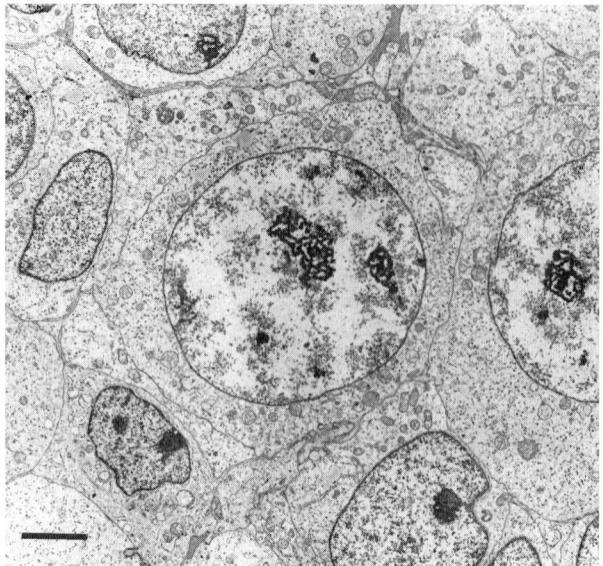

FIGURE 2 Fine structure of an oogonium from a human fetus (crown–rump length, 41 mm). These cells have a high nucleocytoplasm ratio and are larger than the surrounding somatic cells, which may include pregranulosa cells. Scale bar = 3 μm (courtesy of Dr. Daniel Szollosi, INRA, France; reproduced by permission from Gosden, 1995).

rested at the diplotene (dictyate) stage. Diplotene oocytes are larger than oogonia, have more cytoplasmic organelles, and, most significantly, have undergone genetic recombination of maternally and paternally derived DNA. Coincident with the initiation of meiosis, oocytes in the medullary region of the ovary lose any intercellular bridges and become enclosed in a single layer of flattened or polyhedral pregranulosa cells which rest on a delicate basement membrane. These very early stages in follicle development, which include the so-called primordial follicles, are recognizable in the human fetal ovary at 22 weeks of gestation. The primordial follicles represent the lifeboats of the ovary because once growth is initiated the somatic cells that they contain nourish and regulate the development of oocytes. Any oocytes remaining naked are destined to die. The third somatic cell type in ovarian follicles, the theca, is not morphologically recognizable until follicles have started to grow.

The scheduling of the mitotic phase of the oogonia has a crucial bearing on the long-term functional capacity of the ovary. Depending on the species, mitotic activity of the oogonia ceases before or shortly after birth, preventing the addition or replacement of lost oocytes. Only a fraction of the original germ cell population survives and fewer still successfully progress to ovulation in adult life—the great majority are destined to undergo programmed cell death (apoptosis) as naked germ cells or in atretic follicles. In humans ovaries, for example, germ cell number peaks around midgestation at approximately 7 million, decreases to 1 or-2 million by birth, and declines to approximately 250,000 by puberty; of these survivors, only 400 or-500 follicles will ovulate during the reproductive life span. In the ovary, it is therefore a case of many are called but few are chosen.

III. OOCYTE GROWTH

Oocytes are not fertile at first but have to grow in follicles to become competent to resume nuclear maturation and undergo fertilization and cleavage divisions. During the growth phase they acquire a complex cytoplasmic organization dependent both on the production of new gene products and organelles and on the modification and redistribution of existing ones. The fidelity of replication of cytoplasmic organelles during oogenesis, especially mitochondria and their DNA molecules, is crucial because cytoplasmic inheritance of the zygote is mostly, if not exclusively, derived from the egg.

A. Cytological Changes Occurring during Oocyte Growth

1. Cytoplasm

Mammalian oocytes are voluminous compared with most somatic cells due to the bulky ooplasm, which contains the stored molecules and organelles needed for preimplantation embryo development. This cellular enlargement occurs because oocyte growth takes place over a long period and is uninterrupted by mitosis and cleavage of the cytoplasm. In parallel with expanding cell volume, oocytes accumulate water, ions, and lipids and both the rate of protein synthesis and total cellular protein content rise. The mouse oocyte, for example, grows from 15 to 80 mm over a period of 2 or-3 weeks, whereas

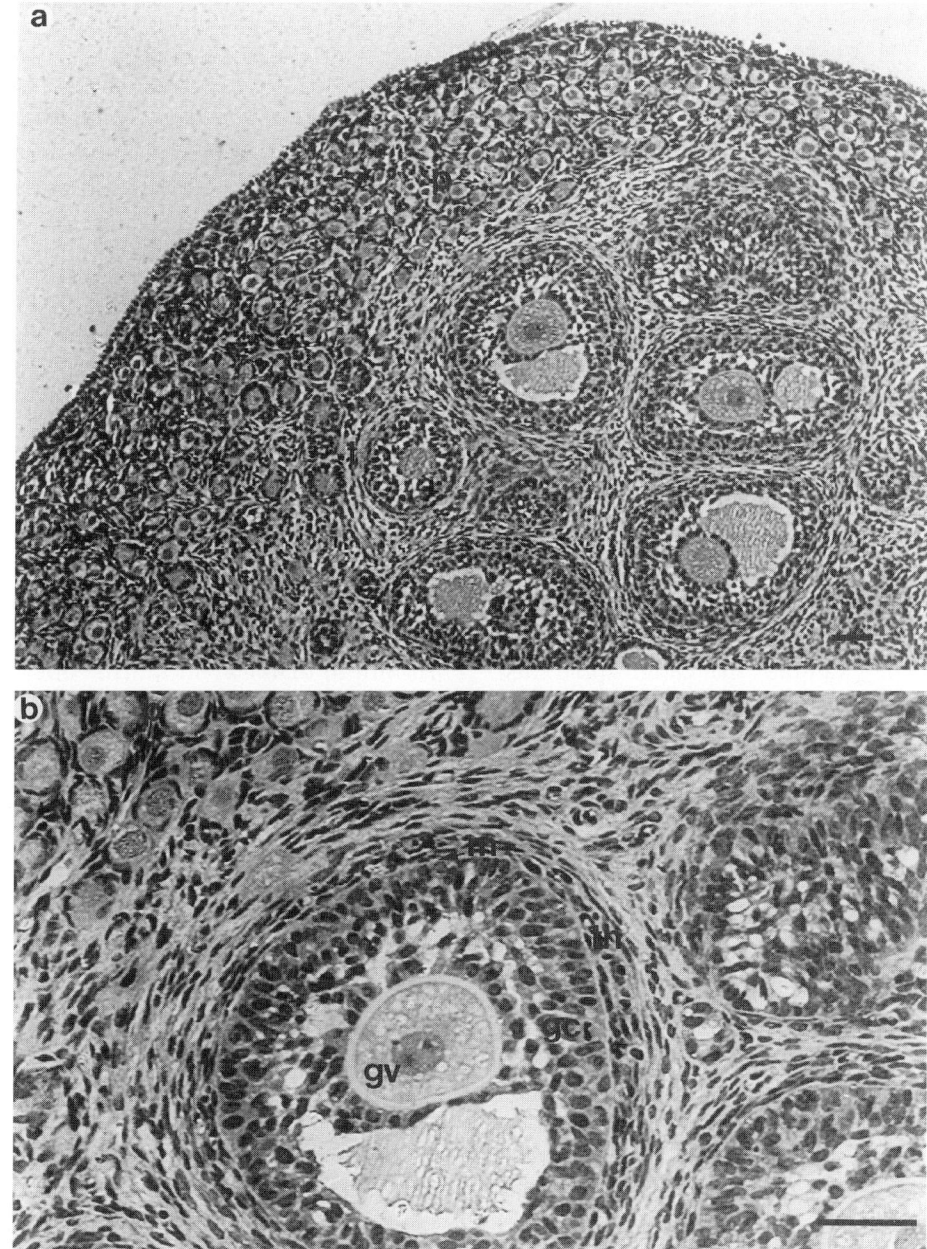

FIGURE 3 Morphology of sheep ovary on Day 134 of gestation showing (a) the gradation of follicular and oocyte development in the transition from the cortex to the medulla of the ovary. Germ cells are present as naked oocytes, primordial follicles (p), or transitional (t) growing follicles. (b) The most advanced follicles are present as small antral follicles of 0.25–0.8 mm diameter in which the theca interna (th) and basement membrane (m) enclose the granulosa cells (gc) and GV oocyte (gv). Scale bar = 50 μm.

the human oocyte begins at 35 mm and requires several weeks longer to reach a final size of 120 mm. In the mouse, most of the growth of the oocyte has occurred by the time the first two layers of granulosa cells have been laid down and is complete when the follicular antrum starts to form (approximately 2 or 3 weeks),-in contrast to farm species and humans in which oocytes require up to 6 months to reach full size.

Cytoplasmic organelles become far more abundant during oocyte growth— the notable exception being centrioles, which disappear at the midgrowth stage and are not found again until after fertilization because in many species, including humans, the centriole is paternally inherited from the sperm. The numerous organelles in small oocytes are initially clustered around the nucleus to form the so-called Balbiani body or "yolk" nucleus. As oocyte growth proceeds the organelles disperse toward the periphery of the cell. For example, the Golgi apparatus enlarges and transforms from a few flattened sacs into numerous units in the cortex of the cell, where it is active in exporting glycoproteins to the zona pellucida (ZP) and in forming the cortical granules which are required at fertilization. As growth proceeds, the extensive endoplasmic reticulum also becomes progressively more cortical in distribution and it may be involved in calcium release for cortical granule exocytosis. During growth, the number of ribosomes multiply, but polyribosomes are rare.

The ooplasm contains numerous membrane-bound vesicles, multivesicular and crystalline bodies, many of which represent the products of molecular transport across the cell membrane. Additionally, mature mammalian oocytes contain variable amounts of yolk in the form of glycogen granules, small lipid droplets, and protein accumulations though apparently without the specific yolk proteins characteristic of oviparous species. In farm species there is so much lipid that the nucleus is obscured and in mature rodent oocytes fibrous lattices are a dominant feature, accumulating to almost 10% of the ooplasm but disappearing during cleavage. The significance of the lattices has been controversial because they have been variously regarded as sites of ribosomal storage or protein yolk but are probably a form of intermediate filament.

Ultrastructural studies have shown that the abundance and morphology of mitochondria change during oogenesis and early embryogenesis. They are initially elongated with numerous transversely orientated cristae but become more spherical and vacuolated during growth and correspondingly less active as indicated by the presence of fewer and concentrically arched cristae. Additionally, mitochondrial copy number increases to more than 105 in mice with the increase in ooplasmic volume.

The appearance of the large oocyte nucleus is unlike that of any interphase somatic cell and has been given the special term germinal vesicle (GV). The rodent oocyte GV is unusually pale and featureless, in contrast to primates in which chromosomal threads are more obvious. The nucleus usually contains a single nucleolus which enlarges until the germinal vesicle breaks down (GVBD) at meiosis, when it disappears. This increase in size (from 2 to almost 10 mm in diameter in murine oocytes) represents a shrinking proportion of cell volume as the cytoplasm enlarges faster than the nucleus. During oocyte growth the nuclear membrane becomes undulated with pores, reflecting the greater nucleocytoplasmic traffic of substrate and informational molecules. In large oocytes capable of resuming meiosis, a rim of chromatin forms around the nucleolus.

2. Zona Pellucida

One of the most conspicuous changes occurring during oogenesis in mammals is the secretion of a fibrillar glycoprotein coat, the ZP (Fig. 4). This protective coat, which is approximately 7 mm thick in murine oocytes, is secreted by the oocyte into the perivitelline space and is impermeable to molecules larger than 170 kDa. The functions of the ZP are well established and include presentation of species-specific receptors to sperm, the induction of the acrosome reaction before fertilization, provision of a block to polyspermic fertilization, protection of the embryo after fertilization, and exchange of molecules with follicular fluid. The ZP, which is only secreted by growing oocytes, initially appears as patches of uniform, fine filaments and develops into a dense meshwork of interconnected filaments which completely surround the oocyte and largely separate it from the follicular cells. During growth the oolemma

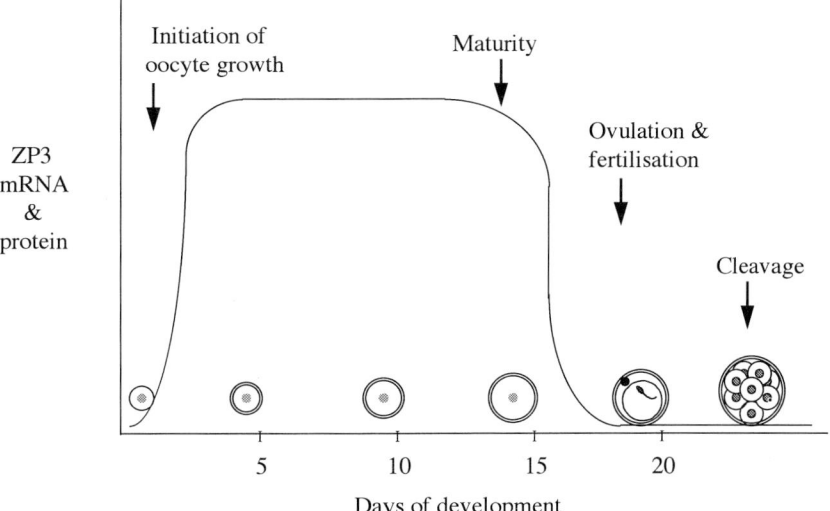

FIGURE 4 Oocyte-specific expression of *Zp3* is confined to oogenesis, when the zona pellucida is forming around the growing oocyte (reproduced by permission from Gosden et al., 1997).

becomes increasingly folded and eventually a uniform cover of microvilli extends a short way from the oolemma into the ZP and junctional contact is made with foot processes extending from the corona granulosa cells surrounding the oocyte. These membrane specializations (gap junctions) produce a syncytium by which the oocyte and follicle cells are metabolically coupled to one another and which is required for oocyte growth.

B. Production and Storage of RNA

In mammals, growing oocytes have characteristically high levels of RNA synthesis which cease after the reinitiation of meiosis and GVBD. The synthesis and uptake of macromolecules into the oocyte serve two purposes. The first is for growth, development, and maturation of the oocyte itself. The second is for the synthesis and storage of RNA and proteins which provide much of the information and structural materials required to support early postfertilization development until the embryonic genome is expressed (two- cell stage in mice, four cells in pigs, and four to eight cells in humans). Successful postfertilization development is critically dependent on the information that has accumulated during oogenesis.

Oocytes in primordial follicles are relatively quiescent compared to the burst of transcription and translation which occurs after the onset of follicle growth and which is sustained until the oocyte is fully grown. By the time meiosis resumes the storage of the maternal program is almost complete and transcription falls to barely detectable levels. In mouse eggs, for example, by the time the oocyte is three-quarters grown, total RNA content has increased 300-fold at an average rate of synthesis of 15 ng min^{-1}. A fully mature mouse egg contains 0.3–0.55 ng of RNA *in toto*. The proportions of the different RNA classes are approximately 60–65% ribosomal RNA, 20–25% transfer RNA, and 10–15% polyA$^+$. Most of this RNA is extremely stable and ~75% of the nascent molecules formed early in growth are retained for 10–20 days.

The transcriptional control of the stored mRNA may be targeted to specific stages of oocyte maturation and early postfertilization development. For example, GVBD signals the beginning of degradation of much of the stable and dormant untranslated mRNAs that have accumulated during oogenesis. Beginning about 3 hr after GVBD in mice, tissue plasminogen activator mRNA undergoes progressive cytoplasmic polyadenylation followed by translation and then degradation. Polyadenylation destabilizes the mRNA because it is much more stable when it has a shorter polyA$^+$ tail. Precocious translation of the mRNA may be prevented by association with specialized proteins which package the RNA in such a way as to exclude the ribosomes from gaining access to the regulatory elements. Further control for timing the release of stored mRNA is by U-rich *cis*-acting adenylation control elements which, like the TATA box of transcription, are highly conserved throughout evolution. Following the addition of a long polyA$^+$ tail the transcripts become vulnerable to degradation.

C. Synthesis and Storage of Proteins

The growing oocyte is very active in the synthesis of proteins from nuclear and mitochondrial transcripts. Some of these proteins are required for the differentiation of the oocyte itself, others are for interactions with the surrounding somatic cells, and yet others will be significant for formation of the ZP, fertilization, and early embryo development. Additionally, some of the proteins which are detected in the oocyte may not be synthesized there but rather are taken up by endocytosis from follicular fluid or granulosa cell secretions.

On commencing growth, the oocyte does not halt until it reaches full size—unless the follicle becomes atretic. During this period the rate of protein synthesis per oocyte rises from 1 to 40 pg hr^{-1} in murine oocytes and the total protein accumulating amounts to about 25 ng, excluding 3 ng for the zona pellucida. Since cell volume is expanding faster than the amount of protein, the absolute rate of protein synthesis actually declines. Despite this, several hundred species of polypeptide have been shown to accumulate steadily in the oocyte and their relative proportions change during oocyte growth and maturation. At GVBD, protein synthesis and the free methionine pool (which remained constant hitherto) decline. This could be due to concomitant degradation of RNA or to changes in the translational control mechanisms which are triggered by the resumption of the meiotic cycle.

Some oocyte proteins are far more abundant than expected from their conventional role. There is, for example, ~200–400 pg of LDH-β in mouse oocytes, representing 2–5% of the total protein synthesized and greatly exceeding the requirements for carbohydrate metabolism. The protein may serve as an amino acid reserve during early embryonic cleavage and, indeed, embryos can develop *in vitro* in media completely lacking in amino acids. Surplus proteins could therefore serve as an amino acid energy reserve and represent a modest substitute for true yolk proteins that have been lost during the evolution of diminutive mammalian oocytes.

IV. OOCYTE-SPECIFIC GENE EXPRESSION

A number of genes that are expressed exclusively in the oocyte or germ cell line have recently come to light. These include *Zp1*, *-2*, and *-3* (Fig. 4), growth/differentiation factor-9 (*gdf-9*), required by the granulosa cells, and *Oct-4* (also known as *Oct-3*), a POU factor which is also expressed in pluripotential stem cells of the embryo.

A. The Zona Pellucida

The ZP consists of three relatively well-conserved proteins in the mouse oocyte (cf. two to four proteins in the pig, rabbit, monkey, and human). The coding sequences of the murine and human genes are 74% identical. The heavily glycosylated murine ZP proteins have been designated ZP1 (molecular weight, 200 kDa), ZP2 (120 kDa), and ZP3 (83 kDa) and amount to 5 ng of protein or 17% of the total protein content of the cell. A complete set of these proteins is necessary because the zona matrix consists of polymeric filaments of ZP2 and ZP3 that are cross-linked noncovalently by ZP1 dimers.

Murine *Zp1*, *Zp2*, and *Zp3* genes are present as single copies on chromosomes 19, 7, and 5, respectively. With the exception of *Zp2*, gene expression does not occur in primordial follicles and only commences when the oocyte begins to grow. Less is known about ZP1, which is less abundant than the other two, which are present in equimolar concentrations. *Zp1*, *Zp2*, and *Zp3* are coordinately expressed and may have common transcriptional regulatory elements. At peak activity during the midgrowth phase, they represent 1.5% of the total polyA$^+$ RNA, and ZP3, which is the primary sperm receptor in the zona, amounts to 7 or 8% of total protein synthesis. In murine oocytes they are expressed for just the 2 weeks required for the cell to reach full size, and transcripts fall to 5% of peak values at ovulation and are virtually undetectable in embryos. Thus, like other long-lived maternal mRNAs, they are rapidly deadenylated and degraded when transcription ceases during the resumption of meiosis.

V. CELL INTERACTIONS

The entire follicle is a functional syncytium with component cells metabolically coupled by gap junctions. In the absence of this close association, oocyte and follicle growth will not progress. From the primordial stage, oocytes are invested in follicular cells which not only produce growth factors and hormones but also provide the oocyte with physical support, nutrients (such as pyruvate), metabolic precursors (such as amino acids and nucleotides), and other small molecules which can equilibrate between the compartments without affecting the distinctive macromolecular phenotype of either cell. Granulosa cells can therefore liberate the oocyte from responsibility for producing some of its nutritional requirements. Furthermore, specific molecules produced by the somatic cells have been identified as taking part in the mechanism for the maintenance of meiotic arrest in the oocyte.

The relationship between granulosa cells and oocytes is one of reciprocal influences rather than unbalanced dependence because the oocyte actively steers proliferation, morphogenesis, and differentiation of the granulosa cells. Persuasive evidence has revealed that granulosa cell mitosis is stimulated by a soluble factor from the oocyte, whereas granulosa cell differentiation is dependent on the presence of the oocyte. This oocyte factor acts in a concentration-dependent manner to increase estradiol and decrease progesterone secretion by affecting specific steroidogenic enzymes in granulosa cells stimulated with follicle-stimulating hormone (FSH) and testosterone. Furthermore, removal of the oocyte results in the premature luteinization and elevated progesterone secretion by the follicular cells.

Another oocyte factor, GDF-9, a member of the TGF-β superfamily, is the first growth factor found to be exclusively expressed by growing oocytes. GDF-9 is synthesized throughout oocyte growth and is thought to be essential for initiating the early stages of follicle growth. Without GDF-9, follicle development halts at the unilaminar stage, though the oocyte grows and is well developed.

It is important to emphasize that a reciprocal relationship exists between the oocyte and its granulosa cells. Although denuded oocytes cannot grow, some development is possible when in coculture with soluble factors from granulosa cells. Both homologous gap junctions between granulosa cells and heterologous ones with the oocyte are maintained throughout growth until shortly after meiotic maturation has resumed, when cellular processes are withdrawn and metabolites and informational molecules can no longer pass. The causes and consequences of this loss of functional contact between the oocyte and somatic cells are discussed in more detail in the following section.

VI. OOCYTE MEIOTIC MATURATION

A. General Features

Oocytes acquire meiotic competence and fertile potential in a stepwise manner during Graafian follicle development and meiotic maturation is triggered by the preovulatory surge of gonadotrophins. Meiotic maturation specifically involves development of the capacity for nuclear membrane breakdown and for progression from the dictyate state of the first meiotic prophase to metaphase II (first reduction division). In meiotically competent oocytes, the chromatin forms a dense ring around the nucleoli and microtubular organizing centers congregate because there is a reduction in cytoplasmic microtubules. Chromosomes condense and the germinal vesicle breaks down and disperses. These changes are accompanied by a reduction in overall mitochondrial number, an increase in mitochondrial volume, and translocation of mitochondria to the perinuclear region during the formation of the meiotic spindle. This movement enables mitochondrial sequestration to areas that require high concentrations of adenosine triphosphate (ATP). The first meiotic division proceeds through metaphase I with the formation of the first meiotic spindle and expulsion of the first polar body. Completion of meiosis I is immediately followed without an interphase or S phase by entry into the second meiotic division, which is arrested again at metaphase II at which stage the cell is fertile for a few hours. After ovulation and fertilization of the oocyte within the oviduct, meiosis is completed and the second polar body is emitted. Strictly speaking, therefore, a true haploid ovum never exists in mammals because the reduction divisions are not completed until after sperm entry.

B. Control of Meiotic Maturation

The control of meiotic maturation involves a complex interplay between somatic and germ cells, with the participation of numerous metabolic pathways. Direct communication between the follicle cells and the oocyte facilitates the transfer of both inhibitory and stimulatory meiotic signals to the oocyte. Resumption of meiosis appears to be associated with a reduction in cyclic adenosine monophosphate (cAMP) concentrations within the oocyte and is negatively regulated by an activated cAMP-dependent protein kinase A (PKA). Furthermore, meiotic arrest can be explained in terms of activation and inactivation of cell cycle proteins and an increasing body of evidence supports a positive stimulus for maturation at the time of the preovulatory gonadotrophin surge.

It has been hypothesized that a threshold level of cAMP, originating in the granulosa cells and to a limited extent in the oocyte itself, is responsible for the maintenance of meiotic arrest. Stimulation of granulosa cell cAMP by the binding of the gonadotrophins to their receptors will therefore lead to the maintenance of meiosis-arresting levels of cAMP in the oocyte. Because oocytes and granulosa cells are metabolically coupled to their companion cells by gap junctions, molecules of a similar size and structure to cAMP such as the purines, adenosine, uridine, and hypoxanthine, or their metabolites produced by the mural granulosa cells can diffuse to the oocyte and vice versa. Hypoxanthine and adenosine maintain the GV status of oocytes isolated in culture.

Although the precise mechanism of action remains to be elucidated, it has been proposed that adenosine stimulates oocyte adenylate cyclase via a receptor on the surface of the oolemma, whereas hypoxanthine prevents hydrolysis of cAMP. Together these actions would sustain meiosis-arresting levels of cAMP. Alternatively, adenosine could participate in meiotic arrest either by conversion to ATP or as a substrate for adenyl cyclase within the oocyte.

It is now well established that one of the follicular responses to the preovulatory surge of luteinizing hormone (LH) is the production of hyaluronic acid which leads to the mucification and expansion of the cumulus granulosa cells with the attendant termination of gap junctional contact between the cumulus cells and the oocyte. This loss of intercellular communication may serve as a trigger for the resumption of meiosis *in vivo*. The action of high levels of LH in shifting steroidogenesis from predominately estrogen to progesterone during luteinization may also helps to modulate the meiotic status of the oocyte.

Preovulatory levels of LH and FSH may actively promote the resumption of meiosis by stimulating the production of a GVBD-inducing signal. Hormone-mediated signals may induce a granulosa cell increase in free inositol 1,4,5-triphosphate (IP$_3$), calcium ions, or both. Transferal of IP$_3$/calcium to the oocyte via gap junctions would mobilize the intracellular calcium stores in the egg, thus activating phosphodiesterase. The level of cAMP would in turn fall below the threshold level required to maintain meiotic arrest in the oocyte, thus initiating the cascade of events that finally cause GVBD.

Regardless of the signaling system which initiates GVBD, nuclear maturation in mammalian species appears to be mediated by the production of active maturation-promoting factor (MPF) in the ooplasm. The direct actions of MPF during GVBD may involve dissolution of the nucleoli, chromosomal condensation, and reorganization of the microtubular complex to form a functional spindle apparatus. MPF consists of two components: a 34-kDa protein, homologous to the product of the *cdc2*$^+$ gene in fission yeast, and cyclin B. The p34^{cdc2} component requires dephosphorylation for activation. Cyclin B, in contrast, requires phosphorylation for activation and is probably a substrate for p39^{mos}. The latter is a product of the *mos* oncogene which is expressed early in oocyte maturation and disappears immediately after fertilization and may itself participate in the activation of MPF. Although mammalian oocytes from different species do not undergo maturation with the same kinetics, MPF activity is consistently low in GV-stage oocytes, increases during GVBD, and high at both metaphase I and metaphase II. Mammalian oocytes require p39^{mos} as well as the accumulation of the other components and substrates of MPF to become meiotically competent. However, GVBD is known to be arrested by PKA. We can hypothesize, therefore, that the decrease in intracellular cAMP occurring in response to the LH surge, and the attendant loss of gap junction contact, reduces PKA activity, thus allowing the association of cyclin B and p34^{cdc2}. p39^{mos} then participates in the activation of MPF, possibly by phosphorylation of the cyclin.

Finally, p39^{mos} may itself act as a cytostatic factor which stabilizes MPF, presumably by maintaining the phosphorylated state of cyclin B and thereby maintaining metaphase II arrest. p39^{mos} synthesis may be stimulated by progesterone produced as the somatic cells luteinize in response to the preovulatory LH surge. p39^{mos} is therefore a key component during oocyte maturation and it is possible that the second meiotic arrest is due to the transcription of Mos as the oocyte undergoes meiotic maturation.

See Also the Following Articles

Apoptosis (Cell Death); Meiosis; Meiotic Cell Cycle, Oocytes; Oogenesis, in Nonmammalian Vertebrates; Primordial Germ Cells

Bibliography

Bachvarova, R. (1985). Gene expression during oogenesis and oocyte development in mammals. In *Developmental Biology. A Comprehensive Synthesis, Volume 1, Oogenesis* (L. W. Browder, Ed.), pp. 453–524.. Plenum, New York.

Buccione, R., Schroeder, A. C., and Eppig, J. J. (1990). Interactions between somatic cells and germ cells throughout mammalian oogenesis. *Biol. Reprod.* 43, 543–547.

Dekel, N. (1996). Protein phosphorylation/dephosphorylation in the meiotic cell cycle of mammalian oocytes. *Rev. Reprod.* 1, 82–88.

Dong, J., Albertini, D. F., Nishimori, K., Kumar, T. R., Lu, N., and Matzuk, M. M. (1996). Growth differentiation factor-9 is required during early ovarian folliculogenesis. *Nature* 383, 531–535.

Eppig, J. J., and O'Brien, M. J. (1996). Development in vitro of mouse oocytes from primordial follicles. *Biol. Reprod.* 54, 197–207.

Faddy, M. J., and Gosden, R. G. (1995). A mathematical model for follicle dynamics in human ovaries. *Hum. Reprod.* 10, 770–775.

Gosden, R. G. (1995). Ovulation 1: Oocyte development throughout life. In *Cambridge Reviews in Human Reproduction, Gametes—The Oocyte* (J. G. Grudzinskas and J. L. Yovich, Eds.), pp. 119–149. Cambridge Univ. Press, Cambridge, UK.

Gosden, R. G., and Bownes, M. (1995). Cellular and molecular aspects of oocyte development. In *Cambridge Reviews in Human Reproduction, Gametes—The Oocyte* (J. G. Grudzinskas and J. L. Yovich, Eds.), pp. 23–53. Cambridge Univ. Press, Cambridge, UK.

Gosden, R. G., Krapez, J., and Briggs, D. (1997). Growth and development of the mammalian oocyte. *Bioessays,* 19, 857–882.

Gougeon, A. (1996). Regulation of ovarian follicular development in primates—Facts and hypotheses. *Endocr. Rev.* 17, 121–155.

Hirshfield, A. N. (1991). Development of follicles in the mammalian ovary. *Int. Rev. Cytol.* 124, 43–100.

Sagata, N. (1997). What does Mos do in oocytes and somatic cells? *BioEssays* 19, 13–21.

Wassarman, P. M. (1996). Oogenesis. In *Reproductive Endocrinology, Surgery and Technology* (E. Y. Adashi, J. A. Rock, and Z. Rosenwaks, Eds.), Vol. 1, pp. 341–359. Lippincott-Raven, Philadelphia.

Wassarman, P. M., Liu, C., and Litscher, E. S. (1996). Constructing the mammalian egg zona pellucida: Some new pieces of an old puzzle. *J. Cell Sci.* 109, 2001–2004.

Whitaker, M. (1996). Control of meiotic arrest. *Rev. Reprod.* 1, 127–135.

Oogenesis, in Nonmammalian Vertebrates

Charles A. Lessman

The University of Memphis

I. Introduction
II. Cyclostome Oogenesis
III. Chondrichthyes Oogenesis
IV. Teleost Oogenesis
V. Amphibian Oogenesis
VI. Reptilian Oogenesis

GLOSSARY

atresia The process of oocyte resorption; generally this phenomenon may occur at any stage of oogenesis.

cortical granules Membrane-bound structures rich in mucopolysaccharides or glycoproteins, formed in the early vitellogenic phase; important in the fertilization reaction and in formation of the perivitelline space; analogous to cortical alveoli found in teleost oocytes.

gonadotropin Hormones derived from the pituitary which stimulate oogenesis and steroidogenesis in the ovary.

oogonia Stem cells of the ovary, derived from primordial germ cells, which undergo mitosis in oogonial proliferation and then become committed to the meiotic cell cycle as primary oocytes.

oolemma The limiting plasma membrane of oocytes.

primordial germ cells Cells destined to become oocytes, formed early in embryogenesis.

prophase I The first phase of meiosis in which genetic recombination and crossing over occurs; consists of leptotene, pachytene, zygotene, and diplotene stages.

vitellogenesis The general phase of oogenesis involving production and incorporation of yolk in the growing oocyte.

vitellogenin Yolk precursor synthesized and released into the blood by the liver; incorporated by the vitellogenic phase oocyte.

Oogenesis may be defined as the development of the stem germ cell into the fertilizable egg or female gamete. The process begins with the stem cell (oogonium) becoming committed to enter meiosis, a special type of cell division peculiar to germ cells. The cell, now called an oocyte, produces the cellular materials needed to complete the previtellogenic phase of development. These may include gene amplification and organelle production. The cell then enters the vitellogenic phase, during which the cell may grow to considerable size due to the formation and incorporation of yolk and other nutrients necessary for subsequent embryonic development. Lastly, the oocyte undergoes a final meiotic maturation and ovulation which prepares the female gamete for fertilization.

I. INTRODUCTION

The major regulatory system of oogenesis is the hypothalamic–pituitary–gonadal axis. Releasing factors from the hypothalamus, a portion of the brain, stimulate production and release of gonadotropins from the anterior pituitary. These in turn act on the ovary, at the level of the follicle wall, to elicit steroid release which controls a variety of processes in oogenesis. The role of this regulatory system has been well documented, especially for the vitellogenic and maturation phases of oogenesis.

The oogenesis of the teleosts and amphibians have been studied more thoroughly than the other groups of nonmammalian vertebrates discussed in this article. This is probably a reflection of the relative ease of rearing some species of fish and amphibians throughout their reproductive cycle in controlled laboratory conditions. In addition, technical problems arise in the preparation of very large, yolky oocytes, such as some reptilian oocytes, for ultrastructural analysis.

During oogenesis, the oocyte is endowed with information molecules that are often spatially segregated in the cell such as messenger RNA (mRNA) which act as embryonic determinants, i.e., important in producing the embryo body plan. Thus, oogenesis may be considered as a crucial step in the development of the individual, even though the oocyte still resides in the maternal ovary.

II. CYCLOSTOME OOGENESIS

The cyclostomes or agnaths are primitive vertebrates and their oogenesis has been little studied compared with other vertebrate groups. Many of the species are specialized as parasites. Pituitary control seems to be limited to the vitellogenic phase because hypophysectomy (pituitary ablation) does not prevent oogonial proliferation or previtellogenic phases.

A. Lampreys

The lamprey has a synchronous ovary, i.e., all of the stem oogonial cells become committed to become oocytes and no stem oogonial cells remain in the fully developed ovary. Oogonial proliferation occurs in the larval or ammocoete stage prior to adult metamorphosis. Prior to metamorphosis, primary oocytes are arrested in the diplotene of meiosis; after metamorphosis, oocytes enter the final period of growth and vitellogenesis. Oocytes all develop as one large clutch or batch of oocytes and thus are homogeneous in size during oogenesis. Similar to the Pacific salmon, the lamprey spawns only once and then dies. The largest anadromous species release up to 170,000 eggs, whereas nonparasitic brook lampreys produce between 500 and 2500 eggs at spawning.

B. Hagfish

The hagfish is a continuous breeder; thus, its ovary is of the asynchronous type in which oogonial stem cell proliferation occurs at the ventral end of the organ (Fig. 1). Many different stages of oocyte development are present at any given time in these ovaries. Since these animals live at depth in the ocean, where few seasonal cues occur, breeding becomes independent of the season and eggs may be produced throughout the year. Atresia, or resorption of oocytes, occurs in this type of ovary. The ovary is unusual since only the outer region or cortex is present in both sexes, whereas the medulla, or middle region found in many types of vertebrate gonad, is absent.

III. CHONDRICHTHYES OOGENESIS

The cartilaginous fishes, sharks and rays, have both a cortex and medulla in the early gonad. During embryogenesis, the primordial germ cells, formed in the extraembryonic endoderm, migrate to the primitive gonad. In the ovary, the germ cells reside and proliferate in the cortex. While the ovary is a paired structure initially, in many species only one becomes functional in the adult. The ovary is unusual since it is intimately associated with a lymphomyeloid hemopoietic structure called the epigonal organ. Some species are oviparous, i.e., they lay eggs, and others are viviparous, i.e., they bear live young.

A. Previtellogenic Phase

The previtellogenic oocytes accumulate more 5S rRNA than needed for ribosome assembly. Amplification of rRNA genes is low or even absent, unlike the situation in many other lower vertebrate ovaries. A yolk nucleus is formed as a large juxtanuclear mass containing mitochondria and lipid; this is thought to be a special area for organelle production in oocytes. Under the microscope, the yolk nucleus has a dense core and loose peripheral zone.

B. Vitellogenic Phase

This phase of oogenesis is characterized by incorporation of yolk precursors called vitellogenin by the growing oocyte. The vitellogenin is produced by the liver under the influence of estrogens. In the dogfish shark, injection of 17β-estradiol increases vitellogenin in the blood. Vitellogenic follicles have an extensive network of blood vessels between the theca interna and basement membrane, i.e., layers of the follicle wall which surround the oocyte (Fig. 2). The basement membrane is specialized with channels to allow vitellogenin to pass through to the granulosa cell layer. There, the vitellogenin passes between cells and is taken up by oocyte pinocytosis. Primary yolk, also known as cortical alveoli or cortical granules, is apparently absent in this group and may account for the prevalence of polyspermy.

IV. TELEOST OOGENESIS

The teleosts or bony fishes comprise the largest vertebrate group, with over 20,000 species. It is a very diverse group in terms of habitat and reproduction. Most species are either exclusively freshwater or seawater inhabitants, with some species preferring brackish water as a habitat. Certain species are anadromous, i.e., live primarily in the oceans but return to fresh water to reproduce (e.g., salmon), and others are catadromous, i.e., freshwater dwellers which return to the oceans to reproduce (e.g., American eel). Teleosts may be oviparous (egg layers), ovoviparous (carry eggs internally through development), or viviparous (live bearers which nourish embryos). A unique feature of the teleost egg is the presence of a micropyle, a preformed sperm entry canal which is formed by a specialized follicle cell during oogenesis. There are three major types of oogenesis in the teleost assemblage: (i) synchronous—members of this group reproduce only once then die, thus all oocytes develop simultaneously (e.g., Pacific salmon); (ii) group synchronous—members with this type of oogenesis have at least two populations (clutches) of

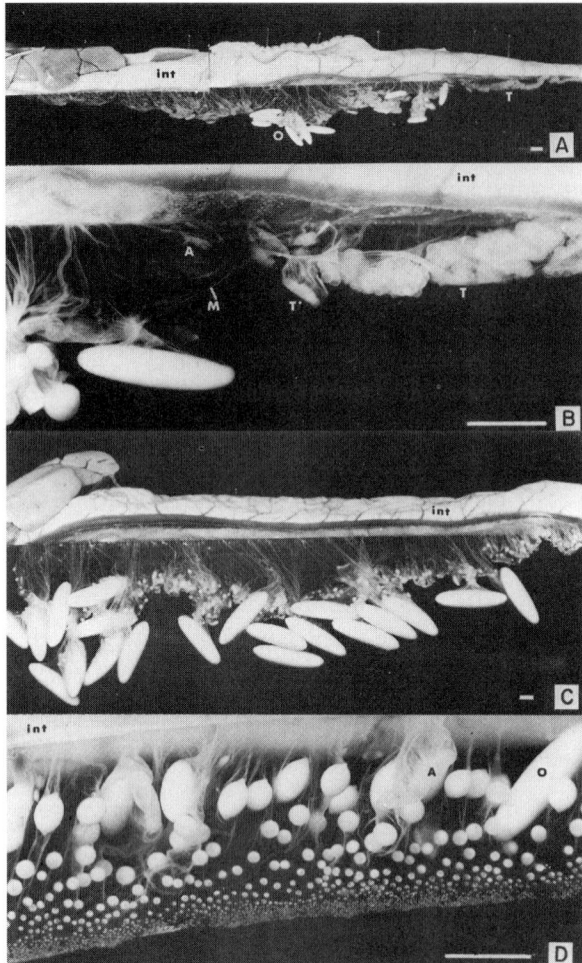

FIGURE 1 Gonads of *Eptatretus stouti*, showing the intestine above in each case, with the mesentery of the gonad joined to the dorsal mesentery of the intestine. (A and B) Hermaphrodite, 54 cm long, two magnifications. In each figure the scale bar = 5 mm. (C) Adult female (58 cm long) with almost mature eggs, 22–24 mm in length. (D) Small female (43 cm long) apparently forming a first clutch of eggs; the largest egg is 9 mm in length. A, atretic follicle; int, intestine; M, mesovarium; O, vitellogenic egg; T, testis. (A and B) In this hermaphrodite the testis (T) occupies a normal posterior position and ovary forms the remainder of the gonad. At higher magnification the sperm follicles are visible in the testis. Between testis and ovary there is a clear interruption of germinal elements although the gonadal mesentery (M) connects the two. The number of large maturing eggs (14–18 mm in length) is unusually small (only nine, divided into two groups). The number of atretic egg follicles is unusually high. (C) In this mature female there are 20 vitellogenic eggs completing their development. Each large egg hangs by a tubular outpocketing of the mesovarium, attached at the more dorsal point from which the egg first developed. There are many small eggs developing at the free ventral edge of the mesovarium, none more than about 4 mm in length. There are no intermediate-sized eggs between 4 mm and the large maturing eggs. In the mesovarium more dorsal than the immature eggs are shadowy atretic follicular structures. (D) In this younger female the largest eggs are 9 mm long. One is shown at the right (O). The smallest oocytes are at the free ventral edge of the membranous ovary and larger ones progress dorsally. An atretic egg follicle (A) about 4 mm long already shows the outpocketing of mesovarium, as do several slightly smaller eggs to the left. Between these and the 9-mm eggs, there are no intermediate-sized eggs (from Gorbman, 1983).

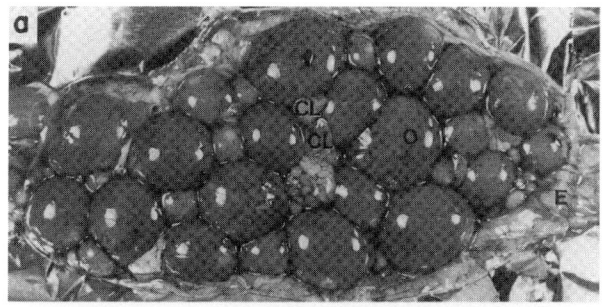

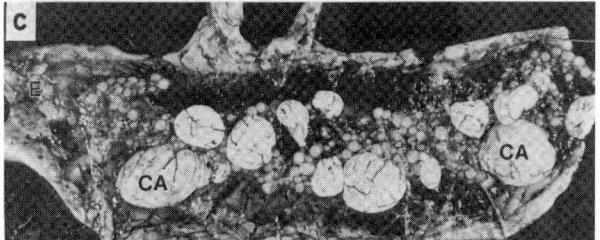
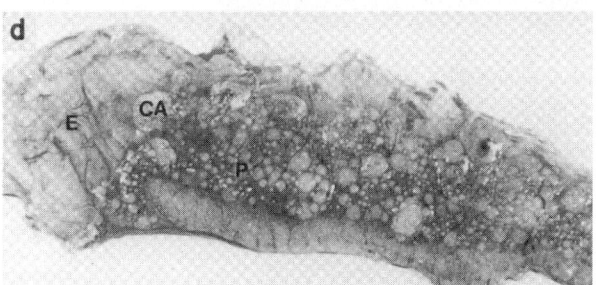

FIGURE 2 The ovary of *S. canicula*. (a) Surface view of partly dissected ovary. (b) Vitellogenic oocytes removed and placed peripherally to demonstrate paired size hierarchy. (c) Ovary after thyroidectomy in May. Note that all large follicles are atretic and none of the smaller follicles are undergoing vitellogenesis. (d) Ovary in late atresia, 11 months after removal of VL. CA, corpus atreticum; CL, corpus luteum; E, epigonal organ; O, large vitellogenic oocyte, ovulatable size: P, previtellogenic follicle (from Dodd, 1983).

oocytes at different stages (e.g., trout) (Fig. 3); and (iii) asynchronous—members of this group have multiple populations (clutches) of oocytes and often breed several times (e.g., goldfish) or continuously throughout the year as adults (e.g., zebra danio). The process of atresia or oocyte resorption may occur at any time during oogenesis in many teleost species. This process involves hypertrophy of the granulosa cells and a progressive deterioration of the oocyte; the purpose and regulation of this phenomenon is not well understood. Yamamoto *et al.* have described eight stages in trout oogenesis: The first three comprise the previtellogenic phase of oogenesis, whereas the next four stages make up the vitellogenic phase and the last stage is involved with final meiotic maturation and ovulation. These eight stages, which are distinguished by the size, amount, and distribution of cell inclusions, especially yolk, and by the morphology of chromosomes, will be described in detail.

A. Previtellogenic Phase

Primordial germ cells are found only in the cortex of the gonad of both sexes since the medulla, which predominates in males of many vertebrates, is absent. This has been suggested as the reason some teleosts are able to undergo sex reversal. In the ovary, oogonial proliferation occurs and the oogonial cells which become committed to meiosis are then surrounded by granulosa (follicle) cells, a basement membrane, a layer of thecal cells, and finally an ovarian epithelium. These layers comprise the follicle wall (Fig. 4) which is important in steroidogenesis.

Once the oogonial cells have become committed to meiosis, they enter the first of the eight stages of oogenesis described by Yamamoto *et al.* Stage 1, the chromatin–nucleolus stage, is characterized by the association of at least one conspicuous nucleolus with chromatin; it is during this stage that synapsis of homologous chromosomes occurs (zygotene of meiotic prophase I). Next, the oocyte enters the early perinucleolus stage (stage 2), which is characterized by enlargement of the nucleus and the formation of multiple nucleoli with the concomitant amplification of rRNA genes. In the late perinucleolus stage (stage 3), the nucleoli take up positions at the extreme margins of the nucleus, the oocyte enlarges, lamp-

FIGURE 3 A portion of the ovary of the brook trout, *Salvelinus fontinalis*. Note the fully grown oocytes with surrounding follicle wall containing blood vessels. Small follicles, representing future clutches, are also visible. Scale bar = 2 mm; LF, large follicle corresponding to a stage 7 oocyte; SF, small follicle corresponding to the previtellogenic phase. Blood vessels are visible as dark lines over the surface of large follicles.

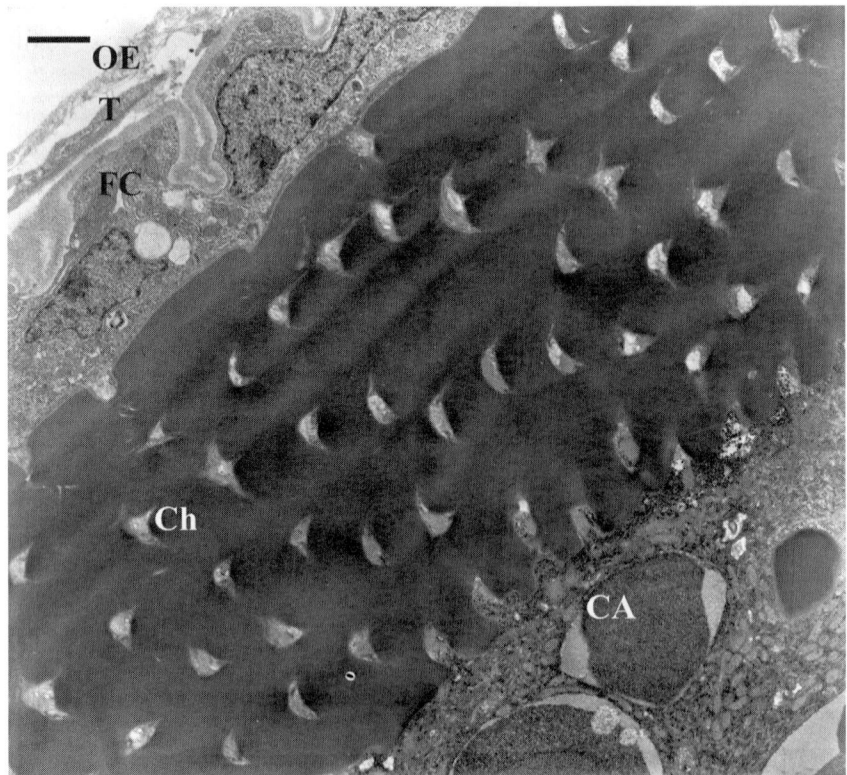

FIGURE 4 A transmission electron micrograph showing the follicle wall, chorion, and cortex of a fully grown oocyte of the goldfish, *Carassius auratus*. Note the thick chorion with its many perforations which allow microvilli from the follicle cells and oocyte to intertwine. Scale bar = 2 mm; OE, ovarian epithelium; T, theca; FC, follicle cells; Ch, chorion; CA, cortical alveoli.

brush chromosomes appear indicating an increase in RNA synthesis, and close to the nucleus or germinal vesicle (GV) the yolk nucleus or Balbiani body forms. This latter structure is composed of mitochondria, Golgi bodies, smooth endoplasmic reticulum, multivesicular bodies, annulate lamellae, and lipid. It is thought that the yolk nucleus acts as a site of organelle production to provide some of the needed cellular machinery for further oogenesis and to provide initial supplies of organelles for early embryonic development. At the end of this stage nucleolar extrusion from the nucleus begins and continues into the vitellogenic phase. Nucleoli are extruded through the nuclear envelope into the cytoplasm (ooplasm), where they disintegrate and the material is then distributed in the ooplasm. Presumably this process provides materials such as 28S and 18S rRNA for ribosome assembly.

The previtellogenic phases are reported to be independent of pituitary control as indicated by progression through these stages by oocytes in hypophysectomized females. In fact, the regulatory processes controlling progression through these stages are not well understood and require further research.

B. Vitellogenic Phase

In this phase of oogenesis, a remarkable increase in size of the oocyte occurs, primarily due to the elaboration and incorporation of yolk. Three major types of yolk are found in teleost oocytes: oil or lipid, cortical alveoli (yolk vesicles or primary yolk), and yolk globules. These three types, the production of which is dependent on pituitary function, also typify the next four stages in oogenesis according to Yamamoto *et al.* In the oil drop stage (stage 4), oil or lipid droplets appear apparently as elaborations of the yolk nucleus; the mitochondria, smooth endoplasmic reticulum, and Golgi bodies of the yolk nucleus are spatially associated with the lipid accumulations and thought to be responsible for their formation. Both phospholipids and neutral triglycerides are formed and make up the obvious oil drops in the ooplasm.

In the primary yolk stage (stage 5), acidic and/or neutral mucopolysaccharides or glycoproteins accumulate to form yolk vesicles or cortical alveoli (Fig. 4). These appear first in the mid- and outer cortical zones; later they will be positioned just below the oolemma or cytoplasmic membrane in the outermost portion of the cortex. The formation of these structures is thought to involve the rough endoplasmic reticulum, in which the protein core would be synthesized and then transported in vesicles to the Golgi bodies for addition of oligo- and polysaccharides. Lectins and membranous structures are also often included in the cortical alveoli interior. In hypophysectomized females, 17β-estradiol can promote cortical alveoli formation. The cortical alveoli are important in the fertilization reaction, during which their contents are exocytosed into the perivitelline space. There they hydrate, causing the space to enlarge, and the contents also modify and harden the chorion (Fig. 4), a tough protective coating around the egg.

The secondary yolk stage (stage 6) is typified by the incorporation of vitellogenin and the formation of protein yolk as membrane-bound yolk globules. Vitellogenin is produced by the liver under the influence of estrogens. Blood-borne vitellogenin is taken up via coated pits of the oolemma and endocytosed as coated vesicles which fuse together to form larger yolk globules. Pituitary extracts have been reported to enhance the micropinocytotic activity of the oolemma. At the electron microscope level, the yolk globules often have a crystalline core of lipovitellin and phosvitin surrounded by a heterogeneous layer of amorphous material. While it is generally accepted that the majority of protein yolk is produced by the liver, there are reports that the oocyte contributes some protein yolk components via endogenous yolk formation. This controversial process is thought to involve the rough endoplasmic reticulum and mitochondria.

In the tertiary yolk stage (stage 7), the yolk globules fuse to form a large homogeneous yolk mass, leaving a relatively thin layer of ooplasm at the periphery. This stage is absent in many teleosts, such as the goldfish fully grown oocyte, in which the yolk globules persist as discrete membrane-bound structures with ooplasm between them.

C. Meiotic Maturation and Ovulation

At this point (stage 8), the oocyte is fully grown and vitellogenin incorporation slows. Under the influence of progestogens, especially 17α,20β-dihydroxyprogesterone, the oocyte nucleus (GV), which is positioned at the center of the oocyte or part way toward the micropyle in the animal pole depending on the species, migrates to the micropyle. There, the GV disassembles its nuclear envelope [GV dissolution (GVD)] and the nucleoplasm mixes with the ooplasm. The cytoskeleton is thought to be involved in the positioning and migration of the GV. The first meiotic spindle is then assembled close to the micropyle and the first polar body is given off as the oocyte goes through metaphase I, anaphase I, and telophase I of meiosis. The oocyte then proceeds to metaphase II as the second meiotic spindle is formed after the reduction division. During this time, depending on the species, lipid droplets may fuse as do yolk globules. Often the oocyte swells through hydration and the ooplasm may become more translucent indicating changes in the yolk. Organelles such as the annulate lamellae disappear and are replaced by a tubular array of endoplasmic reticulum. Concomitant with these changes, the membrane potential (Vm) depolarizes. The oocyte, now a fertilizable egg, is ovulated under the influence of prostaglandins.

V. AMPHIBIAN OOGENESIS

Most amphibians are oviparous with external fertilization; however, some viviparous species exist in both anurans (frogs) and urodeles (salamanders). Since many amphibians must return to water to breed, rainfall is an important regulatory factor in oogenesis.

Oogenesis varies depending on the habitat and whether the species undergoes metamorphosis. Most species are either group synchronous or asynchronous. For example, the gray tree frog (*Hyla chrysoscelis*) has group-synchronous oogenesis and may have up to four clutches or cohorts of fully grown oocytes, which are ovulated about a month apart during the breeding season from May to August. While the leopard frog (*Rana pipiens*) has a single clutch per year, it is considered group synchronous since at least two additional cohorts are also present in the ovary and represent the future spawning of subsequent years (Fig. 5). The African clawed frog (*Xenopus laevis*) is a tropical species which has asynchronous oogenesis and oocytes at many different stages of oogenesis may be found in one individual. A detailed analysis at both the optical and elec-

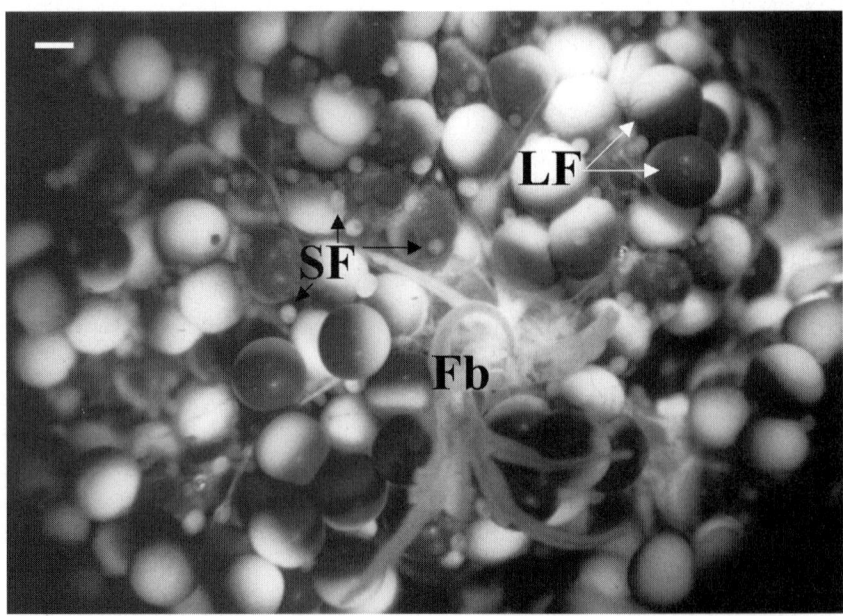

FIGURE 5 A portion of the ovary of the leopard frog, *Rana pipiens*. Scale bar = 1 mm; LF, large follicle corresponding to a stage VI oocyte; SF, small follicle corresponding to a previtellogenic oocyte; Fb, fat body, a lipid storage organ intimately associated with the ovary.

tron microscope level has described six stages of oogenesis in the well-studied African clawed frog, *X. laevis*; these stages are briefly described below.

A. Previtellogenic Phase

Oogenesis is initiated in the larval stage in those amphibians that exhibit metamorphosis. Generally, oogonial proliferation occurs both in the larval ovary and in the adult after spawning to replace the cohort ovulated during breeding. In some species, such as *R. pipiens* and *X. laevis*, intercellular bridges occur between oogonia in nested arrays. These bridges persist into meiotic pachytene and all members of a nest develop synchronously. By late pachytene, oocytes separate and begin asynchronous development. The previtellogenic phase is thought to be gonadotropin independent; however, other hormones such as growth hormone may be important.

Oogonia which have entered meiosis are considered stage I oocytes. They range from 50 to 300 mm in diameter and are transparent. The nucleus or GV occupies most of the cell. A juxtanuclear mass approximately 20–60 mm in size, analogous to the yolk nucleus and termed the mitochondrial mass or cloud in amphibians, is apparent. Surrounding each oocyte is an outer surface epithelium layer, also called the ovarian epithelium. Underlying this is the theca layer containing fibrocytes and blood vessels. Beneath this is the follicle cell layer, which is intimately associated with the oocyte oolemma. These three main layers make up the follicle wall, which is important in steroid production and is thought to contribute to development of the enclosed oocyte during oogenesis (Fig. 6). This is especially true of the follicle cell layer, which develops macrovilli that intertwine with microvilli from the late stage I oocyte. Gap junctions eventually form between these projections of the follicle cells and oocyte, allowing direct chemical communication between the two cell types. The GV

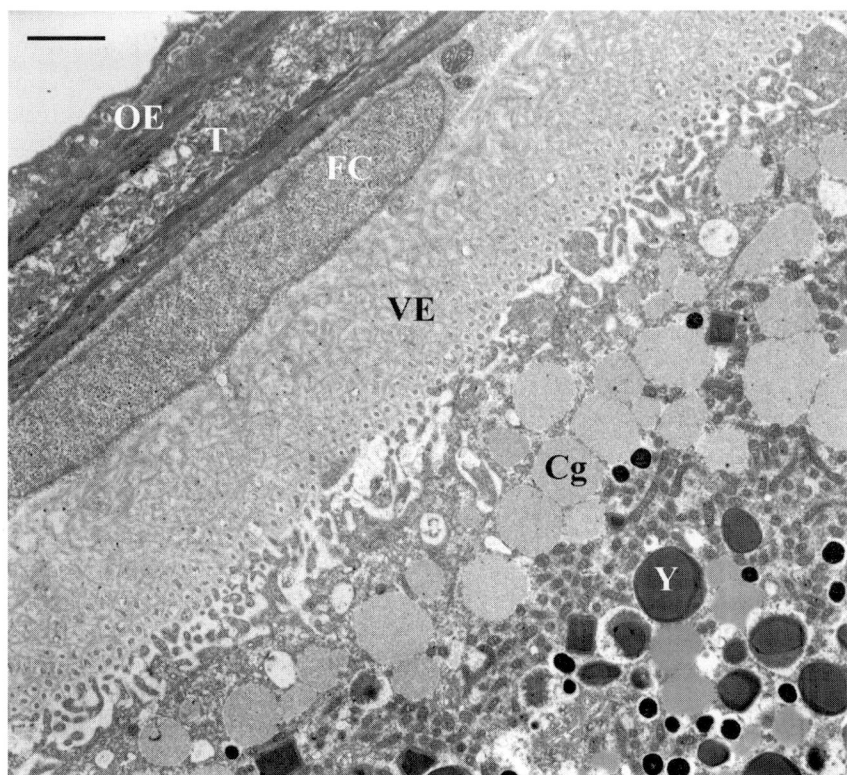

FIGURE 6 A transmission electron micrograph of a thin section through the follicle wall and cortex of a *Rana pipiens* stage VI follicle. Scale = 2 mm; OE, ovarian epithelium; T, theca; FC, follicle cell; VE, vitelline envelope; MV, microvilli; Cg, cortical granules; Y, yolk platelet.

contains multiple nucleoli which lie close to the nuclear envelope.

B. Vitellogenic Phase

The oocytes at stage II are white and opaque with a diameter ranging from 300 to 450 mm. The nucleoli increase in number and are located at the periphery of the GV. At the ultrastructural level, five cytoplasmic components predominate: cortical granules, mitochondria, small yolk platelets, lipid, and premelanosomes without melanin pigment. Components of the follicle wall, including the surface epithilium and theca layer, thicken, while a basement membrane forms between the theca and follicle cell layers. The vitelline envelope begins to form as patches below each follicle cell next to the oolemma.

Oocytes in stage III exhibit uniform brown pigmentation and range from 450 to 600 mm in diameter. Vitellogenin is actively taken up by micropinocytosis, and this has been confirmed by *in vitro* studies. Blood vessels in the theca enlarge. Early in this stage the lampbrush chromosomes are prominent, whereas later in this stage they retract. Most of the rRNA is synthesized at this point and the nucleoli develop vacuoles and migrate to the center of the GV. The mitochondrial mass disperses and the contents are distributed throughout the ooplasm. The cortical granules are located in the periphery of the cell with the melanosomes, now with melanin pigment, located just below. The primordial yolk globules or platelets fuse to form definitive yolk platelets with a distinctive crystalline pattern at the center.

Differentiation of the dark animal and light vegetal poles occurs at stage IV. The oocytes range from 600 to 1000 mm in diameter. The vegetal pole contains large yolk platelets, whereas the GV is offset toward the animal pole. The theca contains many large blood vessels and channels are apparent between follicle cells, presumably to facilitate vitellogenin uptake. In fact, this stage has the most intense vitellogenin uptake as assessed by *in vitro* assay.

At stage V, the oocyte is distinguished by a change in the animal pole pigment from dark brown to brown or beige and a distinct boundary is apparent between the two poles. The oocyte ranges from 1000 to 1200 mm in diameter. Yolk accumulation and micropinocytosis subside in this stage; thus, fewer primordial yolk platelets are apparent. The GV moves closer to the animal pole and becomes distinctly polarized with the vegetal (basal) side, which is notably convoluted. Small nucleoli and the condensed chromosomes form a mass at the GV interior.

C. Meiotic Maturation and Ovulation

Stage VI is the postvitellogenic phase, distinguished by a prominent unpigmented equatorial band separating the animal and vegetal poles. The oocyte ranges from 1200 to 1300 mm in diameter. Some nucleoli remain, especially near the highly folded basal side of the GV. It is at this stage that the oocyte becomes competent to respond to progesterone.

During progesterone-induced maturation a number of events occur, including movement of the GV toward the animal pole (GVM), GV dissolution or breakdown (GVD), and depolarization of the oolemma (V_m). GVM is thought to involve changes in the oocyte cytoskeleton, especially the microtubule system. GVD seems to involve phosphorylation of the nuclear lamins to elicit nuclear envelope vesiculation. Chromosomes condense further and become associated with a meiotic spindle, a product of the cytoskeletal system of the oocyte. The spindle associates with the oolemma at the animal pole and produces the first polar body as the cell goes through anaphase I and telophase I of the first meiotic division. The remaining oocyte chromosomes reassemble onto a second spindle and remain at the oolemma at metaphase II awaiting fertilization. Meanwhile, the oolemma depolarizes from approximately −60 to −15 mV, with a concomitant increase in membrane resistance. This seems to prepare the membrane to undergo the fertilization reaction or activation potential which acts as a fast block to polyspermy.

The overall control of maturation, which also involves cell division, has been reported in some detail. This is particularly significant since several of the regulatory factors have been reported to also control cell division in nonoocyte (somatic) cells. Progesterone is thought to elicit oocyte maturation by causing the translation of mRNA for c-Mos protein (pp39^{mos}), a cyclin protease inhibitor. Cyclin then accumulates in the oocyte and becomes associated with another protein called cyclin-dependent protein kinase. Together, they act as the maturation-promoting factor (MPF), which elicits the cascade of cellular events leading to the oocyte's progression from prophase I and eventually into the metaphase II state. The amphibian oocyte has played a prominent role in elucidating this important regulatory system; for example, MPF was first discovered as a biological activity in *R. pipiens* oocytes.

Ovulation involves the expulsion of the mature oocyte or egg from its follicle wall investments. This process has been compared to an inflammatory reaction in which proteases and prostaglandins are involved to weaken the follicle wall and cause contractile elements in the wall to contract, thereby releasing the gamete from the ovary.

VI. REPTILIAN OOGENESIS

Reptiles are a diverse group and include the crocodilians, turtles, lizards, and snakes. They inhabit an equally diverse array of environments; thus, their oogenesis and its regulation also varies. Species within this group may exhibit oviparity, ovoviparity, or vivaparity. Generally, breeding is seasonal and usually occurs once a year. Temperature, rainfall, and food availability may be important regulatory factors depending on the species.

A. Previtellogenic Phase

In snakes, primordial germ cells from the anterior portion of the extraembryonic area migrate to the presumptive gonad via a passive movement similar to that in birds. In lizards and turtles, the primordial germ cells actively migrate via ameboid motion from the posterior portion of the extraembryonic area. The primordial germ cells are distinguished by their lipid and yolk inclusions and by a juxtanuclear sphere containing mitochondria and Golgi bodies. Oogonial proliferation occurs in the embryonic ovary and in the adult female. Clusters of oogonia develop synchronously in nests called germinal beds. It is thought that the number of germinal beds correlates with the size of the egg clutch produced. The oogonia are connected by intercellular bridges which contain microfilament bundles. Gonadotropins are reported to increase oogonia through either increased mitosis or reduction in oogonial atresia.

Oogonia which enter meiosis become primary oocytes and follow the general meiotic pattern. Ovaries of adult turtles, lizards, and snakes contain oocytes in all stages of meiotic prophase. Embryonic ovaries of lizards contain oocytes in the diplotene stage of prophase I; thus, this group may have rather advanced ovarian development even in the late embryo. Synaptonemal complexes, indicating homologous chromosome pairing, are evident in zygotene of prophase I. By pachytene, the chromosomes form a characteristic bouquet formation. In both the pachytene and zygotene oocytes, mitochondria and Golgi bodies lie in the juxtanuclear zone which includes the protein- and RNA-rich yolk nucleus.

By early diplotene, the oocytes begin asynchronous development. This is probably due to the loss of intercellular bridges between oocytes as the follicle wall forms around each oocyte, effectively isolating it from the others. At this stage the large nucleolus vacuolates and breaks into several portions; however, they remain scattered within the nucleoplasm rather than taking peripheral locations in the GV as occurs in many other vertebrate oocytes. Lampbrush chromosomes are prominent in early diplotene, but by late diplotene the chromosomes condense into rod-shaped bodies and form a central mass in the GV. Two major layers of the follicle wall invest individual oocytes at diplotene. The outermost of these, the theca, is subdivided into two layers: the theca externa, which is rich in fibroblasts, and the theca interna, which has many blood vessels. The other major layer is the granulosa, which underlies the theca. Initially, the granulosa is composed of a single layer of uniform small cells. This condition persists in the crocodilians and turtles, producing a simple epithelium with a single cell type throughout the rest of oogenesis. However, in the lizards and snakes, another cell type emerges called the intermediate cell, which then gives rise to pyriform cells. Thus, at one point the granulosa of this group contains three distinct cells types. Later, the pyriform cells disintegrate, leaving a simple epithelium once again. The function of the pyriform cells has not been definitively elucidated. Microvilli form from both the granulosa cells and the oocyte. In this area of interdigitating microvilli, a perivitelline space appears which becomes filled with an amorphous material forming the zona pellucida.

B. Vitellogenic Phase

Most of the information on this phase of oogenesis has been derived from studies of lizards. Typical of the other vertebrates covered in this article, yolk precursors are formed in the liver under the influence of gonadotropins and estrogens. The yolk precursors, which are released from the liver into the blood, are carried to the blood vessels of the theca interna. The yolk precursors leave these blood vessels and infiltrate spaces between granulosa cells, cross the zona pellucida, and are taken up by oocyte pinocytosis. Some reptile oocytes become very large and contain massive amounts of yolk similar to eggs of many bird species.

See Also the Following Articles

AMPHIBIAN REPRODUCTION, OVERVIEW; FISH, MODES OF REPRODUCTION IN; OOGENESIS, IN MAMMALS; REPTILIAN REPRODUCTION, OVERVIEW

Bibliography

Daniel, J. F. (1934). *The Elasmobranch Fishes*, pp. 303–306. Univ. of California Press, Berkeley.

Dodd, J. M. (1983). Reproduction in cartilaginous fishes (Chondrichthyes). In *Fish Physiology* (W. S. Hoar, D. J. Randall, and E. M. Donaldson, Eds.), Vol. IX, pp. 31–95. Academic Press, New York.

Dumont, J. N. (1972). Oogenesis in *Xenopus laevis* (Daudin) I. Stages of oocyte development in laboratory maintained animals. *J. Morphol.* **136**, 153–180.

Gorbman, A. (1983). Reproduction in the cyclostome fishes and its regulation. In *Fish Physiology* (W. S. Hoar, D. J. Randall, and E. M. Donaldson, Eds.), Vol. IX, pp. 1–29. Academic Press, New York.

Guraya, S. S. (1986). *The Cell and Molecular Biology of Fish Oogenesis*. Karger, Basel.

Hardisty, M. W. (1971). Gonadogenesis, sex determination and gametogenesis. In *The Biology of Lampreys* (M. J. Hardisty and I. C. Potter, Eds.), Vol. 1, pp. 295–400. Academic Press, New York.

Hubert, J. (1985). Origin and development of oocytes. In *Biology of the Reptilia* (C. Gans, F. Billet, and P. F. A. Maderson, Eds.), Vol. 14, pp. 41–74. Wiley, New York.

Nagahama, Y. (1983). The functional morphology of teleost gonads. In *Fish Physiology* (W. S. Hoar, D. J. Randall, and E. M. Donaldson, Eds.), Vol. IX, pp. 223–275. Academic Press, New York.

Yamamoto, K., Oota, I., Takano, K., and Ishikawa, T. (1965). Studies on the maturing process of the rainbow trout, *Salmo gairdneri irideus*. 1. Maturation of the ovary of a one-year old fish. *Bull. Jpn. Soc. Sci. Fish.* **31**, 123–132.

Oogonia

see Oogenesis

Oostatins, Folliculostatins, and Antigonadotropins, Insects

T. S. Adams

U.S. Department of Agriculture Biosciences Research Laboratory

I. Introduction
II. Pathways for Oostatic Activity
III. Gonotropic Cycle Types
IV. Regulation of Follicle Cell Patency
V. Regulation of Vitellogenin Production
V. Species Cross-Reactivity of Oostatins

GLOSSARY

ecdysteroid A generic term for a series of related steroid hormones that induce molting and in some insects are involved in reproduction.

juvenile hormone A group of related sesquiterpenoids involved in metamorphosis and reproduction.

patency The development of spaces between the follicular epithelial cells surrounding the oocyte.

vitellin Yolk protein granules found in the oocyte.

vitellogenin Yolk proteins that are secreted into the hemolymph by the fat body or follicle cells.

Oostatins, folliculostatins, and antigonadotropins are naturally occurring materials that inhibit vitellogenic development in oocytes and will be referred to hereafter as oostatins. The apparent role of oostatins is to regulate egg development cycles. When eggs are produced in cycles, only the ultimate cycle develops and oogenesis in the penultimate cycle is arrested. Insects inhibit oogenesis through two different pathways. In several species, the oostatin inhibits follicle cell patency, whereas in others, vitellogenin production is inhibited. Thus far, five unique peptides with oostatic activity have been identified. Oostatins found in different species may be produced by the ovaries, abdominal neurosecretory organs, or the central nervous system.

I. INTRODUCTION

Insects contain either panoistic or meroistic ovaries. Panoistic ovaries do not contain trophocytes and are found in the more primitive insects such as the silverfish, *Lepisma saccharina* (Thysanura). Meroistic ovaries have trophocytes that supply the oocytes with RNA and are of two types. Meroistic polytrophic ovaries have trophocyte–oocyte cysts, whereas meroistic telotrophic ovaries have oocytes that are connected to the trophocytes with trophic cords. All these ovarian types have oocytes or an oocyte–trophocyte cyst surrounded by follicular epithelial cells that form a barrier between the oocyte and hemolymph. Except for some hymenopteran parasitoids, insects contain yolky eggs. Females expend a significant amount of energy to synthesize vitellogenin, which is then taken up by the oocyte to form vitellin. Depending on the insect, about 40–90% of the total soluble protein in an insect egg is vitellin. The native insect vitellogenic protein is a glycolipoprotein with a species-dependent molecular mass ranging from 160 to 650 kDa. Vitellogenin is synthesized by the fat body and/or ovarian follicle cells and is released into the hemolymph. Vitellogenin synthesis is regulated by either juvenile hormone (JH) or ecdysteroid and depending on the insect may occur before or after eclosion. When hormones are involved, insects with preeclosion vitellogenesis generally regulate vitellogenin synthesis with ecdysteroids, whereas those with posteclosion vitellogenesis use either juvenile hormone or ecdysteroid.

After vitellogenin contacts the oocyte plasma membrane it binds with its receptor and is taken into the oocyte by a process termed receptor-mediated endocytosis. However, before the vitellogenin circulating in the hemolymph gains access to the oocyte plasma membrane it must first pass through the follicle cell barrier separating the oocyte from the hemolymph. Follicle cell patency is the process that allows the large vitellogenic proteins to pass through the interfollicular spaces and contact the oocyte plasma membrane.

Naturally occurring materials that inhibit insect ovarian maturation have been referred to in the literature as oostatic hormones (*Musca* and *Aedes*), folliculostatins (*Neobellaria*), and antigonadotropins (*Rhodnius*). These materials regulate oogeneic cycles within the ovarioles. Oostatins have been extracted from ovaries of *Musca domestica*, *Aedes atropalpus*, *Aedes aegypti*, and *Neobellieria bullata*; from thoracic ganglia of *Musca* and *Locusta migratoria*; from abdominal neurosecretory organs in *Rhodnius prolixus*; and from the corpus cardiacum and pars intercerebralis of *Locusta*.

II. PATHWAYS FOR OOSTATIC ACTIVITY

Oostatins may inhibit oogenesis through different pathways involving vitellogenin synthesis or uptake (Fig. 1). Follicle cell patency could be inhibited by an oostatin, OS_p, that blocks the synthesis of the hormone that induces patency pathway 1 or may act directly on the follicle cell to prevent patency pathway 2. In another scenario the oostatin, Os_r, may compete with vitellogenin for binding sites or inhibit receptor synthesis. Finally, an oostatin, OS_{vg}, could inhibit vitellogenin synthesis. This could be accomplished by (I) inhibiting the digestion of protein so that amino acids required for vitellogenin synthesis are not available, (ii) inhibiting the synthesis or release of the endocrines required for the induction of vitellogenin synthesis, or (iii) inhibiting vitellogenin synthesis directly. Thus far, oostatins acting through

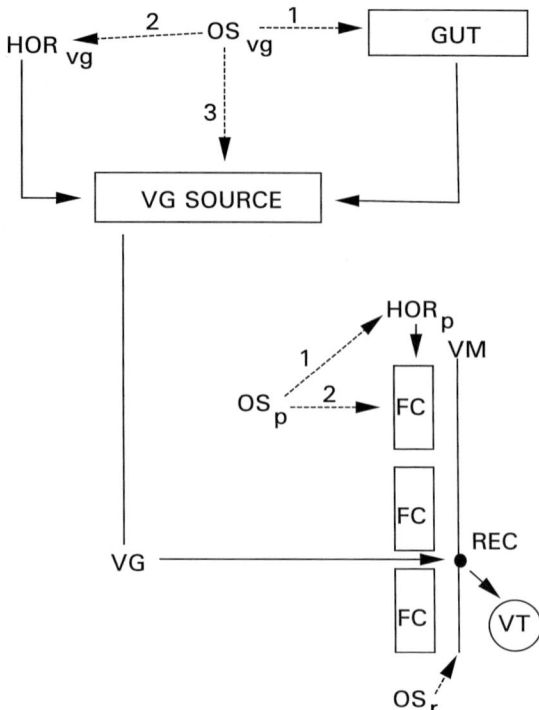

FIGURE 1 The three major pathways through which an oostatin may inhibit ovarian maturation: follicle cell patency (OS_p), vitellogenin synthesis (OS_{vg}), or the vitellogenin receptor (Os_r). The numbers indicate the level in each pathway at which the oostatins may act. FC, ovarian follicular epithelium; HOR_p, a hormone such as JH that induces follicle cell patency; HOR_{vg}, a hormone such as JH or ecdysone that regulates vitellogenin synthesis; Os_r, an oostatin that interferes with the vitellogenin receptor; OS_{vg}, an oostatin that inhibits vitellogenin production; REC, receptor; VG, vitellogenin; VG SOURCE, the biosynthetic tissue for vitellogenin such as the fat body or ovarian follicle cells; VM, vitelline membrane; VT, vitellin.

pathways Os_{vg} 1 and 2 and Os_p 2 have been shown or proposed for several insects.

III. GONOTROPIC CYCLE TYPES

Ovarian follicle development within the ovarioles follows a species-specific pattern that is either asynchronous or synchronous. Ovarioles with asynchronous follicle development have a sequence of follicles from different gonotropic cycles that are all vitellogenic. Some examples of insects with asynchronous development are the large milkweed bug, *Oncopeltus fasciatus* (Hemiptera), the boll weevil, *Anthonomus grandis* (Coleoptera), and the bollworm, *Heliothis zea* (Lepidoptera). Ovaries with asynchronous development produce eggs continuously and oostatins would not be expected. Two types of synchronous follicle development are known. Insects with interovariole synchrony show synchronized follicle development in all ovarioles within both ovaries. The ultimate cycle develops to maturity while the penultimate cycle ceases development at early vitellogenesis or previtellogenesis until the ultimate cycle of eggs is oviposited. Egg production in this instance is cyclical. Some insects with interovariole synchrony are *L. migratoria* (Orthopterta), the yellow fever mosquito, *A. aegypti*, the housefly, *Musca domestica*, and the gray fleshfly, *N. bullata*. Intraovariole synchrony is characterized by synchronous development within an individual ovariole, but follicle development between ovarioles is not synchronized. In this case the follicle in the ultimate cycle in an ovariole must also be oviposited before the penultimate cycle can complete oogenesis. Egg production within ovaries would be acyclical if the insects oviposited continuously. However, if the insect did not oviposit one would expect the ovary to show interovariole synchrony. *Rhodnius prolixus* (Hemiptera) and *Drosophila melanogaster* (Diptera) are representative insects with intraovariole synchrony. Either type of synchronous follicle maturation requires a regulating mechanism. In some insects, an oostatin is the regulator that prevents the penultimate cycle of follicles from developing, whereas the ultimate cycle of follicles is still vitellogenic. Hypothesis concerning the mechanism by which oostatins inhibit oogenesis can be based on the type of ovariole synchrony present and the vitellogenin levels found in a specific insect (Fig. 2). Insects such as *Rhodnius*, with intraovariole synchrony (Fig. 2A), have vitellogenin present continuously and the ovary always contains ovarioles with vitellogenic follicles. If mated and a substrate is available, these insects will oviposit eggs as soon as they are formed in the ovariole. This suggests that the penultimate cycle does not take up vitellogenin because either the follicle cells are not patent or the vitellogenin receptor is unavailable. Insects with interovariole synchrony have distinct peaks of vitellogenin in the hemolymph (Figs. 2B and 2C) and oostatins could have a role in maintaining the cyclical levels of vitellogenin. If the penultimate cycle is negatively correlated with vitellogenin concentration, as in *A. aegypti* (Fig. 2B), the oocyte in the penultimate

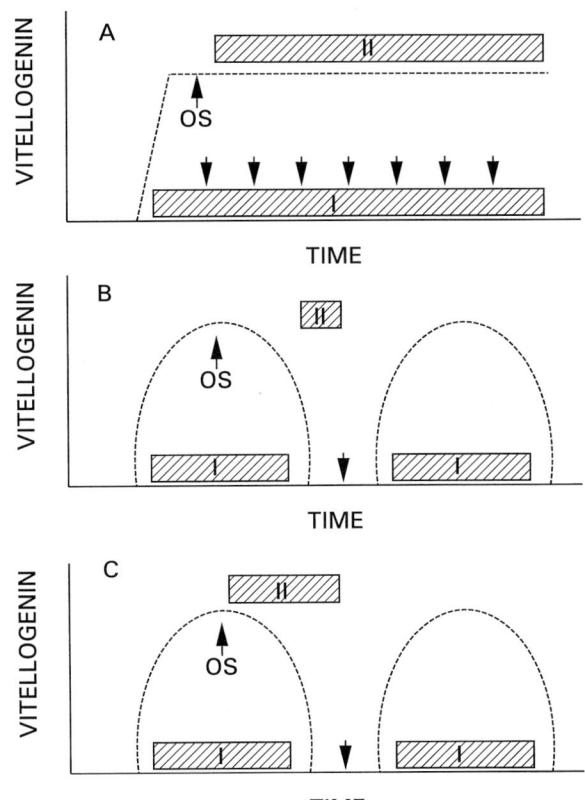

FIGURE 2 Correlation of ovarian development in the first and second gonotropic cycles with vitellogenin levels and purported time of oostatin release. I, developing follicles in the ultimate cycle; II, follicles with arrested development in the penultimate cycle; OS, oostatin release; dashed line, vitellogenin levels. Down arrows indicate oviposition. (A) The oostatin in an insect with intraovariole synchrony, such as *R. prolixus*, would prevent vitellogenin uptake by follicles in II but would have no effect on vitellogenin production. (B) An insect such as *A. aegypti* has interovariole synchrony in which the penultimate cycle is negatively correlated with vitellogenin levels. Decreasing vitellogenin levels could be attributed to the oostatin, but there would be no effect on vitellogenin uptake by follicles in II. (C) An insect such as *M. domestica* has interovariole synchrony in which the penultimate cycle is positively correlated with vitellogenin levels. Both vitellogenin levels and uptake by follicles in II could be inhibited by the oostatin.

cycle is in the resting phase and vitellogenin is absent until the ultimate cycle of eggs is oviposited and the female takes a blood meal. These conditions suggest that the oostatin only regulates vitellogenin synthesis. Insects such as *M. domestica* have the penultimate follicles positively correlated with vitellogenin levels but do not take up vitellogenin when it is present in the hemolymph (Fig. 2C). The penultimate cycle starts to develop after oviposition of the first cycle and the female has fed on protein. In this case, the oostatin could inhibit patency, receptor availability, or vitellogenin synthesis.

IV. REGULATION OF FOLLICLE CELL PATENCY

Rhodnius contains telotrophic ovaries with intraovariole synchrony. Both vitellogenin synthesis and vitellogenin uptake involve JH I. The oostatin, produced by four pairs of abdominal neurosecretory organs which stretch between the dorsal tergites and ventral sternites on segments II–V, inhibit follicle cell patency. Patency requires two exposures to JH I. The first exposure stimulates ATPase production and the second activates ATPase. ATPase is activated in the presence of protein kinase C when a 100-kDa polypeptide α subunit is phosphorylated after JH I binds to its receptor. The activated ATPase in the follicle cell membrane causes fluid loss and interfollicular spaces develop allowing vitellogenin to contact the oocyte plasma membrane. Extracts prepared from the abdominal neurosecretory organ reduced the JH I-induced patency, thereby inhibiting vitellogenin uptake. *Rhodnius* oostatin is proteinaceous in nature with a molecular mass of 1.4 kDa, but it has not been characterized further due to the difficulty in obtaining sufficient material for extraction. *Locusta migratoria* has ovaries with interovariole synchrony and uses JH III for both vitellogenin synthesis and uptake. JH III induces follicle cell patency through the action of a Na^+,K^+-ATPase as in *Rhodnius*. Both *Rhodnius* oostatin and *Locusta* thoracic gland extract inhibit follicle cell patency in *L. migratoria*. Since *Rhodnius* antigonadotropin inhibits patency in systems that use JH I or JH III, but does not inhibit other JH-regulated processes, the antigonadotropin

TABLE 1
Structure of Insect Oostatins

Aea-TMOF (Borovsky *et al.*, 1993)
    Tyr Asp Pro Ala Pro Pro Pro Pro Pro Pro
Neb-TMOF (Bylemans *et al.*, 1994a)
    Asn Pro Thr Asn Leu His
Neb-coloostatin (Bylemans *et al.*, 1994b)
    Ser Ile Val Pro Leu Gly Leu Pro Val Pro Ile Gly Pro Ile Val Val Gly Pro Arg
Neoparsin A (Girardie *et al.*, 1990)[a]
    Asn Pro Ile Ser Arg ↕ Ser Cys Glu Gly Ala Asn Cys Val Val Asp Leu Thr Arg Cys Glu Tyr Gly Asp Val Thr Asp Phe Phe Gly Arg Lys Val Cys Ala Lys Gly Pro Gly Asp Lys Cys Gly Gly Pro Tyr Glu Leu His Gly Lys Cys Gly Val Gly Met Asp Cys Arg Cys Gly Leu Cys Ser Gly Cys Ser Leu His Asn Leu Gln Cys Phe Phe Phe Glu Gly Gly Leu Pro Ser Ser Cys

[a] Cleavage of Neuroparsin A at the double arrow produces Neuroparsin B.

probably acts at a level not involving the JH receptor. Other materials with antigonadotropic activity in the locust are neuroparsin A and neuroparsin B (Table 1). When these materials were injected, ovarian maturation stopped at midvitellogenesis, but no other JH-dependent systems were affected. However, the level at which neuroparsin inhibits oogenesis is not known.

Abdomens of the gray fleshfly, *N. bullata*, produce a 19-amino acid hydrophobic peptide, Neb-colloostatin (Table 1), that blocks vitellogenin uptake in previtellogenic ovarian follicles but not in developing, vitellogenic follicles. *Musca* might also have a similar oostatin because the penultimate follicles do not develop past early vitellogenesis even though vitellogenin is present.

V. REGULATION OF VITELLOGENIN PRODUCTION

A. Digestion of Protein Meal

The mosquito, *A. aegypti*, produces eggs in response to a blood meal in a process requiring both JH and ecdysteroid. *Aedes aegypti* oostatin is a decapeptide (Table 1) produced by the ovarian follicle cells and is released into the hemolymph 24–48 hr after the blood meal. After binding to receptors on the hemolymph side of the middle portion of the midgut, *Aedes* oostatin inhibits midgut trypsin-like enzyme activity, thereby stopping digestion of the blood meal. To reflect its mode of action, the oostatin is called trypsin-modulating oostatic factor (Aea-TMOF). When digestion of the blood meal is inhibited, vitellogenin synthesis stops, with a concurrent drop in vitellogenin levels. Without vitellogenin, ovarian vitellogeninc development is not possible and this may partially explain the cessation of penultimate ovarian follicle development at previtellogenesis and the cycling of vitellogenin levels.

Midgut trypsin activity is also correlated with oogenesis in the gray fleshfly, *N. bullata*, and reaches a peak 6 hr after feeding on protein and then declines. *Neobellieria* ovaries produce a hexapeptide with both oostatic- and trypsin-inhibiting activity, Neb-TMOF (Table 1). Neb-TMOF arises from a 75-kDa precursor protein associated with the cortical layer in vitellogenic oocytes. The 75-kDa protein is found only in the ovary, and it was proposed that degradation of vitelline membrane proteins during remodeling of the oocyte during follicular growth is the source of Neb-TMOF. Furthermore, Aea-TMOF may also be produced in a similar fashion. A region at the N-terminal end of the 15α-2 vitelline membrane gene of *A. aegypti* encodes an amino acid sequence that differs from that of the Aea-TMOF decapeptide only by the order of the first two amino acids, suggesting that Aea-TMOF is cleaved from the translation product of the 15α-2 gene.

B. Ovarian Ecdysteroid Production

Since ecdysteroid is required for vitellogenin synthesis in many Diptera, inhibiting ecdysteroid production may be another means of stopping vitellogenin synthesis. The insect ovary is a primary source of ecdysone in *A. aegypti*, *D. melanogaster*, *M. domestica*, and other insects. Egg development neurosecretory hormone (EDNH), an ecdysteroidogenin, stimulates JH-activated dipteran ovaries to secrete ecdysteroids. EDNH is produced in the brain and is released into the hemolymph from the corpus cardiacum in response to protein feeding. Ecdysteroid synthesis may be blocked either indirectly through the ecdysteroidogenin pathway or directly through the steroid biosynthetic pathway.

Musca oostatin inhibited ovarian ecdysteroid production *in vitro* in ovaries that were EDNH activated *in vivo* or were incubated with EDNH *in vitro*. Ecdysteroid production resumed when the oostatin was removed from the incubation medium and the ovaries were incubated with EDNH alone. This suggests that *Musca* oostatin does not block the EDNH induction of ecdysone biosynthesis but instead has a direct effect on the ecdysteroid biosynthetic mechanism. This could explain the decreasing ecdysteroid levels and vitellogenin levels found after midvitellogenesis. The penultimate ovarian follicles cease growing because vitellogenin levels are low. Oviposition of the ultimate cycle of eggs triggers a mechanism that causes flies to feed selectively on a diet that contains protein, which then reactivates the entire cycle of EDNH release, ecdysteroid synthesis, and vitellogenin synthesis.

VI. SPECIES CROSS-REACTIVITY OF OOSTATINS

Species cross-reactivity of dipteran oostatic factors has been reported. Aea-TMOF decreased yolk deposition in *Musca*. *Musca* ovary extracts inhibited ovarian development in *A. atropalpus* and *D. melanogaster*. *Drosophila melanogaster* ovary extracts had oostatic activity in *A. atropalpus*, *Anopheles albimanus*, and *Culex quinquefasciatus*. There is low cross-reactivity between Aea-TMOF and Neb-TMOF, even though Neb-TMOF has no sequence homology with Aea-TMOF (Table 1). The recognition of *Musca* oostatin by Aea-TMOF antibody in dot-ELISA tests and the oostatic activity of Aea-TMOF in *Musca* bioassays suggests a similarity between *Aedes* TMOF and *Musca* oostatin. Also, injection of *Musca* oostatin into *A. atropalpus* inhibited ecdysteroid production *in vivo*.

See Also the Following Articles

Juvenile Hormone; Rhodnius prolixus; Vitellogenins and Vitellogenesis; Yolk Proteins, Invertebrates

Bibliography

Adams, T. S. (1997). 9-Insecta. In *Progress in Reproductive Endocrinology* (T. S. Adams, Ed.), Vol. 8, Reproductive Biology of the Invertebrates, pp. 277–338. Wiley, New York.

Borovsky, D., Carlson, D. A., Griffin, P. R., Shabanowitz, J., and Hunt, F. (1993). Mass spectrometry and characterization of *Aedes aegypti* trypsin modulating oostatic factor (TMOF) and its analogs. *Insect Biochem. Mol. Biol.* 23, 703–712.

Borovsky, D., Powell, C. A., Nayar, J. K., Blalock, J. E., and Hayes, T. K. (1994a). Characterization and location of mosquito-gut receptors for trypsin modulating oostatic factor using a complementary peptide and immunocytochemistry. *FASEB J.* 8, 350–355.

Borovsky, D., Song, Q., Ma, M. C., and Carlson, D. A. (1994b). Biosynthesis, secretion and immunocytochemistry of trypsin modulating oostatic factor of *Aedes aegypti*. *Arch. Insect Biochem. Physiol.* 27, 27–38.

Bylemans, D., Borovsky, D., Hunt, D. F., Shabanowitz, J., Grauwels, L., and De Loof, A. (1994). Sequencing and characterization of trypsin modulating oostatic factor (TMOF) from the ovaries of the grey fleshfly, *Neobellieria (Sarcophaga) bullata*. *Regul, Peptides* 50, 61–74.

Bylemans, D., Proost, P., Samyn, B., Borovsky, D., Grauwels, L., Huybrechts, R., Van Damme, J., Van Beeumen, J., and De Loof, A. (1995). Neb-colloostatin, a second folliculostatin of the grey fleshfly, *Neobellieria bullata*. *Eur. J. Biochem.* 228, 45–49.

Davey, K. G. (1996). Hormonal control of the follicular epithelium during vitellogenin uptake. *Invertebr. Reprod. Dev.* 30, 249–254.

Davey, K. G., Sevala, V. L., and Gordon, D. R. B. (1993). The action of juvenile hormone and antigonadotropin on the follicle cells of *Locusta migratoria*. *Invertebr. Reprod. Dev.* 24, 39–46.

Girardie, J., Huet, J.-C., and Pernollet, J.-C. (1990). The locust neuroparsin A: Sequence and similarities with vertebrate and insect polypeptide hormones. *Insect Biochem.* 20, 659–666.

Hagedorn, H. H. (1985). The role of ecdysteroids in adult insects. In *Comprehensive Insect Physiology, Biochemistry, and Pharmacology,* (G. A. Kerkut and L. I. Gilbert, Eds.), Vol. 8, Endocrinology 2, pp. 205–262. Pergammon, New York.

Huebner, E. (1983). Oostatic hormones—Antigonadotropin and reproduction. In *Endocrinology of Insects* (R. G. H. Downer and H. Laufer, Eds.), pp. 319–329. A. R. Liss, New York.

Li, Q.-J., Zheng, W.-H., Gong, H., and Adams, T. S. (1996). Oostatic activity in extracts prepared from thoracic ganglia and ovaries of the housefly, *Musca domestica vicina* (Arthropoda: Insecta: Diptera). *Invertebr. Reprod. Dev.* 29, 249–255.

Lin, Y., Hamblin, M. T., Edwards, M. J., Barillas-Mury, C., Kanost, M. R., Knipple, C., Wolfner, M. F., and Hagedorn, H. H. (1993). Structure, expression, and hormonal control of genes from the mosquito, *Aedes aegypti*, which encode proteins similar to the vitellin membrane protein of *Drosophila melanogaster*. *Dev. Biol.* 155, 558–568.

Raikhel, A. S., and Dhadialla, T. S. (1992). Accumulation of yolk proteins in insect oocytes. *Annu. Rev. Entomol.* 37, 217–251.

Opossums

John D. Harder and Leslie M. Jackson

The Ohio State University

I. Introduction
II. Reproductive Anatomy and Sperm Maturation
III. Reproductive Cycles, Ovulation, and Fertilization
IV. Endocrinology of the Estrous Cycle and Gestation
V. Gestation, Development, and Placentation
VI. Birth and Lactation

GLOSSARY

choriovitelline placenta The placental type found in most marsupial taxa; formed from two embryonic membranes, the chorion and the vitelline membrane, or yolk sac.

pheromone A chemical released by one individual that elicits a behavioral or physiological response in another of the same species.

urogenital sinus A common chamber and external opening for the urinary and reproductive tracts.

vomeronasal organ A bilateral, tubular structure dorsal to the anterior palate that contains chemoreceptors for the accessory olfactory system.

The common name "opossum" refers to New World marsupials, members of the family Didelphidae, order Didelphimorphia. Possum is colloquial for American opossums but is the proper common name for opossum-like marsupials of Australia. Didelphidae contains a large number of typically small (<100 g), pouchless opossums with diverse expressions of the basic marsupial mode of reproduction; that is, they give birth to small, fetus-like neonates following a brief pregnancy and then nurse their young for an extended period of time.

I. INTRODUCTION

Marsupials originated in North America about 100 million years ago but ultimately experienced greater diversification in the Neotropical and Australian regions. The oldest fossil marsupials are didelphids, similar in dental morphology to neotropical opossums of today. Therefore, the reproductive patterns seen in Didelphidae can be considered primitive in some features but derived in others, particularly in light of the adaptive radiation among the 63 species in this, the largest of all marsupial families.

The major differences between the mammalian infraclasses Metatheria (marsupials) and Eutheria (all other viviparous mammals) are reproductive, reflecting evolution, since the Cretaceous, of alternative strategies for gestation and lactation. This dichotomy is most apparent in the relative brevity of gestation and the small size and near fetal developmental state of marsupial neonates. This, coupled with a low metabolic rate (about 30% lower than that in eutherians), extends the period of lactation several times longer than that required for eutherians of similar body size. Thus, the marsupial mode of reproduction is characterized by a slow neonatal growth rate and low maternal energy investment per unit time during lactation.

Basic reproductive information (e.g., litters per year and duration of gestation and lactation) is known for fewer than 10 of the 63 species of Didelphidae. However, 3 of the best known species differ markedly in body size, geographic distribution, and ecology, and, therefore, their reproductive patterns and adaptations offer valuable insights into the extent of didelphid adaptive radiation. The Virginia opossum (*Didelphis*), the only marsupial in North America, represents a group of 9 large opossums (Fig. 1). They tend to be more terrestrial, mobile, and omnivorous than smaller species. Seasonal breeding in *Didelphis* is controlled largely by energetics and females invest in a single, large litter when opportunities for second or third litters are reduced

TABLE I
Comparison of Natural History and Reproductive Patterns in Three Representative Species of the Family Didelphidae

Natural history and reproduction	Species			Reference
	Didelphis virginiana	Caluromys philander	Monodelphis domestica	
Representative group	Large opossums	Medium opossums	Small opossums	Emmons (1990), Gardner (1993), Eisenberg (1989)
No. species	9	5	49	
Body size	0.4–3 kg	150–400 g	20–150 g	
Distribution	Northern United States to Mexico	Northeastern South America	Southeastern Brazil	Emmons (1990), Gardner (1997)
Locomotion and habitat	Terrestrial and nomadic; wooded areas interspersed with farmland or suburbia	Arboreal; rain forest canopy	Semiarboreal; semiarid scrub vegetation with mesic rock outcroppings	Emmons (1990), Eisenberg (1989), Streilein (1982)
Diet	Insects, carrion, and fruits; highly omnivorous	Fruit, nectar, gum, and invertebrates	Fruits, insects, and small vertebrates	Emmons (1990), Eisenberg (1989)
Adult body size	1–3 kg	300 g	60–130 g	
Breeding season and ecology	February–July, north, January–September, south; influenced by temperature and food availability	Year-round; influenced by precipitation and food availability in local populations	Year-round; influenced by precipitation and microhabitat	Streilein (1982), Hartman (1928), Perret and Atramentowicz (1989)
Length of estrous cycle (luteal phase)	28 days (11 days)	38 days (20 days)	No estrous cycle; induced estrus (15 days)	Hartman (1928), Perrett and Atramentowicz (1989), Harder and Fleming (1981), Hinds et al. (1992), Fadem (1985)
Length of gestation	13 days	24 days	15 days	Hartman (1928), Harder et al. (1993), Perret and M'Barek (1991)
Length of lactation	100 days	120 days	56 days	Hartman (1928), Perret and Atramentowicz (1989), Harder et al. (1993)
Average litter size	6 in south 9 in north	4	7	Hartman (1928), Perret and Atramentowicz (1989), Harder et al. (1993)
Marsupium	Well-developed	Develops during lactation	Absent	Hartman (1928), Perret and Atramentowicz (1989), Harder et al. (1993)

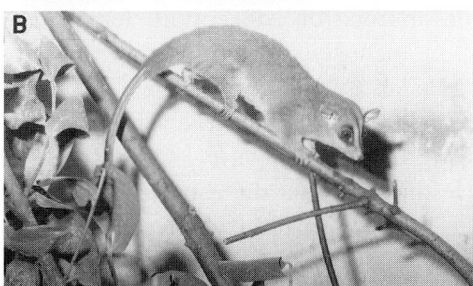

FIGURE 1 *Didelphis* with 87-day-old young (A), *Caluromys* in arboreal habitat (B), and *Monodelphis* showing the pouchless condition with week-old young firmly attached to the teats of the mother (C). Photographs by R. Hossler (A), M. Atramentowicz (B), and D. Denis (C).

by cold weather or dry conditions (Table 1). The bare-tailed woolly opossum (*Caluromys philander*; referred to here as *Caluromys*) (Fig. 1) represents a second group of five highly arboreal opossums from tropical rain forest habitats. Here, reproductive activity and success for *Caluromys* is closely related to forest type, precipitation, and the number of fruiting trees (Table 1). The third group, the remaining 49 species in Didelphidae, is represented by the gray short-tailed opossum (*Monodelphis domestica*; referred to here as *Monodelphis*) (Fig. 1), which is the most widely studied laboratory marsupial. This group of small-bodied, pouchless opossums includes three large genera: *Marmosa*, *Marmosops*, and *Monodelphis* (Table 1).

II. REPRODUCTIVE ANATOMY AND SPERM MATURATION

Sexual dimorphism in opossums is evident in larger (30%) body size, longer canine teeth, and greater skull breadth in males. The scrotum lies anterior to the penis. The testes of *Monodelphis* are large, representing about 2.5% of body mass compared to an average of 0.5% in most other marsupials. The male reproductive tract is similar to that in other marsupials, consisting of paired testes, epididymides, and vas deferentia, a single prostate gland, Cowper's glands, and urethra. All American marsupials have a bifurcate glans penis; its function is unknown, but it might help direct semen into the two lateral vaginal canals at copulation.

Spermatogenesis and sperm maturation in opossums is, except for one feature, similar to that in other marsupials. The nuclei of spermatozoa entering the epididymis from the testes are flattened and oriented at a right angle to the long axis of the midpiece and flagellum (Fig. 2). Maturation of sperm passing through the midsection of the epididymis involves a near 90° rotation of the nucleus and loss of the cytoplasmic droplet so that the acrosome comes to lie on one side of the head. It is at this point that sperm of American marsupials undergo a unique and most remarkable process—sperm pairing—which is unknown in any other group of vertebrates (Fig. 2). Ejaculated spermatozoa remain paired during their passage through the female tract to the upper oviducts where they separate just prior to fertilization. Sperm pairing might serve to protect the acrosomes during this passage or it might enhance motility of sperm through coordinated beating of the two flagella.

The female opossum reproductive tract, similar to other marsupials, is completely paired. The two tracts are separated by ventral medial passage of the ureters to the urinary bladder. Spermatozoa deposited at copulation pass through the two lateral vaginal canals on their way to the two cervixes, uteri, and oviducts, but at the time of parturition, the fetuses pass

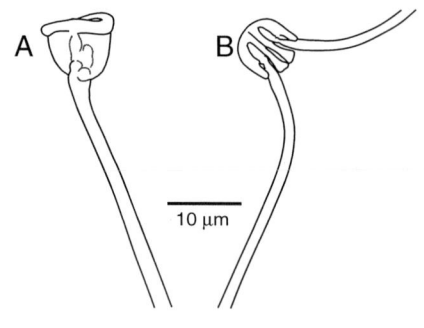

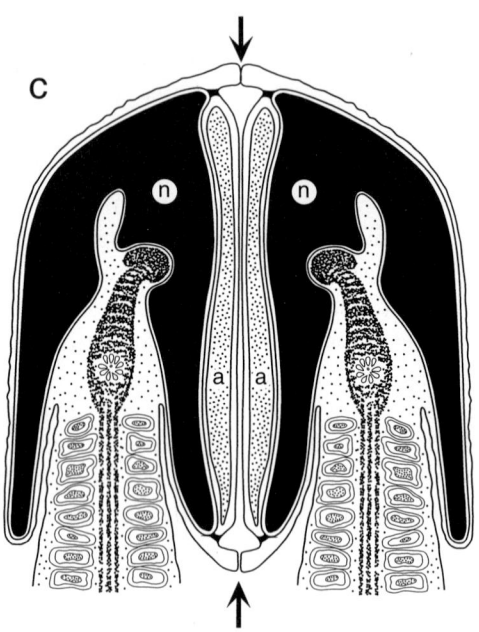

FIGURE 2 Immature spermatozoa prior to 90° rotation of head (A) and paired sperm (B) from *Monodelphis*. Diagram shows two sperm heads with nuclei (n) aligned acrosome to acrosome (a) and the seal (arrows) formed by close association of the plasma membranes (C) (C, reproduced with permission from J. M. Bedford, J. C. Rodger, and W. G. Breed, Why so many mammalian spermatozoa—A clue from marsupials?, *Proc. R. Soc. London B* **221**, 221–233, 1984).

through a central birth canal, formed in loosened connective tissue between the cervixes and the distal end of the urogenital sinus.

The marsupium varies in development among marsupials (Table 1). Larger opossums, such as *Didelphis*, have a deep, well-developed pouch that completely covers the litter. A shallow marsupium or lateral skin folds develop during lactation in a few other species, but in the vast majority of didelphids (about 75%), no pouch develops. The pouchless condition is probably primitive but it is also strongly correlated with body mass; only the larger didelphids (>250 g) possess a well-developed pouch.

III. REPRODUCTIVE CYCLES, OVULATION, AND FERTILIZATION

In the absence of conception, female *Didelphis* cycle spontaneously six or eight times per year, depending on latitude (Table 1). When a male encounters a female in estrus, he makes a smacking sound while cautiously investigating and nuzzling the female. He then mounts the estrous female, grasping her midsection with his front feet and holding on to her hindlegs using the opposable hallux of his hindfeet; the two usually fall on their sides during the course of copulation. *Didelphis* ovulates spontaneously shortly after the onset of behavioral estrus, with or without copulation. The ovulation rate in *Didelphis* is among the highest known for mammals; about 60 eggs are released per cycle. However, many of the ova are not viable or fertilized; the total number of embryos counted *in utero* is normally <30. *Caluromys* is polyestrous and exhibits estrus at 28- to 45-day intervals, except for a hiatus in September–November. Spontaneous ovulation is followed by a relatively long (20 days) luteal phase (Table 1).

Female *Monodelphis* do not have an estrous cycle; they remain reproductively inactive when isolated from males (Table 1). Males deposit scent marks by rubbing their large, androgen-dependent suprasternal (base of neck) gland on hard objects, e.g., food bowls. Females access the nonvolatile pheromone by nuzzling, i.e., gently rubbing their moistened nose and upper lip over the scent mark (Fig. 3). Surgical ablation of the vomeronasal organ (VNO) prevents induction of estrus in females exposed only to male scent marks, but in the absence of the VNO, the main olfactory system is sufficient for induction of estrus if females are exposed directly to males by cohabitation. Pheromonal activation of the hypothalamo–pituitary–ovarian axis is evident (about 5 days after exposure to male scent marks) from estrogenic cytology (cornified epithelial cells) in urogenital sinus smears. Copulation, which follows one or two

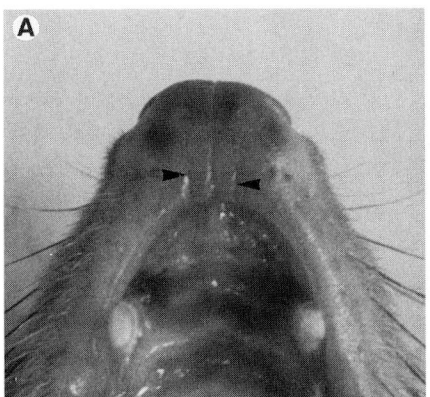 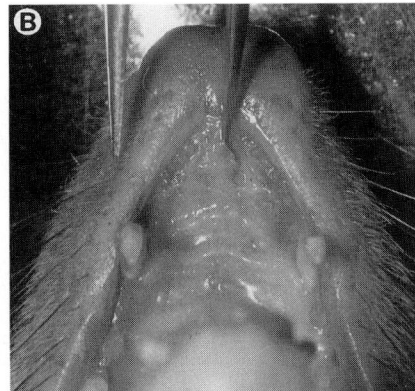

FIGURE 3 Labial–palatine groove (arrowheads in A) and the nasopalatine canal (exposed with probe) (B) through which pheromone-laden mucus from the male scent mark is drawn into the vomeronasal organ of the female (reprinted from *Physiol. Behav.* 53, N. S. Poran, A. Vandoros, and M. Halprin, Nuzzling in the gray short-tailed opossum I: Delivery of odors to the vomeronasal organ, 959–967, 1993, with permission from Elsevier Science).

nights later, is similar in behavior to that in *Didelphis* but is accompanied by brief locking of the penis in the urogenital sinus. Ovulation is induced, i.e., it requires the stimulus of copulation or at least physical contact with a male. Although obligatory, male-induced estrus has not been observed in other didelphids; male urine has a stimulatory effect on the occurrence of estrus in woolly opossums (Table 1).

IV. ENDOCRINOLOGY OF THE ESTROUS CYCLE AND GESTATION

Estrogen levels rise sharply late in the follicular phase of the cycle of Didelphis—1–4 days prior to estrus. This rise in estrogen from growing preovulatory follicles is reflected in a transient gain in body mass, swollen vulva, urogenital sinus smears, and, ultimately, in behavioral estrus and copulation. Spontaneous ovulation in *Didelphis* is presumably stimulated by a surge in circulating levels of luteinizing hormone (LH), as in other mammals. Plasma LH levels peak in *Monodelphis* females (indicative of a preovulatory LH surge) 6–8 days after initial exposure and pairing with males. Following ovulation corpora lutea (CL) develop from the ruptured ovarian follicles and secrete progesterone that stimulates uterine secretory activity during gestation (Fig. 4). The luteal phase of the cycle in *Caluromys*, marked by 20 days of elevated plasma progesterone, is prolonged (Fig. 4 and Table 1), as is gestation; at 24 days it is the longest known for any didelphid marsupial.

Gestation in opossums is sufficiently brief that it is accomplished within the temporal confines of the luteal phase and, by many measures, it appears to be physiologically equivalent to the estrous cycle. For example, progesterone profiles are very similar in pregnant and nonpregnant females (Fig. 4). This equivalence of the pregnant and nonpregnant cycle in opossums differs markedly from the eutherian pattern in which pregnancy extends the life of the CL, interrupts the estrous cycle, and alters the endocrinology and physiology of the mother.

V. GESTATION, DEVELOPMENT, AND PLACENTATION

The length of gestation in *Didelphis* (13 days) is remarkably short for a 2-kg animal—among the shortest gestation periods known for any marsupial. By comparison, the gestation period of the 100-g *Monodelphis* is 2 days longer and that of the 300-g *Caluromys*, which is 24 days. This illustrates a general truth: Length of gestation in marsupials is not closely related to body size.

The marsupial embryo is enclosed in a shell membrane—a transparent, proteinaceous material, which is deposited around the embryo by the first days of uterine life. The shell membrane persists for at least

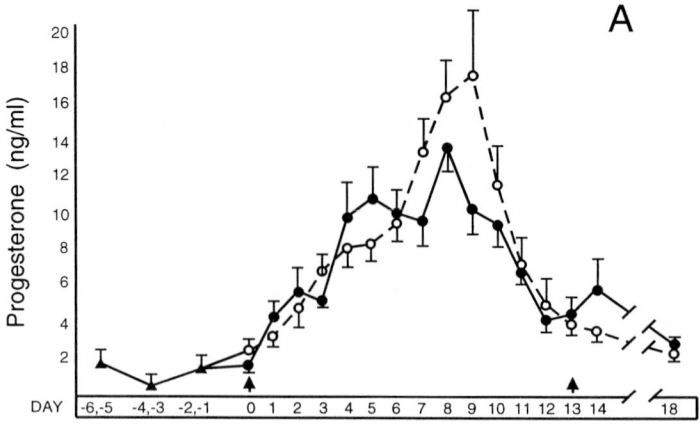
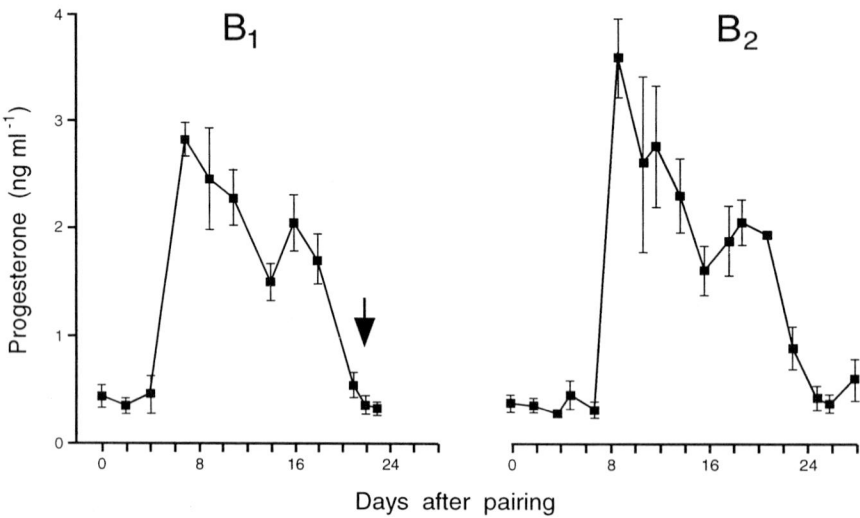
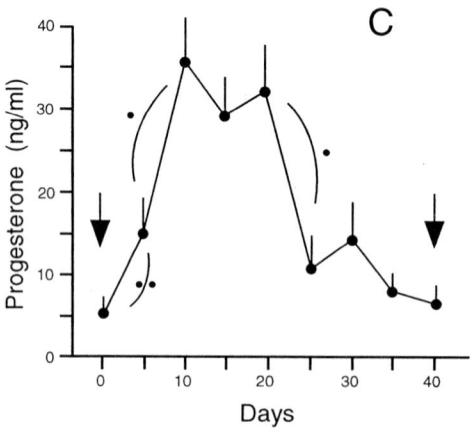

FIGURE 4 Progesterone concentrations in plasma collected during pregnancy (○) and the nonpregnant luteal phase (●) in *Didelphis virginiana* (A) and in *Caluromys philander* (C) and pregnant (B_1) and nonpregnant (B_2) *Monodelphis domestica*. Data have been aligned on the day of copulation in A, day of elevation in progesterone in B, and an indication of ovulation (Days 0 and 40) in C. Progesterone returned to basal levels on or before parturition (arrows). [Figures reprinted with permission from (A) Harder, J. D., and Fleming, M. W. (1981). Estradiol and progesterone profiles indicate a lack of endocrine recognition of pregnancy in the opossum. *Science* 212, 1400–1402. © 1981 American Association of the Advancement of Science; (B) Hinds *et al.*, 1992; and (C) Perret, M., and Atramentowicz, M. (1989). Plasma concentrations of progesterone and testosterone in woolly opossums (*caluromys philander*). *J. Reprod. Fert.* 85, 31–41.]

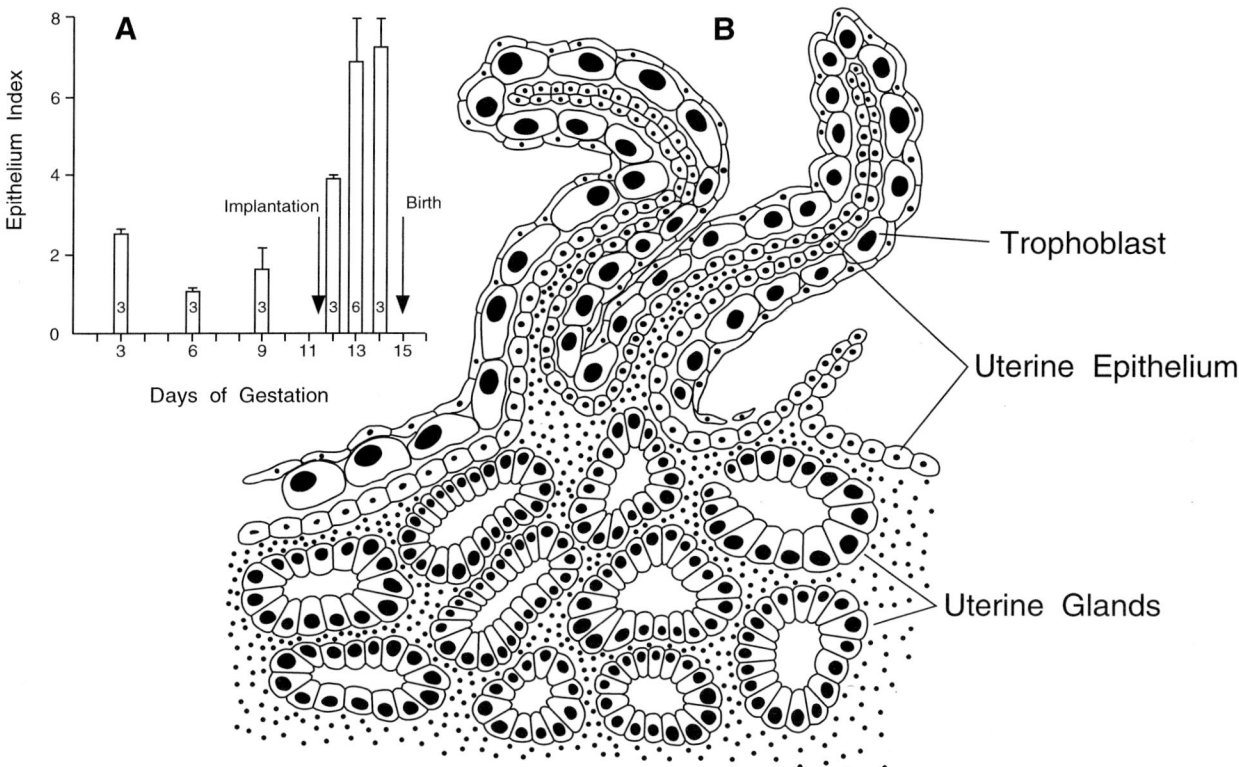

FIGURE 5 An index of the surface area of the uterine epithelium (A) increases markedly after implantation and close association of the fetal trophoblast and the highly folded uterine epithelium (B) (reproduced from Gestation and placentation in two new world opossums: *Didelphis virginiana* and *Monodelphis domestica*, Harder et al., *J. Exp. Zool.*, Copyright © 1993 John Wiley & Sons, Inc.).

the first two-thirds of gestation, and although it is permeable to nutrients in the uterine fluid, it also isolates embryos from direct contact with maternal tissues. *Monodelphis* embryos reach the unilaminar blastocyst stage by Day 6 of gestation and double in size by Day 9 (primitive groove stage). Rupture of the shell membrane and implantation occurs on Day 10 in *Didelphis* and by Day 12 of gestation in *Monodelphis*. Thereafter, the yolk sac, or choriovitelline, placenta of *Monodelphis* spreads rapidly into the crypts of the highly folded uterine mucosa (Fig. 5), which increases fourfold in surface area from Day 9 to Day 13 of gestation. Uterine (endometrial) gland density in *Monodelphis* is high during Days 3–9 of gestation and then gradually declines following implantation as its nutritive role is supplanted by secretions of the uterine epithelium.

VI. BIRTH AND LACTATION

The embryonic appearance of newborn marsupials, one of the hallmarks of the infraclass, actually represents a mosaic of development. The oral cavity and forelimbs are well developed at birth, whereas the hindlimbs and caudal elements of the central nervous system are in a near fetal state of development. At birth, opossum young move without maternal assistance with a wriggling motion from the opening of the urogenital sinus to the mammary area or pouch. If the neonate finds an unoccupied teat, it takes it into its well-developed oral cavity wherein structural modifications and swelling of the teat firmly affix the young to the mother. This grip is not voluntarily released during the continuous teat attachment phase, which varies from less than one-

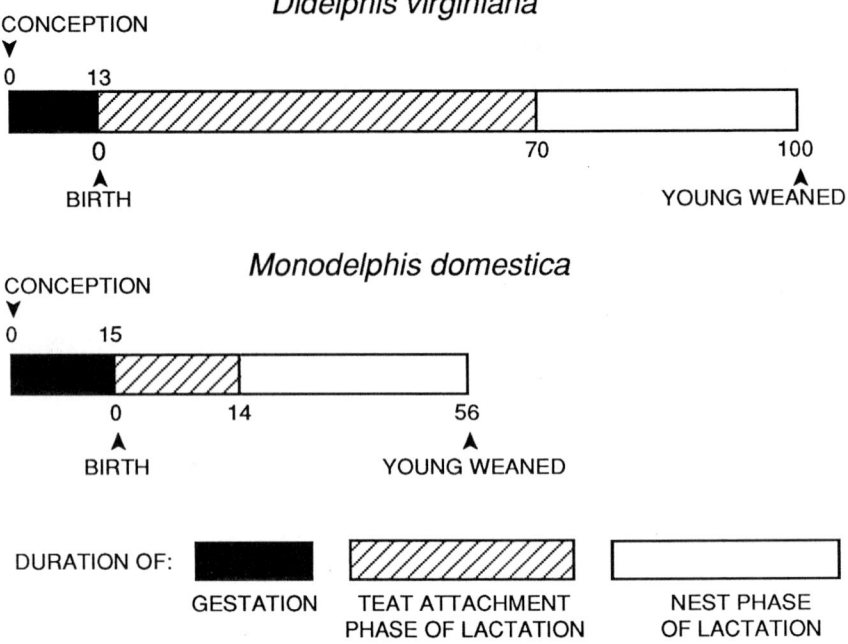

FIGURE 6 Comparison of the temporal features (in days) of the reproductive cycles of *Didelphis* and *Monodelphis* (reproduced from Gestation and placentation in two new world opossums: *Didelphis virginiana* and *Monodelphis domestica*, Harder et al., *J. Exp. Zool.*, Copyright © 1993 John Wiley & Sons, Inc.).

third of the lactation period for small pouchless species to approximately half of the lactation period in larger, pouched species (Fig. 6). At the time such young are left in the nest, they are blind, thinly furred, ectothermic, and in need of active maternal care. The period of postnatal care is relatively long in *Caluromys* compared to other didelphids of similar size (Table 1).

Marsupial young change markedly during the long period of lactation and their changing nutritional requirements are reflected in the changing nutrient composition of the milk. Apparently, opossums exhibit the general marsupial pattern of high lipid and low carbohydrate levels late in lactation.

Relatively few (about one-third) of the ova shed at the time of mating in *Didelphis* are fertilized and develop to term. Even fewer are born, reach the pouch, and locate an unoccupied, functional teat. Litter size in *Didelphis* is correlated with maternal condition, which might determine the number of functional teats (8–13). In any case, the meaning of a nearly 10-fold reduction from the number of ova released at estrus (60) to the number of young weaned is obscure, particularly in light of the fact that the number of ova released (12) in *Monodelphis* does not greatly exceed the number of midgestational embryos (10) or litter size (7). This suggests a fundamental difference between *Monodelphis* and *Didelphis* in patterns of follicular atresia and hormonal selection of oocytes for ovulation.

See Also the Following Articles

Marsupials; Monotremes

Bibliography

Baggott, L. M., and Moore, H. D. M. (1990). Early embryonic development of the grey short-tailed opossum, *Monodelphis domestica*, in vivo and in vitro. *J. Zool.* (London) 222, 623–639.

Eisenberg, J. F. (1989). *Mammals of the Neotropics.* Univ. of Chicago Press, Chicago.

Emmons, L. H. (1990). *Neotropical Rainforest Mammals, A Field Guide.* Univ. of Chicago Press, Chicago.

Fadem, B. H. (1985). Evidence for the activation of female reproduction by males in a marsupial, the gray short-tailed opossum (*Monodelphis domestica*). *Biol. Reprod.* **33**, 112–116.

Gardner, A. L. (1993). Order Didelphimorphia. In *Mammal Species of the World: a Taxonomic and Geographic Reference* (D. E. Wilson and D. M. Reeder, Eds.), pp. 15–23. Smithsonian Inst. Press, Washington, DC.

Green, B., Krause, W. J., and Newgrain, K. (1996). Milk composition in the North American opossum. *Comp. Biochem. Physiol.* **113B**, 619–623.

Harder, J. D. (1992). Reproductive biology of South American marsupials. In *Reproduction in South American Vertebrates* (H. H. Hamlett, Ed.), pp. 211–228. Springer-Verlag, New York.

Harder, J. D., and Fleming, M. W. (1981). Estradiol and progesterone profiles indicate a lack of endocrine recognition of pregnancy in the opossum. *Science* **212**, 1400–1402.

Harder, J. D., Stonerook, M. J., and Pondy, J. (1993). Gestation and placentation in two new world opossums: *Didelphis virginiana* and *Monodelphis domestica*. *J. Exp. Zool.* **266**, 463–479.

Hartman, C. G. (1928). The breeding season of the opossum *Didelphis virginia* and the rate of intrauterine and postnatal development. *J. Morphol.* **46**, 143–215.

Hinds, L. A., Reader, M., Wernberg-Moller, S., and Saunders, N. R. (1992). Hormonal evidence for induced ovulation in *Monodelphis domestica*. *J. Reprod. Fertil.* **95**, 303–312.

Jackson, L. M., and Harder, J. D. (1996). Vomeronasal organ removal blocks pheromonal induction of estrus in gray short-tailed opossums (*Monodelphis domestica*). *Biol. Reprod.* **54**, 506–512.

Marshall, L. G. (1982). Evolution of South American marsupialia. In *Mammalian Biology in South America. Pymatuning Symposia in Ecology* (M. A. Mares and H. H. Genoways, Eds.), Vol. 6, pp. 251–272. Univ. of Pittsburgh, Pittsburgh.

Moore, H. D. M. (1992). Reproduction in the gray short-tailed opossum (*Monodelphis domestica*). In *Reproduction of South American Vertebrates* (H. H. Hamlett, Ed.), pp. 229–241. Springer-Verlag, New York.

Pelengaris, S. A., Abbott, D. H., Barrett, J., and Moore, H. D. M. (1992). Induction of estrus and ovulation in female grey short-tailed opossums, *Monodelphis domestica*, involves the main olfactory epithelium. In *Chemical Signals in Vertebrates* (R. L. Doty and D. Muller-Schwarze, Eds.), Vol. 6, pp. 253–257. Plenum Press, New York.

Perret, M., and Atramentowicz, M. (1989). Plasma concentrations of progesterone and testosterone in captive woolly opossums (*Caluromys philander*). *J. Reprod. Fertil.* **85**, 31–41.

Perret, M., and M'Barek, S. B. (1991). Male influence on oestrous cycles in female woolly opossum (*Caluromys philander*). *J. Reprod. Fertil.* **91**, 557–566.

Renfree, M. B. (1994). Endocrinology of pregnancy, parturition and lactation of marsupials. In *Marshall's Physiology of Reproduction* (G. E. Lamming, Ed.), 4th ed., pp. 677–766. Churchill Livingstone, Edinburgh, UK.

Saunders, N. R., and Hinds, L. A. (Eds.) (1997). *Marsupial Biology: Recent Research, New Perspectives*. Univ. New South Wales Press, Sydney, Australia.

Streilein, K. E. (1982). Behavior, ecology, and distribution of South American marsupials. In *Mammalian Biology in South America* (M. A. Mares and H. H. Genoways, Eds.), Pymatuning Symposia in Ecology, Vol. 6, pp. 231–250. Univ. of Pittsburgh Press, Pittsburgh

Temple-Smith, P. D., and Bedford, J. M. (1980). Sperm maturation and the formation of sperm pairs in the epididymis of the opossum, *Didelphis virginiana*. *J. Exp. Zool.* **214**, 161–171.

Tyndale-Biscoe, C. H., and Renfree, M. B. (1987). *Reproductive Physiology of Marsupials*. Cambridge Univ. Press, Cambridge, UK.

Orchidectomy

see Castration, Effects in Humans (Male)

Orchitis

W. Weidner
University of Giessen

W. Krause
University of Marburg

I. Epidemiology
II. Etiology and Pathogenesis
III. Pathomorphology
IV. Clinical Features and Diagnostic Work-up
V. Therapy and Prognosis

GLOSSARY

classification The categorization of orchitis based on etiology.
diagnostic work-up Medical history, palpation, standardized investigations for microorganisms, and ejaculate analysis are the hallmark of clinical considerations.
therapy Proven benefit of therapy is only given in the acute disease.

Orchitis is an inflammatory lesion of the testicle associated with a predominantly leukocytal exudate inside and outside the seminiferous tubules resulting in (a focal) tubular sclerosis.

I. EPIDEMIOLOGY

Normally, orchitis, the isolated inflammation of the testis, is an uncommon disease. There are no precise data available concerning incidence and prevalence of this infection. Generally, inflammatory disorders of the testis, epididymis, and tunica vaginalis cannot be considered separately. The term epididymoorchitis is reserved for the most common case, in which the structures of both the testis and the epididymis are affected. In an analysis of 400,000 testicular pathological specimens, only 0.42% demonstrated isolated orchitis.

II. ETIOLOGY AND PATHOGENESIS

The classification of orchitis depends on etiology (Table 1). Acute epididymoorchitis of the adult is the most common type. It is reported that acute epididymoorchitis in young males is associated with sexual activity and infection of the consort. The disease caused by sexually transmitted organisms (*Neisseria gonorrhoeae* and *Chlamydia trachomatis*) mainly occurs in males under the age of 35 years. The majority of cases are due to common urinary pathogens, e.g., *Escherichia coli*. Bladder outlet obstruction caused by benign prostate hyperplasia or urogenital malformations are risk factors for this type of infection. The spread of microorganisms is canalicular from the urethra via the spermatic pathways.

Chronic nonspecific epidydymoorchitis occurs in combination with an infiltration of the epididymis, possibly as a degenerative consequence of recurrent inflammatory attacks in elder men. In most of the cases the nature of the disease remains unclear.

Nonspecific (idiopathic) granulomatous orchitis is a chronic inflammatory lesion of uncertain etiology in the elderly patient. An autoimmune process secondary to extravasation of spermatozoa has been considered responsible. The first occurrence was observed after surgical trauma, parallel to the development of sperm autoantibodies. The disease was also observed in the course of other autoimmune diseases. Physiologically, the maturating cells of the spermatogenesis appear only after puberty. Thus, immunocompetence is not achieved for sperm-specific antigens. Mature spermatozoa are the most competent cells for inducing immunologic reactions in the normal organism.

The term "autoimmune orchitis," however, mainly

TABLE 1
Classification of Orchitis

Nonspecific	Specific	Viral
Acute bacterial epididymoorchitis	Specific granulomatous	Mumps orchitis
N. gonorrhoeae	Tuberculosis	Coxsackie-B
C. trachomatis	Lues	
E. coli (and other Enterobacteriaceae)	Brucellosis	
Nonspecific chronic epididymoorchitis		
Granulomatous (idiopathic) orchitis		
Orchitis of the child		
Pneumococci		
Salmonella		
Klebsiella		
Hemophilus influenza		

refers to the disease which can be induced in laboratory animals. Obviously, these animals are more susceptible to the autoimmune process than the human male. There is no other explanation why the autoimmune orchitis in rat and mouse is induced by simple intravascular injection of testicular homogenates, whereas the occurrence of an orchitis in the human is very rare, even following severe testicular trauma. This indicates that the results of animal experiments are only in part transducible to the human male.

Nonspecific acute epididymoorchitis may also occur predominantly in children as a simultaneous involvement of both the testis and the epididymides in the course of systemic diseases such as pneumococcal meningitis, salmonellosis, pneumonia due to klebsiella, and hemophilus influenza infections. The pathogenesis is hematogenous.

Specific granulomatous orchitis is commonly tuberculous in nature. Normally, the infection occurs as a canalicular spread from another genitourinary tract lesion, but a hematogenous infection from the primary lung lesion or a combined infection are also common. Syphilitic orchitis can appear as both congenitally and acquired; the acquired type may occur as gumma of the testis. Another granulomatous orchitis is caused by brucella species.

Mumps orchitis is a major complication of a general infection with the mumps (paramyxo-) virus. After parotitis, orchitis occurs in up to 37% of postpubertal men, and it occurs bilaterally in one-third of these patients. A hematogenous infection of the gland is followed by an immunological reaction with acute necrosis of the tubular structures. Frequency clearly depends on the vaccination status of the population.

III. PATHOMORPHOLOGY

Acute bacterial epididymoorchitis is characterized by an infiltration of granulocytes inside and outside the tubulous seminiferous. Granulomatous orchitis is characterized by a granulomatous type of inflammation; in specific granulomatous infections typical lesions also occur, including Langhans-type giant cells. In mumps orchitis, granulocytal and lymphocytal infiltrations are common. Hallmark in the progression of this lesion is the focal degeneration and atrophy of the seminiferous tubules, with a disappearing of the inflammatory cells and a focal tubular sclerosis. Fifty percent of all patients who have unilateral or bilateral orchitis develop testicular atrophy.

IV. CLINICAL FEATURES AND DIAGNOSTIC WORK-UP

In acute orchitis, painful swelling of the testicle is typical. In epididymoorchitis, the inflammation and swelling usually begin in the tail of the epididymis and spread to the rest of the epididymis and testicular tissue. Concomitant urethritis is an indication of a sexually transmitted origin. A medical history of bladder outlet obstruction in adults or congenital urogenital malformations in children provides evidence for urinary tract infections as a cause of the inflammation.

The microbial etiology of epididymoorchitis can usually be determined easily by examination of Gram-stained urethral smear for urethritis and stain of midstream urine specimen for Gram-negative bacteriuria. The presence of intracellular Gram-negative diplococci on the smear correlates with the presence of *N. gonorrhoeae*. The presence of only white blood cells on urethral smear indicates the presence of nongonococcal urethritis. *C. trachomatis* can be isolated in approximately two-thirds of these patients.

In chronic granulomatous orchitis in adults, testicular swelling with and without pain is the hallmark of the disease. Typically, the clinical work-up including ultrasonography cannot ensure the diagnosis and thus it is necessary to explore the testis surgically to rule out a malignant lesion.

Mumps orchitis develops in 20–30% of postpubertal patients undergoing mumps infection. Mumps orchitis occurs bilaterally in about 30% of patients. Orchitis with fever occurs 3–10 days after parotitis. In clinical suspicion, history of parotitis and evidence of IgM antibodies in serum provide the diagnosis. The clinical features of other types of orchitis are well-known from the clinical features of the underlying diseases.

A. Differential Diagnosis

In every case of acute orchitis, it is necessary to exclude spermatic cord torsion immediately using all information, including scrotal sonography and duplex scanning of the testicular blood flow. In chronic (epididymo-) orchitis, ultrasound does not suffice to rule out a testicular tumor. Thus, in these cases inguinal exploration and biopsy of the lesion is necessary for histological evaluation.

B. Ejaculate Analysis

Ejaculate analysis according to World Health Organization criteria, including leukocyte analysis, may indicate persistent inflammatory activity. In many cases, especially in acute epididymoorchitis, transient decreased sperm counts and reduced forward motility can be evaluated. Acute obstructive azoospermia due to a complete obstruction of the rete testis is a rare complication. In cases of mumps orchitis, the disease may result in a bilateral testicular atrophy and testicular azoospermia. When a granulomatous orchitis is suspected, the presence of sperm-bound autoantibodies should be investigated by the MAR test. If the patient is azoospermic, an indirect MAR test may be used to prove the presence of sperm antibodies in seminal fluid.

C. Further Investigations

In cases with severe oligo- or azoospermia, measurement of follicle-stimulating hormone (FSH) is suggested. If the FSH level is increased and the testicular volume is <15 ml, the constellation is indicative of damage of the seminiferous epithelium. In these cases, testicular biopsy has been suggested in order to confirm the diagnosis. On the other hand, systematic analysis in nonobstructive azoospermia demonstrates ICSI-potent motile spermatozoa even in biopsies of the "hopeless" cases.

V. THERAPY AND PROGNOSIS

Only the therapy of acute bacterial epididymoorchitis and of specific granulomatous orchitis is standardized (Table 2). Several trials are suggested to improve the inflammatory lesion. Corticoids and nonsteroidal antiphlogistic substances, such as diclofenac, indomethacin and acetyl-salicyl acid, have unfortunately not been evaluated as to their andrological outcome. A further nonrandomized therapeutic trial is based on the idea of preventing deleterious effects of orchitis on spermatogenesis by protective gonadotrophin-releasing hormone treatment.

In mumps orchitis, systemic α-2b-interferon therapy is suggested for preventing testicular atrophy and azoospermia. Prospective studies which provide evidence of the benefit of the first communications are missing. In idiopathic granulomatous orchitis, the therapy of choice is surgical removal of the testis.

TABLE 2
Therapy of Orchitis

Classification	Drug treatment
Acute bacterial epididymoorchitis	
N. gonorrhoeae	Tetracyclines
C. trachomatis	
E. coli, Enterobacteriaceae	Fluoroquinolones
Nonspecific chronic epididymoorchitis	Steroidal and nonsteroidal antiphlogistic substances (?)
Granulomatous (idiopathic) orchitis	Semicastration
Orchitis of the child	IV antibiotic therapy
Specific orchitis	According to the therapy of the underlying diseases
Mumps orchitis	α-2b-Interferon

See Also the Following Articles

Epididymis; Male Reproductive Disorders; Testis, Overview

Bibliography

Aitchison, M., Mufti, G. R., Farrell, J., Paterson, P. J., and Scott, R. (1990). Granulomatous orchitis. *Br. J. Urol.* **66**, 312–314.

Bergmann, M., Behre, H. M., and Nieschlag, E. (1994). Serum FSH and testicular morphology in male infertility. *Clin. Endocrinol.* **40**, 133–136.

Chen, C. S., Chu, S. H., Lai, Y. M., Wang, M. L., and Chan, P. R. (1996). Reconsideration of testicular biopsy and follicle-stimulating hormone measurement in the era of intracytoplasmic sperm injection for non-obstructive azoospermia. *Hum. Reprod.* **11**, 2176–2179.

Erpenbach, K., and Derschum, W. (1991). Die systemische α-Interferontherapie: Ein möglicher Weg zur Prävention testikulärer Atrophien und dauerhafter Sterilitäten bei Patienten mit bilateraler Mumpsorchitis. *Urologe A* **30**, 244–248.

Mikuz, G. (1978). *Orchitis.* Thieme, Stuttgart.

Nistal, M., and Paniagua, R. (1984). *Testicular and Epididymal Pathology.* Thieme, Stuttgart.

Perimenis, P., Athanasopoulos, A., Venetsanou-Petrochilou, C., and Barbalias, G. (1991). Idiopathic granulomatous orchitis. *Eur. J. Urol.* **19**, 118–120.

Scholz, M., Graf, N., Steffens, J., Schönkofen, H., Jeannelle, J.-P., Schofer, O., and Sitzmann, F. C. (1996). Mumpsorchitis im Jugend-und Erwachsenenalter. *Dt. Arztbl. A* **93**, 2087–2090.

Scott, R. F., and Bayliss, A. P. (1985). Ultrasound in the diagnosis of granulomatous orchitis. *Br. J. Radiol.* **58**, 907–909.

Suominen, J. J. (1995). Sympathetic auto-immune orchitis. *Andrologia* **27**, 213–216.

Vicari, E., and Mongioi, A. (1995). Effectiveness of long-acting gonadotrophin-releasing hormone agonist treatment in combination with conventional therapy on testicular outcome in human orchitis/epididymo-orchitis. *Hum. Reprod.* **10**, 2072–2078.

Weidner, W., Schiefer, H. G., and Garbe, C. H. (1987). Acute nongonococcal epididymitis. Aetiological and therapeutic aspects. *Drugs* **34**(Suppl. 1), 111–115.

Orgasm

Kevin E. McKenna

Northwestern University School of Medicine

I. Definition
II. Genital Responses
III. Extragenital Responses
IV. Orgasm in Nonhumans
V. Neural Control
VI. Afferent Mechanisms
VII. Refractory Period

GLOSSARY

afferent Moving toward a certain region; often used in neurobiology to refer to sensory nerve fibers projecting from peripheral structures to the spinal cord or brain.

anorgasmia Inability to achieve orgasm.

descending Referring to neural pathways which originate above the spinal cord and project to it, providing control of spinal cord activity by higher portions of the brain.

efferent Moving away from a region; often used to refer to nerve fibers which leave the brain or spinal cord to innervate an organ.

homologous Having a common origin and having a similar organization as another anatomical structure in the opposite sex.

perineal muscles The skeletal muscles of the floor of the pelvis, including the ischiocavernosus, bulbospongiosus, external urethral and anal sphincters, and the levator ani.

Orgasm is defined as the physiological, behavioral, and subjective responses which occur in both sexes at sexual climax. It is characterized by rhythmic pelvic muscle contractions, cardiorespiratory and other autonomic activation, generalized muscle tension, and intensely pleasurable sensations and is followed by a profound sense of relaxation. There is a remarkable physiological similarity between male and female orgasms. The physiological reactions in the pelvis are produced by a spinal pattern generator. The spinal center is under supraspinal inhibitory and excitatory influences. The neural mechanisms responsible for triggering orgasm have not been defined in either sex.

I. DEFINITION

A consensus for an accurate scientific definition of orgasm is problematic. Some writers use it exclusively for the subjective cerebral experience, whereas others define it by the pelvic reactions. A purely subjective definition precludes neurobiological and animal-based studies of the phenomenon. A definition which is based solely on genital responses ignores fundamental psychosocial issues. A common approach is to use the term orgasm to refer to the entire physiological and psychological experience of sexual climax.

II. GENITAL RESPONSES

In the male, the most obvious aspect of orgasm is ejaculation of semen. Sexual stimulation leads to emission of seminal fluids into the urethra. Rhythmic contractions of the urethral smooth muscle and the bulbospongiosus muscle surrounding the urethral bulb lead to the forceful expulsion of semen from the urethral meatus. Contraction of the bulbospongiosus is synchronous with contraction of several other pelvic muscles, such as the ischiocavernosus muscle, the external anal and urethral sphincters, and the levator ani.

In the female, orgasm consists of rhythmic contractions of the vaginal and uterine smooth muscle as well as contractions of the striated pelvic muscles such as the external urethral and anal sphincter and

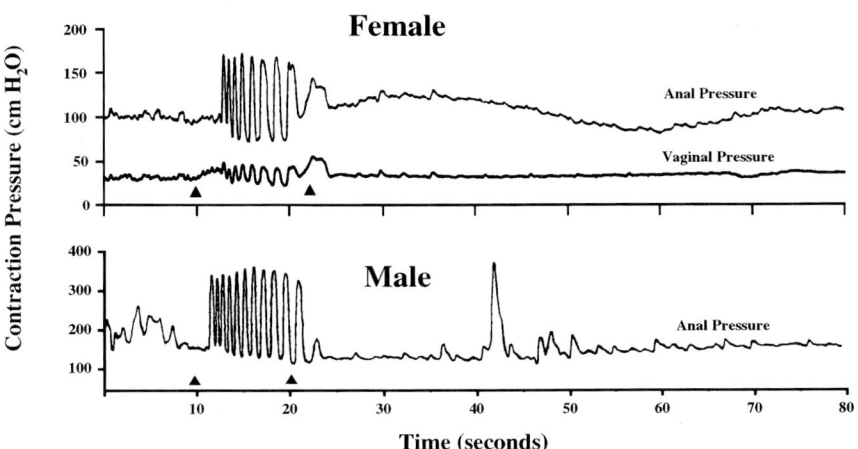

FIGURE 1 Physiological recordings of the most common pattern of orgasm in male and female volunteers induced by masturbation. The pelvic contractions of orgasm were measured by the means of probes inserted into the anus and vagina and the contractions are indicated by increases in pressure around the probes. Note that the contractions are highly regular with a consistent increase in interval between each contraction. The arrows indicate the beginning and end of subjectively recorded orgasm (adapted with permission from J. G. Bohlen, J. P. Held, M. O. Anderson, and A. Ahlgren, The female orgasm: Pelvic contractions, *Arch. Sexual Behav.* **11**, 367–386, 1984; and J. G. Bohlen, J. P. Held, and M. O. Anderson, The male orgasm: Pelvic contractions measured by anal probe, *Arch. Sexual Behav.* **9**, 503–521, 1980).

the circumvaginal muscles. The rhythmic contractions of the pelvic muscles are characteristic of orgasm in both sexes. These can be measured by vaginal or anal pressure measurements or by recording of the electromyographic activity of the muscles. Such recordings demonstrate that the human orgasm is a highly stereotyped pattern of activity. It consists of approximately 8–20 contractions. The interval between contractions starts at approximately 0.6 or 0.7 sec and increases by about 0.1 sec with each successive contraction. Individual subjects tend to show consistent patterns of muscle contractions. The number of contractions, frequency of contractions, and total duration are similar in males and females. Figure 1 demonstrates the essential similarity of male and female orgasm. Female orgasm is also associated with a significant increase in vaginal blood flow and oxygen tension. In both sexes orgasm is preceded by tonic contraction of the perineal muscles.

III. EXTRAGENITAL RESPONSES

In addition to the pelvic activity, orgasm is accompanied by characteristic physiological responses. There is an increase in blood pressure, heart rate, and respiration coincident with orgasm. Pain threshold is greatly increased, but nonpainful tactile sensitivity is not changed. Generalized skeletal muscle tension and facial grimaces are common as orgasm approaches. Erection of the nipples in both sexes may occur. Less regularly, there may be flushing of the skin and sweating. In both sexes, there is a large increase in circulating oxytocin at orgasm. Subjectively, orgasm is accompanied by intense pleasure and a release of tension.

IV. ORGASM IN NONHUMANS

There is considerable evidence that nonhuman mammals experience orgasm. The existence of orgasm in male animals has never been controversial given that ejaculation and male orgasm are so closely identified. The existence of female animal orgasm has been more contentious, largely because human female orgasm has so commonly been defined purely by subjective experiences. Obviously, subjective experiences of animals are problematic to determine. However, if objective physiological criteria are employed, clear evidence of female animal orgasm is available. Female animals commonly demonstrate characteristic elevations of blood pressure and heart rate during copulation which are similar to those seen in humans at orgasm. Uterine contractions induced by copulation have been recorded in several species, including rat, rabbit, cattle, and monkey. Release of oxytocin has been noted in several species of both sexes, again similar to that seen in the human male and female. Female vocalization in nonhuman primates has also been compared to human orgasmic responses. A definitive demonstration would be to record in females the most characteristic aspect of orgasm, regular rhythmic striated pelvic muscle contractions. Such experiments have not yet been performed. However, the available evidence indicates that male and female animals demonstrate a constellation of pelvic, autonomic, and hormonal responses which are indicative of orgasm.

V. NEURAL CONTROL

The neural control of the genital, autonomic, hormonal, and subjective responses of orgasm have not been elucidated. The neural control of the genital responses of male orgasm have been reviewed in this encyclopedia. A similar overview is not yet possible for the female. However, the great similarity in orgasmic responses seen in men and women strongly indicates that the underlying neural substrates must be largely homologous. For example, the release of oxytocin at orgasm indicates an essential role of the paraventricular nucleus (PVN) of the hypothalamus. Stimulation of the PVN and medial preoptic area of the hypothalamus in anesthetized female rats elicits an orgasmic-like pelvic response similar to that seen in males following the same stimulation.

The regular, rhythmic pelvic muscle contractions of orgasm in both sexes are produced by spinal cord reflex pathways. This conclusion is based on studies of orgasm in spinal cord injured men and women and in experimental animal studies. The spinal systems producing this response are influenced by genital sensory stimulation and both excited and inhibited by supraspinal sites. Of particular note is the inhibition of the orgasmic response by descending serotonin pathways. Drugs which increase serotonin levels, such as the serotonin selective reuptake inhibitor antidepressants, are associated with a very high incidence of anorgasmia in both sexes.

VI. AFFERENT MECHANISMS

No area of sex behavior has generated more speculation than the question of the sensory trigger for female orgasm without a corresponding volume of scientific data. It is generally acknowledged for men that penile stimulation is the necessary stimulus for orgasm. Nonetheless, it is not clear what processes determine the amount of penile stimulation necessary to induce orgasm or the mechanisms which are responsible for the integration of sensory stimulation. In female sexuality, even the question of the necessary genital stimulation is unsettled.

Much of the controversy traces to Freudian theories postulating that orgasms triggered by vaginal stimulation were more mature and psychologically satisfying than those induced by clitoral stimulation. Physiologically, there is no basis to conclude that there are any significant differences between copulatory (vaginal) and masturbatory (clitoral) orgasms. Controversies continue on this issue, despite lack of adequate experimental evidence. However, there is a growing consensus from survey data that most women find significant clitoral stimulation necessary for copulatory orgasm, whether it occurs indirectly as proposed by Masters and Johnson or by direct stimulation.

An important consideration is the role of the supraspinal centers. The orgasm response in both men and women is a spinal reflex elicited by undefined stimuli. This spinal reflex pathway is under inhibitory and excitatory control from supraspinal sites. The processes by which inhibitory controls are suppressed are not known. The supraspinal excitatory sites themselves receive genital sensory input and are affected by higher sensory processes. The descending excitatory pathways are powerful enough to induce orgasm even without any genital stimulus, as in the case of sleep-related orgasms, imagery-induced orgasms, and some extreme cases of premature ejaculation. The supraspinal sources of excitatory drive (e.g., paraventricular nucleus of the hypothalamus) are excited by genital and breast stimulation. Thus, orgasm may be induced by genital stimulation via a direct spinal reflex or by activating descending pathways. Strong psychological arousal may facilitate the entire process, increasing the effectiveness of sensory stimuli. Thus, identifying the sensory trigger for orgasm in either sex involves questions of sensory inputs which are effective in activating spinal reflexes, stimuli which suppress inhibitory controls of spinal reflexes, stimuli which activate descending excitatory drive, and stimuli which are psychologically arousing. It is currently unknown how all these mechanisms interact.

VII. REFRACTORY PERIOD

The most important sex difference between male and female orgasm is the refractory period. Following male orgasm, there is a period of time during which the male is incapable of further sexual arousal and orgasm. Immediately after ejaculation, formerly pleasurable stimuli may even become aversive. This refractory period is age dependent, with adolescents and young men able to achieve rearousal in short periods of time. The origin of the refractory period is unknown.

In contrast, in many, and according to some authors all, women have the capacity to remain aroused after orgasm and achieve additional orgasm(s); that is, achieve multiple orgasms. This is an area also filled with controversy and speculation. One issue is whether all women possess multiorgastic capacity. Despite the vigor with which authors have answered this yes and no, a definitive conclusion cannot be offered. Nonetheless, there is good evidence that a significant fraction of women (around 40% in surveys) have experienced multiple orgasms. Multiple orgasms in males have been documented, but they are exceptional cases. No satisfactory explanation for this sex difference is available.

See Also the Following Articles

EJACULATION; PENIS; VAGINA

Bibliography

Fox, C. A., and Fox, B. (1971). A comparative study of coital physiology, with special reference to the sexual climax. *J. Reprod. Fertil.* 24, 319–336.

Kinsey, A. C., Pomeroy, W. B., and Martin, C. E. (1948). *Sexual Behavior in the Human Male.* Saunders, Philadelphia.

Kinsey, A. C., Pomeroy, W. B., Martin, C. E., and Gebhard, P. H. (1953). *Sexual Behavior in the Human Female.* Saunders, Philadelphia.

Levin, R. J. (1980). The physiology of sexual function in women. *Clin. Obstet. Gynecol.* 7, 213–252.

Levin, R. J. (1981). The female orgasm: A current appraisal. *J. Psychosomatic Res.* 25, 119–133.

Marberger, H. (1974). Mechanisms of ejaculation. In *Physiology and Genetics of Reproduction* (E. M. Coutinho and F. Fuchs, Eds.). Plenum, New York.

Masters, W. H., and Johnson, V. E. (1966). *Human Sexual Response.* Little, Brown, Boston.

McKenna, K. E., and Marson, L. (1997). Spinal and brainstem control of sexual function. In *Central Control of Autonomic Function* (D. Jordan, Ed.). Harwood Academic, London.

Rose, J. D. (1990). Brainstem influences on sexual behavior. In *Brainstem Influences on Sexual Behavior* (W. R. Klemm and R. P. Vertes, Eds.). Wiley, New York.

Orthonectida

John S. Pearse
University of California, Santa Cruz

I. Introduction
II. Infection and Growth in Host
III. Sexual Reproduction

GLOSSARY

agametes Orthonectid cells or groups of cells within the hypertrophied host cell ("plasmodium"); these develop into ciliated, vermiform adults that swim free of the host.

plasmodium Hypertrophied host cell infected by orthonectid cells; formerly thought to be a syncytial mass of the orthonectid itself, which enclosed agametes.

Orthonectids are nonfeeding, ciliated, vermiform organisms that live for short periods in the sea; they are filled with gametes—eggs, sperms, or both. Males inject sperms into females, and the fertilized eggs develop into minute larvae that infect a wide variety of marine invertebrate hosts. Once inside the host, cells from the larvae disaggregate and enter host cells, muscle cells underlying the peritoneum in ophiuroids, which have been closely examined. The infected host cells hypertrophy enormously to fill host body spaces, causing considerable damage to the host. Orthonectid cells within the hypertrophied host cell develop into the multicellular male and female worms that swim free of the hosts.

I. INTRODUCTION

Orthonectida includes about 20 species of parasites that infrequently infect marine turbellarians, nemerteans, gastropods, bivalves, ophiuroids, and ascidians. Within the host, the parasite consists of single cells, or groups of cells, within hypertrophied host cells. In the one host studied in detail, the ophiuroid *Amphipholis squamata,* cells of the orthonectid *Rhopalura ophiocomae* infect muscle cells underlying the peritoneum adjacent to the genital bursae and gut, and the muscle cells containing the orthonectid cells grow into large masses that bulge into the body coelom. The mass is enclosed by the coelomic peritoneum, giving it discrete form. These masses were called plasmodia when it was thought that they were part of the orthonectid, a view favored for over a century that is now known to be incorrect. Eventually, some of the orthonectid cells within the hypertrophied host cells develop into ciliated adults, <1 mm long, which escape from the host and swim free.

The unusual organization of the adults—with a central mass of gametes surrounded by a layer of ciliated cells—is somewhat similar to that of the parasitic Rhombozoa, and the two groups have been placed together in the phylum Mesozoa. However, they differ in two ways: (i) Rhombozoans have a central axial cell within which germ cells, axoblasts, develop into both more worms and what appears to be hermaphroditic gonads, processes unlike anything in orthonectids, and (ii) they lack the free-living, sexual form found in orthonectids. Moreover, recent molecular analyses using 18S rDNA indicate a deep separation between the two groups. Although orthonectids appear simple morphologically, these molecular data also suggest that they are not primitive, but instead are aligned with triploblastic animals.

II. INFECTION AND GROWTH IN HOST

Once the free-swimming larva enters its host, it sheds its ciliated outer cells and the inner cells scatter within the host's tissue spaces. In the ophiuroid *A.*

squamata, larvae of *R. ophiocomae* enter through the genital bursae and gut, and the parasite's cells, called agametes or germline cells, infect muscle cells in the connective tissue underlying the peritoneum. The agametes probably proliferate by mitosis, whereas the infected muscle cells hypertrophy greatly to bulge into the adjacent coelomic spaces, displacing and destroying the gonads. Such hypertrophy is not seen in other host species, in which infected individuals are difficult to identify. Little else is known about host–parasite interactions.

III. SEXUAL REPRODUCTION

Under unknown conditions, agametes undergo embryogenesis within the host cell to form multicellular sexual forms. Both males and females are formed, usually within different host cells, suggesting that the cells were infected by individual orthonectid cells. When fully developed, the sexual forms leave the host as tiny, nonfeeding adult worms, characteristically swimming in straight lines (hence the phylum's name).

Both sexes consist of an annulated outer layer of ciliated cells enclosing a central mass of gametes; in some species there are also circular and longitudinal muscles. Females reach 1 mm in length, whereas males are much smaller.

Males inject sperms into females, and fertilization is internal. Details of the development remain undescribed, but the resulting ciliated larvae that escape from the female directly infect more hosts.

See Also the Following Articles

MARINE INVERTEBRATE LARVAE; RHOMBOZOA

Bibliography

Hanelt, B., Van Schyndel, D., Adema, C. M., Lewis, L. A., and Loker, E. S. (1996). The phylogenetic position of *Rhopalura ophiocomae* (Orthonectida) based on 18S ribosomal DNA sequence analysis. *Mol. Biol. Evol.* 13, 1187–1191.

Kozloff, E. N. (1969). Morphology of the orthonectid *Rhopalura ophiocomae*. *J. Parasitol.* 57, 171–195.

Kozloff, E. N. (1994). The structure and origin of the plasmodium of *Rhopalura ophiocomae* (phylum Orthonectida). *Acta Zool.* 75, 191–199.

Rader, D. N. (1982). Orthonectid parasitism: Effects on the ophiuroid *Amphipholis squamata*. In *Echinoderms: Proceedings of the International Conference, Tampa Bay* (J. M. Lawrence, Ed.), pp. 395–401. Balkema, Rotterdam.

Osteoporosis

E. Bruce Roe and Claude D. Arnaud
University of California at San Francisco

I. Magnitude of the Problem
II. Etiology
III. Evaluation
IV. Treatment
V. Conclusion

GLOSSARY

dual-energy X-ray absorptiometry Noninvasive measurement of bone density used to diagnose osteoporosis, predict fractures, and to monitor therapy.

osteoporosis Asymptomatic reduction in the quantity and quality of bone, resulting in an increased susceptibility to fractures.

osteoporotic fracture A fracture occurring with minimal trauma, such as a fall from standing height. The most common sites of osteoporotic fracture are the proximal femur, the distal radius, the vertebrae, the humerus, the pelvis, and the ribs.

Osteoporosis is an asymptomatic reduction in the quantity and quality of bone, resulting in an increased susceptibility to fractures. Osteoporosis can be diagnosed by measuring bone mineral density using dual-energy X-ray absorptiometry or quantitative computerized tomography. However, the presence of osteoporosis is often not recognized until the occurrence of a fracture with minimal trauma. The most common sites of osteoporotic fracture are the proximal femur, the distal radius (Colles' fractures), the vertebrae, the humerus, the pelvis, and the ribs. Osteoporosis occurs most frequently in postmenopausal white women and in the elderly. Osteoporosis in men will not be directly discussed in this article.

I. MAGNITUDE OF THE PROBLEM

Approximately 20% of women suffer one or more osteoporotic fractures by age 65, and as many as 40% sustain fractures after age 65. Osteoporotic fractures are not usually seen in men and black women until after age 70, when fracture rates progressively increase in these groups.

About 260,000 hip fractures occurred in North America in 1986, and as a result of an aging population, this figure is projected to increase to 540,000 by the Year 2050. The vast majority of these occur with minimal trauma, i.e., a fall from standing height or less, in persons with reduced bone mass. The incidence of hip fracture increases almost logarithmically after age 65 so that by age 80, a white woman has a 1 or 2% annual risk of fracturing her hip. The impact of hip fractures is huge; 12–20% more women die during the first year after a hip fracture than would be expected on the basis of age and sex alone, and as many as 15–25% of women living independently before hip fracture need institutionalization in a long-term care facility for at least 1 year after hip fracture.

Vertebral fractures are estimated to have an incidence that is double that of hip fractures, but only one-third of these come to medical attention with acute pain. In fact, by age 65, there is approximately a 20% prevalence of vertebral deformities found on X rays. The consequences of the "nonclinical" fractures diagnosed on X ray have not been known until recently. It is now clear from population studies that these fractures significantly impair functional status and quality of life, as well as causing kyphosis, loss of height, and chronic pain.

The incidence of fractures of the distal radius increases in white women after age 50 but plateaus after age 65. Although a 50-year-old woman has a 15% risk of sustaining a Colles' fracture during her remaining lifetime, the disability caused by this fracture is generally minimal, and the costs of treatment are <10% of those incurred with hip fracture.

Besides the impact of osteoporotic fractures on mortality and quality of life, the economic cost for caring for patients with fractures is staggering. The current cost for hip fractures alone in the United States may be as high as $7 billion per year, and the cost will continue to grow with the increasing number of fractures projected.

II. ETIOLOGY

Osteoporosis is most commonly associated with menopause and aging. Other causes as well as complicating factors in osteoporosis are listed in Table 1.

The dynamics of bone density in women, including skeletal maturation and bone accrual, early bone loss, estrogen-dependent bone loss, and age-related bone loss, are illustrated in Fig. 1. Most of the skeleton's calcium acquisition occurs by the time of menarche in females and is completed by early in the third decade of life. The most important determinant of maximal or peak bone mass is genetic because it is estimated that genetic factors account for as much as 80% of the variation in bone mass in the popula-

TABLE 1
Risk Factors in Osteoporosis

Genetic factors
    Caucasian or Asian ethnicity
    Positive family history
    Small body frame
    Defects in collagen synthesis or structure
Nutritional
    Low calcium intake
    Vitamin D deficiency (sunlight deprivation without supplementation)
    Vegetarian diet
    Long-standing high protein diet
Drugs
    Glucocorticoids
    Excessive doses of thyroid hormone
    Anticonvulsants
    Anticoagulants
    Alcohol
    Caffeine
Lifestyle
    Inactivity
    Smoking
    Excessive exercise leading to amenorrhea
Endocrine
    Hypogonadism
    Early menopause
    Hypercortisolism
    Hyperparathyroidism
    Hyperthyroidism
    Diabetes mellitus

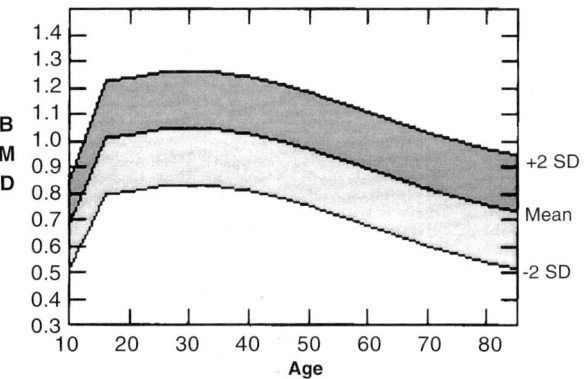

FIGURE 1 Bone density changes in women with age. Lumbar spine bone mineral density (BMD) in g/cm2 plotted against age in years. The middle line is the mean BMD over the age range, and the shaded area includes the mean (2 standard deviations) (reproduced with permission from Hologic, Inc).

tion. The mechanism for the genetic effect is elusive. There have been conflicting results linking bone mass to vitamin D receptor genotypes, and work is being done to evaluate the contribution of various estrogen receptor genotypes to peak bone mass. The contribution of ethnicity is also important because blacks have a higher peak bone mass than whites or Asians. Achieving genetically programmed peak bone mass is dependent on adequate calcium intake, physical activity, and endocrine factors including sex steroids. The optimal calcium intake during adolescence, the major period of bone accretion, is 1200–1500 mg/day (Table 2).

After the consolidation of peak bone mass in the third decade, women and men start to lose bone at a slow rate that continues throughout life. Bone loss accelerates in women when estrogen deficiency starts and results in a 5–15% loss in bone in the first 5 years after natural or surgical menopause. Physiologic doses of estrogen prevent or retard these losses. Estrogen deficiency associated with excessive exercise, eating disorders, hyperprolactinemia, or with gonadotropin analog therapy for endometriosis is also complicated by accelerated bone loss. Estrogen deficiency results in a disproportionate loss of trabec-

TABLE 2
Optimal Daily Calcium Intakes
(NIH Consensus Statement 1994)

	Intake (mg)
Children	
1–5 years	800
6–10 years	800–1200
Adolescents	1200–1500
Adults 25–50 year	1000
Women	
Pregnant or lactating	1200
Postmenopausal, on estrogen	1000
Postmenopausal, not on estrogen	1500
Elderly (age >65 years)	1500

ular bone, which is the predominate type of bone in the spine.

After this period of estrogen-related bone loss, age-related bone loss continues at a rate of 0.5–1%/year and is secondary in part to calcium and vitamin D deficiencies and continued estrogen deficiency. Intestinal calcium absorption and the ability to adapt to low-calcium diets are impaired in many elderly persons. The pathogenesis of these abnormalities is controversial, but evidence suggests that they may be due to a decrease in the ability of the kidney to produce $1,25(OH)_2$ vitamin D and/or an intrinsic age-related defect in the intestine to absorb calcium. The finding that parathyroid hormone levels increase with age implies that those defects in calcium absorption are functionally important, and that elderly persons are forced increasingly to rely on their own bones rather than the external environment as a source of calcium for maintaining normal extracellular fluid calcium. The impact of the calcium and vitamin D deficiencies on bone mass and fracture risk are difficult to quantify, but calcium and vitamin D supplementation in elderly women has been demonstrated in clinical trials to eliminate bone loss and reduce risk of hip fractures. A number of other factors can cause or exacerbate bone loss. They include:

A. Thin Body Build

Thin women have twice the risk of hip and Colles' fractures as obese women. The reason is not known, but obese women may have a greater supply of estrogen during their postmenopausal years than thin women because the major source of estrogen during this time is from the conversion of androstenedione to estrone in adipose tissue. Another possible explanation for the increase in hip fractures is that thin women have less fat "padding" protecting them when they fall.

B. Cigarette Smoking

Most studies show that smoking is associated with reduced bone mass, and this appears to be mediated by altered estrogen metabolism and reduced estrogen levels. Smoking also causes a reduction in the amount of adipose tissue in most smokers, which will result in decreased endogenous estrogen production as described previously.

C. Alcohol

Alcohol use reduces bone formation directly by inhibiting osteoblast function. Alcohol abuse is also associated with nutritional deficiencies, lean body build, and liver disease, which can all contribute to bone loss. The increased risk of falls also contributes to the fracture risk. Alcohol use is less frequently a cause of fractures in women than in men, but the fracture risk is clearly increased when alcohol abuse is severe (>15 g alcohol per day) and when body weight is low. The effects of moderate alcohol use on bone density and fracture risk are less clear.

D. Physical Activity

Immobilization can decrease bone mass. The bone loss produced can be localized (e.g., associated with fracture casting or painful limbs), generalized (associated with prolonged bed rest), or neurologic (associated with paraplegia or quadriplegia). The mechanisms are not known, but it is thought that the absence of stress and muscle pull on bone may be the major etiologic factor.

The results of studies of the influence of increased physical activity on bone mass are mixed. Several controlled trials have shown that postmenopausal bone loss may be inhibited by moderate weight-bearing exercise, but this finding is not universal. Unfortunately, few of these studies used randomized designs. In premenopausal women, it has been well demonstrated that exercise sufficient to produce amenorrhea can result in marked decreases in bone mineral density.

E. Medications

1. Glucocorticoids

Excessive glucocorticoids can cause severe osteoporosis both in persons receiving pharmacologic doses of these drugs and in persons with endogenous hypercortisolism such as Cushing's disease. Biochemical markers of bone metabolism show changes within 48 hr of large doses of glucocorticoids, and bone loss can be measured in many patients within a few months of the initiation of therapy. Glucocorticoids have multiple effects on bone and mineral metabolism. Glucocorticoids act directly on the osteoblast to inhibit bone formation. Intestinal absorption of calcium is inhibited and renal excretion of calcium is increased, resulting in secondary hyperparathyroidism and increased bone resorption. The combined effects of diminished bone formation and increased bone resorption result in marked negative bone balance and a rapid decrease in bone mass, especially at skeletal sites containing primarily trabecular bone such as the spine.

2. Thyroid Hormone

Bone turnover is increased in persons with hyperthyroidism and in persons taking excessive doses of thyroid hormones. As a result, bone density is decreased and fracture risk is increased in individuals with hyperthyroidism, but these appear to be reversible with the successful treatment of hyperthyroidism. In postmenopausal women taking suppressive doses of thyroid hormone, the rate of bone loss is increased but may normalize with estrogen or bisphosphonate therapy.

3. Anticoagulants

Long-term heparin therapy is associated with decreased bone density and fractures. There are little data on the skeletal effects of low-molecular-weight heparin.

4. Anticonvulsant Drugs

Chronic therapy with anticonvulsants induces changes in hepatic vitamin D metabolism which leads to bone loss. Osteoporosis is preventable with vitamin D supplementation in this population.

5. Caffeine

A high caffeine intake exacerbates bone loss in individuals with inadequate calcium intake. In postmenopausal women with adequate calcium intakes, caffeine does not contribute to bone loss.

F. Other Risk Factors for Fracture

While bone density and the level of bone turnover are the most important predictors of osteoporotic fractures, there are a number of other independent risk factors which reflect either qualities of bone not measured by bone density or risk of falling. The Study of Osteoporotic Fractures, a prospective study of almost 10,000 noninstitutionalized elderly women, was instrumental in identifying or confirming many of these factors. These factors include a history of previous fractures, history of maternal hip fracture, physical inactivity, inability to rise from a chair, and use of long-acting benzodiazepines. Assessment of these risk factors is an important adjunct to measuring bone density.

III. EVALUATION

A. Quantitative Measurement of Bone Mass

Bone mineral density (BMD) correlates strongly with bone strength as well as fracture risk. Accurate, highly reproducible measurements of bone density at various sites of the skeleton have been available for more than a decade. This has allowed the use of bone density measurements to diagnose osteoporosis before the complications of fractures, to assess fracture risk in an individual, and to monitor progression of osteoporosis or response to therapy. The development of "dual-energy" technologies, such as dual-photon absorptiometry (DPA) and dual X-ray absorptiometry (DXA), allowed the measurement of "central" skeletal sites, such as the hip and lumbar spine, where there is variable soft tissue composition and thickness. DXA is rapid, accurate, and precise (with a measurement error of 1% for the spine). The dose of radiation is also very low (1–3 μSv) compared with a standard chest X ray (50 μSv). It has become the standard for measurement of bone mineral density. The utility of posteroanterior lumbar spine DXA measurements is decreased in elderly persons due to the prevalence of artifacts such as aortic calcification, osteophytes, and other degenerative changes that will artificially elevate the measured

BMD. Lateral lumbar spine measurements may be more accurate because the projection will exclude these sources of error, but the reproducibility of the lumbar spine measurements has not been optimal. Another technique uses quantitative computer-assisted tomography (QCT) to obtain direct measurements of trabecular bone mass in the central portion of a given vertebra. This is in contrast with DXA, which measures bone mass in entire vertebrae, including cortical and trabecular bone, as well as overlying tissues (e.g., calcified aorta). QCT is therefore a more specific measurement of the trabecular bone in the vertebral bodies, but it is more expensive and exposes the patient to higher doses of radiation.

Measurements of peripheral skeletal sites using DXA or QCT have recently become widely available and have been proposed as screening tools. However, their utility is not established in monitoring disease progression or response to therapy, and their ability to predict osteoporotic fracture is low compared with central sites, such as the hip and lumbar spine. Quantitative ultrasound has recently been used to assess bone at the calcaneus, patella, tibia, or phalanges and may provide additional information regarding bone strength independent of bone density. These peripheral measurements are attractive due to the low cost, portability, and, in the case of ultrasound, elimination of radiation exposure.

Bone density measurements are compared with the mean BMD for young women who have attained peak bone density and expressed as the number of standard deviations above or below this reference point (T score). The World Health Organization has defined osteoporosis as a bone density below a T score of -2.5. Individuals with T scores below -2.5 who have had one or more osteoporotic fractures are defined as having severe osteoporosis. Individuals with T scores between -1.0 and -2.5 are classified as having osteopenia, whereas T scores above -1.0 are consistent with normal bone density. Bone density measurements are also valuable prognostically because the risk of hip fracture increases two or threefold for every standard deviation below the mean for an age- and gender-matched reference population.

While there is some controversy as to the cost-effectiveness of bone density measurements in individuals at risk for the development of osteoporosis, there is little question that they are essential for deciding who needs treatment for established osteoporosis and who needs preventive therapy against menopausal bone loss.

B. Evaluation of Individuals with Low Bone Density or Fracture with Minimal Trauma

The most likely reason for low bone mass detected by DXA or QCT measurement or fracture due to minimal trauma is the failure to achieve genetically programmed peak bone mass during adolescence and young adulthood or bone loss secondary to factors such as estrogen and calcium deficiency. However, there are many factors outlined in Table 1 that mimic or complicate postmenopausal or age-related osteoporosis, and it is essential that they be kept in mind during the clinical and laboratory evaluation. For example, multiple myeloma and diffuse malignant metastases to the vertebrae are commonly misdiagnosed as osteoporosis. Total serum calcium, phosphate, alkaline phosphatase, protein electrophoresis, serum intact PTH, thyroxine, thyrotropin, and urinalysis should be measured in all patients with osteoporosis. Serum 25-hydroxyvitamin D should be measured in persons suspected of sunlight deprivation or malabsorption. In individuals with postmenopausal or age-related osteoporosis, the values for all these indices should be within the normal range. Lateral X rays of the thoracolumbar spine are valuable to determine the number and type of preexisting vertebral fractures, which will serve as a baseline and as a predictor of future fractures. Bone mineral density measurements of both the proximal femur and the vertebrae should be obtained at a facility that can offer long-term accurate and consistent results for comparisons. The rate of bone turnover can be assessed by measuring biochemical markers of bone metabolism. Osteocalcin, bone alkaline phosphatase, and procollagen I peptide can all be used to assess the level of bone formation. Urinary excretion of collagen crosslinks, including deoxypyridinoline and N-telopeptide, can be used to assess levels of bone resorption. These markers may not only be useful in predicting bone loss but also there is evidence that the resorption markers, as an index of bone turnover, may independently predict fracture risk.

IV. TREATMENT

A. General Measures

In addition to addressing risk factors for bone loss, specific risk factors for falls and fractures need to be addressed. The individual's home environment needs to be examined by the individual with the help of a family member to find physical hazards that can be eliminated. Examples of these include loose area rugs, exposed telephone or lamp cords, poorly lit stairs and hallways, and the absence of grab bars in the bathroom. Other risk factors, such as sedatives, muscle weakness, postural hypotension, and uncorrected visual deficits, should also be addressed. Initiation of an exercise program is also valuable with regular weight-bearing exercise to preserve bone mass and to maintain good neuromuscular conditioning. Supervised weight training may be of added benefit to reverse the muscular deconditioning so many of these patients suffer, and this training results in a improvement in their ability to prevent falls as well as improving their sense of well-being and quality of life.

B. Calcium

Calcium requirements typically exceed dietary intake in postmenopausal and elderly women, making supplementation mandatory in order to meet the goals described in Table 2. Concerns about caloric intake and cholesterol limit the intake of dairy products in many women.

Calcium supplementation in the range of 500–1000 mg elemental calcium per day is usually required after an assessment of dietary intake. The most common supplements used are calcium carbonate and calcium citrate. Because calcium carbonate has enhanced availability in an acidic environment, it needs to be taken with meals and is not a reliable source in elderly individuals that are achlorhydric. Calcium carbonate is much less expensive than calcium citrate, but its use is occasionally complicated by abdominal bloating, gaseousness, and constipation. Individuals with a history of nephrolithiasis should have a more detailed evaluation, and if supplemented, calcium citrate is the best choice because the excreted citrate will increase calcium solubility. Vitamin D supplementation (400–800 IU/day) will also ensure adequate calcium absorption. Multivitamins are a convenient source for vitamin D because they typically contain 400 IU per tablet.

C. Estrogens

Estrogen prevents bone loss in oophorectomized and postmenopausal women, and long-term estrogen therapy decreases the fracture rate by approximately 50%. Estrogen needs to be taken in doses equivalent to 0.625 mg of conjugated equine estrogen, but the route, type of estrogen, or regimen does not appear to be important. Estrogen use must be maintained to prevent fractures because bone turnover increases and bone loss resumes upon its discontinuation. The risks of estrogen are described in detail elsewhere. Estrogen replacement therapy remains the prime choice for the prevention and treatment of osteoporosis. The side effects of hormone replacement therapy, including bleeding and breast pain and engorgement, and concerns about breast cancer risks are the largest barriers to compliance and widespread estrogen use in the postmenopausal period. Therefore, the decision to initiate long-term treatment should be made only after the patient has considered its advantages and disadvantages with the sympathetic support and guidance of her physician.

D. Selective Estrogen Receptor Modulators

A new class of drugs under development takes advantage of some of the positive effects of estrogen on bone and lipid profiles but bypasses some of the side effects or risks with estrogen therapy through tissue-selective estrogen receptor agonist and antagonist actions. The first of these, raloxifene, has been approved by the Federal Food and Drug Administration (FDA) for the prevention of osteoporosis. Raloxifene has antiresorptive actions and prevents postmenopausal bone loss similar to estrogen when used at a dose of 60 mg daily. It also lowers total and low-density lipoprotein cholesterol, does not stimulate the endometrium, and may reduce the risk of breast cancer below that of placebo-controlled postmenopausal women. Raloxifene has not yet been approved for treatment of established osteoporosis.

E. Calcitonin

Calcitonin inhibits osteoclastic activity and has been approved by the FDA for the treatment of osteoporosis. Salmon calcitonin is most frequently used because it is 30–50 times more potent than the human form. It has been given parenterally but is now approved in the more convenient and equally effective intranasal spray, to be used at doses of 200 IU daily. It is effective at preventing further bone loss and decreases the risk of vertebral fractures by about 35%. Calcitonin is an extraordinarily safe drug, and intranasal administration appears to decrease the associated side effects seen with subcutaneous use, such as nausea, bloating, and flushing. The long-term efficacy of salmon calcitonin may be limited by desensitization or by the development of neutralizing antibodies with long-term use.

F. Bisphosphonate Compounds

Bisphosphonates are pyrophosphate analogs that adhere to hydroxyapatite but are not metabolized. They inhibit osteoclastic bone resorption and have varying effects on bone mineralization. Bisphosphonates appear safe, are generally well tolerated, and have few extraskeletal effects. Alendronate is a second-generation bisphosphonate that is potent at inhibiting bone resorption and does not inhibit bone mineralization at the doses used clinically. Alendronate has been demonstrated to increase bone density 8% at the spine and 6% at the femoral neck over 3 years of therapy (10 mg daily) in osteoporotic women. It also reduces the risk of hip and vertebral fractures by 50% in this population. Alendronate has also been demonstrated to be effective in preventing osteoporosis in early postmenopausal women at doses of 5 mg daily. Because of its poor intestinal absorption, it must be taken on an empty stomach at least 30 min before food intake. Its main side effect is irritation of esophageal and gastric mucosa and individuals with gastroesophageal reflux should avoid it.

Sodium etidronate, a first-generation bisphosphonate, is also available. Cyclical therapy has been shown to be effective: Two long-term studies demonstrated that etidronate prevented bone loss when given in doses of 400 mg daily by mouth for 2 weeks out of every 3 months. Its effect on fracture incidence was difficult to evaluate because so few fractures occurred in both the treatment and the control groups. Bone biopsies obtained in one of these studies did not reveal mineralization defects. It has also been demonstrated to prevent bone loss in early postmenopausal women. Etidronate does cause bone mineralization defects when given continuously for more than 4 months at doses exceeding 400 mg daily. Etidronate is FDA approved for Paget's disease but has not been approved for use in osteoporosis.

G. Fluoride

Sodium fluoride is one of the few agents that is capable of stimulating bone formation. Unfortunately, two randomized long-term studies have shown that fluoride, at doses of 50 mg/day, had little effect on vertebral fracture incidence and actually increased the occurrence of appendicular fractures, even though vertebral bone mass increased dramatically in most participants. Recently, lower dose, slow-release sodium fluoride has been demonstrated to dramatically reduce vertebral fractures without increasing appendicular fractures. It is clear that the therapeutic window for fluoride is narrow, and more clinical trials are needed using low doses of fluoride before it can be recommended for the treatment of osteoporosis.

H. Parathyroid Hormone

Small intermittent doses of the 1–34 fragment of PTH stimulates bone formation in several animal species and has been demonstrated to increase vertebral bone density in postmenopausal women taking hormone replacement therapy by 13% over 3 years. This is in contrast to its ability to stimulate bone resorption in animals when it is given continuously in large doses. Human PTH (1–34) therefore has been considered as the only available alternative to fluoride as a means of stimulating bone formation in osteoporotic individuals. More definitive and larger clinical trials are under way to evaluate human PTH (1–34) therapy in osteoporosis.

V. CONCLUSION

There have been a number of advances in the past few years in not only the treatment of osteoporosis and prevention of fractures but also the prevention of osteoporosis. While the effects of estrogen on the skeleton have long been known, its widespread use as a prevention for osteoporosis has been limited by its numerous side effects as well as by concern about its effect on breast cancer risk. There are now a number of other options in the prevention of osteoporosis, including raloxifene, alendronate, and etidronate. The decision to start preventive therapy should be made after the evaluation of bone density and a thorough discussion of the risks and benefits of the potential therapies. We have the potential to prevent the painful and crippling complications of osteoporosis by addressing it at the time of menopause. The treatment of established osteoporosis should also be undertaken after a thorough evaluation for causes of secondary osteoporosis and other factors affecting bone metabolism. The treatment options currently include antiresorptive therapies, including estrogen, bisphosphonates, and calcitonin. In the future, we will most likely have the potential to restore large amounts of bone that has been lost using anabolic agents currently being studied, such as PTH.

See Also the Following Articles

ESTROGEN REPLACEMENT THERAPY; MENOPAUSE

Bibliography

Bikle, D. D. (1997). Biochemical markers in the assessment of bone disease. *Am. J. Med.* 103(5), 427–436.

Black, D. M., Cummings, S. R., Karpf, D. B., Cauley, J. A., Thompson, D. E., Nevitt, M. C,, et al. (1996). Randomized trial of effect of alendronate on risk of fracture in women with existing vertebral fractures. *Lancet* 348, 1535–1541.

Cummings, S. R., Nevitt, M. C., Browner, W. S., Stone, K., Fox, K. M., Ensrud, K. E., et al. (1995). Risk factors for hip fracture in white women. Study of Osteoporotic Fractures Research Group. *N. Engl. J. Med.* 332, 767–773.

Genant, H. K., Engelke, K., Fuerst, T., Gluer, C. C., Grampp, S., Harris, S. T., et al. (1996). Noninvasive assessment of bone mineral and structure: State of the art. *J. Bone Miner. Res.* 11, 707–730.

Hol, T., Cox, M. B., Bryant, H. U., and Draper, M. W. (1997). Selective estrogen receptor modulators and postmenopausal women's health. *J. Women's Health* 6(5), 523–531.

Johnston, C. C., Jr., and Slemenda, C. W. (1995). Pathogenesis of osteoporosis. *Bone* 17(Suppl. 2), 19S–22S.

King, M. B., and Tinetti, M. E. (1996). A multifactorial approach to reducing injurious falls. *Clin. Geriatr. Med.* 12(4), 745–759.

Lindsay, R., Nieves, J., Formica, C., Henneman, E., Woelfert, L., Shen, V., Dempster, D., and Cosman, F. (1997). Randomised controlled study of effect of parathyroid hormone on vertebral-bone mass and fracture incidence among postmenopausal women on oestrogen with osteoporosis. *Lancet* 350(9077), 550–555.

Nelson, D. A., and Kleerekoper, M. (1997). The search for the osteoporosis gene. *J. Clin. Endocrinol. Metab.* 82(4), 989–990.

Riggs, B. L., and Melton, L. J., III (1995). *Osteoporosis: Etiology, Diagnosis, and Management*, 2nd ed. Lippincott-Raven, New York.

Taunton, J. E., Martin, A. D., Rhodes, E. C., Wolski, L. A., Donelly, M., and Elliot, J. (1997). Exercise for the older woman: Choosing the right prescription. *Br. J. Sports Med.* 31(1), 5–10.

The Writing Group for the PEPI (1996). Effects of hormone therapy on bone mineral density: Results from the postmenopausal estrogen/progestin interventions (PEPI) trial. *J. Am. Med. Assoc.* 276(17), 1389–1396.

Ovarian Cancer

Mark K. Dodson and Jason L. Johnson

University of Utah School of Medicine

I. Classification of Ovarian Cancers
II. Epidemiology
III. Genetic Predisposition
IV. Screening
V. Presentation and Evaluation
VI. Patterns of Spread
VII. Prognostic Factors
VIII. Staging
IX. Treatment
X. Survival
XI. Future Directions

GLOSSARY

debulking Removal of as much cancer as possible.
epithelium Cells that compromise the surface of an organ.
hereditary cancer syndrome Groups of individuals that carry a genetic defect which can predispose them to developing cancer.
lymphatics A system of ducts and nodules that help filter out invaders and transport fluid and protein in the body.
staging The categorization of ovarian cancer as based on the extent of spread; helps determine treatment and prognosis.

Ovarian cancer is a malignancy which develops as a primary disease in one or both female gonads. This disease continues to be a difficult diagnostic and clinical problem. We have made few strides in the treatment of ovarian cancer and patient survival has changed very little in the past 20 years. This article will attempt to outline the difficulties in diagnosing and treating this disease and the future directions that current research may take us.

I. CLASSIFICATION OF OVARIAN CANCERS

The epithelial (surface cells of the ovary) ovarian cancers are by far the most common and account for approximately 90% of ovarian malignancies. These are further divided into histologic cell types. Serous is the most common and represents 75% of the epithelial ovarian cancers, whereas mucinous tumors account for 20%. The remaining cell types are endometrioid, clear cell, Brenner, and undifferentiated. The bulk of this article will discuss epithelial ovarian cancers.

The remaining 10% of ovarian malignancies are composed of germ cell tumors and sex cord–stromal cell tumors or metastatic tumors to the ovary. All these tumors represent a very interesting contrast to the epithelial tumors because many secrete biologically active substances or markers. For instance, the granulosa–stromal cell ovarian tumors can excrete high levels of estrogen causing endometrial hyperplasia or frank malignancy of the uterus. Conversely, the Sertoli–Leydig ovarian tumors can secrete androgens and produce masculinization. The germ cell tumors typically carry a better prognosis than all other ovarian cancers because they are very sensitive to chemotherapy. Further details of these tumors are beyond the scope of this article.

II. EPIDEMIOLOGY

Ovarian cancer is the fifth leading cause of cancer deaths in women. There are 24,000 new cases and 13,600 deaths each year from ovarian cancer. A women's lifetime risk of developing ovarian cancer is 1 in 70 or a 1.4% chance. This disease typically strikes patients of certain age depending on the type of tumor and possible predisposing risk factors. The germ cell tumors occur most often in the first two decades of life, whereas the epithelial ovarian carcinomas occur in the sixth and seventh decades.

The identified risk factors for ovarian cancer usually relate to uninterrupted ovulation which likely results in ovarian surface epithelial damage. Uninterrupted ovulation occurs in patients who are nulligravid (have never had a pregnancy) or those that have had a pregnancy but did not breast-feed, therefore allowing ovulation to resume earlier in the postpartum period. Additionally, women who take birth control pills will not ovulate and therefore be protected against ovarian cancer.

III. GENETIC PREDISPOSITION

It is not uncommon for a woman to request physician consultation secondary to having a family history of ovarian cancer. The term ovarian cancer may mean many things to the layperson, including ovarian cysts or benign ovarian enlargement and less likely ovarian cancer. Therefore, it would behoove both the physician and the patient with the family history to obtain the pathology report(s) from the family member(s) to determine the actual risk to the patient. The simple act of obtaining pathology reports will help limit expensive testing and even provide reassurance to an otherwise frightened patient.

Three distinct hereditary cancer syndromes have been identified: the BRCA1 (breast and ovarian), site-specific ovary, and Lynch II. Patients in the BRCA families are predisposed to early onset breast cancer and/or ovarian cancer as the result of a germline mutation in a gene on chromosome 17q, also known as BRCA1. Those patients in the site-specific cancer syndrome develop only ovarian cancer and the specific gene leading to the predisposition has not been positively identified, but many investigators believe that BRCA1 may be that gene. The Lynch II syndrome consists of families who predominantly develop hereditary nonpolyposis colorectal cancer (HNPCC). Additionally, those in a Lynch II family are at an increased risk of developing endometrial and ovarian cancer.

If a particular patient has documented familial cases of ovarian cancer a risk assessment can be made for that patient. Those patients with one first-degree (mother, sister, or daughter) or second-degree (grandmother, aunt, first cousin, or granddaughter) relative affected with ovarian cancer harbor a 3.7% lifetime risk of developing ovarian cancer themselves. Patients with two or more relatives with ovarian cancer have a 5.5% lifetime risk of developing the disease. In contrast, patients in a hereditary ovarian cancer syndrome family have up to a 50% lifetime risk of developing the disease because the hereditary syndromes are thought to be inherited in an autosomal-dominant fashion. Individual risks will vary depending on penetrance. Additionally, those with a known mutation at BRCA1 have a 63% risk of developing ovarian cancer and an 85% risk of developing breast cancer by age 70. The risk of developing one of these two cancers by age 70 is 95%.

It must be emphasized that most ovarian cancers develop secondary to a sporadic event. Only 5–10% of cases can be attributed to familial or hereditary causes. If patients do have a genetic predisposition or belong to a family cancer syndrome they will likely manifest ovarian cancer at a younger age. The sporadic cases occur at a mean age of 59 years compared with 52 years for BRCA1, 49 years for site-specific ovary cases, and 45 years for Lynch II cases.

IV. SCREENING

Screening for ovarian cancer remains very difficult and most patients will present to their physicians with advanced disease. Patients with advanced disease have a diminished survival and this, in part, accounts for the lack of significant progress in improving survival during the past 20 years. Several modalities for screening have been, and are being, investigated.

CA125 represents a tumor marker for "peritoneal irritation": It is elevated with ovarian cancer; however, it is also elevated with almost any other intraperitoneal process (i.e., pancreatitis, diverticulitis, endometriosis, etc.). It therefore lacks specificity for ovarian cancer. It is not recommended for routine screening in the general population because of the lack of specificity.

Transvaginal ultrasound is another important tool in the management of patients; however, its role in screening for ovarian cancer is questionable. When this modality is used for ovarian cancer screening it appears to lack specificity and if used alone may require 32 laparotomies to discover one ovarian cancer.

The current recommendation for screening the general population of women for ovarian cancer is an annual pelvic exam. CA125 appears to provide guidance for treatment in patients with known ovarian cancer and can be a marker for recurrent disease. It is most useful in postmenopausal women in whom it is more sensitive and specific and is being investigated in high risk patients. The transvaginal ultrasound is a useful adjuvant for diagnosis of pelvic masses or suspected ovarian enlargement.

V. PRESENTATION AND EVALUATION

The difficulty with ovarian cancer lies in the fact that it presents with symptoms in most cases, but at a point when widespread disease is typically present. A detailed history will often reveal early symptoms that are vague and nonspecific. Patients may have pressure symptoms or bowel irregularities, pain, and possibly dyspareunia (pain with intercourse). Once the disease has progressed the symptoms become characteristic such as bloating, distention, constipation, nausea, vomiting, weight loss, and early satiety (quickly becoming full with meals). Ovarian cancer is referred to as "The cancer commonly bathed in antacids." When these symptoms are present the patient likely has stage III disease.

A detailed general physical examination should be performed and will often show a pelvic mass that is solid, fixed, and irregular in character with occasional palpable nodules on rectovaginal exam. In addition, all lymph nodes should be evaluated thoroughly and a full clinical exam should be performed, paying particular attention to the abdominal exam for the presence of an omental cake (tumor invading the fatty drape that overlies the bowel). Signs of ascites (fluid collected in the abdomen) should be elicited. The chest should be evaluated for the presence of pleural effusions.

Further evaluation should include laboratory studies that will identify any other occult disease. Liver function tests may be elevated with the presence of hepatic parenchymal metastasis. Renal function should be assessed in addition to a complete blood count. Most gynecologic oncologists would also obtain a preoperative CA125 which aids with postsurgical treatment and follow-up.

Imaging studies have an important role in medicine; however, they do not play a major role in the evaluation of patients presenting with ovarian cancer. Many of these patients will have had an ultrasound to characterize any pelvic mass that has been palpated. All patients should have a preoperative chest X ray to evaluate for pleural effusions and possible need for pleuracentesis (drawing fluid out from around the lungs). The fluid would then be evaluated for malignant cells. A magnetic resonance imaging test or computed tomography test are usually not indicated because, regardless of the findings, surgical exploration is necessary in all suspected ovarian cancers.

VI. PATTERNS OF SPREAD

"Like leaves falling from a tree." This analogy describes the most common mode of metastatic spread of ovarian cancer. The epithelial cells are thought to exfoliate from the ovary and are carried around the abdomen in the peritoneal fluid. This fluid travels in a clockwise fashion driven by the patient's respirations. In turn, the ovarian cancer cells are disseminated throughout the abdominal cavity and begin to grow on the surfaces that they touch.

Lymphatic channels are an important mode of tumor dissemination. Commonly with stage III disease lymph nodes will be found to have metastatic cancer cells. Hematologic spread is unusual with ovarian cancer and occurs in only 2 or 3% of patients.

VII. PROGNOSTIC FACTORS

Ovarian cancer commonly presents with widely metastatic disease. Several studies have shown that patients outcome is directly related to the degree of tumor debulking (removing the bulk of tumor) achieved. The goal of the gynecologic oncologist is to remove all the cancer from every patient. When this is not possible, "optimal debulking" is attempted and this represents diminishing the size of any and all tumor nodules to <2.0 cm. If residual tumor remains, the smaller the volume the better the prognosis.

The stage of the ovarian cancer is critical to prognosis and future treatment recommendations. It is imperative that each patient with ovarian cancer undergoes a proper surgical staging procedure by a qualified gynecologic oncologist. A recent study suggests that patients operated on by gynecologic oncologists have a better survival and are more properly staged. Survival by stage will be outlined later.

The tumor grade plays a significant role in patient survival. An important category of ovarian cancer is the borderline tumors. Their histological appearance and biological behavior are intermediate between the benign and frankly malignant ovarian neoplasms. The patients with borderline ovarian tumors usually experience a prolonged survival even without the use of chemotherapy. Frankly invasive low-grade ovarian cancers have a much better prognosis than the higher grade group. If a patient has a high-grade tumor, sometimes referred to as "highly aggressive," her prognosis is poor.

Many tumor markers and genetic markers are being evaluated for prognostic purposes. The ploidy (number of chromosomes within each cell) of a tumor is highly correlated with survival. Protooncogenes have recently been described as prognostic factors. For example, ovarian cancers with overexpression of *HER-2/neu* show a poorer prognosis.

VIII. STAGING

FIGO Staging for Primary Ovarian Carcinoma

Stage I: Growth limited to the ovaries.
    Ia. Confined to one ovary only.
    Ib. Confined to both ovaries only.
    Ic. Either a or b but with capsule rupture, ascites containing malignant cells, or positive peritoneal washings.

Stage II: Growth involving one or both ovaries but extending to the pelvis.
    IIa. Extension and/or metastasis to the uterus and/or tubes.
    IIb. Extension to other pelvic organs.
    IIc. Either a or b with capsule rupture, ascites containing malignant cells, or positive peritoneal washings.

Stage III: Growth involving one or both ovaries but extending outside the pelvis, including superficial liver metastasis, positive retroperitoneal or inguinal lymph nodes, and histologic proof of spread outside of the true pelvis.
    IIIa. Tumor limited to true pelvis with negative lymph nodes but histologically confirmed microscopic seeding of abdominal peritoneal surfaces.
    IIIb. Tumor of one or both ovaries and histologically confirmed visible implants <2 cm in size, negative lymph nodes.
    IIIc. Abdominal implants of >2 cm and/or positive retroperitoneal or inguinal lymph nodes.

Stage IV: Growth involving one or both ovaries and the presence of distant metastasis. Positive pleural fluid for carcinoma cells or parenchymal liver mets are also included in stage IV.

IX. TREATMENT

A. Surgery

All patients with suspected ovarian cancer need surgical exploration. The surgical techniques involved include obtaining ascites or washings for malignant cells, systematic exploration, multiple biopsies, resection of disease, and debulking (which typically includes a hysterectomy and bilateral salpingooophorectomy, which is removal of Fallopian tubes and ovaries). An aggressive approach confers the patient a survival advantage as discussed previously.

B. Chemotherapy

All patients are candidates for chemotherapy except those with stage Ia grade I tumors. The primary agents used for ovarian cancer include platinum-based chemotherapy (i.e., cisplatin) plus taxol. These can be used as single agents in recurrent disease but combination chemotherapy is currently the treatment of choice for primary disease. Many ongoing trials are evaluating the optimal regimen of chemotherapy.

C. Radiation

Radiation therapy is effective against ovarian cancer; however, it has many complications when given in a "whole abdomen" dose. For this reason it is not frequently used. Another novel approach to administering radiation to tumor cells is the use of intraperitoneal radiocolloids (^{32}P). The ^{32}P is infused into the abdomen and allowed to disperse over the peritoneal surfaces where the ovarian carcinoma cells have implanted. The radiocolloid then emits radiation that destroys those cells receiving a sufficient dose.

X. SURVIVAL

5-Year Survival by FIGO Stage

Stages I and II:	80–100%
Stage IIIa:	30–40%
Stage IIIb:	20%
Stages IIIc and IV:	5% or less

XI. FUTURE DIRECTIONS

Intensive research is ongoing in an effort to identify novel treatments and detection regimens for ovarian cancer. Recent work has focused on boosting the body's own immune system to destroy foreign tissues when recognized. Also, cancers need an abundant blood supply to meet the demands of their rapid growth. Researchers are investigating the possibility of turning off the blood vessel proliferation associated with rapidly growing cancers. Gene therapy is also being evaluated in a clinical trial to determine its efficacy in advanced ovarian cancers.

Bibliography

Averette, H., and Nguyen, H. N. (1994). The role of prophylactic oophorectomy in cancer prevention. *Gynecol. Oncol.* **55**, S38–S41.

Baak, J. P., Chan, K. K., Stolk, J. G., *et al.* (1987). Prognostic factors in borderline and invasive ovarian tumours of the common epithelial type. *Pathol. Res. Pract.* **182**, 755.

Barnes, M. N., *et al.* (1997). Gene therapy and ovarian cancer: A review. *Obstet. Gynecol.* **89**(1), 145–155.

Booth, M., Beral, V., and Smith, P. (1989). Risk factors for ovarian cancer: A case control study. *Br. J. Cancer* **60**, 592.

Boring, C. C., *et al.* (1994). Cancer statistics, 1994. *Cancer J. Clin.* **44**(1), 7–26.

Carlson, K. J., *et al.* (1994). Screening for ovarian cancer. *Ann. Intern. Med.* **121**(2), 124–132.

Easton, D. F., *et al.* (1995). Breast and ovarian cancer incidence in BRCA-1 mutation carriers. *Am. J. Hum. Genet.* **56**, 265–271.

Einhorn, N., Sjovall, K., Knapp, R. C., *et al.* (1992). A prospective evaluation of serum CA 125 levels for early detection of ovarian cancer. *Obstet. Gynecol.* **80**, 14.

Folkman, J. (1996). New perspectives in clinical oncology from angiogenesis research. *Eur. J. Cancer* **32A**(14), 2534–2539.

Friedlander, M. L., Headley, D. H., Taylor, I., *et al.* (1984). Influence of cellular DNA content on survival in advanced ovarian cancer. *Cancer Res.* **44**, 397.

Karlan, B. Y., *et al.* (1994). The current status of ultrasound and color doppler imaging in screening for ovarian cancer. *Gynecol. Oncol.* **55**, S28–S33.

Lynch, H. T., *et al.* (1985). Hereditary nonpolyposis colorectal cancer (Lynch syndromes I and II), parts I and II. *Cancer* **55**, 934–951.

Mayer, A. R., *et al.* (1992). Ovarian cancer staging: Does it require a gynecologic oncologist? *Gynecol. Oncol.* **47**, 223–227.

McGuire, W. P. (1993). Primary treatment of epithelial ovarian malignancies. *Cancer* **71**, 1541–1550.

Miki, Y., *et al.* (1994). A strong candidate for the breast and ovarian cancer susceptibility gene BRCA1. *Science* **266**, 66–71.

Omura, G. A., Brady, M. F., Homesley, H. D., *et al.* (1991). Long-term follow-up and prognostic factor analysis in advanced ovarian carcinoma: The Gynecologic Oncology Group experience. *J. Clin. Oncol.* **9**, 1138.

Piver, M. S., Barlow, J. J., Lele, S. B., *et al.* (1982). Intraperitoneal chromic phosphate in peritoneoscopically confirmed Stage I ovarian adenocarcinoma. *Am. J. Obstet. Gynecol.* **144**, 836.

Schildkraut, J. M., and Thompson, W. D. (1988). Familial ovarian cancer: A population-based case-control study. *Am. J. Epidemiol.* **128**, 456–466.

Scully, R. E. (1979). Tumors of the ovary and maldeveloped gonads, Fascicle 16. Armed Forces Institute of Pathology, Washington, DC.

Slamon, D. J., Godolphin, W., Jones, L. A., *et al.* (1989). Studies of the HER-2/neu protooncogene in human breast and ovarian cancer. *Science* **244**, 707.

Steichen-Gersdorf, E., *et al.* (1994). Familial site-specific ovarian cancer is linked to BRCA1 on 17q12–q21. *Am. J. Hum. Genet.* **55**, 870–875.

Ovarian Cycle, Mammals

Michel Ferin
Columbia University

I. Introduction
II. Major Hormones of the Reproductive Cycle
III. The Gonadotropin-Releasing Hormone Pulse Generator
IV. The Gonadotropins
V. The Ovary
VI. The Genital Tract

GLOSSARY

corpus luteum A structure within the ovary which derives from the mature follicle, after it has ovulated. It has a finite life span of 12–14 days in the primate. It secretes both estradiol and progesterone.

estradiol A steroid secreted by the growing follicle and the corpus luteum. Through its feedback signal, it controls the release of gonadotropin-releasing hormone and luteinizing hormone.

follicle The structure within the ovary which surrounds the oocyte. While there is a large stockpile of follicles, only three to six are recruited in each reproductive cycle in the primate, and only one of these will mature. Its main product of secretion is estradiol.

follicle-stimulating hormone (FSH) A gonadotropin. Its main role is to recruit a new cohort of follicles at the beginning of each cycle.

gonadotropin-releasing hormone The major neurohormone which controls the cycle.

luteinizing hormone (LH) A gonadotropin. Its main role is to stimulate estradiol and progesterone secretion from the follicle and corpus luteum. A surge of LH serves as the ovulatory signal.

oocyte The female gamete.

ovulation The release of the oocyte from the mature follicle.

progesterone A steroid secreted by the corpus luteum.

The reproductive cycle (or ovarian cycle) is the result of several events taking place in geographically disparate organs of the female body, such as the brain, the pituitary gland, the ovaries, and the reproductive tract. While the cycle is driven by the hypothalamus and by the resultant gonadotropin secretion, it also depends on an orderly sequence of morphological events within the ovary. The overall sequence of cyclic events requires a remarkable coordination between morphological changes, hormonal secretion, and proper signaling, through hormonal feedback loops, between the ovary and the hypothalamic–pituitary unit.

I. INTRODUCTION

The two major goals of reproductive function in the female are the production of (a) mature oocyte(s) which upon release from the ovary can be fertilized and the preparation of the genital tract so that the fertilized egg can implant into the uterus. Hormonal dynamics are such that in most mammals the reproductive process occurs in a cyclic fashion in a predetermined sequence of events which includes three stages: the growth of the follicle (the follicular phase), the release of the oocyte from the mature follicle (ovulation), and the formation of a corpus luteum (the luteal phase). Species can be differentiated into the reflex ovulators (e.g., the rabbit, cat, mouse, and ferret), in which the ovulatory signal derives from the mating stimulus and the transmission of neural signals from the genital tract to the hypothalamus, and the spontaneous ovulators (e.g., the human, nonhuman primate, rat, guinea pig, and sheep), in which the ovulatory signal is an endogenous hormonal feedback. Corpus luteum formation can be classified as spontaneous, such as in the primate, cow, and sheep, when the luteal phase spontaneously follows ovulation and as induced, such as in the mouse, rat, and hamster, when full-blown luteal function requires cervical stimulation. In many species, cyclic reproductive function is influenced by the environment, primarily the photoperiodic stimulus which cues for seasonal breeding. Most mammals have distinct breeding seasons: During anestrus, photoperiodically controlled inhibitory neurons hold hormonal secretion in check so that normal follicular maturation and ovulation do not occur. The human cycle is most removed from environmental influences and occurs year-round. Cycle length is species specific, varying from 4 or 5 days in the rat to 28 days in the primate. In the latter, the uterine lining undergoes a gradual necrotic and ischemic change upon hormonal withdrawal at the end of the luteal phase resulting in menstruation (the menstrual cycle). This phenomenon is not seen in most lower species (the estrous cycle).

II. MAJOR HORMONES OF THE REPRODUCTIVE CYCLE

Reproductive hormonal markers are easily monitored by modern assay methods, such as radioimmunoassays, immunoradiometric assays, and enzyme-linked immunosorbent assays. Two of the major reproductive hormones are glycoproteins secreted by the anterior pituitary gland. These are luteinizing hormone (LH) and follicle-stimulating hormone (FSH). The most dramatic secretory change in both gonadotropins occurs during the preovulatory period, at which time they both rise abruptly several-fold over baseline. This gonadotropin "surge" acts as the signal for ovulation of the mature follicle, an event which in the primate occurs about 36 hr after the initiation of the surge. In addition, there is a distinct (but modest) rise in FSH in the early follicular phase, signaling the beginning of the cycle. The other two major reproductive hormones are steroids secreted by the ovaries: estradiol and progesterone. Estradiol is secreted both by the growing follicles and by the corpus luteum, whereas progesterone is mostly produced by the corpus luteum. An illustrative example of the cyclic secretory patterns of the gonadotropins and of these two ovarian steroids is shown in Fig. 1.

Proteins secreted by the ovary, such as the inhibins, also appear to play a role in the reproductive cycle. These proteins consist of a dimer, in which two different subunits (α and β) are bound together. Since there are currently two identified forms of the subunit (A and B), there are two identified forms of inhibin (inhibin A and inhibin B) with dissimilar secretory patterns. Inhibin A is secreted mainly during the intermenstrual period, whereas inhibin B is released in the follicular and luteal phase. Their precise role in cyclic events remains to be fully determined, however.

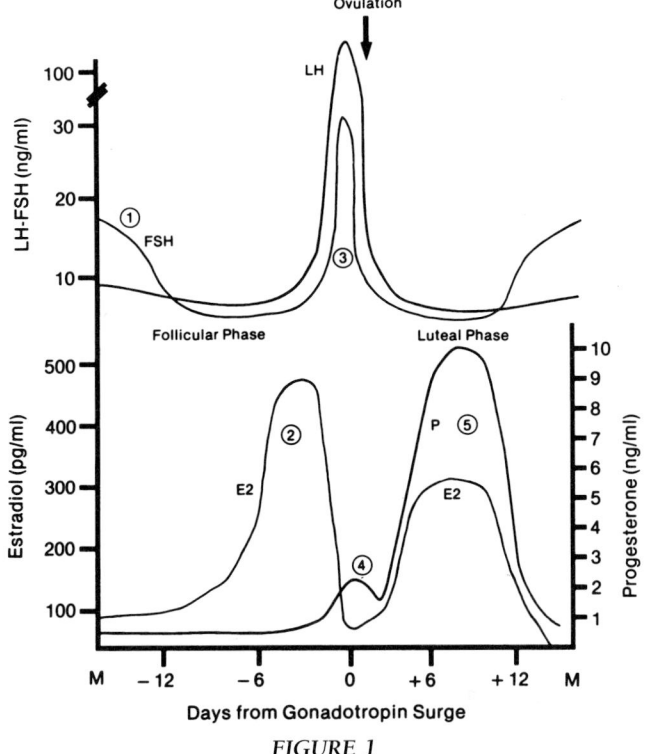

FIGURE 1

Note. Reprinted by permission of Oxford University Press.

III. THE GONADOTROPIN-RELEASING HORMONE PULSE GENERATOR

In all mammals, pituitary gonadotropin secretion is driven by neurons in the hypothalamus. This hypothalamic control is exerted via the release of a neurohormone, gonadotropin-releasing hormone (GnRH), a small peptide consisting of 10 amino acids. GnRH neurons differ from other neurons in that they do not originate in the brain but migrate there during fetal life from the embryonic olfactory placodes. Only after proper migration into the hypothalamus do these GnRH neurons establish a connection with the capillaries and portal veins of the hypothalamic–hypophyseal portal system. In turn, GnRH will be released in these capillaries to be transported undiluted to the anterior pituitary.

A critical advance in our understanding of the hypothalamic control of the ovarian cycle was the observation that GnRH release occurs in a pulsatile mode (the GnRH "pulse generator"). The question as to whether the GnRH neuron is inherently pulsatile or whether other neurosecretory systems are implicated in the generation of GnRH pulses cannot currently be entirely resolved. Although the primary location of the GnRH pulse generator in the hypothalamus may differ from species to species, a proper pulsatile activity of this structure is required for the normal cycle to occur. In the human, abnormalities of the GnRH pulse generator will interfere with the proper function of the cycle. For example, in nutrition-, exercise-, and stress-related amenorrhea (the absence of menstrual cycles), the fundamental insult appears to be to the GnRH pulse generator and results in a dramatic decrease in the GnRH pulse frequency below the critical level required for normal cyclic function.

IV. THE GONADOTROPINS

Each gonadotropin, LH and FSH, is a glycoprotein consisting of two subunits: a common one (α) and one with a different sequence conferring hormonal identity (β). Hormonal activity requires association of both the α and β subunits. These are synthesized and combined within the gonadotrope in the anterior pituitary gland. The gonadotropin response to a GnRH stimulus occurs within minutes. Importantly, the gonadotrope is programmed to respond only to intermittent GnRH stimulation. It is known that a pulsatile GnRH stimulus is required to increase transcription of the gonadotropin subunit genes. On the other hand, continuous GnRH stimulation of the gonadotrope will result in a "downregulation" of the GnRH receptor and in a rapid decrease in gonadotropin secretion. Through changes in the frequency of its pulse signal, the GnRH pulse generator can also selectively regulate gonadotropin subunit gene transcription, such that a fast GnRH pulse frequency favors the synthesis of LH, whereas a slow pulse frequency favors FSH subunit synthesis.

V. THE OVARY

There are two main functional units within the ovary: the follicle and the corpus luteum. They have a different function, are transient, and appear at different phases of the reproductive cycle. The follicle contains the female germ cell, the oocyte; following maturation (the process of folliculogenesis), the oocyte is released from the ovary (ovulation). The corpus luteum is formed from the mature follicle, after it has ovulated; one of its main function is to prepare the uterus for the initial stages of pregnancy in a fertile cycle.

A. Folliculogenesis

Follicles recruited at the beginning of each cycle derive from a stockpile of primordial follicles representing the pool from which all developing follicles will emerge. The process whereby primordial follicles leave the nongrowing stockpile to be converted into primary follicles remains unknown but is already initiated during fetal life. Thus, at birth the number of primordial follicles will have decreased from over 2 million to 500,000. Only a minority of these follicles (400 or less in the primate) will ovulate. Exhaustion of the stockpile of primordial follicles signals the end of reproductive life (menopause in the human).

Gonadotropins do not play a major role in the early stages of folliculogenesis. In their absence, growth from the primordial follicle (consisting of the germ cell and a small layer of surrounding epithelial cells) to a secondary follicle [the epithelial cells have proliferated into multilayers: an inner layer (the granulosa) and an outer layer (the theca)] and to a tertiary or antral follicle (with a well-developed theca layer, gap junctions between the oocyte and granulosa, and accumulation of follicular fluid in a space or antrum) can occur. At all stages of folliculogenesis and at all times of life, most follicles will degenerate through the process of atresia.

Continued growth past the early antral stage requires the stimulation by gonadotropins. FSH appears to provide the fundamental signal for the recruitment of a new cohort of antral follicles (the number of which is species specific) at the start of the follicular phase in each new cycle. In the primate, there is a clear shift in the gonadotropin ratio in the early follicular phase in favor of FSH. Experimental manipulation which results in diminished FSH levels at that time delays the recruitment of the cohort, whereas the administration of supraphysiological levels of FSH results in the recruitment of a larger number of follicles within the cohort.

Although several follicles are recruited at the beginning of each follicular phase in the primate, a process of "selection" occurs early on whereby one of these follicles becomes "dominant" while the others begin atresia. The process of dominance is completed by days 6 or 7 of the follicular phase, at which time there is no surrogate follicle capable to take over if the dominant follicle is experimentally destroyed.

FSH stimulates granulosa cell mitosis and production of the enzyme aromatase, which facilitates the local conversion of androgens (produced by the theca) into estradiol. Increasing local estrogen levels promote mitotic activity by a direct local action and induce the local release of a variety of growth factors which will act to further promote growth of the follicle. FSH and estradiol act to promote the synthesis of the LH receptor in the growing follicle. This allows for LH to stimulate the production of androgens, which are then aromatized into estrogens, thus further promoting the local estradiol microenvironment.

The mature female germ cell, the oocyte, derives from the primordial germ cells. Oocytes become surrounded by granulosa cells and grow within follicles. The growth process of the oocyte is initiated at the time of appearance of the zona pellucida, a glycoprotein coat surrounding it. Oocyte growth, to one of the largest cell in the organism, is completed by the time the follicular structure around it reaches the secondary follicle stage. Thus, by the time it is recruited in a reproductive cycle, the oocyte has completed most of its growth.

B. The Estradiol Feedback Loop

In order to avoid hyperstimulation of the ovary, the output of gonadotropins by the pituitary gland must be controlled. As in other endocrine systems, this is accomplished by an inhibitory or negative feedback loop, whereby hormones secreted by the ovary act as a feedback signal to the hypothalamic–pituitary unit to readjust GnRH and gonadotropin release. This negative feedback loop is active throughout the reproductive cycle. After ovariectomy or at menopause, when ovarian estrogen secretion ceases, gonadotropin concentrations increase rapidly as the negative feedback becomes inactive.

In the female, estradiol may also exert a stimulatory or positive feedback effect on gonadotropins. The existence of this positive feedback loop is critical to cyclicity because it allows for the precise coordination between follicular maturation and the stimulus to ovulation. At the end of the follicular phase, the follicle reaches maturity and its secretion of estradiol is maximal. Under these conditions, the estradiol positive feedback loop is activated and, within 1.5 days in the primate, there follows a large release of gonadotropins—the gonadotropin surge which will act as the signal to ovulation.

C. Ovulation

The gonadotropin surge terminates estradiol secretion by stopping the production of androgen (its precursor). Steroid synthesis in the follicle now switches to progesterone. Vascular changes occur in the preovulatory follicle within minutes of the LH surge, with a large increase in ovarian blood flow occurring later. The LH surge also initiates protein synthesis within the follicle, an increase which may reflect the production of proteolytic enzymes involved in follicle wall degradation. The oocyte responds to the preovulatory LH surge by completing meiosis and entering the second meiotic division. In the primate, ovulation, the release of the mature fertilizable oocyte from the mature follicle, occurs about 36 hr after the initiation of the LH surge. At ovulation, meiosis will be arrested again, to be completed at the time of fertilization.

D. The Corpus Luteum

The expulsion of the oocyte is followed by the spontaneous transformation of the follicle into a corpus luteum. This new structure results from LH-induced morphological changes in both the granulosa and the theca layers of the follicle, which include the vascularization of the previously avascular granulosa and the acquisition by the granulosa cell of *de novo* steroid synthesis capability. Thus, while during the follicular phase, granulosa cells were capable only to aromatize delivered products (androgens), the "luteinized" granulosa cell, now transformed into a structure containing all elements of a typical steroid-producing cell, is now capable of synthesizing both progesterone and estradiol.

The newly formed corpus luteum is a transient endocrine organ. In the human, this structure has an inherent 12- to 14-day life span and its regression (luteolysis) is inevitable unless fertilization occurs. In this case, the corpus luteum is rescued by another hormone, human chorionic gonadotropin. The corpus luteum attains maturity in about 5 days after ovulation, at which point it has become a large structure 15 mm in diameter, easily recognizable on the surface of the ovary. Secretory activity of the corpus luteum is dependent on LH stimulation (a luteotropic agent) throughout the duration of the luteal phase. Regression starts on Days 7–9 of the luteal phase and results in a dramatic decrease in the number of secretory granules and in the secretion of progesterone and estradiol.

VI. THE GENITAL TRACT

During the early follicular phase, with little ovarian steroid activity present, the endometrium (the mucosa lining the inner wall of the uterine cavity) is thin, the endometrial glands are short and narrow, and the stroma is compact. With estradiol increasing with follicle growth, endometrial glands proliferate and the stroma becomes well organized. With the advent of progesterone in the luteal phase, glands stop growing and undergo a secretory differentiation. Glycogen and other secretory products begin to accumulate in the glands. By Day 7 of the luteal phase, these products will be extruded into the glandular space. If fertilization of the oocyte has taken place, this coincides with the time of implantation of the blastocyst.

In the absence of implantation, estradiol and progesterone secretion decrease dramatically during the final days of the luteal phase. Because of withdrawal of the hormonal support, the endometrium undergoes a gradual necrotic ischemia, with progressive exfoliation of all cells and the resultant menstrual bleeding in the primate.

See Also the Following Articles

FOLLICULAR DEVELOPMENT, CONTROL OF; FSH (FOLLICLE-STIMULATING HORMONE; LH (LUTEINIZING HORMONE); MEIOTIC CELL CYCLE, OOCYTES

Bibliography

Adashi E. Y. (1996a). The ovarian follicular apparatus. In *Reproductive Endocrinology, Surgery and Technology* (E. Y. Adashi, J. A. Rock, and Z. Rosenwaks, Eds.), pp. 17–40. Lippincott-Raven, Philadelphia.

Adashi, E. Y. (1996b). The ovarian follicle: Life cycle of a pelvic clock. In *Reproductive Endocrinology, Surgery and Technology* (E. Y. Adashi, J. A. Rock, and Z. Rosenwaks, Eds.), pp. 211–234. Lippincott-Raven, Philadelphia.

Ferin, M., Jewelewicz, R., and Warren, M. (1993). *The Menstrual Cycle: Physiology, Reproductive Disorders, and Infertility*. Oxford Univ. Press, New York.

Freeman M. E. (1994). The neuroendocrine control of the ovarian cycle of the rat. In *The Physiology of Reproduction* (E. Knobil and J. D. Neill, Eds.), 2nd ed., pp. 613–658. Raven Press, New York.

Hotchkiss, J., and Knobil, E. (1996). The hypothalamic pulse generator: The reproductive core. In *Reproductive Endocrinology, Surgery and Technology* (E. Y. Adashi, J. A. Rock, and Z. Rosenwaks, Eds.), pp. 123–162. Lippincott-Raven, Philadelphia.

Ovarian Cycle, Teleost Fish

Izhar A. Khan and Peter Thomas

The University of Texas at Austin

I. Introduction
II. Environmental Control of Ovarian Cycle
III. Neuroendocrine Control of Ovarian Cycle
IV. Oogonial Proliferation and Primary Oocyte Growth
V. Secondary Oocyte Growth
VI. Oocyte Maturation
VII. Ovulation
VIII. Spawning and Fertilization

GLOSSARY

final oocyte maturation The process of transformation of postvitellogenic oocytes into fully mature oocytes covering the period between resumption of meiosis from prophase I up to metaphase II. Meiosis remains arrested in metaphase II until fertilization.

germinal vesicle breakdown The breakdown of germinal vesicle (nucleus) after it migrates to the periphery on resumption of meiosis during final oocyte maturation.

maturation-inducing steroid The steroid produced in response to the GTH II (maturational gonadotropin) surge and responsible for the control of the final stages of oocyte maturation.

oogonia The small, rounded female germ cells that give rise to primary oocytes.

ovulation The process of separation of a mature oocyte from the surrounding follicular cells, rupture of the follicular wall, and expulsion of the oocyte.

vitellogenin The yolk precursor protein synthesized in response to estrogen induction in the liver.

A striking feature of egg production in teleost fish in comparison to most other vertebrates is their enormous fecundity (potential to produce a large number of offspring). Most fishes produce from a few thousand to millions of eggs during each ovarian cycle. The ovarian weight also increases dramatically as the oocytes grow and mature, sometimes reaching up to 50% of the body weight. The duration of ovarian cycles in fishes varies, ranging from a few days to years. A wide variety of reproductive strategies to ensure spawning at the most appropriate time of the seasonal or diurnal cycle for optimum survival of the offspring have been observed among over 20,000 extant species of fish. The ovarian cycles of relatively few of them have been investigated, so it is possible that generalizations based on these limited studies may not apply to all fishes. However, the available evidence suggests that, despite differences in the dynamics and patterns of follicular development in fish ovaries, certain basic features of ovarian cycles in fish, such as the various stages of oocyte and follicular development and endocrine control, are remarkably similar. The sequence of events during the ovarian cycle, beginning with oogonial proliferation, primary oocyte growth, and secondary oocyte growth, and culminating in final oocyte maturation, ovulation, spawning, and fertilization will be described as will their endocrine and environmental control.

I. INTRODUCTION

The oocytes in fish ovaries undergo marked morphological and biochemical changes during their growth and maturation, although the number of oocytes in the ovary that undergo these changes at a given time varies considerably. In addition, the size of eggs in fishes ranges from a fraction of a millimeter to a few centimeters in diameter. Ovulated eggs up to 9 cm in diameter and weighing over 300 g have been recorded in specimens of the coelacanth, *Latimeria chalumnae*. Fish that scatter their eggs (pelagic spawners) produce a large number of small buoyant eggs (up to several million); whereas those that guard their offspring and show parental care tend to produce fewer and larger eggs. Most teleost fishes are oviparous (egg laying) and the majority of oviparous fishes fertilize their eggs externally. Some fishes carry the embryos or young internally (internal bearers) and produce only a few large offspring. Internal bearers employ various methods to provide nourishment for the developing embryo. In viviparous species fertilization of eggs usually occurs within the follicle (e.g., *Cymatogaster aggregata*), although it may occur in the ovarian cavity, such as in *Zoarces viviparous*. A highly vascularized connection to the embryo develops in many species (pseudoplacenta) to transport nutrients to the offspring. The sole source of nourishment for marine rockfish (Scorpaenidae) embryos and some sculpins is their egg yolk (ovoviviparity), whereas other internal bearers provide additional nutrition for their offspring (viviparity). Although the ovarian follicles are modified for gestation in viviparous fishes, the stages of oocyte growth within the ovary in general are similar to those of oviparous teleosts. Similarly, the stages of oocyte growth during the ovarian cycles in sequential and simultaneous hermaphrodite fishes and parthenogenic species also do not differ significantly from those of bisexual fishes.

Three basic patterns of oocyte growth have been described in teleosts, i.e., synchronous, group-synchronous, and asynchronous. A synchronous oocyte growth pattern is characteristic of some of the anadromous Pacific salmonid species and catadromous eels which spawn once in their lifetime and die (semelparity). Species that reproduce more than once in their lifetime (iteroparous) have at least two distinct populations (clutches) of oocytes at some time during their ovarian cycles and display group-synchronous, asynchronous, or some combination of these growth patterns. In fish having only two clutches of oocytes, i.e., primary oocytes and synchronously developing secondary oocytes, the primary oocytes remain in the resting stage until the next ovarian cycle (e.g., the rainbow trout, *Oncorhynchus mykiss*, and Indian catfish, *Heteropneustes fossilis*). A more or less synchronous secondary growth pattern is followed up to the late vitellogenic phase in several marine species that produce large numbers of pelagic eggs. However, oocyte maturation and ovulation in these species occur in batches (only a fraction of oocytes mature at one time) leading to multiple spawning events during their breeding cycles. Three distinct oocyte clutches (vitellogenic, cortical alveoli stage, and primary oocytes) are simultaneously present during the spawning season in many species displaying a group-synchronous pattern (e.g., the three-spined stickleback, *Gasterosteus aculeatus*). Recruitment of the second oocyte clutch from cortical alveoli into the vitellogenic stage does not usually occur until initiation of maturation in the leading clutch. Most species displaying asynchronous patterns of oocyte growth spawn several times over a protracted period of breeding season. The pipefish (*Syngnathus scovelli*), killifish (*Fundulus heteroclitus*), and spotted seatrout (*Cynoscion nebulosus*) show asynchronous growth patterns with all oocyte stages present in a single ovary. Several clutches of oocytes can be seen in the ovaries of some daily spawners or various stages of oocytes may be present in an individual without distinct clutches, for example, in the blenny (*Blennius pholis*). Thus, fish employ a variety of modifications of the three basic patterns of ovarian and oocyte growth as part of their successful reproductive strategies.

II. ENVIRONMENTAL CONTROL OF OVARIAN CYCLE

Ovarian cycles in most fish species are influenced by external environmental stimuli as well as their endogenous reproductive rhythms. The relative importance of each of these mechanisms in controlling the ovarian cycle varies among fishes. One part of the ovarian cycle may be under an endogenous control, whereas the other may be regulated by external factors in the same species. Most studies on the control of ovarian cycles in fishes have investigated exogenous factors, whereas relatively few have examined endogenous circannual rhythms. To demonstrate that an ovarian cycle is endogenously driven, fish must be maintained under a constant environment for long periods of time (up to 2 years or more). The studies conducted with the Indian catfish, rainbow trout, and stickleback convincingly demonstrate that underlying endogenous rhythms are important components of their seasonal ovarian cycles.

Photoperiod and temperature are the two primary environmental variables that influence ovarian cycles in most fish species, but pH, salinity, and other physicochemical characteristics of water and biological factors such as food availability and social interactions are also often important. In general, alterations of the photoperiod initiate the fish ovarian cycles, whereas temperature changes are important for completion of the later stages of the cycle in species inhabiting temperate regions. Tropical species which experience only minor changes in photoperiod and water temperature often have an extended breeding season and spawning peaks may instead be associated with increasing water levels and reduced conductivity (ionic concentration) during the rainy season. Ovarian growth is stimulated by long or increasing photoperiods in combination with warm temperatures in fish that spawn in spring or early summer, such as the stickleback (*G. aculeatus*), medaka (*Oryzias latipes*), golden shiner (*Notemigonus crysoleucas*), and goldfish (*Carassius auratus*). Decreasing photoperiod promotes ovarian recrudescence in species that spawn in autumn or early winter, for example, rainbow trout and several species of bitterlings. Temperature may play a dominant role in some species such as the goby (*Gillichthys mirabilus*) in which low temperature stimulates sexual maturation. On the other hand, temperature requirements may vary during different phases of the ovarian cycle, for example, in the marsh killifish (*Fundulus confluentus*), low temperature promotes early phases of oocyte growth and high temperature induces the late phases. Effects of photoperiod and temperature on fish ovarian cycles often vary depending on the phase of the annual reproductive season, i.e., they act in concert or in phase with the endogenous rhythms to control the ovarian cycle and successful breeding.

Our knowledge of the neurochemical signals and neural pathways in fishes by which environmental information influences the neuroendocrine system controlling the ovarian cycle is rudimentary. Whether the photoperiodic information is relayed via the suprachiasmatic nucleus (SCN) similar to the situation in birds and mammals is unknown, although retinal projections do reach the SCN in fishes. Similarly, the role of the pineal in the control of ovarian cycle has also not been fully established. Rhythmic secretion of melatonin is probably involved in the photoperiodic control of reproductive cycles in at least some teleosts. Available data from goldfish and the Atlantic croaker (*Micropogonias undulatus*) suggest that the pineal and its hormone melatonin may be involved in the control of daily gonadotropin rhythms and/or in the timing of spawning. Melatonin receptors have been localized in brain areas known to be involved in the control of gonadotropin secretion in several teleosts (e.g., goldfish, the catfish *Silurus asotus*, and some salmonid species). However, more direct evidence is needed to link these receptors to the control of gonadotropin secretion and thus the ovarian cycle in teleosts.

III. NEUROENDOCRINE CONTROL OF OVARIAN CYCLE

Environmental information perceived by the sense organs activates neuroendocrine pathways that control the synthesis and release of the gonadotropins. The neuroendocrine control of gonadotropin secretion involves complex interactions of several regulatory factors, including monoamine and amino acid neurotransmitters, neuropeptides, and gonadal steroids. These regulatory factors exert both stimulatory and inhibitory influences on gonadotropin synthesis and release via their actions on the gonadotropin-releasing hormone (GnRH) neurons in the preoptic–anterior hypothalamic area and/or by acting directly at the pituitary level. The teleostean pituitary is unique in that both inhibitory and stimulatory neuroendocrine pathways directly innervate the anterior pituitary and have their nerve endings on or near the gonadotrops to control gonadotropin release. Thus, unlike tetrapods, they lack a median eminence and a hypothalamus–pituitary portal system to deliver releasing hormones to the adenohypophysial cells. The GnRH system is the major stimulatory pathway in fish, similar to other vertebrates, and its actions on gonadotropin secretion are modulated by a variety of neurotransmitters, including serotonin, dopamine, norepinephrine, and γ-aminobutyric acid (GABA).

Two gonadotropins, GTH-I and GTH-II, homologous to follicle-stimulating hormone (FSH) and luteinizing hormone (LH) of tetrapods, have been isolated and purified from the pituitaries of all the teleost species examined so far except the African catfish (*Clarias gariepinus*), in which GTH-I could not be detected. However, their functions may not be as distinguishable in fishes as those of FSH and LH in tetrapods. Relatively few studies have investigated the neuroendocrine control of GTH-I secretion, the hormone which likely controls the earlier stages of ovarian development and growth in salmonids and probably other teleost species. For example, during the period of ovarian recrudescence, gonadectomy in coho salmon (*Oncorhynchus kisutch*) elicits an increase in plasma GTH-I levels, although the nature of inhibitory gonadal factors and their mechanism of action remain to be elucidated. Neuroendocrine mechanisms controlling GTH-II release, the gonadotropin that controls final oocyte maturation and ovulation, have been investigated in greater detail. GTH-II secretion is under a variable degree of inhibitory dopaminergic control in teleosts, being dominant in carps and some catfishes, of minor importance in salmonids, and completely absent in Atlantic croaker and other perciform species. Serotonin stimulates GTH-II release in goldfish and Atlantic croaker by acting directly on the pituitary gland and also via its interactions with GnRH neurons in the preoptic area. GABA stimulates GTH-II release in the regressed and recrudescing goldfish and in regressed Atlantic croaker. In addition, GABA can inhibit GTH-II release in croaker with fully developed ovaries, thereby displaying a dual action depending on the stage of ovarian cycle similar to the situation in several mammals. A variety of other neuromodulatory factors have also been shown to either stimulate or potentiate the effect of GnRH on GTH-II release in some teleosts.

Multiple GnRH forms (two or more) are present in the brains of all the fish investigated to date. Chicken GnRH-II (cGnRH-II) is the most conserved form of GnRH which is present in all vertebrates except in some placental mammals. In teleosts, one or two additional GnRHs are present, for example, catfish GnRH in African catfish, salmon GnRH (sGnRH) in goldfish and salmonid species, and seabream GnRH (sbGnRH) together with sGnRH in several perciform fish species. The GnRH-containing neurons have distinct distribution within the brains of fishes, with cGnRH-II perikarya being present in the midbrain and the native fish GnRH forms in the forebrain. In the gilthead seabream, sbGnRH is present in the preoptic area and is the predominant form in the pituitary (500-fold higher concentration than sGnRH; cGnRH-II is nondetectable). Therefore, sbGnRH appears to be the primary form responsible for gonadotropin release from the pituitary in seabream and possibly in other perciform species. On the other hand, both sGnRH and cGnRH-II are present in goldfish pituitary and are potent releasers of gonadotropin. Whether the two GnRHs interact to control gonadotropin release from goldfish pituitary or differentially regulate the two gonadotropins is not known. The number of GnRH receptors in the pituitary varies with the ovarian cycle, the receptors being most abundant during late ovarian recrudescence. GnRH binding to its receptors on the gonadotrops activates the Ca^{2+}/protein kinase C-dependent signaling pathway and possibly other signal transduction systems to stimulate gonadotropin release.

The gonadotropins are released in the circulation and bind to the specific receptors in the ovary to stimulate the production of steroids. Growth hormone can also stimulate ovarian steroidogenesis in some teleosts (e.g., killifish, rainbow trout, and spotted seatrout), similar to the observation in mammals, and may have a gonadotropic function during the ovarian cycle. Ovarian steroids and other factors (e.g., activin and inhibin) exert both positive and negative feedback control on gonadotropin release. For example, estradiol and testosterone (primarily via its aromatization to estradiol) exert a positive feedback effect on gonadotropin release in immature (e.g., salmonids and eels), recrudescing (e.g., goldfish and Atlantic croaker), or regressed (e.g., goldfish) individuals. In addition, estradiol and testosterone can exert a negative feedback influence when fully recrudesced ovaries are surgically removed, for example, in goldfish, African catfish, and Atlantic croaker. Although certain details of the steroid feedback control of gonadotropin release may differ in different fish models, the essential features of these feedback mechanisms are the same and are integral components of the neuroendocrine mechanisms controlling gonadotropin release. In fact, most of the other reproductive neuroendocrine control mechanisms appear to be modulated by the stage of ovarian cycle or the endogenous steroid hormone milieu. The primary function of the neuroendocrine system is to induce a surge of GTH-II secretion at the appropriate time for successful final oocyte maturation, ovulation, and spawning. The accumulation of GTH-II in the gonadotrops, blockage of the dopaminergic inhibitory control of GTH-II release, potentiation of the stimulatory effect of GnRH, release of GnRH, and upregulation of GnRH receptors are all coordinated by the neuroendocrine system to achieve the preovulatory GTH-II surge.

Circulating GTH-I and GTH-II levels show distinct seasonal profiles in salmonids which correlate with the expression of GTH-I and GTH-II mRNA in some species in which it has been examined. Plasma GTH-I levels are high during ovarian recrudescence and decline during oocyte maturation and spawning in the coho salmon (Fig. 1). A more or less similar pattern of circulating gonadotropin levels has been reported in common carp. In rainbow trout, plasma GTH-I levels are elevated during the early part of oocyte growth, return to basal levels during late vitellogenesis, and increase again during ovulation. On the other hand, the GTH-II levels remain low throughout the period of ovarian growth and show a prominent surge prior to oocyte maturation and ovulation in all the fish species investigated. Thus, GTH-I most likely controls earlier stages of ovarian cycle, whereas GTH-II controls final stages of oocyte maturation, ovulation, and spawning, at least in salmonids and carp.

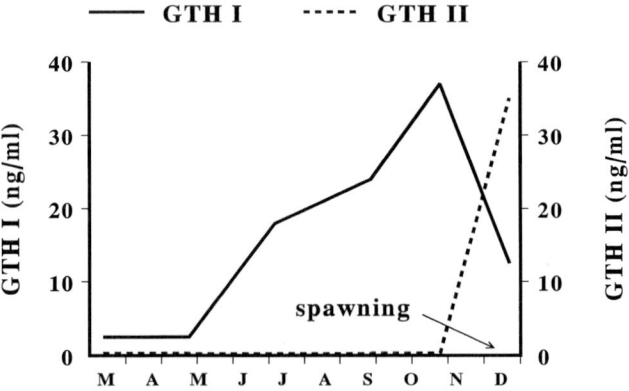

FIGURE 1 Plasma levels of GTH-I and GTH-II during the ovarian cycle in coho salmon (ovarian recrudescence from May to December) [hormonal profiles adapted with permission from P. Swanson, In *Reproductive Physiology of Fish* (A. P. Scott, J. P. Sumpter, D. E. Kime, and M. S. Rolfe, Eds.), pp. 2–7, FishSymp 91, Sheffield, UK, 1991].

IV. OOGONIAL PROLIFERATION AND PRIMARY OOCYTE GROWTH

Differentiated teleost ovaries consist of stroma and oogonia. Fish oogonia are small, rounded female germ cells with a high nucleus to cytoplasm ratio and usually occur in small nests. Oogonia undergo periodic mitosis (oogonial proliferation) correlating with the beginning of the seasonal ovarian cycle. Earlier studies using hypophysectomy and injections of pituitary extract or implantation of pituitary glands in the coelomic cavity suggest pituitary involvement in the control of oogonial mitosis. However, there is no direct evidence for a role of gonadotropins in the control of oogonial proliferation. In fish with recurring ovarian cycles, oogonia divide and proliferate to give rise to primary oocytes. Their number reaches a peak again during the postspawning phase in preparation for the next ovarian cycle. Many tropical species have a continuous series of short cycles with oocyte production throughout the year. Small waves of oogonial mitosis occur in these fish, with several peaks of oogonial proliferation corresponding with each short cycle of ovarian recrudescence.

Oogenesis, i.e., the transformation of oogonia into oocytes (Fig. 2), starts during or immediately after the peak of oogonial proliferation. The beginning of oogenesis is characterized by the arrest of chromosomes in the prophase (diplotene stage) of the first meiotic division. Hypophysectomy experiments in several fish species have shown that pituitary hormone input is not required for the transformation of oogonia into oocytes or for the growth of primary oocytes.

Folliculogenesis, i.e., the formation of a follicular wall around the growing oocyte, begins soon after transformation of oogonia into primary oocytes. Prefollicular cells of the primordial follicle most likely arise from the coelomic epithelium. Mitosis of germinal epithelium adds new prefollicular cells as the oocytes enlarge and migrate away from the oogonial nests. A primordial follicle is single layered, and processes from the follicular wall and microvilli from

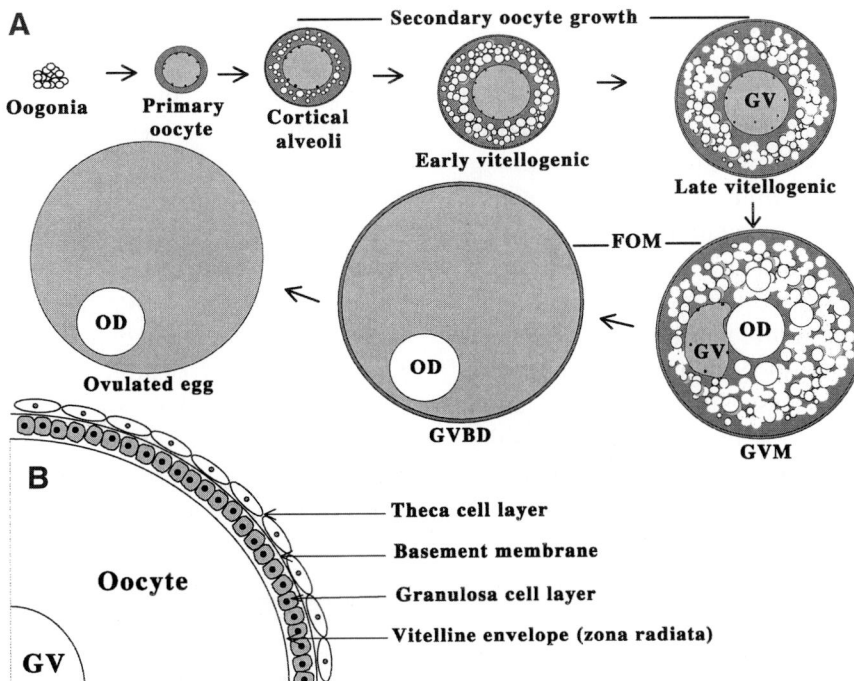

FIGURE 2 Diagrammatic representation of the different stages of oocyte growth and maturation during the ovarian cycle of fishes (A) and follicular layers enclosing a growing oocyte (B). FOM, final oocyte maturation; GV, germinal vesicle (nucleus); GVM, GV migration; GVBD, GV breakdown; OD, oil droplet, typically present in pelagic eggs.

the oocyte are already formed during the diplotene stage. These processes and microvilli facilitate transfer of nutrients and chemical messengers to and from the oocytes. Little is known about the hormonal control of folliculogenesis during the primary growth phase in fishes. Growth of follicles during this period is most likely controlled by the extraovarian and/or intraovarian factors that govern growth and metabolism in general. The process of folliculogenesis continues throughout the growth and maturation of the oocytes and is strikingly similar in all nonmammalian vertebrates.

V. SECONDARY OOCYTE GROWTH

Most of the oocyte and ovarian growth occurs during this period primarily due to accumulation of yolk from its precursor vitellogenin. The yolk proteins aggregate in yolk granules and the size of oocytes increases with the increase in the number and size of yolk granules as vitellogenesis proceeds. The oocytes remain arrested in prophase I throughout this growth phase. The appearance of round mucopolysaccharide or glycoprotein-rich structures in fish oocytes, called cortical alveoli, marks the beginning of secondary oocyte growth (Fig. 2). The acellular vitelline envelope (zona radiata) is deposited between the oocyte plasma membrane and surrounding follicle cells and is composed of two to four major proteins which are synthesized in the liver. The follicle cells consist of an inner layer of granulosa cells and one or sometimes two outer layers of theca cells. The outer margins of granulosa cells are in contact with a rich capillary plexus supported by a basement membrane which separates it from the theca layer. The theca layer comprising flattened fibroblast-like theca cells is bounded by a peritoneal epithelium. The cytoplasmic connections between the oocyte and granulosa cells form where oocyte microvilli and granulosa cells make contact via specialized membrane junctions known as "gap junctions." A gap junction is formed by two hemichannels or connexons, where each connexon is a hexamer of connexin protein subunits. These components of the follicular wall play important roles throughout the growth, maturation, and ovulation, including protection of the oocyte, synthesis of a variety of hormones and their receptors, and transfer of nutrients and chemical messengers across and within the follicular layers. In many teleosts, especially marine species, oil droplets begin to accumulate in the cytoplasm during this period.

A. Endocrine Control

Steroidogenesis in the follicle cells of the growing oocytes is regulated primarily by GTH-I in salmonids and probably many other species because GTH-II levels remain extremely low (close to the detection limit of most assays; Fig. 1) during this period, although both GTH-I and GTH-II can stimulate ovarian steroidogenesis *in vitro* by binding to type I gonadotropin receptors present on the theca and granulosa cells. Gonadotropin binding to the receptors activates a cascade of intracellular events leading to steroid biosynthesis. An increase in cAMP levels activates protein kinase A to modulate the activity of the proteins involved in steroidogenesis. Calcium and possibly other second messenger systems are also involved in mediating gonadotropin action on steroidogenic enzymes that regulate steroid biosynthesis. Ovarian steroids play important roles throughout the secondary oocyte growth. In salmonids, testosterone is synthesized in the theca cells and is transferred to granulosa layer where it is aromatized to estradiol. The enzymatic pathways for testosterone synthesis in theca cells include conversion of cholesterol to pregnenolone by cholesterol side chain cleavage P450 enzyme and to progesterone by 3β-hydroxysteroid dehydrogenase, 17α-hydroxylation, formation of androstenedione by C_{17-20} lyase, and finally its conversion to testosterone by 17β-hydroxysteroid dehydrogenase. Production of testosterone by theca cells *in vitro* is stimulated by gonadotropins, but its conversion to estradiol in the granulosa cells does not seem to require gonadotropin action. However, this two-cell model may not be valid for all teleosts because in *Fundulus heteroclitus*, production of testosterone and estradiol *in vitro* does not seem to require the presence of theca cells.

Circulating estradiol (the major estrogen in fish) levels increase during oocyte growth and remain high during vitellogenesis in rainbow trout (Fig. 3) and

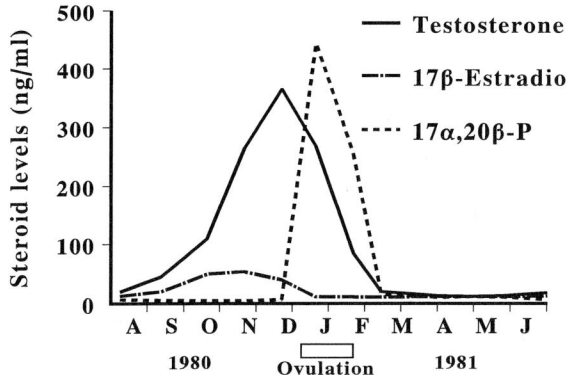

FIGURE 3 Changes in plasma steroid hormones during the ovarian cycle in a winter-spawning strain of rainbow trout (ovarian recrudescence from May and June to December and January). 17α,20β-P, 17α-hydroxy-20β-dihydroprogesterone (hormonal profiles adapted with permission from A. P. Scott and J. P. Sumpter, Gen. Comp. Endocrinol. 52, 79–85, 1983).

other teleosts. Estrogens induce vitellogenesis in the liver by binding to hepatic nuclear estrogen receptors (ERs). Significant correlations between plasma estradiol and hepatic cytosolic ER concentrations, and between hepatic nuclear ERs and plasma vitellogenin levels, have been observed during the period of ovarian recrudescence in spotted seatrout. Cytosolic and nuclear ER concentrations increased fourfold and eightfold, respectively, in vitellogenic versus nonvitellogenic females. Hepatic ER mRNA and protein are both induced by estradiol in salmonids. The increase in receptor concentration in turn increases the sensitivity of the hepatocytes to estrogens. Preliminary recent evidence suggests that genes encoding multiple ERs are present in Atlantic croaker and rainbow trout similar to the recent discovery of ERα and ERβ in mammals; however, their physiological significance and possible differential regulation remain to be investigated.

A major function of estrogens is to induce vitellogenin synthesis in the liver. The process of vitellogenesis includes synthesis of vitellogenin in the liver hepatocytes in response to estradiol stimulation, its secretion and transport through blood to the growing oocytes, its uptake via vitellogenin receptor-mediated endocytosis, and its cleavage into yolk proteins, mainly lipovitellin and phosvitin. Oocyte surface receptors for vitellogenin which facilitate its uptake have been characterized in several teleost species.

There is evidence for the accumulation of lipoproteins other than vitellogenin in the growing oocytes in a few teleosts. Plasma levels of very low-density lipoprotein increase during ovarian recrudescence in rainbow trout and appear to be under the control of estrogens. Estradiol also regulates the hepatic synthesis and release of eggshell (vitelline envelope) proteins, a novel group of glycoproteins in teleosts. Plasma levels of these proteins correlate positively with GTH-I, estradiol, and the gonadosomatic index during the ovarian recrudescence in Atlantic salmon.

Plasma levels of testosterone also increase during ovarian recrudescence in many species. The androgen may indirectly promote early stages of secondary oocyte growth via changes in gonadotropin secretion. This contention is supported by the fact that testosterone exerts a positive feedback effect on gonadotropin release during ovarian recrudescence in Atlantic croaker and goldfish, immature rainbow trout, and European eel. Testosterone and/or other androgens also probably act directly on the ovary via their binding to the androgen receptors that have been characterized in a few teleost species (e.g., goldfish, Atlantic croaker, and Nile tilapia).

Estradiol and testosterone are transported in the blood bound to specific sex steroid-binding proteins. Concentration of the testosterone and estradiol-binding protein in spotted seatrout plasma increases during ovarian recrudescence, reaching a peak during late vitellogenesis and oocyte maturation. Although the physiological importance of these proteins remains unclear, they probably act to buffer plasma steroid levels, preventing their rapid metabolic clearance and modulating steroid availability at the target tissues. Receptors for these proteins have been identified on steroid target tissues in several vertebrate species.

The mechanisms of intraovarian regulation of steroidogenesis and follicular growth by other factors (e.g., activin, inhibin, GnRH, oxytocin, vasopressin, and growth factors) have not been fully elucidated in teleosts. Preliminary evidence supports a role for insulin, insulin-like growth factors (IGFs; especially IGF-I), and their receptors during the oocyte growth. The levels of insulin and IGF-I binding in the carp ovary are high during primary oocyte growth and vitellogenesis, decline in postvitellogenic ovaries, and increase again just prior to ovulation. Involvement of IGF-I in the control of steroid biosynthesis, vitellogenin uptake, and follicular proliferation and differentiation has been suggested in salmonids and carp.

Estradiol levels decline prior to oocyte maturation and ovulation in rainbow trout (Fig. 3), whereas the levels remain high even after ovulation in goldfish. This difference can be attributed to the asynchronous oocyte growth in goldfish in which batches of vitellogenic oocytes are present even after the first round of ovulation. Testosterone levels peak just before oocyte maturation and decline during final oocyte maturation (FOM) and ovulation in most teleost species examined, including rainbow trout (Fig. 3).

Androstenedione and 11-ketotestosterone levels also vary during the fish ovarian cycle in a few species, for example, in the spring chinook salmon (*O. tshwytscha*), although their roles in the control of ovarian cycle are unknown. Sex steroids in fish are often conjugated to other molecules such as glucuronides and the conjugates are disposed of as inactive metabolites and/or may function as pheromones.

VI. OOCYTE MATURATION

At the end of the period of oocyte growth, the postvitellogenic oocytes of teleosts cannot be fertilized because the first meiotic division is incomplete and is arrested at prophase I. A surge in gonadotropin (GTH-II) secretion induces the resumption of meiosis and structural changes in the ooplasm which enable the oocytes to be fertilized after they have been ovulated. This process, termed FOM, involves breakdown of the germinal vesicle or nucleus (GVBD), chromosome condensation, formation of the first meiotic spindle and polar body, and continuation of the meiotic division up to metaphase II. The appearance of the ooplasm usually changes dramatically during FOM due to proteolytic cleavage and mobilization of yolk reserves.

FOM is highly synchronized in the ovaries of most teleosts and is often completed within a short time frame, <24 hr in many species. The first visible sign of FOM in postvitellogenic spotted seatrout oocytes is coalescence of the yolk granules around the nucleus (lipid coalescence) which begins around dawn of the day of spawning. Lipid coalescence is complete and the ooplasm has become clear by midmorning, and the nucleus has begun to migrate to the animal pole (germinal vesicle migration). GVBD has occurred by midafternoon and the oocytes have begun to swell due to massive influx of water (hydration) which is dependent on active ion transport by oubain-sensitive Na^+,K^+-ATPase, thereby increasing the osmotic pressure inside the oocyte. Increases in the internal concentrations of organic solutes due to proteolysis of yolk proteins also contribute significantly to the osmotic gradient generated across the oocyte membrane. At dusk the oocyte has swelled to eight times its original volume, and an oil droplet, which is also important for buoyancy, has formed. Shortly afterwards the follicular layer surrounding the oocyte ruptures and large numbers of oocytes (about 300,000) are released into the ovarian cavity (ovulation). A micropyle has formed at the animal pole by this time through which the sperm will later enter and fertilize the egg. Spawning occurs within a 2- or 3-hr period and is completed by midnight.

A. Endocrine Control

The entire process of FOM can readily be induced by gonadotropins in cultured follicle-enclosed fish oocytes *in vitro* and this has led to the development of *in vitro* bioassays to assess the potencies of various hormones using GVBD as an endpoint of FOM. The availability of reliable FOM bioassays and large numbers of mature oocytes in the ovaries of most fishes has made them excellent vertebrate models for investigating the hormonal control of FOM. Extensive studies have shown that gonadotropin (GTH II) induces FOM in teleosts indirectly, as it does in amphibians, by stimulating the production of certain 21-carbon steroids, called maturation-inducing steroids (MISs), by the ovarian follicular cells. A variety of progestins and 11-deoxycorticosteroids have been proposed as MISs in teleosts based on their potencies in the *in vitro* bioassay and their synthesis during FOM; but convincing evidence has only been obtained for two of them. The progestin 17,20βdihydroxy-4-pregnen-3-one (17,20β-P) has been positively identified as the MIS in the amago salmon and is the likely MIS in other salmonids, some carp species, atherinids, catfishes, and fishes belonging to several other teleost orders. The 11-deoxycorticosteroid, 17,20β-21-trihydroxy-4-pregnen-3-one (20β-S) has been positively identified as the MIS in two marine perciform species belonging to the family Sciaenidae, Atlantic croaker, and spotted seatrout, and appears to be major MIS in many perciform fishes and several species of flatfish. Recent studies have shown that in addition to their functions as MISs, these steroids or their conjugated metabolites act as primary pheromones in several species of cyprinids and representatives of other teleost families, signaling to the male conspecific that final maturation is proceeding and spawning is imminent.

The mechanism by which the GTH-II surge causes a switch in the steroidogenic pathway from the production of estrogens and androgens to the synthesis of C21 MISs during FOM remains poorly understood, even in the extensively studied salmonid model. The decline of estradiol levels in postvitellogenic salmonid follicles appears to be due to decreases in C_{17-20} lyase and possibly aromatase activities, whereas there is a rapid increase in 20β-hydroxysteroid dehydrogenase activity in the granulosa cells due to synthesis of the enzyme in response to gonadotropin which is mediated by a G protein, adenylate cyclase, and protein kinase signaling pathway. The involvement of the theca cells in MIS production is unclear and may vary among teleosts. Granulosa cells of *Fundulus* and medaka are capable of producing 17,20β-P, the MIS in these species, when cultured with gonadotropin. In contrast, coculture of the granulosa cells with theca cells is necessary for 17,20β-P synthesis in amago salmon and rainbow trout. The theca cells in salmonids possess the steroidogenic enzyme, 17-hydroxylase ($P450_{C17}$), which converts progesterone to the immediate precursor of 17,20β-P, 17-hydroxyprogesterone. Recent studies indicate that the genes encoding $P450_{C17}$ and earlier steps in the steroidogenic pathway (3β-HSD and P450scc) are also present in the granulosa cells in these species although their functions are unknown. The involvement of both theca and granulosa cells in MIS production (two-cell type model) in salmonids suggests that both possess gonadotropin receptors. Evidence has recently been obtained for the existence of two types of gonadotropin receptors in salmonids. Type 1 receptors recognize both GTH-I and GTH-II, but have a higher affinity for GTH-I, and are present on both cell types; type II receptors primarily recognize GTH-II, are only present in granulosa cells, and increase in number in the mature postvitellogenic follicles. Confirmation that receptors specific for GTH-II are present on granulosa cells would provide a plausible explanation for how GTH-II causes upregulation of 20β-HSD and a switch in steroid synthesis from estrogen to MIS production.

A novel feature of MIS induction of FOM in fish and amphibians is that it does not conform to the classical model of hormone action; that is, it is not mediated by binding to nuclear steroid receptors to alter gene transcription. Early experiments in which inhibitors of transcription, but not translation, did not prevent MIS-induced FOM in a variety of fish species provided the first evidence that the action of the MIS is nongenomic. It was subsequently shown that the MIS, 17,20β-P, does not induce FOM when it is microinjected into goldfish oocytes but is effective when applied externally, which indicates that its site of action is on or near the oocyte plasma membrane. The finding that pharmacologically induced increases in cAMP levels block MIS stimulation of FOM *in vitro* also suggested that the action of the MIS is nongenomic and instead involves a second messenger signal transduction pathway. Subsequently, specific high-affinity, low-capacity plasma membrane receptors for 20β-S have been identified in the ovaries of spotted seatrout and striped bass and for 17,20β-P in defolliculated ovaries of rainbow trout. Binding to the spotted seatrout receptor is highly specific for 20β-S and structurally related C21 steroids. Moreover, there is a close correlation between receptor binding and their agonist or antagonist activities in the spotted seatrout FOM bioassay. Interestingly, the presence of hydroxyl groups at both the 17 and 20β positions on the progesterone nucleus appears to be essential for agonist activity in the seatrout FOM bioassay, whereas hydroxyls at both the 20β and 21 positions are required for high binding affinity for the 20β-S receptor. The presence of hydroxyls at all three positions (17, 20β, and 21 positions, i.e., 20β-S) results in greatest affinity for the receptor and the most potent induction of FOM in the bioassay. The action of 20β-S is rapid—a 1-min exposure to 20β-S is sufficient to induce FOM—which is consistent with its rapid rate of association with the receptor.

Other studies have shown that MIS induction of FOM occurs in Atlantic croaker, kisu, and dragonet only if their full-grown intrafollicular oocytes have previously been exposed to gonadotropin. The ability of croaker oocytes to respond to 20β-S and complete GVBD *in vitro* (oocyte maturational competence) was examined after incubation with gonadotropin [human chorionic gonadotropin (hCG)] for various periods *in vitro*. Follicle-enclosed oocytes not exposed to hCG (0 hr) did not undergo GVBD in the bioassay, but after 6 hr of exposure to hCG the oocytes had become maturationally competent (Fig. 4). This action of gonadotropin is not dependent on steroid synthesis since it was not blocked by coincubation with a 3β-hydroxysteroid dehydrogenase inhibitor, cyanoketone (Fig. 4), but it does require new mRNA and protein synthesis. It was concluded that there are two distinct stages of gonadotropin control of FOM in croaker and several other species: a "priming phase," which is largely steroid independent and induces the oocytes to become responsive to the MIS, followed by the "GVBD phase," which involves MIS synthesis and the completion of FOM in response to the MIS. Recent studies have shown that spotted seatrout 20β-S membrane receptor concentrations increase several-fold during gonadotropin treatment of mature oocytes *in vivo* and *in vitro* and that this increase in receptor concentrations is coincident with the acquisition of oocyte maturational competence. Moreover, similar to the development of maturational competence, gonadotropin upregulation of the receptor does not require synthesis of 20β-S or other steroids but is dependent on mRNA and protein synthesis. These experiments suggest that gonadotropin upregulation of MIS receptor concentrations is a physiologically important regulatory step in the hormonal control of oocyte maturational competence and FOM in these species. Gonadotropin also activates the connexin (*Cx 32.2*) gene in fully grown

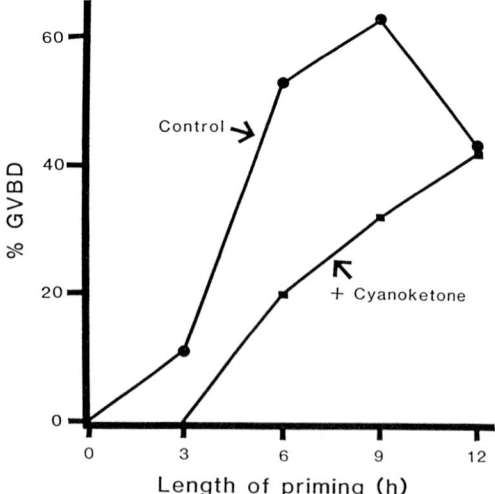

FIGURE 4 Effect of duration of priming *in vitro* with gonadotropin on the development of maturational competence of croaker oocytes in the presence and absence of cyanoketone. At the end of incubations with gonadotropin, tissues were rinsed and incubated with the maturation-inducing steroid, 20β-S, and oocytes were scored for completion of GVBD 10 hr later (adapted with permission from R. Patiño and P. Thomas, *Biol. Reprod.* 43, 818–827, 1990).

Atlantic croaker oocytes via a cAMP-dependent transcriptional process. An increase in *Cx 32.2* mRNA levels is associated with the increased heterocellular (granulosa cells–oocyte) gap junction contacts during the acquisition of oocyte maturational competence.

The signal transduction pathways and intracellular mechanisms of MIS action have been investigated in rainbow trout. A decrease in cAMP is required for MIS induction of FOM which appears to involve a guanine nucleotide binding G protein in the signal transduction pathway across the oocyte plasma membrane to the cytoplasm. A cytoplasmic factor, maturation-promoting factor composed of cdc2 kinase and cyclin B, is formed and is the intracellular mediator of FOM.

VII. OVULATION

Once oocyte maturation is complete the oocyte is released from the ovarian follicle by a process termed ovulation and is transported to an extraovarian site, where fertilization occurs. The ovulatory process includes several preparatory changes in the oocyte and ovarian follicle prior to expulsion of the oocyte. Microvillar connections between the oocyte and surrounding granulosa cells are broken (follicular separation), parts of the follicular wall decrease in thickness, and an opening forms in the follicle wall (follicular rupture) due to digestion of tissue by proteolytic enzymes. The oocyte is released by contraction of smooth muscles in the follicular wall. This process is regulated in part by prostaglandins. The action of prostaglandins in the brook trout appears to be mediated by an increase in ovarian cAMP levels.

A. Endocrine Control

These structural and biochemical changes in the ovary during ovulation are controlled by GTH-II, which is elevated in the blood during FOM prior to ovulation. The MISs, 17,20β-P and 20β-S, which are elevated at this time, have been shown to induce ovulation *in vitro* of yellow perch and spotted seatrout oocytes, respectively, by a genomic mechanism. One likely action of the MISs in teleosts is to regulate ovarian prostaglandin F synthesis (PGF). An increase in PGF levels has been observed in brook trout and yellow perch ovaries around the time of ovulation. The extrafollicular tissue, which includes the stroma, appears to be a major site of prostaglandin synthesis. An interesting finding is that removal of the extrafollicular tissue from yellow perch ovarian follicles abolishes 17,20β-P induction of PGF synthesis and ovulation, which is not restored when the two tissues are coincubated. This suggests that the close association between the follicles and extrafollicular tissue must be maintained for 17,20β-P induction of ovulation. Recently, a nuclear progestogen receptor has been identified in spotted seatrout ovaries which has high binding affinities for 17,20β-P and 20β-S. Thus, in spotted seatrout, and probably other teleost species, there are two ovarian progestogen receptors that recognize the MIS: a membrane receptor on the oocyte that mediates the MIS induction of FOM and a nuclear receptor presumably in the extrafollicular cells that is involved in MIS regulation of ovulation.

VIII. SPAWNING AND FERTILIZATION

Spawning in most fish species involves the release of ovulated eggs for eventual external fertilization by sperm. In goldfish, prostaglandins (especially PGF2α) released at ovulation trigger spawning behavior in males. Spawning behavior commences at dawn on the day of ovulation in goldfish, provided a sexually active male and aquatic vegetation are available, and continues for several hours until all ovulated eggs are released. Entry of sperm into the oocyte occurs via the micropyle. Fertilization induces hardening of the vitelline envelope (cortical reaction). The cortical reaction causes structural changes in the micropyle, which prevents additional sperm from entering the egg. The hardened vitelline envelope forms a protective barrier around the egg and the developing embryo.

See Also the Following Articles

Female Reproductive System, Fish; Fish, Modes of Reproduction in; Ovarian Cycle, Mammals; Ovarian Cycles and Follicle Development in, Birds; Seasonal Reproduction, Fish

Bibliography

Dodd, J. M., and Sumpter, J. P. (1984). Fishes. In *Marshall's Physiology of Reproduction, Vol. 1, Reproductive Cycles of Vertebrates* (G. E. Lamming, Ed.), pp. 1–126. Churchill Livingstone, London.

Goetz, F. W., and Thomas, P. (Eds.) (1995). *Reproductive Physiology of Fish*. Univ. of Texas Press, Austin.

Goetz, F. W., Berndtson, A. K., and Ranjan, M. (1991). Ovulation: Mediators at the ovarian level. In *Vertebrate Endocrinology: Fundamental and Biomedical Implications* (P. K. T. Pang and M. Schreibman, Eds.), pp. 127–203. Academic Press, San Diego.

Munro, A. D., Scott, A. P., and Lam, T. J. (Eds.) (1990). *Reproductive Seasonality in Teleosts: Environmental Influences*. CRC Press, Boca Raton, FL.

Nagahama, Y., Yoshikuni, M., Yamashita, M., and Tanaka, M. (1994). Regulation of oocyte maturation in fish. In *Fish Physiology, Vol. XIII, Molecular Endocrinology of Fish* (N. M. Sherwood and C. L. Hew, Eds.), pp. 393–439. Academic Press, San Diego.

Patiño, R., and Thomas, P. (1990). Characterization of membrane receptor activity for 17α,20β,21-trihydroxy-4-pregnen-3-one in ovaries of spotted seatrout (*Cynoscion nebulosus*). *Gen. Comp. Endocrinol.* **78**, 204–217.

Peter, R. E., and Yu, K. L. (1997). Neuroendocrine regulation of ovulation in fishes: Basic and applied aspects. *Rev. Fish Biol. Fish.* **7**, 173–197.

Smith, J. S., and Thomas, P. (1991). Changes in hepatic estrogen receptor concentrations during the annual reproductive and ovarian cycles of a marine teleost, the spotted seatrout, *Cynoscion nebulosus*. *Gen. Comp. Endocrinol.* **81**, 234–245.

Thomas, P. (1994). Hormonal control of final oocyte maturation in sciaenid fishes. In *Perspectives in Comparative Endocrinology* (K. G. Davey, R. E. Peter, and S. S. Tobe, Eds.), pp. 619–625. National Research Council of Canada, Ottawa.

Wallace, R. A., and Selman, K. (1990). Ultrastructural aspects of oogenesis and oocyte growth in fish and amphibians. *J. Electron Microsc. Technique* **16**, 175–201.

Ovarian Cycles and Follicle Development in Birds

A. L. Johnson
University of Notre Dame

I. Seasonal Reproduction
II. The Avian Ovary and Dynamics of Follicle Growth
III. Patterns of Ovulation
IV. Ovulation and Oviposition
V. Follicle Atresia and Postreproductive Regression of the Ovary

GLOSSARY

atresia The death of ovarian follicles not yet developed to the ovulatory stage. The process of atresia is mediated by apoptosis and is initiated within the granulosa cell layer.

granulosa layer The innermost layer of follicle cells (of epithelial cell origin) that nurtures and interacts with the oocyte.

oocyte A single cell containing the germinal vesicle and surrounded by the plasma (vitelline) membrane.

ovarian follicle A single germ cell (oocyte), granulosa layer, and theca tissue.

ovulation The rupture of the fully developed ovarian follicle and release of the oocyte plus yolk into the reproductive tract for subsequent fertilization, membrane deposition, and shell formation.

photorefractory Referring to a physiological state in which the gonads are incapable of remaining, or initially becoming, functional in the presence a stimulatory photoperiod.

photosensitive Referring to a physiological state in which the reproductive system responds to a stimulatory environmental cue (usually the increasing photoperiod during spring) with growth and development.

previtellogenic follicles Follicles in the slow-growth stage of development.

primordial follicles The oocyte arrested at the diplotene stage of meiotic development and surrounded only by a single layer of granulosa cells; they are embedded within the ovarian stromal tissue.

theca The heterogeneous outermost tissue of the follicle consisting of steroidogenic cells, connective tissue, smooth muscle, blood, and immune cells. The theca is functionally divided into theca interna and theca externa.

vitellogenic follicles Rapidly growing follicles due to the uptake of the lipoproteins, vitellogenin, and very low-density lipoprotein; such follicles constitute the preovulatory hierarchy.

For the majority of avian species (excluding as a notable example the domestic hen, *Gallus gallus*), seasonal ovarian cycles entail the precisely timed and highly regulated development of the ovary (and its ovarian follicles) followed by its subsequent postreproductive regression. A prerequisite for seasonal reproductive activity in virtually all birds, with the possible exception of some equatorial species, is the ability to respond to photoperiod (e.g., acquisition of photosensitivity). Ovarian activity with each new reproductive season generally entails the initial growth and differentiation of tens to perhaps hundreds of ovarian follicles, only a small fraction of which will successfully develop to the ovulatory stage. At ovulation, the oocyte, together with its sustaining yolk, is released from the follicle and collected by the proximal region (infundibulum) of the oviduct. Following fertilization and the deposition of egg membranes, the oocyte is encased in a calcareous shell by the shell gland prior to oviposition. The ovarian follicular hierarchy represents an adaptation evolved to enable birds to ovulate an egg every or every other day for a finite period of time (several days to weeks) before the initiation of incubation. Eventually, termination of the annual reproductive cycle is precipitated by a loss of sensitivity to a stimulatory photoperiod. This condition of photorefractoriness is characterized by a reduction of gonadotropin-releasing hormone within the hypothalamus and decreased circulating gonadotropin concentrations. At the level of the ovary, this results in the rapid death of all growing follicles (atresia) and the complete regression of the ovary to its preseasonal inactive state.

I. SEASONAL REPRODUCTION

Very few avian species exhibit breeding behavior throughout the year. Most, if not all birds are considered to be photoperiodic and possess, to varying degrees, an endogenous (circannual) rhythm of reproduction that is synchronized and modified by external environmental cues. The yearly reproductive cycle consists of a predictable alternation between photosensitivity and photorefractoriness, and the development of photosensitivity is critical for timing the onset of the breeding season. The photosensitive state is one in which the hypothalamus and pituitary (hypothalamo–hypophyseal axis) becomes capable of producing and secreting gonadotropin-releasing hormone (GnRH) and gonadotropins, respectively, in response to an appropriate photoperiod. Thus, the onset of ovarian development is associated with a steady increase in circulating concentrations of the gonadotropins, follicle-stimulating hormone (FSH) and luteinizing hormone (LH).

Wingfield and Farner outlined four categories of information that describe the influence of various environmental factors on seasonal reproduction: initial predictive, essential supplementary, synchronizing and integrating, and modifying information. The most frequent form of initial predictive information critical for the induction of gonadal development is the highly predictable annual cycle of day length. Growth of the ovary and follicle development are initiated in most avian species in response to increasing photoperiod, though there are a few species that begin reproductive activity during the period of decreasing daily photoperiod (e.g., emperor penguin, *Aptenodytes forsteri*). While such predictive information initiates development of a reproductively competent state, supplementary information, such as weather, food availability, and an available breeding territory, are often required to induce final stages of the nesting phase, mating and egg laying. In turn, synchronizing and integrating information represents the fine-tuning of temporal events which is perhaps best achieved by behavioral interactions between mates. In the wood pigeon (*Columba palumbus*), photoperiod-induced increases in gonadotropin secretion stimulate the development of ovarian follicles to a size range of 1–5 mm, but final growth and maturation of follicles occur only following stimulation of the female by male courtship and nest-building behavior. Sometimes supplementary or synchronizing information alone can initiate reproductive activity. For instance, the presence of a mate during the winter months can stimulate gonadotropin secretion in the kestrel (*Falco tinnunculus*) well before the onset of predictive (photoperiodic) cues. In general, once courtship is initiated, there occurs a series of behavioral and physiological events that leads to ovarian follicle growth, differentiation, and ovulation of one or more eggs.

In most species, the termination of egg laying is followed by incubation behavior. It is generally accepted that this behavior is promoted by increased circulating concentrations of prolactin, and prolactin levels almost always decline with the termination of parental care. There are, however, examples of species that show a poor correlation between prolactin levels and nesting behavior (e.g., several species of sparrows). Furthermore, brood parasitic species such as the cowbird (*Molothrus ater*) show elevated levels of prolactin during the breeding season despite the fact that they do not express parental behavior. In species in which the male is the primary attendant of eggs (e.g., spotted sandpiper, *Actitis macularia*), the female routinely has lower concentrations of prolactin than the male. Elevated levels of prolactin have also been implicated as a hormonal regulator of migration at the end of the breeding season, though again this may not hold true for all migrating species (e.g., European quail, *Coturnix coturnix*).

Unfortunately, the precise neural, endocrine, and cellular mechanisms that mediate many of the events mentioned previously, if known at all, have been described in only a few avian species. Thus, much of the remaining discussion is necessarily derived from experimental data accumulated from primarily domesticated species [e.g., domestic hen, Japanese quail (*Coturnix coturnix japonica*), and turkey hen (*Meleagris gallopavo*)]. Nevertheless, it is useful to consider such data in an effort to develop testable hypotheses which can subsequently be investigated in wild species.

II. THE AVIAN OVARY AND DYNAMICS OF FOLLICLE GROWTH

In most species of birds the right ovary and oviduct regress during embryonic development, and only the left ovary and reproductive tract are functional in the adult. There are, however, exceptions to this arrangement, including many hawk and falcon species that have two functional ovaries and oviducts. Such species are thought to alternate ovulations between ovaries throughout the breeding season. Perhaps more curious is the reproductive anatomy of the kiwi (*Apteryx australis*), which typically has two functional ovaries that alternately ovulate a follicle but only one functional oviduct to transport the egg.

A. Early Organization of the Ovary

Unlike some teleost and many reptilian species, the ovary of the adult bird is not known to contain mitotically active germ cells; instead, all oocytes are arrested at the diplotene stage of the first meiotic division by or shortly after the time of hatch. Oocyte viability and meiotic arrest are maintained by paracrine factors, such as stem cell factor, which are produced by mesenchymal cells from the ovarian cortex and by presumptive granulosa cells. The extent of blood flow to the ovary during the posthatch interval is relatively minimal but later increases dramatically during the immediate prepubertal period. Beginning early in the posthatch period, the ovary is already innervated by both cholinergic and adrenergic fibers. Neurotropins produced within neuron cell bodies and secreted from the nerve terminals are proposed to promote the organization of primordial follicles at a time when the ovary is not fully responsive to gonadotropins. The process of primordial follicle organization consists of a single layer of presumptive granulosa cells of epithelial cell origin completely surrounding a "naked" germ cell. Germ cells that do not become organized during the early posthatch period are destined to die via the process of apoptosis. Primordial follicles formed during this period serve as the only source of germ cells available for the remainder of the female's life.

B. Cellular Changes Associated with Follicle Growth and Differentiation

During each nonbreeding season, the ovary of postpubertal females contains primarily primordial follicles embedded within the stromal tissue but also a few slow-growing follicles at various early stages of development. Due to low levels of gonadotropins and other trophic factors which exist during the nonbreeding season, these follicles will eventually become atretic without reaching the preovulatory stage. Increased mass of the ovary occurs with the onset of each reproductive season coincident with the formation of new blood vessels plus the initial slow growth of tens to a hundred or more follicles. The factors that initially stimulate the growth of only a select (and species-dependent) number of primordial follicles per reproductive season are unknown. Once a cohort of follicles begins the slow-growth phase they protrude from the ovary, and each follicle forms a pedicle (stalk), through which the sole source of blood supply and nervous innervation is provided.

It is useful to summarize here only some of the more relevant data that describe hen follicle growth and differentiation. More extensive reviews of this topic can be found elsewhere (Etches, 1990; Johnson, 1990, 1996, 1997). Primordial follicles may remain in a resting stage within the ovarian stroma for many reproductive seasons before being selected to begin the slow-growth phase. In the domestic hen, slow-growing follicles increase in diameter from <1 to 6–8 mm over a period of several weeks to months. At any time there may be a total of 4–12 follicles 6–8 mm in diameter, and the size of this group is determined both by the number of follicles starting slow growth and by their rate of premature death (atresia). It is from this cohort of 6- to 8–mm-diameter follicles that a single follicle per day is selected into the vitellogenic phase in final preparation for ovulation. Significantly, it is this process of selection that allows a single egg per day to be ovulated, and this process differentiates birds from mammals (which typically ovulate a cohort of follicles simultaneously).

Perhaps one of the most important functions of the ovarian follicle, aside from providing a nurturing environment for the oocyte, is the production of steroid hormones. Sufficient amounts of progestins, androgens, and estrogens synthesized at the appropriate times are critical for the positive and negative feedback actions on the hypothalamo–hypophyseal axis, growth and maintenance of the reproductive tract, expression of reproductive behaviors, and even maintenance of the follicle itself. Clearly, however, sex steroids are not the only hormones synthesized by the avian ovary because the ever-expanding list of endocrine/paracrine/autocrine factors currently includes growth factors (e.g., transforming growth factors-α and -β, stem cell factor, and insulin-like growth factor), cytokines (tumor necrosis factor-α and interleukins), prostaglandins, inhibin and activin, relaxin and oxytocin, vasoactive substances (e.g., vasoactive intestinal peptide and arginine vasotocin), and neurotropins. Many of these factors act within the ovary itself to regulate follicle development, whereas others serve as agents that provide feedback information to the hypothalamus and pituitary.

C. Granulosa Cells

Increased follicle diameter during initial growth is accommodated by active granulosa cell mitosis. The germinal disc region within the follicle is both the center of active granulosa cell mitosis and the site where the germinal vesicle is localized. In the hen, granulosa cells from follicles <8 mm in diameter form a multicell (two or three deep) layer. By comparison, during the final stages follicle development (i.e., in vitellogenic follicles 9–40 mm in diameter), granulosa cells become mitotically inactive, except to a limited extent within the germinal disc region. Thus, the increasing follicle diameter in rapidly growing follicles occurs within granulosa primarily by reorganization into a single, cuboidal cell layer (Fig. 1).

The previtellogenic phase of follicle development is associated with the highest rate of granulosa cell proliferation during follicle development but the lowest capacity for steroidogenesis. Granulosa cells from hen follicles express FSH receptor messenger ribonucleic acid (mRNA) at all stages of development, but levels are highest in granulosa from follicles <8 mm diameter (Fig. 2C). Only after the follicle has been selected into the rapid growth phase does the granulosa layer become capable of producing steroids, and this competence is associated with the first detectable cytochrome P450 cholesterol side chain cleavage enzyme activity, progesterone production, and expression of LH receptor mRNA (Figs. 2A and 2B). The granulosa cell layer of preovulatory follicles is the primary source of secreted progesterone, and it also produces lesser amounts of testosterone. Steroid production in granulosa cells from vitellogenic follicles is primarily under the stimulatory control of LH.

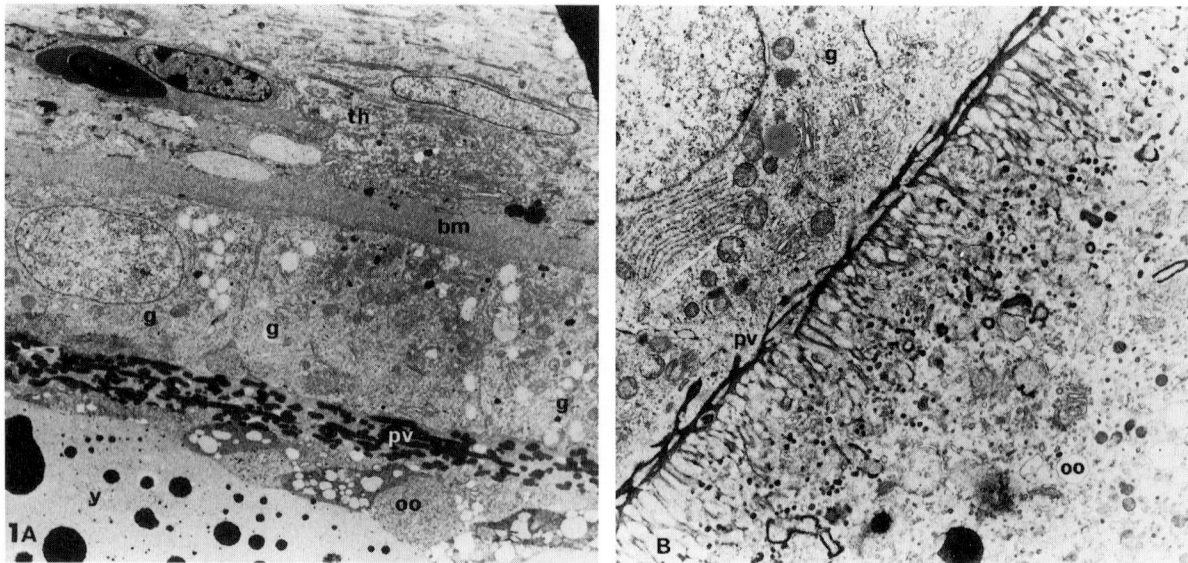

FIGURE 1 (A) Electron micrograph (magnification, ×4600) of a cross section through the cell layers of a preovulatory (~30-mm diameter) chicken follicle. Theca externa (th ex) is visible only at the uppermost portion of the figure. The basement membrane (bm) separates the theca layer from the single cell granulosa (g) layer. pv, perivitelline membrane; oo, oocyte cytoplasm; y, yolk spheres. (B) Higher magnification (×9900) of a 7-mm-diameter follicle showing the radially oriented invaginations of the granulosa cell layer (not visible in A). These invaginations are lost just prior to ovulation, allowing the granulosa layer to remain part of the postovulatory follicle. Note that this stage of follicle development is previtellogenic (i.e., there are no large lipid spheres) [reprinted with permission from Shen et al. (1993). Chicken oocyte growth: Receptor-mediated yolk deposition. *Cell Tissue Res.* 272, 459–471].

D. Theca Tissue

Another critical developmental stage in hen follicles <1 mm diameter is the organization of the theca layer (from mesodermal origin) external to the granulosa layer. Theca tissue consists of steroidogenic cells, blood cells, connective tissue, and smooth muscle and serves as the primary support in the face of increasing intrafollicular tension. Sometime during early follicle development, the theca differentiates into morphologically and functionally distinct theca externa and theca interna layers. As each follicle grows, the development of new blood vasculature and the innervation by adrenergic and cholinergic nerve fibers increases. Blood and nervous tissues are localized exclusively within the theca layer and predominantly within the theca interna. Accommodation of the rapid growth phase by the theca is accomplished by decreased cohesion of theca cells and increased elasticity due to the dissociation of connective tissue.

The theca layer is steroidogenically active at all stages of follicle development. Luteinizing hormone receptor and FSH receptor mRNA are expressed within the theca at all stages of follicle development, but steroid production is primarily under the stimulatory control of LH. Progesterone produced within the theca is metabolized to androgens (primarily androstenedione) in the theca interna and is subsequently converted to estrogen (primarily estradiol-17β) by groups of aromatase-expressing cells located within the theca externa. Interestingly, the steroidogenic pathway to androstenedione synthesis changes from the Δ^5 pathway in slow-growing follicles to the Δ^4 pathway in preovulatory follicles. The physiological significance of this metabolic switch has yet to be determined. The interaction of the granulosa, theca interna, and theca externa layers is critical for the production and local distribution and secretion of paracrine/endocrine steroids and growth factors required for the growth and function of the developing follicle.

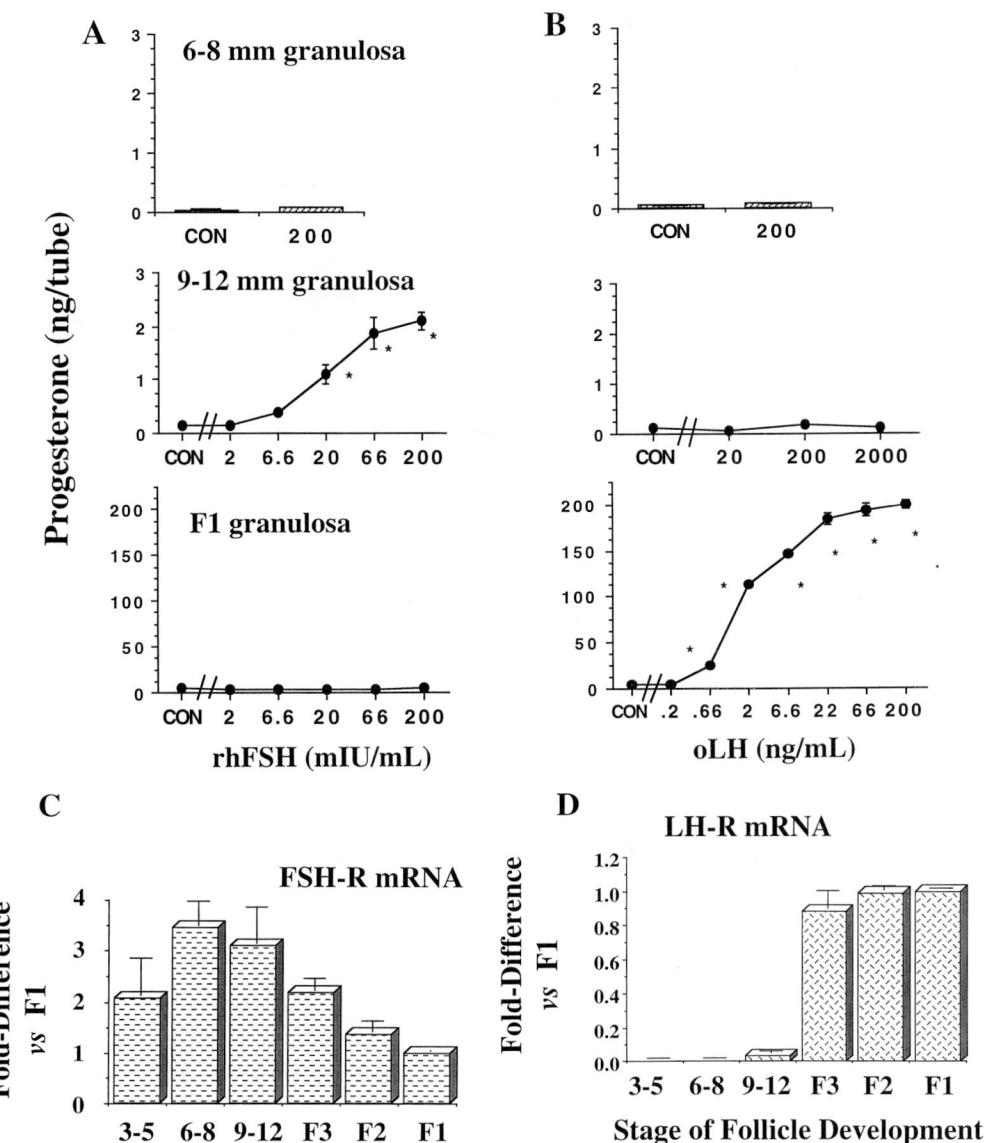

FIGURE 2 In vitro production of progesterone in granulosa cells from 6- to 8-mm (previtellogenic), 9- to 12-mm (the earliest phase of vitellogenesis), and the largest preovulatory (F1) follicles in response to recombinant human FSH (A) or ovine LH (B). Relative levels of FSH receptor and LH receptor mRNA are depicted in C and D, respectively. Note that granulosa cells from previtellogenic follicles do not produce progesterone even though they express FSH receptor mRNA. By comparison, steroid production is observed in response to FSH, but not LH, during early vitellogenesis, whereas LH becomes the predominantly active gonadotropin during the final stages of follicle maturation [reprinted with permission from Johnson, A. L., Bridgham, J. T., Witty, J., and Tilly, J. L. Susceptibility of avian granulosa cells to apoptosis is dependent upon stage of follicle development and is related to endogenous levels of bcl-xlong gene expression. Endocrinology 137, 2059–2066, 1996, © The Endocrine Society].

E. Vitellogenesis

At all stages of follicle development the granulosa layer is separated from the theca by an acellular basement membrane (basal lamina). On the side adjacent to the oocyte, granulosa cells during early follicle growth come in direct contact with the oocyte plasma (vitelline) membrane (Fig. 1). Sometime prior to the beginning of the rapid growth (vitellogenic) phase the perivitelline membrane (equivalent to the mammalian zona pellucida) is deposited outside of the vitelline membrane by granulosa cells. A final acellular matrix surrounding the perivitelline membrane is eventually deposited by the oviduct following ovulation.

Following selection and during the rapid growth phase, follicle growth is accomplished almost exclusively by the accumulation of very low-density lipoprotein and vitellogenin from the blood. The deposition of yolk is facilitated by extensive blood vascularization developed within the follicle prior to the vitellogenic stage. Lipoproteins escape from blood capillaries in the theca interna, passively cross the basement membrane through gaps between granulosa cells and across the perivitelline matrix, and finally are actively transported through the oocyte plasma membrane by receptor-mediated endocytosis. The yolk completely fills the interior of the oocyte, and there is no fluid-filled antrum comparable to that in the mammalian preovulatory follicle. During the final 8–10 days prior to ovulation, a hen follicle initially weighing <0.5 g can increase its mass by as much as 2.5 g per day to attain an ovulatory size of 15 g (40 mm diameter) or more. Uptake of yolk occurs evenly across the entire oocyte surface with the exception of the area around the germinal disc. Presumably, the presence of large yolk spheres within this region would eventually interfere with the process of fertilization.

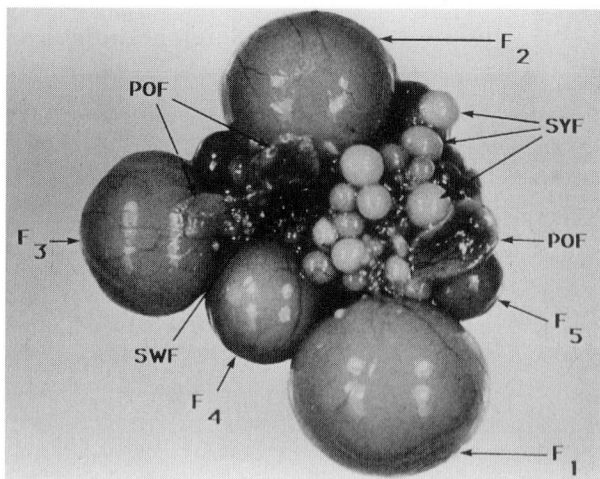

FIGURE 3 Ovary of the domestic hen illustrating the distinct preovulatory follicle hierarchy. F1 denotes the largest preovulatory follicle which will be ovulated next, whereas F2–F5 represent successively smaller and less mature hierarchical follicles. Small yellow follicles (SYF; 6–8 mm diameter) comprise a pool of 6–12 follicles, from which a single follicle per day is recruited into the rapid growth phase. There may be tens to 100 or more slow-growing small white follicles (SWF; ~1–5 mm diameter) as well as several postovulatory follicles (POF) in various stages of being resorbed [reproduced with permission from Johnson, A. L. (1990). Reproduction in the Female. In *Avian Physiology*, 4th ed. (P. D. Sturkie, Ed.). Springer-Verlag, New York.

III. PATTERNS OF OVULATION

The arrangement of growing follicles in an orderly and morphologically distinguishable fashion (the ovarian hierarchy) is an adaptation of birds to enable the production of eggs on successive days (Fig. 3). It has also been hypothesized that the avian follicular hierarchy represents a weight-reducing adaptation evolved for flight.

During the breeding season, the ovary is capable of ovulating a single egg once every or every other day until a species-specific clutch size is reached. The number of eggs in the clutch can vary dramatically among species (from 1 to >15 eggs). In those species with a clutch size of 1 (e.g., puffins, several swifts, some of the larger doves, and penguins) it appears that only a single ovarian follicle per clutch develops to the vitellogenic (preovulatory) stage. It has been proposed that a relatively large number of eggs per clutch in a given species is related to an overall lower survival success rate for the offspring.

For a given clutch, two alternative patterns of ovulation may be distinguished. Determinate layers (e.g., brant, *Branta bernicla*; snow goose, *Chen caerulescens*) produce a single clutch, and the number of eggs ovulated does not increase should eggs be destroyed. By comparison, indeterminate layers are capable of exceeding the normal clutch size or will duplicate the clutch should the progression of a normal nesting phase be disrupted. Such an event might occur following predation and would constitute a form of modifying information. For instance, the mallard duck (*Anas platyrhynchos*), song sparrow (*Melospiza melodia*), and flicker (*Colaptes auratus*) will produce a second clutch if eggs from the first clutch are removed from the nest.

The physiological mechanisms which suppress further daily ovulations once the species-specific clutch size has been obtained are not known but are at least in part related to the tactile (and possibly thermal stimulation) of the brood patch during incubation and the resulting elevated secretion of prolactin. High levels of circulating prolactin suppress pituitary secretion of gonadotropins and can directly inhibit ovarian steroidogenesis and continued follicle development.

Finally, many species that breed at mid- to low latitudes will produce two, and sometimes more, broods per season. In general, the initiation of a subsequent clutch does not occur until the first brood becomes independent from the parents.

IV. OVULATION AND OVIPOSITION

Ovulation of the largest preovulatory follicle is ultimately induced by a transient increase in blood levels of LH produced by the pituitary. The preovulatory LH surge is thought to be initiated and subsequently potentiated by steroid hormone produced predominantly by the largest preovulatory follicle. In all avian species studied to date, the primary ovulation-inducing ovarian steroid is progesterone; preovulatory surges of both progesterone and LH occur in most species about 4–8 hr prior to ovulation. Germinal vesicle breakdown and the completion of the first meiotic division, stimulated in response to the LH surge, are completed about 2 hr before ovulation, whereas the second meiotic division takes place subsequent to ovulation and fertilization. Some final changes in follicle cell morphology also occur 1 or 2 hr prior to ovulation. For instance, the single layer of cuboidal granulosa cells becomes progressively more flattened, and the number of cytoplasmic processes extending from the granulosa layer toward the oocyte plasma membrane is decreased. These changes eventually enable the granulosa cell layer to detach from the oocyte at ovulation and remain a part of the postovulatory follicle.

Rupture of the preovulatory follicle at ovulation occurs from the stigma, a narrow finger-like region along the surface of the follicle easily discernible by the relative absence of blood vasculature. The process of oocyte expulsion is most likely the result of many physiological factors working in parallel: activation of proteases that results in the reduction of tensile strength in the follicle wall, increased intrafollicular pressure due to increased blood flow to the follicle, initiation of smooth muscle contractions in the follicle wall, a weakening of granulosa cell support following the withdrawal of microvilli from the surface of the oocyte, and a selective loss of cells within the stigma region via apoptosis.

Unlike mammalian species, in which the remains of the follicle after ovulation are rapidly transformed into an active progesterone-producing structure (the corpus luteum), the avian postovulatory follicle (POF) in most species remains a viable structure for less than a day or two before it begins to undergo active resorption. Exceptions to this include the pheasant (*Phasianus colchicus*) and mallard duck, in which the POF appears to persist for several months; it is not clear, however, whether the structure is physiologically active at any time during this period. Relaxin and oxytocin, together with relatively low levels of steroids and prostaglandins, are a few of the products that have been localized to the avian POF. It is possible that one or more of these hormones is involved with regulating the timing of egg laying because surgical removal of the most recent POF in the chicken delays oviposition of the egg by 1–7 days.

The exact time of ovulation (and, in turn, oviposition) within the day is species dependent and is undoubtedly adapted to the bird's behavior and environment. For instance, many passerines lay their eggs early in the morning, pigeons and pheasants tend to lay in the afternoon to evening, whereas the American coot (*Fulica americana*) and many ducks lay during the night. Some galliform birds, including the chicken, turkey, and bobwhite quail (*Colinis virginianus*), lay their eggs at intervals in excess of 24 hr (24–28 hr). Thus, oviposition of the first eggs in the sequence occurs in the early morning, and each subsequent egg in the sequence is laid progressively later in the day. The neural, neuroendocrine, and endocrine mechanisms that regulate this unique and complex variation of the ovulatory cycle have been the subject of considerable investigation over the years.

V. FOLLICLE ATRESIA AND POSTREPRODUCTIVE REGRESSION OF THE OVARY

Termination of egg laying must occur with sufficient time remaining in the season for the resulting young to obtain adequate resources before the resources are lost to inclement weather or changes in seasonal availability and/or to adequately mature prior to molt and/or migration. The termination of breeding frequently occurs before the summer solstice, while the photoperiod is still increasing and is actually longer than that which initiated gonadal growth. This physiological state results in a lower hypothalamic content of GnRH and decreased circulating gonadotropins to levels that are unable to sustain follicle development. The result is the death of all developing ovarian follicles via the process of atresia; the loss of yolk-filled, preovulatory follicles is sometimes referred to as "bursting atresia." Moreover, the mass of ovarian stromal tissue itself decreases, due in large part to the concomitant decrease in the amount of blood perfusing the ovary. The photorefractory state generally requires exposure to some duration of short photoperiod, such as the decreasing day length of autumn, before photosensitivity can be restored. Although all the physiological mediators of photorefractoriness have yet to be defined, it has been experimentally determined in several species that thyroidectomy will prevent long photoperiod-induced photorefractoriness and thyroxine treatment can replicate the effects of long photoperiod in reducing hypothalamic levels of gonadotropins.

The loss of follicles by atresia is by no means limited to the end of the reproductive season. Indeed, follicle atresia normally occurs in a large percentage of follicles that begin the slow growth phase but fail to survive to the preovulatory stage. In the chicken it is estimated that >90% of such developing follicles succumb to atresia, and that atresia occurs most frequently in follicles measuring approximately 3–7 mm in diameter. By contrast, follicles that have begun the rapid growth phase rarely become atretic unless the bird is perturbed (i.e., food withdrawn or photoperiod decreased).

While it is not known what physiological signal(s) determines when and which follicles will die, or why there occurs such an apparent wastage of female germ cells, it is now recognized that the process of follicle atresia is mediated via apoptosis. During the early stages of follicle atresia apoptosis occurs almost exclusively in the granulosa layer, and more specifically in cells randomly dispersed throughout the layer. The total number of affected granulosa cells increases as the atretic process progresses. Characteristics of granulosa cells undergoing apoptosis include initial cell shrinkage and membrane blebbing, formation of pyknotic nuclei (Fig. 4), and the cleavage

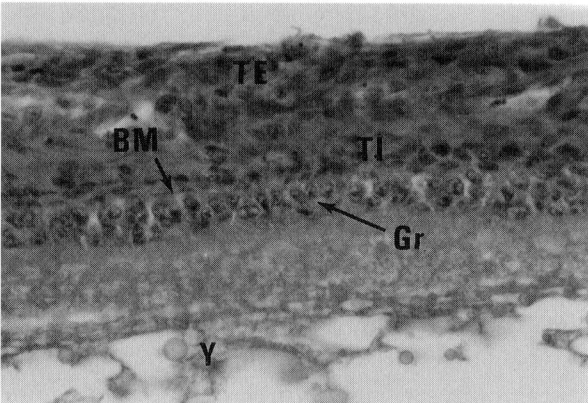

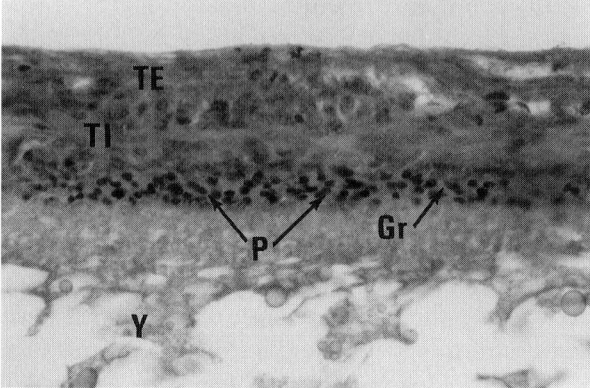

FIGURE 4 Morphological evaluation of tissues from a ~6-mm-diameter, morphologically normal follicle (A) and a follicle actively undergoing atresia (B). Note the presence of numerous pyknotic (P) nuclei (stained with hematoxylin and picric acid methyl blue) almost exclusively within the multicellular layer of granulosa (Gr) from atretic follicles. BM, basement membrane; TE, theca externa; TI, theca interna; Y, yolk. Magnification, ×400 [reproduced with permission from Johnson, A. L., Bridgham, J. T., Witty, J., and Tilly, J. L. Susceptibility of avian granulosa cells to apoptosis is dependent upon stage of follicle development and is related to endogenous levels of *bcl*-xlong gene expression. *Endocrinology* 137, 2059–2066, 1996, © The Endocrine Society].

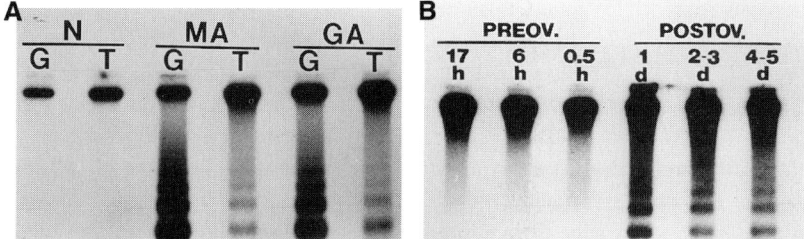

FIGURE 5 (A) Analysis of oligonucleosome formation, a hallmark of apoptosis, in granulosa (G) and theca (T) tissue from normal (N), early atretic (MA), and grossly atretic (GA) 4- to 6-mm-diameter follicles. Note that evidence of apoptosis occurs only in follicles undergoing atresia, and that the vast majority of oligonucleosome formation occurs within the granulosa. (B) Evaluation of oligonucleosome formation in preovulatory (PREOV) and postovulatory (POSTOV) follicles collected approximately 17, 6, and 0.5 hr before a predicted ovulation and 1, 2 or 3, and 4 or 5 days following ovulation (reproduced with permission from J. L. Tilly et al., Involvement of apoptosis in ovarian follicular atresia and postovulatory regression, Endocrinology 129, 2799–2801, 1991; © The Endocrine Society).

of DNA which results in the characteristic ladder (oligonucleosome) formation (Fig. 5). Eventually, small portions of the cell bud off forming apoptotic bodies, and these are resorbed by adjacent cells. Of significance is that this entire process is accomplished rapidly (within a matter of hours) and that it occurs in the absence of an inflammatory response. Significantly, it is this latter characteristic that enables large numbers of follicles to die simultaneously (i.e., at the termination of breeding) or relatively fewer follicles to be removed on a continuing basis (i.e., during follicle development) without severely disrupting ovarian function and the surrounding environment within the abdomen. Not unexpectedly, it has been determined that resorption of the POF is also accomplished by apoptosis (Fig. 5).

Although the specific factors responsible for inducing or preventing the onset of granulosa cell apoptosis, and thus follicle atresia, have yet to be unequivocally identified, there is some evidence that the absence of a viable germinal vesicle results in follicle death. While the presence of a defective germinal vesicle may provide an explanation for the occurrence of at least some percentage of atretic follicles during the breeding season, it cannot explain the synchronous loss of virtually all developing follicles at the termination of the breeding season. On the other hand, it has been determined that endocrine/paracrine factors capable of cell signaling via the adenylyl cyclase/cAMP second messenger system (e.g., LH, FSH, and vasoactive intestinal polypeptide) prevent or attenuate the onset of apoptotic cell death in hen granulosa cells. Thus, it is logical to propose that the sudden decline in circulating gonadotropins that accompanies photorefractoriness and the termination of breeding in wild species is sufficient to initiate widespread granulosa cell apoptosis and follicle atresia. Finally, while virtually all the experimental data pertaining to follicle atresia mediated via the process of apoptosis have been derived from domesticated birds, it is speculated with confidence that the mechanisms responsible for follicle atresia and ovarian regression in wild species will eventually be determined to be similar or identical.

See Also the Following Articles

Follicular Atresia; Ovarian Cycle, Mammals; Photoperiodism, Vertebrates; Seasonal Reproduction, Birds

Bibliography

Cockrem, J. F. (1995). Timing of seasonal breeding in birds, with particular reference to New Zealand birds. Reprod. Fertil. Dev. 7, 1–19.

Etches, R. J. (1990). The ovulatory cycle of the hen. Crit. Rev. Poultry Biol. 2(4), 293–318.

Johnson, A. L. (1990). Steroidogenesis and actions of steroids in the hen ovary. Crit. Rev. Poultry Biol. 2(4), 319–346.

Johnson, A. L. (1996). The avian ovarian hierarchy: A balance between follicle differentiation and atresia. Poultry Avian Biol. Rev. 7(2/3), 99–110.

Johnson, A. L. (1997). Apoptosis-susceptible versus -resistant granulosa cells from hen ovarian follicles. In *Cell Death in Reproductive Physiology* (J. L. Tilly, J. Strauss, and M. Tenniswood, Eds.). Springer-Verlag, New York.

Johnson, A. L., Bridgham, J. T., Witty, J., and Tilly, J. L. (1996). Susceptibility of avian granulosa cells to apoptosis is dependent upon stage of follicle development and is related to endogenous levels of *bcl*-xlong gene expression. *Endocrinology* **137**, 2059–2066.

Shen, X., Steyrer, E., Retzek, H., Sanders, E. J., and Schneider, W. J. (1993). Chicken oocyte growth: receptor-mediated yolk deposition. *Cell & Tissue Res.* **272**, 459–471.

Wingfield, J. C., and Farner D. S. (1993). Endocrinology of reproduction in wild species. In *Avian Biology*, Vol. IX. Academic Press, New York.

Yoshimura, Y., Tischkau, S. A., and Bahr, J. M. (1994). Destruction of the germinal disc region of an immature preovulatory follicle suppresses follicular maturation and ovulation. *Biol. Reprod.* **51**(2), 229–233.

Ovarian Function in the Perimenopause

Elizabeth B. Connell
Emory University School of Medicine

I. Introduction
II. The Stages of Decreasing Ovarian Function
III. Implications of Ovarian Deficiency
IV. Clinical Considerations

GLOSSARY

cardiovascular Pertaining to or comprising the heart and blood vessels.

estrogen Any of a family of steroid hormones synthesized and secreted by the ovarian follicle.

estrogen receptor A cellular structure possessing a specific affinity for the binding of estrogenic hormones.

menopause The immediate postreproductive phase of a woman's life.

osteoporosis A reduction in the quantity and quality of bone by the loss of both bone mineral and protein content.

ovaries Paired internal genital organs of the female.

During the reproductive era, a woman's ovaries are the primary source of her sex steroids. Key among these is estrogen, which is elaborated primarily by the ovarian follicles. This hormone is responsible for the development and proper functioning of the female genital tract. It also plays a major role in the maintenance and functioning of numerous tissues throughout the body, all of which have now been found to contain estrogen receptors. During the perimenopausal years, most women experience a gradual decline in the amount of circulating estrogen. This, in turn, produces numerous anatomical and functional changes.

I. INTRODUCTION

The phenomenon of waning ovarian function ending in menopause is a relatively recent one in the evolution of the human female. Not too many generations ago, the majority of women did not survive long enough to experience a menopause, most of them dying in their mid- to late 30s. As more and more women lived beyond their reproductive era, it gradually became apparent that the declining amounts of estrogen that were being produced by their aging ovaries had profound effects on many parts of their bodies.

Today, these changes are being observed far more commonly as increasing numbers of women reach their 80s, 90s, and beyond. Indeed, in developed countries many will spend one-third to one-half of their lives as postmenopausal women. Thus, there is growing concern about the medical and social implications of prolonged estrogen deficiency.

Androgens are produced not only in the male but also in the female. Even after the menopause, low levels of androgen continue to play a role in women. Although not conclusively proven, the addition of an androgen, usually in conjunction with estrogen replacement therapy (ERT) or hormone replacement therapy (HRT), has been reported to have certain beneficial effects. Among these are an improved sense of well-being, an increase in libido, less depression, and fewer estrogen-induced headaches. Some studies have also indicated an increase in bone mass, particularly when used with estrogen. Androgens can be administered as oral tablets, by injection, or by implantation.

II. THE STAGES OF DECREASING OVARIAN FUNCTION

A. Early

The earliest changes noted by most women relate to their reproductive organs. As their ovaries decline in functional activity, ovulation becomes more and more irregular. Their menstrual periods may also lose their cyclicity, often changing to episodes of irregular bleeding and ultimately ceasing. Concurrent with this, many women begin to experience a variety of other symptoms. The majority of these are in the neurologic area, notably hot flushes and flashes, depression, sleeplessness, insomnia, irritability, and mood changes. Until relatively recently, these symptoms were believed to be just part of a woman's lot in life. In fact, they were often shrouded in an aura of disbelief and even varying degrees of ridicule. As a result of the growing attention which is being given to research in this area, evidence is accumulating that the central nervous system is, indeed, very sensitive to declining levels of estrogen.

B. Intermediate

As time goes on, women often note a number of changes in their genital tracts. Among the most common of these are vulvar thinning, flattening, and pruritis. Vaginal symptoms may also begin to appear. Some women develop a vaginal discharge, and with thinning of their vaginal tissues experience episodes of infection and bleeding, particularly with sexual intercourse. As their vaginal atrophy becomes more pronounced, it frequently results in dyspareunia (painful intercourse).

Not surprisingly, since the genital and urinary tracts develop together embryologically, estrogen deficiency also begins to produce problems with the urinary tract. Some women experience urinary frequency and various types of incontinence, stress incontinence being most troublesome to large numbers of women as they grow older. Anal incontinence may also be experienced, which is a most disturbing and often embarrassing situation. These events are becoming explainable as numerous studies have identified estrogen receptors throughout the female pelvis. They have now been found not only in the urogenital tract but also in the pelvic supporting structures, the periurethral tissues, and the anal sphincter.

With continuing estrogen deficiency, other changes in the genital tract may occur. Lack of support to the pelvic floor may lead to prolapse. A decrease in the amount of collagen in the supporting musculature of the pelvic organs has been identified and may provide another reason for this problem.

Women also notice a number of other changes as they produce decreasing amounts of estrogen. Breasts in older women become smaller and softer and have a tendency toward droopiness. The loss of estrogen is also observable in the skin. Progressive wrinkling occurs in the skin of the face, particularly if a woman is a heavy smoker. Her skin, in general, tends to become dry and less pliable and is more easily traumatized. In addition, pubic and axillary hair begin to disappear, and facial hair may start to appear.

In addition to causing many disease states, smoking has been documented to have profound effects on the human female and her reproductive functions. There is early evidence that both oocytes and ovarian follicles are depleted by smoking. Menopause occurs 1–5 years earlier in smokers than would be anticipated (the actual time appears to be dose dependent). Progressive wrinkling of the skin of the face also appears to be caused by smoking. The mechanism of action of these changes has been suggested by findings of decreased levels of both luteinizing hormone and estrogen in smokers.

C. Late

While all the early and intermediate postmenopausal changes are very important to women, none of them are life-threatening. The areas of greatest interest today are related to the progressive damage which is observed in the skeletons and cardiovascular systems of women as they become more and more estrogen deficient.

Studies have shown a gradual loss of bone which leads, over time, to an increased incidence of fractures. These tend to occur at about 5 year intervals, being noted first in the wrist, second in the spine, and third, and most lethally, in the hip. As the population ages, osteoporotic hip fractures assume a greater role in the causation of morbidity and mortality in older women.

Even more significant are the problems related to cardiovascular disease (CVD) in older and elderly women. While the number of deaths caused by breast, lung, and endometrial cancer are surpassed by the deaths due to osteoporosis, they are all dwarfed in comparison to those due to CVD. As estrogen levels decline, the incidence of angina and coronary insufficiency increases. In fact, myocardial infarction is now recognized to be the leading cause of death of women in this age group.

There is also growing evidence that some of the changes in blood vessels which lead to atherosclerosis may be induced by estrogen deficiency. For years, there has been a popular assumption that estrogen administration to older women would increase their risk of CVD. It now appears that precisely the opposite may be true. Population studies have shown that women taking ERT have 50% less coronary artery disease than untreated controls. Two mechanisms have been proposed for this. First, estrogen produces favorable changes in lipoproteins, raising HDL and lowering LDL. Second, and probably more important, estrogen has been shown to improve cardiac output and coronary blood flow, either by preventing vascular spasm or by somehow preventing the buildup of arterial plaques.

Similarly, estrogen has also been found to play a major role in the maintenance of an adequate blood supply to the brain. The same two reasons have been cited to cause less risk of plaque development and relief of vascular spasm. In addition, the brain shares in the increase in the systemic arterial blood flow found with estrogen administration, including that of the internal carotid arteries.

Finally, there is a new and intriguing evidence that estrogen deficiency may somehow be involved in the neurodegenerative changes associated with aging and dementia, specifically Alzheimer's disease (AD). It has been found that there are estrogen receptors in several areas of the central nervous system, changes in which are now linked to the signs and symptoms of AD. Given the progressive demographic shift upward in the age of American women, these findings may some day prove to be of major significance.

III. IMPLICATIONS OF OVARIAN DEFICIENCY

The progressive loss of estrogen which follows the waning of ovarian function clearly poses many problems with regard to the health and well-being of older women. The personal and public health importance of the changes which have been described is particularly evident when one considers the nature of the population dynamics among American women. The "graying" of American has been well documented, and the results of estrogen deficiency become more clear with each passing year.

A. Medical

Whereas estrogen was formerly believed to affect the breasts and the female genital tract almost exclusively, there is increasing evidence that estrogen receptors can be found in virtually every tissue of a woman's body. Thus, a long-standing and progressive hormone deficiency inflicts widespread damage over time, being most significantly seen in bone and the cardiovascular system.

B. Personal

The loss of ovarian function produces adverse effects on all women, with some women being affected to a far greater degree than others. The early signs and symptoms of the menopause may interfere with a woman's daily activities as well as her sexual life. Changes coming somewhat later may affect her appearance and her social interactions and limit the overall quality of her life. The final stages of estrogen deficiency not only threaten her long-term health but also may substantially limit her ability to continue to function as a productive member of society.

IV. CLINICAL CONSIDERATIONS

Inevitably, a careful consideration of all of these issues leads to a consideration of whether or not HRT should be instituted in some or all women. If so, should it be given in low doses for a relatively short period of time for the relief of early symptoms, as some physicians are still advising, or should it be continued for the lifetime of the individual in view of the newer evidence related to the prevention of osteoporosis and cardiovascular disease? There appears to be a growing consensus that, particularly in the high-risk woman, the later approach should be far more often considered. The treatment of other hormonal deficiencies is now well accepted. Perhaps the time has now come to apply the same approach to the prevention and/or treatment of estrogen deficiency.

See Also the Following Articles

ESTROGEN ACTION ON THE FEMALE REPRODUCTIVE TRACT; ESTROGEN REPLACEMENT THERAPY; MENOPAUSE

Bibliography

Berg, G., Hammar, M., et al. (1994). *The Modern Management of the Menopause. Perspective for the 21st Century.* Parthenon, New York.
Grady, M. H. T., et al. (1995). *Benign Postreproductive Gynecologic Surgery.* McGraw-Hill, New York.
Kase, N., Weingold, A. B., Gershenson, D. M., et al. (1990). *Principles and Practice of Clinical Gynecology,* 2nd ed. Churchill Livingston, New York.
Mishell, D. R., Jr., et al. (1993). *Menopause: Physiology and Pharmacology.*: Year Book Med. Pub., Chicago.
Notelovitz, M., and Tonnessen, D. (1993). *Menopause and Midlife Health.* St. Martin's, New York.
Selzer, V. L., and Pearse, W. H. (1995). *Women's Primary HealthCare. Office Practice and Procedures.* McGraw-Hill, New York.
Zuspan, F. P., and Quilligan, E. J. (1994). *Current Therapy in Obstetrics and Gynecology.* Saunders, Philadelphia.

Ovarian Hormones, Overview

Shao-Yao Ying and Zhong Zhang

University of Southern California

I. Steroidal Hormones in the Ovary
II. Nonsteroidal Hormones in the Ovary

GLOSSARY

autocrine actions Hormones produced by a cell which act on the cell itself to produce an effect.

growth factors Polypeptides that often take a longer time than hormones to act and regulate cell growth, apoptosis, and/or differentiation through paracrine and/or autocrine actions.

nonsteroid hormones Widely distributed, water-soluble polypeptides that are growth factors or growth factors related and often regulate cellular proliferation and/or differentiation through paracrine and/or autocrine actions.

paracrine actions Hormones produced by a cell which act on adjacent cells to produce an effect.

steroid hormones Nonpolar, fat-soluble hormones generally derived from cholesterol and having a cyclopentanoperhydrophenanthrene ring core; consequently, they have intracellular receptors, are not readily soluble in blood and are transported bound to proteins, and synthetic forms can often be administered orally.

The ovary produces both steroidal and nonsteroidal hormones. Steroidal hormones are fat-soluble and generally derived from cholesterol; they bind to sex-binding proteins and are metabolized in the liver and kidney. Nonsteroidal hormones are water-soluble, peptidergic in nature, and may function as growth factors or be growth factor related; they may be involved in modulating cell proliferation and/or cell differentiation. Whereas the steroids are synthesized in thecal, granulosa, and luteal cells, nonsteroidal hormones are primarily produced in granulosa and luteal cells.

I. STEROIDAL HORMONES IN THE OVARY

A. Androgen

Androgen is the name for a class of sex hormones secreted principally by the testes in males but also by the adrenal cortex. However, in the female, the ovary also secretes androgen. In the ovary, androgen is produced primarily by the theca interna (thecal cells) in the preovulatory follicle. The thecal cells are the connective tissue-like cells surrounding the developing follicles of the ovary. The major ovarian androgens are androstenedione and testosterone. The thecal cells have luteinizing hormone (LH) receptors, and LH acts to stimulate the production of androstendione and testosterone (both are estrogen precursors), which are converted to estrogen. Small quantities of dihydrotestosterone are produced in the ovaries. The serum testosterone level of a cycling woman is 0.2–0.7 ng/ml, which fluctuates with her reproductive cycle; about half of it is produced by the ovary and the other half by the adrenal cortex. Androgens, primarily of adrenal origin, are responsible for pubic and axillary hair development in the female because the hair starts to grow before the ovaries begin to produce androgens. If androgens are present in abnormally high quantities, they interfere with the proper functioning of the ovaries and induce the development of male secondary sex characteristics.

B. Estrogen

Estrogens are produced predominantly by granulosa cells, utilizing androstenedione produced by the thecal cells as a precursor. Granulosa cells surround the ovum and regulate the availability of nutrients to it. Granulosa cells have follicle-stimulating hormone (FSH) receptors, and FSH acts to stimulate granulosa cell aromatization of thecal androgen to produce estrogen. Early in the follicular phase, the granulosa cells contain only FSH receptors. As the follicle grows in response to the action of FSH and estrogen production increases as a result of the action of LH on thecal cells and FSH on granulosa cells, serum estrogen levels rise, stimulating further secretion of FSH and consequently inducing the development of granulosa cell LH receptors. Once LH receptors develop, granulosa cells begin to secrete progesterone. After ovulation, granulosa cells change to luteal cells; LH stimulates luteal cells to secrete both progesterone and estrogen. Both LH and FSH bind to their specific receptors and trigger a cyclic adenosine monophosphate-mediated mechanism for a rate-limiting estrogen production through cyclic AMP-activated protein kinases. Reciprocally, estradiol feeds positively and negatively back to stimulate and inhibit LH and FSH synthesis and secretion at the hypothalamic and pituitary levels, respectively. Approximately 60% of the estrogen secreted is transported bound to sex hormone-binding globulin (SHBG), 20% is bound to albumin, and 20% is in the free form. Four estrogens (α-estradiol, β-estradiol, estriol, and estrone) found in women have identical biological activities with different potencies, with β-estradiol being the most potent one normally secreted by the ovaries. The serum levels of estrogen in a cycling woman range from undetectable to 700 pg/ml, with peaks at 1 day before and Day 8 after ovulation. Estrogens are degraded in the liver and kidney to inactive metabolites, conjugated with sulfate or glucuronide, and excreted in the bile and the urine. Major metabolites of estradiol are estrone, estriol, and catecholestrogens. Estrogens readily penetrate cell membranes and bind to intracellular receptors; in turn, the activated receptors bind to estrogen response elements in the vicinity of promoter regions in target genes, resulting in synthesis of ribosomal and messenger RNAs which is followed by various types of biological manifestations. Estradiol stimulates the growth and development of the uterus, Fallopian tubes, cervix, vagina, labia, and breasts at puberty and controls secondary sex characteristics in the female, including body contour and skeletal development. It also stimulates uterine endometrial proliferation and increases spontaneous myometrial electrical activity and the consequent uterine contraction as well as uterine sensitivity to oxytocin. Estrogens are thought to regulate the flow and thickness of the mucous secretion of the cervix, the growth of the vagina to its adult size, the thickening of the vaginal wall, and the increase in vaginal acidity that reduces bacterial infections. Estrogen is responsible for estrous behavior in animals. Estrogens promote deposition of subcutaneous fat and suppress the activities of sebaceous glands and thereby reduce the likelihood of acne in the female. Estrogens also affect the metabolism of water, salt, and calcium. In addition to the ovary, estrogens are secreted by the placenta and the adrenal cortex. Estrogens are also secreted by ovaries of non-mammals, including mollusks, starfish, and dogfish, and also by some plants. The male testes produce relatively small amounts of estrogens. Many estrogens, including synthetic estrogens such as diethylstilbestrol, have been used as drugs; for example, 17β-estradiol and progesterone are frequently components of birth control pills.

C. Progestins

Progestins (progesterone and 17α-hydroxyprogesterone) are produced predominantly by luteal cells. The estrogen produced by granulosa cells synergizes with FSH to stimulate development of granulosa cell LH receptors. In these granulosa cells, LH acts to stimulate progesterone production. After ovulation, granulosa cells and thecal cells become luteal cells. Luteal cells have LH receptors and secrete estrogens and progesterone in response to LH stimulation. The major progestin is progesterone, which is a precursor for androgens and estrogens, and it is produced in all ovarian endocrine cells except early granulosa cells. After secretion, progesterone binds primarily to transcortin and albumin. The serum levels of progesterone in a cycling women ranges from undetectable to 10 ng/ml, with the peak at Day 8 after ovulation. Similar to estrogens, progestins bind firmly to intranuclear receptors and the receptors are released from the complex with heat shock protein 90 and other proteins and bind to their hormone response element upstream of the target gene for cellular activities. Progestins are degraded in the liver and kidney as sulfate or glucuronide conjugates and excreted in the urine. The major metabolite of progesterone is pregnanediol, which is conjugated with glucuronide and excreted in the urine. Progesterone stimulates conversion of a proliferative-type endometrium into the secretory-type endometrium, which is optimal for the implantation of the developing embryo. It increases the myometrial transmembrane potential, thereby stabilizing the membrane and decreasing uterine motility and uterine response to oxytocin. Progesterone stimulates the production of a scanty, viscous, acidic cervical mucus that inhibits sperm penetration. Progesterone also increases the set point for thermoregulation; it increases body temperature, which is the basis for determining the occurrence of ovulation. Together with estrogen, progesterone stimulates the lobular–alveolar growth of the breast. Progesterone is also produced by the placenta, thus maintaining pregnancy, and the adrenal glands. Synthetic progestins inhibit egg growth and release in the ovaries, preventing fertilization, and have been used as contraceptives.

Progesterone prepares the wall of the uterus so that the lining is able to accept a blastocyst for implantation and development. It also inhibits muscular contractions of the uterus that would probably cause the wall to reject the adhering blastocyst. During pregnancy, progesterone also stimulates development of the glands in the breasts that are responsible for milk production.

II. NONSTEROIDAL HORMONES IN THE OVARY

A. Inhibin

The concept of inhibin's presence as a nonsteroidal modulator of gonadotropin has been accepted for more than 60 years. Inhibin's activity was defined later as a polypeptide of gonadal origin responsible for suppression of FSH secretion by the pituitary. In 1985, inhibin was isolated from ovarian follicular fluid based on its ability to suppress FSH secretion by cultured rat pituitary cells. Inhibins ($\alpha\beta$A and $\alpha\beta$B) were characterized as heterodimeric glycoproteins consisting of a common α chain and one of two highly homologous β chains (βA and βB). The major sites of inhibin synthesis are granulosa cells in preovulatory follicles in the ovary. However, synthesis of inhibin has also been found in the corpus luteum and theca cells. Removal of the ovaries results in no detectable serum inhibin. Ordinarily, inhibin can be detected in serum throughout the estrous cycle. Inhibin rises in the late follicular phase and during midcycle and its level is even higher during the luteal phase. The level of inhibin parallels the serum progesterone level. The production of inhibin is primarily controlled by FSH and is associated with development of follicles. Women who have received *in vitro* fertilization have increased peripheral levels of inhibin. LH and androgen have also been reported to stimulate the production of inhibin. Inhibin selectively suppresses both basal and gonadotropin-releasing hormone-stimulated FSH synthesis and secretion without influencing LH secretion. Inhibin has paracrine and autocrine functions in the ovary. Although inhibin has been found to bind to the granulosa cell surface, the receptor for inhibin has not been isolated. Inhibin stimulates the proliferation of luteinized granulosa cells and suppresses FSH-mediated estrogen production by granulosa cells. It also stimulates androgen production and synergizes with LH and IGF-I to increase androgen production by theca cells. Current data suggest that inhibin plays a role in the regulation of follicular development. Injection of inhibin into the ovary increases follicular diameter similar to the effects observed following systemic administration of pregnant mares' serum gonadotropin. Recently, inhibin has been suggested as a tumor suppressor gene product with gonadal specificity because deletion of the gene for inhibin α subunit led to the development of granulosa cell tumors in mice.

B. Activin

In 1986, a molecule capable of stimulating pituitary FSH secretion was identified from follicular fluid as a dimeric protein composed of two inhibin β subunits; it was named activin. Three forms of activin, activin A (βA/βA), activin (βA/βB), and activin B (βB/βB), have been isolated and shown to have similar biological functions. For quite some time, it was believed that only three forms of activin were present. Recently, three new activin β subunits, $\beta C-\beta E$, have been cloned. These three subunits have the same general structure as activin βA and βB subunits with conservation of the nine cysteine residues. General information about these three newly discovered activin subunits, such as dimer formation with other inhibin/activin subunits, receptor binding, and bioactivity, is not available. Activin elicits its biological effects via binding to specific cell surface receptors. There are two classes of membrane proteins, with approximate molecular weights of 50 and 70 kDa, which bind to activin with high affinity: activin type I and type II receptors. These receptors are composed of an extracellular domain, a single membrane-spanning domain, and an intracellular protein kinase domain. It is believed that both activin and transforming growth factor-β (TGF-β) type II receptors are specific for a given ligand family, whereas type I receptors determine the signaling specificity of kinase receptor complexes.

The granulosa cell, but not the corpus luteum, is the major site for activin production. Inhibin/activin β subunit expression is highest in immature follicles. Levels of activin, unlike inhibin, do not seem to fluctuate during the estrous cycle. The functions of activin are more diverse than originally anticipated. In addition to stimulating the production of FSH by the pituitary (endocrine actions), activin has long been known to regulate the differentiation of the ovarian granulosa cell. Activin reduces human chorionic gonadotropin (hCG)-stimulated dehydroepiandrosterone accumulation and IGF-I- and LH-stimulated androgen production by theca cells, but it increases the conversion of pregnenolone and dehydroepiandrosterone to testosterone (paracrine actions). Activin enhances FSH-stimulated progesterone and estrogen production as well as FSH-induced aromatase activity in granulosa cells. Activin alone stimulates basal inhibin secretion and has no effect on progesterone or estradiol production. Activin also induces FSH receptor expression and augments FSH-induced LH receptor expression in granulosa cells. Administration of activin to rat ovary causes extensive follicular atresia, whereas injection of activin into monkey ovary interrupts follicular maturation. It is clear that activin is involved in folliculogenesis and reproductive hormone regulation.

Activin is structurally related to inhibin and a member of the TGF-β superfamily; it often acts as a functional antagonist of inhibin in pituitary as well as in many other tissues. Since inhibin has been proposed to be a tumor suppressor, it is tempting to speculate that activin may have a function antagonistic to tumor suppression, namely, tumor promotion. Inhibin α subunit-deficient mutant mice have been generated through homologous recombination. All of the homozygous mutant mice developed gonadal stromal tumors. In these mice, serum activin levels are highly increased and this has led people to speculate that the tumor formation is due to the action of high-level, unopposed activin. Indeed, activin does have an autocrine stimulatory effect on the growth of cell lines derived from gonadal stromal tumors.

Activin was first isolated from ovarian fluid for its action on anterior pituitary cells. Subsequently, activin was independently isolated based on a number of other bioactivities, including erythroid and megakaryocyte differentiation, nerve cell survival, and mesoderm induction. Its broad range of activities include effects on the endocrine system and reproduction, embryogenesis and differentiation, and cell growth and tumorigenesis (autocrine actions).

C. Follistatin

Follistatin is a single-chain polypeptide originally isolated from ovarian fluid as another pituitary FSH-secretion inhibitor. The major sites for follistatin synthesis are granulosa, luteal, and thecal cells. Although mRNA for follistatin is expressed in healthy and atretic follicles, follistatin protein is only detected in tertiary follicles and newly formed corpus luteum. Many early experiments showed that follistatin has opposite bioactivity to activin in the ovary. It was later found that follistatin is a high-affinity binding protein for activin and inhibin. Follistatin binds to both activin and inhibin through their common β subunit and neutralizes some activities of activin but not those of inhibin. In fact, follistatin exerts its effects primarily through binding and neutralization of activin. Site-specific mutational analysis of follistatin showed that destroying the activin-binding domain of follistatin abolished follistatin activity on FSH suppression. The neutralization of activin activity by follistatin has been observed in many biological systems, including osteoblasts, granulosa cells, and embryonal carcinoma cells. It must be noted that not all activin functions were neutralized by follistatin. The interaction between activin and follistatin is probably important to the local regulation of activin's function.

D. Insulin-like Growth Factor

Several growth factor families [EGF, FGF, TGF-β, and insulin-like growth factor (IGF)] have been shown to be produced and have autocrine and paracrine function in the ovary. Among these, IGF has been the best studied. Two IGFs, IGF-I and IGF-II, have been described. Both are dimeric peptides consisting of one A chain and one B chain linked by disulfide bonds. The IGFs elicit their biological effects by binding to the specific cell surface receptors (type I and type II) on the target. In addition, IGFs also bind to six serum IGF-binding proteins (IGFBPs) with comparable high affinity to IGF receptor. IGFBPs can either inhibit or potentiate IGF action at the level of target cells by sequestration or releasing of IGF through the change of IGF-binding affinity.

IGF-I is mainly expressed in the theca–interstitial cells, whereas IGF-II expression is localized to the granulosa cells in human. The expression of IGFs is stimulated by gonadotropin and estrogen. FSH also stimulates the expression of IGF receptors and increases binding of IGF to granulosa cells. IGF concentrations in follicular fluid are slightly increased during late follicular growth and decreased in atresia. However, the concentrations of IGFBPs are dramatically changed at the same time. The major function of IGFs seems to be to potentiate the action of gonadotropins during follicular development. IGFs synergize with FSH to increase estrogen, progesterone, and inhibin production. IGFs also induce LH receptor expression and increase ovarian androgen production. Both growth and steridogenesis of granulosa cells are stimulated by IGFs. Transgenic mice without IGF-II are fertile, whereas knockout of the IGF-I gene results in infertility.

E. Relaxin

Relaxin is a dimeric peptide hormone produced in the ovary and placenta. It has an α and a β chain connected by two disulfide bridges. Structurally, it is similar to insulin and insulin-like growth factors, suggesting that they derive from the duplication of a common ancestral gene. The hormone action and concept of relaxin were proposed more than 50 year ago. However, the genes were not cloned until the early 1970s. Relaxin induces relaxation of the pelvic ligaments and softens the cervix to facilitate childbirth as well as inhibits uterine motility. Two relaxins have been discovered, namely, relaxin H1 and relaxin H2. Corpora lutea of menstrual cycle and pregnancy are the main sites for relaxin H2 production. Relaxin H1 expression is identified in the decidua and placenta but not in the ovary. The serum relaxin levels consistently rise after the LH surge during the menstrual cycle. Although the absolute level of relaxin is low, relaxin reaches its maximum concentration during the first trimester of normal pregnancy and then gradually declines to term. The production of relaxin is stimulated by LH during the menstrual cycle and hCG during pregnancy.

See Also the Following Articles

Activin and Activin Receptors; Androgens; Estrogens, Overview; Follistatin; IGF (Insulin-like Growth Factors); Inhibins; Progestins; Relaxin, Mammalian

Bibliography

Giudice, L. C. (1992). Insulin-like growth factors and ovarian follicular development. *Endocr. Rev.* 13, 641–669.

Mathews, L. S. (1994). Activin receptors and cellular signaling by the receptor serine kinase family. *Endocr. Rev.* 15, 310–325.

Sherwood, O. D. (1993). Relaxin. In *The Physiology of Reproduction*. Raven Press, New York.

Ying, S. Y. (1988). Inhibin, activin, and follistatins: Gonadal proteins modulating the secretion of follicle-stimulating hormone. *Endocr. Rev.* 9, 267–293.

Ovarian Innervation

Gregory A. Dissen and Sergio R. Ojeda

Oregon Regional Primate Research Center, Oregon Health Sciences University

I. Introduction
II. Ovarian Innervation
III. Innervation of Primate Ovaries
IV. Ovarian Nerves and Follicular Steroidogenesis
V. Ovarian Nerves and Follicular Development
VI. Conclusions

GLOSSARY

folliculogenesis The process by which ovarian follicles are formed.

granulosa cells Cuboidal-shaped cells of epithelial origin which surround the oocyte and line the inside of the follicle basal membrane. These cells produce progesterone and estrogens.

interstitial cells Cells of mesenchymal origin that fill the spaces between follicles and receive innervation from the peripheral nervous system. Some of them are steroidogenic.

luteinizing hormone-releasing hormone The hypothalamic neuropeptide that controls sexual development and adult reproductive function. Also known as gonadotropin hormone releasing hormone.

primordial, primary, and secondary follicles The first three stages of follicular development. Primordial follicles are characterized by a single layer of flattened cells surrounding the oocyte. Primary follicles are those in which the single layer of cells becomes cuboidal. Secondary follicles exhibit two layers of cuboidal cells surrounding the oocyte.

steroidogenesis The synthesis of steroid hormones by gonadal or nongonadal tissues.

thecal cells Elongate cells of mesenchymal origin that surround the basal membrane of the follicle. They produce androgens and receive innervation from the peripheral nervous system.

Most endocrine glands, including the ovary, are innervated by neurons of the peripheral nervous system. The early demonstrations that this innervation reaches both the vasculature of the gland and the ovarian follicles provided the bases for the concept that the extrinsic ovarian nerves may not only be involved in the regulation of blood flow but also participate directly in the control of steroidogenesis and follicular development. A series of subsequent studies demonstrated the overall validity of this view by showing the ability of neurotransmitters contained in ovarian nerves to stimulate steroid release via specific recognition molecules located on steroidogenic cells. Although more difficult to obtain, evidence has also been provided supporting a facilitatory role of ovarian nerves in follicular development. Thus, in addition to the well-known control effected by pituitary gonadotropins, the ovary appears to be under the direct neural regulatory influence of its extrinsic innervation.

I. INTRODUCTION

Mammalian ovaries are innervated by both sympathetic and sensory neurons, whose fibers reach most structural components of the gland, including the vasculature, the interstitial tissue, and developing follicles. A substantial body of evidence exists describing the intraovarian distribution of the projecting fibers and the chemical identity of the neurotransmitters they contain. Significant progress has also been made toward elucidating the functional impact of the ovarian innervation on both the steroidogenic activity of the gland and follicular development. It is now clear that norepinephrine (NE) and vasoactive intestinal peptide (VIP), two neurotransmitters contained in ovarian nerves, stimulate the production of ovarian steroids. While both enhance the secretion of progesterone and androgens, only VIP stimulates estradiol production. Importantly, NE also facilitates the stimulatory effect of gonadotropins on ovarian steroidogenesis. Recent experiments have demonstrated that VIP and NE contribute to the biochemical differentiation of newly formed follicles by inducing the formation of follicle-stimulating hormone (FSH) receptors, thereby endowing the young follicle with the ability to respond to the gonadotropin.

The importance of the extrinsic ovarian nerves in follicular development has been shown by experiments in which the sympathetic innervation of the neonatal ovary was prevented by immunoneutralization of nerve growth factor (NGF). In these animals, follicular development was retarded, puberty was delayed, and estradiol output from the gland was reduced. It is thus evident that the nervous system affects specific processes of ovarian function and, in doing so, facilitates the stimulatory action of gonadotropins on both follicular growth and ovarian steroidogenesis.

II. OVARIAN INNERVATION

The ovary's extrinsic innervation is mostly composed of sympathetic and sensory fibers, with a small parasympathetic component. The sympathetic innervation is composed of catecholaminergic and neuropeptide Y (NPY)-containing neurons, whereas the sensory innervation is provided by substance P (SP) and calcitonin gene-related peptide (CGRP)-containing fibers. The ovary is also innervated by nerve fibers of both a sympathetic and a sensory nature that contain VIP and its distant relative, pituitary adenylate cyclase activating polypeptide (PACAP).

There are two major nerves that innervate the mammalian ovary. One is the ovarian plexus nerve (OPN), which runs along the ovarian artery; in the rat, the second is referred to as the superior ovarian nerve (SON). The SON is associated with the suspensory ligament, which attaches to the ovary, oviduct, and uterus. In the human, however, the second nerve supply is provided by the superior hypogastric plexus (or hypogastric nerve), which divides into a small number of bundles that run along the broad ligament to reach the ovary. The ovarian nerves arise from the celiac ganglia, the mesenteric ganglia, and the lumbar splanchnic nerves, which originate in ganglia located along the aorta in the vicinity of the renal arteries. The sympathetic portion of the ovarian innervation originates in the thoracic-11 to lumbar-4 segments of the sympathetic chain ganglia and synapses in the celiac and mesenteric ganglia (Fig. 1). The postganglionic fibers then innervate the ovary via one of the two ovarian nerves. Regarding the parasympathetic innervation, there is an anatomical difference between humans and rats. In the human, the cell bodies that project to the ovary are located in the spinal cord (sacral segments 2–4). Axons from these neurons run in the pelvic nerves and form

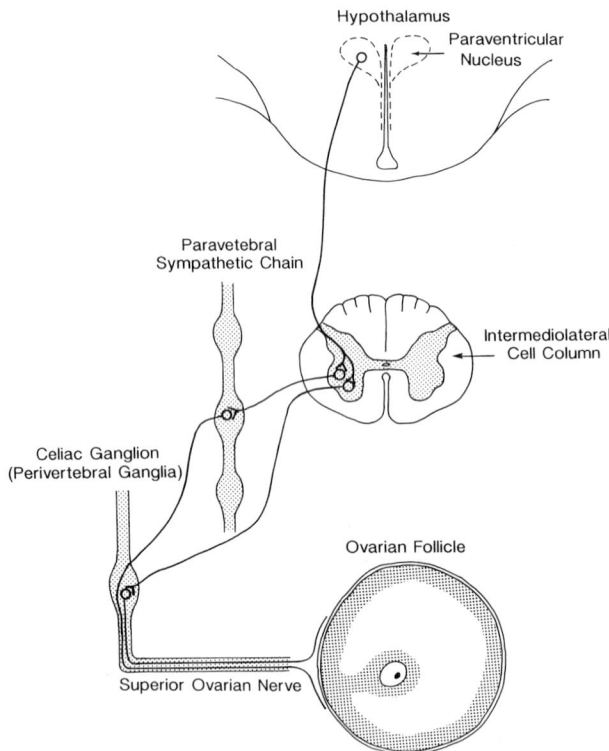

FIGURE 1 The sympathetic innervation of the ovary reaching the gland via the superior ovarian nerve. The perikarya of the first-order neurons projecting to the ovary are located in the celiac ganglion and ganglia of the paravertebral sympathetic chain. They receive projections from neurons located in the intermediolateral cell column of the spinal cord which, in turn, appear to be synaptically connected to central oxytocin-containing neurons located in the paraventricular nucleus of the hypothalamus.

synapses in the uterovaginal ganglion. The route followed by the postganglionic fibers to reach the ovary is not known. No such projection reaches the rat ovary, which instead appears to receive vagal innervation, the path of which is also unknown. The sensory innervation of the ovary derives from both the nodose ganglia and dorsal root ganglia located between the caudal thoracic segment (T9–T11) and the cranial lumbar segment of the spinal cord (L2–L4). The projecting fibers arrive at the ovary via the OPN.

Several studies have suggested the existence of a bidirectional neural connection between the hypothalamus and the ovary, mediated by the extrinsic innervation of the gland. For instance, electrical stimulation of the anterior hypothalamic area of rats was shown to increase ovarian steroid secretion in the absence of the pituitary and adrenal gland. On the other hand, electrolytic lesions of the same region resulted in ipsilateral changes in the ovarian concentration of VIP and receptors for luteinizing hormone-releasing hormone (LHRH). In turn, that the ovary can affect hypothalamic function via an afferent route was suggested by the unilateral changes in LHRH content detected in the hypothalamus following hemiovariectomy. The existence of this hypothalamic–ovarian neural connection was recently demonstrated using a viral transneuronal tracing method (Lee et al., 1996). Upon injection of a pseudorabies virus that infects only synaptically connected neuronal networks, a predominance of labeled neurons was found in diencephalic nuclei, and among them the paraventricular nucleus was the most intensively labeled. Most of the labeled neurons were found in the magnocellular portion of the nucleus, suggesting that the most important groups of hypothalamic neurons that project to the ovary are those that produce oxytocin. This assumption was confirmed by triple-label experiments in which the retrograde tracer, cholera toxin conjugated to horseradish peroxidase, was stereotaxically injected into the neurohypophysis (the main site to where oxytocin neurons project), and the magnocellular neurons of the paraventricular nucleus that project to the neurohypophysis and to the ovary were identified by immunohistochemistry for oxytocin and a pseudorabies viral protein, respectively. As ovarian activity is suppressed during breast feeding, but the activity of oxytocin neurons is increased, a neural connection between oxytocin neurons and the ovary may serve to provide a direct transsynaptic mechanism by which the central nervous system maintains the state of infertility that accompanies lactation in mammals.

Studies in rats and cows have shown that the innervation of the ovary is an early event that either precedes (rat) or accompanies (bovine) the initiation of folliculogenesis. While some of the early fibers reaching the rat ovary have been identified as catecholaminergic based on their content of tyrosine hydroxylase (TH), the rate-limiting enzyme in catecholamine synthesis, some of the nerve fibers reaching the fetal bovine ovary have been found to contain VIP. Interestingly, folliculogenesis in the rat is initiated near the medulla of the gland, and this is the region that, at the outset, contains most—if not all—of the developing innervation. In the bovine ovary folliculogenesis occurs in the cortex of the ovary (not the medullary portion) between 4.5 and 6 months of gestation. Nerve fibers containing VIP were detected in the bovine ovary at this same time and were confined to the cortex of the ovary. Thus, while the specific developmental patterns are different in the rat and the cow, the initial distribution of nerve fibers suggests a relationship between the presence of the fibers and folliculogenesis. As it will become apparent, this may be a feature of considerable physiological significance because in the rat the inner part of the ovary is precisely the region where the rete ovary enters the gland and follicular formation is initiated, whereas in the bovine the ovarian cortex is the region where follicular formation is initiated.

III. INNERVATION OF PRIMATE OVARIES

The ovary of humans and rhesus monkeys is innervated by the same types of nerve fibers present in other species, with the exception of nerves containing SP, which appear to be absent. In an ontogenic study of the rhesus monkey ovary, SP-containing fibers were not detected at any of the postnatal intervals examined, ranging from the neonatal period (2 month of age) to senescence (27 years of age). An interesting feature observed in the nonhuman primate ovary is the innervation of a subpopulation of primordial follicles by VIPergic fibers, which appear to target only some follicles and enclose them almost completely, leaving neighboring follicles uninnervated. There seems to be some specificity for this distribution because it was not observed with NPY or TH fibers. Because VIP has been recently shown to induce FSH receptors in primary follicles, the possibility exists that the ovarian VIPergic innervation of higher primates contributes to the selection of follicles that will enter the proliferative pool by conferring them the ability to respond to gonadotropins at early stages of development.

The view that ovarian nerves may facilitate prepubertal ovarian development, born from the findings that NE and VIP stimulate ovarian steroidogenesis, is inferentially supported by the demonstration that the density of catecholaminergic and VIPergic nerve fibers in the monkey ovary increases significantly between the neonatal period and the expected time of puberty (approximately 3 years of life). No further changes were noted during adulthood and no developmental changes in the density of the sensory innervation were detected, suggesting that the changes observed are specific for nerve fibers containing neurotransmitters able to affect ovarian steroid output.

In addition to its extrinsic innervation, the monkey ovary has been found to contain an intrinsic network of neurons. The cells have a small perikarya (5–10 μm) and are immunoreactive for the cytoplasmic neuronal protein neuron-specific enolase and the neuronal cytoskeletal protein neurofilament 68K. The neurons also have immunoreactive NGF receptors of the p75 class (i.e., the low-affinity receptors that bind all members of the NGF family with similar affinity). Striking changes in the distribution and neurotransmitter phenotype of these neurons have been observed during sexual development. Shortly after birth, the neurons form ganglion-like structures in the medulla of the ovary and spread toward the cortex as the animal approaches puberty, becoming undetectable at senescence. The function(s) of this intrinsic neuronal network has not been elucidated, but it is interesting to note that although the neurons are also present in the human and pig, only a few of them can be detected in the rat ovary. There is perhaps some evolutionary significance in this species-dependent expression so that the acquisition of an intrinsic neuronal network by the ovary of nonrodent species may provide the gonad with an independent neural regulatory component able to influence the development and/or secretory activity of the gland.

IV. OVARIAN NERVES AND FOLLICULAR STEROIDOGENESIS

As already mentioned, ovarian steroidogenesis is stimulated by both VIP and NE, two of the neurotransmitters contained in ovarian nerves. The sympathetic fibers containing NE reach the ovary via both the SON and OPN. While the fibers originating from the OPN mostly innervate the ovarian vasculature, the majority of those derived from the SON innervate the secretory compartments of the gland, i.e., interstitial cells and the thecal layer of follicles. Catecholamines affect ovarian steroidogenesis via activation of β_2-adrenergic receptors. They stimulate release of progesterone (P) from granulosa and luteal cells and androgens from thecal cells. They also facilitate the steroidogenic response of ovarian cells to low concentrations of gonadotropins, suggesting that in physiological circumstances the catecholaminergic sympathetic nerves amplify the effects of circulating gonadotropins on ovarian steroidogenesis.

VIP-containing fibers reach the ovary via the SON. Very much like the noradrenergic nerves, VIPergic fibers innervate all three main compartments of the prepubertal ovary, i.e., the interstitial tissue, follicles, and the ovarian vasculature. VIP is a potent stimulus of P, estradiol (E_2), and androgen secretion. It does so by enhancing the synthesis of all three components of the cholesterol side chain cleavage enzyme complex, the rate-limiting enzyme in steroid biosynthesis, as well as by stimulating aromatase activity. Experiments with rat granulosa cells in culture have shown that VIP targets a population of immature granulosa cells that is unresponsive to FSH, implying that the maturation of these cells comprises an initial phase of responsiveness to the neurotransmitter, followed by the acquisition of responsiveness to the gonadotropin. As recently shown in neonatal ovaries that only contain primary and secondary follicles, VIP may also play a role in inducing granulosa cell differentiation in larger follicles via activation of cyclic AMP synthesis. Acquisition of FSH receptors is required for granulosa cells to respond to the gonadotropin, and VIP has been shown to induce not only the expression of the FSH receptor gene in neonatal ovaries but also the development of functional receptors able to cause follicular growth upon FSH stimulation.

The mammalian ovary is innervated by a network of sympathetic fibers containing NPY. As shown in other sympathetic neurons, many of the ovarian fibers containing NPY also contain NE. However, noradrenergic fibers reach the ovary via both the SON and the OPN, whereas NPYergic nerves are restricted to the OPN. The profuse NPYergic innervation of the ovarian vasculature suggests a role for NPY in the control of blood flow. Although NPY does not appear to participate in the ovulatory process, it may contribute to modulating the response of granulosa cells to gonadotropins and catecholamines, as evidenced by an inhibitory effect of the neuropeptide on hCG-induced P release from pig granulosa cells and on NE release from nerve terminals of the rat ovary, respectively.

The sensory innervation of the ovary, represented by SP and CGRP fibers, reaches the ovary via the OPN. As in other organs innervated by the peripheral nervous system, SP and CGRP coexist in ovarian nerves. The fibers innervate all ovarian compartments but predominantly the vasculature. Since neither SP nor CGRP have any discernible effect on ovarian steroidogenesis, SPergic and CGRPergic fibers are thought to be predominantly involved in regulating blood flow and in providing the sensory afferents from the ovary. The nature of the information conveyed by these fibers to the central nervous system remains speculative, but it may consist of "sensing" the individual stage of follicular development.

With regard to cholinergic nerves, they have been detected around follicles, and nicotinic cholinergic agonists have been shown to inhibit gonadotropin-induced biosynthesis in granulosa cells, suggesting that cholinergic nerves may have a modulatory function in ovarian steroidogenesis. As indicated before, the source of these fibers has not been established. In addition to the aforementioned neurotransmitters, the ovary has been shown to contain other putative neurotransmitters/neuromodulators such as γ-aminobutyric acid (GABA), gastrin-releasing peptide, the VIP relatives peptide histidine isoleucine and PACAP, enkephalin, somatostatin, and cholecystokinin. The function of most of these peptides is unknown, but some evidence exists that GABA may increase ovarian blood flow, stimulate E_2 secretion, and decrease P secretion. Whether these actions of GABA have physiological relevance remains to be determined.

V. OVARIAN NERVES AND FOLLICULAR DEVELOPMENT

Many experiments, conducted over the years, to determine if the ovarian nerves influence follicular development have, for the most part, supported the notion that the ovarian sympathetic innervation exerts a facilitatory influence on follicular growth. The majority of these studies involved transection of the ovarian nerves followed by morphological and morphometric analysis of follicular development. The changes observed, though clear-cut, have never been dramatic, and this outcome has led to the conclusion that the extrinsic innervation of the ovary plays, if any, a small, nonessential role in follicular development. The problem that one faces when interpreting the long-term consequences of nerve transection is the absence of information about the degree and rate of reinnervation that may occur after the transection; the development of compensatory mechanisms, such as hypersensitivity to circulating catecholamines by denervation; and the impact of severing partially or completely the vascular supply to the gland. In recent experiments, we ablated the ovarian sympathetic innervation during neonatal life of the rat by using two approaches. In one of them, development of the sympathetic innervation was prevented by administration of antibodies to NGF. In the other, the sympathetic nerves were destroyed by treatment with guanethidine. In both cases, the resulting loss of the sympathetic innervation was associated with delayed follicular development, reduced ovarian steroid output, and marked irregularities of the estrous cycle. Of special interest are the results of experiments in which the animals were passively immunosympathectomized via administration of NGF antibodies. This approach was chosen based on the finding that the developing ovary synthesizes NGF, which is essential for the survival of peripheral sympathetic neurons. Since the ovary is a target for sympathetic neurons, blockade of NGF action during development of the ovarian innervation would be effective in preventing the innervation of the gland. As expected, the treatment almost completely eliminated the sympathetic nerves of the ovary, as assessed by immunohistochemistry in older prepubertal animals. Directly relevant to the purpose of the experiment, the ovaries of sympathectomized rats showed a significant reduction in large antral follicles compared with the normally innervated ovaries of control animals.

These findings are consistent with the notion that the loss of the innervation abolishes a facilitatory sympathetic input to the developing follicles, and thus leads to stunted follicular growth. It is also possible that the removal of NGF itself may have contributed to the deficit in follicular development. Thecal and interstitial cells not only produce NGF but also contain the recognition molecules able to transduce NGF-mediated signals. While this possibility remains open, recent experiments have provided direct evidence for a facilitatory role of NE and VIP in follicular development. Using organotypic culture of neonatal rat ovaries, which mainly contain primary and secondary follicles, the effects of administration of the neurotransmitter VIP and the β-adrenergic agonist isoproterenol (ISO) were examined. Both agents increased the ovarian content of the mRNA encoding the FSH receptor (FSHR), mimicking the effect observed following forskolin-induced activation of cAMP formation. The FSHR mRNA was localized to follicles as determined by *in situ* hybridization. The increase in FSHR mRNA was accompanied by formation of functional receptor molecules, as demonstrated by the ability of FSH to stimulate cAMP formation in ovaries preexposed to either ISO or VIP but not in untreated ovaries. Treatment of VIP-primed ovaries with FSH resulted in follicular growth, demonstrating that exposure of the gland to the neurotransmitter led to the formation of a functional complement of FSH receptors. These results suggest that ovarian nerves, acting via neurotransmitters coupled to the cAMP-generating system, contribute to the differentiation process by which newly formed primary follicles acquire FSH receptors and responsiveness to FSH. These findings and the previously reported ability of FSH-insensitive granulosa cells to respond to VIP suggest that an important function of neurotransmitters working via the cAMP-generating system is to facilitate the initial differentiation of granulosa cells and thus promote the acquisition of gonadotropin dependency by the growing follicle.

VI. CONCLUSIONS

The mammalian ovary is innervated by sensory and sympathetic neurons of the peripheral nervous system, with innervation occurring before the initiation of folliculogenesis. While the sensory neurons contain SP and CGRP, the sympathetic neurons use NE and VIP as neurotransmitters. The neural control of the ovary involves not only neurons of the peripheral nervous system but also brain neurons, as evidenced by the recent demonstration that oxytocin-producing neurons of the hypothalamus are synaptically connected to the ovary. Little is known about the ovarian functions affected by SP and CGRP, but a role in controlling blood flow is strongly suggested by their perivascular distribution and their inability to affect steroid secretion. Whether the SP and CGRP fibers seen around follicles provide afferent information to the nervous system on the status of follicular development remains a matter of speculation. In contrast to this paucity of information, strong evidence exists indicating that NE and VIP, the other two neurotransmitters contained in ovarian nerves, stimulate ovarian steroidogenesis. At least part of the steroidogenic effect of these neurotransmitters is related to their ability to increase the gene expression of key steroidogenic enzymes involved in the synthesis of progesterone (NE and VIP) and estrogen (VIP). Ovarian sympathetic nerves also facilitate follicular development. A potential mechanism by which NE and VIP may facilitate follicular development is by inducing the synthesis of FSH receptor in newly formed follicles. Thus, the noradrenergic and VIPergic innervation of the ovary appear to act both sequentially and concomitantly with gonadotropins in facilitating ovarian steroidogenesis and promoting follicular development.

Acknowledgments

We appreciate the collaboration of Drs. Artur Mayerhofer (Department of Molecular Anatomy, Anatomical Institute, Technical University, Munich, Germany), Bonghee Lee (Department of Anatomy, School of Medicine, Gyeongsang National University, Gyeongnam, Korea), Les Dees (Department of Veterinary Anatomy and Public Health, Texas A&M University, College Station), Sasha Malamed (Department of Neurosciences and Cell Biology, Robert Wood Johnson School of Medicine, UMDNJ, Piscataway, NJ), and Anne Hirshfield (Department of Anatomy, University of Maryland School of Medicine, Baltimore) in some of the studies discussed in this chapter. We thank Janie Gliessman for editorial and typing assistance, Diane Hill for technical and editorial assistance, and Maria Costa for technical assistance. This work was supported by National Institutes of Health Grants HD24870, HD18185, and RR00163.

See Also the Following Articles

FOLLICULAR DEVELOPMENT, CONTROL OF; FOLLICULAR STEROIDOGENESIS; NEUROTRANSMITTERS

Bibliography

Baljet, B., and Drukker, J. (1979). The extrinsic innervation of the abdominal organs in the female rat. *Acta Anat.* 104, 243–267.

Burden, H. W. (1978). Ovarian innervation. In *The Vertebrate Ovary: Comparative Biology and Evolution* (R. E. Jones, Ed.), pp. 615–638. Plenum, New York.

Burden, H. W. (1985). The adrenergic innervation of mammalian ovaries. In *Catecholamines as Hormone Regulators* (N. Ben-Jonathan, J. M. Bahr, and R. I. Weiner, Eds.), pp. 261–278. Raven Press, New York.

Hsueh, A. J. W., Adashi, E. Y., Jones, P. B. C., and Welsh, T. H., Jr. (1984). Hormonal regulation of the differentiation of cultured ovarian granulosa cells. *Endocr. Rev.* 5, 76–127.

Lawrence, I. E., Jr., and Burden, H. W. (1980). The origin of the extrinsic adrenergic innervation to the rat ovary. *Anat. Rec.* 196, 51–59.

Lee et al. (1996). *Soc. Neurosci. Abstr.* 22, 1576.

Mayerhofer, A., Dissen, G. A., Costa, M. E., and Ojeda, S. R. (1997). A role for neurotransmitters in early follicular development: Induction of functional FSH receptors in newly formed follicles of the rat ovary. *Endocrinology*, in press.

Ojeda, S. R., Lara, H., and Ahmed, C. E. (1989). Potential relevance of vasoactive intestinal peptide to ovarian physiology. *Sem. Reprod. Endocrinol.* 7, 52–60.

Owman, C., Kannisto, P., Liedberg, F., Schmidt, G., Sjöberg, N.-O., Stjernquist, M., and Walles, B. (1992). Innervation of the ovary. In *Local Regulation of Ovarian Function* (N.-O. Sjöberg, L. Hamberger, P. O. Janson, and C. Owman, Eds.), pp. 149–170. Parthenon, Park Ridge, NJ.

Schultea, T. D., Dees, W. L., and Ojeda, S. R. (1992). Postnatal development of sympathetic and sensory innervation of the rhesus monkey ovary. *Biol. Reprod.* 47, 760–767.

Ovariectomy

see Castration, Effects in Nonhumans (Female)

Ovary, Overview

Humphrey H. C. Yao and Janice M. Bahr

University of Illinois at Urbana–Champaign

I. Development of the Ovary
II. Anatomy of the Ovary
III. Functions of the Ovary
IV. Regulation of Ovarian Functions

GLOSSARY

corpus luteum An endocrine gland formed from the wall of an ovulated follicle.
follicle A structure in the ovary consisting of the oocyte and surrounding granulosa and theca cell layers.
genital ridges Embryonic ridges on the dorsal wall of the abdominal cavity of an embryo.
granulosa cells Somatic cells directly surrounding the oocyte.
meiosis A type of cell division found in gonads, in which the chromosome number is reduced from two copies to one.
oocyte The female gamete.
ovary The female gonad.
steroids Molecules with a basic structure similar to that of cholesterol.
theca cells The layer of steroidogenic cells and connective tissue covering the follicle.

Ovaries are female gonads responsible for generation of female gametes (oocytes) and synthesis of hormones necessary for the regulation of reproductive functions. Since the first description of the ovary reported by Aristotle more than 2000 years ago, information about the ovary has expanded significantly. Knowledge of formation of the ovary and its endocrine function is essential to understand the mystery of the regeneration of life.

I. DEVELOPMENT OF THE OVARY

Development of the gonad is dependent on the genetic makeup of the individual. The testis-determining gene of the Y chromosome is necessary for development of the mammalian testis. Without the Y chromosome or the testis-determining gene, testes fail to develop and ovaries form. The postulated ovary-determining gene (Z gene) would inhibit testis differentiation and stimulate formation of ovaries. In male mammals, the product of the testis-determining gene would suppress the Z gene (or its product), thereby allowing testicular development. As a result, the Z gene in female mammals would then be active due to the absence of the Y chromosome and the ovarian phenotype would be expressed. However, the ovary-determining Z gene has not yet been cloned and the molecular basis for formation of the ovary remains an unsolved mystery of biology.

Sexual differentiation of gonads starts early during embryonic development. Development of vertebrate ovaries can be divided into four major stages (Fig. 1). During the first stage (Fig. 1a), primordial germ cells or undifferentiated germ cells from the endoderm of the yolk sac of the embryo travel to the genital ridges by ameboid movement. Primordial germ cells migrate through the gut endoderm and into the mesoderm of the mesentery, finally residing in the genital ridges. The second stage (Fig. 1b), which occurs after the arrival of the primordial germ cells at the genital ridges, consists of the proliferation of primordial germ cells and the coelomic epithelium on the genital ridges. The epithelium on the surface of the genital ridges infiltrates the loose connective mesenchymal tissue and forms the primitive sex cords surrounding the germ cells. This development leads to the formation of indifferent gonads that are identical in both sexes. During the third stage (Fig. 1c), the initial female sex cords degenerate. New sets of sex cords (cortical sex cords) develop immediately and reside near the outer zone (cortex) of the organ. The fourth and final stage of ovarian formation (Fig.

1d) is characterized by development of the cortex and involution of the medulla (inner cord) of the ovary. Sex cords in the cortex are split into clusters, with each cluster surrounding a germ cell. Germ cells will become oogonia and the surrounding epithelial sex cords will differentiate into granulosa cells. Mesenchymal cells of the ovary develop into theca cells. Together, granulosa and theca cells form follicles that envelop the oogonia.

II. ANATOMY OF THE OVARY

Most vertebrates develop a pair of ovaries with the exception of some birds, reptiles, and a few mammals that only have one ovary. Ovaries lie on either side of the upper pelvic cavity and against the pelvic wall. They are held in place by a mesentery (mesovarium) connected to a broad ligament. Ovaries are one of the most vascular organs. The ovarian artery (or uteroovarian artery) which arises from the abdominal aorta reaches the ovary along with the mesovarium. Branches of the ovarian artery enter the ovary through the hilus, the same site at which the venous blood exits. Adrenergic and cholinergic nerves also enter the ovary through the hilus.

Even though the size of the ovary varies, the structure of the ovary is similar among mammalian species (Fig. 2). The ovary consists of an inner medulla, containing a rich vascular bed within loose connective tissue, and an outer cortex, where the ovarian follicles are located. The outermost layer of the cortex is a single squamous or cuboidal germinal epithelium derived from the peritoneum. Under the germinal epithelium lies the tunica albuginea, a poorly delineated layer of dense connective tissue that gives the ovary a whitish color. The cortex of the ovary is made up of numerous follicles embedded in the stroma. The stroma is composed of at least three different cell types: connective tissue cells (fibroblasts) performing support functions, smooth muscle cells regulating the contraction of blood vessels, and interstitial cells including undifferentiated theca cells and degenerated follicular cells from atretic follicles

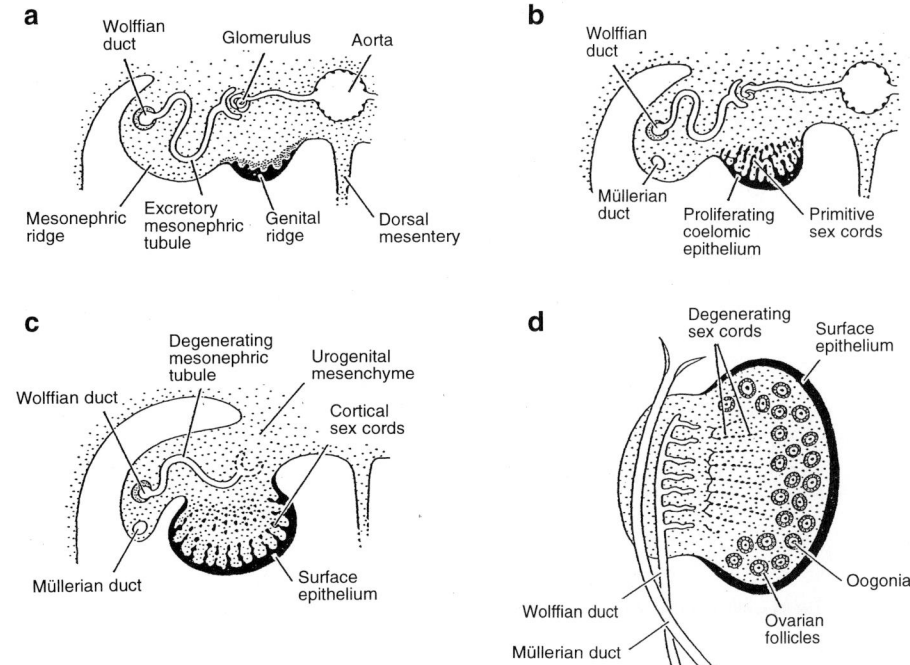

FIGURE 1 Differentiation of the ovary. (a) First stage of formation of the ovary showing the genital ridge. (b) Genital ridge at the second stage of formation showing primitive sex cords. (c) Ovary development as primitive sex cords degenerate at the third stage. (d) The last stage of ovary formation as new cortical sex cords surround germ cells and form follicles (adapted from Gilbert, 1994, p. 758).

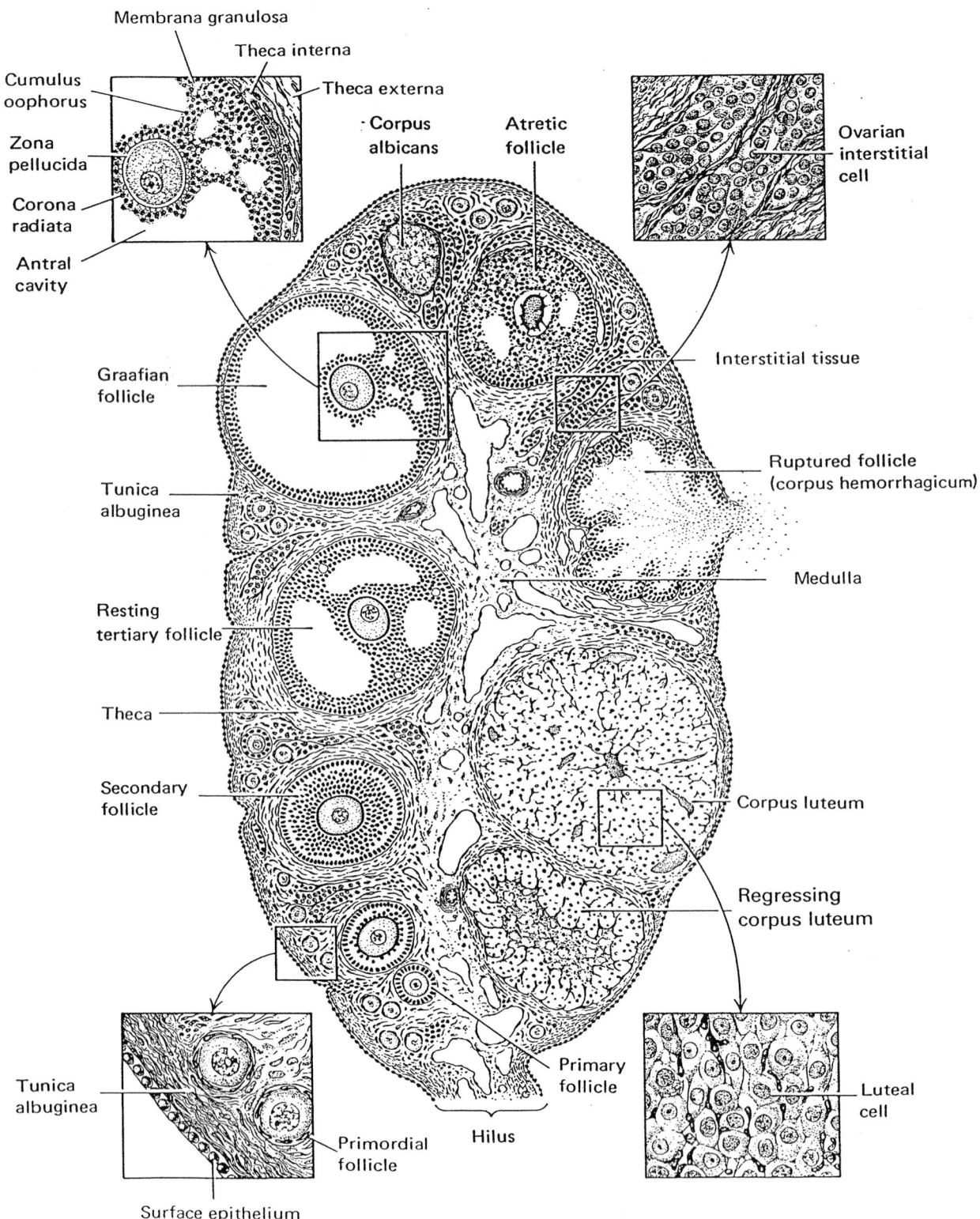

FIGURE 2 A cross section of the ovary illustrating follicles at different stages of development (from primordial to Graafian follicles), corpus hemorrhagicum, corpus luteum, and corpus albicans. The microscopic structures of follicles are also shown (adapted from Jones, 1991, p. 42).

and regressed corpus lutea. The follicles (follicle is Latin for "little bag") are structurally the most conspicuous and functionally the most important units in the ovarian cortex. The microscopic appearance of follicles is different depending on the stage of follicular development, whereas the basic cellular organization of follicles is the same. A follicle consists of an oocyte and surrounding follicular wall. Between the oocyte and the follicular wall is a thin transparent membrane, the zona pellucida. The follicular wall contains an inner granulosa layer and an outer theca layer. The granulosa layer surrounds the oocyte and is separated from the theca layer by the basement membrane. In mature follicles, the theca layer can be further divided into the theca interna, containing differentiated steroid-producing cells, and the theca externa, consisting mainly of connective tissue. The boundary between the theca interna and externa is not clear: neither is the boundary between the theca externa and the ovarian stroma. The blood and nerve supply terminate in the theca interna. There are no blood vessels in the granulosa layer during any stage of follicular growth.

Once ovulation has occurred, blood derived from torn blood vessels of the follicular wall infiltrates the collapsed follicle and results in the formation of the corpus hemorrhagicum. Luteinized granulosa and theca cells (luteal cells) begin to divide and invade the old antral cavity, forming the corpus luteum (Latin for "yellow body"). Blood vessels from the theca layer grow and penetrate the luteal cell mass. If pregnancy does not occur, the corpus luteum degenerates. The connective tissue replaces the luteal cells and forms the corpus albicans (Latin for "white body"). The ovarian medulla, devoid of follicles, contains large, spirally arranged blood vessels, lymphatic vessels, and nerves.

III. FUNCTIONS OF THE OVARY

A. Generation of the Female Gametes

1. Oogenesis

Female gametes, or oocytes, provide the maternal genetic material and nutrients for early development of the embryo. The ovary nurtures thousands of oocytes and functions as an incubator for their development. The development of oocytes (oogenesis) starts with primordial germ cells, the precursor cells for oocytes. Primordial germ cells residing in sex cords divide mitotically producing oogonia. Oogonia then become primary oocytes and undergo the first meiosis. The primary oocytes are arrested at the diplotene stage of the first meiosis until shortly after ovulation. The first meiosis is reinitiated after ovulation and the membrane of the oocyte nucleus (germinal vesicle) disintegrates, which is a process called germinal vesicle breakdown. Meiosis of the oocyte is unequal and produces a large haploid secondary oocyte and a tiny haploid first polar body. This polar body can divide again or remain single; in either case, it degenerates. Then, the secondary oocyte begins the second meiotic division, but this division is arrested at the metaphase until after sperm penetration of the oocyte, which occurs in the oviduct. Completion of the second meiosis results in a haploid ovum and the second polar body.

2. Folliculogenesis

Oocytes reside in their individual follicles and interact intimately with the surrounding follicular wall. Maturation of oocytes (oogenesis) is closely associated with development of follicles (folliculogenesis (Fig. 2). Folliculogenesis always begins in the innermost part of the ovarian cortex in mammals. Primordial follicles consist of primary oocytes surrounded by a single layer of flattened granulosa, the membrana granulosa. As primordial follicles develop into primary follicles, the membrana granulosa gradually transforms from a flat into a cuboidal-shaped layer. In addition, primary follicles acquire the theca layer that encloses the membrana granulosa. As a primary follicle continues to grow, granulosa cells divide mitotically so that secondary follicles have a membrana granulosa with two to six layers of granulosa cells. The theca remains a single layer in secondary follicles. During formation of tertiary follicles, granulosa cells secrete fluid that accumulates between cells. Large amounts of additional fluid diffuse out of the thecal blood vessels and are added to the fluid. This fluid-filled space is the antrum or antral cavity, and the fluid is called follicular fluid. Follicular fluid contains steroid and protein hormones, anticoagulants, enzymes, and electrolytes and is similar to blood serum in appearance and contents. Tertiary follicles have a membrana granulosa of more than four cell layers, and the theca layer is now differentiated into an inner theca interna and outer theca externa. Oocytes in tertiary follicles are suspended in follicular fluid by a stalk of granulosa cells, the cumulus oophorus. Immediately surrounding oocytes is a thin ring of granulosa cells, the corona radiata. At this stage, the follicle is called the Graafian follicle and appears as a transparent vesicle that bulges from the surface of the ovary.

Even though one of the functions of the ovary is to produce oocytes, the majority of oocytes never ovulate. The number of oocytes reaches its maximum soon after the ovaries are formed. After that time, oocyte number decreases dramatically. At birth a female has all the follicles she will have in her life; no new follicles are made after birth. The vast majority of the follicles, ranging between 70 and 99.9%, are eliminated before reaching ovulation. This degenerative process is termed atresia. Atresia is a universal phenomenon, characteristic of both mammalian and nonmammalian vertebrates. Follicles can become atretic at any stage of development. Some follicles undergo atresia when they are very small (primordial follicles) or even when they are larger (tertiary follicles). Recent studies have demonstrated that apoptosis, or programmed cell death, is the molecular mechanism underlying follicular atresia. The overall hormonal regulation of follicular atresia is still not understood.

B. Production of Hormones

Another function of the ovary is to secrete hormones which act on the hypothalamus and pituitary to regulate the secretion of hormones by these two tissues, thus establishing the hypothalamic–pituitary–ovarian axis. The ovarian hormones also affect the functions of the reproductive tract. This action of ovarian hormones is important because the success of follicle development, ovulation, fertilization, and eventually embryonic development depends on correct functioning of the hypothalamus, pituitary, and reproductive tract.

1. Protein and Peptide Hormones

i. Inhibin and Activin Inhibin and activin were first isolated from gonadal fluids because of their effects on production of follicle-stimulating hormone (FSH) by the pituitary in mammals. Inhibins consist of two disulfide-bridged subunits, the α and β subunits, whereas activins consist of two β subunits. The primary source of inhibin and activin in the ovary is the maturing follicles and corpus luteum. The function of inhibins is to modulate FSH secretion at the level of the pituitary, whereas the function of activins is not yet determined. Inhibins and activins may also act as intraovarian regulators of follicular development.

ii. Relaxin Relaxin is produced by the corpus luteum. Relaxin's structure is very similar to that of insulin but has <20% amino acid homology. Relaxin concentration in the blood reaches the highest level before parturition. Relaxin functions to soften the cervix and vagina for the passage of the fetus during parturition and to promote the growth of nipples. Relaxin also acts on nonreproductive tissues, such as skin and gastrointestinal tract.

iii. Growth Factors The ovary not only secretes endocrine hormones to regulate functions of other reproductive organs but also produces growth factors to coordinate the activities of different ovarian compartments. Many growth factors, such as insulin-like growth factors, transforming growth factors, and epidermal growth factor, are produced by the germ cells and somatic cells in the ovary. This complex intraovarian regulation system is no less important than the extraovarian regulation by pituitary hormones. These growth factors form a delicate interactive communication web inside the ovary. Without them, the ovarian cells cannot interact with each other and the growth of the ovary is halted.

2. Steroid Hormones

The ovary uses cholesterol as the precursor for steroid synthesis. Cholesterol is metabolized into progestins, androgens, and estrogens by different compartments of the follicles (Fig. 3).

i. Progestins Pregnenolone is the most important progestin (C21 pregnane family) produced by

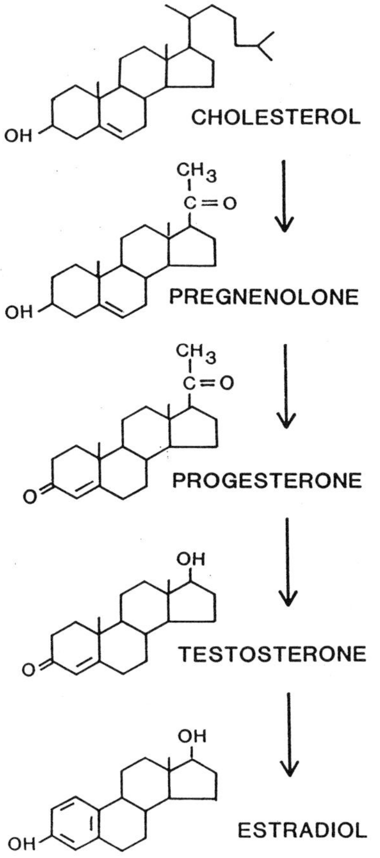

FIGURE 3 Biosynthesis of steroid hormones from cholesterol. This scheme provides a simplistic view of a highly organized and complicated process that requires multiple enzymes (adapted from Hafez, 1993, p. 79).

follicles because of its key position as the precursor of all steroid hormones. The most abundant progestin is progesterone, produced as a biosynthetic intermediate by follicles at all growing stages of development and as a secretory endproduct of the corpus luteum. In developing follicles, the theca layer is the primary site of progestin production. After ovulation, the corpus luteum synthesizes large amounts of progesterone.

ii. Androgens The follicle is a significant source of ovarian androgens (C19 androstane family). Pregnenolone and progesterone are converted into androgen metabolites, dehydroepiandrosterone and androstenedione, respectively. These two metabolites are then transformed into testosterone. The theca layer of the follicle is the primary source of ovarian androgens.

iii. Estrogens Physiologically, the estrogens (C18 estrane family), especially estrone and estradiol-17β, are the most important of the ovarian steroids. Androstenedione and testosterone are the immediate biosynthetic precursors of estrone and estradiol-17β, respectively. Their names reflect their roles in the induction of sexual receptivity (estrus) in female mammals. Estrone was the first sex steroid to be isolated and identified. The granulosa layer is the major site of estrogen synthesis in the ovary.

IV. REGULATION OF OVARIAN FUNCTIONS

A. Regulation of Folliculogenesis

Growth of primordial follicles to the preantral stage is independent of gonadotropins and is controlled by an unknown intraovarian factor(s). Growth of follicles after the preantral stage depends on appropriate patterns of secretion, sufficient concentrations, and adequate ratios of FSH and luteinizing hormone (LH) in the blood. FSH plays a major role in early follicular development. FSH stimulates granulosa cell mitosis and accumulation of follicular fluid. Granulosa cells synthesize estrogens in response to FSH which further enhance the mitotic effect of FSH. Moreover, FSH induces granulosa cell sensitivity to LH by increasing LH receptor expression. Abundant LH receptors in granulosa cells prepare for the luteinization of granulosa cells in response to the LH ovulatory surge in mammals. In contrast, theca cells are stimulated only by LH, and LH receptors are present from the beginning of the formation of the theca layer.

B. Regulation of Steroidogenesis

The steroidogenic output of the ovary is a function of coordinated actions of theca and granulosa cells. Differences in gonadotropin receptors on the membrane, in steroidogenic enzyme activities, and in compartmentalization in the follicle result in an

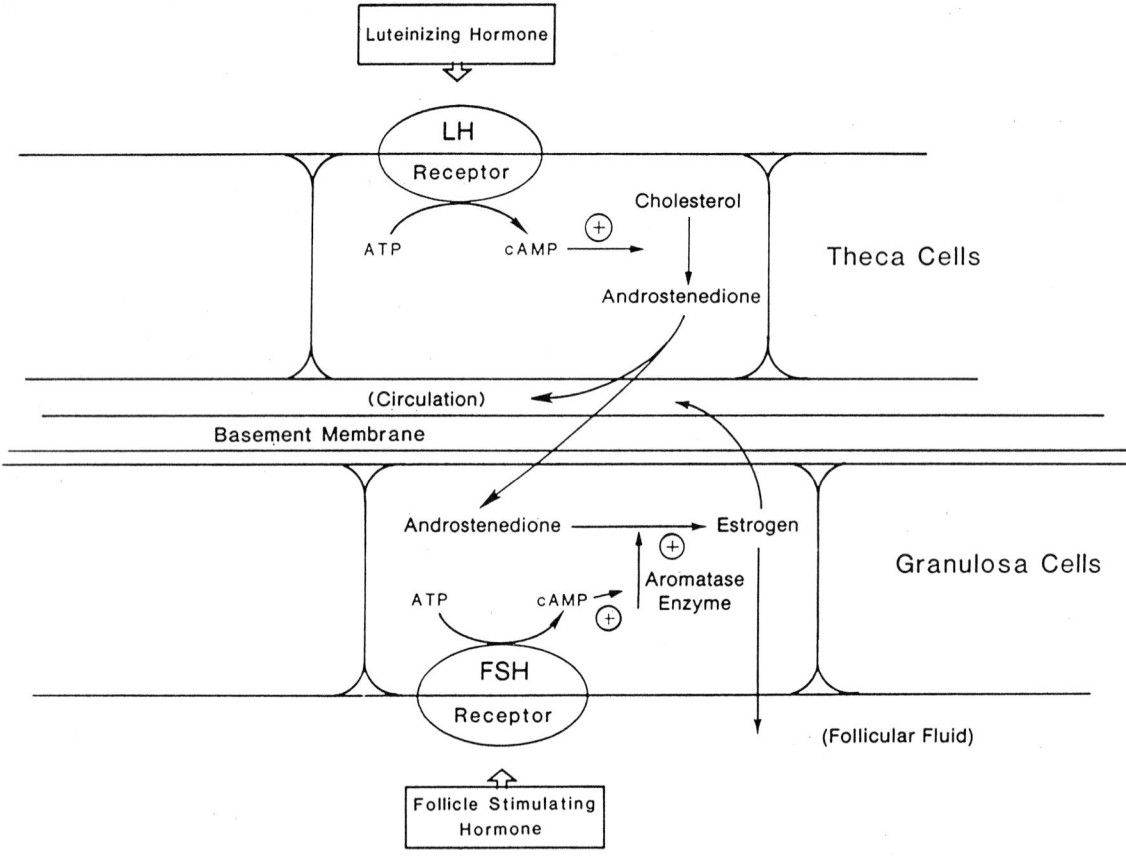

FIGURE 4 "Two-cell, two-hormone" theory of follicular steroidogenesis. LH binds to specific membrane receptors on theca cells and stimulates cyclic AMP production and the conversion of cholesterol to androgens, primarily androstenedione and testosterone. These androgens diffuse into the circulation and across the basement membrane into granulosa cells. FSH binds to specific membrane receptors on granulosa cells and stimulates cyclic AMP production, which leads to increased aromatase enzyme and the conversion of theca androgens to estrogens (adapted from Yen and Jaffe, 1986, p. 124).

unique partnership in steroid synthesis between theca and granulosa cells. The principal site of estrogen synthesis in the ovary is granulosa cells under the control of FSH. FSH stimulates not only estrogen production of granulosa cells at all stages of follicular development but also progesterone synthesis by mature follicles before ovulation. Androgen production appears to be the primary steroidogenic function of theca cells in response to LH. Androgens from theca cells provide substrates for granulosa cells to synthesize estrogens. The action of LH on theca androgen production, together with the action of FSH in granulosa estrogen synthesis, forms the basis of the "two-cell, two-hormone" theory for the control of steroidogenesis in the ovary (Fig. 4).

See Also the Following Articles

Female Reproductive System, Humans; Follicular Steroidogenesis; Granulosa Cells; Meiosis; Theca Cells

Bibliography

Burkitt, H. G., Young, B., and Heath, J. W. (1993). Female reproductive system. In *Wheater's Functional Histology*, 3rd ed., pp. 335–337. Churchill Livingstone, New York.

Findlay, J. K., Xiao, S., Shukovski, L., and Michel, U. (1993). Novel peptides in ovarian physiology: Inhibin, activin and follistatin. In *The Ovary* (E. Y. Adashi and Leung, Eds.), pp. 413–432. Raven Press, New York.

Gilbert, S. F. (1994). Sex determination. In *Developmental Biology*, 4th ed., pp. 754–768. Sinauer, Sunderland, MA.

Gillet, J. Y., Maillet, R., and Gautier, C. (1980). Blood and lymph supply of the ovary. In *Biology of the Ovary* (Motta and Hafez, Eds.), pp. 86–98. Kluwer, Boston.

Gore-Langton, R. E., and Armstrong, D. T. (1993). Follicular steroidogenesis and its control. In *The Physiology of Reproduction* (E. Knobil and J. D. Neil, Eds.), 2nd ed., pp. 571–628. Raven Press, New York.

Hafez, E. S. E. (1993). Folliculogenesis, egg maturation, and ovulation. In *Reproduction in Farm Animals,* 6th ed., pp. 114–143. Lea & Febiger, Philadelphia.

Hsueh, A. J. W., Billig, H., and Tsafriri, A. (1994). Ovarian follicle atresia: A hormonal controlled apoptotic process. *Endocrinol. Rev.* 15, 707–724.

Jones, R. E. (1991). The ovaries. In *Human Reproductive Biology*, pp. 39–53. Academic Press, San Diego.

Ross, G. T., and Schreiber, J. R. (1986). The ovary. In *Reproductive Endocrinology* (S. S. C. Yen and R. B. Jaffe, Eds.), 2nd ed., pp. 115–139. Saunders, Philadelphia.

Yen, S. S. C., and Jaffe, R. B. (1986). *Reproductive Endocrinology,* 2nd ed. Saunders, Philadelphia.

Zuckerman, S., and Baker, T. G. (1977). The development of the ovary and the process of oogenesis. In *The Ovary* (S. Zuckerman and Weir, Eds.), pp. 42–59. Academic Press, New York.

Ovidae

see Sheep and Goats

Oviduct, Mammalian

see Fallopian Tube

Oviparity

see Viviparity and Oviparity

Oviposition in Molluscs

Jeffrey L. Ram

Wayne State University

I. Reproductive Behavior in Molluscs
II. Hormonal Regulation in Gastropods
III. Neural Regulation in Bivalves

GLOSSARY

Bivalvia A class of molluscs that has two shells ("valves") joined together at a flexible hinge; includes freshwater and marine mussels, clams, scallops, and oysters.

Cephalopoda A class of molluscs that includes octopus, squid, cuttlefish, and *Nautilus*.

dioecious Having separate sexes (male and female). Most molluscan species are dioecious.

egg-laying hormone In gastropods, peptides of approximately 4.5 kDa in size found in clusters of neurons in the nervous system. Upon secretion into the circulation, these peptides activate egg laying and related egg-release behavior.

Gastropoda A class of molluscs that includes marine snails (Prosobranchia), sea slugs (Opisthobranchia), freshwater and terrestrial snails, and terrestrial slugs (Pulmonata).

hermaphrodite An animal capable of producing gametes of both male and female type, often at the same time.

serotonin A neurotransmitter (chemical name, 5-hydroxytryptamine) that causes reinitiation of oocyte maturation and spawning in many bivalves. The gonads of bivalves are directly innervated by serotonin-containing neurons.

Although a small number of molluscan species internally brood their eggs, most molluscs release or deposit them into the environment in a process known as oviposition or spawning. Spawning is affected by environmental conditions and may be initiated by specific stimuli, such as chemicals or tactile input. The spawning responses are controlled by the nervous system in all molluscs; however, bivalves and gastropods differ in the neural control mechanisms they utilize. Gastropod oviposition is activated by the neurohormonal release of a peptide egg-laying hormone into the circulation. In contrast, spawning in bivalves is activated by serotonin released in the gonad by a direct neuronal innervation. Cephalopods have complex oviposition behaviors but the internal regulatory systems regulating them have not yet been determined.

I. REPRODUCTIVE BEHAVIOR IN MOLLUSCS

Reproductive behavior and the handling of gametes varies considerably among the molluscs. Although most species are dioecious (having separate sexes), some species are hermaphrodites. Some species discharge their eggs directly into surrounding waters where they must be fertilized externally and develop initially as free-floating larvae. This is common among the most primitive species, such as in Amphineura, Archaeogastropoda, Scaphapoda, and most Bivalvia. In other species, the eggs are fertilized internally and packaged into more or less elaborate egg capsules that are attached by adhesive proteins to an external substrate. Development to juvenile form is completed in the capsules in some species (e.g., in *Busycon*, a large marine snail found along the eastern coast of the United States), whereas in others, larval stages are released from the capsules to complete development in the plankton. For example, the planktonic larval stage of the Chilean marine snail *Concholepas concholepas* (common name, "loco") lasts up to 3 months. Other species may brood their developing larvae internally or release them to develop parasitically in fish.

A. Bivalves

As noted previously, the most common reproductive pattern among bivalves is to release oocytes into surrounding waters for external fertilization. The release of free-floating gametes is influenced by a variety of external factors, including temperature, chemicals from potential food sources and other spawning animals, and possibly light. For example, the zebra mussel (*Dreissena polymorpha*) is found in temperate regions and is predominantly dioecious, and both sexes release their gametes into surrounding water for external fertilization. Prior to spawning, ovary and testes grow and sperm and eggs mature as the water warms from midwinter through early summer. Animals rarely spawn below 12°C (54°F); however, above that temperature animals in nearby locations usually spawn more or less at the same time. For bivalves and other molluscs with external fertilization and separate sexes, such as the zebra mussel, synchronous spawning facilitates encounters between sperm and eggs. It is thought that chemicals play a critical role in synchronizing bivalve spawning since it has been shown in zebra mussels, marine mussels (*Mytilus*), and oysters that chemicals extracted from algae (potential food for larvae) or from gametes can trigger spawning. In addition to chemicals and temperature, another factor might be cyclic patterns or levels of light since some of the peaks of zebra mussel spawning seem to be correlated with the lunar cycle.

Not all bivalves have external fertilization. In some freshwater bivalves, internally brooded larvae known as glochidia may be released to attach to passing fish on which they develop parasitically attached to the fish's gills.

B. Gastropods

Gastropods tend to have a complex pattern of behavior associated with deposition of their eggs. The sea hare *Aplysia* (Fig. 1a) prepares to lay eggs with circular head and neck movements ("undulations"). The eggs emerge encapsulated in a gelatinous string from a groove near the animal's mouth and are placed in a compact clump with characteristic weaves and tamps of its head and anterior foot. At the same time

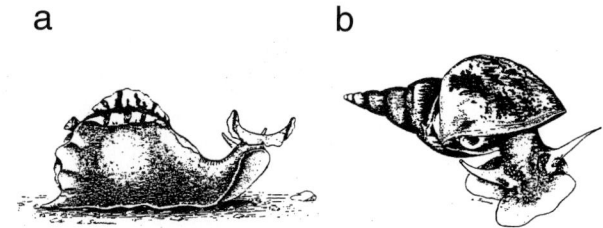

FIGURE 1 (a) *Aplysia californica*. Mature specimens range from 10 cm to more than 20 cm in length. (b) *Lymnaea stagnalis*. Reproductive animals have shell heights of 23–35 mm (from Gerearts *et al.*, 1988; reprinted by permission of Wiley-Liss, Inc., a subsidiary of John Wiley & Sons, Inc.).

the animal stops locomoting and becomes unwilling to eat for the duration of the egg-laying episode (a few hours), perhaps to prevent itself from mistakenly eating its own eggs. In this hermaphroditic animal, egg laying often is preceded by the animal mating as a female and followed by the animal mating as a male if suitable mates are available. Egg laying occasionally begins spontaneously with no apparent triggering stimulus; however, a stimulus that has consistently been observed to activate laying in *Aplysia* is the freshly laid egg string of another animal. Within 30–60 min (depending on the temperature and other factors) of contacting a fresh egg string with its mouth, lips, and tentacles, ripe *Aplysia* usually begin laying eggs themselves and placing their egg string on the previously laid clump of eggs. Peptides ("attractins") released from *Aplysia* egg strings have been shown to attract other *Aplysia* and could be involved in this response.

Among the freshwater gastropods, the most thoroughly studied animal is the snail *Lymnaea stagnalis* (Fig. 1b). Egg laying can be triggered by transfer of *L. stagnalis* from stagnant water to clean aerated water. Typically, prior to egg laying *L. stagnalis* ceases locomoting for about an hour and during the next hour begins slow turning movements, rasping the surface with feeding structures, and finally begins depositing its gelatinous oblong mass of eggs on the scraped surface while taking small steps, continuing the rasping movements, and pressing the egg mass to the substrate.

Other gastropods also have complex stereotyped egg-laying behaviors. In the marine slug *Pleuro-*

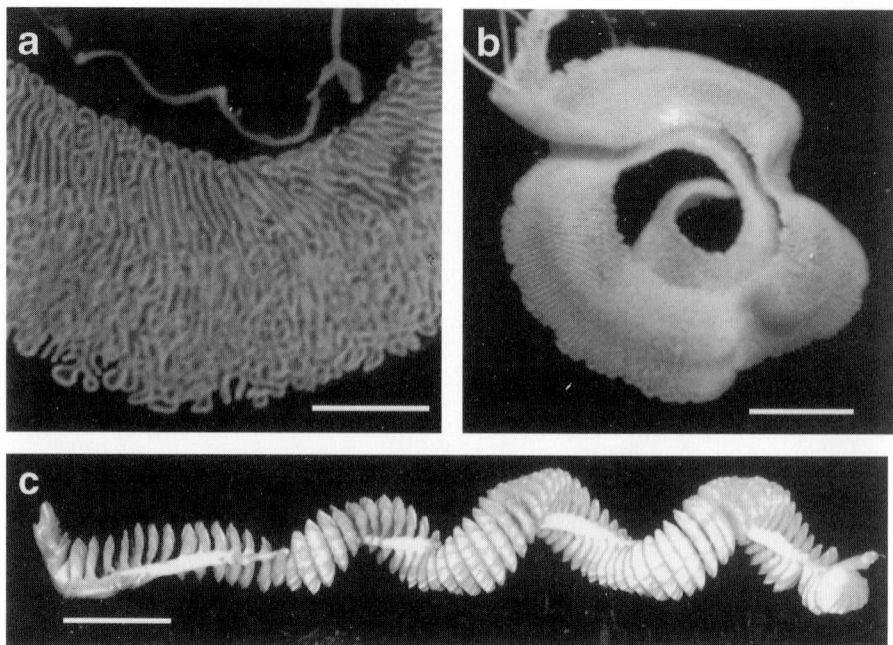

FIGURE 2 Gastropod egg ribbons and capsules. (a) Egg ribbon with glue strip from *Pleurobranchaea californica*. (b) Spiral-shaped ribbon from *Anisodoris*. (c) String of egg capsules from *Busycon canaliculatum*. This string contained 79 capsules and took approximately 10 days to lay at a rate of about one capsule every 3 hr. Scale bars = (a) 1 cm, (b) 1 cm, and (c) 3 cm.

branchaea, feeding is inhibited during egg laying; however, in contrast to *Aplysia*, which uses its mouth to manipulate eggs, *Pleurobranchaea* has a specialized ovipositor that extends out of its right side to deposit its eggs in a ribbon. The egg ribbon has a strip of "glue" along one edge to adhere it to the substrate (Fig. 2a). *Anisodoris* also deposits eggs in a ribbon but moves slowly in a tight spiral during the process, producing spiral-shaped egg masses (Fig. 2b).

The longest egg-laying episodes have been observed in prosobranch gastropods (marine snails). *Busycon* (Fig. 3) lays its eggs in leathery disciform capsules that are attached to one another in long strings (Fig. 2c). Each capsule emerges as a soft bulb-shaped whitish bag from the female gonopore and is passed through a groove along the side of the foot into a gland in the bottom of the foot where it is hardened and given its species-specific shape. At the same time, the capsule is glued to the substrate or to a previously laid egg capsule to extend the string (Fig. 3). Generally, the first few capsules in a string do not contain eggs, and subsequently laid capsules contain 30–80 eggs. The average time between laying capsules in *Busycon* is 1–3 hr, and capsule strings may contain 50–110 capsules. Thus, egg-laying episodes may last 1 or 2 weeks! *Concholepas concholepas* has been reported to lay more than 200 capsules in

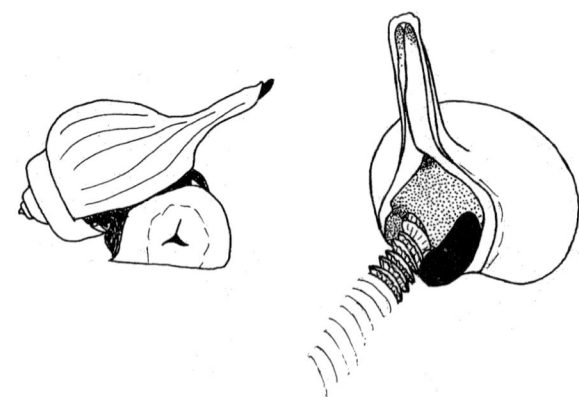

FIGURE 3 *Busycon canaliculatum* laying eggs. (Left) An animal has a single capsule in its pedal gland, from which the capsule will emerge adhering to the substrate. (Right) Capsules are added onto an egg capsule string that emerges from the foot of another animal. Mature females range from 12 to 18 cm in length from siphon to apex of shell.

an episode lasting 11 days. Similar to *Busycon*, the capsules are hardened and cemented to the underlying surface while enveloped in a pedal gland. During laying and short periods between laying episodes *C. Concholepas* refuses to eat.

C. Cephalopods

Among cephalopods, the octopus has been observed to lay its eggs in strings on the roofs of caves or in crevices of rocks. Females deposit their eggs shortly after mating, with oviposition following preparation of the surface by blowing away detritus, plucking the surface with suckers, and applying adhesive material to attach the eggs to the surface or to one another by a stalk to form the egg strings. In *Octopus vulgaris* the entire process of laying a typical clutch of 150,000 eggs in numerous strings of several hundred eggs apiece may take a week or more; however, other species, such as *O. bimaculoides,* may lay only a few dozen eggs. The female octopus guards her eggs, aerating and cleaning them with her suckers and jets of water and not leaving the eggs or eating while the embryos develop. The female thus loses much weight while brooding her eggs and typically dies shortly after they hatch.

In *Nautilus*, eggs are enclosed in capsules and are usually deposited singly or in small batches. Egg laying in *Nautilus* has been reported to require temperatures of at least 18–21°C and to occur only at night. In contrast to octopus, female *Nautilus* appear capable of producing eggs almost continuously and to have longevities of decades compared to only about a year or two for octopus.

II. HORMONAL REGULATION IN GASTROPODS

The internal mechanisms regulating egg-laying behavior in molluscs are understood best in gastropods, in which oviposition is controlled by specific peptide neurohormones. Hormonal activation of egg laying in gastropods was first demonstrated in *Aplysia* in 1967 by Irving Kupfermann, who showed that injection of homogenates of the bag cells, two clusters of neurons in the abdominal ganglion, elicited egg-laying behavior. Subsequently, the active factor was identified as a 36-amino acid peptide, designated egg-laying hormone (ELH). The behavior elicited by the pure peptide is identical in almost all respects (undulations, head weaving and tamping, and inhibition of locomotion and feeding) to the behavior that occurs following activation of the bag cell neurons *in vivo*. In *Pleurobranchaea* peptides in the medial lobes of the pedal ganglia (Fig. 4) activate egg laying and inhibit feeding upon injection into ripe recipients. In freshwater snails, groups of neurons in the cerebral ganglia known as the caudodorsal cells (Fig. 5) similarly elicit complete spawning episodes upon injec-

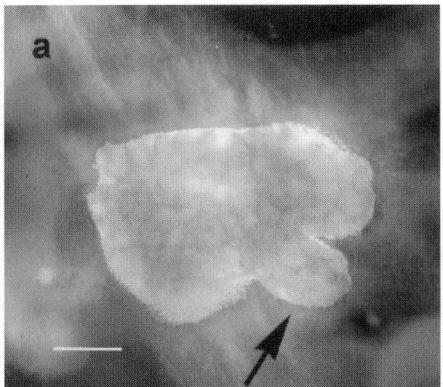 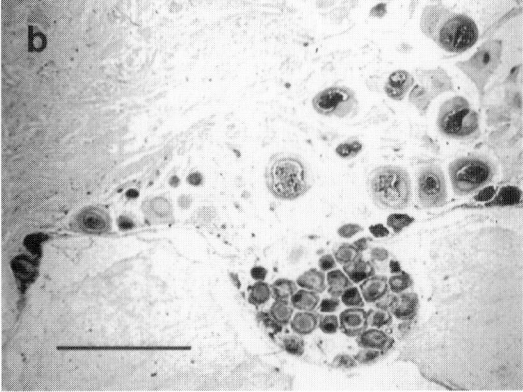

FIGURE 4 Location of ELH in *Pleurobranchaea californica*. (a) Whole mount of left pedal ganglion. ELH is found in medial lobe (arrow). Scale bar = 500 μm. (b) Medial lobe and adjacent area stained with phloxine, also known to stain ELH-containing neurons in *Lymnaea* and *Aplysia*. Scale bar = 400 μm.

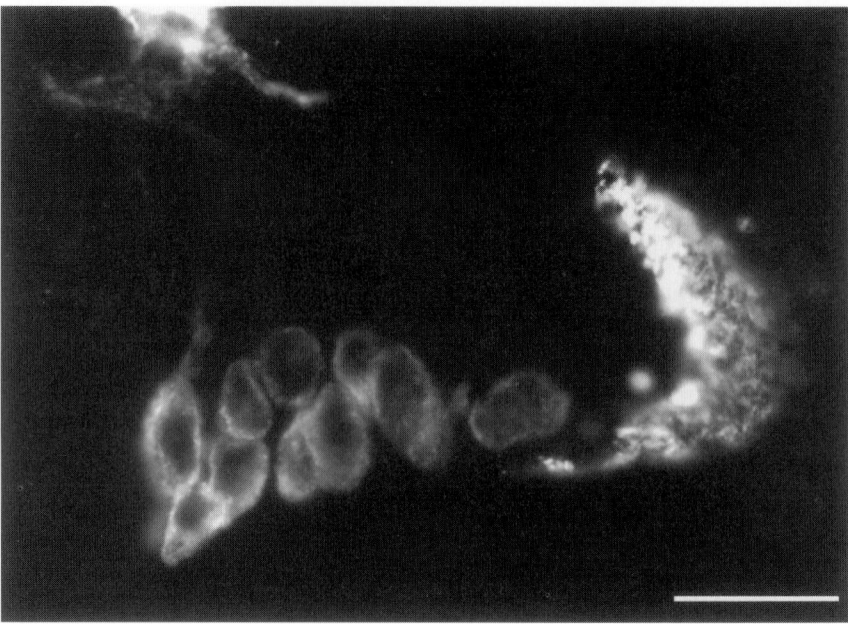

FIGURE 5 Caudodorsal cells and adjacent neurohemal area in the cerebral commissure in *Lymnaea*. These are stained with an antibody to *Lymnaea* alpha peptide, developed with an FITC-labeled second antibody. Alpha peptide is synthesized from the same precursor peptide as ELH (see Fig. 6). Scale bar = 100 μm.

tion into ripe recipients. As in *Aplysia*, the active factor in *Lymnaea* is a 36-amino acid peptide. The *Lymnaea* ELH is approximately 45% identical to *Aplysia* ELH; however, the differences in ELH from one species to another are large enough that interspecies injections between these genera do not result in egg laying.

Among prosobranchs, injection of extracts of cerebral and several other ganglia into *Busycon* elicits the laying of egg capsules. *Busycon* ELH is sensitive to proteolytic enzymes and is approximately the same molecular weight as the hormones in *Aplysia*, *Pleurobranchaea*, and *Lymnaea*; however, ELH sequences in prosobranchs have not been determined. Unlike *Aplysia*, *Pleurobranchaea*, and *Lymnaea*, the egg-laying episode elicited in *Busycon* by hormone-containing extracts is incomplete: Only a single egg-less capsule is laid in response to a single injection. Nevertheless, upon repeated injection of hormone-containing extracts at 2-hr intervals, a capsule is laid following each injection, and eggs begin to appear in the capsules after several empty capsules have been laid, similar to the distribution of eggs in capsule strings laid naturally. It has been suggested that the ELH-secreting neurons in *Busycon* are either active continuously for days at a time or cyclically reexcited at intervals to maintain laying over the 1 or 2 weeks that animals lay their eggs in typical egg-laying episodes.

In *Aplysia* and *Lymnaea*, in which ELH synthesis and secretion has been studied most intensively, ELH is synthesized as part of a large precursor which is cleaved into several peptides (Fig. 6). These cosynthesized peptides include the "alpha peptides," which can be demonstrated immunohistologically in ELH-containing neurons (Fig. 5). Immunohistological methods have demonstrated that ELH and alpha peptides are also present in adjacent neurohemal secretory regions (Fig. 5) and in individual neurons or small clusters of neurons elsewhere in the nervous systems in addition to their main clusters. The alpha peptides have been suggested to act as excitatory neurotransmitters together with other peptides co-synthesized with ELH. Autoexcitation from these peptides may be part of the mechanism that causes the bag cell neurons in *Aplysia* and the caudodorsal

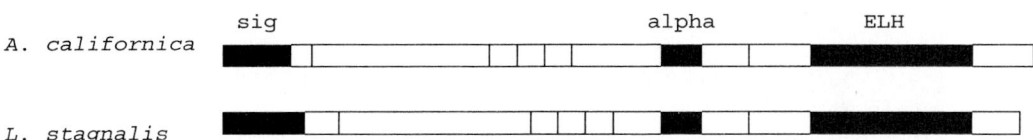

FIGURE 6 Proposed fragmentation patterns of the ELH/alpha peptide precursors from *Aplysia* and *Lymnaea*, based on positions of basic amino acid residues and known peptide products. sig, signal peptide; alpha, alpha peptide; ELH, egg-laying hormone.

cells in *Lymnaea* to produce long-lasting (15–70 min) trains of action potentials when sufficiently activated to initiate action potentials.

Initial activation of ELH-secretory neurons involves sensory input. In *Aplysia*, stimulation of nerves from the head region can initiate the activity, and contact of ripe *Aplysia* with a newly laid egg mass results in activation of the bag cell neurons with latencies of up to 20 min until onset of action potentials. Neurons containing alpha peptides in the pleural ganglia may mediate these responses. Similarly, the clean water stimulus that activates egg laying in *Lymnaea* has been shown to activate the caudodorsal cells prior to appearance of ELH in the circulation and the initiation of egg-laying behavior.

ELH released into the circulation acts on the gonad to release eggs. ELH has been detected in the blood in *Lymnaea* after caudodorsal cell activity. ELH is secreted from the bag cells of *Aplysia* abdominal ganglia *in vitro* in sufficient amounts to trigger egg laying in intact animals. Direct egg-releasing effects of ELH on the gonad have been demonstrated in *Aplysia* by application of ELH to gonad pieces *in vitro*.

Behavioral changes in feeding and locomotion that accompany egg laying may also be mediated by ELH since ELH has been shown in *Aplysia* to activate feeding retractor motoneurons and to modulate central pattern generator activity in the pedal–pleural ganglia, respectively. In *Lymnaea*, activation of the caudodorsal cells is similarly accompanied by changes in activity in neurons that mediate feeding and locomotion. These central neuronal responses to ELH may be mediated either hormonally or by paracrine secretion of ELH not only from the main clusters of ELH-containing neurons but also from neurons outside of the main clusters. In addition, some behavioral changes associated with egg laying are due to reflex responses to the egg strings, clumps, or capsules themselves. It seems especially important to decipher the mechanisms by which feeding is inhibited in association with egg laying since this behavioral change is present in many molluscan species.

III. NEURAL REGULATION IN BIVALVES

In contrast to gastropods, in which egg release is triggered by a peptide hormone acting through the circulation, egg release in bivalves is mediated by direct innervation of the gonad utilizing the biogenic amine serotonin. Serotonin is a neurotransmitter found both in invertebrate and in vertebrate nervous systems. Injection of serotonin into a variety of bivalve species elicits spawning in both males and females. Spawning responses in either or both sexes have been observed in scallops, clams, oysters, and some mussels (e.g., *Dreissena* spp. but not *Mytilus edulis*).

The evidence that serotonin is the physiological mediator of spawning in bivalves is most complete for scallops and zebra mussels. Serotonin has been detected by immunohistology in neuronal fibers innervating the testes and ovaries of both scallops and zebra mussels (Fig. 7). In the hermaphroditic scallop *A. purpuratus*, the levels of serotonin measured in both male and female regions of the gonad decrease by more than 50% when animals are stimulated to spawn by an external stimulus. Application of serotonin to ovarian tissue of scallops *in vitro* causes the release of eggs. In zebra mussels, serotonin reinitiates meiosis prior to spawning of oocytes, a process that also occurs in isolated ovaries *in vitro* in response to direct application of serotonin.

The pharmacology of spawning in intact animals

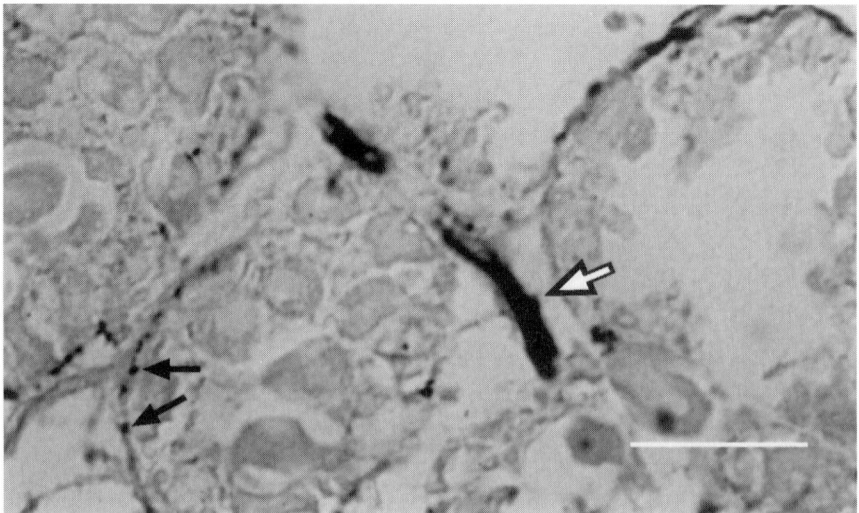

FIGURE 7 Serotonin-like immunoractivity in the gonad of a female zebra mussel, *Dreissena polymorpha*. Open arrow, portions of a nerve; solid arrows, varicosities in the walls of follicles surrounding oocytes. Scale bar = 100 μm.

and *in vitro* preparations provides additional evidence for the physiological role of serotonin. In scallops the serotonin antagonist methysergide inhibits both the spawning response of intact animals and the release of eggs by serotonin *in vitro*. In zebra mussels, spawning of oocytes can be inhibited by low concentrations (1 μmol/liter) of the serotonin antagonist methiothepin. 8-OH-DPAT, a serotonin receptor activator, has been shown both to activate spawning in intact zebra mussels and to reinitiate meiosis in isolated ovaries *in vitro*. Thus, anatomical, biochemical, and pharmacological evidence all support a physiological role of serotonergic innervation of the gonad in regulating spawning of eggs in bivalves. Serotonin may play a similar role even in brooding bivalves, in which serotonin has been shown to cause parturition in the brooding freshwater bivalve *Sphaerium transversum*.

The reproductive role of serotonin in bivalves clearly contrasts with its function in gastropods. Although serotonin is widely distributed in gastropod nervous systems, no serotonergic innervation of snail gonads has been observed. Furthermore, in *Aplysia* serotonin inhibits spawning in response to an ELH neuron activator, in contrast to the activation of spawning serotonin elicits in bivalves. Conversely, ELH-like peptides have been observed immunohistologically in bivalve nervous systems, but no reproductive functions of these ELH-containing neurons have yet been identified in bivalves. Thus, the spawning regulators of gastropods and bivalves appear to be based on two very different systems: a neuroendocrine system in gastropods utilizing a peptide as the activator of oocyte release, and neuronal innervation of the gonad in bivalves using serotonin as the activator of oocyte release. Furthermore, it is not known whether either of these systems mediate the complex behaviors of cephalopod spawning.

See Also the Following Articles

Hermaphroditism; Mollusca

Bibliography

Bavendam, F. (1991, March). Eye to eye with the giant octopus. *Natl. Geogr.* 179(3), 86–97.

Begnoche, V. L., Moore, S. K., Blum, N., Van Gils, C., and Mayeri, E. (1996). Sign stimulus activates a peptidergic neural system controlling reproductive behavior in *Aplysia*. *J. Neurophysiol.* 75, 2161–2166.

Bernheim, S. M., and Mayeri, E. (1995). Complex behavior induced by egg-laying hormone in *Aplysia*. *J. Comp. Physiol. A* 176, 131–136.

Gerearts, W. P. M., Ter Maat, A., and Vreugdenhil, E. (1988). The peptidergic neuroendocrine control of egg-laying behavior in *Aplysia* and *Lymnaea*. In *Invertebrate Endocrinology, Vol. 2, Endocrinology of Selected Invertebrate Types* (H. Laufer and G. H. Downer, Eds.), pp. 141–231. A. R. Liss, New York.

Ram, J. L. The zebra mussel page. *http://www.science.wayne.edu/~jram/zmussel.htm*.

Ram, J. L., and Ram, M. L. (1990). Gastropod egg-laying hormones. In *Advances in Invertebrate Reproduction 5* (M. Hoshi and O. Yamashita, Eds.), pp. 257–264. Elsevier, Amsterdam.

Ram, J. L., Fong, P. P., and Garton, D. (1996). Physiological aspects of zebra mussel reproduction: Maturation, spawning, and fertilization. *Am. Zool.* 36, 326–338.

Saunders, W. B., and Landman, N. H. (1987). *Nautilus, The Biology and Paleobiology of a Living Fossil*. Plenum, New York.

Smith, S. A., and Croll, R. P. (1997). Molluscs. In *Reproductive Biology of Invertebrates, Volume VI, Progress in Reproductive Endocrinology* (K. G. Adiyodi and R. G. Adiyodi, series Eds.; T. Adams, Vol. Ed.). Wiley, Chichester, UK.

Wells, M. J. (1978). *Octopus, Physiology and Behavior of an Advanced Invertebrate*. Chapman & Hall, London.

Ovulation

Lawrence L. Espey
Trinity University

I. Introduction
II. Anatomy of Ovulation
III. Biochemistry of Ovulation
IV. Current Hypothesis on Ovulation

GLOSSARY

differential display A series of molecular procedures involving extraction of RNA from tissues, conversion of the mRNA into cDNA, amplification of radiolabeled cDNA by the polymerase chain reaction, and electrophoresis of the cDNA on an acrylamide gel for the purpose of detecting differential expression of genes between control and experimental (or pathological) tissues.

gonadotropin Any of a number of glycoprotein hormones, such as luteinizing hormone (LH), follicle-stimulating hormone (FSH), and human chorionic gonadotropin (hCG), that stimulate ovarian follicles to develop and ovulate.

hypothalamo/hypophyseal axis The neuronal and vascular associations between the hypothalamus at the base of the brain and the hypophysis (pituitary gland) that regulate gonadotropin secretion and the sexual cycle.

inflammation A complex sequence of metabolic changes leading to vasodilatation, hyperemia, exudation, edema, proteolysis, and eventual tissue remodeling in response to microbial invasion, radiation, friction, chemical irritation, or other factors such as acute stimulation of target tissues by glycoprotein hormones.

luteinization The physical and metabolic transformation of a mature ovarian follicle into a corpus luteum, principally characterized by a marked increase in follicular progesterone synthesis in response to an ovulatory surge in LH.

mature follicle An ovarian follicle that has acquired an adequate concentration of gonadotropin receptors to undergo an ovulatory response when stimulated by a surge in endogenous LH (and/or FSH) or by an adequate amount of exogenous chorionic gonadotropin such as hCG.

ovulatory Describing or referring to events that follow initiation of the ovulatory process by gonadotropic hormone(s).

ovulatory process The complete sequence of physical and chemical events that begin when a mature ovarian follicle is stimulated by an ovulatory surge in gonadotropic hormones and end when the follicle releases an egg; ovulation.

postovulatory Describing or referring to events that follow expulsion of the oocyte from the follicle.

preovulatory Describing or referring to events that precede initiation of the ovulatory process by gonadotropic hormone(s).

Ovulation is the release of fertile eggs from the adult ovary. Mammalian ovulation is a clearly defined biological process that begins when gonadotropic hormone(s) stimulates mature ovarian follicles, and it ends when the follicles rupture and release fertile eggs into the oviduct. The duration of the ovulatory process is a species-specific interval of time, ranging from 10 hr in the rabbit to possibly as much as 30–36 hr in the human.

I. INTRODUCTION

Mammalian ovaries have two principal functions. They produce sex steroids to prepare the adult female for reproduction, and they release eggs at appropriate intervals during the fertile years of the organism. It is generally thought that the underlying mechanisms of hormone action leading to fertility and ovulation are homologous in all mammals.

A. The Sexual Cycle and Ovulation

At sexual maturity (puberty), the female begins a sexual cycle (i.e., a menstrual cycle in humans) that is based on rhythmic interaction between the hypothalamo–hypophyseal axis and the ovaries. A given cycle is initiated when the hypothalamus begins secreting gonadotropin-releasing hormone (GnRH) into the hypothalamohypophyseal portal system where it travels to the anterior hypophysis (pituitary) and stimulates gonadotropes to secrete follicle-stimulating hormone (FSH) and luteinizing hormone (LH). Under the influence of these two gonadotropic hormones from the pituitary, primordial follicles begin growing in the ovary. During this developmental process, known as folliculogenesis, a large antral cavity forms in the center of the follicle and a thick layer of collagenous connective tissue forms around its perimeter. As the follicle grows, it begins secreting androgens and estrogens. Ovarian β-estradiol promotes the expression of gonadotropin receptors on the plasma membranes of follicular cells. A follicle is said to be mature when it is endowed with an adequate population of gonadotropin receptors that are responsive to LH and/or FSH. At this stage of the sexual cycle, the elevated level of circulating β-estradiol induces a sudden increase in GnRH secretion from the neurosecretory cells of the hypothalamus, and this releasing hormone causes a surge in LH and FSH secretion from the pituitary gland. This surge in gonadotropins initiates the ovulatory process. During the next several hours, androgen and estrogen secretion is replaced by a marked increase in ovarian progesterone synthesis. The rise in this progestin signals the onset of luteinization of the ovarian follicle. In addition, the elevation in circulating progesterone inhibits further secretion of GnRH, LH, and FSH. The hypothalamo–hypophyseal axis begins to secrete these hormones once again to initiate the next sexual cycle only after the corpus luteum begins to deteriorate (i.e., undergoes luteolysis) and progesterone secretion is diminished.

B. Rupture of the Ovarian Follicle

Mammalian ovulation is a unique biological phenomenon in that it requires the physical disruption of healthy tissue at the surface of the ovary. Initially, during the first several hours after a mature ovarian follicle has been stimulated by an ovulatory surge in pituitary gonadotropins, there is no conspicuous change in the appearance of an ovulatory follicle. However, 4–6 hr into the ovulatory process, a follicle will begin to blush. There is clear evidence that the capillaries in the follicle wall have dilated, and the tissue has become hyperemic. There is negligible other macroscopic, or microscopic, evidence of pending rupture until 1 or 2 hr before the follicle wall will actually burst. As the time of rupture nears, the apex of a mature follicle protrudes more and more above the surface of the ovary and the follicle wall itself gradually becomes thinner. Eventually, the apical-most portion of the follicle becomes translucent and rapidly balloons above the normal curvature of the follicle wall to form a stigma. This nipple-like bleb may not form in all species of mammals and will not occur if the vascular supply to the ovary has been impaired. However, a follicle will usually rupture within several minutes after the stigma forms. The eventual rupture of a follicle is dependent on adequate degradation of the collagenous connective tissue in the follicle wall and on a modest, but essential, intrafollicular pressure of about 20 mm Hg that arises from capillary hydrostatic pressure. After the follicle wall bursts, the oocyte and surrounding cumulus cells are usually extruded within 1 or 2 min. Ovulation is complete when the egg-bearing cumulus mass is expelled from the ovary.

II. ANATOMY OF OVULATION

At the apex of a mature follicle, where a stigma forms and the follicle ruptures, there are five different layers of cells (Fig. 1). The outermost layer is the surface epithelium, a single-cell layer of cuboidal epithelial cells. The second layer is the tunica albuginea, consisting of fibroblasts and collagen that form a tenacious sheath around the entire ovary. The third layer is the theca externa, the follicle's own capsule of collagenous connective tissue which delineates its boundary. The fourth layer consists of the secretory cells of the theca interna, just inside the theca externa. The fifth and innermost layer is the stratum granulosum, from which extends the cumulus mass and its oocyte.

This section examines the morphophysiological changes that occur in these layers of the follicle wall during the hours preceding the release of the egg. Although the details of the changes are based on ultrastructural studies of ovulatory follicles from the rabbit, most reproductive physiologists agree that the basic anatomy of mammalian follicles is comparable from one species to the next, and it is generally thought that the mechanism of follicular rupture is basically the same in all species of mammals. In the case of the rabbit, the ovulatory process normally requires about 10 hr. Therefore, this analysis will examine the morphophysiological aspects of the follicle wall at 10 hr before follicular rupture (i.e., at the beginning of the ovulatory process) (Fig. 1), at approximately 30 min to 1 hr before rupture (Fig. 2), and at about 1–5 min before rupture (Fig. 3).

A. 10 hr before Follicular Rupture

1. Surface Epithelium

In mature ovarian follicles, the surface epithelium is a single layer of cuboidal cells that are loosely attached to a thin basal lamina at the surface of the connective tissue (the tunica albuginea) surrounding the ovary (Fig. 1). The most conspicuous feature of these cells is the dense cytoplasmic spheres that are common on the basal side. The composition of these dense granules is unknown, but it is unlikely that they are involved in the ovulatory process because the surface epithelium can be gently scraped from the surface of a mature follicle, but the follicle will still ovulate in response to adequate stimulation by gonadotropin(s). The other interesting feature of the surface epithelium is that the cells contain polymorphous nuclei that somewhat resemble the nuclei of polymorphonuclear leukocytes. It is possible that the surface epithelium functions as a first line of defense to protect the vital procreative elements of the ovary.

2. Tunica Albuginea

Beneath the ovarian surface epithelium is the layer of dense collagenous connective tissue known as the tunica albuginea. This layer, which surrounds the entire ovary, consists almost entirely of fibroblasts, along with extracellular collagen and related ground substance (Fig. 1). The collagen usually is not readily visible by transmission electron microscopy unless the tissue is treated with 1% phosphotungstic acid or some other stain that increases the electron density of collagen. The fibroblasts in this thecal tissue give the appearance of spindle-shaped smooth muscle cells if the follicle is sectioned on a plane perpendicular to the apical follicle wall. However, if one cuts thin sections of a follicle on a plane that is tangential to the surface of the ovary, then the cells in this layer appear round, or ovoid, and it is quite obvious that they are platter-shaped fibroblasts that produce substantial amounts of collagen.

3. Theca Externa

The follicle itself is surrounded by its own layer of collagenous connective tissue called the theca externa. This tissue is quite similar to the tunica albuginea, and these two layers of thecal tissue are so contiguous at the apex of a follicle that it is difficult to distinguish them from one another (Fig. 1). The theca externa usually contains a few more fibroblasts than the tunica albuginea, but the outer tunic of connective tissue contains more collagenous extracellular material. There is not a conspicuous differ-

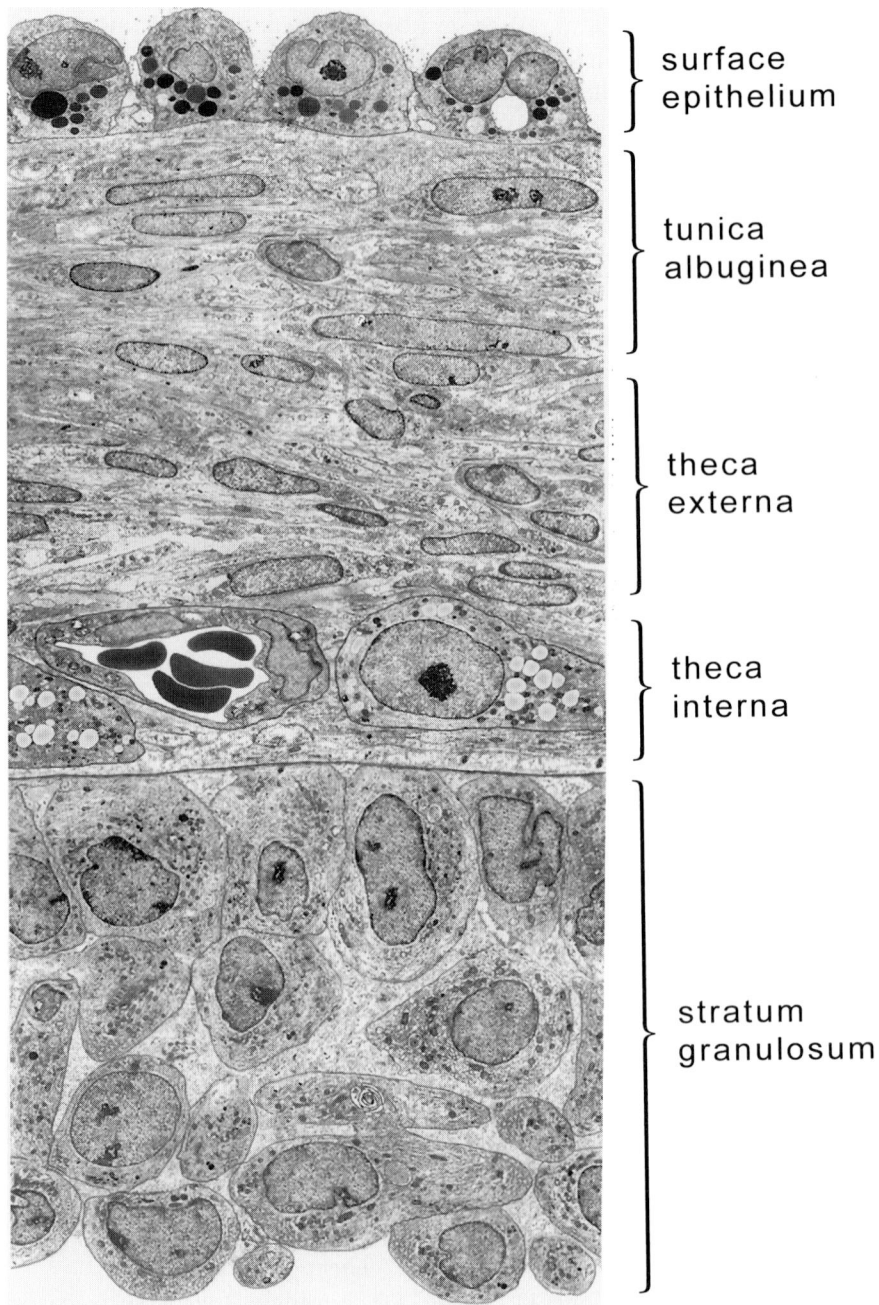

FIGURE 1 Ultrastructure of the apex of a rabbit follicle 10 hr before rupture. See Section II for further details. (Width of the electron micrograph is approximately 56 μm.)

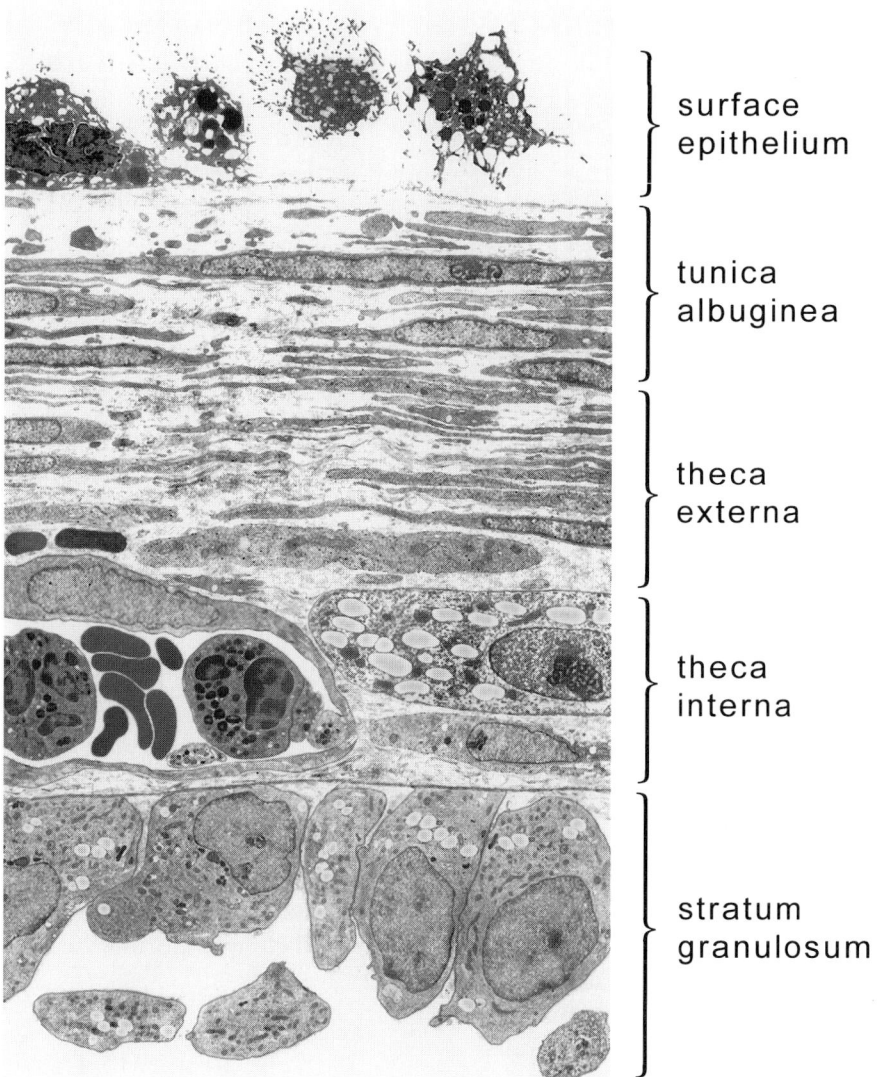

FIGURE 2 Ultrastructure of the apex of a rabbit follicle 30 min to 1 hr before rupture. See Section II for further details. (Width of the electron micrograph is approximately 56 µm.)

ence in the cellular composition of the theca externa at the apex of a follicle (where rupture will occur) versus the base of the follicle (which is surrounded by ovarian stromal tissue).

4. Theca Interna

This highly differentiated thecal tissue is a thin layer of steroidogenically active cells that are supplied by a number of large capillaries which collectively comprise the bulk of the ovarian circulation (Fig. 1). Fibroblasts and collagen are sparse in this thin layer. The secretory cells of the theca interna are characterized by large oval nuclei with prominent nucleoli and, like most steroid-secreting cells, their cytoplasm is dominated by lipid droplets, numerous mitochondria, and Golgi networks that are distributed throughout their smooth endoplasmic reticulum. These cells are sometimes referred to as "interstitial cells" in the current literature. The interior border of the theca interna is clearly delineated by a thin, but conspicuous, basal lamina called the membrana propria. This basal lamina has been erron-

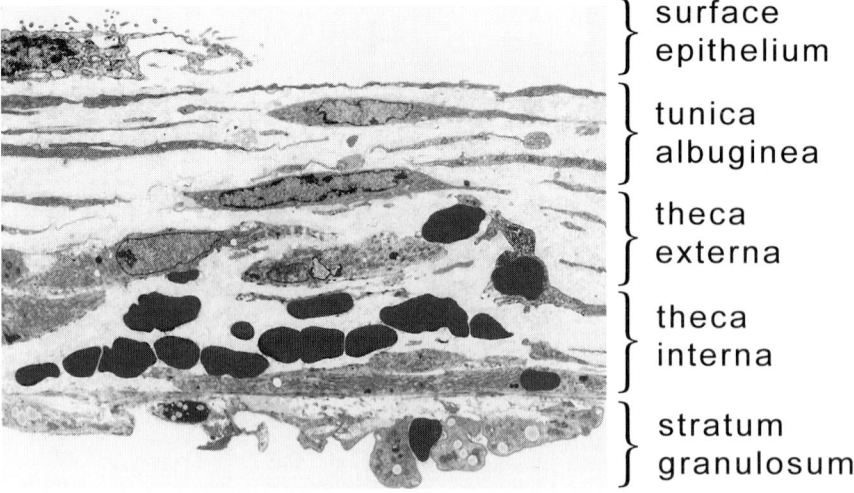

FIGURE 3 Ultrastructure of the apex of a rabbit follicle 1–5 min before rupture. See Section II for further details. (Width of the electron micrograph is approximately 56 μm.)

eously referred to as a double membrane because of its close association to the plasma membranes of the granulosa cells that adhere to its inner border.

5. Stratum Granulosum

The granulosa layer at the innermost surface of the follicle wall arises from a single layer of epithelial cells which surround the oocytes of primordial follicles. The granulosa cells that are attached to the membrana propria extend in a columnar pattern from this basement membrane (Fig. 1). The remaining cells toward the follicular antrum are more cuboidal and are distributed inward for a total depth of 3–10 cells, depending on the species of animal. The cells of the granulosa layer are metabolically integrated by an extensive labyrinth of gap junctions that couple this layer into a syncytium with the cumulus oophorus. In the vicinity of the tight junctions between granulosa cells it is common to observe invaginations from one cell to the other that become pinched off and form phagocytic-like vesicles within one or the other of the abutting cells. These vesicles constitute the transfer of cytoplasm from one granulosa cell to another, and their contents can include mitochondria, lipid droplets, or other large areas of cytoplasm. The cumulus mass, which includes the oocyte, consists of granulosa-like cells that protrude inward from any portion of the stratum granulosum, i.e., from either the apical or the basal region of this innermost layer of the follicle. This random morphological arrangement positions the oocyte toward the center of the follicular antrum and probably facilitates its dislodgement and expulsion from the follicle at the time of ovulation. Also, this central location of the fragile germ cell may serve to protect it from the degradative events that occur within the thecal layers of the follicle wall during the ovulatory process.

B. 30 min to 1 hr before Follicular Rupture

1. Surface Epithelium

Within the last several hours of the ovulatory process, conspicuous morphological changes take place in the epithelial cells on the apical surface where rupture is destined to occur. The cells of the surface epithelium develop numerous vacuoles within their cytoplasm, and the cells appear to be necrotic (Fig. 2). There is no evidence that the mucin-like dense granules in these cells release their contents into the thecal layers of the follicle or contribute in any way to the mechanism of ovulation. During the final hour of the ovulatory process, these cells begin to slough from the stigma area of the follicle, and they are usually absent at the time of rupture.

2. Tunica Albuginea

The tunica albuginea that covers the ovarian surface of most preovulatory follicles consists of quiescent fibroblasts. However, conspicuous changes occur in this collagenous connective tissue during the final several hours prior to follicular rupture. The fibroblasts begin to project long cytoplasmic process from their central mass, and on their longitudinal plane these cells become as much as 100 μm in length (Fig. 2). The tips of these cytoplasmic processes often consist of unusual multivesicular structures that probably contain bioactive agents that contribute to the decomposition of the extracellular collagenous elements during the final stages of the ovulatory process. In addition to these cellular changes, the extracellular matrix of collagen and ground substance begins to dissociate during the hour preceding follicular rupture.

3. Theca Externa

As a follicle approaches the moment of rupture, the collagenous connective tissue of the theca externa undergoes changes similar to those of the overlying tunica albuginea (Fig. 2). The fibroblasts become much more elongated, their cytoplasmic processes exhibit the same type of multivesicular structures, and the extracellular matrix of this layer is less integrated. The fibroblasts of both the theca externa and tunica albuginea are transformed from quiescent, resting cells into active, proliferating fibroblasts. As these activated cells become motile and begin moving around within the local area of the follicle, they probably secrete proteolytic enzymes that soften the extracellular matrix and facilitate movement of the fibroblasts. In this weakened state, the tissue in the apical area of a follicle begins to separate under the force of a relatively low, but steady, intrafollicular pressure of 15–20 mm Hg. The result of this dissociation is a gradually thinning of the follicle wall at the site where rupture will eventually occur.

4. Theca Interna

The principal ovulatory changes in the theca interna occur in the extensive network of capillaries that is characteristic of this area of the follicle wall. Within 4–6 hr after initiation of the ovulatory process there is a measurable increase in ovarian blood flow and the follicles become hyperemic. Follicles become visibly redder as their capillaries dilate, and their blood content increases as much as fivefold. In addition, the permeability of the thecal capillaries increases significantly during ovulation. These marked changes in the vasculature result in occasional extravasation of blood and the formation of petechiae in the walls of some follicles. Also, there is an increase in the number of polymorphonuclear leukocytes in the patent blood vessels (Fig. 2). However, macrophages, or other derivatives of leukocytes, are rarely observed in the thecal tissues outside the vascular compartment prior to follicular rupture. Therefore, although some investigators believe that leukocytic cells may contribute the proteolytic enzymes that are thought to degrade the follicular connective tissue during ovulation, it is probably more likely that leukocytes begin to accumulate in ovulatory follicles in response to leukotactic agents that are generated by an acute inflammatory reaction that activates the thecal fibroblasts and initiates ovulatory decomposition of the follicular wall prior to infiltration of the area by leukocytes.

5. Stratum Granulosum

During the last hour preceding follicular rupture, the cells of the granulosa layer become less firmly attached to one another. As the apical area dissociates and the follicle wall becomes thinner, the innermost granulosa cells begin to slough from the wall and become dispersed in the follicular fluid (Fig. 2). Those cuboidal cells that remain attached to the membrana propria usually contain an increasing number of lipid droplets that were not present a few hours earlier. Thus, the granulosa cells become steroidogenically active during the hours preceding follicular rupture. Otherwise, there are negligible changes in the ultrastructure of the granulosa cells at the apex of an ovulatory follicle.

C. 1–5 min before Follicular Rupture

It has been possible to obtain electron micrographs of rabbit follicles only a few minutes before rupture (Fig. 3). Shortly before the follicle wall breaks, it balloons out to form a stigma at the apicalmost area where it will rupture. Only traces of the surface epithelium remain clinging to the disintegrated tunica albuginea and theca externa. Extravasated erythrocytes appear more frequently in the vicinity of the theca interna, and most of the stigmal region is void of capillaries. Essentially all the granulosa cells have sloughed into the follicular fluid or have retracted toward the base of the stigma as the membrana propria disintegrates and the follicle wall undergoes its final thinning before rupture. Rupture ultimately occurs at the apicalmost area of the follicle simply because this is, morphologically, the thinnest (i.e., the weakest) site along the ovarian surface.

III. BIOCHEMISTRY OF OVULATION

A. Historical Background

1. The 1960s

In the 1960s, it became apparent that ovarian follicles do not rupture as a consequence of any significant increase in intrafollicular pressure. A variety of evidence revealed that rupture is probably due, instead, to the action of proteolytic enzymes that decompose the collagenous connective tissue in the thecal layers of the follicle wall and the ovarian tunic. By the end of the 1960s, it was also apparent that ovarian steroid metabolism changes markedly in response to an ovulatory surge in gonadotropin. Progesterone synthesis increases within several hours after initiation of the ovulatory process, whereas ovarian estrogens and androgens decline in a reciprocal pattern.

2. The 1970s

In the early 1970s, it was discovered that there was a significant increase in ovarian prostanoid synthesis during ovulation. This information, together with considerable evidence that antiinflammatory agents such as indomethacin can inhibit ovulation, led to the general assumption that prostaglandins E_2 and $F_2\alpha$ are essential for ovulation, but other data raise questions about the role of ovarian prostanoids in the mechanism of ovulation. In this decade it became apparent that ovarian plasminogen activator also increases in response to most gonadotropic hormones. It has been hypothesized that this serine protease might contribute to the ovulatory process by digesting connective tissue components of the follicle wall or by activating a procollagenase. However, targeted deletion of the genes for several types of plasminogen activator has not yielded an anovulatory phenotype. Therefore, the precise role of plasminogen activator in ovulation remains unclear.

3. The 1980s

By the 1980s, more attention was being given to the fact that ovulatory follicles are hyperemic, and that such a vascular response is a cardinal sign of inflammation. In addition, it became evident that a wide variety of nonsteroidal antiinflammatory drugs can inhibit ovulation. This information, along with other supporting data, led to the hypothesis that an ovulatory dose of gonadotropin initiates an inflammatory-like response in mature ovarian follicles. Subsequently, it was demonstrated that ovarian kallikrein activity increases, and that kinin formation might contribute to the ovarian vascular changes during ovulation. As this decade ended, there were reports that cytokines, platelet-activating factor, growth factors, and metalloproteases might also influence the inflammatory response that occurs in ovarian follicles during ovulation.

B. Current Knowledge about the Biochemistry of Ovulation

To augment the previously discussed knowledge about the biochemical events of ovulation, several laboratories have characterized the hormonal regulation of genes for enzymes involved in the synthesis of steroids, eicosanoids, proteases, and other agents that have been implicated in the ovulatory process.

1. Steroid Metabolism

Several members of the cytochrome P450 family of enzymes are now known to be expressed in ovarian tissue in a pattern that is consistent with what is currently known about ovarian progestin and estrogen synthesis during ovulation. Transcription of the gene for cytochrome P450 side chain cleavage enzyme, which increases progesterone synthesis by increasing the rate of conversion of cholesterol to pregnenalone, is upregulated in ovarian follicles (mainly in the theca interna and stratum granulosum) several hours after the ovulatory process has been initiated by gonadotropins. Conversely, the gene for cytochrome P450 aromatase, which converts testosterone into 17-estradiol, is concomitantly down-regulated in a pattern parallel to the decline in ovarian estrogen synthesis at the time of ovulation.

2. Eicosanoid Metabolism

Two prostaglandin synthase genes have been identified in ovarian follicular tissues. Prostaglandin synthase-1 is constituitively expressed and has been localized to thecal cells and luteal cells. Prostaglandin synthase-2 is rapidly and transiently induced by the ovulatory surge in luteinizing hormone, and it is localized inclusively in the granulosa cells of those follicles destined to ovulate. The expression of these genes leads to enzymatic activity that causes 50- to 100-fold increases in prostaglandins E_2 and $F_2\alpha$ in follicular tissue during ovulation. In addition to these two eicosanoids, there is recent evidence that lipoxygenase enzymes cause marked increases in 12- and 15-hydroxyeicosatetranoic acids associated with bioactive agents such as the lipoxins.

3. Expression of Other Bioactive Factors

Recent work at the molecular level has revealed ovarian increases of mRNAs for nerve growth factor (NGF), oxytocin, tissue inhibitor of metalloproteinase (TIMP), kallikrein, and vascular endothelial growth factor/vascular permeability factor (VEG/PF) after the stimulation of follicles by gonadotropin. Regarding NGF, both the growth factor itself and the receptor for NGF are expressed by thecal fibroblasts of ovulatory follicles. Most of the evidence indicates that oxytocin mRNA is expressed in the granulosa layer, but the function of this neuropeptide in the ovary is uncertain. TIMP is also expressed in the granulosa layer, and this inhibitor may modulate ovarian proteolytic activity in the vicinity of the oocyte during ovulation. Ovarian kallikrein activity produces kinins that promote vasodilatation and contribute to the hyperemic reaction in ovulatory follicles. The increase in VEG/PF causes follicular capillaries to become more permeable and promotes angiogenesis during the luteinization of ovulatory follicles.

4. Detection of Novel Biochemicals in Ovulation by Differential Display

It is likely that innumerable other mediators of the ovulatory process remain to be elucidated. The new molecular protocol known as "differential display" is a valuable method that is now being used to isolate and identify mRNAs of genes that are uniquely expressed in the ovary during ovulation. This molecular technique, which was developed by P. Liang and A. B. Pardee at Harvard University in 1992, is based on the display of differentially expressed mRNA/cDNA by electrophoresis on an acrylamide gel following RT-PCR amplification of subpopulations of gene transcripts from different groups of experimental tissues. This method has been used recently to discover the unique expression during ovulation of genes for a carbonyl reductase with 20α-hydroxysteroid dehydrogenase activity, a long interspersed nucleotide element that is highly repeated in mammalian genomes, and a nerve growth factor-induced substance (NGFI-A). NGFI-A is usually expressed concomitantly with the protooncogene c-*fos* and the metalloproteinase stromelysin-1, and therefore it is quite likely that the transcription of genes for these factors is also upregulated during ovulation. Thus, in the future, the differential display procedure has the potential of elucidating many other biochemical agents that are involved in the ovulatory process.

IV. CURRENT HYPOTHESIS ON OVULATION

The previous background information on ovulation, along with the current literature on this topic, serve as the basis for the following "working hypothesis" on the mechanism of mammalian ovulation: The ovulatory surge in LH (or exogenous hCG) initiates acute changes in steroid and eicosanoid metabolism in the granulosa cells of mature follicles. The ensuing increases in local prostaglandins, lipoxins, kinins, platelet-activating factor, VEG/PF, and other vasoactive agents collectively cause substantial dilatation of the capillaries in the theca interna of the follicle wall. This significant change in the capillaries of ovulatory follicles results in a fourfold increase in the volume of the ovarian vascular compartment.

Concomitant with this hyperemic response, the permeability of the thecal capillaries increases to the extent that serum proteins are exuded into the interstitial spaces of the follicle in a manner characteristic of inflamed tissues. Since blood serum is well-known for its ability to activate fibroblasts, it can be predicted that the exuded serum stimulates the quiescent fibroblasts in the theca externa and tunica albuginea of an ovulatory follicle and causes them to transform into proliferating cells. The activated thecal fibroblasts begin secreting a metalloproteinase (perhaps stromolysin-1, which is regulated by NGF) that degrades the extracellular matrix of collagenous connective tissue in the follicle wall. Ultimately, the follicle wall loses its tensile strength and eventually ruptures under the force of a modest, but effective, intrafollicular pressure.

Acknowledgment

This work was supported in part by NIH Grant HD31634 and by NSF Grant 9870793.

See Also the Following Articles

Cytokines; Follicular Development, Control of; Luteinization; Theca Cells

Bibliography

Adashi, E. Y. (1990). The potential relevance of cytokines to ovarian physiology: The emerging role of resident ovarian cells of the white blood cell series. *Endocr. Rev.* 11, 454–464.

Curry, T. E., Jr., Mann, J. S., Estes, R. S., and Jones, P. B. (1990). Alpha 2-macroglobulin and tissue inhibitor of metalloproteinases: Collagenase inhibitors in human preovulatory ovaries. *Endocrinology* 127, 63–68.

Dissen, G. A., Hill, D. F., Costa, M. E., Dees, W. L., Lara, H. E., and Ojeda, S. R. (1996). A role for *TrkA* nerve growth factor receptors in mammalian ovulation. *Endocrinology* 137, 198–209.

Espey, L. L. (1980). Ovulation as an inflammatory reaction—A hypothesis. *Biol. Reprod.* 22, 73–106.

Espey, L. L. (1991). Ultrastructure of the ovulatory process. In *Ultrastructure of the Ovary* (G. Familiari, S. Makabe, and P. Motta, P., Eds.), pp. 143–159. Kluwer, Norwell.

Espey, L. L. (1992). A review of factors that could influence membrane potentials of follicular cells during mammalian ovulation. *Acta Endocrinol.* 126(Suppl. 2), 1–32.

Espey, L. L. (1994). Current status of the hypothesis that mammalian ovulation is comparable to an inflammatory reaction. *Biol. Reprod.* 50, 233–238.

Espey, L. L., and Lipner, H. (1994). Ovulation. In *The Physiology of Reproduction* (E. Knobil and J. D. Neill, Eds.), pp. 725–780. Raven Press, New York.

Hartman, C. G. (1932). Ovulation and the transport and viability of ova and sperm in the female genital tract. In *Sex and Internal Secretions* (E. Allen, Ed.), pp. 647–688. Williams & Wilkins, Baltimore.

Koos, R. D. (1995). Increased expression of vascular endothelial growth/permeability factor in the rat ovary following an ovulatory gonadotropin stimulus: Potential roles in follicle rupture. *Biol. Reprod.* 52, 1426–1435.

Liang, P., and Pardee, A. B. (1995). Recent advances in differential display. *Curr. Opin. Immunol.* 7, 274–280.

Richards, J. S. (1994). Hormonal control of gene expression in the ovary. *Endocr. Rev.* 15, 725–751.

Wathes, D. C., and Denning-Kendall, P. A. (1992). Control of synthesis and secretion of ovarian oxytocin in ruminants. *J. Reprod. Fertil.* 45(Suppl.), 39–52.

Ovulation and Oviposition, Insects

Marc J. Klowden
University of Idaho

I. Introduction
II. Control of Ovulation
III. Fertilization of the Egg
IV. Oviposition Site Location
V. Control of Oviposition

GLOSSARY

chorion The eggshell of the insect egg.
common oviduct An ectodermal invagination that connects the genital opening with the lateral oviducts.
lateral oviduct Branches from the ovary connecting to the common oviduct.
micropyle An opening in the chorion for the entrance of sperm.
myotropin A chemical agent that causes muscles to contract.
neurosecretory cells Modified nerve cells that secrete hormones into general circulation.
ovariole Individual tapering units that produce eggs and together comprise the ovary.
oviposition Passage of the egg from the genital tract to outside the body of the female.
ovulation Passage of the egg from the ovariole of the ovary into the oviduct.
spermatheca A sac in the female reproductive tract that stores sperm after mating and releases them while the egg passes by in the common oviduct.

Ovulation in insects involves the passage of mature eggs from the ovary of the female into the oviduct, after which the process of oviposition moves the eggs to the outside of the female where they are deposited.

I. INTRODUCTION

Once egg development has been completed in insects, the eggs must be ovulated, or moved out of the ovaries, and then oviposited, or deposited onto a site where the immatures can develop once they hatch. In some insects that simply drop their eggs when they have matured, the two processes of ovulation and oviposition are successive and appear to be part of a single mechanism. However, in some species there is a period of egg storage between ovulation and oviposition that underscores the differences between them. For example, some species of ovoviviparous cockroaches retain their eggs in a brood sac for a variable period of time after they are ovulated until the female finds a suitable site on which the eggs can be laid. Other viviparous insects, like the tsetse fly, retain the immatures after they hatch from the egg, provide them with food during the larval stage, and larviposit them just before they pupate. Thus, depending on the ecological habits of the insect, ovulation and oviposition may either be part of a continuum or exist as separate events that are not completed until environmental conditions are suitable for the survival of progeny. There is probably as much variation in oviposition habits as there are insect species, but the ones most studied are a few species of medical and agricultural importance.

During egg maturation, the oocyte is surrounded by a monolayer of follicle cells that is derived from somatic prefollicular cells in the germarium of the ovariole (Fig. 1A). These follicle cells are involved in the transport of substances from the hemolymph into the oocyte during vitellogenesis, and among their final acts is the secretion of the chorion, or eggshell, inwards toward the surface of the oocyte.

The chorion provides protection and waterproofing after the egg is oviposited. Following chorion formation, the follicle cells are no longer functional nor necessary and they degenerate. Without the follicle cells to restrain it, the oocyte is free to move within the oviduct and muscular contractions within the ovariole cause ovulation to occur (Fig. 1B). Ovulation is therefore a prerequisite for oviposition, which follows as a result of the contractions of the oviducts. In insects that show a concordance between blood feeding and egg development, such as mosquitoes, it has been possible to estimate the age of the insect by the number of ovariole dilatations that remain after ovulation and oviposition occur. Just prior to oviposition, the egg is fertilized by sperm stored in the spermathecae, and in the majority of species embryogenesis of the fertilized egg commences outside the mother after it is laid.

II. CONTROL OF OVULATION

Ovulation occurs when contractions of the walls of the ovarioles and lateral oviducts move the egg down toward the common oviduct. Activation of the muscles and the initiation of ovulation occurs in the few insects that have been studied as a result of endocrine signals from both the female and the male with which she mates.

In tsetse flies within the genus *Glossina*, ovulation and the first reproductive cycle are initiated when a least two conditions are met: A mature oocyte is present in the ovariole and the female has mated. A myotropic hormone synthesized in the medial neurosecretory cells of the brain is released from either the corpus cardiacum or the neuromuscular junctions in the reproductive tract once these conditions are satisfied and causes the muscles on the ovariole wall and the oviducts to contract. The release of the hormone does not occur until mating takes place. Females that are prevented from mating until after the egg chorion is formed will ovulate within 24 hr of copulation, and ablation of either the neurosecretory cells or the corpus cardiacum prevents the ovulation from occurring. Cyclic AMP appears to be involved as a second messenger in hormone action. In *Drosophila*, another dipteran, an ovulation-stimulating substance has been identified as a peptide with a molecular weight of 3990.

The female tsetse receives the information that mating has occurred from prolonged mechanical stimulation associated with copulation and also from substances transferred from the male accessory glands. Several short copulatory experiences that fail to transfer sperm can initiate ovulation as long as their total duration is as long as a single successful copulation. Also, a prolonged mating with a male whose accessory reproductive glands have been removed, and thus do not transfer accessory gland substances, can initiate ovulation. Even the insertion of a glass bead that is slightly smaller than a sperma-

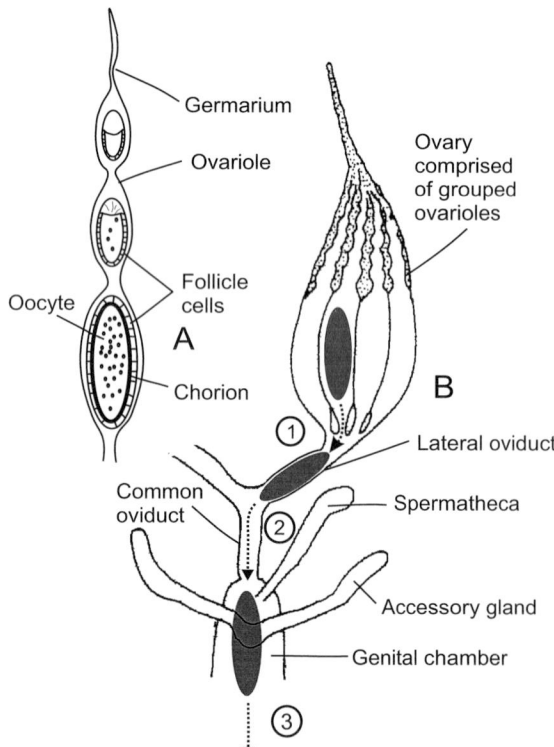

FIGURE 1 A typical insect ovariole (A) and female reproductive system (B). An ovary generally consists of several ovarioles that produce oocytes originating in the germarium (A). During vitellogenesis, the oocytes deposit yolk that is stored within their cytoplasm. They are initially surrounded by follicle cells that ultimately secrete the chorion inwards. Once the chorion is produced, the follicle cells degenerate. The egg then is ovulated (B, 1 and 2) and oviposited (B, 3) (modified from Snodgrass, 1935).

tophore into the vagina stimulates ovulation in unmated females, which is further evidence for the role of mechanical stimulation in informing the female she has mated. There is also some evidence that male accessory gland substances stimulate ovulation.

Similarly, in the blood-sucking bug *Rhodnius prolixus*, ovulation is initiated by a FMRFamide-like myotropin with a molecular weight of approximately 8500 that is released from 10 neurosecretory cells in the brain. As in the tsetse fly, this release occurs only when the female is mated and has a maturing egg within the ovariole. In *Rhodnius,* however, the signal that the female is mated is chemical instead of mechanical as it is in the tsetse. When filled with sperm and secretions of the male accessory gland, the spermathecae produce a chemical factor responsible for the physiological changes that mating induces. Observations of the oviducts through a plastic window inserted into the abdominal wall disclose two peaks of muscular activity in mated females but only the first occurs in the absence of mating. The signal that a maturing egg is also present comes from the steroid hormone 20-hydroxy-ecdysone, produced by the ovaries during oogenesis. Because *Rhodnius* does not show the delay between ovulation and oviposition, the processes appear to be more continuous.

III. FERTILIZATION OF THE EGG

Although largely covered by a waterproof chorion, a small area of the egg is modified for the receipt of sperm that fertilize the egg toward the end of ovulation. There is a variable number of these micropyles on the chorion of each egg; dipteran eggs only have a single micropyle, but those of grasshoppers may have as many as 40 circularly arranged on the posterior of the egg. The female stores sperm acquired during mating in a capsule-like spermathecae, where they remain until released as the egg passes down the common oviduct. The sperm enter the egg through the micropyle as it passes the spermathecal duct. In the locust, *Schistocerca*, sensory receptors in the oviduct are mechanically stimulated by the presence of the egg and transmit a message to the terminal abdominal ganglion, which then activates motor neurons that innervate the muscular walls of the spermathecal duct. Peristaltic contractions of the muscles then force the sperm out into the genital chamber where the egg is located. In hymenopterans, in which sex determination occurs by haplodiploidy, the process of sperm release is controlled more directly by the female. For example, in the honeybee, eggs that are fertilized develop into female workers, whereas those that are not fertilized develop into male drones. The workers construct cells for rearing male drones, queens, or other workers, and before she lays the egg in a particular cell, the queen measures its size with sensory receptors at the tip of the abdomen. If the cell is larger, she releases sperm from her spermatheca during ovulation and produces a female that may become a queen or worker.

IV. OVIPOSITION SITE LOCATION

The oviposition habits of insects are varied. Some insects, like dragonflies, are less selective in where they place their eggs and oviposit while in flight over habitats such as water that are suitable for their larvae to survive. Some terrestrial insects such as phasmids may simply drop their eggs to the ground while they are resting or feeding. In contrast, parasitic hymenopterans often lay their eggs not only on a particular host but sometimes even identify a particular organ within that host, such as on the ventral ganglion, that may allow the egg to elude the host's immunological defenses.

In phytophagous insects, host plant selection by the female plays a critical role in determining the survival of progeny. Because the larval stage generally has limited mobility and nutritional reserves, it is important that the female place her eggs where the larvae are most likely to find food and survive. The identification of a particular host plant is governed by sequences of phases that are mediated by the central nervous system. There are usually two major phases, the first involving the selection of the plant from a distance that involves visual and olfactory cues, followed by closer range selection where plant chemicals and physical factors such as texture, stem thickness, and color determine whether eggs are laid. The so-called secondary plant substances, which are minor products or metabolic intermediaries, are probably most important in determining host specificity. The quality of the oviposition site can affect the rate at which eggs are laid, either by modulating egg production or by causing the female to retain the eggs that have matured.

A circadian rhythmicity may determine when eggs are laid, often by controlling overall activity patterns that normally expose the female to oviposition site stimuli and her responses once the stimuli are perceived. The mosquito, *Aedes aegypti*, normally lays its eggs just prior to the scotophase, but these oviposition rhythms are abolished when the females are maintained under constant light or constant dark. Stick insects are defenseless and are active at night when most predators are least active, and their oviposition is also confined to nocturnal hours. The oviposition behavior of the milkweed bug, *Oncopeltus*, consists of a rhythm based on the ovarian cycle as well as the circadian rhythmicity of oviposition itself.

In order to avoid the overutilization of resources used for oviposition, some females mark them with chemicals to signal to other individuals that the site has already been used. Parasitic wasps may mark the host eggs in which they lay their own eggs, informing others of their species that the host egg is already occupied. The apple maggot fly, *Rhagoletis*, drags its ovipositor over fruit after laying its eggs as a sign to other females to prevent the competition for food and space that might jeopardize larval development.

In those insects that select specific sites to lay their eggs, the female often produces substances from her accessory gland that may bind them or protect them. The blood-sucking bug, *Rhodnius,* produces an adhesive from its accessory gland that causes the egg to stick to the oviposition substrate. The horse bot fly, *Gasterophilus*, attaches her eggs to the inner surface of the knees of horses, where they can be removed when the animal grooms itself. The higher temperature of the horse's tongue causes the eggs to hatch and larvae then burrow into it. In an amazing suggestion of foresight, the female fly, *Dermatobia*, captures another blood-sucking fly, such as a mosquito, onto which it glues its eggs. The egg hatches when the mosquito subsequently feeds on a mammalian host and the larvae burrow into its skin. Several cockroaches produce proteins from their glands that, upon oviposition, enclose the eggs in a molded ootheca, or egg case, for protection. Mantids produce a frothy secretion from their accessory glands that hardens to form a protective covering for eggs. The sting apparatus of many Hymenoptera consists of a modified ovipositor, and the accessory glands produce the venom.

V. CONTROL OF OVIPOSITION

Ovulated eggs move down the common oviduct by waves of peristalsis and out of the body through the ovipositor. The origin of the ovipositor differs in some insects. It may be derived from modified abdominal appendages, as in the elongated ovipositor of grasshoppers and hymenopterous parasites, or from the modified terminal segments of the abdomen, as in the telescoping ovipositor of the housefly. In many insects there may not be a specialized ovipositor present at all. The inside of the genital valves that form the ovipositor often contain scales that are backwardly directed to provide a unidirectional "ratchet" that moves eggs outward by the oscillatory movements of the valves and peristaltic contractions of the muscles that surround the oviduct. Mechanoreceptors and contact chemoreceptors are often located at the posterior tip to enable the female to evaluate the site before the eggs are laid. Several sequential oviposition behavior programs have been identified in the cricket, all requiring sensory feedback as the programs are performed. These include the examination of the surface for suitability, positioning the ovipositor relative to the surface, the penetration of the substrate, a short lift of the ovipositor, followed by a resting phase and finally egg deposition. In locusts, the superextension of the abdomen during oviposition occurs through the action of specialized intersegmental abdominal muscles that can elongate by up to 10 times their contracted length.

The peristaltic contractions of the oviduct that propel the eggs outward generally result from myotropins released from the central nervous system, but they also may be initiated by nervous pathways. The terminal abdominal ganglion innervates the oviducts and the removal of the ganglion in many insects prevents oviposition. In some insects, decapitation causes the immediate oviposition of mature eggs, which may occur from the removal of a nervous inhibition of oviposition by the brain. In other insects, decapitation inhibits oviposition, suggesting a positive control. The extent of the involvement of the nervous system in oviposition appears to be related to the degree to which the female searches for a proper oviposition site, with a greater degree of nervous participation in insects that choose specific places to lay their eggs.

A control system may be in place to prevent the inopportune laying of eggs. For example, locusts prefer oviposition sites of moist uncompacted soil so the eggs can be buried and not exposed to desiccation or predation. Until these sites are located, the ovulated eggs are stored in the lateral oviducts and are prevented from being expelled. Two coordinated neural networks, or central pattern generators, are involved in this oviposition. One, the digging central pattern generator, coordinates the opening and closing of the ovipositor valves that are necessary for digging the oviposition hole. This motor pattern generator is located in the eighth abdominal ganglion and is initiated by the release of neural inhibition from the brain and thoracic ganglia as well as the sensory input from sensilla at the tips of the ovipositor valves that maintain motor output. Proctolin, a neuropeptide that enhances neuromuscular transmission and muscle contraction, binds to receptors on the oviduct and appears to be involved in the expression of the motor pattern. The second central pattern generator is located in the seventh abdominal ganglion, but its activity inhibits oviposition by blocking the passage of eggs. This second motor pattern generator shows activity just before and during the digging of the oviposition hole and when egg-laying is interrupted and is activated by the brain and thoracic ganglia to prevent oviposition from occurring when it would be inappropriate. Similarly, in the stick insect, *Carausius*, which oviposits only at night, the movement of eggs from the lateral into the common oviduct is prevented during the day when the muscles of the common oviduct contract from stimulation by nerves originating in the seventh abdominal ganglion and block the progression of the egg into the oviduct. Another stick insect, *Clitumnus*, restricts its egg-laying by producing a hormone from the thoracic and abdominal ganglia that acts on the genital tract and stimulates this nocturnal oviposition.

The female tsetse fly incubates a larva within its uterus, where it is fed from specialized milk glands. During this larval gestation, the female releases a parturition-inhibiting hormone that maintains the larva within the uterus. The subsequent growth of the larva stimulates maternal stretch receptors, which terminates the parturition-inhibiting hormone and initiates the production of parturition-stimulating hormone from neurosecretory axons in the uterus that causes the muscles in the uterine wall to contract. Just prior to parturition, the overall activity patterns of the female increase, presumably to increase her chance of locating a suitable site to deposit the larva.

In many insects, oviposition will only take place after mating occurs. The ultimate result of this mechanism may be to increase the period of egg retention by the unmated female and thus increase her chances of encountering a male before laying the unfertilized, nonviable eggs that may have already been ovulated. The effect of males on females can be mediated in several ways. The male cricket transfers a spermatophore containing sperm, fatty acids, and the enzyme prostaglandin synthetase, which catalyzes the formation of prostaglandins from precursors already present in the female. The prostaglandins in the female stimulate oviposition by acting on the genital chamber. In other grasshoppers and many dipterans, products of the male accessory gland that are transferred during mating stimulate oviposition by the female. The protein from the male accessory gland that stimulates oviposition in *Drosophila* has the structural features of a prohormone and contains a region of amino acid sequence similar to that of an egg-laying hormone identified from the snail, *Aplysia*. In the blood-sucking bug *Rhodnius*, the contributions from the male stimulate the female spermathecae to produce a factor that causes the release of a myotropin from neurosecretory cells in the brain that stimulates both ovulation and oviposition. The presence of sperm in females of the ovoviviparous cockroach, *Pycnoscelus*, is detected by stretch receptors in the spermathecae that stimulate oviposition and ootheca formation.

See Also the Following Articles

DROSOPHILA; EGG COVERINGS, INSECTS; INSECT ACCESSORY GLANDS; LOCUSTS; OVIPOSITION IN MOLLUSCS; RHODNIUS PROLIXUS; TSETSE FLIES

Bibliography

Belanger, J. H., and Orchard, I. (1993). The locust ovipositor opener muscle: Properties of the neuromuscular system. *J. Exp. Biol.* **174,** 321–342.

Chapman, R. F. (1982). *The Insects: Structure and Function.* Hodder & Stoughton, London.

Davey, K. G. (1985). The female reproductive tract. *Comprehensive Insect Physiol. Biochem. Pharmacol.* **1,** 15–36.

Engelmann, F. (1970). *The Physiology of Insect Reproduction.* Pergamon, New York.

Hinton, H. E. (1994). *Biology of Insect Eggs*, Vol. 1. Pergamon, Oxford, UK.

Renwick, J. A. A., and Chew, F. S. (1994). Oviposition behavior in Lepidoptera. *Annu. Rev. Entomol.* **39,** 377–400.

Sugawara, T., and Loher, W. (1986). Oviposition behaviour of the cricket *Teleogryllus commodus*: Observation of external and internal events. *J. Insect Physiol.* **32,** 179–188.

Oxytocics

Ramkrishna Mehendale and Laird Wilson, Jr.

University of Illinois at Chicago

I. Oxytocin
II. Prostaglandins
III. Other Oxytocics
IV. Oxytocics and Parturition

GLOSSARY

eicosanoids Generic term for a series of 20-carbon unsaturated fatty acids. An example is eicosatetraenoic acid, which is the chemical name for arachidonic acid, the precursor for prostaglandins, thromboxanes, leukotrienes, and epoxides containing two double bonds in the aliphatic side chains.

G-protein Guanosine triphosphate-binding proteins that associate with hormonal receptors and transduce the stimulus in the cell. There are many different types of G-proteins including G_s and G_i, associated with stimulation and inhibition of adenyl cyclase, respectively, and G_q, associated with phospholipase C activity. The G-proteins are heterotrimers consisting of α, β, and γ subunits.

neurohypophysis The posterior pituitary gland or *pars nervosa* containing axons from hypothalamic magnocellular neurons which secrete oxytocin and vasopressin.

NSAID Commonly used acronym which stands for nonsteroidal antiinflammatory drug. Examples include aspirin, indomethacin, naproxen, and ibuprofen.

oxytocin A nine-amino acid peptide hormone secreted by the neurohypophysis that stimulates milk letdown and uterine contractions.

prostaglandins Twenty-carbon unsaturated fatty acids with a unique cyclopentane ring; produced in most, if not all, cells and have autocrine, paracrine, and sometimes endocrine activity.

Oxytocics are by definition compounds that induce uterine contractions during pregnancy. The best known oxytocics, and the ones that have important physiologic and/or pharmacologic roles during pregnancy in primates, are oxytocin, prostaglandins, and ergot alkaloids. The oxytocic used most frequently for clinical induction of labor is oxytocin, with prostaglandins as a second choice. For postpartum hemorrhage oxytocin, prostaglandins and ergot alkaloids are potential uterotonic treatments. Besides these medically relevant oxytocics, many other compounds have the ability to induce uterine contractions during pregnancy either directly or indirectly, such as cytokines, endothelins, estrogens, and platelet-activating factors.

I. OXYTOCIN

In 1906 Sir Henry Dale reported that an extract of the neuohypophysis had oxytocic effects on mammalian uterine tissue. The application of such an extract to induce labor was reported as early as 1911. The name given to this substance was oxytocin, derived from the Greek word meaning "swift-birth." Oxytocin (OT) is a neuropeptide hormone produced by magnocellular cells in the hypothalamus, transported to the posterior pituitary via axons, and stored and secreted by the posterior pituitary. The one unequivocal physiologic function of OT in mammals is the stimulation of milk letdown from the mammary glands. In addition, in primates, and perhaps certain nonprimates, OT is important in inducing uterine contraction during pregnancy. Recently, OT has been discovered to have other roles in the central nervous system and steroidogenic tissues.

A. Structure and Metabolism

1. Structure

In the 1950s Du Vigneaud and co-workers established the structure and synthesized OT. OT was the first peptide to be synthesized. OT is a nine-amino acid peptide consisting of a six-member ring formed by disulfide bonds between positions 1 and 6 and a three-amino acid tail. It is structurally similar to vasopressin (VP), differing by only two amino acids, and thought to evolve from a common ancestral gene (Fig. 1).

2. Synthesis

Genes for both OT and VP reside on chromosome 20 in opposite transcriptional orientation separated by intergenic sequences. OT is initially synthesized as a 125-amino acid preprohormone in the magnocellular neurons of the supraoptic (SON) and paraventricular (PVN) nuclei of the hypothalamus. The prohormone is first cleaved by a serine endoprotease to

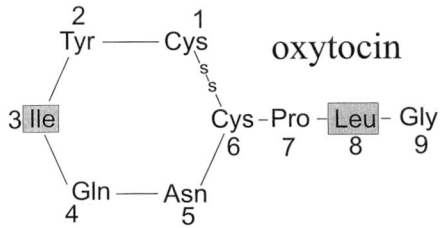
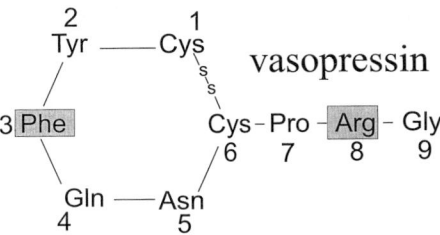

FIGURE 1 Structure of oxytocin and vasopressin. Oxytocin differs from vasopressin at two amino acids: Ile$_3$Phe$_3$ and Leu$_8$Arg$_8$. Both peptides have a disulfide bond between the first and sixth Cys residue.

form oxytocin-Gly-Lys-Arg. This is further cleaved by convertases and carboxypeptidase E in the secretory granules to oxytocin-Gly which is -amidated to oxytocin by peptidylglycine -amidating monooxygenase (PAM). The rate-limiting step is the conversion of oxytocin–glycine to oxytocin by the enzyme PAM. The cleavage of the precursor to oxytocin occurs in the secretory vesicle during axonal transport to the neurohypophysis.

3. Catabolism

Oxytocin is rapidly metabolized in the liver, kidney, and placenta. In addition, the plasma from pregnant women metabolizes OT to biologically inactive peptides due to the action of an enzyme produced by the placenta and released into the blood, cystine aminopeptidase (oxytocinase or CAP). Plasma CAP appears to be unique to humans, being absent in rat, sheep, and nonhuman primates (baboon and rhesus monkey) plasma. CAP cleaves the Cys^1-Tyr^2 bond and then cleaves Tyr. Other enzymes known to metabolize oxytocin are postproline endopeptidase, which cleaves the Pro^7-Leu^8 bond and produces the metabolite oxytocin-(1-7), chymotrypsin (Leu^8-Gly^9), and disulfide-cleaving enzyme (Cys^1-Cys^6).

4. Pharmacokinetics

The metabolic clearance rate (MCR) of oxytocin varies between 15 and 25 ml/min in men and pregnant and nonpregnant women. Although there are differences in opinion in the literature on whether the MCR for OT differs between pregnant and nonpregnant women, recent data suggest it is greater during pregnancy. This seems reasonable since two important sources of OT metabolism, CAP in plasma and the placenta, are only present during pregnancy. The MCR should reflect the sum total of all sources for clearance. Unlike the MCR, the volume of distribution and the half-lives of dispersion (HLα) and clearance (HLβ) do not differ between pregnant and nonpregnant states. The HLα and HLβ are about 1 and 8–10 min, respectively.

B. Mode of Action

Oxytocin is known to act through specific oxytocin receptors (Fig. 2). These are G-protein-coupled receptors with seven transmembrane domains. The major second messenger system that mediates uterine contraction is the phospholipase C (PLC), inositol triphosphate pathway. Oxytocin receptor binding results in activation of G-proteins, in particular the G_q and G_i families. These activate phospholipase C, which results in the production of inositol triphosphate (IP_3) and diacylglycerol. IP_3 stimulates the release of calcium from intracellular stores. PLC has three isomers (PLCβ, PLCγ, and PLCδ), and each of these is subdivided by subscripts 1, 2, and 3. OT action through G_q is thought to be mediated via PLC$\beta_{1\text{ and }3}$, whereas G_i activity is mediated via PLC$\beta_{2\text{ and }3}$.

Oxytocin receptors have been identified in other tissues besides the myometrium, including the endometrium/decidua and amnion. Although OT can stimulate prostaglandin (PG) production in these tissues, the physiologic function of OT is only speculative.

C. Physiology

Release of oxytocin into the circulation requires the stimulation of the magnocellular neurons through neuronal reflexes. The 5' upstream promoter region of the oxytocin gene contains response elements for estrogen, cAMP, and activating protein 2. Estrogen promotes OT synthesis and lowers the threshold for its release. Oxytocin gene (mRNA) expression has also been found in the uterus suggesting an autocrine/paracrine role for the hormone in parturition. However, not much OT peptide is present but there are rather high concentrations of OT precursors (i.e., OT-Gly, OT-Gly-Lys-Arg, etc.). The OT precursors are referred collectively as OTX since they are extended forms of OT. In addition to the hypothalamus and uterus, OT gene expression is also found in steroid-producing tissues such as the ovaries, testes, adrenal glands, and placenta.

1. Lactation

Oxytocin has long been known to be involved in lactation. Suckling induces OT release from the pituitary and results in milk letdown in mammals. The stimulus is thought to be more mechanical than thermal and is sensed by free nerve endings in the nipple. The impulses are carried to the PVN and SON of the hypothalamus. Stimulation of the magnocellular neurons in these nuclei leads to OT secretion from the posterior pituitary. The secretion of OT

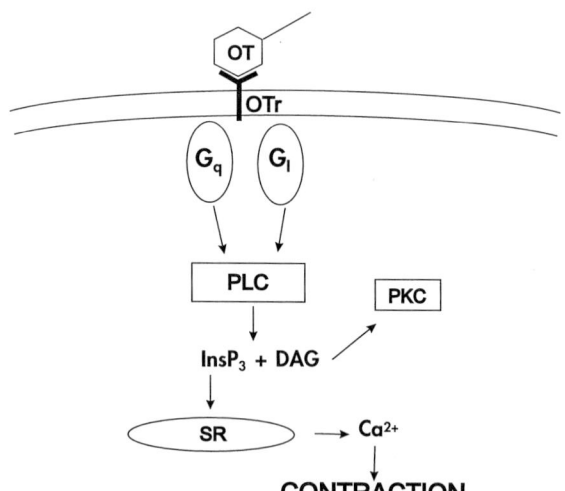

FIGURE 2 Oxytocin (OT) second messenger systems. OT induces uterine contraction by increasing intracellular Ca^{2+} levels via its action on PLC. OTr, oxytocin receptor; PLC, phospholipase C; PKC, protein kinase C; InsP$_3$, inositol triphosphate; DAG, diacylglycerol; SR, sarcoplasm reticulum; G_q/G_i, G-proteins.

from the SON and PVN is also under the control of higher brain centers such as the forebrain and limbic system, which afford emotional and cerebral input in the control of OT release during the suckling reflex.

The alveoli of the mammary gland are surrounded by a network of myoepithelial cells. Both the cell number and myofibrillar content increase during pregnancy and they attain maturity toward the end of pregnancy and the onset of lactation. OT receptors are expressed by myoepithelial cells during the lactational period and they disappear during weaning. OT binding results in the contraction of these cells and milk ejection.

2. Parturition

Oxytocin is the most potent known parturient. The action of oxytocin is mediated by binding to specific oxytocin receptors (OTR) present in the uterus and other target tissues. The binding of oxytocin to OTR ultimately causes an increase in intracellular calcium resulting in contraction of myometrial cells from the uterus and myoepithelial cells of the mammary gland. There has been a controversy regarding the role of oxytocin in the initiation and progress of labor, at least in humans. The controversy is based on the diversity of reports as to whether plasma OT levels increase at the time of labor. As a consequence of this controversy, the interest in OT diminished until Soloff and co-workers reported that the OT receptor number increases in the myometrium at the time of labor and suggested, therefore, that plasma OT levels do not have to increase, but rather the uterus could simply become more sensitive to the steady-state plasma levels of OT. In addition, as indicated previously, local synthesis of OT might obviate the need for elevated systemic OT. The ability of OT antagonists, which block the action of OT at the receptor, to inhibit nocturnal uterine contractions and contractions of early labor adds support to the role of OT in parturition in primates. In nonprimate species the role of OT in parturition is less clear. For example, OT knockout mice (lacking the OT gene) appear to have normal deliveries but they do not lactate. Thus, at least in the mouse, OT would not appear to be important in the parturition process.

Clinically, the primary drug used to induce labor in women is synthetic OT. OT is infused iv beginning at about 1 or 2 μU/min (2 pg OT = 1 μU) and the dose is increased every 20–30 min until effective uterine contractions are achieved. The uterus is more sensitive to OT during the last trimester of pregnancy due to the increase in the number of myometrial OT receptors.

3. Other Effects

Central and behavioral effects of OT have been implicated in a multitude of local effects in the brain including but not limited to release of prolactin, inhibition of ACTH secretion, autonomic effects such as tachycardia, and increased blood pressure. OT is also known to be involved in pair bonding in prairie voles, maternal behavior in sheep and mice, and nest defense.

OT gene expression (mRNA) is also found in corpora lutea of the ovary, testes, and adrenal gland. The function of OT in these tissues is controversial and seems to vary from species to species. For example, in sheep luteal OT appears to be important in regression of the corpus luteum but in humans this is less clear.

Finally, OT is the drug of choice for preventing and treating postpartum hemorrhage. This is discussed in more detail in the following section.

II. PROSTAGLANDINS

Prostaglandins were discovered in the early 1930s as substances in human seminal plasma that induced, and sometimes relaxed, uterine muscle activity *in vitro*. Von Euler named the substances prostaglandins believing that they came from the prostate gland. However, it was later determined that the seminal vesicles was the major source of PGs in human seminal plasma, thus prostaglandins is somewhat of a misnomer. In 1957 Bergstrom and Sjovall reported the first isolation of PGs and subsequently determined their structures. Beginning in the 1960s Sammuelsson and co-workers began describing the PG metabolic pathways. In 1971 Vane reported that aspirin inhibited PG production and later showed that most nonsteriodal antiinflammatory drugs (NSAIDs) acted via PG inhibition. Bergstrom, Sammuelsson, and Vane were awarded the Nobel prize in 1982 for their work with PGs. Eicosanoids metabolites and PGs are ubiquitous, and one type or the other is found in most, if not all cells in the body. They usually act as autocrine or paracrine factors although they can behave as endocrine hormones (i.e., corpus luteum regression in some nonprimate species).

A. Structure and Metabolism

1. Structure

Prostaglandins are 20-carbon fatty acids with the unique feature of a cyclopentane ring from carbons 8 to 12, an α-carboxylic acid, and two aliphatic side chains. The theoretical parent structure for PGs is prostanoic acid (Fig. 3). With the discovery of additional products derived from presursors of PGs whose parent structure was not prostanoic acid, such as thromboxane A_2 (TXA_2), a more generic name was required to encompass these other compounds; thus the name prostanoid. Prostanoids consist of PGs and TXA_2. There are many different type of PGs, depending on the number of double bonds in the side chains and functional groups in the cyclopentane ring (Fig. 3). The numerical subscript 1, 2, and 3 refers to the number of double bonds in the aliphatic side chains. The alphabetical nomenclature began when seminal plasma in a phosphate buffer was extracted with ether. The PGs in the ether phase were the E PGs, whereas those remaining in the phosphate buffer were the F PGs (from the Swedish word fosfat). When the E PGs were placed in an *acid* with a pH < 3.0 the A PGs were formed and when placed in a *base* with a pH > 8 the B PGs formed. Thus, the alphabetical nomenclature is related to the different types of functional groups in the cyclopentane ring. The PGF and PGE with the 1, 2, and 3 subscripts are referred to as the primary PGs. Since those early studies many different PGs have been discovered as

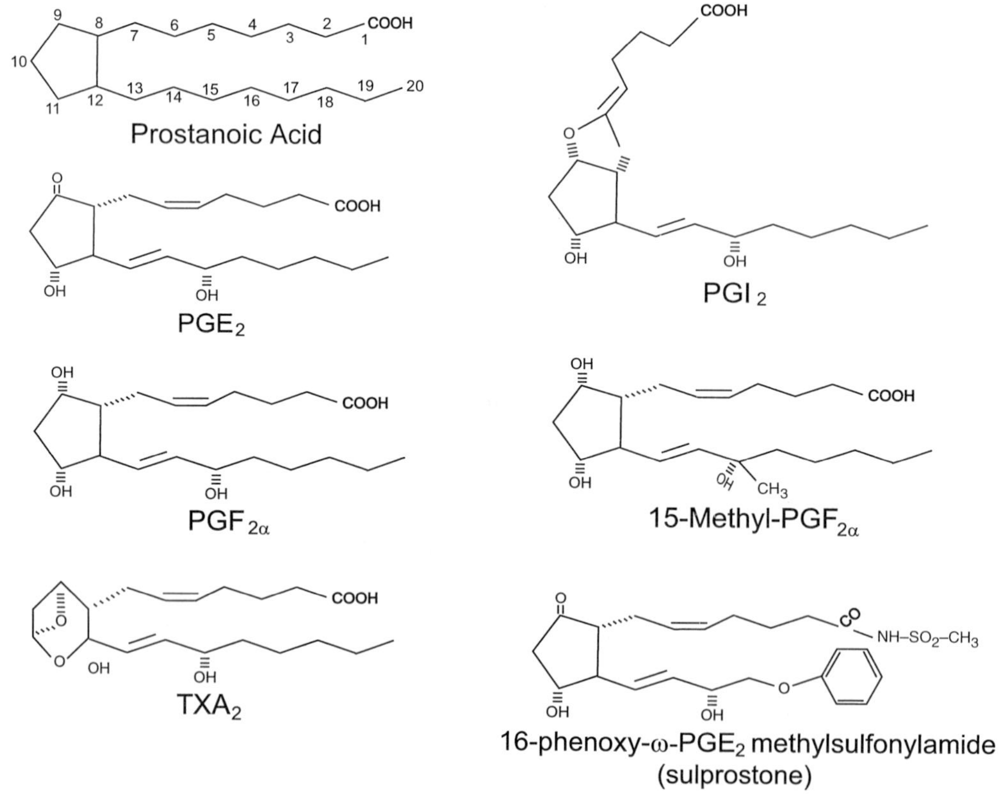

FIGURE 3 The chemical structures of the hypothetical PG parent compounds, prostanoic acid, several commonly referenced PGs ($PGF_{2\alpha}$, PGE_2, TXA_2, and PGI_2), and examples of two PG analogs (sulprostone and 15-methyl-$PGF_{2\alpha}$) produced by drug companies. The analogs are 10–100× more active than the parent compounds mainly due to reduced metabolic clearance rates.

well as new pathways of eicosanoid metabolism. The predominant prostanoids in animal tissues have two double bonds present and thus the subscript 2 (i.e., PGE_2 and $PGF_{2\alpha}$). The precursor for PG with two double bonds is arachidonic acid (AA), also referred to as eicosatetraenoic acid (ETE). AA is also a precursor for leukotrienes (LTA_4, LTB_4, and LTC_4) and 5-, 12-, and 15-hydroxy-ETE (HETEs), epoxides, and TXA_2.

2. Synthesis

The synthetic pathway for PGs is shown in Fig. 4.

i. Phospholipases A_2 and C Fatty acids are stored in the body esterified to triglycerides, cholesterol, and phospholipids, but the highest concentration is in phospholipids. Phospholipids (PLs) are present in cell membranes and fatty acids, such as AA, with multiple double bonds always stored in the middle position (sn-2 position) of the PL. The enzyme that releases fatty acids in the sn-2 position of the three-carbon glyerol backbone of PL is PLA_2. Thus, many studies have focused on the regulation of this enzyme since the release of AA is thought to be a rate-limiting step in PG production. Recent studies have also implicated a role for PLC, which, in conjunction with di- and monoglycerases, will release AA from PLs.

During the past decade it has been determined that there are several types of PLA_2 referred to as cytosolic ($cPLA_2$) and secretory ($sPLA_2$). The $sPLA_2$ is further divided into group I (associated with digestive function) and group II (secreted or membrane bound and implicated in inflammatory responses). The PLA_2's relevant to parturition appear to be $sPLA_2$-II and $cPLA_2$. The latter is particularly interesting because it is thought to be a cellular mediator of hormonal induction and shows preference for arachidonic acid. However, at least as relates to uterine contractions and regulation of PG production, the relative importance of $cPLA_2$ vs $sPLA_2$-II is not clear.

ii. Prostaglandin Endoperoxide H Synthase The conversion of AA to PGs occurs via an enzyme originally referred to as cycloxygenase since it was responsible for the formation of the cyclopentane ring from the linear AA. It was later determined that ephemeral intermediates existed (PGG_2 and PGH_2) and that a peroxidase enzyme was involved. It was debated whether there was one protein with two activities or actually two enzymes. It was finally determined there was one bifunctional protein with two activities. The nomenclature for the enzyme was then changed to take into account these observations. The correct nomenclature is now prostaglandin endoperoxide H synthase (PGHS). The rate-limiting step in PG synthesis has often been stated as the release of the

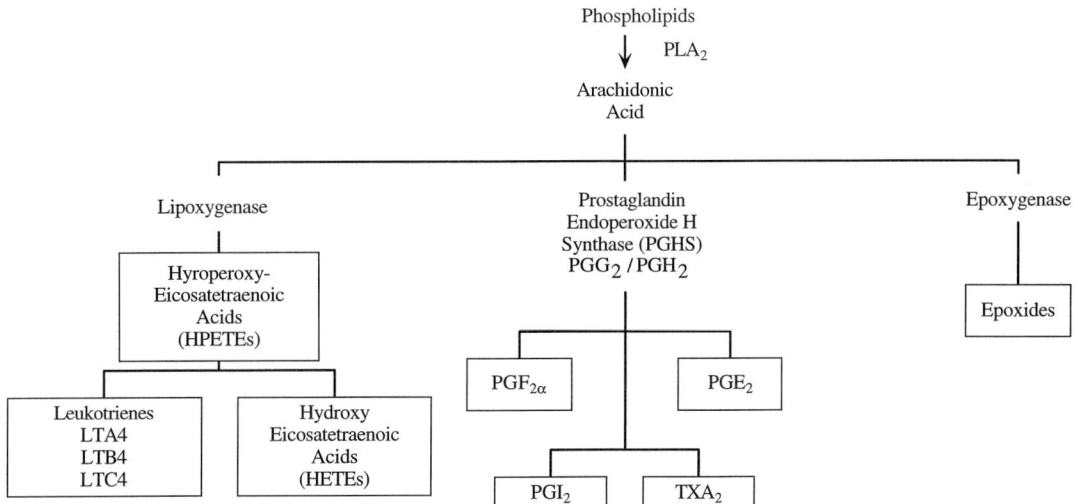

FIGURE 4 Metabolic pathways of arachidonic acid metabolism. The prostaglandin endoperoxide H synthase is the pathway for PG and thromboxane synthesis. It is at this enzyme that NSAIDs such as aspirin and indomethacin block PG production.

precusor (e.g., arachidonic acid) by PLA_2. However, the conversion of AA to PGG_2 and PGH_2 by PGHS is also a rate-limiting step. Recently it was shown that there are two isoenzymes referred to as PGHS-1 and PGHS-2 (also appears in the literature as COX-1 and COX-2). PGHS-1 is thought to be constitutively expressed in most mammalian cells and maintains PG production for stable physiologic conditions. PGHS-2, which shares about 60% homology with PGHS-1, is induced by proinflammatory agents and thus might be hormonally regulated.

The effects of aspirin and NSAIDs are via inhibition of cycloxygenase activity; aspirin actually inactivates the enzyme by acetylation, whereas compounds such as indomethacin are reversible inhibitors. With the identification of PGHS-1 and PGHS-2, pharmaceutical companies have developed specific inhibitors that discriminate between these two enzymes. For example, nimesulide is a preferential inhibitor of PGHS-2, whereas indomethacin and ibuprofen are relatively preferential PGHS-1 inhibitors.

3. Catabolism

Most prostanoids are catabolized near or within the cells that produce them. Two enzymes are involved: PG dehydrogenase, which reduces the 15-hydroxy group to 15-keto, and Δ^{13}-reductase, which reduces the double bond in the 13-14 position. Once the 15-keto group is formed the PG is biologically inactive. The metabolism of $PGF_{2\alpha}$ by these enzymes results in a metabolite referred to as 13-14-dihyro, 15-keto $PGF_{2\alpha}$. The metabolites are released into the blood where they travel to the liver and undergo β and ω oxidation forming 14- or 16-carbon compounds that are released into the urine. If biologically active PGs are released into the blood they are metabolized by enzymes in the liver, kidney, and particularly the lung, which has high concentrations of the two catabolic enzymes. After one passage of $PGF_{2\alpha}$ or PGE_2 through the lung, over 90% of the PG is metabolized. This is one reason most PGs have very short half-lives in blood (<1 min). The one exception appears to be PGI_2, which is not readily metabolized by the lungs. In addition to the lung, the placenta also has very high concentrations of PG dehydrogenase. Thus, it is very unlikely the biologically active PGs can cross the placenta and affect the fetus.

C. Mechanism of Action

The prostanoid receptors are classified into five basic types based on their sensitivity to the five primary prostanoids: FP, TP, EP, IP, and DP sensitive to $PGF_{2\alpha}$, TXA_2, PGE_2, PGI_2, and PGD_2, respectively. In addition, four receptors have been identified for EP (EP_1–EP_4). Unique G-protein-coupled receptors have been cloned for each of these receptors. Finally, splice variant isoforms have been cloned for EP_3, TP, and FP. According to smooth muscle assay systems the receptors have been divided into two branches which coincide with relaxant and excitatory receptors. The former consists of EP_2, EP_4, IP, and DP, whereas the latter consists of EP_1, EP_3, TP, and FP. The homology of receptor structure within these groups is about 30–40%.

The receptors can be grouped into three categories based on the types of signal transduction: (i) adenylcyclase stimulation (EP_2, EP_4, IP, and DP), (ii) Ca^{2+} mobilization (EP_1, EP_3, and TP), and (iii) inhibition of adenylcyclase (EP_3). Therefore, the relaxant branch is coupled to adenylcyclase, whereas the stimulant branch is coupled to PI hydrolysis and Ca^{2+} increase or inhibition of adenylcyclase. These effects are consistent with the observed biologic actions of the different PGs since stimulation of AC relaxes smooth muscle, whereas increased intracellular Ca^{2+} induces muscle contraction.

The sequence of the receptors suggests they are all members of the G-protein-linked receptor family having seven membrane spanning domains. The G-protein linked to the EP_1 receptor is apparently a member of the pertussis toxin-sensitive G_q family, which acts via the phosphoinositol–phospholipase C receptor system and enhances intracellular PKC activity and intracellular calcium (iCa) via inositol triphosphate. Therefore, one would anticipate that activity through this receptor would stimulate smooth muscle contraction by elevating iCa levels. The EP_2 and EP_4 receptors are linked to the G_s family, which acts via adenylcyclase to elevate the levels of intracellular cAMP and PKA activity. Thus, one would expect activity through these receptors to relax smooth muscle via elevated cAMP, which induces sequestration of iCa. Activity through the EP_3 signal transduction receptor is by G-proteins, which lowers the level of cellular cAMP and thus stimulates smooth

TABLE 1
Comparison of Prostanoid Receptor Intracellular
Mediators and Smooth Muscle Bioactivity

Receptor	Intracellular mediator	G-protein	Smooth muscle activity
FP	PLC	G_q	Stimulates
EP_1	Ca^{2+} channel	?	Stimulates
EP_2	Adenyl cyclase	G_s	Inhibits
EP_3	Adenyl cyclase	G_i	Stimulates
EP_4	Adenyl cyclase	G_s	Inhibits
TP	PLC	G_q	Stimulates
IP	Adenyl cyclase, PLC	G_s, G_q	Inhibits
DP	Adenyl cyclase	G_s	Inhibits

Note. FP, $PGF_{2\alpha}$; EP_{1-4}, PGE_2; TP, TXA_2; IP, PGI_2; DP, PGD_2.

muscle contraction. These data are summarized in Table 1.

D. Physiological and Pharmacological Effects

PGs and PG inhibitors have important applications in treating medical conditions in human pregnancy. These include (i) induction of labor, (ii) induction of abortion, (iii) preventing preterm labor with PG synthesis inhibitors, and (iv) suppression of postpartum hemorrhage.

1. Induction of Labor

PGs can be used to induce parturition. They have been approved for this use by the FDA in many countries, but not in the United States. The clinical testing for the use of PGs to induce labor was completed in the 1970s. The studies indicated they were efficacious and easy to use since they were taken orally. However, the FDA in the United States did not want to place on the market a drug that would make it easier to perform induction of labor, which the FDA believed was already being performed in excess. In addition, unlike in other countries, in the United States induction of labor is only permitted for medical reasons. Thus, approval of the drug was denied. PGs appear to have a physiologic role in the process of parturition. The evidence that supports this is threefold. First, levels of PGs increase in uterine tissue, amniotic fluid, and blood at the time of labor. Second, inhibitors of PG production, such as indomethacin, or antibodies to PGs can delay labor. Third, infusion of PGs such as PGE_2 and $PGF_{2\alpha}$ can induce labor. Thus, the application of PGs for labor induction was reasonable since they physiologically played a role in this process. The potency of PGE_2 is about 10 times greater than that of $PGF_{2\alpha}$ for inducing uterine contractions during pregnancy. One of the concerns with PGs is that the dose–response range is rather narrow and one has to be careful not to hyperstimulate the uterus.

In addition to the uterotonic effects, PGs also ripen the cervix. To induce labor or abortion the cervix must be ripened for the uterine contractions to effectively deliver its contents. This is an important advantage PGs have over other oxytocics, such as oxytocin. Currently in the United States, PGs are used in conjunction with OT to induce labor in individuals with an unripened cervix. Although PGs are not approved by the FDA for inducing labor, they are approved for inducing cervical ripening. The protocol for ripening the cervix is to place PGE_2 in a gel in the cervical os on the evening before induction of labor and the next morning begin the induction with OT. About 50% of the women given PGE_2 intracervically to ripen the cervix go into labor spontaneously and do not require OT. Thus, indirectly PGs are used to induce labor in the United States.

2. Abortion

PGs have also been used to induce abortions, particularly in the second trimester because of their potent uterotonic effects. However, since PGs are not toxic to the fetus their usage to induce abortion has diminished significantly due to the ethical, medical, and legal issues involved in delivering a live fetus. For this reason, hypertonic saline, which is toxic to the fetus, is the method most commonly used to induce an abortion after 16 weeks of pregnancy. However, PGs are used for inducing an abortion in the case of a fetal demise, a time when the mother does not handle hypertonic saline very well.

Abortions in early pregnancy can be induced with the antiprogestational pill RU 486 (mifepristone). Progesterone is required to maintain uterine quiescence and withdrawal of progesterone action induces dramatic uterine contractions within 24 hr. However, these contractions are often not effective; thus, PG analogs (misoprostol) are used in combination with mifepristone to abort the fetus.

3. Preventing Preterm Labor with PG Synthesis Inhibitors

PGs appear to have an important physiologic role in term and preterm labor (PTL). PTL is the number one cause of fetal mortality and morbidity in developed countries. There is currently no tocolytic agent that has proven to be efficacious for more than 48 hr. The most commonly used tocolytics are β-sympathomimetics and magnesium sulfate, with indomethacin as a third choice. However, a recent review concluded that the only effective agent appeared to be indomethacin. The problem with indomethacin, and the reason it is not used more frequently, is the potential side effects on the fetus. The major side effects are premature closure of ductus arteriosus (PGs have an important role in maintaining a patent ductus in the fetus), oligohydramnios, intraventicular hemorrahge, and necrotizing enterocolitis. However, the evidence in humans suggest that when given before 34 weeks of pregnancy and for less than 72 hr, the effects on the ductus arterious and oligohydramnios are transient and thus not a major problem.

4. Postpartum Hemorrhage

Following vaginal delivery of the fetus the uterus normally contracts which stops the bleeding that occurs when the placenta disassociates from the uterine lining. When bleeding becomes excessive, several choices are available to induce uterine contractions to stop the hemorrhaging. The first line of defense is oxytocin. Usually 20 U of oxytocin is given intravenously in lactated Ringer's or normal saline and infused over a 1- or 2-hr period. If the uterus does not respond to oxytocin, the second line of defense is either ergot alkaloids or prostaglandins. Historically, ergot alkaloids have been used as the second choice but in the mid-1980s prostaglandins were approved for this purpose and in many institutions prostaglandins have become the second choice. Two ergot alkaloids can be used, ergonovine (Ergotrate) or methylergonovine (Methergine), administered at 0.2 or 0.3 mg intramuscularly or intravenously, the latter for immediate uterotonic effect. The mode of action of ergot alkaloids appears to be via serotoninergic and α-adrenergic receptors. Hypertension is a potential deleterious side effect of the ergot alkaloids; thus, they should not be given to individuals with hypertension.

The prostaglandin approved in the mid-1980s for treating postpartum hemorrhage is Hemobate (15-methyl-$PGF_{2\alpha}$-tromethamine) and it is very effective. If a patient is hypertensive, then Hemobate is the choice over ergot alkaloids. Oxytocin is routinely given following a normal cesarean section since the uterus is often atonic due to the anesthesia. Interestingly, oxytocin is also routinely given following a normal vaginal delivery, although the uterus of most patients spontaneously contracts after delivery of the placenta.

III. OTHER OXYTOCICS

A variety of agents have the ability to induce uterine contractions in pregnant animals. However, their physiological and clinical importance is unclear. These compounds include platelet-activating factor, cytokines such as interleukin-1β (IL-1β) and IL-6, growth factors such as epidermal growth factor, insulin-like growth factors-I and -II, and endothelins.

In addition, the withdrawal of certain factors can bring about uterine contractions. In particular, inhibition of progesterone action by administering progesterone receptor anatgonist such as RU 486 induces dramatic uterine activity within 24 hr during pregnancy. In addition, nitric oxide (NO) induces uterine relaxation and inhibition of NO synthesis in combination with subthreshold doses of RU 486 will precipitate preterm labor in rats.

IV. OXYTOCICS AND PARTURITION

A number of articles have proclaimed that PGs are the trigger of labor and/or PTL in women, whereas others have suggested that OT, endothelins, cytokines, and NO withdrawal might also be triggers. It is likely that many of these factors contribute to parturition and PTL. It has become apparent that, at least in primates, the events that lead to labor occur over several weeks and that conceptually it is better to consider parturition as a "process." Evidence suggests that estrogen is important in the labor process, enhancing the production and/or enabling factors such as OT and PGs to act on the uterus. A key to understanding labor is to appreciate that changes in the cervix are occurring in parallel with the uterus and that the cervix influences uterine activity. For purposes of conceptualization, the parturition process can be divided into two phases: the endocrine phase and the paracrine phase. The endocrine phase begins with an increase in the estrogen/progesterone ratio that brings about the release of OT and nocturnal uterine contractions. Nocturnal uterine contractions appear to precede normal labor by 2–4 weeks in women. As these nocturnal contractions become stronger and last longer, they contribute to the pushing of the fetal head into the ripening cervix and this aids in ripening and opening the cervical os. As the cervix opens, this exposes the chorioamnion membranes to vagina bacteria and this, along with the trauma of the head pushing into the cervix, induces an inflammatory-like reaction that brings blood elements, such as macrophages and neutrophils, into the lower uterine segment and cervix. Cytokines are released locally and stimulate PG production and the control of the uterine contractions switches from those dominated by OT to paracrine and autocrine factors such as PGs. Thus, depending on the phase of the delivery process, different tocolytics may be needed to stop labor. Based on this concept antagonists to OT will be most effective in the endocrine phase, whereas inhibitors to PG production or inflammatory reactions will be more effective during the paracrine phase of parturition. Thus, the current research focus is on selecting factors in the early stages of the parturition process that identify patients at risk and treating them with agents such as OT antagonists that will stop uterine activity in the endocrine phase. Factors that are being examined as early indicators of patients at risk are (i) elevated salivary estriol levels, (ii) vaginal fetal fibronectin, and (iii) cervical length as evaluated by ultrasound.

See Also the Following Articles

EICOSANOIDS; LABOR AND DELIVERY, HUMAN; UTERINE CONTRACTION

Bibliography

Bergstrom, S., and Sjovall, J. (1957). The isolation of prostaglandin. *Acta Chem. Scand.* 11, 1086.

Gainer, H., and Wray, S. (1994). Cellular and molecular biology of oxytocin and vasopressin. In *The Physiology of Reproduction* (E. Knobil and J. D. Neill, Eds.), pp. 1099–1129. Raven Press, New York.

Higby, K., Xanakis, E. M., and Pauerstein, C. J. (1993). Do tocolytic agents stop preterm labor? A critical and comprehensive review of efficacy and safety. *Am. J. Obstet. Gynecol.* 168, 1247–1256.

Negishi, M., Sugimoto, Y., and Ichikawa, A. (1995). Molecular mechanisms of diverse actions of prostanoid receptors. *Biochim. Biophys. Acta* 1259, 109–120.

Niebyl, J. R., and Witter, F. R. (1986). Neonatal outcome after indomethacin treatment for preterm labor. *Am. J. Obstet. Gynecol.* 155, 747–749.

Norman, J. E., Thong, K. J., and Baird, K. J. (1991). Uterine contractility and induction of abortion in early pregnancy by misoprostol and mifepristone. *Lancet* 338, 1233–1236.

Mitchell, M. D. (1996). Reproductive roles of eicosanoids. In *Reproductive Endocrinology, Surgery, and Technology* (E. Y. Adashi, J. A. Rock, and Z. Resenwaks, Eds.), p. 841. Lippincott-Raven, Philadelphia.

Phaneuf, S., Europe-Finner, G. N., Carrasco, M. P., Hamilton, C. H., and Lopez Bernal, A. (1995). Oxytocin signalling in human myometrium. *Adv. Exp. Biol. Med.* 395, 453–467.

Soloff, M. S., Alexandrova, M., and Fernstrom, M. J. (1979). Oxytocin receptors: Triggers for parturition and lactation. *Science* 204, 1313–1315.

Thorton, S., Davison, J. M., and Baylis, P. H. (1990). Effect of human pregnancy on metabolic clearance rate of oxytocin. *Am. J. Physiol.* 259, F21–F24.

Wakerley, J. B., Clarke, G., and Summerle, A. J. S. (1994). Milk ejection and its control. In *The Physiology of Reproduction* (E. Knobil and J. D. Neill, Eds.), pp. 1131–1178. Raven Press, New York.

Wilson, L., Jr., and Parsons, M. T. (1996). Endocrinology of pregnancy. In *Reproductive Endocrinology, Surgery and Technology* (E. Y. Adashi, J. A. Rock, and Z. Rosenwaks, Eds.), pp. 451–475. Raven Press, New York.

Wilson, L., Jr., Flouret, G., Parsons, M., Fejgin, M., and Pak, S. C. (1995). Oxytocin antagonists. In *First World Congress on Labor and Delivery* (D. Wienstein and S. Gabbe, Eds.), pp. 155–162. Monduzzi Editore, Bologna, Italy.

Vane, J. R., and Botting, R. M. (1995). New insights into the mode of action of anti-inflammatory drugs. *Inflammation Res.* 44, 1–10.

Zingg, H. H., Rozen, F., Chu, K., Larcher, A., Arslan, A., Richard, S., and Lefebvre, D. (1995). Oxytocin and oxytocin receptor gene expression in the uterus *Recent Prog. Horm. Res.* 50, 255–273.

Oxytocin

A. Courtney DeVries and C. Sue Carter
University of Maryland

I. General Features of Oxytocin
II. Lactation
III. Parturition
IV. Sexual Physiology and Behavior
V. Maternal Behavior
VI. Social Contact and Bonding

GLOSSARY

antisense oligonucleotide A small piece of artificially generated DNA that prevents the production of a particular protein by binding to the mRNA sequence that codes for the protein.

glucocorticoids The class of steroid hormones produced by the adrenal gland that is commonly associated with stress responses.

in vivo microdialysis A neurochemical technique that takes advantage of the properties of a dialysis membrane to allow the measurement of substances in extracellular fluid in awake animals.

lordosis An immobile posture in which the female concavely arches her back and deflects her tail to allow the mounting male to achieve penile insertion.

neurophysin The carrier protein for oxytocin and vasopressin that increases their half-life in blood.

paracrine Cells that secrete hormones that affect adjacent cells, whereas endocrine cells secrete hormones into the bloodstream so that they can have an effect on distant target cells.

prairie voles Small (30–50 g) monogamous rodents that live throughout the midwestern United States.

Oxytocin (OT) was the first peptide hormone to have its structure completely determined and the first peptide to be chemically synthesized. OT consists of nine amino acid residues, with a disulfide cysteine–cysteine bond that gives the molecule a ring and tail structure. Although the effects of OT on uterine contractibility and milk ejection have been known for nearly a century, it has been primarily within the past two decades that an assemblage of other physiological and behavioral effects of OT have been described. The majority of the behavioral pharmacology studies have been conducted in rats, but similar effects of OT have been reported in several mammalian species, including humans.

I. GENERAL FEATURES OF OXYTOCIN

OT is synthesized by the magnocellular neurons in the supraoptic nucleus (SON) and paraventricular nucleus (PVN) of the hypothalamus. The axons of the oxytocinergic neurons from both these regions project through the median eminence, extend down the infundibulum, and terminate in the neurohypophysis (posterior pituitary). OT is produced in the neuronal cell body and packaged into secretory vesicles which are transported by axonal flow to axon terminals located in the posterior pituitary. The biologically active peptide and its carrier protein, neurophysin, are then released via exocytosis into peripheral circulation, following appropriate stimulation.

During parturition and lactation, the release of OT into the periphery occurs in a pulsatile fashion, even when the stimulus eliciting the release of OT is fairly constant. The phasic nature of the OT release pattern is due to the unusual morphological and electrical nature of oxytocinergic cells in the PVN and SON. Ultrastructural analyses indicate that OT-containing magnocellular neurons originating in these two regions occur in clusters. Prior to parturition and lactation, the astroglial processes that normally isolate these neurons from each other retract, allowing the formation of direct contacts between the neurons (gap junctions). The formation of gap junctions permits the synchronous firing of several OT-containing neurons, which results in the pulsatile release of OT into the periphery. In a parturient female, these structural changes are maintained throughout lactation and can be induced in nonparturient rats via chronic central infusion of OT. Thus, the OT-containing magnocellular neurons in the PVN and SON act as both endocrine cells and typical neurons that are capable of conducting action potentials.

Most of the centrally located OT is found in the magnocellular neurons of the PVN and SON; however, the smaller, parvocellular neurons of the PVN also synthesize OT. Oxytocinergic projections of parvocellular neurons originating in the PVN extend to other parts of the central nervous system (CNS), including the forebrain, caudal brain stem, and spinal cord, where they form synapses with neurons in these regions. In these circumstances OT acts as a neurotransmitter. OT synthesis also has been demonstrated in the adrenal gland, ovary, uterus, testis, and in fetal membranes; in these tissues OT is expressed in such low concentrations that it probably is not released into the blood and therefore functions in a paracrine rather than endocrine fashion. As a peptide, OT does not pass easily through the blood–brain barrier, but there is evidence of bidirectional transport of small amounts of OT between the general circulation and cerebrospinal fluid.

The pattern of distribution and density of OT receptors within the CNS varies between sexes, across development, and among species. Steroid hormones, including estrogen, progesterone, testosterone, and glucocorticoids, also can affect the density and distribution of OT receptors. Within the CNS, OT receptors are particularly abundant in regions of the brain that mediate reproduction and autonomic function, including the anterior olfactory nucleus, ventromedial hypothalamus, bed nucleus of the stria terminalis, central amygdala, lateral septum, nucleus tractus solitarius, and motor nucleus of the vagus nerve.

OT is found exclusively in mammals, although homologs of OT have been identified in amphibians, reptiles, fish, and birds. OT also is very similar in structure to another mammalian hormone, arginine vasopressin (AVP; also known as antidiuretic hormone). OT and AVP differ from each other by two amino acids and may have arisen from a common ancestral peptide. In common with OT, AVP is produced in the magnocellular neurons of the SON and PVN and is released into peripheral circulation via axon terminals in the posterior pituitary. However, OT and AVP are rarely colocalized in the same neurons, and the release of these peptides can occur independently. AVP increases blood pressure and is responsible for water reabsorption in the kidneys; it also has been implicated in learning, memory, and territorial or self-defensive behaviors. Pharmacological studies indicate that OT can bind to AVP receptors, and that AVP can bind to OT receptors; however, the physiological significance of this cross-signaling remains unspecified.

II. LACTATION

Lactation is a uniquely mammalian event. Somatosensory stimulation of the mammary glands results in an increase in the release of OT from the posterior pituitary that subsequently causes the contraction of the mammary myoepithelial cells and the expulsion of milk. Although suckling may provide a relatively constant stimulus, milk ejection occurs only intermittently. The periodic nature of the milk ejection response may be due to morphological modifications in the magnocellular neurons of the SON and PVN which accompany pregnancy and lactation (described previously) and permit synchronous firing in the oxytocinergic neurons and the subsequent pulsatile release of OT. *In vivo* microdialysis (a technique that allows the continual measurement of substances in extracellular space) reveals that there is an increase in OT release in the SON during parturition and in the SON and PVN during suckling. *In situ* hybridization histochemistry also indicates that suckling increases OT gene expression in magnocellular regions of PVN and SON, which in turn project to posterior pituitary. In lactating female rats, central injection of OT causes an increase in neuronal excitatory activity and increases the frequency of milk ejection. The failure of milk ejection to occur in OT-deficient rats and mice supports the hypothesis that OT is essential for successful lactation.

III. PARTURITION

OT is a potent stimulus for uterine contractions in several mammalian species, including humans. For more than 90 years, posterior pituitary extracts and, recently, OT and its synthetic analogs (e.g., Pitocin) have been used to induce or augment labor in women. However, despite the application of OT in labor induction, there is continuing controversy about the role of circulating OT in normal parturition. Some researchers report an increase in OT pulse amplitude in blood prior to parturition and a surge associated with active labor and delivery, whereas others do not. The discrepancy in these findings may be methological since the pulse of OT release that is assumed to occur during parturition could be missed if the interval between blood samples is too long. However, an increase in uterine OT receptor density which precedes the onset of parturition in several species, including humans, also could increase uterine sensitivity to OT. An increase in OT receptors, even in the absence of an increase in plasma OT concentrations, would make the uterus more sensitive to OT. Preterm labor in women is associated with an increase in plasma OT concentrations and uterine receptor binding. The subsequent administration of OT antagonists decreases uterine contractility in these women.

In rats there is a more than 100-fold increase in OT gene expression in the uterine epithelium during the last 3 days of gestation compared to diestrus. These findings suggest that the uterus is a site of OT gene expression during the last part of gestation and that locally produced OT may act in a paracrine manner in the uterus to facilitate contractions. However, not all data support a physiological role for OT in parturition. For example, administration of OT antiserum impairs milk ejection but does not prevent parturition in rats. Also, genetically engineered mice, deficient in OT, deliver pups successfully. Taken together, these data suggest that OT may not be essential for parturition, although OT is capable of enhancing the intensity of uterine contraction and thus can facilitate birth. The role of OT in the initiation of birth under normal conditions remains uncertain and may be species specific.

IV. SEXUAL PHYSIOLOGY AND BEHAVIOR

Until recently, OT was considered a "female" reproductive hormone because of its role in labor induction and lactation. However, early reports that milk ejection sometimes occurred during orgasm in lactating women suggested a link between OT and sexual behavior. Hormone assays have subsequently confirmed that serum OT concentrations rise during sexual arousal and peak during orgasm in both women and men. An increase in serum OT during the ejaculatory phase also has been documented in bulls, rams, and rabbits.

OT gene expression also occurs locally in the ovary and testis. In females, the concentration of OT is particularly high in the corpus luteum, where it is presumed to play a role in luteolysis in several species. In male rats, OT has been localized in the Leydig cells of the testes, where it is able to influence steroidogenesis. In both males and females, OT also causes smooth muscle contractions in the reproductive tract which may assist in gamete transport.

Research on the role of OT in female sexual behavior has focused primarily on lordosis in rats. Lordosis has been used as an index of female sexual receptivity in several rodent species. Infusion of OT into the medial preoptic area increases the frequency of lordosis, and infusion into the ventromedial hypothalamus increases the duration of lordosis in rats that have been treated with a subthreshold dose of estrogen. The effects of OT on lordosis are most likely mediated through OT-specific receptors because treatment of estrogen and progesterone-primed female rats with an OT receptor antagonist (OTA) inhibits lordosis (provided that the OTA and progesterone are administered concurrently). Additional evidence for OT receptor-mediated facilitation of female sexual behavior has been obtained using a technique from molecular biology which is commonly referred to as "antisense." Intrahypothalamic infusions of antisense oligonucleotides (small pieces of DNA) directed against OT receptor mRNA reduce lordosis behavior in estrogen-primed rats, presumably by preventing the production of OT receptors.

OT also appears to have an effect on sexual behavior in males. For example, central administration of OT facilitates penile erection in a variety of species, including rats, rabbits, and monkeys. In rats, the effects of OT on penile erection can be reversed by either OTA administration or electrolytic lesions of the PVN that deplete central OT concentrations. Penile erections induced by N-methyl-D-aspartic acid and various dopamine agonists are also mediated via increased central OT transmission and are blocked by OTA. In addition, OT-induced penile erection is inhibited by nitric oxide synthase inhibitors injected centrally, suggesting a role for nitric oxide (a gaseous neurotransmitter) in OT-induced penile erection.

In sexually experienced 5-month-old rats, intraperitoneal and ICV injection of OT does not affect the frequency or latency to mount or intromit; however, OT does shorten the ejaculation latency and postejaculatory interval. In older, sexually experienced male rats (20 months) OT shortens mount, intromission, and ejaculation latencies as well as postejaculatory intervals. However, OT does not reverse sexual impotence in males of any age. The effects of OT on penile erection appear to follow a bell-shaped dose–response curve. Maximal facilitation of penile erection in rats occurs following central administration of between 10 and 60 ng of OT. At higher doses of OT (>100 ng) penile erection is inhibited and males emit ultrasonic vocalizations similar to those characteristic of the postejaculatory refractory period. These data suggest that in rats low to moderate doses of OT facilitate male sexual behavior, whereas high doses of OT may create a physiological state analogous to sexual satiety.

V. MATERNAL BEHAVIOR

Most mammalian species require exposure to the hormonal changes associated with pregnancy to show parental behavior. For example, the onset of spontaneous maternal behavior occurs just prior to parturition in rats and coincides with an increase in OT fibers and gene expression in the CNS. Exogenous OT can facilitate the onset of maternal behavior in estrogen-primed female rats. However, the effects of OT on maternal behavior are modulated by environmental factors, including stress. Also, OT plays a role in the onset, but not the maintenance, of maternal behavior. Lesioning the PVN (a major hypothalamic source of OT) during late pregnancy disrupts the onset of maternal behavior, whereas lesioning the PVN 5 days postpartum does not inhibit ongoing maternal behavior. Pregnant female rats treated with OTA or OT antiserum also exhibit a longer latency to the initiation of maternal behavior, thereby suggesting that endogenous OT plays a facilitatory role in the natural onset of maternal behavior in rats.

The effect of OT on maternal behavior in mice varies among strains. In virgin and pregnant wild mice, OT increases maternal behavior and inhibits infanticide of unrelated pups. However, genetically engineered mice deficient in OT are not infanticidal and exhibited a level of maternal care that is indistinguishable from animals with a functional OT gene. Thus, in some strains of mice the induction of maternal behavior, like parturition, can occur in the absence of OT.

The hormonal regulation of maternal behavior has also been particularly well studied in sheep. Maternal behaviors in sheep include approach and licking of the lamb, the emission of low-pitch bleats, and suckling. In this species, the newborns must follow their mothers, and individual recognition between lambs and mothers is vital. In natural circumstances, ewes are only maternally responsive to lambs after the onset of parturition. Within 2 hr of parturition, suckling occurs and the ewe and her lamb form an exclusive bond. Following the formation of the bond, the mother will actively reject any unfamiliar lamb that attempts to suckle.

In vivo microdialysis in sheep reveals that OT is released during suckling in several regions of the CNS that have been implicated in the control of maternal behavior in other species and may be important to the onset of maternal behavior in sheep as well. By elevating the concentration of OT in the CNS, through treatment with exogenous OT or exposure to vaginocervical stimulation (VCS), researchers can induce both parturient- and estrogen-primed nonparturient sheep to accept and form a bond with an unfamiliar lamb. The natural formation of a bond between an ewe and a lamb, as well as bonds induced via OT treatment or VCS, can be blocked by opiate receptor antagonists which prevent the actions of OT. Thus, OT appears to be crucial to the formation of mother–infant bonds and the initiation of maternal behavior in ewes.

VI. SOCIAL CONTACT AND BONDING

OT has been implicated in several species-specific behavioral patterns including both positive social behaviors and agonistic behaviors in rats and other species. In prairie voles, which are monogamous rodents, the formation of adult social bonds is facilitated by OT and inhibited by OTA. In several species of voles and deer mice, the distribution of OT receptors correlates with monogamy or polygyny, suggesting that sociality, or more specifically social organization, may be influenced by patterns of OT receptors and peptide release. OT is also capable of increasing tactile contact and other indications of sociality in species that are not monogamous, such as rats and squirrel monkeys.

See Also the Following Articles

Lordosis; Mating Behaviors, Mammals; Microtinae (Voles); Milk Ejection; Parturition, Nonhuman Mammals

Bibliography

Carter, C. S. (1992). Oxytocin and sexual behavior. *Neurosci. Biobehav. Rev.* 16, 131–144.

Carter, C. S., Lederhendler, I. I., and Kirkpatrick, B. (Eds.) (1997). Integrative neurobiology of affiliation. *N. Y. Acad. Sci.* 807, 1–614.

Ivell, R., and Russel, J. (Eds.) (1995). Oxytocin. In *Advances in Experimental Medicine and Biology,* Vol. 395. Plenum, New York.

Pedersen, C., Caldwell, J., Jirikowski, G., and Insel, T. (1992). Oxytocin in maternal sexual and social behaviors. *N. Y. Acad. Sci.* 652, 1–492.

Richard, P., Moos, F., and Freund-Mercier, M.-J. (1991). Central effects of oxytocin. *Physiol. Rev.* 71, 331–370.

Pampiniform Plexus

B. P. Setchell
University of Adelaide

I. Structure
II. Functions

GLOSSARY

arteriovenous anatomosis A direct channel linking an artery to a nearby vein.

countercurrent heat exchanger An arrangement of two sets of tubes carrying fluids in opposite directions, in which one lot of fluid is cooled while the other is warmed by heat transfer from one side to the other.

lipophilic Highly soluble in fats; the opposite of hydrophilic.

pulse pressure The difference between systolic and diastolic blood pressure in a vessel.

spermatic cord The structure formed by the pampiniform plexus and the coils of the internal spermatic artery between the inguinal canal and the testis.

The pampiniform plexus is a complex of veins surrounding the coils of the internal spermatic or testicular artery to form the spermatic cord in mammals with scrotal testes. There are numerous arteriovenous anatomoses between the artery and the veins in the cord which allow a significant fraction of the blood to flow directly from the artery to the veins without passing through the testis. The plexus and the coiled artery act as an efficient countercurrent heat exchanger, cooling the arterial blood before it reaches the testis and warming the venous blood from scrotal to near body temperature. Pulse pressure in the arterial blood is dramatically reduced before it reaches the testis. While lipophilic substances can pass readily from vein to artery and vice versa, there appears to be little transfer of testosterone from the venous blood to the arterial blood in the cord.

The veins carrying blood from a scrotal testis divide as they leave the testis to form the pampiniform plexus, so-called because of its supposed resemblance to the pampinus, the new growth including the tendrils, which comes from a bud on a year-old grape vine cane. This plexus surrounds the coils of the internal spermatic or testicular artery to form the spermatic cord, and on its outside there are several prominent lymphatic vessels. The plexus and the spermatic cord are most highly developed in animals such as the ram and bull, with pendulous scrota, and are less obvious in humans and laboratory rodents.

I. STRUCTURE

A. Veins

In eutherian mammals, up to 200 venous branches can be found in a cross section of the plexus. They are formed when the small veins leaving the testis divide again to form an intercommunicating plexus; near the inguinal canal, the venous branches reunite to form the internal spermatic or testicular vein, which then carries the blood to the vena cava, the renal vein, the hypogastric vein, or the iliac vein, depending on the species and the side of the animal. In marsupials, there is a similar arrangement of veins, except that they all arise from one face of the testis. In bulls, the veins in the plexus can be classified into three types on the basis of size; they intercommunicate freely with one another, and none have obvious valves.

B. Artery

As the testis descends during fetal life from its origin near the mesonephros, its arterial supply becomes elongated, and beyond the inguinal canal the artery becomes coiled so that in rams up to 7 m of artery is tightly coiled into about 10 cm of spermatic cord. These arterial coils are surrounded by the branches of the venous plexus. Around the artery, there are abundant nerve fibers containing neuropeptide Y and dopamine β-hydroxylase, and a few contain substance P. In rams, the artery near the inguinal canal is much more sensitive than the artery at the testicular end of the cord to the vasoconstrictor activity of norepinephrine, and the artery on the surface of the testis is virtually unresponsive. In marsupials, instead of a single, long, coiled artery, the internal spermatic artery divides as it leaves the inguinal canal and forms an arterial rete, with a similar number of branches to the venous complex. The arterial and venous branches run interspersed parallel to one another, and at the testis the arterial branches reunite to form two or three vessels, which run along the opposite face of the testis from the veins.

C. Arteriovenous Anastomoses

Evidence for the presence of arteriovenous anastomoses between the artery and veins in the spermatic cord has been sought since the time of de Graaf in the seventeenth century. He could not find any, but later studies have produced both anatomical and functional evidence for their existence. Blood flow though the testicular artery above the cord is substantially higher than that in the same vessel below the cord, and the testosterone concentration is much higher and oxygen saturation of the hemoglobin is lower in veins on the testis than above the cord in scrotal testes in rams, rats, and cryptorchid pig testes, but not in normal pig testes. It has been calculated that up to 50% of the blood entering the artery at the top of the spermatic cord crosses into the veins without passing through the capillary bed of the testis.

II. FUNCTIONS

A. Countercurrent Heat Exchange

The closely associated venous network and the single coiled artery in eutherian mammals with scrotal testes and the intermingled venous and arterial retia in marsupials act as efficient countercurrent heat exchangers. Arterial blood entering the spermatic cord in rams at 39°C is cooled to about 35°C, which is close to scrotal skin temperature, whereas venous blood leaving the testis at 33°C is warmed up to 38.6°C by the time it reaches the inguinal canal (Fig. 1). Body temperature is 39.6°C and testis temperature is 34.1°C in this species. Corresponding values for tammar wallabies are 35.9 to 30.5°C for the artery and 31.0 to 36.1°C for the veins, with a body temperature of 36.5°C and a testis temperature of 31.2°C. Thus, the testis is kept at a uniformly

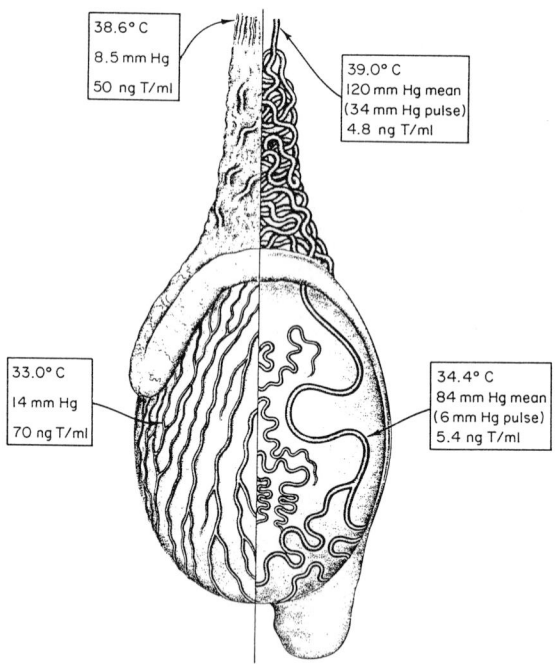

FIGURE 1 A diagram showing the venous drainage and the pampiniform plexus of a ram testis (left) and the arterial supply (right). The temperatures, mean blood pressure, pulse pressure, and testosterone concentrations of testosterone inside the testicular artery at the inguinal canal, on the testis and veins on the testis, and at the top of the pampiniform plexus are shown in the boxes (reproduced with permission from Setchell, 1991).

lower temperature than is the body, provided that the scrotal skin is able to maintain its temperature. Furthermore, if the temperature of the testis is raised even by as little as 2 or 3°C for period of several hr by exposing the animal to a hot environment or by scrotal insulation, or to 43°C for as little as 15 min by immersing the testis in warm water, spermatogenesis is severely disrupted, the spermatozoa that are formed have reduced motility and fertilizing ability, and the embryos they produce in normal females are less viable.

B. Pulse Elimination

As the arterial blood passes through the spermatic cord, its mean pressure falls slightly, but the pulse is virtually eliminated. This phenomenon has been demonstrated in eutherian mammals and marsupials, but its significance is unclear.

C. Transfer of Substances from Vein to Artery in the Cord

It has been suggested that testosterone may pass down its concentration gradient from the veins to the artery in the spermatic cord. There is no doubt that this occurs with highly permeable lipophilic substances, such as krypton and xenon, and with tritiated water, but the extent of transfer of testosterone appears to be negligible, raising the concentration of this steroid by about 10–20% compared with the increase of 50-fold or greater as the blood passes through the capillary bed of the testis. The increased concentration of testosterone in arterial blood below the cord may have significance for the epididymis because the arterial supply to and venous drainage from this organ also form part of the spermatic cord. The possibility remains that some other substance more lipophilic than testosterone might be transferred in significant quantities from vein to artery, and this would have the effect of producing a higher concentration in the testis for a given rate of production.

See Also the Following Article

TESTIS, OVERVIEW

Bibliography

Setchell, B. P. (1978). *The Mammalian Testis*. Elek/Cornell Univ. Press, London/Ithaca, NY.

Setchell, B. P. (1991). Male reproductive organs and semen. In *Reproduction in Domestic Animals* (P. T. Cupps, Ed.), 4th ed., pp. 221–249. Academic Press, San Diego.

Setchell, B. P. (1998). Heat and the testis. *J. Reprod. Fertil.*, in press.

Setchell, B. P., Maddocks, S., and Brooks, D. E. (1994). Anatomy, vasculature, innervation and fluids of the male reproductive tract. In *Physiology of Reproduction* (E. Knobil and J. D. Neill, Eds.), pp. 1063–1175. Raven Press, New York.

Paracine Control

see Local Control Systems

Parasites and Reproduction

Jack J. O'Brien

University of South Alabama

I. Introduction
II. Decreased Reproduction
III. Enhanced Reproduction
IV. Evolutionary Impact of Parasites on Sexual Reproduction

GLOSSARY

brood parasite Some species of birds and insects which lay their eggs among the clutch of the host species which then cares for the young of the parasite as if it were its own.

egg predator A symbiont that lives within and feeds on the egg mass of its host; many nemertean worms and some copepods have adapted this manner of existence in crustacean hosts.

parasitic castrator A parasite that in addition to acquiring its nutritional requirements from the host actively inhibits the reproductive development of the host usually by either destroying or inhibiting maturation of the gonads.

parthenogenesis Asexual reproduction in which reproductive adults develop from unfertilized gametes; common in arthropods, rotifers, and some groups of molluscs.

Rhizocephala An order of barnacles (class Cirripedia) containing at least five families that are parasitic castrators of decapod crustaceans (crabs and shrimp).

Trematoda A class of parasitic flatworms (phylum Platyhelminthes) exhibiting complex life cycles involving multiple hosts; adult worms are typically intestinal parasites of vertebrates, whereas the subsequent parasitic stages (rediae and sporocysts) of the life cycle are parasitic castrators of molluscs.

I. INTRODUCTION

Imagine an extraterrestrial sent to Earth with the assignment to enumerate how many species of Earth animals displayed what type of lifestyle. Such a being would observe that there were many species that walked around and ate plants, noticeably fewer species that ran around and ate the plant-eaters, and a third group that mostly sat around waiting for the plant-eaters and the animal-eaters to drop dead so that they could eat them. However, if the extraterrestrial were a competent observer, he/she/it would notice that there was an easily overlooked fourth group consisting of small creatures that lived on and in organisms in the aforementioned categories. If our visitor were to sum up his/her/its findings, he/she/it would discover that there were twice as many animals in the latter category than all the plant-eaters, animal-eaters, and carrion-eaters combined. In other words, two-thirds of all the animals on Earth are parasites and, as one might expect, within this extremely diverse group of organisms are species that exert profound influences on the reproduction of their hosts.

A. What Is a Parasite?

It is difficult to provide a good short definition of parasitism because types of interactions among organisms lie along a continuum rather than occupy well-defined categories. As a starting point, a parasite is an organism (plant, animal, fungus, or bacterium) that establishes a symbiotic association with a second organism, the host, from whom it acquires its nutritional requirements and who is harmed by the activities of the parasite. The concept of symbiosis or "living together" is crucial to this definition because it helps to distinguish parasitism from other interactions (such as predation and herbivory) wherein creatures may also harm the organisms from whom they acquire their nutritional needs. Indeed, many parasitologists would argue that, as a group, parasites have suffered from a bad press in the mind of the public when, as the following comparisons demonstrate, they actually exemplify characteristics that are admired in fictional politically correct heroes and heroines.

First, although there may be much squealing and screaming (which some erroneously associate with passion), predator–prey interactions are very brief relative to the lifetime of the organisms. Parasites, on the other hand, engage in long-term, intimate relationships with their hosts; they personify lifetime commitment. Second, as is typical of bullies, predators rarely pick on animals their own size. What chance does a timid, furry little mouse have against a silent, nighttime, sneak attack (often from behind) by a horned owl armed with two huge talons and a rapacious beak? In comparison, a larval hookworm, barely visible to the naked eye, risks being crushed underfoot in order to slip between the toes and burrow into the skin of its enormous human host. Once inside, the hookworm must then battle legion upon legion of cells of the human immune system. Finally, while there are parasitic–host associations that may result in the eventual death of the host, many, if not most, do not. However, every victim in a successful predatory–prey interaction is killed. Simply put, predators murder to get their food, whereas parasites steal.

B. Trophic Interactions Similar to Parasites

Although parasitology texts devote chapters to mosquitoes, fleas, ticks, and their ilk, these animals occupy an intermediate position in the continuum stretching from parasitism to predation. These organisms have been termed micropredators or sublethal predators to distinguish them from predators and parasites. Micropredators are similar to parasites in that they are smaller than the organism on which they feed and they do not kill their victims. As do predators, however, micropredators are not committed to one individual prey organism but will move from victim to victim. Kuris gives an excellent discussion of the differences and similarities of predators, parasites, micropredators, and related trophic interactions.

II. DECREASED REPRODUCTION

A. Egg Predators

There are organisms that acquire nutrition by eating eggs of other organisms. Opportunistic birds, such as jays and crows, will consume unguarded eggs of other birds, but the organisms under discussion in this section are small organisms that eat eggs before they have been released from the mother. Many aquatic organisms do not lay eggs in clutches separate from the mother but rather brood them in either internal cavities or attached to specialized external appendages. Female decapod (crabs, shrimp, and lobsters) crustaceans, for example, carry tens of thousands of eggs in external egg masses which also serve as habitats for a plethora of fungi, bacteria, protozoans, copepods, and nemeretean worms. These symbionts are commonly overlooked because they are hidden among the closely packed developing embryos. In addition, they often closely resemble the color, size, and shape of the egg that they feed on (Fig. 1).

Some researchers have argued that these symbionts exert a greater impact on the reproductive output of their hosts than is commonly recognized. During the 1980s, the commercial yields of the Dungeness crab dropped significantly along the north Pacific Coast. Female crabs often carried egg masses with significantly fewer eggs than in previous years as well as relatively large numbers of the nemertean worm, *Carcinonemertes epialti*. Whether it was an abundance of worms that caused the crab population to crash or whether the worms were a symptom of a

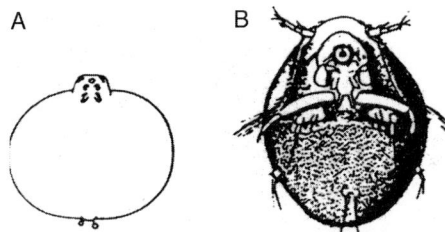

FIGURE 1 Mimicry of decapod host eggs by copepod egg predators. (A) Ventral view of a female *Sphaeronella rotunda* (×32) exhibiting a rotund body and atrophied body appendages. (B) Ventral view of a male *S. aeginae* (×87) showing appendages at higher magnification (reproduced with permission from Hansen, 1923).

fishery stressed by overfishing is still debated. Not much is known about the ecological interactions within the egg mass habitat. The mere presence of an organism within an egg mass is not enough evidence to conclude that it is harmful to the eggs. For example, there is evidence to suggest that some bacteria may protect eggs from other fouling organisms, whereas other organisms in the egg mass consume only eggs that die of "natural" causes (i.e., are scavengers). Even whether these symbionts should be classified as predators or parasites is tricky. Are the developing crab embryos independent organisms that are being preyed on by egg predators or do they constitute an organ of the female crab that is being attacked by egg parasites? The consensus among people who study these relationships is that organisms that are larger than and do not form an intimate association with the individual eggs that they consume are egg predators. Technically, an egg parasite would be smaller than the egg and would not kill it.

B. Brood Parasites

Previously, reference was made to the predatory activities of jays and crows on the eggs of other birds. There exists yet another threat to the reproductive activity of birds—that of nest parasitism or brood parasites. Some female bluebirds, for example, watch the nests of other bluebirds and deposit eggs in those nests when they are left unguarded. There are risks inherent in such behavior. Keeping a nest under surveillance requires time, during which the watcher's nest is unguarded. Indeed, investigators have documented parasitic bluebirds caring for one or more offspring of other parasitic females indicating that poetic justice does exist in the bluebird world. Although bluebirds only parasitize nests of their own species, some ducks and cuckoos will, in addition to laying eggs in their own nest, lay eggs in nests of similar species.

None of the aforementioned species are considered to have a significant negative impact on their victims. Such is not the case with the brown-headed cowbird. Female cowbirds do not even make nests; cowbird young are always raised by foster-parent birds (Fig. 2). Studies have shown that cowbirds will deposit eggs in nests of over 200 bird species (Fig. 3) and since cowbird eggs hatch sooner than eggs of most

FIGURE 2 Foster parenting of a large fledgling brown-headed cowbird by a smaller red-eyed vireo adult (T. Angell, illustrator, reproduced with permission of the University of Washington Press from Orians, 1985).

other birds, the parasitic nestlings often outcompete the true offspring for food and space within the nest. What concerns conservationists is that the cowbirds have undergone an extensive range expansion and population increase over the past 50 years and there has been a reduction recently in the numbers of songbirds. Although the reduction of nesting habitat has undoubtedly had a negative impact on songbird reproduction, there is mounting evidence that brood parasitism by cowbirds may also be pushing some species toward extinction.

Brood parasitism is not limited to birds; there are numerous species of organisms (usually other insects) that have a negative impact on the broods of

FIGURE 3 Cowbird egg (right) in brown towhee nest (T. Angell, illustrator, reproduced with permission of the University of Washington Press from Orians, 1985).

social insects. Some ant species do not forage but live in the nests of other species, consuming food brought into the nest by their hosts. As a result, colonies that maintain such social parasites have fewer offspring than unparasitized colonies because less energy is available for reproduction. The impact of so-called "slave-making" ants is much more direct. These ants conduct raids on colonies of other ant species and physically carry developing cocoons of their neighbors back to their own nests. "Slaves" that hatch from transplanted cocoons work in the nest of their captors as though it were their own. An interaction whereby one species completely eliminates a victimized colony occurs with certain ant species and cuckoo wasps. In these cases, a parasitic queen invades the colony of another species and through either physical combat or release of chemicals stops the reproductive activities of the host queen. Eggs laid by the usurper are cared for by workers in the host colony and through attrition all the original workers are eventually replaced by workers of the invading species. Because cuckoo wasps cause the reproductive death of the host colony, they are similar in many respects to parasitic castrators.

C. Parasitic Castrators

Another type of interaction, known as parasitic castration, consists of symbionts that selectively destroy host gonads. Parasitic castration has evolved numerous times and parasitic castrators are found in many phyla. One explanation for the success of this lifestyle emphasizes that parasitic castrators do not utilize energy resources nor affect defense mechanisms that the hosts use for survival. In other words, by targeting only those metabolic reserves that the host would have allocated to reproduction, parasitic castrators limit their cost to the host to that of normal host reproduction. This argument assumes that parasitic castrators do not reapportion the energy reserves of their hosts. Indeed, at least one detailed study found that unparasitized shrimp used 4.4% of their energy intake for shrimp reproduction compared to 4.3% used by the gill parasite that castrates this same shrimp species.

The ecological implications of parasitic castration are profound. Though they do not reproduce, hosts harboring parasitic castrators continue to live, feed, and grow in the ecosystem occupied by uninfected hosts. Predators kill their prey, leaving more resources (such as food and space) available for the survivors; thus, predators reduce intraspecific competition within prey species. Although not well studied, the situation is probably more complex with parasitic castrators. Castrated hosts do not contribute to the gene pool of the next host generation because they are reproductively dead and in this way are similar to the victims of predation. However, hosts of parasitic castrators remain in the habitat competing for food and space with unparasitized hosts; thus, it is conceivable that in situations in which food resources are scarce, parasitic castration could be "a fate worse than death" to the population of hosts.

The definition of species is based on the concept of reproductive isolation—the ability (or inability) of organisms to produce sexually functional offspring. For example, mules result from crosses between horses and donkeys. Even though mules are alive and often kicking, they are sterile; consequently, horses and donkeys are without question different species. Clearly, hosts parasitized by parasitic castrators are reproductively isolated from the rest of the host population. Metabolic reserves of hosts carrying parasitic castrators will go toward the production of larval parasites and not host offspring. In a reproductive sense, hosts of parasitic castrators are no longer members of the host species. Consequently, parasitic castration results in a situation similar to those in which two different genotypes (host and parasite) possess the same phenotype (that of the host).

Because many species of trematode worms pose a serious human health threat, whereas others are economically important, parasitic castration caused by trematodes has been studied most often. As a group, trematodes possess incredibly complex life cycles. During its life cycle, any one trematode species commonly parasitizes three, sometimes four, hosts belonging to different phyla. Adult trematodes do not castrate, but they are typical parasitic worms that reproduce sexually either in the digestive tracts of vertebrates or occasionally, as in the case of schistosomes, in vertebrate blood vessels. It is the next host in the life cycle (usually a gastropod snail) that is castrated. Two morphologically different trematode stages (rediae and/or sporocysts) can be found in snails; these occupy host gonads in which they undergo prolific asexual reproduction. Host snails continue to feed and commonly attain a larger size than unparasitized snails while daily shedding hundreds if not thousands of cercariae, the next stage in the life cycle. With the important exception of schistosomes (whose cercariae burrow into the skin of waterfowl or humans working in rice paddies), cercariae usually encyst within a third species of host that is not castrated. Typically, the trematode life cycle is completed when the latter is eaten by the appropriate vertebrate.

Gastropods exhibit plasticity in their methods of reproduction. Some reproduce sexually as well as parthenogenically, whereas others, such as *Biomphalaria glabrata* which is an important intermediate host of *Schistosoma mansoni*, are simultaneous hermaphrodites capable of cross and self-fertilization. These species possess female and male reproductive tissue in an ovotestes. One response of the host that occurs shortly after invasion by trematode infection has been termed fecundity compensation. Parasitized *B. glabrata*, for example, undergo a pulse of egg deposition during the initial stages of the trematode's development prior to disruption of host ovarian tissue and can continue to donate sperm for some time after their egg-laying capabilities have been destroyed. That a host of a parasitic castrator may continue to reproduce may seem paradoxical; however, such an effect is arguably beneficial to the parasite as well as the host. In populations wherein the parasite may encounter host strains that are resistant to infection, reproduction of susceptible hosts increases the likelihood that the offspring of that parasite will also encounter vulnerable hosts.

How trematodes induce parasitic castration is not completely understood; evidence suggests that both direct and indirect influences may occur. Early investigators noted that daily shedding of cercariae certainly placed a nutritional burden on the host, that the presence of mouths on rediae would enable them to ingest host gonadal tissue, and that the physical presence of so many asexually reproducing trematodes in host gonads left little room for host gametes. Results from recent investigations strongly suggest that there are factors in parasitized snail hemolymph that inhibit gonad development. Transplants of immature ovotestes into unparasitized adult controls rapidly matured, whereas similar transplants into parasitized snails demonstrated blocked or impaired gonadal development without direct contact with the parasites. Although these results could be the result of depleted nutrients and/or host reproductive hormones in parasitized transplant recipients, other investigators have found evidence of molecules in snail hemolymph that inhibit host gonadal development.

In addition to the trematodes, there are numerous examples of parasitic castrators within the Crustacea. Kentrogonid rhizocephalan barnacles (Fig. 4) and epicaridean isopods parasitize decapod crustaceans (crabs and shrimp). Interestingly, male crabs parasitized by sacculinid rhizocephalans acquire a broad abdomen typical of females (i.e., they are feminized). In addition, hosts stop molting. A sacculinid in the Gulf of Mexico is an economic pest as parasitized blue crabs do not attain harvestable size. Members of the epicaridean families Bopyridae and Dajidae are external parasites that feed on the hemolymph of their hosts. At least with bopyrids, the degree to which host reproduction is inhibited does vary because hosts parasitized by bopyrids have occasionally

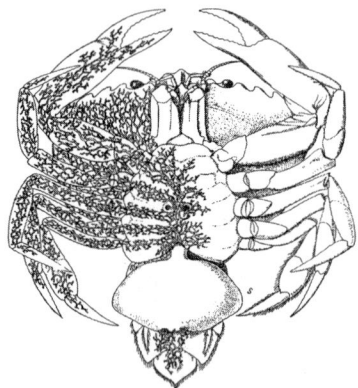

FIGURE 4 A green crab infected by the parasitic castrator, *Sacculina carcini*. The root system of the rhizocephalan barnacle is illustrated on the left side of the figure to show the extensive ramification of the root system of the parasite throughout the host's body cavity. The external portion of the rhizocephalan contains the gonads and occupies the same position as the egg mass of an unparasitized host. After Boas from Kaestner, 1970. © 1970 John Wiley and Sons, Inc.

been found bearing a small brood of eggs. Ascothoracican barnacles castrate starfish and other echinoderms. *Anselma* is a barnacle found attached behind the pectoral fins of dogfish. Unlike whale barnacles which feed on the zooplankton, *Anselma* feeds on host fluids and infected dogfish have undeveloped gonads. Females of the copepod *Cardiodectes* feed on host blood by burying their heads into the hearts of castrated lanternfish.

III. ENHANCED REPRODUCTION

As mentioned previously, in theory, enhanced reproduction of susceptible hosts is adaptive for parasites because it would increase the likelihood that parasite progeny would encounter vulnerable hosts. Although examples in the animal kingdom of enhanced host reproduction directly attributable to parasites are rare, there are some. Wade and Chang found that the gametes of male flour beetles infected with the intracellular rickettsia parasite, *Wolbachia*, had a fertility advantage over unparasitized gametes. Since this parasite can spread from parents to offspring through the cytoplasm of egg and sperm cells, outcompeting unparasitized gametes leads to accelerated spread of the parasite through host populations. The reproductive isolation effects caused by *Wolbachia* on its hosts are discussed in further detail in Section IV. Care must be taken in interpreting results such as a correlation between high levels of parasitism and high fecundity. One must be sure to distinguish between the cause and the effect. Richner and colleagues manipulated brood size in birds and found that the prevalence of bird malaria in males foraging for artificially enlarged broods was more than two times that of males tending unmanipulated nests. The authors concluded that the high parasite prevalence in overworked males arose from either increased exposure to the malaria vectors due to extensive foraging or the stress of overwork somehow reducing the ability of their immune system to protect them from infection. In other words, the high level of parasitism was caused by caring for a large brood rather than a large brood having resulted from being parasitized.

Enhanced asexual reproduction is commonly found with host plants parasitized by fungi that selectively attack host flowers, effectively castrating the host plants. Parasitized plants continue to propagate vegetatively, however, and often exceed unparasitized conspecifics in growth rate, size, and vigor. Long-lived hosts have been observed to outcompete unparasitized plants that devote significant energy resources to seed production. This suggests that enhanced asexual reproduction might be found in animal groups such as sponges, sea anemones, and corals that reproduce sexually as well as asexually if they harbor parasites that selectively infect host gonadal tissue.

IV. EVOLUTIONARY IMPACT OF PARASITES ON SEXUAL REPRODUCTION

Evolution is the gradual change in the genetic composition of a species over generations. According to the theory of evolution, the most important influences on the direction evolutionary changes take are those that determine what genes are passed on to the next generation. Two mechanisms have been demonstrated to affect this process: natural selection (individuals with maladaptive traits die before they can reproduce) and sexual selection (individuals mate with conspecifics that have certain impressive traits, such as bright plumage, large antlers, or baby-blue eyes). Evidence is accumulating indicating that parasitism exerts a profound influence on both these processes.

One fundamental question is the following: Why do almost all organisms reproduce sexually? If it is advantageous to pass on as many of one's genes into the next generation as possible, why not reproduce asexually because there are costs incurred by sexual reproduction. (One model predicts that for species in which the male provides no assistance in caring for its progeny, the benefits derived from sexual reproduction must, at times, be two times greater than asexual reproduction for sexuality to last over time.) For example, one's genes are diluted by 50% when mixed with the genes of one's sexual partner. Additionally, time and energy are spent in finding a mate. In species that are widely dispersed, there is the tangible risk that one will not be able to find a sexual partner. Finally, there is the problem of wastage. Using *Homo sapiens* as an example of an organism utilizing internal fertilization, a female never comes close to fertilizing her approximately 400,000 primary oocytes, and the percentage of male spermatozoa that successfully fertilize an ovum is almost infinitesimally small relative to the number of sperm cells produced in a male's lifetime. Broadcast spawners such as fish and amphibians suffer even higher levels of wastage. Therefore, wouldn't it be easier to reproduce asexually? The answer appears to be, "In the short term, yes, but over the long term, no." This is because offspring produced by sexual reproduction are more variable over the long term and variability enhances the chances for survival. If one individual of a population of genetically identical individuals were to acquire a transmittable fatal disease, the entire host population would succumb. Genetic variability increases the likelihood that some individuals within a population will be resistant to any particular pathogen. As discussed later, evidence is accumulating to indicate that, at least in some populations, parasites are responsible for the maintenance of sexual reproduction.

A. Red Queen Hypothesis

Quite often changes in one species will induce changes in another. This process, termed coevolution, is perhaps most easily observed in predator–prey (fast cheetahs must catch fast gazelles) and insect–plant pollination (pollinators transfer pollen when "mating" with flowers whose shape induces the mating response) interactions, but parasites and their hosts coevolve as well. The red queen hypothesis (named after an episode in Lewis Carroll's *Through the Looking Glass* in which the Queen of Hearts exhorted Alice to run as fast as she can in order to stay in the same place) predicts that in order to survive (i.e., stay in the same place) each of two species will accrue significant energetic costs (i.e., run hard) counteracting changes that occur in the other. In genetically diverse populations of parasites and hosts, it is expected that some hosts will have a defense mechanism, such as an effective immune reaction, that will offer some degree of protection from infection. As this trait becomes common within the host population, those individuals within the parasite population that can avoid that defense mechanism will acquire a reproductive advantage. What was an advantageous trait in one generation may be a liability in the next and the most successful strategy over the long term is to continue to produce variable offspring. Red queen interactions result in a dynamic standoff similar to the arms race that existed during the Cold War.

Parasitic castration of snails caused by trematodes provides an opportunity to test the validity of the red queen hypothesis with host snails that exhibit both sexual and parthenogenic reproduction. In associations wherein parasitization is the major influence on natural selection, the red queen hypothesis predicts that parthenogenesis would be common where trematode prevalence is low because asexually reproducing females would have an advantage over sexually reproducing females because they use energy for reproduction much more efficiently. In populations in which trematode prevalence is high, most females would be reproducing sexually (although energetically wasteful, greater genetic variation will increase the likelihood that some offspring will be resistant to infection). In a series of papers, Lively and coworkers accumulated data consistent with the predictions of the red queen hypothesis: Parasites prevent parthenogenically produced clones from replacing sexually reproducing populations. It is not that the effect of parasitic castrators is to directly select for sexual reproduction, but rather that it selects for genetic diversity which is supplied by sexual reproduction.

B. Sexual Selection: The Hamilton–Zuk Hypothesis

It has been noted that the females (but for the theory it does not really matter which sex does the choosing) of many species act as though they are choosing the males with whom they will mate. Males engage in such activities as performing "courtship dances," singing songs, spraying urine, displaying bright plumage, and fighting with competing males. In a classic paper, Hamilton and Zuk gathered evidence to suggest that these displays allow choosing organisms to evaluate potential mates for their resistance to parasitic diseases. The authors used the metaphor of a physician conducting a physical examination who checks reflexes, listens, performs a urine test, evaluates skin color, and eventually assesses vigor. The implication is that organisms that are heavily parasitized will not perform these rituals as well as healthy ones and that by choosing healthy mates one's offspring are more likely to acquire genes conferring resistance to parasitic diseases.

C. Speciation and Wolbachia

If one were to describe multicellular life on this planet, one would unquestionably conclude that it is dominated by arthropods. Approximately 80% of all species are arthropods. It is not intuitively obvious why speciation has been so pronounced in this one phylum, but there is a growing body of evidence to indicate that a very interesting intracellular bacterial parasite, *Wolbachia*, may have played a role.

Wolbachia are cytoplasmically transmitted/inherited rickettsia bacteria found in the reproductive tissues of arthropods. Mitochondria, the powerhouse of the eukaryotic cell, are believed to have descended from an ancestor of this group. *Wolbachia* influence host reproduction in at least three ways. First, they can cause reproduction isolation through cytoplasmic incompatibility. For some parasitized arthropod species, neither sperm nor eggs will form viable zygotes with unparasitized gametes. In other species the isolation is not as complete; eggs of parasitized hosts can be fertilized by either unparasitized or unparasitized sperm, but parasitized sperm is only compatible with parasitized eggs. Second, in another reproductive isolation mechanism, *Wolbachia* can induce some species to reproduce solely by parthenogenesis. Third, the bacteria have been shown to convert male hosts into reproductively active females. The duration of the these effects over generations is not well understood, but the effects of *Wolbachia* are completely reversible through the application of antibiotics suggesting that some host species may be able to resist infections. Whether *Wolbachia* is partly responsible for the great numbers of arthropods species by increasing the occurrence of reproductive isolation in arthropods relative to other phyla remains to be seen, but the possibility is intriguing.

Bibliography

Anderson, G. (1977). The effects of parasitism on energy flow through laboratory shrimp populations. *Mar. Biol.* 42, 239–251.

Clay, K. (1991). Parasitic castration of plants by fungi. *Trends Ecol. Evol.* 6, 162–166.

Cooper, L. A., Larson, S. E., and Lewis, F. A. (1996). Male reproductive success of *Schistosoma mansoni*-infected *Biomphalaria glabrata* snails. *J. Parasitol.* 82, 428–431.

De Jong-Brink, M. (1992). Interference of schistosome parasites with neuroendocrine mechanisms in their snail host causes physiological changes. *Adv. Neuroimmunol.* 2, 199–233.

De Jong-Brink, M., Hoek, R. M., Smit, A. B., Bergamin-Sassen, M. J. M., and Lageweg, W. (1995). Schistosoma parasites evoke stress responses in their snail host by a cytokine-like factor interfering with neuro-endocrine mechanisms. *Netherlands J. Zool.* 45, 113–116.

Hamilton, W. D., and Zuk, M. (1982). Heritable true fitness and bright birds: A role for parasites? *Science* 218, 384–386.

Hamilton, W. D., Axelrod, R., and Tanese, R. (1990). Sexual reproduction as an adaptation to resist parasites (A review). *Proc. Natl. Acad. Sci. USA* 87, 3566–3573.

Hansen, H. J. (1923). *Crustacea Copepoda II. The Danish Ingolf Expedition*, Vol. III. Bianco Luno, Copenhagen

Kaestner, A. (1990). *Invertebrate Zoology, Crustacea*, Vol. III. Wiley, New York.

Kuris, A. M. (1974). Trophic interactions: Similarity of parasitic castrators to parasitoids. *Q. Rev. Biol.* 49, 129–148.

Lively, C. M. (1987). Evidence from a New Zealand snail for the maintenance of sex by parasitism. *Nature* 328, 519–521.

Lively, C. M. (1992). Parthogenesis in a freshwater snail: Reproductive assurance versus parasitic release. *Evolution* 46, 907–913.

Lively, C. M. (1996). Host–parasite coevolution and sex. *Bioscience* 46, 107–114.

Lively, C. M., Craddock, C., and Vrijenhoek, R. C. (1990). Red queen hypothesis supported by parasitism in sexual and clonal fish. *Nature* 344, 864–866.

Manger, P. H., Li, B. M., Christensen, M., and Yoshino, T. P. (1996). Biogenic monoamines in the freshwater snail, *Biomphalaria glabrata*: Influence of infection by the human blood fluke, *Schistosoma mansoni*. *Comp. Biochem. Physiol.* A 114, 227–234.

Minchella, D. J. (1985). Host life-history variation in response to parasitism. *Parasitology* 90, 205–216.

O'Brien, J., and van Wyk, P. (1985). Effects of crustacean parasitic castrators (epicaridean isopods and rhizocephalan barnacles) on growth of crustacean hosts. In *Crustacean*

Issues 3: Factors in Adult Growth (A. Wenner, Ed.), pp. 191–218. Balkema Press, Rotterdam, Amsterdam.

Orians, G. H. (1985). *Blackbirds of the Americas*. Univ. of Washington Press, Seattle.

Richner, H., Christe, P., and Oppliger, A. (1995). Paternal investment affects prevalence of malaria. *Proc. Natl. Acad. Sci. USA* 92, 1192–1194.

Sullivan, J. T., Lares, R. R., and Galvan, A. G. (1998). *Schistosoma mansoni* infection inhibits maturation of ovotestis allografts in *Biomphalaria glabrata* (Mollusca: Pulmonata). *J. Parasitol.* 84, 82–87.

Van Wyk, P. M. (1982). Inhibition of the growth and reproduction of the porcellanid crab *Pachycheles rudis* by the bopyrid isopod, *Aporobopyrus muguensis*. *Parasitology* 85, 459–473.

Wade, M. J., and Chang, N. W. (1995). Increased male fertility in *Tribolium confusum* beetles after infection with the intracellular parasite *Wolbachia*. *Nature* 373, 72–74.

Werren, J. H. (1997). Biology of *Wolbachia*. *Annu. Rev. Entomol.* 42, 587–609.

Wilson, E. O. (1971). *The Insect Societies*. Harvard Univ. Press, Cambridge MA.

Parasitoids

Nancy E. Beckage
University of California, Riverside

I. Introduction
II. Reproductive Adaptations of Parasitoids
III. Parasitoids as Inhibitors of Host Reproduction
IV. Viruses Associated with the Reproductive Tract of Parasitoids
V. Reproductive Cues in Parasitoids
VI. Sex Ratio Determination in Parasitoids

GLOSSARY

arrhenotokous Describing or referring to a process of reproduction whereby an unfertilized egg gives rise to a male haploid individual and a fertilized egg gives rise to a female diploid individual.

ecdysteroids The class of arthropod molting hormones.

egg resorption The process whereby eggs are recycled and are broken down during periods when parasitoids are unable to oviposit and fail to find hosts to parasitize.

juvenile hormone The hormone produced in the larval or nymphal stages of insects which preserves the larval state prior to metamorphosis.

teratocyte A cell derived from the serosa of endoparasitic wasps which persists in the hemolymph of the host.

vitellin Egg yolk protein derived from vitellogenin.

vitellogenin The hemolymph-borne protein destined to become vitellin following uptake by a developing oocyte.

Parasitoids are parasites of insect hosts and usually cause the premature death of their host prior to the onset of its reproductive stage. The fact that they cause death of their host distinguishes them from parasites, which debilitate but do not kill their host. Due to the fact they cause host death, many parasitoids are widely exploited in the biological control of insect pests since host mortality is invoked. The reproduction of the host is stopped even if its current feeding activity is not. In contrast, most other parasites fail to kill their host, and indeed parasitism may prolong the life of the host so that both partners intimately comingle for an extended period. Being attacked by a parasitoid has a marked influence on host reproduction, and parasitic castration frequently occurs, even if the host dies before metamorphosis occurs. Hosts that survive until the adult stage suffer fecundity reduction. The parasitoids themselves exhibit several reproductive adaptations allowing them to exploit the host resources available to them. In the Braconidae and Ichneumonidae, viruses are produced in the reproductive tract and transmitted to the host during parasitization. Several reproductive cues associated with hosts are exploited by parasitoids, including odors of the host plant and odors of the host itself, that trigger host searching and oviposition behaviors. Hymenopterous parasitoids utilize arrhenotokous mechanisms of sex determination. The sex ratio is skewed toward males or females depending on certain characteristics of the host, including its size as assessed by the female at the time of oviposition.

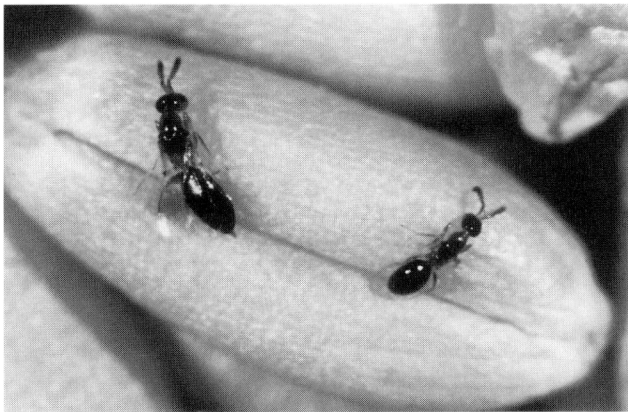

FIGURE 1 Photo showing a female (left) and male (right) of the pteromalid *Choetospilla elegans* photographed on a wheat kernal. The wasps are seeking hosts that feed within the wheat kernel. This species parasitizes *Rhyzoperhta dominica* (lesser grain borer) and *Sitophilus oryza* (rice weevil); the wasp stings the host and then deposits one egg externally (photo provided by Paul Flinn).

I. INTRODUCTION

Most species utilized in biological control of insect pests are Hymenoptera, and these species are emphasized in this review. Dipteran parasitoids such as tachinid flies may also figure prominently in the natural control of pests but are not often reared for control purposes. As a result, compared to the body of literature on fly parasitoids, much more is known about the biology of the parasitic Hymenoptera, particularly the Braconidae and Ichneumonidae. Most of the comments outlined in this review are particularly relevant to these species.

II. REPRODUCTIVE ADAPTATIONS OF PARASITOIDS

Hymenoptera exhibit arrhenotokous reproduction in which haploid progeny, developing from unfertilized eggs, are destined to become males and fertilized eggs give rise to females. Thus, even in the absence of mating, reproduction may occur, with the nonmated female producing an all-male brood. Following mating, the female may preferentially deposit fertilized eggs on larger hosts and unfertilized eggs on smaller prospective hosts. How females with this haplodiploid strategy of sex determination regulate the fertilization process based on sensory input they receive about host quality and size remains unknown. Female insects are often larger than males and hence may be more heavily parasitized due to their larger, preferred size.

In bethylid parasitoids, which are aculeate hymenopterans related to the social bees and wasps, the female stings the host larva with a potent venom to paralyze the host and then deposits its progeny externally. Several pteromalids also often live in stored products (i.e., grain or processed materials; Fig. 1) and the females first find and then parasitize beetles or moth larvae in this cryptic environment. The number of parasitoids deposited per host often will vary positively with host size; often the clutch laid is entirely female except for a single male deposited as the last egg of the clutch. With particularly large hosts, ectoparasitoids may deposit multiple clutches of eggs on a single host (e.g., *Euplectrus plathypenae* developing on tobacco hornworm larvae; Fig. 2).

With the external parasitoids, the developing larvae may also secrete potent paralyzing salivary secretions while they imbibe nutrients contained in the host. Pupation occurs externally adjacent to the host remains. The male (which often ecloses before the females) may then mate with his sisters following adult emergence of the wasps. Many facultatively

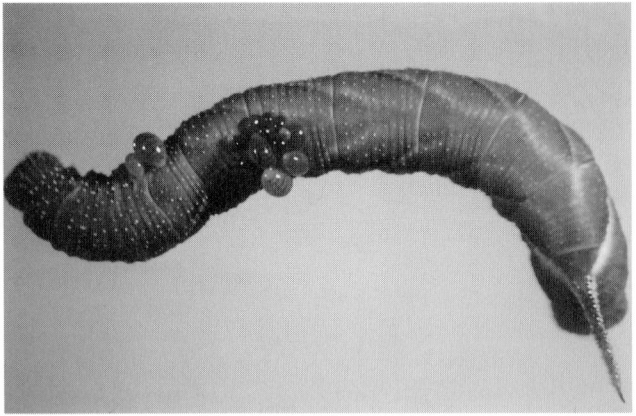

FIGURE 2 Photo showing a third-instar larva of the tobacco hornworm, Manduca sexta, with attached Euplectrus plathypenae. The host has been parasitized with two clutches of eggs, which are now developing as larvae on the dorsum of the host. The parasitoid larvae are blue-green, similar to the integument and hemolymph of the host.

gregarious parasitoids will deposit variable numbers of eggs within the body cavity (endoparasitoids) of the host in accordance with the size of the host. Even parasitoids that attack host eggs may deposit variable numbers of eggs within the host egg, which varies with egg diameter and circumference (e.g., Trichogramma).

As seen in other insects, the size of parasitoid wasps, particularly females, determines the level of their fecundity. Several size parameters, including the tibia, thorax, or wing length (or ovipositor length in females), are often used to assess adult size in very small parasitoids, which may be too small to weigh individually. A positive relationship between body size and fecundity has been demonstrated in many parasitoids. Male reproductive potential is more difficult to assess compared to female fecundity as monitored by assessing egg load and the number of ovarioles containing vitellogenic eggs and the number which are chorionated (ready to be laid). Total lifetime fecundity is assessed by exposing a female to as many hosts as she will parasitize and counting the number of progeny she lays.

The wasp's nutritive status is also an important determinant of fecundity; some species must have a protein source (i.e., these species may normally feed on host hemolymph) in order to mature chorionated eggs or continue producing them. Hosts which are fed upon are usually rejected as being suitable for oviposition. Some species select small prospective hosts for host feeding preferentially, whereas larger hosts are utilized for oviposition. Others need only sugars (contained in flower nectar) in order to produce mature eggs. Continuous access to a flower nectar source (or sugars) appears important to maintain the highest levels of fecundity and longevity in many parasitoid species. In the absence of adequate nutritive resources, few or no eggs are produced, and longevity is dramatically reduced. Other species can live for long periods without access to carbohydrate, water, or a protein source.

Hymenopteran parasitoids that develop as internal parasites of other insects and are endoparasitic (i.e., species such as Cotesia congregata) produce microtype oocytes in which little yolk is incorporated into the egg and the chorion is extremely thin, allowing hydration of the egg to occur immediately upon deposition of the egg in the host hemocoel (Figs. 3 and 4). Some species allow their host to continue to grow and molt (koinobionts), whereas others cause an immediate cessation of host growth and development (idiobionts).

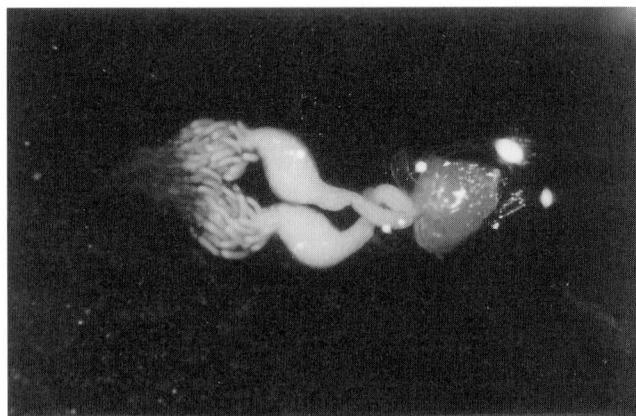

FIGURE 3 Photo showing dissected ovaries of Hyposoter exiguae. The ovarioles are seen at the far left, and the chorionated oocytes are visible at the proximal ends of the ovarioles. The venom gland is seen as a transparent thin filamentous sac attached to the reproductive tract at the common oviduct (photograph by Frances Tan; wasp provided by Jennifer Thaler).

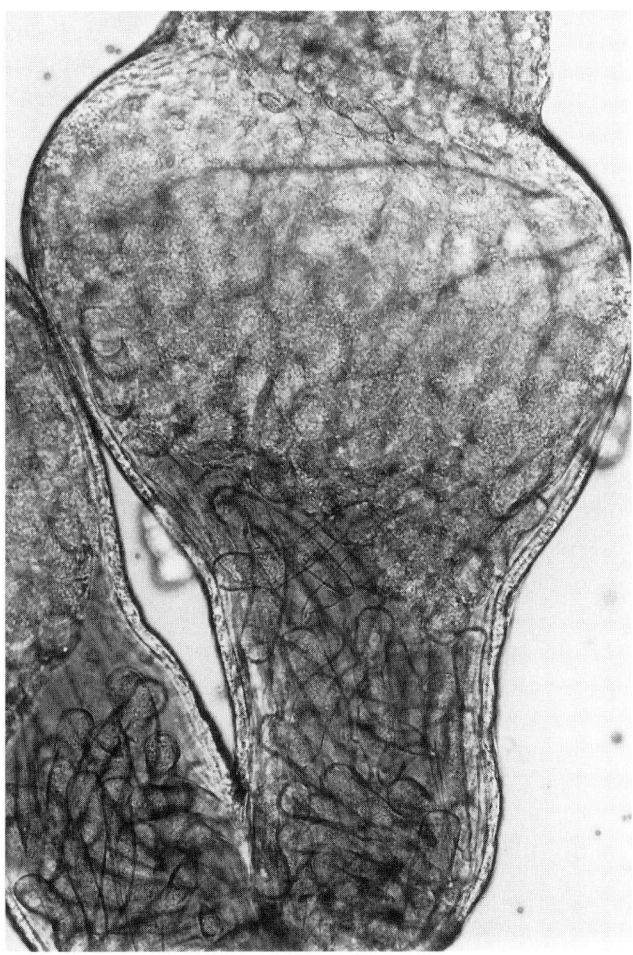

FIGURE 4 Light micrograph of the ovary of the endoparasitic wasp *Cotesia congregata* showing that multiple microtype eggs are ready to be oviposited following their maturation in the calyx of the ovary. In the calyx the eggs are mixed with calyx fluid containing polydnavirus particles (Fig. 6) (photograph provided by Isaure de Buron).

Complex developmental and endocrine interactions involving juvenile hormone, ecydsteroids, and neuropeptides serve to coordinate the development of the parasitoid(s) with that of the host. Developing parasitoids also secrete hormones including juvenile hormone and ecdysteroids into the hemolymph of their host. Parasite-derived juvenile hormone may prevent host metamorphosis, whereas ecdysteroids produced by the wasp may have an effect on the molting status of the host. Thus, a bidirectional endocrine communication pathway exists between host and parasitoid, with the parasitoid being responsive to hormones derived from the host and vice versa. Alterations in the metabolic pathways of hormone degradation may also occur. For example, many endoparasitoids interfere with or precociously induce the expression of juvenile hormone esterase in their host.

As mentioned earlier, ectoparasitoid wasps (e.g., *Bracon hebetor*) usually paralyze their host with a potent venom or arrestment agent, ejected from the venom or poison gland into the host, preparatory to deposition of the eggs externally on the cuticle. The venom may be injected at a specific site (e.g., near the subesophageal ganglion) or randomly along the body surface at multiple sites. Once the wasp detects that the host is semiparalyzed, elaborate grooming of the host (removal of setae, etc.) may or may not be involved prior to egg deposition, and then the host may be moved to a concealed site before parasitization. Once eggs are deposited, the adult female wasp may (depending on the species) defend her brood from other intruding females that approach the newly parasitized host. The parasitoids hatch and the larvae grow on the external surface while the host itself shrinks as its body fluids are consumed by the developing parasitoids. Not surprisingly, the parasitoids often turn blue-green as they consume host hemolymph of the same color. Nothing may remain of the host in the end except its cuticle or head capsule (Fig. 5).

Still another pattern of reproduction is seen in tachinid flies, which deposit eggs or larvae (depending on the species) on the external surface of the head capsule or anterior thoracic segments of the host, where the host is unable to groom, that then burrow as first-instar larvae into the body cavity of the host. Alternatively, the tachnids may be larvaposited on vegetation and then ingested by the host, burrow through the gut wall, and take up residence in the hemocoel. Their mode of reproduction differs from that of Hymenoptera in that all progeny are diploid. In addition, the larvae may be encased within a capsule of host hemocytes. They may develop initially as free-floating larvae then attach to a network of host trachae emanating from the spiracles, obtaining oxygen directly from the respiratory system of the host. The larvae may remain attached to the trachae for the duration of parasitism or make peri-

FIGURE 5 Photo showing the remains of the beetle *Trogoderma granarium* with two developing larvae of the bethylid parasitoid *Laelius pedatus*. Note that little remains of the host except its cuticle, whereas the parasitoids are quite large relative to the host (photograph provided by Janet Klein)

odic migrations into the ecdysial space during molting of the host, as does *Gonia cinerescens* in host larvae of the wax moth, *Galleria mellonella*. Tachinid larvae may be lost during this migration process. Depending on the species, the early instars may be strictly hemolymph borne with the larvae only attaching to the spiracles in the final instars.

In hymenopteran parasitoids, egg resorption occurs when no appropriate hosts are found and the contents of the mature eggs are eventually recycled to provide nutrients for new developing oocytes. The fully vitellogenic eggs that normally would complete the maturation process are resorbed to conserve the nutrient constituents (i.e., vitellin) normally incorporated into the egg in large amounts as the oocyte descends through the ovariole. Assessing the reproductive status of parasitoid females caught in the wild thus may be complicated by the simultaneous occurrence of oocyte resorption as well as oocyte maturation. The phenomenon of resorption has not been well studied at the physiological level compared to the process of vitellogenesis, for which several models now exist in different taxa. In the absence of hosts and/or sources of nutrients such as carbohydrates, resorptive processes are favored over vitellogenic pathways.

The reproductive potential of some parasitoids is truly enormous, and parasitoids have evolved several strategies to compensate for the fact that their progeny often suffer high rates of mortality prior to maturity. Finding a host, followed by successful development within a host, is often a high-risk proposition for a parasitoid, and hence many have compensated for high mortality at the population level by producing very large numbers of offspring. In gregarious species, multiple eggs, varying from just a few to several hundred, are deposited within the host. A mean of about 150 eggs is deposited into hosts of *Cotesia congregata*; preliminary evidence suggests that more eggs are deposited into older larger hosts (tobacco hornworms) compared to early instar larvae. The ovaries of these females contain many mature eggs ready for oviposition soon after the wasps eclose. Other gregarious species have been shown to exhibit a similar phenomenon, in which more parasites are deposited into larger hosts. These gregarious species deposit multiple small eggs instead of a single large egg, and the individual adult parasitoids are much smaller compared to a solitary species (*Hyposoter exiguae*) which attacks the same host, *Manduca sexta*. Parasitized *M. sexta* larvae containing *C. congregata* wasps grow larger; thus, metabolism of the host appears to be altered such that more nutrients become available to feed a larger collective mass of parasitoids. In the tobacco hornworm, a total of about 200 parasitoids appears to be the maximum number of successfully emerging parasitoids even though several hundred may be present within a single host. This number approximates the host's carrying capacity in terms of parasitoid output per host, at least when the host is parasitized at the beginning of the fourth instar.

Another strategy evolved by parasitoids is polyembryony, a process in which several hundred (or even thousands) of genetically identical progeny develop from a single fertilized egg. This strategy arose independently in four different parasitic groups: the Platygasteridae, Braconidae, Encyrtidae, and Dryinidae. The cells comprising the primary morula separate from each other to form multiple morulae, each of which develops into a complete adult wasp which is genetically identical to its siblings. As many as a thousand or more wasps may emerge from a single host. Some species have evolved characteristic sterile precocious or "defender morphs" with large mandibles which protect the developing brood from invading wasp parasitoids within the host. These defender morphs are fated to die within the host while the main brood of reproductive larvae survives. When a host is attacked by polyembryonic species, nearly 100% of the host's tissues may be transformed into a "mummy" composed of developing parasitoids. In contrast, when parasitized tobacco hornworms are attacked by the gregarious species *C. congregata,* no more than 20–30% of the mass of the host–parastoid complex may be represented by the parasitoids despite the presence of several hundred parasitoids in the hemocoel. Thus, the polyembryonic species appear more efficient in terms of converting host biomass to parasitoid biomass on a gram-for-gram basis.

Yet another anomaly unique to parasitoids is the formation of teratocytes, or "giant cells," arising from the serosa enclosing first the embryo and then the developing first-instar larva (Fig. 6). Teratocytes are released into the host hemocoel upon hatching of the first instar and persist in the hemolymph for the duration of parasitism by braconid species. In tobacco hornworm larvae parasitized by *C. congregata,* the cells grow rapidly from 15 to 150–200 μm in diameter within a short period of time. During the process, they secrete a variety of proteins and develop large secretory vesicles on their surface. Their precise role remains unknown but is hypothesized to be nutritive or to be regulative in nature, facilitating successful parasitism by the wasps.

III. PARASITOIDS AS INHIBITORS OF HOST REPRODUCTION

Like other parasites, parasitoids have been demonstrated to be effective parasitic castrators, inhibiting the reproduction of their host organism. Parasites such as mermithid nematodes (worm-like parasites) cause sex reversal in mayflies, affecting the secondary sexual morphology (evident in the head, eyes, and genitalia) and behavior of the adult mayfly host; males are transformed to female intersexes and some infected males even undergo a complete sex reversal. The behavior of the host also undergoes a transformation in that genetic males behave similarly to females in flying upstream similar to nonparasitized female counterparts. Thus, the nervous system and behavioral programming are also involved in the sexual transformation process in parasitized mayflies.

In addition to influencing the reproduction of their hosts, parasitoids also affect other aspects of their host's physiology, including its immune responses (many "knock out" the host immune responses, at least transitorily), hemolymph proteins, metabolic physiology, and behavior (feeding, locomotion, and mating). Virtually no aspect of the host's physiology remains aloof to endoparasitism.

Even in species in which the host dies prior to metamorphosis, development of the gonads frequently is adversely affected. For example, in egg–larval parasitoids such as *Chelonus* or *Ascogaster,* the presence of the parasitoid stops development of the testes. These parasitoids oviposit into egg-stage hosts, and the resultant progeny emerge during the larval stage of the host. The earlier during embryonic development that parasitization occurs, the more drastic is the observed effect, with hosts parasitized early during embryonic development having virtually nonexistent testes. Effects on females are more

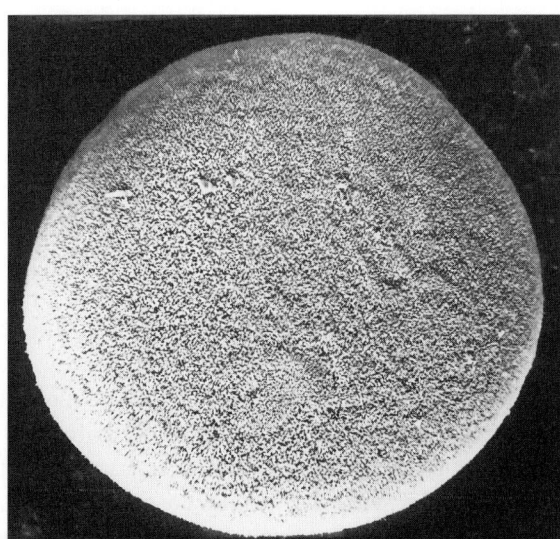

FIGURE 6 Scanning electron micrograph showing a teratocyte of *Cotesia congregata* isolated from the hemolymph of a host *M. sexta* larva. Note that the surface is covered with dense microvilli, and the cells are thought to have an adsorptive, nutritive role which facilitates successful parasitism (photograph provided by Isaure de Buron)

difficult to discern since ovarian development normally does not begin until the pupal stage.

In larval–larval parasitoids, testicular inhibition may also occur in hosts initially parasitized in the larval stage; for example, in tobacco hornworm larvae parasitized by *C. congregata*, atrophy and disintegration of the testes occur if parasitization occurs during the first instar. The later during larval life the host is parasitized, the lesser are the effects induced, but still the testes fail to proliferate at the normal rate even if parasitization does not occur until the fourth or fifth instar. Whether effects occur on the ovary is not clear because the female gonad normally does not differentiate until the pupal stage, and the host larvae invariably die prematurely following induction of developmental arrest prior to wandering and pupation.

Fecundity of adult female insects is adversely affected by the presence of a variety of parasites, including cestodes, nematodes, and tachinid parasitoids. Vitellogenin production by the fat body is sometimes diminished. In other cases, vitellogenin synthesis continues unabated (as in beetles of *Tenebrio molitor* carrying tapeworm *Hymenolepis* metacestodes) and instead the process of vitellogenesis and incorporation of the protein into the oocyte as vitellin is inhibited. A lack of juvenile hormone (JH), which is required for normal vitellogenic processes, explains the inhibition of vitellogenesis during parasitism; in other cases, as in the tapeworm-carrying beetles, the levels of JH and JH esterase appear relatively normal while the ovary fails to accumulate vitellin. The main lesion appears to be the failure of the ovary to develop follicular patency so that the follicle cells move apart from each other as needed to allow vitellogenin to move between the cells to enter the oocyte.

IV. VIRUSES ASSOCIATED WITH THE REPRODUCTIVE TRACT OF PARASITOIDS

A variety of virus-like elements are associated with the reproductive tract of parasitoids (Fig. 7), including polydnaviruses, baculovirus-like filamentous particles, poxviruses, and other agents. Their replication proceeds either in the outermost calyx epithelium, within inner nonepithelial layers lining the ovary, or in nonovarian tissues such as the venom or poison glands associated with the reproductive tract. The sheer variety of types of particles produced is intriguing and suggestive that parasitoids initiated associations with virus-like particles, derived either from pathogens of the wasps or from their hosts, on more than one occasion during their evolutionary history. The biological function of the virus particles appears to be, in those species in which this has been studied, physiological regulation of the host insect larva.

The genomes of the polydnaviruses are incorporated into the genomes of the respective parasitoids that carry them, and they replicate solely in the reproductive tissues of the adult female parasitoid wasp. How replication occurs is not known, but it is initiated during the pupal stage during the onset of adult development prior to eclosion of the adult wasp. Ecdysteroid production during adult development appears to trigger virus replication, suggesting the viruses may have hormonally activated replication mechanisms. Replication of the viruses may (bracoviruses) or may not (ichnoviruses) lyse the ovarian cells, in which they are assembled into mature virions. The bracoviruses have a single envelope and undergo nuclear replication; the ichnoviruses replicate in the cytoplasm, bud from the cell in which they replicate, and are enclosed by a double envelope.

In addition to braco- or ichnoviruses, other viral forms may also be present. For example, in *C. congregata*, filamentous long particles (which appear to be baculovirus-like) are present in addition to the polydnaviruses. Whether the baculovirus-like or poxvirus-like viral particles likewise have integrated forms of their genomes localized within the genomes of the wasps that "vector" the viruses remains to be established. How the parasitoids have managed to capture viral elements for the purpose of utilization of the viruses for physiological regulation of the host remains an intriguing enigma. In most systems examined, the viruses have been found to cause immunosuppression of the host and suppression of the encapsulation of the invading parasitoids; this phenomenon, linked to onset of apoptosis or other

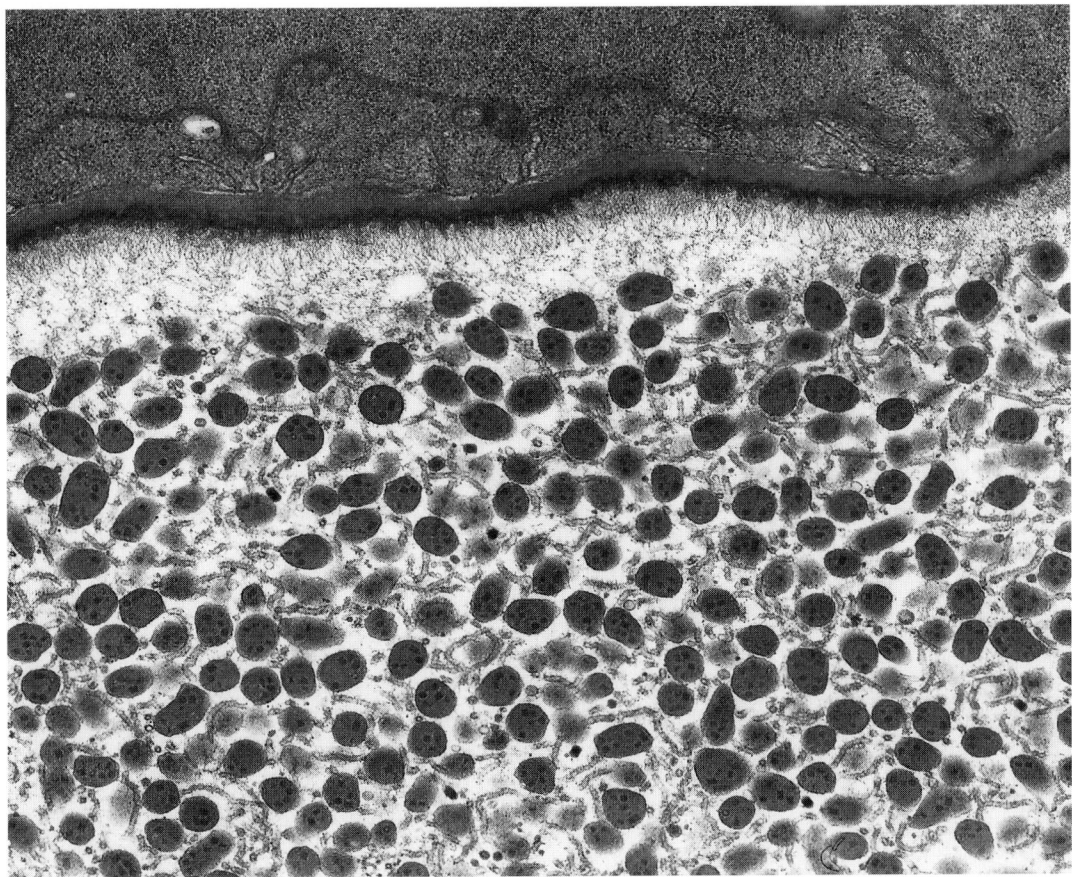

FIGURE 7 Transmission electron micrograph showing the surface of a *Cotesia congregata* egg in the calyx (above) and the abundant polydnavirus virions in the calyx fluid (below) that are injected during oviposition into the host larva. Each polydnavirus virion contains multiple nucleocapsids and has a prominent tail. Some naked nucleocapsids can be seen in the calyx fluid. Note that virions are not embedded in the chorion of the egg but are injected apart from the egg, during parasitization (photograph provided by Isaure de Buron)

dysfunctions (i.e., failure to spread and adhere to a substrate) in host hemocytes, may either be transitory in its nature or last for the duration of endoparasitism. Other polydnavirus roles appear to be the expression of polydnavirus-encoded hemolymph proteins seen only in parasitized insects (so-called "parasitism-specific proteins") and the induction of developmental arrest of the host larva such that its metamorphosis is inhibited. In other species, namely, in noctuid larvae parasitized by wasps belonging to the genus *Chelonus*, precocious metamorphosis is induced by the presence of the parasitoid combined with the polydnavirus and venom injected by the female parasitoid into the host.

In one instance of a poxvirus transmitted by *Diachasmamimorpha longicaudatus*, the virus has been shown to replicate in host hemocytes following parasitization of host dipteran (Caribbean fruit fly) larvae. Hence, the viruses may or may not replicate following transfer to the host insect carrying the parasitoid.

V. REPRODUCTIVE CUES IN PARASITOIDS

How does a parasitoid locate a prospective host? Many behavioral ecology studies have shown that parasitoids use a complex combination of chemical and visual cues to locate their host. Some species even exploit vibration of host larvae within plant tissues to find their respective host. However, the neurophysiological basis of host finding behavior remains largely unexplored.

A combination of antennal and ovipositor receptors are utilized by the female in locating her host and determining its parasitized (or nonparasitized) status. For example, a nonmotile but nonetheless chemically attractive host may be rejected by ovipositing females, which detect such larvae as previously parasitized by other females. Vibrational cues are often detected by the ovipositor receptors, especially in parasitoid species which attack hosts that construct galleries, mines, or other internal conduits in plant tissues. If damaged plant tissue (or pitch, for example, in the case of parasitoids seeking beetles which construct galleries within tree trunks) is identified, but then an immobile host is localized in the tissue, no egg is laid by the searching female parasitoid. Parasitoids of leaf miners show this type of assessment of the mobile versus immobile status of their prospective host and reject an otherwise appropriate but immobile host.

Plant cues are very important in the host finding process for many parasitoids. Plant chemicals combined with those biochemical factors emitted by the host may provide the most potent searching cues. Host kairomones (i.e., volatile compounds) emanating from the host clearly are important. Antennal searching is the first response elicited when the parasitoid female comes into contact with host integument, exuviae (cast skins), salivary secretions, or frass produced by the prospective host larva. Ovipositor probing may follow, assuming an appropriate host is found, to allow the female to assess the parasitization status of the host.

The solitary species appear quite deliberate in their oviposition behavior. In contrast, gregarious species, including female wasps of the species *C. congregata*, may display rapid oviposition decision-making behaviors, either accepting or rejecting a host very quickly and choosing to frequently superparasitize previously parasitized hosts while exhibiting a rapid hypodermic oviposition behavior. Nonetheless, some gregarious species do lay more eggs in larger hosts, or alter their clutch size according to host species, to most efficiently utilize host resources without causing undue stress to the host involved, which would jeopardize the survival of both parasitoid and host.

When prospective hosts are reared on plants, the odor(s) of host-conditioned damaged plant foliage is particularly attractive to many parasitoids. Volatiles are emitted which clearly attract parasitoids to the insect-damaged foliage, providing the first cues to draw a parasitoid to the search arena. Chemicals emitted following the mixing of host insect salivary components with plant chemicals during plant wounding are particularly attractive. Caterpillar-induced damage may be sensorily distinguished from mechanically induced plant damage by searching females.

Following parasitization, the females of many solitary species mark the host by dragging their ovipositor across the surface of the newly parasitized host, depositing kairomones to warn conspecifics of the already-parasitized status of the hosts. For example, the egg-recognition kairomone detected by the scelonid parasitoid *Trissolcus basalis* is a mucopolysaccharide produced by the ovariole follicular cells and is a component of the adhesive used by the wasp for attaching eggs to the substrate and to each other. This compound is detected by the female's antennae while searching for new hosts to parastize. Other species are thought to mark the host internally by depositing chemicals detected by later arriving females via their ovipositor. Later they discriminate and reject previously parasitized hosts when searching for prospective hosts to parasitize.

In contrast, gregarious species are less discerning about marking their host so that the host larvae often are superparasitized by other females of the same species. Moreover, a single parasitoid may oviposit more than once into the same host larva. This behavioral difference in solitary versus gregarious species is not unexpected given the life histories of many gregarious species and the fact that more than a single individual parasitoid can emerge from each host. Gregarious parasitoids may optimize their chances of reproduction by parasitizing all available hosts of an appropriate species without regard to whether they have been previously parasitized simply to optimize their chances of successful reproduction.

VI. SEX RATIO DETERMINATION IN PARASITOIDS

Wolbachia is an intracellular rickettsia which drastically affects the sex ratio of the *Trichogramma* parasitoids that carry it. Specifically, those populations of *Trichogramma* carrying *Wolbachia* become thelotokous and produce an all-female brood. Like other Hymenoptera, the host *Trichogramma* has a haplodiploid method of sex determination but in individuals in which *Wolbachia* is present in the cytoplasm of an egg, it fails to undergo the first cleavage division and becomes diploid. Therefore, a diploid female develops. Thelytokous strains of *Trichogramma* may be converted to arrhenotokous wasps by antibiotic treatment, suggesting a microbe is responsible for production of all-female broods. *Wolbachia* have also been isolated from *Nasonia vitripennis*, as well as other species in this genus, and may be responsible for reproductive isolation of species within this genus. In other insects, such as flies and mosquitoes, infected males are unable to fertilize uninfected eggs, a phenomenon which has been termed cytoplasmic incompatibility. Thus, *Wolbachia* has a variety of effects on reproduction of the insect species carrying this agent. The presence of *Wolbachia* and the occurrence of cytoplasmic incompatibility has been confirmed in the parasitic wasps, flour beetles, weevils, planthoppers, moths, mosquitoes, and fruit flies, indicative of a widespread distribution among insect species.

See Also the Following Articles

ECDYSTEROIDS; JUVENILE HORMONE; OVULATION AND OVIPOSITION, INSECTS

Bibliography

Beckage, N. E. (1997). New insights: How parasites and pathogens alter the endocrine physiology and development of insect hosts. In *Parasites and Pathogens: Effects on Host Hormones and Behavior* (N. E. Beckage, Ed.), pp. 3–36. Chapman & Hall, New York.

Beckage, N. E., Thompson, S. N., and Federici, B. A. (Eds.) (1993). *Parasites and Pathogens of Insects. Vol. 1. Parasites.* Academic Press, New York.

Brodeur, J., Geervliet, J. B. F., and Vet, L. E. M. (1997). The role of host species, age and defensive behavior on ovipositional decisions in a solitary specialist and gregarious generalist parasitoid (*Cotesia* species). *Ent. Exp. Appl.* **81**, 125–132.

Geervliet, J. B. F., Vet, L. E. M., and Dicke, M. Innate responses of the parasitoids *Cotesia glomerata* and *C. rubecula* (Hymenoptera: Braconidae) to volatiles from different plant–herbivore complexes. *J. Insect Behav.* **9**, 525–538.

Godfray, H. C. J. (1994). *Parasitoids: Behavioral and Evolutionary Ecology.* Princeton Univ. Press, Princeton, NJ.

Hurd, H., and Webb, T. (1997). The role of endocrinological versus nutritional influences in mediating reproductive changes in insect hosts and insect vectors. In *Parasites and Pathogens: Effects on Host Hormones and Behavior* (N. E. Beckage, Ed.), pp. 179–197. Chapman & Hall, New York.

King, B. H. (1993). Sex ratio manipulation by parasitoid wasps. In *Evolution and Diversity in Sex Ratios in Insects and Mites* (D. L. Wrensch and M. S. Ebbert, Eds.), pp. 404–417. Chapman & Hall, New York.

Mackauer, M., Machaud, J. P., and Volkl, W. (1996). Host choice by aphidiid parasitoids (Hymenoptera:Aphidiidae): Host recognition, host quality, and host value. *Can. Entomol.* **128**, 959–968.

Quicke, D. L. J. (1997). *Parasitic Wasps.* Chapman & Hall, New York.

Pfister-Wilhelm, R., and Lanzrein, B. Precocious induction of metamorphosis in *Spodoptera littoralis* (Noctuidae) by the parasitic wasp *Chelonus inanitus* (Braconidae): Identification of the parasitoid larva as the key regulatory element and the host corpora allata as the main targets. *Arch. Insect Biochem. Physiol.* **33**, 511–525.

Terrasse, C., Nowbahari, B., and Rojas-Rousse, D. (1996). Sex ratio regulation in the wasp *Eupelmus vuilleti*, an ectoparasitoid on bean weevil larvae (Hymenoptera: Pteromalidae). *J. Insect Behav.* **9**, 251–263.

Vance, S. A. (1996). Morphological and behavioural sex reversal in mermithid-infected mayflies. *Proc. R. Soc. London B* **263**, 907–912.

Werren, J. H. (1997). Biology of *Wolbachia*. *Annu. Rev. Entomol.* **42**, 587–609.

Paraspermatozoa

Alan N. Hodgson

Rhodes University

I. Introduction
II. The Structure of Paraspermatozoa
III. The Formation of Paraspermatozoa
IV. The Functions of Paraspermatozoa
V. Other Forms of Invertebrate Sperm Dimorphism

GLOSSARY

acrosome Membrane-bound vesicle in the head of an animal sperm which is first to fuse with the egg membrane and enter the egg cell. The acrosome is usually positioned anterior to the nucleus.

axoneme The central structure of a cilium or flagellum, composed of microtubules (and cross-bridges) which are usually arranged as a ring of nine outer pairs (doublets) and a central pair of single microtubules.

chromatin The nucleic acid–protein complex found in eukaryotic chromosomes.

euspermatozoa Sperm which fertilize the egg and therefore contribute to the zygote genome.

seminal vesicle Regions of the male reproductive system, usually diverticula of the vas deferens, which store sperm.

spermatheca A sac or receptacle in a female or hermaphrodite animal in which sperm cells received from a male or another individual are stored until required to fertilize the eggs.

spermatophore A mass of spermatozoa held together by mucilage (or similar substance) which enables the animal to manipulate or direct it to the receiving parts of the female.

spermatozeugmata Sperm aggregates implanted in the spermatheca by the concopulant, characterized by repetitive order of the spermatozoa and the presence of some kind of cementing agent but lacking a proper capsule.

Paraspermatozoa are sperm cells produced in the testis of invertebrates which may aid, but do not contribute genetically to fertilization. Such sperm cells usually possess one or more axonemes which may be free (such as a flagellum or flagella) or partly or completely incorporated into the cell body. The possession of flagella and the release of them by the male along with fertilizing sperm are features which distinguish parasperm from "nurse cells" and Sertoli cells, which are also found in the testes of invertebrates.

I. INTRODUCTION

Some invertebrates form more than one type of sperm within the testis and are therefore said to produce dimorphic (or even polymorphic) sperm. Sperm dimorphism was first observed in the prosobranch gastropod *Paludina vivipara* by von Siebold in 1836. He described a flagellate sperm with a head and a tail and a multiflagellate form with very little chromatin. Since this first description, dimorphic sperm have been discovered to be a feature not only of some gastropod taxa but also of some annelids, arthropods (Myriopoda, Arachnida, and Lepidoptera), and one echinoderm.

The first classification of sperm types was performed by Meves in 1903, whose system was based on chromatin content. Sperm which fertilized the egg and had a nucleus with normal amounts of chromatin were termed eupyrene; sperm with very little chromatin were termed oligopyrene; those with an excess of chromatin were termed hyperpyrene; and sperm with no chromatin were termed apyrene. A number of workers subsequently referred to the sperm with unusual amounts of chromatin as "atypical" and eupyrene sperm as "typical." However, the term atypical carries the implication of abnormality. In many animals, sperm which are genetically or morphologically abnormal are often formed in the testis. Atypical sperm, however, are not abnormal and they can play an important role in the reproductive biology of those species which produce them (see Section IV). For this reason, the term atypical was replaced by the use of paraspermatozoon (para = near or close), and this term is now widely accepted. Most parasperm are either oligopyrenic or apyrenic and, with the possible exception of a single observation in the echinoderms, are produced only by animals with internal fertilization.

II. THE STRUCTURE OF PARASPERMATOZOA

Before considering the structure of paraspermatozoa it is important to consider briefly the fundamental structure of the euspermatozoon (fertilizing sperm). Although the morphology of the eusperm varies between and within invertebrate phyla, the majority of invertebrates have sperm cells which consist of the following regions arranged in an anterior to posterior sequence: a head, which is composed of a nucleus with highly condensed chromatin capped by an acrosomal complex; a midpiece consisting of the centrioles, accessory structures, and mitochondria; and a tail, which usually (but not always) has an axoneme with a 9 + 2 arrangement of microtubules. Glycogen may be associated with various regions of the sperm, but it is most commonly associated with the midpiece and axoneme or as a glycogen piece surrounding the tail. Many of the previously discussed details are illustrated in Fig. 1.

In some invertebrate taxa (e.g., Annelida), the parasperm are morphologically similar to the eusperm, whereas in other taxa (e.g., Gastropoda) the two sperm types are distinctly different. To highlight any structural similarities and differences, a summary of the morphology of both eusperm and parasperm is presented for each taxon.

A. Annelida

Paraspermatozoa have been described from three very distantly related annelid taxa, the polychaete families Protodrilidae and Siboglinidae (formerly known as the Pogonophora) and tubificid oligochaetes of the subfamilies Tubificinae and Limnodriloidinae.

Tubificids transfer spermatozoa as sperm bundles called spermatozeugmata. In the Tubificinae, the rod-shaped spermatozeugmata form in the spermathecae and are composed of both the euspermatozoa and the paraspermatozoa. The eusperm lie parallel to one another in the center of the spermatozeugmata and are surrounded by a cortex of parasperm (Figs. 2A and 2B). In some species (e.g., *Tubifex tubifex*) each spermatozeugma contains more parasperm than eusperm, whereas in others (e.g., *Limnodrilus hoffmeisteri*) both sperm types are produced in equal numbers.

The head of the eusperm of tubificids has an acrosomal complex which consists of the acrosomal vesicle, an acrosomal tube (and often secondary tube), and acrosomal rod, all of which lie anterior to the elongated nucleus, which is essentially cylindrical in shape (Figs. 1, 2G, and 2H). The midpiece is composed of a ring of mitochondria between the nucleus and centriolar apparatus (Figs. 1 and 2H). Finally, the tail consists of the axoneme which is surrounded by β-glycogen particles (Fig. 1).

The parasperm of the Tubificinae also have a head, midpiece, and tail (Figs. 1 and 2C). When compared to that of the eusperm, the nucleus is smaller, irregular in shape, and contains incompletely condensed chromatin. In some species the nuclear contents may degenerate into myelinated figures. Thus, the parasperm are oligopyrenic and in *T. tubifex* the DNA content is eight times less than that of the eusperm. In addition, a small amount of head cytoplasm remains. Although the parasperm possess an acrosome, its contents are more electron-lucent and there is no acrosome rod or secondary tube (Figs. 1 and 2D). The midpiece of parasperm has fewer, larger mitochondria (often twice the size of those of the eusperm), and the axoneme is surrounded by cytoplasm (Figs. 1, 2C, 2E, and 2F). The tails are helically wound to form the cortex of the spermatozeugma and in some the tails are joined to neighbors by junctional complexes (Fig. 2F).

In *Smithsonidrilus hummelincki* (Tubificidae and Limnodriloidinae) the parasperm and eusperm form separate spermatozeugmata in the spermathecae, both of which are transferred to the copulating part-

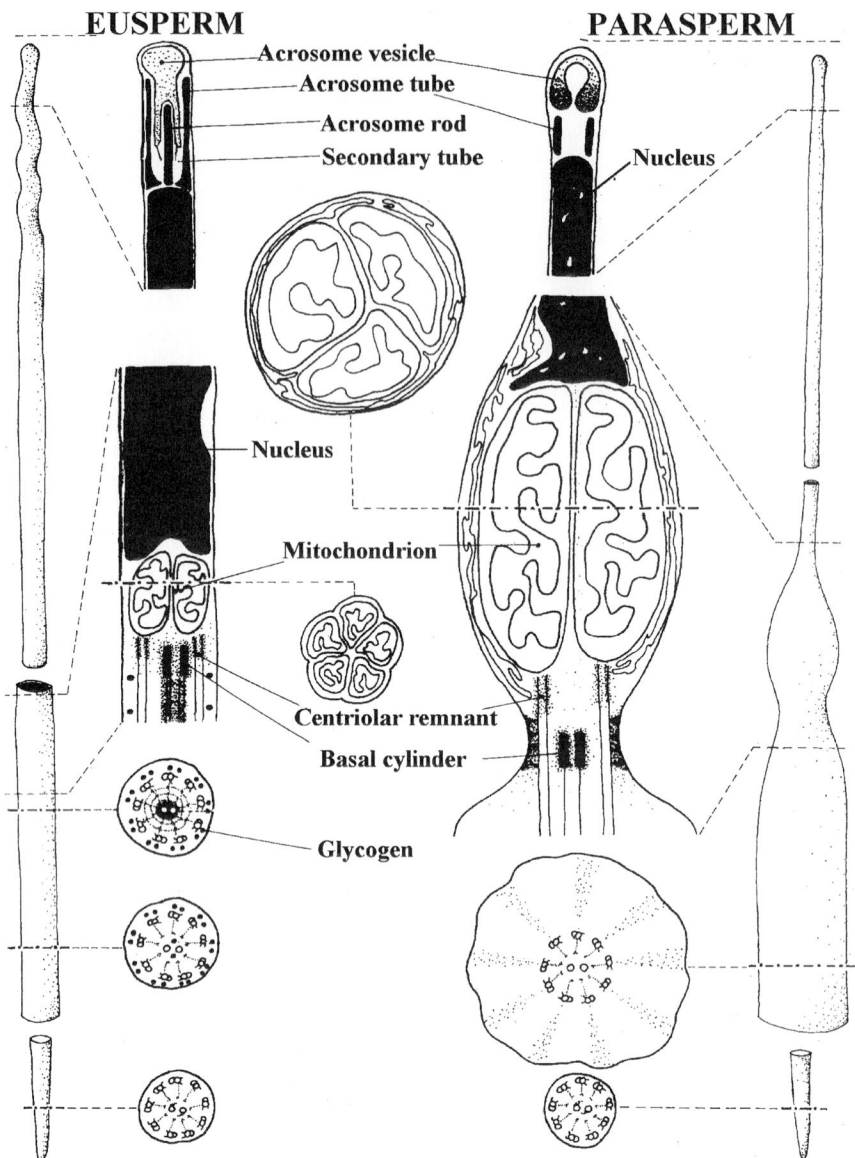

FIGURE 1 Diagrammatic representations of the eusperm and parasperm of the tubificid oligochaete *Clitellio arenarius*. Both sperm types are drawn to the same scale (modified from Ferraguti and Ruprecht, 1992).

ner. However, the euspermatozoa are more numerous. The parasperm are oligopyrenic with a much smaller acrosome. The number of mitochondria are reduced (two to four in the parasperm; four or five in the eusperm), and the plasma membrane of the tail is swollen. The axoneme of the parasperm appears to degenerate in the spermatheca.

In polychaetes of the genus *Protodrilus* (Protodrilinae), the parasperm constitute about 5–20% of the total number of male germ cells. Although having a similar fundamental structure to the eusperm, the parasperm are usually thinner and shorter. However, for each species of *Protodrilus*, the nucleus of the parasperm, which has incompletely condensed chromatin, is longer and thinner. The acrosome, which is anterior to the nucleus, is smaller and does not have an acrosomal rod. The midpiece is simpler. When observed in the females (*Protodrilus* has internal fertilization) the paraspermatozoa have lost the acrosome and have reduced motility. It has been

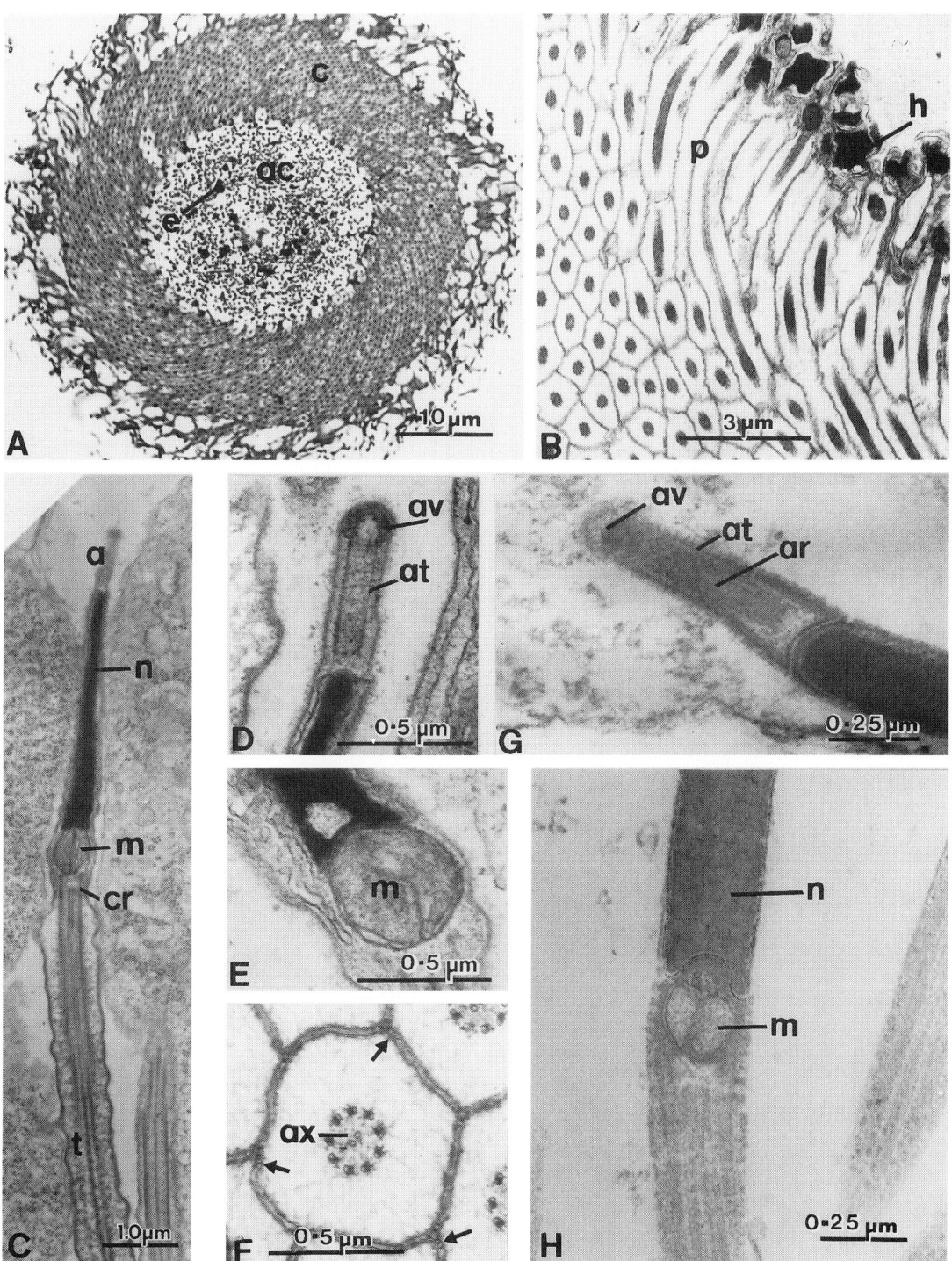

FIGURE 2 The spermatozeugma, euspermatozoon, and paraspermatozoon of the oligochaete *Tubifex tubifex*. A, light micrograph; B–H, transmission electron micrographs. (A) Cross section through a spermatozeugma showing the cortex (c) of parasperm and axial cylinder (ac) with eusperm (e). (B) Higher magnification of a portion of the cortex of the spermatozeugma showing the parasperm (p) with their heads (h) orientated toward the center. (C) Longitudinal section of a paraspermatozoon. (D) Longitudinal section of the acrosome of the parasperm. (E) Longitudinal section of a mitochondrion from the midpiece of the parasperm. (F) Transverse section of the tail of the parasperm showing the large amount of cytoplasm around the axoneme (ax) and junctional complexes between parasperm (arrows). (G) Longitudinal section of the acrosome of a eusperm. (H) Longitudinal section of the posterior nucleus (n) and midpiece of the eusperm. av, acrosomal vesicle; ar, acrosome rod; at, acrosomal tube; cr, centriolar remnant; m, mitochondrion; n, nucleus; p, parasperm; t, tail. (courtesy of M. Ferraguti).

suggested that the paraspermatozoa participate in the construction of the spermatophores, with the parasperm forming the external layer.

In many respects, tubificid and *Protodrilus* parasperm have a morphology which is similar to that of euspermatids. This has led some authors to propose that parasperm in these taxa have evolved from the "arrest" of spermatogenesis in the paraspermatogenic line.

Two sizes of sperm have been described in two species of Siboglinidae (*Siboglinum ekmani* and *Riftia pachyptila*) and it has been proposed that the larger are eusperm and smaller parasperm. Both sizes of sperm are transferred to the females in spermatophores, the smaller sperm being about half as abundant as the larger. Because it has not been established which sperm fertilize the eggs it is perhaps presumptive to classify the larger sperm as eusperm and the smaller as the parasperm. In *S. ekmani* the larger spermatozoa are filiform with a nucleus (about 30 μm long) capped by a corkscrew-shaped acrosome. The basal region of the nucleus is surrounded by three mitochondrial derivatives. Posterior to the nucleus is a pair of centrioles which give rise to the axoneme. The smaller (para?)spermatozoon has a shorter acrosome and the oligopyrene nucleus is shorter and thinner. The mitochondria are bigger than those of the larger sperm, but the arrangement of the centrioles and length of the flagellum is similar to that of the larger sperm.

B. Lepidoptera

In addition to the eupyrene sperm, Lepidoptera (moths and butterflies) develop apyrene parasperm. These sperm are produced in large numbers in the testis (but not in the same testicular cyst as the eupyrene euspermatozoa) and can constitute half the sperm complement. In the tobacco hornworm, *Manduca sexta*, an average of 96% of the ejaculate is parasperm!

The eupyrene sperm are elongate cells, varying from 300 μm to 1 mm in length in different species of Lepidoptera. The eusperm have an elongate nucleus which has highly condensed chromatin, anterior to which is the acrosomal complex. The nucleus has a lateroposterior fossa which houses a centriole or centriole-like structure from which the axoneme emerges. The axoneme is composed of an outer ring of microtubule singlets (which are electron-dense) to the inside of which is a ring of nine doublets and a central pair of singlets (also electron-dense) (Fig. 3A). Posterior to the nucleus are two mitochondrial derivatives (usually unequal in size) with poorly defined cristae. The testicular eusperm of higher Lepidoptera are surrounded by two types of structures, the laciniate and reticular appendages (Fig. 3A). Both of these are absent in the lower Lepidoptera (e.g., Zeugloptera). The laciniate appendages extend from the eusperm cell membrane and in cross section appear to be composed of alternating electron-lucent and electron-dense zones with a periodicity of 900 nm. They are usually larger anteriorly, gradually decreasing in size along the length of the eusperm. The shape of the laciniate appendages, and the number of bands of which they are composed, differs between species. The reticular appendage has a granular substructure. Euspermatozoa from the vas deferens lack the laciniate appendages. There is some evidence that material from the dissolution of these accessory structures in the vas deferens is used to form a sheath around both eusperm and parasperm (Fig. 3C).

Within a species, the dimensions of the apyrene parasperm differ from those of the eusperm. The parasperm are generally shorter and about half the diameter and length of the eusperm (e.g., in the silkworm, *Bombyx mori*, the mean length of the eusperm is 633 μm, whereas that of the parasperm is 350 μm). The apyrene sperm consist of a flagellum which arises from a centriole. Anterior to the centriole is a small (about 1 μm long × 0.3 μm diameter) electron-dense cone, the apical cap (Fig. 3C), with the nucleus and acrosome being absent. The mitochondrial derivatives of the midpiece are small and of equal diameter (the lengths of the derivatives may or may not be equal). The derivatives extend from the posterior end of the apical cap to almost the posterior end of the parasperm. The axoneme structure is essentially similar to that of the eusperm (Fig. 3B). Unlike the eupyrene sperm, apyrene sperm do not have laciniate and reticular appendages surrounding them, although in the species of hawkmoth examined here (*Hippotion celerio*) an accessory structure is present in testicular parasperm (Fig. 3B).

In the vas deferens and female tract, the parasperm are surrounded by a sheath of electron-dense material

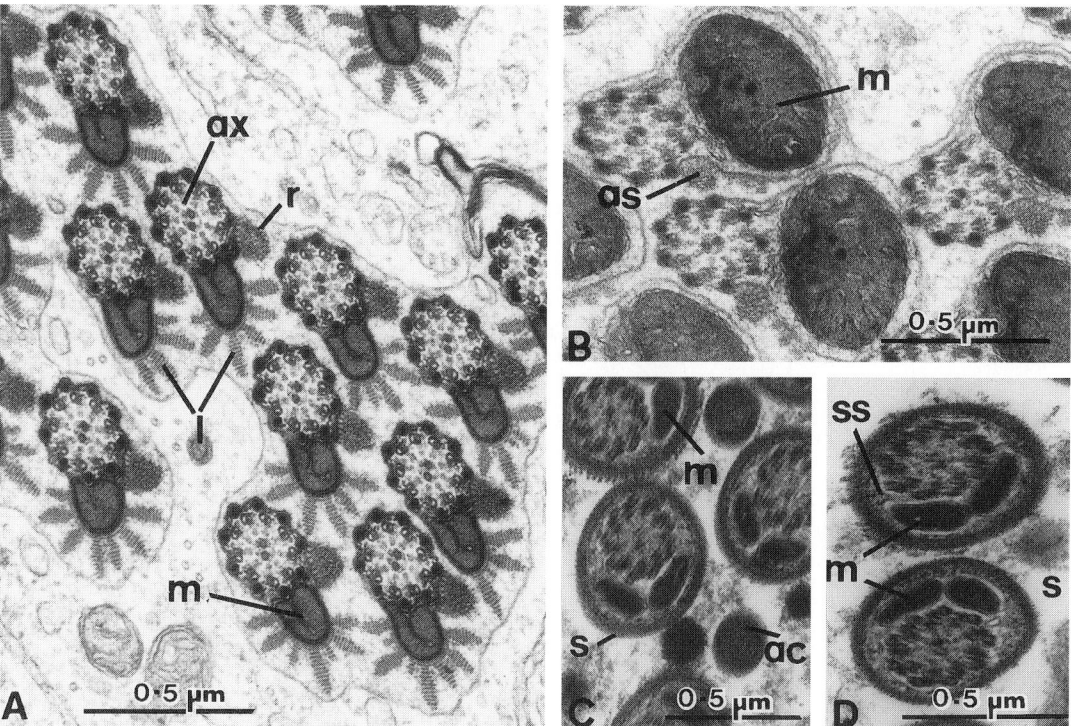

FIGURE 3 Transmission electron micrographs of the eupyrene and apyrene sperm of the lepidopteran *Hippotion celerio*. (A) Transverse sections through the midpiece regions of euspermatozoa from the testis showing lacinate (l) and reticulate appendages (r). (B) Midpiece sections of testicular apyrene sperm. (C and D) Apyrene sperm from the vas deferens showing sections through the apical cap (ac) and midpiece regions. Note the electron-dense sheath (s) surrounding the parasperm. as, accessory structure; ax, axoneme; m, mitochondrion, ss, subsheath material.

(Figs. 3C and 3D) which may display a substructure of concentric electron-dense rings. The number of rings varies not only between species but also along the length of the sheath within a species, becoming more numerous posteriorly. A relatively large amount of subsheath material lies between the external sheath and the parasperm (Figs. 3C and 3D), and it is thought that this represents some type of male accessory material which is transported by the parasperm to the spermatheca in the female.

C. Gastropoda

Within the class Gastropoda, dimorphic sperm (i.e., eusperm and parasperm) have been described from taxa from the subclass Prosobranchia only (specifically the Caenogastropoda, Neritimorpha, and Skeneidae). Although there is enormous variability in prosobranch eusperm morphology, all are uniflagellate, have a midpiece with mitochondria (or mitochondrial derivatives) and centrioles (or centriolar apparatus), and a head with a nucleus and acrosome. Some of these features are illustrated in Fig. 5C.

The paraspermatozoa of prosobranchs display the greatest morphological variability of all invertebrate taxa, and some of this variability is represented diagrammatically in Fig. 4. The apyrene parasperm of the Neritimorpha are uniflagellate, the flagellum being surrounded by a central cytoplasmic body containing mitochondria. In some caenogastropod taxa (e.g., Cyclophoroidea, Viviparoidea, and Cerithioidea) the parasperm have a posterior tuft of flagella. In the Triphoroidea, Cerithiopsoidea, and Epitonioidea the parasperm are large with multiple axonemes. Eusperm attach to these parasperm, forming spermatozeugmata. Spermatozeugmata are also formed in the Littorinoidea, in which the eusperm are attached to more rounded aflagellate parasperm. In the Vermetoidea the parasperm have bipolar tufts

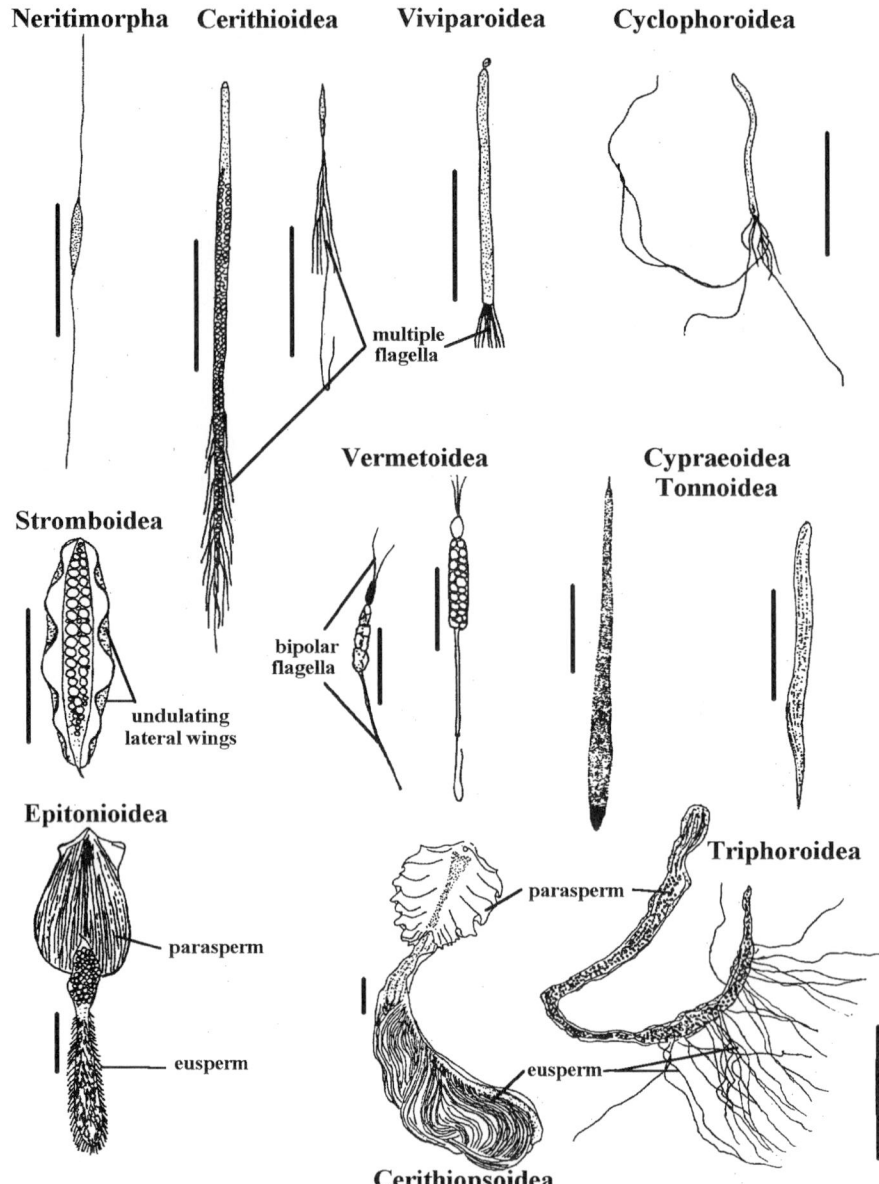

FIGURE 4 Diagrammatic representations of parasperm from some prosobranch gastropods Scale bars = 50 μm except for Vermetoidea, Cypraeoidea, and Tonnoidea (scale bars = 20 μm) and Epitonioidea and Cerithiopsoidea (scale bars = 100 μm) (modified from Healy, 1988).

of flagella, whereas in the Stromboidea, the parasperm have a fusiform body with lateral undulating membranes. The parasperm of the Tonnoidea, Xenophoroidea, and Cypraeoidea is vermiform and in one species of tonnoidean, *Fusitriton oregonensis*, each male produces two types of parasperm (Figs. 5A and 5B). In the higher caenogastropods (e.g., Buccinacea, Conacea, and Volutacea) the parasperm are usually vermiform with internal axonemes.

Despite this variability, most prosobranch parasperm are either apyrene or oligopyrene (e.g., Fig. 6), and the majority have a number of structural features in common.

The majority of prosobranch parasperm have multiple axonemes (which may be either partially free or completely enclosed in the cell body; Figs. 4 and 5) which arise from numerous basal bodies or centrioles. The number of axonemes (and therefore fla-

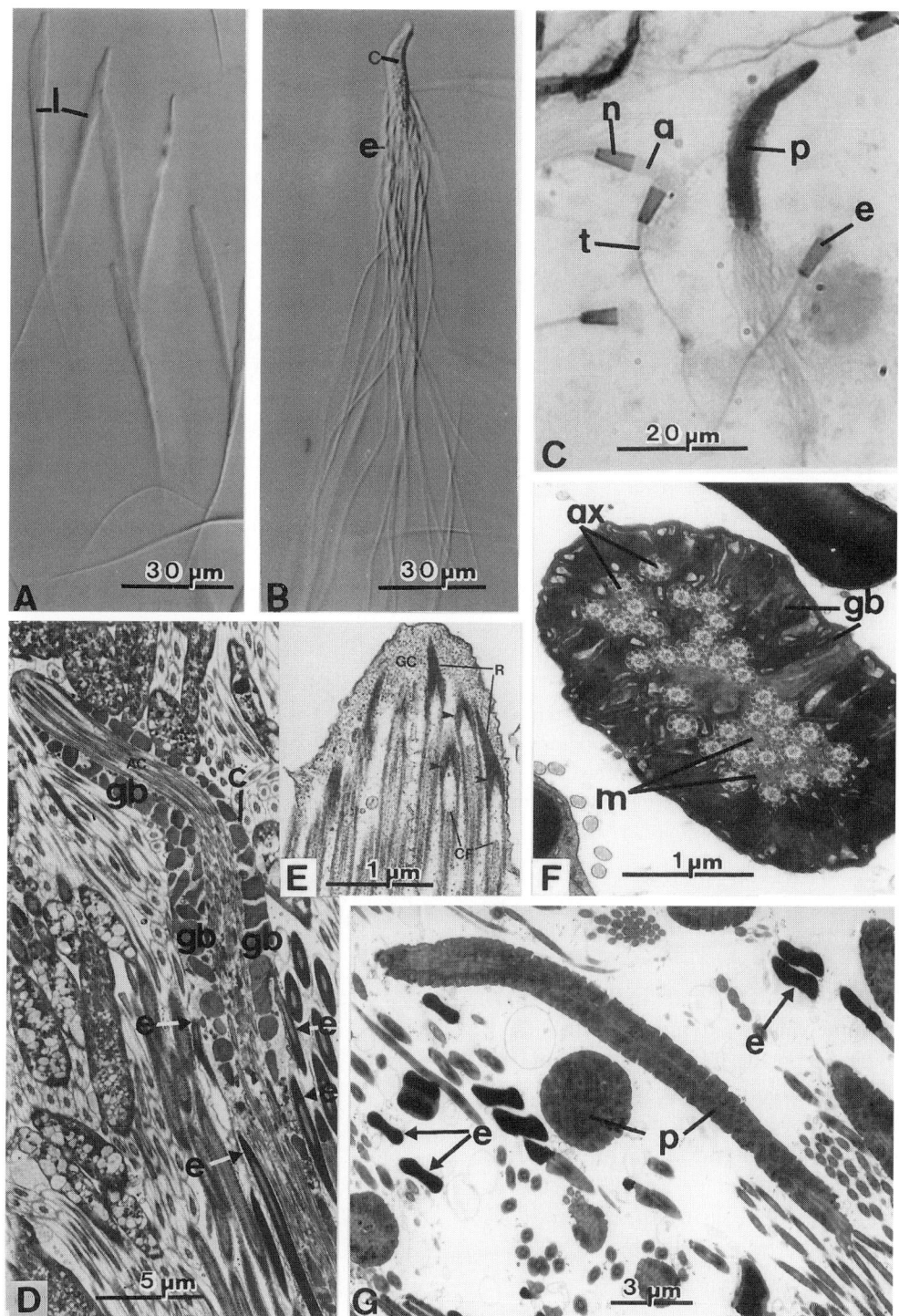

FIGURE 5 (A and B) Nomarski diffraction images of the lancet (l) and carrier (c) parasperm of *Fusitriton oregonensis*. Note the euspermatozoa (e) attached to the carrier sperm in 5B. (C) Testicular smear of *Melanopsis* sp. stained in Spermac showing the multiflagellate parasperm (p) and uniflagellate eusperm (e). (D) Transmission electron micrograph of the carrier sperm (c) with attached eusperm (e) from the seminal vesicle of *F. oregonensis*. (E) Higher magnification of the anterior of the carrier sperm of *F. oregonensis* showing the basal bodies of the axonemal core (arrowheads). Note the unstriated rootlets (R) of the basal bodies [which give rise to central fibers (CF) embedded in a granular cap (GC)]. (F) Transmission electron micrograph of a transverse section through the parasperm of *Melanopsis* sp. showing the core of axonemes (ax) and mitochondria (m) surrounded by electron-dense glycoprotein blocks (gb). (G) Transmission electron micrograph of a longitudinal and transverse section of parasperm (p) and transverse sections through the nucleus of the eusperm (e) of *Melanopsis* sp. a, acrosome; ac, axonemal core; n, nucleus; t, tail (A, B, D, and E from Buckland-Nicks *et al*., 1982).

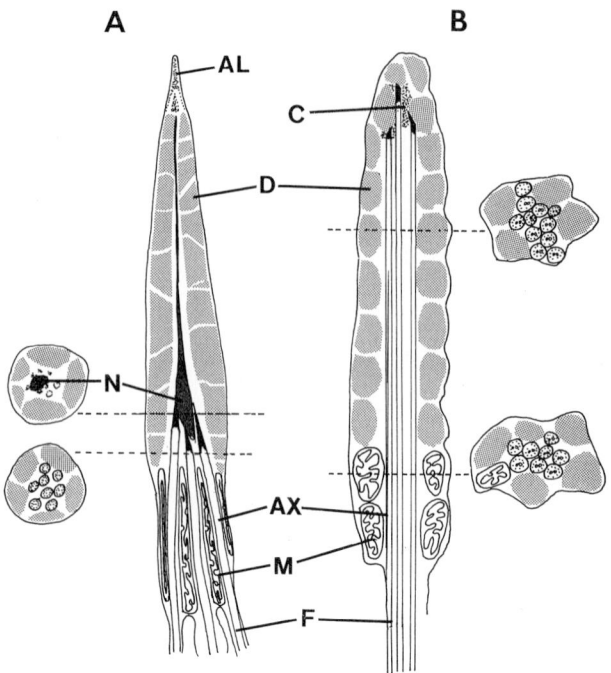

FIGURE 6 Diagrammatic representations of oligopyrene (A) and apyrene (B) parasperm. AL, acrosome-like structure; AX, axonemes; C, centriolar cap; D, dense glycoprotein blocks; F, flagella; M, mitochondria; N, nuclear core [modified from Healy and Jamieson (1981) and Hodgson (1997)].

gella) varies between species, with up to 3000 having been observed in some parasperm. Often closely associated with the flagella in the body of the paraspermatozoon are numerous small mitochondria (Fig. 5F) and glycogen deposits. The cytoplasm of the parasperm is filled with large electron-dense blocks which are composed of glycoprotein (Figs. 5D, 5F, and 5G). These surround the nuclear remnant when it is present. Finally, the parasperm do not possess an acrosome. Examples of ultrastructure of two types of prosobranch parasperm are given in Fig. 6.

D. Echinodermata

Nearly all echinoderms have external fertilization, and these animals are noted for producing structurally simple sperm which are conservative in morphology within taxa. However, Eckelbarger et al. have reported the presence of two sperm types in the abyssal echinoid *Phrissocystis multispina*. In addition to the normal type of echinoid sperm, this species produces an approximately equal number of sperm which are bipolar tailed (Fig. 7) with both a posteriorly and anteriorly directed flagellum. Both flagella have a length of about 75 μm. The microtubules within the anterior flagellum, however, do not have the usual 9 + 2 arrangement and appear disorganized. Eckelbarger et al. suggested that such sperm are parasperm, although they were unable to observe whether the sperm are released during spawning and are motile. Consequently, it is not known what role (if any) they play in reproduction. Development of both euspermatozoa and parasperm was found to be identical until the spermatid stage, when the nucleus of the parasperm was noted to undergo chromatin reduction.

III. THE FORMATION OF PARASPERMATOZOA

All paraspermatozoa develop from undifferentiated spermatogonia by spermatogenesis. The subject of paraspermatogenesis is too broad to be covered

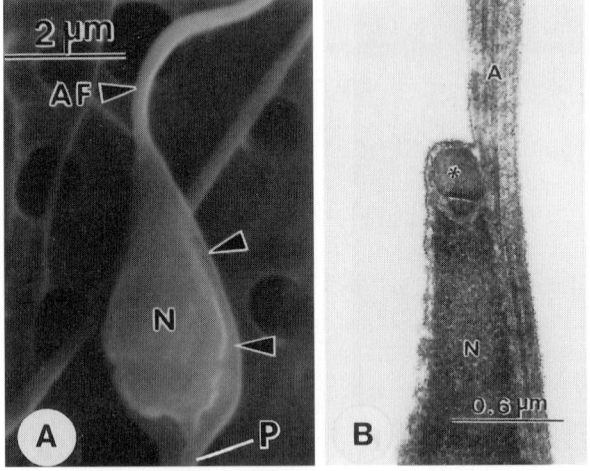

FIGURE 7 (A) Scanning electron micrograph of the parasperm of the deep-sea echinoid *Phrissocystis multispina* showing anteriorly (AF) and posteriorly (P) directed components of the flagellum. Arrowheads indicate primary axoneme lateral to nucleus (N). (B) Transmission electron micrograph of the apical region of the nucleus (N) showing acrosome (*) and anteriorly directed axoneme (A) of the parasperm (from Eckelbarger et al., 1989).

in detail here, and therefore a brief overview only is presented.

A. Annelida

Spermatogenesis in tubificids involves the production of groups of cells by the testes which are released into the seminal vesicles. Each group of cells consists of a central cytoplasmic mass (or cytophore) to which developing sperm cells are attached. In the seminal vesicles the cells divide. At the spermatid stage it is possible to recognize those cysts which will produce parasperm from those which form the eusperm.

In some species (e.g., *Tubifex tubifex*), the parasperm cysts are more numerous. In addition, these cysts are formed by several hundreds of small cells, whereas the eusperm cysts have far fewer larger spermatids (e.g., 128 in *T. tubifex*). The cysts display a number of morphological differences. The cytophore of the parasperm cysts contain pycnotic (degenerate) nuclei. Early in spermatogenesis the nuclei of the parasperm become irregular in shape and chromatin condensation is also irregular. Although differences in the eusperm and parasperm cysts are readily recognized, the mechanism or types of cell division which produce the numerous paraspermatids are unknown.

B. Lepidoptera

In Lepidoptera, spermatogenesis commences in the larvae. Initially, larvae produce only eupyrene euspermatozoa. Just prior to pupation the cysts of the testis begin to produce parasperm, their formation continuing during pupation. The numerous parasperm develop rapidly and are formed by asynaptic meiosis. As they develop and elongate, the nuclear material fragments and is degraded by DNAase. Experimentation has shown that the formation of both types of sperm is under hormonal control.

C. Gastropoda

In prosobranch gastropods the eusperm and parasperm are produced simultaneously in the same testicular acini. Parasperm develop from spermatogonia which are similar in morphology to those which produce the eusperm. Formation of the parasperm proceeds by atypical meiotic divisions. During paraspermiogenesis multiple flagella develop from numerous basal bodies which in turn originate from two parent and their satellite centrioles. In developing apyrene parasperm, the nucleus fragments into a number of vesicles in which the chromatin gradually degenerates. In developing oligopyrene parasperm the nucleus undergoes a gradual decrease in size as the chromatin condenses. The glycoprotein blocks are produced by the Golgi bodies.

IV. THE FUNCTIONS OF PARASPERMATOZOA

The precise functions of the paraspermatozoa are better understood in some taxa than in others. However, the fundamental role of parasperm is to assist euspermatozoa in the fertilization of eggs.

A. Annelida

In the annelids the parasperm protect and bind together the euspermatozoa, carrying them to the spermathecal opening. The enlarged mitochondria of the parasperm indicates that these cells have an enhanced metabolic activity. The free tails of the parasperm are motile and beat with metachronal waves which pass helically along the spermatozeugmata with a wavelength of 2.8 μm and a beat of 10 Hz. The parasperm may also play an important role in the selection of nutrients from the spermatheca for the eusperm. Finally, it is also possible that the parasperm, by filling up the spermatheca, may prevent the female from participating in further copulation, a function which has been suggested for the apyrene sperm of some Lepidoptera.

B. Lepidoptera

There are at least six hypotheses to explain the function of the apyrene parasperm of Lepidoptera. Some are based on observation only, whereas other hypotheses are supported or derived from experimental evidence. The hypotheses fall into one of two groups: those which suggest physiological functions for the parasperm and those which propose that the parasperm play a role in sperm competition.

1. Physiological Functions

i. Facilitation of Emigration from the Testis Apyrene sperm move out of the testis before the eupyrene sperm. It has therefore been proposed that the apyrene sperm act as "trailblazers" for the eupyrene sperm by creating holes in the layer of cells which separate the lumina of the follicles of the testis from the vas deferens.

ii. Transport of Eupyrene Sperm in the Female Tract Motility of the apyrene sperm is activated by a protein (a serine endopeptidase of molecular weight 30 kDa) secreted by the glandula prostactica (the most distal gland of the male reproductive tract). At ejaculation of the spermatophores, the apyrene sperm are therefore vigorously motile. The fact that the eupyrene sperm have little motility has led to the suggestion that the parasperm help in propelling the eusperm through the female tract. Unfortunately, no direct association of eusperm and parasperm has been observed.

iii. Nutritive Role The final destination of the apyrene sperm is the spermatheca of the female, where they degenerate. In some species the apyrene sperm reach the spermatheca several hours before the eupyrene sperm. It is therefore possible that by their breakdown they supply nutrients for the eupyrene sperm.

iv. Facilitation of Euspermatozoan Maturation As previously mentioned, the parasperm in the spermatophores are very motile. However, this motility does not result in their forward movement. A feature of eusperm maturation is that it proceeds in the spermatophore. Initially, the eusperm are arranged in bundles within the spermatophore, but as maturation proceeds the bundles begin to break up. It has been observed that the vigorously swimming parasperm mechanically facilitate the dissociation of the eusperm bundles. Furthermore, the stirring action of the parasperm in the spermatophore accelerates the metabolic reactions (e.g., arginine degradation cascade) which are necessary for eusperm maturation.

2. Sperm Competition

Many female Lepidoptera mate more than once, and therefore the sperm of more than one male may be deposited into the spermatheca of the female. This in turn may result in sperm competition. The fact that the apyrene sperm are activated first and reach the spermatheca first (in some species) has led to the idea that their function is to eliminate previously deposited eusperm or to displace them (the elimination hypothesis). Simply put, their function is to "seek and destroy" another male's sperm.

Alternatively, parasperm may function in preventing or delaying further mating by females (the prevention hypothesis). After mating, many female Lepidoptera have a refractory period (several days) during which time they do not mate again. Refractoriness to mating is correlated to the presence of sperm in the spermatheca. It is therefore possible that the apyrene sperm (which may be energetically inexpensive to produce) simply help to fill the spermatheca and therefore assist in inducing the refractory period of the females. This in turn ensures that the sperm of other males cannot fertilize any eggs.

C. Gastropoda

Several functions have been proposed for the parasperm of prosobranch molluscs. In some species numerous eusperm become attached to the parasperm (Fig. 5B) which clearly functions to transport the fertilizing sperm to, or within, the female. In other gastropods the eusperm and parasperm have no direct association (Fig. 5C) and therefore the function of the parasperm is not clear. In a few species it has been noted that the parasperm degenerate in the female tract. It is therefore possible that the breakdown products of the parasperm provide valuable nutrients to the female or that they stimulate eusperm motility (presumably by the release of enzymes) in the female. Unfortunately, there is no experimental evidence to support these theories. It is possible that some parasperm are multifunctional.

D. Echinodermata

In the deep-sea sea urchin in which parasperm have been recorded, it was observed that eusperm–parasperm mixtures tended to clump due to the axonemes becoming entangled. Eckelbarger et al. proposed that such entanglements may facilitate fertilization in this animal through the reduction of eusperm diffusion during spawning.

V. OTHER FORMS OF INVERTEBRATE SPERM DIMORPHISM

Brief mention must be made of those invertebrates which produce dimorphic sperm whereby it is unclear as to whether the sperm can be classified as eu- or parasperm. Such sperm dimorphism has been observed in some myriopods (Chilopoda and Symphyla), the mite-like opilionid *Siro* (Arachnida: Opiliones), and the dipteran family Drosophilidae. In chilopods two sizes of sperm (termed micro- and macrosperm) are formed, and in some male fruit flies of the genus *Drosophila* (e.g., *D. azteca*) up to three sizes of sperm can be present. In chilopods and *Drosophila*, all sperm possess a nucleus and a fully formed acrosome, but there is uncertainty as to whether all sperm are capable of, or participate in, fertilization. Recent studies on *D. subobscura* have shown that different-sized sperm are produced in different numbers, with 66% of the ejaculate being composed of short sperm and 34% of long sperm. Both sizes of sperm are transferred to females, in which they are stored in the spermatheca. However, long sperm appear to be preferentially selected for fertilization.

Until the function of dimorphic sperm is clarified in these taxa, it would be incorrect to classify any of their sperm as parasperm.

Acknowledgments

This article would not have been possible without the significant input of my friends and colleagues. I am particularly grateful to Drs. Marco Ferraguti, John Buckland-Nicks, Greg Rouse, Elizabeth Hauschteck-Jungen, Kevin Eckelbarger, and Guido Melone. Dr. Martin Villet kindly provided the lepidopteran material. The technical assistance of Val Hodgson, Robin Cross, and Marvin Randall is also gratefully acknowledged.

Bibliography

Alberti, G. (1995). Comparative spermatology of Chelicerata: Review and perspective. *Mäm. Musäum Natl. d'Histoire Nat.* **166**, 203–230.

Bressac, C., and Hauschteck-Jungen, E. (1996). *Drosophila subobscura* females preferentially select long sperm for storage and use. *J. Insect Physiol.* **42**, 323–328.

Buckland-Nicks, J., Williams, D., Chia, F.-S., and Fontaine, A. (1982). The fine structure of the polymorphic spermatozoa of *Fusitriton oregonensis* (Mollusca: Gastropoda), with notes on the cytochemistry of the internal secretions. *Cell Tissue Res.* **227**, 235–255.

Eckelbarger, K. J., Young, C. M., and Cameron, J. L. (1989). Ultrastructure and development of dimorphic sperm in the abyssal echinoid *Phrissocystis multispina* (Echinodermata: Echinoidea): Implications for deep sea reproductive biology. *Biol. Bull.* **176**, 257–271.

Ferraguti, M. (1994). An ultrastructural overview of tubificid spermatozoa. *Hydrobiologia* **278**, 165–178.

Ferraguti, M., and Ruprecht, D. (1992). The double sperm line in the tubificid *Clitellio arenarius* (Annelida, Oligochaeta). *Bollettino Zool.* **59**, 349–362.

Friedlander, M. (1983). A comparative study on the spermatogenesis of Trichoptera and Lepidoptera. In *The Sperm Cell* (J. André, Ed.), pp. 436–439. Nijhoff, The Hague.

Healy, J. M. (1988). Sperm morphology and its systematic importance in the Gastropoda. *Malacol. Rev. Suppl.* **4**, 251–266.

Healy, J. M. (1996). Molluscan sperm ultrastructure: Correlation with taxonomic units within the Gastropoda, Cephalopoda and Bivalvia. In *Origin and Evolutionary Radiation of the Mollusca* (J. D. Taylor, Ed.), pp. 99–113. Oxford Univ. Press, Oxford, UK.

Healy, J. M., and Jamieson, B. G. M. (1981). An ultrastructural examination of developing and mature paraspermatozoa in *Pyrazus ebeninus* (Mollusca, Gastropoda, Potamididae). *Zoomorphology* **98**, 101–119.

Hodgson, A. N. (1997). Paraspermatogenesis in gastropod molluscs. *Invertebr. Reprod. Dev.* **31**, 31–38.

Jamieson, B. G. M. (1987a). *The Ultrastructure and Phylogeny of Insect Spermatozoa*. Cambridge Univ. Press, Cambridge, UK.

Jamieson, B. G. M. (1987b). A biological classification of

sperm types, with special reference to annelids and molluscs, and an example of spermiocladistics. In *New Horizons in Sperm Cell Research* (H. Mohri, Ed.), pp. 311–322. Gordon & Breach, New York.

Nishiwaki, S. (1964). Phylogenetic study on the type of the dimorphic spermatozoa in Prosobranchia. *Sci. Rep. Tokyo Kyoiku Daigaku B* **11**, 237–275.

Silberglied, R. E., Shepherd, J. G., and Dickinson, J. L. (1984). Eunuchs: The role of apyrene sperm in Lepidoptera. *Am. Nat.* **123**, 255–265.

Parental Behavior, Arthropods

Gary A. Polis, Joseph D. Barnes, Andrew M. Beld, and C. Todd Jackson

Vanderbilt University

I. Introduction
II. Order Araneae: The Spiders
III. Parental Care in Other Arachnids
IV. Myriapods
V. Terrestrial Crustacea

GLOSSARY

instar In arthropods, any subadult life stage between molts. For example, a first instar would be the stage between hatching or birth and the next molt.

oostegite Plate-like structures on the ventral thoracic surface of some crustaceans used in the incubation of developing eggs or young.

oviposition Selective placement (laying) of eggs by the female at a particular locality.

viviparity Direct nourishment, from after fertilization to birth, of developing embryos by the mother; seen in mammals, some reptiles, and many chelicerate arthropods.

I. INTRODUCTION

A. What Are the Terrestrial Arthropods?

Terrestrial arthropods include insects, arachnids, myriapods, and a few crustaceans. The best examples of parental care in the insects are to be found in the social insects, and that entry in this book should be consulted for further information. In other insect groups, the nature of parental care is too varied and widespread over too many taxa to be addressed in this brief section. Excluding the insects, we can rank the order of importance of terrestrial arthropod groups in terms of their relative abundance, diversity, and ecological significance as follows: subclass Acari (mites and ticks, class Arachnida), order Aranea (spiders, class Arachnida), order Scorpionida (scorpions, class Arachnida), class Diplopoda (millipedes), and class Chilopoda (centipedes). Less common and/or diverse are the eight other orders of arachnids, two minor classes of myriapods, and the various terrestrial Crustacea. The depth of our knowledge of the basic biology and parental care of these groups generally mirrors their importance. Regarding parental care, we know a fair amount for a subset of spider, scorpion, millipede, and isopod species. Except for scattered reports, we are largely ignorant of how parents interact with their offspring in the remaining groups. In the following section, the basic reproductive biology of terrestrial arthropods is sketched with particular emphasis on how this biology may influence parent offspring association.

B. Basic Reproductive Biology

Fertilization is internal in all terrestrial arthropods. This is necessary because sperm would desiccate if exposed to air. In the vast majority, males transfer sperm directly via insertion of their copulatory organs into the female's genital track. In a few groups, sperm transfer is indirect via spermatophores (packages of sperm) that the male places on the ground and from which females inseminate themselves. After internal fertilization, as the developing egg passes down the females oviduct, it is placed into a protective case, the chorion. The chorion provides a very effective barrier against water loss to air and is a major factor in the success of terrestrial arthropods. The chorion is also impenetrable by sperm and is thus the reason why fertilization must be internal. Fertilized eggs are then oviposited (laid) in the environment. Only in scorpions and a few insects do eggs or embryos develop internally.

C. General Types of Parental Care

Eggs are either placed into a moist natural environment (e.g., soil) or a specially constructed chamber (e.g., an egg sac) or guarded and/or carried by the female. Females that guard or carry eggs inevitably provide protection from predators, parasites, and pathogens such as fungus and mold. If the female is present when the young hatch, she may or may not stay with the young. If the young do not disperse and the mother does not leave, the mother continues to provide protection for her offspring. In a few cases (e.g., some species of spiders, scorpions, and isopods), the young remain with the mother for an extended period. In a very few cases, fairly complex social groups of kin may characterize a species, e.g., in some of the social spiders, a couple species of social scorpions, and the desert isopod, *Hemilepistus reaumuri*. These species engage in cooperative activities such as prey capture and feeding, construction and maintenance of a homesite, communal brood care, and/or defense against predators.

II. ORDER ARANEAE: THE SPIDERS

A. Egg Sac Production

Spiders are common faunal components of nearly every terrestrial landscape. The type of parental care provided to their young is quite variable among the 34,000 described species. The simplest form, present in nearly all species, is simply enclosing the eggs in a protective silken sac. The egg sac is usually constructed with a tough outer layer of parchment-like silk and an inner layer of fluffy silk in which eggs are laid. This egg sac provides protection from the elements, mechanical disturbance, and many parasites and predators, and it provides a favorable microclimate for development. Furthermore, the egg sac is often camouflaged with bits of old prey and debris, making it inconspicuous to visually oriented predators and parasites. Some spiders hide the egg sac. For example, several species of large orb weavers (e.g., *Nephila maculata*, Araneidae) leave their web and travel to the forest floor to construct and bury their egg sac under leaf litter and debris.

B. Guarding the Eggs

Many species actively guard their egg sacs. For example, green lynx spiders (*Peucetia viridans*, Oxyopidae) suspend their egg sacs from silk strands in a retreat in which they remain. They guard the egg sac from enemies and actively attack intruders, even spitting venom at them. If an egg sac is disturbed too much, the mother will move it to a new location. Egg sac guarding occurs in a wide variety of other species of non-web-building spiders, such as crab spiders (Thomisidae), jumping spiders (Salticidae), and various sac spiders (Clubionidae and Gnaphosidae). Species that inhabit burrows guard egg sacs laid within their retreat. As long as the mother is alive, she will defend her eggs. Many web-building spiders place egg sacs in their web or retreat (e.g., Araneidae and Theridiidae). These species actively monitor and defend them against intruders. Wolf spiders (Lycosidae), a common spider that actively forages on the ground, use yet another method. Since they have no web in which to suspend their egg sacs, they carry them attached to their spinnerets.

C. Parental Care Immediately after Birth

1. Mothers Guard Young
In some species of burrowing spiders, such as purse-web weavers (Atypidae), the young live with

the mother for a short time before dispersing. Similarly, when young wolf spiders emerge, they climb onto their mothers back, where they stay for a week or two before dispersing. A related group, the nursery web spiders (Pisauridae), carry their egg sacs in their mouthparts. When the young are ready to emerge, they construct a tangled "nursery web" in vegetation. The mother remains with her young to defend against enemies until the young disperse.

2. Mothers Feed Young

The next level of parental care by spiders occurs when the mother feeds her young. This occurs in three ways. First, young spiders remain with their mother in her web; when she catches prey, she wraps it in silk, gives it a paralyzing bite, and leaves it for her young to consume (e.g., *Coelotes terrestris*, Agelenidae). Some species (e.g., *Achaearanea saxitile*, Theridiidae) even communicate with their young to "stay away" if the prey item is not yet subdued and to "come and get it" when the prey is rendered harmless. Second, mothers may regurgitate prey they consume to their young. In a few species, the spiderlings that remain in the web with their mother (e.g., some *Theridion*, Theridiidae; some *Stegodyphus*, Eresidae) beg for food by tapping on her mouthparts with their legs. This stimulates the mother to regurgitate predigested food and, in some cases, even some of her own intestinal cells, which the young spiderlings consume. Third, in at least nine species (e.g., in some Eresidae), the mother sequesters herself away with her young, becomes inactive, and allows the young to feed on her moribund yet still living body. The mother becomes a usable food depot when she directs digestive juices inward to liquefy her body, with her heart and respiratory system functioning to the end.

D. Social Spiders: Mother and Young Cooperate over Extended Periods

In many instances, extended parental care has been the origin of social groupings in spiders. In at least four families (Agelenidae, Eresidae, Lycosidae, and Theridiidae), young spiders remain with their mother in her web until or near maturity, at which time they disperse and build webs of their own. Many disperse only locally such that a colony of closely related individuals forms. Individuals construct their own webs but are interconnected to conspecifics by a matrix of silken threads. Experimental evidence shows that this type of aggregation increases the efficiency of prey capture because insects escaping one web are usually caught in a neighboring one.

A more complex sociality characterizes spiders that inhabit their mothers web throughout their lives. Spiders from at least six families (Agelenidae, Araneidae, Dictynidae, Eresidae, Theridiidae, and Uloboridae) exhibit this behavior. These species live communally on the same web complex, share food and prey capture, and cooperate in brood care and protection. Some species (e.g., *Mallos gregalis*, Dictynidae) form colonies of thousands of related individuals, all living in a communal network of silken tunnels and webs.

III. PARENTAL CARE IN OTHER ARACHNIDS

In comparing parental care of spiders with other arachnid groups, there are many differences and a few similarities. In fact, a great deal of variation exists among arachnid orders, although parental behaviors are generally similar within each group. Very little is known about the parental care or even the basic biology of many orders (e.g., Palpigradi and Schizomida). For other groups, parental care is better known (e.g., scorpions).

A. Order Scorpionida

In many terrestrial habitats, scorpions are a minor component and relatively uncommon. In deserts, however, scorpions may be very important as major predators and may exhibit more standing biomass per unit area than all vertebrates combined. Parental care is one aspect of scorpion biology that may contribute to this success in arid areas.

1. Viviparity

Scorpions differ from all other arachnids and most other invertebrates in that all 1500 species are viviparous, i.e., there is live birth and the young are nourished directly by the mother during development, much as in placental mammals. The mother supplies the embryos with nutrition and they develop over a period of 5–18 months. However, the young are resorbed when the mother is food limited.

2. Birth

Litter size varies from 1 to over 100 young, with most species producing 20–30 babies per gestation. At birth, the young exit the female's reproductive tract via the ventral gonopore, a structure directly posterior to the sternum. Females of many species curve one or two pairs of legs under the body, producing a "birth basket" for newborns (first instars). Within minutes of birth, all newborns crawl to the mother's back. The young remain on the mother's back but do not feed. They remain until the first molt, a period that lasts from 2 to 4 weeks. After this molt, the young can fend for themselves and, in most species, disperse to assume an independent existence.

3. Early Mother–Young Association

First instars scarcely resemble adults: They are relatively soft, white, and fat-bodied. Until their first molt, their cuticle is unsclerotized, i.e., it is very soft and susceptible to water loss. Two lines of evidence suggest that the young acquire moisture from the mother. First, newborns removed from the mother's back quickly die from dehydration. Second, experiments with tritiated water (water labeled with a heavy, radioactive isotope of hydrogen) show direct transfer of water from the mother to the young. The exact mechanism of such transfer through the mother's cuticle is unknown. Furthermore, without maternal protection, first instars have no defenses and would be prone to predation. Studies indicate that when faced with an enemy, females carrying young are quicker to engage in defensive and/or avoidance behavior than females without young.

It is particularly important for scorpions, known as "inveterate cannibals," for mother and young to recognize one another. How is this accomplished? Apparently, species-specific chemical cues are used for recognition. Transfer experiments indicate that first-instar young will only remain on the back of a conspecific female. Furthermore, if young are treated with chemicals to disguise or alter their scent, the mother will abandon or eat them.

4. Extended Mother–Young Association

A few species of scorpions exhibit some social behavior beyond the newborn–maternal interaction. Two distinct evolutionary pathways find individuals aggregating to form groups: Along the "familial route," young remain with their mother for extended periods, whereas unrelated individuals aggregate along the "communal route." For example, in many species from the family Buthidae, conspecific individuals form temporary communal groups in suitable (and often limited) microhabitats (e.g., fallen wood), usually during the colder months.

Most sociality occurs along the familial route as a natural extension of the universal parental care. For example, in the North African *Scorpio maurus* (family Scorpionidae), young may stay in the mother's burrow for 3 months before dispersing, well after the end of the first instar. In some species, this extended association may last longer. In two South American species in the family Bothriuridae, *Urophonius iheringi* and *U. brachycentrus*, mothers expand their burrows into "gestation chambers" in which the mother and young remain in an inactive state for up to 6.5 months.

A further elaboration of this association is seen in species in which the mother may share food with the young or even forage communally. In the Mexican *Didymocentrus caboensis* (family Diplocentridae), young remain with the mother for extended periods and communally capture large prey items. When the young do disperse, it is only in the immediate area; thus, patches containing closely associated burrows of related individuals can be found.

5. Social Scorpions: Mother and Young Cooperate over Extended Periods

More permanent (sub)social groups can also be found. These are usually associated with extended parental care. The offspring of one to several broods (and possibly some unrelated individuals) remain in a central place with one or more mothers. Scorpions from several species of the Indian *Heterometrus* (Scorpionidae) often live together in a complex burrow system in groups of 2 to over 30 individuals of all ages. In the West African emperor scorpion *Pandinus imperator* (Scorpionidae), aggregations of 10–25 related and nonrelated individuals live in expanded chambers beneath fallen logs. Sociality among *Pandinus* scorpions is quite advanced, with group defense, prey capture, and homesite maintenance. For many months, young *Pandinus* depend heavily on larger animals to dismember prey into pieces that can be handled. This represents one of the most extreme cases of parental care by scorpions.

B. Other Arachnid Orders

In the other orders of arachnids, parental care is limited to guarding eggs and, in many cases, allowing the defenseless newborn to stay with the mother until after their first molt. In almost all cases, second-instar individuals, with a sclerotized cuticle and the ability to capture food, disperse from the mother to assume a fully independent existence.

1. Order Amblypygi (Tailless Whip Scorpions)

Amblypygids are scorpion-like arachnids which can be fairly common in the tropics. These organisms, unlike scorpions, lay eggs and carry them in a ventral brood pouch. After hatching, the mother carries the young on her back until after the first molt. Little more is known of parental care in these animals.

2. Order Uropygi (Whip Scorpions)

Uropygids, another mainly tropical group, also lay eggs. Eggs are carried in a brood pouch on the mother's underside until they hatch. After hatching, the young climb onto the mother's dorsum and reside there through the first molt.

3. Order Pseudoscorpionida (Pseudoscorpions)

Pseudoscorpions are small predators in soil and leaf litter. They lay eggs and secrete a brood pouch in which the developing eggs incubate until hatching. In some species, females guard the eggs.

4. Order Solpugida (Wind Scorpions)

Solpugids are voracious predators common to many deserts. Generally, eggs are laid in a burrow and abandoned. For some large African species in the genus *Galleodes*, the female remains in the burrow to guard the eggs until they hatch. Other aspects of parental care are unknown.

5. Order Opiliones (Harvestmen)

Harvestmen are common arachnids in moist habitats worldwide. Originally, it was thought that parental care in this group was minimal to nonexistent. Generally, females oviposit eggs in the soil and leave them. Recent work has shown three cases of parental care. In two, the eggs are guarded after laying; in one species, *Zygopachylus albomarginus*, the male guards. In the third case, *Leytpodoctis oviger*, eggs are attached to one of the legs of the female and carried until the young emerge.

6. Order Ricinulei

Ricinuleids are rare, tick-like arachnids which live in (sub)tropical areas. Usually, females produce only one large egg at a time, which is guarded and carried around using the pedipalps and chelicerae. Little more is known about these cryptic organisms.

7. Subclass Acarina (Mites and Ticks)

This group comprises the largest order of arachnids; to date, only a tiny fraction of the species have been identified. In most of the groups that have been studied, little evidence of parental care exists. One known example involves a parasite of honeybees, *Varroa jacobsoni*. These mites feed upon developing bee larvae and pupae and may have multiple generations within a single chamber (cell) of a bee colony. Females choose oviposition sites on bee pupae that are least likely to be disturbed by worker bees. The female then aggregates her feces at one site on the chamber's wall; these droppings act as a cue to show the next generation where to aggregate and mate.

IV. MYRIAPODS

Myriapods consist of the centipedes (class Chilopoda), millipedes (class Diplopoda), and two minor groups, the symphylans (class Symphyla) and pauropods (class Pauropoda). Parental care is typically very limited. Most species invest nothing more than the effort of selecting a suitable oviposition site.

The chilopod orders Lithobiomorpha and Scutigeromorpha and the classes Symphyla and Pauropoda show no parental behavior beyond attempting to conceal the egg cluster. Symphylan reproductive behavior is so simplified that the female discovers and picks up the spermatophore which the male had left behind some time before; thus, males and females never actually meet.

A. Class Chilopoda

Brooding behavior is consistently shown in the orders Geophilomorpha and Scolopendromorpha. The 12–75 eggs are laid in humid soil, under bark, or in litter and held tightly in a spherical mass by coils of the females body. She keeps approximately two-thirds of her body segments wrapped around the egg mass at all times, even while defending it from enemies. Females continue to guard their eggs after hatching; the young undergo several molts before leaving the mother. The female does not feed during the 40–50 days that she broods. Females of many species continuously probe the eggs with their antennae and maxillae, turn them with their legs, and coat them with a glandular secretion that protects them from fungal and bacterial infections. Eggs separated from their mother are frequently infected with fungi, showing the importance of maternal care. In laboratory experiments, females of various species usually consumed their brood if disturbed but were less inclined to consume larvae than eggs.

B. Class Diplopoda

With the exception of males in the order Platydesmida, millipedes have never been observed to display brooding behavior (actively guarding eggs until they hatch). However, millipedes do construct an energetically expensive egg chamber. Eggs are typically laid in a crevice of a rotting log or a small hole in the soil which the female has dug and covered. Some species line the inside of the sealed egg chamber with silk. Once the eggs are laid, the female seals the chamber with a mixture of feces, anal gland secretions, and soil. Thus, the eggs are protected from temperature and humidity fluctuations, yet they remain vulnerable to fungal decay or nest predation from various soil invertebrates.

V. TERRESTRIAL CRUSTACEA

A few groups of Crustacea have invaded land successfully. These include a few types of land crabs, some hermit crabs, some crayfish, a few amphipods, and several species of isopods. Most are restricted to relatively moist habitats and microhabitats. The isopods are the most successful terrestrial crustacean and are common inhabitants in a diversity of habitats, ranging from forests to deserts.

All Crustacea copulate and eggs are fertilized internally. Almost all brood their eggs until they hatch. Many terrestrial Crustacea also brood the developing larvae, although some land crabs return to place the young in water. In most cases, brooding ceases when the fully formed juveniles disperse from the female. The most advanced parental care is shown by terrestrial isopods.

A. Isopods (Class Isopoda)

Terrestrial isopods keep their young in a brood pouch similar to other Crustacea. The ventral brood pouch or marsupium is formed by membranes connected to five pairs of sternal oostegites; mucous filled, it provides the eggs with oxygen and nutrients from the mother. The eggs are carried by the mother until the larvae have molted several times, typically 1 or 2 weeks after hatching. Larva that fail to develop are eaten by the viable portion of the brood.

Greater parental care is demonstrated by some isopods, such as the well-studied *Hemilepistus reaumuri*. This species is a common inhabitant of the northern Sahara Desert, a surprising habitat for a creature with such high vulnerability to desiccation. Individuals of this group form family units of 30–50 closely related individuals that share communal burrows and cooperatively dig through soil. Communal digging and burrows decrease water loss of individuals. When the young reach 9 months, they leave the group to search for a new colony site. Selection of a proper burrow site by a pioneering female is critical; the established colony will die in a matter of weeks if she chooses a site which is too dry. The female is also under high risk of predation during her search. However, once the colony is established, the mortality for the family unit as a whole is low.

Bibliography

Brusca, R. C., and Brusca, G. J. (1990). *Invertebrates*. Sinaur, Sunderland, MA.
Foelix, R. F. (1982). *Biology of Spiders*. Harvard Univ. Press, Cambridge, MA.
Gertsch, W. J. (1979). *American Spiders*. Van Nostrand, New York.
Kaston, B. J. (1972). *How to Know the Spiders*. Brown, Dubuque, IA.
Lewis, J. G. E. (1981). *The Biology of Centipedes*. Cambridge Univ. Press, Cambridge, UK.
Polis, G. A. (1990). *The Biology of Scorpions*. Stanford Univ. Press, Stanford, CA.

Parental Behavior, Birds

John D. Buntin
University of Wisconsin at Milwaukee

I. Mating Systems, Modes of Nestling Development, and Parental Care Patterns
II. Hormonal Regulation of Parental Behavior
III. Neural Mechanisms Underlying Parental Behavior

GLOSSARY

altricial young Nestlings that are helpless at birth and completely dependent on their parents for warmth, protection, and food.
brood parasitism The practice of laying eggs in the nests of birds of another species, thereby transferring the parental responsibilities of egg incubation and care of young to the host.
brood patch A thickened, edematous, and highly vascularized patch of featherless skin that facilitates the transfer of heat from the incubating or brooding parent to the eggs or nestlings.
broody behavior The act of sitting on or crouching over young, which serves a thermoregulatory and protective function.
crop milk A nutritive substance consisting of epithelial cells sloughed from the crop sac wall of incubating and brooding pigeons and doves that is regurgitated to the nestlings at hatching.
crop sac An outpocketing of the esophagus that is used as a seed storage organ in many birds and as a source of crop milk in pigeons and doves.
incubation behavior The act of sitting on eggs, which serves a thermoregulatory and protective function.
monogamy A mating system in which one male mates with one female for at least one reproductive cycle.
passerines Representatives of the largest avian order, Passeriformes, which includes the perching birds.
polyandry A mating system in which one female mates with two or more males.
polygny A mating system in which one male mates with two or more females.
precocial young Chicks that are able to leave the nest, follow their parents, and forage for food on their own soon after hatching.
semialtricial young Nestlings that are capable of leaving the nest soon after hatching but are dependent on their parents for food for several days or weeks.
squab A nestling pigeon or dove.

In birds, parental behaviors are traditionally defined as those that enhance the survival of eggs or young. Included in this category are incubation of eggs, brooding and feeding of young, and behaviors that serve to protect the eggs and young from predators and weather. While parental care is necessary for successful reproduction in birds, there are instances in which the parents themselves do not provide this care. For example, brown-headed cowbirds lay their eggs in the nests of interspecific hosts and rely on the hosts to incubate their eggs and hatch and rear their chicks. Such examples of brood parasitism are rare because over 99% of the approximately 9000 species of birds exhibit some form of parental behavior. Nevertheless, the type and amount of care delivered varies markedly across species, as does the nature and extent of parental contributions by the male and female breeding partners. These differences correlate most strongly with interspecific variations in mating systems and the maturity and mobility of the young at hatching, although they may also relate to other factors, such as habitat, number and size of eggs laid, and diet.

I. MATING SYSTEMS, MODES OF NESTLING DEVELOPMENT, AND PARENTAL CARE PATTERNS

The amount and type of parental care displayed depends strongly on the mode of development of the young. Over 70% of avian species rear altricial young, which are born naked, or nearly so, and are incapable of effective thermoregulation, locomotion, or independent feeding. In contrast, species with precocial young rear chicks that are capable of leaving the nest, following their parents, and feeding on their own soon after hatching. Gulls and terns are examples of the 10% of avian species that rear semialtricial young, which are more mobile than altricial young at hatching but are still dependent on their parents for food for an extended period. Guarding of nests and young is common in all avian species that exhibit parental behavior, but other aspects of the parental behavior repertoire may vary with nestling capabilities. For example, parents of precocial young do not feed their chicks but they typically lead them toward food and away from danger by emitting distinctive vocalizations.

Following hatching, most young birds are unable to fully regulate their body temperatures, and the brooding behavior shown by the parent serves both a thermoregulatory and protective function. While newly hatched altricial young require almost constant brooding, precocial young require only intermittent brooding. Brooding typically occurs over a longer period of time in altricial species than in precocial species. However, brooding time declines sooner in large broods than in small broods because nestlings in large broods conserve heat more efficiently.

Parental provisioning of young also varies with mode of nestling development. Once mobile, most precocial young feed on their own, even though they are typically led to food sources by one or both parents. Most altricial nestlings, in contrast, depend on food delivered to the nest or directly to their mouths by their parents, which obtain it by foraging or hunting. In columbiform birds (pigeons and doves), however, parents manufacture their own "nestling food" in the form of "crop milk," which is regurgitated to their altricial young.

Parental care patterns also vary with the mating system employed. As a general rule, biparental care, uniparental care by males, and uniparental care by females are associated with monogamy, polyandry, and polygyny, respectively. Biparental care is seen in approximately 90% of all avian species, a feature that sets birds apart from other vertebrates. Sharing of parental duties between the breeding partners is most extensive in species rearing altricial young in which both sexes are monogamous. In these species, it is common for both breeding partners to engage in incubation of eggs and care of young. However, females typically make a larger parental contribution than males, and some division of labor among the breeding partners is common. In many passerines (perching birds), the male and the female contribute relatively equally to feeding the young. However, males often spend less time incubating and brooding than females even though they may provision the female while she sits on eggs. While biparental care is most common in monogamous species with altricial young, it is also seen in some monogamous species with precocial young such as swans and geese. Biparental care is also seen in a few polyandrous species such as dunnocks, the Galapagos hawk, and the Tasmanian native hen. It has also been reported in some species, such as the starlings and icterine blackbirds, which exhibit polygny to some degree. However, recent studies in the facultatively polygnous European starling indicate that the male's parental care contribution is more extensive in monogous pairs than in birds practicing polygny.

Fewer than 90 species have been reported in which parental care is provided exclusively by the female. Typically, these are polygynous species with precocial young, such as turkeys and pheasants. However, this pattern is also exhibited by some monogamous ducks and grouse, which rear precocial young, and by polygynous hummingbirds, bowerbirds, manakins, and birds of paradise, which rear altricial young.

Less than 10% of all avian subfamilies and <1% of living avian species display a pattern in which the male is the predominant or exclusive parental caregiver (Fig. 1). Many of the birds that fall into this category are shorebirds that nest on the ground and rear precocial or semialtricial young. Some of these species exhibit sequential polyandry, such as the phalaropes, emus, and cassowaries, whereas others, such as the jacanas, exhibit simultaneous polyandry in which the larger and more brightly colored females compete for access to males. In the ostriches, rheas, and tinamous, males are largely if not exclusively responsible for incubating a communal clutch of eggs laid by multiple females. Although rare, exclusive male parental care or male-dominated biparental care is also seen in a few monogamous species including the kiwis.

Cooperative breeding, in which more than two birds provide parental care, is practiced by at least 3% of all avian species. Two forms of cooperative breeding have been distinguished. In communal breeding systems, incubation and care of young are shared among several breeding individuals of both sexes that also share parentage of the offspring. In the "helper-at-the-nest" system, however, breeding birds are assisted in their parental duties by nonbreeding auxilliaries that are usually related to the parents as nuclear or extended family members. The Mexican jay and the Florida scrub jay are examples of species that exhibit the helper-at-the-nest pattern, whereas the groove-billed ani, a member of the cuckoo family, is an example of communal breeder. Other species, such as the acorn woodpecker, may combine elements of both strategies.

II. HORMONAL REGULATION OF PARENTAL BEHAVIOR

The physiological basis of avian parental behavior has only been investigated in detail in two domesticated members of the order Galliformes and one domesticated member of the order Columbiformes. Accordingly, it is not possible to correlate differences in parental care strategies among birds with differences in underlying physiological mechanisms. The two galliform species, the domestic chicken and turkey, are polygnous species that exhibit uniparental female care of precocial young. They are therefore atypical in the sense that they do not exhibit the pattern of biparental care of altricial young that is most common in birds. While the ringdove is more typical in this respect, it exhibits a highly specialized parental feeding pattern that is unique to columbiform birds in which parents synthesize and regurgitate crop milk to their newly hatched young. While these galliform and columbiform species may not represent the avian norm, similarities in the physiological underpinnings of parental behavior in these

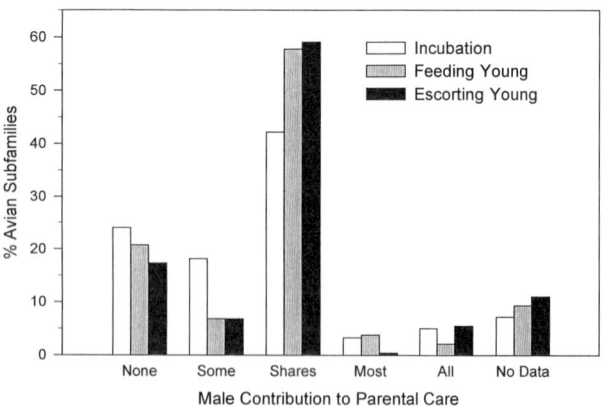

FIGURE 1 Prevalence of different parental care patterns in males, as reflected by the percentage of avian subfamilies exhibiting the pattern. Note that male-dominated parental care (i.e., male performs most or all the parental behaviors shown) is exhibited in a small minority of avian subfamilies (modified from Silver et al., 1985, with permission from the American Zoologist).

two widely different groups may suggest features that are common to the expression of parental care in variety of birds.

A. Common Themes in Hormonal Mechanisms Regulating Avian Parental Care

Although important species, sex, and breeding stage differences have been documented, several general themes emerge from a survey of the available literature on hormonal determinants of avian parental care: (i) Increased secretion of the pituitary hormone prolactin (PRL) is closely associated with parental care expression in a wide variety of avian species and, at least in some, it has been shown to directly facilitate parental responsiveness; (ii) steroid hormones from the gonads can exert both facilitatory and inhibitory influences on parental behavior expression and may interact with PRL in complex ways to influence these behaviors; (iii) hormone-dependent changes in parental behavior rarely occur in the absence of appropriate environmental cues, and social/environmental signals may eventually replace hormonal signals as the principal regulators of parental behavior as the breeding cycle proceeds; and (iv) previous breeding experience may act as a powerful modulator of the birds' responsiveness to the hormonal and environmental signals that regulate parental activity.

B. Onset of Incubation Behavior

Although sitting on eggs is the most common method of providing heat for embryonic development in birds, it is not the only method of doing so. An alternative strategy is seen in the mallee fowl and other mound-building megapodes of the order Galliformes. In these birds, eggs are deposited in a mound consisting of soil and vegetation that is built by one or both parents. The eggs are typically warmed by the sun and by fermentation of the decaying plant matter, and nest temperature is closely monitored and regulated by active parental interventions such as mound ventilation and additional mound building.

In the remaining species that transfer their own heat to their eggs during incubation, the sitting behavior involved is often a natural extension of the nest attachment that is established during nest building and egg laying. Studies in ovariectomized chickens, turkeys, and ringdoves have clearly shown that the initiation of female nesting behavior requires estrogen and progesterone. These gonadal steroid hormones, which are secreted from the developing ovarian follicles, also prepare the oviduct for egg production and stimulate the release of luteinizing hormone for ovulation. In ovariectomized ringdoves, the increased nest attentiveness induced by estrogen and progesterone is sufficiently robust to support the induction of full incubation behavior when eggs are provided. In ovariectomized chickens and turkeys, however, the sustained sitting that is characteristic of full incubation requires the addition of the pituitary hormone PRL to the steroid hormone regimen. A similar process of PRL–steroid synergism appears to be involved in the induction of incubation behavior in female budgerigars. In contrast to the ringdove, which shows no increase in plasma prolactin concentrations at the onset of incubation, PRL levels rise abruptly in chickens and turkeys and closely parallel the increase in time spent on the nest as incubation is established (Fig. 2). Prolactin also synergizes with gonadal steroid hormones to form the brood patch in chickens and several other avian species. This thickened, defeathered patch of ventral surface skin increases tactile sensitivity and enhances heat transfer between the incubating bird and its eggs (Fig. 2).

The hormonal basis (if any) for the onset of incubation behavior in male birds has not been well-defined. In many species, plasma testosterone levels decrease as incubation begins. In the spotted sandpiper, a polyandrous species in which the male is the predominant caregiver, this decline appears to be an important signal for initiation of sitting, since sitting onset is accelerated when antiandrogenic drugs are administered while androgen levels are still elevated. In male ringdoves, testosterone levels are high during the courtship phase and then decrease at the onset of incubation. This courtship-phase elevation in plasma androgen may be important since testosterone has been shown to enhance the ability of progesterone to induce incubation behavior in nonbreeding males. Although plasma progesterone levels do not change in males at the onset of incubation, testosterone in-

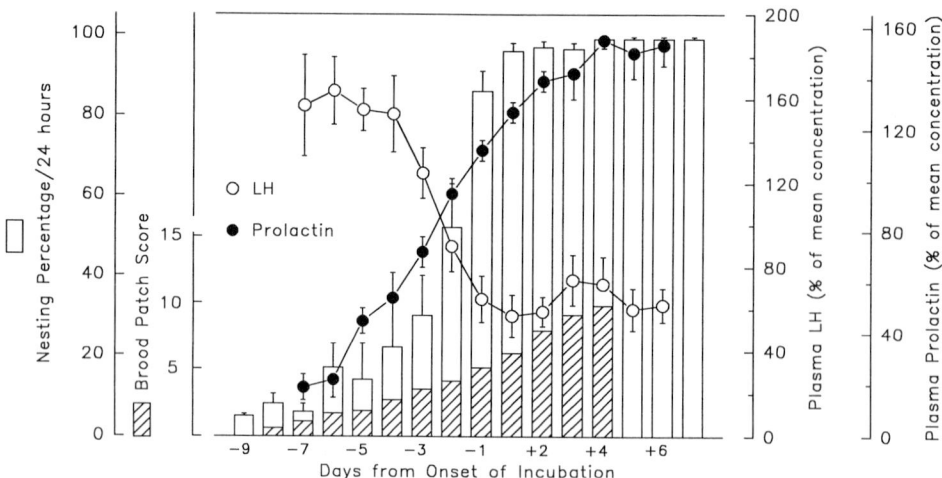

FIGURE 2 Changes in plasma PRL, luteinizing hormone (LH), time in the nest, and brood patch development in domestic chickens (bantam hens) in relation to incubation onset. Values shown means ± standard error of the mean (modified from R. W. Lea, A. S. Dods, P. J. Sharp, and A. Chadwick, *J. Endocrinol.* **91**, 89–97, 1981, with permission from the Journal of Endocrinology LTD).

fluences sensitivity to progesterone by increasing progesterone receptors in brain areas that support incubation behavior. Despite the demonstrated effectiveness of progesterone in promoting sitting behavior in male doves, incubation can also be expressed in the absence of adrenal or gonadal sources of steroids, provided that the breeding males receive adequate exposure to social and environmental signals from the mate and nest that normally occur at this stage. This suggests that hormonal and environmental stimuli act together in complementary fashion to promote the onset of incubation in male doves.

C. Maintenance of Incubation

In most species, plasma levels of pituitary gonadotropic hormones and gonadal steroid hormones decrease after egg laying and remain low throughout incubation and into the early posthatching period. Reproductive suppression is due in part to the high level of PRL in the blood during incubation and, in some species, negative feedback effects of gonadal steroids on gonadotropin secretion. Stimuli generated while sitting on eggs and cues received from young after hatching may also inhibit gonadotropin secretion, most probably through a hormone-independent suppression of gonadotropin-releasing hormone secretion from the hypothalamus. The functional significance of this decline in gonadal activity during incubation is not entirely clear. In the male spotted sandpiper and female starling, a decrease in gonadal steroid secretion is necessary to maintain interest in sitting on eggs. However, plasma estradiol or testosterone can be maintained at high concentrations without disrupting sitting behavior in males or females of several other species. Although gonadal steroids are important in initiating incubation behavior, the low levels of gonadal steroids present during incubation apparently play no major role in maintaining the behavior since incubation persists in chickens, turkeys, and ringdoves after gonad removal.

The hormone that is most closely linked to incubation behavior in birds is PRL, and evidence for a causal relationship between PRL and sitting behavior is persuasive. Virtually all species of birds that have been studied show an increase in plasma PRL concentrations during incubation, although there are differences among species in the timing of this rise in relation to egg laying (Fig. 3). Plasma PRL levels

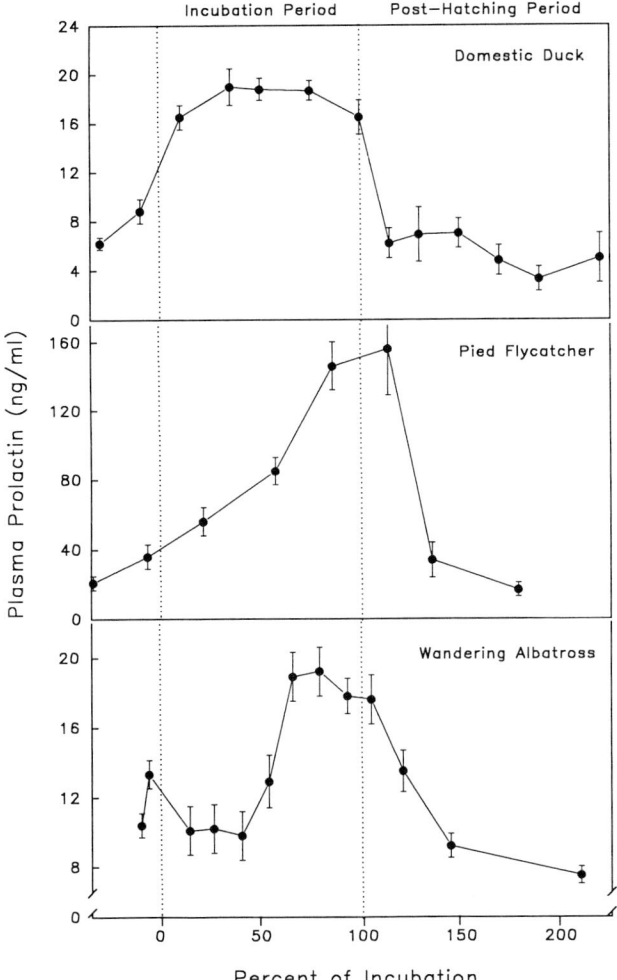

FIGURE 3 Three patterns of plasma PRL changes in relation to parental behavior during the breeding cycle. In the domestic duck (top), plasma PRL levels increase abruptly to near-maximal levels in association with incubation onset. This pattern is also typical of domestic galliform species such as chickens and turkeys. The pied flycatcher exhibits a second pattern seen in several passerine species in which prolactin levels begin to increase at the onset of incubation but do not reach peak concentrations until later in this breeding stage. The wandering albatross, like the ringdove (see Fig. 4), exhibits a third pattern in which plasma PRL levels remain low during the first half of incubation. Plasma PRL levels typically fall abruptly at hatching in species with precocial young such as the domestic duck. In contrast, PRL levels tend to remain elevated for some period after hatching in species with altricial young (pied flycatcher) or semialtricial young (wandering albatross) [reprinted from Buntin, 1996, with permission from Academic Press; original source data reprinted with per-

tend to be elevated in both sexes during incubation in species in which both breeding partners make a substantial sitting contribution. In species in which incubation is predominantly or exclusively performed by one breeding partner, however, PRL levels tend to be higher in the incubating sex. Prolactin levels in blood decline when sitting behavior is interrupted or if the brood patch is anesthetized and they rebound when incubation is reinstated after a relatively brief (1–3 days) period of nest deprivation. PRL injections maintain interest in sitting on eggs in ringdoves and bantam hens during extended periods of nest deprivation and they extend the period over which ringdoves will incubate infertile eggs.

Despite strong evidence that PRL facilitates sitting and that sitting, in turn, stimulates PRL secretion, these two events can be dissociated. One confounding influence in many temperate zone passerine species is increasing day length, which stimulates PRL secretion during the breeding season even in birds such as cowbirds that do not engage in parental activities. In incubating male and female ringdoves, PRL levels remain elevated in both sexes during normal nest recesses, which typically last for several hours. PRL secretion during these nest recesses could be maintained in part by visual stimuli from the incubating mate since male pigeons and doves deprived of the opportunity to sit on eggs continue to secrete PRL if they are allowed to view their sitting partners. In other species, such as in the king penguin, however, PRL remains elevated during routine sitting recesses lasting several days or weeks, during which time the birds are foraging at sea. This suggests a possible endogenous timing mechanism regulating PRL secretion. Sitting will resume following a period of nest deprivation in canaries, ringdoves, and turkeys even though PRL levels have markedly declined during the deprivation period, and at least in turkeys

mission from Academic Press and Oxford University Press (M. R. Hall and A. R. Goldsmith, *Gen. Comp. Endocrinol.* **49**, 270–276, 1983; J. A. Hector and A. R. Goldsmith, *Gen. Comp. Endocrinol.* **60**, 236–243, 1985: B. Silverin and A. R. Goldsmith, *J. Zool. (London)* **200**, 119–130, 1983; B. Silverin and A. R. Goldsmith, *Gen. Comp. Endocrinol.* **55**, 239–244, 1984)].

and doves, the resumption of sitting occurs before an increase in plasma PRL is detected. High plasma levels of PRL are apparently not necessary for the maintenance of incubation behavior in ringdoves since incubating birds given antibodies to vasoactive intestinal polypeptide, a potent prolactin-releasing neuropeptide, continue to sit on eggs even though the incubation-associated rise in plasma PRL is largely eliminated by antibody treatment. Similar manipulations in bantam hens cause incubation to terminate, thereby highlighting the importance of species differences in this regard. The relationship between PRL and sitting activity is therefore complex, and although PRL appears to be an important hormonal facilitator of sitting behavior during the incubation period in those species that have been extensively investigated, it is not the only factor that promotes incubation. Nonhormonal cues from the nest, eggs, and mate may also sustain interest in sitting by exerting direct stimulatory effects on the behavior or, more indirectly, by stimulating PRL release. The effectiveness of these nonhormonal factors in promoting incubation is likely to vary with species and with the experiential history of the breeding pair.

D. Care of Precocial Young

Plasma prolactin levels drop sharply after hatching in domestic fowl, turkey, and duck and remain low while the mothers are actively caring for their young (Fig. 3). This decline appears to result from the disruptive effect of contact stimuli from the chicks on sustained brooding activity of the hen. In some strains of chickens, this decline in PRL, although rapid, is not as precipitous as the decrease in PRL induced by nest deprivation alone, thereby suggesting that signals from chicks retard this process. However, it is not clear if this attenuated decline in PRL is important for parental behavior expression during the posthatching period. On the one hand, there is ample evidence from a variety of galliform species that prolactin injections promote parental responses toward chicks. On the other hand, maternal behavior can be induced in hens in the absence of changes in plasma PRL levels by simply confining them with chicks. This observation, together with the fact that plasma PRL levels normally decline during the posthatching period, suggests that PRL is not essential for the display of parental behavior at this stage. The low levels of gonadal steroids that are normally present after hatching favor the expression of parental responsiveness since maternal behavior is delayed or disrupted when plasma levels of these hormones are artificially elevated. Nevertheless, like PRL, gonadals steroids are not necessary since normal onset and maintenance of maternal behavior is seen in ovariectomized hens. In general terms, therefore, hatching signals a transition from a hormone-dominated mode of parental behavior regulation, as seen during incubation, to a mode in which nonhormonal stimuli from the developing chicks assume preeminence as regulators of behavioral expression.

E. Care of Altricial Young

In general, plasma PRL levels remain elevated for a longer period after hatching in altricial birds and semialtricial birds (Fig. 3, middle and bottom) than in precocial birds (Fig. 3, top). This in turn correlates with the extended period of brooding and feeding young that occurs in these species. Several mechanisms may be responsible for the protracted period of PRL secretion after hatching. In male and female ringdoves and in the female pied flycatcher, there is good evidence that stimuli from the young or stimuli generated by parent–young interactions promote PRL release. In the scrub jay, a species in which nonbreeding helpers at the nest assist the parents in feeding young, plasma PRL levels in helpers are positively correlated with the number of nestings being attended, thereby suggesting a similar mechanism of PRL regulation. Nevertheless, the period of elevated PRL release cannot be prolonged indefinitely in either flycatchers or doves by exposing the parent to maximally effective stimuli from young. This suggests that endogenous timing mechanisms may also be at work to modulate the effectiveness of these cues. Even more rigid timing mechanisms may be involved in regulating the period of elevated PRL secretion in other species, such as the wandering albatross and king penguin. In the wandering albatross, for example, pairs given chicks prematurely show a prolonged period of elevated PRL secretion and a prolonged period of chick brooding, whereas those in which the incubation period is artificially prolonged display a shortened posthatching phase of elevated PRL and abandon their chicks prematurely.

The correlation between elevated plasma PRL and the display of parental behavior in many species with altricial young suggests that PRL stimulates these activities. Studies in the ringdove provide the strongest support for this hypothesis. In both male and female doves, rising plasma PRL levels during the latter half of incubation stimulate the development of the crop sac, which culminates in the formation of crop milk at the time of hatching (Fig. 4). Prolactin is not only responsible for synthesis of this nestling food but also important for transferring it to the young by promoting a variety of parental activities. First, PRL facilitates the transition from incubation of eggs to care of young that occurs at hatching, as evidenced by the observation that PRL injections cause incubating birds to shift their preference from eggs to young in a choice test. Second, PRL promotes parental regurgitation feeding behavior itself. This is supported by studies in nonbreeding doves with previous breeding experience which have documented the effectiveness of PRL injections in promoting parental regurgitation behavior toward foster young. Nonbreeding doves that lack previous breeding experience do not respond to PRL in this way. Nevertheless, they are capable of exhibiting regurgitation feeding if they receive injections of both progesterone and PRL. Progesterone apparently stimulates these birds to sit on the nest containing the foster young, thereby enhancing exposure to the young and providing the contextual cues that might be necessary for PRL to effectively stimulate regurgitation in this testing situation. Finally, PRL may stimulate the foraging behavior needed to provision the young with increasing amounts of seed and other food as the young grow and develop. In nonbreeding males and females, PRL stimulates a marked increase in food intake when it is injected subcutaneously or into the cerebral ventricles. This parallels the increase in food intake seen in breeding pairs during the posthatching period (Fig. 4). In contrast to the situation in nonbreeding doves, the increase in food intake is not accompanied by an increase in body weight in the breeding birds since the food is brought

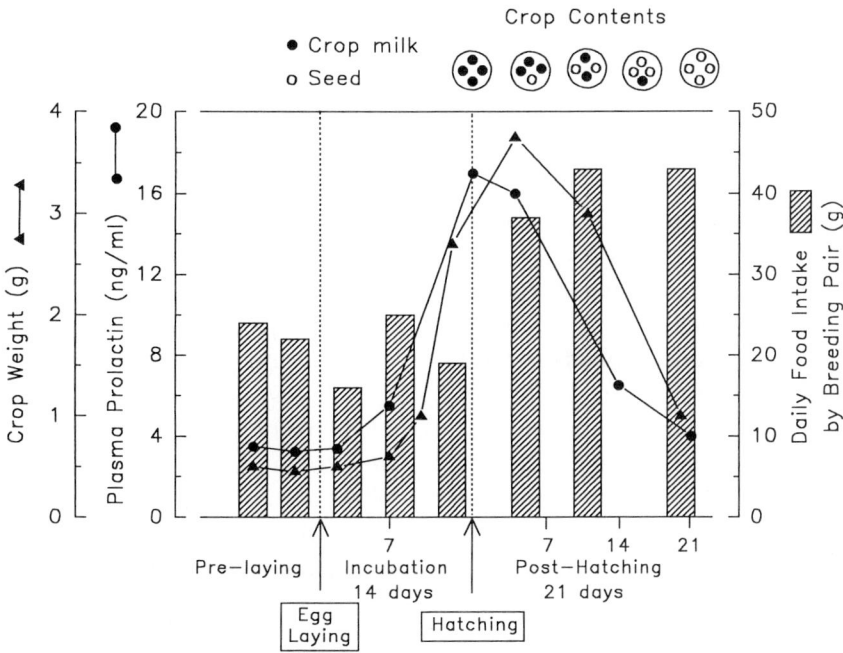

FIGURE 4 Changes in plasma PRL, crop sac weight, crop contents, and parental foraging behavior during the ringdove reproductive cycle. Sex differences in plasma PRL and crop sac development are minimal in this species (reprinted with permission from N. D. Horseman and J. D. Buntin, 1995, with permission from Annual Reviews of Nutrition, Vol. 15, © 1995 by Annual Reviews, Inc.)

back to the nest and regurgitated to the young. As the posthatching period proceeds, however, parental food intake remains elevated despite a decrease in plasma PRL, thereby suggesting that provisioning may eventually become independent of hormonal support (Fig. 4).

As during incubation, steroid hormones influence the expression of parental behaviors seen after hatching in species with altricial young. Many male passerines actively feed young during the posthatching period, and a necessary condition for this behavior to be displayed appears to be maintenance of low androgen levels. Implants of testosterone have been shown to disrupt parental feeding of nestlings by males in at least four different passerine species and, in one species (the house sparrow), the decline in feeding that is normally seen as testosterone levels rise is eliminated by the antiandrogenic drug flutamide. Although reduced secretion of PRL could conceivably mediate these androgen-induced deficits in parental behavior, this is not the case in at least one species, the song sparrow. Based on behavioral observations, it is likely that these deficits result from an increase in time spent engaging in other activities that are incompatible with parental provisioning, such as territorial defense and mate guarding.

Adrenal steroid hormones may also modulate parental behavior expression, albeit in complex ways. Corticosterone, the major glucocorticoid in birds, is normally present in low concentrations during the posthatching period in pied flycatchers, but plasma levels may rise during times of stress associated with the rearing of large broods or in females rearing young without the aid of the male partner. Studies of flycatchers given corticosterone implants suggest that modest elevations in plasma corticosterone of the type that would occur under these conditions stimulate an increase in parental foraging activity to help meet the increased food demands from the nestlings. At higher corticosterone concentrations, however, provisioning of the young decreased even though foraging increased. This is presumably due to an increase in the amount of food consumed by the parent. At very high levels, corticosterone caused birds to abandon their nests, leave the breeding area, and increase their foraging rate even further, thereby signaling a shift away from reproductive effort altogether. Augmentation of feeding by corticosterone has also been reported in PRL-treated pigeons and in white-crowned sparrows that had been previously food deprived. However, it is not yet clear if corticosterone modulates parental provisioning of nestlings in these or other species.

F. Defense of the Nest and Young

Most birds exhibit some form of defensive behavior when their eggs or young are threatened by intruders. Defensive behaviors displayed while on the nest may involve feather erection, wing raising, and aggressive counterattacks against the intruder using the bill, feet, or wings. Some species exhibit more elaborate defensive reactions, such as the injury-feigning distraction displays seen in many species that nest on or near the ground. Little is known of the physiological basis of these behaviors, although it is assumed that the mechanisms are similar to those involved in promoting other parental activities. In ringdoves, both progesterone and PRL facilitate aggressive and defensive responses when incubating birds are challenged with a model spider "intruder," and although their effects may be sex specific, both of these hormones may be involved in modulating nest defense expression during the normal breeding cycle. In females, aggressive and defensive reactions to an intruder are highest at the time of egg laying and early incubation when progesterone levels are elevated. Male doves, in contrast, exhibit a sharp rise in these behaviors around the time of hatching and during the early posthatching period when PRL levels are maximal. Hormonal modulation of defensive responses is less clearly understood in other species. Chronic infusion of PRL has been reported to increase the incidence of distraction displays induced by the approach of an intruder in free-living willow ptarmigan hens with newly hatched chicks. Because plasma PRL levels normally remain elevated in this species during the first week after hatching, this effect may be of functional significance. In domestic fowl with chicks, however, increased aggression toward intruding conspecifics occurs even though PRL levels drop precipitously after hatching. In this case, therefore, other regulatory mechanisms would appear to be involved.

III. NEURAL MECHANISMS UNDERLYING PARENTAL BEHAVIOR

A. Sensory Mechanisms

Denervation studies in the domestic turkey have revealed that somatosensory signals from the brood patch area are essential for the rise in plasma PRL at egg laying and the resulting onset of incubation behavior. Anesthetization or denervation of the brood patch area has also been shown to depress PRL secretion in incubating ducks, although the effects on incubation behavior itself are inconclusive. In contrast, denervation of the apterial skin that contacts the eggs has little effect on established incubation in pigeons. Whether these differences reflect species variation or differences in the stimulus requirements for onset and maintenance of incubation cannot be discerned from the fragmentary evidence that is currently available. However, nontactile cues from young have been shown to maintain PRL secretion in parent ringdoves during the incubation and posthatching periods, thereby demonstrating that other sensory modalities are capable of supplementing the tactile signals received from the eggs and young in maintaining the physiological state associated with parental behavior expression.

B. Brain Sites Mediating Parental Responses

Based on studies in the domestic chicken, turkey, and ringdove, the preoptic area (POA) of the avian brain, which lies immediately anterior and dorsal to the optic chiasm, appears to be a key component of the neural substrate underlying incubation, brooding of young, and feeding of young. Lesions of the POA prevent the onset of incubation in turkey hens and PRL implants in this region reportedly promote incubation onset in domestic fowl. In doves, POA lesions severely disrupt PRL-induced parental regurgitation feeding in nonbreeding ringdoves with previous breeding experience. Microinjections of PRL into the POA also elevate food intake in nonbreeding ringdoves, thus simulating the hyperphagia that occurs in normally breeding doves that are provisioning their nestlings during the posthatching period. The possibility that the POA is an important site of PRL action in promoting parental behavior is strengthened by evidence that the region contains relatively high concentrations of PRL receptors in several avian species, including the ringdove, Wilson's phalarope, redwing blackbird, European starling, and dark-eyed junco. It is also strengthened by the observation that the level of PRL binding activity in the POA of the brown-headed cowbird, a brood parasite that does not exhibit parental behavior, is less than that of two related songbird species, the redwing blackbird and the European starling, which do care for their young. In addition to its possible role as a target of PRL action, the POA, together with the adjacent anterior hypothalamus, is an effective site of progesterone action for initiating incubation behavior in nonbreeding doves.

Receptor mapping studies have shown that the POA is not the only brain area that exhibits high PRL binding activity. Receptors are also concentrated in several forebrain and midbrain sites, including the paraventricular and ventromedial hypothalamic regions, the tuberal region of the hypothalamus, and the lateral septum. Behavior studies have implicated some of these regions in the expression of PRL-induced parental behavior. For example, the ventromedial hypothalamus is the most effective site of PRL action in elevating food intake in nonbreeding doves, which presumably mimics the increased parental provisioning of older nestings that is displayed by breeding pairs.

Although the blood–brain barrier prevents direct access of brain cells to PRL and other proteins, there is evidence that PRL in the blood does enter the brain. A possible route by which PRL gains access to the brain is through the choroid plexus, which makes the cerebrospinal fluid (CSF). Prolactin receptors are present in the choroid plexus in several avian and mammalian species, and evidence in the rat suggests that this structure may be part of a receptor-mediated blood-to-CSF transport mechanism. Once in the CSF, PRL could presumably gain access to target neurons with little difficulty since most PRL receptors in the brain are concentrated in regions that lie in close proximity to the ventricular spaces.

See Also the Following Articles

ALTRICIAL AND PRECOCIAL BIRDS; AVIAN REPRODUCTION, OVERVIEW; BROOD PARASITISM IN BIRDS; NESTING, BIRDS; PARENTAL BEHAVIOR, BIRDS; PROLACTIN, IN NONMAMMALS

Acknowledgment

The author is grateful to Linda Whittingham for her comments and suggestions on a earlier draft. The author's work in this area was supported by NIMH Grant MH41447.

Bibliography

Buntin, J. D. (1996). Neural and hormonal control of parental behavior in birds. *Adv. Study Behav.* **25**, 161–213.

Clutton-Brock, T. H. (1991). *The Evolution of Parental Care.* Princeton Univ. Press, Princeton, NJ.

El Halawani, M. E. (1993). Incubation behavior in the turkey: Molecular and endocrinological implications. In *Avian Endocrinology* (P. J. Sharp, Ed.). Journal of Endocrinology Limited, Bristol, UK.

Gowaty, P. A. (1996). Field studies of parental care in birds. *Adv. Study Behav.* **25**, 477–531.

Horseman, N. D., and Buntin, J. D. (1995). Regulation of pigeon cropmilk secretion and parental behaviors by prolactin. *Annu. Rev. Nutr.* **15**, 213–238.

Kendeigh, S. C. (1952). Parental care and its evolution in birds. *Ill. Biol. Monogr.* **22**, 1–356.

Lea, R. W. (1987). Prolactin and avian incubation: A comparison between Galliformes and Columbiformes. *Sitta* **1**, 117–141.

Rosenblatt, J. S. (1992). Hormone-behavior relations in the regulation of parental behavior. In *Behavioral Endocrinology* (J. B. Becker, S. M. Breedlove, and D. Crews, Eds.), pp. 219–260. MIT Press, Cambridge, MA.

Silver, R. (1984). Prolactin and parenting in the pigeon family. *J. Exp. Zool.* **232**, 617–625.

Silver, R., Andrews, H., and Ball, G. F. (1985). Parental care in an ecological perspective: A quantitative analysis of avian subfamilies. *Am. Zool.* **25**, 823–840.

Skutch, A. F. (1976). *Parent Birds and Their Young.* Univ. of Texas Press, Austin.

Parental Behavior, Mammals

Michael Numan
Boston College

I. Introduction
II. Physiological Control Mechanisms of Parental Behavior in Nesting Mammals
III. Paternal Behavior in Nesting Mammals
IV. Parental Behavior in Sheep
V. Conclusions

GLOSSARY

amygdala A neural region located in the limbic system which relays olfactory information from the olfactory bulbs to the hypothalamus.

fos family of genes Genes that produce the Fos proteins, which serve as transcription factors that activate or inhibit the expression of other genes; the production of Fos proteins within neurons has been used as a measure of neural activation.

lactogenic hormones Peptide hormones such as pituitary prolactin and placental lactogens which not only stimulate lactogenesis but also act on the brain, in conjunction with steroid hormones (estradiol and progesterone), to promote maternal responsiveness.

medial preoptic area A neural region located in the rostral hypothalamus that is critical for the expression of maternal behavior in nesting mammals; it is a site at which estradiol, *lactogenic hormones*, and *oxytocin* act to stimulate maternal responsiveness.

mesotelencephalic dopamine system A dopamine-containing neural system that originates in the ventral tegmental area and retrorubral field of the midbrain and projects to such forebrain regions as the nucleus accumbens and septal area; it is involved in the regulation of motivational processes.

oxytocin A peptide that is produced in the brain which can serve as either a hormone (when released from the posterior pituitary) or a neurotransmitter/neuromodulator (when released at synapses in the brain); oxytocin action in the brain is involved in stimulating the onset of maternal responsiveness at parturition.

sensitization The process by which a nonpregnant/nonlactating female rodent can be induced to show maternal behavior as a result of exposing her to pups over a period of days.

Physiological control mechanisms regulate parental responsiveness in mammals. Most information on these mechanisms comes from research on small mammals, primarily rodents, that give birth to altricial young which are cared for in nests (nesting mammals). An additional source of important information comes from research on sheep. Therefore, the focus of this article will be on rodents and sheep, but it will end with a discussion of the relevance of the reviewed findings for parental behavior in primates.

I. INTRODUCTION

Many rodent species, as typified by rats and mice, are short-gestation species that give birth to multiple altricial young. Such young are helpless at birth, essentially immobile, poikilothermic, and completely dependent on the mother for care, protection, and nurturance. The young are kept in a nest that the mother builds prior to parturition. This nest is in a secluded place and serves to insulate the young in the mother's absence. Materials used to build nests can include paper, straw, grass, and, in the case of rabbits, fur plucked from the mother's ventral surface. In the nest area the mother crouches over the young in order to expose her mammary region, thus allowing the young to suckle. In certain species the young appear to be attracted to a pheromone on the mother's nipple region which allows them to nipple attach. While crouching over the young, the mother also warms them by transferring her body heat to them. If the nest site is disrupted, or if a pup becomes displaced from the nest, the rodent mother will engage in retrieval or transport behavior during which the mother carries the pup in her mouth to a new nest site or back to the original nest.

It is interesting to note that mothers of most rodent species do not recognize their own young. That is, one can cross-foster litters between two mothers, and the females will care for the foster young. Under natural conditions, of course, altricial young cannot wander from one nest to another. Therefore, the female only needs to recognize the location of her nest site, and there should not be any evolutionary pressure for her to learn to recognize her particular litter. Importantly, in some rodent species (e.g., ground squirrels) the mother continues to care for her young after they have become fully mobile and, therefore, at a time when the young can leave and then return to the nest site. In these cases, the mother does learn to recognize her young, but such recognition develops only after the young have become mobile.

An additional aspect of maternal care in rodents is the occurrence of maternal aggression toward intruders at the nest site, which protects the young from possible injury or infanticide. Therefore, when describing the maternal condition in nesting mammals, one can refer to pup-directed activities (retrieving, grooming, and nursing) and nonpup directed activities (nest building and aggression toward intruders), both of which contribute to the survival of the young.

Sheep are at the opposite end of the continuum from nesting mammals with respect to infant development: The young are precocial at birth and are able to follow their mother. Since sheep are herding animals, and since the herd contains unrelated individuals, it is not surprising that postpartum sheep discriminate their own young from alien young. Maternal sheep nurse, lick, and protect their young from danger, while rejecting the suckling attempts of alien young.

II. Physiological Control Mechanisms of Parental Behavior in Nesting Mammals

A. Basic Substrate

A certain basic level of parental responsiveness appears to exist in many rodent species in the absence of the physiological conditions associated with late pregnancy and lactation. Estrous cycling virgin female rats, when first presented with alien young, usually avoid them but may sometimes attack them. However, if the investigator supplies the female with freshly nourished pups (provided by a donor lactating female) on a daily basis, after about a week of constant cohabitation with the pups the virgin will come to show maternal behavior: She will build a nest, retrieve the pups to it, and adopt a nursing posture over the young (although she will be unable to lactate). This process of pup-induced parental responsiveness has been called sensitization, and it has been shown that sensitized maternal behavior in rats can occur in the absence of hormonal mediation, in that it occurs in ovariectomized and hypophysectomized females.

The behavior of the virgin female rodent toward young should be contrasted with that of the parturient female giving birth for the first time. Unlike the virgin, the parturient female will respond with immediate maternal interest to her own or alien young. Therefore, the virgin brain appears to be refractory to pup-elicited maternal responsiveness in comparison to the maternal brain, and this difference suggests that the physiological events of pregnancy promote maternal responsiveness. However, the fact that the virgin female can show maternal behavior indicates that maternal motivation is not rigidly and strictly tied to the physiological states associated with late pregnancy and lactation.

Possibly related to this basic level of parental responsiveness in nonpregnant/nonlactating females is the occurrence of alloparental behavior shown by certain juveniles. That is, in certain rodent species juveniles may remain at the nest site after they are weaned and after the birth of subsequent litters. The older juveniles may help care for the younger litters (huddling over the young pups and retrieving them), and it has been suggested that such behavior may contribute to the effective development of maternal responsiveness so that when the juvenile matures and has its own offspring it might be a better parent. Therefore, the neural substrate underlying maternal behavior is not only influenced by the physiological state associated with pregnancy and lactation but also affected by experience.

B. Hormonal Mechanisms

The classic studies of Terkel and Rosenblatt showed that humoral factors stimulate maternal responsiveness in rats. They cross-transfused blood between a parturient female and an estrous cycling virgin female and then presented the virgin with pups. The virgin responded with a short-latency onset of maternal behavior toward the foster pups, retrieving them within the first few hours after presentation, which contrasts with the long latency of the normal virgin response. Although parturient blood was effective, blood transfused from one virgin to another was ineffective as a maternal stimulant. Subsequent research has attempted to identify the factors present in the blood of the parturient female which stimulate maternal behavior. The onset of maternal behavior at parturition is coincident with a sharp decline in serum progesterone levels superimposed on high levels of estradiol, pituitary prolactin, and placental lactogens (pituitary prolactin and placental lactogens stimulate lactogenesis). Research has shown that it is this hormonal profile which stimulates maternal behavior in rats. High levels of estradiol and lactogenic hormones superimposed on progesterone withdrawal may also be involved in stimulating maternal responsiveness in mice and rabbits.

The effects of progesterone on maternal behavior in rats are complex. It appears that the fall from high levels (progesterone withdrawal) potentiates the ability of estradiol and lactogenic hormones to stimulate maternal behavior, and that if the high levels of progesterone are experimentally prevented from falling, then maternal behavior is inhibited even in the presence of effective levels of estradiol and lactogens.

Although hormones activate immediate maternal responsiveness at parturition, once the behavior has become established during the early postpartum period hormones no longer appear to be essential, at least for the rat. This has given rise to the view that there is a sensitive period at parturition during which hormones allow females to respond to young, and that once maternal responsiveness has been initiated there is a shift in control mechanisms so that infant stimuli alone can regulate continued maternal behavior.

C. Sensory Factors

A major change in olfactory processing appears to underlie the onset of maternal responsiveness at parturition in rodent species. For example, in mice olfactory bulbectomy, or the production of anosmia (inability to smell) by peripheral destruction of the olfactory receptors, disrupts maternal behavior in parturient females. In rats and hamsters, however, anosmia does not disrupt maternal behavior but instead appears to facilitate maternal behavior in estrous cycling virgin females: Anosmic females will show maternal behavior after a brief exposure to young even in the absence of hormonal treatment. Since parturient rodent females, or females primed with maternal hormones, find litter-related odors attractive, whereas virgin females are not attracted to such odors, the view has been offered that in the absence of hormone priming, females find the odor of pups aversive, whereas after hormone priming pup odors become attractive and this facilitates maternal responding. For mice, the attractiveness of pup odors may be primary for the occurrence of maternal responding, whereas for rats and hamsters, the important factor may be a reduction in the aversive qualities of pup odors.

Once a female approaches and interacts with her young, tactile somatic-sensory input originating from the perioral region (as the female sniffs, licks, and nuzzles the pups) and from the female's ventral surface (during suckling) provides a positive proximal stimulus for the occurrence of maternal behavior.

D. Neural Control Mechanisms

What are the neural processes that are influenced by maternal hormones and infant-related stimuli? As suggested previously, several investigators have recently argued that maternal behavior is facilitated when the tendency to approach infant stimuli and engage in maternal behavior is greater than the tendency to avoid or withdraw from such stimuli. As shown in Fig. 1, this perspective suggests that the hormonal events of late pregnancy act on the brain to decrease fear/aversion of infant stimuli, increase attraction/approach toward such stimuli, or both. Research on the neural substrate of maternal behavior will be presented in the context of such approach/avoidance models.

Figure 2 shows the location of the medial preoptic area (MPOA) and bed nucleus of the stria terminalis (BST). These two regions are part of an excitatory system for maternal behavior, critical for attraction to infant-related stimuli and for the performance of maternal behavior: Destruction of the MPOA/BST disrupts all components of maternal behavior in rats. The MPOA/BST region contains both intracellular

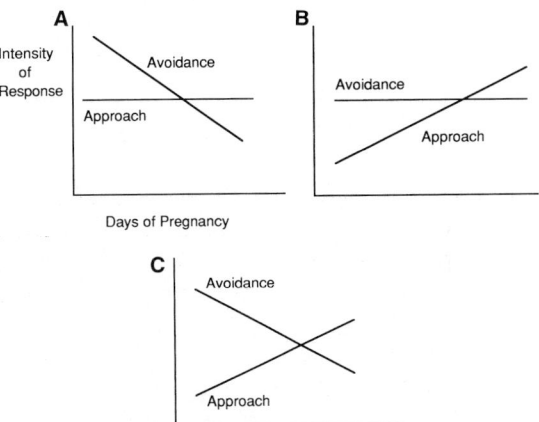

FIGURE 1 Three alternative approach–avoidance models of the onset of maternal behavior at parturition. Maternal behavior occurs when central neural approach system are more active than central neural avoidance systems with respect to infant-related stimuli. In A, the physiological events of pregnancy primarily decrease avoidance, in B these physiological factors promote approach responses, and in C avoidance systems are depressed and approach systems are activated toward the end of pregnancy (reproduced with permission from M. Numan and T. P. Sheehan, Neuroanatomical circuitry for mammalian maternal behavior, Ann. N. Y. Acad. Sci. 807, 101–125, 1997. Copyright 1997 by New York Academy of Sciences).

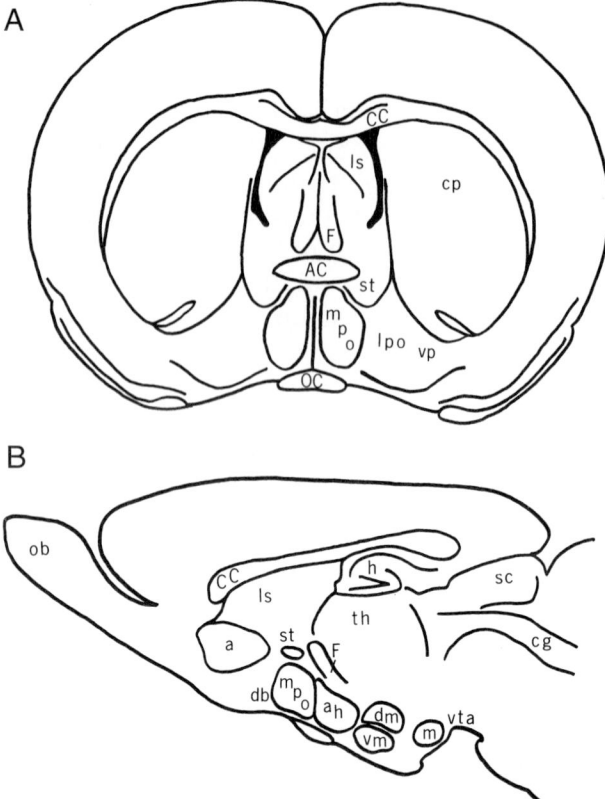

FIGURE 2 Frontal (A) and sagittal (B) sections of the rat brain at the level of the medial preoptic area. a, nucleus accumbens; AC, anterior commissure; ah, anterior hypothalamus; CC, corpus callosum; cg, central gray; cp, caudate-putamen; db, nucleus of the diagonal band; dm, dorsomedial nucleus of the hypothalamus; F, fornix; h, hippocampus; ls, lateral septum; lpo, lateral preoptic area; mpo, medial preoptic area; ob, olfactory bulb; OC, optic chiasm; sc, superior colliculus; st, bed nucleus of stria terminalis; th, thalamus; vm, ventromedial hypothalamic nucleus; vp, ventral pallidum; vta, ventral tegmental area (reproduced with permission from Numan, 1994. Copyright 1994 by Raven Press).

estrogen and progesterone receptors (steroids produce intracellular changes by affecting gene transcription) and cell membrane-bound lactogenic hormone receptors. Therefore, it might be one of the neural sites at which hormones act to promote maternal behavior. Research supports this view: Injections of small amounts of estradiol or lactogenic hormones into the MPOA have been found to stimulate maternal behavior in female rats that have been suboptimally hormone primed. These findings concur with other work which shows that the number of intracellular estrogen receptors increases in the MPOA/BST as parturition approaches. Importantly, a recent study with a genetically engineered transgenic mouse line that was deficient in the estrogen receptor gene showed that such mice had deficits in maternal behavior. Such a finding emphasizes the importance of estrogen for maternal behavior, and, along with the brain work just reviewed, suggests the importance of genomic effects of estrogen acting on its receptor at the level of the MPOA/BST. With respect to lactogenic hormones, studies have shown that the expression of prolactin receptors within the brain can be influenced by both steroid hormones and contact with infant-related stimuli.

In addition to pituitary prolactin and placental lactogens, there is also evidence for a brain prolactin system: Prolactin appears to be produced by neurons to serve as either a neurotransmitter or a neuromodulator. The degree to which the brain prolactin system, as opposed to the endocrine prolactin/lactogen system, influences parental responsiveness remains to be determined.

Given the importance of the MPOA/BST region as part of an excitatory neural system for maternal behavior, if it was known where the critical neurons projected we would be in a better position to understand the particular functions that the MPOA/BST serves with respect to maternal responsiveness. Recent research on the *fos* family of genes and their respective proteins has provided insights into this issue. The *fos* family consists of four genes, *c-fos*, *fosB*, *fra-1*, and *fra-2*, and these genes are referred to as immediate early genes. With respect to the function of Fos proteins (the products of activation of the *fos* gene family) within neurons, the following should be noted. When a neuron is stimulated, in many cases its biosynthetic machinery is activated, which includes the activation of early and late responding genes. That is, neural activation causes intracellular changes which first act on the genome to stimulate immediate early gene expression. The protein products of immediate early gene activation then serve as transcription factors which, in turn, activate late-responding structural genes, the protein products of which alter neuronal function, for example, by causing an increase in neurotransmitter receptors, peptide neurotransmitters, or enzymes involved in neurotransmitter synthesis. An increase in the production of Fos proteins within neurons has been taken as evidence that those neurons have been activated. Within this context, recent studies have shown

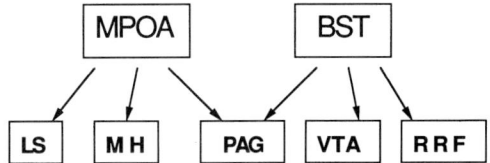

FIGURE 4 Diagrammatic representation of the neural regions that receive input from medial preoptic area (MPOA) and bed nucleus of stria terminalis (BST) neurons that express Fos during maternal behavior. LS, lateral septum; MH, medial hypothalamus; PAG, periaqueductal gray (also referred to as central gray); RRF, retrorubral field; VTA, ventral tegmental area.

that the production of c-Fos protein increases in the MPOA/BST of maternal females (Fig. 3), and that this increase is closely tied to the performance of maternal behavior.

Since the production of Fos proteins within neurons during maternal behavior serves as an anatomical marker for neurons which are active during maternal behavior, recent work has set out to determine where these neurons project. As shown in Fig. 4, results have indicated that MPOA neurons which are activated during maternal behavior project most strongly to the lateral septum (LS) and to the medial hypothalamus (MH) [in the vicinity of the ventromedial nucleus (VMN)] caudal to the MPOA] and that BST neurons that are similarly activated project most strongly to two midbrain sites: the ventral tegmental area and the retrorubral field (VTA/RRF). In addition to these projections, MPOA/BST Fos-containing neurons also have strong projections to the periaqueductal gray (PAG) in the midbrain.

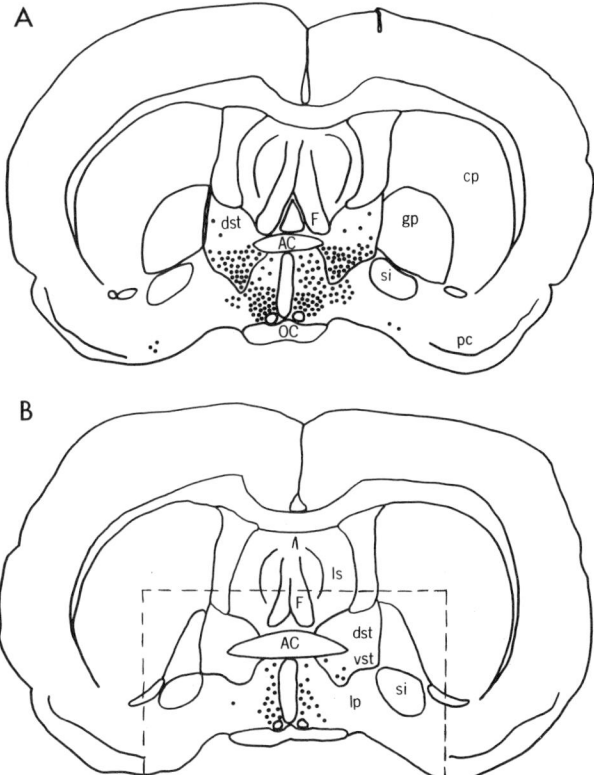

FIGURE 3 Distribution of cells labeled with Fos-like immunoreactivity on a single frontal section through the preoptic area of a postpartum female rat that was exposed to pups and showed maternal behavior (A) and from a postpartum female that was exposed to candy and therefore did not show maternal behavior (B). Each dot represents five labeled cells. The area analyzed in both sections is represented in the lower section by dashed lines. cp, caudate-putamen; dst, dorsal bed nucleus of stria terminalis; F, fornix; gp, globus pallidus; lp, lateral preoptic area; ls, lateral septum; OC, optic chiasm; pc, piriform cortex; si, substantia innominata; vst, ventral bed nucleus of stria terminalis (reproduced with permission from M. Numan and T. P. Sheehan, Neuroanatomical circuitry for mammalian maternal behavior, Ann. N. Y. Acad. Sci. 807, 101–125, 1997. Copyright 1997 by New York Academy of Sciences).

This functional–neuroanatomical approach to the brain circuits underlying maternal responsiveness in rodents has led to some interesting hypotheses which will be tested by future work. First, BST projections to the VTA/RRF may be important for the motivational aspects of maternal behavior, increasing the attractive or incentive value of pup-related stimuli, increasing the persistence and vigor of maternal responses, and increasing the rewarding aspects of maternal behavior. This view is probable because the VTA/RRF gives rise to the mesotelencephalic dopamine (DA) systems, which are known to be involved in the motivational control of a variety of behaviors.

In addition to influencing the motivation to engage in maternal behavior, MPOA/BST projections may also influence the ability to perform particular maternal responses. Projections to the PAG may serve this purpose since the PAG projects to the trigeminal sensory complex, and therefore MPOA/BST projections to PAG may regulate oral and perioral sensorimotor integration related to retrieval behavior.

Although MPOA/BST projections to VTA/RRF and PAG may be part of a positive control system, MPOA projections to LS and MH may be part of a negative control system. That is, there is evidence that LS and MH are part of the brain's neural machinery for fear, defensive aggression, and avoidance behavior. Perhaps MPOA projections to these regions alter their function so as to depress fear and avoidance responses toward pup-related stimuli.

Once the neurochemical makeup of each projection circuit has been determined, we will be better able to test the previous hypotheses. Then, by applying the relevant neurochemical agonists and antagonists to the appropriate terminal region, one will be able to examine the nature of each projection with respect to maternal behavior. Some answers to this neurochemical story may be forthcoming once the particular late-responding genes, and their protein products, which are activated by Fos proteins within MPOA/BST neurons during maternal behavior, are determined. With respect to these issues, it is worth pointing out that a recent study has reported that the *fos* family of immediate early genes is critical for the normal expression of parental behavior in rodents. A mouse line which contained a knockout mutation of the *fosB* gene, although normal in many respects, showed a lack of maternal responsiveness. The authors suggested that the deletion of the *fosB* gene from MPOA neurons may have participated in the observed disruption of maternal behavior. Future research with antisense oligonucleotides to *fos* gene products injected *in vivo* directly into the MPOA will help test this assertion. Clearly, determining the function of *fos* genes within MPOA/BST neurons will answer some important neurochemical questions.

Studies have shown that the amygdala plays a role in parental responsiveness in rodents. In rats, destruction of the medial amygdala (MA) facilitates maternal behavior in virgin females, suggesting an inhibitory influence. Figure 5 shows some of the neural circuitry involving MA. This region receives and integrates olfactory input from both main olfactory bulb and the vomeronasal system and, in turn, projects strongly to the MPOA/BST and MH/VMN. Perhaps, in the estrous cycling virgin female the MA-to-MH projection is dominant, and since MH is part of the brain's aversion system, this might be the route over which olfactory input from pups causes avoidance and withdrawal responses. As a result of

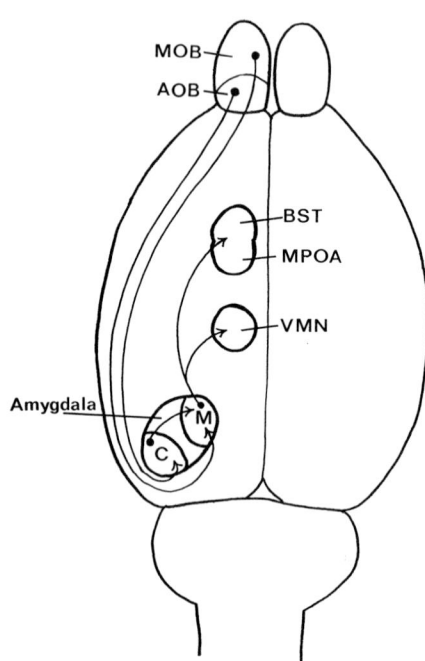

FIGURE 5 Diagrammatic representation of olfactory bulb connections to the corticomedial amygdala and amygdaloid projections to the hypothalamus and bed nucleus of stria terminalis (BST). Projections are shown on only one side of the brain. Not all known projections are shown. The main olfactory bulb (MOB) has projections to the anterior cortical amygdaloid nucleus (C) and the accessory olfactory bulb (AOB) projects to the medial amygdaloid nucleus (M). C projects to M. The medial amygdala projects to the BST, medial preoptic area (MPOA), and ventromedial nucleus of the hypothalamus (VMN) (reproduced with permission from M. Numan and T. P. Sheehan, Neuroanatomical circuitry for mammalian maternal behavior, *Ann. N. Y. Acad. Sci.* **807**, 101–125, 1997. Copyright 1997 by New York Academy of Sciences).

central neural biochemical changes (e.g., alterations in neurotransmitters and their receptors) caused by the hormonal events of late pregnancy, MA projections to MPOA/BST may become dominant and this may change the valence of pup-related odors from negative to positive.

Another region which has recently been shown to play an important role in maternal responsiveness is the lateral habenula. How the lateral habenula (LHb) fits into the neural circuitry underlying maternal behavior remains to be determined. It is known that Fos-expressing neurons in the MPOA and BST do not project to the LHb. Perhaps the LHb exerts its influence through its interactions with the VTA/RRF.

E. Neurochemical Factors

1. Monoamines

Very little work has been done on the role of serotonin and norepinephrine in maternal behavior. Most of the research has concentrated on DA. If BST projections to VTA/RRF DA systems influence the motivational aspects of maternal behavior, one might expect that disruption of the mesotelencephalic DA system (VTA/RRF dopaminergic projections to the nucleus accumbens) would interfere with maternal behavior, and this is indeed the case. In addition, dopamine acts postsynaptically on D1 and D2 dopamine receptors, and drugs which preferentially block D2 receptors disrupt maternal behavior. Finally, intracerebral injections of antisense oligonucleotides aimed specifically at D2 receptor mRNA disrupt maternal behavior in rats. These findings fit well with the observation that quantitative changes occur in the distribution of D2 receptors in the rat telencephalon as parturition approaches.

2. Oxytocin

Oxytocin has long been known as a neurohormone that is produced by hypothalamic neurosecretory cells and released into the blood from the posterior pituitary to influence both the uterine contractions associated with parturition and the milk ejection reflex. However, oxytocin also serves as a neurotransmitter in the brain, where it acts to promote maternal responsiveness. Central neural oxytocin is primarily produced by neurons in the paraventricular nucleus of the hypothalamus, and these neurons have widespread intracerebral projections. Importantly, oxytocin receptors exist in the MPOA/BST and in the VTA, and the number of these receptors increases as parturition approaches. In suboptimally hormone-primed female rodents, intracerebral injections of oxytocin facilitate maternal behavior. Conversely, the microinjection of oxytocin antagonists into either the VTA or the MPOA/BST blocks the normal onset of maternal behavior at parturition. One view, therefore, is that the hormonal events of late pregnancy act to upregulate oxytocin receptors, making the MPOA/BST and the VTA more sensitive to the positive influences of oxytocin, thus facilitating maternal responsiveness. Additionally, the same hormonal events appear to activate the synthesis and release of neural oxytocin near term. A recent finding has caused some confusion with respect to oxytocinergic neural systems and maternal behavior. A transgenic mouse line with a knockout mutation of the oxytocin gene showed deficits in lactation due to disruption of the milk ejection reflex, but maternal behavior appeared normal. This research suggested that central oxytocin systems are not important for mouse maternal behavior. These findings can possibly best be explained as follows: Unlike other rodents, many laboratory mouse strains show extremely short sensitization latencies. That is, the hormonal events of late pregnancy play a very minor role in facilitating maternal behavior in such strains and, therefore, hormonally primed oxytocinergic neural systems may not be critical for the display of maternal behavior in these strains.

3. Opiates

Opioid neural systems appear to exert dual effects on maternal responsiveness. A β-endorphin system arising in the arcuate nucleus and projecting to the MPOA appears to inhibit maternal behavior. This concurs with work which shows that as parturition approaches β-endorphin levels in the MPOA decrease. In contrast, opiates acting at the level of the VTA/RRF (presumably an enkephalinergic system of unknown origin) stimulate maternal responsiveness, possibly by exciting the mesotelencephalic DA systems.

4. Tachykinins

Tachykinin peptides (substance P and related peptides) serve as neurotransmitters/neuromodulators in amygdaloid projections to the hypothalamus. Since other research has shown that tachykinins are involved in pain transmission and fearful behaviors, these peptides might also be involved in relaying odor-related aversive inputs from pups between the amygdala and MH (Fig. 5). In support of this model, it has recently been shown that microinjection of the tachykinin neuropeptide K into MH disrupts maternal behavior. It has been hypothesized that the hormonal events of late pregnancy act to downregulate this inhibitory tachykininergic system.

5. Neurotensin

Recent work has shown an increase in the number of MPOA/BST neurons that express neurotensin immunoreactivity in postpartum females. Since other work has shown that Fos proteins influence neurotensin synthesis, these preliminary findings suggest that neurotensin may be utilized by MPOA/BST neurons involved in maternal behavior control.

III. PATERNAL BEHAVIOR IN NESTING MAMMALS

Since it is the female that lactates, she is the primary parental figure in mammals. However, paternal behavior does sometimes occur, particularly in monogamous species. It should be clear that since males are not exposed to the hormonal fluctuations associated with pregnancy and lactation, other mechanisms must exist which promote their parental responsiveness. Although not much work has been done on this problem, recent findings in rodents have suggested that the MPOA/BST may be involved in paternal behavior and that a vasopressinergic system, possibly arising from the MPOA/BST and terminating in the lateral septum, may be particularly important. This finding is interesting because vasopressin is a neuropeptide closely related to oxytocin. Indeed, it has been suggested that vasopressin neural systems have evolved to play important roles in male prosocial and paternal behaviors, whereas oxytocinergic systems play similar roles in females.

IV. PARENTAL BEHAVIOR IN SHEEP

Estrous cycling nonpregnant ewes will not respond maternally toward young, whereas a ewe that has just given birth will show maternal behavior toward its own or alien young, showing no discriminatory behavior: She will allow any lamb to approach and suckle without rejection, and she will lick and clean the young. Discriminatory behavior does develop, however, over the first 24 hr postpartum. As a result of exposure to a particular lamb, the parturient female learns the olfactory characteristics of that lamb, and on subsequent tests she will only respond with maternal acceptance to that young and reject the advances of lambs to which she has not been exposed. Importantly, if a recently parturient ewe is not exposed to any lamb during the first 24 hr postpartum (her own young are immediately removed by the experimenter at birth), then maternal motivation quickly wanes, and on subsequent tests she will reject all young.

With this background, research on the physiological basis of maternal behavior in sheep has attempted to answer two questions: What are the factors which enhance maternal motivation at parturition and what are the physiological events which underlie the ability of the parturient female to learn the olfactory characteristics of the lamb(s) to which she has been exposed?

Many of the players that regulate maternal motivation in sheep are the same as those which operate in rodents: steroid hormones and neural oxytocin. Currently, however, there is no evidence linking prolactin or other lactogens to maternal behavior in sheep. In addition, the vaginal and cervical stimulation (VCS) which accompanies parturition is also critically involved in potentiating maternal behavior in sheep. Research suggests that as parturition approaches, declining progesterone and rising estradiol levels prepare the brain for oxytocin action: Oxytocin synthesis within central neurons increases, as does the synthesis of oxytocin receptors in diverse brain regions. The VCS that occurs at parturition then causes a dramatic release of oxytocin from the axon terminals of paraventricular hypothalamic nucleus neurons which project to diverse brain regions. It has been suggested that oxytocin action at multiple central neural sites is necessary for the complete display of maternal responsiveness in sheep: decreased avoidance and rejection of young and increased approach and acceptance of young. There is evidence that oxytocin action on MPOA/BST neurons decreases avoidance and rejection behavior, but that oxytocin action at other, as yet unspecified sites, is necessary for the positive aspects of maternal behavior in sheep.

With respect to the development of discriminatory ability, the VCS that occurs at parturition appears to cause the release of both norepinephrine and oxytocin in the olfactory bulb, and these events, in conjunction with exposure to lamb odors, cause modifications in olfactory bulb synaptic circuitry which allow the female to respond selectively to olfactory signals emanating from the young to which she had been exposed.

V. CONCLUSIONS

A basic neural substrate regulates maternal behavior in nesting mammals, and the MPOA/BST and its projections seem to play a key integrative role. Critical sensory factors are olfactory and tactile inputs. As parturition approaches, hormonal factors and neuropeptides (e.g., oxytocin) act on the basic neural substrate to ensure that the female is more likely to be attracted by pup-related stimuli and to engage in maternal behavior and is less likely to withdraw from such stimuli. Modifications in limbic mechanisms (amygdala and septum), and in other parts of the hypothalamus (MH), all come into play to foster a high level of maternal responsiveness. Similar mechanisms regulate maternal responsiveness in sheep.

An evolutionary perspective on the neural basis of parental behavior has been presented by MacLean. In comparing mammals with reptiles, he noted the high level of maternal responsiveness in mammals and the almost complete absence of such behavior in reptiles. Since the limbic system is more developed in mammals, MacLean suggested that limbic structures, particularly the amygdala, septum, cingulate cortex, and hippocampus, are critically important for maternal responsiveness. This review of the neural control of maternal behavior in mammals clearly suggests the involvement of such limbic regions as the amygdala, BST, and septum. However, the MPOA and MH are important hypothalamic regulatory structures. It must be stressed that although reptiles hardly show any parental behavior (there are some exceptions), fish and amphibia do show strong parental responses, but they have undeveloped limbic structures, although homologs of the amygdala and septum appear to exist. Therefore, hypothalamic–brain stem connections may form a core substrate that is capable of influencing parental responsiveness in a variety of vertebrate classes, with limbic structures exerting important modulatory influences, particularly in mammals.

Although the few studies that exist suggest that hormones do exert an influence on parental responsiveness in primates, a dominant position is that parental behavior in primates has been emancipated from hormonal control. Clearly, humans are capable of adopting and caring for young without being exposed to the physiological events of pregnancy and lactation. Within this context, what is the value, with respect to understanding human maternal behavior and its abnormalities, of exploring the hormonal basis of maternal behavior in nonprimate mammals? First, even in the rat it should be recalled that a basic level of maternal responsiveness exists which is free of endocrine control. Second, studies on nesting mammals and sheep have utilized hormones as probes to explore the central neural, neurochemical, and molecular mechanisms underlying maternal behavior. It is highly likely that a common basic neural substrate underlies parental responsiveness in most mammals, and that what varies between species is the degree to which this core substrate is activated or affected by hormonal, sensory, and/or experiential factors. For example, an influential position is that for rodents and sheep hormones are of primary importance for maternal responsiveness, whereas in primates, early life experiences, such as preadult experiences with infants ("play mothering"), are more important for the development of maternal responsiveness. However, independent of whether maternal behavior is influenced more by experiential or hormonal factors, the underlying neural processes that are affected may be the same. That is, experiential and/or hormonal factors may promote maternal behavior by both increasing the attractive value of infant stimuli and decreasing the aversive qualities of such stimuli.

See Also the Following Articles

NESTING, BIRDS; OXYTOCIN; RODENTIA; SHEEP AND GOATS

Bibliography

Ben-Jonathan, N., Mershon, J. L., Allen, D. L., and Steinmetz, R. W. (1996). Extrapituitary prolactin: Distribution, regulation, functions, and clinical aspects. *Endocr. Rev.* 17, 639–669.

Brown, J. R., Ye, H., Bronson, R. T., Dikkes, P., and Greenberg, M. E. (1996). A defect in nurturing in mice lacking the immediate early gene fosB. *Cell* 86, 297–309.

Carter, C. S., Lederhendler, I. I., and Kirkpatrick, B. (Eds.) (1997). The integrative neurobiology of affiliation. *Ann. N. Y. Acad. Sci.* 807.

Insel, T. R. (1997). A neurobiological basis of social attachment. *Am. J. Psych.* 154, 726–735.

Keverne, E. B. (1996). Psychopharmacolgy of maternal behaviour. *J. Psychpharmacol.* 10, 16–22.

Keverne, E. B., Martel, F. L., and Nevison, C. M. (1996). Primate brain evolution: Genetic and functional considerations. *Proc. R. Soc. London B* 262, 689–696.

Krasnegor, N. A., and Bridges, R. S. (Eds.) (1990). *Mammalian Parenting.* Oxford Univ. Press, New York.

MacLean, P. D. (1990). *The Triune Brain in Evolution.* Plenum, New York.

Morrell, J. I., Wagner, C. K., Malik, K. F., and Lisciotto, C. A. (1995). Estrogen receptor mRNA: Neuroanatomical distribution and regulation in three behaviorally relevant physiological models. In *Neurobiological Effects of Sex Steroid Hormones* (P. E. Micevych and R. P. Hammer, Eds.), pp. 57–84. Cambridge Univ. Press, New York.

Numan, M. (1994). Maternal behavior. In *The Physiology of Reproduction, Volume 2* (E. Knobil and J. D. Neill, Eds.), 2nd ed., pp. 221–302. Raven Press, New York.

Ogawa, S., Taylor, J. A., Lubahn, D. B., Korach, K. S., and Pfaff, D. W. (1996). Reversal of sex roles in genetic female mice by disruption of estrogen receptor gene. *Neuroendocrinology* 64, 467–470.

Pedersen, C. A., Caldwell, J. D., Jirikowski, G. F., and Insel, T. R. (Eds.) (1992). Oxytocin in maternal, sexual, and social behaviors. *Ann. N. Y. Acad. Sci.* 652.

Pryce, C. R. (1992). A comparative systems model of the regulation of maternal motivation in mammals. *Anim. Behav.* 43, 417–441.

Rosenblatt, J. S., and Snowdon, C. T. (Eds.) (1996). Parental care: Evolution, mechanisms, and adaptive significance. In *Advances in the Study of Behavior, Volume 25.* Academic Press, San Diego.

Terkel, J., and Rosenblatt, J. S. (1972). Humoral factors underlying maternal behavior at parturition: Cross transfusion between freely moving rats. *J. Comp. Physiol. Psychol.* 80, 365–371.

Thompson, A. C., and Kristal, M. B. (1996). Opioid stimulation in the ventral tegmental area facilitates the onset of maternal behavior in rats. *Brain Res.* 743, 184–201.

Young, W. C., Shepard, E., Amico, J., Hennighausen, L., Wagner, K., LaMarca, M. E., McKinney, C., and Ginns, E. I. (1996). Deficiency in mouse oxytocin prevents milk ejection, but not fertility or parturition. *J. Neuroendocrinol.* 8, 847–853.

Parthenogenesis and Natural Clones

Robert C. Vrijenhoek

Rutgers University

I. Parthenogenesis and Asexual Reproduction
II. Origins of Parthenogens
III. Ecological Considerations
IV. Evolutionary Considerations
V. Parthenogens as Study Organisms

GLOSSARY

amphimixis Sexual reproduction; the mixing of the genes from two distinct individuals; involving the recombinational effects of meiotic reduction and fusion of gametes.

apomixis Asexual reproduction without chromosome reduction or fusion of gametes; ameiotic parthenogenesis; retains parental heterozygosity.

automixis Asexual reproduction with chromosomal reduction but without fusion of gametes; meiotic parthenogenesis; rapidly leads to complete homozygosity.

endoduplication Duplication of the entire chromosomal set without cell division prior to meiosis.

gynogenesis Sperm-dependent parthenogenesis; sperm are used to activate embryogenesis but fusion of egg and sperm nuclei does not occur; pseudogamy.

hemiclone A haploid clonal genome that is transmitted without recombination by hybridogenetic females.

hybridogenesis The perpetuation of a hybrid genotype (AB) by hemiclonal inheritance in which the maternal genome (A) is transmitted to eggs; the paternal genome (B) is discarded during oogenesis and restored by true fertilization with sperm from males of a sexual host species B.

pseudogamy Sperm-dependent parthenogenesis in plants; pollen is required to activate seed development, but the seed nucleus is produced clonally.

tychoparthenogenesis Occasional or accidental parthenogenetic development in unfertilized eggs.

Parthenogenesis (virgin birth) is reproduction via eggs but without sex. Eggs develop into new individuals without fertilization by sperm. Parthenogenetic lineages occur in many plant and animal taxa, and they may flourish under a variety of ecological conditions. Nevertheless, individual clones are believed to be evolutionary dead ends, because they lack the ability to respond genetically to changes in their physical and biotic environments.

I. PARTHENOGENESIS AND ASEXUAL REPRODUCTION

Reproduction does not require sex, or amphimixis, a complex process that involves two basic elements: (i) meiotic reduction—chromosomal segregation, assortment, and crossing over that generate an immense variety of haploid gametes; and (ii) syngamy—fusion of gametes that produces unique new individuals in each generation. Mixing the genotypes from different individuals (recombination) is the essential characteristic of sex in eukaryotic organisms, and circumvention of these processes leads to parthenogenesis and cloning.

Vegetative reproduction (budding, fragmentation, fission, etc.) is common in plants and some invertebrate animals. Although comparable to parthenogenesis in producing clones, vegetative modes of reproduction should be distinguished because they do not involve egg production and meiotic processing of chromosomes. Chromosome processing may be necessary to reset imprinted DNA methylation patterns and restore developmental totipotency in some organisms. Additionally, fertilized seeds and eggs (and subsequent larvae) are often the essential dispersal phase of many plants and animals. In most cases, vegetative propagules tend to remain close to the parent organism. Corals ordinarily reproduce by budding, but they employ sexual reproduction to produce planula larvae, the dispersal phase of the life cycle. In an ecological sense, vegetative reproduction is more appropriately compared with growth than reproduction.

Cyclical parthenogenesis alternates between sexual and asexual egg production. Because cyclical parthenogens engage in periodic recombination, they are facultatively sexual. The cladoceran waterflea, *Daphnia pulex*, produces a new assemblage of clones after each cycle of sexual reproduction. Sexual reproduction generally is stimulated by high density or other forms of stress and is used to produce the overwintering eggs. However, some populations occurring at high latitudes and in more permanent bodies of water have given rise to obligately parthenogenetic lineages that no longer reproduce sexually.

True parthenogenesis is a strictly clonal form of reproduction that transmits the female's diploid (or polyploid) genome to eggs, which develop spontaneously into genetically identical daughters. The terminology favored by botanists is more precise in its distinction among cytological mechanisms involved in the production of eggs. The term apomixis (ameiotic parthenogenesis) is used to describe zygote production without chromosomal reduction (some researchers include vegetative reproduction under apomixis). Some apomicts eliminate the reductional division (meiosis I) and produce nonrecombinant eggs with a single equational division (meiosis II). Other ameiotic methods of egg production are known and the primary genetic consequences are strict clonal inheritance and retention of the maternal level of heterozygosity.

In contrast, automixis (meiotic parthenogenesis) restores diploidy by fusing various meiotic products. For example, some free-living *Rhabditis* nematodes fuse the second polar body with the egg nucleus. In most cases, automixis is comparable to self-fertilization and quickly leads to complete homozygosity. Some automicts produce normal haploid ova and then duplicate the generative nucleus in a subsequent mitotic division. Fusion of these mitotic products restores diploidy but leads to complete homozygosity in one step. Once automicts are completely homozygous, inheritance is effectively clonal.

Most parthenogenetic animals are functionally apomictic. They retain elements of meiosis while circumventing chromosomal recombination and reduction. For example, parthenogenetic whiptail lizards of the genus *Cnemidophorus* duplicate the entire chromosomal complement prior to meiosis, a process known as endoduplication. Because synapsis occurs between the duplicated pairs of chromosomes, meiotic recombination is genetically inconsequential. Eggs contain a functionally nonrecombinant version of the maternal genotype. A great variety of functionally apomictic mechanisms are known. Their common theme is the circumvention of reduction and recombination. Many parthenogenetic animals arose as hybrids, and functional apomixis effectively preserves their hybrid genotypes. Why functionally apomictic animals are more common than true apomicts is not understood. Perhaps, chromosomal processing during prophase of meiosis I is necessary for normal embryonic development.

Sperm-dependent modes of parthenogenetic reproduction also are known. Dandelions in North America (they were introduced from Europe) are pseudogamous apomicts: Pollination is necessary to activate development of endosperm tissue in the seed, but the generative nucleus develops apomictically. Pseudogamy is more commonly called gynogenesis in animals (Fig. 1). Despite the need for sperm, pseudogamous inheritance is strictly maternal and clonal. The fall cankerworm moth, *Alsophila pometaria*, has pseudogamous lineages that use sperm from males of a coexisting sexual lineage, but gynogenetic fish such as the Amazon molly, *Poecilia formosa*, use sperm from males of closely related sexual species. The need for sperm produces a kind of host–parasite relationship between sexually reproducing sperm donors and all-female gynogens. However, pseudogamous planarians are hermaphrodites, and they can use their own sperm. Although pseudogamous forms are not parthenogenetic in the strict sense (i.e., virgin birth), genetic consequences are the same: Syngamy does not occur and inheritance is clonal. Nevertheless, sperm-dependent versus sperm-independent forms of parthenogenesis function under very different ecological constraints.

Hybridogenesis, an unusual form of matrilineal inheritance that perpetuates a hybrid genotype, combines elements of parthenogenesis and sexual reproduction. The hybridogenetic fish *Poeciliopsis*

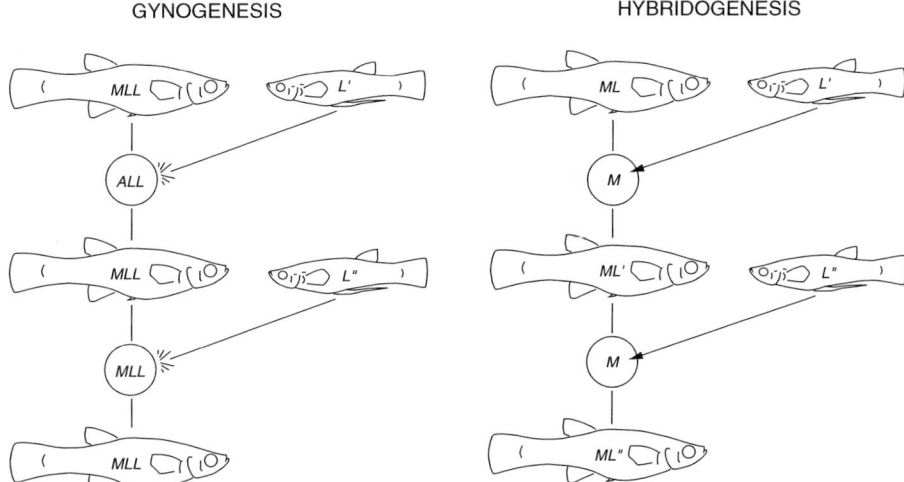

FIGURE 1 Gynogenetic and hybridogenetic reproduction in all-female fish (genus *Poeciliopsis*) of hybrid origin. The letters M and L represent whole chromosome sets from the sexually reproducing progenitors, *P. monacha* and *P. lucida*. The triploid gynogen, *P. monacha-2 lucida* (or MLL), has one set of monacha chromosomes and two sets of lucida chromosomes; and the diploid hybridogen, *P. monacha-lucida* (or ML) has one set of chromosomes from each species. Both the gynogen and hybridogen are pictured mating with males of *P. lucida*. During gynogenesis, the entire triploid genome, MLL, is transmitted between generations without recombination. Different markers associated with the sperm source (L, L', L", etc.) are not incorporated or expressed in the offspring. During hybridogenesis, only the haploid M genome (hemiclone) is transmitted to eggs. The paternal L genome is replaced in each generation.

monacha-lucida is a hybrid between the sexual species *P. monacha* and *P. lucida*. It is easier to describe hybridogenesis if we substitute the letters M and L for monacha and lucida chromosome sets of the hybrid (Fig. 1). Just before meiosis, these ML hybrids discard the L chromosomes. Functional eggs contain only a nonrecombinant M set that must fuse with sperm provided by *P. lucida* males, producing a new hybrid, ML'. New paternal genomes (L, L', L", etc.) are (i) drawn anew from the sexual gene pool in each generation, (ii) paired with the M genome, (iii) fully expressed in ML hybrids, and then (iv) discarded. The M genome is called a hemiclone because it comprises only half of the organism's chromosomal complement, and it is cloned. Populations of *P. monacha-lucida* usually contain several hemiclones, marked by distinct M genomes that were independently derived from *P. monacha*. The European water frog, *Rana esculenta*, also is hybridogenetic. Hybridogenesis is also found in some insects, but overall it is a rare form of clonal reproduction.

Numerous variations exist on these basic themes of clonal reproduction and parthenogenesis in plants and animals. The reference by Suomalainen and co-workers (1987) provide a useful summary of what is known about cytogenetic mechanisms.

II. ORIGINS OF PARTHENOGENS

Most plant and animal parthenogens (agamospecies or parthenoforms) have arisen relatively recently from sexual progenitors. Additionally, a large proportion of parthenogens are polyploids and many are interspecific hybrids. In the majority of cases, the sexual progenitors are extant and living sympatrically or parapatrically with the parthenogens (see Section III).

A. Spontaneous Origins

Meiotic parthenogens arise spontaneously in many plant and animal species. *Tychoparthenogenesis* (occasional development of unfertilized eggs) may be favored in colonizing species that often find themselves at low density and without mates. Artificial selection can improve the rate of tychoparthenogenesis in Drosophila mercatorum, which suggests that automictic species such as D. mangabierai may have arisen spontaneously from tychoparthenogenetic ancestors. Nevertheless, the transition to automixis may be difficult if the sexual ancestors carry deleterious recessive mutations. Selection will rapidly eliminate automictic lineages that are homozygous for such mutations and fix the "lucky" lineages that lack them. Colonization, founder events, and small population sizes can purge the genetic load of the sexual progenitors and facilitate the transition to automixis.

B. Hybrid Origins

Many apomictic and functionally apomictic parthenogens arose as interspecific hybrids. All known asexual vertebrates are hybrids, as are many insects. The strong association between asexuality and hybrid origins led some researchers to suggest that cloning fixes heterosis (hybrid vigor) that may confer broad ecological tolerance. Although evidence exists for broad tolerance to physical stresses in some asexual plants, fish, and frogs, the phenomenon may be a consequence of interclonal selection for the best hybrid combinations rather than heterosis per se. Experimental studies with laboratory-synthesized hybridogenetic fish (P. monacha-lucida; Fig. 1) revealed that most hybrids were inferior to the parental forms; however, a small proportion of hybrid combinations had relatively high fitness. Fitness was not a consequence of heterosis; it was a consequence of the combining properties of parental genomes. Inferences about heterosis and fitness from comparative studies of natural parthenogens and their sexual counterparts are likely to be biased because we only see the successful genomic combinations in nature and not the failures that were purged by selection.

The association between parthenogenesis and hybridization may be a consequence of hybrid dysgenesis. Interspecific hybridization often leads to disruption of meiosis and sterility. Natural selection will preserve the lucky cytogenetic accidents that rescue egg production and restore or retain diploidy. Hybridization is one of a number of dysgenic processes that can produce windows of opportunity for the selection of ameiotic or functionally apomictic reproduction.

C. Parthenogenesis and Polyploidy

The majority of unisexual vertebrates, insects, and plants are polyploids. Although some researchers have suggested that elevated ploidy may produce superior genetic combinations, the association between polyploidy and parthenogenesis may also result from dysgenic processes. Accidental fertilization of a diploid (unreduced) egg will produce triploid progeny that typically are sterile. Such events create another window of opportunity for the selection of lucky cytological accidents that rescue egg production.

Prior establishment of functionally apomictic diploids can facilitate the elevation of ploidy because it removes the sterility barrier. For example, the triploid gynogenetic fish P. monacha-2 lucida ($3n = 72$; Fig. 1) arose by addition of a second lucida genome ($1n = 24$) to a P. monacha-lucida ($2n = 48$) hybrid. For most polyploids, we do not know whether unisexuality or polyploidy came first or if they arose together. If some of these polyploids outperform their diploid counterparts, enhanced performance may be a product of interclonal selection and fixation of the best genomic combinations from sexual ancestors rather than a direct consequence of elevated ploidy.

III. ECOLOGICAL CONSIDERATIONS

All other things being equal (i.e., survival, fecundity, niche requirements, etc.), an all-female lineage should rapidly replace its sexual relatives because a parthenogenetic female produces two daughters for every one produced by an equivalent sexual female. This twofold "cost of sex" may be exacerbated by numerous additional liabilities, such as the risks and energetic costs associated with finding a mate, courtship, and mating itself. Despite the costs of sex, asexual lineages generally have not completely replaced their sexual counterparts in animal taxa that regularly produce clones. Williams (1975) referred to this ecological and evolutionary problem as the "paradox of sex." Why does biparental sexuality predominate so overwhelmingly despite its costs? Ecological studies that attempt to address this question have focused on the primary assumption behind this paradox—that all else is equal between sexual progenitors and derived asexual lineages.

A. Primary Fitness (Fecundity and Survival)

No investigator has succeeded in comparing the lifetime fertility and survival schedules of closely related sexual and asexual lineages in their natural environments, so it is impossible to say that everything else is equal with respect to primary fitness (fertility and survival). Some field and laboratory investigations have obtained data on components of fitness, although few generalizations can be drawn from the current studies. Gynogenetic and hybridogenetic *Poeciliopsis* have fecundities that are similar to those of their sexual counterparts. All-female reproduction is limited, however, by the availability of sperm from the sexual hosts. Parthenogenetic flies (*Drosophila*) and lizards (*Lacerta*) exhibit lower hatching rates than comparable sexual species. Finally, automictic lineages tend to have low hatching success, perhaps due to expression of deleterious recessive genes and inbreeding depression.

Survival differences have been observed in field and laboratory studies. Some unisexual fish (*Phoxinus eos-neogaeus*) and frogs (*R. esculenta*) may be more tolerant of thermal stresses than their sexual counterparts, but the differences do not appear to be generalizable. The roles of hybridity and selection for resistant clones are confounded in these organisms. Studies of survival under stress in *Poeciliopsis* revealed considerable variation among clones and no consistent advantage over the sexual counterparts for the various kinds of stress tested.

B. Geographical Parthenogenesis and General-Purpose Genotypes

Parthenogens should have superior colonizing abilities because they do not have to find mates when they initially occur at low density. Some researchers argue that parthenogens are general-purpose genotypes (jack-of-all-trades) that have wider ecological tolerances than their sexual counterparts. Other researchers argue that parthenogens are narrowly adapted fugitive species that escape from competition with their sexual ancestors. Biogeographical studies reveal that parthenogens are more frequent at the margins of a species range, at extreme latitudes, at higher altitudes, and in regularly disturbed communities—a pattern known as geographical parthenogenesis. It is unclear in most cases, however, whether this pattern is due to enhanced colonization abilities of parthenogens, an inability to compete with sexual progenitors in ecologically central areas, or an increased tolerance of ecologically marginal conditions.

Many widespread apomictic weeds appear to have general-purpose genotypes that can tolerate a wide range of environmental conditions. Selection in a varying environment should favor clones that fluctuate least in fitness. General-purpose clones may not be the best genotype in a particular set of circumstances but, more important, they avoid being the worst during many circumstances. Although the wide geographical distribution of many asexual plants and animals is often cited as supporting the general-purpose genotype hypothesis, such taxa may be composed of numerous cryptic (hidden) clones, each with different environmental tolerances, as found in some asexual waterfleas, brine shrimp, snails, and topminnows. Furthermore, a wide geographical distribution alone may not be sufficient evidence for general-purpose genotypes because a single widespread clone might occupy a narrow but universally available niche. For example, humans introduced dandelion (*Taraxacum officinale*) clones to North America and their success is a consequence of human habitat disruption (grassy lawns).

The fugitive species aspect of geographical parthenogenesis does not apply to sperm-dependent parthenogens. Their colonization and competitive abilities are constrained by the need for sperm from coexisting sexual hosts. Outcompeting or geographically escaping the sexual host will lead to their own reproductive failure. Hybridogenetic and gynogenetic fish (*Poeciliopsis*) have relatively limited ranges encompassed within the geographical limits of their sexual relatives and hosts, whereas some parthenogens, such as the cockroach *Pycnoscelus surinamensis*, have immense distributions, all outside the range of the putative sexual ancestors.

C. Niche Requirements

The niches of parthenogenetic clones and their sexual counterparts appear to differ in many cases. A sexual population should have greater niche breath than a single clone if the differences between genotypes contribute to a wider use of resources. For example, it is difficult to imagine a single jack-of-all-trades human clone (if humans were to be cloned) that has the breadth of talents of the entire human population from which it was drawn. The difference in niche breadth between a sexual population and a single clone will result in asymmetrical competition, in which the sexual lineage has a greater competitive impact on the clone than vice versa. However, an assemblage of ecologically divergent clones may equal or exceed the niche breadth of the sexual ancestors, leading to symmetrical competition and, perhaps, competitive exclusion of the sexuals.

Computer simulations of these ideas revealed that clonal invasion of the sexual niche proceeds from the margins to the center of the resource distribution. According to the frozen niche-variation model, a diverse array of clonal genotypes is frozen from the sexual gene pool. Interclonal selection will eliminate clones that overlap substantially with one another and the sexual ancestors and fix an assemblage of clones that maximally exploits the range of available resources. Sexual and clonal forms can coexist as long as competition remains asymmetrical and the combined niche of the clones is less than that of the sexuals.

Some hybrid parthenogens appear to occupy a weakly contested intermediate niche between the parental forms. However, hybrids are not necessarily intermediate for all niche-related characters. For example, some clones of the hybridogenetic fish *P. monacha-lucida* exhibit dominant phenotypes and extreme trophic behaviors. Hybridity does not necessarily constrain unisexual organisms to ecological intermediacy. Evidence also exists for niche separation between diploid and polyploid parthenogens in several taxa.

IV. EVOLUTIONARY CONSIDERATIONS

Asexual lineages may flourish briefly in some environments, but most appear to be dead ends with limited adaptive potential. From a phylogenetic perspective, obligately asexual plants and animals are little more than buds at the ends of branches that are fundamentally sexual. The rotifer class Bdelloidea is a notable exception. Although they appear to be strictly asexual, bdelloids have diversified into hundreds of morphologically distinct species that are classified into several families. We know of few other asexual taxa that have diversified in a similar way.

Bdelloids notwithstanding, numerous theories exist concerning the genetic, ecological, and evolutionary benefits of sex. Theories about the origin of recombination and meiosis in eukaryotic organisms are poorly understood and beyond the scope of this article. However, factors that favored the origin of sex (e.g., recombinational repair of DNA damage) need not be the same as those that currently maintain sex in higher organisms. Critical reviews of current hypotheses are provided in several of the listed references. Some major ideas related to the maintenance of sex in higher organisms are outlined in the following sections.

A. The Fisher–Muller Hypothesis (Sex Accelerates Evolution)

Adaptation by natural selection requires heritable genetic variation, and sexuality generates a new array of genotypes in each generation. Having more variation, sexual species should be able to adapt more quickly in a changing environment than asexual species. In the early 1930s, Ronald Fisher and Hermann Muller restated this hypothesis in genetic terms. Good mutations occur rarely (e.g., let the mutation rate, μ, be 10^{-8}). The probability of two good mutations arising simultaneously in the same asexual lineage is the vanishingly small product of these numbers (μ^2 or 10^{-16}). It is more likely that two good mutations will come together in the same clone if the first mutation spreads to near fixation before the second mutation arises in the same lineage. In a sexual population, however, the mutations can arise simultaneously in different individuals, and mixis will bring them together as each spreads to fixation.

Although the idea that sex is good for evolution seems intuitively satisfying, it suffers from several fundamental problems. It provides an advantage to sexual populations but not to the individuals that participate in sex. Sexual individuals will not spread at the expense of clones, unless the individuals also gain an advantage that compensates for the cost of males or meiosis. Furthermore, it is hard to see how sex could spread for the purpose of accelerating evolution of the species if evolution itself has no purpose. Evolution is a consequence of heritable variation among individuals and natural selection; it has no goals. Furthermore, evolving rapidly does not necessarily guarantee evolutionary success. Some "living fossils" such as *Limulus,* the horseshoe crab, and *Lingula,* an articulated brachiopod, have changed very little morphologically for hundreds of millions of years.

B. Muller's Ratchet (Sex Is a Way to Get Rid of Bad Mutations)

In 1960, Muller recognized another problem with the Fisher–Muller theory: The vast majority of expressed mutations are slightly deleterious. Recombination uncouples mutations and facilitates purging the bad ones. Muller suggested that slightly deleterious mutations will accumulate in asexual lineages and hitchhike along with the rare good mutations. Clones with the lowest genetic load may be lost due to genetic drift in finite populations. Except for the exceedingly rare back-mutation, the expected fate of an asexual population is to ratchet forward with deteriorating fitness. Other researchers have examined this problem in greater mathematical detail and refer to the mutational meltdown of clones. Despite the attractiveness of this argument, the evolutionary time scale for Muller's ratchet makes it difficult to imagine how it can compensate for the twofold cost of sex on an ecologically relevant time scale (but see Section IV,D).

C. The Tangled Bank (Sex Increases Niche Breadth)

Genotypic differences among individuals of a sexual species may contribute to more effective utilization of natural resources. In a heterogeneous environment, a sexual parent that produces diverse progeny may leave more offspring than a clonal parent that produces only one specialized type of offspring. Competition should be lower among the diverse sexual offspring than among clonal offspring. Thus, sexuals may gain a slight advantage over individual clones in a heterogeneous environment, but they may be eclipsed and replaced by an ecologically diverse assemblage of clones. Without considerable demographic stochasticity that leads to the random loss of clones, it is hard to see how this model can compensate for the twofold cost of sex.

D. The Red Queen (Sex Is Needed to Stay in Coevolutionary Race with Biological Enemies)

A consensus seems to be emerging that coevolutionary pressures from biological enemies (parasites, predators, and competitors) may provide sufficient ecological compensation for the costs of sex. Rapidly evolving microparasites (bacteria, viruses, etc.), because of their short generation times and vast numbers, will rapidly evolve means to avoid immune surveillance and exploit the most common host phenotypes. This provides rare host phenotypes a temporary advantage, until they rise in frequency and become the targets of newly evolved mechanisms of parasitic attack. Fitness of the host is frequency dependent, always favoring rare and different phenotypes, a cycle that maintains genetic polymorphism. Such a process would favor the parents of diverse offspring by spreading the risks of survival. This benefit is even more evident for species that brood their young and thereby increase the risk of contagion.

Red Queen processes may also facilitate the advance of Muller's ratchet. Frequency-dependent fitness will cause clones to cycle in abundance. Clones are more susceptible to random extinction when they are rare, and these losses may also remove clones with the smallest load of deleterious mutations. Working together, the Red Queen and Muller's ratchet may result in a rapid decay of fitness that may account for the maintenance of sex on ecological time scales.

V. PARTHENOGENS AS STUDY ORGANISMS

Comparative studies of sexual and asexual organisms have provided considerable insight into the adaptive benefits of sex. Just as a physician studies deficiencies and diseases to understand the functioning of normal health, evolutionary biologists and ecologists study parthenogenetic clones as deviations from the normal sexual processes. Understanding the conditions under which asexuals prosper has provided insight into the short-term limitations of biparental sex. The overall biogeographical patterns of asexual organisms have likewise allowed biologists to reject some of models for the benefits of sex.

Efforts are also under way to compare the evolutionary longevity of closely related sexual and asexual taxa. Analyses of mitochondrial and nuclear genes provide a general picture that most asexual taxa, except bdelloid rotifers perhaps, arose recently and are relatively short-lived. Few asexual taxa have diversified to the extent that a taxonomist would be tempted to erect new species, genera, or families. For the most part, clonal diversity in asexual populations can be explained by recurrent origins of new clones from extant sexual progenitors. This observation leads to a surprising conclusion that the ecological success of many asexual taxa may depend on periodic recruitment of new genotypes from the sexual gene pool. Thus, sex, and periodic recombination, may also be essential for the ecological success and persistence of asexual populations.

See Also the Following Articles

ASEXUAL REPRODUCTION; CLONING; HYBRIDIZATION; MEIOSIS

Bibliography

Bell, G. (1982). *The Masterpiece of Nature: The Evolution and Genetics of Sexuality*. Univ. of California Press, Berkeley.

Beukeboom, L., and Vrijenhoek, R. C. (1998). Evolutionary genetics and ecology of sperm-dependent parthenogenesis. *J. Evol. Biol*. in press.

Lynch, M. (1984). Destabilizing hybridization, general-purpose genotypes and geographical parthenogenesis. *Q. Rev. Biol.* **59**, 257–290.

Lynch, M., Bürger, R., Butcher, D., and Gabriel, W. (1993). The mutational meltdown in asexual populations. *J. Heredity* **84**, 339–344.

Maynard Smith, J. (1978). *The Evolution of Sex*. Cambridge Univ. Press, Cambridge, UK.

Michod, R. M., and Levin, B. R. (1988). *The Evolution of Sex: An Examination of Current Ideas*. Sinauer, Sunderland, MA.

Suomalainen, E., Saura, A., and Lokki, J. (1987). *Cytology and Evolution in Parthenogenesis*. CRC Press, Boca Raton, FL.

Templeton, A. R. (1982). The prophecies of parthenogenesis. In *Evolution and Genetics of Life Histories* (H. Dingle and J. P. Hegmann, Eds.), pp. 75–102. Springer-Verlag, Berlin.

Vrijenhoek, R. C. (1994). Unisexual fish: Models for studying ecology and evolution. *Annu. Rev. Ecol. Syst.* **25**, 71–96.

Williams, G. C. (1975). *Sex and Evolution*. Princeton Univ. Press, Princeton, NJ.

Parturition, Nonhuman Mammals

Anna-Riitta Fuchs
Cornell University Medical College

Michael J. Fields
University of Florida at Gainesville

I. Introduction
II. Maintenance and Termination of Pregnancy
III. Maternal Preparation for Parturition
IV. Endocrine and Physiologic Mechanisms of Parturition
V. Summary and Future Directions

GLOSSARY

cervix Neck of the uterus consisting of connective tissue admixed with bundles of smooth muscle, lined with mucus-secreting epithelium.

cytokines A group of mitogenic compounds synthesized by cells of the immune system as well as in certain other tissues, including endometrium and placenta. Cytokines have numerous actions and function as signals between different cells.

endometrium/decidua The lining of the uterus which consists of luminal epithelium and stroma. Under the influence of progesterone stromal cells of many species undergo decidual transformation, which involves cytoplasmic enlargement and changes in nuclear and cellular shape.

estrogens, progesterone Steroid hormones produced by the ovary and in many species also by the placenta. After binding to their receptors both steroids translocate into the cell nucleus where they induce transcription of genes for several enzymes, structural proteins, and growth factors that regulate growth and function of the reproductive tract.

fetal membranes Membranes that envelope and protect the fetus in a fluid-filled space. Chorion is the outermost of the membranes and consists of an outer trophoblast epithelium and inner sheath of mesoderm; next are allantois and yolk sac and innermost is the amnion, which surrounds the fetus in a sac filled with amniotic fluid.

hypothalamo–pituitary–adrenal axis Hypothalamus is the basal part of a subdivision of the brain. Specialized blood vessels, the portal vessels, connect it to the anterior pituitary, the "master endocrine gland" that is located just below the base of the brain. A stalk provides nervous connection to the posterior lobe of the pituitary. Hypothalamus integrates both nervous impulses and hormonal signals from the periphery and regulates anterior pituitary hormone secretion by emitting releasing factors into the portal vessels. The hormones of anterior pituitary, luteinizing hormone and follicle-stimulating hormone, prolactin, and adrenocorticotrophin, are released into the circulation. Their target glands are the gonads, the adrenals, and the thyroid gland, which secrete sex steroids, gluco- and mineralo-corticoids, or thyroid hormones, respectively, in response to stimulation by anterior pituitary hormones.

myometrium The muscular outer layer of the uterus (womb).

oxytocin An octapeptide neurohormone synthesized in the hypothalamus and released into the bloodstream from the posterior lobe of the pituitary gland. Oxytocin is a potent uterine stimulant, causes milk ejection, and induces COX-2 expression in endometrium.

placenta Organ that arises from fetal trophoblast and maternal endometrium in which maternal and fetal blood vessels are brought to close apposition, thus providing a site for physiologic exchange between mother and fetus.

prostaglandins Ubiquitous unsaturated fatty acids formed from arachidonic acid. Prostaglandin $F_{2\alpha}$ and prostaglandin E_2, prostacyclin, and tromboxane are produced in tissues of the reproductive tract.

Parturition (from Latin *parturire* = to have the pains of labor) is the process by which intrauterine life of the fetus(es) is terminated in a manner that ensures extrauterine viability of the offspring and survival of the mother with minimal damage to her reproductive tract. Expulsion of the fetus(es) is achieved in three phases: During the first phase (labor) uterine contractions occur with increasing frequency and severity, resulting in dilatation of the cervix; during the second phase (delivery) uterine contractions, aided by abdominal straining efforts, push the fetus(es) through the dilated cervix and vagina; during the third phase (afterbirth) the placenta(s) and fetal membranes are expelled. A preparatory phase is not part of the classic obstetric definition of parturition although it is essential for successful parturition. During this phase various endocrine and developmental processes bring about changes in the maternal, fetal, and placental tissues that prepare the maternal reproductive tract for the act of giving birth.

I. INTRODUCTION

Length of gestation varies greatly among species but is confined within narrow limits in each species. The developmental stage of the fetus at birth varies considerably among species; in some the young are born in very immature state, as in marsupials or rodents; in others the young are able to survive on their own within a few days of birth, as in guinea pigs. Currently, no explanation exists for these genetic differences in reproductive performance. From the times of Hippocrates it has been assumed that the fetus initiates the onset of labor by some unknown mechanism. Strong support for this hypothesis has been provided during the past 30 years by studies in sheep. The stimulus for these studies was the finding that congenital absence of fetal pituitary gland or experimental removal of fetal adrenals or pituitary gland prevented parturition in pregnant sheep. The converse was also proved, namely, that stimulation of fetal adrenal secretion by adrenocorticotrophin (ACTH) infusions or injections of glucocorticoids into fetal lambs precipitated preterm delivery. In other species of the family Artiodactyla fetal cortisol levels were also shown to increase in late gestation and preterm delivery could be induced by infusions of ACTH or glucocorticoids. These observations suggested that maturation of the fetal hypothalamo–pituitary axis is important for the timing of parturition. Further support for this notion is provided by the fact that fetal genotype influences the length of gestation in several ungulates. On the other hand, fetal factors do not appear to be important in many species, including marsupials, rodents, rabbits, cats, and dogs, in which fetal decapitation *in utero* does not affect onset of labor. Maternal factors have the dominant control over the timing of parturition in these species. In primates both fetal and maternal factors appear to be significant. Thus, a continuum exists among species from almost total fetal control in the ewe to total maternal control in marsupials over the timing of parturition.

The interests of mother and fetus are essentially opposite in regard to the timing of parturition, with the fetus favoring a long intrauterine life and the mother preferring a short period of gestation. The balance established between the best interests of mother and fetus during evolution differs greatly among species. The ability to exercise maternal control over gestation length has probably been advantageous for species with small body size because they have had to rely on agility and speed to avoid predators. Similarly, carnivores that depend on speed and agility for catching food have been better off with short gestations, whereas ungulates that graze for food need to be mobile from an early age and therefore prolonged fetal development *in utero* was an advantage. Domestication and hence lack of predators may also have contributed to evolution in this direction.

II. MAINTENANCE AND TERMINATION OF PREGNANCY

Knowledge of the factors that maintain pregnancy is a necessary prerequisite for understanding the mechanisms of parturition. A brief description of such factors is therefore included here.

A. Estrogens and Progesterone

The uterus is unique in that its growth and function are regulated entirely by hormones. By and large, estrogen promotes growth and progesterone promotes differentiation of tissues of the reproductive tract. Estrogens are necessary for the action of progesterone because they regulate the formation of progesterone receptors, and progesterone in turn downregulates estrogen receptors. Neither hormone has a direct action on myometrial contractions. Estrogens promote contractility because they induce the synthesis of contractile proteins and enzymes necessary for energy production and because they induce synthesis of myometrial receptors for uterotonic agonists, such as oxytocin, α-adrenergic compounds, and some prostanoids, and suppress receptors for inhibitory agonists, such as β-adrenergic compounds. Estrogens induce gap junction formation between myometrial cells, which promotes coordination of myometrial contractions, and in progesterone-exposed animals estrogens increase endometrial prostaglandin synthesis. Among many other actions, progesterone opposes many effects of estrogens and it suppresses the synthesis of oxytocin receptors and gap junctions. Under the influence of progesterone the innervation of the uterus becomes very sparse, and the myometrium is predominantly under the influence of inhibitory β-receptors. In contrast to estrogen, which promotes the softening action of relaxin and collagenases on the cervix, progesterone inhibits the softening by relaxin. In many corpus luteum-dependent species administration of progesterone prolongs pregnancy, but in species that depend on placental progesterone this is not the case.

B. Luteotrophic Hormones: Luteinizing Hormone, Prolactin, Estrogen, Chorionic Gonadotropins, and Placental Lactogens

The corpus luteum is the sole source of progesterone throughout gestation in a number of species. Gestation length is usually longer than the duration of luteal life span in nonpregnant cycles, therefore additional luteotrophic factors are necessary for maintenance of progesterone production during pregnancy. A luteotrophic complex secreted by maternal anterior pituitary consisting of luteinizing hormone (LH), prolactin (PRL), and estradiol from the ovary maintains luteal function beyond the nonpregnant luteal life span in rodents, lagomorphs, carnivores, and some artiodactyls. Depending on the species anterior pituitary hormones are required for maintenance of corpus luteum function throughout pregnancy or only during the first half of pregnancy, after which luteotrophic hormones produced by the placenta maintain luteal function. Trophoblast cells of fetal origin are the source of placental gonadotropins in all instances. In many species corpus luteum becomes superfluous beyond the early days of pregnancy, and placenta becomes the main source of progesterone. Bats, guinea pig, sheep, horses, subhuman primates, and humans belong to this group. A system for the rescue of the corpus luteum from luteolysis after conception that does not depend on gonadotropins has evolved in cows, goats, pigs, and mares. In these species the embryo produces antiluteolytic factors, which act locally within the uterus and prevent generation of luteolytic factors of uterine origin.

C. Uterine Luteolytic Factor

Uterine luteolytic factor consists of prostaglandin $F_{2\alpha}$; ($PGF_{2\alpha}$), which is synthesized in the endometrium of nonpregnant females at the end of each luteal cycle and in pregnant females at the end of pregnancy.

D. Conceptus-Produced Factors

The embryo acquires at an early stage the capacity to synthesize proteins or steroids that diffuse into the maternal endometrium; some of these are capable of inhibiting production of $PGF_{2\alpha}$ in the surrounding endometrium. Best known among these are ovine and bovine interferons-τ and estrogens secreted by pig and horse embryos. Chorionic gonadotropins produced by the trophoblastic cells of the placenta are also conceptus-secreted protein that influence maternal endocrine function.

E. Fetoplacental Unit

In several species fetal liver, pituitary gland, gonads, and adrenals begin to secrete compounds, which serve as precursors for placental steroidogenic enzymes and thereby influence or alter placental function. Interactions of maternal and fetal factors within the placenta, which result in production of hormones that neither organism alone is capable of producing, constitute a fetoplacental unit. Humans, some subhuman primates, and ungulates have a fetoplacental unit that produces estrogens from fetal precursors. As the fetus matures and grows, the source of precursors increases and the production of estrogens increases in advancing pregnancy. The fetus is protected against inappropriate stimulation by ovarian hormones, particularly estrogens, by sex steroid-binding proteins produced by fetal liver, such as α-fetoprotein in the rodents and primates, or by conversion of the steroids to water-soluble sulfates that diffuse freely into maternal circulation. Placenta, endometrium, and some other maternal tissues possess sulfatase activity by which steroid sulfates are converted to free steroids, whereas fetal organs lack sulfatase activity.

III. MATERNAL PREPARATION FOR PARTURITION

A. Structural Alterations

Maternal preparation for parturition involves softening of the connective tissue of the cervix, relaxation of the pubic symphysis and pelvic ligaments, loosening of the placental attachment, induction of receptors for uterine activating agents, and formation of gap junctions between uterine smooth muscle fibers to promote the spread of excitation and coordination of myometrial contractions. These changes are brought about by an interaction of several maternal and fetal factors, among which estrogens are important. Estrogens promote the activity of collagenase and certain metalloproteinases thought to be important in the remodeling of uterine connective tissues. Rising estrogen levels also attract invasion of eosinophiles and neutrophiles into uterine and cervical tissues. These lymphocytes are rich sources of collagenase activity and may participate actively in cervical softening. In late gestation the growth of the fetus surpasses the growth of the uterine wall, which becomes distended; distention has been shown to increase the number of gap junctions and thus promote contractility, and up to a certain point distention also increases the force of myometrial contractions.

B. Relaxin

This hormone is released from the corpus luteum in association with luteal demise at the end of gestation, and its release can be induced preterm by administration of $PGF_{2\alpha}$. In several species, including horse, dog, and cat, relaxin is also produced in the placenta, but the stimulus for its release from the placenta is not known. Relaxin receptors are found in uterus, cervix, mammary gland and nipple, and several other tissues. Relaxin acts on connective tissue causing softening of the cervix and relaxation of pubic ligaments. At least in rats and pigs, parturition cannot progress normally in the absence of relaxin.

C. Prostaglandins

Prostaglandins have many actions, including luteolysis ($PGF_{2\alpha}$), stimulation of myometrium and other smooth muscles [$PGF_{2\alpha}$, prostaglandin E_2 (PGE_2), and tromboxane (TXA_2)], constriction ($PGF_{2\alpha}$ and TXA_2) or dilatation [prostacyclin (PGI_2) and PGE_2] of blood vessels, and remodeling of uterine and cervical connective tissue (PGE_2). $PGF_{2\alpha}$ by itself elicits myometrial contractions only at high concentrations but potentiates contractions induced by oxytocin at low concentrations. Prostaglandins are not stored in tissues but are formed upon stimulation from free arachidonic acid by the enzyme cyclooxygenase. Arachidonic acid is in many tissues rate limiting and is formed from tissue phospholipids by either phospholipase A_2 (PLA_2) or phospholipase C (PLC). Figure 1 shows diagrammatically the enzymatic pathways involved in arachidonic acid mobilization in uterine tissues. Cyclooxygenase (COX) exists in two isoforms, COX-1, which is constitutively expressed, and COX-2, which is inducible. Endometrial and placental tissues express mainly the inducible COX-2 gene, whereas the constitutively expressed COX-1 is expressed in vascular tissue, myometrium, and other tissues that require prostanoids for general "housekeeping" functions. The action of agonists that induce transcription of the COX-2 gene is required for the synthesis and release of

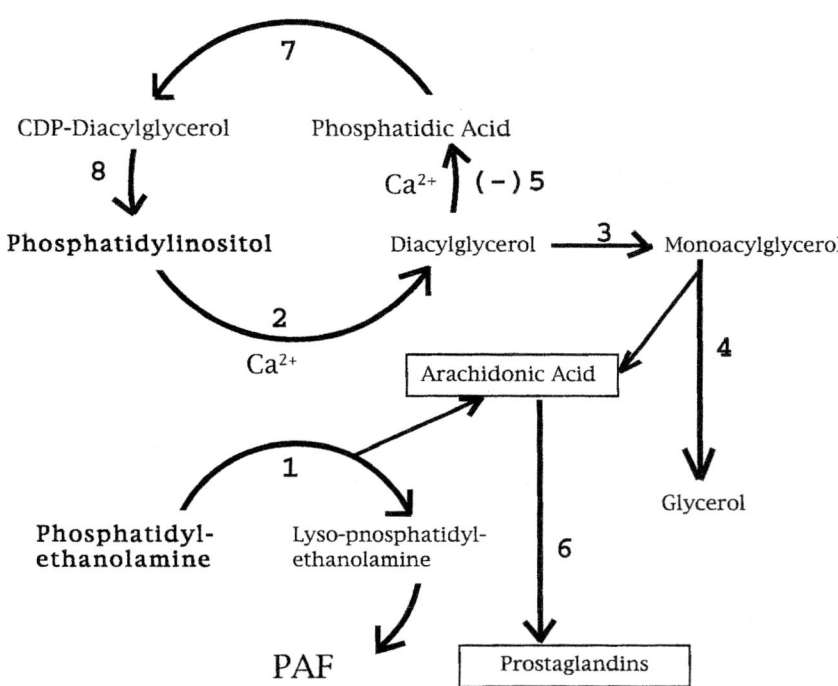

FIGURE 1 The enzymatic pathways involved in mobilization of arachidonic acid from membrane phospholipids. Reactions are catalyzed by (1) phospholipase A_2, (2) phosphatidyl inositol-specific phospholipase C, (3) diacylglycerol lipase, (4) monoacylglycerol lipase, (5) diacylglycerol kinase, (6) cyclooxygenase–peroxidase complex, (7) CTP (phosphatidate cytidyltransferase), (8) inositol 3-phosphatidyltransferase. Platelet-activating factor (PAF) is formed by lysoPAF (acetyl-coenzyme A acetyltransferase from alkylacyl-glycerophosphocholine). Intracellular Ca^{2+} plays an important role. Increase in Ca^{2+} inhibits diacylglycerol kinase activity and hence the recycling of diacylglycerol; Ca^{2+} promotes the activity of diacylglycerol lipase and consequently the mobilization of arachidonic acid. (after J. E. Bleasdale and J. M. Johnston, Reviews in Perinatal Medicine, Vol. 5, pp. 151–191, A. R. Liss, New York, 1984).

prostaglandins from endometrium, placenta, and fetal membranes. Inflammatory and infectious agents induce expression of COX-2. Oxytocin is capable of inducing COX-2 gene expression in endometrium and cervical mucosa of late pregnancy. Platelet-activating factor (PAF), which is produced near term in increasing amounts by fetal lungs, is capable of induction of COX-2 expression in fetal membranes.

D. Oxytocin, Oxytocin Receptors, and Their Interaction with Prostaglandins

Oxytocin plays an important role in parturition. Oxytocin has a threefold action on the uterus: It initiates myometrial contractions, causes release of $PGF_{2\alpha}$ from endometrium, and causes release of PGE_2 from cervical mucosa. Because of its great potency, oxytocin concentrations in circulation are usually very low, which makes measurements difficult; moreover, oxytocin is often secreted in a pulsatile fashion and the short half-life of oxytocin further complicates determination of plasma concentrations. The secretory pattern of oxytocin during gestation and parturition is therefore known in only a few species. In rats, cows, sheep, rhesus monkeys, and humans oxytocin is secreted from the posterior pituitary in a pulsatile fashion during the latter part of gestation. The oxytocin receptor gene is expressed in myometrium, endometrium, and cervical mucosa. A marked increase in uterine and cervical oxytocin receptor concentrations occurs in all species at term, which increases uterine responsiveness to circulating oxytocin levels. Uterine oxytocin receptors bind vasopressin with almost as high affinity as oxytocin, which may explain why parturition in oxytocin knockout mice is reportedly not impeded. The expression of oxytocin receptors in rabbits, rodents, and carnivores depends on an increase in estrogen and fall in progesterone secretion, but in ruminants, mares, guinea pigs, and primates a fall in plasma progesterone concentration is not necessary; rising levels of estrogen can induce oxytocin receptor formation in the face of high gestational progesterone concentrations.

E. Central Nervous System

At the level of the hypothalamus some changes are evident before parturition. The activity of the neurons that synthesize oxytocin is increased and oxytocin content of the posterior pituitary lobe reaches a peak at term. Maternal behavioral changes, such as nest building and hair plucking, are observed shortly before parturition. Timing of parturition exhibits circadian variations in many species, which testifies to the participation of the central nervous system. In pregnant sheep and rhesus monkeys continuous recordings of uterine contractile activity have shown that for several days before parturition uterine activity exhibits circadian variations in intensity and degree of coordination. This nocturnal activity increases progressively as onset of parturition approaches. Administration of an oxytocin antagonist to late pregnant monkeys abolished the nocturnal increases in uterine activity, indicating that pituitary oxytocin elicited these contractions. Concentrations of adrenal glucocorticoids, estrogens, progesterone, oxytocin, and antidiuretic hormone in maternal blood also exhibit circadian variations. Circadian variations in maternal endocrine secretions imprint a "clock" upon the fetal hypothalamus that in late gestation initiates circadian variations in fetal endocrine activity.

IV. ENDOCRINE AND PHYSIOLOGIC MECHANISMS OF PARTURITION

The anatomy and physiology of the uterus varies markedly among mammals resulting in variations in the mechanism of parturition. Endocrine aspects of parturition in mammalian species with different uterine and placental features are briefly described in the following sections. The diversity of the shape of placentas and uteri in different species is shown in Figs. 2 and 3.

A. Marsupials

Marsupials have the most primitive form of reproductive tract and placenta of all mammals. The Müllerian ducts do not fuse at the embryonic stage as in other mammals, and mature females have two completely separate uteri and vaginas (Fig. 3A). The placenta is choriovitelline (yolk sac placenta) with only superficial apposition to maternal endometrium. Maternal endometrium is not cast off after birth but is slowly absorbed. Females often mate and are impregnated at the postpartum estrus but enter diapause (a period of physiologically imposed inactivity) until the pouch young leaves the teat; implantation then takes place and gestation is resumed. The duration of gestation from implantation to parturition is approximately the same as the length of a nonpregnant cycle, and the pup is born at a very immature state. Parturition is associated with a simi-

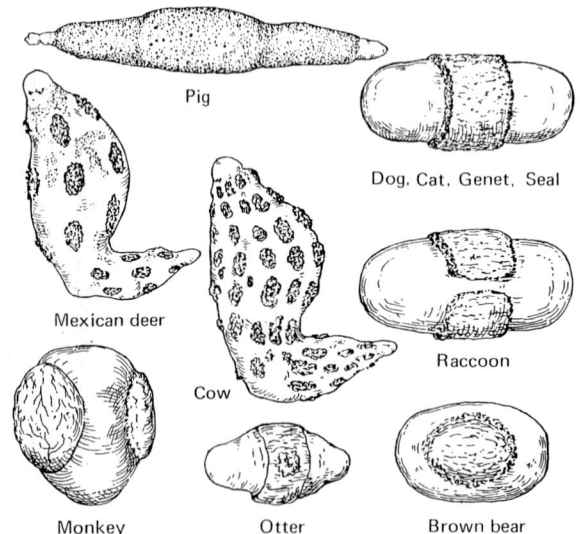

FIGURE 2 The gestation sacs of representative animals showing the gross shape of the placentas. Pig, diffuse; deer and cow, cotyledonary; dog, otter, and raccoon, annular or zonary; monkey and bear, discoid [after W. J. Hamilton and H. W. Mossman (Eds.), *Human Embryology*, 3rd ed., Williams & Wilkins, Baltimore, 1962].

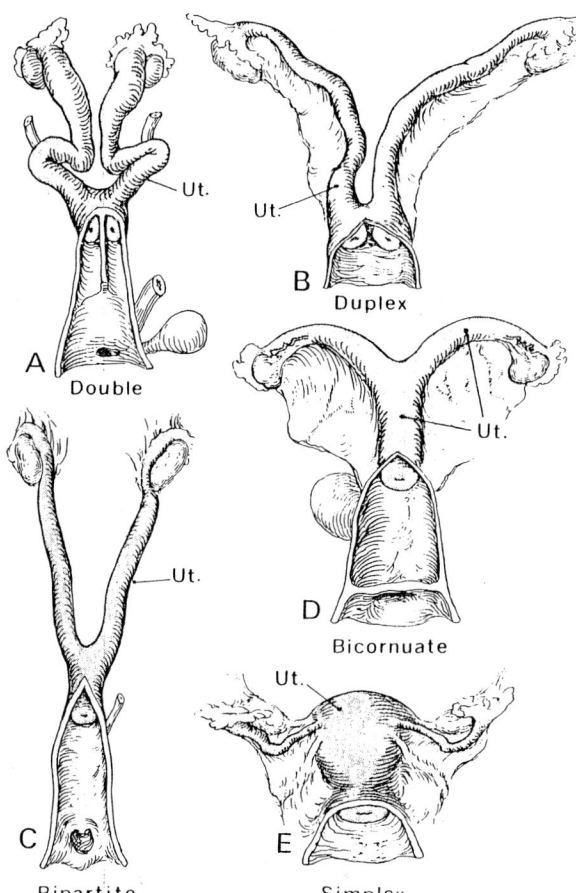

FIGURE 3 Representative types of uteri in various animals. (A) Double, in marsupials; (B) duplex, in rodents and rabbits; (C) bipartite, in carnivores; (D) bicornuate, in ungulates; (E) simplex, in primates. Ut, uterus (after S. R. M. Reynolds, *Physiology of the Uterus*, 2nd ed., 1949, Hafner, New York).

lar fall in plasma progesterone as at the end of the luteal cycle, and the pup is expelled by a series of strong uterine contractions that are elicited by a sudden surge of oxytocin and oxytocin-induced release of $PGF_{2\alpha}$. Both uteri, pregnant and nonpregnant, exhibit similar changes in oxytocin receptor concentrations and myometrial sensitivity to oxytocin during pregnancy, but at parturition oxytocin receptors increase only in the pregnant uterus. The immature pup manages to climb into the pouch immediately after birth and gets attached to the teat, which swells in its mouth preventing the fetus from falling off. Prolactin secretion in response to suckling is responsible for the ovarian inactivity during the diapause that follows the attachment of the pouch young.

B. Lagomorphs: Rabbit

Lagomorphs and rodents have a similar uterus and placenta. The reproductive tract is duplex (Fig. 3B), the two uterine horns open into a single vagina through separate cervices, and the litter consists of several fetuses. The chorioallantoic placenta is usually discoid in shape and labyrinthine in structure, and it produces placental lactogen but no progesterone. Rabbit is a reflex ovulator and usually ovulates in response to the mating stimulus. When the male is infertile, mating results in pseudopregnancy with formation of corpora lutea of significantly shorter life span than corpora lutea of pregnancy (16–18 vs 28–30 days). Concentration of plasma progesterone reaches a plateau at about 14 days of gestation and begins to decline at about 21 or 22 days and falls to 1 or 2 ng/ml on Days 30 or 31. Plasma estradiol concentration is relatively high and stable throughout pregnancy. Uterine contractions are virtually absent during gestation and delivery can be inhibited by administration of progesterone. These observations gave rise to the progesterone "block" theory, according to which progesterone acts directly on myometrium to inhibit contractile activity and cessation of progesterone secretion suffices for the development of labor-like uterine contractions. It is now known that labor contractions are elicited by the release of uterotonic agonists. In rabbits they are initiated by activation of the hypothalamic nuclei that stimulate oxytocin secretion from the posterior pituitary. A large surge of oxytocin sets off a train of strong contractions that expel the litter of up to 14 pups in about 10–15 min. The first stage of labor lasts only a few minutes because the cervix of rabbits is soft and easily extensible. Oxytocin also causes the release of a surge of $PGF_{2\alpha}$ from the endometrium which peaks postpartum and may prevent bleeding from the placental sites. Inhibitors of oxytocin release delay onset of labor and bolus injections of oxytocin induce parturition when given within 24 hr before spontaneous onset of labor. The inhibitory effects of progesterone on uterine contractility are exerted indirectly by suppression of oxytocin receptor and gap junction formation. Hypophysectomy of fetuses *in utero* by decapitation has no effect on gestation length. Pregnant rabbits exhibit a very characteristic behavior shortly before the onset of parturition. They are restless and suddenly begin plucking abdominal hair in a frenzy; this hair is later used for a nest. The stimulus for the sudden release of oxytocin has not been identified but may be associated with the central nervous system excitation demonstrated by the prepartum behavior.

C. Rodents: Rats and Mice

With the exception of guinea pigs, rodents depend on progesterone from corpora lutea during gestation. They are spontaneous ovulators with a very short cycle. Copulation induces release of anterior pituitary hormones that prolong corpora lutea life span. Infertile matings result in pseudopregnancy of 14–16 days duration, and fertile matings result in gestation of 22–24 days in rats and 19 or 20 days in mice. A luteotrophic polypeptide from the placenta maintains luteal function in conjunction with ovarian estradiol from midgestation onwards; progesterone concentrations in plasma peak a few days before term and then fall precipitously during the last 1 or 2 days before parturition. Estradiol concentrations in plasma are low most of gestation and rise steeply just before parturition. Decapitation of fetuses does not significantly affect length of gestation. Uterine contractile activity is frequent during gestation but remains localized and nonpropagated. Near term the pattern changes to intermittent series of rhythmic contractions alternating with quiescent periods. During the last day of gestation the duration of active periods lengthens and contractions become propagated along the uterine horns; gradually the quiescent intervals disappear when the rat enters the first stage of labor. In rats the cervix is hard and closed and requires a few hours of labor before it permits passage of the fetuses. The first stage of labor lasts about 1 or 2 hr; expulsion of the pups is achieved with the aid of strong abdominal straining efforts at the peak of each contraction. The intervals between the birth of individual pups are longer than those in rabbits, and the expulsion of the litter takes 1–1.5 hr. The placentas are more firmly attached than they are in rabbits, and the mother frequently grabs the cord and pulls the placenta out after the pup is born. Uterine contractions subside gradually about 1 or 2 hr after delivery. Labor and delivery can be mimicked with administration of an oxytocin infusion of 4 or 5 hr duration or with small doses of oxytocin given every 10 min over 4–6 hr on average. Myometrial oxytocin receptors are sparse during gestation; their concentration rises exponentially during the last 24 hr before parturition. Normal delivery can therefore not be induced with oxytocin earlier than about 8 hr before the expected time of spontaneous parturition. Temporal relationships among plasma levels of oxytocin, myometrial oxytocin receptor concentrations, and $PGF_{2\alpha}$ release during parturition in rodents are illustrated in Fig. 4.

In contrast to oxytocin, infusions of $PGF_{2\alpha}$ or PGE_2 to rats on the day of expected parturition fail to induce labor, and uterine contractions that were elicited initially wane during the infusions, suggesting that myometrial prostaglandin receptors are downregulated. Administration of substantial doses of $PGF_{2\alpha}$ on Days 18–20 of gestation results in abortion within about 24 hr. The mechanism of the abortifacient action of $PGF_{2\alpha}$ is induction of luteolysis. Decidual oxytocin receptor concentrations in rats peak around Days 18–20 and then decline to low levels at term. It is not known whether maternal oxytocin elicits $PGF_{2\alpha}$ release from the decidua causing luteolysis at that stage of gestation. Maternal behavior is very pronounced in rodents and is expressed before parturition in the form of nest building. After parturition the mothers retrieve the young, lick them, and bring them to a nest, where the mothers spend hours at a time hovering over them and nursing them.

D. Guinea Pig and Bats

Guinea pig is a rodent and the bats belong to Chiroptera, but one aspect of the physiology of gestation is common to both: Relative to their size these species have longer pregnancies and deliver proportionally larger young than any other species. A guinea pig pup at birth weighs up to 20% and individual

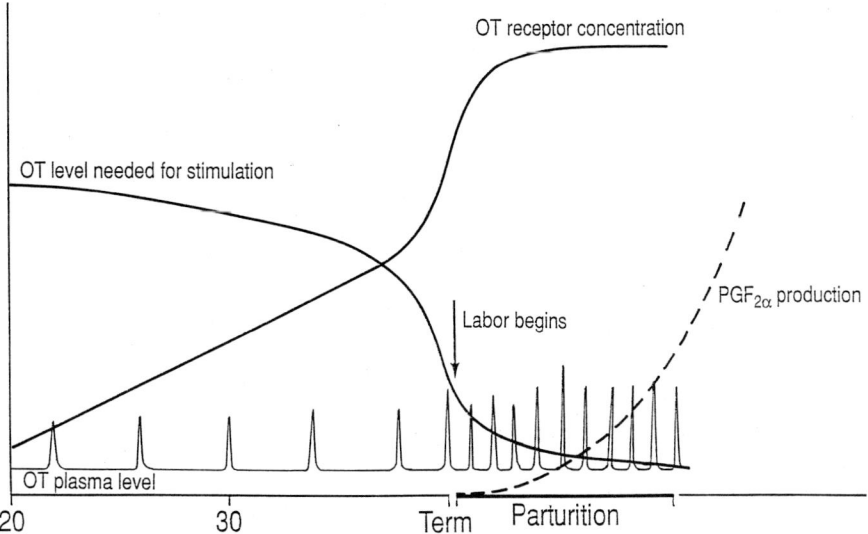

FIGURE 4 Temporal relationships among myometrial oxytocin (OT) receptor concentration, circulating oxytocin (OT) concentration, $PGF_{2\alpha}$ release, and onset of labor [after A. R. Fuchs, F. Fuchs, and P. G. Stubblefield (Eds.), *Preterm Birth*, 2nd ed., p. 84, McGraw-Hill, New York, 1993].

newborn bats weigh 30–50% of maternal weight. Bats carry only one fetus at a time but guinea pigs have litters of two to four fetuses. Not surprisingly, both species have very highly developed, discoid, and labyrinthine placentas that secrete both progesterone and estrogens. Ovariectomy at an early embryonic stage does not interfere with maintenance of pregnancy. Both species depend on relaxin for relaxation of the pubic symphysis, pelvic ligaments, and cervix to permit the passage of the bulky fetuses and ensure live birth. In bats the ligament is stretched to about 15 times its nonpregnant length. Bats have the ability to prolong the length of gestation when they are under stress, presumably by entering diapause, but little is known about the endocrinology of parturition in bats. Guinea pigs have been more thoroughly investigated. In contrast to bats, guinea pigs tend to abort or deliver premature young when they are stressed. Plasma progesterone concentrations do not decline before parturition as in other rodents but remain high throughout delivery; plasma estrogen levels increase moderately before parturition. The fetuses are very mature and their anterior pituitary regulated endocrine systems are fully functional at term. At birth glucocorticoids in fetal plasma are elevated, as are maternal glucocorticoids, but injections of ACTH or glucocorticoids into the fetuses near term do not initiate parturition. The release of uterine prostaglandins increases during parturition, but the stimulus is not known with certainty and may include maternal or fetal oxytocin, fetal PAF, or EGF. The event that initiates parturition in guinea pigs has not been established.

E. Carnivores: Dog and Cat

Carnivores have a bipartite uterus which has two separate horns opening into the vagina through a single cervix (Fig. 3C); they have a zonary placenta in which fetal chorionic villi encircle as a girdle the inside of the uterus (Fig. 2). Corpus luteum is essential throughout pregnancy in these species. A placental luteotrophin is secreted only for a short period at the end of gestation. The duration of pregnancy (dogs, 63–67 days; cats, 55 or 56 days) is only slightly longer than the nonpregnant luteal phase in each species; concentrations of progesterone in plasma are also similar or slightly higher than those in nonpregnant cycles. After a peak of about 50–60 pg/ml during proestrus, in dogs plasma estrogen concentration falls and remains low throughout pregnancy, and in cats a moderate rise in plasma estradiol occurs at term. The corpora lutea of pregnancy secrete relaxin and in association with luteolysis a surge of relaxin enters the bloodstream. Low concentrations of relaxin are also synthesized in the placenta. Pseudopregnant bitches have no circulating relaxin. Prolactin levels rise very markedly near term in pregnant bitches but do not seem to have a role in the mechanism of parturition. Uterine contractile activity increases during the last week before whelping and is quite strong for a few days before visible signs of labor are evident. Labor lasts about 4 hr and intervals between successive pups are rather long, up to 30 min or more. Oxytocin administered in several small doses can speed up the process considerably, but prostaglandins are ineffective stimulants of uterine contractions in dogs. Large doses of $PGF_{2\alpha}$ induce luteolysis and abortion when given earlier in gestation. Gestation and placentation of the cat are similar to those in the dog anatomically, morphologically, and endocrinologically. Circulating progesterone and estrogen concentrations are comparable to those in pseudopregnant cats. Little is known about endocrine changes in the fetuses of cats, dogs, and other carnivores at the time of parturition. Only a few studies have been published on plasma progesterone and estrogen concentrations in blue and red fox, mink, and weasel. These animals are reflex ovulators and depend on the corpora lutea for progesterone production; duration of gestation and nonpregnant cycle are similar, as are the patterns of ovarian steroids.

F. Artiodactyls: Sheep, Cow, Goat, and Pig

The artiodactyls and other ungulates have a bicornuate uterus in which the horns are joined caudally to form an uterine body before entering the vagina (Fig. 3D). The ruminants have a multiplex or cotyledonary placenta, whereas the pig (and hippopotamus) have a diffuse placenta, in which fetal chorionic villi are distributed fairly uniformly over the whole fetal sac (Fig. 2). In animals having a cotyledonary placenta the nonpregnant uterine lining possesses localized thickenings (caruncles) to which the fetal villi attach after implantation. In sheep the placenta becomes the main organ for progesterone production around Day 50 after mating, whereas cows, goats, and pigs depend on the corpus luteum for progesterone production throughout gestation. Near term the placenta of these species acquires the ability to synthesize low amounts of progesterone. Fetal membranes are capable of estrogen synthesis and produce mainly the less active estrone and estradiol-17α. They are conjugated into water-soluble sulfates in fetal liver, pass the placenta, and are excreted in maternal urine. The ruminant placenta is largely impermeable to free steroids as well as large-molecular-weight anterior pituitary hormones. Fetal maturation therefore depends on functioning fetal endocrine system. Early fetal development is independent of endocrine influences but late development and maturation depend on endocrine secretions, particularly insulin, thyroid, and adrenal hormones.

The ewe has been widely used as a model in studies of fetal endocrinology and fetomaternal interactions following pioneering studies conducted 30 years ago that provided strong support for the concept that maturation of fetal adrenal glands initiates the process of parturition. The endocrine pathways are illustrated in Fig. 5. The placenta of the ewe produces

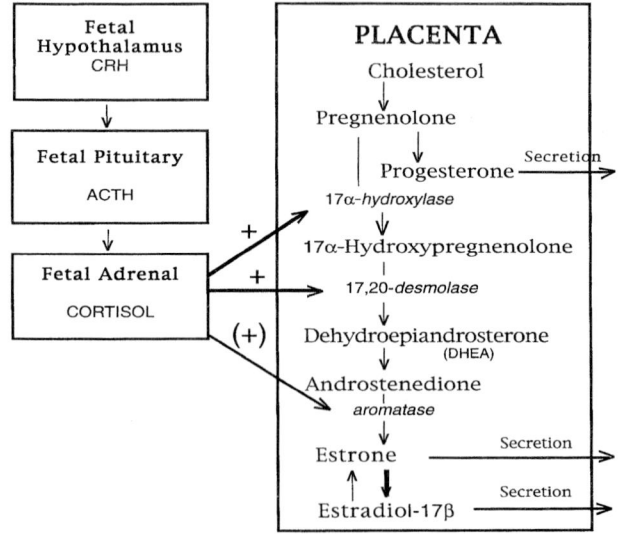

FIGURE 5 Endocrine secretions and interactions among fetal hypothalamus, fetal anterior pituitary gland, fetal adrenals, and placenta at initiation of parturition in sheep. +, Induction of activity; (+), enhancement of existing activity. Secretion of oxytocin from the posterior lobe of the pituitary is not shown.

progesterone from cholesterol but lacks a key enzyme in the synthetic pathway from progesterone to estrogens, namely, 17α-hydroxylase. During the last week of pregnancy fetal adrenals mature and respond with cortisol production to ACTH secreted by fetal anterior pituitary. Cortisol induces 17α-hydroxylase enzyme in the fetal placenta, which begins to convert progesterone to estrogen causing plasma progesterone levels to fall and estradiol-17β levels to rise. The shift in hormonal balance results in increased posterior pituitary oxytocin content; increased numbers of gap junctions; increased density of oxytocin receptors in myometrium, endometrium, cervical mucosa, and fetal membranes; and induction of COX-2 gene expression and prostaglandin production in these tissues. These events gradually lead to more powerful and coordinated myometrial contractions and labor. $PGF_{2\alpha}$ release from endometrium increases in response to oxytocin and other COX-2 inducers causing luteolysis and release of relaxin from the corpus luteum; relaxin in turn softens cervical connective tissues so that the cervix yields to the pressure exerted by the advancing fetus and dilates. Oxytocin secretion increases slowly at first and occurs in a pulsatile manner during the first stage, and later in labor a sustained surge of oxytocin stimulates powerful contractions that expel the fetus.

Parturition can also be induced preterm with ACTH or cortisol infusions into the fetus in cows and goats. The prepartum hormonal changes in cows involve more gradual increases in plasma concentrations of estradiol-17β and estrone than in sheep. The main site of progesterone production in cows and goats is the corpus luteum; it has therefore been difficult to understand how fetal adrenal activation signals the maternal corpus luteum to stop secreting progesterone. As in sheep, fetal cortisol activates the enzymes 17α-hydoxylyase and 17,20-lyase in the placentas of goats and cows, resulting in conversion of placental progesterone to estradiol-17β. Although placental progesterone production in goats and cows amounts to only about 20% of total progesterone, estradiol-17β produced from placental progesterone may suffice for the induction of $PGF_{2\alpha}$ release from the endometrium, thus constituting the signal by which fetus indicates its readiness to be born in cows and goats.

G. Perissodactyls: Horse

The horse has, like other ungulates, a bicornuate uterus. The placenta is diffuse as in pigs (Fig. 2) and villous in structure. The villi are not uniformly distributed as in the pig but occur in tufts scattered over the entire fetal sac. The definite chorioallantoic placenta develops late, the villi start to penetrate endometrial folds on the seventh or eighth week, and attachment is not complete until the 14th week. Until then there is a loose attachment of the embryo to the endometrium via a choriovitelline (yolk sac) placenta and the embryo obtains much of its nutrition in the form of uterine milk produced by endometrial glands and from deteriorating uterine tissue. The maternal side of the placenta of the horse has a unique feature because it forms the so-called endometrial cups, which are the source of pregnant mare serum gonadotropin (PMSG). PMSG initiates the formation of secondary corpora lutea at approximately Day 60, after which the cups disappear.

The endocrine profile of the pregnant mare is also unusual. The corpora lutea secrete progesterone and plasma levels rise to a peak of 15 ng/ml at about Day 60, but between 180 and 220 days of gestation luteal secretion ceases and progesterone disappears from maternal blood. The placenta secretes a number of progesterone metabolites from about Day 100 onwards. From Days 200 to 300 the total concentration of these progestagens is low (below 10 ng/ml) but increases during the last month of gestation to about 25 ng/ml and then decreases during the last few days before parturition to about 10 ng/ml. The biological significance of these progestagens has not been clarified. The adrenal glands of horse fetuses do not respond to ACTH with cortisol production until at birth and in the early postpartum period. On the other hand, fetal gonads develop at an early stage and secrete significant amounts of dehydroepiandrosterone (DHEA) in midgestation. DHEA is converted to estrone by the placenta. Maternal ovary also secretes estrone but removal of fetal gonads significantly decreases maternal plasma estrone concentrations, indicating that fetal gonads are the major source of estrogens. The concentrations of plasma estrone sulfate and equilin, an estrogenic compound of unknown origin, are highest around 240 days and decrease toward term. Delivery cannot be induced by fetal infusion of ACTH or glucocorticoids, whereas a bolus injection of oxytocin at term results in foaling within 1 hr; an abrupt rise in plasma PGF metabolite concentration accompanies the oxytocin injection. The cervix of pregnant horses is soft and easily extensible. The fetal hypothalamo–adrenal axis does not seem to be important for the initiation of parturition in horses but the fetus nevertheless has an influence on the timing of birth as evidenced by the different lengths of gestation in mares and donkeys and the crosses between them.

H. Subhuman Primates

Primates have a simplex uterus consisting of a single cavity that opens into a single cervix and vagina (Fig. 3E); the chorioallantoic placenta is discoid in form and villous in structure. Most primates have a bony pelvis which limits the size of the fetal head that can pass through the birth canal. In this situation both mother and fetus have a vital stake in the timing of parturition. Primate placenta has developed an extensive endocrine repertoire producing numerous polypeptide hormones, releasing factors, and growth factors, as well as steroid hormones. It permits the passage of steroid hormones so that the fetus is under the influence of maternal hormones throughout pregnancy. ACTH and glucocorticoids do not induce preterm delivery in primates. The corpus luteum is required only for a limited time after the end of normal luteal phase when it is rescued in subhuman primates by monkey chorionic gonadotropin. Plasma progesterone concentrations are low in comparison to humans, but as in women, they do not fall at term. The concentrations of estrogens in plasma of pregnant rhesus monkeys are also considerably lower than those in pregnant women and consist of estrone, estrone sulfate, and estradiol. Near term plasma levels of estrogens increase. Primate placenta does not have 17α-hydroxylase activity and therefore cannot convert progesterone to estrogens. The fetoplacental unit of rhesus monkey converts fetal androgens, DHEAS, and androstenedione sulfate into estrone and estradiol-17β as shown in Fig. 6. The androgens are secreted by fetal adrenals in response to stimulation by fetal ACTH. Maternal endocrine activity modulates the secretion of fetal ACTH because in late gestation maternal cortisol is converted into the less active cortisone by placental 11β-hydroxylase; this lowers the feedback inhibition of fetal ACTH secretion by cortisol. As a consequence fetal adrenals are

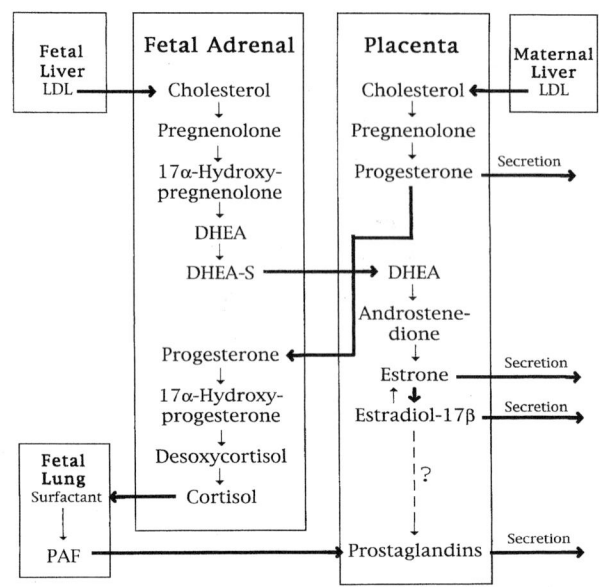

FIGURE 6 The endocrine interactions among fetal liver, fetal adrenals, fetal lungs, placenta, and maternal liver at initiation of parturition in subhuman primates. (For clarity, the activities of anterior and posterior pituitary glands are omitted.) Plasma low-density lipoproteins (LDL) produced by fetal and maternal liver, respectively, serve as precursors for fetal adrenal and placental production of pregnenolone. Placenta is rich in 3β-hydroxysteroiddehydrogenase (3β-HSD) activity and converts pregnenolone to progesterone, which is secreted into the circulation, whereas 17α-hydroxylase/17,20-desmolase activity in fetal adrenals converts pregnenolone to 17α-hydroxypregnenolone and DHEA, which is sulfated and secreted as DHEAS. Placenta has sulfatase activity and converts DHEAS to DHEA, which is a substrate for 3β-HSD and is converted to androstenedione; this in turn is aromatized to estrone and eventually reduced to estradiol-17β, both of which are secreted into the circulation. Pregnenolone is a better substrate for 17α-hydroxylase than progesterone, but in late gestation some progesterone is converted in fetal adrenals to 17α-hydroxyprogesterone, desoxycortisol, and cortisol, which is secreted into fetal circulation. Important among the actions of cortisol is stimulation of surfactant production in fetal lungs, platelet-activating factor (PAF) is produced as a by-product. PAF is one of the compounds that elicits prostaglandin release from fetal membranes and placenta.

stimulated to produce more androgenic precursors for placental estradiol-17β synthesis. The effects of rising estradiol-17β concentrations in pregnant primates have not been established with certainty but estradiol-17β may induce endometrial COX-2 synthesis, increase oxytocin receptor concentrations, and stimulate secretion of oxytocin. In primates the preparturient changes are more gradual than in the lower animals and take place over 1 or 2 weeks in monkeys. The circadian rhythm in maternal adrenal steroid secretion is imprinted on the fetal hypothalamic–adrenal axis and its epiphenomena, placental estradiol production, and posterior pituitary oxytocin secretion. Oxytocin-induced nocturnal uterine activity increases and becomes progressively stronger and more coordinated over the last 7–10 days before parturition and eventually evolves into the first stage of labor. Neither progesterone receptor antagonists, estradiol injections, $PGF_{2\alpha}$ infusions, or oxytocin infusions alone can reproduce the process of normal, spontaneous parturition. The interaction and coordination of several of these factors is required over a certain length of time to mimic the physiology of parturition in primates. Figure 7 depicts the three main factors involved in eliciting uterine contractions during labor.

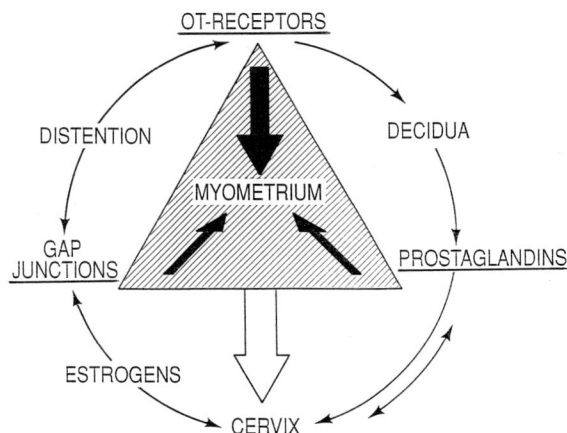

FIGURE 7 Three main factors in activation of uterine contractions at parturition: oxytocin (OT) receptors, prostaglandins ($PGF_{2\alpha}$ and PGE_2), and gap junctions. An increase in OT receptor concentration lowers the threshold for initiation of myometrial contractions by circulating OT, and release of $PGF_{2\alpha}$ from decidua/endometrium by OT potentiates OT-induced contractions, which act on the cervix causing shortening and dilatation. OT receptors in cervical mucosa mediate PGE_2 release which acts on the surrounding connective tissue causing softening. Estrogens produced by the placenta stimulate metalloproteinases that act to soften the cervix, and estrogens increase the formation of gap junctions that potentiate OT-induced contractions by increasing conduction of excitation along the myometrium. Distention of the uterine wall by the rapidly growing fetus in late gestation increases the number of open gap junctions and the number of OT receptors enhancing the action of OT on myometrial contractility [after A.-R. Fuchs, in *Uterine Contractility* (R. Garfield, Ed.), Sereno Symposia, Norwell, MA, 1992].

V. SUMMARY AND FUTURE DIRECTIONS

Ovarian or placental estrogens and progesterone, adrenal glucocorticoids and androgenic precursors of estrogens produced by the adrenals, lipid precursors produced by the liver, $PGF_{2\alpha}$ and PGE_2 produced by uterine tissues, and oxytocin from the central nervous system are the main factors that control the duration of pregnancy and initiation of parturition across species. Sterols and steroids are evolutionary old molecules, as are the posterior pituitary hormones. They are found in all vertebrate classes and in some invertebrates. During evolution various species have adapted and modified them for different uses and in different combinations. In some species maturation of the fetus determines the onset of parturition, and in others the maternal luteal life span is the determining factor. In between these extremes are numerous examples of mutual dependence between mother and fetus in providing the necessary conditions required for the successful act of birth. Because of its function as the organ providing a site for physiological exchange between mother and fetus, the placenta has evolved as the coordinator that integrates the signals from maternal and fetal organisms. Developmental factors and programmed cell death, which are currently under intense scrutiny, may also play an important role in the design of these signals. Various cytokines may play a greater role than is currently assumed, and the possible role of the immune system in initiation of parturition is poorly understood. The exact mechanisms by which the production of various luteotrophic factors are turned on and off during pregnancy are not known and need further study. The mechanisms by which steroid hormones control the gene expression of plasma membrane receptors for excitatory and inhibitory agents, such as posterior pituitary hormones and prostaglandins, need to be elucidated. The role of the central nervous system in coordinating the function of endocrine glands involved in the circadian variations of uterine contractions that become prominent as parturition approaches needs to be investigated, as does the nature of the signal for the surge of oxytocin at parturition.

See Also the Following Articles

Cats (Felidae); Cattle (Bovidae); Dogs (Canidae); Horses (Equidae); Marsupials; Oxytocin; Pigs (Suidae); Primates, Nonhuman; Rodents; Sheep and Goats

Bibliography

Amoroso, E. C. (1959). Comparative anatomy of the placenta. *Ann. N. Y. Acad. Sci.* **75**, 855–872.

Currie, W. B., Gorewit, C., and Michel, F. J. (1989). Endocrine changes, with special emphasis on oestradiol-17β, prolactin and oxytocin, before and during labour and delivery in goats. *J. Reprod. Fertil.* **82**, 299–308.

Fuchs, A.-R. (1985). Oxytocin in animal parturition. In *Oxytocin; Laboratory and Clinical Studies* (J. A. Amico and A. G. Robinson, Eds.), International Congress Series Vol. 666, pp. 207–236. Elsevier, Amsterdam.

Liggins, C. G. (1988). Endocrinology of parturition. In *Fetal Endocrinology* (J. M. Novy and J. A. Resko, Eds.), pp. 211–237. Academic Press, New York.

Renfree, M. (1994). Endocrinology of pregnancy, parturition and lactation in marsupials. In *Marshall's Physiology of Reproduction* (G. E. Lamming, Ed.), 4th ed., Vol. 3, pp. 677–766.

Steinetz, B. G., Goldsmith, L. T., Harvey, H. J., and Lust, G. (1989). Serum relaxin and progesterone concentrations in pregnant, pseudopregnant, and ovariectomized, progestin treated pregnant bitches: Detection of relaxin as a marker of pregnancy. *Am. J. Vet. Res.* **50**, 68–71.

Thorburn, G. D., and Challis, J. R. G. (1979). Endocrine control of parturition. *Physiol. Rev.* **59**, 863–919.

Wood, C. E., and Cudd, T. A. (1997). Development of the hypothalamus–pituitary–adrenal axis in the equine fetus: A comparative review. *Equine Vet. J. Suppl.* **24**, 74–82.

Pelvic Inflammatory Disease (PID)

Thomas E. Snyder
Bowman Gray School of Medicine

I. Epidemiology
II. Microbiology and Pathogenesis
III. Diagnosis
IV. Treatment
V. Prevention
VI. Summary

GLOSSARY

acute pelvic inflammatory disease (PID) Infection originating from cervicovaginal flora that ascends along mucosal surfaces to the uterus and fallopian tubes.

Centers for Disease Control and Prevention Agency for research of and formulating guidelines and treatment norms for infectious diseases.

Chlamydia trachomatis Identified as causative organism in majority of cases of the sexually transmitted disease cervicitis in United States.

dyspareunia Painful intercourse.

mucopurulent cervicitis Endocervicitis characterized by a mucopurulent (green or yellow) endocervical discharge and endocervix that is erythematous, edematous, and friable.

neisseria gonorrhea Gram-negative intracellular diplococci, causative organism of cervicitis and PID.

sexually transmitted disease A disease process transferred from one individual to another via sexual intercourse.

tuboovarian abscess Inflammation and abscess formation involving the ovary and the fallopian tube.

Pelvic inflammatory disease (PID) is an infection originating from the cervicovaginal flora which ascends along the mucosal surfaces of the uterus and fallopian tubes. The infection may spread to the peritoneal cavity causing peritonitis and/or intraabdominal abscesses. The disease may be acute with marked symptoms, chronic, causing ongoing mild or moderate pain, or silent. Essentially all PID occurs in nonpregnant, generally young females. Infection can occur after instrumentation, such as abortion, dilatation and curettage, or intrauterine device insertion, but is uncommon. The patient generally has a period of gonorrheal or chlamydial cervicitis which then spreads to the fallopian tubes. Chronic or recurrent PID is a poorly defined symptom complex of pelvic pain/dyspareunia over a long period of time. The sequelae of single or multiple episodes include chronic pelvic pain, ectopic pregnancy, involuntary sterility, abscess formation, and, occasionally, sepsis and death.

I. EPIDEMIOLOGY

It is estimated that 10% of reproductive-age women in the United States have been treated for pelvic inflammatory disease (PID). The highest incidence occurs in the teenager, with two-thirds of all cases expected in women <25 years of age. One million women are treated for PID each year, of which 180,000 are hospitalized. The direct costs for PID approached $4.2 billion a year in 1990 and are projected to reach $10 billion a year by 2000 with a sizable percentage covered by public sources. The total number of cases is probably underestimated because the disease is asymptomatic in many patients and many patients are treated with outpatient therapy and go unreported. Three percent of sexually active teenagers are estimated to be affected each year, resulting in increasing rates of ectopic pregnancy four to six times that of the unaffected population. Twenty-five percent of patients with a diagnosis of PID have serious sequelae, including ovarian abscess, ectopic pregnancy, dyspareunia, pelvic adhesive disease, and infertility. The risk of infertility is related to the severity of the episode and the number of times the patient is infected, with 10, 25, and 60% of women becoming infertile with one, two, or three exposures, respectively. Patients present with symptoms characteristic of the effect of the causative organism. Patients seen in urban emergency rooms often exhibit acute abdominal pain and fever characteristically associated with gonococcal PID. Women seen in sexually transmitted disease (STD) clinics, private offices, or college clinics often have more subtle signs of infection associated with chlamydia.

A. Risk Factors

Risk factors for PID that affect the risk of transmission on progression of disease are variable (Table 1).

1. Age

One-third of the 1 million women affected annually are <19 years old and often nulliparous. These women are more susceptible because of lower prevalence of protective antibodies, larger zones of cervical ectopy (columnar epithelium in the ectocervix), and

TABLE 1
Factors Associated with Pelvic Inflammatory Disease

Age
Sexual behavior (multiple partners)
Method of contraception
Health care behaviors
Douching
Menses
Mucopurulent cervicitis
Substance abuse

Note. Adapted from Washington *et al.* (1991). *JAMA* **266**, 2581–2586. © 1991 American Medical Association.

greater penetrability of the cervical mucous. The rate of PID among 15- to 19-year-old women is approximately three times the rate of 20–24 year olds and twice the rate of 25- to 29-year-old women. PID is the most common serious infectious disease of young women.

2. Sexual Behavior

Except for occasions when women have been instrumented, PID almost exclusively affects sexually active females. Young age at first intercourse, frequency of intercourse, multiple sexual partners, substance abuse, and history of other STDs are associated with an increased risk of PID.

3. Contraception

Mechanical and chemical contraception, such as condoms, diaphragms, vaginal spermicides, and foams, when used properly decrease the risk of PID. Intrauterine device (IUD) use is associated with increased risk of 1.5–2.6 times. However, infection associated with PID usage is usually limited to certain groups of women and is highest in the first 4 months following insertion and decreases to baseline by 5 months after insertion. Women at low risk of STD exposure have low rates of IUD-associated PID, whereas women at high risk of STD exposure are not good candidates for IUD usage. Risk of PID may be decreased by administration of 200 mg doxycycline daily for 2 days following insertion.

Most studies show a two- or threefold increase in the prevalence of chlamydial infection in women using oral contraceptives (OCs). However, this may be confined to cervicitis since rates of PID requiring hospitalization may be as much as 50% lower in women on OCs compared to sexually active women not using contraceptives. Oral contraceptives increase the zone of cervical ectopy, creating a larger surface for potential infection; however, the progesterone component of most OCs makes cervical mucous less penetrable by both sperm and bacteria. Also, decreased menstrual flow may lead to less retrograde menstruation and less optimal environment for bacterial proliferation. Women using OCs have the same rate of tubal infertility as those using other methods of birth control, which suggests that they may have an increased incidence of chlamydial (silent/asymptomatic) PID.

4. Douching and Menses

Women who douche are more likely to have PID than those who do not, although a causal relationship has not been established. The risk is possibly associated with the frequency of douching by (i) altering vaginal flora or (ii) flushing vaginal and causal organisms into the upper genital tract. Symptoms of PID occur more often within 7 days of menses than 7–14 days prior to onset of flow.

5. Socioeconomic Variables

Rates of chlamydia in white women continue to increase, whereas rates in other ethnic groups remain stable. Married women have a decreased incidence compared to never married or divorced women. The total number of cases hospitalized has decreased since 1980, whereas office-treated cases have remained unchanged. Office exposure for white women and those <24 years old have increased, perhaps reflecting more asymptomatic nongonococcal cases. Lesbian women appear to be at low risk; however, studies are lacking. HIV and PID may obviously coexist secondary to similar modes of transmission. Patients with HIV are more likely to have tuboovarian abscesses and require surgical intervention but may exhibit lower white cell counts.

II. MICROBIOLOGY AND PATHOGENESIS

Agents causing PID include sexually transmitted bacteria, endogenous vaginal and bowel flora, and other organisms (Table 2). The most common bacterial STD in the United States is *Chlamydia trachomatis*, with an estimated 4 million cases annually. If chlamydia is detected on endocervical culture, 10–30% of patients will develop PID. Chlamydia causes damage to tubal epithelia through immunologic mechanisms rather than cytotoxicity; therefore, these infections may go unnoticed more often secondary to lack of acute symptoms.

TABLE 2
Microbial Origin of Pelvic Inflammatory Disease

Exogenous	Endogenous	Other
Sexually transmitted diseases	Vaginal bacteriosis	Escherichia coli
Neisseria gonorrhoeae	Gardnerella vaginalis	Haemophilus spp.
Chlamydia trachomatis	Bacteroides spp.	Group B streptococci
Mycoplasmas[a]	Prevotella spp.	Staphylococci
	Peptostreptococci	Pneumonococci
	Mobiluncus	
	Streptococci	
	Mycoplasmas	

Note. From Washington and Berg (1996). Originally published in R. L. Sweet, Changing etiology of PID. In *Genitourinary Infections in Women: Updated on Urinary Tract Infections and Pelvic Inflammatory Disease*, Consensus Conference Proceedings, 1994.

[a] Role in origin is undetermined.

In the past *Neisseria gonorrhea* was thought to be the predominate organism causing PID, with positive cervical cultures in 30–80% of cases. Recent studies utilizing culdocentesis, laparoscopy, and endometrial biopsy have demonstrated the role of anaerobic bacteria in the etiology of PID. Studies performed in the 1970s using transvaginal culdocentesis suggested a polymicrobial nature to PID. However, it is now apparent that vaginal contamination yielded many false-positive results. Current studies from Scandinavia and the United States utilizing laparoscopic and endometrial specimens have deemphasized the role of *N. gonorrhea* and demonstrated chlamydia to be present in 23% of patients hospitalized for PID. Mixed anaerobic and aerobic infections constituting normal vaginal flora also play a major role in PID. In a summary of six published studies utilizing laparoscopy for diagnosis, anaerobic and facultative organisms, *C. trachomatis,* and *N. gonorrhea* were recovered in 61, 31, and 27% of patients, respectively. However, in two other studies, *N. gonorrhea* was recovered 22.2 and 1.2% and mixed anaerobic flora in 31 and 30% of cases. Anaerobic flora were also isolated in 50% of patients who had positive cultures for *N. gonorrhea* and *C. trachomatis.* Therefore, a mixed anaerobic/aerobic flora plays a significant role in infection of the upper genital tract.

The mechanisms providing spread of the organism from the cervix to the upper genital tract are poorly understood. Studies showing that particulate matter and dyes may be transported to the fallopian tubes after placement in the vagina suggest a vehicle is unnecessary for bacteria to gain access to the upper tracts. However, other direct and indirect evidence suggests sperm also play an important role. In addition, organisms may be spread via retrograde menstruation. Once within the fallopian tube, *N. gonorrhea* produces damage via a cytotoxic effect and secondary, complement-mediated, inflammatory reaction resulting in cell death and tissue damage. The end result is production of inflammation and purulent exudate noted clinically in these patients.

A. Fitz–Hugh–Curtis Syndrome

Fitz–Hugh–Curtis syndrome is an extrapelvic manifestation consisting of an acute peritonitis and perihepatitis caused by inflammation of the subdiaphragmatic space. The patients may have inflammatory exudate on the surface of the liver. In the chronic phase, "violin string" adhesions are found between the anterior surface of the liver and anterior abdominal wall secondary to scarring. Usually, Fitz–Hugh–Curtis syndrome is an incidental finding since the patient's pelvic symptoms are much more prominent clinically. A presumptive diagnosis is entertained in those with pleuritic and/or upper right quadrant abdominal pain radiating to the back. Liver enzymes are usually normal.

B. "Silent PID"

Patients presenting for infertility evaluations commonly exhibit antibody to chlamydia or gonorrhea or both, indicating previous infection. However, examination of the fallopian tubes of these patients reveals no differences from those without a history of PID, suggesting damage may occur without overt clinical symptoms.

III. DIAGNOSIS

Diagnosis of PID is often problematic since the patient may present a range of symptoms from mild pelvic pain of many days duration to septic shock. PID often mimics other diseases such as ectopic pregnancy, endometriosis, ovarian cysts, appendicitis, and inflammatory bowel disease.

A thorough history should include recent sexual contact, menstrual history, number of sexual partners, genital tract surgical procedures, and method of contraception. History regarding other pelvic diseases, such as urinary tract or inflammatory bowel disease, is often helpful in developing a differential diagnosis.

Classically, the patient will have lower abdominal pain of moderate to severe intensity and usually of short duration. A history of chronic pelvic pain or intermittent pain lessen the likelihood of the diagnosis. Gram stain should be done, and if positive for gram-negative, diplococci is highly suggestive of diagnosis; however, it will not be useful for diagnosis of chlamydia. Symptoms usually occur around menstruation, are exacerbated by intercourse, and remain bilateral and unrelenting in the pelvis.

Dependent on severity of the disease, the patient may variably have only lower abdominal pain or pain above the navel. The pain is usually produced on deep palpation of the lower quadrants and the patient may exhibit rebound. Upper abdominal pain may be present if spread to the perihepatic region is present.

Examination of vaginal discharges may be helpful secondary to leukorrhea, defined as the presence of more than one polymorphonuclear leukocyte per epithelial cell observed on vaginal wet prep. Mucopurulent cervicitis (MCP) characterized by a greenish and yellow discharge from an erythematous, friable cervix is often present. Cultures for *N. gonorrhea* and *C. trachomatous* should be done.

Pelvic examination should be done systematically evaluating for cervical motion tenderness without abdominal wall palpation. The classic "chandelier sign" refers to pain produced on cervical motion and is nonspecific and unreliable for exact diagnosis. The size, mobility, and tenderness of the uterus are determined followed by similar evaluation of the adnexa.

Endometrial biopsy may be useful at least retrospectively, confirming the diagnosis of endometritis. While not considered standard of care, it provides evidence of upper tract infection without the cost or potential risk of laparoscopy and causes the patient only transient discomfort. Ultrasound will be negative in cases of mild PID; however, it may be useful in defining characteristics of adnexal masses in patients with tuboovarian abscess. In addition, ultrasound or computerized tomography guided drainage of these masses is possible and may be useful in following resolution of masses after treatment.

Laparoscopy provides the "gold standard" for diagnosis of acute salpingitis. Criteria include (i) hyperemia of the fallopian tube surface, (ii) edema, and (iii) purulent exudate on the surface and fimbriated ends of the tube. Severity of disease is not necessarily related to findings at laparoscopy. Patients with marked rebound may have lesser findings at laparoscopy than the patient with a tuboovarian abscess (TOA).

In 1993 the Centers for Disease Control (CDC) established diagnostic criteria for PID secondary to lack of consistent findings in these patients (Table 3). Minimum criteria include (i) abdominal and (ii) adnexal tenderness and (iii) cervical motion tenderness. Additional criteria to narrow the diagnosis include (i) elevated temperature, (ii) cervical discharge, (iii) elevated erythrocyte sedimentation rate, (iv) C-reactive protein, and (v) positive laboratory cultures. The CDC recommends a low threshold for diagnosis secondary to potential severe, long-term health consequences. It is emphasized that these are guidelines and not firm rules. Most causes of PID may be presumptively diagnosed with clinical criteria; however, the more elaborate tests may be utilized in more obscure cases or in those not responding to initial therapy.

TABLE 3
Diagnostic Criteria for PID

Minimum[a]
   Lower abdominal tenderness
   Adnexal tenderness
   Cervical rectum tenderness
Additional criteria
   Routine
      Oral temperature >38.3°C
      Abnormal cervical or vaginal discharge
      Elevated erythrocyte sedimentation rate
      Elevated C-reactive protein
      Laboratory documentation of cervical infection with
         N. gonorrhea or C. trachomatis
Elaborate[b]
   Histopathologic evidence of endometritis on biopsy
   Tuboovarian abscess on sonography or other radiologic test
   Laparoscopic abnormalities consistent with PID

Note. From Centers for Disease Control and Prevention (1993b).
[a] All three must be present.
[b] Data may provide additional help but are more costly.

IV. TREATMENT

Current guidelines for treatment of PID emphasize the probable polymicrobial nature of the disease. Regimens recommended by the CDC (Table 4) provide a broad spectrum of aerobic and anaerobic coverage in addition to coverage of N. gonorrhea and C. trachomatous. The major decision to be made by the clinician is whether to hospitalize the patient or utilize ambulatory management. Some experts emphasize hospital management in an attempt to prevent tubal sequelae. There are no studies which document differential outcomes in treatment regimens. CDC criteria (Table 5) for hospitalization follow standard clinical patterns. If the diagnosis is insecure, severe disease is suspected, or if the patient is not thought reliable for adequate outpatient follow-up and treatment, inpatient therapy is appropriate. Outpatient therapy is not usually recommended for the teenage population secondary to concerns regarding compliance and possible prevention of sequelae.

Inpatient regimens A and B are time tested and have been evaluated in multiple protocols with clinical cure rates of 92–94% demonstrated. The regimens dictate total duration of therapy of 14 days secondary to feeling that this will result in an approximately 10-day compliance, whereas a recommended 10-day regimen may result in 7 or less days of therapy and a less favorable outcome.

Outpatient regimens A and B both are associated with a clinical cure of 95%; however, studies on long-term outcomes are lacking. One concern is whether a single dose of cefoxitin provides adequate coverage of anaerobic organisms.

Patients who are hospitalized should meet criteria of decreased to no abdominal pain and normalization of temperature and white blood cell count prior to discharge. Outpatients should be reevaluated 72 hr

TABLE 4
Centers for Disease Control and Prevention Treatment Recommendations for Acute Pelvic Inflammatory Disease

Inpatient regimens
   A. Cefoxitin 2 g iv q6h or cefotetan 2 g iv q12h
      Plus
     Doxycycline 100 mg iv or po q12h[a]
   B. Clindamycin 900 mg iv q8h
      Plus
     Gentamicin 2 mg/kg iv or im (loading dose), followed by 1.5 mg/kg (maintenance dose) q8h[b]
Alternative inpatient regimens
   Ampicillin/sulbactam plus doxycycline
   or
   Ofloxacin iv plus clindamycin or metronidazole
Outpatient regimens
   A. Cefoxitin 2 g im plus probenecid 1 g po concurrently, or ceftriaxone 250 mg im, or other third-generation cephalosporin im once
      Plus
     Doxycycline 100 mg po bid for 14 days
   B. Ofloxacin 400 mg po bid for 14 days
      Plus
     Clindamycin 450 mg po qid, metronidazole 500 mg po bid or for 14 days

Note. From Centers for Disease Control and Prevention (1993b). iv, intravenous; q, every; h, hour; po, by mouth; im, intramuscularly; bid, twice daily; qid, four times daily.
[a] Continue regimen for at least 48 hr after patient demonstrates substantial clinical improvement and then follow with doxycycline 100 mg po bid for a total of 14 days.
[b] Continue regimen for at least 48 hr after patient demonstrates substantial clinical improvement and then follow with doxycycline 100 mg po bid or clindamycin 450 mg po qid for a total of 14 days.

TABLE 5
Criteria for Hospitalization

Diagnosis is uncertain and surgical emergencies such as appendicitis and ectopic pregnancy cannot be excluded
Pelvic abscess is suspected
The patient is pregnant
The patient is an adolescent (compliance issue)
HIV infection
Severe illness or nausea and vomiting preclude outpatient management
The patient is unable to follow or tolerate an outpatient regimen
The patient has failed to respond clinically to outpatient therapy
Clinical follow-up within 72 hr of starting antibiotics cannot be arranged

Note. From Centers for Disease Control and Prevention (1993b).

after therapy is initiated to determine if changes in therapy or hospitalization are needed.

A. Complications

1. Tuboovarian Abscess

Tuboovarian abscess is the most severe consequence of PID and occurs in approximately 15% of cases. Organisms appear to gain access to ovarian stroma and the subsequent inflammatory response leads to abscess formation. The process may involve the ovary, both tube and ovary, and/or adjacent structures such as the bowel. Diagnosis is based on abdominal findings similar to those in uncomplicated PID plus the finding of a palpable or sonographically identifiable adnexal mass. Outpatient therapy is not an option with findings of TOA since rupture of these abscesses is a life-threatening event. If the patient fails to respond to the initial antibiotic regimens, the diagnosis remains insecure, or rupture of abscess is suspected, surgical intervention is needed. In the past, removal of the uterus and both adnexa was thought necessary to affect cure. Currently, either drainage of the abscess or unilateral adnexectomy is considered appropriate if future fertility is desired. Total abdominal hysterectomy and bilateral salpingo-oophorectomy is reserved for patients with extraordinary disease or those not desirous of future fertility.

2. Sequelae

PID results in sequelae such as involuntary infertility, chronic pelvic pain, dyspareunia, and ectopic pregnancy. Infertility appears proportional to the severity as well as the number of episodes of disease in 0.6, 6.2, and 21.4% of patients with a single episode of mild, moderate, and severe disease, respectively. Oral contraceptives appear to favorably affect the prognosis; however, infertility may occur in 8, 19.5, and 40% of patients after one, two, or three or more episodes, respectively.

Ectopic pregnancy is a common and potentially life-threatening sequelae of PID. Westrum demonstrated a fourfold increase in occurrence compared to control patients. Pelvic pain may last 6 months or more after hospitalization in 24% of patients with documented PID and approximately 18% may have chronic pelvic pain of many months or years duration. Causation of adhesive disease commonly associated with PID remains controversial. In addition, 50% of patients may experience dyspareunia, resulting in complicated social relations.

V. PREVENTION

Since damage is already done to the pelvic organs of those who present with PID, initially preventing the disease is obviously important. If the patient has already had one episode of disease, treatment of her current partner and information to avoid repeat exposure should be provided. Reproductive-age females bear the consequences of these diseases, including infertility, pain, and ectopic pregnancy as described previously. Therefore, they need knowledge of behaviors that place them at risk and effective methods to prevent disease when intercourse occurs. The CDC in 1991 (Table 6) provided recommendations for prevention; however, only limited data regarding the effectiveness of these recommendations exist. Since a high percentage of PID is asymptomatic, education regarding symptoms will be ineffective in preventing sequelae. Therefore, measures aimed at decreasing exposure are of utmost importance.

Prevention programs include both primary and secondary strategies. Primary strategies are aimed at prevention and include (i) behavioral changes to

TABLE 6
Recommendations for Health Providers to Prevent Sexually Transmitted Disease/Pelvic Inflammatory Disease (STD/PID)

General preventive measure	Specific recommendations[a]
Maintain up-to-date knowledge about STD/PID prevention and management	Develop an accurate base of information on the diagnosis, treatment, and prevention of STD/PID
	Complete continuing education courses periodically to update knowledge of STD/PID prevention and management
Provide patient education and preventive medicine services	Educate patients about STD/PID and potential complications
	Encourage healthy sexual behavior, use of barrier methods, and seeking medical care
	Provide epidemiologic treatment for STD/PID when appropriate
	Screen women for chlamydial and gonococcal infection routinely when indicated
Treat infections appropriately	Diagnose STD/PID promptly
	Treat STD/PID promptly and with effective antibiotics
	Encourage women with STD/PID to adhere to management instructions
Ensure examination of sex partners	Encourage infected women to refer sex partners for medical assessment
	Evaluate and treat sex partners appropriately

Note. Based on Centers for Disease Control and Prevention (1991).
[a] More detail is provided in text.

reduce risk of acquiring or transmitting infection, including delaying first intercourse, decreasing number of partners, careful selection of partners, and use of barrier contraception; and (ii) that effort should be made to identify and treat carriers before they infect either partners or, in the case of pregnant females, babies.

Secondary strategies are aimed at preventing complications once a patient is infected. This may be accomplished by (i) screening patients to identify and treat asymptomatic infection, (ii) treating female partners of males with infection, and (iii) recognizing and treating clinical signs of disease.

The target population for prevention should emphasize teenagers since they have the highest incidence of disease, are most susceptible to infection, and potentially bear the most significant sequelae.

Prevention strategies include community and health care provider strategies. Community-based strategies should make teachers, parents, and health care providers aware of the disease process and its consequences and provide access to diagnosis and treatment facilities. Programs to reduce the risk of chlamydia/gonorrhea and related PIDs also impact on transmission of HIV and other STDs. Therefore, the presence of subclinical chlamydia discovered on culture should trigger the health care provider to educate the patient about initiation of sexual activity, number of partners, and avoidance of high-risk partners.

Schools and other educational organizations should be responsible for basic sex education, including (i) rates of gonorrhea and chlamydial infection, (ii) consequences of PID, (iii) signs and symptoms of PID and other STDs, (iv) signs of asymptomatic infections, (v) treatment of partners, and (vi) locations and methods of contacting health care facilities. In addition to information, schools may logically provide school-based clinics, testing of asymptomatic males, and access to health care providers and facilities. Adolescents who have dropped out of school may be at even greater risk of infection than those in school. Organizations such as Job Corps, Vo-Tech centers, and other community-based programs may help reach this group of higher risk individuals.

Health care providers must be acutely aware of the high incidence of disease and be ready to aggressively screen persons at risk and their partners. Knowledge of current screening and management strategies is essential.

Screening is essential, especially for chlamydia secondary to its commonly asymptomatic status. Specimens may easily be obtained at the time of routine health care examination. Other opportunities for screening include adolescent care providers, induced-abortion clinics, and detention facilities.

Screening at family planning and prenatal clinics is perhaps most cost-effective because of the large number of persons examined. The following are CDC guidelines for who should be screened: (i) women with MPC, (ii) sexually active women <20 years of age, and (iii) women 20–24 years old or >24 who do not use barrier contraception or have a new or have had more than one partner during the past 3 months. Women <20 years old should be tested at the time of routine pelvic exam unless sexual activity is limited to a single, mutually monogamous partner. Other patients meeting criteria should be screened annually. Men should be screened in the same circumstances.

Treatment of sexual partners of patients at risk is extremely important to keep the pool of potentially infected persons at the minimum size. Many males are asymptomatic carries and, if not treated, will reinfect their partners.

Finally, in addition to conventional screening, health care providers should provide risk-reduction counseling, including (i) education of sexually active patients about HIV and STDs, (ii) assessment of patient's risk factors, (iii) advice about behavioral changes to reduce risk of infection, and (iv) encouraging the use of barriers to STDs such as condoms.

VI. SUMMARY

Pelvic inflammatory disease is a sexually transmitted disease mainly affecting young, reproductive-age women, requiring many hospitalizations and marked expense in the United States and other countries. It is caused by exposure of the upper genital tract to *N. gonorrhea, C. trachomatis,* or other aerobic and anaerobic organisms endogenous to the vagina. The disease causes symptoms of abdominal pain, fever, and laboratory abnormalities. It may mimic other intraabdominal pathologies such as appendicitis, torsion of an ovary, ectopic pregnancy, or inflammatory bowel disease. Treatment involves broad-spectrum antibiotics either in the hospital or on an outpatient basis. Sequelae include involuntary infertility, chronic abdominal pain, and dyspareunia. Education and prevention are mandatory to prevent the significant social and economic consequences of this disease.

See Also the Following Article

SEXUALLY TRANSMITTED DISEASES

Bibliography

Centers for Disease Control and Prevention (1991). Pelvic inflammatory disease: Guidelines for prevention and management. *MMWR* **40**(RR-5), 1–25.

Centers for Disease Control and Prevention (1993a, August 6). Recommendations for the prevention and management of Chlamydia trachomatis infection. *MMWR*(RR-12).

Centers for Disease Control and Prevention (1993b). Sexually transmitted diseases treatment guidelines. *MMWR* **42**(RR-14), 77–80.

Eschenbach, D. A. (1996). Acute pelvic inflammatory disease. In *Gynecology and Obstetrics* (J. J. Sciarra, Ed.,) Vol. 1, Chap. 44, pp. 1–16. Lippencott, Philadelphia.

Kottmann, L. M. (1995). Pelvic inflammatory disease: Clinical overview. *J. Obstet. Gynecol. Nurs.* **24**(8), 759–766.

Soper, D. E. (1993). Upper genital tract infections. In *Textbook of Gynecology* (J. Copeland, Ed.), pp. 517–530. Saunders, Philadelphia.

Soper, D. E. (1994). Pelvic inflammatory disease. *Infect. Dis. Clin. North Am.* **8**(4), 821–840.

Sweet, R. L. (1995). Role of bacterial vaginosis in pelvic inflammatory disease. *Clin. Infect. Dis.* **20**(Suppl. 2), S271–S275.

Washington, A. E., and Aral, S. O. (1991). Assessing risk for pelvic inflammatory disease and its sequelae. *J. Am. Med. Assoc.* **266**, 2581–2586.

Washington, A. E., Cates, W., and Wasserheit, J. N. (1991). Preventing pelvic inflammatory disease. *J. Am. Med. Assoc.* **266**, 2574–2580.

Washington, E., and Berg, A. O. (1996). Preventing and managing pelvic inflammatory disease: Key questions, practices, and evidence. *J. Family Pract.* **43**(3), 283–293.

The Pelvic Nerve

Karen J. Berkley
Florida State University Program in Neuroscience

I. The Pelvic Nerve and Other Nerves Supplying the Pelvic Region
II. Problems in Dissociating Innervation Patterns of Nerves Supplying the Pelvic Region
III. Organs Innervated by the Pelvic Nerve
IV. Functions of the Pelvic Nerve
V. Summary and Future Issues

GLOSSARY

Aδ fibers Small-diameter primary afferent nerve fibers that are thinly myelinated.
analgesia The absence of pain in response to a stimulus that would normally be painful.
C fibers Small-diameter peripheral nerve fibers (primary afferents or efferents) that are unmyelinated.
efferent fibers (efferents) Nerve fibers that convey information by means of action potentials (or axonal transport) from the central nervous system to the periphery.
enteric nervous system An extensive endogenous and partially autonomous neural supply of the gastrointestinal tract composed of myenteric and submucosal plexes.
parasympathetic nerve fibers A division of the autonomic nervous system composed of those nerve fibers innervating internal organs whose preganglionic nerve cell bodies are located in the medulla (dorsal motor nucleus of the vagus nerve) or in the sacral (S2–S4) spinal segments (humans) or L6/S1 spinal cord segments (rats) and that synapse with postganglionic neurons in ganglia located adjacent to or in the walls of the target organ. The postganglionic fibers then travel only a short distance to innervate the peripheral organs, often carrying out their functions there using acetylcholine, although many other neuroactive agents, including purines, nitric oxide, and various peptides, are also involved.
perineum (adjective: perineal) The area between the thighs from the coccyx to pubis, below the pelvic diaphragm.
plasma extravasation The leakage of plasma from capillaries in peripheral tissues that can be induced by retrograde (distally directed) action potentials in afferent C-fibers.
plasticity (or neural plasticity) Long-term changes in peripheral or central innervation density, synaptic efficacy, or phenotypic action of neurons (e.g., neurotransmitter characteristics) that can be induced by pathophysiology and other means.
primary afferent fibers (or afferents) Nerve fibers that convey information about events occurring in peripheral tissues by means of action potentials (or axonal transport) traveling from the tissue to the central nervous system.
sympathetic nerve fibers A division of the autonomic nervous system composed of those nerve fibers innervating internal organs whose preganglionic nerve cell bodies are located in thoracolumbar spinal cord segments and synapse with postganglionic neurons in sympathetic chain ganglia or various para- and prevertebral ganglia. The postganglionic fibers then travel a relatively long distance to innervate the peripheral organs ("long adrenergic neurons"). However, exceptions to this pattern exist for some of the fibers innervating pelvic viscera; these fibers synapse on postganglionic neurons in ganglia located closer to the target organ ("short adrenergic neurons"). The sympathetic postganglionic fibers often carry out their functions on target organs using noradrenaline, although many other neuroactive agents, as for the parasympathetic nerve fibers, are also involved.

The pelvic nerve is one of three major nerves exiting the spinal cord that innervates the pelvis. Actions by the three nerves are intricately coordinated, together possibly with those of the vagus nerve, by a combination of peripheral, intraspinal, supraspinal (brain), and hormonal interactions to affect many pelvic reproductive, urinary, and gastrointestinal functions. Thus, understanding pelvic nerve contributions to these functions requires understanding its relation to the other nerves. In general, more is known about urological functions than about gastrointestinal and reproductive ones (Fig. 1).

I. THE PELVIC NERVE AND OTHER NERVES SUPPLYING THE PELVIC REGION

The pelvic nerve, known also in humans as the nervi erigentes or pelvic splanchic nerves, innervates various pelvic visceral structures with both sensory afferent fibers and efferent preganglionic fibers that are derived from neurons located in spinal segments S2–S4 in humans (L6–S2 in rats). These preganglionic fibers terminate in the pelvic ganglion (or pelvic plexus), known in males as the major pelvic gangion and in females variously as the paracervical ganglion, uterine cervical ganglion, or Frankenhäusen's ganglion. The ganglion is located adjacent to the prostate in males and the cervix in females.

The second of the other two nerves emerging from the spinal cord (segments T10–L2) to innervate the

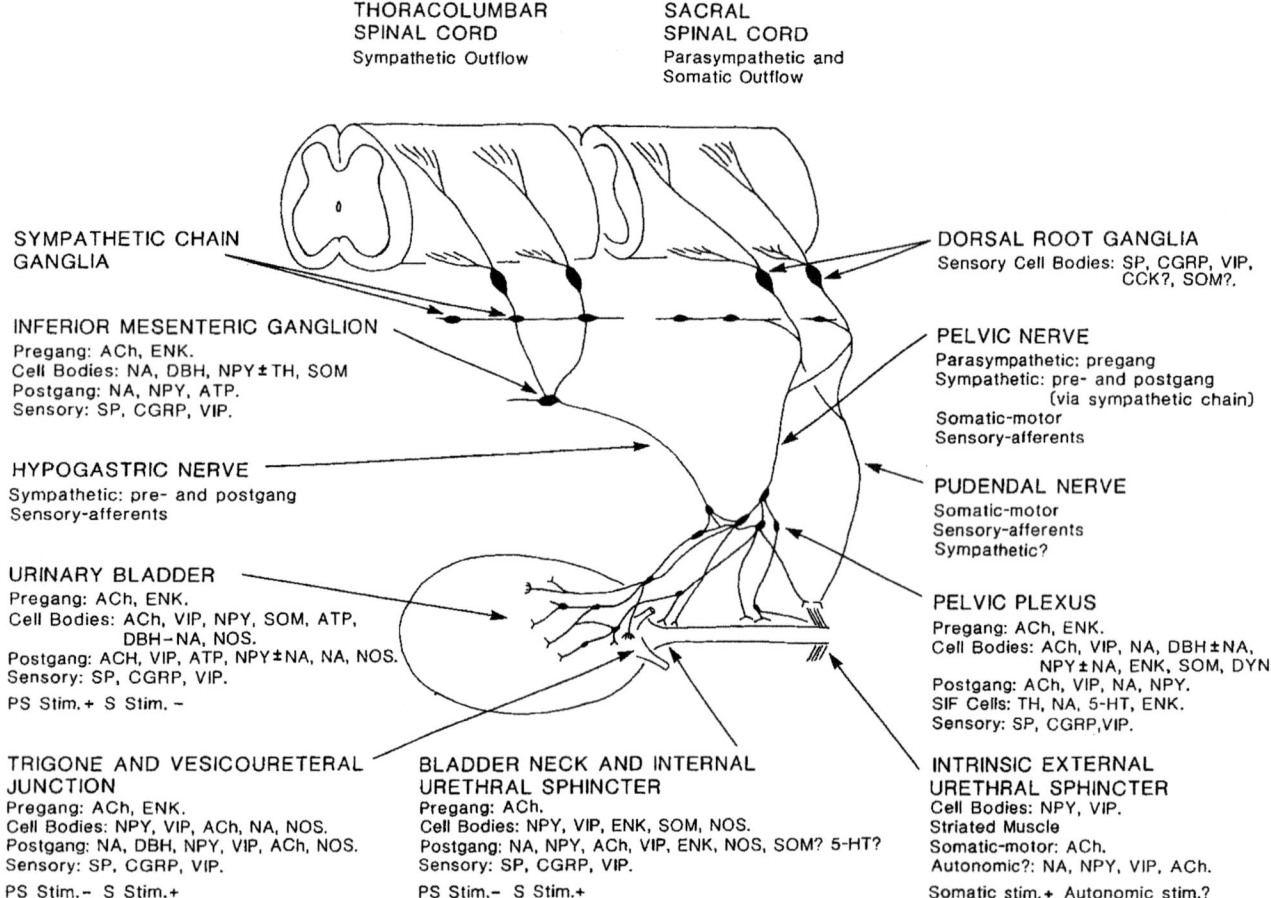

FIGURE 1 Schematic representation of sympathetic (S), parasympathetic (PS), sensory, and somatomotor pathways to the bladder and urethral sphincters, together with some of the neurotransmitters and neuromodulators they contain. The effect of stimulation (stim) of each nerve pathway is shown by the following: +, excitatory response (contraction); − inhibitory response (relaxation). ---, Neuromuscular junction; Λ, sensory endings; ?, findings not yet fully established. ACh, acetylcholine; ATP, adenosine 5′-triphosphate; CCK, cholecystokinin; CGRP, calcitonin gene-related peptide; DBH, dopamine β-hydroxylase; DYN, dynorphin; ENK, enkephalin; 5-HT, serotonin (5-hydroxytryptamine); NA, noradrenaline; NOS, nitric oxide synthase; NPY, neuropeptide Y; SOM, somatostatin; SP, substance P; TH, tyrosine hydroxylase; VIP, vasoactive intestinal polypeptide (reproduced by permission from Lincoln and Burnstock, 1993; © 1993 by Harwood Academic).

pelvic region is the hypogastric nerve, which supplies various pelvic visceral structures with sensory afferent fibers and efferent postganglionic sympathetic fibers, most of which are derived from neuronal cell bodies located in the inferior mesenteric ganglion that had received their preganglionic fibers from lower thoracic and upper lumbar spinal cord segments (T10–L2). However; some parasympthetic fibers also travel in the hypogastric nerve; these fibers, instead of terminating on postganglionic neurons in the inferior mesenteric ganglion, travel through the ganglion to terminate in the pelvic ganglion along with pelvic nerve preganglionic fibers.

The third nerve is the pudendal nerve. Derived from spinal segments S2 (or S1) to S4 in humans (L5, L6, and S1 in rats), it innervates perineal skin (including clitoris, vulva, penis, and scrotum), accessory sex glands associated with the urethra, muscles of the pelvic floor, and the external urethral and anal sphincters with both somatic sensory afferent fibers and somatomotor efferent fibers. The pudendal nerve also innervates the entire perineum with sympathetic postganglionic fibers derived from the sympathetic chain. Most of the research on the anatomy, neurochemistry, physiology, and pelvic organ functions of these nerve fibers has been carried out in rodents and cats.

II. PROBLEMS IN DISSOCIATING INNERVATION PATTERNS OF NERVES SUPPLYING THE PELVIC REGION

Part of the difficulty in understanding pelvic nerve innervation patterns and their functions derives from the situation described previously: The hypogastric nerve contains a substantial proportion of preganglionic fibers that travel through the inferior mesenteric ganglion without synapsing and synapse instead in the pelvic ganglion where they converge with pelvic nerve preganglionic fibers. This situation makes it difficult to assess which organs are served by the pelvic nerve-derived versus the hypogastric nerve-derived postganglionic neurons that exit the pelvic ganglion. This distinction has been made in the past by assuming that cholinergic actions in reproductive and urinary target tissues signify parasympathetic involvement by way of pelvic and hypogastric nerve fibers, whereas adrenergic actions signify sympathetic involvement by way of hypogastric nerve fibers. However, this assumption is now no longer valid because of the existence of noncholinergic–nonadrenergic mechanisms, involving numerous purines and peptides, not only within the pelvic ganglion and other autonomomic ganglia in the pelvic region but also within the peripheral pelvic reproductive (and other) organs where the postganglionic fibers exert their effects (Fig. 1). This situation is even more complicated for the gastrointestinal system because it has its own extensive internal neural supply (i.e., the enteric nervous system) that is only partially influenced by activity in autonomic fibers.

Another factor that hinders understanding of pelvic nerve innervation patterns and functions is that as the pelvic nerve afferent and efferent fibers course toward the periphery, they are variously accompanied by both hypogastric and pudendal nerve fibers. Thus, as the pelvic nerve fibers exit the spinal cord and head toward the pelvic ganglion, they are first joined by fibers that will form the pudendal nerve. As the pudendal nerve fibers then branch away toward their perineal targets, the pelvic nerve fibers continue caudally to join the middle hypogastric plexus of nerve fibers (not shown in Fig 1), where at this level the entire group of fibers is also called "the hypogastric nerve". Moving further caudally, the two groups of fibers then travel together through the inferior hypogastric plexus (not shown in Fig. 1). This plexus contains numerous small ganglia where some of the preganglionic fibers of both the pelvic and the hypogastric nerves synapse, a feature that further complicates the innervation pattern described previously and shown in Fig. 1. As this new mix of hypogastric nerve pre- and postganglionic (T10–L2 derived) and pelvic nerve pre- and postganglionic fibers (S2–S4 derived) continues caudally, most of the remaining preganglionic pelvic nerve fibers then synapse on postganglionic neurons located the pelvic ganglion. The remainder of the preganglionic fibers continue through the ganglion to synapse on postganglionic neurons located in small ganglia embedded in the walls of various pelvic viscera. This anatomical situation has not surprisingly led to considerable confusion in nomenclature.

III. ORGANS INNERVATED BY THE PELVIC NERVE

A. Afferent Innervation

Given this intertwining of nerve fibers, particularly those of the pelvic and hypogastric nerves on their course to and from the periphery and the problems of distinguishing them by the biochemical mechanisms of their postganglionic neurons, it has been difficult to sort out which pelvic structures each nerve innervates. One strategy has been to use various anatomical labeling techniques, mainly in rodents and cats, to determine the innervation targets of the afferent fibers that travel in each.

The first step is to apply the labeling techniques to the pelvic nerve proper, i.e., that portion of the nerve closest to the spinal cord where it is composed of all the afferent fibers (along with all the efferent preganglionic fibers), so as to identify the dorsal root ganglia and dorsal roots through which all the pelvic nerve afferent fibers enter the spinal cord. Using this strategy, the pelvic nerve was found in both male and female rats to label neurons in the rat L6–S2 dorsal root ganglia (which would be the S2–S4 ganglia in humans). Of importance, however, is that these methods have also shown that, while most of the primary afferent fibers entering at L6–S2 segments form synapses on neurons in those spinal segments, branches of the primary afferents extend rostrally and caudally within the white matter of the cord to terminate in distant segments, with some extending even as far as the cervical spinal cord.

The next step is to determine the peripheral source of the L6–S2 primary afferent fibers either by using axonal labeling techniques or by stimulating the pelvic nerve electrically to produce plasma extravasation of intravenously applied Evans blue dye, thereby revealing the peripheral targets of the afferent fibers. Although there appear to be important species differences, these determinations have shown that the pelvic nerve afferents are composed mainly of A and C nerve fibers that supply the bladder, urethra, distal sigmoid colon, and anorectum in both males and females, with the main entrance of the urinary afferents located more rostrally than that of the gastrointestinal afferents. Other A and C fiber afferents in the pelvic nerve innervate the vagina and cervix in female, concentrated heavily in the cervix–vaginal area. Less work has been carried out on pelvic nerve afferents from the male reproductive organs with these methods, but so far the evidence suggests that they probably supply the prostate, seminal vesicles, vas deferens (uncertain), penis, epididymis (possibly), and bulbourethral accessory sex glands.

B. Efferent Innervation

Given the current knowledge of the afferent innervation territory of pelvic nerve fibers, one might wish to assume that the efferent targets are the same. However, in part because of the convergence of hypogastric and pelvic preganglionic fibers within the pelvic ganglion, this assumption is unlikely to be valid. One possible means of identifying the efferent targets might be to stimulate the pelvic nerve electrically and observe the effects of stimulation on the peripheral organs, but here again, unfortunately, converging influences within the pelvic ganglion and in the periphery confound interpretations. On the other hand, one recent clever anatomical approach that may produce better answers involves using a combination of retrograde transneuronal and nontransneuronal labeling (tracing) methods. Here, preganglionic neurons in the spinal cord that give rise to all the pelvic nerve fibers are identified by applying nontransganglionic tracer to the pelvic nerve (as was done to identify the afferents) while the peripheral targets of those preganglionic neurons are simultaneously identified by applying a differently colored transneuronal tracer into potential target organs (see Papka et al., 1995, for an example). Currently, however, the precise efferent targets of the pelvic nerve remain uncertain.

IV. FUNCTIONS OF THE PELVIC NERVE

Much of the information about functions of the pelvic nerve (i.e., of the preganglionic fibers traveling toward the pelvic ganglion and afferent fibers from the targets mentioned previously) is derived from either the effects of pelvic neurectomy or electrical stimulation, carried out experimentally in animals or clinically in humans (Fig. 1). There are obvious interpretative problems associated with such methods, and these problems are further complicated by the confluence, (discussed previously) of hypogastric and pelvic nerve influences not only within the pelvic ganglion but also within their peripheral targets (where there are additional interactions in some regions with fibers of the pudendal nerve; e.g., external anal and urethral sphincters and some accessory sex glands). Nevertheless, some general conclusions are possible.

One generality is that the pelvic nerve serves functions within the three physiological domains contained within the pelvis; that is, urological, gastrointestinal, and reproductive. Many details of these functions (presented briefly below) may be found in the references cited in the Bibliography. Currently, the pelvic nerve provides information to and from the central nervous system that, in cooperation with the other two nerves, is important for (i) coordinating actions via spinal reflexes and supraspinal loops of the detrusor muscle of the bladder and the bladder outlet (mainly via the hypogastric nerve) and the external urethral sphincter (mainly via the pelvic and pudendal nerves) for bladder continence and micturition; (ii) coordinating actions via spinal reflexes and supraspinal mechanisms between the sigmoid colon and anorectum for defecation (during which the pelvic nerve contracts the colon/anorectum and relaxes the internal anal sphincter); (iii) coordinating many functions associated with vaginal and/or cervix stimulation important for events during the ovarian cycle, pregnancy, and parturition, such as changes in vaginal tone (increases or decreases), facilitation of body positional adjustments during coitus, changes in cervix composition and cervix secretions, changes in blood pressure, and initiation of the hormonal conditions of pregnancy (in rats); (iv) participating along with fibers in the pudendal and hypogastric nerve to coordinate smooth muscle contractile activity associated with secretions from the prostate (and seminal vesicles) and changes in penile vascular permeability and pressures, all important for sperm movements and both erectile (penile tumescence) and ejaculatory and orgasmic functions; (v) providing afferent information which, in both pathophysiological and healthy circumstances, is relevant to pain, analgesia, other sensations (e.g., vaginal orgasm) and vaginal tone changes associated with stimulation of the vagina and cervix (little is known about the sensory consequences of such afferent information arriving from the male reproductive organs), as well as pain and other sensations (e.g., fullness) associated with stimulation of the sigmoid colon and anorectum and the bladder and urethra in both sexes.

V. SUMMARY AND FUTURE ISSUES

Pelvic nerve afferent fibers provide sensory information to the central nervous system from caudal components of the female internal reproductive tract (vagina and cervix) as well as from the bladder and urethra and the sigmoid colon and anorectum in both sexes (less is known about afferents from the male reproductive organs). These afferent fibers synapse mainly on neurons in L6–S1 segments of the rat (S2–S4 in humans), with branches extending many segments rostrally and caudally. Pudendal nerve afferents arrive at the same segments (with similar distant branching patterns within the cord), but their innervation territory is mainly of somatic structures (skin and muscle) of the perineum, including clitoris, vulva, scrotum, penis, and external urethral and anal sphincters. The hypogastric afferent fibers provide sensory information to thoracolumbar spinal segments from components of the reproductive and urinary tracts (uterus, testis, epididymis, and ureter) rostral to those supplied by the pelvic afferents as well as from some of the same components (cervix and bladder). On the other hand, there is nearly complete overlap of the pelvic and hypogastric afferents from the anorectum.

This afferent innervation pattern generates a situation in which primary afferents arriving from the uterus, cervix, testis, epididymis bladder, and ureter by way of the hypogastric nerve are concentrated in thoracolumbar spinal cord segments (mainly T10–L3) that are separated considerably from the spinal segments (S2–S4 in humans; L6–S1 in rats) that receive afferents by way of the pelvic nerve from the cervix, vagina, bladder, urethra, urethral accessory sex glands, vas deferens, seminal vesicles, prostate, and anorectum and by way of the pudendal nerve from the vulva, clitoris, penis, scrotum, accessory sex glands, and urethral and anal sphincters. The significance of this segmental separation for intraspinal and supraspinal intercoordination and regulation of pelvic organ sensations and the many functions described previously, especially in light of the convergence of the efferent hypogastric nerve and pelvic/pudendal nerve influences exiting from the separated segments on postganglionic neurons supplying all these regions, is poorly understood. This situation becomes relevant and important, of course, in cases of spinal injury as well as in many other conditions of spinal and pelvic pathophysiology. Research on these questions is only in its infancy, in large part because most of the studies so far have focused on peripheral mechanisms of control.

Also poorly understood is the significance of the extensive branching and divergence of primary afferent fibers after entering the spinal cord. Recent evidence indicates another emerging complication: The high probability that the vagus nerve innervates many pelvic organs (e.g., cervix, uterus, and anorectum), adding yet another confounding neural influence to the mix of pelvic, hypogastric, and pudendal innervation. Even less well understood are the plastic changes in neural function (neuronal plasticity) that are induced by pathophysiology of peripheral tissues, peripheral nerves (e.g., diabetes and nerve injury), and the central nervous system (e.g., strokes and central neurodegenerative diseases such as multiple sclerosis and parkinsonism).

One strategic situation that hinders progress in our understanding is that information about functions within the domains of the reproductive, urinary, and gastrointestinal systems is largely separated in the experimental and clinical literature (to wit, this encyclopedia), whereas for the individual organism information is intricately interlinked (e.g., pelvic visceral pain, changes in bowel activity during micturition, and changes in both bowel and urinary activity during coitus). More cross-talk is needed between both scientists and clinicians in these three arenas. Furthermore, huge sex differences in reproductive structures create huge differences in the relative arrangements of various pelvic and abdominal organs and therefore in their interacting controls. Additional sex differences in supraspinal and intraspinal processing, the existence of which has been ignored until recently, are also beginning to be recognized. A more active pursuit of this issue is needed; comparison of male and female mechanisms can enhance information pertinent to both sexes.

I hope that the next edition of this encyclopedia will provide more information in all the following arenas: (i) supraspinal/intraspinal coordination mechanisms; (ii) the significance of primary afferent divergence; (ii) the significance of vagal innervation; (iv) the afferent innervation patterns of male reproductive organs; (v) pathophysiologically induced plastic changes within both the central nervous system and the periphery; (vi) cross-linkages between reproductive, gastrointestinal, and urinary domains: and (vii) sex differences.

See Also the Following Articles

Erection; Male Reproductive System, Human; Ovarian Innervation

Bibliography

Berkley, K. J., and Hubscher, C. H. (1995). Visceral and somatic sensory tracks through the neuroaxis and their relation to pain: Lessons from the rat female reproductive system. In *Visceral Pain, Progress in Pain Research and Management* (G. F. Gebhart, Ed.), Vol. 5, pp. 195–216. IASP Press, Seattle

Bonica, J. J. (1990). General considerations of pain in the pelvis and perineum. In *The Management of Pain* (J. J. Bonica, Ed.), 2nd ed, pp. 1283–1312. Lea & Febiger, Philadelphia

Dail, W. G. (1993). Autonomic innervation of male reproductive genitalia. In *Nervous Control of the Urogenital System* (C. A. Maggi, Ed.), pp. 69–101. Harwood Academic, The Netherlands

De Groat, W. C., and Booth, A. M. (1993). Neural control of penile erection. In *Nervous Control of the Urogenital System* (C. A. Maggi, Ed.), pp. 467–524. Harwood Academic, The Netherlands.

De Groat, W. C., Booth, A. M., and Yoshimura, N. (1993). Neurophysiology of micturition and its modification in animal models of human disease. In *Nervous Control of the Urogenital System* (C. A. Maggi, Ed.), pp. 227–290. Harwood Academic, The Netherlands.

Gee, W. F., and Ansell, J. S (1990). Pelvic and perineal pain of urological origin. In *The Management of Pain* (J. J. Bonica, Ed.), 2nd ed, pp. 1368–1382. Lea & Febiger, Philadelphia.

Jänig, W., and McLachlan, E. M. (1987). Organization of lumbar spinal outflow to distal colon and pelvic organs. *Physiol. Rev.* **67**, 1332–1403.

Komisaruk, B. R., and Whipple, B. (1995). The suppression of pain by genital stimulation in females. *Annu. Rev. Sex Res.* **6**, 151–186.

Lincoln, J., and Burnstock, G. (1993). Autonomic innervation of the urinary bladder and urethra. In *Nervous Control of the Urogenital System* (C. A. Maggi, Ed.), pp. 33–68. Harwood Academic, The Netherlands.

McKenna, K. E., and Marson, L. (1997). Spinal and brainstem control of sexual function. In *Central Nervous Control of Autonomic Function* (D. Jordan, Ed.), pp. 151–187. Harwood Academic, The Netherlands.

Newton, B. W., and Hammill, R. W. (1996). Sexual differentiation of the autonomic nervous system. In *Autonomic–Endocrine Interactions* (K. Unsicker, Ed.), pp. 425–463. Harwood Academic, The Netherlands.

Papka, R. E., and Traurig, H. H. (1993). Autonomic efferent and visceral sensory innervation of the female reproductive system: Special reference to neurochemical markers in nerves and ganglionic connections. In *Nervous Control of the Urogenital System* (C. A. Maggi, Ed.), pp. 423–466. Harwood Academic, The Netherlands.

Papka, R. E., McCurdy, J. R., Williams, S. J., Mayer, B., Marson, L., and Platt, K. B. (1995). Parasympathetic preganglionic neurons in the spinal cord involved in uterine innervation are cholinergic and nitric oxide-containing. *Anat. Rec.* **241**, 554–562.

Rose, J. D. (1990). Forebrain influences on brainstem and spinal mechanisms of copulatory behavior: A current perspective on Frank Beach's contribution. *Neurosci. Biobehav. Rev.* **14**, 207–215.

Smith, T. K., and Sanders, K. M. (1995). Motility of the large intestine. In *Textbook of Gastroenterology* (T. Yamada, Ed.), 2nd ed., pp. 234–261. Lippincott, Philadelphia.

Pelvimetry

Samuel Parry and Mark A. Morgan
University of Pennsylvania Medical Center

I. Pelvic Anatomy: Obstetric Considerations
II. Methodologies for Measuring Maternal Pelvic Diameters and Fetal Indices Associated with Fetal–Pelvic Disproportion
III. Defining the Appropriate Population to Screen: Risks versus Benefits
IV. Conclusions

GLOSSARY

Colcher–Sussman technique (of X-ray pelvimetry) Standard two-image X-ray pelvimetry technique, utilizing a lateral image to measure the anteroposterior diameters of the pelvic inlet and midpelvis and an axial image to measure the transverse diameters of the pelvic inlet and midpelvis.

diagonal conjugate The distance between the promontory of the sacrum and the lower margin of the pubic symphysis which can be measured clinically.

dystocia Difficult labor and delivery, generally caused by an abnormality of the mother or fetus or fetal–pelvic disproportion.

fetal–pelvic disproportion Condition in which the relative size of the fetus exceeds maternal pelvic diameters, often precluding vaginal delivery and associated with increased risks of labor dystocia and birth trauma.

fetal–pelvic index A method that combines maternal X-ray pelvimetry with fetal sonography to predict fetal–pelvic disproportion.

interspinous diameter The transverse diameter of the midpelvis, between the ischial spines, that represents the narrowest pelvic diameter.

Leopold's maneuvers Four maneuvers performed during abdominal examination to determine fetal position: the first to determine which fetal part is present at the fundus; the second to evaluate the location of the fetal back and small parts; the third to palpate the presenting part above the pubic symphysis; and the fourth to determine the direction and degree of flexion of the fetal head.

midpelvis One of two clinically important pelvic planes which is bounded by the sacrum posteriorly, the inferior margins of the ischial spines laterally, and the lower margin of the pubic symphysis anteriorly.

obstetric conjugate The shortest anteroposterior diameter of the pelvic inlet, estimated by measuring the diagonal conjugate and subtracting 2 cm from the result.

pelvic inlet One of two clinically important pelvic planes that is bounded by the sacral promontory posteriorly, the linea terminalis laterally, and the horizontal rami of the pubis anteriorly.

pelvic shapes Caldwell and Moloy classification: android, the male type of pelvis, with a wedge-shaped inlet, convergent pelvic sidewalls, prominent ischial spines, and a shortened posterior sagittal diameter at the inlet; anthropoid, pelvic type that is elongated anterioposteriorly at the inlet; gynecoid, the so-called normal female type of pelvis, found in almost 50% of women—the sidewalls of the gynecoid pelvis are straight, the pubic arch is wide, and the transverse diameter at the ischial spines exceeds 10 cm; platypelloid, pelvic type that is elongated in tranverse diameter at the inlet.

pelvimetry Measurement of the diameters of the pelvis: clinical pelvimetry, estimation of the length of the diameters of the pelvis obtained by pelvic examination; radiographic pelvimetry, measurement of pelvic diameters by roentgenogram, computerized tomography, or magnetic resonance imaging.

Obstetricians have understood for many centuries that successful vaginal delivery is dependent on the relative size of the fetus and the maternal pelvis. The concept of fetal–pelvic disproportion was originally proposed by Savonarola in 1560. Subsequently, a great deal of data have accumulated that demonstrate increased maternal and perinatal morbidity and mortality for those patients in whom fetal–pelvic disproportion is present. Therefore, the purpose of present-day pelvimetry, which encompasses the manual and radiographic measurements of the diameters of the female pelvis, is to identify those patients at greatest risk for fetal–pelvic disproportion. In order to accurately predict fetal–pelvic disproportion, an assessment of fetal size must be included.

I. PELVIC ANATOMY: OBSTETRIC CONSIDERATIONS

The adult pelvis is formed by the fusion of four bones: the sacrum, the coccyx, and the two innominate bones (which in turn are formed by the fusion of the ilium, the ischium, and the pubis) (Fig. 1). The linea terminalis, which demarcates the line of fusion between the iliac and ischial segments of the innominate bone, separates the false pelvis (above) from the true pelvis (below). The size of the false pelvis varies considerably in women without obstetric significance. However, two clinically relevant planes within the true pelvis determine the space

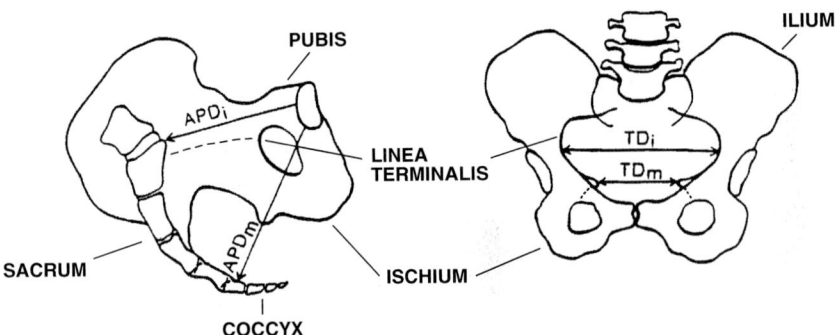

FIGURE 1 Anteroposterior and transverse diameters of the pelvic inlet and midpelvis are illustrated. APDi, anteroposterior diameter of the pelvic inlet; APDm, anteroposterior diameter of the midpelvis; TDi, transverse diameter of the pelvic inlet; TDm, transverse diameter of the midpelvis (modified with permission from Thurnau et al., 1992).

limitations through which the fetus must pass during parturition. The plane of the pelvic inlet, through which the fetal vertex must pass for head engagement, is bounded by the promontory of the sacrum posteriorly, the linea terminalis laterally, and the horizontal rami of the pubis anteriorly. The shape of the pelvic inlet has been classified into four types by Caldwell and Moloy, although intermediate types of pelves are much more frequent than pure types (Fig. 2). The second plane of clinical importance is that of the midpelvis, which is bounded by the sacrum posteriorly, the inferior margins of the ischial spines laterally, and the lower margin of the pubic symphysis anteriorly. The midpelvis is of particular importance because the transverse interspinous diameter is usually the smallest diameter of the pelvis.

In 1948, Mengert reported to the American Medical Association his findings regarding the average values of essential pelvic measurements in 935 women at Parkland Hospital (Table 1). By a process of trial and error, Mengert determined that 85% of normal capacity (the product of the transverse and anteroposterior diameters) of either inlet or midplane represented the threshold between adequacy and contraction. Pelvic contracture (<85% normal capacity) was strongly associated with the need for operative vaginal and cesarean delivery. Based on Mengert's findings, critical limit values for pelvic measurements have been described and are listed in Table 1.

II. METHODOLOGIES FOR MEASURING MATERNAL PELVIC DIAMETERS AND FETAL INDICES ASSOCIATED WITH FETAL–PELVIC DISPROPORTION

A. Clinical Pelvimetry

Valuable information concerning the obstetric capacity of the bony pelvis may be obtained by physical

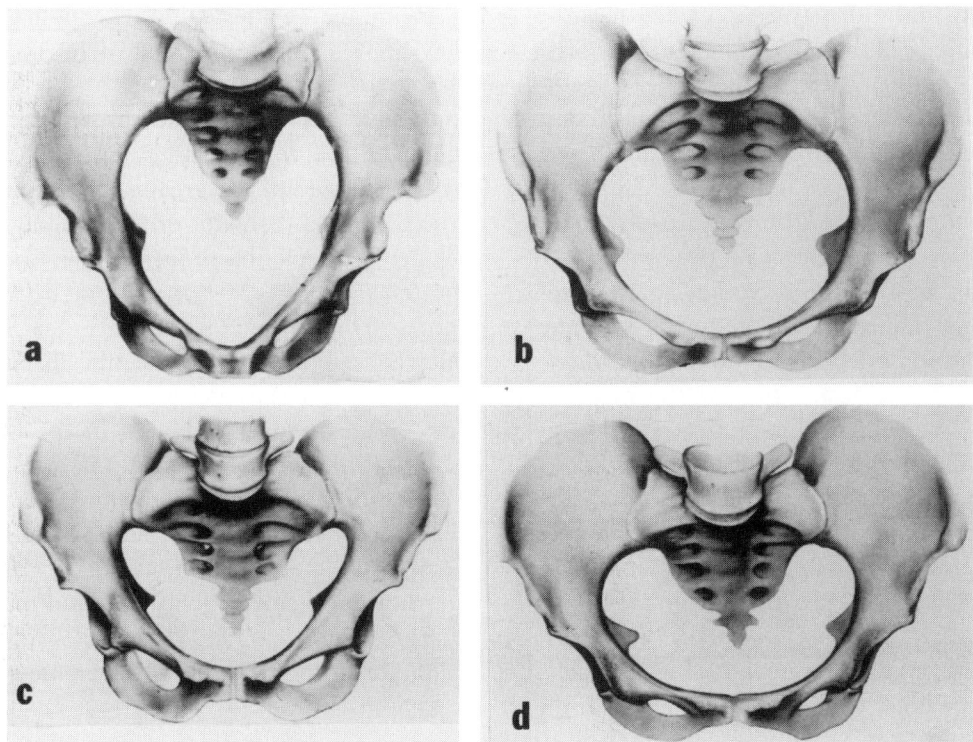

FIGURE 2 Caldwell–Moloy classification of pelvic types. a, anthropoid pelvis; b, gynecoid pelvis; c, android pelvis; d, platypelloid pelvis. It should be noted that intermediate or mixed types of pelves are much more frequent than pure types (reproduced with permission from Danforth, 1986).

TABLE 1
Essential Pelvic Measurements

	Average values (n = 935)[a] (cm)	85% of average values[a] (cm)	Commonly used critical values[b] (cm)
Pelvic inlet			
Transverse diameter	12.5	10.6	12
AP[c] diameter	11.6	9.9	10
Total	24.1	20.5	22
Midpelvis			
Transverse diameter	10.3	8.8	10
AP[c] diameter	12.1	10.3	10
Total	22.4	19.1	20

[a] Based on data originally reported by Mengert (1948).
[b] Values adopted from Mengert's data (85% of average) (O'Brien and Cefalo, 1996).
[c] AP, anteroposterior.

examination. The most easily obtained index of the pelvic inlet is the diagonal conjugate, which is an indirect measure of the obstetric conjugate (the anteroposterior diameter of the pelvic inlet; Fig. 1). Pelvic examination is performed with the patient in the dorsal lithotomy position, with the buttocks at the end of the operating table and the hips and knees fully flexed with feet strapped in position. The examiner measures the distance from the tip of his/her finger (which is palpating the sacral promontory) to the point on the examining hand that contacts the symphysis pubis. The obstetric conjugate is calculated by subtracting 2 cm from the value of the diagonal conjugate. The critical value for the obstetric conjugate is 10 cm (Table 1); hence, the critical value for the diagonal conjugate is 12 cm.

Direct measurement of the midpelvis is not possible by clinical pelvimetry, although the prominence of the ischial spines should be noted, and the experienced obstetrician may estimate the interspinous diameter. Importantly, the interspinous diameter is the transverse diameter of the midpelvis (Fig. 1) and is the narrowest diameter that must be traversed by the fetus during parturition. A value exceeding 9 or 10 cm suggests adequate size for vaginal delivery of a normal sized fetus.

The obstetrician may describe other features of the true pelvis, particularly the capacity of the posterior pelvis, which is the distinguishing feature of favorable versus unfavorable pelvic types (Caldwell and Moloy classification; Fig. 2). The space within the posterior pelvis is best defined by the posterior sagittal diameter, the distance from the sacral promontory to the greatest transverse diameter of the pelvis. Unfortunately, the posterior sagittal diameter cannot be directly measured by clinical or X-ray pelvimetry. At the pelvic outlet, the capacity of the posterior pelvis can be inferred from the biischial diameter. With the patient in the dorsal lithotomy position, the ischial tuberosities are palpated and the distance between them is measured, usually with a closed fist. The critical value is 8 cm per Mengert's data (not shown in Table 1). Other characteristics of the maternal pelvis that may be evaluated include the angulation of the pubic rami beneath the pubic arch and the curvature of the sacrum and coccyx. Collectively, these measurements and qualitative assessments supply significant information regarding the obstetric capacity of the maternal pelvis, but clinical pelvimetry does not provide information about fetal size.

Abdominal examination (Leopold's maneuvers) may be used to estimate fetal size, and engagement should be assessed to utilize the fetal head as an internal pelvimeter and demonstrate that the pelvic inlet is ample for that fetus. Engagement is defined by the passage of the greatest fetal cranial diameter (the biparietal diameter of the flexed head) through the pelvic inlet and is thought to have occurred when the fetal vertex is stationed at or below the ischial spines. The distance from the ischial spines to the inlet is approximately 5 cm; the distance from the vertex to the biparietal diameter is generally 3 or 4 cm. Thus, when engagement of the fetal head has occurred, the vertex is usually palpable at or below the ischial spines. Although the incidence of pelvic contracture is probably higher in primigravidas with unengaged fetal heads at the onset of labor, assessment of engagement, fetal size, and maternal pelvic characteristics cannot conclusively predict fetal–pelvic disproportion.

B. X-Ray Pelvimetry

The Colcher–Sussman technique of X-ray pelvimetry requires two films, one taken with the patient laterally positioned to obtain anteroposterior diameters of the inlet and midpelvis and the other taken with the patient in a dorsal supine position to obtain transverse diameters of the inlet and midpelvis. With each view, an appropriately positioned ruler is included on the respective radiograph to compensate for magnification. This technique exposes the fetus to approximately 0.08 rad of radiation, or 0.04 rad per film.

Mengert validated the accuracy of X-ray pelvimetry by evaluating interobserver variation and by direct measurement with calipers and a ruler during the course of gynecologic surgery. He concluded that no errors >0.5 cm were found. Later investigators have confirmed the accuracy of X-ray pelvimetry using various models.

Pelvic adequacy is determined by comparing the sum of the patient's anteroposterior and transverse diameters at the inlet and midpelvis with critical values based on Mengert's findings (85% of average values in his series from Parkland Hospital). The critical values described in most obstetrics textbooks slightly overestimate Mengert's raw data (Table 1). At the inlet, the critical value (the sum of the anteroposterior and transverse diameters) is 22 cm; at the midpelvis, the critical value is 20 cm. Pelvic measurements which exceed these critical values describe pelves that are adequate for delivery of a normal-sized neonate. Unfortunately, no estimation of fetal size is determined by X-ray pelvimetry.

C. Computed Tomography

Since Federle's original description of computed tomography (CT) pelvimetry in 1982, many institutions have preferred CT pelvimetry to conventional X-ray pelvimetry when considering accuracy and radiation dose to the fetus. Using anteroposterior and lateral scout films, Federle and subsequent investigators estimated the total fetal radiation dose with CT pelvimetry to be 0.04 rad. An additional axial section at the level of the ischial spines is required to measure the interspinous diameter. The axial CT section produces an absorbed dose of 0.38 rad; however, the fetus may be shielded because of the low position at which the CT scanner is directed.

Recently, an Australian group criticized other investigators for comparing state-of-the-art CT pelvimetry with outmoded X-ray pelvimetry techniques. These investigators concluded that both CT and X-ray pelvimetry have similar accuracy and radiation doses. Additionally, they reported novel CT techniques, utilizing low radiographic factors for their axial sections and eliminating the axial scan altogether in 67% of their patients, to minimize fetal radiation exposure. When transverse and anteroposterior diameters of the inlet and midpelvis could be measured by scout films only (67% of their patients), the fetal radiation dose was 0.04 rad. In patients requiring an axial section to measure the transverse diameter of the midpelvis, the total fetal radiation exposure was 0.14 rad. Another group, at the University of Virginia, reduced the radiation dose to 0.06 rad by angling the axial section to exclude the fetal vertex.

In summary, CT pelvimetry does not seem to offer any significant advantage when compared to conventional X-ray pelvimetry. In most institutions the costs are similar, and each method has acceptable accuracy and radiation exposure. CT pelvimetry may be disadvantageous at institutions that are not expert at modifying the standard axial section and at institutions in which CT pelvimetry is not readily available on a 24-hr basis.

D. Magnetic Resonance Imaging

The use of magnetic resonance imaging (MRI) has been shown to be safe during pregnancy and compares favorably with X-ray pelvimetry. Anteroposterior and transverse diameters of the inlet and midpelvis can be measured using sagittal and transverse images. Tukeva and associates reported that MRI scanning time was less than 6 min and was well accepted by their patients. MRI and X-ray pelvimetry measurements were closely correlated. Because the fetus is not exposed to ionizing radiation, MRI is an attractive method for pelvimetry.

In addition to imaging the maternal pelvis, novel fast-scan MRI techniques which limit artifact caused by fetal movement have been utilized to image the fetus *in utero*. Fetal MRI is gaining widespread acceptance as an adjunctive technique to ultrasound for diagnosing fetal anomalies. MRI also has been used to accurately measure fetal shoulder width and biparietal diameter. Although no one has used MRI to compare maternal and fetal measurements, the ability of MRI to image bony structures and soft tissues in the mother and fetus suggests that MRI may be an ideal method for predicting fetal–pelvic disproportion. Currently, however, the prohibitive cost of MRI and its limited availability temper enthusiasm for this method of pelvimetry.

E. Fetal–Pelvic Index

The fetal-pelvic index combines maternal X-ray pelvimetry (Colcher–Sussman technique) with fetal sonographic biometry measurements to predict fetal–pelvic disproportion. The theoretic advantage of the fetal–pelvic index is that it compares maternal pelvic measurements with fetal size to determine if the mother's pelvis is adequate to deliver that particular fetus. Because most macrosomic fetuses can be delivered vaginally and most cases of shoulder dystocia and/or fetal–pelvic disproportion occur in the noncontracted maternal pelvis, information regarding fetal biometry or maternal pelvic diameters without measuring the other parameter cannot accurately predict fetal–pelvic disproportion.

The fetal–pelvic index was originally described by Morgan, Thurnau, and associates at the University of Oklahoma in 1986. Measurement of fetal head (HC) and abdominal circumferences (AC) are obtained by standard ultrasound techniques. Anteroposterior and transverse diameters of the maternal pelvic inlet and midpelvis are measured according to the X-ray pelvimetry technique described by Colcher and Sussman. The circumferences of the inlet and midpelvis are calculated by multiplying the sum of the diameters of each plane by one-half. Four circumference differences between the fetus and maternal pelvis (HC − maternal inlet circumference, HC − maternal midpelvis circumference, AC − maternal inlet circumference, and AC − maternal midpelvis circumference) are calculated. The differences are negative values when maternal circumferences are larger than fetal HC or AC but positive values when fetal circumferences exceed maternal circumferences. The fetal–pelvic index is derived from the sum of the two most positive circumference differences (Fig. 3). A positive fetal–pelvic index identifies a fetus that is larger than the maternal pelvis and predicts fetal–pelvic disproportion. A negative fetal–pelvic index value identifies a fetus that is smaller than the maternal pelvis and predicts the absence of fetal–pelvic disproportion.

Morgan, Thurnau, and associates initially used the fetal–pelvic index to predict the incidence of fetal–pelvic disproportion and delivery outcome (cesarean section vs vaginal delivery) in patients whose physi-

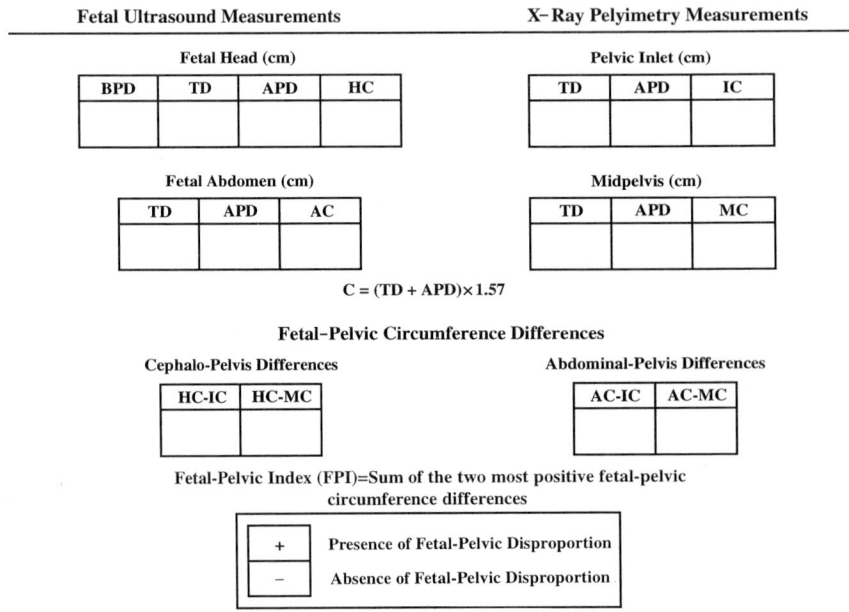

FIGURE 3 Fetal–pelvic index form (reproduced with permission from Thurnau *et al.*, 1992).

cians were blinded to the fetal–pelvic index values. Subsequently, the same methodologies were employed to determine the predictive value of the fetal–pelvic index in higher risk groups, including patients carrying macrosomic fetuses (>4000 g), patients requiring induction of labor, patients requiring augmentation of labor, patients attempting vaginal birth after a previous cesarean delivery, and nulliparous patients. In their six reports, 406 patients were studied. Overall, the sensitivity of the fetal–pelvic index is 0.80, its specificity is 0.98, its positive predictive value is 0.97, and its negative predictive value is 0.86. Thirty-seven patients with negative fetal–pelvic indices required cesarean delivery (false negatives), but among these, 34 fetuses presented with persistent occiput posterior position. When these cases are considered as a special category, the overall negative predictive value is 0.98. Of course, these numbers compared much more favorably than the techniques of using fetal weights or X-ray pelvimetry alone to predict fetal–pelvic disproportion.

Despite the obvious advantages of directly comparing maternal pelvic circumferences with fetal circumferences in order to predict fetal–pelvic disproportion, there are several criticisms of the fetal–pelvic index that merit consideration. Fetal biometric measurements are difficult to obtain in the third trimester, and the accuracy of the fetal–pelvic index has not yet been reproduced by clinicians at other institutions. Additionally, as with other techniques utilizing X-ray pelvimetry, maternal soft tissue dystocias cannot be evaluated. Finally, in the case of the nulliparous patient whose fetus' head is already engaged, inclusion of the fetal HC − maternal inlet circumference seems superfluous.

III. DEFINING THE APPROPRIATE POPULATION TO SCREEN: RISKS VERSUS BENEFITS

Clinical pelvimetry, which entails the assessment of maternal pelvic adequacy by physical examination, in combination with abdominal examination to estimate fetal size and presentation (Leopold's maneuvers), provides useful information for the obstetrician in making labor management decisions. Because these findings may be obtained without risk to the patient, clinical pelvimetry and Leopold's maneuvers should be performed in every laboring patient. However, the accuracy of physical examination in predicting fetal–pelvic disproportion has not been evaluated in a prospective fashion, and the decision to perform cesarean delivery based on clinical pelvimetry and Leopold's maneuvers must be individualized.

The primary concern governing the use of radiographic pelvic measurement is the potential radiation hazard to the fetus. Third-trimester exposure to radiation has been associated with an increased risk of leukemia and other childhood cancers. Collectively, retrospective data demonstrate approximately a 50% increased risk for childhood leukemia in infants exposed to ionizing radiation antenatally. Among U.S. Caucasian children not exposed to radiation before birth, the risk of childhood leukemia is 1:2880 per 10 years. Among those children exposed to X-ray pelvimetry *in utero*, the risk is 1:2000, with a relative risk of 1.4. Harvey *et al.* compared the rates of childhood cancers in twins with prenatal exposure to X rays versus unexposed twins. Rather than using studies in singletons that evaluate medical complications that may have predisposed the infants to cancer, twins were selected because their exposure to X rays likely resulted from studies to diagnose twin gestation or fetal presentation in labor. After adjusting for low birth weight, which was associated with cancer risk, the relative risk for all childhood cancers in exposed twins was 2.4 (95% confidence interval, 1.0–5.9). The relative risk for leukemia was 1.6 (95% confidence interval, 0.4–6.8). The estimated X-ray dose to the exposed twins ranged from 0.16 to 4 rad, with an average dose of 1 rad. According to Colcher and Sussman, however, the standard two-film pelvimetry technique exposes the fetus to only 0.08 rad. Thus, the data that associate X-ray pelvimetry with childhood cancer are limited by their retrospective nature, their inclusion of fetuses exposed to doses of radiation not typically associated with X-ray pelvimetry, and, in many cases, their inability to achieve statistical significance.

If performed accurately, X-ray pelvimetry may benefit the patient by reducing the incidence of complications associated with fetal–pelvic disproportion. The Collaborative Perinatal Study (1972) documented a perinatal death rate of 17.9 per 1000 in the presence of labor dystocia. Furthermore, a prolonged second stage of labor has been associated with increased perinatal morbidity. These adverse outcomes appear to be much more frequent than childhood cancers.

Although the risk–benefit profile favors the use of X-ray pelvimetry, these studies are inappropriate if they are not used to guide clinical management. Over the past 20–30 years, many authorities have advocated a trial of labor for all patients. In 1979, the American Colleges of Radiology and Obstetrics and Gynecology each recognized that X-ray pelvimetry is not a prerequisite to clinical decisions concerning obstetric management but instead should be requested only when the physician believes that X-ray pelvimetry will contribute to decision making on an individual basis. Our challenge, then, is to identify which patients X-ray pelvimetry may benefit. Mengert enumerated criteria of suspicion indicating the need for employment of radiographic pelvimetry. These included patients with a history of difficult labor, those with abnormal findings by clinical pelvimetry, and primigravidas presenting at term with nonengagement of the fetal head. Another indication not elaborated by Mengert is the woman with a previous injury or disease likely to affect the bony pelvis. Finally, and perhaps most important, the hazards of breech delivery may be sufficient enough to warrant X-ray pelvimetry. In the presence of abnormal pelvimetry findings, perinatal mortality subsequent to vaginal breech delivery ranges from 2.2 to 8.5%. Although many authorities advise performing X-ray pelvimetry before all vaginal breech deliveries, we cannot report that outcome data support the universal use of X-ray pelvimetry alone in any clinical setting.

IV. CONCLUSIONS

Radiographic pelvimetry is not an entirely benign procedure because it exposes the fetus to low levels of ionizing radiation that increase the risk of childhood cancers by 50%. The challenge for obstetricians is to identify groups of patients in whom pelvimetry may affect labor management decisions. Many authorities support the use of X-ray pelvimetry before attempted vaginal breech delivery, citing the increased perinatal mortality of breech delivery in the presence of fetal–pelvic disproportion. Additionally, Mengert has described other high-risk patients, including those with histories of difficult labor, nulliparous patients with a nonengaged fetal head, and patients with abnormal findings by clinical pelvimetry, who may be appropriately screened by X-ray pelvimetry. The most accurate technique for prospectively predicting fetal–pelvic disproportion is the fetal–pelvic index, but the safest technique is MRI pelvimetry, in which ionizing radiation is avoided. In the future, fast-scan MRI imaging techniques may be used to measure fetal and maternal bony and soft tissue parameters, which may be incorporated into a calculation similar to the fetal–pelvic index to predict fetal–pelvic disproportion.

See Also the Following Articles

Cesarean Delivery; Ultrasound

Bibliography

Benson, W. L., Boyce, D. C., and Vaughn, D. L. (1972). Breech delivery in the primigravida. *Obstet. Gynecol.* **40**, 417–428.

Brent, R. L. (1974). The effects of radiation on embrogenesis. *J. Reprod. Med.* **12**, 6–17.

Caldwell, W. E., and Moloy, H. C. (1933). Anatomical variations in the female pelvis and their effect in labor with suggested classification. *Am. J. Obstet. Gynecol.* **26**, 479–505.

Colcher, A. E., and Sussman, W. (1944). A practical technique for roentgen pelvimetry with a new positioning. *Am. J. Radiol.* **51**, 207–214.

Danforth, D. N. (1986). Mechanism of normal labor. In *Obstetrics and Gynecology* (D. N. Danforth and J. R. Scott, Eds.), 5th ed., pp. 631–633. Lippincott, Philadelphia.

Federle, M. A., Cohen, H. A., Rosenwein, M. F., Brant-Zawadzki, M. N., and Cann, L. E. (1982). Pelvimetry by digital radiography: A low dose examination. *Radiology* **143**, 733–735.

Ferguson, J. E., DeAngelis, G. A., Newberry, Y. G., Finnerty, J. J., and Agarwal, S. (1996). Fetal radiation exposure is minimal after pelvimetry by modified digital radiography. *Am. J. Obstet. Gynecol.* **175**, 260–269.

Goodwin, J. W., and Reid, D. (1963). Risk to the fetus in prolonged and trial labor. *Am. J. Obstet. Gynecol.* **85**, 209–222.

Harvey, E. B., Boice, J. D., Honeyman, M., and Flannery, J. T. (1985). Prenatal x-ray exposure and childhood cancer in twins. *N. Engl. J. Med.* **312**, 541–545.

Kastler, B., Gangi, A., Mathelin, C., *et al.* (1993). Fetal shoulder measurements with MRI. *J. Computer Assisted Tomogr.* **17**, 777–780.

Mengert, W. F. (1948). Estimation of pelvic capacity. *J. Am. Med. Assoc.* **138**, 169–174.

Morgan, M. A., and Thurnau, G. R. (1988a). Efficacy of the fetal–pelvic index for delivery of neonates weighing 4000 grams or greater: A preliminary report. *Am. J. Obstet. Gynecol.* **158**, 1133–1137.

Morgan, M. A., and Thurnau, G. R. (1988b). Efficacy of the fetal–pelvic index in patients requiring labor induction. *Am. J. Obstet. Gynecol.* **159**, 621–625.

Morgan, M. A., and Thurnau, G. R. (1992). Efficacy of the fetal–pelvic index in nulliparous women at high risk for fetal–pelvic disproportion. *Am. J. Obstet. Gynecol.* **166**, 810–814.

Morgan, M. A., Thurnau, G. R., and Fishburn, J. I., Jr. (1986). The fetal–pelvic index as an indicator of fetal–pelvic disproportion: A preliminary report. *Am. J. Obstet. Gynecol.* **155**, 608–613.

Morris, C. W., Heggie, J. C. P., and Acton, C. M. (1993). Computed tomography pelvimetry: Accuracy and radiation dose compared with conventional pelvimetry. *Aust. Radiol.* **37**, 186–191.

O'Brien, W. F., and Cefalo, R. C. (1996). Labor and delivery. In *Obstetrics: Normal and Problem Pregnancies* (S. G. Gabbe, J. R. Niebyl, and J. L. Simpson, Eds.), 3rd ed., p. 377. Churchill Livingstone, New York.

Thurnau, G. R., and Morgan, M. A. (1988). Efficacy of the fetal–pelvic index as a predictor of fetal–pelvic disproportion in women with abnormal labor patterns that require labor augmentations. *Am. J. Obstet. Gynecol.* **159**, 1168–1172.

Thurnau, G. R., Scates, D. H., and Morgan, M. A. (1991). The fetal–pelvic index: A method of identifying fetal–pelvic disproportion in women attempting vaginal birth after previous cesarean delivery. *Am. J. Obstet. Gynecol.* **165**, 353–358.

Thurnau, G. R., Hales, K. A., and Morgan, M. A. (1992). Evaluation of the fetal–pelvic relationship. *Clin. Obstet. Gynecol.* **35**, 570–581.

Tukeva, T. A., Aronen, H. J., Karjalainen, P. T., and Makela, P. J. (1997). Low-field MRI pelvimetry. *Eur. Radiol.* **7**, 230–234.

Penis

Gerald H. Jordan
Eastern Virginia Medical School and The Devine Center for Genitourinary Reconstruction

Paul F. Schellhammer
Eastern Virginia Medical School

I. Surgical Anatomy of the Penis
II. Vascular Anatomy
III. Physiology of Erection

GLOSSARY

Buck's fascia The fascial support structure of the deep structures of the penis.

cavernosal venoocclusion The terminology applied to the process by which blood flow from the lacunar spaces becomes obstructed. Cavernosal venoocclusion is accomplished via physical factors (i.e., pressure compression of the subtunical venules, shuttering as the emissary vein perforates the tunica albuginea, and stretching compression of the circumferential deep dorsal veins). Additionally, it is postulated that there may be a neurologic component to venoocclusion.

circumflex veins The veins that travel circumferentially around the corpora cavernosa, joining the emissary veins to the deep dorsal vein of the penis.

common penile artery The continuation of the deep internal pudendal artery after the departure of the perineal artery and the labial scrotal artery.

corpus cavernosum One of the two erectile bodies responsible for tumescence and the creation of rigidity.

corpus spongiosum The third erectile body; this erectile body tumesces but offers nothing with regard to rigidity.

Dartos fascia The fascial support structure intimately associated with the skin and vascularity to the skin.

deep dorsal vein The major component of the intermediate venous drainage of the penis, beginning subcoronally as the retrocoronal plexus extending proximally to join the preprostatic plexus.

dorsal arteries The superficial extension of the common penile artery, the main arterial supply to the glans penis.

emissary veins The veins joining the subtunical venules with the circumflex veins. The emissary veins perforate the tunica albuginea.

helicine artery The terminal arborization of the intracavernosal artery; the helicine arteries enter the lacunar spaces.

intracavernosal artery The extension of the common penile artery; the intracavernosal artery traverses through the substance of the corpus cavernosum.

nitergic system One of the three systems governing smooth muscle relaxation. The nitergic system is governed primarily by nitric oxide and mediates the conversion of GTP to cyclic GMP.

prostacycline system One of the three systems modulating smooth muscle relaxation. The prostacycline system is mediated essentially by prostaglandin and mediates the conversion of ATP to cyclic AMP.

tumescence The property of the erectile bodies in which the vasculature spaces dilate causing engorgement.

tunica albuginea of the corpora cavernosa The lining of the corpora cavernosa, bilaminar throughout most of the circumference, becoming monolaminar in the ventral midline.

vipergic system One of the three systems modulating smooth muscle relaxation. Under the influence of the vasoactive intestinal polypeptide, ATP is converted to cyclic-AMP.

I. SURGICAL ANATOMY OF THE PENIS

The anatomy of the penis, scrotum, and perineum is complex, with multiple fascial layers surrounding the corpora cavernosa, the corpus spongiosum, and urethra. The urethra acts as a conduit for voiding and the products of ejaculation. The corpus spongiosum, while having the property of tumescence, essentially serves the function of protection to the urethra and also, probably due to its thickness, enhances complete emptying of the urethra following voiding. Tumescence and its role in causing increased efficiency of the conduit with ejaculation is proposed but not proven. The major function performed by the penis, however, is that of erection, which is a complex phenomenon mediated through the interaction of psychological, hormonal, neurologic, arterial, venous, and muscular factors.

If one views the penile shaft in cross section, immediately deep to the skin, one finds the dartos fascia (Fig. 1). In the penis, the dartos fascia is predominantly fascial in composition. The dartos fascia becomes the tunica dartos of the scrotum, which is more musculofascial in composition. In the perineum, the tunica dartos becomes Colles' fascia, whereas over the abdomen, the dartos fascia becomes Scarpa's fascia. Immediately deep to the dartos fascia is Buck's fascia. The nature of Buck's fascia is very different than that of the dartos fascia. The dartos fascia is thick and areolar, whereas Buck's fascia is thin but relatively stronger. Both Buck's fascia and the dartos fascia have the usual fascial vasculature—a plexus on the deep aspect and the superficial aspect of the fascia, with perforators connecting the two. The Buck's fascia surrounds the dorsal neurovascular bundle, the corpora cavernosa, and the corpus spongiosum. On the dorsum, the Buck's fascia contains the dorsal neurovascular bundle in an envelope-type configuration, being fully bilaminar, with fascia encompassing those structures on both the dorsal and ventral aspects. Likewise, Buck's fascia divides as it encounters the corpus spongiosum, with a layer of Buck's fascia extending beneath the corpus spongiosum becoming attenuated immediately dorsal to the midline of the corpus spongiosum and a more prominent layer extending superficial/ventral to the corpus spongiosum. The corpus spongiosum is attached to the ventral aspect of the corpora cavernosa. The urethra lies within the corpus spongiosum (Fig. 2). In the pendulous portion of the corpus spongiosum, the urethra lies roughly in the middle. As the corpus spongiosum extends into the perineum, the urethra

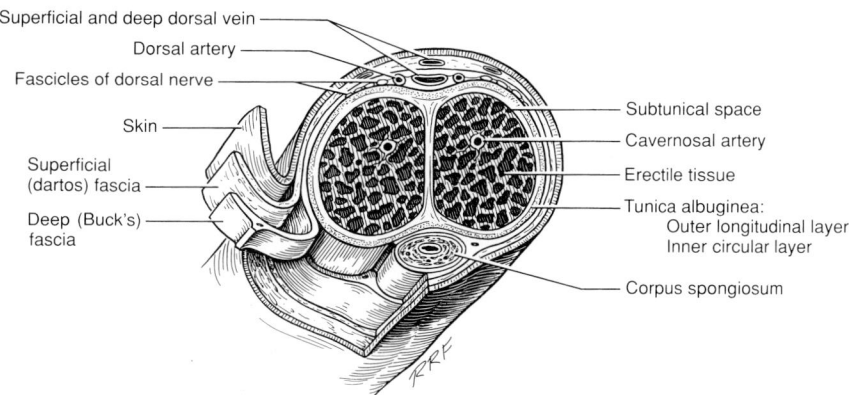

FIGURE 1 Diagram illustrating the layers of the penile shaft.

becomes eccentrically placed lying closer to the corpora cavernosa. The corpus spongiosum attaches in the perineum to the perineal body. The urethra departs the corpus spongiosum in advance of that attachment to the perineal body, with the corpus spongiosum extending to its attachment to the perineal body as the bulbospongiosum. Distally, the corpus spongiosum is contiguous with the spongy erectile tissue of the glans penis. The urethra extends through the glans penis, with the fossa navicularis being that portion of the urethra contained in the glans penis. The glans penis caps the tip of the corpora cavernosa.

The corpora cavernosa are classically thought of as dual bodies (Fig. 3). In the pendulous portion of the urethra, the corpora cavernosa actually function as a single body, with intimate midline fusion. The corpora are so intimately fused that there is free traverse of blood between the chambers of the corpora cavernosa and the midline septum actually becomes midline septal fibers. The covering of the corpora cavernosa is the tunica albuginea of the corpora cavernosa. The tunica albuginea is bilaminar

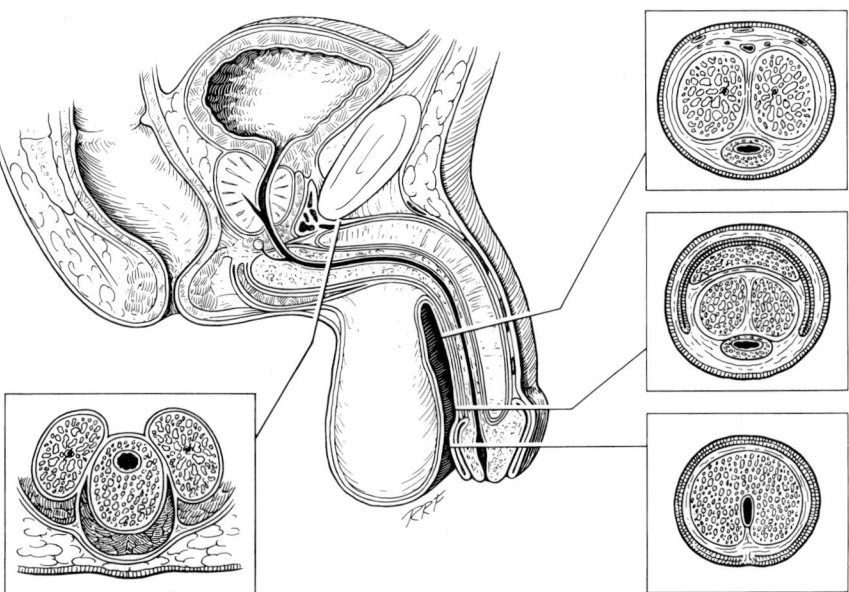

FIGURE 2 Diagram illustrating male pelvis. Notice the cross sections demonstrating the progressively eccentric location toward the dorsum of the corpus spongiosum as one goes proximal.

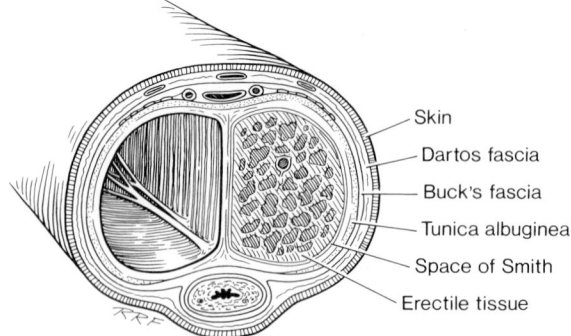

FIGURE 3 Diagram illustrating a cross section of the penis with particular emphasis on the anatomy of the tunica albuginea, midline septal fibers, and oblique support fibers.

and the bulbospongiosum, in composite create a layer which is diaphragm-like.

II. VASCULAR ANATOMY

A. Venous System

The penis contains multiple superficial veins usually running on the dorsal lateral surface of the penis in the dartos fascia or between the dartos and Buck's fascias. In many men, these veins unite at the penopubic area to form a single superficial dorsal vein which drains into the ileofemoral system, and the drainage obligatorily seems to be predominantly toward the left side. These veins in most men consolidate to become in essence the vena comitans of the superficial external pudendal artery (Fig. 4). The intermediate drainage system is composed of the deep dorsal vein, the circumflex veins, and the emissary veins (Fig. 5). The deep dorsal vein emerges from the glans as multiple small veins forming the retrocoronal plexus (Fig. 6). This plexus then joins to become the deep dorsal vein of the penis. The deep dorsal vein is contained in Buck's fascia. In most patients, the deep dorsal vein exists as a single trunk, running in a "groove" immediately over the septum of the corpora cavernosa. This vein then

throughout most of the circumference of the penis. The inner layer is composed of circular fibers and this layer is evident around the entire circumference. The midline septal fibers interweave with the fibers of this layer, as do obliquely oriented support fibers. The outer layer is composed of longitudinal fibers and is prominent dorsally and laterally but attenuates as the layer approaches the ventral midline. This bilaminar architecture seems to be important not only for support of the corporal bodies because of the high pressure developed with erection and intercourse but also as a component of the physical compression and obstruction believed to be a major cause of venous occlusion. Additionally, the bilaminar architecture explains the variety of presentations seen with buckling trauma to the penis.

The tunica albuginea is a very compliant layer, as are the other fascial layers of the penis. Immediately deep to the tunica albuginea is a space described by Smith. This is an areolar space between the tunica albuginea and the deep spongy erectile tissue of the corpora cavernosa. The spongy erectile tissue is composed of endothelial cell-lined lacunar spaces, small arterials and venules, a collagen matrix, and smooth muscle. Proximally, the corpora cavernosa are attached to the inferior pubic ramus and affixed to the ischial tuberosities. Buck's fascia in the perineum is in continuity with the superficial genitourinary perineal membrane. While a true diaphragm does not exist in the male, the plane created by the crura of the corpora cavernosa, the triangular ligament,

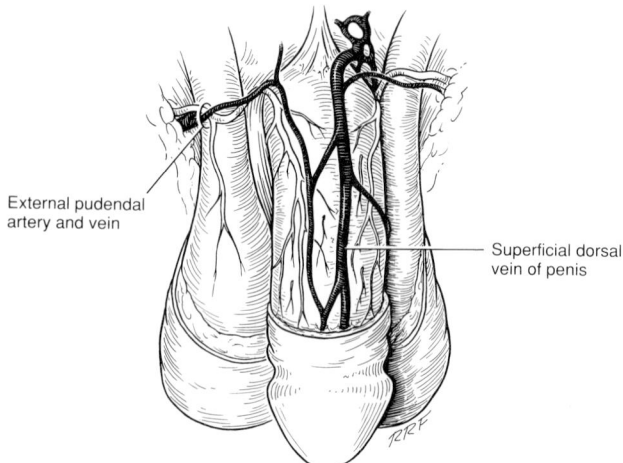

FIGURE 4 Diagram illustrating the superficial venous system. In most individuals the superficial venous system essentially becomes the vena comitans associated with the superficial external pudendal artery.

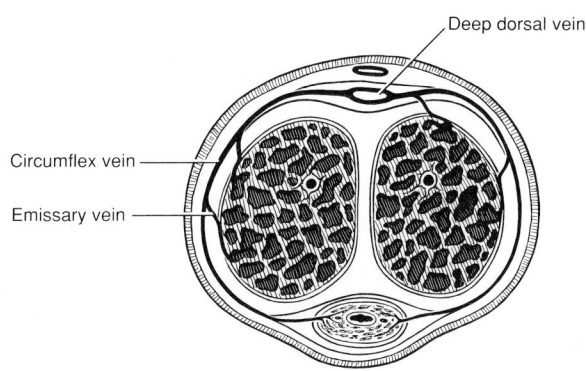

FIGURE 5 Cross section of the shaft of the penis demonstrating the intermediate venous system.

drains beneath the pubis into the preprostatic plexus. However, more than one deep dorsal vein can often be found. In many cases, on the pendulous portion of the penis, the vein is a single structure but then becomes multiple branches in the area of the crus, thus descending beneath the pubis, already divided, to join the preprostatic plexus. The circumflex veins arise from the corpora cavernosa or from the corpus spongiosum. On the ventral surface, frequently the circumflex veins unite to form periurethral veins. These veins course along the inferolateral surface of the corpora cavernosa and proximally seem to join the deep dorsal venous system into the deep dorsal vein. The emissary veins are tiny veins which penetrate the tunica albuginea in either a perpendicular or oblique course. These veins emerge from the corpora cavernosa from the lateral and dorsal surface of the corpora cavernosa and empty into the circumflex veins. In some patients, there is a confluence of the circumflex veins lateral to the course of the dorsal artery on the dorsum of the penis; these have been termed the lateral dorsal veins. In most patients, there are connections between the deep and superficial dorsal venous systems. The deep drainage system of the penis is composed of the cavernosal and crural veins. The veins which arise in the hilum of the penis on the dorsal medial surface of the corpora cavernosa are termed the crural veins and usually join to form a large venous channel which drains into the preprostatic plexus. In some patients the drainage is into the internal pudendal veins. Additionally, the cavernosal veins run most medially and deeply in the penile hilum. Also, at the insertion of the corpora cavernosa to the ischial tuberosities, there can be lateral cavernosal veins that drain laterally to the ileofemoral system. The venous anatomy of the proximal penis is quite variable.

B. Arterial System

The arterial supply to the penis begins at the common penile artery (Fig. 7), which is a branch of the internal pudendal artery. The internal pudendal artery pierces the "urogenital diaphragm" and courses along the medial margin of the inferior ramus

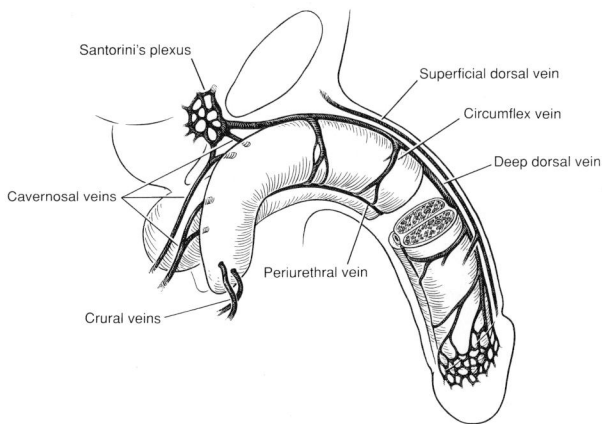

FIGURE 6 Diagrammatic representation of the venous drainage of the deep structures of the penis.

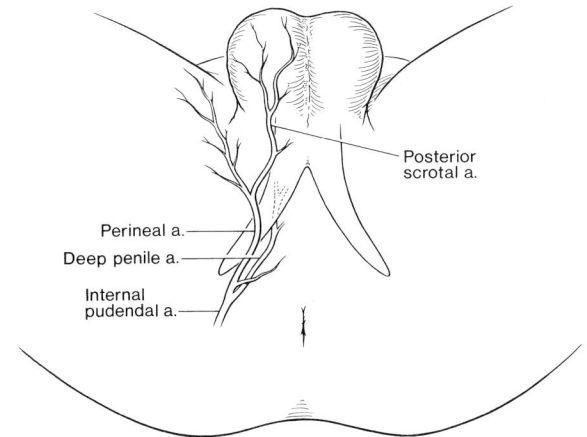

FIGURE 7 Diagrammatic illustration of the distribution of the perineal artery. Note that the perineal artery is a branch of the deep internal pudendal artery that continues after that branch becomes the common penile artery.

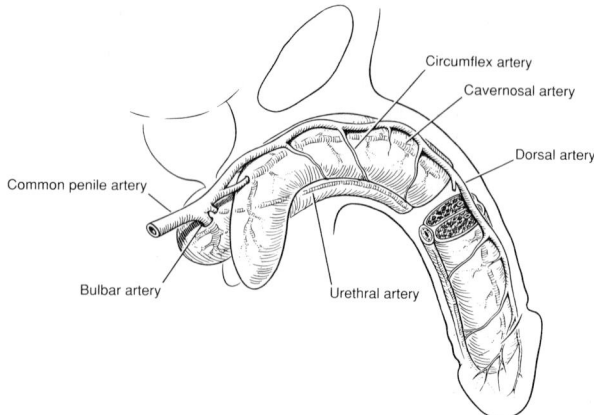

FIGURE 8 Diagrammatic representation of the arborization of the common penile artery.

of the pubis. The artery gives off the perineal artery, which further divides into the medial posterior scrotal branch and extends laterally on the medial thigh as the perineal artery. From that point, the internal pudendal artery becomes the common penile artery. Near the bulbospongiosum, the common penile artery divides into its ultimate branches (Fig. 8). One branch of the penile artery is the artery to the bulb or the bulbourethral artery. Generally, this arises as a single branch from each common penile artery. Obviously, there are two common penile arteries. However, the bulbourethral artery can further branch. The bulbourethral artery penetrates into the bulbospongiosum, usually just caudal to the departure of the urethra from the bulbospongiosum. The common penile artery then further divides into the dorsal artery of the penis and the deep central artery or cavernosal artery. The dorsal artery of the penis courses distally along the dorsum of the penis between the deep dorsal vein medially and the dorsal nerves laterally. Along its course it gives off from 3 to 10 circumflex branches. Proximally the circumflex cavernosal arteries join the bulbospongiosum and comprise the second of the proximal blood supplies to the corpus spongiosum. Patients demonstrate marked variations in both origin and number of circumflex cavernosal arteries. Classically, a single cavernosal artery penetrates the crura of the corpora cavernosa and extends distally in the center of the corpus cavernosum. The cavernosal arteries arborize, forming helicine arteries or arterioles which pene-

trate the lacunar spaces. Emerging from the lacunar spaces are venules which become confluent and eventually emerge from the tunica albuginea as the emissary veins (Fig. 5).

C. Nervous System

The dorsal nerve of the penis arises in the pudendal canal or Alcock's canal as the first branch of the pudendal nerve. It travels ventral in relation to the main pudendal trunk, above the obturator internus, and under the levator ani, thus perforating the transverse perinei muscle to course along the dorsum of the penis and to extend distally along the dorsal lateral penile surface. At the level of the penopubic area, the dorsal nerve fascicles spread and the nerve extends as multiple branches which terminate on the glans penis. The origin and course of the dorsal nerves is highly consistent in most patients. The cavernous nerves are found in the hilum of the penis. The cavernous nerve begins as a plexus coursing along the posterior lateral aspect of the prostate. This plexus then begins to form trunks which course along the lateral surface of the membranous urethra after penetrating the transverse perinei muscles. These bundles then run into the hilum of the penis to penetrate the corpora cavernosa, where the nerves then again form a network of fine fibers spreading over the cavernous arteries, veins, and lacunar spaces. The cavernous nerve is the distal extension of nerves referred to as the "nervi erigentes." Walsh's dissections demonstrated the nervi erigentes as a plexus coursing the dorsal lateral aspect of the bladder and prostate and then penetrating the perineal membranes to course into the penis. The sensory enervation on the shaft of the penis, which is less well-defined, is believed to occur via branches of the genitofemoral nerve. While the glans is markedly sensate, the skin of the shaft is remarkably poorly sensate.

III. PHYSIOLOGY OF ERECTION

The lacunar spaces within the spongy erectile tissue of the corpora cavernosa are surrounded by smooth muscle. The lacunar spaces are lined by en-

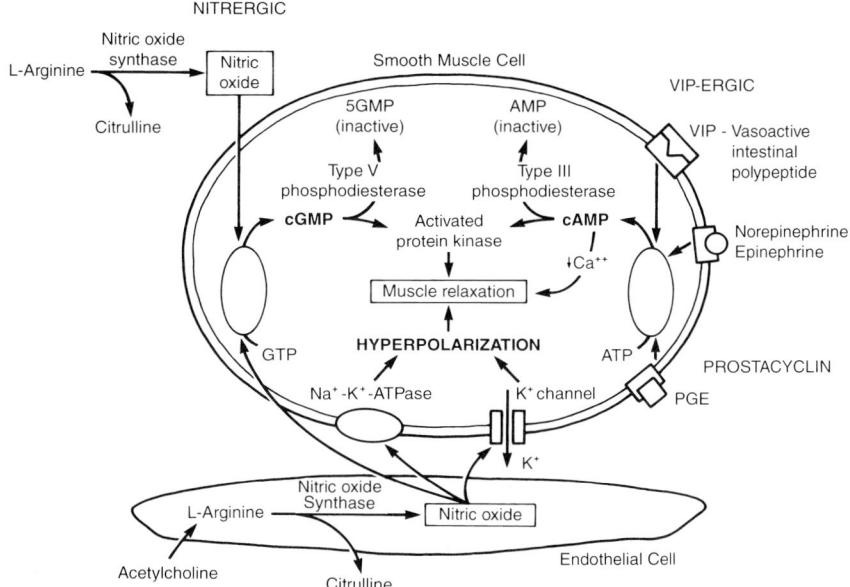

FIGURE 9 Diagrammatic representation of mechanisms of smooth muscle relaxation.

dothelial cells. Stimulation of the cavernosus nerve either by erotic stimuli via the brain's limbic system or by tactile stimulation of the genitalia results in active relaxation of the cavernosal smooth muscle. The parasympathetic neurotransmitters involved include acetylcholine released by the cholinergic autonomic nerves, vasoactive intestinal polypeptide (VIP) (the vipergic system), and nitric oxide (the nitrergic system) (Fig. 9). These neurotransmitters act on the endothelial cells of the lacunar spaces, the intralacunar smooth muscle, as well as the muscle of the walls of the arteries and perhaps the venules. Muscle relaxation is mediated by cyclic GMP, cyclic AMP, and hyperpolarization. Arginine is converted to nitric oxide by the enzyme nitric oxide synthase. Nitric oxide converts GTP to cyclic GMP. Cyclic GMP causes smooth muscle relaxation. VIP and prostaglandin both affect the ATP/cyclic AMP system, with the net result being smooth muscle relaxation. Flaccibility is maintained by constant tonus/contraction of the intralunar smooth muscle mediated via norepinephrine, the sympathetic system. With manual and/or psychological stimulation and hormonal mediation, the sympathetics turn off and the parasympathetics via the nitrergic, VIP-ergic, and prostacyclin systems turn on. The helicine arteries dilate, the lucnar spaces dilate, and there is marked increase in arterial flow to the corpora cavernosa. In short, the lucnar spaces become large vascular sinks. Initially, there is an increase in venous outflow; however, the veins occlude quickly and venous outflow becomes almost nonexistent. Venous occlusion is clearly a mechanical phenomenon. The subtunical venules are compressed by the rising pressure within the corpora, the emissary veins probably "shutter" as they perforate the tunica albuginea, the circumflex veins are stretched around the enlarging corporal bodies, and the longitudinal veins are stretched by the lengthening penis. There also appears to be strong evidence of an early neurologic component to venoocclusion that is unknown. As pressure rises, arterial flow diminishes. Soon the intracorporal cavernosal space equilibrates at mean arterial pressure and remains at that level due to the trapping phenomena previously described. Pressures within the corpora at times exceed mean arterial pressure, but these abnormally high pressures occur during intercourse, due to buckling forces, and ejaculation with muscular compression of the proximal corpora.

Following ejaculation, the parasympathetic stimuli decrease. Increased sympathetic tone is noted, the veins begin to flow, and the system eventually equilibrates at baseline. During tumescence, rigidity is achieved by the accumulation of pressure within the corpora cavernosa due to the vascular events which were previously described. The pressures within the corpora cavernosa distend the tunica albuginea to the point of nondistensability. Then, the midline septal fibers are tightly stretched between the dorsal and ventral corpora, thus creating, in effect, an "I" beam. The I-beam arrangement of the midline septal fiber accounts for the anteroposterior rigidity of the penis seen with erection. The relative indistensability of the paired corpora/lateral columns adds lateral stability to the penis during erection. Hence, patients can notice relatively impressive tumescence and yet not accumulate sufficient pressures to cause the tunica albuginea and septal fibers to become nondistensable and thus provide rigidity. Additionally, Goldstein *et al.* identified an intracorporal fibrous collagen matrix. The relationship of these fibers to erectile stability and rigidity is not known. It may be that this fibrous network functions in the same fashion as the struts within the Von Hendenburg.

Impotence is defined as the inability to obtain or maintain an erection which is suitable for the completion of intercourse to the satisfaction of both partners. Erectile dysfunction is nothing more than dysfunctional erections and does not tacitly imply inability to have intercourse. Erectile dysfunction is now the preferred term since term impotence was confining regarding what it defined. Erectile dysfunction may be organic, psychogenic, or mixed, and in the vast majority of patients, there is a mix of organic and psychogenic factors. The art of evaluating erectile dysfunction is to determine the amount of the blend between organic dysfunction and functional dysfunction, thus determining the disabling factor for a given patient. In some patients, there is minimal organic dysfunction; however, the patient loads himself with a number of functional issues and is hence disabled more by the functional aspects of erectile dysfunction than the organic aspects. In other patients, however, the organic dysfunction is so profound that the functional aspects loaded on top of the organic dysfunction are mute in that the organic dysfunction would never allow for satisfactory intercourse. The blend for an individual patient needs to be determined in order to determine which patients will benefit from initial sex therapy and which patients, in addition to some sex therapy counseling, would be candidates for further therapy.

The approach to the work-up of erectile dysfunction must be systematic and pragmatic. Much information is available from a detailed history. Patients suffering predominantly from organic factors will note an insidious onset of their erectile dysfunction. They will notice diminished erectile function at night, with morning erections/ "pee hard-ons," Additionally, the degree of erectile dysfunction remains relatively constant, with patients noticing diminished erections both for masturbation and for attempts at intercourse. In many cases, a concomitant cause of systemic disease exists. There seems to be a sequence of gradual deterioration of sexual function with an initial decrease in both rigidity and frequency of erections. Not uncommonly, patients will notice tumescence but will also notice diminished rigidity. Additionally, with organic factors, erections may be achieved but not maintained. Eventually, the patient may complain of complete loss of tumescence. Generally, with pure organic impotence, without other complicating factors such as hypogonadism, there is no loss of libido.

In contradistinction, when psychological factors predominate, there is often an acute onset of erectile dysfunction which may be temporally related to a particular emotional stress. Once functional factors predominate, erectile function or dysfunction may be intermittent in nature. At times there may be limited or no impairment of erectile function, whereas at other times erectile function may be severely diminished. There may be a history of no impairment of erectile capacity with masturbation and nocturnal penile tumescence should not be affected. It is emphasized that in the older adult patient, it is extremely unusual for a patient to demonstrate only psychogenic erectile dysfunction unless there has been a long chronic history of such dysfunction in the past. The vast majority of mature adult patients, particularly those who have functioned successfully prior to the onset of their erectile dysfunction, will demonstrate a blend and the nature of this blend must be determined.

The history must include a complete review of symptoms with particular emphasis on neurologic, genitourinary, endocrine, and cardiovascular systems. Special emphasis should be placed on history suggesting diabetes mellitus, lower extremity claudication, medication ingestion, as well as smoking and other ingestions, previous abdominal or pelvic surgery, previous back surgery, or history of traumatic pelvic fracture or back injury. Excessive alcohol intake can result in decreased libido. Additionally, in some individuals, chronic alcohol ingestion can cause primary hypogonadism. Chronic cigarette smoking has been demonstrated, in a rat model, to be directly toxic to the endothelial cells and hence result in organic impotence. As mentioned, most patients have some psychological factors complicating their erectile dysfunction. In some patients, however, detailed psychological screening may be useful.

Physical examination should include a detailed examination of the peripheral pulses and an examination for bruits of the neck, abdomen, or groin which may be indicative of other systemic cardiovascular disease. The examination must include an evaluation of the patient's endocrine status looking for signs of reduced testosterone or elevated estrogen, i.e., gynecomastia, testicular atrophy, or a prostate which is unexpectedly small for the patient's age. The genitalia must be examined in detail looking for signs of diminished neurologic function or mechanical factors which may be interfering with the patient's ability to have intercourse. Testicular volume and consistency must be evaluated in addition to the rectal examination to assess the prostate.

Beyond the physical examination, further evaluation must be dictated by the patient's ultimate goals with regard to treatment and must also be tempered by realism with regard to efficacies of particular therapies in the given patient. Most patients with erectile dysfunction are endocrinologically normal. Without specific indications, expensive tests such as those for testosterones, gonadotropins, or gonadotropin-releasing hormones are not necessary or indicated. Likewise, if a patient is not motivated for surgery, then detailed evaluation of venoocclusive function or pudendal/common penile arterial disease are also not indicated. Additionally, a patient might be highly motivated for surgery and specifically have "an operation in mind that he has heard about," but for that patient the efficacy of the given operation is poor.

For that patient, instead of tests, he needs candid discussion of realistic expectations.

The wide range of testing that is available as well as the level of knowledge that we now have regarding erectile function allow us to consider patients' motivation and realistic outcomes and thus arrive at a suitably complete yet individualized evaluation and offer treatment that, it is hoped, restores the couple to acceptable sexual function. In reality, however, given the prevalence of erectile dysfunction, the treatment for the most part is ineffectual. Long-term efficacy is low. The effectiveness of therapy is perception mediated. If the treatment is believed inappropriate or the outcome perceived as less than reasonable, the patient will drop out.

See Also the Following Article

MALE REPRODUCTIVE SYSTEM, OVERVIEW

Bibliography

Aboseif, S. R., Breza, T. F., Lue, T. F., and Tanagho, E. A. (1989). Penile venous drainage in erectile dysfunction. Anatomical, radiological and functional considerations. *Br. J. Urol.* **64**, 183–190.

Bastuba, M. D., DeTejada, I. S., Dinlenc, C. Z., Sarazen, A., Krane, R. J., and Goldstein, I. (1994). Arterial priapism: Diagnosis, treatment and long-term follow up. *J. Urol.* **151**, 1231–1237.

Breza, J., Aboseif, S. R., Orvis, B. R., Lue, T. F., and Tanagho, E. A. (1989). Detailed anatomy of penile neurovascular structures: Surgical significance. *J. Urol.* **141**, 437.

Burnett, A. L., Tillman, S. L., Chang, T. S., Epstein, J. I., Lowenstein, C. J., Bredt, D. S., Snyder, S. H., and Walso, P. C. (1993). Immunohistochemnical localization of nitric oxide synthase in the autonomic innervation of the human penis. *J. Urol.* **150**, 73–76.

Bush, P. A., Aronson, W. J., Buga, G. M., Rajfer, J., and Ignarro, L. J. (1992). Nitric oxide is a potent relaxant of human and rabbit corpus cavernosum. *J. Urol.* **147**, 1650–1655.

De Tejada, I. S. (1995). Commentary on mechanism for the regulation of penile smooth muscle contractility. *J. Urol.* **153**, 1762.

Giuliano, F. A., Rampin, O., Benoit, G. and Jardin, A. (1995). Neural control of penile erection. *Urol. Clin. North Am.* **22**(4), 747–766.

Hauri, D., Spycher, M., and Bruhlmann, W. (1983). Erection and priapism: A new physiopathological concept. *Urol. Int.* **38**, 138–145.

Hellstrom, W. J. G., Monga, M., Wang, R., Domer, F. R., Kadowitz, P. J., and Roberts, J. A. (1994). Penile erection in the primate: Induction with nitric-oxide donors. *J. Urol.* **151**, 1723–1727.

Ignarro, L. J., Bush, P. A., Buga, G. M., *et al.* (1993). NO2 and cyclic GMP formation upon electrical field stimulation cause relaxation of corpus cavernosum smooth muscle. *Biochem. Biophys. Res. Commun.* **843**, 1990.

Juskieweiski, S., Vayss, P. H., Moscovia, J., *et al.* (1982). A study of the anatomic blood supply to the penis. *Anat. Chir.* **4**, 101.

Kim, N., Azadzoi, K. M., Goldstein, I., *et al.* (1991). A nitric oxide-like factor mediates nonadrenergic, noncholinergic neurogenic relaxation of penile corpus cavernosum smooth muscle. *J. Clin. Invest.* **88**, 112–118.

Kodos, A. B. (1967). The vascular supply of the penis. *Arkh. Anat. Embriol.* **43**, 525.

Lepor, H., Greggerman, M., Crosby, R., Mostofi, F., and Walsh, P. C. (1995). Precise localization of the autonomic nerves from the pelvic plexus to the corpora cavernosa: A detailed anatomical study of the adult male pelvis. *J. Urol.* **133**, 207.

Lue, T. F., and Tanagho, E. A. (1987a). Physiology of erection and pharmacological management of impotence. *J. Urol.* **137**, 829–836.

Lue, T. F., and Tanagho, E. A. (1987b, December). Surgical anatomy of the penis. Paper presented at the American Urological Association Conference on Impotence.

Lugg, J. A., Gonzalez-Cadavid, N. F., and Rajfer, J. (1995). The role of nitric oxide in erectile function. *J. Androl.* **16**, 2.

Miller, M. A. W., Morgan, R. J., Thompson, C. S., Mikhalidis, D. P., and Jeremy, J. Y. (1995). Effects of papaverine and vasointestinal polypeptide on penile and vascular cAMP and Cgmp in control and diabetic animals: An in vitro study. *Int. J. Impotence Res.* **7**, 91–100.

Quartey, J. K. M. (1997). Microcirculation of penile and scrotal skin. *Atlas Urol. Clin. North Am.* **5**(1), 1–9.

Pheromones, Fish

Norm Stacey
University of Alberta

Peter Sorensen
University of Minnesota

I. Introduction
II. Hormonal Sex Pheromones in Goldfish
III. Patterns of Olfactory Sensitivity to Hormones in Fish
IV. Biological Evidence for Hormonal Pheromones in Species Other Than Goldfish
V. Function and Evolution of Hormonal Pheromones

GLOSSARY

final oocyte maturation Resumption of meiosis in response to the preovulatory surge of gonadotropin.
gonadotropin II Maturational gonadotropin; shown in salmonids to be homologous to mammalian luteinizing hormone.
ovulation Rupture of the ovarian follicle and release of the oocyte.
primer effect Physiological response to a pheromone.
releaser effect Rapid behavioral response to a pheromone.

Pheromones are chemicals that are released by organisms and elicit adaptive, rapid, and instinctual responses in conspecifics. Living in an aqueous medium, which is a universal solvent but a poor transmitter of light, fish have evolved highly developed pheromonal signaling systems. These systems have many functions, one of the most important being the promotion of reproductive synchrony. Much recent evidence indicates that the reproductive pheromones of many fish species are composed of sex hormones (steroids and prostaglandins) and their metabolites. The best understood of these hormonally derived pheromone systems is that of the goldfish (*Carassius auratus*): Ovulatory female goldfish release at least five hormones and metabolites which have dramatic effects on the physiology and behavior of conspecifics. Less complete studies suggest similar hormonally derived pheromone systems are used by a variety of other teleosts. The question of signal specificity remains poorly understood.

I. INTRODUCTION

Living in a medium that readily transports compounds released as a result of metabolism, excretion, and osmoregulation, fish are routinely exposed to hormones and metabolites that can carry important information about the internal physiological state of conspecifics. Accordingly, fish have evolved the ability to detect particular components of the "hormonal soup" released by conspecifics and to use them as social cues (pheromones). The sensory basis of this ability is the olfactory system, an ancient and highly conserved sensory system renowned for its ability to discriminate large numbers of odors with extreme sensitivity. Although the physiological basis for odor discrimination in vertebrates is poorly understood, it is thought to be associated with the numerous transmembrane receptors found on the olfactory sensory neurons of the olfactory epithelium. The discovery that fish use hormonal products as pheromones extends the classical concept that the actions of hormones are restricted to coordinating reproductive events within the individual.

Although the concept that hormones and their metabolites function as pheromones is appealing, there is definitive evidence of such function for only a few species. This is so because proof of pheromonal function in a species requires three types of information: (i) biochemical evidence that a compound is synthesized and released, (ii) physiological evidence that it is discriminated by the olfactory (nervous) system with high sensitivity, and (iii) biological evidence that it evokes specific and relevant responses at appropriate concentrations. Only in goldfish have these three criteria for demonstrating pheromonal function been met. Nonetheless, it is clear that a very large number of fish (perhaps all) use pheromones, that fish release complex mixtures of hormones and their metabolites, that olfactory systems of at least some fish discriminate low concentrations of hormonal products, and that various species respond to waterborne hormonal compounds in an apparently adaptive manner.

II. HORMONAL SEX PHEROMONES IN GOLDFISH

Goldfish have become a model species for pheromone studies not because their hormonal or pheromonal systems are unusual but simply because they readily ovulate and spawn in the laboratory and are an established fish reproductive model for neuroendocrine and gonadal studies. The mating system of goldfish is typical of many other nonterritorial species in the family Cyprinidae: ovulation occurs late in scotophase, at which time small groups of males compete for spawning access to the female, which enters aquatic vegetation to oviposit many times over a period of several hours. The absence of pair bonding and territoriality and the resulting intense male–male sperm competition of the goldfish mating system undoubtedly have been major factors shaping the evolution of goldfish pheromone function, which is strongly characterized by male responses to female cues.

Female goldfish release at least two distinct hormonal pheromones during the periovulatory period: a steroid pheromone which is released during the preovulatory surge of gonadotropin II (GtH II) and a prostaglandin pheromone which is released during the several hours that ovulated eggs are in the reproductive tract. Both the steroid and prostaglandin pheromones have primer (slow physiological) and releaser (rapid behavioral) effects that together increase male reproductive success during competitive spawning.

A. The Preovulatory Steroid Pheromone

Female goldfish release dozens of steroid hormones and derivatives in substantial and changing quantities during the ovulatory surge of GtH II. This steroidal mixture, whose release drops at the time of ovulation and spawning, functions as a preovulatory pheromone whose primary function appears to be synchronizing the endocrine state of the male with that of the female.

1. Synthesis and Release

Among the dozens of steroid hormones and derivatives released by preovulatory goldfish, at least three are specifically discriminated by the olfactory system and have pheromonal function. Of these, the steroid hormone 17,20ß-dihydroxy-4-pregnen-3-one (17,20ßP), which induces final oocyte maturation, appears to be a primary component. The other components are the sulfated derivative of 17,20ßP (17,20ßP-20-S) and androstenedione (AD). Like other fish species, female goldfish release unconjugated steroids (17,20ßP and AD) across the gills and conjugated forms of 17,20ßP in the urine and/or bile.

2. Olfactory Sensitivity of the Male

Electroolfactogram (EOG) recording has been an important technique for assessing the sensitivity of the olfactory system of goldfish and other fish to putative pheromones. EOG measures multiunit extracellular voltage transients across the olfactory epithelium, which are thought to reflect summated receptor generator potentials. EOG recording shows that the goldfish olfactory system is remarkably sensitive (picomolar detection thresholds) to 17,20ßP, 17,20ßP-20-S, and AD. EOG and accompanying bioassays also show this sensitivity to be highly specific because steroids which differ from these compounds at only one or two positions fail to elicit electrical responses at nanomolar concentrations. Furthermore, EOG cross-adaptation studies (which examine olfactory response to one compound during adaptation to another) indicate the presence of independent receptor sites which discriminate distinctive molecular attributes of all three steroids. 17,20ßP binds specifically and reversibly with membrane preparations from the goldfish olfactory epithelium in a manner which indicates the presence of a single class of high-affinity, low-capacity receptors for this steroid. Based on the rate of steroid release to the water, the olfactory specificity and sensitivity determined by EOG, and the tendency of goldfish to form bisexual aggregations, males likely detect and discriminate the mixture of these steroids throughout the preovulatory period.

3. Organismal Responses

Although the actions of a tertiary mixture of 17,20ßP, 17,20ßP-20-S, and AD have yet to be studied, exposure to 17,20ßP alone can mimic the ability of ovulatory female water to stimulate hormonal change in males. If mature male goldfish are exposed to picomolar concentrations of 17,20ßP, their blood concentration of GtH-II increases within 15 min, and the volume of milt (sperm and seminal fluid) that can be stripped from their sperm ducts increases within 4 hr. The increased milt volume is associated with decreased sperm concentration in the milt, increased total number of ductal sperm, and increases in both the proportion of motile sperm and the duration of sperm motility. 17,20ßP-20-S also increases GtH II and milt volume, but these effects have not been studied in detail.

The increase in GtH II and milt volume normally induced by exposure to 17,20ßP is blocked if males are exposed simultaneously to 17,20ßP and AD. Although the functional significance of this inhibition by AD is not understood, it seems likely that males are responding not simply to 17,20ßP but rather to the mixture of 17,20ßP, 17,20ßP-20-S, and AD released by females. Thus, when the female is nonovulatory, and the relative quantity of AD is high, a low but potentially detectable concentration of released 17,20ßP would be "masked" by AD, preventing the male from testicular responses when there is no opportunity for spawning.

In addition to exhibiting dramatic endocrine and testicular responses, males that are exposed to 17,20ßP immediately decrease spontaneous feeding behavior and increase their interactions with females. Following overnight exposure to 17,20ßP, males in competition with nonexposed males display more courtship behaviors, perform more spawning acts, and, as revealed by microsatellite DNA fingerprinting, fertilize dramatically more eggs. Pheromonal enhancement of fertility during spawning appears to result at least partially from enhanced sperm function because, in competitive *in vitro* fertilizations, sperm from 17,20ßP-exposed males fertilize more eggs than do sperm from control males. Notably, whereas release of 17,20ßP decreases dramatically at ovulation, the reduction in 17,20ßP-20-S is more gradual; suggesting a continued function in the postovulatory period.

B. Postovulatory Prostaglandin Pheromone

At ovulation, release of the preovulatory steroid pheromone decreases precipitously, and the female becomes sexually attractive to males and sexually receptive to male courtship. These sexual interactions are triggered by a prostaglandin pheromone derived from circulating prostaglandin $F_{2\alpha}$ (PGF) that is responsible for hormonal regulation of female sexual behavior.

1. Synthesis and Release

Ovulation in goldfish is closely associated with the synthesis of PGF, which then travels to the brain to trigger female sexual activity. Circulating PGF is then rapidly metabolized and released to the water where it functions as a potent olfactory stimulant with pheromonal activity. Thus, PGF and its derivatives synchronize both female and male behavior with ovulation.

PGF concentration in female blood increases approximately 50-fold during the several hours that ovulated eggs are in the goldfish oviduct. If nonovulated females receive an intraovarian injection of ovulated eggs or egg substitutes, their circulating PGF concentration increases and they become sexually active. Evidence that the PGF increase is responsible for female sexual activity comes from studies showing that prostaglandin synthesis inhibitors block sexual activity in ovulated or egg-injected females, whereas PGF injection induces sexual activity (without egg release) in nonovulated fish.

As with the preovulatory steroid pheromone, the PGF pheromone is a mixture of only a few of the metabolites released. Based on EOG recording, the two most likely candidates for pheromonal activity are PGF and 15-keto-PGF (15-K-PGF). 15-K-PGF is released almost exclusively in the urine, whereas PGF is released both in the urine and apparently across the gills; release at these sites might account for the tendency of courting males to nudge the female's cloacal and opercular areas with their nares.

2. Olfactory Sensitivity of the Male

EOG studies indicate that PGF and 15-K-PGF are detected by at least two classes of olfactory receptor mechanisms whose specificity is not mutually exclusive. Despite this evidence for separate receptors and differential release, there is no evidence that PGF and 15-K-PGF have different pheromonal functions. As measured by EOG recording, the olfactory epithelium of mature males is more responsive to prostaglandins than is that of immature males or females. Sexual dimorphism in response to prostaglandins, which is observed in other cyprinids and which can be induced by treating fish with androgenic steroids, is one of the few examples of sex hormones influencing olfactory function in a vertebrate. The physiological basis for the dimorphism is unknown.

3. Organismal Responses

Males show no evidence that they can discriminate between ovulated females and nonovulated females injected with PGF and readily court and spawn with either type of female unless their olfactory system is blocked. If PGF or 15-K-PGF is added to tank water, males immediately increase locomotory activity and begin to investigate conspecifics while exhibiting following and chasing behaviors that are typical of those displayed toward ovulated females.

In addition to eliciting behavioral responses typical of normal sexual activity, exposure to prostaglandins rapidly increases blood GtH-II concentration and milt volume. As discussed in Section II,C, these hormonal and testicular responses to the prostaglandin pheromone are physiologically distinct from those induced by pheromonal 17,20ßP. In particular, pheromonal 17,20ßP increases GtH-II concentration and milt volume directly (i.e., it will act on isolated individuals), whereas pheromonal prostaglandin acts indirectly by inducing sexual interactions among male conspecifics, which then respond by increasing hormone concentration and milt volume.

C. Neuroendocrine Mechanisms of Pheromone Action

Lesion and recording studies show that the medial subdivisions of the paired olfactory tracts convey neural activity induced by pheromones, whereas the lateral subdivisions convey activity induced by food odors. The medial tracts project directly to forebrain areas controlling reproductive behavior. Despite speculation that pheromonally induced neural activity might be conducted by the terminal nerve (nervus terminalis; cranial nerve 0), which is anatomically associated with the olfactory nerve in vertebrates and is immunoreactive for gonadotropin-releasing hormone (GnRH), neither goldfish hormonal pheromones nor other compounds detected by the goldfish olfactory system affect terminal nerve activity.

Although pheromonal 17,20ßP and PGF both increase GtH II concentration and milt volume, their mechanisms of action differ in two important ways. First, the two pheromones stimulate GtH II release through different neuroendocrine mechanisms. Pheromonal 17,20ßP evidently induces GtH II increase by reducing tonic dopaminergic inhibition of pituitary GtH II release, whereas pheromonal prostaglandin appears to increase GtH II through a mechanism involving GnRH. Second, the effect of 17,20ßP exposure on milt volume appears to be mediated by a 17,20ßP-induced increase in GtH II concentration, whereas the effect of pheromonal PGF on milt volume is not. The clearest evidence for this differential control of milt volume is that whereas acute hypophysectomy completely blocks the stimulatory effect of pheromonal 17,20ßP on milt volume, it does not block the increase in milt volume that normally occurs when males spawn with PGF-injected females.

Taken together, the available information indicates that pheromonal 17,20ßP increases milt volume through increased GtH II release and with a latency sufficient to increase milt volume in time for spawning. In contrast, interaction with PGF-injected females increases milt volume through a GtH II-independent mechanism (likely the stimulation of neuromuscular components within the male reproductive tract) with a latency short enough to increase milt during the ongoing spawning. The function of the GtH II increase in response to prostaglandin pheromone is not known; however, it may serve to replenish sperm stores depleted by spawning.

III. PATTERNS OF OLFACTORY SENSITIVITY TO HORMONES IN FISH

EOG studies that test olfactory responsiveness to a large number of steroids show that many fishes detect hormonal compounds with great specificity and sensitivity. Several orders of fish have been found to contain species which detect hormonal products: Cypriniformes (goldfish and carp), Characiformes (tetra), Siluriformes (catfish), Elopiformes (tarpon), Salmoniformes, Perciformes (goby), and Gymnotiformes (knifefish). The olfactory systems of the majority of species studied discriminate at nanomolar concentrations at least one of the 100 or more hormone-related products which have been used for EOG testing. Where sufficient species from one order have been tested, the results reveal a strong relationship between phylogeny and the kinds of hormonal compounds detected (Table 1). That is, the more closely related the species, the more similar the compounds detected. For several reasons, caution must be exercised in interpreting the results of such studies. First, the fact that a compound is detected does not necessarily mean that it has pheromonal activity. Second, because hormone metabolism and release is poorly understood in fish, and because the fish olfactory system exhibits such great specificity for steroids and prostaglandins, many apparently unresponsive species in fact may detect hormonal compounds that have not been used for EOG testing. Despite these weaknesses, the findings indicate widespread use of hormonal pheromones in certain groups of fish.

Our understanding of hormones as olfactory signals is best for the cypriniform fish. All cypriniform species tested detect PGF and related F prostaglandins (Table 1), indicating that a prostaglandin pheromone system similar to that seen in goldfish may be ubiquitous in this order. However, steroid detection appears restricted to the family Cyprinidae, in which species from three of the four subfamilies tested are responsive. Within cyprinidae, variance among species in the spectra of detected steroids decreases progressively from higher (superfamily and family) to lower taxa (tribe and genus).

IV. BIOLOGICAL EVIDENCE FOR HORMONAL PHEROMONES IN SPECIES OTHER THAN GOLDFISH

In addition to EOG evidence that waterborne hormones and their metabolites are detected by many

TABLE 1
EOG Responses to Prostaglandins and Steroids in the Order Cypriniformes

Superfamily, family, and subfamily	Number of species tested in taxon	Prostaglandins	21-Carbon steroids			19-Carbon steroids		
			F^a	G^b	S^c	F	G	S
Cyprinoidea								
Cyprinidae								
Acheilognathinae (bitterlings)	4	$+^d$	$-^e$	−	−	−	−	−
Danioninae (danios)	17	+	$(+)^f$	(+)	(+)	−	(+)	−
Leuciscinae (North American species)	8	+	+	+	+	−	−	−
Cyprininae (goldfish, carp)	41	+	(+)	(+)	+	(+)	(+)	−
Cobitoidae								
Gyrinocheilidae (algae eaters)	1	+	−	−	−	−	−	−
Catostomidae (suckers)	5	+	−	−	−	−	−	−
Cobitidae (loaches)	5	+	−	−	−	−	−	−
Balitoridae (river loaches)	1	+	−	−	−	−	−	−

[a] Unconjugated steroid.
[b] Glucuronated steroid.
[c] Sulfated steroid.
[d] Compounds in this chemical group detected by all species tested in taxon.
[e] Compounds in this chemical group not detected by any species tested in taxon.
[f] Compounds in this chemical group detected only by some species tested in taxon.

fish, there also is limited evidence for biological responses in a variety of species, several of which are briefly reviewed.

A. Common Carp (*Cyprinus carpio*) and Crucian Carp (*Carassius carassius*)

Common and crucian carp are closely related to goldfish (all are in the subtribe Cyprini), have similar mating systems, and detect a similar set of prostaglandins and steroids. Although pheromonal prostaglandin function has been investigated only in goldfish, males of all three species increase blood GtH II concentration and milt volume in response to 17,20ßP exposure.

B. African Catfish (*Clarias gariepinus*)

Although the mating system of African catfish (order Siluriformes) is similar to that of goldfish (males compete for ovulated females), research in the catfish demonstrates that males release pheromones that induce ovulation and attract the ovulated female. Much evidence from biochemical and EOG studies indicates that the attractant pheromone is produced by the seminal vesicles and is at least partially composed of steroid glucuronides, the most potent of which are 5ß-pregnan-3α,17α-diol-20-one-3-glucuronide and 5ß-androstan-3α,11ß-diol-17-one-3-glucuronide.

C. Atlantic Salmon (*Salmo salar*)

As measured by EOG recording, the EOG sensitivity of precociously maturing male parr to testosterone (T), testosterone–sulfate (T–S), and F prostaglandins increases for a brief period during testicular maturation, a time when they display a positive rheotactic response to T. Another, unconfirmed, report suggests that following brief exposure to urine of ovulated females, EOG responsiveness to 17,20ßP-20-S increases rapidly in male *S. salar*. The biological function of 17,20ßP-20-S and the mechanism and significance of these changes in olfactory responsiveness is not known. Interestingly, none of these changes in EOG responsiveness to prostaglandins and steroids have been observed in males of the brown trout (*Salmo trutta*), which occasionally hybridizes with Atlantic salmon. Nevertheless, olfactory responsiveness to prostaglandins is reported in several salmonid species, suggesting that prostaglandin pheromones may be common in this group. Their function is unclear and possibly complex: in *Salvelinus* prostaglandins have been suggested to be a male releaser; in *Salmo*, in which exposure to F prostaglandins increases plasma 17,20ßP concentration in maturing male parr, F prostaglandins have been suggested to be a priming pheromone released in urine of ovulated females.

D. Gobies

The first reported hormonal pheromone in fish was etiocholanolone–glucuronide (Etio-G), released by male black gobies, *Gobius jozo* (order Perciformes) to attract ovulated females to their nests. Etio-G is produced in a nongametogenic portion of the testis (mesorchial gland) rich in Leydig cells, one of the few examples of specialization for signaling in a hormonal pheromone system. These early findings are supported by recent EOG studies in the roundhead goby, *Neogobius melanostomus*, showing that this species detects a number of steroids through classes of olfactory receptor sites that are most responsive to Etio-G, estrone, estradiol-G, and dehydroepiandrosterone-G. The function(s) of these odorants, and whether they are released by *Neogobius*, is not known.

V. FUNCTION AND EVOLUTION OF HORMONAL PHEROMONES

Much information indicates released hormones and their metabolites are reproductive pheromones in many fish. However, it is generally unclear what specific functions these hormonal pheromones perform and whether they are specialized, species-specific signals. For example, despite the considerable information on hormonal pheromones of goldfish, it is not known if the female benefits from the male's responses or if the hormonal compounds the female releases are specialized for pheromonal function and distinct from those of related species. A parallel can be drawn with the insects, in which pheromone systems are well understood and often composed of rather common components whose precise mixture determines their species-typical identity. Even though related insect species frequently use the same pheromonal compounds and detect them with similar sensitivities, careful examination reveals that these compounds are released in specialized, species-specific mixtures. Although the difficult question of whether fish hormonal pheromones are specialized remains to be explored, it is instructive to consider how these cues might have evolved from released hormones.

Because endogenous physiological functions of steroids and prostaglandins are highly conserved among vertebrates, it is reasonable to assume that these compounds originally functioned only as hormones (a "preadaptation" stage; Fig. 1), and that their functions as pheromones have evolved recently. The first step in the evolution of pheromone function might occur when mutation(s) in olfactory function enables detection of, and response to, hormonal compounds. We term this hypothetical stage, which likely benefits only the receiver, "chemical spying" (Fig. 1). The steroid pheromone system of goldfish may be an example of this initial step in pheromone evolution. Hypothetically, chemical spying might then progress to chemical communication if the receiver's response also benefits the sender, which con-

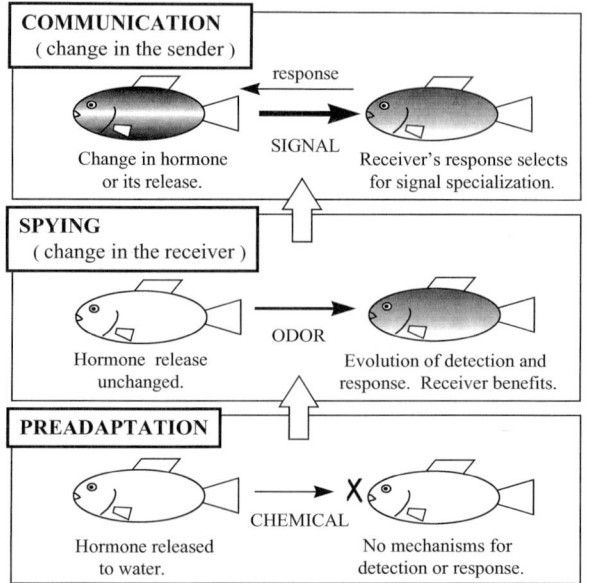

FIGURE 1 Proposed scenario for the evolution of hormonal pheromones.

sequently evolves the ability to produce a specialized signal by changing hormone production, metabolism, or release. The apparently specialized production of pheromonal Etio-G in *G. jozo* might represent this communicatory stage of pheromone function and might have evolved because the ability of females to locate males through released Etio-G favored those males producing the strongest chemical signal. Because much is known about the hormonal compounds detected by fish, and because biochemical techniques for measuring their production and release are readily available, the stage is set for exploring the fundamental questions regarding the function and evolution of fish reproductive pheromones.

See Also the Following Articles

PHEROMONES, INSECTS; PHEROMONES, MAMMALS

Bibliography

Colombo, L., Belvedere, P. C., Marconato, A., and Bentivegna, F. (1982). Pheromones in teleost fish. In *Proceedings of the Second International Symposium on the Reproductive Physiology of Fish* (C. J. J. Richter and H. J. T. Goos, Eds.), pp. 84–94. Pudoc, The Netherlands.

Liley, N. R. (1982). Chemical communication in fish. *Can. J. Fish. Aquat. Sci.* **39**, 22–35.

Sorensen, P. W. (1992). Hormones, pheromones and chemoreception. In *Fish Chemoreception* (T. J. Hara, Ed.), pp. 199–228. Chapman & Hall, London.

Sorensen, P. W., and Caprio, J. (1997). Chemoreception. In *The Physiology of Fishes* (D. Evans, Ed.), 2nd ed., pp. 375–405. CRC Press, Boca Raton, FL.

Stacey, N. E., and Cardwell, J. R. (1995). Hormones as sex pheromones in fish: Widespread distribution among freshwater species. In *Proceedings of the Fifth International Symposium on the Reproductive Physiology of Fish* (F. W. Goetz and P. Thomas, Eds.), pp. 244–248. Fish Symposium 95, Austin, TX.

Stacey, N. E., and Sorensen, P. W. (1991). Function and evolution of fish hormonal pheromones. In *Biochemistry and Molecular Biology of Fishes* (P. W. Hochachka and T. P. Mommsen, Eds.), Vol. 1, pp. 109–135. Elsevier, Amsterdam.

Stacey, N. E., Kyle, A. L., and Liley, N. R. (1986). Fish reproductive pheromones. In *Chemical Signals in Vertebrates* (D. Duvall, D. Muller-Schwarze, and R. M. Silverstein, Eds.), Vol. 4, pp. 117–133. Plenum, New York.

Van Den Hurk, R., and Resink, J. W. (1992). Male reproductive system as sex pheromone producer in teleost fish. *J. Exp. Zool.* **261**, 204–213.

Pheromones, Insects

Jeremy N. McNeil
Université Laval

Johanne Delisle
Natural Resources Canada

Claude Everaerts
CNRS

I. Which Sex Produces Sex Pheromones?
II. Location of Pheromone Glands
III. Chemical Composition of Pheromones
IV. Ecological Conditions That Affect Pheromone Emission and Reception
V. Physiology of Pheromone Production in Virgins
VI. Pheromone Physiology Following Mating

GLOSSARY

corpus bursa A component of the female reproductive tract where the spermatophore is deposited.

hair pencils Specialized reversible structures associated with male pheromones.

infochemical A chemical substance involved in the transfer of information between two individuals (of the same or different species) that results in a behavioral and/or physiological response in the receiver that is adaptive for one or the other (or both) of the individuals.

monandrous Species in which females generally mate only once during their lifetime.

pheromone An infochemical that modulates interactions between two individuals of the same species and may be beneficial for the emitter, the receiver, or both.

pheromonostasis The temporary or permanent inhibition of pheromone production following mating.

polyandrous Species in which females mate several times throughout the reproductive period.

refractory period The time that pheromone production is inhibited following mating.

spermatheca The section of the female reproductive tract housing the sperm following their migration from the bursa after mating has terminated.

spermatophore A male-produced capsule, containing sperm and accessory gland secretions, transferred to the female at the time of mating.

Many intraspecific interactions in insects may be modulated through chemical cues, and examples of sex pheromones playing important roles in reproduction have been reported for most orders of insects. However, the most detailed studies have been carried out on Lepidoptera, Coleoptera, Diptera, Hymenoptera, and Orthoptera of importance to agricultural, forestry, and health-related problems. We will first examine the different roles pheromones play in mating and consider which sex produces them, discuss the locations of the pheromone producing glands, and consider some of the chemical aspects of pheromone composition that help provide species-specific communication channels. Then we will examine how biotic and abiotic factors may influence the emission and reception of these chemical signals and examine the physiological mechanisms that govern pheromone production in virgin and mated individuals. Because of space limitations we have restricted most of our examples to the moths, which have unquestionably been the most intensively studied group with respect to pheromone-mediated mating.

I. WHICH SEX PRODUCES SEX PHEROMONES?

Sex pheromones may be involved in two aspects of reproduction: (i) to facilitate encounters between the sexes and (ii) to modulate mate choice once potential mates come into close proximity. In the great majority of species, it is the female who releases the long-distance sex pheromone to attract potential mates (Box 1), although in some species, such as the cockroach *Eurycotis floridana*, it is the receptive male

BOX 1

In the armyworm, *Pseudaletia unipuncta*, it is the female that emits the long-distance sex pheromone (solid arrow). When sexually mature, the receptive female exhibits an evident calling behavior, similar to that seen in many Lepidoptera. She extrudes her ovipositor, thereby exposing the pheromone gland, the modified intersegmental membrane located between abdominal segments 8 and 9. Her wings are raised slightly and there is intermittent wing fluttering during the calling period. In this species the sex pheromone is a blend of Z11–16 acetate and Z11–16 alcohol, with the acetate being the major component (more than 99.0% of the blend). Once a male is in close proximity to the calling female, he extrudes a pair of bilateral hair pencils, housed in lateroventral pouches on the second to fifth abdominal segments (open arrow). These hair pencils contain two active compounds, benzaldehyde and acetic acid. Removal of these hair pencils markedly reduces male acceptance by females, supporting the hypothesis that male pheromones serve in female mate choice.

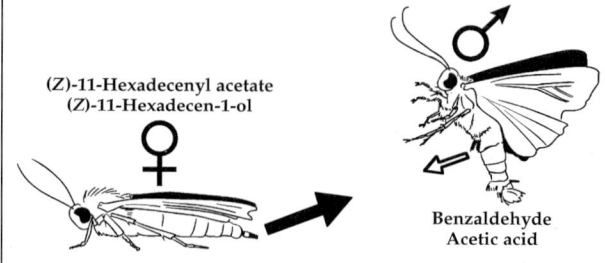

BOX 2

In the cockroach, *Eurycotis floridana*, males produce a long-distance sex pheromone from epidermal glands located under the tergites of the seventh and eighth abdominal segments (solid arrow). The pheromone is a blend of 4-hydroxy-5-methyl-2[3H]furanone, produced by the gland on segment 7, and benzyl 2-hydroxybenzoate, ($2R^*$, $3R^*$) butanediol, and dodecanol originating from the gland on segment 8. Mature males rock from side to side with a typical body posture that exposes the pheromone glands. It is of interest to note that while the pheromone blend attracts females it also repels conspecific males. Whether females produce a close-distance pheromone (open arrow) is unclear.

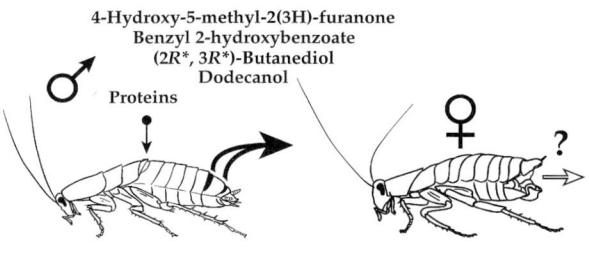

(Box 2). In a few species, such as the cabbage looper *Trichoplusia ni*, there is clear evidence that both males and females produce sex pheromones to attract the opposite sex over considerable distances (Box 3).

In the majority of insect species females are the choosy sex and, as a result, the pheromones involved in mate choice are generally of male origin. These are released by the male when in close proximity to a potential mate and may provide the female information concerning male quality. Eliminating the structures that contain the male pheromone, or the antennal receptors of females, results in males being refused by females and leads to a reduced incidence of mating, providing support for the idea that male pheromones play a role in mate choice. Male sex pheromones are often produced prior to adult moth emergence, so the amount of pheromone a male has should decline with each mating attempt and information on previous male mating history could be detected by females. Male quality generally declines with successive matings and females accepting previously mated males as mates often have reduced fecundity and fertility. Thus, the ability of females to discriminate between different-quality mates is of major importance for their overall reproductive success. The quantity of male pheromone may also be related to the quality of the larval host plants, especially if, as seen in a number of arctiid species, male pheromones are derived from host plant defense compounds. These compounds may be transferred to the female at the time of mating, providing both her and her egg complement additional defense against predators. Thus, the level of defense compounds in the plant influences male pheromone titer and provides females an index of male quality with respect to male-derived resources.

BOX 3

In the cabbage looper, *Trichoplusia ni*, both sexes produce long-distance pheromones (solid arrows). The female pheromone gland has a similar structure and location as that described for the true armyworm in Box 1 and females assume a typical calling posture. Males release a three-component pheromone blend from hair pencils located on the eighth and ninth abdominal segments. This blend serves as both long- and short-distance (open arrow) pheromones. The peak of male emission that induces upwind movement of females occurs in the first hour of the scotophase, whereas the calling period of females to attract males occurs later in the night.

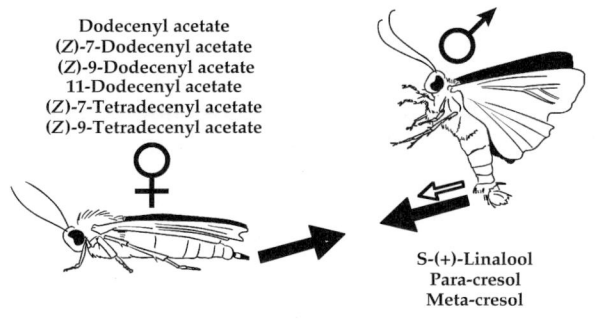

II. LOCATION OF PHEROMONE GLANDS

Pheromones may be produced in a wide variety of structures, located on different parts of the body. The production and release of long-distance pheromones may be associated with distinct behaviors (referred to as calling behavior) that expose specific pheromone glands (Boxes 1–3), whereas in other species they may be released from much of the body surface without any evident calling behavior. The short-distance male pheromones, which generally play a role in the last phases of the courtship behavior, are released from specific structures located on different parts of the abdomen (Boxes 1 and 3), the legs, or the wings. The diversity in both site of production and manner of release of long- and short-distance pheromones is observed both within and between insect orders.

III. CHEMICAL COMPOSITION OF PHEROMONES

The pheromones in any species are usually composed of a very precise mixture of chemicals rather than a single chemical (Boxes 1–3) and provide species-specific communication channels. The specificity of a blend may be attained in a number of ways other than the obvious use of different constituents. Closely related species may use the same components but retain species-specific communication channels by varying the relative proportions within the blend, e.g., a three-component blend of A + B + C in 80:10:10 and 40:30:30 ratios. In a similar fashion, different stereoisomers of the same compound may determine blend specificity. For example, there are two races of the European corn borer, *Ostrinia nubilalis*, and the female pheromone in both is composed of the Z and E isomers of 11-tetradecenyl acetate. In one strain (the Z strain) females emit, and males respond to, a pheromone blend composed of a 97:3 ratio of Z to E isomers, whereas in the other (the E strain) females emit, and males respond to, a blend of 99:1 ratio of E to Z. It has been suggested that two sympatric species may use the same pheromone but that specificity is obtained by the use of different concentrations. While this is possible, the example most frequently cited is no longer valid because different minor components have been identified in the pheromone blends.

IV. ECOLOGICAL CONDITIONS THAT AFFECT PHEROMONE EMISSION AND RECEPTION

There is generally a distinct diel periodicity in the emission of, and response to, the long-distance sex pheromone, with each species having a specific temporal window during the 24-hr period. These temporal differences have been postulated as one possible reproductive isolating mechanism for closely related, sympatric species. However, for any given species the temporal window for both the release of and response to the long-distance pheromone can change from one day to the next because prevailing biotic and abiotic conditions may markedly affect a number of different variables associated with pheromone-mediated mating systems.

A. Biotic Factors

1. Age and Mating

There may be a marked change in the patterns of pheromone production and emission by virgin female moths with age. Older females usually have lower pheromone titers than younger individuals and they initiate calling, the behavior associated with the release of the pheromone, earlier during the calling window than younger conspecifics. Furthermore, older females spend more time calling than do younger ones. It has been postulated that these differences serve to increase the probability of mating by reducing competition with younger females because these individuals tend to call late in the calling period, a hypothesis that is supported by a number of field experiments. Females may resume calling after an initial mating, but because these individuals tend to call later in the calling window they should not represent direct competition when older virgin females first initiate calling.

Male receptivity to the female sex pheromone is affected by age, and in many species there is an increase in receptivity during the first few days following emergence. This is often related to the phenomenon of protandry, in which males emerge several days before females and reach sexual maturity as the newly emerged, receptive females become available. The impact of mating on subsequent male responsiveness is not particularly pronounced in males. This is not surprising because the best male reproductive strategy is to acquire as many mates as possible and any postmating decline in receptivity could prove costly.

2. Population Density

In a number of species, females not only emit but also can detect the long-distance pheromone released by conspecifics and there is evidence that they may alter their calling behavior in response to the levels of ambient pheromone present. For example, spruce budworm females start calling earlier and spend considerably more time calling when there is pheromone present compared to when there is no pheromone. Furthermore, there is a marked increase in female flight activity when atmospheric levels of pheromone are high. It has been postulated that the response to increased levels of conspecific pheromone plays an important role in emigration which, in this species, is generally undertaken by mated females moving from habitats of declining quality. However, in several tea tortrix species (in the same family as the spruce budworm) the presence of conspecific pheromones results in a reduction of calling behavior. Currently, there is not enough information to postulate why such interspecific differences have evolved and their importance as reproductive strategies remains to be clarified.

Males obviously detect ambient levels of female sex pheromone, but under natural ecological conditions it seems unlikely that levels would reach concentrations to significantly disrupt male searching behavior. However, the basis of mating disruption as a control means is to increase the background pheromone to such levels that males can no longer locate potential mates. This reduction in mate location may result from the saturation of antennal pheromone receptors and/or the masking of plumes produced by receptive feral females.

In certain species, males may be able to detect the pheromone released by conspecific males when courting a receptive female. It was postulated that male pheromone could serve to reduce male–male competition if the male approaching a calling female arrested upwind flight upon detecting the pheromone of another male. This hypothesis was not supported by subsequent experiments and when considered from an evolutionary perspective this is not particularly surprising. If the female rejects the first male then the second male would miss a potential mating opportunity if he stopped upwind flight upon detecting male pheromone in the plume.

3. Host Plants

The presence of host plants, suitable for oviposition and larval development, is an important component of the pheromone-mediated mating of insects. While the most elaborate studies have been carried out on different species of bark beetle, there is an ever-growing body of evidence that similar plant–insect interactions may occur in the Lepidoptera. For example, the sunflower moth, which uses host plant pollen as an oviposition stimulant, is believed to migrate in search of resources that fluctuate both in time and in space. In this species, females become sexually mature within 2 days of emergence when pollen is present, whereas in the absence of pollen the average time for the same process is nearly 1 week. Furthermore, the time spent calling is significantly longer and the production of chorionated eggs is greater when females are in the presence of pollen. Host plant volatiles may also increase the quantity of pheromone synthesized, as recently reported in the corn earworm. The combined effect of these changes in pheromone biology and ovarian development will increase the probability of attracting a mate and subsequently permit the female to exploit the suitable oviposition sites available, thereby improving her reproductive success.

4. Pathogens

Many insects often have sublethal parasitic or pathogenic infections and these are known to affect parameters, such as growth, longevity, and fecundity, that have an impact on reproductive success. However, despite the frequency of pathogenic and parasitic infections in natural populations there is a noticeable dirth of studies examining their possible effects on pheromone production and female mating success. A recent study has shown that pheromone production in the beetle, *Tenebrio molitor*, an intermediate host of the rat tapeworm, is significantly reduced when females are infected with metacestodes.

As with females, very little work has been done on the role of pathogens on male responsiveness to pheromones. Infection by metacestodes of the rat tapeworm depressed the copulatory response of yellow mealworm males to the female sex pheromone at the peak of normal responsiveness, about a week after emergence. However, this effect was not observed in older males, possibly because the response of control males also declined significantly with age. Protozoan infections may reduce the ability of western spruce budworm males to express the complete sequence of behaviors associated with upwind flight to a pheromone source, and the degree to which behavior is modified increases with increased pathogen load. It seems reasonable to assume that a closer examination of sublethal parasitic and pathogenic infections on pheromone communication systems of both sexes will uncover a number of different effects influencing the reproductive success of infected individuals.

B. Abiotic Factors

1. Temperature

As noted earlier, in many species there is a distinct diel periodicity of calling. However, temperature conditions can have a marked effect on the timing and duration of the calling window associated with the emission of female sex pheromone. For example, oblique banded leafroller females are normally nocturnal, calling at the onset of the scotophase, but if ambient temperatures are cool, calling begins at the end of the photophase. In the true armyworm, while calling activity is always nocturnal, this behavior is expressed much earlier in the scotophase at low air temperatures than when they are high. Clear wing moths are generally diurnal and on cool days the onset of calling occurs later in the day than it does on warm days. Similar trends have been observed in many species in which calling behavior has been carefully examined, and these changes in response to temperature are considered an adaptation to ensure receptive females release pheromone at a time when air temperatures are conducive for male flight. As one would expect, parallel temperature-modulated shifts are seen in male responsiveness to the female pheromone, thereby ensuring that the activity periods of the sexes are well synchronized.

In species that migrate in response to predictable seasonal deterioration of the habitat, temperature may be a major cue determining the age at which females reach sexual maturity. For example, females of the true armyworm have fully developed ovaries and exhibit calling behavior about a week following emergence at 25°C, whereas at 10°C these same processes require approximately 3 weeks. A similar delay is seen in the responsiveness of males to the female sex pheromone. Migration is generally undertaken by sexually immature individuals, and it has been argued that this delay in maturation provides a suitable time window for both sexes to migrate and locate a suitable habitat for reproduction.

2. Relative Humidity

Relative humidity may also play an important role in pheromone-mediated mating. The European corn borer is a species that normally mates in grassy areas outside the agroecosystem, where relative humidity is nearly 100%. Females held under lower relative humidity conditions generally (i) spent less time calling than controls at high humidity, (ii) did not show a pronounced advance in the calling window seen in controls, and (iii) on certain nights failed to call at all. The lepidopteran pheromone gland represents a very large membranous surface area from which water may be lost, so a reduction in calling behavior under dry conditions is believed to be an adaptation to minimize desiccation. Low-humidity conditions also reduce male corn borer responsiveness to the female sex pheromone. The exact causal mechanism for this is unknown, but it has been postulated that the efficiency of the binding at the pheromone receptor sites on the antenna is affected. The combined effect on both sexes results in a significant delay in mating and thus may reduce the reproductive success of males and females.

3. Light Intensity and Day Length

The effect of temperature on the periodicity of female calling and male responsiveness mentioned previously may be significantly modified by light intensity. For example, when light intensity is high, a lower temperature is required for pheromone-mediated mating behaviors to occur during the photophase in crepuscular species than when light intensities are low.

In multivoltine species, adults active in early spring may be subjected to very different day length conditions than those flying in midsummer and significant temporal differences in the mating period could be observed between generations. However, within any given generation changes in day length would probably not be a major factor affecting either the calling behavior of receptive females or males' responsiveness. Day length may, however, be of considerable importance for migratory species that use abiotic conditions as indicators of habitat quality. Both sexes of several migratory armyworm species delay the onset of sexual maturation by several days under "short-day" conditions (12 hr light : 12 hr dark) compared to "long-day" ones (16 hr light : 8 hr dark) under identical temperature conditions. This response is again considered an adaptation to provide a greater time window for emigrating adults to locate suitable breeding sites before they reach sexual maturity.

4. Wind Speed and Atmospheric Conditions

Wind velocity and turbulence obviously have an impact on flight, with each species having a wind speed threshold (generally size determined) above which sustained, oriented flight is impossible. In addition, these same factors will play a determining role in the diffusion and stability of pheromone plumes and consequently the efficiency of the communication system. At very low wind speeds the pheromone will only be carried over short distances and thus provides very few directional olfactory cues to potential mates. Furthermore, the plume stability will be low, thereby reducing the probability that effective mate location occurs. However, within the range in which flight is possible, higher wind velocities will increase the dispersal of the female's pheromone plume and generate a more stable plume structure which will augment the probability of males locating the source. Despite the obvious importance of wind speed there are very few detailed studies examining how this parameter impacts on calling and upwind flight, although work on the cabbage looper has shown that there may be significant behavioral changes. Cabbage looper males fly less at wind velocities more than 2 or 3 m/sec and females spend more time calling at velocities between 0.3 and 1 m/sec than at higher and lower wind speeds. Calling is totally inhibited when wind speeds reach 4 m/sec. Thus, pheromone emission changes with wind speed and is maximal at velocities most conducive for male flight.

There is a small body of data that supports the idea that changes in barometric pressure may modify pheromone biology. In the spruce budworm the capture of males in pheromone traps baited with virgin females appears greatest during periods of marked increases or decreases in pressure. In contrast, the male upwind flight of the small parasitic wasp, *Aphidius nigripes*, to virgin females is maximum under stable pressure conditions and declines if there are major pressure changes in either direction. These findings are not surprising when one considers how wind speed and turbulence affect both flight capacity and the efficacy of pheromone communication systems. The ability to use short-term changes in barometric pressure as cues about future atmospheric conditions would permit individuals to modify their behavior in order to maximize reproductive success. However, as noted in the previous two examples, the responses vary with species, so a greater database is necessary before any general, ecologically valid scenarios can be developed.

V. PHYSIOLOGY OF PHEROMONE PRODUCTION IN VIRGINS

Synthesis of the female sex pheromone in Lepidoptera normally follows a diel pattern of production and degradation, and the initiation of pheromone biosynthesis is generally associated with the release of a pheromone biosynthesis-activating neuropeptide (PBAN) produced in the brain–subesophageal complex. These polypeptides occur in several insect

BOX 4

In several *Helicoverpa* species pheromone production follows pathway No. 1, with PBAN being produced in the subesophageal ganglion (SOG), released via the corpora cardiaca (CC) into the hemolymph and transported by the blood directly to the pheromone gland. The red-banded leafroller, *Arygrotaenia velutinana*, also releases the PBAN into the hemolymph but the target site is the corpus bursa (pathway No. 2). The bursa subsequently releases a "bursa factor" and this peptide acts on the pheromone gland to stimulate pheromone biosynthesis. Pathway No. 3, where PBAN moves along the ventral nerve cord (VNC) to the terminal abdominal ganglion (TAG) and causes the release of octopamine or other neurotransmitters to activate pheromone synthesis, was first proposed for *H. zea*. While there is doubt as to the validity of the model for the corn earworm, it is currently considered the appropriate pathway for pheromone synthesis in the gypsy moth.

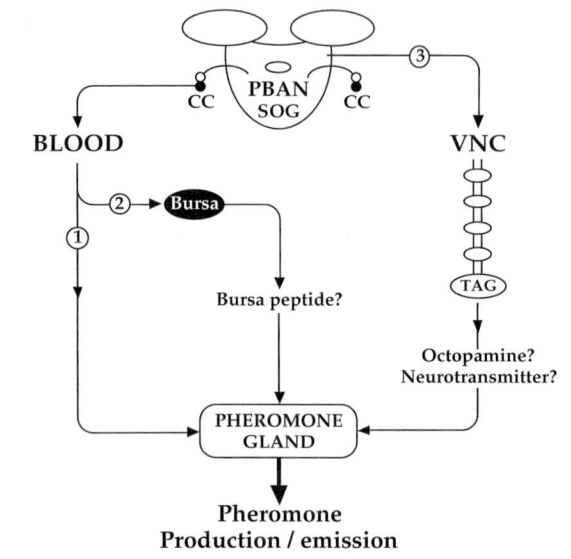

able that between-order similarities will be found. PBAN is not directly implicated in pheromone biosynthesis by the cabbage looper moth in which, unlike most other species studied, there is no diel pattern of pheromone production. However, there is evidence that in this species PBAN-like substances play a role in transporting the pheromone to the gland surface at the time of release. These findings suggest PBAN-like substances may modulate a number of processes associated with the emission of pheromone in females.

Studies on a number of different species have shown that several different pathways exist for this peptide to initiate the process of pheromone biosynthesis (Box 4). It may be released into the bloodstream and (i) transported directly to the pheromone gland or (ii) stimulate the bursa to release a bursa peptide which subsequently activates biosynthesis in the pheromone gland. It is also possible that PBAN passes directly along the ventral nerve cord to the terminal abdominal ganglion, where the release of second messengers subsequently activates pheromone production. Whether the various pathways observed reflect differences in the life histories of the species studied or whether different pathways may be used by the same species under different ecological/physiological conditions remains to be elucidated.

When adults of the true armyworm are reared under low-temperature, short-day conditions—cues announcing impending habitat deterioration—there is a significant delay in sexual maturation and this change is directly associated with low levels of juvenile hormone (JH), the hormone implicated in both ovarian development and migration in sexually immature adults. JH is also required for both pheromone synthesis and the expression of calling behavior in this species. It has been proposed that the release of PBAN is modulated by JH, a process inhibited by the lower JH levels present during the migratory phase but stimulated by the higher JH titers observed during the sexual maturation process. It is unknown exactly how PBAN, once released, functions in the true armyworm but preliminary data suggest that it follows pathway No. 1 (Box 4).

PBAN-like substances are present in the brains of both sexes but little is known about the possible roles of these polypeptides in males. Work on the black cutworm, *Agrotis ipsilon*, suggested that there

orders and there is a high degree of cross-specific activity both within and between orders. There is a considerable degree of homology in the different lepidopteran PBANs sequenced to date and it is prob-

> **BOX 5**
>
> Pheromonostasis in tortricids, such as the red-banded leafroller and the light brown apple moth, is initiated by neural messages to the brain. In the corn earworm the decline in pheromone production is a result of the combined effect of neural message (responsible for stopping pheromone biosynthesis) and a humoral signal of male origin that serves to deplete the pheromone present in the gland at the time of mating. There are no clear case humoral messages of female origin, but given the small number of experimental studies carried out to date, the possibility cannot be excluded.
>
>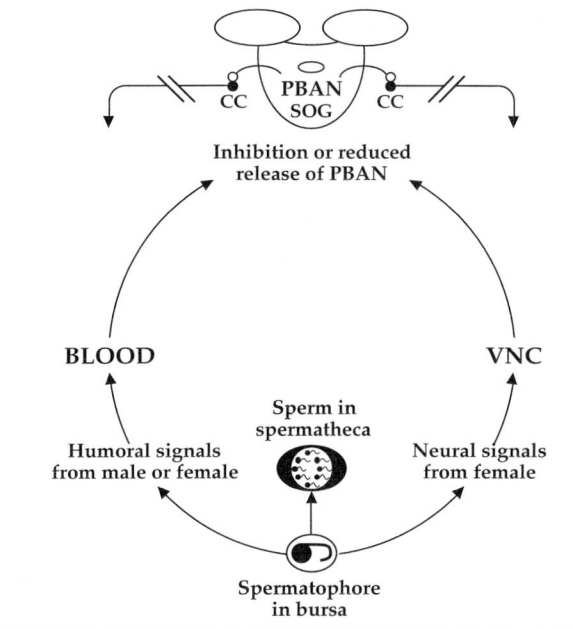

cesses involved in pheromonostasis vary between species (Box 5). The presence of the spermatophore in the corpus bursa or sperm in the spermatheca may result in humoral factors of female origin being released and carried to the brain, where they inhibit the production and/or release of PBAN. In other cases, the humoral factors may be of male origin (e.g., the accessory glands) and transferred to the female at the time of mating. Pheromonostasis may also be induced through neural signals from the bursa or spermatheca to the brain. It should be noted that neural and humoral pathways are not necessarily mutually exclusive.

The duration of the refractory period following mating may vary for a number of reasons. In monandrous species, remating generally occurs in response to the lack of viable sperm to fertilize the full egg complement. However, in polyandrous species, remating may also serve to acquire male-derived resources that actually increase the total number of eggs that a female produces, even if the first mate provided sufficient high-quality sperm to fertilize all eggs. The quality of the previous male, as determined by factors such as age, previous mating history, and larval host plants, will certainly affect whether females remate as well as influence the duration of the refractory period between matings. Reduced spermatophore size, low production of sperm, and/or accessory gland secretions could all result in a reduced pheromonostasis and thus increase the probability of additional matings.

is an effect of PBAN on male receptivity to the female pheromone. If true, because *A. ipsilon* is a migrant species, it is possible that different JH titers during migratory and reproductive phases modulate the production and release of PBAN in males, as proposed for true armyworm females.

VI. PHEROMONE PHYSIOLOGY FOLLOWING MATING

The production and emission of pheromone terminate following mating and the physiological pro-

See Also the Following Articles

Pheromones, Fish; Pheromones, Mammals

Bibliography

Agosta, W. C. (1992). *Chemical Communication: The Language of Pheromones.* Scientific American Library, New York.

Bell, W. J., and Cardé, R. T. (Eds.) (1984). *Chemical Ecology of Insects.* Sinauer, Sunderland, MA.

Birch, M. C., Poppy, G. M., and Baker, T. C. (1990). Scents and eversible scent structures of male moths. *Annu. Rev. Entomol.* **35,** 25–58.

Brossut, R. (1996). *Phéromones: La Communication Chimique Chez Les Animaux.* CNRS Editions, Paris.

Cardé, R. T., and Bell, W. J. (Eds.) (1995). *Chemical Ecology of Insects 2*. Chapman & Hall, New York.

Cardé, R. T., and Minks, A. K. (Eds.) (1997). *Insect Pheromone Research: New Directions*. Chapman & Hall, New York.

Landolt, P. J., and Phillips, T. W. (1991). Host plant influences on sex pheromone behaviour of phytophagous insects. *Annu. Rev. Entomol.* 42, 407–430.

McNeil, J. N. (1990). Behavioral ecology of pheromone-mediated communication in moths and its importance in the use of pheromone traps. *Annu. Rev. Entomol.* 36, 25–58.

Payne, T. L., Birch, M. C., and Kennedy, C. E. J. (Eds.) (1986). *Mechanisms in Insect Olfaction*. Clarendon, Oxford, UK.

Prestwich, G. D., and Blomquist, G. J. (Eds.) (1987). *Pheromone Biochemistry*. Academic Press, New York.

Roitberg, B. D., and Isman, M. B. (Eds.) (1992). *Insect Chemical Ecology: An Evolutionary Approach*. Chapman & Hall, New York.

Pheromones, Mammals

John G. Vandenbergh
North Carolina State University

I. Introduction
II. Signaling Pheromones
III. Maternal–Offspring Interaction
IV. Reproduction-Related Signals
V. Priming Pheromones
VI. Mechanism of Pheromone Reception

GLOSSARY

accessory olfactory bulb The caudal portion of the olfactory bulb that receives neural input from the vomeronasal organ.

estrous synchronization Correspondence in time of the estrous cycle in social groups of some female mammals; also known as the Whitten effect.

estrus Originally a behavioral term that indicates a period of heightened activity at the time of peak fertility in female mammals. It now refers to a series of behavioral and physiological changes that occur at or just before ovulation in female mammals.

flehmen A characteristic behavioral posture (including flaring of nostrils and curling of upper lip) shown by many male mammals to detect chemical signals from estrous females.

olfactory bulb The most rostral part of the brain that receives neural input from the main olfactory epithelium.

pheromone A chemical substance released into the environment that influences the behavioral and/or physiological actions of other members of the same species.

pregnancy blockage The termination of pregnancy in female rodents exposed to the presence of an unfamiliar male of their species or to pheromones from the unfamiliar male; also known as the Bruce effect.

priming pheromone A chemical signal produced by one or more individuals resulting in a developmental or physiological change, usually over the long term, in the recipient animals of the same species.

puberty acceleration The acceleration of puberty in females of many mammalian species through direct contact with the adult male or by exposure to a male urinary pheromone; also known as the Vandenbergh effect.

puberty inhibition The delay of puberty among female mammals housed in groups due to direct physical interactions or to a female urinary pheromone.

signaling pheromone A chemical substance produced by one or more individuals which results in a behavioral, and usually almost immediate, response in the recipients of the same species.

vomeronasal organ A tube-shaped structure found in most mammals bilaterally located on either side of the septum in the nasal cavity. This is the primary receptor organ for priming pheromones.

I. INTRODUCTION

Organisms as ancient as bacteria and slime molds produce chemical substances that are released into the environment and influence the behavioral and physiological actions of other members of the same species. This is the basic definition of a pheromone, a term coined by Karlson and Luscher in 1959. Pheromonal communication occurs throughout the animal kingdom and, during the course of study of this form of communication, pheromones have been classified into two types: (i) signaling, or releasing pheromones, which are produced by one individual and which have a behavioral, and usually almost immediate, response in the recipient, and (ii) priming pheromones, which are chemical signals produced by one or more individuals that result in a developmental or physiological change in a receiving animal and affect changes over longer periods of time. This classification system is not all-or-none because some pheromonal systems seem to fall into an intermediary category having attributes of both signaling and priming effects. Pheromonal communication requires a producer and at least one recipient of the same species to result in a change in physiology, development, or behavior. This excludes a number of other chemicals that serve as signals in the environment such as the aroma of flowers or the scents of other species.

Using chemical, rather than visual or auditory signals has some benefits for the animals. Chemosignals are effective in the dark, are relatively long-lasting, and can be highly specific.

II. SIGNALING PHEROMONES

A. Mate Attraction

A silkworm moth male, fluttering through the air, when encountering just a few molecules of an odor called bombykol alters its flight pattern so that it continually moves toward higher and higher concentrations of bombykol molecules in the air. It uses highly sensitive chemoreceptors on its antennae to detect remarkably small quantities of the bombykol. The pheromone is produced by specialized glands in the abdomen of the female. By following ever greater concentrations, the male moth (and other males) "home in" on the female, and mating can occur. This is a classic example of a signaling pheromone. The female moth produces a specific compound, which is transmitted through the air, is detected by a recipient, and induces a behavioral response that results in a positive benefit for either the producer or the receiver of the signal or both. The main cue, in this case bombykol, may be part of a blend of chemicals that serve as a sex attractant. The use of blends or minor differences in concentration of compounds in the blend serving as sex attractants may permit moths to recognize their own species from closely related moths that overlap their range.

Most insects use chemical cues to organize their behavior and to communicate among members of the species. Signaling pheromones serve as sex attractants, to repel competing males, and for dispersion associated with high population density. Temporal patterning of the pheromone in the air may be important. This is especially true for the complex upwind orientation behavior of a male moth seeking a receptive, pheromone-emitting female. Some of this temporal patterning is the result of pulsatile release, some due to turbulence in the air, and some due to the male moth's zigzag pattern of movement through a plume of pheromone-containing air. Such temporal patterning reduces the risk of olfactory accommodation and permits more accurate homing in on the female.

Pheromones are not limited to airborne molecules. Aquatic organisms have evolved chemical signaling systems that are transmitted through the water. Fishes often live their lives in murky water or are nocturnal and depend to a great extent on chemical signals to attract conspecifics, recognize kin, or synchronize their reproductive physiology and behavior.

Among mammals, the hamster provides an excellent example of the role of signaling pheromones in mate attraction. The female hamster has a very regular ovarian cycle 4 days in length. Males can detect the stage of the female's cycle by changes in vaginal scent. At the time of proestrus the female scent marks frequently by pressing the vulva of the vagina to the substrate. This leaves behind an olfactory signal to which the males attend by "sexual trailing." The signal, which was first thought to be a very common chemical, dimethyl disulfide, remains unknown. Nevertheless, experimental application of a "trail" of the vaginal secretion will be followed by the male, and in the presence of another chemical signal (a peptide) the males will attempt to mount the female. The scent deposited by the female also stimulates the male to scent mark his environment and thus a two-way communication between the sexes is facilitated.

The hamster is but one of many mammals in which chemical signals play an important role in sex attraction and in mating. Among the animals using scent for sex attraction are other rodents such as mice and gerbils, ungulates such as black-tailed deer and several antelope, and carnivores such as lions, cheetah, and domestic cats and dogs.

Male ungulates and many other animals routinely investigate the urine or anogenital area of females presumably to determine the reproductive state of the female. Investigatory behavior in many species often elicits urination by the female. Males place their noses directly into the urine and show a stylized sniffing behavior with the head up, upper lip curled, and the nostrils flared. This is termed "flehmen." During flehmen the fluids from the female enter the vomeronasal organ, a specialized olfactory organ which perceives the chemical message and relays the signal on to the brain.

B. Mate Selection

The first task an animal faces in selecting a mate is to identify a member of the opposite sex. Sex recognition is often facilitated by olfactory signals, either a specific compound or a subtle difference in the blend of compounds. Almost all species tested show differences in responsiveness to male and female odors. In many, males prefer the odor of females over the odor of males and their response varies with the reproductive state of the female. The odors from the female that are most attractive typically peak at the time of highest sexual receptivity. Females also show highest preference for male vs female odors at the time of estrus. The intensity of the female's response to odors can vary with the social status of the male, the quality of the nutrients the male has consumed, and the seasonal status of the male. An important internal mediating factor for this varying response by the female seems to be the androgen status of the male. The higher the androgen level, the higher the production of direct metabolites or indirect compounds from glands and body organs that serve as male pheromones.

Mate selection can have obvious genetic consequences for the offspring and the population. In many species the individual's complex blend of odors is sufficiently unique to be used for individual recognition. In house mice and rats, for example, the unique odor of the individual is related to a specific part of the genome, the major histocompatibility complex (MHC). Male mice can be trained to selectively respond to the odors of individual females who are genetically identical (congenic) except for the genes at the MHC locus. If given a choice they will preferentially mate with the females that differ from their own genome at that locus over those with which they are congenic. The finding that they will mate with the female that is different at the MHC locus is important because the MHC locus in the mouse (homologous with the HLA locus in humans) provides the individual with its immunological identity. Heterozygosity among the genes at this locus is very beneficial at the population level for disease resistance and other traits. Thus, having the genes that convey self in an immunological sense also provide cues for mate selection is highly adaptive.

III. MATERNAL–OFFSPRING INTERACTION

Chemical communication is also an important channel of communication between mother and young in mammals. The first requirement is that the mother can discriminate her own young from offspring in general. This is accomplished largely by olfactory cues in most mammals. In sheep there is a short period during which an ewe will accept another's lamb, but after a few hours she becomes imprinted upon her own and will reject others. In goats and many other species identification of a mother's own young seems instantaneous. Tests on the source of the individual recognition odor point to diet-based differences. Under wild conditions, diet can provide a great deal of variability. In the laboratory, standard diets are provided and, as one would expect, mothers are much more likely to accept cross-fostered young. Intensive work on rat maternal behavior has shown that mother rats at first find their pup's odor offensive but, after continued exposure, the pups become attractive and receive maternal care. As is true of other mammals, human mothers can also discriminate between their own infants and those of others based on smell alone.

In addition to using scents to identify their young, mother rats use them to orient to their pups and to retrieve them when out of the nest. Rat pups also show attraction to their mothers and to the nest site. Anal excreta emitted by the dams is the principal source of the attractant pheromone. The female rat produces a special type of fecal pellet, a cecotroph, as the source of this odor. The odor is a consequence of bacterial action since its presence can be suppressed by antibiotics and its attractiveness to the pup depends on concentration.

IV. REPRODUCTION-RELATED SIGNALS

A number of chemical signals are indirectly related to reproduction through enhancing the potential for reproductive success. An example is the deterrent quality of the castor scent of the beaver. The beaver has a specialized gland near its anus that secretes an oily substance. After building small mounds of mud and debris around the perimeter of its territory, the beaver places a small quantity of this castor on the mound. Intruders are thus warned that the territory is occupied. Animals use many other signals to enhance their chances of survival and reproduction. Here the focus is on those pheromones directly related to developmental and physiological reproductive processes.

V. PRIMING PHEROMONES

A. Puberty Regulation

Juvenile female mice, like juvenile females of all species, must reach puberty before they can become reproductively active. In the house mouse, the onset of puberty can occur as early as 30 days of age or as late as 60 days of age, even if the mice are fed a standard diet and maintained under excellent housing conditions. Much of the difference in the age of puberty onset is attributable to social factors—specifically, the presence of signals from the male and from other females.

The presence of an adult male or a pheromone from the male can accelerate the onset of puberty in a number of species. Adult male house mice produce a chemical signal, probably a peptide, in urine that induces earlier onset of puberty among females. This compound, or blend of compounds, is under androgen control because if the mouse is castrated, his urine loses its ability to accelerate the onset of puberty within 1 week. If the castrated male is injected with testosterone, the puberty-accelerating effect of his urine is restored. Similarly, female mice do not normally produce puberty-accelerating pheromones in their urine, but if injected with testosterone for several days, their urine gains this ability to accelerate puberty in juvenile females. This puberty-acceleration effect (also called the Vandenbergh effect) is just one of many reproductively related pheromonal effects that have been described.

It is interesting to note that puberty inhibition due to exposure to females or their urine also occurs in mice and several other species of mammals. This inhibiting effect balances out the male acceleratory effect; in fact, it can override the male's effect. Female mice produce the urinary inhibitory pheromone when crowded together for several days. This suggests that the crowding due to high population density in the wild could feedback on the population by slowing the rate of puberty acquisition. The puberty-inhibition effect has been demonstrated under field conditions. Crowded females in a seminatural population produce the puberty-inhibiting pheromone, whereas those in a sparse population do not. Thus, some support is available for the notion that there could be a density-dependent factor operating through a pheromonal mechanism to regulate mouse populations.

B. Synchronization of Ovarian Cycles

Among the first pheromonal effects on reproduction to be discovered was the estrus synchronizing effect of the male described by Whitten in 1959. Female house mice, when reared in groups, show suppressed ovarian cycles. When paired with a male or exposed to urine collected from an adult male, a proportion of the females ovulate synchronously on the third night after male stimulation. Since Whitten's original discovery, studies have shown that the ovarian cycles in a number of species can be synchronized, including the golden hamster, goat, and sheep. In rats, females seem to be able to drive each others cycles to result in synchronous estrus. A similar effect has been reported for humans and some evidence exists suggesting that an axillary pheromone is responsible.

C. Pregnancy Block

During the course of breeding mice, Bruce noticed that if she mated a pair of mice, then removed the male and replaced him with a different ("strange") male, within 3 days of the first mating the pregnancy of that female was terminated in about 80% of the cases. The more different the genetic constitution of the male the more potent is his effect in blocking the pregnancy of the female. The blockage of pregnancy in the female seems to be secondary to the induction of estrus and related to a surge in the female's production of prolactin. If prolactin is blocked by bromocryptine, the blocking effect of the strange male disappears. The Bruce effect is a fascinating laboratory phenomenon which allows us to understand some of the pheromonal and physiological mechanisms involved in the females' response to the male, the regulation of the ovarian cycle, and the progress of pregnancy. The blocking effect can have important evolutionary consequences in that a new male, when entering a territory, may terminate a resident female's pregnancy, thereby releasing her to come into estrus again to be inseminated by the new male, thus expediting the entry of his genes into the next generation.

VI. MECHANISM OF PHEROMONE RECEPTION

In mammals the two primary sites of reception for chemical signals are the main olfactory system and the vomeronasal system. The sensory epithelium of the main olfactory system is specialized for perceiving airborne chemicals, some of which are pheromones. Upon reception the signals are conveyed via axons that pass back to the olfactory bulb of the brain, where initial processing occurs in the second-order neurons prior to distribution to a variety of brain areas. The vomeronasal system consists of a paired tubular structure on either side of the nasal septum that communicates with the nasal cavity and/or the oral cavity depending on the species. The sensory epithelium is similar to the main olfactory epithelium but it is lacking in cilia. Axons from the sensory epithelium of the vomeronasal organ form nerves that course back to the accessory olfactory bulb, immediately caudal to the main olfactory bulb. The second-order neurons in the accessory olfactory bulb connect through a synapse in the amygdala to the hypothalamic portions of the brain known to be involved with regulation of pituitary function, particularly as it relates to reproduction. In general, the main olfactory system is the receptor organ for signaling pheromones as well as other chemicals and distributes its information to many regions of the brain, including those with cognitive function. The vomeronasal organ is the receptor of priming pheromones and its distribution in the central nervous system bypasses the cognitive processing areas. Thus, it provides an almost direct channel for chemical signals to affect portions of the brain involved with the regulation of reproduction. The main olfactory system is found throughout the mammalian order but a functional vomeronasal system seems not to be present in higher primates, including humans. Recent anatomical studies suggest that a remnant of the vomeronasal organ remains in humans, but no clear neural connection to the central nervous system has been demonstrated to date.

Three other, less well-understood potential receptors of chemical signals exist: the septal organ, which is a patch of olfactory epithelium-like tissue on the septum of the nasal cavity; the nervus terminalis, which is found near the tip of the nose in some species; and the trigeminal nerve with its distribution to specific facial areas. Further study is needed to clarify the role of these potential receptors.

How odors are encoded has long perplexed researchers. Recent advances using molecular techniques have revealed that there are G-protein-coupled odorant receptors, encoded by an array of about 1000 genes, on the cells of the olfactory epithelium. The binding of a receptor by a specific odorant activates metabolic changes in the cell, i.e., activation of a heterotrimeric G-protein leading to activation of adenyl cyclase which in turn generates the neuronal electrical potential. To discriminate among the thousands of odors available to an animal, each of these receptors must respond to several odor molecules and each of these molecules must bind to several receptors. To permit discrimination among specific odors, specific neurons project to specific parts of the olfactory bulb allowing neural processing to occur in the brain.

Reception of pheromones in the vomeronasal organ differs from that of the main olfactory system. The intracellular response in the vomeronasal organ involves induction of an inositol-(1,4,5)-trisphosphate (IP_3) pathway involving G-proteins. The apparent use of the IP_3 pathway by the receptors of the vomeronasal organ may be homologous to that used in insect chemoreception and may explain why remarkable coincidences occur in pheromone structure used by such different organisms as mammals and insects. For example, the compound *exo*-brevicomin, which is produced by female western pine beetles to attract male beetles, is apparently the same substance found in male mouse urine that elicits male aggression and female attraction. Another, even more remarkable coincidence is the finding that (Z)-7-dodecenyl acetate, a common attractant among Lepidoptera. also serves as a urinary pheromone produced by female Asian elephants to signal sexual receptivity.

See Also the Following Articles

Mating Choice; Olfaction and Reproduction; Puberty Acceleration; Sexual Attractants; Whitten Effect

Bibliography

Axel, R. (1995). The molecular logic of smell. *Sci. Am.* October, 154–159.

Bruce, H. M. (1959). An exteroceptive block to pregnancy in the mouse. *Nature (London)* **184**, 105.

Kelly, D. R. (1996). When is a butterfly like an elephant? *Chem. Biol.* **3**, 595–602.

McClintock, M. K. (1971). Menstrual synchrony and suppression. *Nature* **229**, 244–245.

Preti, G., Cutler, W. B., Garcia, C. R., Huggins, G. R., and Lawley, H. J. (1986). Human axillary secretions influence women's menstrual cycles: The role of donor extract of females. *Horm. Behav.* **20**, 474–482.

Schneider, D. (1992). 100 years of pheromone research: An essay on Lepidoptera. *Nat. Wissenschaften* **79**, 241–250.

Sorensen, P. W. (1996). Biological responsiveness to pheromones provides fundamental and unique insight into olfactory function. *Chem. Senses* **21**, 245–256.

Stoddart, D. M. (1990). *The Scented Ape.* Cambridge Univ. Press, Cambridge, UK.

Vandenbergh, J. G. (1969). Male odor accelerates female sexual maturation in mice. *Endocrinology* **84**, 658–660.

Vandenbergh, J. G. (Ed.) (1983). *Pheromones and Reproduction in Mammals.* Academic Press, New York.

Vandenbergh, J. G. (1994). Pheromones and mammalian reproduction. In *The Physiology of Reproduction* (E. Knobil and J. Neill, Eds.), 2nd ed., pp. 343–362. Raven Press, New York.

Whitten, W. K. (1956). Modifications of the oestrus cycle of the mouse by external stimuli associated with the male. *J. Endocrinol.* **13**, 399–404.

Phoronida

Russel L. Zimmer
University of Southern California, Los Angeles

I. Description of the Taxon
II. Modes of Reproduction: Sexual versus Asexual
III. Modes of Sex: Gonochorism versus Hermaphoditism
IV. Anatomy of the Reproductive Systems
V. Reproductive Physiology and Endocrine Control
VI. Modes of Fertilization
VII. Modes of Development
VIII. Larvae and Metamorphosis

GLOSSARY

actinotroch The characteristic larva of the phylum Phoronida.

heterochrony A term describing changes in the timing of appearance of features during development (or in rate of development of an organism). For example, the precocious formation of adult structures in a larval stage.

lophophore The circumoral ring of ciliated tentacles which arise from the mesosome in brachiopods, bryozoans, and phoronids; sometimes used for the entire mesosome in these groups.

metanephridia A common excretory structure in invertebrates, characterized by a ciliated funnel opening to the visceral coelom.

nidamental gland Epidermal glands whose secretions serve to retain early developmental stages in phoronid species that brood their young.

protonephridia Excretory structures characterized by having either flame cells or solenocytes as their collecting units.

semelparous A reproductive pattern in which there is a single spawning season after which the adults normally die.

spermatophoral gland Epidermal glands within which sperm are fashioned into spermatophores.

trimerous A three-part body plan that is characteristic of lophophorates and some other deuterostome animals. The generalized terms for the three body parts are protosome (a small preoral region), mesosome (a small region at the level of the mouth, typically provided with tentacles that encircle the mouth but not the anus), and metasome (a large "trunk" region that constitutes most of the body). These three body parts each have their own coelom (proto-, meso-, and metacoel, respectively) and often have specialized names depending on the animal group (e.g., epistome, lophophore, and trunk in phoronid adults) and/or life stage (e.g., preoral hood, collar, and trunk in phoronid larvae).

vasoperitoneal tissue Peritoneal tissue that serves both as the lining of a coelom and as the lining of certain blood vessels.

This article reviews the reproductive and developmental biology of the phylum Phoronida. This small, exclusively marine phylum is of interest because there are radically different interpretations of its phylogenetic relationships, the biology of fertilization is very unusual, the developmental modes are diverse, and the rapid metamorphosis, which is facilitated by extensive heterochrony, involves exceptionally dramatic rearrangements (and loss) of parts.

I. DESCRIPTION OF THE TAXON

The phylum Phoronida contains but two genera and about a dozen species worldwide. Most species are either intertidal or shallow subtidal in distribution, but one ranges from the intertidal to the deep sea. All are elongate, tube-dwelling marine "worms," with different species ranging in length from about 0.2 to 20 cm. The term worm may be misleading because phoronids are not segmented, lacking serial repetition of parts. However, phoronid adults (and larvae) have three distinctive regions to their bodies and are considered tri- or oligomerous. The three parts of the adult body are a small preoral epistome or protosome, a circumoral tentacle-bearing lophophore or mesosome, and a large postoral trunk or metasome that contains the U-shaped gut and most other parts of the body. Although bilateral, phoronid adults have an extremely short dorsal surface and a correspondingly long ventral one, as revealed during the dramatic metamorphosis from larva to adult. Each of the three body regions contains a coelomic cavity, although only that of the trunk is paired in the adult and even this is unpaired embryologically. As complex coelomates, they are of necessity triploblastic.

The relationship of phoronids to other phyla is controversial. Most zoologists recognize this phylum as one of three related phyla (the other two are Bryozoa and Brachiopoda) in a supraphyletic group the lophophorates. Alternative classifications unite only the Brachiopoda and Phoronida, separating the Bryozoa as a distinctive phylum. Most morphological and developmental data support the affinity of Phoronida with the Deuterostomia (which includes Hemichordata, Echinodermata, and Chordata) rather than the Protostomia (including, among others, Annelida, Mollusca, and Arthropoda), but most current molecular data favor the opposite alignment.

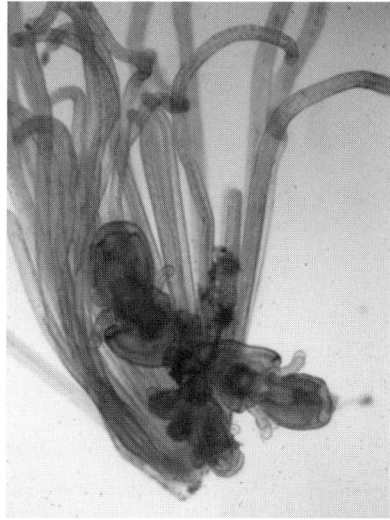

FIGURE 2 *Phoronis hippocrepia*. Tentacles with attached embryos have been removed from a brooding adult. In most species, the brooded young escape as planktotrophic larvae as soon as they are capable of feeding, but the three larger individuals in this example have utilized food brought in by maternal currents to reach advanced "larval" stages. The largest stage is about 0.5 mm in length.

FIGURE 1 Adult of *Phoronis hippocrepia* (Sebastian Inlet, Florida). The lophophore has been extended for feeding and measures about 4 mm in width. Attached to some of the tentacles of its inner whorl are a small number of opaque white embryos. A single file of eggs is seen in the trunk. This brooding species is unusual in that an unusually small number of embryos, some of which reach advanced stages, are retained at any one time (see Fig. 2).

The term lophophorate derives from the ring of tentacles in the adult that surrounds the mouth and is used in feeding, respiration, and, in many species, as a site for brooding of early embryos (Figures 1, 2, 3). A lophophore is distinguished from the tentacle rings of protostomes such as annelids and sipunculids in that it is often indented or horseshoe shaped, surrounds the mouth but not the adjacent anus of the U-shaped gut, contains extensions of a circumoral coelom, and has ciliation that collects food using an "upstream" or "single-band" collection system.

II. MODES OF REPRODUCTION: SEXUAL VERSUS ASEXUAL

Asexual reproduction is well documented in only one species, *Phoronis ovalis*. In this smallest species, division of the adult body into two approximately equal parts (achieved by contraction of circular muscles in the body wall of the trunk) is followed by regeneration of the missing parts in both fragments. This species is also reported to be capable of budding.

FIGURE 3 *Phoronis australis* (Sidney Harbor). The spiral lophophore of this species consists of one or more thousands of black tentacles and measures more than a centimeter across. Hundreds of small white embryos are being brooded while attached to the bases of certain tentacles.

Experimental bisection of adults of other species has demonstrated that two complete organisms can result provided each part contains some of the gonadal region. It is not well documented that such asexual multiplication occurs naturally in species other than *P. ovalis*.

Although regeneration resulting in asexual reproduction may be limited, regeneration to restore lost parts is an important trait. Phoronids commonly cast off (autotomize) the anterior part of their body during adverse conditions or after injury. The small anterior part (including the entire epistome and lophophore as well as the distal part of the trunk with the mouth, anus, nephridia, circumesophageal nerve ring, and main ganglion) typically degenerates so that asexual proliferation is not achieved following autotomy (an exception may occur in *P. ovalis*). All species appear capable of replacing the distal parts lost during autotomy, but whether regeneration involves undifferentiated stem cells or dedifferentiation of specialized cells has not been studied. Regenerative potential has not been documented or assessed in phoronid larvae.

All phoronids are capable of sexual reproduction. The life cycle of all species is indirect, involving a larva the morphology of which is radically different from that of the adult, thus requiring a complex

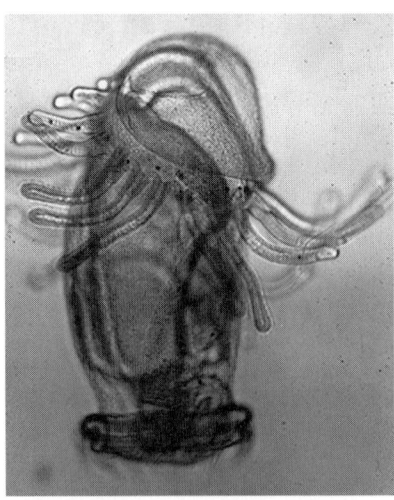

FIGURE 4 Actinotroch larva of unidentified species in right lateral view (Los Angeles, California). The preoral hood with the brain lies above the ring of tentacles which serve in feeding and locomotion. The larval trunk, which is posterior to the tentacles, contains a central stomach and intestine around which the metasomal sac is folded. At the posterior end of the trunk is a powerful band of locomotory cilia. The larva is about 1.1 mm in length.

metamorphosis. The characteristic larva for the phylum is the planktotrophic actinotroch (Figure 4), but one species has a slug-shaped lecithotrophic larva. Additional details concerning sexual reproduction and life histories are provided in the following sections.

III. MODES OF SEX: GONOCHORISM VERSUS HERMAPHODITISM

There are about equal numbers of hermaphroditic and gonochoristic species, but the mechanisms of sex determination is not known. Because sperm production begins before that of eggs in both gonochoristic and hermaphroditic species, the latter are commonly identified as protandrous hermaphrodites, although for much of the breeding season they produce eggs and sperm concurrently.

Males and females of gonochoristic species are not strongly dimorphic. During the breeding season (and in anticipation of it), males can be recognized by the presence of spermatophoral glands which develop within the lophophore. These glands are absent in females and immature forms and regress in mature males after the breeding season. In the only gonochoristic species that broods, *Phoronis psammophila*, reproductive females can be distinguished by the presence of brooded embryos (and nidamental glands which are involved in retaining the embryos); as with the spermatophoral glands, nidamental glands are developed only during the breeding season. Even though sexual dimorphism is limited, one can often sex gonochoristic adults by identifying male or female gametes within the trunk coeloms or metanephridia.

IV. ANATOMY OF THE REPRODUCTIVE SYSTEMS

A. Structures Common to Both Sexes

The common components of male and female reproductive systems include the gonads, vasoperitoneal tissue or fat body, visceral or trunk coelom, and paired metanephridia.

Although both gonad types have at least some cells that are not germline ones, the testis and ovary are extremely simple and the gametes of both sexes are directly shed into the surrounding visceral coelom. The testicular and ovarian tissues develop from the coelomic lining of blind-ending capillaries that radiate from a hemal sinus associated with the stomach of the adult. It is not known when germline cells are established (primordial germ cells have not be identified), but some of the stomachic capillaries develop during larval life and it is widely assumed that gonial cells are modified peritoneal ones. Although phoronids are clearly bilateral animals, the gonads themselves are usually highly asymmetrically developed on the left and right sides of the adult.

Between breeding seasons, the vasoperitoneal cells (those cells of the coelomic lining which clothe the blood vessels) of the stomachic capillaries become swollen with various inclusions. Because there is an inverse relationship between its size and that of the gonad proper, the gonadal vasoperitoneum (or fat body) probably fuels gamete production with its nutrient reserves and would thus serve as an accessory sex structure.

After they are released from the gonad into the proximal part of the trunk coelom, gametes of both sexes may be stored here for various periods of time but are eventually transported to the distal end of the trunk for release. Peristaltic actions of the dermomuscular trunk wall and currents created by the ciliated coelomic funnels of the metanephridia are primarily responsible for the proximal-to-distal transport of the gametes within the trunk cavity.

At the distal or lophophoral end of the trunk, a pair of metanephridia connect the visceral coeloms with the exterior. Presumed to function primarily in excretion, the metanephridia also serve as male and/or female gonoducts in all phoronids except *P. ovalis*. In this diminutive species, the eggs are too large to exit via the metanephridia and their release requires autotomy of the adult's distal end, including the first part of the visceral coelom.

B. Male Structures

Testicular tissue is developed from the peritoneal lining of capillaries arising from the stomachic sinus and as such is somewhat diffuse. Spermatozoa are highly modified and of a V shape. One limb consists of a simple flagellum that originates from a centriole at the angle of the V. The free end of the other limb consists of two laterally aligned mitochondrial rods. An elongate, poorly condensed nucleus, interposed between the mitochondria and flagellum, forms the other half of the nonflagellar limb. A narrow electron-dense rod which parallels that part of the nucleus near the apex of the V may represent an acrosome. Coelomic sperm isolated from males appear filiform rather than V shaped because the two limbs are closely adherent initially.

Functional males of all but one species are thought to produce spermatophores within a pair of complex, pocket-like spermatophoral glands adjacent to the nephridiopores (the exception is *P. ovalis*, in which spermatophoral glands have not been identified). Masses of sperm are concentrated within the metanephridia and, rather than being shed directly into the seawater, are first passed into the cavity of the spermatophoral glands where they are packaged into spermatophores. Spermatophores have species-specific sizes and shapes but are of only two major types. The simpler forms consist of a consolidated sperm mass sheathed in mucus, whereas the more complex are also provided with a spiral, sail-like appendage also composed of mucus. Spermatophores are released freely into the seawater rather than being transferred to or near a recipient by copulation or pseudocopulation. The "sail" probably facilitates suspension of the spermatophore in the water column, thus permitting more extensive transport by currents. The fate of spermatophores will be considered later.

C. Female Structures

As in male-acting individuals, female-acting ones have simple diffuse ovaries associated with stomachic capillaries and release gametes into the visceral coelom or metacoel. At the time of their release the female gametes are primary oocytes at metaphase of first meiosis. It is thought that fertilization occurs as the oocytes leave the ovary since coelomic, but not ovarian, oocytes regularly have a male pronucleus (as demonstrated with DAPI staining).

The zygotes are collected from the visceral coelom by the metanephridia and released to the exterior via the nephridiopores as fertilized primary oocytes at metaphase of first meiosis. Outside of the female, the zygotes alternatively are passed into the ambient seawater where all development occurs or are temporarily brooded to a larval stage either within the tube of the mother or attached to some of her lophophoral tentacles. In the latter, the fertilized oocytes are retained by mucous secretions of nidamental glands found on certain tentacles and adjacent tissues of the lophophore (Figures 1, 2, 3).

In the absence of male intromittent organs and of female openings or ducts to receive sperm, it was long assumed that fertilization had to be external. That carefully collected female gametes develop without exposure to sperm, that the sperm are highly modified, and that sperm are fashioned into spermatophores suggest that fertilization is probably highly modified; the recognition of male pronuclei in coelomic eggs of both hermaphroditic and gonochoristic species dictates that it is internal. The following information is integral to the fertilization process and is reported here because the structures through which sperm reach the site of fertilization are properly accessory female elements.

Although spermatophores are freely broadcast into the seawater, their abundant production and the facts that phoronids commonly occur in dense aggregations and have strong cilia-generated feeding currents make it possible for spermatophores to reach the exposed lophophore of potential mates. Only one pathway by which foreign sperm gain the metacoel of mates has been documented and this has been determined for only three species, but superficial evidence for a second avenue exists; both involve lytic penetration into the coelom. In the three species for which their fate has been followed, spermatophores, on contact with one or several tentacles, lose the enveloping mucous envelopes. The sperm are released as a naked, coherent mass that is somewhat "ameboid" and adheres closely to several tentacles. After a few minutes, the wall of several (rarely one) tentacles is lysed (presumably by enzymes either from the sperm themselves, although whether they have an acrosome is uncertain, or packaged with them inside the spermatophore) and thick columns of sperm enter the coelomic lumen of each tentacle and stream toward the proximal end of the tentacle. Here, at the base of the lophophore, the several columns of sperm aggregate as a single mass adjacent to the transverse septum that separates the visceral or trunk coelom from that of the lophophore. After several minutes, the sperm disperse into the trunk coelom, apparently through a hole that has been lysed through the transverse septum. Either by their own motility or aided by coelomic currents, the sperm eventually reach the opposite end of the trunk coelom where the gonads lie, and here oocytes are fertilized as they are dehisced from the ovarian surface. In at least two species, spermatophores do not break down on contact with tentacles but are ingested intact as if normal food items. The fate of these spermatophores is unknown but it is reasonable to speculate that, in these cases, the sperm reach the visceral coelom through a pore lysed in the wall of the stomach.

Coelomic spermatozoa of males of gonochoristic species appear filiform due to the close adherence of their two limbs, whereas sperm released from spermatophores are clearly V shaped. It has been suggested that exposure to secretions of the spermatophoral glands may be required for morphological completion and/or physiological capacitation of the male gametes. Self-fertilization in hermaphroditic species could theoretically be avoided if this is true.

V. REPRODUCTIVE PHYSIOLOGY AND ENDOCRINE CONTROL

Virtually nothing is known about either the proximate or the ultimate factors that are involved in sexual (or asexual) reproduction. As with many invertebrates, the reproductive seasons of most species occur over periods of several months and reproduction typically occurs during the spring and summer months. Exceptions include *Phoronopsis californica* and *P. ovalis*, which breed in fall and winter, respectively.

It is widely assumed that individuals of most phoronid species live (and are reproductively active) for several years, but there is little evidence to confirm this. In all but two species, individuals probably release eggs every day for several weeks to several months. Although some species have been reported to release gametes preferentially at night, this has not been confirmed in recent studies. The number of eggs released each day ranges from one to a few dozen in brooding species (see exceptions below, however) to hundreds or even thousands in nonbrooding ones. Two exceptions to repeated daily spawning are found in *P. ovalis* and *P. psammophila*. *Phoronis psammophila* females release several hundred eggs at one time and brood these within the lophophore; a given individual probably produces several clutches of eggs over the reproductive season. *Phoronis ovalis* probably spawns only once each season since the lophophore is autotomized for spawning and the maternal individuals cannot feed until the long brooding period is completed. Considering its small size and mode of reproduction, this species may be semelparous.

VI. MODES OF FERTILIZATION

Despite reports to the contrary, it is doubtful that egg–sperm union has ever been observed in phoronids since it now seems probable that there is internal, cross-fertilization in all species. (For information on how sperm are transferred from male-functioning to female-functioning ones, see Section IV,C.) Where examined, coelomic oocytes, but not ovarian ones, have a sperm pronucleus, so fertilization probably occurs as or shortly after the female gametes rupture through the ovarian surface. About this time, the oocytes undergo germinal vesicle breakdown and then are arrested at metaphase I of meiosis so fertilization is probably at late prophase I or metaphase I. In the absence of observations at the time of syngamy, little is known of a possible acrosome reaction (although sperm have a structure that could be acrosomal), egg activation, or other aspects of egg–sperm interaction.

The remaining stages of first meiosis, the second meiotic division, and the subsequent union of the two pronuclei normally occur only after the fertilized primary oocytes are released into seawater. However, cleavage stages and even early embryos have occasionally been observed within the maternal coeloms of several species (e.g., *Phoronis muelleri* and *Phoronopsis harmeri*). The factors controlling arrest and reinitiation of meiotic events are unexplored.

VII. MODES OF DEVELOPMENT

There are three patterns of development in the phylum, each correlated with a specific egg size. The largest eggs are about 125 μm in diameter and are produced only by *P. ovalis*; whether this species is hermaphroditic or gonochoristic remains unclear. Only about two dozen eggs are produced at one time. These are simultaneously released to the exterior after autotomy of the distal (lophophoral) end of the adult and are brooded within the maternal tube where all embryogenesis occurs. The length of the brooding period is not known and the short-lived, lecithotrophic, slug-shaped larvae have no identifying name.

About half of the remaining species produce, over a season, hundreds to thousands of eggs which average about 100 μm in diameter. The eggs in these species become attached to the maternal tentacles by secretions of nidamental glands as they are released (Figures 1, 2, 3). While still attached, embryogenesis to the stage of an early larva—the actinotroch—occurs. In most species, larvae escape from the mother occurs shortly after they are capable of feeding. The larvae are planktotrophic and spend a considerable time in the plankton, feeding, growing, and eventually reaching their definitive morphology and competence to metamorphose. In at least one species (*Pho-

ronis hippocrepia), some but not all of the larvae remain attached to the parent, capturing food from her feeding current until they are actinotrochs competent to metamorphose (Figures 1, 2). Phoronids with this second developmental pattern are all hermaphroditic with one exception (*P. psammophila*).

The third pattern is similar to the second one in that long-lived, planktotrophic actinotrochs are produced. However, these develop from 60-μm eggs which are not brooded even briefly, so all development is planktonic. All phoronids with this third pattern of reproduction are gonochoristic with one exception (*Phoronis pallida*) and tend to be larger species which are capable of producing many thousands of eggs over a season.

VIII. LARVAE AND METAMORPHOSIS

A. Larvae

All phoronids except *P. ovalis* produce long-lived, planktotrophic larvae called actinotrochs. In the one exception, the progeny escape the maternal tube (where they were retained through embryogenesis) as short-lived, lecithotrophic, slug-shaped larvae. The ciliated slug-like larvae largely crawl along the bottom on a flattened sole rather than swimming freely in the water column. After perhaps a week of demersal life during which no obvious morphological changes occur, the larva attaches to the substratum and assumes a simple hemispherical shape. Despite observation over an extended period of weeks, the conversion of this mass to the elongate tubiculous adult remains unknown. Although extremely different externally and ecologically, actinotrochs and the larvae of *P. ovalis* share several important internal characteristics.

Actinotrochs are thought to spend 4–8 weeks swimming freely in the plankton where they feed on plankton and undergo dramatic changes in size and complexity. Metamorphosis will be described later, but it is catastrophic and rapidly leads to a functional adult-like imago.

The actinotroch (Figure 4), like the adult, is considered to be trimerous. The preoral hood, which overhangs the larval mouth as a moveable flap, corresponds to the preoral epistome of the adult. The hood has an unpaired coelomic compartment, but it is doubtful that this becomes the protocoel of the adult.

The second larval body region is called the collar; this bears a ring of ciliated tentacles that extend radially from the larval body and give the actinotroch its name. Late in larval life, the collar region gains a small C-shaped coelom which will be retained through metamorphosis as the lophophoral coelom or mesocoel. The larger part of the collar region remains as a remnant of the embryonic blastocoel. The blastocoelic portion will be the site of development of one or more clumps of erythrocytes and eventually will contribute the ring and tentacular blood vessels of the adult circulatory system during the complex metamorphosis. As might be anticipated, the circumoral collar region of the larva corresponds to the lophophore of the adult and its tentacles are usually retained in total or in part as the first tentacles of the adult's lophophore. As do the adult tentacles, the larval tentacles serve in collection of phytoplankton, but they also serve in slow propulsion of the larva.

The remainder of the larval body is called the trunk. Initially very small, the larval trunk will grow rapidly to become the largest portion of the larva. The larval trunk contains most of the digestive tract, which is now straight rather than U shaped as in the adult. The posterior end of the cylindrical trunk is girdled by a powerful ciliated band, the perianal ciliated ring or telotroch; this plays no role in feeding but is capable of moving the larva quite rapidly. Paired solenocytic protonephridia are located at the boundary between the collar and trunk. As within the collar, components of the adult circulatory system are preformed in the larval trunk, including the afferent and efferent longitudinal vessels, the stomachic sinus, and some of the capillaries that extend from this sinus.

The trunk has an unpaired coelom provided with a midventral but not a middorsal mesentery. The midventral mesentery initially attaches the larval gut to the ventral midline of the larval trunk, but this arrangement will be modified with the development of the metasomal sac, a diagnostic feature of the actinotroch larva. This blind-ending tubular sac originates as a midventral invagination of the epidermis near the anterior limit of the larval trunk. First appearing when the larva has relatively few tentacles, the metasomal sac grows rapidly, pushing between the right and left leaves of the midventral mesentery of the trunk coelom. Thus, the midventral mesentery connects the metasomal sac and the larval gut; this is a critical feature at the time of metamorphosis. The metasomal sac at the time of metamorphosis will be equal to or greater in length than the larval body, but until then it remains coiled within the larval trunk coelom, occupying virtually all of its space.

Actinotrochs possess a complex musculature and an extensive intraepidermal nervous system. The larval brain is near the apex of the preoral hood at the anterior extremity of the larva. Catecholamine-, serotonin- and FMRFamide-reactive neurons have been described on the basis of immunocytochemistry and electron microscopy. In late larval life a conspicuous apical sensory organ develops on the hood near the larval brain with which it is united by three nerve tracts. This complex, protrusible, glandular structure almost certainly plays some role in site selection at metamorphosis, although how favorable stimuli are transduced has not been explored.

B. Metamorphosis

As noted earlier, the metamorphosis of actinotrochs is cataclysmic and very rapid; it is also global in that virtually every larval feature is dramatically rearranged (or lost). Factors that mediate metamorphosis (in at least some species) include various gram-positive bacteria (if in log-growth phase) and certain ions including cesium. Settlement is gregarious and results in populations as dense as 28,000 individuals/m^2. Neither intraspecific factors, which mediate this gregarious settlement, nor interspecific ones, which initiate the symbiotic relationship in the two phoronid species which seemingly are obligate commensals (with thalassinid shrimp in one case and cerianthid anemones in the other), have been analyzed.

After exploration of benthic substrata in which the dramatically extended larval sense organ is protruded against (or even into soft) substrates, actinotroch larvae initiate a series of irreversible actions that result in metamorphosis to a tubiculous benthic imago that is a miniature replica of the adult. In the first step of this transformation, the metasomal sac is everted due to shortening of the larval trunk by muscles (Figure 5). After eversion, the metasomal sac is recognized as the trunk of the incipient adult. Secretions of gland cells in the trunk epidermis form a tubular chitinous sheath to which sand grains and other particles of the sediment adhere, forming a tube within which the juvenile will live.

As the metasomal sac rapidly protrudes at right angles to the larval axis, the midventral mesentery of the larval trunk coelom pulls the once straight larval gut out of the larval trunk into the everted sac. The gut—now that of the imago—thus assumes a U shape, with the middle of the gut deep within the evaginated sac and the two ends of the larval digestive tract (mouth and anus) positioned very close to each other due to the dramatic shortening of the long axis of the larva (Figure 5). The short distance between mouth and anus must be recog-

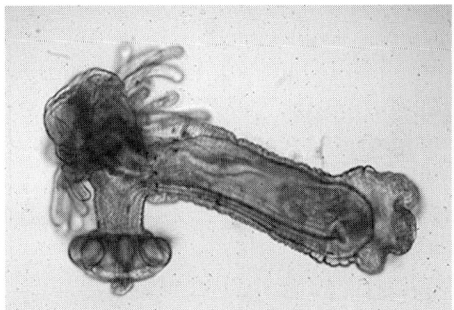

FIGURE 5 Metamorphosing actinotroch of unidentified species shown in the same orientation as the larva in Figure 4, but at lower magnification. (Los Angeles, California,) The metasomal sac, which had been within the larval trunk, has been everted at right angles to the larval body as the rudiment of the adult trunk (under natural conditions this would have been protruded into the substratum). The larval trunk is both shorter and thinner as it no longer contains either the metasomal sac or most of the larval gut; much of the latter has been drawn in a U-shape into the everted sach which measures about 1 mm in length. The preoral hood and most of the larval tentacles are undergoing histolysis and will be swallowed as the imago's first meal.

nized as the dorsal midline, whereas the rest of the sagittal profile is ventral.

As eversion of the metasomal sac occurs, most of the larva's preoral flap and often parts of its tentacles are cast off and ingested as the imago's first meal. The epidermis of the larval trunk is also lost, but this is usually slowly resorbed *in situ*. The larval tentacles or their shortened remnants (or occasionally a second set formed during larval life) are erected around the mouth to form the circumoral lophophore. A complex circulatory system is established from precursors preformed during larval life. The ducts of the larval protonephridia are retained as those of the metanephridia of the adult, but the solenocytes, which extended into the collar blastocoel, will be lost and subsequently replaced by coelomic funnels that open into the trunk coelom.

In a period as short as 30 min, actinotrochs are thus transformed from larvae fully adapted for a free-living, planktonic life to benthic juveniles, encased in a tube of their own secretion and provided with a complex vascular system. The rapidity of this dramatic conversion, which obviated the extended existence of ill-adapted intermediate forms, was made possible by the precocious formation of numerous adult features during larval development (a form of heterochrony—adultation).

Postmetamorphic growth is quite rapid, but massive predation on the juveniles has been reported. The time needed to reach reproductive maturity has not been determined, but juveniles are probably sexually active the following season.

See Also the Following Articles

BRACHIOPODA; BRYOZOA; ECHINODERMATA; HEMICHORDATA; HERMAPHRODITISM; MARINE INVERTEBRATES, MODES OF REPRODUCTION IN

Bibliography

Emig, C. C. (1977). Embryology of the Phoronida. *Am. Zool.* 17, 21–38.

Franzén, A., and Ahlfors, K. (1980). Ultrastructure of the spermatids and spermatozoan in *Phoronis pallida*. *J. Submicrosc. Cytol.* 12, 585–597.

Hay-Schmidt, A. (1990). Catecholamine-containing, serotonin-like, and FMRFamide-like immunoreactive neurons and processes in the nervous system of the early actinotroch larva of *Phoronis vancouverensis* (Phoronida): Distribution and development. *Can. J. Zool.* 68, 1525–1536.

Herrman, K. (1995). Induction and regulation of metamorphosis in planktonic larvae—*Phoronis muelleri* (Tentaculata) as archetype. *Helgol. Meeresunters.* 49, 255–281.

Zimmer, R. L. (1997). Phoronids. In *Reproduction of Marine Invertebrates, Vol. VI: Lophophorates and Echinoderms* (A. C. Giese, J. S. Pearse, and V. B. Pearse, Eds.), pp. 1–35. Boxwood Press, Pacific Grove, CA.

Zimmer, R. L. (1997). Phoronids, brachiopods, and bryozoans, the lophophorates. In *Embryology—Constructing the Organism* (S. F. Gilbert and A. M. Raunio, Eds.), pp. 279–305. Sinauer, Sunderland, MA.

Photoperiodism, Vertebrates

Randy J. Nelson

Johns Hopkins University

I. Introduction
II. Mammals
III. Birds
IV. Reptiles, Amphibians, and Fishes

GLOSSARY

circadian rhythm A self-sustained, endogenous biological rhythm that persists in constant conditions with a period of approximately 24 hr.

entrainment The synchronization of biological rhythms to a periodic environmental cue.

external coincidence model According to this model, photoperiodic time measurement depends on an endogenous circadian rhythm of 12 hr of photo-nonresponsiveness followed by 12 hr of photoresponsiveness. When light is coincident with the photoresponsive phase, day lengths are physiologically interpreted as long; when light does not coincide with the photoresponsive stage, day lengths are physiologically interpreted as short. This model appears to explain photoperiodic time measurement in most vertebrates.

free-running rhythm A biological rhythm that is not synchronized to any environmental cue and that expresses its own endogenous rhythm.

hourglass model This model of photoperiodic time measurement depends on the accumulation (or depletion) of some physiological product during the light or dark portion of the light–dark cycle. When a critical threshold of the product is attained, then a photoperiodic response is observed. This model appears to account for photoperiodic time measurement in many insects as well as a few vertebrate species.

internal coincidence model This model of photoperiodic time measurement is based on entrainment theory and assumes that the photoperiodic time measurement system depends on the internal coincidence of two or more endogenous circadian rhythms whose relative phase relationships change with the annual fluctuations in day length. Thus, during one time of year (e.g., winter) the two circadian rhythms do not coincide, but when day lengths change (e.g., summer), the phase relationships of the multiple circadian rhythms may coincide or phase lock, and a photoperiodic response is observed.

melatonin An indole amine hormone secreted from the pineal gland at night that is important in the regulation of daily and seasonal biological rhythms.

phase–response curve A graphic representation of the different effects that a periodic environmental signal (usually light) has on the timing of biological rhythms; these temporal effects depend on the phase relationship of the temporal cue with the circadian organization of the individual in question. Depending on the timing of exposure to the temporal signal, the unsynchronized biological rhythm may advance, delay, or remain unchanged.

photoperiod Day length or the amount of light per day.

photoperiodism A term first proposed in 1922 by the botanists, W. W. Garner and H. A. Allard, to describe the response of plants to the length of day and night. Currently, the term photoperiodism includes the ability to determine day length (usually by measuring night length) in both plants and animals.

photorefractoriness The temporary loss of biological responsiveness to changes in photoperiod.

pineal gland An endocrine gland (also called the epiphysis) that secretes melatonin usually located between the telencephalon and diencephalon in vertebrates.

Photoperiodism is the ability of plants and animals to measure environmental day length (photoperiod). The biological ability to measure day length permits organisms to ascertain the time of year and engage in seasonally appropriate adaptations. Although the precise mechanisms underlying photoperiodism differ among taxa, individuals that respond to day length can precisely, and reliably, ascertain the time of year with just two bits of data: (i) the length of the daily photoperiod and (ii) whether day lengths are increasing or decreasing.

I. INTRODUCTION

Photoperiodism has evolved in virtually all taxa of plants and animals that experience seasonal changes in their habitats. Among vertebrate animals, photoperiodism is linked to a number of seasonal adaptations, including reproductive, metabolic, immunological, and morphological adaptations to cope with seasonal changes in ambient conditions. The first demonstration of photoperiodic time measurement regulating vertebrate reproduction was established in birds by Rowan. During the winter, juncos (*Junco hyemalis*) were maintained in outdoor aviaries in Edmonton, Alberta, and exposed to several minutes of artificial light after the onset of dark each day (lights were illuminated at sunset). Under these artificial lighting conditions, these birds came into reproductive condition despite the harsh Canadian winter temperatures. In comparison, juncos living in the wild in the relatively mild "Riviera-like" climate of Berkeley, California, but exposed to normal winter day lengths remained in nonreproductive condition. Thus, Rowan concluded that the number of hours of light per day, not ambient temperature or food availability, regulated the annual breeding cycle of juncos. The initial demonstration of photoperiod regulating mammalian reproduction was reported for European field voles, *Microtus agrestis*. To date, the role of photoperiod in mediating seasonal adaptions has been documented for hundreds of vertebrate species.

Since those early studies, environmental photoperiod has been established to be an important factor used by temperate and boreal vertebrates to time reproduction. Changes in day length, while probably of little direct importance to most animals, provide the most error-free indication of time of year, and thus enable individuals to anticipate seasonal conditions. Because the same photoperiod occurs twice a year (i.e., March 21 and September 21), it is advantageous for animals to be able to discriminate between these two dates. Many photoperiodic vertebrates have solved this problem by developing an annual alteration between two physiological states termed photoresponsiveness and photorefractoriness.

Syrian, or golden, hamsters (*Mesocricetus auratus*) represent the most common mammalian model used in laboratory investigations of photoperiodism. Hamsters, in common with most small mammals, are "long-day breeders." Gestation is relatively brief in these animals; mating, pregnancy, and lactation occur during the long days of late spring and early summer. Adult male hamsters undergo gonadal regression when day lengths fall below 12.5 hr of light. The minimum day length that supports reproduction is called the critical day length (Fig. 1). Critical day length is not a fixed variable but differs among populations of animals living at different altitudes and latitudes. For example, Siberian (*Phodopus sungorus*) hamsters occupy very different habitats than Syrian hamsters; Siberian hamsters live in high-latitude habitats, and in the lab they display a critical day length for reproductive function of 16 hr of light/day.

When Syrian hamsters are maintained in day lengths <12.5 hr of light, blood concentrations of gonadotropins and sex steroid hormones decrease, accessory organ mass diminishes to about 10% of

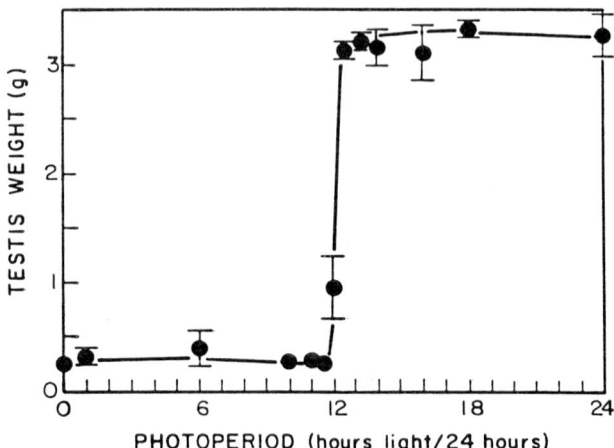

FIGURE 1 Testicular development in response to photoperiods ranged from 0 to 24 hr light/day. Each circle represents the mean (± SEM) paired testes mass (g) of groups of Syrian hamsters housed for approximately 3 months in these photoperiods. The minimum day length that supports reproduction is called the critical day length and is between 12 and 12.5 hr light/day for these hamsters (reproduced with permission from Elliott, 1976).

the original size, and reproductive behaviors stop. Male hamsters remain reproductively quiescent for approximately 16–20 weeks, a period of time that roughly corresponds to the duration of short days experienced in the wild during autumn and winter. Hamsters are photorefractory during the recrudescent phase; i.e., gonadal condition becomes unlinked from photoperiodic inhibition. Photorefractoriness permits attainment of fully functional gonads in the spring before environmental photoperiods attain 12.5 hr, the day length necessary for gonadal maintenance in the autumn. This is an important adaptation that allows burrowing animals (that live in constant dark conditions) to anticipate spring conditions with the development of fully functional reproductive systems without long-day exposure. Chronic maintenance in the long days of summer is necessary to "reset" or reestablish the photoperiodic mechanism so that short days can again attain inhibitory status.

Sheep (*Ovis aries*) are another important model species in studies of mammalian photoperiodism. In common with other ungulates and most large mammals, sheep are "short-day breeders." Ewes have relatively long periods of gestation; mating typically occurs in the autumn and lambs are born and nursed in the spring when food and other conditions are most conducive for survival. As day lengths decrease in late summer, the rate of hypothalamic pulses of gonadotropin-releasing hormone (GnRH) secretion increases, eventually stimulating increased gonadotropin secretion that initiates reproductive function. In Suffolk ewe, seasonal reproductive transitions primarily appear to reflect changes in the responsiveness of the GnRH neurosecretory system to the negative feedback influence of estradiol. In common with other mammals, GnRH neurons in the sheep lack estrogen receptors, and the influence of estradiol on GnRH neurosecretory activity is probably conveyed via afferents. The ultrastructure and synaptic inputs of GnRH neurons form in the preoptic area of ewes during the breeding season and seasonal anestrus. GnRH neurons were examined in both ovary-intact ewes and ovariectomized ewes bearing implants that produced constant levels of estradiol. Preoptic GnRH neurons in ewes during the breeding season received more than twice the mean number of synaptic inputs per unit of plasma membrane compared to GnRH neurons in anestrous animals. Although GnRH dendrites received more synaptic input than GnRH cell bodies, significant seasonal differences were observed in both axodendritic and axosomatic inputs. Seasonal changes in synaptic inputs onto GnRH neurons were reported in both intact animals and ovariectomized ewes bearing estradiol implants. Consequently, these seasonal alterations are unlikely to reflect fluctuating endogenous sex steroid concentrations but may instead reflect unspecified changes in the environmental photoperiod, the expression of an endogenous circannual rhythm, or both. After mating, sheep become photorefractory; i.e., short days lose their stimulatory effects and mating behavior wanes. Exposure to the long day lengths of summer is not necessary to reestablish responsiveness to short days, suggesting the presence of an underlying circannual cycle of photosensitivity.

A. Hourglass Model of Photoperiodism

The critical problem for any photoperiodic individual is to discriminate between long and short day lengths. Two broad hypotheses have been advanced to explain how animals respond to changes in day length: (i) The response depends on the total number of hours of light per day or (ii) the response depends on the phase of the light exposure relative to some internal rhythm of photoresponsiveness. According to the first model, animals monitor the accumulation (or depletion) of some physiological agent during one part of the light–dark cycle; this process is reversed during another portion of the cycle. This time measurement hypothesis is referred to as the hourglass model. The hourglass model appears to be used by several invertebrate species but only by a very few vertebrate species.

The hourglass model presumes that the absolute duration of either the light or dark period is monitored, and when some threshold value is attained, a photoperiodic "adjustment" is made. Alternatively, the absolute ratio between the amount of light and dark per day could also be monitored. To determine if animals use an hourglass timing mechanism to measure photoperiod, the first step is to determine the critical day length necessary to mediate the photoperiodic response. Male anole lizards (*Anolis caro-*

linensis) undergo gonadal regression during the autumn, then display recrudescence during early spring. Housing male anole lizards in a variety of photoperiods revealed that day lengths <13.5 hr of light per day evoked gonadal regression. This critical day length corresponds to the day lengths of mid-August in the natural habitat of these lizards. To determine if anole lizards monitor and respond to the length of the daily dark period (i.e., night lengths >11.5 hr/day should induce gonadal regression), male anole lizards were maintained in light–dark (LD) cycles of LD 8:16, 16:8, and 16:20. If anole lizards turn off reproductive function in response to night lengths >11.5 hr, then males housed in both LD 8:16 and 16:20 should display regressed reproductive systems. However, only male lizards maintained in LD 8:16 photoperiods underwent gonadal regression. Males housed in LD 16:8 or 16:20 photoperiods maintained reproductive function, suggesting that these lizards were monitoring absolute day length.

Alternative models exist to explain how organisms respond to day length. These so-called "coincidence models" rely on phase relations between internal circadian oscillators or phase relations between external factors and endogenous circadian cycles of receptive states. There are two types of coincidence models of photoperiodic time measurement: (i) the external coincidence model and (ii) the internal coincidence model.

B. Coincidence Models of Photoperiodism

1. External Coincidence Model

The second hypothesis of photoperiodic time measurement was originally formulated to explain flowering in plants. According to Bünning (1973), "The physiological basis of photoperiodism ... lies with the endogenous daily rhythms." The crux of Bünning's hypothesis is the assumption of an endogenous circadian rhythm of subjective day (photoinsensitivity/noninducibility, with a duration of about 12 hr) and subjective night (photosensitivity/inducibility, also with a duration of about 12 hr). Light was postulated to serve two functions: (i) Light entrains (i.e., synchronizes) the endogenous circadian rhythm of subjective day and subjective night, and (ii) light stimulates photoperiodic responses if it is coincident with the subjective night (i.e., the photoinducible phase). Thus, when first exposed to light, the 12-hr subjective day is set; light intruding during the next 12 hr will not maintain the reproductive system. For example, a photoperiod of 6 hr of light and 18 hr of darkness (LD 6:18) will not maintain the reproductive system of a hamster because light will only coincide with the subjective night of the cycle. The inverse photoperiod (i.e., LD 18:6) would maintain the reproductive system because 6 hr of light would coincide with the photosensitive/inducible part of the cycle. Many studies across different taxa have produced results consistent with Bünning's hypothesis. In principle, a few seconds of light per day, appropriately timed to coincide with an individual's photosensitive/inducible phase, would evoke a "long-day" response.

There have been several tests of Bünning's hypothesis in vertebrates. One type of test of circadian involvement in photoperiodic time measurement makes use of the resonance paradigm. In resonance studies, groups of animals are maintained on photoperiod cycles that couple a fixed light phase (e.g., 6 or 8 hr) with various durations of darkness, resulting in LD cycles of varying lengths. For example, hamsters can be maintained in LD 6:18, 6:24, 6:30, 6:42, and 6:54 photoperiods. If hamsters use an hourglass timer for photoperiodic time measurement, then animals maintained in each of these photoperiods should have regressed reproductive systems; the duration of light in each instance is less than 12 hr per day. Alternatively, if hamsters employ a circadian timer for photoperiodic time measurement, then only animals housed in the LD 6:18 and 6:42 photoperiods should have regressed reproductive systems because only during these photo-regimens is light restricted to the putative photoinsensitive/noninducible phase. Animals exposed to LD 6:30 and 6:54 photoperiods should have comparably sized reproductive systems to those of long-day males because light would coincide with the subjective night every other day or every third day, respectively. The results of these types of studies are consistent with Bünning's hypothesis of photoperiodic time measurement (Fig. 2).

Skeleton photoperiods have also been used to test Bünning's hypothesis. In skeleton photoperiods, the

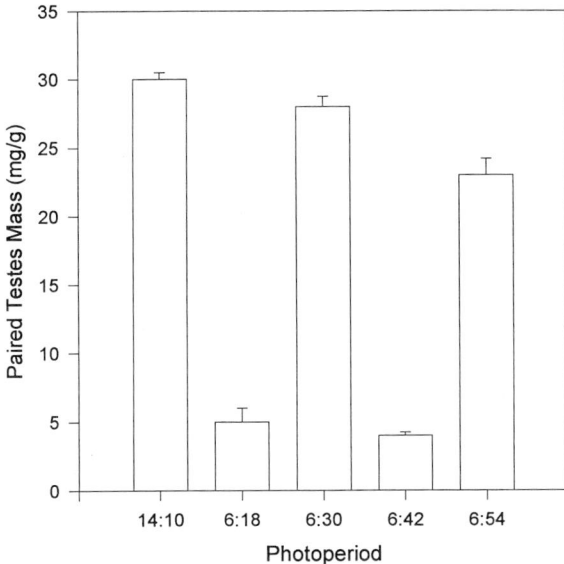

FIGURE 2 Testicular development in response to resonance photoperiods shows that the absolute ratio of light to dark does not mediate photoperiodic responsiveness in hamsters. Mean (± SEM) paired testes mass corrected for body mass (mg/g) of hamsters housed in LD 14:10 (long-day control), 6:18, 6:30, 6:42, or 6:54 photoperiods. Note that these photoperiods provided 6 hr of light every 24, 36, 48, or 60 hr, and that testicular function regressed only when the photocycle was 24 or 48 hr (after Elliott, 1976).

full light cycle is usually replaced by appropriately timed light pulses. For example, the reproductive system of male hamsters maintained in LD 16:8 photoperiods could be sustained in LD 2:22 photoperiods if 1 hr of light occurred from 8:00 to 9:00 hr and another hour of light occurred at 23:00 to 24:00 hr. This skeleton photoperiod would provide a pulse of light in the morning to entrain the cycle of photononinducible/photoinducible phases and a second pulse of light near the middle of the photoinducible phase. Thus, an appropriately timed, very short photoperiod of LD 2:22 could "trick" the reproductive system into a long-day response. Of course, it is possible for the individual to entrain in such a way that the 1-hr pulse of light between 23:00 and 24:00 hr would entrain the cycle of photo-noninducible/ photoinducible phases and the second pulse 8 hr later would coincide with the photo-noninducible phase resulting in short-day responses. Thus, predictions about specific skeleton photoperiods depend on entrainment patterns; monitoring locomotor activity or other circadian cycles is important in explaining photoperiodic responsiveness to skeleton photoperiods.

2. Internal Coincidence Model

This model of photoperiodic time measurement assumes the existence of two or more internal oscillators that change in their phase relationship during the annual change in day length. For example, if the peak blood concentration of glucocorticoids was coupled to a "dawn" oscillator and the peak blood concentration of prolactin was coupled to a "dusk" oscillator, the difference in the timing of the peak values of these two hormones might range from 15 to 9 hr throughout the year among temperate zone animals. According to the internal coincidence model, a photoperiodic response should be observed when certain internal phase relations are attained. Although the existence of internal coincidence processes in vertebrate photoperiodic time measurement systems has not been firmly established, it continues to be an attractive hypothesis because this model is based on entrainment theory and does not rely on a special photoinducible oscillator.

C. Entrainment

The process of synchronization of endogenous biological rhythms to a periodic cue in the environment is called entrainment. Light is an important entraining agent for circadian rhythms, including the rhythm of photoresponsiveness/inducibility–photononresponsiveness/noninducibility. An understanding of phase–response curves is useful to understand the process of entrainment. Phase–response curves are graphic representations of the differential effects that light has on the timing of biological rhythms; these temporal effects depend on the phase relationship of the light to the circadian organization of the individual in question (Fig. 3). Exposure to light at different times throughout the day does not result in uniform responses of free-running biological rhythms. For example, if hamsters are maintained in constant dark conditions, then they begin to free-run, with the period of the onset of their daily loco-

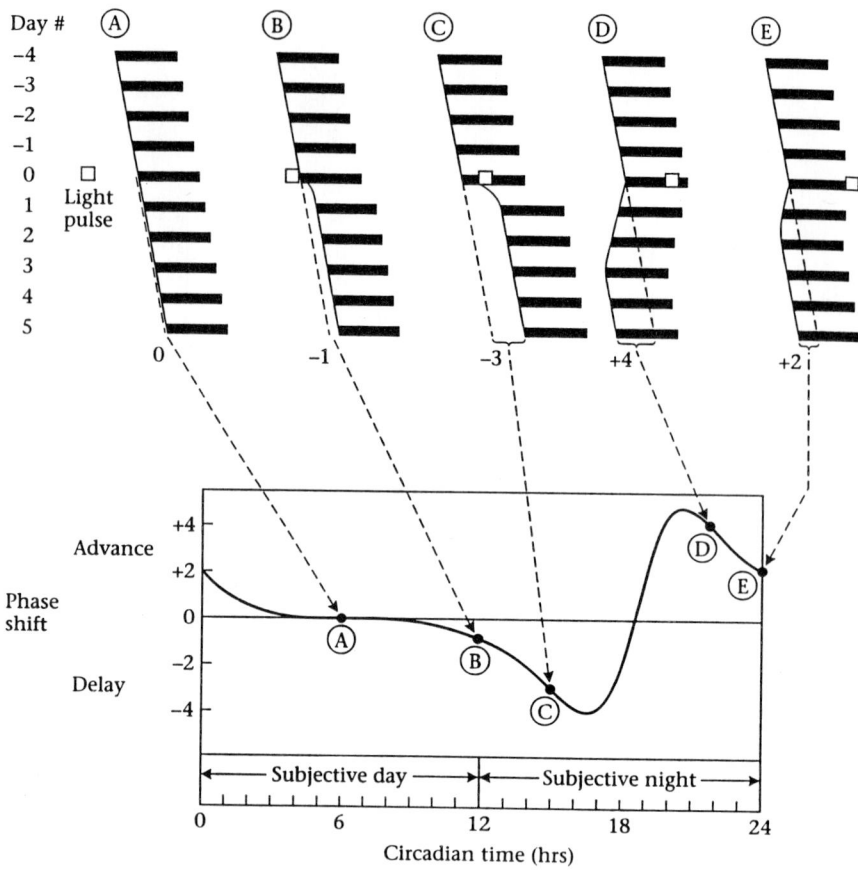

FIGURE 3 Phase–response curves are graphic representations of the differential effects that light has on the timing of biological rhythms. A free-running nocturnal individual maintained in constant dark conditions was exposed to a 1-hr light pulse at various times during the subjective day and night. (A) When light was given in the middle of the subjective day, the subsequent activity onset was unaffected; the middle of the subjective day is thus called the "dead zone." (B) A light pulse at the end of the subjective day phase-delayed activity the next day by about 1 hr. (C) With a light pulse at the start of the subjective night, a substantial phase delay (3 hr) was observed. (D) Light given later during the subjective night caused a substantial advance (4 hr) of activity onset the next day. (E) Finally, a light pulse at the start of the subjective day caused a 2-hr phase advance in activity the following day. The effect of light on the endogenous rhythm is involved in synchronizing circadian rhythms to exactly 24 hr each day (after Moore-Ede et al., 1982).

motor activity rhythms deviating slightly from 24 hr each day.

The circadian period can be divided into a subjective day and subjective night. Because hamsters are nocturnal, the beginning of activity usually coincides with the onset of their subjective night, whereas the rest period begins at the onset of the animal's subjective day. For diurnal animals, the onset of activity coincides with the beginning of the animal's subjective day, and the rest period begins at the onset of the animal's subjective night. If free-running hamsters that are housed in constant dark conditions are exposed to a 1-hr pulse of light at any time during the middle of the subjective day, the time of onset of the next bout of locomotor activity is unaffected; that is, for a hamster with a free-running period of 24.2 hr, wheel running still begins 24 hr and 12 min after the onset of the previous bout of activity. However, if the 1-hr pulse of light is given early in the subjective night (or late in the subjective day), then the hamster does not begin its activity 24.2 hr later, but rather delays its activity onset for 1–4 hr, depending on when exactly the pulse of light occurs. If the 1-hr pulse of light is given late in the subjective night (or early in the subjective day), then again the hamster does not begin its activity 24.2 hr later, but rather advances its activity onset by 1–4 hr, depending on exactly when the pulse of light occurs. Thus, the largest phase delays can be produced early during the subjective night, when a nocturnal animal has just awakened and a diurnal animal has just retired, and the largest phase advance can be produced late during the subjective night. Light appears to have little effect on the circadian system during the subjective day of either nocturnal or diurnal animals. Although these phase responses to light are best observed in free-running animals, these same effects of light also operate on individuals that are entrained to a light–dark cycle. The function of an animal's daily cycle of responsiveness to light is to reset the biological rhythms to exactly 24 hr. Phase–response curves have been generated for many species, including humans, and it is also apparent that many species measure day length. However, the extent to which some species, including humans, couple reproductive function to day length remains unspecified.

II. MAMMALS

A. Pineal Gland and Photoperiodism

The pineal gland and its hormone, melatonin, mediate photoperiodic time measurement in mammals. Pinealectomy blocks responsiveness to photoperiod in every mammalian species studied. Information about environmental light arrives to the brain via the lateral eyes in mammals. A nonvisual neuronal pathway called the retinohypothalamic tract carries light information directly to the suprachiasmatic nuclei (SCN) of the hypothalamus (Fig. 4). The SCNs are the primary mammalian biological clocks. From the SCN, photoperiod information actually leaves the brain and synapses in the superior cervical ganglion in the spinal cord. Postganglionic noradrenergic fibers eventually project back into the brain and innervate the pineal gland, in which neural information is transduced into a hormonal message. Melatonin, an indole amine hormone secreted rhythmically

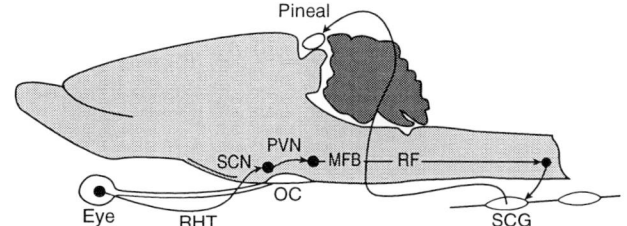

FIGURE 4 Photoperiodic time measurement requires an intact circuit among the pineal gland, SCN, and the eyes. Environmental light enters the SCN from the lateral eyes via the retinohypothalamic tract (RHT). Projections from the SCN travel to the paraventricular nuclei (PVN), from the PVN to the medial forebrain bundle (MFB), and from the MFB to the superior cervical ganglion (SCG). Postganglionic noradrenergic fibers project back into the brain and innervate the pineal gland, where neural information is transduced into a hormonal message (after Klein *et al.*, 1983).

by the pineal, can induce photoperiodic responses when administered in a number of ways. Nevertheless, particular aspects of the responses observed following constant release implants, daily injections, or daily infusions of melatonin have led to some controversy regarding the significance of various temporal characteristics of melatonin secretion. Melatonin is normally secreted in a circadian fashion, with an extended peak occurring at night, and basal secretion occurs during the day. The duration of this nocturnal "peak" varies inversely with day length in several species, including humans. There has been significant controversy regarding the relative importance of the phase versus the duration of the nocturnal elevation of melatonin secretion from the pineal gland in the mediation of this gland on photoperiodic responsiveness. Results obtained in sheep and in Siberian hamsters strongly favor the overriding importance of the duration of the melatonin peak. When pinealectomized hamsters or sheep were infused with melatonin on a daily basis, the types of responses elicited depended on the duration, but not on the phase, of the daily infusion. For example, when male hamsters received melatonin for 6 hr or less each day, they exhibited functional testes (a long-day response); when melatonin was administered in daily infusions of an 8-hr duration or more, the animals inhibited reproductive function (a short-day response). Recent data have been obtained from similar studies in several other mammalian species; in each instance, the results support the conclusions reached in the studies using Siberian hamsters and sheep; that is, the duration of melatonin is the critical physiological parameter providing photoperiod information.

B. Melatonin Influences on the Hypothalamus–Pituitary–Gonadal Axis

One of the mechanisms by which the nightly duration of melatonin secretion affects the reproductive system is by altering the steroid negative feedback mechanisms in the hypothalamus and pituitary gland. In long-day breeders, reduced secretion of pituitary luteinizing hormone (LH) and follicle-stimulating hormone (FSH) during short days results from increased sensitivity of the hypothalamic–pituitary axis to the negative feedback effects of gonadal steroid hormones. In males, low concentrations of blood androgens are more effective at inhibiting postcastration elevations in pituitary gonadotropin secretion in hamsters and rams when males are housed in short or long days, respectively. A return to the lower level of sensitivity returns the animals to a state of reproductive activity via increased pituitary hormone secretion. A similar phenomenon has been implicated in puberty, during which a prepubertal decrease in sensitivity to steroid negative feedback leads to increased secretion of LH and FSH and activation of the reproductive system.

In addition to changes in the sensitivity of the gonadotropin secretion system to gonadal steroid hormones, a steroid-independent mechanism has also been implicated in the regulation of seasonal changes in the rate of gonadotropin secretion. A steroid-independent effect of photoperiod on gonadotropin secretion is particularly evident in female hamsters. In long days, female hamsters exhibit an approximately 8- to 10-fold increase in baseline serum LH concentrations following ovariectomy, and LH concentrations can be returned to baseline by administration of estrogens. After several weeks of short-day exposure, female hamsters become anovulatory and serum LH concentrations are very low during most of the 24-hr cycle; however, the anovulatory females display daily LH surges during the afternoon. This pattern of LH secretion continues following ovariectomy or adrenalectomy. In other words, removal of the sources of steroid hormones does not cause any detectable elevation in the baseline serum LH concentrations in short-day female hamsters, and the daily LH surges continue in the absence of steroid hormones. Seasonal variations in circulating and pituitary concentrations of gonadotropins have been reported for several mammalian species.

With the use of 2-[^{125}I]iodo-melatonin, melatonin-binding sites have been discovered throughout the periphery and nervous system in vertebrates. Among mammals, high melatonin binding is commonly observed in the pars tuberalis and pars distalis of the pituitary, the SCN, and dorsalmedial nucleus of the hypothalamus, preoptic area, and the area postrema. However, there are some important exceptions to this general pattern, e.g., melatonin receptors have not been reported in the pars tuberalis of humans, and neither sheep nor mustelids (e.g., mink and ferrets) possess melatonin receptors in the SCN. The SCN melatonin receptors are not critical for photoperiodic responses in hamsters because appropriately timed melatonin infusions induce reproductive regression in SCN-lesioned animals. Lesions of the 2-[^{125}I]iodo-melatonin-binding sites of the mediobasal hypothalamus block the inhibitory effects of infused melatonin on gonadotropin secretion. Low densities of 2-[^{125}I]iodo-melatonin binding have been reported elsewhere throughout the CNS and periphery, but their functional significance remains unspecified.

All the available evidence suggests that melatonin inhibits cAMP through a G-protein-coupled mechanism at some sites. In the pars distalis, melatonin appears to inhibit calcium influx and membrane potential. Successful cloning of melatonin receptors from various vertebrate target tissues has revealed three receptor subtypes (i.e., MEL-1A, MEL-1B, and MEL-1C). Recent evidence suggests that the MEL-1A melatonin receptor is coupled to parallel signal transduction pathways that inhibit adenylyl cyclase and enhance phospholipase activation. The genetic sequence encoding the various melatonin receptors is currently being mapped; presumably, mice with targeted disruption of melatonin receptor genes will be available soon and will provide another important tool to study the mechanisms of melatonin action.

III. BIRDS

Long day lengths maintain, and short day lengths inhibit, reproductive function in mammalian long-day breeders such as hamsters. Short day lengths stimulate, and then inhibit, breeding in mammalian short-day breeders such as sheep. Photoperiodic regulation of avian breeding differs fundamentally from photoperiodic regulation of mammalian breeding. In all species of temperate zone birds studied to date, long day lengths stimulate the reproductive system. After a few weeks or months, the reproductive system "switches off" despite exposure to long day lengths. In other words, the individual becomes photorefractory, and long days no longer provoke reproductive function. Birds require several weeks of short-day exposure to reestablish sensitivity to stimulatory effects of long days.

Several elegant studies of photoperiodic time measurement, using non-24-hr days (T-cycles) and skeleton and resonance photoperiods, suggest that birds measure day length with a coincidence model of photoperiodic time measurement. The pineal gland is an important circadian oscillator and is essential in entrainment of locomotor and other biological rhythms in birds. Melatonin is secreted during the dark but not during the light portion of the day, as in mammals. In common with mammals, the SCN is necessary for the pineal to maintain the daily cycle of melatonin secretion in chickens (*Gallus domesticus*). However, in other bird species [e.g., Japanese quail (*Coturnix coturnix japonica*)] the integrity of the SCN is not necessary to maintain the daily pattern of melatonin secretion, suggesting the presence of photoreceptors in the pineal or communication with the pineal through a non-SCN route. However, unlike mammals, the pineal gland is unnecessary for photoperiodic time measurement in birds. For example, pinealectomized American tree sparrows (*Spizella arborea*) undergo the complete cycle of photostimulation, photorefractoriness, and photostimulation when exposed to the appropriate photoperiods. Although endogenous melatonin administration can affect gonadal size in some bird species, generally melatonin is thought not to mediate photoperiodic information in birds. Deep hypothalamic photoreceptors are hypothesized to transduce photoperiodic information into either a hormonal or neurochemical signal that affects photoperiod-mediated responses in birds. Input from these photoreceptors stimulates hypothalamic GnRH secretion, the first step in the cascade of photostimulation of the avian reproductive system. However, in common with mammals, the physiological mechanisms by which photorefractoriness is mediated remains unknown. In fact, to a large extent, "We remain largely ignorant of the processes, at least at the highest neural levels, by which the vertebrate reproductive system switches on and off" (Nicholls *et al.*, 1988).

IV. REPTILES, AMPHIBIANS, AND FISHES

Seasonal reproductive cycles and photoperiodic time measurement have been reported for various species of reptiles, amphibians, and fishes. However, the mechanisms underlying photoperiodic time measurement are not very well documented among poikilothermic (heterothermic) vertebrates. As described earlier, the best studied reptilian species is the anole lizards. These lizards appear to use an hourglass mechanism to measure photoperiod. Although exogenous melatonin treatment can affect reproductive function in anole lizards, a series of well-designed experiments determined that neither the duration nor the amplitude of the nocturnal melatonin pulse mediates photoperiodic time measurement in this species. However, it remains possible that the phase relationship of the melatonin rhythm to other circadian rhythms may mediate photoperiodic time measurement in this species. In poikilothermic vertebrates, temperature, in addition to day length, may affect pineal melatonin secretion. Again, few studies have established the physiological mechanisms by which photoperiodism is mediated among poikilothermic vertebrates.

Long days stimulate reproductive function among amphibians. Whether amphibians use an hourglass model or a coincidence mechanism to measure day length has not been established. The role of the pineal gland and melatonin in photoperiodic time measurement also remains unspecified. For example, exogenous administration of melatonin to some species of frogs (*Hyla cinerea, Rana ridibunda, R. cyanophlyctis,* and *R. temporaria*) and toads (*Bufo arenarium*) caused gonadal regression; however, in other amphibian species (*Rana tigrina, r. Perezi,* or *Discoglossus pictus*) melatonin treatment either enhanced or had no effect on reproductive function.

More research has been conducted on the mechanisms of seasonality in fishes than on reptiles and amphibians, possibly reflecting the commercial importance of managing fish reproduction and development. With the exception of goldfish (*Carassius auratus*), much of this research has been focused on the role of photoperiod in mediating molt. The mechanisms by which photoperiod affects molt appear to involve nonretinal photoreceptors that affect prolactin and thyroid hormone secretion. Photoperiod and melatonin affect circadian rhythms of activity and hormone secretion in fishes, but the role of pineal melatonin on fish (teleost) reproduction remains controversial. In goldfish, melatonin secretion mimics the duration of the dark period of the daily light–dark cycle. The melatonin receptor in goldfish and trout (*Salmo gairdneri*) appears to be a transmembrane receptor belonging to the family of G-protein-coupled receptors with approximately 85% homology with the mammalian melatonin brain receptor. It also appears that a similar melatonin signal transduction system exists among all vertebrates.

The evolution of melatonin appears to reflect a change in its uses rather than a change in hormonal structure or receptor dynamics. Indeed, melatonin receptors are present in nonoptic brain regions of deep-sea fish living in complete darkness. Importantly, despite the lack of light in their habitats, these deep-sea fish breed seasonally. Seasonal breeding has many advantages. In order to time breeding appropriately, animals must determine the time of year. Photoperiodic time measurement has evolved among vertebrates, and this trait permits precise determination of time of year. However, the mechanisms underlying photoperiodism appear to have changed over the course of vertebrate evolution. Further studies are required to determine the various mechanisms that have evolved that permit vertebrates to measure day length.

See Also the Following Articles

Circadian Rhythms; Circannual Rhythms; Cricetidae (Hamsters and Lemmings); Pineal Gland, Birds and Others; Pineal Gland, Mammals; Seasonal Reproduction, Mammals; Sheep and Goats

Bibliography

Baker, J. R., and Ranson, R. M. (1932). Factors affecting the breeding of the field mouse (*Microtus agrestis*). Part I. Light. *Proc. R. Soc. London* 110, 313–323.

Bartness, T. J., Powers, J. B., Hastings, M. H., Bittman, E. L., and Goldman, B. D. (1993). The timed infusion paradigm for melatonin delivery: What has it taught us about the melatonin signal, its reception, and the photoperiodic control of seasonal responses? *J. Pineal Res.* 15, 161–190.

Bittman, E. L., and Karsch, F. J. (1984). Nightly duration of pineal melatonin secretion determines the reproductive response to inhibitory day length in ewe. *Biol. Reprod.* 30, 585–593.

Bünning, E. (1973). *The Physiological Clock, Cicadian Rhythms and Biological Chronometry.* Springer-Verlag, New York.

Carter, D. S., and Goldman, B. D. (1983). Antigonadal effects of timed melatonin infusion in pinealectomized male Djungarian hamsters (*Phodopus sungorus sungorus*): Mediation by melatonin. *Endocrinology* 113, 1261–1267.

Elliott J. A. (1976). Circadian rhythms and photoperiodic time measurement in mammals. *Fed. Proc.* 35, 2339–2346.

Filadelfi, A. M. C., and Castrucci, A. M. (1996). Comparative aspects of the pineal/melatonin system of poikilothermic vertebrates. *J. Pineal Res.* 20, 175–186.

Follett, B. K., and Sharp, P. J. (1969). Circadian rhythmicity in photoperiodically induced gonadotrophin release and gonadal growth in the quail. *Nature* 223, 968–971.

Garner, W. W., and Allard, H. A. (1922). Photoperiodism, the response of the plant to relative length of day and night. *Science* 60, 582–583.

Godson, C., and Reppert, S. M. (1997). The Mel-1A melatonin receptor is coupled to parallel signal transduction pathways. *Endocrinology* 138, 397–404.

Goldman, B. D., and Nelson, R. J. (1993). Melatonin and seasonality in mammals. In *Melatonin: Biosynthesis, Physiological Effects and Clinical Applications* (H. S. Yu and R. J. Reiter, Eds), pp. 225–252. CRC Press, New York.

Hyde, L. L., and Underwood, H. (1993). Effects of nightbreak, T-cycle, and resonance lighting schedules on the pineal melatonin rhythm of the lizard *Anolis carolinensis*: Correlations with the reproductive response. *J. Pineal Res.* **15**, 70–80.

Iigo, M., Kezuka, H., Suzuki, T., Tabata, M., and Aida, K. (1994). Melatonin signal transduction in the goldfish, *Carassius auratus*. *Neurosci. Biobehav. Rev.* **18**, 563–569.

Klein, D. C., Reppert, S. M., and Moore, R. Y. (1991). *The Suprachiasmatic Nucleus: The Mind's Clock*. Elsevier, New York.

Kumar, V., Jain, N., and Follett, B. K. (1996). The photoperiodic clock in blackheaded buntings (*Emberiza melanocephala*) is mediated by a self-sustaining circadian system. *J. Comp. Physiol. A* **179**, 59–64.

Matthews, C. D., Guerin, M. V., and Deed, J. R. (1993). Melatonin and photoperiodic time measurement: Seasonal breeding in the sheep. *J. Pineal Res.* **14**, 105–116.

Maywood, E. S., and Hastings, M. H. (1995). Lesions of the iodomelatonin-binding sites of the mediobasal hypothalamus spare the lactotropic, but block the gonadotropic response of male Syrian hamsters to short photoperiod and to melatonin. *Endocrinology* **136**, 144–153.

Moore-Ede, M. C., Sulzman, F. M., and Fuller, C. A. (1982). *The Clocks That Time Us*. Harvard Univ. Press, Cambridge, MA.

Nanda, K. K., and Hamner, K. C. (1958). Studies on the nature of the endogenous rhythm affecting photoperiodic response of Biloxi soybean. *Bot. Gazette* **120**, 14–28.

Nicholls, T. J., Goldsmith, A. R., and Dawson, A. (1988). Photorefractoriness in birds and comparison with mammals. *Physiol. Rev.* **68**, 133–176.

Pittendrigh, C. S., and Minis, D. H. (1964). The entrainment of circadian oscillations by light and their role as photoperiodic clocks. *Am. Nat.* **98**, 261–299.

Roca, A. L., Godson, C., Weaver, D. R., and Reppert, S. M. (1996). Structure, characterization, and expression of the gene encoding the mouse Mel1a melatonin receptor. *Endocrinology* **137**, 3469–3477.

Rowan, W. (1925). Relation of light to bird migration and developmental changes. *Nature* **115**, 494–496.

Smith, A., Trudeau, V. L., Williams, L. M., Martinoli, M. G., and Priede, I. G. (1996). Melatonin receptors are present in non-optic regions of the brain of a deep-sea fish living in the absence of solar light. *J. Neuroendocrinol.* **8**, 655–658.

Underwood, H., and Goldman, B. D. (1987). Vertebrate circadian and photoperiodic systems. Role of the pineal gland and melatonin. *J. Biol. Rhythms* **2**, 279–315.

Xiong, J. J., Karsch, F. J., and Lehman, M. N. (1997). Evidence for seasonal plasticity in the gonadotropin-releasing hormone (GnRH) system of the ewe: Changes in synaptic inputs onto GnRH neurons. *Endocrinology* **138**, 1240–1250.

Pigeons

Richard F. Johnston
The University of Kansas

I. Introduction
II. The Pair Bond
III. Nests
IV. Eggs and Incubation
V. Squabs

GLOSSARY

endothermic Having body temperature determined by heat energy generated internally.

extrapair fertilization Result of sexual union outside the pair bond.

monophyletic Derived from one ancestral taxon.

pair bond The persistent social relationship between a male and a female whereby the sexual and parental aspects of reproduction are enabled.

photoperiodic Response of an organism to changes in day length or a light–dark cycle.

squab The young of pigeons, applied indiscriminately to hatchlings, nestlings, and individuals ready to leave the nest.

Pigeons belong to the Columbidae, a monophyletic family consisting of pigeons and doves. Reproductively, all members of the family have much in common, but this article focuses on two species. The rock dove (rock pigeon), *Columba livia*, and the ringed turtledove (barbary dove and ringdove), *Streptopelia risoria*, have been subjects of study of reproduction for more than a century and have provided insight into some of the fundamental aspects of vertebrate reproduction.

I. INTRODUCTION

Pigeons have been used in biomedical research more than would be expected of birds. Part of the reason for this is that the rock dove and ringdove have long been domesticated, for at least 5000 years in the case of the rock dove, and therefore much information on the care and handling of pigeons had been accumulated by the time experimental biology was begun. The importance of this cannot be overemphasized—pigeons are user-friendly, adjust readily to confinement, and have simple dietary needs and a nearly interminable reproductive schedule. These characteristics were responsible for the species being brought into domestication at such an early time in human history; their current state, however, is largely a consequence of thousands of generations of selective breeding by humans.

The diet of rock and ringdoves consists mostly of seeds, except in the first week of life. The young squabs are fed by both parents on crop milk, an energy-rich product of crop gland activity. One of the transcendent discoveries in the history of vertebrate reproduction was that production of crop milk was a result of the hormone prolactin. Because prolactin is the hormone generating lactation in mammals, pigeons for a time assumed a position of significance in laboratories doing research on human endocrinology. The response of pigeon crop sacs to prolactin has been the basis on which much of the research on prolactin in other vertebrates has proceeded. Endocrinologic coincidences between pigeons and other vertebrates do not foreshadow many parallels in other aspects of reproductive biology, although some nevertheless exist.

Many species of pigeons are capable of undertaking reproduction at an early age, perhaps at 6 or 7 months and occasionally earlier. Reproductive performance of juvenile birds can be variously inadequate or unfocussed at the outset. Confined mated pairs can maintain a high level of reproductive activity for several years, but after age 6 signs of sexual senescence usually appear. The reproductive cycle can be examined under conventional categories of behavior.

II. THE PAIR BOND

Pigeons are monogamous, and in free-living birds the pair bond is rarely broken. The bond is associated with crop milk feeding of squabs because both parents are necessary for satisfactory provisioning of a pair of squabs, or even of singleton squabs in time of food limitation. The bond is generated by a complex of behavioral interactions of the two birds ("courtship"), brought on by sex steroids stimulated by pituitary gonadotropins, generated by hypothalamic hormones under the control of photoperiodic variation and other factors. The nexus of endocrinologic and behavioral events is generally characteristic of endothermic vertebrates, but the range and format of behavior itself is entirely species specific.

Rock and ringdoves are sexually monochromatic for plumage and detect sex by behavior. Males with sufficient androgen levels display to individuals near them with the "bow-coo," which involves strutting and bowing and a vocalization. Females under a complex of hormonal stimulation that includes estrogen may respond if unmated, and subsequent behavior is partly dependent on a succession of endocrinologic changes and partly on the presence of the displaying partner.

Both sexes discriminate in choice of mate, and at least seven variables are used: age, body size, color of plumage, pattern of plumage, previous breeding experience, condition of the feather coat, and dominance status. Some of these are redundant. Color and pattern of plumage are exercised by females only. After initial choice, successive phases of pair formation include mutual preening, courtship feeding (symbolic feeding of females by males), and copulation. Bonded pairs are close partners, and their lives together are largely concerned with reproduction.

Plumage color and pattern covary with variation in reproductive output. Melanics (checker, T-pattern, and spread) produce more offspring in dense urban centers and in winter than do birds in other plumages; bluebar individuals (wild type) have higher productivity in suburbs and rural habitats.

Because of the nearly permanent pair bond (divorce is rare) mistakes in mate choice may entail severe penalties. For instance, a sterile mate prevents its partner from contributing to future populations. Irregular means of circumventing such a penalty include extrapair fertilization, a solution that appears to be used more frequently than divorce.

III. NESTS

Coincident with first copulations, the pair builds a nest. Males gather twigs and other matter and bring them to their females, who sit where they have decided to make the nest. Males approach and stand on the backs of females and give them what they have brought. Females then tuck the material under flank or breast feathers, while males fly off to get additional supplies. First nests can be relatively simple, but with continued use they grow to considerable size, owing to accumulation of droppings of squabs and the additional nesting material brought by the male for each successive clutch of eggs. Nest sites may be difficult to find, so continued use of a site, despite the nest itself gradually becoming attractive to ectoparasites, may be a condition for reproduction.

IV. EGGS AND INCUBATION

Normal clutches are of two eggs in rock and ringdoves. The first egg in rock doves tends (70%) to be male and the second female, and the sex ratio at ovulation is 1:1. Females thus control sex of offspring, but the manner in which this is achieved has not been studied.

In rock doves, incubation begins a few hours after laying of the first egg, and each egg receives about 18 days of parental incubation. Eggs are covered nearly 100% of the time; the pair exchange places just twice each day, the female sitting all night and the male from midmorning to late afternoon.

Mature pigeons in good condition in midsummer will initiate successive clutches prior to the time that squabs from the earlier clutch leave the nest. Overlapping of broods with eggs can occur when squabs are only 18–20 days old so that the last 10–12 days of squab nest life may be coincident with early incubation of new eggs. Two to four such overlaps may occur for mature pairs, increasing their potential number of fledged juveniles. The endocrinologic background to clutch overlaps has not been directly studied.

V. SQUABS

Emergence from the shell requires nearly a day for a single squab, and it is a critical period for the young birds. Mortality of hatching squabs can be considerable and may average 10%. Reasons for this are probably related to both thermoregulation (squabs regulate body temperature poorly in the first week but adequately after 10 days) and the shift to taking food through the alimentary canal. Effectiveness of parental care varies and is partly dependent on previous experience.

The foundations for crop milk production appear in adults after the first week of incubation. The mucosal lining of the crop thickens, vascularity increases, and the lining itself begins to slough into the crop interior. The cheesy material is available in the crop around a day before hatching. Crop milk contains about 33% fats and nearly 60% protein. It is this remarkable quality that emancipates parents from the need to search for high-protein food in the form of insects or other invertebrates.

For the first 4 days posthatch, each rock dove parent produces crop milk at a rate sufficient for the growth of one squab. Both parents are thus necessary for the satisfactory growth of two squabs. Other kinds of pigeons, usually those feeding largely on fruits, lay just one egg and can rear only one squab at a time, with both parents feeding it.

Rock dove squabs from first eggs tend to be male and are 24 hr older than second squabs, which tend to be female. The sex and age differences result in squabs of significantly different sizes. Second squabs nevertheless have a more rapid growth rate than first squabs, and if food is plentiful the size differential will be reduced in a week or 10 days. Under food shortage, however, the second squab usually dies of starvation.

Some squabs leave the nest site at age 30 days, but most stay near the parental nest until age 35 days (and some even longer). Juveniles leaving the nest may be only 84% of adult weight and have wings only 77% as long as those of adults. Juveniles accompany parents to feeding sites and learn feeding behavior by observation. By Days 45–50, juveniles are essentially independent, although they still benefit from living among adults in foraging and loafing flocks.

See Also the Following Articles

AVIAN REPRODUCTION, OVERVIEW; FEMALE REPRODUCTIVE SYSTEM, BIRDS; MALE REPRODUCTIVE SYSTEM, BIRDS; NESTING, BIRDS; PHOTOPERIODISM, VERTEBRATES

Bibliography

Abs, M. (1983). *Physiology and Behaviour of the Pigeon.* Academic Press, London.

Johnston, R., and Janiga, M. (1995). *Feral Pigeons.* Oxford Univ. Press, New York.

Levi, W. (1974). *The Pigeon.* Levi, Sumter, SC.

Murton, R., and Westwood, N. (1977). *Avian Breeding Cycles.* Oxford Univ. Press, Oxford, UK.

Murton, R., Thearle, R., and Coombs, C. (1974). Ecological studies of the feral pigeon Columba livia var. III. Reproduction and plumage polymorphism. *J. Appl. Ecol.* **11**, 841–854.

Riddle, O., Bates, R., and Dykshorn, S. (1932). A new hormone of the anterior pituitary. *Proc. Soc. Exp. Biol. Med.* **29**, 1211–1212.

Pigs

Rodney D. Geisert
Oklahoma State University

I. Introduction
II. Anatomy of the Male and Female Reproductive System
III. Puberty in the Male and Female
IV. Spermatogenesis and Ejaculate Characteristics
V. Characteristics and Endocrine Patterns of the Estrous Cycle and Early Pregnancy
VI. Early Embryonic Development, Establishment of Pregnancy, and Placental Attachment
VII. Parturition and Lactation

GLOSSARY

bicornuate uterus A uterus in which horns are convoluted and may reach lengths of 4 to 5 ft, whereas the uterine body is only 1 or 2 in.

boar taint A musk-smelling odor present in boar salvia, semen, and meat caused by testis production of 16-androstenes.

conceptus The embryo and its extraembryonic placental membranes.

epitheliochorial placentation A noninvasive type of placental attachment in which microvilli of placenta and uterine epithelium interlock superficially at the maternal–fetal interface in the uterus.

estrus The portion of females' reproductive cycle when they are sexually active, receptive, and fertile.

lordosis The arching of back and locking of legs for presentation of female to the male during mating at estrus.

luteolysis Lysis of a functional corpus luteum causing a dramatic decrease in plasma progesterone.

maternal recognition of pregnancy A chemical signal(s) produced by the developing conceptus to extend pregnancy beyond the normal length of an estrous cycle.

polytocous Producing many offspring or ova at a single time.

puberty Period of adolescence when a male or female is first able to release gametes.

I. INTRODUCTION

The pig is a nonseasonal, polytocous species that has reoccurring estrous cycles (polyestrous) of approximately 21 days (range, 17–24) in length when not mated. The estrous cycle is characterized by four basic phases: proestrus, estrus, metestrus, and diestrus. During proestrus, the sow becomes restless, the vulva may be red and swollen, mucous discharge is sometimes evident, and she may attempt to mount other females. Proestrus occurs 24–48 hr prior to the female exhibiting lordosis (standing for mounting by the boar, other females, or hand back pressure), which is the true evidence of estrus in the pig. Characteristics of reproduction in the gilt are presented in Table 1. The length of standing heat is usually between 24 and 72 hr, with gilts (young nonbred females) showing a slightly shorter expression of estrus compared to mature sows. Ovulation of 10–24 mature Graafian follicles (10 mm in diameter) occurs over 1–3 hr at 30–40 hr from the onset of behavioral estrus.

Following ovulation, the granulosa and theca cells of the follicle wall become luteinized to initiate formation of functional corpora lutea. Initially, blood fills the lumen of the ruptured follicle, forming a corpus hemorrhagicum. Metestrus is the phase of the estrous cycle in which the luteal cells proliferate, fill the ruptured follicle cavity, and increase progesterone synthesis over the next 1 or 2 days. Formation of multiple corpora lutea (8–11 mm in diameter) on each ovary of the sow results in a steady increase of progesterone during the start of diestrus. Diestrus continues for approximately 14 or 15 days until regression of the corpora lutea and reinitiation of proestrus.

II. ANATOMY OF THE MALE AND FEMALE REPRODUCTIVE SYSTEM

A. Boar

Testes of the boar are contained outside the body cavity in the scrotum, which is positioned just below the anus. The testis, which can weigh approximately 400 g in a mature boar, must remain 3–5°C below body temperature for normal spermatogenesis to occur. Cooling of the testis is accomplished through positioning testicles outside the body cavity, abundance of sweat glands in the scrotum, and cooling of arterial blood entering the testis through countercurrent heat exchange of the coiled artery and veins which form the pamipiniform plexus.

The boar has three major accessory glands for which the seminal vesicles provide the major fluid volume of the seminal plasma in the ejaculate. The prostate gland contributes a minor portion to the fluid volume of the semen. Cowper's glands of the boar are quite pronounced and unique it that they produce gel plugs which serve to seal the sows cervix following emission of the sperm-rich fraction of the boar's ejaculate. The fibroelastic penis of the boar is characterized by an S-shape curve called the sigmoid flexure, which is allowed to straighten with relaxation of the retractor penis muscle during erection. The anterior end of the glans penis has a spiral shape which the boar uses to lock into the interdigitating pads of the sow's cervix during copulation.

B. Sow

Paired ovaries of the female are oval in shape but have a mulberry-like appearance when follicles and corpora lutea are present. Ovulated eggs are picked up by the oviductal infundibulum, which surrounds the ovary but is not firmly attached.

The sow has a bicornuate type of uterus in which each uterine horn is extremely long, measuring approximately 150–200 cm in length. The endome-

TABLE 1
Characteristics of Reproduction in the Female

Puberty	6–7 months
Estrous cycle length	18–21 days
Estrus	48–72 hr
Ovulation[a]	30–40 hr
Number of corpora lutea	10–25
Gestation length	114 days
Average litter size	12

[a] Hours from initiation of estrus.

trium within the uterus is very vascular and dense. Uterine horns join at the uterine body, which is only 5 cm in length. Cervix of the sow (10 cm in length) is continuous with the vagina because the pig has no fornix vagina as is seen in other species. The cervix contains a number of interdigitating rounded projections (pads) which serve to tighten around the corkscrew end of the boar's glans penis during copulation. The vagina and vestibule of the pig are similar to those of other species.

III. PUBERTY IN THE MALE AND FEMALE

Spermatozoa can be present within the testes of boars by 120 days of age; however, mating interest and ejaculation of fertile sperm in the majority of boars usually develop between 5 and 8 months. Puberty can be affected by breed, nutrition, and possibly season of the year. Most boars develop sexually after reaching 110 kg. Sperm production and seminal volume increase with increasing age usually to 18 months of age.

Gilts express their first pubertal estrus by 6–9 months of age, when they reach 80–120 kg in body weight. Puberty can be effected by breed, nutrition, season, social environment, and exposure to the boar. A delay in puberty has been observed in the summer months of the year and can also occur during crowding and confinement of young females housed in indoor pens. Mixing of gilts from different pens or transporting prepubertal gilts can advance puberty in gilts given adequate nutritional development.

As a species, the pig is a very social animal in which olfactory signaling through pheromones has highly evolved in regulating not only expression of estrus but also attainment of puberty. Daily exposure of gilts to boars has a direct effect on advancing puberty in gilts which are between 135 and 160 days of age. Stimulation of puberty by the boar (boar effect) occurs through physical contact, vocalization, and secretion of musk-smelling compounds, 16-androstenes, which give the typical odor to mature boars and boar taint to the meat. Boar odor provides signals to the female for inducing cyclicity and also the strong lordosis response during estrus. Puberty can be advanced hormonally through the treatment of 5.5-month-old prepuberal gilts with a combination of 400 IU pregnant mare serum gonadotropin and 200 IU human chorionic gonadotropin which is under the trade name P.G. 600 (Inervet America, Inc.).

The first pubertal estrus in the gilt is generally less fertile than subsequent estrous cycles because larger litters are obtained from gilts bred on their second or third heats. Increase in litter size is due to the increasing ovulation rate between the first and fourth estrus.

IV. SPERMATOGENESIS AND EJACULATE CHARACTERISTICS

Generation of mature spermatozoa from base cells to the lumen of the seminiferous tubule takes 34.5 days in the boar. Testicular spermatozoa are not immediately capable of fertilization and therefore must mature as they are transported through the epididymus, which takes approximately 10 days in the pig. Sperm can be stored in the cauda epididymus for 4–7 days prior to ejaculation. Spermatogenesis is very sensitive to elevated environmental temperatures in the boar. A 1 or 2°C rise in body temperature caused by either a fever or heat stress during the summer months can have dramatic effects on the percentage of females which conceive and litter size. Even a short-term stress can result in fertility problems for 3 or 4 weeks in the boar. As a management practice, boars are cooled during the hot summer months to avoid a drop in conception rate.

A boar ejaculates in three phases: A presperm fraction, which is a clear, sperm-free seminal fluid, is followed by the sperm-rich fraction that has a creamy appearance due to the high concentration of sperm. These two phases can alternate several times before the gel fraction is ejaculated to seal the sow's cervix. Small gel plugs secreted from the Cowper's glands and adhere to one another, forming a plug in the cervix. The plug serves to prevent the 200–300 ml of semen from retrograde flow back through the cervix.

The boar is stimulated to ejaculate upon locking his corkscrewed shape glans penis in the cervical protrusions of the sow or by hand pressure on the glans penis during semen collection for artificial in-

TABLE 2
Characteristics of Boar Ejaculate

Fluid volume (ml)	100–500
Gel volume (ml)	50–100
Sperm concentration (million/ml)	100–600
Total sperm/ejaculate (billion)	45–120
Normal motility (%)	60–80
Normal morphology (%)	70–90
Number of collections/week for AI	2–5

semination (AI). Complete ejaculation during mating or semen collection can last 1–5 min. Characteristics of boar ejaculate are presented in Table 2.

Hand-mating of boars to sows or AI is usually done 12 and 24 hr following first detection of estrus with twice/day heat detection (Fig. 1). For AI, extended semen must contain a minimal volume of 50 ml and a total of 3 or 4 billion sperm/breeding.

Boar semen can be frozen, but the freezing process causes a substantial loss of sperm viability and shortens the life span of the sperm in the female reproductive tract. The poor freezing quality of boar semen is due to the lipid content in the plasma membrane compared to that of other species. With frozen semen, it is critical to breed females very close (4–8 hr) to the time of ovulation in order to achieve acceptable pregnancy rates.

V. CHARACTERISTICS AND ENDOCRINE PATTERNS OF THE ESTROUS CYCLE AND EARLY PREGNANCY

A. Estrous Cycle

Endocrine changes during the estrous cycle of the sow are presented in Fig. 2. During proestrus, growth of Graafian follicles increases the concentration of estrogen in plasma to 40 pg/ml approximately 48 hr prior to the surge release of luteinizing hormone (LH) from the anterior pituitary which induces ovulation. Peak concentrations of LH (6–10 ng/ml) occur near the time of estrus onset and decline to basal levels (1 ng/ml) after about 10 hr. Concentration of estrogen declines prior to ovulation and remains basal during diestrous. Follicle-stimulating hormone increases and peaks 2 or 3 days from the initiation of estrus, with the decline caused by increased release of inhibin from the ovary.

Development of the corpora lutea increases plasma progesterone concentrations rapidly from <1 ng/ml on the day of estrus to 30–40 ng/ml on Day 12 of the estrous cycle. Peripheral plasma progesterone concentrations rapidly decline after Day 15 of the estrous cycle following the episodic release of prostaglandin $F_{2\alpha}$ ($PGF_{2\alpha}$) from the uterine endometrium to lysis of the corpora lutea and reinitiate follicular growth for the next estrous cycle. Exogenous $PGF_{2\alpha}$ can be utilized to induce corpora lutea (CL) regression in the pig; however, CL do not respond well to

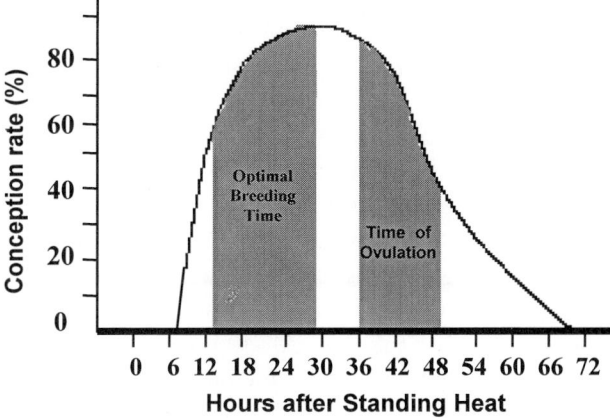

FIGURE 1 Optimal time for breeding and time of ovulation relative to first expression of estrus in the sow.

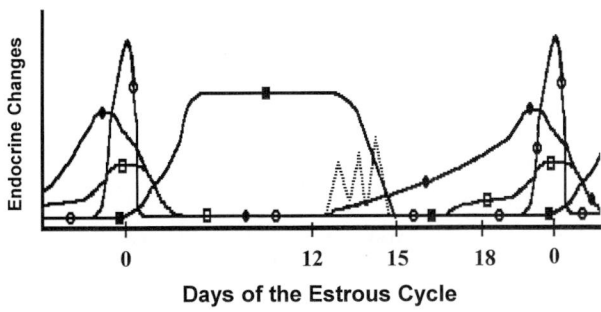

FIGURE 2 Endocrine changes in peripheral plasma during the estrous cycle of the sow. Patterns of hormone changes during the cycle are illustrated. ■, Progesterone; ●, estrogen; □, follicle-stimulating hormone; ○, LH; dashed line, $PGF_{2\alpha}$.

PGF$_{2\alpha}$-induced regression until after Day 12 of the estrous cycle or pregnancy.

Because CL are not readily sensitive to lysis by PGF$_{2\alpha}$ until after Day 12 of the estrous cycle, PGF$_{2\alpha}$ is not effective in synchronizing cyclic gilts for mating. However, gilts can be synchronized by feeding the progestin, altrenogest (20 mg/head/day), for 14–18 days. Gilts will usually express estrus 4–7 days following progestin withdrawal.

B. Pregnancy

Corpora lutea do not regress during pregnancy because the conceptus prevents release of endometrial PGF$_{2\alpha}$ into the uterine vasculature. Although corpora lutea are maintained during pregnancy, plasma concentrations of progesterone decline after Day 14 of gestation but remain relatively constant (10–20 ng/ml) until term. The decline of plasma progesterone results from the increased steroid metabolism of the uterus and conceptus during pregnancy. Maintenance of pregnancy throughout the 114-day gestation is totally dependent on CL progesterone production because regression of the CL induces abortion during any stage of pregnancy in the pig.

Plasma concentrations of unconjugated and conjugated estrogens such as estrone sulfate increase in peripheral circulation from Day 16 to peak levels of >3 ng/ml between Days 23 and 30 of gestation. The plasma concentrations then decline to low levels (35 pg/ml) by Day 45 before slowly increasing to attain a second peak on the day before parturition. Increases in estrogen follow the steriodogenic activity of the pig placenta and changes in placental allantoic fluid volume throughout gestation.

VI. EARLY EMBRYONIC DEVELOPMENT, ESTABLISHMENT OF PREGNANCY, AND PLACENTAL ATTACHMENT

A. Embryonic Development

Fertilized ova develop to the four-cell stage in the ampullary–isthmic junction until progesterone-

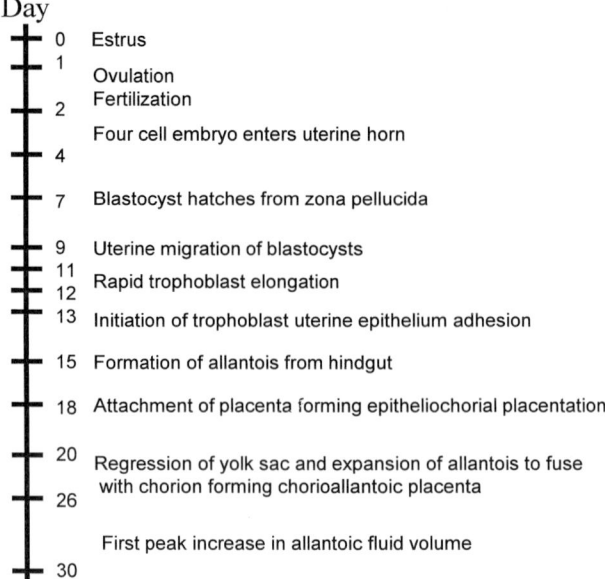

FIGURE 3 Summary of early conceptus development during early pregnancy in the pig.

stimulated relaxation of the isthmus allows movement into the uterine on Days 3 or 4 of gestation. Early embryos continue to cleave, developing to the morula stage of development by Days 5 or 6 of gestation. Fully formed blastocysts develop by Days 7 or 8 of gestation, with hatching from the zona pellucida occurring on Days 8 or 9 (Fig. 3). From Days 7 to 12 of pregnancy, blastocysts begin to intermix by migrating through and between the uterine horns to space themselves equidistantly apart within the uterine lumen. Blastocyst migration is stimulated by prostaglandin and estrogen synthesis. During uterine migration, the trophoblast and inner cell mass continue to expand as the diameter of the spherical conceptus (inner cells mass and trophoblast) reaches 5 mm by Day 11 of gestation.

Spherical conceptuses increase in diameter at a rate of 1 mm/4 hr over the next 24 hr. However, upon reaching a spherical diameter of 10 mm, the conceptuses undergo a rapid morphological change from spherical to tubular (20–40 mm in length) and finally thin filamentous threads measuring 100–150 mm in length (Fig. 4). Transformation from a 10-mm sphere to a filamentous thread occurs in less than 3 hr. This dramatic change in morphology occurs

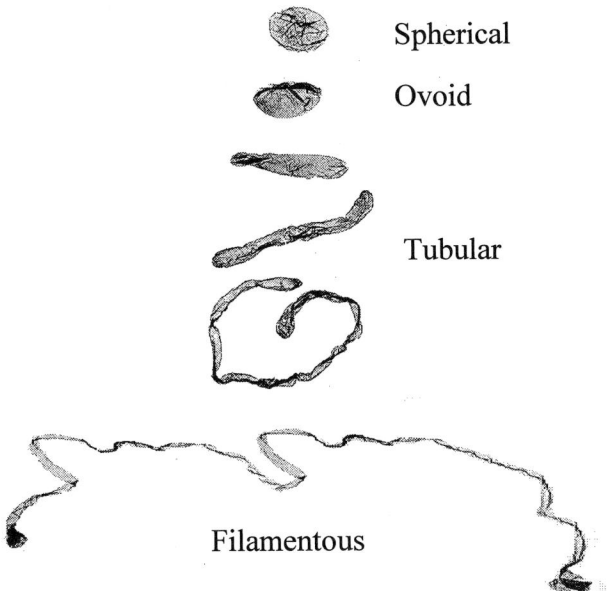

FIGURE 4 Morphological appearance of early pig conceptuses from Days 11 or 12 of gestation. Conceptuses rapidly elongate (3 or 4 hr) from 10 mm spherical to tubular and filamentous forms.

through cellular remodeling rather than cellular proliferation.

B. Establishment of Pregnancy

Rapid expansion of the conceptuses throughout the uterine horns is essential because initially each uterine horn must contain at least two embryos to prevent endometrial release of $PGF_{2\alpha}$ into the uterine vasculature to maintain functional corpora lutea. Conceptus synthesis and release of estrogen during rapid trophoblast elongation is the maternal recognition of pregnancy signal for establishment and maintenance of pregnancy in the pig. Estrogen stimulation of the endometrium does not inhibit $PGF_{2\alpha}$ synthesis but induces containment of $PGF_{2\alpha}$ within the uterine lumen rather than release into the uterine capillaries for luteolysis as occurs during the estrous cycle.

Complete maintenance of pregnancy to term in the pig requires two phases of conceptus estrogen secretion. The first phase is the short increase in conceptus estrogen secretion during trophoblast elongation which was previously described. This increase in estrogen maintains the CL until 30 days of gestation; however, a second sustained increase in conceptus estrogen secretion from Days 16 to 20 of gestation is necessary to maintain the CL to term.

C. Placental Attachment

Elongation of the trophoblast provides the initial uterine surface for placental development and nutrient exchange for the embryo. Length of the conceptus membranes can measure 1 m by Day 15 of gestation when allantoic membrane fills with fluid to expand and press the chorion against the entire uterine lumen. Placental membranes of individual embryos do not overlap within the uterus and fusion of placentae is a very rare event. Placentation in the pig is noninvasive, forming the epitheliochorial placenta through interdigitation of the chorion with the mircovilli present on the surface epithelium of the endometrium.

Surface area of uterus is increased by the many primary and secondary folds which permit a meter length of placenta to occupy only one-third this distance in the uterus. With expansion of the allantois on Day 18 and regression of the yolk sac, the allantois begins to fill with fluid on Day 18, reaching a volume of 200–300 ml by Day 30 of gestation. Fluid volume decreases by Day 45, with a second peak in allantoic fluid volume occurring by Day 70 of gestation. The allantois serves as a reservoir for nutrient support because the epitheliochorial type of placentation requires transfer of maternal nutrients through the chorion and the absorptive areolae of the chorion which cover the uterine glands to maintain fetal growth throughout gestation. Fluid expansion of the allantois by Day 27 of gestation permits first detection of pregnancy in gilts and sows with ultrasonography.

Since the maintenance of pregnancy in the sow occurs through conceptus secretion of estrogens, reproduction in the pig is susceptible to many environmental estrogenic compounds which can be consumed in feed. Feed containing a high concentration of alfotoxins, which is usually found in moldy corn, induces an estrogenic effect on the porcine uterus. Cyclic females consuming high levels of alfotoxin become psuedopregnant and will not return to estrus until 60–100 days later. However, females can be induced to come into heat when treated with $PGF_{2\alpha}$ after Day 12 of gestation. Females which consume alfotoxins following mating will have total embryonic loss by Day 18 and maintain functional CL for 70–100 days.

D. Embryonic Mortality

Fertilization of ovulated ova in the pig is usually high (>90%), with minimal embryonic loss before Day 8 of gestation. Estimates indicate that 20–30% of the potential embryos die before Day 30 of gestation, with an additional fetal loss of 10–20% occurring before term.

A large percentage of the early embryonic losses in the pig occurs during the periimplantation period of development from Days 11 to 18 of gestation. This period includes the time of conceptus trophoblast elongation, maternal recognition of pregnancy, and placental attachment. Competition for limited uterine space for placental surface area between littermate embryos results in sequential loss of potential piglets throughout gestation.

There are breed differences in ovulation rate, embryo survival, and litter size, but the Chinese Meishan pig is the breed most noted for high embryonic survival and litter size in swine. Meishan females farrow approximately three to five more piglets/litter compared to U.S. and European breeds of swine. The increased proliferacy of the Meishan breed is attributed to the ability of the maternal uterus and conceptus to control growth of placenta. Placenta of the Meishan fetuses also enhances placental efficiency through increasing the density of placental blood vessels at the maternal–placental interface.

VII. PARTURITION AND LACTATION

A. Parturition

Parturition occurs on Day 114 of gestation in the pig. Plasma concentrations of estrogens peak approximately 2 or 3 days prior to delivery of the piglets. Release of glucocorticoids from the maturing fetal adrenal initiates parturition through luteolysis of the

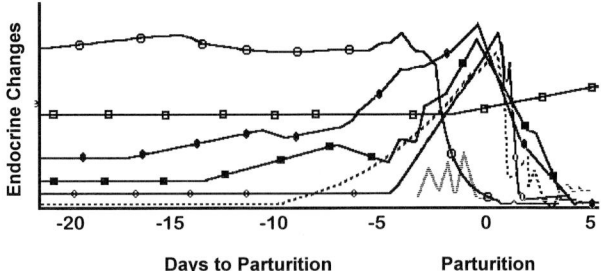

FIGURE 5 Endocrine changes in peripheral plasma during parturition of the sow. Increase in fetal corticosteriods initiates uterine release of prostaglandin $F_{2\alpha}$ which induces regression of the CL and a rapid decrease in progesterone. Decline in progesterone triggers myometrial concentrations through increased oxytocin release and the cervix softens with the release of relaxin from the CL. Prior to parturition, high concentrations of estrogens and relaxin stimulate development of the mammary gland. Prolactin, which is involved in lactogenesis, increases a few days prior to parturition. ○, Progesterone; ●, estrogen; dashed line, glucocorticoids; shaded line, $PGF_{2\alpha}$; □, growth hormone; ■, prolactin; ◇, relaxin.

CL by $PGF_{2\alpha}$ release from the uterine endometrium (Fig. 5).

Uterine release of $PGF_{2\alpha}$ stimulates a rapid decline in maternal plasma concentrations of progesterone 1 or 2 days prior to parturition. Initial rise in $PGF_{2\alpha}$ also induces release of prolactin from the anterior pituitary and relaxin from the CL. Release of prolactin stimulates mammary growth and synthesis of milk along with relaxin, which also serves with estrogen to soften and dilate the cervix for passage of the piglets at delivery. With regression of the CL and basal concentrations of plasma progesterone, increases in $PGF_{2\alpha}$ stimulate oxytocin release which heightens myometrial activity for the delivery of the piglets during parturition. Parturition can be induced with glucocorticoids (dexamethasone); however, $PGF_{2\alpha}$ (lutalyze) is the most effective and predictable method of successfully inducing parturition in the sow.

Sows can be induced to farrow after Day 110 of gestation. However, caution must be used because induction prior to Day 110 of gestation will result in a high percentage of stillborn piglets. Parturition is usually initiated within 48 hr of treatment. Delivery of the piglets in a litter occurs over 2–5 hr, with an average interval of 10–15 min between piglets. Piglets are delivered randomly between the uterine horns, with expulsion of the placentae within 4 hr of delivery of the last piglet. With delivery of each piglet, the uterine horn constricts to reduce the distance the piglet must pass to be expelled from the uterus. However, there is a higher incidence of stillborn piglets delivered last in sows which have a large litter. Risk of stillborns increases when delivery becomes longer than 20 min between piglets. Oxytocin has been utilized to enhance the delivery time of sows with large litters or when showing uterine inertia (weak myometrial contractions) during parturition.

B. Lactation

There is no evidence of placental lactogen synthesis in the pig as occurs in other species. Placental estrogen synthesis may increase prolactin receptors within the mammary tissue to stimulate proliferation. The increase of prolactin, glucocorticoids, growth hormone, and relaxin during the days prior to parturition stimulates lactogenesis in the pig. Milk letdown occurs during parturition with the release of oxytocin as well as oxytocin released during suckling by the piglets. Milk production in the sow increases to peak levels between the third and fourth week of lactation. The nursing interval by piglets during early lactation is <1 hr and gradually increases over time.

Gilts or sows have a lactational anestrous following parturition. Occasionally, a female will come into heat shortly following farrowing, but the heat is not accompanied by ovulation. Lactational anestrus will continue until sows are weaned from the piglets. Under normal management practices, weaning of piglets is done at 3 or 4 weeks of age. Estrus can be easily synchronized when litters are weaned from sows on the same day. Sows return to estrus within 3–7 days following removal of the suckling influence. Sows which do not exhibit estrus within a week of weaning can be treated with P.G. 600 to induce estrus and allow breeding.

See Also the Following Articles

Cattle (Bovidae); Epitheliochorial Placentation; Lordosis; Luteolysis; Pregnancy, Maternal Recognition of

Bibliography

Anderson, L. L. (1993). Pigs. In *Reproduction in Farm Animals* (E. S. E. Hafez, Ed.), 6th ed., pp. 343–360. Lea & Febiger, Malvern, PA.

Cole, D. J. A., and Foxcroft, G. R. (1982). *Control of Pig Reproduction*. Butterworth, Boston.

Cole, D. J. A., Foxcroft, G. R., and Weir, B. J. (1990). Control of pig reproduction III. *J. Reprod. Fertil. Suppl.* **40**.

Ford, S. P. (1997). Control of pig reproduction V. *J. Reprod. Fertil. Suppl.* **52**.

Perry, J. S. (1981). The mammalian fetal membranes. *J. Reprod. Fertil.* **62**, 321–335.

Pope, W. F. (1994). In *Embryonic Mortality in Swine* (M. T. Zavy and R. D. Geisert, Eds.), pp. 53–77. CRC Press, Boca Raton, FL.

Pineal Gland, Melatonin Biosynthesis and Secretion

Stuart E. Dryer

University of Houston

I. Phylogenetic and Ontogenetic Patterns in the Cellular Morphology of the Vertebrate Pineal Organ
II. Role of the Pineal Organ in the Circadian Organization of Avian Physiology and Behavior
III. Biochemical Pathways Regulating Melatonin Biosynthesis and Secretion
IV. Neurohumoral Control of Melatonin Secretion
V. Intrinsic Circadian Control of Pineal Melatonin Secretion
VI. Direct Photic Control of Pineal Melatonin Secretion
VII. Circadian Regulation of Cyclic AMP and Intracellular Free Calcium

GLOSSARY

circadian oscillator A self-sustaining biological clock that drives daily rhythms in the biochemistry, physiology, and behavior of an organism. The period of a circadian oscillator is close to but rarely equal to 24 hr (*circa dian*). A circadian oscillator continues to function in environments devoid of external time cues but can be entrained by external cues known as zeitgebers. The most common zeitgeber is the environmental light–dark cycle.

melatonin The principal hormone secreted by the vertebrate pineal gland. It is also secreted by photoreceptors in the vertebrate retina and also by the Harderian gland of some vertebrates.

phase–response curve In circadian biology, the relationship between the steady-state shift in the phase of a free-running oscillator produced by a given stimulus and the time of day at which the stimulus was applied. For example, light pulses typically evoke a phase delay if applied early in the nighttime, a phase advance if applied late in the nighttime, and have little or no effect if applied during the daytime.

photoperiod The number of consecutive hours in a day during which an animal is exposed to substantial light intensities, *i.e.* photopic conditions. In temperate latitudes the photoperiod will depend on the season, with long photoperiods occurring during the summer and short photoperiods occurring during the winter. Scotoperiod is a corresponding term referring to the daily duration of low light intensities. Conditions of low light intensity are referred to as scotopic.

suprachiasmatic nucleus Hypothalamic site of an important circadian oscillator in mammals and birds. Source of a neuronal pattern generator that drives polysynaptic neuronal pathways projecting to the avian and mammalian pineal gland.

visual pigment A class of G-protein-coupled receptors essential for phototransduction. Functional visual pigments consist of an apoprotein, known as an opsin, and a bound vitamin A-derived chromophore that isomerizes upon absorption of a photon. Examples in birds include rhodopsin and iodopsin.

The vertebrate pineal organ, also known as the epiphysis, is the principal source of circulating melatonin, a methoxyindole hormone that is also produced in the retina and the Harderian gland. The vertebrate pineal gland is rhythmic and nocturnal, as melatonin secretion is markedly enhanced during the nighttime but is reduced during the daytime. The pinealocytes of nonmammalian vertebrates are intrinsically photosensitive and can regulate melatonin secretion in direct response to changes in environmental illumination. In addition, the pinealocytes of most birds and some other nonmammalian vertebrates contain intrinsic circadian oscillators that support daily rhythms in melatonin secretion, even in isolated cells maintained in constant conditions free of environmental time cues. Here, regulatory mechanisms in the nonmammalian pineal organ are discussed, with a special emphasis on birds.

I. PHYLOGENETIC AND ONTOGENETIC PATTERNS IN THE CELLULAR MORPHOLOGY OF THE VERTEBRATE PINEAL ORGAN

The anamniote pineal organ is characterized by a distinct bineuronal association in which a photoreceptor cell with a prominent cone-like outer segment is synaptically coupled to a second-order neuron that gives rise to extensive pinealofugal neuronal projections. In lampreys and teleosts, pineal second-order neurons and retinal ganglion neurons project to many of the same visual centers. In addition to photoreceptors and second-order neurons, anamniote pineal organs contain at least some neuroendocrine pinealocytes that do not exhibit obvious photoreceptor specializations and that secrete hormones into the circulation. Two main evolutionary trends have shaped the cellular organization of the vertebrate pineal organ. The first is a progressive reduction in the number of second-order neurons relative to photoreceptors as one moves from more ancient to more modern species. The second is a concurrent trend toward less elaborate photoreceptor outer segments and a relative increase in the number of neuroendocrine pinealocytes. The pineals of birds and reptiles contain modified photoreceptor cells with outer segments that are either substantially reduced or that have an unusual whorl-like appearance very different from that of retinal photoreceptors. These modified photoreceptors represent an intermediate case between the cone-like pineal photoreceptors of anamniotes and the purely neuroendocrine pinealocytes that are the primary cell type found in mammals. The modified photoreceptors of birds make few if any synapses with second-order neurons and pinealofugal projections are sparse. Instead, the avian pineal gland has a follicular structure that may facilitate secretion of melatonin into the circulation. The mature pinealocytes of mammals lack photoreceptive structures altogether and are purely secretory in function. Pineal second-order neurons are not present in mammals.

The pineal gland and retina are both derived from ventricular germinal zones located in the embryonic diencephalon. In the chick embryo, the epiphysis forms as an evagination of the surface of the diencephalon at about 55–60 hr of incubation (27–30 somites). Differentiated pinealocytes can be observed by the Embryonic Day 8 in the chick. Removal of the entire surface of the diencephalon is required to prevent formation of the pineal organ. The inductive processes involved in the formation of the pineal are not understood, but it is reasonable to speculate that they may be similar to inductive processes that occur in the lateral eyes. Indeed, embryonic avian pineal cells or retinal cells that develop *in vitro* can be pushed toward a neuronal phenotype or toward a photoreceptor phenotype depending on the culture conditions. Cultured quail pineal cells can even produce lens and pigment epithelium during development *in vitro*. In teleosts, photoreceptor-specific proteins can be detected in the pineal at earlier developmental stages then in retina. The avian pineal gland has a follicular structure. In chickens, the follicles become greater in number and more compressed as development proceeds, a process that extends several months into posthatching life. This is associated with an increase in the size and complexity of the photoreceptor outer segments that extend into the follicular lumen. Light-sensitive circadian rhythms in pineal melatonin secretion first appear 2 or 3 days before hatching in the chicken.

II. ROLE OF THE PINEAL ORGAN IN THE CIRCADIAN ORGANIZATION OF AVIAN PHYSIOLOGY AND BEHAVIOR

The avian pineal gland is one of at least three diencephalic structures in birds that contain circadian oscillators and that are important in the overall circadian system of these species. The other structures include the visual suprachiasmatic nucleus and the retina. There is now compelling evidence that deep brain photoreceptors are also important in some birds and reptiles. The various circadian oscillators are coupled under normal conditions but can be made to operate independently under certain experimental conditions. The relative importance of these various components in the overall circadian organization of birds is highly species dependent, so much so that it is difficult to make general statements. For example, pinealectomy in the house sparrow causes an almost complete disruption of circadian rhythms in locomotor activity and feeding. However, in some other species, such as chickens and European starlings, the effects of pinealectomy are less profound because circadian rhythms in locomotor activity usually persist, even if they are less robust. In several species there is reason to believe that melatonin enhances the ability of an organism to resynchronize to a shift in the phase of a low-amplitude environmental zeitgeber. It should be noted that the pineal is not always the sole source of systemic melatonin. For example, in the quail the retina also makes a significant contribution to circulating melatonin.

The pineal gland and melatonin do not appear to play an essential role in the reproductive physiology of birds, even in species in which photoperiod is a critical environmental variable regulating reproduction. There is evidence that the pineal gland is more important in at least some reptiles. Many birds enter into breeding at the onset of long photoperiods, i.e., in the spring. They often end breeding some weeks later at which time they become refractory to long photoperiods. This refractory state subsequently dissipates with the return of short photoperiods in the autumn and winter. Each of these photoperiodic changes is typically associated with corresponding changes in endocrine status, especially in plasma levels of prolactin and leuteinizing hormone (LH). In nearly every species tested, these photoperiodic changes persist after pinealectomy or other manipulations that alter plasma melatonin rhythms. This suggests that most birds do not utilize the duration of elevated nocturnal melatonin levels as the principal index of the scotoperiod. On the other hand, the pineal may play a subtle role in regulating certain types of photoperiodic reproductive behavior. For example, male doves tend to engage in greater nest building behavior under long photoperiods then under short photoperiods. The effect of photoperiod, but not the behavior itself, is abolished after pinealectomy. Pinealectomy has no effect on courtship or copulation in the same species. That the pineal gland plays some role in avian reproductive physiology is further suggested by a few reports indicating that melatonin is antigonadotrophic, or progonadotrophic, and that melatonin receptors are present in the testes and ovaries of several avian species. Thus, it remains possible that some subtle effects of the pineal on gonadal function and reproductive physiology occur independent of changes in LH or other hormones. This subject requires more detailed study.

III. BIOCHEMICAL PATHWAYS REGULATING MELATONIN BIOSYNTHESIS AND SECRETION

Melatonin is extremely lipophilic and is not secreted by exocytosis. Instead, it is secreted by simple diffusion across the plasma membrane and its rate of secretion is simply determined by its rate of synthesis.

A. Biosynthetic Pathways

Melatonin is synthesized from tryptophan by the biosynthetic pathway shown in Fig. 1. All these en-

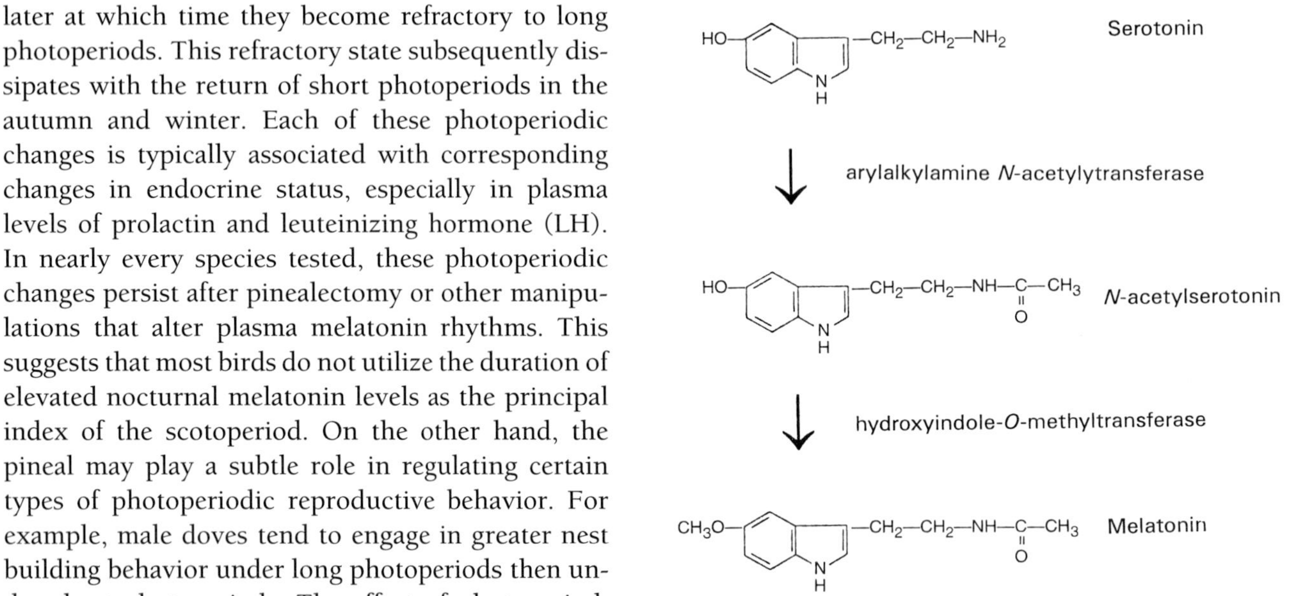

FIGURE 1 Melatonin biosynthetic pathway. Under most conditions the rate-limiting step is catalyzed by arylalkylamine N-acetyltransferase. The activity of this enzyme determines the rate of melatonin secretion and is elevated during the nighttime. As with other acetyltransferases, its activity requires the cofactor acetyl-coenzyme A. The activity of hydroxyindole-O-methyltransferase requires the cofactor S-adenosyl methione. Details of the circadian regulation of this pathway are described in the text.

zymes have been cloned and sequenced in several species including the chicken. Under most conditions, the rate-limiting step is catalyzed by arylalkylamine N-acetyltransferase (AA-NAT) and the activity of this enzyme determines the overall rate of melatonin secretion. AA-NAT is expressed at high levels in the pineal gland and at lower levels in retinal photoreceptors. In both tissues, AA-NAT activity is higher at night and is much lower during the day. These daily rhythms are associated with changes in AA-NAT mRNA levels and posttranslational modification of the enzyme molecules. The relative contribution of transcriptional and posttranslational mechanisms to the overall control of AA-NAT activity is species dependent. The enzyme tryptophan hydroxylase (TPH) also exhibits a nocturnal pattern of activity in pineal and retinal photoreceptors and contributes to the regulation of melatonin secretion. The daily rhythm in TPH activity is associated with robust changes in mRNA levels. The final step of the pathway is catalyzed by hydroxyindole-O-methyltransferase (HIOMT). In the chicken, the levels of this enzyme and its transcripts are highest during the daytime, when melatonin secretion is lowest.

B. Role of Cyclic AMP and Calcium

Agents that increase intracellular cyclic AMP increase melatonin biosynthesis and secretion and increase the activities of NAT and TPH and the levels of their transcripts. Chicken pineal cells contain adenylate cyclases that can be regulated by neurohormones and by the intrinsic circadian oscillator. Agents that increase intracellular free Ca^{2+} also increase melatonin biosynthesis. These two messengers interact at several levels. For example, cyclic AMP production in pineal cells is stimulated by agents that increase intracellular free Ca^{2+}, probably owing to Ca^{2+}/calmodulin-dependent stimulation of type II adenylate cyclase. However, Ca^{2+} can also regulate melatonin synthesis and secretion independent of any effects on cyclic AMP. Chicken pineal cells express voltage-activated L-type Ca^{2+} channels that are blocked by Co^{2+} and Cd^{2+} and by dihydropyridines such as nifedipine and nitrendipine. These channels are stimulated to open by BAY K 8644. The L-type Ca^{2+} channels are responsible for increases in melatonin secretion evoked by KCl depolarization and for at least a portion of the nocturnal increase in melatonin secretion.

The activity of the L-type Ca^{2+} channels is dependent on the gating of other types of plasma membrane ionic channels that probably initiate rhythmic changes in membrane potential. In pineal cells these include cyclic GMP-activated cationic channels similar to those of retinal photoreceptors, as well as an unusual circadian clock-regulated cationic channel known as I_{LOT} (for long open time cationic channel). These nonselective cationic channels are permeable to Ca^{2+}, and their properties are described further below. Chick pineal cells also express several types of K^+ channels. The L-type Ca^{2+} channels may serve as a mechanism to amplify changes in Ca^{2+} influx initiated by changes in the gating of other ionic channels.

Chick pineal cells also contain intracellular Ca^{2+} stores. Influx of Ca^{2+} across the plasma membrane can be evoked by depletion of the intracellular stores. A subpopulation of pineal cells isolated from the chicken and the trout exhibits spontaneous oscillations in intracellular free Ca^{2+} concentration that are dependent in part on Ca^{2+} influx across the plasma membrane. These Ca^{2+} oscillations, which are somewhat irregular with periods of seconds to minutes, are not seen in mammalian pineal cells. The physiological significance of the intracellular Ca^{2+} stores and the spontaneous Ca^{2+} oscillations is unknown.

IV. NEUROHUMORAL CONTROL OF MELATONIN SECRETION

The chicken pineal organ is controlled by at least two distinct neuronal inputs. The first is composed of noradrenergic fibers that originate from cervical sympathetic ganglia. The second consists of a plexus of vasoactive intestinal peptide (VIP)-containing fibers of unknown origin. The significance of the VIP-containing fibers in controlling melatonin secretion *in vivo* is unknown, but a significant role for the sympathetic innervation in suppressing melatonin secretion during the daytime is well established. Note that this is different from mammals, in which sympathetic input stimulates melatonin secretion and is more active during the nighttime. In chickens, an intact sympathetic input is necessary to prevent damping of circadian rhythms in melatonin secretion in constant dark conditions. Norepinephrine pulses can also prevent damping of melatonin rhythms in cultured cells maintained at 37°C. The mechanism of this effect is not known. In quail there is evidence that extrinsic neuronal control of the pineal gland persists in sympathectomized animals. Birds contain a hypothalamic structure analogous to the mammalian suprachiasmatic nucleus that controls the various efferent projections to the pineal gland.

Norepinephrine inhibits melatonin biosynthesis and secretion from cultured chick pineal cells. This effect is mediated by α_2-adrenergic receptors coupled to the guanine nucleotide binding protein G_i, which together act by causing inhibition of adenylate cyclase and a fall in cyclic AMP. Norepinephrine does not shift the phase of the intrinsic circadian oscillator. In cultured chick pineal cells, application of VIP causes stimulation of adenylate cyclase and thereby causes an increase in melatonin biosynthesis and secretion. At least three other neurohormones have been shown to stimulate melatonin secretion from cultured chick pineal cells. These are prostaglandin E, adenosine, and histamine, the latter acting through an unusual type of histamine receptor. Prostaglandins can stimulate melatonin secretion at concentrations 20- to 300-fold lower than are required to stimulate cyclic AMP formation. The stimulatory effects of these low concentrations of prostaglandins can be blocked by several calmodulin inhibitors, suggesting a direct role for intracellular Ca^{2+} in the regulation of melatonin biosynthesis.

V. INTRINSIC CIRCADIAN CONTROL OF PINEAL MELATONIN SECRETION

Chick pineal cells exhibit daily rhythms in melatonin synthesis and secretion even when they are removed from the animal and maintained in cell culture in constant dark conditions free of environmental time cues. This is true even when the cells are grown in low-density cultures on glass beads and superfused continuously. This indicates that circadian rhythms are a property of individual pineal cells and do not require neuronal or paracrine interactions. The molecular mechanisms of vertebrate circadian oscillators are unknown. However, mutations that affect the period of the circadian oscillator in the mammalian suprachiasmatic nucleus also affect circadian oscillators in photoreceptors, suggesting that they share common mechanisms.

Free-running circadian rhythms are also observed in the activity of the melatonin biosynthetic enzymes TPH, AA-NAT, and HIOMT, as well as in the intracellular concentrations of cyclic AMP, cyclic GMP, and Ca^{2+} influx. In the case of AA-NAT, the free-running activity rhythms in cultured chick pineal cells reflect changes in mRNA levels. Posttranslational modifications of the enzyme also contribute to daily rhythms observed *in vivo*. If cultured chick pineal cells are maintained at 37°C the melatonin rhythms dampen out in three to five cycles. However, if pineal cells are cultured at 39 or 40°C (normal body temperature in chickens) the free-running rhythms can be maintained indefinitely. The mechanism of this effect of temperature is unknown. However, the period of the free-running rhythm observed in constant darkness is at least partially temperature compensated. Moreover, changes in temperature can entrain the circadian oscillator.

The amplitudes of free-running rhythms in melatonin secretion are markedly suppressed in constant light conditions. However, the underlying circadian oscillator continues to function and it is still possible to measure other rhythms, such as the one in AANAT activity. Constant light causes the peak in AANAT activity to be shifted by 6 hr (from CT18 to CT24). Such a shift can be predicted from the phase–response relationship for light pulses, described in more detail in the next section.

There are only a few physiological or pharmacological manipulations that have been shown to shift the phase of the intrinsic circadian oscillator of cultured chick pineal cells. For example, a variety of treatments that alter membrane potential, intracellular free Ca^{2+}, and/or cyclic AMP do not shift the phase of the circadian oscillator, although they can produce acute effects on AA-NAT activity and melatonin secretion. Similarly, neurohormones such as norepinephrine do not shift the phase of the circadian oscillator. Protein synthesis inhibitors, such as the reversible translational inhibitor anisomycin, can shift the phase of the circadian oscillator of pineal cells. In addition, application of ouabain or K^+-free salines (both of which inhibit Na^+,K^+-ATPase) shift the phase of the circadian oscillator, with a phase–response curve similar to that produced by dark pulses in constant red light (see below). This effect is mimicked by treatment with hypotonic salines. Ouabain and low salt also cause cell swelling. By contrast, treatment with hypertonic salines, which cause cells to shrink, produces opposite phase shifts similar to those evoked by pulses of white light applied to cells maintained in red light. It is possible that the effects of ouabain, hypotonic, and hypertonic salines are mediated by changes in intracellular pH, which is known to affect other circadian oscillators.

VI. DIRECT PHOTIC CONTROL OF PINEAL MELATONIN SECRETION

The avian skull is translucent and a considerable amount of light can reach the pineal gland and the rest of the brain *in vivo*. As a result, light–dark cycles can entrain circadian rhythms in enucleated birds and lower vertebrates. Light produces at least two different effects on cultured chicken pineal cells. Light always causes an acute inhibition of melatonin secretion. Light (and dark) can also entrain the intrinsic circadian oscillator such that the free-running rhythm retains a constant relationship to an initial light–dark cycle. Entrainment can also be seen as a shift in the phase of subsequent cycles of the melatonin rhythm evoked by a single pulse of light applied in otherwise constant conditions. The amplitude and direction of the phase shift depend on the time of day at which the light pulse is applied (the so-called phase–response relationship). For example, a single pulse of white light applied to cells free -running in dim red light will cause a phase delay if applied early in the nighttime but will cause a phase advance if applied late in the nighttime. By contrast, application of a single dark pulse in constant dim red light will cause a phase advance if applied early in the nighttime and a phase delay if applied later in the nighttime.

The two effects of light can be separated pharmacologically because the acute inhibitory effects of light on melatonin secretion are blocked by pretreatment of pineal cells with pertussis toxin, whereas the entraining effects of light persist after pertussis toxin treatment. The two effects of light are also differentially sensitive to vitamin A depletion. These results suggest the existence of multiple phototransduction cascades. The precise nature of either of these cascades is unknown, but chicken pineal cells clearly express all the essential components of the phototransduction cascades previously described in retinal photoreceptors. This includes the visual pigment pinopsin, also known as P-opsin, which is 43–48% similar at the amino acid level to retinal cone opsins. There is evidence that avian pineal cells also express rhodopsin and iodopsin, the latter based on both immunochemistry and Northern blot analysis. Chicken pineal cells also express the G-protein transducin, a light-sensitive cyclic nucleotide phosphodiesterase, two types of cyclic GMP-activated channels, arrestin, recoverin, visinin, and calretinin. Application of light pulses to chicken pineal cells causes isomerization of the visual pigment chromophore 11-*cis*-retinal to all-*trans*-retinal. Electrophysiological responses to light have not been described in chicken pineal cells. However, light has been shown to hyperpolarize pineal photoreceptors in lampreys and teleosts. As with retinal rods and cones, those responses to light are mediated by changes in Na^+ influx. It should be noted that the pineal photoreceptors of trout do not contain intrinsic circadian oscillators. In those cells the principal effect of light is acute inhibition of melatonin secretion associated with membrane hyperpolarization. This suggests that the acute effects of light on pineal melatonin secretion are mediated by a cyclic GMP-dependent phototransduction mechanism similar to that of retinal rods and cones. On the other hand, there is evidence that the entraining effects of light on chick pineal cells are mediated by a distinctly different mechanism that does not require changes in Na^+ influx or membrane potential and that is resistant to blockade by pertussis toxin. There are preliminary reports that this effect is mediated by a pathway leading to activation of MAP kinase. Distinct entraining and acute effects of light have also been observed in zebrafish pinealocytes and in retinal photoreceptors of *Xenopus*.

VII. CIRCADIAN REGULATION OF CYCLIC AMP AND INTRACELLULAR FREE CALCIUM

Chick pineal cells exhibit a free-running rhythm in cyclic AMP production and efflux. This rhythm is both in phase and highly correlated with the secretion of melatonin. Because cyclic AMP protagonists stimulate melatonin secretion but do not produce steady-state phase shifts, it was initially proposed that the circadian oscillator acts through cyclic AMP to regulate melatonin synthesis and secretion. That theory has been disproved in its simplest form. Thus, application of high concentrations of forskolin or 8-Br-cyclic AMP produced a saturating effect on melatonin secretion such that higher concentrations of these agents could not evoke additional melatonin secretion. Nevertheless, the circadian oscillator continued to drive 24-hr rhythms in melatonin secretion under those conditions. Therefore, the circadian oscillator cannot be acting exclusively through cyclic AMP to drive melatonin synthesis. It has also been proposed that the circadian rhythms in intracellular Ca^{2+} contribute to the free-running rhythms in melatonin biosynthesis. There is indirect evidence that this takes place. Thus, 8-Br-cyclic AMP is more potent and effective during the nighttime in pineal cells maintained in constant darkness if Ca^{2+} is present in the external saline. Day–night differences in the 8-Br-cyclic AMP dose–response curve disappear in Ca^{2+}-free external salines, and the curve is shifted down and to the right. This indicates that Ca^{2+} influx is greater during the nighttime even when there is no change in environmental lighting. Only a portion of this effect is inhibited by blockers of L-type Ca^{2+} channels, suggesting the existence of other pathways responsible for enhanced nocturnal Ca^{2+} influx. At least one plausible mechanism has been identified. Chick pineal cells express an unusual cationic channel known as I_{LOT} that is under direct circadian control. This channel is permeable to Ca^{2+} and is only active during the nighttime, even in pineal cells maintained in constant darkness. I_{LOT} is not voltage or stretch activated and it remains active after patch excision (implying that I_{LOT} gating is not dependent on continued contact with a soluble cytosolic messenger). I_{LOT} is not activated in quiescent patches by melatonin, cyclic AMP protagonists, depletion of intracellular Ca^{2+} stores, or patch excision. The molecular mechanisms that control I_{LOT} gating remain unknown. However, this channel provides a mechanism for increased nocturnal Ca^{2+} influx because of its intrinsic Ca^{2+} permeability and because its activity should lead to depolarization and secondary activation of L-type Ca^{2+} channels. At this time it is not possible to rule out that the circadian oscillator also acts to mobilize intracellular Ca^{2+} stores and associated store-operated plasma membrane Ca^{2+} channels or to regulate other Ca^{2+}-binding proteins or transport systems.

See Also the Following Articles

Melatonin; Photoperiodism, Vertebrates; Seasonal Reproduction, Birds

Bibliography

Cassone, V. M., and Menaker, M. (1984). Is the avian circadian system a neuroendocrine feedback loop? *J. Exp. Zool.* 232, 539–549.

Chabot, C. C., and Menaker, M. (1992). Circadian feeding and locomotor rhythms in pigeons and house sparrows. *J. Biol. Rhythms* 7, 287–299.

Collin, J. P., and Oksche, A. (1981). Structural and functional relationships in the non-mammalian pineal gland. In *The Pineal Gland* (R. Reiter, Ed.), Vol. 1, pp. 27–67. CRC Press, Boca Raton, FL.

D'Souza, S. W., and Dryer, S. E. (1996). A cationic channel regulated by a vertebrate intrinsic circadian oscillator. *Nature* 382, 165–167.

Follett, B. K., Foster, R. G., and Nicholls, T. J. (1985). Photoperiodism in birds. *Ciba Foundation Symp.* 117, 93–105.

Foster, R., Korf, H. W., and Schalken, J. (1987). Immunocytochemical markers revealing retinal and pineal but not hypothalamic photoreceptor systems in the Japanese quail. *Cell Tissue Res.* 248, 161–169.

Nikaido, S. S., and Takahashi, J. S. (1996). Calcium modulates circadian variation in cAMP-stimulated melatonin in chick pineal cells. *Brain Res.* 716, 1–10.

O'Brien, P. J., and Klein, D. C. (Eds.) (1986). *Pineal and Retinal Relationships*. Academic Press, New York.

Okano, T., Yoshizawa, T., and Fukada, Y. (1994). Pinopsin is a chicken pineal photoreceptive molecule. *Nature* 372, 94–97.

Pineal Gland, Regulatory Function

Fred W. Turek

Northwestern University

I. Introduction
II. Role in the Regulation of Seasonal Rhythms
III. Role in the Regulation of Circadian Rhythms
IV. Role in the Regulation of Sleep
V. Other Possible Functions for Melatonin

GLOSSARY

chronobiotic An agent (pharmacological or nonpharmacological) which can induce phase shifts or period changes in biological rhythms—primarily used in conjunction with alterations in circadian rhythms.

circadian clock An internal timing device which regulates the expression of circadian rhythms. In mammals, a master circadian clock is located in the hypothalamic suprachiasmatic nucleus.

circadian rhythms Biochemical, physiological, and behavioral events that reoccur in a cyclic fashion with a period of about 24 hr.

melatonin The primary secretory product produced by the pineal gland. This hormone is released almost exclusively into the circulation during the nighttime in both day-active (diurnal) and night-active (nocturnal) species.

photoperiodism The regulation of seasonal rhythms (in both plants and animals) by the seasonal change in day length which occurs in all nonequatorial regions on earth.

seasonal (or *annual*) *rhythms* Behavioral and physiological rhythms that vary throughout the year such that specific functions only occur during certain seasons of the year (e.g., reproduction, hibernation, and migration).

Even without a wristwatch or a calendar, plants and animals can tell the time of day as well as the time of year. In vertebrates, the pineal gland, through its secretion during the nighttime of the hormone, melatonin, plays a central role in enabling animals to keep track of time. Indeed, in lower vertebrates (e.g., birds and lizards) the pineal gland is part of the "circadian clock system" that regulates the timing of 24-hr rhythms. In mammals, the 24-hr rhythm in melatonin production can influence the timing of other physiological and behavioral daily rhythms. A truly clever use of the 24-hr rhythm of melatonin production in mammals is the measuring of how long melatonin levels are high during the nighttime hours to determine how long the night or day is and to use this information to regulate a variety of seasonal rhythms including seasonal reproduction. By enabling the organism to know the "time of day" and/or the "time of year," melatonin may have indirect effects on many physiological and behavioral processes, including the timing of sleep and wakefulness.

I. INTRODUCTION

The pineal gland of mammals lies between the two cerebral hemispheres at various depths in the brain. The only established output of this gland with any clear function is the hormone melatonin. The salient feature of melatonin production is that it is secreted into the circulation primarily during the dark period. Furthermore, the longer the period of darkness, the longer the duration of time when circulating melatonin levels are high. This control of melatonin production and release by the light–dark cycle has the potential of serving as a clock and a calendar, capable of informing the rest of the organism of whether it is day or night (i.e., functions as a clock) and the season of the year (i.e., functions as a calendar). Indeed, there is now substantial evidence in mammals that the pineal gland plays a central role in mediating the effects of day length or seasonal or annual rhythms, and it is the duration of the high nighttime melatonin levels which informs the organism of the length of the night (and thus the day). In lower vertebrates, the pineal gland and melatonin play a central role in the regulation of circadian (i.e., 24-hr) rhythms and recent evidence indicates that it may play a modulating role in the regulation of circadian rhythms in mammals. Such "chronobiotic" effects of melatonin are just one way it could influence the sleep–wake cycle. In addition, melatonin may have direct hypnotic effects. By providing information to the brain and the rest of the body as to the time of day and time of year, the pineal melatonin signal may influence a variety of physiological systems that use this signal to adaptively change to meet the demands placed on the organism due to changes in daylight and day length.

II. ROLE IN THE REGULATION OF SEASONAL RHYTHMS

Seasonality comes in many varieties. Some animals hibernate, some migrate, some change color, some gain or lose weight, etc. One common seasonal rhythm for many mammals is that of reproduction. Most animals only breed at specific times of the year so that the young are born during those seasons when their chances of survival are optimum. As first demonstrated in the 1920s, it is the seasonal change in day length that is used by many plants and animals to control the timing of their breeding season. The effects of day length are truly dramatic. For example, regardless of the season of the year, if golden hamsters are maintained on long days (e.g., 14 hr of light per day) the paired testes weight of the adult male will be about 3000 mg and the female will ovulate every 4 days if pregnancy does not occur. In contrast, exposure to winter-like short days (e.g., 10 hr of light per day) will induce complete testicular regression in about 10 weeks such that the testes will weigh about 400 mg and be devoid of any advanced stages of spermatogenesis (Fig. 1). The female of this species will stop ovulating during exposure to short days. Many experiments have established in the golden hamster, as well as in many other species, that the

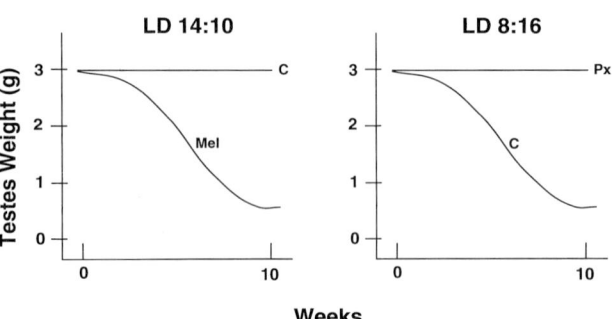

FIGURE 1 Schematic representation of the testicular response of golden hamsters maintained on long days (i.e., LD 14:10) or short days (i.e., LD 8:16) for a 10-week period. Control animals (C) maintained on long days will maintain normal testicular function, whereas exposure to short days will lead to complete testicular regression in about 10 weeks. Infusion of melatonin (Mel) on a daily basis such that circulating melatonin levels are high for a period of time each day, simulating the endogenous melatonin profile during exposure to short days, will lead to complete testicular regression in this species even when exposed to stimulating long days. In contrast, the removal of the pineal gland (Px) completely blocks the inhibitory effects of short days on the reproductive axis. Thus, melatonin treatment can induce testicular regression in animals on long days, whereas removal of the source of melatonin prevents the normal short-day inhibition of reproductive function.

same internal circadian clock located in the hypothalamic suprachiasmatic nucleus (SCN) that regulates the expression of daily rhythms is involved in measuring the seasonal change in the length of the day. The reproductive response to changes in day length is abolished in SCN-lesioned animals.

In mammals, the SCN controls the timing of the reproductive season in photoperiodic animals through its regulation of the pineal melatonin rhythm. Indeed, in many mammalian species, removal of the pineal gland abolishes the photoperiodic response to a change in day length (Fig. 1). Since melatonin levels are only high during the nighttime, as the length of the day changes the duration of nighttime pineal melatonin production changes such that on short days (long nights) melatonin is produced and released into the circulation for a longer period of time than it is during exposure to long days (short nights). This melatonin duration signal "informs" the brain as to the season of the year. Somehow (this is the major mystery in the field of seasonality today) the brain is able to "decode" the melatonin duration signal which leads to a cascade of biochemical, physiological, and behavioral changes that are appropriate for a given season of the year. Note that melatonin itself is not inhibitory or stimulatory to reproductive function. Instead, it supplies information about the length of the day to the brain—information which is used in a species-specific manner to enhance the survival of the species. For example, in sheep the short days of fall, and the long-duration melatonin signal, inform the brain that it is time to reproduce so that the lambs will be born in the spring. For smaller rodent species, with a shorter gestation period, it is the longer days of spring and summer, and the short-duration melatonin signal, that are stimulatory to reproductive function (Fig. 1). Where and how melatonin is acting as a calendar in the brain is an active area of research.

III. ROLE IN THE REGULATION OF CIRCADIAN RHYTHMS

Pinealectomy and treatment with melatonin have been shown to abolish and severely disrupt or alter the expression of circadian rhythms in a number of vertebrate species, particularly in birds. Indeed, there are substantial data in lower vertebrates to support the hypothesis that the pineal gland can function as a "circadian biological clock." However, although claims have been made that the pineal gland is the biological clock in mammals, including humans, it is clearly not necessary or sufficient for the expression of 24-hr rhythms in mammals. Indeed, pinealectomy has little, if any, effect on the expression of 24-hr rhythms in mammals, and circadian rhythms can be abolished in mammals with an intact pineal gland following lesions in the anterior hypothalamus. The "master" circadian clock in mammals was discovered just about the same time that the importance of the pineal gland in the circadian organization of lower vertebrates was elucidated. The clock was found to be located in the bilaterally paired SCN of the hypothalamus. It has been firmly established that the circadian rhythms of pineal melatonin synthesis and release are regulated via neural signals from the SCN and that the pineal melatonin rhythm is abolished, as are seemingly all endogenous circadian rhythms, following the destruction of the SCN in mammals.

Recent studies indicate that the SCN-regulated rhythmic release of pineal melatonin may influence the expression of other 24-hr rhythms and that exogenous treatment with melatonin can lead to phase shifts in the circadian clock system of mammals (Fig. 2), including humans. The finding that the SCN contains a dense concentration of melatonin receptors suggests that there may be a feedback loop between the SCN and the pineal gland. Therefore, while the pineal melatonin rhythm is clearly not the ultimate source of 24-hr rhythmicity, it may play an important modulatory role in the regulation of circadian rhythmicity in mammals.

IV. ROLE IN THE REGULATION OF SLEEP

It has been speculated for many years that melatonin may play a role in inducing sleep in humans. Indeed, soon after melatonin was discovered it was found to have hypnotic properties, at least at pharmacological doses. There are now numerous reports that the administration of melatonin at both physiological and nonphysiological doses can induce a feeling of

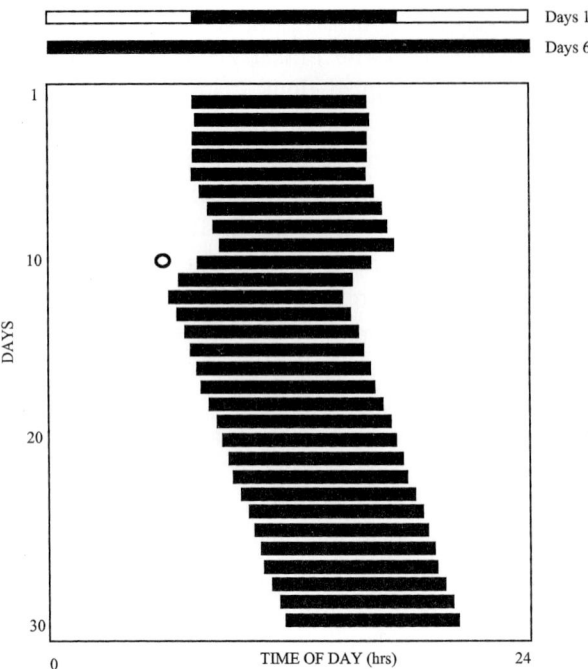

FIGURE 2 Computer simulation of a hamster's circadian running-wheel behavior. Time of day is plotted across the horizontal axis; successive days are plotted on the vertical axis. Dark areas indicate bouts of activity. The light–dark schedules are indicated at the top of the figure. During the first 5 days, the animal is entrained to a 14:10 hr light–dark cycle. On the sixth day, the animal is placed in constant darkness; the circadian clock free-runs with a period slightly greater than 24 hr, as indicated by the daily delay in time of activity onset. The delivery of exogenous melatonin prior to activity onset on Day 10, indicated by the open circle, advances the phase of the free-running rhythm of locomotor activity. After reaching a steady state on Day 12, the rhythm continues to free-run with the same period as before the melatonin administration.

fatigue or sleep and decrease alertness in humans, particularly when it is administered during the normal daytime waking hours. These hypnotic effects of melatonin, coupled with the fact that melatonin is released into the circulation primarily during the normal time of sleep, have led to recent interest in its possible use as a sleep aid in humans.

As noted in the previous section, there is increasing evidence that melatonin can act as a chronobiotic since its administration can induce phase shifts in human rhythms. Such phase shifts might be useful in phasing the sleep–wake cycle to the desired time for sleep, particularly in situations in which humans desire to alter the time of sleep as occurs in shift workers or in individuals moving rapidly across time zones (i.e., jet lag).

While both the chronobiotic and hypnotic properties of melatonin raise the possibility that it may be useful in the treatment of various sleep disorders, well-designed placebo-controlled studies have yet to be performed to determine if it is an effective and useful sleeping aid for any particular disorder of the sleep–wake cycle. The high cost of such clinical studies, and the fact that melatonin cannot be patented, makes it unlikely that such large-scale clinical studies will be performed. However, a number of pharmaceutical companies are now testing various melatonin-related drugs for both their chronobiotic and hypnotic properties.

V. OTHER POSSIBLE FUNCTIONS FOR MELATONIN

There is now a good deal of evidence to indicate that treatment with melatonin can influence the physiological and cellular properties of a variety of other systems. As noted previously, when one examines the vertebrates one sees that melatonin, depending on the species, is acting to provide information to the rest of the organism as to the time of day as well as the time of year. Such information may be useful to many physiological systems and/or cellular processes that need to change over the course of the day or year to meet the challenging demands of a changing environment. It is perhaps constructive to think in terms of the daily rhythm of melatonin as providing temporal information to the brain and body rather than in terms of melatonin being inhibitory or stimulatory to a particular physiological/cellular process. For example, nocturnal organisms are awake and active primarily during the time of high nighttime melatonin levels. Thus, the hypnotic effects of melatonin in humans do not represent a universal response to melatonin. How the brain and body respond to the melatonin signal may depend on the evolutionary history of the organism and how it successfully met the challenges faced by the ever-changing physical environment on a daily and a seasonal basis.

One final cautionary note about melatonin should be added. Many unsupported claims have been made for the beneficial effects of melatonin for the treatment of a wide variety of human illnesses and health problems including cancer, AIDS, heart disease, infertility, diabetes, etc., and claims have been made that melatonin is an anti-aging hormone. This latter claim is based in part on the finding that in advanced age there is a decrease in melatonin production which itself may contribute to aging. However, the evidence for many of these claims is well below meeting the level of acceptable scientific evidence. Interestingly, claims that melatonin can have effects on so many physiological and cellular processes are usually accompanied by claims that melatonin is safe and that it has no unwelcome side effects. However, the very claim that melatonin can have effects on so many biological processes is itself a red flag for urging caution for the indiscriminate use of melatonin, particularly its use on a chronic basis. The fact that the increased daily production of melatonin (in particular the duration of high levels) can induce complete testicular atrophy in male hamsters, and an anovulatory condition in the female, should make that point clear.

See Also the Following Articles

CIRCADIAN RHYTHMS; MELATONIN; SEASONAL REPRODUCTION, MAMMALS

Bibliography

Arendt, J. (1995). *Melatonin and the Mammalian Pineal Gland*. Chapman & Hall, London.

Cassone, V. M., Warren, W. S., Brooks, D. S., and Lu, J. (1993). Melatonin, the pineal gland, and circadian rhythms. *J. Biol. Rhythms* 8(Suppl.), S73–S81.

Goldman, B. D., and R. J. N. (1993). Melatonin and seasonality in mammals. In *Melatonin: Biosynthesis, Physiological Effects, and Clinical Applications* (H. Yu and R. Reiter, Eds.), pp. 225–252. CRC Press, Boca Raton, FL.

Karsch, F. J., Woodfill, C. J. I., Malpaux, B., Robinson, J. E., and Wayne, N. L. (1991). Melatonin and mammalian photoperiodism: Synchronization of annual reproductive cycles. In *Suprachiasmatic Nucleus: The Mind's Clock* (D. C. Klein, R. Y. Moore, and S. M. Reppert, Eds.), pp. 217–232. Oxford Univ. Press, New York.

Reppert, S. M., and Weaver, D. R. (1995). Melatonin madness. *Cell* 83, 1059–1062.

Tang, P. L., Pang, S. F., and Reiter, R. J. (1997). *Melatonin: A Universal Photoperiodic Signal with Diverse Actions*. Karger, Basel.

Turek, F. W. (1996). Melatonin hype hard to swallow. *Nature* 379, 295–296.

Turek, F. W., and Van Cauter, E. (1995). Rhythms in reproduction. In *The Physiology of Reproduction* (E. Knobil and J. D. Neill, Eds.), Vol. 2, pp. 487–540. Raven Press, New York.

Pinnipedia

see Seals

Pituitary Gland, in Fish

Martin P. Schreibman
City University of New York

Lucia Magliulo-Cepriano
State University of New York

I. Development (Embryology)
II. Structural Association of the Adenohypophysis, Neurohypophysis, and the Brain
III. Microscopic Anatomy/Cytology
IV. Pituitary Gland Morphology

GLOSSARY

adenohypophysis The epithelial component of the hypophysis composed of the pars tuberalis, pars distalis, and pars intermedia. In fishes, the pars tuberalis is absent and the pars distalis is further divided into a rostral and caudal region.

fishes A group composed of three major classes of vertebrates: Agnatha (hagfish and lamprey); the Chondrichthyes, the cartilaginous fish, which are further divided into the elasmobranchii (sharks, skates, and rays) and the holocephali (rabbitfishes and chimeroids); and the Osteichthyes, fish with a bony skeleton, which are further divided into two subclasses, Actinopterygii (chondrostei, holostei, and teleostei) and Sarcopterygii (Dipnoi and Crossopterygii). The term "fish" (as opposed to "fishes") is used when referring to organisms within a single species.

gonadotrope Cell type of the adenohypophysis of the pituitary gland that produces and secretes gonadotropin hormones.

hypophysis Also called the pituitary gland; an endocrine gland lying in the sella turcica of the sphenoid bone and attached to the hypothalamus by the infundibular stalk. It consists of a neural component (neurohypophysis), composed of the axonal terminals of hypothalamic neurons that serve as storage and release sites for pituitary-regulating hypothalamic hormones, and an epithelial component (adenohypophysis), which produces and secretes a variety of hormones that regulate other endocrine glands and diverse physiological events.

hypothalamus The inferior region of the diencephalon consisting of numerous, discrete brain nuclei that serve to maintain homeostasis by regulating vital physiological phenomena such as temperature, heart rate, and blood pressure. Additionally, the hypothalamus coordinates the endocrine and nervous systems by producing the hormones of the neurohypophysis and the regulating factors that control the functioning of the adenohypophysis.

infundibulum Also called the infundibular stalk. It is the narrow, stalk-like region of the neurohypophysis that serves to suspend the hypophysis from the floor of the diencephalon. It is composed largely of the neuronal fibers of neurons that have their perikarya in hypothalamic nuclei and their axonal terminals in the pituitary gland.

median eminence A capillary plexus located in the infundibular stalk which links the superior hypophysial artery to the hypophysial portal vessel. The portal vessels act to link the median eminence to a secondary plexus in the adenohypophysis of the pituitary gland, forming a portal circulatory system that serves to convey hypothalamic-regulating factors to adenohypophysial endocrine cells.

neurohypophysis The neural component of the hypophysis, often referred to as the pars nervosa. Hormones that are produced in hypothalamic perikarya are transported down axonal fibers to terminals in the neurohypophysis, where they are stored until released.

Rathke's pouch The region of buccal epithelium that develops into the adenohypophysis of the pituitary gland during embryogenesis.

somatotrope Cell type of the adenohypophysis of the pituitary gland that produces and secretes growth hormone.

thyrotrope Cell type of the adenohypophysis of the pituitary gland that produces and secretes thyroid-stimulating hormone.

The pituitary gland (or hypophysis) is considered by many to be the most complex organ (both structurally and functionally) of the endocrine system. All vertebrates, beginning with the two extant classes of cyclostomes, have a pituitary gland. Interest in comparative endocrinology and in the pituitary gland, in particular, has provided exciting new information and new questions have been posed which challenge or redefine well-accepted tenets of hypophysial structure and function. These include such basic phenomena as the direction of blood flow between the brain and pituitary gland, the number of hormones synthesized by a single endocrine cell, the discovery of novel pituitary products and variant forms of well-known pituitary hormones, and the suggestion of new roles for pituitary cells, products, and regions. This article summarizes the morphological features of the pituitary gland of the various classes of fishes. The accumulation of data on the fish pituitary gland has served to stimulate biomedical and basic researchers, clinicians, and students of comparative endocrinology to view these nonmammalian animals as important experimental models to complement, supplement, and/or replace traditional models in future research and clinical application.

I. DEVELOPMENT (EMBRYOLOGY)

There is an underlying pattern of development for the pituitary gland that is common to all vertebrates, despite the diversity that is found in the structure of the adult gland. In all vertebrates, the hypophysis has a dual origin. The neurohypophysis, which comprises the neural component of the pituitary gland and serves as a means of suspending and connecting the adenohypophysis in close proximity to the base of the brain, develops from a downgrowth of the diencephalon. The adenohypophysis has its origin in the primitive buccal epithelium (stomodeum), known as "Rathke's pouch," which is generally hollow in some vertebrates but may be solid, as it is in fishes and amphibians. An epithelial stalk that connects the adenohypophysial anlage with the buccal epithelium commonly disappears during development; however, in some teleosts (e.g., *Polypterus* and *Calamoichthys*), it may persist to form an open ciliated duct. A recently formulated evolutionary hypothesis proposes that the modern vertebrate pituitary gland evolved from a chemoreceptive olfactory organ that gradually lost its sensory function and developed, in its place, links of varying intensity to the hypothalamus and neural structures. Direct connections between the environment and the pituitary gland of some fish species, such as the presence of an orohyphophysial duct, lined with prolactin cells in some species of primitive teleosts, and the existence of the nasohypophysial duct linking the adenohypophysis to the olfactory organ of lampreys in larval stages, lend credence to this hypothesis.

The formation of specific lobes from Rathke's pouch during development accounts for class-related differences in the size and position of pituitary regions in the adult gland. The lateral lobes always form the pars tuberalis, whereas the pars intermedia develops from the portion of the aboral lobe that first makes contact with the infundibulum. The third region of the adenohypophysis, the pars distalis, forms from both the oral and aboral lobes of Rathke's pouch and it is this developmental event that gives rise to the zonation characteristic of the adult gland. The pars distalis of adults is further characterized as having a rostral portion and a caudal portion. Pituitary gland development is a dynamic phenomenon that involves Rathke's pouch formation and migration and the intimate contact of neural and buccal components. The development of the neurohypophysis by active growth of neural tissue occurs concomitantly with the proliferation of the secretory epithelium to establish the morphology of the adult gland.

II. STRUCTURAL ASSOCIATION OF THE ADENOHYPOPHYSIS, NEUROHYPOPHYSIS, AND THE BRAIN

The structural intimacy of the neurohypophysis and adenohypophysis that is established early during the embryology of the pituitary gland reflects the direct functional interaction between the central nervous system and the endocrine system. Vascular and/or neuronal pathways provide the means of exchanging chemical signals, thus enabling hypophysiotropic neurons of the hypothalamus to exert control over the synthesis and release of adenohypophysial hormones. The neurohypophysis also serves as a holding station for neurohypophysial hormones synthesized in the brain before they are released into the general circulation to act at endocrine and non-endocrine sites.

The extent of anatomical intimacy between neuropituitary and adenopituitary components ranges considerably among vertebrate classes and is particularly significant in teleosts in which there is an interdigitation between the two (Fig. 1). Neurohormones, which are synthesized in the specific regions of the brain, are conveyed to the neurohypophysis by way of axonal tracts, where they may be stored in distended endings termed Herring bodies. Axons may also contact blood vessels and discharge neurosecretory products into the systemic circulation or into a

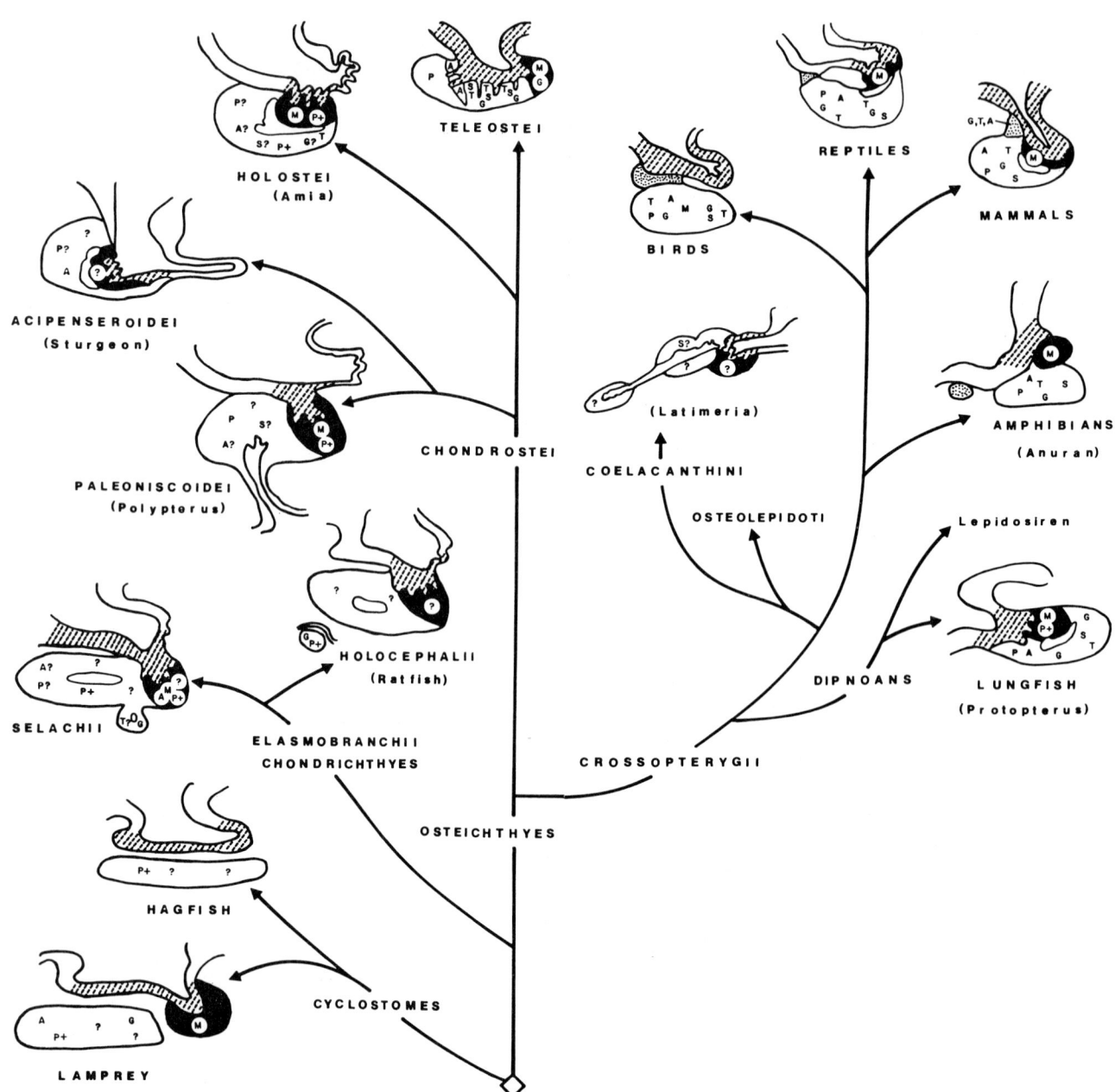

FIGURE 1 A phylogenetic tree of generalized diagrams of vertebrate pituitary glands as seen in midsagittal section. Anterior is on the left. Cell distribution, where indicated, is represented by tintictorial affinity for PAS (P+) or by function. P, prolactin; S, somatotropin; G, gonadotropin; T, thyrotropin; A, adrenocorticotropin; M, melanocyte-stimulating hormone; ?, unidentified function. Neurohypophysis is indicated by dashed lines, pars tuberalis is represented by dots; pas intermedia is colored black. Drawing is not to scale (from Schreibman, p. 17, 1986).

portal system leading to the adenohypophysis, or they may directly innervate pituitary cells. The proximity of the hypothalamic fibers to the adenohypophysial endocrine cells allow for the diffusion of regulating factors across the small intervening spaces to regulate adenohypophysial cells. Neural fibers have traditionally been categorized as type A (peptidergic) or type B (aminergic).

The histology of the neurohypophysis reflects the preponderance of neuronal tracts conveying neurosecretory products along with networks of vascular elements. The neurohormones of the fish neurohypophysis usually include both arginine vasotocin (AVT) and either oxytocin or a second peptide related to oxytocin. The cyclostomes are known to possess only the single peptide AVT. Pituicytes in the neurohypophysis are a class of neuroglia elements whose cytoplasmic processes envelop secretory axon terminals and, along with the nerve terminals, may make contact with the perivascular space. Further study is needed to determine whether pituicytes serve in some metabolic capacity in the secretory process or are merely supportive.

Ependymal cells, which form a boundary of the neurohypophysis by lining the cavity of the third ventricle, are in a strategic position for affecting the transfer of materials between the central nervous system and pituitary gland cells. For example, tanycytes are modified ependymal cells that were first described in elasmobranchs. Their potential importance is strongly suggested by the functional cytological bridge they form between the ventricular systems and portal and pituitary elements as well as by changes in their cytology that occur in association with specific endocrine-regulated processes. The role of these cells in pituitary gland function and in information transfer between the central nervous and endocrine systems requires clarification.

III. MICROSCOPIC ANATOMY/CYTOLOGY

The adenohypophysis of all vertebrates is essentially a conglomerate of cells that range in order from masses highly segregated by cell type (teleosts) to the more randomized mixtures of endocrinotropic cells (birds and mammals). The most widely studied fish species are the teleosts due, in large part, to the almost complete segregation of physiological cell types to distinct zones of the gland (Fig. 2). It is in these fish species that the most abundant information concerning the hormones of the pituitary gland and the regulatory mechanisms exerted upon them can be found. Identification of cell types according to the function they serve is accomplished either by (i) the application of stains that identify the cells

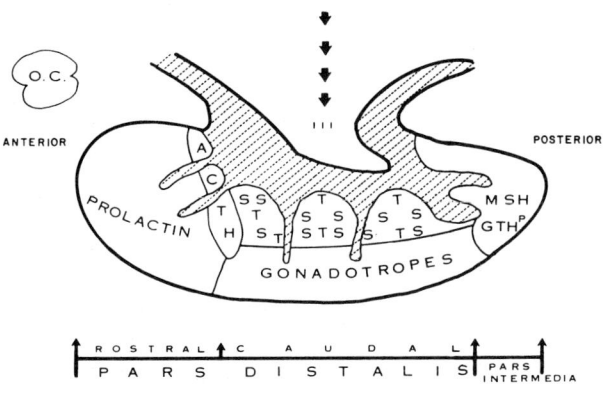

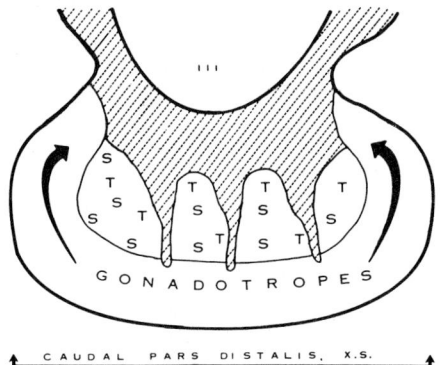

FIGURE 2 A diagrammatic representation of the pituitary gland from a sexually mature platyfish as seen in midsagittal section (top) and in a cross section through the caudal pars distalis (bottom). The major regions of the gland and the hormone-producing cells that populate the various zones are indicated. The arrows in the top diagram indicate the plane of section for the cross section depicted below. O.C., optic chiasma; ACTH, adrenocorticotropes; T, thyrotropes; S, somatotropes; III, third ventricle; MSH, melanocyte-stimulating hormone; GTHP, gonadotropes (PAS$^+$ cells) of the pars intermedia; dashed lines, neurohypophysis (from Schreibman, p. 31, 1986).

according to their tinctorial affinity (acidophilic, basophilic, or chromophobe) or to the more specific chemical characteristics of the hormones contained in such cells, such as periodic acid-Schiff (PAS) aldehyde fuchsin (AF), alcian blue (AB), and lead hematoxylin (PbH), or (ii) immunological procedures utilizing antibodies generated against specific pituitary hormones and visualized with a florescent dye or chromogen. While these methods adequately demonstrate the presence of a particular antigen (hormone) in a specific cell, the question remains whether the hormone was elaborated at the site of localization or was transported there after synthesis in some other region. This question is being addressed by *in situ* hybridization techniques that localize the messenger RNA of the hormone under investigation.

IV. PITUITARY GLAND MORPHOLOGY

A. Class Agnatha

Agnathans are the most primitive living vertebrates and are represented by two orders: Petromyzontiformes (lamprey eels) and Myxiniformes (hagfish). The great disparity between the way hagfishes and lampreys regulate reproductive functions may be reflected in part by the structure and function of their pituitary glands.

The hagfish pituitary gland appears to be the most primitive of the two. Its neurohypophysis is highly developed, flattened, sac-like, and contains many AF^+ neurosecretory endings. The hypophysial–portal system of hagfish is markedly reduced but this may be attributed to species variation. The suggestion that neurosecretions may reach the adenohypophysis by diffusion through connective tissue rather than by a vascular route has gained a good deal of support due to recent studies.

The adenohypophysis is separate from the neurohypophysis and is not differentiated into zones except, perhaps, in the posterior region of older and larger animals (*Myxine*). Generally, islets of adenohypophysial cells are embedded in a poorly vascularized, loose connective tissue which is continuous with the layer separating the neurohypophysis from the adenohypophysis. Cell types have been difficult to characterize but several types have been classified by their affinity for standard pituitary stains. Some cells form follicles containing PAS^+ colloid. Acidophils and PAS^+ cells have been noted, but their function remains obscure. A thyrotropic substance may be present. The presence of a prolactin-like substance in the hagfish has not been established and neither has a gonadotropic factor, although the presence of a gonadotropin-releasing hormone (GnRH)-like substance has been detected in the hagfish brain. Total hypophysectomy in hagfish has not been shown to alter gonadal function. These data reflect particular physiological conditions of hagfish hypothalamic regulation of pituitary function and reproduction.

In lampreys, the neurohypophysis consists of a thin anterior portion and a thickened posterior part, which, because it possesses neurosecretory neurons ending in a neurohemal structure, has been termed a "pars nervosa." The adenohypophysis is compact and is differentiated into two zones, a pars distalis and a pars intermedia. A well-developed pars intermedia forms a neurointermediate lobe with the pars nervosa; however, there are no direct nervous or vascular connections between the two components.

The lamprey pars distalis is separated from the infundibulum by connective tissue, and its cells display a regional distribution. The rostral pars distalis (RPD) contains either chromophobes or PAS^+ and AF^+ basophils. Cells containing immunoreactive adrenocorticotropin (ACTH) have been identified, and ACTH activity in the pars distalis has been reported. A newly characterized pituitary protein has been identified in the lamprey adenohypophysis known as nasohypophysial factor; it is found in the olfactory organ and nasohypophysial duct of some lampreys. Its function remains unclear. The caudal pars distalis (CPD) contains both acidophils and fewer basophils. The functions of the pars distalis cells have not been clarified, despite attempts to either relate them to metamorphosis and spawning or to subject them to hypophysectomy, inhibiting or blocking agents, and immunochemical analysis. The presence of cells in the CPD that cross-react with antisera to several molecular forms of gonadotropin-releasing hormone suggests that typical control mechanisms may operate in lampreys. Reports for the pars intermedia vary, but generally one cell type is present that is carminophilic: either PAS$^+$ or PAS$^-$ and PbH$^+$. These cells undergo cytological changes with varying illumination, migration, and spawning, and presumably they secrete melanocyte-stimulating hormone (MSH) since antiserum to α-MSH cross-reacts with all cells in the pars intermedia.

Other than the existence of hypothalamic regulating factors in the brain and pituitary gland of lampreys, there is little evidence to suggest that the hypothalamus exerts any influence on the pars distalis. It has been shown that even with an ectopic adenohypophysis, complete gonadal maturation may take place in male lampreys. Ultrastructural studies failed to demonstrate the presence of nervous or vascular connections between the anterior neurohypophysis and the pars distalis. Diffusion appears to be the primary method of communication between hypothalamic structures and the pituitary gland. This could represent the most primitive of hypothalamus–pituitary associations.

Unlike hagfish, adult marine lampreys generally lead an active predatory life. Changes in the amount of neurosecretory material in their neural lobe may reflect accommodations to varying environments brought about by their migrations between rivers and the sea, variations in daylight, and a variety of other seasonally related phenomena. What remains a mystery is how, in the absence of anatomical links between sensory receptors and reproductive structures, seasonal changes and reproductive functions are related in lampreys. It is interesting to speculate that changes in reproductive processes could be related to patterns in feeding. Variations in quantity and type of food consumed may reflect alterations in environmental conditions. The pituitary may be affected directly when food passes through the buccal cavity or, perhaps, indirectly by gut hormones that are released upon feeding to affect brain, pituitary, and/or gonad activity.

B. Class Chondrichthyes

The cartilaginous fish diverged early and pursued an evolution independent of all other fish groups. There are two extant subclasses: the Elasmobranchii, composed of the sharks, skates, and rays, and the Holocephali, composed of the ratfishes or rabbitfishes and chimeras. In elasmobranch fishes, the neurohypophysis consists of a thin-walled anterior portion that is a true median eminence, which is connected to the pars distalis by a hypothalamo–hypophysial portal system. The median eminence is divided into anterior and posterior portions in at least some species. The posterior region receives both hypothalamic aminergic and peptidergic axons and is linked by the portal system to the CPD and perhaps indirectly to the ventral lobe; the anterior region contains less neurosecretory material and is linked to the RPD. It has often been suggested that this arrangement could reflect a more rapid and efficient method for controlling pituitary cell types. This situation is similar to that of the median eminence in birds, which is also divided into anterior and posterior portions and serves in a "point-to-point" delivery of vessels to the adenohypophysis. Since, in contrast to teleosts, there is no direct innervation of the pars distalis, the portal system is the only route for hypothalamic control. There is also a portal supply to the neurointermediate lobe. A saccus vasculosus is present in elasmobranchs, but it is not as developed as its homolog in teleosts. It develops from the hypothalamus as a balloon-like structure just above the neurointermediate lobe, but it is considered to be sensory and nonendocrine.

The adenohypophysis is divided into four areas: an elongated pars distalis (the "dorsal lobe"), with its characteristic RPD and CPD; a ventral lobe attached by a stalk to the CPD; and a large pars intermedia that is heavily penetrated by neurohypophysial tissue to form a typical neurointermediate lobe. A pars tuberalis is not seen.

Follicles and spaces are characteristic of the chondrichthyian pituitary gland and range from a highly developed system of vesicles and tubules that communicates with a hollow pars distalis in sharks and dogfish to the situation in skates and rays in which a hypophysial cavity is small or lost entirely. The ventral lobe, which forms from the fusion of the lateral lobes of Rathke's pouch, is also hollow and often vesicular. All of the cavities and spaces, including a persistent hypophysial cleft, are apparently remnants of Rathke's pouch. (Elasmobranchii are the only fishes that have a hollow Rathke's pouch during development.) The cavity frequently contains a PAS$^+$, AF$^+$, and AB$^+$ colloid. It is likely that this colloid is secreted from nonendocrine, chromophobic cells that line the walls of the cavity. The structure of these cells and the chemistry of their secretions suggest that they are similar to the mucus-secreting epithelial cells of the buccal cavity. It has also been suggested that they are homologous to the stellate cells that are found in the adenohypophysis of most vertebrates.

Although certain tropic hormone functions have been clearly defined and have been associated with specific areas of the gland, others remain elusive. Attempts to link hormones to specific cells have spawned contradictions. ACTH and prolactin are presumably produced in the RPD. However, there is one cell type in the elasmobranch RPD that stains with PAS and orange G but not with AF or AB, making it unlike either the corticotropin- or prolactin-producing cells of other fish. Early reports that localized ACTH in the pars intermedia as well as in the RPD were eventually confirmed by immunofluorescence.

Cells of the CPD are acidophils that have a varied response to PAS and a relatively unknown function. The cells of the ventral lobe are large and stain with AF, PAS, and AB and are presumed to represent gonadotropes and thyrotropes. Immunofluroescence confirms luteinizing hormone (LH) activity in this region. In addition, removal of the ventral lobe from the dogfish results in the degeneration and phagocytosis of a specific stage of spermatogenesis.

The holocephalian neurohypophysis (e.g., in the ratfish) has a pars nervosa and a prominent median eminence that sends many short blood vessels into the RPD and CPD. A typical neurointermediate lobe is seen. In holocephalians, the entire pars distalis is hollow and is not as clearly divisible into rostral and caudal regions as in the elasmobranchs. A ventral lobe, characteristic of the elasmobranchs, is not present. In adult holocephalians, however, there is a large compact follicular structure called the Rachendachhypophyse (the pharyngeal lobe) that lies outside the cranium and has an independent blood supply. It is derived from the embryonic pars distalis rudiment and is ultimately separated from it when a connecting stalk disappears. The Rachendachhypophyse contains two kinds of basophilic cells; both are PAS$^+$ but only one is AF$^+$. Gonadotropic activity has been identified in the Rachendachhypophyse of the rabbitfish, which would make this buccal lobe comparable to the ventral lobe of elasmobranchs. However, it is not likely that the two are homologous because their embryogenesis differs.

C. Class Osteichthyes

1. Subclass Actinopterygii

The subclass Actinopterygii (fish with a bony skeleton and paired ray fins) includes the infraclasses Teleostei, Chondrostei, and Holostei (the latter two taxa are often referred to as the ganoid fishes).

i. Infraclass Chondrostei The Chondrostei contain two orders, Polypteriformes (*Polypterus* and *Calamoichthys*) and Acipenseriformes (paddlefish and sturgeons). The pituitary glands of *Polypterus* and *Calamoichthys* are apparently similar in structure. Type A fibers originate in the preoptic nucleus and terminate in the pars nervosa. The origin of the type B fibers is unknown (though it is suspected that it is also the preoptic nucleus) but they terminate in the median eminence. A typical intermediate lobe is present. *Polypterus* and *Calamoichthys* both possess a typical median eminence and portal system and, in this respect, are different from the phylogenetically more recent teleosts. There is no direct innervation of the pars distalis cells, a feature common in teleosts. The gland subdivisions are easily discernible and resemble those seen in teleosts. Basophils, chromophobes, and acidophils have been identified but studies ascribing specific hormones to them have been few.

The most remarkable feature of the Chondrostei is the presence of a duct or canal that connects a cavity in the ventral region of the RPD to the roof of the mouth (the buccohypophysial or orohypophysial duct). In tissue sections, this sometimes appears as follicles within the RPD. The duct presumably arises from Rathke's pouch. Duct cells do not stain with fluorescent-labeled antibodies to ovine prolactin, as do other cells scattered throughout the RPD, and so are not homologous with the prolactin cells that surround the follicles of the Salmonidae and Clupeidae. They are generally chromophobic, containing PAS$^+$ and AB$^+$ granules, mucus-producing, and non-endocrine.

In the order Acipenseriformes, the most notable anatomical feature is a large, central hypophysial cavity with many tubular extensions. Unlike the palaeoniscoids, however, there is no hypophysial duct. A thin-walled neural lobe penetrates into the deepest regions of a large pars intermedia. There is no direct innervation of the pars distalis, but there is a well-developed median eminence and portal system.

ii. Infraclass Holostei The Holostei are the closest living primitive relatives of the teleosts. Neither genus of the two living holostean genra, *Amia* (bowfin) and *Lepisosteus* (gars), has a hypophysial cavity or duct as adults. The gland is attached along most of its length to the infundibular floor, and a vascular connective tissue is found in between. There is a large saccus vasculosus.

In contrast to teleosts, the pars distalis is not penetrated by large strands of neurohypophysial tissue. It has been shown, however, that there are small numbers of fibers with neurosecretory material (AB$^+$) that extend into and abut against the cell cords of the CPD. Innervation of the pars intermedia of *Amia* is similar to that of other ganoids and primitive teleosts in that type A fibers contact the basement membrane of the intervascular space that separates the neural processes from endocrine cells. Occasionally, type A fibers appear to synapse with pars intermedia cells. This type of novel contact is rare in other ganoids but common in teleosts. Holosteans also maintain a median eminence that is somewhat better developed than their phylogenetic predecessors, the acipenseroids. Thus, it appears from an examination of the pituitary in *Amia* that innervation of the adenohypophysial cells, a characteristic that typifies teleosts, may have preceded the change in the median eminence/portal system that is seen in the bony fishes.

The adenohypophysis consists of an RPD, a CPD, and a pars intermedia. The RPD contains follicles that enclose PAS$^+$, AB$^+$, and AF$^+$ material and are lined with the acidophils that may be the source of prolactin. These cells are in close association with an erythrosin-, aniline blue-, and PbH$^+$-cell that may be the corticotrope. At the boundary of RPD and CPD are the so-called basophils (PAS$^+$, AF$^+$, PbH$^+$, and AB$^+$), which also stain with erythrosin and could be (although no experimental evidence is available) the gonadotropes. In the CPD, there is an acidophil that stains differently from acidophils of the RPD and is suspected of being a somatotrope. Two types of basophils, one of which is a thyrotrope and the other unknown, are also present in the CPD. In the pars intermedia, a prominent PbH$^+$ cell thought to produce MSH and a less common PAS$^+$ cell of unknown function are present.

iii. Infraclass Teleostei Reviews of the structure and function of the teleostean pituitary gland have appeared with a regular periodicity. Teleosts represent the most diverse group of vertebrates, with more than 20,000 extant species recorded. Inasmuch as they occupy every conceivable habitat, it would not be surprising to find this variability reflected in a number of different patterns of pituitary organization and function. However, this is generally not the case; within all this diversity there are common anatomical, histological, and cytological denominators. The most notable of these is that the adenohypophysis is clearly divided into three regions: the RPD, CPD, and pars intermedia. (There is no pars tuberalis.) The clarity of this separation is essentially due to a restriction of specific cell types to distinct regions of the gland (Fig. 2). In teleosts, there is a structural intimacy of the neurohypophysis with both RPD and CPD as well as with the pars intermedia, and a unique feature is that the adenohypophysial cells are innervated directly. A few exceptions to the general rule have been noted, and in these forms an elaborate system of extravascular circulation serves to convey neurosecretions to pituitary cells. There is generally no true median eminence or portal system present in teleosts.

Variations in the structure of the pituitary gland appear in the axis formed between the pituitary gland and the brain and, thus, in the orientation of the various adenohypophysial regions. Specific cell types may not be found in the same region of the gland in all fishes. This is especially true for the thyrotropes. The arrangement of cells may also vary. For example, follicles that are common in the Salmonidae (salmon and trout), Clupeidae (herrings), and Anguillidae (eels) generally are not found in other fish. A persistent orohypophysial duct may be seen in adult *Chanos chanos* (milkfish) and in some Clupeidae. In a number of primitive teleosts, this duct has been shown to be lined with prolactin cells. Different hypothalamic control mechanisms also have been noted among the teleosts.

The neurohypophysis receives neuronal endings from hypothalamic and extrahypothalamic nuclei. Axons may originate from perikarya as close as the nucleus lateralis tuberis or as distant as the nucleus olfactoretinalis (nervus terminalis). The neurohypophysis typically can be divided into two parts: an anterior region with AF^-, type B fibers (and a modicum of pituicytes) in contact with the pars distalis, and a posterior region composed mainly of AF^+, type A fibers (but with many pituicytes) in direct association with the pars intermedia. It has been suggested from functional, anatomical, and embryological evidence that these two regions correspond to the median eminence and the neural lobe, respectively. However, there has been considerable discussion as to whether a true median eminence exists in teleosts that is similar in structure and function to that seen in other groups, and opposing conclusions have been reached. Some maintain that the median eminence is represented by the vascular system resulting from the close anatomical association of the anterior neurohypophysis and the pars distalis and by evidence of synaptoid contacts on perivascular spaces. Other evidence to support the presence of a median eminence comparable to tetrapods is frequently based on india ink injections that are difficult to evaluate and on reconstructions from serial histological sections. Current thinking is inclined to agree with Gorbman that although it is possible that neurosecretions could be released into blood vessels in the neurohypophysis that course through the pars distalis and thus reach pituitary secretory cells, there is no vascular pattern seen that is comparable to that of other vertebrates. A true median eminence system is composed of a primary capillary plexus, which drains to a secondary plexus by way of a portal vein. From a functional point of view, it would be a redundant, unnecessary system since pituitary cells are directly (or almost directly) innervated by these nerve fibers.

In general, hormone-producing cells of a particular type are restricted to specific regions of the adenohypophysis (Fig. 2). Nomenclature for pituitary cells has been derived from studies involving the application of standard staining methods to fish whose physiological homeostasis has been challenged (castration, interference with thyroid gland function, alteration of ionic content, administration of exogenous humoral agents, etc.). The functional accuracy of these terms has been examined using immunocytochemical methods with antisera from homologous and heterologous organisms, including antisera generated against human pituitary hormones. The RPD has a preponderance (essentially a single mass) of prolactin cells (erythrosinophilic). A narrow band of corticotropes forms the posterior boundary of the RPD. In the CPD, thyrotropes and somatotropes are intermingled in islets formed by pervading neurohypophysial tissue. The external boundary of the CPD is made up of several cell layers of gonadotropes. In midsagittal section these gonadotropes form a distinct ventral border in the CPD, which does not appear in poeciliids until the process of sexual maturation commences. In platyfish, the age at which this gonadotropic zone develops is under genetic control and is dependent on the prior maturation of specific GnRH-containing neuronal systems in the brain. It is fairly well established that, in fish, pituitary gonadotropes of the CPD produce two forms of gonadotropin hormone (GTH), known as GTH-I and GTH-II. GTH-I is a follicle-stimulating hormone-like gonadotropin hormone that is genetically distinct from GTH-II, an LH-like gonadotropin. In most species, GTH-I appears earlier in the process of sexual development, with GTH-II not appearing until the onset of, or just prior to, reproductive function. The physiological distinction between these two hormones varies among the different species and has yet to be determined.

The pars intermedia, a highly variable structure among different species, generally contains two cell types. One is PbH^+, is known to produce POMC-related peptides. These cells cross-react weakly with antiserum to ACTH and are presumed to be the source of MSH. The other cell type is strongly PAS^+

and is believed to produce a recently characterized pituitary protein, belonging to the growth hormone-prolactin family of protein hormones, known as somatolactin. This protein exists in both a glycosylated and nonglycosylated form and has also been identified in species which lack PAS$^+$ cells in the pars intermedia. Although the biochemical and physicochemical properties of somatolactin have been the subject of much study, its physiological significance is still unclear.

2. Subclass Sarcopterygii

The subclass Sarcopterygii consists of the orders Dipnoi and Crossopterygii. The single Coelacanth genus, *Latimeria,* and the half dozen species of lungfish are interesting to consider in the study of the evolution of the tetrapod pituitary gland.

i. Order Dipnoi The neurohypophysis of lungfish is divisible into a median eminence, infundibular stalk, and neural lobe. The preoptic nucleus is the major hypothalamic nucleus projecting to the neurohypophysis, and it stains with the usual neurosecretory material methods. The projecting neurons are peptidergic and terminate around the primary portal vessels, the plexus intermedius, the neural lobe, and in the pars intermedia. In Lepidosiren, many small aminergic neurons have been described whose perikarya lie just below the ependyma of the median eminence and whose axons end in the pars distalis, pars intermedia, and the neural lobe close to the pars intermedia. These neurons resemble the nucleus lateralis tuberis of teleosts and the infundibular and arcuate nuclei of tetrapods. Thus, this is a system of type A fibers terminating on capillaries anteriorly but on an avascular connective tissue posteriorly.

The adenohypophysis of lungfish is more similar in structure to that of amphibians than it is to that of other fishes. Among the amphibian characteristics of the lungfish pituitary are a well-formed neural lobe, a prominent median eminence, the absence of a saccus vasculosus, and a less obvious regional distribution of cell types. Nevertheless, certain fish-like characteristics persist. Most notable are the direct aminergic innervation of the pars distalis cells, the interdigitation of the neurohypophysis and pars intermedia, the prominent hypophysial cleft between the pars distalis and the pars intermedia, the absence of a pars tuberalis, and the presence of follicles, especially in Neoceratodus.

The five tinctorial cell types that have been defined in the pars distalis are intermingled and lack the distinct zonation seen in teleosts; however, differences in relative proportions of cells permit delineating pars distalis regions. The absence of immunocytochemical evidence and the lack of physiological experiments preclude assigning specific hormonal roles to pituitary cells. Nevertheless, investigators have extrapolated from the data gathered in amphibians to assign function to pituitary cells in the dipnoans on the grounds that cells in these two groups show similarities of staining responses, ontogenesis, and distribution. One type of acidophil that stains with alizarine BT is distributed throughout the gland and secretes prolactin. A second acidophil type, orangeophilic and faintly PAS$^+$, is restricted to the CPD and is purported to be a somatotrope. Three types of basophils, distinguishable by size and distribution, are the putative sources for thyrotropin, gonadotropin, and adrenocorticotropin. The size of the pars intermedia, the arrangement of its cells, and its structural association with the neural lobe processes vary among the lungfish; however, in all cases the pars intermedia is separated from the pars distalis by the hypophysial cleft. Two cell types are found in the pars intermedia; one is weakly PAS$^+$ and alcian blue- and aniline blue-positive. The second, which is more abundant in young fish, is strongly PAS$^+$, alcian blue-positive, and orangeophilic. Both types are directly innervated by type A and a few type B fibers. The results reviewed here relating tinctorial cell types to function are not to be construed as being confirmatory, but rather they are merely meant to stimulate thought and to serve as a basis for further, more sophisticated cytological analyses.

There is still much to be learned about the lungfish pituitary. We need to know the variations in hypophysial structure, aside from the relative number of follicles, size of the pars intermedia, and innervation of cells, that would account for differences in the physiological activities of the three living genera (*Protopterus,* African; *Neoceratodus,* Australian; and *Lepidosiren,* South American) and the six species of the Dipnoi. This would be especially interesting in regard to the differences in the ability of lungfish to estivate.

ii. Order Crossopterygii *Latimeria chaulumnae* is the only living representative of the suborder Coelacanthini. Relatively few specimens have been collected and described in the past two decades. The orthodox features of the coelacanth pituitary include a neurointermediate lobe (typically fish-like) ventral to the brain, direct contact between pars distalis and pars intermedia, follicles and tubules containing glycoprotein colloid seen in many fish, a well-developed hypothalamo–hypophysial portal system, and a probable median eminence. Other features that are more "fish-like" include a small saccus vasculosus, penetration of neurohypophysial tissue into the pars distalis and regional differentiation of the gland based on cell type.

Among the exceptional features is a greatly extended pars distalis that is tripartite in nature. An orthodox proximal portion lies close to (even interdigitating with) a neurointermediate lobe and can be separated on the basis of vascular supply and cell type into dorsal (rostral) and posterior (caudal) regions. The rostral portion contains orangeophils and erythrosinophils, and orangeophils and basophils populate the caudal region. The presence of growth hormone has been reported; however, its cellular origin is unknown. An elongate extension of the pars distalis comprises its most rostral lobe (buccal portion). Relatively few basophils were found in the buccal rostral lobe in the one immature female studied. Unique is a long (up to 12 cm), cylindrical, tubular hypophysial cavity that extends from the rostral lobe of the pars distalis and contains vascularized masses of adenohypophysial cells, the so-called rostral islets or pars buccalis. These structures are comparable with the Rachendachhypophyse of holocephalians rather than with the ventral lobe of elasmobranchs with regard to their anatomical connection; however, cell types of the islets are similar to those in the ventral lobe. The need to study additional specimens is obvious.

Thus, the pituitary gland of *Latimeria* shows many specializations and peculiarities. It possesses many features of the teleost and elasmobranchiomorph gland rather than those of dipnoans and amphibians.

The patent diversity of the hypophysial system in vertebrates, in general, and in fishes in particular, offers a wealth of elegant material to study the evolution of neuroendocrine systems and the structure–function relationships involving the pituitary gland. Additionally, by gaining knowledge of the species-dependent features of the pituitary gland, we are provided the opportunity to utilize relatively simple systems to explore the more complex questions of general vertebrate endocrine significance.

See Also the Following Article

Pituitary Gland, Overview

Bibliography

Ball, J. N., and Baker, B. I. (1969). The pituitary gland: Anatomy and histophysiology. In *Fish Physiology* (W. S. Hoar and D. J. Randall, Eds.), Vol. 2, pp. 1–110. Academic Press, New York.

Gorbman, A. (1995). Olfactory origins and evolution of the brain–pituitary endocrine systems: Facts and speculation. *Gen. Comp. Endocrinol.* 97, 171–178.

Schreibman, M. P. (1986). The pituitary gland. In *Vertebrate Endocrinology: Fundamentals and Biomedical Implications* (P. K. T. Pang and M. P. Schreibman, Eds.), Vol. I, pp. 11–55. Academic Press, New York.

Schreibman, M. P., Leatherland, J. F., and McKeown, B. A. (1973). Functional morphology of the teleost pituitary gland. *Am. Zool.* 13, 719–742.

Schreibman, M. P., Holtzman, S., and Cepriano, L. (1990). The life cycle of the brain–pituitary–gonadal axis in teleosts. In *Progress in Comparative Endocrinology* (A. Epple, C. G. Scanes, and M. H. Stetson, Eds.), pp. 399–408. Wiley-Liss, New York.

Wingstrand, K. G. (1966). Comparative anatomy and evolution of the hypophysis. In *The Pituitary Gland: Anterior Pituitary* (G. W. Harris and B. T. Donovan, Eds.), Vol. 1, pp. 58–126. Univ. of California Press, Berkeley.

Pituitary Gland, Overview

Béla Halász
Semmelweis University Medical School

I. Introduction
II. Size, Weight, and Location
III. Terminology
IV. Development
V. Blood Supply

GLOSSARY

adenohypophysis A major part of the pituitary arising from the ectodermal outgrowth of the primary oral cavity called stomodeum.
anterior pituitary A major part of the adenohypophysis; also called pars distalis; it contains glandular cells secreting different anterior pituitary hormones.
endocrine gland A ductless gland whose secretion enters small blood vessels and is conveyed by the bloodstream to the site of action.
hormone A product of endocrine glandular cells.
hypophysis Another term for the pituitary gland.
neurohypophysis A major part of the pituitary gland arising from the downgrowth of the floor of the third ventricle of the brain.
posterior pituitary Part of the neurohypophysis; also called infundibular process or pars nervosa; it contains nerve endings of hypothalamic supraoptic and paraventricular neurons synthesizing posterior pituitary hormones which are released from the nerve endings in the posterior pituitary.

The pituitary gland (or hypophysis cerebri) is an unpaired endocrine gland connected with the hypothalamus. It lies on the inner surface of the base of the skull in the hypophysial fossa of the sphenoid bone and contains two major parts, adenohypophysis and neurohypophysis, which have different embryonic origin, structure, and function. The adenohypophysis develops from an ectodermal outgrowth of the roof of the primary oral cavity and contains glandular cells secreting different hormones, some of which act on target endocrine glands (thyroid gland, adrenal cortex, and gonads), whereas others exert their influence without the intervention of other endocrine glands. The neurohypophysis is a downgrowth of the floor of the third ventricle and consists largely of nerve fibers and terminals arising from supraoptic and paraventricular neurons of the hypothalamus synthesizing and releasing posterior pituitary hormones, which are released from the nerve terminals in the neurohypophysis and enter blood capillaries and veins.

I. INTRODUCTION

The pituitary gland or hypophysis cerebri is an unpaired endocrine gland and is continuous with the brain, the ventral part of the hypothalamus. It consists of two major parts: adenohypophysis and neurohypophysis. The term pituitary is related to the idea of Galen, who regarded the pituitary as a sump for waste products (phlegm = pituita) derived in the brain from distillation of "animal spirit." He supposed that the phlegm would then filter through openings in the ethmoid bone into the nasal passages. That notion held without question until it was discovered that the openings in the cribriform plate of the ethmoid bone are for the olfactory nerves and that fluids cannot pass from the cranial cavity into the nose. An important step in the development of our knowledge about the pituitary was made by Rathke, who recognized the dual embryonic origin of the pituitary: the neurohypophysis arising from the diencephalon and the adenohypophysis from the ectodermal saccule

(Rathke's pouch) of the roof of the primary oral cavity (stomodeum). Knowledge of the histological structure of the respective components was derived for many years in the future. The identification of ductless glands structurally specialized for internal secretion required the microscopic observations which were made possible in the second part of the nineteenth century.

The structure and function of the two parts of the gland are also different (Fig. 1). The neurohypophysis consists largely of nerve fibers and terminals whose cell bodies are outside the hypophysis in the supraoptic and paraventricular nuclei of the hypothalamus. Neurosecretory material (containing hormones and carrier proteins) manufactured in the cell bodies of these nuclei migrates along their axons and ends in the distal part of the neurohypophysis, from which the hormones (posterior pituitary hormones) are released into the general circulation. The adenohypophysis is a highly vascular structure and contains large numbers of different glandular cells synthesizing and releasing various hormones, some of which act on target endocrine glands (such as thyroid gland, adrenal cortex, and gonads), whereas

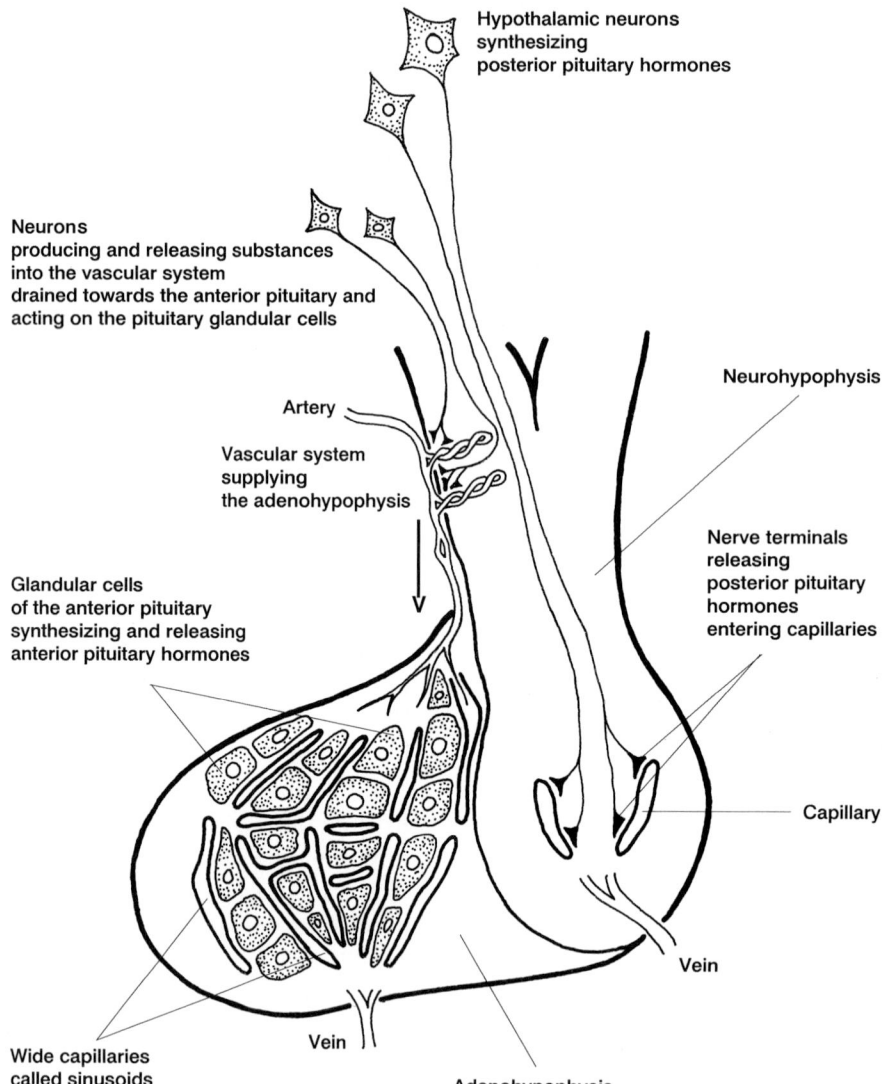

FIGURE 1 Illustration of the basic structural characteristics of the neurohypophysis and adenohypophysis. For details, see text.

others exert their influence without the intervention of such glands. Thus, the pituitary is an endocrine gland with fairly unique characteristics: Some of the hormones released from the gland are synthesized outside the organ (i.e., by neurons in the hypothalamus), whereas other pituitary hormones are produced by secretory cells of the gland, and some of these latter substances control the structure and function of other endocrine glands. The pituitary, at least its adenohypophysis part, is also unique in the sense that the central nervous system, which controls to a great extent this part of the gland, exerts its regulatory influence via a neurohumoral mechanism (Fig. 1): Pituitary hormone-releasing substances are produced and released by hypothalamic neurons. The regulatory substances enter the special vascular system supplying the adenohypophysis and are carried by the bloodstream to the pituitary.

II. SIZE, WEIGHT, AND LOCATION

The pituitary gland is a reddish-gray, more or less round or ovoid body continuous with the infundibulum, a hollow conical inferior process from the tuber cinereum of the hypothalamus between the optic chiasma and the mamillary bodies. The size and weight of the pituitary shows great variation. In humans, the anteroposterior diameter of the gland is about 8 mm, the transverse diameter is about 12 mm, and its weight is approximately 500–700 mg. In animals such as the cow and horse, the anteroposterior diameter of the pituitary is about 25 mm and it weighs approximately 3 g. The pituitary weight of dogs is approximately 65–70 mg and that of rats approximately 8–10 mg. It should be noted that the weight also varies according to sex and age. The gland is heavier in females than in males and weighs more during pregnancy and lactation. The weight decreases with age.

The pituitary lies in the intracranial cavity in the midline of the inner surface of the base of the skull, within the hypophysial fossa of the sphenoid bone (Fig. 2). It is covered superiorly by a membrane of the dura mater, called diaphragma sellae, which arises from the margins of the hypophysial fossa. There is a small opening on the diaphragm for the infundibular stem of the neurohypophysis and the pars tuberalis of the adenohypophysis (collectively called pituitary stalk) to pass through, providing the connection of the pituitary gland with the hypothalamus. On the two sides of the hypophysial fossa are the cavernous sinuses connected rostrally and caudally by the intercavernous sinuses. The cavernous sinuses contain not only venous blood but also a segment of the internal carotid artery, and some cranial nerves (oculomotor, trochlear, ophthalmic branch of the trigeminal nerve and the abducens nerve) pass through it or are closely related to the wall of the sinus. In humans and in some animals, the optic chiasma is just in front of the upper surface of the dural diaphragm separating the pituitary from the brain. Therefore, pituitary tumors may press the optic chiasma, leading to defects in the visual field. There are significant species variations in the length and direction of the pituitary stalk as well as in the direction of its course; therefore, the topographical relationship of the pituitary gland to the brain varies to some extent. The gland may be just below the tuber cinereum of the hypothalamus or somewhat more caudally, under the mamillary bodies.

III. TERMINOLOGY

Both parts of the pituitary, adeno- and neurohypophysis having different embryonic origin, are divided into three subdivisions (Table 1).

The subdivisions are illustrated in Fig. 3. The pars distalis is the major part of the adenohypophysis. It is placed rostral or rostroventral to the neurohypophysis, but it may also surround the neurohypophysis. The pars intermedia is a rather small subdivision between the pars distalis and the infundibular process, but it may also surround to various extent the whole infundibular process. The pars tuberalis is a small dorsal extension of the pars distalis along the infundibular stem, covering the stem like a cuff. The median eminence is a slight centerline prominence on the ventral surface of the hypothalamus. It is the proximal part of the infundibulum, a hollow conical inferior process from the tuber cinereum, and continues into the infundibular stem. On the inside of the median eminence is the infundibular

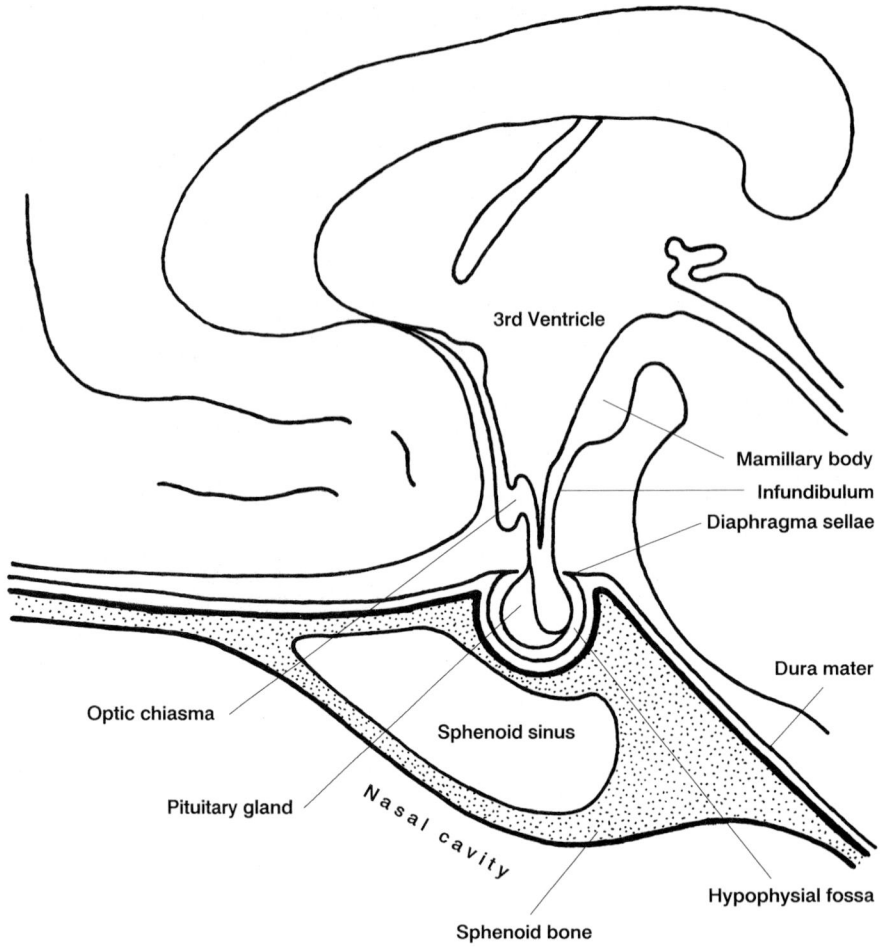

FIGURE 2 Schematic drawing of the pituitary gland and adjacent brain structures in midsagittal section illustrating the location of the gland.

TABLE 1
Adenohypophysis and Neurohypophysis Subdivisions

Adenohypophysis	Pars distalis	Anterior lobe or anterior pituitary
	Pars tuberalis	
Neurohypophysis	Pars intermedia (intermediate lobe)	Posterior lobe or posterior pituitary
	Infundibular process (neural lobe or pars nervosa)	
	Infundibular stem	
	Median eminence	

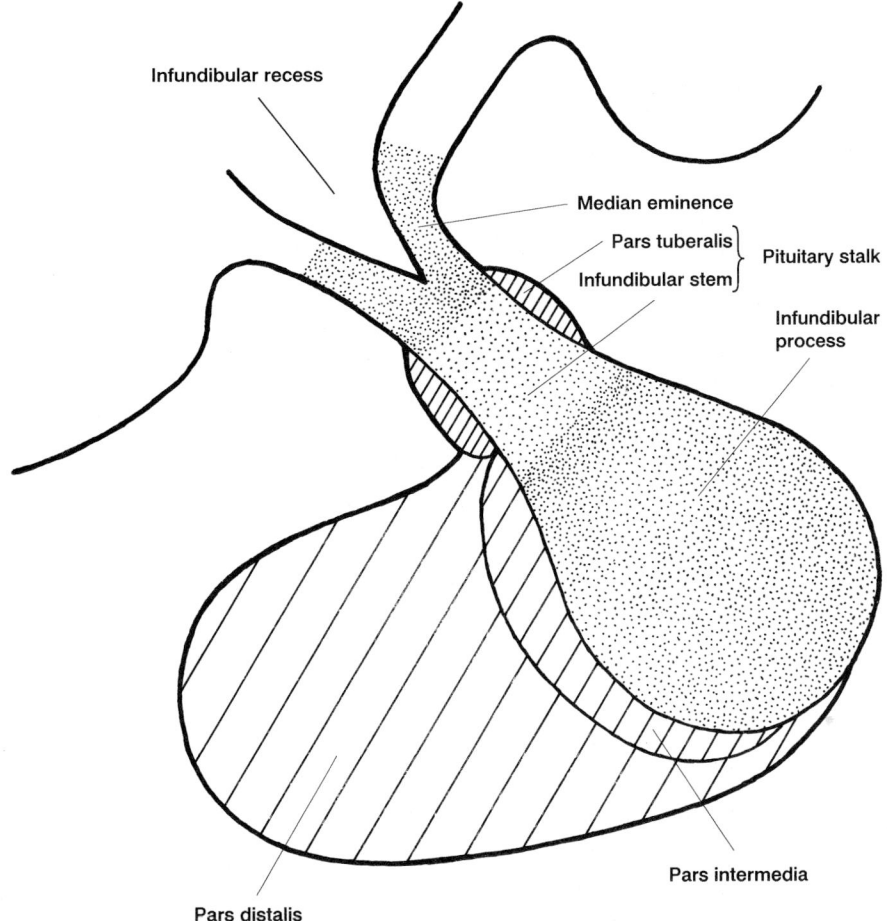

FIGURE 3 Subdivisions of the adenohypophysis (striped area) and neurohypophysis (shaded area) in sagittal section of the pituitary.

recess of the third ventricle. This recess may extend into the infundibular stem or even into the infundibular process. The infundibular stem connects the median eminence and through the median eminence the hypothalamus with the infundibular process, which lies behind or more or less above the pars distalis.

The pars tuberalis of the adenohypophysis and the infundibular stem of the neurohypophysis are called the pituitary stalk (hypophysial stalk). The infundibular process and the pars intermedia are often collectively termed the posterior lobe. However, the infundibular process per se without the pars intermedia is also sometimes considered as a posterior lobe or posterior pituitary. There is no general agreement whether the pars tuberalis of the adenohypophysis belongs to the anterior lobe. Some people include it, whereas others use anterior lobe (anterior pituitary) as just another term of the pars distalis.

IV. DEVELOPMENT

The neurohypophysis and the adenohypophysis have different embryonic origins (Fig. 4). The neurohypophysis is an outgrowth of the floor of the third ventricle. This outgrowth gives rise to the median eminence, the infundibular stem, and the infundibular process. Initially, the wall of the outgrowth is thin, like the floor plate of the diencephalon. During

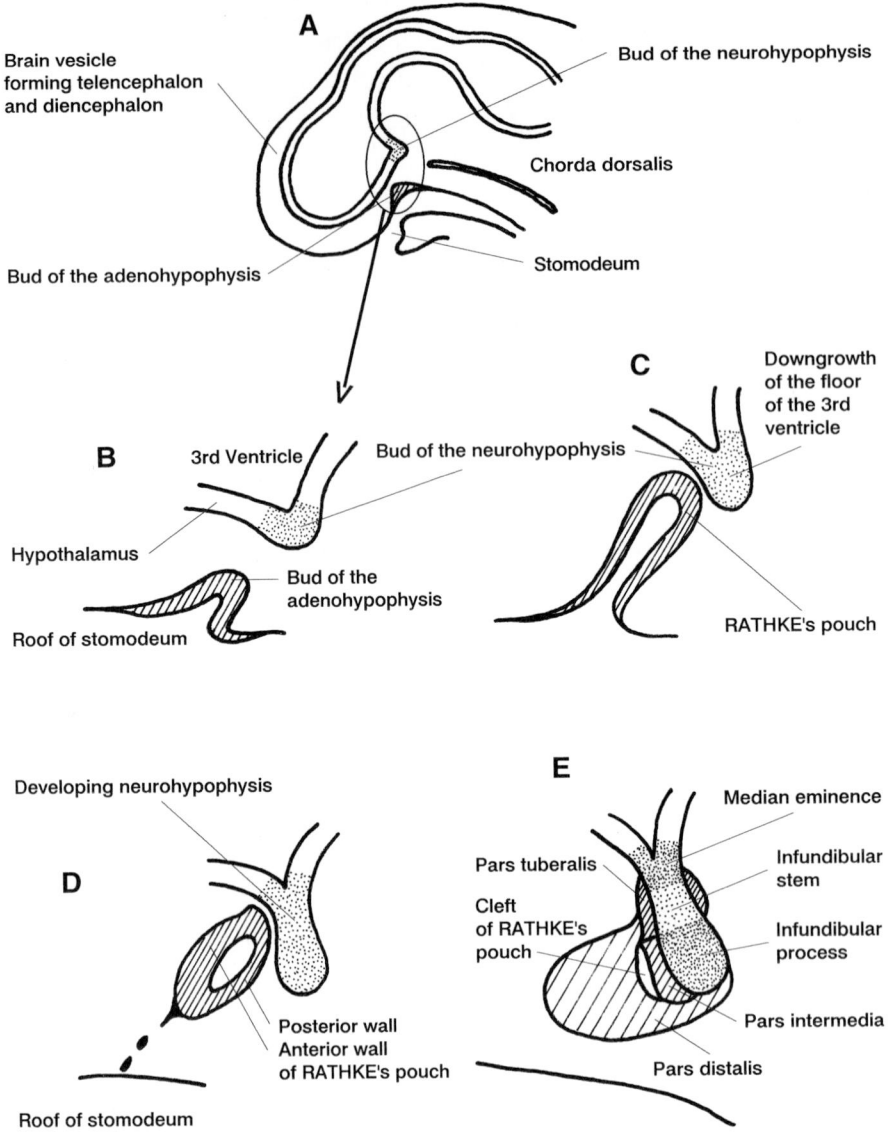

FIGURE 4 Illustrations of the development of the pituitary showing the origin of the bud of the adeno- and neurohypophysis (A) and the early (B, C) and later stages of the development of the two major parts and their subdivisions (D, E).

subsequent development the distal end of the outgrowth becomes solid as the neuroepithelial cells proliferate. These cells later differentiate into pituicytes resembling neuroglial cells. Nerve fibers arising from the neurons of the supraoptic and paraventricular nucleus of the hypothalamus grow into the infundibular process, to which the infundibular stem is attached.

The adenohypophysis arises from an evagination of the ectodermal epithelium covering the vault of the stomodeum. At first, there is a thickening in the roof of the mouth opening just external to the oropharyngeal membrane. This thickening forms a pit known as Rathke's pouch. This dorsal outpocketing from the roof of the stomodeum grows toward the brain and gradually approaches the diencephalic outgrowth, the ventral process from the diencephalic floor. Rathke's pouch elongates and becomes constricted at its attachment to the oral epithelium. The stalk of Rathke's pouch passes between the chondrification centers of the developing bones of the skull. The connection of Rathke's pouch with the oral cavity disappears. A remnant of this stalk may persist and give rise to a pharyngeal hypophysis in the pharyngeal roof. Very rarely, accessory masses of anterior lobe tissue may occur outside the capsule of the gland. Cells of the anterior wall of Rathke's pouch proliferate actively during subsequent development and give rise to the pars distalis of the pituitary. Later, a small extension grows around the infundibular stem, forming the pars tuberalis of the adenohypophysis. The extensive proliferation of the anterior wall of Rathke's pouch reduces the lumen; it becomes a narrow residual cleft, which later disappears and usually is not recognizable in the adult gland. It may be represented by a zone of cysts. Cells of the posterior wall of Rathke's pouch proliferate differentially and give rise to the pars intermedia.

V. BLOOD SUPPLY

The arteries of the pituitary arise from the internal carotid arteries, from the anterior and posterior cerebral arteries, and from their communications. Usually there are a single inferior and several superior hypophysial arteries on each side. In addition, there is also a third source of arterial supply, of which the vessel (variously named trabecular artery, peduncular artery, middle hypophysial artery, etc.) arises from the internal carotid artery between the origins of the inferior and superior hypophysial arteries.

All three kinds of hypophysial arteries are involved in supplying the neurohypophysis (Fig. 5). The infundibular process is supplied by the inferior hypophysial arteries arising from the intracavernous part

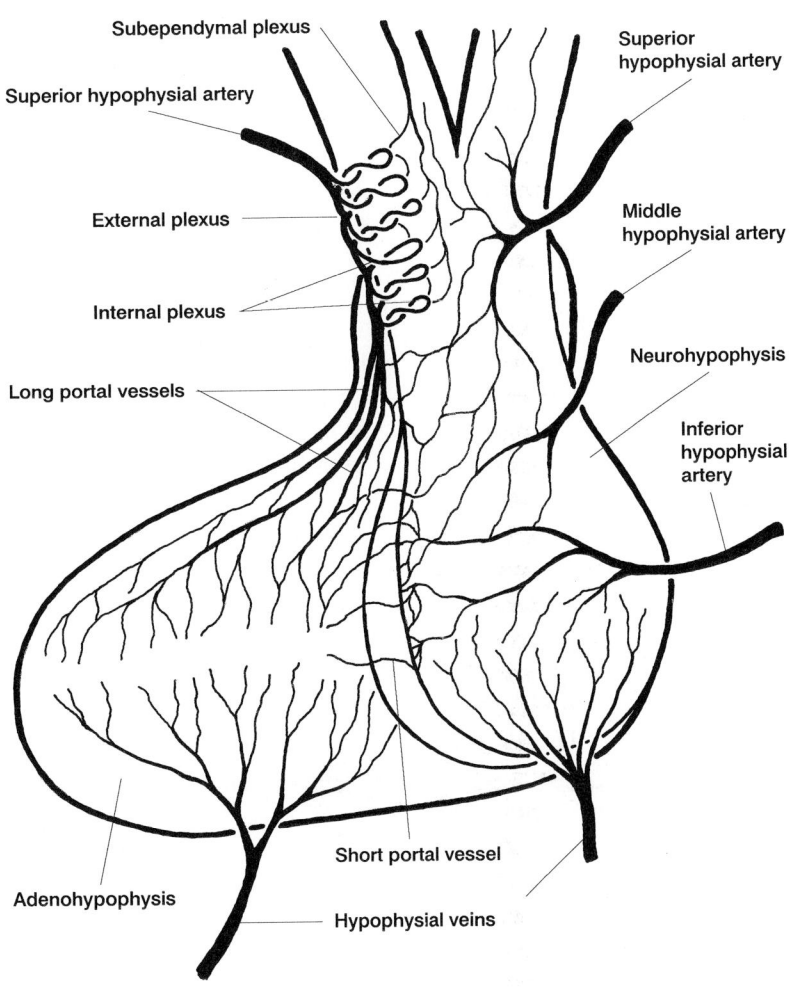

FIGURE 5 Schematic drawing illustrating the blood supply of the adeno- and neurohypophysis. For details, see text.

of the internal carotid arteries. These inferior arteries frequently unite prior to supplying the infundibular process. The middle hypophysial arteries supply the infundibular stem. They also frequently unite to form a single artery before entering the infundibular stem. The median eminence part of the neurohypophysis is supplied by superior hypophysial arteries arising from the intracranial carotid arteries and from the circle of Willis. These latter arteries form a ring about the median eminence before entering it. The capillaries of the neurohypophysis supplied by the three sets of hypophysial vessels form a confluent capillary net extending through the three subdivisions of the neurohypophysis. Reversal of flow can occur in capillary beds lying between the three supplies and has been suggested in the neurohypophysial network.

Inferior hypophysial veins drain the blood from the infundibular process to the cavernous sinuses and posterior intercavernous sinus. There are also vessels from the infundibular process to the adjacent pars distalis (capillaries and so-called short portal vessels). There are no direct venous drainage routes from the infundibular stem, except for the infundibular stem with the pars distalis. Venous drainage routes from the median eminence to the systemic circulation are not known. The blood is drained toward the adenohypophysis (see below).

The adenohypophysis has no direct arterial supply. Blood for the adenohypophysis first passes through the vessels of the median eminence and infundibular stem. Veins (called portal vessels) and capillaries bring blood to the pars distalis and pars tuberalis.

The arteries of the median eminence and infundibular stem form a characteristic capillary network on the surface of these structures (external plexus) and another plexus within the median eminence and infundibular stem (internal or deep plexus) (Fig. 5). The external plexus, fed by the superior hypophysial arteries, is drained by so-called long portal vessels descending to the anterior pituitary. The internal plexus, which invades the median eminence carrying with it the perivascular structures, is supplied by the external plexus, forms capillary loops, and is continuous with the infundibular capillary bed. The capillary loops of the internal plexus are made up of an ascending limb, an apex, and a descending limb. The ascending limb arises from the external plexus and passes into the median eminence to various depths. It may pass as deep as the subependymal layer, where an extensive arborization may be at the apex of the loops. The descending limb passes back to the surface of the median eminence, where it joins either the external plexus or a portal capillary. A large number of nerve terminals are evident in the surface zone of the median eminence and infundibular stem close to the capillaries of the external and internal plexus. This close anatomical relationship between nerve terminals and the vessels supplying the pars tuberalis and pars distalis represents the structural basis of the neurovascular connections existing between the central nervous system and the anterior pituitary, through which the central nervous system acts on the anterior pituitary. Observations of blood flow in the long portal vessels of living animals clearly indicate that blood flows downwards, i.e., from the median eminence to the pars distalis.

In some species there is also a subependymal system of capillaries in the median eminence which forms a third capillary network in the median eminence of mammals. The capillaries unite at the tip of long capillary loops and pass to the hypothalamic arcuate nucleus. In other species, there are just connections between the internal plexus of the median eminence and the subependymal capillaries at about the third ventricle in the region of the hypothalamic arcuate nucleus. It appears that the capillary beds of the arcuate nucleus and the neurohypophysis should be considered as a single unit, which means that the hypophysial vascular system is not isolated from the hypothalamus. In accordance with this are observations indicating that blood from the external plexus may pass into capillary loops of the internal plexus, then into subependymal vessels, and then to the hypothalamus. Furthermore, other findings suggest that there are deep portal vessels in which blood from the pars distalis may flow upwards to the median eminence, providing access for anterior pituitary hormones to directly reach the hypothalamus. Short portal vessels arising mainly from the branches of the inferior hypophysial arteries within the infundibular process (infundibular plexus) run from the infundibular process to the anterior pituitary, passing through the pars intermedia. Both types of portal vessels open into vascular sinusoids (dilated capillaries) lying between the secretory cords in the adenohypophysis, providing all its blood.

The venous drainage of the blood from the pars distalis is toward the paired cavernous and posterior intercavernous sinuses. The veins which drain the pars distalis may unite with the veins that drain the infundibular process and drain through a common stem to the cavernous sinus.

See Also the Following Articles

ANTERIOR PITUITARY; HYPOTHALAMIC–HYPOPHYSIAL COMPLEX; MEDIAN EMINENCE

Bibliography

Barrington, E. J. W. (1975). *An Introduction to General and Comparative Endocrinology*, 2nd ed. Clarendon, Oxford, UK.

Everett, J. V. (1994). Pituitary and hypothalamus: Perspectives and overview. In *The Physiology of Reproduction* (E. Knobil and J. D. Neill, Eds.), 2nd ed., Vol. 1, pp. 1509–1526 Raven Press, New York..

Harris, G. W., and Donovan, B. T (1966). *The Pituitary Gland*, Vols. I, II, and III. Butterworth, London.

Imura, H. (Ed.) (1994). *The Pituitary Gland. Comprehensive Endocrinology* (L. Martini, Series Ed.),. 2nd ed. Raven Press, New York.

Page, R. B. (1994). The anatomy of the hypothalamo–hypophysial complex. In *The Physiology of Reproduction* (E. Knobil and J. D. Neill, Eds.), 2nd ed., Vol. 1, pp. 1527–1619 Raven Press, New York.

Williams, P. L., Bannister, L. H., Berry, M. M., Collins, P., Dyson, M., Dussek, J. E., and Ferguson, M. W. J. (Eds.) (1995). Pituitary gland. In *Gray's Anatomy. The Anatomical Basis of Medicine and Surgery*, 38th ed., pp. 1883–1888. Churchill Livingstone, London.

Placenta and Placental Analogs in Elasmobranchs

William C. Hamlett
Indiana University School of Medicine

I. Introduction
II. Viviparity in Elasmobranchs
III. Conclusions

GLOSSARY

appendiculae Vascularized ectodermal and mesodermal villous extensions of the umbilical cord in some placental sharks that absorb uterine secretions.

histotroph Often called uterine milk. Nutritive substances derived directly from maternal tissues other than blood, including glandular secretions and cell fragments.

placental analog Counterpart to placenta, includes uterine villi in some sharks and trophonemata in stingrays.

placental viviparity Bearing of young nourished within the maternal uterus by transfer of nutrients via an epitheliochorial placenta.

trophonemata Vascularized, villous extensions of the uterine mucosa in stingrays used for nutrient delivery and respiration.

yolk sac placenta Ectoderm, mesoderm, and endoderm of the yolk sac persist after yolk contents have been absorbed as the fetal portion of the placenta. The contribution of the old yolk sac plus modified regions of the maternal uterus thus form the functional placental unit. The tertiary egg envelope is incorporated between maternal and fetal tissues in most species.

Elasmobranch fishes possess a remarkable array of reproductive specializations to provision their developing young with nutrients, eliminate wastes, and serve respiratory demands. Reproduction is either oviparous (egg laying) or viviparous (live bearing). Fertilization is internal and the initial stages of embryogenesis occur within the female genital system, regardless of the mode of reproduction. Viviparous development is either aplacental (no morphologically definitive vascular exchange organ formed) or placental (a maternal–fetal exchange organ is formed).

I. INTRODUCTION

All elasmobranch embryos undergo an initial stage of development during which they are reliant on yolk reserves sequestered in the fetal yolk sac. Yolk precursors are synthesized in the maternal liver and transported via the systemic circulation to follicle cells of the ovary and thence to the egg. Following internal fertilization, embryogenesis ensues. The embryo is attached to the yolk sac by a yolk stalk containing a yolk sac artery, vein, and ciliated ductus vitellointestinalis, which conveys yolk to the embryonic alimentary canal for digestion. Yolk is also digested in the yolk syncytial–endoderm complex of the yolk sac. As yolk reserves are depleted, maternal and fetal tissues are structurally and functionally modified to perform nutrient delivery functions.

II. VIVIPARITY IN ELASMOBRANCHS

Viviparity in elasmobranchs may be categorized as (i) aplacental yolk sac variety, (ii) aplacental oophagous with or without intrauterine cannibalism, (iii) aplacental with uterine villi or trophonemata, and (iv) placental viviparity with or without appendiculae.

Various forms of aplacental viviparity operate in sharks and rays. Most species ovulate relatively large eggs. Nearly all viviparous species encapsulate eggs along with egg jelly in tertiary egg envelopes produced by the oviducal gland, also referred to as the shell or nidamental gland. The initial phase of embryogenesis occurs inside these egg investments. What has traditionally been termed ovoviviparity, now called aplacental yolk sac viviparity, is utilized by approximately one-quarter of contemporary shark species. In this mode, embryos rely primarily on the substantial yolk in the ovulated egg for the organic material required by development. Embryonic development in these species is similar to that in oviparous species. Yolk is mobilized from the yolk sac and transported via the ductus vitellointestinalis of the yolk stalk to the embryonic intestine, and it is digested in the yolk syncytial–endoderm complex. In aplacental yolk sac species such as *Squalus acanthias*, the embryos break out of the so-called "candle cases" early in gestation to complete development free in the uterus, which provides no supplemental nutrients.

Another form of aplacental viviparity, oophagy, occurs in a few species of the lamnoid sharks. Early development in these sharks is essentially identical to that in aplacental yolk sac species. Embryos are initially reliant on a relatively small quantity of yolk in the yolk sac. Thereafter, development relies on the continued supply of yolk in the form of ovulated eggs. The embryos actively feed on these eggs and store the yolk for processing in the cardiac portion of the stomach. The sand tiger, *Carcharias taurus*, also practices intrauterine cannibalism in addition to oophagy. An egg capsule initially encloses each embryo. The more gestationally advanced embryos exhibit precocious development of dentition and escape from their egg capsules to feed on uterine egg capsules containing eggs and embryos and free embryos. The uterus provides no supplemental nutrients.

Stingrays, such as *Rhinoptera bonasus* and *Dasyatis americana*, display aplacental development with uterine trophonemata (Fig. 1). Embryos are initially dependent on yolk reserves. As yolk stores are depleted, the maternal uterus develops vascularized villi termed trophonemata that secrete a nutrient histotroph or "uterine milk" to further nourish the developing young and provide for respiratory demands.

In placental species, such as *Rhizoprionodon terraenovae*, the tertiary egg envelope surrounds the embryos throughout gestation and is incorporated into the placenta. Initially, yolk stores fuel early development. When the yolk sac has been depleted, it differentiates into a placenta rather than being resorbed into the body wall of the embryo. The uterus provides maternally derived nutrients which are delivered to the yolk sac portion of the placental complex by transport through the intervening egg envelope. The placental complex mediates all metabolic transport,

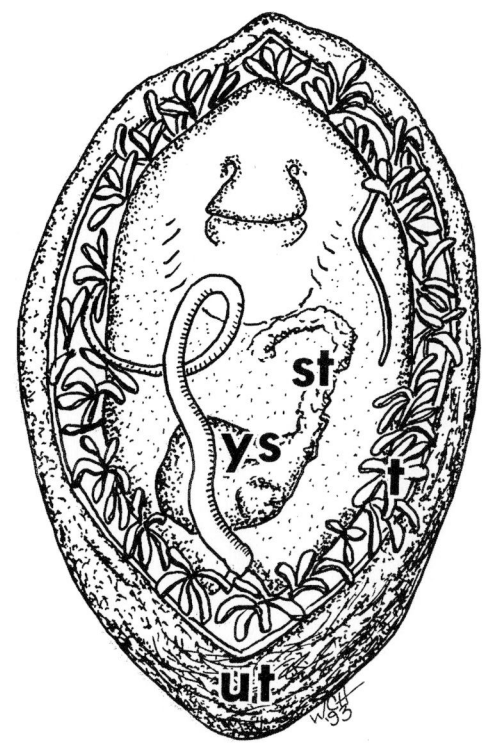

FIGURE 1 A Southern stingray, with yolk stalk (st) and yolk sac (ys) still attached, rests in the uterus (ut) adorned with trophonemata (from Hamlett *et al.*, 1993c).

excretory demands, and respiratory exchange between the mother and embryo.

A. External Gills

External gill or branchial filaments have historically been considered as respiratory structures in early embryos of sharks, skates, and rays. Early workers suspected these transitory structures to also function as sites of nutrient absorption, particularly in viviparous forms. During the initial yolk-reliant phase, external gill filaments are present and have been demonstrated to be capable of endocytosing macromolecular protein tracer, horseradish peroxidase, after 10 min of *in vitro* incubation. As yolk stores are depleted, the yolk sac develops into an epitheliochorial yolk sac placenta. External gills may thus serve as a nutrient absorptive membrane before the establishment of the yolk sac placenta as well as perform its respiratory function. Similar absorptive functions of the gills may occur in aplacental development of stingrays prior to the secretion of uterine histotroph.

B. Aplacental Viviparity with Uterine Villi or Trophonemata

In stingrays, aplacental viviparity involves production and secretion of an organically rich uterine-derived histotroph which is then ingested by the embryo and/or absorbed by external gill filaments. The uterine epithelium forms elongate villi, termed trophonemata, that significantly increase the surface area for histotroph secretion and respiratory exchange. Gestation in stingrays is shorter than that in most sharks. Ray embryos develop from 2 to 4 months, whereas viviparous sharks produce young in 10–12 months. Rays also have smaller litters and produce smaller eggs, the yolk content of which is insufficient for growth to term.

Trophonemata are closely apposed to the embryo which may form the basis of earlier reports that trophonemata enter the spiracles where they secrete directly into the embryonic gut. These observations have not been confirmed by contemporary observations.

Early studies reported 13.3% organic substance in the uterine milk of *Dasyatis violacea* and 1.2% for *Torpedo ocellata*. In the Southern stingray, *D. americana*, there is a 3750% increase in wet weight from the egg to the term fetus. Trophonemata are 1.5 cm long, more narrow at the base and spatulate at the tip. Surface epithelial cells form a pattern of surface cables, each with a small blood vessel at its core. In females containing fertilized eggs, the epithelium is simple and cuboidal (Fig. 2). In contrast, in uteri containing late-term fetuses, the epithelium is squamous, which reduces the diffusion distance to augment respiratory exchange. Epithelial cells, with periodic acid Schiff-positive cytoplasmic vesicles, form invaginated crypts. Epithelial cells produce proteinaceous, mucous, and lipid secretions which constitute the histotroph. Thus, the term uterolactation has been coined to describe this phenomenon.

C. Placental Viviparity

Placental viviparity occurs only in sharks and occurs in about 10% of extant species. The initial phase

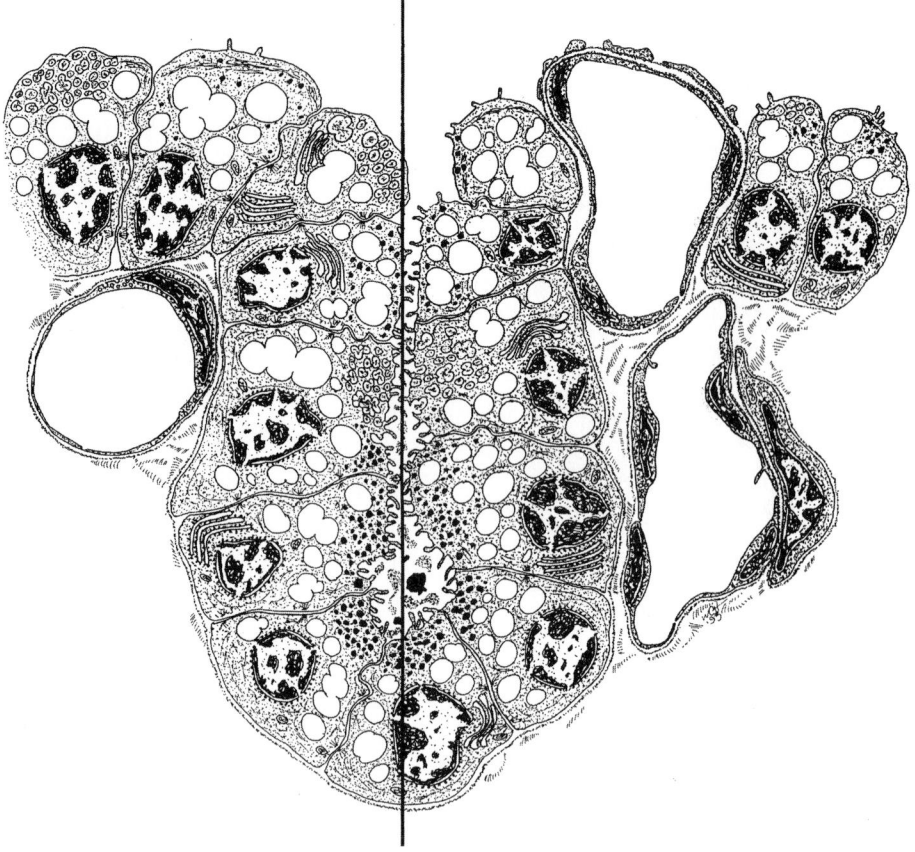

FIGURE 2 Composite line drawing illustrating a Southern stingray uterus containing fertilized eggs, characterized by cuboidal surface epithelium and capillaries. The right side shows a uterus containing term fetuses which is characterized by squamous epithelium overlying dilated sinusoids (from Hamlett et al., 1996a).

of development is similar to the aplacental strategy in which eggs are encapsulated in some form of egg covering and embryos develop from yolk contained in the ovulated egg. During the middle of gestation, uterine secretions augment the declining yolk stores and only during the latter phases of pregnancy does the placenta become functional. The egg covering is transient in *Prionace glauca* and *Scoliodon laticaudus*. In all other species examined, the egg envelope persists and is incorporated into the placenta. In all cases the placenta is of the yolk sac or epitheliochorial variety. The placenta is noninvasive and nondeciduate. Hemotrophic transport is the major route of nutrient transfer from mother to fetus. The placental unit consists of (i) an umbilical stalk with or without appendiculae; (ii) the smooth, proximal portion of the placenta; (iii) the distal, rugose portion; (iv) the egg envelope; and (v) the maternal uterine tissues. Exchange of metabolites is effected through the intervening egg envelope (Fig. 3). In some species, such as *Rhizoprionodon terraenovae*, the umbilical cord develops villous extensions, appendiculae, that may serve as a paraplacental nutrient absorptive site.

1. Maternal Attachment Site

During gestation, the uterus develops specialized sites for exchange of metabolites between the mother and embryo. Uterine attachment sites are highly vascular, rugose elevations of the maternal uterine lining that interdigitate with the fetal placenta. The mater-

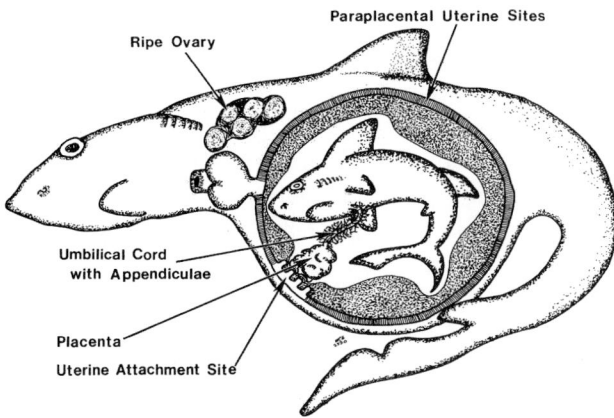

FIGURE 3 At term, the yolk stalk of *Rhizoprionodon terraenovae* is transformed into an umbilical cord with appendiculae. Specialized sites of the uterus and fetal yolk sac mediate metabolic exchange (from Hamlett, 1993).

nal epithelium remains intact and there is no erosion. The epithelium is bilayered and overlays a profuse vascular bed. The epithelium actively produces secretion vesicles that are periodic acid Schiff and alcian blue positive.

2. Paraplacental Uterine Sites

Ultrastructural characteristics of the paraplacental portion of the uterus at term indicate that it functions as a lubricating structure by elaborating mucous from prominent columnar cells. Wandering phagocytic cells also populate this region. Light microscopic studies of paraplacental regions of uteri containing midterm fetuses suggest that secretory products other than mucous may be produced during midgestation, namely, nutrient histotroph or uterine milk.

3. Egg Envelope

In viviparous shark species with a placenta, the tertiary egg envelope is greatly reduced in thickness, reflecting altered function of the oviducal gland. In all placental species examined, the egg envelope is retained throughout gestation, with the exception of the blue shark, *P. glauca*, and the Telok Anson shark, *Scoliodon laticaudus*, in which the egg envelope is transient. In most placental sharks, all metabolic exchange between the uterus and fetus must, therefore, be effected through or across the egg envelope. A gap-like separation exists between the uterus and the egg envelope, whereas the placental cells form an intimate apposition with the envelope. The envelope consists of layers which are compact on the placental side but it has delaminations on the uterine side.

4. Fetal Yolk Sac Portion

The fetal portion of the yolk sac placenta is composed of two portions: a proximal, saccular region and a heavily vascularized, rugose, distal portion (Fig. 4). The proximal portion has ultrastructural characteristics of a steroid hormone-producing tissue, including massive smooth endoplasmic reticulum frequently forming whorled arrays. Although the human placenta produces steroid hormones, it lacks the complete repertoire of enzymes necessary for synthesis of all the hormones it produces. An exchange between the fetus and placenta by the so-called fetoplacental unit is responsible for the final hormone production. It is feasible that similar mechanisms might be operating in the shark placenta. Whether the shark placenta functions as an endocrine organ awaits investigation; however, prelimi-

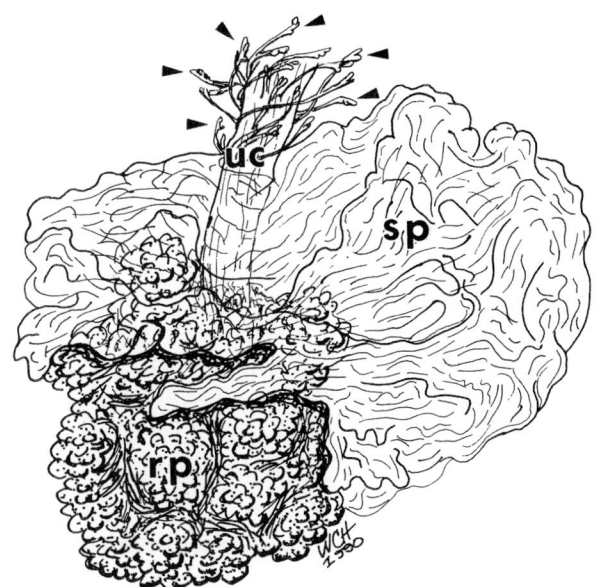

FIGURE 4 Appendiculae (arrowheads) emerge from the term umbilical cord (uc) of *Rhizoprionodon terraenovae*. The proximal portion is smooth (sp) and billowing. The distal portion is rough in texture (rp) and mulberry-like (from Hamlett, 1993).

nary studies employing radioimmunoassay procedures have identified measurable levels of some steroids in the placenta of the blacknose shark, *C. acronotus.*

The distal portion of the fetal yolk sac is composed of a squamous epithelial bilayer that is separated from the underlying vascular network by a continuous basal lamina. The endothelium of the vessels is fenestrated. Cytoplasmic characteristics of these cells include an extensive Golgi complex, smooth-walled caveolae, vesicles with electron-dense contents which are presumably endocytotic in nature, and dense bodies that are suggested to be lysosomes involved in the digestion of material, perhaps yolk metabolites. *In vitro* exposure of full-term placentae to solutions of trypan blue and horseradish peroxidase (HRP) reveals little uptake by the smooth portion of the placenta but rapid absorption by the surface epithelial cells of the distal, rugose portion. HRP enters these cells by an extensive apical system of smooth-walled membranous anastomosing canaliculi and tubules. Prominent whorl-like inclusions that occupy the basal cytoplasm of the surface cells, adjacent to the pinocytotically active endothelium of the vitelline capillaries, are proposed to be yolk proteins that are transferred from the mother to the embryo throughout gestation (Fig. 5).

Crystalline inclusions in the ovary, uterus, preimplantation embryo, and fetal membranes of mammals have been considered as storage forms of material that is to be used by the developing embryo of other vertebrates. There is a striking similarity in the ultrastructural appearance of nonmembrane-bound inclusions in the distal portion of the term placenta of the sandbar shark, *C. plumbeus,* and the teleost and amphibian yolk protein precursors. The inclusions may conceivably represent residual bodies derived from partial lysosomal digestion of yolk. Residual bodies are frequently characterized by membrane fragments, myelin figures, or whorled inclusions. Based on ultrastructural analyses of the shark yolk sac and placenta, it has been suggested that yolk protein precursors of hepatic origin continue to be produced during pregnancy in placental sharks. Initial yolk stores in the egg mass are depleted and used for embryogenesis and subsequent growth. As these stores are diminishing, the maternal liver produces yolk protein precursors that are secreted into the blood, where they are taken up by the new crop of developing oocytes for the next reproductive cycle. The same yolk precursors could then be transported across the uterus and fetal placenta to nourish the fetuses during the latter stages of gestation. In addition, the uterus provides material via merocrine secretion.

5. Appendiculae

In most carcharhinid species the umbilical cord is a smooth, glistening structure that contains the umbilical artery, umbilical vein, extra embryonic coelom, and the ductus vitellointestinalis, all of which persist from the yolk stalk (Fig. 6). The term appendiculae refers to villiform extensions of the umbilical cord in some placental sharks, including members of the genera *Rhizoprionodon, Zygaena, Scoliodon, Hemigaleus, Sphyrna,* and *Paragaleus.*

Considerable morphological diversity in appendiculae exists between and within genera. Four types of appendiculae and three types of placentae in sharks from the Indian Ocean have been described. The hypothesis is that species with a less efficient placenta develop appendiculae to augment the metabolic activities of the placenta.

In early stages of development in the Atlantic sharpnose shark, the yolk stalk is a smooth, cylindrical structure. By the time the embryo is 25 mm in total length (TL), modest longitudinal folds form on the surface of the yolk stalk alternating with punctate depressions. These are first indications of appendiculae. By the time the embryo is 4 cm TL, appendiculae are present as rounded, longitudinal protuberances. By the time the embryo attains a length of 6 cm TL, yolk stalk is adorned with a bushy covering of cylindrical and modestly branched appendiculae. Scanning electron microscopy reveals that the surface epithelium to be columnar and a profusion of small blood vessels populate the connective tissue core in close proximity to the surface cells. Closer examination reveals two distinct cell types—one with elongate microvilli and one with low-relief microvilli. Adjacent cells are joined by continuous apical tight junctions. The yolk stalk of 10-cm TL embryos is covered with a dense mat of branched appendiculae. The bases of the appendiculae are

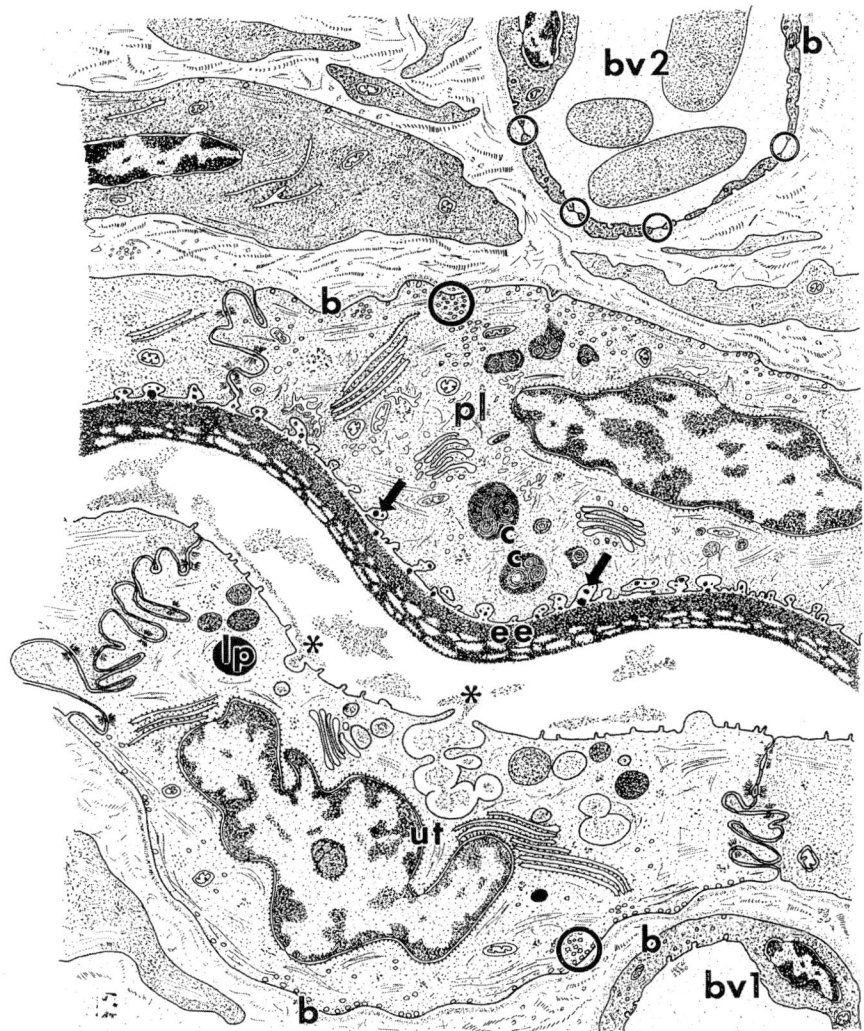

FIGURE 5 At term the placenta of *Rhizoprionodon terraenovae* is characterized by a squamous uterine epithelium (ut) that has basal pinocytotic vesicles (large circles), lipid inclusions (lp), and secretory vesicles that release their contents to the uterine lumen (asterisks). Subjacent capillaries (bv1) are continuous. The egg envelope (ee) separates the fetal and maternal portions of the placenta. Electron-dense vesicles (arrows), of uterine origin, are endocytosed by the distal portion of the fetal placenta (pl). The epithelium has basal pinocytotic vesicles (large circles) and crystalline inclusions (c) that are interpreted as yolk precursors. Subjacent capillaries (bv2) are fenestrated (small circles). b, basal lamina (from Hamlett, 1993).

rounded but flatten distally as they branch. The yolk stalk of a 18-cm TL embryo shows appendiculae basically characteristic of term fetus.

In appendiculae of the Atlantic sharpnose shark, microvillar cells form pyriform processes insinuated between adjacent cells (Fig. 7). Their cytoplasm contains lipid and apical micropinocytotic vesicles. Uptake experiments suggest that these cells are capable of limited endocytosis *in vitro*. Dilated intercellular spaces frequently occur between adjacent cells. These may represent functional transport channels or may be produced as a stress response associated with capture of animals. Granulated cells undergo synthetic and secretory cycles. Cells seemingly devoid of most cytoplasmic organelles have, in reality, just undergone exocytosis. A large aggregation of mitochondria occupies the base of the cells, having been displaced by secretion granules. The rough endoplas-

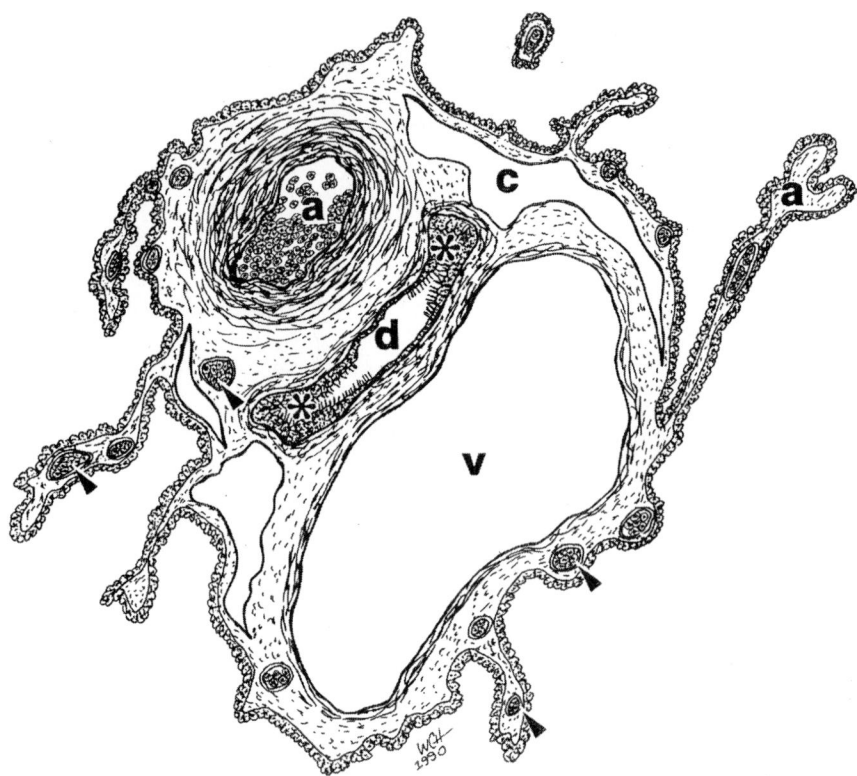

FIGURE 6 Cross section of a mature umbilical cord from *Rhizoprionodon terraenovae* reveals a muscular umbilical artery (a), umbilical vein (v), ductus vitellointestinalis (d) with rows of cilia (asterisks) and extraembryonic coelom (c). Small blood vessels (arrowheads) occupy the connective tissue stroma of the appendiculae (ap) as well as the body of the cord (from Hamlett, 1993).

mic reticulum is active and the supranuclear Golgi complex is prominent. As the synthesis of material proceeds, the apical cytoplasm becomes engorged with secretion granules. Following exocytosis, the synthesis-secretion cycle repeats. The chemical nature of the secretion vesicles is unknown and the function of the secretory cells is also unresolved. What has emerged is a concept of what functions appendiculae might perform. The microvillar cells are able to absorb exogenous material the size of protein; thus, they may serve as a paraplacental nutrient-absorptive fetal membrane. They could absorb material in the periembryonic fluid that is produced by the maternal uterus. Granulated cells may be the source of material that is absorbed by the microvillar cells or their secretion may serve other, as yet unelucidated, functions. Their elaborations may be a lubricant, an antibacterial factor, an immunological component, or any of several other possibilities.

III. CONCLUSIONS

All elasmobranchs provision their developing young with yolk stores. Oviparous embryos rely exclusively on yolk, as do aplacental yolk sac species. In addition to reliance on yolk stores, aplacental oophagous species also ingest uterine eggs. Stingrays utilize placental analogs in the form of trophonemata to produce uterine histotroph to augment yolk stores. These structures also function as respiratory exchange structures. In placental species, following a yolk-dependent stage and as yolk stores are depleted, the uterus produces uterine secretions prior to the establishment of a functional placenta. Elasmobranchs thus display an impressive variety of successful reproductive modes involving yolk reliance and assorted devices to supplement yolk stores, including placental analogs such as trophonemata and definitive placentae.

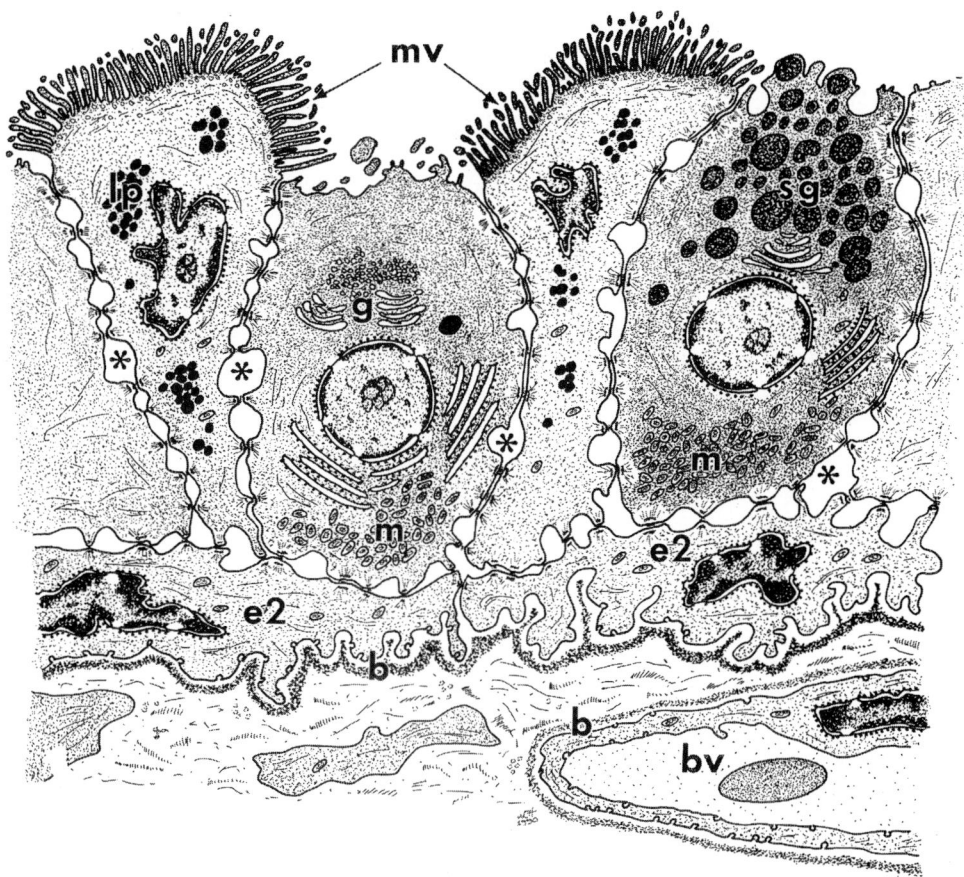

FIGURE 7 Microvillar cells (mv) cover the apex of some columnar appendicular cells of *Rhizoprionodon terraenovae*. These cells are also characterized by lipid inclusions (lp) and are joined by a luminal complex of zonular tight junctions, zonula adherens, and macula adherens. Intercellular spaces (asterisks), often dilated, occur between adjacent surface cells and the underlying epithelial layer (e2). b, basal lamina; m, mitochondria; g, Golgi; bv, blood vessel; sg, secretory granules (from Hamlett, 1993).

See Also the Following Article

PLACENTA AND PLACENTAL ANALOGS IN REPTILES AND AMPHIBIANS

Bibliography

Hamlett, W. C. (1987). Comparative morphology of the elasmobranch placental barrier. *Arch. Biol. (Bruxelles)* **98**, 135–162.

Hamlett, W. C. (1989). Evolution and morphogenesis of the placenta in sharks. *J. Exp. Zool. Suppl.* **2**, 35–52.

Hamlett, W. C. (1993). Ontogeny of the umbilical cord and placenta in the Atlantic sharpnose shark, *Rhizoprionodon terraenovae*. *Environ. Biol. Fishes* **38**, 253–267.

Hamlett, W. C., Wourms, J. P., and Hudson, J. S. (1985a). Ultrastructure of the full term shark yolk sac placenta. I. Morphology and cellular transport at the fetal attachment site. *J. Ultrastruct. Res.* **91**, 192–206.

Hamlett, W. C., Wourms, J. P., and Hudson, J. S. (1985b). Ultrastructure of the full term shark yolk sac placenta. II. The smooth, proximal segment. *J. Ultrastruct. Res.* **91**, 207–220.

Hamlett, W. C., Wourms, J. P., and Hudson, J. S. (1985c). Ultrastructure of the full term shark yolk sac placenta. III. The maternal attachment site. *J. Ultrastruct. Res.* **91**, 221–231.

Hamlett, W. C., Miglino, M. A., and DiDio, L. J. A. (1993a). Subcellular organization of the placenta in the Atlantic sharpnose shark, *Rhizoprionodon terraenovae*. *J. Submicrosc. Cytol. Pathol.* **25**, 535–545.

Placenta and Placental Analogs in Reptiles and Amphibians

Daniel G. Blackburn

Trinity College

I. Reproductive Modes and Embryonic Nutrition
II. Placentae in Viviparous Squamate Reptiles
III. Placentae and Placental Analogs in Amphibians
IV. Evolution of Placentae and Matrotrophy

GLOSSARY

chorioallantois The vascularized respiratory membrane of the amniote egg, which contributes to placental formation in all viviparous squamates and eutherian mammals.

extraembryonic membranes Specific tissues lying outside of the body of the embryo proper (such as the yolk sac, amnion, and chorioallantois) which can function in embryo protection, physiological exchange, and placental formation.

lecithotrophy A developmental pattern in which the yolk of the ovum provides nutrients to the embryo.

matrotrophy A developmental pattern in which the mother provides nutrients during gestation by a means other than the yolk of the ovum (e.g., oviductal secretions, sibling yolks, and placental tissues).

oviparity A reproductive mode in which females lay eggs, whether they be developing or unfertilized; also known as "egg-laying" reproduction.

placenta Any apposition or fusion of the fetal organs to the maternal (or paternal) tissues for physiological exchange.

placentotrophy A type of matrotrophy in which nutrients for development are supplied by placental organs.

viviparity A reproductive mode in which embryos develop inside the female reproductive tract and are born as viable offspring; also known as "live-bearing" reproduction.

yolk sac Extraembryonic tissues that surround and digest the yolk and that contribute to several distinct types of placentation in viviparous reptiles and mammals.

Placentae are organs formed from extraembryonic and parental tissues that help sustain embryos physiologically during their development. In viviparous squamate reptiles, in which the fetus develops to term inside the maternal oviduct, the placental membranes typically accomplish gas exchange and provide water and small quantities of nutrients to the embryo. In certain lizards, placental membranes supply virtually all the nutrients for embryonic development. Morphological attributes of squamate chorioallantoic placentae correlate with the degree of placentotrophy. Some viviparous species of amphibians exhibit specializations for maternal–fetal nutrient provision that are functionally analogous to placentae. Simple placenta-like structures that function in gas exchange are occasionally found among egg-laying amphibians.

I. REPRODUCTIVE MODES AND EMBRYONIC NUTRITION

In most species of animals, females reproduce by laying unfertilized or fertilized eggs, a reproductive pattern known as "oviparity." In contrast, in species with "viviparity" (e.g., most mammals, a few amphibians, and many lizards, snakes, and fishes), females retain their developing eggs to term in their reproductive tracts and eventually give birth to their offspring. Many viviparous species exhibit placentae or placental analogs by which embryos are sustained physiologically during their development in the female reproductive tract.

The structure, function, and evolution of placentae are matters of considerable empirical and theoretical interest. Although placentae supply oxygen and water to the developing embryo, they also can provide organic and inorganic nutrients. Maternal provision of nutrients in this fashion is known as "placentotrophy." This pattern contrasts with "lecithotrophy," in which nutrients are supplied by the yolk of the ovum. Placentotrophy is a special case of "matrotrophy," which also includes various analogous ways of providing extravitelline nutrients for embryonic development (Table 1).

Most reptiles and amphibians exhibit lecithotrophic oviparity. Viviparity occurs in about 20% of the squamates, most of which are relatively lecithotrophic; however, placentotrophy characterizes three or four lineages of viviparous lizards. Among amphibian species, viviparity is rare and placentotrophy (strictly speaking) is absent. However, some viviparous amphibians are matrotrophic, having evolved analogous means of providing nutrients to the developing embryos.

II. PLACENTAE IN VIVIPAROUS SQUAMATE REPTILES

A. Physiological Relationships in Oviparity and Viviparity

In oviparous squamates, as in other egg-laying amniotes, the eggs are fertilized inside the maternal reproductive tract, an eggshell is deposited by uterine glands, and the egg is laid on land. The oviposited eggs of lizards and snakes exchange gases with the surrounding air through the eggshell and absorb water from the substrate (Fig. 1). In fact, the eggs may

TABLE 1
Forms of Substantial Matrotrophy among Viviparous Reptiles and Amphibians

Type	Nutrient source	Animal
Placentotrophy	Placental membranes	A few viviparous lizards;[a] typhlonectid caecilians
Oophagy[b]	Sibling ova	*Salamandra atra* (Urodela)
Histotrophy, histophagy[c]	Oviductal secretions	*Nectophrynoides occidentalis* (Anura); viviparous caecilians

[a] Genera *Chalcides*, *Mabuya*, and *Pseudemoia* (family Scincidae); many (and perhaps all) viviparous squamates show incipient placentotrophy.
[b] Embryonic ingestion of sibling egg yolks.
[c] Absorption or ingestion of oviductal secretions and tissues.

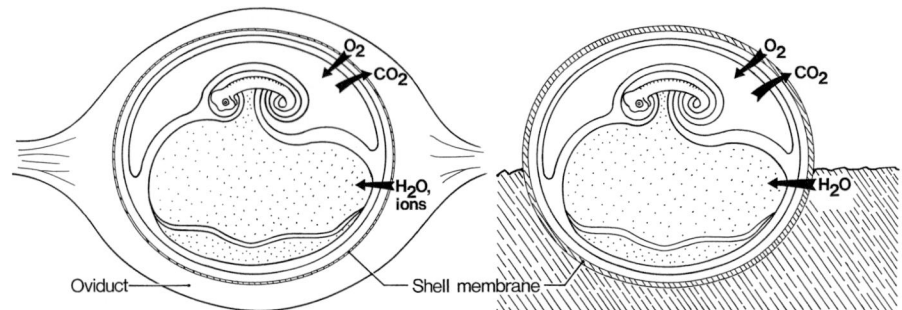

FIGURE 1 Physiological relationship between a squamate egg and its environment under conditions of viviparity (left) and oviparity (right). In the viviparous condition, exchange occurs with the uterine oviduct, and in the oviparous condition it occurs with the air and substrate. In each case, gas exchange occurs via the chorioallantois. The yolk sac is inferred to function in water uptake, although experimental evidence is not yet available (reproduced from Blackburn, 1993).

double or triple in wet mass after oviposition through water uptake while decreasing in dry mass (Table 2). Gas exchange is accomplished by the chorioallantois, a vascularized extraembryonic membrane that lines the inner surface of the eggshell at the dorsal (embryonic) hemisphere of the egg. Uptake of water from the substrate is probably accomplished by the yolk sac, although the mechanism is not well understood. Embryo nutrition is lecithotrophic; nutrients are supplied via the egg yolk. In many species, the eggshell supplies calcium to the embryo.

The physiological relationship between an egg and its environment in typical (lecithotrophic) viviparous squamates is similar to that in oviparous forms (Fig. 1). The developing egg increases in wet mass and decreases in dry mass approximately to the same extent as it does in oviparous squamates (Table 3). The main difference from the oviparous situation is that in the viviparous condition, oxygen and water are supplied (and carbon dioxide removed) by the uterine oviduct, where the egg resides during gestation. In addition, in viviparous squamates the eggshell is vestigial; in many species this remnant of the eggshell becomes progressively thinner during gestation. As a result, the maternal and extraembryonic tissues lie in close proximity, enhancing the potential for physiological exchange (Fig. 1). In viviparous, matrotrophic species of at least two lizard genera (*Chalcides* and *Mabuya*), the vestigial shell

TABLE 2
Developmental Changes in Mass of the Egg between Oviposition and Hatching in Various Oviparous Squamates

Species	Change in mass (%)	
	Dry mass	Wet mass
Anolis auratus	—	+82–204
Callisaurus draconoides	—	+53–100
Coluber constrictor	−14	+53
Crotaphytus collaris	—	+82
Eumeces fasciatus	−33	+14
Liolaemus tenuis	—	+125
Pogona barbata	−30	+66
Sceloporus undulatus	—	+106–170

TABLE 3
Developmental Changes in Mass of the Oviductal Conceptus in Selected Viviparous Squamates[a]

Species	Change in mass (%)	
	Dry mass	Wet mass
Elgaria coerulea	−18	+142
Eulamprus quoyii	−10	+96
Nerodia rhombifera	−22	+59
Notechis scutatus	−35	+96
Virginia striatula	−31	+48
Vipera berus	—	+51
Pseudemoia entrecasteauxii[b]	+68	+296
Chalcides chalcides[b]	—	+683
Mabuya heathi[b]	+38,400	+53,800

[a] In the relatively lecithotrophic species, dry mass of the conceptus decreases, as in oviparous forms.
[b] Species with substantial placentotrophy.

membrane ruptures and is shed early in development, with the result that extraembryonic and maternal tissues are in direct contact for most of gestation.

The close association of extraembryonic membranes with the uterine lining that exists in viviparous squamates meets Harland Mossman's (1937) criteria for recognition of a placenta: "Any intimate apposition or fusion of the fetal organs to the maternal (or paternal) tissues for physiological exchange." Given that biologists in past decades have applied the concept of placentation to reptiles in discrepant ways, three points should be noted. First, organs need not transfer organic nutrients to be recognized as placental; even oxygen provision qualifies as physiological exchange. Second, placentae are not necessarily vascularized. As mammalian studies have shown, some placentae transfer nutrients from mother to offspring through secretion and absorption rather than between bloodstreams. Third, the presence of a remnant of the eggshell between extraembryonic and maternal tissues does not disqualify the arrangement as placental. After all, acellular layers lying between fetal and maternal tissues are a common feature of mammalian placentae.

B. Types of Placentae

Squamate placentae are defined anatomically according to the particular extraembryonic membrane that contributes to them. For example, apposition of the chorioallantois to the uterine lining forms the chorioallantoic placenta (Fig. 2). Similarly, the avascular chorion (formed from ectoderm and mesoderm) can contribute to the chorionic placenta, a transitory structure of early development. Both placental types have equivalents among mammals. Squamates differ from other amniotes in the variety of extraembryonic tissues that develop from the yolk sac, a fact that is overlooked in embryology textbooks and all but a handful of reviews. Each of these membranes contributes to a distinct type of yolk sac placenta. Thus, viviparous squamates can exhibit a choriovitelline placenta, a bilaminar yolk sac placenta, an omphaloplacenta (formed from the avascular omphalopleure and isolated yolk mass), and an omphalallantoic placenta (the omphalopleure as vascularized internally by the allantois).

Of these six placental types, the chorioallantoic

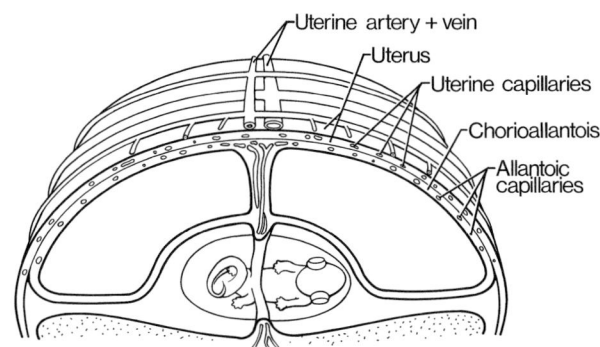

FIGURE 2 Apposition of the chorioallantois to the uterine lining, forming the squamate chorioallantoic placenta. The vestigial shell membrane is not illustrated (reproduced from Blackburn, 1993).

placenta is the only one that invariably persists until the end of gestation in all species. Either an omphaloplacenta or an omphalallantoic placenta also persists to the end of development. Although the remaining placental types have been described only occasionally, they may well be universal among viviparous squamates, having been overlooked due to their transitory nature. Our definition of the concept "placenta" has both a structural and a functional component. Unfortunately, because information on placental function is limited, we do not know whether all the placental types recognized on morphological grounds function in physiological exchange. We do have excellent reason to believe that the chorioallantoic placenta functions in gas exchange and nutrient transfer, as discussed later. We also know that yolk sac placentae can be sites of striking cellular specializations. Clarification of the particular roles of the various placental arrangements is one of the most important and difficult tasks facing placentologists who study reptiles.

C. Placentae in Lecithotrophic Squamates

1. Gas Exchange and the Chorioallantoic Placenta

Because the lumen of the oviduct is a relatively hypoxic environment, a major problem facing an egg retained in the uterus is that of gas exchange. Embryonic needs for oxygen progressively increase during development and accelerate in the postembryogenic growth phase. In the oviparous condition, gas exchange is accomplished by the chorioallantois. Because it is the only vascularized extraembryonic membrane that invariably persists until the end of gestation, this membrane probably accomplishes this same function in all viviparous squamates.

Examination of the chorioallantoic placenta reveals an organ with considerable potential for gas exchange. The entire dorsal hemisphere of the egg is devoted to chorioallantoic placentation (Fig. 2), and in some squamates this placenta continually expands at the expense of the yolk sac during later development. Both the uterus and the chorioallantois are vascularized via an extensive network of capillaries. Capillary density increases during development. In addition, the uterine and chorionic epithelia become progressively thinner as capillaries migrate toward the luminal surfaces. As a consequence, by late development a very thin layer of epithelium separates the uterine and allantoic capillaries, presenting a narrow barrier to gas diffusion (Fig. 3). Although previous studies inferred that the uterine epithelium erodes to expose the underlying capillaries, recent work has not confirmed such erosion. Nevertheless, the diffusion distance between allantoic and uterine capillaries can be <0.5 μm. In a peculiar (and largely unappreciated) feature of squamate development, the allantois vascularizes the inner face of the omphalopleure in many species, forming an omphalallantoic membrane. The extent to which the omphalallantoic placenta contributes to gas exchange is not known.

2. Water and Nutrient Provision

Although the yolk provides most of the nutrients for development in lecithotrophic viviparous squamates, radiotracer studies have revealed that amino acids and inorganic ions are transferred from maternal to embryonic tissues in several species. Quantitative analyses have shown that placental provision can account for considerable amounts of sodium and calcium, among other ions. Thus, even lecithotrophic viviparous species exhibit an incipient form of placentotrophy. In addition, the developing viviparous egg absorbs water from maternal tissues.

The precise site of placental transfer of water, ions, and organic molecules in lecithotrophic squamates has not been determined. However, a few physiological studies have implicated yolk sac placentae in nutrient transfer in these species. In addition, anatomical studies have revealed cellular specializations suggestive of secretion and absorption across the yolk sac placenta in some viviparous lizards and snakes.

D. Specializations for Placentotrophy

In at least three lineages of viviparous lizards, placental membranes transfer substantial quantities of organic nutrients. For example, in South American *Mabuya* as well as the Mediterranean skink *Chalcides chalcides*, the ovulated egg is very small and virtually all of the nutrients for development are supplied by placental means. In these lizards, the dry mass of the conceptus between ovulation and birth increases markedly, in contrast to the decrease in mass that characterizes lecithotrophic forms (Table 3).

The chorioallantoic placentae in such species are correspondingly specialized. Each exhibits a "placentome"—a region of interdigitation between hypertrophied uterine and chorionic tissues. At this site, the chorionic epithelial cells are enlarged and bear mi-

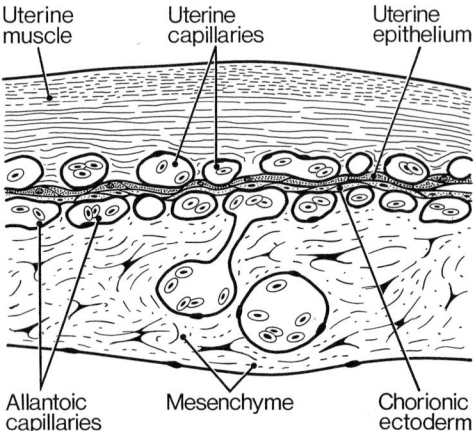

FIGURE 3 Interface of uterine and chorioallantoic tissues in the generalized chorioallantoic placenta of lizards and snakes. This placental type is characteristic of relatively lecithotrophic, viviparous squamates, in which it functions as a respiratory organ. Only thin layers of uterine and chorionic epithelia separate the fetal and maternal vascular systems (reproduced from Blackburn, 1993).

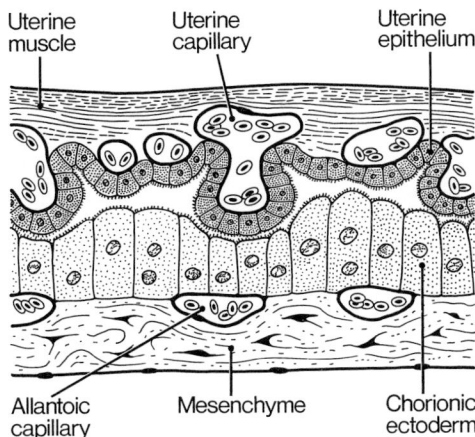

FIGURE 4 Interface of uterine and chorioallantoic tissues in the placentome *Chalcides chalcides*. The uterine epithelium is secretory, and the chorionic epithelium bears microvilli. Outside the placentome, the placenta looks like that shown in Fig. 3 (reproduced from Blackburn, 1993).

crovilli, a morphology typical of absorptive cells (Fig. 4). The apposed uterine epithelium consists of large secretory cells. Similar specializations are found in some unrelated Australian lizards of the genus *Pseudemoia*. In *Mabuya*, the uterus contains enlarged glands whose secretions appear to be absorbed by the chorionic cells. Outside of the placentome, the chorioallantoic placenta in these lizards appears much like that of generalized (lecithotrophic) species. Thus, the chorioallantoic placenta shows regional differentiation in placentotrophic forms, with one region being specialized for nutrient transfer and the other for gas exchange.

III. PLACENTAE AND PLACENTAL ANALOGS IN AMPHIBIANS

The vast majority of amphibians reproduce by laying eggs, which are either fertilized externally [most frogs and toads (anurans)] or internally (caecilians and most salamanders). However, viviparity occurs in one genus of salamanders, two lineages of frogs, and among the caecilians. Representatives of each of these amphibian groups have placental analogs (i.e., mechanisms by which matrotrophys is accomplished), as do a few oviparous anurans.

A. Specializations for Gas Exchange

In the viviparous amphibians, the eggs are fertilized internally and the larvae develop within the maternal oviduct. Respiration via the skin and gills may suffice for the small larvae of some of these species. However, some developing caecilians exhibit enormously elongated gills during development that function in gas exchange with the oviduct. In the viviparous frog *Eleutherodactylus jasperi*, a lecithotropic species, the tail of the developing tadpole may accomplish respiratory exchange; it is broad and highly vascularized and adpressed to the oviductal epithelium. Analogous situations occur in various oviparous anurans. For example, in certain *Gastrotheca*, after fertilization the eggs are carried in integumentary pouches on the back of the mother frog, and the enlarged gills of the developing tadpole closely appose the pouch lining. Although the term "placenta" has not traditionally been applied to such specializations, they do lie within the bounds of the definition.

B. Specializations for Nutrient Provision

Although few amphibians exhibit matrotrophic viviparity, their mechanisms of nutrient provision are diverse and unusual compared to those of squamate reptiles (Table 1). In the East African frog *Nectophrynoides occidentalis*, females produce tiny (0.6-mm) eggs and nourish their fetuses by oviductal secretions. The fetuses have specialized oral papillae that may aid in ingestion of the secretions. In the urodele *Salamandra atra*, the gestation period lasts from 2 to 5 years, during which time the embryos consume sibling yolks as well as uterine secretions. A specialized region of the pregnant oviduct sheds epithelial cells that are ingested by the embryos.

About 50% of those caecilians whose reproductive modes are known are both viviparous and matrotrophic. The developing fetuses ingest nutritious oviductal secretions as well as epithelial cells, which they scrape away with specialized teeth. Abrasion of the oviduct lining is said to stimulate release of the secretions. Embryos of typhlonectid caecilians exhibit highly unusual sac-like gills that are thought to function not only in gas exchange but also in absorption of oviductal nutrients.

Nutrient provision following hatching may occur in Darwin's frog, *Rhinoderma darwinii*, a species in which males carry their young in their vocal sacs though the period of metamorphosis. Experimental study indicates that labeled amino acids injected into the paternal circulation are transferred to the larvae—a pattern which legitimately could be labeled "patrotrophy."

IV. EVOLUTION OF PLACENTAE AND MATROTROPHY

A. Reptiles

Since the early 1900s, biologists have assumed that placentae evolve subsequent to viviparity in squamates. Much effort has been devoted toward estimating whether particular viviparous squamates have structures worthy of being called placentae. However, by modern criteria, functional placentae are universal among viviparous squamates. Placentae apparently evolve concomitantly with viviparity because thinning of the eggshell allows apposition of the extraembryonic membranes to the uterine lining.

The evolutionary implications are profound. Given that viviparity has evolved on over 100 separate occasions among squamates, placental organs must have also originated as frequently. No other animal organ is known to have originated on so many independent occasions. Squamate placentae therefore offer a powerful resource for studies of the interplay of structure and function in an evolutionary context.

From comparisons of oviparous and viviparous squamates, we can trace structural and functional changes through which placentation evolves. Retention of the viviparous egg inside the maternal oviduct and thinning of the eggshell places the chorioallantois in close proximity to the vascularized uterine mucosa (Fig. 1). Associated evolutionary modifications include diminution of uterine gland secretion (such that only a remnant of the eggshell is deposited) and, according to some studies, an increase in vascularity of the uterus. As noted previously, the epithelia of both the uterus and the chorioallantois are greatly thinned in viviparous forms, minimizing the diffusion distance between fetal and maternal bloodstreams (Fig. 3). Whether oviparous and viviparous squamates differ in this regard has not been determined.

In any case, the result is a chorioallantoic placenta that functions in gas exchange, as does the chorioallantois of oviparous species. If yolk sac placentae function in water uptake, as some biologists have speculated, the yolk sac of viviparous forms may also have retained its original oviparous function. Interestingly, intermediate evolutionary stages in the evolution of viviparity and placentation are exceedingly rare among extant species, perhaps because they are evolutionarily unstable. The great majority of squamates either lay fully shelled eggs at an early stage of development or retain eggs to term and form functional placentae.

Modification of the chorioallantoic placenta for nutrient transfer has occurred in three separate lineages of squamates, all members of the lizard family Scincidae. Specializations of the chorioallantoic placentae are similar in these three lineages, providing a striking example of evolutionary convergence at the cellular level.

B. Amphibians

Matrotrophy clearly has evolved independently in *Salamandra*, *Nectophrynoides*, and the caecilians. Because the embryos of these amphibians are not surrounded by extraembryonic membranes, ingestion and absorption of products of the maternal reproductive tract can occur. True placentotrophy appears to be confined to typhlonectid caecilians, in which it occurs via specialized gills of the embryo. Thus, although the anamniote condition has precluded forms of placentotrophy such as those of squamates, it has facilitated alternative mechanisms of maternal–fetal nutrient transfer.

Placentae also have evolved to serve respiratory functions in both oviparous and viviparous amphibians through the evolutionary recruitment of embryonic gills and tails. The integumentary placental structures found in some oviparous frogs have no equivalent among amniotes, and their evolution is made possible by the glandular skin and small unshelled eggs.

Our understanding of the structure and function of placentae and their analogs in reptiles and amphibians is based on incomplete information on a miscellaneous sampling of species. Without question, much more needs to be learned. Unfortunately, several key viviparous species are now threatened in large parts of their ranges (e.g., certain South American *Mabuya*), and at least one (the frog *E. jasperi*) recently appears to have gone extinct. Thus, studies of reproduction in viviparous reptiles and amphibians gain special urgency from the prospects that what we do not learn soon may be forever lost to science and humankind.

See Also the Following Articles

ALLANTOCHORION; AMPHIBIAN REPRODUCTION, OVERVIEW; PLACENTA AND PLACENTAL ANALOGS IN ELASMOBRANCHS; PLACENTAL GAS EXCHANGE; PLACENTAL NUTRIENT TRANSPORT; REPTILIAN REPRODUCTION, OVERVIEW; VIVIPARITY AND OVIPARITY

Bibliography

Blackburn, D. G. (1992). Convergent evolution of viviparity, matrotrophy, and specializations for fetal nutrition in reptiles and other vertebrates. *Am. Zool.* **32**, 313–321.

Blackburn, D. G. (1993). Chorioallantoic placentation in squamate reptiles: Structure, function, development, and evolution. *J. Exp. Zool.* **266**, 414–430.

Blackburn, D. G. (1994). Standardized criteria for the recognition of developmental nutritional patterns in squamate reptiles. *Copeia* **1994**, 925–935.

Blackburn, D. G. (1995). Saltationist and punctuated equilibrium models for the evolution of viviparity and placentation. *J. Theor. Biol.* **174**, 199–216.

Blackburn, D. G. (1998). Structure, function, and evolution of the oviducts of squamate reptiles, with special reference to viviparity and placentation. *J Exp. Zool.*, in press.

Greven, H. (1977). Comparative ultrastructural investigations of the uterine epithelium in the viviparous *Salamandra atra* Laur. and the ovoviviparous *Salamandra salamandra* (L.) (Amphibia, Urodela). *Cell Tissue Res.* **181**, 215–237.

Jones, R. E., and Baxter, D. C. (1991). Gestation, with emphasis on corpus luteum biology, placentation, and parturition. In *Vertebrate Endocrinology: Fundamentals and Biomedical Implications*, Vol. 4, Part A, pp. 205–302. Academic Press, New York.

Mossman, H. W. (1937). Comparative morphogenesis of the fetal membranes and accessory uterine structures. *Carnegie Inst. Contrib. Embryol.* **26**, 129–246.

Stewart, J. R. (1992). Placental structure and nutritional provision to embryos in predominantly lecithotrophic viviparous reptiles. *Am. Zool.* **32**, 303–312.

Stewart, J. R. (1993). Yolk sac placentation in reptiles: Structural innovation in a fundamental vertebrate fetal nutritional system. *J. Exp. Zool.* **266**, 414–449.

Stewart, J. R., and Blackburn, D. G. (1988). Reptilian placentation: Structural diversity and terminology. *Copeia* **1988**, 838–851.

Wake, M. H. (1977). Fetal maintenance and its evolutionary significance in the Amphibia: Gymnophiona. *J. Herpetol.* **11**, 379–386.

Wake, M. H. (1982). Diversity within a framework of constraints. Amphibian reproductive modes. In *Environmental Adaptation and Evolution* (D. Mossakowski and D. G. Roth, Eds.), pp. 87–106. Fischer, Stuttgart.

Wake, M. H. (1993). Evolution of oviductal gestation in amphibians. *J. Exp. Zool.* **266**, 394–413.

Yaron, Z. (1985). Reptile placentation and gestation: Structure, function, and endocrine control. In *Biology of the Reptilia*, Vol. 15 (C. Gans and F. Billet, Eds.), pp. 527–603. Wiley, New York.

Placenta: Implantation and Development

Kurt Benirschke
University of California, San Diego

I. Introduction
II. Gross Anatomy
III. Implantation and Early Stages
IV. The Amnion
V. The Yolk Sac
VI. The Mesoderm
VII. Microscopic Anatomy

GLOSSARY

amnion The innermost layer of the placental membranes.
chorion That layer of the placental surface that carries the blood vessels.
placenta Afterbirth.
trophoblast The outermost surface of the placental villi, regulating transport.

The placenta or afterbirth is predominantly a fetal organ. The maternal part consists of blood, which circulates about the fetal tissue in the intervillous space, and a very thin maternal floor, modified uterine lining, to which the placenta attaches to the uterus. The placenta is the largest organ of the fetus. It connects with the baby through the umbilical cord. Normally, because of uterine contractions which persist beyond the delivery of the baby, the placenta is detached from the uterus within a few minutes of the baby's birth.

I. INTRODUCTION

The appearance and structure of the placenta varies greatly among different mammalian species. Some, such as the human placenta, have a single disk, whereas many species have multiple "cotyledons" (lobules); others have ring-shaped placentas, and so on. The depth of invasion into the maternal tissue is also variable, leading Grosser to classify the organ from different animals depending on the number and types of tissue layers that separate mother from the fetus. There are thus the "hemochorial" placenta as seen in humans, "epitheliochorial" placentas (pig), and several other divisions, depending on the number of cells between maternal blood (hemo-) or surface endometrium (epithelio-) and the trophoblast (chorio-). Trophoblast is the cellular epithelial tissue that surrounds the supporting tissue "stroma" of the entire placenta. This epithelium differentiates into various specific cell types: cytotrophoblast, syncytiotrophoblast, and intermediate trophoblast which is also known as the X cell. The word trophoblast is of Greek origin and means nourishment; *blastos* means germ.

The placenta regulates the exchange of nutrients from mother to fetus. Nutrients are brought via the maternal blood and absorbed by trophoblast and then carried via the placental (fetal) circulation to the fetus. The placenta similarly disposes of fetal waste products in a reverse exchange. Oxygen transport between the two organisms is also accomplished here and the placenta dictates the water exchange. The placenta also produces certain hormones that vary greatly from species to species. Since the placenta is no longer needed after birth, it is detached from the newborn baby. In animals, the cord is severed by biting or it spontaneously ruptures. In most species, the mother eats the placenta. This may be beneficial, serving as welcome nutrient and it may stimulate milk production; moreover, its presence on the ground attracts predators unnecessarily.

II. GROSS ANATOMY

The placenta consists of a disk-shaped bloody mass to which is attached a large, water-filled fibrous sac, often referred to as the fetal membranes (Fig. 1). Terms such as amnion, chorion, trophoblast, and villi are used to describe various portions of the organ. The mature human placenta, trimmed of membranes and umbilical cord, weighs between 450 and 550 g (about 1 lb). Prenatal influences, such as maternal diabetes, preeclampsia, or virus infection, markedly alter this weight. The human placenta is an oval or round disk measuring 20 cm (8 in.) in diameter and 2.5 cm (1 in.) in thickness and almost all of it is of fetal origin. Surrounding it is the only maternal tissue, the decidua, representing endometrium changed by hormones of pregnancy, especially progesterone. There is now good evidence that the paternal genes "imprint" on the placenta, thereby expressing an overriding influence on its development; maternal genes imprint on embryonic genetic expression.

On the fetal surface of the placenta lies the detachable, thin amnion; beneath it, the chorion carries the fetal blood vessels and the villi also originate here. The arteries invariably cross over veins on the placental surface. At their terminations, these vessels dip or penetrate into the placental parenchyma, the villous tissue. Each ramification supplies one placental cotyledon, a subdivision of the placenta. The villous tissue is the most abundant component. It is composed of numerous ramifications of blood vessels which are carried in connective tissue. These are covered by the trophoblast, the epithelial surface of villi. This then lines the intervillous space in which the maternal blood circulates.

The umbilical cord normally inserts near the center of the placenta; a marginal insertion, also called a "Battledore" placenta, is found in 7% and a membranous (velamentous) cord insertion occurs in 1% of singleton placentas. In this latter condition, the umbilical cord inserts onto the membranes rather than onto the placental disk. Both are abnormal and their frequencies are greater with multiple pregnancies. The cord has an average length of 54 cm. It is regularly spiraled in a counterclockwise direction. I believe that the length of the umbilical cord is largely determined by the fetal movements. Excessively long cords often lead to fetal entanglement or knotting. There are normally two arteries and one vein, and the two arteries have an anastomosis near the placental surface. The absence of one artery, which occurs in 1% of placentas, is distinctly abnormal. It is associated with higher fetal mortality and malformations. The arteries are perfused by the fetal heart. They bring the deoxygenated blood to the placenta; there, oxygen and nutrients are collected and returned to the fetus via the umbilical vein. The umbilical cord contains neither nerves nor lymphatics.

The extraplacental membranes of the fetal sac (the "chorion laeve") attach to the edge of the placenta. The amnion is on the inside of the membranes; the next layer is the chorion, with remnants of fetal vessels and villi. On the outside, the membranes are covered by decidua capsularis, representing maternal tissue.

Because of the fetal blood content of the villous tissue, the maternal surface of the placenta is dark red. In mature placentas one also finds finely distributed yellow flecks of calcification. Sonographers have "graded" placental maturation and degeneration according to the amount of calcification. However, it represents only dystrophic calcification in the normal fibrin deposits and has no major importance for fetal development.

The cotyledonary subdivisions of the placenta are readily apparent from the maternal side. When long,

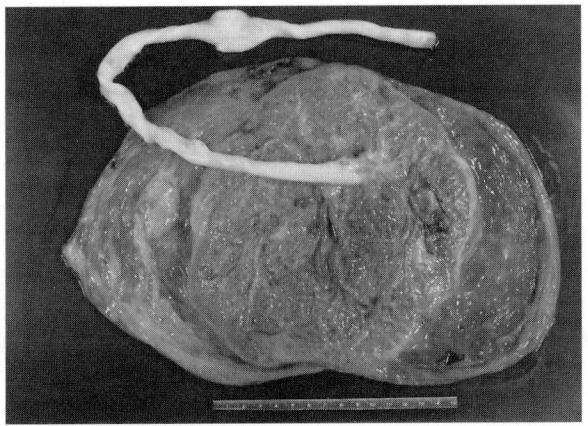

FIGURE 1 Normal mature human placenta at term. The umbilical cord contains a false knot, a common vascular redundancy.

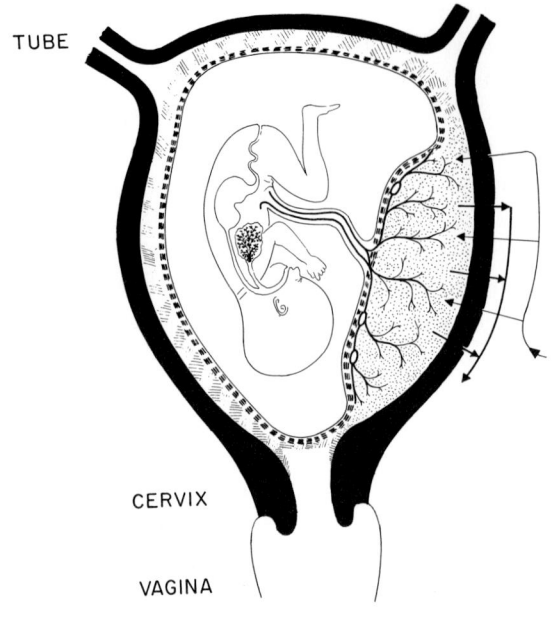

FIGURE 2 Schematics of uterus with attached placenta and fetus within its membranes.

clean slices are made through the placental substance, one encounters distinctive holes in the central portions of cotyledons toward the maternal surface. They represent spaces formed by the initial jet-like injection of blood from the maternal spiral arterioles (the terminal arteries of the endometrium). *In vivo*, this can be seen by sonographers.

The placenta is most frequently attached to the anterior or posterior fundal portions of the uterus (Fig. 2). Sonographic evidence indicates that placenta previa (lying low and covering the cervix) is more frequent in early gestation than at term. From this it is inferred that the placenta "wanders" during gestation. Obviously, it does not detach and reattach itself at a higher point; rather, the lower margin often atrophies, whereas the placenta expands laterally. This is reflected by the often eccentric insertion of the umbilical cord and also by abnormalities of membrane insertion.

III. IMPLANTATION AND EARLY STAGES

Implantation and development of the placenta are complex processes. Most of the principal features of these early changes at implantation have been described from systematic studies by Hertig and Rock. They correlated the endometrial cycle with the histology of the earliest implantation of the blastocyst. Later stages of placental development and the anatomy of the placenta have been illustrated in a monumental monograph by Boyd and Hamilton. One major deficiency still exists, however: We do not know what genes control placental development and to what extent the maternal (uterine) environment influences this process.

A. Fertilization, Transport

For this discussion, the endometrial cycle is divided into two halves, each 14 days long. The proliferative phase lasts from Days 1 to 14, with its first portion made up of the menstrual phase and the majority being the reparative, proliferative phase. In this idealized cycle, ovulation occurs at the end of the proliferative phase, on Day 14; the progesterone of the corpus luteum (the remainder of the maternal ovarian follicle) then causes the endometrium to undergo changes that are referred to as the "secretory," progestational phase. In this 14-day span, the glands secrete a thin fluid, sometimes referred to as uterine milk. Its composition is unexplored in the human uterus but it plays an important role in animals. Profound changes affect the endometrial stroma, with the "predecidua" beginning to form around the spiral arterioles of the endometrium. Hallmarks of this include cytoplasmic enlargement of stromal cells, cessation of mitoses, and dilatation with hyperactivity of secretory glands. Shortly before the beginning of the next menses, "endometrial granular cells," leukocytes, and edema are conspicuous in the stroma.

After fertilization of the ovum and the formation of the blastocyst it is the objective of the blastocyst to first reach the endometrial cavity, to then attach itself to its surface, and thereafter to invade the endometrium. Next it seeks a connection to the maternal blood supply by eroding the endometrial blood vessels. By so doing it delivers human chorionic gonadotropin, a protein hormone produced by placental syncytiotrophoblast, into the maternal bloodstream. This is essential to encourage the ovarian corpus luteum to continue progesterone production, thus preventing the next menstruation. Timing of implantation and placental expansion are critical, lest the endometrium is shed with the loss of its early conceptus.

Ovulation occurs on the 14th day in this idealized 28-day cycle, with fertilization taking place in the ampulla of the Fallopian tube a few hours later. Reconstitution of a nucleus and division of the egg follow rapidly. A well-studied 58-cell ovum, washed out of the uterus 96 hr following coitus, shows the first delineation of the embryo. This conceptus contained a 5-cell "inner cell mass" (the future embryo), separate from the 53-cell trophoblastic shell, that is destined to become the placenta. Thus, at 96 hr a "blastocyst" has developed with a cavity in which the embryo-to-be is located. Hereafter, the initial development is primarily confined to the outer trophoblastic shell, whereas the "embryo" has to wait for most of its future growth until implantation is complete and a trophoblast–maternal blood relationship has been established. It is presumed that the trophoblastic shell takes some nourishment from the tubal secretions and, subsequently, from the "uterine milk" during this interval.

Tubal contractions transport the blastocyst into the endometrial cavity on Day 17 of the endometrial cycle. The exact time of its implantation, however, remains unknown. The next stage recorded occurs on the 21st day of the cycle, the ovum now 7.5 days old. By this time, the surface of the endometrium has been penetrated by the trophoblast and a portion of the blastocyst has become embedded. The nuclei of this implanting portion of the shell are much larger and seem to be polyploid (having more than a single set of chromosomes) when compared to the cells away from the implantation site. We have reason to believe that this trophoblast incorporates ("eats") portions of the superficial endometrium, not merely dissociating its cells in order to implant. Now the blastocyst invades into the decidua until it is completely embedded. It thus attains a truly "interstitial" implantation. This aspect differs markedly from species to species.

The trophoblast now proliferates circumferentially and seeks a direct connection with the endometrial vessels which it invades. The peripheral trophoblast forms columns in which spaces or "lacunae" appear, i.e., the development of the future intervillous space.

From now on we refer to the "days" of embryonic development; 14 days are then added to the calculation to arrive at the "menstrual" age. The maternal spiral arterioles are next invaded and maternal blood or serum then flows into the lacunar spaces which establishes the primitive intervillous circulation. By this means, a hemochorial placenta is formed, in which maternal blood directly surrounds the trophoblast. In many other species the maternal blood may be separated from the trophoblast by several layers of maternal elements, usually epithelium and connective tissue. In the epithelial–chorial placenta of pigs and whales, for instance, the trophoblast only touches the endometrial epithelium—no true invasion occurs.

During this interval the embryonic cavity has enlarged and become more cystic, and the future embryo has grown slightly. It has become elongated and begun to differentiate. The cavity in which it lies is filled with a thin, gelatinous fluid called magma reticulare. The embryo develops rapidly and its layers (ectoderm, mesoderm, and endoderm) participate importantly in shaping the future of the placenta—an aspect that is crucial for understanding the placenta.

The conceptus' cavity first expands rapidly. This causes the surface of the endometrium to bulge and that area of blastocyst wall becomes the "membranes," eventually the chorion laeve of the future placenta. The part of the placenta which remains at the base of the endometrial implantation site becomes the placenta, or chorion frondosum. The membranes are covered on their outer surface by the decidua capsularis. The portion of decidua which remains beneath the site of implantation is called the decidua basalis. The other endometrium, not invaded by trophoblastic cells, is called the decidua vera. It is important to recognize that the decidua capsularis is vascularized by maternal vessels from the area of implantation rather than from the opposite side of the uterus. While there remains some slit between the expanding placental membranes and the opposite uterine wall for some time, ultimately the membranes touch the other side of the uterus but never truly grow together.

The trophoblast differentiates into an outer syncytiotrophoblast from the inner cytotrophoblast. This inner layer is also called the Langhans cell layer. The syncytium (synonymous with syncytiotrophoblast) is a remarkable cell type. It is unable to undergo mitosis and lacks cytoplasmic separations between its nuclei. Cytoplasm and nuclei are contributed constantly by the cytotrophoblast, whose only function appears to be the regeneration of syncytium. The nuclei of the syncytium are diploid ($2n = 46$) but unable to divide. The syncytium is essentially a sheet of one huge cell which, at term, makes up about 10 m^2 of surface. Its exchange capacity is enhanced by a myriad of microvillous projections. As it matures, being influenced by the expansion and contraction of villi, the syncytium may buckle up to form "sprouts" or "knots." These are often swept away in the intervillous circulation and end up in the mother's lung. Here, they are destroyed. The syncytium is the working end of the placenta; it is filled with enzymes and transport vesicles and produces the various placental hormones. It also reigns over the complex placental exchange. The more primitive, as yet undifferentiated, trophoblast invades the decidua and creeps up within the vascular lumina of the maternal vessels. It transforms these vessels by invasion, changing them into more rigid structures at the site of implantation. A third type of trophoblast later differentiates and is often incorrectly referred to as "intermediate" trophoblast. It forms a part of the placental floor and of the septa that separate the cotyledons ("X cells," so named because their origin had remained undefined). The cysts it produces contain a mucoid fluid with a protein that is identical to that contained in the granules of eosinophilic leukocytes, the major basic protein.

IV. THE AMNION

The amnionic membrane is avascular and translucent. It consists of a single layer of epithelial cells that sit on a thin membrane of fibrous tissue. The amnion has an ectodermal origin and its development is best appreciated by reconstructing the early developing ovum. It is then seen to be a derivative of the embryo rather than of the trophoblastic shell. As the embryo lengthens it forms three main layers: an ectoderm that is directed toward the base of the placenta, an endoderm that looks toward the center, and in between is the mesoderm. As the embryo further differentiates it folds in all directions so as to surround the endoderm. At the same time, the embryonic layers proliferate and differentiate. The amnion, being attached to the ectoderm on all ends of its plate, forms a vesicle and fills with fluid. It resides on the back of the folding embryo. If one envisages that the embryo, while forming, also herniates backwards into this expanding bag of amnion, then one can understand how the amnion not only comes to line the entire placental cavity but also, later, to cover the developing umbilical cord. It is not until the 12th week of fetal life that the amnion completely touches the entire chorionic surface. The initial amnionic fluid presumably is a filtrate from the magma reticulare. Later it is mostly made from fetal urine.

V. THE YOLK SAC

The endoderm develops its own cavity, the primitive yolk sac, in whose walls the hematopoietic cells and perhaps germ cells differentiate. Its future progeny are the liver and the gastrointestinal tract. The yolk sac epithelial cells also produce α-fetoprotein, an embryonic albumin. The yolk sac is partially enclosed by the folding embryo, although a portion remains extraembryonic and is connected to the embryo by the evanescent omphalomesenteric (vitelline) duct of the umbilical cord. Ultimately, the extraembryonic yolk sac of human placentation atrophies. In human placentas, the vitelline duct and the few yolk sac vessels also atrophy as the gut rotates and pulls the initial endoderm into the embryo to become its intestine.

VI. THE MESODERM

In between ectoderm and endoderm develops the mesoderm, the future connective tissue of embryo and placenta. The mesoderm proliferates toward the tail end of the embryo and streaks toward the inner surface of the trophoblastic shell. It proliferates rapidly to line the entire trophoblastic cavity and then invades the trophoblastic columns in a manner that

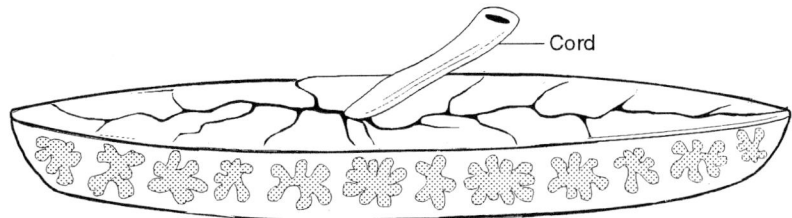
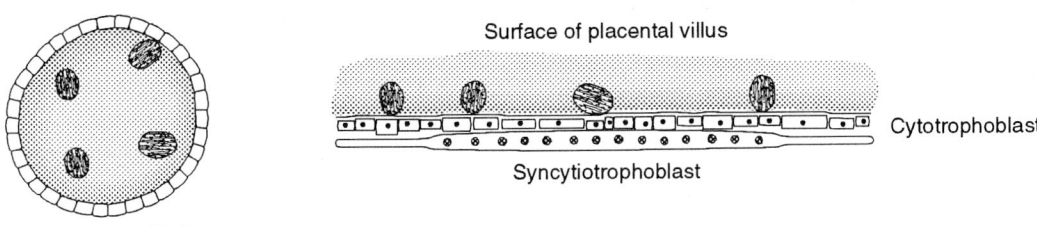

FIGURE 3 The finer structure of the human placenta.

can be visualized as fingers (mesoderm) pushing into the dough of trophoblast. There, the mesoderm continues to proliferate peripherally, and vessels follow its path from the embryo. Thus, chorion with the fetal vasculature and villi are formed. Near the embryo, the mesoderm condenses as it is being invaded by vessels to become the umbilical cord.

Now the basic layers of the placenta are complete and maternal–fetal nutrient, gas, and water exchanges can take place. Hereafter, the fetus develops relatively more rapidly than the placenta, which only grows in size. The ultimate site of umbilical cord insertion is thus influenced by the position of the embryo at the time of embryo positioning or in relation to the original outgrowth of the mesoderm (Fig. 3).

VII. MICROSCOPIC ANATOMY

A. Decidua

Under the influence of progesterone, the endometrium gradually changes from predecidua to decidua. Because of the trophoblastic invasion, some necrosis occurs and fibrin is deposited in an irregular fashion, its quantity increasing with gestation. Two areas of fibrin deposits, Nitabuch's fibrin layer and Rohr's striae, merge frequently in the floor of the placenta, the attachment plane. It was once considered that these layers of fibrin may restrict transplacental cell transport or that they provide some immunologic barrier between mother and fetus. It is no longer held that this fibrin helps in preventing the immunologically "foreign" placental cells from being rejected by the mother. Also, both fibrin layers are not the site of placental detachment at delivery because they are invariably found in the delivered placenta.

Trophoblastic cells infiltrate the decidua at the site of implantation. The initial cell columns that attach the blastocyst to the endometrium differentiate rapidly. They lose their original peripheral polyploidy (increased DNA content) and the inner cytotrophoblast develops the syncytiotrophoblastic (syncytial) shell. Trophoblast infiltrates the maternal vascular spaces and creeps up within their walls for some distance into the myometrium. The cells enter the walls of these vessels, evoke fibrin deposits, and destroy the muscular coat of arterioles. They thus cause profound change in the structure and physiologic reactivity of these vessels. The ultimate fate and the evolution of these placental site trophoblastic giant cells is poorly explored. Some produce various placental proteins.

B. Umbilical Cord

The cord consists largely of connective tissue with a substantial amount of mucopolysaccharides and mast cells in addition to the three fetal blood vessels it carries (one vein and two arteries). The Wharton's jelly of the cord is readily compressible and varies remarkably in quantity. The vessels lack typical elastic membranes of similar-sized vessels in the body and differ from other large vessels by their remarkable ability to allow leukocytes to traverse their walls in infectious conditions. The surface of the cord is amnionic epithelium.

C. Amnion

The amnion is composed of a single layer of flat to columnar epithelial cells resting on a sheet of connective tissue. Amnion has no blood vessels. It subsists on the nutrient and gaseous supplies found within amnionic fluid and/or the underlying chorionic vasculature. A large population of macrophages is present in the connective tissue of the amnion. When these are challenged by pigments or infection, they enlarge and become vacuolated.

D. Chorion

The tissue of the chorionic membrane is a tough connective tissue layer that carries the branches of the fetal blood vessels, the so-called chorioallantoic vessels. The vascular walls of a delivered placenta are much thinner toward the embryonic cavity than toward the placental villous tissue. Smaller allantoic vessels, as seen in ruminants, or vitelline vessels, as in rodents, are not found in the human placenta. The chorion contains a similar macrophage population as that of the amnion and, on its undersurface, it has a layer of trophoblastic cells which are frequently enmeshed in dense fibrin deposits.

E. Villous Tissue

The bulk of the mature placenta is composed of tertiary villi. Secondary villi are generally considered the villous structures found when the mesodermal invasion of the trophoblastic columns (primary villi) has taken place; when capillaries have entered, they become the "tertiary villi." The villi are the major structural component of the placenta; all nutrient exchange is accomplished here. The villi take off in all directions from their primary ramifications and branch a number of times until, at the periphery, the terminal villus consists only of fetal capillaries and a trophoblastic surface that sits on a thin basement membrane. The capillaries are supported by delicate connective tissue. In this matrix there is only one other cell type, the Hofbauer cell. Hofbauer cells comprise a population of macrophages similar to the phagocytes in amnion and chorion. These cells have a common origin from the fetal circulation and contain various debris, including some iron. In the normal mature placenta, hematopoiesis is absent, and nucleated red blood cell are pathologic.

Capillaries are clearly the dominating feature of mature placentas and it is here where the most important exchange takes place. In some areas these small vessels nearly bulge into the trophoblast and only a thin cover of syncytium lies between the fetal vascular lumen and the maternal blood. This is the so-called vasculosyncytial membrane, through which maximum nutrient exchange is possible. Some capillaries are sinusoidally dilated, reaching diameters of 50 μm, areas that may serve to reduce blood flow resistance.

The trophoblast that covers the villi consists of cytotrophoblast and syncytiotrophoblast or syncytium. The former is nearer the fibrous core of the villi, uninuclear, and forms the single cellular Langhans layer of early pregnancy. This trophoblast has little cytoplasmic differentiation and is capable of undergoing mitosis until late in pregnancy. With villous growth and capillary expansion, however, this layer becomes discontinuous. Enders provided evidence that supports that some cytoplasm of Langhans cells fuses with that of the syncytium, perhaps with a brief intermediate stage, after which the cell borders disappear. The syncytium is the most complex of placental cells. It is a remarkable cellular membrane in which the nuclei float freely, without cell borders. The cytoplasm has an enormous variety of cytoplasmic structures. The maternal and fetal surfaces are studded with microvilli, extensions of the cytoplasm, to greatly increase the overall surface of the cell and thus enhance maternal–fetal exchange of nutrients, gases, and water. The syncytium is a very invasive cell, as can be seen in very early implantation stages and from the tumors it occasionally produces; however, it is a nonreproductive cell type. As it matures, small knots regularly break off and are deported into the maternal bloodstream, to vanish in the mother's lung. Breaks also often occur on the surface of villi, thus focally denuding the villus. At these sites "healing" maternal fibrin is deposited in ever-increasing quantity. The syncytium contains smaller mitochondria than the cytotrophoblast and has numerous lipid droplets and innumerable transport vacuoles. However, the syncytium of one villus does not invade the surface of other villi from which it is separated by very small sinus-like spaces. Thus, the intervillous space is not a big "lake" as was so frequently envisaged.

See Also the Following Articles

AMNIOTIC FLUID; DECIDUA; FETAL-PLACENTAL UNIT; TROPHOBLASTS AND THE HUMAN PLACENTA; UMBILICAL CORD; YOLK SAC

Bibliography

Benirschke, K., and Kaufmann, P. (1995). *The Pathology of the Human Placenta.* Springer-Verlag, New York.

Boyd, J. D., and Hamilton, W. J. (1970). *The Human Placenta.* Heffer, Cambridge, UK.

Enders, A. C. (1965). Formation of syncytium from cytotrophoblast in the human placenta. *Obstet. Gynecol.* **25**, 378–386.

Grosser, O. (1927). *Frühentwicklung, Eihautbildung und Placentation des Menschen und der Saeugetiere.* Bergmann, München.

Hertig, A. T. (1968). *Human Trophoblast.* Thomas, Springfield, IL..

King, D. L. (1973). Placental migration demonstrated by ultrasonography. A hypothesis of dynamic placentation. *Radiology* **109**, 167–190.

Placental and Decidual Protein Hormones, Human

Stuart Handwerger

University of Cincinnati College of Medicine and Children's Hospital Medical Center

I. Introduction
II. The hPL/hGH/hPRL Gene Family
III. Chorionic Gonadotropin
IV. Summary

GLOSSARY

chorionic gonadotropin A glycoprotein hormone released by the human placenta that has striking homologies to pituitary luteinizing hormone.

decidual prolactin A protein hormone synthesized and released by human endometrial stromal cells during the luteal phase of the menstrual cycle and during pregnancy that is identical in structure to pituitary prolactin.

growth hormone-variant A protein hormone synthesized and released by the human placenta that has biologic actions similar to those of pituitary growth hormone.

placental lactogen A protein hormone synthesized and released by the human placenta that has striking homologies to pituitary growth hormone and prolactin; important in regulation of fetal growth.

placental lactogen/growth hormone/prolactin gene family A family of six closely related hormones with growth-promoting and lactogenic activities that evolve from a common ancestral gene. The placental lactogen and growth hormone genes are located on chromosome 17; the prolactin gene is on chromosome 6.

I. INTRODUCTION

During pregnancy, the human placenta and uterine decidua synthesize and release protein hormones with striking homologies in structure and function to pituitary and hypothalamic hormones. This chapter will focus on four hormones: the placental hormones, placental lactogen, placental growth hormone, and chorionic gonadotropin, and the decidual hormone decidual prolactin.

The placental hormones, placental lactogen (hPL) and growth hormone (hGH-V), and the decidual hormone decidual prolactin (dPRL) are members of the growth hormone/placental lactogen/prolactin gene family. The hormone, chorionic gonadotropin (hCG), is a glycoprotein hormone with striking homologies in structure and function to the pituitary glycoprotein hormone luteinizing hormone (hLH). In addition to these protein hormones, the human placenta synthesizes and secretes other proteins and polypeptide hormones, many of which have striking homologies to pituitary and hypothalamic hormones: corticotropin-releasing hormone (CRH), growth hormone-releasing hormone (GHRH), gonadotropin-releasing hormone (GnRH), somatostatin, activin, inhibin, relaxin, and parathyroid hormone-related peptide.

II. THE hPL/hGH/hPRL GENE FAMILY

Molecular studies strongly suggest that the hGH, hPL and hPRL genes evolved from a common ancestral precursor in which the five genes of the hPL/hGH gene cluster arose from recombination events involving moderately repeated sequences. The hGH/hPL and hPRL precursor genes then segregated onto two different chromosomes, with the hGH/hPL genes on chromosome 17 (q22–q24) (Fig. 1) and the hPRL gene on chromosome 6. The hGH/hPL locus contains five structurally related genes spanning 47 kb that are organized in the same transcriptional orientation and are each composed of five exons and four introns. The genes from 5′ to 3′ are hGH-N, hPL-L, hPL-A, hGH-V, and hPL-B. The entire cluster and about 19.8 kb of flanking sequences have been sequenced. The genes are expressed in two mutually exclusive tissue-specific patterns, with the hGH-N gene expressed solely in the pituitary and the other genes expressed solely in the syncytiotrophoblast layer of the placenta. The hGH-N gene encodes two alternatively spliced mRNAs that are translated into 22- and 20-kDa GH proteins. The placental genes hPL-A and

GENES	hGH-N	hPL-L	hPL-A	hGH-V	hPL-B
PRODUCTS	22K hGH / 20K hGH	20K hPL-L	hPL-A	22K hGH variant (20K hGH variant)	hPL-B
TISSUE	PITUITARY	PLACENTA	PLACENTA	PLACENTA	PLACENTA
% mRNA	3	0.01	3	<0.001	0.5

FIGURE 1 Structure of the hGH/hPL gene cluster. The gene assignment, the proteins produced by the genes, and the tissues in which the mRNAs are expressed are shown below the gene map (modified with permission from Chen *et al.*, The human growth hormone locus: Nucleotide sequence, biology, and evolution, *Genomics* **4**, 479–497, 1989).

hGH-V are alternatively spliced and encode 22- and 26-kDa gene products, whereas hPL-B encodes a single 22-kDa protein product. The expression of the hPL-L protein product(s) has not been identified in maternal blood, although the gene produces several alternatively spliced mRNA transcripts with leader sequences. The members of the gene family share 91–99% sequence identity throughout the coding regions and are located within a 500-bp region immediately upstream of the genes. The hPL-A and hPL-B mRNAs are 98% homologous and encode identical mature proteins but have one amino acid difference in the signal peptides. The mRNAs for hPL-A and hPL-B comprise approximately 3.5% of the total placental mRNA. During normal pregnancies, the hPL-A gene is expressed three to six times more than the hPL-B gene, probably due to differences in stability of the two mRNAs.

The nucleotide sequence of the cDNA for decidual prolactin is identical to the cDNA for human pituitary prolactin except for the addition of a noncoding exon. Furthermore, the promoter for the decidual prolactin gene is distinct from the pituitary prolactin gene and located 5.3 kb upstream of the human pituitary prolactin gene. The "decidual type" prolactin promoter is also present in other nonpituitary cells that express prolactin (e.g., T lymphocytes, brain, myometrium dermal fibroblasts, and lacrimal gland), strongly suggesting that the expression of prolactin in these cells is regulated by the decidual type promoter.

A. Placental Lactogen

hPL, which is synthesized by the syncytiotrophoblast cells of the placenta, is first detected in the maternal circulation at about 6 weeks of gestation. The concentration of the hormone then increases linearly to about Week 13 of gestation, when peak concentrations of 5000–7000 ng/ml are reached. During gestation, the concentration in maternal plasma is positively correlated with placental mass and is greater in multiple than singleton gestations. In contrast, the plasma concentration of pituitary growth hormone in the mother throughout pregnancy remains in the range of 4–6 ng/ml. Comparisons of the concentrations of hPL, pituitary growth hormone, hGH-V, and pituitary prolactin in maternal and fetal plasma and in amniotic fluid at term are shown in Table 1.

hPL has anabolic and growth-promoting actions in both the mother and the fetus that are similar to those of pituitary growth hormone (Fig. 2). In the mother, hPL enhances insulin secretion, impairs glucose tolerance, and promotes nitrogen retention. The antagonism of insulin action and stimulation of lipolysis and proteolysis promotes the transfer of glucose and amino acids to the fetus. Both of these substrates are critical for the growth and metabolism of the fetus. hPL also appears to be responsible in part for

TABLE 1
The Concentrations of hPL, hGH-N, hGH-V, and hPRL in Maternal and Fetal Plasma and in Amniotic Fluid at Term

Hormone	ng/ml		
	Maternal plasma	Fetal plasma	Amniotic fluid
hPL	5000–7500	80–125	55–70
Pituitary hGH	4–6	28–38	1200
Placental hGH	5–18	0	0
hPRL	130–200	150–250	15

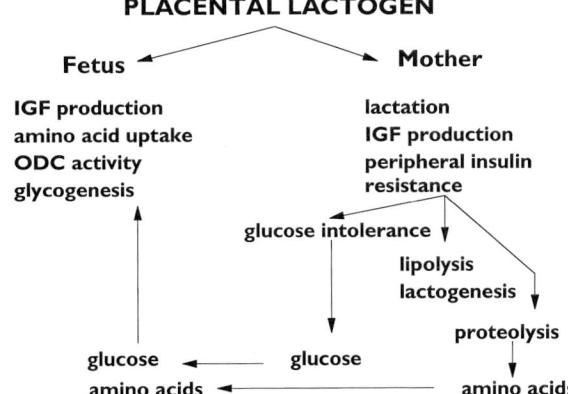

FIGURE 2 The biologic actions of hPL in the mother and fetus during pregnancy [modified with permission from S. Handwerger and M. Freeman, Role of placental lactogen and prolactin in human pregnancy, in *Regulation of Ovarian and Testicular Function* (V. B. Mahesh, D. H. Dhindsa, E. Anderson, and S. Kalra, Eds.), pp. 399–420, Plenum, New York, 1987].

the progressive increase in maternal plasma insulin-like growth factor-I (IGF-I) concentrations that normally occurs during pregnancy. The elevated IGF-I concentrations in maternal plasma may act as a growth factor for the uterus and other maternal tissues. hPL, acting through the prolactin receptor, may also contribute to the stimulation of breast development during pregnancy.

hPL, in addition to increasing the availability of substrates to the fetus, also affects fetal growth by acting directly on fetal tissues. The hormone stimulates ornithine decarboxylase activity in fetal liver, amino acid transport and uptake in fetal muscle, and DNA synthesis and IGF production in fetal tissues. The observation that hPL stimulates ornithine decarboxylase (ODC) activity and IGF production in fetal tissues has important implications for understanding the mechanisms by which hPL directly affects fetal growth. ODC is the rate-limiting enzyme in the synthesis of polyamines, which are critical in the regulation of protein, RNA, and DNA synthesis. The IGFs have been shown to be important growth factors in the regulation of fetal growth. The increase in maternal IGF-I concentrations, however, does not directly affect the growth of the fetus since IGF-I does not cross the placenta.

Placental lactogen has also been shown to stimulate glycogen synthesis in isolated hepatocytes and to inhibit the glycogenolytic actions of glucagon on fetal hepatocytes by suppressing glucagon-induced activation of phosphorylase A. This observation suggests that placental lactogen promotes glycogen storage in the fetal liver by antagonizing the actions of glucagon as well as by stimulating the synthesis of glycogen. Accumulation of hepatic glycogen in late pregnancy is of critical importance since hepatic glycogen is the major source of glucose in the brain and red cells of the neonate. The abrupt decrease in hPL concentrations after delivery may contribute to mobilization of hepatic glycogen immediately after birth.

Although placental lactogen has many growth hormone-like actions in fetal tissues, growth hormone itself has no effect on fetal tissues due to the lack of growth hormone receptors. While hepatic placental lactogen receptors are detected in fetal liver by midgestation and increase markedly in number throughout the latter half of gestation, hepatic growth hormone receptors are not detected until the first week of life. This situation is in sharp contrast to that in postnatal tissues, in which placental lactogen competes with growth hormone and prolactin for binding to somatotrophic and lactogenic receptors; the actions of placental lactogen in these tissues are felt to be mediated by binding to one or both of these receptors. The absence of growth hormone action in the fetus cannot be attributed to low growth hormone concentrations since plasma concentrations of growth hormone in the fetus exceed plasma growth hormone concentrations in postnatal animals. The absence or marked decrease in growth hormone receptors explains why human fetuses born with congenital absence of the pituitary gland or anencephaly do not have impaired fetal growth and why transgenic mice that markedly overexpress growth hormone are normal size at birth. Since growth hormone has little or no biological actions in the fetus and hPL has actions in the mother and fetus similar to those of growth hormone in nonpregnant humans and animals, it is clear that hPL functions as a "growth hormone" of pregnancy.

Although placental lactogen has striking homologies in its chemical and biological properties to growth hormone and prolactin, the factors that regulate the synthesis and secretion of placental lactogen are different than those that regulate the pituitary hormones. For example, glucocorticoids, estrogen, oxytocin, epinephrine, thyroid-releasing hormone (TRH), GnRH, L-DOPA, and somatostatin, which regulate the secretion of growth hormone and prolactin in the pituitary, have no consistent effects on the synthesis or secretion of placental lactogen. Moreover, acute and chronic changes in extracellular glucose concentrations, which are know to modulate growth hormone secretion, have no consistent effects on placental lactogen secretion *in vivo* or *in vitro*. Angiotensin II and IGF-I have been reported to stimulate placental lactogen release, but the physiologic significance of these findings is unknown. Recent studies strongly suggest a novel physiologic role for apolipoprotein A-I (apoA-I), the major apoprotein constituent of high-density lipoprotein, in the regulation of placental lactogen gene expression during pregnancy. Clinical studies have shown that plasma apoA-I and placental lactogen concentrations increase concomitantly during pregnancy and that the patterns of plasma apoA-I and placental lactogen concentrations during pregnancy are nearly identical. Plasma apoA-I concentrations have also been reported to be significantly lower than normal in preeclampsia, a pathologic condition of pregnancy characterized by low plasma placental lactogen concentrations. hPL gene expression is also regulated by retinoic acid, vitamin D3, thyroid hormone, and several cytokines.

Both the adenylate cyclase and phospholipase C-mediated phosphoinositide hydrolysis pathways appear to be involved in the signal transduction of apoA-I-mediated placental lactogen release. Since earlier studies have shown that phospholipase A2 and arachidonic acid also stimulate placental lactogen release, it is also possible that an arachidonic acid-dependent pathway is also involved in the regulation of apoA-I-mediated placental lactogen release. hPL release therefore appears to be mediated by a complex interaction of multiple signal transduction pathways.

Aberrations of placental lactogen secretion have been detected in many pathologic conditions of pregnancy associated with abnormal fetal growth, including preeclampsia, diabetes mellitus, intrauterine growth retardation, and erythroblastosis. In one large series of patients, a single placental lactogen concentration <4000 ng/ml in the last 5 weeks of pregnancy was associated with 30% risk of fetal distress or neonatal asphyxia. Low placental lactogen concentrations on two separate occasions during the last 5 weeks were associated with a fetal risk of 50% and low concentrations on three occasions with a risk of 71%. In another study, placental lactogen concentrations <4000 ng/ml were detected in 47 of 98 (48%) preeclamptic patients. Perinatal mortality in neonates born to the mothers with low placental lactogen concentrations was 13%, and intrauterine growth retardation was noted in 57% of the neonates.

Several pregnancies have been reported in which the mothers had absent or extremely low placental lactogen concentrations during gestation resulting from deletions of the hPL-A and hPL-B genes. In each instance, the pregnancies were uneventful, and the neonates were normal size at birth. Although extremely low placental lactogen concentrations were detected in the circulation, placental lactogen-like immunoreactivity was detected in the placentas. It appears that expression of the hPL-L gene may account for the placental lactogen-like immunoreactivity found in the placentas of these mothers and that expression of the hPL-L gene can compensate for deletions of the other two hPL genes.

B. Placental Growth Hormone

The amino acid sequence of hGH-V differs from that of hGH in 15 positions, 13 of which are in the mature protein and are distributed throughout the sequence. hGH-V mRNA expression is first detected in the syncytiotrophoblast layer of the placenta at 9 weeks of gestation and then increases progressively during the remainder of pregnancy. The tissue-specific expression of hGH-V and the pattern of plasma hGH-V concentrations during pregnancy are similar to those of hPL. Immunocytochemical studies of hGH-V also indicate that the protein is located in the placental basal plate, a region that contains fetal and maternal cells as well as the syncytiotrophoblast cells of placental villi.

HGH-V is first detected in maternal plasma at about 10 weeks of pregnancy and reaches a maximum of about 25 ng/ml in the third trimester. In contrast, plasma concentrations of pituitary growth hormone decrease to undetectable levels during the second half of pregnancy, probably due to suppression at the level of the pituitary. Pituitary growth hormone, however, is detected in relatively high concentrations in the fetus, whereas placental growth hormone is not detected in the fetal circulation. hGH-V is also not detected in amniotic fluid.

Although little is known about the regulation of the synthesis and release of placental growth hormone, it appears that the regulation of placental growth hormone is markedly different than that of pituitary growth hormone. For example, pituitary growth hormone is secreted episodically, whereas placental growth hormone is secreted tonically. In addition, GHRH does not stimulate placental growth hormone release *in vitro* from human trophoblast cells. The release of hGH-V may be stimulated by thyroid hormone.

Like pituitary growth hormone, placental growth hormone binds to both somatotrophic and lactogenic plasma membrane receptors and has biological activities similar to those of prolactin and growth hormone. However, the specificity of placental growth hormone for the somatotrophic receptor is approximately sevenfold greater than that of pituitary growth hormone, whereas the lactogenic activities of the two growth hormones are identical or nearly identical. This finding suggests that placental growth hormone may be a more potent growth-promoting factor than pituitary growth hormone. The finding of a significant correlation between placental growth hormone and IGF-I concentrations during late pregnancy suggests that placental growth hormone may also stimulate IGF-I production. Presumably, placental growth hormone, in concert with placental lactogen and pituitary prolactin, also plays a role in preparing the breast for lactation.

C. Decidual Prolactin

During the luteal phase of the cycle, endometrial stromal cells under the influence of progesterone undergo a "decidual" reaction in which the cells differentiate to a phenotype with different morphologic and biochemical characteristics. This decidualization process is critical for favorable uterine receptivity of the blastocyst and survival of the embryo. If pregnancy ensues, the decidualized endometrial stromal cells line the fetal placental membranes and comprise the maternal component of the placenta. During decidualization, the endometrial cells express several proteins that are specific markers for endometrial cell differentiation, including prolactin and insulin-like growth factor-binding protein-1 (IGFBP-1). Prolactin is also detected in endometrial tissues of women receiving progesterone or combined progesterone/estrogen therapy and in endometrial tissue exposed *in* vitro to progesterone.

Numerous studies strongly suggest that the decidua is the source of the large amounts of prolactin in amniotic fluid. In *in vitro* studies, prolactin synthesized by decidual tissue is transported across the fetal membranes to the luminal side of the amnion (i.e., toward the amniotic cavity). Although prolactin is transported from decidual tissue to amniotic fluid, there is no evidence for the transport of decidual prolactin to either the maternal or fetal circulations. The amounts of prolactin in decidual tissue and amniotic fluid reach maximal concentrations during the first half of gestation, whereas maternal and fetal plasma prolactin concentrations reach maximal concentrations near term. Maximal amniotic fluid prolactin concentrations of approximately 4000 ng/ml are attained at 16–20 weeks of gestation when maternal and fetal prolactin concentrations are approximately 50 and 10 ng/ml, respectively. Abnormalities of amniotic fluid prolactin concentrations and decidual prolactin production have been noted in several pathologic conditions of pregnancy, including diabetes mellitus, intrauterine growth retardation, moderately severe toxemia, moderately severe erythroblastosis, chronic hypertension, and idiopathic polyhydramnios.

Numerous investigations suggest physiologic roles for decidual prolactin during pregnancy. Since prolactin affects water and ion transport in lower vertebrates and binds specifically to amniotic membranes with high affinity, amniotic fluid prolactin may play a role in the regulation of amniotic fluid volume and ion content. *In vitro* and *in vivo* studies suggest that amniotic fluid prolactin may protect the fetus from sudden changes in the extracellular fluid content of water and electrolytes that might occur as a consequence of altered amniotic fluid tonicities. Amniotic fluid prolactin may also be one of the factors involved in the regulation of surfactant synthesis in the fetus. Prolactin has also been shown to inhibit uterine contractility, suppress the immune response, and be involved in the maintenance of T cell immunocompetence. Since prolactin is present in decidual tissue throughout gestation, decidual prolactin may therefore act locally on the uterus to inhibit myometrial contractility and prevent immunologic rejection of the blastocyst and fetus. The observation that human decidual tissue contains specific membrane receptors for prolactin further supports an autocrine/paracrine role for decidual prolactin.

Although decidual prolactin has chemical and biological properties that are identical to those of pituitary prolactin, the mechanisms involved in the release of decidual prolactin are different than those involved in the release of pituitary prolactin. Comprehensive reviews of the factors involved in the regulation of the synthesis and release of decidual prolactin have been recently published. TRH, dopamine, and bromocriptine, each of which affects the synthesis and release of pituitary prolactin, do not affect either the synthesis or the release of decidual prolactin—even at concentrations considerably greater than the half-maximal concentrations that affect pituitary prolactin release. The observation that bromocriptine does not affect the release of decidual prolactin is consistent with clinical studies that indicate that bromocriptine therapy during pregnancy has no effect on amniotic fluid prolactin concentrations even though maternal and fetal plasma prolactin concentrations are markedly inhibited.

The release of decidual prolactin is stimulated by decidual prolactin-releasing factor, a 23.5-K Mr protein that has been purified to homogeneity from conditioned medium of human placental explants and from extracts of human placenta. The releasing factor, however, has no effect on the release of pituitary prolactin or other pituitary hormones. Recent studies have also demonstrated that decidual prolactin release is acutely stimulated by physiologic concentrations of free α molecules, forms of a glycoprotein subunits that no longer combine with gonadotropin β subunits. IGF-I, which is synthesized and released by the placenta, stimulates decidual prolactin release. Insulin also stimulates the synthesis and release of decidual prolactin release. However, the stimulation by insulin does not result from the binding of insulin to the IGF-I receptor. Relaxin, which is synthesized by both decidual and placental tissues and has homologies in chemical structure to both IGF-I and insulin, also stimulates prolactin release. Since relaxin has not been observed to bind to insulin or IGF-I receptors in other cells, the stimulation of prolactin release by the three hormones appears to be due to binding to three distinct receptors. The release of decidual prolactin is inhibited by lipocortin I, one of six members of the lipocortin/annexin group of calcium- and phospholipid-binding proteins that are present in the placenta, as well as by TNF-α and several other cytokines.

Both the protein kinase A and protein kinase C signal transduction pathways are involved in the regulation of decidual prolactin release. In addition, arachidonic acid, which is released from membrane phospholipids by the activation of phospholipase A2, may act as a second messenger in the release of decidual prolactin. Currently, little is known about the transcription factors involved in transcription of the decidual prolactin gene. Although the transcription factor PIT-1 is important in the regulation of pituitary prolactin gene expression, PIT-1 has no role in the regulation of decidual prolactin gene expression.

III. CHORIONIC GONADOTROPIN

hCG, like the pituitary glycoprotein hormones, is composed of two subunits, an α subunit and a β subunit. The α subunit is identical to that of the pituitary glycoprotein hormones. The β subunit, which is larger than the β subunit of LH and the other pituitary glycoprotein hormones, contains a 24-amino acid extension at the carboxyl-terminal end that is not present in β-hLH. The α and β subunits are regulated separately and are derived from separate genes. The α subunit of hCG and the glycoprotein hormones is coded by a single gene located on chromosome 6 (q12–q21). β-hCG and β-LH are coded by eight distinct genes located on chromosome 19. Seven of the genes code for β-hCG, but only three of the genes are expressed. The amino acid sequence of the free α subunit in the placenta and blood during pregnancy is identical to that of the α subunit of intact hCG except for a difference in the oligosaccharide structure. The N-linked oligosaccharides of free α subunit are more highly branched and more extensively fucosylated and sialylated than those of the α subunit of intact hCG. The carbohydrate moieties of hCG are not essential for receptor binding but appear to protect the hormone from degradation. The individual subunits do not bind to the LH/hCG receptor and do not have biological actions.

During the first 6 weeks of gestation, hCG is synthesized by cytotrophoblast cells. However, after the sixth week the hormone is synthesized by syncytiotrophoblast cells. The intact hCG molecule as well as the β subunit are synthesized by several fetal tissues as well as by choriocarcinoma cells. Intact hCG has been detected in blastocysts 7 days after fertilization and in maternal blood between 7.5 and 9.5 days after the midcycle LH surge that precedes ovulation. hCG concentrations then increase rapidly, reaching a maximum of 10–15 mg/ml at about 8–10 weeks of gestation. Free α subunit is detected in maternal blood by the sixth week of gestation and its concentration increases gradually thereafter to a maximum of about 0.5 mg/ml. During the third trimester of gestation, free α subunit constitutes approximately 30–50% of total α-hCG in maternal blood. The concentration of hCG in maternal urine parallels closely that in plasma. hCG is also present in fetal blood although at a significantly decreased amount. Multiple isoforms for hCG are present in maternal circulation and are excreted in urine. The circulation also contains free α subunits that do not combine with the β subunit. Elevated plasma concentrations of hCG and β-hCG have been detected in women bearing fetuses with trisomy 21 (Down's syndrome). Vaccines against hCG have been used successfully as a contraceptive.

The most studied function of hCG is its role in regulating ovarian progesterone synthesis in early pregnancy. hCG stimulates steroid precursor availability by increasing the synthesis of low-density lipoprotein (LDL) receptors and increasing the internalization, degradation, and intracellular utilization of LDL as well as by stimulating the synthesis of components of the cholesterol side chain cleavage complex. hCG may also play a role in regulating placental growth and the production of other placental hormones and growth factors. hCG also appears to be important in regulating adrenal and testicular steroidogenesis in the fetus. In pathologic conditions associated with markedly elevated plasma concentrations of hCG, such as choriocarcinomas and hydatidiform moles, hCG may also have a thyrotopic action resulting in hyperthyroidism.

The regulation of hCG release from cultures of trophoblast cells has been found to be affected by many different factors, including GnRH, inhibin, TRH, CRH, somatostatin, opioids, oxytocin, vasopressin, several cytokines, epidermal growth factor, and numerous other growth factors and hormones. However, the physiological role of most of these factors in the regulation of hCG secretion during pregnancy is unclear. Since many of these factors are synthesized and released by the placenta, the factors may act, at least in part, by an autocrine–paracrine mechanism. Both the protein kinase A and protein kinase C signal transduction pathways appear to be involved in the regulation of hCG secretion.

IV. SUMMARY

During pregnancy the placenta and decidua synthesize and secrete many hormones with chemical and biological properties similar to those of pituitary and hypothalamic hormones. hPL, which has striking homologies to pituitary growth hormone and prolactin, functions as a "growth hormone" of pregnancy, regulating both maternal and fetal metabolism. The hormone affects fetal growth by increasing the availability of substrates to the fetus and stimulating the utilization of substrates by fetal tissues. In addition, hPL stimulates the synthesis of IGFs, which are critical growth factors for fetal growth and development. hGH-V, which has striking homologies in its biological properties to pituitary growth hormone, is not present in the fetal circulation but is present in relatively large amounts in the maternal circulation. Decidual prolactin, which is identical in chemical and biological properties to pituitary prolactin, probably plays a role in immunocompetence during pregnancy and in the stimulation of fetal pulmonary surfactant synthesis. hCG, which is similar in its chemical and biological properties to pituitary LH, is critical for the maintenance of the corpus luteum in early pregnancy and the stimulation of fetal steroidogenesis in the testes and adrenal.

Acknowledgments

This work was supported by NIH Grants HD-07447 and HD-15201.

See Also the Following Articles

Decidua; IGF (Insulin-like Growth Factors); LH (Luteinizing Hormone); Placenta, Implantation and Development; Prolactin, Actions of

Bibliography

Anthony, R. V., Pratt, S. L., Liang, R., and Holland, M. D. (1995). Placental–fetal hormonal interactions: Impact on fetal growth. *J. Anim. Sci.* 73(6), 1861–1871.

Eberhardt, N. L., Jiang, S. W., Shepard, A. R., Arnold, A. M., and Trujillo, M. A. (1996). Hormonal and cell-specific regulation of the human growth hormone and chorionic somatomammotropin genes. *Prog. Nucleic Acid Res. Mol. Biol.* 54, 127–163.

Evain-Brion, D. (1994). Hormonal regulation of fetal growth. *Horm. Res.* 42(4–5), 207–214.

Forsyth, I. A. (1991). The biology of the placental prolactin/growth hormone gene family. *Oxford Rev. Reprod. Biol.* 13, 97–148.

Handwerger, S., and Brar, A. (1992). Placental lactogen, placental growth hormone, and decidual prolactin. *Sem. Reprod. Endocrinol.* 10, 106–115.

Iles, R. K., and Chard, T. (1993). Molecular insights into the structure and function of human chorionic gonadotrophin. *J. Mol. Endocrinol.* 10(3), 217–234.

Ogren, L., and Talamantes, F. (1994). The placenta as an endocrine organ: Polypeptides. In *The Physiology of Reproduction* (E. Knobil and J. D. Neill, Eds.), 2nd ed., pp. 875–945. Raven Press, New York.

Placental Gas Exchange

Lawrence D. Longo
Loma Linda University

I. Introduction
II. Factors Affecting Placental Oxygen Transfer
III. Acid Base Regulation in the Fetal Placental Unit
IV. High Altitude and Respiratory Gas Exchange
V. Comparison of Gas Exchange in the Placenta and Lung
VI. Conclusions

GLOSSARY

acclimatization Physiological adjustment to a new environment; occurs over hours to weeks.

adaptation A characteristic of structure, function, or behavior that enables an organism to live and reproduce in a given environment; occurs over years to generations.

diffusing capacity A quantitative measure of respiratory gas exchange.

hypoxemia Deficient oxygenation of arterial blood.

hypoxia Relative lack of oxygen (e.g., low oxygen content or tension) in the ambient air or tissues.

ischemia Insufficiency of blood flow to a given organ or tissue.

oxygen affinity Expression of tendency of O_2 to combine with hemoglobin by chemical bonding.

oxygen capacity Expression of the amount of O_2 in blood combined with hemoglobin.

oxygen transport system The lungs, heart, and blood vessels which conduct oxygen from the atmosphere to the mitochondria of cells, the site of oxidative phosphorylation.

oxyhemoglobin Hemoglobin chemically bound to oxygen.

Development of the embryo and fetus demands appropriate exchange of the respiratory gases oxygen and carbon dioxide across the placenta between maternal and fetal blood. For the fetus, the placenta, which couples substrate delivery from the mother to its needs, serves as its lung as well as fulfilling the functions of many other organs to extrauterine existence.

I. INTRODUCTION

The placenta supplies about 8 ml O_2 per minute per kilogram fetal mass (about twice that of an adult per weight basis, e.g., 24 ml/min for a 3-kg term

TABLE 1
Normal Values of O_2, CO_2, and pH in Human Maternal and Fetal Blood

	Maternal uterine		Fetal umbilical	
	Artery	Vein	Vein	Artery
P_{O_2} (Torr)	95	38	30	22
HbO_2 (% saturation)	98	72	75	50
O_2 content (ml/dl)	16.4	11.8	16.2	10.9
O_2 content (mM)	7.3	5.3	7.2	4.5
Hb (g/dl)	12.0	12.0	16.0	16.0
O_2 capacity (ml/dl)	16.4	16.4	21.9	21.9
O_2 capacity (mm)	7.3	7.3	9.8	9.8
P_{CO_2} (Torr)	32	40	43	48
CO_2 content, mM	19.6	21.8	25.2	26.3
HCO_3^-	18.8	20.7	24.0	25.0
pH	7.42	7.35	7.38	7.34

Note. P_{O_2} and P_{CO_2}, partial pressures of O_2 and CO_2, respectively; Hb, hemoglobin.

fetus), and because fetal blood O_2 stores are only sufficient for 1 to 2 min, this must be continuous on a moment-to-moment basis. Table 1 gives normal values of blood gases and pH in maternal and fetal placental exchange vessels.

II. FACTORS AFFECTING PLACENTAL OXYGEN TRANSFER

Placental O_2 exchange is altered by varying the properties of maternal or fetal blood, such as O_2 capacity or affinity, or by variations in maternal or fetal placental blood flow. Respiratory gas transfer is also dependent on the spatial configuration of the blood vessels and on the diffusion characteristics of the placental membranes. Table 2 lists some of these factors and their components. Many of the variables which affect placental gas exchange are interdependent. Uterine and umbilical venous outflows, rather than representing blood from a single exchange unit, consist of blood from numerous compartments with differing O_2 and CO_2 tensions and contents. This results from a combination of nonuniform distribution of maternal and fetal placental blood flows, vascular shunts, and so forth.

A. Placental Diffusing Capacity

As in the lung, the quantity of O_2 crossing the placenta is a function of the so-called "diffusing capacity" and the partial pressure gradient. The diffusion characteristics of the placental membrane may

TABLE 2
Principal Factors Affecting Placental Oxygen Transfer

Variable	Associated components
Placental diffusing capacity	Membrane diffusing capacity (area, thickness, O_2 solubility, diffusivity of tissues); capillary blood volume; diffusing capacity of blood (O_2 capacity, hemoglobin reaction rates, concentration of reduced hemoglobin)
Maternal arterial P_{O_2}	Inspired P_{O_2}; alveolar ventilation: mixed venous P_{O_2}; pulmonary blood flow; pulmonary diffusing capacity
Fetal arterial P_{O_2}	Umbilical venous P_{O_2}; fetal O_2 consumption; peripheral blood flow; maternal arterial P_{O_2}; maternal placental hemoglobin flow; placental diffusing capacity
Maternal placental hemoglobin flow rate	Arterial pressure; placental resistance to blood flow; venous pressure; blood O_2 capacity
Fetal placental hemoglobin flow rate	Umbilical arterial blood pressure; umbilical venous blood pressure (or maternal vascular pressure under conditions of sluice flow); placental resistance to blood flow; blood O_2 capacity
Spatial relation of maternal to fetal flow	—
Amount of CO_2 exchange	—

be described by Ficks' first law:

$$\frac{dQ}{dt} = AD\,\Delta X \qquad (1)$$

where dQ/dt is the quantity of a given substance (e.g., O_2) crossing the membranes per unit time, A is the exchange area, D is the diffusion constant (cm²/sec), ΔC is the concentration difference (by volume) across the membrane, and ΔX is the diffusion distance.

The placental membrane is a complex structure with varying thickness and permeability. Nonetheless, overall, permeability, diffusibility, and thickness may be treated as constants and combined into a single term which expresses the membrane characteristics

$$\frac{dQ}{dt} = D_p P_z \qquad (2)$$

where D_p is the placental diffusing capacity in ml/min/Torr partial pressure difference for gas z. For respiratory gases, the Bunsen solubility coefficient (α) and the partial pressure difference (ΔP) are used rather than the concentration difference (ΔC). Thus, placental diffusing capacity is commonly expressed as

$$D_p = \dot{V}/P_m - P_f \qquad (3)$$

where $\dot{V}$ is the quantity of respiratory gas exchanging across the placental membrane per unit time, and $P_m - P_f$ is the mean partial pressure difference between maternal and fetal placental exchange vessels.

Ideally, one would like to study O_2 or CO_2 exchange using those gases; however, this is not practical because significant amounts of the total O_2 exchanging are consumed by the placenta (probably close to one-third at term). In addition, uterine and umbilical mixed venous blood samples represent a mixture of blood from compartments with differing ratios of maternal to fetal blood flow and probably differing ratios of diffusing capacity to blood flows. In almost all circumstances, O_2 exchange is limited by blood flow rather than diffusion. Thus, to quantify exchange, a metabolically inert gas whose exchange is limited by diffusion and which combines with hemoglobin must be used.

Carbon monoxide in low concentrations has been shown to be the most practical gas for studies of transplacental diffusion. The placental diffusing capacity for carbon monoxide (D_{PCO}) can be calculated by use of the Haldane relation:

$$[HbCO] = P_{CO} \times M[HbO_2]/P_{O_2} \qquad (4)$$

where [HbCO] is the carboxyhemoglobin concentration, [HbO_2] is the oxyhemoglobin saturation, P_{CO} is the CO partial pressure in Torr, and M is the relative affinity of hemoglobin for CO compared with O_2.

Placental CO diffusing capacity equals about 0.6 ml/min/Torr/kg fetal weight in several species (sheep, dog, and macaque monkeys); however, in the rabbit and guinea pig it is several-fold greater. Such studies suggest that the mean maternal to fetal partial pressure difference for O_2 equals only about 6 Torr, a value similar to that of the pulmonary alveolar to capillary P_{O2} difference. In turn, this suggests that the placenta does not constitute a significant barrier to respiratory gas diffusion, and that placental O_2 exchange is limited by blood flow rather than by diffusion.

During the course of gestation, the placental mass and exchange area increase to meet the demands of the developing conceptus. Nonetheless, during the last trimester, while fetal mass increases three- or fourfold, and placental mass doubles (so that the ratio of placental to fetal mass is halved), and placental diffusing capacity (per kilogram fetal weight) remains constant.

Unfortunately, no reliable measurements of D_{PCO} in humans are available. The value would be predicted to decrease in conditions in which the placental membranes are thickened (e.g., diabetes mellitus and syphillus, edema), in association with intrauterine growth retardation and in association with decreased blood volume or hemoglobin concentrations in the placental exchange vessels. Such clinical associations have not been established, however.

B. Maternal and Fetal Arterial O_2 Partial Pressures

Both theoretical and experimental studies suggest that placental O_2 exchange is sensitive to changes in maternal or fetal arterial O_2 tensions. Above ~70 Torr the oxyhemoglobin saturation curve is relatively flat, oxyhemoglobin remains saturated, and placental

O_2 exchange and fetal oxygenation are normal. However, a decrease in maternal arterial P_{O_2} below ~70 Torr results in decreased amounts of O_2 crossing to the fetus. On the other hand, raising maternal arterial P_{O_2} to ~600 Torr by breathing 100% O_2 increases the amount of O_2 in maternal blood slightly and increases fetal umbilical venous O_2 tension 3–5 Torr. Although probably of little value in normal circumstances, such an increase in fetal blood O_2 tension may be of great benefit in instances of fetal hypoxemia.

Fetal arterial O_2 tensions also influence placental O_2 exchange, the amount of such exchange varying inversely with the umbilical P_{O_2} value. Of course, fetal arterial P_{O_2} in turn is a function of transplacental O_2 exchange, umbilical venous P_{O_2}, and the rate of fetal O_2 consumption.

Umbilical venous O_2 tension normally is 10–15 Torr less than that of the uterine venous blood. The blood gas values shown in Table 1 are based on studies in chronically catheterized sheep and monkeys as well as on human data obtained by puncture of the umbilical cord under ultrasonic guidance (cordocentesis). A number of factors could theoretically affect placental O_2 exchange and account for the O_2 tension difference between uterine and umbilical venous blood. Such factors include the geometric relation of fetal vessels to maternal blood in the exchange area, placental shunts in which uterine or umbilical arterial blood enter the venous circulation without traversing the exchange areas, and nonuniform or uneven distribution of maternal and fetal blood flow in localized regions of the placenta.

C. Maternal and Fetal Blood O_2 Affinity and Capacity

Hemoglobin in maternal blood contributes considerably to placental O_2 transfer. The reduced form of hemoglobin binds with O_2 to form oxyhemoglobin. This binding is reversible so that as the O_2 partial pressure decreases hemoglobin unloads O_2 to diffuse across the placenta. The ability of hemoglobin to bind oxygen depends not only on the P_{O_2} but also on the hemoglobin–O_2 affinity, as indicated by the sigmoid-shaped oxyhemoglobin curve (Fig. 1).

The P_{50} describes the O_2 partial pressure required to half-saturate hemoglobin. Under standard conditions (pH = 7.40, P_{CO_2} = 40 Torr, 37°C) the P_{50} for normal adult human blood, including that of the pregnant mother, is 26.5 Torr (Fig. 1). The curve is shifted to the right (i.e., lowered O_2 affinity) in association with increased concentrations of CO_2, hydrogen ion (H^+), 2,3-diphosphoglycerate, adenosine triphosphate, and/or chloride ion. In many species, but not all, the fetal oxyhemoglobin saturation curve is shifted to the left compared to that of maternal blood. The P_{50} for fetal blood near term is about 20 Torr (Fig. 1). Under physiologic conditions *in vivo* the maternal curve is shifted to the left (pH = 7.42, P_{CO_2} = 34 Torr), whereas that of the fetus is shifted to the right (pH = 7.35, P_{CO_2} = 45 Torr, 37.5°C) so that they are, in fact, almost superimposed (Fig. 1).

Blood oxygen capacity is the maximum amount of O_2 which can reversibly bind with hemoglobin. With a hemoglobin concentration of 14 g/dl the nonpregnant woman has a blood O_2 capacity of ~19 g/dl ([Hb] × 1.36). During the course of gestation a physiological hemodilution occurs as plasma volume increases ~50% while erythrocyte mass increases ~25%. Thus, near-term maternal hemoglobin concentration decreases to ~11.5 g/dl with an O_2 capacity of 15.6 g/dl. In humans, fetal hemoglobin concentration increases from ~8.5 g/dl at 10 weeks gestation to a mean value of ~16.5 g/dl at term. Thus, during the last third of gestation the fetal blood O_2 capacity exceeds that of the mother (Fig. 2).

As noted previously, *in vivo* the maternal and fetal O_2 saturation curves are probably superimposed. Figure 2 shows maternal and fetal blood O_2 content as a function of O_2 partial pressure. This illustrates that a normal fetal umbilical venous P_{CO_2} of only ~28 Torr is associated with an O_2 content of 15.5 ml/dl, a value as great as the maternal O_2 content of 15.4 ml/dl. Thus, despite the fetal hemoglobin being only ~75% saturated compared to ~98% in the adult, its greater hemoglobin concentration allows for a higher O_2 content. The maternal and fetal blood oxyhemoglobin saturations have important implications for placental O_2 transfer; an increase of either maternal or fetal O_2 capacity promotes placental O_2 exchange. Other factors remaining constant, the larger the sum of maternal and fetal blood O_2 capacity, the more O_2

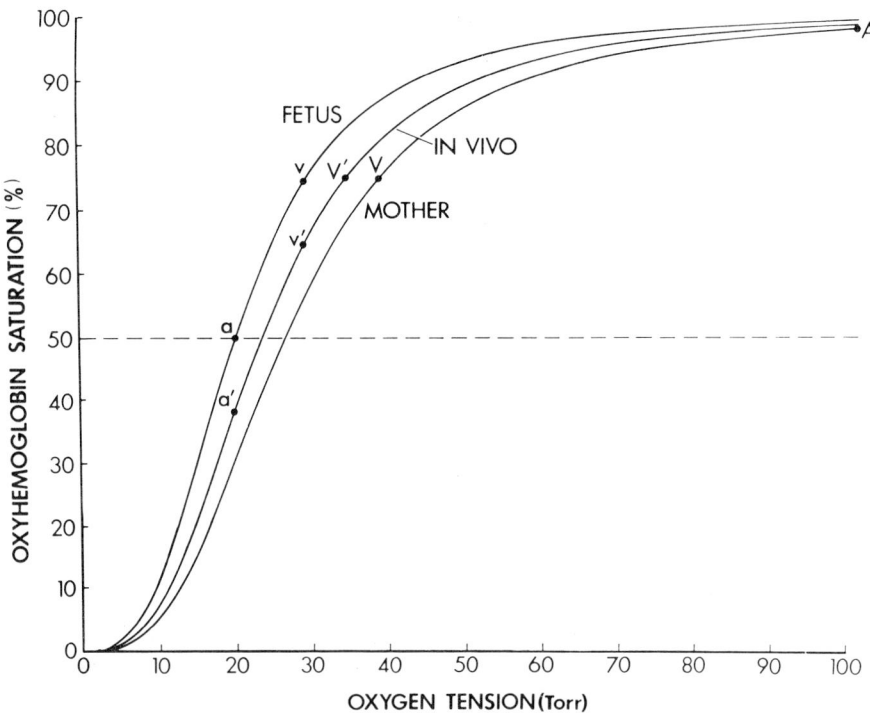

FIGURE 1 HbO$_2$ saturation curves for human maternal and near-term fetal blood. Maternal and fetal HbO$_2$ affinities (P$_{50}$) are 26.5 and 20 Torr, respectively. A and V, maternal arterial and venous values, respectively, under standard conditions; a and v, umbilical arterial and venous values, respectively; V', a', and v', probable *in vivo* maternal venous, umbilical arterial, and umbilical venous values, respectively.

will be exchanged before equilibration of P$_{O_2}$ values in these bloodstreams are reached.

D. Bohr and Haldane Effects

As maternal and fetal blood course through placental exchange vessels, H$^+$ and CO$_2$ diffuse from fetal blood across the placenta so that the fetal curve is shifted to the left promoting O$_2$ uptake by the fetal erythrocytes. At the same time, maternal blood becomes more acidotic and hypercarbic, shifting the oxyhemoglobin saturation curve to the right and thus increasing the O$_2$ available for transfer. Theoretical studies suggest that this mechanism, the so-called "Bohr effect," accounts for ~8% of placental O$_2$ exchange.

As a consequence of this exchange process, deoxyhemoglobin concentration increases in the maternal placental blood while decreasing in that of the fetus. Because deoxyhemoglobin binds CO$_2$ to a greater extent than oxyhemoglobin, CO$_2$ exchange from fetal to maternal blood is augmented. This so-called "Haldane effect" is calculated to account for ~46% of placental CO$_2$ exchange.

E. Maternal and Fetal Placental Blood Flows

To a great extent, placental O$_2$ and CO$_2$ exchange depend on the rates of uterine and umbilical blood flows. During the course of gestation uteroplacental blood flow increases from very little to ~1000 ml/min at term. Eighty to 90% of this flow supplies the placenta and thus is available for O$_2$ and nutrient transfer to the fetus. In the human, maternal blood from uterine spiral arteries enters the intervillous space and is diverted toward the placental villi. After bathing fetal vessels within the villi, uterine blood spreads laterally into the uterine venous sinuses. Studies in the monkey, sheep, and other species indi-

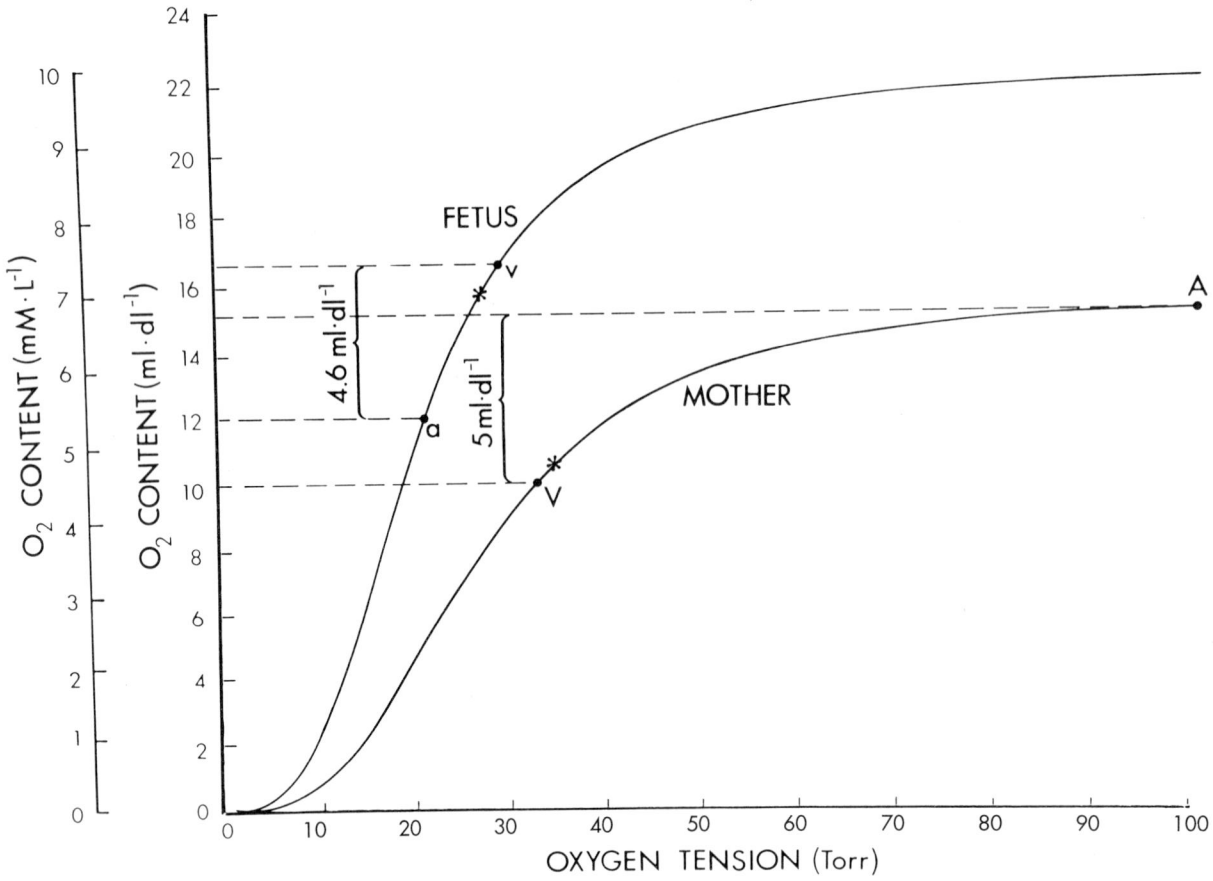

FIGURE 2 Blood O_2 content as function of P_{O_2} for maternal hemoglobin = 12 g/dl and P_{50} = 26.5 Torr, and near-term fetal hemoglobin = 16.5 g/dl and P_{50} = 20 Torr blood. Asterisks, mean maternal and fetal P_{O_2}; A and V, maternal arterial and venous values; a and v, umbilical arterial and venous values.

cate that placental O_2 exchange varies as a function of uterine blood flow (Fig. 3).

The role of uterine blood flow on placental O_2 exchange and fetal oxygenation has obvious clinical implications. For example, uteroplacental blood flow is reduced during the uterine contractions of labor, and this is associated with transient decreases in O_2 transfer to the fetus. In addition, blood flow may be decreased in women with vascular disease or hypertensive disorders of pregnancy. Fetal oxygenation is highly dependent on the rate of uterine blood flow. Fetal umbilical flow is also an important determinant of placental O_2 exchange.

In most vascular beds blood flow is proportional to the hydrostatic pressure difference between arterial and venous vessels. However, in the placenta, because of the close association of maternal and fetal circulations, evidence suggests that increases in maternal placental blood volume may impinge upon fetal vessels in placental villi, increasing the resistance of umbilical flow. Under such conditions fetal placental blood flow would be proportional to the fetal inflow pressure minus that of surrounding maternal placental blood pressure. Such a "sluice" or "waterfall" relation, in which surrounding pressure affects vascular resistance, appears to operate in the placenta as well as in the lung. In addition, evidence suggests that such a mechanism may also operate so that increases in fetal placental blood volume affect maternal placental flow. In a pregnant women near term, in the supine position the gravid uterus may compress the inferior vena cava, impeding uterine

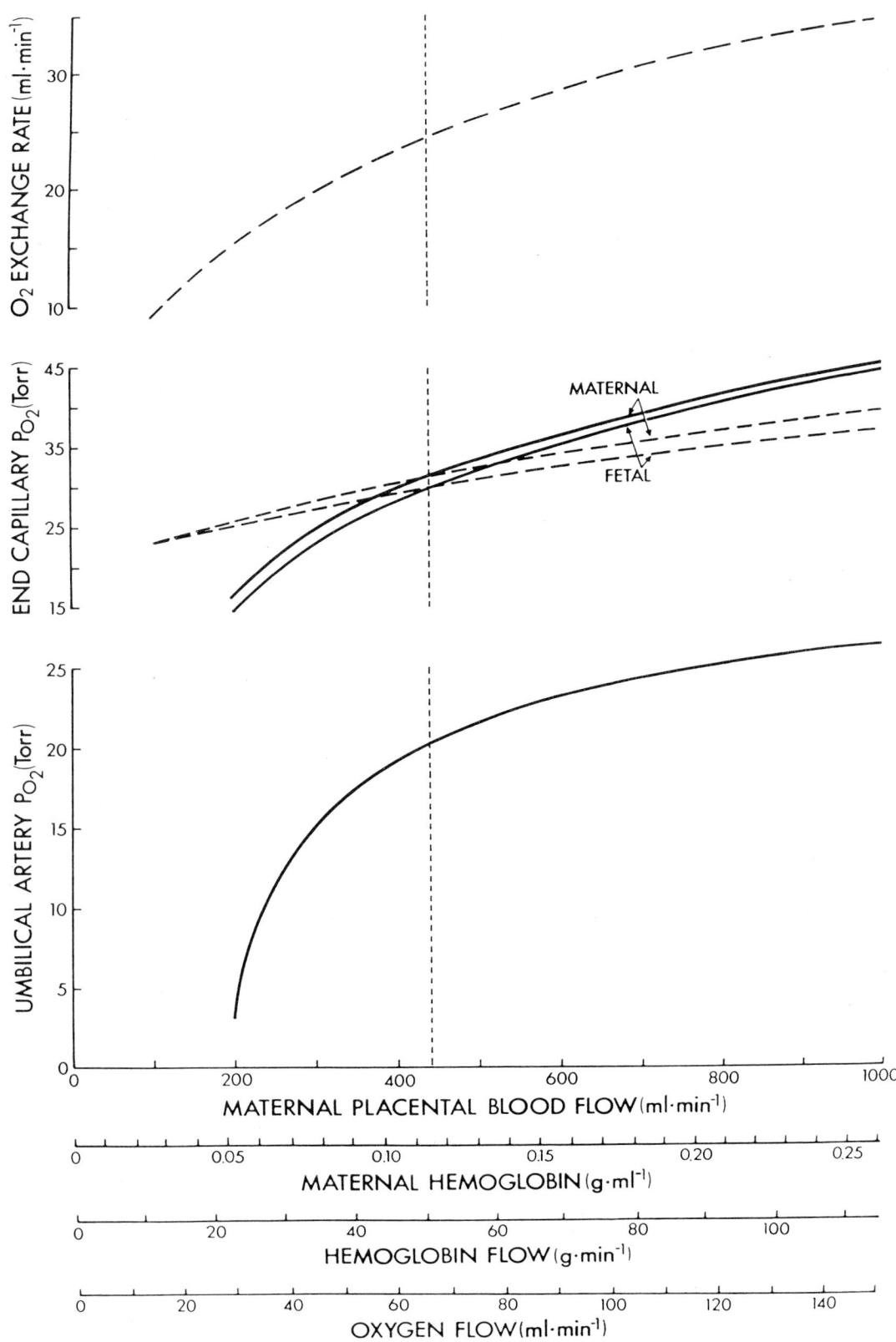

FIGURE 3 Calculated effects of changes in maternal placental blood flow, hemoglobin flow, and O_2 flow on transient maternal and fetal placental end capillary P_{O_2} values and V_{O_2} (dashed lines) and on steady-state end capillary P_{O_2} values and fetal arterial P_{O_2} (solid lines).

venous outflow and causing intervillous space pressure to rise. Thus, fetal placental vessels would be compressed, increasing vascular resistance and decreasing umbilical flow. Of course, fetal placental flow would be restored once fetal arterial pressure increased insufficiently.

F. Maternal and Fetal Placental Oxygen Flow

Oxygen delivery to an organ equals the product of blood flow and blood O_2 content. In many respects, although not all, a decrease in hemoglobin concentration has an effect similar to that of decreasing blood flow on O_2 delivery and exchange. In the maternal circulation, decreases in hemoglobin concentration may be compensated for by increases in uteroplacental blood flow. Figure 3 depicts how changes in uteroplacental blood flow and O_2 affect placental O_2 exchange. For the fetus, similar principles apply (Fig. 4). However, the fetus has a limited ability to increase its already relatively high cardiac output and thus its uteroplacental blood flow.

III. ACID BASE REGULATION IN THE FETAL PLACENTAL UNIT

A consideration of placental gas exchange would be incomplete without attention to the impact of respiration on acid base balance. In general, oxidative metabolism in the mother and fetus produces carbon dioxide, which is rapidly hydrated to carbonic acid. The carbonic acid then dissociates to form hydrogen and bicarbonate ions, such that

$$CO_2 + H_2O \rightleftharpoons H_2CO_3 \rightleftharpoons H^+ + HCO_3 \quad (5)$$

The Henderson–Hasselbach equation demonstrates that the pH of plasma is dependent on the ratio of the concentrations of bicarbonate and carbon dioxide such that

$$pH = 6.1 + \log([HCO_3^-]/\alpha\, PCO_2) \quad (6)$$

where α is the solubility coefficient for CO_2. Under normal physiologic conditions, the serum bicarbonate concentrations exceeds that of CO_2 by 20-fold.

Because of the concentration gradient required to transfer CO_2 across the placenta, the fetal pH is normally 0.1 unit lower than that of the mother (Table 1).

When placental oxygen transfer is compromised, there is impaired oxidative metabolism of carbohydrate to carbon dioxide and water, resulting in the accumulation of lactic acid. Incomplete oxidation of fatty acids and altered amino acid metabolism results in excess formation of ketoacid and uric acid, respectively. These nonvolatile acids traverse the placenta slowly in comparison to the rapid transit of CO_2.

Carbon dioxide is carried to the blood in three forms: ~62% is carried in the erythrocyte as bicarbonate, ~30% is carried by hemoglobin as carbamate, and the remaining 8% is dissolved. Furthermore, fetal bicarbonate transported to the placenta is converted by carbonic anhydrase to CO_2. The diffusion constant for CO_2 is approximately 20 times that of O_2, and because it is so diffusible its transport is chiefly affected by uteroplacental and umbilicoplacental blood flow rates. The placenta is poorly permeable to electrically charged bicarbonate. Nonetheless, bicarbonate appears to equilibrate across the placenta, not solely as a result of rapid CO_2 equilibration.

Buffers act to maintain both maternal and fetal acid base homeostasis, with hydrogen ions being buffered intracellularly and extracellularly. Carbon dioxide produced metabolically diffuses into the erythrocyte and is hydrated by carbonic anhydrase, and the hydrogen ion produced is buffered by erythrocyte hemoglobin, bicarbonate, and inorganic phosphate. Additional buffering is provided by the extracellular carbonic acid–bicarbonate buffer system, forming volatile CO_2, which is rapidly eliminated by the placenta and eventually the maternal lung.

Maternal respiratory changes occur in the first few weeks of pregnancy due to an increase in tidal volume. This results in a 60–70% increase in alveolar ventilation and a decrease in arterial CO_2 tension to the range of 26–34 Torr (Table 1). The fetal CO_2 tension is maintained in the 35–40 Torr range. Despite decreases in maternal CO_2, maternal pH remains normal secondary to compensatory increases in renal excretion of bicarbonate. Evidence suggests that increased maternal progesterone during pregnancy facilitates hyperventilation. This relative hy-

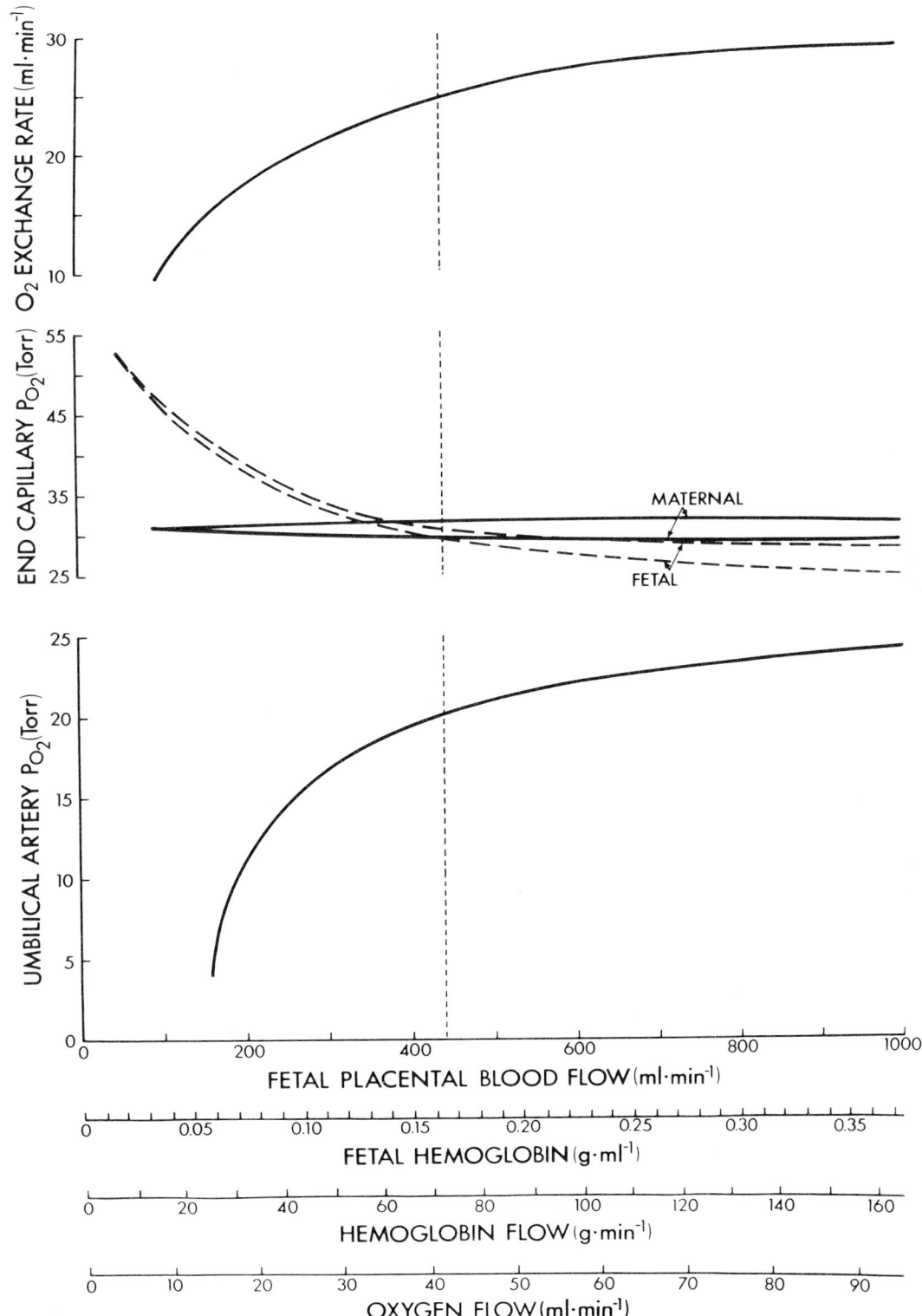

FIGURE 4 Calculated effects of changes in fetal placental blood flow, hemoglobin flow, and O_2 flow on transient maternal and fetal placental end capillary P_{O_2} values and V_{O_2} (dashed lines) and on steady-state end capillary P_{O_2} values and fetal arterial P_{O_2} (solid lines).

perventilation is exaggerated at high altitude and during labor, the latter being affected by such factors as pain, fear, and emotional excitement. Acute hyperventilation, although enhancing placental CO_2 exchange by increasing the transplacental CO_2 gradient, may endanger the fetus if extreme.

IV. HIGH ALTITUDE AND RESPIRATORY GAS EXCHANGE

Despite the hypoxia associated with high altitude, many individuals live at elevations over 3000 m, and in the Andes permanent residents live at higher than 4600 m. Because at sea level fetal arterial O_2 tension approximates that of an adult at approximately 5000 m, one would anticipate that the PO_2 of the fetus at high altitude would be much lower than normal. Surprisingly, this is not the case. Not only do the mother and fetus adapt to this hypobaric hypoxic environment but also the placenta displays compensatory mechanisms to ensure adequate fetal O_2 delivery. Acclimatization responses observed in pregnant women at 3100 m include hyperventilation and increased hemoglobin concentration. Fetal acclimatization responses include increases in fetal hemoglobin and redistribution of cardiac output. This redistribution of fetal blood flow maintains oxygen delivery to the fetal brain and heart at the expense of the kidney, gastrointestinal tract, and carcass. With respect to fetoplacental blood flow, umbilical artery resistance, as measured by Doppler, does not change significantly at altitudes up to 1500 m above sea level.

Examination of human placenta from pregnancies at high altitude reveals histologic features suggestive of placental hypoxia. Long-term exposure of the pregnant mother to high altitude results in placental changes which appear to optimize placental gas exchange to maintain fetal oxygenation. Changes observed in placentas from pregnancies at high altitude include increased volume of the fetal capillaries, dilation of the fetal capillary sinusoids, increased fetal arteriole and venule branching and capillary coiling, increased fetal and maternal luminal size per vessel, shorter and more tightly clustered peripheral villi, and thinning of the villous membranes.

The end result of these maternal, fetal, and placental acclimatization responses is illustrated by blood gas values and intrauterine growth statistics. At an altitude of 3100 m, newborn infants, before the onset of respiration, have umbilical arterial oxyhemoglobin saturations averaging 50% compared with 58% at 1525 m. The high-altitude fetus is also born with a mixed acid base disturbance, represented by a slight metabolic acidosis and respiratory alkalosis. The latter correlates with the decreased maternal P_{CO_2} observed with maternal hyperventilation accompanying acclimatization to high altitude.

Despite acclimatization responses to optimize placental respiratory gas exchange, newborn infants at high altitude show retardation of intrauterine growth and decreased birth weight. Additionally, there is an increased infant rate of morbidity and mortality observed at high altitude, although this occurs in primarily preterm infants.

V. COMPARISON OF GAS EXCHANGE IN THE PLACENTA AND LUNG

The question arises as to how respiratory gas exchange in the placenta compares with that of the lung. Table 3 presents comparative values for several measures of placental and pulmonary respiratory gas exchange and blood flow. Despite the similar weights of these organs, 10- to 20-fold more O_2 is exchanged per minute in the lungs than in the term placenta, in rough accord with the mass of organism supplied. Although in the lungs, O_2 consumption by parenchymal tissue is an insignificant fraction of the total quantity exchanged, at term ~30% of the O_2 derived from maternal blood is consumed by placental tissue before it reaches the fetus. In both organs O_2 and CO_2 exchange mutually enhance one another, and the placenta shows double Bohr and Haldane effects because these reactions occur in both maternal and fetal blood.

Both organs receive a generous blood supply. However, in the placenta 20–40% of flow functionally bypasses gas exchange sites, a fraction much larger than that in the lung except in newborns with marked uneven ventilation to perfusion. Although some of the placental shunt may be anatomical, probably most is physiologic, analogous to nonuniform distribution of ventilation to perfusion in the lung. Both

TABLE 3
Comparisons of Blood Flow and Gas Exchange in Placenta and Lungs

	Placenta	Lungs
CO diffusing capacity (ml/min·Torr)	1.81	25
CO diffusing capacity (ml/min·Torr·kg body wt)	0.60	0.42
CO diffusing capacity (ml/min·Torr·kg organ wt)	3.6	42
O_2 diffusing capacity (l/min·Torr)	2.3	30
Mean alveolar-to-pulmonary capillary P_{O_2} difference (Torr)		6
Mean maternal–fetal placental P_{O_2} difference (Torr)	6	
O_2 transfer rate (ml/min)	24	300
O_2 transfer per unit blood flow (ml/min)	6	54
Tissue O_2 consumption and CO_2 production	Significant	Insignificant
Interaction of O_2 and CO_2	Double Bohr and double Haldane effects	Bohr and Haldane effects
Fixed acid transfer	Significant	Insignificant
Blood flow (% of cardiac output)	45	100
Distribution	Uneven maternal flow/fetal flow	Uneven ventilation/blood flow
Shunt (%)	20	2
Type of flow	Sluice (maternal vascular pressure surrounding fetal capillaries)	Sluice (alveolar pressure surrounding pulmonary capillaries)
Regulation	Unknown	Active and precise

organs have flow characterized by a sluice or waterfall phenomenon. Although the pulmonary circulation displays active and precise regulation, the regulation of maternal and fetal placental blood flows remain poorly understood.

Another question concerns the relative efficiency of placental O_2 exchange. A large exchange surface with a small barrier (as expressed by the Dp) is advantageous for substrate exchange. In addition, both uniform distribution of maternal to fetal placental flow and a countercurrent flow pattern optimize exchange. Placental "efficiency" has been considered from several points of view, e.g., the magnitude of the degree of arterialization of umbilical venous blood, the maternal-to-fetal venous P_{O_2} or P_{CO_2} differences, the umbilical arterial to venous P_{O_2} difference, and the percentage of O_2 extracted from maternal arterial blood. In general, although the placenta is designed to facilitate respiratory gas exchange, the vascular architecture is not arranged most efficiently, O_2 consumption and CO_2 production occur in the regions of gas transfer, and inhomogeneities of several types introduce further inefficiency. Although it was previously thought that the membranes separating maternal and fetal blood were a significant barrier to diffusion, we now realize that, as in the lung, these tissues constitute only a minor resistance to exchange.

VI. CONCLUSIONS

Formerly, the placenta was considered as a glorified sieve separating the fetus from the mother. It now is understood to perform complex metabolic syntheses in the interplay of hormones between the two organisms and to serve other metabolic and immunologic functions as vital as respiratory gas exchange. Finally, it is designed for rapid growth during a brief life span. In view of the diversity of placental morphologic types and vascular arrangements in the various mammalian species, it is evident that a wide divergence of architecture is compatible with similar physiological functions. None of this new information denies the fact that the respiratory function of the placenta is of critical importance for optimal fetal development. It does, however, suggest that the respiratory function may be a subsidiary consideration in its design.

Problems for the future include elucidating the regulation of both maternal and fetal placental blood flows and their matching, as well as the regulation of vascular development in the placenta in response to long-term hypoxia and other stress.

See Also the Following Articles

ALTITUDE, EFFECTS ON REPRODUCTION; FETAL GROWTH AND DEVELOPMENT; FETAL HORMONES; FETAL-PLACENTAL UNIT; FETUS, OVERVIEW; PLACENTA, HUMAN

Bibliography

Bartels, H. (1970). *Prenatal Respiration*. North Holland, Amsterdam.

Carter, A. M. (1989). Factors affecting gas transfer across the placenta and the oxygen supply to the fetus. *J. Dev. Physiol.*; 12, 305–322.

Dawes, G. S. (1968). *Foetal and Neonatal Physiology. A Comparative Study of the Changes at Birth*. Year Book Pub., Chicago.

Faber, J. J., and Thornburg, K. L. (1983). *Placental Physiology*. Raven Press, New York.

Longo, L. D. (1987). Respiratory gas exchange in the placenta. In *Handbook of Physiology, Section 3: The Respiratory System, Vol. IV, Gas Exchange* (A. P. Fishman, L. E. Farhi, and S. M. Tenney, Eds.), pp 351–401. American Physiological Society, Washington, DC.

Wilkening, R. B., and Meschia, G. (1992). Current topic: Comparative physiology of placental oxygen transport. *Placenta* 13, 1–15.

Placental Lactogens

Daniel I. H. Linzer
Northwestern University

I. Introduction
II. Evolution of the Placental Lactogen Genes
III. Structure of the Placental Lactogens
IV. Synthesis of Placental Lactogens
V. Actions of Rodent Placental Lactogens
VI. Actions of Ruminant Placental Lactogens
VII. Future Directions in Placental Lactogen Research

GLOSSARY

corpus luteum The structure in the ovary that forms from a follicle that has ruptured to release a mature egg; a major source of progesterone.
lactogen A factor that stimulates milk production.
placental Synthesized by the placenta.
prolactin A protein hormone made by the pituitary gland that acts on numerous tissues in the body to stimulate, for example, mammary gland development and milk production and steroid hormone production by the ovary.
receptor A protein on or in a cell that is activated by the binding of a hormone or growth factor; receptors for protein hormones are on the surface of cells, where they contact the protein hormone in the circulation.
trophoblasts Extraembryonic cells that derive from the fertilized egg and are the major cell type that forms the placenta.

The placental lactogens are pregnancy-specific protein hormones that regulate maternal and fetal physiology. In rodents and ruminants, the placental lactogens are very similar to prolactin in structure and function. A primary function of these hormones is to stimulate the development of the mammary glands during pregnancy to prepare the mother for postpartum lactation. Additional functions may vary among species but likely include the regulation of steroid hormone production, maternal metabolism, and fetal growth.

I. INTRODUCTION

The placenta is a transient organ that plays numerous roles in mammalian reproduction. The placenta attaches the embryo to the mother's uterus, where the embryo grows and develops; it forms a barrier that prevents the mother's immune system from attacking the embryo (the embryo makes proteins specified by both the mother's and father's genetic information and would therefore be seen by the mother's immune system as foreign); it mediates the transport and exchange of nutrients and waste products between the mother and the fetus; and, it is an endocrine organ—a factory that produces hormones.

Some of the hormones produced in the placenta are steroid hormones, such as progesterone, and some of the hormones are proteins. Steroid hormones are able to penetrate the outer membrane of the cell and bind to specific receptor proteins within the cell that then turn on or turn off target genes. Protein hormones, in contrast, bind to specific receptor proteins on the surface of the cell. These receptors convert the hormone-binding event into a signal, much like the push of a finger on a button on one side of a door can trigger the ringing of a bell on the other side. Some of the placental protein hormones are identical to hormones produced elsewhere in the adult, but the placenta is also the source of hormones that are uniquely produced by this organ. These latter proteins constitute a set of pregnancy-specific hormones that may, for example, act on the mother to change her metabolism and behavior, on the fetus to stimulate growth, or on the placenta itself to regulate the amount of hormone being released into the circulation.

Although the set of these placental-specific protein hormones that are produced varies among mammals, the placenta of many rodents, ruminants, and primates produces protein hormones that are closely related to prolactin and growth hormone. These proteins are made at very high levels (the concentration of one of these placental hormones in the mother's bloodstream may be 100 times higher than the concentration of pituitary-derived prolactin or growth hormone) and generally represent the major placental-specific proteins that are released into the maternal circulation. The placental relatives of prolactin and growth hormone can be divided into two groups—those that have biological effects that are the same as prolactin's and those with distinct (or uncharacterized) activities. The hormones that act like prolactin are called placental lactogens (PLs), even though the term "lactogen" refers to only one function of prolactin—the ability to stimulate mammary development and milk production (lactation).

The PL group of hormones will be the focus of this chapter; recent reviews may be consulted for additional information on both the PLs and the other placental hormones related to prolactin. Since human PL (also referred to as chorionic somatomammotropin) is described in another article in this volume, this article will concentrate on the nonprimate PLs, in particular the PLs of rodents (mice and rats) and ruminants (sheep and cows). Mice and rats provide the major laboratory models for the study of the PLs; mice represent the primary laboratory model for studies in mammalian genetics, including investigations into the effects of the introduction or removal of genetic information, whereas rats are still preferred for studies on mammalian reproduction because of their short and very regular reproductive (estrous) cycle. Sheep and cows offer investigators the opportunity to isolate placental hormones in large quantities for biochemical and functional analyses, and studies with these species directly address important features in reproduction, growth, and milk production in commercially important animals.

II. EVOLUTION OF THE PLACENTAL LACTOGEN GENES

The amino acid sequences of the rodent and ruminant PLs are much more similar to the sequence of prolactin than of growth hormone. Consistent with the close relationship between the PLs and prolactin, the PL genes in rodents are not only located on the same chromosome as the prolactin gene (the growth hormone gene is on a different chromosome) but also found in a very closely spaced cluster of prolactin-related genes that includes the prolactin gene. Similarly, the bovine PL gene lies near the prolactin gene.

Since prolactin, but not the PLs, is found in fish, amphibians, and birds, the rodent and ruminant PL genes appear to be derived from the duplication and subsequent divergence of the prolactin gene after the appearance of mammals on earth. However, not all mammals have a similar cluster of prolactin-related genes, and in some species no placental hormones in this family have yet been identified. In humans the PL gene arose from the growth hormone gene, which is on a different chromosome than the prolactin gene, even though human PL is functionally more similar to prolactin than growth hormone. This diversity in the prolactin family of genes in mammals indicates that the PL genes arose after the first mammals branched off into different orders. The fact that different orders of mammals have separately evolved genes that code for placental hormones related to prolactin and growth hormone, and that function like prolactin, suggests that these hormones provide profound advantages for reproductive success.

III. STRUCTURE OF THE PLACENTAL LACTOGENS

The PLs are synthesized as precursor proteins of approximately 230 amino acids, including an extension at the beginning of each protein that serves as a signal for secretion of the mature hormone from the cell. The signal sequence of about 30 amino acids is cleaved during synthesis, so the mature hormones are generally 200 amino acids in length. In rodents, two PL proteins are produced, with PL-I further modified by extensive addition of carbohydrate, whereas PL-II does not have any attached sugars. Bovine PL is heavily glycosylated (like rodent PL-I), whereas ovine PL is not glycosylated (like rodent PL-II).

Despite the difference between PL-I and PL-II in terms of attachment of sugar groups, both proteins are predicted to form very similar three-dimensional

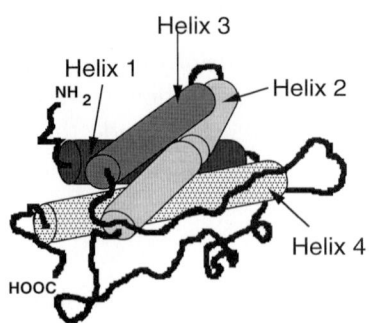

FIGURE 1 Schematic diagram of the three-dimensional structure of growth hormone. The structure of the PLs is expected to be very similar to the structure of growth hormone, which contains four helical regions (cylinders) connected by loops of protein (heavy lines). Note that the four helices come together in the center of the protein to form a bundle. The two ends are marked as the amino (NH_2) and carboxy (COOH) termini (drawing based on Abdel-Meguid et al., Proc. Natl. Acad. Sci. USA 84, 6434–6437, 1987).

structures. The structures of these proteins have not yet been determined experimentally, but the structure of growth hormone has been solved by a group of scientists at Monsanto Corporation. Given the sequence similarity among all the proteins in this family, it is likely that the PLs will be found to have a structure very similar to that of growth hormone. The growth hormone protein chain contains four helical regions that come together to form a bundle (Fig. 1), and the sequence of amino acids in the PL chains is consistent with the formation of similar four-helix structures. In prolactin and related hormones, the four-helix bundle structure is reinforced by bonds between two pairs of cysteine amino acids that are found in the same positions in each of these proteins. However, PL-I appears to be uniquely unable to make both of these cysteine–cysteine bonds because of the presence of another, interfering cysteine in the protein chain.

IV. SYNTHESIS OF PLACENTAL LACTOGENS

Two peaks of lactogenic hormone activity during pregnancy were discovered in the rat by Henry Friesen's group and in the mouse by Frank Talamantes and coworkers. These two peaks were later found to correspond to two distinct PLs: PL-I and PL-II. PL-I is made in the placenta at the time of implantation of the embryo into the uterus and continues to be made until the middle of pregnancy (pregnancy lasts approximately 3 weeks for mice and rats). At midpregnancy, a switch occurs from the synthesis of PL-I to the synthesis of PL-II.

Both these hormones are made exclusively in one cell type of the placenta, and at midpregnancy these cells briefly synthesize both hormones simultaneously. The PL-I- and PL-II-producing cells are called trophoblast giant cells, so named for their enormous size. Trophoblasts derive from the fertilized egg and give rise to the extraembryonic tissue that forms the placenta; the giant cells represent a final stage in trophoblast differentiation. The giant cells become very large as a result of duplicating their genetic material over and over without undergoing cell division. As a consequence, giant cells are no longer able to give rise to more cells, but they still retain the capacity to synthesize proteins. The giant cells also form the border between the maternal and fetal tissues, and these cells come into direct contact with the mother's bloodstream; thus, the giant cells are perfectly positioned to release hormones that can be rapidly dispersed throughout the mother and that can then act on many maternal organs.

The ruminant PLs are also produced in a bizarre cell type of the placenta. In sheep and cows, the source of this hormone is the binucleate trophoblast cell, a cell that contains two nuclei, the compartment in which the genetic information of the cell is stored. These cells form near the borders of the finger-like extensions, or villi, that the placenta extends into the maternal tissue. After arising from cells with the normal single nucleus, the binculeate cells migrate toward the border of the villi and fuse with other cells, an unusual method of delivering the contents of the cell (including hormones) into the maternal system.

The precise temporal pattern of PL-I and PL-II synthesis during pregnancy suggests that the production of these hormones is tightly controlled. The mechanisms that turn the PL-I and PL-II genes on and off are only now beginning to be characterized at the molecular level, so it is not known how the switch from the synthesis of PL-I to PL-II is accomplished. Concurrent investigations have found that several circulating factors can stimulate or inhibit PL production, suggesting that the switch may not be the result of an internal placental "clock," but instead may be initiated by events that occur at a specific stage of pregnancy elsewhere in the mother, with these events communicated to the placenta by factors that travel through the bloodstream.

In the sheep and the cow, PL derives from a single gene, and the amount of PL circulating in the mother's bloodstream gradually increases until midpregnancy and then remains at that level or slowly declines. An important clue about PL function is that the amount of this hormone released into the circulation increases if the mother is fasting, suggesting that this hormone may be involved in mobilizing stored resources when nutrients are not being provided by the digestive system so that the fetus remains well-fed.

V. ACTIONS OF RODENT PLACENTAL LACTOGENS

The rodent PLs are able to mimic prolactin because they bind to the same cell surface receptor through which prolactin acts, the prolactin receptor. Indeed, PL-I and PL-II actually replace prolactin during pregnancy because prolactin levels in the bloodstream fall dramatically as PL-I levels rise. Furthermore, PL-I and PL-II are equivalent to prolactin in their affinity for the prolactin receptor: Equal amounts of each of these three hormones bind the prolactin receptor equally well. Also, because PL-I and PL-II are present at concentrations far exceeding the maximum level that prolactin reaches in the nonpregnant adult, the prolactin receptors on the surface of cells are likely to be filled with bound PL hormone rather than prolactin during pregnancy. The prolactin receptor is present on most if not all tissues of the mother and fetus, so PL-I and, later, PL-II have the potential to act on many different organ systems, thereby regulating many different aspects of maternal and fetal physiology. The best studied targets for these hormones are the maternal ovary, mammary gland, pancreas, and brain, along with the fetus, although almost certainly other tissues also represent functionally important sites of action.

The rodent ovary is the primary source of progesterone, and this steroid hormone is required to maintain the uterus in a receptive state for embryo implantation and development. Should progesterone levels drop during pregnancy (or the action of the progesterone receptor be blocked, for example, with RU 486), the uterine lining, along with the embryo, will be shed. Within the ovary, progesterone is produced in the structure known as the corpus luteum, which forms from the ruptured follicle (the structure in which the egg develops) once the egg is released. The corpus luteum is also short-lived, but one of the first identified functions of prolactin was its ability to increase the life span of the corpus luteum in rats (and therefore prolactin was also called "luteotropin"). Similarly, PL-I and PL-II are able to bind to prolactin receptors on cells within the corpus luteum, resulting in preservation of this structure and continued production of progesterone throughout pregnancy.

Mammals have in common the ability to nurse their young by producing milk, and prolactin is the key hormone that regulates milk production. Nevertheless, throughout the latter half of pregnancy, a period when the mammary glands must develop for the mother to be ready to lactate soon after giving birth, prolactin levels are extremely low. The PLs are therefore also responsible for replacing prolactin to stimulate this phase of mammary gland development. The process of mammary development involves more than just the mammary gland growing larger; instead, new structures in the mammary gland (alveoli) develop much like dormant tree limbs at the beginning of spring sending out new buds that will open into leaves and flowers. Prolactin and the PLs are responsible for this differentiation (formation of new structures) in the mammary gland as well as for the increase in mammary gland size.

Changes in metabolism are evident in the mother throughout pregnancy, including changes in the balance of fat and sugar uptake from the circulation and storage versus release of fats and sugars from these stores back into the circulation. In general, during pregnancy maternal stores are broken down to release nutrients that can pass through the placenta to the fetus, thereby providing the fetus with the necessary nutrition for growth. PLs may not be required to mobilize nutrients for delivery to the fetus when the mother is well-fed because under this condition adequate nutrients would be available as the products of digestion; however, if the mother is unable to eat enough (or enough of the appropriate foods), the effect of the PLs on maternal metabolism may become much more important. Robert Sorenson and colleagues found that in rats PL-I and PL-II are key regulators of insulin production by the β cells of the pancreas, with insulin directly controlling the levels of glucose in the circulation. Additional metabolic effects may result from PL binding to prolactin receptors in the liver, which has the highest concentration of prolactin receptors of any tissue. The effects initiated by PL binding to the liver have not been elucidated in rodents, but a reasonable hypothesis is that PLs act on the liver to regulate the production of insulin-like growth factor, which in turn may act on many other target tissues to regulate growth and metabolism.

Prolactin receptors in the brain may also be targets for PL-I and PL-II. Prolactin is likely to be able to reach these brain receptors because it is produced in the pituitary, a structure that is nestled into the base of the brain, but PLs must travel from the placenta through the circulatory system to reach the brain. However, recent evidence has been obtained that both PL-I and PL-II enter the cerebrospinal fluid. Thus, PLs may be able to initiate certain maternal responses attributed to prolactin, such as nesting behavior (building a nest and keeping the newborns in the nest). PL-I, in particular, may also play an important role in communicating between the placenta and the brain to regulate hormone release. James Voogt and Michael Soares obtained evidence that suggests that one of the effects of high levels of PL-I may be to drive sufficient amounts of this hormone to a region of the brain that controls prolactin release from the pituitary. Thus, they proposed that PL-I is responsible for the precipitous decline in the concentration of prolactin in the bloodstream at midpregnancy, when PL-I levels are maximal.

In addition to actions of PLs on the mother, these proteins almost certainly act directly on the fetus as well. PL-II enters the fetal circulation, but whether or not PL-I similarly reaches this compartment has been difficult to determine given the minimal amount of fetal blood at midpregnancy in mice and rats. The fetus also displays an abundant expression of prolactin receptors on numerous organs and tissues, yet the fetus has not yet begun to produce its own prolactin. Thus, the hormone most likely to act on fetal prolactin receptors during the latter half of pregnancy is PL-II. The effect of an interaction of PL-II with fetal prolactin receptors is not known, and the importance of such an interaction for normal fetal development can be questioned because Paul Kelly and coworkers demonstrated that viable mice that lack the prolactin receptor can be born. Nevertheless, fetuses and newborns lacking the prolactin receptor display subtle defects, for example, in the formation of the skeleton.

In many ways, PL-I and PL-II may act as the prolactins of pregnancy. Although rodent PL-I and PL-II are equivalent to prolactin in all the tests conducted in the laboratory, these tests typically compare the effects of the same concentration of each hormone. In the animal, though, PL-I at its peak is present in the bloodstream at approximately 10-fold higher levels than PL-II at its maximum and at more than 100-fold higher levels than prolactin at its highest concentration in the nonpregnant adult. The very high concentration of PL-I may simply mean that the prolactin receptors are more efficiently occupied during pregnancy, and therefore that the receptors are kept "on" a greater percentage of the time. However, the mechanism by which hormone binding activates the prolactin receptor raises another surprising possibility. Scientists at Genentech found that one molecule of growth hormone simultaneously binds two molecules of the growth hormone receptor—a remarkable and (at the time of the discovery) unique feat because the same region of the receptor can be occupied by either of two completely different regions of growth hormone. One of these regions of growth hormone binds more avidly to the receptor than the other region; in other words, growth hormone has both a high-affinity receptor binding region and a low-affinity receptor binding region. Because growth hormone and the growth hormone receptor are so similar to prolactin (and the PLs) and the prolactin receptor, activation of the prolactin receptor is also very likely to occur by one hormone molecule bringing two separate receptor molecules close together (much like connecting the ends of two wires

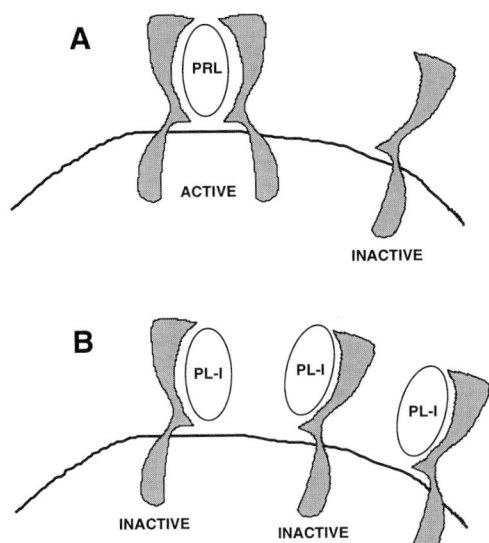

FIGURE 2 Model of hormone activation and inactivation of the prolactin receptor (gray structure). (A) When concentrations of hormone are low relative to the number of available receptors [as is typically the case for prolactin (PRL)], one prolactin molecule will simultaneously bind to two different prolactin receptor molecules, resulting in activation. Note that under these conditions, additional free (unoccupied) receptor molecules are available. (B) When concentrations of hormone are high relative to the number of available receptors (as may be the case for PL-I), each prolactin receptor molecule may be occupied, and therefore no free receptor would be available to add to the complex. The result would be many hormone–receptor pairs, but all these will be inactive This model is based on the finding that one molecule of growth hormone binds simultaneously to two molecules of growth hormone receptor, reported by Cunningham et al., Science 254, 821–825, 1991, and by de Vos et al., Science 255, 306–312, 1992. Modified from Linzer, D. I. H. and Arey, B. J. (1995). Ovarian prolactin receptors and their placental ligands. In *Somatotrophic Axis and the Reproductive Process in Health and Disease* (eds. E. Y. Adashi and M. O. Thorner) Springer-Verlag, NY, pp. 28–39.

to complete an electrical circuit). Paradoxically, at very high hormone concentrations the receptor might actually be turned "off" because the relatively limited number of receptors will each have a hormone molecule bound (through its high-affinity region), and no free receptor will be available to form a larger complex containing one hormone and two receptors (Fig. 2). The one hormone/one receptor complexes are not active (the ends of the two "wires" are not connected), so the presence of too much hormone can effectively mimic the absence of hormone. Thus, high concentrations of PL-I may act at certain target organs (especially those with lower amounts of receptor) to turn off the responses that require the prolactin receptor. The ability of PL-I to act in this manner in the animal is still speculative, but this scenario highlights a potential concern for therapeutic uses of similar hormones if the levels of these proteins become too elevated.

VI. ACTIONS OF RUMINANT PLACENTAL LACTOGENS

In contrast to the clear evidence that rodent PLs act exclusively through the prolactin receptor, the identity of the receptor for the sheep and cow PL is still not entirely established. Michael Freemark and coworkers presented the initial evidence that points to the presence of a unique PL receptor in the sheep fetus; they demonstrated that sheep PL binds more strongly to a fetal liver receptor compared to the binding of either prolactin or growth hormone. Similarly, Robert Collier and colleagues detected distinct receptors for the cow PL, prolactin, and growth hormone in the maternal uterus. Current evidence suggests, though, that the specific PL receptor may be a modified form of the growth hormone receptor rather than a completely unique protein.

No matter which receptor protein mediates the response to PL in sheep and cows, what is clear is that the ruminant PLs (like their rodent counterparts) have the ability to stimulate mammary gland development and milk production, and hence they are appropriately named lactogens. However, other functions are less certain. In comparison to laboratory mice and rats, which can be maintained under identical conditions and can be used to compare different treatments on genetically identical individuals, sheep and cows present greater challenges for conducting reproducible and standardized experiments. Thus, the types of animals, the conditions under which they are maintained, and the procedures of the experiments often vary among investigators, and these variations may well contribute to different results and conclusions.

Most studies have found significant effects of ruminant PLs on maternal metabolism and suggest that sheep and cow PL regulate the conversion of nutrient stores into forms that can be transported and utilized by the fetus, especially during periods of fasting. However, not all reports agree on which metabolic effects are induced by PL. Earlier work reported that ruminant PLs are unable to act on the ovary to induce progesterone production, but recent results suggest that the corpus luteum is indeed a target of PL, leading to elevated progesterone release. The ability of PL to stimulate fetal growth directly is also open to question because evidence has been presented both for and against this function. However, PL does enter the fetus at very high levels and can bind to fetal receptors (especially receptors on the fetal liver), suggesting that this hormone does have direct effects on the fetus by regulating fetal metabolism.

VII. FUTURE DIRECTIONS IN PLACENTAL LACTOGEN RESEARCH

The PLs in rodents and ruminants are hormones with a multitude of functions, and with the broad distribution of receptors for these proteins on tissues in the mother and the fetus, it seems likely that many of the functions of these hormones remain to be established. A major goal in PL research will therefore be to define these functions and to clarify the activities that have been reported but remain controversial. Not only will characterization of these activities lead to a new appreciation for the regulation of pregnancy but also it may well uncover regulatory processes that are currently not even recognized. Identification of these physiological processes and how they are regulated should, in turn, lead to new approaches to managing reproduction in animals and to new clinical applications that will enhance fertility and healthy fetal and maternal development.

Acknowledgment

I thank Linda Ogren for helpful comments on this chapter.

See Also the Following Articles

Corpus Luteum; Lactation; Mammary Gland Development; Placenta: Implantation and Development; Placental and Decidual Protein Hormones; Prolactin, Overview

Bibliography

Anthony, R. V., Liang, R., Kayl, E. P., and Pratt, S. L. (1995). The growth hormone/prolactin gene family in ruminant placentae. *J. Reprod. Fertil. Suppl.* **49**, 83–95.

Byatt, J. C., Warren, W. C., Eppard, P. J., Staten, N. R., Krivi, G. G., and Collier, R. J. (1992). Ruminant placental lactogens: Structure and biology. *J. Anim. Sci.* **70**, 2911–2923.

Forsyth, I. A. (1994). Comparative aspects of placental lactogens: Structure and function. *Exp. Clin. Endocrinol.* **102**, 244–251.

Goffin, V., Shiverick, K. T., Kelly, P. A., and Martial, J. A. (1996). Sequence–function relationships within the expanding family of prolactin, growth hormone, placental lactogen, and related proteins in mammals. *Endocr. Rev.* **17**, 385–410.

Ogren, L., and Talamantes, F. (1988). Prolactins of pregnancy and their cellular source. *Int. Rev. Cytol.* **112**, 1–65.

Soares, M. J., Faria, T. N., Roby, K. F., and Deb, S. (1991). Pregnancy and the prolactin family of hormones: Coordination of anterior pituitary, uterine, and placental expression. *Endocr. Rev.* **12**, 402–423.

Sorenson, R. L., and Brelje, T. C. (1997). Adaptation of islets of Langerhans to pregnancy: β-cell growth, enhanced insulin secretion and the role of lactogenic hormones. *Horm. Metab. Res.* **29**, 301–307.

Talamantes, F. (1990). Structure and regulation of secretion of mouse placental lactogens. *Prog. Clin. Biol. Res.* **342**, 81–85.

Placental Nutrient Transport

Colin P. Sibley
University of Manchester

I. Introduction
II. General Structural and Physiological Considerations
III. Flow-Limited and Membrane-Limited Diffusion
IV. Transport Protein-Mediated Transfer
V. Endocytosis/Exocytosis
VI. Summary and Future Directions

GLOSSARY

endocytosis/exocytosis A process by which large molecules may move across cell layers, caught in an invagination of the plasma membrane on one side, which then moves across the cell as a vesicle that then fuses with the opposite plasma membrane, expelling its content of macromolecules, etc.

Fick's law of diffusion A law that describes the diffusional flux of a solute between two compartments in terms of the surface area available for exchange, the path length down which diffusion must occur, the diffusion coefficient of the solute in question, and the gradient of its concentration in the two compartments.

flow-limited diffusion Diffusion of a molecule which is not limited by the properties of the placental barrier and so is dependent on its rate of delivery to and removal from the exchange barrier, i.e., the blood flow.

membrane-limited diffusion Diffusion of a molecule which is limited by the properties of the placental barrier; rate of delivery to or removal from the exchange barrier is not important.

syncytiotrophoblast The multinucleated transporting epithelial cell of the placenta; a true syncytium.

transport protein A protein which catalyzes the movement of a molecule across the plasma membrane. These may be channels, which allow diffusion across the plasma membrane, or carriers, which may allow active transport against electrochemical gradients by utilization of ATP.

Placental nutrient transport is the process by which there is net transfer of molecules, essential to fetal growth and development, from maternal to fetal blood across the cellular exchange barrier separating the two circulations in the placenta. Of equal importance is the net transfer from fetal to maternal blood of waste products of metabolic processes in the fetus. The two transfer processes occur by essentially identical mechanisms and are primary functions of the placenta. In this role the placenta provides for the fetus the homeostatic functions carried out *post partum* by the kidney, lungs and gut.

I. INTRODUCTION

More than 90% of the mass of the newborn baby is derived from molecules which have been transferred across the placenta. The mechanisms involved are similar to those utilized by transporting epithelia in the kidney, lung, and gut but are more poorly understood. This lack of knowledge may be, at least partially, attributed to two factors:(i) the considerable heterogeneity in placental structure and function between different species, making it more difficult than normal to extrapolate data obtained in animals to the human situation, and (ii) the short life span of the organ and the marked changes it undergoes during the course of pregnancy. However, in some counterbalance to this, placental research in humans is aided by the ready availability of normal placentas following term delivery. For this reason most of what follows focuses on the term human placenta.

There are very few molecules (perhaps only glucose) for which the complete mechanism of transfer from maternal to fetal blood is clearly understood. This article will therefore provide an overview of mechanisms of nutrient transfer across the placenta, only mentioning particular molecules to provide examples. More detailed information about specific molecules may be found in other articles in this encyclopedia and in the references cited in the Bibliography.

II. GENERAL STRUCTURAL AND PHYSIOLOGICAL CONSIDERATIONS

The functional unit of the human placenta, with regards to maternofetal exchange, is the villous tree which is bathed by maternal blood and which has the blood vessels of the fetoplacental circulation running through its core. The cellular and connective tissue layers of the villus, forming the barrier to exchange between maternal and fetal blood, are shown in Fig. 1. The outer cell layer is the syncytiotrophoblast. This is a very unusual epithelium in that it is a true syncytium, without any obvious extracellular space or plasma membranes intervening between nuclei; it may therefore be considered as one giant cell covering the villi. The syncytiotrophoblast has two plasma membranes: the microvillous plasma membrane, directly bathed in maternal blood, and the basal plasma membrane, facing the core of the villus. Underneath the syncytiotrophoblast is a layer of cytotrophoblast cells. These mononucleate cells form a continuous layer in the first trimester but from then on become more sparse as they differentiate and fuse with the syncytiotrophoblast, which is thereby enlarged.

Working further toward the core of the villus, one finds the basement membrane of the trophoblast, followed by connective tissue, the basement membrane of the fetal capillary endothelium, and the endothelium itself, enclosing the fetal blood.

All the cells and tissues outlined previously will provide a barrier, to greater or lesser degrees, to nutrient transfer. However, only the fetal capillary endothelium and the syncytiotrophoblast have been considered experimentally in this regard.

The fetal capillary endothelium is a normal continuous endothelium with large (5–10 nm diameter) spaces between the cells. For this reason it is not thought to provide a major barrier to small hydrophilic solutes but, as in any capillary bed with such an endothelium, the diffusion of proteins, such as α-fetoprotein and immunoglobulin G, through the intercellular spaces will be sterically hindered. Therefore, the endothelium cannot be ignored when considering transfer of these large molecules.

The syncytiotrophoblast is the transporting epithelium of the placenta and is thought to have the major role in enhancing or restricting nutrient transfer. It is this layer on which the rest of this article will focus.

In general terms there are four main mechanisms by which solute might cross the syncytiotrophoblast (shown diagramatically in Fig. 2 and considered in detail in the following sections).

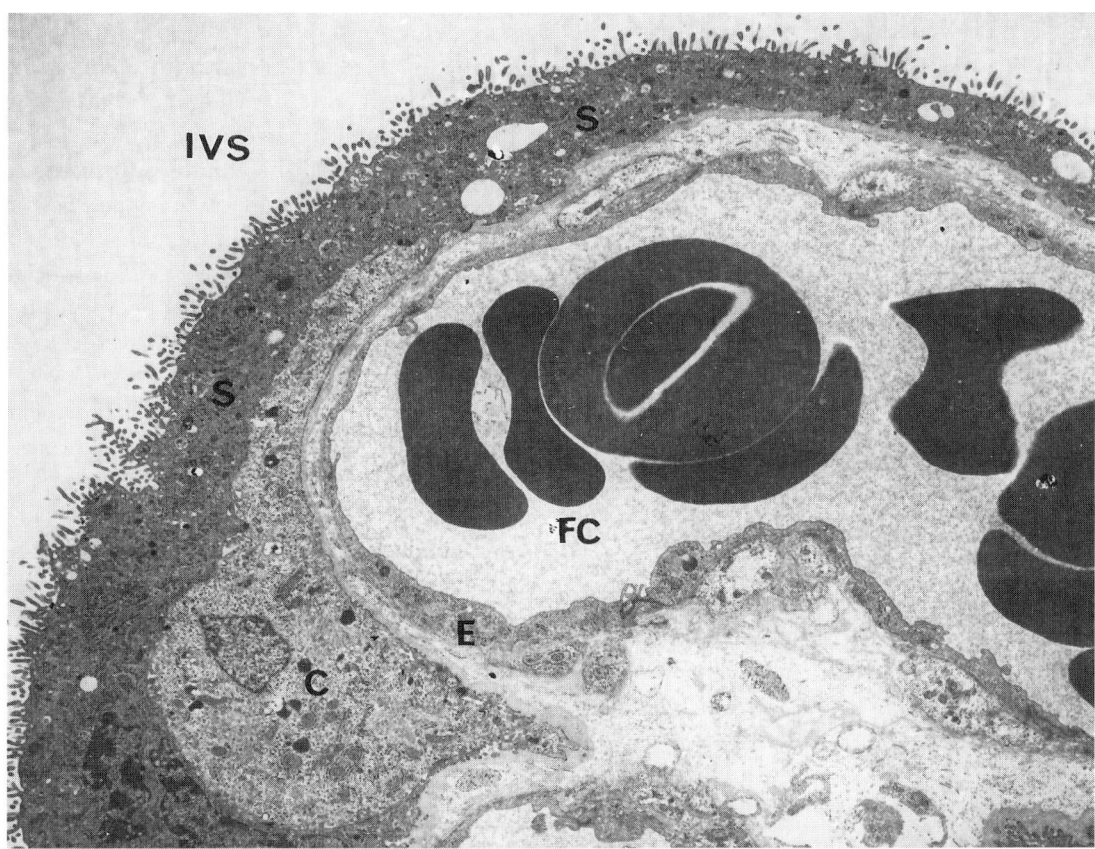

FIGURE 1 Electron micrograph showing the exchange barrier across which placental nutrient transport occurs. IVS, intervillus space (containing maternal blood); S, syncytiotrophoblast; C, cytotrophoblast; E, endothelium; FC, fetal capillary lumen (containing fetal blood). Magnification, ×5000 (the micrograph was kindly provided by Dr. C. J. P. Jones).

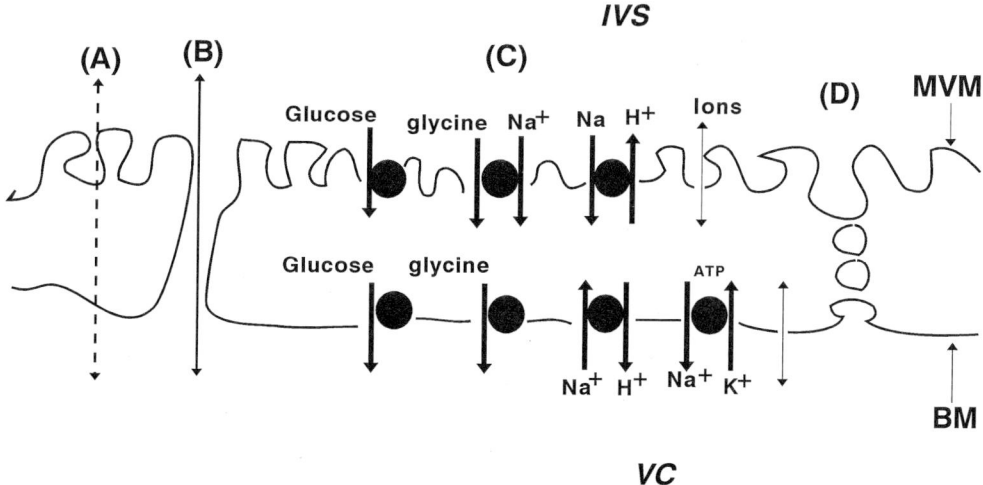

FIGURE 2 The four main mechanisms of solute transport across the syncytiotrophoblast: A, flow-limited diffusion; B, membrane-limited diffusion; C, Transport protein-mediated transfer; D, endocytosis/exocytosis. IVS, intervillus space; VC, villus core; MVM, microvillous plasma membrane; BM, basal plasma membrane (adapted with permission from Sibley et al., 1997).

III. FLOW-LIMITED AND MEMBRANE-LIMITED DIFFUSION

A. Fick's Law and Diffusion across the Placenta

Diffusion is a process by which molecules, moving randomly in solution, expand to fill the volume available to them. Each individual molecule is equally likely to move into a region of high concentration or a region of low concentration. However, because there are more molecules in the region of high concentration, there will be a greater number moving to the region of low concentration than in the reverse direction, i.e., net flux is from the region of high concentration to the region of low concentration. All molecules will be able to move between maternal and fetal plasma across the placenta by means of diffusion. However, the rate at which they will be able to diffuse will vary greatly between different molecules, depending on their physicochemical characteristics, particularly size and degree of water/fat solubility. The net rate of diffusion of an uncharged molecule between maternal and fetal plasma may be determined from an adaptation of Fick's law of diffusion:

$$J = AD/l(C_m - C_f),$$

where J is the net flux (mol/sec), A is the total cross-sectional surface area available for diffusion (m^2), D is the diffusion coefficient of the molecule in question (m^2/sec), l is the thickness of the barrier over which diffusion is occurring (m), and $(C_m - C_f)$ is the geometric mean concentration gradient in maternal and fetal plasma at the site of exchange (mol/liter), i.e., in the intervillous space on the maternal side of the placenta and in the capillaries on the fetal side. Bear in mind that the direction of net transfer by diffusion will depend on the prevailing concentration gradient.

The diffusion coefficient is dependent on the temperature (37°C in the context here), the size of the molecule (large molecules will diffuse more slowly), and the solubility of the molecule in the solvent in question (a water-soluble molecule will diffuse more slowly through a hydrophobic barrier, such as the lipid bilayer of the plasma membrane, than it will through water). Because of this latter consideration, the AD/l term in the previous equation, which represents the placental permeability for the solute in question, will tend to be large for hydrophobic molecules and small for hydrophilic molecules. The former molecules will be able to diffuse rapidly across the plasma membranes of the syncytiotrophoblast, whereas the hydrophilic molecules will not.

For this reason the transfer of hydrophobic solutes will be more limited by their concentration gradients in maternal and fetal plasma (the $C_m - C_f$ term in the equation). Because the concentration gradient is, in turn, dependent on the rate of delivery of solute on one side of the barrier and rate of removal on the other (i.e., the blood flows), the transfer of hydrophobic molecules is said to be flow limited. On the other hand, the low permeability to hydrophilic molecules means that their rate of transfer is much more dependent on the properties of the barrier than on the concentration gradients. Consequently, these molecules are said to be membrane or diffusion limited. These two situations are depicted in Fig. 3. It should be borne in mind that what is described are the extreme cases; many molecules will have physicochemical characteristics that result in their rate of transfer by diffusion being partially flow or membrane limited, depending on the set of circumstances prevailing at the time. For example, the transfer of a small hydrophilic compound may become essentially flow limited if blood flow through one or both circulations of the placenta is considerably reduced.

B. Geometric Arrangement of Blood Flows

The efficiency of transfer of flow-limited molecules is dependent on the relative directions of blood flow in the uteroplacental and fetoplacental circulations. The most efficient system would be the countercurrent one, in which the two blood flows are moving in opposite directions on either side of the placental barrier. The least efficient system would be the concurrent one, in which the two blood flows are moving in the same direction. In the human placenta, maternal blood spurts fountain-like into the intervillus space and so the pattern of blood flow does not fit a neat geometric arrangement. The actual efficiency is probably intermediate between the two extremes listed previously.

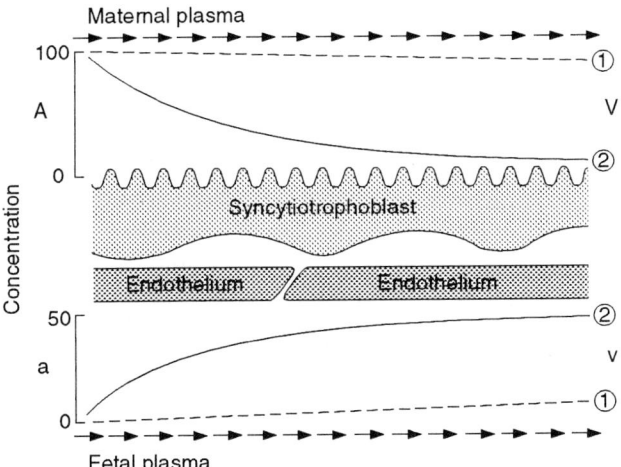

FIGURE 3 Theoretical plot of the maternal and fetal plasma concentration of two substances as they pass through the capillary bed on the maternal side (where capillary bed is equivalent to the intervillous space) and fetal side of the placenta from the arterial end to the venous end. Substance 1 is a small hydrophobic molecule such as O_2 which freely diffuses across plasma membranes and rapidly equilibrates on maternal and fetal sides. Its rate of transfer is limited only by the blood flow supplying or removing it and so it is said to be flow limited. Substance 2 is a hydrophilic molecule such as creatinine which cannot easily diffuse across plasma membranes so that its maternal plasma concentration does not approach the fetal plasma concentration. Its rate of transfer is affected only by the surface area of the placenta available for its transfer and the width of the barrier separating maternal and fetal blood and so it is said to be diffusion or membrane limited (reproduced with permission from Sibley et al., 1988).

C. Permeability to Hydrophilic Solutes and Paracellular Routes

Measurements of the permeability of the human placenta (as well as those of other species) to inert, uncharged, hydrophilic molecules with a range of sizes, both *in vivo* and *in vitro*, have shown an inverse linear correlation between permeability and diffusion coefficient in water. Although there are a number of explanations for these data, the simplest is that there is an extracellular water-filled paracellular route across the placenta along which these molecules diffuse. However, this is very difficult to reconcile with the morphology of the syncytiotrophoblast which, as already noted, is a true syncytium without any obvious paracellular route. Three possible explanations have been advanced: (i) The plasma membranes of the syncytiotrophoblast, because of their particular phospholipid composition, have a relatively high permeability to small hydrophilic solutes; (ii) there are transtrophoblastic channels which are normally too narrow or too rare to be reliably detected by electron microscopy but which can be visualized if they are expanded by increasing hydrostatic pressure in one of the circulations of the placenta; and (iii) areas of syncytial denudation, often plugged by fibrin containing fibrinoid deposits, are a normal feature of all placentas and undoubtedly constitute a paracellular route. The true situation remains to be resolved; it is likely that the paracellular route is heterologous and that all three routes contribute to the overall permeability to hydrophilic molecules.

An important final point on this topic is that the data described previously, as well as other experiments, have shown that the permeability of the human placenta to hydrophilic molecules is very high so that diffusion will contribute a large proportion of their flux in each direction across the placenta. This also means that hydrophilic compounds introduced into the maternal circulation will have ready access to the fetal circulation.

D. Electrical Gradients

Fick's law only takes into account concentration gradients, but diffusion of charged species will also be driven by any electrical gradient (potential difference; pd). The existence or otherwise of such a pd has been, and remains, controversial. However, recent *in vitro* and *in vivo* experiments suggest that there is normally a small fetal-side negative pd of, at maximum, 15 mV across the exchange barrier of the placenta. Such a pd would increase the rate of transfer of cations by diffusion toward the fetus but retard diffusional transfer of anions in this direction.

IV. TRANSPORT PROTEIN-MEDIATED TRANSFER

Transport proteins are inserted into plasma membranes and essentially catalyze the transfer of molecules between the extracellular and intracellular space. They increase the rate of transcellular transfer of hydrophilic molecules above that which would be expected by diffusion alone. There are broadly two different categories: channels and carriers.

A. Channels

Channels are proteins which form water-filled pores in the plasma membrane through which molecules may diffuse dependent on electrochemical gradients. They are highly selective for specific molecules, usually ions. There is no doubt that both plasma membranes of the syncytiotrophoblast possess an array of such ion channels which are involved in both cellular homeostasis and maternofetal exchange. However, very little is known about their specific roles in these processes.

B. Carriers

These proteins are highly selective for specific molecules, are saturable (i.e., increasing the concentration of the specific molecules will only increase the rate of transfer on the carrier up to a certain point), and show competition for transport between similar molecules. They can be further characterized into facilitated diffusion types, exchangers and cotransporters, and active transporters (Fig. 2).

1. Facilitated Diffusion

These carriers can only promote transfer across the plasma membrane in the direction of the prevailing electrochemical gradient. A good example in the placenta is the glucose transporter. There are several types of facilitated glucose transporter but the predominant one in both the microvillous and the basal plasma membranes of the syncytiotrophoblast is GLUT-1. This high-capacity, low-affinity transporter can catalyze transcellular glucose transfer to the fetus because, as noted, it is in both plasma membranes and because the maternal plasma concentration of glucose is almost always higher than that in the fetal plasma.

2. Cotransporters and Exchangers

These may be regarded as a subset of the facilitated diffusion carriers. Cotransporters transport two molecules in the same direction. Exchangers transport two molecules in opposite directions. Cotransport provides a means of driving the transport of one molecule against its concentration gradient by utilizing a favorable electrochemical gradient for the cotransported molecule. Na^+/glycine cotransport is depicted in Fig. 2. In this example, the electrochemical gradient for Na^+ across the microvillous plasma membrane, maintained by the Na^+ pump (see below) in the basal plasma membrane, drives the carrier in the direction of the cell cytosol and results in transport of glycine against its concentration gradient (the concentrations of amino acids in the syncytiotrophoblast are considerably higher than those in maternal or fetal plasma).

The transport on exchangers will be dependent on the relative electrochemical gradients for the two molecules. The Na^+/H^+ exchanger is shown in Fig. 2 and is on both microvillous and basal plasma membranes. The electrochemical gradient for Na^+ in both situations will invariably be into the cell, but that for H^+ will depend on the metabolic condition, and therefore pH, of fetus, placenta, and mother; the possibility exists that in some situations it might be able to drive Na^+ extrusion against its electrochemical gradient. For this reason, it is difficult to determine the specific role of this and other exchangers such as the Cl^-/HCO_3^- exchanger, which is also found on both syncytiotrophoblast plasma membranes in maternofetal transfer. It could be that they have a more important role in syncytiotrophoblast cell homeostasis.

3. Active Transporters

These carriers utilize ATP to drive transport, usually of ions, against their electrochemical gradients. The classic example is the Na^+ pump (Na^+,K^+-ATPase). This is predominantly localized to the basal plasma membrane of the human syncytiotrophoblast as is another similar pump, the Ca^{2+}-ATPase, which extrudes Ca^{2+} from the cell into the villus core.

4. Carriers and Vectorial Transplacental Transfer

The mechanisms by which carriers can bring about vectorial transplacental transfer can be exemplified with reference to amino acid and calcium.

TABLE 1
Examples of Amino Acid Transport Systems in the Human Placenta

Name of transporter system	Plamsa membrane location	Amino acids (aa) transported
A	MVM + BM[a]	neutral aa, MeAIB
ASC	BM	Ala, Ser, Cys, anionic aa
N	MVM	His, Gln
β	MVM	Tau
X_{AG}^-	MVM + BM	Asp, Glu
l	MVM + BM	Leu, BCH
y^+	MVM + BM	Lys, Arg

[a] MVM, microvillous plasma membrane; BM, basal plasma membrane.

i. Amino Acid Transport There are an array of amino acid transporters in both microvillous and basal plasma membranes, each highly selective for specific groups of amino acids (see Table 1). The concentrations of amino acids in fetal plasma are higher than those in maternal plasma so maternofetal transport is likely to be an active process. As already mentioned, the concentration of free amino acids in the syncytiotrophoblast is higher than that in either plasma. Transfer is therefore said to involve a pump/leak mechanism whereby amino acid is driven into the cell against the electrochemical gradient on the maternal side, utilizing Na^+-dependent carriers as already described or, perhaps, by exchange with other amino acids. Extrusion across the basal plasma membrane is then down the electrochemical gradient and may utilize facilitated diffusion-type carriers or leak across the plasma membrane. Although net maternofetal transfer may be brought about as described, net transfer in the fetomaternal direction is also possible by a similar means dependent on the prevailing gradients and the particular asymmetry in localization of carriers in the microvillous and basal plasma membrane.

ii. Calcium Transport The concentration of ionized Ca^{2+} in fetal plasma is higher than that in maternal plasma and intracellular concentrations of Ca^{2+} in the syncytiotrophoblast are at least 1000-fold lower than those in either plasma. Active Ca^{2+} transport to the fetus is therefore said to occur by a leak/pump mechanism. The cation diffuses across the microvillous plasma membrane down its electrochemical gradient utilizing either a channel (for which there is currently no evidence) or a facilitated diffusion carrier (for which there is some evidence). It then diffuses across the cytosol, probably bound to a specific calcium-binding protein which allows a large transcellular flux without any major alteration in intracellular ionized Ca^{2+} concentration. Finally, the Ca^{2+} pump extrudes the cation into the villous core against the electrochemical gradient, with expenditure of ATP.

V. ENDOCYTOSIS/EXOCYTOSIS

Large hydrophilic molecules may gain access to the syncytiotrophoblast by endocytosis. This is a process by which an area of plasma membrane invaginates and then becomes pinched off, forming a vesicle in which extracellular water and its contents are entrapped (Fig. 2). The process as described is called fluid-phase endocytosis, but the process can be more selective in a similar process called receptor-mediated endocytosis. Here, a molecule is bound to specific receptors in the area of plasma membrane which invaginates.

Theoretically, the vesicle formed by endocytosis, following random or directed diffusion through the cytosol, might eventually come into association with the basal plasma membrane, fuse with it, evaginate, and release its contents into the villus core. However, while the evidence that endocytosis occurs is reasonable, data showing exocytosis are sparse. Nevertheless, it is thought that this is the process by which immunoglobulin G is transported across the syncytiotrophoblast.

VI. SUMMARY AND FUTURE DIRECTIONS

The main cellular barriers to maternofetal exchange across the placenta are the syncytiotrophoblast and the fetal capillary endothelium, the former being the key transporting epithelium. The human placenta is a highly permeable organ and so diffusion will make the major contribution to transfer of most nutrients. However, transport protein-mediated systems and endocytosis/exocytosis provide additional mechanisms of transfer for specific molecules and may make the most important contribution to net transfer as well as provide loci for regulation of nutrient transport.

A considerable amount of work is still required for a full understanding of nutrient transfer across the placenta. The following are particular areas which require attention:

1. Gestational changes in placental transport mechanisms: The information provided here comes almost exclusively from studies of the term placenta.
2. Mechanisms by which placental nutrient transport is regulated: Virtually nothing is known about this.
3. Further molecular characterization of syncytiotrophoblast transporters.
4. Intracellular concentrations in the syncytiotrophoblast of ions and other solutes need to be known.
5. A better understanding of how placental transport is altered by and contributes to pathophysiological situations such as intrauterine growth retardation and preeclampsia.

See Also the Following Articles

FETAL GROWTH AND DEVELOPMENT; PLACENTAL AND DECIDUAL PROTEIN HORMONES

Bibliography

Kaufmann, P., and Burton, G. (1994). Anatomy and genesis of the placenta. In *The Physiology of Reproduction* (E. Knobil and J. D. Neil, Eds.), Vol. 1, 2nd ed., pp. 441–484. Raven Press, New York.

Morriss, F. H., Boyd, R. D. H., and Mahendran, D. (1994). Placental transport. In *The Physiology of Reproduction* (E. Knobil and J. D. Neil, Eds.), Vol. 1, 2nd ed., pp 813–862. Raven Press, New York.

Sibley, C. P. (1994). Mechanisms of ion transport by the rat placenta: A model for the human placenta? *Placenta* 15, 675–691.

Sibley, C. P., and Boyd, R. D. H. (1988). In *Oxford Reviews of Reproductive Biology* (J. R. Clarke, Ed.), Vol. 10. Oxford Univ. Press, Oxford, UK.

Sibley, C. P., and Boyd, R. D. H. (1998). Mechanisms of transfer across the human placenta. In *Fetal and Neonatal Physiology* (R. A. Polin and W. W. Fox, Eds.), 2nd ed. Saunders, Philadelphia.

Sibley, C. P., Glazier, J. D., and D'Souza, S. W. (1997). Placental transporter activity and expression in relation to fetal growth. *Exp. Physiol.* 82, 389–402.

Stein, W. D. (1990). *Channels, Carriers, and Pumps: An Introduction to Membrane Transport.* Academic Press, San Diego.

Placental Steroidogenesis in Primate Pregnancy

Eugene D. Albrecht
The University of Maryland School of Medicine

Gerald J. Pepe
Eastern Virginia Medical School

I. Introduction
II. Estrogen Production
III. Progesterone Production
IV. Regulation of Placental Corticosteroid Metabolism by Estrogen
V. Summary

GLOSSARY

fetal adrenal An organ located at the top of the kidney that, in the primate, is composed of a medulla and a cortex made up of a fetal zone which produces steroid precursors for estrogen biosynthesis and a definitive "adult-type" zone which produces cortisol important for fetal maturation.

fetal anencephaly A congenital condition occurring in the fetus in which there is a developmental absence of portions of the cranial structures, including the pituitary gland located at the base of the brain

hypothalamic–pituitary–adrenocortical axis The functionally integrated components of the endocrine axis involved in the production of adrenocorticotropic hormone and cortisol.

low-density lipoprotein A complex macromolecule composed of apolipoprotein and lipid subunits, including cholesterol, a substrate for steroidogenesis.

placenta An organ composed of cytotrophoblasts and syncytiotrophoblasts of fetal embryonic origin and the decidua of maternal uterine origin and which is the site of steroid and protein hormone biosynthesis and exchange of gases and substrates between mother and fetus.

steroid A member of the lipid class of compounds that is composed of a four-ring structure, the cyclopentano-perhydro-phenanthrene nucleus, and which is the basic structural component of the progestogens, corticosteroids, androgens, and estrogens.

trophoblast Cells derived from the trophectoderm cells of the developing blastocyst and which undergo differentiation from stem cell-like cytotrophoblasts to the mature functional syncytiotrophoblasts of the hemochorial placenta of the primate.

Placental steroidogenesis is the process by which the placenta forms the steroid hormones, estrogen and progesterone, and metabolizes the corticosteroids of maternal origin as they cross into the fetus. The formation of estrogen in humans and nonhuman primates requires functional communication between the fetal adrenal gland and the placenta. A change in the metabolism of maternal corticosteroids within the placenta during the second half of primate gestation results in the activation of the hypothalamic–pituitary–adrenocortical axis of the fetus and consequently the maturation of the fetal adrenal gland. After their biosynthesis, estrogen and progesterone have important roles in the maintenance of pregnancy, whereas the fetal adrenal produces cortisol essential to maturation of the lung and other organs in the developing fetus.

I. INTRODUCTION

A. Physiological Roles of Estrogen and Progesterone

Estrogen and progesterone are the principal steroid hormones produced by the placental syncytiotrophoblasts during primate pregnancy, and these hormones have important actions in preparing the uterine endometrium for implantation of the blastocyst and in the maintenance of pregnancy thereafter (Pepe and Albrecht, 1995). Progesterone has anti-inflammatory and immunosuppressive properties which may, along with other hormones, protect the conceptus from immunologic rejection by the mother. Estrogen and progesterone also have a role, along with prolactin and other hormones, in mammary gland development in preparation for lactation. Estrogen enhances blood flow within the uterus and placenta and regulates key components of the maternal cardiovascular system for purposes of adaptation to the increased physiologic demands of the developing fetus. Finally, estrogen enhances and progesterone suppresses the contractile nature of the uterus, and thus these hormones seemingly have important roles in the cascade of hormonal events leading to the onset of labor and delivery of the newborn.

II. ESTROGEN PRODUCTION

A. Biosynthetic Pathway

1. Fetal–Placental Unit

Although the ovary produces estrogen and progesterone immediately after ovulation and conception, the placenta becomes the principal source of these hormones at approximately 8 weeks of human pregnancy (length of gestation is 39.5 weeks). There is a progressive rise in the maternal plasma concentrations of the three principal estrogens—estrone, estradiol, and estriol—with advancing stages of human gestation (Fig. 1). The steroid biosynthetic pathway involved in the conversion of substrate cholesterol to the estrogens in a typical steroidogenic organ is shown in Fig. 2. The primate placenta, however, is very unique in that it does not express the P450 17-hydroxylase/17-20 lyase enzyme (abbreviated P-450_{C17} in Fig. 2) which converts de novo those C_{21} steroids (i.e., possessing 21 carbon units such as pregnenolone and progesterone) to the C_{19} steroids such as dehydroepiandrosterone (DHA) and androstenedione. Consequently, the placental syncytiotrophoblast cannot produce estrogen on its own but rather obtains the C_{19} steroid precursors from the fetal and to some extent maternal adrenal glands (Fig. 3). A functional fetal–placental unit for the formation of estrogen therefore exists during primate pregnancy. As a result of this functional interaction between the adrenals and placenta, more than 90% of all the estriol that is formed each day by the human placenta is derived from fetal adrenal steroid precursors, whereas the fetal and maternal adrenals contribute about equally to the production of estrone and estradiol. Consequently, situations that result in subnormal fetal adrenal function, including fetal anen-

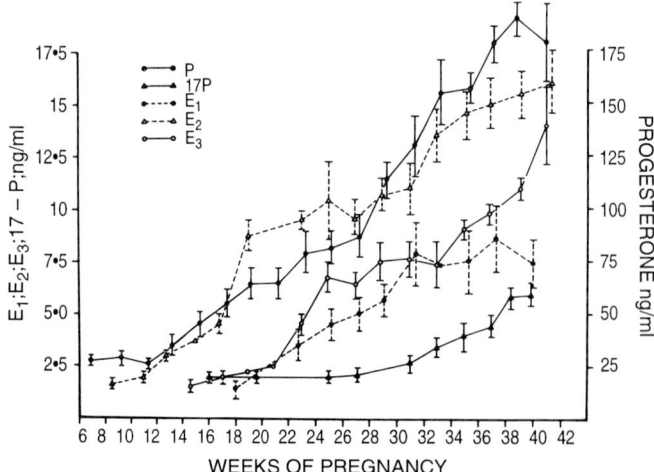

FIGURE 1 Mean (SE) maternal plasma concentrations of progesterone (P), 17-hydroxyprogesterone (17P), unconjugated estrone (E_1), unconjugated estradiol (E_2), and unconjugated estriol (E_3) during human pregnancy (reproduced with permission from D. Tulchinsky et al., Am. J. Obstet. Gynecol. **112**, 1095–1100, 1972).

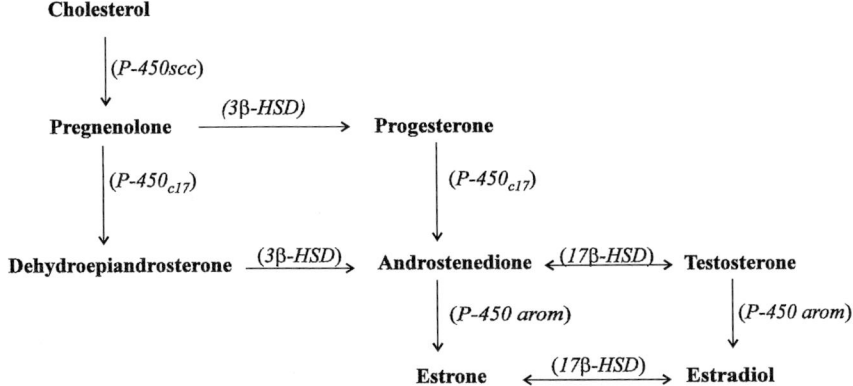

FIGURE 2 The steroid biosynthetic pathway for the conversion of substrate cholesterol to the progestogens, androgens, and estrogens.

cephaly in which there is little or no secretion of pituitary adrenocorticotropic hormone (ACTH), fetal death, or suppression of the fetal hypothalamic–pituitary–adrenocortical axis (HPAA) by the administration of synthetic glucocorticosteroids, are associated with very low levels of estrogen. Because almost all the estriol produced during human pregnancy is derived from fetal adrenal steroid precursors, this particular estrogen can be valuable in assessing fetal well-being.

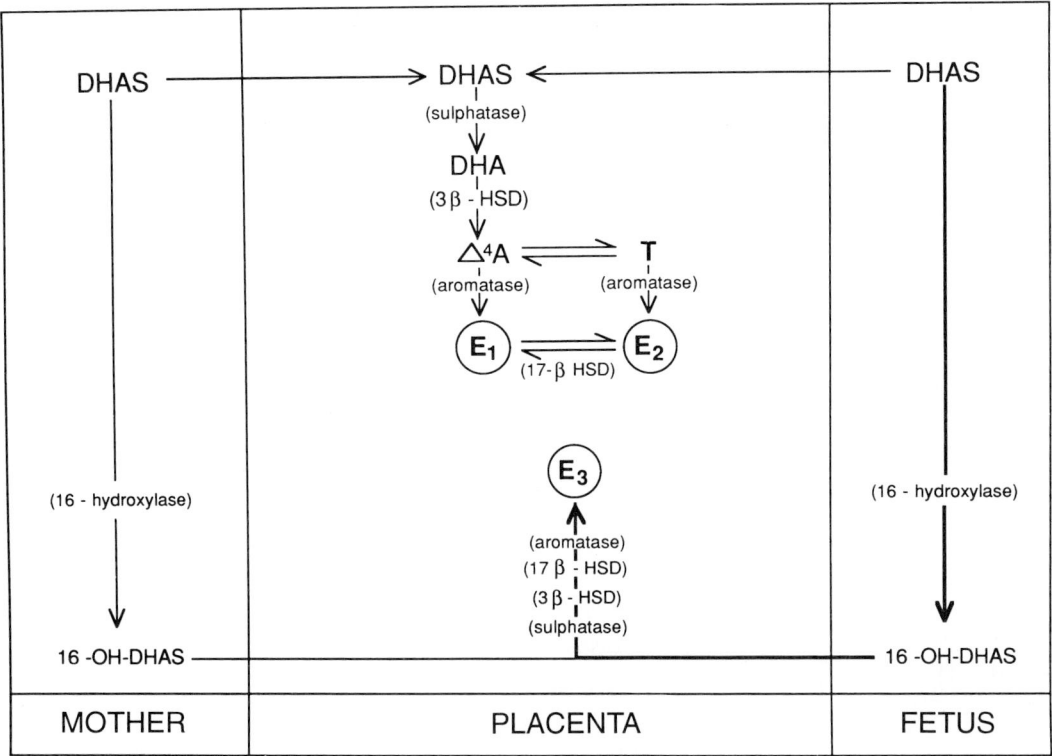

FIGURE 3 The biosynthesis of estrone (E_1), estradiol (E_2), and estriol (E_3) by the human fetal–placental unit. DHAS and 16-hydroxy DHAS of fetal and maternal origin are converted via androstenedione (^{4}A) and testosterone (T) to E_1/E_2 and E_3, respectively (reproduced with permission from E. D. Albrecht and G. J. Pepe, Endocr. Rev. 11, 124–150, 1990).

2. Enzymes for Estrogen Biosynthesis

The biosynthesis of estrogen within the placenta from the C_{19} steroid precursors requires the following enzymes: the sulfatase, Δ^5-3β-hydroxysteroid dehydrogenase/Δ^{5-4} isomerase (3β-HSD); the P450 aromatase; and 17β-hydroxysteroid oxidoreductase (17β-HSD; Fig. 3).

i. Sulfatase A major portion of the DHA secreted by the fetal adrenal is sulfated in the fetal liver to form DHAS, or 16-hydroxy DHAS in the case of the precursor for estriol formation. Upon arrival in the placenta these sulfated steroids are quickly acted on by a sulfatase enzyme to form free DHA and 16-hydroxy DHA. The latter Δ^5 steroids (i.e., double bond located between carbons 5 and 6) are then converted by the 3β-HSD enzyme (which will be described later) to the Δ^4 steroids (i.e., double bond located between carbons 4 and 5) such as androstenedione and testosterone.

ii. P450 Aromatase The conversions of androstenedione and testosterone to estrone and estradiol and 16-hydroxyandrostenedione to estriol (Fig. 3) require the action of a P450 aromatase (P450arom) enzyme complex. The P450arom is a member of the cytochrome P450 superfamily of enzymes, which is composed of more than 220 members (reviewed in Simpson et al., 1994). The enzyme binds C_{19} steroid substrate and catalyzes a series of reactions leading to the very unique phenolic A ring formation characteristic of the estrogens. The aromatase complex also consists of cytochrome P450 reductase, a ubiquitous protein which transfers reducing equivalents from NADPH to the P450arom. In the process of aromatization, 1 mol of C_{19} steroid, 3 mol of molecular oxygen, and 3 mol of NADPH are utilized. Aromatization is a very complex and incompletely understood enzymatic process in which the first two oxygen molecules are involved in the oxidation of the C_{19} angular methyl group, whereas the third oxidative reaction may involve a peroxidative attack on the C_{19} methyl group. However, a single active site for aromatization appears to exist, whereby the entire three-step process is catalyzed by a single cytochrome P450arom.

The gene encoding human cytochrome P450arom, CYP19, is located on chromosome 15 and is composed of nine exons and two polyadenylation sites in the last coding exon downstream from the terminating stop codon which give rise to 3.4- and 2.9-kb transcripts which encode human P450arom. Using site-directed mutagenesis to delete or structural modify potentially important nucleotides, Simpson and co-workers (1994) began to characterize the functional domains and structure–function relationships of the human cytochrome P450arom. In the human placenta, almost all the CYP19 transcripts include sequences encoded by exon 1.1, whereas a minor proportion contain sequences encoded by exon 1.2. Although the entire intron sequences have not been mapped to this point, the CYP19 gene is at least 70 kb long, the largest cytochrome P450 gene characterized to date.

The human P450arom gene is highly unique because it appears to be the first cytochrome P450 that utilizes alternative promoters in the regulation of tissue-specific expression within the placenta as well as the ovary and adipose tissue, which are also sites of estrogen biosynthesis within the body. Thus, transcripts specific for proximal promoter II are exhibited in the ovary, whereas transcripts specific for the distal promoter 1.1 are confined to the placenta. Simpson et al. (1994) have suggested that the P450arom may have resorted to the use of alternative promoters to allow for greater versatility in determining regulation of expression in the various tissue sites. Although there has been rather limited study of the regulation of CYP19 gene expression, a transcriptional enhancer element, possibly involved in cell-specific expression, has been identified upstream of the promoter transcriptional start site.

iii. 17-Hydroxysteroid Oxidoreductase The interconversions of androstenedione and testosterone and of estrone and estradiol are catalyzed by a 17-hydroxysteroid oxidoreductase enzyme in the placenta. The human 17-hydroxysteroid oxidoreductase has been sequenced and localized on chromosome 17. Estrone and estradiol are extensively interconverted within the placenta and the transfer constants in each direction are close to unity. However, there is a highly significant unequal distribution of placental secretion of these two estrogens into the maternal and fetal compartments. Thus, estradiol, the much more biological active steroid, is secreted primarily into the maternal circulation, whereas the relatively inactive counterpart, estrone, is released primarily into the fetal circulation. The relatively low level of bioactive estradiol in the primate fetus, therefore, can be explained by this selective placental secretion, potentially as a means to prevent exposure of the developing fetus to excessive estrogenic activity. Although a carrier system specific for estradiol has been suggested to be responsible for this selective secretion, differential expression of the oxidase and reductase activities of the 17-hydroxysteroid oxidoreductase by basal and apical portions of the trophoblast cell may also potentially be involved.

B. Regulation

Because the sulfatase and aromatase are present in abundant amounts in the primate placenta, these enzymes do not seem to be rate limiting in the production of estrogen. Therefore, the rate of estrogen production during pregnancy appears to primarily be dictated by the level of fetal adrenal precursor production, uteroplacental blood flow, and mass of the placental trophoblast.

1. Fetal Adrenal C_{19} Steroids

Because the fetal adrenal is a major source of the C_{19} steroids (e.g., DHAS and 16-hydroxy DHAS) required as substrate for estrogen production within the primate placenta, the overall rate of estrogen production is in part dictated by the relative level of fetal adrenal function. Thus, acute stress induced by hypoxemia in fetal baboons causes a marked increase in fetal adrenal C_{19} steroid secretion and in a corresponding rise in placental estrogen production. Moreover, suppression of the fetal HPAA by the administration of synthetic glucocorticosteroids to humans, rhesus monkeys, and baboons results in a marked suppression of estrogen levels. Consequently, those factors which stimulate fetal adrenal development and function, such as ACTH, indirectly have a major role in regulating placental estrogen production.

2. Role of Uteroplacental Blood Flow

Because there is a decrease in placental utilization of DHAS for estrogen formation in pregnancies complicated by a decrease in uteroplacental perfusion (e.g., pregnancy-induced hypertension), it has been proposed that placental clearance of steroid precursors through estrogen formation is directly related to uteroplacental blood flow. In support of this concept, a reduction in uteroplacental blood flow induced in nonhuman primates by graded reductions in maternal distal aortic blood flow causes a corresponding decrease in the placental clearance of DHA through estradiol formation. Thus, uteroplacental perfusion has a role in regulating placental estrogen production in primates.

3. Role of cAMP

Although the P450arom and sulfatase do not appear to be rate-limiting enzymatic steps for estrogen biosynthesis within the primate placenta, there seems to be a progressive increase in the capacity for placental aromatization with advancing stages of human pregnancy. In malignant trophoblast cells in culture, cAMP increases the levels of aromatase enzyme and NADPH–cytochrome P450 reductase, and thus estrogen formation. The messenger ribonucleic acid (mRNA) levels for P450arom are also enhanced in malignant trophoblast cells by cholera toxin, an agent that increases cAMP. However, the stimulatory effects of cAMP on aromatization in transformed trophoblast have not been consistently observed in normal placental tissue. Thus, depending on the culture conditions, aromatase activity and estradiol production were increased, unaltered, or decreased by cAMP in normal human placental minces or monolayer cultures. The reasons for these apparently disparate results are unknown, and thus cAMP is considered only potentially important in the regulation of placental estrogen biosynthesis.

III. PROGESTERONE PRODUCTION

A. Biosynthetic Pathway

1. The Low-Density Lipoprotein Pathway

There is a progressive increase in the plasma concentrations of progesterone and 17-hydroxyprogesterone with advancing human pregnancy (Fig. 1). The primate placenta expresses abundant quantities of the P450 cholesterol side chain (P450scc) and 3-HSD enzymes required to convert substrate cholesterol into progesterone (Fig. 2). Thus, unlike the situation for estrogen formation, progesterone production does not require, at least directly, the participation of the fetus. However, in contrast to other steroid-secreting organs the human and nonhuman primate placenta exhibits a very limited capacity to form cholesterol and thus progesterone *de novo* from acetate. Earlier studies conducted *in vivo* in pregnant women and by *in vitro* perfusion showed that the placenta utilizes cholesterol from the maternal circulation to form progesterone. It was subsequently demonstrated that human trophoblasts in primary culture take up low-density lipoprotein (LDL) cholesterol via a saturable high-affinity binding process, and LDL degradation occurs as a consequence of cellular LDL uptake and internalization. Therefore, the utilization and metabolism of LDL, and possibly VLDL, within the primate placenta (Fig. 4) appears to occur in a manner similar to the classical LDL pathway originally described in fibroblasts. Simpson and MacDonald (1981) have suggested that as a result of LDL uptake kinetics, there is sufficient LDL cholesterol substrate within the maternal circulation to saturate placental trophoblast receptors and thus the rate of LDL uptake and placental progesterone formation by the human placenta are determined by the number of LDL receptors. Because cholesterol provided via the LDL pathway has the capacity to suppress the rate-limiting enzyme 3-hydroxy-3-methylglutaryl coenzyme A (HMG CoA) reductase for *de novo* cholesterol synthesis, this may explain the low rate of activity of this pathway in the term placenta. When maternal serum LDL concentrations are suppressed in pregnant baboons, however, there is an increase in placental HMG CoA reductase mRNA expression, a result which supports the concept that the *de novo* pathway has the capacity to be upregulated in this system potentially to provide an alternative source of cholesterol substrate for steroidogenesis. Nevertheless, the LDL pathway typically provides the major source of substrate for progesterone biosynthesis within the primate placenta, a concept consistent with the earlier *in vivo* studies in women which showed that the primary source of progesterone precursor was the maternal circulation.

2. P-450scc

The formation of pregnenolone from cholesterol represents a rate-limiting step in the biosynthesis of steroids in many tissues. During this process, cholesterol carbons 20 and 22 are hydroxylated and the cholesterol side chain between these carbon atoms is cleaved via the P450scc enzyme located in the

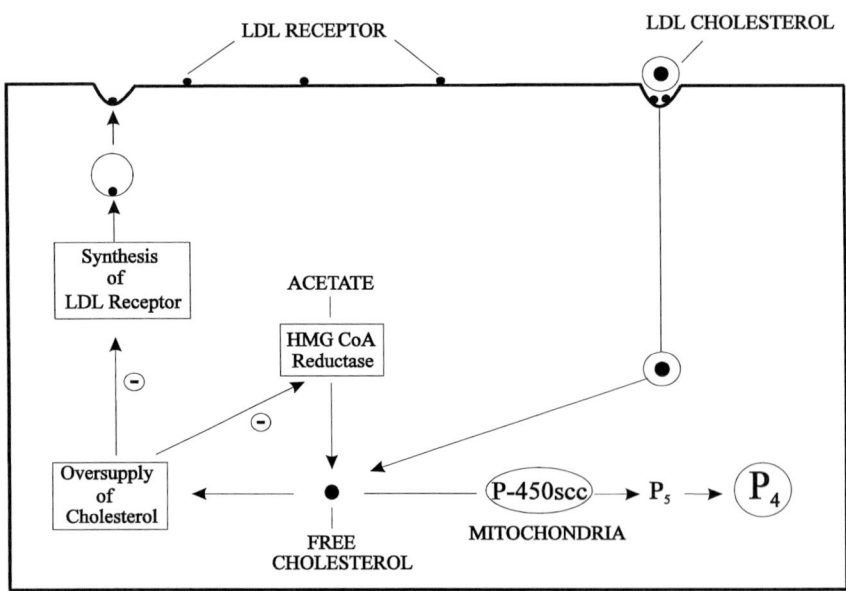

FIGURE 4 Receptor-mediated uptake of LDL cholesterol for the formation of pregnenolone (P_5) and progesterone (P_4) within the human placental trophoblast. An oversupply of free cholesterol within the cell downregulates the HMG CoA reductase enzyme, the *de novo* formation of cholesterol, and the synthesis of LDL receptor.

inner mitochondrial membrane (reviewed in Miller, 1988). The hydroxylation and lyase reactions require a pair of electrons and molecular oxygen. P450scc functions as a terminal oxidase in a mitochondrial electron transport system. Electrons from NADPH are accepted by an adrenodoxin reductase flavoprotein also associated with the inner mitochondrial membrane. Adrenodoxin reductase transfers the electrons to adrenodoxin, an iron–sulfur protein located in the mitochondrial matrix, which then denotes them to cytochrome P450scc.

P450scc exists as a single gene, P450XIA, which is located on human chromosome 15 and is expressed as early as Week 10 of gestation in the human placental trophoblast. This interval approximates the time in human pregnancy when the placenta assumes the role for progesterone production from the ovary. Adrenodoxin reductase also exists as a single gene on the long arm of chromosome 17 and appears to be expressed in many human nonsteroidogenic tissues, suggesting that it serves as a general mitochondrial ferredoxin reductase. Adrenodoxin is synthesized as a larger precursor protein and is encoded by a single functional gene lying on chromosome 11 in the human.

3. 3β-HSD

The conversions of pregnenolone to progesterone and DHA to androstenedione are catalyzed by the 3β-HSD enzyme, which has been localized in the microsomal and mitochondrial fractions of the human placenta. Placental 3β-HSD exists as a monomer peptide of 41 kDa, and dehydrogenase and isomerase activities are inseparable and responsible for both the C_{19} and C_{21} steroid reactions. However, different isoforms of the 3β-HSD are expressed in a tissue-specific manner within the placenta, testis, and adrenal gland. The 3β-HSD is expressed in very high level within the human and baboon placenta and thus does not appear to represent a rate-limiting step in the biosynthesis of progesterone.

B. Regulation

1. Role of the Fetus

Although the progesterone biosynthetic pathway does not directly require the fetus for substrate, the potential role which the fetus may have indirectly upon placental progesterone production during primate pregnancy has been investigated under various conditions with apparently disparate results. Thus, intrauterine fetal death or fetal anencephaly during human or rhesus monkey pregnancy have been reported to either have no effect upon or result in a decrease in maternal progesterone levels. Fetectomy, in which the fetus but not the placenta is surgically removed, provides another experimental approach to study the role of the fetus in nonhuman primates. After fetectomy at midgestation the placenta is maintained *in situ*, viable, and functional with respect to the capacity for aromatization. Fetectomy in baboons results in a progressive decline in serum estradiol as well as progesterone concentrations, findings which indicate that the fetus is required for the maintenance of placental estrogen as well as progesterone secretion.

Collectively, the various studies indicate that although the fetus has a well-defined role in placental estrogen biosynthesis, the requirement for placental progesterone production is less clear. The apparently disparate results may reflect the very different experimental and clinical situations that have been considered when addressing the functional role of the fetus during human and nonhuman primate pregnancy.

2. Effect of cAMP

Strauss *et al.* (1995, 1996) has shown that cAMP regulates the expression of P450scc and consequently progesterone production *in vitro* by the placental trophoblast. Thus, P450scc mRNA expression and progesterone secretion were increased in a dose-dependent manner by 8-bromo-cAMP in cultures of human trophoblasts. Because the cAMP-dependent increase in progesterone secretion occurred in cytotrophoblasts prevented from transforming into syncytiotrophoblasts by culturing cells in the absence of fetal serum, it appears that this process does not depend on syncytia formation. The stimulatory action of cAMP, however, required the catalytic unit of protein kinase because a kinase inhibitor blocked the increase in cytochrome P450scc mRNA expression and progesterone secretion. Strauss has suggested, therefore, that cAMP activates a protein kinase that phosphorylates a specific protein(s), which may be critical to gene transcription and consequently expression of the mRNAs for key regulatory elements in progesterone biosynthesis. It appears, therefore, that the level of adenylate cyclase activity is one determinant of endocrine function of the differentiating trophoblast.

3. Role of Estrogen

It is well established in the laboratory rat and rabbit, species in which the corpus luteum is the principal source of progesterone during pregnancy, that estrogen, which is also formed within the ovary, has a pivotal role in luteal maintenance and the biosynthesis of progesterone. During primate pregnancy, when the placenta is the primary site of progesterone production, estrogen appears to have a similar regulatory role in acting on critical steps in the biosynthetic pathway for progesterone (Albrecht and Pepe, 1990; Pepe and Albrecht, 1995). Thus, coinciding with the increase in estrogen during advancing baboon pregnancy there is a developmental increase in expression of the mRNAs for the P450scc enzyme and LDL receptor, as well as LDL uptake, by placental syncytiotrophoblasts. Moreover, placental syncytiotrophoblast LDL uptake, P450scc activity, and progesterone production are decreased in baboons in which the levels or action of estrogen are suppressed by fetectomy or administration of an estrogen receptor antagonist, respectively, and these effects are reversed by restoring estrogen. Therefore, as illustrated in Fig. 5, the expression of the LDL receptor and thus receptor-mediated LDL cholesterol uptake and expression of the mitochondrial P450scc, and consequently the conversion of cholesterol to pregnenolone and progesterone thereafter, are developmentally regulated by estrogen within the primate placenta. Because placental estrogen formation is dependent on C_{19} steroid precursors, such as DHA/DHAS, of fetal adrenal origin, the fetus ultimately appears to be important in regulating placental progesterone formation during primate pregnancy.

4. Other Factors

The translocation of cholesterol from the outer to the inner mitochondrial membrane, which accounts for the rapid increase in steroidogenesis within the adrenals and gonads in response to tropic stimulation, seems to be mediated by a recently discovered protein, steroidogenic acute regulatory protein (StAR). Mutations in the StAR gene result in lipoid

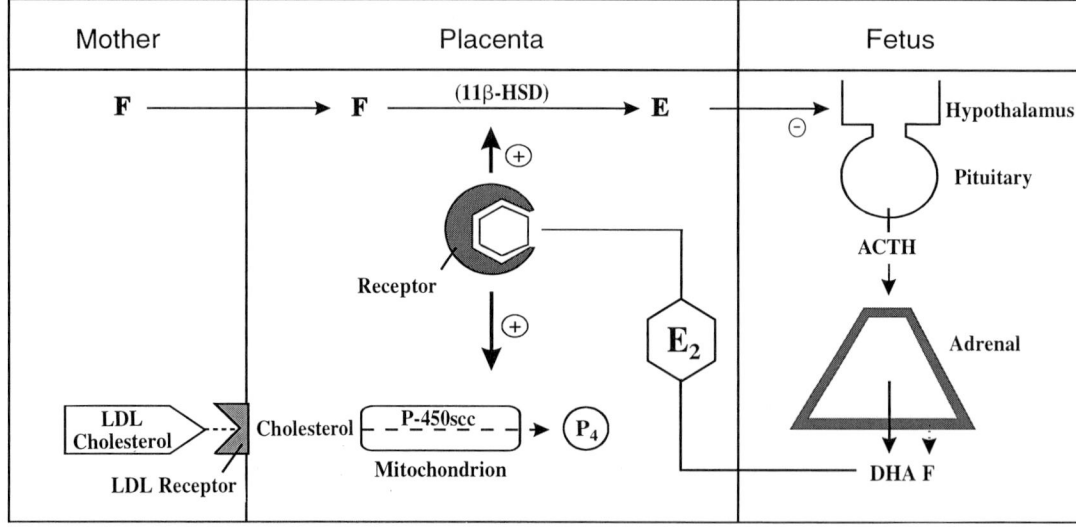

FIGURE 5 The central integrative role of estrogen in modulating the communication between the placenta and fetus in primate pregnancy. Estradiol (E_2), formed from fetal adrenal DHA/DHAS, regulates the receptor-mediated uptake of LDL cholesterol and the P450scc enzyme steps in the progesterone (P_4) biosynthetic pathway in placental syncytiotrophoblasts. Estrogen also regulates the 11β-HSD enzyme which dictates the developmental change in transplacental cortisol (F) and cortisone (E) metabolism, which results in the release of fetal pituitary ACTH and *de novo* F formation by the definitive zone of the fetal adrenal.

congenital adrenal hyperplasia, in which steroidogenesis is impaired and cholesterol accumulates in the cytoplasm of these cells because it cannot be effectively transported to the inner mitochondrial membrane. Other factors [e.g., steroidogenic factor-1 (SF-1)] which bind to the promoters of cytochrome P450 enzymes such as P450scc also seem to play a role in steroidogenesis in the adrenals, ovaries, and testis. However, human trophoblasts do not express StAR or SF-1, suggesting that other mechanisms may be involved in the intracellular trafficking of cholesterol within the placenta. Recent studies by Strauss show that the morphology of the mitochondria of human syncytiotrophoblasts is very different from that of cytotrophoblast, adrenal, or gonadal cells, and it has been suggested that this unique structure may facilitate cholesterol entry and steroidogenesis within syncytiotrophoblast mitochondria despite the absence of factors such as StAR.

IV. REGULATION OF PLACENTAL CORTICOSTEROID METABOLISM BY ESTROGEN

A. Activation of the Fetal HPAA

Although steroid hormones are typically catabolized to their biologically inactive metabolites as they cross the placenta, throughout much of human and nonhuman primate pregnancy the biologically inactive corticosteroid, cortisone, is preferentially metabolized to its biologically active counterpart, cortisol, as it crosses the placenta from the mother to the fetus. However, between mid- and late gestation in the baboon, and apparently also in the human, the pattern of transplacental corticosteroid metabolism changes so that the oxidation of cortisol to cortisone exceeds that of the reverse reaction. Moreover, the increase in estrogen typical of advancing gestation regulates the 11β-HSD enzyme responsible for this change in transplacental corticosteroid metabolism, which would seem to result in a decline in the level of bioactive cortisol coming across the placenta from the mother into the fetus at some point between mid- and late gestation. Because the fetal HPAA is responsive to cortisol feedback at this time in gestation, it has been proposed (Pepe and Albrecht, 1995) that the estrogen-induced developmental increase in transplacental oxidation of cortisol to cortisone would disinhibit the fetal HPAA, resulting in the release of fetal pituitary ACTH and maturation of the fetal adrenal gland (Fig. 5). This appears to be the case because when estrogen levels are prematurely elevated in pregnant baboons at midgestation, there is an increase in transplacental oxidation of cortisol to cortisone, fetal pituitary ACTH expression, and activities of the 3β-HSD and 17β-hydroxylase-17/20 lyase enzymes in and *de novo* cortisol formation by the fetal adrenal. Cortisol, produced by the definitive zone of the fetal adrenal, then matures the lung and other organs critical to neonatal self-sufficiency. It is evident that estrogen, indirectly through its regulatory action on placental metabolism, has a critical role in development of the fetal adrenal gland and consequently maturation of the fetus.

V. SUMMARY

The functional interaction which exists between the primate fetal adrenal gland and placenta for the biosynthesis of estrogen has been well established. However, it is evident from recent studies that more extensive communication occurs between the placenta and fetus with regard to the regulation of steroidogenesis and development during primate pregnancy. Steroid hormones then ensure the maintenance of pregnancy and maturation of the fetus.

See Also the Following Articles

FETAL ADRENALS; FETAL–PLACENTAL UNIT; PROTEIN HORMONES OF PRIMATE PREGNANCY

Bibliography

Albrecht, E. D., and Pepe, G. J. (1990). Placental steroid hormone biosynthesis in primate pregnancy. *Endocr. Rev.* **11**, 124–150.

Gwynne, J. T., and Strauss, J. F., III (1982). The role of lipoproteins in steroidogenesis and cholesterol metabolism in steroidogenic glands. *Endocr. Rev.* **3**, 299–329.

Miller, W. L. (1988). Molecular biology of steroid hormone synthesis. *Endocr. Rev.* **9**, 295–318.

Pepe, G. J., and Albrecht, E. D. (1995). Actions of placental and fetal adrenal steroid hormones in primate pregnancy. *Endocr. Rev.* **16**, 608–648.

Simpson, E. R., and MacDonald, P. C. (1981). Endocrine physiology of the placenta. *Annu. Rev. Physiol.* **43**, 163–188.

Simpson, E. R., Mahendroo, M. S., Means, G. D., Kilgore, M. W., Hinshelwood, M. M., Graham-Lorence, S., Amarneh, B., Ito, Y., Fisher, C. R., Michael, M. D., Mendelson, C. R., and Bulun, S. E. (1994). Aromatase cytochrome P450, the enzyme responsible for estrogen biosynthesis. *Endocr. Rev.* **15**, 342–355.

Strauss, J. F., III, Gafvels, M., and King, B. F. (1995). Placental hormones. In *Endocrinology* (L. J. Degroot, ed.), Vol. 3, pp. 2171–2206. Saunders, Philadelphia.

Strauss, J. F., III, Martinez, F., and Kiriakidou, M. (1996). Placental steroid hormone synthesis: Unique features and unanswered questions. *Biol. Reprod.* **54**, 303–311.

Placozoa

Vicki Buchsbaum Pearse
University of California, Santa Cruz

I. General Features
II. Asexual Reproduction
III. Regeneration
IV. Sexual Reproduction
V. Field Populations

GLOSSARY

cover cells Flattened cells of the upper epithelium, which faces away from the substrate.

cylinder cells Elongate cells of the lower epithelium, which faces toward the substrate.

fiber cells Contractile, stellate cells forming a syncytial network in the interspace between the two epithelia.

interspace Fluid-filled compartment between the two epithelia that contains the fiber cells.

swarmers Small, hollow, floating forms that arise as buds.

Members of the recently established phylum Placozoa Grell 1971 are the simplest metazoans known. The taxon includes only one (or possibly two) species: *Trichoplax adhaerens* Schulze 1883 is abundant and widely distributed in nearshore tropical and subtropical marine habitats, swimming or creeping over the substrate. The plate-like individuals are microscopic or no larger than 2 or 3 mm across and only about 20 μm thick. Hence, they are seldom noticed and have been little studied. Free-living, they appear to feed on algal films, flagellates, and detritus; they cause no known disease and probably have little direct impact on humans or other animals. Their interest lies in their uniquely simple organization and apparently primitive phylogenetic position among animals.

I. GENERAL FEATURES

Placozoans must be compared to other animals mostly in terms of what they lack; they are the "have nots" of the animal realm. They have neither radial nor bilateral symmetry and none of the organ systems typical of most animals (digestive, circulatory, excretory, neurosensory, or muscular). Even the most closely related groups—cnidarians and ctenophores—have a mouth or other body openings; a digestive cavity or gastrovascular network; sensory, nerve, and epitheliomuscular cells; a basal lamina or other collagenous layer underlying the epithelia; totipotent stem cells; and clearly defined female and male gametes. Sponges also display many of these characters. Placozoans lack all of them.

Though the discoid placozoan body is multicellular and its cells contain the usual organelles (nucleus, mitochondria, Golgi, etc.), only four somatic cell types are present. Flattened cover cells and elongate cylinder cells constitute an upper and a lower epithelium, facing away from and toward the substrate, respectively. These epithelia are sometimes designated "dorsal" and "ventral" by analogy with bilaterally symmetrical animals. Both epithelial cell types are monoflagellated and contribute to swimming; the more numerous flagella of the lower epithelium effect a sort of gliding locomotion over the substrate. The lower epithelium is always kept against the substrate, to which the animal adheres strongly, and a dislodged and overturned placozoan will right itself to restore this orientation. Among the cylinder cells are also a smaller number of gland cells, and between the epithelia is a fluid-filled interspace containing a syncytial network of mesenchymatous fiber cells. Contractions of the fiber cells, which contain filaments of actin, are responsible for the animal's continually changing outline and superficial resemblance to an ameba.

II. ASEXUAL REPRODUCTION

Placozoans replicate rapidly by asexual means and can thus form large clonal assemblages. Laboratory populations, fed on cultured flagellates, have been maintained in this way for decades, and field populations may be perpetuated in the same way.

A. Fission

The most commonly seen type of asexual replication is more or less equal fission. The two portions of a dividing individual may remain connected for some time by a thin cellular strand, which eventually breaks and is quickly incorporated into the two newly independent products. Factors stimulating fission have not been studied, but a few data suggest that fission may occur at fairly regular intervals, not closely related to food supply; with less food, the animals are smaller but do not necessarily fission less often.

In another pattern observed, the animal assumes a doughnut shape and then fissions rapidly into multiple products that creep away. (A doughnut-shaped individual may also revert to a solid, discoid form without dividing.) Large numbers of solid spheres (mostly 20–50 μm diameter), which may similarly result from fragmentation, are sometimes seen on the bottom of culture dishes; the surface is half cover cells and half cylinder cells, with fiber cells in the interior. These solid balls appear to assume the discoid form by simply flattening.

B. Budding

Hollow spherical forms, or swarmers, arise by budding from the upper surface of a flattened individual. In swarmers (mostly 40–60 μm diameter), the outer surface is composed of cover cells, the inner cavity is lined by cylinder cells with their flagella projecting to the interior, and fiber cells lie in the interspace. After floating for about a week, a swarmer settles to the bottom, and the wall breaks through, exposing the interior of the cavity, which opens out; the sphere then flattens until the disk-like shape with differentiated upper and lower epithelia is restored.

III. REGENERATION

Placozoans that have been wounded or cut through will quickly close over the cut edges and glide away. However, neither a narrow piece of the rim nor a piece cut from the center of an individual will regenerate; both marginal and central portions of the disk must be present. Also, as has been shown for a number of sponges and hydrozoan cnidarians, placozoan cells that have been dissociated (e.g., by mechanical disruption in Ca/Mg-free seawater) will adhere in masses, sort themselves out into the usual differentiated layers, and develop into normal discoid forms.

IV. SEXUAL REPRODUCTION

The evidence for sexual reproduction in placozoans is equivocal. If no sexual reproduction occurs, this is the only animal phylum to be completely without it.

A. Gametes

What may be gametogenesis sometimes occurs in the laboratory under crowded conditions, in cultures that have entered a "degenerative phase" and perhaps also when two different cultures are mixed. The normally discoid animals become swollen and detached from the substrate, and two new kinds of cells appear in the interspace. One is a large cell resembling an oocyte, reaching 70–120 μm diameter and containing what appear to be numerous yolk granules and cortical granules. Under the same conditions, numerous small (4 μm) round cells, nonflagellated but possibly representing sperms, may also be seen in the interspace. No meiosis or interaction between the two cell types has been documented.

B. Development

The nucleus of the oocyte-like cell enlarges grossly in size and DNA content and becomes fragmented. The cell may later display what looks like a raised fertilization membrane and perivitelline space, but no egg with a fertilization membrane has been found to include an intact nucleus. Early cleavages (up to 32–64 blastomeres) may proceed, but without normal concurrent nuclear mitosis. No development beyond early cleavage stages has been seen.

V. FIELD POPULATIONS

The forms of placozoans seen in laboratory cultures are not easy to relate to those obtained in the field. Glass slides hung in the sea near tropical shores (especially near mangroves or coral reefs) will usually soon bear placozoans. The smallest individuals seen are ~100–120 μm in diameter, a size that best matches the "oocyte" diameters seen in laboratory studies; except for this coincidence of size, however, these settlers could easily represent asexually produced fragments or buds. Feeding on the slides where they settle, placozoans may digest visible paths, e.g., through red algal films, themselves turning pink as they feed and growing and dividing repeatedly until a clone of many individuals results. Oocyte-like cells have never been seen on my field slides, even when these were subsequently kept in the laboratory through degenerative phase or when placozoans from different areas were placed together. Thus, the enigma of sexual reproduction in placozoans remains unresolved.

See Also the Following Articles

CNIDARIA; CTENOPHORA; PORIFERA

Bibliography

Grell, K. G., and Ruthmann, A. (1991). Placozoa. In *Microscopic Anatomy of Invertebrates, Vol. 2, Placozoa, Porifera, Cnidaria, and Ctenophora* (F. W. Harrison and J. A. Westfall, Eds.), pp. 13–27. Wiley-Liss, New York.

Platyhelminthes

Seth Tyler

University of Maine

I. Description of the Platyhelminthes
II. Modes of Reproduction
III. Anatomy of Reproductive Organs
IV. Reproductive Physiology and Endocrine Control
V. Modes of Fertilization
VI. Development
VII. Larvae and Metamorphosis

GLOSSARY

ectolecithal, entolecithal Conditions of eggs of platyhelminths and relating to position of yolk; yolk is packaged either within the oocyte (entolecithal) or in yolk cells that are separate from the oocyte but packaged with it in the egg (ectolecithal).

germarium That part of the female gonad that produces oocytes; distinguished from vitellarium.

neoblasts Quiescent undifferentiated cells capable of differentiating into other cell types, including gametes.

Neodermata The taxon of platyhelminths encompassing the major parasitic groups: the trematodes [aspidobothreans (parasites of molluscs) and digeneans (parasites of vertebrates)], the monogeneans (mostly ectoparasites of fish), and tapeworms and their relatives (parasites of vertebrates).

ootype Egg-packaging region of the oviduct in neodermatan platyhelminths; site at which eggshell develops and encloses oocytes and yolk cells.

polyembryony Asexual reproductive process whereby multiple individuals are produced from a single ovum.

turbellarians All platyhelminths that are not classified in the Neodermata; a paraphyletic group previously considered a distinct class of the Platyhelminthes.

vitellarium That part of the female gonad that produces yolk cells; distinguished from germarium.

I. DESCRIPTION OF THE PLATYHELMINTHES

The Platyhelminthes is a phylum of worm-shaped, acoelomate, bilaterally symmetrical animals that are considered primitive in most schemes of phylogeny of the Metazoa, falling between the coelenterates and the remaining bilaterians such as coelomate worms and molluscs. Larger-bodied platyhelminths have a dorsoventrally flattened body, which is why the group is commonly referred to as flatworms (Platyhelminthes means "flat worms"); actually, many of the small free-living platyhelminths are cylindrical. Platyhelminths have a sack-like intestine with only one opening, the mouth; some parasitic platyhelminths, notably the tapeworms, lack intestine and mouth. Lower platyhelminths, known as turbellarians, have a ciliated epidermis and use the cilia in locomotion; the remainder constitute the Neodermata, comprising the major parasitic taxa of the platyhelminths (including flukes and tapeworms), and these have a syncytial, nonciliated epidermis, evidently an adaptation to parasitism. The Platyhelminthes is an ancient group, with a long evolutionary history and considerable diversity, including a remarkable diversity in their reproductive biology.

II. MODES OF REPRODUCTION

A. Sexual Reproduction

Platyhelminths are by and large hermaphroditic and rely on internal fertilization. A few presumably primitive turbellarians deposit sperm on the epidermis of a partner; the greater majority inject sperm into the body of the partner, either into ducts of the female reproductive tract by copulation or into the body by hypodermic impregnation. For such otherwise simple and presumably primitive animals, the reproductive system is surprisingly complex, with an elaborate male copulatory organ and specialized parts for storing sperm and packaging eggs.

Very few examples are known of gonochoristic turbellarians; some other turbellarians may appear to be dominantly male or female at any given time because of the tendency to protandry and later degeneration of the male reproductive organs as the female system is prominently replete with eggs. Among the trematodes are a few famous examples of gonochorism, notably the blood flukes or schistosomes, which display marked sexual dimorphism; a few tapeworms are gonochoristic. In general, however, hermaphroditism prevails among platyhelminths.

B. Asexual Reproduction

Asexual reproduction is quite common and probably a basic feature of the group. Given their high powers of regeneration, turbellarians suffering accidental fragmentation of the body can likely produce new individuals through regrowth of the pieces, provided totipotent cells known as neoblasts are present in the pieces. Distinctive processes of paratomy (by which new individuals differentiate in a chain-like fashion from a parent worm before separating from it), architomy (in which the body spontaneously fragments and each fragment then differentiates a new individual), and budding are known in given groups of turbellarians, notably acoels, catenulids, macrostomids, and triclads (Fig. 1).

Many of the parasitic platyhelminths use asexual multiplication of stages in the life cycle to produce multiple individuals from a single egg. Specifically, digenean trematodes can produce multiple stages in the life cycle by a kind of internal budding (polyembryony) in which totipotent cells (propagatory cells) develop into whole new individuals that appear as successive stages, each in turn with propagatory cells that generate the next stage (Fig. 2); some monogeneans have what appears to be a kind of polyembryony by which a juvenile develops within the uterus of a parent and has a juvenile developing in its uterus, with yet another juvenile developing in the uterus of that juvenile, etc.—an arrangement like Russian dolls. Some juvenile tapeworm stages, such as cysticercoids and cysticerci, also asexually produce multiple infective individuals within them by a kind of budding; a few other tapeworm juvenile stages are known to multiply by longitudinal fission. Adult tapeworms form chain-like arrangements of multiple segment-like copies of the body, each with its own set of reproductive organs; these are proglottids, and though these reproductive units should not be considered separable individuals in the usual sense of asexual reproduction, they are essentially asexually produced copies of the sexual generative machinery of the tapeworm body.

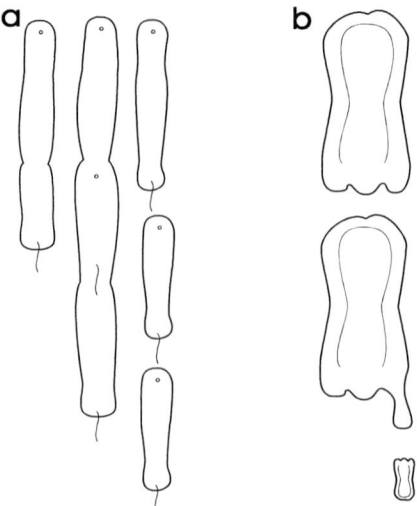

FIGURE 1 Two modes of asexual reproduction in turbellarians. (a) Paratomy in the acoel *Paratomella*, whereby a worm forms a chain of zooids that eventually separate into new individuals. (b) Budding in the acoel *Convolutriloba*: The posterior margin of the body is stretched by reversed swimming of this portion, and a new worm is eventually released with polarity opposite to that of the parent.

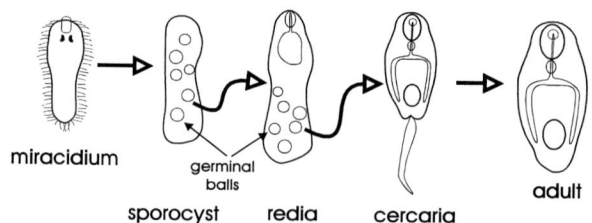

FIGURE 2 Asexual reproduction of juvenile stages in the life cycle of digenetic trematodes. The miracidium is a larva that metamorphoses into the sporocyst; germinal balls develop within the sporocyst from propagotory cells and each develops into the next stage, here represented as the redia. Within the redia the same process asexually produces the cercarial stage which eventually becomes the adult stage. All the asexual reproduction takes place within a first intermediate host, typically a snail.

Parthenogenesis is probably the mode of reproduction of catenulid turbellarians. In polyploid races of other platyhelminths, notably triclads and tapeworms, parthenogenesis—specifically pseudogamous parthenogenesis in which the role of the sperm is to simply activate the egg to initiate parthenogenetic development—is probably the mode. Parthenogenesis may operate in other turbellarians as well or possibly self-fertilization—little documentation exists. Tapeworms and some digeneans are reported to be capable of self-fertilization, but while it is likely that many platyhelminths can facultatively use self-fertilization, cross-fertilization is probably the rule.

III. ANATOMY OF REPRODUCTIVE ORGANS

Being the most complicated and varied part of most platyhelminths, the reproductive organs are used for taxonomy, for delimiting major taxa as well as species from one another.

A. Gonads

The gonads are typically distinct from surrounding tissue, delimited by a tunic. Some of the more primitive members of the group, however, such as acoel and catenulid turbellarians (Fig. 3), have gametes developing in the tissues between gut and body wall without any lining setting them off from other tissues; these taxa have no female canal system and simply release the eggs through rupture of the body wall or possibly through the mouth. The sperm are released through some sort of copulatory structure except in some of the most primitive turbellarians, which have only a simple male pore. Mixed gonads, in which oocytes and spermatozoa are produced in the same general tissue, are known in some turbellarians.

B. Ducts of the Reproductive System

Male gonads typically sit anterior to the female gonads; members of certain groups, including most among the major parasitic taxa such as the trematodes and monogeneans, have them posterior. Some species have many testes, whereas others have only one or two. The ducts from the male gonad comprise the sperm ducts (vasa deferentia), which can be swollen at their distal ends to form spermiducal bulbs or vesicles, and a copulatory organ, consisting of a seminal vesicle, prostatic glandular vesicle, and various penial structures serving as intromittent organs (Figs. 3 and 4). The male canal may open through a separate gonopore or jointly with the female canal into a common atrium that opens to the outside. The penial structure may be quite elaborate, consisting of a sclerotized tubular stylet or a group of needles or a spine-studded cirrus; it may also be a simpler glandular–muscular fold of the wall of the male atrium.

Stemming from each female gonad is an oviduct, joining a common oviduct in cases in which there are two ovaries, and usually a female canal that connects the female system to a separately opening female atrium or vagina. A uterus may be developed, appearing as a separate evagination of the female atrium or, as in the major parasitic groups, as a long expansion of the female canal which extends to the female genital pore. Higher platyhelminths have a vitellarium distinct from the ovary, and it produces yolk cells; it joins the oviduct through a vitelline duct so that yolk cells and ova (or zygotes) are packaged together in the egg (Fig. 4). Such a two-part female gonad is said to be heterocellular, and the egg is termed ectolecithal; lower turbellarians, such as acoels and macrostomids, which have an ovary that produces yolk-rich ova are called homocellular and their eggs are called entolecithal (Fig. 3). Among the heterocellular systems are ovovitellaria in which vitellarium and ovary are contiguous and release both yolk cells and ova into a common duct, the oviduct. Vitellaria are typically quite large and extensively ramifying or follicular in the body, even alongside a small ovary. They are presumably derived from the ovary, and the yolk cells they produce are thus modified oocytes. Because of this two-part nature of the female gonad, the terms "vitellarium" and "germarium" are used by some to refer, respectively, to yolk-cell-producing and oocyte-producing portions.

Typical of the heterocellular female system of the Neodermata is a characteristic egg-forming apparatus, sometimes referred to as oogenotop (Fig. 4). At its center is the ootype, a gland-rich expansion of the oviduct near the point at which it joins the common

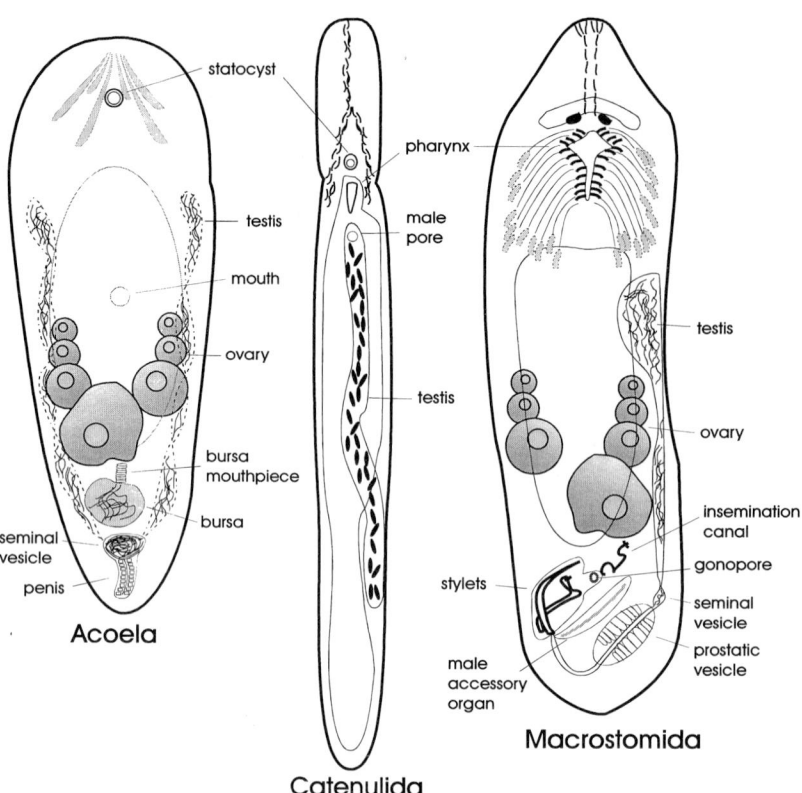

FIGURE 3 Reproductive systems in lower platyhelminths with homocellular female gonad. In the Acoela, represented by *Convoluta*, the gametes develop in the body tissues without being enveloped by a tunic in a distinct gonad, and there are no female ducts; the male copulatory organ in Convoluta is a tubular muscular penis. In the Catenulida, represented by *Retronectes*, the male pore is simple and sits in middorsal position; the sperm are globular. No female canals or bursal organ are known in this group, and even an oocyte has been seen only twice. In the Macrostomida, represented by *Cylindromacrostomum* (*cylindromacrostomum* from unpublished drawing by R. M. Rieger), female canals are developed, and the copulatory organ has a sclerotized stylet (and also, in this particular species, an accessory stylet on a glandular accessory organ and a sclerotized insemination canal, all near a common gonopore into which both female and male canals lead).

vitelline duct. Each oocyte released from the ovary is fertilized as it enters and is then packaged with a cluster of yolk cells in a shell formed by secretions of the yolk cells and probably secretions of a complex of glands surrounding the ootype (Mehlis's glands). Sperm reach the ootype from a seminal receptacle or other parts of the female duct (uterus or vagina). A uterus is well developed in these parasites, correlated with their prodigious production of eggs. It may be short and straight or long and coiled, folded, reticular, or lobulated.

Many turbellarians have a bursal organ, which variously receives and stores sperm, controls their release to the oocyte, or digests excess sperm; it may be separate from the other ducts or linked to the common atrium, the female duct, or even the male duct. It may open to the outside via a separate vaginal or bursal canal (for receiving the penis) and connect to the female gonad or female canal system via an insemination canal which may have sclerotized valves (bursa mouthpiece) controlling release of sperm to the ova. The bursa itself may have sclerotized walls.

C. Germ Cells

Oocytes produced in homocellular gonads, those in which the yolk is incorporated into the oocyte, are on the order of 100–250 μm in diameter; those from the more advanced heterocellular gonads are

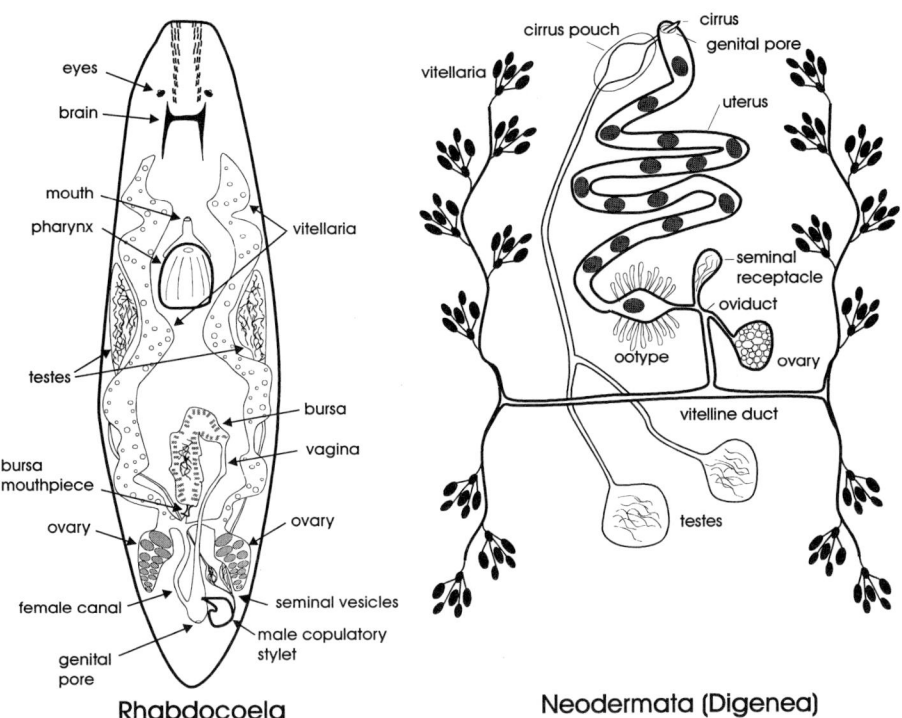

FIGURE 4 Reproductive systems in higher platyhelminths with heterocellular female gonad; yolk-producing vitellaria are separate from oocyte-producing ovary. In the Rhabdocoela, represented by *Proxenetes*, vitellaria are usually paired and ovaries are paired or unpaired (paired in *Proxenetes*), and the canal systems usually end at a common gonopore, although many have a separate opening for the bursa (after Reisinger, 1928, p. 33). In the Neodermata, represented by a generalized scheme for the Digenea, the ovary is single and its oviduct joins the vitelline duct at an egg-packaging center, the ootype, and projects from there as an expanded region serving as uterus to hold the embryonated eggs.

smaller, commonly <100 μm. The oocytes have characteristic inclusions: the eggshell granules, which are produced first, and the yolk granules, which appear next.

Spermatozoa in most platyhelminths are characteristically biflagellate and highly elongate, lacking the head, middle piece, and tail typical of sperm in most animals. The two axonemes in many are incorporated for most of their length in the body of the sperm, whereas others have free flagella. Uniflagellate sperm are known in specific groups of turbellarians (notably the primitive Nemertodermatida and, through secondary loss of one flagellum, in some rhabdocoel turbellarians and certain tapeworms); certain other turbellarian taxa have aflagellate sperm, with ovoid shape or with two stiff bristle-like projections instead of flagella. The major parasitic groups and the higher turbellarians have a characteristic pattern in the microtubules of the axoneme in their two flagella, with the nine doublets typical of motile flagella but with a central banded solid core, not microtubular.

IV. REPRODUCTIVE PHYSIOLOGY AND ENDOCRINE CONTROL

Maturation of the gonads is probably under neurosecretory control in platyhelminths, but little is known about such control. In triclad turbellarians, for example, maturation of the gonads is seen to correlate with the number of neurosecretory vesicles in nerve plexuses around those gonads. Hypothetical gradients of gonad-inducing factors along the length of the body in triclads have also been invoked to explain the differentiation of more anteriorly positioned neoblasts into ovaries and more posterior ones into testes, and these factors are thought to correlate with the abundance of neurosecretory vesicles in the ventral nerve cords. Other platyhelminths are also thought to derive the gonads from such stem cells and presumably their differentiation is through hormonal control. Differentiation of the copulatory organs has also been shown to be inducible by testosterone in triclads; similarly, tapeworms are known to respond, at least in growth of the individual and reproductive organs, to the testosterone of their hosts. Much less is known about other flatworms.

Pheromones may come into play in coordinating copulatory behavior among platyhelminths in that individuals of the same species will aggregate and the timing of copulation and production of eggs appears conditional on release of substances into the environment by mature worms.

Tapeworms are known to secrete substances that have hormone-like effects on their host; for instance, substances that inhibit gonad development in fish hosts or that mimic growth hormone in mammals.

V. MODES OF FERTILIZATION

Fertilization is always internal, apparently necessitated among marine species by the small size of the animal and relatively large size of the gametes; internal fertilization is clearly also adaptive in the freshwater, terrestrial, and parasitic habitats that higher flatworms have invaded. The sperm are introduced into a partner usually by copulation or hypodermic impregnation; a few lower turbellarians simply deposit sperm onto the epidermis of the partner in loose bundles or as spermatophores. Evidence for insemination through cannibalism of another individual is also known. Sperm are mutually exchanged usually in copulatory pairings, but each partner can function as one sex only.

Self-fertilization is probably rare despite the hermaphroditic nature of the reproductive system. It has been reported for turbellarians—including marine triclads, a few freshwater triclads, and for the production of summer eggs (subitaneous eggs) in a rhabdocoel (whereas production of resting overwintering eggs requires cross-fertilization)—and in a few representatives of the monogeneans, trematodes, and tapeworms. Some turbellarians raised individually in isolation can produce young, whereas others cannot.

Fertilization takes place typically in the female canal system at a site proximal to that at which the egg envelopes are produced; in the parasitic groups, the Neodermata, this fertilization occurs in the oviduct just proximal to the ootype. In some turbellarians, fertilization apparently takes place even in the ovary.

VI. DEVELOPMENT

The embryo of certain lower turbellarians cleaves by a spiralian pattern. Particularly in the polyclad turbellarians, the cleavage follows a typical quartet spiral pattern quite like that of annelids and molluscs; in acoel turbellarians, the spiral cleavage is in duets rather than the quartets typical of spiralian taxa; and in others the spiral nature of cleavage can be seen only in early stages, later stages being modified. The heterocellular nature of the eggs in the higher platyhelminths, with small oocytes accompanied by abundant yolk cells, results in cleavage that is highly modified and seemingly chaotic; the blastomeres are often scattered among yolk cells and only later regroup to form, for instance, blastemas of specific organs.

The remarkable capacity for asexual production of stages in the life cycles of parasitic flatworms such as the trematodes is manifest even in the first cleavage of the zygote: The first two cells produced are a stem cell (propagatory cell) and a somatic cell. The somatic cell develops into the tissues of the embryo, whereas the stem cell divides to produce a line of more stem cells and more somatic cells, which are the cells of the tissues of future stages. The stem cells are thus the germinal cells that differentiate into the later asexual stages, and they also produce the gonads of the adult.

VII. LARVAE AND METAMORPHOSIS

Turbellarians, by and large, develop directly, without larval stages; the major parasitic taxa of flatworms, however, have indirect development with distinctive larvae. The few turbellarians known to have larvae include some polyclads, which develop a Müller's larva or a Götte's larva, and a catenulid, which develops a Luther's larva. The polyclad larvae have ciliated lobes (eight in the Müller's larva and two in the Götte's larva) by which they swim in the plankton and gather nanoplankton for food. Metamorphosis involves shedding of the lobes and assumption of a more worm-like shape. The Luther's larva has circumferential ciliary bands by which it swims and a statocyst; at metamorphosis the bands are reduced and the statocyst is lost. Another planktonic form thought to be the larva of a nemertodermatid turbellarian has been found in nearshore ocean waters.

Parasitic flatworms of the Neodermata, on the other hand, all have larval stages and rely on them to reach a first host. Trematodes, monogeneans, and tapeworms with aquatic life cycles have a ciliated larva that hatches from the egg, namely, the cotylocidium in aspidobothrean trematodes, the miracidium in digenean trematodes, the oncomiracidium in monogeneans, and the coracidium or lycophora in tapeworms; these types are distinguished by, among other features, the presence or differences in sclerotic attachment or host-penetration hooks. Cotylocidia and oncomiracidia have mouth, pharynx, and gut; miracidia, coracidia, and lycophora do not, but all these larvae are nonfeeding and quite short-lived, dying if they cannot find a host. They have sensory organs appropriate for locating a host and they have penetration organs. As it penetrates the host, such a ciliated larva undergoes a metamorphosis in which it sheds the ciliated cells and develops a syncytial epidermis to replace them from beneath. Tapeworms with terrestrial life cycles have a larva (oncosphere) that lacks the cilia but still undergoes a metamorphosis involving replacement of larval epidermis as it invades the first intermediate host.

In parasites with single-host life cycles, such as the aspidobothrean trematodes and monogeneans, the larva metamorphoses directly into an adult. Multihost life cycles, typical of digenean trematodes and tapeworms, may be quite complex and involve a series of additional juvenile stages that develop through asexual multiplication. As many as four morphologically different juvenile stages follow the larva in the life cycle of a single species of trematode. Tapeworms have a distinct juvenile stage, generally referred to as a metacestode, between the larva and adult. This may involve two successive forms, the procercoid and the plerocercoid, typical of aquatic life cycles and which directly transform into a single adult. Alternatively, it may also be a single stage, generically called cysticercoid or cysticercus, some forms of which may asexually produce multiple progeny internally, each capable of developing into an adult in the next suitable host.

The most complex life cycles are those of the Digenea (Fig. 2). Hatching from the egg is the larva miracidium which, upon finding an appropriate host, metamorphoses (in most species) into a sporocyst—a sack-like stage within which germinal balls develop into the next stage, daughter sporocysts or rediae (rediae may develop earlier in the miracidium and be released immediately as that larva metamorphoses in the first host). Rediae are feeding stages, with a pharynx, and the germinal balls within them develop into cercariae or more rediae (daughter rediae). Cercariae are typically free-swimming and may metamorphose into the adult directly or, more usually, encyst in an intermediate host as metacercariae. A few uncommon forms produce miracidia from sporocysts or produce cercariae from an adult-like animal in an intermediate host (such cercariae in turn produce another generation of cercariae).

Bibliography

Adiyodi, K. G., and Adiyodi, R. G. (Eds.) (1983, 1988, 1989, 1992, 1993, 1998). *Reproductive Biology of Invertebrates*. Wiley, Chichester, UK. [A 7-volume treatise including 13 chapters on the Platyhelminthes]

Baguna, J., and Boyer, B. C. (1990). Descriptive and experimental embryology of the Turbellaria: Present knowledge, open questions and future trends. In *Experimental Embryology in Aquatic Plants and Animals* (H.-J. Marthy, Ed.), pp. 95–128. Plenum, New York.

Coil, W. H. (1991). Platyhelminthes: Cestoidea. In *Microscopic Anatomy of Invertebrates. Vol. 3: Platyhelminthes and Nemertinea* (F. W. Harrison and B. J. Bogitsh, Eds.), pp. 211–283. Wiley-Liss, New York.

Fried, B., and Haseeb, M. A. (1991). Platyhelminthes: Aspidogastrea, Monogenea, and Digenea. In *Microscopic Anatomy of Invertebrates. Vol. 3: Platyhelminthes and Nemertinea*

PMSG

see Equine Chorionic Gonadotropin

PMS (Premenstrual Syndrome)

Ellen W. Freeman

University of Pennsylvania

I. What Is Premenstrual Syndrome?
II. Who Has Premenstrual Syndrome?
III. What Causes Premenstrual Syndrome?
IV. Diagnosis
V. Medical Treatments
VI. Future Directions

GLOSSARY

follicular Describing or relating to the first half of the menstrual cycle, from menses to ovulation.

gonadotropin-releasing hormone A neurohormone of the hypothalamus that controls gonadotropins.

luteal Describing or relating to the second half of the menstrual cycle, from ovulation to menses.

postmenstrual Describing or relating to the days following menses and before ovulation, in clinical assessments usually about Days 6–12 (Day 1 is the first day of menses).

premenstrual Describing or relating to the days preceding the menstrual flow, usually about 1 week in clinical assessments, but symptomatic days may range from several days to 2 weeks.

premenstrual dysphoric disorder The diagnostic term for premenstrual syndrome in the *Diagnostic and Statistical Manual of Mental Disorders*.

selective serotonin reuptake inhibitor A class of antidepressant medications.

Premenstrual syndrome (PMS) is a cluster of emotional, behavioral, and physical symptoms that occur for several days to several weeks before menses and subside following the menstrual period. Most women experience symptoms of PMS at some time in their reproductive years but do not perceive these menstrually related changes as either distressing or debilitating. A small minority of women experience severe premenstrual symptoms that diminish productivity, result in lost work time, and disrupt relationships. The etiology of PMS remains unknown. The mood and behavioral changes that characterize PMS may derive from central nervous system effects, possibly an altered sensitivity to normal hormonal changes of the menstrual cycle. Diagnostic criteria are evolving, and information on the effectiveness of treatments has increased in recent years. Serotonergic antidepressants show the most promising treatment results at this time.

TABLE 1
Common Symptoms of PMS

Irritability	Difficulty concentrating
Nervous tension	Insominia
Anxiety	Swelling, bloating
Feeling overwhelmed	Breast tenderness
Mood swings	Headache
Depressed mood	Muscle aches
Decreased interest	Poor coordination
Fatigue	Food cravings, appetite changes

I. WHAT IS PREMENSTRUAL SYNDROME?

Premenstrual syndrome (PMS) has appeared in the medical literature for nearly 70 years, but only recently have the causes and treatments of this disorder had rigorous scientific scrutiny. The term itself lacks precision because it is commonly used by both the laypublic and clinicians to describe a broad range of menstrually related changes, ranging from the normal and nondistressing to severe and debilitating.

The most frequently reported symptoms of moderate to severe PMS include irritability, nervous tension, mood swings, depression, feeling overwhelmed or out of control, fatigue, physical symptoms of swelling or bloating of the abdomen or extremities, appetite changes, food cravings, aches, and breast discomfort. In severe PMS, the mood and behavioral symptoms of the syndrome are almost invariably the primary complaint.

II. WHO HAS PREMENSTRUAL SYNDROME?

PMS can occur throughout the reproductive years, from menarche to menopause. However, the ages of women seeking treatment for PMS are overwhelmingly from the early twenties to the late thirties. PMS does not occur during pregnancy or after menopause but does continue after a hysterectomy as long as ovarian production remains. Demographic background factors are not associated with PMS, which appears to occur across all socioeconomic levels without regard for education, occupation, family size, household composition, or ethnic background. Many women diagnosed with other physical or psychiatric disorders experience exacerbated symptoms premenstrually, as for example in depressive disorders, epilepsy, migraine, and endometriosis. These observations suggest that PMS is not simply associated with ovarian steriod production in the reproductive years, but that the effects of ovarian hormones on brain regions may be important in this disorder (Table 1).

III. WHAT CAUSES PREMENSTRUAL SYNDROME?

The etiology of PMS is unknown. The dominant current hypotheses are that an underlying neurobiological vulnerability or altered central neuroregulation may be causal in PMS. Historically, the gonadal steroids estrogen and progesterone were hypothesized as the cause of PMS symptoms, but there is no consistent evidence of abnormal circulating levels of gonadal steroids in PMS. Since PMS occurs in the context of normal endocrine function, the reproductive endocrine factor may be necessary but is not causally sufficient. Nonetheless, the effects of ovarian steroids on neurotransmitters appear likely to be important in PMS symptomatology. Current evidence supports serotonergic, γ-aminobutyric acid (GABA), adrenergic, and opioid involvement.

A. Serotonergic Involvement

Various preliminary studies of markers and activity of the serotonin system are consistent with hypotheses of central serotonergic activity in PMS symptomatology. The serotonin system is an important central modulator of mood and behavior. Brain serotonin neurotransmission is influenced by alterations in plasma concentrations of progesterone and estradiol, suggesting a possible pathway by which gonadal steroids may by involved in PMS. Indications of abnormalities in markers of serotonergic transmission in women with severe PMS include evidence of a lowered platelet imipramine binding (a peripheral marker of 5-HT function) in the early luteal phase, decreased platelet 5-HT content and 5-HT uptake during the luteal phase, and significantly decreased whole blood 5-HT levels premenstrually. PMS patients have shown a lower 5-HT response to tryptophan (a 5-HT precursor) during the luteal phase compared to other phases of the menstrual cycle. Challenge tests depleting tryptophan can provoke PMS symptoms, whereas tryptophan supplementation relieves PMS symptoms, as observed in open-label treatment. D-Fenfluramine, which releases serotonin and blocks its uptake, can block food craving and improve depressed mood in PMS subjects. Treatment trials of the new serotonergic antidepressants, such as fluoxetine, sertraline, and clomipramine, show significant benefit over a placebo in PMS patients, further suggesting that serotonin may be involved in the symptoms of PMS.

B. Relationship to Depression

Because depressive symptoms, such as low mood, fatigue, appetite changes, sleep difficulties, and decreased concentration, are among the most common symptoms of PMS, a biological relationship between PMS and depression has been postulated. The overlap between PMS and atypical depression, in which emotional hypersensitivity, increased anxiety, irritability, and food cravings predominate, is particularly notable. A relationship with depression is further supported by a high incidence of lifetime history of depression in PMS patients, ranging from 30 to 76% in reported studies, compared to a prevalence of depressive illness in the general female population of about 25%. A family history of depressive illness is also common in women with moderate to severe PMS. Preliminary evidence that the GABA system may be altered in women with premenstrual dysphoric disorder supports the possibility of a biological continuum of premenstrual dysphoria and major depressive disorders.

C. Differences from Depression

PMS patients report feeling "down, blue, or depressed" to the same degree in the premenstrual phase as patients with major depression. However, when clinical ratings of other depression items are examined, PMS patients have less loss of interest, diurnal mood variation, guilt, early morning waking, and hypochondriasis (all core symptoms of depression) than depressed patients. PMS symptoms are intermittent, occurring only in the luteal phase of the menstrual cycle, and the symptoms increase and diminish relatively swiftly. Thus, neither the expression nor the timing of depressive symptoms in PMS is identical to major depression or dysthymia.

Although PMS frequently mimics depressive disorders premenstrually, women with severe PMS also report anxiety and other behavioral symptoms. Studies show increased sensitivity of PMS women to lactate infusion, carbon dioxide inhalation, and cholecystokinin-tetra-peptide, responses that are also seen in persons with panic disorder. Notably, the increased sensitivity to lactate infusion is evident whether or not PMS is associated with affective disorders. Possibly, PMS symptoms, like those of panic patients, are associated with an overinterpretation of internal cues.

Suppressing ovulation offers further evidence of a biological distinction between PMS and depression. Treatment with a gonadotropin-releasing hormone (GnRH) agonist, which downregulates and desensitizes the pituitary to produce a hypogonadotropic and hypogonadal state, results in significant reduction or remission of premenstrual depressive symptoms in women with clearly diagnosed PMS. Ovulation suppression with a GnRH agonist does not reduce the mood and behavioral symptoms in women whose symptoms are not limited to the luteal phase of the cycle.

The strongest biological evidence of a distinction between PMS and depression is the cortisol response of PMS patients to corticotropin-releasing hormone

(CRH) stimulation. In contrast to the response to CRH stimulation of depressed patients [blunted adrenocorticotropic hormone (ACTH) and normal cortisol], PMS patients exhibit an increased response to CRH stimulation. Importantly, the excessive response of PMS patients to cortisol can occur in the absence of any comorbid affective disorder. These results suggest a possible hypersensitivity of the pituitary corticotrophs to CRH in PMS, in contrast to the HPA hyperactivation-induced hyposensitivity to CRH in depression.

Cortisol levels are consistently in the normal range in PMS studies. However, decreased plasma levels of β-endorphin and ACTH, which is cosecreted with β-endorphin and originates from the same precursor, have been identified in PMS subjects. Other research has failed to find significant differences in basal levels of ACTH in women with and without PMS but has reported lower argonine vasopressin levels in PMS women throughout the cycle as well as positive significant correlations between atrial natriuretic peptide and ACTH levels in PMS women but not in asymptomatic controls. These preliminary findings of menstrual cycle-independent effects suggest a possible underlying neurobiological vulnerability in PMS subjects that differs from major depression.

Decreased β-endorphin levels in PMS patients are reported by several research groups, although replications in the small study samples have not been consistent. A loss of central opioid tone during the midluteal phase in women with PMS as indicated by the loss of luteinizing hormone (LH) response to naloxone is consistent with the hypothesis that PMS symptoms are associated with an altered central neuroregulation, although alternative explanations cannot be excluded in view of inconsistent findings.

IV. DIAGNOSIS

There is no laboratory test of hormones or other physiologic measure that identifies PMS. Currently, a diagnosis is made from the patient's daily rating of symptoms for two or more menstrual cycles. These prospective daily symptom reports must demonstrate a cyclic symptom pattern, with the greatest symptom severity in the week preceding menses and minimal or no symptoms in the week following menses. The severity of the symptoms is further determined by the extent to which they interfere with usual functioning at work, at home, and in personal relationships. Additional physical examination and psychiatric evaluation is needed to determine that the symptoms are not accounted for by another physical or emotional disorder.

A number of instruments are reported in the medical literature for rating PMS symptoms, but there is considerable variability in their scope and utility for clinical practice. Fewer than 20 symptoms appear to be the most common in PMS, and shorter rating scales are appropriate for primary clinical care. One simple approach is to list the woman's 5 worst symptoms with a grid for rating the severity of each symptom daily. This is done each evening for at least two menstrual cycles to determine the specific symptoms, their severity, and changes over the cycle (Fig. 1).

Notably, less than half the women who report PMS will have prospective symptom reports that corroborate a PMS diagnosis. Symptoms that occur other than premenstrually are not PMS. Depending on the stringency of the criteria for the severity and number of symptoms and the required degree of postmenstrual symptom remission, as few as 2–10% may be diagnosed as having severe PMS with no concomitant physical or mood disorder (Table 2).

The leading attempt to develop a uniform assessment of severe PMS is the evolving diagnosis of premenstrual dysphoric disorder (PMDD), which is in-

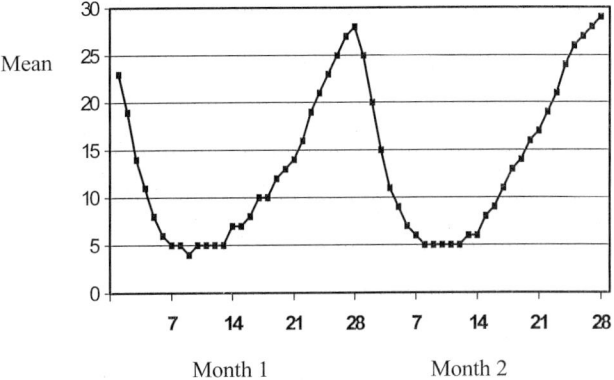

FIGURE 1 Daily symptom scores for two menstrual cycles (standardized to 28 days) for 170 women who met clinical criteria for PMS.

TABLE 2
Daily Symptom Rating

DAILY SYMPTOM RATING

UNIVERSITY OF PENNSYLVANIA MEDICAL CENTER
PMS PROGRAM

NAME _____

Date (Mo/Day/Yr)																																
Check, if Menstruating																																
Study Medication: Number Taken																																
Check if other Meds Taken—List on other side																																
Fatigue, Lack of Energy																																
Poor Coordination																																
Feeling Out of Control, Overwhelmed																																
Feeling Hopeless, Worthless or Guilty																																
Headache																																
Anxiety, Tension, "On Edge"																																
Aches																																
Irritability, Persistent Anger																																
Mood Swings																																
Swelling, Bloating, Weight Gain																																
Craving Foods, Increased Appetite, Overeating																																
Decreased Interest in Usual Activities																																
Cramps																																
Depression, Feeling Sad, Down or Blue																																
Breast Tenderness																																
Insomnia or Hypersomnia																																
Difficulty Concentrating																																

0 = Not present at all
1 = Mild: only slightly apparent to you
2 = Moderate: aware of symptom, but don't affect daily activity at all
3 = Severe: continuously bothered by symptoms and/or it interferes with daily activity
4 = Very Severe: symptom is overwhelming and/or unable to carry out daily activity

TABLE 3
Criteria and Symptoms for
Premenstrual Dysphoric Disorder

Criteria

- Five or more of the symptoms listed below (including at least one of the first four symptoms) are present for most of the time during the week before menses for at least 1 year. The symptoms diminish after the onset of menses and are absent in the week following menses.
- The symptoms markedly interfere with usual functioning.
- The symptoms are not merely an exacerbation of another disorder.
- The symptoms are confirmed by prospective daily ratings for at least two consecutive menstrual cycles.

Symptoms

Depressed mood, hopelessness
Anxiety, tension
Affective lability, mood swings
Irritability, persistent anger
Decreased interest in usual activities
Difficulty concentrating
Fatigue
Appetite changes, overeating, food cravings
Insomnia, sleep difficulties
Feeling overwhelmed or out of control
Physical symptoms such as breast tenderness, swelling, headaches, aches, feeling bloated

---

Note. Reprinted with permission from the *Diagnostic and Statistical Manual of Mental Disorders*, 4th ed. © 1994 American Psychiatric Association.

---

cluded as a provisional diagnosis in the *Diagnostic and Statistical Manual of Mental Disorders* (*DSM-IV*). PMDD replaces and differs little from the earlier diagnosis of late luteal-phase dysphoric disorder, which was included in the *DSM-III-R*.

The PMDD diagnosis requires that 5 of 11 specified symptoms be "marked or severe" premenstrually and "low or absent" postmenstrually. The symptoms must markedly interfere with usual functioning, not be due to another current physical or psychiatric disorder, and be confirmed by daily symptom ratings for at least two consecutive menstrual cycles.

The PMDD symptoms include depressed mood, anxiety/tension, mood swings, irritability, decreased interest in usual activities, difficulty concentrating, fatigue, appetite changes, sleep difficulties, feeling out of control, and physical symptoms. One of the first four symptoms is required for the diagnosis; all physical symptoms are considered as a single symptom.

The PMDD diagnosis is important as an effort to develop a systematic diagnosis of the disorder. PMDD emphasizes the severity of PMS (requiring at least five severe symptoms that interfere with work or relationships), but it also stresses dysphoric symptoms and may exclude other women whose premenstrual distress is predominantly characterized by anxiety. A continuing difficulty is that the methods of determining "severe" and "remission" are undefined and vary widely in both clinical practice and research. It is demonstrated that among women requesting treatment for PMS, the number diagnosed as having PMS can range from 14 to 45%, depending on the methods used to define symptom severity both pre- and postmenstrually and the degree of symptom changes over the cycle (Table 3).

V. MEDICAL TREATMENTS

After many years of unexamined or ineffective treatments, there are now three classes of medications that have evidence of efficacy for treating severe PMS or PMDD: antidepressants, anxiolytics, and GnRH agonists. These medications have demonstrated effectiveness in rigorous scientific studies, although no medication is successful for all patients in the heterogeneous PMS population and no medication is approved by the U.S. Food and Drug Administration for the indication of PMS or PMDD.

A. Antidepressants

The recent development of the serotonergic antidepressants (commonly termed SSRIs) has provided a new class of medication that shows promising results for treatment of severe PMS or PMDD. Clinical studies to date suggest that approximately 60–70% of clearly diagnosed PMS subjects report significant reduction of premenstrual symptoms with serotonergic antidepressants. Studies of efficacy for PMS that are reported in the medical literature include fluoxetine, sertraline, paroxetine, clomipramine, and nefazodone.

In the studies reported thus far, the dosing for the serotonergic antidepressants starts at the standard starting dose for depression, and the medication is taken daily throughout the cycle. In the absence of major improvement or dose-limiting side effects, the dose is increased at the beginning of the second treatment cycle and again at the beginning of the third treatment cycle. Efficacy is generally observed in the first or second cycle of treatment, i.e., at relatively low doses, but it is important to ensure that a poor response is not due to an inadequate dose. If the response remains insufficient, it is generally useful to try another SSRI. It is important to taper when discontinuing the medication. Discontinuation symptoms have been observed in some patients, but the likelihood is not known.

Side effects are common with the onset of the serotonergic antidepressants but are typically transient and subside with time or dose adjustments. Discontinuation due to side effects (other than loss of libido, which may not decrease with time if it occurs) is infrequent. The most common side effects include nausea, insomnia, headache, fatigue, diarrhea, decreased concentration, dizziness, and sexual dysfunction. Of the SSRIs, only nefazodone does not have sexual dysfunction side effects in PMS studies.

Tricyclic antidepressants may be helpful for some PMS patients, but there is currently insufficient information. Preliminary reports of nortriptylene and desipramine suggest efficacy for depressive symptoms of PMS, but PMS symptoms are not limited to depression, and patients with nondepressive PMS symptoms may not be helped with this treatment. More important, tolerability of the tricyclic antidepressants by PMS patients appears to be very low, and discontinuation of treatment precludes improvement.

Because PMS symptoms are intermittent rather than continuous, there is clinical demand for treatment that can be taken symptomatically and limited to the premenstrual phase of the cycle. Clinical experience and several case reports in the literature support intermittent dosing with fluoxetine or sertraline, at least for a subgroup of PMS patients. However, well-designed, placebo-controlled studies are needed to support or refute these observations.

B. Anxiolytics

Alprazolam, a triazolobenzodiazepine, is an anxiolytic and antipanic medication that may have some antidepressant properties. Alprazolam shows efficacy for PMS symptoms in several but not all placebo-controlled double-blind studies. The conflicting results may be due to differences in the characteristics of small samples. The largest study found that alprazolam was significantly better than a placebo, whereas another study showed that alprazolam was effective for pure PMS in which symptoms occurred only premenstrually but not for women who had mild symptoms of depression and anxiety continuing after the menstrual period.

The effective dose level of alprazolam for PMS patients appears to be 0.25 mg taken three or four times/day. Dosing starts several days before symptoms in the luteal phase and is tapered for several days beginning at the onset of menses to preclude rebound or withdrawal symptoms.

Overall, alprazolam appears to have a modest effect in the PMS population. The advantage of alprazolam is that it can be taken only in the symptomatic premenstrual phase (rather than daily). Also, its mechanism of action is clearly distinct from that of the serotonergic antidepressants, thus providing another therapeutic option. The limitation of alprazolam is the risk of possible dependence when doses are not strictly limited to the luteal phase.

C. Gonadotropin-Releasing Hormone Agonists

Gonadotropin-releasing hormone agonists (GnRHas) downregulate and desensitize the pituitary to produce a hypogonadotropic and hypogonadal state. This treatment has significantly reduced premenstrual symptoms in women in a number of small studies. It appears to be most effective for women with pure PMS, i.e., the symptoms are clearly limited to the premenstrual phase. Women with depressive symptoms throughout the menstrual cycle, despite a clear premenstrual exacerbation, do not improve with GnRHa treatment.

Efficacy does not occur until after several months of GnRHa treatment. This may in part be due to the "flare" effect, an initial agonist action, which is associated with the increased circulating levels of the gonadotropins, follicle-stimulating hormone and LH. However, there is little evidence that PMS symptoms worsen with the onset of treatment, and it may be that improvement occurs when the circulating levels of estrogen and progesterone decrease to menopausal levels.

Side effects with GnRHa treatment are common and include hot flashes, aches, headache, vaginal dryness, sweating, tingling, and decreased libido. The health risks of the hypoestrogenic state are essential to consider in long-term treatment. Steroid "add-back" therapy to protect against bone mineral loss has shown only a small increase in premenstrual symptoms, but the long-term utility of add-back regimens for PMS treatment is not yet demonstrated.

D. Other Medical Treatments

Progesterone was long the treatment of choice for PMS, but numerous studies have reported a lack of efficacy for progesterone suppository treatment. A large randomized treatment trial of oral micronized progesterone likewise reported negative results for the overall symptoms of PMS. Although progesterone may be more effective than a placebo for physical symptoms of breast tenderness and swelling, severe PMS usually includes mood and behavioral symptoms that are not improved with progesterone treatment.

Other medical treatments advocated for PMS include evening primrose oil, spironolactone (an aldosterone receptor agonist), bromocriptine, prostaglandin synthesis inhibitors, pyridoxine, synthetic progestational agents, and oral contraceptives. None of these treatments are supported by scientific evidence of beneficial effects exceeding a placebo in rigorous studies of clearly defined PMS patients. Each of these treatments may be beneficial for some women in reducing specific symptoms, and the possibility of small effects cannot be excluded. Such small effects have heuristic value but little clinical utility.

E. Nonpharmacologic Treatments

Dietary modifications, cognitive behavior therapy, relaxation training, group therapy, biofeedback, light therapy, and sleep deprivation have been investigated as treatments for PMS. These noninvasive therapies appear to have many benefits and may be the most appropriate for mild or moderate symptoms, although definitive scientific studies that demonstrate their efficacy are generally lacking.

F. Self-Help Approaches

It is appropriate to try nonmedical approaches to PMS symptoms before initiating pharmacologic treatments, particularly when symptoms are not severe. Many women find that they can control their premenstrual symptoms by making changes in diet, exercise, or lifestyle. If such approaches do not help significantly within several months, medication should then be considered.

There is no evidence that nutritional deficiencies cause PMS. However, poor dietary habits may exacerbate symptoms, and dietary changes that enhance nutritional status and promote good health are often recommended for women with PMS symptoms.

A PMS diet has been described as 60% complex carbohydrates, 20% protein, and 20% fat. Caffeine, salt, sugar, and alcohol consumption should be reduced or eliminated. Consuming meals that are high in carbohydrate-rich foods and low in protein, particularly during the premenstrual symptomatic time, may improve mood symptoms of PMS. This counters advice to resist carbohydrate cravings, but it is known that carbohydrate-rich foods that are low in protein increase release of serotonin, which is involved in both dysphoric mood and appetite.

Understanding the symptoms and the regularity of their occurrence helps many women gain control over their symptoms. A primary tool for this is the daily symptom report, and the process of rating symptoms each day is a cognitive learning experience. Women frequently report improvement after rating their symptoms for several months. A support group in which women with the same problem share their experiences is a helpful approach for some women. In clinical treatment, a positive therapeutic relationship may augment the response to medication.

VI. FUTURE DIRECTIONS

PMS is a complex, chronic, menstrually related mood disorder that affects large numbers of women in their reproductive years. It is well documented that depressive mood disorders as a class are associated with increased accident rates, increased rates of substance abuse, increased medical hospitalization, somatic illnesses, and outpatient medical utilization. Despite the disruptions associated with severe PMS, it has historically been on the medical fringe, in part because of the absence of effective treatments and the scientific and technical limitations in studying brain center actions. Only in recent years have there been well-designed scientific studies of the treatments and possible causal factors of PMS.

The serotonergic antidepressants may control symptoms for about 60% of PMS patients, but further studies are needed to identify treatments for those who are not helped by these medications. In addition to increasing information on the serotonergic system, other potential areas are the GABA system, the adrenergic system, and the linkages between brain center actions and ovarian hormones as suggested by GnRHa studies that suppress ovulation.

Although there has been progress in developing diagnostic criteria for PMS, the syndrome remains highly heterogeneous in combining a wide range of mood, behavioral, and physical symptoms and in its degree of severity. Research to define specific signs and the core symptoms that constitute the syndrome and to more clearly delineate PMS from other disorders is needed.

Basic research to increase understanding of relationships in the hypothalamic–pituitary–ovarian and hypothalamic–pituitary–adrenal axes that effect mood and behavior have the greatest importance for understanding the pathophysiology of PMS. It is been observed that neurotransmitter research has focused on steroid exposure rather than steroid exposure followed by withdrawal, which may be more relevant to the possible etiology of PMS, and promising target sites for such studies are now being identified.

See Also the Following Articles

GnRH (Gonadotropin-Releasing Hormone); Menstrual Cycle; Neurotransmitters

Bibliography

ACOG Committee Opinion: Premenstrual Syndrome (1995). *Int. J. Gynecol. Obstet.* **50**, 80–84.

American Psychiatric Association (1994). *Diagnostic and Statistical Manual of Mental Disorders–IV*, 4th ed. American Psychiatric Association, Washington, DC.

Fink, G., Sumner, B. E. H., Rosie, R., *et al.* (1996). Estrogen control of central neurotransmission: Effect on mood, mental state, and memory. *Cell. Mol. Neurobiol.* **16**, 325–344.

Freeman, E. W., Rickels, K., Sondheimer, S. J., *et al.* (1995). A double-blind trial of oral progesterone, alprazolam, and placebo in treatment of severe premenstrual syndrome. *J. Am. Med. Assoc.* **274**, 51–57.

Freeman, E. W., Kielich, A. M., and Sondheimer, S. S. (1996). PMS: New treatments that really work. *Contemp. Ob./Gyn.* **41**, 25–44.

Mezrow, G., Lobo, R., Shoupe, D., *et al.* (1994). Depot leuprolide acetate with estrogen and progestin add-back for long-term treatment of premenstrual syndrome. *Fertil. Steril.* **62**, 932–937.

Pearlstein, T. (1996). Nonpharmacologic treatment of premenstrual syndrome. *Psychiatr. Ann.* **26**, 590–594.

Rubinow, D. R., and Schmidt, P. J. (1995). The neuroendocrinology of menstrual cycle mood disorders. *Ann. N. Y. Acad. Sci.* **29**, 648–659.

Smith, S., and Schiff, I. (1993). *Modern Management of Premenstrual Syndrome*. Norton, New York.

Steiner, M., Steinberg, S., Stewart, D., *et al.* (1995). Fluoxetine in the treatment of premenstrual dysphoria. *N. Engl. J. Med.* **332**, 1529–1534.

Wieck, A. (1996). Ovarian hormones, mood and neurotransmitters. *Int. Rev. Psychiatr.* **8**, 17–25.

Yonkers, K. A., Halbreich, U., Freeman, E. W., *et al.* (1996). Sertraline in the treatment of premenstrual dysphoric disorder. *Psychopharmacol. Bull.* **32**, 41–46.

Poecilogony

Glenys D. Gibson

Acadia University

I. Introduction
II. Types and Consequences of Poecilogony
III. Determination of Development Mode
IV. Conclusions

GLOSSARY

adelphophagy A form of lecithotrophy in which offspring feed on extraembryonic yolk.
lecithotrophy A mode of larval development in which offspring rely on maternally derived yolk to support development to the juvenile stage.
nurse egg A nondeveloping egg which is ingested by developing larvae.
planktotrophy A mode of larval development in which offspring feed on extrinsic sources (e.g., phytoplankton, bacteria, and detritus).

Poecilogony is the production of more than one type of offspring within a single species. The term is generally used in reference to the sexual reproduction of marine invertebrates.

I. INTRODUCTION

The production of more than one type of offspring by a single species of marine invertebrate is not common. Most species produce only one type of offspring. Since Giard's (1905) original description of poecilogony, dozens of examples have been described spanning several phyla of marine invertebrates, but closer investigation revealed that the majority of these species were cryptic species, each exhibiting a single development mode. Despite the prevalence of these taxonomic mishaps, a few poecilogonous species have been well described and all are opisthobranch molluscs and polychaete annelids. Poecilogonous species, although uncommon, are of great interest because they allow for intraspecific comparison of life history traits typical of distinct modes of larval development, they have enormous potential for investigation of the mechanisms that trigger shifts in development mode, and they provide models with which to consider the evolution of alternate development modes.

Development mode is generally defined by larval type because most species of benthic marine invertebrates produce a single type of offspring. Larval development is classified by three major categories: planktotrophy, lecithotrophy, and benthic development. Species with planktotrophic larval development produce large numbers of small, feeding larvae that have a relatively long planktonic period before becoming competent to metamorphose into benthic juveniles. Lecithotrophy involves the production of fewer, larger larvae which rely on yolk reserves and are not obliged to feed during a short planktonic phase. Species with benthic development spend their entire embryonic and larval period within a benthic egg mass or brood and hatch as nondispersing juveniles. The major differences between these three patterns of development are larval dispersal and larval trophic mode. Within each of these ecological patterns there is an impressive variety of larval morphologies, each considered representative of specific taxa.

Poecilogonous species differ in that they produce offspring that develop in more than one of these modes. For example, within a single poecilogonous species, development may include both planktotrophy and lecithotrophy (e.g., *Elysia chlorotica* and *Streblospio benedicti*), lecithotrophy and benthic development (e.g., *Haminaea callidegenita* and *Tenellia adspersa*), or even all three (e.g., *Boccardia proboscidea*). Development may vary among offspring from a single brood or between broods produced by conspecific females. Where development mode varies among females or among populations, one must ensure that descriptions are made of a single species. Congeneric species are often very similar in both opisthobranchs and polychaetes and sibling species are sometimes difficult to recognize. Evidence to support poecilogony includes interfertility of different development morphs, allozyme or DNA analysis, and testing for lab-induced artifacts.

Although poecilogony is relatively uncommon in marine invertebrates, the production of different types of offspring by a single species is not unusual in insects and plants. Many insect species have flight polymorphisms that vary both spatially within a generation and temporally over several generations and are influenced by genetic and environmental factors. Flowering plants are known to produce seeds with dormancy or dispersal polymorphisms. These species are extremely sensitive to maternal environment and in consequence may simultaneously produce different types of offspring during a single reproductive event.

The objective of this article is to summarize some of the major themes that underlie poecilogony and to suggest some considerations for future research. Table 1 summarizes details and references for the specific examples discussed here. I will focus on a few well-studied species with the caveat that many other species of polychaete and opisthobranch are thought to be poecilogonous but have not yet been thoroughly examined and therefore are difficult to categorize in the general discussion presented here. Excellent reviews of these additional cases are available elsewhere.

II. TYPES AND CONSEQUENCES OF POECILOGONY

Poecilogony is manifested in a variety of ways in different species (Table 1). In all known cases, variability in larval development brings about a range of dispersal potentials among offspring. In most species, development is additionally linked to differing levels of maternal investment and larval trophic mode. Dispersal and reproductive investment are instrumental in shaping life history patterns in general and are the focus of most of the current research on larval evolution in marine invertebrates. A dispersive larva maintains continuity among populations when adults are sessile or of limited mobility. Dispersal is also tightly linked with maternal investment because low per-offspring investment necessitates a long feeding period while in the plankton. The trade-off with long-range dispersal is high larval mortality because planktonic offspring are particularly vulnerable to starvation, predation. and advection.

In poecilogonous species, dispersal polymorphisms exist either among siblings of a single brood or between offspring produced by different females (Table 1; Fig. 1). Some species produce both planktonic and benthic offspring (e.g., the sea slugs *H. callidegenita* and *E. chlorotica*), whereas others produce planktonic larvae with either a short or long swimming period (e.g., <7 days or 2–7 weeks in the polychaete *S. benedicti*). The production of offspring that collectively encompass a wide range in dispersal potentials would allow for local recruitment into presumably favorable (or at least known) sites, mixing among populations, and potential protection against local extinction should parental habitats deteriorate.

Reproductive investment has a major influence on larval development. Females have finite resources available for reproduction which may be divided among many yolk-poor propagules (characteristic of planktotrophy) or fewer yolk-rich offspring (lecithotrophy or benthic development). In many poecilogonous species, differential partitioning of resources among offspring leads to differences in egg size, development rates, larval trophic mode, and length of the planktonic dispersive period (Table 1; Fig. 1). For example, in *S. benedicti*, some females allocate only small amounts of yolk to each egg and produce planktotrophic larvae. Other *S. benedicti* females make few, large eggs that contain enough yolk to sustain larval development through metamorphosis. Increasing investment per offspring decreases female fecundity, increases offspring hatching size, decreases the length of the planktonic period, and increases time to maturity.

TABLE 1
Examples of Poecilogonous Marine Invertebrates

Species	Occurs within or among broods	Variability in larval dispersal	Variability in larval trophic mode	Investment differences	Developmental events influenced by peocilogony
Opisthobranch molluscs					
Berghia verrucicornis[a]	With brood (nonaerated cultures only)	Short-term planktonic larvae (1–3 days) and benthic juveniles	None (all lecithotrophic)	No	Planktonic period, time of metamorphosis
Elysia chlorotica[b]	Among females (separate populations)	Long-term planktonic larvae (14 days) or benthic juveniles	Planktotrophic or lecithotrophic	Egg size among broods, clutch size among females	Time to hatching, planktonic period, time of metamorphosis
Haminaea callidegenita[c]	Within brood	Planktonic larvae (1–30 days) and benthic juveniles	None (all lecithotrophic)	No	Planktonic period, time of metamorphosis
Spurilla neopolitana[d]	Among broods of one female (switch if starved)	Switch from short- to long-term dispersers	Switch (lecithotrophy to planktotrophy)	Egg size and clutch size decrease with starvation	Planktonic period
Tenellia adspersa[e]	Among broods of one female (switch if starved)	Planktonic larvae or benthic juveniles	None (all lecithotrophic)	Egg size and clutch size change with starvation	Planktonic period, time of metamorphosis
Tenellia pallida[f]	Among females	Planktonic larvae or benthic juveniles	None (all lecithotrophic)	Egg size, fecundity	Hatching time, planktonic period, time of metamorphosis
Polychaete annelids					
Boccardia semibranchiata[g]	Within brood	Release both short- and long-term swimming larvae	Planktotrophic and/or adelphophagic	Varies among sibs with nurse egg consumption	Planktonic period, time of metamorphosis
Boccardia proboscidea[h,i]	Within brood, among females	Release long-term and short-term swimming larvae and benthic juveniles	Planktotrophic or adelphophagic	Egg size, fecundity	Time to hatching, planktonic period, time of metamorphosis
Capitella sp.[j]	Within one brood, among females	Short or long planktonic period	Planktotrophic or lecithotrophic	Egg size, organic content, fecundity	Planktonic period, time of metamorphosis
Streblospio benedicti[k,l]	Among females	Short or long planktonic period	Planktotrophic or lecithotropic	Egg size, fecundity	Planktonic period, time of metamorphosis

[a] Carroll and Kempf (1990).
[b] West et al. (1984).
[c] Gibson and Chia (1989, 1995).
[d] Clark and Goetzfried (1978).
[e] Chester (1996).
[f] Eyster (1979).
[g] Guérin (1990).
[h] Blake and Kudenov (1981).
[i] Gibson (1997).
[j] Qian and Chia (1991, 1992).
[k] Levin (1984).
[l] Levin and Bridges (1994).

In other poecilogonous species, investment varies through the production and differential utilization of nurse eggs, or nondeveloping eggs that are ingested by developing siblings in a process called adelphophagy. Adelphophagy has similar consequences to the production of larger eggs in other species, including rapid prehatching larval development, larger hatching size, decreased dispersal, and shorter time to metamorphosis (e.g., *Boccardia* spp.; Table 1). Larval development may vary among broods in species in which certain females produce nurse eggs and others do not (e.g., *B. proboscidea*) or within a single brood in species in which only some siblings ingest nurse eggs, whereas others are nonadelphophagic and hatch as planktotrophic larvae (*B. proboscidea* and *B. semibranchiata*).

1. Variation in the time of hatching or release (varies among broods)

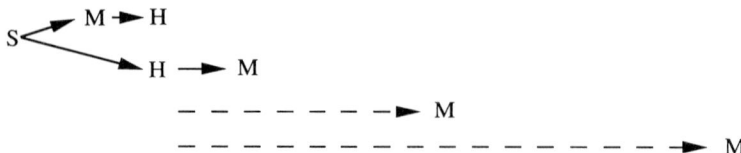

Consequences: length of the embryonic period, length of the planktonic period, dispersal

Examples: *Boccardia proboscidea* [1,2], *Elysia chlorotica* [3], *Tenellia pallida* [4]

2. Variation in the time metamorphic competence is reached (within or among broods)

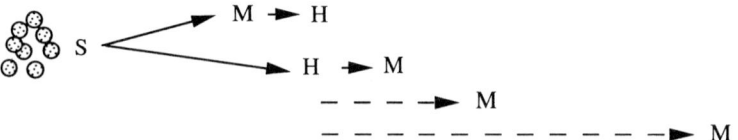

Consequences: presence and/ or length of the planktonic period, dispersal

Examples: *Berghia verrucicornis* [5], *Cirriformia tentaculata* [6], *Haminaea callidegenita* [7]

3. Offspring trophic mode and patterns of maternal investment in reproduction (among broods)

a. Planktotrophy

b. Lecithotrophy

c. Adelphophagy

Consequences: length of the planktonic period, dispersal, necessity of larval feeding, fecundity

Examples: *Capitella* sp, [8], *Elysia chlorotica* [3], *Streblospio benedicti* [9]

Examples: *Boccardia semibranchiata* [10], *B. proboscidea* [1,2]

FIGURE 1 General patterns of poecilogony in marine invertebrates, summarizing shifts in the timing of ontogenetic events among conspecific offspring. S, time of spawning; H, hatching; M, metamorphosis. Solid arrows represent the relative duration of the embryonic (prehatching) period, and dashed arrows represent the duration of the planktonic period. In part 3, differences in yolk investment per offspring are approximated as small eggs (planktotrophy), large eggs (lecithotrophy), and nurse eggs (adelphophagy). 1, Blake and Kudenov (1981); 2 Gibson (1997); 3 West *et al.* (1984); 4, Eyster (1979); 5, Carroll and Kempf (1990); 6, George (1963); 7, Gibson and Chia (1989); 8, Qian and Chia (1992); 9, Levin (1984); 10, Guérin (1990).

Dispersal polymorphisms and variance in maternal investment are tightly correlated and in some cases have similar consequences (Fig. 1). The outcome is the production of offspring that fall into different ecological categories. Reproductive resources are also spread out among different ecological types of offspring, which presumably will not all be successful given a particular set of environmental conditions. However, some offspring would be likely to survive, even in unpredictable or ephemeral environments.

One fascinating aspect of poecilogony is that the different modes of development result in the same or almost identical adult morphology. In most poecilogonous species, reproductive type is not based on adult features but rather on brood or offspring characteristics. In many species, both types of offspring may be produced by a single female during a single spawning event (e.g., *H. callidegenita*, *Boccardia* spp., *T. adspersa*, *Capitella* sp., and *Berghia verrucicornis*; Table 1). Where development mode varies among females, morphological differences are not observed or are slight and may be environmentally mediated. The most commonly reported difference is female size and, in general, females with planktotrophic development are larger than those producing benthic or lecithotrophic offspring (e.g., *T. pallida*). Size differences are linked to nutritional (e.g., *Capitella* sp.) or heritable differences (*S. benedicti*). Size differences are usually small and allometric differences have not been reported.

Larval morphology also seems to be similar overall among conspecifics of different development types, but the timing of certain developmental events may be accelerated or delayed and the "type" of offspring (larva or juvenile) is frequently determined by differences in the time of hatching or metamorphosis.

Variation in time of hatching leads to the release of offspring at different ontogenetic stages (Fig. 1). Metamorphosis, in some species, can occur before hatching, immediately after hatching, or several weeks after hatching. For example, in *H. callidegenita* morphological differences between hatchlings (larva and juvenile) are those typical of metamorphosis in this species, primarily loss of the velar cilia and resorption of the velar lobes (Fig. 2). Offspring size and trophic mode are the same.

Small differences in larval morphology exist in some poecilogonous species in which larval development is similar to variable maternal provisioning. Size differences are common in these species because lecithotrophic offspring tend to be larger at hatching than do planktotrophic conspecifics (e.g., *E. chlorotica*, *S. benedicti*, and *B. proboscidea*; Figs. 1 and 3). Structures associated with a planktonic lifestyle may also vary. For example, lecithotrophic offspring tend to have yolkier, more poorly developed intestines and digestive glands than do their planktotrophic

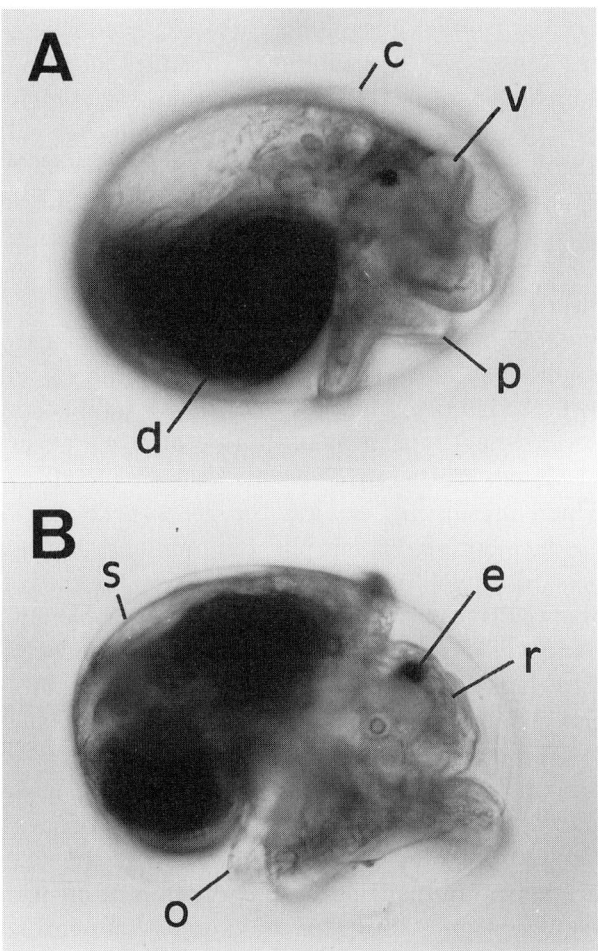

FIGURE 2 Photomicrographs of the opisthobranch *H. callidegenita* at hatching, including (A) veliger just prior to hatching as a planktonic larva and (B) metamorphosed juvenile, also just before hatching. Both types of offspring are simultaneously released from the same egg mass. The only morphological differences between the two offspring shown here are those associated with metamorphosis, namely, loss of velar cilia and resorption of the velar lobes (shell is retained throughout life). c, capsule; d, digestive gland; e, eye; o, operculum; p, propodium; r, resorbed velum; s, shell; v, velum.

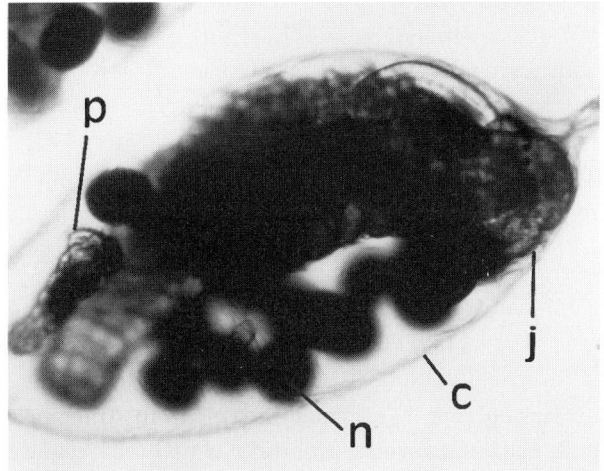

FIGURE 3 Photomicrograph of the polychaete *B. proboscidea* before hatching. In this species, poecilogony occurs within a single brood. Eggs within a single capsule (c) are categorized as follows: n, nurse eggs, or nondeveloping eggs that are ingested by siblings; p, zygotes that develop into planktotrophic larvae with a 3-week planktonic feeding period before metamorphosing; and j, adelphophagic offspring that ingest nurse eggs, have accelerated development, and hatch as juveniles. Note the size difference between the two siblings: The larva is only a few segments long and the juvenile fills almost the entire capsule.

conspecifics (e.g., *T. pallida*) and swimming chaetae are reduced in lecithotrophic or adelphophagic polychaetes (e.g., *B. proboscidea* and *S. benedicti*). Although most of these differences may be yolk related, reduction in number of swimming chaetae is heritable in *Streblospio*. In some species, differences in morphology reflect changes in rates of morphogenesis of certain structures relative to specific ontogenetic events. In the sea slug *T. pallida*, eyes develop at about the same age in all siblings, but this event has been decoupled from time of hatching such that planktonic larvae develop eyes after hatching (hatch 3 days after oviposition) but benthic developers form eyes before hatching (hatch 6 days after oviposition).

A change in the timing of a particular ontogenetic event (e.g., hatching and metamorphosis) can provide a relatively simple means of varying offspring traits, without invoking major morphological differences. Even very small changes in timing, spanning hours, days, or weeks, can have major ecological consequences (Fig. 1). It is not surprising that these events would be so plastic in poecilogonous species because they are highly variable among marine invertebrates in general. Stage at hatching or release varies from gamete to juvenile among even closely related species. Age of metamorphosis is extremely variable both among taxa and within species in terms of when competence is reached and how long larvae can delay metamorphosis once competent.

III. DETERMINATION OF DEVELOPMENT MODE

Although we are far from understanding the mechanisms that determine a particular path of larval development, poecilogony in known to be mediated by a number of factors, including heritable differences, reproductive investment, maternal effects (e.g., specific maternal signals), and environmental factors (parental and embryo environment). In some species, development is constant within a particular strain (e.g., *E. chlorotica* and *S. benedicti*), whereas in others, females may change the type of offspring they produce if exposed to or stressed by specific environmental factors (*Capitella* sp., *H. callidegenita*, and *T. adspersa*).

Mode of larval development is known to be heritable in two poecilogonous species. In *S. benedicti*, some females produce planktotrophic larvae, whereas others are lecithotrophic and a single female will produce several broods of the same type throughout her life. The type of larva produced is highly heritable as defined by several genetically correlated traits, including egg size and larval number. In the opisthobranch *H. callidegenita* , both swimming larvae and benthic juveniles are released from a single egg mass. The percentage of juvenile hatchlings, ranging from 4 to 100% among egg masses, is highly heritable and is positively genetically correlated with length of the planktonic phase in sibs. Female *E. chlorotica* have either planktotrophic or benthic development and lab crosses have shown that F1s exhibit the maternal development mode and F2s have intermediate characteristics, indicative of heritable and/ or maternal effects. Lab crosses of planktotrophic and adelphophagic offspring of the spionid *B. proboscidea* produce F1s that have the maternal development type, also suggesting maternal effects. These lab crosses involved the analysis of full sibs which presumably possess the genes necessary to produce both larval morphs, but only one set is expressed. It would be interesting to examine the role of specific genes and gene interactions in generating particular developmental pathways in these species as well as to investigate the potential regulation of gene activity by maternal and/or environmental cues.

Reproductive investment is the most prevalent factor known to influence poecilogony. In most poecilogonous species, investment varies per offspring through the production of either larger eggs (e.g., lecithotrophy vs planktotrophy) or nurse eggs (e.g., adelphophagy). Increased investment influences a number of larval and fitness traits, as discussed previously. In three opisthobranch species (*Spurilla neapolitana*, *H. callidegenita* , and *T. adspersa* ; Table 1), females responded to decreased food availability by increasing the percentage of their offspring that dispersed. Larval trophic mode may also change as in *S. neapolitana,* in which it switched from lecithotrophy to planktotrophy under the same conditions. In polychaetes, parental food quality is known to influence fecundity in lecithotrophic *Streblospio benedicti* and both egg size and egg number in lecithotrophic *Capitella* sp., although larval trophic mode was not affected.

Maternal signals may also trigger specific developmental events which, if timing of the response varies among sibs, may also lead to poecilogony. The opisthobranch *H. callidegenita* metamorphoses in response to a maternally derived compound found in the jelly that covers the egg mass. Only some sibs (~50%) are competent to metamorphose (i.e., physiologically able to respond) in response to this compound before hatching and are released as juveniles, whereas the rest will hatch as nonfeeding larvae. Siblings that are not competent to respond to egg mass jelly before hatching may do so, in the lab, between 1 and 30 days after hatching. This leads to the release of both dispersive larvae and metamorphosed juveniles (nondispersive) from a single clutch.

Parental environment also influences poecilogony in some species, but in others no effect could be demonstrated. In species in which poecilogony is related to differences in maternal investment, parental food supply may trigger shifts in development, although this does not occur in all species tested. Poecilogony may be induced by culturing eggs in aerated or flowing conditions (e.g., *Berghia verrucicornis*) or not (*H. callidegenita*). Poecilogony is highly conservative when exposed to different parental environments in some species such as *S. benedicti,* in which females could not be induced to switch development modes, despite testing with a wide variety of environmental parameters (food quantity, food quality, temperature, and photoperiod).

IV. CONCLUSIONS

Poecilogony is highly variable in different species, but dispersal and investment differences are themes common to most examples. These two factors are phenotypically highly correlated and are also associated with a number of adult traits (size and fecundity) and offspring traits (egg size, development time, and sometimes small morphological differences). Polymorphic development appears to occur in species in which development varies among females but is the same within one brood or in species in which an individual produces the same proportion of different morphs of offspring, regardless of environmental conditions. Poecilogony can also be phenotypically plastic in some species, such as those that are sensitive to change in parental environment. In some species, poecilogony could simply be an opportunistic association between increased investment and larval trophic mode. Alternatively, females may produce specific signals which may directly trigger an alternate development mode in some of their offspring, although this has not been investigated. Heritable and maternal effects have a strong influence on poecilogony but the mechanisms that determine a particular mode are not known. This form of bet-hedging spreads resources over different morphs of offspring with differing ecological potentials, which perhaps is suited to species found in unpredictable or ephemeral habitats. Exposure to fluctuating selection pressures within or among generations may contribute to maintenance of poecilogony over time, although long-term stability of variable development seems unlikely because poecilogonous species are rare.

Species which exhibit phenotypically plastic poecilogony are interesting models in which to study mechanisms underlying developmental change and deserve further consideration. Current work has focused on life history traits and ecological parameters that are variable in poecilogonous species and have widespread relevance to other species. Several avenues of research would contribute to this knowledge: How do genes and gene interactions determine a particular development mode, and how is gene activity influenced by maternal effects and parental environment?

See Also the Following Articles

MARINE INVERTEBRATE LARVAE; MARINE INVERTEBRATES, MODES OF REPRODUCTION IN

Bibliography

Blake, J., and Kudenov, J. (1981). Larval development, larval nutrition and growth for two *Boccardia* species (Polychaeta: Spionida) from Victoria, Australia. *Mar. Ecol. Prog. Ser*:, 175–182.

Carroll, D., and Kempf, S. (1990). Laboratory culture of the aeolid nudibranch *Berghia verrucicornis* (Mollusca: Opisthobranchia): Some aspects of its development and life history. *Biol. Bull.* 179, 243–253.

Chester, C. (1996). The effect of adult nutrition on the reproduction and development of the estuarine nudibranch, *Tenellia adspersa* (Nordmann, 1845). *J. Exp. Mor. Biol. Ecol.* 198, 113–130.

Chia, F. S., Gibson, G., and Qian, P. Y. (1996). Poecilogony as a reproductive strategy of marine invertebrates. *Oceano Acta* 19(3/4), 203–308.

Clark, K., and Goetzfried, A. (1978). Zoogeographic influences on development patterns of North Atlantic Ascoglossa and Nudibranchia, with a discussion of factors affecting egg size and number. *J. Moll. Stud.* 44, 283–294.

Eyster, L. (1979). Reproduction and developmental variability in the opisthobranch *Tenellia pallida*. *Mar. Biol.* 51, 133–140.

George, J. (1963). Behavioural differences between the larval stages of *Cirriformia tentaculata* (Montagu) from Drake's Island (Plymouth Sound) and from Southampton water. *Nature* 199, 195.

Giard, A. (1905). La Poecilogonie. *Comp. Rend. Sixth Congr s. Int. Zool. Berne 1904*, 617–646.

Gibson, G. (1997). Variable development in the spionid *Boccardia proboscidea* (Polychaeta) is linked to nurse egg production and larval trophic mode. *Invertebr. Biol.* 116(3), 213–226.

Gibson, G., and Chia, F. S. (1989). Developmental variability (pelagic and benthic) in *Haminoea callidegenita* (Opisthobranchia: Cephalaspidea) is influenced by egg mass jelly. *Biol. Bull.* 176, 103–110.

Gibson, G., and Chia, F. S. (1995). Developmental variability in the poecilogonous opisthobranch *Haminaea callidegenita*: Life-history traits and effects of environmental parameters. *Mar. Ecol. Prog. Ser.* 121, 139–155.

Gurin, J.-P. (1990). Description d'une nouvelle espce de spionid (Annlides, Polychtes) *Boccardia semibranchiata*. *Ann. Inst. Ocanogr. Paris* 66(1/2), 37–45.

Hoagland, E., and Robertson, R. (1988). An assessment of poecilogony in marine invertebrates: Phenomenon or fantasy? *Biol. Bull.* 174, 109–125.

Levin, L. (1984). Multiple patterns of development in *Streblospio benedicti* Webster (Spionidae) from three coasts of North America. *Biol. Bull.* 166, 494–508.

Levin, L., and Bridges, T. (1994). Control and consequences of alternate developmental modes in a poecilogonous polychaete. *Am. Zool.* 34, 323–332.

Levin, L. A., Zhu, J., and Creed, E. (1991). The genetic basis of life-history characteristics in a polychaete exhibiting planktotrophy and lecithotrophy. *Evolution* 45(2), 380–397.

Qian, P. Y., and Chia, F. S. (1992). Effects of diet type on the demographics of *Capitella* sp. (Annelida: Polychaeta): Lecithotrophic development vs. planktotrophic development. *J. Exp. Mar. Biol. Ecol.* 157, 159–179.

West, H., Harrigan, J., and Pierce, S. (1984). Hybridization of two populations of a marine opisthobranch with different patterns of development. *Veliger* 26(3), 199–206.

Polyandry

see Mating Behaviors

Polycystic Ovary Syndrome

Richard S. Legro

Pennsylvania State University College of Medicine

I. Definition and Diagnostic Criteria
II. Reproductive Sequelae of PCOS
III. Metabolic and Malignant Sequelae of PCOS
IV. PCOS: The Next Century

GLOSSARY

insulin resistance A subnormal biologic response to a given amount of insulin.

oligomenorrhea: A condition characterized by menstrual bleeding intervals that exceed 35 days.

ovarian hyperstimulation syndrome A potential complication of medical ovulation induction. This is a syndrome of massive enlargement of the ovaries and transudation of ascites into the abdominal cavity that can lead to rapid and symptomatic enlargement of the abdomen, intravascular contraction, hypercoaguability, and systemic organ dysfunction.

ovulation induction The medical or surgical techniques used to cause ovulation.

pilosebaceous unit Hair follicles and sebaceous glands.

polycystic ovaries Enlarged ovaries due to increased central stroma; the subcapsular area is strewn with multiple small (2–8 mm) follicles ("necklace of pearls").

polycystic ovary syndrome (PCOS) A syndrome of unexplained hyperandrogenic chronic anovulation.

wedge resection of the ovary A surgical procedure in which a wedge-sized portion of the ovary is removed in women with PCOS to cause spontaneous ovulation.

Polycystic ovary syndrome is probably the most common, least understood endocrinopathy that affects women in the developed world. Approximately 5% of women may have this disorder. It is a syndrome of unexplained hyperandrogenic chronic anovulation and most likely represents a heterogeneous disorder. Its etiology remains unknown, its long-term sequelae poorly documented, and the quest for further understanding of this complex disorder may at times seem like that for the Holy Grail.

I. DEFINITION AND DIAGNOSTIC CRITERIA

Part of the difficulty in understanding polycystic ovary syndrome (PCOS) is that there is no universal definition of it. While chronic anovulation may be the sine qua non of the syndrome, only a small percentage of women with PCOS are completely amenorrheic. The majority are oligomenorrheic and experience varying intervals of vaginal bleeding. The cause of this vaginal bleeding is not always a postovulatory withdrawal bleed (after an infrequent ovulation)—it may also be caused by anovulatory events (due to prolonged exposure of the endometrium to weak estrogens or to a relative lack of any potent estrogen). How infrequent should the menstrual bleeding be to qualify as "chronic anovulation"? There is no consensus. Thus, the most readily available historical variable in the syndrome, menstrual irregularity, is itself a source of confusion.

This confusion is further confounded by defining hyperandrogenism. Some will focus on hyperandrogenic stigmata such as hirsutism as evidence of hyperandrogenism, whereas others will focus on elevated circulating androgens (testosterone, free testosterone, androstendione, DHEA-S-alone or in combination, and single elevated or multiple elevated values). Both the ovary and adrenal gland contribute to the increased androgens in the syndrome. Hirsutism is not an invariable component of the phenotype because just as many cycling women may manifest hirsutism, women with PCOS (especially Asian women) may have minimal hirsutism. Clearly, hirsutism has its own separate and distinct etiologies. The magnitude of circulating androgens in PCOS women, while elevated compared to other women, rarely approaches the lower limits of normal for men. For instance, in our lab the normal level of total testosterone for women is 20–60 ng/dl, most women with PCOS are in the range of 80–150 ng/dl, and the normal range for males begins at 250 ng/dl. Therefore, we only infrequently encounter signs of virilization such as temporal balding, clitoromegaly, voice deepening, and changes in body topography in these women.

Some investigators examine the appearance or size of the ovary by bimanual exam, laparoscopy, and, recently, ultrasound. The ovaries tend to be enlarged due to increased central stroma and the subcapsular area is strewn with multiple small (2–8 mm) follicles ("necklace of pearls"). It is felt that these represent arrested follicles due to the increased intraovarian androgenic environment. Polycystic ovaries are distinct from polycystic ovary syndrome; up to 20% of an unselected population may have polycystic ovaries and many of these are normoandrogenic cycling females. Given's aphorism that polycystic ovaries are a sign, not a diagnosis, is applicable here. Others want to see evidence of abnormal hypothalamic–pituitary activity by elevated luteinizing hormone:follicle-stimulating hormone ratios. It is uncertain whether abnormalities in gonadotropin pulsatility and amplitude are a primary cause of the syndrome or a secondary response to the hormonal milieu of chronic anovulation.

What are the phenocopies of hyperandrogenic chronic anovulation that should be excluded before we can diagnose PCOS? Most would agree that an androgen-secreting tumor, exogenous androgens, Cushing's syndrome, and nonclassical congenital adrenal hyperplasia should be excluded. Hyperprolactinemia is on many lists, but a substantial number of women with PCOS have mild elevations of circulating prolactin and this appears to be an epiphenomenon rather than a cause of the excess circulating androgens.

Although there is no consensus to the definition of PCOS, the diagnostic criteria from the 1990 NIH-NICHD conference on PCOS have been frequently cited as a starting point: hyperandrogenism and/or hyperandrogenemia, oligoovulation, and exclusion of other potential causes such as congenital adrenal hyperplasia, hyperprolactinemia, Cushings's syndrome, and androgen-secreting tumors. In our practice and research studies we base our definition of hyperandrogenism on hyperandrogenemia elevation of circulating testosterone and/or free testosterone that is two standard deviations above a control group of cycling reproductive-age women. In addition we look for six or fewer bleeding episodes a year as evidence for oligomenorrhea. Where other definitions PCOS are used, mention will be made.

II. REPRODUCTIVE SEQUELAE OF PCOS

PCOS tends to develop shortly after menarche and last for most of the reproductive life, although questions persist about its natural history during the reproductive years. It has been suggested that the syndrome is mollified by destructive ovarian interventions such as wedge resection or ovarian drilling. There is aprocryphal data to suggest pregnancy may improve the phenotype.

A. Infertility

The most common reason that women with PCOS present to the gynecologist is because of infertility, most likely due to chronic anovulation. These women are not absolutely infertile but subfertile due to the infrequency and unpredictability of their menstrual cycles. These women frequently require ovulation induction. As a general rule, women with PCOS represent one of the most difficult groups to induce ovulation successfully and safely. Many are unresponsive to clomiphene citrate and human menopausal gonadotropins. On the other end of the spectrum are women with PCOS who overrespond to both of these medications. Women with PCOS are at especially increased risks of ovarian hyperstimulation syndrome (a syndrome of massive enlargement of the ovaries and transudation of ascites into the abdominal cavity that can lead to rapid and symptomatic enlargement of the abdomen, intravascular contraction, hypercoaguability, and systemic organ dysfunction). They are also at increased risk for multiple pregnancy. The tendency of women who respond to the medications to overrespond is probably due to the multiple "arrested" small follicles that are recruited simultaneously. Low dose or step down gonadotropin regimens have been utilized in women with PCOs to avoid complications. Good success rates with a low complication rate of ovarian hyperstimulation syndrome or multiple gestation have been reported using pulsatile gonadotropin-releasing hormone agonist (GnRHa) therapy.

Other forms of therapy that may successfully induce ovulation include wedge resection of the ovary, which has been modified to consist of laparoscopic ovarian "drilling," destruction of ovarian tissue, and presumably androgen-producing stroma via electrocautery or laser. This technique has been very successful in inducing ovulation, although the effects may not last beyond 1 or 2 years. Weight loss in obese women with PCOS may restore menstrual cyclicity. Agents which improve insulin sensitivity, such as metformin and troglitazone, have been reported to cause spontaneous ovulation in women with PCOS, but large-scale clinical trials are still pending.

B. Menstrual Irregularity

This can be a major source of frustration and in severe cases of menorrhagia, a cause of anemia. Many of the previous therapies have been utilized in these women, such as weight loss, or are undergoing investigation, as in the case of the insulin-sensitizing agents. Frequently, menstrual regularity is achieved through oral contraceptive pill or cyclic progestin administration.

C. Hirsutism and Acne

Both hirsutism and acne have been attributed to an increased circulating pool of androgens and increased androgen production at the pilosebaceous unit. There may be increased activity of 5-α reductase which can convert androgen precursors into more potent androgens at the skin level. Both hirsutism and acne can be treated with a combination regimen of suppression of the ovaries (most commonly with the oral contraceptive pill), to decrease the precursor pool, and androgen antagonism at the pilosebaceous unit.

III. METABOLIC AND MALIGNANT SEQUELAE

There is a gap in our knowledge of the natural course of PCOS. Women with PCOS are subject to menopause as are all women and with senescence of the ovaries comes decreased androgen production (at least compared to their reproductive-age counterparts) and eventual cessation of the menses. Thus, the criteria for PCOS—hyperandrogenic chronic anovulation—may resolve with age. However, as the curtain falls for the reproductive abnormalities, the curtain may begin rising for the metabolic/malignant complications. Most of these studies are retrospective and thus subject to many forms of bias, especially recall bias about hirsutism and menstrual irregularity. Many of these retrospective studies come from a Scandinavian population and it is uncertain how applicable they are to other ethnicities, or they are based on morphologic appearance of the ovaries and not hyperandrogenic chronic anovulation. Clearly, there are ethnic differences in the reproductive phenotype of women with PCOS. The lack of a well-defined prospective cohort study of women with PCOS has led to a large amount of speculation and little data about the long-term effects of PCOS.

A. Gynecological Cancers

Many gynecological cancers have been reported to be more common in women with PCOS, including endometrial cancer, ovarian cancer, and breast cancer. The best case of an association between PCOS and cancer can be made for endometrial cancer. Although menstrual infrequency and infertility are risk factors for endometrial cancer, there is less certainty that PCOS is an independent risk factor. Most studies are case-series or case-control studies that assign affected status on a basis of hirsutism and menstrual irregularity. These articles suggest that many of the young women of reproductive age who develop endometrial cancer (usually a well-differentiated adenocarcinoma with a favorable prognosis) have a history consistent with PCOS. However, the women in this age group represent only a small number of women who develop endometrial cancer. The vast majority of these women are postmenopausal and assigning a diagnosis of PCOS in retrospect is often flawed.

There are also case-control studies suggesting that women with PCOS are at increased risk for ovarian cancer—which would seem to be the one cancer that PCOS women would be protected against due to their chronic anovulation (uninterrupted ovulation being one of the risk factors for ovarian cancer). There are also studies that suggest PCOS women have an increased risk of breast cancer and studies suggesting that women with breast cancer have elevated circulating androgens compared to age-matched controls. These latter two associations are less well established in the literature.

B. Type II Diabetes Mellitus

Women with PCOS have been found to be profoundly insulin resistant. The mechanism is unknown. This baseline insulin resistance combined with the worsening effect of obesity (which may affect up to 50% of the American PCOS population) places these women at increased risk for impaired glucose tolerance and most likely diabetes. The level of insulin resistance found in these women based on dynamic measures of insulin action has in other populations (i.e., children of parents with diabetes) led to a marked increase in their chance for developing type II diabetes mellitus. About 30–40% of obese reproductive-age women with PCOS have been found to have impaired glucose tolerance. Retrospective studies from a Scandinavian population have found a fivefold higher prevalence of diabetes (about 15%) among women with PCOS compared to controls. Of note, the diagnosis of PCOS was based on ovarian morphology and not hyperandrogenic chronic anovulation. There are currently no good prospective studies of a cohort of women with PCOS to confirm this trend.

One of the many vicious circles in women with PCOS is the relationship between hyperinsulinemia and hyperandrogenism. There is evidence to suggest that the ovary may be sensitive to the excess insulin, which can function as a cogonadotropin to increase steroid production (whereas the periphery, mainly skeletal muscle is resistant to its action). Additionally, sex hormone-binding globulin, the major carrier protein of androgens and estrogens, is suppressed by elevated insulin levels, such that insulin-resistant women have increased free (thus, bioavailable) androgens. Screening for insulin resistance in PCOS is difficult because there is currently no good clinical screening. It is prudent to assume that all women with PCOS, both obese and thin, are insulin resistant. Obese women especially need to monitor for impaired glucose tolerance or frank diabetes. Weight loss and diet will improve insulin sensitivity in these women. The role of insulin-sensitizing agents in preventing diabetes has not yet been established for women with PCOS.

C. Cardiovascular Disease

Many of the studies suggesting an increased incidence of cardiovascular disease are inferential based on risk factor models. Women with PCOS appear to form a subset of the insulin-resistance syndrome first described by Reaven consisting of insulin resistance, hypertension, and lipid abnormalities. Women with PCOS tend to have elevated triglycerides (perhaps the best lipid marker of insulin resistance) and an unfavorable elevated low-density lipoproteins:high-density lipoproteins ratio. Reproductive-aged PCOS women have not consistently been found to be hypertensive. Recent studies have documented other interim risk factors (e.g., increased carotid artery intimal thickness or decreased peak systolic cardiac blood flow) to suggest that these women have multiple risk factors for cardiovascular disease. Confirmation of this based on actual cardiovascular events, such as stroke or myocardial infarction, in PCOS women has not yet appeared.

IV. PCOS: THE NEXT CENTURY

The debate about the etiology of the syndrome may find its solution before the millennium. There is the age-old argument that is also applicable to PCOS about the question of nature versus nurture. PCOS tends to aggregate in families with a high number of sisters and mothers affected. Whether this is genetic, due to the inheritance of a certain allele or a combination of alleles, or whether this is environmental, due to a common diet and lifestyle in these families, is a point of contention. Large-scale linkage studies are ongoing. It is provocative to speculate about a potential male phenotype, which some have suggested is premature male pattern baldness if there is a genetic component. Preliminary results have suggested positive associations with several candidate genes, including the insulin gene, dopamine receptor, and steroidogenic enzymes, but these studies have been based on varying definitions of the affected phenotype. The common thread to these associations has not yet been discovered.

See Also the Following Articles

HIRSUTISM; INFERTILITY; MENSTRUAL DISORDERS

Bibliography

Ben-Shlomo, I., Franks, S., and Adashi, E. Y. (1995). The polycystic ovary syndrome: Nature or nurture? *Fertil. Steril.* 63, 953–954.

Dahlgren, E., and Janson, P. O. (1994). Polycystic ovary syndrome—Long-term metabolic consequences. *Int. J. Gynaecol. Obstet.* 44, 3–8.

Franks, S. (1995). Polycystic ovary syndrome. *N. Engl. J. Med.* 333, 853–861.

Givens, J. R., and Kurtz, B. R. (1986). Understanding the polycystic ovary syndrome. *Prog. Clin. Biol. Res.* 225, 355–376.

Legro, R. S. (1995). The genetics of polycystic ovary syndrome. *Am. J. Med.* 98, 9S–16S.

Legro, R. S., and Dunaif, A. (1996). The role of insulin resistance in polycystic ovary syndrome. *Endocrinologist* 6, 307–321.

Lobo, R. A. (1995). Polycystic ovary syndrome/hyperandrogenic chronic anovulation. *Adv. Endocrinol. Metab.* 6, 167–191.

Reaven, G. M. (1988). Banting lecture 1988. Role on insulin resistance in human disease. *Diabetes* 37, 1595–1607.

Udoff, L., and Adashi, E. Y. (1995). Polycystic ovarian disease: A new look at an old subject. *Curr. Opin. Obstet. Gynecol.* 7, 340–343.

Polygamy

see Mating Behaviors

Polyspermy

R. H. F. Hunter

Royal Veterinary and Agricultural University

I. Introduction
II. Evolutionary Considerations
III. Block to Polyspermy
IV. Experimental Circumstances Generating Polyspermy
V. *In Vitro* Fertilization
VI. Stability of Block to Polyspermy
VII. Molecular Aspects of Block to Polyspermy
VIII. Consequences of Polyspermy

GLOSSARY

cumulus investment The several layers of granulosa cells surrounding an oocyte and supporting it in the antrum of a Graafian follicle. The innermost layer of cumulus cells, in contact with the zona pellucida, is referred to as the corona radiata. Following the preovulatory surge of gonadotrophic hormones that stimulates resumption of meiosis, the cumulus oophorus undergoes a substantial remodeling, expansion, and mucification before dispersal at fertilization.

Graafian follicle A multilayered ovarian structure containing an oocyte and a fluid-filled antrum. A Graafian follicle responds to the preovulatory surge of pituitary gonadotrophic hormones by a series of biochemical, endocrine, and structural changes culminating in shedding of the oocyte (ovulation).

perivitelline space The fluid-filled space in an ovulated egg (secondary oocyte) between the vitelline (plasma) membrane and zona pellucida that contains the first polar body. Upon activation of the oocyte by a fertilizing spermatozoon, contraction of the vitellus significantly increases the volume of the perivitelline space and its accumulated fluid.

vitellus Strictly speaking, this is the egg proper in an unfertilized condition, enclosed by its plasma membrane (the vitelline membrane) and encompassed by the zona pellucida.

zona pellucida A relatively thick acellular glycoprotein coat that surrounds the mammalian oocyte and young embryo. It is usually shed at the blastocyst stage of development, referred to as hatching from the zona pellucida.

I. INTRODUCTION

Abnormalities of fertilization have been extensively reported in eutherian mammals and are a significant cause of early embryonic death. Indeed, such loss may occur so precociously that it fails to influence the life span of the corpus luteum and thus the duration of the estrous or menstrual cycle.

Perhaps the most widely reported abnormality of fertilization in mammals is penetration of the oocyte cytoplasm—the vitellus—by more than one spermatozoon. This is termed polyspermic fertilization. In order to restore the diploid condition, fertilization of a secondary oocyte requires to be monospermic, i.e., by a single (haploid) spermatozoon. Polyspermy may be dispermic, the most common condition, or trispermic, tetraspermic, and so on. Complete failure of the defense mechanism against polyspermy may lead to multiple, even massive, sperm penetration of the vitellus. The egg becomes literally overwhelmed by sperm heads and tails, with severe derangement of the cytoplasm.

II. EVOLUTIONARY CONSIDERATIONS

Before discussing the phenomenon of polyspermy in a systematic manner, brief reference should be made to the evolutionary circumstances associated with the transition from external fertilization under aquatic conditions in ancestral species to internal fertilization in highly specialized portions of the paramesonephric (Müllerian) ducts in modern animals. External fertilization frequently involves shedding of enormous numbers of male gametes in the immediate vicinity of the eggs; such liberated spermatozoa usually have an extremely short life span. Fertilization is therefore prompt, and the eggs of many species have accordingly developed a rapid block against further sperm penetration.

Internal fertilization in eutherian mammals, by contrast, involves many strategies to reduce the risk of polyspermic fertilization. First, there is a steeply declining gradient in sperm numbers between the site of semen deposition—usually the vagina but sometimes the uterus—and the site of fertilization at the ampullary–isthmic junction of the Fallopian tubes. Second, there is a region of the Fallopian tubes, the caudal portion of the isthmus, in which viable spermatozoa are arrested before ovulation and then released close to the time of ovulation in discrete numbers. Initial sperm:egg ratios at the onset of fertilization may be close to 1:1. Third, the egg is shed as a secondary oocyte encompassed by a mass of cumulus (granulosa) cells from the Graafian follicle. This cellular investment acts to reduce the number of spermatozoa arriving simultaneously at the egg surface, and a proportion of such follicular cells may themselves incorporate spermatozoa. Fourth, the egg is surrounded by an acellular envelope or coat rich in glycoprotein material, the zona pellucida, which has to be penetrated by the fertilizing spermatozoon and usually impedes passage of subsequent spermatozoa. Together, these features function powerfully to reduce the chances of two or more competent spermatozoa arriving simultaneously at the egg plasmalemma after a spontaneous mating, thereby avoiding the potential risk of polyspermic penetration. There is, however, a more specific defense mechanism termed the block to polyspermy.

III. BLOCK TO POLYSPERMY

Although there is certainly an electrical component in terms of membrane hyperpolarization, the basis of a block to polyspermy in mammalian eggs resides principally in the action of so-called cortical granules. These are vesicular organelles of approximately 1 μm diameter located in the newly ovulated secondary oocyte just beneath the plasmalemma (Fig. 1). Cortical granules are formed in the Golgi apparatus and migrate toward the surface of the oocyte, especially during resumption of meiosis shortly

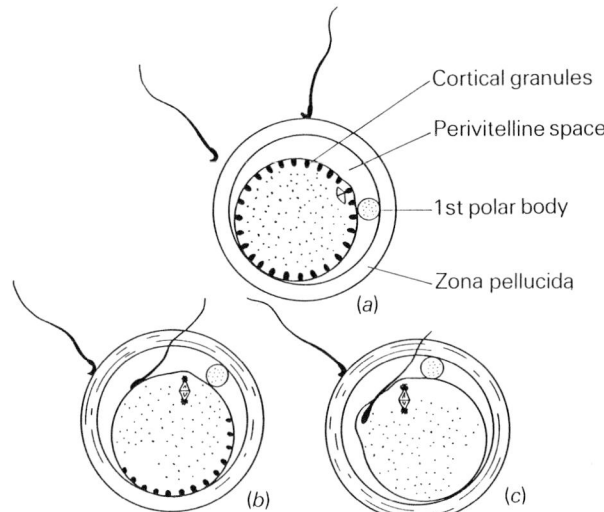

FIGURE 1 Three mammalian eggs demonstrating loss of the cortical granules that occurs as an activation response to sperm penetration. The cortical granule material diffuses across the fluid-filled perivitelline space to make the zona pellucida impermeable to further sperm penetration—the block to polyspermy. The block is progressive from the point of initial entry of the fertilizing spermatozoon.

before ovulation. In circumstances of postovulatory aging of the egg, the granules lose this well-defined location and drift in a disorganized manner to deeper regions of the cortex. This loss of peripheral location has serious consequences for the egg and is one explanation for the deleterious influence of postovulatory aging of eggs in the Fallopian tubes–a phenomenon that may frequently commence within 6–8 hr of ovulation.

Cortical granules or, more correctly stated, the granular-like contents of the vesicular organelles are enzymatic in nature. The enzymatic material (proteases) has an ability to alter the characteristics of the zona pellucida, rendering it impenetrable to all supplementary spermatozoa after entry of the fertilizing spermatozoon. Release of cortical granule material is triggered upon activation of the egg by a fertilizing spermatozoon. The peripheral (outermost) portion of the membrane of suitably located cortical granules undergoes a fusion reaction with the overlying plasma membrane, forming escape ports and enabling the hydrolytic enzymatic contents of individual vesicles to be discharged into the fluid-filled perivitelline space. This is a classical exocytosis reaction involving a dramatic mobilization of intracellular Ca^{2+} ions. Diffusion of enzymatic material across the still quite limited perivitelline space causes the zona pellucida to be rendered impenetrable by further spermatozoa. Structural changes in the zona have not been distinguished in the light or transmission electron microscope, so molecular modifications are inferred. One interpretation is that the putative molecular changes render the lytic enzymes of the sperm acrosome nonfunctional within the substance of the zona pellucida. In other words, cortical granule-prompted modifications act to antagonize or inhibit acrosomal enzymes. This interpretation is especially interesting since both acrosomal enzymes and cortical granule enzymes are Golgi derivatives.

In the physiological situation, the change brought about in the zona pellucida by the cortical reaction clearly develops from the interior surface due to the influence of enzymes liberated into fluid in the perivitelline space (Fig. 2). However, treatment of the exterior surface of the zona pellucida of unfertilized hamster eggs with cortical granule material obtained by sonication and centrifugation also enables treated hamster secondary oocytes to be rendered unfertilizable. Indeed, in an as yet unspecified manner, exposure of human eggs to cortical granule material has been proposed to have contraceptive potential.

The influence of cortical granule exocytosis—the cortical reaction—is primarily on the substance of the zona pellucida in most eutherian mammals studied. The heads of supplementary spermatozoa may continue to enter the outermost portion of the zona pellucida—as in the domestic farm species—but the inner portion (not distinguishable by specific anatomical detail) is rendered impenetrable. Rabbit eggs, by contrast, develop a block to polyspermy at the vitelline surface (plasma membrane) rather than in the zona pellucida, enabling supplementary spermatozoa to accumulate in considerable numbers in the perivitelline space. Rodents such as rats are said to have a weak zona but strong vitelline block. In fact, there is growing evidence for a supplementary vitelline block to polyspermy in species classically thought of as having a strong zona block, such as pigs and man.

Under physiological conditions, activation of the

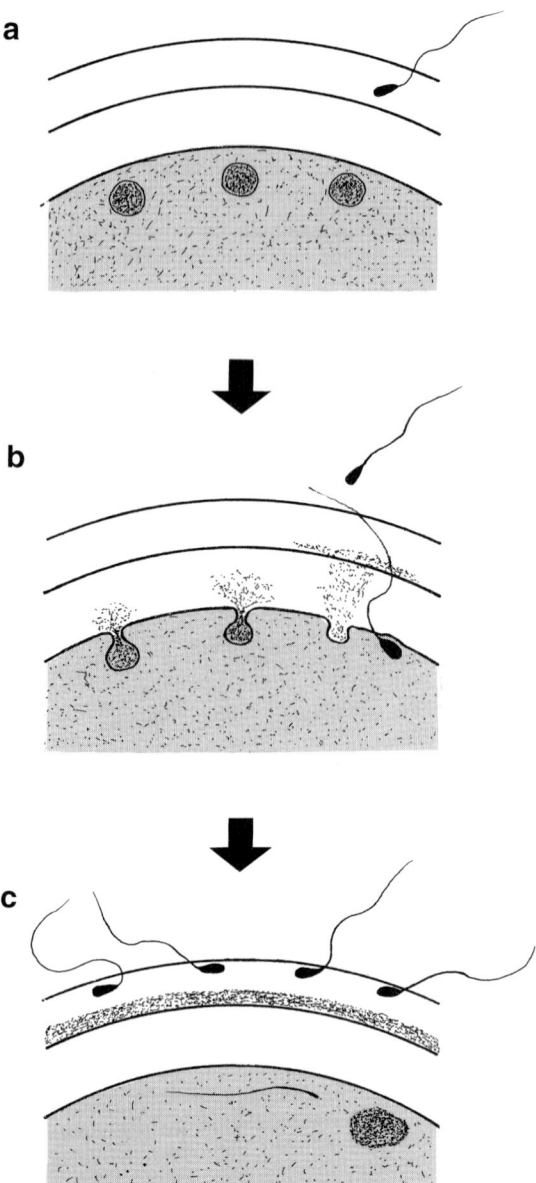

FIGURE 2 A semidiagrammatic representation of the response of peripheral cortical granules to a fertilizing spermatozoon. (a) Cortical granules are still intact as a spermatozoon penetrates the zona pellucida. (b) Activation of the cortical response after the fertilizing sperm head undergoes fusion with the egg plasma membrane. Enzymatic material is discharged into the perivitelline space. (c) Establishment of a block to polyspermy in the innermost portion of the zona pellucida. Supplementary sperm heads can still penetrate the outermost portion of the zona pellucida. Note: Cortical granules are drawn many times larger than their actual size (see text).

cortical reaction, i.e., release of the granular enzymatic substance into the perivitelline space, occurs as soon as fusion commences between the fertilizing spermatozoon and the vitelline membrane. A sperm cytosolic factor—a soluble sperm protein—termed oscillogen introduced into the vitellus at fusion has been invoked as the activating factor that triggers repetitive Ca^{2+} oscillations in the egg cytoplasm; these act to prompt cortical granule exocytosis. This initial functional contact between male and female gametes occurs in the equatorial region of the sperm head where microvillous processes of the egg plasma membrane promptly extend around and incorporate this portion of the head before extending laterally. The equatorial segment has become fusigenic as a consequence of completion of the acrosome reaction and thus modification of the anterior sperm head surface.

In a context of polyspermy, it is relevant that the cortical reaction is propagated from the point of sperm–egg fusion progressively around the surface of the sphere in three dimensions. Accordingly, the chances of a second sperm entering the egg before completion of the block to polyspermy are greatest at a point in the opposite hemisphere of the egg and furthest from entry of the fertilizing spermatozoon. Careful examination of dispermic eggs soon after activation has shown this invariably to be the case: Swollen sperm heads in a peripheral location in the vitellus are usually almost opposite each other in the two hemispheres. Whether an accessory fertilizing sperm has some means of "sensing" the vulnerable hemisphere is a moot point: Perhaps the surface characteristics of the zona give some guidance while the egg is rotating in the Fallopian tube due to the action of endosalpingeal cilia. However, in circumstances of multiple sperm penetration, the various accessory spermatozoa may simply overwhelm the block to polyspermy rather than beating it in a chronological sense.

IV. EXPERIMENTAL CIRCUMSTANCES GENERATING POLYSPERMY

Polyspermic penetration of mammalian eggs has been produced in a variety of experimental situations. Usually, but not always, these reflect an increased number of spermatozoa at the site of fertilization in the Fallopian tubes at the time of initial egg penetration. Almost simultaneous penetration of an egg by more than one spermatozoon is therefore thought to occur rather than belated penetration of an already fertilized egg. It is worth mentioning two exceptions; they are a consequence of the location of the cortical granules.

Primary oocytes are seldom ovulated spontaneously: One study in domestic animals recorded a total of 3 in 1677 recently ovulated eggs, an incidence of 0.2%. However, if Graafian follicles are induced to ovulate prematurely during the luteal or follicular phases of the estrous cycle, a significant proportion of primary oocytes will be shed. After artificial insemination of such treated animals, a majority of primary oocytes will show polyspermic penetration of the cytoplasm but the sperm head chromatin does not decondense. The principal explanation for polyspermy is that the cortical granules of these primary oocytes have not yet migrated from the Golgi apparatus up to the egg plasma membrane, so they are not in a position to undergo a fusion reaction with the egg surface at the time of initial sperm penetration. Cortical granule enzymes are therefore not released and a block to polyspermy is not instigated.

A second and, in a sense, corresponding exception concerns postovulatory aging of eggs in the Fallopian tubes prior to sperm penetration. Once again, inadequate cortical granule function is a primary explanation for the polyspermy that occurs in these circumstances. As mentioned, cortical granules drift away from the egg surface deeper into the cortex with postovulatory aging, thereby preventing membrane fusion with the plasmalemma and establishment of an effective block to polyspermy. An associated problem, frequently overlooked in descriptions of the postovulatory syndrome, may also be an increased number of spermatozoa at the site of fertilization due to lack of regulation. This arises because the Fallopian tubes lose their muscular tone and edematous mucosa as progesterone secretion from the elaborating corpora lutea increases, leading to a larger tubal lumen and a poorer quantitative control of sperm transport. Therefore, both immature and overripe oocytes are at risk of polyspermic penetration.

Experimental circumstances that generate polyspermic penetration of secondary oocytes include the following:

1. Artificial insemination of animals induced to ovulate secondary oocytes during the luteal phase of the estrous cycle (Fig. 3)
2. Systemic injections or local microinjections of a solution of progesterone in oil into the wall of the Fallopian tubes some hours before ovulation
3. Reconstructive surgery that involves resection of the isthmus portion of the Fallopian tubes
4. Surgical insemination directly into the Fallopian tubes of large numbers of spermatozoa at various times before ovulation

The first two categories of experiment, performed in the 1960s and early 1970s, suggested that the steroid hormone progesterone might be implicated in the polyspermic situation, either by directly influencing the gametes themselves or by modifying the nature of the fluids in the lumen of the Fallopian tubes in which the gametes were bathed. Indeed, further evidence for such an interpretation came from reciprocal egg transplantation experiments: Spontaneously ovulated secondary oocytes were transplanted into the Fallopian tubes of inseminated luteal-phase animals and showed a high incidence of polyspermy. By contrast, eggs induced to ovulate during the luteal phase and then transplanted to the Fallopian tubes of spontaneously mated animals in estrus showed negligible polyspermy. However, it is now appreciated that the principal factor leading to polyspermy in these experiments was an enhanced number of capacitated spermatozoa at the site of fertilization due to the relaxing influence of progesterone on the wall of the Fallopian tubes.

An increased number of competent spermatozoa confronting the recently ovulated secondary oocytes is also the explanation for polyspermic penetration after tubal reconstructive surgery. The caudal portion of the Fallopian tube imposes a steeply declining gradient in sperm numbers after mating not only due to the extremely restricted lumen containing a viscous mucus but also because of the binding of sperm heads to the endosalpinx of the caudal isthmus. Surgical resection of the isthmus and reanastomosis of the remaining portions effectively removes the sperm gradient and permits massive numbers of spermatozoa to confront the newly ovulated eggs. Polyspermy is an inevitable sequel.

Similarly, surgical introduction of sperm suspensions directly into the Fallopian tube lumen invariably overrides the regulatory functions of the isthmus. In experiments in which surgical insemination is performed at least several hours before ovulation, large numbers of capacitated spermatozoa are in a position to meet the freshly ovulated eggs. At least a proportion of them will become polyspermic, frequently highly polyspermic with adjoining sperm heads forming aggregates of chromatin. All these experiments have been summarized by Hunter.

V. IN VITRO FERTILIZATION

It is appropriate to consider the results of *in vitro* fertilization. Polyspermy has consistently been the principal anomaly during procedures of *in vitro* fertilization, especially because of the relatively vast numbers of motile spermatozoa present in the microdrop of tissue culture medium used for the fertilizing procedure. Perhaps the most surprising finding is that only a proportion of secondary oocytes so exposed become polyspermic: frequently about 5–10% in man, 10–20% in ruminants, and 50% or more in pigs. Doubtless the expanded and mucified cumulus investment helps to reduce the number of competent spermatozoa reaching the surface of the zona pellucida simultaneously. Perhaps an associated reason is that only a small proportion of cells in the sperm suspension are competent to penetrate at the time of fertilization. No form of gamete selection has been imposed that corresponds to that found in the female genital tract.

It is worth noting that polyspermic penetration of zona-free hamster eggs *in vitro* has been used as a means of finding suitable electron microscopic images of the earliest stages of fertilization, a technique developed in 1970. The most recent technique for studying sperm–egg interactions *in vitro* is intracytoplasmic sperm injection, which offers the possibility of a precisely controlled degree of polyspermy: Known numbers of spermatozoa can be microinjected into specific regions of the vitellus and subsequent cellular events monitored.

VI. STABILITY OF BLOCK TO POLYSPERMY

Because the block to polyspermy functions to prevent penetration of an activated egg by supplementary spermatozoa, the question arises as to whether this defense mechanism is stable and permanent. The question has relevance since, in various species of mammal that are not induced ovulators, animals may remain in estrus and be mated for several hours after ovulation and thus after the initial stages of fertilization. In addition, there would be considerable reserves of spermatozoa at different levels in the female genital tract, frequently permitting a sustained release of viable spermatozoa toward the site of fertilization for several hours after ovulation. If applied mistakenly too late after ovulation or after an unrecorded mating or insemination, the procedure of artificial insemination may also expose recently fertilized eggs to the attentions of considerable numbers of competent spermatozoa. For all these reasons, the stability of the block to polyspermy has been worth examining.

In the domestic pig (Fig. 3), surgical instillation of suspensions of freshly ejaculated or washed boar spermatozoa directly into the Fallopian tubes in the presence of pronucleate and two-celled embryos failed to disturb the ploidy of the embryos or lead to supplementary sperm penetration through the zona pellucida into the perivitelline space. This is not surprising because the pig zona shows structural resilience to supplementary sperm penetration from the Fallopian tube reservoirs: As many as 200–400 sperm heads may penetrate into, but not traverse, the zona of individual eggs. Similarly, cattle embryos exposed at the stages of zona-enclosed or zona-free (hatched) blastocysts to suspensions of capacitated bull spermatozoa *in vitro* were not penetrated or visibly damaged by the presence of large numbers of supplementary spermatozoa. Although this finding would be worth verifying *in vivo*, there are nonetheless reasons for believing that the block to polyspermy in mam-

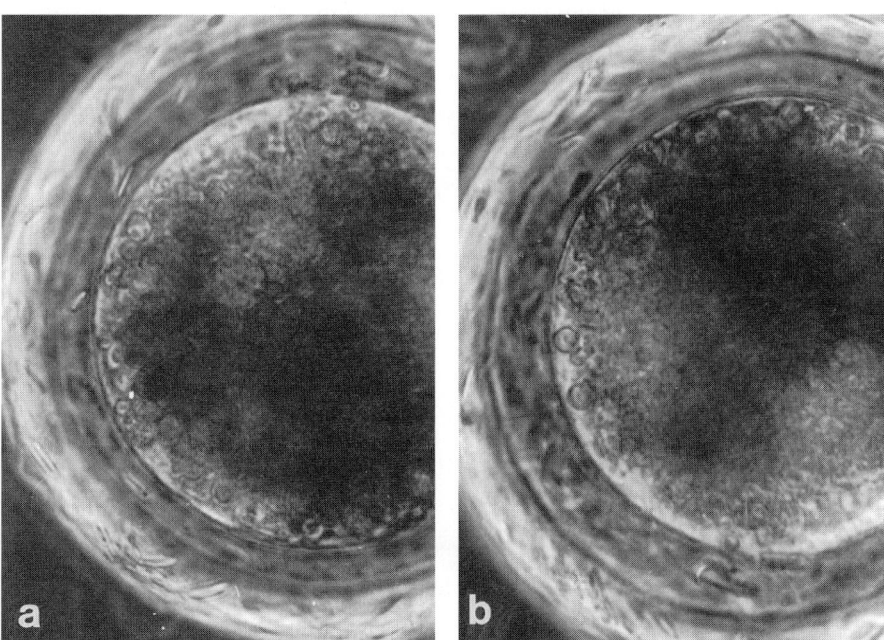

FIGURE 3 (a and b) Whole mount preparations of pig eggs seen under phase-contrast microscopy with many spermatozoa visible in both the perivitelline space and zona pellucida. After fixing and staining, these eggs were also found to exhibit polyspermic penetration of the vitellus, with unswollen sperm heads and pronuclei in the same preparation.

malian embryos is permanent and irreversible—a situation that would make good biological sense.

VII. MOLECULAR ASPECTS OF BLOCK TO POLYSPERMY

In recent years, elegant molecular studies of the process of fertilization have been performed by Wassarman and colleagues using a mouse model. These have focused on interactions between spermatozoa and the zona pellucida and also on the constituent glycoproteins of the zona substance. In particular, they have expanded the earlier studies of Gwatkin's group on sperm binding to the surface of the zona pellucida. In brief, Wassarman's group has proposed what is referred to as the ZP_3–ZP_2 sequence of events, whereby mouse sperm head binding to the mouse zona by the ZP_3 receptor induces the acrosome reaction and thus transfers receptor (binding) activity to the ZP_2 glycoprotein; the latter becomes a secondary ligand, interacting with molecules on the inner acrosomal membrane. The loss of surface ZP_3 receptors, in theory at least, is supposed to prevent the binding of further spermatozoa.

Two points must be noted. First, polyspermic penetration of mouse eggs has been recorded *in vivo* by various authors, so either (i) the molecular changes require a significant period of time during which almost simultaneous penetration of two or more spermatozoa enables polyspermy or (ii) initial sperm head binding may not always be absolutely essential for sperm penetration of the zona. In other words, a spermatozoon undergoing a spontaneous rather than induced acrosome reaction just at the surface of the zona pellucida may be able to bypass the ZP_3–ZP_2 system of receptors and simply use its lytic enzymes and propulsive force for direct penetration of the zona. It should be noted that the surface of the zona pellucida is not smooth but extremely uneven with steep undulations, furrows, and crevices; therefore, this unevenness could easily act to trap a highly motile spermatozoon if the granulosa (cumulus) cell investment was no longer present. In any event, the mouse model does not have close parallels in other mammals such as the domestic farm species because, as already stated, many accessory sperm heads are able to penetrate into the zona pellucida. Another factor which would influence sperm binding and zona penetrability is the extent of glycoprotein secretion such as oviductin onto the egg surface in the Fallopian tubes. This secretion may be associated with what is termed as zona hardening.

In theory, polyspermy could arise due to an inadequate quantity or quality of cortical granule material released belatedly or not uniformly at the time of egg activation or due to inappropriate responses in the zona pellucida. The concept of zona pellucida heterogeneity is currently gaining favor, i.e., the notion that the zonae pellucidae around newly ovulated eggs do not all have equivalent functional competence. Consideration should also be given to the fact that the cortical granule exocytotic reaction requires diffusion of the enzyme message across the perivitelline fluid. Presumably this fluid may also vary in composition between oocytes, resulting in differing degrees of enzymatic dispersion. Indeed, using modern microtechniques, the topic of perivitelline fluid composition urgently requires sensitive and sustained research. After all, perivitelline fluid provides the fluid microenvironment not only of the secondary oocyte but also of the embryo until the time of hatching from the zona pellucida. It would therefore not only be fascinating but also extremely valuable to monitor the dynamic state of this fluid both with the stage of embryonic development and with location of embryos in the genital tract.

VIII. CONSEQUENCES OF POLYSPERMY

Polyspermy does not necessarily infer polyploidy. Dispermic penetration of the vitellus invariably leads to the formation of two male pronuclei and usually to the approximation of these and the female pronucleus in the center of the egg prior to condensation of chromosomes and organization of the first mitotic spindle. A zygote of triploid constitution (3n) would thus be anticipated. Trispermic penetration may frequently lead to a similar sequence of events and perhaps to a tetraploid zygote (4n). However, as the number of accessory spermatozoa in the vitellus increases, the development of each individual pronucleus seems to be restricted in size and the tendency toward centripetal migration seems to diminish, mostly due to disruption of the cytoskeleton. Regarding the growth of individual pronuclei, their reduced size (volume) in a tetraspermic or pentaspermic situation led Austin to propose the presence of an hypothetical cytoplasmic factor available only in limited quantities that restricts growth of individual male pronuclei. The speculation has proved correct, and the substance is generally referred to as male pronucleus growth factor.

Polyspermic eggs *in vivo* almost always undergo degeneration and/or fragmentation, although the possibility of elimination of blastomeres containing two nuclei remains in question. If polyploid embryos are formed, these invariably die precociously, usually by the time of blastocyst formation and thus before implantation. Triploids are an exception. They may continue up to conceptus formation and in primates are associated with early spontaneous abortion. Therefore, polyspermy in mammals is a pathological condition, usually heralding an early demise of the zygote or embryo and standing in marked contrast to the so-called physiological polyspermy found in fish and birds. As indicated in the Introduction, an embryo derived from this abnormal form of fertilization may die so precociously that it frequently fails to influence the duration of estrous or menstrual cycles.

Acknowledgments

Writing of this article was supported by a Carlsberg Guest Professorship in Veterinary Physiology, for which grateful acknowledgment is made. Frances Anderson kindly prepared the manuscript.

See Also the Following Articles

FALLOPIAN TUBE; FERTILIZATION, CELL BIOLOGY; GRAAFIAN FOLLICLE; IN VITRO FERTILIZATION

Bibliography

Austin, C. R. (1961). *The Mammalian Egg.* Blackwell, Oxford, UK.

Barros, C., and Yanagimachi, R. (1971). Induction of the zona reaction in golden hamster eggs by cortical granule material. *Nature (London)* **233**, 268–269.

Braden, A. W. H., Austin, C. R., and David, H. A. (1954). The reaction of the zona pellucida to sperm penetration. *Aust. J. Biol. Sci.* **7**, 391–409.

Chang, M. C., and Pincus, G. (1951). Physiology of fertilisation in mammals. *Physiol. Rev.* **31**, 1–26.

Harper, M. J. K. (1973). Stimulation of sperm movement from the isthmus to the site of fertilisation in the rabbit oviduct. *Biol. Reprod.* **8**, 369–377.

Hunter, R. H. F. (1991). Oviduct function in pigs, with particular reference to the pathological condition of polyspermy. *Mol. Reprod. Dev.* **29**, 385–391.

Hunter, R. H. F., and Nichol, R. (1988). Capacitation potential of the Fallopian tube: A study involving surgical insemination and the subsequent incidence of polyspermy. *Gamete Res.* **21**, 255–266.

Parrington, J., Swann, K., Shevchenko, V. I., Sesay, A. K., and Lai, F. A. (1996). Calcium oscillations in mammalian eggs triggered by a soluble sperm protein. *Nature (London)* **379**, 364–368.

Rothschild, Lord (1956). *Fertilisation.* Methuen, London.

Szollosi, D. (1975). Mammalian eggs ageing in the Fallopian tubes. In *Ageing Gametes* (R. J. Blandau and A. G. Karger, Eds.), pp. 98–121. Karger, Basel.

Wassarman, P. M. (1990). Profile of a mammalian sperm receptor. *Development* **108**, 1–17.

Porifera

Paul E. Fell
Connecticut College

I. Introduction
II. Sexual Reproduction
III. Asexual Reproduction
IV. Reproductive Cycles

GLOSSARY

amphiblastula Hollow larva with anterior flagellated cells and posterior nonflagellated cells.
carrier cell Modified choanocyte that transports spermatozoon to oocyte during internal fertilization.
coeloblastula Simple blastula, like larva.
gemmule Asexually produced propagule consisting of a mass of nutrient-laden cells enclosed within a protective capsule.
nurse cell Cell that contributes nutrients and possibly organelles to growing oocytes, embryos, or developing gemmules.
parenchymella A usually solid flagellated larva.
trichimella Solid flagellated larva of hexactinellid sponges.

Sponges are relatively simple multicellular animals that inhabit water bodies ranging in salinity from fresh water to seawater. They pump environmental water through canals and chambers within their bodies; and the flow of water functions in feeding, respiratory gas exchange, and the removal of wastes. The current also transports reproductive elements. Sponges exhibit a wide range of size and complexity as adults, as well as many different reproductive patterns. This article discusses reproduction by members of the major groups of sponges that have been relatively well studied.

I. INTRODUCTION

All sponges evidently reproduce sexually; however, many of them may also reproduce by asexual means. The relative importance of sexual and asexual reproduction varies according to species and perhaps also in relation to environmental conditions. Most studies of reproduction in sponges are descriptive (Table 1). Very little is known concerning regulatory mechanisms.

II. SEXUAL REPRODUCTION

Gonads and reproductive tracts do not occur in sponges. Oocytes, frequently enclosed within a flattened follicular epithelium, are distributed individually throughout much of the body of the sponge (Fig. 1). In some species, they occur in loose aggregates. Similarly, spermatozoa differentiate from masses of cells surrounded by a thin epithelium, called spermatic cysts, that are scattered within the sponge. In many species, individuals simultaneously produce oocytes and spermatozoa. However, in other species, individuals either produce only one type of gamete (gonochorism) or generate first one type of gamete and then the other (successive hermaphroditism). In many cases, it has not been possible to distinguish with certainty between these latter patterns in gamete production. Nothing is known about sex determination in sponges. Among sponges with separate sexes, the apparent sex ratios are highly variable.

There is very little information about the longevity of sponges. Some species normally live for little more than a year; others may live for decades. Life spans

TABLE 1
Major Groups of Sponges with Some Genera within Each Group

Subdivision	Examples
Class Calcarea	
Subclass Calcinea	*Ascandra, Clathrina*
Subclass Calcaronea	*Grantia, Sycon, Scypha, Leucandra*
Class Demospongiae	
Subclass Homoscleromorpha	*Octavella*
Subclass Tetractinomorpha	*Tetilla, Suberites, Cliona*
Subclass Ceractinomorpha	*Haliclona, Halichondria, Microciona, Spongilla*
Class Hexactinellida	*Euplectella, Oopsacas*

of a few years may be common. Individuals of *Halichondria* sp. often begin to produce gametes when they are only a few weeks old and <1 cm in dimension. On the other hand, certain long-lived species may not start to reproduce sexually until they are at least several years old.

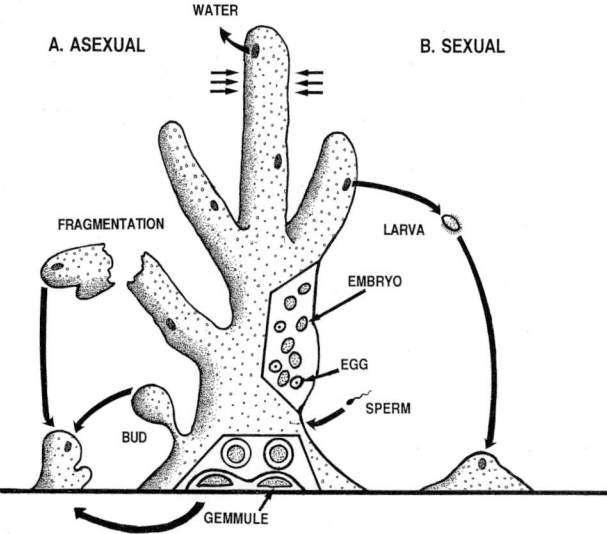

FIGURE 1 Schematic representation of reproduction in sponges. (A) Asexual reproduction by fragmentation, budding, and gemmules. (B) Sexual reproduction with internal development of larvae that are released from the excurrent canal system.

A. Gametogenesis

The origin of the gametes has not been well documented. Morphological studies indicate that in some sponges gametes arise from archeocytes (totipotent ameboid cells), whereas in other sponges they derive from choanocytes (flagellate cells involved in feeding and in generating the movement of water through the sponge). For still other species, it appears that oocytes develop from archeocytes and spermatogenic cells arise directly from choanocytes. However, no experimental studies have been undertaken to test the conclusions drawn from morphological studies.

In most sponges that incubate embryos, oocytes and spermatic cysts develop asynchronously within individuals during the reproductive season. On the other hand, among oviparous species, many exhibit a single cycle of synchronized oocyte development both within individual sponges and throughout entire populations and some also display a synchronized differentiation of spermatic cysts. In oviparous species, oogenesis frequently requires one to several months, whereas spermatogenesis may occur in less than a week.

Within the spermatic cysts (Fig. 2), cells progress from spermatogonia to primary spermatocytes, to secondary spermatocytes, and then to spermatids as they undergo the meiotic divisions. Sister secondary spermatocytes and spermatids often are joined by cytoplasmic bridges formed as a result of incomplete cytokinesis. Finally, the spermatids differentiate into spermatozoa. The fully developed spermatozoa are characterized by a compact nucleus, one to several mitochondria, and a flagellum with a pair of centrioles at its base. The flagellum may already be present in the primary spermatocytes. An acrosomal granule has been observed in the spermatozoa of *Oscarella lobularis*, but this organelle apparently does not occur in the spermatozoa of many sponges.

Small oocytes are frequently ameboid. However, as the oocytes increase in size, they become more or less spherical and may be enclosed by flattened follicle cells. The early growth of the oocytes is usually achieved without the production of yolk; but during the period of major growth, numerous yolk granules accumulate within the cytoplasm. In certain oviparous sponges with small eggs, it appears that the

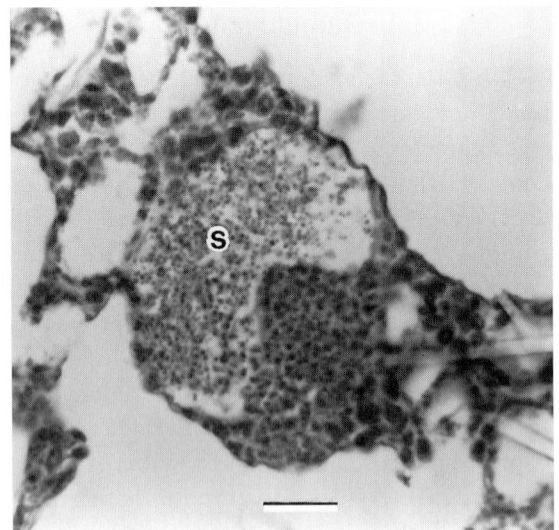

FIGURE 2 A spermatic cyst (s) in *Halichondria* sp. Scale bar = 25 μm.

oocytes may synthesize much or all of the yolk. The oocytes contain well-developed rough endoplasmic reticulum and Golgi complexes. On the other hand, in many species, much of the yolk apparently is supplied by nurse cells (trophocytes) that provide the oocytes with organelles and/or macromolecules. The nurse cells often resemble large archeocytes; but among the calcareous sponges, cells derived from choanocytes frequently function as nurse cells. When the oocytes attain a certain size, large numbers of nurse cells may aggregate around them (Fig. 3), presumably in response to a factor(s) produced by the oocytes. In many sponges, the nurse cells move between the follicle cells and are engulfed by the oocyte. Transfer of materials from nurse cells to oocytes may also occur in other ways, including passage through cytoplasmic bridges joining the two types of cells and endocytotic ingestion by the oocyte of substances secreted by the nurse cells. However, conclusive evidence for such mechanisms is lacking. Currently, little is known about the synthetic activities of the oocytes, the composition of the yolk, or the nature of the materials provided by nurse cells.

Meiotic divisions and the production of polar bodies have been seen in the oocytes of some calcareous sponges and a few demosponges. The failure to observe these events in more species may be related to the small size of the chromosomes and the large amount of yolk contained in the oocytes at the end of their growth.

B. Spawning and Fertilization

When gametes are released, they exit the sponge by way of the excurrent canal system. In many sponges, fertilization and embryonic development occur within the oocyte follicles so that only spermatozoa are shed. Following spawning the spermatozoa are dispersed by currents and may enter egg-bearing individuals through their incurrent canal systems (Fig. 1). Some oviparous species release eggs that are fertilized externally; but for many oviparous sponges, it has not been determined whether fertilization is internal or external. In some cases the eggs/zygotes are released in a mucous sheet that binds them to the parent sponge and the surrounding substratum. For certain oviparous species, the mass release of gametes is highly predictable with respect to period of year, lunar phase, and time of day.

Stages of internal fertilization have been observed in fixed tissue sections of a few sponges, mostly belonging to the class Calcarea. Spermatozoa apparently are transported from the canal system to eggs by modified choanocytes, called carrier cells. After

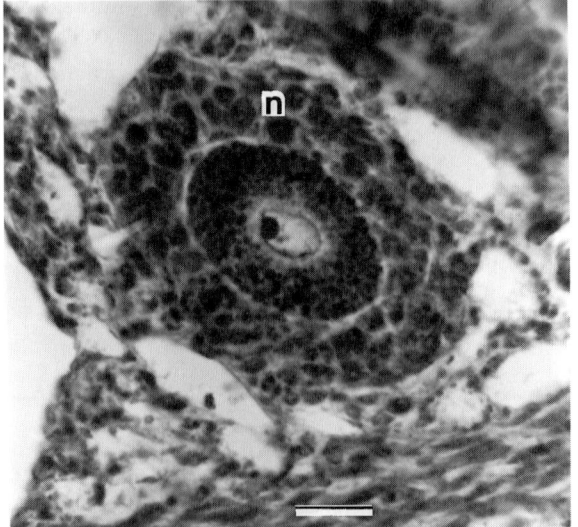

FIGURE 3 A growing oocyte surrounded by nurse cells (n) in *Halichondria* sp. Scale bar = 25 μm.

engulfing a spermatozoon, the choanocyte loses its collar and flagellum and becomes closely associated with an egg. The spermatozoon, within a membrane-bound vesicle in the cytoplasm of the carrier cell, also changes. Its head and mitochondrial body enlarge and its flagellum is lost. Finally, the carrier cell depresses the surface of the egg and transfers the spermiocyst (vesicle containing the modified spermatozoon) to it. The male and female pronuclei meet near the center of the egg and fuse.

Not only is research on internal fertilization in sponges limited to a few morphological studies but also there have been virtually no detailed studies of external fertilization. Clearly, experimental investigations of both internal and external fertilization are needed.

C. Embryonic and Larval Development

Although in most cases it appears that zygotes initiate cleavage soon after fertilization, the zygotes of commercial bath sponges (*Spongia* and *Hippospongia*) grow from about 50 μm to approximately 300 μm in diameter before mitosis begins. Each zygote is surrounded by nurse cells from which it apparently receives materials by phagocytosis and possibly transport through cytoplasmic bridges. In all species in which it has been observed, cleavage is total (holoblastic); however, a typical cleavage sequence is not apparent in some sponges presumably because abundant yolk reserves obscure cytological details. Oocytes, late-stage embryos, and larvae are recognizable in these latter species, but the early stage embryos appear as dense masses of yolk granules.

The embryos of calcareous sponges and some demosponges develop into hollow flagellated blastulae. In *Ascandra* and *Clathrina* (subclass Calcinea), the blastocoel of the coeloblastula larva is progressively filled with cells that lose their flagellum and leave the epithelial layer in a process known as ingression (Fig. 4A). *Grantia* and *Sycon* (subclass Calcaronea) possess an amphiblastula larva (Fig. 4B) with flagellated epithelial cells occupying the anterior hemisphere and large nonflagellated cells making up the posterior hemisphere. This larval stage develops from a blastula in which the flagella of the epithelial

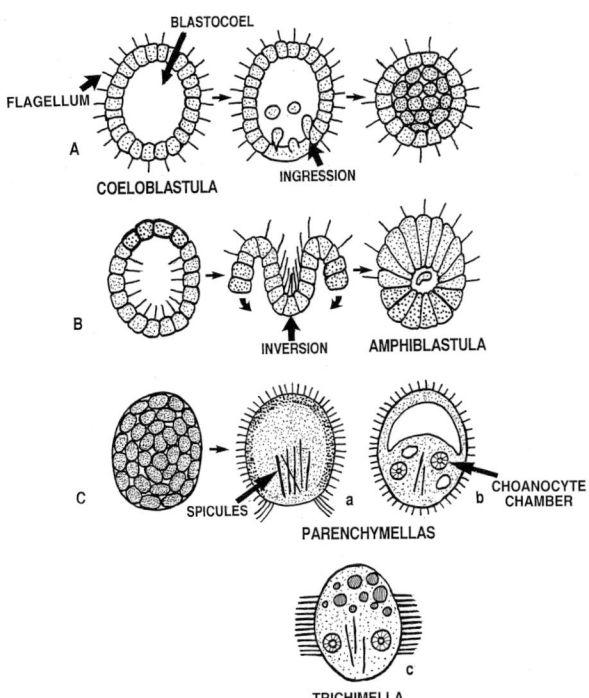

FIGURE 4 Different patterns of larval development in sponges. (A) Transformation of the coeloblastula larva of *Clathrina* into a solid mass of cells by means of ingression. (B) Inversion of the blastula of *Sycon* leading to the formation of an amphiblastula larva. (C) Development of a solid embryo into a parenchymella lava in *Haliclona* (a) or *Spongilla* (b) or a trichimella larva in *Oopsacas* (c).

cells are initially directed into the blastocoel. Later, an opening appears among the nonflagellated cells and the blastula turns inside out. Inversion produces the amphiblastula with external flagella and a small central cavity that may contain a small number of maternal nurse cells.

Although *Octavella* and certain tetractinomorph sponges (e.g., *Polymastia* and *Chondrosia*) possess blastular larvae, many demosponges produce solid parenchymella larvae (Fig. 4C). In most cases, cleavage leads to the formation of an aggregated mass of cells that differentiate progressively into larval structures. Small cells at the surface of the developing larva form a flagellated epithelium in which each cell possesses a single flagellum. In some species the entire surface of the parenchymella is flagellated; in others one or both ends of the larva lack flagella.

The central region of the parenchymella is filled by amebocytes and often other types of cells that are present in the adult sponge. Many parenchymellas carry a bundle of siliceous spicules and some possess choanocyte chambers. The larvae of freshwater sponges (e.g. *Ephydatia* and *Spongilla*) are also characterized by a large anterior cavity as well as smaller rudiments of the canal system.

Hexactinellid sponges produce a solid trichimella larva (Fig. 4C). This type of larva possesses a broad flagellated equatorial zone in which the cells are multiflagellated and covered by a thin epithelium through which the flagella project. Inside the larva, large lipid granules fill the anterior end and distinctive larval spicules and choanochambers are present within the posterior hemisphere.

Although most sponges possess a free-living larval stage, *Tetilla japonica* is an oviparous sponge with direct development. The zygote attaches to the substratum and develops into a small sponge.

The eggs of many sponges with internal development possess abundant nutrient stores and there appears to be little, if any, transfer of materials from the parent to the developing embryos. Incubation of embryos and larvae apparently serves primarily a protective function. On the other hand, in some sponges with internal development, including many calcareous sponges, the embryos evidently ingest parental cells and/or cellular products. Furthermore, certain oviparous sponges release eggs or zygotes that are enclosed by parental cells which are subsequently taken into the embryos.

D. Free-Living Larva and Metamorphosis

Larvae are released from the parent by way of the excurrent canal system (Fig. 1). In some species such release occurs during the morning in response to light signals.

Sponge larvae are nonfeeding and typically short-lived. Most are planktonic, but those of a few species (e.g., *Polymastia robusta* and *Halichondria moorei*) are benthic and creep over the substratum. During the exploratory phase that precedes settlement, planktonic larvae also move over the substratum. The larvae of many sponges exhibit behavioral responses to light and/or gravity that appear to be important for dispersal and for determining where they are likely to settle. However, to a large degree planktonic larvae are passively distributed by currents. Larval settlement often takes place within a few hours to several days following release from the parent sponge; however, in some cases, it may be delayed for 2 or more weeks.

Upon finding a suitable settling site, a larva attaches to the substratum and begins to spread over it. Then in the process of metamorphosis, the newly attached larva develops into a small sponge. Currently, little is known concerning the fates of various types of larval cells. Knowledge of sponge metamorphosis is based primarily on morphological studies. More experimental studies are needed.

At the time of settlement, the coeloblastula larva of certain calcareous sponges (subclass Calcinea) consists of morphologically similar cells. It is hypothesized that the cells are developmentally uncommitted and that they differentiate according to their positions within the cellular mass. The amphiblastula of other calcareous sponges (subclass Calcaronea) usually attaches to the substratum by the flagellated anterior pole, and within a few hours the flagella are lost. The anterior larval cells apparently give rise to most of the structures of the young sponge, with the exception of the external epithelium, which is formed by the nonflagellated cells of the posterior half of the larva.

As already noted, the parenchymella larvae of certain freshwater sponges have a structure that in many respects resembles the structure of the adult. Metamorphosis involves replacement of the flagellated epithelium with a thin nonflagellated epithelium and continued development of the canal system. On the other hand, the parenchymellas of most demosponges are less highly developed, lacking choanocyte chambers, rudiments of the canal system, and, in some cases, spicules. In some species (e.g., *Microciona prolifera*) it appears that all the components of the adult sponge are derived from the central cell mass of the larva and that the flagellated epithelial cells are phagocytized. However, evidence suggests that in other species (e.g., *Mycale contarenii* and *Haliclona permollis*) the larval flagellated cells may give rise to the choanocytes of the young sponge.

III. ASEXUAL REPRODUCTION

Sponges exhibit a number of mechanisms for asexual propagation, including fragmentation, budding, and gemmule production (Fig. 1). Gemmules have been extensively studied, but other forms of asexual reproduction also appear to be important.

A. Gemmules

Many freshwater and some estuarine/marine sponges produce gemmules. A gemmule consists of a mass of nutrient-laden cells enclosed by a collagenous capsule that in many cases contains spicules (Fig. 5). The cells, called thesocytes, are morphologically homogeneous; and in many freshwater sponges they are binucleate. An individual sponge frequently produces hundreds to thousands of gemmules that may be attached to the substratum and/or distributed throughout much of the body of the sponge. In some cases, most of the sponge tissue is converted into gemmules.

Gemmules develop from aggregations of cells, consisting of archeocytes, nurse cells, and capsule-secreting cells (spongocytes). As the archeocytes phagocytize the nurse cells in the interior of an aggregate, they are transformed into thesocytes. Peripherally, the spongocytes constitute a columnar epithelium that produces the capsule. Following gemmule production, the parent sponge frequently dies. The gemmules of many species are highly tolerant of environmental stresses and tend to ensure sponge survival during unfavorable periods. Gemmules may be tolerant of extreme temperatures, drying, anoxia/hypoxia, and/or large salinity fluctuations. In some cases, they persist for long periods under adverse conditions.

Gemmules typically undergo a period of dormancy. In some cases, dormancy is imposed by unfavorable environmental conditions, a state known as quiescence. When favorable conditions are reestablished, the gemmules hatch and develop into a functional sponge(s). The gemmules of many sponges exhibit diapause, a dormant state that is maintained by an unknown endogenous mechanism even under favorable conditions. The breaking of diapause frequently is promoted by exposure to low temperature and may be followed by a period of gemmule quiescence. Diapause tends to ensure that gemmules will hatch at a time when favorable environmental conditions persist.

Although the cells from many gemmules, hatching at about the same time, often develop into a large individual, single gemmules are capable of giving rise to functional sponges. Gemmules are usually discrete entities held together only by a meshwork of parental skeleton. Therefore, the gemmules produced by one individual may become separated from one another, potentially resulting in asexual propagation and perhaps dispersal. Because of their resistance to harsh environmental conditions, gemmules appear to be well suited for dispersal within and between habitats. Unfortunately, little information concerning the dispersal of sponges by gemmules exists.

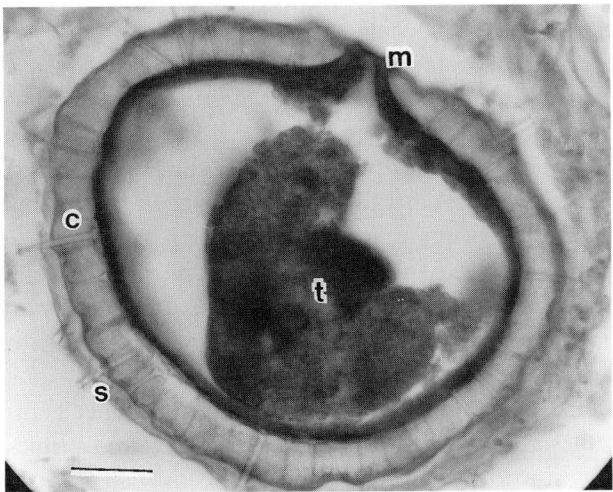

FIGURE 5 Gemmule of *Anheteromeyenia ryderi* with inner mass of thesocytes (t) enclosed by a thick capsule (c) in which spicules (s) are embedded. The micropyle (m) is the site where the cells emerge during gemmule hatching. Scale bar = 88 μm.

B. Fragmentation and Budding

Fragmentation appears to be a major means of propagation in some sponges, especially those of branching habit. Because sponges exhibit little regional differentiation and possess a highly plastic morphology, even small fragments often can give rise to new individuals. The torn surface of each fragment is covered and some reorganization of the canal system evidently takes place. Attachment of the fragment to a stable substratum appears to be crucial to its survival.

Some sponges produce external buds that separate from the parent. The buds of certain species (e.g., *Radiospongilla cerebellata*) are essentially protrusions of the body and possess a functional organization from their inception. Detachment of the buds is analogous to fragmentation. The buds of other species (e.g., *Tethya aurantium* and *Axinella damicornis*) consist of archeocyterich cellular masses with no evidence of choanocyte chambers or canals. After such buds separate from the parent sponge, they attach to the substratum and develop into functional sponges. In some cases, asexual budding appears to play an important role in maintaining sponge populations.

IV. REPRODUCTIVE CYCLES

Some sponges reproduce sexually throughout the year, especially in regions where annual fluctuations in environmental parameters are small. In such cases, the level of reproductive activity may vary to some degree with the seasons. However, most sponges reproduce sexually during a limited, predictable period(s) of the year at any particular location. Where large annual changes in water temperature occur, this factor may be of major importance in determining the reproductive period. For each species there appears to be a range of temperatures in which reproduction can occur as long as other conditions are favorable. Many sponges initiate gametogenesis when water temperature increases during the spring; in some cases, reproduction ceases when water temperature declines. In contrast, other species initiate/sustain gametogenesis when water temperature is falling or near its yearly low extreme. For certain species, it has been shown that reproduction commences at different times in different localities, but at each place it starts when the water temperature reaches a critical level. Similarly, at one site, the reproductive period of a sponge may begin and/or end at somewhat different times during different years in relation to variations in water temperature. In a number of instances, it appears that the reproductive period of a sponge is shorter than the period when temperatures favorable for reproduction occur. Evidently, a variety of factors, both exogenous and endogenous, act to control reproductive activity, but these factors remain largely undefined.

Some species of sponge allocate a large portion of their biomass to sexual reproduction. Much of the body is filled with reproductive elements and there may be a long reproductive period during which gametes and embryos are continuously produced. For other species, gametes/embryos represent a small fraction of the sponge biomass and the reproductive period may be relatively short. A growing body of evidence suggests that some sponges, which exhibit low levels of sexual reproductive activity, may rely to a large degree on asexual propagation. Both sexual and asexual reproduction appear to be of major significance in the life cycles of certain species.

Asexual reproduction by sponges also appears to be controlled by environmental factors, including temperature; however, as with sexual reproduction, regulation is poorly understood. More is known about gemmules than about other forms of asexual propagation. Some species in certain habitats begin to produce gemmules a short time before the sponges degenerate. In some of these cases, the formation of gemmules may be triggered directly or indirectly by changes in water temperature and/or falling water levels. Other species typically produce gemmules during a relatively long period when environmental conditions favor growth. As individuals increase in size, they invest more biomass in gemmules. It seems likely that in many cases factors such as the nutritional state of the sponge and its sexual reproductive condition may be important in controlling gemmule production. Various species respond differently to similar environmental conditions. Within the same habitat, different species may begin to produce gemmules at different times of the year and some species may not form gemmules.

Particular species may exhibit different patterns of gemmule production and hatching in different types of habitat within a restricted area as well as in habitats in different portions of their geographical ranges. For example, in Louisiana *Eunapius fragilis* living in temporary water bodies forms gemmules during the spring before the habitats dry up, whereas in permanent streams of the same region, it produces gemmules during the fall. In Connecticut, *Haliclona loosanoffi* occurs only as gemmules during the winter. On the other hand, in North Carolina this sponge is active during the winter and persists as gemmules during the middle of the summer.

Acknowledgments

The author is grateful to Audrey Goldstein for help in preparing the manuscript and to Susan Stone for the drawn illustrations.

Bibliography

Amano, S., and Hori, I. (1996). Transdifferentiation of larval flagellated cells to choanocytes in the metamorphosis of the demosponge *Haliclona permollis*. *Biol. Bull.* **190**, 161–172.

Boury-Esnault, N., and Vacelet, J. (1994). Preliminary studies on the organization and development of a hexactinellid sponge from a Mediterranean cave, *Oopsacas minuta*. In *Sponges in Time and Space* (R. W. M. van Soest, Th. M. G. van Kempen, and J. C. Braekman, Eds.), pp. 407–415. Balkema, Rotterdam.

Corriero, G.; Sarà, M., and Vaccaro, P. (1996). Sexual and asexual reproduction in two species of *Tethya* (Porifera: Demospongiae) from a Mediterranean coastal lagoon. *Mar. Biol.* **126**, 175–181.

Fell, P. E. (1974). Porifera. In *Reproduction of Marine Invertebrates* (A. C. Giese and J. S. Pearse, Eds.), pp. 51–132. Academic Press, New York.

Fell, P. E. (1983/1989/1993). Porifera. In *Reproductive Biology of Invertebrates I. Oogenesis, Oviposition and Oosorption* pp. 1–29, 1983; *IV. Fertilization, Development and Parental Care*, pp. 1–41, 1989; and *VI. Asexual Propagation and Reproductive Strategies*, pp. 1–44, 1993 (K. G. Adiyodi and R. G. Adiyodi, Eds.). Wiley/IBH, New York/New Delhi.

Hoppe, W. F., and Reichert, M. J. M. (1987). Predictable annual mass release of gametes by the coral reef sponge *Neofibularia nolitangere* (Porifera: Demospongiae). *Mar. Biol.* **94**, 277–285.

Reiswig, H. M. (1983). Porifera. In *Reproductive Biology of Invertebrates II. Spermatogenesis and Sperm Function* (K. G. Adiyodi and R. G. Adiyodi, Eds.), pp. 1–21. Wiley, New York.

Sarà, M. (1992). Porifera. In *Reproductive Biology of Invertebrates V. Sexual Differentiation and Behaviour* (K. G. Adiyodi and R. G. Adiyodi, Eds.), pp. 1–29. Wiley, New York.

Simpson, T. L. (1984). *The Cell Biology of Sponges*. Springer-Verlag, New York.

Porpoises

see Whales and Porpoises

Postdate (Postterm) Pregnancy

Jennifer Claman and Brian Koos

UCLA School of Medicine

I. Diagnosis
II. Parturition
III. Complications
IV. Evaluation
V. Management
VI. Conclusions

GLOSSARY

amniotic fluid index (AFI) Ultrasound is used to determine the sum of the vertical measurement of the largest amniotic fluid pocket in each of the four abdominal quadrants. The AFI normally ranges between 8 and 26 cm. An AFI < 8 represents reduced amniotic fluid volume; values < 5 cm are associated with significantly increased perinatal morbidity.

biophysical profile The biophysical profile is an ultrasonographic evaluation of fetal movements, amniotic fluid volume, and heart rate activity. This 30-min evaluation credits two points each to each of the following: (i) reactive nonstress test, (ii) 20 sec of fetal breathing, (iii) flexion and extension of fetal extremities, (iv) three distinct fetal movements, and (v) adequate amniotic fluid volume (AFI > 5 cm). Management is generally based on the composite score.

cervical ripening A process that softens the cervix resulting in less resistance to dilatation.

contraction stress test (CST) An external monitor records fetal heart rate responses to spontaneous or induced contractions. A CST is negative if no late decelerations occur with at least three contractions of moderate intensity lasting 40–60 sec. The CST is positive if late decelerations occur with more than 50% of the uterine contractions. A CST is equivocal if late decelerations are present with fewer than 50% of contractions.

gestational age This dating of pregnancy is based on menstrual age and is calculated from the first day of the last menstrual period which actually precedes conception by about 2 weeks.

macrosomia This term describes a large infant and is commonly defined as an infant who weighs 4500 g or more after 38 weeks of gestation.

meconium This is a dark green material consisting of gastrointestinal secretions, cellular debris, bile acids, bilirubin, and water and appears in the fetal ileum by 16 weeks of gestation. The passage of meconium into amniotic fluid occurs in up to 22% of live births and is more commonly seen in postdate pregnancies.

meconium aspiration syndrome A syndrome in the neonate characterized by severe respiratory distress resulting from aspiration of meconium. The meconium produces a mechanical obstruction of airways as well as a diffuse chemical pneumonitis.

membrane stripping An index finger is inserted as far through the internal os as possible and rotated 360°. This procedure separates the membranes from the lower uterine segment.

nonstress test (NST) Fetal heart rate (FHR) is recorded with an external monitor. The NST is reactive if at least two FHR accelerations of at least 15 beats in amplitude and lasting at least 15 sec occur within a 20-min period. The NST is nonreactive if such accelerations are not observed during at least 40 min of continuous FHR recording.

oligohydramnios Excessive reduction in amniotic fluid volume; commonly defined as an amniotic fluid index of 5 cm or less.

shoulder dystocia Gentle downward traction fails to dislodge the anterior shoulder from behind the pubic bone, thereby requiring additional maneuvers to facilitate delivery of the anterior shoulder.

Postterm or postdate pregnancy is commonly defined as a gestation exceeding 42 weeks (294 days) from the onset of the last menstrual period. Based on a menstrual cycle length of 28 days, this condition corresponds to a postconceptional age ≧40 weeks or 280 days.

I. DIAGNOSIS

Accurate identification of the postdate pregnancy depends on how closely the estimated gestational age reflects the postconceptional age. That the incidence of postdate pregnancy determined retrospectively varies from 3 to 12% reflects the difficulty in accurately determining the due date. Up to 45% of pregnant women may be assigned incorrect dates because of uncertainty as to when the last menstrual period occurred, irregular cycles, or use of oral contraceptives near the time of conception. Approximately 95% of patients deliver spontaneously within 2 weeks of their estimated date of delivery as predicted by early ultrasound, but only about 75% of patients go into spontaneous labor within 2 weeks of their expected dates of delivery based on menstrual dates. This disparity suggests that a significant number of gravidas are considered to be postterm as a result of inaccurate dating.

Because menstrual dating alone can be a poor predictor of the due date, a careful history must be obtained to uncover factors that may affect the reliability of menstrual dates. Because ovulation occurs 2 weeks prior to menses, the length and regularity of the menstrual cycle affect the accuracy of menstrual dating. The cycle length is also difficult to estimate if the patient has not established her natural cycle after stopping the use of oral contraceptives. Implantation bleeding, which is usually lighter than that for normal menses, can also be mistaken for the last menstrual period. Maternal perception of fetal movement or quickening, which begins at 16–20 weeks of gestation, can help date the pregnancy.

Several physical findings are also useful in establishing the duration of pregnancy:

1. Fetal heart auscultation with Doppler instruments: The fetal heart is normally heard at about 10 weeks of gestation using this method.
2. Uterine size: The size of the uterus correlates with gestation age, with first-trimester assessments being more accurate predictors of pregnancy duration. An obstetrical ultrasound should be performed when uterine size does not agree with gestational age to determine whether the discrepancy is caused by incorrect dates or some other factor.
3. Obstetrical ultrasound: Ultrasound assessment of gestational age is accurate to within 3–5 days during the first trimester, within 1.5 weeks during the second trimester, and within 3 weeks during the third trimester. Ultrasound to determine gestational age should be performed when the menstrual dates may be unreliable.

Prolonged pregnancy refers to gestations in which there is strong corroborating evidence that the postconception age exceeds 40 weeks. Unfortunately, patients often present for health care after 26 weeks' gestation, at a time when physical findings and ultrasound assessment are less useful for dating.

II. PARTURITION

Prolonged gestation most likely results from a delayed maturation of hormonal systems and other factors involved in the labor process.

A. Sheep

In sheep, the trigger for parturition apparently begins in the hypothalamus with increased secretion of corticotrophin-releasing hormone and arginine vasopressin. The resulting increased release of adrenocorticotrophin hormone (ACTH) from the posterior pituitary produces a rise in circulating cortisol concentrations in the fetus. This increase in cortisol induces 17α-hydroxylase synthesis in the placenta, increasing the conversion of progesterone to 17α-hydroxyprogesterone and subsequently to estrogen. As a result, the ewe's plasma progesterone levels fall and estrogen concentrations increase. This decrease in the progesterone/estrogen ratio, which occurs 1 or 2 days prior to labor, promotes prostaglandin synthesis, which in turn facilitates cervical ripening and the onset of labor.

Intravascular infusions of ACTH or cortisol to the pregnant ewes induce premature parturition, whereas ablation of the fetal pituitary prolongs pregnancy. Thus, cortisol plays a critical role in parturition in this species.

B. Primates

The mechanism underlying parturition in primates is somewhat different, although it apparently involves activation of the fetal hypothalamic–pituitary–adrenal axis. Although the primate placenta lacks the 17α-hydroxylase enzyme, it can convert C_{19} steroids, derived from the fetal cortical zone of the adrenals, into estrogen. Androstenedione infusion to rhesus monkeys increases maternal plasma estrogen concentrations and induces uterine contractions, and a rise in dehydroepiandrostenedione sulfate concentrations in fetal blood precedes parturition in this species. Thus, androgens derived from the fetal adrenal appear to be critically involved in initiating parturition.

The gestational length of subhuman primates is increased by fetectomy, fetal decapitation, and glucocorticoid suppression of androgen production by the fetal adrenal, suggesting that the fetus is involved in initiating parturition by modulating estrogen production by the placenta. However, the exact role of estrogen in triggering parturition is unclear because estrogen administration to nonhuman primates does not induce uterine contractions and because a clear reduction in the progesterone/estrogen ratio, either systemic or local, has not been shown to precede the onset of parturition.

Anecephalic fetuses have a variable absence of the diecephalon and the rostral brain stem with hypoplastic adrenals. Although some women carrying anecephalic fetuses have prolonged pregnancies, the average gestational age for the spontaneous parturition for anencephalic fetuses as a whole does not differ significantly from normal. The prolongation of pregnancy associated with a subgroup of anecephalic fetuses presumably results from the lack of hypothalamic and pituitary tissue. However, this interpretation apparently does not apply generally to anecephalic fetuses, and the reason for this disparity remains to be established.

As in sheep and subhuman primates, prostaglandins, derived from amnion (primarily prostaglandin E_2) and decidua (principally $PGF_{2\alpha}$), are critically involved in parturition in women. Prostaglandin concentrations increase in plasma and amniotic fluid during parturition, with elevated levels detected even at the onset of labor. Prostaglandins appear to be directly related to parturition because prostaglandins can induce labor and because inhibitors of prostaglandin synthesis can prolong pregnancy in women with premature labor. Prostaglandin release within the uterus is modulated by inhibitors and promotors of prostaglandin synthesis in amniotic fluid.

The functional expression of prostaglandin endoperoxidase H synthase-2 (PGHS-2), the rate-limiting enzyme of prostaglandin synthesis, increases in amnion and chorion but not in decidua prior to the onset of labor. This increase in PGHS-2 is likely part of the chain of fetal endocrine events that trigger parturition in humans. Thus, there is reason to suspect that a delay in the development of such pathways predisposes gravidas to prolonged pregnancy.

III. COMPLICATIONS

A. Macrosomia

About 25% of postterm gravidas give birth to infants weighing over 4000 g, which is about twice the incidence occurring in normal-term gestations. Fetal macrosomia predisposes the infant to birth trauma related to shoulder dystocia and the mother to perineal trauma and cesarean delivery.

B. Meconium Aspiration

The incidence of meconium passage is also greater in postdate pregnancy. Because postdate gravidas frequently have reduced amniotic fluid, this meconium is often dispersed in less volume, which results in a more particulate fluid. Meconium enters the lungs through gasping efforts associated with fetal asphyxia or through the initiation of air breathing at the time of birth. This aspirated meconium can obstruct bronchi and cause a chemical pneumonitis, which characterizes the meconium aspiration syndrome. Suspicious or ominous fetal heart rate abnormalities have been associated with this life-threatening disorder, which develops in up to 18% of infants delivered with meconium-stained fluid and can cause long-term respiratory morbidity. Although meconium passage occurs in pregnancies of normal length, the risk of developing meconium aspiration syndrome is greater in postdate pregnancy because the newborn is more likely to aspirate thicker meconium.

C. Dysmaturity Syndrome

Newborns with the dysmaturity syndrome, as described by Clifford in 1954, have a loss of subcutaneous fat; dry, wrinkled, and cracked skin; meconium staining of the skin, nails, umbilical cord, and membranes; long nails; and an unusual degree of alertness (Fig. 1). Some characteristics of dysmaturity syndrome are observed in about 20% of postterm infants, an incidence about 10-fold greater than that for normal neonates. Attributing this "postmature syndrome" to placental dysfunction, Clifford alerted clinicians to the associated high rates of perinatal morbidity and mortality.

The increase in postterm perinatal mortality correlates with the development of degenerative changes in placenta. The postterm placenta is more likely to have excessive intervillous thrombosis and fibrin deposition, which can decrease the effective area of exchange between maternal and fetal blood. Other features of placental senescence may also reduce placental transfer of nutrients and contribute to fetal wasting. However, there is very little direct evidence that actually relates these pathological changes to an alteration in mechanisms of placental transfer. Thus, further work is required to determine the role of "placental insufficiency" in dysmaturity.

More is known about the developmental sequelae of dysmaturity. Dysmature infants who escape complications related to perinatal asphyxia or meconium aspiration appear to have normal growth and intelligence up to 7 years of age, although the possibility of more subtle developmental abnormalities cannot be excluded.

D. Oligohydramnios

Oligohydramnios, which commonly occurs in prolonged pregnancy, is a decrease in amniotic fluid volume that has been attributed to fetal hypoxia caused by placental dysfunction. However, clinical evidence suggests that other mechanisms account for the decrease in fluid. For example, the majority of postterm fetuses have normal blood gases as determined from scalp samples obtained before or during labor. Cesarean deliveries for fetal heart rate abnormalities are more commonly performed for patterns (variable decelerations or bradycardias) associated with cord compression rather than from late decelerations related to uteroplacental insufficiency. Thus, the reduction in amniotic fluid commonly observed in prolonged pregnancy apparently results from the poorly understood mechanisms underlying the general decline in amniotic fluid volume that begins after 34–36 weeks of gestation. Oligohydramnios significantly increases the risk of meconium-stained amniotic fluid, growth-restricted fetuses, and cesarean delivery.

E. Fetal Mortality

In 1963, McClure Brown reported that the perinatal mortality rate doubles at 43 weeks and triples at 44 weeks of gestation when compared to that (10.5/1000 live births) at 39–41 weeks' gestation. These mortality rates from England and Wales were derived before routine monitoring of postterm fetuses. Recently, the effect of prolonged pregnancy on fetal

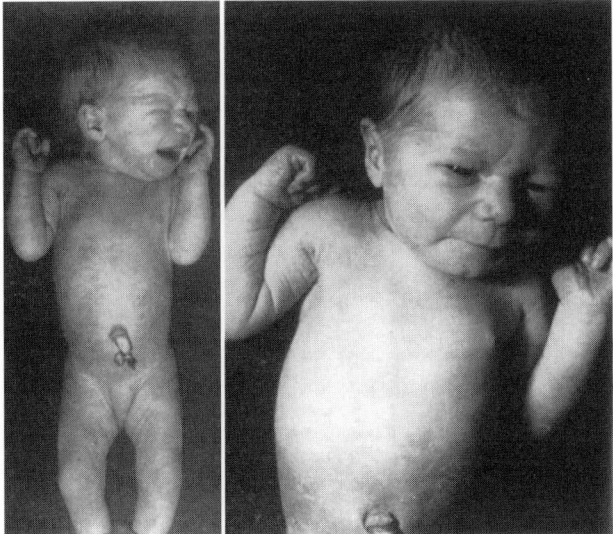

FIGURE 1 A dysmature infant delivered at 305 days of gestation with a long, thin body (left), peeling skin, and an alert appearance (right) (reproduced from Clifford, 1954).

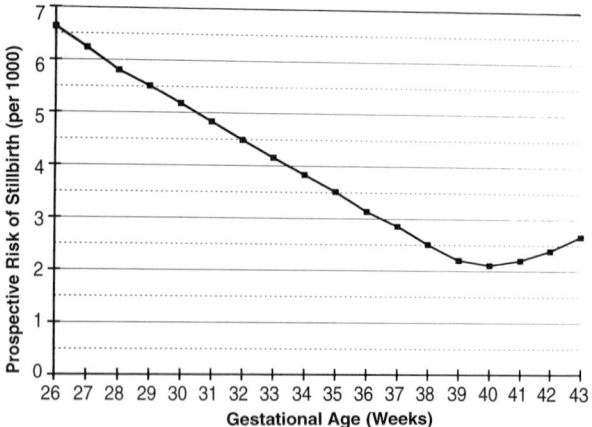

FIGURE 2 Prospective risk of stillbirth as calculated from reported births in New York City from 1987 to 1989 (reproduced from Feldman, 1992).

mortality rate in New York City has been calculated using the "prospective risk of stillbirth" for a given gestational age:

Risk of stillbirth at gestational week N
$$= \frac{\text{Stillbirths} \geqq \text{gestational week } N}{\text{Births} \geqq \text{gestational week } N}$$

In an evaluation of 370,051 births between 1987 and 1989, the prospective risk of stillbirth increased after 40 weeks EGA (Fig. 2). The relative risk was greater in certain ethnic groups, women of advanced maternal age, multiple gestations, and women receiving poor prenatal care.

During most of this century, there has been controversy regarding the risk of prolonged pregnancy and the need for obstetrical intervention. Most studies have not found an appreciable increase in perinatal mortality with expectant management involving fetal surveillance. Because of the low perinatal mortality rate in postdate gestation, a very large number of patients need to be studied to make a plausible case for earlier delivery. Outcome can be more easily estimated by reviewing published randomized trials. In such a meta-analysis involving 5727 women, induction of labor at >41 weeks of gestation was associated with a significantly reduced perinatal mortality rate (0.3/1000) compared to that (2.5/1000) of expectant management.

IV. EVALUATION

Several antenatal tests have been used to assess the well-being of postdate fetuses.

A. Nonstress Test

Fetal heart recording can provide useful information regarding the fetus. A reactive nonstress test (NST) suggests that the fetus is not hypoxic. A nonreactive NST is not reassuring and generally is an indication for delivery in this group of patients. If there is good reason to delay delivery, a contraction stress test or biophysical profile can be performed to further assess fetal status. In postdate pregnancy, particular attention is given to the presence of heart rate decelerations that can occur with reduced amniotic fluid volume, predisposing the fetus to additional risks during pregnancy and labor.

The subjective nature of visual analysis of fetal heart rate records results in a high inter- and intraobserver variation in fetal heart rate. A more reliable analysis of heart rate can be given by computer. Commercially available software for fetal heart rate analysis of heart rate can identify heart rate accelerations, decelerations, and fetal movements. The analysis also provides a measure of short-term fetal heart rate variability, which can be very helpful in assessing fetal health.

B. Contraction Stress Test

Fetal heart rate responses to uterine activity can be determined by inducing contractions through intravenous administration of oxytocin or through nipple stimulation, which promotes the release of endogenous oxytocin release. A negative contraction stress test (CST), which suggests adequate uteroplacental reserve, is associated with a perinatal mortality of 1/1000 or less if performed weekly. An equivocal test is repeated within 24 hr. Because the CST has about a 25% false-positive rate related to ominous fetal heart rate abnormalities in labor, a positive CST with a reactive heart rate usually indicates a need for induction with continuous fetal heart rate monitoring. Cesarean delivery is generally performed if a positive CST is accompanied by a nonreactive fetal heart rate recording. Disadvantages of this method include the need for an intravenous infusion, the time required for study, and the high rate (≈14%) of equivocal tests.

C. Biophysical Profile

Fetal assessment can also be performed using the biophysical profile (BBP). Management of the pregnancy is typically based on the composite score as follows:

10 normal

8 normal

6 If ≧36 weeks, consider induction of labor or repeat testing in 4–6 hr

4 If ≧32 weeks, consider induction of labor or repeat testing in 4–6 hr

0–2 Strongly suggests chronic placental insufficiency. Extend testing to 120 min; if score ≦, deliver regardless of gestational age if the fetus is considered to be viable

The absence of fetal breathing is one of the more sensitive indicators of fetal hypoxia, but prolonged monitoring (up to an hour or more) may be necessary to encounter a breathing episode in normal fetuses. The amniotic fluid index (AFI) can be a useful predictor of perinatal morbidity in postdate pregnancy, with values <5 cm associated with an increased risk of ominous fetal heart rate patterns, meconium aspiration, and cesarean delivery.

D. Doppler

Doppler velocimetry of the umbilical artery and other fetal vessels may be useful in identifying growth-restricted fetuses who require more intensive monitoring. Abnormal Doppler waveforms (i.e., increased S/D ratio or pulsatility index and absent or reversed diastolic flow) of the umbilical artery tend to precede changes in fetal heart rate. Although some have advocated its use in postterm pregnancies, its usefulness in evaluating the postterm fetus has not been critically evaluated.

In prolonged pregnancy, fetal assessment is best performed by combining a test that evaluates fetal health (e.g., nonstress test) with an assessment of amniotic fluid volume. A reduced AFI (<5 cm) is associated with an increased risk of cord compression and dysmaturity syndrome. Although an occasional fetal death will occur despite normal testing, most studies indicate that such testing in postdate fetuses is associated with a perinatal mortality rate approaching that for gestations of normal duration.

Fetal testing is less helpful in preventing the perinatal morbidity of prolonged pregnancy. Although oligohydramnios is often associated with dysmaturity, a normal amniotic fluid volume does not completely exclude the possibility of this syndrome. Thus, these commonly used tests are not entirely reliable in detecting the dysmature fetus and will not eliminate the morbidity related to meconium aspiration, shoulder dystocia, or the need for cesarean delivery.

V. MANAGEMENT

Obstetrical management should take into consideration that the postdate gravida has an increased risk for complications related to labor and delivery. Because these patients have a greater incidence of large fetuses, the obstetrician should be prepared for a shoulder dystocia. Although a cesarean delivery may be performed for all large fetuses with estimated weights of >4500 g, a reasonable approach is to individualize management based on obstetrical history, size of the pelvis, and the labor curve. Midpelvic delivery by forceps or vacuum should be avoided in these patients because of the high incidence of shoulder dystocia.

Special attention should also be given to the management of fetuses with oligohydramnios and meconium-stained fluid. Amnioinfusions have been shown to reduce operative delivery for abnormal fetal heart rate patterns; therefore, these fetuses may benefit from transvaginal infusion of saline into the amniotic sac during labor or transabdominal infusion prior to induction of labor. Upon delivery of the fetal head, the obstetrician should remove meconium by suctioning the nasopharynx and posterior oropharynx. Attendants should also be available to evaluate the newborn and to perform endotracheal suctioning if necessary. Although not entirely preventable, the risk of severe meconium aspiration syndrome can be reduced by such management.

A. Uncertain Dates

If the postconceptional age is uncertain, postterm pregnancies generally should undergo fetal testing while awaiting spontaneous labor.

B. Complicated Pregnancy

In prolonged pregnancy, induction of labor is recommended for gravidas with favorable cervices and for those with medical or obstetrical complications that put them at increased risk for poor outcome.

C. Unfavorable Cervix

The management of prolonged pregnancy in a normal gravida with an unfavorable cervix is more controversial. In this case, the risks of postmaturity must be weighed against the risk of complications relating to induction of labor. Abnormal fetal heart rate testing, oligohydramnios, or a postconceptional age of 41 weeks or more generally indicate the need for labor induction.

Outpatient procedures, such as stripping the membranes and applying prostaglandins to vagina or endocervix, can enhance cervical ripening and the onset of labor. If prostaglandins are given, uterine activity and fetal heart rate should be monitored before and after administration of the drug. The delivery of these patients has also been facilitated by the inpatient use of prostaglandins to ripen the cervix and to induce labor. Prostaglandin preparations are available as a 0.5-mg prostaglandin E_2 (PGE_2) gel (Prepidil, Upjohn, Kalamazoo, MI) for endocervical application or as a vaginal insert (Cervidil, Forest Pharmaceuticals, Inc., St. Louis, MO). The latter releases dinoprostone at controlled rate of 0.3 g/hr over a 12-hr period. Several small studies have also demonstrated clinical safety and efficacy for misoprostol (Cytotec, Searle, Chicago), a prostaglandin E_1 (PGE_1) analog available in 100- or 200-μg tablets. The most commonly used approach is to place 25–50 μg intravaginally every 4 hr until the onset of active labor. When compared with intracervical PGE_2, patients treated with PGE_1 had shorter time from induction of labor to vaginal delivery, and fewer patients required oxytocin augmentation. Mechanical approaches, such as osmotic dilators (laminaria) and the transcervical placement of a Foley catheter bulb, have also been used to ripen the cervix. These methods complement the more traditional approaches of labor induction which have involved artificial rupture of the membranes and oxytocin administration.

As a result of these technical advances, the risk of inducing labor in gravidas with an unfavorable cervix may be less than commonly perceived. For example, in a randomized study of 302 low-risk obstetrical patients with an unfavorable cervix at 291 days of gestation, induction of labor was associated with a lower cesarean delivery rate compared to that for those managed expectantly with fetal monitoring. In this study, labor induction reduced perinatal morbidity by reducing the incidence of ominous fetal heart rate abnormalities, meconium aspiration, postmaturity syndrome, and low Apgar scores.

VI. CONCLUSIONS

Postdate pregnancy is one of the most common disorders requiring fetal assessment. While ultrasound and electronic fetal heart rate monitoring have apparently been useful in reducing the perinatal mortality associated with this condition, these methods of evaluation are not totally reliable in preventing fetal morbidity or mortality. Management should be individualized according to the certainty of the postconceptional age, the state of the cervix, amniotic fluid volume, and the existence of maternal or fetal complications.

See Also the Following Article

PRETERM LABOR AND DELIVERY

Bibliography

Clifford, S. H. (1954). Postmaturity—With placental dysfunction. Clinical syndrome and pathologic findings. *J. Pediatr.* 44, 1–14.

Dawes, G. S., Moulden, M., and Redman, C. (1996). Improvements in computerized fetal heart rate analysis antepartum. *J. Perinat. Med.* 24, 25–36.

Feldman, G. B. (1992). Prospective risk of stillbirth. *Obstet. Gynecol.* 79, 547–553.

Grant, J. M. (1994). Induction of labour confers benefits in prolonged pregnancy. *Br. J. Obstet. Gynaecol.* 101, 99–102.

Lagrew, D. C., and Freeman, R. K. (1986). Management of postdate pregnancy. *Am. J. Obstet. Gynecol.* 154, 8–13.

Leveno, K., Quirk, J. G., Cunningham, F. G., Nelson, S. D., Santos-Ramos, R., Toofanian, A., and De Palma, R. T. (1984). Prolonged pregnancy. I. Observations concerning the causes of fetal distress. *Am. J. Obstet. Gynecol.* 150, 465–473.

Longo, L. D. Disorders of placental transfer. In *Pathophysiology of Gestation* (N. S. Assali, Ed.), Vol. 2, pp. 1–76. Academic Press, New York.

Mannino, F. (1988). Neonatal complications of postterm gestation. *J. Reprod. Med.* 33, 271–276.

Mijovic, J. E., and Olson, D. M. The physiology of human parturition. In *Advances in Organ Biology, Vol. 1*, pp. 89–119. JAI Press, London.,

Mijovic, J. E., Zakar, T., and Olson, D. M. (1997). An exclusive increase in the functional expression of prostaglandin endoperoxide H synthase-2 (PGHS-2) by amnion and chorion but not decidua is associated with the initiation of term labor in humans. *J. Soc. Gynecol. Invest.* 4(Suppl.), 85A.

O'Brien, J. M., Mercer, B. M., Clearly, N. T., and Sibai, B. M. (1995). Efficacy of outpatient induction with low-dose intravaginal prostaglandin E_2: A randomized, double-blind, placebo-controlled trial. *Am. J. Obstet. Gynecol.* 173, 1855–1859.

Phelan, J. P., Platt, L. D., Yeh, S.-Y., Broussard, P., and Paul, R. H. (1985). The role of ultrasound assessment of amniotic fluid volume in the management of the postdate pregnancy. *Am. J. Obstet. Gynecol.* 151, 304–308.

Rayburn, W., Motley, M. E., Stempel, L. E., and Gendreau, R. M. (1982). Antepartum prediction of the postmature infant. *Obstet. Gynecol.* 60, 148–153.

Resnik, R. (1994). Post-term pregnancy. In *Maternal–Fetal Medicine. Principles and Practice* (R. Creasy and R. Resnick, Eds.), 3d ed., pp. 521–526. Saunders, Philadelphia.

Usta, I., Mercer, B. M., and Sibai, B. M. (1995). Risk factors for meconium aspiration syndrome. *Obstet. Gynecol.* 86, 230–234.

Varaklis, K., Gumina, R., and Stubblefield, P. G. (1995). Randomized controlled trial of vaginal misoprostol and intracervical prostaglandin E_2 gel for induction of labor at term. *Obstet. Gynecol.* 86, 541–544.

Vorherr, H. (1975). Placental insufficiency in relation to postterm pregnancy and fetal postmaturity. Evaluation of fetoplacental function; management of the postterm gravida. *Am. J. Obstet. Gynecol.* 123, 67–102.

Postpartum Depression

Joseph F. Mortola

Cook County Hospital

I. Incidence
II. Prenatal Patient Identification
III. Symptoms
IV. Differential Diagnosis
V. Pathophysiology
VI. Treatment

GLOSSARY

baby blues A self-limited episode of unexplained sadness experienced by as many as 60% of women in the postpartum period usually beginning 3 days to a week after delivery.

bipolar disorder A disorder of mood characterized by episodes of both severe depression and mania. The episodes of depression are commonly severe and often of psychotic proportions.

catecholestrogen A compound formed by either 2-hydroxylation or 4-hydroxylation of an estrogen. The compound has weak affinity to both estrogen receptors and catecholamine receptors.

dysthymic disorder A persistent state of feeling sad or unhappy lasting more than 2 years in which the feelings of dejection are present more days than not.

psychotic depression An episode of intense sadness characterized by the presence of delusions or hallucinations.

schizotypal behavior Behavior in a nonpsychotic individual that is highly idiosyncratic and somewhat bizarre. It often includes inability to relate well with others and failure to show a normal range of feelings.

serotonin reuptake inhibitor One of a new class of antidepressants which are highly selective for their effects on serotonin economy. Examples include fluoxetine and paroxetine.

tricyclic antidepressant An older, widely used, and effective class of antidepressant named for their three-ring chemical structures. These antidepressants affect a variety of neurotransmitters, including noradrenelin, dopamine, and acetylcholine; as such, they are not considered "selective." Examples include amitriptyline, imipramine, and nortriptyline.

Postpartum depression is a term applied to a variety of mood changes that occur in women during the time after delivery of a child (the puerperium). These disorders share in common the feelings of sadness and are often manifested by episodes of crying. However, the prevalence of associated symptoms such as anhedonia (loss of interest in almost all activities), psychomotor retardation, sleep disturbance, and appetite disturbance may vary considerably depending on the severity of the depression. There are three degrees of postpartum depression: a minor depressive disorder which is extremely common ("baby blues"), major depressive disorder (nonpsychotic type), and major depressive disorder with psychotic features (Table 1). The latter two forms may be the first occurrence of a serious depressive disorder, an episode that is part of a recurrent major depressive disorder, or a manifestation of bipolar disorder (manic-depressive illness).

I. INCIDENCE

The incidence of the mildest form of the postpartum depressive disorder, the baby blues or maternity blues, is extremely high. Fifty to 80% of women have been reported to experience these symptoms. The high incidence of this disorder in the puerperium strongly suggests a hormonal basis for the disorder. The temporal occurrence of the symptoms of the baby blues closely approximates the decay curves of serum estradiol and serum progesterone in the puerperium. In contrast to the baby blues, the incidence of major depressive disorder, of either the psychotic or nonpsychotic type, is much smaller. Prevalence rates in Britain and North America have been reported to be 6–16%. This prevalence is no

TABLE 1
Classification of Postpartum Depressive Disorders

Disorder	Characteristics
Baby blues	Prevalence 50–80%
	Appears 3–10 days postpartum
	Self-limited (1 or 2 weeks)
	Requires no treatment
	Requires observation for progression to major depression
	No suicidal ideation
Major depression	Previous episodes of depression or family history are risk factors
	Meets *DSM IV* criteria for depression
	Requires treatment
	Tricyclic antidepressants preferred
	May have suicidal thoughts or acts
Nonpsychotic type	No hallucinations or delusions
Psychotic type	Hallucinations or delusions
	Hospitalization usually required
	Risk of infanticide
	History of bipolar disorder a risk factor

higher than that which has been reported for women in general, in which the prevalence is 16–26% using *DSM IV* criteria. The rate of depression in women in general is higher than that in men (approximately 15 vs 7%). However, it has not been demonstrated that this increased prevalence is due to the postpartum state experienced exclusively by women. There is also no conclusive evidence that women with bipolar disorder are more prone to manic or depressive episodes in the puerperium. Psychotic depressions, while extremely dangerous for both mother and child, are relatively rare occurrences in the puerperium, with an incidence of 1 or 2/1000 births.

II. PRENATAL PATIENT IDENTIFICATION

Methods for prepartum diagnosis of patients at risk for the mildest form of postpartum depression, the baby blues or maternity blues, are not currently available. In contrast, since major postpartum depression is likely to be related to an underlying depressive disorder, or biologic predisposition to depression, careful questioning of patients as to prior depressive episodes, or a family history of depression, is the most reliable method of assessing patients at risk. Such patients should be tracked carefully in the puerperal period for the onset of depressive symptoms.

III. SYMPTOMS

The symptoms of the baby blues include sadness, agitation, irritability, loss of appetite, and mood swings. Panic episodes are not infrequently observed. Typically, the symptoms appear 2 or 3 days after delivery, peak within 7 days, and remit substantially by 2 weeks postpartum. The symptoms of the baby blues most often mimic those observed in agitated depression. As such, lethargy is less common than nervousness and fidgeting. The presence of suicidal thinking should alert the clinician to the presence of a major depression rather than the baby blues since suicidal thoughts rarely accompany the latter. Other hallmarks of a major depressive disorder are (i) the prolonged duration of the symptoms and (ii) interference with normal functioning or ability to care for the child (Table 2). The presence of suicidal

TABLE 2
Criteria for Major Depressive Episode[a]

Both of the following symptoms for a 2-week period
   Depressed mood nearly every day by self-report or report of others
   Markedly decreased interest in almost all activities
At least three of the following symptoms lasting for the same 2-week period
   Decrease or increase in appetite nearly every day
   Insomnia or hypersomnia nearly every day
   Psychomotor agitation (more common in postpartum) or psychomotor retardation nearly every day
   Fatigue or loss of energy nearly every day
   Feelings of worthlessness or inappropriate guilt nearly every day
   Diminished ability to concentrate or indecisiveness nearly every day
   Recurrent thoughts of death or suicide without a plan or any suicide plan or attempt
Symptoms not due to substance abuse or medical condition

[a] Adapted from *DSM IV* American Psychiatric Association (1987).

thinking connotes a depression of sufficient severity to be diagnosed as a major depressive disorder, although it does not distinguish between nonpsychotic and the more serious psychotic variant of the illness. The diagnosis of a psychotic postpartum depression is made when there are either delusions or hallucinations.

IV. DIFFERENTIAL DIAGNOSIS

The differential diagnosis of postpartum depression includes anxiety disorders, dysthymic disorder, thyroid disorder, metabolic disturbances, substance abuse, sleep deprivation, and, in the psychotic form of the disorder, schizophrenia. The diagnosis of postpartum depression is aided by the prepregnant symptom profile. Patients with schizophrenia will usually show evidence of thought disorder and bizarre or schizotypal behavior before the postpartum period. Similarly, dysthymic disorder is characterized by long-standing (usually longer than 2 years) periods of sadness and unhappiness. A prior history of depression or anxiety disorder is extremely helpful in differentiating the postpartum episode as one of acute anxiety or depression. In patients in whom substance abuse is suspected, toxicology screening is necessary, and the clinician is mandated in most localities to report the suspected abuse to appropriate child-protective agencies.

The prevalence of thyroid disease in the postpartum period is increased. Up to 6% of women postpartum exhibit evidence of transient thyroid disturbance, which may be either of the hyperthyroid or hypothyroid variety. This has been shown to be correlated with an increase in circulating anti-microsomal antibodies. A thyroid-stimulating hormone (TSH) level is therefore required in the evaluation of all patients in whom major depression is suspected postpartum. Since a serum TSH level is the most sensitive indicator of thyroid disease, no further thyroid function studies are required if the TSH is normal.

The evaluation of the patient with postpartum depression should also include a careful physical examination to exclude metabolic disturbances, such as diabetes or electrolyte imbalance, as well as a complete blood count to evaluate the extent of anemia.

V. PATHOPHYSIOLOGY

The pathophysiology of postpartum depressive disorders is not definitively known. Nonetheless, the abrupt decline in plancental steroid production which occurs immediately postpartum is believed to be a potent stimulus for changes in mood-altering neurotransmitters. Estrogen has long been recognized to increase endogenous opiate peptide levels in the central nervous system; an effect potentiated by progesterone in estrogen-primed animal model systems. An abrupt decline in estrogen, as during the postpartum period, would be expected to produce opiate-withdrawal type symptoms including the type of agitated depression that is characteristic of postpartum depressive disorders. Recently, the effects of estrogen on catecholamine economy have been elucidated. Estrogen influences both adrenergic receptor expression and catecholamine metabolism. Decreased catecholamine synthesis occurs through catecholestrogen competition with tyrosine β-hydroxylase, the rate-limiting enzyme in catecholamine synthesis. In addition, catecholestrogens can compete with catechol-*O*-methyl transferase and inhibit catecholamine degradation. In addition to this mechanism, it is well established that the aforementioned estrogen-induced increases in the endogenous opiate peptide system are responsible for secondary alterations in noradrenergic neurons.

Among the most ubiquitous mood-influencing neurotransmitters in the central nervous system are the GABA-ergic and cholinergic systems. Biochemical, pharmacological, and morphological methods have all confirmed that estrogen-responsive neurons use GABA as a neurotransmitter. The GABA system has been shown to have considerable anxiolytic effects. Presumably, with the sudden postpartum decline in estrogen levels, this system is significantly disrupted. Like GABA, acetylcholine is also influenced by estrogen, particularly through induction of the acetylcholine-synthesizing enzyme acetyl transferase. Decreased acetylcholine during the postpartum period would also be expected to cause depression and anxiety.

Progesterone also has mood-altering effects. In general, administration of high doses of progesterone, such as those seen in pregnancy, serve to have anxiolytic effects. This is believed to be mediated largely through progesterone metabolites, which have been shown to be benzodiazepine-like modulators of the GABA–receptor complex. Overall, there is compelling evidence that the abrupt and dramatic changes in estrogen and progesterone in the postpartum period are sufficient to cause mood alterations through a variety of neurotransmitter mechanisms.

VI. TREATMENT

The usual course of the baby blues is relatively brief and self-limited. It is therefore both unnecessary and unwarranted to begin pharmacologic treatment. Nonetheless, the importance of providing a watchful eye on the patient with any depressive disorder in the postpartum state cannot be overemphasized. Patients may progress from this condition to disabling depression and even psychosis and infanticide. As depression worsens, the ability of the patient to have insight into her disorder and her ability to make an emotional bond with the professional caretaker decrease. Thus, early intervention can be not only beneficial but also life saving.

When patients meet criteria for major depressive disorder, treatment is warranted. The optimal intervention includes both psychological counseling and a strong consideration of the use of antidepressants. There are usually psychological issues superimposed upon, and perhaps contributing to, the biological alterations that accompany the depressive state. Often, the stress of reorganizing the home to accommodate a new baby, fears of being an inadequate parent, and changes in identity that ensue upon leaving the workforce can be assuaged by the therapist. Patients who meet the criteria for major depression should also be offered a trial of antidepressants. The institution of antidepressants may be particularly difficult in the breast-feeding mother because antidepressants are excreted in the breast milk. The long-term effects of these agents on the development of neurotransmitter systems are largely unknown, as are the effects on psychomotor development of the child. However, there is also evidence that the tricyclic antidepressant nortriptyline, which the infant may receive through the breast milk, does not accumulate in the infants serum. For this reason Sichel *et al.* recommend that nortriptyline be used in mothers who continue to breast-feed.

In general, however, the largely unknown long-term effects of psychotropic agents and the reported ill-effects of some tricyclic antidepressants on the infants of breast-feeding women require careful informed consent before prescribing the drug.

Although there is a paucity of randomized data, the use of a tricyclic antidepressant such as nortriptyline is the preferred initial agent for treatment of postpartum depression. Newer agents in, or related to, the serotonin-reuptake inhibitor class of antidepressants, such as fluoxetine and paroxetine, may initially cause agitation in some patients. Since postpartum depression is often an agitated depression, these agents appear less suitable as a first-line agent.

The usual starting dose of nortriptyline is 25 mg at night, with this dose increased every 3 days by 25 mg until a 75-mg daily dose is achieved. At that point, it is reasonable to allow 1 or 2 weeks to achieve efficacy. In cases in which this does not occur, the dose may be increased or an alternative antidepressant chosen.

Patients with major depression in the postpartum state may exhibit treatment resistance. In these cases dual-agent therapy may be required. In the most resistant cases, electroconvulsive therapy may be required simultaneous with, or followed by, antidepressants. If the patient has a history of bipolar disorder or presents with a manic episode, lithium should be instituted.

Hospitalization should be considered in postpartum patients with major depression who lack adequate social support. In addition, hospitalization is required in psychotic depressions accompanied by suicidal or infanticidal thoughts and in nonpsychotic depressions where the degree of suicidal or homicidal ideation is considered to constitute a real threat of harm to the woman or infant. It may be lethal to assume that a postpartum women will not commit suicide because she has a new baby for whom to care. On the contrary, the depressed women may believe that she is such a failure as a parent that the child would be better off without her, and she may act on this idea.

Bibliography

American Psychiatric Association (1987). *Diagnostic and Statistical Manual of Mental Disorders*, 4th ed., pp. 213–223. American Psychiatric Association Press, Washington, DC.

Amino, N., Mori, Y., Iwantani, O., *et al.* (1982). High prevalence of transient postpartum thyrotoxicosis and hypothyroidism. *N. Engl. J. Med.* **306**, 849–852.

Brown, B. J. (1956). Urinary excretions of estrogen during pregnancy, lactation and the reinitiation of menstruation. *Lancet* **1**, 704.

Cox, J. L., Conner, Y., and Kendall, R. E. (1982). Prospective study of the psychiatric disorders of childbirth. *Br. J. Psychol.* **140**, 111–118.

Fein, H. G., Goldman, J., and Weintraub, B. D. (1985). Postpartum lymphocytic thyroiditis in American women: A spectrum of thyroid dysfunction. *Am. J. Obstet. Gynecol.* **138**, 504–509.

Gardner, D. K., and Rayburn, W. F. (1980). Drugs in breast milk. In *Drug Therapy in Obstetrics and Gynecology* (W. F. Rayburn and F. P. Zuspan, Eds.), pp. 175–196. Appleton-Century-Crofts, Norwalk, CT.

Hamilton, R. M. (1962). *Postpartum Psychiatric Problems*. Mosby, St. Louis, MO.

Kendall, R. E., Chalmers, J. C., and Platz, E. (1987). Epidemiology of puerperal psychoses. *Br. J. Psychol.* **150**, 662–666.

Kumar, R., and Ronson, J. M. (1984). A prospective study of emotional disorders in childbearing women. *Br. J. Psychol.* **144**, 35–43.

Mortola, J. F. (1989). Use of psychotropic agents in pregnancy and lactation. *Psychiatr. Clin. North Am.* **12**, 69–87.

Shearer, W. T., Schreiner, R. L., and Marshall, R. E. (1972). Urinary retention in a neonate secondary to nortriptyline. *J. Petitur.* **81**, 570–572.

Sichel, D. A. (1995). Postpartum psychiatric disorders. In *Primary Care of Women* (K. J. Carlson and S. A. Eisenstat, Eds.), pp. 394–399. Mosby, St. Louis, MO.

Postpartum Period

see Puerperium

Prader–Willi Syndrome

Shahab S. Minassian

Allegheny University of the Health Sciences

I. History/Clinical Presentation
II. Etiology and Genetics
III. Diagnosis
IV. Management

GLOSSARY

DNA methylation Process which has been implicated in the control of genomic imprinting.

fluorescent in situ hybridization The hybridization of a specific DNA probe to a standard metaphase preparation on a microscope slide. Site-specific labeling of only that area of chromosome complementary to the probe occurs.

genomic imprinting A process occurring during gametogenesis by which specific genes are differentially marked. The resulting expression of these genes is dependent on the parent of origin.

high-resolution chromosome analysis Laboratory technique intended to yield less condensed mitotic figures. More detail will then be seen after banding techniques.

uniparental disomy Inheritance of two copies of a chromosome from one parent.

In 1956 Prader, Labhart, and Willi were the first to report a syndrome characterized by the findings of obesity, mental retardation, infantile hypotonia, short stature, delayed motor and language development, and cryptorchidism. It is now known that this complex disorder is characterized by two phases during its natural history; the severe hypotonia and feeding problems of infancy are followed by a childhood plagued with insatiable appetite and obesity. Hundreds of studies and several thousand cases have since been published on the Prader–Willi syndrome (PWS). The incidence of PWS has proven to be relatively rare and has been variably estimated at 1:10,000 to 1:25,000. The etiology of this syndrome was poorly understood until the 1980s, when cytogenetic abnormalities responsible for PWS, specifically on the long arm chromosome 15, were discovered. A small deletion of the paternal chromosome is responsible for most of the affected individuals. Recently, more advanced methods of cytogenetic study have revealed new information about the genetic anomalies associated with PWS. In the following review, the etiologies, clinical course, diagnosis, and management of PWS will be covered.

I. HISTORY/CLINICAL PRESENTATION

A. Overview

The diagnosis of Prader–Willi syndrome (PWS) requires a detailed evaluation of the natural history and physical findings. This disorder, like many others, will vary in its presentation, making the diagnosis occasionally difficult. Also, the clinical features and symptoms of PWS during infancy will differ greatly from those of childhood and adulthood. Two clinical phases of the disorder, therefore, have been described.

B. Prenatal Course

Gestations yielding subsequent PWS have been associated with several perinatal complications. The most common is a high incidence of decreased fetal activity, ranging from 54 to 86%. Another frequent finding is that of breech presentation with an overall incidence of 33%. Mild polyhydramnios and joint contractures have also been reported. These complications are thought to be related to the severe hypotonia which characterizes PWS. They are also frequent prenatal indications for the clinician to offer cytogenetic testing, especially in the newborn. A high preterm delivery rate has been reported as common in one publication, but no confirmation of this complication has been reported.

C. Infancy Phase

This first phase is quite dramatically characterized by marked muscle hypotonia in virtually all patients. The hypotonia results in a weak suck and sucking reflex, with subsequent feeding difficulties requiring intravenous or tube feeding in most cases. Diminished arousal is common. The infant fails to thrive. The regulation of body temperature is dysfunctional, resulting in hypothermia or profuse sweating. Delayed psychomotor development occurs in the large majority of individuals and is most likely of central nervous system origin. Milestones such as crawling, walking, and especially speech lag behind controls. Hypogonadism and genital hypoplasia are major physical findings of PWS in infancy. Noted features are cryptorchidism and a small penis in males and a small clitoris and hypoplastic labia in females. Sertoli cell-only patterns have been reported in many testicular biopsies. No comparable series have been reported in females. These findings persist into childhood and adulthood. Distinct craniofacial features can be found in most PWS individuals. Upward slanting or almond-shaped palpebral fissures are the most common. A small forehead and small mouth have been described. Interestingly, PWS individuals with the paternal deletion etiology are frequently hypopigmented, having fair hair and skin color regardless of family characteristics. The 15q11–13 locus has been implicated in the appearance of this finding.

D. Childhood/Adult Phase

After an average of 2 years PWS children undergo a dramatic change in their symptoms as the infantile hypotonia resolves. The onset of hyperphagia is most important and is followed by an uncontrolled weight gain (Fig. 1). The impression of most clinicians is that they do not sense satiety and will continue to eat when presented with food. Gorging, sneaking of food, food foraging, and food preoccupation have

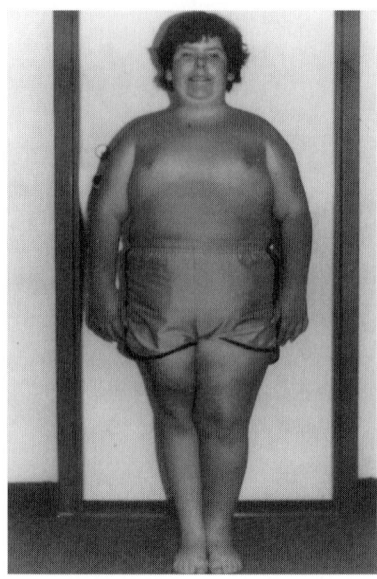

FIGURE 1 Characteristic obesity and facial appearance of a boy with Prader–Willi syndrome (reproduced with permission from Rimoin et al., 1996).

been described. Violent tantrums may occur when dietary modification are attempted. A decreased caloric requirement for PWS patients has been reported to contribute to the obesity. Truncal fat distribution is a major characteristic of PWS, with the distal extremities spared. The massive obesity that afflicts PWS individuals eventually may cause severe cardiopulmonary disease, and indeed this outcome is the major cause of the shorter life span seen in untreated patients. Pickwickian syndrome, with hypoventilation and somnolence, has been reported along with hypercapnea, hypoxia, and right-sided heart failure. No abnormalities of fat metabolism have been found. Diabetes mellitus is eventually diagnosed in up to 20% of PWS patients. It is of the type II class clinically, and insulin resistance is characteristic. This sequela is treatable with weight loss and oral hypoglycemics.

Another important feature of PWS is short stature, which is found in up to 95% of patients. Growth falls below the 50th percentile by the third year, then follows parallel and low to the normal growth curves. This results in most patients reaching young adulthood below the 5th percentile. Bone age is delayed in childhood. Growth hormone (GH) secretion has been evaluated with basal levels and dynamic testing to isolate an etiology for the deficient growth. Generally, no GH deficiency has been shown.

The hypogonadism of PWS infancy continues into childhood. Genital hypoplasia of either gender is common. In boys, cryptorchidism is found in at least 80% of patients. Puberty in both sexes is usually delayed, although normal and even precocious puberty have been reported. Sparse axillary and pubic hair is noted in both sexes. The development of secondary sex characteristics in males is incomplete, with poor hair distribution and lack of voice change. Sexual functioning is essentially nil, with lack of erection and ejaculation. In females, secondary sexual development will occur. Breast development, however, can be small for age. Delayed menarche or amenorrhea is very common, and when menstruation does occur, patients are invariably oligomenorrheic. Fertility has never been proven in confirmed PWS patients. Basal reproductive hormone studies reveal low circulatory follicle-stimulating hormone, luteinizing hormone, testosterone (in males), and estradiol (in females) Acute dynamic testing using gonadotropin-releasing hormone (GnRH) in subjects with pubertal delay results in a blunted gonadotropin response. However, prolonged stimulation with clomiphene citrate has resulted in increased gonadotropin and sex steroid levels in both genders. The overall evidence strongly suggests deficient GnRH secretion resulting in hypogonadotropic hypogonadism. Although no specific anatomic or functional defect has been demonstrated, a recent study has reported a decrease in paraventricular nucleus volume in five PWS autopsies. This finding, in such a vital area of the reproductive process, may lead to greater understanding of PWS hypogonadism.

Behavioral problems and mental retardation are a hallmark of PWS. In contrast to the friendliness and cheerful moods reported in infancy, the childhood phase is plagued with outbursts of temper, mood changes, and irritability. The majority of individuals are prone to automutilation (skin picking or scratching) and have episodes of somnolence. Cognitive disabilities are universal and appear similar to those seen with learning disabilities. IQ testing has found borderline, mild or moderate mental retardation in over 90% of PWS patients; 50% of IQ scores are between 30 and 50.

Other clinical features reported in higher incidence are scoliosis, strabismus, myopia, iris translucency, and dental problems.

II. ETIOLOGY AND GENETICS

The search for the etiology of PWS, and a genetic one in particular, has been long and intensive. This process has led to discoveries which have helped shape a new picture of genetic inheritance. The first reports of a cytogenetic cause associated with PWS were published in the late 1970s. Subsequent high-resolution cytogenetic studies demonstrated that a small interstitial deletion located on the long arm of chromosome 15, in the 15q11–13 region, is responsible (Fig. 2). Approximately 60% of patients have the deletion, and up to 15% of patients with apparently normal chromosomes will show a deletion with the use of molecular studies. These deletions are usually sporadic and *de novo*, although rare familial translocations have been reported. Virtually all deletions have now been shown to occur on the paternally derived chromosome. The PWS deletion most likely occurs during spermatogenesis at the time of meiosis. Additionally, the remaining 25% of PWS patients have only maternal copies of the involved chromosome 15 locus, an event termed uniparental disomy. Nondisjunction is the likely cause for this event. Concurrent studies on patients with Angelman's syndrome (a neurobehavioral disorder of mental retardation, seizures, paroxysms of laughter, and pup-

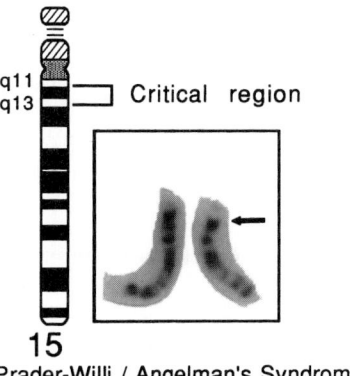

FIGURE 2 Karyotype and ideogram of chromosome 15 displaying the deletion responsible for Prader–Willi syndrome (reproduced with permission from Rimoin *et al.*, 1996).

pet-like movements) found a 15q deletion of the maternally derived chromosome. Thus, both parental copies of this chromosome are required for normal development. This phenomenon has been termed genomic imprinting, a process which temporarily marks a parental gene so that it differs from the other parent's gene. DNA probes and fluorescent *in situ* hybridization (FISH) techniques have isolated the 15q11–13 critical region for PWS. DNA methylation has been implicated as a method by which imprinting occurs; recent studies have found a difference in DNA methylation of parental alleles in the PWS region of chromosome 15. The cytogenetic study of PWS has helped open a new avenue of investigation which may modify the traditional views of Mendelian inheritance.

III. DIAGNOSIS

A. Clinical

The diagnosis of PWS begins clinically. A scoring system by Holm *et al.* (1993) for each of the two historical phases has been published to assist clinicians when the syndrome is suspected. In particular, the profound hypotonia and tube-feeding requirements noted previously are critical findings. Other causes for infantile hypotonia must be considered in the differential diagnosis, including brain injuries, intracranial hemorrhage, and cerebral malformations. Facial characteristics alone cannot be relied on. In the childhood/adult stage, Laurence–Moon–Beidl syndrome may offer a similar presentation, but despite the sharing of some major characteristics, the features of retinitis pigmentosa and polydactyly are uniquely different from those of PWS. Overall, the clinical suspicion for PWS should be reached relatively easily provided the clinician is aware of the salient features.

B. Cytogenetic

Once a clinical suspicion of PWS exists, a laboratory diagnosis must confirm the disorder. Conventional cytogenetic testing is not recommended. High-resolution chromosome analysis, which yields less condensed mitotic figures and thus more detail, combined with FISH will rapidly uncover deletions. If a deletion is found, the paternal karyotype is sampled and compared. It will not, however, diagnose the 25% of patients with maternal uniparental disomy. These patients, who will have negative high-resolution/FISH studies, must be diagnosed by more specific molecular techniques. Should the FISH results be negative, testing to confirm uniparental disomy should be performed. Such techniques to confirm the lack of DNA methylation patterns (and thus the absence of paternal 15q11–13, the PWS deletion) include DNA probe methods and, recently, a reverse transcriptase-polymerase chain reaction test. The latter detects the lack of a gene in the 15q11–13 region which apparently disrupts the methylation/imprinting process.

IV. MANAGEMENT

PWS is a disorder characterized by numerous medical and social problems for the patients and loved ones. Ideally, a multidisciplinary approach with subspecialty care in a dedicated PWS clinic is desirable. Pediatric, psychological, nutritional, behavior management, and endocrine consultants are particularly needed. Social work and educational specialists can also be invaluable team members.

Newborn hypotonia usually requires a prolonged hospitalization and tube feeding. Physical therapy to prevent muscular atrophy and joint contractures is encouraged. Once resolution of the hypotonia occurs, obesity and behavioral problems become great concerns. An early diagnosis helps to prospectively manage the increased food intake and avert the morbid obesity responsible for earlier mortality. Strict parental supervision and restriction of access to food by locking of cabinets and refrigerators is essential. Educating school contacts is important. Once morbid obesity is established weight reduction may be very difficult. Behavioral management and energy restriction by qualified nutritional specialists have had some modest success. Other forms of effective treatment have included gastric bypass surgery in extreme cases. Pharmacologic treatment using appetite suppressants and naltrexone have yielded only mixed results. GH has been administered to treat the short stature of PWS and has been found efficacious in several short-term studies. Long-term studies on final adult heights and drug safety are needed to confirm the benefit of this approach.

Hypogonadism for both genders has been treated by replacement of sex steroids. In males, penile hypoplasia has been corrected with courses of intramuscular testosterone injection every few years. Late spontaneous testicular descent in childhood has been reported in a minority of patients, prompting some clinicians to delay the treatment of cryptorchidism. However, for psychological reasons some individuals may require earlier intervention, usually in the form of orchidopexy or intramuscular hCG. Because of the increased risk of malignant transformation of intraabdominal testes after childhood, orchidectomy may be considered after adolescence. Since puberty itself can be incomplete or delayed, exogenous sex steroid supplementation is a treatment consideration for some patients and especially for males, who usually require testosterone supplementation to complete their normal maturation.

Given the intense medical and emotional struggles with the upbringing of PWS patients, support mechanisms for parents are critical. In addition to medical team management, parent support groups worldwide have filled this need admirably. PWS associations in the United States and Europe offer information and services and can be easily located on the Internet. Increased physician awareness, continued advances in genetic research, and application of successful clinical trials will provide a more promising future for early diagnosis and effective treatment of this devastating syndrome.

See Also the Following Articles

Cryptochordism; GnRH (Gonadotropin-Releasing Hormone); Hypogonadism

Bibliography

ASHG/ACMG Report (1996). Diagnostic testing for Prader–Willi and Angelman syndromes: Report of the ASHG/ACMG Test and Technology Transfer Committee. *Am. J. Hum. Genet.* 58, 1085–1088.

Butler, M. G., Meaney, F. J., and Palmer, C. G. (1986). Clinical and cytogenetic survey of 39 individuals with Prader–Labhart–Willi syndrome. *Am. J. Hum. Genet.* 23, 793–809.

Cassidy, S. B. (1986). Prader–Willi syndrome. *Curr. Prob. Pediatr.* 14(1), 1–55.

Donaldson, M. D. C., Chu, C. E., Cooke, A., Wilson, A., Greene, S. A., and Stephenson, J. B. P. (1994). The Prader–Willi syndrome. *Arch. Dis. Child* 70, 58–63.

Holm, V., Cassidy, S. B., Butler, M. G., Hanchett, J. M., Greenswag, L. R., Whitman, B. Y., and Greenberg, F. (1993). Prader–Willi syndrome: Consensus diagnostic criteria. *Pediatrics* 91, 398–402.

Mascari, M. J., Gottleib, W., Rogan, P. K., Butler, M. G., Waller, D. A., Armour, J. A. L., Jeffreys, A. J., Ladda, R. L., and Nicholls, R. D. (1992). The frequency of uniparental disomy in Prader–Willi syndrome: Implications for molecular diagnosis. *N. Engl. J. Med.* 326, 1599–1607.

Nicholls, R. D. (1994). New insights reveal complex mechanisms involved in genomic imprinting. *Am. J. Hum. Genet.* 54, 733–740.

Rimoin, D. L., Connor, J. M., and Pyeritz, R. E. (Eds.) (1996). Prader–Willi syndrome. In *Emery and Rimoin's Principles and Practice of Medical Genetics*, 3rd ed., Vol. 1, pp. 1007–1008. New York.

Sapienza, C. (1990, October). Parental imprinting of genes. *Sci. Am.*, 52–60.

Swaab, D. F., Purba, J. S., and Hofman, M. A. (1995). Alterations in the hypothalamic paraventricular nucleus and its oxytocin neurons (putative satiety cells) in Prader–Willi syndrome: A study of five cases. *JCEM* 80, 573–579.

Wevrick, R., and Francke, U. (1996). Diagnostic test for the Prader–Willi syndrome by SNRPN expression in blood. *Lancet* 348, 1068–1069.

Wharton, R. H., and Loechner, K. J. (1996). Genetic and clinical advances in Prader–Willi syndrome. *Curr. Opin. Pediatr.* 8, 618–624.

Precocial Birds

see Altricial and Precocial Development

Preeclampsia/Eclampsia

Everett F. Magann and James N. Martin, Jr.

University of Mississippi Medical Center

I. Introduction
II. Pathophysiology
III. Laboratory Assessment
IV. Target Organ Involvement
V. Management

GLOSSARY

antiphospholipid syndrome Antibodies directed against phospholipids in cell membranes associated with an increased risk of fetal loss.

coagulation cascade The sequential process of blood clot formation in the vascular system.

eicosanoids All of the products of the metabolism of arachidonic acid and prostaglandins.

HELLP Hemolysis, elevated liver enzymes, and low platelets.

microangiopathic hemolytic anemia Pathologic process causing anemia from destruction of red blood cells in the microcirculation.

mitogen Protein derived from plants that is used in the laboratory to stimulate cells to divide.

perinatal The period of time between the 20th week of pregnancy through the 28th day following birth.

procoagulant A factor that can assume the role of factor VIII in the coagulation cascade.

Preeclampsia is a complication of pregnancy characterized by hypertension, edema, and/or proteinuria. Hypertension is the most common medical condition associated with pregnancy, with a reported incidence of 5–10%. Preeclampsia as a disorder is unique to human pregnancy since it has not been observed in subprimates. This uniqueness has hindered research and study of this entity. Without the availability of suitable laboratory animals to study this disease, experimentation is limited and answers must be derived from observations of pregnant women and limited human clinical trials. Inquiry about the etiology and management of preeclampsia is important because hypertension during gestation remains an important cause of serious maternal morbidity and mortality. Maternal mortality associated with preeclampsia is primarily due to complications of abruptio placentae, hepatic rupture, or eclamptic seizures. Perinatal morbidity and mortality also are increased due to the direct effect of hypertension on the fetus itself and the complications engendered by preterm delivery. Because the only definitive treatment for preeclampsia is delivery of the placenta, preterm delivery often is necessary to preserve maternal health.

I. INTRODUCTION

Preeclampsia/eclampsia remains, as it has for nearly 50 years, one of the most serious complications of pregnancy. Despite the efforts of hundreds of thousands of investigators, the precise etiology of preeclampsia remains unknown. Preeclampsia entails a wide range of clinical features which are the result of profound disruptions within multiple organ systems. Preeclampsia occurs only during pregnancy and requires a placenta for its initiation. The importance of the placenta to the genesis of preeclampsia is demonstrated clearly in affected pregnancies that have placental components but without a fetus such as the gestation with hydatiform mole. These pregnancies are at an even greater risk for the development of preeclampsia than pregnancies with a fetus and the preeclampsia which develops occurs earlier in pregnancy and is more severe. The amount of placental tissue present also affects the appearance and severity of preeclampsia. As demonstrated in multiple gestations and those with Rh isoimmunization or diabetes mellitus, the amount of placental tissue usually is increased, which adds to the greater risk of developing preeclampsia.

A. Risk Factors for the Development of Preeclampsia

This is a pregnancy complication which is more common in women who are at the extremes of reproductive age, specifically <20 or >35 years of age. The incidence of this disorder in women with their first pregnancy is 10–14%, whereas in women with a previous pregnancy the risk is 5–7%. Women characterized by African American race, obesity, preexisting hypertension and/or renal disease, multiple gestation, elevated blood pressure in a previous pregnancy, elevated systolic blood pressure at the first prenatal visit, or with antiphospholipid syndrome are at increased risk for the development of preeclampsia. First-degree relatives of women who have had preeclampsia, (sisters, mothers, and daughters; not sisters-in-law or daughters-in-law) are also at increased risk for the development of preeclampsia in pregnancy. This increased risk in blood relatives suggests an autosomal recessive or autosomal dominant mode of inheritance with variable penetrance. Cigarette smoking, although it increases the risk of many other complications including early fetal loss, intrauterine growth restriction, and stillborn at term, has been associated with a decreased risk of developing preeclampsia.

B. Terminology

The terms used to identify hypertension in pregnancy are confusing and not uniform. This disorder was known for many years as "toxemia of pregnancy" because it was believed that a toxic substance was circulating in the blood creating this disease. The term "pregnancy-induced hypertension" is now frequently used but fails to differentiate between preeclampsia and gestational hypertension, so more specific terms must be used for better communication.

In order for uniformity in studies and for researchers and clinicians to be able to interpret treatments and results, an attempt has been made to classify the hypertensive disorders of pregnancy. The hypertensive disorders complicating pregnancy were divided into four distinct categories by the National High Blood Pressure Education Working Group in 1990: preeclampsia/eclampsia, chronic hypertension, chronic hypertension with superimposed preeclampsia, and gestational hypertension.

Preeclampsia is the triad of hypertension, proteinuria, and/or edema occurring after the 20th week of pregnancy. Hypertension is defined as an absolute, sustained (≥ 6 hr) stystolic blood pressure of ≥ 140 and/or diastoic blood pressure of ≥ 90 mm Hg. Proteinuria is defined as ≥ 300 mg of protein in a 24-hr specimen. Edema is fluid collection outside the vascular system in the body's third space. In pregnancy, the majority of women will have pretibial edema secondary to restricted venous return from the legs. This impairment in venous return is caused by pressure of the gravid uterus on the vena cava. The edema of preeclampsia is the nondependent, more central edema of the face and hands. This is often characterized as facial puffiness and the inability to wear rings. Rapid weight gain in pregnancy (≥ 5 pounds in 1 week) may also be suggestive of fluid retention and impending preeclampsia.

For a patient to be classified with severe preeclampsia rather than mild preeclampsia, she must show evidence of at least two of the following: blood pressure $\geq 160/110$ on two occasions 6 hr apart; ≥ 5 g of protein in a 24-hr urine collection; right upper quadrant or midepigastric pain; severe headaches; persistent, worsening visual changes; intrauterine fetal growth restriction; oliguria (500 ml or less in 24 hr); and thrombocytopenia with platelets $\leq 100,000/$ml. The addition of seizures in the pregnant woman with mild or severe preeclampsia in the absence of a history of either a neurologic disorder or head trauma is defined as eclampsia.

Chronic hypertension is defined as hypertension that is present in the absence of pregnancy, prior to 20 weeks gestation; hypertension which persists beyond 42 days postpartum; or eye, renal, or laboratory changes that are detected which are consistent with the diagnosis of chronic hypertension.

Chronic hypertension with superimposed preeclampsia is the development of preeclampsia (hypertension, proteinuria, and/or edema) in a pregnant woman with known chronic hypertension. This is a particularly worrisome group of patients because there exists the greatest risk of an adverse maternal or fetal outcome.

Gestational hypertension is the elevation of blood pressure only, in a pregnancy beyond 20 weeks or in the immediate postpartum period without proteinuria or other manifestations of preeclampsia. Gestational hypertension disappears within the first 10 days postpartum and is not considered to place the mother or her unborn child at an increased risk for an unfavorable outcome. A small percentage of women with gestational hypertension may develop preeclampsia postpartum.

This classification is useful for identifying the hypertensive disorders of pregnancy. However, by emphasizing high blood pressure for the diagnosis, other patients who do not manifest hypertension but have underlying preeclampsia may be mismanaged because their disease is not recognized. An example is the patient with an atypical form of severe preeclampsia known as HELLP syndrome (*H*emolysis, *e*levated *l*iver enzymes, and *l*ow *p*latelets). It is characterized by microangiopathic hemoytic anemia, thrombocytopenia, and elevated liver enzymes reflective of hepatic dysfunction in a patient with preeclampsia. Some notable differences have been observed between women who develop HELLP syndrome and women with preeclampsia alone. Typically, HELLP syndrome occurs in older multiparous women, whereas preeclampsia occurs in primagravidas. They may present with mild nausea, vomiting, or a clinical picture more suggestive of a viral syndrome or a gastrointestinal illness than preeclampsia. Also, the usual tests of urine protein and uric acid which assist in determining the severity of preeclampsia may not be helpful in women with HELLP syndrome. Variable blood pressures and a widened pulse pressure are often present in HELLP syndrome and may be overlooked, resulting in a delay in diagnosis or a failure to diagnose this potentially life-threatening derangement of pregnancy.

To prevent the misdiagnosis or the failure to diagnosis the hypertensive disorders of pregnancy, the entire pathophysiological picture must be carefully assessed and understood.

II. PATHOPHYSIOLOGY

In the evaluation of pregnant women who will develop or who have developed preeclampsia, a number of common observations have been made.

A. Sensitivity to Pressor Agents

Prior to the clinical onset of preeclampsia, an increased sensitivity to pressor agents has been observed. Normally pregnant women are very resistant to pressor agents. In women who subsequently develop preeclampsia, however, this insensitivity is lost before the clinical signs of preeclampsia become evident.

B. Prostanoid Production

An alteration in normal eicosanoid production in pregnancy tissues is detected in women who develop preeclampsia. Among the aberrant prostaglandins are the relative amounts and ratios of prostacyclin and thromboxane to each other. Prostacyclin causes vasodilatation and inhibits platelet aggregation, whereas thromboxane causes vasoconstriction and platelet aggregation. In the patient who develops preeclampsia, the normal prostacyclin/thromboxane ratio encountered in normal pregnant patients which promotes vasodilatation and inhibits platelet aggregation becomes reversed and vasoconstriction and platelet aggregation predominate.

C. Placental Implantation

Even by the beginning of the second trimester of pregnancy, an altered implantation of the placenta is noted in patients who later exhibit preeclampsia. Placentation entails two waves of invasion of trophoblastic cells into the uterine vasculature. The first wave of invasion into the superficial vasculature of the spiral arteries is similar between women who develop preeclampsia and those who do not. However, in the former group the second wave of vasculature invasion does not occur into the larger branches of the spiral arteries which reside under the placenta. Normally, the two waves of invasion create a low-resistance, high-flow system which is optimal for placental perfusion. When this high-flow, low-resistance vascular system is not formed, blood flow becomes responsive to vasoconstriction created by pressor agents. In times of altered or impaired uterine blood flow, placental perfusion is further decreased by vasoactive amines as the protective high-flow, low-resistance system is lost. The reduced placental perfusion that results from abnormal implantation may be the stimulus for multiple manifestations of the systemic disease of preeclampsia. The prevailing hypothesis for the abnormal placentation is that it is immunologically mediated by a new or first exposure to paternal antigen. This alliance with immunology in the etiology of preeclampsia is suggested by multiple observations including the increased likelihood of preeclampsia in first pregnancies, in first pregnancies with a new partner, in pregnancies following prolonged barrier contraception use, or in pregnancies with donor insemination.

D. Endothelial Dysfunction

Endothelial cells serve complex functions to preserve vascular integrity and to mediate immune and inflammatory responses. With injury or activation, endothelial cells produce vasoconstrictors, procoagulants, and mitogens. Injured endothelial cells can activate the coagulation cascade and the "leaking" of fluid caused as the vascular integrity is lost creates extravascular fluid extravasation. This fluid extravasation takes place into the peripheral tissues and can leak into the alveoli of the lungs. The "shock lung" or adult acute respiratory distress syndrome with the loss of capillary–alveolar membrane integrity can rapidly lead to the need for intubation and assisted ventilatory support. The systemwide loss of endothelial integrity and health to some degree is thought to be the primary pathophysiologic mechanism in the development of the signs and symptoms of hypertension, proteinuria, and nondependent edema which is characteristic of preeclampsia.

III. LABORATORY ASSESSMENT

One of the markers used in identifying preeclampsia is ≥ 300 mg of urine protein in a 24-hr specimen. Antithrombin III has been reported to be decreased in women with preeclampsia but not in women with chronic hypertension—a useful observation to help differentiate preeclampsia from chronic hypertension. Levels of $\alpha 2$-antiplasmin are reduced in preeclampsia, consistent with endothelial injury and fibrinolysis. Laboratory tests useful in determining the severity of the preeclampsia include uric acid, liver enzymes, (transaminases) platelet count, coagulation profile, and lactic dehydrogenase. A rising hematocrit is frequently observed in women developing preeclampsia as fluid is moved outside the vascular space, plasma volume is reduced, and hemoconcentration ensues. Promising future tests to identify the patient with preeclampsia include α-2-antiplasmin, soluble thrombomodulin, and α-human chorionic gonadotropin.

Other markers which identify pregnancies at risk to develop preeclampsia include a decreased excretion of calcium in the urine in the second trimester of pregnancy and the presence of cellular fibronectin in the plasma. Fibronectin becomes elevated in early pregnancy in women who subsequently develop preeclampsia. These markers, although not currently being used clinically, are present weeks before the clinical signs of preeclampsia develop and could be used to identify women in need of more intensive surveillance.

IV. TARGET ORGAN INVOLVEMENT

Preeclampsia can cause multiorgan dysfunction, but the four organ systems which are primarily targeted are the liver, kidney, central nervous system, and coagulation system. The consequences of this disease are the result of vasoconstriction, platelet aggregation, and endothelial damage.

A. Liver

Liver injury is thought to result from disseminated deposition of fibrin and fibrinogen within the periportal areas which leads to necrosis. Areas of hemorrhage caused by disseminated intravascular coagulation or microangiopathic hemolytic anemia can coalesce to form intrahepatic or a subcapsular hemorrhage. The most serious consequence of hemorrhage underneath the capsule of the liver is a hepatic rupture, a surgical emergency that can be fatal unless prompt surgical intervention or interventional embolization is undertaken. Even with appropriate management, the maternal mortality approaches 30%. Liver enzymes may be elevated, but that elevation is variable and the degree of elevation is not reflective of the severity of the disease process or how quickly the patient will recover from the preeclampsia.

B. Kidneys

Renal involvement includes swelling of and hyaline droplets in the endothelial cells, capillary narrowing, and mesangial thickening. These changes contribute to the protein loss which is characteristic of this disorder. When preeclampsia is severe, these lesions can progress, particularly when blood volume decreases suddenly, and result in acute tubular necrosis with resultant oliguria/anuria or in the more permanent and irreversible cortical necrosis. Decreased glomerular filtration is commonly observed. The decreased clearance of uric acid results in elevated serum uric acid levels which have been linked to perinatal outcome. The higher the serum uric acid level, irrespective of the elevation of blood pressure, the worse the perinatal outcome.

C. Central Nervous System

The central nervous system effects include hyperreflexia and seizures. Clinically, a great deal of attention has been given to hyperreflexia because this is increased in many women prior to seizures, but not all preeclamptic women will be hyperreflexic. The degree of hyperreflexia seen in preeclampsia has not been shown to correlate with the severity of the preeclampsia. Cerebral vasospasm with resultant cerebral ischemia may lower the threshold for seizure in certain areas of the brain, although the relationship is poorly understood.

D. Coagulation System

The most common hematological abnormality observed in a patient with preeclampsia is thrombocytopenia. The platelet count can be reliably used to evaluate the coagulation system without the measurement of any other clotting parameter until that count is <100,000 μl. The coagulation system then may undergo a systemwide consumption of clotting factors, resulting in disseminated intravascular coagulopathy. This can also occur suddenly subsequent to other obstetric remature crises such as separation of the placenta from the uterine wall.

V. MANAGEMENT

The only definitive treatment of preeclampsia is removal of all pregnancy tissue from the mother. It is only with delivery of the fetus and removal of the trophoblastic tissue that this disorder can be stopped. Persistence of this disorder can and does result in significant risk of serious maternal and perinatal morbidity and mortality.

A. Antepartum

In the pregnancy in which the fetus is premature, and the maternal–fetal condition is stable, the preecamptic pregnancy can be prolonged in order to administer steroids to accelerate fetal lung maturity and decrease the risk of intraventricular hemorrhage, necrotizing enterocolitis, and fetal death. The benefits of continued *in utero* existence of the fetus must always be weighed against the risk of pregnancy prolongation. In any pregnancy with preecampsia in which the fetus is mature or in any pregnancy in which there is evidence of maternal or fetal compromise, the risks of continued pregnancy clearly outweigh any benefits and the fetus should be promptly delivered.

Selective patients with mild preeclampsia can be managed on an outpatient basis after inpatient stabilization and confirmation that their disease is mild. Outpatient monitoring optimally requires blood pressure monitoring four times a day, frequent home visits, or at least twice weekly office visits with antenatal testing. Worsening of the preeclampsia or nonreassuring antenatal testing mandates hospitalization.

Chronic hypertension is usually managed with antihypertensive agents that have been shown to be safe in pregnancy. Currently, α-methyldopa is the drug with the greatest published experience in pregnancy. Blood pressures are titrated in pregnancy to keep the systolic blood pressure between 90 and 100 and the systolic blood pressure between 140 and 150. Chronically hypertensive patients are seen frequently with antenatal fetal surveillance testing begun by 32 weeks of pregnancy. These women are monitored closely for superimposed preeclampsia.

B. Intrapartum

Mild or severe preeclampsia in a term pregnancy is managed by magnesium sulfate administration to prevent eclamptic seizures and control blood pressure with hydralazine and induction of labor. If the cervix is not favorable, cervical ripening agents can be used to dilate and efface the cervix prior to labor induction so that the chance of a successful induction of labor can be increased. Blood pressure is managed by the administration of hydralazine or β-adrenergic blocking agents to maintain a blood pressure of 150/90–100 mm Hg. Both mother and fetus are continuously monitored during labor. The fetus is monitored with an external monitoring device until membrane rupture and then monitored internally with a scalp electrode. The mother is monitored with blood pressures every 15 min, hourly urine outputs, and assessments of platelets, hemoglobin, and hematocrit about every 6 hr. Liver enzymes and lactic dehydrogenase are monitored every 12–24 hr. Maternal deterioration may require more frequent monitoring or the addition of other tests of the coagulation system if the maternal platelet count falls below 100,000/μl. Additional monitoring may be required for decreased urinary output or the development of multiorgan failure.

C. Postpartum

Postpartum care is in an intensive care unit or in an obstetric recovery room with hourly blood pressure measurements, hourly assessments of urinary output, and frequent neurologic and reflex checks. Frequent evaluations of the platelet count, liver enzymes, and lactic dehydrogenase are made particularly in the patients with HELLP syndrome. These patients are managed in the units with magnesium sulfate administered to prevent seizures and antihypertensives as needed to control blood pressure. These are discontinued only after blood pressure stabilization, adequate urinary output, and clinical improvement in the patient's condition.

D. Future

Pregnancies complicated by preeclampsia are at an increased risk of developing preeclampsia in a subsequent pregnancy. The earlier in gestation that preeclampsia is detected clinically and the greater its severity, the greater is the risk of developing preeclampsia in a future pregnancy. For example, in patients whose HELLP syndrome developed prior to 30 weeks in a preceding gestation, the chance of developing some hypertensive disorder in the next pregnancy is 61%.

Invasive monitoring in selected cases, newer medications, the emergence of obstetric intensive care units, and the availability of subspecialists skilled in the management of these high-risk women have led to much improved maternal outcome. The use of corticosteroids and the sophistication of neonatal intensive care units has significantly reduced the perinatal morbidity and mortality related to preterm deliveries. Despite these advances, pregnancies complicated by preeclampsia still are at risk for an adverse outcome. A thorough understanding of preeclampsia by the clinician relative to the diagnosis, pathophysiology, and management of the disease is currently the best way to approach this pregnancy challenge in order to minimize its consequences to the mother and her fetus.

Bibliography

American College of Obstetricians and Gynecologists (ACOG) (1986, February). Management of Preeclampsia, Tech. Bull. No. 91. ACOG.

National High Blood Pressure Education Program Working Group (1990). Report on high blood pressure in pregnancy. *Am. J. Obstet. Gynecol.* **163**, 1689.

Magann, E. F., and Martin, J. N., Jr. (1995). New-onset hypertension in the pregnant patient. *Clin. Obstet. Gynecol. North Am.* **22**, 157–171.

Martin, J. N., Jr., and Magann, E. F. (1966). HELLP syndrome. Current principles and recommended practices. In *Current Obstetric Medicine* (R. V. Lee, P. R. Garner, W. M. Barron, and D. R. Coustan, Eds.), 2nd ed., Vol. 2, pp. 129–175. Mosby/Year Book, Chicago.

Roberts, J. M. (1994). Pregnancy-related hypertension. In *Maternal–Fetal Medicine: Principles and Practice* (R. K. Creasy and R. Resnik, Eds.), 3rd ed., pp. 804–843. Saunders, Philadelphia.

Sibai, B. M. (1996). Hypertension in pregnancy. In *Obstetrics Normal and Problem Pregnancies* (S. G. Gabbe, J. R. Niebyl, and J. L. Simpson, Eds.), 3rd ed., pp. 935–996. Churchill Livingstone, New York.

Pregnancy in Dogs and Cats

Patrick W. Concannon
Cornell University College of Veterinary Medicine

John Verstegen
University of Liege College of Veterinary Medicine

I. Background
II. Ovulation and Fertilization
III. Implantation, Placentation, and Timing of Pregnancy Events
IV. Pregnancy Physiology and Endocrinology
V. Luteal Function and Maintenance of Pregnancy
VI. Parturition
VII. Lactation and Neonatal Period
VIII. Postpartum Anestrus and Interestrus Intervals

GLOSSARY

anestrus A period of apparent ovarian quiescence which occurs postpartum in dogs and some cats, after the end of the luteal phase in nonpregnant dogs, and seasonally in cats.

bitch The term for female domestic dog used by veterinarians and breeders.

diestrus Veterinary terminology used for canine metestrus and used consistently across veterinary species to designate the period of ovarian cycles dominated by corpus luteum secretion of progesterone.

metestrus The term classically used to designate the period following estrus until the absence of evidence of ovarian activity in the ovarian cycle. Proestrus, estrus, and metestrus are primarily behavioral terms related to the sexual receptivity of estrus. Clinically, the term metestrus or preferably vaginal metestrus refers to the period which begins at the time that vaginal cornification associated with the period of peak estrus behavior is reduced, as reflected in vaginal cytology or vaginoscopic examination.

pseudopregnancy A condition in which changes similar to those of pregnancy occur during the luteal phase of a nonpregnant cycle, ranging from a subclinical physiological pseudopregnancy to an overt, clinical pseudopregnancy.

queen The term for female domestic cat used by veterinarians and breeders.

The biology of pregnancy has not been investigated in dogs and cats as vigorously as it has been in other species of common domestic and laboratory animals. However, the information available from veterinary observations and limited directed research suggests that the underlying physiology is most likely similar to that of many other species. In both dogs and cats, this includes periovulatory priming of the reproductive tract first by estrogen and then progesterone, an absolute requirement for progesterone for pregnancy maintenance, pregnancy-specific mechanisms to ensure adequate secretion of progesterone, and a prepartum reduction in progesterone to initiate or facilitate parturition.

I. BACKGROUND

A. Seasonality and Litters

Pregnancy in dogs and cats is remarkably similar in some aspects, including a gestation length of 64–66 days in nearly all instances, although the underlying ovarian cycles differ markedly. The bitch is monoestrous, with little or no evidence of seasonality in most breeds, and ovulates spontaneously at intervals of 5–12 months. She can produce and raise one or two litters a year. Litter size is 1–12 and the average is 6–8 in many breeds. Pregnancy can occur any time of the year in most breeds but appears limited to the autumn in Basenji, which have cycles entrained to photoperiod. In cats, ovulation is most often induced by copulation. The queen is thus a reflex ovulator. Queens are usually seasonally polyestrous in the absence of pregnancy or pseudopregnancy and have repeated estrous cycles every 2 or 3 weeks except for a seasonal anestrus typically occurring from midautumn to early winter. Most pregnancies occur between late winter and summer, although some breeds can become pregnant at any time of year. Cats can produce and raise one to three litters a year. Litter size is typically 1–8 and averages 3–5 in most breeds. The potential effects of artificial lighting on the seasonality of cat breeding are not well characterized, but continuous exposure to 14 hr of light per day causes cats to cycle year-round.

B. Pregnant versus Nonpregnant Ovulatory Cycles

In dogs, the 2–8 months of obligatory anestrus are followed by a 1- to 3-week follicular phase and estrogen secretion that produces vaginal cornification, vulval swelling, and behavioral signs termed proestrus which usually last 1 or 2 weeks. It culminates in an estradiol peak and a resulting preovulatory surge in luteinizing hormone (LH) triggered by the preovulatory decline in estradiol and rise in progesterone associated with preovulatory luteinization. The decline in the estrogen:progesterone ratio initiates behavioral estrus, which lasts typically for 1 week. It also results in a decrease in vaginal cornification which is observed at 6–11 days after the LH surge (Day 0) and referred to as the end of vaginal estrus, onset of metestrus, or onset of diestrus. The luteal phase involves peak progesterone secretion around Days 15–25 and a slow decline during a progressive functional arrest of the corpora lutea which lasts until Days 55–70 in nonpregnant cycles. Its termination, marked by serum progesterone below 1 ng/ml, is considered to mark the end of metestrus or diestrus and onset of anestrus. In pregnant bitches, the luteal phase is remarkably similar to that of the ovarian cycle but there are often secondary increases in circulating progesterone between Days 25 and 35, suggesting a pregnancy-specific stimulation of progesterone production. Luteal function is terminated by an abrupt prepartum luteolysis.

In cats, ovulations which either occur spontaneously or are induced by infertile matings are followed by a luteal phase of progesterone secretion which typically lasts 25–35 days and is termed a physiological pseudopregnancy. It is followed by 1–4 weeks of anestrus and a return to polyestrous cycles except during autumn–winter anestrus. Cats vary in the occurrence of spontaneous ovulation in the absence of males. Under laboratory conditions, up to 35% of singly housed queens and up to 60% of group-housed queens can ovulate one or more times a year in the absence of a male. In feline pregnancy, luteal progesterone secretion is extended to the duration of gestation, is supplemented by placental progesterone production, and is terminated at parturition. Both dogs and cats can experience overt mammary enlargement and behavioral changes referred to as clinical pseudopregnancy during or following the latter half of the luteal phase of the nonpregnant ovarian cycle. It is common in dogs and can be accompanied by lactation and maternal behavior. It is uncommon in cats and rarely involves lactation.

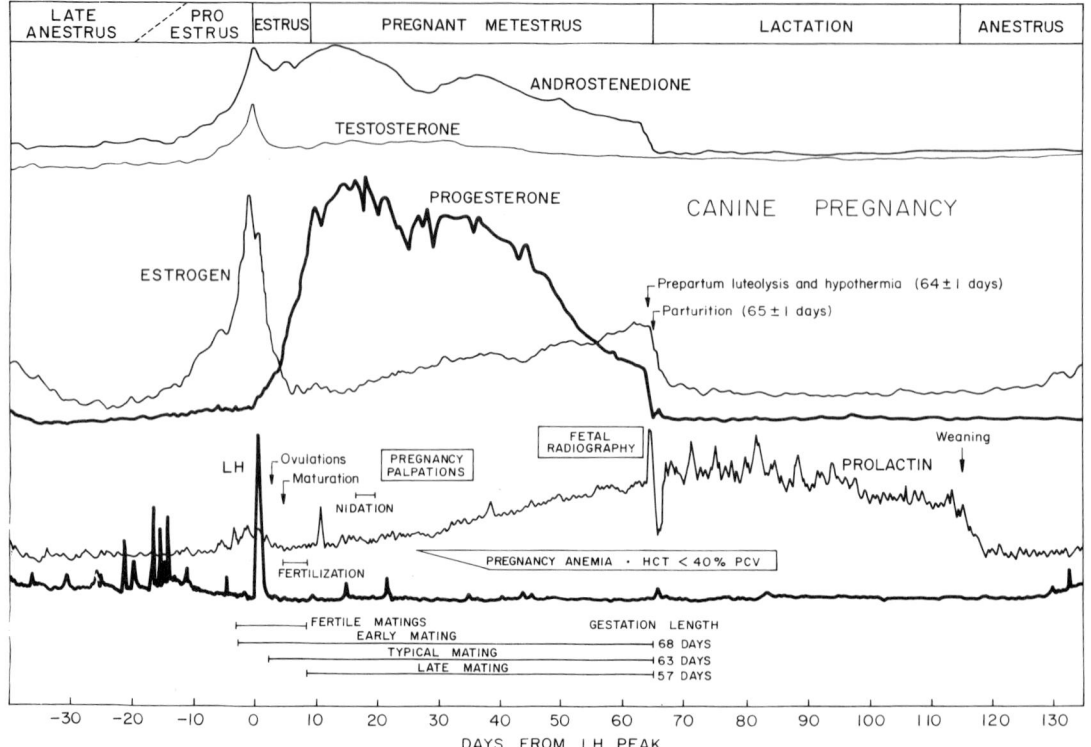

FIGURE 1 A schematic representation of typical changes in serum or plasma concentrations of various hormones reported or presumed to occur during pregnancy and lactation in the bitch and their relation to the events indicated and considered important for breeding programs and clinical management of pregnancy. Typical basal and peak serum levels of steroids are, respectively; 5–10 and 50–100 pg estradiol/ml, 0.2–0.5 and 15–90 ng progesterone/ml, 0.1 and 1 ng testosterone/ml, and 0.2 and 2 ng androstenedione/ml. Basal and preovulatory peak levels of LH based on the dog LH standard LER 1685-1 are 0.8 and 25 ng/ml, respectively. Prolactin concentrations based on the dog prolactin standard AFP-2451-B range from basal levels of 1.5 ng/ml to peak prepartum and lactation levels of 35 ng/ml (from Concannon, 1986).

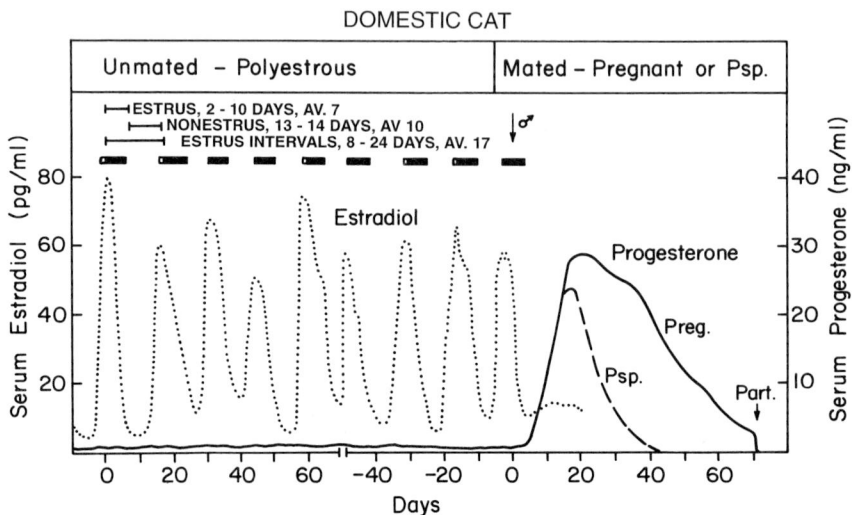

FIGURE 2 Schematic representation of endocrine changes typically observed during polyestrus cyclic activity in unmated queens and during pregnancy or pseudopregnancy (Psp) following fertile matings or vaginal stimulation resulting in ovulation. Solid bars indicate periods of estrus; open portions indicate potential proestrous periods (from Concannon and Lein, 1989).

II. OVULATION AND FERTILIZATION

A. Dogs

In dogs, initiation of estrus behavior may be as early as 5 days before ovulation or as late as 3 days after ovulation. Ovulation occurs spontaneously about 2 days after the preovulatory LH surge. Oocytes are ovulated as primary oocytes which undergo maturation slowly, over 2 or 3 days in the oviduct. Following oocyte maturation, many bitches remain fertile for 3 days or longer. The lengthy fertile life spans of both the spermatozoa and the mature oocytes may explain the high fertility rates (>90%) often obtained in commercial dog breeding facilities. Dog sperm may survive in the female tract for up to 8 days and still retain motility. In some cases, sperm can survive in the bitch for 5 days before ovulation and still result in fertilization and pregnancy. The opportunity to have multiple sires for a single litter is great if mating opportunities are not strictly controlled. In dogs, individual matings are usually followed by a coital lock for 5–20 min. Ovulation is timed clinically by using serum progesterone assays to detect the rapid preovulatory luteinization which occurs during the preovulatory LH surge or by monitoring the vulval detumescence and vaginal wrinkling that accompany the preovulatory decline in estradiol. Ovulation has also been timed using ultrasound to detect the loss of the anechoic appearance of preovulatory follicular fluid. Ovulation can be timed retrospectively as having occurred 5–7 days before the decline in vaginal cornification at the end of estrus.

B. Cats

In cats, LH increases within minutes of copulation and ovulation occurs about 30 hr after a mating-induced release of LH. Queens may allow brief 5- to 30-sec matings up to 30 times in a 36-hr period and may mate over a period of 2–6 days. Oocytes are ovulated as secondary oocytes. The duration of the fertile period for superfecundation by multiple sires, which is documented, has not been determined. Luteinization results in a rise in progesterone beginning 1 or 2 days after ovulation.

III. IMPLANTATION, PLACENTATION, AND TIMING OF PREGNANCY EVENTS

A. Dogs

In dogs, gestation almost invariably lasts 64–66 days after the LH surge (Day 0), whereas the duration following a single mating can range from 56 to 69 days (Table 1). Implantation and earlier events are apparently timed by maternal endocrine changes initiated around the time of ovulation. Furthermore, the actual period in which a viable syngamy of nuclear material can occur is limited and begins 2 or 3 days after ovulation. Oocyte maturation into secondary oocytes does not occur until Day 4 or 5 and oocytes begin to degenerate 2 or 3 days later. The fact that mature oocytes may survive for 4 or 5 days, i.e., until Days 9 or 10, in some cases, and then become fertilized by a mating in late estrus accounts for instances of pregnancy only lasting 56 or 57 days after mating. Blastocysts enter the uterus around Days 9–12, preimplantation uterine swellings are grossly detectable around Day 16, trophoblast attachment occurs around Day 18, and implantation occurs around Day 20. Using ultrasound, embryonic vesicles are detectable at Days 18–20, embryonic masses at Days 21–23, fetal heart movement at Days 24 or 25, the zonary placenta at Day 28, and fetal movement at Day 35. As the placenta and embryo continue to grow, the embryo is shorter than the length of the placental band until Day 38 and is obviously longer after Day 43.

B. Cats

In cats, events of pregnancy can be timed accurately if mating has been limited to a 1- or 2-day period and spontaneous ovulation has not occurred. Then gestation length is 63–67 days, with an average of 66 days. Following mating, blastocysts enter the uterus on Days 5 or 6, uterine enlargements are seen at Days 12 or 13, attachment occurs on Days 14 or 15, and implantation occurs on Day 16. Using ultrasound, embryonic vesicles are detectable by Days 11–13, heartbeats at Day 18–22, and fetal movement at Day 33. There is transmigration of em-

TABLE 1
Events of Typical Pregnancy in Dogs and Cats

Events and parameters	Dog	Cat
Day 0	LH surge	Mating and LH surge
Fertile matings		
Earliest	Day −3	Day 0
Latest	Day +8	—
First rise in progesterone	Day 0	Days 3–5
Peak progesterone		
Day	Days 15–25	Days 10–30
Value	15–80 ng/ml	10–60 ng/ml
Ovulation	Day 2	Day 1
Oocyte maturation	Days 4–5	Day 1
Completed fertilization		
Earliest	Days 4–5	Day 1
Latest	Day 8	—
Blastocysts in uterus	Day 10	Days 5–6
Uterine swellings	Days 16–18	Day 12
Embryo attachment	Days 16–18	Days 11–12
Implantation	Days 18–20	Days 13–14
Ultrasound (initial detection)		
Embryonic vesicle	Days 18–20	Days 11–13
Embryo mass 2–4 mm	Days 22–24	Days 15–16
Fetal heart beat	Days 24–25	Days 18–22
Heart rate monitorable	Days 28–32	Days 25–26
Fibrinogen		
Increased	Days 26–30	NA[a]
Peak	Days 30–50	NA
Relaxin		
Increased	Days 26–30	Days 24–26
Peak	Days 40–55	Days 40–55
Prolactin		
Increased	Days 28–32	Days 35–42
Peak	Days 64–65	Days 64–65
Anemia		
Detected	Days 25–30	Day 30
Term hematocrit	29–35%	34–44%
Radioopacity onset		
Skull and spine	Days 44–46	Days 34–38
Limbs and pelvis	Days 53–57	Days 34–38
Teeth	Days 58–61	NA
Decline in progesterone		
Onset	Days 62–64	Days 62–66
Complete	Days 64–66	Days 64–68
Parturition	Days 64–66	Days 63–67
Days from mating to parturition	57–69	63–67

[a] NA, Not available or not applicable.

bryos between uterine horns before implantation in both species.

In both dogs and cats, placentation is endothelial–chorial in composition and zonary and circumferential in morphology. The girdle of fetal trophoblast tissue develops marginal hematomas, whereas the chorioallantoic poles remain thin and transparent. The marginal hematomas develop large pools of stagnant blood from which the extraembryonic circulation absorbs various metabolites, especially iron. Clinically, pregnancy can be diagnosed by manual palpation of discrete uterine enlargements between Days 20 and 35 in dogs and Days 15 and 30 in cats, by ultrasound after Day 25 in dogs and after Day 20 in cats, and by radiography after Day 46 in dogs and after Days 38–42 in cats. Fetal heart rate can be monitored readily by ultrasound, beginning at Day 25 in cats and Day 32 in dogs. The average fetal heart rate in both species is around 230 bpm. Biochemical pregnancy tests are not available, but serum relaxin assay after Day 30 would be diagnostic in both species.

IV. PREGNANCY PHYSIOLOGY AND ENDOCRINOLOGY

A. Dogs

In dogs, several endocrine differences have been reported for pregnancy relative to the nonpregnant cycle. These include (i) a modest secondary increase in progesterone and estradiol secretion after Days 25–30; (ii) an elevation in relaxin concentrations from about Days 26–30 until term; (iii) an elevation in prolactin concentrations from Days 30–35 until term and followed by suckling-related elevations during 6–8 weeks of lactation; (iv) pregnancy-related elevations in acute-phase proteins between Days 30 and 50; and (v) changes associated with parturition including declining progesterone, increased circulating prostaglandin, prepartum surge in prolactin, and brief prepartum increase in cortisol.

Several physiological effects of pregnancy have been reported in the bitch. A progressive, normochromic, normocytic anemia starts between Days 25 and 30, becomes maximal near term, recovers over an 8- to 10-week period postpartum, and can involve 20–40% decreases in PCV to hematocrit values as low as 29–35%. The anemia of pregnancy presumably represents hemodilution due to an increased plasma volume, but actual changes in blood volume have not been determined in the pregnant bitch. Pregnancy causes resistance to exogenous insulin in diabetic bitches and can cause insulin resistance and hyperglycemia in some normal bitches. The likely cause is an increase in growth hormone secretion by the progesterone-stimulated mammary gland, which can occur in nonpregnant cycles as well. Litter size varies from 1 to 12, and the average varies among breeds and is typically 6–8 pups. Within breeds average litter size varies little with age but is maximal at 3 years and declines after 7 years. Postimplantation loss of 1 or more fetuses was documented in 6% of pregnancies in one study and in 25% in another. Spontaneous resorption or abortion of entire litters can occur. Effects of litter size on the extent of anemia and weight gain are not known. Body weight typically increases 20–55%, with an average of 35%. There is an increase in the absolute requirement for both protein and carbohydrate. It is recommended that food be offered at up to 150% of maintenance levels or more during the last 3 weeks of pregnancy and at 200% of maintenance or more during early and peak lactation.

There are postimplantation increases in serum concentrations of acute-phase proteins. Midgestational increases in circulating levels of fibrinogen, C-reactive protein, and haptoglobin have been reported, but the source, if not hepatic, has not been identified. Pregnancy-specific changes in plasma fibrinogen and C-reactive protein in dogs have been suggested as pregnancy tests, but false-positive results would be expected for bitches with any inflammatory disease. Pregnancy elevations in acute-phase proteins probably represent an inflammatory-like response to the presence of the fetal–placental unit which is recognized as containing foreign protein. There is also a moderate leucocytosis in midgestation.

The canine luteal phase is similar to that of the ovarian cycle, with an initial peak in circulating progesterone occurring between Days 15 and 25 followed by a slow decline. However, there are usually secondary increases in circulating progesterone between Days 25 and 40 reflecting a pregnancy-specific stimulation of progesterone production. Secondary, pregnancy-related increases in progesterone and estradiol secretion are masked in the circulation due to increased metabolism and dilution in an increased plasma volume. During the 24-hr prior to parturition, progesterone declines to less than the 2 ng/ml required to support pregnancy. Prolactin becomes distinctly elevated above levels seen in most nonpregnant cycles by Days 30–35, increases to near peak level in late gestation, and surges during parturition. The stimulus for the gestational rise in prolactin is not known. Prolactin concentrations during overt, clinical pseudopregnancy have not been adequately studied to determine whether they are similar or lower than those observed in pregnancy. Relaxin levels in the bitch increase between Days 27 and 30, peak by Days 40–50, and remain elevated but decline toward term. Relaxin concentrations then decline at parturition and reach nondetectable levels in 1–6 weeks. Relaxin levels within the normal range are observed in ovariectomized pregnant bitches in which pregnancy is maintained by an exogenous progestin. Therefore, the placenta is the primary source of relaxin in dogs. The extent of any ovarian production is not known. The initial increase in relaxin at about Days 27–30 appears to coincide with both the increase in fibrinogen and the initial pregnancy-related increase in prolactin. Relaxin assay is the best candidate for a pregnancy test because it is pregnancy specific in bitches.

B. Cats

Pregnancy physiology of cats appears to be similar to that of dogs but is less well documented. Average litter size is about 4 in most breeds and 2.6 in Persians. There is about a 20% decline in hematocrit between Weeks 3 and 9. Litter size increases with parity until the fifth to eighth pregnancy and declines after the seventh or eighth pregnancy. The reported incidences of spontaneous resorption or abortion of entire litters range from 3 to 20%. Primiparous queens are reported to have smaller litters and higher neonatal mortality. Typically, circulating progesterone initially peaks around Days 10–25, with further increases or secondary peaks between Days 15 and 35 followed by a slow decline through to term. However, peak values can occur as early as Day 11 and as late as Day 60. There is an abrupt decline before and during parturition. Relaxin is elevated after Days 25–30, peaks at Days 40–50, and declines to a nondetectable level after parturition. Prolactin is progressively elevated above nonpregnant values after Day 35 and increases rapidly prepartum. Estrus and copulatory behavior during pregnancy are not uncommon, although there are no large increases in estradiol.

V. LUTEAL FUNCTION AND MAINTENANCE OF PREGNANCY

In dogs, the corpora lutea appear to be the only source of circulating progesterone during pregnancy. The source of the modest concentrations of estradiol and androstenedione which parallel those of progesterone is probably also luteal. Both LH and prolactin are required luteotrophins. The pregnancy-related increase in prolactin is presumed to be the stimulus for the pregnancy-specific increase in luteal function. Ovariectomy or induction of luteolysis at any time in gestation causes termination of pregnancy or premature parturition. Exogenous prostaglandin $F_{2\alpha}$ (PGF) is weakly luteolytic in the bitch and can be used to induce complete luteolysis, and thus terminate pregnancy, any time after Day 15 if administered two or more times a day for a sufficient duration, typically 4–9 days. However, much higher doses of PGF are often required before Day 20 than after Day 30. Dopamine agonists such as bromocriptine and cabergoline are also luteolytic and abortifacient by virtue of their ability to suppress prolactin secretion. Combined treatment with dopamine agonist and prostaglandin has been successful for elective termination of postimplantation pregnancy. High-dose glucocorticoid administration for a week or more is also abortifacient, but the mechanism has not been determined.

In cats, luteal secretion of progesterone appears to continue until term but is supplemented by placental progesterone secretion after Day 50, based on maintenance of gestation in some queens following ovariectomy at Day 55. Circulating progesterone is mostly or entirely of ovarian origin and declines to nondetectable in pregnant queens maintained on synthetic progestin after ovariectomy. Placental progesterone is assumed to act locally and may not contribute significantly to circulating concentrations. Prolactin is luteotrophic in cats, and the maintenance of luteal function beyond the duration observed in nonpregnant cycles is presumed, as in dogs, to be due to increased prolactin. The existence of placental gonadotrophin has not been investigated in either cats or dogs. In cats, as in dogs, prostaglandin and dopamine agonists can be luteolytic and abortifacient, particularly in combination.

VI. PARTURITION

A. Dogs

In the bitch, normal parturition occurs as a result of a rapid decline in progesterone during the 24-hr prepartum. The decline is the result of a rapid luteolysis caused by a concurrent increase in circulating levels of prostaglandin $F_{2\alpha}$, as evidenced by a rise in prostaglandin metabolite. It is accompanied by a prepartum transient hypothermia due to rapid withdrawal of the thermogenic effects of progesteronemia. The increase in prostaglandin is presumably the result of a cascade of fetal–placental endocrine changes initiated by fetal maturation. The latter may include maturation of the fetal hypothalamic–pituitary–adrenal axis, as suggested for other domestic species, based on the increased maternal cortisol concentration observed 1 day prepartum. However, unlike the sheep, the dog does not have any observable prepartum increase in estrogen to serve as a stimulus for prostaglandin release. During the 6–24 hr before birth of the first pup, behavior changes typically include seeking of seclusion, digging, scratching at the floor, panting, anorexia, vomiting, and shivering. Typically, there is a potentially copious green mucoid vaginal discharge before, during, and after parturition. The color is due to normal hemoglobin metabolites, termed uteroverdin, from placental blood stores. There is often a prepartum surge in prolactin, followed by somewhat reduced levels for 1 or 2 days before suckling-induced elevations become near maximal. Pups may be delivered within intact membranes or attached to ruptured membranes. Membranes and placenta are typically eaten by the bitch, and the umbilical cord is thus severed. Vomiting of placental material is common. Intervals between pups are normally <30 min but can vary from 15 min to several hours. A litter of six to eight pups can require 4–18 hr.

B. Cats

Parturition in cats is accompanied by a decline in progesterone, but levels do not appear to become basal until after parturition, possibly because parturition largely depends on local withdrawal of the placental source of progesterone. The peripartum luteolytic mechanism involves a large increase in uteroplacental production of prostaglandin $F_{2\alpha}$ just before parturition. Parturition in cats is similar to that in dogs, except that litter size tends to be more consistent and on average smaller, although the range is 1–10 kittens. Coloration of discharge is brown. Intervals between kittens are typically 5–60 min, but interruptions of 11–24 hr can occur. The 2–12 hr immediately prepartum are accompanied by grooming and vocalization as well as restlessness and nesting. The prepartum temperature fall is not consistently observed.

In both dogs and cats, dystocia may involve malposition requiring cesarean section, although manual assistance may suffice in the bitch. In both species, uterine inertia is treated by administration of oxytocin and calcium gluconate at carefully monitored doses and rates.

VII. LACTATION AND NEONATAL PERIOD

A. Dogs

In dogs, obvious mammary development usually occurs by Day 45, and obvious milk secretion normally begins at or after parturition. However, instances of delayed mammary enlargement or of early milk secretion can occur. Suckling induces prolactin elevations to values similar to those of late pregnancy or higher during peak lactation and progressively lower levels later in lactation. Prolactin is required for normal lactation, and suppression by dopamine agonists inhibits lactation. Oxytocin released in response to suckling has been measured. In both the dog and cat lactation lasts about 6 weeks, with the dam encouraging weaning beginning about Week 5. Neonatal deaths are not uncommon for both pet and laboratory dogs. There is a moderate to serious transient metabolic acidosis that occurs in pups during parturition in normal births. The possible consequences, particularly in combination with compromised respiration, have not been evaluated. Average neonatal mortalities of 15–25% have been reported for dogs. Causes include respiratory distress following dystocia and bacterial infection. Puppy fading syndrome is recognized. The contributing factors are not known but may include hypothermia in cases in which bitches do not aggressively retrieve wandering pups as well as unrecognized infection or congenital abnormality and even failure to suckle competitively.

B. Cats

In cats, mammary development is obvious around Day 40. Lactation is initiated at parturition. Lactation is maintained by suckling-induced prolactin release. Weaning occurs at 5–9 weeks in pet cats and at 10–12 weeks in feral cats. Neonatal kitten mortality can average 20%, in addition to 10% born dead, with etiology including dystocia, congenital defects, milking problems, bacterial and viral disease, and unknown causes.

VIII. POSTPARTUM ANESTRUS AND INTERESTRUS INTERVALS

A. Dogs

Hematocrit returns to near normal by 30–90 days postpartum. Factors regulating the duration of anestrus, and thus the interestrus interval, have not been extensively studied in dogs. Likewise, the heritability of anestrus duration has not been addressed. Heritability seems likely since many bitches of at least one breed (German shepherd) often have characteristically short, 4- or 5-month interestrus intervals, and a few breeds appear to have average cycle lengths longer than the 7-month average seen in beagles and most breeds. Cycle length increases with age over 8 years in beagles.

Photoperiod may play a modifying role in at least one breed, the Basenji, in which estrus is usually restricted to late summer and early autumn. Circannual changes in metabolic hormones could also play a role in the duration of anestrus in other breeds. Such a relationship would be in agreement with the observation that in beagles housed outdoors, estrus may occur in any month, but there can be twice as many cycles in winter and spring than in summer and autumn. Furthermore, in dogs housed outdoors, a circannual rhythm in prolactin secretion has been reported. Prolactin is most likely inhibitory to the termination of anestrus and initiation of proestrus. Suppression of prolactin with a dopamine agonist in early anestrus can result in a precocious onset of proestrus within 10–50 days, resulting in an interestrus interval of about 100–170 days which is 40–90 days less than normal. In the lactating bitch, suckling-induced prolactin secretion presumably contributes to a physiological lactational anestrus which has not been studied as a discrete phenomenon. However, there are no consistent differences in the interestrus intervals for pregnant versus nonpregnant cycles. The normal physiology of anestrus includes the presence of elevated levels of serum follicle-stimulating hormone and relatively low levels of LH. Increased LH pulsatility, as a result of increased gonadotropin-releasing hormone pulse frequency, stimulates ovarian activity at the end of anestrus. That the critical element in the transition from anestrus to proestrus is LH is suggested by the fact that administration of highly purified LH alone, three times a day for 1 week, induced proestrus in all treated bitches.

B. Cats

Return to estrus in cats can be in a few days postpartum in some instances but is typically observed in 2–8 weeks (average, 4 weeks) after normal weaning. Cats can become pregnant while still lactating. If kittens are lost or weaned prematurely, return to estrus is commonly in 6–8 days.

See Also the Following Articles

CIRCANNUAL RHYTHMS; ESTRUS; MATING BEHAVIORS, MAMMALS

Bibliography

Banks, D. R. (1986). Physiology and endocrinology of the feline estrous cycle. In *Current Therapy in Theriogenology* (D. A. Morrow, Ed.), Vol. 2, pp 795–800. Saunders, Philadelphia.

Christiansen, I. J. (1984). *Reproduction In the Dog and Cat*. Bailliere Tindall, London.

Concannon, P. W. (1986). Canine pregnancy and parturition. *Vet. Clin. North Am. Small Anim. Pract.* 16, 453–475.

Concannon, P. W. (1991). Reproduction in the dog and cat. In *Reproduction in Domestic Animals* (E. T. Cupps, Ed.), 4th ed., pp. 517–554. Academic Press, San Diego.

Concannon, P. (1995). Reproductive endocrinology, contraception and pregnancy termination in dogs. In *Textbook of Veterinary Internal Medicine* (S. Ettinger and E. Feldman, Eds.), pp. 1625–1636. Saunders, Philadelphia.

Concannon, P. W., and Lein, D. H. (1989). Hormonal and clinical correlates of ovarian cycles, ovulation, pseudopregnancy and pregnancy in dogs. In *Current Veterinary Therapy, Small Animal Practice* (R. Kirk, Ed.), Vol. X, pp. 1269–1282. Saunders, Philadelphia.

Concannon, P. W., McCann, J. P., and Temple, M. (1989a). Biology and endocrinology of ovulation, pregnancy and parturition in the dog. *J. Reprod. Fertil. Suppl.* 39, 3–25.

Concannon, P. W., Morton, D. B., and Weir, B. J. (Eds.) (1989b). Dog and cat reproduction, contraception and artificial insemination. *J. Reprod. Fertil. Suppl.* 39, 1–350.

Concannon, P. W., England, G. C. W., Verstegen, J. P., and Russell, H. A. (Eds.) (1993). Fertility and infertility in dogs, cats and other carnivores. *J. Reprod. Fertil. Suppl.* 47, 1–569.

Concannon, P., Gimple, T., Newton, L., and Castracane, D. (1996). Increase in fibrinogen coincident with rise in relaxin following implantation in dogs. *Am. J. Vet. Res.* 57, 1382–1385.

Concannon, P., England, G., Verstegen, J., Rijnbeck, A., and Doberska, C. (Eds.) (1997). Reproduction in dogs, cats and exotic carnivores. *J. Reprod. Fertil. Suppl.* 51, 1–367.

Goodrowe, K. L. (1992). Feline reproduction and artificial breeding technologies. *Anim. Reprod. Sci.* 28, 389–397.

Johnson, C A. (Ed.) (1986). Reproduction and periparturient care. *Vet. Clin. North Am. Small Anim. Pract.* 16, 9.

Johnston, S. D., and Romagnoli, S. E. (Eds.) (1991). *The Veterinary Clinics of North America, Small Animal Practice, Canine Reproduction,* Vol. 21. Saunders, Philadelphia.

Noden, D., and Delahunta, A. (1984). *The Embryology of Domestic Animals*. Williams & Wilkins, Philadelphia.

Onclin, K., and Verstegen, J. (1996). Practical use of a combination of a dopamine agonist and a synthetic prostaglandin analogue to terminate unwanted pregnancy in dogs. *J. Small Anim. Pract.* 37, 211–216.

Shille, V. (1989). Reproductive physiology and endocrinology of the female and male dog and cat. In *Textbook of Veterinary Internal Medicine* (S. J. Ettinger, Ed.), pp. 1777–1791. Saunders, Philadelphia.

Shille, V., and Sojka, N. (1995). Feline reproduction. In *Textbook of Veterinary Internal Medicine* (S. Ettinger and E. Feldman, Eds.). Saunders, Philadelphia.

Verstegen, J., Onclin, K., Silva, L. D. M., and Concannon, P. (1997). Termination of obligate anoestrus and induction of fertile ovarian cycles in dogs by administration of purified porcine LH. *J. Reprod. Fertil.*, 35–40.

Yeager, A. E., and Concannon, P. W. (1990). Association between the preovulatory LH surge and the early ultrasonographic detection of pregnancy and fetal heart-beats in beagle dogs. *Theriogenology* 34, 655–665.

Yeager, A. E., and Concannon, P. W. (1996). Ovaries (Chap. 16) and Uterus (Chap. 17). In *Small Animal Ultrasound* (R. W. Green, Ed.), pp. 265–303. Lippincott-Raven, Philadelphia.

Pregnancy, in Farm Animals

Troy L. Ott
University of Idaho

I. Introduction
II. Etymology
III. Pregnancy in Farm Animals
IV. Pregnancy Diagnosis

GLOSSARY

caruncle Fixed structures on the maternal uterine endometrium that develop into the maternal half of the placentome in domestic ruminants.

chorionic gonadotropin A globular, heterodimeric glycoprotein consisting of α and β subunits produced by the placentae of equids (formerly pregnant mare serum gonadotropin) and primates with follicle-stimulating hormone and luteinizing hormone bioactivities, respectively.

conceptus Products of conception (fertile mating) including embryonic (fetal) and extraembryonic (placental) tissues.

corpus luteum Structure that develops on the ovary from the ovulated follicle that secretes progesterone as well as protein hormones, which, in domestic ruminants, include oxytocin.

cotyledon Fetal placental structures that form on areas of the chorioallantois overlying maternal caruncles, which ultimately develop into the fetal half of a placentome in domestic ruminants.

epitheliochorial placenta Placental type characterized by six tissue layers separating maternal and fetal blood (fetal vascular endothelium, stroma and chorionic epithelium, maternal uterine epithelium, stroma, and vascular endothelium).

estrous cycle Period between estrus of one cycle and estrus of the subsequent cycle for animals exhibiting distinct periods of estrus behavior. The length of the estrous cycle is characteristic of a given species.

estrus The period of the estrous cycle characterized by high levels of follicular estrogen production during which time the female will accept the male for mating; also termed "heat."

hemochorial placenta Placental type characterized by three tissue layers separating maternal and fetal blood (fetal vascular endothelium, stroma, and fetal chorionic epithelium).

parturition The process of giving birth to offspring.

placentome Placental structure consisting of maternal (caruncular) and fetal (cotyledonary) tissue which functions to bring maternal and placental blood vessels in close proximity for transfer of nutrients and wastes and for production of steroid and protein hormones. Between 50 and 100 placentomes develop on the placenta of domestic ruminants.

pregnancy-specific protein B Glycoprotein hormone produced by the trophoblast/chorion in cattle and assayable in the maternal circulation as a method of pregnancy diagnosis; also know as pregnancy-associated glycoprotein.

progesterone The hormone of pregnancy which prepares the uterus to support embryonic growth, implantation, and placentation; a 21-carbon-containing steroid hormone produced from cholesterol through the intermediate pregnenolone by the corpus luteum (and the placenta of certain domestic farm animals, especially sheep and horses).

I. INTRODUCTION

In eutherian (placental) mammals, pregnancy involves interactions between two independent and interdependent systems, the conceptus and the mother, which result in the birth of a fully developed individual at term. The length of pregnancy or gestation is calculated as the interval from fertile mating to parturition. Length of gestation is species specific (Table 1) and is influenced by genetic (breed effects and fetal genotype), fetal (sex of fetus, litter size, and pituitary and adrenal function), maternal (age

TABLE 1
Characteristics of Pregnancy in Domestic Farm Animals

			Days					
Species	Placental type[a] morphology	CL regression[b]	Conceptus in uterus	Hatching from ZP	Elongation of blastocyst	Pregnancy recognition	Initial placentation	Length of gestation
Sheep (ewe)	Epitheliochorial/cotyledonary	14–16	3–4	8–9	10–11	11–13	15	144–152
Cattle (heifer/cow)	Epitheliochorial/cotyledonary	17–20	3–4	9–10	13–15	16–19	22	270–300
Pig (gilt/sow)	Epitheliochorial/diffuse	16–18	3–4	8–9	10–11	11–12	13	112–115
Horse (filly/mare)	Epitheliochorial/diffuse[c]	14–16	4–5	6–7	10–12[d]	12–14	24	315–360
Goat (doe/nanny)	Epitheliochorial/cotyledonary	17–19	3–4	8–9	13–15	14–17	20–22	146–154

[a] Placental type classification based on the number of tissue layers separating maternal and fetal blood. For epitheliochorial placentas, six tissue layers (fetal vascular endothelium, stroma and fetal chorionic eipthelium, maternal uterine epithelium, stroma, and maternal vascular endothelium) comprise the barrier between maternal and fetal blood.

[b] Represent typical ranges in days postestrus or mating (estrus = Day 0).

[c] Placentae of horses are classified morphologically as diffuse/microcotylededonary.

[d] Horse blastocyst initially undergoes symmetrical expansion, not elongation.

of dam), and environmental (nutrition, temperature, and photoperiod) factors (Fig. 1). In domestic farm animals, pregnancy is established by conceptus-induced mechanisms which prolong luteal function and ensure continued production of progesterone, the hormone of pregnancy. All domestic farm animals require corpus luteum (CL) function for at least a part of gestation. During early gestation the conceptus must block development of a uterine-dependent mechanism responsible for regression of the CL. This effect of the conceptus is termed antiluteolytic (blocking luteal regression) and the action of the conceptus is local, at the level of the uterus. This is quite different from that in primates, in which the conceptus directly supports CL function (luteotrophic) and the action of the conceptus is systemic. Pregnancy is terminated at parturition, an event initiated by the fetus.

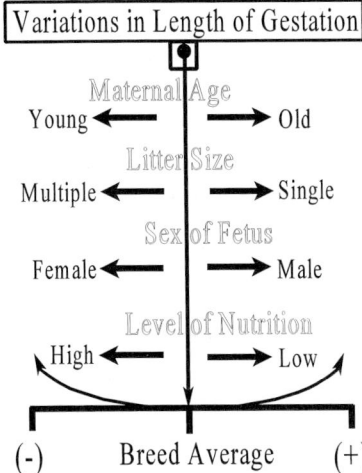

FIGURE 1 Factors which can affect length of gestation in domestic farm animals.

II. ETYMOLOGY

The word pregnancy is derived from the Latin *prae* ("from") + *(g)nasci* ("to be born"). The perfect (or past) participle of this Latin verb, *natus*, may be more familiar because it gives us such words as natal or nation. "Pregnant" has been used in the sense of "fertilized with child" since the fifteenth century and figuratively, as in a "pregnant wit," since the sixteenth century. "Gestation" is derived from the Latin *gerere* or "to carry" from the perfect participle.

III. PREGNANCY IN FARM ANIMALS

During pregnancy the relationship between the conceptus and mother can be described as symbiotic inasmuch as continuation of the species benefits the mother by ensuring that her genes are represented in the next generation. Practically, however, pregnancy is more accurately described as a parasitic relationship between fetus and mother. Pregnancy (and the ensuing lactation) places tremendous metabolic demands on the mother and puts her at increased risk of predation and succumbing to drought, famine, or disease. In addition, the fetal semiallograft behaves like a parasite in its ability to evade maternal immune rejection.

The obvious difference between a cyclic and a pregnant animal is the presence of a conceptus in the uterus. Pregnancy, therefore, can be considered a response of the mother to the conceptus in the uterine lumen. Pregnancy is initiated at fertilization; however, in domestic farm animals, necessary endocrine differences do not occur between cyclic and pregnant animals until the period of maternal recognition of pregnancy (MRP), typically occurring near the end of the luteal phase (Table 1). Embryo transfer experiments in cattle, sheep, goats, and pigs established the timing of MRP and revealed that the early embryo initiates no change in maternal/uterine physiology which is required for pregnancy to term prior to the period of MRP. Clearly, however, the conceptus is affected by the uterine environment prior to MRP, but the nature of the effect of the uterus on the preimplantation conceptus has not been described.

Blocking development of the luteolytic mechanism, a uterine-dependent phenomenon in domestic livestock, is followed by placentation involving apposition, adhesion, and attachment between uterine and chorionic epithelia. In general, placental growth precedes fetal growth, setting the stage to support rapid fetal growth which occurs during the third trimester of gestation. During the last third of gestation, growth of the placenta (in terms of mass) is minimal; however, tremendous functional maturation occurs during this period which supports the large increase in fetal metabolic demands. This functional maturation involves changes in the biochemistry of the placental barrier (particularly the extracellular matrix molecules), allowing for more efficient movement and exchange of nutrients and metabolites. This is accompanied by continued growth of the placental vascular bed. Factors which compromise placental growth early in gestation (i.e., undernutrition or disease) can result in placental insufficiency during late gestation. Placental insufficiency can result in growth retardation of the fetus and offspring, which are born small for their gestational age.

Domestic farm animals possess epitheliochorial placentae and, in general, their young are more fully developed at birth. These species do not support ectopic pregnancies and their conceptuses undergo characteristically prolonged free-living periods *in utero* prior to attachment. Offspring of animals having hemochorial placentation (e.g., primates) are born in an immature and helpless state and require prolonged, intense postnatal care. Conceptus adaptation to prolonged uterine gestation involved development of the ability to produce biologically active agents necessary for maintenance of pregnancy. Therefore, interdependence between conceptus and maternal units might be considered most highly developed in species such as cattle, sheep, goats, pigs, and horses, which possess epitheliochorial placentae, since (i) the uterus is the preferred site for establishment of pregnancy, (ii) conceptuses experience a prolonged free-living period *in utero* during early gestation, (iii) attachment is superficial, (iv) gestation is prolonged, and (v) offspring are born relatively more fully developed.

Extension of luteal life span for at least a portion of gestation is a requirement for all domestic farm animals. In the cow and sow, but not in the ewe and mare, maintenance of luteal function is required for the entire period of gestation. In the ewe and mare ovariectomy during the second half of gestation (after Days 55 and 150–200, respectively) does not result in abortion since placental production of progesterone is sufficient to maintain pregnancy.

Pregnancy is terminated at parturition, with the length of gestation being species specific (Table 1) in domestic ruminants. Parturition is initiated by maturation and activation of the fetal hypothalamic–pituitary–adrenal axis and increased cortisol secretion by the fetal adrenals.

IV. PREGNANCY DIAGNOSIS

Accurate pregnancy diagnosis is an important contributor to efficiency in animal production systems allowing for timely decisions regarding culling (sale) or rebreeding of animals. Methods in current use include visual (detection of estrus or "heat"), physical (palpation and ultrasonography), and immunologic (pregnancy-specific hormone assays). The method of choice most often relates to its reliability, cost, and ease of use versus the value of the animal being diagnosed.

A. Detection of Estrus

Return to estrus is one of the most common methods for identifying nonpregnant females following mating or insemination. Detection of estrus following mating, artificial insemination, or embryo transfer is simple and inexpensive and provides immediate information about which females are not pregnant. The accuracy of this diagnosis depends on the species tested, method and frequency of detection, skill of personnel, and season (for seasonal breeders). However, it is not unusual for pregnant animals to exhibit estrous behaviors. For example, cows will occasionally exhibit estrous behavior during gestation in response to estrogen production by growing follicles on the ovary.

The method of estrus detection depends on the species being tested. Cows in estrus exhibit homosexual behavior and can be observed mounting other cattle as they are coming into estrus and then allowing themselves to be mounted as they are in estrus. Ewes and does do not exhibit these behaviors and estrus detection requires the presence of a ram or buck. Pigs are intermediate in terms of the overt signs of estrus. Gilts exhibiting estrus can be observed mounting other gilts and gilts in estrus will allow themselves to be mounted or will exhibit a lordosis response characterized by a rigid stance with ears erect when firm hand pressure is applied to their back. The presence of a male will greatly facilitate estrus detection in swine. Near the time of estrus in swine, it is typical for the vulva (external genitalia) to become red and swollen and to have a clear mucus discharge. Estrus detection can be accomplished in the mare using a teaser stallion. Mares in estrus will often assume a stance characteristic of urination, with tail head raised and diverted to one side. This is often accompanied by rhythmic contraction of the vulva, termed "winking," during which the clitoris is repeatedly exposed and small amounts of urine are expelled. Duration and intensity of estrous behavior are quite variable in the mare, but can last from 5 to 7 days.

B. Physical

Palpation per rectum is the most common form of physical diagnosis of pregnancy in the cow and mare. Its accuracy depends on the stage of pregnancy and skill of the palpator. Palpation per rectum has been associated with a small but measurable fetal loss (5–10% by some estimates). This loss is probably most associated with the skill of the palpator and stage of gestation (palpation earlier than Day 50 is associated with higher risk). Pregnancy can be accurately diagnosed by rectal palpation as early as the second month for both cattle (35–40 days) and horses (20–30 days). During early pregnancy, diagnosis is made based on the shape and tone of the uterine horn being palpated. In cattle, during the latter third of gestation, the uterus and its contents are pulled over the brim of the pelvis and down into the abdomen, making the fetus increasingly difficult to palpate (also an indicator of pregnancy). Moreover, placentomes (doughnut-shaped structures between 1 and 10 cm in diameter on the placenta) and the middle uterine artery (along the floor of the pelvis) can be palpated and used as predictors of pregnancy after 120 days of gestation. The diameter of the middle uterine artery of the cow ranges in size from 3 or 4 mm in nonpregnant cows to 1–1.5 cm at the end of gestation, and when it is encircled with the thumb and forefinger, a "buzzing" or pulsing sensation can be felt on the side of the uterus containing the pregnancy.

Abdominal palpation (termed "ballotment"), used primarily in sheep and goats after about 75 days of gestation, is accomplished by placing one hand on either side of the abdomen of the ewe in the inguinal area and firmly pressing with one hand while holding the other hand in place. This procedure is done with the ewe resting on its hindquarters and tilted slightly backwards against the legs of the palpator. A rectal–abdominal palpation has also been used in sheep and goats which is highly accurate after about Day 60 of gestation but can be associated with damage to the rectal lining and abortion (especially in the doe).

Ultrasonography (pulse echo; includes both A and B type), either per rectum or through the abdomen, is increasingly being used as an accurate method for pregnancy diagnosis in domestic animals. Once prohibitive due to cost, ultrasound units are now more affordable, especially for larger operations. Ultrasonography measures differences in acoustical impedance between tissues varying in water content by bouncing sound waves off the tissue using a handheld transducer (either rectal or abdominal), with fluid-filled structures (i.e., fetal amniotic and allantoic sacs and maternal bladder) being the least reflective and tissues such as bone and cartilage being the most reflective.

In the swine industry, the less expensive A-mode units (amplitude; one-dimensional measure of echo versus distance) are finding increasing use. The more expensive B-mode units (brightness; two-dimensional picture of soft tissue) allow for real-time imaging of the abdominal contents. Real-time B-mode machines are used commonly on large brood mare farms and are often used by veterinarians and stud managers to diagnose pregnancy or reproductive disorders. Rectal ultrasonography can be used to detect pregnancy in ruminants as early as 20–25 days postmating (10–12 days postmating in the mare) and at 2 or 3 weeks in swine. Accuracy of pregnancy diagnosis by rectal ultrasonography increases significantly after Day 26 in cattle. In addition, ultrasonographic measurements of the fetus (e.g., width of skull) can be used to estimate day of gestation when breeding dates are not known and can accurately determine the number of fetal–placental units in domestic ruminants and horses.

Transabdominal Doppler-effect ultrasonography is also used, particularly in sheep and swine (starting at the third month), to detect the fetal heart beat and/or umbilical blood flow. In addition, radiography can be used to visualize the fetal skeleton, but it is costly and can only be used during the third trimester.

C. Immunologic

Immunologic methods of pregnancy diagnosis involve assays for the presence of steroid or protein hormones specific to, or indicative of, pregnancy in blood, milk, or urine. These assays can be specific for conceptus-produced substances such as chorionic gonadotropin (eCG) in horses or pregnancy-specific protein B (PSPB) in cattle. Other assays rely on detection of pregnancy-specific increases in substances (i.e., progesterone or estrogen) in blood, milk, or urine.

Progesterone is a common hormone assayed to diagnose pregnancy. Progesterone is typically assayed in blood or milk of cows that have been inseminated by sending the sample to a central testing facility or, increasingly, kits are available for on-farm pregnancy diagnosis. The on-farm or "cow-side" tests yield a color reaction when samples contain progesterone at >2 ng/ml and are, in general, good predictors of pregnancy status. Studies have demonstrated that plasma and milk progesterone concentrations are highly correlated. Therefore, estimates of progesterone concentrations in a milk sample obtained at 21–24 days postinsemination are highly predictive as a method of pregnancy diagnosis. Cows exhibiting milk progesterone concentrations of >2 ng/ml at this time are considered pregnant. Cows with values <1 ng/ml should be considered nonpregnant, evaluated for estrous behavior, and rebred. The timing of collecting milk for this method of pregnancy diagnosis is critical since progesterone levels in the milk of luteal phase, nonpregnant cows can range from 10 to 80 ng/ml. The milk progesterone assay can also be used on sheep, goats, and mares; however, its use on these species is currently limited.

The equine placenta develops structures called endometrial cups starting at about Day 35 of gestation which produce the protein hormone eCG. eCG production is maximal between about Days 40 and 120 of gestation in the mare. This protein hormone can be assayed in the blood and serves as a sensitive method for pregnancy diagnosis in this species. After Day 120, eCG levels decline as the endometrial cups degenerate and analysis of eCG concentration is no longer a reliable predictor of pregnancy.

PSPB is produced by the placentae of cattle and other ruminants. It can be measured in the maternal circulation after about Day 24 of pregnancy in cattle. Currently, PSPB cannot be measured in milk or urine and requires blood sampling. In addition, PSPB levels can remain elevated in the maternal circulation for more that 60 days postpartum, which limits its use for pregnancy diagnosis of lactating cows due to excessive number of false positives.

Estrogens, produced by the placenta, can also be assayed in plasma, milk, or urine and elevated levels reflect fetoplacental function. Assays for estrone sulfate are accurate for pregnancy diagnosis in plasma or milk of domestic farm animals around the normal time of luteolysis. In the mare, estrogens in the urine are useful predictors of pregnancy after Day 150 when eCG levels decline.

See Also the Following Articles

Cloning Mammals by Nuclear Transfer; Epitheliochorial Placentation; Estrus; Horses (Equidae); Sheep and Goats

Bibliography

Bazer, F. W., Ott, T. L., and Spencer, T. E. (1998). Endocrinology of the transition from recurring estrous cycles to establishment of pregnancy in sub-primate mammals. In *Endocrinology of Pregnancy* (P. M. Conn, Ed.), Contemporary Endocrinology Series, pp. 1–34. Humana Press, Clifton, NJ.

Chemineau, P. H., Chupin, D., Cognie, Y., and Thimonier, J. (1993). Control of reproduction in domestic mammals. In *Reproduction in Mammals and Man* (C. Thibault, M.-C. Levasseur, and R. H. F. Hunter, Eds.), pp. 673–693. Ellipses, Paris.

Fields, M. J., and Sand, R. S. (Eds.) (1994). *Factors Affecting Calf Crop.* CRC Press, Boca Raton, FL.

Ginther, O. J. (1995a). *Ultrasonic Imaging and Animal Reproduction: Fundamentals. Book 1.* Equiservices, Cross Plains, WI.

Ginther, O. J. (1995b). *Ultrasonic Imaging and Animal Reproduction: Horses. Book 2.* Equiservices, Cross Plains, WI.

Hafez, E. S. E. (Ed.) (1993). *Reproduction in Farm Animals*, 6th ed. Lea & Febiger, Philadelphia.

Heap, R. B., and Flint, A. P. F. (1985). Pregnancy. In *Reproduction in Mammals: 3; Hormonal Control of Reproduction* (C. R. Austin and R. V. Short, Eds.), pp. 153–194. Cambridge Univ. Press, Cambridge, UK.

Roberts, R. M., and Anthony, R. V. (1994). Molecular biology of trophectoderm and placental hormones. In *Molecular Biology of the Female Reproductive System* (J. K. Findlay, Ed.), pp. 329–440. Academic Press, San Diego.

Sharp, D. C., and Bazer, F. W. (Eds.) (1995). *Equine Reproduction VI,* Biology of Reproduction Monograph Series. Edwards Brothers, Ann Arbor, MI.

Zavy, M. T., and Geisert, R. D. (Eds.) (1994). *Embryonic Mortality in Domestic Species.* CRC Press, Boca Raton, FL.

Pregnancy in Humans, Overview

Carmen L. Regan

Hospital of the University of Pennsylvania

I. Ovulation, Fertilization, and Implantation
II. Organogenesis
III. Fetal Maturation and Growth
IV. Maternal Adaptation to Pregnancy
V. Labor and Delivery
VI. The Puerperium
VII. Lactation

GLOSSARY

integrins Transmembrane proteins which are receptors for extracellular linkers at the cell surface.
ontogeny The process whereby the fetus ordinarily develops.
syncytiotrophoblast Differentiated trophoblast lining the maternal villous spaces and responsible for β-human chorionic gonadotrophin hormone production in pregnancy.
trophoblast Fundamental placental cell of epithelial origin.

Pregnancy is made unique by the fact of maternal tolerance of the fetus, which is by definition a semiallograft. Incompletely understood, this immune tolerance is thought to be due in part to unique antigen expression by the fetal trophoblast and to altered maternal immune responses in pregnancy. Following implantation an orchestrated sequence of events occurs local to and distant from the fetoplacental unit leading to the altered physiological state of pregnancy.

I. OVULATION, FERTILIZATION, AND IMPLANTATION

Estrus in the human female is a 28-day cycle. Ovulation characteristically occurs on Day 14 and is followed by menstruation 14 days later if fertilization fails to occur. Average human gestation lasts 280 days or 40 weeks and is divided into three parts, described as trimesters. The first trimester is from conception to 12 weeks' gestation, the second from 12 weeks to 28 weeks, and the third from 28 weeks to term.

The ovum, once fertilized, commences division over a period of hours to days and becomes the morula, which is composed of a cluster of dividing cells. The morula enters the uterus on the fifth day after fertilization. During the following few days it lies free in the uterine cavity bathed by secretions from uterine glands. A fluid-filled cavity appears within the cavity of the morula which is then termed a blastocyst. Placentation in the human involves an invasive phenomenon in which embryo-derived trophoblastic cells progressively integrate into the maternal tissues through production of extracellular matrix degrading enzymes (matrix metalloproteinases), migratory activity, and rapid cell division. The human placenta is hemochorial. The trophoblast cells at the implanting blastocyst invade into the uterus in order to establish a blood supply. During the first 2 weeks of development nutrients are exchanged by diffusion; thereafter, a blood supply is established by the cytotrophoblastic columns which invade the decidua blood vessels by a process of endovascular invasion. In the process of endovascular invasion the epithelial trophoblast acquires endothelial characteristics, including the expression of endothelial-specific integrins. Human chorionic gonadotrophin (hCG) is a glycoprotein secreted by the placental syncytiotrophoblast throughout gestation. Serum levels rise rapidly over the 10 days following implantation, reaching a peak in the ninth week of gestation at 100,000 mIU. Thereafter, it falls to a level of about 10,000 mIU, at which it remains for the duration of pregnancy. hCG acts on the corpus luteum to prevent its regression and to stimulate its production of progesterone and estradiol.

Abnormal placentation can result in loss of the fetus and may cause severe complications for the mother. For example, preeclampsia or pregnancy-induced hypertension, a disease exclusive to human pregnancy, affects 7–10% of all pregnancies and is thought to be due to abnormal placentation. Trophoblast cells from such pregnancies fail to invade endovasculature and show the characteristic changes to an endothelial-like phenotype.

II. ORGANOGENESIS

Fetal organogenesis occurs in the first 12 weeks, although ontogeny continues throughout pregnancy, particularly within the fetal brain. The first trimester is therefore a time of rapid cell division and differentiation. It is during this critical period that drugs exert most of their teratogenic effects. The neural tube closes at 24–28 days after conception, often before the pregnancy is apparent. The primitive fetal heartbeat can be detected by ultrasound from as early as 6 weeks. The fetal kidney is developed at 10 weeks and fetal urine production commences early in the second trimester. By the end of the embryonic period (10 menstrual weeks) the extremity bones, joints, and musculature have differentiated into structures with relative position and form identical to those of an adult.

III. FETAL MATURATION AND GROWTH

A. Rate of Fetal Growth

From about 14 or 15 weeks the fetus gains weight at a rate of 5 g/day increasing to 10 g per day at around 20 weeks of gestation. In the third trimester the average daily weight gain is 30–35 g per day. The mean growth rate peaks at 230 g per week; this occurs between 33 and 36 weeks. The maximum percentage growth rate occurs in the first trimester.

B. Factors Contributing to Fetal Growth

Forty percent of total birth weight variation is due to genetic contributions from the mother and fetus; the remainder is environmental. The paternal influence on growth is limited to the contribution of a Y chromosome. Male fetuses grow faster than females and weigh 150–200 g more at birth.

Numerous studies show that inadequate nutrition in pregnancy can predispose to intrauterine growth restriction (IUGR). If the insult occurs early in pregnancy then the number of cells is decreased; if it occurs later in pregnancy cell size is decreased. Inadequate weight gain in pregnancy (<0.27 kg/week or <10 kg at 40 weeks) may also contribute to low birth weight. Prepregnancy nutritional status is also important, and studies from famine situations indicate that the IUGR is more profound if nutritional deprivation predates and continues through pregnancy.

Alterations in uteroplacental perfusion affect both the growth and the status of the placenta as well as the fetus. Umbilical artery flow studies indicate a reduction of umbilical blood flow in some human growth restricted fetuses. This may be due to increased resistance downstream as a result of impaired placental perfusion secondary to thrombosis and vasoconstriction. Maternal cigarette smoking decreases birth weight by 135–300 g. If smoking is stopped in the third trimester, this effect is not seen. The mechanism is not well established but may be due to carboxyhemoglobin concentrations in the maternal and fetal bloodstream, which displaces oxygen from circulating hemoglobin. Maternal alcohol consumption and cocaine usage are associated with low birth weight. Growth restriction in these cases is global, affecting both fetal weight and head size. In the latter case, the reduction in head circumference is more pronounced.

IV. MATERNAL ADAPTATION TO PREGNANCY

In order to facilitate the growth of the fetus *in utero* the maternal physiology undergoes dramatic change. Multiple organ systems are involved but the following sections discuss the most relevant.

A. Respiratory System

Progesterone from placenta stimulates the respiratory center in the brain to produce hyperventilation. This results in a decreased alveolar CO_2 and arterial pCO_2. The hypocarbia results in a reduced plasma bicarbonate via increased renal excretion and a minimal change in pH. Thus, pregnancy is a state of compensated respiratory alkalosis.

Tidal volume increases from about 450 to 600 ml, representing a 40% increase. In addition, minute ventilation increases by 40%, thus increasing the oxygen available to the fetus and facilitating CO_2 transfer from fetus to the mother.

B. Cardiovascular Changes

Cardiac output increases 30–50% in pregnancy and reaches a maximum at 10 weeks of gestation. It remains elevated until term, when 20% of output is directed to the kidneys, 17% to the uterus, and 10% to the skin. Cardiac output is dependent on position. Supine occlusion of the inferior vena cava occurs in late pregnancy, with an 8% drop in cardiac output from decubitus lateral to back and an 18% drop from standing to lying. Most women do not become hypotensive when standing because the fall in cardiac output is accompanied by a rise in peripheral vascular resistance. Blood pressure in pregnancy is highest when seated, lower when supine, and lowest in the left lateral position. Peripheral vascular resistance falls in pregnancy due in part to the relaxing effects of progesterone on smooth muscle. Systemic arterial blood pressure falls in the first 24 weeks and increases gradually to term.

Echocardiographic studies demonstrate that, despite increases in left ventricular dimensions and volume during pregnancy, most parameters of left ventricular function are generally similar to those in the nonpregnant state. Central hemodynamics are altered in normal pregnancy. Significant increases in cardiac output and heart rate occur, and additionally significant decreases in systemic and peripheral vascular resistance occur, resulting in no net change in mean arterial pressure, pulmonary capillary wedge pressure, central venous pressure, or left ventricular work index during normal pregnancy. There is a reduction in colloid osmotic pressure which may explain the propensity to pulmonary edema in pregnant women with enhanced capillary permeability or cardiac preload.

C. Plasma Volume and Red Cell Mass

Plasma volume in normal pregnancy increases from 6 to 8 weeks and increases progressively until 30–34 weeks, after which time it reaches a plateau. The mean increase is 45–50% and is larger in multiple gestation and in women with bigger babies. Erythrocyte mass increases from 10 weeks and rises steadily toward term. The increase is thought to be due to erythropoietin production. Red blood cell mass increases by about 18% by term in unsupplemented patients and by 30% in those given iron. Because plasma volume increases by 50% and red blood cell mass by 18–30% the hematocrit falls, reaching a nadir at 30–34 weeks. This physiological response enhances placental perfusion by decreasing viscosity.

D. Iron Requirements

Iron requirements increase in pregnancy. Placental transfer of iron occurs in the first trimester as soon as the tertiary villi are formed. Iron is absorbed from the proximal duodenum in the ferrous stat; only 10% is absorbed in the nonpregnant state. In pregnancy 20% of oral iron is absorbed and in deficiency states up to 40% may be absorbed. The total iron requirement in pregnancy is 1000 mg. Five hundred milligrams is required for increased red cell mass, 300 mg for the fetus, and 200 mg to compensate for normal daily losses by the mother. In the third trimester the fetus takes up all the iron available to it. Iron is actively transferred to the fetus by the placenta against a high concentration gradient, and fetal levels do not correlate with maternal levels. Iron requirements increase with advancing gestation to support the increasing red cell mass and the requirements of the fetoplacental unit. Placental volume and weight has been correlated with maternal anemia, and placental hypertrophy may represent a mechanism for improving transfer and supply of oxygen to the fetus.

E. The Renal Tract

Pregnancy is associated with major anatomical and functional changes in the renal tract. Kidney volume, weight, and size increase in pregnancy, with renal length increasing by 1 cm. The renal collecting system undergoes marked dilatation, seen as early as the first trimester and persisting up to 4 months postpartum. These effects are attributed to both mechanical and hormonal effects. The glomerular filtration rate (GFR) and effective renal plasma flow increase by 50–80% above the nonpregnant value. Renal 24-hr creatinine clearance increases at 4 weeks of pregnancy, rises to a maximum at 9 or 10 weeks, and remains elevated until late pregnancy. Serum creatinine and levels of blood urea nitrogen fall in pregnancy secondary to the increase in GFR. Uric acid falls and increases toward term as a result of increased tubular reabsorbtion of urate. Plasma osmolality falls in early pregnancy due to a reduction in sodium and associated anions. A diuretic response does not occur because of a lower osmoreceptor setting in pregnancy. Sodium metabolism is altered in pregnancy. Sodium loss by the kidneys is enhanced by increased glomerular filtration and the natriuretic effect of progesterone. This is balanced by enhanced renal tubular reabsorbtion of sodium as a result of increases in circulating aldosterone, estrogen, and deoxycortisone.

F. The Coagulation System

Pregnancy is a hypercoagulable state. Fibrinogen I increases during pregnancy, which is also associated with elevated levels of factors VII–X. Prothrombin II and factors V and VII remain unchanged during pregnancy, whereas XI and XIII decline somewhat. The risk of thromboembolism is 1.8 times that in the nonpregnant state in pregnancy and 5.5 times that in the puerperium. The naturally occurring anticoagulants antithrombin III and proteins C and S are important in maintaining hemostasis. Protein S falls but protein C and antithrombin III remain stable. The platelet count declines progressively in pregnancy and is associated with a fall in platelet volume. This is thought to reflect increased platelet consumption and augmented production due to a shortened platelet life span.

V. LABOR AND DELIVERY

Term labor is defined as labor occurring after 37 weeks of completed gestation. The stimulus heralding the onset of labor is unknown but is thought to be fetal in origin. A number of mechanisms for the initiation of labor have been proposed, including a shift in the balance of estrogen/progesterone effects toward estrogen, release of oxytocin, and increased uterine synthesis of prostaglandins.

The actual mechanism is not fully understood, although there does appear to be a common biochemical end point of an increased synthesis of prostaglandins. Recently, induction of the prostaglandin endoperoxide synthase isoform, PGHS-2 (also known as cyclooxygenase-2), in maternal reproductive tissues prior to the onset of labor has been described, and this may account for increased prostaglandin biosynthesis. In addition, prostaglandin synthase inhibitors are effective in delaying preterm labor.

Following initiation of labor, oxytocin release from the posterior pituitary gland stimulates rhythmic uterine contractions. Progressive softening and connective tissue remodeling of the cervix results in effacement, a process whereby the cervix is incorporated or "taken up" into the lower uterine segment. Following complete effacement, cervical dilatation commences. The first stage of labor is that time from the initiation of labor to complete cervical dilatation and is of variable duration depending on fetal, maternal, and uterine factors. Delay in the first stage of labor is arbitrarily defined as time taken in excess of 12 hr and is termed dystocia. The second stage of labor is the time from complete cervical dilatation to delivery of the fetus. Classically, this stage involves descent of the fetal head into the maternal pelvis by a sequence of flexion, internal rotation, extension, and, following delivery, restitution. Delay in the second stage may be due to a combination of factors, including fetal size, pelvic anatomy, and inefficenct uterine action. Augmentation of labor is deemed necessary when inadequate progress is made in the first stage of labor as judged by cervical dilatation and in the second stage by failure of descent of the fetal presenting part. Synthetic oxytocin is given intravenously in order to stimulate uterine contractions. The rate of oxytocin infusion varies in different centers and low- and high-dose regimens have been described. A system of labor management for nulliparous women, termed the active management of labor, has been developed and practiced in Ireland, and has resulted in a reduction in prolonged labor and cesarean section rates for dystocia. The third stage of labor is the time from delivery of the fetus to delivery of the placenta. Following delivery of the fetus, the placenta separates from the uterine wall. Separation is heralded by a vaginal gush of blood, lengthening of the umbilical cord, and firming of the uterus as palpated abdominally by the examining hand. Expulsion of the placenta follows shortly thereafter.

VI. THE PUERPERIUM

The puerperium commences immediately following the delivery of the placenta and is arbitrarily defined as a period lasting 6 weeks. Involution of the uterus occurs immediately and within a week is 50% of its size at the end of pregnancy. After 2 weeks of normal involution the uterus cannot be palpated abdominally and at 6 weeks is almost its prepregnant size. The superficial layer of decidualized endometrium is sloughed off as the lochia, whereas regeneration of the underlying endometrium occurs and is complete on the 16th postpartum day. The cervix gradually returns to a nonpregnant state over a period of 3 or 4 months and the vaginal epithelium returns to its nonpregnant state over a period of 6–10 weeks, although varying degrees of mucosal and facial relaxation may remain. Ovulation occurs in nonlactating women at about 10 weeks after delivery and menstruation will occurs at 12 weeks postpartum in 70% of cases. Following term delivery hCG disappears from the circulation by about 12 days.

The systemic changes reflecting the maternal adaptation to pregnancy return to prepregnant levels over varying degrees of time. The major circulatory changes return to baseline over a period of 6 weeks. Renal function returns promptly to prepregnancy levels after delivery, with renal plasma flow being substantially diminished by 5 days postpartum. In contrast, the anatomical changes in the urinary system, such as increased renal size and ureteral dilatation, may persist for months.

VII. LACTATION

During pregnancy there is a gradual increase in serum levels of prolactin. The effects of prolactin on the breast are inhibited by high levels of estrogen which prevents lactation during pregnancy. Following delivery there is a rapid fall in the level of estrogen, and progesterone and prolactin are able to initiate lactation. Prolactin levels decline to nonpregnant values within 4–6 weeks in women who do not breast-feed. In breast-feeding mothers the levels remain elevated for about 2 or 3 months postpartum and thereafter decline. As lactation continues, suckling elicits progressively less prolactin release, although the amount is sufficient to maintain lactation. Initiation of lactation is via the "letdown" reflex. Impulses generated by suckling enter the spinal cord and are relayed to the hypothalamus. The neurosecretory cells in the supraoptic and paraventricular nuclei are stimulated to secrete oxytocin and prolactin via mechanisms which are imprecisely understood. Myoepithelial cells are the effector organ for oxytocin. Contraction of these cells forces milk out of the alveolar lumina. Oxytocin is released in a pulsatile fashion which is responsible for the rhythmic contraction of myoepithelial glands within the mammary gland.

Although suckling is the primary stimulus for the release of oxytocin, this reflex may be conditioned so that the sight or sound of the baby may cause the letdown of milk. Pain, embarrassment, or distraction may inhibit it. Prolactin release following suckling is not a conditioned reflex and release is dependent solely on suckling. The composition of human milk is a mixture of fat in water that is isotonic with plasma, with water being the major constituent. Colostrum, the milk secreted in the first few days of lactation, is higher in protein and lower in carbohydrate than mature breast milk. Human milk is composed of more than 100 constituents. The principal proteins are caseins, α-lactalbumin, lactoferrin, immunoglobulin A (IgA), lysozyme, and albumin. A number of peptide hormones are present in breast milk, including epidermal growth factor and transforming growth factor-α: These may play a role in the growth of the developing infant. Growth hormones have also been identified, as have naturally occurring benzodiazepines which may have sedative properties. The advantages of nursing are evident for both mother and child. Uterine involution is facilitated by the pulsatile release of oxytocin from the posterior pituitary upon suckling. Maternal weight loss is facilitated by transfer of proteins, carbohydrates, and fats to the neonate. Bonding between mother and infant is facilitated.

Nutritionally, breast milk cannot be improved upon by formula. From an immunologic standpoint, neonatal immunity to infection is boosted by the ingestion of maternal IgA antibodies which protect against respiratory and gastrointestinal pathogens; this may be the mechanism in the reduction in sudden infant death syndrome seen in breast-fed babies. Other benefits may result from the ingestion of trophic hormones and other factors in human milk. Little is known about the cause of failing lactation in humans, and a decline in milk production occurs in the first 3 months postpartum. A variety of drugs have effects on lactation and breast feeding, either by effects on the mammary gland or by altering prolactin secretion. Many substances may be transmitted to the neonate through human milk and only drugs necessary for the welfare of the mother should be prescribed.

In conclusion, pregnancy involves major physiological adaptations to allow normal growth and development of the fetus. These adaptations of pregnancy begin early in gestation and are associated with major changes in important organ systems. When the normal physiological response does not take place, poor fetal growth and increased maternal morbidity result.

Bibliography

Bennett, P. R., Henderson, D. J., and Moore, G. E. (1992). Changes in the expression of the human cyclooxygenase gene in human fetal membranes and placenta with labor. *Am. J. Obstet. Gynecol.* 167(1), 212–216.

Clark, S. L., and Cotton, D. B. (1988). Clinical indications for pulmonary artery catheterisation in the patient with severe preeclampsia. *Am. J. Obstet. Gynecol.* 158, 453–458.

Creasy, R. K., and Resnik, R. (1994). *Maternal–Fetal Medicine: Principles and Practice*, 3rd ed. Saunders, Philadelphia.

Dubin, W. H., Johnson, J. W. C., Calhoun, S., et al. (1980). Plasma prostaglandins in pregnant women with term and preterm deliveries. *Obstet. Gynecol.*, 203–306.

McMaster, Librach, C. L., Zhou, et al. (1995). Human placental HLA-G expression is restricted to differentiated cytotrophoblasts. *J. Immunol.* 154, 3771–3778.

O'Driscoll, K., Foley, M., and MacDonald, D. (1992). Active management of labor as an alternative for cesarean section for dystocia. *Obstet. Gynecol.*, 485–490.

Williams, R. L., Creasy, R. K., Cunningham, G. C., et al. (1982). Fetal growth and prenatal viability in California. *Obstet. Gynecol.* 59, 624.

Zhou, Y., Damsky, C. H., and Fisher, S. J. (1997a). Preeclampsia is associated with failure of human cytotrophoblasts to mimic a vascular adhesion phenotype. *J. Clin. Invest.* 99, 2152–2164.

Zhou, Y., Fisher, S. J., Janatpour, M., Genbacev, O., Dejana, E., Wheelock, M., and Damsky, C. H. (1997b). Human cytotrophoblasts adopt a vascular phenotype as they differentiate. *J. Clin. Invest.* 99, 2139–2151.

Pregnancy, in Other Mammals

Lloyd L. Anderson

Iowa State University

I. Introduction
II. Monotremes
III. Marsupials
IV. Nonprimate Eutherians
V. Nonhuman Primates
VI. Reproductive Strategies of Mammals

GLOSSARY

Eutheria An infraclass or division to which all placental mammals belong (viviparous).

Marsupialia Orders of nonplacental mammals (viviparous).

monotocous, polytocous Describing or referring to single- and litter-bearing mammals, respectively.

Monotremata Orders of primitive egg-laying mammals (oviparous).

pseudopregnancy A condition of extended corpus luteum function in nongravid mammals less or equivalent to the duration of normal pregnancy.

I. INTRODUCTION

Mammals are distinguished from all other animals by their production of milk from mammary glands. Lactation is an essential feature for early immunological protection, and for growth and development of the young. The three major functional groups or taxa of mammals are Monotremata, Marsupialia, and Eutheria. Monotremes lay and incubate eggs (oviparous), whereas reproductive duct development in marsupials and eutherians (viviparous) allows embryonic and placental growth in a uterus for incomplete (marsupials) or advanced (eutherians) development of the young at birth. Reproductive processes of these three major groups retain some common characteristics [i.e., preovulatory ovarian (Graafian) follicles, functional corpora lutea, uterine secretions, bilaminar blastocysts, yolk sac placentae, mammary glands, and lactation]. The basic mode of mammalian reproduction likely evolved simultaneously with the origin of mammals in the Triassic but remained almost unchanged until the late Mesozoic because it was the most appropriate mode for small nocturnal insectivorous mammals (Tyndale-Biscoe and Renfree, 1987). Divergence of characteristics in egg-laying monotremes may represent a common ancestral group that evolved to marsupials and eutherians.

Although reproductive strategies of mammalian species differ in many ways in their production of young, lactation, whether brief (4 days in hooded seal) or prolonged (>900 days in great apes), is common to all. Mammary glands of monotremes, marsupials, and eutherians contain clusters of galactophores that drain into teats (marsupials and eutherians) which are absent in monotremes. Mammary glands may have derived from cutaneous (sebaceous and apocrine) secreting glands that specialized into synthesizing a milk rich in carbohydrates, proteins, and lipids with antimicrobial properties. Secretions of the mammary gland precursor may have enhanced survival of the eggs or young by their antimicrobial properties. Blackburn and colleagues (1989) suggested that such protolacteal secretions from these cutaneous glands controlled or destroyed potential pathogens on or near the eggshell surface and, when ingested by hatchlings, enhanced survival by controlling microflora of the pharynx and digestive tract. These secretions may have provided immunity to the young as seen in immunoglobulins of present-day milk. Thus, hypertrophy of the cutaneous glands of the vascularized incubation patch with a more copious secretion may have evolved to maternal secretion of milk as a primary source of energy for development and growth of the young. Maternal secretion of milk from myoepithelial cells and mammary areolae by the suckling hatchlings (monotremes) requires teat development, attachment and suckling stimulus by the young, and a neurogenic response by the maternal system for milk release (marsupials and eutherians).

Comparative aspects of pregnancy representative of the class Mammalia are presented. A descriptive account of aspects of reproductive biology of most families within an order is available from the listed references. The endocrinology of pregnancy and central nervous system regulation of pregnancy, parturition, and lactation are comprehensive only in a few laboratory animals (i.e., rat, mouse, and rabbit), farm animals (i.e., cattle, pig, sheep, and horse), nonhuman primates, and man.

II. MONOTREMES

Echidnas (spiny anteaters) and the duck-billed platypus are living representatives of the mammalian subclass Prototheria. They resemble some anatomical features of reptiles and differ from all other mammals by laying shell-covered eggs that are incubated and hatched outside of the body of the mother. In both sexes of all monotremes the posterior end of the intestine, the ducts of the excretory system, and the genital ducts open into a common chamber known as the cloaca. Thus, convergence of these three systems into a single external opening is the basis for the ordinal name, Monotremata. The penis of the male is attached to the ventral wall of the cloaca and divided at the tip into paired canals for passage of spermatozoa; testes remain abdominal. In the female, the paired ovaries (the left only in the platypus) release eggs into the oviducts where fertilization occurs, and then they are covered with albumen and a sticky, leather-like shell. The oviduct leads through a junctional region into a uterus with a thick mucosa and uterine glands. The paired uteri open anteriorly into the long urogenital sinus that leads into the cloaca where eggs, urine, and feces pass through the external opening. Mature ovarian follicles release yolk-filled ova, up to 5 mm diameter, that acquire a permeable shell membrane and shell by absorption of nutrients in the uterus. Corpora lutea, formed from ovulated ovarian follicles, are secretory and autonomous; progesterone blood concentration during pregnancy was 10 ng/ml compared with 2 ng/ml in nonpregnant specimen (platypus). Ovoid eggs are laid about 2 weeks after breeding when the embryos have 19 or 20 somites. The young hatch after an incubation of about 10 days in a pouch (echidnas); platypus has no pouch but the female curls about the eggs while incubating. Embryogenesis occurs mainly during incubation. Lactation is long (3 or 4 months) and milk composition shows major changes through lactation but is high in total solids, lipids, iron, and copper. Litter size is one to three eggs and interlitter intervals range from 1 to 2 years. They show seasonality of breeding but do not hibernate. Echidnas (short- and long-nosed) frequent a variety of habitats including forests, rocky areas, hilly tracts, and sandy plains. They feed on termites, ants, other insects, and worms and are generally solitary and with a home range of about 800 m. Self-protection is achieved by rolling up into a partial or complete spiny ball. Echidnas are powerful diggers for burrowing and can run swiftly and climb well. Hatchlings are 12–14 mm in diameter and about 400 g when they are ejected from the pouch about 55 days of age and 150–200 mm long. Their life spans in captivity range from 30 to 50 years. The platypus is an aquatic nocturnal predator of crayfish, shrimp, larva of water insects, snails, tadpoles, worms, and small fish in freshwater streams, lakes, and lagoons. A deep amber or black fur is short and dense with bladelike guard hairs that contribute to insulation of the body. Burrows constructed in banks along streams and ponds provide shelter for both sexes and a nest for rearing the young by the female in isolation from the male. When hatched, the young are about 25 mm long, blind, and naked. About 4 months later, they emerge from the burrow fully furred and about 335 mm long. Limbs are short with webbed feet, each with five clawed digits, and there is a poison spur on the hindfeet on the male. The thick fatty tail is a reserve energy source. The snout, resembling a duck's bill, is covered with moist, soft leathery skin and perforated on the surface by openings to sensitive nerve endings. Life span in captivity ranges up to 17 years.

III. MARSUPIALS

In marsupials the ureters pass between the oviducts, whereas in eutherian mammals the oviducts pass between the ureters (Fig. 1). There is fusion of the vaginae anterior to the ureters with development of a median pseudovaginal (birth) canal. The oviducts of marsupials cannot meet in the midline to form a single median uterus and wide median vagina opening to the exterior. This necessarily limits size of the young at birth; the smallest weigh <5 mg and none weighs more than 1 g. Neonatal weights as a percentage of maternal weight range from 0.002% (eastern native cat) and 0.003% (red kangaroos) to 0.05% (terrestrial nocturnal insectivores) compared with 2 or 3% of a single young of eutherian species. Thus, reproductive strategies of marsupials include a relatively brief intrauterine gestation, birth, and transit to a pouch for attachment to a teat for suckling during a long period of lactation in which differentiation and growth of the young occur.

The mother's special endocrine control of lactation allows the suckling young to increase body weight 12-fold between parturition and >100 days later when the young begin to leave the pouch. Marsupials are seasonal breeders. Their reproduction patterns require prolonged parental investment and can be markedly affected by availability of adequate food and rainfall. Of living marsupials, 249 species in 16 families are recognized, but the reproductive biology of only a few major families (7) is well described

TABLE 1
Representative Families of Marsupials

Family	Total species	Representative species and common name
Didelphidae	70	*Didelphis virginiana* (opossum)
Dasyuridae	49	*Dasyurus viverrinus* (eastern native cat)
Peramelidae	16	*Perameles nasuta* (long-nosed bandicoot)
		Isoodon macrourus (short-nosed bandicoot)
Petauridae	22	*Petaurids* (nocturnal, arboreal gliding omnivores)
Phalangeridae	11	*Trichosurus vulpecula* (brush-tailed possum)
Phascolarctidae	1	*Phascolarctos cinereus* (koala)
Macropodidae	56	*Setonix brachyurus* (quokka)
		Macropus eugenii (tammer)
		Macropus rufus (red kangaroo)
		Macropus giganteus (eastern grey kangaroo)

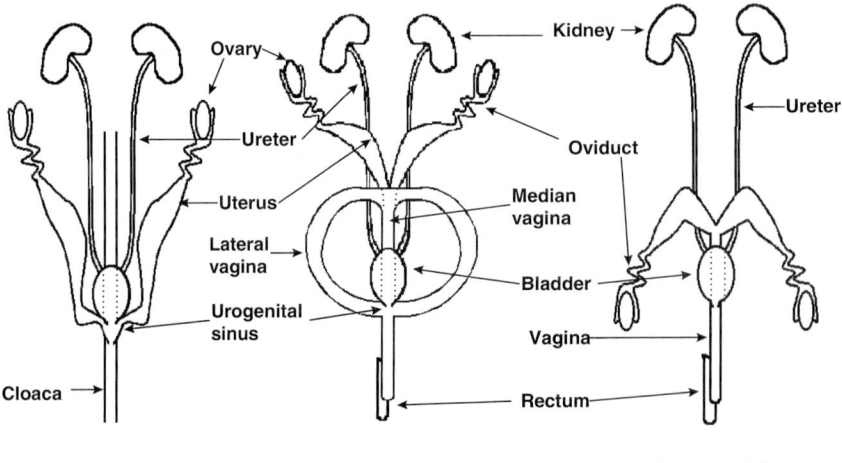

FIGURE 1 Urogenital duct development in three taxa of mammals.

representing 21 species (Table 1). Three phases of lactation are recognized. The first is firm attachment of the young to one teat for milk. Only suckled teats continue to produce milk and thus determine litter size. The second phase is when the young can relinquish the teat but depends on the mother for nourishment. In the third phase the young lives outside the pouch but continues occasional suckling until fully weaned. This is a critical stage for survival of the young for adequate food and a favorable environment.

Living marsupials of the Americas are divided into three families of opossums. Gestation averages 13 days and neonatal weights average 130 mg. Attachment phase from birth is 48 days and pouch emergence from birth averages 70 days with weaning at 110 days. Litter size ranges from 3 to 13.

For species inhabiting Australia, Indonesia, and New Guinea, gestation in the eastern native cat is 16–20 days with a neonatal birth weight of 12 mg. Teat attachment phase from birth lasts 49–56 days and weaning to birth ranges from 110 to 120 days. Litter size ranges from 6 to 8. Gestation in the long- and short-nosed bandicoots is 12.5 days with a neonatal birth weight of 235 mg. Although teat attachment phase is unknown, pouch emergence from birth averages 50 days with weaning at 56–63 days. Litter size averages 2.4–3.4. Gestation in the brush-tailed possum is 17.5 days with birth of a single neonate weighing 200 mg. Teat attachment phase from birth lasts 94 days with pouch emergence at 140–150 days and weaning at 230 days.

Gestation in the koala is 35 days with only one newborn. Although teat attachment phase is unknown, pouch emergence ranges from 209 to 224 days and weaning occurs at 365–380 days. Gestation in the quokka, tammer, red kangaroo, and eastern gray kangaroo ranges from 27 to 33 days. Neonatal birth weights among these species range from 350 to 750 mg. Teat attachment from birth is 110–150 days in the tammer but unknown for the other species. Pouch emergence from birth ranges from 185 to 250 days and weaning occurs at 240–365 days. Removal of the pouch young results in return to estrus within 26–35 days. Although litter size is limited to one, the species retain a reproductive strategy of being able to breed after removal of the pouch young and experience a delayed gestation of 25–31 days. Thus, one offspring can continue suckling the mother while she is initiating another pregnancy.

IV. NONPRIMATE EUTHERIANS

Only brief comments about fecundity, lactation, and number of young can be cited.

Order Insectivora (shrew, mole, and hedgehog; 375 species) is widely distributed and nearly all small animals live on insectivorous or carnivorous diets. Most are seasonally polyestrus with moderate-size litters (up to five). The European common shrew is an induced ovulator with corpus luteum and placental production of progesterone. Gestation as well as lactation lasts 20 or 21 days; mating during lactation induces embryonic diapause. European moles are seasonal breeders (two litters per year) with gestation of 30–40 days and lactation of 30 days. Hedgehog hibernates and produces two litters of five young per year. Gestation is 35–42 days and lactation is 40 days.

Order Chiroptera [nearly 1000 species of bats worldwide of mega- (fruit-eating) and micro- (insectivores) chiroptera suborders]. Many hibernate with torpor beginning in autumn to spring. Asymmetry of the morphology of reproductive organs with unilateral ovarian and uterine function is a common feature with usually only one newborn. Estrus and copulation in autumn is followed by hibernation and birth 4 or 5 months later (breed usually one time per year). Gestation lasts 155–160 days and lactation lasts for several weeks. Reproductive strategies include autumn breeding, implantation followed by embryonic diapause dependent on corpus luteum progesterone, whereas rapid embryonic differentiation and placental progesterone production (February and March) occurs with birth in June in nonhibernating American leaf-nose bat. The common or small brown bat (Pipistrellus) is a seasonal breeder that hibernates. Estrus and copulation occur in late summer and early autumn and reservoirs of spermatozoa are stored in the uterus for more than 100 days with fertilization of a recently ovulated ovum and a 2-month pregnancy and birth in the spring. Another feature is accumulation of a protective coat of glycogen in cumulus cells surrounding ovum during hibernation.

Order Edentata (anteater, sloth, and armadillo; 32 species) is monotocous with gestation ranges of 135 days (armadillos), 190 days (anteaters), and 263 days (sloths). Luteal regression occurs during the later half of gestation with progesterone production then dependent on placenta. Unique features of armadillos are a single corpus luteum, fertilization of the single ovum, and delayed implantation of blastocysts 4 months later. Four young are born (monozygotic quadruplets) and lactation lasts about 2 months.

Order Pholidota (scaly anteater; five species) are monotocous seasonal breeders with estrous cycles of 9–14 days and gestation of 65–139 days. Neonates weigh 235–450 g and lactation lasts 3 or 4 months.

Order Lagomorpha (rabbit, hare, and pika) is widely distributed. Reproductive biology of the domestic rabbit (*Oryctolagus cuniculus*) is comprehensive. Puberty and breeding occur at 6–9 months of age with estrus induced by coital stimulation; ovulation occurs 10–18 hr later. Nonfertile mating results is pseudopregnancy lasting 16–20 days that includes maternal behavior, nest building, mammary gland development, and fur-plucking. Pregnancy lasts 30–32 days with four to eight young born and a lactation of 4–6 weeks. Ovaries contain abundant interstitial tissue with cholesterol stores and luteinizing hormone (LH)-dependent production of 20α-dihydroxyprogesterone. Progesterone production by the newly developing corpora lutea begins 2 or 3 days after mating to maintain pregnancy to term.

Order Rodentia includes several suborders of myomorphs (mouse, rat, hamster, lemming, gerbil, and vole; 1183 species), sciuromorphs (squirrel, chipmunk, gopher, marmot, beaver, and spring hare; 366 species), and hystricomorphs (guinea pig, porcupine, agouti, coypu, chinchilla, viscacha, and naked mole rat; approximately 180 species with a wide but unequal geographic distribution between and within the New and Old Worlds). The reproductive biology of the mutant albino rat (*Rattus norvegicus*) is well described and it is used extensively for laboratory investigations. Reproductive strategies include the ability to conceive when bred at parturition, delay implantation (embryonic diapause) during lactation, and deliver young about 30 days later rather than the normal 22-day pregnancy. Lactation lasts about 14 days.

Order Cetacea (whales, dolphins, and porpoises; 64 species) is widely distributed in all the major oceans and in some large rivers. Mating, parturition, and lactation occur below the surface of the water. Sperm whale pregnancy lasts 14.5 months, with parturition in November–June and birth of a calf weighing about 800 kg. Lactation persists for about 25 months. The corpus luteum of an ovulation in the absence of mating lasts 2 or 3 weeks, but during pregnancy the corpus luteum remains at maximal size (5–10 cm in diameter) until late pregnancy and regresses after parturition. The largest whale (blue whale, 30 m in length and >100,000 kg) reaches puberty at 3–7 years. Gestation lasts about 12 months and is followed by 7 months of lactation and 5 months of anestrus to maintain a biennial reproductive cycle. The corpus luteum of pregnancy attains a diameter of 13–18 cm and weight of 4 kg that is maintained until after parturition. Gestation in the bottle-nose dolphin lasts about 12 months and lactation continues for 12–18 months. The corpus luteum at parturition measures 26–28 mm in diameter and regresses slowly during several years. The common porpoise, found in the Southern Hemisphere, has a gestation of 11 months and lactation of 7 or 8 months followed by anestrus to maintain a biennial reproductive cycle. The corpus luteum of early gestation (2.5 cm in diameter) persists until early lactation, when it regresses, and is rejuvenated until the end of lactation, when it becomes a corpus albicans.

Order Carnivora is grouped into seven families (ferret, weasel, stoat, marten, mink, badger, skunk, sea otter, civet, genet, mongoose, raccoon, panda, bear, hyena, domestic cat, lion, jaguar, domestic dog, wolf, coyote, and fox; 252 species). Most species are seasonal breeders, being either seasonally polyestrus or having ovulation induced by copulation. Sterile mating in the ferret is followed by development of a functional corpus luteum of pseudopregnancy lasting 42 days and equivalent to that of pregnancy. In several species (i.e., stoat, weasel, sable, badger, and skunk) mating is followed by a prolonged preimplantation embryonic diapause and a relatively brief postimplantation period (18–42 days). After implantation the corpora lutea hypertrophy and progesterone blood concentration increases about twofold. During pseudopregnancy circulating progesterone remains slightly above basal concentration for a period equivalent to pregnancy. However, in the ferret, mink, dog, blue fox, and cat, progesterone blood concentration remains similar during pseudopregnancy and pregnancy and implicates the corpus luteum as the primary source of the hormone. Although progesterone blood levels decrease earlier in pseudopregnant compared with pregnant domestic cats, the duration of luteal function of 63 days is the same. In the seven species of bears, the interval from mating to parturition ranges from 175 to 240 days. Delayed implantation (embryonic diapause) is followed by embryonic differentiation lasting only about 60 days. Thus, birth of an extremely small size neonate, in comparison to maternal size, occurs during the later part of maternal starvation (hibernation in black bear). Recent evidence in the spotted hyena emphasizes the importance of relaxin production during late gestation to remodel a penile urethra for successful birth of the young. In the seasonally breeding domestic dog, the transition from proestrus includes a period of ovarian follicular growth with ovulation occurring during the first half of an 8- to 10-day estrous period. The duration (60–65 days) and similar patterns of mean plasma progesterone concentrations obtain in pseudopregnant and pregnant beagle bitches. Strong maternal behavior develops when the corpora lutea of pseudopregnancy regress.

Order Pinnipedia includes 32 species of marine mammals (eared and earless seals, sea lions, and walrus). Gestation in pinnipeds ranges from 280 to 360 days, with an embryonic diapause lasting from 60 to 150 days. Lactation can be brief (4–42 days) in earless seals but range from 3 to 12 months in eared seals and thus affect annual or biennial reproductive cycles.

Order Proboscidea includes two living species of Indian and African elephants. Gestation lasts 21 or 22 months and lactation 24 months. During pregnancy generations of corpora lutea of different sizes persist, but only the largest corpora lutea (30–56 g) are associated with pregnancy. There are marked differences in peripheral plasma progesterone concentrations in nonpregnant (208–215 pg/ml) and pregnant (416–482 pg/ml) African elephants.

Order Hydracoidea includes six living species and 30 subspecies of herbivorous hyrax in Africa. In the rock hyrax, gestation is about 225 days with birth of one or two young. The corpora lutea of pregnancy develop and are maintained until regressing just preceding parturition.

Order Perissodactyla (odd-toed ungulates) includes the family equidae (horse, donkey, and zebra; seven species), four species of tapins, and two genera of rhinoceros. A remarkable feature of equidae is the wide range in chromosome number (44–66) and interspecies breeding with viable young (horse × donkey) to produce nonfertile mule and hinney as well as zebra × horse and zebra × donkey nonfertile offspring. Domestication and selection have produced vast differences in overall size (Shetland vs Shire) and performance traits (draft vs racing). A well-defined sexual season dependent on increasing photoperiod defines most breeds of horses. Estrous cycles of 19–22 days are characterized by estrus lasting 3–7 days with ovulation occurring about 14 hr before termination. Gestation lasts about 360–365 days, followed by lactation and a foal-estrus (heat) that usually is nonfertile. Mares produce peak amounts of placental pregnant mare serum gonadotropin (PMSG) by Day 60 of pregnancy; relaxin of primarily placental origin is produced during the later two-thirds of pregnancy. PMSG production is reduced markedly in horse × donkey crosses. Secondary corpora lutea are a source of progesterone and a coincident demise of the primary corpus luteum. Parturition usually occurs during isolation from handler in the nighttime. Although the incidence of multiple ovulation increases in older mares, twinning rarely occurs. Gestation in the black rhinoceros lasts 438–476 days. Estrous cycles of 21–30 days are characterized by estrus lasting 1–4 days and copulation up to 60 min.

Order Artiodactyla [even-toed (hooves) ungulates] includes 194 living species with suborders suiformes (domestic pig and warthog), tayassuidae (peccaries), hippopotamidae (hippopotamus), tylopoda (camel and alpaca), and ruminatia [deer, giraffe, pronghorn, buffalo, ox, oryx, waterbuck, leche, kob (antelope), gnu, wildebeest, impala, gazelle, dik-dik, sheep, and goat). Notable features of the domesticated pig include gestation of 114–116 days with litter size ranging from 8 to 12 piglets. Ovulation occurs 38–42 hr after onset of estrus. The corpora lutea are the only source of progesterone during pregnancy; they also produce relaxin that is released in peak amounts about 14 hr before parturition. Luteal maintenance of progesterone secretion during early pregnancy requires pituitary LH and then shifts to pituitary prolactin after Day 65 of pregnancy. Lactation suppresses ovarian cyclicity; conception can occur at the postweaning estrus. The hippopotamus is monotocous with a gestation of 240 days. The corpus luteum by midpregnancy reaches a maximum diameter of 5 cm with little decrease in size by parturition. Lactation lasts 10–12 months. Camels are polyestrus throughout the year. Gestation lasts about 406 days. Ovulations range from one to three and implantation occurs in the left uterine horn regardless of side (right/left) of the functional corpus luteum. Lactation does not inhibit postpartum cyclicity. Gestation averages 240 days in alpaca, 280 days in vicuña, and 308 days in llama and guanaco. In the alpaca the right uterine horn has a local luteolytic effect on the corpus luteum and the left uterine horn has a local and systemic luteolytic effect. Reproductive strategies in various species of deer include rut (breeding season) and delayed implantation (embryonic diapause) with an overall gestation of 200–231 days. Progesterone blood concentration peaks at implantation and remains elevated before decreasing within 30 days of parturition. Giraffes experience brief estrous cycles of 12–15 days but long gestations of 420–450 days. Fetal ovaries also contain two to six corpus luteum-like structures that secrete progesterone. Reproduction in the family bovidae has been well characterized, being most extensive in *bos taurus* (domesticated cattle). Estrous cycles average 20 or 21 days with ovulation occurring 12–14 hr after an estrus period of 18–20 hr. Gestation ranges from 278 to 300 days, likely as a result of selection for production traits of milk production, fecundity, meat quality, and environment. The maternal–calf bonding and suckling intensity affect the duration of postpartum anestrus. Although cows of various species usually produce one calf, genetic selection of beef breeds has resulted in twinning rates exceeding 25%. The corpus luteum is the primary source of progesterone during pregnancy; recent evidence indicates isolation of mRNA encoding a relaxin-like factor from granulosa cells of ovarian follicles, but this was not found in luteal cells of the mature corpus luteum. Implantation usually occurs in the uterine horn adjacent to the ovulating ovary. The freemartin condition (genetic female with rudimentary reproductive duct development) occurs with high incidence (>85%) when a male is cotwin with a female fetus. Antimullerian hormone produced by Sertoli cells of the fetal testis suppresses uterine development in the cotwin at 45–50 days of gestation and renders the female infertile. Domesticated sheep and goats (subfamily caprinae; 24 species) are seasonal breeders producing one to five young, but the freemartin condition does not obtain. Onset of the breeding season is initiated by decreasing photoperiod. Most breeds of sheep have estrous cycles of 16 or 17 days, estrous periods about 30 hr duration, and gestation of 144–152 days. The corpus luteum is required for progesterone production during the first 50 days of pregnancy, with placental production of the hormone sufficient during the later two-thirds of pregnancy. The corpus luteum of pregnancy is maintained by the synergistic action of prolactin and LH. Extensive investigation in the endocrinology of gestation and parturition in the sheep has revealed an important role of the maturing fetal hypothalamic–pituitary–adrenal axis for secretion of corticotropin-releasing hormone, adrenocorticotropin, and cortisol in initiating the timing of birth. Goats depend on corpus luteum production of progesterone throughout pregnancy.

V. NONHUMAN PRIMATES

Pregnancy in four groups of primates [prosimians (galago, lemur, and tree shrew), Old World monkeys (talapoin, rhesus, and patas monkeys; chacma baboon; bonnet; and stump-tailed and Taiwan macaques), New World monkeys (marmoset and white-fronted capuchin monkey), and great apes (chimpanzee, gibbon, and gorilla)] is described briefly. Menstrual cycles range from 9 to 40 days; rhesus monkeys average 28 days. The length of pregnancy varies greatly among primate species (from 50 days in common tree shrew to 265 days in gorilla). Elevated circulating concentrations of progesterone required for the maintenance of pregnancy are secreted initially from a corpus luteum, whose rescue depends on continued stimulation by chorionic gonadotropin. The corpus luteum eventually regresses, with the fetoplacental unit assuming the major steroidogenic support for the remainder of pregnancy. Luteolysis may result from estradiol and prostaglandin release.

VI. REPRODUCTIVE STRATEGIES OF MAMMALS

In all mammalian species the basic reproductive cycle is that of pregnancy. Females in the wild need to be pregnant and lactate in order to perpetuate the species. Ovarian cycles reflect these conditions. With onset of puberty, a goal of estrus or recurrent estrous cycles is attraction of the male and conception. Progesterone is required throughout pregnancy in all mammalian species. In many species (i.e., rabbit, mouse, pig, and goat) the corpus luteum is the only source of progesterone production, whereas the placenta takes over this function in others (i.e., sheep and rhesus monkey). Functional corpora lutea prepare the uterus for implantation of the conceptus. If the female does not conceive, the life span of the corpus luteum may be extended to a pseudopregnancy equal to (i.e., ferret and dog) or less than (i.e., rat and rabbit) that of normal pregnancy (Table 2).

In other species (i.e., pig, cow, and guinea pig) the life span of the corpus luteum is finite regardless of whether mating occurred. Fertilization occurs in the oviduct and implantation in the uterus. In monotremes and marsupials embryogenesis continues outside the uterus, whereas it continues to full term in nonprimate eutherians and nonhuman primates.

The ephemeral nature of the corpus luteum in its diverse patterns of progesterone secretion during nongravid and gravid states obtain for all mammalian species. Thus, luteinization of the postovulatory follicle initiates a series of intermediate stages, such as ovoviviparity in monotremes, a limited form of viviparity in marsupials, and progesterone secretion by the corpus luteum for at least the initial stages of pregnancy in eutherians. Characteristics of mammalian corpora lutea include (i) an ability to secrete progesterone autonomously in the absence of any known extrinsic stimulus in several species, (ii) the luteolytic effect of prostaglandins of uterine as well as ovarian origin, and (iii) an ability to extend luteal function by extrinsic factors [gonadotropic support (LH, prolactin, and in some cases follicle-stimulating hormone)] as required for the species. In eutherians, the evolution of the trophoblastic placenta assumed an important role in maintaining pregnancy by its ability to produce progesterone as well as hormones with luteotropic properties while protecting the embryo from immune rejection by the mother. During nongravid cycles the progesterone secretion by the corpora lutea may be deficient, resulting in only intermittent cycles of 4 or 5 days (i.e., mouse, rat,

TABLE 2
Degree of Maintenance of Luteal Function after Total Hysterectomy in Various Species Compared with Duration of Pregnancy

Species	Normal estrous cycle	Cycle after hysterectomy	Normal pseudopregnancy	Pseudopregnancy after hysterectomy	Normal pregnancy
Hamster	4	4	8–10	16–23	16–19
Rat	4–5	4	12–14	18–25	21–22
Rabbit	—	—	15–16	25–29	30–32
Guinea pig	15–18	80–110	—	—	65–70
Sheep	16–18	160–170	—	—	145–150
Pig	20–22	>140	—	—	114–115
Cow	18–22	>270	—	—	275–290

and hamster), but capable of an extended life span (pseudopregnancy) by neurogenic stimuli or finite (i.e., guinea pig, pig, cow, and sheep; Table 2). An intriguing aspect of corpus luteum physiology is the influence of the uterus in control of luteal regression. In some species uterine removal has no effect on luteal life span when hysterectomized during a normal estrous cycle, but it can extend pseudopregnancy equivalent to that of pregnancy. In species in which the functional corpora lutea produce sufficient amounts of progesterone during an estrous cycle, hysterectomy extends luteal life span equivalent to or exceeding that of normal pregnancy (Table 2). Hysterectomy of nongravid primates, however, is without effect on progesterone secretion during subsequent menstrual cycles (Table 3). Thus, in monotremes and marsupials, progesterone secretion by the corpus luteum is brief and not influenced by the fertilized shell-covered egg (monotremes) or small embryo (marsupials) in the uterus. In eutherians, the conceptus(es) plays an important role in negating a luteolytic effect of the uterus, resulting in sustained progesterone secretion throughout pregnancy, and even in determining the time for its own birth. Thus, the reproductive cycle may be defined as that of pregnancy (whether brief or prolonged) and similar to nongravid cycles or extended by removal of the nongravid uterus. Such an entrained reproductive cycle, even in nongravid animals, is indicated by peak release of relaxin and decreased progesterone secretion by the corpora lutea in unmated hysterectomized pigs coincident with prepartum relaxin release and progesterone decrease during the last hours of normal pregnancy in the pig.

Lactation in mammals (milk production and its constituents and duration) is directly related to the degree of embryonic development at the time of birth. For example, in monotremes and marsupials a prolonged lactation is required for completion of embryogenesis and growth of the young. In eutherians, lactation, whether of short (4 days) or long (900 days) duration, is a function of maturity of neonatal development and independence for survival at birth. For example, newborn guinea pigs have a full hair

TABLE 3
Effect of Hysterectomy on Life Span of Corpus Luteum in Various Species

		Reproductive state		
Order	Species	Unmated	Pseudopregnancy	Pregnancy
Marsupialia	Opossum (Americas)	No change		
	Brush possum (Australasia)	No change		
Rodentia	13-lined ground squirrel	No change		No change
	Mouse	No change	Increase	
	Rat	No change	Increase	Decrease
	Golden hamster	No change	Increase	Decrease
	Guinea pig	Increase		
Lagomorpha	Rabbit	No change	Increase	Decrease
Carnivora	Dog	No change		
	Ferret	No change	No change	
	Cat	No change	No change	
	Western spotted skunk			No change
	Badger			No change
Artiodactyla	Swine	Increase		Increase
	Cattle	Increase		
	Sheep	Increase		No change
Primates	Monkey	No change		
	Man	No change		

coat and eat leafy food within minutes after birth and the dam (sow) secretes little milk. In contrast, gorillas provide parental care and protection during a lactation that can exceed 24 months.

See Also the Following Articles

CORPUS LUTEUM; MARSUPIALS; MONOTREMES; PRIMATES, NONHUMAN; PSEUDOPREGNANCY

Bibliography

Anderson, L. L., and Musah, A. I. (1989). Uterine control of ovarian function. In *Biology of the Uterus* (R. M. Wynn and W. P. Jollie, Eds.), 2nd ed., pp. 505–557. Plenum, New York.

Anderson, L. L., Bland, K. P., and Melampy, R. M. (1969). Comparative aspects of uterine–luteal relationships. *Recent Prog. Horm. Res.* **25**, 57–104.

Asdell, S. A. (1964). *Patterns of Mammalian Reproduction*, 2nd ed. Comstock, Ithaca, NY.

Augee, M. L. (1992). *Platypus and Echidnas*. Royal Zoological Society of New South Wales, Sydney.

Blackburn, D. G., Haysen, V., and Murphy, C. J. (1989). The origins of lactation and the evolution of milk: A review with new hypothesis. *Mammal Rev.* **19**, 1–26.

Carrick, F. N., Drinan, J. G., and Cox, R. I. (1975). Progestagens and oestrogens in peripheral plasma of the platypus *Orinithorhynchus anatinus*. *J. Reprod. Fertil.* **43**, 375–376.

Felder, K. J., Molina, J. R., Benoit, A. M., and Anderson, L. L. (1986). Precise timing for peak relaxin and decreased progesterone secretion after hysterectomy in the pig. *Endocrinology* **119**, 1502–1509.

Griffiths, M. (1984). Mammals: Monotremes. In *Marshall's Physiology of Reproduction* (G. E. Lamming, Ed.), 4th ed., Reproductive Cycles of Vertebrates, Vol. 1, pp. 351–385. Churchill Livingstone, New York.

Grzimek, B. (1990). *Encyclopedia of Mammals*. McGraw-Hill, New York.

Hayssen, V. D., Van Tienhoven, A., and Van Tienhoven, A. (1993). *Asdell's Patterns of Mammalian Reproduction: A Compendium of Species-Specific Data*. Cornell Univ. Press, Ithaca, NY.

Nowak, R. M. (1991). *Walker's Mammals of the World*, 5th ed. Johns Hopkins Univ. Press, Baltimore.

Renfree, M. B. (1993). Ontogeny, genetic control, and phylogeny of female reproduction in monotreme and therian mammals. In *Mammal Phylogeny: Mesozoic Differentiation, Multituberculates, Monotremes, Early Therians, and Marsupials* (F. S. Szalay, M. J. Novacek, and M. C. McKenna, Eds.), pp. 4–20. Springer-Verlag, New York.

Rothchild, I. (1981). The regulation of the mammalian corpus luteum. *Recent Prog. Horm. Res.* **37**, 183–298.

Rowlands, I. W., and Weir, B. J. (1984). Mammals: Nonprimate eutherians. In *Marshall's Physiology of Reproduction* (G. E. Lamming, Ed.), 4th ed., Reproductive Cycles of Vertebrates, Vol. 1, pp. 455–658. Churchill Livingstone, New York.

Tyndale-Biscoe, C. H., and Renfree, M. (1987). *Reproductive Physiology of Marsupials*. Cambridge Univ. Press, New York.

Wilson, D. E., and Reeder, D. M. (1993). *Mammal Species of the World: A Taxonomic and Geographic Reference*, 2nd ed. Smithsonian Institution Press, Washington. DC.

Pregnancy, Maintenance of

Fuller W. Bazer

Texas A&M University

I. Pregnancy Maintenance in Ruminants
II. Pregnancy Maintenance in Rodents
III. Pregnancy Maintenance in Swine
IV. Pregnancy Maintenance in Horses
V. Pregnancy Maintenance in Rabbits
VI. Pregnancy Maintenance in Cats
VII. Pregnancy Maintenance in Dogs

GLOSSARY

conceptus The embryo/fetus and associated membranes.

corpus luteum The endocrine gland(s) formed by the granulosa and theca cells of the ovarian follicle after ovulation that secretes progesterone which is necessary for establishment and/or maintenance of pregnancy.

luteolytic hormones Hormones that cause the functional and structural demise of the corpus luteum with prostaglandin $F_{2\alpha}$ of uterine or intraovarian origin considered the major luteolysin.

luteotrophic hormone Luteinizing hormone from the anterior pituitary gland or chorionic gonadotropin from the placenta of most primates.

prostaglandin $F_{2\alpha}$ The luteolytic hormone that causes the functional and structural demise of the corpus luteum.

Pregnancy is established and maintained in subprimate mammals in response to a series of interactions between the conceptus (embryo and associated membranes), uterus, and/or ovarian corpus luteum (CL). These interactions prevent functional and structural regression of the CL, or luteolysis. During the periimplantation period, pregnancy recognition signals from the conceptus to the maternal system are antiluteolytic and/or luteotrophic. The functional life span of the CL is controlled by release of prostaglandin $F_{2\alpha}$ (PGF) from the uterus and/or ovaries, whereas pregnancy recognition signals from the trophoblast act in a paracrine or endocrine manner to interrupt endometrial or intraovarian production of luteolytic PGF (antiluteolytic) or the effect may be directly on the CL (luteotrophic).

I. PREGNANCY MAINTENANCE IN RUMINANTS

An ovulatory surge of luteinizing hormone (LH) coincident with onset of estrus (Day 0) initiates events which culminate in ovulation about 30 hr later. The antiluteolytic signal for pregnancy recognition in ruminants is interferon-τ (IFN-τ) produced by mononuclear cells of the embryonic trophectoderm. The IFN-τ exerts a paracrine, antiluteolytic effect on the endometrium to inhibit endometrial production of luteolytic pulses of prostaglandin $F_{2\alpha}$ (PGF). Other conceptus and/or uterine products secreted during early pregnancy, e.g., PGE and platelet-activating factor, may exert secondary luteal protective effects. The mechanism for pregnancy recognition is similar, if not identical, for ruminants such as sheep, cattle, and goats.

Luteolysis is initiated by increased expression of estrogen receptors (ERs) and subsequently oxytocin receptors (OTRs) by the uterine endometrial epithelium. Oxytocin secreted by the corpus luteum (CL) and posterior pituitary binds to the OTRs and stimulates uterine secretion of a pulse of PGF. The IFN-τ inhibits ER and OTR gene transcription in endometrial epithelium to abrogate oxytocin-induced release of luteolytic pulses of PGF and protects CL function. Nevertheless, interestrous intervals of 30–35 days result when conceptuses are flushed from the uterus of ewes on Day 16 or after intrauterine infusions of either highly purified native oIFN-τ or roIFN-τ on Days 11–15. Thus, an endometrial luteolytic mechanism is activated between 14 and 16 days postestrus which must be abrogated by the conceptus. Although the mechanism has not been established, placental lactogen (oPL) may be a second "signal" from the conceptus that reinforces effects of IFN-τ to allow both establishment and maintenance of pregnancy in ruminants.

Binucleate cells in trophoblast of sheep conceptuses appear by Day 16, migrate into the uterine epithelium, and secrete oPL. Placentae of other ruminant species secrete hormones structurally related to pituitary growth hormone (GH) and prolactin (PRL), which are termed placental lactogens (PLs). oGH and oPL have similar somatogenic activities; however, their circulating levels are regulated differently during the pregnancy. oPL is detected in maternal serum by Day 50 and concentrations peak at 120–130 days of gestation. oPL is lower in fetal serum and peak levels at midgestation may remain stable or decrease to term. oGH is low in maternal serum about 12 days during pregnancy and concentrations do not correlate with gestational age. oGH is higher in fetal serum and values are highest at mid- and late gestation. oPL may be critical to maintenance of pregnancy in sheep since lactogenic hormones influence steroidogenesis in CL and transport of water by placental membranes. Lactogenic hormones also stimulate proliferation, PR gene expression, protein synthesis, and exocrine secretion of PGF (pigs) by uterine endometrium, as well as mammary growth and lactation. oGH from sheep placenta may also have a role(s) in pregnancy.

II. PREGNANCY MAINTENANCE IN RODENTS

Gestation in rodents lasts 20–22 days, and functional CL must be maintained until Day 17. The transition in rodents (i.e., rats, mice, and hamsters) from recurrent estrous cycles to pregnancy is dependent on maintenance of progesterone production by the CL, the main source of progesterone throughout pregnancy. Replacement therapy with progesterone alone is sufficient to maintain pregnancy in ovariectomized rats. In addition to lacking a true luteal phase during the estrous cycle, rodents do not exhibit a change in source of progesterone from the CL to the placenta. Thus, maternal recognition of pregnancy in rodents involves activation of nonfunctional CL of the cycle into functional CL of pregnancy.

Mating of rodents during estrus results in pseudopregnancy or pregnancy, and the activated CL secrete progesterone for 12–14 days. Extension of CL life span past 12 days is dependent on the presence of viable conceptuses within the uterus. A successful pregnancy in rats requires active progesterone secretion from CL until at least Days 17 or 18. Therefore, establishment and maintenance of pregnancy in rodents requires two separate endocrine events. The first endocrine event is mating, which induces both diurnal and nocturnal surges of prolactin that last about 12 days during pregnancy or pseudopregnancy in rats. The increase in prolactin secretion is necessary for maintenance of active CL and progesterone secretion. Therefore, the luteotrophic effects of PRL during early pregnancy are required to convert the nonfunctional CL of the estrous cycle into functional CL of pregnancy or pseudopregnancy.

The second endocrine event required for the maintenance of pregnancy in rodents is dependent on implantation and development of normal conceptuses. In pregnant rodents, the placenta and decidua produce PRL-like hormones which exert luteotrophic effects on CL to ensure production of progesterone during the middle and late stages of pregnancy. The antimesometrial uterine decidua secretes numerous hormones, including PRL-like protein B (PLP-B) and decidual PRL-like protein (dPRP). The main luteotropic hormone of the decidua is a PRL-like protein, although decidual tissue also expresses PLP-B.

PLs are found in a variety of subprimate mammals, including rodents. However, the only established physiological roles for PLs are in rodents. In rats, seven members of the PRL gene family are expressed by trophoblast cells of the placenta: placental lactogen-I (PL-I), PL-I variant (PL-Iv), PL-I mosaic (PL-Im), PL-II, PRL-like protein A (PLP-A), PLP-B, and PLP-C. Both PL-I and PL-II have biological activities similar to those of pituitary PRL, including maintenance of CL functions and stimulation of mammary gland growth.

Factors regulating production of PLs by trophoblast cells are not well-known. The ontogeny of PL-I expression appears to be linked to trophoblast cell differentiation during the periimplantation period. In mice, PL-I gene expression appears to be regulated by the number of conceptuses and by the pituitary gland. Similarly, factors affecting PL-II gene expression include the number of conceptuses, genotype of the conceptuses, pituitary via GH, ovarian steroids, and the nutritional status of the mother. The shift from PL-I to PL-II expression is temporally associated with degeneration of the choriovitelline placenta on Days 13 and 14 of gestation.

In rats, PRL secretion is essential for maintenance of progesterone secretion by CL throughout pseudopregnancy. However, in the pregnant rat, removal of the anterior pituitary after midgestation does not affect luteal function and pregnancy is maintained. Thus, pituitary PRL is not necessary after Day 6 of gestation in rodents. Given their PRL-like activity and their ontogeny during pregnancy, PLs are likely placental luteotrophic factors during mid- to late pregnancy in rodents.

III. PREGNANCY MAINTENANCE IN SWINE

Estrogens produced by conceptuses on Days 11 and 12 of gestation provide the initial signal for maternal recognition of pregnancy in swine. This signal results in a switch in the direction of endometrial PGF secretion from one that is endocrine (i.e., toward the maternal blood) to one that is exocrine (i.e., toward the uterine lumen). A second period of estrogen production occurs between Days 15 and 25–30 of pregnancy. Injection of exogenous estrogen (estradiol valerate, 5 mg/day) on Days 11–15 of the estrous cycle will result in CL maintenance for a period equivalent to or slightly longer than pregnancy. This condition, referred to as pseudopregnancy, persists for about 120 days. Thus, two phases of exogenous estradiol, similar to that produced by conceptuses on Days 11–13 and Days 15 to 25–30, are necessary for prolonged secretion of PGF into the uterine lumen. Estradiol may induce receptors for maternal hormones, e.g., prolactin, or conceptus secretory proteins, which influence exocrine (into the uterine lumen) secretion of PGF. The first estrogen signal may induce those receptors and the second estrogen signal may be required to replenish those receptors. Administration of estradiol on Day 9 advances the uterine secretory response in pregnant gilts which leads to conceptus death by Day 16. An explanation for this "induced" conceptus death is not available, but it may result from asynchrony between the developing conceptus and uterine environment.

Estrogen induces endometrial receptors for prolactin in pigs, and prolactin acts on the endometrium to induce calcium cycling across the epithelium. In pigs, PRL interacts with estrogen and progesterone to increase total recoverable uteroferrin, glucose, and PGF in uterine flushings. Available results strongly indicate that prolactin enhances uterine responsiveness to progesterone during periods critical for maintenance of pregnancy.

IV. PREGNANCY MAINTENANCE IN HORSES

The uterine luteolytic hormone in mares is PGF and the conceptus appears to inhibit production of PGF by the uterine endometrium. In cycling mares, concentrations of PGF in uterine venous plasma and uterine flushings increase between Days 14 and 16 when luteolysis occurs and plasma progesterone levels decline. The amount of PGF bound by luteal receptors is maximal on Day 14 of the estrous cycle and Day 18 of pregnancy. Since CL of mares can respond to circulating PGF during pregnancy, the conceptus must evoke an antiluteolytic mechanism. Pregnant mares have little PGF in uterine fluids, PGF in uterine venous plasma is reduced, and PGFM in peripheral plasma has no episodic pattern of release. In the presence of the conceptus, endometrial production of PGF in response to cervical stimulation and exogenous oxytocin is markedly reduced, indicating the absence/reduction of endometrial receptors for oxytocin in mares during early pregnancy.

The pregnancy recognition signal in mares is not known, but evidence has been presented to indicate that estradiol and/or proteins from the conceptus are the most likely candidates. There is some evidence that PGEs may also play such a role. The equine conceptus migrates between uterine horns 12–14 times per day on Days 12–18 of pregnancy, presumably to inhibit endometrial PGF production and protect the CL. Thus, the equine conceptus suppresses PGF production by the endometrium, but the factor responsible has not been identified. The equine conceptus also produces increasing amounts of estradiol between Days 8 and 20 of gestation. A similar trend, but of greater magnitude, was found for estrone. Attempts to prolong CL life span in mares by injection of estrogens have been inconsistent.

Horse conceptuses secrete three major proteins between Days 12 and 14 of pregnancy with molecular weights of ≧400,000, 50,000, and 65,000. However, the role(s) of these proteins is not known. Estrogens and/or conceptus secretory proteins may provide the maternal recognition of pregnancy signal in the mare by directly or indirectly inhibiting endometrial production of luteolytic pulses of PGF.

V. PREGNANCY MAINTENANCE IN RABBITS

Rabbits are induced ovulators and multiple ova are ovulated from each ovary approximately 10 hr pc and the eggs remain fertilizable for about 6 hr. The fertilized ova arrive in the uterus 3 days postovulation and implantation occurs on Day 7 at the blastocyst stage of conceptus development. The rabbit placenta is not a significant source of progesterone and the CL are required for pregnancy to go to term. Following sterile mating, CL form and persist for 14–16 days without support from conceptus products. For both pseudopregnant and pregnant does, progesterone begins to increase 2 days after mating to maximal levels of 12–20 ng/ml between Days 6 and 8 pc. Between Days 8 and 10 pc, progesterone profiles of pregnant and pseudopregnant does begin to diverge with levels declining rapidly after Day 12 to basal levels between Days 16 and 18 of pseudopregnancy. Pregnant does exhibit elevated progesterone levels which begin to decline 3 or 4 days prior to parturition (kindling).

Maternal recognition of pregnancy occurs after implantation between Days 10 and 12 of gestation. Estrogen and a placental luteotropin interact to maintain progesterone production until term (28–35 days). Estrogen from developing follicles is required for luteal progesterone production for the first 10–12 days of pregnancy but not to term. Rabbit luteal cells contain LH receptors; however, LH does not appear to be the stimulus for progesterone production *in vivo*.

Production of a placental luteotrophin is necessary for luteal progesterone production to term and exogenous estrogen will not support progesterone production in does hysterectomized during late pregnancy. Rabbit placentae secrete immunoreactive gonadotropin-releasing hormone-like activity, but there is no evidence for it being transported from the uterus to the CL. A putative placental luteotrophic factor with a molecular weight >6–8 kDa has been shown to enhance progesterone production by cultured luteal cells. Rabbit placental giant cells also contain immunoreactive chorionic gonadotropin, and cytotrophoblast cells contain immunoreactive PL/PRL. However, the function(s) of these proteins has not been determined.

VI. PREGNANCY MAINTENANCE IN CATS

The cat, also an induced ovulator, ovulates 25–50 hr pc (about 24–36 hr after the LH peak), with frequent matings reducing the time to ovulation. Fertilization takes place in the oviduct up to 48 hr after ovulation. The embryo enters the uterus at the blastocyst stage (4–6 days postovulation), hatches on Day 11, and begins implanting by Days 12 or 13. Following mating, plasma progesterone concentrations increase from about Day 3 to maximal levels (15–90 ng/ml) between Days 10 and 40 of pregnancy or Days 13 and 30 of pseudopregnancy. Pseudopregnancy typically lasts 40 days and gestation ranges between 56 and 71 days, averaging 63–65 days. By Day 30, circulating levels of progesterone are significantly higher in pregnant than in pseudopregnant queens.

The placenta does not appear to be a significant source of progesterone during gestation in cats. Prolactin levels increase during the latter third of gestation, peak just prior to parturition (5–10 ng/ml), and are elevated during lactation in response to suckling. Prolactin is thought to be an important luteotrophin in late gestation. Relaxin is produced by the fetal–placental unit and increases to peak levels of 5–10 ng/ml during the latter half of gestation. Relaxin is thought to work in concert with progesterone to keep the uterus quiescent and to facilitate parturition by softening the connective tissues of the pelvis. Following parturition, queens experience a period of lactational anestrus and resume cycling 2 or 3 weeks after weaning kittens.

VII. PREGNANCY MAINTENANCE IN DOGS

Fertilization takes place 2–5 days after ovulation in the bitch and embryos enter the uterus at the blastocyst stage around Day 10. Embryos remain free-floating in the uterus until hatching and implantation around Day 16. The ovary is the primary source of progesterone, and ovariectomy or hypophysectomy at any stage of pregnancy results in abortion. Since CL of pregnancy and pseudopregnancy have similar life spans there is no known requirement for signaling between the conceptus and maternal system for CL maintenance, pregnancy recognition, or maintenance of pregnancy.

See Also the Following Articles

CORPUS LUTEUM; LUTEOTROPIC HORMONES; PREGNANCY, MATERNAL RECOGNITION OF; PROLACTIN, ACTIONS OF; PSEUDOPREGNANCY

Bibliography

Bazer, F. W., Ott, T. L., and Spencer, T. E. (1994). Pregnancy recognition in ruminants, pigs and horses: Signals from the trophoblast. *Theriogenology* **41**, 79–94.

Bazer, F. W., Spencer, T. E., and Ott, T. L. (1996). Placental interferons. *Am. J. Reprod. Immunol.* **35**, 297–308.

Bazer, F. W., Ott, T. L., and Spencer, T. E. (1997). Endocrinology of the transition from recurring estrous cycles to establishment of pregnancy in subprimate mammals. In *Endocrinology of Pregnancy*, Contemporary Endocrinology Series. Human Press, Totawa, NJ.

Spencer, T. E., Ott, T. L., and Bazer, F. W. (1996). τ-Interferons: Pregnancy recognition signals in ruminants. *Proc. Soc. Exp. Biol. Med.* **213**, 215–229.

Pregnancy, Maternal Recognition of

Thomas E. Spencer
Texas A&M University

I. Introduction
II. Humans and Primates
III. Rodents
IV. Domestic Animals
V. Other Subprimate Mammals
VI. Summary and Future Research

GLOSSARY

antiluteolytic signal Substance which inhibits development of a luteolytic mechanism to, in most cases, prevent luteolysis.

conceptus Embryo and associated extraembryonic membranes (placenta).

luteolysis Physiological process induced by prostaglandin $F_{2\alpha}$, in most cases, that results in functional and structural demise of the corpus luteum.

luteotropic signal Substance which supports corpus luteum function and/or inhibits intraovarian mechanisms of luteolysis.

maternal recognition of pregnancy Physiological process whereby the conceptus produces a signal which acts on the maternal system to prolong the life span of the corpus luteum which secretes progesterone, the hormone of pregnancy.

progesterone Steroid hormone produced by the corpus luteum which is the hormone of pregnancy and acts on the uterus to produce an environment permissive to conceptus growth and development to term.

Maternal recognition of pregnancy is a phrase first coined by Roger Short in 1969; it is the physiological process whereby the conceptus signals its presence to the maternal system and prolongs the life span of the corpus luteum (CL). In most mammalian species, progesterone production by the CL is required for a successful pregnancy. Progesterone acts on the uterus to stimulate and maintain uterine functions that are permissive to early embryonic development, implantation, placentation, and successful fetal and placental development to term. Prolonged life span of the CL is a characteristic feature of mammalian pregnancy in species with a gestation period that exceeds the length of a normal estrous or menstrual cycle. Interestingly, there appears to be no common mechanism for maternal recognition of pregnancy given the wide array of species differences in uterine and embryonic/fetal physiology.

I. INTRODUCTION

Once fertilized in the oviduct, the developing embryo is transported to the uterus. In the uterus, the conceptus continues to develop and must signal its presence to the maternal system. If the embryo fails to signal its presence, the mother prepares for another opportunity to mate and successfully reproduce. Depending on the mammal, the mother may undergo recurring estrous or menstrual cycles until successful maternal recognition of pregnancy. Recurrent estrous and menstrual cycles are interrelated with the ovarian cycle. The ovarian cycle is composed of repeated episodes of follicular growth, ovulation, luteinization of follicles to form a corpus luteum (CL), and regression of the CL. The CL undergoes structural and functional regression or luteolysis, which allows for another subsequent ovulation. The luteolytic signal that is common to most, if not all, mammals is prostaglandin $F_{2\alpha}$ ($PGF_{2\alpha}$). Depending on the mammal, ovarian cycles may be either uterine dependent or uterine independent depending on the source of the luteolytic mechanism.

The luteal phase of the ovarian cycle is when the CL actively synthesizes and secretes progesterone. In mammals, the life span of the CL, which is normally a transient ovarian organ, can be extended by one of several mechanisms employed by the conceptus. The placenta of primates and humans produces a luteotropic signal, usually chorionic gonadotropin, which acts in an antiluteolytic manner directly on the ovarian CL to prolong CL life span during pregnancy. In rodents, a luteotropic signal, consisting of prolactin and luteinizing hormone (LH), is secreted by the pituitary at mating. Moreover, in this species, the endometrium produces prolactin during implantation, and the conceptus produces placental lactogens which are prolactin-related luteotropic signals. In domestic animals, the conceptus produces an antiluteolytic signal that prevents development of the endometrial luteolytic mechanism. However, the life span of the CL need not always be extended. In marsupials the normal life span of the CL in pregnant and nonpregnant animals is essentially the same. Moreover, the gestation period is approximately the length of the estrous cycle. In cats and dogs, the length of the luteal phase of the estrous cycle is also similar to the length of gestation. In these species, pregnancy recognition may not be necessary. Finally, in some species, such as the rabbit, the conceptus does not seem to affect function of the CL until after implantation.

II. HUMANS AND PRIMATES

A. Luteolytic Mechanism

The menstrual cycle in humans and primates is from the onset of menses in one cycle to the onset of menses in the subsequent cycle and averages 28 days in length. Ovulation occurs in response to a surge of LH released on Day 14, and luteolysis precedes the onset of menses. The intraovarian luteolytic mechanism is poorly understood in primate mammals and results from the actions of $PGF_{2\alpha}$, oxytocin, or other undefined molecules acting independently or in concert within the ovary. Luteolysis is uterine independent because hysterectomy does not prolong CL life span. Therefore, the conceptus must act on the CL to block intraovarian luteolytic mechanisms during maternal recognition of pregnancy.

B. Pregnancy Recognition and Chorionic Gonadotropin

Near the time of implantation, the trophoblasts of human and subhuman primate conceptuses begin to synthesize and secrete chorionic gonadotropin (CG). The CG has primarily LH-like activity and acts directly on the CL to inhibit the intraovarian luteolytic mechanism and ensure continued production of progesterone. The β subunit of CG is detectable as early as the six- to eight-cell stage of human embryo development, but secretion of human CG is associated with implantation. In humans, implantation begins on Days 7–9, and CG is detectable in peripheral plasma between Days 9 and 12. A similar sequence of events occurs in other primates. Passive or active immunization of marmoset monkeys against CG results in luteolysis and termination of pregnancy. Thus, CG appears to be the primary mediator of maternal recognition of pregnancy and is a luteotropic signal that directly acts on the CL to prevent luteolysis.

III. RODENTS

A. Luteolytic Mechanism

Rodents (rats, mice, and hamsters) are spontaneously ovulating mammals with an estrous cycle that lasts only 4 or 5 days. In these species the CL secretes progesterone for approximately 2 days, and it never becomes fully functional during the estrous cycle. Therefore, rodents lack a "true" luteal phase that is characteristic of humans, primates, and a number of subprimate mammals. The life span of the rodent CL is purposefully limited to produce short, recurrent estrous cycles that ensure numerous opportunities for mating and pregnancy. The exact nature and underlying mechanisms of intraovarian luteolysis in rodents are not well understood.

Newly formed CLs attain maximal size during diestrus and are maintained throughout metestrus of the following cycle. These CLs abruptly regress during diestrus of the second cycle. The CLs in an unmated animal are referred to as "nonfunctional" CLs because they do not secrete sufficient quantities of progesterone to permit a decidual reaction in the uterus. Progesterone is also metabolized by 20α-hydroxysteroid dehydrogenase (20α-HSD) in the CL to 20α-hydroxyprogesterone, an inactive progesterone metabolite. This metabolite of progesterone is not an active progestin and will not support pregnancy or a decidual reaction in the uterus.

B. Pregnancy Recognition

Gestation in rodents lasts 20–22 days, and a functional CL must be maintained until Day 17, a period virtually equal to that of pregnancy. As in other subprimate mammals, the transition from recurrent estrous cycles to pregnancy is entirely dependent on maintenance of progesterone production by the CL. The main source of progesterone throughout pregnancy is the ovarian CL. In addition to lacking a true luteal phase during the estrous cycle, rodents also lack a shift in progesterone production from the CL to the placenta that is characteristic of many subprimate mammals, such as sheep and cattle, during mid- to late pregnancy.

Maternal recognition of pregnancy in rodents involves rescuing the "nonfunctional" CL of the cycle and subsequent conversion into the "functional" CL of pregnancy. Sterile mating or cervical stimulation of rodents during estrus results in pseudopregnancy, and the CL continues to secrete progesterone for 12–14 days. Function of the CL past 12 days is dependent on the presence of viable embryos within the uterus. Two separate endocrine events appear to be required for pregnancy recognition in rodents. In the first endocrine event, surges of prolactin from the pituitary are produced via activation of a neural reflex arc in response to either mating or cervical manipulation. In rats, cervical stimulation results in both diurnal and nocturnal surges of prolactin (PRL) that last throughout pseudopregnancy, and this phenomenon also occurs in pregnant rats. Likewise, cervical stimulation by either mating or mechanical means on the morning of estrus leads to increased serum concentrations of PRL and progesterone. The increase in PRL is necessary for maintenance of active progesterone secretion. Injections of PRL will maintain the CL in cycling rats, whereas injection of a PRL secretion inhibitor, ergocornine, terminates pseudopregnancy. The PRL luteotropic signal during early pregnancy is required to rescue the nonfunctional CL and convert it into the functional CL of pregnancy. The second endocrine event required for the maintenance of pregnancy in rodents is dependent on implantation and normal development of the embryo. In pregnant rodents, the decidua and placenta produce PRL and placental lactogens, respectively, which are luteotropic signals to the CL and ensure production of progesterone during the middle and late stages of pregnancy to term.

C. Luteotropic Actions of Prolactin

During the first half of pregnancy, PRL and LH are two major components of the luteotropic complex. The mating-induced secretion of PRL appears to increase the number of luteal receptors for LH and estrogen. The LH is necessary to maintain synthesis of estrogen and progesterone by the CL. The estrogen produced by the CL is thought to stimulate progesterone synthesis by an unknown mechanism which probably involves a component of the steroidogenic pathway from cholesterol. The actions of decidual PRL and placental and decidual PRL-related proteins during the latter half of gestation maintain the synthesis of estrogen and progesterone in the CL. This luteotropic signaling mechanism has some striking similarities to that in the rabbit. A reciprocal relationship exists between plasma levels of progesterone and 20α-dihydroprogesterone during pregnancy. Recently, PRL has been demonstrated to suppress the activity of 20α-HSD in the CL, which suppresses the conversion of progesterone to 20α-hydroxyprogesterone. The activity of 20α-HSD in the rat CL is suppressed by PRL during the luteal phase, but 20α-HSD enzyme activity increases at the end of pseudopregnancy. Moreover, the twice-daily surges of PRL in response to mating also cease between Days 12 and 14 postmating in both pregnant and psuedopregnant rodents. Therefore, another luteotropic signal must be involved in maintaining progesterone production by the CL to term.

Implantation in rodents involves the transformation of stromal cells into decidual cells—a process referred to as decidualization. The presence of decidual cells affects the production of progesterone by the CL. The hormones produced the antimesometrial decidua sustain luteal production of progesterone. The main luteotropic action of the decidua is through production of decidual PRL-like proteins (dPRPs), members of the PRL family. Decidual PRPs bind to PRL receptors on luteal cell membranes and apparently have the same effects on luteal cell function to suppress 20α-HSD, thereby stimulating active progesterone secretion. Moreover, dPRP binds receptors in the mesometrial decidua and upregulates expression of several other hormones. Therefore, dPRP appears to have actions on both the ovary and the decidua itself.

D. Luteotropic Actions of Placental Lactogens

Placental lactogens (PLs) are found in a variety of subprimate mammals, including rodents and ruminants. However, clear physiological roles for these PLs have only been established in rodents. In the rat, seven different members of the PRL gene family are expressed in trophoblast cells of the placenta: placental lactogen-I (PL-I), PL-I variant (PL-Iv), PL-I mosaic (PL-Im), PL-II, PRL-like protein A (PLP-A), PLP-B, and PLP-C. The initial members of this family (PL-I and PL-II) were identified by their biological activities, which are similar to that of pituitary PRL, including maintenance of the CL and stimulation of mammary gland growth. As mentioned previously, two members of the PRL gene family, dPRP and PLP-B, are expressed in the uterus. Although dPRP is exclusively expressed by antimesometrial decidual cells, PLP-B appears to be expressed in both the placenta and decidua.

Given that members of the placental PRL-like family display structural and functional similarity to PRL and twice-daily surges of PRL cease when secretion of PLs start in the rat, PLs are proposed to replace the luteotropic actions of pituitary PRL. In support of this contention, PL-I, PL-Iv, PL-Im, and PL-II bind the PRL receptor and may activate a signal transduction pathway similar to other lactogenic hormones. In contrast, PLP-A, PLP-B, and PLP-C do not bind the PRL receptor, and their activities are largely unknown. Given that PLs have PRL-like activity and ontogeny during pregnancy, the PLs are most likely the placental luteotrophic factors responsible for the maintenance of CL function during mid- to late pregnancy in rodents. Recent results have demonstrated the direct luteotropic actions of mouse PL-I and PL-II during midpregnancy.

IV. DOMESTIC ANIMALS

A. Ruminants

Sheep, cattle, and goats are spontaneously ovulating, polyestrous mammals that exhibit recurrent estrous cycles of 17, 21, and 20 days, respectively. In these ruminants, the ovarian cycle is uterine dependent because hysterectomy prolongs CL life span. The estrous cycle of ruminants is dependent on the uterus because the endometrium produces the luteolysin $PGF_{2\alpha}$. The antiluteolytic signal for pregnancy recognition in ruminants is trophoblast interferon-τ (IFN-τ). The IFN-τ exerts a paracrine, antiluteolytic effect on the endometrium to inhibit endometrial production of luteolytic pulses of PGF. Other conceptus and/or uterine products secreted during early pregnancy, such as PGE and platelet-activating factor, may exert secondary luteal protective effects. Although each species has some unique features, general mechanisms of the ovarian cycle, the luteolytic mechanism, and pregnancy recognition appear to be similar, if not identical, for sheep, cattle, and goats.

1. Luteolytic Mechanism in Sheep

In sheep, estrus is denoted as Day 0, when the ewe will first accept the ram for mating. An ovulatory surge of LH initiates events which culminate in ovulation about 30 hr later. With maturation of the CL, concentrations of progesterone in peripheral blood are maximum in diestrus (Days 10–14) and, in cyclic females, luteolysis is induced by pulsatile release of $PGF_{2\alpha}$ from endometrial epithelium during late diestrus (Days 15 or 16). If ewes are hysterectomized during the active life of the CL, luteolysis does not occur and CL life span is prolonged to about 5 months, the duration of normal pregnancy. In ruminants an intimate anatomical relationship between the uterine branch of the ovarian vein and the ovarian artery is required for countercurrent exchange of luteolytic $PGF_{2\alpha}$ from the uterine venous drainage to the ovarian artery. Uterine $PGF_{2\alpha}$ is also transported from the lymphatic drainage to the ovarian artery. Injection of ewes with exogenous $PGF_{2\alpha}$ causes premature luteolysis, and immunization of ewes against $PGF_{2\alpha}$ blocks luteolysis. Uterine-derived $PGF_{2\alpha}$ binds to receptors on luteal cells and initiates intracellular events which terminate production of progesterone, initiate cell death, and culminate in CL regression. Uterine secretion of luteolytic pulses of $PGF_{2\alpha}$ in sheep is dependent on effects of progesterone, estrogen, and oxytocin on the uterine lumenal epithelium (LE) and superficial glandular epithelium (sGE).

John McCracken originally proposed that the ovarian steroid hormones, estrogen and progesterone, regulate the oxytocin receptor (OTR) gene expression in the endometrial epithelium. Progesterone inhibits endometrial OTR synthesis for 10–12 days; a phenomenon termed the "progesterone block" to endometrial OTR formation. Results from *in vivo* studies clearly indicate that progesterone progressively suppresses expression of PR and, therefore, progesterone loses its ability to suppress endometrial expression of ER and OTR during mid- to late diestrus. Moreover, a preponderance of evidence suggests that estrogen induces expression of OTR in the LE and sGE, which is responsible for production and release of luteolytic $PGF_{2\alpha}$. In cyclic ewes, endometrial ER expression is maximal during estrus and metestrus (Days 0–5) due to high levels of estrogen secreted by the ovulatory Graafian follicle and very low levels of circulating progesterone. Interactions of estrogen and ER, in the absence of occupied PR, allows estrogen to induce OTR expression in all cell types of the endometrium. Although OTR expression is high, circulating oxytocin is low or absent and pulsatile secretion of $PGF_{2\alpha}$ by the endometrium is absent. During early to middiestrus (Days 6–10 or 12), the progesterone block prevents expression of endometrial ER and OTR. However, chronic progesterone exposure downregulates expression of PR in the LE and sGE by Days 12 or 13, which permits increases in ER and then OTR gene expression and activation of the luteolytic mechanism. During luteolysis (Days 14–17), estrogen from maturing ovarian follicles further stimulates ER and OTR expression to ensure that oxytocin from the CL and posterior pituitary stimulates pulsatile release of luteolytic $PGF_{2\alpha}$ by the endometrium to terminate CL function which allows the onset of estrus and another opportunity for mating and successful establishment of pregnancy. Thus, progesterone is critical for luteolysis as well as pregnancy. Progesterone ensures the potential for uterine release of luteolytic $PGF_{2\alpha}$ in the absence of a viable conceptus and it also ensures a proper uterine environment for the growth and differentiation of the conceptus during pregnancy.

2. Pregnancy Recognition

Embryo transfer experiments by Bob Moore and Tim Rowson in the late 1960s provided the first evidence that the sheep conceptus in pregnant ewes produces an antiluteolytic signal that prevents development of the endometrial luteolytic mechanism. In pregnant ruminants, the mononuclear cells of the conceptus trophectoderm synthesize and release large amounts of IFN-τ, which is a unique subclass of the 172-amino acid ω interferons and the sole antiluteolytic signal for maternal recognition of pregnancy. Injections of purified native or recombinant ovine IFN-τ into the uterus of cyclic sheep, cattle, and goats prevent the pulsatile production of $PGF_{2\alpha}$, thereby abrogating development of the luteolytic mechanism and extending CL life span. The antiluteolytic effects of IFN-τ are assumed to be local and limited to the endometrium because IFN-τ is not detectable in the uterine venous drainage or lymphatic drainage during early pregnancy. The antiluteolytic effects of IFN-τ in uteri of ewes, cows, and goats may differ slightly, but the primary mechanism(s) of action is very similar.

Ovine IFN-τ is secreted by the sheep conceptus between Days 10 and 21 of pregnancy and exerts a paracrine, antiestrogenic effect on the endometrium to suppress ER and OTR expression in the lumenal epithelium and superficial glandular epithelium. Intrauterine infusions of either purified native or recombinant ovine IFN-τ block formation of epithelial ER and OTR in cyclic ewes which are necessary for secretion of luteolytic pulses of $PGF_{2\alpha}$ in response to oxytocin. Although intrauterine injections of IFN-τ do not prevent endometrial PR downregulation by progesterone, IFN-τ does prevent estrogen-induced luteolysis in cyclic ewes by preventing increases in steady-state levels of ER mRNA that precede OTR formation in the endometrium. Moreover, the conceptus and IFN-τ suppress transcription of the ER and OTR genes in the ovine endometrium *in vivo*. Therefore, IFN-τ activates intracellular mechanisms in the endometrial epithelium which silence transcription of both the ER and OTR genes. Available evidence supports the working hypothesis that IFN-τ suppresses transcription of the ER gene which precludes estrogen induction of OTR gene expression, OTR formation on the endometrial epithelium, and thus pulsatile release of $PGF_{2\alpha}$ in response to luteal oxytocin.

Interestingly, cyclic ewes receiving intrauterine infusions of either native or recombinant ovine IFN-τ will eventually undergo luteolysis between 30 and 50 days postestrus. These observations suggest that there may be secondary luteotropic signals produced by the conceptus that ensure maintenance of the CL. Similar to rodents, the binucleate cells of the ruminant trophoblast do secrete placental lactogens beginning on Days 15–17 of pregnancy. However, it is not known if the biological activities of these PLs are directly or indirectly luteotropic. Moreover, a second period of type I IFN production has been suggested to occur between Days 25 and 35 of pregnancy; however, the nature of this substance has not been determined. Therefore, it is likely that the conceptus also employs other mechanisms for maintenance of the CL that are different from those used

B. Pigs

1. Luteolytic Mechanism

The pig is a spontaneously ovulating mammal with an estrous cycle of 21 days. Luteolysis is uterine dependent and occurs during late diestrus following stimulation of the uterine endometrium by progesterone for 10–12 days. The uterine luteolysin in pigs is $PGF_{2\alpha}$, although endocrine requirements for luteolysis in pigs have not been clearly delineated. Hysterectomy extends the estrous cycle and CL function by removing the source of $PGF_{2\alpha}$. The CLs of pigs are refractory to luteolytic effects of $PGF_{2\alpha}$ until Days 12 or 13 of the estrous cycle because of low numbers of luteal receptors for $PGF_{2\alpha}$. Luteolysis occurs when pulsatile release of $PGF_{2\alpha}$ into the uterine venous drainage begins on Days 15 or 16 of the estrous cycle.

In ruminants, oxytocin from the CL binds uterine OTR to elicit pulses of $PGF_{2\alpha}$, but this mechanism is not well defined for pigs. The CLs of pigs contain very low levels of oxytocin and vasopressin and undetectable levels of oxytocin mRNA, and the role of these neuropeptides of ovarian and/or posterior pituitary origin in luteolysis has not been established. Recent results indicate that the endometrium is a source of oxytocin that may be involved in luteolysis, but the role of endometrial oxytocin remains to be defined. Exogenous oxytocin decreases interestrous interval when administered to cyclic gilts between Days 10 and 16 postestrus but not when administered to ovarian-intact hysterectomized gilts, suggesting that the effect of oxytocin is uterine dependent. The endometrium of pigs contains receptors for oxytocin and lysine vasopressin but only responds to oxytocin with increased secretion of $PGF_{2\alpha}$.

Concentrations of oxytocin increase in the peripheral circulation during luteolysis. In addition, oxytocin-induced increases in circulating concentrations of $PGF_{2\alpha}$ metabolite (PGFM) are reduced in pregnant gilts compared to cyclic gilts or in gilts made pseudopregnant by injection of exogenous estrogen from Days 11 to 15 postestrus. Prostaglandins, however, are thought to be critical for establishment of pregnancy in the pig because inhibition of prostaglandin synthesis results in pregnancy failure and basal peripheral concentrations of PGFM are elevated in pregnant gilts on Day 12.

2. Endocrine–Exocrine Theory of Pregnancy Recognition in Pigs

This theory of maternal recognition of pregnancy in pigs was first published by Fuller Bazer and William Thatcher. The major assumptions are that the uterine endometrium secretes $PGF_{2\alpha}$ and that the conceptuses secrete estrogens which are antiluteolytic. The current theory is that in cyclic gilts, $PGF_{2\alpha}$ is secreted in an endocrine direction, toward the uterine vasculature, and transported to the CL to exert its luteolytic effect. However, in pregnant pigs, the direction of secretion of $PGF_{2\alpha}$ is exocrine into the uterine lumen, where it is sequestered to exert its biological effects *in utero* and/or metabolized to prevent luteolysis. Mean concentrations, peak frequency, and peak amplitude of $PGF_{2\alpha}$ in uteroovarian vein plasma are lower in pregnant and estrogen-induced pseudopregnant gilts than in cyclic gilts. On the other hand, uterine flushings from pseudopregnant and pregnant gilts have significantly higher amounts of PGF than those from cyclic gilts. These results indicate that $PGF_{2\alpha}$ is released primarily into the uterine venous drainage (endocrine) in cyclic gilts but into the uterine lumen (exocrine) in pregnant and pseudopregnant pigs, and that secretion of $PGF_{2\alpha}$ is not inhibited during either pregnancy or pseudopregnancy.

The transition from endocrine to exocrine secretion occurs between Days 10 and 12 of pregnancy and is temporally associated with initiation of estrogen secretion by elongating pig conceptuses. Estrogens, either secreted by the conceptus or injected, induce a transient release of calcium into the uterine lumen within 12 hr. Reuptake of that calcium by endometrial and/or conceptus tissues occurs about 12 hr after concentrations of calcium in uterine secretions reach maximum values. The switch in direction of endometrial secretion of $PGF_{2\alpha}$ from an endocrine to an exocrine orientation is closely associated with this period of release and reuptake of calcium by the endometrium in pregnant and pseudopregnant gilts. When endometrium from Day 14 cyclic gilts was treated with the calcium ionophore A23187 (induces calcium flux across epithelial membranes), secretion of $PGF_{2\alpha}$ changed from an endocrine to an exocrine direction. These results suggest that induction of calcium cycling across endometrial epithelium is associated with redirection of secretion of $PGF_{2\alpha}$.

3. Conceptus Estrogen as an Antiluteolytic Signal for Pregnancy Recognition

Uterine $PGF_{2\alpha}$ is luteolytic and estrogens produced by conceptuses on Days 11 or 12 of gestation provide the initial signal for maternal recognition of pregnancy in swine. A second period of estrogen production occurs between Days 15 and 25–30 of pregnancy. Injection of exogenous estrogen on Days 11–15 of the estrous cycle will result in CL maintenance for a period equivalent to or slightly longer than pregnancy (about 120 days later). This condition is referred to as pseudopregnancy. Whereas a single injection of estradiol on Days 9.5, 11, 12.5, 14, 15.5, or 14–16 resulted in interestrous intervals of about 30 days, estradiol must be administered to gilts on Days 11 and 14–16 or daily from Days 11 to 15 to obtain interestrous intervals >60 days. This suggests that two phases of estradiol, similar to that produced by conceptuses on Days 11–13 and 15 to 25–30, are necessary for prolonged secretion of $PGF_{2\alpha}$ into the uterine lumen. Estradiol may induce receptors for maternal hormones, e.g., prolactin, or conceptus secretory proteins, which influence exocrine (into the uterine lumen) secretion of $PGF_{2\alpha}$. The first estrogen signal may induce the receptors and the second estrogen signal may be required to replenish those receptors. Administration of estradiol on Day 9 advances the uterine secretory response in pregnant gilts, leading to conceptus death by Day 16. An explanation for this "induced" conceptus death is not available, but it may result from ansynchrony between development of the conceptus and uterine environment. Available results indicate that estrogens of blastocyst origin are essential for maternal recognition of pregnancy in pigs and that pCSPs, including interferons, play other roles during early pregnancy in pigs.

C. Horses

1. Luteolytic Mechanism

Horses are spontaneously ovulating mammals with an estrous cycle of 28 days. The uterine luteolytic substance in mares is $PGF_{2\alpha}$, and the conceptus appears to inhibit production of $PGF_{2\alpha}$ by the uterine endometrium. In cycling mares, concentrations of $PGF_{2\alpha}$ in uterine venous plasma and uterine flushings increase between Days 14 and 16 when luteolysis occurs and plasma progesterone levels decline. The amount of $PGF_{2\alpha}$ bound by luteal receptors is maximal on Day 14 of the estrous cycle and Day 18 of pregnancy. Since CLs of mares can respond to circulating $PGF_{2\alpha}$ during pregnancy, the conceptus must evoke an antiluteolytic mechanism. Pregnant mares have little $PGF_{2\alpha}$ in uterine fluids, $PGF_{2\alpha}$ in uterine venous plasma is reduced, and PGFM in peripheral plasma has no episodic pattern of release. In the presence of the conceptus, endometrial production of $PGF_{2\alpha}$ in response to cervical stimulation and exogenous oxytocin is markedly reduced, indicating the absence/reduction of endometrial receptors for oxytocin in mares during early pregnancy.

2. Pregnancy Recognition

The horse is an example of a precocious form of maternal recognition of pregnancy. In the horse, only fertilized eggs are transported into the uterus. Unfertilized eggs remain in the oviduct, trapped at the isthmus, where they often undergo parthenogenetic cleavage. These unfertilized ova slowly degenerate over the next few months. In the event that another egg is ovulated during another estrous cycle and successfully fertilized, the developing morula can bypass its degenerating predecessors and enter the uterus. The physiological mechanism underlying this remarkable process is unknown but could be attributed to differences in surface properties of the two egg types. The ability of the maternal reproductive tract to differentiate between eggs precedes by at least 2 weeks the time when the life of the CL is prolonged by the presence of an embryo.

During Days 12–18 of early pregnancy, the equine conceptus migrates between uterine horns 12–14 times per day, presumably to inhibit endometrial $PGF_{2\alpha}$ production to rescue the CL. The equine conceptus does suppress $PGF_{2\alpha}$ production by the endometrium, but the nature of this antiluteolytic signal has not been identified. The equine conceptus also produces increasing amounts of estradiol and estrone between Days 8 and 20 of gestation. However, attempts to prolong CL life span in mares by injection of estrogens have yielded inconsistent results.

Horse conceptuses secrete three major proteins between Days 12 and 14 of pregnancy with molecular weights >40, 50, and 65 kDa. However, the role(s) of these proteins is not known. Estrogens and/or conceptus secretory proteins may provide the maternal recognition of pregnancy signal in the mare by directly or indirectly inhibiting endometrial production of luteolytic pulses of $PGF_{2\alpha}$.

IV. OTHER SUBPRIMATE MAMMALS

A. Rabbits

The rabbit is an induced ovulator. The rabbit placenta is not a significant source of progesterone, and the CLs are required for pregnancy to go to term. Following a sterile mating, CLs form and persist for 14–16 days without apparent effects of the conceptus products. For both pseudopregnant and pregnant does, progesterone begins to increase 2 days after mating to maximal levels between Days 6 and 8 postcoitum (pc). Between Days 8 and 10 pc, progesterone profiles of pregnant and pseudopregnant does begin to diverge, with levels declining rapidly after Day 12 to basal levels between Days 16 and 18 of pseudopregnancy. Pregnant does exhibit elevated progesterone levels which begin to decline 3 or 4 days prior to parturition.

In the rabbit, maternal recognition of pregnancy occurs after implantation between Days 10 and 12 of gestation. Estrogen and a placental luteotropin interact to maintain progesterone production until term (28–35 days). Estrogen from developing follicles is required for luteal progesterone production for the first 10–12 days of pregnancy but not to term. Rabbit luteal cells contain LH receptors; however, LH does not appear to be the stimulus for progesterone production *in vivo*. Rather, estrogen exerts a luteotrophic effect which results from uncoupling progesterone production from cyclic AMP, allowing progesterone production to proceed autonomously.

Production of a placental luteotrophic signal is necessary for luteal progesterone production to term, and exogenous estrogen will not support progesterone production in does hysterectomized during late pregnancy. The luteotrophic effect of the placenta is apparently not a result of increased concentrations or affinity of luteal estrogen receptors. The rabbit placenta secretes immunoreactive gonadotropin-releasing hormone (GnRH)-like activity; however, this activity is not present in the uterine vein at concentrations greater that those of estrus rabbits, suggesting a local site of action for this placental product. Therefore, a direct effect of a placental GnRH-like factor on luteal progesterone production in the rabbit is not likely.

Several putative placental luteotrophic factors have been partially purified and some stimulate progesterone production by cultured luteal cells. However, the identity and mechanism of action of this luteotropic activity remain to be determined. Recently, rabbit placental giant cells were shown to contain immunoreactive CG, and cytotrophoblast cells were demonstrated to contain immunoreactive PL/PRL. However, the function(s) of these substances as antiluteolytic signals in the rabbit has not been determined.

B. Marsupials, Cats, and Dogs

In marsupials, such as the Tammar wallaby and red kangaroo, the life span of the CL in the pregnant and nonpregnant animal is similar, and the gestation length is about the same length as the estrous cycle. Therefore, no pregnancy recognition signal is needed. Likewise in cats and dogs, the CLs of pregnancy and pseudopregnancy have similar life spans, and there is no known requirement for signaling between the conceptus and maternal system for pregnancy recognition and maintenance of the CL.

VI. SUMMARY AND FUTURE RESEARCH

The transition from recurring estrous or menstrual cycles to pregnancy requires intricately orchestrated interactions between the conceptus, uterus, and ovaries. There appears to be no single theme for maternal recognition of pregnancy that can be applied to all eutherian or placental mammals. The ability or necessity of the conceptus to signal its presence to the mother is dictated by variations in estrous and ovarian cycles, type of implantation and placentation, and gestation length. During the evolution of viviparity, it is quite possible that all adaptive strategies of the conceptus to maternal reproductive physiology can be found in one or more species. However, all mammalian pregnancies require the continued actions of progesterone on maternal physiology. Strategies to accomplish this include direct conceptus support of luteal function (luteotropic) as occurs in humans and primates (hCG) and rodents (PRL, PL); conceptus-mediated signals which abrogate the uterine luteolytic mechanism (antiluteolytic) as occurs in domestic ruminants, horses, and pigs; and tailoring the length of the luteal phase to the length of pregnancy as occurs in marsupials, cats, and dogs.

In the majority of these mammals, effects of the conceptus on uterine gene expression are only beginning to be unraveled. This area of research should prove to be quite exciting as advances provide a clearer picture of intercellular communications between the periimplantation conceptus and the uterus as well as between cell types in the uterus that mediate effects of hormones in the reproductive tract of the female during pregnancy. Discovery of genes which regulate uterine function during the estrous or menstrual cycle and pregnancy may be useful to ameliorate pregnancy loss in both humans and domestic animals.

See Also the Following Articles

IMPLANTATION; LUTEOLYSIS; OXYTOCIN; RABBITS (LAGOMORPHA)

Bibliography

Austin, C. R., and Short, R. V. (1984). *Reproduction in Mammals*. Cambridge Univ. Press, New York.

Bazer, F. W., Spencer, T. E., and Ott, T. L. (1997). Endocrinology of the transition from recurring estrous cycles to establishment of pregnancy in subprimate mammals. In *The Endocrinology of Pregnancy* (F. W. Bazer, Ed.). Humana Press, Clifton, NJ.

Hamberger, L., Hahlin, M., Hillensjo, T., Johanson, C., and Sjogren, A. (1988). Luteotropic and luteolytic factors regulating human corpus luteum function. *Ann. N. Y. Acad. Sci.* **54**, 485–497.

Heap, R. B., and Flint, A. P. FA. P.: Pregnancy. Hormonal Control of Reproduction. New York: Cambridge University Press, 1984.

Knobil, E., and Neill, J. D. (1994). *The Physiology of Reproduction*. Raven Press, New York.

Mossman, H. W. (1987). *Vertebrate Fetal Membranes*. Rutgers Univ. Press, New Brunswick, NJ.

Short, R. V. (1969). Implantation and the maternal recognition of pregnancy. In *Foetal Autonomy*. Churchill, London.

Spencer, T. E., Ott, T. L., and Bazer, F. W. (1996). τ-Interferon: Pregnancy recognition signal in ruminants. *Proc. Soc. Exp. Biol. Med.* **213**, 215–229.

Pregnancy, Metabolic Changes in

William W. Hay, Jr.
University of Colorado School of Medicine

I. Introduction
II. Impact of Pregnancy on Maternal Metabolism
III. Glucose Metabolism: Early Gestation
IV. Glucose Metabolism: Late Gestation
V. Lipid Metabolism during Pregnancy
VI. Protein Metabolism during Pregnancy
VII. Summary

GLOSSARY

anabolism Net synthesis and growth of tissue.
catabolism Net breakdown and loss of tissue.
conceptus The content of the entire pregnant uterus, including the uterine decidual tissues, the placenta and its membranes, and the fetus.
contrainsulin Inhibition of insulin action or producing metabolic actions opposite to those of insulin.
diabetogenic Tending to cause diabetes mellitus, including insulin resistance and glucose intolerance.
euglycemia Normal glucose concentration.
gestation The period of embryonic and fetal growth from conception to birth.
gluconeogenesis The production of glucose from nonglucose precursors such as amino acids and lactate.
hyperphagia Increased food intake, above normal.
ketogenesis Production of ketone bodies by partial fatty acid oxidation.
leptin A peptide hormone produced by adipocytes under the influence of insulin and increased fat mass which acts at the hypothalamus and leads, probably via neuropeptide Y, to decreased appetite and decreased food intake; it also leads to increased metabolic rate.
lipoprotein lipase An enzyme which breaks down lipoproteins into fatty acids and triglycerides.
maternal constraint Nongenomic limitation of fetal growth by maternal (and thus uterine) size and related physiological factors, such as uterine blood flow and uterine capacity to support placental growth and function.
postprandial The period after a meal.
primiparous Females having delivered one offspring.

Maternal metabolic changes during pregnancy produce increased maternal stores of nutrients in early pregnancy, primarily of fat in adipose tissue and amino acids in protein, followed by mobilization of these stored nutrients in later pregnancy. The mobilized nutrients, plus ingested nutrients, are increasingly directed to the uterus in later pregnancy to meet the requirements for metabolism and growth of the placenta and fetus.

I. INTRODUCTION

The most significant changes in maternal metabolism are those that occur in the third trimester. Transfer of nutrient substrates to the placenta and fetus is dependent in part on their concentrations in maternal plasma. Mobilization of maternal lipid stores increases maternal plasma lipid concentrations (free fatty acids and triglycerides) and transport to the fetus. Decreased maternal utilization of glucose and amino acids tends to maintain their maternal plasma concentrations, even though there is an overall decrease in maternal plasma concentrations of these substrates in late gestation due to increasing placental and fetal utilization. The small decrease in maternal plasma concentrations of glucose and amino acids also limits their utilization by maternal tissues as well as their transport to the placenta and fetus. The latter process is counterbalanced by increased nutrient transport capacity of the placenta in late gestation. Such metabolic changes in the mother are linked to supportive physiological and behavioral changes of increased appetite and hyperphagia, perhaps via alterations in leptin and neuropeptide Y activity, and increasing blood flow to the uterus via increased nitric oxide production leading to vasodilation in the uterine circulation. Overall, therefore, nutrient metabolism in the pregnant mother is balanced to support both the mother's own nutrition and that of the placenta and fetus.

II. IMPACT OF PREGNANCY ON MATERNAL METABOLISM

A. Maternal Weight Gain

Early gestation is characterized as an anabolic phase during which maternal energy reserves are built up in preparation for the demands of rapid fetal and placental growth in later gestation. During this period maternal blood volume and uterine blood flow also increase to support nutrient delivery to the conceptus. Maternal weight gain is greater than the simple growth of the pregnant uterus, including the placenta and fetus. In this period, the pregnant mother develops increased appetite and hyperphagia and conserves more exogenous nutrients whenever she eats. This maternal anabolic phase is most striking during the second trimester, when maternal weight gain is approximately 500 g per week. This rate of weight gain produces a total weight gain of about 6500 g, of which the fetal weight increase is only approximately 1000 g. One-half of the maternal weight gain through midpregnancy is accounted for by maternal lipid stores. Based on indirect measures of body fat content, previous studies have indicated that the lipid proportion of maternal weight gain decreases steadily until term, when it represents 25% of the total maternal weight gained (Table 1). Recent measurements of total body water content using deuterium dilution demonstrate a steady increase throughout pregnancy, with a marked increase near term. The fraction of total body weight that is water remains relatively constant, however, indicating that increases in lean body mass, blood volume, and extracellular fluid combine to balance increases in fat stores. Such information and the differences among studies also point out the difficulty in assessing metabolic changes in the mother during pregnancy, such as glucose, lipid, and amino acid kinetics, because the composition of the mother's tissues and fluids, as well as maternal weight, change at the same time. The changes in maternal body composition are widely divergent among animal species, as are diets, food intakes, and degrees of physical activity, adding further difficulty to determining actual changes in maternal metabolic rates.

TABLE 1
Tissues and Fluids That Account for Weight Gain (in Grams) during Pregnancy in Normal Nonedematous Women[a]

Tissue or fluid	Week 10	Week 20	Week 30	Week 40
Fetus	5	300	1500	3,400
Placenta	20	170	430	650
Amniotic fluid	30	350	750	800
Uterus	140	320	600	970
Mammary gland	45	180	360	405
Blood	100	600	1300	1,250
Extracellular, extravascular fluid	0	30	80	1,680
Maternal lipid stores	310	2050	3480	3,345
Total weight gained	650	4000	8500	12,500

[a] From N. Kretchmer, E. J. Quilligan, and J. D. Johnson. *Prenatal and Perinatal Biology and Medicine,* Harwood, New York, 1987.

B. Fetal/Maternal Mass Ratios

The mass of the conceptus has a major impact on maternal metabolism during pregnancy. Conceptus mass and especially fetal mass vary considerably among species, from as little as 0.1% of maternal weight in the marsupials and 0.2–0.3% in bears to as much as 60% of maternal weight in the guinea pig (Table 2). Within a species, increasing fetal mass also imposes increased metabolic demands on the mother to provide nutrients to the conceptus. For example, maternal glucose concentration tends to decrease in pregnant women as their number of fetuses increases. Maternal constraint of fetal growth, primarily by limitation of uterine size and thus the growth of the placenta, prevents the demand for maternal nutrients from becoming excessive and harmful to the mother. As a result, fetal and placental growth restriction develop as fetal number increases. A relatively similar fetal heart rate and thus oxygen

TABLE 2
Total Fetal Mass as a Fraction of Maternal Mass in Late Gestation[a]

	Fetal/maternal mass ratio	Fetal mass, fraction of maternal mass
Human	3.5/65 kg	0.06
Guinea pig	500/700 g	0.70
Sheep	4/50 kg	0.07
Rat	55/300 g	0.18

[a] From Battaglia and Meschia (1986).

consumption rate per fetal mass among species (Fig. 1) limit the impact of a larger conceptus/maternal mass ratio on maternal metabolism. Thus, small mammals with relatively large fetal mass have relatively lower metabolic rates than would be expected for their size after birth, and large mammals have relatively small fetal mass with relatively greater metabolic rate than would be expected for their size after birth. These differences in fetal/maternal metabolic rate allow for larger fetal/maternal mass ratios, particularly of multiple fetuses per mother, in the smaller animals.

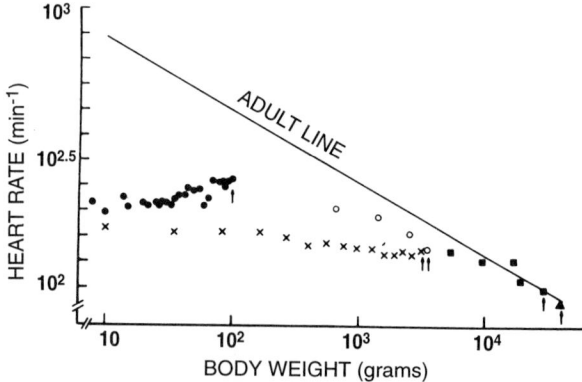

FIGURE 1 Logarithmic plot of heart rate versus body weight in fetuses of different species: guinea pig, ●; human, X; sheep, ○; cow, ■; horse ▲. The "adult line" represents the relationship of heart rate to body mass in adult mammals (reproduced from Battaglia and Meschia, 1986, p. 197).

C. Competition between Maternal and Fetal Metabolic Needs

In the latter part of gestation, the rapid fetal growth is sustained by the placental transfer from the mother of a variety of nutrients—primarily glucose, but also lipids and amino acids. This intense loss of nutrients from the maternal circulation is not compensated by the mother's hyperphagia and seems to contribute to a switch to a net catabolic state that is especially evident in adipose tissue and becomes accelerated when food is withheld. Under this condition, exaggerated ketogenesis and gluconeogenesis actively contribute to the availability of fuels for both the mother and the fetus.

Maternal metabolic demands also can supersede those of the fetus and placenta, particularly when the mother is still growing. In both humans and sheep, evidence from adolescent pregnancies in which the mother is still undergoing the normal adolescent growth spurt show that nutrients are increasingly partitioned to maternal use, resulting in maintained maternal growth but the development of placental and fetal growth restriction. This phenomenon is especially prominent when the mother is fed on a high dietary plane. The mechanisms in these circumstances that might primarily limit placental development (e.g., decreased pregnancy-specific hormones) or divert maternal plasma nutrients to maternal use, thereby limiting nutrient delivery to the placenta and fetus and leading to a reduced rate of placental and fetal growth, are not known.

III. GLUCOSE METABOLISM: EARLY GESTATION

A. Insulin Resistance

The first change in maternal glucose metabolism in early gestation is the development of insulin resistance in the peripheral tissues. Insulin resistance is inversely related to the percentage of body fat in pregnant women, although further increases in body fat in later gestation are less pronounced as insulin resistance increases. Insulin resistance only develops late in the first half of gestation, with little change observed before 12 weeks and only as much as 7–10% increased resistance by 14 weeks. Increasing insulin resistance with advancing gestation has been documented by studies using the glucose–insulin clamp technique that showed that less glucose is consumed and less glucose infusion is required to maintain euglycemia during a steady-state condition of hyperinsulinemia produced by insulin infusion. Exact mechanisms for this change are not known. Insulin resistance, in general, appears to be a postinsulin receptor defect, most likely at the level of signaling of insulin-responsive glucose transporter translocation and/or activity, perhaps via reduced levels of insulin receptor substrate (IRS-1) protein or signal transduction events subsequent to IRS-1 protein activation. Whether these changes are primarily involved in insulin resistance during pregnancy is not known, especially because a variety of contrainsulin hormones with different sites of action are primarily involved. As a result of this insulin resistance, maternal pancreatic insulin secretion increases (both first and second phase secretion are increased), resulting in relative hyperinsulinemia. Specific mechanisms that increase maternal pancreatic insulin secretion are not clearly defined.

B. Hormonal Changes

A variety of hormonal changes during pregnancy have been associated with the decreases in insulin action and increases in insulin secretion (Table 3). Interestingly, those hormones with the strongest contrainsulin (or diabetogenic) tendencies increase most in late pregnancy when the need to provide increased glucose supply to the placenta and fetus is greatest. Fasting plasma insulin concentrations increase slowly over gestation, doubling by the third trimester. Insulin response to oral or intravenous glucose challenge also increases by 1.5- to 2.5-fold. Estrogen, progesterone, and especially prolactin promote pancreatic β-cell hyperplasia and insulin production, even in early gestation, which tends to promote greater rates of maternal glucose utilization, primarily for nonoxidative metabolism of glycogenesis and lipogenesis, resulting in progressive fasting hypoglycemia. In contrast, progesterone also inhibits

TABLE 3
Sequential Rise and Potency of the Diabetogenic (Contrainsulin) Hormones during Human Pregnancy[a]

Hormone	Onset of increase (days)	Peak increase (weeks)	Relative diabetogenic potency[b]
Estradiol	32	26	1
Prolactin	36	10	2
Human placental lactogen	45	26	3
Cortisol	50	26	5
Progesterone	65	32	4

[a] From Jovanovic-Peterson and Peterson (1991).
[b] Scale: 1, weak; 5, strong.

insulin's action to suppress hepatic glucose production. Hyperinsulinemia promotes lipogenesis from triglycerides and from fatty acids, to the extent that glycerol is available from glucose metabolism through glycolysis. Glycolysis is maintained by hyperglycemia that results from the developing glucose intolerance, particularly in the postprandial period.

In summary, the principal changes in glucose metabolism in early gestation in pregnant women include fasting hypoglycemia followed by postprandial hyperglycemia and enhanced hepatic glucose production from insulin resistance, maintained or enhanced glycogen storage in the liver and skeletal muscle, and lipogenesis from increased glycolysis-derived glycerol and relative hyperinsulinemia. There are considerable species-specific differences in these metabolic changes that are further modulated by unique diets and food intakes, fetal number and mass of the conceptus, and maternal behavior.

IV. GLUCOSE METABOLISM: LATE GESTATION

A. Glucose Intolerance

Later gestational changes in glucose metabolism represent exaggerations of the changes that began in early gestation. Fasting glucose concentrations decrease further as rates of fetal and placental glucose consumption increase. The combined utilization of glucose by the placenta and fetus may account for as much as 30–50% of maternal glucose production near term. Peripheral insulin resistance increases, decreasing the action of insulin to promote maternal glucose utilization to as little as 20% of prepregnancy conditions, followed by further increases in plasma insulin concentration. Postprandial glucose intolerance and hyperglycemia increase in direct relation to peak increases in those hormones of pregnancy with the strongest contrainsulin effects—hPL, cortisol, and progesterone. Meal-associated lipogenesis becomes less significant relative to fasting lipolysis and increased plasma concentrations of triglycerides and fatty acids. These occur due to the increasing degree of fasting hypoglycemia, augmented by increased catecholamine secretion, which acts to mobilize fatty acids from adipose tissue triglycerides. As a result, fatty acid inhibition of the normal action of insulin to suppress hepatic glucose production increases, leading to increased meal-associated hepatic glucose production. The specificity of fatty acids to interfere with insulin's normal suppression of hepatic glucose production has been shown in the rabbit by intravenous infusion of lipids into nonpregnant rabbits, which mimics the reduction of the effect of insulin on hepatic glucose production during pregnancy. Such effects are diet and species specific, however, because studies in the sheep indicate an increased capacity for insulin to suppress hepatic glucose production both during pregnancy and during the fasted state in nonpregnant ewes. Together, these changes produce increasingly wider swings in meal-associated glycemia and insulinemia (Fig. 2). Glucose production is also enhanced by gluconeogenesis by several mechanisms, including (i) increased substrate supply, such as glycerol, lactate, and fatty acids, but not amino acids, which in general are decreased in concentration in the pregnant mother; (ii) the action of increased concentrations of cortisol on protein and lipid catabolism and the activity of gluconeogenic enzymes (especially PEPCK); and (iii) increased hepatic β-oxidation of fatty acids that provides NADPH and energy for gluconeogenesis. Overall, there is an increase in glucose production in late gestation, supporting noninsulin-mediated glucose metabolism in the mother and pro-

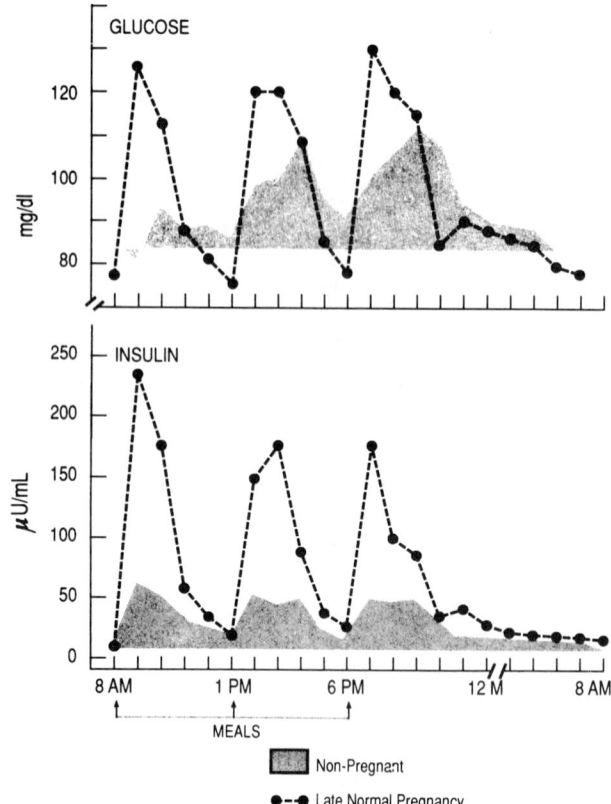

FIGURE 2 Effect of normal late pregnancy on diurnal changes in plasma glucose and insulin [from Z. Hagay and E. A. Reece, Diabetes mellitus in pregnancy. In *Medicine of the Fetus & Mother* (E. A. Reece, J. C. Hobbins, M. J. Mahoney, and R. H. Petrie, Eds.), p. 989, Lippincott, Philadelphia, 1992].

viding more glucose for the growing placenta and fetus. This increased glucose production and disposal rates lead to increased glucose turnover. Glucose turnover measurements in pregnant women studied serially from prepregnancy to late pregnancy show a 30% increase in basal hepatic glucose production in late pregnancy relative to pregravid estimates (Fig. 3).

B. Gestational Diabetes

In late pregnancy, some women develop exaggerated glucose intolerance, which has been labeled gestational diabetes. The main cause of gestational diabetes, or diabetes mellitus that is uncovered during pregnancy, is insulin resistance coupled with a failure

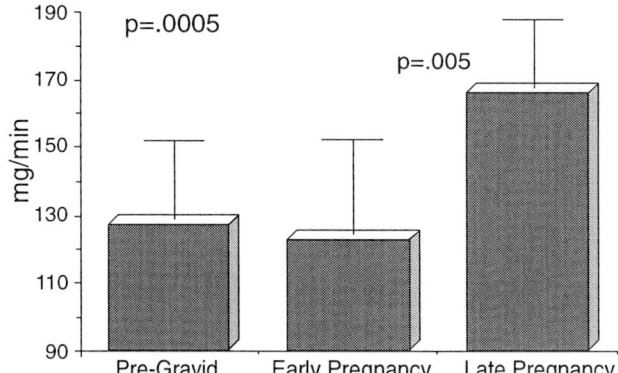

FIGURE 3 Changes in total basal endogenous glucose production from before pregnancy through early and late gestation (mean ± SD) (from P. M. Catalano, E. D. Tyzbir, R. R. Wolfe, N. M. Roman, S. B. Amini, and E. A. Sims, Longitudinal changes in basal hepatic glucose production and suppression during insulin infusion in normal pregnant women, *Am. J. Obstet. Gynecol.* **167**, 913–919, 1992).

of pancreatic insulin secretion. It is the failure of insulin secretion to keep up with the glucose intolerance from insulin resistance that is key. These particular pregnant women also have greater than normal increases in free fatty acid and triglyceride concentrations, which could have the effect of limiting insulin's action to inhibit hepatic glucose production. As a result, these women have exaggerated glucose intolerance and increased glucose production, especially in the postprandial period. Up to 12% of pregnant women will develop glucose intolerance of pregnancy or gestational diabetes mellitus. Gestational diabetes usually presents during the second trimester when the concentrations of the contrainsulin hormones of pregnancy, principally hPL, cortisol, and progesterone, are greatest. It is possible that all gestational diabetic women have undiagnosed type II diabetes mellitus, which is manifest by peripheral insulin resistance in the presence of insufficient pancreatic insulin secretion.

V. LIPID METABOLISM DURING PREGNANCY

A. Fat Accumulation

Maternal weight gain during pregnancy includes a variable but relatively large amount of subcutaneous and intraabdominal adipose tissue. Body fat accumulation in women reaches a maximum at midgestation and does not increase further or may even decline slightly near term (Fig. 4); similar patterns of maternal adipose tissue accumulation in pregnancy occur in most other animals, with slight variation according to species, strain, and diet. In those species studied— human, rat, and hampster—attenuation of fat mass accumulation toward the end of gestation is associated with a reduction of fatty acid synthesis and increased incorporation of glucose into glycerol, indicating increased fatty acid reesterification rather than lipogenesis. This pattern of lipid metabolism indicates an increase in insulin resistance and intracellular contrainsulin effects. Associated with the decreased insulin effects is increased fatty acid mobilization by lipolysis, resulting in increased maternal plasma free fatty acid and ketone concentrations, especially during fasting.

Marked differences among individual pregnant women are observed in the extent to which they enter pregnancy lean or obese and then augment weight gain during the course of pregnancy. In general, obese women tend to gain less fat mass during gestation and may undergo an actual loss in body fat storage, in contrast to lean, primiparous women who usually gain a substantial amount of adipose tissue mass during their first pregnancy. The different mechanisms controlling the augmented accumulation of fat in the lean women and the restricted or

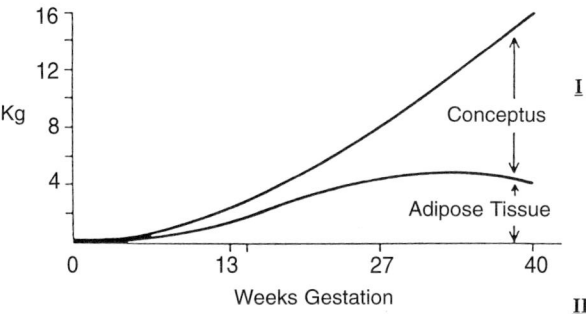

FIGURE 4 Maternal body fat accumulation in pregnancy [after Hytten and Leitch, 1971; From R. H. Knopp, *Univ. Washington Med.* **7**, 20, 1980; and R. H. Knopp, M. S. Magee, B. Bonet, and D. Gomez-Coronado, Lipid metabolism in pregnancy. In *Principles of Perinatal–Neonatal Metabolism* (R. M. Cowett, Ed.), p. 183, Springer-Verlag, New York, 1991].

even negative fat accumulation in an already obese woman are not known. The hypothalamus may become more sensitized to body signals that reflect fat storage accumulation, such as central nervous system insulin concentrations. Alternatively, exaggerated increases in basal and glucose-stimulated insulin concentrations in the obese versus lean pregnant women might reach such levels as to inhibit appetite center receptors insensitive to lower levels of insulin feedback. Other feedback signal alterations, such as adipsin and leptin, are possible. Leptin concentrations increase as much as 67% in late human pregnancy, which is attributed to increased fat mass and increased insulin concentrations. Increased leptin concentrations may suppress appetite in late gestation. These effects may be countered by pregnancy hormones such as progesterone acting through stimulation of the hypothalamic appetite center.

B. Fatty Acid, Triglyceride, and Lipoprotein Metabolism

Gestational changes in the disposition of circulating fuels are shown in Fig. 5. During early gestation,

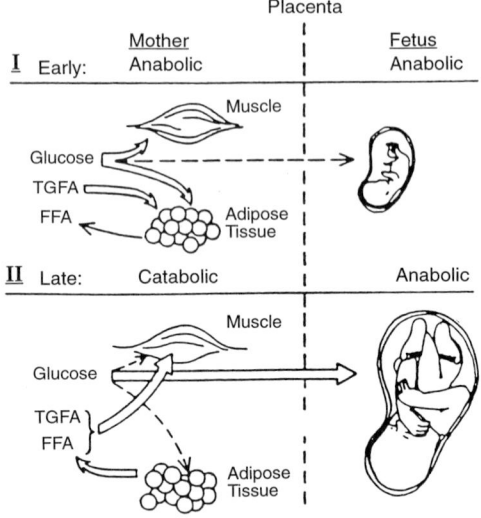

FIGURE 5 Reallocation of body fuels in pregnancy after an overnight fast [from R. H. Knopp, *Contemp. Obstet. Gynecol.* **12**, 83, 1978. Contemporary OB/GYN, Medical Economics Company; and R. H. Knopp, M. S. Magee, B. Bonet, and D. Gomez-Coronado, Lipid metabolism in pregnancy, In *Principles of Perinatal–Neonatal Metabolism* (R. M. Cowett, Ed.), p. 183, Springer-Verlag, New York, 1991].

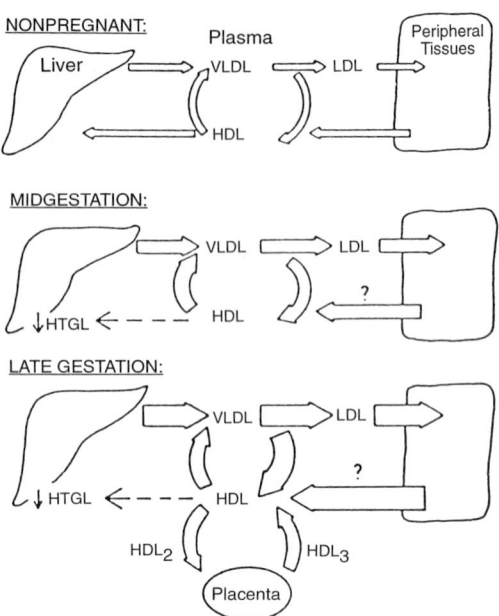

FIGURE 6 Changes in lipoprotein flow in pregnancy [from R. H. Knopp, M. S. Magee, B. Bonet, and D. Gomez-Coronado, Lipid metabolism in pregnancy. In *Principles of Perinatal–Neonatal Metabolism* (R. M. Cowett, Ed.), p. 188, Springer-Verlag, New York, 1991].

maternal caloric intake increases (up to 250–300 kcal day in women). Circulating glucose, triglycerides, and fatty acids are used by muscle and adipose tissue, with little transfer of glucose or lipids to the fetus. In later gestation, fetal requirements for glucose and lipids increase. Increasing maternal insulin resistance, as discussed previously, promotes lipolysis, particularly in the fasting state, due to increased catecholamine secretion and during both fed and fasting conditions by the action of pregnancy-related hormones. Hypertriglyceridemia is the most dominant change in circulating lipids. This is caused by the hepatic production of triglycerides from fatty acids. Lipoprotein metabolism follows this general pattern (Fig. 6), with sustained hyperlipoproteinemia. Enhanced triglyceride production, estrogen stimulation of very low-density lipoprotein (VLDL) production, and decreased extrahepatic lipoprotein lipase (LPL) activity combine to augment the concentration of circulating VLDLs. In the fed state, alimentary chylomicron or VLDL triglyceride can be oxidized by muscle, which retains some LPL activity.

Adipose tissue LPL activity decreases throughout gestation (in contrast to muscle LPL activity, which does not change during pregnancy), especially in the third trimester, with the greatest reduction in adipose tissue LPL activity. The reduced adipose tissue LPL activity may account for the attenuation in growth of fat stores during the second half of pregnancy. With progressive caloric restriction, the tendency to fatty acid mobilization is even higher. Even in the fed state, increased fatty acid mobilization from adipose tissue, along with increased fatty acid reesterification and recycling, is observed. A certain amount of caloric wastage appears to be due to fatty acid recycling within adipose tissue itself during pregnancy, resulting in generation of extra heat. The increased fatty acid metabolism also contributes to increased glycerol and ketone production, primarily in the fasting state. Glycerol and ketones can be used for maternal fuel metabolism. Glycerol also provides substrate for the increased rate of maternal hepatic gluconeogenesis and glucose production, and ketones can be taken up by the placenta and fetus and used for fuel metabolism (Table 4).

In summary, adipose tissue mass in pregnancy is sensitive to hormonal effects that alter lipid metabolism and to hormonally mediated alterations in maternal appetite. Increased caloric intake expands the adipose tissue stores until late gestation when contrainsulin pregnancy hormones as well as leptin, which may interfere with increased appetite stimulated by steroid hormones, oppose lipid anabolism.

VI. PROTEIN METABOLISM DURING PREGNANCY

A. Maternal Nitrogen Gain

Maternal nitrogen gain during pregnancy in the form of lean body mass represents a significant protein cost above that deposited in the fetus and placenta. Methodological differences among studies prevent a firm conclusion about how much nitrogen is deposited in the pregnant mother, but estimates range as high as 250 g nitrogen in humans, deposited at a rate of about 0.17 g/day between 12 and 37 weeks. The mechanisms that lead to protein and nitrogen gain in maternal tissues during pregnancy are not established. Presumably, increased plasma concentrations of insulin in early to midgestation are partly responsible in that there is no evidence for resistance to insulin's effect on amino acid metabolism and protein balance as there is for glucose and lipid metabolism. Increased plasma insulin concentrations would tend to limit proteolysis in the mother and perhaps increase amino acid uptake for protein synthesis, although the latter effect becomes limiting as the plasma amino acid concentrations decrease. In support of this scenario, experimentally exaggerated maternal hyperinsulinemia produces a further reduction in maternal plasma amino acid concentrations.

B. Amino Acid Concentrations

Maternal plasma amino acid concentrations in both humans and animals decrease during pregnancy (Table 5). With prolonged fasting, maternal amino acid concentrations decrease even more than during normal dietary conditions during pregnancy, especially concentrations of the gluconeogenic amino acids alanine—serine, threonine, and glutamine. These changes, along with decreasing glucose concentrations and increasing fatty acid concentrations, occur sooner during fasting in pregnancy; this is termed "accelerated starvation of pregnancy." Recent

TABLE 4
Maternal Plasma Lipid Concentrations in Late Pregnancy Compared with Nonpregnant Women (Mean ± SD)[a]

	Concentration	
	Nonpregnant	36 Weeks gestation
Triglycerides (mg/dl)		
Total	59 ± 19	222 ± 60
VLDL	33 ± 14	107 ± 41
LDL	14 ± 10	72 ± 21
HDL	12 ± 6	29 ± 9
Cholesterol	171 ± 26	251 ± 32
Free fatty acids (μEg/liter)		672 ± 39

[a] Adapted from R. H. Knopp, M. S. Magee, B. Bonet, and D. Gomez-Coronado, Lipid metabolism in pregnancy. In *Principles of Perinatal–Neonatal Metabolism* (R. M. Cowett, Ed.), p. 188, Springer-Verlag, New York, 1991.

TABLE 5
Plasma Amino Acid Concentrations in Nonpregnant and Pregnant Women at 16–20 Weeks Gestation after a 12-hr Fast[a]

Amino acid	Concentration in μmol/liter (mean ± SEM)		P
	Nonpregnant women (n = 11)	Pregnant women (n = 12)	
Taurine	51.5 ± 2.5	36.0 ± 2.9	<0.001
Threonine	138.0 ± 11.3	150.9 ± 7.0	
Serine	126.3 ± 6.0	107.4 ± 4.6	<0.025
Proline	151.3 ± 10.1	93.8 ± 4.9	<0.001
Citrulline	27.2 ± 1.2	18.9 ± 0.8	<0.001
Glycine	197.4 ± 14.6	120.8 ± 7.6	<0.001
Alanine	279.3 ± 15.1	221.6 ± 9.0	<0.005
Valine	201.4 ± 11.7	169.8 ± 3.9	<0.025
Cystine	94.8 ± 5.4	69.1 ± 3.0	<0.001
Methionine	18.0 ± 1.7	16.7 ± 0.7	
Isoleucine	51.4 ± 2.7	47.0 ± 1.8	
Leucine	99.5 ± 2.0	91.0 ± 2.5	<0.025
Tyrosine	40.4 ± 3.3	34.5 ± 1.7	
Phenylalanine	45.8 ± 2.4	42.3 ± 1.1	
Ornithine	71.8 ± 7.6	33.7 ± 5.0	<0.001
Lysine	164.1 ± 12.1	175.8 ± 13.9	
Histidine	81.4 ± 5.5	100.7 ± 5.7	<0.025
Arginine	55.5 ± 6.6	51.0 ± 7.6	

[a] From P. Felig, Y. J. Kim, V. Lynch, and R. Hendler, Amino acid metabolism during starvation in human pregnancy, *J. Clin. Invest.* **51**, 1195–1202, 1972. © The American Society for Clinical Investigation.

studies have shown further reductions in maternal plasma amino acid concentrations in pregnancies carrying a growth-restricted fetus. In these cases, fetal concentrations of amino acids tend to be lower, producing a decreased maternal–fetal amino acid concentration difference. Such conditions tend to conserve amino acids for the mother at the expense of the fetus, with fetal growth restriction a successful though less than optimal reproductive alternative.

C. Urea

Maternal plasma urea nitrogen concentration and urea synthesis are both decreased in pregnancy as early as the first trimester, indicating that these changes are caused by adaptations in maternal metabolism. As a result, nitrogen is spared as urea excretion by the kidney decreases both during fasting and in response to amino acid load. Mechanisms for these changes are not clear. One mechanism could involve decreased hepatic extraction of circulating amino acids, perhaps due to the relative hypoaminoacidemia. For example, decreased incorporation of alanine into glucose production has been observed in pregnant women. Another possibility is decreased activity of the urea cycle. Both progesterone and estrogen, which are increased in pregnancy, have been shown to have suppressive effects on the activity of urea cycle enzymes. Prolonged fasting results in greater catabolism and urea excretion, especially in pregnant rats. Differences in results between animal and human studies may be due to the different impact of fetal mass/maternal mass among species, the duration of fasting relative to normal dietary eating and fasting patterns, and differences in the availability of alternate fuels.

D. Amino Acid and Protein Kinetics

Protein kinetics in pregnancy are best expressed in relation to maternal fat free mass. This is difficult to measure; thus, most data are expressed per weight of combined maternal and conceptus weight (total maternal weight). There is wide variation among studies of amino acid and protein kinetics in different species and by different methods. Studies in pregnant sheep indicate that protein turnover is decreased in later pregnancy, as measured using a leucine tracer. Thus, protein breakdown is probably diminished as well. The relative hyperinsulinemia of pregnancy may account for this decrease in protein breakdown, leading to the relative hypoaminoacidemia. Among small animals such as the rat, conflicting results from different studies are difficult to reconcile, but overall there appears to be an increase in whole body amino acid, and therefore protein, turnover in pregnancy. This condition perhaps reflects the increased fetal/maternal mass ratio and the metabolic impact of the relatively larger conceptus on maternal protein balance. Information from other small animals, such as the guinea pig, with different fetal/maternal mass ratios, diets, and activity levels is not available. In

TABLE 6
Whole Body Protein and Nitrogen Metabolism in Pregnancy (Mean ± SEM)[a]

	Metabolism			
	Nonpregnant women (n = 6)	Pregnant women		
Parameter		12 Weeks (n = 6)	24 Weeks (n = 6)	33 Weeks (n = 6)
Nitrogen flux (mg/kg/day)	525 ± 44	770 ± 89	665 ± 35	562 ± 30
Protein synthesis (g/kg/day)	1.9 ± 0.4	3.8 ± 0.5	2.8 ± 0.3	2.1 ± 0.2
Protein breakdown (g/kg/day)	1.7 ± 0.2	3.3 ± 0.5	2.3 ± 0.4	1.4 ± 0.4
Protein balance (g/kg/day)	0.1 ± 0.02	0.5 ± 0.1	0.5 ± 0.1	0.8 ± 0.1
Protein oxidation (g/kg/day)	1.1 ± 0.2	1.0 ± 0.2	1.0 ± 0.4	0.9 ± 0.3

[a] Adapted from B. de Benoist, A. A. Jackson, J. S. Hall, and C. Persaud, Whole-body protein turnover in Jamaican women during normal pregnancy, Hum. Nutr.–Clin. Nutr. 39, 167–179, 1985.

humans, labeled glycine studies have shown an increase in nitrogen flux, including both protein synthesis and breakdown, in the first and second trimesters, with turnover and particularly breakdown decreasing to less than nonpregnant rates by the third trimester. These studies also showed no change in protein oxidation (Table 6). As a result, protein balance (protein synthesis minus breakdown) was positive and increased progressively throughout pregnancy. In contrast, leucine tracer studies indicate a lower rate of leucine nitrogen turnover, primarily in the third trimester but clearly present in the first and second trimesters. An almost 30% lower rate of urea synthesis, when compared with nonpregnant controls, is evident in the first trimester and a 45% lower rate is evident in the third trimester. These carbon-labeled leucine tracer studies also show no change in leucine carbon flux or the calculated rate of protein turnover (expressed per total body weight) during pregnancy, representing a relatively constant contribution of leucine carbon to oxidation. Presumably the lower rate of nitrogen turnover represents the positive protein balance in early to midpregnancy, with lower rates of protein breakdown. The rate of leucine deamination is also lower and correlated with the decreased rate of urea synthesis. These changes begin as early as the first trimester, before the nutritional and metabolic demands of the fetus and placenta are significant, indicating hormonal changes as causative.

In summary, maternal nitrogen balance and protein balance increase in pregnancy. This reflects total maternal body nitrogen and protein, not just that in the conceptus. This nitrogen and protein anabolism may be produced or enhanced by increased insulin secretion and circulating insulin concentrations that develop in response to peripheral insulin resistance to glucose transport and metabolism. It is unclear whether the increased maternal nitrogen and protein stores are mobilized by metabolic processes in the late second and the third trimester, as for glucose and lipid in response to insulin resistance, to meet the increasing amino acid needs of the placenta and fetus.

VII. SUMMARY

Maternal metabolic changes in pregnancy involve tissue and energy anabolism in early pregnancy, followed by decreased maternal nutrient substrate utilization and storage and increased energy substrate mobilization in later pregnancy. These metabolic changes in the mother are primarily linked to changes in maternal hormonal status that are signaled from the placenta and fetus. As a result, the mother prepares her tissue stores and metabolism in early pregnancy to provide for the increasing metabolic and growth requirements for nutrients of the growing placenta and fetus in later pregnancy.

See Also the Following Articles

FETAL GROWTH AND DEVELOPMENT; FETUS, OVERVIEW; PLACENTA: IMPLANTATION AND DEVELOPMENT; PLACENTAL NUTRIENT TRANSPORT; PREGNANCY IN HUMANS, OVERVIEW

Bibliography

Battaglia, F. C., and Meschia, G. (1986). *An Introduction to Fetal Physiology*. Academic Press, Orlando, FL.

Catalano, P. M., Ishizuka, T., and Friedman, J. E. (1998). Glucose metabolism in pregnancy. In *Principles of Perinatal–Neonatal Metabolism* (R. M. Cowett, Ed.), 2nd ed. Springer-Verlag, New York.

Cowett, R. M. (Ed.) (1998). *Principles of Perinatal–Neonatal Metabolism*, 2nd ed. Springer-Verlag, New York.

Herrera, E., Lasunción, M. A., Martin, A., and Zorzano, A. (1992). Carbohydrate–lipid interactions in pregnancy. In *Perinatal Biochemistry* (E. Herrera and R. H. Knopp, Eds.), pp. 1–18. CRC Press, Boca Raton, FL.

Herrera, E., Bonet, B., and Lasunción, M. A. (1997). Maternal–fetal transfer of lipid metabolites. In *Fetal and Neonatal Physiology* (R. A. Polin and W. W. Fox, Eds.), 2nd ed. Saunders, Philadelphia.

Hytten, F. E., and Leitch, I. (1971). The gross composition of the components of weight gain. In *The Physiology of Human Pregnancy*, 2nd ed., pp. 371–387. Blackwell Scientific, London.

Jovanovic-Peterson, L., and Peterson, C. M. (1991a). Pregnancy and the endocrine pancreas. In *The Endocrine Pancreas* (E. Samois, Ed.), pp. 229–252. Raven Press, New York.

Jovanovich-Peterson, L., and Peterson, C. M. (1991b). Glucose metabolism in pregnancy. In *Principles of Perinatal–Neonatal Metabolism* (R. M. Cowett, Ed.), pp. 149–162. Springer-Verlag, New York.

Kalhan, S. C. (1998). Protein metabolism in pregnancy. In *Principles of Perinatal–Neonatal Metabolism* (R. M. Cowett, Ed.), 2nd ed. Springer-Verlag, New York.

Kalkhoff, R. K., Kissebah, A. H., and Kim, H. J. (1978). Carbohydrate and lipid metabolism during normal pregnancy: Relationship to gestational hormonal action. *Sem. Perinatol.* 2, 291–307.

Knopp, R. H., Bonet, B., Lasunción, M. A., Montelongo, A., and Herrera, E. (1992). Lipoprotein metabolism in pregnancy. In *Perinatal Biochemistry* (E. Herrera and R. H. Knopp, Eds.), pp. 19–52. CRC Press, Boca Raton, FL.

Knopp, R. H., Bonet, B., and Zhu, X. (1998). Lipid metabolism in pregnancy. In *Principles of Perinatal–Neonatal Metabolism* (R. M. Cowett, Ed.), 2nd ed. Springer-Verlag, New York.

Ogata, E. S. (1997). Maternal metabolism during pregnancy. In *Fetal and Neonatal Physiology* (R. A. Polin and W. W. Fox, Eds.), 2nd ed. Saunders, Philadelphia.

Verhaeghe, J., and Van Assche, A. (1992). Maternal amino acid metabolism during pregnancy. In *Perinatal Biochemistry* (E. Herrera and R. H. Knopp, Eds.), pp. 53–68. CRC Press, Boca Raton, FL.

Zorzano, A., Palacín, M, and Testar, X. (1992). Insulin resistance in pregnancy. In *Perinatal Biochemistry* (E. Herrera and R. H. Knopp, Eds.), pp. 69–92. CRC Press, Boca Raton, FL.

Pregnant Mare Serum Gonadotropin
see Equine Chorionic Gonadotropin

Pregnenolone
see Steroidogenesis, Overview

Premature Birth

see Preterm Labor and Delivery

Prenatal Genetic Screening

Deborah A. Driscoll
University of Pennsylvania School of Medicine

I. Genetic Risk Assessment
II. Chromosomal Disorders
III. Single Gene Disorders
IV. Multifactorial and Polygenic Disorders
V. Future of Prenatal Genetic Testing

GLOSSARY

amniocentesis An invasive technique performed during the second and third trimester to obtain fetal cells and amniotic fluid for prenatal testing.

chorionic villus sampling A procedure performed in the late first trimester (10–12 weeks) to obtain fragments of chorionic villi from the placenta for prenatal testing.

cordocentesis A technique for obtaining a sample of fetal blood from the umbilical vein using ultrasound guidance.

fetal echocardiography A noninvasive technique used to identify structural cardiac defects and disturbances of cardiac rhythm.

fluorescence in situ hybridization A technique used to detect complementary sequences of DNA within a chromosome using a specific DNA probe that is tagged to a fluorescent marker.

linkage analysis A process used to trace the inheritance of a disease-causing gene within a family based on the coinheritance of two or more DNA markers because they are in close proximity on the same chromosome.

nondisjunction The failure of two like chromosomes to separate during cell division such that one cell receives both copies and the other cell receives neither.

single gene disorders Disorders resulting from alterations or mutations in the DNA sequence of a specific gene resulting in a disease phenotype, birth defect, growth problems, and/or mental retardation.

translocation An exchange of chromosomal material between two or more chromosomes.

trinucleotide repeat A short, repetitive sequence of DNA composed of three base pairs which occur throughout the genome but can change in size on transmission from parent to offspring.

ultrasonography A noninvasive technique used to date a pregnancy, diagnose fetal structural anomalies, and identify multiple gestation.

Prenatal genetic testing is the use of specific assays to determine the status of a pregnancy suspected to be at risk for a particular genetic condition based on family history, parental age, ethnicity, or clinical symptoms. In contrast, genetic screening is the use of various genetic tests to evaluate populations or groups of individuals independent of a family history. However, these terms are often used interchangeably. Genetic counseling is the process by which individuals are provided with information about the nature and etiology of the condition, the risk of an inheritable condition, testing options, risks and benefits of genetic testing, interpretation of the test results, reproductive options, and appropriate support and medical care.

I. GENETIC RISK ASSESSMENT

The initial step in prenatal genetic screening is to ascertain the risk of an inheritable or genetic condition. Risk assessment is based on family history, medical and reproductive history, parental age, and ethnicity. Table 1 summarizes the indications for prenatal diagnosis. A three-generation family pedigree can help to establish whether a pregnancy and/or other family members are at risk for a genetic disorder or congenital malformation. Family history may be obtained by a health care professional such as a physician, nurse practitioner, nurse, or genetic counselor. The pedigree is constructed based on a couple's family history of Down syndrome and other chromosome abnormalities; birth defects such as cleft palate, congenital cardiac defects, and spina bifida; mental retardation; inherited diseases such as muscular dystrophy, hemophilia, and cystic fibrosis. A reproductive history of multiple spontaneous abortions and/or stillbirths may indicate an increased risk for chromosomal abnormalities and inherited conditions. Parental age is also important for risk assessment. Advanced maternal age (≥35 years) is associated with an increased risk for having a child with Down syndrome and other chromosomal abnormalities, whereas advanced paternal age is associated with an increased risk for new mutations which may result in genetic disorders such as achondroplasia and premature craniosynostosis (i.e., Apert syndrome). Ethnic background is also important since particular populations are at an increased risk for specific genetic disorders such as Tay–Sachs disease or sickle cell anemia.

In addition, the mother's medical history, occupation, and medication exposures are a fundamental part of genetic screening and used to assess the risk for congenital anomalies. For example, maternal insulin-dependent diabetes is associated with a two- or threefold increased risk for congenital malformations, in particular, neural tube defects, congenital heart disease, and caudal regression syndrome. Hence, excellent glucose control prior to conception and during the first trimester is recommended to minimize the risk to the fetus. Maternal phenylketonuria associated with maternal blood levels of phenylalanine >20 mg/dl is more likely to result in fetal microcephaly, mental retardation, congenital heart disease, and intrauterine growth retardation, whereas the initiation of dietary changes that result in lowered phenylalanine levels during the first trimester can result in a normal outcome. Recently, folate intake prior to conception and during the first trimester has been shown to prevent recurrences of neural tube defects. Hence, the U.S. Public Health Service currently recommends that all women of childbearing age consume 0.4 mg of folate per day. In contrast, the daily recommendation for women who have had a prior child or fetus with a neural tube defect is 4.0 mg folate per day, unless contraindicated by a history of pernicious anemia.

Most often, genetic risk assessment occurs during the initial prenatal visit or if a patient is referred for genetic counseling. Many obstetricians utilize prenatal genetic screening questionnaires to identify patients at risk for potential genetic problems. Although several studies suggest that a three-generation pedigree obtained by a trained individual is more sensitive for the accurate assessment of genetic risk, self-administered genetic screening questionnaires may be a useful adjunct to the family pedigree, particularly in a busy obstetrical practice.

Ideally, a genetic risk assessment should be performed prior to conception since this information may influence a couple's reproductive decisions. Identification of many of the genes for disorders such as cystic fibrosis and fragile X mental retardation

TABLE 1
Indications for Prenatal Genetic Testing

Increased risk for Down syndrome or chromosomal abnormality based on maternal age (≥35 years) or results of maternal serum screening

Previous child with numerical or structural chromosome abnormality

Parent is a carrier of a chromosomal rearrangement

Parent is a carrier of an autosomal recessive, autosomal dominant, or X-linked genetic disorder

Prior stillbirth or history of recurrent spontaneous abortions and/or stillbirths

Congenital malformation

Previous child with a neural tube defect

syndrome have forced us to reexamine our current approach to prenatal testing and carrier screening and emphasizes the importance and value of preconception genetic screening. Furthermore, this is an ideal time to educate patients about nutritional needs during pregnancy, folate supplementation, medications, and exposures to avoid which may lead to improved perinatal outcomes. Immunizations may be recommended prior to pregnancy to prevent congenital rubella syndrome. Women at risk for toxoplasmosis should be informed to avoid eating raw meat or cleaning cat litter. Medications such as valproic acid, isoretinoin, coumadin, and lithium are associated with an increased risk of fetal malformations of the central nervous and cardiovascular system as well as craniofacial defects. Hence, a patient may wish to avoid these medications or try an alternative medication prior to conception. Alternatively, they may wish to avoid pregnancy with the full knowledge of the risks and testing options as well as alternative reproductive choices such as adoption, and donor egg and donor insemination programs.

II. CHROMOSOMAL DISORDERS

A. Down Syndrome (Trisomy 21)

The most common indication for prenatal genetic counseling and testing is advanced maternal age. Indications for prenatal cytogenetic testing are summarized in Table 2. Women 35 years and older have an increased risk of chromosomal abnormalities such as trisomy 21 due to nondisjunction of the chromosomes usually during maternal meiosis. Age-specific rates for Down syndrome are available and are used to assess a patient's risk for trisomy 21. Trisomy 21 or Down syndrome is the most common chromosomal abnormality, occurring in 1 in 800 liveborns, but the risk increases to 1 in 350 at 35 years and 1 in 100 at 40 years. Individuals with Down syndrome have characteristic craniofacial features, including a flat occiput, epicanthal folds, broad nasal bridge, protruding tongue, small low-set ears with an overlapping helix and prominent antihelix; broad short fingers with an incurving of the fifth finger, a single palmar crease, and wide space between the first two toes; and mental retardation. Approximately 40% have a congenital heart defect and, less commonly, duodenal atresia.

The majority of cases of Down syndrome (95%) result from an extra chromosome 21. A small number result from segregation of an unbalanced translocation which may be inherited from a parent with a balanced translocation or may be a new event in the fetus. The risk of Down syndrome is increased when a parent is a carrier of a translocation involving chromosome 21. Hence, prenatal testing is routinely offered to parents who are known translocation carriers. Furthermore, in the United States it is standard obstetrical practice to offer genetic counseling and prenatal chromosomal testing to women age 35 or older at delivery. Amniocentesis at 15–20 weeks of gestation and chorionic villus sampling at 10–12 weeks of gestation are well established, relatively safe techniques to obtain cells for cytogenetic testing.

In the past decade maternal serum screening has been used to adjust a mother's age-based risk for Down syndrome. This screening test was initially designed for women under the age of 35 since approximately 80% of children with Down syndrome are born to women in this age group. The majority of screening programs utilize three markers: α-fetoprotein (AFP), human chorionic gonadotropin (hCG), and unconjugated estriol. Women with an adjusted risk equal to or greater than the risk of a 35-year-old at midtrimester are offered amniocentesis for cytogenetic testing. Recent studies suggest that maternal serum screening may be used to revise the risk for Down syndrome in women 35 years and older. Although this sounds attractive since it is a noninvasive test, patients need to be counseled that the frequency of a positive maternal serum screening

TABLE 2
Indications for Prenatal Cytogenetic Analysis

Advanced maternal age (≥35 years)
Positive multiple marker screen
Previous child with trisomy or chromosome abnormality
Parent is a carrier of a chromosomal rearrangement
Prior stillbirth or history of recurrent spontaneous abortions and/or stillbirths
Congenital malformation

test is high and, in contrast to amniocentesis, maternal serum screening is not a diagnostic test for Down syndrome and other chromosomal disorders that result from nondisjunction such as trisomy 18.

Although ultrasonography is a useful technique for the detection of congenital anomalies and may detect features suggestive of Down syndrome such as an atrioventricular canal defect or nuchal thickening, it is not a diagnostic test nor should it be used as the sole screening test for Down syndrome. In contrast, when a congenital malformation is demonstrated by sonographic examination, a fetal karyotype is often recommended as part of the evaluation.

B. Other Autosomal Trisomies

Women 35 years and older are also at risk for other chromosomal abnormalities as a result of nondisjunction; however, these are less frequent than trisomy 21. Trisomy 18 occurs in 1 in 8000 live births and in the majority of cases is lethal. Very few cases of trisomy 18 survive beyond days or months of life. This disorder causes severe mental retardation, growth retardation, and cardiovascular, genitourinary, central nervous system and skeletal anomalies. Many of these anomalies may be detected sonographically. In addition, trisomy 18 may be suspected based on the maternal serum marker screening when all three values (AFP, hCG, and unconjugated estriol) are low. Less frequent is trisomy 13 (1 in 20,000 live births), which results in growth retardation, severe mental retardation, cardiac defects, cleft lip and palate, and opthamologic abnormalities. Less than 5% survive beyond the first year; many result in spontaneous miscarriages. Approximately 80% of cases of trisomy 13 and 18 are the result of nondisjunction. Trisomy for other chromosomes is rare and may result in miscarriage or, in some cases such as trisomy 8 and 22, anomalies and mental retardation of variable severity.

C. Sex Chromosomal Abnormalities

Prenatal cytogenetic testing may identify abnormalities involving the sex chromosomes X and Y. These occur in approximately 1 in 250 amniocenteses performed for advanced maternal age.

1. Turner's Syndrome

Turner's syndrome is caused by absence of one X chromosome (45,X) or loss of part of the X chromosome. This chromosomal disorder is not related to maternal age, and in the majority of cases results in spontaneous pregnancy loss. Typically, females with Turner's syndrome have short stature and may have a webbed neck and cardiac and renal anomalies. In addition, the ovaries undergo premature attrition resulting in lack of development of secondary sex characteristics, amenorrhea, and infertility. In rare instances, a patient with Turner's syndrome may menstruate for a short time. However, most will benefit from growth hormone and hormone replacement therapy at the time of puberty. In addition, with assisted reproductive technology currently available, patients may carry a pregnancy. In contrast to the autosomal chromosomal disorders, intelligence is within the normal range.

2. Klinefelter's Syndrome

Klinefelter's syndrome occurs in 1 in 600 males, resulting from the presence of at least one extra X chromosome (47,XXY) due to nondisjunction during either male or female meiosis. Males with this disorder have tall stature, small testes, normal sexual development, low or absent sperm, infertility, and have an increased risk for learning disabilities and developmental problems. However, mental retardation is uncommon. Males with Klinefelter's syndrome may benefit from testosterone supplementation.

3. 47,XYY

The presence of more than one Y chromosome is found in 1 in 1000 males. These individuals are more likely to be tall but otherwise appear phenotypically male. Intelligence is within the normal range but IQ scores may be 10–15 points below those of siblings. Approximately 50% have neurodevelopmental and learning disabilities; however, recent prospective studies of males ascertained through prenatal diagnosis indicate that 47,XYY males in a supportive family are better adjusted and are behaviorally no different from controls.

4. 47,XXX

Females with an extra X chromosome (47,XXX) are phenotypically female, fertile, and tend to be tall. This occurs in 1 in 1200 female newborns, usually secondary to nondisjunction. Mental retardation is uncommon but IQ scores tend to be 10–15 points less than those of siblings and they may have neurodevelopmental and language delays. When a female has more than three X chromosomes mental retardation and minor phenotypic abnormalities are common.

D. Autosomal Chromosomal Structural Abnormalities

Prenatal genetic testing is recommended when a parent has a structural chromosomal abnormality such as a translocation, deletion, or duplication.

1. Translocation

A translocation results from the exchange of chromosomal material between two different chromosomes. Translocations are considered balanced when there is no apparent loss of chromosomal or genetic material. Balanced translocations occur in approximately 1 in 500 individuals. In general, carriers of balanced translocations are phenotypically normal, although there are several rare reported exceptions associated with genetic disorders. Translocations may occur as a new event in the fetus or may be inherited. Carriers of a balanced translocation are at increased risk for recurrent pregnancy losses, stillbirths, and offspring with congenital anomalies and/or mental retardation as a result of transmitting the translocation to their offspring as an unbalanced translocation, in which case a segment of one chromosome is absent and the segment from the reciprocal chromosome is duplicated.

2. Deletion

Deletion or loss of a segment of a chromosome can result in a variety of anomalies and growth and learning problems depending on the size, location, and the function of the genes within the deleted region. There are several well-known syndromes associated with deletions, including DiGeorge/velocardiofacial syndrome with deletions of 22, Prader–Willi and Angelman with 15, and Miller–Dieker with 17. Deletions may occur as sporadic events or can be inherited from a parent with a deletion, in which case there is a 50% chance the parent will transmit the deletion to his or her offspring. Although the recurrence risk for parents of a child with a sporadic deletion is low, due to possible germline mosaicism many couples elect to have prenatal testing for reassurance. Detection of these deletions is best accomplished using fluorescence *in situ* hybridization (FISH) as an adjunct to routine cytogenetic analysis. Testing for these deletions is offered when a parent has the deletion, the couple have had a previous child with the deletion, or there is sonographic evidence of a malformation suggestive of a particular deletion syndrome. For example, testing for a chromosome 22 deletion is recommended when a conotruncal cardiac defect, such as an interrupted aortic arch or truncus arteriosus, is detected since a significant percentage of these types of cardiac defects are seen in association with this deletion syndrome. Prenatal confirmation of a deletion provides the parents, obstetrician, and pediatricians with important prognostic information.

3. Duplication

Duplications of a segment of a chromosome may occur within a chromosome as an interstitial duplication or as an extra or supernumerary chromosome, also referred to as a marker chromosome. Duplications can result in various congenital malformations, growth disturbances, and cognitive impairment. There are several well-described genetic syndromes associated with supernumerary chromosomes involving duplications of chromosome 22, 15, and 18. In general, the risk for an abnormality is dependent on the origin of the marker chromosome and whether it is familial or *de novo*. The origin may be determined based on the cytogenetic features or may be characterized further using molecular–cytogenetic techniques such as FISH. It is not always possible to determine the etiology nor to accurately predict the phenotypic outcome antenatally.

III. SINGLE GENE DISORDERS

Single gene disorders are caused by mutations or changes in genes which result in an altered phenotype or condition. These conditions are usually inherited but can occur as a result of a new mutation. In general, these disorders are rare, but some occur more commonly in specific populations such as Tay–Sachs disease in individuals of Eastern European Jewish descent. Advances in molecular genetics during the past 20 years have had a profound impact on our ability to detect these disorders antenatally as well as identify carriers of some genetic disorders such as cystic fibrosis. The rapid identification of disease-causing genes as a result of the Human Genome Project and the advances in biotechnology has resulted in an increase in prenatal testing of at-risk pregnancies and may result in population-based screening programs for some of these disorders in the future. Although molecular-based assays are available for a large number of genetic disorders, prenatal genetic testing is, in general, limited to pregnancies at risk based on family history. In the absence of a known gene defect or mutation, biochemical assays or linkage analysis may be possible. Linkage analysis requires knowledge of the relative location of the gene, availability of informative DNA markers in close proximity to the gene, and DNA from the affected individual and family members. In some cases, prenatal diagnosis may not be possible or may rely on the sonographic finding of specific defects associated with the disorder.

A. Population-Based Carrier Testing

Carrier testing for inherited disorders may be recommended based on the family history and ethnicity. Currently, population-based carrier testing is only offered based on ethnicity, whereas carrier testing for disorders such as cystic fibrosis and fragile X mental retardation is currently recommended for individuals with a positive family history. Ethnicity determines whether carrier testing for diseases such as Tay–Sachs, sickle cell, and thalassemia is necessary. Carrier testing programs for these disorders are well established and successfully used to screen specific populations of patients at risk.

1. Tay–Sachs Disease

Tay–Sachs disease is a lysosomal storage disease caused by a deficiency of hexosaminidase A. This results in the abnormal accumulation of gangliosides in the central nervous system which leads to progressive neurological impairment and, ultimately, death in early childhood. Tay–Sachs disease is transmitted in an autosomal recessive fashion and is more prevalent in individuals of Eastern European Jewish, French Canadian, and Cajun descent. The carrier frequency is 1 in 30 in individuals of Ashkenazi Jewish descent in contrast to a carrier frequency of 1 in 300 in the general population. Measurement of serum hexosaminidase levels is used to determine carrier status. However, this testing is not accurate in women who are pregnant or taking oral contraceptive pills, in which case leukocyte testing is recommended. Molecular studies of the gene for the α subunit of hexosaminidase have identified mutations; however, this type of testing is usually reserved for individuals in which the biochemical testing is ambiguous or inconclusive. When both partners are carriers of Tay–Sachs disease, prenatal genetic counseling is recommended and prenatal testing is offered.

Recently, carrier testing has been offered for several other genetic disorders which occur more commonly in the Ashkenazi Jewish population. Canavan disease is a lethal autosomal recessive disorder caused by a deficiency of aspartoacyclase in the brain which results in spongy degeneration of the central nervous system. The clinical features include hyperextension of the legs and flexion of the arms, blindness, a large head, and severe mental retardation. The gene for Canavan disease was identified in 1993, and subsequently three mutations were found in 98.8% of individuals with Canavan disease of Ashkenazi Jewish descent. Based on a population-based mutation analysis of Ashkenzi Jews, the carrier frequency for Canavan disease is 1 in 59. DNA-based assays are now available to assess carrier status for this disease.

Similarly, a small number of mutations have been found in the majority (96%) of individuals with Gaucher's disease, an autosomal recessive disorder more frequent but not limited to individuals of Eastern European Jewish descent. Gaucher's disease is caused by glucocerebrosidase deficiency, which can result in anemia, thrombocytopenia, enlarged spleen, bone lesions, increased skin pigmentation, neurological problems, and pingueculae. There are three types of Gaucher's disease with variable severity and age of onset. The carrier frequency is 1 in 13 in individuals of Ashkenazi Jewish descent; carrier testing for the four most common mutations is available.

Niemann–Pick disease is also an autosomal recessive disorder which occurs more frequently in individuals of Eastern European Jewish descent. The main features are hepatosplenomegaly, neurological problems, and growth retardation. There are several types of varying severity and age of onset. The classical or infantile form (type A) occurs within 6 months of life and is lethal. This disorder is caused by a deficiency of sphingomyelinase which leads to abnormal accumulation of sphingomyelin within cells throughout the body, in particular the central nervous system and reticuloendothelial cells. In the past, prenatal testing was based on the finding of this metabolic defect in the amniotic fluid cells; however, currently, carrier and prenatal testing relies on the knowledge of the gene structure.

2. Sickle Cell Disease

Sickle cell disease is an autosomal recessive disorder characterized by anemia, episodes of severe pain or "crises," increased risk for stroke, and cardiovascular and renal impairment. The incidence of sickle cell disease is 1 in 500 among African Americans, and therefore, 1 in 12 African Americans is a carrier of the gene for sickle cell disease. This risk may be higher in individuals with a positive family history. For example, if a first-degree relative (a parent or sibling) has sickle cell disease then the risk of carrying the gene increases to 50–66%. Carrier testing is routinely performed in most obstetric practices during the initial prenatal visit using a sickle prep and/or hemoglobin electrophoresis. When both partners are carriers, they have a 1 in 4 chance of having a child with sickle cell disease in each pregnancy, and prenatal genetic counseling and testing are offered. Prenatal testing is possible because the gene mutation for sickle cell disease is known. The disease is caused by a mutation in the β-globin gene; a single base pair substitution in the DNA sequence results in an amino acid change from glutamic acid to valine. Prenatal testing can be performed using either amniotic fluid cells or chorionic villi obtained by either amniocentesis or chorionic villus sampling, respectively.

3. Thalassemia

Thalassemia is a form of anemia caused by either mutations or deletions in the hemoglobin genes, α- and β-globin. α-Thalassemia is more common in individuals of Asian descent, whereas β-thalassemia is more common among individuals of African and Mediterranean descent. Carriers of thalassemia are usually asymptomatic but are at risk for having more severely affected offspring if their partner is also a carrier. Carrier status is initially determined by the mean corpuscular volume (MCV), a test routinely done as part of a complete blood count. Patients with a low MCV should have a hemoglobin electrophoresis test to determine whether they are indeed carriers of one form of thalassemia. In the absence of iron-deficiency anemia, a positive hemoglobin electrophoresis should prompt genetic counseling and possible molecular testing if both partners are carriers of either thalassemia or sickle cell disease.

B. Prenatal Genetic Testing for Single Gene Disorders

Prenatal testing for specific genetic disorders is based on family history and is only feasible when either a biochemical assay is available or the gene structure and/or function are known. Most of these diseases are rare, and often couples at risk are not identified until the birth of an affected child. Historically, inborn errors of metabolism were the first genetic disorders amenable to prenatal testing using biochemical techniques. Rarely, fetal biopsies of skin, liver, and muscle have been performed to establish a diagnosis when the gene defect is unknown. However, this has been largely replaced by DNA analysis using either linkage analysis or direct detection of gene mutations. The list of diseases which can be diagnosed using DNA analysis is rapidly growing and includes Duchenne muscular dystrophy, cystic fibrosis, Huntington's disease, fragile X syndrome, myotonic dystrophy, spinal muscular atrophy, neurofibromatosis, sickle cell disease, thalassemias, and Tay–Sachs disease. Two of the more common genetic disorders, cystic fibrosis and fragile X syndrome, are discussed in the following sections.

1. Cystic Fibrosis

Cystic fibrosis (CF) is the most common autosomal recessive disorder of Caucasians. This disease affects approximately 1 in 3300 newborns and 1 in 29 Caucasians carries the gene for CF. CF is rare in Africans and Asians. Children with CF develop chronic obstructive pulmonary disease, pancreatic insufficiency, and failure to thrive. Males also have bilateral absence of the vas deferens, resulting in infertility. The diagnosis of CF is usually made by a "sweat test," which demonstrates elevated levels of chloride in the sweat, and confirmed by DNA-based testing since the gene, CF transmembrane conductance regulator, was identified in 1989. Since 1989 over 600 mutations have been reported. The frequency of mutations varies among populations. For example, ΔF508 accounts for 75–88% of CF mutations in Caucasians of northern European descent and only 50–60% in southeastern European populations. In contrast, mutation W1282X appears to be more common in individuals of Ashkenazi Jewish descent. Hence, the accuracy of mutation screening is dependent on ethnic background. Commercial and hospital-based DNA diagnostic laboratories have designed mutation screening tests which will identify approximately 95% of Ashkenazi Jewish and 85% of non-Jewish northern European CF carriers. Prenatal DNA testing is available for couples with a previously affected child and for couples in which both are carriers of a known mutation and hence have a 1 in 4 risk of having a child with CF. If one or both mutations cannot be identified linkage analysis may be utilized. Currently, carrier testing is recommended when there is a family history of CF and when a partner is a carrier of a CF mutation. Population-based carrier testing for CF is controversial and currently is not been recommended; however, testing is available to couples who wish to know their CF carrier status.

2. Fragile X Syndrome

Fragile X syndrome is the most common form of inherited mental retardation. In addition to mental retardation, males with fragile X syndrome may have behavioral problems, autism, and physical features such as large ears, long and narrow face, large testes, and hyperextensible joints. Fragile X syndrome occurs in approximately 1 in 1200 males and 1 in 2500 females. Females are less severely affected. The gene for fragile X syndrome was characterized in 1991; it contains a trinucleotide repeat. The normal number of repeats is approximately 6–50. Expansions of this repeated trinucleotide sequence to 50–200 repeats are called premutations which can expand further to full mutations of >200 repeats. Males and females with premutations are phenotypically normal but the females are at risk for having offspring with full mutations. In contrast, males with premutations transmit the premutation unchanged in size to their female offspring. Most males with a full mutation are mentally retarded and show some of the physical and behavioral features of fragile X syndrome. However, the clinical features are of variable severity. Although females with the full mutation may be of normal intelligence, approximately one-third are mentally retarded. DNA analysis to determine the size of the trinucleotide repeat is accomplished by the polymerase chain reaction (PCR) and/or Southern blot analysis. Currently, DNA testing is recommended for individuals of either sex with mental retardation, developmental delay, or autism, especially when they have physical and/or behavioral features characteristic of fragile X syndrome, a family history of fragile X syndrome, or relatives with undiagnosed mental retardation. Carrier testing is recommended for individuals with a family history of fragile X syndrome or undiagnosed mental retardation. Population screening is currently not recommended. Prenatal testing of cultured amniocytes is possible when a mother is a carrier.

IV. MULTIFACTORIAL AND POLYGENIC DISORDERS

The majority of birth defects or congenital malformations which occur as isolated defects are presumed to be multifactorial, the result of both genetic and environmental factors. Examples of common malformations which are presumably inherited in a multifactorial fashion include neural tube defects, congenital heart defects, cleft lip and palate, pyloric stenosis, and omphalocele. However, other causes, including chromosomal abnormalities, single gene defects, and teratogens, need to be excluded. The recurrence risk for these anomalies ranges from 1 to 5% for first-degree relatives of an affected sibling or parent. As our understanding of the genetic basis for these common birth defects increases, we will continually need to reassess the recurrence risk and prenatal testing options to establish an etiology as well as a better estimate of the outcome and prognosis. One of the best examples of this is the finding that a significant proportion of newborns and children with congenital heart defects involving the outflow tract and aortic arch occur as a result of submicroscopic deletions of chromosome 22. While the majority of these deletions occur sporadically, approximately 10% are familial, and in these cases the recurrence risk can be as high as 50% in contrast to 2 or 3%. Furthermore, individuals with the deletion are at a significant risk for other medical and learning problems and hence may benefit from comprehensive medical and learning evaluations.

A. Detection of Fetal Malformations

1. Ultrasonography

Many major anatomic defects may be diagnosed antenatally by ultrasonography and fetal echocardiography during the second trimester. In general, these studies are recommended when a first- or second-degree relative has a congenital malformation. In many countries it is standard practice to perform a routine ultrasound between 16 and 20 weeks of gestation. In the United States, ultrasonography may be performed for a variety of indications, including establishment of gestational age, assessment of fetal growth, position, amniotic fluid volume, placental location and grade, diagnosis of multiple gestation, and structural defects. Transvaginal ultrasound during the first trimester may allow for earlier detection of certain anomalies such as anencephaly. However, in some instances malformations such as hydrocephalus and renal anomalies may not be apparent until late in gestation.

2. Maternal Serum Screening

In addition to ultrasonography, maternal serum AFP screening programs are well established and very effective at identifying a pregnancy at risk for an open neural tube defect (NTD), an abnormality in the development of the neural tube resulting in spina bifida or anencephaly. It is standard practice to offer AFP testing between 15 and 20 weeks of gestation. AFP levels are usually elevated in the amniotic fluid when a fetus has an open NTD or abdominal wall defect such as omphalocele or gastroschisis. In some instances, maternal serum AFP measurements and ultrasound examination are sufficient for the diagnosis of open spina bifida and anencephaly. Measurement of amniotic fluid AFP is considered in women with an elevated maternal serum AFP, sonographic evidence of a NTD, family history of a previous child or parent with a NTD, maternal diabetes, and exposure to medications such as valproic acid which are associated with an increased risk for NTDs.

B. Approach to the Fetus with a Congenital Malformation

When a structural anomaly is suspected, further diagnostic studies are indicated, including a comprehensive level II ultrasound examination and echocardiogram to exclude associated defects and either an amniocentesis or cordocentesis to obtain a fetal sample of tissue or blood for cytogenetic and/or infectious studies. In general, cordocentesis or percutaneous umbilical blood sampling is utilized when a rapid diagnosis is essential for obstetrical and/or fetal management. Numerical and structural chromosomal abnormalities are not an infrequent cause of fetal anomalies and intrauterine growth restriction, and the finding of a chromosomal abnormality may influence the obstetrical and neonatal management.

Following an extensive evaluation of the fetus with a congenital anomaly, the parents are counseled about the nature of the defect, the possible etiologies, potential outcomes, and management issues. Decisions regarding fetal therapy, resuscitative efforts, mode of delivery, and management of the infant may differ depending on the severity of the defect, the fetal karyotype, and the associated prognosis. In many cases, it is helpful to involve the obstetrician, geneticist, pediatrician, and pediatric specialists in these discussions. In selected situations, the possibility of *in utero* medical or surgical intervention may be appropriate and may be reviewed with the family. Fetal surgery for congenital diaphragmatic hernia, cystic adenomatoid malformation of the lung, and obstructive uropathies has been successfully performed at a few centers that specialize in fetal therapy.

In general, the management of a pregnancy complicated by a fetal malformation or genetic disorder depends on the gestational age of the fetus, fetal viability, nature and severity of the disorder, the prognosis, as well as consideration of the parent's wishes. In the event that a couple do not continue the pregnancy, an effort should be made to establish a diagnosis in order to provide the family with an accurate assessment of the recurrence risk. Cytogenetic studies, an autopsy, and genetic evaluation are recommended. Follow-up counseling is provided to review the findings, the diagnosis, the recurrence risk in subsequent pregnancies, the availability of prenatal diagnostic testing, and reproductive options. In addition, this is an excellent opportunity to address the emotional needs of the couple. When a couple chooses to continue a pregnancy, close monitoring is recommended with possible delivery at a tertiary care hospital prepared to provide the appropriate postnatal care for the infant. In some cases, a timed delivery is necessary to ensure that the essential personnel are available, such as the perinatologist, pediatric specialist, and/or surgeon. The time preceding the delivery is used to prepare a couple for the birth of a child, who may require intensive medical and/or surgical care.

V. FUTURE OF PRENATAL GENETIC TESTING

The availability of DNA-based prenatal genetic testing and carrier testing will continue to grow as a result of the identification of the genes for single gene disorders as well as susceptibility genes for complex diseases. Furthermore, advances in biotechnology such as DNA microchip technology will allow for rapid, accurate, and comprehensive screening for genetic mutations. The GeneChip (Affymetrix, Santa Clara, CA) system is currently being used for research in gene expression, genetic mapping, and genotyping. In the future, DNA chip-based assays may allow for the simultaneous analysis of thousands of genes for evidence of genetic alterations.

Concurrently, efforts have been directed at developing techniques for earlier, noninvasive methods of prenatal testing. In the late 1980s, techniques to obtain cells from the developing embryo prior to implantation were developed and used for human preimplantation genetic diagnosis. Single cells obtained from a biopsy of eight-cell embryos of pregnancies conceived through *in vitro* fertilization have been used for the genetic diagnosis of point mutations associated with specific genetic disorders such as cystic fibrosis and for fetal sex determination for X-linked diseases. This technique allows for selection of unaffected embryos for transfer into the uterus and, ultimately, the birth of a child not affected by the specific genetic disorder. Preimplantation genetic testing is available at several centers in the United States, Australia, and Europe on an investigational basis. Testing has been limited to a select group of genetic diseases amenable to PCR-based diagnosis of a known mutation or chromosomal abnormalities which may be detected using FISH techniques. However, the centers currently offering preimplantation genetic diagnosis have had limited experience and continue to refine the molecular technology to provide couples with an affordable and reliable testing option.

Another possibility for earlier and noninvasive prenatal genetic screening is the analysis of fetal cells isolated from the maternal circulation. Research studies have demonstrated that fetal cells can be successfully separated from the maternal circulation and used to detect numerical chromosomal abnormalities using FISH with chromosome-specific probes. The diagnosis of single gene disorders may prove to be more difficult since it will require a pure population of fetal cells although sex determination may be possible for couples at risk for an X-linked recessive disorder. However, large collaborative studies will be necessary to evaluate the sensitivity and specificity for chromosomal and genetic disorders before this technique can be adopted for genetic screening.

See Also the Following Articles

Amniocentesis; Fetal Anomalies; Fetal Monitoring and Testing; Intrauterine Growth Retardation; Klinefelter's Syndrome; Prader–Willi Syndrome; Turner's Syndrome; Ultrasound

Bibliography

American College of Obstetricians and Gynecologists (ACOG) (1986, September). Sample prenatal genetic screen, Tech. Bull. No. 108. ACOG, Washington, DC.

Beaudet, A. L. (1992). Genetic testing for cystic fibrosis (review). *Pediatr. Clin. North Am.* **39**, 213–228.

Cefalo, R. C., and Moos, M. K. (1995). *Preconceptional Health Care: A Practical Guide*, 2nd ed. Mosby, St. Louis, MO.

Centers for Disease Control and Prevention (1992). Recommendations for the use of folic acid to reduce the number of cases of spina bifida and other neural tube defects. *MMWR* **41**(RR-14), 1–7.

Cystic Fibrosis Genetic Analysis Consortium (1994). Population variation of common cystic fibrosis mutations. *Hum. Mutat.* **4**, 167–177.

Haddow, J. E., Palmaki, G. E., *et al.* (1992). Prenatal screening for Down's syndrome with use of maternal serum markers. *N. Engl. J. Med.* **327**, 588–593.

Handyside, A. H., Lesko, J. G., Tarin, J. J., Winston, R. M. L., and Hughes, M. R. (1992). Birth of a normal girl after in vitro fertilization and preimplantation diagnostic testing for cystic fibrosis. *N. Engl. J. Med.* **327**, 905–909.

Kaback, M., Lim-Steele, J., Dabholkar, D., Brown, D., Levy, N., and Zeiger, K. (1993). Tay–Sachs disease-carrier screening, prenatal diagnosis, and the molecular era. An international perspective, 1970 to 1993. The International TSD Data Collection Network. *J. Am. Med. Assoc.* **270**, 2307–2315.

New England Regional Genetics Group Prenatal Collaborative Study of Down Syndrome Screening (1989). Combining maternal serum alpha-fetoprotein measurements and age to screen for Down syndrome in pregnant women under age 35. *Am. J. Obstet. Gynecol.* **160**, 575–581.

Rimoin, D. L., Connor, J. M., and Pyeritz, R. E. (Eds.) (1996). *Emery and Rimoin's Principles and Practice of Medical Genetics*, 3rd ed. Churchill Livingston, New York.

Romero, R., Pilu, G., Jeanty, P., Ghidini, A., and Hobbins, J. C. (1988). *Prenatal Diagnosis of Congenital Anomalies*. Appleton & Lange, Norwalk, CT.

Scriver, C. R., Beaudet, A. L., Sly, W. S., and Valle, D. (Eds.) (1995). *The Metabolic and Molecular Basis of Inherited Disease*, 7th ed. McGraw-Hill, New York.

Shulman, L. P., and Elias, S. (1990). Percutaneous umbilical blood sampling, fetal skin sampling, and fetal liver biopsy. *Sem. Perinatal.* **14**, 456–464.

Warren, S. T., and Nelson, D. L. (1994). Advances in molecular analysis of fragile X syndrome. *J. Am. Med. Assoc.* **271**, 536–542.

Preterm Labor and Delivery

Douglas Woelkers and Steve Caritis
University of Pittsburgh

I. Introduction and Definitions
II. Epidemiology
III. Risk Factors
IV. Infant Outcome
V. Physiology and Pathophysiology
VI. Diagnosis and Prediction
VII. Treatment and Prevention

GLOSSARY

decidua Glandular (endometrial) lining of the uterus during pregnancy.

iatrogenic Describing an unintended adverse condition induced by medical intervention in the course of treatment.

labor Forceful uterine contractions resulting in progressive cervical dilatation, fetal descent, and delivery of the conceptus.

neonatal The period of time from birth through the first 28 days of life.

oxytocin Endogenous hormonal peptide released from posterior pituitary gland which stimulates uterine contractions.

parturition The ultimate phase of pregnancy in which fetal and maternal biochemical and physical processes bring about delivery of the conceptus.

prostaglandins A family of bioactive metabolites of arachidonic acid which are involved in the initiation and progression of labor.

surfactant Phospholipid and protein containing secretions of alveolar epithelium having surface tension lowering properties essential for early pulmonary function.

tocolysis Pharmacologic inhibition of uterine contractions.

Birth is the consummation of the reproductive process. It symbolizes the successful culmination of all the biological, behavioral, and social imperatives to regenerate a species. The preterm delivery of immature offspring can be a reproductive failure, subverting the great expenditure of energy and resources dedicated to reproduction. Among humans, who are uniquely susceptible to this condition, the neonatal consequences of preterm birth may include death, illness, or lifelong disability, and the costs to society are enormous, being further compounded by the lost potential productivity of its youngest members. Despite intensive research and clinical efforts, the rate of preterm delivery has remained relatively stable worldwide, continuing to complicate approximately 7–10% of the 140 million births per year. The study of preterm labor and delivery has recently undergone a paradigm shift: It is now appreciated by researchers not as a symptom or event but as a multifactorial syndrome of premature activation of the parturitional process.

I. INTRODUCTION AND DEFINITIONS

The study of preterm birth has been confounded by the use of nonstandardized terms describing gestational length and fetal maturity. It is important, therefore, to set forth definitions of the terms used in this article. The length of the ideal human gestation is 252–280 days from conception. Because the exact date of conception is frequently unknown, gestational age is calculated from the first day of the last menstrual period (which usually precedes conception by 14 days). Using this convention, a term infant is born between days 266 and 294 (38–42 weeks) after the last menstrual period (Fig. 1). A preterm birth is any delivery prior to 38 weeks but after 20 weeks of gestation. Delivery or evacuation of the uterus prior to 20 weeks is generally termed an abortion and may be spontaneous (miscarriage) or induced. The delivery of a dead fetus after 20 weeks is a stillbirth. The neonatal period encompasses the first 28 days of infant life, and neonatal mortality often reflects lethal congenital anomalies and diseases of prematurity. The word premature (meaning ill-timed or not meeting some physiological criteria of maturity) is inexact and should not be used to describe the labor process. It is important to differentiate between the concepts of labor and delivery:

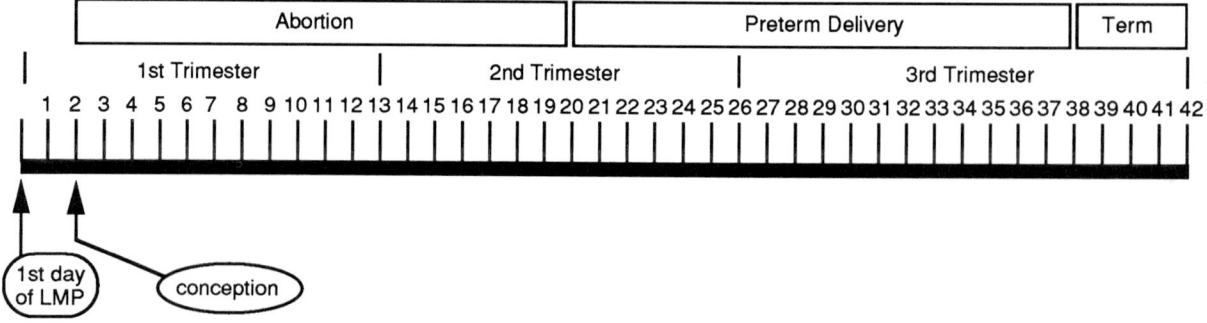

FIGURE 1 Time line of gestational events. Gestational age is calculated as weeks from the first day of the last menstrual period (LMP).

Labor is the painful contractile effort of the uterus to expel the fetus, and delivery is the actual birth. Clearly, preterm labor will not always eventuate in delivery. However, both labor and delivery may be viewed as merely the final overt manifestations of the parturitional process, a complex process terminating pregnancy in which fetal, maternal, and placental signals prepare for and initiate the onset of labor and subsequent birth.

Because of the difficulty in accurately assigning gestational age, epidemiologic studies often equate low birth weight (<2500 g) with preterm delivery. Although closely correlated, birth weight and gestational age are not synonymous, and a low-birth-weight infant may in fact be the full-term product of a pregnancy complicated by fetal, placental, or maternal disease. Indeed, only 30–40% of infants weighing <2500 g are preterm, the remainder being small for gestational age. Conversely, only 40% of preterm infants are of low birth weight. Finally, one must distinguish between the iatrogenic preterm delivery of an infant for maternal or fetal indications (accounting for 25% of preterm births) and the spontaneous preterm activation of the parturitional process leading to preterm birth.

II. EPIDEMIOLOGY

In the United States, preterm birth continues to be a major health concern, affecting 11% of the 3.9 million births in 1995. This rate represents a 17% increase since 1981, a rise that reflects improved reporting of very preterm deliveries, but also underscores the resistant nature of this public health dilemma (Fig. 2). Complications of prematurity dis-

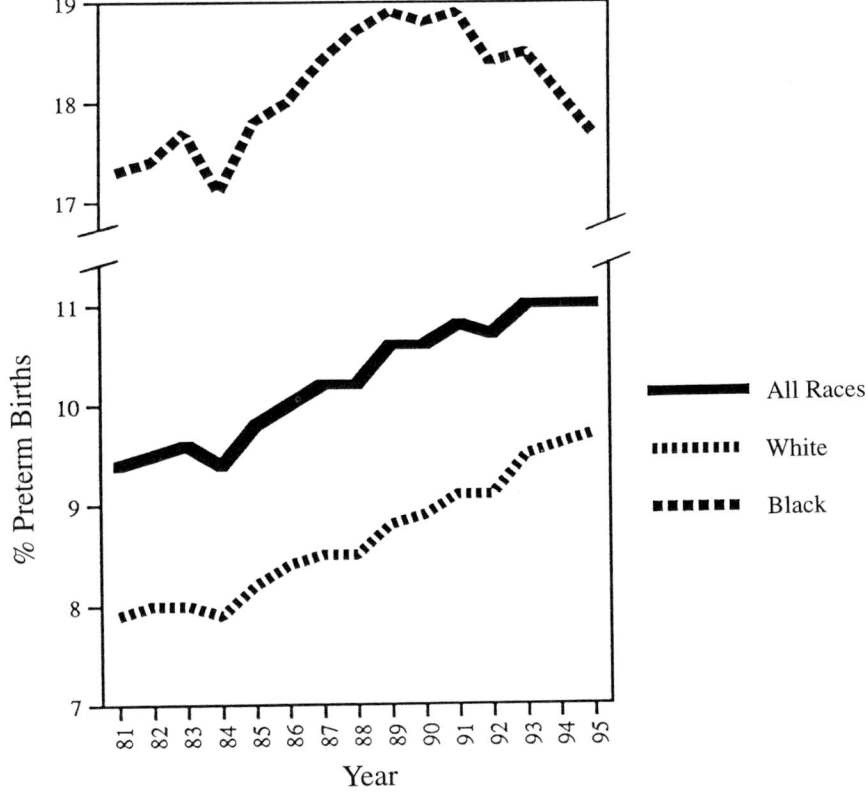

FIGURE 2 Preterm birth rates in the United States. Percentage of live births <37 completed weeks by race of mother: United States, 1981–1995. Notice the 17% increase in overall preterm birth during this 14-year period of time. Black Americans suffer almost twice the rate of preterm birth compared to white Americans (reproduced with permission from the National Center for Health Statistics, *Monthly Vital Statistics Rep.* 45(Suppl. 11), 69, 1997).

proportionately account for 60–80% of all neonatal deaths in normal infants, second only to congenital anomalies as the leading cause of infant mortality. Remarkable improvements in the survival of preterm infants are largely attributable to costly technical advances in neonatal intensive care, and the United States spends upwards of $5 billion dollars annually for the acute care and rehabilitation of these infants. The burden of prematurity is not shared equally among all populations. For reasons largely, but not fully, explained by socioeconomic differences, the rate of preterm delivery among black American women (17.7%) is nearly twice the rate among white American women (9.7%) (Fig. 2). Despite salutary achievements in neonatal care, the United States' infant mortality rate ranks 22nd in the world—largely due to an excess of very preterm deliveries.

III. RISK FACTORS

The epidemiologic study of preterm birth identifies a long list of medical, demographic, and behavioral risk factors for preterm delivery (Table 1). Some risk factors are intrinsic; that is, they will complicate successive pregnancies (i.e., maternal race or chronic disease). Other risk factors are sporadic, impacting only the index pregnancy (i.e., rupture of membranes or preeclampsia). Conceptually, it is helpful to further divide these risk factors into two groups: Those which lead to an indicated (iatrogenic) delivery, and those which predict spontaneous preterm parturition. For example, hypertensive diseases often necessitate early obstetrical intervention and are therefore identified as risk factors for preterm birth but are not likely in and of themselves to initiate spontaneous labor. Conversely, genital tract infection and race

TABLE 1
Risk Factors for Preterm Birth

Constitution risk factors	RR	Sporadic risk factors	RR
Spontaneous PTD		Spontaneous PTD	
Maternal race (black)^c	1.1–3.3	Preterm labor^b	3.1
Low socioeconomic status^c	1.2–3.4	Preterm premature ROM^a	2.0
Age (<18 or >34 years)^a	1.8–2.6	Preterm contractions^c	1.6–5.5
Prior preterm delivery (ea)	1.6–2.5	Early cervical dilation^c	1.5–6.5
One	3.9	Vaginal bleeding^c	1.1–2.3
Two	6.5	Bacterial vaginosis	1.5–1.8
Incompetent cervix^b	2.6–4.0	Working during pregnancy^c	1.2–3.5
Uterine anomaly^a	1.8–3.1	Personal stress^a	3.0
Poor nutrition^a	2.3	Spontaneous or induced	
Alcohol and drug abuse^a	1.1–2.3	Multiple gestation^a	2.0–5.5
Smoking^a	1.5–1.6	Polyhydramnios^a	2.6
Spontaneous or induced		Maternal illness	
Chronic disease		Pyelonephritis^b	1.5
Heart disease		Acute lung disease^c	1.2–5.1
Liver disease		Induced	
Lung disease^c	1.2–5.1	Preeclampsia/eclampsia^a	5.8
		Placenta previa/abruption^a	6.0–8.0

Note. This table identifies risk factors for preterm delivery (PTD). Risks are classified as constitutional or sporadic and accompanied by their estimated relative risk (RR) where available. Factors are further divided into those which may induce spontaneous labor and those which would likely necessitate induced preterm delivery. ROM, rupture of membranes.
^{a} Wen, S. W., et al. (1990). Am. J. Obstet. Gynecol. 162, 213.
^{b} Holbrook, R. H., et al. (1989). Am. J. Perinat. 6(1), 62.
^{c} Mercer, B. M., et al. (1995). Am. J. Obstet. Gynecol, 174, 1885.

have been consistently identified as risk factors for spontaneous preterm birth. A recent multicenter study identified black race, poor social environment, work during pregnancy, prior spontaneous preterm delivery, acute or chronic lung disease, low body mass index, vaginal bleeding, and early cervical ripening as the most significant risk factors for spontaneous preterm birth. Having had a prior preterm delivery is perhaps the single most powerful predictor of future preterm birth, nearly quadrupling the risk of recurrence from 4.4 to 17.2% (Table 1). Identification of these diverse risk factors provides clues as to the mechanisms of spontaneous preterm parturition and helps to direct efforts at treatment and prevention. It is important to remember, however, that more than one-half of spontaneous preterm births occur among woman who have no apparent risk factors, and that no single risk factor consistently predicts preterm birth.

Many attempts have been made to prospectively identify those women and pregnancies at risk of preterm delivery by quantitatively assessing risk factors. This risk-scoring approach should allow a more selective and cost-effective application of surveillance and intervention, with the ultimate goal of reducing the incidence of preterm delivery. The most critical function of a screening test such as this is the positive predictive value—the proportion of patients identified as high risk who actually deliver preterm. The sensitivity (proportion of patients delivering preterm who were identified as high risk) and negative predictive value (proportion of patients identified as low risk who did not delivery preterm) also determine the clinical usefulness of these screening tests. Unfortunately, trials of these systems in both low- and high-risk populations have been disappointing, largely because of the high prevalence of risk factors in the general population and the current lack of effective preventative intervention. Of those women labeled "high risk," only 4–30% actually deliver preterm, along with 5–15% of the "low-risk" patients. It is hoped that improved sensitivity, specificity, and predictive power will be achieved by combining epidemiological risk factors with clinical parameters (i.e., cervical length) into comprehensive risk scoring systems.

IV. INFANT OUTCOME

A. Mortality

Due largely to advances in neonatal intensive care, survival rates for preterm infants have improved dramatically over the past two decades, despite an overall increase in preterm delivery rates (Fig. 3). Survival rates for preterm infants are directly related to the gestational age at delivery and range from approximately 50% at 24 weeks to >95% at 36 weeks. There is no absolute threshold of viability, but neonatal survival at less than 23 weeks is unlikely. Interestingly, neonatal survival is related by unknown mechanisms to gender (girls fare better than boys) and race (black infants do better than white). The advances in infant survival over the past decades have occurred

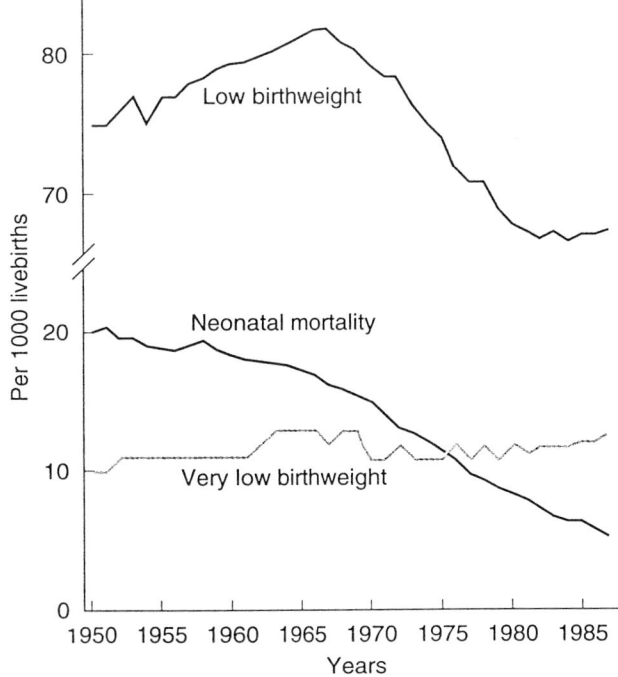

FIGURE 3 Rates of neonatal mortality and low birth weight. This graph demonstrates a steady decline in neonatal mortality rates (death in the first 28 days of life), despite stabilization of the incidence of low birth weight (<2500 g) and very low birth weight (<1500 g) infants [reproduced with permission of Churchill Livingstone, from D. Main and E. Main, In *Gynecology and Obstetrics: A Longitudinal Approach* (T. R. Moore, R. C. Reiter, R. W. Rebor, and V. V. Baker, Eds.), Churchill Livingstone, New York, 1991].

among all subgroups and without a significant accumulation of neonatal morbidities.

B. Morbidity

The major proximate causes of morbidity and mortality among preterm infants are related to the immaturity of their organ systems and their inability to adapt to the extrauterine environment. Hence, the incidence and severity of various diseases of the preterm infant are inversely related to gestational age at delivery. Among the major contributors to morbidity are (i) surfactant deficient respiratory distress syndrome, (ii) intraventricular hemorrhage, (iii) sepsis, (iv) patent ductus arteriosis, and (v) necrotizing enterocolitis (Fig. 4). Long-term morbidities include bronchopulmonary dysplasia, retinopathy of prematurity, cerebral palsy, developmental delay, and learning and behavioral difficulties. The treatment and prevention of these conditions begins even before delivery with the maternal administration of medications and antepartum transport to tertiary care centers prior to anticipated preterm delivery. Thus, the ability to anticipate or delay preterm delivery by as little as 1 or 2 days can yield significant benefit for the preterm infant.

V. PHYSIOLOGY AND PATHOPHYSIOLOGY

A. Overview

It is very likely that the ultimate understanding of preterm parturition will have to await the elucidation of normal term labor, a field which is still fragmentary and evolving. Until then, much insight into the mechanisms of preterm parturition can be gained from the study of suitable animal models and *in vitro* experimental systems—with the recognized caveat that human pathways may not replicate animal physiology. Our current understanding of parturition portrays an intricate, coordinated system of communication between the mother and fetus featuring steroid and peptide hormones, prostaglandins, cytokines, and neurotransmitters working in endocrine, paracrine, and autocrine fashion. The following review of normal term parturition will facilitate the discussion of preterm labor pathophysiology and treatment.

B. Anatomy and Physiology of the Reproductive Tract

From a simple anatomical perspective, the tissues participating in labor include the posterior pituitary, which secretes oxytocic hormones (oxytocin and vasopressin), the muscular portion of the uterus (the myometrium), the hormonally active glandular tissue lining the uterine cavity (the decidua), the placenta, the cervix, and the fetus. The myometrium is composed of intricate bundles of smooth muscle cells embedded in a collagen and ground substance matrix. Similar to all smooth muscle, the myocytes contain actin and myosin (thin and thick) filaments which contract in response to propagated action potentials as well as to local paracrine (prostaglandin) and endocrine (oxytocin) stimulation (Fig. 5). Following stimulation, calcium flows into the cytoplasm from the extracellular fluid or is released from vesicles within the myocyte. Calcium is bound by calmodulin, which then activates myosin light chain

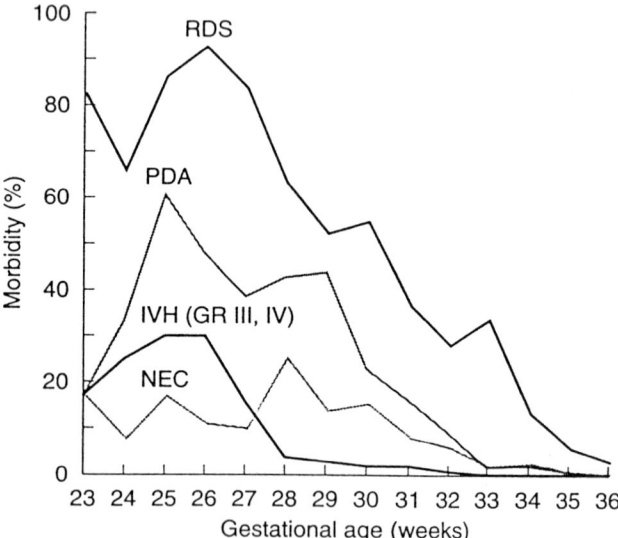

FIGURE 4 This graph shows the incidence of the major neonatal morbidities facing preterm infants (1983–1986). RDS, respiratory distress syndrome; IVH, intraventricular hemorrhage, grades III and IV; NEC, necrotizing enterocolitis; PDA, patent ductus arteriosus (adapted with permission from P. Robertson, *et al.*, *Am. J. Obstet. Gynecol.* **166**, 1626, 1989).

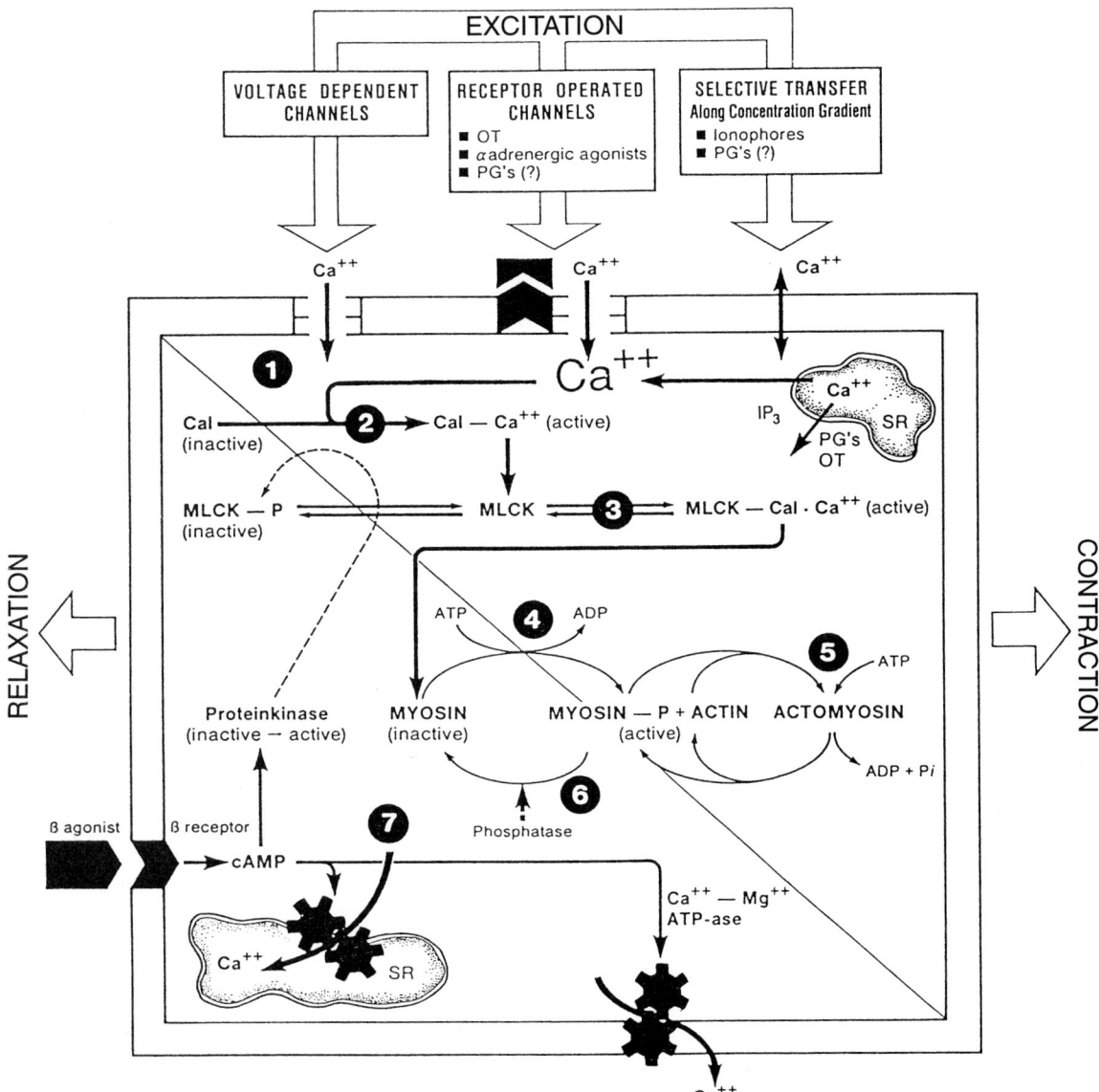

FIGURE 5 Uterine smooth muscle excitation. This figure illustrates the mechanisms of myocyte activation and relaxation. Contraction begins with excitation of the myocyte cell membrane by oxytocin (OT), prostaglandins (PGs), or other agonists which increase intracellular calcium (1). Calcium binds with calmodulin (Cal) (2) to activate myosin light-chain kinase (MLCK) (3). Myosin is then phosphorylated by the MLCK–Cal–Ca^{2+} complex (4) to permit actin/myosin interaction (5) and muscle contraction. When intracellular Ca2+ levels fall, MLCK is inactivated and myosin is dephosphorylated by phosphatase (6) which relaxes the muscle. Relaxation is maintained or induced through cAMP-mediated extrusion of Ca2+ to the cell exterior or into sarcoplasmic reticulum (SR) (7). β-Agonists (i.e., ritodrine) relax myocytes by upregulating cAMP, which activates protein kinase and thus inactivates MLCK (reproduced with permission from E. Braunwald, N. Engl. J. Med. 207, 1618, 1982).

kinase (MLCK) to phosphorylate the globular enzymatic head of the myosin protein. This phosphorylation permits the interaction of myosin with actin filaments and simultaneously activates myosin ATPase. Dephosphorylation of ATP provides the energy to induce conformational changes in the flexible region of myosin, thus pulling against the actin filaments and leading to cell contraction. A key regulatory step is the inactivation of MLCK by protein kinase, an enzyme activated by cAMP and inhibited by phosphodiesterase. Many pharmacological attempts to suppress preterm labor focus on the inhibition of protein kinase by increasing intracellular cAMP concentrations. Isolated, uncoordinated myocyte contractions occur throughout pregnancy—this is the nature of a muscular organ such as the uterus. In fact, as early as 20 weeks' gestation a circadian pattern of uterine activity can be demonstrated, with peak contractility occurring in the early morning hours. Not until the early stages of parturition are the muscular forces potentiated by increasing sensitivity to oxytocic substances and coordinated by increasing numbers of gap junctions between cells. Oxytocin, a nonapeptide produced in the hypothalamus and released from the posterior pituitary gland, is a potent uterine stimulant that exerts its action by binding to specific cell surface receptors and mobilizing myocyte calcium influx.

Positioned as the interface between mother and fetus, the decidua, placenta, and fetal membranes are active metabolic centers engaged in the reception and production of signaling molecules involved in parturition. The decidua and placenta contain cells from many lineages (including immune cells) which participate in the parturitional process. In essence, these tissues integrate and amplify signals from the fetus, the local environment, and the maternal host, and their activation appears to be necessary for labor initiation.

The cervix serves two opposing functions in pregnancy: (i) isolating and supporting the developing fetus within the uterine cavity and (ii) allowing delivery via ripening and permissive dilatation during labor. The cervix is composed mainly of collagen, ground substance, and a few muscle fibers (6–25%). Bacteriostatic agents in cervical mucous (i.e., cytokines) have been shown to act as a barrier to microbial invasion. The function of the cervix during labor is usually integrated with that of the uterus in that both tissues respond to arachidonic acid metabolites (prostaglandins) and oxytocin.

C. Initiation of Term Labor

The initiation of normal parturition in humans is a very complex process which may be thought of as the convergence of multiple stimuli breaching a threshold of uterine quiescence and thus leading to active labor. Indeed, it is controversial whether term parturition results mainly from the withdrawal of a state of labor inhibition or from the *de novo* release of substances which stimulate the labor process. Detailed animal studies (in sheep) demonstrated that parturition begins primarily as a fetal signal emanating from the hypothalamus. The subsequent cascade of hormonal activation leads to fetal adrenal cortisol production which decreases placental progesterone secretion and increases circulating maternal estrogen concentrations (Table 2). This "progesterone with-

TABLE 2
Parturitional Functions of Steroid Hormones

Progesterone functions	*Estrogen functions*
Uncoupling of excitation–contraction	Increases prostaglandin synthesis
Decreases oxytocin receptors	Increases oxytocin release
Inhibits gap junction formation	Increases smooth muscle response to oxytocin
Enhances muscle relaxation	
Suppresses synthesis of prostaglandins	Promotes gap junction formation

drawal" directly initiates the myometrial contractility and cervical ripening of labor.

In humans, parturition simultaneously involves (i) the withdrawal of labor-inhibiting factors, (ii) the elaboration of labor-promoting signals, and (iii) the augmentation of cell-to-cell and tissue-to-tissue communication. Contrary to the sheep model, labor initiation in humans is not clearly dominated by the fetus, and progesterone withdrawal is not a uniform characteristic of human parturition. Instead, studies demonstrate a more significant role for oxytocin, prostaglandins, and estrogen. Diurnal variation in maternal estrogen levels precedes labor and correlates with periods of peak uterine activity in the early morning hours. Oxytocin function prior to labor is enhanced by increasing pulsatile release from the pituitary gland and dramatic increases in myocyte membrane receptor concentration (80- to 100-fold more than nonpregnant levels). These transcriptional increases in receptor concentration are probably mediated by estrogen levels and local signals from uterine distention. Consequently, the uterine sensitivity to oxytocin increases significantly in the last 1 or 2 weeks preceding labor. Oxytocin is the most potent uterine stimulant known, and its activity during labor may be augmented by local decidual and fetal production.

The prostaglandins (PGs) are metabolites of arachidonic acid, a key constituent of phospholipid membranes. $PGF2\alpha$ and $PGE2$ are the two predominant metabolites involved in human parturition, and their activity within the genital tract varies from tissue to tissue. Unlike oxytocin, a systemic increase in prostaglandin concentrations (measured in urine or amniotic fluid) appears to occur with the onset of labor, although local production may occur earlier. It appears that *de novo* synthesis is mediated by a newly discovered inducible form of prostaglandin H synthase. $PGF2\alpha$ and $PGE2$ are produced in different ratios in fetal membranes, umbilical cord, placenta, and decidua and act in a paracrine fashion to stimulate both myometrial contractility and cervical ripening. Prostaglandin production is stimulated by many agents found in the laboring uterus, including oxytocin, cytokines, and growth factors. Interleukin-1 (IL-1) and tumor necrosis factor-α (TNF-α) are two especially potent proinflammatory cytokines expressed in decidua, membranes, and placenta which activate production of prostaglandins. These mediators are themselves subject to modulation by suppressor cytokines. Furthermore, the placenta and chorionic membranes can regulate production of prostaglandins by endogenous agents capable of metabolizing arachidonic acid metabolites and inhibiting their synthesis. Downregulation of these suppressor functions may enhance the positive balance of prostaglandin production at term.

In summary, the early phases of normal term parturition involve the following key preparatory steps:

1. Cervical ripening via the endogenous release of PGE2, relaxin, and circulating estrogen;
2. Upregulation of oxytocin and vasopressin receptors and the subsequent increased sensitivity of the myometrium to oxytocic substances;
3. Increased pulsatility of oxytocin release, leading to Braxton–Hicks contractions and greater mechanical stress at the cervix;
4. Formation of gap junctions and membranous voltage channels which further increase the strength and coordination of contractions; and
5. Enhanced fetal production of oxytocin, cytokines, and growth factors.

D. Preterm Labor

Spontaneous preterm birth results from the untimely activation of the parturitional process. Similar to labor at term, some threshold of labor initiation is surpassed by a combination of endogenous and exogenous factors which conspire to activate parturition and remove labor inhibition (Fig. 6). The earlier in gestation, the less prepared is the uterus for labor, and the greater the exogenous stimulus must be to overcome inherent resistance to labor. Because many risk factors for prematurity (especially infection) are mediated by the same biological signals as those involved in term labor, it is suspected that preterm labor may have an analogous physiology. This model of spontaneous preterm birth has gained clinical and laboratory support from recent studies demonstrating increased production of prostaglandins (and other bioactive lipid products), cytokines, and steroid hormones in the uteroplacental unit preceding

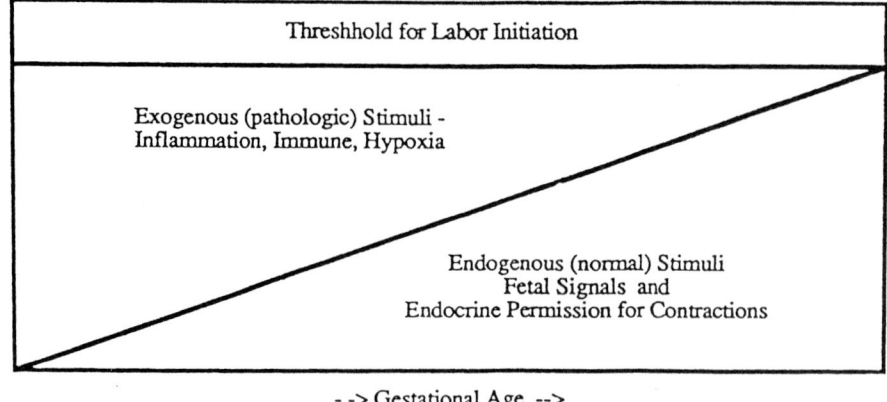

FIGURE 6 Hypothetical model of labor initiation. This schematic represents the composition of exogenous and endogenous signals exceeding a threshold for initiation of labor (reproduced with permission from Iams, 1996).

and during preterm labor (Fig. 7). While most risk factors for spontaneous prematurity remain unexplained (i.e., race and maternal weight), recent studies have helped to formulate rational mechanisms for the pathophysiology of preterm parturition associated with infection, preterm premature rupture of membranes, cervical factors, multiple gestation, and other correlates of preterm birth.

1. Infection

The close epidemiologic association of infection with preterm birth is understandable in light of the

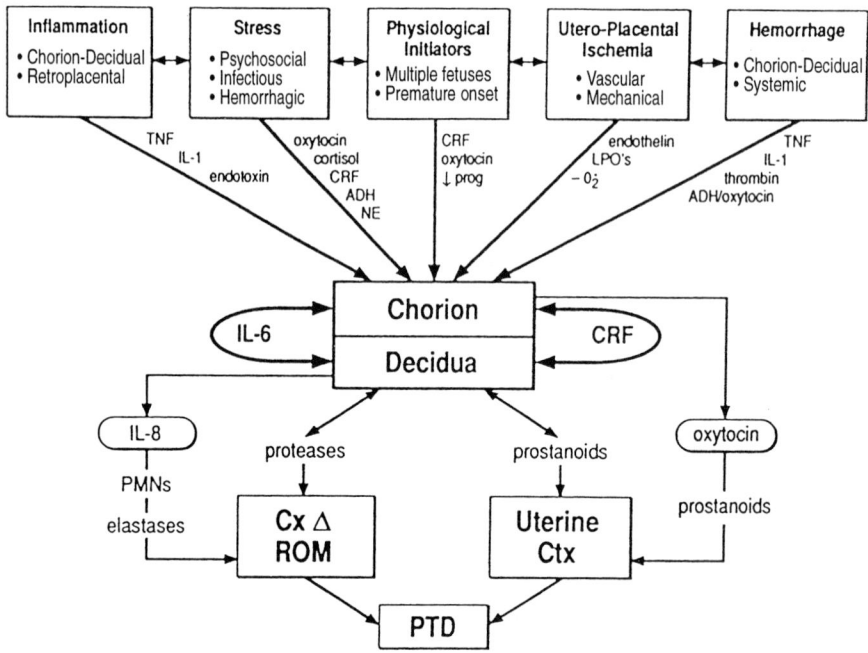

FIGURE 7 Schematic depiction of the biochemical pathways of preterm labor initiation. TNF, tumor necrosis factor; IL-1, interleukin-1; CRF, corticotropin-releasing factor; ADH, antidiuretic hormone; NE, norepinephrine; prog, progesterone; LPOs, lipid peroxides; $^{-}O_2$, oxygen radical; IL-6, interleukin-6; PMNs, polymorphonuclear cells; Cx Δ, cervical change; ROM, rupture of membranes; uterine Ctx, uterine contractions; PTD, preterm delivery (reproduced with permission from C. J. Lockwood, *Clin. Obstet. Gynecol.* 38(4), 675, 1995).

role that cytokines and prostaglandins play in both of these processes. Microbiological or histological evidence of infection is identified in 30% of cases of spontaneous preterm delivery. Initial clinical evidence linked maternal systemic infections (pneumonia, pyelonephritis, etc.) to preterm deliveries, with further evidence derived from the observation that effective antibiotic treatment tended to lessen the association. Early animal experiments demonstrated that exposure to bacteria or bacterial products induced abortion or preterm delivery. Recent microbiological data show that bacteria can be isolated from the amniotic fluid of 20–30% of woman in preterm labor and that histologic examination of the chorioamnion and placenta may identify acute inflammation (chorioamnionitis) in up to 45% of cases. Additionally, there are statistically significant increases in the rates of postpartum endometritis and neonatal sepsis following preterm delivery. Importantly, these associations are strengthened as gestational age at delivery decreases.

What is the source of these infections? Of the four potential routes of intrauterine inoculation (transcervical, hematogenous, transtubal, or iatrogenic), recent evidence suggests that pathologic species of bacteria migrate through the cervix to the extraamniotic space, where inflammation of the membranes and decidua ensues. Bacterial pathogens incite activation of local decidual immune cells and the release of prostaglandins and cytokines. Moreover, transmembranous invasion of the amniotic cavity by bacteria leads to chorioamnionitis and subsequent inoculation of the fetus (Fig. 8). Invasive studies have documented a fetal systemic inflammatory response

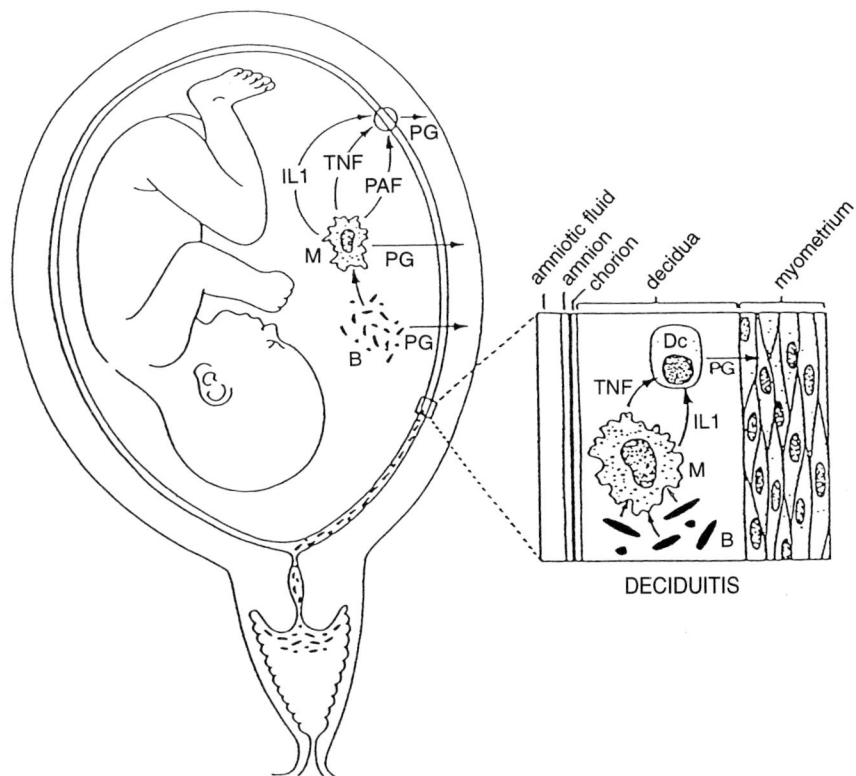

FIGURE 8 Initiation of preterm parturition in the presence of intrauterine infection. Bacteria (B) gain access to the decidua (Dc) or the amniotic cavity, where they activate macrophages (M) to release cytokines such as interleukin-1 (IL-1), tumor necrosis factor (TNF), and platelet-activating factor (PAF). Further activation of decidua or membranes results in release of prostaglandins (PG) which act on myometrium to initiate labor (reproduced with permission from R. Romero and M. Mazor, *Clin. Obstet. Gynecol.* **31**, 553, 1988).

characterized by elevations of fetal plasma IL-6 and TNF-α. There is now clear evidence linking bacterial vaginosis, group B streptococcus, *Chlamydia*, and *Mycoplasma* with preterm parturition. Inflammatory mediators are found in increased concentrations in genital tract secretions during preterm labor associated with infection and these products are known to elicit the elaboration of uterotonic arachidonic acid metabolites which can initiate labor.

2. Preterm Premature Rupture of Membranes

Preterm premature rupture of the membranes (PPROM) is defined as rupture of the chorioamnionic membranes before 37 weeks and before the onset of labor. PPROM complicates 1–4% of all pregnancies and precedes 20–50% of preterm deliveries, one of the strongest predictors of preterm birth. The latency period between ROM and labor varies inversely with gestational age and may range from a few hours to many weeks. Consistent with the multifactorial model of spontaneous preterm birth, PPROM or preterm labor may be thought of as two possible clinical consequence of decidual activation and host immune response. Chronic membrane stress or injury may predispose to structural failure when superimposed on an acute inflammatory response in the decidua or overlying the cervical os. For example, smoking, repetitive stretching, bleeding, multiple gestation, polyhydramnios, and malnutrition are chronic factors which can weaken membranes through induction of thinning (strain hardening), impairment of collagen synthesis, or downregulation of protease and collagenase inhibitors. On the other hand, acute exposure to microbial or host factors such as metalloproteinases and elastases may lead directly to impaired membrane integrity and PPROM.

As with preterm labor, infection is strongly associated with PPROM and may play a primary or secondary role in its pathogenesis. Epidemiological, histological, and clinical data all support the involvement of infectious agents in PPROM. Bacteria can be isolated from amniotic fluid in up to 35% of cases of PPROM, and clinical chorioamnionitis will occur in 10–40%, and endometritis in 10–30% of cases. Whether or not labor ensues may depend on the host–microbial response and the degree of inflammation at the time of ROM. Prolonged latency is unlikely in the face of infection. Prompt treatment with broad-spectrum antibiotics has been conclusively shown to prolong the latency period by an average of 7 days, thus lending further support to an infectious etiology of labor induction. Regardless of the cause of rupture, once the membranes have failed, vaginal flora gain potential access to the usually sterile uterine cavity, thus increasing the chances that acute choriodeciduitis or fetal systemic inflammatory response will initiate parturition.

3. Cervical Factors

The role of cervical function in preterm birth has long been recognized and debated. Traditionally, the cervix was viewed in dichotomous terms—either competent or incompetent. Incompetent cervix is best diagnosed in light of a clinical history of passive, painless dilatation and pregnancy loss without antecedent ROM or uterine activity. Classically diagnosed, incompetent cervix contributes <10% to the overall rate of preterm birth. It should be remembered, however, that every birth is necessarily preceded by cervical dilatation and that the newly discovered roles of cervical production of prostaglandins and cytokines means that the cervix usually acts in concert with the uteroplacental unit and not as an independent factor. Moreover, recent studies based on serial vaginal ultrasound examinations of cervical length demonstrate a spectrum of cervical competence which is reflected in the continuous and inversely related risk of preterm delivery. For example, the relative risk of preterm delivery is 6.19 at the 10th percentile of cervical length and 13.99 at the first percentile . Because every shortened cervix does not lead to delivery, length is but a marker predictive of cervical function and the activation of parturitional pathways. It may also be hypothesized that decreased cervical length predisposes to bacterial invasion of the endocervix and decidua which leads to production of mediators of cervical ripening and parturition.

4. Decidual Ischemia

The preceding discussion of the pathophysiology of preterm labor has largely ignored the potential role of the fetus in signaling the onset of parturition.

As mentioned, there is less evidence for a significant role of fetal signaling in normal term human labor than in the sheep. However, a documented association of intrauterine growth restriction among supposedly "spontaneous" preterm infants hints that intrauterine fetal stress, or the conditions causing it, may in fact participate in the onset of preterm parturition. Abnormal uterine artery blood flow, placental vascular lesions, and hemosiderin-stained fetal membranes are found in significantly more preterm gestations than in term pregnancies. Each of these factors may directly or indirectly lead to placental or decidual hypoxia, which in turn may elicit local hypoxia-inducible factors or fetal stress response factors that interact with the mechanisms responsible for uterine quiescence and labor initiation.

5. Multiple Gestation, Polyhydramnios, and Uterine Anomalies

Multiple gestation, polyhydramnios, and uterine anomalies are causes of both membrane and myometrial stress. Myometrial stretch is known to increase the force and frequency of spontaneous contractions. Furthermore, prostaglandin release is augmented by myometrial stretch, which may increase sensitivity to oxytocin, promote cervical ripening, and increase expression of gap junctions. Uterine monitoring in twins demonstrates a significant increase in spontaneous uterine contractions throughout gestation. Clinical studies show that approximately 50% of all multiple gestations are delivered before 37 weeks, accounting for 15–20% of all preterm deliveries and a disproportionate percentage of preterm morbidity. Weakening of the membranes combined with uterine activity leads to ROM in 50% of preterm multiple gestations. It is noteworthy, however, that many of the risk factors for prematurity in singletons do not predict preterm delivery in multiple gestations, suggesting that unique factors are at work.

VI. DIAGNOSIS AND PREDICTION

The diagnosis of preterm labor is relatively straightforward, requiring regular, painful contractions and progressive cervical dilatation or effacement. This clinical definition is problematic for two reasons: (i) It will invariably include a large proportion of patients in false labor as well as exclude some who are indeed in early labor, and (ii) it is based on relatively late events in the parturitional process which makes intervention less likely to succeed. In clinical trials of tocolytic therapy, for example, up to 40% of women receiving placebo alone will deliver at term. Conversely, up to 20% of patients evaluated for symptoms of preterm labor and deemed not to be in labor will subsequently deliver preterm. The consequences of such an ambiguous diagnosis are the overtreatment of women not in preterm labor and the difficulty of establishing strategies for truly effective treatment and prevention of preterm birth. Part of the ambiguity of the diagnosis is related to the high prevalence of the signs and symptoms of preterm labor in the general gravid population (Table 3). If preterm labor could be accurately predicted earlier in the parturitional process, then patients could be more intensively targeted for intervention. Currently, even the most effective pharmacologic inhibitors of labor cannot be expected to delay preterm delivery for more than a few days.

Because the prediction of preterm birth based on demographic risk factors has been unsuccessful, researchers have sought physical and biochemical markers to enrich our predictive ability. Digital cervical exams can detect precocious ripening but are fraught with interexaminer variation and ultimately provide low sensitivity and predictive value. Transvaginal ultrasonography is a more reproducible and precise method of cervical assessment and may find greater utility for its ability to exclude patients thought to be in preterm labor (negative predictive value of 97% and positive predictive value <20%).

TABLE 3
Signs and Symptoms of Preterm Labor

Pelvic pressure
Vaginal discharge
Backache
Diarrhea
Vaginal bleeding
Cramping
Contractions
Changed perception of fetal movement

Electronic home uterine activity monitoring has been proposed as a means of predicting preterm labor/delivery, but extensive testing has raised controversy without definitively demonstrating a reduction in the rate of preterm delivery. Certain biochemical markers of parturition, such as fetal fibronectin and C-reactive protein, have also been investigated. Fetal fibronectin is a cell matrix protein commonly thought of as the "glue" which affixes fetal membranes to the decidua. Studies strongly support an association between increased cervicovaginal fetal fibronectin and preterm labor. Although this assay is able to reliably exclude patients not in preterm labor (negative predictive value of 97%), the positive predictive value ranges from only 13 to 46%. Future efforts to predict preterm parturition may well make use of a combined panel of sociodemographic, clinical, and biochemical factors to provide the predictive power necessary to selectively test and apply interventions.

VII. TREATMENT AND PREVENTION

The ideal treatment for preterm labor does not exist. In principle, the goal of therapy is the delivery of a mature, healthy infant without jeopardizing maternal well-being. The strategy may include, alone or in combination, treatment to (i) delay delivery until fetal maturity is reached and transport to high risk centers is accomplished (tocolysis), (ii) accelerate fetal maturity and reduce gestational age-dependent morbidities (glucocorticoids), and (iii) prevent maternal and fetal complications of preterm labor and delivery (birth trauma and sepsis). Unfortunately, the unique fetal–maternal relationship often pits the well-being of the fetus against that of the mother. In some cases, expedited delivery of the fetus is necessary for maternal indications, whereas in other cases, further delay of a preterm delivery may place the fetus at higher risk of *in utero* death or injury.

A. Tocolysis

Tocolysis is the pharmacologic inhibition of uterine contractions thought to represent preterm labor. The rational use and development of tocolytic medicines has paralleled our improved understanding of smooth muscle physiology and the events initiating labor. It is important to recognize that without properly controlled trials, almost any intervention will appear effective because of the high false-positive rate of a clinical diagnosis of preterm labor. The first agents employed included ethanol and progestational agents, but subsequent data did not support their effectiveness. Despite the widespread use of tocolytics in the United States, however, the rate of preterm delivery has not only failed to fall, but has actually increased (Fig. 2).

The most extensively studied class of tocolytic agents are the ß-agonists. Their mode of action is presumed to be through stimulation of ß2 receptors on the uterine smooth muscle which activates adenylate cyclase and increases intracellular cAMP, thus inactivating myosin light chain kinase (Fig. 5). The pharmacological effect is a disruption of the actin/myosin interaction, thereby inhibiting smooth muscle contractility. Commonly used agents include ritodrine (the only drug approved for tocolysis in the United States) and terbutaline, although many drugs with differing ß1 and ß2 receptor selectivity have been studied. Critical analysis of these agents shows a modest ability to delay delivery by up to 48 hr in the presence of intact membranes, which may be long enough to allow transportation of the patient to a high-risk facility and administration of adjunctive therapy (such as corticosteroids to enhance fetal pulmonary maturation and antibiotics). Side effects include maternal and fetal tachycardia, hyperglycemia, hypokalemia, and a myriad of other metabolic and cardiac consequences of ß-agonist stimulation. The most severe complications include pulmonary edema, myocardial ischemia, and ketoacidosis.

In the United States, magnesium sulfate is the most commonly used tocolytic agent despite a paucity of well-designed clinical trials. Conflicting reports of efficacy have been generated by small, often nonrandomized studies comparing magnesium to ritodrine or placebo. By competing with calcium for specific ion channels, magnesium is suspected to inhibit smooth muscle contraction by preventing depolarization and calcium influx into the myocyte (Fig. 5). Its side effect profile makes it better tolerated by most patients, although pulmonary edema and accidental overdosage remain significant risks. Interestingly, recent studies have indicated that maternal magnesium administration prior to preterm birth reduces the risk of subsequent cerebral palsy. This effect has yet to be substantiated in a randomized, prospective fashion, however.

The integral role of arachidonic acid metabolites in the parturitional process has led to studies of prostaglandin synthase inhibitors as tocolytic agents. The prostaglandin metabolites of cyclooxygenase (PGE2 and PGF2α) increase intracellular calcium in myometrial cells and enhance myometrial gap junction formation prior to labor. Inhibition of cyclooxygenase by nonsteroidal antiinflammatory drugs (NSAIDs) reduces prostaglandin concentrations. Indomethacin, a potent inhibitor of cyclooxygenase, has been shown to effectively delay delivery up to 7 days in several small randomized studies. Maternal risks of NSAIDs are few but may include gastrointestinal bleeding, alterations in hemostasis, and nephrotoxicity. Contrary to their relative safety in the mother, NSAIDs may have profound negative effects on fetal physiology, including constriction of the ductus arteriosis, oligohydramnios/renal failure, and neonatal pulmonary hypertension. Because of these rare but significant concerns, Indomethacin use is generally restricted to short courses <48 hr in duration and to gestational ages <32 weeks.

Calcium channel blocking agents have also been investigated as tocolytic agents. Again, a few studies have suggested some limited ability to prolong pregnancy in the face of suspected preterm labor comparable to the effect of ritodrine. By preventing the voltage-dependent influx of calcium ions into the myocyte (Fig. 5), these agents indirectly prevent actin/myosin interaction and thus weaken contractile forces. Other than their mild cardiovascular effects, they are well tolerated by mother and fetus. Finally, a novel antagonist of oxytocin, Atosiban, has recently been studied in humans and demonstrates short-term efficacy in arresting uterine contractions.

B. Steroids

If tocolytic drugs can only reasonably be expected to delay preterm delivery by a few days, how then can they afford significant benefit to the neonate? In 1972, Liggins and Howie reported that antenatally administered glucocorticoids significantly reduced the incidence of respiratory distress syndrome (by 50%) when given more than 48 hr prior to preterm delivery. Since then, multiple studies have confirmed these findings and demonstrated additional benefits, including reduced incidence and severity of intraventricular hemorrhage (50% decrease) and, most important, decreased mortality (61% decrease). Meta-analyses have concluded that up to 34 weeks' gestational age and in most circumstances, including extreme prematurity and ruptured membranes, glucocorticoids improve the neonatal condition. Steroids are generally thought to enhance cell maturation and differentiation. In the fetal lung, they have been shown to increase production of certain surfactant lipids and proteins, to increase lung compliance, and to decrease protein leakage, whereas in the brain they are thought to promote maturation of the germinal matrix, thereby reducing the risk of intraventricular hemorrhage.

C. Antibiotics

There is ample evidence demonstrating the role of bacterial infection/colonization in the etiology and pathology of preterm birth. It is therefore somewhat surprising that multiple trials of adjunctive treatment of patients in preterm labor with intact membranes have failed to demonstrate significant benefit. Perhaps treatment at such a stage in the parturitional process is "too little, too late." Nonetheless, accumulating evidence does suggest a benefit of antibiotic therapy in patients with preterm premature rupture of the membranes, who demonstrate approximately 7 days prolongation and decreased rates of maternal and fetal infection. Furthermore, antibiotic treatment of local genitourinary tract infections (such as pyelonephritis, asymptomatic bacteriuria, sexually transmitted diseases, and bacterial vaginosis) has been clearly shown to decrease the risk of preterm labor and delivery.

D. Primary Prevention

The primary prevention of preterm labor and delivery is a long-sought goal of obstetrical and pediatric caregivers. Primary prevention involves the identification of patients at risk and the institution of safe, appropriate medical and social interventions. Reductions in prematurity have been noted after the implementation of comprehensive prenatal care programs, most notably in France, where the preterm birth rate has fallen from 8.2 to 5.6%. These programs include, among other strategies, a means to delay childbearing, provide nutrition, provide stress and job relief, diagnose and treat genitourinary infections, reduce smoking, and educate mothers.

See Also the Following Articles

DECIDUA; OXYTOCIN; POSTDATE PREGNANCY; TOCOLYTIC AGENTS; UTERINE ANOMALIES

Bibliography

Challis, J. R. G., and Lye, S. J. (1994). Parturition. In *The Physiology of Reproduction* (E. Knobil and J. D. Neill, Eds.), 2nd ed. Raven Press, New York.

Creasy, R. K. (1994). Preterm labor and delivery. In *Maternal–Fetal Medicine* (R. K. Creasy and R. Resnik, Eds.), 3rd ed. Saunders, Philadelphia.

Cunningham, F. G., MacDonald, P. C., Gant, N. F., Leveno, K. J., Gilstrap, L. C., Hankins, G. D. V., and Clark, S. L. (Eds.) (1997). Preterm birth. In *William's Obstetrics*, 20th ed. Appleton & Lange, Stamford, CT.

Fuchs, A. R., Fuchs, F., and Stubblefield, P. G. (1993). *Preterm Birth: Causes, Prevention, and Management*, 2nd ed. McGraw-Hill, New York.

Iams, J. D. (1995). Preterm labor. In *Clinical Obstetrics and Gynecology* (R. M. Pitkin and J. R. Scott, Eds.), Vol. 38(4). Lippincott, Philadelphia.

Iams, J. D. (1996). Preterm birth. In *Obstetrics: Normal and Problem Pregnancies* (S. G. Gabbe, J. R. Niebyl, and J. L. Simpson, Eds.), 3rd ed. Churchill Livingston, New York.

Mercer, B. M. (1996), Premature rupture of the membranes. *Sem. Perinat.* 20(5), 343–461.

National Center for Health Statistics, Centers for Disease Control and Prevention (1997, June 10). Report of final natality statistics. *Monthly Vital Statistics Rep.* 45(Suppl. 11).

Papiernik, E., Keith, L. G., Bouyer, J., Dreyfus, J., and Lazar, P. (Eds.) (1989).. *Effective Prevention of Preterm Birth: The French Experience Measured at Haguenau*. March of Dimes Birth Defects Foundation, White Plains, NY.

Priapism
see Erection

Priapulida

Christian Lemburg
University of Göttingen

Andreas Schmidt-Rhaesa
University of South Florida in Tampa

I. Introduction
II. Morphology of the Reproductive System
III. Morphology of the Gametes
IV. Reproduction
V. Development

GLOSSARY

body cavity A spacious, fluid-filled cavity that in Priapulida is not bordered by an epithelium and therefore is a primary body cavity or pseudocoel.

extracellular matrix A matrix secreted from cells that serves mainly as an attachment for cells.

lorica A number of cuticularized plates in priapulid larvae covering the abdomen and probably having a protective function.

mesentery A tissue consisting of flattened longitudinal muscle cells that connects the urogenital system to the body wall.

primitive spermatozoon A special type of spermatozoon with a round body and a long cilium which is thought to be phylogenetically plesiomorphic and connected to free spawning.

protonephridia Excretory organs consisting of a cylindrical terminal cell with pores or clefts covered with extracellular matrix through which fluid is filtered into a central lumen and moved to the body surface through a protonephridial canal.

solenocyte tree Clusters of protonephridia (solenocytes) that merge into one common duct per solenocyte tree.

solenocytes A synonym for protonephridia.

urogenital system Ducts of the reproductive and the excretory system are fused and release their products through the same pore.

Priapulida is a small group of 17 extant species known to date. Although only a few species are known, priapulids differ greatly in morphology and reproductive mode, coupled with differing gamete morphology and developmental pathways. All priapulids are burrowers in marine sediments; some are large (macrobenthic) and fairly common (e.g., *Priapulus caudatus* and *Halicryptus spinulosus*), and others are small (1–3 mm, meiobenthic) interstitial animals (*Meiopriapulus fijiensis*, *Tubiluchus* spp., and *Maccabeus tentaculatus*).

I. INTRODUCTION

Priapulids are dioecious, but sexual dimorphism is only present in species of *Tubiluchus*. Sporadic hermaphroditic individuals have been observed in a population of *Priapulus caudatus*, but this seems to be an exception. In *Maccabeus*, only females have been found, leading to the assumption that this species may reproduce by protogynous hermaphroditism or parthenogenesis, but neither mode has been observed.

Summaries on reproduction, gamete structure and development are given by Franzén (1983), van der Land (1975), and Nørrevang and van der Land (1983, 1989).

II. MORPHOLOGY OF THE REPRODUCTIVE SYSTEM

The gonads in priapulids are closely associated with the excretory system and therefore represent a urogenital system (Fig. 1). Generally, the gonads are paired (only females of *Meiopriapulus fijiensis* have one ovary) and positioned laterally or ventrolaterally on either side of the intestine. The gonads of the large, free-spawning species are voluminous and contain large quantities of gametes, whereas in intersti-

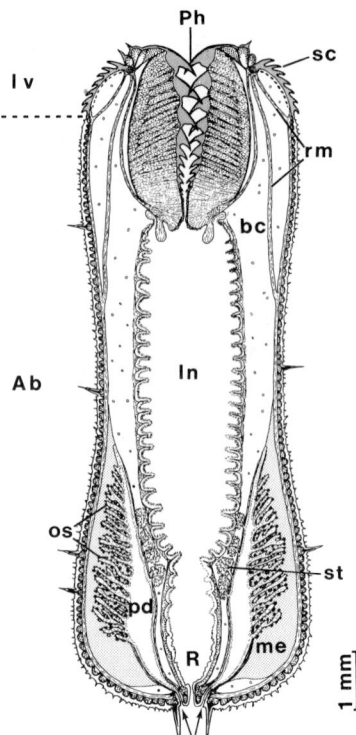

FIGURE 1 Schematic horizontal section of an adult female of *Halicryptus spinulosus* showing ovarial sacs (os) extending into the mesentery (me). Ovarial sacs and solenocyte trees (st) open into the primary urogenital duct (pd). Ab, abdomen; bc, body cavity; In, intestine; Iv, introvert with scalids; Ph, pharynx; R, rectum; rm, retractor muscles; sc, scalids. Arrows point to urogenital pores.

tial priapulids the gonads are smaller and produce only a few gametes. The urogenital system leads to a pair of pores that open on either side of the anus. In *M. fijiensis* the urogenital ducts appear to lead into the posterior hindgut. If not otherwise stated, the following information is based on observations on *Priapulus caudatus* and *Halicryptus spinulosus* summarized by van der Land (1975) and Nørrevang and van der Land (1983).

The gonads develop from the anterior part of the main excretory duct, named the primary duct. The protonephridia (often called solenocytes in priapulids) are clustered in so-called solenocyte trees. There is one duct from the solenocyte trees (*Tubiluchus*) or several ducts that either lead separately into the primary duct (*H. spinulosus*; Fig. 1) or merge and lead together into the primary duct (*P. caudatus* and *Meiopriapulus*). In all species (except *H. spinulosus*), the solenocyte trees fuse with the primary duct in its posterior region and so the anterior part of the primary duct functions as a gonoduct. A bladder-like extension of the primary ducts is present in macrobenthic priapulids, whereas in male *Meiopriapulus* the duct of the solenocyte trees is extended and spermatozoa which are immotile and probably mature have been observed in this "bladder," suggesting it functions as a sperm storage organ.

The primary duct is attached to the lateral body wall by a mesentery or ligament that mainly consists of flattened longitudinal muscle cells which also completely surround the primary urogenital duct. In the earliest stage of organogenesis, the anterior region of the primary duct is solid tissue. This tissue proliferates at the start of gonad growth, developing a central lumen which extends in the macrobenthic species with numerous secondary ducts into the mesentery. In females these secondary ducts widen and become ovarial sacks. At least in *P. caudatus*, gland cells are associated with the secondary ducts where they merge into the primary duct. The whole primary urogenital duct is surrounded by circular and longitudinal musculature. Each gonad is surrounded by a layer of extracellular matrix (ECM) that separates it from the body cavity and from the mesentery cells (Figs. 2a and 2b). The gonad itself consists of a germinal epithelium that gives rise to the gametes. The epithelial cells are tightly interdigitated with each other and have apical microvilli and a long cilium (Fig. 2b) (the cilium is not described for ovarial cells in *Meiopriapulus*. The oocytes develop from epithelial cells and remain in association with the epithelium until maturity. Some epithelial cells become supporting cells. Generally, different stages of gametogenesis can be found within the gonads.

III. MORPHOLOGY OF THE GAMETES

Two types of spermatozoa are present in priapulids: the so-called "primitive" type has been found in *P. caudatus*, *H. spinulosus*, and *M. fijiensis*, whereas in species of *Tubiluchus* a derived and unique type has been observed (for references see Storch, 1991). The primitive spermatozoon consists of an ovoid head, midpiece, and a long flagellum (Figs. 3a and

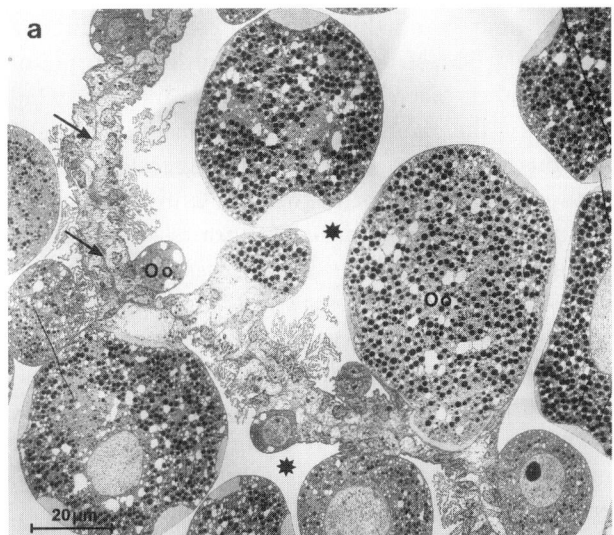

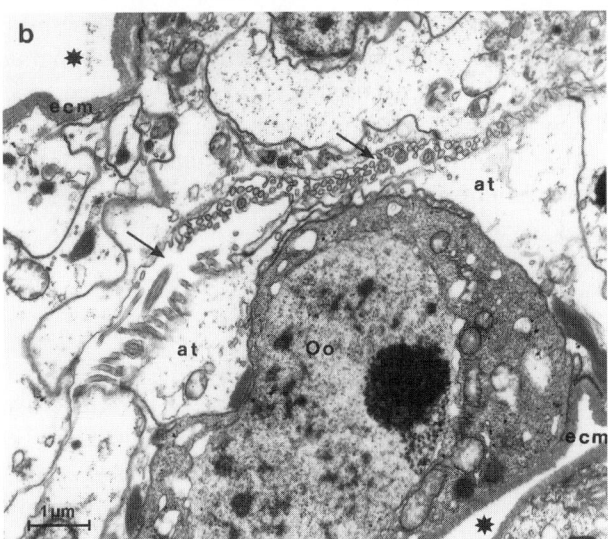

two (*P. caudatus*) centrioles are present at the base of the flagellum. The flagellum has the "typical" ciliary pattern of $9 \times 2 + 2$ microtubules in cross section. The mature spermatozoa of *T. corallicola* and *T. philippinensis* are elongated and divided into three regions (Fig. 3b). In the head the acrosome and nucleus are present, but in contrast to the primitive spermatozoa the acrosomal vesicle is paired and the two halves are spirally coiled around each other. Additionally, the anterior region of the nucleus is bifurcated and the two "arms" are also wound spirally around each other (Fig. 3b). In the midpiece three or four mitochondria encircle a central axoneme. Besides the usual pattern of $9 \times 2 + 2$ tubuli which is present along the whole flagellum, in the midpiece are 27 accessory tubules, arranged in nine triplets. In the tailpiece, only 9 single accessory tubules remain, whereas the distal end of the flagellum is free of accessory tubules. The unique structure of the spermatozoa develops during spermatogenesis from large spermatids that contain a rounded nucleus and an acrosomal vesicle that is most likely derived from a Golgi complex. The spiral pattern in the anterior end is produced by a bilobed growth of the acrosome

FIGURE 2 (a) Overview of ovarial sacs in a female *Priapulus caudatus* with different developmental stages of oocytes (Oo). Large oocytes bulge into the body cavity (*), and arrows show the ovarial lumen. (b) A very young oocyte (Oo) lying in a basal position of the ovarial sac. Attachment cells (at) cover the oocyte apically; basally it is only surrounded by extracellular matrix (ecm). Microvilli and cilia of the apical cells of the ovarial sac reach into the lumen (arrows). The asterisk indicates the spacious body cavity.

4a). The head contains a large nucleus and an anteriorly located acrosomal vesicle, and posterior of the nucleus in the midpiece are a small number (usually four or five) of mitochondria. The acrosomal vesicle is cup shaped and supplemented by a subacrosomal structure (= perforatorium). One (*M. fijiensis*) or

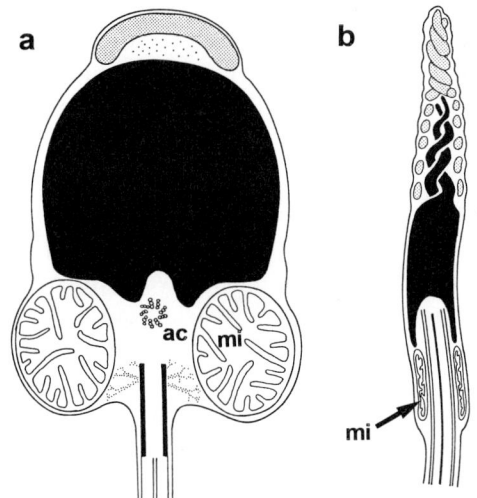

FIGURE 3 (a) "Primitive" spermatozoon of *Priapulus caudatus* with acrosomal vesicle (gray), subacrosomal structure (= perforatorium), large central nucleus (black), accessory centriole (ac), mitochondria (mi), and the base of the flagellum. (b) Spermatozoon of *Tubiluchus* with anteriorly wound acrosomal vesicles (gray), nucleus (black), and mitochondria around the base of the flagellum (a, after Afzelius and Ferraguti, 1978; b, after Storch *et al.*, 1989).

and a paired anterior outgrowth of the anterior region of the nucleus.

In all priapulids spermatogonia develop from epithelial cells of the testicular follicles. During spermatogenesis, mitochondria are concentrated around the base of the cilium and the acrosome is formed from a Golgi complex.

Oogenesis has been investigated in detail in *P. caudatus*. Oogonia are interspersed among the epithelial cells of the ovary. During development, the oogonia lose their apical cilium and microvilli and then shift into a basal position within the epithelium. Neighboring epithelial cells overgrow them apically. One to three cells form a cup-like process that rests on the apical side of the developing oocyte and they function as "attachment cells" (Fig. 2b). The main part of the oocyte bulges out basally into the body cavity (Figs. 2a and 2b). During oocyte growth the ECM becomes thinner but still surrounds each oocyte completely. This gives the appearance that the oocytes lie on the outside of the ovarial sac. In contrast to *P. caudatus*, oocytes have a peripheral microvillar border in *Meiopriapulus* and *Tubiluchus*. Nurse cells are not found in priapulids. The oocytes contain large amounts of mitochondria, yolk platelets, rough endoplasmic reticulum, lipid droplets, and glycogen. The absence of nurse cells and the mode of bulging into the body cavity suggests that nutrients for oogenesis are absorbed from the fluid in the primary body cavity.

Egg size is 60–80 μm in *H. spinulosus* and *P. caudatus*, 80–100 μm in *Maccabeus tentaculatus*, 60–70 μm in *Tubiluchus*, and up to 250 μm in *M. fijiensis*. The macrobenthic priapulids have numerous oocytes, whereas meiobenthic species have only a few oocytes present (about 20 in *Tubiluchus*, 3–8 in *Maccabeus*, and up to 8 in *Meiopriapulus*). In *Maccabeus*, *Tubiluchus*, and *Meiopriapulus* the most mature oocyte is in the anterior region of the gonad.

IV. REPRODUCTION

The macrobenthic priapulids (*Priapulus* and *Halicryptus*) with primitive spermatozoa release large quantities of gametes into the water. Spawning seems to be induced by the presence of individuals of the opposite sex in *P. caudatus* and usually the male spawns first. In northern Europe, the reproductive period may be restricted to winter, although gametogenesis takes place year-round. It is likely that all species of the meiobenthic taxon *Tubiluchus* have internal fertilization, which was suggested by the finding of spermatozoa in the female urogenital system of *T. philippinensis*. In *M. fijiensis*, only a small number of oocytes have been observed in females with just one mature oocyte at a time. This finding, together with the observation of one embryo being half released by the female, suggests that *Meiopriapulus* also has internal fertilization and is viviparous. *Priapulus*, *Halicryptus*, and *Tubiluchus* fit the theory that large-bodied organisms tend to have external fertilization with a primitive type of spermatozoon, whereas small organisms reproduce by some kind of internal fertilization, having modified types of spermatozoa. *Meiopriapulus* does not fit into this scheme because it is a small organism with a primitive type of spermatozoon and probably internal fertilization.

After fertilization, eggs of *P. caudatus* form a fertilization membrane and sink to the sediment.

V. DEVELOPMENT

The development of priapulids is still insufficiently known. Summaries on the early development can be found in Nørrevang and van der Land (1989) and Adrianov and Malakhov (1996), including the only original results of Lang, Zhinkin, and Zhinkin and Korsakova. Cleavage is total and radial. The third cleavage from the 4- to the 8-cell stage is equatorial. A coeloblastula is formed at the 32-cell stage and gastrulation is most likely by invagination. Mesoderm seems to be derived by ingression of cells in the blastopore region.

Later development and hatching from eggs is only imperfectly known, but a peculiar type of larva has been observed in priapulids (Fig. 4b). It is characterized by a division of the body in three regions: introvert, neck, and abdomen. The introvert and neck can be completely retracted into the abdomen, which is covered by a more or less thickened cuticle forming a lorica. In *Tubiluchus* and at least the earliest stages of *P. caudatus*, the lorica is round in cross section and consists of 20 longitudinally arranged flexible

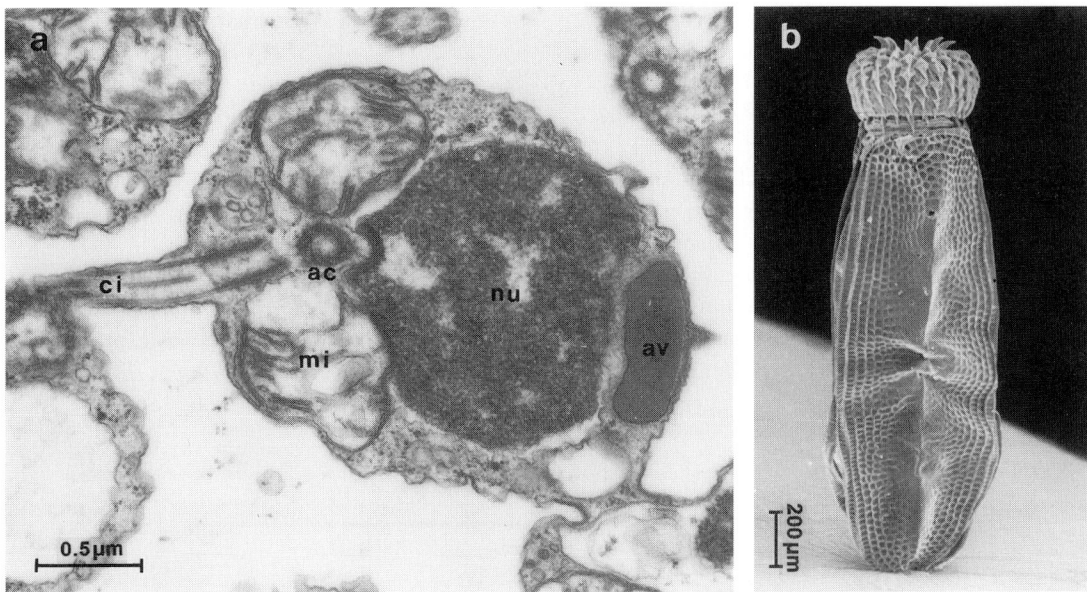

FIGURE 4 (a) Almost mature spermatozoon of *H. spinulosus* with acrosomal vesicle (av), nucleus (nu), mitochondria (mi), accessory centriole (ac), and flagellum (ci). (b) Larva of *H. spinulosus* with lorica and retractable, scalid-bearing introvert (reproduced from Lemburg, 1995, with kind permission of Springer-Verlag, Berlin).

plates. In other species and later stages of *P. caudatus*, the lorica is dorsoventrally flattened with strongly cuticularized dorsal and ventral plates that show a species-specific sculpturing (Fig. 4b). The dorsal and ventral plates are connected by several accordion-like folded lateral plates. The most recent overview of known larval forms is given by Adrianov and Malakhov (1996). The larva grows by a series of molts. Metamorphosis to the postlarval stage takes place within the last larval cuticle. After being released, the tiny postlarva resembles the adult but still grows by molts until it reaches the final size and maturity.

In contrast to all other priapulids, *Meiopriapulus* develops directly without larval stages. The female releases a postembryonic stage from the urogenital opening which already resembles the adult. It differs only in not having a functional introvert, a smaller number of anal hooks, and no gonads, and it is therefore comparable to postlarval stages of other species.

See Also the Following Article

MARINE INVERTEBRATES, MODES OF REPRODUCTION IN

Bibliography

Adrianov, A. V., and Malakhov, V. V. (1996). *Priapilida (Priapulida): Structure, Development, Phylogeny, and Classification.* KMK Scientific, Moscow.

Afzelius, B. A., and Ferraguti, M. (1978). The spermatozoon of *Priapulus caudatus* Lamarck. *J. Submicrosc. Cytol.* **10**, 71–79.

Alberti, G., and Storch, V. (1988). Internal fertilization in a meiobenthic priapulid worm: *Tubiluchus philippinensis* (Tubiluchidae, Priapulida). *Protoplasma* **143**, 193–196.

Alberti, G., and Storch, V. (1989): Zur Feinstruktur des weiblichen Geschlechstraktes von *Tubiluchus philippinensis* (Tubiluchidae, Priapulida). *Zool. Anz.* **222**, 12–26.

Franzén, Å. (1983). Priapulida. In *Reproductive Biology of Invertebrates. Vol. 2: Spermatogenesis and Sperm Function* (K. G. Adiyodi and R. G. Adiyodi, Eds.), pp. 269–274. Wiley, Chichester, UK.

Higgins, R. P., and Storch, V. (1991). Evidence for direct development in *Meiopriapulus fijiensis* (Priapulida). *Trans. Am. Microsc. Soc.* **110**, 37–46.

Lemburg, C. (1995). Ultrastructure of the introvert and associated structures of the larva of *Halicryptus spinulosus* (Priapulida). *Zoomorphology* **115**, 11–29.

Nørrevang, A., and van der Land, J. (1983). Priapulida. In *Reproductive Biology of Invertebrates. Vol. 1: Oogenesis, Ovi-*

position, and Oosorption* (K. G. Adiyodi and R. G. Adiyodi, Eds.), pp. 269–282. Wiley, Chichester, UK.

Nørrevang, A., and van der Land, J. (1989). Priapulida. In *Reproductive Biology of Invertebrates. Vol. 4A: Fertilization, Development, and Parental Care* (K. G. Adiyodi and R. G. Adiyodi, Eds.), pp. 259–262. Wiley, Chichester, UK.

Paulay, G., and Holthuis, B. V. (1994). Biology and morphology of living *Meiopriapulus fijiensis* Morse (Priapulida). In *Reproduction and Development of Marine Invertebrates* (W. H. Wilson, S. A. Stricker, and G. L. Shinn, Eds.), pp. 158–165. Johns Hopkins Univ. Press, Baltimore.

Por, F. D., and Bromley, H. J. (1974). Morphology and anatomy of *Maccabeus tentaculatus* (Priapulida: Seticoronaria). *J. Zool. London* 173, 173–197.

Storch, V. (1991): Priapulida. In *Microscopic Anatomy of Invertebrates. Vol. 4: Aschelminthes* (F. W. Harrison and E. E. Ruppert, Eds.), pp. 333–350. Wiley-Liss, New York.

Storch, V., Higgins, R. P., and Morse, M. P. (1989). Internal anatomy of *Meiopriapulus fijiensis* (Priapulida). *Trans. Am. Microsc. Soc.* 108, 245–261.

van der Land, J. (1975). Priapulida. In *Reproduction of Marine Invertebrates. Vol. 2: Entoprocts and Lesser Protostomes* (A. C. Giese and J. S. Pearse, Eds.), pp. 55–65. Academic Press, New York.

Primates, Nonhuman

Bill Lasley and Susan Shideler
University of California at Davis

I. Introduction
II. Geographic Distribution
III. Catarrhini Reproduction
IV. Endocrine Comparisons to the Human
V. Reproductive Strategies

GLOSSARY

Catarrhini A group (tribe) of the Anthropiodea or "higher primates" that includes Old World monkeys, apes, and humans.

corpus luteum The organelle that produces progesterone prior to and during pregnancy in most mammals.

endometrium The lining of the uterus in which the embryo first implants.

estrous cycle An ovarian cycle type in which the fertile period is associated with an increased desire to mate or the behavioral manifestations of "heat."

luteal Related to the function of the corpus luteum.

menstrual cycle An ovarian cycle type in which the endometrium is sloughed (menstruation) at the end of each nonconceptive lutea phase.

placenta The organ derived from the embryo that attaches the embryo to the maternal blood supply.

Platyrrhini A group (tribe) of the Anthropoidea or higher primates that includes New World monkeys.

polyandry A mating system in which females have multiple matings involving several males.

polygamy A mating system involving multiple matings between males and females.

polygyny A mating system in which males have multiple matings involving several females; contrasts with monogamy, in which there is a single mate.

The Catarrhini possess reproductive characteristics that most closely resemble those of the human. The hallmark of these reproductive characteristics is an ovarian cycle, which is similar in hormonal patterns and temporal sequence to that of the human female menstrual cycle. The primate ovarian cycle type has led to members of this group becoming the animal models of choice in studies of human reproduction. The catarrhines (Old World monkeys and great apes) are distinguished reproductively from the platyrrhines (New World monkeys) found

and social adaptations, such as mating or parental care systems, but this categorization combines all mammalian species possessing reproductive characteristics that closely resemble those of another primate: the human. The hallmark of these reproductive characteristics is an ovarian cycle similar in hormonal patterns and temporal sequence to that of the human female menstrual cycle. The primate ovarian cycle type has led to members of this group becoming the animal models of choice in studies of human reproduction. In terms of all species considered to be "higher" primates, the Old World monkeys and great apes are distinguished from the New World monkeys found in Southern Mexico, Central America, and South America. The female ovarian cycle of the Old World monkeys and great apes is referred to as a menstrual cycle in which there is overt sloughing of the endometrium, or menstruation, as the result of progesterone withdrawal following a nonconceptive ovulation. New World monkeys generally lack any overt sloughing of the endometrium following a nonconceptive ovarian cycle. This difference in Old World monkey and the great ape ovarian cycles is in contrast to not only most New World monkeys but also the ovarian cycles of virtually all other female mammals which have estrous cycles. Distinctions between these two ovarian cycle types will be described.

II. GEOGRAPHIC DISTRIBUTION

The largest number of primate genera are found in this group of primates, exist in the Africa, Europe, and Asia, and include approximately 35 genera and 125 species, including man. The largest number of species with a single genus of Catarrhini, and some of the best described, are the macaques which include the laboratory macaques (spp. rhesus, speciosa, cynomolgus, and radiata). The rhesus macaque was recognized early in this century as having reproductive qualities similar to those of humans. This recognition led to the adoption of the rhesus, and closely related species, as the animal model of choice for investigations which had specific relevance to human reproduction. Recently, the smaller but closely related cynomolgus macaque (long-tailed macaque)

in Southern Mexico, Central America, and South America by an overt sloughing of the endometrium, or menstruation, as the result of progesterone withdrawal following a nonconceptive ovulation. This difference in Old World monkey and the great ape ovarian cycles is in contrast to not only most New World monkeys but also to the ovarian cycles of virtually all other female mammals which have estrous cycles. The largest number of species with a single genus of Catarrhini, and some of the best described, are the macaques, which include the laboratory macaques (spp. rhesus, speciosa, cynomolgus, and radiata). The rhesus macaque was recognized early in this century as having reproductive qualities similar to those of humans. This led to the adoption of the rhesus monkey and closely related species as the animal models of choice for investigations having specific relevance to human reproduction. This group of primates has adapted to equatorial savannas, dense jungles, islands, coastlines, and temperate and mountains regions and one species of macaque still exists on the southernmost tip of the Iberian Peninsula on Gibralter. Gorillas and chimpanzees represent the African great apes, whereas gibbons and siamangs are limited to Asia and orangutans to Indonesia. In general catarrhine species are polygynous, monotocous, and provide large amounts of parental investment to their offspring. Geologic and paleontologic studies indicate that the principal evolutionary development of the primates took place during the Cenozoic, or during the past 65 million years. Clear ancestors of modern Old World monkeys appeared in Africa as early as 30–35 million years ago.

I. INTRODUCTION

From a reproductive point of view, the Catarrhini (which includes the Old World monkeys, i.e., macaques, langurs, mangabeys, colobus, baboons, drills, mandrills, vervets, guenons, etc., and the great apes, i.e., gorillas, chimpanzees, orangutans, and gibbons-siamangs) can be considered as a group which has separate features from those which characterize the taxonomic group such as dentition, nostrils, and ischial callosities. Catarrhines are a diverse group of primate species in terms of geographic distribution

has been used synonymously with the rhesus, particularly in studies relating to reproduction.

With arguably 12 (possibly more) extant macaque species, this group has adapted to equatorial savannas, dense jungles, islands, coastlines, and temperate and mountains regions. One species of macaque still exists on the southernmost tip of the Iberian Peninsula on Gibraltar. Gorillas and chimpanzees represent the African great apes, whereas gibbons and siamangs are limited to Asia and orangutans to Indonesia. Some catarrhines have retained the ancestral arboreal lifestyle, whereas others spend all their lives on the ground. Old World monkey species, such as colobus and proboscis monkeys and langurs, have adapted to a leaf-eating diet that allows them to process food that occurs in abundance but that is low in energy value and hard to digest, whereas other species have adapted to more terrestrial and omnivorous habits. A general tendency toward foraging for food, as a group of species, has influenced types of social groupings and reproductive strategies. Geologic and paleontologic studies indicate that the principal evolutionary development of the primates took place during the Cenozoic, or during the past 65 million years. Earliest primates ranged over the north American and western European continents. Clear ancestors of modern Old World monkeys appeared in Africa as early as 30–35 million years ago. Extant catarrhine species exist in Africa, India, southern Asia, Malaysia, Indonesia, and the Philippines, and one species exists in Japan.

III. CATARRHINI REPRODUCTION

The most significant unique physical reproductive trait of the Old World monkeys and great apes is the organization of the ovarian cycle to include a prolonged follicular phase in which estrogen rises more slowly than in the estrous cycle and which is comparable in duration to the following progesterone-dominated luteal phase. Ovulation, which usually results in a single ovulation, occurs just after the midcycle estrogen peak. Proliferation of the endometrium resulting from the preovulatory estrogen rise requires the sloughing of the endometrium (menstruation) if pregnancy does not occur following ovulation. Mating can, and often does, occur throughout the ovarian cycle; thus, the overt hallmark of the menstrual cycle is menstruation, not a behavioral change. While the adaptive benefit of the menstrual cycle over the "estrous cycle" is not known, it has been hypothesized that the menstrual cycle evolved at the same time that some primates moved from an arboreal lifestyle to the savanna or to ground living. The menstrual cycle is also associated with a trend toward decreasing hormonal and increasing neurocortical control over behavior as well as prolonged infant dependency, maintenance of the mother–infant unit, continuous female receptivity, and the consolidation of the male–female pair bond.

Female monkeys and great apes have paired pectoral mammary glands, simplex cervix, and intrapelvic ovaries. Some Old World monkey females have distinct sex skin swelling and/or coloration which affects the vulva and perineal region and presumably serves as a visual display. Under the influence of estrogen, this hairless skin becomes engorged with fluid and swells to large proportions, especially in certain macaques, baboons, mangabeys, and chimpanzees. Other regions of hairless skin may be similarly affected, such as the pink skin on the chest of the female gelada that also becomes engorged and brightly colored at midcycle and during pregnancy. Sex skin seems to be most common in female catarrhines living in polygynous, multimale groups and is not seen in monogamous species. The distribution of female sex skin varies from species to species, individual to individual, and even in the same individual with increasing age or parity.

In a few species, the sexually mature males have brightly colored faces and/or genitalia. Males frequently have pendulous genitalia which may be incorporated into visual displays. Most species have an os penis, with the exception of a few New World monkey species and *Homo*, and all have scrotal testes. In general, most of the Catarrhini exhibit sexual dimorphism in terms of body size. Males are 20–60% larger that females in most species. In a few cases dimorphic coloration also distinguishes males from females (e.g., siamangs and drills) and, in even fewer species, actual body contours are overtly dimorphic (gorillas, orangutans, baboons, etc.). There is a large variation in the size and appearance of the external reproductive organs and degree of sexual display, which ranges from the extreme to the subtle; therefore, generalizations are difficult to make.

Seasonality in primates is highly variable and is related to reproductive cycles themselves, day length, and/or food and water availability. Some species within a genera may be seasonal, whereas others may not be seasonal. Regardless of the seasonality found in the wild, animals adapted to artificial conditions can lose the annual pattern. Seasonal flexibility is a compromise between several factors; for examples, when the lactational load is great on the mother, requiring adequate nutrition to sustain that load and when weaning must be timed to the appropriate forage for infant primates.

In general, the hormone profile of the menstrual cycle is similar to that of other female mammals, i.e., a rise of estrogen prior to ovulation in association with developing follicles and the production of progesterone following ovulation in association with formation of the corpus luteum. The differences between the menstrual and estrus cycles are subtle but critically important: (i) The estrogen rise is prolonged in menstruating females catarrhines compared to females of other mammals of the same size, and (ii) the period of progesterone production in the menstrual cycle is predetermined. These two differences make the menstrual cycle substantially different from the estrous cycle and preclude one from being mistaken for the other. The prolonged rise of unopposed estrogen in the menstrual cycle leads to an equally prolonged period of sexual attractiveness and permits a prolonged and complex mate selection process. Furthermore, this more gradual rise of estrogen just prior to ovulation does not lead to the acute onset of sexual receptivity or "estrus," for which the ovarian cycles of all other mammals are named. Instead, the menstrual cycle is named for the cyclic sloughing of the endometrium, which is its most overt physical characteristic, and the synchrony between mating and ovulation is controlled by unknown factors.

The gradual, prolonged, and unopposed rise of estrogen in the follicular phase of the menstrual cycle may also cause a greater amount of endometrial tissue to develop than in estrous cycle females and thereby require the subsequent sloughing of the endometrium. It is clear that all female mammals have a preovulatory rise of unopposed estrogen, albeit not as that of pronounced as Old World monkeys and great apes. It is possible that the difference between the estrous cycle and the menstrual cycle is one of degree rather than a categorical one. Since the adaptive benefit of menstruation has never been adequately explained, it may be convenient to understand it as a cost resulting from the benefits of a prolonged follicular phase and enhanced opportunity for mate selection by the female since social rank is important in most social groupings associated with Old World monkeys and great apes. There is no doubt that the loss of tissue and blood is an energetic expense which should have some adaptive benefit. Theories that menstruation represents a cleansing of pathogens from the female tract are attractive but lack scientific foundation.

The predetermined life span of the corpus luteum necessitates the conceptus to "rescue" the maternal ovary when a primate pregnancy does occur. This need, in turn, requires the production of chorionic gonadotropin, which is another trait that is not found in most other animal species. The requirement of the conceptus to control female reproduction through the production of chorionic gonadotropin occurs only once outside of the higher primates—in the horse family (horses, zebras, and asses). All other species that have been studied have a completely different mechanism for the initiation and support of pregnancy involving control by the maternal reproductive tract. Again, as with menstruation, the adaptive benefit of placing the control of early pregnancy in primates with the conceptus is not clear.

The characteristics of pregnancy and parturition are strikingly similar among the catarrhines. While pregnancy length varies from 6–9 months, the temporal events are essentially the same. The placenta is capable of steroid hormone production and obviates the need for the maternal ovaries after the first 3 or 4 weeks. The major source of steroid precursors is the fetal adrenal, which has adapted a unique feature—the fetal zone. The fetal zone disappears shortly after birth. It produces a weak androgen which is converted to estrogen by the placenta that gives rise to some of the highest levels of estrogen production seen in mammals. Most Old World monkeys exhibit some type of seasonality in their reproduction. Implantation, pregnancy, and placentation are relatively uniform among the Old Word monkeys and great apes. The perceived differences between species are generally considered differences of degree rather than differences in adaptation. Typically, implantation of a single embryo occurs approximately 1 week postfertilization and is associated with the release of chorionic gonadotropin to rescue the corpus luteum of the ovarian cycle which is necessary for the maintenance of early pregnancy. The simplex uterus and chorioallantoic placenta in catarrhines replace the bicornuate uterus and choriovitelline placenta of the haplorines, e.g., lemurs, lorises, and tarsiers, which appear to be older phylogenetically. This anatomical adaptation is associated with a tendency toward monotocous pregnancies, increasing invasiveness of the trophoblast, and a more distinct discoid placenta. Some Old World monkeys, such as the macaques, develop two placental discs, one at the site of initial implantation and a second on the opposing uterine wall where contact is made with the developing trophoblast. Parturition, or "delivery," in Old World monkeys occurs prior to a decline in circulating progesterone and is under partial control of the developing fetus through increasing adrenal production of estrogen precursors. The maternal compartment appears to have limited sympathetic control because births usually occur at night and can be delayed 24 hr when conditions are less than optimal.

IV. ENDOCRINE COMPARISONS TO THE HUMAN

Since hormone patterns are best described in the human female, comparisons are often made between the human and other primate species. Of the more than 100 species of Old World monkey and great ape, only about a dozen have been characterized in terms of circulating hormones. For obvious reason, daily blood samples have not been analyzed from the larger, less tractable or the smaller, more excitable nonhuman primate species. Comparing the profiles from those that have been reported, the larger great apes (gorillas, chimpanzees, and orangutans) appear to be most like the human female in terms of ovarian and placental hormones. All the apes except the orangutan appear to have lower estrogen production during pregnancy and some subtle differences in ovarian hormone production. The Old World monkeys have much less circulating estrogen in the luteal phase of the menstrual cycle as well as much less estrogen in circulation during pregnancy compared to the human. Daily urinary profiles have been collected and analyzed in many nonhuman primates and more comparisons to the human can be made in this regard compared to comparisons of circulating hormones. Taking this approach, the orangutan is virtually indistinguishable from the human, whereas the other apes have clearly lower estrogen production during pregnancy compared to the human female. Steroid metabolism is similar in humans and most of the apes; however, progesterone is metabolized to a much greater extent in all the Old World monkeys studied to date compared to the great apes.

V. REPRODUCTIVE STRATEGIES

In some catarrhine species, males learn how to be aggressive and dominant during their maturation, after which they will compete for social status and female resources in later life. In general, adult males compete for access to females for mating. Male competitiveness is thought to be related to sexual development and current levels of testosterone. Immature, very old, and subdominant males are thought to have lower levels of testosterone and are therefore less able to compete successfully for the female resources. Young males are physiologically capable of breeding much earlier than they are socially allowed to do so. The gibbon male essentially has a subadulthood comparable to human adolescence: it is fully capable of reproduction but requires a mate, a territory, a higher level in the social hierarchy, or all three to be successful. For female primates, age at first reproduction indicates sexual maturity and is expected to occur in the second or third year of life if social barriers do not preclude it. In contrast, reproduction for males may not occur until they are 4–6 years old. This time period is even longer in the larger great ape species, such as chimpanzees, gorillas, and orangutans. Dominant males have access to the greatest number of females and females tend to prefer mating with dominant males. Testosterone levels are only marginally influenced by season and are instead thought to reflect sexual maturation and social stature. Subordinate males can and do copulate using a wide range of strategies, such as in the case of the "subordinate follower" which has been described in the hamadrayas baboon, gelada, and gorilla. Social influence may indirectly affect reproductive success by having a direct effect on gonadal function.

Female Old World monkeys and great apes can mate at any time during the ovarian cycle. Females frequently undergo a period of adolescent sterility during which time they are attractive to males and mate with them but do not conceive. Unlike males, social limitations less commonly restrict female access to breeding males. In some primate species, females suppress other female's sex cycles and, in baboons, subordinate females conceive less frequently than their dominant counterparts. Social synchrony of sexual cycles can also occur within a group but does not appear to extend between female groups. Most females are sexually active as soon as they are physiologically capable and pregnancy can occur at the end of the second year in some of the smaller species. In most cases females are active in mate selection. While there is a strong dogma that sex hormones have a minor role in female sexual behavior of animals that menstruate, females appear to be most attractive to males near the time of ovulation and mating primarily takes place at midcycle. There is strong evidence that in some catarrhine species, sexual activity is either limited to (gorilla) or is increased (macaques) at the time of ovulation compared to other times of the menstrual cycle.

In general, Catarrhini exhibit polygamy in which polygyny is the primary form. Polygyny is present in multimale systems in which males compete directly for access to reproducing females such as vervets, baboons, common chimpanzees, and bonobos (pygmy chimpanzees). Polygyny also occurs in single male social systems, such as in the harem polygyny of hamadrayas baboons and the patas monkey, as well as in the multimale social groups of gelada, drills, macaques, colobus monkeys, and langurs and in the solitary orangutan. In many of the species in which polygyny occurs, strong matrilines exist. Polyandry is rare in primates, and, where it is reported to occur, it is associated with simultaneous or serial mating. Monogamy is also rare and occurs in approximately 15% of all primate species. In mammals, monogamy is associated with limited resources that require defense, harsh environments, early reproductive effort, widely dispersed females, and with females that require (bioenergetically) male help to successfully rear offspring. Some of these general associations exits in monogamous primates species such as the gibbon and siamang, among the Catarrhini, and among the titis, tamarins, and marmosets of the Platyrrhini (New World monkey species). Thus, a large number of mating strategies exist with diversity in adaptations to different habitats, social systems, and reproductive strategies.

Bibliography

Dunbar, R. I. M. (1988). *Primate Social Systems.* Comstock/Cornell Univ. Press, Ithaca, NY.

Luckett, W. P. (1974). Comparative development and evolution in the placenta of primate. In *Contributions to Primatology*, (W. P. Luckett, Ed.), Vol. 3, pp. 142–234. Karger, Basel.

Mossman, H. W. (1937). Comparative morphogenesis of the fetal membranes and accessory uterine structures. *Contrib. Embryol. Carnegie Inst.* 26, 129–246.

Napier, J. R., and Napier, P. H. (1994). *The Natural History of the Primates.* MIT Press, Cambridge.

Rowe, N. (1996). *The Pictorial Guide to the Living Primates.* Pogonias Press, East Hampton, NY.

Short, R. V., and Balaban, E. (1994). *The Difference between the Sexes.* Cambridge Univ. Press, Cambridge, UK.

Primordial Germ Cells

Peter J. Donovan

Kimmel Cancer Institute, Thomas Jefferson University

I. Introduction
II. Determination and Segregation of the Germline
III. Germ Cell Migration
IV. PGC Survival and Proliferation
V. PGC Differentiation
VI. Conclusions and Prospects

GLOSSARY

alkaline phosphatase An enzyme detected histochemically and found to be expressed by the primordial germ cells (PGCs) of some vertebrates, especially mammalian PGCs.

embryonic germ cells Pluripotent stem cells derived *in vitro* from PGCs cultured in a combination of growth factors including Steel factor, basic fibroblast growth factor, and leukemia inhibitory factor. Embryonic germ cells can differentiate into a variety of cell types *in vitro* and can reenter the germline when introduced into a host embryo.

genital ridge The embryonic structure that is the gonad anlagen or the embryonic precursor of the gonad of the adult animal. Before the time of sexual differentiation the genital ridges of male and female embryos are indistinguishable and may be referred to as the indifferent gonads. Upon sexual differentiation the testis is easily distinguishable by the formation of testis cords.

germ plasm Cytoplasmic structures containing ribonucleoprotein complexes laid down in the oocyte and segregated into the PGCs during early development in some vertebrate species. Germ plasm is thought to determine, or specify, cells to become PGCs.

stage-specific antigen-1 (SSEA-1) A carbohydrate differentiation antigen recognized by monoclonal antibodies raised against embryonal carcinoma cells. It is expressed in a stage-specific manner in the early mouse embryo and on the PGCs of a variety of species including the mouse and chick. Monoclonal antibodies to the SSEA-1 antigen can be used to identify PGCs *in vivo* as well as *in vitro* and can be used to isolate PGCs by fluorescence-activated cell sorting or related techniques.

teratomas/teratocarcinomas Respectively, benign and malignant tumors that develop from PGCs in the mammalian embryonic testis. Teratomas contain many differentiated cell types representative of all the three primary embryonic germ layers. The differentiated cells are derived from a stem cell called the embryonal carcinoma (EC) cell which is probably derived from PGCs. Teratocarcinomas form when the EC cell loses the ability to differentiate and continues to proliferate.

Primordial germ cells (PGCs) are the embryonic precursors of the gametes of the mature adult animal and, collectively with the gametes, are considered cells of the germline or germ cell lineage. The germline, uniquely among the cell lineages of the embryo, will carry the genome on to the next generation, whereas the somatic cell lineages will form the body of the animal and will eventually die.

I. INTRODUCTION

PGCs are set aside from the somatic cell lineages early in development but arise some distance from the gonad, the site at which they will eventually differentiate into the gametes. To reach the gonad the PGCs are translocated by the morphogenetic movements of the embryo and also undergo a period of active migration. Once in the embryonic gonad, the PGCs proliferate to establish the population of cells that will eventually enter into meiosis and differentiate into eggs and sperm. Failure of PGCs to survive or proliferate in the embryo can lead to infertility, whereas increased proliferation of PGCs in the mammalian embryonic testis is associated with the development of testicular tumors, teratomas, and teratocarcinomas. The development of PGCs in embryos has been studied using classical histological techniques as well as using immunological, histochemical, and molecular biological markers. Cell culture systems have also been used to study PGC behavior and growth requirements, whereas sterile mouse mutants have allowed the characterization of some of the genes involved in PGC development in mammals. The analysis of PGC development is central to understanding of the regulation of developmental totipotency in vertebrates.

II. DETERMINATION AND SEGREGATION OF THE GERMLINE

The fusion of an egg and a sperm at fertilization creates a zygote with all the genetic information to form an entire embryo and in mammals the extraembryonic membranes as well. Therefore, the zygote is referred to as developmentally totipotent. As development proceeds the cells of the embryo gradually differentiate, losing the ability to completely recapitulate development and become developmentally restricted. Some cells may retain the ability to generate cells in multiple lineages and may be referred to as developmentally pluripotent. The stem cells found in many tissue compartments of the adult animal can be considered pluripotent. However, most cells in the late embryo and in the adult have completely differentiated into highly specialized cell types and have lost developmental pluripotency and are nullipotent. By some as yet undetermined mechanism, the PGCs must maintain the genetic material in a condition which allows it to recapitulate the entire developmental program. An important question about PGC development is the following: How are PGCs formed in the early embryo and how are they segregated from the somatic cell lineages? Both the timing and the molecular mechanisms underlying the segregation of the PGCs from the somatic cell lineages in the embryo are distinct, even among the vertebrates. The analysis of PGC development has been most extensively studied in three vertebrate groups: anuran amphibians (including *Xenopus laevis* and *Rana pipiens*), birds (chick and quail), and the laboratory mouse. This article will focus on these three vertebrate groups.

In *X. laevis*, determinants present in the vegetal pole cytoplasm of the egg are segregated during the early cleavage divisions of the embryo and inherited by a small number of blastomeres. The cells inheriting these cytoplasmic determinants will become the presumptive PGCs. The cytoplasmic determinants are observed as aggregates of basophilic cytoplasm and may also be called germ plasm. Germ plasm begins to accumulate in the stage I oocyte in the structure known as the mitochondrial cloud. In the mature oocyte, germ plasm is identified as islands of yolk-free cytoplasm in the vegetal cortex. After fertilization of the egg, these yolk-free islands aggregate into large masses at the vegetal pole (Fig. 1). This process is known to be dependent on the kinesin-like protein, Xklp-1. Subsequent cell divisions bring the germ plasm into association with the cleavage furrows. In blastula stage embryos, germ plasm continues to be associated with the plasma membrane but during mitosis it becomes localized to only one spindle pole. Consequently, germ plasm is segregated into only one daughter of each cell division. By following the fate of vegetal pole plasm in the developing embryo, it was determined that this cytoplasmic material could be traced to the population of cells that colonized the gonad and that were the definitive PGCs. The cells that do not inherit the germ plasm are thought to enter other lineages. During gastrulation, germ plasm moves from its peripheral location to a perinuclear location so that at mitosis each daughter inherits germ plasm. Germ plasm contains germinal granules and electron-dense material of about 10- to 20-nm diameter, and it is associated with dense clusters of mitochondria and cytoskeletal elements. Ultraviolet (UV) irradiation of eggs, which results in the production of animals with reduced numbers of PGCs, suggests that germ plasm also contains an RNA component. Several lines of evidence suggest that, in anuran amphibians, germ plasm is required for development of PGCs. First, physical removal of germ plasm or destruction by UV light irradiation reduces or destroys the PGC population. Second, injection of germ plasm can, in some cases, reverse the effects of UV light irradiation of embryos on PGC numbers. Third, injection of

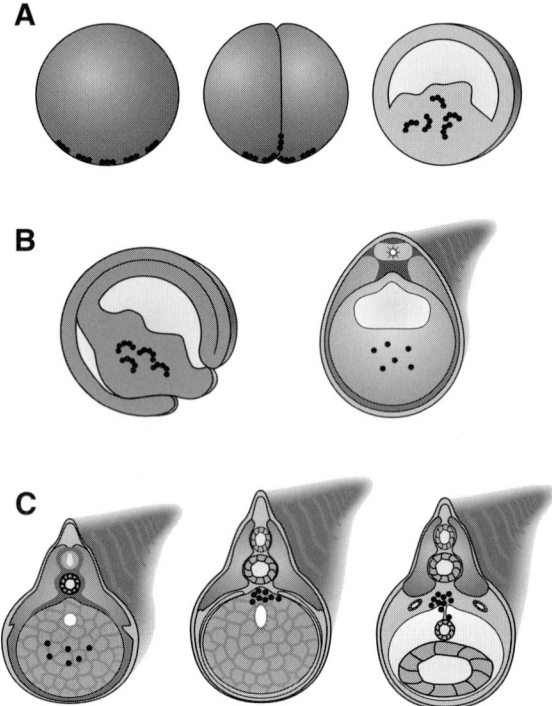

FIGURE 1 Primordial germ cell development in an amphibian, *Xenopus laevis*. (A) The oocyte (left hand panel) contains cytoplasmic aggregates termed pole plasm that lie at the vegetal cortex (black dots). During subsequent development these cytoplasmic determinants become segregated into cells that will become the primordial germ cell progenitors. These cells will eventually end up in the endoderm of the gut (B). From here they are moved by the morphogenetic movements of the embryo towards the dorsal surface of the coelomic cavity. They then actively migrate up the mesentery of the hindgut towards the developing genital ridges (C). Not to scale.

the germline Pole cells or P lineage. In *Drosophila*, the ordered assembly of protein and RNA components is required for the formation of Pole plasm. Although few of the components of germ plasm in *Xenopus* have been identified, a number of RNAs accumulate in, and localize to, the mitochondrial cloud of stage I oocytes. They then translocate to and become concentrated in the vegetal cortex. Interestingly, one of the RNAs, termed *Xcat-2*, is related to the *Drosophila* Pole plasm component *nanos* and another is related to the Drosophila *boule* genes and the mammalian *Deleted in Azoopersmia* (*DAZ*) and *Daz-like* (*DAZL*) genes. *Nanos* is required for germline development in *Drosophila*, whereas *boule* is required for G_2/M transition in fly spermatocytes. The *DAZ* gene, localized on the human Y chromosome, is deleted in patients with azoopermia and severe oligospermia, suggesting that *DAZ* may have functions other than completion of the meiotic cell cycle. A common feature of these genes is that they encode RNA-binding proteins. These data suggest that the molecular mechanisms governing germ plasm assembly and/or function may be evolutionarily conserved.

In mammals such cytoplasmic determinants have not been identified and PGCs are thought to be determined relatively late in development. Individual blastomeres of cleavage-stage mouse embryos retain developmental totipotency, suggesting that mammalian eggs are unlikely to contain cytoplasmic germline determinants. In mice and a number of other vertebrates species, PGCs have been classically identified by alkaline phosphatase histochemistry. Mouse PGCs express a specific izozyme of alkaline phosphatase, tissue nonspecific alkaline phosphatase (TNAP), on their cell surface. TNAP has been used as the classical marker by which PGCs have been traced in mammalian embryos. This has allowed both the number of PGCs and their location to be accurately determined during embryonic development. The reduction in the numbers of TNAP-positive cells in sterile mice confirms the identity of the TNAP-positive cells as PGCs. In mice, PGCs are first identified by TNAP staining in the 7 days postcoitus (dpc) embryo when a small number of PGCs (8–10) are identified at the caudal end of the primitive streak in the extraembryonic mesoderm (Fig. 2). Fate-mapping studies demonstrate that these cells arise from

additional germ plasm into an egg can increase the numbers of germ cells. These data have led to the hypothesis that inheritance of the germ plasm causes cells to enter the germline. However, PGCs transplanted ectopically into the blastocoel differentiated into somatic cells suggesting that germ plasm specifies cells to become germ cells rather than causing irreversible determination of cells as PGCs.

The association of germ plasm with PGC development is clearly analogous to the situation in *Drosophila* and *Caenorhabditis*, in which cytoplasmic determinants present in the egg (Pole plasm in *Drosophila* and P-granules in *Caenorhabditis*) are inherited by

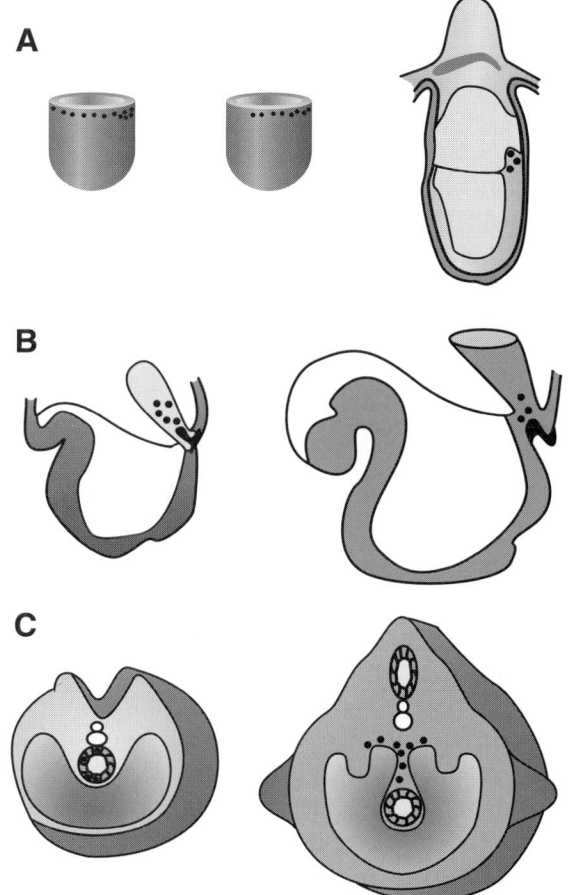

FIGURE 2 Primoridal germ cell development in the mouse embryo. In early mouse development (6.0 to 6.5 dpc) the progenitor population of the PGCs (black dots) are scattered in a ring in the cylindrical epiblast close to the extraembryonic ectoderm (A, left). These cells are then translocated as part of the expanding epiblast to the posterior end of the primitive streak (A, middle). By the next day of development they come to lie in an extraembryonic location in the extraembryonic mesoderm (cross section of embryo; A, right). During the subsequent two days of development, the PGCs, which lie close to the allantois at the caudal end of the primitive streak, are drawn into the embryo proper as the hind gut forms by invagination (B). By 9.5 dpc the PGCs are found within the hindgut (C, left). As the embryo expands, the ventral mesentery of the hindgut disappears and the gut becomes hung in the coelomic cavity, suspended by the dorsal mesentery. The PGCs migrate up the hindgut mesentery towards the two genital ridges (C, right). Not to scale.

a small number of progenitors in the most proximal part of the presumptive extraembryonic mesoderm before gastrulation. At this time of development, these progenitors are not yet restricted to a germline fate. In other words, the progenitors of the PGCs can also give rise to other cell lineages. The PGC progenitors are scattered in a ring in the epiblast close to the extraembryonic ectoderm before the formation of the primitive streak. The PGC progenitors are then translocated, as part of the expanding epiblast, to the posterior end of the primitive streak as the epiblast expands toward and through the primitive streak early in gastrulation. During the subsequent few days of embryonic development (7.0–8.5 dpc) the PGCs are found to be located in an extraembryonic location at the caudal end of the primitive streak near the allantois. At 8.5 dpc, TNAP staining reveals that there are approximately 50–100 PGCs in the embryo. The reason why PGCs are located outside of the embryo proper at this time of development is not understood.

In the chick embryo, cytoplasmic determinants, such as those found in anuran amphibians, have not been identified and it is presumed that PGC development and differentiation in birds occurs by a distinct mechanism. Nevertheless, PGCs can be traced for part of their development by staining for glycogen with periodic acid-Schiff reagent and with monoclonal antibodies recognizing cell surface differentiation antigens such as stage-specific antigen-1 (SSEA-1). Moreover, by culturing fragments of embryos and staining for glycogen it has been possible to determine the regions of the chick embryo that give rise to PGCs. The cells capable of giving rise to the PGCs are first identified in stage X chick embryos at the uterine stage of embryogenesis. The PGC progenitors are identified in the ventral surface of the area pellucida (Fig. 3). During subsequent development, this area expands outwards and then the PGCs migrate as single cells down from the epiblast through the embryo to the hypoblast layer. During gastrulation, morphogenetic movements of the embryo push the hypoblast anteriorly and the PGCs are carried with the hypoblast to an extraembryonic location, cranial and lateral to the primitive streak. This region is called the germinal crescent. PGCs are found in the germinal crescent from the primitive

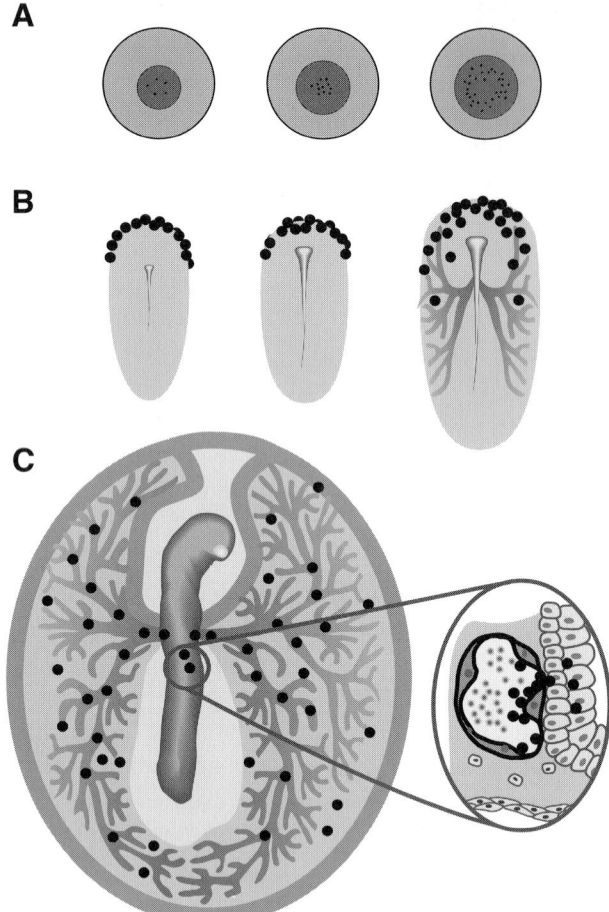

FIGURE 3 PGC development in the chick embryo. In the uterine stage embryos (VII-IX) (A) the PGC progenitors (black dots) can be detected by periodic acid Schiff (PAS) staining in the area pelucida. The PGCs start to segregate from the area pelucida at around stage X and enter the hypoblast. As the hypoblast moves anteriorly, the PGCs are then translocated anteriorly towards the extraembryonic germinal crescent at around stages 4–6 (Hamburger and Hamilton staging) of embryonic development (B, left and middle). From here the PGCs will enter the capillaries of the developing blood stream (B, right). The PGCs will exit the blood stream in the vicinity of the developing gonad in a process that resembles lymphocyte homing (C and inset). Not to scale.

streak stage to the 10-somite stage. As in both mammals and anuran amphibians, the PGCs of avian embryos find themselves at a distance from the gonad and must undertake a period of migration to reach the site at which they will differentiate into the gametes.

III. GERM CELL MIGRATION

In order to reach the gonad, the PGCs must undergo a period of directed migration. In mammals and amphibians, the mechanisms of PGC migration, while distinct, have many common features. On the other hand, PGC migration in avian embryos is quite different and is more similar to lymphocyte homing than to the migration of PGCs of mammals or amphibians or to the migration of other embryonic cell types such as neural crest cells. PGC migration must be controlled very strictly spatially and temporally. Migration must commence at a specific time and place and end at a specific time and place. Failure of PGCs to reach the gonad (or to stop there once they arrive) could lead to sterility and, in mammals, could lead to the development of PGC-derived tumors.

In both *Xenopus* and mice, migration into the gonad anlagen is brought about in part passively, through the morphogenetic movements of the embryo, and in part actively, through locomotion by the PGCs. In anuran embryos, such as *Xenopus*, the normal fate of cells at the vegetal pole of the blastula is to be carried by the morphogenetic movements of the gastrulating embryo into the endodermal gut tube. In mice, as in *Xenopus*, the first stage of migration is thought to be passive, brought about by the morphogenetic movements of the embryo. In this stage, the germ cells, located at the caudal end of the primitive streak, are thought to be drawn into the embryo as the hindgut forms by invagination. In both amphibian and mammalian embryos, PGCs then migrate out the gut and up the hindgut mesentery toward two thickened ridges of tissues (the genital ridges) that lie on either side of the hindgut mesentery. The genital ridge is the gonad anlagen and is the precursor of the adult gonad. The migration and proliferation of PGCs in mice outlined previously has been confirmed using other PGC markers such as the POU transcription factor, Oct 4, and by using monoclonal antibodies to the carbohydrate cell surface differentiation antigen, SSEA-1. Most likely, PGC migration is regulated by complex interactions with surrounding somatic cells that involve extracellular matrix components (fibronectin, laminin, etc.) as well as growth factors and growth factor receptors.

PGC migration could be guided by both haptotactic and chemotactic mechanisms which are not mutually exclusive mechanisms of cell guidance. In *Xenopus*, the PGCs migrate at an angle along the mesentery to the dorsal body wall. Then they migrate laterally until they reach the thickened ridges that form on the dorsal body wall that are the gonad anlage or genital ridges. PGC migration in *Xenopus* may be controlled by the shape of the coelomic epithelial cells that line the hindgut mesentery. Scanning electron microscopy reveals that the PGCs align themselves with the coelomic epithelial cells. Transmission electron microscopic (TEM) examination of PGCs *in vivo* shows them to be elongated and extending filopodia containing microfilament-containing stress fibers. When isolated from the embryo and placed in culture, migrating PGCs extend filopodia and align themselves with the cells of the substratum. Taken together these data suggest that *Xenopus* PGCs reach the gonad partly through active cell locomotion.

In the mouse embryo, PGC migration may not be as extensive as previously thought but, nevertheless, mouse PGCs also undergo a period of active locomotion into the gonad anlagen. TEM reveals that between 8.5 and 9.5 dpc the PGCs undergo dramatic changes in morphology and undergo a transition from a rounded to a polarized shape. During the period in which they are migrating out of the hindgut as well as up the dorsal mesentery, they have prominent filopodia which extend between the surrounding somatic cells. These filopodia can be identified by TEM and when the PGCs are stained with monoclonal antibodies in whole mount embryos and observed by laser scanning confocal microscopy. In many cases the filopodia extend between PGCs which appear to be physically attached to each other during migration. When PGCs are isolated into culture and placed on feeder layers of certain established fibroblast cell lines they are actively motile and display some of the characteristics of invasive cells. Alternatively, when PGCs that have colonized the gonad anlagen are isolated into culture they remain immotile on the feeder layer suggesting that they have undergone some differentiation event that switches off their motile activity. The changes in PGC behavior *in vivo* and *in vitro* can be correlated with the appearance and disappearance of cell surface differentiation antigens such as the SSEA-1 antigen. However, it seems likely that the changes that occur upon colonization of the gonad are complex and may involve multiple cell surface molecules, including both integrins and growth factor receptors. The mechanism(s) by which PGCs find their way to the gonad anlagen remains obscure. Some evidence suggests that a member of the transforming growth factor-β superfamily may play a role in chemotactic guidance of PGCs to the gonad.

In the chick embryo, PGCs also arise outside of the gonad and undergo migration into the embryo. In this case PGCs migrate posteriorly and laterally from the germinal crescent toward the forming blood islands. Here the vitelline circulation will form to carry blood between the embryo and the yolk sac. PGCs enter the bloodstream of the embryo and are carried around in it. They then adhere to the capillary walls in the area of the gonad anlagen. From this site they exit the bloodstream to colonize the gonad anlagen. The molecular mechanisms regulating this process are not understood but the factors effecting this process are evolutionarily conserved since turkey (and even mouse) PGCs introduced into the chick bloodstream will colonize the gonad anlagen.

IV. PGC SURVIVAL AND PROLIFERATION

During the course of PGC migration toward the genital ridges, and for a few days after PGCs reach them, the PGCs proliferate to establish the population of cells that will enter into meiosis. Little is known about the factors that regulate this process in any vertebrate species except the mouse. The identification and characterization of mouse mutants that cause sterility has allowed some of the genes that regulate this process in mammals to be identified.

In the 8.5 dpc mouse embryo, the 50–100 cells identified as PGCs will give rise to approximately 25,000–35,000 PGCs in the fully colonized genital ridge at 13.5 dpc. Mutations at the *Dominant White Spotting* (*W*), *Steel* (*Sl*), *hertwig's anemia* (*an*), and *germ cell deficient* (*gcd*) loci all affect early PGC development, thereby identifying genes involved in early germline development. The *W* and *Sl* mutations affect not only PGC development but also development of the hemopoietic lineages and melanoblasts, and they also affect other aspects of gametogenesis in mammals. Both mutations have been characterized at the molecular level. The *W* locus, on mouse chromosome 5, encodes a receptor tyrosine kinase (c-kit) which is related to the platelet-derived growth factor receptor. It is characterized, in part, by the presence of five immunoglobulin-like repeats in the extracellular domain and by the presence of an insert within the catalytic kinase domain. The *Sl* locus, on mouse chromosome 10, was found to encode the ligand for c-kit, variously called kit-ligand, stem cell factor, Steel factor, and mast cell growth factor. Steel factor is a transmembrane growth factor encoded in two forms generated by alternate splicing. One form has a proteoltyic cleavage site present in the ectodomain of the protein and can be cleaved to generate a soluble factor. Another form lacks the proteolytic cleavage site and is thought to remain as a membrane-bound growth factor. Compelling evidence suggests that the *W/Sl* signal transduction pathway is required for PGC survival in mice. The c-kit receptor is expressed in PGCs, whereas Steel factor is expressed in the surrounding somatic cells. Mice carrying W mutations that delete or inactivate the catalytic domain of the c-kit receptor have severe deficiencies in PGCs and these animals (if viable) are usually completely sterile. The early period of germ cell development in mice (up to 9.5 dpc) seems to be independent of the c-kit signaling pathway since W mutants have no deficit in PGC numbers up until 9.5 dpc. Paralleling mutations at the *W* locus, mutations at the *Sl* locus also severely affect PGC development and can cause sterility. Steel factor seems to be required in a membrane-bound form for long-term survival of PGCs. The Sl^d mutation is caused by a 4.0-kb intragenic deletion that removes sequences encoding the cytoplasmic tail of Steel factor. Therefore, these mice can only produce a soluble form of Steel factor. Since mice carrying the Sl^d mutation are sterile, this suggests that the membrane-bound form of the factor is required for proper PGC survival in mice. In the absence of signaling via the c-kit receptor, PGCs undergo programmed cell death or apoptosis. In addition to a role in PGC survival, c-kit and its ligand may play a role in PGC migration since both W and Sl mutants have been reported to show defects in PGC migration. This may reflect the fact that Steel factor is a transmembrane growth factor that is required in its membrane-bound form for long-term survival of PGCs. The c-kit/Sl signal transduction pathway may therefore provide an exquisite mechanism for controlling PGC survival and migration in the developing mouse embryo.

Most likely, PGC survival and proliferation are regulated by multiple polypeptide growth factors. One of these factors is likely to be a member of the interleukin-6/leukemia inhibitory factor (IL-6/LIF) family. The IL-6/LIF family of growth factors have heterodimeric receptors. A common subunit of these receptors is gp130, which is required for signal transduction in response to ligand binding. PGC survival is blocked when signaling via gp130 is inhibited *in vitro* and, importantly, mice lacking gp130 have a severe deficiency of PGCs. The physiological ligand that activates gp130 on the PGC surface remains to be determined. While other growth factors have been shown to affect PGC survival and/or proliferation in *in vitro* culture systems, the physiological relevance of these factors remains unclear.

A second source of information regarding the factors that regulate PGC development in mammals comes from the analysis of PGC-derived tumors. In mammals, such as mice and humans, aberrant proliferation or differentiation of PGCs (particularly in males) can lead to the development of testicular tumors, termed teratomas and teratocarcinomas. Teratomas are benign tumors consisting of cell types derived from all three primary embryonic germ layers. These differentiated cell types are derived from a stem cell, the embryonal carcinoma (EC) cell. Studies in mice have demonstrated that the EC cell is most likely derived from PGCs. In some cases the EC cells lose the ability to differentiate and continue to proliferate, forming a malignant teratocarcinoma. Some of the genetic loci affecting teratocarcinogenesis in mice have been identified and these loci are likely to encode genes that regulate PGC proliferation or differentiation. One of the identified loci is the *Teratoma* (*Ter*) locus on mouse chromosome 8, and *Ter/Ter* mice have severe deficiencies in PGC numbers demonstrating the role of this gene in PGC development. *In vitro* culture of mouse PGCs has identified many growth factors that affect their survival and/or proliferation. A combination of Steel factor, basic fibroblast growth factor, and leukemia inhibitory factor effectively immortalizes PGCs in culture. The resultant cell lines are called embryonic germ cells or EG cells. These cells are similar in property to pluripotent EC cells and embryonic stem cells, and they can differentiate into a variety of cell types *in vitro*. When EG cells are introduced into a host blastocyst they can colonize many lineages and, importantly, reenter the germline. The mechanisms causing the formation of EG cells *in vitro* may be related to the formation of EC cells in spontaneously arising teratomas.

The growth factors regulating the development of PGCs in other vertebrates have not been elucidated. However, some of the factors found to regulate the survival and proliferation of mouse PGCs have also been found to act on PGCs isolated from chick embryos. For example, Steel factor can affect the survival and proliferation of isolated chick PGCs. These data suggest that some of the mechanisms regulating PGC growth in the embryo may be conserved during evolution.

V. PGC DIFFERENTIATION

Following mitotic expansion of the PGC population in the embryonic gonad, the PGCs commence the differentiation events that will finally lead to the production of the functional gametes of the mature adult animal. The differentiation of PGCs is dependent on the sex of the embryo which also becomes apparent at this time of development. In the mouse embryo at 12.5 dpc, the first signs of sexual differentiation are observed as the sex cords begin to form in the developing testis. In male embryos, the PGCs, now called gonocytes, enter a cell cycle arrest. They will remain arrested in G_0 or G_1 of the cell cycle until a few days after birth when they resume mitosis forming the mitotic stem cell of the adult testis, the spermatogonium. The molecular mechanisms causing such cell cycle arrest are unknown. In female embryos, PGCs exit from the mitotic cell cycle and enter directly into the meiotic cell cycle. They then arrest at meiotic prophase, at which point they can remain for the entire reproductive life span of the organism. Following entry into meiosis many female gonocytes will die in a process called atresia. The molecular mechanisms regulating cell cycle arrest in the male PGCs or entry into meiosis in the female PGCs remain unknown.

VI. CONCLUSIONS AND PROSPECTS

Historically the germline has been one of the most studied lineages in the vertebrate embryo. The molecular mechanisms regulating PGC development are gradually being elucidated in a variety of species including vertebrates. Understanding the mechanisms by which PGCs are formed in the embryo and by which they regulate developmental totipotency is one of the most important questions in modern developmental biology. Moreover, factors which affect PGC development are likely to be important in determining fertility and in influencing gonadal tumor development in males and females. Advances in this area of research are likely to come from a variety of avenues. First, characterization of sterile mouse mutants will identify some of the genes involved in PGC development. Furthermore, a large number of mice carrying targeted gene mutations (generated by homologous recombination in embryonic stem cells) have been produced over the past few years. Many of these "knockout" mice have been found to have defects in germline development. This has provided a windfall of information about the molecular regulation of germline development in mammals and is likely to continue to do so for some time. However, some important information may be missed if it is presumed that knockout mice, if fertile, have no defects in germline development. The plasticity of the germline can make up for deficiencies during PGC development. For example, some *Sl* mutants have severe deficiencies in PGC development but are fully fertile because spermatogonial proliferation compensates for the reduced numbers of PGCs.

It will be important, therefore, to examine germline development in many knockout mice even if an obvious reproductive defect is not apparent. Second, advances in research concerning germline development in vertebrates are likely to come as a result of new genomic and high-throughput sequencing technologies. The ability to rapidly determine the chromosomal localization of new genes will continue to provide important new candidates for genetic loci affecting fertility and teratocarcinogenesis in mice and humans. Furthermore, the ability to rapidly construct YAC and BAC contigs of a region and to isolate candidate genes means that many loci affecting fertility and teratocarcinogenesis will likely be molecularly cloned in the near future, even if no good candidate gene already exists. Over the past few years high-throughput sequencing of expressed sequence tags has provided a valuable insight into gene expression in a variety of cell types. The application of this technology to the analysis of germline development is already under way and is likely to generate a wealth of information on the genes expressed in, and regulating the development of, germ cells. Third, the analysis of germline development in *Drosophila* and *Caenorhabditis* will likely continue to provide important clues as to the mechanisms regulating germline development in vertebrates. Significant information regarding the regulation of the meiotic cell cycle may also come from studies on meiosis in yeast. A new player has also entered this field, namely, the zebrafish, an experimental organism with a combination of the superb genetics of flies and worms and the experimental manipulability of *Xenopus*. Analysis of PGC development in the zebrafish would undoubtedly advance our understanding of germline development in the vertebrates.

Acknowledgments

This research was sponsored in part by the National Cancer Institute, DHHS, under contract with ABL-Basic Research Program. I am grateful to Richard Frederickson, Carolyn Whistler, and Ellen Frazier for preparing the figures.

Bibliography

Donovan, P. J. (1994). Growth factor regulation of mouse primoridal germ cell development. In *Current Topics in Developmental Biology, Vol. 29* (R. A. Pederson, Ed.). Academic Press, San Diego.

McLaren, A. (1981). *Germ Cells and Soma: A New Look at an Old Problem*. Alpine Press, Stoughton, MA.

McLaren, A., and Wylie, C. C. (Eds.) (1983). *Current Problems in Germ Cell Differentiation*. Cambridge Univ. Press, Cambridge, MA.

Niewkoop, P. D., and Sutasurya, L. A. (1979). *Primordial Germ Cells in the Chordates*. Cambridge Univ. Press, Cambridge, MA.

ISBN 0-12-227023-1